Preparation Strategy & Analysis Videos

AF889473

# 24 Years

2025-2013 JEE Main & 2012-2002 AIEEE

# JEE Main

## Chapterwise Topicwise Solved Papers

# CHEMISTRY

**Authors**

Pooja Yadav

Shalini Singh

arihant

SUCCESS AND NOTHING LESS

arihant
SUCCESS AND NOTHING LESS

**ARIHANT PRAKASHAN (Series), MEERUT**



卐 **Administrative & Production Offices**

**Regd. Office**

'Ramchhaya' 4577/15, Agarwal Road, Darya Ganj, New Delhi -110002
Tele: 011- 47630600, 43518550

卐 **Head Office**

Kalindi, TP Nagar, Meerut (UP) - 250002
Tel: 0121-7156203, 7156204

卐 **Sales & Support Offices**

Agra, Ahmedabad, Bengaluru, Bareilly, Chennai, Delhi, Guwahati, Hyderabad, Jaipur, Jhansi, Kolkata, Lucknow, Nagpur & Pune.

卐 **PRICE** ₹625.00

卐 **PO No :** TXT-XX-XXXXXXX-X-XX

Published by Arihant Publications (India) Ltd.

For further information about the books published by Arihant, log on to www.arihantbooks.com or e-mail at info@arihantbooks.com

Follow us on

# Preface

JEE Main is a gateway examination for candidates expecting to seek admission in Bachelor in Engineering (BE), Bachelor of Technology (B.Tech) and Bachelor of Architecture (B.Arch) at Indian Institutes of Information Technology (IIITs), National Institutes of Technology (NITs), Delhi Technological University and other Centrally Funded Technical Institutes (CFTIs).

JEE Main is also an examination which is like screening examination for JEE Advanced (The gateway examination to India's most reputed Technical Institutes, Indian Institutes of Technology— IITs).

To make the students well-versed with the pattern as well as the level of the questions asked in the exam, this book contains Chapterwise Topicwise Solutions of Questions asked in Last 24 Years' Examinations of JEE Main (formerly known as AIEEE).

**The Unique and exclusive features of this book are:**

- Questions are further sorted according to their difficulty level in each topic.
- Questions on frequently asked concepts are marked with concept name to get edge over others.

Solutions to all the questions have been kept very detailed and accurate with ingredient of special features like Key idea, Concept Enhancer & Alternate/Time saver techniques.

We hope this book would be highly beneficial for the students. We would be grateful if any discrepancy or mistake in the questions or answers is brought to our notice so that these could be rectified in subsequent editions.

Publisher

# Contents

## Class 11


1. **SOME BASIC CONCEPTS IN CHEMISTRY** 1-25
   - *Topic-1* Fundamental Concepts and Laws of Chemical Combinations
   - *Topic-2* Atomic & Molecular Masses, Percentage, Composition, Mole Concept, Empirical & Molecular Formula
   - *Topic-3* Chemical Equations, Stoichiometry and Concentration Terms

2. **ATOMIC STRUCTURE** 26-50
   - *Topic-1* Preliminary Models and Dual Nature of Light
   - *Topic-2* Bohr's Model and Hydrogen Spectrum
   - *Topic-3* Dual Nature of Matter and Uncertainty Principle
   - *Topic-4* Quantum Mechanical Model

3. **CLASSIFICATION OF ELEMENTS AND PERIODICITY IN PROPERTIES** 51-67
   - *Topic-1* Modern Periodic Table
   - *Topic-2* Periodic Properties and their Trends 1(Atomic Radius, IE, EA, EN, Metallic Character)
   - *Topic-3* Periodic Properties and their Trends 2(Lattice & Hydration Energy, Oxidation State & Chemical Properties)

4. **CHEMICAL BONDING AND MOLECULAR STRUCTURE** 68-91
   - *Topic-1* Ionic and Covalent Bonding, Fajan's Rule, Bond Parameters
   - *Topic-2* Hybridisation, VSEPR, VBT
   - *Topic-3* MOT , Metallic & Hydrogen Bonding

5. **THERMODYNAMICS** 92-120
   - *Topic-1* Fundamentals of Thermodynamics
   - *Topic-2* First Law of Thermodynamics
   - *Topic-3* Reactions Related to Enthalpies and Hess's Law
   - *Topic-4* Second Law of Thermodynamics, Entropy, Free Energy Change and Spontaneity

6. **CHEMICAL EQUILIBRIUM** 121-135
   - *Topic-1* Chemical Equilibrium, Law of Mass Action and Equilibrium Constant
   - *Topic-2* Le-Chatelier's Principle and Factors Affecting Chemical Equilibria

7. **IONIC EQUILIBRIUM** 136-152
   - *Topic-1* Ostwald's Dilution Law and Concept of Acids and Bases
   - *Topic-2* Solubility Product and Common Ion Effect
   - *Topic-3* pH, Buffer and Indicators, Salt Hydrolysis

**8. REDOX REACTIONS** **153-158**

*Topic-1* Oxidation - Reduction and Oxidation Number

*Topic-2* Balancing of Chemical Equations

**9. *p*-BLOCK ELEMENTS -1** **159-170**

*Topic-1* Group-13 Elements

*Topic-2* Group- 14 Elements

**10. PURIFICATION AND CHARACTERISATION OF ORGANIC COMPOUNDS** **171-187**

*Topic-1* Methods of Purification

*Topic-2* Qualitative Analysis

*Topic-3* Quantitative Analysis

**11. SOME BASIC PRINCIPLES OF ORGANIC CHEMISTRY** **188-209**

*Topic-1* Nomenclature, Hybridisation and Classification

*Topic-2* Isomerism of Organic Compounds

*Topic-3* Bond Fission, Electronic Displacement in Organic Molecules, Attacking Reagent and Reaction Intermediates

*Topic-4* Types of Organic Reactions

**12. HYDROCARBONS** **210-242**

*Topic-1* Alkanes

*Topic-2* Alkenes

*Topic-3* Alkynes

*Topic-4* Aromatic Hydrocarbons

## Class 12

**13. SOLUTIONS** **243-271**

*Topic-1* Expression of Concentrations of Solutions

*Topic-2* Vapour Pressure, Henry's Law and Raoult's Law

*Topic-3* Colligative Properties

*Topic-4* Abnormal Molecular Masses and van't Hoff Factor

**14. ELECTROCHEMISTRY** **272-296**

*Topic-1* Conductance and Electrolysis

*Topic-2* Electrochemical Series, Cells and Their EMF

*Topic-3* Batteries, Fuel Cells and Corrosion

**15. CHEMICAL KINETICS** **297-327**

*Topic-1* Rate of Chemical Reaction and Rate Expression

*Topic-2* Order, Molecularity and Half- life Period

*Topic-3* Arrhenius Theory, Activation Energy, Collision and Related Theories

**16. *p*-BLOCK ELEMENTS -2** **328-346**

*Topic-1* Group 15 Elements

*Topic-2* Group 16 Elements

Topic-3 Group 17 Elements
Topic-4 Group 18 Elements

17. d- AND f- BLOCK ELEMENTS 347-368
Topic-1 Properties of Transition Elements
Topic-2 Important Compounds of Transition Elements
Topic-3 Inner Transition Elements (Lanthanoids and Actinoids)

18. COORDINATION COMPOUNDS 369-409
Topic-1 Werner's Theory, Coordination Number, Nomenclature and Isomerism of Coordination Compounds
Topic-2 Valence Bond Theory and CFT
Topic-3 Application of Coordination Compounds and Organometallic Compounds

19. ORGANIC COMPOUNDS CONTAINING HALOGENS 410-445
Topic-1 Haloalkanes
Topic-2 Haloarenes
Topic-3 Some Important Polyhalogen Compounds

20. ALCOHOLS, PHENOLS AND ETHERS 446-477
Topic-1 Preparation, Properties and Uses of Alcohols
Topic-2 Preparation, Properties and Uses of Phenols
Topic-3 Preparation, Properties and Uses of Ethers

21. ALDEHYDES, KETONES & CARBOXYLIC ACIDS 478-531
Topic-1 Preparation, Properties and Uses of Aldehydes
Topic-2 Preparation, Properties and Uses of Ketones
Topic-3 Preparation, Properties and Uses of Carboxylic Acids

22. ORGANIC COMPOUNDS CONTAINING NITROGEN 532-574
Topic-1 Aliphatic Amines
Topic-2 Aromatic Amines
Topic-3 Diazonium Salts and Other Nitrogen Containing Compounds

23. BIOMOLECULES 575-598
Topic-1 Carbohydrates
Topic-2 Proteins and Enzymes
Topic-3 Vitamins and Nucleic Acids

24. PRINCIPLES RELATED TO PRACTICAL CHEMISTRY 599-613
Topic-1 Detection of Extra Elements and Functional Groups
Topic-2 Chemistry involved in the Preparation of Organic and Inorganic Compounds
Topic-3 Chemistry involved in Titrimetric Experiments
Topic-4 Qualitative Salt Analysis

# SYLLABUS

## Chemistry

### Physical Chemistry

**UNIT 1 Some Basic Concepts in Chemistry**

Matter and its nature, Dalton's atomic theory: Concept of atom, molecule, element, and compound:: Laws of chemical combination; Atomic and molecular masses, mole concept, molar mass, percentage composition, empirical and molecular formulae: Chemical equations and stoichiometry.

**UNIT 2 Atomic Structure**

Nature of electromagnetic radiation, photoelectric effect; Spectrum of the hydrogen atom. Bohr model of a hydrogen atom - its postulates, derivation of the relations for the energy of the electron and radii of the different orbits, limitations of Bohr's model; Dual nature of matter, de Broglie's relationship. Heisenberg uncertainty principle. Elementary ideas of quantum mechanics, quantum mechanics, the quantum mechanical model of the atom, and its important features. Concept of atomic orbitals as one-electron wave functions: Variation of y and $y^2$ with r for 1s and 2s orbitals; various quantum numbers (principal, angular momentum, and magnetic quantum numbers) and their significance; shapes of s, p, and d - orbitals, electron spin and spin quantum number: Rules for filling electrons in orbitals – Aufbau principle. Pauli's exclusion principle and Hund's rule, electronic configuration of elements, and extra stability of half-filled and completely filled orbitals.

**UNIT 3 Chemical Bonding and Molecular Structure**

Kossel-Lewis approach to chemical bond formation, the concept of ionic and covalent bonds.

**Ionic Bonding:** Formation of ionic bonds, factors affecting the formation of ionic bonds; calculation of lattice enthalpy.

**Covalent Bonding:** Concept of electronegativity. Fajan's rule, dipole moment: Valence Shell Electron Pair Repulsion (VSEPR ) theory and shapes of simple molecules.

Quantum mechanical approach to covalent bonding: Valence bond theory - its important features, the concept of hybridization involving s, p, and d orbitals; Resonance.

**Molecular Orbital Theory** - Its important features. LCAOs, types of molecular orbitals (bonding, antibonding), sigma and pi-bonds, molecular orbital electronic configurations of homonuclear diatomic molecules, the concept of bond order, bond length, and bond energy. Elementary idea of metallic bonding. Hydrogen bonding and its applications.

**UNIT 4 Chemical Thermodynamics**

**Fundamentals of thermodynamics:** System and surroundings, extensive and intensive properties, state functions, entropy, types of processes.

**The first law of thermodynamics** - Concept of work, heat internal energy and enthalpy, heat capacity, molar heat capacity; Hess's law of constant heat summation; Enthalpies of bond dissociation, combustion, formation, atomization, sublimation, phase transition, hydration, ionization, and solution.

**The second law of thermodynamics** - Spontaneity of processes; $\Delta S$ of the universe and $\Delta G$ of the system as criteria for spontaneity. DG° (Standard Gibbs energy change) and equilibrium constant.

**UNIT 5 Solutions**

Different methods for expressing the concentration of solution - molality, molarity, mole fraction, percentage (by volume and mass both), the vapour pressure of solutions and Raoult's Law - Ideal and non-ideal solutions, vapour pressure - composition, plots for ideal and non-ideal solutions; Colligative properties of dilute solutions - a relative lowering of vapour pressure, depression of freezing point, the elevation of boiling point and osmotic pressure; Determination of molecular mass using colligative properties; Abnormal value of molar mass, Van't Hoff factor and its significance.

**UNIT 6 Equilibrium**

Meaning of equilibrium is the concept of dynamic equilibrium.

**Equilibria involving physical processes:** Solid-liquid, liquid-gas - gas and solid-gas equilibria, Henry's law. General characteristics of equilibrium involving physical processes.

Equilibrium involving chemical processes: Law of chemical equilibrium, equilibrium constants ($K_p$ and $K_c$) and their significance, the significance of $\Delta G$ and $\Delta G°$ in chemical equilibrium, factors affecting equilibrium concentration, pressure, temperature, the effect of catalyst; Le Chatelier's principle.

Ionic equilibrium: Weak and strong electrolytes, ionization of electrolytes, various concepts of acids and bases (Arrhenius. Bronsted - Lowry and Lewis) and their ionization, acid-base equilibria (including multistage ionization) and ionization constants, ionization of water. pH scale, common ion effect, hydrolysis of salts and pH of their solutions, the solubility of sparingly soluble salts and solubility products, and buffer solutions.

### UNIT 7 Redox Reactions and Electrochemistry

Electronic concepts of oxidation and reduction, redox reactions, oxidation number, rules for assigning oxidation number and balancing of redox reactions.

Electrolytic and metallic conduction, conductance in electrolytic solutions, molar conductivities and their variation with concentration: Kohlrausch's law and its applications.

Electrochemical cells - Electrolytic and Galvanic cells, different types of electrodes, electrode potentials including standard electrode potential, half-cell and cell reactions, emf of a Galvanic cell and its measurement: Nernst equation and its applications; Relationship between cell potential and Gibbs' energy change: Dry cell and lead accumulator; Fuel cells.

### UNIT 8 Chemical Kinetics

Rate of a chemical reaction, factors affecting the rate of reactions: concentration, temperature, pressure, and catalyst; elementary and complex reactions, order and molecularity of reactions, rate law, rate constant and its units, differential and integral forms of zero and first-order reactions, their characteristics and half-lives, the effect of temperature on the rate of reactions, Arrhenius theory, activation energy and its calculation, collision theory of bimolecular gaseous reactions (no derivation).

## Inorganic Chemistry

### UNIT 9 Classification of Elements and Periodicity in Properties

Modem periodic law and present form of the periodic table, *s*, *p*, *d* and *f* block elements, periodic trends in properties of elements, atomic and ionic radii, ionisation enthalpy, electron gain enthalpy, valence, oxidation states, and chemical reactivity.

### UNIT 10 *p*-block Elements

Group-13 to Group 18 Elements

General Introduction: Electronic configuration and general trends in physical and chemical properties of elements across the periods and down the groups; unique behaviour of the first element in each group.

### UNIT 11 *d*- and *f*- Block Elements

Transition Elements

General introduction, electronic configuration, occurrence and characteristics, general trends in properties of the first-row transition elements - physical properties, ionization enthalpy, oxidation states, atomic radii, colour, catalytic behaviour, magnetic properties, complex formation, interstitial compounds, alloy formation; Preparation, properties, and uses of $K_2Cr_2O_7$, and $KMnO_4$.

Inner Transition Elements

Lanthanoids - Electronic configuration, oxidation states, and lanthanoid contraction.

Actinoids - Electronic configuration and oxidation states.

### UNIT 12 Co-ordination Compounds

Introduction to coordination compounds. Werner's theory; ligands, coordination number, denticity. chelation; IUPAC nomenclature of mononuclear co-ordination compounds, isomerism; Bonding-Valence bond approach and basic ideas of Crystal field theory, colour and magnetic properties; Importance of co-ordination compounds (in qualitative analysis, extraction of metals, and in biological systems).

## Organic Chemistry

### UNIT 13 Purification and Characterisation of Organic Compounds

**Purification** - Crystallization, sublimation, distillation, differential extraction, and chromatography - principles and their applications.

**Quanlitative analysis** Detection of nitrogen, sulphur, phosphorus, and halogens.

**Quantitative analysis** (basic principles only) - Estimation of carbon, hydrogen, nitrogen, halogens, sulphur, and phosphorus. Calculations of empirical formulae and molecular formulae: Numerical problems in organic quantitative analysis.

### UNIT 14 Some Basic Principles of Organic Chemistry

**Tetravalency of carbon:** Shapes of simple molecules - hybridization (s and p): Classification of organic compounds based on functional groups: and those containing halogens, oxygen, nitrogen, and sulphur; Homologous series: Isomerism - structural and stereoisomerism.

Nomenclature (Trivial and IUPAC)

Covalent bond fission - Homolytic and heterolytic: free radicals, carbocations and carbanions; stability of carbocations and free radicals, electrophiles and nucleophiles.

Electronic displacement in a covalent bond - Inductive effect, electromeric effect, resonance and hyperconjugation.

Common types of organic reactions - Substitution, addition, elimination, and rearrangement

### UNIT 15 Hydrocarbons

Classification, isomerism, IUPAC nomenclature, general methods of preparation, properties, and reactions.

**Alkanes** - Conformations: Sawhorse and Newman projections (of ethane): Mechanism of halogenation of alkanes.

**Alkenes** - Geometrical isomerism: Mechanism of electrophilic addition: addition of hydrogen, halogens, water, hydrogen halides (Markownikoffs and peroxide effect): Ozonolysis and polymerization.

**Alkynes** - Acidic character: Addition of hydrogen, halogens, water, and hydrogen halides: Polymerisation.

**Aromatic hydrocarbons** - Nomenclature, benzene - structure and aromaticity: Mechanism of electrophilic substitution: halogenation, nitration. Friedel-Craft's alkylation and acylation, directive influence of the functional group in monosubstituted benzene.

### UNIT 16 Organic Compounds Containing Halogens

General methods of preparation, properties, and reactions: Nature of C—*X* bond; Mechanisms of substitution reactions. Uses: Environmental effects of chloroform, iodoform, freons and DDT.

### UNIT 17 Organic Compounds Containing Oxygen

General methods of preparation, properties, reactions, and uses.

ALCOHOLS, PHENOLS, AND ETHERS

**Alcohols:** Identification of primary, secondary, and tertiary alcohols: mechanism of dehydration.

**Phenols:** Acidic nature, electrophilic substitution reactions: halogenation. nitration and sulphonation. Reimer - Tiemann reaction.

**Ethers:** Structure.

**Aldehyde and Ketones:** Nature of carbonyl group; Nucleophilic addition to C > O = group, relative reactivities of aldehydes and ketones; Important reactions such as - Nucleophilic addition reactions (addition of HCN. $NH_3$, and its derivatives), Grignard reagent; oxidation: reduction (Wolf Kishner and Clemmensen); the acidity of a-hydrogen. aldol condensation, Cannizzaro reaction. Haloform reaction, Chemical tests to distinguish between aldehydes and ketones.

**Carboxylic Acids**

Acidic strength and factors affecting it.

### UNIT 18 Organic Compounds Containing Nitrogen

General methods of preparation. Properties, reactions, and uses.

**Amines:** Nomenclature, classification structure, basic character, and identification of primary, secondary, and tertiary amines and their basic character.

**Diazonium Salts:** Importance in synthetic organic chemistry.

## UNIT 19 Biomolecules

General introduction and importance of biomolecules.

CARBOHYDRATES - Classification; aldoses and ketoses: monosaccharides (glucose and fructose) and constituent monosaccharides of oligosaccharides (sucrose, lactose, and maltose).

PROTEINS - Elementary Idea of a-amino acids, peptide bond, polypeptides. Proteins: primary, secondary, tertiary, and quaternary structure (qualitative idea only), denaturation of proteins, enzymes.

VITAMINS – Classification and functions.

NUCLEIC ACIDS – Chemical constitution of DNA and RNA.

Biological functions of nucleic acids.

Hormones (General introduction)

## UNIT 20 Principles Related To Practical Chemistry

Detection of extra elements (Nitrogen, Sulphur, halogens) in organic compounds; Detection of the following functional groups; hydroxyl (alcoholic and phenolic), carbonyl (aldehyde and ketones) carboxyl, and amino groups in organic compounds.

- The chemistry involved in the preparation of the following:
  Inorganic compounds; Mohr's salt, potash alum.
  Organic compounds: Acetanilide, *p*-nitro acetanilide, aniline yellow, iodoform.
- The chemistry involved in the titrimetric exercises – Acids, bases, and the use of indicators, oxalic-acid vs $KMnO_4$, Mohr's salt vs $KMnO_4$.
- Chemical principles involved in the qualitative salt analysis:

**Cations** — $Pb^{2+}$, $Cu^{2+}$, $Al^{3+}$, $Fe^{3+}$, $Zn^{2+}$, $Ni^{2+}$, $Ca^{2+}$, $Ba^{2+}$, $Mg^{2+}$, $NH^{4+}$.

**Anions** – $CO_3^{2-}$, $S^{2-}$, $SO_4^{2-}$, $NO_2$, $NO_3$, $Cl^-$, $Br^-$, $I^-$ (Insoluble salts excluded).

Chemical principles involved in the following experiments:

1. Enthalpy of solution of $CuSO_4$.
2. Enthalpy of neutralisation of strong acid and strong base.
3. Preparation of lyophilic and lyophobic sols.
4. Kinetic study of the reaction of iodide ions with hydrogen peroxide at room temperature.

# C-01

JEE Main Chapterwise Topicwise Chemistry

In JEE Main there is usually 1-2 questions per paper are visible from this chapter. In past 3 years 37 conceptual and 59 numerical questions were asked of moderate difficulty level from the chapter.

**For detailed analysis scan QR code**

# SOME BASIC CONCEPTS IN CHEMISTRY

## Topic 1 Fundamental Concepts and Laws of Chemical Combinations

### Level 1

**01** Which of the following have same number of significant figures? *[2023, 8 April Shift-II]*

A. 0.00253 B. 1.0003
C. 15.0 D. 163

Choose the correct answer from the options given below

(a) A, C and D only (b) A, B and C only
(c) C and D only (d) B and C only

**02** Using the rules for significant figures, the correct answer for the expression $\frac{0.02858 \times 0.112}{0.5702}$ will be *[2022, 29 June Shift-II]*

(a) 0.005613 (b) 0.00561
(c) 0.0056 (d) 0.006

### Level 2

**03** The incorrect postulates of the Dalton's atomic theory are *[2024, 5 April Shift-I]*

(A) Atoms of different elements differ in mass.
(B) Matter consists of divisible atoms.
(C) Compounds are formed when atoms of different element combine in a fixed ratio.
(D) All the atoms of given element have different properties including mass.
(E) Chemical reactions involve reorganisation of atoms.

Choose the correct answer from the options given below

(a) (A), (B), (C) only (b) (C), (D), (E) only
(c) (B), (D) only (d) (B), (D), (E) only

**04** Choose the incorrect statement about Dalton's atomic theory. *[2024, 4 April Shift-II]*

(a) All the atoms of a given element have identical properties including identical mass.
(b) Matter consists of indivisible atoms.
(c) Chemical reactions involve reorganisation of atoms
(d) Compounds are formed when atoms of different elements combine in any ratio.

### Level 3

**05** Choose the correct statements. *[2025, 29 Jan Shift-I]*

A. Weight of a substance is the amount of matter present in it.
B. Mass is the force exerted by gravity on an object.
C. Volume is the amount of space occupied by a substance.
D. Temperatures below 0°C are possible in Celsius scale, but in Kelvin scale negative temperature is not possible.
E. Precision refers to the closeness of various measurements for the same quantity.

Choose the correct answer from the options given below

(a) C, D and E only (b) B, C and D only
(c) A, B and C only (d) A, D and E only

**06** 0.27 g of a long chain fatty acid was dissolved in 100 $cm^3$ of hexane. 10 mL of this solution was added dropwise to the surface of water in a round watch glass. Hexane evaporates and a monolayer is formed. The distance from edge to centre of the watch glass is 10 cm. What is the height of the monolayer? [Density of fatty acid $= 0.9 \text{ g cm}^{-3}$; $\pi = 3$] *[2019, 8 April Shift-II]*

(a) $10^{-6}$ m (b) $10^{-4}$ m
(c) $10^{-8}$ m (d) $10^{-2}$ m

## Numerical Value Type Questions

**07** The number of significant figures in $50000.020 \times 10^{-3}$ is ............ . [2021, 26 Feb Shift-I]

**08** The number of significant figures in 0.00340 is ...... . [2021, 25 July Shift-II]

## Topic 2 Atomic & Molecular Masses, Percentage Composition, Mole Concept, Empirical & Molecular Formula

### Level 1

**01** Mass of magnesium required to produce 220 mL of hydrogen gas at STP on reaction with excess of dil. HCl is

(Given, molar mass of Mg is 24 g $mol^{-1}$)

[2025, 3 April Shift-II]

(a) 236 mg (b) 2.444 g
(c) 0.24 mg (d) 235.7 g

**02** $2.8 \times 10^{-3}$ mol of $CO_2$ is left after removing $10^{21}$ molecules from its '*x*' mg sample. The mass of $CO_2$ taken initially is

(Given, $N_A = 6.02 \times 10^{23}$ $mol^{-1}$) [2025, 23 Jan Shift-I]

(a) 150.4 mg (b) 196.2 mg
(c) 48.2 mg (d) 98.3 mg

**03** A solution is prepared by adding 2 g of "*X*" to 1 mole of water. Mass per cent of "*X*" in the solution is [2022, 11 April Shift-II]

(a) 20% (b) 10% (c) 5% (d) 2%

**04** The average molar mass of chlorine is 35.5 g $mol^{-1}$. The ratio of $^{35}Cl$ to $^{37}Cl$ in naturally occuring chlorine is close to [2020, 6 Sep Shift-II]

(a) 4 : 1 (b) 3 : 1 (c) 2 : 1 (d) 1 : 1

**05** The most abundant elements by mass in the body of a healthy human adult are Oxygen (61.4%); Carbon (22.9%), Hydrogen (10.0 %); and Nitrogen (2.6%). The weight which a 75 kg person would gain if all $^1H$ atoms are replaced by $^2H$ atoms is [JEE Main 2017]

(a) 15 kg (b) 37.5 kg
(c) 7.5 kg (d) 10 kg

**06** The ratio of masses of oxygen and nitrogen in a particular gaseous mixture is 1 : 4. The ratio of number of their molecules is [JEE Mains 2014]

(a) 3:16 (b) 1:4
(c) 7:32 (d) 1:8

**Imp.** Empirical Formula of Hydrocarbon

**07** A gaseous hydrocarbon gives upon combustion, 0.72 g of water and 3.08 g of $CO_2$. The empirical formula of the hydrocarbon is [JEE Main 2013]

(a) $C_2H_4$ (b) $C_3H_4$
(c) $C_6H_5$ (d) $C_7H_8$

### Level 2

**08** On complete combustion 1.0 g of an organic compound (*X*) gave 1.46 g of $CO_2$ and 0.567 g of $H_2O$. The empirical formula mass of compound (*X*) is .......... g.

(Given, molar mass in g $mol^{-1}$ C : 12, H : 1, O : 16)

[2025, 2 April Shift-I]

(a) 15 (b) 45 (c) 60 (d) 30

**09** Among $10^{-9}$ g (each) of the following elements, which one will have the highest number of atoms? (Element: Pb, Po, Pr and Pt) [2025, 3 April Shift-I]

(a) Pr (b) Po (c) Pt (d) Pb

**10** The elemental composition of a compound is 54.2% C, 9.2% H and 36.6% O. [2025, 24 Jan Shift-II]

If the molar mass of the compound is 132 g $mol^{-1}$, the molecular formula of the compound is

[Given, the relative atomic mass of C : H : O = 2 : 1 : 16]

(a) $C_6H_{12}O_6$ (b) $C_4H_8O_2$
(c) $C_4H_9O_3$ (d) $C_6H_{12}O_3$

**11** A metal chloride contains 55.0% of chlorine by weight 100 mL vapours of the metal chloride at STP weigh 0.57 g. The molecular formula of the metal chloride is (Given, atomic mass of chlorine is 35.5 u)

[2023, 12 April Shift-I]

(a) $MCl_4$ (b) $MCl_2$
(c) $MCl$ (d) $MCl_3$

**12** The number of molecules and moles in 2.8375 litres of $O_2$ at STP are respectively [2023, 10 April Shift-I]

(a) $7.527 \times 10^{23}$ and 0.125 mol
(b) $1.505 \times 10^{23}$ and 0.250 mol
(c) $7.527 \times 10^{22}$ and 0.250 mol
(d) $7.527 \times 10^{22}$ and 0.125 mol

**13** 120 g of an organic compound that contains only carbon and hydrogen, gives 330 g of $CO_2$ and 270 g of water on complete combustion. The percentage of carbon and hydrogen, respectively are

[2022, 24 June Shift-II]

(a) 25 and 75 (b) 40 and 60
(c) 60 and 40 (d) 75 and 25

**14** 5 moles of $AB_2$ weigh $125 \times 10^{-3}$ kg and 10 moles of $A_2B_2$ weigh $300 \times 10^{-3}$ kg. The molar mass of $A (M_A)$ and molar mass of $B (M_B)$ in kg $mol^{-1}$ are

[2019, 12 April Shift-I]

(a) $M_A = 10 \times 10^{-3}$ and $M_B = 5 \times 10^{-3}$
(b) $M_A = 50 \times 10^{-3}$ and $M_B = 25 \times 10^{-3}$
(c) $M_A = 25 \times 10^{-3}$ and $M_B = 50 \times 10^{-3}$
(d) $M_A = 5 \times 10^{-3}$ and $M_B = 10 \times 10^{-3}$

**15** A 10 mg effervescent tablet containing sodium bicarbonate and oxalic acid releases 0.25 mL of $CO_2$ at $T = 298.15$ K and $p = 1$ bar. If molar volume of $CO_2$ is 25.0 L under such condition, what is the percentage of sodium bicarbonate in each tablet? [Molar mass of $NaHCO_3 = 84$ g $mol^{-1}$]

*[2019, 11 Jan Shift-I]*

(a) 8.4 (b) 0.84 (c) 16.8 (d) 33.6

**16** At 300 K and 1 atm, 15 mL of a gaseous hydrocarbon requires 375 mL air containing 20% $O_2$ by volume for complete combustion. After combustion, the gases occupy 330 mL. Assuming that the water formed is in liquid form and the volumes were measured at the same temperature and pressure, the formula of the hydrocarbon is

*[JEE Main 2016]*

(a) $C_3H_8$ (b) $C_4H_8$ (c) $C_4H_{10}$ (d) $C_3H_6$

## Level 3

**17** 0.1 M solution of KI reacts with excess of $H_2SO_4$ and $KIO_3$ solutions. According to equation

$$5I^- + IO_3^- + 6H^+ \longrightarrow 3I_2 + 3H_2O$$

Identify the correct statements.

(A) 200 mL of KI solution reacts with 0.004 mol of $KIO_3$.
(B) 200 mL of KI solution reacts with 0.006 mol of $H_2SO_4$.
(C) 0.5 L of KI solution produced 0.005 mol of $I_2$.
(D) Equivalent weight of $KIO_3$ is equal to $\left(\frac{\text{Molecular weight}}{5}\right)$

*[2025, 29 Jan Shift-II]*

Choose the correct answer from the options given below

(a) (A) and (D) only (b) (C) and (D) only
(c) (B) and (C) only (d) (A) and (B) only

**18** A sample of $CaCO_3$ and $MgCO_3$ weighed 2.21 g is ignited to constant weight of 1.152 g. The composition of mixture is
(Given, molar mass in g $mol^{-1}$ $CaCO_3$ : 100, $MgCO_3$ : 84)

*[2024, 31 Jan Shift-II]*

(a) 1.187 g $CaCO_3$ + 1.187 g $MgCO_3$
(b) 1.023 g $CaCO_3$ + 1.187 g $MgCO_3$
(c) 1.187 g $CaCO_3$ + 1.023 g $MgCO_3$
(d) 1.023 g $CaCO_3$ + 1.023 g $MgCO_3$

**19** Match List I with List II.

| List I | List II |
|---|---|
| A. 16 g of $CH_4(g)$ | I. Weighs 28 g |
| B. 1 g of $H_2(g)$ | II. $60.2 \times 10^{23}$ electrons |
| C. 1 mole of $N_2(g)$ | III. Weighs 32 g |
| D. 0.5 mol of $SO_2(g)$ | IV. Occupies 11.4 L volume at STP |

Choose the correct answer from the options given below

*[2023, 10 April Shift-II]*

(a) A-I, B-III, C-II, D-IV
(b) A-II, B-IV, C-I, D-III
(c) A-II. B-III, C-IV, D-I
(d) A-II, B-IV, C-III, D-I

**20** Compound *A* contains 8.7% hydrogen, 74% carbon and 17.3% nitrogen. The molecular formula of the compound is (Given, atomic masses of C, H and N are 12, 1 and 14 amu respectively).
The molar mass of the compound *A* is 162 g mol⁻.

*[2022, 28 June Shift-II]*

(a) $C_4H_6N_2$ (b) $C_2H_3N$
(c) $C_5H_7N$ (d) $C_{10}H_{14}N_2$

**21** Complete combustion of 1.80 g of an oxygen containing compound $(C_xH_yO_z)$ gave 2.64 g of $CO_2$ and 1.08 g of $H_2O$. The percentage of oxygen in the organic compound is *[2021, 25 Feb Shift-I]*

(a) 50.33 (b) 53.33
(c) 63.53 (d) 51.63

**22** The ratio of mass per cent of C and H of an organic compound $(C_xH_yO_z)$ is 6 : 1. If one molecule of the above compound $(C_xH_yO_z)$ contains half as much oxygen as required to burn one molecule of compound $C_xH_y$ completely to $CO_2$ and $H_2O$. The empirical formula of compound $C_xH_yO_z$ is *[JEE Main 2018]*

(a) $C_3H_6O_3$ (b) $C_2H_4O$
(c) $C_3H_4O_2$ (d) $C_2H_4O_3$

**23** An organic compound contains C, H and S. The minimum molecular weight of the compound containing 8% sulphur is (atomic weight of S = 32 amu) *[JEE Mains 2016]*

(a) 600 g $mol^{-1}$ (b) 200 g $mol^{-1}$
(c) 400 g $mol^{-1}$ (d) 300 g $mol^{-1}$

**24** The ratio of masses of oxygen and nitrogen in a particular gaseous mixture is 1:4. The ratio of number of their molecule is *[JEE Main 2014]*

(a) 1 : 4 (b) 7 : 32
(c) 1 : 8 (d) 3 : 16

## Numerical Value Type Questions

**25** *X* g of nitrobenzene on nitration gave 4.2 g of *m*-dinitrobenzene. *X* = ......... g. (nearest integer)
[Given, molar mass (in g $mol^{-1}$) C : 12, H : 1, O : 16, N : 14] *[2025, 3 April Shift-II]*

**26** Fortification of food with iron is done using $FeSO_4 \cdot 7H_2O$. The mass in grams of the $FeSO_4 \cdot 7H_2O$ required to achieve 12 ppm of iron in 150 kg of wheat is ......... (Nearest integer)
(Given, Molar mass of Fe, S and and O respectively are 56, 32 and 16 g $mol^{-1}$.) *[2025, 4 April Shift-I]*

**27** 0.1 mol of the following given antiviral compound (*P*) will weigh ......... $\times 10^{-1}$ g (Nearest integer)

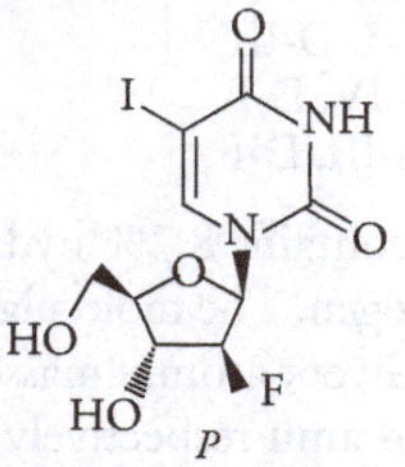

*P* **[2025, 2 April Shift-I]**

(Given, molar mass in g $mol^{-1}$ H : 1, C : 12, N : 14, O : 16, F : 19, I : 127)

**28** 0.01 mole of an organic compound (*X*) containing 10% hydrogen, on complete combustion produced 0.9 g $H_2O$. Molar mass of (*X*) is ....g $mol^{-1}$. **[2025, 23 Jan Shift-II]**

**29** Some $CO_2$ gas was kept in a sealed container at a pressure of 1 atm and at 273 K. This entire amount of $CO_2$ gas was later passed through an aqueous solution of $Ca(OH)_2$. The excess unreacted $Ca(OH)_2$ was later neutralised with 0.1 M of 40 mL HCl. If the volume of the sealed container of $CO_2$ was *x*, then *x* is ......... $cm^3$ (nearest integer).

[Given, the entire amount of $CO_2(g)$ reacted with exactly half the initial amount of $Ca(OH)_2$ present in the aqueous solution.]

$A \rightarrow B$ **[2025, 22 Jan Shift-I]**

**30** Quantitative analysis of an organic compound (*X*) shows following % composition. **[2025, 28 Jan Shift-I]**

C : 14.5% Cl : 64.46% H : 1.8%

(Empirical formula mass of the compound (*X*) is ...... $\times 10^{-1}$

(Given, molar mass in g $mol^{-1}$ of C : 12, H : 1, O : 16, Cl : 35.5)

**31** In the Claisen-Schmidt reaction to prepare, dibenzalacetone from 5.3g of benzaldehyde, a total of 3.51 g of product was obtained. The percentage yield in this reaction was........%. **[2025, 29 Jan Shift-II]**

**32** A solution is prepared by adding 1 mole ethyl alcohol in 9 mole water. The mass percent of solute in the solution is ....... (integer answer) (Given : Molar mass in g $mol^{-1}$, ethyl alcohol : 46, water : 18) **[2024, 8 April Shift-II]**

**Imp.** Reaction of Aniline in Blend with Percentage Yield

**33** 9.3 g of pure aniline is treated with bromine water at room temperature to give a white precipitate of the product '*P*'. The mass of product '*P*' obtained is 26.4 g. The percentage yield is ......... % **[2024, 5 April Shift-I]**

**34** Molar mass of the salt from NaBr, $NaNO_3$, KI and $CaF_2$ which does not evolve coloured vapours on heating with concentrated $H_2SO_4$ is........ g $mol^{-1}$. (Molar mass in g $mol^{-1}$ : Na : 23, N : 14, K : 39, O : 16, Br : 80, I : 127, F : 19, Ca : 40) **[2024, 31 Jan Shift-I]**

**35** Number of moles of methane required to produce 22g $CO_2(g)$ after combustion is $x \times 10^{-2}$ moles. The value of *x* is ............ **[2024, 31 Jan Shift-I]**

**36** 0.05 cm thick coating of silver is deposited on a plate of 0.05 $m^2$ area. The number of silver atoms deposited on plate are ......... $\times 10^{23}$. (Atomic mass of Ag = 108, $d$ = 7.9 g $cm^{-3}$) **[2024, 30 Jan Shift-I]**

**37** 1 mole of PbS is oxidised by "*X*" moles of $O_3$ to get "*Y*" moles of $O_2$. $X + Y$ = ........ **[2024, 27 Jan Shift-II]**

**38** A solution of sugar is obtained by mixing 200 g of its 25% solution and 500 g of its 40% solution (both by mass). The mass percentage of the resulting sugar solution is ........... (Nearest integer) **[2023, 11 April Shift-I]**

**39** Molar mass of the hydrocarbon (*X*) which on ozonolysis consumes one mole of $O_3$ per mole of (*X*) and gives one mole each of ethanal and propanone is .......... g $mol^{-1}$. (Molar mass of C : 12 g $mol^{-1}$, H : 1 g $mol^{-1}$) **[2023, 8 April Shift-I]**

**40** A solution of phenol in chloroform when treated with aqueous NaOH gives compound *P* as a major product. The mass percentage of carbon in *P* is ......... . (to the nearest integer)

(Atomic mass: C = 12; H = 1; O = 16) **[2022, 6 Sep Shift-II]**

**41** In the estimation of bromine, 0.5 g of an organic compound gave 0.40 g of silver bromide. the percentage of bromine in the given compound is .........% (Nearest integer)

(Relative atomic masses of Ag and Br are 108 u and 80 u, respectively). **[2022, 28 June Shift-I]**

**42** A protein '*A*' contains 0.30% of glycine (molecular weight 75). The minimum molar mass of the protein '*A*' is ......... $\times 10^3$ g $mol^{-1}$. [Nearest integer] **[2022, 25 June Shift-II]**

**43** The mass percentage of nitrogen in histamine is ......... **[2022, 9 Jan Shift-I]**

**44** Complete combustion of 750 g of an organic compound provides 420 g of $CO_2$ and 210 g of $H_2O$. The percentage composition of carbon and hydrogen in organic compound is 15.3 and ...... respectively (Round off to the nearest integer). **[2021, 16 March Shift-I]**

45 ........ grams of 3-hydroxypropanal (MW = 74) must be dehydrated to produce 7.8 g of acrolein (MW = 56) ($C_3H_4O$) if the percentage yield is 64. (Rounded off to the Nearest Integer). *[2021, 18 March Shift-I]*

46 A 20.0 mL solution containing 0.2 g impure $H_2O_2$ reacts completely with 0.316 g of $KMnO_4$ in acid solution. the purity of $H_2O_2$ (in %) is ........ (mol. weight of $H_2O_2$ = 34 ; mol weight of $KMnO_4$ = 158) *[2020, 04 Sep Shift-I]*

## Topic 3 Chemical Equations, Stoichiometry and Concentration Terms

### Level 1

01 Consider the reaction,

$$4HNO_3(l) + 3KCl(s) \longrightarrow Cl_2(g) + NOCl(g) + 2H_2O(g) + 3KNO_3(s)$$

The amount of $HNO_3$ required to produce 110.0 g of $KNO_3$ is

(Given, atomic masses of H, O, N and K are 1, 16, 14 and 39 respectively.) *[2022, 29 July Shift-II]*

(a) 32.2 g (b) 69.4 g (c) 91.5 g (d) 162.5 g

02 $SO_2Cl_2$ on reaction with excess of water results into acidic mixture $SO_2Cl_2 + 2H_2O \longrightarrow H_2SO_4 + 2HCl$

16 moles of NaOH is required for the complete neutralisation of the resultant acidic mixture. The number of moles of $SO_2Cl_2$ used is *[2022, 25 July Shift-I]*

(a) 16 (b) 8 (c) 4 (d) 2

03 The hardness of a water sample (in terms of equivalents of $CaCO_3$) containing $10^{-3}$ M $CaSO_4$ is (Molar mass of $CaSO_4$ = 136 g $mol^{-1}$) *[2019, 12 Jan Shift-I]*

(a) 100 ppm (b) 10 ppm (c) 50 ppm (d) 90 ppm

04 1 g of a carbonate ($M_2CO_3$) on treatment with excess HCl produces 0.01186 mole of $CO_2$. The molar mass of $M_2CO_3$ in g $mol^{-1}$ is *[JEE Main 2017]*

(a) 1186 (b) 84.3 (c) 118.6 (d) 11.86

05 The molarity of a solution obtained by mixing 750 mL of 0.5 M HCl with 250 mL of 2M HCl will be *[JEE Main 2013]*

(a) 0.875 M (b) 1.00 M (c) 1.75 M (d) 0.0975 M

06 The density of a solution prepared by dissolving 120 g of urea (mol. mass = 60 u) in 1000 g of water is 1.15 g/mL. The molarity of this solution is *[AIEEE 2012]*

(a) 0.50 M (b) 1.78 M (c) 1.02 M (d) 2.05 M

07 How many moles of magnesium phosphate, $Mg_3(PO_4)_2$ will contain 0.25 mole of oxygen atoms? *[AIEEE 2006]*

(a) 0.02 (b) $3.125 \times 10^{-2}$
(c) $1.25 \times 10^{-2}$ (d) $2.5 \times 10^{-2}$

### Level 2

08 On combustion 0.210 g of an organic compound containing C, H and O gave 0.127 g $H_2O$ and 0.307 g $CO_2$. The percentages of hydrogen and oxygen in the given organic compound respectively are *[2025, 8 April Shift-II]*

(a) 53.41, 39.6 (b) 6.72, 53.41
(c) 7.55, 43.85 (d) 6.72, 39.87

09 When a concentrated solution of sulphanilic acid and 1-naphthylamine is treated with nitrous acid (273 K) and acidified with acetic acid, the mass (g) of 0.1 mole of product formed is

(Given, molar mass in g $mol^{-1}$ H : 1, C : 12, N : 14, O : 16, S : 32) *[2025, 2 April Shift-II]*

(a) 330 (b) 33 (c) 343 (d) 66

10 $CaCO_3(s) + 2HCl(aq) \longrightarrow CaCl_2(aq) + CO_2(g) + H_2O(l)$

Consider the above reaction, what mass of $CaCl_2$ will be formed if 250 mL of 0.76 M HCl reacts with 1000 g of $CaCO_3$?

(Given, molar mass of Ca, C, O, H and Cl are 40, 12, 16, 1 and 35.5 g mol-1, respectively) *[2025, 2 April Shift-I]*

(a) 10.545 g (b) 5.272 g
(c) 3.908 g (d) 2.636 g

11 Density of 3 M NaCl solution is 1.25 g/mL. The molality of the solution is *[2025, 22 Jan Shift-II]*

(a) 2.79 m (b) 1.79 m
(c) 3 m (d) 2 m

12 Given below are two statements: *[2025, 22 Jan Shift-I]*

**Statement I** One mole of propyne reacts with excess of sodium to liberate half a mole of $H_2$ gas.

**Statement II** Four grams of propyne reacts with $NaNH_2$ to liberate $NH_3$ gas which occupies 224 mL at STP.

In the light of the above statements, choose the most appropriate answer from the options given below:

(a) Both Statement I and Statement II are correct.
(b) Both Statement I and Statement II are incorrect.
(c) Statement I is correct but Statement II is incorrect.
(d) Statement I is incorrect but Statement II is correct.

13 Combustion of glucose ($C_6H_{12}O_6$) produces $CO_2$ and water. The amount of oxygen (in g) required for the complete combustion of 900 g of glucose is [Molar mass of glucose in g $mol^{-1}$ = 180] *[2024, 8 April Shift-I]*

(a) 32 (b) 480 (c) 800 (d) 960

14 Molality (*m*) of 3 M aqueous solution of NaCl is (Given : Density of solution = 1.25 g $mL^{-1}$, Molar mass in g $mol^{-1}$: Na-23, Cl-35.5) *[2024, 6 April Shift-II]*

(a) 2.90 m (b) 2.79 m (c) 1.90 m (d) 3.85 m

**15** The density of '$x$' M solution ('$x$' molar) of NaOH is $1.12 \text{ g mL}^{-1}$, while in molality, the concentration of the solution is 3 m(3 molal). Then $x$ is
(Given : Molar mass of NaOH is 40 g/mol) *[2024, 6 April Shift-I]*

(a) 3.5 (b) 3.0 (c) 3.8 (d) 2.8

**16** The number of moles of methane required to produce 11 g $CO_2(g)$ after complete combustion is
(Given, molar mass of methane in g $mol^{-1}$ : 16) *[2024, 5 April Shift-II]*

(a) 0.75 (b) 0.5
(c) 0.35 (d) 0.25

**17** Given below are two statements: one is labelled as Assertion (A) and the other is labelled as Reason (R).

**Assertion (A)** 3.1500 g of hydrated oxalic acid dissolved in water to make 250.0 mL solution will result in 0.1 M oxalic acid solution.

**Reason (R)** Molar mass of hydrated oxalic acid is $126 \text{ g mol}^{-1}$.

In the light of the above statements, choose the correct answer from the options given below. *[2023, 10 April Shift-II]*

(a) A is true but R is false
(b) Both A and R are true and R is the correct explanation of A
(c) A is false but R is true
(d) Both A and R are true but R is not the correct explanation of A

**18** If a rocket runs on a fuel ($C_{15}H_{30}$) and liquid oxygen, the weight of oxygen required and $CO_2$ released for every litre of fuel respectively are (Given, density of the fuel is 0.756 g/mL) *[2022, 24 June Shift-I]*

(a) 1188 g and 1296 g (b) 2376 g and 2592 g
(c) 2592g and 2376 g (d) 3429 g and 3142 g

**19** Haemoglobin contains 0.34% of iron by mass. The number of Fe-atoms in 3.3 g of haemoglobin is
(Given, atomic mass of Fe is 56 u,
$N_A = 6.022 \times 10^{23} \text{ mol}^{-1}$) *[2022, 26 July Shift-II]*

(a) $1.21 \times 10^5$ (b) $12.0 \times 10^{16}$
(c) $1.21 \times 10^{20}$ (d) $3.4 \times 10^{22}$

**20** Production of iron in blast furnace follows the following equation,

$$Fe_3O_4(s) + 4CO(g) \longrightarrow 3Fe(l) + 4CO_2(g)$$

when 4.640 kg of $Fe_3O_4$ and 2.520 kg of CO are allowed to react then the amount of iron (in g) produced is
[Given, molar atomic mass (g $mol^{-1}$) Fe = 56 molar atomic mass (g $mol^{-1}$) O = 16 and molar atomic mass (g $mol^{-1}$) C = 12.] *[2022, 29 June Shift-I]*

(a) 1400 (b) 2200
(c) 3360 (d) 4200

**21** The strengths of 5.6 volume hydrogen peroxide (of density 1 g/mL) in terms of mass percentage and molarity ($M$), respectively, are
(Take molar mass of hydrogen peroxide as 34 g/mol) *[2020, 3 Sep Shift-II]*

(a) 0.85 and 0.25 (b) 0.85 and 0.5
(c) 1.7 and 0.5 (d) 1.7 and 0.25

**22** An unknown chlorohydrocarbon has 3.55% of chlorine. If each molecule of the hydrocarbon has one chlorine atom only, chlorine atoms present in 1 g of chlorohydrocarbon are
(Atomic weight of Cl = 35.5 u ; Avogadro constant $= 6.023 \times 10^{23} \text{ mol}^{-1}$) *[JEE Mains 2018]*

(a) $6.023 \times 10^{21}$
(b) $6.023 \times 10^{23}$
(c) $6.023 \times 10^{20}$
(d) $6.023 \times 10^{9}$

**23** 3 g of activated charcoal was added to 50 mL of acetic acid solution (0.06 N) in a flask. After an hour it was filtered and the strength of the filtrate was found to be 0.042 N. The amount of acetic acid adsorbed (per gram of charcoal) is *[JEE Main 2015]*

(a) 18 mg (b) 36 mg
(c) 42 mg (d) 54 mg

## Level 3

**24** What amount of bromine will be required to convert 2 g of phenol into 2, 4, 6-tribromophenol?
(Given, molar mass in g $mol^{-1}$ of C, H, O, Br are 12, 1, 16, 80 respectively.) *[2025, 23 Jan Shift-I]*

(a) 6.0 g (b) 10.22 g
(c) 4.0 g (d) 20.44 g

**25** 1 L, 0.02 M solution of $[Co(NH_3)_5SO_4]Br$ is mixed with 1 L, 0.02 M solution of $[Co(NH_3)_5Br]SO_4$.

The resulting solution is divided into two equal parts ($X$) and treated with excess of $AgNO_3$ solution and $BaCl_2$ solution respectively as shown below.

1 L solution ($X$) + $AgNO_3$ solution (excess) $\longrightarrow Y$
1 L solution ($X$) + $BaCl_2$ solution (excess) $\longrightarrow Z$

The number of moles of $Y$ and $Z$ respectively are *[2023, 30 Jan Shift-II]*

(a) 0.02, 0.01
(b) 0.01, 0.02
(c) 0.02, 0.02
(d) 0.01, 0.01

**26** 250 g solution of D-glucose in water contains 10.8% of carbon by weight. The molality of the solution is nearest to (Given: Atomic weights are, H, 1u; C, 12u; O, 16u) *[2022, 27 July Shift-I]*

(a) 1.03 (b) 2.06
(c) 3.09 (d) 5.40

**Imp.** **Limiting Reagent**

**27** $N_2(g) + 3H_2(g) \longrightarrow 2NH_3(g)$
(20 g) (5 g)

Consider the above reaction, the limiting reagent of the reaction and number of moles of $NH_3$ formed respectively are **[2022, 29 July shift-I]**

(a) $H_2$, 1.42 mol
(b) $H_2$, 0.71 mol
(c) $N_2$, 1.42 mol
(d) $N_2$, 0.71 mol

**28** The exact volume of 1 M NaOH solution required to neutralise 50 mL of 1 M $H_3PO_3$ solution and 100 mL of 2 M $H_3PO_2$ solution respectively, are **[2021, 16 March Shift-II]**

(a) 100 mL and 100 mL
(b) 100 mL and 200 mL
(c) 100 mL and 50 mL
(d) 50 mL and 50 mL

**29** A solution of two components containing $n_1$ moles of the 1st component and $n_2$ moles of the 2nd component is prepared. $M_1$ and $M_2$ are the molecular weights of component 1 and 2 respectively. If $d$ is the density of the solution in g $mL^{-1}$, $C_2$ is the molarity and $x_2$ is the mole fraction of the 2nd component, then $C_2$ can be expressed as **[2020, 06 Sep Shift-I]**

(a) $C_2 = \dfrac{1000\, x_2}{M_1 + x_2(M_2 - M_1)}$

(b) $C_2 = \dfrac{d\, x_2}{M_2 + x_2(M_2 - M_1)}$

(c) $C_2 = \dfrac{1000 d\, x_2}{M_1 + x_2(M_2 - M_1)}$

(d) $C_2 = \dfrac{d\, x_1}{M_2 + x_2(M_2 - M_1)}$

## Numerical Value Type Questions

**30** Butane reacts with oxygen to produce carbon dioxide and water following the equation given below.

$$C_4H_{10}(g) + \frac{13}{2}O_2(g) \rightarrow 4CO_2(g) + 5H_2O(l)$$

If 174.0 kg of butane is mixed with 320.0 kg of $O_2$ the volume of water formed in liters is .............. . (Nearest integer)

[Given, (a) Molar mass of C, H, O are 12, 1, 16 g $mol^{-1}$ respectively, (b) Density of water = 1 g $mL^{-1}$] **[2025, 7 April Shift-II]**

**31**

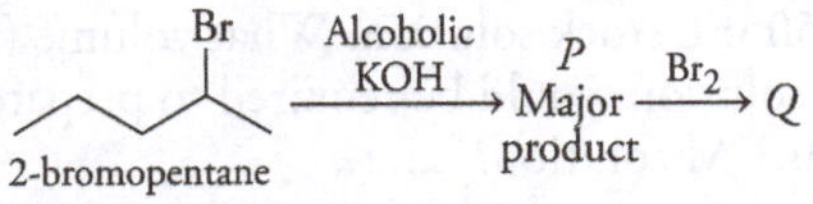


Consider the above sequence of reactions. 151 g of 2-bromopentane is made to react.

Yield of major product $P$ is 80% whereas $Q$ is 100%. Mass of product $Q$ obtained is ........ g.

(Given, molar mass in g $mol^{-1}$ H : 1, C : 12, O : 16, Br : 80) **[2025, 2 April Shift-II]**

**32** An organic compound weighing 500 mg produced 220 mg of $CO_2$, on complete combustion. The percentage composition of carbon in the compound is ........%. (Nearest integer)

(Given, molar mass in g $mol^{-1}$ of C : 12, O : 16) **[2025, 7 April Shift-I]**

**33** 0.5 g of an organic compound on combustion gave 1.46 g of $CO_2$ and 0.9 g of $H_2O$. The percentage of carbon in the compound is ...... (Nearest integer)

[Given, molar mass (in g $mol^{-1}$) C : 12, H : 1, O :16] **[2025, 3 April Shift-I]**

**34** The amount of calcium oxide produced on heating 150 kg limestone (75% pure) is ........... kg. (Nearest integer)

[Given, molar mass (in g $mol^{-1}$) of Ca-40, O-16, C-12] **[2025, 4 April Shift-II]**

**35** 20 mL of 2 M NaOH solution is added to 400 mL of 0.5 M NaOH solution. The final concentration of the solution is ......... $\times 10^{-2}$ M. (Nearest integer) **[2025, 22 Jan Shift-II]**

**36** $X$g of benzoic acid on reaction with $aq$ $NaHCO_3$ released $CO_2$ that occupied 11.2L volume at STP. $X$ is .......g. **[2025, 24 Jan Shift-I]**

**37** When 81.0 g of aluminium is allowed to react with 128.0 g of oxygen gas, the mass of aluminium oxide produced in grams is......(Nearest integer)

[Given; Molar mass of Al is 27.0 g $mol^{-1}$;
Molar mass of O is 16.0 g $mol^{-1}$] **[2025, 23 Jan Shift-II]**

**38** Consider the following reaction occurring in the blast furnace **[2025, 24 Jan Shift-I]**

$$Fe_3O_4(s) + 4CO(g) \rightarrow 3Fe(l) + 4CO_2(g)$$

'$x$' kg of iron is produced when $2.32 \times 10^3$ kg $Fe_3O_4$ and $2.8 \times 10^2$ kg CO are brought together in the furnace. The value of '$x$' is .........(Nearest integer)

[Given,
molar mass of $Fe_3O_4$ = 232 g $mol^{-1}$
molar mass of CO = 28 g $mol^{-1}$
molar mass of Fe = 56 g $mol^{-1}$]

**39**

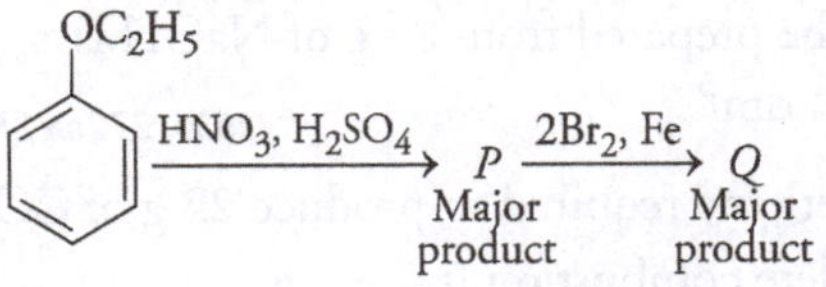


**[2024, 6 April Shift-II]**

The ratio of number of oxygen atoms to bromine atoms in the product $Q$ is .......... $\times 10^{-1}$.

**Imp.** Diazotisation Reaction in Blend with Stoichiometry

**40** 9.3 g of pure aniline upon diazotisation followed by coupling with phenol gives an orange dye. The mass of orange dye produced (assume 100% yield conversion) is ......... g. (Nearest integer) *[2024, 6 April Shift-I]*

**41** In the Claisen-Schmidt reaction to prepare 351 g of dibenzalacetone using 87 g of acetone, the amount of benzaldehyde required is ........... g. (Nearest integer) *[2024, 5 April Shift-II]*

**42** *X* g of ethanamine was subjected to reaction with $NaNO_2/HCl$ followed by hydrolysis to liberate $N_2$ and HCl. The HCl generated was completely neutralised by 0.2 moles of NaOH. *X* is ......... g. *[2024, 5 April Shift-II]*

**43** Only 2 mL of $KMnO_4$ solution of unknown molarity is required to reach the end point of a titration of 20 mL of oxalic acid (2 M) in acidic medium. The molarity of $KMnO_4$ solution should be...... M. *[2024, 4 April Shift-I]*

**44** *X* g of ethylamine is subjected to reaction with $NaNO_2$ / HCl followed by water; evolved dinitrogen gas which occupied 2.24 L volume at STP. *X* is...... $\times 10^{-1}$ g. *[2024, 4 April Shift-I]*

**45** Consider the following reaction,

$$3PbCl_2 + 2(NH_4)_3PO_4 \longrightarrow Pb_3(PO_4)_2 + 6NH_4Cl$$

If 72 m mol of $PbCl_2$ is mixed with 50 m mol of $(NH_4)_3PO_4$, then the amount of $Pb_3(PO_4)_2$ formed is ......... mmol (Nearest integer). *[2024, 1 Feb Shift-I]*

**46** Number of moles of $H^+$ ions required by 1 mole of $MnO_4^-$ to oxidise oxalate ion to $CO_2$ is ......... . *[2024, 31 Jan Shift-II]*

**47** If 50 mL of 0.5 M oxalic acid is required to neutralise 25 mL of NaOH solution, the amount of NaOH in 50 mL of given NaOH solution is ......... g. *[2024, 29 Jan Shift-II]*

**48** Volume of 3 M NaOH (formula weight 40 g $mol^{-1}$) which can be prepared from 84 g of NaOH is ....... $\times 10^{-1}$ $dm^3$. *[2024, 27 Jan Shift-II]*

**49** Mass of methane required to produce 22 g of $CO_2$ after complete combustion is ........ g.

(Given, molar mass in g $mol^{-1}$ C = 12.0, H = 1.0, O = 16.0) *[2024, 27 Jan Shift-I]*

**50** The volume of hydrogen liberated at STP by treating 2.4 g of magnesium with excess of hydrochloric acid is ........... $\times 10^{-2}$L.

Given, molar volume of gas is 22.4L at STP.

Molar mass of magnesium is 24 g $mol^{-1}$. *[2023, 11 April Shift-II]*

**51** If 5 moles of $BaCl_2$, is mixed with 2 moles of $Na_3PO_4$, the maximum number of moles of $Ba_3(PO_4)_2$ formed is ......... . (Nearest integer) *[2023, 6 April Shift-I]*

**52** The molality of a 10% (*V* / *V*) solution of dibromine solution in $CCl_4$ (carbon tetrachloride) is '*x*', *x* = .......... $\times 10^{-2}$ m. (Nearest integer) *[2023, 1 Feb Shift-II]*

[Given, molar mass of $Br_2$ = 160 g $mol^{-1}$
atomic mass of C = 12 g $mol^{-1}$
atomic mass of Cl = 35.5 g $mol^{-1}$
density of dibromine = 3.2 g $cm^{-3}$
density of $CCl_4$ = 1.6 g $cm^{-3}$]

**53** The density of 3 M solution of NaCl is 1.0 g $mL^{-1}$. Molality of the solution is .......... $\times 10^{-2}$ m. (Nearest integer).
[Given, molar mass of Na and Cl is 23 and 35.5 g $mol^{-1}$ respectively.] *[2023, 1 Feb Shift-I]*

**54** Zinc reacts with hydrochloric acid to give hydrogen and zinc chloride. The volume of hydrogen gas produced at STP from the reaction of 11.5 g of zinc with excess HCl is ....... L. (Nearest integer)

(Given, molar mass of Zn is 65.4 g $mol^{-1}$ and molar volume of $H_2$ at STP = 22.7 L) *[2023, 31 Jan Shift-I]*

**55** Assume carbon burns according to following equation.

$$2C(s) + O_2(g) \longrightarrow 2CO(g)$$

When 12 g carbon is burnt in 48 g of oxygen, the volume of carbon monoxide produced is ....... $\times 10^{-1}$ L at STP. [Nearest integer]
[Given, assume CO as ideal gas, mass of C is 12 g $mol^{-1}$, mass of O is 16 g $mol^{-1}$ and molar volume of an ideal gas at STP is 22.7 L $mol^{-1}$] *[2023, 31 Jan Shift-II]*

**56** 5 g of NaOH was dissolved in deionised water to prepare a 450 mL stock solution. What volume (in mL) of this solution would be required to prepare 500 mL of 0.1 M solution? ........

[Given, molar mass of Na, O and H is 23, 16 and 1 g $mol^{-1}$ respectively.] *[2023, 24 Jan Shift-I]*

**57** A 1.84 mg sample of polyhydric alcoholic compound '*x*' of molar mass 92.0 g/mol gave 1.344 mL of $H_2$ gas at STP. The number of alcoholic hydrogens present in compound '*x*' is ......... *[2022, 29 July Shift-II]*

**58** 2 L of 0.2 M $H_2SO_4$ is reacted with 2 L of 0.1 M NaOH solution, the molarity of the resulting product $Na_2SO_4$ in the solution is ........... millimolar. (Nearest integer) *[2022, 28 July Shift-II]*

**59** In the given reaction,

$$X + Y + 3Z \rightleftharpoons XYZ_3$$

if one mole of each of $X$ and $Y$ with 0.05 mole of $Z$ gives compound $XYZ_3$. (Given : Atomic masses of $X$, $Y$ and $Z$ are 10, 20 and 30 amu, respectively.) the yield of $XYZ_3$ is ...... g. (Nearest integer) *[2022, 28 July Shift-I]*

**60** Chlorophyll extracted from the crushed green leaves was dissolved in water to make 2 L solution of Mg of concentration 48 ppm. The number of atoms of Mg in this solution is $x \times 10^{20}$ atoms. The value of $x$ is .......... (Nearest integer)
(Given, atomic mass of Mg is 24 g $mol^{-1}$; $N_A = 6.02 \times 10^{23}$ $mol^{-1}$) *[2022, 26 July Shift-I]*

**61** When 800 mL of 0.5 M nitric acid is heated in a beaker, its volume is reduced to half and 11.5 g of nitric acid is evaporated. The molarity of the remaining nitric acid solution is $x \times 10^{-2}$ M. (Nearest integer) (Molar mass of nitric acid is 63 g $mol^{-1}$) *[2022, 26 July Shift-I]*

**62** The normality of $H_2SO_4$ in the solution obtained on mixing 100 mL of 0.1 M $H_2SO_4$ with 50 mL of 0.1 M NaOH is ........ $\times 10^{-1}$ N. (nearest integer) *[2022, 27 July Shift-I]*

**63** 56.0 L of nitrogen gas is mixed with excess of hydrogen gas and it is found that 20 L of ammonia gas is produced. The volume of unused nitrogen gas is found to be ........ L. *[2022, 25 July Shift-II]*

**64** Number of grams of bromine that will completely react with 5.0 g of pent-1-ene is ........ $\times 10^{-2}$ g. (Atomic mass of Br = 80 g/mol.) (Nearest integer) *[2022, 25 June Shift-I]*

**65** The moles of methane required to produce 81 g of water after complete combustion is ...... $\times 10^{-2}$ mol. [Nearest integer] *[2022, 26 June Shift-II]*

**66** 0.01 M $KMnO_4$ solution was added to 20.0 mL of 0.05 M Mohr's salt solution through a burette. The initial reading of 50 mL burette is zero. The volume of $KMnO_4$ solution left in the burette after the end point is.......... mL. (Nearest integer) *[2022, 28 June Shift-II]*

**67** CNG is an important transportation fuel. When 100 g CNG is mixed with 208g oxygen in vehicles, it leads to the formation of $CO_2$ and $H_2O$ and produced large quantity of heat during this combustion, then the amount of carbon dioxide, produced in grams is ...... [Nearest integer] [Assume CNG to be methane] *[2022, 26 June Shift-II]*

**68** The neutralisation occurs when 10 mL of 0.1 M acid '$A$' is allowed to react with 30 mL of 0.05 M base $M(OH)_2$. The basicity of the acid '$A$' is ........... [$M$ is a metal] *[2022, 25 June Shift-II]*

**69** 100 mL of $Na_3PO_4$ solution contains 3.45 g of sodium. The molarity of the solution is ........ $\times 10^{-2}$ mol $L^{-1}$. (Nearest integer)
[Atomic masses - Na : 23.0 u, O : 16.0 u, P : 31.0 u] *[2021, 26 Aug Shift-II]*

**70** 10.0 mL of 0.05 M $KMnO_4$ solution was consumed in a titration with 10.0 mL of given oxalic acid dihydrate solution. The strength of given oxalic acid solution is ..... $\times 10^{-2}$ g/L.

(Round off to the nearest integer) *[2021, 27 July Shift-II]*

**71** When 10 mL of an aqueous solution of $Fe^{2+}$ ions was titrated in the presence of dil. $H_2SO_4$ using diphenylamine indicator, 15 mL of 0.02 M solution of $K_2Cr_2O_7$ was required to get the end point. The molarity of the solution containing $Fe^{2+}$ ions is $x \times 10^{-2}$ M. The value of $x$ is ...... (Nearest integer) *[2021, 25 July Shift-I]*

**72** If the concentration of glucose ($C_6H_{12}O_6$) in blood is 0.72 $gL^{-1}$, the molarity of glucose in blood is ..... $\times 10^{-3}$ M. (Nearest integer)
[Given : Atomic mass of C = 12, H = 1, O = 16 u] *[2021, 22 July Shift-II]*

**73** 4g equimolar mixture of NaOH and $Na_2CO_3$ contains $x$ g of NaOH and $y$ g of $Na_2CO_3$. The value of $x$ is ..... g. (Nearest integer) *[2021, 20 July Shift-II]*

**74** 250 mL of 0.5 M NaOH was added to 500 mL of 1 M HCl. The number of unreacted HCl molecules in the solution after complete reaction is ...... $\times 10^{21}$. (Nearest integer) ($N_A = 6.022 \times 10^{23}$) *[2021, 20 July Shift-I]*

**75** The density of NaOH solution is 1.2 g $cm^{-3}$. The molality of this solution is ........ m. (Round off to the Nearest Integer)

[Use : Atomic masses : Na : 23.0 u, O : 16.0 u, H : 1.0 u density of $H_2O$ : 1.0 g $cm^{-3}$] *[2021, 27 July Shift-I]*

**76** 15 mL of aqueous solution of $Fe^{2+}$ in acidic medium completely reacted with 20 mL of 0.03 M aqueous $Cr_2O_7^{2-}$. The molarity of the $Fe^{2+}$ solution is ...... $\times 10^{-2}$ M (Round off to the nearest integer). *[2021, 17 March Shift-I]*

**77** The mole fraction of a solute in a 100 molal aqueous solution ...... $\times 10^{-2}$

(Round off to the nearest integer).

[Given, atomic masses H : 1.0 u, O : 16.0 u] *[2021, 17 March Shift-I]*

**78** 10.0 mL of $Na_2CO_3$ solution is titrated against 0.2 M HCl solution. The following titre values were obtained in 5 readings. 4.8 mL, 4.9 mL, 5.0 mL, 5.0 mL and 5.0 mL based on these readings and convention of titrimetric estimation of concentration of $Na_2CO_3$ solution is ........ mM (Round off to the nearest integer). *[2021, 18 March Shift-II]*

**79** The number of chlorine atoms in 20 mL of chlorine gas at STP is ........ $\times 10^{21}$. (Round off to the Nearest integer) *[2021, 17 March Shift-II]*

[Assume chlorine is an ideal gas at STP]

$R = 0.083$ L bar $mol^{-1}K^{-1}$, $N_A = 6.023 \times 10^{23}$

**80** When 35 mL of 0.15 M lead nitrate solution is mixed with 20 mL of 0.12 M chromic sulphate solution ........ $\times 10^{-5}$ moles of lead sulphate precipitate out. *[2021, 16 March Shift-II]*

(Round off to the Nearest Integer).

**81** A reaction of 0.1 mole of Benzylamine with bromomethane gave 23 g of Benzyl trimethyl ammonium bromide. The number of moles of bromomethane consumed in this reaction are $n \times 10^{-1}$, when $n =$ ......... .

(Rounded off to the Nearest Integer).

[Given : Atomic masses are C : 12.0 u, H : 1.0 u, N : 14.0 u, Br : 80.0 u] *[2021, 18 March Shift-I]*

**82** The $NaNO_3$ weighed out to make 50 mL of an aqueous solution containing 70.0 mg $Na^+$ per mL is ........ g. (Rounded off to the nearest integer)

[Given : Atomic weight in g $mol^{-1}$, Na : 23 ; N : 14 ; O : 16]. *[2021, 26 Feb Shift-II]*

**83** The formula of a gaseous hydrocarbon, which requires 6 times of its own volume of $O_2$ for complete oxidation and produces 4 times its own volume of $CO_2$ is $C_xH_y$. The value of $y$ is ......... . *[2021, 24 Feb Shift II]*

**84** 1.86 g of aniline completely reacts to form acetanilide. 10% of the product is lost during purification. Amount of acetanilide obtained after purification (in g) is ........$\times 10^{-2}$. *[2021, 24 Feb Shift-II]*

**85** 4.5 g of compound $A$ (MW = 90) was used to make 250 mL of its aqueous solution. the molarity of the solution is $x \times 10^{-1}$. The value of $x$ is ........

(Rounded off to the nearest integer) *[2021, 24 Feb Shift-I]*

**86** 0.4 g mixture of NaOH, $Na_2CO_3$ and some inert impurities was first titrated with $\frac{N}{10}$ HCl using phenolphthalein as an indicator, 17.5 mL of HCl was required at the end point. After this, methyl orange was added and titrated. 1.5 mL of same HCl was required for the next end point. The weight percentage of $Na_2CO_3$ in the mixture is ......... . (Rounded-off to the nearest integer) *[2021, 25 Feb Shift-I]*

**87** $6.023 \times 10^{22}$ molecules are present in 10 g of a substance '$x$'. The molarity of a solution containing 5 g of substance '$x$' in 2 L solution is ..... $\times 10^{-3}$. *[2020, 3 Sep Shift-II]*

**88** A 100 mL solution was made by adding 1.43 g of $Na_2CO_3 \cdot xH_2O$. The normality of the solution is 0.1 N. The value of $x$ is ......... . *[2020, 4 Sep Shift-II]*

**89** The ratio of the mass percentages of 'C' and H' and 'C and O' of a saturated acyclic organic compound '$X$' are 4 : 1 and 3 : 4 respectively. Then, the moles of oxygen gas required for complete combustion of two moles of organic compound '$X$' is ......... . *[2020, 2 Sep Shift-II]*

**90** The volume (in mL) of 0.1 N NaOH required to neutralise 10 mL of 0.1 N phosphonic acid is .......... . *[2020, 3 Sep Shift-II]*

**91** The mass of ammonia in grams produced when 2.8 kg of dinitrogen quantitatively reacts with 1 kg of dihydrogen is ......... . *[2020, 4 Sep Shift-I]*

**92** The minimum number of moles of $O_2$ required for complete combustion of 1 mole of propane and 2 moles of butane is ...... . *[2020, 5 Sep Shift-I]*

**93** The volume, in mL, of 0.02 M $K_2Cr_2O_7$ solution required to react with 0.288 g of ferrous oxalate in acidic medium is ......... .

(Molar mass of Fe = 56 g $mol^{-1}$) *[2020, 5 Sep Shift-II]*

**94** Ferrous sulphate heptahydrate is used to fortify foods with iron. The amount (in grams) of the salt required to achieve 10 ppm of iron in 100 kg of wheat is ......... .

Atomic weight : Fe = 55.85; S = 32.00; O = 16.00 *[2020, 8 Jan Shift-I]*

**95** The molarity of $HNO_3$ in a sample which has density 1.4 g/mL and mass percentage of 63% is ......... (Molecular weight of $HNO_3 = 63$)
*[2020, 9 Jan Shift-I]*

**96** The volume (in mL) of 0.125 M $AgNO_3$ required to quantitatively precipitate chloride ions in 0.3 g of $[Co(NH_3)_6]Cl_3$ is .........

$$M_{[Co(NH_3)_6]Cl_3} = 267.46 \text{ g/mol}$$
$$M_{AgNO_3} = 169.87 \text{ g/mol}$$
*[2020, 8 Jan Shift-I]*

**97** $NaClO_3$ is used, even in spacecrafts, to produce $O_2$. The daily consumption of pure $O_2$ by a person is 492 L at 1 atm 300 K. How much amount of $NaClO_3$, in grams, is required to produce $O_2$ for the daily consumption of a person at 1 atm, 300 K ......... ?

$$NaClO_3(s) + Fe(s) \longrightarrow O_2(g) + NaCl(s) + FeO(s)$$
$R = 0.082$ L atm mol$^{-1}$ K$^{-1}$
*[2020, 8 Jan Shift-II]*

## Key Index

### Topic 1

| | | | | | | | |
|---|---|---|---|---|---|---|---|
| 1 (a) | 2 (b) | 3 (c) | 4 (d) | 5 (a) | 6 (a) | 7 (8) | 8 (3) |

### Topic 2

| | | | | | | | | | |
|---|---|---|---|---|---|---|---|---|---|
| 1 (a) | 2 (b) | 3 (b) | 4 (b) | 5 (c) | 6 (c) | 7 (d) | 8 (d) | 9 (a) | 10 (d) |
| 11 (b) | 12 (d) | 13 (d) | 14 (d) | 15 (a) | 16 (a) | 17 (a) | 18 (c) | 19 (b) | 20 (d) |
| 21 (b) | 22 (d) | 23 (c) | 24 (b) | 25 (3) | 26 (9) | 27 (372) | 28 (100) | 29 (45) | 30 (1655) |
| 31 (60) | 32 (22) | 33 (80) | 34 (78) | 35 (50) | 36 (11) | 37 (8) | 38 (36) | 39 (70) | 40 (69) |
| 41 (34) | 42 (25) | 43 (37.84) | 44 (3) | 45 (16) | 46 (85) | | | | |

### Topic 3

| | | | | | | | | | |
|---|---|---|---|---|---|---|---|---|---|
| 1 (c) | 2 (c) | 3 (a) | 4 (b) | 5 (a) | 6 (d) | 7 (b) | 8 (b) | 9 (b) | 10 (a) |
| 11 (a) | 12 (c) | 13 (d) | 14 (b) | 15 (b) | 16 (d) | 17 (b) | 18 (c) | 19 (c) | 20 (c) |
| 21 (c) | 22 (c) | 23 (a) | 24 (b) | 25 (d) | 26 (b) | 27 (c) | 28 (b) | 29 (c) | 30 (138) |
| 31 (184) | 32 (12) | 33 (80) | 34 (63) | 35 (57) | 36 (61) | 37 (153) | 38 (420) | 39 (15) | 40 (20) |
| 41 (318) | 42 (9) | 43 (8) | 44 (45) | 45 (24) | 46 (8) | 47 (4) | 48 (7) | 49 (8) | 50 (224) |
| 51 (1) | 52 (139) | 53 (364) | 54 (4) | 55 (227) | 56 (180) | 57 (3) | 58 (25) | 59 (2) | 60 (24) |
| 61 (54) | 62 (1) | 63 (46) | 64 (1143) | 65 (225) | 66 (30) | 67 (143) | 68 (3) | 69 (50) | 70 (1575) |
| 71 (18) | 72 (4) | 73 (1) | 74 (226) | 75 (5) | 76 (24) | 77 (64) | 78 (50) | 79 (1) | 80 (525) |
| 81 (3) | 82 (13) | 83 (8) | 84 (243) | 85 (2) | 86 (4) | 87 (25) | 88 (10) | 89 (5) | 90 (10) |
| 91 (3400) | 92 (18) | 93 (50) | 94 (5) | 95 (14) | 96 (27) | 97 (2130) | | | |

# Detailed **Explanations**

## Topic 1

**01.** *(a)* The number of significant figures of the given numbers are as follows:

| | |
|---|---|
| (A) 0.00253 | 3 |
| (B) 1.0003 | 5 |
| (C) 15.0 | 3 |
| (D) 163 | 3 |

Thus, (A), (C) and (D) have same number of significant figures.

**02.** *(b)* Answer should be reported with 3 significant figures.

$$\frac{0.02858 \times 0.112}{0.5702} = 0.00\underbrace{561}$$

↓ 3 significant figures

> • Concept **Enhancer**
>
> In multiplication and division, the result must be reported with number of significant figures equal to those present in the expression with least significant number involved in the calculation.

**03.** *(c)* Statement given in options B and D are incorrect, while in option A, C and E are correct. The correct form of incorrect statements are

Matter consists of indivisible atoms.

All the atoms of a given element have same properties including mass.

**04.** *(d)* Statement given in option (d) is incorrect. Its correct form is as follows according to Dalton's atomic theory, compounds are formed when atoms of different elements combine in fixed ratio.

**05.** *(a)* Among the given statements, statement given in C, D and E are correct while A and B are incorrect.

Their correct forms are

Weight of substance is the force applied by gravity on body.

Mass of a substance is the amount of matter present in it.

**06.** *(a)* In 10 mL solution, mass of the fatty acid,

$$m = \frac{0.27}{100} \times 10 = 0.027 \text{ g}$$

Density of fatty acid, $d = 0.9$ g cm$^{-3}$

$\therefore$ Volume of the fatty acid over the watch glass,

$$V = \frac{m}{d} = \frac{0.027}{0.9} = 0.03 \text{ cm}^3$$

$\because$ Volume of the cylinder = Volume of fatty acid

$$\Rightarrow \quad V = \pi r^2 \times h$$

$$\Rightarrow \quad h = \frac{V}{\pi r^2} = \frac{0.03 \text{ cm}^3}{3 \times (10)^2 \text{ cm}^2}$$

$$= 1 \times 10^{-4} \text{ cm}$$

$$= 1 \times 10^{-6} \text{ m}$$

**07.** *(8)* Zero's between 5 and 2 are all significant. Trailing zeroes after decimal are also significant.

$\underbrace{50000.020}\times 10^{-3}$

↑

(Number of significant figures = 8)

**08.** *(3)* Total number of significant figure = 3.

Here,

$\underbrace{0.00}\ \underbrace{34}\ \underbrace{0}$

↓ Leading zero ↓ Non-zero digits → Trailing zero

Non-zero digits and trailing zero are considered as significant figures.

## Topic 2

**01.** *(a)* The reaction involved is as follows

$$Mg + 2HCl \longrightarrow MgCl_2 + H_2$$

Volume of $H_2$ evolved = 220 mL

$$\text{Mole of } H_2 = \frac{220 \times 10^{-3}}{22.4} \times 24$$

$$= 235.7 \times 10^{-3} \text{ g} \cong 236 \text{ mg}$$

**02.** *(b)* Let the mass of $CO_2$ taken initially $= x$

No. of moles of $CO_2$ left

$$= 2.8 \times 10^{-3} \text{ mol}$$

No. of molecules removed

$$= 10^{21} \text{ molecules}$$

$$\text{Now, no. of initial moles} = \frac{x \times 10^{-3}}{44}$$

$$\text{No. of moles removed} = \frac{10^{21}}{6.02 \times 10^{23}}$$

No. of moles left

= initial moles − moles removed

$$2.8 \times 10^{-3} = \frac{x \times 10^{-3}}{44} - \frac{10^{21}}{6.02 \times 10^{23}}$$

$$x = 196.2 \text{ mg}$$

**03.** *(b)* Given,

Mass of solute $(X) = 2$g

Mass of solvent $(H_2O) = 1$ mole = 18 g

So, total mass of the solution

$= 18 + 2 = 20$ g

Mass % of $X$

$$= \frac{\text{Mass of solute}}{\text{Total mass of the solution}} \times 100$$

$$= \frac{2}{20} \times 100 = 10\%$$

**04.** *(b)* Suppose the mole ratio of $^{35}$Cl be $x$ and that of $^{37}$Cl be $y$.

$$\text{Average molar mass} = \frac{n_1 m_1 + n_2 m_2}{n_1 + n_2}$$

$$35.5 = \frac{35x + 37y}{x + y}$$

$$\Rightarrow \quad 1.5y = 0.5x \quad \Rightarrow \frac{x}{y} = \frac{3}{1}$$

**05.** *(c)* Mass due to $^1$H $= \dfrac{75 \times 10}{100} = 7.5$ kg

$^1$H atoms are replaced by $^2$H atoms,

Mass due to $^2$H $= (7.5 \times 2)$ kg $= 15$ kg

According to the question, all $^1$H are replaced by $^2$H.

So, the increase in weight

$= 15 - 7.5 = 7.5$

$\therefore$ Mass gain by person = 7.5 kg

**06.** *(c)* Suppose the mass of $O_2 = x$

Mass of $N_2 = 4x$

$$\text{Number of molecules of } O_2 = \frac{x}{32} \times N_A$$

Number of molecules of

$$N_2 = \frac{4x}{28} N_A = \frac{x}{7} N_A$$

Ratio of molecules of $O_2$ and $N_2$

$$= \frac{x}{32} : \frac{x}{7} \text{ or } 7:32$$

**07.** *(d)* 18 g $H_2O$ contain 2g H.

$\therefore$ 0.72 g $H_2O$ contain 0.08 g H

44 g $CO_2$ contain 12 g C.

$\therefore$ 3.08 g $CO_2$ contain 0.84 g C.

$$\therefore C : H = \frac{0.84}{12} : \frac{0.08}{1}$$

$$= 0.07 : 0.08 = 7 : 8$$

$\therefore$ Empirical formula = $C_7H_8$

**08.** *(d)* Number of moles of carbon

$(C) = n_{CO_2} = \frac{1.46}{44} = 0.033$

Moles of 'C' $= W_C = 0.033 \times 12$

Moles of hydrogen

$(H) = W_H = 2 \times n_{H_2O} = \frac{2 \times 0.567}{18} = 0.063$

Mass of hydrogen $= 0.0063$

Mass of oxygen $= 1 - (W_C + W_H)$

$= 1 - (0.033 \times 12 + 0.063 \times 1) = 0.541\,g$

Mole of 'O' $= \frac{0.541}{16} = 0.033$

Empirical formula $= CH_2O$

Empirical formula mass = 30g

**09.** *(a)* Number of atoms can be calculated by using the formula,

Number of atoms

$= \frac{\text{Mass in g}}{\text{Molar mass (g/mol)}} \times N_A$

So, for the same given mass, element with least molar mass will have highest number of atoms.

Now, molar mass of Pr = 141 g/mol, Po = 209 g/mol, Pt = 195 g/mol and Pb = 207 g/mol

Thus, Pr has highest number of atoms.

**10.** *(d)* C : H : O

$\frac{54.2}{12} : \frac{9.2}{1} : \frac{36.6}{16}$

$4.516 : 9.2 : 2.287$

$\frac{4.516}{2.287} : \frac{9.2}{2.287} : \frac{2.287}{2.287}$

$1.97 : 4.02 : 1 \approx 2 : 4 : 1$

$C_2H_4O \Rightarrow$ empirical formula

Empirical formula mass = 24 + 4 + 16 = 44 g/mol

and molar mass = 132 g/mol

Hence, molecular formula is

$(C_2H_4O)_3 = C_6H_{12}O_3$

**11.** *(b)* Weight of metal chloride at STP is

$= \frac{0.57}{100} \times 22400 = 127.68$

Weight or mass of Cl = 127.68 × 0.55 in metal chloride = 70.224

Now, number of moles of Cl

$= \frac{\text{Mass of Cl}}{\text{Molecular mass of Cl}} = \frac{70.224}{35.5} \approx 2$

Thus, the molecular formula is $MCl_2$.

**12.** *(d)* No. of moles $(n) = \frac{\text{Given volume}}{\text{Volume at STP}}$

$\therefore$ No. of moles of $O_2 = \frac{2.8375}{22.7} = 0.125$

Now, number of molecules = Moles × Avagadro's number

$= 0.125 N_A$

$= (0.125) \times (6.022 \times 10^{23})$

$= 7.527 \times 10^{22}$

**13.** *(d)* The general equation for complete combustion of organic compound is

$$C_xH_y + \left(x + \frac{y}{4}\right) O_2 \longrightarrow x CO_2 + \frac{y}{2} H_2O$$

Moles of $CO_2 = \frac{330}{44} = 7.5$ moles

Moles of water $= \frac{270}{18} = 15$ moles

Moles of $CO_2$ = moles of carbon

Mass of carbon = number of moles × atomic mass

$= 7.5 \times 12 = 90$ g

% of carbon in organic compound

$= \frac{90}{120} \times 100 = 75\%$

% of hydrogen in organic compound

$= \left(\frac{30}{120}\right) \times 100 = 25\%.$

**14.** *(d)*

> **key idea** To find the mass of *A* and *B* in the given question concept of mole is used.
>
> Number of moles $(n) = \frac{\text{Given mass } (w)}{\text{Molecular mass } (M)}$

| Compound | Mass of *A* (g) | Mass of *B* (g) |
|---|---|---|
| $AB_2$ | $M_A$ | $2M_B$ |
| $A_2B_2$ | $2M_A$ | $2M_B$ |

We know that,

Number of moles

$(n) = \frac{\text{Given mass } (w)}{\text{molecular mass } (M)}$ ...(A)

Using equation (A), it can be concluded that

$5(M_A + 2M_B) = 125 \times 10^{-3}$ kg ...(i)

$10(2M_A + 2M_B) = 300 \times 10^{-3}$ kg ...(ii)

Dividing Eq. (i) and (ii),

$\frac{1}{2} \frac{(M_A + 2M_B)}{(2M_A + 2M_B)} = \left(\frac{125}{300}\right)$

On solving the equation, we obtain

$M_A = 5 \times 10^{-3}$ and $M_B = 10 \times 10^{-3}$

**15.** *(a)* $\underset{2\text{ mol}}{2NaHCO_3} + \underset{1\text{ mol}}{H_2C_2O_4} \longrightarrow \underset{2\text{ mol}}{2CO_2} + Na_2C_2O_4 + 2H_2O$

$\Rightarrow$ In the reaction, number of mole of $CO_2$ produced.

$n = \frac{pV}{RT}$

$= \frac{1\text{ bar} \times 0.25 \times 10^{-3}\text{ L}}{0.082\text{ L atm K}^{-1}\text{mol}^{-1} \times 298.15\text{ K}}$

$= 1.02 \times 10^{-5}$ mol

Number of mole of $NaHCO_3$

$= \frac{\text{Weight of } NaHCO_3}{\text{Molecular mass of } NaHCO_3}$

$\therefore w_{NaHCO_3} = 1.02 \times 10^{-5} \times 84 \times 10^3$ mg

$= 0.856$ mg

$\Rightarrow \%NaHCO_3 = \frac{0.856}{10} \times 100$

$= 8.56\% \approx 8.4$

**16.** *(a)* $\underset{15\text{ mL}}{C_xH_y(g)} + \underset{75\text{ mL}}{\left(x + \frac{y}{4}\right) O_2(g)} \longrightarrow x CO_2(g) + \frac{y}{2} H_2O(l)$

1 mL $C_xH_y$ requires $= \left(\frac{4x + y}{4}\right)$ mL $O_2$

According to the question

$15\left(\frac{4x + y}{4}\right) = 75$

$4x + y = 20$

$\Rightarrow$ $x = 3$

$y = 8$

$\therefore$ The formula of hydrocarbon is $C_3H_8$.

**17.** *(a)* Correct statements regarding given equation is given in statements A and D. i.e., 200 mL of KI solution reacts with 0.004 mol of $KIO_3$. Equivalent weight of $KIO_3$ is equal to $\left(\frac{\text{molecular weight}}{5}\right)$.

**18.** *(c)* The reaction involved are

$CaCO_3(s) \xrightarrow{\Delta} CaO(s) + CO_2(g)$

$MgCO_3(s) \xrightarrow{\Delta} MgO(s) + CO_2(g)$

Let the weight of $CaCO_3$ be *x* g.

$\therefore$ Weight of $MgCO_3 = (2.21 - x)$ g

Moles of $CaCO_3$ decomposed = Moles of CaO formed

$\frac{x}{100}$ = Moles of CaO formed

$\therefore$ Weight of CaO formed $= \frac{x}{100} \times 56$

Moles of $MgCO_3$ decomposed = Moles of MgO formed

$\frac{2.21 - x}{84}$ = Mole of MgO formed

$\therefore$ Weight of MgO formed

$$= \frac{2.21 - x}{84} \times 40$$

$$\frac{2.21 - x}{84} \times 40 + \frac{x}{100} \times 56 = 1.152$$

$$\therefore \quad x = 1.1886 \text{ g} \approx 1.187 \text{ g}$$
$$= \text{Weight of } CaCO_3$$

$$\text{Weight of } MgCO_3 = 1.0214 \text{ g} \approx 1.023 \text{ g}$$

Hence, option (c) is correct.

**19.** *(b)* The correct match is

A-II, B-IV, C-I, D-III.

(A) 16 g of $CH_4(g) = 1$ mole $CH_4$ contains $10 \times 6.02 \times 10^{23}$ electrons $= 60.2 \times 10^{23}$

(B) 1 g $H_2(g) = 0.5$ mole $H_2$ gas occupy 1 mole contains = 22.7 L
$\therefore$ 0.5 mole contains 11.35 (approx) litre volume at STP.

(C) 1 mole of $N_2(g) = 28$ g

(D) Mole of $SO_2(g) = 64$ g
0.5 mole of $SO_2(g) = 32$ g

**20.** *(d)* Empirical formula can be found out as follows

| Elements | % composition | Atomic mass | Relative number of moles | Simple ratio of mole |
|---|---|---|---|---|
| Carbon (C) | 74 | 12 | $\frac{74}{12} = 6.17$ | $\frac{6.17}{1.24} \approx 5$ |
| Hydrogen (H) | 8.7 | 1 | $\frac{8.7}{1} = 8.7$ | $\frac{8.7}{1.24} \approx 7$ |
| Nitrogen (N) | 17.3 | 14 | $\frac{17.3}{14} = 1.24$ | $\frac{1.24}{1.24} = 1$ |

So, empirical formula is $C_5H_7N$.

$$\text{Empirical mass} = 5 \times 12 + 7 \times 1 + 14 \times 1 = 60 + 7 + 14 = 81$$

Molecular mass = 162 (given)

$$\text{Thus,} \quad n = \frac{\text{Molecular mass}}{\text{Empirical mass}}$$

$$n = \frac{162}{81} = 2$$

$\therefore$ Molecular formula = imperical formula $= (C_5H_7N)_2 = C_{10}H_{14}N_2$.

**21.** *(b)* $\underset{1.8\text{ g}}{C_xH_yO_z} + O_2 \longrightarrow \underset{2.64\text{ g}}{CO_2} + \underset{1.08\text{ g}}{H_2O}$

$$n_C = \underset{\text{(Moles)}}{n_{CO_2}} = \frac{2.64 \text{ (Given mass)}}{44 \text{ (Molecular mass)}} = 0.06$$

$$n_H = 2 \times n_{H_2O} = \frac{1.08}{18} \times 2 = 0.12$$

Weight of oxygen in $C_xH_yO_z$

$$= 1.80 - 12 \times 0.06 - 0.06 \times 2 = 1.80 - 0.72 - 0.12 = 0.96 \text{ g}$$

% of oxygen by weight

$$= \frac{0.96}{1.80} \times 100 = 53.33\%$$

**22.** *(d)* We can calculate the simplest whole number ratio of C and H from the data given, as

| Element | Relative mass | Molar mass | Relative mole | Simplest whole no. ratio |
|---|---|---|---|---|
| C | 6 | 12 | $\frac{6}{12} = 0.5$ | $\frac{0.5}{0.5} = 1$ |
| H | 1 | 1 | $\frac{1}{1} = 1$ | $\frac{1}{0.5} = 2$ |

We can calculate number of oxygen atoms from this by considering the equation.

$$C_xH_y + \left[x + \frac{y}{4}\right] O_2 \longrightarrow xCO_2 + \frac{y}{2} H_2O$$

Number of oxygen atoms required

$$= 2 \times \left[x + \frac{y}{4}\right] = \left[2x + \frac{y}{2}\right]$$

As 1 molecule of the above compound $(C_xH_yO_z)$ contains half number of oxygen atoms as required to burn 1 mole of compound $C_xH_y$.

$$\text{Now given, } z = \frac{1}{2}\left[2x + \frac{y}{2}\right] = \left[x + \frac{y}{4}\right]$$

So now putting the values of $x$ and $y$ in the above equation.

$$z = \left[x + \frac{y}{4}\right] = \left[1 + \frac{2}{4}\right] = 1.5$$

Now, put these values in the formula given i.e.

$$C_xH_yO_z = C_1H_2O_{1.5}$$

So, empirical formula will be

$$[C_1H_2O_{1.5}] \times 2 = C_2H_4O_3$$

**23.** *(c)* Percentage of sulphur

$$= \frac{\text{Atomic weight of sulphur}}{\text{Mol. weight of compound}} \times 100$$

$$8 = \frac{32}{\text{Mol. weight of compound}} \times 100$$

Therefore, molecular weight of compound = 400 g $\text{mol}^{-1}$

**24.** *(b)* The number of number of their molecules is given by

$$n = \frac{\text{weight } (w)}{\text{molecular weight } (M)}$$

Thus, ratio of number of their molecules of $O_2$ and $N_2$ is given by

$$\frac{N_{O_2}}{N_{N_2}} = \frac{n_{O_2}}{n_{N_2}} = \left(\frac{W_{O_2}/M_{O_2}}{W_{N_2}/M_{N_2}}\right)$$

$$= \left(\frac{1}{4}\right) \times \left(\frac{28}{32}\right) = \frac{7}{32}$$

Hence, ratio of $N_{O_2}$ and $N_{N_2}$ is 7 : 32.

**25.** *(3)* The complete reaction is as follows.

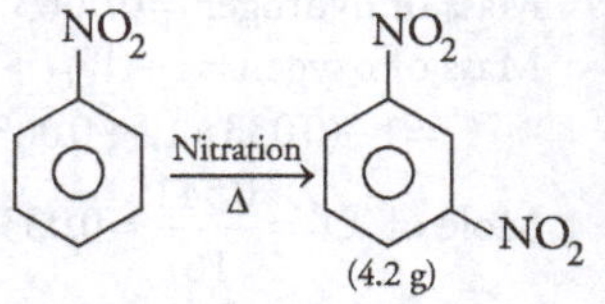

| | $C_6H_5NO_2$ , | $C_6H_4N_2O_4$ |
|---|---|---|
| Molecular weight | 123, | 168 |

$$\therefore \frac{4.2}{168} = 0.025 \text{ mol}$$

$\therefore$ Required gram of nitrobenzene

$$= 123 \times 0.025 = 3.075$$

$\therefore$ Nearest integer is 3.

**26.** *(9)* Let the mass of iron be $x$ g.

$$\frac{x}{150 \times 10^3} \times 10^6 = 12$$

$$\Rightarrow \quad x = 1.8 \text{ g}$$

Now, let mass of $FeSO_4 \cdot 7H_2O = y$ g

$$\text{So, moles of Fe} = \frac{1.8}{56} = \frac{y}{56 + 96 + (7 \times 18)}$$

$$\Rightarrow \quad y = 8.935 \text{ g} \approx 9$$

**27.** *(372)* Considering the given compound

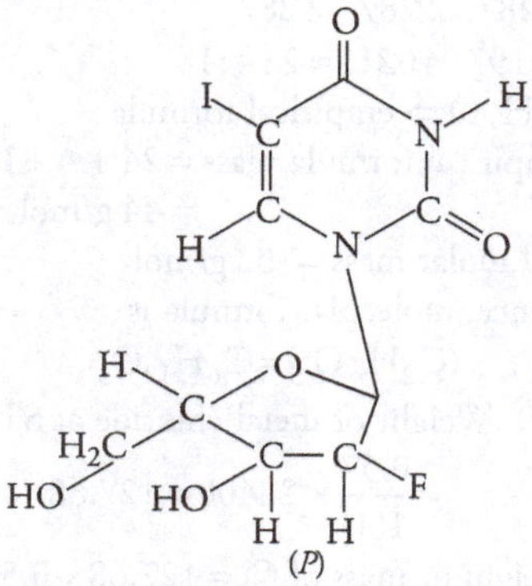

*(P)*

Molar mass = 372 g

1 mole = 372 g

0.1 mole has $= 372 \times 10^{-1}$ g

**28.** *(100)* Given, mass of $H_2O$ on combustion = 0.9 g

Number of moles of organic compound $(X) = 0.01$ mol

Molar mass of $X$ = ?

$$\underset{(X)}{\text{Organic compound}} \xrightarrow{\text{Combustion}} \underset{0.9\text{ g}}{H_2O}$$

$\therefore$ Mole of $H_2O = \frac{0.9}{18} = 0.05$ mole

$\therefore$ Mole of H in $H_2O = 0.05 \times 2 = 0.1$ mol
= mole of H in 0.01 mole organic compound

$\therefore$ Weight of H in 0.01 mole compound $= 0.1 \times 1 = 0.1$ g

$\therefore$ Weight of H in one mole compound

$= \frac{0.1}{0.01} = 10$ g

$\because$ Weight % of H

$$= \frac{\text{Weight of H in one mole compound}}{\text{Molar mass of compound}} \times 100$$

$$10 = \frac{10}{M} \times 100$$

$\therefore$ $M = 100$ g/mol

So, the molar mass of $x$ is 100 g/mol.

**29.** *(45)* Let the number of moles of $CO_2 = y$.
Total initial moles of $Ca(OH)_2 = 2y$
Excess $Ca(OH)_2 = y$
Gram equivalent of $Ca(OH)_2$
= Gram equivalent of HCl

$$y \times 2 = \frac{0.1 \times 40}{1000} \times 1$$

$$y = 2 \times 10^{-3}$$

Volume of $CO_2$

$= x = 2 \times 10^{-3} \times 22400$

$= 44.8\ cm^3$

$\approx 45\ cm^3$.

So, the value of $x$ is 45.

**30.** *(1655)*

**Key idea**

To solve this problem, first of all find the molar ratio of all the elements, then find the minimum integral ratio and put this value as a subscript with respective element to obtain empirical formula.

| | C : | Cl : | H : | O |
|---|---|---|---|---|
| % mass | 14.5 | 64.46 | 1.8 | 19.24 |

Molar ratio

$$= \frac{14.5}{12}, \frac{64.46}{35.5}, \frac{1.8}{1}, \frac{19.24}{16}$$

$= 1.2 \quad 1.8 \quad 1.8 \quad 1.2$

Minimum integral ratio 2 3 3 2

$\therefore$ Empirical formulae $= C_2H_3Cl_3O_2$

So, the empirical mass is $= 165.5$

Mass $= 1655 \times 10^{-1}$

**31.** *(60)* The complete reaction is as follows

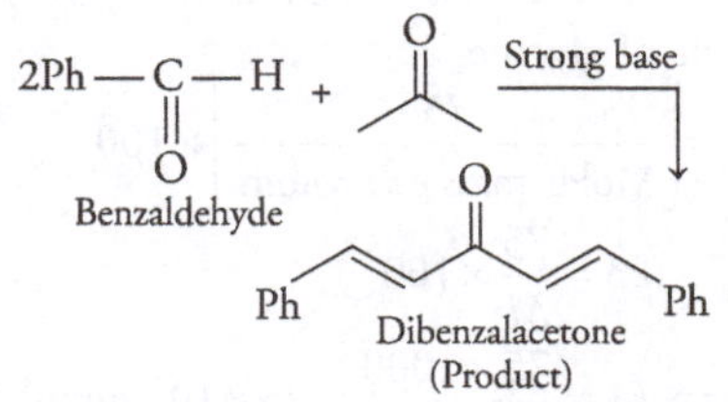


Now, 5.3 g of benzaldehyde
$= 3.51$ g of dibenzalacetone

$\therefore \frac{5.3}{106}$ mol of benzaldehyde

$= \frac{3.51}{234}$ mol of dibenzalacetone

or, $\frac{1}{20}$ mol (Actual) of benzaldehyde

$= 0.015$ mol of dibenzalacetone

Theoretical $= \frac{1}{40}$ mol

[$\because$ 2 mole sof benzaldehyde give 1 mol]

$$\% \text{ yield} = \frac{0.015}{1/40} \times 100 = 60\%$$

**32.** *(22)* Mass percent of alcohol

$$= \frac{\text{Mass of ethyl alcohol}}{\text{Total mass of solution}} \times 100$$

$$= \frac{1 \times 46}{1 \times 46 + (9 \times 18)} \times 100$$

$$= \frac{4600}{208} = 22.11$$

Nearest integer = 22

**33.** *(80)* The reaction that take place is given by

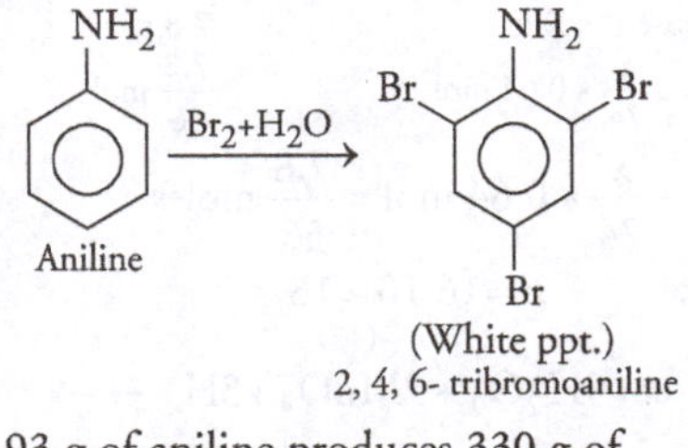


93 g of aniline produces 330 g of 2, 4, 6-tribromoaniline
Hence, 9.3 g of aniline will give 33 g of 2, 4, 6-tribromoaniline

$$\text{Percentage yield} = \frac{26.4 \times 100}{33} = 80\%$$

**34.** *(78)*

- $2NaBr + \underset{(\text{Conc.})}{2H_2SO_4} \xrightarrow{\Delta} \underset{(\text{Reddish brown vapour})}{Br_2} + SO_2 + 2H_2O + Na_2SO_4$

- $4NaNO_3 + \underset{(\text{Conc.})}{2H_2SO_4} \xrightarrow{\Delta} \underset{(\text{Reddish brown vapour})}{4NO_2} + O_2 + 2Na_2SO_4 + 2H_2O$

- $2KI + \underset{(\text{Conc.})}{H_2SO_4} \xrightarrow{\Delta} \underset{(\text{Violet vapour})}{I_2} + SO_2 + 2H_2O + K_2SO_4$

- $CaF_2 + \underset{(\text{Conc.})}{H_2SO_4} \xrightarrow{\Delta} CaSO_4 + \underset{(\text{Colourless})}{2HF}$

So, $CaF_2$ does not evolved coloured vapours. Hence its molar mass is 78 g/mol.

**35.** *(50)* $CH_4(g) + 2O_2(g) \longrightarrow CO_2(g) + 2H_2O(l)$

$$n = \frac{22}{44} = \frac{1}{2} \text{ mol of } CO_2$$

as 1mol $CH_4$ form 1 mol $CO_2$

So, $n_{CH_4} = \frac{1}{2}$ mol $= 0.5$ mol

$n_{CH_4} = 50 \times 10^{-2}$ mol

So, the value of $x$ is 50.

**36.** *(11)* Given, thickness $(d) = 0.05$ cm
Area of plate $(A) = 0.05\ m^2$
Volume of silver coating
$V = A \times d = 0.05 \times 0.05 \times 10000$
$= 25\ cm^3$

Mass of silver deposited $= 25 \times 7.9$ g

$$\text{Mole of Ag atoms} = \frac{25 \times 7.9}{108}$$

Number of silver atoms

$$= \frac{25 \times 7.9}{108} \times N_A$$

$$= \frac{25 \times 7.9}{108} \times 6.023 \times 10^{23}$$

$$= 11.01 \times 10^{23} \approx 11 \times 10^{23}$$

Hence, the number of Ag atoms are $11 \times 10^{23}$.

**37.** *(8)* The reaction involved is as follows,
$PbS + 4O_3 \longrightarrow PbSO_4 + 4O_2$
Here $X = 4$ and $Y = 4$
$\therefore X + Y = 4 + 4 = 8$
Hence, $X + Y$ is 8.

**38.** *(36)* Total mass of sugar in the given mixture is
$= 0.25 \times 200 + 0.40 \times 500$
$= 250$ g

Now, total mass of solution
$= 200 + 500 = 700$ g

So, mass % of sugar $= \frac{250}{700} \times 100$
$= 35.7\% \approx 36\%$.

**• Concept Enhancer**

Empirical formula of a chemical compound is the simplest whole number ratio of atoms present in a compound.

**39.** *(70)* Compound $(X)$ on ozonolysis consume one mole of $O_3$ and will give one mole of ethanal and one mole of propanone. Thus, $(X)$ can be 2-methylbut-2-ene

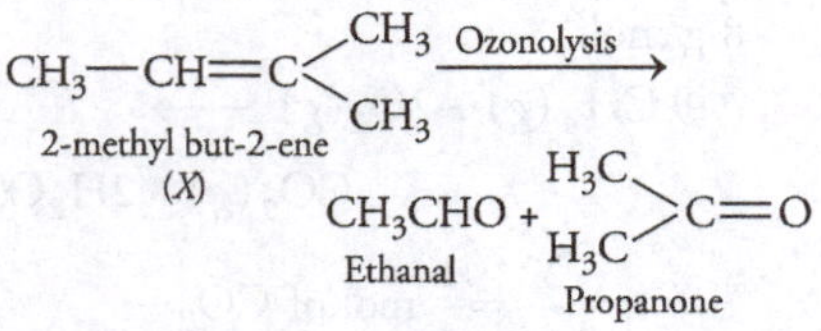


Molar mass of $X = 12 \times 5 + 1 \times 10$
$= 60 + 10 = 70$ g/mol

**40.** *(69)*

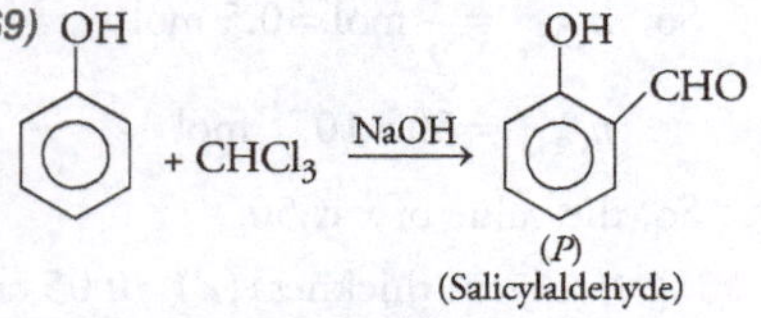


$\therefore$ Mass % of C in $P$ (Compound)

$= \frac{\text{Mass of C}}{\text{Mass of compound}} \times 100$

Here, mass of C $= 12 \times 7 = 84$ g

Mass of compound $(P)$
$= 12 \times 7 + 6 \times 1 + 16 \times 2 = 122$ g

$= \frac{84}{122} \times 100 = 68.85\% \approx 69\%$

**• Concept Enhancer**

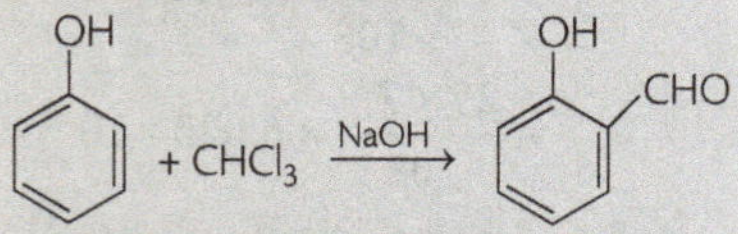


This reaction is known as Reimer Tieman reaction where phenol is converted to orthohydroxy benzaldehyde, also known as salicylaldehyde using chloroform and a strong base.

**41.** *(34)* Molecular mass of AgBr
$= 108 + 80 = 188$ g/mol

188 g of AgBr contains 80 g bromine

0.4 g of AgBr contains $\frac{80 \times 0.4}{188}$ g bromine

% of bromine $= \frac{80 \times 0.4 \times 100}{188 \times 0.5}$
$= 34.04\% \approx 34\%$

**42.** *(25)* Given, mass % of glycine = 0.30%

Molecular weight of glycine $= 75$ g mol$^{-1}$

% of glycine

$= \left[\frac{75}{\text{Molar mass of protein}}\right] \times 100$

$0.3 = \frac{75}{M} \times 100$

$\Rightarrow M = \frac{75 \times 1000}{3} = 25 \times 10^3$ g mol$^{-1}$

**43.** *(37.84)* The structure of histamine

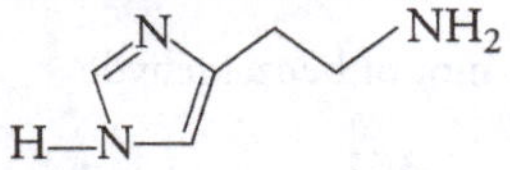


Molecular formula $= C_5H_9N_3$.

Molecular mass = 111

Number of nitrogen atoms per molecule = 3

Mass due to N $= 3 \times 14 = 42$

% of N $= \frac{42}{111} \times 100 = 37.84\%$

**44.** *(3)* Mass of water formed = 210 g

18 g of $H_2O$ contains = 2 g of hydrogen

210 g of $H_2O$ contains

$= \left(\frac{2}{18} \times 210 = \frac{70}{3}\right)$ g of hydrogen.

Given, mass of organic compound = 750 g

Percentage composition of hydrogen

$= \frac{70}{3 \times 750} \times 100 = 3.11\%$

Nearest integer = 3

**45.** *(16)* When 3-hydroxy propanal is heated,

$$(HO)CH_2CH_2CHO \xrightarrow[64\%]{\Delta} C_3H_4O + H_2O$$

Suppose $x$ g ... 7.8 g

$\frac{x}{74} \times 0.64$ mol ... $\frac{7.8}{56}$ mol

$\therefore \quad \frac{x}{74} \times 0.64 \text{ mol} = \frac{7.8}{56}$ moles

$\Rightarrow \quad x = 16.10 \approx 16$

**46.** *(85)* $5H_2O_2 + 2\overset{(+7)}{Mn}O_4^- + 3H^+ \longrightarrow 2Mn^{2+} + 5O_2 + 8H_2O$

Moles of $KMnO_4 = \frac{0.316}{158} = 2 \times 10^{-3}$

(Molar mass of $KMnO_4 = 158$)

Equivalents of $H_2O_2$ = Equivalents of $KMnO_4$

Equivalents of $KMnO_4$
$= 2 \times 10^{-3} \times 5 = 0.01$

Moles of $H_2O_2 = \frac{\text{Equivalents of } H_2O_2}{n\text{-factor of } H_2O_2}$

$= \frac{0.01}{2} = 0.005$

Mass of pure $H_2O_2 = 0.005 \times 34$
$= 0.170$ g

% purity $= \frac{0.17}{0.2} \times 100 = 85\%$

## Topic 3

**01.** *(c)* The given reaction is

$$4HNO_3(l) + 3KCl(s) \longrightarrow Cl_2(g) + NOCl(g) + 2H_2O(g) + 3KNO_3(s)$$

Number of moles of $KNO_3 = \frac{110}{101} = 1.089$

As 3 moles of $KNO_3$ is produced from 4 moles or $HNO_3$

$\therefore$ 1.089 moles of $KNO_3$ is produced from $= \frac{4}{3} \times 1.089 = 1.452$ mol

Mass of $HNO_3$ required is
$1.452 \times 63 = 91.5$ g

**02.** *(c)*

$$SO_2Cl_2 + 2H_2O \longrightarrow H_2SO_4 + 2HCl$$

(given)

Number of moles of NaOH required for complete neutralisation = 16

Now, as given chemical equation is balanced, so stoichiometric coefficients are equal to their moles respectively.

Number of moles of $H^+$ = 2 moles [from $H_2SO_4$] + 2 moles [HCl] = 4 moles

Now, 1 mole of $SO_2Cl_2$ gives 4 moles of $H^+$ and 16 moles of NaOH are used in neutralisation of these 4 moles.

So, moles of $SO_2Cl_2$ used

$= \frac{\text{Moles of NaOH used in neutralisation}}{\text{Moles of acid } [H^+] \text{ formed from 1 mole of } SO_2Cl_2}$

$= \frac{16}{4} = 4$

**03.** *(a)* Hardness of water sample can be calculated in terms of ppm concentration of $CaCO_3$.

Given, molarity $= 10^{-3}$ M

i.e. 1000 mL of solution contains $10^{-3}$ mole of $CaCO_3$.

$\therefore$ Hardness of water = ppm of $CaCO_3$

$= \frac{10^{-3} \times 100}{1000} \times 10^6 = 100$ ppm

**04.** *(b)* $\underset{1\text{ g}}{M_2CO_3} + \underset{\text{Excess}}{2HCl} \longrightarrow 2MCl + H_2O + \underset{0.01186 \text{ mole}}{CO_2}$

From balanced chemical equation

Number of moles of $M_2CO_3$ reacted = Number of moles of $CO_2$ evolved

Number of mole of $M_2CO_3$

$$= \frac{1}{M} = 0.01186$$

$[M = \text{molar mass of } M_2CO_3]$

$$M = \frac{1}{0.01186} = 84.3 \text{ g mol}^{-1}$$

**05.** *(a)* The molarity of a resulting solution is given by

$$M = \frac{M_1V_1 + M_2V_2}{V_1 + V_2}$$

$$= \frac{750 \times 0.5 + 250 \times 2}{750 + 250} = \frac{875}{1000} = 0.875 \text{ M}$$

**06.** *(d)* Total mass of solution = 1000 g water + 120 g urea = 1120 g

Density of solution = 1.15 g / mL

Thus, volume of solution $= \dfrac{\text{mass}}{\text{density}}$

$$= \frac{1120 \text{ g}}{1.15 \text{ g / mL}}$$

$= 973.91$ mL

$= 0.974$ L

Moles of solute $= \dfrac{120}{60} = 2$ mol

Molarity $= \dfrac{2 \text{ mol}}{0.974 \text{ L}} = 2.05 \text{ mol L}^{-1}$

**07.** *(b)* 8 moles of O-atom are contained by 1 mole $Mg_3(PO_4)_2$.

Hence, 0.25 moles of O-atom are contained by

$$= \frac{1}{8} \times 0.25 \text{ mol } Mg_3(PO_4)_2$$

$$= 3.125 \times 10^{-2}$$

**08.** *(b)* In the combustion of organic compound, all carbon in $CO_2$ and all hydrogen in $H_2O$ comes from organic compound.

The reaction is as follows,

$$\underset{0.210 \text{ g}}{C_XH_YO_Z} + O_2 \longrightarrow \underset{0.307\text{g}}{CO_2} + \underset{0.127 \text{ g}}{H_2O}$$

Weight of 'C' in $CO_2$ is

$$= \frac{12}{44} \times 0.307 = 0.0837 \text{ g}$$

Weight of 'H' in $H_2O$ is

$$= \frac{2}{18} \times 0.127 = 0.0141 \text{ g}$$

% of H in compound

$$= \frac{0.0141}{0.21} \times 100 = 6.72\%$$

Weight of 'O' in compound

$= 0.210 - (0.0837 + 0.0141) = 0.1122$

% of 'O' in compound

$$= \frac{0.1122}{0.21} \times 100 = 53.41\%$$

**09.** *(b)* The complete reaction is as follows,

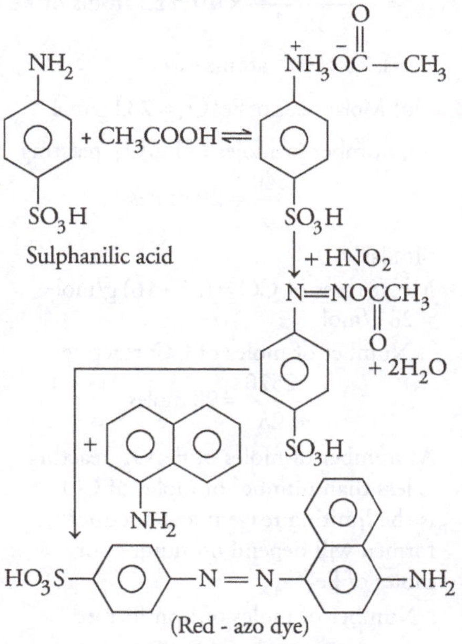


0.1 mole of red-azo dye (Molar mass = 327 g/mol) will have 32.7 g mass.

Nearly 33 g

**10.** *(a)* The reaction involved is as follows,

$$CaCO_3 + 2HCl \longrightarrow CaCl_2 + CO_2 + H_2O$$

Moles of $CaCO_3 = \dfrac{1000}{100} = 10$

Mole of HCl $= 0.76 \times \dfrac{250}{1000} = 0.19$

Moles of $CaCl_2$ formed $= \dfrac{0.19}{2}$

Mass of $CaCl_2 = \dfrac{0.19}{2} \times 111$

$= 10.545$ g

**11.** *(a)* Given, Molarity = 3M

Density = 1.25 g/mL

Molality = ?

Mass of solution = Density × Volume

Mass of solution = 1.25 × 1000 = 1250 g

Now, number of moles of NaCl

= Molarity × Volume

= 3 × 1 = 3 moles

Molar mass of NaCl = 58.5 g/mol

Mass of NaCl = 3 × 58.5 = 175.5 g

Now mass of solvent = 1250 − 175.5

= 1074.5 g

So, molality $= \dfrac{\text{Number of moles} \times 1000}{\text{Mass of solvent (in g)}}$

$$= \frac{3 \times 1000}{1074.5} = 2.79 \text{ m}$$

**Alternate Solution**

$$\text{Molality} = \frac{M \times 1000}{1000d - M \times Mw}$$

$$= \frac{3000}{1250 - 175.5} = 2.79 \text{ m}$$

**12.** (c) $H_3C—C \equiv CH + \underset{(4g)}{NaNH_2}$

$$\longrightarrow H_3CC \equiv C^-Na^+ + NH_3$$

Number of moles of propyne

$$= \frac{4}{40} = 0.1 \text{ mol}$$

Number of moles of $NH_3$ produced = 0.1

Volume occupied by 0.1 mol of $NH_3$ at STP = 2240 mL

Thus, Statement I is correct but Statement II is incorrect.

**13.** *(d)* The complete reaction is as follows,

$$C_6H_{12}O_6 + 6O_2 \longrightarrow 6CO_2 + 6H_2O$$

Mole of glucose $= \dfrac{900 \text{ g}}{180 \text{ g / mol}} = 5$ mol

1 mol glucose reacts with 6 mol of $O_2$, so 5 mol glucose will react with 30 mol $O_2$.

Mole of $O_2$ required = 30 mol

Mass of $O_2$ required = 32 × 30 g = 960 g

**14.** *(b)* Given,

density = 1.25 g $mL^{-1}$, $M = 3$ M

molar mass in g $mol^{-1}$ :

Na = 23, Cl = 35.5

Molality = ?

Molality (*m*)

$$= \frac{1000 \times M}{1000 \times d - (M \times W_{\text{Molecular weight of solute}})}$$

$$m = \frac{3 \times 1000}{1.25 \times 1000 - [3 \times 58.5]}$$

$m = 2.79$ m

**15.** *(b)* Given, molality = 3

Density of *x* M of NaOH

= 1.12 g $mL^{-1}$

Using the formula for molality

$$\text{Molality} = \frac{1000 \times M}{1000 \times d - (M \times W_{\text{solute}})}$$

(Molecular weight = *W*)

$$3 = \frac{1000 \times x}{1000 \times 1.12 - (x \times 40)} \quad \text{...(i)}$$

[40 = Molecular weight of NaOH]

By solving equation (i)

$x = 3$

**16.** *(d)* The reaction given in question is,

$$C_nH_{2n+2} + \frac{3n+1}{2}O_2 \longrightarrow nCO_2 + (n+1)H_2O$$

$$\underset{\substack{4\,g \\ 0.25\,mol}}{CH_4} + 2O_2 \longrightarrow \underset{\substack{11\,g \\ 0.25\,mol}}{CO_2} + 2H_2O$$

0.25 mol $CH_4$ gives 0.25 mol 11 g $CO_2$.

Hence, option (d) is correct.

> **• Concept Enhancer**
>
> Generally the combustion reactions are exothermic and $\Delta H$ is negative but combustion of $N_2$ is endothermic because a large amount of energy is required to break $N{\equiv}N$ bond.

**17.** *(b)* Both A and R are true and R is the correct explanation of A.

$$\text{Molarity} = \frac{\text{Number of moles hydrated oxalic acid}}{\text{Volume of the solution}}$$

$$0.1 = \frac{\frac{3.1500}{\text{molar mass}}}{250} \times 1000$$

$$0.1 = \frac{3.1500 \times 1000}{250 \times mm}$$

$$mm = \frac{3.1500 \times 1000}{250 \times 0.1} = 126\ g/mol$$

**18.** *(c)* The chemical reaction involved is as follows

$$C_{15}H_{30} + \frac{45}{2}O_2 \longrightarrow 15CO_2 + 15H_2O$$

Mass of the fuel ($C_{15}H_{30}$)

$$= 0.756 \times 1000 = 756\ g$$

Number of moles of fuel

$$= \frac{756}{210} = 3.6\ mol$$

Now, weight of oxygen required

$$= 3.6 \times \frac{45}{2} \times 32 = 2592\ g$$

and weight of $CO_2$ released

$$= 3.6 \times 15 \times 44 = 2376\ g$$

**19.** *(c)* Since, 100 g of haemoglobin has $= 0.34$ g of Fe

Therefore, 3.3 g haemoglobin

$$= \frac{0.34}{100} \times 3.3\ \text{g of Fe}$$

$$= 0.01122\ \text{g of Fe}$$

Since, 56 g (1 mole) of Fe has

$$= 6.022 \times 10^{23}\ \text{atoms of Fe}$$

Therefore, 0.01122 g of Fe has

$$= \frac{6.022 \times 10^{23}}{56} \times 0.01122\ \text{atoms of Fe}$$

$$= 1.21 \times 10^{20}\ \text{atoms of Fe}$$

**20.** *(c)* Molar mass of $Fe_3O_4 = 232$ g/mol

So, number of moles of $Fe_3O_4$ reacting

$$= \frac{4640}{232} = 20\ \text{moles}$$

Similarly,

Molar mass of CO $= (12 + 16)$ g/mol $= 28$ g/mol

$\therefore$ Number of moles of CO reacting

$$= \frac{2520}{28} = 90\ \text{moles}$$

As number of moles of $Fe_3O_4$ reacting is less than number of moles of CO, it is the limiting reagent and product formed will depend on number of moles of $Fe_3O_4$.

$\therefore$ Number of moles of iron formed

$$= 3 \times \text{moles of } Fe_3O_4$$

$$= 3 \times 20 = 60\ \text{moles}$$

Weight of iron produced

$$= 60 \times 56\ (\text{in g}) = 3360\ g$$

**21.** *(c)* Volume of hydrogen $= 5.6$

Volume strength $= 11.2 \times$ molarity

$$\Rightarrow \text{Molarity} = \frac{\text{Volume strength}}{11.2}$$

$$= \frac{5.6}{11.2} = 0.5$$

Suppose, that the solution taken is $= 1$ L

Mass of solution $= 1000$ mL $\times$ 1 g/mL

$$= 1000\ g$$

Mass of solute = Moles × Molar mass

$$= 0.5 \times 34$$

[$\because$ molar mass of $H_2O_2 = 34$]

$$= 17\ g$$

$$\Rightarrow \text{Mass } \% = \frac{17}{1000} \times 100 = 1.7\%$$

**22.** *(c)* 1 g of chlorohydrocarbon will contain

$$= \frac{3.55}{100}\ \text{g of chlorine}$$

$\therefore$ Number of moles of chlorine atoms

$$= \frac{\text{Mass}}{\text{Atomic mass}} = \frac{3.55/100}{35.5} = 1 \times 10^{-3}$$

1 mole of chlorine contains $6.023 \times 10^{23}$ chlorine atoms. Therefore, $1 \times 10^{-3}$ mole of chlorine will have

$$= 1 \times 10^{-3} \times 6.023 \times 10^{23}$$

$$= 6.023 \times 10^{20}\ \text{chlorine atoms}$$

**23.** *(a)* Given, initial strength of acetic acid $= 0.06$N

Final strength = 0.042 N

Volume given = 50 mL

$\therefore$ Initial number of millimoles of $CH_3COOH = 0.06 \times 50 = 3$ millimoles

Final number of millimoles of $CH_3COOH$

$$= 0.042 \times 50 = 2.1\ \text{millimoles}$$

$\therefore$ Number of millimoles of $CH_3COOH$ adsorbed

$$= 3 - 2.1 = 0.9\ \text{millimoles}$$

Hence, mass of $CH_3COOH$ absorbed per gram of charcoal

$$= \frac{0.9 \times 60}{3}$$

($\because$ Molar mass of $CH_3COOH = 60\ gmol^{-1}$)

$$= \frac{54}{3} = 18\ mg$$

**24.** *(b)* The reaction involved in conversion of phenol to 2, 4, 6-tribromophenol is as follows

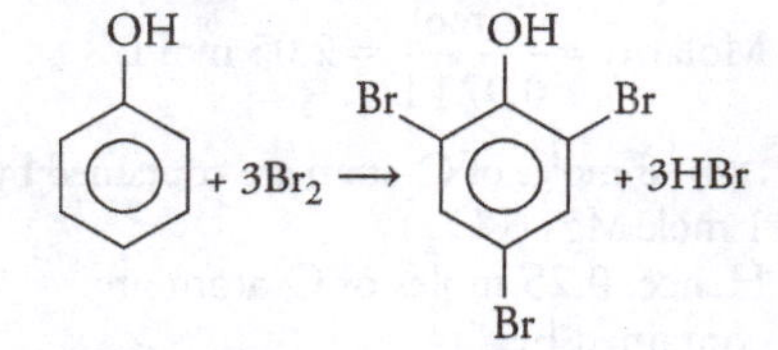


Molar mass of phenol $= 94$ g/mol

Molar mass of $Br_2 = 80$ g/mol

$\therefore$ Mass of $3Br_2 = 80 \times 3 \times 2 = 480$ g

So, 94 g phenol required 480 g bromine

$\therefore$ 2 g phenol required

$$= \frac{480}{94} \times 2\ \text{g of bromine.}$$

$$= 10.22\ \text{g of bromine}$$

**25.** *(d)* Both complexes have equal volume and half concentration due to mixing.

$$\underset{0.01\,M}{[Co(NH_3)_5SO_4]Br} + \underset{0.01\,M}{[Co(NH_3)_5Br]SO_4}$$

(I) $\underset{\left(\frac{0.02}{2}\right)mol}{[Co(NH_3)_5SO_4]Br} + \underset{(Excess)}{AgNO_3} \longrightarrow \underset{\substack{0.01\,mol \\ (Y)}}{AgBr\downarrow}$

(II) $\underset{\left(\frac{0.02}{2}\right)mol}{[Co(NH_3)_5Br]SO_4} + \underset{Excess}{BaCl_2} \rightarrow \underset{\substack{0.01\,mol \\ (Z)}}{BaSO_4\downarrow}$

Thus, option (d) is correct.

**26.** *(b)* Percentage of carbon in glucose

$$10.8 = \frac{\text{Mass of C}}{250} \times 100$$

$$\Rightarrow \text{Mass of C} = \frac{10.8 \times 250}{100} = 27 \text{ g}$$

Molar mass of glucose = 180 g

72 g of carbon is present in 180 g of glucose.

$\therefore$ 27 g of carbon present in

$$= \frac{180}{72} \times 27 = 67.5 \text{ g of glucose}$$

Mass of solvent = 250 g − 67.5 g

$= 182.5$ g

$$\text{Molality} = \frac{\text{Moles of solute}}{\text{Mass of solvent}} \times 1000$$

$$= \frac{67.5 \times 1000}{180 \times 182.5} = 2.054 \text{ mol/kg}$$

**27.** *(c)* $\underset{(20\text{ g})}{N_2(g)} + \underset{(5\text{ g})}{3H_2(g)} \rightleftharpoons 2NH_3(g)$

Number of moles of $N_2 = \frac{20}{28}$ mol

Number of moles of $H_2 = \frac{5}{2}$ mol

Limiting reagent can be calculated using a factor

$$= \frac{\text{Number of moles}}{\text{Stoichiometric coefficient}}$$

For $NH_3 = \frac{20}{28}$

For $H_2 = \frac{5}{2 \times 3} = \frac{5}{6}$

Clearly, $\frac{20}{28} < \frac{5}{6}$ so, $N_2$ is a limiting reagent and thus decides the formation of products.

1 mole of $N_2$ forms 2 mole of $NH_3$.

So, $\frac{20}{28}$ mole of $N_2$ forms

$$= 2 \times \frac{20}{28} = \frac{20}{14} = \frac{10}{7} = 1.42 \text{ mol.}$$

**28.** *(b)* When $H_3PO_3$ reacts with sodium hydroxide

$$H_3PO_3 + 2NaOH \longrightarrow Na_2HPO_3 + 2H_2O$$

Number of moles of $H_3PO_3$ = 0.05 mol

Number of moles of NaOH required $= 0.05 \times 2 = 0.1$ mol

Volume of NaOH = 0.1 L = 100 mL

For $H_3PO_2$,

$$NaOH + H_3PO_2 \longrightarrow NaH_2PO_2 + H_2O$$

Number of moles of $H_3PO_2$ = 0.2 mol

Number of moles of NaOH required $= 0.2 \times 1 = 0.2$ mol

Volume of NaOH = 0.2 L

= 200 mL

**29.** *(c)*

| | 1st component | 2nd component |
|---|---|---|
| mole | $n_1$ | $n_2$ |
| M. weight | $M_1$ | $M_2$ |
| mass | $n_1M_1$ | $n_2M_2$ |

Mass of solution $= n_1M_1 + n_2M_2$

Mole fraction of the 2nd component

$$(x_2) = \frac{n_2}{n_1 + n_2}$$

$$\Rightarrow \quad n_1 = \frac{n_2(1 - x_2)}{x_2}$$

Mass of solution $= n_1M_1 + n_2M_2$

$$= \frac{n_2M_1(1 - x_2)}{x_2} + n_2M_2$$

$$= \frac{n_2}{x_2}[M_2x_2 - x_2M_1 + M_1]$$

Volume of solution

$$= \frac{n_2\,[M_2x_2 - x_2M_1 + M_1]}{1000dx_2}$$

(where $d$ is the density)

Thus, $C_2$ is represented as

$$\Rightarrow \quad C_2 = \frac{1000dx_2}{M_1 + x_2(M_2 - M_1)}$$

**30.** *(138)* The reaction is as follows,

$$\underset{3 \times 10^3}{C_4H_{10}} + \underset{10 \times 10^3}{\frac{13}{2}O_2} \longrightarrow 4CO_2 + 5H_2O$$

Mole of $H_2O$ formed

$$= n_{H_2O} = 5 \times \frac{2}{13} \times 10 \times 10^3$$

Then, $w_{H_2O} = \frac{10^5}{13} \times 18$

$= 1.3846 \times 10^5$ g

Volume of $H_2O$ will be = 138.46 L

**31.** *(184)* The complete reaction occurs as follows.

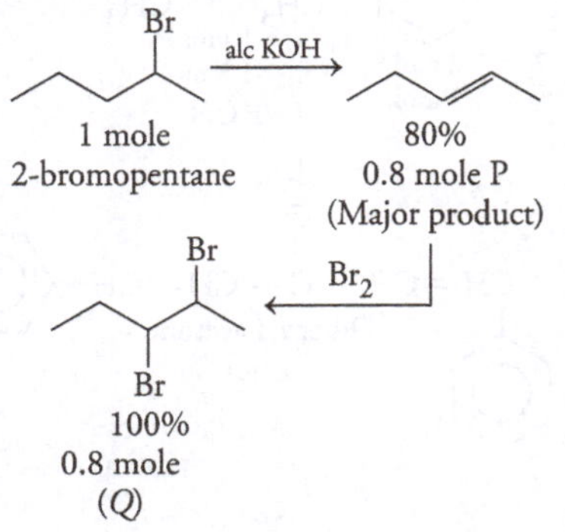


Molecular mass of $Q = 230$ g mol$^{-1}$

Mass of $Q = 0.8 \times 230$

$= 184$ g

**32.** *(12)* The reaction involved in complete combustion is as follows.

$$\text{Organic compound} \xrightarrow[\Delta]{\text{CuO}} CO_2 + H_2O$$

$$n_{CO_2} = \frac{220 \times 10^{-3}}{44} = 5 \times 10^{-3} \text{ moles}$$

$$M_C = 5 \times 10^{-3} \times 12$$

$$\% M_C = \frac{5 \times 10^{-3} \times 12}{500 \times 10^{-3}} \times 100 = 12\%$$

**33.** *(80)* $\underset{0.50\text{ g}}{\text{Organic compound}} \rightarrow \underset{1.46\text{ g}}{CO_2} + \underset{0.9\text{ g}}{H_2O}$

Mole of C in organic compound

$= n_{CO_2} \times 1$

By using principle of POAC,

So, mass of C in compound

$$= \frac{1.46}{44} \times 12$$

Now, the % of C in compound

$$= \frac{1.46}{44} \times \frac{12}{0.5} \times 100$$

$= 79.63 \approx 80$

**34.** *(63)* The reaction involved is

$$CaCO_3 \longrightarrow CaO + CO_2$$

$$\text{Mass of } CaCO_3 = \frac{150 \times 75}{100} = 112.5 \text{ kg}$$

$= 112500$ g

$n_{CaCO_3} = 1125$

So, moles of CaO = 1125

$$\text{So, mass of CaO} = \frac{1125 \times 56}{1000} = 63 \text{ kg}$$

**35.** *(57)* $M_F = \frac{M_1V_1 + M_2V_2}{V_1 + V_2}$

$$= \frac{2 \times 20 + 0.5 \times 400}{420}$$

$= 0.571$ M

$= 57.1 \times 10^{-2}$ M $\approx 57$

**36.** *(61)* The complete reaction

$$\underset{X\text{ g}}{C_6H_5COOH} + NaHCO_3 \longrightarrow C_6H_5COO^-Na^+ + H_2O + CO_2$$

$$\text{Mole of } CO_2 = \frac{11.2}{22.4} = 0.5$$

Mass of $C_6H_5COOH = X = 0.5 \times 122$

$= 61$ g

**37.** *(153)* Given,

Mass of Al = 81 g

Mass of oxygen = 128 g

Mass of $Al_2O_3$ = ?

The reaction involved is

$$4Al + 3O_2 \longrightarrow 2Al_2O_3$$

$$\frac{81}{27} = 3 \text{ mol} \qquad \frac{128}{32} = 4 \text{ mol}$$

Limiting reagent,

$\therefore$ Mole of $Al_2O_3$ formed $= \frac{1}{2} \times 3$ mole

$\therefore$ Weight of $Al_2O_3$ formed $= \frac{3}{2} \times 102$

$= 153$ g

**38.** *(420)*

**Key idea**
To solve this question first of calculate the number of moles of $Fe_3O_4$ and CO. Then with the help of this find out limiting reagent. After that use stoichiometry to calculate the mass of Fe.

$$\text{Moles of } Fe_3O_4 = \frac{2.32 \times 10^3 \times 10^3}{232}$$

$= 10000$ mol

$$\text{Moles of CO} = \frac{2.8 \times 10^2 \times 10^3}{28}$$

$= 10000$ mol

$$Fe_3O_4 + 4CO \longrightarrow 3Fe + 4CO_2$$

$10^4$mol $\quad 10^4$mol

CO is limiting reagent

$$\text{Mole of Fe} = \frac{3}{4} \times 10^4$$

$$\text{Mass of Fe} = \frac{3}{4} \times \frac{10^4 \times 56}{1000}$$

$= 420 \text{ kg} = x$ kg

**39.** *(15)* The complete reaction is as follow,

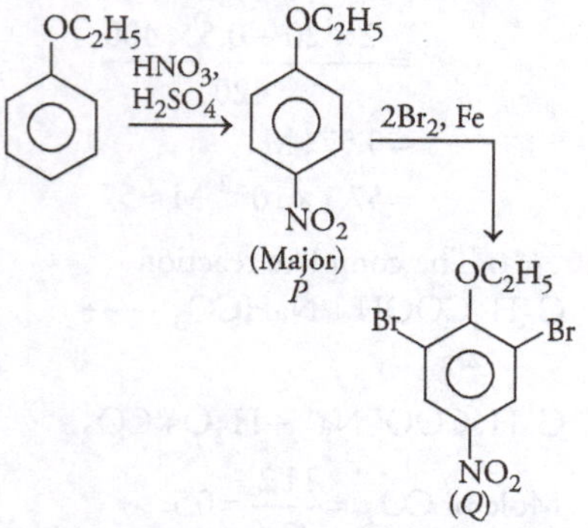


Number of O-atoms in product $Q = 3$

Number of Br atoms in product $Q = 2$

$$\text{Required ratio} = \frac{3}{2} = 1.5 = 15 \times 10^{-1}$$

Hence, required value is 15.

**40.** *(20)*

**key idea** To solve this question first of all write the complete balanced o To solve this question first of all write the complete balanced equation and then compare the number of moles of aniline with number of moles of orange dye to calculate the mass of orange dye.

The complete reaction of aniline to produce to orange dye is as follows

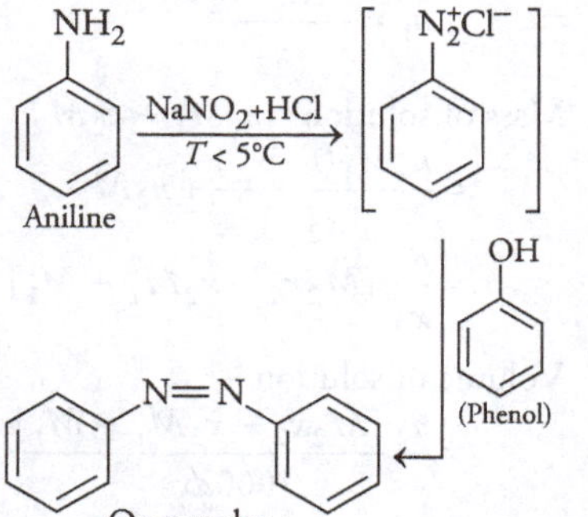


Orange dye

Above reaction indicates that 1 mole of aniline gives 1 mole of orange dye.

So, $(\text{Mole})_{\text{Aniline}} = (\text{Mole})_{\text{Orange dye}}$

$$\left(\frac{\text{Mass}}{\text{Molar mass}}\right)_{\text{Aniline}} = \left(\frac{\text{Mass}}{\text{Molar mass}}\right)_{\text{Orange dye}}$$

$$\frac{93 \text{ g}}{93 \text{ g mol}^{-1}} = \frac{\text{Mass of orange dye}}{199 \text{ g mol}^{-1}}$$

Mass of orange dye = 19.9 g $\approx$ 20 g

**41.** *(318)* The complete Claisen-Schmidt reaction is,

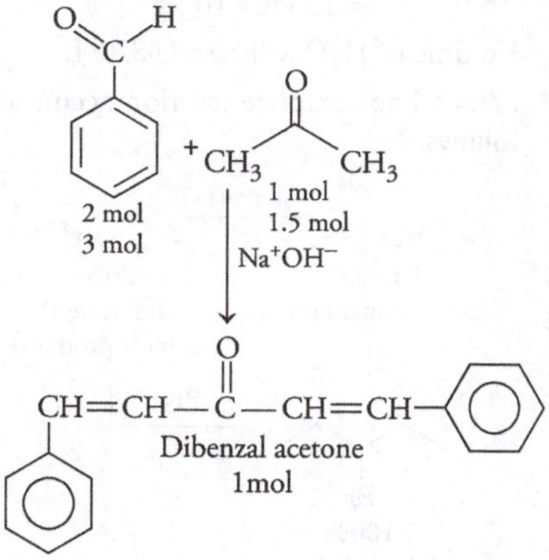


$\Rightarrow$ 351 g = 1.5 mol

Molecular weight of benzaldehyde

= 106 g/mol

$106 \times 3 = 318$ g required to give 1.5 mole (or 351 g) product.

**42.** *(9)* The reaction is as follow,

$$\underset{\substack{\text{Ethanamine} \\ 0.2 \text{ mol}}}{CH_3-CH_2-NH_2} \xrightarrow{NaNO_2 + HCl} CH_3-CH_2-N_2Cl$$

$$\xrightarrow[-N_2]{HOH} CH_3-CH_2OH + \underset{0.2 \text{ mol}}{HCl(g)}$$

Molecular weight of ethanamine = 45 g/mol

Mass = molecular weight × mole

$= 45 \times 0.2 = 9$ g

Hence, 9 g of ethanamine was subjected to reaction.

**43.** *(8)* Equivalent of $KMnO_4$ = Equivalent of $H_2C_2O_4$

So, $M \times 2 \times 5 = 2 \times 20 \times 2$

$\Rightarrow M = 8$ M

**44.** *(45)*

$$\underset{45 \text{ g (1 mol)}}{CH_3CH_2NH_2} \xrightarrow[H_2O]{NaNO_2 + HCl} CH_3CH_2OH + \underset{28 \text{ g (1 mol)}}{N_2}$$

Given that $N_2$ evolved in 2.24 L or 0.1 mole.

So, for 0.1 mole the ethylamine $(CH_3CH_2NH_2)$ will be 4.5 g or $45 \times 10^{-1}$ g.

**45.** *(24)*

$$3PbCl_2 + 2(NH_4)_3PO_4 \longrightarrow Pb_3(PO_4)_2 + 6NH_4Cl$$

Limiting reagent is $PbCl_2$.

mmol of $Pb_3 \cdot (PO_4)_2$ formed

$$= \frac{\text{mmol of } PbCl_2 \text{ react}}{3}$$

$$= \frac{72 \text{ mmol}^{-1}}{3}$$

$= 24$ mmol

Hence, the amount of $Pb_3(PO_4)_2$ formed is 24 mmol.

**46.** *(8)* The complete reaction to oxidise oxalate ion to $CO_2$ is,

$$2MnO_4^- + 5C_2O_4^{2-} + 16H^+ \longrightarrow 2Mn^{2+} + 10CO_2 + 8H_2O$$

$\therefore$Number of moles of $H^+$ ions required by 1 mole of $MnO_4^-$ to oxidise oxalate ion to $CO_2$ is 8.

**47.** *(4)* $C_2H_2O_4 + 2NaOH \longrightarrow Na_2C_2O_4 + 2H_2O$

Equivalents of oxalic acid = Equivalents of NaOH

Volume × Molarity × $n$-factor = Volume × Molarity × $n$-factor

$50 \times 0.5 \times 2 = 25 \times M_{NaOH} \times 1$

$M_{NaOH} = 2$

Now, mass of NaOH is 50 mL solution, Molarity $= \dfrac{\text{Mass} \times 1000}{\text{Molar mass} \times V_{(mL)}}$

$\text{Mass} = \dfrac{50}{1000} \times 2 \times 40 = 4\text{ g}$

**48.** *(7)* Given,

Weight of NaOH = 84 g

Molar mass of NaOH = 40

$M = \dfrac{n_{NaOH}}{V_{sol}\text{ (in L)}}$

$\Rightarrow \quad 3 = \dfrac{\left(\dfrac{84}{40}\right)}{V}$

$V = \dfrac{84}{40 \times 3} = 0.7\text{ L}$

or $V = 7 \times 10^{-1}\text{ L}$

$= 7 \times 10^{-1}\text{ dm}^3$

Hence, missing value is 7.

**49.** *(8)* The combustion reaction of methane is,

$CH_4 + 2O_2 \longrightarrow CO_2 + 2H_2O$

Moles of $CO_2 = \dfrac{22}{44} = 0.5$ mol

1 mole of $CO_2$ is produced from 1 mole of $CH_4$.

So, 0.5 mole of $CO_2$ will be produced from 0.5 mole of $CH_4$.

Mass of $CH_4$ formed = Moles × Molar mass

$= 0.5 \times 16 = 8\text{ g}$

**50.** *(224)*

$\underset{1\text{ mol}}{Mg} + \underset{2\text{ mol}}{2HCl} \longrightarrow \underset{1\text{ mol}}{MgCl_2} + \underset{1\text{ mol}}{H_2\uparrow}$

No. of moles $= \dfrac{\text{Mass of Mg}}{\text{Molar mass of Mg}} = \dfrac{2.4}{24}$

$= 0.1$ mol

So, 1 mole of Mg will give 1 mole of $H_2$

1 mole of gas at STP = 22.4 L

∴ 0.1 mole of gas = 0.1 × 22.4

$= 2.24\text{ L or } 224 \times 10^{-2}\text{ L}$

Thus, volume of hydrogen liberated is $224 \times 10^{-2}$ L.

**51.** *(1)* $3BaCl_2 + 2Na_3PO_4 \longrightarrow Ba_3(PO_4)_2 + 6NaCl$

The coefficient of the balanced equation tell you the ratio of the reactant and the product.

3 mol of $BaCl_2$ react with 2 mol of $Na_3PO_4$ to yield 1 mol of $Ba_3(PO_4)_2$

So, required ratio of $BaCl_2 : Na_3PO_4 = 3:2$

5 mol $BaCl_2$ require $2/3 \times 5 = 3{\cdot}3$ $Na_3PO_4$

We have only 2 mol $Na_3PO_4$. So, it is limiting reagent and $BaCl_2$ is in excess.

2 mol of $Na_3PO_4$ will yeild 1 mol of $Ba_3(PO_4)_2$.

**52.** *(139)* 10 mL solute in 90 mL solvent.

Mass of solute = 10 × 3.2 = 32 g

Mass of solvent = 90 × 1.6 g

$m = \dfrac{32 \times 1000}{160 \times 90 \times 1.6} = 1.388$

$= 138.8 \times 10^{-2} = 139 \times 10^{-2}$ m

**53.** *(364)*

> **key idea** In this type of question, whenever the density and molarity is given, then use this formula of molality
>
> i.e., $m = \dfrac{1000 \times M}{1000d - M \times \text{molecular weight}}$

$m = \dfrac{1000 \times M}{1000d - M \times \text{molecular weight}}$

$= \dfrac{1000 \times 3}{1000 \times 1 - (3 \times 58.5)} = 3.6385$ m

$\approx 3.64\text{ m} = 364 \times 10^{-2}$ m

**54.** *(4)* $Zn + 2HCl \longrightarrow ZnCl_2 + H_2\uparrow$

Amount of Zn = 11.5 g

Number of moles of Zn $= \dfrac{11.5}{65.4} = 0.176$

Number of moles of $H_2$ = Number of moles of Zn

Volume of $H_2$ liberated = 0.176 × 22.7 L

$= 3.99\text{ L} \approx 4\text{ L}$

**55.** *(227)* $\underset{\substack{12\text{ g}\\1\text{ mol}}}{2C(s)} + \underset{\substack{48\text{ g}\\1.5\text{ mol}}}{O_2(g)} \longrightarrow 2CO(g)$

Limiting reagent is carbon. One mole of carbon produces one mole of CO.

Thus, volume at STP is $227 \times 10^{-1}$ L.

**56.** *(180)* Molarity $= \dfrac{5 \times 1000}{40 \times 450}$

Let $x$ mL NaOH is required from $\dfrac{5 \times 1000}{40 \times 450}$ M NaOH solution to prepare 500 mL 0.1 M solution.

$\therefore \left(\dfrac{5 \times 1000}{40 \times 450}\right) x = 0.1\text{ M} \times 500\text{ mL}$

$(\because M_1V_1 = M_2V_2)$

$x = \dfrac{0.1 \times 500\text{ mL} \times 40 \times 450}{5 \times 1000}$

$= 180$ mL

**57.** *(3)* $R(OH)_y \longrightarrow H_2$

∴ Number of moles of 'OH' present in $x$

$= \dfrac{1}{2}$ (Number of moles of $H_2$ formed)

$\Rightarrow \quad y\left(\dfrac{1.84 \times 10^{-3}\text{ g}}{92\text{g mol}^{-1}}\right) = \dfrac{1}{2} \times \dfrac{1.344\text{ mL}}{22.4 \times 10^3\text{ mL}} \times 2$

[Volume of 1 mole of gas at STP = 22.42]

$\Rightarrow \quad y = \dfrac{1.344 \times 2 \times 92 \times 1000}{2 \times 22.4 \times 1000 \times 1.84}$

$\Rightarrow \quad y = 3$

So, number of alcoholic hydrogens present in compound '$x$' is 3.

**58.** *(25)*

| | $H_2SO_4$ + | $2NaOH \longrightarrow$ | $Na_2SO_4$ + | $2H_2O$ |
|---|---|---|---|---|
| Initial moles | 0.4 | 0.2 | 0 | 0 |
| Final moles | 0.3 | 0 | 0.1 | 0.2 |

Limiting reagent → NaOH

$n_{Na_2SO_4} = 0.1$ mol

Total volume,

$V = V_{H_2SO_4} + V_{NaOH} = 2 + 2 = 4\text{ L}$

Molarity, $M_{Na_2SO_4} = \dfrac{0.1}{4}$

$= 0.025$ M

$= 25 \times 10^{-3}$ M

$= 25$ millimolar

**59.** *(2)* 3 moles of $Z$ are required to give 1 mole of $XYZ_3$.

So, 0.05 moles of $Z$ are required to give

$= \dfrac{1}{3} \times 0.05$ mole of $XYZ_3$

$= 0.016$ mole of $XYZ_3$

Molecular weight of $XYZ_3 = 10 + 20 + (3 \times 30) = 120$ amu

∴ Number of moles of

$XYZ_3 = \dfrac{\text{Weight of } XYZ_3}{\text{Molecular weight of } XYZ_3}$

$0.016 = \dfrac{\text{Weight of } XYZ_3}{120}$

∴ Weight of $XYZ_3 = 0.016 \times 120$

$= 1.92\text{ g} \approx 2\text{ g}$

**60.** *(24)* Given, volume of solution = 2 L

Concentration = 48 ppm

48 mg $\longrightarrow$ in 1 L

$\therefore$ 2 L contains 48 × 2 = 96 mg of Mg

24 g of Mg contains $6.02 \times 10^{23}$ atoms

$\therefore$ 96 mg of Mg will contain

$$= \frac{96 \times 6.02 \times 10^{23} \times 10^{-3}}{24}$$

$$= 24 \times 10^{20}$$

$\therefore \quad x = 24$

**61.** *(54)* Moles of nitric acid

= Volume (in L) × Molarity

$= 800 \times 0.5 \times 10^{-3}$

$= 400 \times 10^{-3} = 0.4$ moles

Weight of $HNO_3$ = Number of moles × Molar mass

$= 0.4 \times 63 = 25.2$ g

Amount of $HNO_3$ remaining in the solution

$= 25.2 - 11.5 = 13.7$ g

$$\text{Molarity} = \frac{13.7}{63} \times \frac{1000}{400} = 54 \times 10^{-2} \text{ M}$$

**62.** *(1)* Reaction involved

$$H_2SO_4 + 2NaOH \longrightarrow Na_2SO_4 + H_2O$$

Number of equivalents of $H_2SO_4$

$= 100 \times 0.1 \times 2 = 20$

Number of equivalent of NaOH

$= 50 \times 0.1 = 5$

Normality

$$= \frac{\text{Number of gram equivalent}}{\text{Volume of the solution}}$$

Volume of the solution = 100 + 50

= 150 mL

$$\therefore \quad N = \frac{15}{150} = 0.1 = 1 \times 10^{-1} \text{ N}$$

**63.** *(46)* $N_2 + 3H_2 \longrightarrow 2NH_3$

Since the reactants and product are at same condition of temperature and pressure.

So, by Avagadro's law, $V \propto n$

56.0 L of $N_2$ gas = 56 moles of $NH_3$ gas

20.0 L of $NH_3$ gas = 20.0 moles of $NH_3$ gas

To produce 20 moles of $NH_3$, 10 mole of $N_2$ required.

So, to produce 20 L of $NH_3$, 10 L of $N_2$ required.

So, unused volume of $N_2$ will be

$= 56 - 10 = 46$ L

46 L of $N_2$ will remain unused.

**64.** *(1143)* When pent-1-ene reacts with bromine, we get

$$\underset{70\text{ g}}{CH_3CH_2CH_2CH{=}CH_2} + \underset{160\text{g}}{Br_2} \longrightarrow CH_3CH_2CH_2\underset{Br}{CH}\underset{Br}{CH_2}$$

70 g of pent-1-ene reacts with 160 g $Br_2$

5 g of pent-1-ene will react with

$$\frac{160}{70} \times 5 \approx 11.43 \text{ g or } 1143 \times 10^{-2} \text{ g}$$

**65.** *(225)*

$$CH_4 + 2O_2 \longrightarrow CO_2 + 2H_2O$$

**Mole-mole analysis**

$$\frac{n_{CH_4}}{1} = \frac{n_{H_2O}}{2}$$

Mass of water to be produced = 81 g

$$\text{Mass of } CH_4 = \frac{1}{2} \times \frac{81}{18} \times 16 = 36 \text{ g}$$

$$\text{So, moles of } CH_4 = \frac{36 \text{ g}}{16 \text{ g}} = 2.25 \text{ mol}$$

or $225 \times 10^{-2}$ mol

> **Alternate Solution**
>
> $CH_4 + 2O_2 \longrightarrow CO_2 + 2H_2O$
>
> Number of moles of $H_2O = \frac{81}{18} = 4.5$ moles
>
> From balanced equation
>
> 2 moles of $H_2O$ are produced 1 mole of $CH_4$.
>
> 4.5 moles of $H_2O$ are produced by 1 mole of $CH_4$
>
> $= \frac{1}{2} \times 4.5 = 2.25$
>
> $\approx 225 \times 10^{-2}$ mol

**66.** *(30)* $M_{eq}$ of oxidising agent = $M_{eq}$ of reducing agent

$$(M \times V \times n_f)_{KMnO_4} = (M \times V \times n_f)_{\text{Mohr's salt}}$$

$0.01 \times 20 \times 5 = 0.05 \times V \times 1$

Volume needed = 20 mL

Initial volume of $KMnO_4$ in burette is 50 mL

Thus, vol. of $KMnO_4$ left in the burette, after end point is 30 mL.

**67.** *(143)* $CH_4 + 2O_2 \longrightarrow CO_2 + 2H_2O$

| | $CH_4$ | $O_2$ |
|---|---|---|
| Moles | $\frac{100}{16} = 6.25$ | $\frac{208}{32} = 6.5$ |
| Moles / Stoichiometric coefficient | $= \frac{6.25}{1} = 6.25$ | $= \frac{6.5}{2} = 3.25$ |

So, $O_2$ is limiting reagent.

Using mole-mole analysis

$$\frac{n_{O_2}}{2} = \frac{n_{CO_2}}{1} \Rightarrow \frac{6.5}{2} = n_{CO_2}$$

$n_{CO_2} = 3.25$

$\therefore$ Mass of $CO_2 = 3.25 \times 44 = 143$ g.

**68.** *(3)* Let basicity of acid is $x$.

$$\underset{\substack{0.1\text{ M} \\ 10\text{ mol}}}{A} + \underset{\substack{0.05\text{ M} \\ 30\text{ mL}}}{M(OH)_2} \longrightarrow \text{Salt} + H_2O$$

On complete neutralisation,

Milliequivalent of acid

= Milliequivalent of base

$x(0.1 \times 10) = 2(0.05 \times 30)$

$x = 2 \times 1.5 = 3$

So, the basicity of acid, $A$ is 3.

**69.** *(50)* Moles of sodium in 100 mL

$$= \frac{3.45}{23} = 0.15$$

Moles of sodium in 1000 mL (1 L) = 1.5

1 mole of $Na_3PO_4$ which contains 3 moles of Na

1.5 moles of Na will be present in $\frac{1}{2}$ moles of $Na_3PO_4$

($M$) Molarity

$$= \frac{\text{Number of moles of solute}}{\text{Volume of solution in L}}$$

$$= \frac{1/2}{1} = 0.5 = 50 \times 10^{-2} \text{ mol/L}$$

**70.** *(1575)* For titration of $KMnO_4$ with oxalic acid,

$$n_{eq}(KMnO_4) = n_{eq}(H_2C_2O_4 \cdot 2H_2O)$$

$$\text{or} \quad \frac{10 \times 0.05}{1000} \times 5 = \frac{10 \times M}{1000} \times 2$$

$\Rightarrow \quad M = 0.125$ M

Molar mass of $H_2C_2O_4 \cdot 2H_2O$

$= 2 \times 1 + 2 \times 12 + 4 \times 16 + 2(18)$

= 126 g/mol

Strength of oxalic acid = $(126 \times 0.125)$ g/L

= 15.75 g/L = $1575 \times 10^{-2}$ g/L

**71.** *(18)* For the given reaction,

$$Fe^{2+} + Cr_2O_7^{2-} \longrightarrow Fe^{3+} + Cr^{3+}$$

Volume of aqueous solution of $Fe^{2+}$ ion ($V_1$) = 10 mL.

Volume of $K_2Cr_2O_7$ solution ($V_2$) = 15 mL

$$\underset{(Fe^{2+})}{n_1 M_1 V_1} = \underset{(K_2Cr_2O_7)}{n_2 M_2 V_2}$$

$M_1 \times 10 \times 1 = 6 \times 0.02 \times 15$

$$\Rightarrow \quad M_1 = \frac{6 \times 0.02 \times 15}{10} = 0.18$$

$M_1 \approx 18 \times 10^{-2}$

So, value of $x$ is 18.

**72.** *(4)* Molar mass of glucose $(C_6H_{12}O_6)$

$=12\times6+12\times1+16\times6=180\text{ g mol}^{-1}$

$\therefore$ Molarity $(M)$

$$=\frac{\text{Concentration of glucose}}{\text{Molar mass of glucose}}=\frac{W_B}{M_B}$$

$$M=\frac{0.72}{180}=0.004=4\times10^{-3}\text{ M.}$$

**73.** *(1)* Total mass = 4 g (given)

Mass of NaOH $= x$ g

Mass of $Na_2CO_3 = y$ g

Total mass $= x + y = 4$

$$\frac{\text{Mass of NaOH}}{\text{Molar mass of NaOH}}=\frac{\text{Mass of }Na_2CO_3}{\text{Molar mass of }Na_2CO_3}$$

$$\frac{x}{40}=\frac{y}{106}\quad\text{(Equimolar)}$$

$$y=\left(\frac{106}{40}\right)x$$

$$\Rightarrow\quad x+\frac{106}{40}x=4$$

$$x+2.65x=4$$

$$\Rightarrow\quad 3.65x=4$$

$$\Rightarrow\quad x=1.096\text{ g}\Rightarrow x\approx1$$

**74.** *(226)* We know that, number of millimoles $=V_{\text{mL}}\times$ molarity

So, millimoles of

NaOH = 250 × 0.5 = 125

Millimoles of HCl = 500 × 1 = 500

Now, reaction is

| | NaOH | + HCl ⟶ | NaCl | + $H_2O$ |
|---|---|---|---|---|
| At $t = 0$, | 125 | 500 | 0 | 0 |
| At $t = t$, | 0 | 375 | 125 | 125 |

125 millimoles of NaOH reacts with 125 millimoles of HCl.

So, millimoles of HCl left = 375

Moles of HCl $=375\times10^{-3}$

Number of HCl molecules

= Avogadro's constant $(N_A)\times$ moles of HCl

$=6.022\times10^{23}\times375\times10^{-3}$

$=225.8\times10^{21}=226\times10^{21}$

Therefore, answer is 226.

**75.** *(5)* Take 1 L solution

mass of solution = Vol. × density

$=1.2\times1000\text{ g}=1200\text{ g}$

Mass of 1 L solvent (water) = 1000 g

($\because$ density $=1\text{ g cm}^{-3}$)

$\Rightarrow$ Mass of NaOH $=(1200-1000)$ g

$=200$ g

$$\Rightarrow\text{Moles of NaOH}=\frac{200\text{ g}}{40\text{ g/mol}}=5\text{ mol}$$

$$\text{Molality}=\frac{\text{Number of moles of solute}}{\text{Mass of solvent (in kg)}}$$

$$\Rightarrow\quad\text{Molality}=\frac{5\text{ mol}}{1\text{ kg}}=5\text{ m}$$

**76.** *(24)* $\underset{+6}{Cr_2O_7^{2-}}+\underset{+2}{Fe^{2+}}\longrightarrow\underset{+3}{Cr^{3+}}+\underset{+3}{Fe^{3+}}$

Dichromate $(Cr_2O_7^{2-})$ oxidised $Fe^{2+}$ to $Fe^{3+}$ and itself get reduced to $Cr^{3+}$.

Valency factor of $Cr_2O_7^{2-}=6$ as Cr gets reduced from +6 to +3.

Valency factor of $Fe^{2+}=1$ as it gets oxidised from +2 to +3.

From law of equivalence,

$$\underset{(Cr_2O_7^{2-})}{M_1V_1n_1}=\underset{(Fe^{2+})}{M_2V_2n_2}$$

$$(M_1\times V_1)\times6=(M_2\times V_2)\times1$$

$$(0.03\times20)\times6=(M_2\times15)\,1$$

$$\Rightarrow\quad M_2=0.24\text{ molar}=24\times10^{-2}\text{ molar}$$

**77.** *(64)* Given, molality = 100

$\Rightarrow$ 100 moles of solute in 1 kg of solvent

Mole fraction of solute $(\chi_{\text{Solute}})$

$$=\frac{\text{Moles of solute}}{\text{Total moles}}=\frac{n_{\text{Solute}}}{n_{\text{Solute}}+n_{\text{Solvent}}}$$

$$n_{\text{Solvent}}=\frac{1000}{18}=55.5$$

$$\chi_{\text{Solute}}=\frac{100}{100+55.5}$$

$$=\frac{100}{155.5}=0.643$$

$$\chi_{\text{Solute}}=64.3\times10^{-2}=64\times10^{-2}$$

**78.** *(50)* 10.0 mL of $Na_2CO_3$ solution is titrated against 0.2 M HCl.

$$\underset{\substack{10\text{mL}\\M_{Na_2CO_3}=?}}{Na_2CO_3}+\underset{\substack{0.2\text{M HCl}\\5\text{ mL}}}{2HCl}\longrightarrow2NaCl+H_2O+CO_2$$

$M_{eq}$ of $Na_2CO_3=M_{eq}$ of HCl

$M_{Na_2CO_3}\times10\times2=0.2\times5\times1$

$M_{Na_2CO_3}=5\times10^{-2}\text{M}=50\times10^{-3}\text{ M}$

$=50$ mM

**79.** *(1)* Volume of $Cl_2$ at STP = 20 mL

Number of moles of chlorine gas

$$=\frac{20\text{ mL}}{22400\text{ mL/mol}}$$

Number of molecules of $Cl_2$ gas

$$=\frac{20}{22400}\times6.023\times10^{23}$$

Thus, number of chlorine atoms

$$=2\times\frac{20}{22400}\times6.023\times10^{23}$$

$$=1.08\times10^{21}\approx1\times10^{21}$$

**80.** *(525)* When lead nitrate solution is reacts with chromic sulphate solution then following reaction takes place.

$$3Pb(NO_3)_2+Cr_2(SO_4)_3\rightarrow3PbSO_4+2Cr(NO_3)_3$$

$m$ moles of $Pb(NO_3)_2=35\times0.15=5.25$

$m$ moles of $Cr_2(SO_4)_3=20\times0.12=2.4$

Thus, $Pb(NO_3)_2$ is limiting reagent,

$m$ moles of $PbSO_4$ formed = 5.25

$=525\times10^{-5}$ moles

**81.** *(3)* When benzylamine reacts with bromomethane then

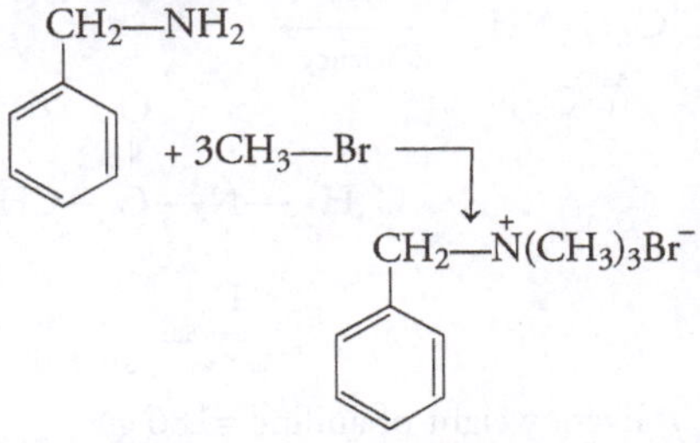


Number of moles of benzyl trimethyl ammonium bromide formed

$$=\frac{23}{230}=0.1$$

$\therefore$ Number of moles of bromomethane consumed in the reaction

$$=3\times0.1=3\times10^{-1}$$

**82.** *(13)* 70 mg $Na^+$ is present in 1 mL of $NaNO_3$ solution.

$\therefore$ 50 mL of $NaNO_3$ will contain

$$=70\times50\text{ mg}=\frac{70\times50}{1000}=3.5\text{ gm}$$

Moles of $Na^+$ in solution = Moles of $NaNO_3$ in solution

$[\because NaNO_3\longrightarrow Na^++NO_3^-]$

$$=\frac{3.5}{23}\text{ mol}$$

$[\because$ Molar mass of $Na^+=23\text{g mol}^{-1}]$

Mass of $NaNO_3$ = mole × molar mass

$[\because$ molar mass of $NaNO_3=85\text{ gmol}^{-1}]$

$$=\frac{3.5}{23}\times85=12.934\text{ g}\approx13\text{ g}$$

**83.** *(8)* Combustion reaction :

$$C_xH_y(g)+\left(x+\frac{y}{4}\right)O_2(g)\longrightarrow xCO_2(g)+\frac{y}{z}H_2O(l)$$

Suppose, volume of $C_xH_y$ is $V$ and volume of $O_2$ is 6 times greater than $C_xH_y = 6V$

then volume of $xCO_2$ produced during complete oxidation.

$\Rightarrow \quad Vx = 4V \Rightarrow x = 4$

Since, $V_{O_2} = 6V$

$V\left(x + \frac{y}{4}\right) = 6V$

$\left(x + \frac{y}{4}\right) = 6 \quad \dots$ (i)

Put value of $x = 4$ in Eq. (i)

We get, $4 + \frac{y}{4} = 6 \Rightarrow y = 8$

**84.** *(243)* Reaction

$$\underset{\substack{\text{Aniline}\\ \text{Mol. wt} = 98}}{C_6H_5NH_2} \xrightarrow[\text{Efficiency}]{90\%}$$

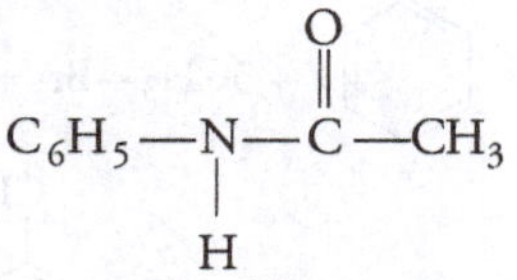

Acetanilide (mol. wt. = 135)

Given, weight of aniline = 1.86 g

Here, 1 mole of aniline gives 1 mole of acetanilide

∴mole of aniline = mole of acetanilide

$\Rightarrow \quad \frac{1.86}{93} = \frac{W_{\text{Acetanilide}}}{135}$

$W_{\text{Acetanilide}} = \frac{1.86 \times 135}{93}\text{ g} = 2.70\text{ g}$

But efficiency of reaction is 90% only. Hence, mass of acetanilide produced

$= 2.70 \times \frac{90}{100}\text{ g}$

$= 2.43\text{ g}$

$= 243 \times 10^{-2}\text{ g}$

$x = 243$

**85.** *(2)* Molarity

$= \frac{4.5}{90} \times \frac{1000}{250} = 0.2 = 2 \times 10^{-1}$

So, $x = 2$

**86.** *(4)* For first end point :

$$\underset{(a)}{NaOH} + \underset{(b)}{Na_2CO_3} + \underset{(c)}{HCl} \longrightarrow \underset{(b)}{NaHCO_3} + NaCl + H_2O$$

Both NaOH and $Na_2CO_3$ are of basic character. Hence, they will not react with each other. However both of them will react with HCl.

According to law of equivalence Equivalents of NaOH $(a)$ + Equivalents of $Na_2CO_3$ $(b)$ = Equivalents of HCl $(c)$ and Equivalents of $Na_2CO_3$ $(b)$ = Equivalents of $NaHCO_3$ $(b)$

For second end point

$$\underset{(b)}{NaHCO_3} + \underset{(d)}{HCl} \longrightarrow H_2CO_3 + NaCl$$

According to law of equivalence

Equivalents of $NaHCO_3$ $(b)$ = Equivalents of HCl $(d)$

$\Rightarrow \quad b = \frac{1}{10} \times 1.5 = 0.15$

From first end point, equivalents of $Na_2CO_3 = 0.15$

$\Rightarrow$ Mass of $Na_2CO_3$ = Number of eq. × Eq. mass

$= 0.15 \times \frac{106}{1} \times 10^{-3}\text{ g}$

$= 0.016\text{ g}$

Weight percentage of

$Na_2CO_3 = \frac{0.016}{0.4} \times 100 = 4\%.$

**87.** *(25)* Given, $6.023 \times 10^{22}$ molecules are present in 10 g substance.

Number of moles

$= \frac{\text{Number of molecules}}{6.023 \times 10^{23}}$

$= \frac{\text{Mass (given)}}{\text{Molar mass}}$

Molar mass $= \frac{10 \times 6.023 \times 10^{23}}{6.023 \times 10^{22}}$

$= 100\text{ g/mol}$

Molarity $= \frac{\text{Number of moles of solute}}{\text{Volume of solution in litres}}$

$= \frac{(5/100)}{2} = \frac{5}{100 \times 2}$

$= 0.025 = 25 \times 10^{-3}$

**88.** *(10)* Let the water of crystallisation be $x$.

Molar mass of $Na_2CO_3 \cdot xH_2O$.

$= (106 + 18x)$

Equivalent weight of $Na_2CO_3 \cdot xH_2O$

$= \frac{\text{Molar mass}}{n\text{ factor}} = \frac{M}{2} = (53 + 9x)$

[Here, $M$ = molar mass and $n$ factor = 2]

Gram equivalent $= \frac{\text{Weight}}{\text{Equivalent weight}}$

[Given, weight of $Na_2CO_3 \cdot xH_2O = 1.43$ g]

Hence, gram equivalent of

$Na_2CO_3 \cdot xH_2O = \frac{1.43}{53 + 9x}$

Normality $= \frac{G_{\text{meq}}}{V_{\text{litre}}}$; $0.1 = \frac{\left(\frac{1.43}{53 + 9x}\right)}{0.1}$

{As, volume = 100 mL = 0.1 L

So, $\quad 10^{-2} = \frac{1.43}{53 + 9x} \Rightarrow x = 10.00$

**89.** *(5)* For the molecular formula, $C_nH_{2n+2}O_2$ ($M$ = Molar mass)

$C\% = \frac{12n}{M} \times 100$, $H\% = \frac{2n + 2}{M} \times 100$,

$O\% = \frac{32}{M} \times 100$

To be saturated and alicyclic, $n = 2$, i.e.,

$C\% = \frac{24}{M} \times 100$, $H\% = \frac{6}{M} \times 100$ and

$O\% = \frac{32}{M} \times 100$

$\Rightarrow \quad C\% : H\% = 24 : 6 = 4 : 1$

$C\% : O\% = 24 : 32 = 3 : 4$

So, $X$ is $C_2H_6O_2$ and its combustion equation is

$C_2H_6O_2 + \left(2 + \frac{6}{4} - \frac{2}{2}\right)O_2 \longrightarrow 2CO_2 + \frac{6}{2}H_2O$

$\Rightarrow \underset{\substack{1\text{ mol}\\ 2\text{ mol}}}{C_2H_6O_2} + \underset{5/2\text{ mol}}{\frac{5}{2}O_2} \rightarrow 2CO_2 + 3H_2O$

$\frac{5}{2} \times 2 = 5$ mol of $O_2$ will be required.

**90.** *(10)*

$$\underset{\text{Base}}{NaOH} + \underset{\substack{\text{Phosphonic}\\ \text{acid}}}{H_3PO_3} \longrightarrow \underset{\text{Salt}}{NaH_2PO_3} + \underset{\text{Water}}{H_2O}$$

Given $\begin{cases} 0.1\text{ N} & 0.1\text{ N} \\ V = ? & 10\text{ mL} \end{cases}$

On dilution

$(N_1V_1)_{NaOH} = (N_2V_2)_{H_3PO_3}$

$0.1\,V_1 = 0.1 \times 10\text{ mL}$

$\Rightarrow \quad V_1 = \frac{0.1 \times 10\text{ mL}}{0.1}$

$V_1 = 10\text{ mL}$

Volume of NaOH = 10 mL

**91.** *(3400)* $\underset{1\text{ mol}}{N_2} + \underset{3\text{ mol}}{3H_2} \longrightarrow \underset{2\text{ mol}}{2NH_3}$

Molecular weight of $N_2$ (kg)

$= 0.028$ kg/mol

Molecular weight of $H_2 = 2$ g
Here, $3H_2$ is $= 3 \times 2 = 6\text{ g} = 0.006$ kg/mol
Molecular weight of $2NH_3$ (kg)
$= 0.034$ kg/mol
0.028 kg $N_2$ require hydrogen $= 0.006$ kg

$$1 \text{ kg } N_2 \text{ require} = \frac{0.006}{0.028}$$

2.8 kg $N_2$ (limiting reagent) require

$$= \frac{0.006}{0.028} \times 2.8 = 0.6 \text{ kg } H_2 \quad \text{(Excess)}$$

Then, 0.6 kg $H_2$ is present in excess and $N_2$ will be limiting reagent.

$$0.028 \text{ kg } N_2 \xrightarrow{\text{Excess}} 0.034 \text{ kg of } NH_3$$

$$1 \text{ kg } N_2 \longrightarrow \frac{0.034}{0.028}$$

$$2.8 \text{ kg } N_2 \longrightarrow \frac{0.034}{0.028} \times 2.8$$

$$= 3.4 \text{ kg or } 3400 \text{ g}$$

Amount of $NH_3 = 3400$ g

**92.** *(18)* **Combustion of propane**

$$\underset{\text{Propane}}{C_3H_8(g)} + 5\,O_2(g) \longrightarrow 3CO_2(g) + 4H_2O(l)$$

For combustion 1 mole propane 5 mole $O_2$ is required.

**Combustion of butane**

$$\underset{\substack{\text{Butane}\\ 2\text{ mol}}}{C_4H_{10}(g)} + \frac{13}{2}O_2(g) \longrightarrow 4CO_2(g) + 5H_2O(l)$$

For 2 moles of butane 13 moles of $O_2$ is required.

Hence, minimum number of moles of $O_2$ required to oxidise 1 mole of propane and 2 moles of butane

$$= 5 + 2 \times \frac{13}{2} = 18$$

**93.** *(50)*

$$K_2Cr_2O_7 + 7H_2SO_4 + FeC_2O_4 \longrightarrow Fe_2(SO_4)_3 + K_2SO_4 + Cr_2(SO_4)_3 + CO_2$$

$m$ eq. of $K_2Cr_2O_7 = m$ eq. $FeC_2O_4$

$$V \times 0.02 \times 6 = \frac{0.288 \times 3 \times 1000}{144}$$

$$\Rightarrow \quad V = 50 \text{ mL}$$

**94.** *(5)* Ferrous sulphate heptahydrate $FeSO_4 \cdot 7H_2O$.

If the amount of the salt is '$x$' gram.

Mass of iron contained in it

$$= \frac{x \times \text{atomic mass of Fe (55.85)}}{\text{Molar mass of } FeSO_4 \cdot 7H_2O (277.85)}$$

$$= \frac{55.85\,x}{277.85}$$

10 ppm of iron in 100 kg of wheat, can be acheived as,

$$= \frac{\text{mass of iron} \times 10^6}{\text{mass of wheat } (100 \text{ kg} = 10^5 \text{ g})} = 10$$

$$\Rightarrow \quad \frac{55.85\,x \times 10^6}{277.85 \times 10^5} = 10$$

$$\Rightarrow \quad x = \frac{277.85}{55.85} = 4.97 \text{ g} \approx 5$$

Practically this value may vary from 4.95 to 4.99.

**95.** *(14)* For 100 g sample of the solution, mass of $HNO_3 = 63$ g,

$$\therefore \quad n_{HNO_3} = \frac{63}{63} = 1 \text{ mol}$$

$$\text{Volume of solution} = \frac{\text{Mass}}{\text{Density}}$$

$$= \frac{100 \text{ g}}{1.4 \text{ g/mL}}$$

$$= \frac{100}{1.4} \text{ mL} = \frac{1}{14} \text{ L}$$

$$\text{Molarity} = \frac{n}{V(\text{L})} = \frac{1}{(1/14)} = 14 \text{ M}$$

**96.** *(27)* Given, Mass of $[Co(NH_3)_6]Cl_3$
$= 0.3$ g

Molar mass $= 267.46$ g $\text{mol}^{-1}$

$$\therefore \quad n = \frac{0.3}{267.46} \text{ mol}$$

From the formula, 1 mol complex will give 3 mol $Cl^-$.

$$\therefore \quad n_{Cl^-} = \frac{0.3}{267.46} \times 3 = \frac{0.9}{267.46}$$

The precipitation reaction takes place as follows :

$$\underset{(0.3\text{g})}{[Co(NH_3)_6]Cl_3} + \underset{(V\text{mL}, 0.125\text{g})}{3AgNO_3} \longrightarrow 2AgCl$$

Let volume of $AgNO_3$ (in mL) be $V$.

Number of moles of $Cl^-$ = number of moles of $N$

$$\therefore \quad \frac{0.3 \times 3}{267.46} = 0.125 \times V \times 10^{-3}$$

$$= \frac{0.9 \times 1000}{0.125 \times 267.46}$$

$$= 26.92 \text{ mL.}$$

Depending upon the accurancy of calculation this value may vary from 26.60 to 27.00.

**97.** *(2130)*

> **key idea** To find the No. of moles of $O_2$ in the given question, use the formula $n = \frac{pV}{RT}$.
> Then, use the value of $n$ to find out the mass of $NaClO_3$.

Given, volume of $O_2 = 492$ L

$$\therefore \quad n = \frac{pV}{RT}$$

$$= \frac{1 \text{ atm} \times 492 \text{ L}}{0.082 \text{ L atm K}^{-1} \text{ mol}^{-1} \times 300 \text{ K}}$$

$$= \frac{492}{8.2 \times 3} \text{ mol}$$

From the reaction,

$$NaClO_3(s) + 5Fe(s) \longrightarrow O_2(g) + NaCl(s) + 5FeO(s)$$

$$n_{NaClO_3} = n_{O_2} = \frac{492}{8.2 \times 3} = 20 \text{ mol}$$

$$m_{NaClO_3} = n \times \text{molar mass of } NaClO_3$$

$$= 20 \times 106.5 = 2130 \text{ g}$$

The value in different situations may range from 2120 to 2140.

# C-02

JEE Main Chapterwise Topicwise Chemistry

In JEE Main there is usually 0-1 question per paper are visible from this chapter. In past 3 years, 26 conceptual and 19 numerical questions were asked of moderate difficulty level from this chapter.

**For detailed analysis scan QR code**

# ATOMIC STRUCTURE

## Topic 1 Preliminary Models and Dual Nature of Light

### Level 1

**01** The minimum energy that must be possessed by photons in order to produce the photoelectric effect with platinum metal is
[Given, the threshold frequency of platinum is $1.3 \times 10^{15} s^{-1}$ and $h = 6.6 \times 10^{-34}$ Js] *[2022, 25 June Shift-II]*

(a) $3.21 \times 10^{-14}$ J (b) $6.24 \times 10^{-16}$ J
(c) $8.58 \times 10^{-19}$ J (d) $9.76 \times 10^{-20}$ J

**02** Hydrogen has three isotopes (*A*), (*B*) and (*C*). If the number of neutron(s) in (*A*), (*B*) and (*C*) respectively, are ($x$), ($y$) and ($z$), the sum of ($x$), ($y$) and ($z$) is *[2020, 8 Jan Shift-II]*

(a) 4 (b) 3 (c) 2 (d) 1

**03** Ejection of the photoelectron from metal in the photoelectric effect experiment can be stopped by applying 0.5 V when the radiation of 250 nm is used. The work function of the metal is *[JEE Main 2018]*

(a) 5 eV (b) 4 eV (c) 5.5 eV (d) 4.5 eV

**04** If $\lambda_0$ and $\lambda$ be threshold wavelength and wavelength of incident light, the velocity of photoelectron ejected from the metal surface is *[JEE Main 2014]*

(a) $\sqrt{\frac{2h}{m}(\lambda_0 - \lambda)}$ (b) $\sqrt{\frac{2hc}{m}(\lambda_0 - \lambda)}$

(c) $\sqrt{\frac{2hc}{m}\left(\frac{\lambda_0 - \lambda}{\lambda\lambda_0}\right)}$ (d) $\sqrt{\frac{2h}{m}\left(\frac{1}{\lambda_0} - \frac{1}{\lambda}\right)}$

### Level 2

**05** Match List-I with List-II. *[2024, 6 April Shift-II]*

| List-I (Alkali metal) | | List-II (Emission wavelength in nm) | |
|---|---|---|---|
| A. | Li | I. | 589.2 |
| B. | Na | II. | 455.5 |
| C. | Rb | III. | 670.8 |
| D. | CS | IV. | 780.0 |

Choose the correct answer from the options given below

| | A | B | C | D | | A | B | C | D |
|---|---|---|---|---|---|---|---|---|---|
| (a) | IV | II | I | III | (b) | II | IV | III | I |
| (c) | III | I | IV | II | (d) | I | IV | III | II |

**06** Given below are two statements, one is labelled as Assertion (A) and the other is labelled as Reason (R).

**Assertion** (A) In the photoelectric effect, the electrons are ejected from the metal surface as soon as the beam of light of frequency greater than threshold frequency strikes the surface.

**Reason** (R) When the photon of any energy strikes an electron in the atom, transfer of energy from the photon to the electron takes place.

In the light of the above statements, choose the most appropriate answer from the options given below. *[2023, 11 April Shift-I]*

(a) Both A and R are correct but R is not the correct explanation of A
(b) Both A and R are correct and R is the correct explanation of A
(c) A is correct but R is not correct
(d) A is not correct but R is correct

**07** The energy of one mole of photons of radiation of wavelength 300 nm is
(Given, $h = 6.63 \times 10^{-34}$ Js, $N_A = 6.02 \times 10^{23}$ mol$^{-1}$, $c = 3 \times 10^8$ ms$^{-1}$) *[2022, 24 June Shift-II]*

(a) 235 kJ mol$^{-1}$ (b) 325 kJ mol$^{-1}$
(c) 399 kJ mol$^{-1}$ (d) 435 kJ mol$^{-1}$

**Imp.** Photoelectric Effect

**08** What is the work function of the metal, if the light of wavelength 4000 Å generates photoelectron of velocity $6 \times 10^5$ ms$^{-1}$ from it?
(Mass of electron $= 9 \times 10^{-31}$ kg
Velocity of light $= 3 \times 10^8$ ms$^{-1}$
Planck's constant $= 6.626 \times 10^{-34}$ Js
Charge of electron $= 1.6 \times 10^{-19}$ JeV$^{-1}$) *[2019, 12 Jan Shift-I]*

(a) 4.0 eV (b) 2.1 eV (c) 0.9 eV (d) 3.1 eV

**09** Which of the graphs shown below does not represent the relationship between incident light and the electron ejected from metal surface? *[2019, 10 Jan Shift-I]*

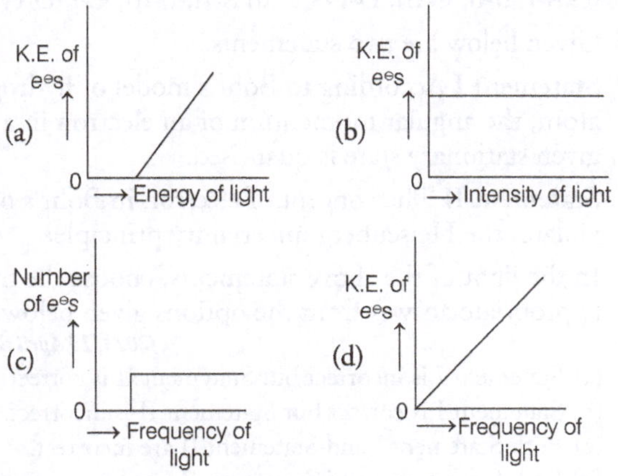


## Level 3

**10** If $p$ is the momentum of the fastest electron ejected from a metal surface after the irradiation of light having wavelength $\lambda$, then for 1.5 $p$ momentum of the photoelectron, the wavelength of the light should be

(Assume kinetic energy of ejected photoelectron to be very high in comparison to work function) *[2019, 8 April Shift-II]*

(a) $\frac{4}{9}\lambda$ (b) $\frac{3}{4}\lambda$

(c) $\frac{2}{3}\lambda$ (d) $\frac{1}{2}\lambda$

**11** Ionisation energy of gaseous Na atoms is 495.5 kJ $mol^{-1}$. The lowest possible frequency of light that ionizes a sodium atom is *[JEE Main 2014]*

($h = 6.626 \times 10^{-34}$ Js, $N_A = 6.022 \times 10^{23}$ $mol^{-1}$)

(a) $7.50 \times 10^4$ $s^{-1}$

(b) $4.76 \times 10^{14}$ $s^{-1}$

(c) $3.15 \times 10^{15}$ $s^{-1}$

(d) $1.24 \times 10^{15}$ $s^{-1}$

## Numerical Value Type Questions

**12** A hypothetical electromagnetic wave is shown below.

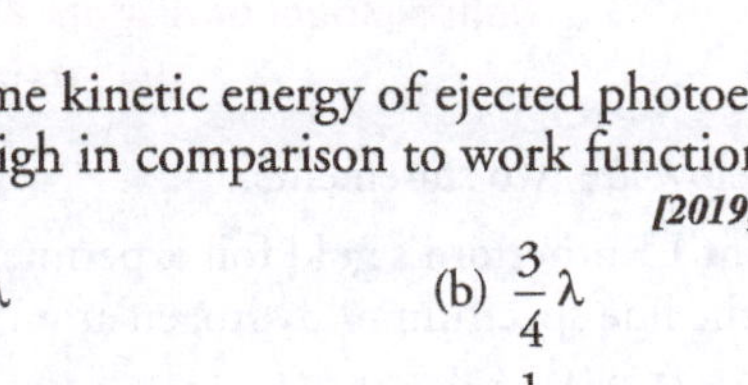


The frequency of the wave is $x \times 10^{19}$ Hz.

$x$ = ..... (nearest integer) *[2024, 8 April Shift-I]*

**Imp.** Ionisation Energy in Blend with Electromagnetic Radiation

**13** The ionisation energy of sodium in kJ $mol^{-1}$, if electromagnetic radiation of wavelength 242 nm is just sufficient to ionise sodium atom is ........ *[2024, 31 Jan Shift-I]*

**14** The energy of one mole of photons of radiation of frequency $2 \times 10^{12}$ Hz in J $mol^{-1}$ is ......... (Nearest integer) [Given, $h = 6.626 \times 10^{-34}$ Js, $N_A = 6.022 \times 10^{23}$ $mol^{-1}$] *[2023, 30 Jan Shift-I]*

**15** Values of work function ($W_0$) for a few metals are given below *[2023, 12 April Shift-I]*

| Metal | Li | Na | K | Mg | Cu | Ag |
|---|---|---|---|---|---|---|
| $W_0$ / eV | 2.42 | 2.3 | 2.25 | 3.7 | 4.8 | 4.3 |

The number of metals which will show photoelectric effect when light of wavelength 400nm falls on it is ..........

Given, $h = 6.6 \times 10^{-34}$ Js, $c = 3 \times 10^8$ $ms^{-1}$

$e = 1.6 \times 10^{-19}$ C

**16** The number of incorrect statement/s about the black body from the following is .......... *[2023, 10 April Shift-I]*

(A) Emit or absorb energy in the form of electromagnetic radiation.

(B) Frequency distribution of the emitted radiation depends on temperature.

(C) At a given temperature, intensity *vs* frequency curve passes through a maximum value.

(D) The maximum of the intensity *vs* frequency curve is at a higher frequency at higher temperature compared to that at lower temperature.

**17** Consider an imaginary ion $^{48}_{22}X^{3-}$. The nucleus contains '$a$'% more neutrons than the number of electrons in the ion. The value of '$a$' is ........... . [nearest integer] *[2022, 26 July Shift-II]*

**18** If the work function of a metal is $6.63 \times 10^{-19}$ J, the maximum wavelength of the photon required to remove a photoelectron from the metal is ...... nm. [Nearest integer] [Given, $h = 6.63 \times 10^{-34}$ Js, and $c = 3 \times 10^8$ m $s^{-1}$] *[2022, 28 June Shift-I]*

**19** The longest wavelength of light that can be used for the ionisation of lithium atom (Li) in its ground state is $x \times 10^{-8}$ m. The value of $x$ is ...... [Nearest integer] (Given, energy of the electron in the first shell of the hydrogen atom is $-2.2 \times 10^{-18}$ J; $h = 6.63 \times 10^{-34}$ Js and $c = 3 \times 10^8$ $ms^{-1}$) *[2022, 25 June Shift-I]*

**20** A 50 watt bulb emits monochromatic red light of wavelength of 795 nm. The number of photons emitted per second by the bulb is $x \times 10^{20}$. The value of $x$ is ......... [Given, $h = 6.63 \times 10^{-34}$ Js and $c = 3.0 \times 10^{8}$ ms$^{-1}$] *[2021, 1 Sep Shift-II]*

**21** The number of photons emitted by a monochromatic (single frequency) infrared range finder of power 1 mW and wavelength of 1000 nm, in 0.1 second is $x \times 10^{13}$. The value of $x$ is ......... (Nearest integer) ($h = 6.63 \times 10^{-34}$ Js, $c = 3.00 \times 10^{8}$ ms$^{-1}$) *[2021, 27 Aug Shift-II]*

**22** A metal surface is exposed to 500 nm radiation. The threshold frequency of the metal for photoelectric current is $4.3 \times 10^{14}$ Hz. The velocity of ejected electron is ........ $\times 10^{5}$ ms$^{-1}$. (Nearest integer) [Use $h = 6.63 \times 10^{-34}$ Js, $m_e = 9.0 \times 10^{-31}$ kg] *[2021, 26 Aug Shift-II]*

**23** The work function of sodium metal is $4.41 \times 10^{-19}$ J. If photons of wavelength 300 nm are incident on the metal, the kinetic energy of the ejected electrons will be ($h = 6.63 \times 10^{-34}$ J s; $c = 3 \times 10^{8}$ m/s) ........... $\times 10^{-21}$ J. *[2020, 2 Sep Shift-II]*

## Topic 2 Bohr's Model and Hydrogen Spectrum

### Level 1

**01** If $a_0$ is denoted as the Bohr radius of hydrogen atom, then what is the de-Broglie wavelength (λ) of the electron present in the second orbit of hydrogen atom? [$n$ : any integer] *[2025, 29 Jan Shift-I]*

(a) $\frac{2a_0}{n\pi}$ (b) $\frac{8\pi a_0}{n}$

(c) $\frac{4\pi a_0}{n}$ (d) $\frac{4n}{\pi a_0}$

**02** Radius of the first excited state of helium ion is given as: $a_0 \rightarrow$ radius of first stationary state of hydrogen atom. *[2025, 22 Jan Shift-I]*

(a) $r = \frac{a_0}{2}$ (b) $r = 2a_0$ (c) $r = \frac{a_0}{4}$ (d) $r = 4a_0$

**03** Match List-I with List-II. *[2024, 29 Jan Shift-II]*

| | List I (Spectral series for hydrogen) | | List II (Spectral region \ higher energy state) |
|---|---|---|---|
| A. | Lyman | I. | Infrared region |
| B. | Balmer | II. | UV region |
| C. | Paschen | III. | Infrared region |
| D. | Pfund | IV. | Visible region |

Choose the correct answer from the options given below

(a) A-I, B-III, C-II, D-IV (b) A-II, B-IV, C-III, D-I

(c) A-I, B-II, C-III, D-IV (d) A-II, B-III, C-I, D-IV

**04** Given below are two statements.

**Statement I** According to Bohr's model of hydrogen atom, the angular momentum of an electron in a given stationary state is quantised.

**Statement II** The concept of electron in Bohr's orbit, violates the Heisenberg uncertainty principle.

In the light of the above statements, choose the most appropriate answer from the options given below. *[2023, 15 April Shift-I]*

(a) Statement I is incorrect but Statement II is correct.
(b) Statement I is correct but Statement II is incorrect.
(c) Both Statement I and Statement II are incorrect.
(d) Both Statement I and Statement II are correct.

**05** Which transition in the hydrogen spectrum would have the same wavelength as the Balmer type transition from $n = 4$ to $n = 2$ of $He^+$ spectrum *[2023, 31 Jan Shift-I]*

(a) $n = 1$ to $n = 3$ (b) $n = 1$ to $n = 2$

(c) $n = 3$ to $n = 4$ (d) $n = 2$ to $n = 1$

Imp. Bohr's Model of H-Atom & Heisenberg Uncertainty Principle

**06** Given below are two statements.

**Statement I** Rutherford's gold foil experiment cannot explain the line spectrum of hydrogen atom.

**Statement II** Bohr's model of hydrogen atom contradicts Heisenberg's uncertainty principle.

In the light of the above statements, choose the most appropriate answer from the options given below : *[2021, 27 July Shift-I]*

(a) Statement I is false but statement II is true.
(b) Statement I is true but statement II is false.
(c) Both statement I and statement II are false.
(d) Both statement I and statement II are true.

**07** The region in the electromagnetic spectrum where the Balmer series lines appear is *[2020, 4 Sep Shift-I]*

(a) infrared (b) ultraviolet

(c) microwave (d) visible

**08** The radius of the second Bohr orbit in terms of the Bohr radius, $a_0$ in $Li^{2+}$ is *[2020, 8 Jan Shift-II]*

(a) $\frac{2a_0}{3}$ (b) $\frac{4a_0}{3}$ (c) $\frac{4a_0}{9}$ (d) $\frac{2a_0}{9}$

**09** The ratio of the shortest wavelength of two spectral series of hydrogen spectrum is found to be about 9. The spectral series are *[2019, 10 April Shift-II]*

(a) Lyman and Paschen (b) Brackett and Pfund

(c) Paschen and Pfund (d) Balmer and Brackett

10 The radius of the second Bohr orbit for hydrogen atom is (Planck's constant ($h$) $=6.6262\times10^{-34}$ Js; mass of electron $=9.1091\times10^{-31}$ kg; charge of electron ($e$) $=1.60210\times10^{-19}$ C; permitivity of vacuum ($\varepsilon_0$) $=8.854185\times10^{-12}\text{kg}^{-1}\text{m}^{-3}\text{A}^2\text{s}^4$) **[JEE Main 2017]**

(a) 1.65 Å (b) 4.76 Å
(c) 0.529 Å (d) 2.12 Å

11 If $m$ and $e$ are the mass and charge of the revolving electron in the orbit of radius $r$ for hydrogen atom, the total energy of the revolving electron will be **[JEE Main 2014]**

(a) $\frac{1}{2}\frac{e^2}{r}$ (b) $-\frac{e^2}{r}$
(c) $\frac{me^2}{r}$ (d) $-\frac{1}{2}\frac{e^2}{r}$

12 Energy of an electron is given by $E=-2.178\times10^{-18}\text{ J}\left(\frac{Z^2}{n^2}\right)$. Wavelength of light required to excite an electron in an hydrogen atom from level $n=1$ to $n=2$ will be ($h=6.62\times10^{-34}$ J s and $c=3.0\times10^8\text{ ms}^{-1}$) **[JEE Main 2013]**

(a) $1.214\times10^{-7}$ m (b) $2.816\times10^{-7}$ m
(c) $6.500\times10^{-7}$ m (d) $8.500\times10^{-7}$ m

13 The limiting line in Balmer series will have a frequency of (Rydberg constant, $R=3.29\times10^{15}$ cycles/s) **[AIEEE 2012]**

(a) $8.22\times10^{14}\text{ s}^{-1}$ (b) $3.29\times10^{15}\text{ s}^{-1}$
(c) $3.65\times10^{14}\text{ s}^{-1}$ (d) $5.26\times10^{13}\text{ s}^{-1}$

14 Ionisation energy of $He^+$ is $19.6\times10^{-18}$ J atom$^{-1}$. The energy of the first stationary state ($n=1$) of $Li^{2+}$ is **[AIEEE 2010]**

(a) $8.82\times10^{-17}$ J atom$^{-1}$
(b) $4.41\times10^{-16}$ J atom$^{-1}$
(c) $-4.41\times10^{-17}$ J atom$^{-1}$
(d) $-2.2\times10^{-15}$ J atom$^{-1}$

15 According to Bohr's theory, the angular momentum of an electron in 5th orbit is **[AIEEE 2006]**

(a) $25\frac{h}{\pi}$ (b) $1.0\frac{h}{\pi}$
(c) $10\frac{h}{\pi}$ (d) $2.5\frac{h}{\pi}$

16 The wavelength of the radiation emitted, when in a hydrogen atom electron falls from infinity to stationary state 1, would be (Rydberg constant $=1.097\times10^7\text{ m}^{-1}$) **[AIEEE 2004]**

(a) 91 nm (b) 192 nm
(c) 406 nm (d) $9.1\times10^{-8}$ nm

## Level 2

17 Which of the following postulate of Bohr's model of hydrogen atom is not in agreement with quantum mechanical model of an atom? **[2025, 3 April Shift-I]**

(a) An atom in a stationary state does not emit electromagnetic radiation as long as it stays in the same state.
(b) The electron in a H atom's stationary state moves in a circle around the nucleus.
(c) When an electron makes a transition from a higher energy stationary state to a lower energy stationary state, then it emits a photon of light.
(d) An atom can take only certain distinct energies $E_1$, $E_2$, $E_3$, etc. These allowed states of constant energy are called the stationary states of atom.

18 Heat treatment of muscular pain involves radiation of wavelength of about 900 nm. Which spectral line of H atom is suitable for this?
(Given, Rydberg constant $R_H=10^5\text{ cm}^{-1}$, $h=6.6\times10^{-34}$ J s, $c=3\times10^8$ m/s) **[2025, 23 Jan Shift-I]**

(a) Balmer series, $\infty\rightarrow2$
(b) Paschen series, $5\rightarrow3$
(c) Lyman series, $\infty\rightarrow1$
(d) Paschen series, $\infty\rightarrow3$

19 The energy of an electron in the first Bohr orbit of hydrogen atom is $-2.18\times10^{-18}$ J. Its energy in the third Bohr orbit is ........... . **[2023, 13 April Shift-I]**

(a) three times of this value (b) $\frac{1}{9}$th of this value
(c) one third of this value (d) $\frac{1}{27}$ of this value

20 If the radius of the 3rd Bohr's orbit of hydrogen atom is $r_3$ and the radius of 4th Bohr's orbit is $r_4$. Then **[2022, 26 June Shift-I]**

(a) $r_4=\frac{9}{16}r_3$ (b) $r_4=\frac{16}{9}r_3$ (c) $r_4=\frac{3}{4}r_3$ (d) $r_4=\frac{4}{3}r_3$

21 Given below are two statements.

**Statement I** According to Bohr's model of an atom, qualitatively the magnitude of velocity of electron increases with decrease in positive charges on the nucleus as there is no strong hold on the electron by the nucleus.

**Statement II** According to Bohr's model of an atom, qualitatively the magnitude of velocity of electron increases with decrease in principal quantum number.

In the light of the above statements, choose the most appropriate answer from the options given below. **[2021, 26 Aug Shift-I]**

(a) Both statement I and statement II are false.
(b) Both statement I and statement II are true.
(c) Statement I is false but statement II is true.
(d) Statement I is true but statement II is false.

Imp.

### Limitations of Bohr's Model of an Atom

**22** Given below are two statements.

**Statement I** Bohr's theory accounts for the stability and line spectrum of $Li^+$ ion.

**Statement II** Bohr's theory was unable to explain the splitting of spectral lines in the presence of a magnetic field.

In the light of the above statements, choose the most appropriate answer from the options given below : *[2021, 18 March Shift-II]*

(a) Both statements I and II are true.
(b) Statement I is false but statement II is true.
(c) Both statements I and II are false.
(d) Statement I is true but statement II is false.

**23** The difference between the radii of 3rd and 4th orbits of $Li^{2+}$ is $\Delta R_1$. The difference between the radii of 3rd and 4th orbits of $He^+$ is $\Delta R_2$. Ratio $\Delta R_1 : \Delta R_2$ is *[2020, 5 Sep Shift-I]*

(a) 3 : 2 (b) 8 : 3 (c) 2 : 3 (d) 3 : 8

**24** For the Balmer series in the spectrum of H atom, $\bar{\nu} = R_H \left\{\frac{1}{n_1^2} - \frac{1}{n_2^2}\right\}$, the correct statements among (I) to (IV) are :

I. As wavelength decreases, the lines in the series converge
II. The integer $n_1$ is equal to 2
III. The lines of longest wavelength corresponds to $n_2 = 3$
IV. The ionisation energy of hydrogen can be calculated from wave number of these lines *[2020, 8 Jan Shift-I]*

(a) I, II, IV (b) II, III, IV
(c) I, III, IV (d) I, II, III

**25** For emission line of atomic hydrogen from $n_i = 8$ to $n_f = n$, the plot of wave number ($\bar{\nu}$) against $\left(\frac{1}{n^2}\right)$ will be (The Rydberg constant, $R_H$ is in wave number unit) *[2019, 9 Jan Shift-I]*

(a) non linear (b) linear with slope $-R_H$
(c) linear with slope $R_H$ (d) linear with intercept $-R_H$

**26** The ground state energy of hydrogen atom is −13.6 eV. The energy of second excited state of $He^+$ ion in eV is *[2019, 10 Jan Shift-II]*

(a) −54.4 (b) − 3.4
(c) − 6.04 (d) − 27.2

**27** Which of the following statements is false? *[JEE Main 2018]*

(a) Photon has momentum as well as wavelength.
(b) Splitting of spectral lines in electrical field is called Stark effect.
(c) Frequency of emitted radiation from a black body goes from a lower wavelength to higher wavelength as the temperature increases.
(d) Rydberg constant has unit of energy.

**28** Which of the following is the energy of a possible excited state of hydrogen? *[JEE Main 2015]*

(a) + 13.6 eV (b) − 6.8 eV
(c) − 3.4 eV (d) + 6.8 eV

**29** The ionisation enthalpy of hydrogen atom is $1.312 \times 10^6$ J mol$^{-1}$. The energy required to excite the electron in the atom from $n = 1$ to $n = 2$ is *[AIEEE 2008]*

(a) $9.84 \times 10^5$ J mol$^{-1}$ (b) $8.51 \times 10^5$ J mol$^{-1}$
(c) $6.56 \times 10^5$ J mol$^{-1}$ (d) $7.56 \times 10^5$ J mol$^{-1}$

## Level 3

**30** Which one of the following about an electron occupying the Is orbital in a hydrogen atom is incorrect? (Bohr's radius is represented by $a_0$) *[2025, 4 April Shift-I]*

(a) The total energy of the electron is maximum when it is at a distance $a_0$ from the nucleus.
(b) The probability density of finding the electron is maximum at the nucleus.
(c) The electron can be found at a distance $2a_0$ from the nucleus.
(d) The 1s orbital is spherically symmetrical.

**31**

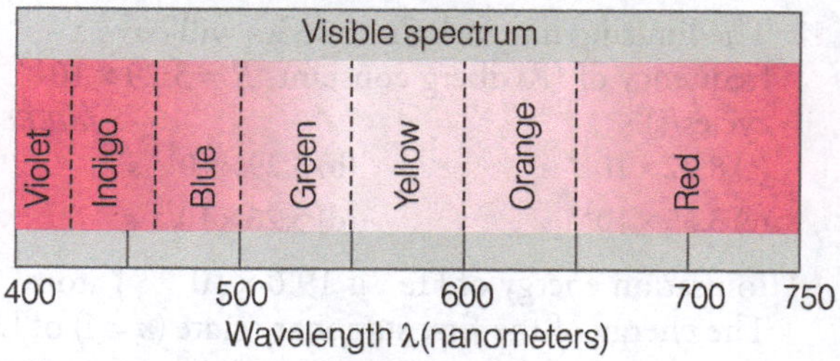


Which of the following statements are correct, if the threshold frequency of caesium is $5.16 \times 10^{14}$ Hz?

A. When Cs is placed inside a vacuum chamber with an ammeter connected to it and yellow light is focused on Cs, the ammeter shows the presence of current.
B. When the brightness of the yellow light is dimmed, the value of the current in the ammeter is reduced.
C. When a red light is used instead of the yellow light, the current produced is higher with respect to the yellow light.
D. When a blue light is used, the ammeter shows the formation of current.
E. When a white light is used, the ammeter shows formation of current.

Choose the correct answer from the options given below: *[2025, 7 April Shift-I]*

(a) B, C and D Only (b) A, D and E Only
(c) A, C, D and E Only (d) A, B, D and E Only

**32** According to Bohr's model of hydrogen atom, which of the following statement is incorrect? *[2025, 2 April Shift-I]*
(a) Radius of 8th orbit is four times larger than that of 4th orbit
(b) Radius of 6th orbit is three times larger than that of 4th orbit
(c) Radius of 4th orbit is four times larger than that of 2nd orbit
(d) Radius of 3rd orbit is nine times larger than that of 1st orbit

**33** For hydrogen like species, which of the following graphs provides the most appropriate representation of $E$ *vs* $Z$ plot for a constant $n$?
[$E$ = energy of the stationary state, $Z$ = atomic number, $n$ = principal quantum number] *[2025, 29 Jan Shift-II]*

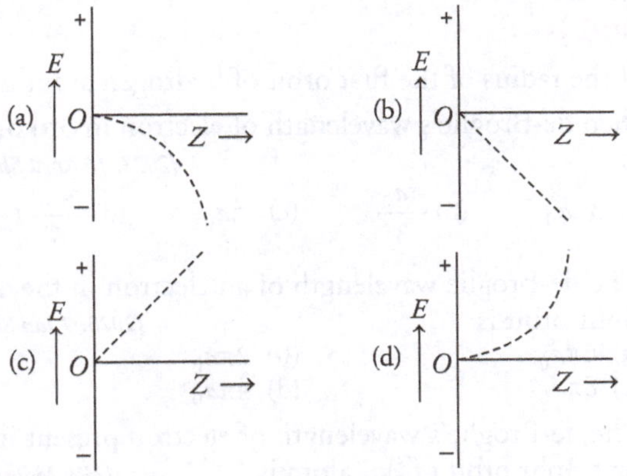


**34** The shortest wavelength of hydrogen atom in Lyman series is $\lambda$. The longest wavelength in Balmer series of $He^+$ is *[2023, 29 Jan Shift-I]*
(a) $\frac{5}{9\lambda}$ (b) $\frac{9\lambda}{5}$
(c) $\frac{5\lambda}{9}$ (d) $\frac{36\lambda}{5}$

**35** According to Bohr's atomic theory,
I. kinetic energy of electron is $\propto \frac{Z^2}{n^2}$
II. the product of velocity ($v$) of electron and principal quantum number ($n$), '$vn$' $\propto Z^2$.
III. frequency of revolution of electron in an orbit is $\propto \frac{Z^3}{n^3}$
IV. coulombic force of attraction on the electron is $\propto \frac{Z^3}{n^4}$

Choose the most appropriate answer from the options given below. *[2021, 24 Feb Shift-II]*
(a) Only III (b) Only I
(c) I, III and IV (d) I and IV

**36** For any given series of spectral lines of atomic hydrogen, let $\Delta\bar{v} = \bar{v}_{max} - \bar{v}_{min}$ be the difference in maximum and minimum frequencies in $cm^{-1}$. The ratio $\Delta\bar{v}_{Lyman} / \Delta\bar{v}_{Balmer}$ is *[2019, 9 April Shift-I]*
(a) 27 : 5 (b) 5 : 4 (c) 9 : 4 (d) 4 : 1

## Numerical Value Type Questions

**37** The energy of an electron in first Bohr orbit of H-atom is – 13.6 eV. The magnitude of energy value of electron in the first excited state of $Be^{3+}$ is ............ eV (Nearest integer value) *[2025, 8 April Shift-II]*

**38** For hydrogen atom, energy of an electron in first excited state is −3.4 eV, KE of the same electron of hydrogen atom is $x$ eV. Value of $x$ is ..........$\times 10^{-1}$ eV. (Nearest integer) *[2024, 6 April Shift-II]*

**39** The value of Rydberg constant ($R_H$) is $2.18 \times 10^{-18}$ J. The velocity of electron having mass $9.1 \times 10^{-31}$ kg in Bohr's first orbit of hydrogen atom = ..... $\times 10^5$ $ms^{-1}$. (Nearest integer). *[2024, 5 April Shift-I]*

**40** Number of spectral lines obtained in $He^+$ spectra, when an electron makes transition from fifth excited state to first excited state will be ...... . *[2024, 30 Jan Shift-I]*

**41** The number of following statement/s which is/are incorrect is *[2023, 8 April Shift-I]*
(A) line emission spectra are used to study the electronic structure.
(B) the emission spectra of atoms in the gas phase show a continuous spread of wavelength from red to violet.
(C) an absorption spectrum is like the photographic negative of an emission spectrum.
(D) the element helium was discovered in the sun by spectroscopic method.

**42** The electron in the $n^{th}$ orbit of $Li^{2+}$ is excited to ($n$ + 1) orbit using the radiation of energy $1.47 \times 10^{-17}$ J (as shown in the diagram). The value of $n$ is ......... [Given: $R_H = 2.18 \times 10^{-18}$ J]

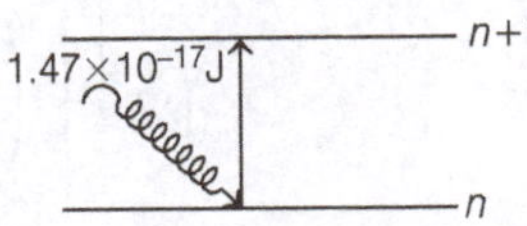


*[2023, 10 April Shift-II]*

**43** If wavelength of the first line of the Paschen series of hydrogen atom is 720 nm, then the wavelength of the second line of this series is ......... nm. (Nearest integer) *[2023, 24 Jan Shift-I]*

**44** Assume that the radius of the first Bohr orbit of hydrogen atom is 0.6 Å. The radius of the third Bohr orbit of $He^+$ is .......... picometer. (Nearest integer) *[2023, 29 Jan Shift-II]*

**45** When the excited electron of a H-atom from $n = 5$ drops to the ground state, the maximum number of emission lines observed are ....... *[2022, 25 July Shift-II]*

**46** The kinetic energy of an electron in the second Bohr orbit of a hydrogen atom is equal to $\frac{h^2}{xma_0^2}$. The value of $10x$ is .........
($a_0$ is radius of Bohr's orbit) (Nearest integer)
[Given, $\pi = 3.14$] *[2021, 27 Aug Shift-I]*

## Topic 3 Dual Nature of Matter and Uncertainity Principle

### Level 1

**01** Given below are two statements. *[2025, 29 Jan Shift-II]*

**Statement I** It is impossible to specify simultaneously with arbitrary precision, both the linear momentum and the position of a particle.

**Statement II** If the uncertainty in the measurement of position and uncertainty in measurement of momentum are equal for an electron, then the uncertainty in the measurement of velocity is

$$\geq \sqrt{\frac{h}{\pi}} \times \frac{1}{2m}$$

In the light of the above statements, choose the correct answer from the options given below

(a) Statement I is false but Statement II is true.
(b) Both Statement I and Statement II are false.
(c) Both Statement I and Statement II are true.
(d) Statement I is true but Statement II is false.

**Imp.** Graphical Relation between Momentum and Wavelength of Matter

**02** According to the wave-particle duality of matter by de-Broglie, which of the following graph plot presents **most appropriate** relationship between wavelength of electron ($\lambda$) and momentum of electron ($p$)? *[2024, 1 Feb Shift-I]*

(a) 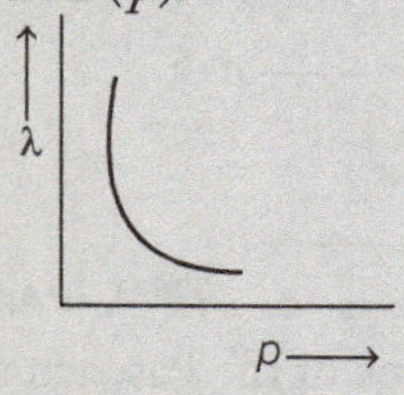

(b) 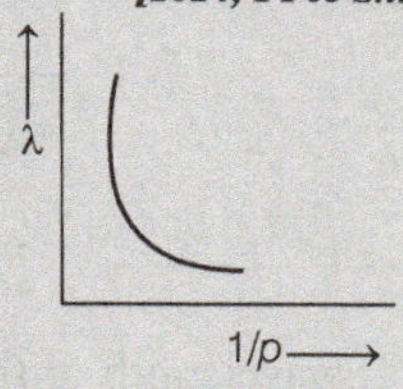

(c) 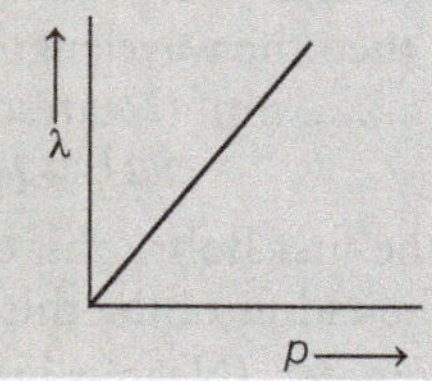

(d) 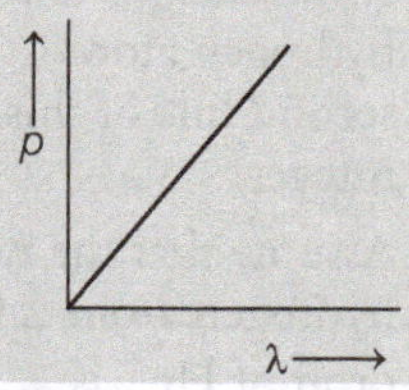


**03** Calculate the wavelength (in nanometre) associate with a proton moving at $1.0 \times 10^3$ m s$^{-1}$. *[AIEEE 2009]*
(Mass of proton $= 1.67 \times 10^{-27}$ kg and $h = 6.63 \times 10^{-34}$ J s)

(a) 0.032 nm (b) 0.40 nm (c) 2.5 nm (d) 14.0 nm

### Level 2

**04** The de-Broglie wavelength ($\lambda$) associated with a photoelectron varies with the frequency ($\nu$) of the incident radiation as, [$\nu_0$ is threshold frequency] *[2019, 11 Jan Shift-II]*

(a) $\lambda \propto \frac{1}{(\nu - \nu_0)^{1/4}}$ (b) $\lambda \propto \frac{1}{(\nu - \nu_0)^{3/2}}$

(c) $\lambda \propto \frac{1}{(\nu - \nu_0)}$ (d) $\lambda \propto \frac{1}{(\nu - \nu_0)^{1/2}}$

### Level 3

**05** If the radius of the first orbit of hydrogen atom is $a_0$, then de-Broglie's wavelength of electron in 3rd orbit is *[2023, 16 April Shift-II]*

(a) $3\pi a_0$ (b) $\frac{\pi a_0}{3}$ (c) $6\pi a_0$ (d) $\frac{\pi a_0}{6}$

**06** The de-Broglie wavelength of an electron in the 4th Bohr orbit is *[2020, 9 Jan Shift-I]*

(a) $6\pi a_0$ (b) $2\pi a_0$
(c) $8\pi a_0$ (d) $4\pi a_0$

**07** The de-Broglie's wavelength of electron present in first Bohr orbit of 'H' atom is *[JEE Main 2018]*

(a) $\frac{0.529}{2\pi}$ Å
(b) $2\pi \times 0.529$ Å
(c) 0.529 Å
(d) $4 \times 0.529$ Å

## Numerical Value Type Questions

**08** Based on Heisenberg's uncertainty principle, the uncertainty in the velocity of the electron to be found within an atomic nucleus of diameter $10^{-15}$ m is ............ $\times 10^9$ ms$^{-1}$. (Nearest integer) *[2024, 9 April Shift-II]*

**09** Frequency of the de-Broglie wave of electron in Bohr's first orbit of hydrogen atom is ....... $10^{13}$ Hz (Nearest integer).
[Given : $R_H$ (Rydberg constant) $= 2.18 \times 10^{-18}$ J, $h$(Plank's constant) $= 6.6 \times 10^{-34}$ Js] *[2024, 6 April Shift-I]*

**10** The de-Broglie's wavelength of an electron in the 4th orbit is...... $\pi a_0$. ($a_0$ = Bohr's radius) *[2024, 4 April Shift-I]*

**11** The wavelength of an electron of kinetic energy $4.50 \times 10^{-29}$ J is ......... $\times 10^{-5}$ m. (Nearest integer) *[2023, 6 April Shift-I]*

**12** Electrons in a cathode ray tube have been emitted with a velocity of 1000 m $s^{-1}$. The number of following statements which is/are true about the emitted radiation is ............. .
[Given, $h = 6 \times 10^{-34}$ J s, $m_e = 9 \times 10^{-31}$ kg.] **[2023, 1 Feb Shift-I]**
(A) The de-Broglie wavelength of the electron emitted is 666.67 nm.
(B) The characteristic of electrons emitted depend upon the material of the electrodes of the cathode ray tube.
(C) The cathode rays start from cathode and move towards anode.
(D) The nature of the emitted electrons depends on the nature of the gas present in cathode ray tube.

**13** The minimum uncertainty in the speed of an electron in an one dimensional region of length $2a_0$ (where, $a_0$ = Bohr radius 52.9 pm) is ...... $kms^{-1}$.
(Given, mass of electron $= 9.1 \times 10^{-31}$ kg. Planck's constant, $h = 6.63 \times 10^{-34}$ Js) **[2022, 29 July Shift-I]**

**14** If the wavelength for an electron emitted from H-atom is $3.3 \times 10^{-10}$ m, then energy absorbed by the electron in its ground state compared to minimum energy required for its escape from the atom, is .............. times. (Nearest integer)
[Given, $h = 6.626 \times 10^{-34}$ Js]
[Mass of electron $= 9.1 \times 10^{-31}$ kg] **[2022, 28 July Shift-II]**

**15** The wavelength of an electron and a neutron will become equal when the velocity of the electron is $x$ times the velocity of neutron. The value of $x$ is ........... . (Nearest integer) (Mass of electron is $9.1 \times 10^{-31}$ kg and mass of neutron is $1.6 \times 10^{-27}$ kg) **[2022, 26 July Shift-I]**

**16** An accelerated electron has a speed of $5 \times 10^6 ms^{-1}$ with an uncertainty of 0.02%. The uncertainty in finding its location while in motion is $x \times 10^{-9}$ m. The value of $x$ is ......... . (Nearest integer)
[Use mass of electron $= 9.1 \times 10^{-31}$ kg, $h = 6.63 \times 10^{-34}$ Js, $\pi = 3.14$] **[2021, 25 July Shift-II]**

**17** The wavelength of electrons accelerated from rest through a potential difference of 40 kV is $x \times 10^{-12}$ m. The value of $x$ is ...... (Nearest integer)
Given : Mass of electron $= 9.1 \times 10^{-31}$ kg
Charge on an electron $= 1.6 \times 10^{-19}$ C
Planck's constant $= 6.63 \times 10^{-34}$ Js **[2021, 20 July Shift-II]**

**18** When light of wavelength 248 nm falls on a metal of threshold energy 3.0 eV, the de- Broglie wavelength of emitted electrons is ...... Å.
[Round off to the nearest integer]
[Use : $\sqrt{3} = 1.73$, $h = 6.63 \times 10^{-34}$ Js, $m_e = 9.1 \times 10^{-31}$ kg, $c = 3.0 \times 10^8$ $ms^{-1}$, 1 eV $= 1.6 \times 10^{-19}$ J] **[2021, 16 March Shift-I]**

**19** A ball weighing 10 g is moving with a velocity of 90 $ms^{-1}$. If the uncertainity in its velocity is 5%, then the uncertainty in its position is ........ $\times 10^{-33}$ m (Rounded off to the nearest integer). [Given, $h = 6.63 \times 10^{-34}$ J-s] **[2021, 26 Feb Shift-II]**

## Topic 4 Quantum Mechanical Model

### Level 1

**01** The correct set of four quantum numbers for the valence electron of rubidium atom ($Z = 37$) is **[2024, 29 Jan Shift-I, JEE Main 2014]**
(a) $5, 1, 1, +\frac{1}{2}$ (b) $5, 1, 0, +\frac{1}{2}$
(c) $5, 0, 0, +\frac{1}{2}$ (d) $5, 0, 1, +\frac{1}{2}$

**02** The number of radial nodes for 3 $p$-orbital is **[2024, 1 Feb Shift-II]**
(a) 1 (b) 2 (c) 4 (d) 3

**03** The wave function ($\psi$) of $2s$ is given by

$$\psi_{2s} = \frac{1}{2\sqrt{2\pi}}\left(\frac{1}{a_0}\right)^{1/2}\left(2 - \frac{r}{a_0}\right)e^{-\frac{r}{2a_0}}$$

At $r = r_0$, radial node is formed. Thus, $r_0$ in terms of $a_0$ **[2023, 30 Jan Shift-II]**
(a) $r_0 = a_0$ (b) $r_0 = 2a_0$ (c) $r_0 = \frac{a_0}{2}$ (d) $r_0 = 4a_0$

**04** Arrange the following orbitals in decreasing order of energy. **[2023, 31 Jan Shift-II]**
A. $n = 3, l = 0, m = 0$ B. $n = 4, l = 0, m = 0$
C. $n = 3, l = 1, m = 0$ D. $n = 3, l = 2, m = 1$
The correct option for the order is
(a) D > B > A > C (b) A > C > B > D
(c) B > D > C > A (d) D > B > C > A

**05** The number of radial and angular nodes in $4d$ orbital are respectively **[2022, 26 June Shift-II]**
(a) 1 and 2 (b) 3 and 2 (c) 1 and 0 (d) 2 and 1

**06** Given below are the quantum numbers for 4 electrons.
A. $n = 3, l = 2, m_1 = 1, m_s = +1/2$
B. $n = 4, l = 1, m_1 = 0, m_s = +1/2$
C. $n = 4, l = 2, m_1 = -2, m_s = -1/2$
D. $n = 3, l = 1, m_1 = -1, m_s = +1/2$
The correct order of increasing energy is **[2022, 29 July Shift-II]**
(a) D < B < A < C (b) D < A < B < C
(c) B < D < A < C (d) B < D < C < A

**07** Identify the incorrect statement from the following. *[2022, 28 July Shift-I]*

(a) A circular path around the nucleus in which an electron moves is proposed as Bohr's orbit.
(b) An orbital is the one electron wave function ($\psi$) in an atom.
(c) The existence of Bohr's orbits is supported by hydrogen spectrum.
(d) Atomic orbital is characterised by the quantum numbers $n$ and $l$ only.

**08** Which of the following sets of quantum numbers is not allowed? *[2022, 25 July Shift-I]*

(a) $n = 3, l = 2, m_l = 0, s = +\frac{1}{2}$

(b) $n = 3, l = 2, m_l = -2, s = +\frac{1}{2}$

(c) $n = 3, l = 3, m_l = -3, s = -\frac{1}{2}$

(d) $n = 3, l = 0, m_l = 0, s = -\frac{1}{2}$

**09** Given below are two statements. One is labelled as Assertion (A) and the other is labelled as Reason (R).

**Assertion (A)** Energy of 2*s* orbital of hydrogen atom is greater than that of 2*s* orbital of lithium.

**Reason (R)** Energies of the orbitals in the same subshell decrease with increase in the atomic number.

In the light of the above statements, choose the correct answer from the options given below: *[2022, 27 July, Shift-I]*

(a) Both A and R are true and R is the correct explanation of A.
(b) Both A and R are true but R is not the correct explanation of A.
(c) A is true but R is false.
(d) A is false but R is true.

**10** The electrons are more likely to be found *[2019, 12 April Shift-I]*

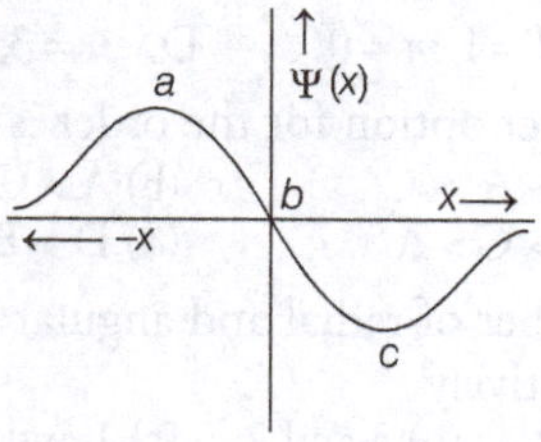


(a) in the region *a* and *c*
(b) in the region *a* and *b*
(c) only in the region *a*
(d) only in the region *c*

**11** The total number of orbitals associated with the principal quantum number 5 is *[JEE Main 2016]*

(a) 20 (b) 25
(c) 10 (d) 5

**12** In a multi-electron atom, which of the following orbitals described by the three quantum numbers will have the same energy in the absence of magnetic and electric fields?

(i) $n = 1, l = 0, m = 0$ (ii) $n = 2, l = 0, m = 0$
(iii) $n = 2, l = 1, m = 1$ (iv) $n = 3, l = 2, m = 1$
(v) $n = 3, l = 2, m = 0$ *[AIEEE 2005]*

(a) (i) and (ii) (b) (ii) and (iii)
(c) (iii) and (iv) (d) (iv) and (v)

## Level 2

**13** For hydrogen atom, the orbital/s with lowest energy is/are *[2025, 24 Jan Shift-II]*

(A) 4*s* (B) $3p_x$ (C) $3d_{x^2-y^2}$ (D) $3d_{z^2}$
(E) $4p_z$

Choose the correct answer from the options given below

(a) (B), (C) and (D) only (b) (A) and (E) only
(c) (B) only (d) (A) only

**14** For electrons in '2*s*' and '2*p*' orbitals, the orbital angular momentum values, respectively are *[2025, 3 April Shift-II]*

(a) $\sqrt{2}\frac{h}{2\pi}$ and 0 (b) 0 and $\sqrt{6}\frac{h}{2\pi}$

(c) $\frac{h}{2\pi}$ and $\sqrt{2}\frac{h}{2\pi}$ (d) 0 and $\sqrt{2}\frac{h}{2\pi}$

**15** Consider the ground state of chromium atom ($Z = 24$). How many electrons are with azimuthal quantum number $l = 1$ and $l = 2$ respectively? *[2025, 4 April Shift-II]*

(a) 12 and 5 (b) 16 and 5 (c) 16 and 4 (d) 12 and 4

**16** Given below are two statements. *[2025, 22 Jan Shift-II]*

**Statement I** A spectral line will be observed for a $2p_x \rightarrow 2p_y$ transition.

**Statement II** $2p_x$ and $2p_y$ are degenerate orbitals.

In the light of the above statements, choose the correct answer from the options given below:

(a) Both Statement I and Statement II are true.
(b) Statement I is false but Statement II is true.
(c) Both Statement I and Statement II are false.
(d) Statement I is true but Statement II is false.

**17** Which of the following is/are not correct with respect to energy of atomic orbitals of hydrogen atom?

(A) $1s < 2p < 3d < 4s$
(B) $1s < 2s = 2p < 3s = 3p$
(C) $1s < 2s < 2p < 3s < 3p$
(D) $1s < 2s < 4s < 3d$

Choose the correct answer from the options given below *[2025, 28 Jan Shift-II]*

(a) (A) and (B) only (b) (C) and (D) only
(c) (B) and (D) only (d) (A) and (C) only

**18** Given below are two statements. *[2025, 23 Jan Shift-II]*

**Statement I** For a given shell, the total number of allowed orbitals is given by $n^2$.

**Statement II** For any subshell, the spatial orientation of the orbitals is given by $-l$ to $+l$ values including zero.

In the light of the above statements, choose the correct answer from the options given below
(a) Statement I is false but Statement II is true.
(b) Statement I is true but Statement II is false.
(c) Both statement I and Statement II are true.
(d) Both statement I and Statement II are false.

**19** Given below are two statements. *[2024, 30 Jan Shift-I]*

**Statement I** The orbitals having same energy are called as degenerate orbitals.

**Statement II** In hydrogen atom, $3p$ and $3d$ orbitals are not degenerate orbitals.

In the light of the above statements, choose the **most appropriate** answer from the options given below
(a) Statement I is false but Statement II is true.
(b) Both Statement I and Statement II are false.
(c) Statement I is true but Statement II is false.
(d) Both Statement I and Statement II are true.

**20** Compare the energies of following sets of quantum numbers for multielectron system.
(A) $n = 4, l = 1$ (B) $n = 4, l = 2$
(C) $n = 3, l = 1$ (D) $n = 3, l = 2$
(E) $n = 4, l = 0$

Choose the correct answer from the options given below: *[2024, 9 April Shift-I]*
(a) (E) < (C) < (D) < (A) < (B)
(b) (C) < (E) < (D) < (A) < (B)
(c) (B) > (A) > (C) > (E) > (D)
(d) (E) > (C) > (A) > (D) > (B)

**21** The four quantum numbers for the electron in the outermost orbital of potassium (atomic no. 19) are *[2024, 31 Jan Shift-II]*
(a) $n = 4, l = 2, m = -1, s = +\frac{1}{2}$
(b) $n = 2, l = 0, m = 0, s = +\frac{1}{2}$
(c) $n = 4, l = 0, m = 0, s = +\frac{1}{2}$
(d) $n = 3, l = 0, m = 1, s = +\frac{1}{2}$

**22** Consider the following pairs of electrons : *[2022, 24 June Shift-I]*
(A) (a) $n = 3, l = 1, m_1 = -1, m_s = +1/2$
(b) $n = 3, l = 2, m_1 = 1, m_s = +1/2$
(B) (a) $n = 3, l = 2, m_1 = -2, m_s = -1/2$
(b) $n = 3, l = 2, m_1 = -1, m_s = -1/2$
(C) (a) $n = 4, l = 2, m_1 = 2, m_s = +1/2$
(b) $n = 3, l = 2, m_1 = 2, m_s = +1/2$

The pairs of electron present in degenerate orbitals is/are
(a) Only A (b) Only B
(c) Only C (d) Both (B) and (C)

**23** Which of the following is the correct plot for the probability density $\psi^2(r)$ as a function of distance '$r$' of the electron from the nucleus for $2s$ orbital? *[2022, 29 June Shift-II]*

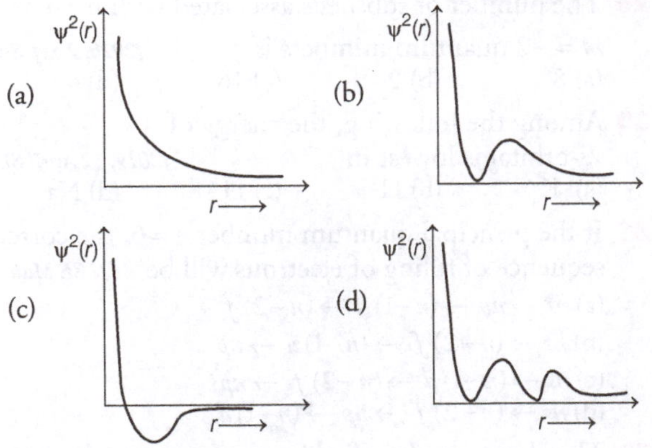


**24** Which of the following statements are correct?
(A) The electronic configuration of Cr is $[Ar]3d^5 4s^1$.
(B) The magnetic quantum number may have a negative value.
(C) In the ground state of an atom, the orbitals are filled in order of their increasing energies.
(D) The total number of nodes are given by $n - 2$.

Choose the most appropriate answer from the options given below *[2022, 29 June Shift-I]*
(a) (A), (C) and (D) only (b) (A) and (B) only
(c) (A) and (C) only (d) (A), (B) and (C) only

**Imp.** Quantum Numbers

**25** Consider the following statements.
(A) The principal quantum number '$n$' is a positive integer with values of '$n$' = 1, 2, 3, ....
(B) The azimuthal quantum number '$l$' for a given '$n$' (principal quantum number) can have values as '$l$' = 0, 1, 2,... $n$.
(C) Magnetic orbital quantum number '$m_l$' for a particular '$l$' (azimuthal quantum number) has $(2l + 1)$ values.
(D) $\pm 1/2$ are the two possible orientations of electron spin.
(E) For $l = 5$, there will be a total of 9 orbital.

Which of the above statements are correct? *[2022, 28 June Shift-II]*
(a) (A), (B) and (C) (b) (A), (C), (D) and (E)
(c) (A), (C) and (D) (d) (A), (B), (C) and (D)

**26** The orbital having two radial as well as two angular nodes is *[2021, 26 Feb Shift-I]*
(a) $3p$ (b) $4f$ (c) $4d$ (d) $5d$

**27** A certain orbital has no angular nodes and two radial nodes. The orbital is *[2021, 18 March Shift-I]*
(a) $2s$ (b) $3s$ (c) $3p$ (d) $2p$

**28** The correct statement about probability density (except at infinite distance from nucleus) is *[2020, 5 Sep Shift-II]*
(a) it can be zero for $1s$ orbital
(b) it can be negative for $2p$ orbital
(c) it can be zero for $3p$ orbital
(d) it can never be zero for $2s$ orbital

**29** The number of subshells associated with $n = 4$ and $m = -2$ quantum numbers is *[2020, 2 Sep Shift-II]*
(a) 8 (b) 2 (c) 16 (d) 4

**30** Among the following, the energy of $2s$-orbital is lowest in *[2019, 12 April Shift-II]*
(a) K (b) H (c) Li (d) Na

**31** If the principal quantum number $n = 6$, the correct sequence of filling of electrons will be *[JEE Main 2015]*
(a) $ns \rightarrow np \rightarrow (n-1)d \rightarrow (n-2)f$
(b) $ns \rightarrow (n-2)f \rightarrow (n-1)d \rightarrow np$
(c) $ns \rightarrow (n-1)d \rightarrow (n-2)f \rightarrow np$
(d) $ns \rightarrow (n-2)f \rightarrow np \rightarrow (n-1)d$

**32** The electrons identified by quantum numbers $n$ and $l$
(1) $n = 4, l = 1$ (2) $n = 4, l = 0$
(3) $n = 3, l = 2$ (4) $n = 3, l = 1$
can be placed in order of increasing energy as *[AIEEE 2012]*
(a) (4) < (2) < (3) < (1) (b) (2) < (4) < (1) < (3)
(c) (1) < (3) < (2) < (4) (d) (3) < (4) < (2) < (1)

**33** Which of the following sets of quantum numbers represent the highest energy of an atom? *[AIEEE 2007]*
(a) $n = 3, l = 0, m = 0, s = +\frac{1}{2}$
(b) $n = 3, l = 1, m = 1, s = +\frac{1}{2}$
(c) $n = 3, l = 2, m = 1, s = +\frac{1}{2}$
(d) $n = 4, l = 0, m = 0, s = +\frac{1}{2}$

## Level 3

**34** In a multielectron atom, which of the following orbitals described by three quantum numbers will have same energy in absence of electric and magnetic fields? *[2025, 28 Jan Shift-I]*
A. $n = 1, l = 0, m_l = 0$ B. $n = 2, l = 0, m_l = 0$
C. $n = 2, l = 1, m_l = 1$ D. $n = 3, l = 2, m_l = 1$
E. $n = 3, l = 2, m_l = 0$

Choose the correct answer from the options given below.
(a) A and B only (b) D and E only
(c) C and D only (d) B and C only

**35** Correct statements for an element with atomic number 9 are *[2025, 8 April Shift-II]*
A. There can be 5 electrons for which $m_s = +\frac{1}{2}$ and 4 electrons for which $m_s = -\frac{1}{2}$.
B. There is only one electron in $p_z$ orbital.
C. The last electron goes to orbital with $n = 2$ and $l = 1$.
D. The sum of angular nodes of all the atomic orbitals is 1.

Choose the correct answer from the options given below.
(a) A and C Only
(b) A and B Only
(c) C and D Only
(d) A, C and D Only

**36** The extra stability of half-filled subshell is due to
(A) symmetrical distribution of electrons.
(B) smaller coulombic repulsion energy.
(C) the presence of electrons with the same spin in non-degenerate orbitals.
(D) larger exchange energy.
(E) relatively smaller shielding of electrons by one another.

Indentify the correct statements: *[2025, 7 April Shift-II]*
(a) (A), (B), (D) and (E) only
(b) (B), (C) and (D) only
(c) (A), (B) and (D) only
(d) (B), (D) and (E) only

**37** Which of the following statements are true? *[2025, 2 April Shift-II]*
(A) The subsidiary quantum number $l$ describes the shape of the orbital occupied by the electron.
(B)

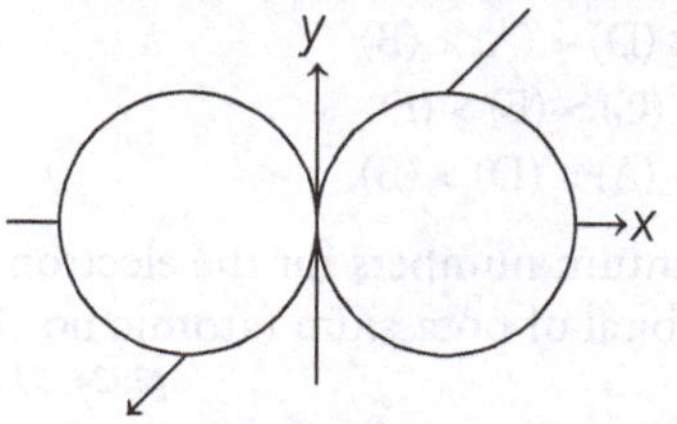


is the boundary surface diagram of the $2p_x$ orbital.
(C) The positive and negative signs in the wave function of the $2p_x$ orbital refer to charge.
(D) The wave function of $2p_x$ orbital is zero everywhere in the $xy$ plane.

Choose the correct answer from the options given below :
(a) (C) and (D) only
(b) (B) and (D) only
(c) (A), (B) and (C) only
(d) (A) and (B) only

**Imp.** Radial Distribution Function of Hydrogen

**38** The plots of radial distribution functions for various orbitals of hydrogen atom against '$r$' are given below. *[2021, 25 Feb Shift-I]*

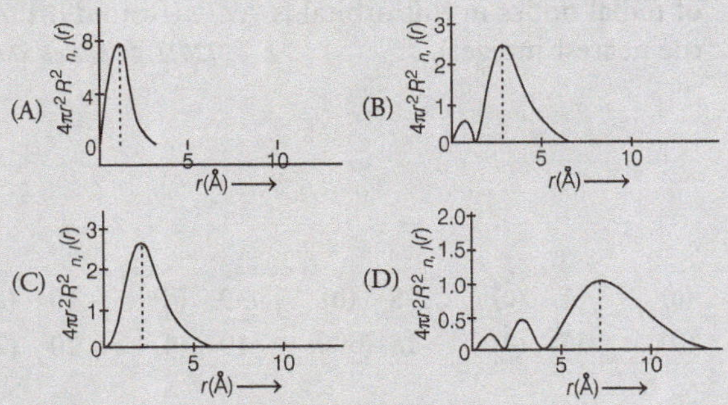


The correct plot for $3s$-orbital is
(a) (A) (b) (B) (c) (C) (d) (D)

**39** The figure that is not a direct manifestation of the quantum nature of atoms is *[2020, 2 Sep Shift-I]*

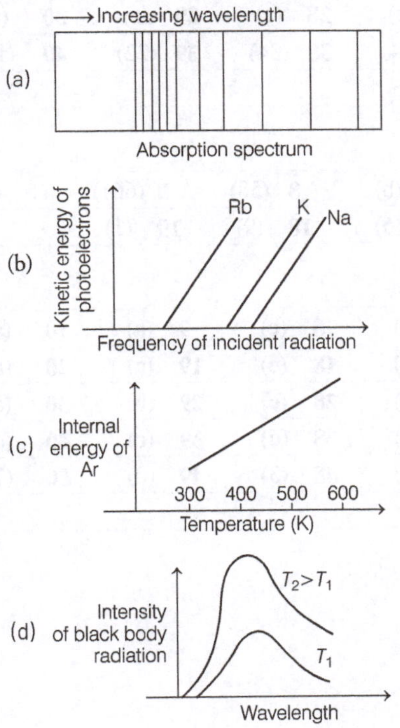


**40** Which of the following combination of statements is true regarding the interpretation of the atomic orbitals?

I. An electron in an orbital of high angular momentum stays away from the nucleus than an electron in the orbital of lower angular momentum.

II. For a given value of the principal quantum number, the size of the orbit is inversely proportional to the azimuthal quantum number.

III. According to wave mechanics, the ground state angular momentum is equal to $\frac{h}{2\pi}$.

IV. The plot of $\psi$ *vs* $r$ for various azimuthal quantum numbers, shows peak shifting towards higher $r$ value. *[2019, 9 Jan Shift-II]*

(a) I, III (b) II, III
(c) I, II (d) I, IV

## Numerical Value Type Questions

**41** The maximum number of orbitals which can be identified with $n = 4$ and $m_l = 0$ is .... . *[2024, 4 April Shift-II]*

**42** In an atom, total number of electrons having quantum numbers $n = 4$, $|m_l| = 1$ and $m_s = -\frac{1}{2}$ is .......... *[2024, 5 April Shift-I]*

**43** The number of given orbitals which have electron density along the axis is ......... $p_x, p_y, p_z, d_{xy}, d_{yz}, d_{z^2}, d_{x^2-y^2}$ *[2023, 25 Jan Shift-II]*

**44** The orbital angular momentum of an electron in $3s$ orbital is $\frac{xh}{2\pi}$. The value of $x$ is ........... (nearest integer) *[2023, 13 April Shift-II]*

**45** The number of correct statements from the following is *[2023, 11 April Shift-II]*

A. For $1s$ orbital, the probability density is maximum at the nucleus.

B. For $2s$ orbital, the probability density first increases to maximum and then decreases sharply to zero.

C. Boundary surface diagrams of the orbitals encloses a region of 100% probability of finding the electron.

D. $p$ and $d$-orbitals have 1 and 2 angular nodes respectively.

E. Probability density of $p$-orbital is zero at the nucleus.

**46** The number of atomic orbitals from the following having 5 radial nodes is ........
$7s, 7p, 6s, 8p, 8d$ *[2023, 8 April Shift-II]*

**47** Consider the following set of quantum numbers.

| | $n$ | $l$ | $m_l$ | | $n$ | $l$ | $m_l$ |
|---|---|---|---|---|---|---|---|
| A. | 3 | 3 | −3 | B. | 3 | 2 | −2 |
| C. | 2 | 1 | +1 | D. | 2 | 2 | +2 |

The number of correct sets of quantum numbers is ......... . *[2022, 27 June Shift-II]*

**48** The number of orbitals with $n = 5$, $m_l = +2$ is ...... . (Round off to the nearest integer). *[2021, 16 March Shift-II]*

**49** The value of magnetic quantum number of the outermost electron of $Zn^+$ ion is *[2021, 31 Aug Shift-II]*

**50** Ge ($Z = 32$) in its ground state electronic configuration has $x$ completely filled orbitals with $m_l = 0$. The value of $x$ is *[2021, 31 Aug Shift-I]*

**51** The Azimuthal quantum number for the valence electrons of $Ga^+$ ion is ........
(Atomic number of Ga = 31) *[2021, 20 July, Shift-I]*

**52** A certain orbital has $n = 4$ and $m_l = -3$. The number of radial nodes in this orbital is ...... (Round off to the nearest integer). *[2021, 17 March Shift-I]*

## • Key Index

### Topic 1

| | | | | | | | | | |
|---|---|---|---|---|---|---|---|---|---|
| 1 (c) | 2 (b) | 3 (d) | 4 (c) | 5 (c) | 6 (c) | 7 (c) | 8 (b) | 9 (d) | 10 (a) |
| 11 (d) | 12 (5) | 13 (494) | 14 (798) | 15 (3) | 16 (0) | 17 (4) | 18 (300) | 19 (4) | 20 (2) |
| 21 (50) | 22 (5) | 23 (222) | | | | | | | |

### Topic 2

| | | | | | | | | | |
|---|---|---|---|---|---|---|---|---|---|
| 1 (b) | 2 (b) | 3 (b) | 4 (d) | 5 (d) | 6 (d) | 7 (d) | 8 (b) | 9 (a) | 10 (d) |
| 11 (d) | 12 (a) | 13 (a) | 14 (c) | 15 (d) | 16 (a) | 17 (b) | 18 (d) | 19 (b) | 20 (b) |
| 21 (c) | 22 (b) | 23 (c) | 24 (d) | 25 (c) | 26 (c) | 27 (c) | 28 (c) | 29 (a) | 30 (a) |
| 31 (d) | 32 (b) | 33 (a) | 34 (b) | 35 (d) | 36 (c) | 37 (54) | 38 (34) | 39 (22) | 40 (10) |
| 41 (1) | 42 (1) | 43 (492) | 44 (270) | 45 (10) | 46 (3155) | | | | |

### Topic 3

| | | | | | | | | | |
|---|---|---|---|---|---|---|---|---|---|
| 1 (c) | 2 (a) | 3 (b) | 4 (d) | 5 (c) | 6 (c) | 7 (b) | 8 (58) | 9 (661) | 10 (8) |
| 11 (7) | 12 (2) | 13 (548) | 14 (2) | 15 (1758) | 16 (58) | 17 (6) | 18 (9) | 19 (1) | |

### Topic 4

| | | | | | | | | | |
|---|---|---|---|---|---|---|---|---|---|
| 1 (c) | 2 (a) | 3 (b) | 4 (d) | 5 (a) | 6 (b) | 7 (d) | 8 (c) | 9 (a) | 10 (a) |
| 11 (b) | 12 (d) | 13 (a) | 14 (d) | 15 (a) | 16 (b) | 17 (b) | 18 (c) | 19 (c) | 20 (b) |
| 21 (c) | 22 (b) | 23 (b) | 24 (d) | 25 (c) | 26 (d) | 27 (b) | 28 (c) | 29 (b) | 30 (a) |
| 31 (b) | 32 (a) | 33 (c) | 34 (b) | 35 (a) | 36 (a) | 37 (d) | 38 (d) | 39 (c) | 40 (d) |
| 41 (4) | 42 (6) | 43 (5) | 44 (0) | 45 (3) | 46 (3) | 47 (2) | 48 (3) | 49 (0) | 50 (7) |
| 51 (0) | 52 (0) | | | | | | | | |

# Detailed **Explanations**

## Topic 1

**01.** *(c)* Minimum kinetic energy required to produce photoelectric effect is

$$E = E_0 = h\nu_0$$
$$= 6.6\times 10^{-34} \times 1.3\times 10^{15}$$
$$= 8.58\times 10^{-19}\ \text{J}$$

**02.** *(b)* Number of neutrons in protium ($^1_1$H) is zero ($x$).

Number of neutrons in deuterium ($^2_1$H or $^2_1$D) is 1 ($y$)

Number of neutrons in tritium ($^3_1$H or $^3_1$T) is 2 ($z$)

So, the sum of $x$, $y$ and $z$ is

$$x + y + z = 3$$

**03.** *(d)* The kinetic energy of the ejected electron is given by the equation :

$$\text{K.E.} = h\nu - h\nu_0$$

Where work function, $w_0 = h\nu_0$

For the given situation stopping potential = KE = 0.5 eV

$$0.5 = \frac{6.626\times 10^{-34} \times 3\times 10^8}{250\times 10^{-9}} - w_0$$

$$\Rightarrow w_0 = 4.46 \approx 4.5\ \text{eV}$$

**• Concept Enhancer**

The ejection of photoelectron from metal in photoelectric effect takes place only when the following condition is satisfied i.e., $\nu > \nu_0$ and $\lambda < \lambda_0$. Where $\nu_0$ is the threshold frequency and $\lambda_0$ is the threshold wavelength.

**04.** *(c)* The kinetic energy of the ejected electron is given by the equation,

$h\nu - h\nu_0 = \text{K.E.}$

$$\frac{1}{2}mv^2 = \frac{hc}{\lambda} - \frac{hc}{\lambda_0} = hc\left(\frac{\lambda_0 - \lambda}{\lambda\lambda_0}\right)$$

$$v = \sqrt{\frac{2hc}{m}\left(\frac{\lambda_0 - \lambda}{\lambda\lambda_0}\right)}$$

**05.** *(c)* The correct match is

A-III, B-I, C-IV, D-II

**06.** *(c)* A is correct but R is not correct.

Emission of electron from metal surface take place only, when the photon with energy greater than work function ($w_0$) strikes the metal surface. The ejected electrons leave with a specific amount of kinetic energy at a frequency $\nu > \nu_0$.

**07.** *(c)* Energy of a photon is given by

$$E = \frac{hc}{\lambda} = \frac{6.63\times 10^{-34}\ \text{Js}\times 3\times 10^8\ \text{ms}^{-1}}{300\times 10^{-9}\ \text{m}}$$

$$\Rightarrow E = 6.63\times 10^{-19}\ \text{J}$$

Energy of 1 mole of photon

$$= 6.63\times 10^{-19} \times 10^{-3} \times 6.02\times 10^{23}\ \text{kJ mol}^{-1}$$
$$= 6.63\times 6.02\times 10$$
$$= 39.9\times 10 = 399\ \text{kJ mol}^{-1}$$

**08.** *(b)* The kinetic energy of the ejected electron is given by

$$\text{KE} = h\nu - h\nu_0$$

where, $h\nu_0 (W_0)$ = Work function

$$\frac{1}{2}m_e v^2 = \frac{hc}{\lambda} - w_0 \qquad \ldots\text{(i)}$$

On substituting all the given values in Eq. (i), we get

$$\frac{1}{2}\times 9\times 10^{-31}\ \text{kg}\times(6\times 10^5\ \text{ms}^{-1})^2$$
$$= \frac{6.626\times 10^{-34}\ \text{Js}\times 3\times 10^8\ \text{ms}^{-1}}{4000\times 10^{-10}\ \text{m}} - w_0$$
$$\therefore w_0 = -1.62\times 10^{-19}\ \text{kgm}^2\text{s}^{-2} + 4.96\times 10^{-19}\ \text{J}$$
$$= 3.34\times 10^{-19}\ \text{J} \quad [1\ \text{kg m}^2\text{s}^{-2} = 1\ \text{J}]$$
$$= 2.1\ \text{eV}$$

**09.** *(d)* For photoelectric effect, the graph of K.E. of electrons versus frequency of light is given as :

$$\text{KE} = h\nu - h\nu_0$$

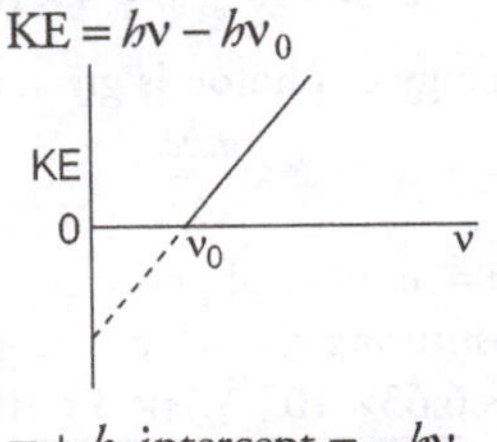


Slope = + $h$, intercept = $-h\nu_0$

So, option (d) is not correct. While the remaining options represents the relationship between incident light and the electron ejected from metal surface.

**10.** *(a)* We know that,

$$\text{K.E.} = \frac{p^2}{2m} = \frac{hc}{\lambda_{\text{incident}}}$$

Now, as per the given condition,

$$\frac{\lambda_2}{\lambda_1} = \left(\frac{p_1}{p_2}\right)^2$$

$$\Rightarrow \frac{\lambda_2}{\lambda} = \left(\frac{p}{1.5\times p}\right)^2 = \left(\frac{2}{3}\right)^2 = \frac{4}{9}$$

$$\Rightarrow \lambda_2 = \frac{4}{9}\lambda \qquad \left[\because \lambda_1 = \lambda,\ p_1 = p\right]$$

**11.** *(d)* Ionisation energy $= N_A h\nu$

$$495.5 = 6.023\times 10^{23}\times 6.6\times 10^{-34}\times \nu$$

On rearranging, we will get

$$\nu = \frac{495.5\times 10^3}{6.023\times 10^{23}\times 6.6\times 10^{-34}}$$
$$= 12.4\times 10^{14}\ \text{s}^{-1} = 1.24\times 10^{15}\ \text{s}^{-1}$$

**12.** *(5)* Wavelength of EM wave $= d\times 4$

$$\lambda = 1.5\times 4 = 6\ \text{pm}$$

$$\nu = \frac{c}{\lambda} = \frac{3\times 10^8}{6\times 10^{-12}}$$
$$= 0.5\times 10^{20}\ \text{Hz} = 5\times 10^{19}\ \text{Hz}$$

Hence, the frequency of the wave is $5\times 10^{19}$ Hz.

**13.** *(494)*

**key idea** To obtain ionisation energy of sodium in kJ/mol. First of all use the formula $E = \dfrac{1240\ (\text{eV})}{\lambda\ (\text{nm})}$.

Then multiply the obtained value with $1.6\times 10^{19}$ to obtain energy in J/atom. After that multiply the calculated value in 2nd step with Avogadro number to get the final answer in kJ/mol.

Given, $\lambda = 242$ nm

using $E = \dfrac{1240\ \text{eV}}{\lambda\ (\text{nm})}$

$$E = \frac{1240}{242}\ \text{eV} = 5.12\ \text{eV}$$
$$E = 5.12\times 1.6\times 10^{-19}\ \text{J/atom}$$
$$E = 8.192\times 10^{-19}\ \text{J/atom}$$

or $$E = 8.192\times 10^{-19}\times 6.022\times 10^{23}\ \text{atom/mol}$$
$$= 493.32\ \text{kJ/mol}$$
$$\approx 494\ \text{kJ/mol}$$

**14.** *(798)* Given, $\nu = 2 \times 10^{12}$ Hz

$E = ?$

For one photon, $E = h\nu$

For one mole photon, $E = nh\nu$

where, $h$ = Planck's constant

$n = 1$ mole

$\therefore E = 6.022 \times 10^{23} \times 6.626 \times 10^{-34} \times 2 \times 10^{12}$

$= 798.03 \text{ J} \approx 798 \text{ J}$

**15.** *(3)* Photoelectric effect will take place, only when the work function of given metals is lower than the energy of incident light with wavelength 400 nm.

Energy of incident light $= h\nu = \dfrac{hc}{\lambda}$

$$E(\text{eV}) = \frac{1240}{\lambda} = \frac{1240}{400} = 3.1 \text{ eV}$$

Thus, Li, Na, K will show photoelectric effect.

**16.** *(0)*

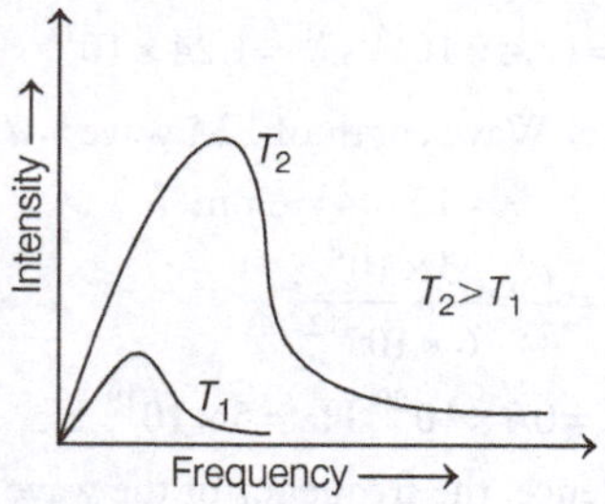


From the above graph we can see that, all statements (A), (B), (C) and (D) are also correct.

A black body can emit and absorb all the wavelength in electromagnetic spectrum. Hence, statement (A) is correct.

(B) The frequency distribution of the emitted radiation by a blackbody indeed depends on its temperature. This phenomenon is described by Planck's law. hence statement (B) is correct.

(C) The intensity *vs* frequency curve for a black body known as the planck's curve, pass through a maximum value for a given temperature. Statement (C) is correct.

(D) The frequency at which maximum intensity occurs increases with the temperature of the black body. Statement (D) is correct.

**17.** *(4)* The symbol ${}^{48}_{22}X^{3-}$ shows,

Atomic number = 22; mass number = 48

Total electrons = 22 + 3 = 25

Total neutrons = Mass number − Atomic number

$= 48 - 22 = 26$

Neutron percentage which is more than number of electrons $(a)$

$$= \frac{\text{Number of neutrons} - \text{Number of electrons}}{\text{Number of electrons}} \times 100$$

$$= \frac{26 - 25}{25} \times 100 = 4$$

**18.** *(300)* Using equation, $E = w_0 + \text{KE}$

[$E$ = energy given, $w_0$ = work function, KE = kinetic energy]

As electron is just ejected, so

$\text{KE} = 0;\ E = w_0 = \dfrac{hc}{\lambda}$

$= 6.63 \times 10^{-19}$

$\Rightarrow \lambda = \dfrac{6.63 \times 10^{-34} \times 3 \times 10^{8}}{6.63 \times 10^{-19}}$

$\therefore \lambda = 3 \times 10^{-7}$ m

$= 300 \times 10^{-9}$ m

Hence, $\lambda = 300$ nm

**19.** *(4)* The potential energy of electron in $n$th orbit is given by,

$E = -22 \times 10^{-19} \times Z^2 / n^2$ J

For Li, $Z = 3$ and $n = 2$

$E = -22 \times 10^{-19} \times \dfrac{3 \times 3}{2 \times 2}$ eV

$= 49.5 \times 10^{-19}$ J

$E = \dfrac{hc}{\lambda}$

$\Rightarrow \lambda = \dfrac{6.63 \times 10^{-34} \times 3.0 \times 10^{8}}{49.5 \times 10^{-19}}$

$= 0.40 \times 10^{-7} \text{ m} = 4 \times 10^{-8}$ m

**20.** *(2)* Energy of photon is given as

$$E = \frac{nhc}{\lambda} \qquad \ldots\text{(i)}$$

where $n$ = number of photons.

On substituting the values, we get

$$50 \text{ J} = \frac{n \times 6.63 \times 10^{-34} \text{ Js} \times 3 \times 10^{8} \text{ m/s}}{795 \times 10^{-9} \text{ m}}$$

$\Rightarrow n = \dfrac{50 \times 795 \times 10^{-9}}{6.63 \times 10^{-34} \times 3 \times 10^{8}}$

$= 1998.49 \times 10^{17} = 1.998 \times 10^{20}$

$\Rightarrow \approx 2 \times 10^{20} \Rightarrow x = 2$

**21.** *(50)* Power of the source

$= 1 \text{ mW} = 10^{-3}$ W

Energy of one photon, $E = h\nu = \dfrac{hc}{\lambda}$

where, $\lambda = 1000$ nm or $10^{-6}$ m

$E = \dfrac{6.63 \times 10^{-34} \times 3 \times 10^{8}}{10^{-6}}$

$\approx 20 \times 10^{-20}$ J

Power = number of photons emitted in one second × energy of 1 photon.

$10^{-3} = n \times 20 \times 10^{-20}$

$\Rightarrow n = \dfrac{10^{-3}}{20 \times 10^{-20}} = 0.5 \times 10^{16}$

Number of photons emitted in 1 s $= 0.5 \times 10^{16}$

Number of photons emitted in 0.1 is given by $= 0.5 \times 10^{16} \times 0.1 = 50 \times 10^{13}$

$x \times 10^{13} = 50 \times 10^{13} \Rightarrow x = 50$

**22.** *(5)* According to photoelectric effect

$\dfrac{hc}{\lambda} = h\nu_0 + \dfrac{1}{2} mv^2$

$\Rightarrow \dfrac{6.63 \times 10^{-34} \times 3 \times 10^{8}}{500 \times 10^{-9}}$

$= 6.63 \times 10^{-34} \times 4.3 \times 10^{14} + \dfrac{1}{2} mv^2$

$\dfrac{6.63 \times 30 \times 10^{-19}}{50} = 6.63 \times 4.3 \times 10^{-20} + \dfrac{1}{2} mv^2$

$11.271 \times 10^{-20} \text{ J} = \dfrac{1}{2} \times 9 \times 10^{-31} \times v^2$

$v = 5 \times 10^{5}$ m/s

$\therefore v = 5$

**23.** *(222)* For photoelectrons, its kinetic energy will be

$\text{KE} = E - E_0 = \dfrac{hc}{\lambda} - E_0$

$= \dfrac{6.63 \times 10^{-34} \times 3 \times 10^{8}}{300 \times 10^{-9}} - 4.41 \times 10^{-19}$ J

$= 222 \times 10^{-21}$ J

## Topic 2

**01.** *(b)* Given, Bohr radius $= a_0$

de-Broglie wavelength $= \lambda$

Radius of $n$th orbit; $r_n = n^2 a_0 = (2)^2 a_0$

$= 4a_0 \quad [\because n = 2]$

Using the formula,

$$2\pi r_n = n\lambda \Rightarrow 2\pi(4a_0) = n\lambda$$

$$\Rightarrow \quad \lambda = \frac{8\pi a_0}{n}$$

**02.** *(b)* For first excited state, $n = 2$

So, substituting the value of $n$ in equation $r = \frac{a_0 n^2}{Z}$ we get

$$r = \frac{(2)^2 a_0}{2} = 2a_0 \quad (\because Z = 2)$$

So, $r = 2a_0$

**03.** *(b)* The correct match is A-II, B-IV, C-III, D-I.

A. Lyman series lies in UV region.
B. Balmer series lies in visible region.
C. Paschen and Pfund series lies in infrared region.

**Time Saver Tech**

Use elimination technique as time saver. Pfund with infrared region (D-I) is given only in option (b). Hence, no need to cross check other option.

**04.** *(d)* Both Statement I and Statement II are correct.

According to Bohr's model, angular momentum is quantised for stationary orbits. In a given stationary state, it can be expressed as

$$mvr = n\frac{h}{2\pi}, n = 1, 2, 3, \ldots$$

It does not follow Heisenberg's uncertainty principle.

**05.** *(d)* Correct option is (d) i.e., $n = 2$ to $n = 1$.

For $He^+$ ion

$$\frac{1}{\lambda(H)} = R(1)^2\left[\frac{1}{n_1^2} - \frac{1}{n_2^2}\right]$$

$$\frac{1}{\lambda(He^+)} = R(2)^2\left[\frac{1}{2^2} - \frac{1}{4^2}\right]$$

Given, Balmer series $\lambda(H) = \lambda(He^+)$.

$$\therefore \quad R(1)^2\left[\frac{1}{n_1^2} - \frac{1}{n_2^2}\right] = R(4)\left[\frac{1}{2^2} - \frac{1}{4^2}\right]$$

$$\frac{1}{n_1^2} - \frac{1}{n_2^2} = \frac{1}{1^2} - \frac{1}{2^2}$$

So, on comparing we will get, $n_1 = 2$ and $n_2 = 1$.

**06.** *(d)* Both Statement I and Statement II are true. It is because definite line in H-spectrum are explained by Bohr and not by the Rutherford.

Bohr's model of an atom is contradicted by Heisenberg uncertainty principle.

**07.** *(d)* Balmer series lies in the visible region of the electromagnetic spectrum.

**08.** *(b)* According to Bohr's model, radius of orbit is given by

$$r_n = \frac{a_0 \times n^2}{Z}, \text{ where } a_0 = \text{Bohr's radius}$$

(Radius of 1st orbit of H-atom)

For, $Li^{2+}$, $Z = 3$ and $n = 2$

$$\therefore \quad r = \frac{2^2 \times a_0}{3} = \frac{4a_0}{3}$$

**09.** *(a)* According to Rydberg's equation,

$$\frac{1}{\lambda} \propto \left(\frac{1}{n_1^2} - \frac{1}{n_2^2}\right)$$

For shortest wavelength, i.e. highest energy spectral line, $n_2$ will be $(\infty)$.

For the given spectral series, ratio of the shortest wavelength of two spectral series can be calculated as follows

$$\frac{\lambda_{\text{Paschen}}}{\lambda_{\text{Lyman}}} = \frac{\frac{1}{1^2} - \frac{1}{\infty^2}}{\frac{1}{3^2} - \frac{1}{\infty^2}} = \frac{\frac{1}{1} - 0}{\frac{1}{9} - 0} = 9$$

Hence, the ratio of wavelength of Paschen and Lyman series of hydrogen spectrum is found to be about 9.

**10.** *(d)* Radius of $n^{th}$ Bohr orbit for H-atom

$$r_n = \frac{n^2 \times 053}{Z} \text{Å}$$

$$= 0.53\, n^2 \text{ Å} \, [Z = 1 \text{for H-atom}]$$

$\therefore$ Radius of $2^{nd}$ Bohr orbit for H-atom $= 0.53 \times (2)^2 = 2.12$ Å

**11.** *(d)* Total energy = K.E. + P.E.

$$\therefore \text{ T.E.} = \frac{Ke^2}{2r} + \left(-\frac{Ke^2}{r}\right)$$

$$= -\frac{Ke^2}{2r}$$

(In CGS system $k = 1$)

$$\text{T.E.} = -\frac{1}{2}\frac{e^2}{r}$$

**12.** *(a)* Energy change

$$\Delta E = 2.178 \times 10^{-18}\left(\frac{1}{1^2} - \frac{1}{2^2}\right) = \frac{hc}{\lambda}$$

$$\text{or} \quad 2.178 \times 10^{-18}\left(\frac{1}{1^2} - \frac{1}{2^2}\right)$$

$$= \frac{6.62 \times 10^{-34} \times 3.0 \times 10^8}{\lambda}$$

$$\Rightarrow \lambda \approx 1.214 \times 10^{-7} \text{ m}$$

**13.** *(a)* The frequency of lines in balmer is given as

$$\nu = R\left[\frac{1}{n_1^2} - \frac{1}{n_2^2}\right] \text{Hz}$$

where $n_1 = 4$

$$\nu = 3.29 \times 10^{15}\left[\frac{1}{4} - \frac{1}{\infty}\right]$$

$$= 8.22 \times 10^{14} \text{ s}^{-1}$$

**14.** *(c)* I.E. of $He^+ = 19.6 \times 10^{-18}$ J/atom

$$= E_H \times Z^2 = E_H \times 4 = -19.6 \times 10^{-18}$$

$$E_1 \text{ (for } Li^{2+}) = E \text{ (for H)} \times 9$$

$$\frac{-19.6 \times 10^{-18} \times 9}{4}$$

$$= -4.41 \times 10^{-17} \text{ J/atom}$$

**15.** *(d)* Angular momentum of the electron according to Bohr's theory is given by $mvr = \frac{nh}{2\pi}$

(given, $n = 5$)

$$\therefore \text{ Angular momentum} = \frac{5h}{2\pi} = \frac{2.5\,h}{\pi}$$

**16.** *(a)* $\frac{1}{\lambda} = R\left[\frac{1}{n_1^2} - \frac{1}{n_2^2}\right]$

$$= 1.097 \times 10^7\left[\frac{1}{1^2} - \frac{1}{\infty^2}\right]$$

$$\Rightarrow \lambda = 0.91 \times 10^{-7} \Rightarrow \lambda = 91 \text{ nm}$$

**17.** *(b)* Postulates given in option (b) is not in agreement with quantum mechanical model of an atom. In Bohr model the electron in a H-atoms stationary state moves in a spherical orbit while electrons are not confined to specific orbits around the molecules in a quantum mechanical model.

**18.** *(d)* Given, $\lambda = 900$ nm

$$= 9 \times 10^{-5} \text{ cm}$$

$$R_H = 10^5 \text{ cm}^{-1}$$

$$h = 6.6 \times 10^{-34} \text{ Js}$$

$$c = 3 \times 10^8 \text{ m/s}$$

Using equation,

$$\frac{1}{\lambda} = R_H Z^2 \times \left[\frac{1}{n_1^2} - \frac{1}{n_2^2}\right]$$

On substituting the given value of we get

$$\frac{1}{9\times10^{-5}}=10^5(1)^2\left[\frac{1}{n_1^2}-\frac{1}{n_2^2}\right]$$

($\because Z=1$ for hydrogen atom)

$$\Rightarrow \frac{1}{n_1^2}-\frac{1}{n_2^2}=\frac{1}{9}$$

So, the given condition is possible only when $n_1=3$, $n_2=\infty$.

**19.** *(b)* Energy of hydrogen atom can be given as :

$$E_n=-2.18\times10^{-18}\left[\frac{Z^2}{n^2}\right]\text{J/atom}$$

$$E_1=-2.18\times10^{-18}\text{ J/atom} \quad \ldots(i)$$

For 3rd orbit : $(n=3)(Z=1)$

$$E_3=-2.18\times10^{-18}\left[\frac{(1)^2}{(3)^2}\right]$$

$$E_3=-2.18\times10^{-18}\left[\frac{1}{9}\right] \quad \ldots(ii)$$

Hence, the energy in the 3rd Bohr orbit is (1/9th) of the energy of electron in the first Bohr orbit.

**20.** *(b)* Radius of $n$th Bohr's orbit of hydrogen atom is given as

$$r=0.529\times\frac{n^2}{Z}\text{ Å} \quad (Z=1\text{ for hydrogen})$$

So, for 3rd Bohr's orbit

$$r_3=0.529\times\frac{(3)^2}{1}$$

$$r_3=0.529\times 9 \quad \ldots(i)$$

For 4th Bohr's orbit $(n=4)$

$$r_4=0.529\times\frac{(4)^2}{1}$$

$$r_4=0.529\times16 \quad \ldots(ii)$$

From Eqs. (i) and (ii),

$$\frac{r_4}{r_3}=\frac{0.529\times16}{0.529\times9}$$

$$\frac{r_4}{r_3}=\frac{16}{9}\Rightarrow r_4=\frac{16}{9}r_3$$

**21.** *(c)* According to Bohr's atom, velocity of electron is given by,

$$v\propto\frac{Z}{n} \quad \ldots(i)$$

From Eq. (i), velocity of electron is directly proportional to atomic number of atom, corresponding to positive charge. So, as $Z$ increases, velocity also increases. Hence, statement I is false, while statement II is true.

**22.** *(b)* Statement I is false, because Bohr's theory accounts for the stability and spectrum of single electronic species (e.g. $He^+$, $Li^{2+}$ etc.) but $Li^+$ has two electrons.

Bohr's theory fails to explain splitting of spectral lines in presence of magnetic field i.e. Zeeman effect.

$\therefore$ Statement II is true.

**23.** *(c)* Bohr's radius is given by,

$$r=0.529\frac{n^2}{Z}\text{ Å}$$

For $Li^{2+}$ $(r_{Li^{2+}})_{n=4}-(r_{Li^{2+}})_{n=3}$

$$\Rightarrow \frac{0.529}{3}(16-9)=\Delta R_1 \quad \ldots(i)$$

For $He^+$ $(r_{He^+})_{n=4}-(r_{He^+})_{n=3}$

$$\Rightarrow \frac{0.529}{2}(16-9)=\Delta R_2 \quad \ldots(ii)$$

$$\frac{\Delta R_1}{\Delta R_2}=\frac{\frac{0.529}{3}(16-9)}{\frac{0.529}{2}(16-9)}=\frac{2}{3}$$

Hence, the ratio of $\Delta R_1:\Delta R_2$ is 2 : 3.

**24.** *(d)* Statements I, II and III are correct. While the statement IV is incorrect. It's corrected form is as follows: Ionisation energy of hydrogen can be calculated from Lyman series lines $(n=1)$ only by using the following formula,

$$\bar{v}=R_H\left[\frac{1}{(1)^2}-\frac{1}{\infty^2}\right]$$

**25.** *(c)* According to Rydberg's formula,

$$\text{wave number }(\bar{v})=R_HZ^2\left[\frac{1}{n_f^2}-\frac{1}{n_i^2}\right]$$

Given, $\quad n_i=8$, $n_f=n$

[$\because$ it is the case of emission]

$$\bar{v}=R_H\times(1)^2\left[\frac{1}{n^2}-\frac{1}{8^2}\right]$$

$$\bar{v}=R_H\left[\frac{1}{n^2}-\frac{1}{64}\right]=\frac{R_H}{n^2}-\frac{R_H}{64}$$

On comparing with equation of straight line, $y=mx+c$, we get

Slope $=R_H$, intercept $=\frac{-R_H}{64}$.

Thus, plot of wave number $(\bar{v})$ against $\frac{1}{n^2}$ will be linear with slope $(+R_H)$.

**26.** *(c)* The ground state energy of H-atom is $-13.6$ eV.

For second excited state, $n=2+1=3$

$$\therefore E_3(He^+)=-13.6\times\frac{Z^2}{n^2}\text{ eV}$$

[$\because$ for $He^+$, $Z=2$]

$$=-13.6\times\frac{2^2}{3^2}\text{ eV}=-6.04\text{ eV}$$

**27.** *(c)* Frequency of emitted radiation shifts from a higher wavelengths to lower wavelength as the temperature increases.

**28.** *(c)*

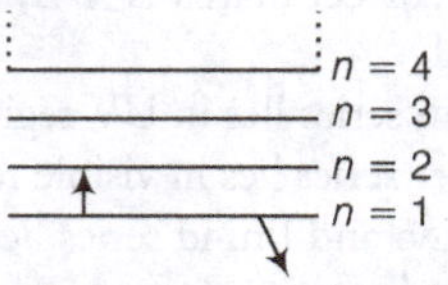


Maximum population of e− of H-atom

Since, at $n=1$, the population of electrons is maximum *i.e.* at ground state. So, maximum excitation will take place from $n=1$ to $n=2$.

Hence, $n=2$ is the possible excited state.

Now, we have the formula for energy of H-atom as

$$(E_n)_H=-13.6\frac{Z^2}{n^2}\text{ eV}$$

$Z$ for H-atom $=1$

$$\therefore \quad (E_n)_H=-13.6\times\frac{1}{2^2}\text{ eV}$$

$$=-\frac{13.6}{4}\text{ eV}=-3.4\text{ eV}$$

**29.** *(a)* I.E. (H-atom) $=E_\infty-E_1$

$$1.312\times10^6=0-E_1$$

or $\quad E_1=-1.312\times10^6$ J/mol

$$E_2=\frac{-1.312\times10^6}{(2)^2}$$

$$=\frac{-1.312\times10^6}{4}$$

$$\Delta E=E_2-E_1$$

$$=\frac{-1.312\times10^6}{4}-(-1.312\times10^6)$$

$$=9.84\times10^5\text{ J/mol}$$

**30.** *(a)* Incorrect statement is given in option (a). Its correct form is energy of electron is maximum at infinite distance from nucleus.

**31.** *(d)* Among the given statement A, B, D and E are correct, while C is incorrect. It's correct form is Red light does not produce photoelectric effect.

**32.** *(b)* According to Bohr's postulate

$r \propto n^2$

(a) $\frac{r_8}{r_4} = \left(\frac{8}{4}\right)^2 = 4$

(b) $\left(\frac{r_4}{r_2}\right) = \left(\frac{4}{2}\right)^2 = 4$

(c) $\frac{r_6}{r_4} = \left(\frac{6}{4}\right)^2 = \frac{9}{4}$

(d) $\frac{r_3}{r_1} = \frac{(3)^2}{1} = \frac{9}{1}$

So, statement given in option (b) is incorrect.

**33.** *(a)* The relation between energy ($E$) and atomic number ($Z$) is

$$E = -\frac{13.6Z^2}{n^2}\text{ eV}$$

$$E \propto -Z^2$$

So, the correct graph is

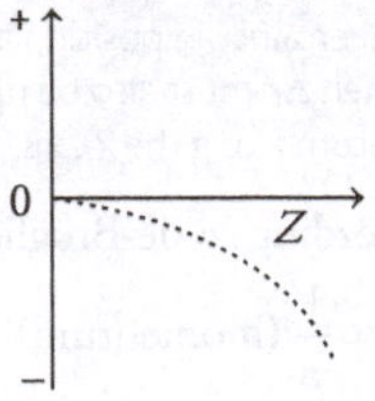


**34.** *(b)* As we know that, $\lambda = \frac{hc}{\Delta E}$

For, $\lambda$ = minimum, i.e. shortest,

$\Delta E$ = maximum

For Lyman series, $n = 1$ and for $\Delta E_{max}$ transition must be form $n = \infty$ to $n = 1$.

So, $\frac{1}{\lambda} = R_H Z^2 \left(\frac{1}{n_1^2} - \frac{1}{n_2^2}\right)$

($\because n_1 = 1$ and $n_2 = \infty$)

$\therefore \frac{1}{\lambda_1} = R_H \left(\frac{1}{1^2} - \frac{1}{\infty^2}\right) \times (1)^2$

$\Rightarrow \lambda_1 = \frac{1}{R_H}$

For longest wavelength,

$\Delta E$ = minimum for Balmer series, $n = 3$ to $n = 2$ will have minimum $\Delta E$.

For $He^+$, $Z = 2$

So, $\frac{1}{\lambda_2} = R_H \times 2^2 \left(\frac{1}{n_1^2} - \frac{1}{n_2^2}\right)$

$\frac{1}{\lambda_2} = R_H \times 4 \left(\frac{1}{4} - \frac{1}{9}\right)$

$\frac{1}{\lambda_2} = R_H \times \frac{5}{9}$

$\Rightarrow \lambda_2 = \lambda \times \frac{9}{5}$

**35.** *(d)* According to Bohr's theory,

I. KE $\propto \frac{Z^2}{n^2}$ or $13.6 \times \frac{Z^2}{n^2}\frac{\text{(eV)}}{\text{(atom)}}$, ($\therefore$ Correct)

II. Speed of electron $\propto \frac{Z}{n}$

(Here, $Z$ = atomic number, $n$ = number of shells)

$\therefore \quad v \times n \propto Z$ ($\therefore$ Incorrect)

III. Frequency of revolution of electron

$= \frac{v}{2\pi r}$

Frequency $\propto \frac{Z^2}{n^3} \left(\because v \propto \frac{Z}{n}, r \propto \frac{n^2}{Z}\right)$

($\therefore$ Incorrect)

IV. Coulombic force of attraction on the electron is given by

$$F = \frac{k q_1 q_2}{r^2} = \frac{kZe^2}{r^2}$$

$$F = \frac{Z}{\left(\frac{n^2}{Z}\right)^2}$$

$$F \propto \frac{Z^3}{n^4} \quad (\therefore \text{ Correct})$$

Hence, only I, and IV are correct.

**36.** *(c)* For any given series of spectral lines of atomic hydrogen.

Let $\Delta\bar{v} = \bar{v}_{max} - \bar{v}_{min}$ be the difference in maximum and minimum frequencies in $cm^{-1}$.

For Lyman series, $\Delta\bar{v} = \bar{v}_{max} - \bar{v}_{min}$

General formula:

$$\bar{v} = 109677\left[\frac{1}{n_i^2} - \frac{1}{n_f^2}\right]$$

For Lyman $n_1 = 1, n_2 = 2, 3, \ldots$

$\bar{v}_{max} = 109{,}677\left(\frac{1}{1^2} - \frac{1}{\infty}\right)$

$= 109{,}677\left(\frac{1}{1} - 0\right) = 109{,}677$

$\bar{v}_{min} = 109{,}677\left(\frac{1}{(1)^2} - \frac{1}{(2)^2}\right)$

$\Delta\bar{v}_{Lyman} = \bar{v}_{max} - \bar{v}_{min}$

$= 109{,}677 - \left[\frac{109{,}677 \times 3}{4}\right]$

$= \frac{109{,}677}{4}$

For Balmer series,

$\bar{v}_{max} = 109{,}677\left(\frac{1}{(2)^2} - \frac{1}{\infty}\right) \Rightarrow \frac{109677}{4}$

$\bar{v}_{min} = 109{,}677\left(\frac{1}{(2)^2} - \frac{1}{(3)^2}\right)$

$= \frac{109677 \times 5}{36}$

$\Delta\bar{v} = \bar{v}_{max} - \bar{v}_{min}$

$\Delta\bar{v}_{Balmer} = \frac{109{,}677}{4} - \left[\frac{109{,}677}{36} \times 5\right]$

$= 109{,}677\left(\frac{1}{9}\right)$

$\frac{\Delta\bar{v}_{Lyman}}{\Delta\bar{v}_{Balmer}} = \frac{109{,}677/4}{109{,}677/9}$

$\Rightarrow \frac{\Delta\bar{v}_{Lyman}}{\Delta\bar{v}_{Balmer}} = \frac{9}{4}$

$\therefore$ The ratio of $\frac{\Delta\bar{v}_{Lyman}}{\Delta\bar{v}_{Balmer}}$ is 9 : 4.

**37.** *(54)* Bohr's energy of $n$th orbit,

$$E_T = \frac{-13.6}{n^2} Z^2 \text{ eV}$$

For energy of H atom, energy of 1st Bohr orbit

$E_1 = -13.6$ eV $\quad [Z = 1, n = 1]$

For $Be^{3+}$ ion, energy of 1st excited state $[Z = 4, n = 2]$

$\frac{E_H}{E_{Be^{3+}}} = \frac{Z_1^2}{n_1^2} \times \frac{n_2^2}{Z_2^2}$

$\Rightarrow \frac{E_H}{E_{Be^{3+}}} = \frac{1}{1} \times \frac{4}{16}$

$E_{Be^{3+}} = -13.6 \times 4 = -54.4$ eV

$\approx 54$ eV

**38.** *(34)* First excited state is $= -3.4$ eV

$E_2 = -3.4$ eV

KE $= -E_2$

KE $= 3.4$ eV

$$KE = 34 \times 10^{-1} \text{ eV}$$

compare it with given equation, value of $x$ will be 34.

**39.** *(22)* Velocity in Bohr's atomic model is given by

$$v = 2.18 \times 10^6 \times \frac{Z}{n}$$

$$= 2.18 \times 10^6 \times \frac{1}{1}$$

$$\approx 22 \times 10^5 \text{ m/s}$$

Hence, missing value is 22.

**40.** *(10)* 5th excited state $\Rightarrow n_2 = 6$

1st excited state $\Rightarrow n_1 = 2$

$$\Delta n = n_2 - n_1 = 6 - 2 = 4$$

Maximum number of spectral lines

$$= \frac{\Delta n(\Delta n + 1)}{2}$$

where $\Delta n = n_2 - n_1 = \frac{4(4+1)}{2} = 10$

Hence, total 10 spectral lines are obtained.

**41.** *(1)* The emission spectra of atoms in gaseous phase do not show a continuous spread of wavelength from red to voilet, rather they emit light at specific wavelength with dark spaces between them. Hence, only statement i.e. (b) is incorrect.

**42.** *(1)* Given,

Radiation of energy $= 1.47 \times 10^{-17}$ J

$$R_H = 2.18 \times 10^{-18} \text{ J}$$

$$\Delta E = R_H Z^2 \left( \frac{1}{n_1^2} - \frac{1}{n_2^2} \right)$$

$$1.47 \times 10^{-17} = 2.18 \times 10^{-18} \times 9 \left( \frac{1}{n^2} - \frac{1}{(n+1)^2} \right)$$

$$\Rightarrow \frac{1.47}{1.962} = \frac{3}{4} = \frac{1}{n^2} - \frac{1}{(n+1)^2}$$

$$\Rightarrow n = 1$$

**43.** *(492)* We know that,

$$\frac{1}{\lambda} = R_H Z^2 \left[ \frac{1}{n_1^2} - \frac{1}{n_2^2} \right]$$

For first line of Paschen series,

$n_1 = 3, n_2 = 4$, on substituting the values, we will get

$$\text{So, } \frac{1}{(\lambda_1)_P} = R_H Z^2 \left[ \frac{1}{(3)^2} - \frac{1}{(4)^2} \right]$$

$$\frac{1}{(\lambda_1)_P} = R_H Z^2 \left[ \frac{1}{9} - \frac{1}{16} \right] \quad \text{...(i)}$$

Similarly, for second line of Paschen series

$n_1 = 3, n_2 = 5$

$$\frac{1}{(\lambda_2)_P} = R_H Z^2 \left[ \frac{1}{(3)^2} - \frac{1}{(5)^2} \right]$$

$$\frac{1}{(\lambda_2)_P} = R_H Z^2 \left[ \frac{1}{9} - \frac{1}{25} \right] \quad \text{...(ii)}$$

From Eqs. (i) and (ii), we get

$$\frac{(\lambda_2)_P}{(\lambda_1)_P} = \frac{\frac{7}{16 \times 9}}{\frac{16}{25 \times 9}} = \frac{25 \times 7}{16 \times 16}$$

$$\Rightarrow (\lambda_2)_P = \frac{25 \times 7}{16 \times 16} \times 720$$

$$(\lambda_2)_P = 492 \text{ nm}$$

**44.** *(270)*

> **key idea** This question includes the concept of calculating the radius of hydrogen or hydrogen like species i.e., $He^+$. For this, we will use the formula $r_n = r_0 \left( \frac{n^2}{Z} \right)$ Å

According to Bohr model, radius of hydrogen like species is,

$$\left[ r_n = r_0 \left( \frac{n^2}{Z} \right) \text{Å} \right]$$

where, $n$ = number of orbit

$Z$ = atomic number

$\Rightarrow$ For 3rd Bohr orbit ($n = 3$)

For $He^+$ ($Z = 2$)

$$r_0 = 0.6 \text{ Å} \quad \text{(Given)}$$

$$\Rightarrow r_3 = 0.6 \left[ \frac{(3)^2}{2} \right] \text{Å} = 2.7 \text{ Å}$$

$$= 2.7 \times 10^{-10} \text{ m}$$

$$= 2.7 \times 10^{-10} \times 10^{12} \text{ pm}$$

$$= 2.7 \times 10^2 \text{ pm} = 270 \text{ pm}$$

$$[\because 1 \text{ Å} = 10^{-10} \text{ m}, 1 \text{ pm} = 10^{-12} \text{ m}]$$

**45.** *(10)* The maximum number of emission lines when the excited electron of the H-atom from $n = 5$ drops to the ground state is given by

$$\Rightarrow \frac{n(n-1)}{2} = \frac{5(5-1)}{2}$$

$$= \frac{5 \times 4}{2} = \frac{20}{2}$$

$$= 10 \text{ lines.}$$

**46.** *(3155)* Bohr's radius of $n$th orbit is given by $r_n = a_0 n^2$.

$$mvr = \frac{nh}{2\pi}$$

For second Bohr orbit $n = 2, r_2 = (2)^2$

$$r_0 = 4a_0$$

$$\therefore \text{Kinetic energy} = \frac{n^2 h^2}{8\pi^2 m r^2}$$

$$= \frac{4h^2}{8\pi^2 m (4a_0)^2} = \left( \frac{4}{8\pi^2 \times 16} \right) \frac{h^2}{ma_0^2}$$

$$= \frac{h^2}{xma_0^2} \quad \text{(Given)}$$

$$x = \frac{16 \times 8\pi^2}{4}$$

$$\Rightarrow x = 315.507$$

$$10x = 3155.07 \approx 3155$$

## Topic 3

**01.** *(c)* Both statement I and statement II are true.

> **Concept Enhancer**
>
> Heisenberg's uncertainty principle is applicable only when the location and momentum are along the same axis. i.e if $\Delta x$ is uncertainty in position along the $x$-axis then $\Delta p$ must also be uncertainty in momentum along the $x$-axis.

**02.** *(a)* According to de-Broglie hypothesis,

$$\lambda \propto \frac{1}{p} \text{ (momentum)}$$

$$\lambda = \frac{h}{p}$$

$$\lambda p = h \text{ (constant)}$$

So, the plot between $\lambda$ and $p$ is rectangular hyperbola.

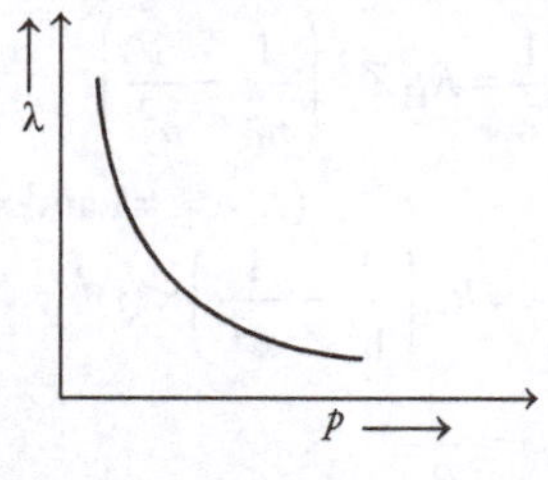


Hence, option (a) is correct.

**03.** *(b)* According to de-Broglie's equation, $\lambda = \frac{h}{mv}$

Given, $\quad v = 1.0 \times 10^3 \text{ m s}^{-1}$

On substituting the given values, we will get

$$\lambda = \frac{6.63 \times 10^{-34}}{1.67 \times 10^{-27} \times 1.0 \times 10^{3}}$$

$$= 3.9 \times 10^{-10} \text{m}$$

$$\approx 0.4 \text{ nm}$$

**04.** *(d)* de-Broglie wavelength ($\lambda$) for electron is given by

$$\lambda = \frac{h}{\sqrt{2m\text{K.E}}} \quad ...(i)$$

Also, according to photoelectric effect

$$KE = h\nu - h\nu_0$$

On substituting the value of KE in Eq (i), we get

$$\lambda = \frac{h}{\sqrt{2m \times (h\nu - h\nu_0)}}$$

$$\Rightarrow \quad \lambda \propto \frac{1}{(\nu - \nu_0)^{1/2}}$$

**05.** *(c)* Using the formula of de-Broglie wavelength, $\lambda = \frac{h}{mv}$

also, angular momentum of an electron by Bohr is : $mvr = \frac{nh}{2\pi}$

or $\quad \frac{2\pi r}{n} = \frac{h}{mv}$

$\therefore$ de-Broglie wavelength $\lambda = \frac{2\pi r}{n}$

de-Broglie wavelength of electron in 3rd orbit, $\quad \lambda_3 = \frac{2\pi r_3}{n} \quad ...(i)$

Also radius of 3rd orbit,

$$r_3 = \text{radius of 1st orbit} \times n^2$$

$$r_3 = a_0 \times n^2$$

$$\frac{r_3}{n} = a_0 \times n \quad ...(ii)$$

From Eqs. (i) and (ii) $\lambda_3 = 2\pi a_0 \times n$

for 3rd orbit, $n = 3$

$\therefore \quad \lambda_3 = 6\pi a_0$

**06.** *(c)* The de-Broglie wavelength of an electron,

$$\lambda = \frac{h}{p}, \text{ or } \frac{h}{mv}$$

But, from the Bohr's postulate of quantisation of angular momentum of electrons, $l = mvr = \frac{nh}{2\pi}$,

$$\therefore \quad mv = \frac{nh}{2\pi r}$$

Upon substitution,

$$\lambda = \frac{h}{\left(\frac{nh}{2\pi r}\right)} = \frac{2\pi r}{n},$$

where, $r$ is the radius of orbits in Bohr model $r = \frac{n^2 a_0}{Z}$

$$\therefore \lambda = \frac{2\pi \times n^2 \times a_0}{n \times Z} = \frac{2\pi n a_0}{Z} = 8\pi a_0$$

$$\left\{\begin{matrix} Z = 1 \text{ for hydrogen} \\ n = 4 \text{ (given)} \end{matrix}\right\}$$

**07.** *(b)* The radius of first Bohr orbit of H-atom is given by

$$r_n = 0.529\,(n)^2 \text{ Å}$$

$$mvr = \frac{nh}{2\pi}$$

$$\Rightarrow \quad mv = \frac{nh}{2\pi r} = \frac{nh}{2\pi\,(0.529)}$$

According to de-Broglie's equation

$$\lambda = \frac{h}{mv} = \frac{h}{nh} \times 2\pi \times 0.529$$

$$= 2\pi \times 0.529 \text{ Å}$$

**08.** *(58)* Using Heisenberg's uncertainty principle

$$\Delta x \cdot m\Delta v = \frac{h}{4\pi}$$

$$\Delta v = \frac{6.626 \times 10^{-34}}{4 \times 3.14 \times (10^{-15}) \times 9.1 \times 10^{-31}}$$

$$\Delta v = 0.0579 \times 10^{12} \text{ m/s}$$

or $\Delta v = 58 \times 10^{9}$ m/s

**09.** *(661)* According to de-Broglie hypothesis

$$\lambda = \frac{h}{mv}$$

$$\Rightarrow \quad \lambda = \frac{h\nu}{mv^2}$$

$$\Rightarrow \quad \frac{mv^2}{h} = \frac{\nu}{\lambda} \quad \left(\frac{v}{\lambda} = \text{Frequency } (\nu)\right)$$

$$\nu = \frac{mv^2}{h} = \frac{2R_H}{h}$$

$$= \frac{2 \times 2.18 \times 10^{-18}}{6.6 \times 10^{-34}}$$

$$= 660.6 \times 10^{13} \text{ Hz}$$

Nearest integer = 661

**10.** *(8)* The de-Broglie's wavelength of an electron is given by

$$2\pi r_n = n\lambda_d$$

or $2\pi a_0 \frac{n^2}{Z} = n\lambda_d \quad \left(\because r_n = \frac{a_0 n^2}{Z}\right)$

Here $n = 4$, $Z = 1$

Substitute the value of $n$ and $Z$ in above equation, we get

$$2\pi a_0 \frac{(4)^2}{1} = 4\lambda_d \Rightarrow \lambda_d = 8\pi a_0$$

**11.** *(7)* Given,

Kinetic energy $= 4.5 \times 10^{-29}$ J

Mass of electron $= 9 \times 10^{-31}$ kg

$h = 6.6 \times 10^{-34}$ Js

Wavelength of an electron = ?

We know that,

$$\text{wavelength } (\lambda) = \frac{h}{\sqrt{2\text{KE}\, m_e}}$$

$$= \frac{6.6 \times 10^{-34} \text{ Js}}{\sqrt{2 \times 4.5 \times 10^{-29} \times 9 \times 10^{-31}}}$$

$$= 7.3 \times 10^{-5} \text{ m}$$

So, wavelength is $\lambda \approx 7 \times 10^{-5}$ m

**12.** *(2)* Statements A and C are correct while statements B and D are incorrect. The characteristic of electrons emitted do not depend upon the material of the electrode of the cathode ray tube. Nature of emitted electrons does not depend on nature of gas present in cathode ray tube.

**13.** *(548)* Uncertainty in position,

$$\Delta x = 2a_0 = 2 \times 52.9 \text{ pm}$$

$$= 2 \times 52.9 \times 10^{-12} \text{ m}$$

Using Heisenberg uncertainity equation,

$$\Delta x \cdot \Delta v = \frac{h}{4\pi m} = \frac{h}{4\pi m \Delta x}$$

$$= \frac{6.63 \times 10^{-34}}{4 \times 3.14 \times 9.1 \times 10^{-31} \times 2 \times 52.9 \times 10^{-12}}$$

$$= 548 \times 10^{3} \text{ m/s}$$

$$= 548 \text{ k m s}^{-1}$$

**14.** *(2)* The de-Broglie wavelength of electron is given by

$$\lambda = \frac{h}{\sqrt{2m\text{KE}}} \Rightarrow \text{KE} = \frac{h^2}{2m\lambda^2}$$

$$= \frac{(6.626 \times 10^{-34})^2}{2 \times 9.1 \times 10^{-31} \times (3.3 \times 10^{-10})^2}$$

$$= 2.215 \times 10^{-18}$$

$$E_{\text{abs}} = E_{\text{req}} + k$$

$$\frac{E_{\text{abs}}}{E_{\text{req}}} = 1 + \frac{k}{E_{\text{req}}}$$

$$= 1 + \frac{2.215 \times 10^{-18}}{13.6 \times 1.602 \times 10^{-19}}$$

$$= 2.01 \simeq 2$$

**15.** *(1758)* We know that wavelength,

$$\lambda = \frac{h}{mv}$$

∴ Wavelength of electron,

$$\lambda_e = \frac{h}{m_e \times v_e}$$

Wavelength of neutron,

$$\lambda_N = \frac{h}{m_N \times v_N}$$

$$\lambda_e = \lambda_N \text{ (given)}$$

when $v_e = xv_N$ ...(i)

$$\frac{h}{m_e \times v_e} = \frac{h}{m_N \times v_N} \quad \text{...(ii)}$$

By substituting Eq. (i) in Eq. (ii), we get

$$\frac{h}{m_e \times x \times v_N} = \frac{h}{m_N \times v_N} \Rightarrow x = \frac{m_N}{m_e}$$

$m_N$ (mass of neutron) $= 1.6 \times 10^{-27}$ kg

$m_e$ (mass of electron) $= 9.1 \times 10^{-31}$ kg

$$\Rightarrow x = \frac{1.6 \times 10^{-27}}{9.1 \times 10^{-31}} = 1758$$

**16.** *(58)* Uncertainity in speed of electron

$$= \Delta v = \frac{0.02}{100} \times 5 \times 10^6 = 10^3 \text{ m/s}$$

Uncertainity in position $= x \times 10^{-9}$ m

According to Heisenberg uncertainty principle,

$$\Delta x \cdot \Delta p = \frac{h}{4\pi m}$$

$$x \times 10^{-9} \times 10^3 = \frac{6.63 \times 10^{-34}}{4 \times 3.14 \times 9.1 \times 10^{-31}}$$

$$x \times 10^{-6} = 0.058 \times 10^{-3}$$

$$\Rightarrow x = \frac{0.058 \times 10^{-3}}{10^{-6}} = 58$$

**17.** *(6)* de-Broglie wavelength of electron

$$\lambda_e = \frac{h}{\sqrt{2m(\text{KE})}}$$

where $\lambda_e$ = wavelength of electron

KE = kinetic energy of electron

also, kinetic energy $= qV$

$q$ = charge of electron,

$m$ = mass of electron,

$V$ = potential.

$$\lambda_e = \frac{6.63 \times 10^{-34}}{\sqrt{2 \times 1.6 \times 10^{-19} \times 9.1 \times 10^{-31} \times 40 \times 10^3}}$$

$$\lambda_e = 6.14 \times 10^{-12} \text{ m}$$

∴ $x = 6$ (nearest integer)

**18.** *(9)* Given, $(\lambda) = 248$ nm $= 2480$ Å

Threshold energy $= 3.0$ eV

de-Broglie wavelength is given by

$$\lambda = \frac{h}{mv}$$

$$\Rightarrow \lambda = \frac{h}{\sqrt{2m(\text{KE})}}$$

$$\text{or} \quad \lambda = \frac{h}{\sqrt{2m(qV)}}$$

Putting mass of electron,

$$m_e = 9.1 \times 10^{-31} \text{ kg}$$

Charge of electron, $q = 1.6 \times 10^{-19}$ C

$$\lambda = \sqrt{\frac{150}{V}} \text{ Å} \quad \text{...(i)}$$

We know,

$$E_{\text{applied}} = \phi + \text{KE}_{\text{max}} \quad \text{...(ii)}$$

$$E_{\text{applied}} = \frac{12400}{\lambda(\text{Å})} = \frac{12400}{2480} = 5 \text{ eV}$$

Putting value in Eq. (ii),

$$5 = 3 + \text{KE}_{\text{max}}$$

$$\Rightarrow \text{KE}_{\text{max}} = 2 \text{ eV}$$

Voltage = 2

$$\lambda_e = \sqrt{\frac{150}{2}} = 8.6 \text{ Å}$$

Nearest integer = 9

**19.** *(1)* According to Heisenberg's uncertainty equation,

$$\Delta x \times m\Delta v \geq \frac{h}{4\pi}$$

$(\therefore m = 10 \text{ g} = 10 \times 10^{-3} \text{ kg}, v = 90 \text{ ms}^{-1})$

Uncertainty in position,

$$\Delta x = \frac{h}{4\pi} \times \frac{1}{m \times \Delta v \text{ (with uncertainty)}}$$

$$= \frac{6.63 \times 10^{-34}}{4 \times 3.14} \times \frac{1}{(10 \times 10^{-3}) \times \left(90 \times \frac{5}{100}\right)} \text{ m}$$

$$= 1.17 \times 10^{-33} \text{ m}$$

$$\simeq 1 \times 10^{-33} \text{ m}$$

## Topic 4

**01.** *(c)* $Z = 37$ (Rubidium)

$$\text{Rb} = [\text{Kr}]5s^1$$

$n = 5$ (Principal quantum number)

$l = 0$ (Azimuthal quantum number)

$m = 0$ (Magnetic quantum number)

$s = +\frac{1}{2}$, or $-\frac{1}{2}$ (spin-quantum number)

Hence, option (c) is correct.

**02.** *(a)* Number of radial nodes is given by

$$n - l - 1$$

For $3p$ orbital, the value of $n = 3, l = 1$

So, number of radial nodes

$$= 3 - 1 - 1 = 1$$

**03.** *(b)* We know that, at node

$$\psi_{2s} = 0$$

So, $\left(2 - \frac{r}{a_0}\right) = 0$

⇒ If, $r = r_0$

$$\Rightarrow 2 - \frac{r_0}{a_0} = 0$$

$$2 = \frac{r_0}{a_0}$$

or $r_0 = 2a_0$

Thus, option (b) is correct.

**04.** *(d)* The orbital with greater value of $n + l$ value will have higher energy.

| | | | | | | | |
|---|---|---|---|---|---|---|---|
| | $n=3$, $l=2$ | > | $n=4$, $l=0$ | > | $n=3$, $l=1$ | > | $n=3$, $l=0$ |
| $n+l$ | 5 | | 4 | | 4 | | 3 |
| | D | > | B | > | C | > | A |

**• Concept Enhancer**

According to $(n + l)$ rule, the orbital with more value of $(n + l)$ has more energy. If $n + l$ value is same for two orbital, then the orbital which has greater value of $n$ will have more energy.

**05.** (a) For $4d$ orbital the value of $n=4$, $l=2$

So, angular nodes $=l=2$

Now, we know that, radial nodes

$$=(n-l-1)$$
$$=4-2-1=1$$

Hence, radial node $=1$, Angular node $=2$

**06.** (b) The energy order of a subshell is determined by $(n+l)$ rule. The subshell with lower value of $(n+l)$ has lower energy. If the value of $n+l$ is same then lower value of $n$ will have lower energy.

Now,

$A: n=3, l=2 \Rightarrow (n+l)=5$

$B: n=4, l=1 \Rightarrow (n+l)=5$

$C: n=4, l=2 \Rightarrow (n+l)=6$

$D: n=3, l=1 \Rightarrow (n+l)=4$

$\therefore$ The order of increasing energy is

$$D < A < B < C.$$

**07.** (d) Option (d) is incorrect. An atomic orbital is characterised by three quantum number $n$ (principal quantum number), $l$ (azimuthal quantum number) and $m$ (magnetic quantum number).

**08.** (c) Given, principal quantum number $=3$

Now, for azimuthal quantum number ($l$) has possible values

$$l=0 \text{ to } n-1$$
$$=0, 1, 2$$

$l$ cannot be equal to the value of $n$.

So, option (c)

$n=3$; $l=3$; $m_l=-3$; $s=-\frac{1}{2}$ is not allowed.

**09.** (a) Energies of the orbitals in the same subshell decrease with increase in the atomic number ($Z$), e.g., energy of $2s$ orbital of hydrogen atom is greater than that of $2s$ orbital of lithium and that of lithium is greater than that of sodium and so on, i.e.,

So, $E_{2s}(\text{H}) > E_{2s}(\text{Li}) > E_{2s}(\text{Na})$
$$> E_{2s}(\text{K}).$$

**10.** (a) The electrons are more likely to be found in the region $a$ and $c$. At $b$, wave function becomes zero and is called radial nodal surface or simply node.

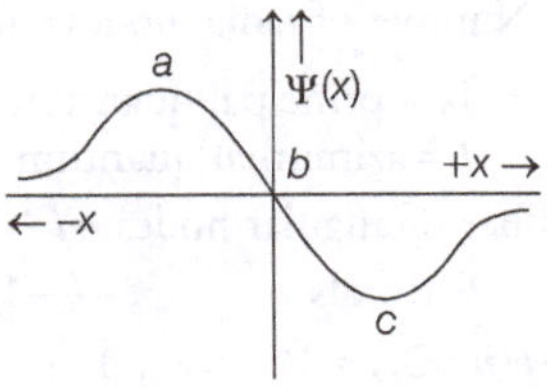


The graph between wavefunction ($\psi$) and distance ($r$) from the nucleus helps in determining the shape of orbital.

**11.** (b) Number of orbitals can be calculated by using the formula $=n^2=(5)^2=25$.

**12.** (d) Orbitals having same value of $n+l$ has same energy.

> **Concept Enhancer**
>
> Thus orbitals having $n=3, l=2, m=1$ and $n=3, l=2, m=0$ will have same energy in the absence of magnetic and electric field.

**13.** (a) For hydrogen atom, the orbital(s) with lowest energy are given in B, C and D.

In hydrogen atom the orbitals in a shell are degenerate, means energy depends only on $n$.

$$E_{3p_x}=E_{3d_{x^2-y^2}}=E_{3d_{z^2}}$$

**14.** (d) Orbital angular momentum is given by

$$=\sqrt{l(l+1)}\frac{h}{2\pi}$$

For 2s orbital; $l=0$

$\therefore$Orbital angular momentum = 0

For $2p$ orbital ; $l=1$

So, orbital angular momentum

$$=\sqrt{1(1+1)}\frac{h}{2\pi}$$
$$=\sqrt{2}\frac{h}{2\pi}$$

**15.** (a) Electronic configuration of Cr is

$$1s^2\, 2s^2\, 2p^6\, 3s^2\, 3p^6\, 3d^5\, 4s^1$$

So, electrons with $l=1$ are 12 (i.e., from $2p^6$ and $3p^6$)

and electrons with $l=2$ are 5 (i.e., from $3d^5$)

So, electrons with $l=1$ and 2 respectively are 12 and 5.

**16.** (b) Statement I is false but Statement II is true.

The correct form of statement I is :

No spectral line will be observed for a $2p_x \longrightarrow 2p_y$ transition because $2p_x$ and $2p_y$ orbitals are degenerate orbitals.

**17.** (b) Energy of atomic orbitals of hydrogen atom is given incorrect in C and D.

For single electron species energy only depends on '$n$' ( principal quantum number)

So, energy of $2s=2p$ and energy of $3d < 4s$

**18.** (c) Both statement I and Statement II are true.

**19.** (c) Statement I is true but Statement II is false.

The correct form of Statement II is :

In hydrogen atom, $3p$ and $3d$ orbitals are degenerate orbitals as in single electron species, energy depends only on principal quantum number ($n$).

**20.** (b) Energy level can be determined by comparing the value of $(n+l)$ value i.e. Bohr-Bury rule.

(A) $n=4, l=1 \Rightarrow (n+l)=5$

(B) $n=4, l=2 \Rightarrow (n+l)=6$

(C) $n=3, l=1 \Rightarrow (n+l)=4$

(D) $n=3, l=2 \Rightarrow (n+l)=5$

(E) $n=4, l=0 \Rightarrow (n+l)=4$

For the same value of $n+l$, the orbital having higher values of $n$, will have more energy

So, the correct order is

$$C < E < D < A < B$$

**21.** (c) Electronic configuration of potassium is,

$${}_{19}K = 1s^2 2s^2 2p^6 3s^2 3p^6 4s^1$$

Outermost orbital of K is $4s$ orbital.

For $4s$,

$n=$ Principal quantum number orbital $=4$

$l=$ Azimuthal quantum number $=0$

$m_l=$ Magnetic quantum number $=0$

$s=$ Spin quantum number $=\pm\frac{1}{2}$

Hence, option (c) is correct.

**22.** *(b)* Electrons pair of the same shells and the same subshell have same energy and these electrons pairs are found in degenerated orbitals. So, only pairs of electron of (B) are present in degenerate orbitals as $n+l$ values in both are cases are same

(a) $n=3, l=2,$

$n+l=3+2=5$

(b) $n=3, l=2$

$n+l=3+2=5$

**23.** *(b)*

**key idea** In this question we will use the concept of nodes.

The point at which the probability of finding an electron is zero is known as node or nodal surface. Number of nodes in any orbital $= n-l-1$

In case of 2*s*-orbital, number of nodes

$=2-0-1=1$

The correct plot for $\psi^2(r)$ as function of distance '*r*'.

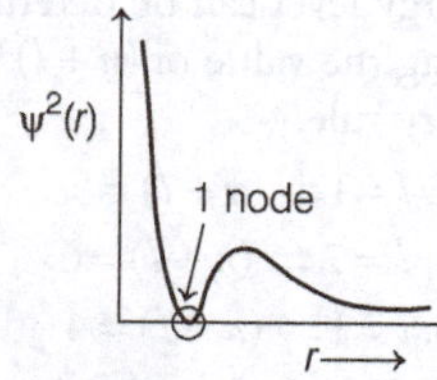


**24.** *(d)* Statement A, B and C are correct while statement D is incorrect. The correct explanation is given as

Total number of nodes present in atomic orbital is given by the formula $n-1$, where $n$ is principal quantum number.

So, the formula "$n-2$" is incorrect.

**25.** *(c)* Statements A, C and D are correct while Statements B and E are incorrect. The correct form of statements B and E are as follows

Azimuthal quantum number '*l*' can have values from 0 to $(n-1)$ for a given '*n*'. It describes the sub-shell and orbital angular momentum.

For any value of '*l*', total number of orbitals are given by $(2l+1)$ formula. So, for $l=5$,

Total number of orbitals

$=2l+1=2\times5+1=11$

**26.** *(d)* Number of radial nodes $=(n-l-1)$

[$n$ = principal quantum number, $l$ = azimuthal quantum number]

Number of angular nodes $= l$

| Orbitals | | $(n-l-1)$ | $l$ |
|---|---|---|---|
| (a) $3p(n=3, l=1)$ | $\Rightarrow$ | 1 | 1 |
| (b) $4f(n=4, l=3)$ | $\Rightarrow$ | 0 | 3 |
| (c) $4d(n=4, l=2)$ | $\Rightarrow$ | 1 | 2 |
| (d) $5d(n=5, l=2)$ | $\Rightarrow$ | 2 | 2 |

So, 5*d*-orbital has two radial as well as two angular nodes (option-d).

**27.** *(b)* As, we know,

Radial nodes $=n-l-1$

For 3*p* orbital, $n=3, l=1$

Angular node $(l)=1$

It has one angular and one radial node.

For 3*s* orbital, $n=3, l=0$

No. of angular node $(l)=0$

No. of radial node $\Rightarrow n-l-1$

$=3-0-1=2=2$

Hence, for 3*s*-orbital has no angular node and 2 radial nodes.

> **• Concept Enhancer**
>
> Radial nodes are spherical regions where the radial wavefunction is equal to zero whereas angular nodes are flat regions where angular wavefunction is zero.

**28.** *(c)*

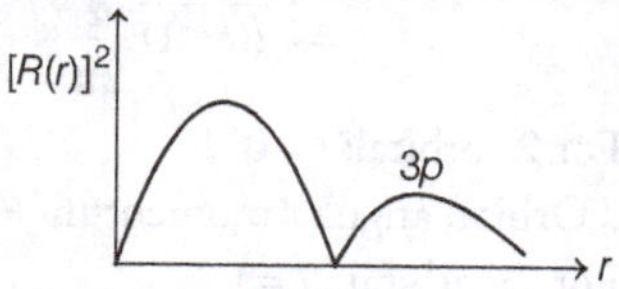


$\psi^2$ (Probability density) can be zero for 3*p* orbital other than infinite distance. Since, $n=3, l=1$ for 3*p* orbital, the number of radial nodes $=3-1-1=1$

**29.** *(b)* When $n=4$

The permissible values of $l$ and subshells are

$l=0 \rightarrow s$-subshell

$=1 \rightarrow p$-subshell

$=2 \rightarrow d$-subshell

$=3 \rightarrow f$-subshell

But, here values of $l$ are restricted by the value of magnetic quantum number, $m=-2$, where, $m$ depends on $l$ as

$m=-l \ldots. 0 \ldots. +l$

So, $m=-2$ value is possible for two subshells, viz, $l=3$ ($f$-subshell) and $l=2$ ($d$-subshell) only.

$l=3 \Rightarrow m=-3, -2-1, 0, +1, +2, +3$

$l=2 \Rightarrow m=-2, -1, 0, +1, +2$

**30.** *(a)* The energy of 2*s*-orbital is lowest in K(potassium). An orbital has smaller energy as the value of atomic number increases.

Correspondingly, the energy of the electron in such an orbital becomes less negative, meaning that the electron is less strongly bound and has less energy. The order energy of 2*s* orbital is H > Li > Na > K.

The graph of principal quantum number with atomic number is

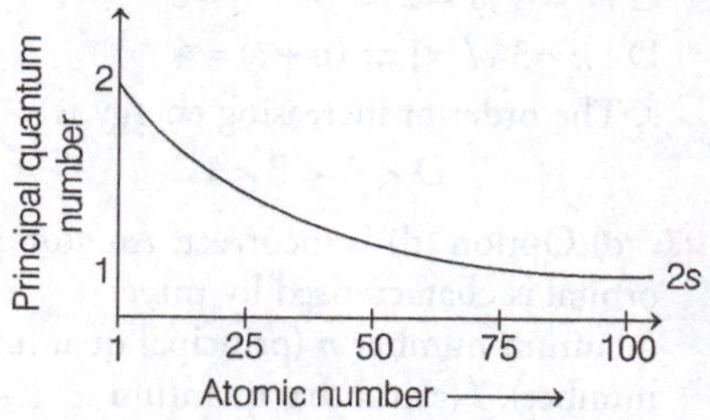


**31.** *(b)* The filling of electrons takes place as per $(n+l)$ rule. So, the orbital with higher $(n+l)$ value is filled first.

When the orbital has same $(n+l)$ value, then the one having lower value of $n$ is filled first. Hence, the sequence of filling is $6s \rightarrow 4f \rightarrow 5d \rightarrow 6p$ i.e. $(ns) \rightarrow (n-2)f \rightarrow (n-1)d \rightarrow np$.

**32.** *(a)* $n+l$ value of given electrons are as follows

(1) $n=4, l=1 \Rightarrow 5$ i.e. 4*p*

(2) $n=4, l=0 \Rightarrow 4$ i.e. 4*s*

(3) $n=3, l=2 \Rightarrow 5$ i.e. 3*d*

(4) $n=3, l=1 \Rightarrow 4$ i.e. 3*p*

$\therefore$ Increasing order of energy is

$3p < 4s < 3d < 4p$ i.e. $4 < 2 < 3 < 1$.

**33.** *(c)* (a) $n+l=3+0=3$ i.e., 3*s* orbital

(b) $n+l=3+1=4$ i.e., 3*p* orbital

(c) $n+l=3+2=5$ i.e., 3*d* orbital

(d) $n+l=4+0=4$ i.e., 4*s* orbital

So, the increasing order is

$3s < 3p < 4s < 3d$.

Hence, option (c) is correct.

**34.** *(b)*

| Set of quantum number | Orbitals |
|---|---|
| $A: n=1, l=0, m_l=0$ | $1s$ |
| $B: n=2, l=0, m_l=0$ | $2s$ |
| $C: n=2, l=1, m_l=1$ | $2p$ |
| $D: n=3, l=2, m_l=1$ | $3d$ |
| $E: n=3, l=2, m_l=0$ | $3d$ |

In absence of electric and magnetic fields, all orbitals of $3d$ are degenerate.

**35.** *(a)* Among the given statements, A and C are correct but B and D are incorrect. Their correct forms are

- Unpaired electron can be in anyone of $p_x, p_y$ or $p_z$ orbital.
- Angular node for $s$-orbital = 0, while of each $p$-orbital is one.

∴ Sum of all angular nodes = 3

**36.** *(a)* Among the given statements, A, B, D, E are reason for extra stability of half-filled subshell.

The stability is due to,

- symmetrical distribution of electrons.
- large exchange of energy
- smaller coulombic repulsion
- smaller shielding of electrons by one another.

**37.** *(d)* Among the given statements, A and B are true, while C and D are incorrect. The correct form of C and D are as follows,

- The positive and negative sign in the more function of $2p_x$ orbital refer to the sign (phase) of the wave function, not the charge.
- The wave function of $2p_x$ orbital will be zero in $YZ$ plane (Nodal plane).

**38.** *(d)* The correct plot for 3s-orbital is

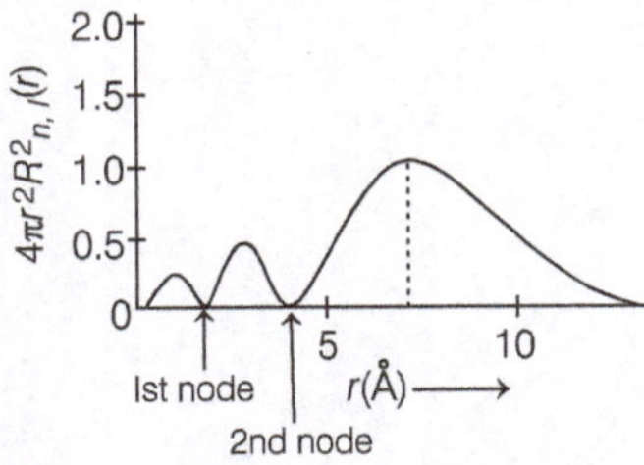


For 3s,

value of $l=0$

value of $n=3$

Number of radial nodes $= n-l-1$
$= 3-2 = 2$

In graph $D$, two nodes are present, so this is the correct plot for 3s-orbital.

**39.** *(c)*

> **key idea** This question includes the phenomenon associated with quantum mechanism. Quantum nature of atoms are associated with following phenomena or processes
> (i) Absorption spectrum
> (ii) Emission spectrum
> (iii) Black body radiation
> (iv) Photoelectric effect

Internal energy ($U$) of particles like atoms depends upon the thermodynamic variables ($p, V, T$) of the system. Thus, $U = f(p, T)$

or $U = f_1(T, V)$ or $U = f_2(p, V)$

So, quantum nature of atom is not associated with its internal energy.

Thus, graph (c) is not a direct manifestation of the quantum nature of atoms.

**40.** *(d)* Statements I and IV are correct while the statements II and III are incorrect.

Statement II

As size of an orbit ∝ Azimuthal quantum number ($l$) (∵ $n$ = constant)

Statement III

As at ground state, $n=1, l=0$

⇒ Orbital angular momentum (wave mechanics)

$$= \sqrt{l(l+1)}\,\frac{h}{2\pi} = 0 \quad [\because l=0]$$

**41.** *(4)* The possible combination of $n=4$ and $m_l=0$ is

$n=4, l=0, m_l=0$

$n=4, l=1, m_l=0$

$n=4, l=2, m_l=0$

$n=4, l=3, m_l=0$

So, total 4 orbitals can be identified with $n=4$ and $m_l=0$.

**42.** *(6)* For principal quantum number $n$, $l=0$ to $n-1$ and for each $l$, $m_l$ will be $-l$ to $+l$.

$n=4$

| $l$ | $m_l$ |
|---|---|
| 0 | 0 |
| 1 | −1, 0, +1 |
| 2 | −2, −1, 0, +1, +2 |
| 3 | −3, −2, −1, 0, +1, +2, +3 |

So, number of orbital associated with $n=4, |m_l|=1$ are 6.

Hence, total number of electrons is 6.

**43.** *(5)* $p_x, p_y, p_z, d_{z^2}$ and $d_{x^2-y^2}$ orbitals have electron density along the axes.

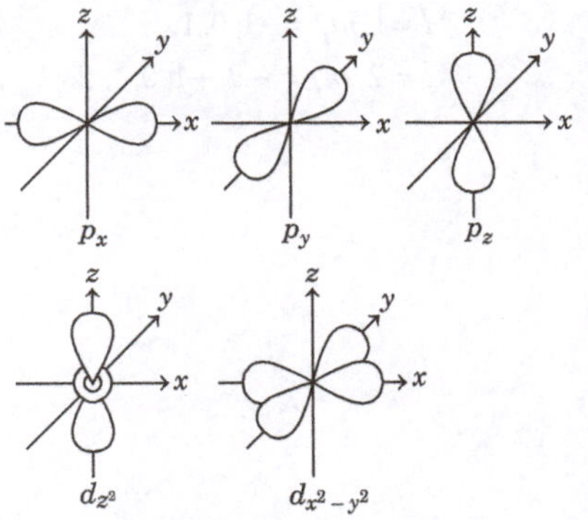


$d_{xy}$ and $d_{yz}$ have electron density between the axis.

**44.** *(0)* Orbital angular momentum ($L$)

$$L = \sqrt{l(l+1)}\,(h/2\pi)$$

For 3s-orbital

[$l = s = 0$]

So, $h = \sqrt{0(0+1)} \times h/2\pi$ or $h = 0$

Thus, the value of $x$ is 0.

**45.** *(3)* Statements given in A, D and E are correct while, B and C are incorrect.

For 2s orbital, the probability density first decreases and then increases.

Boundary surface diagram of orbitals encloses a region of high probability (Not 100%) of finding an electron.

**46.** *(3)* Number of radial nodes in an orbital is given as $n-l-1$

| | | |
|---|---|---|
| $7s$ : | $n=7, l=0$; | $7-0-1=6$ |
| $7p$ : | $n=7, l=1$; | $7-1-1=5$ |

$6s$ : $n=6, l=0$; $6-0-1=5$
$8p$ : $n=8, l=1$; $8-1-1=6$
$8d$ : $n=8, l=2$; $8-2-1=5$

**47.** *(2)* (A) and (D) options are incorrect ones as these two set of quantum numbers are not possible since they both have principle quantum numbers equal to azimuthal quantum number which is not possible.

(B) and (C) are correct set of quantum number which represent electrons in $3d$ and $2p$-orbitals respectively.

**48.** *(3)* Given, $n=5, m_l=+2$

For $n=5$, possible value of $l=0, 1, 2, 3, 4$

For $l=0, m_l=0$
$l=1, m_l=-1, 0, 1$
$l=2, m_l=-2, -1, 0, 1, 2$
$l=3, m_l=-3, -2, -1, 0, 1, 2, 3$
$l=4,$
$m_l=-4, -3, -2, -1, 0, 1, 2, 3, 4$

Possible value of $m_l$ for a given value of $l$

$$=0, \pm 1, \pm 2, \pm 3 \ldots \pm l$$

So, number of orbitals having $n=5$ and $m_l=\pm 2$ are 3.

**49.** *(0)* The configuration of

$Zn^{+} = 1s^2 2s^2 2p^6 3s^2 3p^6 3d^{10} 4s^1$

The outermost electron is in $4s$-orbital. For $4s$-orbital, principal quantum number is 4 and azimuthal quantum number $(l)=0$

Magnetic quantum number, $m=-l$ to $+l$

For $l=0, m=0$

**50.** *(7)* Ge $(Z=32)$ electronic configuration is

$1s^2\ 2s^2\ 2p^6\ 3s^2 3p^6 3d^{10} 4s^2 4p^2$

Thus, it has 7 fully filled orbital with $m_l=0$

**51.** *(0)* The electronic configuration of $Ga^{+}$ is $1s^2 2s^2 2p^6 3s^2 3p^6 3d^{10} 4s^2$

So, the valence electron is in $s$-orbital. Hence, Azimuthal quantum No. $(l)$ for valence shell electron is zero.

**52.** *(0)* Given, $n=4, m_l=-3$, so $l=3$

(Possible value of $l=0, 1, 2, 3$)

As $m=-3$ is possible only for $l=3$

For $l=3, m_l=-3, -2, -1, 0, 1, 2, 3$

Hence, this is $4f$-orbital.

Number of radial nodes $=n-l-1$
$$=4-3-1=0$$

Hence, there is no radial node in $4f$ orbital.

C-03

JEE Main Chapterwise Topicwise Chemistry

In JEE Main there is usually 0-1 question per paper are visible from this chapter. In past 3 years, 38 conceptual and 01 numerical questions were asked of easy to moderate level from this chapter.
**For detailed analysis scan QR code**

# CLASSIFICATION OF ELEMENTS AND PERIODICITY IN PROPERTIES

## Topic 1 Modern Periodic Table

### Level 1

**01** The element that does not belong to the same period of the remaining elements (modern periodic table) is *[2025, 23 Jan Shift-I]*

(a) platinum (b) iridium
(c) palladium (d) osmium

**02** Given below are two statements: *[2024, 27 Jan Shift-I]*

**Statement I** The $4f$ and $5f$-series of elements are placed separately in the periodic table to preserve the principle of classification.

**Statement II** $s$-block elements can be found in pure form in nature.

In the light of the above statements, choose the **most appropriate** answer from the options given below

(a) Statement I is true but Statement II is false.
(b) Both Statement I and Statement II are false.
(c) Both Statement I and Statement II are true.
(d) Statement I is false but Statement II is true.

**03** The IUPAC nomenclature of an element with electronic configuration $[Rn]5f^{14}6d^1 7s^2$ is *[2022, 25 July Shift-I]*

(a) unnilbium (b) unnilunium
(c) unnilquadium (d) unniltrium

**04** Which of the following elements is considered as a metalloid? *[2022, 26 June Shift-II]*

(a) Sc (b) Pb
(c) Bi (d) Te

**05** Which one of the following statements for D.I. Mendeleev, is incorrect? *[2021, 22 July Shift-II]*

(a) Element with atomic number 101 is named after him.
(b) He authored the textbook-Principles of Chemistry.
(c) He invented accurate barometer.
(d) At the time, he proposed periodic table of elements, structure of atom was known.

**06** The atomic number of the element unnilennium is *[2020, 3 Sep Shift-I]*

(a) 109 (b) 102
(c) 108 (d) 119

**07** The group number, number of valence electrons and valency of an element with atomic number 15, respectively, are *[2019, 12 April Shift-I]*

(a) 16, 5 and 2 (b) 15, 5 and 3
(c) 16, 6 and 3 (d) 15, 6 and 2

**08** The IUPAC symbol for the element with atomic number 119 would be *[2019, 8 April Shift-II]*

(a) unh (b) uue
(c) uun (d) une

**Position of Elements in Periodic Table in Blend with Electronic Configuration**

Imp.

**09** In the long form of the periodic table, the valence shell electronic configuration of $5s^2 5p^4$ corresponds to the elements present in *[JEE Main 2015]*

(a) group 16 and period 6 (b) group 17 and period 5
(c) group 16 and period 5 (d) group 17 and period 6

**10** According to the periodic law of elements, the variation in properties of elements is related to their *[AIEEE 2003]*

(a) atomic masses
(b) nuclear masses
(c) atomic numbers
(d) nuclear neutron-proton number ratios

### Level 2

**11** Given below are two statement abouts X-ray spectra of elements : *[2025, 23 Jan Shift-II]*

**Statement I** A plot of $\sqrt{v}$ ($v$ = frequency of X-rays emitted) *vs* atomic mass is a straight line.

**Statement II** A plot of $v$($v$ = frequency of X-rays emitted) *vs* atomic number is a straight line.

In the light of the above statements, choose the correct answer from the options given below
(a) Statement I is true but Statement II is false.
(b) Both Statement I and Statement II are false.
(c) Statement I is false but Statement II is true.
(d) Both Statement I and Statement II are true.

**12** Given below are two statements. *[2025, 28 Jan Shift-II]*

**Statement I** According to the law of octaves, the elements were arranged in the increasing order of their atomic number.

**Statement II** Meyer observed a periodically repeated pattern upon plotting physical properties of certain elements against their respective atomic numbers.

In the light of the above statements, choose the correct answer from the options given below
(a) Statement I is true but Statement II is false.
(b) Both Statement I and Statement II are true.
(c) Statement I is false but Statement II is true.
(d) Both Statement I and Statement II are false.

**13** Number of elements from the following that cannot form compounds with valencies which match with their respective group valencies is...... .
B, C, N, S, O, F, P, Al, Si *[2024, 4 April Shift-I]*
(a) 6 (b) 5 (c) 7 (d) 3

**Imp.** Graph of X-ray Spectra

**14** Henry Moseley studied characteristic X-ray spectra of elements. The graph which represents his observation correctly is
Given $v$ = frequency of X-ray emitted
$Z$ = atomic number *[2023, 8 April Shift-II]*

(a) 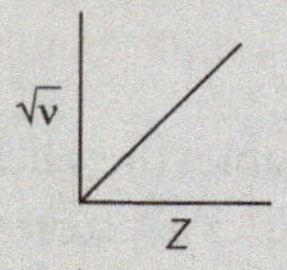


(b) 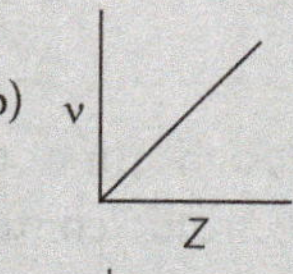


(c) 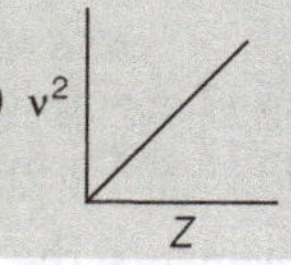


(d) 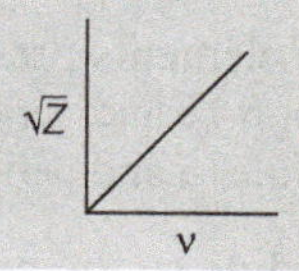


**15** The characteristics of elements $X, Y$ and $Z$ with atomic numbers respectively 33, 53 and 83 are *[2021, 16 March Shift-II]*
(a) $X$ is a metalloid, $Y$ is a non-metal and $Z$ is a metal
(b) $X, Y$ and $Z$ are metals
(c) $X$ and $Y$ are metalloids and $Z$ is a metal
(d) $X$ and $Z$ are non-metals and $Y$ is a metalloid

**16** Similarity in chemical properties of the atoms of elements in a group of the periodic table is most closely related to *[JEE Main 2014, 2003]*
(a) Atomic numbers
(b) Atomic masses
(c) Number of principal energy levels
(d) Number of valence electrons

## Level 3

**17** Match the List-I with List-II

| List-I (Family) | | List-II (Symbol of Element) | |
|---|---|---|---|
| A. | Pnictogen (group 15) | I. | Ts |
| B. | Chalcogen | II. | Og |
| C. | Halogen | III. | Lv |
| D. | Noble gas | IV. | Mc |

Choose the correct answer from the options given below *[2025, 3 April Shift-II]*
(a) A-II, B-III, C-IV, D-I
(b) A-IV, B-III, C-I, D-II
(c) A-III, B-I, C-IV, D-II
(d) A-IV, B-I, C-II, D-III

**18** Which of the following statements are not true about the periodic table? *[2025, 24 Jan Shift-I]*
A. The properties of elements are function of atomic weights.
B. The properties of elements are function of atomic numbers.
C. Elements having similar outer electronic configurations are arranged in same period.
D. An element's location reflects the quantum numbers of the last filled orbital.
E. The number of elements in a period is same as the number of atomic orbitals available in energy level that is being filled.

Choose the correct answer from the options given below
(a) B, C and Eonly
(b) A, C and E only
(c) A and E only
(d) D and E only

**19** Element "$E$" belongs to the period 4 and group 16 of the periodic table. The valence shell electronic configuration of the element, which is just above "$E$" in the group is *[2022, 28 June Shift-I]*
(a) $3s^2, 3p^4$ (b) $3d^{10}, 4s^2, 4p^4$
(c) $4d^{10}, 5s^2, 5p^4$ (d) $2s^2, 2p^4$

**20** Outermost electronic configuration of a group-13 element $E$ is $4s^2\,4p^1$. The electronic configuration of an element of $p$-block period-five placed diagonally to element, $E$ is *[2021, 20 July Shift-II]*
(a) $[Kr]3d^{10}4s^2 4p^2$ (b) $[Ar]3d^{10}4s^2 4p^2$
(c) $[Xe]5d^{10}6s^2 6p^2$ (d) $[Kr]4d^{10}5s^2 5p^2$

## Numerical Value Type Questions

**21** If IUPAC name of an element is "unununnium" then the element belongs to $n$th group of periodic table. The value of $n$ is ........ . *[2024, 30 Jan Shift-I]*

**22** Total number of ions from the following with noble gas configuration is ........
$Sr^{2+}(Z=38)$, $Cs^{+}(Z=55)$, $La^{2+}(Z=57)$, $Pb^{2+}(Z=82)$, $Yb^{2+}(Z=70)$ and $Fe^{2+}(Z=26)$ *[2024, 27 Jan Shift-II]*

**23** The atomic number of unnilunium is ......... *[2020, 6 Sep Shift-II]*

## Topic 2 Periodic Properties and their Trends 1 (Atomic Radius, IE, EA, EN, Metallic Character)

### Level 1

**01** The successive 5 ionisation energies of an element are 800, 2427, 3658, 25024 and 32824 kJ/mol, respectively. By using the above values predict the group in which the above element is present *[2025, 24 Jan Shift-II]*
(a) Group 2 (b) Group 14
(c) Group 4 (d) Group 13

**02** Which of the following electronegativity order is incorrect ? *[2025, 22 Jan Shift-I]*
(a) Al < Mg < B < N
(b) S < Cl < O < F
(c) Mg < Be < B < N
(d) Al < Si < C < N

**03** First ionisation enthalpy values of first four group 15 elements are given below. Choose the correct value for the element that is a main component of apatite family. *[2025, 29 Jan Shift-II]*
(a) 1402 $kJ\ mol^{-1}$ (b) 834 $kJ\ mol^{-1}$
(c) 1012 $kJ\ mol^{-1}$ (d) 947 $kJ\ mol^{-1}$

**04** The incorrect decreasing order of atomic radii is *[2025, 28 Jan Shift-I]*
(a) Be > Mg > Al > Si (b) Si > P > Cl > F
(c) Mg > Al > C > O (d) Al > B > N > F

**05** An element *E* has the ionisation enthalpy value of 374 $kJ\ mol^{-1}$, *E* reacts with elements *A*, *B*, *C* and *D* with electron gain enthalpy values of −328, −349, −325 and −295 $kJ\ mol^{-1}$, respectively. The correct order of the products *EA*, *EB*, *EC* and *ED* in terms of ionic character is *[2025, 29 Jan Shift-I]*
(a) *EB* > *EA* > *EC* > *ED*
(b) *EA* > *EB* > *EC* > *ED*
(c) *ED* > *EC* > *EB* > *EA*
(d) *ED* > *EC* > *EA* > *EB*

**06** The element having the highest first ionisation enthalpy is *[2024, 29 Jan Shift-II]*
(a) Al (b) N
(c) Si (d) C

**Imp.** Order of Electronegativity

**07** The correct order of electronegativity for given elements is *[2023, 8 April Shift-I]*
(a) P > Br > C > At (b) C > P > At > Br
(c) Br > P > At > C (d) Br > C > At > P

**08** The correct order of metallic character is *[2023, 10 April Shift-II]*
(a) K > Be > Ca (b) K > Ca > Be
(c) Be > Ca > K (d) Ca > K > Be

**09** The first ionisation enthalpies of Be, B, N and O follow the order *[2022, 25 July Shift-II]*
(a) O < N < B < Be (b) Be < B < N < O
(c) B < Be < N < O (d) B < Be < O < N

**10** The correct order of electron gain enthalpies of Cl, F, Te and Po is *[2022, 25 June Shift-II]*
(a) F < Cl < Te < Po (b) Po < Te < F < Cl
(c) Te < Po < Cl < F (d) Cl < F < Te < Po

**11** Match List-I with List-II.

| List-I (Electronic configuration of elements) | List-II ($\Delta_i H$ in $kJ\ mol^{-1}$) |
|---|---|
| A. $1s^2 2s^2$ | (i) 801 |
| B. $1s^2 2s^2 2p^4$ | (ii) 899 |
| C. $1s^2 2s^2 2p^3$ | (iii) 1314 |
| D. $1s^2 2s^2 2p^1$ | (iv) 1402 |

Choose the most appropriate answer from the options given below. *[2021, 26 Feb Shift-I]*

| | A | B | C | D | | A | B | C | D |
|---|---|---|---|---|---|---|---|---|---|
| (a) | (ii) | (iii) | (iv) | (i) | (b) | (i) | (iv) | (iii) | (ii) |
| (c) | (i) | (iii) | (iv) | (ii) | (d) | (iv) | (i) | (ii) | (iii) |

**12** The electron gain enthalpy (in kJ/mol) of fluorine, chlorine, bromine and iodine, respectively, are *[2020, 7 Jan Shift-I]*
(a) −333, −325, −349 and −296
(b) −296, −325, −333 and −349
(c) −333, −349, −325 and −296
(d) −349, −333, −325 and −296

**13** The element having greatest difference between its first and second ionisation energy, is *[2019, 9 April Shift-I]*
(a) Ca (b) Sc
(c) Ba (d) K

**14** The size of the iso-electronic species $Cl^-$, Ar and $Ca^{2+}$ is affected by *[2019, 8 April Shift-I]*
(a) Azimuthal quantum number of valence shell
(b) electron-electron interaction in the outer orbitals
(c) principal quantum number of valence shell
(d) nuclear charge

**15** The correct order of atomic radius is *[2019, 12 Jan Shift-II]*
(a) Ce > Eu > Ho > N (b) Ho > N > Eu > Ce
(c) Eu > Ce > Ho > N (d) N > Ce > Eu > Ho

**16** The correct option with respect to the Pauling electronegativity values of the elements is *[2019, 11 Jan Shift-II]*

(a) P > S (b) Si < Al (c) Te > Se (d) Ga < Se

**17** When the first electron gain enthalpy ($\Delta_{eg} H$) of oxygen is –141 kJ/mol, its second electron gain enthalpy is *[2019, 9 Jan Shift-II]*

(a) negative; but less negative than the first
(b) a positive value
(c) a more negative value than the first
(d) almost the same as that of the first

**18** The correct order of electron affinity is *[JEE Main 2018, 15]*

(a) O > F > Cl (b) F > O > Cl
(c) F > Cl > O (d) Cl > F > O

### Electronic Configuration in blend With Ionisation Enthalpy

Imp.

**19** The electronic configuration with the highest ionisation enthalpy is *[JEE Main 2017]*

(a) [Ne] $3s^2 3p^1$ (b) [Ne] $3s^2 3p^2$
(c) [Ne] $3s^2 3p^3$ (d) [Ar] $3d^{10} 4s^2 4p^3$

**20** Which of the following atoms has the highest first ionisation energy? *[JEE Main 2016]*

(a) Na (b) K (c) Sc (d) Rb

**21** The ionic radii (in Å) of $N^{3-}$, $O^{2-}$ and $F^-$ respectively are *[JEE Main 2015]*

(a) 1.36, 1.40 and 1.71 (b) 1.36, 1.71 and 1.40
(c) 1.71, 1.40 and 1.36 (d) 1.71, 1.36 and 1.40

**22** The first ionisation potential of Na is 5.1 eV. The value of electron gain enthalpy of $Na^+$ will be *[JEE Main 2013]*

(a) – 2.55 eV (b) – 5.1 eV
(c) –10.2 eV (d) + 2.55 eV

**23** Which is the correct order of second ionisation potential of C, N, O and F in the following? *[JEE Main 2013, 12]*

(a) O > N > F > C (b) O > F > N > C
(c) F > O > N > C (d) C > N > O > F

**24** The correct order of electron gain enthalpy with negative sign of F, Cl, Br and I, having atomic number 9, 17, 35 and 53 respectively, is *[AIEEE 2011]*

(a) I > Br > Cl > F (b) F > Cl > Br > I
(c) Cl > F > Br > I (d) Br > Cl > I > F

**25** The increasing order of the first ionisation enthalpies of the elements B, P, S and F (lowest first) is *[AIEEE 2006]*

(a) F < S < P < B (b) P < S < B < F
(c) B < P < S < F (d) B < S < P < F

**26** Which one of the following ions has the highest value of ionic radius ? *[AIEEE 2004]*

(a) $Li^+$ (b) $B^{3+}$ (c) $O^{2-}$ (d) $F^-$

## Level 2

**27** Electronic configuration of four elements A, B, C and D are given below *[2025, 2 April Shift-II]*

(A) $1s^2 2s^2 2p^3$ (B) $1s^2 2s^2 2p^4$
(C) $1s^2 2s^2 2p^5$ (D) $1s^2 2s^2 2p^2$

Which of the following is the correct order of increasing electronegativity (Pauling's scale)?

(a) A < C < B < D
(b) A < B < C < D
(c) A < D < B < C
(d) D < A < B < C

**28** The atomic number of the element from the following with lowest 1st ionisation enthalpy is: *[2025, 8 April Shift-II]*

(a) 35 (b) 87
(c) 32 (d) 19

**29** Choose the incorrect trend in the atomic radii ($r$) of the elements. *[2025, 7 April Shift-II]*

(a) $r_{Br} < r_K$ (b) $r_{Rb} < r_{Cs}$
(c) $r_{At} < r_{Cs}$ (d) $r_{Mg} < r_{Al}$

**30** Given below are two statements: *[2025, 2 April Shift-I]*

**Statement I** The metallic radius of Al is less than that of Ga.

**Statement II** The ionic radius of $Al^{3+}$ is less than that of $Ga^{3+}$.

In the light of the above statements, choose the most appropriate answer from the options given below

(a) Statement I is incorrect but Statement II is correct.
(b) Statement I is correct but Statement II is incorrect.
(c) Both Statement I and Statement II are incorrect.
(d) Both Statement I and Statement II are correct.

**31** The electron affinity value are negative for

A. $Be \rightarrow Be^-$ B. $N \rightarrow N^-$
C. $O \rightarrow O^{2-}$ D. $Na \rightarrow Na^-$
E. $Al \rightarrow Al^-$ *[2024, 6 April Shift-I]*

Choose the most appropriate answer from the options given below.

(a) A, B, D and E only (b) A and D only
(c) D and E only (d) A, B and C only

**32** Given below are two statements: *[2024, 5 April Shift-II]*

**Statement I** The metallic radius of Na is 1.86 Å and the ionic radius of $Na^+$ is lesser than 1.86 Å .

**Statement II** Ions are always smaller in size than the corresponding elements.

In the light of the above statements, choose the correct answer from the options given below :

(a) Statement I is correct but Statement II is false.
(b) Statement I is incorrect but Statement II is true.
(c) Both Statement I and Statement II are true.
(d) Both Statement I and Statement II are false.

**33** Given below are two statements. *[2024, 4 April Shift-II]*

**Statement I** The correct order of first ionisation enthalpy values of Li, Na, F and Cl is

$$Na < Li < Cl < F.$$

**Statement II** The correct order of negative electron gain enthalpy values of Li, Na, F and Cl is

$$Na < Li < F < Cl$$

In the light of the above statements, choose the correct answer from the options given below:

(a) Statement I is false but Statement II is true.
(b) Both Statement I and Statement II are true.
(c) Both Statement I and Statement II are false.
(d) Statement I is true but Statement II is false.

**34** The correct order of first ionisation enthalpy values of the following elements is: *[2024, 4 April Shift-I]*

A. O B. N C. Be D. F
E. B

Choose the correct answer from the options given below

(a) A < B < D < C < E (b) B < D < C < E < A
(c) C < E < A < B < D (d) E < C < A < B < D

**35** Given below are two statements. *[2024, 1 Feb Shift-II]*

**Statement I** Both metals and non-metals exist in *p* and *d*-block elements.

**Statement II** Non-metals have higher ionisation enthalpy and higher electronegativity than the metals.

In the light of the above statements, choose the **most appropriate** answer from the options given below

(a) Statement I is true but Statement II is false.
(b) Both Statement I and Statement II are false.
(c) Statement I is false but Statement II is true.
(d) Both Statement I and Statement II are true.

**36** In case of isoelectronic species the size of $F^-$, Ne and $Na^+$ is affected by *[2024, 1 Feb Shift-I]*

(a) electron-electron interaction in the outer orbitals
(b) none of the factors because their size is the same
(c) nuclear charge ($z$)
(d) principal quantum number ($n$)

**37** Consider the following elements. *[2024, 31 Jan Shift-II]*

$$\text{Group}\ \downarrow \begin{matrix} A'B' \longrightarrow \text{Period} \\ C'D' \end{matrix}$$

Which of the following is are true about $A'$, $B'$, $C'$ and $D'$?

A. Order of atomic radii : $B' < A' < D' < C'$
B. Order of metallic character : $B' < A' < D' < C'$
C. Size of the element : $D' < C' < B' < A'$
D. Order of ionic radii : $B'^+ < A'^+ < D'^+ < C'^+$

Choose the correct answer from the options given below

(a) B, C and D only (b) A only
(c) A, B and D only (d) A and B only

**38** Given below are two statements : one is labelled as **Assertion (A)** and the other is labelled as **Reason (R)**.

**Assertion (A)** The first ionisation enthalpy decreases across a period.

**Reason (R)** The increasing nuclear charge outweighs the shielding across the period. *[2024, 29 Jan Shift-I]*

In the light of the above statements, choose the **most appropriate** from the options given below

(a) Both A and R are true and R is the correct explanation of A.
(b) Both A and R are true but R is not the correct explanation of A.
(c) A is false but R is true.
(d) A is true but R is false.

**39** For elements B, C, N, Li, Be, O and F, the correct order of first ionisation enthalpy is *[2023, 11 April Shift-I]*

(a) B > Li > Be > C > N > O > F
(b) Li < B < Be < C < O < N < F
(c) Li < Be < B < C < O < N < F
(d) Li < Be < B < C < N < O < F

**40** For electron gain enthalpies of the elements denoted as $\Delta_{eg} H$, the incorrect option is *[2023 1 Feb Shift-II]*

(a) $\Delta_{eg} H(\text{Cl}) < \Delta_{eg} H(\text{F})$ (b) $\Delta_{eg} H(\text{Te}) < \Delta_{eg} H(\text{Po})$
(c) $\Delta_{eg} H(\text{Se}) < \Delta_{eg} H(\text{S})$ (d) $\Delta_{eg} H(\text{I}) < \Delta_{eg} H(\text{At})$

**41** The correct increasing order of the ionic radius is *[2023, 31 Jan Shift-I]*

(a) $Ca^{2+} < K^+ < Cl^- < S^{2-}$ (b) $S^{2-} < Cl^- < Ca^{2+} < K^+$
(c) $K^+ < S^{2-} < Ca^{2+} < Cl^-$ (d) $Cl^- < Ca^{2+} < K^+ < S^{2-}$

**42** Inert gases have positive electron gain enthalpy. Its correct order is *[2023, 25 Jan Shift-I]*

(a) Xe < Kr < Ne < He (b) He < Ne < Kr < Xe
(c) He < Kr < Xe < Ne (d) He < Xe < Kr < Ne

**43** The first ionisation enthalpy of Na, Mg and Si, respectively, are 496, 737 and 786 kJ $mol^{-1}$. The first ionisation enthalpy (kJ $mol^{-1}$) of Al is *[2022, 29 July Shift-I]*

(a) 487 (b) 768 (c) 577 (d) 856

**44** In which of the following pairs, electron gain enthalpies of constituent elements are nearly the same or identical?

(A) Rb and Cs (B) Na and K
(C) Ar and Kr (D) I and At

Choose the correct answer from the options given below. *[2022, 28 July Shift-I]*

(a) (A) and (B) only (b) (B) and (C) only
(c) (A) and (C) only (d) (C) and (D) only

**45** Given below are two statements.

**Statement I** In $Cl_2$ molecule the covalent radius is double of the atomic radius of chlorine.

**Statement II** Radius of anionic species is always greater than their parent atomic radius.

Choose the most appropriate answer from options given below *[2022, 26 July Shift-I]*
(a) Both Statement I and Statement II are correct.
(b) Both Statement I and Statement II are incorrect.
(c) Statement I is correct but Statement II is incorrect.
(d) Statement I is incorrect but Statement II is correct.

**46** Outermost electronic configurations of four elements *A*, *B*, *C*, *D* are given below. *[2022, 27 July Shift-II]*
(A) $3s^2$ (B) $3s^2 3p^1$ (C) $3s^2 3p^3$ (D) $3s^2 3p^4$
The correct order of first ionisation enthalpy for them is
(a) (A) < (B) < (C) < (D)
(b) (B) < (A) < (D) < (C)
(c) (B) < (D) < (A) < (C)
(d) (B) < (A) < (C) < (D)

**47** Given below are two statements one is labelled as Assertion (A) and the other is labelled as Reason (R).
**Assertion (A)** Metallic character decreases and non-metallic character increases on moving from left to right in a period.
**Reason (R)** It is due to increase in ionisation enthalpy and decrease in electron gain enthalpy, when one moves from left to right in a period.
In the light of the above statements, choose the most appropriate answer from the options given below.
(a) (A) is false but (R) is true. *[2021, 31 Aug Shift-I]*
(b) (A) is true but (R) is false
(c) Both (A) and (R) are correct and (R) is the correct explanation of (A).
(d) Both (A) and (R) are correct but (R) is not the correct explanation of (A).

**48** The correct order of ionic radii for the ions, $P^{3-}$, $S^{2-}$, $Ca^{2+}$, $K^+$, $Cl^-$ is *[2021, 27 Aug Shift-II]*
(a) $P^{3-} > S^{2-} > Cl^- > K^+ > Ca^{2+}$
(b) $Cl^- > S^{2-} > P^{3-} > Ca^{2+} > K^+$
(c) $P^{3-} > S^{2-} > Cl^- > Ca^{2+} > K^+$
(d) $K^+ > Ca^{2+} > P^{3-} > S^{2-} > Cl^-$

**49** The ionic radii of $F^-$ and $O^{2-}$ respectively are 1.33 Å and 1.4 Å while the covalent radius of N is 0.74Å. The correct statement for the ionic radius of $N^{3-}$ from the following is *[2021, 25 July Shift-II]*
(a) it is smaller than $F^-$ and N
(b) it is bigger than $O^{2-}$ and $F^-$
(c) it is bigger than $F^-$ and N, but smaller than of $O^{2-}$
(d) it is smaller than $O^{2-}$ and $F^-$, but bigger than of N

**50** The ionic radii of $K^+$, $Na^+$, $Al^{3+}$ and $Mg^{2+}$ are in the order *[2021, 25 July Shift-I]*
(a) $Na^+ < K^+ < Mg^{2+} < Al^{3+}$
(b) $Al^{3+} < Mg^{2+} < Na^+ < K^+$
(c) $Al^{3+} < Mg^{2+} < K^+ < Na^+$
(d) $K^+ < Al^{3+} < Mg^{2+} < Na^+$

**51** The first ionisation energy of magnesium is smaller as compared to that of elements *X* and *Y*, but higher than that of *Z*. The elements *X*, *Y* and *Z*, respectively, are *[2021, 18 March Shift-II]*
(a) chlorine, lithium and sodium
(b) argon, lithium and sodium
(c) argon, chlorine and sodium
(d) neon, sodium and chlorine

**52** The ionic radius of $Na^+$ ion is 1.02 Å. The ionic radius (in Å) of $Mg^{2+}$ and $Al^{3+}$, respectively are *[2021, 18 March Shift-I]*
(a) 1.05 and 0.99
(b) 0.68 and 0.72
(c) 0.85 and 0.99
(d) 0.72 and 0.54

**53** Consider the elements Mg, Al, S, P and Si, the correct increasing order of their first ionisation enthalpy is *[2021, 24 Feb Shift-I]*
(a) Al < Mg < S < Si < P (b) Mg < Al < Si < S < P
(c) Al < Mg < Si < S < P (d) Mg < Al < Si < P < S

**54** The ionic radii of $O^{2-}$, $F^-$, $Na^+$ and $Mg^{2+}$ are in the order *[2020, 4 Sep Shift-I]*
(a) $O^{2-} > F^- > Mg^{2+} > Na^+$
(b) $Mg^{2+} > Na^+ > F^- > O^{2-}$
(c) $O^{2-} > F^- > Na^+ > Mg^{2+}$
(d) $F^- > O^{2-} > Na^+ > Mg^{2+}$

**55** Among the statements (I-IV), *[2020, 3 Sept Shift-II]*
(I) Be has smaller atomic radius compared to Mg.
(II) Be has higher ionisation enthalpy than Al.
(III) Charge/radius ratio of Be is greater than that of Al.
(IV) Both Be and Al form mainly covalent compounds.
The correct ones are
(a) (I), (II) and (IV) (b) (I), (III) and (IV)
(c) (II), (III) and (IV) (d) (I), (II) and (III)

**56** The third ionisation enthalpy is minimum for *[2020, 8 Jan Shift-I]*
(a) Mn (b) Ni
(c) Co (d) Fe

**57** The process that is not endothermic in nature is
(a) $O^-(g) + e^- \longrightarrow O^{2-}(g)$ *[2020, 4 Sep Shift-II]*
(b) $Na(g) \longrightarrow Na^+(g) + e^-$
(c) $H(g) + e^- \longrightarrow H^-(g)$
(d) $Ar(g) + e^- \longrightarrow Ar^-(g)$

**58** In comparison to boron, beryllium has *[2019, 12 April Shift-II]*
(a) lesser nuclear charge and lesser first ionisation enthalpy
(b) greater nuclear charge and lesser first ionisation enthalpy
(c) greater nuclear charge and greater first ionisation enthalpy
(d) lesser nuclear charge and greater first ionisation enthalpy

**59** The isoelectronic set of ions is *[2019, 10 April Shift-I]*
(a) $F^-$, $Li^+$, $Na^+$ and $Mg^{2+}$
(b) $N^{3-}$, $Li^+$, $Mg^{2+}$ and $O^{2-}$
(c) $Li^+$, $Na^+$, $O^{2-}$ and $F^-$
(d) $N^{3-}$, $O^{2-}$, $F^-$ and $Na^+$

**60** The correct order of the atomic radii of C, Cs, Al and S is *[2019, 11 Jan Shift-I]*
(a) C < S < Al < Cs (b) C < S < Cs < Al
(c) S < C < Cs < Al (d) S < C < Al < Cs

**61** In general, the properties that decrease and increase down a group in the periodic table, respectively are
(a) electronegativity and atomic radius *[2019, 9 Jan Shift-I]*
(b) electronegativity and electron gain enthalpy
(c) electron gain enthalpy and electronegativity
(d) atomic radius and electronegativity

**62** Consider the following ionisation enthalpies of two elements '*A*' and '*B*' *[JEE Main 2018]*

| Elements | Ionisation enthalpy (kJ/mol) | | |
|---|---|---|---|
| | 1st | 2nd | 3rd |
| *A* | 899 | 1757 | 14847 |
| *B* | 737 | 1450 | 7731 |

Which of the following statements is correct?
(a) Both '*A*' and '*B*' belong to group-2 where '*A*' comes below '*B*'.
(b) Both '*A*' and '*B*' belong to group-2 where '*B*' comes below '*A*'.
(c) Both '*A*' and '*B*' belong to group-1 where '*B*' comes below '*A*'.
(d) Both '*A*' and '*B*' belong to group-1 where '*A*' comes below '*B*'.

**63** Which of the following series correctly represents relations between the elements from *X* to *Y*?
$X \rightarrow Y$ *(Online April 11, 2014)*
(a) ${}_3Li \rightarrow {}_{19}K$, Ionisation enthalpy increases
(b) ${}_9F \rightarrow {}_{35}Br$, Electron gain enthalpy (negative sign) increases
(c) ${}_6C \rightarrow {}_{32}Ge$, Atomic radius increases
(d) ${}_{18}Ar \rightarrow {}_{54}Xe$, Noble character increases

**64** The group having isoelectronic species is *[JEE Main 2017]*
(a) $O^{2-}, F^-, Na^+, Mg^{2+}$ (b) $O^-, F^-, Na, Mg^+$
(c) $O^{2-}, F^-, Na, Mg^{2+}$ (d) $O^-, F^-, Na^+, Mg^{2+}$

**65** Which of the following represents the correct order of increasing first ionisation enthalpy for Ca, Ba, S, Se and Ar ? *[JEE Main 2013]*
(a) Ca < S < Ba < Se < Ar
(b) S < Se < Ca < Ba < Ar
(c) Ba < Ca < Se < S < Ar
(d) Ca < Ba < S < Se < Ar

**66** The increasing order of the ionic radii of the given isoelectronic species is *[AIEEE 2012]*
(a) $Cl^-, Ca^{2+}, K^+, S^2$ (b) $S^{2-}, Cl^-, Ca^{2+}, K^+$
(c) $Ca^{2+}, K^+, Cl^-, S^{2-}$ (d) $K^+, S^{2-}, Ca^{2+}, Cl^-$

**67** The set representing the correct order of ionic radius is *[AIEEE 2009]*
(a) $Li^+ > Be^{2+} > Na^+ > Mg^{2+}$
(b) $Na^+ > Li^+ > Mg^{2+} > Be^{2+}$
(c) $Li^+ > Na^+ > Mg^{2+} > Be^{2+}$
(d) $Mg^{2+} > Be^{2+} > Li^+ > Na^+$

**68** The correct sequence which shows decreasing order of the ionic radii of the elements is *[AIEEE 2010]*
(a) $Al^{3+} > Mg^{2+} > Na^+ > F^- > O^{2-}$
(b) $Na^+ > Mg^{2+} > Al^{3+} > O^{2-} > F^-$
(c) $Na^+ > F^- > Mg^{2+} > O^{2-} > Al^{3+}$
(d) $O^{2-} > F^- > Na^+ > Mg^{2+} > Al^{3+}$

## Level 3

**69** The correct orders among the following are
Atomic radius : B < Al < Ga < In < Tl
Electronegativity : Al < Ga < In < Tl < B
Density : Tl < In < Ga < Al < B
1st ionisation energy : In < Al < Ga < Tl < B
Choose the correct answer from the options given below. *[2025, 3 April Shift-II]*
(a) C and D only
(b) A and B only
(c) B and D only
(d) A and C only

**70** Given below are two statements. *[2025, 29 Jan Shift-I]*
**Statement I** The radii of isoelectronic species increases in the order.
$$Mg^{2+} < Na^+ < F^- < O^{2-}$$
**Statement II** The magnitude of electron gain enthalpy of halogen decreases in the order.
$$Cl > F > Br > I$$
In the light of the above statements, choose the most appropriate answer from the options given below.
(a) Statement I is incorrect but Statement II is correct.
(b) Both Statement I and Statement II are correct.
(c) Both Statement I and Statement II are incorrect.
(d) Statement I is correct but Statement II is incorrect.

**71** Consider the following elements In, Tl, Al, Pb, Sn and Ge. The most stable oxidation states of elements with highest and lowest first ionisation enthalpies, respectively, are *[2025, 28 Jan Shift-I]*
(a) +4 and +3 (b) +4 and +1
(c) +1 and +4 (d) +2 and +3

**Imp.** Types of Oxides in Blend with Atomic Radius

**72** The type of oxide formed by the element among Li, Na, Be, Mg, B and Al that has the least atomic radius is *[2025, 29 Jan Shift-II]*
(a) $A_2O$ (b) $A_2O_3$
(c) $AO_2$ (d) $AO$

**73** Match List-I with List-II. *[2025, 22 Jan Shift-I]*

| | List-I | | List-II |
|---|---|---|---|
| (A) | $Al^{3+} < Mg^{2+} < Na^+ < F^-$ | (I) | Ionisation enthalpy |
| (B) | $B < C < O < N$ | (II) | Metallic character |
| (C) | $B < Al < Mg < K$ | (III) | Electronegativity |
| (D) | $Si < P < S < Cl$ | (IV) | Ionic radii |

Choose the correct answer from the options given below.
(a) (A)-(IV), (B)-(I), (C)-(III), (D)-(II)
(b) (A)-(II), (B)-(III), (C)-(IV), (D)-(I)
(c) (A)-(III), (B)-(IV), (C)-(II), (D)-(I)
(d) (A)-(IV), (B)-(I), (C)-(II), (D)-(III)

**74** Given below are two statements. *[2025, 22 Jan Shift-II]*

**Statement I** An element in the extreme left of the periodic table forms acidic oxides.

**Statement II** Acid is formed during the reaction between water and oxide of a reactive element present in the extreme right of the periodic table.

In the light of the above statements, choose the correct answer from the options given below:
(a) Both Statement I and Statement II are true.
(b) Both Statement I and Statement II are false.
(c) Statement I is false but Statement II is true.
(d) Statement I is true but Statement II is false.

**75** Match List-I with List-II.

| List-I (Elements) | List-II (Properties in their respective groups) |
|---|---|
| A. Cl, S | I. Elements with highest electronegativity |
| B. Ge, As | II. Elements with largest atomic size |
| C. Fr, Ra | III. Elements which show properties of both metals and non-metal |
| D. F, O | IV. Elements with highest negative electron gain enthalpy |

Choose the correct answer from the options given below : *[2024, 8 April Shift-I]*
(a) A-III, B-II, C-I, D-IV (b) A-II, B-III, C-IV, D-I
(c) A-IV, B-III, C-II, D-I (d) A-II, B-I, C-IV, D-III

**76** The statement(s) that are correct about the species $O^{2-}, F^-, Na^+$ and $Mg^{2+}$. *[2024, 9 April Shift-I]*
(A) All are isoelectronic
(B) All have the same nuclear charge
(C) $O^{2-}$ has the largest ionic radii
(D) $Mg^{2+}$ has the smallest ionic radii

Choose the most appropriate answer from the options given below:
(a) (C) and (D) only (b) (A), (B), (C) and (D)
(c) (B), (C) and (D) only (d) (A), (C) and (D) only

**77** The correct sequence of electron gain enthalpy of the elements listed below is *[2024, 31 Jan Shift-I]*
A. Ar B. Br C. F D. S
Choose the most appropriate from the options given below.
(a) A > D > C > B (b) D > C > B > A
(c) C > B > D > A (d) A > D > B > C

**78** Identify the elements $X$ and $Y$ using the ionisation energy values given below.

| | Ionisation energy (Ist) | kJ/mol (IInd) |
|---|---|---|
| $X$ | 495 | 4563 |
| $Y$ | 731 | 1450 |

*[2021, 16 March Shift-II]*
(a) $X = Na, Y = Mg$ (b) $X = Mg, Y = F$
(c) $X = Mg, Y = Na$ (d) $X = F, Y = Mg$

**79** Within each pair of elements F and Cl, S and Se, and Li and Na, respectively, the elements that release more energy upon an electron gain are *[2020, 7 Jan Shift-II]*
(a) F, Se and Na (b) F, S and Li
(c) Cl, S and Li (d) Cl, Se and Na

**80** The increasing order of the atomic radii of the following elements is
(A) C (B) O (C) F (D) Cl (E) Br *[2020, 8 Jan Shift-II]*
(a) (A) < (B) < (C) < (D) < (E)
(b) (C) < (B) < (A) < (D) < (E)
(c) (D) < (C) < (B) < (A) < (E)
(d) (B) < (C) < (D) < (A) < (E)

**81** In which of the following arrangements the order is not according to the property indicated against it ? *[AIEEE 2005]*
(a) Li < Na < K < Rb Increasing metallic radius
(b) I < Br < F < Cl Increasing electron gain enthalpy (with negative sign)
(c) B < C < N < O Increasing first ionisation enthalpy
(d) $Al^{3+} < Mg^{2+} < Na^+ < F^-$ Increasing ionic size

**82** The formation of the oxide ion $O^{2-}(g)$ requires first an exothermic and then an endothermic step as shown below

$$O(g) + e^- \longrightarrow O^-(g); \quad \Delta H° = -142 \text{ kJ mol}^{-1}$$
$$O^-(g) + e^- \longrightarrow O^{2-}(g); \quad \Delta H° = 844 \text{ kJ mol}^{-1}$$

This is because *[AIEEE 2004]*
(a) oxygen is more electronegative
(b) oxygen has high electron affinity
(c) $O^-$ ion will tend to resist the addition of another electron
(d) $O^-$ ion has comparatively larger size than oxygen atom

## Numerical Value Type Question

**83** The total number of isoelectronic species from the given set is ............ .
$O^{2-}, F^-, Al, Mg^{2+}, Na^+, O^+, Mg, Al^{3+}, F$

## Topic 3 Periodic Properties and their Trends 2 (Lattice and Hydration Energy, Oxidation State and Chemical Properties)

### Level 1

**01** Given below are two statements *[2024, 9 April Shift-I]*

**Statement I** The oxidation state of an element in a particular compound is the charge acquired by its atom on the basic of electron gain enthalpy consideration from other atoms in the molecule.

**Statement II** $p\pi$-$p\pi$ bond formation is more prevalent in second period elements over other periods.

In the light of the above statements, choose the most appropriate answer from the options given below.

(a) Statement I is correct but Statement II is incorrect.
(b) Both Statement I and Statement II are correct.
(c) Both Statement I and Statement II are incorrect.
(d) Statement I is incorrect and Statement II is correct.

**02** The total number of acidic oxides from the following list is NO, $N_2O$, $B_2O_3$, $N_2O_5$, CO, $SO_3$, $P_4O_{10}$ *[2022, 25 July Shift-II]*

(a) 3 (b) 4
(c) 5 (d) 6

**Imp.** Nature of Oxides

**03** Match List-I with List-II.

| List-I (Oxide) | List-II (Nature) |
|---|---|
| A. $Cl_2O_7$ | (I) Amphoteric |
| B. $Na_2O$ | (II) Basic |
| C. $Al_2O_3$ | (III) Neutral |
| D. $N_2O$ | (IV) Acidic |

Choose the correct answer from the option given below *[2022, 28 June Shift-II]*

(a) (A)–(IV), (B)–(III), (C)–(I), (D)–(II)
(b) (A)–(IV), (B)–(II), (C)–(I), (D)–(III)
(c) (A)–(II), (B)–(IV), (C)–(III), (D)–(I)
(d) (A)–(I), (B)–(II), (C)–(III), (D)–(IV)

**04** Among the following basic oxide is *[2022, 28 June Shift-II]*

(a) $SO_3$ (b) $SiO_2$
(c) CaO (d) $Al_2O_3$

**05** Given below are the oxides :
$Na_2O$, $As_2O_3$, $N_2O$, NO and $Cl_2O_7$
Number of amphoteric oxides is *[2022, 24 June Shift-I]*

(a) 0 (b) 1
(c) 2 (d) 3

**06** Match list-I with list-II.

| List-I | List-II |
|---|---|
| (A) NaOH | (i) Acidic |
| (B) $Be(OH)_2$ | (ii) Basic |
| (C) $Ca(OH)_2$ | (iii) Amphoteric |
| (D) $B(OH)_3$ | |
| (E) $Al(OH)_3$ | |

Choose the most appropriate answer from the options given below. *[2021, 27 July Shift-I]*

(a) (A)-(ii), (B)-(ii), (C)-(iii), (D)-(i), (E)-(iii)
(b) (A)-(ii), (B)-(iii), (C)-(ii), (D)-(i), (E)-(iii)
(c) (A)-(ii), (B)-(ii), (C)-(iii), (D)-(ii), (E)-(iii)
(d) (A)-(ii), (B)-(i), (C)-(ii), (D)-(iii), (E)-(iii)

**07** The set of elements that differ in mutual relationship from those of the other sets is *[2021, 17 March Shift-II]*

(a) Li-Mg (b) B-Si
(c) Be-Al (d) Li-Na

**08** Which pair of oxides is acidic in nature? *[2021, 26 Feb Shift-II]*

(a) $B_2O_3$, CaO (b) $B_2O_3$, $SiO_2$
(c) $N_2O$, BaO (d) CaO, $SiO_2$

**09** The ionic mobility of alkali metal ions in aqueous solution is maximum for *[AIEEE 2006]*

(a) $K^+$ (b) $Rb^+$ (c) $Li^+$ (d) $Na^+$

**10** Which of the following oxides is amphoteric in character? *[AIEEE 2005, 03]*

(a) $SnO_2$ (b) $SiO_2$ (c) $CO_2$ (d) CaO

### Level 2

**11** The nature of oxide ($TeO_2$) and hydride ($TeH_2$) formed by Te, respectively are *[2025, 2 April Shift-II]*

(a) oxidising and basic (b) reducing and basic
(c) reducing and acidic (d) oxidising and acidic

**12** The correct order of hydration enthalpies is
(A) $K^+$ (B) $Rb^+$ (C) $Mg^{2+}$ (D) $Cs^+$ (E) $Ca^{2+}$
Choose the correct answer from the options given below. *[2023, 29 Jan Shift-I]*

(a) C > E > A > B > D (b) C > A > E > B > D
(c) E > C > A > B > D (d) C > E > A > D > B

**13** Metals generally melt at very high temperature. Amongst the following, the metal with the highest melting point will be *[2022, 24 June Shift-II]*

(a) Hg (b) Ag (c) Ga (d) Cs

**14** The set in which compounds have different nature is *[2021, 20 July Shift-I]*

(a) $Be(OH)_2$ and $Al(OH)_3$ (b) $B(OH)_3$ and $Al(OH)_3$
(c) $B(OH)_3$ and $H_3PO_3$ (d) NaOH and $Ca(OH)_2$

**15** Three elements $X, Y$ and $Z$ are in the 3rd period of the periodic table. The oxides of $X, Y$ and $Z$, respectively, are basic, amphoteric and acidic. The correct order of the atomic number of $X, Y$ and $Z$ is *[2020, 2 Sep Shift-II]*

(a) $Z < Y < X$ (b) $X < Y < Z$
(c) $X < Z < Y$ (d) $Y < X < Z$

**16** Which one of the following orders presents the correct sequence of the increasing basic nature of the given oxides? *[AIEEE 2011]*

(a) $Al_2O_3 < MgO < Na_2O < K_2O$
(b) $MgO < K_2O < Al_2O_3 < Na_2O$
(c) $Na_2O < K_2O < MgO < Al_2O_3$
(d) $K_2O < Na_2O < Al_2O_3 < MgO$

**Imp.** Order of Acidic Strength

**17** Among $Al_2O_3$, $SiO_2$, $P_2O_3$ and $SO_2$ the correct order of acid strength is *[AIEEE 2004]*

(a) $SO_2 < P_2O_3 < SiO_2 < Al_2O_3$
(b) $SiO_2 < SO_2 < Al_2O_3 < P_2O_3$
(c) $Al_2O_3 < SiO_2 < SO_2 < P_2O_3$
(d) $Al_2O_3 < SiO_2 < P_2O_3 < SO_2$

## Level 3

**18** Which of the following statements are correct?

A. The process of adding an electron to a neutral gaseous atom is always exothermic.
B. The process of removing an electron from an isolated gaseous atom is always endothermic.
C. The 1st ionisation energy of boron is less than that of beryllium.
D. The electronegativity of C is 2.5 in $CH_4$ and $CCl_4$.
E. Li is the most electropositive among elements of group I.

Choose the correct answer from the options given below: *[2025, 3 April Shift-I]*

(a) A, C and D Only (b) B and D Only
(c) B and C Only (d) B, C and E Only

**19** In which of the following arrangements, the sequence is not strictly according to the property written against it? *[AIEEE 2012, 09, 05]*

(a) $CO_2 < SiO_2 < SnO_2 < PbO_2$ : increasing oxidising power
(b) $NH_3 < PH_3 < AsH_3 < SbH_3$ : increasing basic strength
(c) $HF < HCl < HBr < HI$ : increasing acid strength
(d) $B < C < O < N$ : increasing first ionisation enthalpy

**20** Following statements regarding the periodic trends of chemical reactivity of the alkali metals and the halogens are given. Which of these statements give the correct picture? *[AIEEE 2006]*

(a) The reactivity decreases in the alkali metals but increases in the halogens with increase in atomic number down the group.
(b) In both the alkali metals and the halogens the chemical reactivity decreases with increase in atomic number down the group.
(c) Chemical reactivity increases with increase in atomic number down the group in both the alkali metals and halogens.
(d) In alkali metals, the reactivity increases but in the halogens it decreases with increase in atomic number down the group.

# Key **Index**

### Topic 1

| | | | | | | | | | |
|---|---|---|---|---|---|---|---|---|---|
| 1 (c) | 2 (a) | 3 (d) | 4 (d) | 5 (d) | 6 (a) | 7 (b) | 8 (b) | 9 (c) | 10 (c) |
| 11 (b) | 12 (d) | 13 (d) | 14 (a) | 15 (a) | 16 (d) | 17 (b) | 18 (b) | 19 (a) | 20 (d) |
| 21 (11) | 22 (2) | 23 (101) | | | | | | | |

### Topic 2

| | | | | | | | | | |
|---|---|---|---|---|---|---|---|---|---|
| 1 (d) | 2 (a) | 3 (c) | 4 (a) | 5 (a) | 6 (b) | 7 (d) | 8 (b) | 9 (d) | 10 (b) |
| 11 (a) | 12 (c) | 13 (d) | 14 (d) | 15 (c) | 16 (d) | 17 (b) | 18 (d) | 19 (c) | 20 (c) |
| 21 (c) | 22 (b) | 23 (b) | 24 (c) | 25 (d) | 26 (c) | 27 (d) | 28 (b) | 29 (d) | 30 (a) |
| 31 (d) | 32 (a) | 33 (b) | 34 (d) | 35 (c) | 36 (c) | 37 (c) | 38 (c) | 39 (b) | 40 (c) |
| 41 (a) | 42 (d) | 43 (c) | 44 (c) | 45 (d) | 46 (b) | 47 (b) | 48 (a) | 49 (b) | 50 (b) |
| 51 (c) | 52 (d) | 53 (c) | 54 (c) | 55 (a) | 56 (d) | 57 (c) | 58 (d) | 59 (d) | 60 (a) |
| 61 (a) | 62 (b) | 63 (c) | 64 (a) | 65 (c) | 66 (c) | 67 (b) | 68 (d) | 69 (c) | 70 (b) |
| 71 (a) | 72 (b) | 73 (d) | 74 (c) | 75 (c) | 76 (d) | 77 (d) | 78 (a) | 79 (c) | 80 (b) |
| 81 (c) | 82 (c) | 83 (5) | | | | | | | |

### Topic 3

| | | | | | | | | | |
|---|---|---|---|---|---|---|---|---|---|
| 1 (d) | 2 (b) | 3 (b) | 4 (c) | 5 (b) | 6 (b) | 7 (d) | 8 (b) | 9 (b) | 10 (a) |
| 11 (d) | 12 (a) | 13 (b) | 14 (b) | 15 (b) | 16 (a) | 17 (d) | 18 (c) | 19 (b) | 20 (d) |

# Detailed **Explanations**

## Topic 1

**01.** *(c)* According to modern periodic table, elements platinum, iridium and osmium are placed in 6th period. While palladium is placed in 5th period.
So, option (c) is the correct answer.

> **Concept Enhancer**
> - Principal quantum number or the valence shell corresponds to the period of an element.
> - Group number can be predicted as
>   (i) **For *s*-block elements** Group number is equal to number of valence electrons.
>   (ii) **For *p*-block elements** Group number is equal to 10+ number of electron in the valence shell.
>   (iii) **For *d*-block elements** Group number is equal to the number of electrons in $(n-1)d$ subshell + number of electrons in valence shell.

**02.** *(a)* Statement I is true but Statement II is false.
The correct form of Statement II is as follows, *s*-block element never occurs in pure form in nature. As, they are highly reactive. Thus, they are found as ore or minerals.

**03.** *(d)* The given electronic configuration of an element is

$$=[\text{Rn}]\,5f^{14}6d^{1}7s^{2}$$

Atomic number of Rn $=86$
Atomic number of given element will be

$$86+14+1+2=103$$

IUPAC nomenclature is

$$\begin{matrix} 1 & 0 & 3 \\ \downarrow & \downarrow & \downarrow \\ \text{un} & \text{nil} & \text{tri} \end{matrix}+\text{ium}$$

$=$ Unniltrium

**04.** *(d)* Sc, Pb and Bi are metals while Te is a metalloid.
Elements having characteristics in between metals and non-metals are metalloids. Elements Si, Ge, As, Sb and Te are considered as metalloids.

**05.** *(d)* At the time, Mendeleev proposed periodic table of elements, structure of atom was not known.

**06.** *(a)* The atomic number of the element unnilennium is 109.

$$Z=\underbrace{1}_{\text{un}}\ \underbrace{0}_{\text{nil}}\ \underbrace{9}_{\text{enn}}+\text{ium}$$

$\Rightarrow$ Unnilennium

**07.** *(b)* Electronic configuration of the element having atomic number 15 is

$$1s^2\,2s^2\,2p^6\,\underbrace{3s^2\,3p^3}_{\text{Valence electrons}}$$

Number of valence electrons present in an atom decides the group number. As five electrons are present in valence shell, its group number. Number of valence electrons and valency of an element with atomic number 15 are 15, 5 and 3 respectively. Valency of element having atomic number 15 is +3 i.e., $(8-5=3)$.

**08.** *(b)* Atomic number (119) $=\underset{\text{un}}{\overset{1}{\downarrow}}\ \underset{\text{un}}{\overset{1}{\downarrow}}\ \underset{\text{enn}}{\overset{9}{\downarrow}}$
So, symbol of the element = Uue
IUPAC nomenclature of the element = Ununennium.

**09.** *(c)* $5s^25p^4$ electronic configuration corresponds to the element present in group 16 and period 5.

**10.** *(c)* According to modern periodic law. All physical and chemical properties of elements are periodic function of their atomic number.

**11.** *(b)* Both Statement I and Statement II are false.
The correct form of the statement are as follows.
Statement I. A plot of $\sqrt{\nu}$ *vs* atomic mass is a *zig-zag* plot.
Statement II. A plot of $\sqrt{\nu}$ *vs* atomic number gives a straight line.

**12.** *(d)* Both Statement I and Statement II are false. Their correct forms are
- Law of octaves was arranged in the increasing order of their atomic weight.
- Lothar Meyer plotted the physical properties such as atomic volume, melting point and boiling point against atomic weight.

**13.** *(d)* Among the given elements N, O, F cannot extend their valencies upto their group number due to non-availability of vacant-*d* orbital.

**14.** *(a)* Moseley's observation on characteristic X-ray spectra of elements relates frequency ($\nu$) of X-rays emitted to the atomic number by the following relation : $\sqrt{\nu}=a\,(Z-b)$
The plot of $\sqrt{\nu}$ *vs* $Z$ was found as linear straight line.

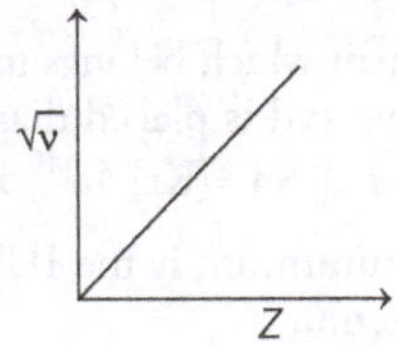


**15.** *(a)* $X={}_{33}\text{As}$ is a metalloid.
$Y={}_{53}\text{I}$ is a non-metal.
$Z={}_{83}\text{Bi}$ is a metal.

**16.** *(d)* Similarity in chemical properties of atoms of elements belonging to same group is most closely related to same number of electrons in their valence shells. This is because chemical properties are electron dependent which involves either loss, gain or sharing of electrons.

**17.** *(b)* The correct match is
A-IV, B-III, C-I, D-II
- Pnictogen $\rightarrow$ Mc (Moscovium) $\rightarrow$ Atomic number = 115
- Chalcogen $\rightarrow$ Lv (Livermorium) $\rightarrow$ Atomic number = 116
- Halogen $\rightarrow$ Ts (Tennessine) $\rightarrow$ Atomic number = 117
- Noble gas $\rightarrow$ Og (Oganesson) $\rightarrow$ Atomic number = 118

**18.** *(b)* Statements given is A, C and E are incorrect while B and D are correct.
The correct form of incorrect statements are as follows
- Properties of elements are periodic function of their atomic number.
- Element with similar outer electronic configuration are arranged in same group.
- Number of elements in a period is not equal to number of atomic orbitals available in energy level that is being filled.

**19.** *(a)* Period number 4 and group 16 of the periodic table indicate that the element '$E$' is selenium (Se). The element

present just above '*E*' is sulphur.
The general electronic configuration of group 16 is $ns^2np^4$. Thus, the electronic configuration of sulphur is [Ne] $3s^2 3p^4$.

So, option (a) is correct.

**20.** *(d)*

| | | 13th | 14th | 15th |
|---|---|---|---|---|
| 2nd period | $2s^2 2p^1$ | B | C | N |
| 3rd period | $3s^2 3p^1$ | Al | Si | P |
| 4th period | $4s^2 4p^1$ | Ga | Ge | As |
| 5th period | $5s^2 5p^1$ | In | Sn | Sb |

The element which belongs to *p*-block period five and is placed diagonally to element is $_{50}Sn = [Kr]\,4d^{10}5s^2 5p^2$

**21.** *(11)* Unununium is the IUPAC name of roentgenium
Atomic number → 111
Group belong → 11th group.
Hence, it belongs to group 11.

**22.** *(2)* Configuration of given ions are as follows.

$[Sr^{2+}] = [Kr]$

$[Cs^+] = [Xe]$

$[Yb^{2+}] = [Kr]\,4d^{10}4f^{14}5s^2 5p^6$

$[La^{2+}] = [Xe]\,5d^1$

$[Pb^{2+}] = [Xe]\,4f^{14}5d^{10}6s^2$

$[Fe^{2+}] = [Ar]\,3d^6$

Thus, only 2 ions i.e., $[Sr^{2+}]$ and $[Cs^+]$ has noble gas configuration.

**23.** *(101)* Unnilunium
IUPAC symbol = Unu
Atomic number (*Z*) = 101

| Digit | Name | Abbreviation |
|---|---|---|
| 1 | un | u |
| 0 | nil | n |
| 1 | un | u |

## Topic 2

**01.** *(d)* The ionisation enthalpy for ($IE_4$) is suddenly very high, therefore element must have 3 valence electrons and it belongs to group 13.

**02.** *(a)* The incorrect order of electronegativity is given in option (a). Its correct order is

$$\underset{(1.2)}{Mg} < \underset{(1.5)}{Al} < \underset{(2.0)}{B} < \underset{(3.0)}{N}$$

**03.** *(c)* The apatite family consists mainly of calcium phosphates, meaning phosphorus (P) is the main component.
The first ionisation enthalpy values are
Nitrogen (N) → 1402 kJ/mol
Phosphorus (P) → 1012 kJ/mol
Arsenic (As) → 947 kJ/mol
Antimony (Sb) → 834 kJ/mol
Since, phosphorus is the main component of apatite, the correct answer is option c.

**04.** *(a)* Atomic radii increases on moving down the group while it decreases along the period.
So, the incorrect decreasing order of atomic radii is given in option (a). Its correct decreasing order is

$$Be < Mg > Al > Si$$

**05.** *(a)* As the difference between ionisation enthalpy and electron gain enthalpy increases, ionic characteristic increases. Thus, the correct order of the ionic characteristic is

$$EB > EA > EC > ED$$

**06.** *(b)* Among the given elements, nitrogen has highest first ionisation enthalpy.
The sequence of ionisation enthalpy of the given elements is
Al < Si < C < N

**07.** *(d)* The correct order of electronegativity is
Br > C > At > P

| Element | Br | C | At | P |
|---|---|---|---|---|
| EN | 2.8 | 2.5 | 2.2 | 2.1 |

**08.** *(b)* On moving from top to bottom in group metallic character increases while on moving from left to right in a period it decreases.
Hence, the correct order of metallic character is K > Ca > Be.

**09.** *(d)* The electronic configuration of given elements are as follows.

$Be = 1s^2 2s^2 \rightarrow$ Fully-filled

$B = 1s^2 2s^2 2p^1$

$N = 1s^2 2s^2 2p^3 \rightarrow$ Half-filled

$O = 1s^2 2s^2 2p^4$

The ionisation enthalpy increases from left to right in a period. It is due to increase in $Z_{eff}$, so order should be Be < B < N < O.
But, due to greater stability nature of fully-filled and half-filled orbitals, the $2p^3$ is more stable configuration than $2p^4$.
Also $2s^2$ is more stable than $2s^2 2p^1$.
So, the correct order of first ionisation energy is B < Be < O < N.

**10.** *(b)* Cl has the highest electron gain enthalpy, even greater than F, due to small size and electron-electron repulsions in fluorine atom. Down the group, electron gain enthalpy decreases and along the period it increases.
Hence, the correct order is Po < Te < F < Cl.

**11.** *(a)* The correct matching is (A)-(ii), (B)-(iii), (C)-(iv), (D)-(i)

**12.** *(c)* Generally, $\Delta_{eg}H$ becomes less negative when comparing elements of the same group from top to bottom.
But among fluorine and chlorine there is an anomaly because inter-electron-electron repulsion is stronger in fluorine due to its extra small size.
∴ $\Delta_{eg}H$ is less exothermic than expected for F-atom.
Thus, the correct values of electron gain enthalpies

$$\underset{(-333)}{F} < \underset{(-349)}{Cl} > \underset{(-325)}{Br} > \underset{(-296)}{I} \quad kJ\,mol^{-1}$$

**13.** *(d)* The second ionisation energy of alkali metals are generally higher because the second electron is to be removed from the stable noble gas configuration. Among the given elements, K belongs to alkali metals thus its first and second ionisation energy has the greatest ionisation difference.

**14.** *(d)* The radius of isoelectronic species is inversely proportional to their nuclear charge or atomic number (*Z*). Thus, greater the value of *Z*, lesser the radii of isoelectronic species.

**15.** *(c)*

| Element | Atomic radius |
|---|---|
| Eu | 199 pm |
| Ce | 183 pm |
| Ho | 175 pm |
| N | 65 pm |

Order of atomic radii is
Eu > Ce > Ho > N

**16.** *(d)* On moving from left to right atomic size decreases from Ga to Se. As a result, the value of electronegativity increases from Ga to Se due to increase in the attraction between the outer electrons and the nucleus. Whereas on moving down the group, (i.e. from Se to Te), the atomic size increases.
As a result, the force of attraction between the outer electron and the nucleus decreases. Hence, the electronegativity decreases.

**17.** (b) Second electron gain enthalpy of O is +850 kJ $mol^{-1}$. This is because energy has to provided in order to overcome the strong electron-electron repulsion.

**18.** (d) On moving across a period from left to right, the electron affinity becomes more negative. On moving from top to bottom in a group, the electron affinity becomes less negative. Fluorine has exceptionally less negative electron affinity than chlorine, because adding an electron to fluorine causes greater repulsion than adding an electron to chlorine which is larger in size. The correct order of electron affinity is Cl > F > O.

**19.** (c) The electronic configuration with the highest ionisation enthalpy is [Ne] $3s^2 3p^3$. On moving down the group, the ionisation enthalpy decreases. On moving from left to right across a period, the ionisation enthalpy increases.

**20.** (c) Order of first ionisation energy is

Sc > Na > K > Rb.

Due to poor shielding effect, removal of one electron from 4*s* orbital of Sc is difficult as compared to 3*s*-orbital of other elements.

**21.** (c) Number of electrons in

$N^{3-} = 7 + 3 = 10$

Number of electrons in

$O^{2-} = 8 + 2 = 10$

Number of electrons in $F^- = 9 + 1 = 10$

Since, all the three species have each 10 electrons hence they are isoelectronic species.

It is considered that, in case of isoelectronic species as the negative charge increases, ionic radii increases and therefore the value of ionic radii are

$N^{3-} = 1.71$

(highest among the three)

$O^{2-} = 1.40$

$F^- = 1.36$

(lowest among the three)

**22.** (b) $Na \longrightarrow Na^+ + e^-$ $IE_2 = 5.1$ eV

$Na^+ + e^- \longrightarrow Na$

Electron gain enthalpy is negative of ionisation enthalpy.

Thus, $\Delta H_{eg} = -5.1$ eV

**23.** (b) $C^+ = 1s^2 2s^2 2p^1$,

$N^+ = 1s^2, 2s^2 2p^2$

$O^+ = 1s^2 2s^2 2p^3$, $F^+ = 1s^2 2s^2 2p^4$

Second ionisation potential means removal of electron from unit charged cation.

Order is O > F > N > C.

**24.** (c) As we go down the group in periodic table, atomic size increases, force of attraction decreases. Hence, electron gain enthalpy decreases down the group.

$$X(g) + e^- \longrightarrow X^-(g)$$

Actual order is Cl > F > Br > I

The fact that fluorine has a less electron gain enthalpy than chlorine is due to greater effectiveness of 2*p*-electron in the small F-atom to repel the additional electron entering the atom than 3*p*-electrons in the larger Cl-atom.

**25.** (d) Ionisation enthalpy increases along a period due to increased nuclear charge and decreased atomic radii. While in a group, it decreases from top to bottom. Along with this trend ionisation enthalpy is highest for the elements possessing half filled and stable electronic configuration. Since, phosphorus is having stable half-filled configuration. Hence, order is B < S < P < F.

**26.** (c) All the ions belong to same period, thus for them cations will be smaller than anions. Now, $O^{2-}$ and $F^-$ are isoelectronic and $r_n \propto \frac{1}{Z}$

Thus, ionic radius of

$O^{2-}$ $(Z = 8) > F^-$ $(Z = 9)$

**27.** (d) The correct order of increasing electronegativity is

D < A < B < C

N : $1s^2 2s^2 2p^3$ (electronegativity = 3)

O : $1s^2 2s^2 2p^4$ (electronegativity = 3.5)

F : $1s^2 2s^2 2p^5$ (electronegativity = 4)

C : $1s^2 2s^2 2p^2$ (electronegativity = 2.5)

**28.** (b) Ge $\rightarrow$ atomic number = 32, Br $\rightarrow$ atomic number = 35

Fr $\rightarrow$ atomic number = 87, K $\rightarrow$ atomic number = 19

Lowest first ionisation enthalpy among the given element will be of Fr[87].

As ionisation enthalpy (1st) is inversely proportional to atomic number.

**29.** (d) In a period from left to right atomic size decreases. So, the correct order of radius is

$r_{Mg} > r_{Al}$

**30.** (a) Statement I is incorrect, but Statement II is correct. The correct form of Statement I is, the metallic radius of Al is more than that of Ga due to poor shielding of *d*-subshell electrons.

**31.** (d) A. $Be + e^- \longrightarrow Be^-$, EA = −ve

B. $N + e^- \longrightarrow N^-$, EA = −ve

C. $O^+ + e^- \longrightarrow O^-$,

$O^- + e^- \longrightarrow O^{2-}$ EA = −ve

D. $Na + e^- \longrightarrow Na^-$, EA = + ve

E. $Al + e^- \longrightarrow Al^-$, EA = + ve

**32.** (a) Statement I is correct but statement II is false. The correct form of Statement II is,

Ions are not always smaller in size than the corresponding element. The size of anions are greater than neutral atom, while the size of cation are smaller than neutral atom.

Hence, option (a) is correct.

**33.** (b) Both Statement I and Statement II are true.

**34.** (d) Generally, ionisation enthalpy increases along the period, due to decrease in size of atoms across a period. Moreover, elements with half filled or fully-filled electronic configuration will have higher ionisation enthalpy. Thus, the correct order of 1st ionisation enthalpy across the period is

Li < B < Be < C < O < N < F < Ne

i.e., E < C < A < B < D

**35.** (c) Statement I is false but Statement II is true. The correct form of Statement I is :

*d*-block contains only metals while *p*-block has metals and non-metals both.

**36.** (c) Electronic configuration of $F^-$, Ne and $Na^+$ are same i.e., $1s^2 2s^2 2p^6$.

They have different size due to the difference in nuclear charge.

Hence, option (c) is correct.

**37.** (c) Statement given in option (c) is correct, and remaining statements are incorrect.

In general along the period from left to right, atomic and ionic size decreases and metallic character decreases. While on moving down the group, atomic and ionic size increases and metallic character increases.

**38.** (c) A is false but, R is true.

The correct form of A is : First ionisation energy increases along the period. This is because, along the

period nuclear charge increases, which out weighs the shielding effect.
Hence, option (c) is correct.

**39.** *(b)* The correct order of first ionisation energy is
$Li < B < Be < C < O < N < F$.
On moving from left to right in a period, atomic size decreases, $Z_{eff}$ increases, so ionisation enthalpy increases. The IE of N is more than O because of presence of half-filled and fully filled configuration which will make an atom more stable thus more will be the ionisation enthalpy.

**40.** *(c)* The incorrect option among the given options is (c).

(a) $\underset{(-345)}{\Delta_{eg}H(Cl)} < \underset{(-328)}{\Delta_{eg}H(F)}$ (correct)

(b) $\underset{(-190)}{\Delta_{eg}H(Te)} < \underset{(-183)}{\Delta_{eg}H(Po)}$ (correct)

(c) $\underset{(-195)}{\Delta_{eg}H(Se)} < \underset{(-200)}{\Delta_{eg}H(S)}$ (incorrect)

(d) $\underset{(-295)}{\Delta_{eg}H(I)} < \underset{(-270)}{\Delta_{eg}H(At)}$ (correct)

**41.** *(a)* The correct increasing order of the ionic radius is

$$Ca^{2+} < K^+ < Cl^- < S^{2-}.$$

Among isoelectronic species, the cation with the greater positive charge will have smaller radius because of the greater attraction of the electrons towards the nucleus, While anion with the greater negative charge will have the larger radius.

**42.** *(d)* Inert gases have stable electronic configuration. Therefore, some external energy is needed (endothermic, $\Delta H > 0$) to add the electron in their subshell. Hence, electron gain enthalpy for inert gases will be positive.
The values of electron gain enthalpy in kJ/mol for He, Ne, Kr, Xe are + 48, + 116, + 96 and + 77 respectively.
Hence, the correct order of positive electron gain enthalpy is

$$He < Xe < Kr < Ne$$

• Concept **Enhancer**

Electron gain enthalpy ($\Delta_{eg}H$) is the enthalpy change accompanying the processes of addition of electron to neutral gaseous atom (X) to convert it into a negative ion. The processes can be either endothemic or exothermic. In general, we can say that electron gain enthalpy becomes more negative with increase in the atomic number across a period.

**43.** *(c)* The value of first ionisation energy of Al is 577 kJ mol$^{-1}$. This is because ionisation energy increases from left to right in the period. But Mg will have greater IE value than Al, as it has fully filled configuration.

**44.** *(c)* Rb and Cs will have nearly same electron gain enthalpy i.e., –47 kJ/mol and – 46 kJ/mol respectively.
Ar and Kr will also have identical electron gain enthalpy as noble gases are inert and they are not found in combined state.

**45.** *(d)* Statement I is incorrect but statement II is correct.
Covalent radius in any molecule is 1/2 the distance between the nuclei of the atoms of the molecule and atomic radius is distance between nuclei of atom and valence shell. So, in molecule like $Cl_2$ it is not exactly double of the atomic radius rather it will be slightly less than that double.

**46.** *(b)* As we move from left to right in a period ionisation enthalpy increases gradually, but ionisation enthalpy of fully and half-filled orbitals are higher due to higher stability.
Hence, the correct order will be
B < A < D < C.

**47.** *(b)* On moving from left to right in periodic table, metallic character decreases while non-metallic character increases. This is due to increase in ionisation enthalpy with increase in atomic number. Hence, (A) is true but (R) is false.
Hence, correct option is (b).

**48.** *(a)* For isoelectronic species more is the negative charge greater will be the radius. So, the order of ionic radii is

$$\underset{15}{P^{3-}} > \underset{16}{S^{2-}} > \underset{17}{Cl^-} > \underset{19}{K^+} > \underset{20}{Ca^{2+}}$$

Therefore, option (a) is correct.

**49.** *(b)* Ionic radii of $N^{3-}$ is greater than $F^-$ and $O^{2-}$, due to higher negative charge. Size of anion increases with increase in magnitude of negative charge.

**50.** *(b)* As positive charge increases ionic size decreases for *iso*-electronic species. Thus, the order of ionic radius is
$Al^{3+} < Mg^{2+} < Na^+ < K^+$.

**51.** *(c)* The order of first ionisation energy of 3$^{rd}$ period is as follows
Na < Al < Mg < Si < S < P < Cl < Ar
On moving along a period from left to right, ionisation enthalpy increases. Ionisation enthalpy of group 2 is greater than group 13 and that of group 15 is greater than group 16. This is due to fully filled and half-filled orbital.
$\therefore$ *X*, *Y* and *Z* are argon, chlorine and sodium respectively.

**52.** *(d)* $Na^+, Mg^{2+}, Al^{3+}$ all are isoelectronic species. So, more is the charge on cation smaller will be the size respectively.

**53.** *(c)* Ionisation enthalpy increases from left to right in a period due to increase in effective nuclear charge. But I.E. (S) < I.E. (P) and I.E. (Al) < I.E. (Mg) due to half-filled and fully filled electronic configuration of P and Mg respectively.
Hence, the order is

$$Al < Mg < Si < S < P.$$

**54.** *(c)* $O^{2-}, F^-, Na^+$ and $Mg^{2+}$ are isoelectronic species.

| **Ionic species** | $O^{2-}$ | $F^-$ | $Na^+$ | $Mg^{2+}$ |
|---|---|---|---|---|
| **Number of $e^-$** | 10 | 10 | 10 | 10 |

For isoelectronic species,

$$\text{Radius} \propto \frac{-\text{ve charge}}{+\text{ ve charge}}$$

So, correct order of ionic radius is

$$O^{2-} > F^- > Na^+ > Mg^{2+}.$$

**55.** *(a)* Statements I, II and IV are correct while Statements III is incorrect. It is because the ratio of charge/radius of Be and Al are same due to diagonal relationship.

**56.** *(d)* Fe has minimum value of third ionisation enthalpy.
$Fe^{2+}$ has $[Ar]3d^6$ configuration, whereas $Fe^{3+}$ has $[Ar]3d^5$ configuration.
The latter is a stable configuration and easier to ionise $Fe^{2+}$ to $Fe^{3+}$ than expected. Hence possess least value.

**57.** *(c)* (a) $O^-(g) + e^- \longrightarrow O^{2-}(g)$ (endothermic)

(b) $Na(g) \longrightarrow Na^+(g) + e^-$ (endothermic)

(c) $H(g) + e^- \longrightarrow H^-(g)$ (exothermic)

(d) $Ar(g) + e^- \longrightarrow Ar^-(g)$ (endothermic)

Reaction (c) is not endothermic. It is exothermic. Electron gain ethalpy of $H(g)$ is negative, due to '$1s^1$' configuration.

**58.** *(d)* In comparison to boron, beryllium has lesser nuclear charge and greater first ionisation enthalpy.

Electronic configuration of $Be(4) = 1s^2, 2s^2$.

It possess completely filled *s*-orbitals. Hence, high amount of energy is required to pull the electron from the gaseous atom.

**59.** *(d)*

**key idea** Isoelectronic species contains same number of electrons.

The species with their atomic number and number of electrons are as follows

| Species (ions) | At. no. (Z) | No. of electrons |
|---|---|---|
| $N^{3-}$ | 7 | $7+3=10$ |
| $O^{2-}$ | 8 | $8+2=10$ |
| $F^-$ | 9 | $9+1=10$ |
| $Na^+$ | 11 | $11-1=10$ |
| $Li^+$ | 3 | $3-1=2$ |
| $Mg^{2+}$ | 12 | $12-2=10$ |

Thus, option (d) contains isoelectronic set of ions.

**60.** *(a)*

| Element | Period No. | Group No. | |
|---|---|---|---|
| C | 2nd | 14 | |
| Al | 3rd | 13 | |
| S | 3rd | 16 | Along the period atomic radius decreases, so, radius of : Al > S. |
| Cs | 6th | 1 | |

With the addition of a new shell is a group, atomic radius increases. So, the order of the atomic radii of the given elements will be:

$$C < S < Al < Cs$$

**61.** *(a)* The summary of variation of periodic properties is given in table below :

| S. No. | Periodic property | Variation Along a period | Variation Along a group |
|---|---|---|---|
| 1. | Atomic radius | Decreases | Increases |
| 2. | Electron gain enthalpy | Increases | Decreases |
| 3. | Electronegativity | Increases | Decreases |

Thus, electronegativity decreases and atomic radius increases down a group in the periodic table.

**62.** *(b)* As the third ionisation energy of *A* and *B* is very high as compared to corresponding second ionisation energy, thus there must be two electrons in their valence shells. Hence, elements *A* and *B* belong to group-2. On moving down the group, the atomic size increases, so force of attraction between valence electron and nucleus decreases. Hence, ionisation energy decreases. So, '*B*' comes below '*A*'.

**63.** *(c)* On moving down in a group atomic radius increases and ionisation enthalpy decreases. While, in Halogens, electron gain enthalpy increases on moving down in group.

**64.** *(a)* Isoelectronic species are those which contains same number of electrons.

| Species | Atomic number | Number of electrons |
|---|---|---|
| $O^{2-}$ | 8 | 10 |
| $F^-$ | 9 | 10 |
| $Na^+$ | 11 | 10 |
| $Mg^{2+}$ | 12 | 10 |
| $O^-$ | 8 | 9 |
| Na | 11 | 11 |
| $Mg^+$ | 12 | 11 |

$\therefore$ Option (a) is correct which contains isoelectronic species $O^{2-}$, $F^-$, $Na^+$, $Mg^{2+}$.

**65.** *(c)* Ionisation energy increases along a period from left to right and decreases down a group. The position of given elements in the periodic table is as follows

| Group No. | 2 | 16 | 18 |
|---|---|---|---|
| | Ca | S | Ar |
| | Ba | Se | |

Thus, the order of increasing $\Delta H_{IE_1}$ is Ba < Ca < Se < S < Ar

**66.** *(c)* For isoelectronic species, $r_n \propto \frac{1}{Z}$

Here, Z is atomic number

| Species | Z | Electrons |
|---|---|---|
| $Cl^-$ | 17 | 18 |
| $Ca^{2+}$ | 20 | 18 |
| $K^+$ | 19 | 18 |
| $S^{2-}$ | 16 | 18 |

Thus, ionic radii is in order

$$Ca^{2+} < K^+ < Cl^- < S^{2-}$$

**67.** *(b)* The ionic radius in general increases on moving from top to bottom and further decreases on moving from left to right. So the correct order is

$$\underset{0.98\,Å}{Na^+} > \underset{0.68\,Å}{Li^+} > \underset{0.65\,Å}{Mg^{2+}} > \underset{0.45\,Å}{Be^{2+}}$$

**68.** *(d)* $O^{2-}, F^-, Na^+, Mg^{2+}$ and $Al^{3+}$ are isoelectronic species.

For isoelectronic species, $r \propto 1/z$

Thus, larger the atomic number (effective nuclear charge) smaller is the ionic radii.

**69.** *(c)* Among the given orders, electronegativity and 1st ionisation enthalpy i.e., B and D are correct while atomic radius and density i.e., A and C are incorrect orders. Their correct orders are

$$\underset{(170)}{Tl} > \underset{(167)}{In} > \underset{(143)}{Al} > \underset{(135)}{Ga} > \underset{(88)}{B}$$

(atomic radius in pm)

$$\underset{(11.85)}{Tl} > \underset{(7.31)}{In} > \underset{(5.9)}{Ga} > \underset{(2.7)}{Al} > \underset{(2.35)}{B} \text{ (density in g/cm}^3)$$

**70.** *(b)* Both Statement I and Statement II are correct. For isoelectronic species. Cation with greater positive charge will have a smaller radius and the anion with greater negative charge will have a larger radius.

$$Mg^{2+} < Na^+ < F^- < O^{2-}$$

Magnitude of E.G.E.

$$Cl > F > Br > I$$

**71.** *(a)* Highest ionisation enthalpy among the given elements = Ge

Most stable oxidation state of Ge = + 4

Lowest first ionisation enthalpy among the given element = In

Most stable oxidation state of In = +3

**72.** *(b)* The type of oxide formed by elements like Li, Na, Be, B, Mg and Al that has least atomic radii is $A_2O_3$.

**73.** *(d)* The correct match is A-IV, B-I, C-II, D-III.

- For isoelectronic species the cation with the greater positive charge will have smaller radius while anion with greater negative charge will have larger radius. So, the order is

$$Al^{3+} < Mg^{2+} < Na^+ < F^-.$$

- Ionisation enthalpy order is B < C < O < N. As it increases along the period and decreases down the group. Further it increases with half or fully-filled orbital.
- Metallic character increases down the group and decreases along the period. So, the correct order is B < Al < Mg < K

- The correct order of electronegativity is

$$\underset{(1.9)}{Si} < \underset{(2.1)}{P} < \underset{(2.5)}{S} < \underset{(3.0)}{Cl}$$

**74.** *(c)* Statement I is false but Statement II is true. It is because elements in extreme left are metallic in nature which generally forms basic oxides.

**75.** *(c)* The correct match is A-IV, B-III, C-II, D-I

Cl, S $\Rightarrow$ Element with highest negative electron gain enthalpy.

Ge, As $\Rightarrow$ Elements which show property of both metal and non-metal.

Fr, Ra $\Rightarrow$ Elements with largest atomic size.

F, O $\Rightarrow$ Element with highest electronegativity.

**76.** *(d)* Statements given in A, C and D are correct while in B is incorrect.

| | $O^{2-}$ | $F^-$ | $Na^+$ | $Mg^{+2}$ |
|---|---|---|---|---|
| Number of electrons | 10 | 10 | 10 | 10 |
| Ionic radius | $O^{2-}$ | $> F^-$ | $> Na^+$ | $> Mg^{+2}$ |
| $Z_{eff}$ | $O^{2-}$ | $< F^-$ | $< Na^+$ | $< Mg^{+2}$ |

**77.** *(d)* The value of electron gain enthalpy of elements given are as follows.

| Elements | Electron gain enthalpy (kJ/mol) |
|---|---|
| F | −333 |
| S | −200 |
| Br | −325 |
| Ar | + 96 |

$Ar > S > Br > F$

So, the correct order is A > D > B > C.

**78.** *(a)* $X$ is Na ($3s^1$) as it has very high second ionisation energy due to removal of electron from highly stable noble gas configuration.

$$\underset{1s^2 2s^2 2p^6 3s^1}{Na} \xrightarrow{IE_1} \underset{1s^2 2s^2 2p^6 \text{ or [Ne]}}{Na^+} + e^-;$$

$$IE_1 = 495 \text{ kJ/mol}$$

$$\underset{1s^2 2s^2 2p^6}{Na^+} \xrightarrow{IE_2} \underset{1s^2 2s^2 2p^5}{Na^{2+}} + e^-;$$

$$IE_2 = 4563 \text{ kJ/mol}$$

(Stable noble gas configuration).

In case of Mg, noble gas configuration is achieved after removal of second electron. So, $IE_2$ is comparatively less.

$$\underset{[Ne]3s^2}{Mg} \longrightarrow \underset{[Ne]3s^1}{Mg^+} + e^-;$$

$$IE_1 = 731 \text{ kJ/mol}$$

$$\underset{[Ne]3s^1}{Mg^+} \longrightarrow \underset{[Ne]}{Mg^{2+}} + e^-;$$

$$IE_2 = 1450 \text{ kJ/mol}$$

**79.** *(c)* Generally, electron gain enthalpy becomes less exothermic (or less negative) when comparing elements of a group from top to bottom.

Therefore, electron gain enthalpy of S > Se and Li > Na.

But there are some exceptions to this. One of them is the case of a group 17 elements where electron gain is most negative for Cl instead of F, due to extra small size of fluorine.

$\therefore$ Upon an electron gain, energy releases in the order :

Cl > F, S > Se and Li > Na

**80.** *(b)* Atomic radius generally decreases as we compare elements in a period from left to right,

$\therefore$ C > O > F

but elements present in next period are larger in size,

$\therefore$ Br > Cl > C > O > F;

So, the correct increasing order of the atomic radii

C < B < A < D < E

**81.** *(c)* (a) Metallic radii increases in a group from top to bottom.

Thus, Li < Na < K < Rb—Correct

(b) Electron gain enthalpy of Cl > F and decreases along a group.

Thus, I < Br < F < Cl is correct.

(c) Ionisation enthalpy increases along a period from left to right but due to presence of half filled orbitals in N, ionisation enthalpy of N > O.

Thus, B < C < N < O is incorrect.

(d) In isoelectronic species Higher the negative charges, larger is their ionic radii $Al^{3+} < Mg^{2+} < Na^+ < F^-$ is also correct.

**82.** *(c)* $O^-(g) + e^- \longrightarrow O^{2-}(g);$

$$\Delta H° = 844 \text{ kJ mol}^{-1}$$

This process is unfavourable in the gas phase because the resulting increase in electron-electron repulsion overweighs the stability gained by achieving the noble gas configuration.

**83.** *(5)* The number of electrons in the given species are as follows:

$O^{2-} = 10$, $O^+ = 7$, $Al^{3+} = 10$, $Al = 13$

$F = 9$, $F^- = 10$, $Mg^{2+} = 10$, $Mg = 12$

$Na^+ = 10$

## Topic 3

**01.** *(d)* Statement I is incorrect but Statement II is correct. The correct form of Statement I is

The oxidation state of an element in a particular compound is the charge acquired by its atom on the basis of electronegativity consideration from other atoms in molecule.

**02.** *(b)* Acidic oxides are $B_2O_3$, $N_2O_5$, $SO_3$, $P_4O_{10}$.

**03.** *(b)* The correct matching is A-IV, B-II, C-I, D-III.

**04.** *(c)* Aluminium and silicon form amphoteric oxides. Non-metals such as sulphur forms acidic oxides. Metals (group 1 or 2) form basic oxides. So, $SiO_2$ and $Al_2O_3$ are amphoteric. $SO_3$ is acidic oxide whereas CaO is basic oxide.

**05.** *(b)* Among the given options $As_2O_3$ is an amphoteric oxide. While $Na_2O$ is basic and $Cl_2O_7$ is acidic in nature $N_2O$ and NO are neutral oxides.

**06.** *(b)* The most appropriate answer is A-(ii), B-(iii), C-(ii), D-(i), E-(iii)

**07.** *(d)* Li-Na pair is different from remaining three options as it does not shows diagonal relationship. Both Li-Na belong to same group and are not placed diagonally.

Li-Mg, B-Si and Be-Al show diagonal relationship with each other.

Li ↘ Mg, Be ↘ Al, B ↘ Si

| Li | Be | B | | (Diagonal relationship) |
|---|---|---|---|---|
| Na | Mg | Al | Si | |

**08.** *(b)* $B_2O_3$ and $SiO_2$ are acidic oxide as they react with base to give salt and water.

$$B_2O_3 + 2NaOH \longrightarrow \underset{\text{Sodium metaborate}}{2NaBO_2} + H_2O$$

$$SiO_2 + 2NaOH \longrightarrow \underset{\text{Sodium silicate}}{Na_2SiO_3} + H_2O$$

**09.** *(b)* $Li^+$ is having largest hydrated ionic size while $Rb^+$ is having smallest. Smaller is the ionic size, greater is the mobility.

**10.** *(a)* CaO is basic as it forms strong base $Ca(OH)_2$ on reaction with water.

$$CaO + H_2O \longrightarrow Ca(OH)_2$$

$CO_2$ is acidic as it dissolves in water forming unstable carbonic acid.

$$H_2O + CO_2 \longrightarrow H_2CO_3$$

Silica ($SiO_2$) acts as a very weak acid. $SnO_2$ is amphoteric in nature as it reacts with both acid and base.

$$SnO_2 + 2H_2SO_4 \longrightarrow Sn(SO_4)_2 + 2H_2O$$

$$SnO_2 + 2KOH \longrightarrow K_2SnO_3 + H_2O$$

**11.** *(d)* $TeO_2$ is oxidising in nature because it can be reduced from + 4 oxidation state to lower oxidation state. $TeH_2$ due to less bond dissociation energy easily break and hence acidic in nature.

**12.** *(a)* Smaller the size of cation, higher will be its hydration enthalpy.

$$\left[\text{Hydration enthalpy} \propto \frac{1}{\text{Size}}\right]$$

Hence, group 2 (alkaline earth metal) cations will be having higher hydration enthalpy than group 1 (alkali metals) cations.

Hence correct answer is

$Mg^{2+} > Ca^{2+} > K^+ > Rb^+ > Cs^+$ i.e., $C > E > A > B > D$.

• Concept **Enhancer**

Hydration enthalpy is the amount of energy produced when one mole of an ion dissolves in the large amount of water.

**13.** *(b)* Mercury is a liquid at room temperature, so it has lowest melting point and Ga and Cs are liquid at slightly above room temperature and hence have melting points around 30°C. So, being solid, Ag has highest melting point. among the given elements.

**14.** *(b)* From the given sets amphoteric oxides are $Be(OH)_2$, $Al(OH)_3$

Acidic oxides are $B(OH)_3$

Basic oxides are NaOH, $Ca(OH)_2$

Among the given sets option (b) has different nature of oxides.

**15.** *(b)* While moving across a period, metallic property of elements changes first into metalloids than into non-metallic nature therefore.

⇒ Metal gives basic oxides.

Metalloid produces amphoteric oxides.

Non-metal gives acidic oxides.

So, $X \rightarrow$ metal, $Y \rightarrow$ metalloid and $Z \rightarrow$ non-metal

Thus, the order of atomic number is

$$X < Y < Z$$

**16.** *(a)* On moving from left to right in a period, the basic character of oxides of elements decreases. So, the order will be $Al_2O_3 < MgO < Na_2O$. While on moving down the group, basic character of oxides increases. So, the order will be $Na_2O < K_2O$. Hence, the overall order will be

$Al_2O_3 < MgO < Na_2O < K_2O$.

**17.** *(d)* On moving down the group group from top to bottom, acidic nature of oxides decreases and along a period from left to right, acidic nature increases.

| | Al | Si | P | S |
|---|---|---|---|---|
| $Z$ | 13 | 14 | 15 | 16 |
| | $Al_2O_3$ | $SiO_2$ | $P_2O_3$ | $SO_2$ |
| | amphoteric | acidic | | max. acidic |

(Arrow pointing right: acidic nature increases →)

Thus, $Al_2O_3 < SiO_2 < P_2O_3 < SO_2$

**18.** *(c)* Statement given in B and C are correct while A, D and E are incorrect. Their correct forms are

(A) The process of adding an electron to a neutral gaseous atom can be endothermic or exothermic.

(D) Electronegativity of C in $CCl_4$ is greater than that of $CH_4$.

(E) Among elements of group I Cs is most electropositive.

**19.** *(b)* Among the given sequence, option (b) is the incorrect order of increasing basic strength. The correct order of increasing basic strength is

$NH_3 > PH_3 > AsH_3 > SbH_3$

**20.** *(d)* The tendency to loose electron by alkali metals increases down the group, therefore, their chemical reactivity increases down the group.

On the other hand, the tendency to gain electron by halogens decreases down the group, hence their chemical reactivity decreases down the group.

# C-04

JEE Main Chapterwise Topicwise Chemistry

In JEE Main there is usually 0-1 question per paper are visible from this chapter. In past 3 years, 53 conceptual and 34 numerical questions were asked of moderate difficulty level from this chapter.

**For detailed analysis scan QR code**

# CHEMICAL BONDING AND MOLECULAR STRUCTURE

## Topic 1 Ionic and Covalent Bonding Fajan's Rule, Bond Parameters

### Level 1

**01** Arrange the following compounds in increasing order of their dipole moment

$HBr, H_2S, NF_3$ and $CHCl_3$

*[2025, 22 Jan Shift-II]*

(a) $NF_3 < HBr < H_2S < CHCl_3$
(b) $CHCl_3 < NF_3 < HBr < H_2S$
(c) $H_2S < HBr < NF_3 < CHCl_3$
(d) $HBr < H_2S < NF_3 < CHCl_3$

**02** Which of the following is least ionic? *[2024, 31 Jan Shift-II]*

(a) KCl
(b) $BaCl_2$
(c) AgCl
(d) $CoCl_2$

**03** Choose the polar molecule from the following *[2024, 27 Jan Shift-I]*

(a) $CH_2{=}CH_2$ (b) $CCl_4$
(c) $CO_2$ (d) $CHCl_3$

**04** Resonance in carbonate ion ($CO_3^{2-}$) is

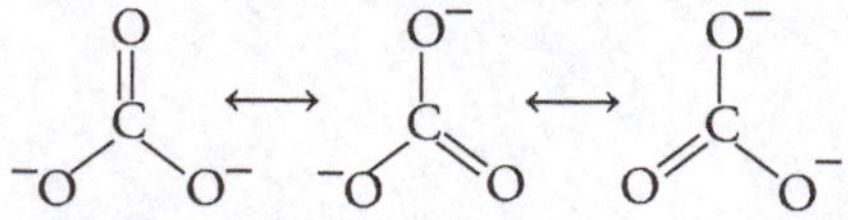

Which of the following is true? *[2023, 1 Feb Shift-I]*

(a) Each structure exists for equal amount of time.
(b) It is possible to identify each structure individually by some physical or chemical method.
(c) All these structures are in dynamic equilibrium with each other.
(d) $CO_3^{2-}$ has a single structure i.e., resonance hybrid of the above three structures.

**05** Given below are two statements.

**Statement I** Dipole moment is a vector quantity and by convention it is depicted by a small arrow with tail on the negative centre and head pointing towards the positive centre.

**Statement II** The crossed arrow of the dipole moment symbolises the direction of the shift of charges in the molecules.

In the light of the above statements, choose the most appropriate answer from the options given below. *[2023, 25 Jan Shift-II]*

(a) Both Statement I and Statement II are incorrect.
(b) Statement I is correct but Statement II is incorrect.
(c) Statement I is incorrect but Statement II is correct.
(d) Both Statement I and Statement II are correct.

**06** Which of the following pair of molecules contain odd electron molecule and an expanded octet molecule? *[2022, 29 July Shift-I]*

(a) $BCl_3$ and $SF_6$ (b) NO and $H_2SO_4$
(c) $SF_6$ and $H_2SO_4$ (d) $BCl_3$ and NO

**07** The potential energy curve for the $H_2$ molecule as a function of internuclear distance is *[2020, 5 Sep Shift-I]*

(a)
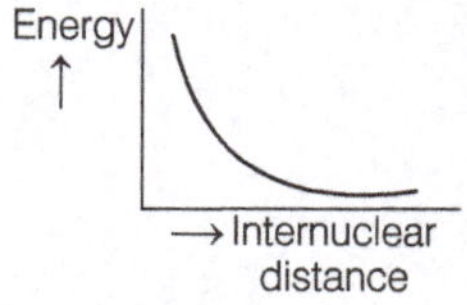


(b)
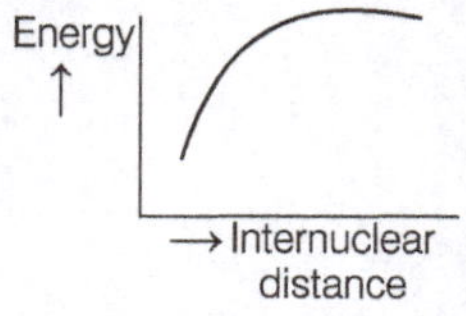


(c)
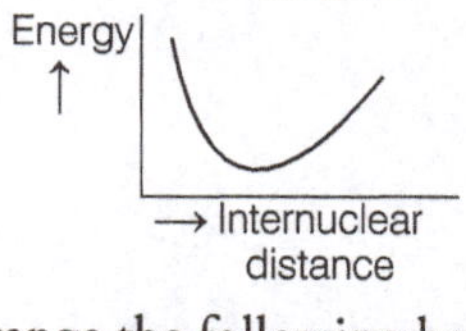


(d)
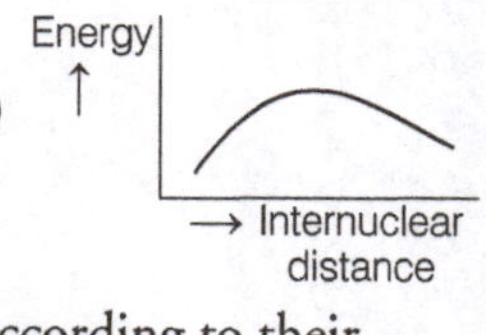


**08** Arrange the following bonds according to their average bond energies in descending order

C—Cl, C—Br, C—F, C—I *[2020, 8 Jan Shift-II]*

(a) C—F > C—Cl > C—Br > C—I
(b) C—Br > C—I > C—Cl > C—F
(c) C—I > C—Br > C—Cl > C—F
(d) C—Cl > C—Br > C—I > C—F

**09** The dipole moments of $CCl_4$, $CHCl_3$ and $CH_4$ are in the order *[2020, 7 Jan Shift-I]*
(a) $CCl_4 < CH_4 < CHCl_3$
(b) $CH_4 < CCl_4 < CHCl_3$
(c) $CH_4 = CCl_4 < CHCl_3$
(d) $CHCl_3 < CH_4 = CCl_4$

**10** Molecule *AB* has a bond length of 1.617 Å and a dipole moment of 0.38 D. The fractional charge on each atom (absolute magnitude) is
$(\varepsilon_0 = 4.802 \times 10^{-10}$ esu) *[JEE Main 2015]*
(a) 0 (b) 0.05
(c) 0.5 (d) 1.0

**11** Among the following, the maximum covalent character is shown by the compound *[AIEEE 2011]*
(a) $FeCl_2$ (b) $SnCl_2$
(c) $AlCl_3$ (d) $MgCl_2$

**12** The charge/size ratio of a cation determines its polarising power. Which one of the following sequences represents the increasing order of the polarising power of the cationic species, $K^+$, $Ca^{2+}$, $Mg^{2+}$, $Be^{2+}$ ? *[AIEEE 2007]*
(a) $Mg^{2+} < Be^{2+} < K^+ < Ca^{2+}$
(b) $Be^{2+} < K^+ < Ca^{2+} < Mg^{2+}$
(c) $K^+ < Ca^{2+} < Mg^{2+} < Be^{2+}$
(d) $Ca^{2+} < Mg^{2+} < Be^{2+} < K^+$

## Level 2

**13** Given below are two statements : one is labelled as Assertion (A) and the other is labelled as Reason (R). *[2024, 5 April Shift-II]*
**Assertion (A)** $NH_3$ and $NF_3$ molecule have pyramidal shape with a lone pair of electrons on nitrogen atom. The resultant dipole moment of $NH_3$ is greater than that of $NF_3$.
**Reason (R)** In $NH_3$, the orbital dipole due to lone pair is in the same direction as the resultant dipole moment of the N—H bond. F is the most electronegative element.
In the light of the above statements, choose the correct answer from the options given below :
(a) (A) is true but (R) is false
(b) (A) is false but (R) is true
(c) Both (A) and (R) are true and (R) is the correct explanation of (A)
(d) Both (A) and (R) are true but (R) is not the correct explanation of (A)

**14** Given below are two statements : one is labelled as Assertion (A) and the other is labelled as Reason (R).
**Assertion (A)** *cis*-form of alkene is found to be more polar than the *trans*-form.
**Reason (R)** Dipole moment of *trans*-isomer of 2-butene is zero. *[2024, 5 April Shift-I]*
In the light of the above statements, choose the correct answer from the options given below.
(a) Both A and R are true but R is not the correct explanation of A.
(b) A is false but R is true.
(c) Both A and R are true and R is the correct explanation of A.
(d) A is true but R is false.

**15** Which one of the following molecules has maximum dipole moment? *[2024, 4 April Shift-I]*
(a) $NF_3$ (b) $NH_3$ (c) $CH_4$ (d) $PF_5$

**Imp.** **Order of Ionic Character**

**16** Arrange the bonds in order of increasing ionic character in the molecules, LiF, $K_2O$, $N_2$, $SO_2$ and $ClF_3$: *[2024, 1st Feb Shift-I]*
(a) $N_2 < SO_2 < ClF_3 < K_2O < LiF$
(b) $LiF < K_2O < ClF_3 < SO_2 < N_2$
(c) $ClF_3 < N_2 < SO_2 < K_2O < LiF$
(d) $N_2 < ClF_3 < SO_2 < K_2O < LiF$

**17** Given below are two statements. *[2024, 30 Jan Shift-II]*
**Statement I** Since fluorine is more electronegative than nitrogen, the net dipole moment of $NF_3$ is greater than $NH_3$.
**Statement II** In $NH_3$, the orbital dipole due to lone pair and the dipole moment of NH bonds are in opposite direction, but in $NF_3$ the orbital dipole due to lone pair and dipole moments of N—F bonds are in same direction.
In the light of the above statements, choose the **most appropriate** from the options given below :
(a) Statement I is false but Statement II is true.
(b) Statement I is true but Statement II is false.
(c) Both Statement I and Statement II are true.
(d) Both Statement I and Statement II are false.

**18** The most polar compound among the following is *[JEE Main 2018]*

(a) 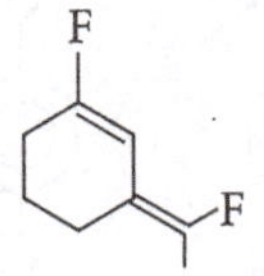


(b) 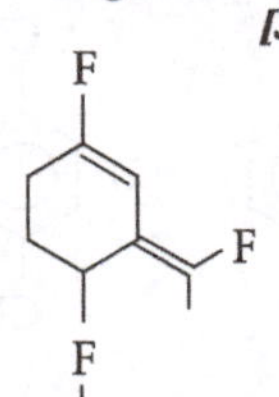


(c) 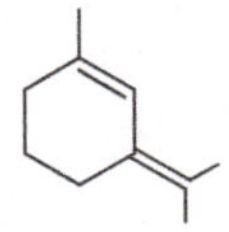

(d) 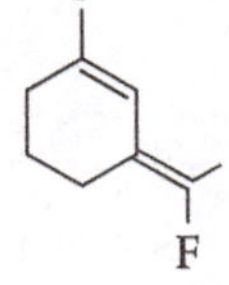

**19** Which compound exhibits maximum dipole moment among the following? *[JEE Main 2015]*

(a) 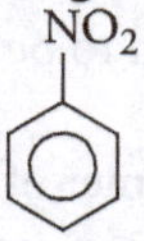

(b) 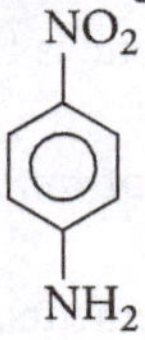

(c) 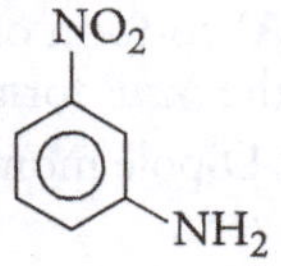

(d) 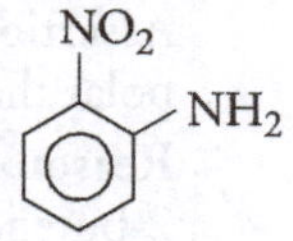


## Level 3

**20** Match the List-I with List-II. *[2025, 23 Jan Shift-I]*

| List-I (Classification of molecules based on octet rule) | | List-II (Example) |
|---|---|---|
| A. Molecules obeying octet rule | I. | NO, $NO_2$ |
| B. Molecules with incomplete octet | II. | $BCl_3$, $AlCl_3$ |
| C. Molecules with incomplete octet with odd electron | III. | $H_2SO_4$, $PCl_5$ |
| D. Molecules with expanded octet | IV. | $CCl_4$, $CO_2$ |

Choose the correct answer from the options given below:

(a) A-II, B-IV, C-III, D-I (b) A-III, B-II, C-I, D-IV
(c) A-IV, B-II, C-I, D-III (d) A-IV, B-I, C-III, D-II

**21** Order of covalent bond;

A. KF > KI; LiF > KF B. KF < KI; LiF > KF
C. $SnCl_4 > SnCl_2$; CuCl > NaCl
D. LiF > KF; CuCl < NaCl
E. KF < KI; CuCl > NaCl

Choose the correct answer from the options given below. *[2023, 24 Jan Shift-I]*

(a) Only B, C (b) Only A, B
(c) Only C, E (d) Only B, C, E

**22** Number of electron deficient molecules among the following

$PH_3$, $B_2H_6$, $CCl_4$, $NH_3$, LiH and $BCl_3$ is *[2022, 25 June Shift-I]*

(a) 0 (b) 1
(c) 2 (d) 3

**23** For which of the following molecules significant $\mu \neq 0$? *[JEE Main 2014]*

(1) 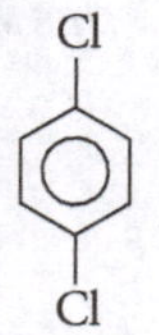

(2) 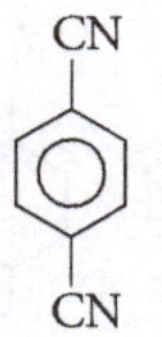

(3) 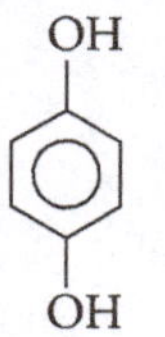

(4) 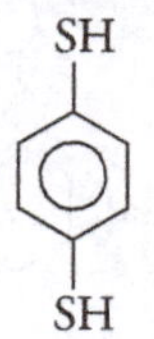


(a) (3) and (4)
(b) Only (1)
(c) (1) and (2)
(d) Only (3)

## Numerical Value Type Questions

**24** Number of molecules from the following which are exceptions to octet rule is ..........

$CO_2$, $NO_2$, $H_2SO_4$, $BF_3$, $CH_4$, $SiF_4$, $ClO_2$, $PCl_5$, $BeF_2$, $C_2H_6$, $CHCl_3$, $CBr_4$ *[2024, 8 April Shift-II]*

**25** Number of compounds from the following with zero dipole moment is ............

HF, $H_2$, $H_2S$, $CO_2$, $NH_3$, $BF_3$, $CH_4$, $CHCl_3$, $SiF_4$, $H_2O$, $BeF_2$ *[2024, 5 April Shift-II]*

**26** In the Lewis dot structure for $NO_2^-$, total number of valence electrons around nitrogen is ......... *[2024, 5 April Shift-I]*

**27** Number of compounds/ species from the following with non-zero dipole moment is...... .

$BeCl_2$, $BCl_3$, $NF_3$, $XeF_4$, $CCl_4$, $H_2O$, $H_2S$, HBr, $CO_2$, $H_2$, HCl *[2024, 4 April Shift-II]*

**28** The total number of lone pairs of electrons on oxygen atoms of ozone is ........ *[2023, Jan Shift-I]*

**29** The total number of molecules with zero dipole moment among $CH_4$, $BF_3$, $H_2O$, HF, $NH_3$, $CO_2$ and $SO_2$ is .......... . *[2022, 29 Jan Shift-II]*

**30** The number of non-polar molecules from the following is ......... .

HF, $H_2O$, $SO_2$, $H_2$, $CO_2$, $CH_4$, $NH_3$, HCl, $CHCl_3$, $BF_3$ *[2022, 27 Jan Shift-II]*

**31** Amongst $BeF_2$, $BF_3$, $H_2O$, $NH_3$, $CCl_4$ and HCl, the number of molecules with non-zero net dipole moment is ............ . *[2022, 25 June Shift-II]*

**32** The Born-Haber cycle for KCl is evaluated with the following data

$\Delta_f H°$ for KCl = –436.7 kJ $mol^{-1}$

$\Delta_{sub} H°$ for K = 89.2 kJ $mol^{-1}$

$\Delta_{ionisation} H°$ for K = 419.0 kJ $mol^{-1}$

$\Delta_{electron\ gain} H°$ for Cl(g) = – 348.6 kJ $mol^{-1}$

$\Delta_{bond} H°$ for $Cl_2$ = 243.0 kJ $mol^{-1}$

The magnitude of lattice enthalpy of KCl (in kJ $mol^{-1}$) is ........ . *[2021, 26 Aug Shift-I]*

## Topic 2 Hybridisation, VSEPR, VBT

### Level 1

**01** Match List-I with List-II

| List-I (Hybridisation) | | List-II (Orientation in space) | |
|---|---|---|---|
| A. | $sp^3$ | I. | Trigonal bipyramidal |
| B. | $dsp^2$ | II. | Octahedral |
| C. | $sp^3d$ | III. | Tetrahedral |
| D. | $sp^3d^2$ | IV. | Square planar |

Choose the correct answer from the options given below. *[2024, 6 April Shift-I]*

(a) A-III, B-I, C-IV, D-II (b) A-IV, B-III, C-I, D-II
(c) A-III, B-IV, C-I, D-II (d) A-II, B-I, C-IV, D-III

**02** In the structure of $SF_4$, the lone pair of electrons on S is in *[2022, 28 June Shift-II]*

(a) equatorial position and there are two lone pair-bond pair repulsions at 90°.
(b) equatorial position and there are three lone pair-bond pair repulsions at 90°.
(c) axial position and there are three lone pair –bond pair repulsion at 90°.
(d) axial position and there are two lone pair–bond pair repulsion at 90°.

**03** Number of lone pair(s) of electrons on central atom and the shape of $BrF_3$ molecule respectively, are *[2022, 29 June Shift-II]*

(a) 0, triangular planar (b) 1, pyramidal
(c) 2, bent T-shape (d) 1, bent T-shape

**04** Amongst the following, the linear species is *[2021, 17 March Shift-II]*

(a) $NO_2$ (b) $Cl_2O$ (c) $O_3$ (d) $N_3^-$

**05** A central atom in a molecule has two lone pairs of electrons and forms three single bonds. The shape of this molecule is *[2021, 17 March Shift-I]*

(a) *see-saw* (b) planar triangular
(c) T-shaped (d) trigonal pyramidal

**06** Given below are two statements : one is labelled as Assertion (A) and the other is labelled as Reason (R).

**Assertion (A)** The H—O—H bond angle in water molecule is 104.5°.

**Reason (R)** The lone pair - lone pair repulsion of electrons is higher than the bond pair - bond pair repulsion. *[2021, 16 March Shift-II]*

(a) A is false but R is true.
(b) Both A and R are true, but R is not the correct correct explanation of A.
(c) A is true but R is false.
(d) Both A and R are true, and R is the correct explanation of A.

**07** The correct shape and I—I—I bond angles respectively in $I_3^-$, ion are *[2021, 24 Feb Shift-II]*

(a) distorted trigonal planar, 135° and 90°
(b) T-shaped, 180° and 90°
(c) Trigonal planar, 120° (d) Linear, 180°

**08** In $\overset{1}{C}H_2=\overset{2}{C}=\overset{3}{C}H-\overset{4}{C}H_3$ molecule, the hybridisation of carbon 1, 2, 3 and 4 respectively are *[2021, 26 Feb Shift-II]*

(a) $sp^3, sp, sp^3, sp^3$ (b) $sp^2, sp^2, sp^2, sp^3$
(c) $sp^2, sp, sp^2, sp^3$ (d) $sp^2, sp^3, sp^2, sp^3$

**09** If $AB_4$ molecule is a polar molecule, a possible geometry of $AB_4$ is *[2020, 2 Sep Shift-I]*

(a) square pyramidal (b) square planar
(c) rectangular planar (d) tetrahedral

**10** The molecular geometry of $SF_6$ is octahedral. What is the geometry of $SF_4$ (including lone pair(s) of electrons, if any)? *[2020, 2 Sep Shift-II]*

(a) Tetrahedral (b) Trigonal bipyramidal
(c) Pyramidal (d) Square planar

### Level 2

**11** In $SO_2$, $NO_2^-$ and $N_3^-$ the hybridisations at the central atom are respectively: *[2025, 7 April Shift-II]*

(a) $sp, sp^2$ and $sp$ (b) $sp^2, sp^2$ and $sp$
(c) $sp^2, sp$ and $sp$ (d) $sp^2, sp^2$ and $sp^2$

**12** Given below are two statements *[2025, 4 April Shift-II]*

**Statement I** For $\ddot{Cl}F_3$, all three possible structures may be drawn as follows.

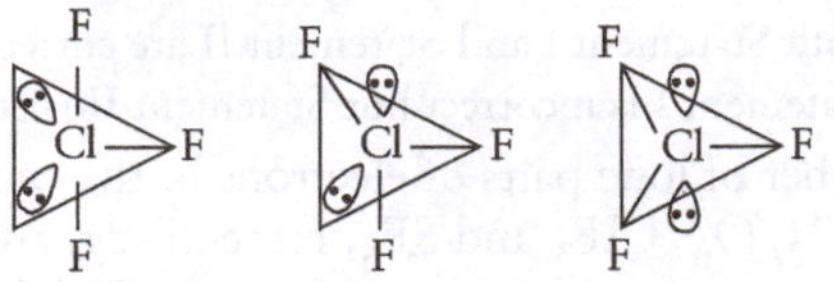


I II III

**Statement II** Structure III is most stable, as the orbitals having the lone pairs are axial, where the lp – bp repulsion is minimum.

In the light of the above statements, choose the most appropriate answer from the options given below:

(a) Statement I is correct but Statement II is incorrect
(b) Statement I is incorrect but Statement II is correct
(c) Both Statement I and Statement II are correct
(d) Both Statement I and Statement II are incorrect

**Imp.** Geometry of Molecules/Ions

**13** The number of molecules/ions that show linear geometry among the following is ...... .

$SO_2$, $BeCl_2$, $CO_2$, $N_3^-$, $NO_2$, $F_2O$, $XeF_2$, $NO_2^+$, $I_3^-$, $O_3$ *[2025, 22 Jan Shift-I]*

**14** Total number of molecules/species from following which will be paramagnetic is....
$O_2, O_2^+, O_2^-$, NO, $NO_2$, CO,
$K_2[NiCl_4]$, $[Co(NH_3)_6]Cl_3$, $K_2[Ni(CN)_4]$ *[2025, 28 Jan Shift-II]*

**15** The correct increasing order for bond angles among $BF_3$, $PF_3$ and $ClF_3$ is *[2024, 9 April Shift-II]*
(a) $PF_3 < BF_3 < ClF_3$ (b) $ClF_3 < PF_3 < BF_3$
(c) $BF_3 = PF_3 < ClF_3$ (d) $BF_3 < PF_3 < ClF_3$

**16** In which one of the following pairs the central atoms exhibit $sp^2$ hybridisation? *[2024, 9 April Shift-I]*
(a) $H_2O$ and $NO_2$ (b) $NH_2^-$ and $H_2O$
(c) $BF_3$ and $NO_2^-$ (d) $NH_2^-$ and $BF_3$

**17** The molecule/ion with square pyramidal shape is *[2024, 30 Jan Shift-II]*
(a) $BrF_5$ (b) $PCl_5$ (c) $[Ni(CN)_4]^{2-}$ (d) $PF_5$

**18** Identify from the following species in which $d^2sp^3$ hybridisation is shown by central atom. *[2024, 27 Jan Shift-II]*
(a) $[Pt(Cl)_4]^{2-}$ (b) $[Co(NH_3)_6]^{3+}$
(c) $BrF_5$ (d) $SF_6$

**19** Given below are two statements.
**Statement I** $SO_2$ and $H_2O$ both posses V-shaped structure.
**Statement II** The bond angle of $SO_2$ is less than that of $H_2O$.
In the light of the above statements, choose the most appropriate answer from the options given below. *[2023, 13 April Shift-II]*
(a) Both Statement I and Statement II are incorrect.
(b) Statement I is correct but Statement II is incorrect.
(c) Both Statement I and Statement II are correct.
(d) Statement I is incorrect but Statement II is correct.

**20** Number of lone pairs of electrons in the central atom of $SCl_2$, $O_3$, $ClF_3$ and $SF_6$, respectively, are *[2022, 29 July Shift-I]*
(a) 0, 1, 2 and 2 (b) 2, 1, 2 and 0
(c) 1, 2, 2 and 0 (d) 2, 2, 2 and 0

**21** Given below are two statements : One is labelled as Assertion (A) and the other is labelled as Reason (R).
**Assertion (A)** Zero orbital overlap is an out of phase overlap.
**Reason (R)** It results due to different orientation/direction of approach of orbitals.
In the light of the above statements, choose the correct answer from the options given below: *[2022, 28 July Shift-II]*
(a) Both A and R are true and R is the correct explanation of A.
(b) Both A and R are true and R is not the correct explanation of A.
(c) A is true but R is false.
(d) A is false but R is true.

**22** Consider the species $CH_4$, $NH_4^+$ and $BH_4^-$. Choose the correct option with respect to these species. *[2022, 29 June Shift-II]*
(a) They are isoelectronic and only two have tetrahedral structures.
(b) They are isoelectronic and all have tetrahedral structures.
(c) Only two are isoelectronic and all have tetrahedral structures.
(d) Only two are isoelectronic and only two have tetrahedral structures.

**23** The molecule in which hybrid MOs involve only one *d*-orbital of the central atom is *[2020, 04 Sep Shift-II]*
(a) $[Ni(CN)_4]^{2-}$ (b) $BrF_5$ (c) $XeF_4$ (d) $[CrF_6]^{3-}$

**24** The compound that has the largest H—*M*—H bond angle ($M$ = N, O, S, C) is *[2020, 5 Sep Shift-II]*
(a) $H_2O$ (b) $NH_3$ (c) $H_2S$ (d) $CH_4$

**25** The decreasing order of bond angles in $BF_3$, $NH_3$, $PF_3$ and $I_3^-$ is *[JEE Main 2018]*
(a) $I_3^- > BF_3 > NH_3 > PF_3$ (b) $BF_3 > NH_3 > PF_3 > I_3^-$
(c) $I_3^- > NH_3 > PF_3 > BF_3$ (d) $BF_3 > I_3^- > PF_3 > NH_3$

**26** Identify the pair in which the geometry of the species is T-shape and square-pyramidal, respectively *[JEE Main 2018]*
(a) $IO_3^-$ and $IO_2F_2^-$ (b) $XeOF_2$ and $XeOF_4$
(c) $ICl_2^-$ and $ICl_5$ (d) $ClF_3$ and $IO_4^-$

**27** The incorrect geometry is represented by *[JEE Main 2018]*
(a) $BF_3$ — trigonal planar (b) $NF_3$ — trigonal planar
(c) $AsF_5$ — trigonal bipyramidal
(d) $H_2O$— bent

**28** Which of the following conversions involves change in both shape and hybridisation? *[JEE Main 2018]*
(a) $BF_3 \rightarrow BF_4^-$ (b) $H_2O \rightarrow H_3O^+$
(c) $CH_4 \rightarrow C_2H_6$ (d) $NH_3 \rightarrow NH_4^+$

**29** $sp^3d^2$ hybridisation is not displayed by *[JEE Main 2017]*
(a) $SF_6$ (b) $PF_5$ (c) $[CrF_6]^{3-}$ (d) $BrF_5$

**30** In which of the following molecules/ions, all the bonds are not equal? *[AIEEE 2006]*
(a) $SF_4$ (b) $SiF_4$ (c) $XeF_4$ (d) $BF_4^-$

**31** The molecular shapes of $SF_4$, $CF_4$ and $XeF_4$ are
(a) different with 1, 0 and 2 lone pair of electrons on the central atom, respectively
(b) different with 0, 1 and 2 lone pair of electrons on the central atom, respectively
(c) the same with 1, 1 and 1 lone pair of electrons on the central atom, respectively
(d) the same with 2, 0 and 1 lone pair of electrons on the central atom, respectively *[AIEEE 2005]*

**32** The maximum number of 90° angles between bond pair-bond pair of electrons is observed in *[AIEEE 2004]*
(a) $dsp^3$ hybridisation (b) $sp^3d$ hybridisation
(c) $dsp^2$ hybridisation (d) $sp^3d^2$ hybridisation

**33** Which one of the following does not have $sp^2$ hybridised carbon? *[AIEEE 2004]*
(a) Acetone (b) Acetic acid
(c) Acetonitrile (d) Acetamide

**34** The pair of species having identical shapes for molecules of both species, is *[AIEEE 2003]*
(a) $CF_4$, $SF_4$ (b) $XeF_2$, $CO_2$
(c) $BF_3$, $PCl_3$ (d) $PF_5$, $IF_5$

## Level 3

**35.** Among $SO_2$, $NF_3$, $NH_3$, $XeF_2$, $ClF_3$ and $SF_4$, the hybridisation of the molecule with non-zero dipole moment highest number of lone-pairs of electrons on the central atom is *[2025, 2 April Shift-I]*
(a) $sp^3$ (b) $dsp^2$
(c) $sp^3d^2$ (d) $sp^3d$

**36.** Match the List-I with List-II. *[2025, 3 April Shift-I]*

| List-I (Molecules/ion) | | List-II (Hybridisation of central atom) | |
|---|---|---|---|
| A. | $PF_5$ | I. | $dsp^2$ |
| B. | $SF_6$ | II. | $sp^3d$ |
| C. | $Ni(CO)_4$ | III. | $sp^3d^2$ |
| D. | $[PtCl_4]^{2-}$ | IV. | $sp^3$ |

Choose the correct answer from the options given below:
(a) A-IV, B-I, C-II, D-III (b) A-I, B-II, C-III, D-IV
(c) A-III, B-I, C-IV, D-II (d) A-II, B-III, C-IV, D-I

**37** The molecules having square pyramidal geometry are *[2025, 28 Jan Shift-I]*
(a) $SbF_5$ and $PCl_5$ (b) $BrF_5$ and $PCl_5$
(c) $SbF_5$ and $XeOF_4$ (d) $BrF_5$ and $XeOF_4$

**38** Which of the following statements is true with respect to $H_2O$, $NH_3$ and $CH_4$? *[2025, 24 Jan Shift-I]*
A. The central atoms of all the molecules are $sp^3$ hybridised.
B. The H—O—H, H—N—H and H—C—H angles in the above molecules are 104.5°, 107.5° and 109.5° respectively.
C. The increasing order of dipole moment is $CH_4 < NH_3 < H_2O$.
D. Both $H_2O$ and $NH_3$ are Lewis acids and $CH_4$ is a Lewis base.
E. A solution of $NH_3$ in $H_2O$ is basic. In this solution $NH_3$ and $H_2O$ act as Lowry-Bronsted acid and base respectively.

Choose the correct answer from the options given below
(a) A, D and E only (b) C, D and E only
(c) A, B, C and E only (d) A, B and C only

**Imp.** Geometry of Molecule

**39** Match List-I with List-II.

| List-I (Molecule) | | List-II (Shape) |
|---|---|---|
| A. | $NH_3$ | I. Square pyramidal |
| B. | $BrF_5$ | II. Tetrahedral |
| C. | $PCl_5$ | III. Trigonal pyramidal |
| D. | $CH_4$ | IV. Trigonal bipyramidal |

Choose the correct answer from the options given below: *[2024, 8 April Shift-I]*
(a) A-IV, B-III, C-I, D-II (b) A-II, B-IV, C-I, D-III
(c) A-III, B-I, C-IV, D-II (d) A-III, B-IV, C-I, D-II

**40** Given below are two statements. *[2024, 6 April Shift-II]*
**Statement I** $PF_5$ and $BrF_5$ both exhibit $sp^3d$ hybridisation.
**Statement II** Both $SF_6$ and $[Co(NH_3)_6]^{3+}$ exhibit $sp^3d^2$ hybridisation.
In the light of the above statements, choose the correct answer from the options given below.
(a) Both Statement I and Statement II are true.
(b) Both Statement I and Statement II are false.
(c) Statement I is true but Statement II is false.
(d) Statement I is false but Statement II is true.

**41** Match List-I with List-II.

| List-I (Compound/species) | | List-II (Shape/Geometry) | |
|---|---|---|---|
| A. | $SF_4$ | I. | Tetrahedral |
| B. | $BrF_3$ | II. | Pyramidal |
| C. | $BrO_3^-$ | III. | See-saw |
| D. | $NH_4^+$ | IV. | Bent T-Shape |

Choose the correct answer from the options given below. *[2024, 6 April Shift-I]*
(a) A-III, B-IV, C-II, D-I (b) A-II, B-IV, C-III, D-I
(c) A-III, B-II, C-IV, D-I (d) A-II, B-III, C-I, D-IV

**Imp.** Magnetic Property of Molecule in Blend with Shape of Molecule

**42** Match List-I with List-II.

| List-I (Molecule/species) | | List-II (Property/shape) | |
|---|---|---|---|
| A. | $SO_2Cl_2$ | I. | Paramagnetic |
| B. | NO | II. | Diamagnetic |
| C. | $NO_2^-$ | III. | Tetrahedral |
| D. | $I_3^-$ | IV. | Linear |

Choose the correct answer from the options given below. *[2024, 6 April Shift-I]*
(a) A-II, B-III, C-I, D-IV (b) A-IV. B-I. C-III, D-II
(c) A-III, B-I, C-II, D-IV (d) A-III, B-IV, C-II, D-I

**43** Match List-I with List-II.

| | List-I | | List-II |
|---|---|---|---|
| A. | ICl | I. | T-shape |
| B. | $ICl_3$ | II. | Square pyramidal |
| C. | $ClF_5$ | III. | Pentagonal bipyramidal |
| D. | $IF_7$ | IV. | Linear |

Choose the correct answer from the options given below *[2024, 5 April Shift-II]*

(a) A-IV, B-I, C-II, D-III (b) A-I, B-IV, C-III, D-II
(c) A-IV, B-III, C-II, D-I (d) A-I, B-III, C-II, D-IV

**44** The number of species from the following that have pyramidal geometry around the central atom is……

$S_2O_3^{2-}, SO_4^{2-}, SO_3^{2-}, S_2O_7^{2-}$ *[2024, 4 April Shift-II]*

(a) 3 (b) 4 (c) 1 (d) 2

**45** Number of molecules/ions from the following in which the central atom is involved in $sp^3$ hybridisation is…… . *[2024, 4 April Shift-I]*

$NO_3^-, BCl_3, ClO_2^-, ClO_3$

(a) 3 (b) 2 (c) 1 (d) 4

**46** Match List-I with List-II.

| | List-I (Species) | | List-II (Geometry/Shape) |
|---|---|---|---|
| A. | $H_3O^+$ | I. | Tetrahedral |
| B. | Acetylide anion | II. | Linear |
| C. | $NH_4^+$ | III. | Pyramidal |
| D. | $ClO_2^-$ | IV. | Bent |

Choose the correct answer from the options given below. *[2023, 11 April Shift-I]*

(a) A-III, B-IV, C-II, D-I (b) A-III, B-I, C-II, D-IV
(c) A-III, B-IV, C-I, D-II (d) A-III, B-II, C-I, D-IV

**47** Match List-I with List-II.

| | List-I (Species) | | List-II (Shape) |
|---|---|---|---|
| A. | $XeF_4$ | I. | *See-saw* |
| B. | $SF_4$ | II. | Square planar |
| C. | $NH_4^+$ | III. | Bent T-shaped |
| D. | $BrF_3$ | IV. | Tetrahedral |

Choose the correct answer from the options given below *[2023, 31 Jan Shift-I]*

(a) A-II, B-I, C-IV, D-III (b) A-IV, B-I, C-II, D-III
(c) A -IV, B-III, C-II, D-I (d) A-II, B-I, C-III, D-IV

**48** Match List-I with List-II.

| List-I (Compound) | List-II (Shape) |
|---|---|
| (A) $BrF_5$ | (I) bent |
| (B) $[CrF_6]^{3-}$ | (II) square pyramidal |
| (C) $O_3$ | (III) trigonal bipyramidal |
| (D) $PCl_5$ | (IV) octahedral |

Choose the correct answer from the options given below : *[2022, 26 July Shift-I]*

| | A | B | C | D | | A | B | C | D |
|---|---|---|---|---|---|---|---|---|---|
| (a) | I | II | III | IV | (b) | IV | III | II | I |
| (c) | II | IV | I | III | (d) | III | IV | II | I |

**49** Based upon VSEPR theory, match the shape (geometry) of the molecules in List-I with the molecules in List-II and select the most appropriate option. *[2022, 27 June Shift-I]*

| List-I (Shape) | List-II (Molecules) |
|---|---|
| A. T-shaped | I. $XeF_4$ |
| B. Trigonal planar | II. $SF_4$ |
| C. Square planar | III. $ClF_3$ |
| D. *See-saw* | IV. $BF_3$ |

| | A | B | C | D | | A | B | C | D |
|---|---|---|---|---|---|---|---|---|---|
| (a) | I | II | III | IV | (b) | III | IV | I | II |
| (c) | III | IV | II | I | (d) | IV | III | I | II |

**50** Match List-I with List-II.

| List-I (Compounds) | List-II (Hybrid orbitals) |
|---|---|
| (A) $[PtCl_4]^{2-}$ | (I) $sp^3d$ |
| (B) $BrF_5$ | (II) $d^2sp^3$ |
| (C) $PCl_5$ | (III) $dsp^2$ |
| (D) $[Co(NH_3)_6]^{3+}$ | (IV) $sp^3d^2$ |

Choose the most appropriate answer from the options given below. *[2022, 24 June Shift-I]*

(a) (A) → (II), (B) → (IV), (C) → (I), (D) → (III)
(b) (A) → (III), (B) → (IV), (C) → (I), (D) → (II)
(c) (A) → (III), (B) → (I), (C) → (IV), (D) → (II)
(d) (A) → (II), (B) → (I), (C) → (IV), (D) → (III)

**Imp.**

## Hybridisation of Molecule

**51** Match List-I and List-II.

| List-I (Species) | List-II (Hybrid orbitals) |
|---|---|
| (A) $SF_4$ | (i) $sp^3d^2$ |
| (B) $IF_5$ | (ii) $d^2sp^3$ |
| (C) $NO_2^+$ | (iii) $sp^3d$ |
| (D) $NH_4^+$ | (iv) $sp^3$ |
| | (v) $sp$ |

*[2021, 22 July Shift-II]*

| | A | B | C | D | | A | B | C | D |
|---|---|---|---|---|---|---|---|---|---|
| (a) | (i) | (ii) | (v) | (iii) | (b) | (ii) | (i) | (iv) | (v) |
| (c) | (iii) | (i) | (v) | (iv) | (d) | (iv) | (iii) | (ii) | (v) |

**52** The hybridisations of the atomic orbitals of nitrogen in $NO_2^-$, $NO_2^+$ and $NH_4^+$ respectively are *[2021, 20 July Shift-II]*

(a) $sp^3$, $sp^2$ and $sp$ (b) $sp$, $sp^2$ and $sp^3$
(c) $sp^3$, $sp$ and $sp^2$ (d) $sp^2$, $sp$ and $sp^3$

53 Which of the following are isostructural pairs ?
A. $SO_4^{2-}$ and $CrO_4^{2-}$ B. $SiCl_4$ and $TiCl_4$
C. $NH_3$ and $NO_3^-$ D. $BCl_3$ and $BrCl_3$
*[2021, 24 Feb Shift-I]*
(a) Only A and B (b) Only A and C
(c) Only B and C (d) Only C and D

54 Which among the following species has unequal bond lengths ? *[2021, 25 Feb Shift-II]*
(a) $XeF_4$ (b) $SiF_4$ (c) $SF_4$ (d) $BF_4^-$

55 The structure of $PCl_5$ in the solid state is *[2020, 5 Sep Shift-I]*
(a) tetrahedral $[PCl_4]^+$ and octahedral $[PCl_6]^-$
(b) trigonal bipyramidal
(c) square planar $[PCl_4]^+$ and octahedral $[PCl_6]^-$
(d) square pyramidal

56 The bond angle H-*X*-H is the greatest in the compound *[JEE Main 2016]*
(a) $PH_3$ (b) $CH_4$ (c) $NH_3$ (d) $H_2O$

57 In which of the following pairs, the two species are not isostructural? *[AIEEE 2012]*
(a) $CO_3^{2-}$ and $NO_3^-$ (b) $PCl_4^+$ and $SiCl_4$
(c) $PF_5$ and $BrF_5$ (d) $AlF_6^{3-}$ and $SF_6$

58 Which one of the following has the regular tetrahedral structure? *[AIEEE 2004]*
(Atomic number : B = 5, S = 16, Ni = 28, Xe = 54)
(a) $XeF_4$ (b) $SF_4$ (c) $BF_4^-$ (d) $[Ni(CN)_4]^{2-}$

## Numerical Value Type Questions

59 Total number of species from the following with central atom utilising $sp^2$ hybrid orbitals for bonding is............
$NH_3$, $SO_2$, $SiO_2$, $BeCl_2$, $C_2H_2$, $C_2H_4$, $BCl_3$, HCHO, $C_6H_6$, $BF_3$, $C_2H_4Cl_2$. *[2024, 6 April Shift-II]*

60 The number of molecules/ion/s having trigonal bipyramidal shapes is .........
$PF_5$, $BrF_5$, $PCl_5$, $[PtCl_4]^{2-}$, $BF_3$, $Fe(CO)_5$
*[2024, 1 Feb Shift-I]*

61 The number of species from the following in which the central atom uses $sp^3$ hybrid orbitals in its bonding is .......... .
$NH_3$, $SO_2$, $SiO_2$, $BeCl_2$, $CO_2$, $H_2O$, $CH_4$, $BF_3$
*[2024, 31 Jan Shift-I]*

62 The maximum number of lone pairs of electrons on the central atom from the following species is .........
$ClO_3^-$, $XeF_4$, $SF_4$ and $I_3^-$ *[2023, 11 April Shift-II]*

63 The number of molecules from the following which contain only two lone pair of electrons is ........
$H_2O$, $N_2$, CO, $XeF_4$, $NH_3$, NO, $CO_2$, $F_2$
*[2023, 10 April Shift-II]*

64 The number of species from the following carrying a single lone pair on central atom xenon is ...... .
$XeF_5^+$, $XeO_3$, $XeO_2F_2$, $XeF_5^-$, $XeO_3F_2$, $XeOF_4$, $XeF_4$ *[2023, 8 April Shift-II]*

65 The number of bent-shaped molecule/s from the following is .....
$N_3^-$, $NO_2^-$, $I_3^-$, $SO_2$, $O_3$ *[2023, 10 April Shift-I]*

66 The sum of lone pairs present on the central atom of the interhalogen $IF_5$ and $IF_7$ is ........ .
*[2023, 10 April Shift-I]*

67 The number of species having a square planar shape from the following is ..............
$XeF_4$, $SF_4$, $SiF_4$, $BF_4^-$, $BrF_4^-$, $[Cu(NH_3)_4]^{2+}$, $[FeCl_4]^{2-}$, $[PtCl_4]^{2-}$
*[2023, 6 April Shift-II]*

68 The number of species from the following which have square pyramidal structure is ............ $PF_5$, $BrF_4^-$, $IF_5$ $BrF_5$, $XeOF_4$, $ICl_4^-$ *[2023, 6 April Shift-I]*

69 Amongst the following, the number of species having the linear shape is .......
$XeF_2$, $I_3^+$, $C_3O_2$, $I_3^-$, $CO_2$ , $SO_2$, $BeCl_2$ and $BCl_2^\ominus$
*[2023, 31 Jan Shift-II]*

70 Consider, $PF_5$, $BrF_5$, $PCl_3$, $SF_6$, $[ICl_4]^-$, $ClF_3$ and $IF_5$.
Amongst the above molecule(s)/ ion(s), the number of molecule(s)/ ion(s) having $sp^3d^2$ hybridisation is ............ . *[2022, 29 July Shift-II]*

71 The number of molecule(s) or ion(s) from the following having non-planar structure is ........... .
$NO_3^-$, $H_2O_2$, $BF_3$, $PCl_3$, $XeF_4$, $SF_4$, $XeO_3$, $PH_4^+$, $SO_3$, $[Al(OH)_4]^-$ *[2022, 27 July Shift-II]*

72 The hybridisation of P exhibited in $PF_5$ is $sp^x d^y$.
The value of $y$ is ...... . *[2022, 28 June Shift-I]*

73 Amongst $SF_4$, $XeF_4$, $CF_4$ and $H_2O$, the number of species with two lone pair of electrons is ...... .
*[2022, 26 June Shift-II]*

74 The number of species having non–pyramidal shape among the following is *[2021, 27 Aug Shift-II]*
(i) $SO_3$ (ii) $NO_3^-$ (iii) $PCl_3$ (iv) $CO_3^{2-}$

75 $AB_3$ is an interhalogen T-shaped molecule. The number of lone pairs of electrons on $A$ is .........
*[2021, 26 Aug Shift -I]*

76 The number of lone pairs of electron on the central I atom in $I_3^-$ is ...... . *[2021, 20 July Shift-I]*

77 The number of species below that have two lone pairs of electrons in their central atom is ......... . (Round off to the nearest integer)
$SF_4$, $BF_4^-$, $ClF_3$, $AsF_3$, $PCl_5$, $BrF_5$, $XeF_4$, $SF_6$
*[2021, 18 March Shift-II]*

**78** Complexes ($ML_5$) of metals Ni and Fe have ideal square pyramidal and trigonal bipyramidal geometrics, respectively. The sum of the 90°, 120° and 180° *L-M-L* angles in the two complexes is ......... . *[2020, 8 Jan Shift-II]*

## Topic 3 MOT, Metallic & Hydrogen Bonding

### Level 1

**01** When $\psi_A$ and $\psi_B$ are the wave functions of atomic orbitals, then $\sigma^*$ is represented by *[2024, 9 April Shift-II]*
(a) $\psi_A - \psi_B$ (b) $\psi_A + \psi_B$
(c) $\psi_A - 2\psi_B$ (d) $\psi_A + 2\psi_B$

**02** Match List-I with List-II.

| List-I | | List-II | |
|---|---|---|---|
| (A) | $\psi_{MO} = \psi_A - \psi_B$ | (I) | Dipole moment |
| (B) | $\mu = Q \times r$ | (II) | Bonding molecular orbital |
| (C) | $\frac{N_b - N_a}{2}$ | (III) | Anti-bonding molecular orbital |
| (D) | $\psi_{MO} = \psi_A + \psi_B$ | (IV) | Bond order |

Choose the correct answer from options given below: *[2022, 27 July Shift-II]*
(a) (A)-(II), (B)-(I), (C)-(IV), (D)-(III)
(b) (A)-(III), (B)-(IV), (C)-(I), (D)-(II)
(c) (A)-(III), (B)-(I), (C)-(IV), (D)-(II)
(d) (A)-(III), (B)-(IV), (C)-(II), (D)-(I)

**03** The bond order and magnetic behaviour of $O_2^-$ ion are, respectively *[2021, 26 Aug Shift-II]*
(a) 1.5 and paramagnetic (b) 1.5 and diamagnetic
(c) 2 and diamagnetic (d) 1 and paramagnetic

**04** Given below are two statements: One is labelled as Assertion (A) and the other is labelled as Reason (R).
**Assertion (A)** Dipole-dipole interactions are the only non-covalent interactions, resulting in hydrogen bond formation.
**Reason (R)** Fluorine is the most electronegative element and hydrogen bonds in HF are symmetrical.
In the light of the above statements, choose the most appropriate answer from the options given below.
(a) A is false but R is true. *[2021, 26 Feb Shift-I]*
(b) Both A and R are true and R is the correct explanation of A.
(c) A is true R is false.
(d) Both A and R are true but R is not the correct explanation of A.

**05** According to molecular orbital theory, the species among the following that does not exist is *[2021, 25 Feb Shift-I]*
(a) $O_2^{2-}$ (b) $He_2^-$
(c) $Be_2$ (d) $He_2^+$

**06** The bond order and the magnetic characteristics of $CN^-$ are *[2020, 7 Jan Shift-II]*
(a) 3, diamagnetic (b) $2\frac{1}{2}$, paramagnetic
(c) 3, paramagnetic (d) $2\frac{1}{2}$, diamagnetic

**07** The correct order of increasing intermolecular hydrogen bond strength is *[2020, 27 June Shift-II]*
(a) $HCN < H_2O < NH_3$ (b) $HCN < CH_4 < NH_3$
(c) $CH_4 < HCN < NH_3$ (d) $CH_4 < NH_3 < HCN$

**08** Among the following species, the diamagnetic molecule is *[2019, 9 April Shift-II]*
(a) CO (b) $B_2$ (c) NO (d) $O_2$

**09** According to molecular orbital theory, which of the following is true with respect to $Li_2^+$ and $Li_2^-$?
(a) Both are unstable *2019, 9 Jan Shift-I]*
(b) $Li_2^+$ is unstable and $Li_2^-$ is stable
(c) Both are stable
(d) $Li_2^+$ is stable and $Li_2^-$ is unstable

**10** According to molecular orbital theory, which of the following will not be a viable molecule? *[JEE Main 2018]*
(a) $He_2^{2+}$ (b) $He_2^+$ (c) $H_2^-$ (d) $H_2^{2-}$

**11** Which of the following best describes the diagram below of a molecular orbital? *[JEE Main 2018]*

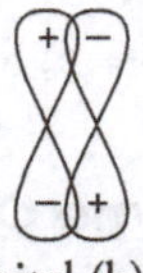


(a) An antibonding $\pi$-orbital (b) An antibonding $\sigma$-orbital
(c) A non-bonding orbital (d) A bonding $\pi$-orbital

**12** Which of the following is paramagnetic? *[JEE Main 2017]*
(a) CO (b) $O_2^{2-}$ (c) $NO^+$ (d) $B_2$

**13** After understanding the assertion and reason, choose the correct option.
**Assertion (A)** In the bonding molecular orbital (MO) of $H_2$, electron density is increased between the nuclei.
**Reason (R)** The bonding MO is $\psi_A + \psi_B$, which shows destructive interference of the combining electron waves. *[JEE Main 2015]*
(a) Assertion and reason are correct and reason is the correct explanation for the assertion.
(b) Assertion and reason are correct, but reason is not the correct explanation for the assertion.
(c) Assertion is correct, reason is incorrect.
(d) Assertion is incorrect, reason is correct.

**14** Which of the following species exhibits diamagnetic behaviour? *[AIEEE 2007]*
(a) $O_2^{2-}$ (b) $O_2^+$ (c) $O_2$ (d) NO

**15** Which of the following hydrogen bond is the strongest? *[AIEEE 2007]*
(a) O—H···N (b) F—H···F
(c) O—H···O (d) O—H···F

**16** Which of the following molecules/ ions does not contain unpaired electrons? *[AIEEE 2006]*
(a) $O_2^{2-}$ (b) $B_2$
(c) $N_2^+$ (d) $O_2$

## Level 2

**17** Which of the following linear combination of atomic orbitals will lead to formation of molecular orbitals in homonuclear diatomic molecules [internuclear axis in z- direction] ? *[2025, 24 Jan Shift-I]*
(A) $2p_z$ and $2p_x$ (B) $2s$ and $2p_x$
(C) $3d_{xy}$ and $3d_{x^2-y^2}$ (D) $2s$ and $2p_z$
(E) $2p_z$ and $3d_{x^2-y^2}$
Choose the correct answer from the options given below
(a) C and D only (b) A and B only
(c) D only (d) E only

**18** The correct statement/s about hydrogen bonding is/are
A. Hydrogen bonding exists when H is covalently bonded to the highly electronegative atom.
B. Intermolecular H-bonding is present in *o*-nitro phenol.
C. Intramolecular H-bonding is present in HF.
D. The magnitude of H-bonding depends on the physical state of the compound.
E. H-bonding has powerful effect on the structure and properties of compounds.
Choose the correct answer from the option given below : *[2024, 4 April Shift-II]*
(a) A, D, E only (b) A, B, C only
(c) A, B, D only (d) A only

**19** H—N ...I... N ...II... N
In hydrogen azide (above) the bond orders of bonds (I) and (II) are *[JEE Main 2018]*

| | (I) | (II) | | (I) | (II) |
|---|---|---|---|---|---|
| (a) | > 2 | < 2 | (b) | < 2 | < 2 |
| (c) | < 2 | > 2 | (d) | > 2 | > 2 |

**20** If the magnetic moment of a dioxygen species is 1.73 B.M, it may be. *[2020, 9 Jan Shift-II]*
(a) $O_2, O_2^-$ or $O_2^+$ (b) $O_2^-$ or $O_2^+$
(c) $O_2$ or $O_2^-$ (d) $O_2$ or $O_2^+$

**21** Among the following, the molecule expected to be stabilised by anion formation is $C_2$, $O_2$, NO, $F_2$. *[2019, 9 April Shift-I]*
(a) $C_2$ (b) $F_2$
(c) NO (d) $O_2$

**22** During the change of $O_2$ to $O_2^-$, the incoming electron goes to the orbital. *[2019, 10 April Shift-I]*
(a) $\pi 2p_x$ (b) $\pi^* 2p_x$
(c) $\pi 2p_y$ (d) $\sigma^* 2p_z$

**23** The bond order and magnetic property of acetylide ion are same as that of *[2023, 12 April Shift-I]*
(a) $O_2^+$ (b) $N_2^+$
(c) $O_2^-$ (d) $NO^+$

**24** In the molecular orbital diagram for the molecular ion $N_2^+$, the number of electrons in the $\sigma 2p_z$ molecular orbital is *[JEE Main 2018]*
(a) 3 (b) 1 (c) 0 (d) 2

**25** Which one of the following molecules is expected to exhibit diamagnetic behaviour? *[JEE Main 2013]*
(a) $C_2$ (b) $N_2$
(c) $O_2$ (d) $S_2$

**26** In which of the following pairs of molecules/ions both the species are not likely to exist? *[JEE Main 2013]*
(a) $H_2^+$, $He_2^{2-}$ (b) $H_2^-$, $He_2^{2-}$
(c) $H_2^{2+}$, $He_2$ (d) $H_2^-$, $He_2^{2+}$

**27** Stability of the species $Li_2$, $Li_2^-$ and $Li_2^+$ increases in the order of *[JEE Main 2013]*
(a) $Li_2 < Li_2^+ < Li_2^-$ (b) $Li_2^- < Li_2^+ < Li_2$
(c) $Li_2 < Li_2^- < Li_2^+$ (d) $Li_2^- < Li_2 < Li_2^+$

**28** In which of the following ionisation processes, the bond order has increased and the magnetic behaviour has changed? *[AIEEE 2007]*
(a) $C_2 \rightarrow C_2^+$ (b) $NO \rightarrow NO^+$
(c) $O_2 \rightarrow O_2^+$ (d) $N_2 \rightarrow N_2^+$

## Level 3

**29.** Which of the following molecule(s) show/s paramagnetic behaviour? *[2025, 4 April Shift-I]*
A. $O_2$ B. $N_2$
C. $F_2$ D. $S_2$
E. $Cl_2$
Choose the correct answer from the options given below:
(a) A and E only (b) B only
(c) A and D only (d) A and C only

**30** What is the number of unpaired electrons in the highest occupied molecular orbital of the following species $N_2$; $N_2^+$; $O_2$; $O_2^+$? *[2023, 24 Jan Shift-II]*
(a) 2, 1, 2, 1 (b) 0, 1, 0, 1
(c) 2, 1, 0, 1 (d) 0, 1, 2, 1

**31** According to MO theory, the bond orders for $O_2^{2-}$, CO and $NO^+$ respectively, are *[2023, 29 Jan Shift-II]*
(a) 2, 3 and 3 (b) 1, 2 and 3
(c) 1, 3 and 3 (d) 1, 3 and 2

**32** Bonding in which of the following diatomic molecule(s) become(s) stronger, on the basis of MO theory, by removal of an electron?

(A) NO (B) $N_2$ (C) $O_2$ (D) $C_2$ (E) $B_2$

Choose the most appropriate answer from the options given below. *[2022, 25 June Shift-I]*

(a) (A), (B), (C) only (b) (B), (C), (E) only
(c) (A), (C) only (d) (D) only

**33** Consider the ions/molecules: $O_2^+, O_2, O_2^-, O_2^{2-}$

For increasing bond order the correct option is *[2022, 26 June Shift-I]*

(a) $O_2^{2-} < O_2^- < O_2 < O_2^+$ (b) $O_2^- < O_2^{2-} < O_2 < O_2^+$
(c) $O_2^- < O_2^{2-} < O_2^+ < O_2$ (d) $O_2^- < O_2^+ < O_2^{2-} < O_2$

**34** The compound/s which will show significant intermolecular H–bonding is/are

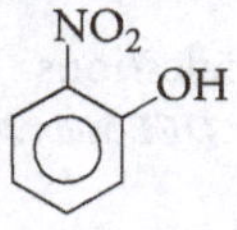
(A)

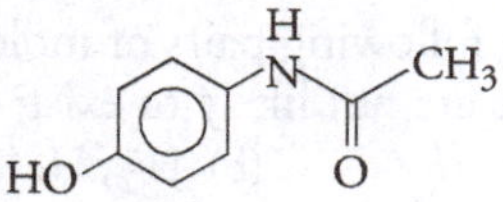
(B)

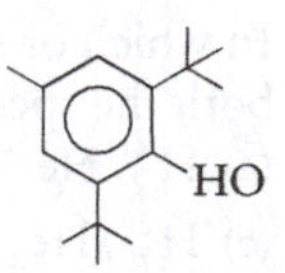
(C)

*[2021, 27 Aug Shift-II]*

(a) Only (B) (b) Only (C)
(c) (A) and (B) (d) (A), (B) and (C)

**35** Match List-I with List-II.

| | List-I (Molecule) | | List-II (Bond order) |
|---|---|---|---|
| A. | $Ne_2$ | (i) | 1 |
| B. | $N_2$ | (ii) | 2 |
| C. | $F_2$ | (iii) | 0 |
| D. | $O_2$ | (iv) | 3 |

Choose the correct answer from the options given below. *[2021, 26 Feb Shift-II]*

| | A | B | C | D |
|---|---|---|---|---|
| (a) | (iii) | (iv) | (i) | (ii) |
| (b) | (i) | (ii) | (iii) | (iv) |
| (c) | (ii) | (i) | (iv) | (iii) |
| (d) | (iv) | (iii) | (ii) | (i) |

**36** Of the species, NO, $NO^+$, $NO^{2+}$ and $NO^-$, the one with minimum bond strength is *[2020, 5 Sep Shift-I]*

(a) $NO^+$ (b) NO (c) $NO^{2+}$ (d) $NO^-$

**37** Among the following molecules/ions,

$C_2^{2-}, N_2^{2-}, O_2^{2-}, O_2$

Which one is diamagnetic and has the shortest bond length? *[2019, 8 April Shift-II]*

(a) $C_2^{2-}$ (b) $O_2$ (c) $O_2^{2-}$ (d) $N_2^{2-}$

**38** Two pi and half sigma bonds are present in *[2019, 10 Jan Shift-I]*

(a) $O_2^+$ (b) $N_2$ (c) $N_2^+$ (d) $O_2$

**Imp.** Magnetic Nature in Blend with MOT

**39** In which of the following processes, the bond order has increased and paramagnetic character has changed to diamagnetic? *[2019, 9 Jan Shift-II]*

(a) $O_2 \rightarrow O_2^+$ (b) $O_2 \rightarrow O_2^{2-}$
(c) $N_2 \rightarrow N_2^+$ (d) $NO \rightarrow NO^+$

**40** Using MO theory, predict which of the following species has the shortest bond length? *[AIEEE 2009]*

(a) $O_2^{2+}$ (b) $O_2^+$
(c) $O_2^-$ (d) $O_2^{2-}$

**41** Which one of the following pairs of species have the same bond order? *[AIEEE 2008]*

(a) $CN^-$ and $NO^+$ (b) $CN^-$ and $CN^+$
(c) $O_2^-$ and $CN^-$ (d) $NO^+$ and $CN^+$

## Numerical Value Type Questions

**42** Total number of electrons present in $(\pi^*)$ molecular orbitals of $O_2$, $O_2^+$ and $O_2^-$ is ........... . *[2024, 9 April Shift-II]*

**43** The total number of species from the following in which one unpaired electron is present, is .......

$N_2, O_2, C_2^-, O_2^-, O_2^{2-}, H_2^+, CN^-, He_2^+$ *[2024, 9 April Shift-I]*

**44** Number of molecules having bond order 2 from the following molecules is ........

$C_2, O_2, Be_2, Li_2, Ne_2, N_2, He_2$ *[2024, 8 April Shift-II]*

**45** Number of molecules from the following, which can exhibit hydrogen bonding is .......... (Nearest integer)

$CH_3OH$, $H_2O$, $C_2H_6$, $C_6H_6$, 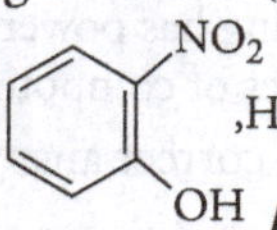 ,HF, $NH_3$ *[2024, 6 April Shift-I]*

**46** Number of molecules/species from the following having one unpaired electron is...... .

$O_2, O_2^{-1}$, NO, $CN^{-1}$, $O_2^{2-}$ *[2024, 4 April Shift-I]*

**47** The total number of molecular orbitals formed from 2*s* and 2*p* atomic orbitals of a diatomic molecule is ...... . *[2024, 30 Jan Shift-I]*

**48** The total number of anti-bonding molecular orbitals, formed from 2*s* and 2*p* atomic orbitals in a diatomic molecule is .......... . *[2024, 29 Jan Shift-II]*

**49** The number of species from the following which are paramagnetic and with bond order equal to one is .........

$H_2, He_2^+, O_2^+, N_2^{2-}, O_2^{2-}, F_2, Ne_2^+, B_2$ *[2024, 29 Jan Shift-I]*

**50** Sum of bond order of CO and $NO^+$ is ...... *[2024, 27 Jan Shift-I]*

**51** Among the following species
$N_2, N_2^+, N_2^-, N_2^{2-}, O_2, O_2^+, O_2^-, O_2^{2-}$
the number of species showing diamagnetism is ……… . **[2022, 25 July Shift-I]**

**52** The number of paramagnetic species among the following is ……… .
$B_2, Li_2, C_2, C_2^-, O_2^{2-}, O_2^+$ and $He_2^+$ **[2022, 28 July Shift-I]**

**53** According to MO (molecular orbital) theory, number of species/ions from the following having identical bond order is …… $CN^-, NO^+, O_2, O_2^+, O_2^{2+}$
**[2022, 27 July Shift-I]**

**54** The spin-only magnetic moment value of $B_2^+$ species is ……… $\times 10^{-2}$ BM. (Nearest integer) [Given, $\sqrt{3} = 1.73$] **[2021, 1 Sep Shift-II]**

**55** According to molecular orbital theory, the number of unpaired electron(s) in $O_2^{2-}$ is …… .
**[2021, 31 Aug Shift-II]**

**56** The total number of electrons in all bonding molecular orbitals of $O_2^{2-}$ is ………. .
(Round off to the nearest integer) **[2021, 27 July Shift-II]**

**57** The difference between bond orders of CO and $NO^{\oplus}$ is $x/2$. where, $x$ = ……… .
(Round off to the nearest integer). **[2021, 27 July Shift-I]**

**58** $AX$ is a covalent diatomic molecule where $A$ and $X$ are second row elements of periodic table. Based on molecular orbital theory, the bond order of $AX$ is 2.5. The total number of electron in $AX$ is ……… .
(Round off to the Nearest Integer).
**[2021, 18 March Shift-I]**

## • Key Index

### Topic 1

| | | | | | | | | | |
|---|---|---|---|---|---|---|---|---|---|
| 1 (a) | 2 (c) | 3 (d) | 4 (d) | 5 (b) | 6 (b) | 7 (c) | 8 (a) | 9 (c) | 10 (b) |
| 11 (c) | 12 (c) | 13 (c) | 14 (c) | 15 (b) | 16 (a) | 17 (d) | 18 (a) | 19 (b) | 20 (c) |
| 21 (d) | 22 (c) | 23 (a) | 24 (6) | 25 (6) | 26 (8) | 27 (5) | 28 (6) | 29 (3) | 30 (4) |
| 31 (3) | 32 (718) | | | | | | | | |

### Topic 2

| | | | | | | | | | |
|---|---|---|---|---|---|---|---|---|---|
| 1 (c) | 2 (a) | 3 (c) | 4 (d) | 5 (c) | 6 (d) | 7 (d) | 8 (c) | 9 (a) | 10 (b) |
| 11 (b) | 12 (a) | 13 (6) | 14 (6) | 15 (b) | 16 (c) | 17 (a) | 18 (b) | 19 (b) | 20 (b) |
| 21 (a) | 22 (b) | 23 (a) | 24 (d) | 25 (a) | 26 (b) | 27 (b) | 28 (a) | 29 (b) | 30 (a) |
| 31 (a) | 32 (d) | 33 (c) | 34 (b) | 35 (d) | 36 (d) | 37 (d) | 38 (d) | 39 (c) | 40 (b) |
| 41 (a) | 42 (c) | 43 (a) | 44 (c) | 45 (b) | 46 (d) | 47 (a) | 48 (c) | 49 (b) | 50 (b) |
| 51 (c) | 52 (d) | 53 (a) | 54 (c) | 55 (a) | 56 (b) | 57 (c) | 58 (c) | 59 (6) | 60 (3) |
| 61 (4) | 62 (3) | 63 (3) | 64 (4) | 65 (3) | 66 (1) | 67 (4) | 68 (3) | 69 (5) | 70 (4) |
| 71 (6) | 72 (1) | 73 (1) | 74 (3) | 75 (2) | 76 (3) | 77 (2) | 78 (20) | | |

### Topic 3

| | | | | | | | | | |
|---|---|---|---|---|---|---|---|---|---|
| 1 (a) | 2 (c) | 3 (a) | 4 (a) | 5 (c) | 6 (a) | 7 (c) | 8 (a) | 9 (d) | 10 (d) |
| 11 (a) | 12 (d) | 13 (c) | 14 (a) | 15 (b) | 16 (a) | 17 (c) | 18 (a) | 19 (c) | 20 (b) |
| 21 (a) | 22 (b) | 23 (d) | 24 (b) | 25 (a,b) | 26 (c) | 27 (b) | 28 (b) | 29 (c) | 30 (d) |
| 31 (c) | 32 (c) | 33 (a) | 34 (a) | 35 (a) | 36 (d) | 37 (a) | 38 (c) | 39 (d) | 40 (a) |
| 41 (a) | 42 (6) | 43 (4) | 44 (2) | 45 (5) | 46 (2) | 47 (8) | 48 (4) | 49 (1) | 50 (6) |
| 51 (2) | 52 (4) | 53 (3) | 54 (173) | 55 (0) | 56 (10) | 57 (0) | 58 (15) | | |

# Detailed **Explanations**

## Topic 1

**01.** *(a)* The correct order of increasing dipole moment of given compounds is

$NF_3 < HBr < H_2S < CHCl_3$
(0.24 D) (0.79 D) (0.95 D) (1.04 D)

**02.** *(c)* According to Fajan's rule, Ag and Cl bond will be covalent in nature not ionic.
The ionic character of the given compounds is in the order,
$AgCl < CoCl_2 < BaCl_2 < KCl$
Hence, AgCl is least ionic.

**03.** *(d)* Polar molecules are those molecules in which there exists an electronegativity difference between the bonded atom. They have non-zero dipole moment.
Hence, $CHCl_3$ is a polar molecule.
Non-polar molecules are those molecules in which electrons are shared equally. They have zero dipole moment.
$CH_2{=}CH_2, CCl_4, CO_2$ are non-polar molecules.
Hence, option (d) is correct.

**04.** *(d)* $CO_3^{2-}$ has a single structure i.e., resonance hybrid of given three canonical structures.
The given structures are resonating structures of carbonate ion.

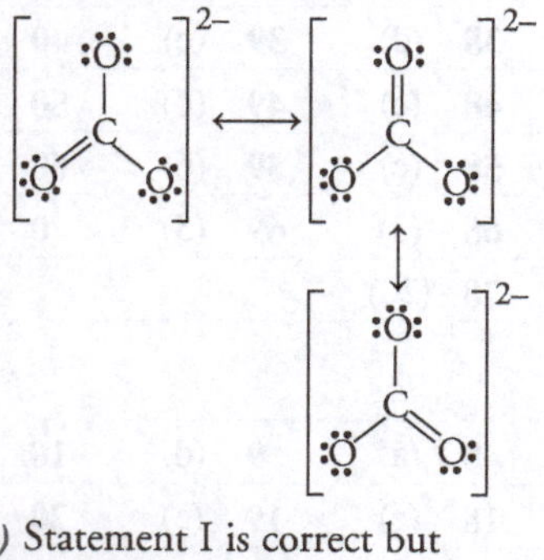

**05.** *(b)* Statement I is correct but Statement II is incorrect.

Dipole moment is a vector quantity i.e., it has magnitude as well as definite directions.
In chemistry, the dipole moment is represented by cross on the positive center and arrowhead on the negative center. This arrow symbolises the shift of electron density in the molecules.

**06.** *(b)* NO contains 15 electrons, thus it is an odd electron molecule.
Expanded octet molecule contains more than eight electrons in their valence shell. $SF_6$, $PCl_5$ and $H_2SO_4$ are examples of expanded octet molecules.
$BCl_3$ is an electron deficient molecule, i.e. it contains less than 8 electrons in its valence shell.

**07.** *(c)* When the distance of hydrogen nucleus is decreasing at a particular point, energy is minimum. At this point potential energy will be least. It is the point when bond formation taken place.

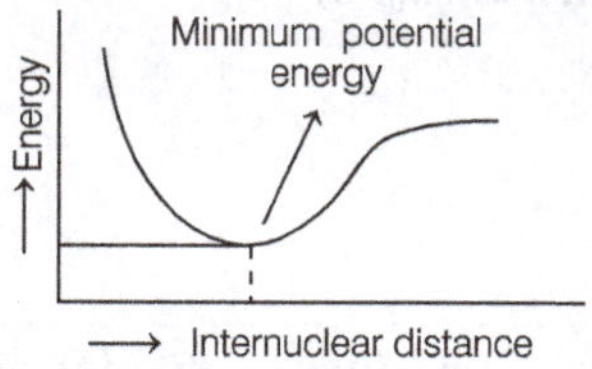


Hence, option (c) is correct.

**08.** *(a)* The correct descending order of bonds according to their average bond energies is as follows :
C—F > C—Cl > C—Br > C—I
Bond length is inversely proportional to bond energy. So, bond length decreases as bond energy increases.

**09.** *(c)* The correct order of dipole moment is $CH_4 = CCl_4 < CHCl_3$

**10.** *(b)* $\mu = q \times d$

$$d = 1.617\ \text{Å} = 1.617 \times 10^{-8}\ \text{cm}$$
$$\mu = 0.38\ \text{D} = 0.38 \times 10^{-18}\ \text{esu cm}$$
$$q = \frac{\mu}{d} = \frac{0.38 \times 10^{-18}}{1.617 \times 10^{-8}}$$

Therefore, fractional charge

$$= \frac{\text{Particle charge}}{\text{Total charge}} = \frac{q}{\varepsilon_0}$$
$$= \frac{0.38 \times 10^{-18}}{1.617 \times 10^{-8} \times 4.802 \times 10^{-10}}$$
$$= 0.0489 \approx 0.05$$

**11.** *(c)* In all the given compounds, anion is same ($Cl^-$), hence polarising power is decided by size and charge of cation.
$Al^{3+}$ with maximum charge and smallest size has maximum polarising power hence, $AlCl_3$ has maximum covalent.

**12.** *(c)* Higher the charge/size ratio, larger is the polarising power. Therefore, order of polarising power of given cationic species will be

$$K^+ < Ca^{2+} < Mg^{2+} < Be^{2+}$$

**13.** *(c)* Both A and R are true and R is the correct explanation of A.
The bond structure of $NF_3$ and $NH_3$ with dipole moment direction are;

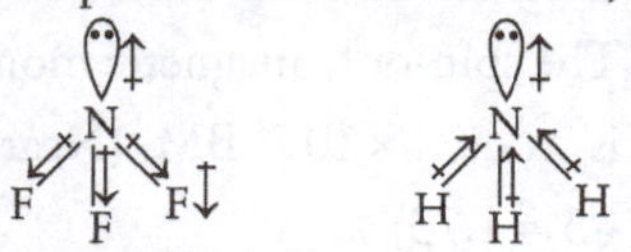


Resultant dipole moment $= 0.80 \times 10^{-30}$ Cm  Resultant dipole moment $= 4.9 \times 10^{-30}$ Cm
Hence, option (c) is correct.

**14.** *(c)* Both A and R are true and R is correct explanation of A.
Dipole moment is a vector quantity and for compound net dipole moment is the vector sum of all dipoles. Hence dipole moment of *cis*-form is greater than *trans*-form.

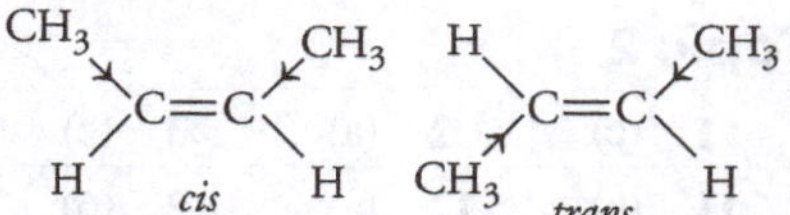


*cis* ($\mu > 0$)  *trans* ($\mu = 0$)

**15.** *(b)* Among the given molecules $CH_4$ and $PF_5$ are non-polar. Hence they have net zero dipole moment while among $NH_3$ and $NF_3$, the dipole moment of $NH_3$ is greater than $NF_3$. It is because in $NH_3$ vector addition of bond moment and lone pair takes place while in $NF_3$ vector subtraction of bond moment and lone pair takes place.

**16.** *(a)* The ionic character in a molecule is dependent upon electronegativity difference between the constituting atoms.
Greater the electronegativity difference, greater will be the ionic character of molecules.
Therefore, the correct sequence of increasing order of ionic character is
$N_2 < SO_2 < ClF_3 < K_2O < LiF$.
Hence, option (a) is correct.

**17.** *(d)* Both Statement I and II are false. Their correct form are as follows

**Statement I** The dipole moment of $NH_3$ is higher than $NF_3$. The reason is that flourine is more electronegative than nitrogen and thus pulls the electron towards itself. As number of NF bonds is 3, the three F-atoms pull electrons and hence resulting in decrease of dipole moment.

**Statement II** In $NH_3$, the orbital dipole due to lone pair and dipole moment of NH bonds are in same direction, but in $NF_3$ it will be in opposite direction.

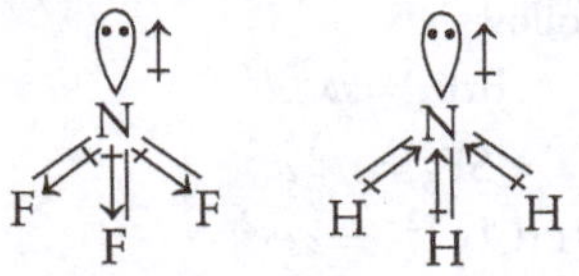

Hence, option (d) is correct.

**18.** *(a)*

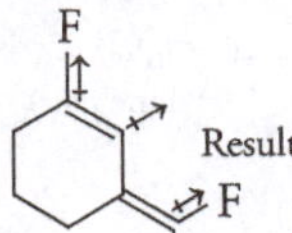


**19.** *(b)* *Para*-nitro aniline has a maximum dipole moment.

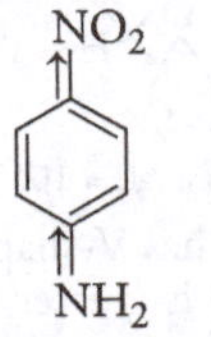

**20.** *(c)* The correct match is A-IV, B-II, C-I, D-III.

**21.** *(d)* Greater the difference of electronegativity between the cation and anion, greater will be the ionic character and lesser will be the covalent character.

Hence, correct order is

KI > KF, LiF > KF; CuCl > NaCl

Smaller the size of the cation higher will be its polarising capacity. Since, $Sn^{4+}$ is smaller in size hence it will have more covalent character. So, $SnCl_4 > SnCl_2$.

> **Alternate Solution**
>
> According to Fajan's rule, covalent character is higher for large size anion and small size cation. Thus,
>
> (A) KF > KI is incorrect, LiF > KF is correct.
>
> (B) KF < KI is correct, LiF > KF is correct.
>
> (C) $SnCl_4 > SnCl_2$ is correct, CuCl > NaCl is correct.
>
> (D) LiF > KF is correct, CuCl < NaCl is incorrect.
>
> (E) KF < KI is correct, CuCl > NaCl is correct.
>
> So, B, C and E are correct.

**22.** *(c)* $B_2H_6$ and $BCl_3$ are electron deficient species with incomplete octets.

**23.** *(a)* Structure given in 1 and 2 are linear and hence,

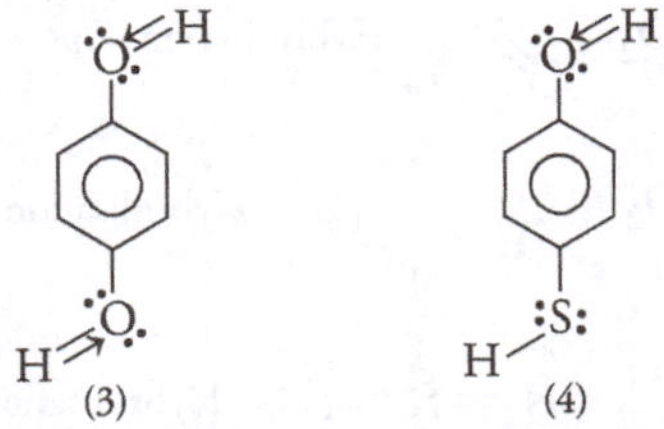

(3) Unsymmetrical, $\mu \neq 0$ shape around O-atom is angular, non-planar

(4) Unsymmetrical, $\mu \neq 0$ shape around S-atom is angular, non-planar

Thus, molecules  and 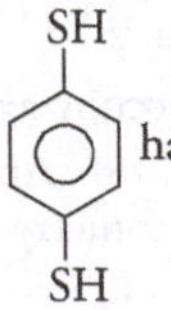 have $\mu \neq 0$.

**24.** *(6)* $NO_2, H_2SO_4, BF_3, ClO_2, PCl_5, BeF_2$. These are exception of octet rule. Hence, among the given molecules total 6 molecule are exception of octet rule.

**25.** *(6)*

> **key idea** The molecules which are symmetrical in nature have zero dipole moment.

Among the given, $H_2, CO_2, BF_3, CH_4, SiF_4, BeF_2$ are symmetrical molecules.

Hence, total 6 molecules will show zero dipole moment.

**26.** *(8)* The bond structure of $NO_2^-$ is,

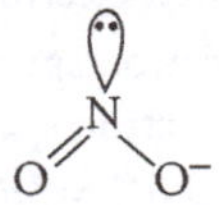

Number of valence $e^-$ around N-atom $=8$.

**27.** *(5)* Polar molecules will have non-zero dipole moments while non-polar molecules will have zero dipole moments. Thus, among the given species/ compounds $NF_3, H_2O, H_2S$, HBr and HCl will have non-zero dipole moment

**28.** *(6)*

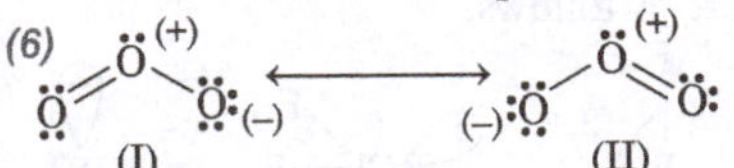

(I) (Ozone) Resonating structures of ozone ($O_3$) (II)

Ozone has 6 lone pair of electrons.

**29.** *(3)* Molecule with net zero dipole moment are $CO_2, CH_4$ and $BF_3$ i.e., 3 Remaining all molecules have non-zero dipole moment.

**30.** *(4)* The number of non-polar molecules are 4. These are $CO_2, H_2, CH_4$ and $BF_3$.

Remaining all compound are polar molecules.

HF, $H_2O, SO_2, NH_3, CHCl_3$, HCl.

**31.** *(3)* Out of given molecules; $H_2O, NH_3$ and HCl are polar molecules and $BeF_2, BF_3$ and $CCl_4$ are non-polar molecules.

**32.** *(718)* $\Delta H_{formation}$ is given as

$$\Delta_f H^\circ_{(KCl)} = \Delta_{sub} H^\circ_{(K)} + \Delta_{ion} H^\circ_{(K)} + \frac{1}{2}\Delta_{bond} H^\circ_{(Cl_2)} + \Delta e_g H^\circ_{(Cl)} + \Delta_{lattice} H^\circ_{(KCl)}$$

$$\Delta_{lattice} H^\circ_{(KCl)} = \Delta_f H^\circ_{(KCl)} - \Delta_{sub} H^\circ_{(K)} - \Delta_{ion} H^\circ_{(K)} - \frac{1}{2}\Delta_{bond} H^\circ_{(Cl_2)} - \Delta e_g H^\circ_{(Cl)}$$

$$= -436.7 - 89.2 - 419.0 - \frac{1}{2}(243) - (-348.6)$$

$$= -717.8 \approx 718 \text{ kJ/mol}$$

## Topic 2

**01.** *(c)* The correct match is A-III, B-IV, C-I, D-II.

| **Hybridisation** | **Orientation** |
|---|---|
| $sp^3$ | → Tetrahedral |
| $dsp^2$ | → Square planar |
| $sp^3d$ | → Trigonal bipyramidal |
| $sp^3d^2$ | → Octahedral |

**02.** *(a)* The hybridisation of $SF_4$ is $sp^3d$ and geometry is trigonal bipyramidal. At axial position, there are 3 *lp-bp* at 90°. But at equatorial position, there are only two lone pair-bond pair repulsions at 90°. So, shape of $SF_4$ molecule is *see-saw*.

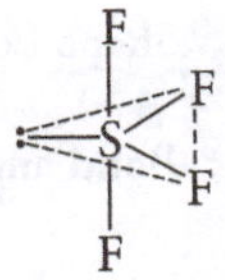

**03.** *(c)*

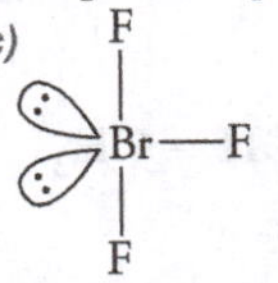

$BrF_3$ has 2 lone pairs of electron and it has bent T-shape.

**04.** *(d)* $N_3^-$ is linear specie.

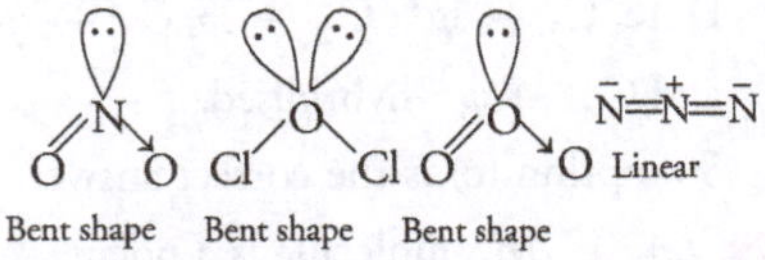


Hybridisation of $N_3^-$ is *sp* and structure is linear.

**05.** *(c)* 2 lone pair + 3 bond pair

Steric number = 5

Hybridisation = $sp^3d$.

So, the shape is T- shape.

> **Concept Enhancer**
>
> Geometry of the compound having $sp^3d$ hybridisation have tetragonal bipyramidal (TBP) geometry according to VSEPR theory. In TBP geometry lone pair occupy equatorial position.
>
> 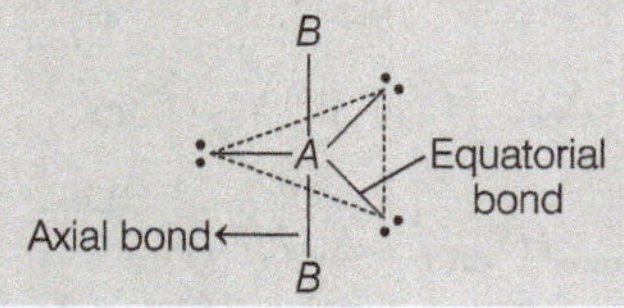
> 

**06.** *(d)* Bond angle of $sp^3$-hybridised compound should be 109.5°.

But in case of water, it is only 104.5° due to repulsion between the lone pairs. Also, lone pair-lone pair repulsion is higher than bond pair-bond pair repulsion. (VSEPR Theory)

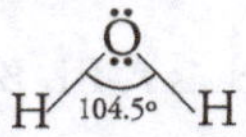


lp-lp Repulsion > bp-bp Repulsion

So, both A and R are true and R is the correct explanation of A.

**07.** *(d)* Hybridisation of central atom I in $I_3^-$ is $sp^3d$ with 3 lone pair and 2 bond pair.

Shape Linear 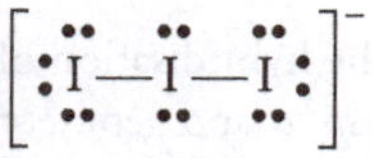

Bond angle 180° (for linear molecule)

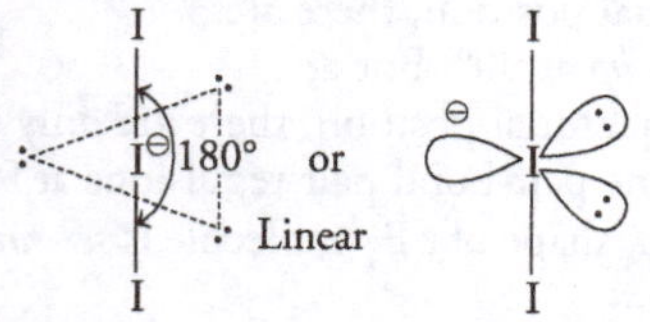


**08.** *(c)* $\overset{1}{C}H_2=\overset{2}{C}=\overset{3}{C}H-\overset{4}{C}H_3$

It is buta -1, 2 - diene,

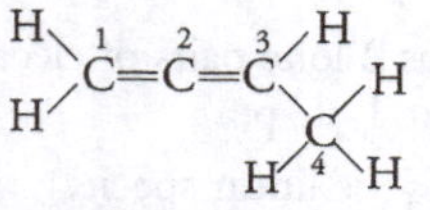

Here, $C_1 \rightarrow sp^2, C_2 \rightarrow sp, C_3 \rightarrow sp^2$ and $C_4 \rightarrow sp^3$-hybridised.

So, option (c) is the correct answer.

**09.** *(a)* If $AB_4$ molecule is a polar molecule, a possible geometry of $AB_4$ is square pyramidal.

**10.** *(b)* The geometry of $SF_6$ is octahedral and the geometry of $SF_4$ is trigonal bipyramidal.

**11.** *(b)*

$SO_2$ 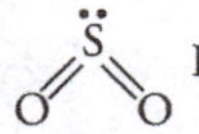 Hybridisation = $sp^2$

$NO_2^-$ 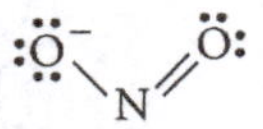 Hybridisation = $sp^2$

$N_3^-$ 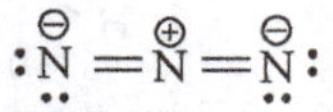 Hybridisation = $sp$

**12.** *(a)* Statement I is correct but Statement II is incorrect. It is because in $sp^3d$ hybridisation lone pair cannot occupy axial position.

**13.** *(6)* Among the given species the geometry of given molecules are linear i.e.,

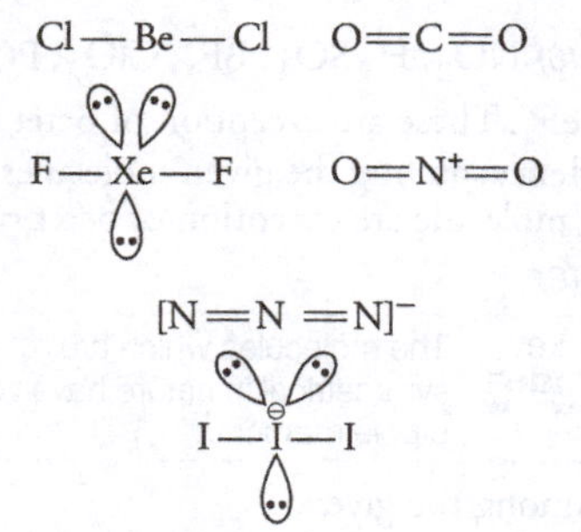


**14.** *(6)* According to molecular orbital theory.

$O_2 \rightarrow 2$ unpaired electron

$O_2^+ \rightarrow 1$ unpaired electron

$O_2^- \rightarrow 1$ unpaired electron

NO → odd electron species.

$NO_2 \rightarrow$ odd electron species.

$K_2[NiCl_4] \rightarrow Ni^{2+} = 3d^8$

(Weak field ligand)

Coordination number = 4

∴Tetrahedral, paramagnetic with 2 unpaired electrons.

So, total 6 paramagnetic species are present.

**15.** *(b)* The structure of given compounds are as follows.

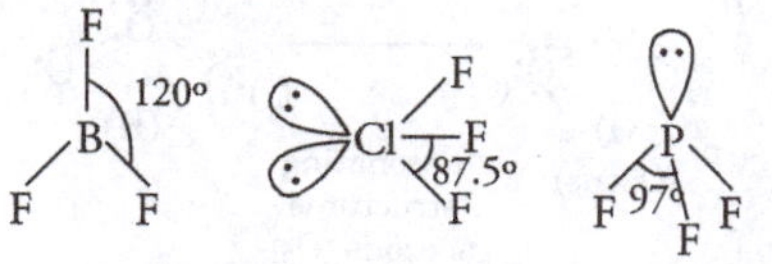


So, the correct order of bond angle is $ClF_3 < PF_3 < BF_3$.

**16.** *(c)* The hybridisation of central atom of given species in different options are

$BF_3, NO_2^-, NO_2 \Rightarrow sp^2$-hybridisation

$H_2O$ and $NH_2^- \Rightarrow sp^3$-hybridisation

So, the pair of central atoms that exhibit $sp^2$-hybridisation is $BF_3$ and $NO_2^-$.

**17.** *(a)* Among the given options, $BrF_5$ has square pyramidal shape.

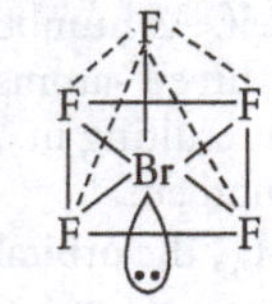


**18.** *(b)* Among the given options, only $[Co(NH_3)_6]^{3+}$ shows $d^2sp^3$ hybridisation.

Hybridisation of rest of the members are as follows

$BrF_5 = sp^3d^2$

$SF_6 = sp^3d^2$

$[PtCl_4]^{2-} = dsp^2$

**19.** *(b)* Shape of $SO_2$ and $H_2O$ are given as

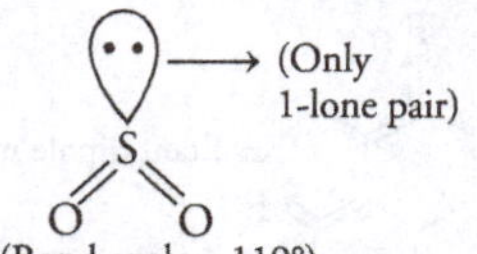


(Bond angle = 119°)

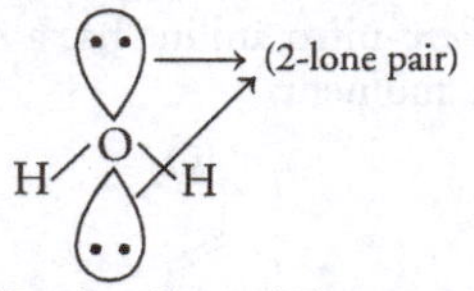


(Bond angle = 1045°)

Hence, both has V-shape, but bond angle in $SO_2$ is greater than $H_2O$, due to repulsion between lone pairs at oxygen. So, statement I is correct but statement II is incorrect.

**20.** *(b)* Structure of $SCl_2$,

(Number of lone pairs of electron on central atom, S = 2)

Structure of $O_3$,

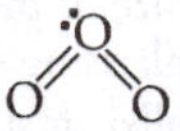

Number of lone pairs on O = 1

Structure of $ClF_3$,

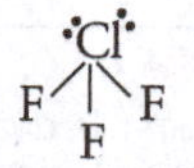

Number of lone pairs on Cl = 2

Structure of $SF_6$,

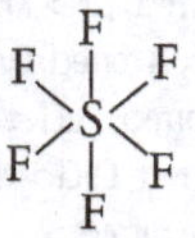

Number of lone pairs on S = 0

**21.** *(a)* Zero orbital overlap is out of phase overlap and it is because of the different orientation / direction of approach of orbitals.

**22.** *(b)* Number of electron in given species

$CH_4 = 6 + 4 = 10$

$NH_4^+ = 7 + 3 = 10$

$BH_4^- = 5 + 4 + 1 = 10$

They all have same number of electron, i.e. are isoelectronic (species with same number of electron) and also have same structure, i.e. tetrahedral.

**23.** *(a)* (a) $[Ni(CN)_4]^{2-} = dsp^2$

(b) $BrF_5 = sp^3d^2$

(c) $XeF_4 = sp^3d^2$

(d) $[CrF_6]^{3-} = d^2sp^2$

**24.** *(d)*

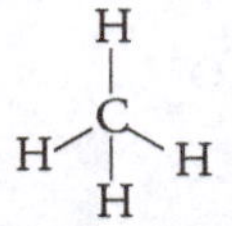

[$CH_4$ = hybridisation is $sp^3$, lone pair of electrons is 0, so that bond angle is maximum = 109°28′ ]

**25.** *(a)*

| Species | Bond angle |
|---|---|
| $BF_3$ | 120° |
| $NH_3$ | 107° |
| $PF_3$ | 100° |
| $I_3^-$ | 180° |

Hence, the correct order of bond angle is

$I_3^- > BF_3 > NH_3 > PF_3$

**26.** *(b)* $XeOF_2 \rightarrow$ 3 bond pairs + 2 lone pairs

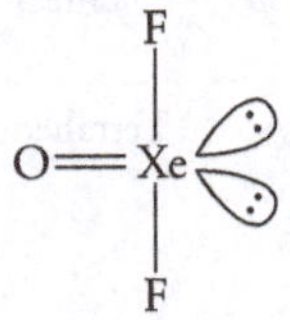

*T*-shape

$XeOF_4 \rightarrow$ 5 bond pairs + 1 lone pair

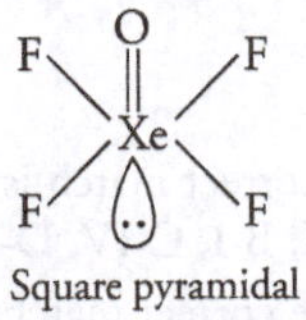

Square pyramidal

**27.** *(b)* $NF_3$ has a trigonal pyramidal geometry and has $sp^3$-hybridisation.

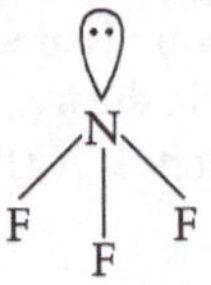

**28.** *(a)* $BF_3 \rightarrow BF_4^-$

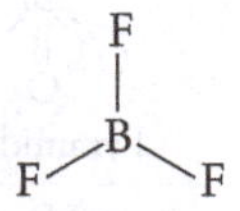

Trigonal planar ($sp^2$-hybridisation)

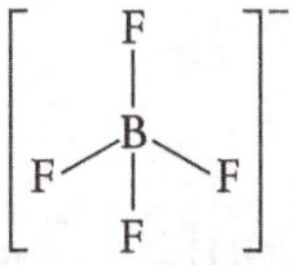

Tetrahedral ($sp^3$-hybridisation)

In this, hybridisation is changes from $sp^2$ to $sp^3$ and shape from trigonal planar to tetrahedral.

**29.** *(b)* Hybridisation

$$= \frac{1}{2}[VE + MA - c + a]$$

$$PF_5 \Rightarrow X = \frac{1}{2}[5 + 5 - 0 + 0]$$

$$= 5 = sp^3d$$

Hence, $PF_5$ exhibits $sp^3d$-hybridisation, not $sp^3d^2$.

**30.** *(a)* $SF_4$ has trigonal bipyramidal geometry, lone pair of electrons repels the axial bond pair and decreases the bond angle to 173°. So all bonds are not equal.

**31.** *(a)*

| Molecule | Structure | Hybridisation of central atom | Lone pair |
|---|---|---|---|
| $SF_4$ | 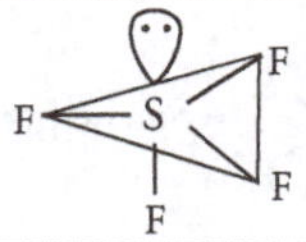 | $sp^3d$ | one |
| $CF_4$ |  | $sp^3$ | zero |
| $XeF_4$ | 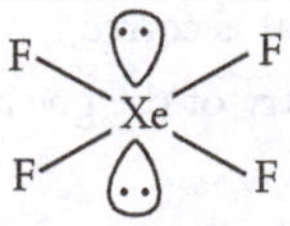 | $sp^3d^2$ | two |

**32.** *(d)* $sp^3d^2$ hybridisation has octahedral structure such that four hybrid orbitals are at 90° with respect to each other and others two at 90° with first four.

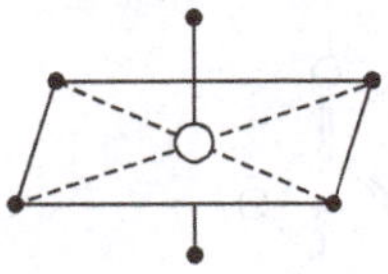

**33.** *(c)* $\underset{sp^3}{CH_3}—\underset{sp}{C}≡N$

Acetonitrile does not contain $sp^2$ hybridised carbon.

**34.** *(b)* $XeF_2$ contains $sp^3d$ hybridised Xe atom with three lone pairs of electrons. While C-atom of $CO_2$ is *sp* hybridised and thus, both possesses a linear shape.

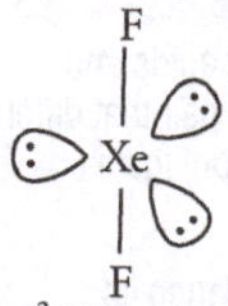

$sp^3d$ hybridised Xe-atom

O═C═O

*sp* hybridised C-atom

**35.** *(d)*

- $SO_2$ has $sp^2$ hybridisation and has non zero dipole moment. It has 1 lone pair.
- $NF_3$ has $sp^3$ hybridisation and has non-zero dipole moment. It has 1 lone pair.
- $NH_3$ ($sp^3$) Non-zero dipole moment - 1 lone pair
- $XeF_2$ ($sp^3d$) Non-zero dipole moment - 3 lone pair
- $ClF_3$ ($sp^3d$) Non-zero dipole moment - 2 lone pair
- $SF_4$ ($sp^3d$) Non-zero dipole moment - 1 lone pair.

**36.** *(d)* The correct match is A-II, B-III, C-IV, D-I

**37.** *(d)* Among the given options, $BrF_5$ and $XeOF_4$ has square pyramidal geometry.

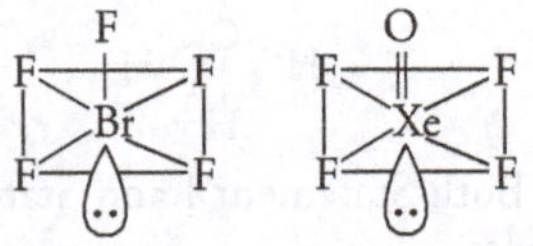

$SbF_5$ and $PCl_5$ has trigonal bipyramidal geometry.

**38.** *(d)* Statements given in *A*, *B* and *C* are correct, while *D* and *E* are incorrect.

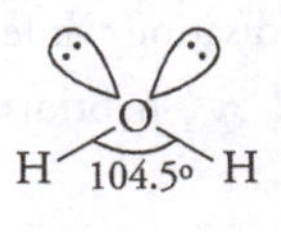


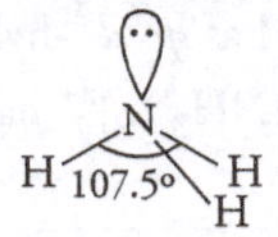


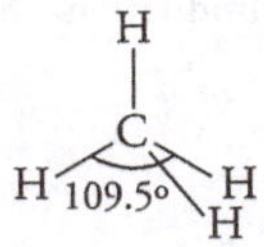


- $O \rightarrow sp^3$ $N \rightarrow sp^3$ $C \rightarrow sp^3$
- Dipole moment order : $CH_4 < NH_3 < H_2O$
- $H_2O$ and $NH_3$ are Lewis base.

- $NH_3$ act as Lowry Bronsted base. In solution of $NH_3$ in $H_2O$.

**Concept Enhancer**

- According to Bronsted-Lowry acids and bases concept, the acid-base pair that differs only by one proton is called conjugate acid-base pair.
- If Bronsted acid is strong acid then its conjugate base is a weak base and *vice-versa*.

**39.** *(c)* The correct match is
A-III, B-I, C-IV, D-II

$NH_3$ = Trigonal pyramidal

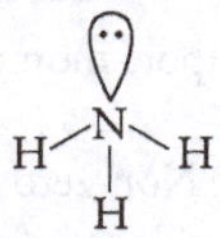

$BrF_5$ = Square pyramidal

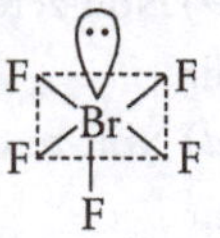

$PCl_5$ = Trigonal bipyramidal

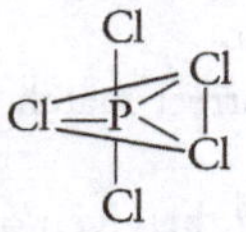

$CH_4$ = Tetrahedral

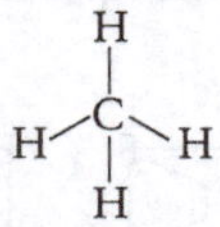

**40.** *(b)* Both Statement I and Statement II are false.
The correct form of these statements are,
$PF_5$ has $sp^3d$-hybridisation while $BrF_5$ has $sp^3d^2$ hybridisation.
$SF_6$ has $sp^3d^2$-hybridisation while $[Co(NH_3)_6]^{3+}$ has $d^2sp^3$-hybridisation.

**41.** *(a)*

| Molecules | Hybridisation | Shape/geometry |
|---|---|---|
| A. $SF_4$ | $sp^3d$ | 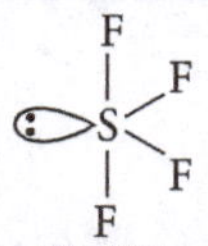 see-saw |
| B. $BrF_3$ | $sp^3d$ | 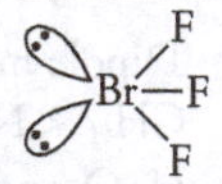 Bent T-shape |
| C. $BrO_3^-$ | $sp^3$ | 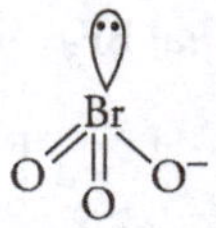 Pyramidal |
| D. $NH_4^+$ | $sp^3$ | 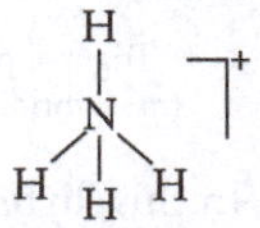 Tetrahedral |

Hence, correct match is A-III, B-IV, C-II, D-I.

**42.** *(c)* The correct match is A-III, B-I, C-II, D-IV.

A. $SO_2Cl_2$ has $sp^3$-hybridisation and has tetrahedral geometry.
B. NO is paramagnetic in nature.
C. $NO_2^-$ is diamagnetic in nature.
D. $I_3^-$ has $sp^3d$-hybridisation and has linear geometry.

**43.** *(a)* The correct match is
A-IV, B-I, C-II, D-III.
ICl = I—Cl = Linear

$ICl_3$ 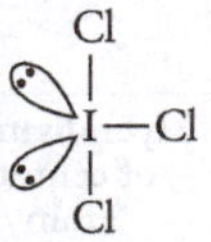 = T-shape

$ClF_5$ 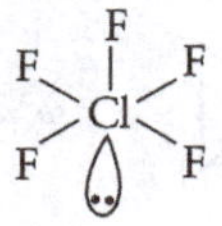 = Square pyramidal

$IF_7$ 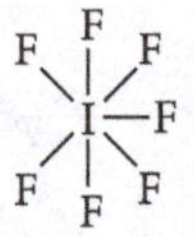 = Pentagonal bipyramidal

Hence, option (a) is correct.

**44.** *(c)* The geometry of the given species are as follows

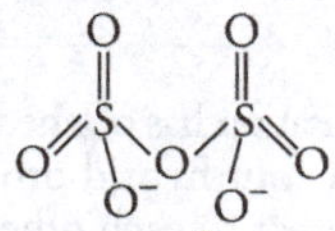

$S_2O_7^{2-}$
Tetrahedral

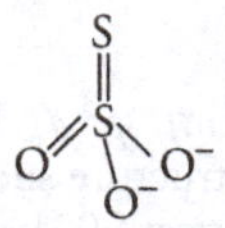

$S_2O_3^{2-}$
Tetrahedral

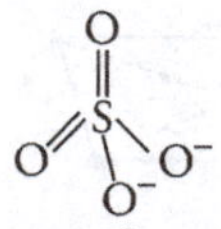

$SO_4^{2-}$
Tetrahedral

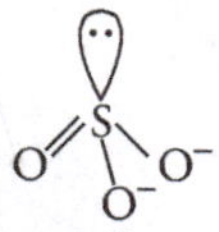

$SO_3^{2-}$
Pyramidal

Thus, among the given species only $SO_3^{2-}$ has pyramidal geometry.

**Time Saver Tech**

The species that does not contain any lone pair of electron will have regular geometry i.e., tetrahedral while the species that contains lone pair of electron will not have regular geometry i.e., pyramidal ($SO_3^{2-}$).

**45.** *(b)* The hybridisation of central atom in the given molecules/ions are as follows

$NO_3^- \rightarrow sp^2$ 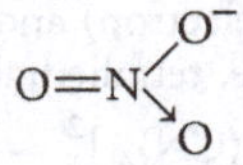

$BCl_3 \rightarrow sp^2$ 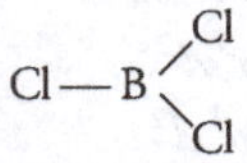

$ClO_2^- \rightarrow sp^3$ 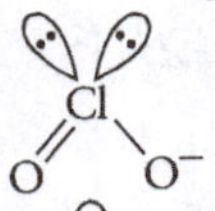

$ClO_3^- \rightarrow sp^3$ 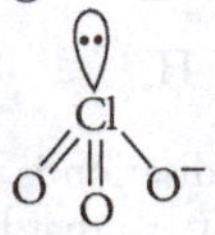

Thus, two molecules/ions i.e., $ClO_2^-$ and $ClO_3^-$ has $sp^3$ hybridisation.

**46.** *(d)* The correct match is
A-III, B-II, C-I, D-IV.

| Species | Hybridisation | Shape | Structure |
|---|---|---|---|
| A. $H_3O^+$ | $sp^3$ | Pyramidal | 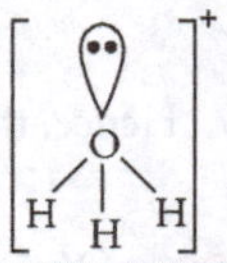 |
| B. Acetylide anion | $sp$ | Linear | $^{\ominus}C{\equiv}C^{\ominus}$ |
| C. $NH_4^+$ | $sp^3$ | Tetrahedral | 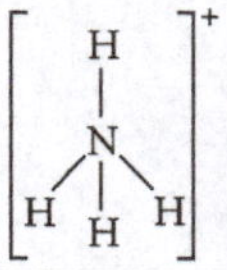 |
| D. $ClO_2^-$ | $sp^3$ | Bent | 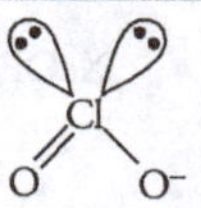 |

**47.** *(a)* The correct match is
A-II, B-I, C-IV, D-III.

**48.** *(c)* ∴ The correct match is A-(II), B-(IV), C-(I), D-(III).

**49.** *(b)* The shapes of the covalently bonded compounds are assigned on the basis of VSEPR theory.
So correct match is A-(III), B-(IV), C-(I), D-(II)

**50.** *(b)* The correct order is (A) → (III), (B) → (IV), (C) → (I), (D) → (II)

**51.** *(c)* The correct matches are : (A)-(III), (B)-(I), (C)-(V) and (D)-(IV).

**52.** *(d)* In $NO_2^-$

In N is $sp^2$-hybridised.

$NO_2^-$ 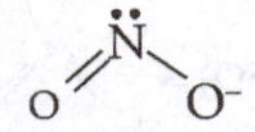

In $NO_2^+$, N *sp*-hybridised.

$NO_2^+$ $O=\overset{\oplus}{N}=O$

In $NH_4^+$, N is $sp^3$-hybridised.

$NH_4^+$ 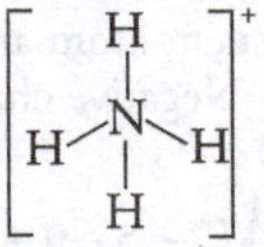

Therefore, hybridisation of N in $NO_2^-$, $NO_2^+$, $NH_4^+$ are $sp^2$, *sp*, $sp^3$ respectively.

**53.** *(a)* If number of sigma bond (σ), co-ordinate bond and lone pair are same for given pairs then, they are isostructural. Hence, $SO_4^{2-}$, $CrO_4^{2-}$, $SiCl_4$ and $TiCl_4$ have tetrahedral shape and are isostructural.

**54.** *(c)* $SF_4$ ⇒

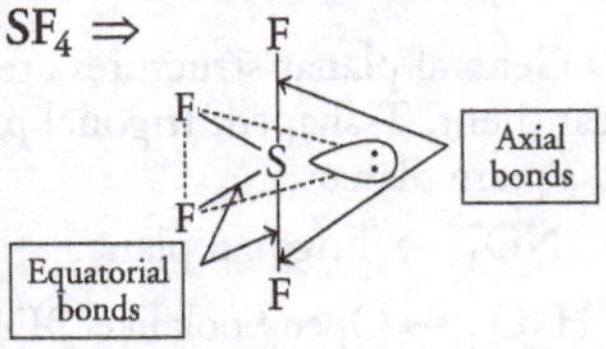


S ($sp^3d$-hybridised)

**Geometry** Trigonal bipyramidal

**Shape** *see-saw*

Axial S—F bonds are longer than equatorial bonds.

Hence, among given species only $SF_4$ has unequal bond lengths.

**55.** *(a)* In solid state, $PCl_5$ prefer to exist as oppositely charged ions i.e. tetrahedral $[PCl_4]^+$ and octahedral $[PCl_6]^-$ as the ionic bonding enhances the crystalline nature. These structures $[PCl_4]^+$ and $[PCl_6]^-$ provide extra stability to solid.

**56.** *(b)*

| S.No. | Molecule | Hybridisation | Bond angle |
|---|---|---|---|
| a. | $PH_3$ | $sp^3$ | 93.6° |
| b. | $CH_4$ | $sp^3$ | 109°28' |
| c. | $NH_3$ | $sp^3$ | 107.8° |
| d. | $H_2O$ | $sp^3$ | 104.5° |

**57.** *(c)* (a) $CO_3^{2-}$, $NO_3^-$ Triangular planar

(b) $PCl_4^+$, $SiCl_4$ Tetrahedral

(c) $PF_5$ Trigonal bipyramidal

$BrF_5$ Square pyramidal

(d) $AlF_6^{3-}$, $SF_6$ Octahedral

**58.** *(c)* Tetrahedral structure is associated with $sp^3$ hybridised central atom without any lone pair.

• $BF_4^-$ 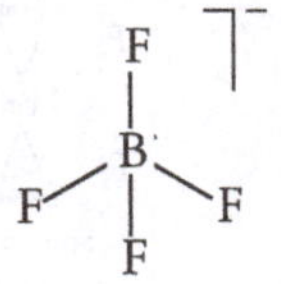 Tetrahedral

**59.** *(6)* Among the given species, total 6 species have $sp^2$ hybrid orbital for bonding with central atoms.

These are

$SO_2$, $C_2H_4$, $BCl_3$, HCHO, $C_6H_6$, $BF_3$

**60.** *(3)*

| Molecules/Ions | Shape |
|---|---|
| $PF_5$, $PCl_5$, $Fe(CO)_5$ | Trigonal bipyramidal |
| $BrF_5$ | Square pyramidal |
| $[PtCl_4]^{2-}$ | Square planar |
| $BF_3$ | Trigonal planar |

Hence, total 3 molecules have trigonal bipyramidal shape.

**61.** *(4)* The number of species that uses $sp^3$ hybrid orbitals in bonding are 4 i.e., $NH_3$, $SiO_2$, $H_2O$, $CH_4$.

While other species uses $sp^2$ and *sp*-hybrid orbitals which are as follows.

$sp^2$ — $SO_2$, $BF_3$

*sp* — $BeCl_2$, $CO_2$

**62.** *(3)*

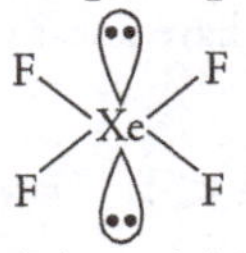

(2 lone pairs)

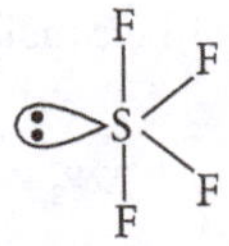

(1 lone pair)

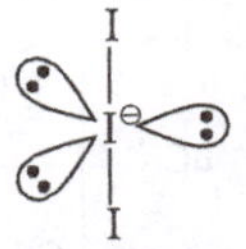

(3 lone pairs)

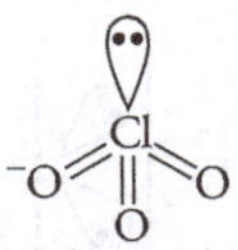

(1 lone pair)

Among the given species $I_3^-$ contains maximum number of unpaired electrons i.e., 3.

**63.** *(3)*

| Molecules | Shape | Lone pairs |
|---|---|---|
| $H_2O$ | 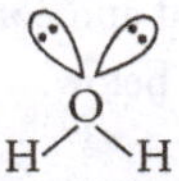 | 2 |
| $N_2$ |  | 2 |
| CO | 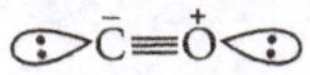 | 2 |
| $XeF_4$ | 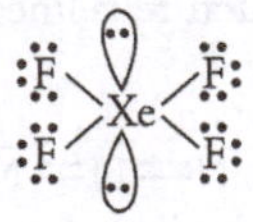 | 14 |
| $NH_3$ | 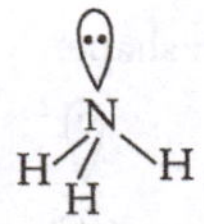 | 1 |
| NO | 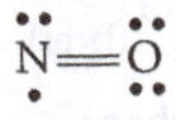 | 3 |
| $CO_2$ | 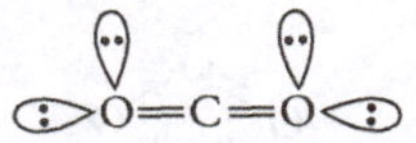 | 4 |
| $F_2$ | 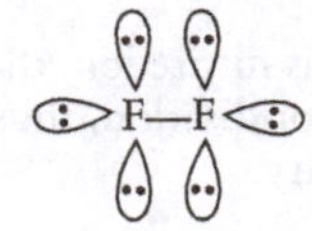 | 6 |

Thus, $H_2O$, $N_2$ and CO i.e., 3 molecules have only 2 lone pair of electrons.

**64.** *(4)*

$XeF_5^+$ 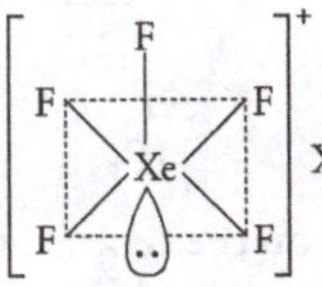 $XeF_5^-$ 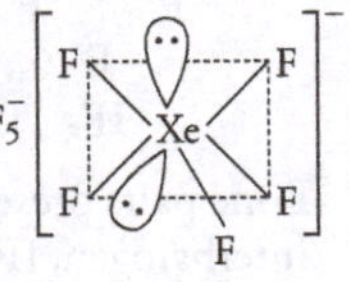

$XeO_3$ 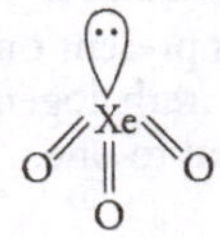 $XeO_3F_2$ 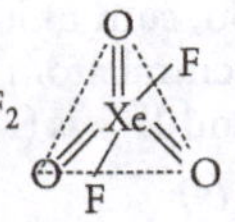

$XeO_2F_2$ 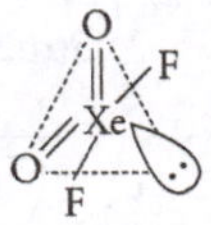 $XeOF_4$ 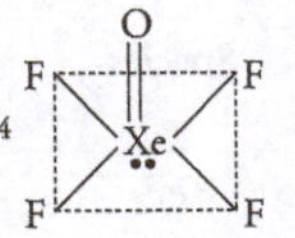

$XeF_4$ 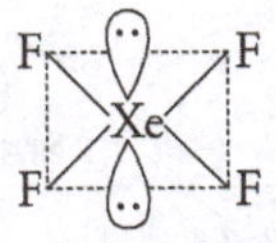

Thus, $XeF_5^+$, $XeO_3$, $XeO_2F_2$ and $XeOF_4$ have single lone pair on central atom.

**65.** *(3)* $N_3^-$ and $I_3^-$ are linear whereas $NO_2^-, O_3$ and $SO_2$ are bent

$I_3^-$ is linear because its central atom is $sp^3d$ hybridised and contains 3 lone pairs as shown below.

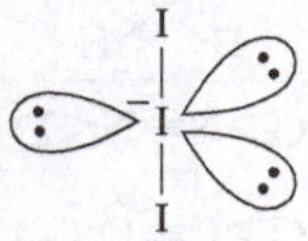

Also, $N_3^-$ structure is linear and has $sp$ hybridisation.

$\bar{N}=\overset{+}{N}=\bar{N}$

(Linear)

$SO_2$ is bent in shape.

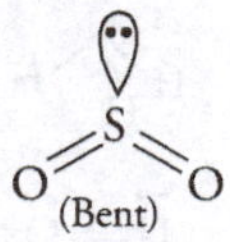

(Bent)

$O_3$ is bent in shape.

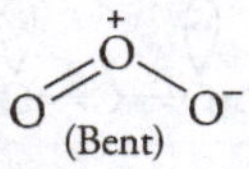

(Bent)

$NO_2^-$, that is nitrite ion, the N-atom have $sp^2$-hybridisation, thus it adopts bent geometry.

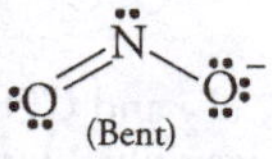

(Bent)

**66.** *(1)*

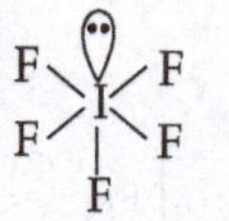

$IF_5$

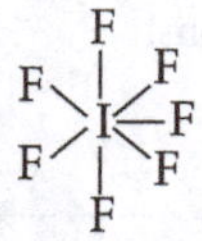

$IF_7$

Lone pairs present on the central atom of interhalogen, $IF_5$ is one and $IF_7$ is zero. So, sum of lone pairs present on the central atom of the interhalogen $IF_5$ and $IF_7$ is $(1+0)$ equal to one.

**67.** *(4)*

| Species | Hybridisation | Structure |
|---|---|---|
| $XeF_4$ | $sp^3d^2$ | 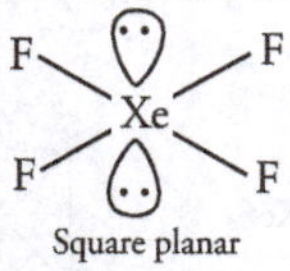 Square planar |
| $SF_4$ | $sp^3d$ | 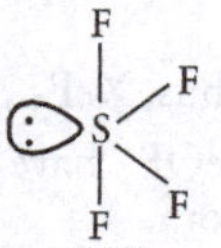 see-saw |
| $SiF_4$ | $sp^3$ | 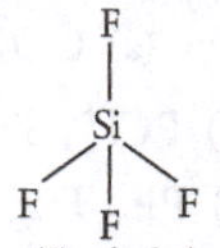 Tetrahedral |
| $BF_4^-$ | $sp^3$ | 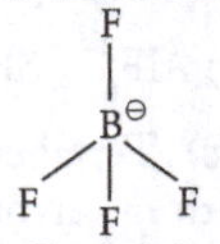 Tetrahedral |
| $BrF_4^-$ | $sp^3d^2$ | 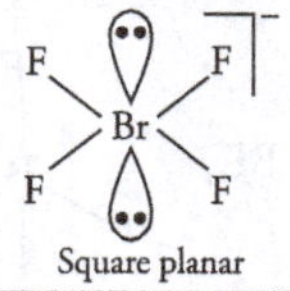 Square planar |
| $[Cu(NH_3)_4]^{2+}$ | $dsp^2$ | 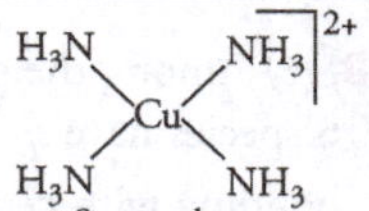 Square planar |
| $[FeCl_4]^{2-}$ | $sp^3$ | 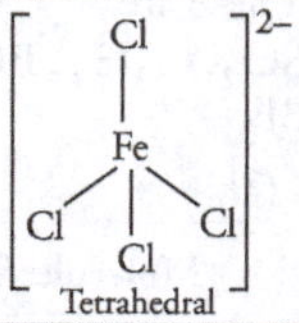 Tetrahedral |
| $[PtCl_4]^{2-}$ | $dsp^2$ | 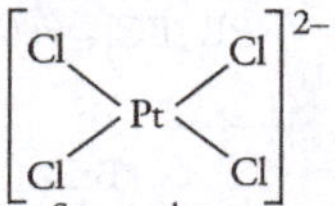 Square planar |

So, total 4 species have square planar shape.

**68.** *(3)*

| | | |
|---|---|---|
| $PF_5$ | $sp^3d$ | Triangular pyramidal |
| $BrF_4^-$ | $sp^3d^2$ | Square planar |
| $IF_5$ | $sp^3d^2$ | Square pyramidal |
| $BrF_5$ | $sp^3d^2$ | Square pyramidal |
| $XeOF_4$ | $sp^3d^2$ | Square pyramidal |
| $ICl_4^-$ | $sp^3d^2$ | Square planar |

In above molecules, 3 molecules ($IF_5, BrF_5, XeOF_4$) have square pyramidal structure.

**69.** *(5)* Five molecules have linear structure.

$XeF_2 =$ 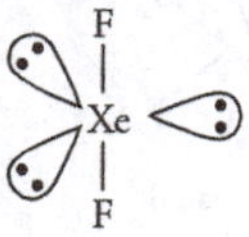 (Linear)

$I_3^- =$ 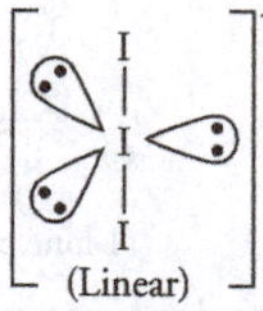 (Linear)

$BCl_2^{\ominus} =$ 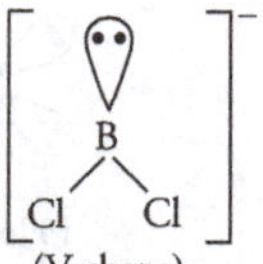 (V-shape)

$C_3O_2 = [O{=}C{=}C{=}C{=}O]$ Linear

$I_3^+ =$ 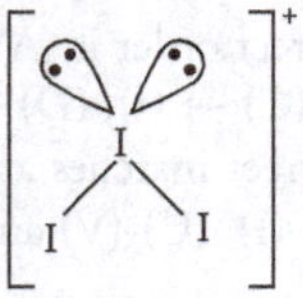 (V-shape)

$CO_2 = O{=}C{=}O$ (Linear)

$SO_2 =$ 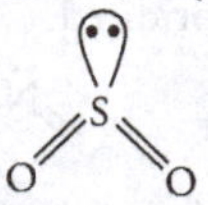 (V-shape)

$BeCl_2 = Cl{-}Be{-}Cl$ (Linear)

**70.** *(4)* Hybridisation $= \frac{1}{2}$ (Valence electron in central atom + Number of monovalent atom attached to central atom + Negative charge − Positive charge)

$BrF_5 : \frac{1}{2}[7+5-0+0] = 6(sp^3d^2)$

$SF_6 : \frac{1}{2}[6+6-0+0] = 6\ (sp^3d^2)$

$ICl_4^- : \frac{1}{2}[7+4-0+1] = 6(sp^3d^2)$

$IF_5 : \frac{1}{2}[7+5-0+0] = 6\ (sp^3d^2)$

So, number of molecules ions having $sp^3d^2$ hybridisation is 4.

**71.** *(6)* General planar structures are linear, bent, T-shaped, trigonal planar and square planar.

$NO_3^- \rightarrow$ Trigonal planar

$H_2O_2 \rightarrow$ Open book like (3D) (non-planar)

$BF_3 \rightarrow$ Trigonal planar

$PCl_3 \rightarrow$ Pyramidal

$XeF_4 \rightarrow$ Square planar

$SF_4 \rightarrow$ *See-saw*

$XeO_3 \rightarrow$ Trigonal pyramidal

$PH_4^+ \rightarrow$ Tetrahedral

$SO_3 \rightarrow$ Trigonal planar

$[Al(OH)_4]^- \rightarrow$ Tetrahedral

Hence, only 6 have non-planar geometry.

**72.** *(1)* $PF_5$ molecule

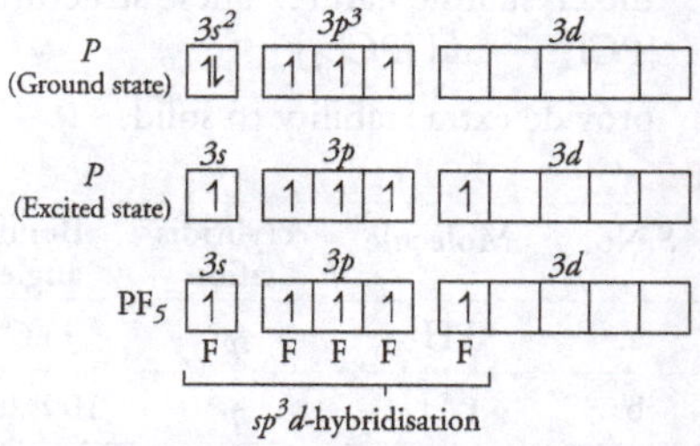


So, hybridisation of P in $PF_5$ is $sp^3d$ and the value of $y$ in $sp^xd^y$ is 1.

**73.** *(1)* In $H_2O$,

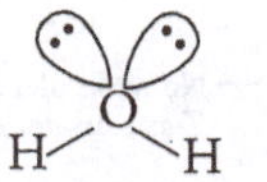

Total lone pairs = 2

**74.** *(3)* The structure of the given compounds are as follows :

(i)

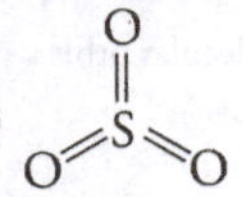

Trigonal planar

(ii)

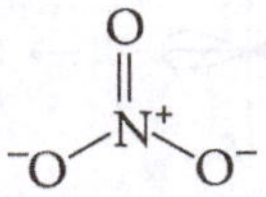

Trigonal planar

(i)

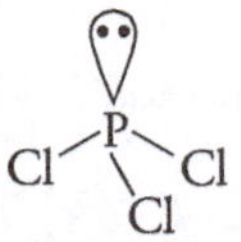

Pyramidal

(iv)

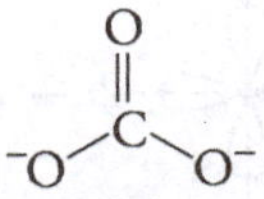

Trigonal planar

3 molecules/ions i.e. (i), (ii) and (iv) have non-pyramidal shape.

**75.** *(2)* $AB_3$ is a T-shaped molecule.

Hence, it can be $ClF_3$ with 2 lone pairs

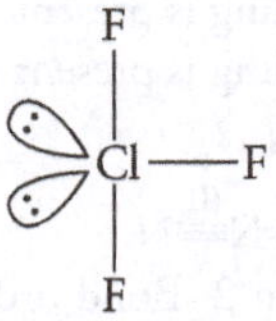

(T-shaped)

**76.** *(3)* Shape of $I_3^-$ is :

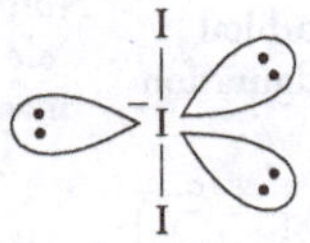

The number of lone pairs of electron on the central atom in $I_3^-$ is 3.

**77.** *(2)* Number of lone pairs on central atom in given compounds are as follows

$ClF_3$ =

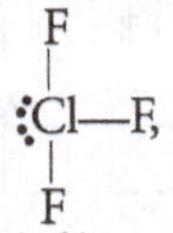

(2 *lp*)

$XeF_4$ =

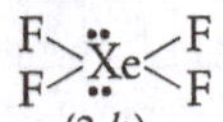

(2 *lp*)

Hence, two (lone pair) lp on central atom is $ClF_3$ and $XeF_4$.

**78.** *(20)*

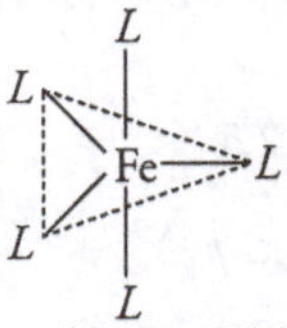

$1L\text{-}M\text{-}L = 180°$
$6L\text{-}M\text{-}L = 90°$
$3L\text{-}M\text{-}L = 120°$

$10L\text{-}M\text{-}L$ angles

Trigonal bipyramidal

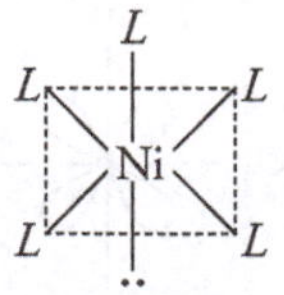

$8L\text{-}M\text{-}L = 90°$
$2L\text{-}M\text{-}L = 180°$

$10L\text{-}M\text{-}L$ angles

Square bipyramidal

$\therefore$ Sum of $L\text{-}M\text{-}L$ angles = 20

# Topic 3

**01.** *(a)* Antibonding molecular orbital (ABMO) are formed by destructive interference of wave function.

(ABMO) $\sigma^* = \psi_A - \psi_B$

**02.** *(c)* The correct match is

A $\rightarrow$ III, B $\rightarrow$ I, C $\rightarrow$ IV, D $\rightarrow$ II.

**03.** *(a)*

> **key idea** Magnetic nature can be detected by molecular orbital theory. Presence of unpaired electrons means paramagnetic and absence of unpaired electrons means diamagnetic in nature.

$O_2^-$ (Total electrons = 17)

$= \sigma 1s^2, \sigma^* 1s^2\ \sigma 2s^2\ \sigma^* 2s^2\ \sigma 2p_z^2,$

$\pi 2p_x^2 = \pi 2p_y^2, \pi^* 2p_x^2 = \pi^* 2p_y^1,$

$\sigma^* 2p_z^0$

Number of unpaired electron = 1

So, it is paramagnetic.

$$\text{Bond order} = \frac{N_b - N_a}{2} = \frac{10-7}{2} = 1.5$$

**04.** *(a)* A is false but R is true.

Corrected form of A is dipole-dipole interactions are purely covalent interactions of dipolar covalent bonds. Ends of dipoles of a polar covalent bonds possess partial charges ($\delta^+$ and $\delta^-$) which are less than unit electronic charge, $e = 1.6 \times 10^{-19}$ C.

**05.** *(c)* $Be_2 = \sigma 1s^2, \sigma^* 1s^2, \sigma 2s^2, \sigma^* 2s^2$

$$\text{Bond order} = \frac{4-4}{2} = 0$$

If bond order of chemical species is zero then that chemical species does not exist.

Therefore, $Be_2$ does not exist.

**06.** *(a)* $CN^-$ has 14 electrons (6 from C, 7 from N and 1 for negative charge).

Its MO configuration must be similar to

$N_2 : \sigma 1s^2\ \sigma^* 1s^2\ \sigma 2s^2\ \sigma^* 2s^2$

$\pi 2p_x^2\ \pi 2p_y^2\ \sigma 2p_z^2$

$$\therefore \quad \text{Bond order} = \frac{10-4}{2} = 3$$

As there is no unpaired electron in $CN^-$, thus it is diamagnetic.

**07.** *(c)* $CH_4$ doesn't show intermolecular hydrogen bonding while out of HCN and $NH_3$, extent of hydrogen bonding is more in $NH_3$ as ammonia could form 'three' hydrogen bond per molecule while HCN could form only one hydrogen bond per molecule.

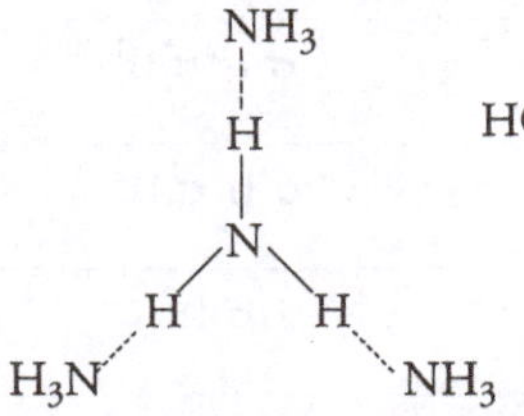

HCN------ HCN

$H_2O$ have even more stronger hydrogen bonding due to the presence of highly electronegative oxygen atom.

Hence, the correct order of increasing intermolecular H-bond strength is : $CH_4 < HCN < NH_3$.

**08.** *(a)* Among the given options, CO is a diamagnetic molecule. It can be proved by molecular orbital (MO) theory. The electronic configuration CO is given below.

CO (Number of electrons = 14)

Electronic configuration

$= \sigma 1s^2, \sigma^* 1s^2, \sigma 2s^2, \sigma^* 2s^2,$

$\sigma 2p_z^2, \pi 2p_x^2 \approx \pi 2p_y^2$

Since, there is no unpaired electron in the CO molecule, so it is diamagnetic.

NO contains 1, while $B_2$ and $O_2$ contains 2 unpaired electrons each hence, all 3 of them are paramagnetic.

**09.** *(d)* Considering molecular orbital theory (MOT) :

The electronic configuration of

$Li_2^+ (Z=5) = \sigma 1s^2, \sigma^* 1s^2, \sigma 2s^1$

Bond order (BO)

$$= \frac{N_b - N_a}{2} = \frac{3-2}{2} = \frac{1}{2}$$

The electronic configuration of

$Li_2^- (Z=7)$

$= \sigma 1s^2, \sigma^* 1s^2, \sigma 2s^2, \sigma^* 2s^1$

Bond order (BO)

$$= \frac{N_b - N_a}{2} = \frac{4-3}{2} = \frac{1}{2}$$

For the species having the same value of BO, the specie having lesser number of antibonding electrons $[N_a]$ will be more stable.

Here, $N_a$ of $Li_2^+$ (2) $< N_a$ of $Li_2^-$ (3).

So, their order of stability will be $Li_2^+ > Li_2^-$.

**10.** *(d)*

**key idea** According to M.O.T, the viability of any molecule can be judged through the calculation of bond order.

| | Electronic Configuration | Bond order |
|---|---|---|
| $He_2^+$ | $\sigma\,1s^2\sigma^*1s^1$ | $\frac{2-1}{2}=0.5$ |
| $H_2^-$ | $\sigma\,1s^2\sigma^*1s^1$ | $\frac{2-1}{2}=0.5$ |
| $H_2^{2-}$ | $\sigma\,1s^2\sigma^*1s^2$ | $\frac{2-2}{2}=0$ |
| $He_2^{2+}$ | $\sigma\,1s^2$ | $\frac{2-0}{2}=1$ |

The molecule having zero bond order will not be viable hence, $H_2^{2-}$ (option d) is the correct answer.

**11.** *(a)* As given in the diagram the opposite wave functions are overlapping each other hence it is a pi antibonding orbital.

**12.** *(d)* $B_2:\sigma 1s^2\sigma^*1s^2\sigma 2s^2\sigma^*2s^2\pi 2p_x^1=\pi 2p_y^1$

Two unpaired electrons are present.

Rest, given species does not contain unpaired electrons.

Hence, $B_2$ is paramagnetic.

**13.** *(c)* Assertion is correct, Reason is incorrect.

The corrected form of R is : Bonding molecular orbital involves constructive interference of the combining electron waves.

**14.** *(a)* The correct option is $O_2^{2-}$. This species has 18 electrons, which are filled in such a way that all molecular orbitals are fully filled, so diamagnetic.

$$\sigma 1s^2, \overset{*}{\sigma}1s^2, \sigma 2s^2, \overset{*}{\sigma}2s^2, \sigma 2p_z^2, \pi 2p_x^2 \approx \pi 2p_y^2, \overset{*}{\pi}2p_x^2 \approx \overset{*}{\pi}2p_y^2$$

**15.** *(b)* Fluorine being the most electronegative element forms strongest H–bonds. On the other hand, H—F possesses larger electronegative difference than O—H.

Therefore, H—F molecule is more polarised than O—H, due to which H-atom of H—F bears larger amount of partial charge than that of O—H. This suggests H–atom of H–F attracts more strongly F–atom than that of O—H.

Hence, H—F forms more stronger H–bonds with fluorine than that of O—H.

**16.** *(a)* Molecular orbital configuration of given molecules are $O_2^{2-}$ (Total number of electrons = 18)

$$\sigma 1s^2, \overset{*}{\sigma}1s^2, \sigma 2s^2, \overset{*}{\sigma}2s^2, \sigma 2p_z^2, \pi 2p_x^2 \approx \pi 2p_y^2, \overset{*}{\pi}2p_x^2 \approx \overset{*}{\pi}2p_y^2$$

Thus, $O_2^{2-}$ does not contain any unpaired electron.

**17.** *(c)*

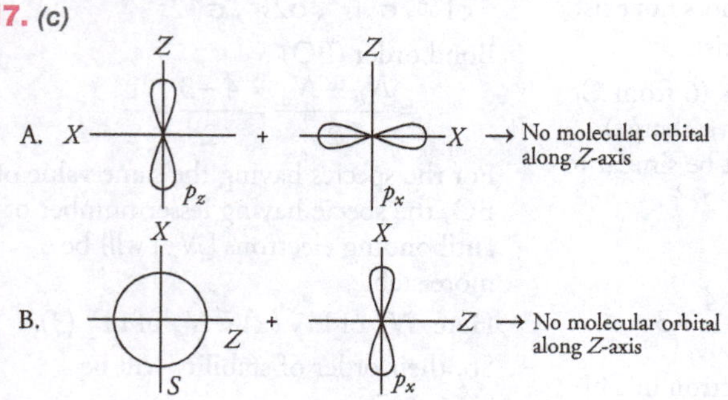


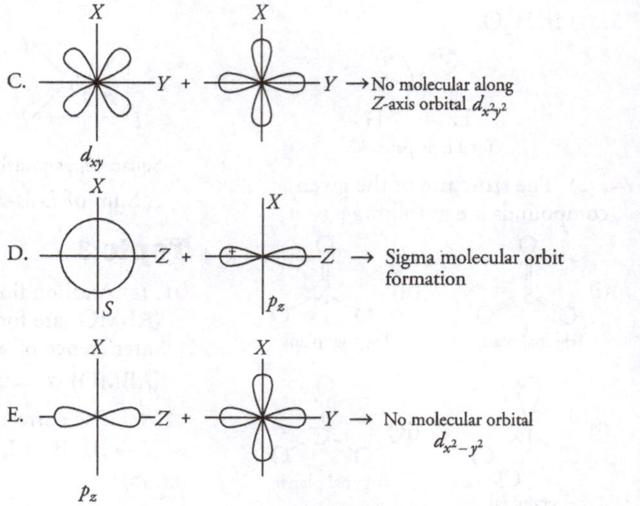


So, only $2s$ and $2p_z$ will lead to formation of molecular orbital in a homonuclear diatomic molecule.

**18.** *(a)* Statements given in A, D and E are correct while statements given in B and C are incorrect. Their correct form are as follows

- Intramolecular hydrogen bonding is present in *o*-nitrophenol.
- Intermolecular hydrogen bonding is present in HF.

**19.** *(c)* Hydrogen azide structure is

$$\mathrm{H{-}N\overset{(I)}{=}N\overset{(II)}{\equiv}N}$$

Bond order of bond I is less than 2. Bond order of bond II is greater than 2.

**20.** *(b)*

| Species and number of electrons | Molecular orbital electronic configuration | Number of unpaired electrons ($n$) and magnetic moment |
|---|---|---|
| 1. $O_2$-16 electrons | σ1s [↑↓] σ*1s [↑↓] σ2s [↑↓] σ*2s [↑↓] σ2$p_z$ [↑↓]<br>π2$p_x$= π2$p_y$ [↑↓][↑↓] π*2$p_x$= π*2$p_y$ [↑][↑] σ*2$p_z$ [ ] | $n=2$,<br>$\mu=\sqrt{n(n+2)}$<br>$\mu=\sqrt{2\times4}$ BM<br>$=2.82$ BM |
| 2. $O_2^+$-15 electrons | σ1s [↑↓] σ*1s [↑↓] σ2s [↑↓] σ*2s [↑↓] σ2$p_z$ [↑↓]<br>π2$p_x$= π2$p_y$ [↑↓][↑↓] π*2$p_x$= π*2$p_y$ [↑][ ] σ*2$p_z$ [ ] | $n=1$, $\mu=\sqrt{1\times3}$<br>$=1.73$ BM |
| 3. $O_2^-$-17 electrons | σ1s [↑↓] σ*1s [↑↓] σ2s [↑↓] σ*2s [↑↓] σ2$p_z$ [↑↓]<br>π2$p_x$= π2$p_y$ [↑↓][↑↓] π*2$p_x$= π*2$p_y$ [↑↓][↑] σ*2$p_z$ [ ] | $n=1$, $\mu=1.73$ BM |

**21.** *(a)* $C_2^-:(\sigma 1s)^2(\sigma^*1s)^2(\sigma 2s)^2(\sigma^*2s)^2(\pi 2p_x^2=\pi 2p_y^2)(\sigma 2p_z^1)$ or

$KK(\sigma 2s)^2(\sigma^*2s)^2(\pi 2p_x^2=\pi 2p_y^2)\sigma 2p_z^1$

$$\text{Bond order}=\frac{1}{2}(N_b-N_a)=\frac{1}{2}(9-4)=2.5$$

$$C_2:\mathrm{KK}(\sigma 2s)^2(\sigma^*2s)^2(\pi 2p_x^2=\pi 2p_y^2)$$

$$\text{Bond order}=\frac{1}{2}(N_b-N_a)=\frac{1}{2}(8-4)=2$$

The BO value of other molecules is

| $O_2$ | NO | $F_2$ |
|---|---|---|
| 2 | 2.5 | 1 |

These values will decrease with addition of electron as incoming electron will occupy AMO. As increase in BO value indicates increase in stability. Hence, $C_2$ will be stabilised by anion formation.

**22.** *(b)* The change of $O_2$ to $O_2^-$ can be as follows:

$O_2$ (16 $e^-$) $\xrightarrow{+e^-}$ $O_2^-$ (17 $e^-$)

[Dioxygen] [Super-oxide]

$\rightarrow$ Energy $\rightarrow$ Energy

$[8e^-]\ \sigma 2p_z^2$ $\quad$ $[8e^-]\ \sigma 2p_z^2$

$\pi 2p_x = \pi 2p_y^2\ \overset{*}{\pi} 2p_x^1 = \overset{*}{\pi} 2p_y^1$ $\quad$ $\pi 2p_x^2 = \pi 2p_y^2\ \overset{*}{\pi} 2p_x^2 = \overset{*}{\pi} 2p_y^1$

Half-filled anti-bonding $\pi^*$-$MO_s$

So, in the formation of $O_2^-$ from $O_2$, the 17th electron goes to the $\overset{*}{\pi} 2p_x$ or $\overset{*}{\pi} 2p_y$ molecular orbital (anti-bonding) which is half-filled in $O_2$.

**23.** *(d)* The bond order of acetylide ion $C_2^{2-}$ ( $^-C{\equiv}C^-$ ) is 3 and this ion is diamagnetic in nature.

- Bond order of $O_2^- = \frac{10-7}{2} = 1.5$ and paramagnetic
- Bond order of $O_2^+ = \frac{10-5}{2} = 2.5$ and paramagnetic
- Bond order of $N_2^+ = \frac{9-4}{2} = 2.5$ and paramagnetic
- Bond order of $NO^+ = \frac{10-4}{2} = 3$ and diamagnetic

Thus, bond order and magnetic property of acetylide ion are same as that of $NO^+$.

**24.** *(b)* Electronic configuration of $N_2^+$ is

$$\sigma 1s^2 \sigma^* 1s^2 \sigma 2s^2 \sigma^* 2s^2 \pi 2p_x^2 = \pi 2p_y^2 \sigma 2p_z^1$$

Hence, the number of electrons in $\sigma 2p_z$ MO $= 1$

**25.** *(a,b)* $C_2 (6+6=12) = \sigma 1s^2, \overset{*}{\sigma} 1s^2, \sigma 2s^2,$

$$\sigma^* 2s^2, \pi 2p_x^2 \approx \pi 2p_y^2$$

Since, all the electrons are paired, it is a diamagnetic species.

$N_2 (7+7=14) = \sigma 1s^2, \overset{*}{\sigma} 1s^2, \sigma 2s^2, \overset{*}{\sigma} 2s^2, \pi 2p_x^2$

$$\approx \pi 2p_y^2, \sigma 2p_z^2$$

It is also a diamagnetic species because of the absence of unpaired electrons.

Since, $O_2$ or $S_2$ contain 2 unpaired electrons, therefore they are paramagnetic.

**26.** *(c)* Species having zero or negative bond order do not exist.

$H_2^{2+} (1+1-2=0) = \sigma 1s^0$

Bond order $= 0$

$He_2\ (2+2=4) = \sigma 1s^2, \overset{*}{\sigma} 1s^2$

Bond order $= \frac{N_b - N_a}{2} = \frac{2-2}{2} = 0$

So, both $H_2^{2+}$ and $He_2$ do not exist.

**27.** *(b)* $Li_2\ (3+3=6) = \sigma 1s^2, \overset{*}{\sigma} 1s^2, \sigma 2s^2$

Bond order $= \frac{N_b - N_a}{2} = \frac{4-2}{2} = 1$

$Li_2^+\ (3+3-1=5) = \sigma 1s^2, \overset{*}{\sigma} 1s^2, \sigma 2s^1$

Bond order $= \frac{3-2}{2} = \frac{1}{2} = 0.5$

$Li_2^-\ (3+3+1=7) = \sigma 1s^2, \overset{*}{\sigma} 1s^2, \sigma 2s^2, \overset{*}{\sigma} 2s^1$

Bond order $= \frac{4-3}{2} = \frac{1}{2} = 0.5$

Stability order is $Li_2^-$ (More $N_a$) $< Li_2^+$ (Less $N_a$) $< Li_2$

**28.** *(b)* NO $\longrightarrow NO^+ + e^-$

($NO^+$) Total electrons $= 14$

$$\sigma 1s^2, \sigma^* 1s^2, \sigma 2s^2, \sigma^* 2s^2, \pi 2p_y^2 \approx \pi 2p_x^2, \sigma 2p_z^2$$

Diamagnetic, bond order $= \frac{10-4}{2} = 3$

(NO) Total electrons $= 15$

$$\sigma 1s^2, \overset{*}{\sigma} 1s^2, \sigma 2s^2, \overset{*}{\sigma} 2s^2, \sigma 2p_z^2, \pi 2p_x^2 \approx \pi 2p_y^2, \overset{*}{\pi} 2p_x^1 \approx \overset{*}{\pi} 2p_z$$

Paramagnetic, bond order $= \frac{10-5}{2} = 2.5$

Electron is taken away from non-bonding molecular orbital, that's why bond order increases.

**29.** *(c)* Paramagnetic behaviour are shown by the species containing unpaired electrons.

Number of unpaired electrons of each species are

| | | |
|---|---|---|
| (A) | $O_2$ | 2 |
| (B) | $N_2$ | 0 |
| (C) | $F_2$ | 0 |
| (D) | $S_2$ | 2 |
| (E) | $Cl_2$ | 0 |

So, A and D are paramagnetic.

**30.** *(d)* $N_2 = \sigma 1s^2\ \sigma^* 1s^2\ \sigma 2s^2\ \sigma^* 2s^2 \pi 2p_x^2 = \pi 2p_y^2, \sigma 2p_z^2$

Number of unpaired electron present in HOMO $= 0$

$N_2^+ = \sigma 1s^2, \sigma^* 1s^2, \sigma 2s^2, \sigma^* 2s^2,$

$\pi 2p_x^2 = \pi 2p_y^2, \sigma 2p_z^1$

Number of unpaired electron present in HOMO $= 1$

$O_2 = \sigma 1s^2, \sigma^* 1s^2, \sigma 2s^2, \sigma^* 2s^2,$

$\sigma 2p_z^2, \pi 2p_x^2 \approx \pi 2p_y^2, \pi^* 2p_x^1 \approx \pi^* 2p_y^1$

Number of unpaired electron present in HOMO $= 2$

$O_2^+ = \sigma 1s^2, \sigma^* 1s^2, \sigma 2s^2, \sigma^* 2s^2,$

$\sigma 2p_z^2, \pi 2p_x^2 = \pi 2p_y^2, \pi^* 2p_x^1 = \pi^* 2p_y^0$

Number of unpaired electron present in HOMO $= 1$

Thus, the number of unpaired electrons are 0, 1, 2, 1.

**31.** *(c)* According to MO theory, the bond orders for $O_2^{2-}$, CO and $NO^+$ are 1, 3 and 3.

**32.** *(c)* NO : $\sigma 1s^2 \sigma^* 1s^2 \sigma 2s^2 \sigma^* 2s^2 \sigma 2p_z^2 \pi 2p_x^2 = \pi 2p_y^2 \pi^* 2p_x^1$

Bond order $= \frac{10-5}{2} = 2.5$

$NO^+$ : $\sigma 1s^2 \sigma^* 1s^2 \sigma 2s^2 \sigma^* 2s^2 \sigma 2p_z^2 \pi 2p_x^2 = \pi 2p_y^2$

Bond order $= \frac{10-4}{2} = 3$

$O_2$: $\sigma 1s^2 \sigma^* 1s^2 \sigma 2s^2 \sigma^* 2s^2 \sigma 2p_z^2 \pi 2p_x^2 = \pi 2p_y^2 \pi^* 2p_x^1 = \pi^* 2p_y^1$

Bond order $= \frac{10-6}{2} = 2$

$O_2^+$: $\sigma 1s^2 \sigma^* 1s^2 \sigma 2s^2 \sigma^* 2s^2 \sigma 2p_z^2 \pi 2p_x^2 = \pi 2p_y^2 \pi^* 2p_x^1$

Bond order $= \frac{10-5}{2} = 2.5$

Increase in bond order indicates increased strength of bond. Hence, correct answer is option (c).

**33.** *(a)*

| Species | $O_2^+$ | $O_2$ | $O_2^-$ | $O_2^{2-}$ |
|---|---|---|---|---|
| No. of $e^-$ | 15 | 16 | 17 | 18 |
| Bond order | 2.5 | 2 | 1.5 | 1 |

Increasing bond order is $O_2^{2-} < O_2^- < O_2 < O_2^+$.

**34.** *(a)* Extensive H-bonding is observed in 2-nitrophenol as two H-atoms bonded to strongly electronegative atoms (nitrogen, oxygen) are present. H-bonding occurs when H-bonded to strongly electronegative atoms exists in the vicinity of other electronegative atoms possessing lone pairs of electron.

**35.** *(a)* A-(iii), B-(iv), C-(i), D-(ii)

**36.** *(d)* Energy order of MOs B.O. $= \frac{N_b - N_a}{2}$

Bond length $\propto \frac{1}{\text{BO (Bond order)}} \propto \frac{1}{\text{Bond strength}}$

So, bond $\propto$ BO strength the BO of given 4 species are :

| Species | NO | $NO^+$ | $NO^{2+}$ | $NO^-$ |
|---|---|---|---|---|
| Bond order | 2.5 | 3 | 2.5 | 2 |

So,the order of bond strength : $NO^+ > NO = NO^{2+} > NO^-$

So, $NO^-$ has minimum bond strength.

**37.** *(a)*

| Species | MO energy order | Bond order (BO) | *n*, no. of unpaired $e^-$ | Magnetic character |
|---|---|---|---|---|
| $C_2^{2-}(14e^-)$ | $[8\bar{e}]\pi 2p_x^2 = \pi 2p_y^2 \sigma 2p_z^2$ | $\frac{6-0}{2} = 3$ | 0 | Diamagnetic |
| $O_2(16e^-)$ | $[8\bar{e}]\sigma 2p_z^2 \pi 2p_x^2 = \pi 2p_y^2 \pi^* 2p_x^1 = \pi^* 2p_y^1$ | $\frac{6-2}{2} = 2$ | 2 | Paramagnetic |
| $O_2^{2-}(18e^-)$ | $[8\bar{e}]\sigma 2p_z^2 \pi 2p_x^2 = \pi 2p_y^2 \pi 2p_x^2 = \pi 2p_y^2$ | $\frac{6-4}{2} = 1$ | 0 | Diamagnetic |
| $N_2^{2-}(16e^-)$ | $[8\bar{e}]\pi 2p_x^2 = \pi 2p_y^2 \sigma 2p_z^2 \overset{*}{\pi} 2p_x^1 = \overset{*}{\pi} 2p_y^1$ | $\frac{6-2}{2} = 2$ | 2 | Paramagnetic |

Bond length $\propto \frac{1}{\text{BO (Bond order)}}$

So order of bond length $\underset{(BO=3)}{C_2^{2-}} < \underset{(BO=2)}{O_2 = N_2^{2-}} < \underset{(BO=1)}{O_2^{2-}}$

The diamagnetic species with shortest bond length is $C_2^{2-}$ (option-a).

**38.** *(c)* $N_2^+ (13e^-\text{'s}) = \sigma 1s^2 \sigma^* 1s^2 \ \sigma 2s^2 \ \sigma^* 2s^2$

$\pi 2p_x^2 = \pi 2p_y^2 \sigma 2p_z^1$

Thus, in case of $N_2^+$, two π-bonds and half σ-bond are present in the bonding MOs.

**39.** *(d)* $NO \longrightarrow NO^+$

Bond order of NO = 2.5 (Paramagnetic)

Bond order of $NO^+$ = 3 (Diamagnetic)

**40.** *(a)* Higher the bond order, shorter is the bond length $O_2^{2+}$ has the shortest bond length (BO = 3)

Bond order of remaining species are

$O_2^+(2.5), O_2^-(1.5)$ and $O_2^{2-}(1)$

**41.** *(a)* $CN^-$ and $NO^+$ both have same number of electrons and hence same bond order (3).

**42.** *(6)* The electronic configuration of $O_2(16e^-)$ is

$\sigma 1s^2 \ \sigma^* 1s^2 \ \sigma 2s^2 \ \sigma^* 2s^2 \ \sigma 2p_z^2$

$\pi 2p_x^2 = \pi 2p_y^2, \pi^* 2p_x^1 = \pi^* 2p_y^1$

So, number of $e^-$ in $\pi^*$ of $O_2 = 2$

number of $e^-$ in $\pi^*$ of $O_2^+ = 1$

number of $e^-$ in $\pi^*$ of $O_2^- = 3$

So, total number of $e^-$ in $\pi^* = 2 + 1 + 3 = 6$

**43.** *(4)*

| Species | Unpaired electrons |
|---|---|
| $N_2$ | 0 |
| $O_2$ | 2 |
| $C_2^-$ | 1 |
| $O_2^-$ | 1 |
| $O_2^{2-}$ | 0 |
| $H_2^+$ | 1 |
| $CN^-$ | 0 |
| $He_2^+$ | 1 |

So, one unpaired electron is present in $C_2^-, O_2^-, H_2^+$ and $He_2^+$.

**44.** *(2)* Formula for Bond order (B.O) $= \frac{N_B - N_A}{2}$

Where, $N_B$ = Number of electron in bonding orbital.

$N_A$ = Number of electron in antibonding orbital.

For $C_2$; B.O. $= \frac{8-4}{2} = 2$

For $O_2$; B.O. $= \frac{10-6}{2} = 2$

For $Be_2$; B.O. $= \frac{4-4}{2} = 0$

For $Li_2$; B.O. $= \frac{4-2}{2} = 1$

For $Ne_2$; B.O. $= \frac{10-10}{2} = 0$

For $N_2$; B.O. $= \frac{10-4}{2} = 6$

$He_2$; B.O. $= \frac{2-2}{2} = 0$

Hence, total 2 molecules have bond order of 2.

**45.** *(5)*

**key idea** This problem is based on the hydrogen bondig. It occurs when H atom is bonded with oxygen, nitrogen and fluorine.

Among the given compounds, following compounds exhibit hydrogen bonding,

$CH_3OH$, $H_2O$, 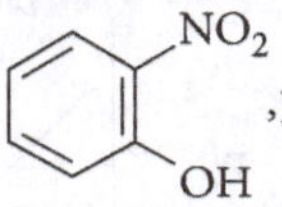

, HF, $NH_3$

Hence, total 5 compounds are there.

**46.** *(2)* According to molecular orbital theory, number of unpaired electrons in the given molecules/species are as follows

$O_2$ = Number of unpaired electrons is 2

$O_2^{-1}$ = Number of unpaired electron is 1

NO = Number of unpaired electron is 1

$CN^{-1}$ = Number of unpaired electron is zero

$O_2^{2-}$ = Number of unpaired electron is zero

**47.** *(8)* Molecular orbital formed by $2s$ atomic orbital = 2 ($\sigma 2s$ and $\sigma^* 2s$)

Molecular orbital formed by $2p$ atomic orbital

$= 6$ ($\sigma 2p_z$ and $\sigma^* 2p_z$, $\pi 2p_x$ $\pi 2p_y$ and $\pi^* 2p_x$, $\pi^* 2p_y$)

So, total molecular orbitals $= 2 + 6 = 8$

**48.** *(4)* Antibonding molecular orbital from $2s = 1$

Antibonding molecular orbital from $2p = 3$

Therefore, total number of antibonding molecular orbitals are $= 1 + 3 = 4$

**49.** *(1)*

| Ion/molecule | Magnetic behaviour | Bond order |
|---|---|---|
| $H_2$ | Diamagnetic | 1 |
| $He_2^+$ | Paramagnetic | 0.5 |
| $N_2^{2-}$ | Paramagnetic | 2 |
| $O_2^{2-}$ | Diamagnetic | 1 |
| $O_2^+$ | Paramagnetic | 2.5 |
| $F_2$ | Diamagnetic | 1 |
| $Ne_2^+$ | Paramagnetic | 0.5 |
| $B_2$ | Paramagnetic | 1 |

Only $B_2$ has paramagnetic magnetic behaviour and bond order of 1.

**50.** *(6)* $NO^+$ and CO both are isoelectronic species and each of them having bond order 3.

$\therefore$ Sum of bond order of $NO^+$ and CO = Bond order ($NO^+$) + Bond order (CO) $= 3 + 3 = 6$

**51.** *(2)* $N_2 : (\sigma 1s)^2 (\sigma^* 1s)^2 (\sigma 2s^2)(\sigma^* 2s^2)$
$(\pi 2p_x^2 = \pi 2p_y^2)(\sigma 2p_z)^2$;

Diamagnetic

$O_2^{2-}$:

$(\sigma 1s)^2 (\sigma^* 1s)^2 (\sigma 2s)^2 (\sigma^* 2s)^2 (\sigma 2p_z)^2$
$(\pi 2p_x^2 = \pi 2p_y^2)(\pi^* 2p_x^2 = \pi^* 2p_y^2)$;

Diamagnetic

The number of species showing diamagnetism are 2.

**52.** *(4)* Molecular orbital configuration of

$B_2 \Rightarrow$
$\sigma 1s^2 \sigma^* 1s^2 \sigma 2s^2 \sigma^* 2s^2 \pi 2p_x^1 = \pi 2p_y^1$
(Paramagnetic)

$C_2^- \Rightarrow$
$\sigma 1s^2 \sigma^* 1s^2 \sigma 2s^2 \sigma^* 2s^2 \pi 2p_x^2 = \pi 2p_y^2 \sigma 2p_z^1$
(Paramagnetic)

$O_2^+ \Rightarrow \sigma 1s^2 \sigma^* 1s^2 \sigma 2s^2 \sigma^* 2s^2 \sigma 2p_z^2 \pi 2p_x^2$
$= \pi 2p_y^2 \pi^* 2p_x^2 = \pi^* 2p_y^1$
(Paramagnetic)

$He_2^+ \Rightarrow \sigma 1s^2 \sigma^* 1s^1$ (Paramagnetic)

The number of paramagnetic species are 4.

**53.** *(3)* Bond order $= \frac{N_B - N_A}{2}$

where, $N_B$ = total $e^-$ in bonding molecular orbitals

$N_A$ = total $e^-$ in antibonding molecular orbitals.

| Species/Ion | No. of $e^-$ | Bond order |
|---|---|---|
| $CN^-$ | 14 | $(10-4/2) = 3$ |
| $NO^+$ | 14 | $(10-4/2) = 3$ |
| $O_2$ | 16 | $(10-6/2) = 2$ |
| $O_2^+$ | 15 | $(10-5/2) = 2.5$ |
| $O_2^{2+}$ | 14 | $(10-4/2) = 3$ |

$\therefore$ 3 species have identical bond order.

**54.** *(173)* Electronic configuration of $B_2^+$ is as follows

$B_2^+ = \sigma 1s^2, \sigma^* 1s^2, \sigma 2s^2, \sigma^* 2s^2, \pi 2p_x^1$

$\therefore$ $B_2^+$ has one unpaired electron,

$\mu = \sqrt{1(1+2)} = \sqrt{3} = 1.73$ BM

$= 173 \times 10^{-2}$ BM

$\therefore \quad x \times 10^{-2} = 173 \times 10^{-2}$

$\therefore \quad x = 173$

Hence, answer is 173.

**55.** *(0)* In $O_2^{2-}$, 18 electrons are present.

Configuration is given by

$\sigma 1s^2, \sigma^* 1s^2, \sigma 2s^2, \sigma^* 2s^2, \sigma 2p_z^2,$
$\pi 2p_x^2 = \pi 2p_y^2, \pi^* 2p_x^2 = \pi^* 2p_y^2$

So, number of unpaired electrons in $O_2^{2-} = 0$.

**56.** *(10)* Configuration of $O_2^{2-}$ is

$(O_2^{2-}) = \sigma 1s^2, \sigma^* 1s^2, \sigma 2s^2, \sigma^* 2s^2,$
$\sigma 2p_z^2, \pi 2p_x^2 = \pi 2p_y^2 \pi^* 2p_x^2 = \pi^* 2p_y^2$

Number of electrons in all bonding molecular orbitals = 10

Number of electrons in all anti-bonding molecular orbitals = 8

**57.** *(0)* Electronic configuration of CO (14 electrons).

$(\sigma 1s)^2 (\sigma^* 1s)^2 (\sigma 2s)^2 (\sigma^* 2s)^2$
$(\pi 2p_x)^2 = (\pi 2p_y)^2, \sigma 2p_z^2$

Bond order

$= \frac{[\text{No. of bonding electron} - \text{No. of antibonding electron}]}{2}$

$= \frac{10-4}{2} = 3$

Electronic configuration of $NO^+$ (14 electrons)

$(\sigma 1s)^2 (\sigma^* 1s)^2 (\sigma 2s)^2 (\sigma^* 2s)^2$
$(\pi 2p_x)^2 = (\pi 2p_y)^2 (\sigma 2p_z)^2$

$BO = \frac{10-4}{2} = 3$

Bond order of both CO and $NO^{\oplus} = 3$

Difference $= 0 = \frac{x}{2}$; $x = 0$

**58.** *(15)* NO is the covalent diatomic molecule which has bond order of 2.5.

Hence, total electrons in NO $= 7 + 8 = 15$

# C-05

JEE Main Chapterwise Topicwise Chemistry

In JEE Main there is usually 0-2 question per paper are visible from this chapter. In past 3 years, 29 conceptual and 46 numerical questions were asked of moderate difficulty level from this chapter.
**For detailed analysis scan QR code**

# THERMODYNAMICS

## Topic 1 Fundamentals of Thermodynamics

### Level 1

**01** Among the following, the set of parameters that represents path function, is *[2019, 09 April Shift-I]*

(A) $q+W$ (B) $q$
(C) $w$ (D) $H-TS$

(a) (B) and (C) (b) (A) and (D)
(c) (B), (C) and (D) (d) (A), (B) and (C)

**02** An ideal gas expands in volume from $1\times 10^{-3}$ $m^3$ to $1\times 10^{-2}$ $m^3$ at 300 K against a constant pressure of $1\times 10^5$ $Nm^{-2}$. The work done is *[AIEEE 2004]*

(a) $-900$ J (b) $-900$ kJ (c) 270 kJ (d) 900 kJ

### Level 2

**03** Ice and water are placed in a closed container at a pressure of 1 atm and temperature 273.15 K. If pressure of the system is increased 2 times, keeping temperature constant, then identify correct observation from following. *[2025, 28 Jan Shift-I]*

(a) The amount of ice decreases.
(b) The solid phase (ice) disappears completely.
(c) Volume of system increases.
(d) Liquid phase disappears completely.

**04** What happens when methane undergoes combustion in systems *A* and *B* respectively? *[2023, 13 April Shift-II]*

| | System A (Adiabatic system) | System B (Diathermic container) |
|---|---|---|
| (a) | Temperature remains same | Temperature rises |
| (b) | Temperature falls | Temperature remains same |
| (c) | Temperature falls | Temperature rises |
| (d) | Temperature rises | Temperature remains same |

**05** The combination of plots which does not represent isothermal expansion of an ideal gas is *[2019, 12 Jan Shift-II]*

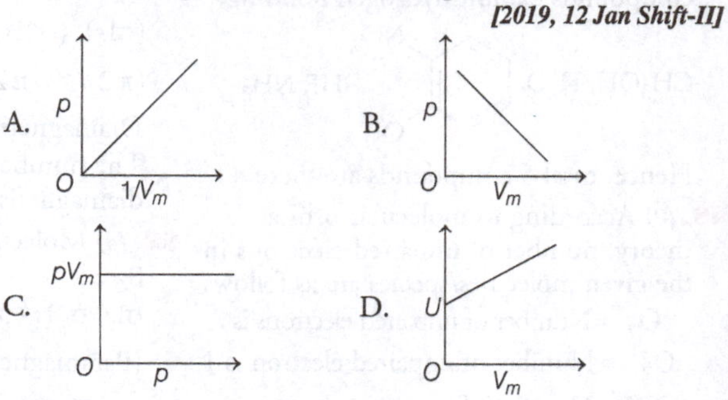


Codes
(a) A and C (b) A and D
(c) B and D (d) B and C

**06** An ideal gas is allowed to expand both reversibly and irreversibly in an isolated system. If $T_i$ is the initial temperature and $T_f$ is the final temperature, then which of the following statements is correct? *[AIEEE 2006]*

(a) $(T_f)_{irrev} > (T_f)_{rev}$
(b) $T_f > T_i$ for reversible process but $T_f = T_i$ for irreversible process
(c) $(T_f)_{rev} = (T_f)_{irrev}$
(d) $T_f = T_i$ for both reversible and irreversible processes

### Level 3

**07.** Which of the following properties will change when system containing solution 1 will become solution 2?

| (Solution 1) | | (Solution 2) |
|---|---|---|
| 10 mol of solute $x$ + 10 L of water | ⟶ | 1 L of solution 1 + 1 mol of solute $x$ + 1 L of water |

*[2025, 3 April Shift-I]*

(a) Density (b) Concentration
(c) Gibbs free energy (d) Molar heat capacity

**08** $C(s)+\frac{1}{2}O_2(g) \longrightarrow CO_2(g)+400\text{ kJ}$

$C(s)+\frac{1}{2}O_2(s) \longrightarrow CO(g)+100\text{ kJ}$

When coal of purity 60% is allowed to burn in presence of insufficient oxygen, 60% of carbon is converted into 'CO' and the remaining is converted into '$CO_2$'. The heat generated when 0.6 kg of coal is burnt is ............ . *[2022, 29 july Shift-II]*

(a) 1600 kJ (b) 3200 kJ (c) 4400 kJ (d) 6600 kJ

**09** Consider the reversible isothermal expansion of an ideal gas in a closed system at two different temp. $T_1$ and $T_2$ $(T_1 < T_2)$. The correct graphical depiction of the dependence of work done ($W$) on the final volume ($V$) is *[2019, 9 Jan Shift-I]*

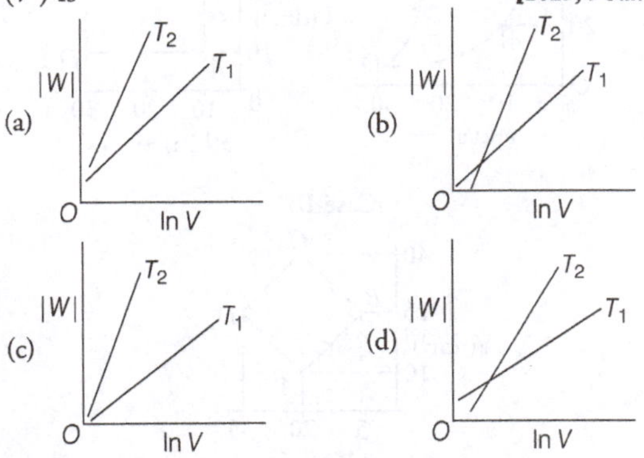


## Numerical Value Type Questions

**Imp.** Work done in Reversible Process

**10**

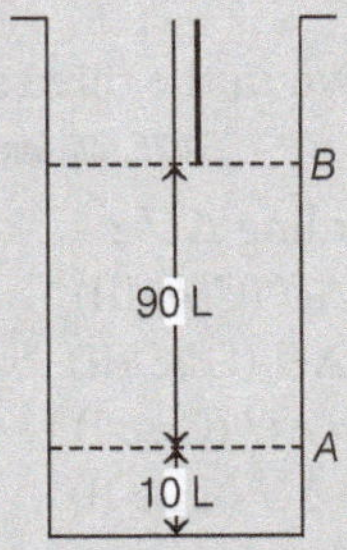


Consider the figure provided.

1 mol of an ideal gas is kept in a cylinder, fitted with a piston, at the position $A$, at 18°C. If the piston is moved to position $B$, keeping the temperature unchanged, then '$x$' L atm work is done in this reversible process.

$x$ = ...... L atm (nearest integer)

[Given : Absolute temperature = °C + 273.15, $R = 0.08206$ L atm mol$^{-1}$ K$^{-1}$]. *[2024, 8 April Shift-I]*

**11** If 5 moles of an ideal gas expands from 10 L to a volume of 100 L at 300 K under isothermal and reversible condition then work, $w$, is $-x$ J. The value of $x$ is ........ .

(Given, $R = 8.314$ JK$^{-1}$ mol$^{-1}$) *[2024, 31 Jan Shift-II]*

**12**

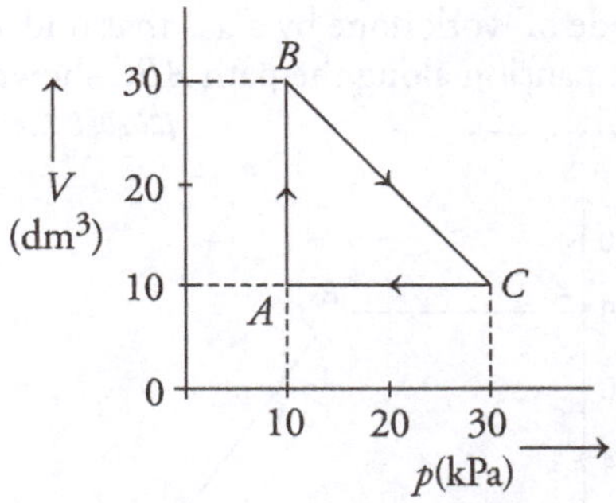


An ideal gas undergoes a cyclic transformation starting from the point $A$ and coming back to the same point by tracing the path $A \rightarrow B \rightarrow C \rightarrow A$ as shown in the diagram above. The total work done in the process is ........ J. *[2024, 30 Jan Shift-I]*

**13** If three moles of an ideal gas at 300 K expand isothermally from 30 dm$^3$ to 45 dm$^3$ against a constant opposing pressure of 80 kPa, then the amount of heat transferred is .......... J. *[2024, 27 Jan Shift-I]*

**14** The total number of intensive properties from the following is ..............

Volume, molar heat capacity, molarity, $E^\circ_{cell}$, Gibbs free energy change, molar mass, mole. *[2023, 11 April Shift-II]*

**15** One mole of an ideal monoatomic gas is subjected to changes as shown in the graph. The magnitude of the work done (by the system or on the system) is ....... J. (Nearest integer)

Given : $\log 2 = 0.3$

$\ln 10 = 2.3$ *[2023, 24 Jan Shift-II]*

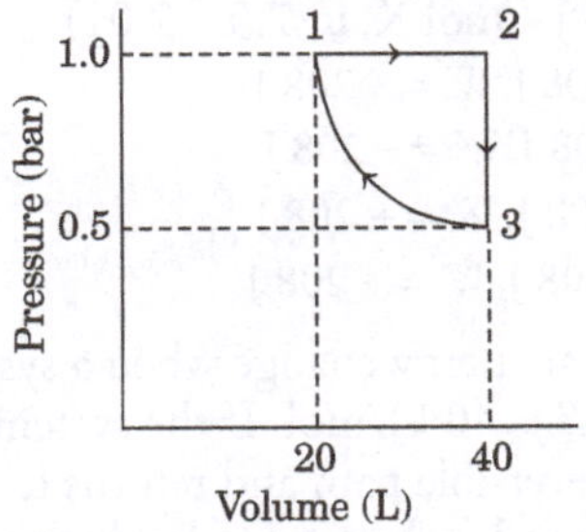


**16** When 5 moles of He gas expand isothermally and reversibly at 300 K from 10 litre to 20 litre, the magnitude of the maximum work obtained is ........... J. [Nearest integer] (Given, $R = 8.3$ J K$^{-1}$ mol$^{-1}$ and $\log 2 = 0.3010$) *[2022, 27 June Shift-II]*

**17** At 25°C, 50 g of iron reacts with HCl to form $FeCl_2$. The evolved hydrogen gas expands against a constant pressure of 1 bar. The work done by the gas during this expansion is ...... J. (Round off to the nearest integer) [Given, $R = 8.314$ J mol$^{-1}$ K$^{-1}$. Assume, hydrogen is an ideal gas]

[Atomic mass of Fe is 55.85 u] *[2021, 16 March Shift-II]*

**18** The magnitude of work done by a gas that undergoes a reversible expansion along the path $ABC$ shown in the figure is ......... . ***[2020, 8 Jan Shift-I]***

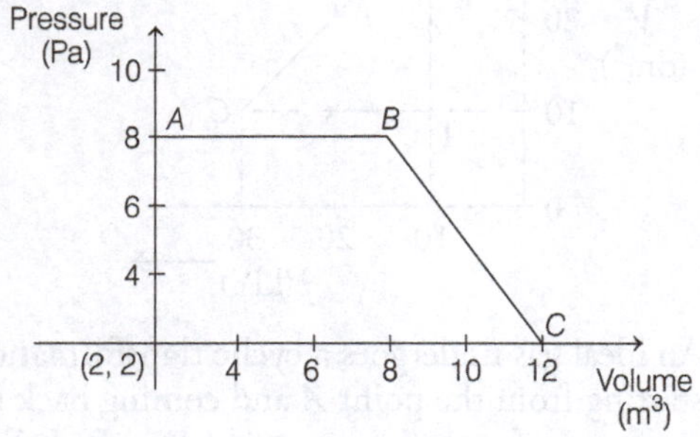


## Topic 2 First Law of Thermodynamics

### ■ Level 1

**01** A liquid when kept inside a thermally insulated closed vessel at 25°C was mechanically stirred from outside. What will be the correct option for the following thermodynamic parameters? ***[2025, 22 Jan Shift-I]***

(a) $\Delta U = 0, q < 0, W > 0$ (b) $\Delta U = 0, q = 0, W = 0$
(c) $\Delta U > 0, q = 0, W > 0$ (d) $\Delta U < 0, q = 0, W > 0$

**02** $\Delta U$ is equal to ***[JEE Main 2017]***

(a) isochoric work (b) isobaric work
(c) adiabatic work (d) isothermal work

**03** A piston filled with 0.04 mole of an ideal gas expands reversibly from 50.0 mL to 375 mL at a constant temperature of 37.0°C. As it does so, it absorbs 208 J of heat. The values of $q$ and $W$ for the process will be (R= 8.314 J / mol K, ln 7.5= 2.01) ***[JEE Main 2013]***

(a) $q = +208$ J, $W = -208$ J
(b) $q = -208$ J, $W = -208$ J
(c) $q = -208$ J, $W = +208$ J
(d) $q = +208$ J, $W = +208$ J

**04** The internal energy change when a system goes from state $A$ to $B$ is 40 kJ/mol. If the system goes from $A$ to $B$ by a reversible path and returns to state $A$ by an irreversible path, what would be the net change in internal energy? ***[AIEEE 2003]***

(a) 40 kJ (b) > 40 kJ (c) < 40 kJ (d) zero

### ■ Level 2

**05.** One mole of an ideal gas expands isothermally and reversibly from 10 $dm^3$ to 20 $dm^3$ at 300 K. $\Delta U$, $q$ and work done in the process respectively are
Given: $R = 8.3\ JK^{-1}\ mol^{-1}$ ***[2025, 4 April Shift-I]***
ln 10 = 2.3
log 2 = 0.30
log 3 = 0.48

(a) 0, 21.84 kJ, 21.84 kJ (b) 0, 21.84 kJ, – 1.726 J
(c) 0, – 17.18 kJ, 1.718 J (d) 0, 1.718 kJ, – 1.718 kJ

**06.** Arrange the following in order of magnitude of work done by the system/on the system at constant temperature. ***[2025, 2 April Shift-II]***

(A) $|W_{reversible}|$ for expansion in infinite stages.
(B) $|W_{irreversible}|$ for expansion in single stage.
(C) $|W_{reversible}|$ for compression in infinite stages.
(D) $|W_{irreversible}|$ for compression in single stage.

Choose the correct answer from the options given below.

(a) C = A > D > B (b) D > C = A > B
(c) A > C > B > D (d) A > B > C > D

**07**

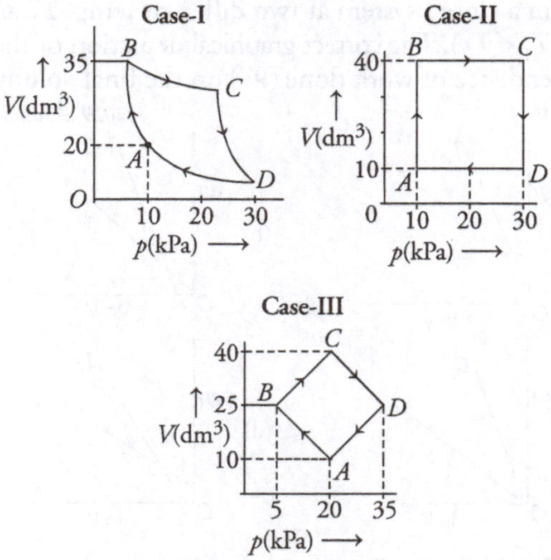


An ideal gas undergoes a cyclic transformation starting from the point $A$ and coming back to the same point by tracing the path $A \rightarrow B \rightarrow C \rightarrow D \rightarrow A$ as shown in the three cases above. ***[2025, 28 Jan Shift-II]***

Choose the correct option regarding $\Delta U$ :

(a) $\Delta U$ (Case-I) = $\Delta U$ (Case-II) = $\Delta U$ (Case-III)
(b) $\Delta U$ (Case-I) > $\Delta U$ (Case-II) > $\Delta U$ (Case-III)
(c) $\Delta U$ (Case-III) > $\Delta U$ (Case-II) > $\Delta U$ (Case-I)
(d) $\Delta U$ (Case-I) > $\Delta U$ (Case-III) > $\Delta U$ (Case-II)

**08** 500 J of energy is transferred as heat to 0.5 mol of Argon gas at 298 K and 1.00 atm. The final temperature and the change in internal energy respectively are (Given : $R = 8.3\text{J K}^{-1}\ \text{mol}^{-1}$ ***[2025, 29 Jan Shift-I]***

(a) 368 K and 500 J (b) 378 K and 300 J
(c) 378 K and 500 J (d) 348 K and 300 J

**09** Choose the correct option for free expansion of an ideal gas under adiabatic condition from the following: ***[2024, 1 Feb Shift-I]***

(a) $q = 0, \Delta T = 0, w = 0$
(b) $q \neq 0, \Delta T = 0, w = 0$
(c) $q = 0, \Delta T < 0, w \neq 0$
(d) $q = 0, \Delta T \neq 0, w = 0$

**10** 5 moles of an ideal gas at 100 K are allowed to undergo reversible compression till its temperature becomes 200 K. If $C_V = 28\ \text{JK}^{-1}\ \text{mol}^{-1}$, calculate $\Delta U$ and $\Delta pV$ for this process. ($R = 8.0\ \text{JK}^{-1}\ \text{mol}^{-1}$) *[2019, 8 April Shift-II]*
(a) $\Delta U = 2.8$ kJ; $\Delta(pV) = 0.8$ kJ
(b) $\Delta U = 14$ J; $\Delta(pV) = 0.8$ J
(c) $\Delta U = 14$ kJ; $\Delta(pV) = 4$ kJ
(d) $\Delta U = 14$ kJ; $\Delta(pV) = 18$ kJ

**11** Which one of the following equations does not correctly represent the first law of thermodynamics for the given processes involving an ideal gas? (Assume non- expansion work is zero) *[2019, 8 April Shift-I]*
(a) Cyclic process : $q = -W$
(b) Adiabatic process : $\Delta U = -W$
(c) Isochoric process : $\Delta U = q$
(d) Isothermal process : $q = -W$

## Level 3

**12.**

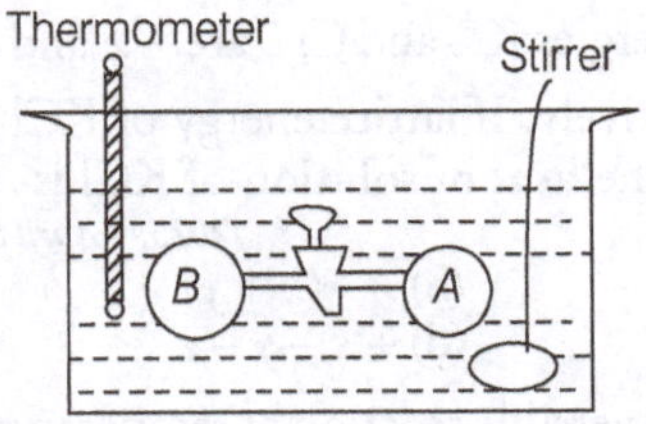


Two vessels $A$ and $B$ are connected *via* stopcock. The vessel $A$ is filled with a gas at a certain pressure. The entire assembly is immersed in water and is allowed to come to thermal equilibrium with water. After opening the stopcock the gas from vessel $A$ expands into vessel $B$ and no change in temperature is observed in the thermometer. Which of the following statement is true? *[2025, 2 April Shift-I]*
(a) $dW \neq 0$ (b) $dU \neq 0$
(c) $dq \neq 0$
(d) The pressure in the vessel $B$ before opening the stopcock is zero

## Numerical Value Type Questions

**13.**

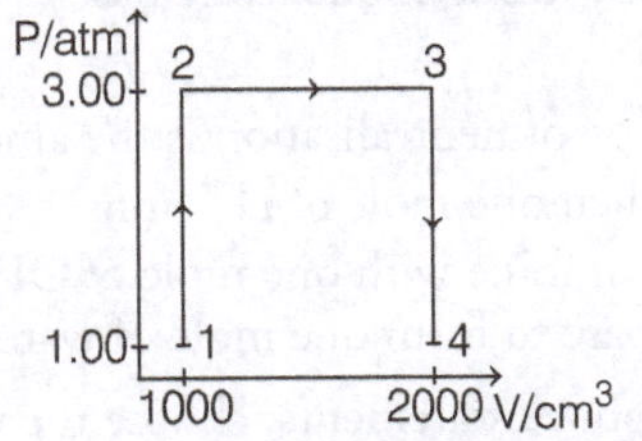


A perfect gas (0.1 mol) having $\overline{C}_v = 1.50R$ (independent of temperature) undergoes the above transformation from point 1 to point 4. If each step is reversible, the total work done ($W$) while going from point 1 to point 4 is (–) ........ J (Nearest integer) [Given : $R = 0.082$ L atm $\text{K}^{-1}\ \text{mol}^{-1}$] *[2025, 3 April Shift-II]*

**14** Three moles of an ideal gas are compressed isothermally from 60 L to 20 L using constant pressure of 5 atm. Heat exchange $Q$ for the compression is – ...... L atm. *[2024, 4 April Shift-II]*

**15** 1 mole of ideal gas is allowed to expand reversibly and adiabatically from a temperature of 27°C. The work done is 3 kJ $\text{mol}^{-1}$. The final temperature of the gas is ........... K. (Nearest integer). [Given : $C_V = 20\ \text{J mol}^{-1}\ \text{K}^{-1}$.] *[2023, 30 Jan Shift-II]*

**16** When a 60 W electric heater is immersed in a gas for 100s in a constant volume container with adiabatic walls, the temperature of the gas rises by 5°C. The heat capacity of the given gas is .......... $\text{JK}^{-1}$ (Nearest integer) *[2023, 8 April Shift-I]*

**17** When 2 L of ideal gas expands isothermally into vacuum to a total volume of 6 L, the change in internal energy is ......... J. (Nearest integer) *[2023, 30 Jan Shift-I]*

**18** 4.0 L of an ideal gas is allowed to expand isothermally into vacuum until the total volume is 20 L. The amount of heat absorbed in this expansion is ...... L atm. *[2022, 28 June Shift-I]*

**19** A system does 200 J of work and at the same time absorbs 150 J of heat. The magnitude of the change in internal energy is ......... J. (Nearest integer) *[2021, 25 July Shift-II]*

**Imp.** Heat Transfer

**20** Five moles of an ideal gas at 293 K is expanded isothermally from an initial pressure of 2.1 MPa to 1.3 MPa against at constant external pressure 4.3 MPa. The heat transferred in this process is ........ kJ $\text{mol}^{-1}$. (Rounded off to the nearest integer) [$R = 8.314\ \text{J mol}^{-1}\text{K}^{-1}$] *[2021, 25 Feb Shift-II]*

# Topic 3 Reactions Related to Enthalpies and Hess's Law

## Level 1

**01** $S(g) + \frac{3}{2}O_2(g) \longrightarrow SO_3(g) + 2x$ kcal

$SO_2(g) + \frac{1}{2}O_2(g) \longrightarrow SO_3(g) + y$ kcal

The heat of formation of $SO_2(g)$ is given by *[2025, 24 Jan Shift-II]*
(a) $y - 2x$ kcal (b) $x + y$ kcal
(c) $2x + y$ kcal (d) $\frac{2x}{y}$ kcal

**02** If C (diamond) $\longrightarrow$ C(graphite) $+X\ \text{kJ mol}^{-1}$

C (diamond) $+ O_2(g) \longrightarrow CO_2(g) + Y\ \text{kJ mol}^{-1}$

$C \text{ (graphite)} + O_2(g) \longrightarrow CO_2(g) + Z \text{ kJ mol}^{-1}$

at constant temperature. *[2025, 29 Jan Shift-II]*

Then

(a) $X = -Y + Z$ (b) $-X = Y + Z$

(c) $X = Y + Z$ (d) $X = Y - Z$

**03** The difference between $\Delta H$ and $\Delta U$ $(\Delta H - \Delta U)$, when the combustion of one mole of heptane ($l$) is carried out at a temperature $T$, is equal to *[2019, 10 April Shift-II]*

(a) $-4\,RT$ (b) $3\,RT$ (c) $4\,RT$ (d) $-3\,RT$

**04** The heats of combustion of carbon and carbon monoxide are $-393.5$ and $-283.5$ kJ mol$^{-1}$, respectively. The heat of formation (in kJ) of carbon monoxide per mole is *[JEE Main 2016]*

(a) 676.5 (b) –676.5

(c) –110.5 (d) 110.5

**05** Oxidising power of chlorine in aqueous solution can be determined by the parameters indicated below

$$\frac{1}{2}Cl_2(g) \xrightarrow{\frac{1}{2}\Delta_{diss} H^{\ominus}} Cl\,(g) \xrightarrow{\Delta_{EA}H^{\ominus}}$$

$$Cl^-(g) \xrightarrow{\Delta_{hyd} H^{\ominus}} Cl^-(aq)$$

The energy involved in the conversion of $\frac{1}{2}Cl_2(g)$ to $Cl^-(aq)$ (using the data, $\Delta_{diss} H^{\ominus}_{Cl_2} = 240$ kJ mol$^{-1}$

$\Delta_{EA} H^{\ominus}_{Cl} = -349$ kJ mol$^{-1}$,

$\Delta_{hyd} H^{\ominus}_{Cl} = -381$ kJ mol$^{-1}$) will be *[AIEEE 2008]*

(a) $+152$ kJ mol$^{-1}$ (b) $-610$ kJ mol$^{-1}$

(c) $-850$ kJ mol$^{-1}$ (d) $+120$ kJ mol$^{-1}$

**06** Assuming that water vapour is an ideal gas, the internal energy change ($\Delta E$) when 1 mole of water is vaporised at 1 bar pressure and 100°C, (Given : molar enthalpy of vaporisation of water at 1 bar and 373 K = 41 kJ mol$^{-1}$ and $R = 8.3$ J mol$^{-1}$ K$^{-1}$) will be *[AIEEE 2007]*

(a) 4.100 kJ mol$^{-1}$ (b) 3.7904 kJ mol$^{-1}$

(c) 37.904 kJ mol$^{-1}$ (d) 41.00 kJ mol$^{-1}$

**07** $(\Delta H - \Delta E)$ for the formation of carbon monoxide (CO) from its elements at 298 K is ($R = 8.314$ JK$^{-1}$ mol$^{-1}$) *[AIEEE 2006]*

(a) $-1238.78$ J mol$^{-1}$ (b) 1238.78 J mol$^{-1}$

(c) $-2477.57$ J mol$^{-1}$ (d) 2477.57 J mol$^{-1}$

**08** Heat required to raise the temperature of 1 mole of a substance by 1° is called *[AIEEE 2002]*

(a) specific heat (b) molar heat capacity

(c) water equivalent (d) specific gravity

## Level 2

**09** Given below are two statements: *[2025, 3 April Shift-II]*

**Statement I** When a system containing ice in equilibrium with water (liquid) is heated, heat is absorbed by the system and there is no change in the temperature of the system until whole ice gets melted.

**Statement II** At melting point of ice, there is absorption of heat in order to overcome intermolecular forces of attraction within the molecules of water in ice and kinetic energy of molecules is not increased at melting point.

In the light of the above statements, choose the correct answer from the options given below.

(a) Statement I is false but Statement II is true.

(b) Both Statement I and Statement II are false.

(c) Both Statement I and Statement II are true.

(d) Statement I is true but Statement II is false.

**10** The hydration energies $K^+$ and $Cl^-$ are $-x$ and $-y$ kJ/mol respectively. If lattice energy of KCl is $-z$ kJ/mol, then the heat of solution of KCl is *[2025, 7 April Shift-II]*

(a) $x + y + z$ (b) $z - (x + y)$

(c) $-z - (x + y)$ (d) $+x - y - z$

**11** Let us consider a reversible reaction at temperature, $T$. In this reaction, both $\Delta H$ and $\Delta S$ were observed to have positive values. If the equilibrium temperature is $T_e$, then the reaction becomes spontaneous at: *[2025, 4 April Shift-I]*

(a) $T > T_e$ (b) $T = T_e$

(c) $T_e = 5T$ (d) $T_e > T$

**Imp.** Enthalpy of Neutralisation

**12** Given below are two statements : one is labelled as **Assertion (A)** and other is labelled as **Reason (R)**. *[2024, 5 April Shift-I]*

**Assertion (A)** Enthalpy of neutralisation of strong monobasic acid with strong monoacidic base is always $-57$ kJ mol$^{-1}$.

**Reason (R)** Enthalpy of neutralisation is the amount of heat liberated when one mole of $H^+$ ions furnished by acid combine with one mole of $OH^-$ ions furnished by base to form one mole of water.

In the light of the above statements, choose the **most appropriate** answer from the options given below.

(a) A is false, but R is true.

(b) Both A and R are true and R is the correct explanation of A.

(c) A is true, but R is false.

(d) Both A and R are true but R is not the correct explanation of A.

**13** At 298.2 K the relationship between enthalpy of bond dissociation (in kJ $mol^{-1}$) for hydrogen ($E_H$) and its isotope, deuterium ($E_D$), is best described by *[2021, 25 July Shift-I]*

(a) $E_H = \frac{1}{2} E_D$ (b) $E_H = E_D$
(c) $E_H \simeq E_D - 7.5$ (d) $E_H = 2E_D$

**14** Lattice enthalpy and enthalpy of solution of NaCl are 788 kJ $mol^{-1}$ and 4 kJ $mol^{-1}$, respectively. The hydration enthalpy of NaCl is *[2020, 5 Sep Shift-II]*

(a) −780 kJ $mol^{-1}$ (b) 780 kJ $mol^{-1}$
(c) −784 kJ $mol^{-1}$ (d) 784 kJ $mol^{-1}$

**15** For one mole of an ideal gas, which of these statements must be true? *[2020, 04 Sept Shift-I]*

(1) $U$ and $H$ each depends only on temperature
(2) Compressibility factor $Z$ is not equal to 1
(3) $C_{P,m} - C_{V,m} = R$
(4) $dU = C_V dT$ for any processes

(a) (1) and (3) (b) (2), (3) and (4)
(c) (3) and (4) (d) (1), (3) and (4)

**16** If enthalpy of atomisation for $Br_2(l)$ is $x$ kJ/mol and bond enthalpy for $Br_2$ is $y$ kJ/mol, the relation between them *[2020, 9 Jan Shift-II]*

(a) is $x < y$ (b) does not exist
(c) is $x > y$ (d) is $x = y$

**Imp.** Calculation of Enthalpy of Sublimation

**17** Enthalpy of sublimation of iodine is 24 cal $g^{-1}$ at 200°C. If specific heat of $I_2(s)$ and $I_2$(vap.) are 0.055 and 0.031 cal $g^{-1}$ $K^{-1}$ respectively, then enthalpy of sublimation of iodine at 250°C in cal $g^{-1}$ is *[2019, 12 April Shift-I]*

(a) 2.85 (b) 5.7 (c) 22.8 (d) 11.4

**18** The combustion of benzene ($l$) gives $CO_2(g)$ and $H_2O(l)$. Given that heat of combustion of benzene at constant volume is −3263.9 kJ $mol^{-1}$ at 25° C; heat of combustion (in kJ $mol^{-1}$) of benzene at constant pressure will be ($R = 8.314$ $JK^{-1}$ $mol^{-1}$) *[JEE Main 2018]*

(a) 4152.6 (b) −452.46 (c) 3260 (d) −3267.6

**19** Given, $C_{(Graphite)} + O_2(g) \longrightarrow CO_2(g);$

$$\Delta_r H^\circ = -393.5 \text{ kJ mol}^{-1}$$

$$H_2(g) + \frac{1}{2}O_2(g) \longrightarrow H_2O(l);$$

$$\Delta_r H^\circ = -285.8 \text{ kJ mol}^{-1}$$

$$CO_2(g) + 2H_2O(l) \longrightarrow CH_4(g) + 2O_2(g);$$

$$\Delta_r H^\circ = +890.3 \text{ kJ mol}^{-1}$$

Based on the above thermochemical equations, the value of $\Delta_r H^\circ$ at 298 K for the reaction, *[JEE Main 2017]*

$C_{(graphite)} + 2H_2(g) \longrightarrow CH_4(g)$ will be

(a) + 78.8 kJ $mol^{-1}$ (b) +144.0 kJ $mol^{-1}$
(c) − 74.8 kJ $mol^{-1}$ (d) −144.0 kJ $mol^{-1}$

**20** The value of enthalpy change ($\Delta H$) for the reaction

$$C_2H_5OH(l) + 3O_2(g) \longrightarrow 2CO_2(g) + 3H_2O(l)$$

at 27°C is − 1366.5 kJ $mol^{-1}$. The value of internal energy change for the above reaction at this temperature will be *[AIEEE 2011]*

(a) − 1371.5 kJ (b) − 1369.0 kJ
(c) − 1364.0 kJ (d) − 1361.5 kJ

**21** Consider the reaction,

$4NO_2(g) + O_2(g) \longrightarrow 2N_2O_5(g)$, $\Delta_r H = -111$ kJ.

If $N_2O_5(s)$ is formed instead of $N_2O_5(g)$ in the above reaction, the $\Delta_r H$ value will be

(Given, $\Delta H$ of sublimation for $N_2O_5$ is 54 kJ $mol^{-1}$) *[AIEEE 2011]*

(a) −165 kJ (b) +54 kJ
(c) +219 kJ (d) − 219 kJ

**22** The standard enthalpy of formation of $NH_3$ is −46.0 kJ $mol^{-1}$. If the enthalpy of formation of $H_2$ from its atoms is − 436 kJ $mol^{-1}$ and that of $N_2$ is −712 kJ $mol^{-1}$, the average bond enthalpy of N—H bond in $NH_3$ is *[AIEEE 2010]*

(a) − 964 kJ $mol^{-1}$ (b) +352 kJ $mol^{-1}$
(c) +1056 kJ $mol^{-1}$ (d) −1102 kJ $mol^{-1}$

**23** On the basis of the following thermochemical data $[\Delta_f G^\circ H^+(aq) = 0]$

$$H_2O(l) \longrightarrow H^+(aq) + OH^-(aq); \Delta H = 57.32 \text{ kJ}$$

$$H_2(g) + \frac{1}{2}O_2(g) \longrightarrow H_2O(l); \Delta H = -286.02 \text{ kJ}$$

The value of enthalpy of formation of $OH^-$ ion at 25°C is *[AIEEE 2009]*

(a) −22.88 kJ (b) −228.88 kJ
(c) +228.88 kJ (d) −343.52 kJ

**24** The enthalpy changes for the following processes are listed below

$Cl_2(g) = 2Cl(g)$, 242.3 kJ $mol^{-1}$
$I_2(g) = 2I(g)$, 151.0 kJ $mol^{-1}$
$ICl(g) = I(g) + Cl(g)$, 211.3 kJ $mol^{-1}$
$I_2(s) = I_2(g)$, 62.76 kJ $mol^{-1}$

Given that the standard states for iodine and chlorine are $I_2(s)$ and $Cl_2(g)$, the standard enthalpy of formation of ICl is *[AIEEE 2006]*

(a) − 14.6 kJ $mol^{-1}$ (b) − 16.8 kJ $mol^{-1}$
(c) + 16.8 kJ $mol^{-1}$ (d) + 244.8 kJ $mol^{-1}$

**25** The standard enthalpy of formation ($\Delta H_f^\circ$) at 298 K for methane, $CH_4(g)$ is – 74.8 kJ $mol^{-1}$. The addition information required to determine the average energy for C—H bond formation would be *[AIEEE 2006]*
(a) the dissociation energy of $H_2$ and enthalpy of sublimation of carbon
(b) latent heat of vaporisation of methane
(c) the first four ionisation energies of carbon and electron gain enthalpy of hydrogen
(d) the dissociation energy of hydrogen molecule, $H_2$

**26** Consider the reaction,
$N_2 + 3H_2 \longrightarrow 2NH_3$ carried out at constant temperature and pressure. If $\Delta H$ and $\Delta E$ are the enthalpy and internal energy changes for the reaction, which of the following expressions is true ? *[AIEEE 2005]*
(a) $\Delta H > \Delta E$ (b) $\Delta H < \Delta E$
(c) $\Delta H = \Delta E$ (d) $\Delta H = 0$

**27** If the bond dissociation energies of $XY$, $X_2$ and $Y_2$ (all diatomic molecules) are in the ratio of 1 : 1 : 0.5 and $\Delta H_f$ for the formation of $XY$ is – 200 kJ $mol^{-1}$. The bond dissociation energy of $X_2$ will be *[AIEEE 2005]*
(a) 400 kJ $mol^{-1}$ (b) 300 kJ $mol^{-1}$
(c) 200 kJ $mol^{-1}$ (d) None of these

**28** If at 298 K, the bond energies of C—H, C—C, C=C and H—H bonds are respectively 414, 347, 615 and 435 kJ $mol^{-1}$, the value of enthalpy change for the reaction,
$H_2C{=}CH_2(g) + H_2(g) \longrightarrow H_3C{-}CH_3(g)$ at 298 K will be *[AIEEE 2003]*
(a) + 250 kJ (b) – 250 kJ (c) + 125 kJ (d) – 125 kJ

## Level 3

**29** The correct statement amongst the following is *[2025, 7 April Shift-II]*
(a) The term 'standard state' implies that the temperature is 0°C.
(b) $\Delta_f H^\theta_{298}$ is zero for O(g).
(c) The standard state of a pure gas is the pure gas at a pressure of 1 bar and temperature 273 K.
(d) $\Delta_f H^\theta_{500}$ is zero for $O_2(g)$.

**30** Total enthalpy change for freezing of 1 mol of water at 10°C to ice at −10°C is ...........
(Given, $\Delta_{fus}H = x$ kJ/mol
$C_p[H_2O(1)] = y$ J $mol^{-1}$ $K^{-1}$
$C_p[H_2O(s)] = z$ J $mol^{-1}$ $K^{-1}$ *[2025, 7 April Shift-I]*
(a) $x - 10y - 10z$ (b) $10(100x + y + z)$
(c) $-10(100x + y + z)$ (d) $-x - 10y - 10z$

**31** Consider the given data:
A. $HCl(g) + 10H_2O(l) \rightarrow HCl.10\ H_2O$
$\Delta H = -69.01$ kJ $mol^{-1}$
B. $HCl(g) + 40H_2O(l) \rightarrow HCl \cdot 40\ H_2O$
$\Delta H = -72.79$ kJ $mol^{-1}$
Choose the correct statement: *[2025, 4 April Shift-I]*
(a) The heat of dilution for the HCl ($HCl \cdot 10\ H_2O$ to $HCl \cdot 40\ H_2O$) is 3.78 kJ $mol^{-1}$.
(b) The heat of solution depends on the amount of solvent.
(c) Dissolution of gas in water is an endothermic process.
(d) The heat of formation of HCl solution is represented by both (A) and (B).

**32** Reactant $A$ converts to product $D$ through the given mechanism (with the net evolution of heat) :
$A \rightarrow B$ slow; $\Delta H =$ +ve
$B \rightarrow C$ fast; $\Delta H =$ –ve
$C \rightarrow D$ fast; $\Delta H =$ –ve
Which of the following represents the above reaction mechanism? *[2025, 2 April Shift-II]*

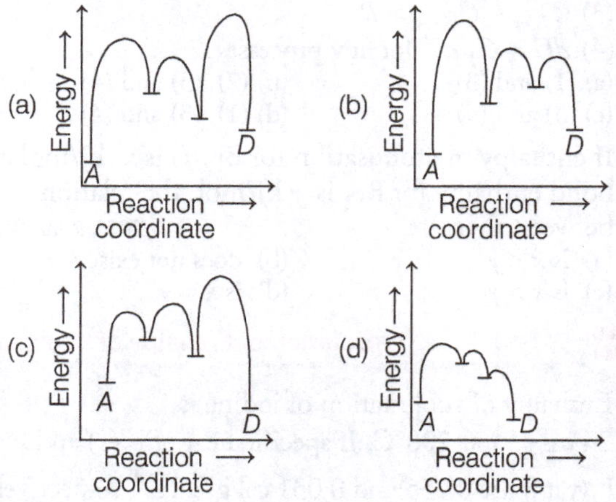


**33** Given,
(A) $2CO(g) + O_2(g) \longrightarrow 2CO_2(g)$
$\Delta H_1^\circ = -x$ kJ $mol^{-1}$
(B) C (graphite) + $O_2(g) \longrightarrow CO_2(g)$
$\Delta H_2^\circ = -\ y$ kJ $mol^{-1}$
The $\Delta H^\circ$ for the reaction
C (graphite) + $\frac{1}{2}O_2(g) \longrightarrow CO(g)$ is *[2023, 10 April Shift-I]*
(a) $\frac{x+2y}{2}$ (b) $\frac{x-2y}{2}$ (c) $\frac{2x-y}{2}$ (d) $2y - x$

**34** At 25°C and 1 atm pressure, the enthalpies of combustion are as given below

| Substance | $H_2$ | C (graphite) | $C_2H_6(g)$ |
|---|---|---|---|
| $\Delta_c H^\ominus$ (kJ $mol^{-1}$) | −286.0 | −394.0 | −1560.0 |

The enthalpy of formation of ethane is *[2022, 24 June Shift-II]*
(a) + 54.0 kJ $mol^{-1}$ (b) – 68.0 kJ $mol^{-1}$
(c) – 86.0 kJ $mol^{-1}$ (d) + 97.0 kJ $mol^{-1}$

**Imp.** Calculation of Enthalpy of Combustion

**35** For the complete combustion of ethanol,
$C_2H_5OH(l) + 3O_2(g) \longrightarrow 2CO_2(g) + 3H_2O(l)$
the amount of heat produced as measured in bomb calorimeter is 1364.47 $kJmol^{-1}$ at 25°C. Assuming ideality, the enthalpy of combustion, $\Delta_C H$ for the reaction will be [$R = 8.314\ JK^{-1}mol^{-1}$] *[JEE Main 2014]*

(a) $-1366.95\ kJmol^{-1}$ (b) $-1361.95\ kJmol^{-1}$

(c) $-1460.50\ kJmol^{-1}$ (d) $-1350.50\ kJmol^{-1}$

## Numerical Value Type Questions

**36** Resonance in $X_2Y$ can be represented as

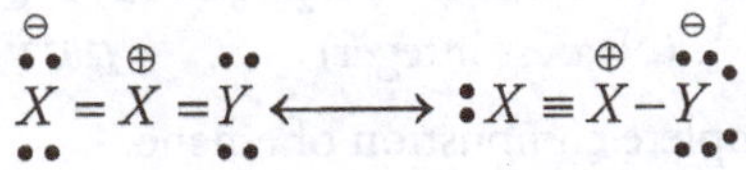

The enthalpy of formation of $X_2Y\left(X \equiv X(g) + \frac{1}{2}Y = Y(g) \rightarrow X_2Y(g)\right)$ is 80 kJ $mol^{-1}$. The magnitude of resonance energy of $X_2Y$ is ........... kJ $mol^{-1}$ (Nearest integer value)

Given: Bond energies of $X \equiv X$, $X = X$, $Y = Y$ and $X = Y$ are 940, 410, 500 and 602 kJ $mol^{-1}$ respectively. Valence $X:3, Y:2$ *[2025, 8 April Shift-II]*

**37** Given:

$\Delta H^{\ominus}_{sub}$ [C (graphite)] = 710 kJ $mol^{-1}$

$\Delta_{C-H}H^{\ominus} = 414$ kJ $mol^{-1}$

$\Delta_{H-H}H^{\ominus} = 436$ kJ $mol^{-1}$

$\Delta_{C=C}H^{\ominus} = 611$ kJ $mol^{-1}$

The $\Delta H_f^{\ominus}$ for $CH_2{=}CH_2$ is ........ kJ $mol^{-1}$ (Nearest integer value) *[2025, 3 April Shift-I]*

**38** A sample of *n*-octane (1.14 g) was completely burnt in excess of oxygen in a bomb calorimeter. whose heat capacity is 5 kJ $K^{-1}$. As a result of combustion reaction, the temperature of the calorimeter is increased by 5 K. The magnitude of the heat of combustion of octane at constant volume is ......... kJ $mol^{-1}$ (Nearest integer). *[2025, 3 April Shift-II]*

**39** Consider the following cases of standard enthalpy of reaction ($\Delta H_r^{\circ}$ in kJ $mol^{-1}$) *[2025, 22 Jan Shift-II]*

$$C_2H_6(g) + \frac{7}{2}O_2(g) \longrightarrow 2CO_2(g) + 3H_2O(l);\ \Delta H_1^{\circ} = -1550$$

$$C\ (graphite) + O_2(g) \longrightarrow CO_2(g);\ \Delta H_2^{\circ} = -393.5$$

$$H_2(g) + \frac{1}{2}O_2(g) \longrightarrow H_2O(l);\ \Delta H_3^{\circ} = -286$$

The magnitude of $\Delta H_f^{\circ}(C_2H_6(g))$ is ......... kJ $mol^{-1}$ (Nearest integer).

**40** The formation enthalpies, $\Delta H_f^{\ominus}$ for H(*g*) and O(*g*) are 220.0 and 250.0 kJ $mol^{-1}$, respectively, at 298.15 K, and $\Delta H_f^{\ominus}$ for $H_2O(g)$ is $-242.0$ kJ $mol^{-1}$ at the same temperature. The average bond enthalpy of the O—H bond in water at 298.15 K is ......... kJ $mol^{-1}$ (Nearest integer). *[2025, 28 Jan Shift-I]*

**41** Consider the following data :

Heat of formation of $CO_2(g) = -393.5$ kJ$mol^{-1}$

Heat of formation of $H_2O(l) = -286.0$ kJ$mol^{-1}$

Heat of combustion of benzene = $-3267.0$ kJ$mol^{-1}$

The heat of formation of benzene is .........kJ$mol^{-1}$ (Nearest integer) *[2025, 28 Jan Shift-II]*

**Imp.** Calculation of Bond Dissociation Enthalpy

**42** The bond dissociation enthalpy of $X_2$, $\Delta H^{\circ}_{bond}$ calculated from the given data is......... kJ $mol^{-1}$. (Nearest integer) *[2025, 23 Jan Shift-II]*

$$M^+X^-(s) \longrightarrow M^+(g) + X^-(g);\ \Delta H^{\circ}_{lattice} = 800\ kJmol^{-1}$$

$$M(s) \longrightarrow M(g);\ \Delta H^{\circ}_{sub} = 100\ kJmol^{-1}$$

$$M(g) \longrightarrow M^+(g) + e^-;\ \Delta H_i^{\circ} = 500\ kJmol^{-1}$$

$$X(g) + e^-(g) \longrightarrow X^-(g);\ \Delta H^{\circ}_{eg} = -300\ kJmol^{-1}$$

$$M(s) + \frac{1}{2}X_2(g) \longrightarrow M^+X^-(s);\ \Delta H_f^{\circ} = -400\ kJmol^{-1}$$

[Given, $M^+X^-$ is a pure ionic compound and $X$ forms a diatomic molecule $X_2$ in gaseous state].

**43** When equal volume of 1 M HCl and 1 M $H_2SO_4$ are separately neutralised by excess volume of 1 M NaOH solution *x* and *y* kJ of heat is liberated respectively. The value of *y*/*x* is .......... *[2024, 9 April Shift-I]*

**44** The heat of solution of anhydrous $CuSO_4$ and $CuSO_4 \cdot 5H_2O$ are $-70$ kJ $mol^{-1}$ and $+12$ kJ $mol^{-1}$ respectively.

The heat of hydration of $CuSO_4$ to $CuSO_4 \cdot 5H_2O$ is $-x$ kJ. The value of *x* is ........ (Nearest integer). *[2024, 9 April Shift-I]*

**45** $\Delta_{vap}H^{\ominus}$ for water is +40.79 kJ $mol^{-1}$ at 1 bar and 100 °C. Change in internal energy for this vapourisation under same condition is ........ kJ $mol^{-1}$. (Integer answer) (Given $R = 8.3\ JK^{-1}\ mol^{-1}$) *[2024, 8 April Shift-II]*

**46** An ideal gas, $C_V = \frac{5}{2}R$, is expanded adiabatically against a constant pressure of 1 atm until it doubles in volume. If the initial temperature and pressure is 298 K and 5 atm, respectively then the final temperature is ....... K (Nearest integer).
[$C_V$ is the molar heat capacity at constant volume] *[2024, 6 April Shift-I]*

**47** The heat of combustion of solid benzoic acid at constant volume is $-321.30$ kJ at 27°C. The heat of combustion at constant pressure is $(-321.30 - xR)$ kJ, the value of $x$ is .......... *[2024, 5 April Shift-II]*

**48** Combustion of 1 mole of benzene is expressed at

$$C_6H_6(l) + \frac{15}{2}O_2(g) \longrightarrow 6CO_2(g) + 3H_2O(l)$$

The standard enthalpy of combustion of 2 mole of benzene is –'$x$' kJ, $x$ = ............. *[2024, 5 April Shift-I]*

Given :

1. Standard enthalpy of formation of 1 mol of $C_6H_6(l)$, for the reactions
   6C (graphite) $+ 3H_2(g) \longrightarrow C_6H_6(l)$ is 48.5 kJ $mol^{-1}$
2. Standard enthalpy of formation of 1 mol of $CO_2(g)$ for the reaction
   C (graphite) + $O_2(g) \longrightarrow CO_2(g)$ is –393.5 kJ $mol^{-1}$
3. Standard and enthalpy of formation of 1 mol of $H_2O(l)$, for the reaction
   $H_2(g) + \frac{1}{2}O_2(g) \longrightarrow H_2O(l)$ is – 286 kJ $mol^{-1}$.

**49** The enthalpy of formation of ethane ($C_2H_6$) from ethylene by addition of hydrogen where the bond-energies of C—H, C—C, C = C, H—H are 414 kJ, 347 kJ, 615 kJ and 435 kJ respectively is – ...... kJ. *[2024, 4 April Shift-I]*

**50** Standard enthalpy of vaporisation for $CCl_4$ is 30.5 kJ $mol^{-1}$. Heat required for vaporisation of 284 g of $CCl_4$ at constant temperature is .......... kJ.
(Given, molar mass in g $mol^{-1}$; C = 12, Cl = 35.5) *[2024, 29 Jan Shift-II]*

**51** Two reactions are given below. *[2024, 30 Jan Shift-II]*

$$2Fe(s) + \frac{3}{2}O_2(g) \longrightarrow Fe_2O_3(s), \quad \Delta H° = -822 \text{ kJ/mol}$$

$$C(s) + \frac{1}{2}O_2(g) \longrightarrow CO(g), \Delta H° = -110 \text{ kJ/mol}$$

Then enthalpy change for the following reaction,
$3C(s) + Fe_2O_3(s) \longrightarrow 2Fe(s) + 3CO(g)$ is ... kJ/mol.

**52** The number of endothermic processes from the following is ........... *[2023, 10 April Shift-II]*
A. $I_2(g) \longrightarrow 2I(g)$
B. $HCl(g) \longrightarrow H(g) + Cl(g)$
C. $H_2O(l) \longrightarrow H_2O(g)$
D. $C(s) + O_2(g) \longrightarrow CO_2(g)$
E. dissolution of ammonium chloride in water

**53** Consider the following data.
Heat of combustion of $H_2(g) = -241.8$ kJ $mol^{-1}$
Heat of combustion of C(s) $= -393.5$ kJ $mol^{-1}$
Heat of combustion of $C_2H_5OH(1) = -1234.7$ kJ $mol^{-1}$
The heat of formation of $C_2H_5OH(l)$ is (–) ......... kJ $mol^{-1}$. (Nearest integer) *[2023, 6 April Shift-II]*

**54** For complete combustion of ethene,
$C_2H_4(g) + 3O_2(g) \longrightarrow 2CO_2(g) + 2H_2O(l)$
the amount or heat produced as measured in bomb calorimeter is 1406 kJ $mol^{-1}$ at 300K. The minimum value of $T\Delta S$ needed to reach equilibrium is (–) ........ kJ. (Nearest integer)
Given: $R = 8.3$ J $K^{-1}$ $mol^{-1}$ *[2023, 8 April Shift-II]*

**55** $A_2 + B_2 \longrightarrow 2AB, \Delta H_f^0 = -200$ kJ $mol^{-1}$
$AB$, $A_2$ and $B_2$ are diatomic molecules. If the bond enthalpies of $A_2$, $B_2$ and $AB$ are in the ratio 1 : 0.5 : 1, then the bond enthalpy of $A_2$ is ........... kJ $mol^{-1}$ (Nearest integer) *[2023, 13 April Shift-I]*

**Imp.** **Bomb Calorimetry**

**56** 0.3 g of ethane undergoes combustion at 27°C in a bomb calorimeter. The temperature of calorimeter system (including the water) is found to rise by 0.5°C. The heat evolved during combustion of ethane at constant pressure is ......... kJ $mol^{-1}$. (Nearest integer)
[Given, the heat capacity of the calorimeter system is 20 kJ $K^{-1}$, $R = 8.3$ J $K^{-1}$ $mol^{-1}$.
Assume ideal gas behaviour. Atomic mass of C and H are 12 and 1 g $mol^{-1}$ respectively] *[2023, 01 Feb Shift-II]*

**57** 28.0 L of $CO_2$ is produced on complete combustion of 16.8 L gaseous mixture of ethene and methane at 25°C and 1 atm. Heat evolved during the combustion processes is – ........... kJ. *[2023, 25 Jan Shift-II]*
[Given, $\Delta H_C(CH_4) = -900$ kJ $mol^{-1}$
$\Delta H_C(C_2H_4) = -1400$ kJ $mol^{-1}$]

**58** When 600 mL of 0.2 M $HNO_3$ is mixed with 400 mL of 0.1 M NaOH solution in a flask, the rise in temperature of the flask is ......... $\times 10^{-2}$°C. *[2022, 29 July Shift-I]*

(Enthalpy of neutralisation = 57 kJ $mol^{-1}$ and Specific heat of water = 4.2 $JK^{-1}g^{-1}$)
(Neglect heat capacity of flask)

**59** A gas (molar mass = 280 g $mol^{-1}$) was burnt in excess $O_2$ in a constant volume calorimeter and during combustion the temperature of calorimeter increased from 298.0 K to 298.45 K. If the heat capacity of calorimeter is 2.5 kJ $K^{-1}$ and enthalpy of combustion of gas is 9 kJ $mol^{-1}$ then amount of gas burnt is .............. g. (nearest integer) *[2022, 27 July Shift-II]*

**60** The molar heat capacity for an ideal gas at constant pressure is 20.785 J $K^{-1}$ $mol^{-1}$. The change in internal energy is 5000 J upon heating it from 300 K to 500 K. The number of moles of the gas at constant volume is ........... . (Nearest integer)
(Given, $R = 8.314$ $JK^{-1}$ $mol^{-1}$) *[2022, 27 July Shift-I]*

**61** For the reaction,

$$H_2F_2(g) \longrightarrow H_2(g) + F_2(g)$$

$$\Delta U = -59.6 \text{ kJ mol}^{-1} \text{ at } 27°\text{C}.$$

The enthalpy change for the above reaction is (–) .............. kJ $mol^{-1}$ [nearest integer]
Given : $R = 8.314$ $JK^{-1}mol^{-1}$ *[2022, 26 July Shift-II]*

**62** 2.4 g coal is burnt in a bomb calorimeter in excess of oxygen at 298 K and 1 atm pressure. The temperature of the calorimeter rises from 298 K to 300 K. The enthalpy change during the combustion of coal is $-x$ kJ $mol^{-1}$. The value of $x$ is ........... .
(Nearest integer)
(Given, heat capacity of bomb calorimeter is 20.0 kJ $K^{-1}$. Assume coal to be pure carbon). *[2022, 26 July Shift-I]*

**63** While performing a thermodynamics experiment, a student made the following observations.

$HCl + NaOH \longrightarrow NaCl + H_2O$; $\Delta H = -57.3$ kJ $mol^{-1}$

$CH_3COOH + NaOH \longrightarrow CH_3COONa + H_2O$; $\Delta H = -55.3$ kJ $mol^{-1}$

The enthalpy of ionisation of $CH_3COOH$ as calculated by the student is ......... kJ $mol^{-1}$. [nearest integer] *[2022, 25 July Shift-II]*

**64** The enthalpy of combustion of propane, graphite and dihydrogen at 298 K are – 2220.0 kJ $mol^{-1}$, – 393.5 kJ $mol^{-1}$ and –285.8 kJ $mol^{-1}$ respectively. The magnitude of enthalpy of formation of propane ($C_3H_8$) is ........ kJ $mol^{-1}$. (Nearest integer) *[2022, 25 July Shift-I]*

**65** For combustion of one mole of magnesium in an open container at 300 K and 1 bar pressure, $\Delta_C H^\ominus = -601.70$ kJ $mol^{-1}$, the magnitude of change in internal energy for the reaction is ... kJ. (Nearest integer) (Given, $R = 8.3$ $JK^{-1}$ $mol^{-1}$) *[2022, 28 June Shift-II]*

**66** 2.2 g of nitrous oxide ($N_2O$) gas is cooled at a constant pressure of 1 atm from 310 K to 270 K causing the compression of the gas from 217.1 mL to 167.75 mL. The change in internal energy of the process, $\Delta U$ is '$-x$' J. The value of '$x$' is [Nearest integer]
(Given, atomic mass of N = 14 g $mol^{-1}$ and of O = 16 g $mol^{-1}$. Molar heat capacity of $N_2O$ is 100 $JK^{-1}$ $mol^{-1}$) *[2022, 29 June Shift-II]*

**67** 17.0 g of $NH_3$ completely vapourises at −33.42°C and 1 bar pressure and the enthalpy change in the process is 23.4 kJ $mol^{-1}$. The enthalpy change for the vaporisation of 85 g of $NH_3$ under the same conditions is ...... kJ. *[2022, 29 June Shift-I]*

**68** A fish swimming in water body when taken out from the water body is covered with a film of water of weight 36 g. When it is subjected to cooking at 100°C, then the internal energy for vaporisation in kJ $mol^{-1}$ is ...... . [Nearest integer]
[Assume steam to be an ideal gas. Given, $\Delta_{vap}H^\ominus$ for water at 373 K and 1 bar is 41.1 kJ $mol^{-1}$; $R = 8.31$ $JK^{-1}$ $mol^{-1}$] *[2022, 26 June Shift-II]*

**69** For complete combustion of methanol

$$CH_3OH(l) + \frac{3}{2}O_2(g) \longrightarrow CO_2(g) + 2H_2O(l)$$

the amount of heat produced as measured by bomb calorimeter is 726 kJ $mol^{-1}$ at 27°C. The enthalpy of combustion for the reaction is $-x$ kJ $mol^{-1}$, where $x$ is ......... (Nearest integer) (Given : $R = 8.3$ $JK^{-1}mol^{-1}$) *[2022, 26 June Shift-I]*

**70** For water $\Delta_{vap}H = 41$ kJ $mol^{-1}$ at 373 K and 1 bar pressure. Assuming that water vapour is an ideal gas that occupies a much larger volume than liquid water, the internal energy change during evaporation of water is ........ kJ $mol^{-1}$.
[Use $R = 8.3$ J $mol^{-1}$ $K^{-1}$] *[2021, 26 Aug Shift-II]*

**71** 200 mL of 0.2 M HCl is mixed with 300 mL of 0.1 M NaOH. The molar heat of neutralisation of this reaction is −57.1 kJ. The increase in temperature

in °C of the system on mixing is $x \times 10^{-2}$. The value of $x$ is ........ (Nearest integer)
[Given, specific heat of water $= 4.18 \text{ J g}^{-1} \text{ K}^{-1}$
Density of water $= 1.00 \text{ g cm}^{-3}$]
(Assume no volume change on mixing)
*[2021, 27 Aug Shift-I]*

**72** When 400 mL of 0.2 M $H_2SO_4$ solution is mixed with 600 mL of 0.1 M NaOH solution, the increase in temperature of the final solution is .............. $\times 10^{-2}$ K. (Round off to the nearest integer).
[Use : $H^+(aq) + OH^+(aq) \longrightarrow H_2O$;
$\Delta_y H = -57.1 \text{ kJ mol}^{-1}$]
Specific heat of $H_2O = 0.18 \text{ J K}^{-1} \text{g}^{-1}$, density of $H_2O = 1.0 \text{ g cm}^{-3}$.
Assume no change in volume of solution on mixing.
*[2021, 27 July Shift-II]*

**73** For water at 100°C and 1 bar,
$\Delta_{vap} H - \Delta_{vap} U = \text{........} \times 10^2 \text{J mol}^{-1}$.
(Round off to the nearest integer)
[Use : $R = 8.31 \text{ J mol}^{-1} \text{ K}^{-1}$]
[Assume volume of $H_2O(l)$ is much smaller than volume of $H_2O(g)$. Assume $H_2O(g)$ treated as an ideal gas]
*[2021, 27 July Shift-I]*

**74** At 298 K, the enthalpy of fusion of a solid ($X$) is 2.8 kJ $\text{mol}^{-1}$ and the enthalpy of vaporisation of the liquid ($X$) is 98.2 kJ $\text{mol}^{-1}$. The enthalpy of sublimation of the substance ($X$) in kJ $\text{mol}^{-1}$ is ..... (Nearest integer)
*[2021, 25 July Shift-I]*

**75** If the standard molar enthalpy change for combustion of graphite powder is $-2.48 \times 10^2 \text{kJ mol}^{-1}$, the amount of heat generated on combustion of 1 g of graphite powder is ..... kJ. (Nearest integer)
*[2021, 22 July Shift-II]*

**76** For the reaction, $C_2H_6 \longrightarrow C_2H_4 + H_2$
the reaction enthalpy $\Delta_r H = \text{.........}$ kJ $\text{mol}^{-1}$
(Round off to the nearest integer).
[Given : Bond enthalpies in
kJ $\text{mol}^{-1}$ : C—C = 347, C=C = 611;
C—H = 414; H—H = 436]
*[2021, 18 March Shift-I]*

**77** The standard enthalpies of formation of $Al_2O_3$ and CaO are $-1675$ kJ $\text{mol}^{-1}$ and $-625$ kJ $\text{mol}^{-1}$ respectively.
For the reaction,
$3CaO + 2Al \longrightarrow 3Ca + Al_2O_3$ the standard reaction enthalpy $\Delta_r H°$ ........ kJ.
(Round off to the nearest integer). *[2021, 17 March Shift-I]*

**78** The average S—F bond energy in kJ $\text{mol}^{-1}$ of $SF_6$ is ........... (Rounded off to the nearest integer)
[Given, the values of standard enthalpy of formation of $SF_6(g)$, $S(g)$ and $F(g)$ are $-1100$, 275 and 80 kJ $\text{mol}^{-1}$ respectively. ]
*[2021, 26 Feb Shift-II]*

**79** The reaction of cyanamide, $NH_2CN(s)$ with oxygen was run in a bomb calorimeter and $\Delta U$ was found to be $-742.24$ kJ $\text{mol}^{-1}$. The magnitude of $\Delta H_{298}$ for the reaction,

$$NH_2CN(s) + \frac{3}{2}O_2(g) \longrightarrow N_2(g) + CO_2(g) + H_2O(l)$$

is ......... kJ. (Rounded off to the nearest integer)
[Assume ideal gases and $R = 8.314 \text{ J mol}^{-1}\text{K}^{-1}$]
*[2021, 25 Feb Shift-I]*

**80** The internal energy change (in J) when 90 g of water undergoes complete evaporation at 100°C is ........,
(Given : $\Delta H_{vap}$ for water at 373 K = 41kJ/mol,
$R = 8.314 \text{ JK}^{-1} \text{ mol}^{-1}$)
*[2020, 5 Sep Shift-I]*

## Topic 4 Second Law of Thermodynamics, Entropy, Free Energy Change and Spontaneity

### Level 1

**01** Which of the following relations are correct?
A. $\Delta U = q + p\Delta V$
B. $\Delta G = \Delta H - T\Delta S$
C. $\Delta S = \dfrac{q_{rev}}{T}$
D. $\Delta H = \Delta U - \Delta nRT$
*[2023, 29 Jan Shift-II]*
(a) A and C
(b) B only
(c) B and C
(d) A and D

**02** Which of the following relation is not correct?
*[2022, 28 July Shift-I]*
(a) $\Delta H = \Delta U - p\Delta V$
(b) $\Delta U = q + W$
(c) $\Delta S_{sys} + \Delta S_{surr} \geq 0$
(d) $\Delta G = \Delta H - T\Delta S$

**03** A process will be spontaneous at all temperature if
*[2019, 10 April Shift-I]*
(a) $\Delta H > 0$ and $\Delta S < 0$
(b) $\Delta H < 0$ and $\Delta S > 0$
(c) $\Delta H < 0$ and $\Delta S < 0$
(d) $\Delta H > 0$ and $\Delta S > 0$

**04** For which of the following processes, $\Delta S$ is negative?
*[JEE Main 2018]*
(a) C(diamond) $\longrightarrow$ C(graphite)
(b) $N_2(g, 1 \text{ atm}) \longrightarrow N_2(g, 5 \text{ atm})$
(c) $N_2(g, 273 \text{ K}) \longrightarrow N_2(g, 300 \text{ K})$
(d) $H_2(g) \longrightarrow 2H(g)$

**05** The correct relationship between free energy change in a reaction and the corresponding equilibrium constant $K_c$ is *[AIEEE 2003]*
(a) $\Delta G = RT \ln K_c$ (b) $-\Delta G = RT \ln K_c$
(c) $\Delta G° = RT \ln K_c$ (d) $-\Delta G° = RT \ln K_c$

## Level 2

**06** Ice at −5°C is heated to become vapour with temperature of 110°C at atmospheric pressure. The entropy change associated with this process can be obtained from *[2025, 23 Jan Shift-I]*

(a) $\int_{268K}^{383K} C_p\, dT + \frac{\Delta H_{melting}}{273} + \frac{\Delta H_{boiling}}{373}$

(b) $\int_{268K}^{383K} C_p dT + \frac{q_{rev}}{T}$

(c) $\int_{268K}^{273K} C_{p,m}\, dT + \frac{\Delta H_m, \text{fusion}}{T_f} + \frac{\Delta H_m, \text{vaporsation}}{T_b} + \int_{273K}^{373K} C_{p,m} dT + \int_{373K}^{383K} C_{p,m}\, dT$

(d) $\int_{268K}^{273K} \frac{C_{p,m}}{T}\, dT + \frac{\Delta H_m, \text{fusion}}{T_f} + \frac{\Delta H_m, \text{vaporisation}}{T_b} + \int_{273K}^{373K} \frac{C_{p,m} dT}{T} + \int_{373K}^{383K} \frac{C_{p,m}\, dT}{T}$

**07** The effect of temperature on spontaneity of reactions are represented as

| | $\Delta H$ | $\Delta S$ | Temperature | Spontaneity |
|---|---|---|---|---|
| (A) | + | − | any $T$ | Non-spontaneous |
| (B) | + | + | low $T$ | Spontaneous |
| (C) | − | − | low $T$ | Non-spontaneous |
| (D) | − | + | any $T$ | Spontaneous |

The incorrect combinations are *[2025, 23 Jan Shift-I]*
(a) (B) and (D) only (b) (A) and (C) only
(c) (B) and (C) only (d) (A) and (D) only

**08** Let us consider an endothermic reaction which is non-spontaneous at the freezing point of water. However, the reaction is spontaneous at boiling point of water. Choose the correct option. *[2025, 24 Jan Shift-I]*
(a) Both $\Delta H$ and $\Delta S$ are (−ve)
(b) Both $\Delta H$ and $\Delta S$ are (+ve)
(c) $\Delta H$ is (+ve) but $\Delta S$ is (−ve)
(d) $\Delta H$ is (−ve) but $\Delta S$ is (+ve)

**09** Which of the following is not correct? *[2024, 29 Jan Shift-I]*
(a) $\Delta G$ is positive for a spontaneous reaction
(b) $\Delta G$ is positive for a non-spontaneous reaction
(c) $\Delta G$ is zero for a reversible reaction
(d) $\Delta G$ is negative for a spontaneous reaction

**10** Match List-I with List-II.

| List-I | List-II |
|---|---|
| A. Spontaneous process | I. $\Delta H < 0$ |
| B. Process with $\Delta p = 0, \Delta T = 0$ | II. $\Delta G_{T,p} < 0$ |
| C. $\Delta H_{reaction}$ | III. Isothermal and isobaric process |
| D. Exothermic process | IV. [Bond energies of molecules in reactants] − [Bond energies of product molecules] |

Choose the correct answer from the options given below. *[2022, 27 June Shift-I]*

| | A | B | C | D |
|---|---|---|---|---|
| (a) | III | II | IV | I |
| (b) | II | III | IV | I |
| (c) | II | III | I | IV |
| (d) | II | I | III | IV |

**11** The incorrect expression among the following is *[2021, 31 Aug Shift-II]*
(a) $\frac{\Delta G_{System}}{\Delta S_{Total}} = -T$ (at constant $p$)
(b) $\ln K = \frac{\Delta H° - T\Delta S°}{RT}$
(c) $K = e^{-\frac{\Delta G°}{RT}}$
(d) For isothermal process, $W_{reversible} = -nRT \ln \frac{V_f}{V_i}$

**12** The true statement amongst the following is *[2020, 9 Jan Shift-II]*
(a) $S$ is not a function of temperature but $\Delta S$ is a function of temperature.
(b) Both $\Delta S$ and $S$ are functions of temperature.
(c) Both $S$ and $\Delta S$ are not functions of temperature.
(d) $S$ is a function of temperature but $\Delta S$ is not a function of temperature.

**13** The process with negative entropy change is
(a) synthesis of ammonia from $N_2$ and $H_2$
(b) dissociation of $CaSO_4(s)$ to $CaO(s)$ and $SO_3(g)$
(c) dissolution of iodine in water
(d) sublimation of dry ice *[2019, 10 Jan Shift-II]*

**14** The incorrect match in the following is *[2019, 12 April Shift-II]*
(a) $\Delta G° < 0, K > 1$ (b) $\Delta G° = 0, K = 1$
(c) $\Delta G° > 0, K < 1$ (d) $\Delta G° < 0, K < 1$

**15** The reaction, $MgO(s) + C(s) \longrightarrow Mg(s) + CO(g)$, for which $\Delta_r H° = +491.1$ kJ mol$^{-1}$ and $\Delta_r S° = 198.0$ JK$^{-1}$mol$^{-1}$, is not feasible at 298 K. Temperature above which reaction will be feasible is *[2019, 11 Jan Shift-II]*
(a) 2040.5 K (b) 1890.0 K
(c) 2380.5 K (d) 2480.3 K

**16** For the chemical reaction, $X \rightleftharpoons Y$, the standard reaction Gibbs energy depends on temperature $T$ (in $K$) as $\Delta_r G°$ (in kJ mol$^{-1}$) $= 120 - \frac{3}{8}T$

The major component of the reaction mixture at $T$ is *[2019, 11 Jan Shift-I]*

(a) $Y$ if $T = 280$ K (b) $X$ if $T = 350$ K
(c) $X$ if $T = 315$ K (d) $Y$ if $T = 300$ K

**17** A process has $\Delta H = 200$ J mol$^{-1}$ and $\Delta S = 40$ JK$^{-1}$ mol$^{-1}$. Out of the values given below, choose the minimum temperature above which the process will be spontaneous *[2019, 10 Jan Shift-II]*

(a) 20 K (b) 4 K
(c) 5 K (d) 12 K

**18** The entropy change involved in the isothermal reversible expansion of 2 moles of an ideal gas from a volume of 10 dm$^3$ to a volume of 100 dm$^3$ at 27°C is *[AIEEE 2011]*

(a) 38.3 J mol$^{-1}$ K$^{-1}$ (b) 35.8 J mol$^{-1}$ K$^{-1}$
(c) 32.3 J mol$^{-1}$ K$^{-1}$ (d) 42.3 J mol$^{-1}$ K$^{-1}$

**19** For a particular reversible reaction, at temperature $T$, $\Delta H$ and $\Delta S$ were found to be both +ve. If $T_e$ is the temperature at equilibrium, the reaction would be spontaneous when *[AIEEE 2010]*

(a) $T_e > T$ (b) $T > T_e$
(c) $T_e$ is 5 times $T$ (d) $T = T_e$

**20** Standard entropy of $X_2, Y_2$ and $XY_3$ are 60, 40 and 50 JK$^{-1}$ mol$^{-1}$, respectively. For the reaction, $\frac{1}{2}X_2 + \frac{3}{2}Y_2 \longrightarrow XY_3$, $\Delta H = -30$ kJ, to be at equilibrium, the temperature will be *[AIEEE 2008]*

(a) 1250 K (b) 500 K
(c) 750 K (d) 1000 K

**21** Identify the correct statement regarding a spontaneous process. *[AIEEE 2007]*

(a) For a spontaneous process in an isolated system, the change in entropy is positive
(b) Endothermic processes are never spontaneous
(c) Exothermic processes are always spontaneous
(d) Lowering of energy in the reaction process is the only criteria for spontaneity

**22** In conversion of limestone to lime,

$$CaCO_3(s) \longrightarrow CaO(s) + CO_2(g)$$

the values of $\Delta H°$ and $\Delta S°$ are $+179.1$ kJ mol$^{-1}$ and 160.2 J / K, respectively at 298 K and 1 bar. Assuming that $\Delta H°$ and $\Delta S°$ do not change with temperature, the temperature above which conversion of limestone to lime will be spontaneous is *[AIEEE 2007]*

(a) 1008 K (b) 1200 K
(c) 845 K (d) 1118 K

**23** In an irreversible process taking place at constant $T$ and $p$ and in which only pressure-volume work is being done, the change in Gibbs free energy ($dG$) and change in entropy ($dS$), satisfy the criteria *[AIEEE 2003]*

(a) $(dS)_{V,E} < 0, (dG)_{T,p} < 0$
(b) $(dS)_{V,E} > 0, (dG)_{T,p} < 0$
(c) $(dS)_{V,E} = 0, (dG)_{T,p} = 0$
(d) $(dS)_{V,E} = 0, (dG)_{T,p} > 0$

**24** A reaction is non-spontaneous at the freezing point of water but is spontaneous at the boiling point of water, then at freezing point *[AIEEE 2002]*

| | $\Delta H$ | $\Delta S$ | | $\Delta H$ | $\Delta S$ |
|---|---|---|---|---|---|
| (a) | + ve | + ve | (b) | – ve | – ve |
| (c) | – ve | + ve | (d) | + ve | – ve |

## Level 3

**25.** Which of the following graphs correctly represents the variation of thermodynamic properties of Haber's process? *[2025, 2 April Shift-II]*

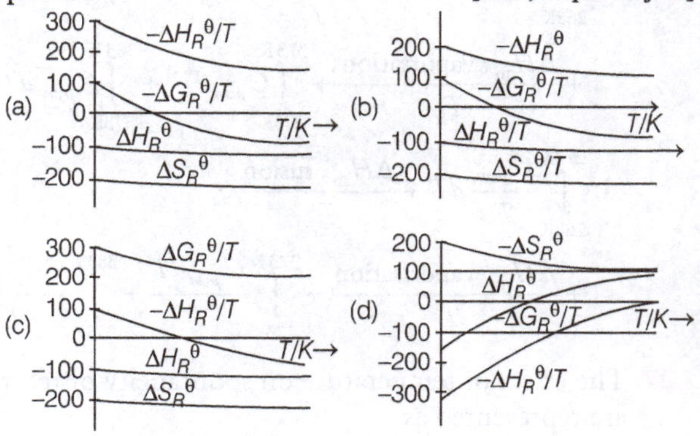


**Imp.** Partial Derivatives of Thermodynamic Quantity

**26** Match List-I with List-II.

| | List-I (Partial derivatives) | | List-II (Thermodynamic quantity) |
|---|---|---|---|
| (A) | $\left(\frac{\partial G}{\partial T}\right)_p$ | (I) | $C_p$ |
| (B) | $\left(\frac{\partial H}{\partial T}\right)_p$ | (II) | $-S$ |
| (C) | $\left(\frac{\partial G}{\partial p}\right)_T$ | (III) | $C_V$ |
| (D) | $\left(\frac{\partial U}{\partial T}\right)_V$ | (IV) | $V$ |

Choose the correct answer from the options given below *[2025, 22 Jan Shift-II]*

(a) (A)-(I), (B)-(II), (C)-(IV), (D)-(III)
(b) (A)-(II), (B)-(III), (C)-(I), (D)-(IV)
(c) (A)-(II), (B)-(I), (C)-(IV), (D)-(III)
(d) (A)-(II), (B)-(I), (C)-(III), (D)-(IV)

**27** Consider the graph of Gibbs free energy G vs extent of reaction. The number of statements from the following which are true with respect to points (*a*), (*b*) and (*c*) is .......... **[2023, 6 April Shift-I]**

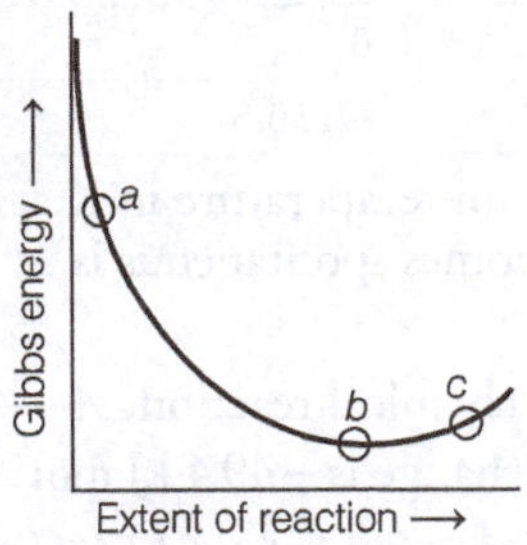


A. reaction is spontaneous at (*a*) and (*b*)
B. reaction is at equilibrium at point (*b*) and non-spontaneous at point (*c*)
C. Reaction is spontaneous at (*a*) and non-spontaneous at (*c*)
D. Reaction is non-spontaneous at (*a*) and (*b*)

**28** During which of the following processes, does entropy decrease ?
A. Freezing of water to ice at 0°C.
B. Freezing of water to ice at −10°C.
C. $N_2(g) + 3H_2(g) \longrightarrow 2NH_3(g)$
D. Adsorption of CO(*g*) and lead surface.
E Dissolution of NaCl in water. **[2021, 17 March Shift-II]**
(a) A, B, C and D (b) B and C
(c) A and E (d) A, C and E

**Imp.** Standard Free Enthalpy of Formation

**29** The following reaction is performed at 298K
$$2NO(g) + O_2(g) \rightleftharpoons 2NO_2(g)$$
The standard free energy of formation of NO(*g*) is 86.6 kJ/mol at 298 K. What is the standard free energy of formation of $NO_2(g)$ at 298 K? $(K_p = 1.6 \times 10^{12})$ **[JEE Main 2015]**
(a) $R(298)\ln(1.6\times 10^{12}) - 86600$
(b) $86600 + R(298)\ln(1.6\times10^{12})$
(c) $86600 - \dfrac{\ln(1.6\times10^{12})}{R(298)}$
(d) $0.5\,[2\times 86600 - R\,(298)\ln(1.6\times10^{12})]$

**30** In a fuel cell, methanol is used as fuel and oxygen gas is used as an oxidiser. The reaction is
$$CH_3OH(l) + \frac{3}{2}O_2(g) \longrightarrow CO_2(g) + 2H_2O(l)$$
At 298 K standard Gibb's energies of formation for $CH_3OH(l)$, $H_2O(l)$ and $CO_2(g)$ are −166.2, −237.2 and − 394.4 kJ $mol^{-1}$, respectively. If standard enthalpy of combustion of methanol is −726 kJ $mol^{-1}$, efficiency of the fuel cell will be **[AIEEE 2009]**
(a) 80% (b) 87% (c) 90% (d) 97%

## Numerical Value Type Questions

**31** Standard entropies of $X_2, Y_2$ and $XY_5$ are 70, 50 and 110 J $K^{-1}mol^{-1}$ respectively. The temperature in Kelvin at which the reaction **[2025, 24 Jan Shift-I]**
$$\frac{1}{2}X_2 + \frac{5}{2}Y_2 \rightleftharpoons XY_5;\ \Delta H^{\ominus} = -35\ \text{kJ mol}^{-1}$$
will be at equilibrium is...........(Nearest integer)

**32** The standard enthalpy and standard entropy of decomposition of $N_2O_4$ to $NO_2$ are 55.0 kJ $mol^{-1}$ and 175.0 J/K/mol respectively. The standard free energy change for this reaction at 25°C in J $mol^{-1}$ is ...... (Nearest integer) **[2025, 23 Jan Shift-I]**

**33** When $\Delta H_{vap} = 30$ kJ/mol and $\Delta S_{vap} = 75$ J $mol^{-1}$ $K^{-1}$, then the temperature of vapour, at one atmosphere is ........... K. **[2024, 9 April Shift-II]**

**34** For the reaction at 298 K, $2A + B \to C$. $\Delta H = 400$ kJ $mol^{-1}$ and $\Delta S = 0.2$ kJ $mol^{-1}$ $K^{-1}$. The reaction will become spontaneous above .............. K. **[2024, 6 April Shift-II]**

**35** For a certain reaction at 300 K, $K = 10$, then $\Delta G°$ for the same reaction is ........ $\times 10^{-1}$ kJ $mol^{-1}$ (Given, $R = 8.314$ J$K^{-1}$ $mol^{-1}$) **[2024, 1 Feb Shift-II]**

**36** For a certain thermochemical reaction, $M \to N$ at $T = 400$ K, $\Delta H^{\ominus} = 77.2$ kJ $mol^{-1}$, $\Delta S = 122$ J$K^{-1}$, log equilibrium constant (log *K*) is −......$\times 10^{-1}$. **[2024, 27 Jan Shift-II]**

**37** The value of log *K* for the reaction $A \rightleftharpoons B$ at 298 K is ......... . (Nearest integer)
Given: $\Delta H° = -54.07$ kJ $mol^{-1}$
$\Delta S° = 10$ J$K^{-1}$ $mol^{-1}$
(Take $2.303 \times 8.314 \times 298 = 5705$) **[2023, 6 April Shift-I]**

**38** One mole of an ideal gas at 350 K is in a 2.0 L vessel of thermally conducting walls, which are in contact with the surroundings. It undergoes isothermal reversible expansion from 2.0 L to 3.0 L against a constant pressure of 4 atm. The change in entropy of the surroundings ($\Delta S$) is .......... $JK^{-1}$. (Nearest integer)
Given, $R = 8.314\ JK^{-1}\ mol^{-1}$ *[2023, 12 April Shift-I]*

**39** Consider the following reaction approaching equilibrium at 27°C and 1 atm pressure.

$$A + B \underset{k_b = 10^2}{\overset{k_f = 10^3}{\rightleftharpoons}} C + D$$

The standard Gibb's energy change ($\Delta_r G^\circ$) at 27°C is (−) .......... kJ $mol^{-1}$. (Nearest integer).
(Given: $R = 8.3\ J\ K^{-1}\ mol^{-1}$ and $\ln 10 = 2.3$)
*[2023, 29 Jan Shift-I]*

**40** The standard free energy change ($\Delta G^\circ$) for 50% dissociation of $N_2O_4$ into $NO_2$ at 27°C and 1 atm pressure is $-x$ J $mol^{-1}$. The value of $x$ is .......... (Nearest integer) [Given : $R = 8.31\ J\ K^{-1} mol^{-1}$, $\log 1.33 = 0.1239$ $\ln 10 = 2.3$] *[2022, 25 June Shift-I]*

**41** The standard entropy change for the reaction, $4Fe(s) + 3O_2(g) \longrightarrow 2Fe_2O_3(s)$ is $-550\ JK^{-1}$ at 298 K.
[Given, the standard enthalpy change for the reaction is − 165 kJ $mol^{-1}$]. The temperature in K at which the reaction attains equilibrium is ... [Nearest integer] *[2022, 25 June Shift-I]*

**42** For the reaction,
$2NO_2(g) \rightleftharpoons N_2O_4(g)$, when
$\Delta S = -176.0\ JK^{-1}$ and $\Delta H = -57.8$ kJ $mol^{-1}$, the magnitude of $\Delta G$ at 298 K for the reaction is .......... kJ $mol^{-1}$. (Nearest integer) *[2021, 1 Sep Shift-II]*

**43** Data given for the following reaction is as follows.

$$FeO(s) + C_{(graphite)} \longrightarrow Fe(s) + CO(g)$$

| Substance | $\Delta H^\circ$ (kJ $mol^{-1}$) | $\Delta S^\circ$ (J $mol^{-1}$ $K^{-1}$) |
|---|---|---|
| FeO(*s*) | −266.3 | 57.49 |
| $C_{(graphite)}$ | 0 | 5.74 |
| Fe(*s*) | 0 | 27.28 |
| CO(*g*) | −110.5 | 197.6 |

The minimum temperature in K at which the reaction becomes spontaneous is .......... (Integer answer) *[2021, 27 Aug Shift-II]*

**44** For a given chemical reaction, $A \longrightarrow B$ at 300 K the free energy change is −49.4 kJ $mol^{-1}$ and the enthalpy of reaction is 51.4 kJ $mol^{-1}$. The entropy change of the reaction is ...... $JK^{-1} mol^{-1}$. *[2021, 20 July Shift-II]*

**45** Assuming ideal behaviour, the magnitude of log *K* for the following reaction at 25°C is $-x \times 10^{-1}$. The value of $x$ is .......... (Integer answer)

$$3HC{\equiv}CH(g) \rightleftharpoons C_6H_6(l)$$

[Given : $\Delta_f G^0(HC{\equiv}CH) = -2.04 \times 10^5\ J\ mol^{-1}$ ; $\Delta_f G^0(C_6H_6) = -1.24 \times 10^5\ J\ mol^{-1}$; $R = 8.314\ J\ K^{-1}mol^{-1}$] *[2021, 24 Feb Shift-II]*

**46** For the reaction $A_{(g)} \rightarrow B_{(g)}$, the value of the equilibrium constant at 300 K and 1 atm is equal to 100.0. The value of $\Delta_r G$ for the reaction at 300 K and 1 atm in J $mol^{-1}$ is $-xR$ where $x$ is ........... (Rounded off to the nearest integer)
[$R = 8.31\ J\ mol^{-1}K^{-1}$ and $\ln 10 = 2.3$] *[2021, 01 Feb Shift-I]*

**47** For a dimerisation reaction

$$2A(g) \longrightarrow A_2(g),$$

at 298 K, $\Delta U^\ominus = -20$ kJ $mol^{-1}$,
$\Delta S^\ominus = -30\ JK^{-1}\ mol^{-1}$, then the $\Delta G^\ominus$ will be − .......... J. *[2020, 5 Sep Shift-II]*

**48** For the reaction; $A(l) \longrightarrow 2B(g)$
$\Delta U = 2.1$ kcal, $\Delta S = 20$ cal $K^{-1}$ at 300 K.
Hence, $\Delta G$ in kcal is − ...... . *[2020, 8 Jan Shift-I]*

## Key Index

### Topic 1

| | | | | | | | | | |
|---|---|---|---|---|---|---|---|---|---|
| 1 (a) | 2 (a) | 3 (b) | 4 (d) | 5 (c) | 6 (d) | 7 (c) | 8 (d) | 9 (c) | 10 (55) |
| 11 (28721) | 12 (200) | 13 (1200) | 14 (4) | 15 (620) | 16 (8630) | 17 (2218) | 18 (48) | | |

### Topic 2

| | | | | | | | | | |
|---|---|---|---|---|---|---|---|---|---|
| 1 (c) | 2 (c) | 3 (a) | 4 (d) | 5 (d) | 6 (b) | 7 (a) | 8 (d) | 9 (a) | 10 (c) |
| 11 (b) | 12 (d) | 13 (304) | 14 (200) | 15 (150) | 16 (1200) | 17 (0) | 18 (0) | 19 (50) | 20 (15) |

### Topic 3

| | | | | | | | | | |
|---|---|---|---|---|---|---|---|---|---|
| 1 (a) | 2 (d) | 3 (a) | 4 (c) | 5 (b) | 6 (c) | 7 (b) | 8 (b) | 9 (c) | 10 (b) |
| 11 (a) | 12 (b) | 13 (c) | 14 (c) | 15 (d) | 16 (c) | 17 (c) | 18 (d) | 19 (c) | 20 (c) |
| 21 (a) | 22 (b) | 23 (b) | 24 (c) | 25 (a) | 26 (b) | 27 (d) | 28 (d) | 29 (d) | 30 (c) |
| 31 (b) | 32 (d) | 33 (b) | 34 (c) | 35 (a) | 36 (98) | 37 (25) | 38 (2500) | 39 (95) | 40 (466) |
| 41 (48) | 42 (200) | 43 (2) | 44 (82) | 45 (38) | 46 (274) | 47 (150) | 48 (6535) | 49 (125) | 50 (56) |
| 51 (492) | 52 (4) | 53 (278) | 54 (1411) | 55 (800) | 56 (1006) | 57 (925) | 58 (54) | 59 (35) | 60 (2) |
| 61 (57) | 62 (200) | 63 (2) | 64 (104) | 65 (600) | 66 (195) | 67 (117) | 68 (38) | 69 (738) | 70 (38) |
| 71 (82) | 72 (82) | 73 (31) | 74 (101) | 75 (21) | 76 (128) | 77 (230) | 78 (309) | 79 (741) | 80 (189494) |

### Topic 4

| | | | | | | | | | |
|---|---|---|---|---|---|---|---|---|---|
| 1 (c) | 2 (a) | 3 (b) | 4 (b) | 5 (d) | 6 (d) | 7 (c) | 8 (b) | 9 (a) | 10 (b) |
| 11 (b) | 12 (b) | 13 (a) | 14 (d) | 15 (d) | 16 (c) | 17 (c) | 18 (a) | 19 (b) | 20 (c) |
| 21 (a) | 22 (d) | 23 (b) | 24 (a) | 25 (a) | 26 (c) | 27 (2) | 28 (a) | 29 (d) | 30 (d) |
| 31 (700) | 32 (2850) | 33 (400) | 34 (2000) | 35 (57) | 36 (37) | 37 (10) | 38 (3) | 39 (6) | 40 (710) |
| 41 (300) | 42 (5) | 43 (964) | 44 (336) | 45 (855) | 46 (1380) | 47 (13537.57) | 48 (2.70) | | |

# Detailed **Explanations**

## Topic 1

**01.** *(a)* From change in internal energy $[\Delta U = q + W]$ = state functions
Heat $(q)$ = path function
Work done $(W)$ = path function
Gibbs energy, $G = H - TS$
= state function.

**02.** *(a)* Work done due to change in volume against constant pressure is

$$W = -p(V_2 - V_1)$$
$$= -1\times10^5 \text{ Nm}^{-2} (1\times10^{-2} - 1\times10^{-3}) \text{ m}^3$$
$$= -900 \text{ Nm} = -900 \text{ J} \quad [1 \text{ Nm} = 1 \text{ J}]$$

**03.** *(b)* The correct answer is option (b). The solid phase (ice) disappears completely.

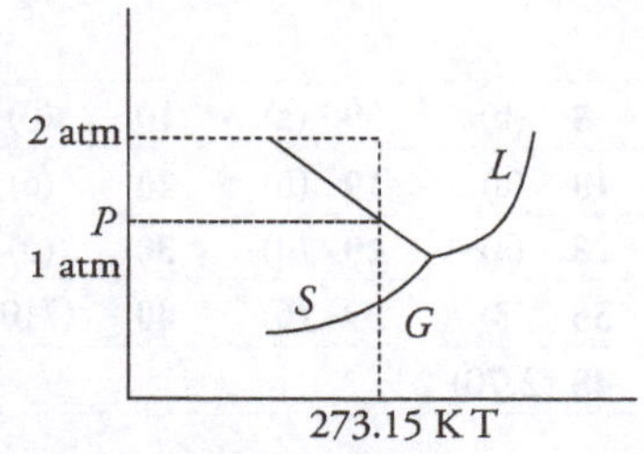


Phase diagram of $H_2O$.
If pressure is made two times, then mixture of ice and water will completely convert into water (liquid) form.

**04.** *(d)* Since in adiabatic system, no heat exchanges between system and surrounding, hence when methane undergo combustion, due to exothermic reaction, the temperature rises in system *(A)*.

$$CH_4 + 2O_2 \longrightarrow CO_2 + 2H_2O + \text{heat}$$

In diathermic container, heat can change between system and surrounding, hence heat releases in to the surrounding and temperature of system *(B)* remains same.

**05.** *(c)* In isothermal expansion,

$$pV_m = K \text{ (constant)}$$

This relation is plotted in graph '*C*'

Likewise, $p = \dfrac{K}{V_m}$

This relation is plotted in graph "*A*".
Thus, graph *B* and *D* are incorrect.

**06.** *(d)* In an ideal gas, there are no intermolecular forces of attraction. Hence, $T_f = T_i$ for both reversible and irreversible processes.

**07.** *(c)* From the given systems of solutions, both solution have same composition, which is 1 mole of *x* in 1 *l*, so all the intensive property will remain same, but as total amount is greater in solution 1 as compared to solution 2. So, all the extensive properties will be different. Hence, Gibbs free energy will be different.

**08.** *(d)* 0.6 kg with 60% purity of coal means

$$0.6 \text{ kg} \times \frac{60}{100} = 0.36 \text{ kg of coal}$$

Out of 0.36 kg of coal, 60% is converting to CO means :

$$0.36 \text{ kg} \times \frac{60}{100} = 0.216 \text{ kg} = 216 \text{ g}$$

Rest will be converting to $CO_2$, means : $0.36 - 0.216 = 0.144$ kg = 144 g
From given equation:

$$C + \frac{1}{2}O_2 \longrightarrow CO + 100 \text{ kJ}$$

means 12g of carbon produces 100 kJ of energy on incomplete combustion
So, 216 g of carbon will produces

$$= \frac{100}{12} \times 216 = 1800 \text{ kJ}$$

From equation:

$$C + O_2 \longrightarrow CO_2 + 400 \text{ kJ}$$

12 g of carbon produces 400 kJ energy on complete combustion.
So, 144 g of carbon will produce

$$\frac{400}{12} \times 144 = 4800 \text{ kJ}$$

Total energy produces

$$= 1800 \text{ kJ} + 4800 \text{ kJ} = 6600 \text{ kJ}$$

**09.** *(c)* For isothermal reversible expansion,

$$|W| = nRT \ln \frac{V_f}{V_i} = nRT \ln \frac{V}{V_i}$$

where, $V$ = final volume,
$V_i$ = initial volume

or $|W| = nRT \ln V - nRT \ln V_i$

On comparing with equation of straight line, $y = mx + c$, we get

slope $= m = +nRT$
intercept $= -nRT \ln V_i$

Thus, plot of $|W|$ with $\ln V$ will give straight line in which slope of $2(T_2)$ is greater than slope of $1(T_1)$ which is given in all options.
Now, if we extend the line of $T_1$ and $T_2$ given in option (c) it seems to be touching the origin. If they touch the origin then *y*-intercept becomes zero which is not possible. Thus, it is not exactly correct answer but among the given options it is the most appropriate one.

**10.** *(55)* Given, $V_1 = 100$ L
and $V_2 = 10$ L
Work done is given,

$$W = -nRT \ln \frac{V_2}{V_1}$$
$$= -1 \times 0.08206 \times 291.15 \times 2.303 \log \frac{10}{100}$$
$$= 55 \text{ L atm}$$

Hence, final answer is 55 L atm.

**11.** *(28721)* It is isothermal reversible expansion, so work done is negative.
Work done is given by

$$W = -2.303 nRT \log\left(\frac{V_2}{V_1}\right)$$
$$= -2.303 \times 5 \times 8.314 \times 300 \log\left(\frac{100}{10}\right)$$
$$= -28720.713 \text{ J}$$
$$\approx -28721 \text{ J}$$

Hence, the value of *x* is 28721.

**12.** *(200)* Work done is given by area enclosed in the *p vs V* cyclic graph.
Sign is +ve for clockwise cyclic process for *V vs p* graph.

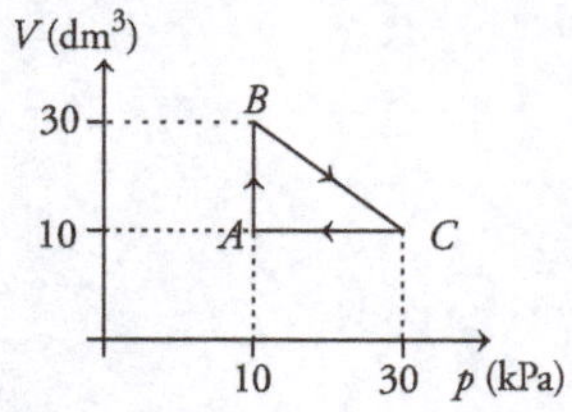


Work done = Area of triangle

$$= \frac{1}{2} \times \text{base} \times \text{height}$$
$$= \frac{1}{2} \times 20 \times 20$$
$$= 200 \text{J}$$

Hence, work done is 200J.

**13.** *(1200)* For isothermal expansion at constant pressure, heat is given by

$Q = -p_{ext}(V_2 - V_1)$

$= -80 \times 1000 \times (45 - 30) \times 10^{-3}$

$= -15 \times 80 = -1200$ J

Negative sign indicates heat is lost. Hence, amount of heat transfer is 1200 J.

**14.** *(4)* **Extensive properties** Mole, volume, Gibb's free energy change

**Intensive properties** Molar mass, molar heat capacity, molarity, $E^\circ_{cell}$

Thus, total number of intensive properties are 4.

**15.** *(620)* $1 \to 2 \Rightarrow$ Isobaric processes

$2 \to 3 \Rightarrow$ Isochoric processes

$3 \to 1 \Rightarrow$ Isothermal processes

Work done

$(W) = W_{1\to2} + W_{2\to3} + W_{3\to1}$

$$= \left(-p(V_2 - V_1) + 0 + \left[-p_1V_1 \ln\left(\frac{V_2}{V_1}\right)\right]\right)$$

$$= \left[-1 \times (40 - 20) + 0 + \left[-1 \times 20 \ln\left(\frac{20}{40}\right)\right]\right]$$

$= -6.2$ bar L

$|W| = 6.2$ bar L $= 620$ J

**16.** *(8630)* Work done $= -nRT \ln \frac{V_2}{V_1}$

$= -5 \times 8.3 \times 300 \ln \frac{20}{10}$

$= -12450 \times 2.303 \log 2$

$= -8630.38$ J

The magnitude of maximum work obtained $= 8630$ J (nearest integer).

**17.** *(2218)*

$Fe(s) + HCl(aq) \to FeCl_2(aq) + H_2(g)$

50g

(Moles of Fe = Moles of $H_2$)

moles $= \frac{50}{55.85}$ 0.8952 mol

$= 0.8952$ mol

Work done can be given by

$W = -p_{ext}\Delta V = -\Delta n_g RT$

$\Delta_{n_g}$ = (gaseous moles on product side) – (gaseous moles on reactant side)

$= 0.8952 - 0$

$W = -0.8952 \times 8.314 \times 298$

$= -2217.92$ J

Closest integer = 2218

**18.** *(48)* Work done is equal to the area under $pV$-curve.

Work done $(W) = W_{AB} + W_{BC}$

$W_{AB}$ = Area of square

$= 6 \times 6 = 36$ Pa. m$^3$

$W_{BC}$ = Area of triangle

$= \frac{1}{2} \times 4 \times 6 = 12$ Pa. m$^3$

$\therefore$ Work done $(W) = (36 + 12)$ Pa. m$^3$

$= 48.00$ Pa. m$^3$ $= 48.00$ J

## Topic 2

**01.** *(c)* For thermally insulated system

$q = 0$

Also from 1st law of thermodynamics

$\Delta U = q + W$

So, $\Delta U = W$

$\Rightarrow W > 0, \Delta U > 0$

So, the correct thermodynamic parameter is

$\Delta U > 0, q = 0, W > 0$

**02.** *(c)*

> **key idea** In this question, use the first law of thermodynamics, to calculate the value of $\Delta U$. Also remember that for isochoric processes, $\Delta V = 0$, for isobaric $\Delta p = 0$, for adiabatic $q = 0$ and for isothermal $\Delta T = 0$.

According to first law of thermodynamics,

$\Delta U = q + W = q - p\Delta V$

In isochoric process $(\Delta V = 0)$, $\Delta U = q$

In isobaric process $(\Delta p = 0)$, $\Delta U = q$

In adiabatic process $(q = 0)$, $\Delta U = W$

In isothermal process $(\Delta T = 0)$ and $\Delta U = 0$

$\therefore \Delta U$ is equal to adiabatic work.

**03.** *(a)* As the process is carrying out at constant temperature, therefore this type of expansion is called **isothermal reversible expansion,** for which, $\Delta U = 0$. Hence, $q = -W$

Thus, $q = +208$ J, $W = -208$ J

**04.** *(d)* $\Delta U = 0$, in a cyclic process.

**05.** *(d)* Given, $T = 300$K

$V_1 = 10$ dm$^3$

$V_2 = 20$ dm$^3$

$\Delta U = ?, q = ?, W = ?$

We know that, for isothermal reversible process

$$-q = W = -nRT \ln \frac{V_2}{V_1}$$

$$W = -8.3 \times 300 \times \ln\left(\frac{20}{10}\right)$$

$W \approx -1.718$ kJ

$\Rightarrow q \approx 1.718$ kJ

$\Delta U = 0$ $(\because \Delta T = 0)$

**06.** *(b)* For isothermal process.

$|W_{reversible}|_{expansion}$

$= |W_{reversible}|_{compression}$

$= -nRT \ln \frac{V_f}{V_i}$

$|W_{irreversible}|_{expansion}$

$< |W_{irreversible}|_{compression}$

$d > c = a > b$

$|W_{irreversible}|_{expansion} = -p_{ext}(V_f - V_i)$

**07.** *(a)* As internal energy '$U$' is a state function, its cyclic integral must be zero in a cyclic process.

$\therefore \Delta U$(case I) $= \Delta U$ (case II)

$= \Delta U$ (case III)

**08.** *(d)* Given, $\Delta q = 500$ J, $n = 0.5$ mol

$T = 298$ K $p = 1$ atm,

$R = 8.3$ JK$^{-1}$ mol$^{-1}$, $\Delta U = ?$ $T_f = ?$

Heat is given by the formula,

$q_p = n \times C_p \times \Delta T$

$500 = 0.5 \times \frac{5}{2} \times 8.3(T_f - 298)$

$\left[\because C_p = \frac{5R}{2}\right]$

On solving, we get,

$T_f = 346.2$ K $\simeq 348$ K

Also, we know that

$$\frac{\Delta H}{\Delta U} = \frac{C_p}{C_V} = \frac{5/2}{3/2} = \frac{5}{3}$$

$\Delta U = \frac{3}{5} \times 500 = 300$ J

So, the final temperature and change in internal energy, respectively are 348 K and 300 J.

**09.** *(a)* For free expansion, $p_{external} = 0$

$\therefore$ Work done, $w = -p_{ext}\Delta V = 0$

$\because$ Adiabatic process, $q = 0$

Also from 1st law of thermodynamics.

$\Delta U = q + w$

$\Delta U = 0$

Internal energy is a function of temperature.

If internal energy is constant

$\Delta T = 0$

So, for free expansion, $q = 0$, $\Delta T = 0$, $w = 0$

Hence, option (a) is correct.

**10.** *(c)* $\Delta U = nC_V\Delta T = nC_V(T_2 - T_1)$

$= 5$ mol $\times 28$ JK$^{-1}$ mol$^{-1}$ $\times (200 - 100)$ K

$= 14{,}000$ J $= 14$ kJ

$\Delta pV = nR\Delta T = nR(T_2 - T_1)$

$= 5\text{ mol} \times 8\text{ JK}^{-1}\text{ mol}^{-1} \times (200 - 100)\text{ K}$
$= 4000\text{ J} = 4\text{ kJ}$

**11.** *(b)* From the 1st law of thermodynamics,
$\Delta U = q + w$

(a) **Cyclic process** For cyclic process,
$\Delta U = 0$
$\therefore \quad q = -w$
Thus, option (a) is correct.

(b) **Adiabatic process** For adiabatic process,
$q = 0 \Rightarrow \Delta U = W$
Thus, option (b) is incorrect.

(c) **Isochoric process** For isochoric process,
$\Delta V = 0$. Thus, $W = 0$ $(\because W = p\Delta V)$.
$\therefore \quad \Delta U = q$
Thus, option (c) is correct.

(d) **Isothermal process** For isothermal process,
$\Delta U = 0$
$\therefore \quad q = -W$
Thus, option (b) is correct.

**12.** *(d)* From the condition mention in question, the gas follows free expansion.
$\therefore$In free expansion of gas, $p_{ext} = 0$
where $W = 0$, $q = 0$ and $\Delta U = 0$

**13.** *(304)* Work done from 1 to 2
$W_{1\to 2} = 0$
(As $W = p\Delta V$ and $\Delta V = 0$)
$W_{2\to 3} = -p\Delta V$
$= -3[2-1] = -3$ atm l
$W_{3\to 4} = 0$
Total work done $= -3$ atm l
$= -3 \times 101.3$ Joule
$= -304$ Joule

**14.** *(200)* For isothermal process, $\Delta U = 0$
So; $Q = -W$
$= -[-p_{ext}(V_2 - V_1)]$
$Q = 5[20 - 60]$
$= -200$ atm L

**15.** *(150)* Given,
Number of moles = 1 mole
Temperature $(T_i) = 27°\text{C} = 300$ K
Since, the processes is adiabatic, hence, $q = 0$
$\Rightarrow \quad W = \Delta U$
On substituting values, we will get
$W = nC_V\Delta T$
$-3 \times 10^3 = 1 \times 20\,[T_2 - 300]$
$T_2 - 300 = -150$ or $T_2 = 150$ K

**16.** *(1200)* Power of heater = 60 W
= 60 J/sec
Total energy emitted $= 60 \times 100$
= 6000 J
heat capacity × temperature difference
= total energy emitted
heat capacity × 5°C = 6000
heat capacity $= \dfrac{6000}{5} = 1200$ J/K

**17.** *(0)* For an isothermal processes, the temperature remains constant. So, the change in internal energy is zero.
$\therefore \quad pV = \text{constant}$

**18.** *(0)* Change in volume,
$\Delta V = 20 - 4 = 16$ L
Work done, $W = -p_{ext}\Delta V$
but in vacuum, $p_{ext} = 0$
$\therefore \quad W = -0 \times 16 = 0$
For an isothermal expansion, $\Delta U = 0$
Using first law of thermodynamics,
$\Delta U = q + W \Rightarrow 0 = q + 0 \Rightarrow q = 0$

**19.** *(50)* Work done by the system,
$W = -200$ J
Heat absorbed, $q = +150$ J
According to first law of thermodynamics
Internal energy, $\Delta U = q + W$
$\Delta U = 150 + (-200)$
{$\Delta U$ = Change in internal energy}
$= -50$ J
Magnitude $= |-50\text{ J}| = |\Delta U| = 50$ J

**20.** *(15)* The gas performs isothermal irreversible work ($W$).
where, $\Delta U = 0$ (change in internal energy)
From, 1st law of thermodynamics,
$\Rightarrow \quad \Delta U = q + W$
$\Rightarrow \quad 0 = q + W \Rightarrow q = -W$
Now, $\quad W = -P_{ext}(V_2 - V_1)$

$$= -P_{ext}\left(\frac{nRT}{p_2} - \frac{nRT}{p_1}\right)$$

$$= -P_{ext} \times nRT\left(\frac{1}{p_2} - \frac{1}{p_1}\right)$$

Given, $P_{ext} = 4.3$ MPa, $p_1 = 2.1$ MPa,
$p_2 = 1.3$ MPa, $n = 5$ mol,
$T = 293$ K
and $\quad R = 8.314\text{ J mol}^{-1}\text{ K}^{-1}$

$$= -4.3 \times 5 \times 8.314 \times 293\left(\frac{1}{1.3} - \frac{1}{2.1}\right)$$

$= -15347.70\text{ J mol}^{-1}$
$= -15.347\text{ kJ mol}^{-1} \simeq -15\text{ kJ mol}^{-1}$
$\Rightarrow q = 15\text{ kJ mol}^{-1}$

## Topic 3

**01.** *(a)* The heat of formation of $SO_2(g)$.

$$SO_2(g) + \frac{1}{2}O_2(g) \longrightarrow SO_3(g), \quad \Delta H = -y$$

$\Delta_r H = (\Delta_f H)_{SO_3} - (\Delta_f H)_{SO_2}$
$-y = -2x - (\Delta H_f)_{SO_2}$
$(\Delta_f H)_{SO_2} = y - 2x$ kcal
Hence, option (a) is correct.

**02.** *(d)* C (diamond) $\rightarrow$ C(graphite) $+ X$ kJ/mol
C(diamond) + $O_2(g) \longrightarrow CO_2(g)$ $+ Y$ kJ/mol
C(graphite) + $O_2(g) \longrightarrow CO_2(g)$ $+ Z$ kJ/mol
then $X = Y - Z$ is the correct relation.

**03.** *(a)* The general combustion reaction of a hydrocarbon is as follows :

$$C_xH_y + \left(x + \frac{y}{4}\right)O_2 \longrightarrow xCO_2 + \frac{y}{2}H_2O$$

For heptane, $x = 7$, $y = 16$

$$\Rightarrow C_7H_{16}(l) + 11O_2(g) \longrightarrow 7CO_2(g) + 8H_2O(l)$$

$\therefore \quad \Delta n_g = 7 - 11 = -4$
Now, from the principle of thermochemistry,
$\Delta H = \Delta U + \Delta n_g RT$
$\Rightarrow \quad \Delta H - \Delta U = \Delta n_g RT = -4RT$

**04.** *(c)*

> **key idea** The relation between $\Delta H$ and $\Delta U$ is
> $\Delta H = \Delta U + \Delta n_g RT$
> where, $\Delta n_g = \Sigma n_p - \Sigma n_R$
> = number of moles of gaseous products
> – number of moles of gaseous reactants.

Combustion reaction of C and CO are as follows

$$C(s) + O_2(g) \longrightarrow CO_2(g) \quad \ldots(i)$$

$$CO(g) + \frac{1}{2}O_2 \longrightarrow CO_2(g) \quad \ldots(ii)$$

On subtracting Eq. (ii) from Eq. (i), we get

$$C(s) + \frac{1}{2}O_2(g) \longrightarrow CO(g);$$

$\Delta H = (-393.5 + 283.5)\text{ kJ mol}^{-1}$
$= -110.5\text{ kJ mol}^{-1}$ (approx.)

**05.** *(b)* $\frac{1}{2}Cl_2(g) \longrightarrow Cl^-(aq)$

$$\Delta H = \frac{1}{2}\Delta H_{diss}(Cl_2) + \Delta H_{EA}Cl + \Delta H_{hyd}(Cl^-)$$

$$= \frac{240}{2} - 349 - 381 = -610\text{ kJ mol}^{-1}$$

**06.** *(c)* $H_2O\,(l) \longrightarrow H_2O\,(g)$

$$\Delta n_g = 1 - 0 = 1$$
$$\Delta E = \Delta H - \Delta n_g\, RT$$
$$= 41 - (1 \times 8.3 \times 373 \times 10^{-3})$$
$$[R = 8.3 \times 10^{-3}]$$
$$= 37.9 \text{ kJ mol}^{-1}$$

**07.** *(b)* Calculate $\Delta H - \Delta E$ by using the formula,

$$\Delta H - \Delta E = \Delta n_g RT$$
$$= \frac{1}{2} \times 8.314 \times 298$$
$$= 1238.78 \text{ J mol}^{-1}$$
$$\left[\text{for } C(s) + \frac{1}{2} O_2(g) \longrightarrow CO(g) \quad \Delta n_g = 1 - \frac{1}{2} = \frac{1}{2}\right]$$

**08.** *(b)* The amount of heat required to raise the temperature of one mole of substance through 1°C is called molar heat capacity.

$$C = \frac{q}{T_2 - T_1}$$

**• Concept Enhancer**

Molar heat capacity should not be confused with heat capacity. It is because heat capacity is the amount of heat required to change the temperature of a given amount of matter by 1°C. While the molar heat capacity is the heat required to raise the temperature of 1 mole of a substance by 1°C.

**09.** *(c)* Both Statement I and II are true. At melting point when ice melts, supplied heat is utilised to overcome intermolecular attraction within the molecules so temperature remains constant.

**10.** *(b)* The reaction is as follows

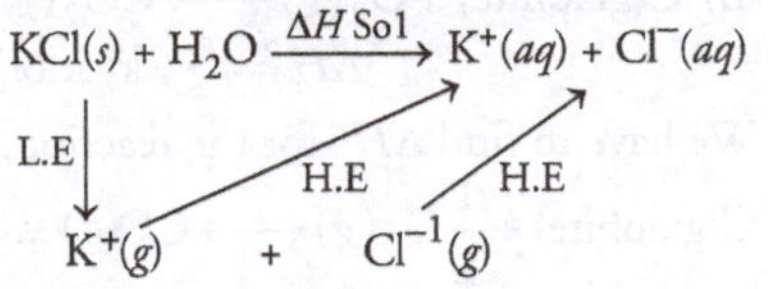


$$\Delta H_{Sol.} = \text{L.E} + (\text{H.E.})_{K^+(g)} + (\text{H.E})_{Cl^{-1}(g)}$$
$$= z - x - y = z - (x + y)$$

**11.** *(a)* According to 2nd law of thermodynamics a reaction is spontaneous, if $\Delta G < 0$

or $\quad \Delta H - T\Delta S < 0$

$$\Rightarrow \quad T > \frac{\Delta H}{\Delta S} = T_e \Rightarrow T > T_e$$

**12.** *(b)* The enthalpy of neutralisation of strong acid and strong base is always −57 kJ/mol. The reason for that is, strong monoacid gives one mole of $H^+$ and strong mono base gives one mole of $OH^-$ which form one mole of water.

**13.** *(c)* Enthalpy of bond dissociation (kJ/mol) at 298.2 K

For hydrogen = 435.88 kJ/mol

For deuterium = 443.35 kJ/ mol

Hence, difference in energy of deuterium and hydrogen is approx. 7.50 kJ/mol

Therefore, equation is $[E_H = E_D - 7.5]$

**14.** *(c)* The enthalpy of solution of an ionic solid is numerically equal to the sum of its hydration and lattice energies.

i.e., $\quad \Delta H^\circ_{sol} = \Delta H^\circ_{hydration} + \Delta H^\circ_{lattice}$

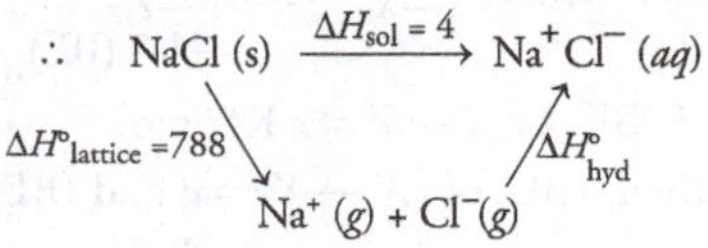


$$\Delta_{sol} H^\circ = \Delta_{lattice} H^\circ + \Delta_{hyd} H^\circ$$
$$4 = 788 + \Delta_{hyd} H^\circ$$
$$\Delta H_{hyd} H^\circ = -784 \text{ kJ/mol}$$

**15.** *(d)* (1) From enthalpy, $H = U + pV$, and for ideal gas, $pV = nRT$

So, $\quad [H = U + nRT]$

and ($H$) depends only on temperature.

(2) For an ideal gas, compressibility factor ($Z = 1$)

(3) $C_p - C_V = R$

(4) For all processes, $[\Delta U = C_V dT]$

**16.** *(c)* Atomisation of $Br_2(l)$

$$Br_2(l) \longrightarrow 2Br(g);$$
$$\Delta H = x \text{ kJ mol}^{-1}$$

which can be considered as sum of following :

$$Br_2(l) \longrightarrow Br_2(g);$$
$$\Delta H = \text{a positive value} \quad ...(i)$$
(endothermic process)

$$Br_2(g) \longrightarrow 2Br(g);\ \Delta H$$
(bond enthalpy of $Br_2$) $= y$ ...(ii)

$\therefore \quad x = y$ + (Positive value), and $x > y$.

**17.** *(c)*

**key idea** When $q$ is the amount of heat involved in a system then at constant pressure

$q = q_p$ and $C_p \Delta T = \Delta H$

Given reaction :

$$I_2(s) \longrightarrow I_2(g)$$

Specific heat of $I_2(s) = 0.055 \text{ cal g}^{-1} \text{ K}^{-1}$.

Specific heat of

$$I_2(\text{vap}) = 0.031 \text{ cal g}^{-1} \text{ K}^{-1}.$$

Enthalpy ($H_1$) of sublimation of iodine = 24 cal $g^{-1}$

If $q$ is the amount of heat involved in a system then at constant pressure $q = q_p$ and

$$\Delta H = C_p \Delta T$$
$$H_2 - H_1 = C_p (T_2 - T_1)$$
$$H_2 = H_1 + \Delta C_p (T_2 - T_1)$$
$$H_2 = 24 + (0.031 - 0.055)(250 - 200)$$
$$H_2 = 24 + (-0.024)(50)$$
$$= 24 - 1.2 = 22.8 \text{ cal/g}$$

**18.** *(d)* From the equation,

$$C_6H_6(l) + \frac{15}{2} O_2(g) \longrightarrow 6CO_2(g) + 3H_2O\,(l)$$

Change in the number of gaseous moles i.e.

$$\Delta n_g = 6 - \frac{15}{2} = -\frac{3}{2} \text{ or } -1.5$$
$$\Delta H_p = \Delta V + \Delta n_g RT$$

So, $\Delta H_p = (-3263.9) + (-1.5) \times 8.314 \times 10^{-3} \times 298$

$$= -3267.6 \text{ kJ mol}^{-1}$$

**19.** *(c)* Required thermochemical reaction is for $\Delta_f H^\circ$ of $CH_4$

Thus, from equation III 890.3

$$890.3 = [\Delta_f H^\circ (CH_4) + 2\Delta_f H^\circ (O_2)] - [\Delta_f H^\circ (CO_2) + 2\Delta_f H^\circ (H_2O)]$$
$$= \Delta_f H^\circ (CH_4) + 0] - [-393.5 - 2 \times 285.8]$$
$$\therefore \Delta_f H^\circ (CH_4) = -74.8 \text{ kJ / mol}$$

**20.** *(c)* Relation between $\Delta H$ (enthalpy change) and $\Delta E$ (internal energy change) is

$$\Delta H = \Delta E + \Delta n_g RT$$

For the given reaction,

$$\Delta n_g = 2 - 3 = -1$$
$$\Rightarrow \quad -1366.5 = \Delta E - 1 \times 8.314 \times 10^{-3} \times 300$$
$$\therefore \quad \Delta E = -1364.0 \text{ kJ mol}^{-1}$$

**21.** *(a)*

$$4NO_2(g) + O_2(g) \xrightarrow{\Delta H = -111 \text{ kJ}} 2N_2O_5(g),$$
$$\Delta H_{sub} = 54 \text{ kJ mol}^{-1}$$

From Hess's law,

$$\Delta H_f + \Delta H_{sub} = \Delta H_{reaction}$$

$$\Delta H_f = \Delta H_{reaction} - \Delta H_{sub}$$
$$= -111\ kJ - (54\ kJ)$$
$$= -111 - 54\ kJ = -165\ kJ$$

Thus, the enthalpy of formation, $\Delta H_f$ for $N_2O_5\ (s)$ is $-165$ kJ.

**22.** *(b)*

$$\frac{1}{2}N_2(g) + \frac{3}{2}H_2(g) \longrightarrow NH_3(g),$$
$$\Delta_r H = -46\ kJ\ mol^{-1}$$
$$H_2(g) \longrightarrow 2H(g)$$

Bond enthalpy of $H_2 = 436\ kJ\ mol^{-1}$

$$N_2(g) \longrightarrow 2N(g)$$

Bond enthalpy of $N_2 = +712\ kJ\ mol^{-1}$

$$\Delta_r H = \left[\frac{1}{2}BE(N_2) + \frac{3}{2}BE(H_2)\right] - 3\ BE(N—H)$$
$$-46 = \left[\frac{1}{2}\times 712 + \frac{3}{2}\times 436\right] - 3\ BE(N—H)$$
$$-46 = (356 + 654) - 3\ BE(N—H)$$
$$3\ BE(N—H) = (1010) + 46$$
$$3\ BE(N—H) = 1056$$
$$BE(N—H) = 1056/3$$
$$= 352\ kJ\ mol^{-1}$$

**23.** *(b)*

$$\Delta H_r = \Delta H_f(H_2O, l) - \Delta H_f(H_2, g) - \frac{1}{2}\Delta H_f(O_2, g)$$
$$= -286.20 = \Delta H_f(H_2O, l) - 0 - 0$$
$$\Delta H_f(H_2O, l) = -286.20$$

Now, consider the ionisation of $H_2O$

$$\Delta H_r = \Delta H_f(H^+, aq) + \Delta H_f(OH^-, aq) - \Delta H_f(H_2O, l)$$
$$57.32 = 0 + \Delta H_f(OH^-, aq) - (-286.20)$$

Thus,

$$\Delta H_f(OH^-, aq) = 57.32 - 286.20$$
$$= -228.88\ kJ$$

**24.** *(c)* $\frac{1}{2}I_2(s) + \frac{1}{2}Cl_2(g) \longrightarrow ICl(g)$

$$\Delta H = \left[\frac{1}{2}\Delta H_{s\to g} + \frac{1}{2}\Delta H_{diss}(Cl_2) + \frac{1}{2}\Delta H_{diss}(I_2)\right] - \Delta H_{ICl}$$
$$= \left(\frac{1}{2}\times 62.76 + \frac{1}{2}\times 242.3 + \frac{1}{2}\times 151.0\right) - 211.3$$
$$= 228.03 - 211.3$$
$$\Delta H = 16.73\ kJ\ mol^{-1} \approx 16.8\ kJ\ mol^{-1}$$

**25.** *(a)* Carbon is found in solid state. The state of substance affects the enthalpy change.

$C(s) \longrightarrow C(g)$ sublimation,

and $H_2(g) \longrightarrow 2H(g)$ dissociation are required for C—H bond formation.

**26.** *(b)* According to relationship of $\Delta H$ and $\Delta E$,

$$\Delta H = \Delta E + \Delta n_g RT$$

For the reaction

$$N_2(g) + 3H_2(g) \longrightarrow 2NH_3(g)$$
$$\Delta n_g = 2 - 4 = -2$$

Thus, $\Delta H = \Delta E - 2RT$

$\Rightarrow$ $\Delta H < \Delta E$

**Note** Numerical value of $\Delta H < \Delta E$ in exothermic reaction and when $\Delta n_g < 0$.

**27.** *(d)* Formation of $XY$ is shown as

$$X_2 + Y_2 \longrightarrow 2XY$$
$$\Delta H = [(BE)_{X—X} + (BE)_{Y—Y}] - [2\,(BE)_{X—Y}]$$

If (BE) of $X—Y = a$ KJ/mol

then, (BE) of $(X—X) = a$ and (BE) of $(Y—Y) = \frac{a}{2}$

$$\therefore -400\ (\text{for 2 mol } XY) = \left(a + \frac{a}{2} - 2a\right)$$
$$-400 = -\frac{a}{2};\ a = +800\ kJ$$

The bond dissociation energy of

$$X_2 = 800\ kJ\ mol^{-1}.$$

**28.** *(d)*

$$CH_2{=}CH_2 + H_2 \longrightarrow CH_3CH_3$$
$$\Delta H = (BE)_{reactants} - (BE)_{products}$$
$$= 4(BE)_{C—H} + (BE)_{C=C} + (BE)_{H—H} - [6\,(BE)_{C—H} + (BE)_{C—C}]$$
$$(4\times 414) + 615 + 435 - (6\times 414) + 347$$
$$= -125\ kJ$$

**29.** *(d)* Among the given statements, option (d) is correct.

For standard state, pressure is 1 bar and temperature is specified only

$$= (\Delta H_f^\circ)_{O_2(g)} = 0$$

**30.** *(c)*

$$\begin{array}{ccc} H_2O(l) & \longrightarrow & H_2O(s) \\ 10°C & & \\ nC_p(l)\Delta T \downarrow & & \uparrow nC_p(s)\Delta T \\ H_2O(l) & \xrightarrow[\text{(fusion)}]{-\Delta T} & H_2O(s) \\ 0°C & & 0°C \end{array}$$

$$\Delta_H = 1\times y\,(0° - 10°) - x\times 100 + 1\times z\,(10° - 0°)$$
$$\Delta_H = -10\,[100x + y + z]\ \text{Joule}$$

**31.** *(b)* From the data given, $\Delta H$ is negative which implies that dissolution of HCl gas is exothermic.

$$HCl(g) + 10H_2O(l) \longrightarrow HCl\cdot 10H_2O \quad ...(i)$$
$$\Delta H_1 = -69.01\ kJ/mol$$
$$HCl(g) + 40H_2O(l) \longrightarrow HCl\cdot 40H_2O \quad ...(ii)$$
$$\Delta H_2 = -72.79\ kJ/mol$$

From above two equations, it is clear that heat of solution depends upon amount of solvent.

Now, subtract Eq. (ii) from (i), we get

$$HCl\cdot 10H_2O + 30H_2O(l) \longrightarrow HCl\cdot 40H_2O$$

so, heat of dilution $= -72.79 - (-69.01)$
$= -3.78$ kJ/mol

So, option (a) is incorrect. For heat of formation reactant should be in elemental form hence option (d) is incorrect.

**32.** *(d)* Consider graph (d)

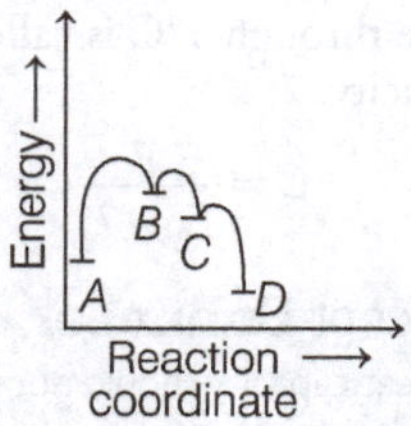


$A \rightarrow B$ (slow)

$\Delta H = +ve$, $E_{a_1}$ = high

$B \rightarrow C$ (fast)

$\Delta H = -ve$, $E_{a_2}$ = low

$C \rightarrow D$ (fast)

$\Delta H = -ve$, $E_{a_3} \rightarrow$ low

This graph correctly matches with given mechanism.

**33.** *(b)* Given equation,

(A) $2CO(g) + O_2(g) \longrightarrow 2CO_2(g)$, $\Delta H_1^\circ = -x\ kJ\ mol^{-1}$

(B) C(graphite) $+ O_2(g) \longrightarrow CO_2(g)$, $\Delta H_2^\circ = -y\ kJ\ mol^{-1}$

We have to find $\Delta H^\circ$ for the reaction,

$$\text{C(graphite)} + \frac{1}{2}O_2(g) \longrightarrow CO(g)\ ...(i)$$
$$\text{C(graphite)} + O_2(g) \longrightarrow CO_2(g)\ ...(ii)$$
$$\Delta H_1 = -y\ kJ/mol^{-1}$$
$$CO_2(g) \longrightarrow CO(g) + \frac{1}{2}O_2(g)\quad ...(iii)$$
$$\Delta H_2 = \frac{x}{2}\ kJ/mole$$

Eq. (i) = Eq. (ii) + Eq. (iii)

$$\therefore \quad \Delta H = -y + \frac{x}{2}$$
$$\Delta H = \frac{x}{2} - y \Rightarrow \Delta H = \frac{x - 2y}{2}$$

**34.** *(c)* The reaction between carbon and hydrogen can be written as

$$2C + 3H_2 \longrightarrow C_2H_6 \quad ...(R)$$

Combustion equation of ethane

$$C_2H_6\,(g) + \frac{7}{2}O_2(g) \longrightarrow 2CO_2 + 3H_2O \quad ...(i)$$

$$\Delta H^\circ_{comb} = -1560 \text{ kJ/mol (given)}$$

$$C(s) + O_2(g) \longrightarrow CO_2(g) \quad ...(ii)$$

$$\Delta H^\circ_{comb} = -394 \text{ kJ/mol}$$

$$H_2(g) + \frac{1}{2}O_2 \longrightarrow H_2O(g) \quad ...(iii)$$

$$\Delta H^\circ_{comb} = -286 \text{ kJ/mol}$$

2 × Eq. (ii) + 3 × Eq. (iii) – Eq (i) = Eq. (*R*)

$$= (-394 \times 2) + (-286 \times 3) - (-1560)$$

$$\Delta H^\circ_f = -86.0 \text{ kJ/mol}$$

**35.** *(a)*

**key idea** Calculate the heat of combustion with the help of following formula

$$\Delta H_p = \Delta U + \Delta n_g RT$$

where, $\Delta H_p$ = Heat of combustion at constant pressure

$\Delta U$ = Heat at constant volume (It is also called $\Delta E$)

$\Delta n_g$ = Change in number of moles (In gaseous state).

$R$ = Gas constant; $T$ = Temperature.

$$C_2H_5OH(l) + 3O_2(g) \longrightarrow 2CO_2(g) + 3H_2O(l)$$

Amount of heat produced in bomb calorimeter,

$$\Delta U = -1364.47 \text{ kJmol}^{-1}$$

Enthalpy of a combustion reaction is

$$\Delta H = \Delta U + \Delta n_g RT$$

As per equation,

$$\Delta n_g = 2 - 3 = -1$$

$$T = 25°C = 25 + 273 \text{ K}$$

$$\Rightarrow \quad T = 298 \text{ K}, R = \frac{8.314}{1000} \text{ kJ mol}^{-1}$$

Thus,

$$\Delta H = -1364.47 + \left[\frac{(-1) \times 8.314 \times 298}{1000}\right]$$

$$\Delta H = -1364.47 - 2.477$$

$$= -1366.947 \text{ kJmol}^{-1}$$

Hence, the enthalpy of combustion, $\Delta_c H$ for the given reaction will be $-1366.947 \text{ kJmol}^{-1}$.

**36.** *(98)* $\Delta H_{\text{Resonance energy}} = \Delta H_{f(\text{experimental})} - \Delta H_{f(\text{theoretical})}$

$\Delta H_{f(\text{exp})}$ for $X_2Y(g) = 80$ kJ / mol

For $\Delta H_{f\,(\text{theoretical})}$

$$X_2(g) + \frac{1}{2}Y_2(g) \longrightarrow X_2Y(g);\ \Delta Hf = ?$$

$$\Delta H_{f(\text{theoretical})} = \left[\text{BE}_{x\equiv x} + \frac{1}{2}\text{BE}_{y=y}\right] - [\text{BE}_{x=x} + \text{BE}_{x=y}]$$

$$= \left(940 + \frac{1}{2} \times 500\right) - (410 + 602)$$

$$= 178 \text{ kJ/ mol}$$

$$\Delta H_{\text{Resonance energy}} = 80 - 178 = -98 \text{ kJ / mol}$$

$$|\Delta H_{\text{Resonance energy}}| = 98$$

**37.** *(25)* Given that,

$$\Delta H^\circ_{sub}\,[(\text{graphite})] = 710 \text{ kJ/mol}$$

$$\Delta_{C-H}H^\circ = 414 \text{ kJ/mol}$$

$$\Delta_{H-H}H^\circ = 436 \text{ kJ/mol}$$

$$\Delta_{C=C}H^\circ = 611 \text{ kJ/mol}$$

$2C(s) + 2H_2(g) \longrightarrow CH_2{=}CH_2(g)$ : $[\Delta H^\circ_F]C_2H_4(g)$

↓

$2C(g) + 4H(g)$ ↗

$$[\Delta H^\circ_F]_{C_2H_4(g)} = 2 \times [\Delta H^\circ_{sub}]_{C(s)} + 2\Delta H^\circ_{H-H} - 1 \times \Delta H^\circ_{C=C} - 4\Delta H^\circ_{C-H}$$

$$[\Delta H^\circ_F]_{C_2H_4\,(g)} = (2 \times 710) + (2 \times 436) - 611 - 4 \times 414$$

$$[\Delta H^\circ_F]_{C_2H_4(g)} = 25 \text{ kJ/mol}$$

**38.** *(2500)*

$$\text{Mole of octane} = \frac{1.14}{114} = 0.01 \text{ mol}$$

$$\text{Heat evolved} = C \times \Delta T = 5 \times 5 = 25 \text{ kJ}$$

∴Magnitude of heat of combustion

$$= \frac{25}{0.01} = 2500 \text{ kJ / mol}$$

**39.** *(95)*

**Key idea**

This question is based upon the concept of Hess's law to calculate the required $\Delta H$ value for the given reaction. First obtain the required equation by multiplying Eq. (ii) by 2, (iii) by 3 then from this value subtract Eq. (i). After that calculate the value by using

$$2\Delta H_2 + 3\Delta H_3 - \Delta H_1$$

Given,

$$C_2H_6(g) + \frac{7}{2}O_2(g) \longrightarrow 2CO_2(g) + 3H_2O(l);$$

$$\Delta H_1^\circ = -1550 \text{ kJ/mol} \quad ...(i)$$

$$\text{C(Graphite)} + O_2(g) \longrightarrow CO_2(g) \quad ...(ii)$$

$$\Delta H_2^\circ = -393.5 \text{ kJ/mol}$$

$$H_2(g) + \frac{1}{2}O_2(g) \longrightarrow H_2O(l) \quad ...(iii)$$

$$\Delta H_3^\circ = -286 \text{ kJ/mol}$$

$$2\text{C(graphite)} + 3H_2(g) \longrightarrow C_2H_6(g);$$

$$\Delta H_f(C_2H_6(g)) = ?$$

$$\Delta H_f = 2\Delta H_2^\circ + 3\Delta H_3^\circ - \Delta H_1^\circ$$

$$= 95 \text{ kJ/mole}$$

**40.** *(466)* The reaction involved is as follows

$$\frac{1}{2}H_2(g) \longrightarrow H(g);$$

$$\Delta_f H(H(g)) = 220 \text{ kJ/ mol} \quad ...(i)$$

$$\frac{1}{2}O_2(g) \longrightarrow O(g);$$

$$\Delta_f H(O_g) = 250 \text{ kJ/mol} \quad ...(ii)$$

[Eq (i) × 2 + Eq (ii)] –2 [BE(O—H)]

$$H_2(g) + \frac{1}{2}O_2(g) \longrightarrow H_2O(g)$$

$$\Delta H_f(H_2O(l)) = -242 = 440 + 250 - 2(\text{BE(O—H)})$$

$$\text{BE(O—H)} = 466 \text{ kJ/ mol}$$

**41.** *(48)* Given, $\Delta H_f\,[CO_2(g)] = -393.5$ kJ/mole

$$\Delta H_f\,[H_2O(l)] = -286 \text{ kJ/mole}$$

$$\Delta H_C\,[C_6H_6] = -3267.0 \text{ kJ/mole}$$

$$\Delta H_f C_6H_6 = ?$$

$$C_6H_6 + \frac{15}{2}O_2(g) \rightarrow 6CO_2(g) + 3H_2O(l)$$

$$\Delta H_R = \Delta H_C = \Sigma\Delta H_f(P) - \Sigma\Delta H_f(R)$$

$$-3267 = 6 \times (-393.5) + 3 \times (-286) - \Delta H_f(C_6H_6)$$

$$\therefore \Delta H_f(C_6H_6) = 48 \text{ kJ/mole}$$

**42.** *(200)* Bond dissociation enthalpy of $X_2$ = ?

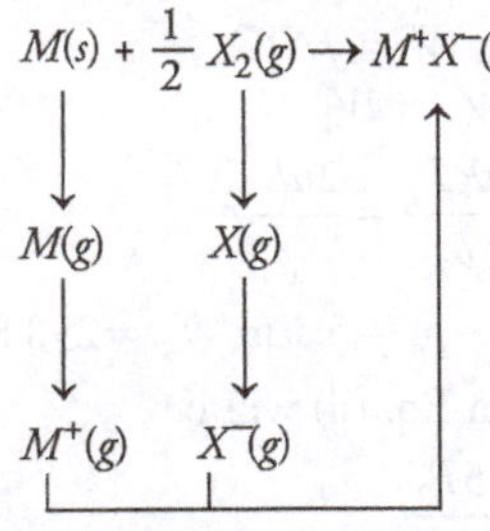


So, the value of $\Delta H_f(MX)$ can be calculated by using

$$\Delta H_f(MX) = \Delta H_{sub}(M) + \text{IE}(M) + \frac{1}{2}[\text{BE}(X - X)] + \text{EG}(X) + \text{LE}(MX)$$

Substitute the values from given data

$$-400 = 100 + 500 + \frac{1}{2}(BE) + (-300) + (-800)$$

$\therefore$ $BE = 200$ kJ/mole

**43.** *(2)* According to question

$H^+ + OH^- \longrightarrow H_2O \Rightarrow x$ (for HCl)

$2H^+ + 2OH^- \longrightarrow 2H_2O$

$\Rightarrow 2x = y$ (for $H_2SO_4$) $\Rightarrow \frac{y}{x} = 2$

**44.** *(82)* $CuSO_4(s)\ aq \longrightarrow CuSO_4(aq)$;

$\Delta H = -70$ kJ mol$^{-1}$ ...(i)

$CuSO_4.5H_2O(s) + aq \longrightarrow CuSO_4(aq) + 5H_2O$; $\Delta H = +12$ kJ mol$^{-1}$ ...(ii)

On reversing Eq. (ii), we get

$CuSO_4(aq) + 5H_2O \longrightarrow CuSO_4 \cdot 5H_2O(s) + aq$ ...(iii)

$\Delta H = -12$ kJ mol$^{-1}$

On adding Eq. (i)and (iii), we get

$CuSO_4(s) + 5H_2O \longrightarrow CuSO_4 \cdot 5H_2O(s)$

$\therefore$ $\Delta H = -70 - 12 = -82$ kJ

Thus, the value of $x$ is 82.

**45.** *(38)* Given,

$\Delta_{vap}H^{\ominus} = +40.79$ kJ/mol

$T = 100°C = 373$ K

Using enthalpy equation,

$\Delta_{vap}H^{\ominus} = \Delta U° + \Delta n_g RT$

$$\Delta U° = 40.79 - \frac{8.3 \times 373}{1000}$$

$= 37.67$ kJ/mol $\approx 38$ kJ/mol

**46.** *(274)* In adiabatic condition, $q = 0$

Then, equation for change in internal energy will be given by

$\Delta U = q + W$ $(q = 0)$

$nC_V \Delta T = -p_{ext}(V_2 - V_1)$ ...(i)

as $(W = p\Delta V)$

$V_2 = 2V_1$ (given)

$$\frac{nRT_2}{p_2} = \frac{2nRT_1}{p_1} \quad ...(ii)$$

$p_1 = 5$ atm, $T_1 = 298$ K

$p_2$ from Eq. (ii) will be

$$p_2 = \frac{5T_2}{2 \times 298}$$

Now from Eq. (i)

(Put $p_{ext} = 1$ atm)

$$n\frac{5}{2}R(T_2 - T_1) = -1\left[\frac{nRT_2}{p_2} - \frac{nRT_1}{p_1}\right]$$

$$\frac{n5}{2}R(T_2 - T_1) = -nR\left[\frac{T_2}{p_2} - \frac{T_1}{p_1}\right]$$

Put, $T_1 = 298$ and $p_2 = \frac{5T_2}{2 \times 298}$

$$\frac{5}{2}(T_2 - 298) = -1\left[\frac{T_2}{\frac{5T_2}{2 \times 298}} - \frac{298}{5}\right] \quad ...(iii)$$

On solving Eq. (iii)

$T_2 = 274.16$ K

Hence, nearest integer value of temperature is 274 K.

**47.** *(150)*

> **key idea** To solve this question, first of all write the balanced chemical equation, then by using this equation, calculate the value of $\Delta n_g$ after that use the equation $\Delta H = \Delta U + \Delta n_g RT$ and substitute the given values. Then compare the equation obtained with the given equation in question to obtain the value of $x$.

The complete reaction is as follow,

$$C_6H_5COOH(s) + \frac{15}{2}O_2(g) \longrightarrow 7CO_2(g) + 3H_2O(l)$$

$\Delta H = \Delta U + \Delta n_g RT$

$$= -321.30 - \frac{1}{2}R \times 300$$

$= (-321.30 - 150R)$ kJ

Matching with equation given in question the value of $x$ is 150.

**48.** *(6535)* Given,

$6C(\text{graphite}) + 3H_2(g) \longrightarrow C_6H_6(l)$;

$\Delta H = 48.5$ kJ/mol ...(i)

$C(\text{graphite}) + O_2(g) \longrightarrow CO_2(g)$;

$\Delta H = -393.5$ kJ/mol ...(ii)

$H_2(g) + \frac{1}{2}O_2(g) \longrightarrow H_2O(l)$;

$\Delta H = -286$ kJ/mol ...(iii)

Eq. (i) $\times 1$ + (ii) $\times 6$ + (iii) $\times 3$

$= -48.5 - 6 \times 393.5 - 3 \times 286$

$= -3267.5$ kJ for 1mol

$= -6535$ kJ for 2 mol

**49.** *(125)* Given, Bond enthalpies of

C—H = 414 kJ; C—C = 347 kJ

H—H = 435 kJ; C=C = 615 kJ

Now, the formation of ethane takes place as follows

$C_2H_4(g) + H_2(g) \rightarrow C_2H_6(g)$

$\Delta H$ = B.E. (C=C) + 4B.E. (C—H) + B.E. (H—H) − B.E. (C—C) − 6B.E. (C—H)

$\Delta H$ = B.E. (C=C) + B.E. (H—H) − B.E. (C—C) − 2B.E. (C—H)

$\Rightarrow \Delta H = 615 + 435 - 347 - 2(414)$

$\Delta H = -125$ kJ

**50.** *(56)* Given,

$\Delta H°_{vapour}$ $CCl_4 = 30.5$ kJ/mol

Mass of $CCl_4 = 284$ g

Molar mass of $CCl_4 = 154$ g/mol

Moles of $CCl_4 = \frac{284}{154} = 1.844$ mol

$\Delta H°_{vapour}$ for 1 mole $= 30.5$ kJ/mole

$\Delta H°_{vapour}$ for 1.844 mole

$= 30.5 \times 1.844 = 56.242$ kJ

Hence, final answer is 56 kJ (nearest integer).

**51.** *(492)*

> **key idea** This question is based upon the concept of Hess's law to calculate the required $\Delta H$ value for the given reaction. First obtain the required equation by multiplying Eq. (ii)with 3 then subtract Eq. (i) from the equation. After that calculate the value through the formula.
> $\Delta H_3 = 3\Delta H_2 - \Delta H_1$

Two given reactions are

$2Fe(s) + \frac{3}{2}O_2(g) \longrightarrow Fe_2O_3(s)$;

$\Delta H° = -822$ kJ/mol ...(i)

$C(s) + \frac{1}{2}O_2(g) \longrightarrow CO(g)$;

$\Delta H° = -110$ kJ/mol ...(ii)

Multiply Eq. (ii) by 3 and then subtract Eq. (i) from this equation to get the required equation.

$3C(s) + Fe_2O_3(s) \longrightarrow 2Fe(s) + 3CO(g)$ ...(iii)

Then,

$\Delta H_3 = 3 \times \Delta H_2 - \Delta H_1$

$= 3(-110) - (-822)$

$= 492$ kJ/mol

Hence, enthalpy change is 492 kJ/mol.

**52.** *(4)* Addition of electron to neutral atom only is exothermic process.

All others corresponding to removal of electron from a neutral atom, unipositive ion and addition of electron to uninegative ion are endothermic processes.

A $\rightarrow$ Endothermic (atomisation)

B $\rightarrow$ Endothermic (atomisation)

C $\rightarrow$ Endothermic (vapourisation)

D $\rightarrow$ Exothermic (combustion)

E $\rightarrow$ Endothermic (dissolution)

**53.** *(278)* $H_2 + \frac{1}{2}O_2 \longrightarrow H_2O$

$\Delta H = -241.8$ kJ/mol ...(i)

$C + O_2 \longrightarrow CO_2$

$\Delta H = -393.5$ kJ/mol ...(ii)

$C_2H_5OH + 3O_2 \longrightarrow 3H_2O + 2CO_2$

$\Delta H = -1234.7$ kJ/mol ...(iii)

Formation of $C_2H_5OH$ will be

$2C(s) + 3H_2(g) + \frac{1}{2}O_2(g) \longrightarrow C_2H_5OH(l)$

Heat of formation of $C_2H_5OH(l)$

$= 2\times$ Eq. (ii) $+ 3\times$ Eq. (i) $-$ Eq. (iii)

$= 2\times(-393.5) + 3\times(-241.8) - (-1234.7)$

$= -787 - 725.4 + 1234.7$

$= -277.7 \approx -278$ kJ/mol

**54.** *(1411)* Consider the following reaction

$C_2H_4(g) + 3O_2(g) \longrightarrow 2CO_2(g) + 2H_2O(l)$

$\Delta U = -1406$ kJ mol$^{-1}$

$\Delta H = \Delta U + \Delta n_g RT$

$= -1406 + \frac{(-2)\times 8.3\times 300}{1000}$

$= -1406 - 4.98$

$\approx -1411$ kJ mol$^{-1}$

$[\because \Delta G = \Delta H - T\Delta S$ Also, $\Delta G = 0]$

The minimum value of $T\Delta S$ at equilibrium, $T\Delta S = -1411$ kJ

**55.** *(800)* $A_2 + B_2 \longrightarrow 2AB$

$\Delta H_f^\circ = -200$ kJ mol$^{-1}$

$\Delta H_f^\circ(AB) = -200$ kJ mol$^{-1}$

$\therefore \Delta H_r^\circ$ for reaction $A_2 + B_2 \longrightarrow 2AB$ is $-400$ kJ mol$^{-1}$

Given that, bond enthalpy of $A_2$, $B_2$ and $AB$ is $1:0.5:1$

Assume that bond enthalpy of $A_2$ is $x$ kJ mol$^{-1}$

$\therefore$ Bond enthalpy of $B_2 = 0.5x$ kJ/mol

$\therefore$ Bond enthalpy of $AB = x$ kJ/mol

$A_2 + B_2 \longrightarrow 2AB$

$-2x$

$\Delta^\circ H_r = -400$ kJ mol$^{-1}$

$-400 = x + 0.5x - 2x$

$-400 = -0.5x$

$\Rightarrow x = 800$ kJ mol$^{-1}$

**56.** *(1006)* Heat released by combustion of 1 mole of

$C_2H_6(\Delta U) = \frac{-20\times 0.5}{0.3}\times 30$

$= -1000$ kJ

$C_2H_6(g) + \frac{7}{2}O_2(g) \longrightarrow 2CO_2(g) + 3H_2O(l)$

$\Delta n_g = 2 - \left(1 + \frac{7}{2}\right) = -\frac{5}{2}$

$\Delta H = \Delta U + \Delta n_g RT$

$= -1000 - \frac{5}{2}\times 8.3\times 300\times 10^{-3}$ kJ

$= -1000 - 6.225 = -1006$ kJ

Thus, heat released $= 1006$ kJ/mol

**57.** *(925)*

> **key idea** This problem is based on the concept of stoichiometry and mole concept. The step involved to solve this question are, firstly write the balanced chemical equation, from that calculate the value of volume. After that calculate the number of moles. Using the value of number of moles, calculate the heat evolved.

Let volume of $C_2H_4$ is $x$ litre,

| | $C_2H_4$ + $3O_2$ $\longrightarrow$ | $2CO_2 + 2H_2O$ |
|---|---|---|
| Initial | $x$ | — |
| Final | — | $2x$ |

| | $CH_4 + 2O_2 \longrightarrow$ | $CO_2 + 2H_2O$ |
|---|---|---|
| Initial | $(16.8 - x)$ | — |
| Final | — | $(16.8 - x)$ |

Total volume of $CO_2 = 2x + 16.8 - x$

$\Rightarrow 28 = 16.8 + x$ or $x = 11.2$ L

Volume of $C_2H_4 = 11.2$ L

Volume of $CH_4 = (16.8 - 11.2)$ L $= 5.6$ L

No. of moles of $CH_4 = \frac{5.6}{22.4} = 0.25$ mol

No. of moles of $C_2H_4 = \frac{11.2}{22.4} = 0.5$ mol

So, the heat evolved will be

$\Rightarrow 0.25(-900) + 0.5(-1400)$

$= -925$ kJ

**58.** *(54)* 600 mL of 0.2 M $HNO_3$ means,

$600\times 0.2 = 120$ m mol

400 mL of 0.1 M NaOH solution means, $400\times 0.1 = 40$ m mol

The reaction between, $HNO_3$ and NaOH can be written as,

| | $HNO_3$ + | NaOH $\longrightarrow$ | $NaNO_3 + H_2O$ |
|---|---|---|---|
| Start : | 120 | 40 | |
| End : | 80 | 0 | [40 m mol] |

$\Delta_r H = 40$ m mol $\times 57\times 10^3$ J/mol

or, $40\times 10^{-3}$ mol $\times 57\times 10^3$ J/mol

or, 2280 J

$mS\Delta T = 2280$

or $1000$ mL $\times \frac{1\text{ gm}}{\text{mL}}\times 4.2\times \Delta T = 2280$

or, $1000$ g $\times 4.2\times \Delta T = 2280$

or, $\Delta T = 54.28\times 10^{-2}$ K

or $54.28\times 10^{-2}$ °C

**59.** *(35)* Given, molar mass of gas

$= 280$ g mol$^{-1}$

Increase in temperature of calorimeter

$= 298.45$ K $- 298.0$ K

$\Delta T = 0.45$ K

Enthalpy of combustion

$\Delta H_{\text{Comb}} = 9$ kJ mol$^{-1}$

$q = C\,\Delta T = 2.5\times 0.45$

$= 1.125$ kJ

Considering $q = \Delta H$ (enthalpy change)

So, total enthalpy change $= q = 1.125$ kJ

Enthalpy change for 1 mole of gas,

$\Delta H = 9$ kJ mol$^{-1}$

Thus, $n_{\text{gas}} = \frac{q}{\Delta H} = \frac{1.125\text{ kJ}}{9\text{ kJ mol}^{-1}}$

and mass of gas taken

$= n_{\text{gas}}\times$ molar mass

$= \frac{1.125}{9}\times 280 \approx 35$ g

**60.** *(2)* $C_V = C_p - R = 20.785 - 8.314$

$= 12.471$ JK$^{-1}$ mol$^{-1}$

$\Delta U = nC_V\Delta T$

$\Rightarrow n = \frac{5000}{12.471\times 200} = \frac{25}{12.471} \approx 2$

**61.** *(57)* For reaction,

$H_2F_2(g) \longrightarrow H_2(g) + F_2(g)$

$\Delta n_{(g)} = n_{(g)p} - n_{(g)r}$

where,

$n_{(g)p}$ = number of moles of product, and

$n_{(g)r}$ = number of moles of reactants.

$\Delta n_{(g)} = 2 - 1 = 1$

$\Delta H = \Delta U + n_g RT$

$= -59.6 + \frac{1\times 8.314\times 300}{1000}$

$= -59.6 + 2.494 = 57.10$ kJ mol$^{-1}$

**62.** *(200)* $Q = \frac{C\Delta T}{n}$ ...(i)

Assuming coal to be pure carbon,

$n_{\text{coal}} = \frac{2.4}{\text{Molar mass of carbon}} = \frac{2.4}{12}$

By substituting the values in Eq. (i), we get,

$$Q = \frac{\frac{-20(300-298)}{2.4}}{12} = -200 \text{ kJ/mol}$$

$\therefore \quad x = 200$

**63.** *(2)* $HCl + NaOH \longrightarrow NaCl + H_2O$

$\Delta H = -57.3$ kJ / mol

Since, HCl, NaOH and NaCl are strong acid, strong base and salt respectively. So, are completely ionised.

$H^+ + Cl^- + Na^+ + OH^- \longrightarrow Na^+ + Cl^- + H_2O$

$\Delta H = -57.3$ kJ / mol

$\Rightarrow \quad H^+ + OH^- \longrightarrow H_2O$

$\Delta H = -57.3$ kJ/mol

$\Rightarrow \quad H_2O \longrightarrow H^+ + OH^- \quad \ldots(i)$

$\Delta H = +57.3$ kJ / mol

$\underset{\text{Weak acid}}{CH_3COOH} + \underset{\text{Strong base}}{NaOH} \longrightarrow CH_3CO\overset{-}{O}\overset{+}{Na} + H_2O$

$\Delta H = -55.3$ kJ/mol

We can also write above equation as

$CH_3COOH + OH^- \longrightarrow CH_3COO^- + H_2O \ldots(ii)$

$\Delta H = -55.3$ kJ/mol

On adding Eq. (i) and Eq. (ii), we get,

$CH_3COOH + OH^- + H_2O \longrightarrow CH_3COO^- + H_2O + H^+ + OH^-$

$CH_3COOH \rightleftharpoons CH_3COO^{\ominus} + H^{\oplus}$

$\Delta H = +57.3 - 55.3 = 2$ kJ/mol

So, ionisation of acid requires 2 kJ/mol of energy.

**64.** *(104)* Reaction for formation of propane is

3C (graphite) + $4H_2(g) \longrightarrow C_3H_8(g)$

Reaction for combustion of propane, graphite and dihydrogen are respectively.

$C_3H_8(g) + 5O_2(g) \longrightarrow 3CO_2(g) + 4H_2O(l), \ldots(i)$

$\Delta H_{(i)} = -2220 \text{ kJ mol}^{-1}$

C(graphite) + $O_2(g) \longrightarrow CO_2(g) \ldots(ii)$

$\Delta H_{(ii)} = -393.5 \text{ kJ mol}^{-1}$

$H_2(g) + \frac{1}{2}O_2(g) \longrightarrow H_2O(l) \ldots(iii)$

$\Delta H_{(iii)} = -285.8 \text{ kJ mol}^{-1}$

$$\Delta H_{\text{reaction}} = 3\Delta H_{(ii)} + 3\Delta H_{(iii)} - \Delta H_{(i)}$$
$$= 3(-393.5) + 4(-285.8) - (-2200)$$
$$= -103.7 \text{ kJ mol}^{-1}$$
$$\approx 104 \text{ kJ/mol}$$

**65.** *(600)* $Mg(s) + \frac{1}{2}O_2(g) \longrightarrow MgO(s)$

Given, $\Delta n_g = 0 - \frac{1}{2} = -\frac{1}{2}$

$\Delta_C H° = -601.70 \text{ kJ mol}^{-1}$

$= -601700 \text{ JK}^{-1} \text{ mol}^{-1}$

We know that,

so, $\quad \Delta U = \Delta H - \Delta n_g RT$

$$= -601700 - \left(-\frac{1}{2}\right)8.3 \times 300$$
$$= -601700 + 1245$$
$$= -600455 \text{ J mol}^{-1}$$

Thus, $|\Delta U| = 600.455 \text{ kJ mol}^{-1}$

The nearest integer is 600.

**66.** *(195)* $q = nC\Delta T$

$$= \frac{2.2}{44} \times 100 \times (270 - 310) = -200 \text{ J}$$

$w = -p\Delta V$

$$= -1 \times (101.3) \times (167.75 - 217.1) \times 10^{-3}$$
$$= 4.999 \text{ J}$$

$\Delta E = q + w = -200 + 4.999 \cong -195$ J

$\therefore \quad x = 195$

**67.** *(117)* Given, mass of $NH_3$ = 17.0 g

$\therefore$ Number of moles of $NH_3$

$$= \frac{\text{Given mass of } NH_3}{\text{Molar mass of } NH_3} = \frac{17}{17} = 1$$

number of moles equivalent to 85 g

$$= \frac{85}{17} = 5$$

$\therefore$ Enthalpy change for vaporisation of 5 moles of $NH_3$

$= 5 \times 23.4$ kJ/mol $= 117$ kJ

**68.** *(38)* $H_2O(l) \longrightarrow H_2O(g)$

Given, weight of water = 36 g

Moles of $H_2O = \frac{36 \text{ g}}{18 \text{ g}} = 2$ moles

Internal energy, $\Delta U = \Delta H - \Delta n_g RT$

$(\Delta n_g = 1 - 0 = 1)$

$$= 41.1 - \frac{1 \times 8.31 \times 373}{1000} \text{ kJ / mol}$$
$$= 38 \text{ kJ/mol}$$

**69.** *(738)* The relation between change in enthalpy and change in internal energy can be given by :

$[\Delta H = \Delta U + \Delta n_g RT]$

From given equation :

$$\left[CH_3OH(l) + \frac{3}{2}O_2(g) \longrightarrow CO_2(g) + 2H_2O(l)\right]$$

$\Delta n_g = 1 - \frac{3}{2} = -\frac{1}{2}$

So, $\Delta H = -726 - \frac{1}{2} \times \frac{8.3 \times 300}{100}$

or, $\Delta H = -738.45$ kJ/mol

**70.** *(38)* $H_2O(l) \longrightarrow H_2O(g)$

$\because \quad \Delta H = \Delta U + \Delta n_g RT$

$\therefore \quad 41 \text{ kJ} = \Delta U + RT$

$$\Delta U = 41 - 8.3 \times 373 \times 10^{-3}$$
$$= 41 - 3.095$$
$$= 37.90 \text{ kJ mol}^{-1} = 38 \text{ kJ mol}^{-1}$$

**71.** *(82)* Millimoles of HCl = 200 × 0.2 = 40

Millimoles of NaOH = 300 × 0.1 = 30

Heat released $(q) = n \times$ molar heat

$$= \frac{30}{1000} \times 57.1 \times 1000 = 1713 \text{ J}$$

Mass of solution = 500 × 1 = 500 g

We know that,

$$\Delta T = \frac{q}{mc} = \frac{1713 \text{ J}}{500 \text{ g} \times 4.18 \text{ J/g-K}}$$
$$= 0.8196 \text{ K} = 81.96 \times 10^{-2} \text{ K}$$

$x \approx 82$

**72.** *(82)* Molarity $= \frac{n \times 1000}{V(\text{mL})}$

$$\text{Moles} = \frac{\text{molarity} \times \text{volume (mL)}}{1000}$$

1 mole of $H_2SO_4$ produces 2 moles of $H^+$

$$n_{H^+} = \frac{400 \times 0.2}{1000} \times 2 = 0.16 \text{ moles}$$

and 1 mole of NaOH produces 1 mole of $OH^-$.

$$n_{OH^-} = \frac{600 \times 0.1}{1000} = 0.06$$

$n_{OH^-}$ is low.

$\underset{1}{H_2SO_4} + \underset{2}{2NaOH} \longrightarrow \underset{1}{Na_2SO_4} + \underset{2}{2H_2O}$ (Mol :)

Now, 1 mole $H_2SO_4$ requires 2 moles NaOH, then 0.06 mole of NaOH will require = 0.03 moles of $H_2SO_4$

$\therefore$ 0.06 mole of $OH^-$ reacts with 0.06 mole of $H^+$.

$m = 1 \times 1000$

Now, heat liberated from reaction = heat gained by solution.

$$n \times \Delta H = m \times C_p \times \Delta T$$

$$0.06 \times 57.1 \times 10^3 = (1000 \times 1) \times 4.18 \times \Delta T$$

$\Rightarrow \quad \Delta T = 0.8196 \text{ K} = 82 \times 10^{-2}$ K

**73.** *(31)* $H_2O(l) \rightleftharpoons H_2O(g)$

Using formula, $\Delta H = \Delta U + \Delta n_g RT$

For 1 mole water,

$\Delta n_g = 1$ mol

$\therefore \Delta n_g RT = 1 \text{ mol} \times 8.31 \text{ J/mol K} \times 373 \text{ K}$

$= 3099.63 \text{ J} \approx 31 \times 10^2 \text{J mol}^{-1}$

**74.** *(101)* Given,

Enthalpy of fusion,

$\Delta H_{fus} = 2.8 \text{ kJ mol}^{-1}$

Enthalpy of vaporisation,

$\Delta H_{vap} = 98.2 \text{ kJ mol}^{-1}$

Using Hess's law,

$\Delta H_{sub} = \Delta H_{fus} + \Delta H_{vap}$

$= 2.8 + 98.2 = 101 \text{ kJ/mol}$

**75.** *(21)* Heat generated by 1g of graphite powder

$= \frac{2.48 \times 10^2}{12} = 20.66 = 21 \text{ kJ}.$

Hence, amount of heat generated is 21 kJ.

**76.** *(128)* This reaction involves the breaking of 2C—H bonds and formation, of 1 C=C bond and 1 H—H bond.

$\Delta_r H$ = [Sum of bond ethalpies of reactants] − [Sum of bond enthalpies of products]

$= [1\,\Delta_{C—C}H + 6\Delta_{C—H}H] - [1\,\Delta_{C=C}H + 4\Delta_{C—H}H + 1\Delta_{H—H}H]$

$= (347 + 6 \times 414) - [611 + 4 \times 414 + 436]$

$= 2831 - 2703 = 128 \text{ kJ/mol}.$

**77.** *(230)*

$\Delta_r H^\circ = \Sigma \Delta_f H^\circ_{Product} - \Sigma \Delta_f H^\circ_{Reactant}$

$= \{\Delta_f H^\circ_{Al_2O_3} + 3\Delta_f H^\circ_{Ca}\} - \{3\Delta_f H^\circ_{CaO} + 2\Delta_f H^\circ_{Al}\}$

$= \Delta H_f^\circ(Al_2O_3) - 3 \times \Delta H_f^\circ(CaO)$

We know, $\Delta_f H^\circ$ for elemental state = 0

$\Rightarrow \Delta_f H^\circ_{Ca} = 0, \Delta_f H^\circ_{Al} = 0$

Putting the value,

$\Delta_r H^\circ = \{-1675 + 0\} - \{3(-625) + 0\}$

$= -1675 + 1905 = 230 \text{ kJ}$

**78.** *(309)* $\Delta_f H^\circ[S, g] + 6 \times \Delta_f H^\circ[F, g] = \Delta_f H^\circ[SF_6, g] + 6 \times E_{S-F}$

[$\because E_{S-F}$ = Average S—F bond energy in $SF_6$]

$275 + 6 \times 80 = -1100 + 6 \times E_{S-F}$

$\Rightarrow \quad E_{S-F} = \frac{275 + 6 \times 80 + 1100}{6}$

$= 309.16 \text{ kJ mol}^{-1}$

$= 309 \text{ kJ mol}^{-1}$

**79.** *(741)* From given equation:

$NH_2CN(s) + \frac{3}{2}O_2(g) \longrightarrow N_2(g) + CO_2(g) + H_2O(l)$

The relation between change in enthalpy and change in internal energy is given by $\Delta H = \Delta U + \Delta n_g RT$

$\left[\Delta n_g = 2 - \frac{3}{2} = 0.5\right]$

So, $\Delta H = -742.24 + \frac{0.5 \times 8.314 \times 298}{1000}$

$= -741 \text{ kJ}$

**80.** *(189494)* Given, $\Delta H = 41$ kJ/mol

$= 41 \times 1000 \times 5$ J

[For 90 g or 5 mol $H_2O(l)$]

$H_2O(l) \longrightarrow H_2O(g)$

$t = 0 \quad \frac{90}{18} \text{ mol} \quad —$

$t = \infty \quad 0 \quad 5 \text{ mol}$

$= \Delta H - \Delta n_g$ RT

$= 41 \times 1000 \times 5 - 5 \times 8.314 \times 373$

$= 189494.39 \text{ J} \approx 189494.00 \text{ J}$

## Topic 4

**01.** *(c)* The most appropriate relations are given in B and C only.

(B) Gibbs free energy is equal to the enthalpy of the system minus the product of the temperature and entropy.

$\Delta G = \Delta H - T\Delta S$ at constant temperature.

(C) Entropy is the measure of disorderness or randomness of the particular system. It can be expressed as

$\left[\Delta S = \frac{q_{rev}}{T}\right]$

**02.** *(a)* Relation given in option (a) is not correct.

According to first law of thermodynamics, the change in heat energy of system is equal to the sum of change in internal energy of system and work done by the system i.e.,

$\Delta H = \Delta U + p\Delta V.$

**03.** *(b)* A process will be spontaneous when its free energy (Gibb's energy) change will be negative, i.e. $\Delta G < 0$.

$\Delta G = \Delta H - T\Delta S$ for positive/negative signs of $\Delta H$ and $\Delta S$ at any/higher/lower temperature as :

| $\Delta H$ | $\Delta S$ | Comment on temperature ($T$) | $\Delta G$ | Comment on the process |
|---|---|---|---|---|
| $<0$ | $>0$ | at any temp. | $<0$ | spontaneous |
| $>0$ | $<0$ | at any temp. | $>0$ | non-spontaneous |
| $<0$ | $<0$ | at lower temp. | $<0$ | spontaneous |
| $>0$ | $>0$ | at higher temp. | $<0$ | spontaneous |

**04.** *(b)* When pressure increases, the intermolecular distance between the molecule decreases, hence entropy also decreases. ($\Delta S < 0$)

**05.** *(d)* Relationship between free energy and equilibrium constant,

$-\Delta G^\circ = RT \ln K_c$

**06.** *(d)*

**Key idea**

To solve this problem, first of all write all the phase changes involved in the reaction. Then calculate individual entropy change for each phase. After that take sum of all the individual entropy change.

The phase change involved as per the question is given by

Ice → Ice ⇌ Water → Water ⇌ Water vapour → Water vapour

268 K (A) 273 K (B) 273 K (C) 373 K (D) 373 K (E) 383 K

So, the overall entropy change is given by

$\Delta S_{overall} = \Delta S_A + \Delta S_B + \Delta S_C + S_D + S_E$

$\Delta S_A = \int_{268 K}^{273 K} C_{p,n} \cdot dT$

$\Delta S_B = \frac{\Delta H_{m\,fusion}}{T_f}, \quad \Delta S_C = \int_{273}^{373} \frac{C_{p,m} dT}{T}$

$\Delta S_D = \frac{\Delta H_{m\,vaporisation}}{T}$

$\Delta S_E = \int_{373}^{383} \frac{C_{p,m}\,dT}{T}$

So, option (d) is the correct answer.

**07.** *(c)* The incorrect combination is given in $B$ and $C$ i.e option (c).

The correct match is given as

| | $\Delta H$ | $\Delta S$ | Temperature | Spontanity |
|---|---|---|---|---|
| (B) | + | + | Low $T$ | Non-spontaneous |
| (C) | – | – | Low $T$ | Spontaneous |

**08.** *(b)* Reaction is spontaneous at relatively high temperature and non spontaneous at low temperature.

$\Delta G = \Delta H - T\Delta S$

$\Delta G < 0$ is only possible when $\Delta H$ and $\Delta S$ both are positive.
Hence, option (b) is correct.

**09.** (a) Among the given options (a) is incorrect.
Its correct form is : $(\Delta G)$ is positive for non-spontaneous reaction.

**10.** (b) The correct match is
A-II, B-III, C-IV, D-I.

**11.** (b) All the expression of thermodynamics are correct except
$$\ln K = \frac{\Delta H^\circ - T\Delta S^\circ}{RT}$$
As we know,
$$\Delta G = \Delta H^\circ - T\Delta S^\circ \quad \ldots(i)$$
Also $\Delta G = -RT \ln K \quad \ldots(ii)$
$\therefore$ From (i) and (ii),
$$-RT \ln K = \Delta H^\circ - T\Delta S^\circ$$
$$\therefore \quad \ln K = \frac{-\Delta H^\circ}{RT} + \frac{\Delta S^\circ}{R}$$
The corect expression is
$$\ln K = \frac{-\Delta H^\circ}{RT} + \frac{\Delta S^\circ}{R} = -\frac{\Delta H^\circ + T\Delta S^\circ}{RT}$$

**12.** (b) Statement (b) is a true statement whereas all other statements are incorrect.
By definition,
$$dS = \frac{dQ}{T} \text{ and } \Delta S = \frac{\int dQ}{T}$$
$\therefore$ $\Delta S$ is a function of temperature.
$S$ is also a function of temperature.

**13.** (a) $N_2(g) + 3H_2(g) \longrightarrow 2NH_3(g)$,
$\Delta n_g = 2 - (1+3) = -2$
So, $\Delta S$ is also negative (entropy decreases)
In rest of the processes entropy change is positive.

**14.** (d) The incorrect match is $\Delta G^\circ < 0$, $K < 1$.
For an ideal gas $\Delta G^\circ = -RT \ln K$.
$$\therefore \ln K = -\frac{\Delta G^\circ}{RT} \text{ and } K = e^{-\Delta G^\circ/RT}$$
The above equation is helpful in predicting the spontaneity of the reaction.

**15.** (d) According to Gibbs-Helmholtz equation,
$$\Delta_r G^\circ = \Delta_r H^\circ - T\Delta_r S^\circ$$
For a reaction to be feasible (spontaneous)
$$\Delta_r G^\circ < 0$$
$$\Delta_r H^\circ - T\Delta_r S^\circ < 0$$
Given, $\Delta_r H^\circ = +491.1 \text{ kJ mol}^{-1}$,
$$\Delta_r S^\circ = 198 \text{ JK}^{-1} \text{ mol}^{-1}$$
$$\therefore \quad 491.1 \times 10^3 - T \times 198 < 0$$
$$T > \frac{491.1 \times 10^3}{198} = 2480.3 \text{ K}$$
$\therefore$ Above 2480.3 K reaction will become spontaneous.

**16.** (c) For a given value of $T$,
(i) If $\Delta_r G^\circ$ becomes < 0, the forward direction will be spontaneous and then the major and minor components will be $Y$ and $X$ respectively.
(ii) If $\Delta_r G^\circ$ becomes > 0, the forward direction will be non-spontaneous and then the major and minor components will be $X$ and $Y$ respectively.
$$\Delta_r G^\circ = 120 - \frac{3}{8} \times 315 = 1.875$$
i.e. $\Delta_r G^\circ > 0$, major component = $X$

**17.** (c) $\Delta G = \Delta H - T\Delta S$
The process will be spontaneous, when
$$\Delta G = -\text{ve}, \text{ i.e. } |T\Delta S| > |\Delta H|$$
Given : $\Delta H = 200 \text{ J mol}^{-1}$
and $\Delta S = 40 \text{ JK}^{-1} \text{ mol}^{-1}$
$$\Rightarrow \quad T > \frac{|\Delta H|}{|\Delta S|} = \frac{200}{40} = 5 \text{ K}$$
So, the minimum temperature for spontaneity of the process is 5 K.

**18.** (a) Entropy change for $n$ moles of isothermal expansion of an ideal gas from volume $V_1$ to volume $V_2$ is
$$\Delta S = 2.303\, nR \log \frac{V_2}{V_1}$$
Here, $n = 2$, $V_2 = 100 \text{ dm}^3$,
$V_1 = 10 \text{ dm}^3$
$$= 2.303 \times 2 \times 8.3143 \log \frac{100}{10}$$
$$= 38.296 \text{ J mol}^{-1}\text{K}^{-1}$$

**19.** (b) $\Delta G = \Delta H - T\Delta S$
At equilibrium, $\Delta G = 0$. Thus, $T\Delta S = \Delta H$, as $\Delta H$ and $\Delta S$ are +ve, for a reaction to be feasible (spontaneous), $\Delta G$ should be –ve. This can be possible only when
$$T\Delta S > T_e \Delta H$$
i.e., $T\Delta S > T_e \Delta S$ or $T > T_e$
[$T_e$ = temperature at equilibrium]

**20.** (c) $\frac{1}{2}X_2 + \frac{3}{2}Y_2 \longrightarrow XY_3$
$$\Delta S_{\text{reaction}} = \Sigma S_{\text{products}} - \Sigma S_{\text{reactants}}$$
$$\Delta S_{\text{reaction}} = 50 - \left(\frac{3}{2} \times 40 + \frac{1}{2} \times 60\right)$$
$$= -40 \text{ J mol}^{-1}$$
$$\Delta G = \Delta H - T\Delta S$$
At equilibrium, $\Delta G = 0$
$$\therefore \quad \Delta H = T\Delta S$$
$$\Rightarrow \quad T = \frac{\Delta H}{\Delta S}$$
$$= \frac{30 \times 10^3}{40} = 750 \text{ K}$$

**21.** (a) In an isolated system where either mass and energy are not exchanged with surrounding for that spontaneous process, the change in entropy is positive.

**22.** (d) The entropy change can be given as
$$\Delta S^\circ = \frac{\Delta H^\circ}{T}$$
$$\Delta S^\circ = 160.2 \text{ J/K}$$
$$\Delta H^\circ = 179.1 \times 10^3 \text{ J/mol}$$
$$T = \frac{179.1 \times 10^3 \text{ J/mol}}{160.2 \text{ J/K}}$$
$$= 1117.97 \text{ K} \approx 1118 \text{ K}$$

**23.** (b) An irreversible process
$\Rightarrow$ spontaneous process
$\Rightarrow (dS)_{V,E}$ (change in entropy) $= +\text{ve} > 0$
$\Rightarrow (dG)_{T,p}$ (change in Gibbs free energy) $- \text{ve} < 0$

**24.** (a) For spontaneity, $\Delta G = -\text{ve}$
$$\Delta G = \Delta H - T\Delta S$$
$\Delta H$, for endothermic process + ve
At lower temperature, $\Delta S$ + ve
Hence, $\Delta G = +$ ve
But at high temperature, $T\Delta S$ will be greater than $\Delta H$.
Hence, $\Delta G = -$ve, spontaneous.

**25.** (a) From Haber's process,
$N_2(g) + 3H_2(g) \longrightarrow 2NH_3$
$\Delta H^\circ = -$ ve, $\Delta S^\circ = -$ ve (As gaseous moles decreases)
(i) As temperature increases $-\frac{\Delta H^\circ}{T}R$, decreases.
(ii) $\Delta G^\circ = -RT \ln K_{eq}$
$$R \ln K_{eq} = \frac{-\Delta G^\circ}{T}$$
On increasing temperature in exothermic reaction $K_{eq}$ decreases.
$\Delta H^\circ$ and $\Delta S^\circ$ are almost constant with temperature.
So, graph (a) correctly represents the variation of thermodynamic properties of Haber's process.

**26.** *(c)* The correct match is A-II, B-I, C-IV, D-III.

- We know that,

$$dG = Vdp - SdT$$

At constant pressure $Vdp = 0$

So, $dG = -SdT$

or $\left(\frac{\partial G}{\partial T}\right)_p = -S$

- $dH = (dq)_p = C_p dT$

$$\left(\frac{\partial H}{\partial T}\right)_p = C_p$$

- $dG = Vdp - SdT$

At constant temperature, $dG = Vdp$

$$\left(\frac{\partial G}{\partial p}\right)_T = V$$

- $dU = C_V dT = (q)_V$

$$\left(\frac{\partial U}{\partial T}\right)_V = C_V$$

**27.** *(2)* Graph between Gibbs free energy *vs* extent of reaction.

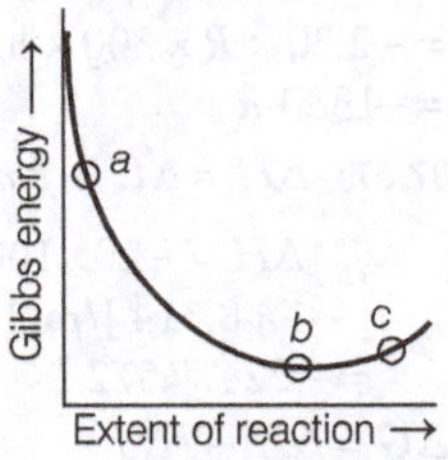


At *a* point we observe $\Delta_r G < 0$

So, reaction is spontaneous.

At *b* point $\Delta G = 0$ (equilibrium)

At *c* point Gibbs free energy $\Delta_r G > 0$

So, reaction is non-spontaneous

two statements B and C are correct (2).

**28.** *(a)* Entropy will decrease in A, B, C and D processes.

(A) Water $\xrightarrow{0°C}$ ice; $\Delta S = -$ ve

(B) Water $\xrightarrow{-10°C}$ ice; $\Delta S = -$ ve

(C) $N_2(g) + 3H_2(g) \longrightarrow 2NH_3(g)$; $\Delta S = -$ ve

Number of moles are decreasing

$$n = 2 - (3+1) \Rightarrow n = -2$$

So, entropy decreases.

(D) Adsorption; $\Delta S = -$ ve

Adsorption will lead to a decrease in the randomness of gaseous particles. So, entropy decreases.

**29.** *(d)* For the given reaction,

$$2NO(g) + O_2(g) \rightleftharpoons 2NO_2(g)$$

Thus, Given, $\Delta G_f^\circ\,(NO) = 86.6$ kJ/mol

$\Delta G_f^\circ\,(NO_2) = ?$ $K_p = 1.6 \times 10^{12}$

Now, we have,

$$\Delta G_f^\circ = 2\Delta G_{f(NO_2)}^\circ - [2\Delta G_{f(NO)}^\circ + \Delta G_{f(O_2)}^\circ]$$

$$= -RT \ln K_p = 2\Delta G_{f(NO_2)}^\circ - [2 \times 86{,}600 + 0]$$

$$\Delta G_{f(NO_2)}^\circ = \frac{1}{2}[2 \times 86{,}600 - R \times 298 \ln(1.6 \times 10^{12})]$$

$$\Delta G_{f(NO_2)}^\circ = 0.5\,[2 \times 86{,}600 - R \times (298) \ln(1.6 \times 10^{12})]$$

**30.** *(d)* Percentage efficiency of the fuel cell

$$= \frac{\Delta G}{\Delta H} \times 100$$

The concerned reaction is

$$CH_3OH(l) + \frac{3}{2}O_2(g) \longrightarrow CO_2(g) + 2H_2O(l)$$

$$\Delta G_r = [\Delta G_f(CO_2, g) + 2\Delta G_f(H_2O, l)] [-\Delta G_f(CH_3OH, l) - \frac{3}{2}\Delta G_f(O_2, g)]$$

$$= -394.4 + 2(-237.2) - (-166.2) - 0$$

$$= -394.4 - 474.4 + 166.2$$

$$= -702.6 \text{ kJ mol}^{-1}$$

Percentage efficiency

$$= \frac{702.6}{726} \times 100 = 96.78\% \approx 97\%$$

**31.** *(700)* The reaction is

$$\frac{1}{2}X_2 + \frac{5}{2}Y_2 \rightleftharpoons XY_5$$

$$\Delta H^- = -35 \text{ kJ/mol} = \Delta_r H^\circ$$

$$\Delta_r S^\circ = 110 - \left[\left(\frac{1}{2} \times 70\right) + \left(\frac{5}{2} \times 50\right)\right]$$

$$= 110 - 160 = -50 \text{ JK}^{-1}\text{mol}^{-1}$$

$\Delta G^\circ = 0$ at equilibrium

At equilibrium

$$T = \frac{|\Delta_r H^\circ|}{|\Delta_r S^\circ|} = \frac{35 \times 1000}{50}$$

$$T = 700 \text{ K}$$

**32.** *(2850)* Given, $\Delta H^\circ = 55$ kJ/mol

$\Delta S^\circ = 175$ J/K/mol

$T = 25$ °C $= 298$ K, $\Delta G^\circ = ?$

Using equation,

$$\Delta G^\circ = \Delta H^\circ - T\Delta S^\circ$$

On substituting the value in above equation, we get

$$\Delta G^\circ = 55000 - 298 \times 175$$

$$= 55000 - 52150 = 2850 \text{ J/mol}$$

**33.** *(400)* Given,

$\Delta H_{vap} = 30$ kJ/mol

$\Delta S_{vap} = 75 \text{ J mol}^{-1}\text{ K}^{-1}$

$T = ?$

At equilibrium,

$$\Delta G = 0$$

$$\Delta H_{vap} = T\Delta S_{vap}$$

So, $30 \times 1000 = T \times 75$

$$T = 400 \text{ K}$$

**34.** *(2000)* From Gibb's free energy equation,

$$\Delta G = \Delta H - T\Delta S$$

For reaction to be spontaneous

$\Delta G = -$ve

For limiting case, $\Delta G = 0$

$$\Delta H = T\Delta S$$

$$T = \frac{\Delta H}{\Delta S} = \frac{400 \text{ kJ/mol}}{0.2 \text{ kJ/mol-K}} = 2000 \text{ K}$$

Hence, reaction will become spontaneous above 2000 K.

**35.** *(57)* $\Delta G^\circ = -RT \ln K_{eq}$

$$= -2.303RT \log K_{eq}$$

$$= -2.303 \times 8.314 \times 300 \times \log 10$$

$$= -5744.14 \text{ J/mol}$$

$$= -5.744 \text{ kJ/mol}$$

$$= -57.44 \times 10^{-1} \text{ kJ/mol}$$

Thus, the value is 57.

**36.** *(37)* Given, $T = 400$ K

$$\Delta H^\ominus = 77.2 \text{ kJ/mol}$$

$$\Delta S = 122 \text{ JK}^{-1}$$

Free energy is given by

$$\Delta G^\circ = \Delta H^\circ - T\Delta S^\circ$$

$$= 77.2 \times 10^3 - 400 \times 122$$

$$= 28400 \text{ J}$$

$$\Delta G^\circ = -2.303RT \log K$$

$$28400 = -2.303 \times 8.314 \times 400 \times \log K$$

$\log K = -3.708$

or $\log K = -37.08 \times 10^{-1}$

Hence, the missing value is 37.

**37.** *(10)* Reaction, $A \rightleftharpoons B$ at 298 K

Given, $\Delta H^\circ = -54.07$ kJ/mol

$\Delta S^\circ = 10$ J/K/mol

We know that

$$\Delta G^\circ = \Delta H^\circ - T\Delta S^\circ$$

or $-2.303\,RT \log K = -54070 - 298 \times 10$

$$\log K = \frac{57050}{5705}$$

So, $\log K = 10$

**38.** *(3)* Given $n=1$, $R=8.314$

$$v_2=3\text{ L}, v_1=2\text{ L}$$

$$\Delta S_{\text{system}}=nR\ln\left(\frac{v_2}{v_1}\right)$$

$$\Delta S_{\text{system}}=1\times 8.314\ln\left(\frac{3}{2}\right)$$

$$\Delta S_{\text{system}}=3.37\text{ JK}^{-1}$$

$$\Delta S_{\text{surrounding}}=3.37\approx 3\text{ JK}^{-1}$$

**39.** *(6)*

> **key idea** To calculate the value of Gibbs free energy, first of all calculate $K_{eq}$ by using the formula $K_{eq}=\frac{K_f}{K_b}$. After that substitute the value obtained, in the equation $\Delta G°=-RT\ln K_{eq}$

$$A+B \underset{K_b=10^2}{\overset{K_f=10^3}{\rightleftharpoons}} C+D$$

Since, equilibrium constant can be calculated as

$$K_{eq}=\frac{K_f}{K_b} \text{ or, } K_{eq}=\frac{10^3}{10^2}=10$$

Again, the relation between $\Delta G°$ and $K_{eq}$ can be calculated from the formula

$$\Delta G°=-RT\ln K$$

So, $\Delta G°=-8.3\times 300\times\ln(10)$

or $\Delta G°=-5.7\text{ kJ/ mol}$

$$\approx -6\text{ kJ/mol}$$

**40.** *(710)* From the given equilibrium equation:

| | $N_2O_4$ | $\rightleftharpoons$ $2NO_2$ |
|---|---|---|
| initially ($t=0$) | [1 mol] | |
| after (time $=t$) | [1 − 0.5] mol = 0.5 mol | 0.5 × 2 mol = 1 mol |

Total pressure = 1 atm

$$\therefore\ K_p=\frac{[NO_2]^2}{[N_2O_4]}=\frac{\left[\frac{1}{1.5}\times 1\right]^2}{\left[\frac{0.5}{1.5}\times 1\right]}$$

$$=\frac{100}{75}=[1.33]$$

$(\Delta G°)$ for the equilibrium reaction can be given by :

$$\Delta G°=-2.303\,RT\log K$$

$$=-2.303\times 8.31\times 300\times\log(1.33)$$

$$=-2.303\times 8.31\times 300\times 0.1239$$

$$\Delta G°=-710.43\text{ J/mol}$$

$\therefore\ x=710$

**41.** *(300)* The relation between enthalpy, entropy and Gibb's free energy is given by equation

$$\Delta G=\Delta H-T\Delta S$$

At equilibrium, $\Delta G=0$

$$0=-165\times 10^3-T(-550)$$

$$\Rightarrow\ T=\frac{165\times 10^3}{550}=300\text{ K}$$

**42.** *(5)* Given, $\Delta H=-57.8\text{ kJ mol}^{-1}$

$$\Delta S=-176\text{ JK}^{-1}\text{ mol}^{-1}$$

$$T=298\text{ K}$$

Using Gibb's free energy relation

$$\Delta G=\Delta H-T\Delta S$$

$$\Delta G=57.8\text{ kJ / mol}-[298\text{ K}\times(-176\text{ JK}^{-1}\text{mol}^{-1})]$$

$$=-57.8\text{ kJ / mol}-\left(298\times\frac{-176}{1000}\text{ kJ}\right)$$

$$[\because 1\text{ kJ}=1000\text{ J}]$$

$$=-5.352\text{ kJ / mol}$$

$$\Delta G=5.352$$

Hence, answer is 5.

**43.** *(964)*

$$\Delta H°=[\Delta H°_{f\,(Fe)}+\Delta H°_{f\,(CO)}]-[\Delta H°_{f\,(FeO)}-\Delta H°_{f\,(C)}]$$

$$=[0+(-110.5)]-[-266.3-0]$$

$$=155.8\text{ kJ}^{-1}\text{ mol}^{-1}$$

$$\Delta S°=[\Delta S°_{(Fe)}+\Delta S°_{(CO)}]-[\Delta S°_{(FeO)}-\Delta S°_{(C)}]$$

$$=[27.28+197.6]-[57.49+5.74]$$

$$=161.6\text{ J K}^{-1}\text{ mol}^{-1}$$

The reaction becomes spontaneous when $\Delta G°$ is atleast zero or negative.

$$0=\Delta H°-T\Delta S°$$

$$\Rightarrow T\Delta S°=\Delta H°$$

$$\Rightarrow\ T=\frac{\Delta H°}{\Delta S°}$$

$$=\frac{155.8\text{ kJ mol}^{-1}}{161.6\text{ J K}^{-1}\text{ mol}^{-1}}$$

$$=\frac{155800\text{ J mol}^{-1}}{161.6\text{ J K}^{-1}\text{ mol}^{-1}}=964\text{ K}$$

The temperature at which reaction becomes spontaneous is 964 K.

**44.** *(336)* For the reaction, $A\xrightarrow{300\text{K}}B$

Given, $\Delta G=-49.4\text{ kJ/mol}$

Change in enthalpy

$$\Delta H_{\text{reaction}}=51.4\text{ kJ/mol}$$

Change in entropy, $\Delta S_{\text{reaction}}=?$

According to Gibbs free energy,

$$\Delta G=\Delta H-T\Delta S$$

$$\Rightarrow \Delta S=\frac{\Delta H-\Delta G}{T}$$

$$=\frac{51.4-(-49.4)}{300}\times 1000\frac{\text{J}}{\text{mol K}}$$

$$\Delta S_{\text{reaction}}=336\text{ J/mol K}$$

**45.** *(855)* $\Delta G°=-2.303\,RT\log K$ ...(i)

Also,

$\Delta G°=\Sigma(\Delta G°_f)_p-\Sigma(\Delta G°_f)_R$ ...(ii)

Equate Eq. (i) and (ii)

$\therefore\ -2.303\times 8.314\times 298\times\log K$

$$=-4.88\times 10^5$$

or $\log K=\frac{4.88\times 10^5}{2.303\times 8.314\times 298}$

$$\log K\approx -85.5\approx -855\times 10^{-1}$$

or, $x=855$

**46.** *(1380)* For equilibrium reaction,

$$\Delta G°=-2.30\,RT\log K$$

$$=-2.30\times R\times 300\times\log 100$$

$$\Delta G°=-1380\,R$$

**47.** *(13537.57)* $\Delta H=\Delta U+\Delta n_g RT$

$$\Delta H=-20\times 1000-1\times 8.314\text{ J/mol.K}\times 298\text{ K}$$

$$=-22477.572\text{ J}$$

$$\Delta G=\Delta H-T\Delta S$$

$$\Delta G=-22477.572+298\times 30$$

$$\Delta G=-13537.57\text{ J}$$

**48.** *(2.70)* Given : $\Delta U=2.1$ kcal

$$\Delta S=20\text{ cal K}^{-1}=\frac{20}{1000}\text{ kcal K}^{-1}$$

$$A(l)\rightleftharpoons 2B(g)$$

$\therefore\ \Delta n_g=2$ mol

$$R=2\text{ cal K}^{-1}\text{ mol}^{-1}$$

$$=\frac{2}{1000}\text{ kcal K}^{-1}\text{ mol}^{-1}$$

As we know,

$$\Delta G=\Delta H-T\Delta S$$

and $\Delta H=\Delta U+\Delta n_g RT$

$\therefore\ \Delta G=\Delta U+\Delta n_g RT-T\Delta S$

$\therefore\ \Delta G=2.1+\frac{2\times 2\times 300}{1000}-\frac{300\times 20}{1000}$

$$=2.1+1.2-6=-2.70$$

# C-06

JEE Main Chapterwise Topicwise Chemistry

In JEE Main there is usually 0-1 question per paper are visible from this chapter. In past 3 years, 12 conceptual and 16 numerical questions were asked of moderate difficulty level from this chapter.

**For detailed analysis scan QR code**

# CHEMICAL EQUILIBRIUM

## Topic 1 Chemical Equilibrium, Law of Mass Action and Equilibrium Constant

### Level 1

**01** For the reaction, *[2025, 24 Jan Shift-II]*

$$H_2(g) + I_2(g) \rightleftharpoons 2HI(g)$$

Attainment of equilibrium is predicted correctly by

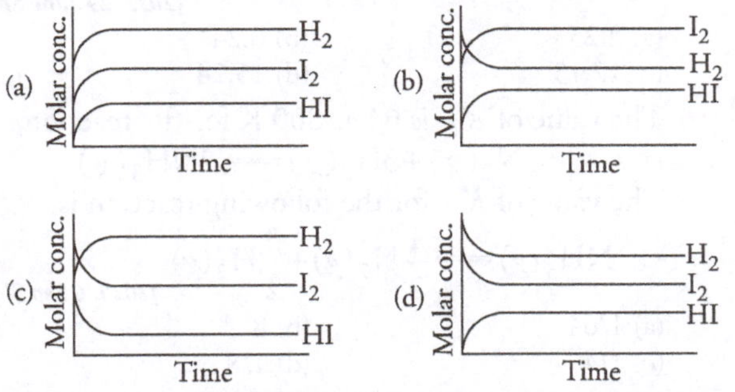


**02** A vessel at 1000 K contains $CO_2$ with a pressure of 0.5 atm. Some of $CO_2$ is converted into CO on addition of graphite. If total pressure at equilibrium is 0.8 atm, then $K_p$ is *[2025, 22 Jan Shift-I]*

(a) 0.18 atm (b) 0.3 atm
(c) 3 atm (d) 1.8 atm

**03** For the given reaction, choose the correct expression of $K_C$ from the following: *[2024, 31 Jan Shift-I]*

$$Fe^{3+}(aq) + SCN^-(aq) \rightleftharpoons (FeSCN)^{2+}(aq)$$

(a) $K_C = \dfrac{[FeSCN^{2+}]}{[Fe^{3+}][SCN^-]}$ (b) $K_C = \dfrac{[FeSCN^{2+}]^2}{[Fe^{3+}][SCN^-]}$

(c) $K_C = \dfrac{[FeSCN^{2+}]}{[Fe^{3+}]^2[SCN^-]^2}$ (d) $K_C = \dfrac{[Fe^{3+}][SCN^-]}{[FeSCN^{2+}]}$

**04** For the equilibrium $A \rightleftharpoons B$, the variation of the rate of the forward (*a*) and reverse (*b*) reaction with time is given by *[2020, 4 Sep Shift-I]*

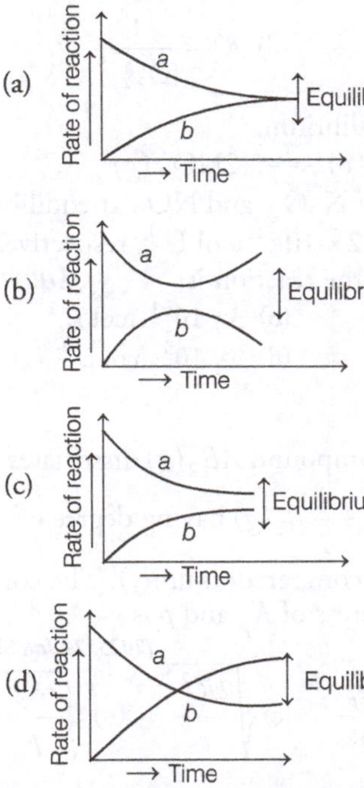


**05** In the figure shown below, reactant *A* (represented by square) is in equilibrium with product *B* (represented by circle). The equilibrium constant is

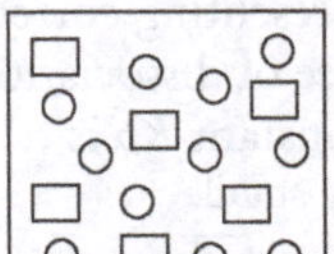

*[2020, 9 Jan Shift-II]*

(a) 4 (b) 2
(c) 8 (d) 1

**06** The incorrect match in the following is *[2019, 12 April Shift-II]*

(a) $\Delta G^\circ < 0, K > 1$ (b) $\Delta G^\circ = 0, K = 1$
(c) $\Delta G^\circ > 0, K < 1$ (d) $\Delta G^\circ < 0, K < 1$

**Imp.** Relation between Equilibrium Constant

**07** For the following three reactions I, II and III, equilibrium constants are given **[AIEEE 2008]**

I. $CO(g) + H_2O(g) \rightleftharpoons CO_2(g) + H_2(g); K_1$

II. $CH_4(g) + H_2O(g) \rightleftharpoons CO(g) + 3H_2(g); K_2$

III. $CH_4(g) + 2H_2O(g) \rightleftharpoons CO_2(g) + 4H_2(g); K_3$

Which of the following relations is correct?

(a) $K_1\sqrt{K_2} = K_3$ (b) $K_2 K_3 = K_1$

(c) $K_3 = K_1 K_2$ (d) $K_3 K_2^3 = K_1^2$

**08** What is the equilibrium expression for the reaction

$P_4(s) + 5O_2(g) \rightleftharpoons P_4O_{10}(s)$? **[AIEEE 2004]**

(a) $K_c = \frac{[P_4O_{10}]}{[P_4][O_2]^5}$ (b) $K_c = \frac{[P_4O_{10}]}{5[P_4][O_2]^5}$

(c) $K_c = [O_2]^5$ (d) $K_c = \frac{1}{[O_2]^5}$

**09** For the reaction equilibrium,

$N_2O_4(g) \rightleftharpoons 2NO_2(g)$

the concentrations of $N_2O_4$ and $NO_2$ at equilibrium are $4.8 \times 10^{-2}$ and $1.2 \times 10^{-2}$ mol $L^{-1}$, respectively. The value of $K_c$ for the reaction is **[AIEEE 2003]**

(a) $3.3 \times 10^2$ mol $L^{-1}$ (b) $3 \times 10^{-1}$ mol $L^{-1}$

(c) $3 \times 10^{-3}$ mol $L^{-1}$ (d) $3 \times 10^3$ mol $L^{-1}$

## Level 2

**10** At temperature $T$, compound $AB_2(g)$ dissociates as $AB_2(g) \rightleftharpoons AB(g) + \frac{1}{2}B_2(g)$ having degree of dissociation $x$ (small compared to unity). The correct expression for $x$ in terms of $K_p$ and $p$ is **[2025, 29 Jan Shift-I]**

(a) $\sqrt{K_p}$ (b) $\sqrt[4]{\frac{2K_p}{p}}$ (c) $\sqrt[3]{\frac{2K_p^2}{p}}$ (d) $\sqrt[3]{\frac{2K_p}{p}}$

**11** Consider the reaction, **[2025, 23 Jan Shift-II]**

$X_2Y(g) \rightleftharpoons X_2(g) + \frac{1}{2}Y_2(g)$

The equation representing correct relationship between the degree of dissociation ($x$) of $X_2Y(g)$ with its equilibrium constant $Kp$ is ......... .

Assume $x$ to be very small.

(a) $x = \sqrt[3]{\frac{2K_p^2}{p}}$ (b) $x = \sqrt[3]{\frac{K_p}{p}}$ (c) $x = \sqrt[3]{\frac{2K_p}{p}}$ (d) $x = \sqrt[3]{\frac{K_p}{2p}}$

**12** For the given hypothetical reactions, the equilibrium constants are as follows: **[2024, 8 April Shift-I]**

$X \rightleftharpoons Y; K_1 = 1.0$

$Y \rightleftharpoons Z; K_2 = 2.0$

$Z \rightleftharpoons W; K_3 = 4.0$

The equilibrium constant for the reaction $X \rightleftharpoons W$ is

(a) 12.0 (b) 7.0 (c) 6.0 (d) 8.0

**13** The ratio $\frac{K_p}{K_c}$ for the reaction, **[2024, 6 April Shift-II]**

$CO(g) + \frac{1}{2}O_2(g) \rightleftharpoons CO_2(g)$ is

(a) 1 (b) $\frac{1}{\sqrt{RT}}$

(c) $(RT)^{1/2}$ (d) $RT$

**14** The equilibrium constant for the reaction

$SO_3(g) \rightleftharpoons SO_2(g) + \frac{1}{2}O_2(g)$

is $K_c = 4.9 \times 10^{-2}$. The value of $K_c$ for the reaction given below is **[2024, 4 April Shift-II]**

$2SO_2(g) + O_2(g) \rightleftharpoons 2SO_3(g)$

(a) 49 (b) 41.6 (c) 4.9 (d) 416

**15** 4.0 moles of argon and 5.0 moles of $PCl_5$ are introduced into an evacuated flask of 100 litre capacity at 610 K. The system is allowed to equilibrate. At equilibrium, the total pressure of mixture was found to be 6.0 atm. The $K_p$ for the reaction is [Given, $R = 0.082$ L atm $K^{-1}$ $mol^{-1}$] **[2022, 29 June Shift-II]**

(a) 2.25 (b) 6.24

(c) 12.13 (d) 15.24

**16** The value of $K_C$ is 64 at 800 K for the reaction,

$N_2(g) + 3H_2(g) \rightleftharpoons 2NH_3(g)$

The value of $K_C$ for the following reaction is

$NH_3(g) \rightleftharpoons \frac{1}{2}N_2(g) + \frac{3}{2}H_2(g)$ **[2020, 6 Sep Shift-II]**

(a) 1/64 (b) 8

(c) 1/4 (d) 1/8

**17** The variation of equilibrium constant with temperature is given below **[2020, 06 Sep Shift-I]**

| Temperature | Equilibrium constant |
|---|---|
| $T_1 = 25°C$ | $K_1 = 10$ |
| $T_2 = 100°C$ | $K_2 = 100$ |

The values of $\Delta H°$, $\Delta G°$ at $T_1$ and $\Delta G°$ at $T_2$ (in kJ $mol^{-1}$) respectively, are close to

[use $R = 8.314$ J $K^{-1}$ $mol^{-1}$]

(a) 28.4, –7.14 and –5.71

(b) 0.64, –7.14 and –5.71

(c) 28.4, –5.71 and –14.29

(d) 0.64, –5.71 and –14.29

**18** For the reaction, $Fe_2N(s) + \frac{3}{2}H_2(g) \rightleftharpoons Fe(s) + NH_3(g)$ **[2020, 6 Sep Shift-I]**

(a) $K_C = K_p(RT)$ (b) $K_C = K_p(RT)^{-1/2}$

(c) $K_C = K_p(RT)^{1/2}$ (d) $K_C = K_p(RT)^{3/2}$

**19** If the equilibrium constant for $A \rightleftharpoons B+C$ is $K_{eq}^{(1)}$ and that of $B+C \rightleftharpoons P$ is $K_{eq}^{(2)}$, the equilibrium constant for $A \rightleftharpoons P$ is *[2020, 4 Sep Shift-II]*

(a) $K_{eq}^{(1)}/K_{eq}^{(2)}$ (b) $K_{eq}^{(1)}+K_{eq}^{(2)}$
(c) $K_{eq}^{(1)}K_{eq}^{(2)}$ (d) $K_{eq}^{(2)}-K_{eq}^{(1)}$

**20** An open beaker of water in equilibrium with water vapour is in a sealed container. When few grams of glucose are added to the beaker of water, the rate at which water molecules *[2020, 2 Sep Shift-I]*

(a) leaves the solution decreases
(b) leaves the solution increases
(c) leaves the vapour increases
(d) leaves the vapour decreases

**21** In which one of the following equilibria, $K_p \neq K_c$? *[2019, 12 April Shift-II]*

(a) $2C(s)+O_2(g) \rightleftharpoons 2CO(g)$
(b) $2HI(g) \rightleftharpoons H_2(g)+I_2(g)$
(c) $NO_2(g)+SO_2(g) \rightleftharpoons NO(g)+SO_3(g)$
(d) $2NO(g) \rightleftharpoons N_2(g)+O_2(g)$

**22** Consider the following reversible chemical reactions,

$$A_2(g)+B_2(g) \overset{K_1}{\rightleftharpoons} 2AB(g) \quad \ldots\text{(i)}$$

$$6AB(g) \overset{K_2}{\rightleftharpoons} 3A_2(g)+3B_2(g) \quad \ldots\text{(ii)}$$

The relation between $K_1$ and $K_2$ is *[2019, 9 Jan Shift-II]*

(a) $K_2 = K_1^3$ (b) $K_1K_2 = 3$
(c) $K_2 = K_1^{-3}$ (d) $K_1K_2 = \frac{1}{3}$

**23** For the following reactions, equilibrium constants are given :

$$S(s)+O_2(g) \rightleftharpoons SO_2(g);\ K_1 = 10^{52}$$

$$2S(s)+3O_2(g) \rightleftharpoons 2SO_3(g);\ K_2 = 10^{129}$$

The equilibrium constant for the reaction,

$$2SO(g)+O_2(g) \rightleftharpoons 2SO_3(g) \text{ is}$$

*[2019, 8 April Shift-II]*

(a) $10^{25}$ (b) $10^{77}$ (c) $10^{154}$ (d) $10^{181}$

**Imp.** Temperature Dependence of Equilibrium Constant

**24** Which of the following lines correctly show the temperature dependence of equilibrium constant, $K$, for an exothermic reaction? *[JEE Main 2018]*

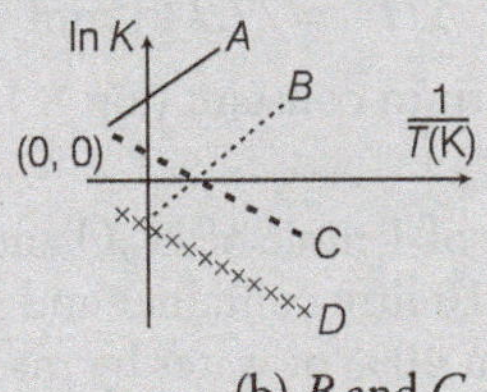


(a) $A$ and $B$ (b) $B$ and $C$
(c) $C$ and $D$ (d) $A$ and $D$

**25** The standard Gibbs energy change at 300K for the reaction, $2A \rightleftharpoons B+C$ is 2494. 2 J. At a given time, the composition of the reaction mixture is $[A]=\frac{1}{2}, [B]=2$ and $[C]=\frac{1}{2}$. The reaction proceeds in the $[R=8.314 \text{ JK/mol}, e=2.718]$ *[JEE Main 2015]*

(a) forward direction because $Q<K_c$
(b) reverse direction because $Q>K_c$
(c) forward direction because $Q>K_c$
(d) reverse direction because $Q<K_c$

**26** For the reaction,

$$SO_2(g)+\frac{1}{2}O_2(g) \rightleftharpoons SO_3(g),$$

if $K_P = K_C(RT)^x$ where the symbols have usual meanings then, the value of $x$ is (assuming ideality). *[JEE Main 2014]*

(a) $-1$ (b) $-\frac{1}{2}$ (c) $\frac{1}{2}$ (d) 1

**27** In a chemical reaction, $A+2B \overset{K}{\rightleftharpoons} 2C+D$, the initial concentration of $B$ was 1.5 times of the concentration of $A$, but the equilibrium concentrations of $A$ and $B$ were found to be equal. The equilibrium constant ($K$) for the aforesaid chemical reaction is: *[JEE Main 2013]*

(a) 4 (b) 16 (c) $\frac{1}{4}$ (d) 1

**28** The equilibrium constant ($K_c$) for the reaction $N_2(g)+O_2(g) \longrightarrow 2NO(g)$ at temperature $T$ is $4\times10^{-4}$. The value of $K_c$ for the reaction $NO(g) \longrightarrow \frac{1}{2}N_2(g)+\frac{1}{2}O_2(g)$ at the same temperature is *[AIEEE 2012]*

(a) 0.02 (b) $2.5\times10^2$
(c) $4\times10^{-4}$ (d) 50.0

**29** A vessel at 1000 K contains $CO_2$ with a pressure of 0.5 atm. Some of the $CO_2$ is converted into CO on the addition of graphite. If the total pressure at equilibrium is 0.8 atm, the value of $K_p$ is *[AIEEE 2011]*

(a) 1.8 atm (b) 3 atm
(c) 0.3 atm (d) 0.18 atm

**30** For the reaction,

$$2NO_2(g) \rightleftharpoons 2NO(g)+O_2(g)$$

$[K_c = 1.8\times10^{-6}$ at 184°C, $R=0.00831$ kJ/(mol K)$]$

When $K_p$ and $K_c$ are compared at 184°C it is found that *[AIEEE 2005]*

(a) whether $K_p$ is greater than, less than or equal to $K_c$ depends upon the total gas pressure
(b) $K_p = K_c$
(c) $K_p$ is less than $K_c$
(d) $K_p$ is greater than $K_c$

## Level 3

**31** $A(g) \rightleftharpoons B(g) + \frac{C}{2}(g)$, the correct relationship between $K_p$, α and equilibrium pressure $p$ is *[2024, 31 Jan Shift-II]*

(a) $K_p = \frac{\alpha^{3/2} p^{1/2}}{(2+\alpha)^{1/2}(1-\alpha)}$ (b) $K_p = \frac{\alpha^{1/2} p^{1/2}}{(2+\alpha)^{1/2}}$

(c) $K_p = \frac{\alpha^{1/2} p^{1/2}}{(2+\alpha)^{3/2}}$ (d) $K_p = \frac{\alpha^{1/2} p^{3/2}}{(2+\alpha)^{3/2}}$

**32** Two solids dissociate as follows

$$A(s) \rightleftharpoons B(g) + C(g); K_{p_1} = x \text{ atm}^2$$

$$D(s) \rightleftharpoons C(g) + E(g); K_{p_2} = y \text{ atm}^2$$

The total pressure when both the solids dissociate simultaneously is *[2019, 12 Jan Shift-I]*

(a) $\sqrt{x+y}$ atm (b) $x^2 + y^2$ atm

(c) $(x+y)$ atm (d) $2(\sqrt{x+y})$ atm

**33** Consider the reaction,

$$N_2(g) + 3H_2(g) \rightleftharpoons 2NH_3(g)$$

The equilibrium constant of the above reaction is $K_p$. If pure ammonia is left to dissociate, the partial pressure of ammonia at equilibrium is given by (Assume that $p_{NH_3} << p_{total}$ at equilibrium) *[2019, 11 Jan Shift-I]*

(a) $\frac{3^{3/2} K_p^{-1/2} p^2}{4}$ (b) $\frac{3^{3/2} K_p^{1/2} p^2}{16}$

(c) $\frac{K_p^{1/2} p^2}{16}$ (d) $\frac{K_p^{1/2} p^2}{4}$

**34** 5.1 g $NH_4SH$ is introduced in 3.0 L evacuated flask at 327°C. 30% of the solid $NH_4SH$ decomposed to $NH_3$ and $H_2S$ as gases. The $K_p$ of the reaction at 327°C is ($R = 0.082$ atm mol$^{-1}$K$^{-1}$, molar mass of S = 32 g mol$^{-1}$, molar mass of N = 14 g mol$^{-1}$) *[2019, 10 Jan Shift-I]*

(a) $0.242 \times 10^{-4}$ atm$^2$ (b) 0.242 atm$^2$

(c) $4.9 \times 10^{-3}$ atm$^2$ (d) $1 \times 10^{-4}$ atm$^2$

**35** The equilibrium constant at 298 K for a reaction, $A + B \rightleftharpoons C + D$ is 100. If the initial concentrations of all the four species were 1 M each, then equilibrium concentration of $D$ (in mol L$^{-1}$) will be *[JEE Main 2016]*

(a) 0.818 (b) 1.818 (c) 1.182 (d) 0.182

**36** At −20°C and 1 atm pressure, a cylinder is filled with equal number of $H_2$, $I_2$ and HI molecules for the reaction $H_2(g) + I_2(g) \rightleftharpoons 2HI(g)$, the $K_p$ for the process is $x \times 10^{-1}$. $x$ = ...... *[2024, 6 April Shift-I]*

[Given : $R = 0.082$ L atm K$^{-1}$mol$^{-1}$]

(a) 1 (b) 0.01 (c) 10 (d) 2

### Numerical Value Type Questions

**37.** Consider the following equilibrium.

$$CO(g) + 2H_2(g) \rightleftharpoons CH_3OH(g)$$

0.1 mol of CO along with a catalyst is present in a 2 dm$^3$ flask maintained at 500 K. Hydrogen is introduced into the flask until the pressure is 5 bar and 0.04 mol of $CH_3OH$ is formed. The $K_p^\circ$ is .... $\times 10^{-3}$ (Nearest integer).

Given: $R = 0.08$ dm$^3$ bar K$^{-1}$ mol$^{-1}$

Assume only methanol is formed as the product and the system follows ideal gas behaviour. *[2025, 2 April Shift-I]*

**38** 37.8 g $N_2O_5$ was taken in a 1L reaction vessel and allowed to undergo the following reaction at 500K

$$2N_2O_5(g) \rightleftharpoons 2N_2O_4(g) + O_2(g)$$

The total pressure at equilibrium was found to be 18.65 bar.

Then, $K_p$ = .........$\times 10^{-2}$[nearest integer]

Assume $N_2O_5$ to behave ideally under these conditions. (Given, $R = 0.082$ bar L mol$^{-1}$K$^{-1}$) *[2025, 24 Jan Shift-I]*

**39** For a certain reaction at 300 K, $K = 10$, then $\Delta G°$ for the same reaction is ......... $\times 10^{-1}$ kJ mol$^{-1}$ (Given, $R = 8.314$ JK$^{-1}$ mol$^{-1}$) *[2024, 1st Feb Shift-II]*

**40** Consider the following reaction at 298 K.

$\frac{3}{2}O_2(g) \rightleftharpoons O_3(g)$, $K_p = 2.47 \times 10^{-29}$, $\Delta_r G°$ for the reaction is ....... kJ (Given, $R = 8.314$ JK$^{-1}$ mol$^{-1}$) *[2024, 31 Jan Shift-I]*

**41** The following concentrations were observed at 500 K for the formation of $NH_3$ from $N_2$ and $H_2$. At equilibrium : $[N_2] = 2\times 10^{-2}$ M, $[H_2] = 3\times 10^{-2}$ M and $[NH_3] = 1.5 \times 10^{-2}$ M. Equilibrium constant for the reaction is .......... . *[2024, 29 Jan Shift-II]*

**42** For the reaction, $N_2O_4(g) \rightleftharpoons 2NO_2(g)$.

$K_p = 0.492$ atm at 300 K. $K_C$ for the reaction at same temperature is .......... $\times 10^{-2}$.

(Given, $R = 0.082$ L atm mol$^{-1}$K$^{-1}$) *[2024, 29 Jan Shift-I]*

**43** For a certain thermochemical reaction, $M \rightarrow N$ at $T = 400$ K, $\Delta H^\ominus = 77.2$ kJ mol$^{-1}$, $\Delta S = 122$ JK$^{-1}$, log equilibrium constant (log $K$) is − ...... $\times 10^{-1}$. *[2024, 27 Jan Shift-II]*

**44** A mixture of 1 mole of $H_2O$ and 1 mole of CO is taken in a 10 litre container and heated to 725 K. At equilibrium 40% of water by mass reacts with carbon monoxide according to the equation

$$CO(g) + H_2O(g) \rightleftharpoons CO_2(g) + H_2(g).$$

The equilibrium constant $K_C \times 10^2$ for the reaction is ........... (Nearest integer) **[2023, 11 April Shift-I]**

**45** 4.5 moles each of hydrogen and iodine is heated in a sealed ten litre vessel. At equilibrium, 3 moles of HI were found. The equilibrium constant for $H_2(g) + I_2(g) \rightleftharpoons 2HI(g)$ is ........... **[2023, 11 April Shift-II]**

**46** $A(g) \rightleftharpoons 2B(g) + C(g)$

For the given reaction, if the initial pressure is 450 mm Hg and the pressure at time $t$ is 720 mm Hg at a constant temperature $T$ and constant volume $V$. The fraction of $A(g)$ decomposed under these conditions is $x \times 10^{-1}$. The value of $x$ is ........... (Nearest integer) **[2023, 10 April Shift-II]**

**47** The equilibrium composition for the reaction, $PCl_3 + Cl_2 \rightleftharpoons PCl_5$ at 298 K is given below $[PCl_3]_{eq} = 0.2 \text{ mol L}^{-1}$, $[Cl_2]_{eq} = 0.1 \text{ mol L}^{-1}$, $[PCl_5]_{eq} = 0.40 \text{ mol L}^{-1}$. If 0.2 mol of $Cl_2$ is added at the same temperature, the equilibrium concentrations of $PCl_5$ is .........$\times 10^{-2}$ $\text{molL}^{-1}$. Given: $K_c$ for the reaction at 298 K is 20. **[2023, 6 April Shift-II]**

**48** (i) $X(g) \rightleftharpoons Y(g) + Z(g), K_{p_1} = 3$

(ii) $A(g) \rightleftharpoons 2B(g), K_{p_2} = 1$

If the degree of dissociation and initial concentration of both the reactants $X(g)$ and $A(g)$ are equal, then the ratio of the total pressure at equilibrium $\left(\frac{p_1}{p_2}\right)$ is equal to $x : 1$. The value of $x$ is ............ (Nearest integer) **[2023, 1 Feb Shift-I]**

**49** At 298 K

$N_2(g) + 3H_2(g) \rightleftharpoons 2NH_3(g); K_1 = 4 \times 10^5$

$N_2(g) + O_2(g) \rightleftharpoons 2NO(g); K_2 = 1.6 \times 10^{12}$

$H_2(g) + \frac{1}{2}O_2(g) \rightleftharpoons H_2O(g); K_3 = 1.0 \times 10^{-13}$

Based on above equilibria, the equilibrium constant of the reaction,

$2NH_3(g) + \frac{5}{2}O_2(g) \rightleftharpoons 2NO(g) + 3H_2O(g)$ is

........ $\times 10^{-33}$. (Nearest integer) **[2023, 29 Jan Shift-II]**

**50** At 600 K, 2 moles of NO are mixed with 1 mole of $O_2$.

$$2NO(g) + O_2(g) \rightleftharpoons 2NO_2(g)$$

The reaction occurring as above comes to equilibrium under a total pressure of 1 atm. Analysis of the system shows that 0.6 mole of oxygen are present at equilibrium. The equilibrium constant for the reaction is ........ (Nearest integer) **[2022, 28 July Shift-II]**

**51** At 298 K, the equilibrium constant is $2 \times 10^{15}$ for the reaction

$$Cu(s) + 2Ag^+(aq) \rightleftharpoons Cu^{2+}(aq) + 2Ag(s)$$

The equilibrium constant for the reaction

$$\frac{1}{2}Cu^{2+}(aq) + Ag(s) \rightleftharpoons \frac{1}{2}Cu(s) + Ag^+(aq)$$

is $x \times 10^{-8}$. The value of $x$ is ........ (nearest integer) **[2022, 26 July Shift-I]**

**52** 40° of HI undergoes decomposition to $H_2$ and $I_2$ at 300 K. $\Delta G°$ for this decomposition reaction at one atmosphere pressure is ........ $\text{J mol}^{-1}$. [Nearest integer]

(Use $R = 8.31 \text{ J K}^{-1}\text{mol}^{-1}$ ; log 2 = 0.3010. In 10 = 2.3, log 3 = 0.477) **[2022, 26 June Shift-II]**

**Imp.** Calculations of Equilibrium Constant $K_p$

**53** $PCl_5$ dissociates as

$$PCl_5(g) \rightleftharpoons PCl_3(g) + Cl_2(g)$$

5 moles of $PCl_5$ are placed in a 200 L vessel which contains 2 moles of $N_2$ and is maintained at 600 K. The equilibrium pressure is 2.46 atm. The equilibrium constant $K_p$ for the dissociation of $PCl_5$ is ...... $\times 10^{-3}$. (Nearest integer)

(Given, $R = 0.082 \text{ L atm K}^{-1} \text{mol}^{-1}$; assume ideal gas behaviour) **[2022, 24 June Shift-II]**

**54** A box contains 0.90 g of liquid water in equilibrium with water vapour at 27°C. The equilibrium vapour pressure of water at 27°C is 32.0 Torr. When the volume of the box is increased, some of the liquid water evaporates to maintain the equilibrium pressure. If all the liquid water evaporates to maintain the equilibrium pressure, then the volume of the box must be ........ litre. [Nearest integer]

(Given : $R = 0.082 \text{ L atm K}^{-1} \text{mol}^{-1}$)

(Ignore the volume of the liquid water and assume water vapours behave as an ideal gas). **[2022, 29 June Shift-II]**

**55** $2NOCl(g) \rightleftharpoons 2NO(g) + Cl_2(g)$

In an experiment, 2.0 moles of NOCl was placed in a one-litre flask and the concentration of NO after equilibrium established, was found to be 0.4 mol/L. The equilibrium constant at 30°C is ...... $\times 10^{-4}$. **[2022, 27 June Shift-I]**

**56** The reaction rate for the reaction

$$[PtCl_4]^{2-} + H_2O \rightleftharpoons [Pt(H_2O)Cl_3]^- + Cl^-$$

was measured as a function of concentrations of different species. It was observed that

$$\frac{-d[[PtCl_4]^{2-}]}{dt} = 4.8 \times 10^{-5}[[PtCl_4]^{2-}] - 2.4 \times 10^{-3}[[Pt(H_2O)Cl_3]^-][Cl^-].$$

where square brackets are used to denote molar concentrations. The equilibrium constant, $K = \ldots\ldots\ldots$ [2021, 26 Aug Shift-II]

**57** The equilibrium constant for the reaction

$$A(s) \rightleftharpoons M(s) + \frac{1}{2}O_2(g)$$

is $K_p = 4$. At equilibrium, the partial pressure of $O_2$ is ........ atm. (Round off to the nearest integer) [2021, 27 July Shift-II]

**58** Value of $K_p$ for the equilibrium reaction

$N_2O_4(g) \rightleftharpoons 2NO_2(g)$ at 288 K is 47.9.

The $K_C$ for this reaction at same temperature is ...... (Nearest integer)

($R = 0.083$ L bar $K^{-1}mol^{-1}$) [2021, 22 July Shift-II]

**59** $2SO_2(g) + O_2(g) \rightleftharpoons 2SO_3(g)$

In an equilibrium mixture, the partial pressures are $p_{SO_3} = 43$kPa; $p_{O_2} = 530$ Pa and $p_{SO_2} = 45$kPa. The equilibrium constant $K_p = \ldots\ldots\ldots \times 10^{-2}$. (Nearest integer) [2021, 20 July Shift-I]

**60** The gas phase reaction

$$2A(g) \rightleftharpoons A_2(g)$$

at 400 K has $\Delta G° = +25.2$ kJ $mol^{-1}$.

The equilibrium constant $K_C$ for this reaction is ........ $\times 10^{-2}$. (Round off to the nearest integer).

[Use : $R = 8.3$ J $mol^{-1}$ $K^{-1}$, ln 10 = 2.3]

$\log_{10} 2 = 0.30$, 1 atm = 1 bar] [antilog $(-0.3) = 0.501$] [2021, 18 March Shift-II]

**61** Consider the reaction, $N_2O_4(g) \rightleftharpoons 2NO_2(g)$.

The temperature at which $K_C = 20.4$ and $K_p = 600.1$, is ........... K. (Round off to the nearest integer).

[Assume all gases are ideal and $R = 0.0831$ L bar, $K^{-1}$ $mol^{-1}$]. [2021, 17 March Shift-II]

**62** For the reaction, $A(g) \rightleftharpoons B(g)$ at 495 K, $\Delta_r G° = -9.478$ kJ $mol^{-1}$.

If we start the reaction in a closed container at 495 K with 22 millimoles of $A$, the amount of $B$ in the equilibrium mixture is ........ millimoles (Round off to the nearest integer).

[$R = 8.314$ J $mol^{-1}$ $K^{-1}$, ln 10 = 2.303] [2021, 16 March Shift-I]

**63** A homogeneous ideal gaseous reaction, $AB_2(g) \rightleftharpoons A(g) + 2B(g)$ is carried out in a 25 L flask at 27°C. The initial amount of $AB_2$ was 1 mole and the equilibrium pressure was 1.9 atm. The value of $K_p$ is $x \times 10^{-2}$. The value of $x$ is ........... . [2021, 26 Feb Shift-I]

**64** At 1990 K and 1 atm pressure, there are equal number of $Cl_2$ molecules and Cl atoms in the reaction mixture. The value of $k_p$ for the reaction $Cl_2(g) \rightleftharpoons 2Cl(g)$ under the above conditions is $x \times 10^{-1}$. The value of $x$ is ........ .

(Rounded off to the nearest integer) [2021, 24 Feb Shift-I]

**65** For the reaction, $A(g) \longrightarrow B(g)$, the value of the equilibrium constant at 300 K and 1 atm is equal to 100.0. The value of $\Delta_r G$ for the reaction at 300 K and 1 atm in J $mol^{-1}$ is $-xR$, where $x$ is ........ .

(Rounded off to the nearest integer)

($R = 8.31$ J $mol^{-1}$ $K^{-1}$ and ln10 = 2.3) [2021, 24 Feb Shift-I]

## Topic 2 Le-Chatelier's Principle and Factors Affecting Chemical Equilibria

### Level 1

**01** Consider the equilibrium, [2025, 29 Jan Shift-II]

$$CO(g) + 3H_2(g) \rightleftharpoons CH_4(g) + H_2O(g)$$

If the pressure applied over the system increases by two fold at constant temperature then

(A) Concentration of reactants and products increases.
(B) Equilibrium will shift in forward direction.
(C) Equilibrium constant increases since concentration of products increases.
(D) Equilibrium constant remains unchanged as concentration of reactants and products remain same.

Choose the correct answer from the options given below

(a) (A), (B) and (C) only
(b) (A) and (B) only
(c) (A),(B) and (D) only
(d) (B) and (C) only

**02** The following reaction occurs in the blast furnace where iron ore is reduced to iron metal

$$Fe_2O_3(s) + 3CO(g) \rightleftharpoons Fe(l) + 3CO_2(g)$$

Using the Le-chatelier's principle, predict which one of the following will not disturb the equilibrium. [2024, 5th April Shift-I]

(a) Removal of CO
(b) Removal of $CO_2$
(c) Addition of $CO_2$
(d) Addition of $Fe_2O_3$

**03** The exothermic formation of $ClF_3$ is represented by the equation

$$Cl_2(g)+3F_2(g) \rightleftharpoons 2ClF_3(g); \Delta H_r=-329 \text{ kJ}$$

Which of the following will increase the quantity of $ClF_3$ in an equilibrium mixture of $Cl_2$, $F_2$ and $ClF_3$? *[AIEEE 2005]*

(a) Adding $F_2$
(b) Increasing the volume of the container
(c) Removing $Cl_2$
(d) Increasing the temperature

**04** One of the following equilibria is not affected by change in volume of the flask *[AIEEE 2002]*

(a) $PCl_5(g) \rightleftharpoons PCl_3(g)+Cl_2(g)$
(b) $N_2(g)+3H_2(g) \rightleftharpoons 2NH_3(g)$
(c) $N_2(g)+O_2(g) \rightleftharpoons 2NO(g)$
(d) $SO_2Cl_2(g) \rightleftharpoons SO_2(g)+Cl_2(g)$

## Level 2

**05** Consider the following chemical equilibrium of the gas phase reaction at a constant temperature :

$$A(g) \rightleftharpoons B(g)+C(g)$$

If $p$ being the total pressure, $K_p$ is the pressure equilibrium constant and $\alpha$ is the degree of dissociation, then which of the following is true at equilibrium? *[2025, 2 April Shift-II]*

(a) If $p$ value is extremely high compared to $K_p$, $\alpha=1$.
(b) When $p$ increases $\alpha$ decreases.
(c) If $K_p$ value is extremely high compared to $p$, $\alpha$ becomes much less than unity.
(d) When $p$ increases $\alpha$ increases.

**06** In the following system, $PCl_5(g) \rightleftharpoons PCl_3(g)+Cl_2(g)$ at equilibrium, upon addition of xenon gas at constant $T$ and $p$, the concentration of *[2025, 3 April Shift-I]*

(a) $PCl_5$ will increase
(b) $PCl_3$ will increase
(c) $Cl_2$ will decrease
(d) $PCl_5$, $PCl_3$ and $Cl_2$ remain constant

**07** For a reaction, *[2025, 24 Jan Shift-I]*

$N_2O_5(g) \longrightarrow 2NO_2(g)+\frac{1}{2}O_2(g)$ in a constant volume container, no products were present initially. The final pressure of the system when 50% of reaction gets completed is

(a) 7/2 times of initial pressure
(b) 7/4 times of initial pressure
(c) 5/2 times of initial pressure
(d) 5 times of initial pressure

**08** Consider the following reaction :

$N_2O_4(g) \rightleftharpoons 2NO_2(g); \Delta H^\circ=+58 \text{ kJ}$

For each of the following cases ($A$, $B$), the direction in which the equilibrium shifts is

($A$) temperature is decreased.
($B$) pressure is increased by adding $N_2$ at constant $T$. *[2020, 5 Sep Shift-I]*

(a) ($A$) towards product, ($B$) towards reactant
(b) ($A$) towards reactant, ($B$) no change
(c) ($A$) towards reactant, ($B$) towards product
(d) ($A$) towards product, ($B$) no change

**Imp.** Factors Affecting Equilibrium Constant

**09** For the reaction,

$$2SO_2(g)+O_2(g) \rightleftharpoons 2SO_3(g),$$

$\Delta H=-57.2 \text{ kJ mol}^{-1}$ and $K_c=1.7\times10^{16}$. Which of the following statement is incorrect? *[2019, 10 April Shift-II]*

(a) The equilibrium constant decreases as the temperature increases
(b) The addition of inert gas at constant volume will not affect the equilibrium constant
(c) The equilibrium will shift in forward direction as the pressure increases
(d) The equilibrium constant is large suggestive of reaction going to completion and so no catalyst is required

## Numerical Value Type Questions

**10** The number of correct statement/s involving equilibria in physical processes from the following is *[2023, 10 April Shift-I]*

(A) Equilibrium is possible only in a closed system at a given temperature.
(B) Both the opposing processes occur at the same rate.
(C) When equilibrium is attained at a given temperature, the value of all its parameters became equal.
(D) For dissolution of solids in liquids, the solubility is constant at a given temperature.

**11** Consider the following equation:

$2SO_2(g)+O_2(g) \rightleftharpoons 2SO_3(g), \Delta H=-190 \text{ kJ}$

The number of factors which will increase the yield of $SO_3$ at equilibrium from the following is ..........

A. increasing temperature
B. increasing pressure
C. adding more $SO_2$
D. adding more $O_2$
E. addition of catalyst *[2023, 30 Jan Shift-II]*

## • Key **Index**

### Topic 1

| | | | | | | | | | |
|---|---|---|---|---|---|---|---|---|---|
| **1** (d) | **2** (d) | **3** (a) | **4** (a) | **5** (b) | **6** (d) | **7** (c) | **8** (d) | **9** (c) | **10** (c) |
| **11** (a) | **12** (d) | **13** (b) | **14** (d) | **15** (a) | **16** (d) | **17** (c) | **18** (c) | **19** (c) | **20** (c) |
| **21** (a) | **22** (c) | **23** (a) | **24** (a) | **25** (b) | **26** (b) | **27** (a) | **28** (d) | **29** (a) | **30** (d) |
| **31** (a) | **32** (d) | **33** (b) | **34** (b) | **35** (b) | **36** (c) | **37** (74) | **38** (962) | **39** (57) | **40** (163) |
| **41** (417) | **42** (2) | **43** (37) | **44** (44) | **45** (1) | **46** (3) | **47** (48) | **48** (12) | **49** (4) | **50** (2) |
| **51** (2) | **52** (2735) | **53** (1107) | **54** (29) | **55** (125) | **56** (0.02) | **57** (16) | **58** (2) | **59** (172) | **60** (2) |
| **61** (354) | **62** (20) | **63** (73) | **64** (5) | **65** (1380) | | | | | |

### Topic 2

| | | | | | | | | | |
|---|---|---|---|---|---|---|---|---|---|
| **1** (b) | **2** (d) | **3** (a) | **4** (c) | **5** (b) | **6** (b) | **7** (b) | **8** (b) | **9** (d) | **10** (3) |
| **11** (3) | | | | | | | | | |

# Detailed **Explanations**

### Topic 1

**01.** *(d)* For the reaction

$$H_2 + I_2 \rightleftharpoons 2HI$$

Concentration of $H_2$ and $I_2$ decrease and concentration of HI increases till equilibrium condition and after equilibrium concentrations of all the reactants and the product remain constant.

**02.** *(d)* The reaction involved in conversion of $CO_2$ into CO is as follows

$$\underset{0.5-x}{\overset{0.5}{CO_2(g)}} + C(s) \rightleftharpoons \underset{2x}{\overset{-}{2CO(g)}}$$

$$p_{\text{Total}} = 0.5 + x = 0.8$$

$$x = 0.3$$

So, $\quad k_p = \dfrac{(0.6)^2}{0.2} = 1.8 \text{ atm}$

So, the value of $k_p$ is 1.8 atm.

**03.** *(a)* $K_C$ is given by the equation

$$K_C = \frac{\text{Concentration of product ion}}{\text{Concentration of reactant ion}}$$

So, for the reaction

$$Fe^{3+}(aq) + SCN^-(aq) \rightleftharpoons (FeSCN)^{2+}(aq)$$

The value of $K_C$ will be

$$= \frac{[FeSCN^{2+}]}{[Fe^{3+}][SCN^-]}$$

**04.** *(a)* Equilibrium is the condition when a forward chemical reaction and its reverse reaction proceed at equal rates.

i.e., at equilibrium, $r_a = r_b$

For the equilibrium, $A \rightleftharpoons B$

The variation of the rate of the forward and reverse reaction can be illustrated by following graph.

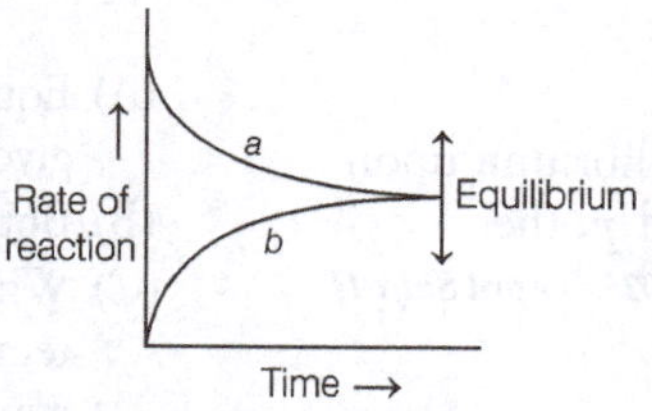


**05.** *(b)* Assuming the reaction to be

$$\underset{\text{(Square)}}{A} \rightleftharpoons \underset{\text{(Circle)}}{B},$$

Equilibrium constant, $K = \dfrac{[B]}{[A]}$

Number of circles in the figure $= 11$

Number of squares in the figure $= 6$

$$\therefore \quad K = \frac{11}{6} = 1.8 \approx 2$$

**06.** *(d)* $\Delta G° = -RT \ln K$

$\therefore$ If $K > 1$ then $\Delta G° < 0$

If $K < 1$ then $\Delta G° > 0$

If $K = 1$ then $\Delta G° = 0$

Therefore, option (d) is incorrect match.

**07.** *(c)* As equation 'III' is obtained on adding equation 'I' and equation 'II', so

$$K_3 = K_1 \cdot K_2.$$

**08.** *(d)* $P_4(s) + 5O_2(g) \rightleftharpoons P_4O_{10}(s)$

Thus, $\quad K_c = \dfrac{1}{[O_2]^5}$

In the expression for equilibrium constant ($K_p$ or $K_c$), species in solid state are not written as their molar concentrations are taken as 1

**09.** *(c)* For the equilibrium reaction

$$N_2O_4(g) \rightleftharpoons 2NO_2(g)$$

$$K_c = \frac{[NO_2]^2}{[N_2O_4]} = \frac{(1.2 \times 10^{-2})^2}{4.8 \times 10^{-2}}$$

$$= 3 \times 10^{-3} \text{ mol L}^{-1}$$

**10.** *(c)* The given reaction is as follows

$$AB_2(g) \rightleftharpoons AB(g) + \frac{1}{2}B_2(g)$$

| | $AB_2(g)$ | $AB(g)$ | $\frac{1}{2}B_2(g)$ |
|---|---|---|---|
| At $t = 0$ | 1 | 0 | 0 |
| At $t = t_{eq}$ | $1-x$ | $x$ | $\frac{x}{2}$ |

$$\therefore K_p = \frac{\left(\dfrac{x}{1+x/2}\right)p\left(\dfrac{x/2}{1+x/2}\right)^{1/2}p^{1/2}}{\left(\dfrac{1-x}{1+x/2}\right)p}$$

$$\because \quad x << 1 \Rightarrow \frac{x}{2} + 1 \text{ and } 1 - x \approx 1$$

$$\therefore \quad K_p = \frac{(xp)\left(\frac{xp}{2}\right)^{1/2}}{p}$$

So, the value of $x$ is terms of $K_p$ and $p$ is $x = \sqrt[3]{\frac{2K_p^2}{p}}$

**11.** *(a)* $X_2Y(g) \rightleftharpoons X_2(g) + \frac{1}{2}Y_2(g)$

$1-x$ mol $\quad x$ mol $\quad x/2$mol

$$\therefore \quad p_{X_2Y} = \frac{1-x}{1+x/2} \times p$$

$$p_{X_2} = \frac{x}{1+x/2} \times p$$

$$p_{Y_2} = \frac{x/2}{1+x/2} \times p$$

Therefore, for the given reaction,

$$K_p = \frac{\left(\frac{x}{1+x/2}p\right)\left(\frac{x}{2(1+x/2)}p\right)^{1/2}}{\left(\frac{1-x}{1+x/2}\right)p}$$

$$\therefore \quad K_p = \left(\frac{x}{1-x}\right)\left(\frac{x}{2(1+x/2)}\right)^{1/2} \times p^{1/2}$$

Also it is given that, the value of $x$ is very small.

So, $\quad K_p = \frac{x^{3/2}}{(2)^{1/2}} \times p^{1/2}$

or $\quad x^{3/2} = \frac{K_p \times 2^{1/2}}{p^{1/2}}$

or $\quad x = \left(\frac{K_p^2 \times 2}{p}\right)^{1/3}$

**12.** *(d)* $X \rightleftharpoons Y; K_1 = 1.0$

$Y \rightleftharpoons Z; K_2 = 2.0$

$Z \rightleftharpoons W; K_3 = 4.0$

For the equilibrium constant value of $X \rightleftharpoons W; K = K_1 \cdot K_2 \cdot K_3$

$K = 1 \times 2 \times 4$

$K = 8$

Hence, option (d) is correct.

**13.** *(b)* Given,

$$CO(g) + \frac{1}{2}O_2(g) \rightleftharpoons CO_2(g)$$

$$\Delta n_g = 1 - \left(1 + \frac{1}{2}\right) = -\frac{1}{2}$$

$$\frac{K_p}{K_c} = (RT)^{\Delta n_g} = (RT)^{-1/2} = \frac{1}{\sqrt{RT}}$$

**14.** *(d)* The reaction $2SO_2 + O_2 \rightleftharpoons 2SO_3$ will be formed from the given reaction by multiplying it by coefficient 2 and then reversing it.

Thus, $K'_C = \frac{1}{K_C^2}$

$$= \left(\frac{1}{4.9 \times 10^{-2}}\right)^2 = 416$$

**15.** *(a)* $p_{total} = \frac{nRT}{V} = \frac{9 \times 0.082 \times 610}{100}$

$= 4.5$ atm

$p_{PCl_5} = \frac{5 \times 4.5}{9} = 2.5$ atm $[p_A = x_A p_T]$

$p_{Ar} = \frac{4 \times 4.5}{9} = 2$ atm

Now, $PCl_5 \rightleftharpoons PCl_3 + Cl_2$

$2.5 - p \quad p \quad p$

$p_{total} = 2.5 - p + p + p + p_{Ar} = 6$

$p = 1.5$

Now, $K_p = \frac{[PCl_3][Cl_2]}{[PCl_5]}$

$$= \frac{1.5 \times 1.5}{(2.5 - 1.5)} = 2.25$$

**16.** *(d)* Equilibrium constant for equation,

$$N_2(g) + 3H_2(g) \rightleftharpoons 2NH_3(g)$$

$$K_C = \frac{[NH_3]^2}{[N_2][H_2]^3} = 64$$

and for

$2NH_3(g) \rightleftharpoons N_2(g) + 3H_2(g)$,

$$K'_C = \frac{1}{K_C}$$

and for equation,

$$NH_3(g) \rightleftharpoons \frac{1}{2}N_2(g) + \frac{3}{2}H_2(g)$$

$$K'_C = \frac{1}{K_C^{1/2}} = \frac{1}{K_C^{1/2}} = \frac{1}{(64)^{1/2}} = \frac{1}{8}$$

**17.** *(c)* $\log\left(\frac{K_2}{K_1}\right) = \frac{\Delta H°}{2.303R}\left[\frac{1}{T_1} - \frac{1}{T_2}\right]$

$$\log\left(\frac{100}{10}\right) = \frac{\Delta H°}{2.303 \times \left(\frac{8.314}{1000}\right)}\left[\frac{1}{298} - \frac{1}{373}\right]$$

$\Delta H° = 28.4$ kJ mol$^{-1}$

$\Delta G° = -RT \ln K$,

$\Delta G°$ at $T_1 = (25°C)$

$$= -\frac{8.314}{1000} \times 298 \times 2.303 \times \log 10$$

$= -5.71$ kJ/mol

$$\Delta G° \text{ at } T_2(100°C) = -\frac{8.314}{1000} \times 373 \times 2.303 \times \log(100)$$

$= -14.29$ kJ/mol

Hence, $\Delta H° = 28.4$ kJ/mol

**At $T_1$**

$\Delta G° = -5.71$ kJ/mol

**At $T_2$**

$\Delta G° = -14.29$ kJ/mol

**18.** *(c)* $K_p = K_C(RT)^{\Delta n_g}$

$\Delta n_g$ = number of gaseous product molecules – number of gaseous reactant molecules

$$Fe_2N(s) + \frac{3}{2}H_2(g) \rightleftharpoons Fe(s) + NH_3(g)$$

$$\Delta n_g = 1 - \frac{3}{2} = -\frac{1}{2}$$

Now, $K_p = K_C(RT)^{-1/2}$

$$\therefore \quad K_C = \frac{K_p}{(RT)^{-1/2}} \Rightarrow K_p(RT)^{1/2}$$

Hence, the correct option is (c).

**19.** *(c)* $A \rightleftharpoons B + C$;

$$K_{eq}^{(1)} = \frac{[B][C]}{[A]} \quad \text{... (i)}$$

$$B + C \rightleftharpoons P; K_{eq}^{(2)} = \frac{[P]}{[B][C]} \quad \text{...(ii)}$$

Equilibrium constant for $A \rightleftharpoons P$

$$A \rightleftharpoons P; K_{eq} = \frac{[P]}{[A]} \quad \text{...(iii)}$$

To obtain the above equilibrium reaction add the other two equilibrium reaction.

In order to obtain $K_{eq}$

Multiplying equation (i) and (ii)

$$K_{(eq)}^{(1)} \times K_{(eq)}^{(2)} = \frac{[P]}{[A]} = K_{eq}$$

**20.** *(c)* At a certain temperature, water maintains an equilibrium state with water vapour in a sealed container. When glucose is added into water, a completely miscible solution is formed in which glucose molecules occupy the inter-molecular space of solvent ($H_2O$) molecules.

Thus, the addition of glucose increases the total number of molecules in the solution, which increases the collision rate of water molecules with glucose molecules. This leads to an increase in the rate of evaporation of water molecules from the surface of the beaker and hence, the rate at which water molecules leave the vapour increases.

**21.** *(a)*

> **key idea** The relationship between $K_p$ and $K_c$ is
> $$K_p = K_c(RT)^{\Delta n_g}$$
> where $\Delta n_g = \Delta n_{Products} - \Delta n_{Reactants}$
> If $\Delta n_g = 0$ then $K_p = K_c$
> If $\Delta n_g = +$ve then $K_p > K_c$
> If $\Delta n_g = -$ve then $K_p < K_c$

Consider the given equilibria reaction

$$2C(s) + O_2(g) \rightleftharpoons 2CO(g)$$

$$\Delta n_g = n_{product} - n_{reactant} = 2 - (1) = 1$$

$$\Delta n_g \neq 0 \quad \Rightarrow \text{So, } K_p \neq K_c$$

**22.** *(c)* (i) $A_2(g) + B_2(g) \overset{K_1}{\rightleftharpoons} 2AB(g)$

$$K_1 = \frac{[AB]^2}{[A_2][B_2]}$$

(ii) $6AB(g) \rightleftharpoons 3A_2(g) + 3B_2(g);$

$$K_2 = \frac{[A_2]^3[B_2]^3}{[AB]^6}$$

$$= \frac{1}{\left(\frac{[AB]^2}{[A_2][B_2]}\right)^3} = \frac{1}{K_1^3},$$

$$\Rightarrow \quad K_2 = K_1^{-3}$$

**23.** *(a)* $S + O_2 \rightleftharpoons SO_2, \quad K_1$

$\therefore \quad SO_2 \rightleftharpoons S + O_2$

$$K_1' = \frac{1}{K_1}$$

or, $\quad 2SO_2 \rightleftharpoons 2S + 2O_2,$

$$K_1'' = (K_1')^2 = \frac{1}{K_1^2} \quad \text{... (i)}$$

$\Rightarrow \quad 2S + 3O_2 \rightleftharpoons 2SO_3, K_2 \quad \text{... (ii)}$

Now, [(i) + (ii)] gives

$$2SO_2 + O_2 \rightleftharpoons 2SO_3, K_3$$

The value of equilibrium constant,

$$K_3 = K_2 \times K_1'' = K_2 \times \frac{1}{K_1^2}$$

$$= 10^{129} \times \frac{1}{(10^{52})^2} = 10^{129-104} = 10^{25}$$

**24.** *(a)*

> **key idea** In exothermic reaction, increasing the temperature, decreases the value of equilibrium constant. However, if the forward reaction is endothermic, increasing the temperature, increases the value of equilibrium constant.

From thermodynamics,

$$\ln k = \frac{-\Delta H°}{RT} + \frac{\Delta S°}{R} \quad \text{...(i)}$$

Mathematically, the equation of straight line is

$$y = c + mx \quad \text{...(ii)}$$

After comparing Eq. (ii) with (i) we get,

$$\text{slope} = \frac{-\Delta H°}{R} \text{ and intercept} = \frac{\Delta S°}{R}$$

Now, we know for exothermic reaction $\Delta H$ is negative (–)ve. But here,

$$\text{Slope} = \frac{-\Delta H°}{R} \text{ is positive}$$

So, lines $A$ and $B$ in the graph represent temperature dependence of equilibrium constant $K$ for an exothermic reaction as shown below

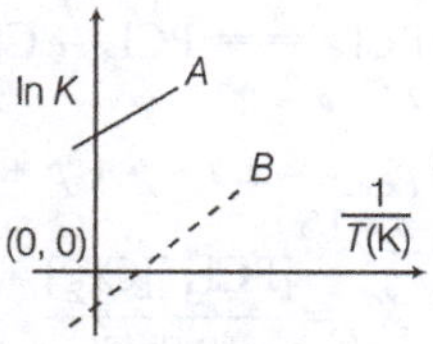


**25.** *(b)* Given, $\Delta G° = 2494.2$J

$$Q = \frac{[B][C]}{[A]^2} = \frac{2 \times (1/2)}{\left(\frac{1}{2}\right)^2} = 4$$

$\therefore$ We know,

$$\Delta G° = -2.303\, RT \log K_c$$

$$2494.2 = -2.303\, RT \log K_c$$

$$\log K_c = \text{antilog}(-0.4342)$$

$$\Rightarrow \quad K_c = 0.3679$$

Therefore, $Q > K_c$ reaction shifts in reverse direction.

> **• Concept Enhancer**
> The direction of the reaction can be predicted with the help of reaction quotient. If $Q = K$, then the reaction is in equilibrium if $Q > K$ the reaction moves in backward direction if $K > Q$ the reaction moves in forward direction.

**26.** *(b)* Equilibrium constant,

$$K_P = K_c\,(RT)^{\Delta n_g}$$

where $\Delta n_g$ = (no. of gaseous moles of products) – (no. of gaseous moles of reactants)

$$\therefore \quad \Delta n_g = x$$

For the given reaction,

$$SO_2\,(g) + \frac{1}{2}O_2\,(g) \rightleftharpoons SO_3\,(g)$$

$$x = \Delta n_g = 1 - \left[\frac{1}{2} + 1\right] = 1 - \left[\frac{3}{2}\right] = -\frac{1}{2}$$

**27.** *(a)*

| | $A\,+$ | $2B$ | $\rightleftharpoons$ | $2C\,+$ | $D$ |
|---|---|---|---|---|---|
| $t = 0$ | 2 | 3 | | 0 | 0 |
| $t_{eq}$ | $2 - x$ | $3 - 2x$ | | $2x$ | $x$ |

Given, $[A] = [B]$

$$3 - 2x = 2 - x \Rightarrow x = 1$$

$$\therefore \quad [C] = 2, [D] = 1, [A] = 1, [B] = 1$$

$$\therefore \quad K_c = \left\{\frac{2^2 \times 1}{1^2 \times 1}\right\} = 4$$

Therefore, the equilibrium constant is 4.

**28.** *(d)* $N_2(g) + O_2(g) \longrightarrow 2NO(g)$

$$K_c = \frac{[NO]^2}{[N_2][O_2]} = 4 \times 10^{-4}$$

For the reverse reaction,

$$2NO(g) \longrightarrow N_2(g) + O_2(g)$$

$$K_c' = \frac{1}{K_c} = \frac{[N_2][O_2]}{[NO]^2}$$

$$= \frac{1}{4 \times 10^{-4}} = \frac{10^4}{4}$$

On dividing the reverse reaction by 2

$$NO(g) \longrightarrow \frac{1}{2}N_2(g) + \frac{1}{2}O_2(g)$$

$$K_c'' = \frac{[N_2]^{1/2}[O_2]^{1/2}}{[NO]} = \sqrt{K_c'}$$

$$= \sqrt{\frac{10^4}{4}} = \frac{100}{2} = 50$$

**29.** *(a)*

| | $CO_2(g) + C(s) \rightleftharpoons$ | $2CO(g)$ |
|---|---|---|
| Initial | 0.5 atm | — |
| At equilibrium | $(0.5 - p)$ | $2p$ atm |

This is a case of heterogeneous equilibrium.

$C(s)$ being solid is not considered.

Total pressure of $CO_2$ and CO gases.

$$p_{CO_2} + p_{CO} = p_{total}$$

$$(0.5 - p) + 2p = 0.8, \quad p = 0.3 \text{ atm}$$

$$\therefore \quad p_{CO_2} = 0.5 - 0.3 = 0.2 \text{ atm}$$

$$p_{CO} = 2p = 0.6 \text{ atm}$$

$$K_p = \frac{p^2_{CO}}{p_{CO_2}} = \frac{0.6 \times 0.6}{0.2} = 1.8 \text{ atm}$$

**30.** *(d)* $2NO_2(g) \rightleftharpoons 2NO(g) + O_2(g)$

$$K_c = 1.8 \times 10^{-6} \text{ at } 184°C\ (= 457\ K)$$

$$R = 0.00831 \text{ kJ mol}^{-1}\ K^{-1}$$

$$K_p = K_c\,(RT)^{\Delta n_g}$$

$$\Delta n_g = (2 + 1) - 2 = 1$$

$$\therefore \quad K_p = 1.8 \times 10^{-6} \times 0.00831 \times 457$$

$$= 6.836 \times 10^{-6} > 1.8 \times 10^{-6}$$

Thus, $K_p > K_c$.

**31.** (a) $A(g) \rightleftharpoons B(g) + \frac{C}{2}(g)$

$t = t_{eq}$ $(1-\alpha)$ $\quad \alpha \quad \frac{\alpha}{2}$

$$p_B = \frac{\alpha}{\left(1+\frac{\alpha}{2}\right)} \cdot p, \; p_A = \frac{(1-\alpha)}{\left(1+\frac{\alpha}{2}\right)} \cdot p,$$

$$p_C = \frac{\frac{\alpha}{2}}{\left(1+\frac{\alpha}{2}\right)} \cdot p$$

$$K_p = \frac{p_B \cdot p_C^{1/2}}{p_A} \quad \text{...(i)}$$

Substitute the value of $p_A$, $p_B$ and $p_C$ in Eq. (i)

$$\therefore \quad K_p = \frac{(\alpha)^{3/2}\,(p)^{1/2}}{(1-\alpha)\,(2+\alpha)^{1/2}}$$

**32.** (d) The equilibrium reaction for the dissociation of two solids is given as:

$A(s) \rightleftharpoons B(g) + C(g)$

At equilibrium $\quad p_1 \quad p_1 + p_2$

$$K_{p_1} = x = p_B \cdot p_C = p_1(p_1 + p_2) \quad \text{...(i)}$$

Similarly,

$D(s) \rightleftharpoons C(g) + E(g)$

At equilibrium $\quad p_1 + p_2 \quad p_2$

$$K_{p_2} = y = p_C \cdot p_E = (p_1 + p_2)p_2 \quad \text{...(ii)}$$

On adding Eq. (i) and (ii), we get.

$$K_{p_1} + K_{p_2} = x + y = p_1(p_1 + p_2) + p_2(p_1 + p_2) = (p_1 + p_2)^2$$

or $\sqrt{x+y} = p_1 + p_2$ ...(iii)

Now, total pressure is given as

$$p_T = p_B + p_C + p_E = p_1 + (p_1 + p_2) + p_2 = 2\,(p_1 + p_2) \quad \text{...(iv)}$$

On substituting the value of $p_1 + p_2$ from Eq. (iii) to Eq. (iv), we get

$$p_T = 2\sqrt{x+y}$$

**33.** (b) $N_2(g) + 3H_2(g) \rightleftharpoons 2NH_3(g)$

At eq. $p_{N_2} = P$, $p_{H_2} = 3P$, $p_{NH_3} = 2P$

$$\Rightarrow p_{(total)} = p_{N_2} + p_{H_2} + p_{NH_3} \approx p_{N_2} + p_{H_2}$$

$[\because p_{(total)} >> p_{NH_3}]$

$$p = P + 3P = 4P$$

$$\text{Now, } K_p = \frac{p^2_{NH_3}}{p_{N_2} \times p^3_{H_2}} = \frac{p^2_{NH_3}}{p \times (3P)^3}$$

$$= \frac{p^2_{NH_3}}{27 \times p^4} = \frac{p^2_{NH_3}}{27 \times \left(\frac{P}{4}\right)^4}$$

$[\because p = 4P]$

$$K_p = \frac{p^2_{NH_3} \times 4^4}{3^2 \times 3 \times p^4}$$

$$\Rightarrow \quad p^2_{NH_3} = \frac{3^2 \times 3 \times p^4 \times K_p}{4^4}$$

$$\Rightarrow \quad p_{NH_3} = \frac{3 \times 3^{1/2} \times p^2 \times K_p^{1/2}}{4^2} = \frac{3^{3/2} \times p^2 \times K_p^{1/2}}{16}$$

**34.** (b) Molar mass of $NH_4SH = 18 + 33 = 51\text{ g mol}^{-1}$

Number of moles of $NH_4SH$ introduced in the vessel

$$= \frac{\text{Weight}}{\text{Molar mass}} = \frac{5 \cdot 1}{51} = 0.1\text{mol}$$

| | $NH_4SH(s) \rightleftharpoons$ | $NH_3(g)$ | $+ H_2S(g)$ |
|---|---|---|---|
| Number of moles at $t = 0$ | 0.1 | 0 | 0 |
| At $t = t_{eq}$ | 0.1(1 − 0.03) | 30% of 0.1 = 0.03 | 30% of 0.1 = 0.03 |
| Active mass (mol $L^{-1}$) | | $\frac{0.03}{3} = 0.01$ | $\frac{0.03}{3} = 0.01$ |

$$K_C = \frac{[NH_3]\,[H_2S]}{[NH_4HS(s)]} = \frac{0.01 \times 0.01}{1} = 10^{-4}\,(\text{mol L}^{-1})^2$$

$$\Rightarrow \quad K_p = K_C(RT)^{\Delta n_g}$$

[where, $\Delta n_g = \Sigma n_{product} - \Sigma n_{reactant}$] $= 2 - 0 = 2$

$$\therefore \quad K_p = K_C(RT)^2 = 10^{-4} \times [0.082 \times (273 + 327)]^2\text{ atm}^2 = 0.242\text{ atm}^2$$

**35.** (b)

| | $A$ | $+ B \rightleftharpoons$ | $C$ | $+ D$ |
|---|---|---|---|---|
| Initially at $t = 0$ | 1 | 1 | 1 | 1 |
| At equilibrium | $1-x$ | $1-x$ | $1+x$ | $1+x$ |

$$K_{eq} = \frac{[C]\,[D]}{[A]\,[B]} = \frac{(1+x)\,(1+x)}{(1-x)\,(1-x)} = \frac{(1+x)^2}{(1-x)^2}$$

$$\text{or} \quad 100 = \left(\frac{1+x}{1-x}\right)^2 \quad \text{or} \quad 10 = \frac{1+x}{1-x}$$

$$\text{or} \quad 10 - 10x = 1 + x$$

$$\Rightarrow \quad 10 - 1 = x + 10x \Rightarrow 9 = 11x$$

$$\Rightarrow \quad x = \frac{9}{11} = 0.818$$

$$\therefore \quad [D] = 1 + x = 1 + 0.818 = 1.818$$

**36.** (c) Given,

$T = -20°\text{C}$, $p = 1$ atm, $n_{H_2} = n_{I_2}$

$K_p = ?$, $R = 0.082\text{ L atm K}^{-1}\text{ mol}^{-1}$

$$\text{Now using } K_P = \frac{(n_{HI})^2}{n_{H_2} n_{I_2}}\left(\frac{p_T}{n_T}\right)^{\Delta n_g}$$

Since $n_{H_2} = n_{I_2} = n_{HI}$

$\Rightarrow \quad \Delta n_g = 0$

So, $\quad K_p = 1$

Compare it with given equation,

$$1 = x \times 10^{-1} \quad \Rightarrow x = 10$$

**37.** (74) Consider the given reaction:

| | $CO(g)$ | $+ 2H_2(g) \rightleftharpoons$ | $CH_3OH(g)$ |
|---|---|---|---|
| $t = 0$ | 0.1 mol | $a$ mol | – |
| $t_{eq}$ | $0.1 - x$ −0.06 | $a - 2x$ 0 − 0.08 = 0.23 − 0.08 = 0.15 mole | $x = 0.04$ |

$V = 2\text{ L}$, $T = 500\text{ K}$, $p_{Total} = 5$ bar

$n_{Total} = 0.25 = 1/4$ mol

$$\text{Using, } p_{Total} = n_{Total} \times \frac{RT}{V} \quad \text{...(i)}$$

Substituting,

$$5 = (0.06 + a - 0.08 + 0.04) \times \frac{0.08 \times 500}{2}$$

$a = 0.23$ mol

$$K_p = \frac{\chi_{CH_3OH}}{\chi_{CO} \times \chi^2_{H_2}} \times \frac{1}{(p_T)^2}$$

$$= \frac{0.04}{0.06 \times (0.15)^2} \times \left[\frac{1/4}{5}\right]^2$$

$$K_p = 0.074 = 74 \times 10^{-3}$$

Hence, $n = 74$

**38.** (962) Initial pressure of $N_2O_5$

$$\frac{\frac{37.8}{108} \times 0.082 \times 500}{1} = 14.35\text{ bar}$$

| | $2N_2O_5 \rightleftharpoons$ | $2N_2O_4$ | $+ O_2$ |
|---|---|---|---|
| at $t = 0$ | 14.35 | | |
| $t$ = eq | 14.35−2$p$ | 2$p$ | $p$ |

$p_{total}$ at eq $= 14.35 + p = 18.65$ bar

$p = 4.3$

$p_{N_2O_5} = 5.75$ bar

$p_{N_2O_4} = 8.6$ bar

$p_{O_2} = 4.3$ bar

$$K_p = \frac{(8.6)^2 \times (4.3)}{(5.75)^2} = 9.619 = X \times 10^{-2}$$

$$X = 961.9 \simeq 962$$

**39.** (57) $\Delta G° = -RT \ln K_{eq} = -2.303RT \log K_{eq}$

$= -2.303 \times 8.314 \times 300 \times \log 10$
$= -5744.14$ J/mol
$= -5.744$ kJ/mol
$= -57.44 \times 10^{-1}$ kJ/mol

Thus, the value is 57.

**40.** *(163)* For the given reaction,

$$\frac{3}{2} O_2(g) \rightleftharpoons O_3(g)$$

$$K_p = 2.47 \times 10^{-29}$$

Using $\Delta_r G^\circ = -RT \ln K_p$

$= -8.314 \times 10^{-3} \times 298 \times \ln(2.47 \times 10^{-29})$

$\Delta_r G^\circ = -8.314 \times 10^{-3} \times 298 \times (-65.87)$

$\Delta_r G^\circ = 163.19$ kJ

So, the value of $x$ is 163.

**41.** *(417)* The formation of $NH_3$ is,

$$N_2 + 3H_2 \rightleftharpoons 2NH_3$$

Equilibrium constant is given by

$$K_C = \frac{[\text{Product}]}{[\text{Reactant}]} = \frac{[NH_3]^2}{[N_2][H_2]^3} = \frac{(1.5 \times 10^{-2})^2}{(2 \times 10^{-2})(3 \times 10^{-2})^3} = 417$$

Hence, equilibrium constant is 417.

**42.** *(2)*

**key idea** To solve this question, first calculate the value of $\Delta n_g$. Then, put this value along with the given values in the formula $K_p = K_C(RT)^{\Delta n_g}$ to calculate the value of $K_C$

Given reaction,

$$N_2O_4(g) \rightleftharpoons 2NO_2(g)$$

$K_p = 0.492$ atm at 300 K

The relation between $K_C$ and $K_p$,

$K_p = K_C(RT)^{\Delta n_g} \Rightarrow \Delta n_g = 1$

$$\Rightarrow K_C = \frac{K_p}{RT} = \frac{0.492}{0.082 \times 300} = 2 \times 10^{-2}$$

Hence, the missing value in the question is 2.

**43.** *(37)* Given, $T = 400$ K

$\Delta H^\ominus = 77.2$ kJ/mol

$\Delta S = 122$ JK$^{-1}$

Free energy is given by

$\Delta G^\circ = \Delta H^\circ - T\Delta S^\circ$

$= 77.2 \times 10^3 - 400 \times 122$

$= 28400$ J

$\Delta G^\circ = -2.303RT \log K$

$28400 = -2.303 \times 8.314 \times 400 \times \log K$

$\log K = -3.708$

or $\log K = -37.08 \times 10^{-1}$

Hence, the missing value is 37.

**44.** *(44)*

$$CO(g) + H_2O(g) \rightleftharpoons CO_2(g) + H_2(g)$$

| | $CO(g)$ | $H_2O(g)$ | $CO_2(g)$ | $H_2(g)$ |
|---|---|---|---|---|
| at $t = 0$ | 1 mol | 1 mol | 0 | 0 |
| at eq. | $1-x$ | $1-x$ | $x$ | $x$ |

Given that 40% by mass, water reacts with CO at equilibrium.

$\Rightarrow x = 0.4$ and $1 - x = 0.6$

$$\text{So, } K_C = \frac{[CO_2][H_2]}{[CO][H_2O]} = \frac{0.4 \times 0.4}{0.6 \times 0.6} = 0.44$$

or $K_C \times 10^2 = 44$

**45.** *(1)* $H_2(g) + I_2(g) \rightleftharpoons 2HI(g)$

| | $H_2(g)$ | $I_2(g)$ | $2HI(g)$ |
|---|---|---|---|
| at $t = 0$ | 4.5 | 4.5 | — |
| at $t$ = eq. | $4.5 - x$ | $4.5 - x$ | $2x$ |

Given, $2x = 3 \Rightarrow x = \frac{3}{2} = 1.5$

$\therefore [H_2] = 4.5 - 1.5 = 3,$
$[I_2] = 4.5 - 1.5 = 3$

$$K_C = \frac{[HI]^2}{[H_2][I_2]} = \frac{(3)^2}{3 \times 3} = 1$$

**46.** *(3)* Given,
Initial pressure = 450 mm Hg
Pressure at time $t$ = 720 mm Hg
Constant temperature $T$ and constant volume $V$.
Fraction of $A(g)$ decomposed $= x \times 10^{-1}$

$$A(g) \rightleftharpoons 2B(g) + C(g)$$

| | $A(g)$ | $2B(g)$ | $C(g)$ |
|---|---|---|---|
| at time $t = 0$, | $p = 450$ | | |
| at time $t$ | $P = 450 - x$ | $2x$ | $x$ |

$p_T = p_A + p_B + p_C$
$720 = 450 - x + 2x + x$
$720 = 450 - 2x$ or, $2x = 270$

$x = \frac{270}{2} \Rightarrow x = 135$

Fraction of $A$ decomposed $= \frac{135}{450} = 0.3$
$= 3 \times 10^{-1}$

So, $x = 3$

**47.** *(48)* $PCl_3 + Cl_2 \rightleftharpoons PCl_5$
0.2 mol L$^{-1}$ 0.1 mol L$^{-1}$ 0.40 mol L$^{-1}$

$$K_C = \frac{[PCl_5]}{[PCl_3][Cl_2]} = \frac{0.40}{0.2 \times 0.1} = 20$$

If 0.2 mol of $Cl_2$ is added then

$$Q_C = \frac{0.40}{0.2 \times (0.1 + 0.2)} = 6.66$$

$\therefore Q_C < K_C$, (Forward reaction)

$\therefore$ $PCl_3 + Cl_2 \rightleftharpoons PCl_5$
At equ. $(0.2 - x)$ $(0.3 - x)$ $0.40 + x$

$$K_C = \frac{0.4 + x}{(0.2 - x)(0.3 - x)}$$

$$\Rightarrow 20 = \frac{0.4 + x}{0.06 - 0.5x + x^2}$$

$20x^2 - 11x + 0.8 = 0$
$x = 0.086$

$\therefore$ Concentration of $PCl_5 = 0.4 + 0.086 = 0.486$
or $48.6 \times 10^{-2} \approx 48 \times 10^{-2}$

**48.** *(12)* For equilibrium equation.

(i) $X(g) \rightleftharpoons Y(g) + Z(g)$, $K_{p_1} = 3$

| | $X(g)$ | $Y(g)$ | $Z(g)$ |
|---|---|---|---|
| at $t = 0$ | 1 | 0 | 0 |
| at $t_{eq}$ | $1-x$ | $x$ | $x$ |

Total number of moles $= 1 - x + x + x = 1 + x$

Partial pressure $\frac{(1-x)p_1}{1+x}$ $\frac{xp_1}{1+x}$ $\frac{xp_1}{1+x}$

$$\Rightarrow Kp_1 = \frac{[p_y][p_z]}{[p_x]} = \frac{\left(\frac{xp_1}{1+x}\right)\left(\frac{xp_1}{1+x}\right)}{\left[\frac{(1-x)p_1}{1+x}\right]} = \frac{x^2}{1-x^2} \times p_1$$

(ii) $A(g) \rightleftharpoons 2B(g)$, $K_{p_2} = 1$

| | $A(g)$ | $2B(g)$ |
|---|---|---|
| at $t = 0$ | 1 | 0 |
| at $t_{eq}$ | $1-x$ | $2x$ |
| Partial pressure | $\frac{(1-x)p_2}{1+x}$ | $\frac{(2x)p_2}{1+x}$ |

$$\Rightarrow Kp_2 = \frac{(p_B)^2}{(p_A)} = \frac{\frac{(2x)^2 p_2^2}{(1+x)^2}}{\left(\frac{1-x}{1+x}\right)p_2}$$

$$= \frac{4x^2}{1-x^2} \times p_2 \Rightarrow \frac{Kp_1}{Kp_2} = \frac{3}{1}$$

$$= \frac{x^2}{(1-x^2)} \times p_1 \times \frac{(1-x^2)}{4x^2} \times \frac{1}{p_2} = \frac{p_1}{4p_2}$$

$$\Rightarrow \frac{p_1}{p_2} = \frac{12}{1} = \frac{x}{1} \quad \text{(Given)}$$

$x = 12$

**49.** *(4)* $N_2(g) + 3H_2(g) \rightleftharpoons 2NH_3(g)$;

$K_1 = 4 \times 10^5$ ...(i)

$$N_2(g)+O_2(g) \rightleftharpoons 2NO(g);$$
$$K_2 = 1.6\times 10^{12} \quad ...(ii)$$
$$H_2(g)+\frac{1}{2}O_2(g) \rightleftharpoons H_2O(g);$$
$$K_3 = 1\times 10^{-13} \quad ...(iii)$$

On solving above equation as
Eq. (ii) + 3 × Eq. (iii) – Eq. (i)
We have,

$$\left[2NH_3(g)+\frac{5}{2}O_2(g) \rightleftharpoons 2NO(g) + 3H_2O(g)\right]$$
... required

Since in Eq. (iii), 3 is multiplied, so

$$3H_2 + 3/2O_2 \rightleftharpoons 3H_2O;$$
$$K'_3 = \frac{[H_2O]^3}{[H_2]^3 [O_2]^{3/2}} = (K_3)^3$$

Eq. (ii) + 3 × Eq. (iii) – Eq. (i)

$$K = \frac{K_2 \times (K_3)^3}{K_1}$$
or $$K = \frac{1.6\times 10^{12} \times (10^{-13})^3}{4\times 10^5}$$
or $$K = 4\times 10^{-33}$$

**50.** (2) $2NO(g) + O_2(g) \rightleftharpoons 2NO_2(g)$

| | | | |
|---|---|---|---|
| At $t = 0$ | 2mol | 1 mol | 0 |
| At $t = t_{eq}$ | 1.2 mol | 0.6 mol | 0.8 mol |

Total moles, $n_T = 1.2 + 0.6 + 0.8 = 2.6$ mol

Equilibrium constant = $K_p$

$$p_T = 1 \text{ atm}$$
$$K_p = \frac{(p_{NO_2})^2}{(p_{NO})^2 \times p_{O_2}} \quad (p_A = p_T x_A)$$
$$= \frac{\left(\frac{0.8}{2.6}\right)^2}{\left(\frac{1.2}{2.6}\right)^2 \times \left(\frac{0.6}{2.6}\right)} = 1.925 \approx 2$$

**51.** (2) Given,

$$Cu(s) + 2Ag^+(aq) \rightleftharpoons Cu^{2+}(aq) + 2Ag(s)$$

Equilibrium constant, $K = 2\times 10^{15}$

$T = 298$ K

For the reaction,

$$\frac{1}{2}Cu^{2+}(aq) + Ag(s) \rightleftharpoons \frac{1}{2}Cu(s) + Ag^+(aq)$$

$K'$ can be calculated as

$$K' = \frac{1}{(K)^{1/2}} = \frac{1}{(2\times 10^{15})^{1/2}}$$
$$= 2.2\times 10^{-8} \quad (K' = x\times 10^{-8})$$
$$\therefore \quad x \approx 2$$

**52.** (2735) $HI \rightleftharpoons \frac{1}{2}H_2 + \frac{1}{2}I_2$

| | | | |
|---|---|---|---|
| $t_i$ | 1 | | |
| $t_{eq}$ | 1 – 0.4 | 0.4/2 | 0.4/2 |

$$K_p = \frac{[H_2]^{1/2}[I_2]^{1/2}}{[HI]}$$
$$K_p = \frac{(0.2)^{1/2}(0.2)^{1/2}}{1-0.4} = \frac{0.2}{0.6} = \frac{1}{3}$$
$$\Delta G = \Delta G° + RT \ln K = 0$$
[∴ at equilibrium $\Delta G = 0$]
$$\Delta G° = -RT \ln K$$
$$= -8.31\times 300\times 2.3 \log\left(\frac{1}{3}\right)$$
$$= 2735 \text{ J/mol}$$

**53.** (1107)

**key idea** We can calculate the equilibrium constant for the process at temperature $T$, by using the relationship between $\Delta G°$ and $K$

$$\Delta G° = RT \ln K_{eq}$$
where $R = 8.314 \text{ J mol}^{-1}\text{ K}^{-1}$
$T$ = Temperature on Kelvin scale.

$$PCl_5 \rightleftharpoons PCl_3 + Cl_2$$

| | | | |
|---|---|---|---|
| Initial mole | 5 | 0 | 0 |
| At equil. | 5(1 – $x$) | 5$x$ | 5$x$ |

Total moles = $5(1-x) + 5x + 5x$
$= 5(1+x) + 2$ (Nitrogen moles)

$$pV = nRT$$
$$n = \frac{2.46\times 200}{0.082\times 600} = 10$$
$$5(1+x) + 2 = 10$$
$$x = \frac{3}{5} = 0.6$$

Partial pressure of

$$p_{PCl_5} = \frac{5(1-x)}{7+5x} \times 2.46$$

Put value of $x = 0.6$

$$p_{PCl_5} = \frac{5\times 0.4}{10}\times 2.46 \Rightarrow \frac{2.46}{5}$$

Partial pressure of $p_{PCl_3}$ and $p_{Cl_2}$

$$p_{PCl_3} = p_{Cl_2} = \frac{5x}{5x+7}\times 2.46$$

Put $x = 0.6$

$$p_{PCl_3} = p_{Cl_2} = \frac{3}{10}\times 2.46$$

$$K_p = \frac{[p_{Cl_3}][p_{Cl_2}]}{p_{PCl_5}}$$
$$= \frac{\frac{3}{10}\times 2.46\times \frac{3}{10}\times 2.46}{\frac{2}{10}\times 2.46}$$
$$= \frac{9}{10}\times\frac{2.46}{2} = 1.107 = 1107\times 10^{-3}$$

**54.** (29) Firstly calculate moles of water

$$\text{Moles of water} = \frac{\text{Given mass}}{\text{Molecular mass}} = \frac{0.90}{18}$$

By ideal gas equation, $pV = nRT$

$$V = \frac{nRT}{p} = \frac{0.90\times 0.082\times 300\times 760}{18\times 32}$$
$$= 29.21 \approx 29$$

**55.** (125) According to the given reaction,

$$2NOCl(g) \rightleftharpoons 2NO(g) + Cl_2(g)$$

| | | | |
|---|---|---|---|
| Initial moles | 2 | 0 | 0 |
| At equili moles | 2 – 0.4 = 1.6 | 0.4 | 0.2 |

Now, equilibrium constant,

$$K_C = \frac{[NO]^2[Cl_2]}{[NOCl]^2} = \frac{(0.4)^2\times(0.2)}{(1.6)^2}$$
$$= 0.0125 = 125\times 10^{-4}$$

**56.** (0.02) $[PtCl_4]^{2-} + H_2O \underset{K_b}{\overset{K_f}{\rightleftharpoons}} [Pt(H_2O)Cl_3]^- + Cl^-$

Rate of a reaction = Rate of forward reaction – rate of backward reaction

$$\frac{-d[[PtCl_4]^{2-}]}{dt} = K_f\,[[PtCl_4]^{2-}] - K_b\,[[Pt(H_2O)Cl_3]^-][Cl^-]$$

Comparing this equation with the given one :

$$\Rightarrow \quad K_{eq} = \frac{K_f}{K_b} = \frac{4.8\times 10^{-5}}{2.4\times 10^{-3}} = 0.02$$

**57.** (16) According to the equation,

$$A(s) \rightleftharpoons M(s) + \frac{1}{2}O_2(g)$$

Equilibrium constant, $K_p = p_{O_2}^{1/2} = 4$

where, $p$ = partial pressure
At equilibrium, the partial pressure of $O_2$ is $p_{O_2} = (4)^2 = 16$ atm.

**58.** (2) Given reaction,

$$N_2O_4(g) \rightleftharpoons 2NO_2(g)$$

So, $\Delta n_g$ = [moles of gaseous products- moles of gaseous reactants]
$= 2 - 1 = 1$

$$K_p = K_C(RT)^{\Delta n_g}$$

where, $K_p$ = equilibrium constant in gaseous state = 47.9

$K_C$ = equilibrium constant

$R$ = gas constant

$T$ = temperature

$$\Rightarrow \quad K_C = \frac{47.9}{0.083 \times 288} = 2.0038$$

∴ Hence, $K_C$ at same temperature (288 K) is 2.

**59.** *(172)* Given reaction,

$$2SO_2(g) + O_2(g) \rightleftharpoons 2SO_3(g)$$

$$\text{Now, } K_p = \frac{[p_{SO_3(g)}]^2}{[p_{SO_2(g)}]^2 \times [p_{O_2}]}$$

On putting given values, we get,

$$\Rightarrow K_p = \frac{(43)^2}{(45)^2 \times 0.53}$$

$$= \frac{1849}{2025 \times 0.53} = \frac{1849}{1073.25}$$

$$= 1.7228$$

$$= 172.28 \times 10^{-2} = 172$$

Hence, the equilibrium constant, $K_p = 172$.

**60.** *(2)* $2A(g) \rightleftharpoons A_2(g)$

Given, $\Delta G° = 25.2 \text{ kJ mol}^{-1}$

$= 25200 \text{ J mol}^{-1}$

$T = 400$ K

According to standard free Gibb's equation,

$$\Delta G° = -RT \ln K_p$$

$$25200 = -2.3 \times 8.3 \times 400 \log(K_p)$$

$$\log K_p = \frac{-25200}{2.3 \times 8.3 \times 400} = -3.3$$

$$K_p = 10^{-3.3} = 10^{-3} \times 0.501$$

$$K_p = 5.01 \times 10^{-4} \text{ bar}^{-1}$$

$$K_p = 5.01 \times 10^{-9} \text{ Pa}^{-1}$$

We know that,

$$K_p = K_c(RT)^{\Delta n_g}$$

$$K_p = K_c(RT)^{-1}$$

$$[\therefore \Delta n_g = 1 - 2 = -1]$$

$$K_p = \frac{K_C}{8.3 \times 400}$$

$$K_c = 5.01 \times 10^{-5} \times 8.3 \times 400$$

$$\Rightarrow \quad K_c = 166 \times 10^{-5} \text{ m}^3/\text{mol}$$

$$= 1.66 \times 10^{-2} \approx 2 \times 10^{-2} \text{ L/mol}$$

According to the answer key of NTA the correct answer is 166 but that is not possible if we consider the units properly. So the correct answer is 2 (nearest integer).

**61.** *(354)* Given reaction is,

$$N_2O_4(g) \rightleftharpoons 2NO_2(g)$$

Given values are : $K_p = 600.1$

$K_C = 20.4$

$\Delta n_g$ = number of moles of product − number of moles of reactant

$= 2 - 1 = 1$

$$K_p = K_C(RT)^{\Delta n_g}$$

where $R$ is the gas constant = 0.0831 L atm/K mol

$\Delta n_g = 1$ (for given reaction)

On putting given values, we will get

$$600.1 = 20.4\,(RT)^1 \Rightarrow T = 354 \text{ K}$$

The temperature at which $K_C = 20.4$ and $K_p = 600.1$, is 354 K.

**62.** *(20)* $A(g) \rightleftharpoons B(g)$

We know,

$$\Delta G° = -2.303\, RT \log K$$

$$\because \quad K = \frac{[B]}{[A]}$$

$$\text{Now, } \log K = \frac{-\Delta G°}{2.303RT}$$

$$\log K = \frac{9.478 \times 1000}{2.303 \times 8.314 \times 495}$$

$$\therefore \quad \log K = 1 \Rightarrow K = 10$$

$$\Rightarrow \quad \frac{[B]}{[A]} = \frac{n_B}{n_A} = 10$$

| Now, | $A(g)$ | $\rightleftharpoons$ | $B(g)$ |
|---|---|---|---|
| $t = 0$ | 22 | | 0 |
| $t = t$ | $22 - x$ | | $x$ |

$$K = \frac{[B]}{[A]} = \frac{x}{22 - x} = 10$$

So, $x = 20$

Milimoles of $B = 20$

**63.** *(73)* $AB_2(g) \rightleftharpoons A(g) + 2B(g)$

| | $AB_2(g)$ | $A(g)$ | $2B(g)$ |
|---|---|---|---|
| $t = 0$ | 1 | 0 | 0 |
| $t = t_{eq}$ | $(1 - x)$ | $x$ | $2x$ |

Total number of moles at equilibrium

$= 1 - x + x + 2x = (1 + 2x)$

| | | | |
|---|---|---|---|
| Partial pressure (atm) | $\frac{1-x}{1+2x}p$ | $\frac{x}{1+2x}p$ | $\frac{2x}{1+2x}p$ |

[$p_T$ = Total pressure at equilibrium = 1.9 atm]

Now, at equilibrium $pV = (1 + 2x)RT$

$$\Rightarrow 1 + 2x = \frac{pV}{RT} = \frac{1.9 \times 25}{0.082 \times 300}$$

$$= 1.93$$

$[V = 25 \text{ L}, R = 0.082 \text{ L atm mol}^{-1}\text{K}^{-1}$

$T = 300 \text{ K}]$

$$\Rightarrow \quad x = \frac{1.93 - 1}{2} = 0.465$$

$$\Rightarrow \quad K_p = \frac{p_A \times p_B^2}{p_{AB_2}}$$

$$\Rightarrow \frac{\left(\frac{x}{1+2x}p\right) \times \left(\frac{2x}{1+2x}p\right)^2}{\left(\frac{1-x}{1+2x}p\right)}$$

$$= \frac{4x^3 \times p^2}{(1 + 2x)^2 \times (1 - x)}$$

$$= \frac{4 \times (0.465)^3 \times (1.9)^2}{(1 + 2 \times 0.465)^2 \times (1 - 0.465)}$$

$$= 0.7285 \text{ atm}$$

$$= 72.85 \times 10^{-2} \text{ atm}$$

$$\simeq 73 \times 10^{-2} = x \times 10^{-2}$$

$$\therefore \quad x = 73$$

**64.** *(5)* $Cl_2(g) \rightleftharpoons 2Cl(g)$

Let, moles of both of $Cl_2$ and Cl molecule be $x$.

(Number of moles of $Cl_2$ = Number of moles of Cl)

Partial pressure of Cl is,

$$p_{Cl} = \frac{x}{2x} \times 1 = \frac{1}{2}$$

Partial pressure of $Cl_2$ is

$$p_{Cl_2} = \frac{x}{2x} \times 1 = \frac{1}{2}$$

$$\text{Now, } K_p = \frac{(p_{Cl})^2}{p_{Cl_2}}$$

$$\Rightarrow \quad K_p = \frac{(1/2)^2}{1/2} = \frac{1}{2} = 0.5$$

or $5 \times 10^{-1}$

Hence, $x \times 10^{-1} \Rightarrow x = 5$

**65.** *(1380)* For a reaction, $A(g) \longrightarrow B(g)$

Formula used

$$\Delta G° = -RT \ln K_p \quad \ldots\text{(i)}$$

Putting given values in Eq (i), we get

$$\Delta G° = -R\,(300) \ln 100$$

$$\Delta G° = -R\,(300)\,(2) \ln (10)$$

$$\because \quad \ln (10) = 2.3$$

$$\Delta G° = -R(300)(2)(2.3)$$
$$\Delta G° = -1380\,R$$

Hence, $\Delta G° = -x\,R$

$$x = 1380$$

## Topic 2

**01.** *(b)* Among the given statements, A and B are correct, while C and D are incorrect.

It is because equilibrium constant remains unchanged as the temperature is constant.

**02.** *(d)* According to Le-Chatelier's principle, if a solid is added, then there will be no effect on the equilibrium. $Fe_2O_3$ is a solid compound. Hence, option (d) is correct.

**03.** *(a)* $Cl_2(g) + 3F_2(g) \rightleftharpoons 2ClF_3$

$$\Delta H_r = -329 \text{ kJ}$$

Since the value of enthalpy is negative, reaction is exothermic. By Le-Chatelier principle, a reaction is spontaneous in forward side (in the direction of formation of more $ClF_3$) if $F_2$ is added, temperature is lowered and $ClF_3$ is removed.

**04.** *(c)* Change in volume affects number of moles per unit volume and the reaction move in the direction which undo the change.

$$N_2(g) + O_2(g) \rightleftharpoons 2NO(g)$$

Number of moles of reactants and products are equal.

**05.** *(b)* The reaction at equilibrium is

| | $A(g)$ | $\rightleftharpoons$ | $B(g)$ | + | $C(g)$ |
|---|---|---|---|---|---|
| $t = 0$ | $a$ | | 0 | | 0 |
| $t = t$ | $a(1-\alpha)$ | | $a\alpha$ | | $a\alpha$ |

'$a$' mole of $A(g)$ taken initially and at time $t = 0$.

Now, mole fraction of $A(g)$, $B(g)$ and $C(g)$.

$$X_A = \frac{a - a\alpha}{a + a\alpha} = \frac{1-\alpha}{1+\alpha}$$

$$X_B = \frac{a\alpha}{a + a\alpha} = \frac{\alpha}{1+\alpha}$$

$$X_C = \frac{a\alpha}{a + a\alpha} = \frac{\alpha}{1+\alpha}$$

Now, if $p$ is total pressure, then partial pressure of each $A$, $B$, $C$ gases are

$$p_A = \left(\frac{1-\alpha}{1+\alpha}\right)p,\ p_B = \left(\frac{\alpha}{1+\alpha}\right)p,$$

$$p_C = \left(\frac{\alpha}{1+\alpha}\right)p$$

$$K_p = \frac{p_B \cdot p_C}{p_A} \Rightarrow \frac{\alpha^2 p}{1-\alpha^2}$$

As $K_p$ is only function of temperature. So, as pressure increases, $\alpha$ decreases.

**06.** *(b)* On addition of xenon at constant $p$ and $T$, reaction will move in forward direction.

So, $[PCl_5]$ will decrease and $[PCl_3]$ and $[Cl_2]$ will increase.

**07.** *(b)* The reaction is

| | $N_2O_5(g) \longrightarrow$ | $2NO_2(g)$ | $+\frac{1}{2}O_2(g)$ |
|---|---|---|---|
| at $t = 0$ | $p_0$ | | |
| $t = t$ | $p_0 - x$ | $2x$ | $\frac{x}{2}$ |

$$p_{\text{total}} = p_0 - \frac{p_0}{2} + p_0 + \frac{p_0}{4} \quad \left[\because x = \frac{p_0}{2}\right]$$

$$= \frac{7}{4}p_0$$

**08.** *(b)* $N_2O_4(g) \rightleftharpoons 2NO_2(g)$

$$\Delta H = +58 \text{ kJ}$$

In case of an endothermic reaction ($\Delta H = +$ ve), the equilibrium constant increases with rise in temperature and hence, the reaction moves in forward direction. On adding $N_2$, pressure is increased at constant $T$, and volume would also be constant, so no change is observed.

**09.** *(d)* Statement given in option (a), (b) and (c) are correct while statement (d) is incorrect.

The larger value of $K_c$ suggest that the reaction should go almost to completion. However, the oxidation of $SO_2$ to $SO_3$ is very low. Hence, the rate of reaction is increased by adding catalyst.

**10.** *(3)* The number of correct statement involving equilibria in physical processes from the given statements are three i.e., A, B and D.

(A) Equilibrium is possible only in a closed system at a given temperature. All measurable properties of the system remain constant.

(B) The opposing processes occur at the same rate and there is dynamic but stable condition.

(C) At equilibrium the value of parameters become constant of a given temperature and not equal.

(D) For a given solid solute and a liquid solvent solubility depends upon temperature only.

**11.** *(3)* $2SO_2(g) + O_2(g) \rightleftharpoons 2SO_3(g)$

$$\Delta H = -190 \text{ kJ}$$

The reaction is exothermic in nature. On increasing pressure, equilibrium will shift in forward direction. Apart from that addition of more $SO_2$ and $O_2$ will also increase the yield of $SO_3$ at equilibrium while other factors decrease the yield of $SO_3$.

# C-07

JEE Main Chapterwise Topicwise Chemistry

In JEE Main there is usually 0-1 question per paper are visible from this chapter. In past 3 years, 10 conceptual and 14 numerical questions were asked of moderate difficulty level from this chapter.

**For detailed analysis scan QR code**

# IONIC EQUILIBRIUM

## Topic 1 Ostwald's Dilution Law and Concept of Acids and Bases

### Level 1

**01** Which of the following is a Lewis acid? ***[JEE Main 2018]***

(a) $PH_3$ (b) $NF_3$
(c) NaH (d) $B(CH_3)_3$

**02** Three reactions involving $H_2PO_4^-$ are given below ***[AIEEE 2010]***

I. $H_3PO_4 + H_2O \longrightarrow H_3O^+ + H_2PO_4^-$

II. $H_2PO_4^- + H_2O \longrightarrow HPO_4^{2-} + H_3O^+$

III. $H_2PO_4^- + OH^- \longrightarrow H_3PO_4 + O^{2-}$

In which of the above does $H_2PO_4^-$ act as an acid?

(a) II only (b) I and II (c) III only (d) I only

**03** The first and second dissociation constants of an acid $H_2A$ are $1.0\times 10^{-5}$ and $5.0\times 10^{-10}$, respectively. The overall dissociation constant of the acid will be ***[AIEEE 2007]***

(a) $5.0\times 10^{-5}$ (b) $5.0\times 10^{15}$
(c) $5.0\times 10^{-15}$ (d) $0.2\times 10^{5}$

**04** What is the conjugate base of $OH^-$? ***[AIEEE 2005]***

(a) $O^{2-}$ (b) $O^-$ (c) $H_2O$ (d) $O_2$

**05** The conjugate base of $H_2PO_4^-$ is ***[AIEEE 2004]***

(a) $PO_4^{3-}$ (b) $P_2O_5$ (c) $H_3PO_4$ (d) $HPO_4^{2-}$

### Level 2

**06** Given below are two statements one is labelled as Assertion (A) and the other is labelled as Reason (R)

**Assertion (A)** The amphoteric nature of water is explained by using Lewis acid/base concept.

**Reason (R)** Water acts as an acid with $NH_3$ and as a base with $H_2S$.

In the light of the above statements choose the correct answer from the options given below: ***[2022, 25 June Shift-II]***

(a) Both A and R are true and R is the correct explanation of A.
(b) Both A and R are true but R is not the correct explanation of A.
(c) A is true but R is false.
(d) A is false but R is true.

**07** Which of the following are Lewis acids? ***[JEE Main 2018]***

(a) $PH_3$ and $BCl_3$ (b) $AlCl_3$ and $SiCl_4$
(c) $PH_3$ and $SiCl_4$ (d) $BCl_3$ and $AlCl_3$

**08** The pH of a 0.1 molar solution of the acid HQ is 3. The value of the ionisation constant, $K_a$ of the acid is ***[AIEEE 2012]***

(a) $3\times 10^{-1}$ (b) $1\times 10^{-3}$
(c) $1\times 10^{-5}$ (d) $1\times 10^{-7}$

**09** An acid $HA$ ionises as $HA \rightleftharpoons H^+ + A^-$

The pH of 1.0 M solution is 5. Its dissociation constant would be ***[AIEEE 2011]***

(a) $1\times 10^{-10}$ (b) 5
(c) $5\times 10^{-8}$ (d) $1\times 10^{-5}$

### Level 3

**10** $K_{a_1}$, $K_{a_2}$ and $K_{a_3}$ are the respective ionisation constants for the following reactions A, B, and C.

A. $H_2C_2O_4 \rightleftharpoons H^+ + HC_2O_4^-$

B. $HC_2O_4^- \rightleftharpoons H^+ + C_2O_4^{2-}$

C. $H_2C_2O_4 \rightleftharpoons 2H^+ + C_2O_4^{2-}$

The relationship between $K_{a_1}$, $K_{a_2}$ and $K_{a_3}$ is given as ***[2022, 25 July Shift-II]***

(a) $K_{a_3} = K_{a_1} + K_{a_2}$ (b) $K_{a_3} = K_{a_1} - K_{a_2}$
(c) $K_{a_3} = K_{a_1} / K_{a_2}$ (d) $K_{a_3} = K_{a_1} \times K_{a_2}$

Imp. Order of Acidic Strength

**11** Four species are listed below

I. $HCO_3^-$ II. $H_3O^+$

III. $HSO_4^-$ IV. $HSO_3F$

Which one of the following is the correct sequence of their acid strength? *[AIEEE 2008]*

(a) IV < II < III < I
(b) II < III < I < IV
(c) I < III < II < IV
(d) III < I < IV < II

## Numerical Value Type Questions

**12** If 1 mM solution of ethylamine produces pH = 9, then the ionisation constant ($K_b$) of ethylamine is $10^{-x}$. The value of $x$ is ........... (Nearest integer).

[The degree of ionisation of ethylamine can be neglected with respect to unity.] *[2025, 23 Jan Shift-I]*

**13** The equilibrium constant for decomposition of $H_2O(g)$

$$H_2O(g) \rightleftharpoons H_2(g) + \frac{1}{2}O_2(g) \; (\Delta G° = 92.34 \text{ kJ mol}^{-1})$$

is $8.0 \times 10^{-3}$ at 2300 K and total pressure at equilibrium is 1 bar. Under this condition, the degree of dissociation (α) of water is ......... $\times 10^{-2}$ (Nearest integer value).

[Assume α is negligible with respect to 1] *[2025, 8 April Shift-II]*

**14** The pH of a 0.01 M weak acid $HX$ ($K_a = 4 \times 10^{-10}$) is found to be 5. Now the acid solution is diluted with excess of water so that the pH of the solution changes to 6. The new concentration of the diluted weak acid is given as $x \times 10^{-4}$ M. The value of $x$ is ........ (Nearest integer) *[2025, 4 April Shift-I]*

**15** Assuming that $Ba(OH)_2$ is completely ionised in aqueous solution under the given conditions the concentration of $H_3O^+$ ions in 0.005 M aqueous solution of $Ba(OH)_2$ at 298 K is ...... $\times 10^{-12}$ mol $L^{-1}$. (Nearest integer) *[2021, 25 July Shift-II]*

**16** 0.01 moles of a weak acid $HA$ ($K_a = 2.0 \times 10^{-6}$) is dissolved in 1.0 L of 0.1 M HCl solution. The degree of dissociation of $HA$ is ...... $\times 10^{-5}$ (Round off to the nearest integer).

[Neglect volume change on adding $HA$. Assume degree of dissociation << 1] *[2021, 17 March Shift-I]*

# Topic 2 Solubility Product and Common Ion Effect

## Level 1

**01** The molar solubility($S$) of zirconium phosphate with molecular formula $(Zr^{4+})_3(PO_4^{3-})_4$ is given by relation *[2025, 22 Jan Shift-II]*

(a) $\left(\frac{K_{sp}}{6912}\right)^{1/7}$ (b) $\left(\frac{K_{sp}}{9612}\right)^{1/3}$

(c) $\left(\frac{K_{sp}}{8435}\right)^{1/7}$ (d) $\left(\frac{K_{sp}}{5348}\right)^{1/6}$

**02** For a sparingly soluble salt $AB_2$, the equilibrium concentrations of $A^{2+}$ ions and $B^-$ ions are $1.2 \times 10^{-4}$ M and $0.24 \times 10^{-3}$ M, respectively. The solubility product of $AB_2$ is *[2024, 9 April Shift-II]*

(a) $0.069 \times 10^{-12}$ (b) $6.91 \times 10^{-12}$
(c) $0.276 \times 10^{-12}$ (d) $27.65 \times 10^{-12}$

**03** The solubility of AgCl will be maximum in which of the following? *[2022, 29 June Shift-I]*

(a) 0.01 M KCl (b) 0.01 M HCl
(c) 0.01 M $AgNO_3$ (d) Deionised water

**04** The $K_{sp}$ for bismuth sulphide ($Bi_2S_3$) is $1.08 \times 10^{-73}$. The solubility of $Bi_2S_3$ in mol $L^{-1}$ at 298 K is *[2022, 25 June Shift-II]*

(a) $1.0 \times 10^{-15}$ (b) $2.7 \times 10^{-12}$
(c) $3.2 \times 10^{-10}$ (d) $4.2 \times 10^{-8}$

**05** The solubility of $Ca(OH)_2$ in water is [Given: The solubility product of $Ca(OH)_2$ in water = $5.5 \times 10^{-6}$] *[2021, 25 Feb Shift-II]*

(a) $1.11 \times 10^{-2}$ (b) $1.11 \times 10^{-6}$
(c) $1.77 \times 10^{-2}$ (d) $1.77 \times 10^{-6}$

**06** The $K_{sp}$ for $Cr(OH)_3$ is $1.6 \times 10^{-30}$. The molar solubility of this compound in water is *[AIEEE 2011]*

(a) $\sqrt[2]{1.6 \times 10^{-30}}$ (b) $\sqrt[4]{1.6 \times 10^{-30}}$
(c) $\sqrt[4]{1.6 \times 10^{-30} / 27}$ (d) $1.6 \times 10^{-30} / 27$

**07** Solid $Ba(NO_3)_2$ is gradually dissolved in a $1.0 \times 10^{-4}$ M $Na_2CO_3$ solution. At what concentration of $Ba^{2+}$, will a precipitate begin to form? ($K_{sp}$ for $BaCO_3 = 5.1 \times 10^{-9}$) *[AIEEE 2009]*

(a) $4.1 \times 10^{-5}$ M (b) $5.1 \times 10^{-5}$ M
(c) $8.1 \times 10^{-8}$ M (d) $8.1 \times 10^{-7}$ M

**08** The solubility product of a salt having general formula $MX_2$, in water is $4 \times 10^{-12}$. The concentration of $M^{2+}$ ions in the aqueous solution of the salt is **[AIEEE 2005]**

(a) $4.0\times 10^{-10}$ M (b) $1.6\times 10^{-4}$ M

(c) $1.0\times 10^{-4}$ M (d) $2.0\times 10^{-6}$ M

**Imp.** Relation between S and $K_{sp}$

**09** The molar solubility (in mol $L^{-1}$) of a sparingly soluble salt $MX_4$ is $S$. The corresponding solubility product is $K_{sp}$. $S$ is given in terms of $K_{sp}$ by the relation. **[AIEEE 2004]**

(a) $S = (K_{sp}/128)^{1/4}$ (b) $S = (K_{sp}/128)^{1/2}$

(c) $S = (K_{sp}/256)^{1/5}$ (d) $S = (K_{sp}/256)^{1/3}$

**10** The solubility in water of a sparingly soluble salt $AB_2$ is $1.0 \times 10^{-5}$ mol $L^{-1}$. Its solubility product will be **[AIEEE 2003]**

(a) $4 \times 10^{-15}$ (b) $4 \times 10^{-10}$ (c) $1 \times 10^{-15}$ (d) $1 \times 10^{-10}$

## Level 2

**Imp.** Order of Solubility Product

**11** Arrange the following in increasing order of solubility product $Ca(OH)_2$, AgBr, PbS, HgS **[2025, 28 Jan Shift-II]**

(a) $Ca(OH)_2 < AgBr < HgS < PbS$

(b) $HgS < PbS < AgBr < Ca(OH)_2$

(c) $PbS < HgS < Ca(OH)_2 < AgBr$

(d) $HgS < AgBr < PbS < Ca(OH)_2$

**12** Which of the following happens when $NH_4OH$ is added gradually to the solution containing [1 M] $A^{2+}$ and [1 M] $B^{3+}$ ions? **[2025, 23 Jan Shift-I]**

(Given, $K_{sp}[A(OH)_2] = 9 \times 10^{-10}$ and $K_{sp}[B(OH)_3] = 27 \times 10^{-18}$ at 298 K.)

(a) $A(OH)_2$ will precipitate before $B(OH)_3$

(b) $B(OH)_3$ will precipitate before $A(OH)_2$

(c) $A(OH)_2$ and $B(OH)_3$ will precipitate together

(d) Both $A(OH)_2$ and $B(OH)_3$ do not show precipitation with $NH_4OH$

**13** $K_{sp}$ for $Cr(OH)_3$ is $1.6 \times 10^{-30}$. What is the molar solubility of this salt in water? **[2025, 24 Jan Shift-I]**

(a) $\sqrt[4]{\dfrac{1.6\times10^{-30}}{27}}$ (b) $\dfrac{1.8\times10^{-30}}{27}$

(c) $\sqrt[5]{1.8\times10^{-30}}$ (d) $\sqrt[2]{1.6\times10^{-30}}$

**14** Given below are two statements One is labelled as Assertion (A) and the other labelled as Reason (R).

**Assertion** (A) During the boiling of water having temporary hardness, $Mg(HCO_3)_2$ is converted to $MgCO_3$.

**Reason** (R) The solubility product of $Mg(OH)_2$ is greater than that of $MgCO_3$.

In the light of the above statements, choose the most appropriate answer from the options given below

(a) Both (A) and (R) are true but (R) is not the correct explanation of (A).

(b) (A) is true but (R) is false.

(c) Both A and R are true and R is the correct explanation of (A).

(d) (A) is false but (R) is true. **[2021, 18 March Shift-I]**

**15** For the following Assertion and Reason, the correct option is

**Assertion** (A) When Cu (II) and sulphide ions are mixed, they react together extremely quickly to give a solid.

**Reason** (R) The equilibrium constant of $Cu^{2+}(aq) + S^{2-}(aq) \rightleftharpoons CuS(s)$ is high because the solubility product is low. **[2020, 2 Sep Shift-I]**

(a) Both (A) and (R) are false.

(b) Both (A) and (R) are true and (R) is the explanation for (A).

(c) Both (A) and (R) are true but (R) is not the explanation for (A).

(d) (A) is false and (R) is true.

**16** The stoichiometry and solubility product of a salt with the solubility curve given below is, respectively : **[2020, 8 Jan Shift-I]**

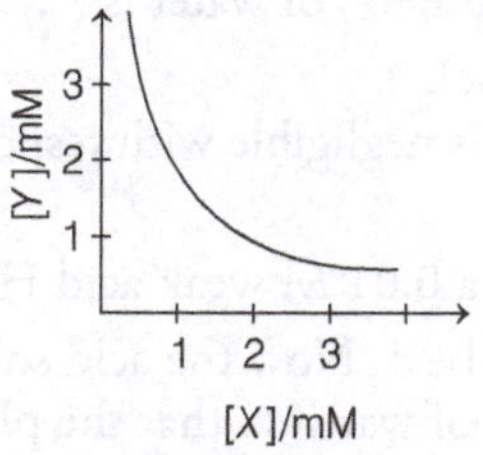


(a) $X_2Y$, $2\times10^{-9}$ $M^3$ (b) $XY$, $2\times10^{-6}$ $M^3$

(c) $XY_2$, $1\times10^{-9}$ $M^3$ (d) $XY_2$, $4\times10^{-9}$ $M^3$

**17** What is the molar solubility of $Al(OH)_3$ in 0.2 M NaOH solution? Given that, solubility product of $Al(OH)_3 = 2.4\times 10^{-24}$ **[2019, 12 April Shift-I]**

(a) $3\times10^{-19}$ (b) $12\times10^{-21}$

(c) $3\times10^{-22}$ (d) $12\times10^{-23}$

**18** If solubility product of $Zr_3(PO_4)_4$ is denoted by $K_{sp}$ and its molar solubility is denoted by $S$, then which of the following relation between $S$ and $K_{sp}$ is correct? **[2019, 8 April Shift-I]**

(a) $S = \left(\dfrac{K_{sp}}{144}\right)^{1/6}$ (b) $S = \left(\dfrac{K_{sp}}{6912}\right)^{1/7}$

(c) $S = \left(\dfrac{K_{sp}}{929}\right)^{1/9}$ (d) $S = \left(\dfrac{K_{sp}}{216}\right)^{1/7}$

**19** Solubility product of silver bromide is $5.0\times10^{-13}$. The quantity of potassium bromide (molar mass taken as 120 g $mol^{-1}$) to be added to 1 L of 0.05 M solution of silver nitrate to start the precipitation of AgBr is *[AIEEE 2010]*

(a) $1.2\times10^{-10}$ g (b) $1.2\times10^{-9}$ g
(c) $6.2\times10^{-5}$ g (d) $5.0\times10^{-8}$ g

**Imp.** Calculation of Mass of Compound in Saturated Solution

**20** In a saturated solution of the sparingly soluble strong electrolyte $AgIO_3$ (molecular mass = 283 g $mol^{-1}$) the equilibrium which sets in is

$$AgIO_3(s) \rightleftharpoons Ag^+(aq) + IO_3^-(aq)$$

If the solubility product constant, $K_{sp}$ of $AgIO_3$ at a given temperature is $1.0\times10^{-8}$, what is the mass of $AgIO_3$ contained in 100 mL of its saturated solution? *[AIEEE 2007]*

(a) $28.3\times10^{-2}$ g (b) $2.83\times10^{-3}$ g
(c) $1.0\times10^{-7}$ g (d) $1.0\times10^{-4}$ g

## Level 3

**21** A solution is 0.1 M in $Cl^-$ and 0.001 M in $CrO_4^{2-}$. Solid $AgNO_3$ is gradually added to it. Assuming that the addition does not change in volume and $K_{sp}(AgCl) = 1.7\times10^{-10}\ M^2$ and $K_{sp}(Ag_2CrO_4) = 1.9\times10^{-12} M^3$

Select correct statement from the following *[2021, 20 July Shift-II]*

(a) AgCl precipitates first because its $K_{sp}$ is high.
(b) $Ag_2CrO_4$ precipitates first as its $K_{sp}$ is low.
(c) $Ag_2CrO_4$ precipitates first because the amount of $Ag^+$ needed is low.
(d) AgCl will precipitate first as the amount of $Ag^+$ needed to precipitate is low.

**22** If $K_{sp}$ of $Ag_2CO_3$ is $8\times10^{-12}$, the molar solubility of $Ag_2CO_3$ in 0.1 M $AgNO_3$ is *[2019, 12 Jan Shift-II]*

(a) $8\times10^{-12}$ M (b) $8\times10^{-13}$ M
(c) $8\times10^{-10}$ M (d) $8\times10^{-11}$ M

**23** An aqueous solution contains 0.10 M $H_2S$ and 0.20 M HCl. If the equilibrium constants for the formation of $HS^-$ from $H_2S$ is $1.0\times10^{-7}$ and that of $S^{2-}$ from $HS^-$ ions is $1.2\times10^{-13}$ then the concentration of $S^{2-}$ ions in aqueous solution is *[JEE Main 2018]*

(a) $5\times10^{-8}$ (b) $3\times10^{-20}$
(c) $6\times10^{-21}$ (d) $5\times10^{-19}$

**24** An aqueous solution contains an unknown concentration of $Ba^{2+}$. When 50 mL of a 1 M solution of $Na_2SO_4$ is added, $BaSO_4$ just begins to precipitate. The final volume is 500 mL. The solubility product of $BaSO_4$ is $1\times10^{-10}$. What is the original concentration of $Ba^{2+}$? *[JEE Main 2018]*

(a) $5\times10^{-9}$ M (b) $2\times10^{-9}$ M
(c) $1.1\times10^{-9}$ M (d) $1.0\times10^{-10}$ M

### Numerical Value Type Questions

**25** 25.0 mL of 0.050 M $Ba(NO_3)_2$ is mixed with 25.0 mL of 0.020 M NaF. $K_{sp}$ of $BaF_2$ is $0.5\times10^{-6}$ at 298 K. The ratio of $[Ba^{2+}][F^-]^2$ and $K_{sp}$ is ............ . (Nearest integer) *[2023, 13 April Shift-I]*

**26** The solubility product of $BaSO_4$ is $1\times10^{-10}$ at 298K. The solubility of $BaSO_4$ in 0.1 M $K_2SO_4(aq)$ solution is ...... $\times10^{-9}$ $gL^{-1}$ (nearest integer). Given, Molar mass of $BaSO_4$ is 233 g $mol^{-1}$ *[2023, 8 April Shift-II]*

**27** At 298 K, the solubility of silver chloride in water is $1.434\times10^{-3}$ g $L^{-1}$. The value of $-\log K_{sp}$ for silver chloride is ........

(Given, mass of Ag is 107.9 g $mol^{-1}$ and mass of Cl is 35.5 g $mol^{-1}$) *[2023, 31 Jan Shift-II]*

**28** If the solubility product of PbS is $8\times10^{-28}$, then the solubility of PbS in pure water at 298 K is $x\times10^{-16}$ mol $L^{-1}$. The value of $x$ is ...... (Nearest integer) [Given, $\sqrt{2} = 1.41$] *[2022, 29 July Shift-I]*

**29** At 310 K, the solubility of $CaF_2$ in water is $2.34\times10^{-3}$ g/100 mL. The solubility product of $CaF_2$ is ...... $\times10^{-8}$ $(mol/L)^3$.
(Given, molar mass of $CaF_2$ = 78 g $mol^{-1}$) *[2022, 27 July Shift-I]*

**30** The molar solubility of $Zn(OH)_2$ in 0.1 M NaOH solution is $x\times10^{-18}$ M. The value of $x$ is ........ (Nearest integer) (Given; The solubility product of $Zn(OH)_2$ is $2\times10^{-20}$). *[2021, 1 Sep Shift-II]*

**31** $A_3B_2$ is a sparingly soluble salt of molar mass $M$ (g $mol^{-1}$) and solubility $x$ g $L^{-1}$. The solubility product satisfies $K_{sp} = a\left(\frac{x}{M}\right)^5$. The value of $a$ is ......... (Integer answer) *[2021, 31 Aug Shift-I]*

**32** The solubility of $CdSO_4$ in water is $8.0 \times 10^{-4}$ mol $L^{-1}$. Its solubility in 0.01 M $H_2SO_4$ solution is ........ $\times 10^{-6}$ mol $L^{-1}$. (Round off to the nearest integer)
(Assume that, solubility is much less than 0.01 M)
*[2021, 18 March Shift-II]*

**33** Two salts $A_2X$ and $MX$ have the same value of solubility product of $4.0 \times 10^{-12}$. The ratio of their molar solubilities i.e $\frac{S(A_2X)}{S(MX)} = ......$
(Round off to the nearest integer) *[2021, 16 March Shift-I]*

## Topic 3 pH, Buffer and Indicators, Salt Hydrolysis

### Level 1

**01** pH of water is 7 at 25°C. If water is heated to 80°C, it's pH will *[2025, 23 Jan Shift-II]*
(a) increase
(b) decrease
(c) $H^+$ concentration increases, $OH^-$ concentration decreases
(d) remains the same

**02** The pH of a 0.02 M $NH_4Cl$ solution will be [given $K_b(NH_4OH) = 10^{-5}$ and log 2 = 0.301]
*[2019, 10 April Shift-II]*
(a) 2.65 (b) 4.35 (c) 4.65 (d) 5.35

**03** Which of the following salts is the most basic in aqueous solution? *[JEE Main 2018]*
(a) $Al(CN)_3$ (b) $CH_3COOK$
(c) $FeCl_3$ (d) $Pb(CH_3COO)_2$

**04** An alkali is titrated against an acid with methyl orange as indicator, which of the following is a correct combination? *[JEE Main 2018]*

| | Base | Acid | End point |
|---|---|---|---|
| (a) | Weak | Strong | Colourless to pink |
| (b) | Strong | Strong | Pinkish red to yellow |
| (c) | Weak | Strong | Yellow to pinkish red |
| (d) | Strong | Strong | Pink to colourless |

**05** The $pK_a$ of a weak acid, $HA$ is 4.80. The $pK_b$ of a weak base, $B$OH, is 4.78. The pH of an aqueous solution of the corresponding salt, $BA$, will be
*[AIEEE 2008]*
(a) 9.58 (b) 4.79 (c) 7.01 (d) 9.22

**06** The $pK_a$ of a weak acid ($HA$) is 4.5. The $p$OH of an aqueous buffered solution of $HA$ in which 50% of the acid ionised is *[AIEEE 2007]*
(a) 4.5 (b) 2.5 (c) 9.5 (d) 7.0

**07** Hydrogen ion concentration in mol/L in a solution of pH = 5.4 will be *[AIEEE 2005]*
(a) $3.98 \times 10^{-6}$ (b) $3.68 \times 10^{-6}$
(c) $3.88 \times 10^{6}$ (d) $3.98 \times 10^{8}$

### Level 2

**08** An aqueous solution of HCl with pH 1.0 is diluted by adding equal volume of water (ignoring dissociation of water). The pH of HCl solution would (Given log 2 = 0.30) *[2025, 7 April Shift-I]*
(a) remain same (b) reduce to 0.5
(c) increase to 1.3 (d) increase to 2

**09** If equal volumes of $AB_2$ and $XY$ (both are salts) aqueous solutions are mixed, which of the following combination will give a precipitate of $AY_2$ at 300 K? (Given $K_{sp}$, (at 300 K) for $AY_2 = 5.2 \times 10^{-7}$)
*[2025, 2 April Shift-I]*
(a) $2.0 \times 10^{-2}$ M $AB_2$, $2.0 \times 10^{-2}$ M $XY$
(b) $2.0 \times 10^{-4}$ M $AB_2$, $0.8 \times 10^{-3}$ M $XY$
(c) $1.5 \times 10^{-4}$ M $AB_2$, $1.5 \times 10^{-3}$ M $XY$
(d) $3.6 \times 10^{-3}$ M $AB_2$, $5.0 \times 10^{-4}$ M $XY$

**10** Which of the following mixing of 1 M base and 1 M acid leads to the largest increase in temperature?
*[2025, 24 Jan Shift-II]*
(a) 30 mL HCl and 30 mL NaOH
(b) 30 mL $CH_3COOH$ and 30 mL NaOH
(c) 50 mL HCl and 20 mL NaOH
(d) 45 mL $CH_3COOH$ and 25 mL NaOH

**11** Given below are two statements: *[2024, 8 April Shift-II]*
**Statement I** A Buffer solution is the mixture of a salt and an acid or a base mixed in any particular quantities.
**Statement II** Blood is naturally occurring buffer solution whose pH is maintained by $H_2CO_3/HCO_3^{\ominus}$ concentrations.
In the light of the above statements, choose the correct from the options given below:
(a) Statement I is false but Statement II is true.
(b) Both Statement I and Statement II are true.
(c) Both Statement I and Statement II are false.
(d) Statement I is true but Statement II is false.

**12** Given below are two statements: *[2024, 27 Jan Shift-I]*
**Statement I** Aqueous solution of ammonium carbonate is basic.
**Statement II** Acidic/basic nature of salt solution of a salt of weak acid and weak base depends on $K_a$ and $K_b$ value of acid and the base forming it.
In the light of the above statements, choose the **most appropriate** answer from the options given below

(a) Statement I is correct but Statement II is incorrect.
(b) Statement I is incorrect but Statement II is correct.
(c) Both Statement I and Statement II are correct.
(d) Both Statement I and Statement II are incorrect.

**13** 200 mL of 0.01 M HCl is mixed with 400 mL of 0.01 M $H_2SO_4$. The pH of the mixture is ………… *[2022, 29 July Shift-II]*

(a) 1.14 (b) 1.78
(c) 2.34 (d) 3.02

**14** A compound '*X*' is a weak acid and it exhibits colour change at pH close to the equivalence point during neutralisation of NaOH with $CH_3COOH$. Compound '*X*' exists in ionised form in basic medium. The compound '*X*' is *[2022, 29 July Shift-II]*

(a) methyl orange (b) methyl red
(c) phenolphthalein (d) erichrome Black T

**15** The solubility of AgCN in a buffer solution of pH = 3 is *x*. The value of *x* is
[Assume : No cyano complex is formed; $K_{sp}(\text{AgCN}) = 2.2 \times 10^{-16}$ and $K_a(\text{HCN}) = 6.2 \times 10^{-10}$] *[2021, 25 Feb Shift-I]*

(a) $1.9 \times 10^{-5}$ (b) $1.6 \times 10^{-6}$
(c) $2.2 \times 10^{-16}$ (d) $0.625 \times 10^{-6}$

**Imp.** pH Graph of Acid-Base Mixture

**16** 100 mL of 0.1M HCl is taken in a beaker and to it 100 mL of 0.1 M NaOH is added in steps of 2 mL and the pH continuously measured. Which of the following graphs correctly depicts the change in pH? *[2020, 3 Sep Shift-II]*

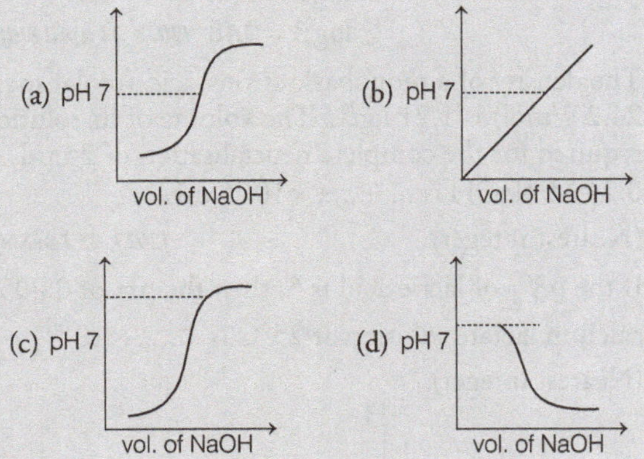


**17** Arrange the following solutions in the decreasing order of pOH.

(A) 0.01 M HCl (B) 0.01 M NaOH
(C) 0.01 M $CH_3COONa$ (D) 0.01 M NaCl
*[2020, 6 Sep Shift-I]*

(a) (A) > (C) > (D) > (B) (b) (A) > (D) > (C) > (B)
(c) (B) > (C) > (D) > (A) (d) (B) > (D) > (C) > (A)

**Imp.** pH of Water in Blend with Dissociation of Water

**18** For the following Assertion and Reason, the correct option is
**Assertion (A)** The pH of water increases with increase in temperature.
**Reason (R)** The dissociation of water into $H^+$ and $OH^-$ is an exothermic reaction. *[2020, 8 Jan Shift-II]*

(a) Assertion is not true, but Reason is true.
(b) Both assertion and Reason are true and the Reason is the correct explanation for the Assertion.
(c) Both Assertion and Reason are false.
(d) Both Assertion and Reason are true, but the Reason is not the correct explanation for the Assertion.

**19** The molar solubility of $Cd(OH)_2$ is $1.84 \times 10^{-5}$ M in water. The expected solubility of $Cd(OH)_2$ in a buffer solution of pH = 12 is *[2019, 12 April Shift-II]*

(a) $1.84 \times 10^{-9}$ M (b) $\frac{2.49}{1.84} \times 10^{-9}$ M
(c) $6.23 \times 10^{-11}$ M (d) $2.49 \times 10^{-10}$ M

**20** The pH of a 0.02 M $NH_4Cl$ solution will be
[Given $K_b(NH_4OH) = 10^{-5}$ and log 2 = 0.301] *[2019, 10 April Shift-II]*

(a) 4.65 (b) 2.65 (c) 5.35 (d) 4.35

**21** Following four solutions are prepared by mixing different volumes of NaOH and HCl of different concentrations, pH of which one of them will be equal to 1? *[JEE Main 2018]*

(a) 55 mL $\frac{M}{10}$ HCl + 45 mL $\frac{M}{10}$ NaOH
(b) 75 mL $\frac{M}{5}$ HCl + 25 mL $\frac{M}{5}$ NaOH
(c) 100 mL $\frac{M}{10}$ HCl + 100 mL $\frac{M}{10}$ NaOH
(d) 60 mL $\frac{M}{10}$ HCl + 40 mL $\frac{M}{10}$ NaOH

**22** $pK_a$ of a weak acid (*HA*) and $pK_b$ of a weak base (*BOH*) are 3.2 and 3.4, respectively. The pH of their salt (*AB*) solution is *[JEE Main 2017]*

(a) 7.2 (b) 6.9 (c) 7.0 (d) 1.0

**23** How many litres of water must be added to 1 L of an aqueous solution of HCl with a pH of 1 to create an aqueous solution with pH of 2? *[JEE Main 2014]*

(a) 0.1 L (b) 0.9 L (c) 2.0 L (d) 9.0 L

## Level 3

**24** A weak acid *HA* has degree of dissociation *x*. Which option gives the correct expression of (pH- $pK_a$)? *[2025, 28 Jan Shift-I]*

(a) 0 (b) $\log(1 + 2x)$
(c) $\log\left(\frac{1-x}{x}\right)$ (d) $\log\left(\frac{x}{1-x}\right)$

**25** Which of the following statement(s) is/are correct?

A. The pH of $1 \times 10^{-8}$ M HCl solution is 8.

B. The conjugate base of $H_2PO_4^-$ is $HPO_4^{2-}$

C. $K_w$ increases with increase in temperature.

D. When a solution of a weak monoprotic acid is titrated against a strong base at half neutralisation point. $pH = \frac{1}{2} pK_a$. *[2023, 15 April Shift-I]*

Choose the correct answer from the options given below.

(a) B, C, D (b) B, C
(c) A, B, C (d) A, D

**26** Class XII students were asked to prepare one litre of buffer solution of pH 8.26 by their chemistry teacher. The amount of ammonium chloride to be dissolved by the student in 0.2 M ammonia solution to make one litre of the buffer is (Given, $pK_b$ $(NH_3) = 4.74$, Molar mass of $NH_3 = 17$ $gmol^{-1}$, Molar mass of $NH_4Cl$ = 53.5 g $mol^{-1}$) *[2022, 26 July Shift-II]*

(a) 53.5 g (b) 72.3 g
(c) 107.0 g (d) 126.0 g

**27** The strength of an aqueous NaOH solution is most accurately determined by titrating (Note consider that an appropriate indicator is used) *[2020, 8 Jan Shift-I]*

(a) Aq. NaOH in a volumetric flask and concentrated $H_2SO_4$ in a conical flask
(b) Aq. NaOH in a pipette and aqueous oxalic acid in a burette
(c) Aq. NaOH in a burette and aqueous oxalic acid in a conical flask
(d) Aq. NaOH in a burette and concentrated $H_2SO_4$ in a conical flask

**28** Consider the following statements.

I. The pH of a mixture containing 400 mL of 0.1 M $H_2SO_4$ and 400 mL of 0.1 M NaOH will be approximately 1.3.

II. Ionic product of water is temperature dependent.

III. A monobasic acid with $K_a = 10^{-5}$ has a pH = 5. The degree of dissociation of this acid is 50%.

IV. The Le-Chatelier's principle is not applicable to common-ion effect.

The correct statements are *[2019, 10 April Shift-I]*

(a) I, II and IV (b) II and III
(c) I and II (d) I, II and III

**29** 20 mL of 0.1 M $H_2SO_4$ solution is added to 30 mL of 0.2 M $NH_4OH$ solution. The pH of the resultant mixture is [$pK_b$ of $NH_4OH = 4.7$] *[2019, 9 Jan Shift-I]*

(a) 9.3 (b) 5.0
(c) 9.0 (d) 5.2

## Numerical Value Type Questions

**30** $x$ mg of $Mg(OH)_2$ (molar mass = 58) is required to be dissolved in 1.0 L of water to produce a pH of 10.0 at 298 K. The value of $x$ is ........... mg. (Nearest integer) [Given $Mg(OH)_2$ is assumed to dissociate completely in $H_2O$] *[2025, 4 April Shift-II]*

**31** One litre buffer solution was prepared by adding 0.10 mol each of $NH_3$ and $NH_4Cl$ in deionised water. The change in pH on addition of 0.05 mol of HCl to the above solution is ......... $\times 10^{-2}$ . (Nearest integer)

Given: $pK_b$ of $NH_3 = 4.745$ and $\log_{10} 3 = 0.477$ *[2025, 7 April Shift-II]*

**32** $K_a$ for $CH_3COOH$ is $1.8 \times 10^{-5}$ and $K_b$ for $NH_4OH$ is $1.8 \times 10^{-5}$. The pH of ammonium acetate solution will be ......... . *[2024, 1st Feb Shift-I]*

**33** The pH of an aqueous solution containing 1 M benzoic acid ($pK_a = 4.20$) and 1 M sodium benzoate is 4.5. The volume of benzoic acid solution in 300 mL of this buffer solution is ........ mL. (Given, log 2 = 0.3) *[2024, 30 Jan Shift-II]*

**34** The pH at which $Mg(OH)_2$ [$K_{sp} = 1 \times 10^{-11}$] begins to precipitate from a solution containing 0.10 M $Mg^{2+}$ ions is *[2024, 30 Jan Shift-I]*

**35** 20 mL of 0.1 M NaOH is added to 50 mL of 0.1 M acetic acid solution. The pH of the resulting solution is ........... $\times 10^{-2}$. (Nearest integer)

Given, $pK_a$ $(CH_3COOH)$ = 4.76
log 2 = 0.30
log 3 = 0.48 *[2023, 13 April Shift-II]*

**36** The density of a monobasic strong acid (molar mass 24.2 g/mol) is 1.21 kg/L. The volume of its solution required for the complete neutralisation of 25 mL of 0.24 M NaOH is ........... $\times 10^{-2}$ mL. (Nearest integer) *[2023, 25 Jan Shift-I]*

**37** If the $pK_a$ of lactic acid is 5, then the pH of 0.005 M calcium lactate solution at 25°C is ........ $\times 10^{-1}$. (Nearest integer)

Lactic acid 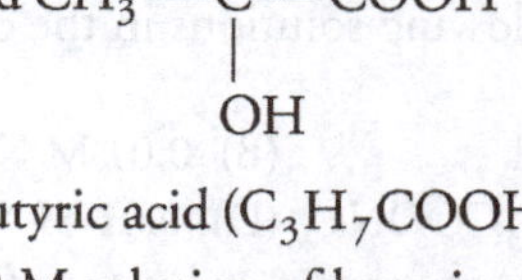


*[2023, 24 Jan Shift-II]*

**38** $K_a$ for butyric acid $(C_3H_7COOH)$ is $2 \times 10^{-5}$. The pH of 0.2 M solution of butyric acid is ........... $\times 10^{-1}$. (Nearest integer) (Given, log 2 = 0.30) *[2022, 28 July Shift-I]*

**39** pH value of 0.001 M NaOH solution is ......... . *[2022, 27 June Shift-II]*

**40** 50 mL of 0.1 M $CH_3COOH$ is being titrated against 0.1 M NaOH. When 25 mL of NaOH has been added, the pH of the solution will be ...... $\times 10^{-2}$. (Nearest integer)
[Given, $pK_a$ ($CH_3COOH$) = 4.76]
$\log 2 = 0.30$, $\log 3 = 0.48$, $\log 5 = 0.69$, $\log 7 = 0.84$, $\log 11 = 1.04$ *[2022, 26 June Shift-I]*

**41** The pH of a solution obtained by mixing 50 mL of 1 M HCl and 30 mL of 1 M NaOH is $x \times 10^{-4}$. The value of $x$ is ........ (Nearest integer)
[log 2.5 = 0.3979] *[2021, 31 Aug Shift-II]*

**42** The $OH^-$ concentration in a mixture of 5.0 mL of 0.0504 M $NH_4Cl$ and 2 mL of 0.0210 M $NH_3$ solution is $x \times 10^{-6}$ M. The value of $x$ is ......... . (Nearest integer)
[Given $K_w = 1 \times 10^{-14}$ and $K_b = 1.8 \times 10^{-5}$] *[2021, 26 Aug Shift-I]*

**43** In order to prepare a buffer solution of pH 5.74, sodium acetate is added to acetic acid. If the concentration of acetic acid in the buffer is 1.0 M, the concentration of sodium acetate in the buffer is ........... M. (Round off to the nearest integer).
[Given, $pK_a$ (acetic acid = 4.74] *[2021, 18 March Shift-I]*

**44** Sulphurous acid ($H_2SO_3$) has $K_{a_1} = 1.7 \times 10^{-2}$ and $K_{a_2} = 6.4 \times 10^{-8}$. The pH of 0.588 M is ......
(Round off to the nearest integer) *[2021, 16 March Shift-II]*

**45** The pH of ammonium phosphate solution, if $pK_a$ of phosphoric acid and $pK_b$ of ammonium hydroxide are 5.23 and 4.75 respectively, is ......... . *[2021, 26 Feb Shift-II]*

**46** 3 g of acetic acid is added to 250 mL of 0.1 M HCl and the solution made up to 500 mL. To 20 mL of this solution $\frac{1}{2}$ mL of 5 M NaOH is added. The pH of the solution is ......... .
[Given, $pK_a$ of acetic acid = 4.75, molar mass of acetic acid = 60 g/mol, log 3 = 0.4771]
Neglect any changes in volume. *[2020, 7 Jan Shift-II]*

**47** Two solutions, *A* and *B*, each of 100 L was made by dissolving 4 g of NaOH and 9.8 g of $H_2SO_4$ in water, respectively. The pH of the resultant solution obtained from mixing 40 L of solution *A* and 10 L of solution *B* is ......... . *[2020, 7 Jan Shift-I]*

## Key Index

### Topic 1

| | | | | | | | | | |
|---|---|---|---|---|---|---|---|---|---|
| 1 (d) | 2 (a) | 3 (c) | 4 (a) | 5 (d) | 6 (d) | 7 (d) | 8 (c) | 9 (a) | 10 (d) |
| 11 (c) | 12 (7) | 13 (5) | 14 (25) | 15 (1) | 16 (2) | | | | |

### Topic 2

| | | | | | | | | | |
|---|---|---|---|---|---|---|---|---|---|
| 1 (a) | 2 (b) | 3 (d) | 4 (a) | 5 (a) | 6 (c) | 7 (b) | 8 (c) | 9 (c) | 10 (a) |
| 11 (b) | 12 (b) | 13 (a) | 14 (d) | 15 (c) | 16 (d) | 17 (c) | 18 (b) | 19 (b) | 20 (b) |
| 21 (d) | 22 (c) | 23 (b) | 24 (c) | 25 (5) | 26 (233) | 27 (10) | 28 (282) | 29 (0) | 30 (2) |
| 31 (108) | 32 (64) | 33 (50) | | | | | | | |

### Topic 3

| | | | | | | | | | |
|---|---|---|---|---|---|---|---|---|---|
| 1 (b) | 2 (d) | 3 (b) | 4 (c) | 5 (c) | 6 (c) | 7 (a) | 8 (c) | 9 (a) | 10 (a) |
| 11 (a) | 12 (c) | 13 (b) | 14 (c) | 15 (a) | 16 (a) | 17 (b) | 18 (c) | 19 (d) | 20 (c) |
| 21 (b) | 22 (b) | 23 (d) | 24 (d) | 25 (b) | 26 (c) | 27 (c) | 28 (d) | 29 (c) | 30 (3) |
| 31 (48) | 32 (7) | 33 (100) | 34 (9) | 35 (458) | 36 (12) | 37 (85) | 38 (27) | 39 (11) | 40 (476) |
| 41 (6021) | 42 (3) | 43 (10) | 44 (1) | 45 (7) | 46 (5.23) | 47 (10.60) | | | |

# Detailed **Explanations**

## Topic 1

**01.** (d)

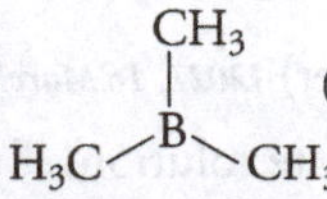


$(6e^-)$

Due to incomplete octet of B, it can act as a Lewis acid.

**02.** (a) Only in reaction (II), $H_2PO_4^-$, gives $H^+$ to $H_2O$. Thus, behaves as an acid.

**03.** (c) The overall dissociation constant of the acid is

$$K = K_1 \times K_2$$
$$= 1.0 \times 10^{-5} \times 5.0 \times 10^{-10}$$
$$= 5 \times 10^{-15}$$

**04.** (a) Conjugate base is formed by loss of $H^+$.

$$OH^- \longrightarrow \underset{\text{Conjugate base}}{O^{2-}} + H^+$$

$O^{2-}$ is the conjugate base of $OH^-$.

> **Concept Enhancer**
>
> The acid-base pair differ by a proton. If the bronsted acid is a strong acid, then its conjugate base is a weak base and vice versa.

**05.** (d) $H_2PO_4^- \rightleftharpoons H^+ + HPO_4^{2-}$

Conjugate base is formed when an acid loses its proton. Thus, $HPO_4^{2-}$ is the conjugate base of $H_2PO_4^-$.

**06.** (d) Assertion is false but Reason is true. The amphoteric nature of water can be explained by Bronsted-Lowry concept.

$$\underset{\text{Acid}}{H_2S} + \underset{\text{Base}}{H_2O} \rightleftharpoons H_3O^+ + HS^-$$

$$\underset{\text{Acid}}{H_2O} + \underset{\text{Base}}{NH_3} \rightleftharpoons NH_4^+ + \overline{O}H$$

**07.** (d) $BCl_3$ has an incomplete octet with following orbital picture.

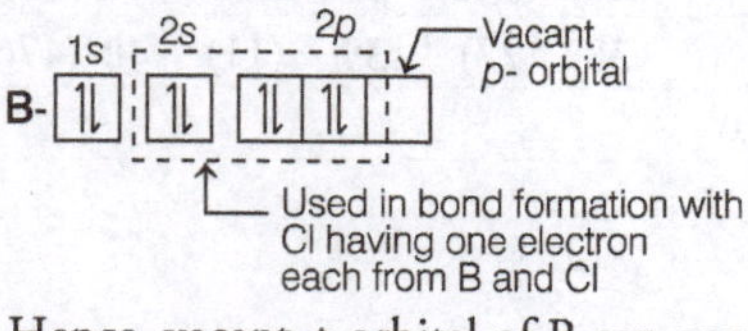


Hence, vacant *p*-orbital of B can accept one lone pair thus it can be considered as Lewis acid.

$AlCl_3$ Similar situation is visible in $AlCl_3$ as well i.e. Al (Valence orbital only) =

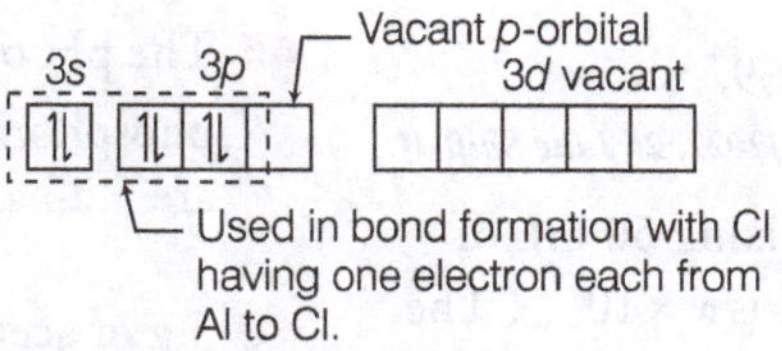


Hence this compound can also be considered as Lewis acid.

Thus both $BCl_3$ and $AlCl_3$ are considered as Lewis acids.

**08.** (c) $HQ = H^+ + Q^-$

$$[H^+] = \sqrt{K_a C}$$

by Ostwald's dilution law

$$[H^+] = 10^{-pH} = 10^{-3}\ M,\ C = 0.1\ M$$

$$\text{Thus, } 10^{-3} = \sqrt{K_a \times 0.1}$$

$$10^{-6} = K_a \times 0.1 \Rightarrow K_a = 10^{-5}$$

> ***Alternate Method***
>
> $$HQ \rightleftharpoons H^+ + Q^-$$
>
> Initial concentration 0.1 M    0    0
>
> Given, pH = 3,
>
> this suggests $[H^+] = 10^{-3}$ M at equilibrium.
>
> Thus, $[Q^-]$ at equilibrium $= 10^{-3}$ M
>
> Hence, $[HQ] = 0.1\ M - 10^{-3}\ M$
>
> $\approx 0.1\ M \quad [\because 10^{-3}\ M <<< 0.1\ M]$
>
> $K_a$ for the above reaction is given by
>
> $$K_a = \frac{[H^+][Q^-]}{[HQ]} = \frac{[10^{-3}][10^{-3}]}{[0.1]} \Rightarrow K_a = 1 \times 10^{-5}$$

**09.** (a) pH of a solution = 5

$$\therefore \quad [H^+] = 10^{-pH} = 10^{-5}$$

*HA* (a weak acid) ionises as

$$HA \rightleftharpoons H^+ + A^-$$

| | $HA$ | $H^+$ | $A^-$ |
|---|---|---|---|
| | 1.0 | 0 | 0 |
| | $(1-10^{-5})$ | $10^{-5}$ | $10^{-5}$ |

$$K_a = \frac{[H^+][A^-]}{[HA]}$$

$$= \frac{10^{-5} \times 10^{-5}}{(1-10^{-5})}$$

$$= \frac{10^{-10}}{1} \quad [\because 1 >>> 10^{-5}]$$

$$= 1 \times 10^{-10}\ M$$

**10.** (d) $H_2C_2O_4 \overset{K_{a_3}}{\rightleftharpoons} 2H^+ + C_2O_4^{2-}$

$$H_2C_2O_4 \overset{K_{a_1}}{\rightleftharpoons} H^+ + HC_2O_4^-$$

$$HC_2O_4^- \overset{K_{a_2}}{\rightleftharpoons} H^+ + C_2O_4^{2-}$$

$$K_{a_3} = \frac{[H^+]^2[C_2O_4^{2-}]}{[H_2C_2O_4]} \quad \ldots\text{(i)}$$

$$K_{a_1} = \frac{[H^+][HC_2O_4^-]}{[H_2C_2O_4]} \quad \ldots\text{(ii)}$$

$$K_{a_2} = \frac{[H^+][C_2O_4^{2-}]}{[HC_2O_4^-]} \quad \ldots\text{(iii)}$$

In order to obtain Eq. (i) we have to add Eq. (ii) and (iii)

$$\therefore K_{a_1} \times K_{a_2} = \frac{[H^+][HC_2O_4^-]}{[H_2C_2O_4]} \times \frac{[H^+][C_2O_4^{2-}]}{[HC_2O_4^-]}$$

$$= \frac{[H^+]^2[C_2O_4^{2-}]}{[H_2C_2O_4]} = K_{a_3}$$

$$K_{a_1} \times K_{a_2} = K_{a_3}$$

So, option (d) is correct.

**11.** (c) Conjugate bases of the given species are

| Species | Conjugate base |
|---|---|
| $HCO_3^-$ | $CO_3^{2-}$ |
| $H_3O^+$ | $H_2O$ |
| $HSO_4^-$ | $SO_4^{2-}$ |
| $HSO_3F$ | $HSO_3^-$ |

Acidic strength depends upon the stability of conjugate base i.e.,

Acidic strength ∝ Stability of conjugate base

Hence, the acidic strength follows the order

$$HSO_3F > H_3O^+ > HSO_4^- > HCO_3^-$$

**12.** (7) Given, pH = 9

$$C_{\text{initial}} = 1\ \text{mM}$$
$$K_b = 10^{-x}$$
$$x = ?$$

Since pH = 9 ⇒ pOH = 5

So, $[\overline{O}H] = 10^{-5}$ M

Chemical reaction involved in ionisation of ethyl amine is as follows

$$C_2H_5NH_2(aq)+H_2O \rightleftharpoons C_2H_2NH_3^+ + \bar{O}H$$

$C=10^{-3}$ M — —

$C(1-\alpha)$ $C\alpha$ $C\alpha$

$\Rightarrow C=10^{-3}$ $=10^{-5}$

$=10^{-5}$

According to question, $1-\alpha \approx 1$

Now we know that,

$$K_b = \frac{[C_2H_5NH_3^+][\bar{O}H]}{[C_2H_5NH_2]}$$

$$K_b = \frac{10^{-5}\times 10^{-5}}{10^{-3}} = 10^{-7}$$

So, the value of $x$ is 7.

**13.** *(5)* Decomposition reaction in equilibrium is,

$$H_2O(g) \rightleftharpoons H_2(g)+\frac{1}{2}O_2(g)$$

$t=0$ 1 mole

$t=t_{eq}$ $1-\alpha$ $\alpha$ $\alpha/2$

$$n_T = 1+\frac{\alpha}{2} = 1 \quad (\alpha <<< 1)$$

$$K_p = \frac{p_{H_2}\cdot p_{O_2}^{1/2}}{p_{H_2O}} = \frac{(\alpha\cdot p)\left(\frac{\alpha}{2}\cdot p\right)^{1/2}}{(1-\alpha)p}$$

as $p=1$

$$8\times 10^{-3} = \frac{\alpha^{3/2}}{\sqrt{2}}$$

$$\alpha^3 = 128\times 10^{-6}$$

$$\alpha = 5.03\times 10^{-2}$$

**14.** *(25)* pH of the new solution $=6$

$\Rightarrow$ $[H^+]=10^{-6}$M

$$HX \rightleftharpoons H^+ + X^-$$

At eqil. $C(1-\alpha)$ $C\alpha$ $C\alpha$

$$K_a = \frac{C\alpha^2}{1-\alpha} \text{ or } K_a = C\alpha^2$$

(for extremely dilute solution)

$$[H^+]=C\alpha=\sqrt{K_aC}$$

$$10^{-6}=\sqrt{4\times 10^{-10}\times C}$$

$$\Rightarrow \quad C=\frac{10^{-12}}{4\times 10^{-10}}=0.25\times 10^{-2}$$

$$=25\times 10^{-4}$$

**15.** *(1)* $Ba(OH)_2$ is completely ionised in aqueous solution.

$$Ba(OH)_2 \longrightarrow Ba^{2+}+2OH^-$$

$$[Ba(OH)_2]=0.005\text{M}$$

$$[OH^-]=2\times 0.005 \Rightarrow 0.01 \Rightarrow =10^{-2}$$

At 298 K, in aqueous solution ;

$$[H_3O^+][OH^-]=10^{-14}$$

$$[H_3O^+]=\frac{10^{-14}}{10^{-2}}=10^{-12}$$

$$[H_3O^+]=1\times 10^{-12} \text{ M}$$

**16.** *(2)* Given, $[HA]=0.01$

$[HCl]=0.1$ M

When strong acid (HCl) is completely dissociated,

$$HCl \longrightarrow H^+(aq)+Cl^-(aq)$$

0.1 0.1 0.1

For weak acid, dissociation is very less,

$$HA \rightleftharpoons H^+(aq)+A^-(aq)$$

$t=0$ 0.01 0.1 0

$t=t_{eq}$ $0.01-0.01\alpha$ $0.1+0.01\alpha$ $0.01\alpha$

$$K_a=\frac{[H^+][A^-]}{[HA]}=2\times 10^{-6}$$

$$\Rightarrow \frac{(0.1+0.01\alpha)(0.01\alpha)}{(0.01-0.01\alpha)}=2\times 10^{-6}$$

As $0.01\alpha << 0.1$, $[H^+]=0.1$

and $0.01\alpha << 0.01$, $[HA]=0.01$

$$\therefore \quad 2\times 10^{-6}=\frac{(0.1)(0.01\alpha)}{0.01}$$

$$\Rightarrow \quad \alpha=2\times 10^{-5} \Rightarrow x=2$$

## Topic 2

**01.** *(a)* The reaction involving dissociation of zirconium phosphate is as follows

$$Zr_3(PO_4)_4(s) \rightleftharpoons 3Zr^{4+}(aq)+4PO_4^{3-}(aq)$$

$3S$ $4S$

So, $K_{sp}$ will be given by

$$K_{sp}=(3S)^3(4S)^4=6912\,S^7$$

$$S=\left(\frac{K_{sp}}{6912}\right)^{1/7}$$

**02.** *(b)* For $AB_2 \rightleftharpoons A^{2+}+2B^-$

$$K_{sp}=[A^{2+}][B^-]^2$$

$$=(1.2\times 10^{-4})(0.24\times 10^{-3})^2$$

$$=0.0691\times 10^{-10}$$

$$K_{sp}=6.91\times 10^{-12}$$

**03.** *(d)* The solubility of AgCl will be maximum in deionised water as rest of all the solutions (0.01 M KCl, 0.01 M HCl and 0.01 M $AgNO_3$) have a common ion with the solute being added (AgCl). Thus, AgCl will be most soluble in deionised water.

**04.** *(a)* Given, $K_{sp}$ for

$Bi_2S_3 = 1.08\times 10^{-73}$

$$Bi_2S_3(s) \rightleftharpoons 2Bi^{3+}(aq)+3S^{2-}(aq)$$

$2S$ $3S$

$$K_{sp}=[Bi^{3+}]^2[S^{2-}]^3$$

So, $K_{sp}=[2S]^2[3S]^3$

$$K_{sp}=4S^2\times 27S^3=108S^5$$

$$\Rightarrow \quad S=\left(\frac{K_{sp}}{108}\right)^{1/5}$$

$$\therefore \quad S=\left(\frac{1.08\times 10^{-73}}{108}\right)^{1/5}=(10^{-75})^{1/5}$$

Hence, solubility,

$$S=1.0\times 10^{-15} \text{ mol L}^{-1}$$

**05.** *(a)* Let, solubility of $Ca(OH)_2$ in pure water $=S$ mol/L

$$Ca(OH)_2 \rightleftharpoons Ca^{2+}+2OH^-$$

$S$ mol/L $2\times S$ (mol/L)

$$K_{sp}=[Ca^{2+}][OH^-]^2=S\times(2S)^2$$

$$=4S^3 \text{ (mol/L)}$$

$$S=\left(\frac{K_{sp}}{4}\right)^{1/3}=\left(\frac{5.5\times 10^{-6}}{4}\right)^{1/3}$$

$$=1.11\times 10^{-2} \text{ mol/L}$$

**06.** *(c)* Let molar solubility of $Cr(OH)_3=S$ mol $L^{-1}$

$$Cr(OH)_3(s) \rightleftharpoons Cr^{3+}(aq)+3OH^-(aq)$$

$S$ $3S$

$$K_{sp}=1.6\times 10^{-30}$$

$$=[Cr^{3+}][OH^-]^3$$

$$=(S)(3S)^3=27S^4$$

$$\therefore \quad S^4=\frac{1.6\times 10^{-30}}{27}$$

$$\Rightarrow \quad S=\sqrt[4]{\frac{1.6\times 10^{-30}}{27}}$$

**07.** *(b)* $Na_2CO_3 \rightleftharpoons 2Na^+ + CO_3^{2-}$

$$[Na_2CO_3]=[CO_3^{2-}]=1\times 10^{-4} \text{ M}$$

$$BaCO_3 \rightleftharpoons Ba^{2+}+CO_3^{2-}$$

$$K_{sp\,(BaCO_3)}=[Ba^{2+}][CO_3^{2-}]$$

$$[Ba^{2+}]=\frac{K_{sp\,(BaCO_3)}}{[CO_3^{2-}]}$$

$$\Rightarrow \quad [Ba^{2+}]=\frac{5.1\times 10^{-9}}{1\times 10^{-4}}$$

$$[Ba^{2+}]=5.1\times 10^{-5} \text{ M}$$

Hence, at $5.1\times 10^{-5}$ M concentration of $Ba^{2+}$, a precipitate will begin to form.

**08.** (c) $MX_2 \rightleftharpoons M^{2+} + 2X^-$
$S \quad 2S$
$$K_{sp} = [M^{2+}][X^-]^2$$
If solubility be $S$, then
$$K_{sp} = (S)(2S)^2 = 4S^3$$
$$4S^3 = 4\times10^{-12} \Rightarrow S = 1\times10^{-4} \text{ M}$$
$$\therefore \quad [M^{2+}] = S = 1\times10^{-4} \text{ M}$$

**09.** (c) $MX_4 \rightleftharpoons M^{4+} + 4X^-$
$$K_{sp} = [M^{4+}][4X^-]^4$$
$$K_{sp} = (S)(4S)^4 \Rightarrow K_{sp} = 256\,S^5$$
$$S^5 = \frac{K_{sp}}{256} \Rightarrow S = \left(\frac{K_{sp}}{256}\right)^{1/5}$$

**10.** (a) $AB_2 \rightleftharpoons A^{2+} + 2B^-$
$S \quad 2S$
$$K_{sp} = [A^{2+}][B^-]^2$$
$$= (S)(2S)^2 = 4S^3$$
$$= 4(1\times10^{-5})^3 = 4\times10^{-15}$$

**• Concept Enhancer**

The value of Q and $K_{sp}$ are helpful in predicting whether ppt would be formed or not. If $Q > K_{sp}$ then ppt will take place. If $Q < K_{sp}$ then ppt. will take place. If $Q < K_{sp}$ then no ppt. will take place.

**11.** (b) Based on $K_{sp}$ values and salt analysis cation identification, we can say that order of $K_{sp}$ value is
$HgS < PbS < AgBr < Ca(OH)_2$
$K_{sp}$ values for given salts are
$HgS \rightarrow 4\times10^{-53}$,
$PbS \rightarrow 8\times10^{-28}$
$AgBr \rightarrow 5\times10^{-13}$,
$Ca(OH)_2 \rightarrow 5.5\times10^{-6}$

**12.** (b) Given, $K_{sp}[A(OH)_2] = 9\times10^{-10}$
$K_{sp}[B(OH)_3] = 27\times10^{-18}$
$[A^{2+}] = 1\text{ M}, [B^{3+}] = 1\text{ M}$
For precipitation $Q_{ip} > K_{sp}$
So, $[A^{2+}][OH^-]^2 > 9\times10^{-10}$
$\Rightarrow [OH^-] > 3\times10^{-5}$ M
For $[B(OH)_3]$
$[B^{3+}][OH^-]^3 > 27\times10^{-18}$
$[OH^-] > 3\times10^{-6}$ M
So, $B(OH)_3$ will precipitate before $A(OH)_2$.

**13.** (a) The complete reaction is
$$Cr(OH)_3(s) \rightleftharpoons Cr^{+3}(aq) + 3OH^-(aq)$$
At equilibrium $\quad SM \quad 3SM$
$$K_{sp} = (S)(3S)^3 = 27S^4$$
$$27S^4 = 1.6\times10^{-30}$$
$$S = \left(\frac{1.6}{27}\times10^{-30}\right)^{1/4} M$$

**14.** (d) During boiling, soluble $Mg(HCO_3)_2$ is converted into insoluble $Mg(OH)_2$ and $Ca(HCO_3)_2$ is converted into insoluble $CaCO_3$. This is because of high solubility product of $Mg(OH)_2$ as compared to $MgCO_3$ hence, $Mg(OH)_2$ is precipitated. These precipitate can be removed by filtration. Thus, filtrate will be obtained in soft water.
$$Mg(HCO_3)_2 \longrightarrow Mg(OH)_2\downarrow + 2CO_2$$
$K_{sp}$ of $Mg(OH)_2 > K_{sp}$ of $MgCO_3$ and hence, $Mg(OH)_2$ precipitated first.

**15.** (c) Both (A) and (R) are true, but (R) is not the correct explanation of (A). Because of high stability of CuS, the value of $K_{eq}$ is high, i.e. the value of $K_{sp}$ of CuS will be low.

**16.** (d) Units given for all options is $M^3$, or $(mol\ L^{-1})^3$, which implies that, the equilibrium has three ions. Therefore,
$$XY_2 \rightleftharpoons X^{2+}(aq) + 2Y^-(aq),$$
$$K_{sp} = [X^{2+}][Y^-]^2$$
or $$X_2Y \rightleftharpoons 2X^+(aq) + Y^{2-}(aq),$$
$$K_{sp} = [X^+]^2[Y^{2-}]$$
Concentation values from graph is
$$[X^{2+}] = 1\text{ mM} = 1\times10^{-3}\text{ M}$$
and $\quad [Y^-] = 2\times10^{-3}$ M
(i) $K_{sp}$ in case of $XY_2$,
$$K_{sp} = (1\times10^{-3})(2\times10^{-3})^2$$
$$= 4\times10^{-9}\text{ M}^3$$
(ii) $K_{sp}$ in case of $X_2Y$,
$$K_{sp} = (1\times10^{-3})^2(2\times10^{-3})$$
$$= 2\times10^{-9}\text{ M}^3$$
The only matching value of $K_{sp}$ from options that satisfies the concentration values from graph is $XY_2$.

**17.** (c)

**key idea** In this question we will use the concepts of solubility. The concentration of a substance in a saturated solution is defined as its solubility ($S$).
For $\quad A_xB_y \rightleftharpoons xA^{y+} + yB^{x-}$
$K_{sp} = [A^{y+}]^x[B^{x-}]^y$

$$NaOH \longrightarrow Na^+ + OH^-$$
$\qquad 0.2 \quad 0.2$

| | $Al(OH)_3 \rightleftharpoons$ | $Al^{3+}$ + | $3OH^-$ |
|---|---|---|---|
| Initially | 1 | 0 | 0 |
| At equilibrium | $1-S$ | $S$ | $3S+0.2$ |

$K_{sp}$ of $Al(OH)_3 = 2.4\times10^{-24}$ (Given)
$$K_{sp} = [Al^{3+}][OH^-]^3$$
$$2.4\times10^{-24} = [S][3S+0.2]^3$$
$[\because 0.2 >> S]$
$$2.4\times10^{-24} = [S][0.008]$$
$$[S] = 3\times10^{-22}$$

**18.** (b) For $A_xB_y \rightleftharpoons xA^{y+} + yB^{x-}$;
$$K_{sp} = [A^{y+}]^x[B^{x-}]^y$$
For, $Zr_3(PO_4)_4$,
$$Zr_3(PO_4)_4(s) \rightleftharpoons 3Zr^{4+}(aq) + 4PO_4^{3-}(aq)$$
$\qquad 3S \qquad 4S$
$$K_{sp} = [Zr^{4+}]^3[PO_4^{3-}]^4$$
$$K_{sp} = (3S)^3(4S)^4 = 6912\,S^7$$
or $S = \left(\frac{K_{sp}}{6912}\right)^{1/7}$

Thus, the relation between molar solubility($S$) and solubility product ($K_{sp}$) will be $S = \left(\frac{K_{sp}}{6912}\right)^{1/7}$

**19.** (b) $[AgBr] = [Ag^+] = 0.05$ M
$$AgBr \rightleftharpoons Ag^+ + Br^-$$
$$K_{sp}[AgBr] = [Ag^+][Br^-]$$
$$\Rightarrow \quad [Br^-] = \frac{K_{sp}(AgBr)}{[Ag^+]}$$
$$= \frac{5.0\times10^{-13}}{0.05}$$
$$= 10^{-11}\text{ M [mol L}^{-1}]$$
Moles of KBr needed to precipitate AgBr
$= [Br^-]\times V = 10^{-11}$ mol $L^{-1} \times 1L$
$= 10^{-11}$ mol
Therefore, amount of KBr needed to precipitate AgBr
$= 10^{-11}$ mol $\times 120$ g $mol^{-1}$
$= 1.2\times10^{-9}$ g

**20.** (b) $AgIO_3(s) \rightleftharpoons Ag^+(aq) + IO_3^-(aq)$
Let the solubility of $AgIO_3$ be $S$
$$K_{sp} = [Ag^+][IO_3^-]$$
$$1.0\times10^{-8} = S^2$$
or $$S = 1\times10^{-4}\text{ mol/L}$$

In 1000 mL moles of $AgIO_3$ dissolved
$=1\times10^{-4}$ moles

In 100 mL moles of $AgIO_3$ dissolved
$=1\times10^{-5}$ moles

Mass of $AgIO_3$ in 100 mL
$=1\times10^{-5}\text{ mol}\times283\text{ g mol}^{-1}$
$=2.83\times10^{-3}$ g

**21.** *(d)* $K_{sp}(AgCl)=1.7\times10^{-10}\ M^2$

$K_{sp}(Ag_2CrO_4)=1.9\times10^{-12}\ M^3$

(i) $[Ag^+]$ required to ppt. AgCl(*s*)

$$K_{sp}=[Ag^+][Cl^-]$$
$$\Rightarrow 1.7\times10^{-10}=[Ag^+][0.1]$$
$$[Ag^+]=1.7\times10^{-9}\ M$$

(ii) $[Ag^+]$ required to ppt. $Ag_2CrO_4(s)$

$$K_{sp}=[Ag^+]^2[CrO_4^{2-}]$$
$$\Rightarrow 1.9\times10^{-12}=[Ag^+]^2[0.001]$$
$$[Ag^+]=\sqrt{1.9\times10^{-9}}=\sqrt{19\times10^{-5}}=4.3\times10^{-5}\ M$$

$\because$ $[Ag^+]$ required to precipitate AgCl is low, so AgCl will precipitate first.

**22.** *(c)* Let the solubility of $Ag_2CO_3$ is $S$. Now, 0.1 M of $AgNO_3$ is added to this solution after which let the solubility of $Ag_2CO_3$ becomes $S'$.

$\therefore$ $[Ag^+]=S+0.1$ and $[CO_3^{2-}]=S'$

$$K_{sp}=(S+0.1)^2(S') \quad \text{...(i)}$$

Given, $K_{sp}=8\times10^{-12}$

$\because$ $K_{sp}$ is very small, we neglect $S'$ against $S$ in Eq. (i)

$\therefore$ $K_{sp}=(0.1)^2S'$

or $8\times10^{-12}=0.01S'$

or $S'=8\times10^{-12}\times10^2=8\times10^{-10}$ M

Thus, molar solubility of $Ag_2CO_3$ in 0.1 M $AgNO_3$ is $8\times10^{-10}$ M.

**23.** *(b)* Given,

$[H_2S]=0.10$ M

$[HCl]=0.20$ M So, $[H^+]=0.20$ M

$$H_2S \rightleftharpoons H^+ + HS^- \quad \text{...(i)}$$
$$K_1=1.0\times10^{-7}$$
$$HS^- \rightleftharpoons H^+ + S^{2-} \quad \text{...(ii)}$$
$$K_2=1.2\times10^{-13}$$

Adding Eqs. (i) and (ii), we get

$$H_2S \rightleftharpoons 2H^+ + S^{2-} \quad \text{...(iii)}$$
$$K=K_1\times K_2=1.0\times10^{-7}\times1.2\times10^{-13}=1.2\times10^{-20}$$

Now $[S^{2-}]=\dfrac{K\times[H_2S]}{[H^+]^2}$ [according to the (iii) equation]

$$=\frac{1.2\times10^{-20}\times0.1\ M}{(0.2M)^2}=\frac{1.2\times10^{-20}\times1\times10^{-1}\ M}{4\times10^{-2}M}=3\times10^{-20}\ M$$

**24.** *(c)* Its given that the final volume is 500 mL and this final volume was achieved when 50 mL of 1 M $Na_2SO_4$ was added to unknown $Ba^{2+}$ solution. So, we can interpret the volume of unknown $Ba^{2+}$ solution as 450 mL i.e.

$$\underset{Ba^{2+}\text{ solution}}{450\text{mL}} + \underset{Na_2SO_4\text{ solution}}{50\text{mL}} \longrightarrow \underset{BaSO_4\text{ solution}}{500\text{mL}}$$

From this we can calculate the concentration of $SO_4^{2-}$ ion in the solution *via*

$$M_1V_1=M_2V_2$$
$$1\times50=M_2\times500$$

(as 1M $Na_2SO_4$ is taken into consideration)

$$M_2=\frac{1}{10}=0.1\ M$$

Now for just precipitation,

Ionic product = Solubility product ($K_{sp}$)

i.e. $[Ba^{2+}][SO_4^{2-}]=K_{sp}$ of $BaSO_4$

Given $K_{sp}$ of $BaSO_4=1\times10^{-10}$

So, $[Ba^{2+}][0.1]=1\times10^{-10}$

or $[Ba^{2+}]=1\times10^{-9}$ M

Remember This is the concentration of $Ba^{2+}$ ions in final solution.

Hence, for calculating the $[Ba^{2+}]$ in original solution we have to use

$$M_1V_1=M_2V_2$$

as $M_1\times450=10^{-9}\times500$

so, $M_1=1.1\times10^{-9}$ M

**25.** *(5)* If equal volume are mixed, concentration becomes half

$$Ba(NO_3)_2 \rightleftharpoons Ba^{2+} + 2NO_3^{\ominus}$$

Before mixing, $[Ba^{2+}]=0.05$ M

After mixing, $Ba^{2+}=\dfrac{0.05}{2}=0.025$ M ... (i)

$$BaF_2 \rightleftharpoons Ba^{2+} + 2F^{\ominus}$$

Before mixing, $[F^{\ominus}]=0.02$ M

After mixing, $[F^-]=\dfrac{0.02}{2}=0.01$ M ...(ii)

From Eqs. (i) and (ii),

$$Q=[Ba^{2+}][F^-]^2=[0.025][0.01]^2=[2.5\times10^{-6}]$$

Since $[Q>K_{sp}]$

So, precipitation takes place and remaining solution is in equilibrium.

$$\therefore \frac{[Ba^{2+}][F^-]^2}{K_{sp}}=\frac{2.5\times10^{-6}}{0.5\times10^{-6}}=5$$

**26.** *(233)*

$$K_2SO_4 \longrightarrow \underset{0.1}{K^+(aq)} + \underset{0.1}{SO_4^{2-}(aq)}$$
$$BaSO_4(s) \rightleftharpoons \underset{S}{Ba^{2+}(aq)} + \underset{S+0.1}{SO_4^{2-}(aq)}$$
$$K_{sp}=[Ba^{2+}][SO_4^{2-}]$$
$$1\times10^{-10}=S(S+0.1)\simeq0.1S$$
$$S=10^{-9}\ M=233\times10^{-9}\ gL^{-1}$$

**27.** *(10)*

> **key idea** To solve this question, the concept of solubility can be used. For this, write the equation involved in the reaction. Then use the formula of $K_{sp}$ in terms of solubility $S$. Then solve the required value i.e., $-\log K_{sp}$.

$$AgCl(s) \rightleftharpoons \underset{S}{Ag^+(aq)} + \underset{S}{Cl^-(aq)}$$
$$K_{sp}=S^2=\left(\frac{1.434}{143.4}\times10^{-3}\right)^2=10^{-10}$$
$$-\log K_{sp}=-(-10)\log10=10$$

**28.** *(282)* Given, $K_{sp}(PbS)=8\times10^{-28}$

$$PbS(s) \rightleftharpoons \underset{S}{Pb^{2+}(aq)} + \underset{S}{S^{2-}(aq)}$$
$$K_{sp}=S^2$$
$$8\times10^{-28}=S^2$$
$$\Rightarrow S=\sqrt{8\times10^{-28}}=2.82\times10^{-14}$$

So, solubility, $S=282\times10^{-16}$ mol $L^{-1}$.

$x=282$

**29.** *(0)* $CaF_2 \rightleftharpoons \underset{(S)}{Ca^{2+}} + \underset{(2S)}{2F^-}$

Solubility product constant,

$$K_{sp}=S\times(2S)^2=4S^3$$

$$\text{Solubility}=\frac{2.34\times10^{-3}\times10}{78}\text{ mol/L}=3\times10^{-4}\text{ mol/L}.$$

$$K_{sp} = 4 \times (3 \times 10^{-4})^3$$
$$= 108 \times 10^{-12}$$
$$= 0.0108 \times 10^{-8} \text{ (mol/L)}^3$$
$$\simeq 0 \times 10^{-8} \text{ (mol / L)}^3$$

**30.** (2)

**key idea** In this question, we will use the concept of common ion effect. Solubility of a sparingly soluble salt is decreased by the addition of a common ion.

$$\underset{S}{Zn(OH)_2} \rightleftharpoons Zn^{2+} + \underset{2S}{2OH^-}$$

Due to common-ion effect (presence of NaOH) the concentration of $OH^-$ will be $(2S + 0.1) \approx 0.1$ ($\because 0.1 >> 2S$)

$\therefore$ Solubility product,

$$K_{sp} = (0.1)^2 \times S$$
$$2 \times 10^{-20} = 0.01 \times S$$
$$\Rightarrow \quad S = \frac{2 \times 10^{-20}}{0.01} = 2 \times 10^{-18}$$
$$\therefore \quad x = 2$$

Hence, answer is 2.

**31.** (108) $A_3B_2 \rightleftharpoons \underset{3S}{3A^{2+}(aq)} + \underset{2S}{2B^{3-}(aq)}$

$$K_{sp} = [A^{2+}]^3 [B^{3-}]^2$$
$$K_{sp} = (3S)^3 (2S)^2 = 108 S^5$$

Also, $S = \frac{x}{M} \Rightarrow K_{sp} = 108\left(\frac{x}{M}\right)^5$

Given that, $K_{sp} = a\left(\frac{x}{M}\right)^5$

$$\therefore \quad a = 108$$

**32.** (64) Given, solubility in water

$(S) = 8.0 \times 10^{-4}$ mol $L^{-1}$

$$CdSO_4(aq) \rightleftharpoons \underset{S}{Cd^{2+}(aq)} + \underset{S}{SO_4^{2-}(aq)}$$

In pure water,

$$K_{sp} = S^2 = (8 \times 10^{-4})^2 = 64 \times 10^{-8}$$

In 0.01 M $H_2SO_4$

$$H_2SO_4(aq) \longrightarrow \underset{0.02}{2H^+(aq)} + \underset{0.01}{SO_4^{2-}(aq)}$$

$$\underset{x}{CdSO_4(s)} \rightleftharpoons \underset{x}{Cd^{2+}(aq)} + \underset{(x+0.01)}{SO_4^{2-}(aq)}$$

$$K_{sp} = S(S + 0.01) = 64 \times 10^{-8}$$
$$S + 0.01 \cong 0.01 \text{ M}$$

So, $S(0.01) = 64 \times 10^{-8}$

$$\Rightarrow \quad S = 64 \times 10^{-6} \text{ M}$$

**33.** (50) Let the solubility of $A_2X$ be 'S'.

$$A_2X(s) \rightleftharpoons \underset{2S}{2A^+(aq)} + \underset{S}{X^{2-}(aq)};$$

$$K_{sp_1} = 4 \times 10^{-12}$$
$$K_{sp_1} = [A^+]^2[X^{2-}]$$
$$4 \times 10^{-12} = (2S)^2(S)$$
$$4 \times 10^{-12} = 4S^3 \Rightarrow S = 10^{-4} \text{ M}$$

Let the solubility of $MX$ be '$S_1$'.

$$MX(s) \rightleftharpoons \underset{S_1}{M^+(aq)} + \underset{S_1}{X^-(aq)},$$

$$K_{sp_2} = 4 \times 10^{-12}$$
$$K_{sp_2} = [M^+][X^-]$$
$$K_{sp_2} = (S_1)^2$$
$$4 \times 10^{-12} = (S_1)^2$$
$$\Rightarrow \quad S_1 = 2 \times 10^{-6} \text{ M}$$
$$\frac{[A_2X]}{[MX]} = \frac{S}{S_1} = \frac{10^{-4}}{2 \times 10^{-6}} = 50$$

## Topic 3

**01.** (b) When there is an increases in temperature, the $k_w$ of water increases. Thus, the degree of dissociation of water also increases. Hence, its pH will decrease.

**02.** (d) For salt of weak base and strong acid.

$$pH = 7 - \frac{1}{2} pK_b - \frac{1}{2} \log C$$
$$pH = 7 - \frac{5}{2} - \frac{1}{2} \log(2 \times 10^{-2}) = 5.35$$

**03.** (b) $CH_3COOK$ is the salt of strong base and weak acid.

Hence, the solution of $CH_3COOK$ will be most basic because of the following reaction.

$$CH_3COOK + H_2O \rightleftharpoons \underset{\text{(Weak acid)}}{CH_3COOH} + \underset{\text{(Strong base)}}{KOH}$$

**04.** (c) Methyl orange show pinkish colour towards more acidic medium and yellow orange colour towards basic or less acidic media.

**05.** (c) For salt of weak acid and weak base

$$pH = \frac{1}{2} pK_a + \frac{1}{2} pK_w - \frac{1}{2} pK_b$$
$$= \frac{1}{2} \times 4.80 + 7 - \frac{1}{2} \times 4.78 = 7.01$$

**06.** (c) When $HA$ is 50% ionised,

$[HA] = [A^-]$

Buffer solution of weak acid $HA \rightarrow$ acidic buffer

$$pH = pK_a + \log\frac{[A^-]}{[HA]}$$

or $\quad pH = pK_a = 4.5$

$$pOH = pK_w - pH$$
$$pOH = 14 - 4.5 = 9.5$$

**07.** (a) pH = 5.4

$$\Rightarrow \quad [H^+] = 10^{-5.4} = 10^{-6} \cdot 10^{0.6}$$

Antilog of 0.6 is $\approx$ 3.98

$$\therefore \quad [H^+] \approx 3.98 \times 10^{-6} \text{ M}$$

**08.** (c) Given,

$HCl(aq)$ pH = 1; $[H^+] = 10^{-1}$

If equal volume of water is added, concentration will become half.

$$[H^+]_{Sol} = \frac{10^{-1}}{2} \quad pH = 1.3$$

So, the pH of the HCl solution would increase to 1.3

**09.** (a) When equal volume are mixed molarity reduced to half.

For precipitation

$$[Q_{sp} > K_{sp}]$$
$$Q_{sp} = [A^{+2}][Y^{-2}]^2 > K_{sp}$$

(i) $Q_{sp} = (1.8 \times 10^{-3})\left(\frac{5}{2} \times 10^{-4}\right)^2 < K_{sp}$

(ii) $Q_{sp} = (10^{-4})(0.4 \times 10^{-3})^2 < K_{sp}$

(iii) $Q_{sp} = (10^{-2})(10^{-2})^2 > K_{sp}$

(iv) $Q_{sp} = \left(\frac{1.5}{2} \times 10^{-4}\right)\left(\frac{1.5}{2} \times 10^{-3}\right)^2 < K_{sp}$

So, among the given combinations $2.0 \times 10^{-2}$ M $AB_2$, $2.0 \times 10^{-2}$ M $XY$ will give a ppt. of $AY_2$ at 300K.

**10.** (a) Complete neutralisation of a strong acid by a strong base leads to the largest increase in temperature. Hence, option (a) is correct.

**11.** (a) Statement I is false but statement II is true. The correct form of statement I is, Buffer solution is the mixture of weak acid or weak base with their conjugate ions.

Hence, option (a) is correct.

**12.** (c) Both the Statement I and Statement II are correct.

**13.** (b) 200 mL of 0.01 M HCl is mixed with 400 mL of 0.01 M $H_2SO_4$.

$\therefore$ Total volume of solution = 200 mL + 400 mL = 600 mL

The moles of $H^+$ given by HCl solution = 0.01 M × 200 mL = 2 moles

The moles of $H^+$ given by $H_2SO_4$ solution = 2 × (0.01M × 400mL)

= 8 moles

$$1H_2SO_4 \longrightarrow 2H^+$$

Total moles of $H^+$ = 2 + 8 = 10 moles

Total concentration of $H^+$,

$$[H^+] = \frac{\text{Total number of moles of } H^+}{\text{Total volume of solution}}$$

$$= \frac{10}{600} = \frac{1}{60}$$

$\therefore$ pH of solution $= -\log[H^+]$

$$= -\log\left[\frac{1}{60}\right] = 1.78$$

**14.** *(c)* The compound $X$ is phenolphthalein. Phenolphthalein have colour change range between pH 8.3 and 10.5 during neutralisation of weak acid and strong base.

**15.** *(a)* $AgCN \rightleftharpoons Ag^+ + CN^-$

Solubility of AgCN is $x$.

$pH = 3 \Rightarrow [H^+] = 10^{-3}$

$HCN \rightleftharpoons H^+ + CN^-$

$$K_a = \frac{[H^+][CN^-]}{[HCN]}$$

$$\Rightarrow \frac{[10^{-3}][CN^-]}{[HCN]} = 6.2 \times 10^{-10}$$

$$\Rightarrow \frac{[CN^-]}{[HCN]} = 6.2 \times 10^{-7}$$

Each CN–hydrolyses to give one HCN as:

$CN^- + H_2O \rightleftharpoons HCN + OH^-$

HCN is a weak acid, thus,

$[HCN] >> [CN^-]$

$x = [Ag^+] = [CN^-] + [HCN]$

$\Rightarrow x = [Ag^+] \approx [HCN]$

$\Rightarrow [CN^-] = x \times 6.2 \times 10^{-7}$

$K_{sp}(AgCN) = [Ag^+][CN^-]$

$\Rightarrow 2.2 \times 10^{-16} = (x)(x \times 6.2 \times 10^{-7})$

$\Rightarrow x^2 = 3.55 \times 10^{-10}$

$\Rightarrow x = 1.88 \times 10^{-5} \approx 1.9 \times 10^{-5}$

**16.** *(a)* Volume of HCl = 100 mL

Molarity of HCl = 0.1 M

Volume of NaOH = 100 mL

Molarity of NaOH = 0.1 M

Initially when only HCl is present in the beaker, then the pH is less than 7, so the graph looks like.

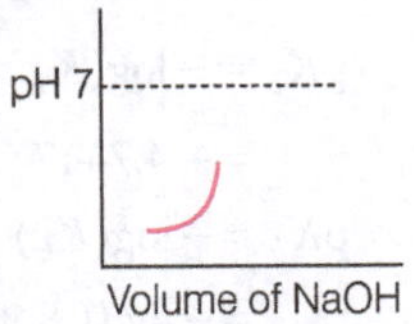


On adding 2 mL of NaOH continuously the pH value gradually increases and reach to equivalence point at pH = 7. On further addition of NaOH the pH value increases further beyond the equivalence point so, the final graph look like.

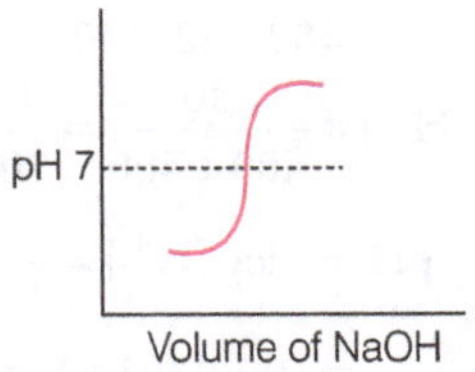


Hence, option (a) is correct.

**17.** *(b)* Given molar concentration are same. So strongest acid (HCl) will have Max. pOH

Strongest base (NaOH) will have Min. pOH

Neutral salt (NaCl) will have pOH equal to 7

$CH_3COONa$ is the salt of weak acid and strong base thus having pOH value less than 7.

Thus, order of pOH value :

(A) > (D) > (C) > (B)

Hence, the correct option is (b).

**18.** *(c)* The dissociation of water into $H^+$ and $OH^-$ is an endothermic reaction.

$\therefore$ As temperature increases, its equilibrium constant also increases.

$[H^+]$ and $[OH^-]$ in water at equilibrium increases with increase in temperature.

$\therefore$ $pH = -\log[H^+]$ decreases, as temperature increases.

So, both Assertion and Reason are false.

**19.** *(d)* $Cd(OH)_2 \rightleftharpoons \underset{S}{Cd^{2+}} + \underset{2S}{2OH^-}$

At equilibrium, $K_{sp} = S(2S)^2 = 4S^3$

$\Rightarrow K_{sp} = 4 \times (1.84 \times 10^{-5})^3$

Solubility in buffer solution having pH = 12

$[OH^-] = 10^{-2}$

$Cd(OH)_2 \rightleftharpoons \underset{S'}{Cd^{2+}} + \underset{\substack{2S+10^{-2} \\ \approx 10^{-2}}}{2OH^-}$

$\therefore K_{sp} = 4 \times (1.84 \times 10^{-5})^3$

$= S'(10^{-2})^2$

$$\Rightarrow S' = \frac{24.9 \times 10^{-15}}{10^{-4}}$$

$= 2.49 \times 10^{-10}$ M

**• Concept Enhancer**

Buffer solution is a solution that resist change in their pH on dilution or an addition of small quantities of acids or alkalies.

**20.** *(c)*

> **key idea** $NH_4Cl$ is a salt of weak base ($NH_4OH$) and strong acid (HCl). On hydrolysis, $NH_4Cl$ will produce an acidic solution (pH < 7) and the expression of pH of the solution is
>
> $$pH = 7 - \frac{1}{2}(pK_b + \log C)$$

Given, $K_b(NH_4OH) = 10^{-5}$

$\therefore pK_b = -\log K_b = -\log(10^{-5}) = 5$

$C$ = concentration of salt solution = 0.02 M = $2 \times 10^{-2}$ M

Now, $pH = 7 - \frac{1}{2}(pK_b + \log C)$

On substituting the given values in above equation, we get

$= 7 - \frac{1}{2}[5 + \log(2 \times 10^{-2})]$

$= 7 - 1.65 = 5.35$

**21.** *(b)* 75 mL $\frac{M}{5}$ HCl + 25 mL $\frac{M}{5}$ NaOH

25 mL $\frac{M}{5}$ NaOH will neutralise

25 mL $\frac{M}{5}$ HCl

75 − 25 = 50 mL $\frac{M}{5}$ HCl will remain.

Total volume will be

75 + 25 = 100 mL

When 50 mL $\frac{M}{5}$ HCl is diluted to 100 mL,

molarity will reduce to half, i.e., $\frac{M}{10}$.

$$[H^+] = [HCl] = \frac{1}{10}$$

$$pH = -\log_{10}[H^+] = -\log_{10}\frac{1}{10} = 1$$

**22.** *(b)* For a salt of weak acid and weak base,

$$pH = 7 + \frac{1}{2}pK_a - \frac{1}{2}pK_b$$

Given, $pK_a(HA) = 3.2$,

$pK_b(BOH) = 3.4$

$\therefore pH = 7 + \frac{1}{2}(3.2) - \frac{1}{2}(3.4)$

$= 7 + 1.6 - 1.7 = 6.9$

**23.** *(d)* $pH = 1 \Rightarrow [H^+] = 10^{-1} = 0.1$ M

$pH = 2 \Rightarrow [H^+] = 10^{-2} = 0.01$ M

For dilution of HCl,

$M_1V_1 = M_2V_2$

$0.1 \times 1 = 0.01 \times V_2 \Rightarrow V_2 = 10$ L

Volume of water to be added = 10 − 1

= 9 L

**24.** *(d)* The dissociation of weak acid is given by

$$HA \rightleftharpoons H^+ + A^-$$

at $t = 0$   $a$   – 

at $t = t$   $a(1-x)$   $ax$   $ax$

$$K_a = (ax)\frac{(x)}{(1-x)}, [H^+] = ax$$

$$-\log(K_a) = -\log(ax) - \log\left(\frac{x}{1-x}\right)$$

$$pK_a = pH - \log\left(\frac{x}{1-x}\right)$$

$$\therefore pH - pK_a = \log\left(\frac{x}{1-x}\right)$$

**25.** *(b)* Statements (B) and (C) are correct while statements (A) and (D) are incorrect.

(A) pH of $10^{-8}$ M HCl is in acidic range i.e. 6.98.

(D) At half neutralisation point, half of the acid is present in the form of salt. $pH = pK_a + \log\frac{1}{1} = pK_a$.

**26.** *(c)* Given, $[Base] = [NH_3] = 0.2$ M

pH of buffer = 8.26

$pK_b (NH_3) = 4.74$

Molar mass of $NH_3$ = 17 g mol$^{-1}$

Molar mass of $NH_4Cl$ = 53.5 g mol$^{-1}$

We know, $pOH = 14 - pH$

$= 14 - 8.26 = 5.74$

Using the expression,

$$pOH = pK_b + \log\frac{[Salt]}{[Base]}$$

or, $$5.74 = 4.74 + \log\frac{[NH_4Cl]}{[0.2]}$$

or $[NH_4Cl] = 2$ moles

Moles of $NH_3$ = Moles of $NH_4Cl = 2$

So, the amount of $NH_4Cl$ required

= Number of moles × Molecular mass

$= 2 \times 53.5 = 107$ g

**27.** *(c)* The strength of aqueous NaOH solution is most accurately determined by titrating aqueous NaOH in burette and aqueous oxalic acid in conical flask with phenolphthalein is used as an indicator. (For weak acid-strong base titration, methyl orange cannot be used).

Thus, solution turns colourless to pink at the end point.

**28.** *(d)* The explanation of given statements are as follows :

In statement (I), millimoles of $H^+$

$= 400 \times 0.1 \times 2 = 80$

Millimoles of $OH^- = 400 \times 0.1 = 40$

$\therefore$ Millimoles of $H^+$ left

$= 80 - 40 = 40$

$$[H^+] = \frac{40}{400+400} = \frac{1}{20} M$$

$$\Rightarrow \quad pH = -\log[H^+] = -\log\left(\frac{1}{20}\right)$$

$= -0 + 0.301 + 1 \Rightarrow 1.30$

Hence, the statement (I) is correct.

In statement (II), ionic product of $H_2O$ is temperature dependent.

$K_w = [H^+][OH^-] \approx 10^{-14} (mol/L)^2$ at 25°C

With increase in temperature, dissociation of $H_2O$ units into $H^+$ and $OH^-$ ions will also increase. As a result, the value of ionic product, $[H^+] \times [OH^-]$ will be increased.

Hence, the statement (II) is correct.

In statement (III), for a weak monobasic acid $HA$

$$HA \rightleftharpoons H^{\oplus} + A^{\ominus}$$

$(1-\alpha)CM$   $\alpha CM$   $\alpha CM$

$\Rightarrow$ pH of the solution is 5, i.e.

$[H^+] = 10^{-5}$ M $= \alpha C$

$$\Rightarrow \quad K_a = \frac{\alpha C \times \alpha C}{(1-\alpha)C} = \frac{10^{-5} \times \alpha}{1-\alpha}$$

$$\Rightarrow \quad 10^{-5} = \frac{10^{-5} \times \alpha}{1-\alpha}$$

$$\Rightarrow \quad \alpha = 0.5 \Rightarrow \alpha\% = 50$$

**29.** *(c)* Volume of $H_2SO_4$ $(V_1) = 20$ mL

Concentration of $H_2SO_4$ $(M_1) = 0.1$ M

Number of moles $(n_{H_2SO_4})$

$= 0.1 \times 2 \times 20 = 4$ mmol

$[\because 1\, H_2SO_4 \rightarrow 2H^+]$

Volume of $NH_4OH$ $(V_2) = 30$ mL

Concentration of $NH_4OH$ $(M_2)$

$= 0.2$ M

Number of moles $(n_{NH_4OH}) = 0.2 \times 30$

$= 6$ mmol

The reaction involved is as follows:

$$H_2SO_4 + 2NH_4OH \longrightarrow (NH_4)_2SO_4 + 2H_2O$$

| | $H_2SO_4$ | $2NH_4OH$ | $(NH_4)_2SO_4$ | $2H_2O$ |
|---|---|---|---|---|
| Initial | 4 mmol | 6 mmol | – | – |
| Final | 0 mmol | 2 mmol | 4 mmol | – |

So, the resulting solution is a basic buffer. According to the Henderson's equation,

$$pOH = pK_b + \log\frac{[Salt]}{[Base]}$$

$$pOH = pK_b + \log\frac{[(NH_4)_2SO_4]}{[NH_4OH]}$$

$$= 4.7 + \log\frac{4}{2} = 4.7 + 0.3 = 5$$

$\Rightarrow \quad pH = 14 - pOH = 14 - 5 = 9$

**30.** *(3)* Given, pH = 10

pOH = 4

So, $[OH^-] = 10^{-4}$

Now, number of moles of $OH^- = 10^{-4}$

number of moles of $Mg(OH)_2 = \frac{10^{-4}}{2}$

$= 5 \times 10^{-5}$

So, mass of $Mg(OH)_2$

$= 5 \times 10^{-5} \times 58 \times 10^{-3}$ mg

$= 2.9 \approx 3$

**31.** *(48)* $pOH = pK_b + \log\frac{[NH_4^+]}{[NH_3]}$

$pOH = 4.745$

On adding 0.05 mole HCl

$$NH_3 + H^{\oplus} \longrightarrow NH_4^{\oplus}$$

| $NH_3$ | $H^{\oplus}$ | $NH_4^{\oplus}$ |
|---|---|---|
| 0.1 | 0.05 | 0.1 |
| 0.05 | 0 | 0.15 |

$pOH' = 4.745 + \log 3$

$pOH' - pOH = 0.477$

$14 - pH' - 14 + pH = 0.477$

$\Delta pH = 0.477 \approx 48 \times 10^{-2}$

**32.** *(7)*

> **key idea** Ammonium acetate is a salt of weak acid and weak base. So, to calculate the pH value of ammonium acetate, we have to first calculate $pK_a$ and $pK_b$ values of acetic acid and ammonium hydroxide. Further apply the formula $pH = 7 + \frac{1}{2}[pK_a - pK_b]$

Given,

$K_a$ of $CH_3COOH = 1.8 \times 10^{-5}$

$K_b$ of $NH_4OH = 1.8 \times 10^{-5}$

Using the formula,

$$pH = 7 + \frac{1}{2}[pK_a - pK_b] \quad ...(i)$$

$pK_a = -\log(K_a) = -\log(1.8 \times 10^{-5})$

$= +4.7447$   ...(ii)

$pK_b = -\log(K_b)$

$= -\log(1.8 \times 10^{-5})$

$= 4.7447$   ...(iii)

Put value of Eq. (ii) and (iii) in Eq. (i)

$$pH = 7 + \frac{1}{2}[4.7447 - 4.7447] = 7$$

**33.** *(100)* Given,

Solution contains = 1 M benzoic acid $(V_s \text{ mL})$ + 1 M sodium benzoate $(V_s \text{ mL})$

Total volume of buffer solution = 300 mL

$$pH = 4.5, \; pK_a = 4.20$$

$$pH = pK_a + \log \frac{[\text{salt}]}{[\text{acid}]} \quad \ldots(i)$$

$$4.5 = 4.2 + \log \left[\frac{V_s}{V_a}\right]$$

$$\frac{V_s}{V_a} = 2 \quad \ldots(ii)$$

$$V_s + V_a = 300 \quad \ldots(iii)$$

From Eqs. (ii) and (iii)

$$V_a = 100 \text{ mL}, V_s = 200 \text{ mL}$$

Hence, volume of benzoic acid is 100 mL.

**34.** *(9)*

> **key idea** When the precipitation just begins, the ionic product of the ions becomes equal to the $K_{sp}$ of the salt. This will give the $[OH^-]$ concentration, and hence the pOH and pH value can be calculated.

When $Mg(OH)_2$ start precipitation, then $[Mg^{2+}][OH^-]^2 = K_{sp}$ of $Mg(OH)_2$

$$[0.1]\,[OH^-]^2 = 1 \times 10^{-11}$$

$$\text{As } K_{sp} = 1 \times 10^{-11} \quad \text{[given]}$$

$$\therefore \quad [OH^-] = 10^{-5} \text{ M}$$

$$pOH = 5$$

$$pH = 14 - pOH = 14 - 5 = 9$$

Hence, pH is 9.

**35.** *(458)* 20 mL of 0.1 M NaOH = 2 m mole of NaOH (base)

50 mL of 0.1 M $CH_3COOH$ = 5 m mole of $CH_3COOH$ (acid)

Now, when acid reacts with base, gives salt and water. This reaction is called neutralisation reaction.

$$\underbrace{\underset{5 \text{ m mol}}{CH_3COOH} + \underset{2 \text{ m mol}}{NaOH}}_{\substack{3 \text{ m mol of} \\ (CH_3COOH \text{ will} \\ \text{left})}} \longrightarrow \underset{\substack{(2 \text{ m mol} \\ \text{of salt} \\ \text{will formed})}}{CH_3COONa} + H_2O$$

Now, with weak acid and the salt of weak acid, a buffer solution will be formed, pH can be calcualted from Henderson's equation

$$pH = pK_a + \log \left[\frac{A^{\ominus} \text{ ion}}{HA \text{ acid}}\right]$$

$$pH = 1.76 + \log \left[\frac{2 \text{ m mol}}{3 \text{ m mol}}\right]$$

$$pH = 4.76 + \{\log[2] - \log(3)\}$$

$$pH = 4.76 + [0.30 - 0.48]$$

$$pH = 4.58 \text{ or } (458 \times 10^{-2})$$

**36.** *(12)*

> **key idea** To find out the volume of solution required, first of all calculate the molarity. It will be equal to the normality of solution due to its monobasic nature. After that use the formula $N_1V_1 = N_2V_2$

$$\text{Molarity} = \frac{\text{Density (g/L)}}{\text{Molar mass (g/mol)}}$$

$$\text{Molarity of acid} = \frac{1.21 \times 10^3}{24.2}$$

$$= \frac{1000}{20} = 50 \text{ M} = 50 \text{ N}$$

$\because$ For both of *HA* and NaOH, *n*-factor = 1

i.e., molarity (*M*) = Normality (*N*).

$\because$ Density = 1.21 kg/L

$$= 1.21 \times 10^3 \text{ g/L}$$

From neutralisation reaction,

$$HA + NaOH \longrightarrow NaA + H_2O$$

$$N_1V_1 = N_2V_2$$

$$\Rightarrow \quad 50 \times V = 0.24 \times 25$$

$$V = 0.12 \text{ mL} = 12 \times 10^{-2} \text{ mL}$$

**37.** *(85)* Concentration of lactate ion

$$= (2 \times 0.005) \text{ M}.$$

As we know that, calcium lactate is a salt of weak acid and strong base so salt hydrolysis will take place.

$$pH = 7 + \frac{1}{2}(pK_a + \log c)$$

$$= 7 + \frac{1}{2}[5 + \log(2 \times 0.005)]$$

$$= 7 + \frac{3}{2} = \frac{17}{2} = 8.5 = 85 \times 10^{-1}$$

Therefore, the pH of 0.005 M calcium lactate solution at 25°C is $85 \times 10^{-1}$.

**38.** *(27)*

$$C_3H_7COOH \rightleftharpoons C_3H_7COO^{\ominus} + H^+$$

| | $C_3H_7COOH$ | $C_3H_7COO^{\ominus}$ | $H^+$ |
|---|---|---|---|
| Initial conc. | $C$ | 0 | 0 |
| Conc. after dissociation | $C - C\alpha$ | $C\alpha$ | $C\alpha$ |

Hence, concentration of $C_3H_7COOH\,(C) = 0.2$ M

So, dissociation constant for weak acid is given as

$$K_a = \frac{[C_3H_7COO^-][H^+]}{[C_3H_7COOH]}$$

$$= \frac{C\alpha \times C\alpha}{C - C\alpha} = \frac{C\alpha^2}{1 - \alpha}$$

Since, butyric acid is weak, so, $1 >> \alpha$

So, $\quad K_a = C\alpha^2 \; [K_a = 2 \times 10^{-5}]$

$$\sqrt{\frac{2 \times 10^{-5}}{0.2}} = \alpha$$

$$10^{-2} = \alpha$$

Concentration of $H^+$ ions $= C\alpha$

$$= 0.2 \times 10^{-2} = 2 \times 10^{-3}$$

$$pH = -\log[H^+]$$

$$= -\log(2 \times 10^{-3})$$

$$= -0.3010 + 3$$

$$pH = 2.7$$

or $\quad pH = 27 \times 10^{-1}$

**39.** *(11)* 0.001 M NaOH will contain 0.001 M $OH^-$ ions.

$\therefore$ Concentration of $OH^-$ ions,

$$[OH^-] = 10^{-3} \text{ M}$$

So, $\quad pOH = -\log[OH^-]$

$$= -\log(10^{-3}) = 3$$

pH, $\; 14 - pOH = 14 - 3 = 11$

Hence, pH value of 0.001 M NaOH solution = 11.

**40.** *(476)* Moles of $CH_3COOH$

= Molarity of $CH_3COOH$ × Volume of $CH_3COOH$(in L)

$$= 0.1 \text{ M} \times \frac{50}{1000} \text{ L} = 0.005 \text{ moles}$$

Similarly, moles of NaOH = 0.0025 moles

$$NaOH + CH_3COOH \longrightarrow CH_3COONa + H_2O$$

| | NaOH | $CH_3COOH$ | $CH_3COONa$ | $H_2O$ |
|---|---|---|---|---|
| Initial conc. | 0.0025 mol | 0.005 mol | 0 | 0 |
| Final conc. | 0 | 0.0025 | 0.0025 mol | 0.0025 mol |

So, buffer is formed, according to Henderson-Hasselbalch equation

$$pH = pK_a + \log\left(\frac{[\text{Salt}]}{[\text{Acid}]}\right)$$

$$= pK_a + \log\left(\frac{\frac{0.0025}{75}}{\frac{0.0025}{75}}\right)$$

$$= pK_a = 4.76 = 476 \times 10^{-2}$$

**41.** *(6021)* Milliequivalents of HCl

$$(N_aV_a) = 50 \times 1 = 50$$

Milliequivalents of NaOH

$$(N_bV_b) = 30 \times 1 = 30$$

Since, $\quad N_aV_a > N_bV_b$

and they neutralise each other

$$[H^+] = \frac{N_aV_a - N_bV_b}{V_a + V_b}$$
$$= \frac{50-30}{80} = 0.25 = 2.5 \times 10^{-1}$$
$$pH = -\log [H^+]$$
$$= -\log (2.5 \times 10^{-1})$$
$$= 1 - 0.3979 = 0.6021$$
$$pH = 6021 \times 10^{-4}$$
$$\therefore \quad x = 6021$$

**42.** *(3)* $NH_3 + H_2O \rightleftharpoons NH_4^+ + OH^-$

$[NH_3] = 2 \times 0.0210 = 0.042$ m mol

$NH_4Cl \longrightarrow NH_4^+ + Cl^-$

$[NH_4Cl] = 5 \times 0.0504$
$= 0.252$ m mol

By Henderson-Hasselbalch equation,

$$pOH = pK_b + \log \frac{[Salt]}{[Base]}$$
$$= -\log (1.8 \times 10^{-5}) + \log \frac{0.252}{0.042}$$
$$= 4.74 + 0.77 = 5.51$$
$$[OH^-] = 10^{-5.51} = 3.09 \times 10^{-6}$$
$$\approx 3 \times 10^{-6}$$

The value of $x$ is $3 \times 10^{-6}$

**43.** *(10)* Concentration of acetic acid in buffer = 1.0 M

Acetic acid and its conjugate base, sodium acetate makes acidic buffer.

$CH_3COOH + CH_3COONa \longrightarrow$ (Acidic buffer)

Using formula,

$$pH = pK_a + \log \frac{[CH_3COONa]}{[CH_3COOH]}$$
$$5.74 = 4.74 + \log \frac{[CH_3COONa]}{[CH_3COOH]}$$
$$\Rightarrow \frac{[CH_3COONa]}{[CH_3COOH]} = 10$$
$$[\therefore [CH_3COOH] = 1]$$
$$[CH_3COONa] = [10 \times 1] = 10$$

Thus, the concentration of sodium acetate in buffer is 10 M.

**44.** *(1)* For $H_2SO_3$, $K_{a_1} >> K_{a_2}$

So, we mainly consider 1st ionisation

| | $H_2SO_3$ | $\rightleftharpoons H^+$ | $+ HSO_3^-$; |
|---|---|---|---|
| $t = 0$ | 0.558 | 0 | 0 |
| $t = t_{eq}$ | $0.558 - 0.558\alpha$ | $0.558\alpha$ | $0.558\alpha$ |

$$K_{a_1} = 1.7 \times 10^{-2}$$
$$K_{a_1} = \frac{[H^+][HSO_3^-]}{[H_2SO_3]}$$
$$= \frac{(0.558\,\alpha)(0.558\,\alpha)}{0.558(1-\alpha)} = \frac{0.558\,\alpha^2}{1-\alpha}$$

$\alpha << 1$ for weak acid

$(1-\alpha) \approx 1$

$$\alpha = \sqrt{\frac{K_{a_1}}{0.558}} = \sqrt{\frac{1.7 \times 10^{-2}}{0.558}}$$
$$\alpha = 1.7 \times 10^{-1} = 0.17$$
$$[H]^+ = 0.558\,\alpha = 9.5 \times 10^{-2}$$
$$pH = -\log[H^+] = -\log(9.5 \times 10^{-2})$$
$$[\because \log 9.9 \approx 1]$$
$$= 2 - \log 9.5 = 2 - 1$$
$$pH = 1$$

**45.** *(7)* Phosphoric acid is a weak tribasic acid ($H_3PO_4$) and $NH_4OH$ is a weak monoacidic base. So, hydrolysis of ammonium phosphate $(NH_4)_3PO_4$ can be shown as,

$(NH_4)_3PO_4 + 3H_2O \rightleftharpoons H_3PO_4 + 3NH_4OH$

[3 : 1 type salt]

$$\text{So, } [H^+] = K_a \times \left(\frac{K_w}{K_a \times K_b}\right)^{1/2}$$
$$pH = pK_a + \frac{1}{2}[pK_w - pK_a - pK_b]$$
$$= 5.23 + \frac{1}{2}(14 - 5.23 - 4.75)$$
$$= 7.24 \approx 7 \quad [\because pK_w = 14\,H_2O]$$

Also, $pK_a = 5.23\ (H_3PO_4)$

$pK_b = 4.75\ (NH_4OH)]$

**46.** *(5.23)* $3g\ CH_3COOH\ (M = 60)$

$$= \frac{3}{60} = \frac{1}{20}\ \text{mol}$$

250 mL of 0.1 M HCl

$$= M \times V = \frac{250}{1000} \times 0.1$$
$$= 25 \times 10^{-3}\ \text{mol}$$

After mixing solution is diluted to 500 mL.

Then from that solution, 20 mL is taken.

Number of moles of both $CH_3COOH$ and HCl will be now only $\frac{20}{500} = \frac{1}{25}$ of original solution.

$$n_{CH_3COOH} = \frac{1}{20} \times \frac{1}{25}$$
$$= \frac{1}{500} = 2 \times 10^{-3}\ \text{mol}$$
$$n_{HCl} = 25 \times 10^{-3} \times \frac{1}{25}$$
$$= 10^{-3}\ \text{mol}$$
$$n_{NaOH}\ \text{added} = M \times V = \frac{1}{2} \times 5 \times 10^{-3}$$
$$= 2.5 \times 10^{-3}$$

After mixing, all HCl and NaOH will be neutralised. Remaining number of moles.

$$n_{NaCl} = 10^{-3}$$
$$n_{CH_3COONa} = 1.5 \times 10^{-3}$$
$$n_{CH_3COOH}\ \text{(unreacted)} = 0.5 \times 10^{-3}$$

$\therefore$ Now the solution is a buffer solution.

$$= pK_a + \log \frac{(n_{CH_3COONa})}{(n_{CH_3COOH})}$$
$$= 4.75 + \log 3 = 4.75 + 0.4771$$
$$= 5.2271 \approx 5.23$$

**47.** *(10.60)* (*A*) 4 g NaOH in 100 L

$$\Rightarrow \quad M = \frac{4/40}{100} = 10^{-3}\ \text{mol L}^{-1}.$$

(*B*) 9.8 g $H_2SO_4$ in 100 L

$$\Rightarrow \quad M = \frac{9.8/98}{100} = 10^{-3}\ \text{mol L}^{-1}$$

40 L of solution *A* contains

$$= 40 \times 10^{-3}\ \text{mol NaOH}$$
$$(\because n = M \times V)$$
$$= 40 \times 10^{-3}\ \text{mol } OH^-$$

10 L of solution *B* contains

$$= 10 \times 10^{-3}\ \text{mol } H_2SO_4$$
$$= 20 \times 10^{-3}\ \text{mol } H^+$$

When they are mixed, total volume = 50 L

and all $H^+$ is neutralised by $OH^-$.

Number of moles ($n_{OH^-}$) of excess $OH^-$.

$$= (40 - 20) \times 10^{-3} = 20 \times 10^{-3}$$
$$[OH^-] = \frac{n_{OH^-}}{V_{solution}}\ \text{(in litre)}$$
$$\text{Or} \quad [OH^-] = \frac{n_{OH^-}}{\text{Total volume}}$$
$$= \frac{20 \times 10^{-3}}{50} = 4 \times 10^{-4}\ \text{M}$$

As we know,

$$pOH = -\log[OH^-]$$
$$\therefore \quad pOH = -\log 4 \times 10^{-4} = 4 - \log 4$$
$$\text{or} \quad pH = 14 - pOH$$
$$= 14 - (4 - \log 4)$$
$$= 10 + \log 4 = 10.602$$

# C-08

JEE Main Chapterwise Topicwise Chemistry

In JEE Main there is usually 0-1 question per paper are visible from this chapter. In past 3 years, 18 conceptual and 2 numerical questions were asked of easy to moderate level.
**For detailed analysis scan QR code**

# REDOX REACTIONS

## Topic 1 Oxidation-Reduction and Oxidation Number

### Level 1

**01** The standard reduction potential values of some of the $p$-block ions are given below. Predict the one with the strongest oxidising capacity. *[2025, 29 Jan Shift-I]*

(a) $E^{\ominus}_{Al^{3+}/Al} = -1.66$ V (b) $E^{\ominus}_{Pb^{4+}/Pb^{2+}} = +1.67$ V

(c) $E^{\ominus}_{Sn^{4+}/Sn^{2+}} = +1.15$ V (d) $E^{\ominus}_{Tl^{3+}/Tl} = +1.26$ V

**02** Based on the data given below *[2025, 24 Jan Shift-II]*

$E^{\circ}_{Cr_2O_7^{2-}/Cr^{3+}} = 1.33$ V; $E^{\circ}_{Cl_2/Cl^{(-)}} = 1.36$ V

$E^{\circ}_{MnO_4^-/Mn^{2+}} = 1.51$ V; $E^{\circ}_{Cr^{3+}/Cr} = -0.74$ V

the strongest reducing agent is

(a) $Cl^-$ (b) Cr

(c) $MnO_4^-$ (d) $Mn^{2+}$

**03** The reaction at cathode in the cells commonly used in clocks involves *[2024, 5 April Shift-I]*

(a) oxidation of Mn from + 2 to + 7
(b) oxidation of Mn from + 3 to + 4
(c) reduction of Mn from + 7 to + 2
(d) reduction of Mn from + 4 to + 3

**04** In alkaline medium, $MnO_4^-$ oxidises $I^-$ to *[2024, 29 Jan Shift-I]*

(a) $I_2$ (b) $IO_4^-$ (c) $IO_3^-$ (d) $IO^-$

**05** Which of the following options are correct for the reaction?

$$2[Au(CN)_2]^-(aq) + Zn(s) \longrightarrow 2Au(s) + [Zn(CN)_4]^{2-}(aq)$$

A. Redox reaction B. Displacement reaction
C. Decomposition reaction D. Combination reaction

Choose the correct answer from the options given below *[2023, 6 April Shift-I]*

(a) C and D only (b) A and B only
(c) A and D only (d) A only

**Imp.** Electrochemical Series

**06** The metal that cannot be obtained by the electrolysis of an aqueous solution of its salts is *[JEE Main 2014]*

(a) Ag (b) Ca (c) Cu (d) Cr

**07** Oxidation number of Cl in $CaOCl_2$ (bleaching powder) is *[AIEEE 2002]*

(a) zero, since it contains $Cl_2$
(b) −1, since it contains $Cl^-$
(c) +1, since it contains $ClO^-$
(d) +1 and −1, since it contains $ClO^-$ and $Cl^-$

### Level 2

**08** The species which does not undergo disproportionation reaction is *[2025, 22 Jan Shift-II]*

(a) $ClO_3^-$ (b) $ClO_2^-$
(c) $ClO^-$ (d) $ClO_4^-$

**09** Match the List-I with List-II. *[2025, 28 Jan Shift-I]*

| List-I (Redox reaction) | List-II (Type of redox reaction) |
|---|---|
| A. $CH_4(g) + 2O_2(g) \xrightarrow{\Delta} CO_2(g) + 2H_2O(l)$ | I. Disproportionation reaction |
| B. $2NaH(s) \xrightarrow{\Delta} 2Na(s) + H_2(g)$ | II. Combination reaction |
| C. $V_2O_5(s) + 5Ca(s) \xrightarrow{\Delta} 2V(s) + 5CaO(s)$ | III. Decomposition reaction |
| D. $2H_2O_2(aq) \xrightarrow{\Delta} 2H_2O(l) + O_2(g)$ | IV. Displacement reaction |

Choose the correct answer from the options given below

(a) A-II, B-III, C-IV, D-I (b) A-III, B-IV, C-I, D-II
(c) A-IV, B-I, C-II, D-III (d) A-II, B-III, C-I, D-IV

**10** Match List-I with List-II.

| List-I (Reaction) | List-II (Type of redox reaction) |
|---|---|
| A. $N_2(g)+O_2(g) \rightarrow 2NO(g)$ | I. Decomposition |
| B. $2Pb(NO_3)_2(s) \rightarrow 2PbO(s)+4NO_2(g)+O_2(g)$ | II. Displacement |
| C. $2Na(s)+2H_2O(l) \rightarrow 2NaOH(aq)+H_2(g)$ | III. Disproportionation |
| D. $2NO_2(g)+2\bar{O}H(aq) \rightarrow NO_2^-(aq)+NO_3^-(aq)+H_2O(l)$ | IV. Combination |

Choose the correct answer from the options given below. *[2024, 6 April Shift-II]*

| | A | B | C | D | | A | B | C | D |
|---|---|---|---|---|---|---|---|---|---|
| (a) | III | II | I | IV | (b) | IV | I | II | III |
| (c) | I | II | III | IV | (d) | II | III | IV | I |

**11** The number of ions from the following that have the ability to liberate hydrogen from a dilute acid is $Ti^{2+}$, $Cr^{2+}$ and $V^{2+}$. *[2024, 5 April Shift-II]*

(a) 1 (b) 0 (c) 2 (d) 3

**Imp.** Disproportionation Reactions

**12** Given below are two statements. *[2024, 31 Jan Shift-II]*

**Statement I** $S_8$ solid undergoes disproportionation reaction under alkaline conditions to form $S^{2-}$ and $S_2O_3^{2-}$.

**Statement II** $ClO_4^-$ can undergo disproportionation reaction under acidic condition.

In the light of the above statements, choose the most appropriate answer from the options given below

(a) Both statement I and Statement II are incorrect.
(b) Statement I is correct but Statement II is incorrect.
(c) Both Statement I and Statement II are correct.
(d) Statement I is incorrect but Statement II is correct.

**13** Which of the following cannot function as an oxidising agent? *[2024, 27 Jan Shift-II]*

(a) $SO_4^{2-}$ (b) $MnO_4^-$ (c) $N^{3-}$ (d) $BrO_3^-$

**14** Cerium (IV) has a noble gas configuration. Which of the following is correct statement about it? *[2022, 25 June Shift-I]*

(a) It will not prefer to undergo redox reactions.
(b) It will prefer to gain electron and act as an oxidising agent.
(c) It will prefer to give away an electron and behave as reducing agent.
(d) It acts as both, oxidising and reducing agent.

**15** Identify the process in which change in the oxidation state is five *[2021, 25 July Shift-II]*

(a) $Cr_2O_7^{2-} \longrightarrow 2Cr^{3+}$ (b) $MnO_4^- \longrightarrow Mn^{2+}$
(c) $CrO_4^{2-} \longrightarrow Cr^{3+}$ (d) $C_2O_4^{2-} \longrightarrow 2CO_2$

**16** The species given below that does not show disproportionation reaction is *[2021, 20 July Shift-I]*

(a) $BrO_4^-$ (b) $BrO^-$ (c) $BrO_2^-$ (d) $BrO_3^-$

**17** The compound that cannot act both as oxidising and reducing agent is *[2020, 9 Jan Shift-II]*

(a) $H_2SO_3$ (b) $H_3PO_4$ (c) $HNO_2$ (d) $H_2O_2$

**18** The oxidation state of chromium in the final product formed by the reaction between KI and acidified potassium dichromate solution is *[AIEEE 2005]*

(a) +3 (b) +2 (c) +6 (d) +4

## Level 3

**19** Thiosulphate reacts differently with iodine and bromine in the reactions given below :

$$2S_2O_3^{2-} + I_2 \longrightarrow S_4O_6^{2-} + 2I^-$$

$$S_2O_3^{2-} + 5Br_2 + 5H_2O \longrightarrow 2SO_4^{2-} + 4Br^- + 10H^+$$

Which of the following statements justifies the above dual behaviour of thiosulphate? *[2024, 8 April Shift-I]*

(a) Thiosulphate undergoes oxidation by bromine and reduction by iodine in these reactions
(b) Bromine is a weaker oxidant than iodine
(c) Bromine is a stronger oxidant than iodine
(d) Bromine undergoes oxidation and iodine undergoes reduction in these reaction.

**20** Which of the following reactions are disproportionation reactions?

A. $Cu^+ \longrightarrow Cu^{2+} + Cu$
B. $3MnO_4^{2-} + 4H^+ \longrightarrow 2MnO_4^- + MnO_2 + 2H_2O$
C. $2KMnO_4 \longrightarrow K_2MnO_4 + MnO_2 + O_2$
D. $2MnO_4^- + 3Mn^{2+} + 2H_2O \longrightarrow 5MnO_2 + 4H^+$

Choose the correct answer from the options given below *[2024, 1 Feb Shift-I]*

(a) A, D (b) A, B, C (c) B, C, D (d) A, B

**21** Which of the given reactions is not an example of disproportionation reaction ? *[2022, 26 July Shift-I]*

(a) $2H_2O_2 \longrightarrow 2H_2O + O_2$
(b) $2NO_2 + H_2O \longrightarrow HNO_3 + HNO_2$
(c) $MnO_4^- + 4H^+ + 3e^- \longrightarrow MnO_2 + 2H_2O$
(d) $3MnO_4^{2-} + 4H^+ \longrightarrow 2MnO_4^- + MnO_2 + 2H_2O$

**22** Which one of the following is an example of disproportionation reaction? *[2022, 26 June Shift-II]*

(a) $3MnO_4^{2-} + 4H^+ \longrightarrow 2MnO_4^- + MnO_2 + 2H_2O$
(b) $MnO_4^- + 4H^+ + 4e^- \longrightarrow MnO_2 + 2H_2O$
(c) $10I^- + 2MnO_4^- + 16H^+ \longrightarrow 2Mn^{2+} + 8H_2O + 5I_2$
(d) $8MnO_4^- + 3S_2O_3^{2-} + H_2O \longrightarrow 8MnO_2 + 6SO_4^{2-} + 2OH^-$

**23** An example of a disproportionation reaction is *[2019, 12 April Shift-I]*

(a) $2MnO_4^- + 10I^- + 16H^+ \longrightarrow 2Mn^{2+} + 5I_2 + 8H_2O$
(b) $2NaBr + Cl_2 \longrightarrow 2NaCl + Br_2$

(c) $2KMnO_4 \longrightarrow K_2MnO_4 + MnO_2 + O_2$
(d) $2CuBr \longrightarrow CuBr_2 + Cu$

**24** The redox reaction among the following is
(a) reaction of $H_2SO_4$ with NaOH *[2020, 7 Jan Shift-II]*
(b) reaction of $[Co(H_2O)_6]Cl_3$ with $AgNO_3$
(c) combination of dinitrogen with dioxygen at 2000 K
(d) formation of ozone from atmospheric oxygen in the presence of sunlight

**25** Which of the following reactions is an example of a redox reaction? *[JEE Main 2017]*
(a) $XeF_4 + O_2F_2 \longrightarrow XeF_6 + O_2$
(b) $XeF_2 + PF_5 \longrightarrow [XeF]^+PF_6^-$
(c) $XeF_6 + H_2O \longrightarrow XeOF_4 + 2HF$
(d) $XeF_6 + 2H_2O \longrightarrow XeO_2F_2 + 4HF$

## Numerical Value Type Questions

**26** The standard reduction potentials at 298 K for the following half cells are given below: *[2024, 9 April, Shift-I]*

$Cr_2O_7^{2-} + 14H^+ + 6e^- \longrightarrow 2Cr^{3+} + 7H_2O,$ $E° = 1.33$ V

$Fe^{3+}(aq) + 3e^- \longrightarrow Fe$ $E° = -0.04$V

$Ni^{2+}(aq) + 2e^- \longrightarrow Ni$ $E° = -0.25$ V

$Ag^+(aq) + e^- \longrightarrow Ag$ $E° = 0.80$ V

$Au^{3+}(aq) + 3e^- \longrightarrow Au$ $E° = 1.40$ V

Consider the given electrochemical reactions,
The number of metal(s) which will be oxidised by $Cr_2O_7^{2-}$, in aqueous solution is .........

**27** From the given list, the number of compounds with +4 oxidation state of sulphur is ........
$SO_3, H_2SO_3, SOCl_2, SF_4, BaSO_4, H_2S_2O_7$ *[2024, 27 April, Shift-I]*

**28** In alkaline medium, the reduction of permanganate anion involves a gain of ........ electrons. *[2023, 10 April, Shift-II]*

**29** In the titration of $KMnO_4$ and oxalic acid in acidic medium, the change in oxidation number of carbon at the end point is ......... . *[2022, 27 July Shift-I]*

**30** The oxidation states of transition metal atoms in $K_2Cr_2O_7$, $KMnO_4$ and $K_2FeO_4$, respectively, are $x$, $y$ and $z$. The sum of $x$, $y$ and $z$ is ......... . *[2020, 2 Sep Shift-II]*

# Topic 2 Balancing of Chemical Equations

## Level 1

**01** $2IO_3^- + xI^- + 12H^+ \longrightarrow 6I_2 + 6H_2O$
What is the value of $x$? *[2023, 8 April Shift-I]*
(a) 6 (b) 2 (c) 10 (d) 12

## Level 2

**02** In acidic medium, $K_2Cr_2O_7$ shows oxidising action as represented in the half reaction:
$$Cr_2O_7^{2-} + XH^+ + Ye^{\ominus} \longrightarrow 2A + Z\,H_2O$$
$X, Y, Z$ and $A$ are respectively are *[2024, 1 Feb Shift-I]*
(a) 8, 4, 6 and $Cr_2O_3$ (b) 8, 6, 4 and $Cr_2O_3$
(c) 14, 7, 6 and $Cr^{3+}$ (d) 14, 6, 7 and $Cr^{3+}$

**03** Consider the following reaction,
$$x\,MnO_4^- + y\,C_2O_4^{2-} + z\,H^+ \longrightarrow x Mn^{2+} + 2y\,CO_2 + \frac{z}{2}H_2O$$
The values of $x$, $y$ and $z$ in the reaction are, respectively *[JEE Main 2013]*
(a) 5, 2 and 16 (b) 2, 5 and 8
(c) 2, 5 and 16 (d) 5, 2 and 8

## Level 3

**04** Chlorine undergoes disproportionation in alkaline medium as shown below
$$a\,Cl_2(g) + b\,OH^-(aq) \longrightarrow c\,ClO^-(aq) + d\,Cl^-(aq) + e\,H_2O(l)$$
The values of $a$, $b$, $c$ and $d$ in a balanced redox reaction are respectively *[2024, 29 Jan Shift-I]*
(a) 1, 2, 1 and 1 (b) 2, 4, 1 and 3
(c) 3, 4, 4 and 2 (d) 2, 2, 1 and 3

## Numerical Value Type Questions

**05** $2MnO_4^- + bI^- + cH_2O \longrightarrow xI_2 + yMnO_2 + z\,\overline{O}H$
If the above equation is balanced with integer coefficients, the value of $z$ is ......... *[2024, 30 Jan Shift-I]*

**06** See the following chemical reaction
$$Cr_2O_7^{2-} + XH^+ + 6Fe^{2+} \longrightarrow YCr^{3+} + 6Fe^{3+} + ZH_2O$$
The sum of $X, Y$ and $Z$ is ...... . *[2023, 13 April Shift-II]*

**07** $2MnO_4^- + bC_2O_4^{2-} + cH^+ \longrightarrow x\,Mn^{2+} + y\,CO_2 + zH_2O$
If the above equation is balanced with integer coefficients, the value of $c$ is ......... . *[2021, 16 March Shift-I]*

**08** Consider the following equations :
$$2Fe^{2+} + H_2O_2 \longrightarrow xA + yB$$
(in basic medium)
$$2MnO_4^- + 6H^+ + 5H_2O_2 \longrightarrow x'C + y'D + z'E$$
(in acidic medium)
The sum of the stoichiometric coefficients $x$, $y$, $x'$, $y'$ and $z'$ for products $A$, $B$, $C$, $D$ and $E$, respectively, is ......... . *[2020, 4 Sep Shift-II]*

## Key Index

### Topic 1

| | | | | | | | | | |
|---|---|---|---|---|---|---|---|---|---|
| 1 (b) | 2 (b) | 3 (d) | 4 (c) | 5 (b) | 6 (b) | 7 (d) | 8 (d) | 9 (d) | 10 (b) |
| 11 (d) | 12 (b) | 13 (c) | 14 (b) | 15 (b) | 16 (a) | 17 (b) | 18 (a) | 19 (c) | 20 (d) |
| 21 (c) | 22 (a) | 23 (d) | 24 (c) | 25 (a) | 26 (3) | 27 (3) | 28 (3) | 29 (1) | 30 (19) |

### Topic 2

| | | | | | | | |
|---|---|---|---|---|---|---|---|
| 1 (c) | 2 (d) | 3 (c) | 4 (a) | 5 (8) | 6 (23) | 7 (16) | 8. (19) |

# Detailed Explanations

## Topic 1

**01.** *(b)* As the standard reduction potential value (positive) increases, oxidising capacity increases.
Hence, option (b) has highest standard reduction potential value.

**02.** *(b)* Among the given options the value of $E°$ for $Cr^{3+}/Cr$ is lowest. i.e,

$$E°_{Cr^{3+}/Cr} = -0.74 \text{ V}$$

Hence, Cr is the strongest reducing agent.

**Time Saver Tech**
Look for the option containing lowest value of reduction potential.
For strongest reducing agent, reduction potential should be lowest.

**03.** *(d)* In the cathode reaction of the cells that are commonly used in clocks. manganese (Mn) is reduced from +4 oxidation state to the +3 state.

**04.** *(c)* The reaction will be as follow,

$$2MnO_4^- + H_2O + I^- \xrightarrow[\text{Medium}]{\text{Alkaline}} 2MnO_2 + 2OH^- + IO_3^-$$

Therefore, $MnO_4^-$ oxidises $I^-$ to $IO_3^-$.
Hence, option (c) is correct.

**05.** *(b)* Consider the following reactions

$$\overset{(+1)}{2[Au(CN)_2]^-}(aq) + \overset{(0)}{Zn}(s) \longrightarrow \overset{(0)}{2Au}(s) + \overset{(+2)}{[Zn(CN)_4]^{2-}}(aq)$$

Reduction (Au: +1 → 0); Oxidation (Zn: 0 → +2)

Above reaction is redox reaction as well as displacement reaction.
Hence, option (b) is correct.

**06.** *(b)* Higher the position of element in the electrochemical series, more difficult is the reduction of its cation. Thus, among all the given elements, only calcium on electrolysis yields $H_2$ gas at cathode while other yields pure metals at cathode. In case of $Ca^{2+}$ ion, water is reduced in preference to calcium.

**For $Ca^{2+}$**
At cathode,

$$H_2O + e^- \longrightarrow \frac{1}{2}H_2 + OH^-$$

**For Cr**
At cathode, $Cr^{3+} + 3e^- \longrightarrow Cr$

**For Ag**
At cathode, $Ag^+ + e^- \longrightarrow Ag$

**For Cu**
At cathode, $Cu^{2+} + 2e^- \longrightarrow Cu$

**07.** *(d)* $CaOCl_2$ is written basically as Ca(OCl)Cl.

$OCl^- \longrightarrow$ Cl has + 1 oxidation state
$Cl^- \longrightarrow$ Cl has −1 oxidation state

**08.** *(d)* Compound containing chlorine in maximum oxidation state will not undergo disproportionation reaction.
So, among the given options, $ClO_4^-$ has chlorine in maximum possible oxidation state i.e. 7. Thus, it will not undergo disproportionation.

**09.** *(d)* The correct match is
A-II, B-III, C-I, D-IV
A : Combustion of hydrocarbon is combination reaction.
B : Decomposition into gaseous product.
C : Displacement of V by Ca atom.
D : Disproportionation of $H_2O_2$.

**10.** *(b)* The correct match is
A-IV, B-I, C-II, D-III

**Concept Enhancer**
All combustion reactions are combination reaction whether there is a use of elemental dioxygen or any other element.

**11.** *(d)* Strong reducing agents will liberate hydrogen from the dilute acid.
In given question, $Ti^{2+}$, $V^{2+}$, $Cr^{2+}$ are strong reducing agents.
e.g.,

$$2Cr^{2+}(aq) + 2H^+(aq) \rightarrow 2Cr^{3+}(aq) + H_2(g)$$

So, total number is 3.
Hence, option (d) is correct.

**12.** *(b)*

**key idea** Disproportionation reaction occurs in element which exhibits at least 3 oxidation states.

Statement I is correct but Statement II is incorrect.
The correct form of Statement II is as follow,
$ClO_4^-$ cannot undergo disproportionation reaction as chlorine is present in its highest oxidation state i.e., +7.

**13.** *(c)* $N^{3-}$ cannot function as an oxidising agent. In $N^{3-}$ ion 'N' is already present in its lowest possible oxidation state, hence it cannot be reduced further because of which it cannot act as an oxidising agent.
Hence, option (c) is correct.

**14.** *(b)* The electronic configuration of Ce $= [Xe]\, 6f^1 5d^1 6s^2$

Ce(IV) has stable noble gas [Xe] electronic configuration. Thus, it will not lose electrons any further so can't acts as reducing agent. However, its stable oxidation state is +3 hence, it instead prefer to gain electrons and act as oxidising agent in the redox reactions.

**15.** *(b)* $\overset{}{MnO_4^-} \longrightarrow Mn^{2+}$

$x - 8 + 1 = 0 \qquad x = +2$

$x = +7 \qquad (\therefore 5e^- \text{ change})$

Hence, in the process of $MnO_4^-$ to $Mn^{2+}$ the change in oxidation state is five.

**16.** *(a)* $BrO_4^-$ will not undergo disproportionation reaction because in it bromine is in its maximum oxidation state, i.e. + 7.

In option (b), we have oxidation state of Br = + 1

In option (c), we have oxidation state of Br = + 3

In option (d), we have oxidation state of Br = + 5

The maximum and minimum possible oxidation state value for Br are + 7 and +1 respectively. So, $BrO_4^-$ cannot oxidise further. Hence, it cannot show disproportionation reaction.

**17.** *(b)* In $H_3PO_4$, P is in its highest oxidation state (+5), it can only act as oxidising agent but not as reducing agents, because it can be reduced but not oxidised.

$H_2SO_3$ : S = +4 can get oxidised or reduced.

$HNO_2$ : N = +3 can get oxidised or reduced.

$H_2O_2$ : O = −1 can get oxidised or reduced.

**18.** *(a)* $Cr_2O_7^{2-} + 14H^+ + 6I^- \longrightarrow 2Cr^{3+} + 7H_2O + 3I_2$

$Cr_2O_7^{2-}$ is reduced to $Cr^{3+}$.

Thus, final state of Cr is + 3.

**19.** *(c)* In the reaction of $S_2O_3^{2-}$ with $I_2$, oxidation state of sulphur change from +2 to +2.5.

In the reaction of $S_2O_3^{2-}$ with $Br_2$, oxidation state of sulphur changes from +2 to +6.

$\therefore$ Both $I_2$ and $Br_2$ are oxidant (oxidising agent) however $Br_2$ is stronger oxidant than $I_2$.

**20.** *(d)* **Disproportionation reaction** is a type of reaction in which, the same element is simultaneously oxidised and reduced. Among the given reactions,

(A) $Cu^+ \longrightarrow Cu^{2+} + Cu$

(B) $3\overset{+6}{Mn}O_4^{2-} + 4H^+ \longrightarrow 2\overset{+7}{Mn}O_4^- + \overset{+4}{Mn}O_2 + 2H_2O$

these two reactions undergoes disproportionation.

**21.** *(c)*

**key idea** Species can undergo disproportionation reaction only, when its oxidation number lies between the maximum and minimum possible oxidation values.

Disproportionation reactions are the reactions in which an element undergoes oxidation and reduction simultaneously, forming two products (one with higher oxidation state and one with lower oxidation state respectively).

(a) $2H_2\overset{-1}{O_2} \longrightarrow 2H_2\overset{-2}{O} + \overset{0}{O_2}$

(b) $2\overset{+4}{N}O_2 + H_2O \longrightarrow H\overset{+5}{N}O_3 + H\overset{+3}{N}O_2$

(c) $\overset{+7}{Mn}O_4^- + 4H^+ + 3e^- \longrightarrow \overset{+4}{Mn}O_2 + 2H_2O$

(d) $3\overset{+6}{Mn}O_4^{2-} + 4H^+ \longrightarrow 2\overset{+7}{Mn}O_4^- + \overset{+4}{Mn}O_2 + 2H_2O$

So, it is only (c) reaction which does not show disproportionation reaction.

**22.** *(a)* A chemical reaction in which the same species undergoes both reduction and oxidation is called disproportionation reaction.

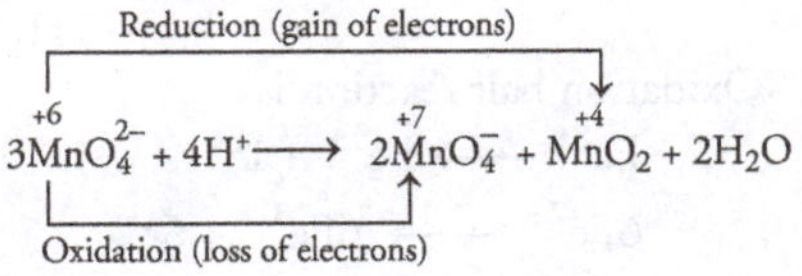


**23.** *(d)*

e.g.,

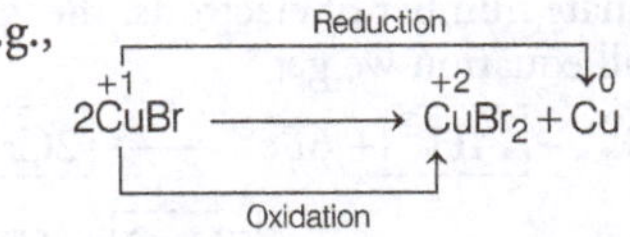


Here, CuBr get oxidised to $CuBr_2$ and also it get reduced to Cu.

**24.** *(c)*

(a) The reactions in the given options are as follows:

$\underset{(+6)}{H_2SO_4} + \underset{(+1)}{2NaOH} \longrightarrow \underset{(+1)\ (+6)}{Na_2SO_4} + 2H_2O$

(No change in oxidation state, so not a redox reaction).

(b) $\underset{(+3)}{[Co(H_2O)_6]Cl_3} + \underset{(+1)}{3AgNO_3} \longrightarrow \underset{(+3)}{[Co(H_2O)_6](NO_3)_3} + \underset{(+1)}{3AgCl\downarrow}$

(No change in oxidation state, so not a redox reduction).

(c) $\underset{(0)}{N_2} + \underset{(0)}{O_2} \xrightarrow{2000\ K} \underset{(+2)(-2)}{2NO}$

($N_2$ oxidised from 0 to +2 oxidation state and $O_2$ reduced from 0 to −2 oxidation state).

$\therefore$ This is redox reaction.

(d) $\underset{(0)}{3O_2} \xrightleftharpoons{h\nu} \underset{(0)}{2O_3}$

(No change in oxidation states, so not a redox reaction).

**25.** *(a)* The reaction in which oxidation and reduction occur simultaneously are termed as redox reaction.

$\overset{+4}{Xe}F_4 + \overset{+1}{O_2}F_2 \longrightarrow \overset{+6}{Xe}F_6 + \overset{0}{O_2}$

Since, Xe undergoes oxidation while O undergoes reduction. So, it is an example of redox reaction.

**26.** *(3)* Among the given metals, Fe, Ni and Ag will be oxidised by $Cr_2O_7^{2-}$ due to lower standard reduction potential values.

**27.** *(3)* Oxidation state of sulphur in given compound is calculated as,

$SO_3 \Rightarrow 0 = x - 6, x = 6$

$H_2SO_3$

$\Rightarrow 0 = 2 \times 1 + x - (2 \times 3), x = +4$

$SOCl_2 \Rightarrow 0 = x - 2 - (2 \times 1), x = +4$

$SF_4 \Rightarrow 0 = x - 4, x = +4$

$BaSO_4 \Rightarrow 0 = 2 + x - 8, x = +6$

$H_2S_2O_7 \Rightarrow 0 = 2 + 2x - 2 \times 7, x = +6$

Hence, in total 3 compounds sulphur have +4 oxidation state.

**28.** *(3)* In alkaline medium,

$MnO_4^- + 3e^- + 2H_2O \longrightarrow MnO_2 + 4OH^-$

Number of electrons gained = 3

**29.** *(1)* Oxidation state of carbon changes from +3 to + 4.

$2KMnO_4 + 5H_2\overset{+3}{C}_2O_4 + 3H_2SO_4\text{(dil.)}$ (Oxalic acid)

$\longrightarrow K_2SO_4 + 2MnSO_4 + 10\overset{+4}{C}O_2 + 8H_2O$

Change in oxidation state of carbon = 1.

**30.** *(19)* $K_2\overset{x}{Cr_2}O_7 \Rightarrow$

$2(+1)+2x+7\times(-2)=0 \Rightarrow x=+6$

$K\overset{y}{Mn}O_4$

$\Rightarrow 1\times(+1)+y+4\times(-2)=0$

$\Rightarrow y=+7$

$K_2\overset{z}{Fe}O_4$

$\Rightarrow 2\times(+1)+z+4\times(-2)=0$

$\Rightarrow z=+6$

So, $(x+y+z)=6+7+6=19$

## Topic 2

**01.** *(c)* Number of iodine on reactant side = Number of iodine on product side

$2+x=6\times 2$

$x=12-2$

$x=10$

Thus, the equation will be

$2IO_3^- + 10I^- + 12H^+ \longrightarrow 6I_2 + 6H_2O$

**02.** *(d)* $Cr_2O_7^{2-} + XH^+ + Ye^- \longrightarrow 2A + ZH_2O$ ...(i)

The balance reaction is,

$Cr_2O_7^{2-} + 14H^+ + 6e^- \longrightarrow 2Cr^{3+} + 7H_2O$ ...(ii)

Thus, $X=14, Y=6,$

$Z=7$ and $A=Cr^{3+}$

Hence, option (d) is correct.

**03.** *(c)* The half equations of the reaction are

$MnO_4^- \longrightarrow Mn^{2+}$

$C_2O_4^{2-} \longrightarrow CO_2$

The balanced half equations are

$(MnO_4^- + 8H^+ + 5e^- \longrightarrow Mn^{2+} + 4H_2O)\times 2$

$(C_2O_4^{2-} \longrightarrow 2CO_2 + 2e^-)\times 5$

Equating number of electrons, we get

$2MnO_4^- + 16H^+ + 10e^- \longrightarrow 2Mn^{2+} + 8H_2O$

$5C_2O_4^{2-} \longrightarrow 10CO_2 + 10e^-$

On adding both the equations, we get

$2MnO_4^- + 5C_2O_4^- + 16H^+ \longrightarrow 2Mn^{2+} + 2\times 5\,CO_2 + \frac{16}{2}H_2O$

Thus, x, y and z are 2, 5 and 16, respectively.

**04.** *(a)* $aCl_2(g) + bOH^-(aq) \longrightarrow cClO^-(aq) + dCl^-(aq) + eH_2O(l)$

$Cl_2 + 2e^- \longrightarrow 2Cl^-$ (Reduction half reaction)

$2H_2O + Cl_2 \longrightarrow 2ClO^- + 4H^+ + 2e^-$ (Oxidation half reaction)

On adding above two equations, we get

$2Cl_2 + 2H_2O \longrightarrow 2Cl^- + 2ClO^- + 4H^+$

$Cl_2 + 2OH^- \longrightarrow Cl^- + ClO^- + H_2O$

$a=1, b=2, c=1, d=1.$

Hence, option (a) is correct.

**05.** *(8)* $2MnO_4^- + bI^- + cH_2O \longrightarrow xI_2 + yMnO_2 + zOH^-$

Reduction half

$2MnO_4^- \longrightarrow 2MnO_2$

$2MnO_4^- + 4H_2O + 6e^- \longrightarrow 2MnO_2 + 8OH^-$

Oxidation half

$2I^- \longrightarrow I_2 + 2e^-$

$6I^- \longrightarrow 3I_2 + 6e^-$

Adding oxidation half reaction and reduction half reaction, net reaction is

$2MnO_4^- + 6I^- + 4H_2O \rightarrow 3I_2 + 2MnO_2 + 8OH^-$ ...(i)

Compare Eq. (i) with equation given in the question, we get

$b=6, c=4, x=3, y=2, z=8$

Hence, the value of z is 8.

**06.** *(23)* Reduction half reaction can be balanced as

$Cr_2O_7^{2-} + 14H^{\oplus} + 6e^{\ominus} \longrightarrow 2Cr^{3+} + 7H_2O$

Oxidation half reaction is

$Fe^{2+} \longrightarrow Fe^{3+} + 1e^{\ominus}$

$6Fe^{2+} \longrightarrow 6Fe^{3+} + 6e^-$

On combining both equation and eliminate number of electrons, the overall equation we get

$Cr_2O_7^{2-} + \boxed{14H^{\oplus}} + 6Fe^{\ominus} \longrightarrow \boxed{2Cr^{3+}} + \boxed{7H_2O} + 6Fe^{3+}$

Hence, $X=14, Y=2, Z=7$

Thus, the sum of

$X+Y+Z=14+2+7=23$

**07.** *(16)* Balancing the given reaction by half-reaction method

$2\overset{+7}{Mn}O_4^- + b\overset{+3}{C_2}O_4^{2-} + cH^+ \longrightarrow xMn^{+2} + y\overset{+4}{C}O_2 + zH_2O$

**Step 1** Separate two half-reactions

Oxidation half (O.H.)

$C_2O_4^{2-} \longrightarrow CO_2$

Reduction half (R.H.)

$MnO_4^- \longrightarrow Mn^{2+}$

**Step 2** Balance all the atoms except oxygen and 'H'.

O.H. $C_2O_4^{2-} \longrightarrow 2CO_2$

R.H. $MnO_4^- \longrightarrow Mn^{2+}$

**Step 3** Balance 'O' by adding $H_2O$ and 'H' by adding $H^+$.

O.H. $C_2O_4^{2-} \longrightarrow 2CO_2$

R.H.

$8H^+ + MnO_4^- \longrightarrow Mn^{2+} + 4H_2O$

**Step 4** Balance charges by adding electron

O.H. $C_2O_4^{2-} \longrightarrow 2CO_2 + 2e^-$

R.H. $5e^- + 8H^+ + MnO_4^- \longrightarrow Mn^{2+} + 4H_2O$

**Step 5** Balance the number of electrons by multiplying O.H. by '5' and R.H. by '2'.

O.H. $5C_2O_4^{2-} \longrightarrow 10CO_2 + 10e^-$

R.H. $10e^- + 16H^+ + 2MnO_4^- \longrightarrow 2Mn^{2+} + 8H_2O$

**Step 6** Add both half-reaction

$16H^+ + 5C_2O_4^{2-} + 2MnO_4^- \longrightarrow 10CO_2 + 2Mn^{2+} + 8H_2O$

So, $c=16$

**08.** *(19)* In basic medium, the balanced equation is given by

$2Fe^{2+} + H_2O_2 \longrightarrow 2Fe^{3+} + 2OH^-$

In acidic medium, the balanced equation is given by

$2MnO_4^- + 6H^+ + 5H_2O_2 \longrightarrow 2Mn^{2+} + 8H_2O + 5O_2$

$\therefore\ x=2, y=2, x'=2, y'=8, z'=5$

$\therefore\ x+y+x'+y'+z'=19$

# C-09

JEE Main Chapterwise Topicwise Chemistry

In JEE Main there is usually 0-1 question per paper are visible from this chapter. In past 3 years, 26 conceptual questions were asked from this chapter of easy to moderate level.

**For detailed analysis scan QR code**

# *p*-BLOCK ELEMENTS-1

## Topic 1 Group 13 Elements

### Level 1

**Imp.** Ionisation Enthalpy Order

**01** The correct order of the first ionisation enthalpy is *[2024, 4 April Shift-II]*

(a) B > Al > Ga (b) Ga > Al > B
(c) Al > Ga > Tl (d) Tl > Ga > Al

**02** The covalency and oxidation state respectively of boron in $[BF_4]^-$, are *[2023, 13 April Shift-II]*

(a) 4 and 3 (b) 3 and 5
(c) 3 and 4 (d) 4 and 4

**03** Given below are two statements.

**Statement I** Boron is extremely hard indicating its high lattice energy.

**Statement II** Boron has highest melting and boiling point compared to its other group members.

In the light of the above statements, choose the most appropriate answer from the options given below. *[2023, 12 April Shift-I]*

(a) Both Statement I and Statement II are correct.
(b) Statement I is incorrect but Statement II is correct.
(c) Statement I is correct but Statement II is incorrect.
(d) Both Statement I and Statement II are incorrect.

**04** Which one of the following elements will remain as liquid inside pure boiling water? *[2023, 6 April Shift-I]*

(a) Ga (b) Li
(c) Cs (d) Br

**05** The Lewis acid character of boron trihalides follows the order *[2023, 31 Jan Shift-II]*

(a) $BBr_3 > BI_3 > BCl_3 > BF_3$
(b) $BF_3 > BCl_3 > BBr_3 > BI_3$
(c) $BI_3 > BBr_3 > BCl_3 > BF_3$
(d) $BCl_3 > BF_3 > BBr_3 > BI_3$

**06** Given below are two statements.

**Statement I** The decrease in first ionisation enthalpy from B to Al is much larger than that from Al to Ga.

**Statement II** The *d*-orbitals in Ga are completely filled.

In the light of the above statements, choose the most appropriate answer from the options given below. *[2023, 29 Jan Shift-II]*

(a) Both the statements I and II are incorrect.
(b) Statement I is correct but statement II is incorrect.
(c) Both the statements I and II are correct.
(d) Statement I is incorrect but statement II is correct.

**07** Given below are two statements, one is labelled as Assertion A and the other is labelled as Reason R.

**Assertion (A)** Boron is unable to form $BF_6^{3-}$.

**Reason (R)** Size of B is very small.

In the light of the above statements, choose the correct answer from the options given below. *[2022, 27 July Shift-II]*

(a) Both A and R are true and R is the correct explanation of A.
(b) Both A and R are true but R is not the correct explanation of A.
(c) A is true but R is false.
(d) A is false but R is true.

**Imp.** Chemical Reaction of Boron in Blend with VSEPR

**08** The geometry around boron in the product '*B*' formed from the following reaction is *[2022, 25 July Shift-I]*

$$BF_3 + NaH \xrightarrow{450\text{ K}} A + NaF$$
$$A + NMe_3 \longrightarrow B$$

(a) trigonal planar (b) tetrahedral
(c) pyramidal (d) square planar

**09** The metal that has very low melting point and its periodic position is closer to a metalloid is *[2022, 26 July Shift-II]*

(a) Al (b) Ga (c) Se (d) In

**10** Choose the correct stability order of group 13 elements in their + 1 oxidation state. *[2022, 26 June Shift-I]*
(a) Al < Ga < In < Tl
(b) Tl < In < Ga < Al
(c) Al < Ga < Tl < In
(d) Al < Tl < Ga < In

**11** In which one of the following molecules strongest back donation of an electron pair from halide to boron is expected? *[2021, 27 Aug Shift-I]*
(a) $BCl_3$ (b) $BF_3$
(c) $BBr_3$ (d) $BI_3$

**12** The number of 2-centre-2-electron and 3-centre-2-electron bonds in $B_2H_6$, respectively, are *[2019, 10 Jan Shift-II]*
(a) 4 and 2 (b) 2 and 4
(c) 2 and 2 (d) 2 and 1

**13** The type of hybridisation of boron in diborane is *[AIEEE 2007]*
(a) $sp$ hybridisation
(b) $sp^2$ hybridisation
(c) $sp^3$ hybridisation
(d) $sp^3d^2$ hybridisation

## Level 2

**14** Given below are the pairs of group 13 elements showing their relation in terms of atomic radius. (B < Al), (Al < Ga), (Ga < In) and (In < Tl)
Identify the elements present in the incorrect pair and in that pair find out the element ($X$) that has higher ionic radius ($M^{3+}$) than the other one. The atomic number of the element ($X$) is *[2025, 4 April Shift-I]*
(a) 49 (b) 31
(c) 13 (d) 81

**15** The elements of Group 13 with highest and lowest first ionisation enthalpies are respectively: *[2025, 4 April Shift-II]*
(a) Tl and B (b) B and Ga
(c) B and In (d) B and Tl

**16** Given below are two statements. One is labelled as **Assertion (A)** and other is labelled as **Reason (R)**.
**Assertion (A)** The stability order of + 1 oxidation state of Ga, In and Tl is Ga < In < Tl.
**Reason (R)** The inert pair effect stabilises the lower oxidation state down the group. *[2024, 8 April Shift-I]*
In the light of the above statements, choose the correct answer from the options given below.
(a) A is false, but R is true.
(b) Both A and R are true but R is not the correct explanation of A.
(c) Both A and R are true and R is the correct explanation of A.
(d) A is true, but R is false.

**17** Given below are two statements : *[2024, 6 April Shift-I]*
**Statement I** Gallium is used in the manufacturing of thermometers.
**Statement II** A thermometer containing gallium is useful for measuring the freezing point (256 K) of brine solution.
In the light of the above statements. choose the correct answer from the options given below
(a) Statement I is true but Statement II is false.
(b) Statement I is false but Statement II is true.
(c) Both Statement I and Statement II are true.
(d) Both Statement I and Statement II are false.

**18** Given below are two statements. *[2024, 5 April Shift-I]*
**Statement I** In group 13, the stability of + 1 oxidation state increases down the group.
**Statement II** The atomic size of gallium is greater than that of aluminium.
In the light of the above statements, choose the most appropriate answer from the options given below:
(a) Both Statement I and Statement II are correct.
(b) Statement I is incorrect but Statement II is correct.
(c) Statement I is correct but Statement II is incorrect.
(d) Both Statement I and Statement II are incorrect.

**19** Given below are two statements. *[2024, 31 Jan Shift-II]*
**Statement I** Group 13 trivalent halides get easily hydrolysed by water due to their covalent nature.
**Statement II** $AlCl_3$ upon hydrolysis in acidified aqueous solution forms octahedral $[Al(H_2O)_6]^{3+}$ ion.
In the light of the above statements, choose the correct answer from the options given below
(a) Both Statement I and Statement II are false.
(b) Both Statement I and Statement II are true.
(c) Statement I is false but Statement II is true.
(d) Statement I is true but Statement II is false.

**20** Given below are two statements : one is labelled as **Assertion (A)** and the other is labelled as **Reason (R)**.
**Assertion (A)** Melting point of boron (245.3 K) is unusually high in group 13 elements. *[2024, 1 Feb Shift-I]*
**Reason (R)** Solid boron has very strong crystalline lattice.
In the light of the above statements, choose the **most appropriate** answer from the options given below
(a) Both (A) and (R) are correct and (R) is the correct explanation of (A).
(b) (A) is true but (R) is false.
(c) Both (A) and (R) are correct and (R) is not the correct explanation of (A).
(d) (A) is false but (R) is true.

**21** The incorrect statement from the following for borazine is **[2023, 13 April Shift-I]**

(a) it is a cyclic compound.
(b) it contains banana bonds.
(c) it has electronic delocalisation.
(d) it can react with water.

**22** For compound having the formula $GaAlCl_4$, the correct option from the following is **[2023, 11 April Shift-I]**

(a) Ga is more electronegative than Al and is present as a cationic part of the salt $GaAlCl_4$.
(b) Ga is coordinated with Cl in $GaAlCl_4$.
(c) Cl forms bond with both Al and Ga in $GaAlCl_4$.
(d) Oxidation state of Ga in the salt $GaAlCl_4$ is +3.

**23** Given below are two statements.

**Statement I** Upon heating a borax bead dipped in cupric sulphate in a luminous flame, the colour of the bead becomes green.

**Statement II** The green colour observed is due to the formation of copper (I) metaborate.

In the light of the above statements, choose the most appropriate answer from the options given below. **[2023, 31 Jan Shift-II]**

(a) Statement I is true but Statement II is false.
(b) Statement I is false but Statement II is true.
(c) Both Statement I and Statement II are true.
(d) Both Statement I and Statement II are false.

**24** Given below are two statements.

**Statement I** The chlorides of Be and Al have Cl-bridged structure. Both are soluble in organic solvents and act as Lewis bases.

**Statement II** Hydroxides of Be and Al dissolve in excess alkali to give beryllate and aluminate ions.

In the light of the above statements, choose the correct answer from the options given below. **[2022, 27 July Shift-I]**

(a) Both Statement I and Statement II are true.
(b) Both Statement I and Statement II are false.
(c) Statement I is true but Statement II is false.
(d) Statement I is false but Statement II is true.

**Imp.** Preparation of Borazine in Blend with Stoichiometry

**25** Borazine, also known as inorganic benzene, can be prepared by the reaction of 3-equivalents of "X" with 6-equivalents of "Y". "X" and "Y", respectively are **[2022, 26 July Shift-I]**

(a) $B(OH)_3$ and $NH_3$
(b) $B_2H_6$ and $NH_3$
(c) $B_2H_6$ and $NH_3$
(d) $NH_3$ and $B_2O_3$

**26** Aqueous solution of which of the following boron compounds will be strongly basic in nature? **[2022, 29 June Shift-II]**

(a) $NaBH_4$
(b) $LiBH_4$
(c) $B_2H_6$
(d) $Na_2B_4O_7$

**27** Given below are the statements about diborane.

(A) Diborane is prepared by the oxidation of $NaBH_4$ and $I_2$.
(B) Each boron atom is in $sp^2$-hybridised state.
(C) Diborane has one bridged 3 centre -2-electron bond.
(D) Diborane is a planar molecule.

The option with correct statement(s) is **[2021, 22 July Shift-II]**

(a) (C) and (D) only
(b) (A) only
(c) (C) only
(d) (A) and (B) only

**28** Given below are two statements: one is labelled as Assertion (A) and the other is labelled as Reason (R). **[2021, 26 Feb Shift-II]**

**Asertion (A)** In $TlI_3$, isomorphous to $CsI_3$, the metal is present in +1 oxidation state.

**Reason (R)** Tl metal has fourteen *f* electrons in its electronic configuration.

In the light of the above statements, choose the most appropriate answer from the options given below:

(a) Both A and R are correct but R is not the correct explanation of A.
(b) Both A and R are correct and R is the correct explanation of A.
(c) A is correct but R is not correct.
(d) A is not correct but R is correct.

**Imp.** Structure of Diborane

**29** The correct statement about $B_2H_6$ is **[2021, 25 Feb Shift-I]**

(a) all B—H—B angles are of 120°
(b) the two B—H—B bonds are not of same length
(c) terminal B—H bonds have less *p*-character when compared to bridging bonds
(d) Its fragment, $BH_3$, behaves as a Lewis base

**30** The reaction of $H_3N_3B_3Cl_3$ (*A*) with $LiBH_4$ in tetrahydrofuran gives inorganic benzene (*B*). Further, the reaction of (*A*) with (*C*) leads to $H_3N_3B_3(Me)_3$. Compounds (*B*) and (*C*) respectively, are **[2020, 9 Jan Shift-II]**

(a) borazine and MeBr
(b) diborane and MeMgBr
(c) boron nitride and MeBr
(d) borazine and MeMgBr

**31** The correct statements among I to III regarding group 13 element oxides are : **[2019, 9 April Shift-II]**

I. Boron trioxide is acidic.
II. Oxides of aluminium and gallium are amphoteric.
III. Oxides of indium and thallium are basic.

(a) I, II and III
(b) I and III only
(c) I and II only
(d) II and III only

**32** Diborane ($B_2H_6$) reacts independently with $O_2$ and $H_2O$ to produce, respectively **[2019, 8 April Shift-I]**

(a) $B_2O_3$ and $H_3BO_3$
(b) $B_2O_3$ and $[BH_4]^-$
(c) $H_3BO_3$ and $B_2O_3$
(d) $HBO_2$ and $H_3BO_3$

**33** Which one of the following order represents the correct sequence of the increasing basic nature of the given oxides? *[AIEEE 2011]*
(a) $MgO < K_2O < Al_2O_3 < Na_2O$
(b) $Na_2O < K_2O < MgO < Al_2O_3$
(c) $K_2O < Na_2O < Al_2O_3 < MgO$
(d) $Al_2O_3 < MgO < Na_2O < K_2O$

**34** Which one of the following is the correct statement ? *[AIEEE 2007]*
(a) Boric acid is a protonic acid.
(b) Beryllium exhibits coordination number of six.
(c) Chlorides of both beryllium and aluminium have bridged chloride structures in solid phase.
(d) $B_2H_6 \cdot 2NH_3$ is known as 'inorganic benzene'.

**35** Heating of an aqueous solution of aluminium chloride to dryness will give *[AIEEE 2005]*
(a) $Al(OH)Cl_2$ (b) $Al_2O_3$
(c) $Al_2Cl_6$ (d) $AlCl_3$

**36** Alum is widely used to purify water since
(a) it forms complex with clay particles
(b) it coagulates the mud particles
(c) it exchanges $Ca^{2+}$ and $Mg^{2+}$ ions present in hard water
(d) its sulphate ion is water purifier *[AIEEE 2002]*

## Level 3

**37** The correct statements from the following are:
(A) $Tl^{3+}$ is a powerful oxidising agent
(B) $Al^{3+}$ does not get reduced easily
(C) Both $Al^{3+}$ and $Tl^{3+}$ are very stable in solution
(D) $Tl^{+}$ is more stable than $Tl^{3+}$
(E) $Al^{3+}$ and $Tl^{+}$ are highly stable
Choose the correct answer from the options given below : *[2025, 7 April Shift-II]*
(a) (A), (B), (C), (D) and (E)
(b) (A), (C) and (D) only
(c) (A), (B), (D) and (E) only
(d) (B), (D) and (E) only

**Imp.** Trends in Properties of Group 13

**38** Match List-I with List-II. *[2024, 9 April Shift-II]*

| List-I | | List-II | |
|---|---|---|---|
| A. | Melting point [K] | I. | Tl > In > Ga > Al > B |
| B. | Ionic radius [$M^{3+}$/pm] | II. | B > Tl > Al ≈ Ga > In |
| C. | $\Delta_i H_1$ [kJ mol$^{-1}$] | III. | Tl > In > Al > Ga > B |
| D. | Atomic radius [pm] | IV. | B > Al > Tl > In > Ga |

Choose the correct answer from the options given below.
(a) A-II, B-III, C-IV, D-I
(b) A-III, B-IV, C-I, D-II
(c) A-I, B-II, C-III, D-IV
(d) A-IV, B-I, C-II, D-III

**39** The correct statements from the following are
(A) the decreasing order of atomic radii of group 13 elements is Tl > In > Ga > Al > B. *[2024, 5April Shift-II]*
(B) down the group 13 electronegativity decreases from top to bottom.
(C) Al dissolves in dil. HCl and liberates $H_2$ but conc. $HNO_3$ renders Al passive by forming a protective oxide layer on the surface.
(D) all elements of group 13 exhibits highly stable +1 oxidation state.
(E) hybridisation of Al in $[Al(H_2O_6)]^{3+}$ ion is $sp^3d^2$.
Choose the correct answer from the options given below.
(a) (C) and (E) only
(b) (A), (C) and (E) only
(c) (A), (B), (C) and (E) only
(d) (A) and (C) only

**40** Identify the correct statement for $B_2H_6$ from those given below :
(A) In $B_2H_6$, all B — H bonds are equivalent.
(B) In $B_2H_6$ there are four 3-centre-2-electron bonds.
(C) $B_2H_6$ is a Lewis acid.
(D) $B_2H_6$ can be synthesised from both $BF_3$ and $NaBH_4$.
(E) $B_2H_6$ is a planar molecule.
Choose the most appropriate answer from the options given below : *[2022, 24 June Shift-I]*
(a) (A) and (E) only
(b) (B), (C) and (E) only
(c) (C) and (D) only
(d) (C) and (E) only

**41** When metal '*M*' is treated with NaOH, a white gelatinous precipitate '*X*' is obtained, which is soluble in excess of NaOH. Compound '*X*' when heated strongly gives an oxide which is used in chromatography as an adsorbent. The metal '*M*' is *[JEE Main 2018]*
(a) Zn (b) Ca
(c) Al (d) Fe

**42** Boron can't form which one of the following anions? *[AIEEE 2011]*
(a) $BF_6^{3-}$ (b) $BH_4^-$
(c) $B(OH)_4^-$ (d) $BO_2^-$

**43** Aluminium chloride exists as dimer, $Al_2Cl_6$ in solid state as well as in solution of non-polar solvents such as benzene. When dissolved in water, it gives *[AIEEE 2004]*
(a) $Al^{3+} + 3Cl^-$
(b) $[Al(H_2O)_6]^{3+} + 3Cl^-$
(c) $[Al(OH)_6]^{3-} + 3HCl$
(d) $Al_2O_3 + 6HCl$

## Topic 2 Group 14 Elements

### Level 1

**01** Which of the following material is not a semiconductor? **[2024, 6 April Shift-I]**
(a) Silicon (b) Graphite (c) Germanium (d) Copper oxide

**02** Which one of the following compounds of group-14 elements is not known? **[2021, 25 July Shift-I]**
(a) $[GeCl_6]^{2-}$ (b) $[Sn(OH)_6]^{2-}$
(c) $[SiCl_6]^{2-}$ (d) $[SiF_6]^{2-}$

**03** The amorphous form of silica is **[2019, 9 April Shift-II]**
(a) tridymite (b) cristobalite (c) kieselguhr (d) quartz

**04** $C_{60}$ an allotrope of carbon contains **[2019, 9 April Shift-I]**
(a) 16 hexagons and 16 pentagons
(b) 20 hexagons and 12 pentagons
(c) 12 hexagons and 20 pentagons
(d) 18 hexagons and 14 pentagons

**05** The correct order of catenation is **[2019, 10 April Shift-I]**
(a) $C > Sn > Si \approx Ge$ (b) $Si > Sn > C > Ge$
(c) $C > Si > Ge \approx Sn$ (d) $Ge > Sn > Si > C$

**06** Which of the following exists as covalent crystals in the solid state? **[JEE Main 2013]**
(a) Iodine (b) Silicon (c) Sulphur (d) Phosphorus

### Level 2

**07** Given below are two statements: **[2025, 4 April Shift-II]**
**Statement (I):** The first ionisation enthalpy of group 14 elements is higher than the corresponding elements of group 13.
**Statement (II):** Melting points and boiling points of group 13 elements are in general much higher than those of corresponding elements of group 14.
In the light of the above statements, choose the most appropriate answer from the options given below:
(a) Both Statement I and Statement II are correct
(b) Statement I is incorrect but Statement II is correct
(c) Statement I is correct but Statement II is incorrect
(d) Both Statement I and Statement II are incorrect

**08** The group 14 elements $A$ and $B$ have the first ionisation enthalpy values of 708 and 715 kJ $mol^{-1}$ respectively. The above values are lowest among their group members. The nature of their ions $A^{2+}$ and $B^{4+}$ respectively is **[2025, 7 April Shift-I]**
(a) both oxidising (b) oxidising and reducing
(c) both reducing (d) reducing and oxidising

**Imp.** Oxidising Property in Blend with Inert Pair Effect

**09** The number of ions from the following that are expected to behave as oxidising agent is
$Sn^{4+}$, $Sn^{2+}$, $Pb^{2+}$, $Tl^{3+}$, $Pb^{4+}$, $Tl^{+}$ **[2024, 6 April Shift-II]**
(a) 1 (b) 3 (c) 4 (d) 2

**10** Given below are two statements. **[2024, 1 Feb Shift-II]**
**Statement I** $SiO_2$ and $GeO_2$ are acidic while SnO and PbO are amphoteric in nature.
**Statement II** Allotropic forms of carbon are due to property of catenation and $p\pi$ - $d\pi$ bond formation.
In the light of the above statements, choose the **most appropriate** answer from the options given below
(a) Both Statement I and Statement II are false.
(b) Statement I is true but Statement II is false.
(c) Statement I is false but Statement II is true.
(d) Both Statement I and Statement II are true.

**11** Consider the oxides of group 14 elements. $SiO_2$, $GeO_2$, $SnO_2$, $PbO_2$, CO and GeO. The amphoteric oxides are **[2024, 31 Jan Shift-I]**
(a) $SnO_2$, CO (b) $SnO_2$, $PbO_2$
(c) $SiO_2$, $GeO_2$ (d) GeO, $GeO_2$

**12** Given below are two statements.
**Statement I** The electronegativity of group 14 elements from Si to Pb, gradually decreases.
**Statement II** Group 14 contains non-metallic, metallic, as well as metalloid elements. **[2024, 29 Jan Shift-I]**
In the light of the above statements, choose the **most appropriate** from the options given below.
(a) Statement I is false but Statement II is true.
(b) Both Statement I and Statement II are false.
(c) Statement I is true but Statement II is false.
(d) Both Statement I and Statement II are true.

**13** Given below are two statements, one is labelled as Assertion (A) and the other is labelled as Reason (R).
**Assertion (A)** Carbon forms two important oxides —CO and $CO_2$, CO is neutral whereas $CO_2$ is acidic in nature. **[2023, 25 Jan Shift-II]**
**Reason (R)** $CO_2$ can combine with water in a limited way to form carbonic acid, while CO is sparingly soluble in water.
In the light of the above statements, choose the most appropriate answer from the options given below.
(a) Both A and R are correct but R is not the correct explanation of A.
(b) Both A and R are correct and R is the correct explanation of A.
(c) A is correct but R is not correct.
(d) A is not correct but R is correct.

**14** Given below are two statements. **[2022, 29 July Shift-II]**
**Statement I** Stannane is an example of a molecular hydride.
**Statement II** Stannane is a planar molecule.
In the light of the above statement, choose the most appropriate answer from the options given below
(a) Both Statement I and Statement II are true.
(b) Both Statement I and Statement II are false.
(c) Statement I is true but Statement II is false.
(d) Statement I is false but Statement II is true.

**15** The incorrect statement regarding the structure of $C_{60}$ is *[2021, 16 March Shift-II]*
(a) the six-membered rings are fused to both six and five-membered rings
(b) each carbon atom forms three sigma bonds
(c) the five-membered rings are fused only to six-membered rings
(d) it contains 12 six-membered rings and 24 five-membered rings

**Imp.** Oxidising Property in Blend with Lead Storage Battery

**16** Compound *A* used as a strong oxidising agent is amphoteric in nature. It is the part of lead storage batteries. Compound *A* is *[2021, 26 Feb Shift-I]*
(a) $Pb_3O_4$ (b) $PbSO_4$
(c) PbO (d) $PbO_2$

**17** Water does not produce CO on reacting with *[2021, 25 Feb Shift-II]*
(a) $CH_4$ (b) C (c) $CO_2$ (d) $C_3H_8$

**18** '*X*' melts at low temperature and is a bad conductor of electricity in both liquid and solid state. *X* is *[2020, 9 Jan Shift-II]*
(a) carbon tetrachloride (b) silicon carbide
(c) mercury (d) zinc sulphide

**19** The basic structural unit of feldspar, zeolites, mica and asbestos is *[2019, 12 April Shift-I]*
(a) $(SiO_3)^{2-}$ (b) $SiO_2$
(c) $(SiO_4)^{4-}$ (d) $-\!\!\left( \overset{R}{\underset{R}{|}}\!\!Si\!-\!O \right)_{\!n}\!\!-$ $(R = Me)$

**20** The correct statement among the following is *[2019, 12 April Shift-I]*
(a) $(SiH_3)_3N$ is planar and less basic than $(CH_3)_3N$.
(b) $(SiH_3)_3N$ is pyramidal and more basic than $(CH_3)_3N$.
(c) $(SiH_3)_3N$ is pyramidal and less basic than $(CH_3)_3N$.
(d) $(SiH_3)_3N$ is planar and more basic than $(CH_3)_3N$.

**21** The number of pentagons in $C_{60}$ and trigons (triangles) in white phosphorus, respectively, are *[2019, 10 April Shift-II]*
(a) 20 and 3 (b) 12 and 4
(c) 20 and 4 (d) 12 and 3

**22** The element that shows greater ability to form $p\pi$-$p\pi$ multiple bonds, is *[2019, 12 Jan Shift-II]*
(a) Ge (b) Si (c) Sn (d) C

**23** Among the following substituted silanes the one which will give rise to cross linked silicone polymer on hydrolysis is *[AIEEE 2007]*
(a) $R_4Si$ (b) $RSiCl_3$
(c) $R_2SiCl_2$ (d) $R_3SiCl$

**24** Graphite is a soft solid lubricant extremely difficult to melt. The reason for this anomalous behaviour is that graphite *[AIEEE 2003]*
(a) is a non-crystalline substance
(b) is an allotropic form of diamond
(c) has molecules of variable molecular masses like polymers
(d) has carbon atoms arranged in large plates of rings of strongly bound carbon atoms with weak interplate bonds

## Level 3

**25** Evaluate the following statements related to group 14 elements for their correctness. *[2024, 6 April Shift-II]*
A. Covalent radius decreases down the group from C to Pb in a regular manner.
B. Eletronegativity decreases from C to Pb down the group gradually.
C. Maximum covalance of C is 4 whereas other elements can expand their covalence due to presence of *d*-orbitals.
D. Heavier elements do not from $p\pi$-$p\pi$ bonds.
E. Carbon can exhibit negative oxidation states.

Choose the correct answer from the options given below.
(a) C and D only (b) C, D and E only
(c) A and B only (d) A, B and C only

**26** Match List-I with List-II.

| List-I (Metal) | List-II (Application) |
|---|---|
| (A) Cs | (I) High temperature thermometer |
| (B) Ga | (II) Water repellent sprays |
| (C) B | (III) Photoelectric cells |
| (D) Si | (IV) Bullet proof vest |

Choose the most appropriate answer from the options given below *[2022, 29 June Shift-I]*
(a) (A)-(III), (B)-(I), (C)-(IV), (D)-(II)
(b) (A)-(IV), (B)-(III), (C)-(II), (D)-(I)
(c) (A)-(II), (B)-(III), (C)-(IV), (D)-(I)
(d) (A)-(I), (B)-(IV), (C)-(II), (D)-(III)

**27** Correct statements among (I) to (IV) regarding silicones are *[2019, 9 Jan Shift-I]*
I. They are polymers with hydrophobic character.
II. They are biocompatible.
III. In general, they have high thermal stability and low dielectric strength.
IV. Usually, they are resistant to oxidation and used as greases.
(a) I and II only
(b) I, II, III only
(c) I, II, III and IV
(d) I, II and IV only

**28** The stability of dihalides of Si, Ge, Sn and Pb increases steadily in the sequence *[AIEEE 2007]*

(a) $GeX_2 < SiX_2 < SnX_2 < PbX_2$
(b) $SiX_2 < GeX_2 < PbX_2 < SnX_2$
(c) $SiX_2 < GeX_2 < SnX_2 < PbX_2$
(d) $PbX_2 < SnX_2 < GeX_2 < SiX_2$

**29** In silicon dioxide *[AIEEE 2005]*

(a) there are double bonds between silicon and oxygen atoms
(b) silicon atom is bonded to two oxygen atoms
(c) each silicon atom is surrounded by two oxygen atoms and each oxygen atom is bonded to two silicon atoms
(d) each silicon atom is surrounded by four oxygen atoms and each oxygen atom is bonded to two silicon atoms

**30** The soldiers of Napolean army while at Alps during freezing winter suffered a serious problem as regards to the tin buttons of their uniforms. White metallic tin buttons got converted to grey powder. This transformation is related to *[AIEEE 2004]*

(a) a change in the crystalline structure of tin
(b) an interaction with nitrogen of the air at very low temperature
(c) a change in the partial pressure of oxygen in the air
(d) an interaction with water vapour contained in the humid air

## • Key Index

### Topic 1

| | | | | | | | | | |
|---|---|---|---|---|---|---|---|---|---|
| 1 (d) | 2 (a) | 3 (a) | 4 (a) | 5 (c) | 6 (c) | 7 (b) | 8 (b) | 9 (b) | 10 (a) |
| 11 (b) | 12 (a) | 13 (c) | 14 (b) | 15 (c) | 16 (c) | 17 (a) | 18 (c) | 19 (b) | 20 (a) |
| 21 (b) | 22 (a) | 23 (d) | 24 (d) | 25 (b) | 26 (d) | 27 (b) | 28 (a) | 29 (c) | 30 (d) |
| 31 (a) | 32 (a) | 33 (d) | 34 (c) | 35 (b) | 36 (b) | 37 (c) | 38 (d) | 39 (a) | 40 (c) |
| 41 (c) | 42 (a) | 43 (b) | | | | | | | |

### Topic 2

| | | | | | | | | | |
|---|---|---|---|---|---|---|---|---|---|
| 1 (b) | 2 (c) | 3 (c) | 4 (b) | 5 (c) | 6 (b) | 7 (c) | 8 (d) | 9 (d) | 10 (b) |
| 11 (b) | 12 (a) | 13 (b) | 14 (c) | 15 (d) | 16 (d) | 17 (c) | 18 (a) | 19 (c) | 20 (a) |
| 21 (b) | 22 (d) | 23 (b) | 24 (d) | 25 (b) | 26 (a) | 27 (d) | 28 (c) | 29 (d) | 30 (a) |

# Detailed **Explanations**

## Topic 1

**01.** *(d)* The values of 1st ionisation enthalpy of Tl, Ga and Al are 589 kJ/mol, 579 kJ/mol and 577 kJ/mol respectively. Thus, the correct order of first ionisation enthalpy is

$$Tl > Ga > Al$$

**02.** *(a)* As 4 fluorine atoms are bonded, hence covalency = 4

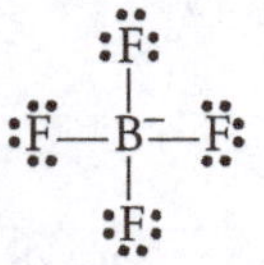


Let oxidation number of boron = $x$

So, $x + 4(-1) = -1$ or, $x = 3$

**03.** *(a)* Both Statement I and Statement II are correct.

Boron is non-metallic in nature. It is extremely hard and black coloured solid. Due to its small atomic size has very strong crystalline lattice. Thus, it has high melting and boiling point.

**04.** *(a)* Gallium (Ga) will remain as liquid inside pure boiling water. It has large liquid range, melting at 29.78°C and boiling at 2400°C, at atmospheric pressure.

**05.** *(c)* Extent of back bonding decreases down the group leading to more Lewis acidic strength. The order of extent of back bonding is

$$\underset{(2p\text{-}2p)}{BF_3} > \underset{(2p\text{-}3p)}{BCl_3} > \underset{(2p\text{-}4p)}{BBr_3} > \underset{(2p\text{-}5p)}{BI_3}$$

Thus, the order of Lewis acidic character is

$$BF_3 < BCl_3 < BBr_3 < BI_3$$

**06.** *(c)* Both the Statements I and II are correct.

The first ionisation enthalpy values (in kJ $mol^{-1}$) of group 13 elements are

| B | Al | Ga | In | Tl |
|---|---|---|---|---|
| 801 | 577 | 579 | 558 | 589 |

On moving down the group, ionisation enthalpy generally decreases due to an increase in the atomic size and decrease in effective nuclear charge. So, from the I.E. values it is clear that decrease in 1st ionisation energy from B to Al is much greater that from Al to Ga. Also, the $d$-orbital in Ga are fully-filled.

**07.** *(b)* Since boron is 2nd period element and cannot expand its octet due to unavailability of $d$-orbitals, it cannot form $[BF_6]^{3-}$. Also the size of boron is very small.

Hence, both A and R are correct but R is not the correct explanation of A.

**08.** *(b)* Complete and balanced reaction is

$$2BF_3 + 6NaH \xrightarrow{450\text{ K}} \underset{\substack{\text{(Diborane)}\\(A)}}{B_2H_6} + 6\,NaF$$

$$B_2H_6 + 2NMe_3 \longrightarrow \underset{(B)}{2BH_3 \cdot NMe_3}$$

$BH_3$ is electron deficient, so $NMe_3$ forms a coordinate bond with it.

Structure =

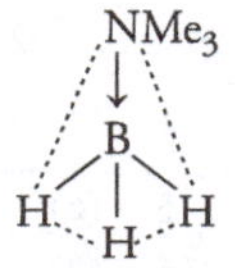


The geometry around boron in the product is tetrahedral.

**09.** *(b)*

| Metal | | Melting point |
|---|---|---|
| Al | → | 933 K |
| Ga | → | 303 K |
| In | → | 430 K |
| Se | → | 490 K |

Gallium has low melting point and its periodic position is closer to metalloids.

**10.** *(a)* On moving down the 13th group from B to Tl, stability of elements showing +1 oxidation state increases. This is due to inert pair effect. This inert pair effect increases down the group due to increase in atomic size. Hence, stability order of group 13 elements in their +1 oxidation state is Al < Ga < In < Tl.

> • Concept **Enhancer**
>
> Inert pair effect is the tendency of the two electron present in the valence $s$-orbital to remain unbonded.

**11.** *(b)* Smaller the size of halide atom, more effective is the back donation of an electron pair from halide atom to Boron.

So, $BF_3$ will show stronger back donation to B.

**12.** *(a)* The structure of $B_2H_6$ can be shown as :

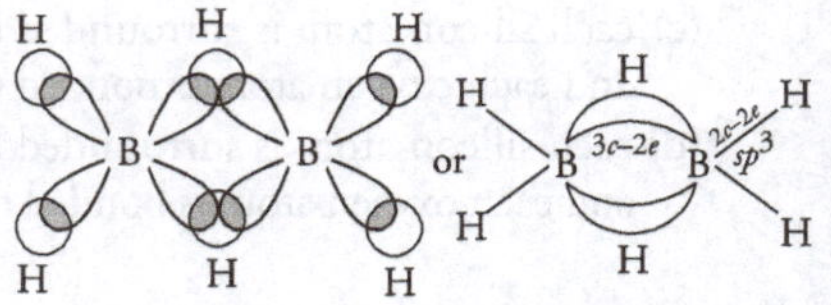


In $B_2H_6$, four 2-centre-2-electron $(2c-2e)$ bonds are present in the same plane and two 3-centre-2-electron $(3c-2e)$ bonds are present in another plane.

**13.** *(c)* Boron atom in $B_2H_6$(diborane) is $sp^3$-hybridised.

**14.** *(b)* Incorrect pair is given is (Al < Ga)

Its correct order is Al > Ga or $Al^{3+} < Ga^{3+}$

So, the atomic number of Ga is 31.

**15.** *(c)* The order of first ionisation enthalpy of group 13 elements are B > Tl > Ga > Al > In.

Thus, highest and lowest first ionisation enthalpies are of B and In respectively.

**16.** *(c)* Both A and R are true and R is the correct explanation of A.

**17.** *(a)* Statement I is true but Statement II is false.

The correct form of Statement II is, Thermometer containing gallium is not used to measure are freezing point of brine solution instead, gallium is used to measure high temperature.

**18.** *(c)* Statement I is correct but Statement II is incorrect.

The correct form of Statement II is, The atomic size of aluminium is greater than that of gallium.

**19.** *(b)* Both Statement I and Statement II are true.

**20.** *(a)* Both (A) and (R) are correct and (R) is the correct explanation of A.

Boron is non-metallic in nature.

Due to very strong crystalline lattice, boron has usually high melting point (245.3 K).

Rest of the members of group 13 element are soft metal with low melting point.

**21.** *(b)* Borazine is $B_3N_3H_6$.

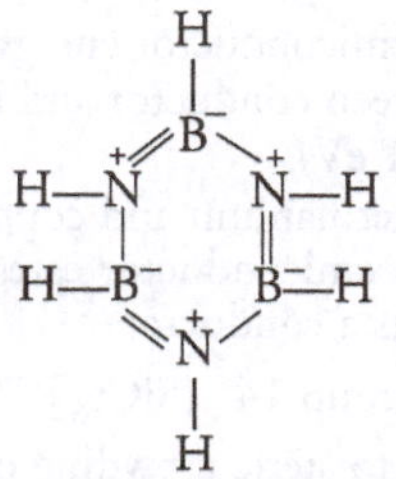

Due to ($p\pi - p\pi$) back bonding, nitrogen donates its lone pair to empty orbital of boron and hence forming double bond, leads to its resemblance with benzene.
It slowly reacts with water and gives $H_3BO_3$.

$$B_3N_3H_6 + H_2O\text{ (excess)} \longrightarrow NH_3 + H_3BO_3 + H_2\uparrow$$

But it does not contain banana bond. Hence option (b) is an incorrect statement.

**22.** *(a)* Structure of $Ga^+AlCl_4^-$ is

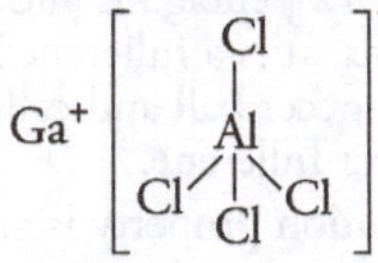

Thus, Ga is cationic part in the given salt. Also it is more electronegative than Al. According to Pauling scale, Al has an electronegativity of 1.61 and Ga has an electronegativity of 1.81.
So, option (a) is correct.

**23.** *(d)* Both Statement I and Statement II are false. Borax ($B_2O_3$) on treatment with metal salt forms metaborate of the metal which gives different colours in oxidising and reducing flame.
With cupric sulphate borax gives blue green flame due to formation of cupric i.e., copper (II) metaborate.

$$CuSO_4 + B_2O_3 \xrightarrow[\text{flame}]{\text{Non-luminous}} \underset{\substack{\text{Cupric metaborate}\\\text{(blue-green)}}}{Cu(BO_2)_2} + SO_3$$

**24.** *(d)* The chlorides of both beryllium and aluminium have Cl - bridged structure in vapour phase.
Both the chlorides are soluble in organic solvents and are strong Lewis acids as both Be and Al are electron defficient centres.
Hence, statement I is false but statement II is true.

**25.** *(b)* The formula of borazine or inorganic benzene is $B_3N_3H_6$. Its structure is as follows

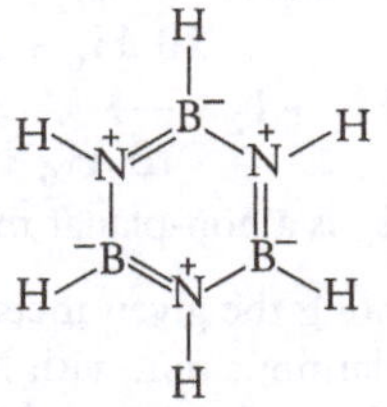

It is prepared by heating diborane ($B_2H_6$) with ammonia at temperature of 250-300°C.

$$3B_2H_6 + 6NH_3 \xrightarrow[250-300^\circ C]{\Delta} 2B_3H_6N_3 + 12H_2$$

**26.** *(d)* $NaBH_4 + H_2O \rightarrow NaBO_2 + H_2\uparrow$

$$LiBH_4 + H_2O \rightarrow LiBO_2\cdot 2H_2O + H_2\uparrow$$

$$B_2H_6 + H_2O \rightarrow \underset{\text{(Weak acid)}}{H_3BO_3}$$

$$Na_2B_4O_7 + H_2O \rightarrow \underset{\text{Strong base}}{NaOH} + \underset{\text{Weak acid}}{H_3BO_3}$$

∴ Aqueous solution of $Na_2B_4O_7$ will be strongly basic in nature due to formation of NaOH which is a strong base.
Hence, option (d) is correct.

**27.** *(b)* Statement (A) is correct whereas all other statements are incorrect.

(A) The oxidation of sodium borohydride with iodine in diglyme gives diborane.

$$I_2 + 2NaBH_4 \xrightarrow{\text{Diglyme}} 2NaI + \underset{\text{(Diborane)}}{B_2H_6} + H_2$$

(B)

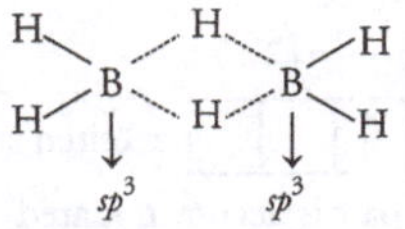

Hybridisation of boron in $B_2H_6$ is $sp^3$.

(C) $B_2H_6$ has two bridged, 3-centre-2-electron bond.

(D) $B_2H_6$ is a non-planar molecule.

**28.** *(a)* Thallium is more stable in +1 oxidation state due to inert pair effect.
So, $TlI_3$ and $CsI_3$ are isomorphous.
Electronic configuration of thallium $= [Xe]\,4f^{14}5d^{10}6s^2 6p^1$.

Hence, both A and R are correct but R is not the correct explanation of A.

**• Concept Enhancer**

Inertness of $ns^2$ electrons of the valence shell to participate in bonding on moving down the group in heavier *p*-block elements is called inert pair effect. It occurs due to poor shielding of the $ns^2$ electrons of the valence shell by the intervening *d* and *f*-electrons.

**29.** *(c)* Statement (c) is correct, whereas all other statements are incorrect.

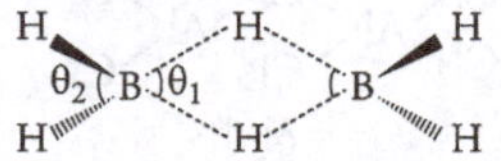

$\theta_2 > \theta_1$

The corrected form of other statements are as follows

- All B—H—B angles are of 90°
- B— H—B bond lengths are same.
- $BH_3$ is electron deficient species and therefore act as Lewis acid.

**30.** *(d)* $\underset{(A)}{H_3N_3B_3Cl_3} + 3LiBH_4 \xrightarrow{\text{THF}} \underset{(B)}{B_3N_3H_6} + 3LiCl + 3BH_3$

$$\underset{(A)}{H_3N_3B_3Cl_3} + \underset{(C)}{3MeMgBr} \longrightarrow H_3N_3B_3(Me)_3 + 3MgBrCl$$

So, $B = B_3N_3H_6$; $C = MeMgBr$

**31.** *(a)* All the given statements are correct. For group 13 elements, the acidic nature of oxides decreases and the basic nature of oxides increases on moving from B to Tl. While, $Al_2O_3$ and $Ga_2O_3$ are amphoteric oxides.

**32.** *(a)* Diborane ($B_2H_6$) reacts independently with $O_2$ and $H_2O$ to produce $B_2O_3$ and $H_3BO_3$ respectively.
The reaction involved is as follows

$$B_2H_6 + 3O_2 \longrightarrow B_2O_3 + 3H_2O + 1976\text{ kJ/mol}$$

It gets hydrolysed readily to give boric acid.

$$\underset{\text{Borane}}{B_2H_6} + 6H_2O \longrightarrow \underset{\text{Orthoboric acid}}{2H_3BO_3} + \underset{\text{Dihydrogen}}{6H_2\uparrow}$$

**33.** *(d)* Metallic character decreases on moving along a period in the sequence as Na > Mg > Al and increases down the group.
As metallic character of the element attached to oxygen atom increases, the basic character of oxides increases.

Hence, the increasing correct order of basic nature is

$$Al_2O_3 < MgO < Na_2O < K_2O.$$

**34.** *(c)* Chlorides of both beryllium and aluminium have bridged structures in solid phase.

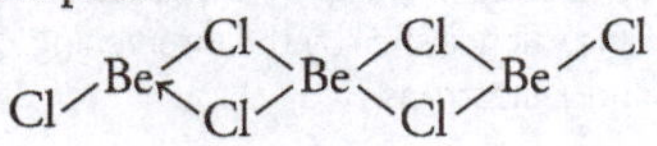

$BeCl_2$

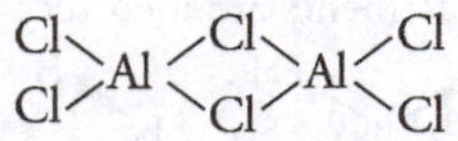

$AlCl_3$

**35.** *(b)* Aqueous solution of $AlCl_3$ is acidic, as it gives HCl on hydrolysis.

$$AlCl_3 + 3H_2O \rightleftharpoons Al(OH)_3 + 3HCl$$

On strongly heating, $Al(OH)_3$ is converted into $Al_2O_3$.

$$2Al(OH)_3 \xrightarrow{\Delta} Al_2O_3 + 3H_2O$$

**36.** *(b)* Mud is colloidal charged solution which on treatment with alum ($Al^{3+}$) gets neutralised or coagulated.

**37.** *(c)* Among the given statements, A, B, D and E are correct, whereas C is incorrect.

The correct form of C is

$Tl^{3+}$ is unstable because it is a powerful oxidising agent.

**38.** *(d)* The correct match is

A-IV, B-I, C-II, D-III

**39.** *(a)* Statements given in C and E are correct, while A, B, D are incorrect. Their correct form are as follows

A : Size order :

$$Tl > In > Al > Ga > B$$

B : Electronegativity order :

$$B > Al < Ga < In < Tl$$

D : B and Al are more stable in +3 oxidation state.

Hence, option (a) is correct.

**40.** *(c)* The structure of $B_2H_6$ is given below

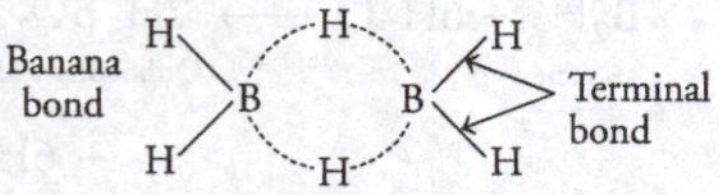


(A) In $B_2H_6$, all B—H bonds are not equivalent.

(B) There are two types of bonds present in $B_2H_6$, four normal covalent bonds and two 3-centre-2-electrons bonds.

(C) $B_2H_6$ is electron deficient and able to accept a pair of electron. Hence, acts as Lewis acid.

(D) $B_2H_6$ can be synthesised from both $BF_3$ and $NaBH_4$ as follows :

$$BF_3 + LiAlH_4 \longrightarrow 2B_2H_6 + 3LiF + 3AlF_3$$

$$NaBH_4 + I_2 \longrightarrow B_2H_6 + 2NaI + H_2$$

(E) $B_2H_6$ is a non-planar molecule.

**41.** *(c)* Among the given metals Al forms white gelatinous ppt. with NaOH. Hence, the probable metal can be Al. This ppt. is dissolved in excess of NaOH due to the formation of sodium metal aluminate. Both the reactions are shown below.

$$Al^{3+} \xrightarrow{NaOH} Al(OH)_3 \downarrow$$

White gelatinous ppt. (*X*) of aluminium hydroxide

$\downarrow$ Excess of NaOH

$NaAlO_2$ (Sodium metaaluminate) soluble

Aluminium hydroxide on strong heating gives alumina ($Al_2O_3$) which is used as an adsorbent in chromatography. This reaction can be seen as :

$$2Al(OH)_3 \xrightarrow{\Delta} Al_2O_3 + 3H_2O$$

Thus, metal *M* is Al.

**42.** *(a)* The filling of electron in the orbital of boron takes place as follows

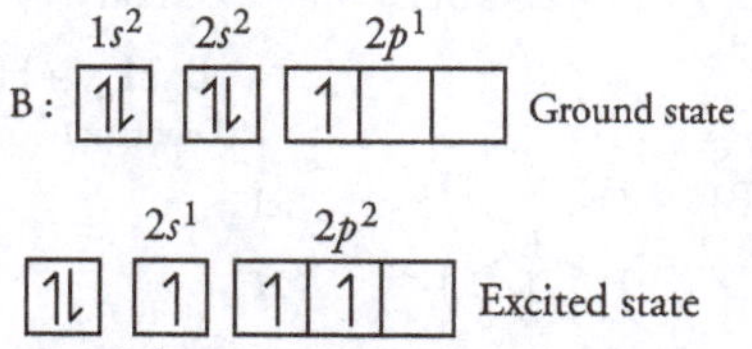


Fourth lone pair is accommodated in this empty orbital

Maximum covalency = 4

Due to absence of 2*d*-orbital, maximum covalency is four.

Therefore, (a) is not formed while $BH_4^-$ ($BH_3 + H^-$), $B(OH)_4^-$, ($B[OH]_3 + OH^-$] and $BO_2^-$ are formed.

**43.** *(b)* $AlCl_3$ is covalent but in water, it become ionic due to large hydration energy of $Al^{3+}$.

$$AlCl_3 + 6H_2O \rightleftharpoons [Al(H_2O)_6]^{3+} + 3Cl^-$$

## Topic 2

**01.** *(b)* In semiconductor energy band gap falls between conductor and insulator (around 1 eV).

Silicon, germanium and copper oxide falls into semiconductor category, while graphite is a conductor.

**02.** *(c)* In group 14, $[SiCl_6]^{2-}$ does not exist due to steric crowding of surrounding atoms.

The main reasons are

(i) Six large chloride ions cannot be accommodated around $Si^{4+}$.

(ii) Interaction between lone pair of chloride ion and $Si^{4+}$ is not very strong.

**03.** *(c)* Silica occurs in nature in several amorphous and crystalline forms. Kieselguhr is the amorphous form of silica. Quartz, tridymite and cristobalite are crystalline forms of silica.

**04.** *(b)* $C_{60}$ is aromatic allotrope of carbon containing 12 pentagons and 20 hexagons. It is a fullerene having a shape like soccer ball and called Buckminster fullerene.

**05.** *(c)* Catenation property is an unique property of group 14 elements. On moving down the group 14, catenation power decreases as :

$$C > Si > Ge \approx Sn$$

Pb does not show catenation.

**06.** *(b)* Silicon exists as covalent crystal in solid state. It exhibits a network structure like diamond.

**07.** *(c)* Statement I is correct but Statement II is incorrect. The correct form of Statement II is melting point of group 14 elements is more than group 13 elements.

**08.** *(d)* Given, 1st ionisation enthalpy of group 14 elements *A* and *B*.

$A = 708$ kJ/mol, element is Sn.

$B = 715$ kJ/mol, element is Pb.

$A^{2+} = Sn^{2+} =$ Reducing agent.

$B^{4+} = Pb^{4+} =$ Oxidising agent.

**09.** *(d)* The number of ions from the following which are expected to behave as oxidising agent are two, namely $Tl^{3+}$ and $Pb^{4+}$. Due to inert pair effect; $Tl^{3+}$ and $Pb^{3+}$ can behave as oxidising agent.

**10.** *(b)* Statement I is true but Statement II is false.

The correct form of Statement II is :

Due to property of catenation and $p\pi - p\pi$ bond formation carbon shows allotropic forms.

**11.** *(b)* Among the given options only $SnO_2$, $PbO_2$ are amphoteric oxides while $SiO_2$, $GeO_2$ and GeO are acidic oxide and CO is a neutral oxide.

**12.** *(a)* Statement I is false, but Statement II is true. The correct form of Statement I is : The electronegativity value of group 14 elements from Si to Pb are almost same. Hence, option (a) is correct.

**13.** *(b)* Both A and R are correct and R is the correct explanation of A.

**14.** *(c)* Stannane $SnH_4$ is a molecular hydride. It has tetrahedral shape and $sp^3$ hybridisation.

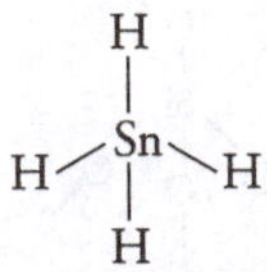

(Stannane)

Thus, statement I is true but statement II is false.

**15.** *(d)* $C_{60}$ molecule has a shape like soccer ball and is called as Buckminster fullerene. It is an allotrope of carbon. It contains twenty-six membered rings and twelve-five membered rings. A six membered ring is fused with both six or five membered rings but a five membered ring can only fuse with six membered rings.

All the carbon atoms are equal and they undergo $sp^2$-hybridisation .

Each carbon atom forms three sigma bonds with other three carbon atoms. So, statement (d) is wrong.

**16.** *(d)* Compound *A* is $PbO_2$. It is a strong oxidising agent because $Pb^{4+}$ is not stable and can be easily reduced to $Pb^{2+}$. $PbO_2$ is amphoteric in nature and is used in lead storage batteries.

**17.** *(c)* Water (steam) can produce CO on reacting with $CH_4$, C and $C_3H_8$ as,

$$CH_4 + H_2O \xrightarrow[\text{Ni}]{1270\text{ K}} CO + 3H_2$$

$$C + H_2O \xrightarrow{1270\text{ K}} CO + H_2$$

$$C_3H_8 + 3H_2O \xrightarrow[\text{Ni}]{1270\text{ K}} 3CO + 7H_2$$

But, water on reaction with $CO_2$ produces carbonic acid ($H_2CO_3$), not CO.

$$CO_2 + H_2O \longrightarrow H_2CO_3$$

**18.** *(a)* '*X*' is carbon tetrachloride. It is a molecular solid. It melts at low temperature and is a bad conductor of electricity in both liquid and solid state. Other options are explained as follows :

| | | |
|---|---|---|
| (1) | Silicon carbide | Network (or covalent) solid. Very high melting point. Bad conductor of electricity in solid or liquid state. |
| (2) | Mercury | Metallic solid with an exceptionally low melting point, conduct electricity in both solid and molten forms. |
| (3) | Zinc sulphide | Ionic crystal. High melting point and conduct electricity in molten state but not in solid state. |

**19.** *(c)* The basic structural unit of feldspar, zeolites, mica and asbestos is $(SiO_4)^{4-}$. These all are silicates.

**• Concept Enhancer**

All silicates involve two types of Si—O bonds.

(i) Terminal Si—O bonds in which oxygen is bonded to a silicon and not other atom.

(ii) Bridging Si—O—Si bonds in which oxygen is bonded to two silicon atoms.

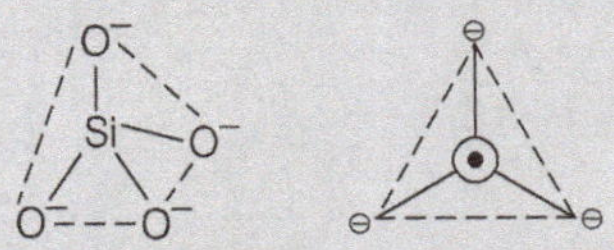

Tetrahedral structure of $SiO_4^{2-}$ anion

In $SiO_4^{4-}$ ion, each Si atom is bonded to four oxygen atoms tetrahedrally.

**20.** *(a)*

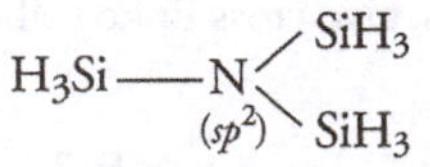

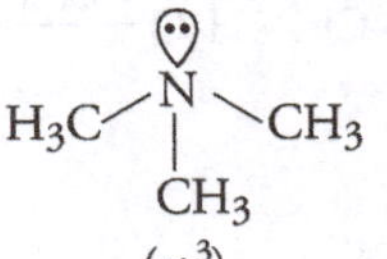

($sp^3$)

$(SiH_3)_3N$ has planar structure. It is less basic than $(CH_3)_3N$ due to backbonding of lone pair of electrons on nitrogen into vacant *d*-orbitals of Si whereas in $(CH_3)_3N$, there is no back bonding and hence, it is more basic.

**21.** *(b)* In $C_{60}$ (Buckminster fullerene) twenty hexagons and twelve pentagons are present which are interlocked resulting in a shape of soccer ball.

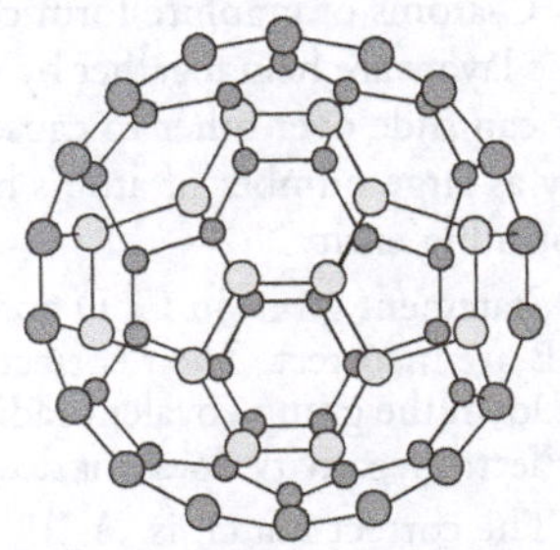

Structure of $C_{60}$ Buckminster fullerene

Phosphorus has large atomic size and less electronegativity, so it forms single bond instead of $p\pi$-$p\pi$ multiple bond. So, it consists of discrete tetrahedral $P_4$ molecule as shown below :

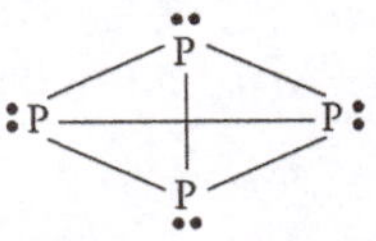

∴ Number of trigons (triangles) = 4

**22.** *(d)* Carbon (C) has greatest ability to form stable $p\pi$-$p\pi$ multiple bonds. 2*p*-orbitals of this element participate in the process. The stability of multiple bonds of C is attributed to their closeness with C-nucleus. Thus, the smaller size of C plays a significant role in the process.

**23.** *(b)* $RSiCl_3$ give cross linked silicone polymer on hydrolysis.

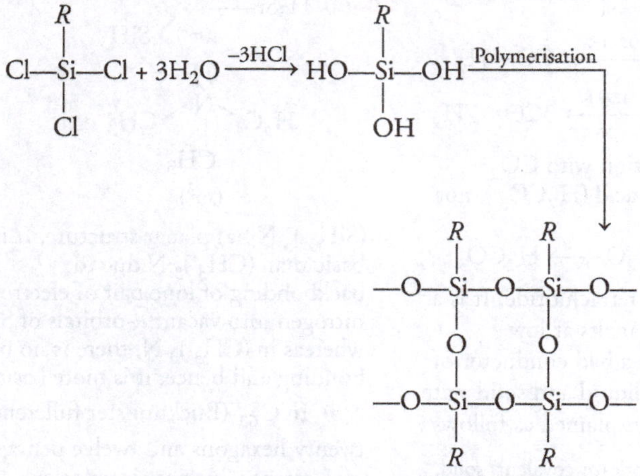

Cross linked silicone

**24.** *(d)* C-atoms of graphite form covalently bonded plates (layers). These layers are held together by weak forces of attraction. i.e., one layer can slide over other to cause lubricacy. It cannot be melted easily as large number of atoms being bonded strongly in the layer to form big entity.

**25.** *(b)* Statement given in C, D and E are correct, while statement A and B are incorrect. Their correct forms are

(A) Down the group covalent radius decreases.

(B) Electronegativity does not decreases gradually from C to Pb.

**26.** *(a)* The correct match is A-III, B-I, C-IV, D-II.

**27.** *(d)* Silicones are polymers with hydrophobic character and form water tight seals. They are bicompatible because they don't support microbiological growth. They are used as greases since they are stable in high temperature not water soluble and is odourless viscous liquids. Thus statements I, II and IV are correct. They have high themal stability, high dielectric strength and resistant to oxidation and chemicals.

**28.** *(c)* Inert pair effect becomes predominant due to which +2 oxidation state becomes more stable on moving down the group and hence, the stability of dihalides increases.

$$SiX_2 < GeX_2 < SnX_2 < PbX_2$$

**29.** *(d)* In $SiO_2$, each silicon atom is bonded to four oxygen atoms, These oxygen atoms are arranged tetrahedrally around Si and each O–atom is attached to two silicon atoms by covalent bonds.

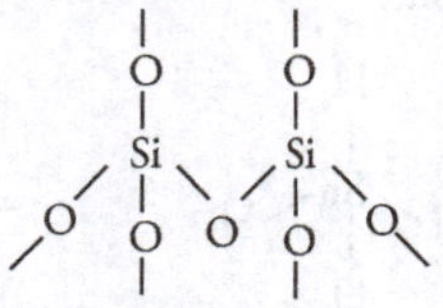

**30.** *(a)* White metallic tin buttons got converted to grey powder. This is because as temperature decreases, white tin ($\beta$-form) changes to grey tin ($\alpha$-form).

$$\alpha\text{-}Sn \xrightleftharpoons{13.2°C} \beta\text{-}Sn$$

In JEE Main there is usually 0-1 question per paper are visible from this chapter. In past 3 years, 32 conceptual and 15 numerical questions were asked of moderate difficulty level from this chapter.
**For detailed analysis scan QR code**

# PURIFICATION AND CHARACTERISATION OF ORGANIC COMPOUNDS

## Topic 1 Methods of Purification

### Level 1

**01** The purification method based on the following physical transformation is *[2025, 28 Jan Shift-II]*

$$\underset{(X)}{\text{Solid}} \xrightarrow{\text{Heat}} \underset{(Y)}{\text{Vapour}} \xrightarrow{\text{Cool}} \underset{(Z)}{\text{Solid}}$$

(a) Extraction (b) Sublimation
(c) Distillation (d) Crystallisation

**02** Which of the following statement is correct for paper chromatography? *[2023, 15 April Shift-I]*

(a) Water present in the pores of the paper forms the stationary phase.
(b) Paper sheet forms the stationary phase.
(c) Water present in the mobile phase gets absorbed by the paper which then forms the stationary phase.
(d) Paper and water present in its pores together form the stationary phase.

**03** Given below are two statements, one is labelled as Assertion A and the other is labelled as Reason R.

**Assertion (A)** Thin layer chromatography is an adsorption chromatography.

**Reason (R)** A thin layer of silica gel is spread over a glass plate of suitable size in thin layer chromatography which acts as an adsorbent.

In the light of the above statements, choose the correct answer from the options given below *[2022, 28 July Shift-II]*

(a) Both A and R are true and R is the correct explanation of A.
(b) Both A and R are true but R is not the correct explanation of A.
(c) A is true but R is false.
(d) A is false but R is true.

**04** Which technique among the following, is most appropriate in separation of a mixture of 100 mg of *p*-nitrophenol and picric acid? *[2022, 26 July Shift-I]*

(a) Steam distillation
(b) 2-5 ft long column of silica gel
(c) Sublimation
(d) Preparative TLC (Thin Layer Chromatography)

**Imp.** Thin Layer Chromatography

**05** Which one of the following techniques is not used to spot components of a mixture separated on thin layer chromatographic plate? *[2022, 28 June Shift-I]*

(a) $I_2$ (Solid) (b) UV light
(c) Visualisation agent as a component of mobile phase
(d) Spraying of an appropriate reagent

**06** Which of the following is structure of a separating funnel? *[2022, 27 June Shift-I]*

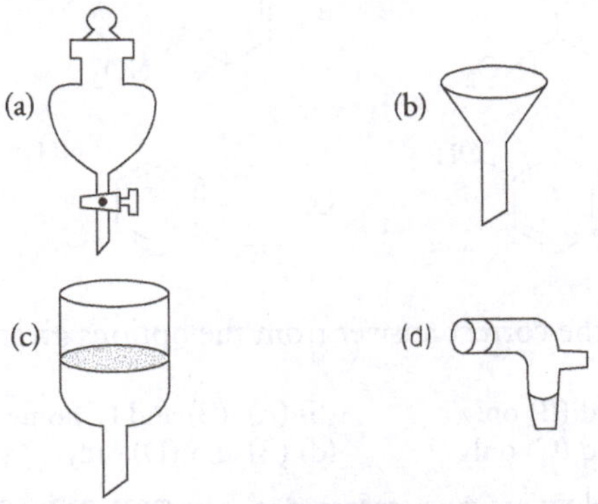


**07** Which purification technique is used for high boiling organic liquid compound (decomposes near its boiling point)? *[2021, 22 July Shift-II]*

(a) Simple distillation (b) Steam distillation
(c) Fractional distillation (d) Reduced pressure distillation

**08** The metal that can be purified economically by fractional distillation method is *[2021, 20 July Shift-I]*
(a) Fe (b) Zn
(c) Cu (d) Ni

**09** In chromatography technique, the purification of compound is independent of *[2021, 16 March Shift-I]*
(a) mobility or flow of solvent system
(b) solubility of the compound
(c) length of the column or TLC plate
(d) physical state of the pure compound

**10** In chromatography, which of the following statements is incorrect for $R_f$? *[2019, 10 April Shift-II]*
(a) $R_f$ value depends on the type of chromatography
(b) Higher $R_f$ value means higher adsorption
(c) $R_f$ value is dependent on the mobile phase
(d) The value of $R_f$ can not be more than one

**Imp.** Calculation of Retention Factor Value by Using Column Chromatography

**11** Two compounds I and II are eluted by column chromatography (adsorption of I > II). Which one of the following is a correct statement? *[JEE Main 2018]*
(a) II moves faster and has higher $R_f$ value than I.
(b) I moves faster and has higher $R_f$ value than II.
(c) II moves slower and has higher $R_f$ value than I.
(d) I moves slower and has higher $R_f$ value than II.

**12** The distillation technique most suited for separating glycerol from spent lye in the soap industry is
(a) fractional distillation *[JEE Main 2016]*
(b) steam distillation
(c) distillation under reduced pressure
(d) simple distillation

## Level 2

**13** The steam volatile compounds among the following are *[2025, 29 Jan Shift-I]*

A. 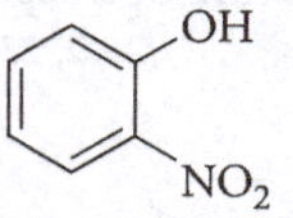


B. 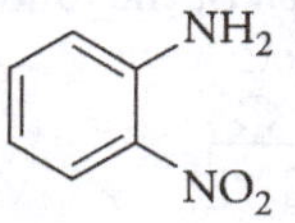


C. 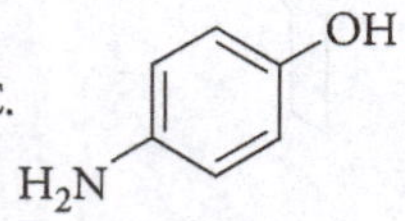


D. 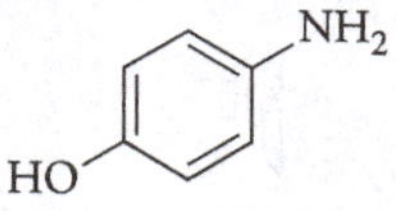


Choose the correct answer from the options given below
(a) (A) and (B) only (b) (A), (B) and (C) only
(c) (A) and (C) only (d) (B) and (D) only

**14** Given below are two statements. *[2025, 29 Jan Shift-I]*
**Statement I** In partition chromatography, stationary phase is thin film of liquid present in the inert support.
**Statement II** In paper chromatography, the material of paper acts as a stationary phase.
In the light of the above statements, choose the correct answer from the options given below:
(a) Statement I is true but Statement II is false.
(b) Statement I is false but Statement II is true.
(c) Both Statement I and Statement II are false.
(d) Both Statement I and Statement II are true.

**15** Methods used for purification of organic compounds are based on *[2024, 9 April Shift-I]*
(a) neither on nature of compound nor on the impurity present.
(b) nature of compound and presence of impurity.
(c) nature of compound only.
(d) presence of impurity only.

**16** The adsorbent used in adsorption chromatography is/are- *[2024, 4 April Shift-II]*
A. silica gel B. alumina
C. quick lime D. magnesia
Choose the most appropriate answer from the options given below:
(a) A only (b) B only
(c) C and D only (d) A and B only

**17** The fragrance of flowers is due to the presence of some steam volatile organic compounds called essential oils. These are generally insoluble in water at room temperature but are miscible with water vapour in vapour phase. A suitable method for the extraction of these oils from the flowers is *[2024, 31 Jan Shift-II]*
(a) distillation
(b) crystallisation
(c) steam distillation
(d) distillation under reduced pressure

**18** 'Adsorption' principle is used for which of the following purification method? *[2024, 31 Jan Shift-I]*
(a) Sublimation (b) Chromatography
(c) Extraction (d) Distillation

**19** Which among the following purification methods is based on the principle of "Solubility" in two different solvents? *[2024, 30 Jan Shift-II]*
(a) Column chromatography
(b) Differential extraction
(c) Distillation
(d) Sublimation

**20** Chromatographic technique\s based on the principle of differential adsorption is\are *[2024, 29 Jan Shift-II]*
A. column chromatography
B. thin layer chromatography
C. paper chromatography
Choose the **most appropriate** answer from the options given below
(a) A only (b) B only
(c) C only (d) A and B only

**21** The technique used for purification of steam volatile water immiscibe substances is **[2024, 27 Jan Shift-II]**
(a) fractional distillation
(b) fractional distillation under reduced pressure
(c) distillation
(d) steam distillation

**22** Thin layer chromatography of a mixture shows the following observation:

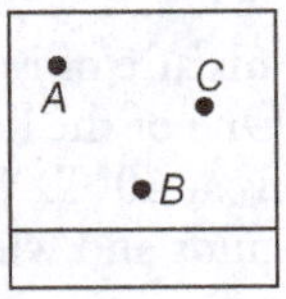


The correct order of elution in the silica gel column chromatography is **[2023, 11 April Shift-I]**
(a) *B, C, A* (b) *A, C, B*
(c) *C, A, B* (d) *B, A, C*

**23** From the figure of column chromatography given below, identify incorrect statements.

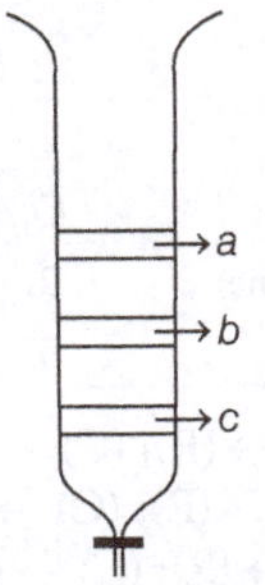


A. Compound '*c*' is more polar than '*a*' and '*b*'.
B. Compound '*a*' is least polar.
C. Compound '*b*' comes out of the column before '*c*' and after '*a*'.
D. Compound '*a*' spends more time in the column.
Choose the correct answer from the options given below: **[2023, 6 April Shift-II]**
(a) B, C and D only (b) A, B and D only
(c) A, B and C only (d) B and D only

**24** Match List-I with List-II.

| List-I (Mixture of compounds) | List-II (Separation technique) |
|---|---|
| A. $H_2O/CH_2Cl_2$ | I. Crystallisation |
| B. 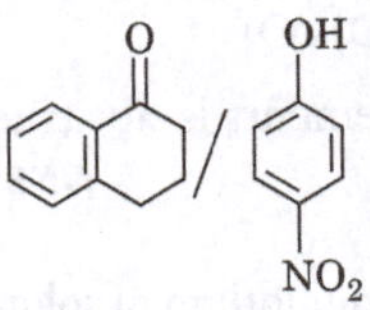 | II. Differential solvent extraction |
| C. Kerosene / naphthalene | III. Column chromatography |
| D. $C_6H_{12}O_6/NaCl$ | IV. Fractional distillation |

Correct match is **[2023, 31 Jan Shift-I]**
(a) A-I, B-III, C-II, D-IV
(b) A-II, B-IV, C-I, D-III
(c) A -III, B-IV, C-II, D-I
(d) A-II, B-III, C-IV, D-I

**25** Match List-I with List-II. **[2023, 30 Jan Shift-II]**

| List-I (Mixture) | List-II (Separation technique) |
|---|---|
| A. $CHCl_3 + C_6H_5NH_2$ | I. Steam distillation |
| B. $C_6H_{14} + C_5H_{12}$ | II. Differential extraction |
| C. $C_6H_5NH_2 + H_2O$ | III. Distillation |
| D. Organic compound in $H_2O$ | IV. Fractional distillation |

(a) A-II, B-I, C-III, D-IV
(b) A-IV, B-I, C-III, D-II
(c) A-III, B-I, C-IV, D-II
(d) A-III, B-IV, C-I, D-II

**26** Match List I with List II.

| List I (Mixture) | List II (Purification process) |
|---|---|
| (A) Chloroform and aniline | (I) Steam distillation |
| (B) Benzoic acid and naphthalene | (II) Sublimation |
| (C) Water and aniline | (III) Distillation |
| (D) Naphthalene and sodium chloride | (IV) Crystallisation |

Choose the correct answer from the options given below **[2022, 27 July Shift-II]**
(a) (A) → (IV), (B) → (III), (C) → (I), (D) → (II)
(b) (A) → (III), (B) → (I), (C) → (IV), (D) → (II)
(c) (A) → (III), (B) → (IV), (C) → (II), (D) → (I)
(d) (A) → (III), (B) → (IV), (C) → (I), (D) → (II)

**27** TLC analysis of a mixture having 3 components (*A, B, C*) using silica gel as the stationary phase gave following $R_f$ values; for $A = 0.72$, $B = 0.48$, $C = 0.20$.
Regarding the above observations, which one of the following statements is not corect for column chromatography of the mixture? **[2022, 30 June Shift-I]**
(a) *A* is the most non-polar component.
(b) *C* is the most soluble component in mobile phase.
(c) *A* will be eluted first.
(d) *C* is the most strongly absorbed component.

**28** Given below are two statements.
One is labelled as Assertion A and the other is labelled as Reason R.
**Assertion (A)** A mixture contains benzoic acid and naphthalene. The pure benzoic acid can be separated out by the use of benzene.

**Reason** (R) Benzoic acid is soluble in hot water.
In the light of the above statements, choose the most appropriate answer from the options given below.
*[2022, 25 June Shift-II]*
(a) Both A and R are true and R is the correct explanation of A.
(b) Both A and R are true but R is not the correct explanation of A.
(c) A is true but R is false.
(d) A is false but R is true.

**29** Phenol on reaction with dilute nitric acid, gives two products. Which method will be most effective for large scale separation? *[2022, 25 June Shift-I]*
(a) Chromatographic separation (b) Fractional crystallisation
(c) Steam distillation (d) Sublimation

**Imp.** Method of Simple Distillation

**30** Given below are two statements.
One is labelled as Assertion (A) and the other is labelled as Reason (R).
**Assertion** (A) A simple distillation can be used to separate a mixture of propanol and propanone.
**Reason** (R) Two liquids with a difference of more than 20°C in their boiling points can be separated by simple distillations.
In the light of the above statements, choose the most appropriate answer from the options given below.
(a) (A) is false but (R) is true. *[2021, 31 Aug Shift-I]*
(b) Both (A) and (R) are correct but (R) is not the correct explanation of (A).
(c) (A) is true but (R) is false.
(d) Both (A) and (R) are correct and (R) is the correct explanation of (A).

**31** Given below are two statements.
**Statement I** A mixture of chloroform and aniline can be separated by simple distillation.
**Statement II** When separating aniline from a mixture of aniline and water by steam distillation aniline boils below its boiling point.
In the light of the above statements, choose the most appropriate answer from the options given below.
*[2021, 26 Feb Shift-I]*
(a) Statement I is false but statement II is true
(b) Both statement I and statement II are false
(c) Statement I is true but statement II is false
(d) Both statement I and statement II are true

**32** Given below are two statements:
**Statement I** *o*-nitrophenol is steam volatile due to intramolecular hydrogen bonding.
**Statement II** *o*-nitrophenol has high melting due to hydrogen bonding.
In the light of the above statements, choose the most appropriate answer from the options given below.
*[2021, 26 Feb Shift-I]*
(a) Statement I is true but statement II is false.
(b) Statement I is false but statement II is true.
(c) Both statement I and statement II are false.
(d) Both statement I and statement II are true.

**Imp.** Technique of Fractional Distillation

**33** A flask contains a mixture of isohexane and 3-methylpentane. One of the liquids boils at 63°C while the other boils at 60°C. What is the best way to separate the two liquids and which one will be distilled out first? *[2020, 8 Jan Shift-I]*
(a) Fractional distillation 3-methylpentane
(b) Fractional distillation, isohexane
(c) Simple distillation, 3-methylpentane
(d) Simple distillation, isohexane

**34** The correct match between items I and II is *[2019, 11 Jan Shift-I]*

| | Item I (Mixture) | | Item II (Separation method) |
|---|---|---|---|
| A. | $H_2O$ : Sugar | P. | Sublimation |
| B. | $H_2O$ : Aniline | Q. | Recrystallisation |
| C. | $H_2O$ : Toluene | R. | Steam distillation |
| | | S. | Differential extraction |

(a) (A) → (Q); (B) → (R); (C) → (S)
(b) (A) → (Q); (B) → (R); (C) → (P)
(c) (A) → (S); (B) → (R); (C) → (P)
(d) (A) → (R); (B) → (P); (C) → (S)

**35** The correct match between items of List-I and List-II is *[JEE Main 2018]*

| | List-I | | List-II |
|---|---|---|---|
| A. | Coloured impurity | P. | Steam distillation |
| B. | Mixture of *o*-nitrophenol and *p*-nitrophenol | Q. | Fractional distillation |
| C. | Crude naphtha | R. | Charcoal treatment |
| D. | Mixture of glycerol and sugars | S. | Distillation under reduced pressure |

(a) (A)-(R), (B)-(S), (C)-(P), (D)-(Q)
(b) (A)-(R), (B)-(P), (C)-(Q), (D)-(S)
(c) (A)-(P), (B)-(S), (C)-(R), (D)-(Q)
(d) (A)-(R), (B)-(P), (C)-(S), (D)-(Q)

**36** Which of the following statements is not true about partition chromatography? *[JEE Main 2017]*
(a) Mobile phase can be a gas.
(b) Separation depends upon equilibrium of solute between a mobile and a stationary phase.
(c) Paper chromatography is an example of partition chromatography.
(d) Stationary phase is a finely divided solid adsorbent.

## Level 3

**37** Match List-I with List-II. *[2025, 4 April Shift-II]*

| List-I (Separation of ) | List-II (Separation Technique) |
|---|---|
| (A) Aniline from aniline-water mixture | (I) Simple distillation |
| (B) Glycerol from spent-lye in soap industry | (II) Fractional distillation |
| (C) Different fractions of crude oil in petroleum industry | (III) Distillation at reduced pressure |
| (D) Chloroform-Aniline mixture | (IV) Steam distillation |

Choose the correct answer from the options given below:

(a) (A)-(IV), (B)-(III), (C)-(II), (D)-(I)
(b) (A)-(III), (B)-(IV), (C)-(I), (D)-(II)
(c) (A)-(II), (B)-(I), (C)-(IV), (D)-(III)
(d) (A)-(I), (B)-(II), (C)-(III), (D)-(IV)

**38** Match List I with list II *[2025, 2 April Shift-II]*

| List I (Purification technique) | List II (Mixture of organic compounds) |
|---|---|
| A. Distillation (simple) | I. Diesel + Petrol |
| B. Fractional distillation | II. Aniline + Water |
| C. Distillation under reduced pressure | III. Chloroform + Aniline |
| D. Steam distillation | IV. Glycerol + Spent-lye |

Choose the correct answer from the options given below.

(a) (A)-(III), (B)-(IV), (C)-(II), (D)-(I)
(b) (A)-(III), (B)-(I), (C)-(IV), (D)-(II)
(c) (A)-(II), (B)-(IV), (C)-(I), (D)-(III)
(d) (A)-(II), (B)-(III), (C)-(IV), (D)-(I)

**39** Which of the following statements are correct? *[2024, 6 April Shift-I]*

A. Glycerol is purified by vacuum distillation because it decomposes at its normal boiling point.
B. Aniline can be purified by steam distillation as aniline is miscible in water.
C. Ethanol can be separated from ethanol water mixture by azeotropic distillation because it forms azeotrope.
D. An organic compound is pure, if mixed M.P. is remained same.

Choose the most appropriate answer from the options given below.

(a) A, C, D only (b) B, C, D only
(c) A, B, C only (d) A, B, D only

**40** Match List-I with List-II. *[2024, 31 Jan Shift-I]*

| List-I (Technique) | List-II (Application) |
|---|---|
| A. Distillation | I. Separation of glycerol from spent-lye |
| B. Fractional distillation | II. Aniline-water mixture |
| C. Steam distillation | III. Separation of crude oil fractions |
| D. Distillation under reduced pressure | IV. Chloroform-aniline |

Choose the correct answer from the options given below:

| | A | B | C | D | | A | B | C | D |
|---|---|---|---|---|---|---|---|---|---|
| (a) | II | III | I | IV | (b) | I | II | IV | III |
| (c) | IV | I | II | III | (d) | IV | III | II | I |

**41** The correct statement among the following for a 'chromatography' purification method is *[2024, 6 April Shift-II]*

(a) non-polar compounds are retained at top and polar compounds come down in column chromatography.
(b) organic compounds run faster than solvent in the thin layer chromatographic plate.
(c) $R_f$ of a polar compound is smaller than that of a non-polar compound.
(d) $R_f$ is an integral value.

## Numerical Value Type Questions

**42**

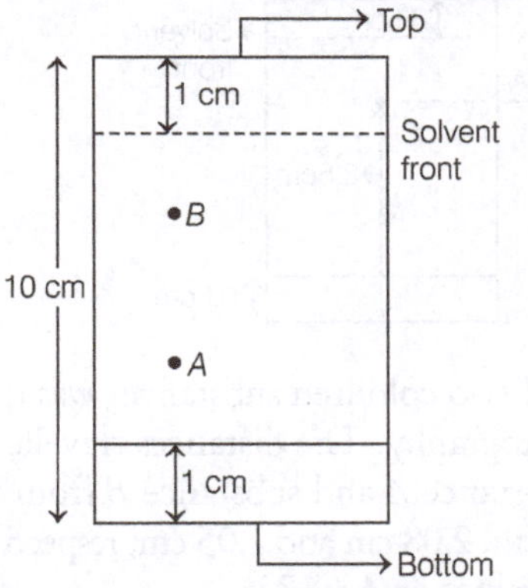


In the given TLC, the distance of spot $A$ and $B$ are 5 cm and 7 cm, from the bottom of TLC plate, respectively.

$R_f$ value of $B$ is $x \times 10^{-1}$ times more than $A$. The value of $x$ is ........... . *[2024, 9 April Shift-II]*

**43** Using the given figure, the ratio of $R_f$ values of sample $A$ and sample $C$ is $x \times 10^{-2}$. Value of $x$ is ............ *[2024, 5 April Shift-II]*

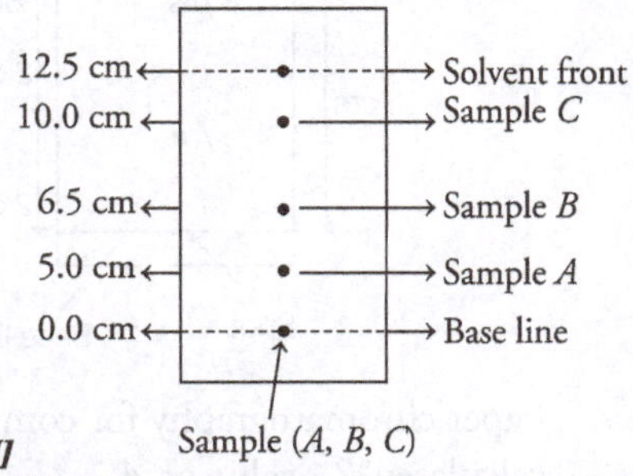


**Fig.** Paper chromatography of samples

**44** On a thin layer chromatographic plate, an organic compound moved by 3.5 cm, while the solvent moved by 5 cm. The retardation factor of the organic compound is ........ $\times 10^{-1}$. *[2024, 30 Jan Shift-I]*

**45** Three organic compounds $A$, $B$ and $C$ were allowed to run in thin layer chromatography using hexane and gave the following result (see figure). The $R_f$ value of the most polar compound is ........ $\times 10^{-2}$.

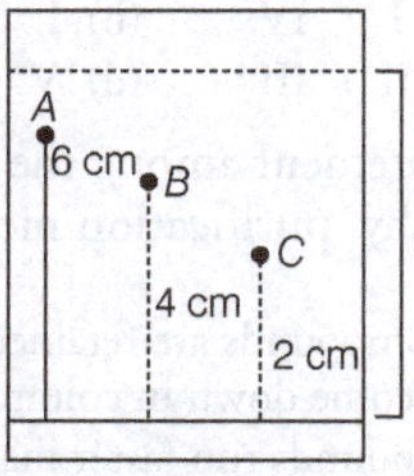


*[2023, 12 April Shift-I]*

**46** Following chromatogram was developed by adsorption of compound 'A' on a 6 cm TLC glass plate. Retardation factor of the compound 'A' is ......... $\times 10^{-1}$. *[2023, 29 Jan Shift-I]*

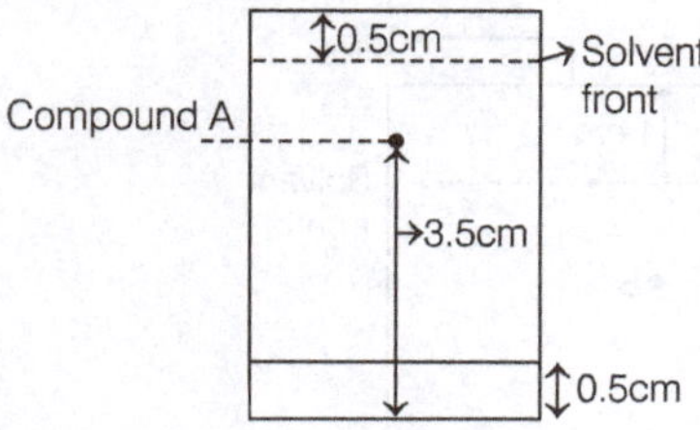


**47** The separation of two coloured substances was done by paper chromatography. The distances travelled by solvent front, substance $A$ and substance $B$ from the base line are 3.25 cm, 2.08 cm and 1.05 cm, respectively. The ratio of $R_f$ values of $A$ to $B$ is......... . *[2022, 25 July Shift-II]*

**48** Using the provided information in the following paper chromatogram *[2021, 25 Feb Shift-I]*

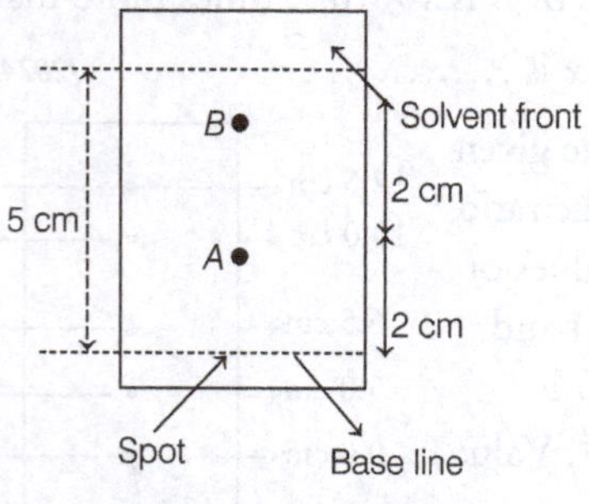


Paper chromatography for compounds $A$ and $B$, then calculated $R_f$ value of $A$ ........ $\times 10^{-1}$.

## Topic 2 Qualitative Analysis

### Level 1

**01** $Fe^{3+}$ cation gives a prussian blue precipitate on addition of potassium ferrocyanide solution due to the formation of *[2022, 27 July Shift-II]*
(a) $[Fe(H_2O)_6]_2[Fe(CN)_6]$
(b) $Fe_2[Fe(CN)_6]$
(c) $Fe_3[Fe(OH)_2(CN)_4]_2$
(d) $Fe_4[Fe(CN)_6]_3$

**02** During halogen test, sodium fusion extract is boiled with concentrated $HNO_3$ to *[2022, 25 June Shift-II]*
(a) remove unreacted sodium
(b) decompose cyanide or sulphide of sodium
(c) extract halogen from organic compound
(d) maintain the pH of extract

**03** Which one of the following set of elements can be detected using sodium fusion extract? *[2021, 26 July Shift-I]*
(a) Phosphorus, oxygen, nitrogen, halogens
(b) Nitrogen, phosphorus, carbon, sulphur
(c) Sulphur, nitrogen, phosphorus, halogens
(d) Halogens, nitrogen, oxygen, sulphur

**04** Which of the following compounds is added to the sodium extract before addition of silver nitrate for testing of halogen? *[2021, 25 Feb Shift-II]*
(a) Nitric acid (b) Ammonia
(c) Hydrochloric acid (d) Sodium hydroxide

### Level 2

**05** Formation of $Na_4[Fe(CN)_5NOS]$, a purple coloured complex formed by addition of sodium nitroprusside in sodium carbonate extract of salt indicates the presence of *[2025, 2 April Shift-II]*
(a) sulphate ion (b) sodium ion
(c) sulphide ion (d) sulphite ion

**Imp.** Duma's Method

**06** Given below are two statements I and II. *[2025, 24 Jan Shift-I]*

**Statement I** Dumas method is used for estimation of "Nitrogen" in an organic compound.

**Statement II** Dumas method involves the formation of ammonium sulphate by heating the organic compound with conc. $H_2SO_4$.

In the light of the above statements, choose the correct answer from the options given below.
(a) Statement I is true but Statement II is false.
(b) Both Statement I and Statement II are false.
(c) Both Statement I and Statement II are true.
(d) Statement I is false but Statement II is true.

**07** Given below are two statements. *[2025, 23 Jan Shift-I]*

**Statement I** In Lassaigne's test, the covalent organic molecules are transformed into ionic compounds.

**Statement II** The sodium fusion extract of an organic compound having N and S gives prussian blue colour with $FeSO_4$ and $Na_4[Fe(CN)_6]$.

In the light of the above statements, choose the correct answer from the options given below

(a) Statement I is false but Statement II is true.
(b) Both Statement I and Statement II is true.
(c) Both Statement I and Statement II are false.
(d) Statement I is true but Statement II is false.

**08** Given below are two statements. *[2025, 22 Jan Shift-II]*

**Statement I** Nitrogen, sulphur, halogen and phosphorus present in an organic compound are detected by Lassaigne's test.

**Statement II** The elements present in the compound are converted from covalent form into ionic form by fusing the compound with magnesium in Lassaigne's test.

In the light of the above statements, choose the correct answer from the options given below:

(a) Statement I is false but Statement II is true.
(b) Both Statement I and Statement II are true.
(c) Statement I is true but Statement II is false.
(d) Both Statement I and Statement II are false.

**Imp.** Lassaigne's Test

**09** Which of the following nitrogen containing compound does not give Lassaigne's test? *[2024, 4th April Shift-I]*

(a) Hydrazine (b) Glycine
(c) Phenyl hydrazine (d) Urea

**10** Compound that will give positive Lassaigne's test for both nitrogen and halogen is *[2023, 29 Jan Shift-I]*

(a) $NH_2OH \cdot HCl$ (b) $NH_4Cl$
(c) $N_2H_4 \cdot HCl$ (d) $CH_3NH_2 \cdot HCl$

**11** Reagent, 1-naphthylamine and sulphanilic acid in acetic acid is used for the detection of *[2021, 18 March Shift-I]*

(a) $N_2O$ (b) $NO_3^-$ (c) NO (d) $NO_2^-$

## Topic 3 Quantitative Analysis

### Level 1

**01** In Duma's method for estimation of nitrogen 0.4 g of an organic compound gave 60 mL of nitrogen collected at 300 K temperature and 715 mm Hg pressure. The percentage composition of nitrogen in the compound is *[2025, 3 April Shift-II]*

(Given: Aqueous tension at 300 K = 15 mm Hg)

(a) 20.95% (b) 17.46%
(c) 15.71% (d) 7.85%

**02** In Duma's method for estimation of nitrogen, 0.5 g of an organic compound gave 60 mL of nitrogen collected at 300 K temperature and 715 mm Hg pressure. The percentage composition of nitrogen in the compound (Aqueous tension at 300 K = 15 mm Hg) is ........ %. *[2025, 2 April Shift-II]*

(a) 20.87 (b) 18.67
(c) 12.57 (d) 1.257

**03** In Duma's method for the estimation of $N_2$, the sample is heated with copper oxide and the gas evolved is passed over *[2023, 31 Jan Shift-II]*

(a) copper gauze (b) Pd
(c) Ni (d) copper oxide

**04** In Carius method, halogen containing organic compound is heated with fuming nitric acid in the presence of *[2021, 20 July Shift-II]*

(a) $HNO_3$ (b) $AgNO_3$ (c) $CuSO_4$ (d) $BaSO_4$

**05** Nitrogen can be estimated by Kjeldahl's method for which of the following compound? *[2021, 17 March Shift-II]*

(a) 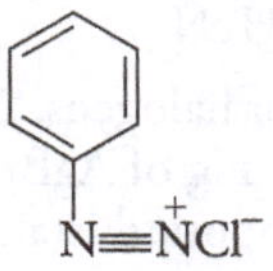


(b) 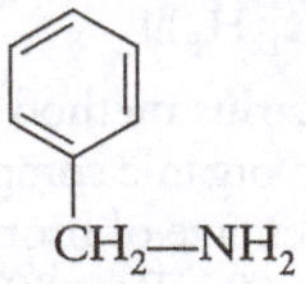


(c) 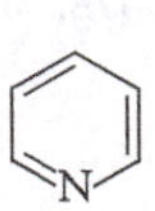


(d) 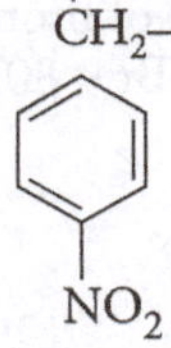


**06** Which of the following compounds will be suitable for Kjeldahl's method for nitrogen estimation? *[JEE Main 2018]*

(a) 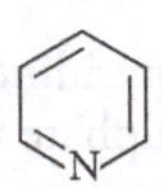


(b) 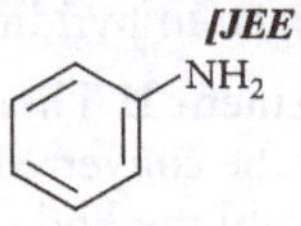


(c) 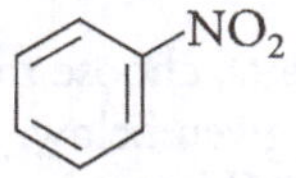


(d) 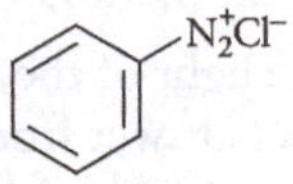


### Level 2

**07** In Kjeldahl's method for estimation of nitrogen, $CuSO_4$ acts as *[2024, 1 Feb Shift-I]*

(a) hydrolysis agent (b) reducing agent
(c) catalytic agent (d) oxidising agent

**08** In Carius method of estimation of halogen, 0.45 g of an organic compound gave 0.36 g of AgBr. Find out the percentage of bromine in the compound.
(Molar masses gBr = 188 g $mol^{-1}$; Br = 80 g $mol^{-1}$) *[2022, 27 July Shift-I]*

(a) 34.04% (b) 40.04% (c) 36.03% (d) 38.04%

**09** 120 g of an organic compound that contains only carbon and hydrogen gives 330 g of $CO_2$ and 270 g of water on complete combustion. The percentage of carbon and hydrogen, respectively are *[2022, 24 June Shift-II]*

(a) 25 and 75 (b) 40 and 60
(c) 60 and 40 (d) 75 and 25

**10** Which of the following is a false statement? *[2021, 26 Feb Shift-I]*

(a) Kjeldahl's method is used for the estimation of nitrogen in an organic compound.
(b) Phosphoric acid produced on oxidation of phosphorus present in an organic compound is precipitated as $Mg_2P_2O_7$ by adding magnesia mixture.
(c) Carius tube is used in the estimation of sulphur in an organic compound.
(d) Carius method is used for the estimation of nitrogen in an organic compound.

**11** An organic compound is estimated through Dumas method and was found to evolved 6 moles of $CO_2$, 4 moles of $H_2O$ and 1 mole of nitrogen gas. The formula of the compound is *[2019, 11 Jan Shift-I]*

(a) $C_6H_8N$ (b) $C_{12}H_8N$
(c) $C_{12}H_8N_2$ (d) $C_6H_8N_2$

**12** In Carius method of estimation of halogens, 250 mg of an organic compound gave 141 mg of AgBr. The percentage of bromine in the compound is(at. mass Ag = 108, Br = 80) *[JEE Main 2015]*

(a) 24 (b) 36
(c) 48 (d) 60

## ■ Level 3

**13** Given below are two statements : *[2024, 8 April Shift-II]*

**Statement I** Kjeldahl method is applicable to estimate nitrogen in pyridine.

**Statement II** The nitrogen present in pyridine can easily be converted into ammonium sulphate in Kjeldahl method.

In the light of the above statements, choose the correct answer from the options given below:
(a) Statement I is false but Statement II is true.
(b) Both Statement I and Statement II are false.
(c) Both Statement I and Statement II are true.
(d) Statement I is true but Statement II is false.

**Imp.** Carius Method

**14** In Carius tube, an organic compound '*X*' is treated with sodium peroxide to form a mineral acid '*Y*'. The solution of $BaCl_2$ is added to '*Y*' to form a precipitate '*Z*'. '*Z*' is used for the quantitative estimation of an extra element. '*X*' could be *[2023, 10 April Shift-II]*

(a) cytosine
(b) methionine
(c) a nucleotide
(d) chloroxylenol

**15** In Carius method of estimation of halogen 0.172 g of an organic compound showed presence of 0.08 g of bromine. Which of these is the correct structure of the compound? *[2020, 2 Sep Shift-I]*

(a) $H_3C—Br$ (b) $H_3C—CH_2Br$

(c) 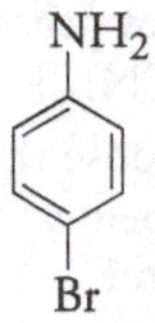

(d) 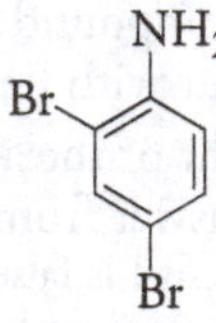

**Imp.** Percentage of Nitrogen by Kjeldahl's Method

**16** For the estimation of nitrogen, 1.4 g of an organic compound was digested by Kjeldahl's method and the evolved ammonia was absorbed in 60 mL of $M/10$ sulphuric acid. The unreacted acid required 20 mL of $M/10$ sodium hydroxide for the complete neutralisation. The percentage of nitrogen in the compound is *[JEE Main 2014]*

(a) 6% (b) 10%
(c) 3% (d) 5%

## Numerical Value Type Questions

**17** In Duma's method 292 mg of an organic compound released 50 mL of nitrogen gas ($N_2$) at 300 K temperature and 715 mm Hg pressure. The percentage composition of 'N' in the organic compound is ........... % (Nearest integer)
(Aqueous tension at 300 K = 15 mm Hg) *[2025, 7 April Shift-II]*

**18** During estimation of nitrogen by Duma's method of compound *X* (0.42 g) *[2025, 3 April Shift-I]*

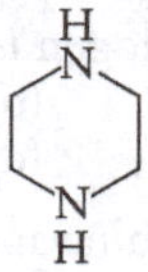

X

......... mL of $N_2$ gas will be liberated at STP. (Nearest integer)
(Given molar mass in g $mol^{-1}$ : C : 12, H : 1, N : 14)

**19** In Duma's method for estimation of nitrogen 1 g of an organic compound gave 150 mL of nitrogen collected at 300 K temperature and 900 mm Hg pressure. The percentage composition of nitrogen in the compound is .......... % (Nearest integer)
(Aqueous tension at 300 K = 15 mm Hg) *[2025, 4 April Shift-I]*

**20** In Carius method of estimation of halogen, 0.25 g of an organic compound gave 0.15 g of silver bromide (AgBr). The percentage of bromine in the organic compound is ............... $\times 10^{-1}$% (Nearest integer)
(Given: Molar mass of Ag is 108 and Br is 80 g $mol^{-1}$) *[2025, 24 Jan Shift-II]*

**21** During "S" estimation, 160 mg of an organic compound gives 466 mg of barium sulphate. The percentage of sulphur in the given compound is ...........%.
(Given, molar mass in g $mol^{-1}$ of Ba : 137, S : 32, O : 16) *[2025, 23 Jan Shift-I]*

**22** In Carius method for estimation of halogens, 180 mg of an organic compound produced 143.5 mg of AgCl. The percentage composition of chlorine in the compound is ...... %.
(Given, molar mass in g $mol^{-1}$ of Ag : 108, Cl : 35.5) *[2025, 22 Jan Shift-I]*

**23** In the sulphur estimation, 0.20 g of a pure organic compound gave 0.40 g of barium sulphate. The percentage of sulphur in the compound is ........ $\times 10^{-1}$%.
(Molar mass : O = 16, S = 32, Ba = 137 in g $mol^{-1}$) *[2025, 29 Jan Shift-II]*

**24** Following Kjeldahl's method, 1g of organic compound released ammonia, that neutralised 10 mL of 2M $H_2SO_4$. The percentage of nitrogen in the compound is ........ %. *[2024, 1st Feb Shift-II]*

**25** 0.400 g of an organic compound $(X)$ gave 0.376 g of AgBr in Carius method for estimation of bromine. % of bromine in the compound $(X)$ is ........... .
(Given : Molar mass AgBr = 188 g $mol^{-1}$
Br = 80 g $mol^{-1}$) *[2023, 10 April Shift-II]*

**26** In sulphur estimation, 0.471 g of an organic compound gave 1.4439 g of barium sulphate. The percentage of sulphur in the compound is ........ . (Nearest integer)
(Given : atomic mass Ba = 137 u, S = 32 u, O = 16 u) *[2023, 25 Jan Shift-I]*

**27** The complete combustion of 0.492 g of an organic compound containing 'C', 'H' and 'O' gives 0.793 g of $CO_2$ and 0.442 g of $H_2O$. The percentage of oxygen composition in the organic compound is ......... (Nearest integer) *[2022, 28 June Shift-II]*

**28** 0.25 g of an organic compound containing chlorine gave 0.40 g of silver chloride in Carius estimation. The percentage of chlorine present in the compound is ........... . [Nearest integer]
(Given, molar mass of Ag is 108 g $mol^{-1}$ and that of Cl is 35.5 g $mol^{-1}$) *[2022, 27 June Shift-II]*

**29** On complete combustion 0.30 g of an organic compound gave 0.20 g of carbon dioxide and 0.10 g of water. The percentage of carbon in the given organic compound is ......... . (Nearest integer) *[2022, 26 June Shift-I]*

**30** 0.2 g of an organic compound was subjected to estimation of nitrogen by Duma's method in which volume of $N_2$ evolved (at STP) was found to be 22.400 mL. The percentage of nitrogen in the compound is ...... .
[Nearest integer]
(Given molar mass of $N_2$ is 28 $mol^{-1}$. Molar volume of $N_2$ at STP 22.4 L) *[2022, 24 June Shift-II]*

**31** A 0.166 g of sample of an organic compound was digested with conc. $H_2SO_4$ and then distilled with NaOH. The ammonia gas evolved was passed through 50.0 mL of 0.5 N $H_2SO_4$. The used acid required 30.0 mL of 0.25 N NaOH for complete neutralisation. The mass percentage of nitrogen in the organic compound is ......... . *[2022, 24 June Shift-I]*

**32** The transformation occurring in Duma's method is given below

$$C_2H_7N + \left(2x + \frac{y}{2}\right) CuO \longrightarrow xCO_2 + \frac{y}{2} H_2O + \frac{z}{2} N_2 + \left(2x + \frac{y}{2}\right) Cu$$

The value of $y$ is ......... . (Integer answer) *[2021, 31 Aug Shift-II]*

**33** The number of moles of CuO, that will be utilised in Dumas method for estimation nitrogen in a sample of 57.5 g of N, N-dimethylaminopentane is ........ $\times 10^{-2}$. (Nearest integer) *[2021, 27 Aug Shift-I]*

**34** In Carius method for estimation of halogens, 0.2 g of an organic compound gave 0.188 g of AgBr. The percentage of bromine in the compound is ........ (Nearest integer)
[Atomic mass; Ag = 108, Br = 80] *[2021, 27 Aug Shift-I]*

**35** In the sulphur estimation, 0.471 g of an organic compound gave 1.44 g of barium sulphate. The percentage of sulphur in the compound is ......... %. (Nearest integer)
(Atomic mass of Ba = 137 u) *[2021, 26 Aug Shift-II]*

**36** In an estimation of bromine by Carius method, 1.6 g of an organic compound gave 1.88 g of AgBr. The mass percentage of bromine in the compound is ....... . (Atomic mass, Ag = 108, Br = 80 g $mol^{-1}$) *[2020, 6 Sep Shift-I]*

# Key Index

## Topic 1

| | | | | | | | | | |
|---|---|---|---|---|---|---|---|---|---|
| 1 (b) | 2 (a) | 3 (a) | 4 (d) | 5 (c) | 6 (a) | 7 (d) | 8 (b) | 9 (d) | 10 (b) |
| 11 (a) | 12 (c) | 13 (a) | 14 (a) | 15 (b) | 16 (d) | 17 (c) | 18 (b) | 19 (b) | 20 (d) |
| 21 (d) | 22 (b) | 23 (c) | 24 (d) | 25 (d) | 26 (d) | 27 (b) | 28 (d) | 29 (c) | 30 (d) |
| 31 (d) | 32 (a) | 33 (b) | 34 (a) | 35 (b) | 36 (d) | 37 (a) | 38 (b) | 39 (a) | 40 (d) |
| 41 (c) | 42 (15) | 43 (50) | 44 (7) | 45 (25) | 46 (6) | 47 (2) | 48 (4) | | |

## Topic 2

| | | | | | | | | | |
|---|---|---|---|---|---|---|---|---|---|
| 1 (d) | 2 (b) | 3 (c) | 4 (a) | 5 (c) | 6 (a) | 7 (d) | 8 (c) | 9 (a) | 10 (d) |
| 11 (d) | | | | | | | | | |

## Topic 3

| | | | | | | | | | |
|---|---|---|---|---|---|---|---|---|---|
| 1 (c) | 2 (c) | 3 (a) | 4 (b) | 5 (b) | 6 (b) | 7 (c) | 8 (a) | 9 (d) | 10 (d) |
| 11 (d) | 12 (a) | 13 (b) | 14 (b) | 15 (c) | 16 (b) | 17 (18) | 18 (111) | 19 (20) | 20 (255) |
| 21 (40) | 22 (20) | 23 (275) | 24 (56) | 25 (40) | 26 (42) | 27 (46) | 28 (40) | 29 (18) | 30 (14) |
| 31 (63) | 32 (7) | 33 (1125) | 34 (40) | 35 (42) | 36 (50) | | | | |

# Detailed **Explanations**

## Topic 1

**01.** *(b)* When solid is heated and it is directly converted to vapour phase without going through liquid phase, then it is called as sublimation.
Same process happen when vapour is directly cooled down to solid phase by passing liquid phase

$$\text{Vapour} \xrightarrow[\text{Sublimation}]{\text{Cool}} \text{Solid}$$

$$\text{Solid} \xrightarrow[\text{Sublimation}]{\text{Hot}} \text{Vapour}$$

**02.** *(a)* Paper chromatography is a type of partition chromatography. In this a special quality paper known as chromatography paper is used that contains water, which gets trapped on it, acts as the stationary phase.
A strip of chromatography paper spotted at the base with the solution of a mixture is suspended in a suitable solvents. Here, the solvent acts as a mobile phase.

**03.** *(a)* Thin layer chromatography (TLC) is a type of adsorption chromatography. It involves separation of substances from a mixture over a thin layer of an adsorbent coated on glass plate. This thin layer of an adsorbent, (silica or alumina) is spread on a glass plate of suitable size. This layer of silica or alumina acts as an adsorbent.
Thus, both A and R are true and R is the correct explanation of A.

**04.** *(d)* *p*-nitrophenol and picric acid is an example of non-volatile mixture. Non-volatile mixtures can be best separated by preparative TLC (thin layer chromatography) technique.

**05.** *(c)* During thin layer chromatography technique spots of coloured compounds are visible on TLC plate. But if colour is not visible, then following techniques can be employed:
(i) By putting TLC plate under ultraviolet light.
(ii) By placing TLC plate in a covered jar containing crystals of iodine.
(iii) By spraying an appropriate reagent, e.g. spraying ninhydrin to detect amino acids.

**06.** *(a)*

| Apparatus | Name |
|---|---|
| 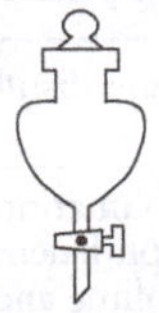 | Separating funnel |
| 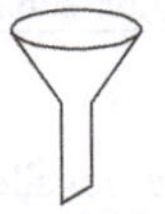 | Normal funnel |
| 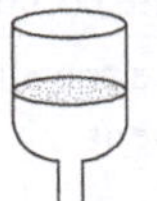 | Buchner funnel |
| 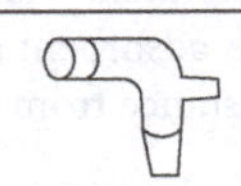 | Distillation adapter |

**07.** *(d)* Distillation under reduced pressure is used to purify liquids having very high boiling points and those, which decompose at or below their boiling points.
Because the boiling point of liquids are decreased at reduced pressure. Organic compounds can be distilled at lower temperatures.
Such liquids are made to boil at a temperature lower than their normal boiling points by reducing the pressure on their surface.
Thus, by reducing pressure, it can be distilled at low boiling point to avoid their decomposition.

**08.** *(b)* Zinc can be purified economically by fractional distillation.
Fractional distillation process utilises the boiling point difference between metal and that of impurity. Using this process, crude zinc containing Cd, Fe and Pb act as impurities can be refined.

**09.** *(d)* In chromatography technique, the purification of compound is independent of physical state of the pure compound (stationary phase). Chromatography is based on the principle of adsorbtion. Different substances are differently adsorbed.
The technique of chromatography uses the difference in the rates at which the components of a mixture move through a porous medium (stationary phase) under the influence of some solvent or gas (moving phase).

**10.** *(b)* In chromatography, the expression of retention factor ($R_f$) is

$$R_f = \frac{\text{Distance travelled by the compound from origin}}{\text{Distance travelled by the solvent from origin}} < 1$$

The value of $R_f$ signifies the relative ratio of migration of each component of the mixture with respect to the developing solvent used. $R_f$ value depends on the type of adsorption chromatography like TLC (Thin-Layer Chromatography), paper chromatography etc. The $R_f$ value is also the characteristic of a compound (sample) for a given developing solvent at a given temperature.
When the compound in the sample (usually less polar) is weakly adsorbed the spot will travel a shorter distance from the origin and hence the $R_f$ value will get increased.

**11.** *(a)* Since the adsorption of (I) > (II), (I) is firmly attached to column (stationary phase). Hence, it will move slowly and will move little distance.
Also, (II) is loosely attached to column (stationary phase). Hence, it will move faster and will move large distance.

**12.** *(c)* Glycerol with high boiling point (290°C) can be separated from spent lye by distillation under reduced pressure.
This process is used to purify liquids having very high boiling points. In this process, liquid is made to boil at lower temperature than its boiling point by lowering the pressure

**13.** *(a)* Among the given compounds, compound present in, A and B are steam volatile.
This is due to intermolecular hydrogen bonding among them.
So, option (a) is the correct answer.

**14.** *(a)* Statement I is true but Statement II is false. The correct form of Statement II is :
In paper chromatography, stationary phase is water.

**15.** *(b)* Methods used for purification of organic compounds are based on nature of compound and presence of impurity.

**16.** *(d)* The most commonly used adsorbent in adsorption chromatography are silica gel and alumina.

**17.** *(c)* Steam distillation technique is applied to separate substances which are steam volatile, and are immiscible with water.
So, extraction of oil from flower is done by steam distillation.

**18.** *(b)* 'Adsorption' principle is used in chromatography method.

**19.** *(b)* Differential extraction is based on principle of solubility. This method is used for immiscible liquid.
Different layers are formed due to different solubility and can be separated in funnel. Hence, option (b) is correct.

**20.** *(d)* Chromatographic techniques based on the principle of differential adsorption are
A. column chromatography
B. thin layer chromatography.

**21.** *(d)* Steam distillation is used for those liquids which are insoluble in water, containing non-volatile impurities and steam volatile.
Hence, option (d) is correct.

**22.** *(b)* The correct order of elution in the silica gel column chromatography is *A*, *C*, *B*.
According to the observation, *A* is more mobile and interacts with mobile phase more than *C*. Similarly *C* is more mobile than *B*.

**23.** *(c)* Incorrect statements are A, B and C. It is because in column chromatography, polar compound will have more interaction with silica gel. Therefore, they are adsorbed by silica gel quickly. So, '*a*' will be most polar while '*c*' will be least polar. Also, polar compound will be eluted lately due to more interaction with silica gel. Hence, order of elution will be $c > b > a$.

**24.** *(d)* The correct match is
A-II, B-III, C-IV, D-I.

**25.** *(d)* The correct match is A-III, B-IV, C-I, D-II.

**26.** *(d)*

| | |
|---|---|
| (A) Chloroform and aniline | (III) can be separated by simple distillation as difference in their boiling points is more than 25°C |
| (B) Benzoic acid and naphthalene | (IV) they can be separated more. conveniently by crystallisation. |
| (C) Water and aniline | (I) steam distillation |
| (D) Naphthalene and NaCl | (II) sublimation (as naphthalene is sublime and NaCl is non-sublime) |

**27.** *(b)*

$$\left[R_f \text{ value} = \frac{\text{Distance travelled by the solute component}}{\text{Distance travelled by the solvent}}\right]$$

Polar molecule have a tendency to get more attracted towards adsorbent and hence cover smaller distance from the baseline.
Since (*A*) has maximum $R_f$ value, so it will get less adsorbed on the surface of adsorbent. It will then travel faster and will be eluded first.
(*C*) has minimum $R_f$ value, so it is more strongly adsorbed and less soluble in mobile phase.

**28.** *(d)* Both benzoic acid and naphthalene are organic compounds. They are soluble in benzene. So, pure benzoic acid cannot separated out by the use of benzene. Hence, A is false.
Aromatic acids are insoluble in water but they become soluble on heating it. So, benzoic acid is soluble in water. Hence, R is true.

**29.** *(c)* Phenol on reaction with dil. nitric acid forms *o*- and *p*-nitrophenol.

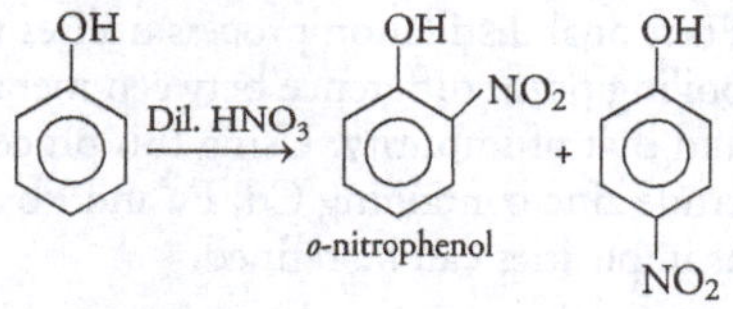

*o*-nitrophenol
*p*-nitrophenol

*p*- nitrophenol has higher boiling point as the molecules get self associated due to intermolecular H- bonding whereas no such association is observed in *o*-nitrophenol. Thus, the two can be separated by steam distillation.

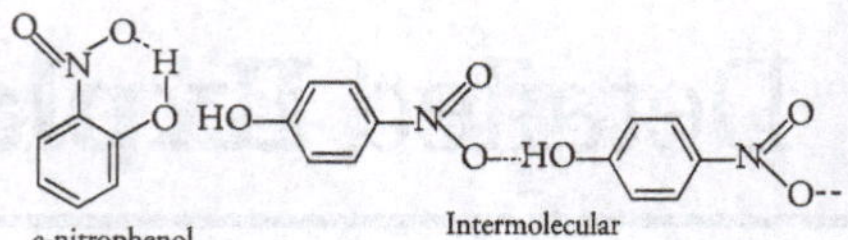

*o*-nitrophenol
(Intramolecular H-bonding)
Intermolecular

**30.** *(d)* Propanol and propanone can be separated by simple distillation technique as difference in boiling point of propanol and propanone is more than 20°C.
Boiling point of propanol = 97°C.
Boiling point of propanone = 56°C
Difference in boiling points
= 41°C > 20°C
Hence, option (d) is correct.

**31.** *(d)* Statement I is true, i.e. a mixture of chloroform and aniline can be separated by simple distillation. Boiling points of chloroform (334 K) and aniline (457 K) differ largely. So, on boiling the mixture, vapours of $CHCl_3$ are formed first which is then condensed to pure liquid $CHCl_3$. Whereas, the vapours of aniline will form later and liquid aniline can be collected separately.
Statement II is also true, i.e. aniline and water can be separated by steam distillation technique. Aniline is steam volatile but immiscible with water. So, a mixture of aniline and water will boil close to but below 373 K. After distillation, the mixture of aniline (bottom layer) and water (top layer) can be separated by separating funnel.
So, both statements I and II are true (option-d).

**32.** *(a)* Statement I is true but Statement II is false.

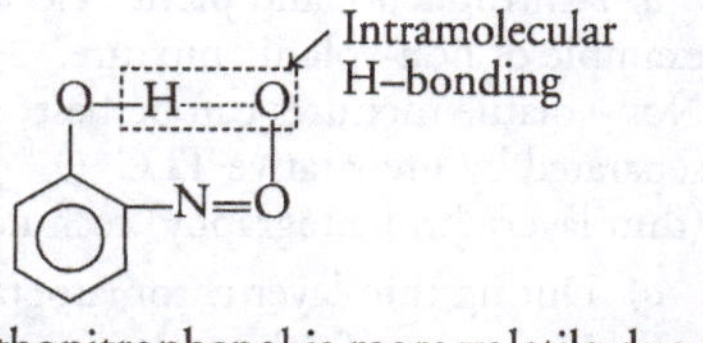


Orthonitrophenol is more volatile due to weak intramolecular hydrogen bonding.
Melting point depends on packing efficiency not on H-bonding, thus statement (II) is false.

**33.** *(b)* The boiling points of the given compounds are very close.
∴ Simple distillation can't be used.
Fractional distillation is only suitable to separate the two liquids.
Further, the liquid with lower boiling point distills out first. Since, 3-methyl pentane has a more symmetrical

structure as compared to isohexane. Therefore, it has a greater area of contact and better interaction between molecules. This accounts for the greater boiling point of 3-methyl pentane as compared to isohexane.

Isohexane 60°C    3-methyl pentane 63°C

So, isohexane will be distilled out first.

**34.** *(a)* The correct option is :
(A) → (Q); (B) → (R); (C) → (S)

**35.** *(b)* The correct match is :
(A) → (R); (B) → (P); (C) → (Q); (D) → (S)

**36.** *(d)* The statement (d) is not true about partition chromatography. Stationary phase may/may not be finely divided solid adsorbent. In paper chromatography, paper is used as stationary phase and is not finely divided.
The separation of the component from the sample mixture is carried out by the process of partition of the components between 2-phases. Both phases are in liquid form.
In this process, the immiscible solid surface coated with the liquid surface on the stationary phase is in the mobile phase.

**37.** *(a)* The correct match is
A-IV, B-III, C-II, D-I

**38.** *(b)* The correct match is
A-III, B-I, C-IV, D-II

**39.** *(a)* Statements given in A, C and D are correct, while in B it is incorrect.
The correct form of incorrect statement is,
Aniline is immiscible in water, hence it can be purified by steam distillation.

**40.** *(d)* The correct match is A-IV, B-III, C-II, D-I.

**Time Saver Tech**
Use elimination technique as time saver. Distillation under reduced pressure with separation of glycerol from spent-lye (D–I) is given only in option (d). Hence, no need to cross check other options.

**41.** *(c)* Statement given in option (c) is correct, while (a), (b) and (d) are incorrect. It's correct form are;
- Polar compound is retained more by the adsorbent than non-polar compound.
- Organic compounds run slower than solvent in thin layer chromatography.
- $R_f$ is not an integral value.

**42.** *(15)*

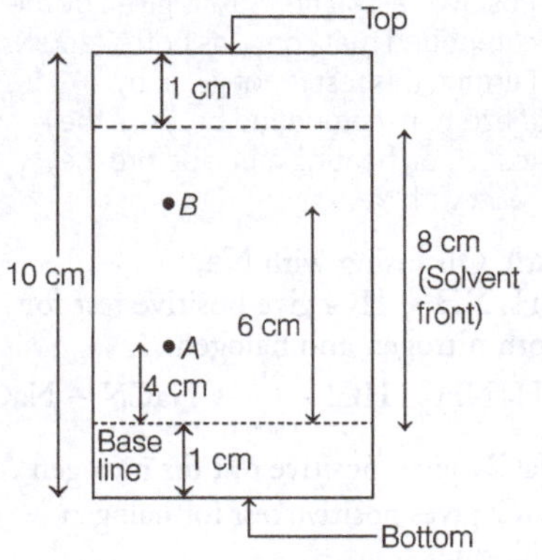


$$R_f = \frac{\text{Distance moved by substance from base line}}{\text{Distance moved by solvent from base line}}$$

$$(R_f)_A = \frac{4}{8}, (R_f)_B = \frac{6}{8}$$

$$\frac{(R_f)_B}{(R_f)_A} = \frac{6}{8} \times \frac{8}{4}$$

$$(R_f)_B = 1.5 (R_f)_A \Rightarrow x = 15$$

**43.** *(50)* $R_f$ of sample

$$= \frac{\text{Distance travel by sample}}{\text{Total distance travel by solvent}}$$

$$R_f \text{ of } A = \frac{5}{12.5}, R_f \text{ of } C = \frac{10}{12.5}$$

$$\text{Ratio} = \frac{R_{f(A)}}{R_{f(C)}} = \frac{5}{12.5} \times \frac{12.5}{10} = \frac{1}{2}$$

$= 0.5$ or $50 \times 10^{-2}$

Hence, missing value is 50.

**44.** *(7)* Retardation factor =

$$\frac{\text{Distance travelled by sample/organic compound } (X)}{\text{Distance travelled by solvent } (Y)}$$

$$= \frac{3.5}{5} = 7 \times 10^{-1}$$

[Given, $X = 3.5$ cm, $Y = 5$ cm]

Hence, retardation factor is $7 \times 10^{-1}$.

Thus, the missing value is 7.

**45.** *(25)* Less $R_f$ value means more polar. Thus among *A*, *B* and *C*, *C* will be most polar. Thus,

$$R_f = \frac{\text{Distance travelled by C from origin}}{\text{Distance travelled by solvent from origin}}$$

$$= \frac{2}{8} = 0.25 = 25 \times 10^{-2}$$

**46.** *(6)*

**key idea** In this question, the concept of thin layer chromatography is used to calculate the retardation factor, use the following formula,

$$R_f = \frac{\text{Distance moved by the substance from origin}}{\text{Distance moved by the solvent from origin}}$$

$$R_f = \frac{\text{Distance moved by the substance from origin}}{\text{Distance moved by the solvent from origin}}$$

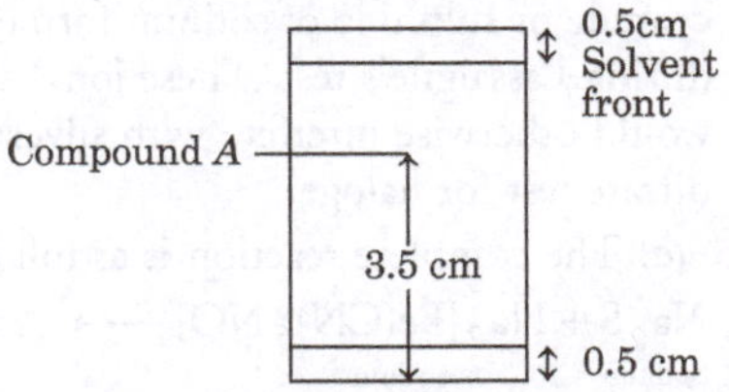


Distance moved by the substance from base line can be calculated as $(3.5 - 0.5) = 3$ cm

Total TLC distance = 6 cm

Distance moved by the solvent from base line can be $6 - (0.5 + 0.5) = 6 - 1 = 5$ cm

$$\therefore R_f = \frac{3 \text{ cm}}{5 \text{ cm}} = 0.6 = 6 \times 10^{-1} \text{ cm}$$

Thus, the missing value is 6.

**47.** *(2)*

$$R_f = \frac{\text{Distance travelled by the substance from the base line}}{\text{Distance travelled by the solvent front}}$$

$$R_f \text{ of } A, (R_f)_A = \frac{2.08}{3.25} = 0.64$$

$$R_f \text{ of } B, (R_f)_B = \frac{1.05}{3.25} = 0.32$$

$$\text{The ratio of } (R_f)_A \text{ to } (R_f)_B = \frac{0.64}{0.32} = 2.$$

**48.** *(4)*
$$\left[R_f = \frac{\text{Distance moved by the substance from the base line}}{\text{Distance moved by the solvent}}\right]$$

Hence, $R_f = \frac{2}{5} = [4 \times 10^{-1}]$

Thus, the missing value is 4.

## Topic 2

**01.** *(d)* $4Fe^{3+} + 3K_4[Fe(CN)_6]$ (Potassium ferrocyanide) $\longrightarrow Fe_4[Fe(CN)_6]_3$ (Prussian blue)

**02.** *(b)* During test for halogen, if nitrogen or sulphur is also present in the compound, then sodium fusion extract is first boiled with concentrated nitric acid to decompoase cyanide or sulphide of sodium formed during Lassaigne's test.

**03.** *(c)* By sodium fusion extract we can detect sulphur, nitrogen, phosphorus and halogens.

**04.** *(a)* If nitrogen or sulphur is also present in the compound, the sodium fusion extract is first boiled with concentrated nitric acid to decompose cyanide or sulphide of sodium formed during Lassaigne's test. These ions would otherwise interfere with silver nitrate test for halogen.

**05.** *(c)* The complete reaction is as follows,

$$\underset{\text{Sodium sulphide}}{Na_2S} + \underset{\text{Sodium nitro prusside}}{Na_2[Fe(CN)_5NO]} \longrightarrow \underset{\text{Violet colour}}{Na_4[Fe(CN)_5NOS]}$$

Hence, salt indicate the presence of sulphide ion.

**06.** *(a)* Statement I is true but II is false. The correct from of statement II is, In Dumas method nitrogen present in organic compound is converted into $N_2$ gas whose volumetric analysis gives the percentage of nitrogen in the organic compounds.

**07.** *(d)* Statement I is true but Statement II is false. The correct form of statement II is : The sodium fusion extract of an organic compound having N and S gives blood red colour with $FeSO_4$ and $Na_4[Fe(CN)_6]$.

**08.** *(c)* Statement I is true but Statement II is false. The correct form of Statement II is
The elements present in the compound are converted from covalent form into ionic form by fusing the compound with sodium in Lassaigne's test.

> • Concept **Enhancer**
> Lassaigne's extract is a chemical extract used to detect the presence of nitrogen, sulphur and halogens in organic compounds. It's also known as the sodium fusion test.

**09.** *(a)* Among the given options, hydrazine $(NH_2 — NH_2)$ will not give Lassiangne's test. As it does not have any carbon.

> • Concept **Enhancer**
> Positive Lassaigne's test is given by the compound that contains both C and N. Further, this test is not given by diazonium compound because they lose $N_2$ on heating much before they react with Na.

**10.** *(d)* On fusing with Na, $CH_3NH_2 \cdot HCl$ give positive test for both nitrogen and halogen.

$$CH_3NH_2 \cdot HCl \xrightarrow[\text{Fusion}]{\text{Na}} NaCN + NaCl$$

NaCN gives positive test for nitrogen
NaCl gives positive test for halogen (chlorine).

> • Concept **Enhancer**
> In Lassaigne's test, a small piece of sodium metal is heated in a fusion tube with the organic compound to form ionic salts. The formed ionic salts are extracted from the fused mass by boiling it with water.

**11.** *(d)* When sulphanilic acid and 1-naphthylamine, is acidified with acetic acid, then red coloured precipitate obtain, indicates the presence of $NO_2^-$ anions.

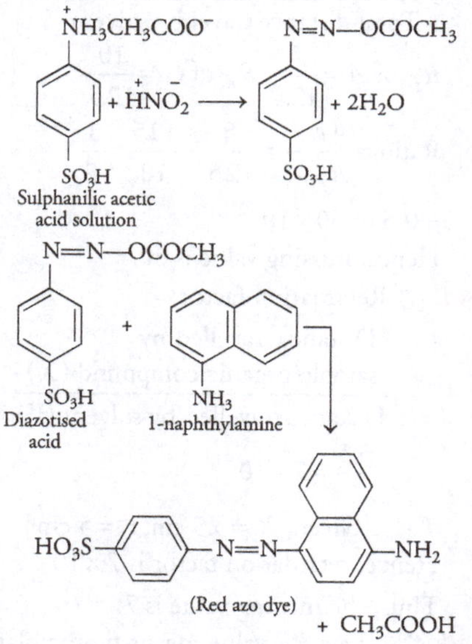


Thus, for the detection of $NO_2^-$ above test is used.

## Topic 3

**01.** *(c)* Pressure of $N_2$ gas evolved

$$= 715 - 15 = 700 \text{ mmHg}$$

$$= \frac{700}{760} \text{ atm}$$

Using ideal gas equation,
Mole of $N_2$ evolved

$$= \frac{pV}{RT} = \frac{700 \times 60 \times 10^{-3}}{760 \times 0.0821 \times 300}$$

$$= 0.0022 \text{ mole}$$

$\therefore$ Weight of $N_2$ evolved

$$= 0.0022 \times 28 = 0.063 \text{ g}$$

$\therefore$ Weight % of nitrogen in compound

$$\frac{\text{Weight of nitrogen}}{\text{Weight of compound}} \times 100$$

$$= \frac{0.063}{0.4} \times 100 = 15.71\%$$

**02.** *(c)* Pressure of $N_2$ gas

$$= (715 - 15) = 700 \text{ mm Hg}$$

$$n_{N_2} = \frac{pV}{RT} \text{ (Using Ideal gas equation)}$$

$$n_{N_2} = \frac{700 \times 60 \times 10^{-3}}{760 \times 0.0821 \times 300}$$

$$= 2.24 \times 10^{-3} \text{ mol}$$

Mass of $N_2 = 2.24 \times 10^{-3} \times 28$ g

$$= 0.06272 \text{ g}$$

$$\% N_2 = \frac{0.06272}{0.5} \times 100 \simeq 12.57$$

**03.** *(a)* Correct option is (a) i.e. copper gauze.
Duma's method involve nitrogen containing organic compound which when heated with copper oxide in presence of $CO_2$ yields, $N_2$ along with $CO_2$ and $H_2O$

$$C_xH_yN_z + \left(2x + \frac{y}{z}\right) CuO \longrightarrow xCO_2 + \frac{y}{2} H_2O + \frac{z}{2} N_2 + \left(2x + \frac{y}{z}\right) Cu$$

Here, if any traces of nitrogen oxide is formed these are reduced to nitrogen by passing the gaseous mixture over heated copper gauze.

**04.** *(b)* Organic compound is heated with fuming nitric acid in the presence of silver nitrate $AgNO_3$ in Carius method. This method used for quantitative determination of halogens in chemical substance. Method is based on the fact that when organic compound containing halogen (Cl, Br, I) is heated in sealed tube with fuming $HNO_3$ in presence of excess of $AgNO_3$, silver halide is formed.

**05.** *(b)* Nitrogen can be estimated by Kjeldahl's method for the benzyl amine,

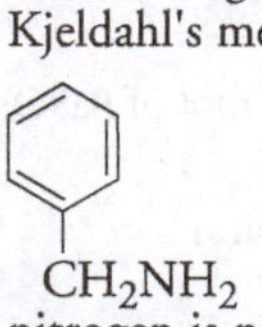

as in this compound nitrogen is not the part of ring and is free to react.

Because this method can be readily applied to the compound in which nitrogen is free to react with the reagent. other compounds which have nitrogen in the ring like given in the question i.e. other pyridine, diazonium salt chloride compound, or in nitro benzene are not readily converted into the ammonium sulphate by the action of sulphuric acid.

**06.** *(b)* Estimation of nitrogen through Kjeldahl's method is not suitable for organic compounds containing nitrogen in ring or nitrogen in nitro or azo groups. It is because of the fact that nitrogen of these compounds does not show conversion to ammonium sulphate $((NH_4)_2SO_4)$ during the process. Hence, among the given compounds only aniline can be used suitably for estimation of nitrogen by Kjeldahl's method.

> • Concept **Enhancer**
> Kjeldahl's method is based on the principle that when nitrogenous compound is heated with conc. $H_2SO_4$ in the presence of potassium sulphate and $CuSO_4$, the nitrogen present in the compound is converted to ammonium sulphate.

**07.** *(c)* The Kjeldhal's method is used to determine the nitrogen content in organic compounds.

In this method $CuSO_4$ acts as catalytic agent. Hence, option (c) is correct.

**08.** *(a)* Given, mass of organic compound = 0.45 g

Mass of AgBr obtained = 0.36 g

$\therefore$ Moles of AgBr $= \frac{0.36}{188}$

$\therefore$ Mass of bromine

$= \frac{0.36}{188} \times 80 = 0.1532$ g

$\therefore$ % bromine in compound

$= \frac{0.1532}{0.45} \times 100 = 34.04\%$.

**09.** *(d)* $C_xH_y + \left[x + \frac{y}{4}\right]O_2 \longrightarrow xCO_2 + \frac{y}{2}H_2O$

$$\left[\% \text{ of C} = \frac{\text{Mass of } CO_2 \text{ produced} \times 12}{\text{Mass of organic compound} \times 44} \times 100\right]$$

$$= \frac{330 \times 12}{120 \times 44} \times 100 = 75\%$$

$$\% \text{ of H} = \frac{\text{mass of } H_2O \text{ formed} \times 2 \times 100}{\text{mass of organic compound} \times 18}$$

$$= \frac{270 \times 100 \times 2}{18 \times 120} = 25\%$$

**10.** *(d)* Carius method is used in estimation of sulphur or halogen in organic compound.

**11.** *(d)*

> **key idea** In this question, the concept of Dumas method is used for the estimation of nitrogen. The estimation is done by heating organic compound with dry cupric oxide in a combustion tube in the atmosphere of $CO_2$. Upon heating, C and H present are oxidised to $CO_2$ and water vapours while $N_2$ is set free.

Let, the molecular formula of the organic compound (1 mol) be $C_xH_yN_z$. In Dumas method,

$$\underset{1\text{ mol}}{C_xH_yN_z} + \left(2x + \frac{y}{2}\right)CuO \longrightarrow \underset{x\text{ mol}}{xCO_2} + \underset{\frac{y}{2}\text{ mol}}{\frac{y}{2}H_2O} + \underset{\frac{z}{2}\text{ mol}}{\frac{z}{2}N_2} + \left(2x + \frac{y}{2}\right)Cu$$

Now, $x = 6, \frac{y}{2} = 4$

$\Rightarrow \quad y = 8$ and $\frac{z}{2} = 1 \Rightarrow z = 2$

$\therefore$ Molecular formula of the compound is $C_6H_8N_2$.

**12.** *(a)* Given, weight of organic compound = 250 mg

Weight of AgBr = 141 mg

$\therefore$ According to formula of % of bromine by Carius method

$$\% \text{ of Br} = \frac{\text{Atomic weight of Br}}{\text{Molecular weight of Ag Br}} \times \frac{\text{Weight of AgBr}}{\text{Weight of organic bromide}} \times 100$$

$$\therefore \% \text{ of Br} = \frac{80}{188} \times \frac{141}{250} \times 100$$

$$= \frac{1128000}{47000} = 24\%$$

**13.** *(b)* Both statement I and Statement II are false. Kjeldahl method is not applicable to compounds containing nitrogen in the ring as nitrogen of these compounds does not change to ammonium sulphate under these conditions. Hence, option (b) is correct.

**14.** *(b)* $X \xrightarrow{Na_2O_2} Y \xrightarrow{BaCl_2} Z\ [BaSO_4]$

Methionine : $C_5H_{11}NO_2S$

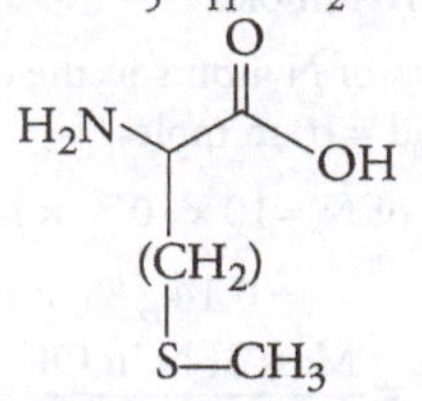

**15.** *(c)* In the organic compound

$$Br\% = \frac{0.08}{0.172} \times 100 = 46.51$$

Now, let us calculate Br % in the given compounds.

(a) $CH_3Br$ ($M$ = 95)

$\Rightarrow$ Br % $= \frac{80}{95} \times 100 = 84.21$

(b) $C_2H_5Br$ ($M$ = 109)

$\Rightarrow$ Br% $= \frac{80}{109} \times 100 = 73.39$

(c) $C_6H_6NBr$ ($M$ = 172)

$\Rightarrow$ Br% $= \frac{80}{172} \times 100 = 46.51$

(d) $C_6H_5NBr_2$ ($M$ = 251)

$\Rightarrow$ Br% $= \frac{160}{251} \times 100 = 63.74$

So, correct structure of compound is (c).

**16.** *(b)* Mass of organic compound = 1.4 g

Suppose $x$ m mole of N-atom is present in organic compound. Thus,

$$\text{Organic compound} \xrightarrow[\text{method}]{\text{Kjeldahl's}} \underset{x \text{ mmole}}{NH_3}$$

Initial mmol of $H_2SO_4 = 60 \times \frac{M}{10}$
$= 6$ mmol

mmoles of NaOH required by

$$H_2SO_4 = 20 \times \frac{M}{10} = 2 \text{ mmol}$$

Involved reactions are

$$2NH_3 + \underset{\left(\substack{6 \text{ mmol} \\ \text{Initially taken}}\right)}{H_2SO_4} \longrightarrow (NH_4)_2SO_4$$

$$\underset{\text{2 mmol reacted}}{H_2SO_4 + 2NaOH} \longrightarrow Na_2SO_4 + 2H_2O$$

Hence, 1 mmole of $H_2SO_4$ reacted with NaOH.

$\Rightarrow$ mmoles of $H_2SO_4$ reacted from first equation $= 6 - 1 = 5$ mmoles

$\Rightarrow$ mmoles of $NH_3$ in first equation $= 2 \times 5 = 10$ mmoles

$\Rightarrow$ mmoles of N atoms in the organic compound $= 10$ m moles

$\Rightarrow$ mass of N $= 10 \times 10^{-3} \times 14$

$= 0.14$ g % of N in

$$OC = \frac{\text{Mass of N in OC}}{\text{Total mass of OC}} \times 100$$

$\therefore$ $$\% \text{ of N} = \frac{0.14}{1.4} \times 100 = 10\%$$

**17.** *(18)* $$\underset{(292\text{ g})}{\text{Organic compound}} \xrightarrow{\text{Duma's}} N_2$$

$V = 50$ mL

$p = 715$ mm Hg

$T = 300$ K

Aqueous tension $= 15$ mm Hg

$p_{N_2} = 715 - 15 = 700$ mm Hg

$$p_{N_2} = \frac{700}{760} \text{ atm}$$

$$n_{N_2} = \frac{p_{N_2} \cdot V}{RT} = \frac{700}{760} \times \frac{50}{1000} \times \frac{1}{0.0821 \times 300}$$

$n_N = 2 \times n_{N_2}$

Mass of N $= 2 \times n_N \times 14$

$$\%N = \frac{\text{Mass of N}}{\text{Mass of organic compound}} \times 100$$

$$\%N = \frac{700}{760} \times \frac{50}{1000} \times \frac{2 \times 14}{0.0821 \times 300} \times \frac{1000}{292} \times 100$$

% N $= 18\%$

**18.** (111) Molecular weight of given compound is 86 g/mol

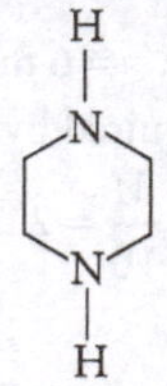

$\longrightarrow N_2$

By applying POAC on N

$n_x \times 2 = n_{N_2} \times 2$

$$n_{N_2} = \frac{0.42}{86}$$

$$(\text{Volume})_{N_2} \text{ at STP} = \frac{0.42}{86} \times 22.4 \text{ L}$$

$= 0.1108$ L or 110.8 ml $\approx 111$ ml

**19.** *(20)* Partial pressure of

$N_2 = (900 - 15) = 885$ mmHg

$$\text{Mole of } N_2 = \frac{\left(\frac{885}{760} \times 0.15\right)}{0.0821 \times 300} = 0.0071 \text{ moles}$$

Now, % of nitrogen in organic

$$\text{compound} = \frac{0.0071 \times 28}{1} \times 10 = 19.85\%$$

**20.** *(255)* Given, mass of organic compound $= 0.25$ g

Mass of AgBr $= 0.15$ g

% of bromine $= ?$

Percentage of bromine

$$= \frac{\text{Molar mass of bromine}}{\text{Molar mass of silver bromide}} \times \frac{\text{Weight of AgBr}}{\text{Weight of sample}} \times 100$$

$$= \frac{80}{188} \times \frac{0.15}{0.25} \times 100$$

$= 25.5319 = 255.319 \times 10^{-1}$

$= 255 \times 10^{-1}$

**21.** *(40)* Given, mass of $BaSO_4 = 466$ mg

mass of organic compound $= 160$ mg

% of sulphur in given compound $= ?$

Using the formula for calculation of % of sulphur

$$= \frac{\text{Milli. moles of } BaSO_4 \times 32}{\text{Mass of organic compound}} \times 100$$

$$= \frac{\frac{466}{233} \times 32}{160} \times 100 = 40\%$$

So, the % of sulphur in given compound is 40%.

**22.** *(20)* Given,

mass of organic compound $= 180$ mg

Mass of AgCl $= 143.5$ mg

% Cl $= ?$

$$n_{Cl} = n_{AgCl} = \frac{143.5 \times 10^{-3}}{143.5} = 10^{-3}$$

$$\text{Now, } \% \text{ Cl} = \frac{10^{-3} \times 35.5}{180 \times 10^{-3}} \times 100 = 19.72$$

$\approx 20$

**23.** *(275)* Given mass of $BaSO_4 = 0.40$ g

Mass of pure organic compound $= 0.20$ g

0.20 g of organic compound gives 0.40 g of $BaSO_4$.

$$\text{Moles of } BaSO_4 = \frac{0.40}{233} \text{ mol of } BaSO_4$$

$$= \frac{0.40}{233} \text{ mol of sulphur}$$

$$= \frac{0.40}{233} \times 32 \text{ g of sulphur}$$

$$\% \text{ of sulphur} = \frac{\frac{0.40}{233} \times 32}{0.20} \times 100$$

$= 27.5\%$ or $275 \times 10^{-1}\%$

**24.** *(56)* Millimoles of $H_2SO_4$ used

$= 10 \text{ mL} \times 2M = 20$

Millimoles of $NH_3$ neutralised

$= 2 \times 20 = 40$

Millimoles of N atom $= 40$ millimoles

$$\text{Weight of N} = \frac{40 \times 14}{1000}$$

Percentage composition of N in organic compound

$$= \frac{40 \times 14}{1000} \times 100 = 56\%$$

**25.** *(40)* Mass of organic compound $(X) = 0.40$ g

Mass of AgBr obtained $= 0.376$ g

$$\therefore \text{Moles of AgBr} = \frac{\text{given mass}}{\text{molar mass}} = \frac{0.376}{188}$$

$$\text{Moles of bromine} = \frac{0.376}{188} \times 80$$

$= 0.16$ gram

$\therefore$% of bromine in organic compound

$$\Rightarrow \frac{0.16}{0.40} \times 100 \Rightarrow 40\%$$

Thus, the % of bromine in the compound $X$ is 40.

**26.** *(42)*

> **key idea** In this question, the concept of Carius method is used which is used for the detection of sulphur and halogen in an organic compound.

Calculation of % sulphur (Carius method)

$$= \frac{32}{233} \times \frac{\text{Weight of } BaSO_4 \text{ formed}}{\text{Weight of organic compound}} \times 100$$

$$= \frac{32}{233} \times \frac{1.4439}{0.471} \times 100 = 42.10 \approx 42$$

**27.** *(46)* $$\underset{[0.49\text{ g}]}{C_xH_yO_z} + O_2 \longrightarrow \underset{[0.793\text{ g}]}{CO_2(g)} + \underset{[0.442\text{ g}]}{H_2O(l)}$$

Weight of carbon

$(W_C) = \left[\left(\frac{0.793}{44}\right) \times 12\right] = [0.216\text{ g}]$

Weight of hydrogen

$(W_H) = \left[\frac{0.442}{18}\right] \times 2 = [0.0491\text{ g}]$

Weight of oxygen

$[W_o] = [0.492 - 0.2651]$
$= [0.2269\text{ g}]$

% of oxygen

$= \left[\frac{0.2269}{0.492} \times 100\right] = 46.11 = (46\%)$

**28.** *(40)* Weight of organic compound $= 0.25$ g

We know, 143.5 g of silver chloride contains 35.5 g of chlorine.

So, 0.40 g of silver chloride contains

$\frac{35.5}{143.5} \times 0.40$ g of chlorine

$= 0.0989$ g of chlorine

Now, mass percentage of Cl in organic compound

$= \frac{\text{Mass of chlorine}}{\text{Mass of organic compound}} \times 100$

$= \frac{0.0989}{0.25} \times 100 = 39.60\% \approx 40\%$

**29.** *(18)*

$\underset{(0.30\text{ g})}{\text{Organic compound}} + O_2 \xrightarrow{\Delta} \underset{(0.20\text{ g})}{CO_2(g)} + \underset{(0.10\text{ g})}{H_2O(l)}$

Moles of $CO_2$ = moles of carbon in organic compound $= \left(\frac{0.2}{44}\right)$ mol

Weight of carbon $= \left[\frac{0.2}{44} \times 12\text{ g}\right]$

$= [0.054\text{ g}]$

% of carbon $= \left[\frac{0.054}{0.3} \times 100\right] = 18\%$

**30.** *(14)* Volume of $N_2$ gas evolved

$= 22.4$ mL at STP $= 22.4 \times 10^{-3}$ L

1 mol = 22.4 L

Therefore, $22.4 \times 10^{-3}$ L occupied

$= \frac{22.4 \times 10^{-3}}{22.4}$ moles of nitrogen gas

Mass of nitrogen gas

$= \frac{22.4 \times 10^{-3}}{22.4} \times 28 = 0.028$ g

% of nitrogen in organic compound

$= \frac{0.028}{0.2} \times 100 = 14\%.$

**31.** *(63)*

> **key idea** In this question we use the concept of Kjeldahl's method which is used for the estimation of nitrogen.

Milliequivalent of $NH_3$
= Milliequivalent of conc. $H_2SO_4$ used
= Milliequivalent of NaOH used
$= 30 \times 0.25 = [7.5]$

Gram of nitrogen

$= 7.5 \times 14 \times 10^{-3} = [0.105\text{ g}]$

% of nitrogen $= \left[\frac{0.105}{0.166} \times 100\right]$

$= (63.25\%) \approx 63\%$

**32.** *(7)* For the reaction,

$C_2H_7N + \left(2x + \frac{y}{2}\right) CuO \longrightarrow$

$x\,CO_2 + \frac{y}{2} H_2O + \frac{z}{2} N_2 + \left(2x + \frac{y}{2}\right) Cu$

On reactant side number of H-atom = 7

On product side number of H-atom

$= \frac{y}{2} \times 2$

No. of H on R.H.S. = No. of H on L.H.S.

$\therefore \quad 7 = \frac{y}{2} \times 2 \quad \therefore \quad y = 7$

**33.** *(1125)* In Dumas method,

$C_xH_yN_z + \left(2x + \frac{y}{2}\right) CuO \longrightarrow xCO_2$

$+ \frac{y}{2} H_2O + \frac{z}{2} N_2 + \left(2x + \frac{y}{2}\right) Cu$

N,N-dimethylaminopentane has formula $C_7H_{17}N$.

So, relating with $C_xH_yN_z$

$x = 7;\ y = 17;\ z = 1$

Molar mass of $C_7H_{17}N = 115$ g

1 mole $C_7H_{17}N$ requires

$= \left(2x + \frac{y}{2}\right) CuO$

= 22.5 moles of CuO

57.5 g i.e. $\left(\frac{57.5}{115}\right) C_7H_{17}N$ will utilise

$= \frac{22.5}{115} \times 57.5$ moles of CuO

= 11.25 g mol

$\approx 1125 \times 10^{-2}$ mol.

**34.** *(40)* Mass of bromine = 80 u

Mass of silver = 108 u

Mass of AgBr $= 108 + 80 = 188$ u

Weight of organic compound = 0.2 g

% of Br $= \frac{\text{Molar mass of Br}}{\text{Molar mass of AgBr}} \times \frac{\text{Weight of AgBr}}{\text{Weight of organic compound}} \times 100$

$= \frac{80}{188} \times \frac{0.188}{0.2} \times 100 = 40\%.$

**35.** *(42)* Atomic mass of sulphur is 32 g.

Molecular weight of $BaSO_4$ is 233 g.

So, weight of sulphur in $BaSO_4$

$= \frac{\text{Atomic mass of sulphur}}{\text{Molecular weight of } BaSO_4} \times \text{Weight of } BaSO_4$

$= \frac{32}{233} \times 1.44$

Percentage of sulphur

$= \frac{\text{Weight of sulphur}}{\text{Weight of organic compound}} \times 100$

$= \frac{32}{233} \times \frac{1.44}{0.471} \times 100$

$= 41.98 \simeq 42\%$

**36.** *(50)* Molar mass of AgBr $= 108 + 80$

$\Rightarrow 188$ g/mol

Weight of organic compound = 1.6 g

Weight of AgBr = 1.88 g

% of Br

$= \frac{\text{Weight of AgBr} \times 80}{\text{Weight of organic compound} \times 188} \times 100$

$= \frac{1.88 \times 80}{1.6 \times 188} \times 100 = 50\%$

Hence, the correct answer is 50.

# C-11

JEE Main Chapterwise Topicwise Chemistry

In JEE Main there is usually 1-2 question per paper are visible from this chapter. In past 3 years, 50 conceptual and 12 numerical questions were asked of moderate difficulty level from this chapter.

**For detailed analysis scan QR code**

# SOME BASIC PRINCIPLES OF ORGANIC CHEMISTRY

## Topic 1 Nomenclature, Hybridisation and Classification

### Level 1

**01** The IUPAC name of the following compound is:

$$HC \equiv C—CH_2—\underset{}{\overset{OH}{\overset{|}{C}}}H—CH_2—CH=CH_2$$

*[2025, 4 April Shift-II]*

(a) 4-hydroxyhept-6-en-1-yne
(b) 4-hydroxyhept-1-en-6-yne
(c) Hept-6-en-1-yn-4-ol
(d) Hept-1-en-6-yn-4-ol

**02** What is the correct IUPAC name of

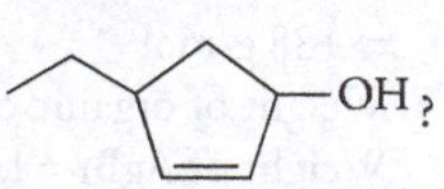


?

*[2025, 8 April Shift-II]*

(a) 4-ethylcyclopent-2-en-1-ol
(b) 4-ethyl-1-hydroxycyclopent-2-ene
(c) 1-ethyl-3-hydroxycyclopent-2-ene
(d) 1-ethylcyclopent-2-en-3-ol

**03** Which of the following is the correct IUPAC name of given organic compound (*X*)? *[2025, 7 April Shift-II]*

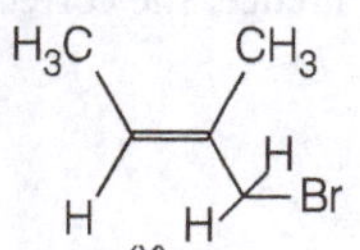


(X)

(a) 1-bromo-2-methylbut-2-ene
(b) 3-bromo-3-methylprop-2-ene
(c) 2-bromo-2-methylbut-2-ene
(d) 4-bromo-3-methylbut-2-ene

**04** What is the correct IUPAC name of

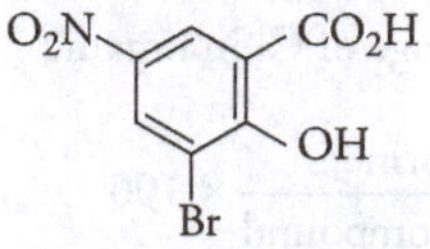


*[2025, 3 April Shift-II]*

(a) 3-bromo-4-hydroxy-1-nitrobenzoic acid
(b) 3-bromo-2-hydroxy-5-nitrobenzoic acid
(c) 5-nitro-3-bromo-2-hydroxybenzoic acid
(d) 2-hydroxy-3-bromo-5-nitrobenzoic acid

**05** In 3,3-dimethylhex-1-en-4-yne, there are ……… $sp^3$, ……… $sp^2$ and ……… $sp$ hybridised carbon atoms respectively. *[2025, 2 April Shift-II]*

(a) 4, 2, 2
(b) 3, 3, 2
(c) 2, 4, 2
(d) 2, 2, 4

**06** The IUPAC name of the following compound is

$$CH_3—\overset{COOH}{\overset{|}{C}}H—CH_2—CH_2—\overset{COOCH_3}{\overset{|}{C}}H—CH_3$$

*[2025, 22 Jan Shift-I]*

(a) Methyl-6-carboxy-2, 5-dimethylhexanoate
(b) Methyl-5-carboxy-2-methylhexanoate
(c) 2-carboxy-5-methoxycarbonylhexane
(d) 6-methoxycarbonyl-2, 5-dimethylhexanoic acid

**07** In the given structure, number of $sp$ and $sp^2$ hybridised carbon atoms present respectively are

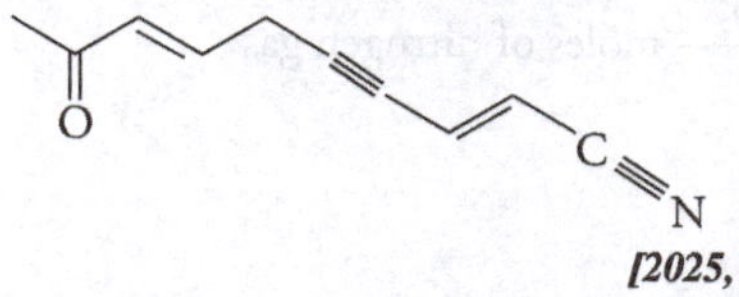


*[2025, 24 Jan Shift-II]*

(a) 3 and 5
(b) 3 and 6
(c) 4 and 6
(d) 4 and 5

**08** IUPAC name of following hydrocarbon ($X$) is

$$\underset{\displaystyle CH_3}{CH_3—\overset{|}{C}H}—CH_2CH_2—\underset{\displaystyle CH_3}{\overset{|}{C}H}—\underset{\displaystyle CH_3}{\overset{|}{C}H}—CH_2CH_3 \quad (X)$$

(a) 2-ethyl-2, 6-diethylheptane *[2024, 9 April Shift-II]*
(b) 3, 4, 7-trimethyloctane
(c) 2-ethyl-3, 6-dimethylheptane
(d) 2, 5, 6-trimethyloctane

**09** The correct nomenclature for the following compound is

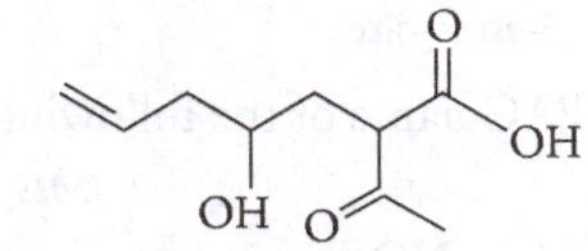

*[2024, 5 April Shift-II]*

(a) 2-carboxy -4-hydroxyhept-7-enal
(b) 2-formyl-4-hydroxyhept-6-enoic acid
(c) 2-formyl-4-hydroxyhept-7-enoic acid
(d) 2-carboxy-4-hydroxyhept-6-enal

**10** Identify structure of 2, 3-dibromo-1-phenylpentane. *[2024, 31 Jan Shift-II]*

(a) 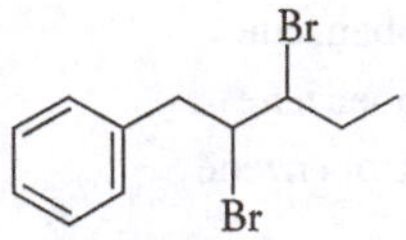
(b) 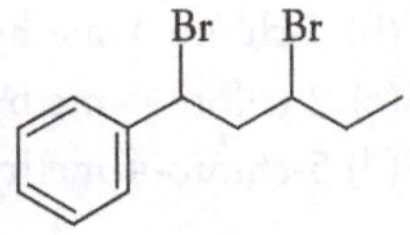
(c) 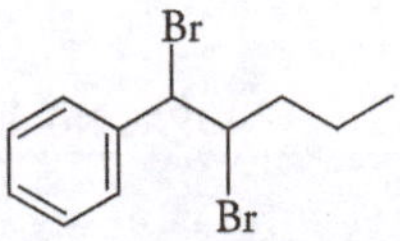
(d) 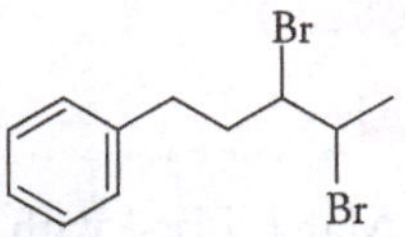

**11** IUPAC name of following compound is

$$CH_3—\underset{\displaystyle NH_2}{\overset{|}{C}H}—CH_2—CN$$

*[2024, 30 Jan Shift-II]*

(a) 2-aminobutanenitrile (b) 2-aminopentanenitrile
(c) 3-aminopropanenitrile (d) 3-aminobutanenitrile

**12** Bond line formula of $HOCH(CN)_2$ is *[2024, 27 Jan Shift-II]*

(a) 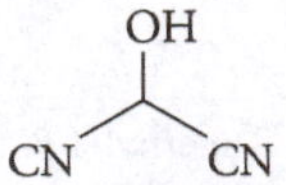
(b) $HO—CH(C{\equiv}N)_2$
(c) $HO—CH(CN)_2$
(d) 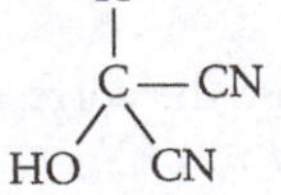

**13** Cyclohexene  is .......... type of an organic compound. *[2024, 27 Jan Shift-I]*

(a) benzenoid aromatic (b) alicyclic
(c) benzenoid non-aromatic (d) acyclic

**14** IUPAC name of following compound ($P$) is

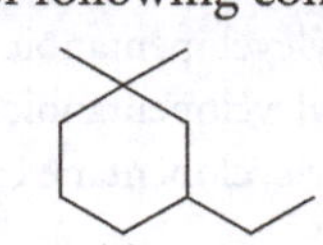

($P$) *[2024, 27 Jan Shift-I]*

(a) 1, 1-dimethyl-3-ethylcyclohexane
(b) 3-ethyl-1, 1-dimethylcyclohexane
(c) 1-ethyl-3, 3-dimethylcyclohexane
(d) 1-ethyl-5, 5-dimethylcyclohexane

**Imp.** Priority Order of Functional Group

**15** The correct decreasing order of priority of functional groups in naming an organic compound as per IUPAC system of nomenclature is *[2022, 26 July Shift-II]*

(a) $—COOH > —CONH_2 > —COCl > —CHO$
(b) $—SO_3H > —COCl > —CONH_2 > —CN$
(c) $—COOR > —COCl > —NH_2 > \rangle C{=}O$
(d) $—COOH > —COOR > —CONH_2 > —COCl$

**16** The correct IUPAC name of the following compound is *[2022, 28 June Shift-II]*

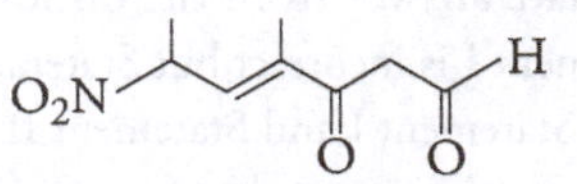

(a) 4-methyl-2-nitro-5-oxohept-3 -enal
(b) 4-methyl-5-oxo-2-nitrohept-3 -enal
(c) 4-methyl-6-nitro-3-oxohept-4 -enal
(d) 6-formyl-4-methyl-2-nitrohex- 3-enal

**17** In the following molecules,

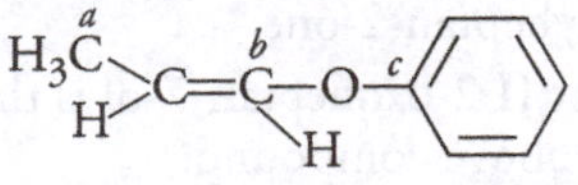

hybridisation of carbon $a$, $b$ and $c$ respectively are *[2021, 18 March Shift-II]*

(a) $sp^3$, $sp$, $sp$ (b) $sp^3$, $sp^2$, $sp$
(c) $sp^3$, $sp^2$, $sp^2$ (d) $sp^3$, $sp$, $sp^2$

**18** In $\overset{1}{C}H_2{=}\overset{2}{C}{=}\overset{3}{C}H—\overset{4}{C}H_3$ molecule, the hybridisation of carbon 1, 2, 3 and 4 respectively, are *[2021, 26 Feb Shift-II]*

(a) $sp^3$, $sp$, $sp^3$, $sp^3$ (b) $sp^2$, $sp^3$, $sp^2$, $sp^3$
(c) $sp^2$, $sp^2$, $sp^2$, $sp^3$ (d) $sp^2$, $sp$, $sp^2$, $sp^3$

**19** The IUPAC name of the following compound is

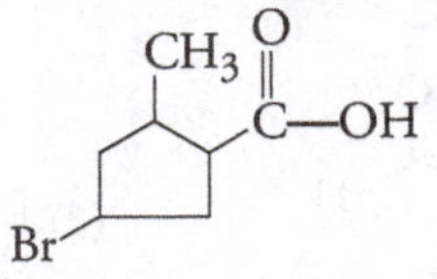

*[2020, 4 Sep Shift-I]*

(a) 3-bromo-5-methylcyclopentane carboxylic acid
(b) 3-bromo-5-methylcyclopentanoic acid
(c) 5-bromo-3-methylcyclopentanoic acid
(d) 4-bromo-2-methylcyclopentane carboxylic acid

## Level 2

**20** Given below are two statements : *[2024, 8 April Shift-I]*

**Statement I**

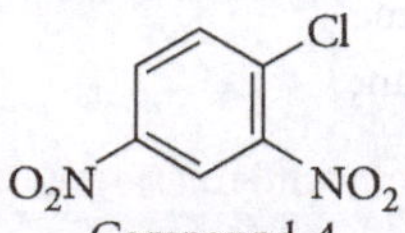

IUPAC name of compound *A* is 4-chloro-1, 3-dinitrobenzene

Compound *A*

**Statement II**

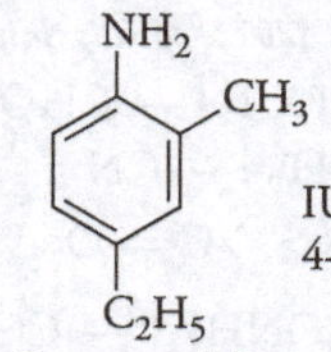

IUPAC name of compound *B* is 4-ethyl-2-methylaniline

Compound *B*

In the light of the above statements. Choose the most appropriate answer from the options given below :
(a) Statement I is incorrect but Statement II is correct.
(b) Both Statement I and Statement II are incorrect.
(c) Statement I is correct but Statement II is incorrect.
(d) Both Statement I and Statement II are correct.

**21** Given below are two statements. *[2024, 31 Jan Shift-I]*

**Statement I** IUPAC name of $HO—CH_2—(CH_2)_3—CH_2—COCH_3$ is 7-hydroxyheptan-2-one.

**Statement II** 2-oxoheptan-7-ol is the correct IUPAC name for above compound.

In the light of the above statements, choose the **most appropriate** answer from the options given below.
(a) Both Statement I and Statement II are incorrect.
(b) Statement I is correct but Statement II is incorrect.
(c) Statement I is incorrect but Statement II is correct.
(d) Both Statement I and Statement II are correct.

**22** Correct structure of γ-methylcyclohexane carbaldehyde is *[2022, 29 July Shift-II]*

(a) 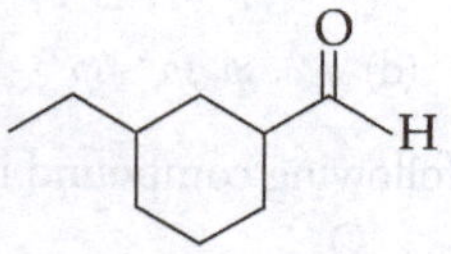

(b) 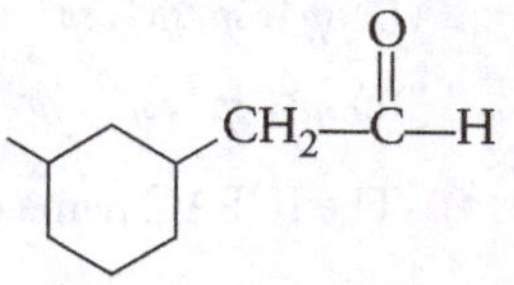

(c) 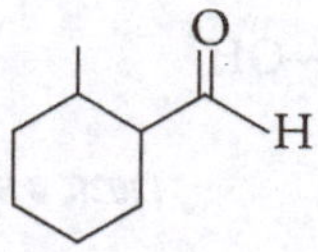

(d) 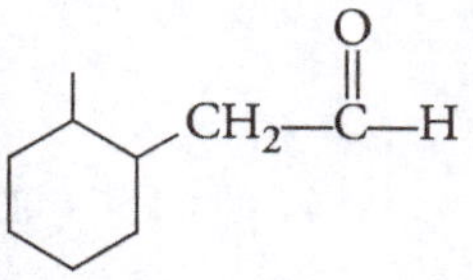

**23** The IUPAC name of ethylidene chloride is *[2022, 25 June Shift-I]*
(a) 1-chloroethene
(b) 1-chloroethyne
(c) 1, 2-dichloroethane
(d) 1, 1-dichloroethane

**24** Mesityl oxide is a common name of *[2021, 17 March Shift-I]*
(a) 2,4-dimethylpentan-3-one
(b) 3-methylcyclohexane carbaldehyde
(c) 2-methylcyclohexanone
(d) 4-methylpent-3-en-2-one

**25** The correct IUPAC name of the following compound is *[2019, 9 April Shift-I]*

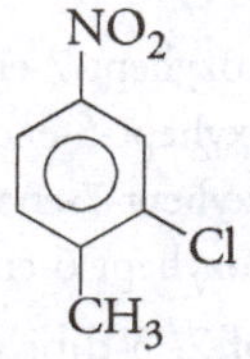

(a) 2-methyl-5-nitro-1 -chlorobenzene
(b) 3-chloro-1-methyl-1-nitrobenzene
(c) 2-chloro-1-methyl-4-nitrobenzene
(d) 5-chloro-4-methyl-1-1-nitrobenzene

## Level 3

**Imp.** IUPAC Nomenclature

**26** Match List-I with List-II.

| | List-I (Structure) | List-II (IUPAC name) |
|---|---|---|
| (A) | $CH_3CH_2CHCH_2CHC_2H_5$ with $C_2H_5$ and $CH_3$ substituents | I. 4-methylpent-1-ene |
| (B) | $(CH_3)_2C\ (C_3H_7)_2$ | II. 3-ethyl-5-methylheptane |
| (C) | 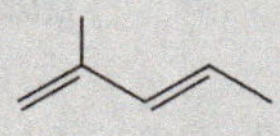 | III. 4,4-dimethylheptane |
| (D) | 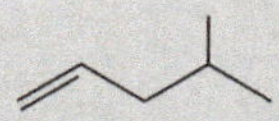 | IV. 2-methyl-1,3-pentadiene |

Choose the correct answer from the options given below : *[2025, 29 Jan Shift-I]*
(a) (A)-(III), (B)-(II), (C)-(I), (D)-(IV)
(b) (A)-(II), (B)-(III), (C)-(I), (D)-(IV)
(c) (A)-(II), (B)-(III), (C)-(IV), (D)-(I)
(d) (A)-(III), (B)-(II), (C)-(IV), (D)-(I)

**27** Match List-I with List-II. *[2022, 28 July Shift-I]*

| List-I | | List-II |
|---|---|---|
| (A) |  | (I) Spiro compound |
| (B) |  | (II) Aromatic compound |
| (C) | 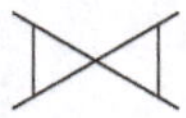 | (III) Non-planar heterocyclic compound |
| (D) |  | (IV) Bicyclo compound |

Choose the correct answer from the options given below:

(a) (A)-(II), (B)-(I), (C)-(IV), (D)-(III)
(b) (A)-(IV), (B)-(III), (C)-(I), (D)-(II)
(c) (A)-(III), (B)-(IV), (C)-(I), (D)-(II)
(d) (A)-(IV), (B)-(III), (C)-(II), (D)-(I)

**28** Choose the correct name for compound given below *[2021, 31 Aug Shift-I]*

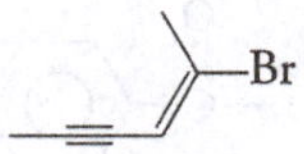


(a) (4*E*)-5-bromohex-4-en-2-yne
(b) (2*E*)-2-bromohex-4-yn-2-ene
(c) (2*E*)-2-bromohex-2-en-4-yne
(d) (4*E*)-5-bromohex-2-en-4-yne

**29** The IUPAC name of the following compounds is

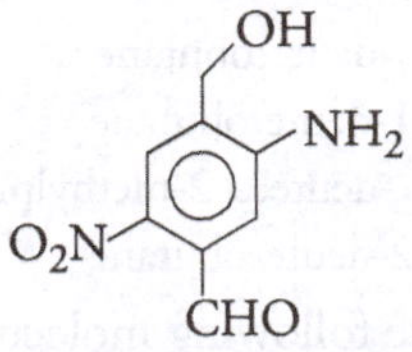


*[2020, 6 Sep Shift-II]*

(a) 2-nitro-4-hydroxymethyl-5-amino benzaldehyde
(b) 3-amino-4-hydroxymethyl-1-5- nitrobenzaldehyde
(c) 5-amino-4-hydroxymethyl-2-nitrobenzaldehyde
(d) 4-amino-2-formyl -5- hydroxymethyl -nitrobenzene

## Numerical Values Type Questions

**30** The sum of sigma ($\sigma$) and pi ($\pi$) bonds in hex-1,3-dien-5-yne is ............... *[2025, 29 Jan Shift-I]*

**31** Two moles of benzaldehyde and one mole of acetone under alkaline conditions using aqueous NaOH after heating gives *X* as the major product. The number of $\pi$-bonds in the product *X* is ........ *[2024, 8 April Shift-II]*

**32** Major product *B* of the following reaction has ........ $\pi$-bond.

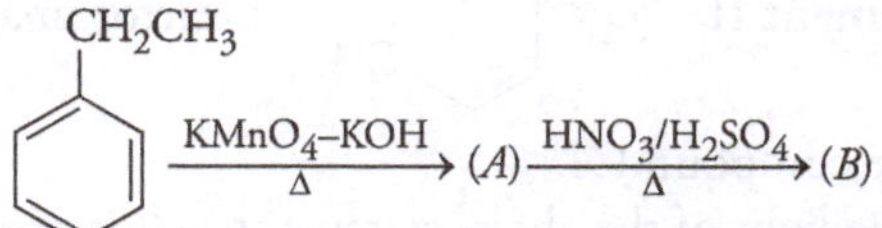


*[2024, 8 April Shift-I]*

**Imp.** Reaction of Alkene in Blend with Sum of $\sigma$ and $\pi$-Bond

**33** The major products from the following reaction sequence are product *A* and product *B*. *[2024, 6 April Shift-I]*

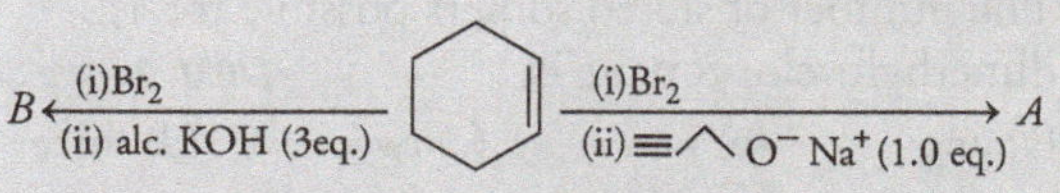


The total sum of $\pi$-electrons in product *A* and product *B* ....... are (Nearest integer)

**34** The total number of 'sigma' and 'Pi' bonds in 2-oxohex-4-ynoic acid is...... . *[2024, 4 April Shift-II]*

**35** The number of $sp^3$ hybridised carbons in an acyclic neutral compound with molecular formula $C_4H_5N$ is ............... . *[2022, 25 July Shift-I]*

**36** The number of sigma bonds in

$$H_3C—\underset{\underset{H}{|}}{C}=CH—C\equiv C—H$$ is ...... .

*[2021, 25 July Shift-I]*

## Topic 2 Isomerism of Organic Compounds

### Level 1

**01** How many different stereoisomers are possible for the given molecule? *[2025, 22 Jan Shift-I]*

$$CH_3—\underset{\underset{OH}{|}}{CH}—CH=CH—CH_3$$

(a) 3 (b) 4 (c) 2 (d) 1

**02** Given below are two statements.

**Statement I** The compound *A* *[2022, 29 July Shift-II]*

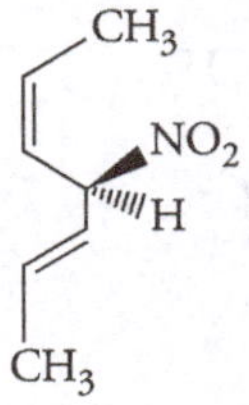


is optically active.

**Statement II** 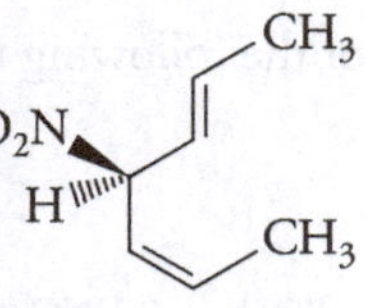 is mirror image of above compound *A*.

In the light of the above statements, choose the most appropriate answer from the options given below.

(a) Both statement I and statement II are correct.
(b) Both statement I and statement II are incorrect.
(c) Statement I is correct but statement II is incorrect.
(d) Statement I is incorrect but statement II is correct.

**03** The number of stereoisomers possible for 1,2-dimethylcyclopropane is **[2021, 26 Aug Shift-II]**

(a) one (b) four (c) two (d) three

**04** Which one of the following pairs of isomers is an example of metamerism? **[2021, 20 July Shift-II]**

(a) $CH_3CH_2CH_2CH_2CH_3$ and 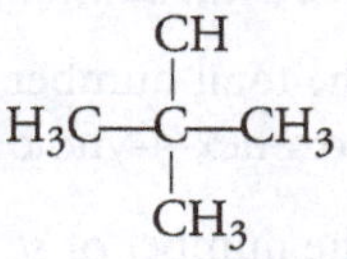

(b) 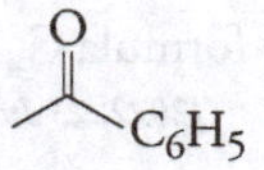 and 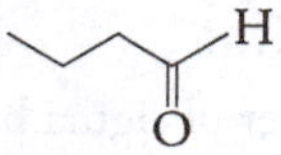

(c) 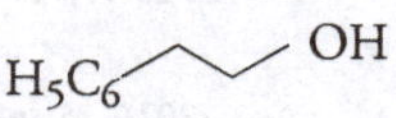 and 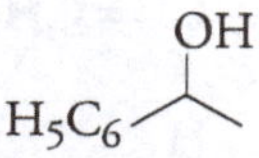

(d) 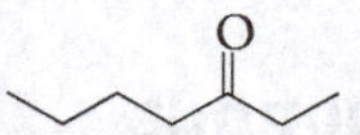 and 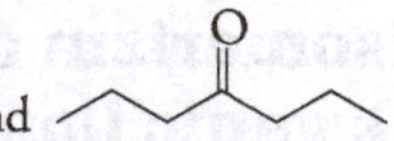

**05** Compound with molecular formula $C_3H_6O$ can show **[2021, 18 March Shift-I]**

(a) positional isomerism
(b) both positional isomerism and metamerism
(c) metamerism
(d) functional group isomerism

**06** Among the following compounds, geometrical isomerism is exhibited by **[2020, 5 Sep Shift-II]**

(a) 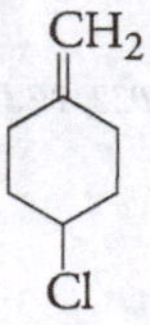 (b) 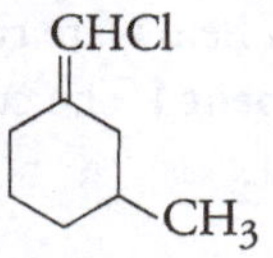

(c) 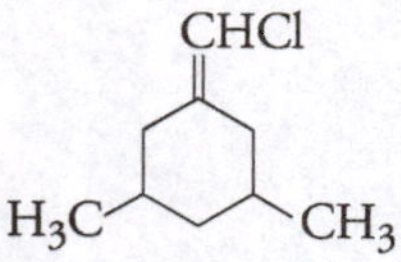 (d) 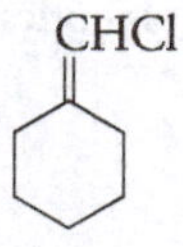

## Level 2

**07** Total number of stereoisomers possible for the given structure **[2024, 9 April Shift-II]**

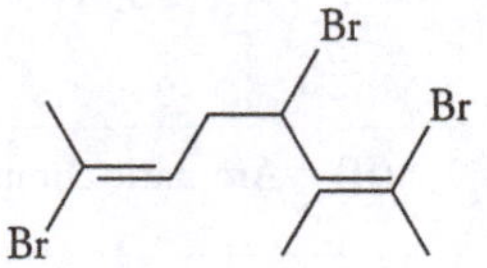

(a) 4 (b) 3
(c) 2 (d) 8

**08** Which of the following is metamer of the given compound (*X*)? **[2024, 6 April Shift-I]**

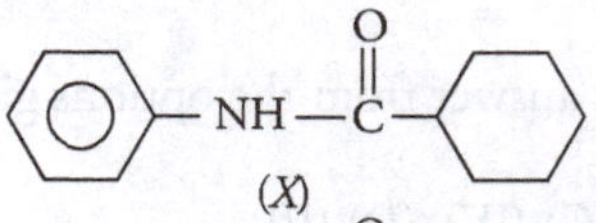
(*X*)

(a) 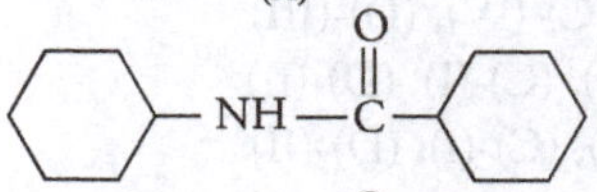

(b) 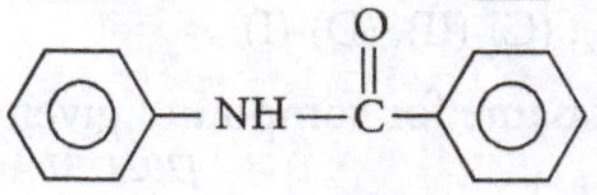

(c) 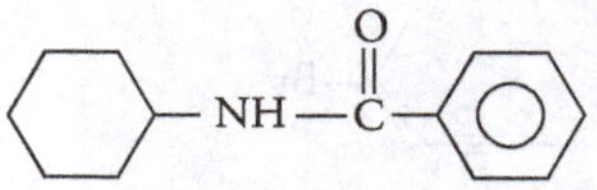

(d) 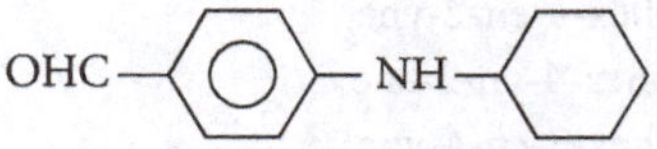

**09** The isomeric deuterated bromide with molecular formula $C_4H_8DBr$ having two chiral carbon atoms is **[2023, 25 Jan Shift-II]**

(a) 2-bromo-3-deuterobutane
(b) 2-bromo-1-deuterobutane
(c) 2-bromo-1-deutero-2-methylpropane
(d) 2-bromo-2-deuterobutane

**10** Which of the following molecules does not show stereoisomerism? **[2021, 22 July Shift-II]**

(a) 3,4-dimethylhex-3-ene (b) 3-methylhex-1-ene
(c) 3-ethylhex-3-ene (d) 4-methylhex-1-ene

**11** Which of the following compounds produces an optically inactive compound on hydrogenation? **[2020, 3 Sep Shift-I]**

(a) 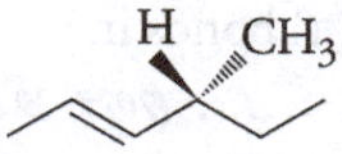 (b) 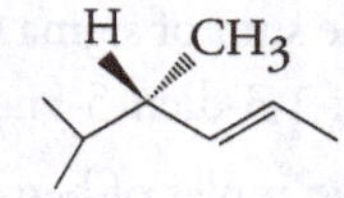

(c)  (d) 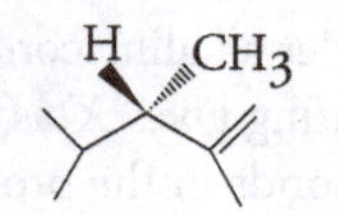

## Level 3

**12** Identify the correct statements from the following.

A 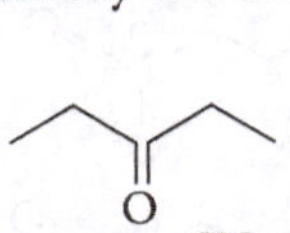 and 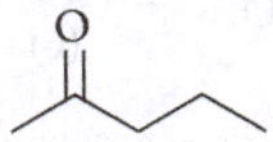 are metamers

B 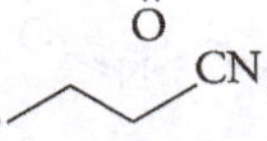 and  are functional isomers

C  and 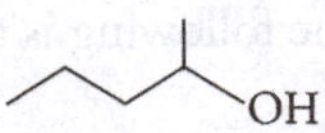 are position isomers

D  and  are homologous

Choose the correct answer from the options given below: *[2025, 3 April Shift-I]*

(a) A, B and C Only (b) A and B Only
(c) C and D Only (d) B and C Only

**13** Match List I with List II.

| List I (Pair of Compounds) | | List II (Isomerism) |
|---|---|---|
| A. | *n*-propanol and iso-propanol | I. Metamerism |
| B. | Methoxypropane and ethoxyethane | II. Chain Isomerism |
| C. | Propanone and propanal | III. Position Isomerism |
| D. | Neopentane and iso-pentane. | IV. Functional Isomerism |

Choose the correct answer from the options given below : *[2024, 5 April Shift-II]*

(a) A-II, B-I, C-IV, D-III (b) A-III, B-I, C-IV, D-II
(c) A-III, B-I, C-II, D-IV (d) A-I, B-III, C-IV, D-II

**14** Match List-I with List-II.

| List-I (Isomeric pairs) | | List-II (Type of isomers) |
|---|---|---|
| (A) | Propanamine and N-methylethanamine | I. Metamers |
| (B) | Hexan-2-one and hexan-3-one | II. Positional isomers |
| (C) | Ethanamide and hydroxyethanimine | III. Functional isomers |
| (D) | *o*-nitrophenol and *p*-nitrophenol | IV. Tautomers |

Choose the correct answer from the options given below. *[2023, 25 Jan Shift-II]*

(a) A-II, B-III, C-I, D-IV
(b) A-III, B-I, C-IV, D-II
(c) A-III, B-IV, C-I, D-II
(d) A-IV, B-III, C-I, D-II

## Numerical Value Type Questions

**15** The total number of structural isomers possible for the substituted benzene derivatives with the molecular formula $C_9H_{12}$ is ......... . *[2025, 3 April Shift-II]*

**Imp.** Isomerism

**16** The possible number of stereoisomers for 5-phenylpent-4-en-2-ol is ......... . *[2025, 24 Jan Shift-II]*

**17** The total number of chiral compounds from the following is ............. . *[2023, 01 Feb Shift-I]*

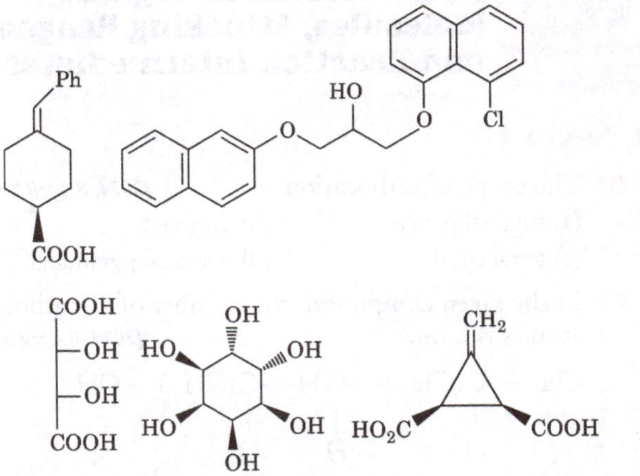

**18** The number of stereoisomers formed in a reaction of (±) Ph(C=O)C(OH)(CN) Ph with HCN is ............ . *[2022, 29 July Shift-II]*

**Imp.** Optical Activity of Enantiomeric Mixture

**19** Optical activity of an enantiomeric mixture is +12.6° and the specific rotation of (+) isomer is +30°. The optical purity is ...... %. *[2022, 27 July Shift-I]*

**20** Observe structures of the following compounds. *[2022, 29 June Shift-I]*

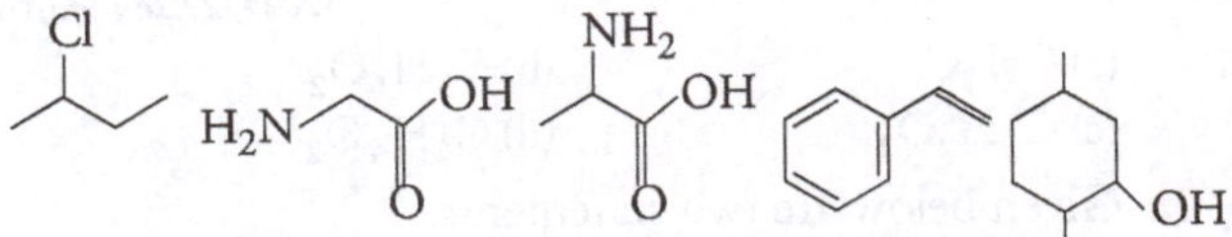

The total number of structures/ compounds which possess asymmetric carbon atoms is ...... .

**21** Total number of possible stereoisomers of dimethyl cyclopentane is ...... . *[2022, 27 June Shift-I]*

**22** The number of acyclic structural isomers (including geometrical isomers) for pentene are ......... . *[2021, 22 July Shift-II]*

**23** The number of chiral carbons present in the molecule given below is ...... . **[2020, 2 Sep Shift-I]**

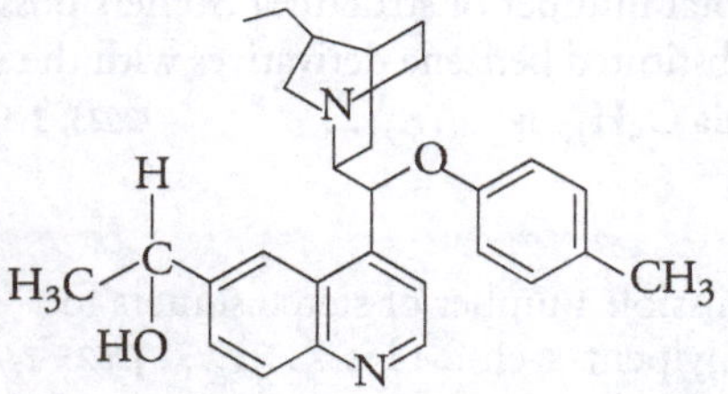


## Topic 3 Bond Fission, Electronic Displacement in Organic Molecules, Attacking Reagent and Reaction Intermediates

### Level 1

**01** The shape of carbocation is **[2024, 8 April Shift-II]**

(a) trigonal planar (b) diagonal
(c) tetrahedral (d) diagonal pyramidal

**02** In the given compound, the number of 2° carbon atom/s is ........ **[2024, 8 April Shift-I]**

$$CH_3—\underset{|\atop H}{C}(CH_3)—\underset{|\atop H}{C}H—\underset{|\atop H}{C}(CH_3)—CH_3$$

(a) Four (b) One
(c) Two (d) Three

**03** The difference in energy between the actual structure and the lowest energy resonance structure for the given compound is **[2024, 29 Jan Shift-I]**

(a) electromeric energy
(b) ionisation energy
(c) hyperconjugation energy
(d) resonance energy

**04** The molecular formula of second homologue in the homologous series of mono carboxylic acids is ....... **[2024, 27 Jan Shift-II]**

(a) $CH_2O$ (b) $C_3H_6O_2$
(c) $C_2H_2O_2$ (d) $C_2H_4O_2$

**05** Given below are two statements.

**Statement I** Hyperconjugation is a permanent effect.

**Statement II** Hyperconjugation in ethyl cation $(CH_3—\overset{+}{C}H_2)$ involves the overlapping of $C_{sp^2}—H_{1s}$ bond with empty $2p$ orbital of other carbon.

Choose the correct option **[2021, 27 July Shift-II]**

(a) Both statement I and statement II are false.
(b) Statement I is incorrect but statement II is true.
(c) Statement I is true but statement II is false.
(d) Both Statement I and statement II are true.

**06** Which one among the following resonating structures is not correct ? **[2021, 25 July Shift-I]**

(a) 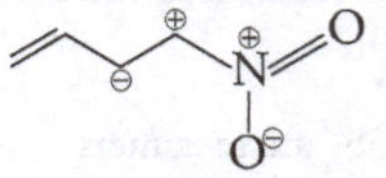 (b) 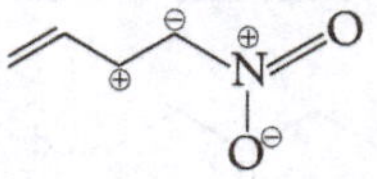

(c) 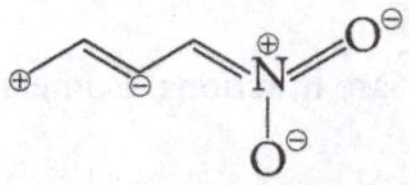 (d) 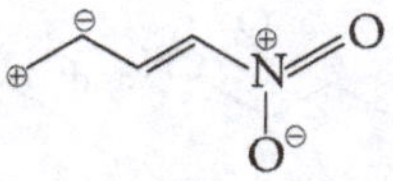

**07** Which among the following is the strongest acid? **[2021, 25 July Shift-II]**

(a) $CH_3CH_2CH_2CH_3$ (b) 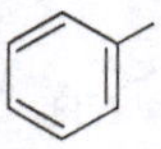

(c)  (d) 

**08** Which of the following has the shortest C—Cl bond? **[2020, 8 Jan Shift-II]**

(a) $ClCH=CH—CH_3$ (b) $ClCH=CH—NO_2$
(c) $ClCH=CH—OCH_3$ (d) $ClCH=CH_2$

**09** Which amongst the following is the strongest acid? **[2019, 9 Jan Shift-I]**

(a) $CHBr_3$ (b) $CHI_3$ (c) $CHCl_3$ (d) $CH(CN)_3$

**10** The correct order for acid strength of compounds $CH\equiv CH$, $CH_3—C\equiv CH$ and $CH_2=CH_2$ is as follows **[2019, 12 Jan Shift-I]**

(a) $CH_3—C\equiv CH > CH_2=CH_2 > HC\equiv CH$
(b) $CH_3—C\equiv CH > CH\equiv CH > CH_2=CH_2$
(c) $HC\equiv CH > CH_3—C\equiv CH > CH_2=CH_2$
(d) $CH\equiv CH > CH_2=CH_2 > CH_3—C\equiv CH$

**11** On the treatment of the following compound with a strong acid, the most susceptible site for bond cleavage is: **[JEE Mains 2018]**

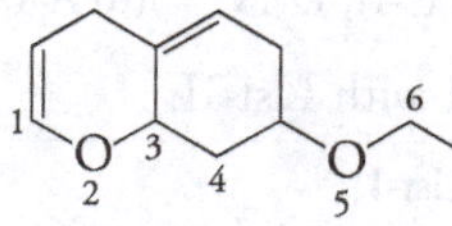


(a) $O_2$-$C_3$ (b) $O_5$-$C_6$ (c) $C_4$-$O_5$ (d) $C_1$-$O_2$

### Level 2

**12.** Match the List-I with List-II. **[2025, 8 April Shift-II]**

| List-I | | List-II | |
|---|---|---|---|
| A. | Carbocation | I. | Species that can supply a pair of electrons. |
| B. | C-Free radical | II. | Species that can receive a pair of electrons. |
| C. | Nucleophile | III. | $sp^2$ hybridised carbon with empty $p$-orbital. |
| D. | Electrophile | IV. | $sp^2/sp^3$ hybridised carbon with one unpaired electron. |

Choose the correct answer from the options given below:

(a) A-III, B-IV, C-II, D-I (b) A-III, B-IV, C-1, D-II
(c) A-II, B-III, C-I, D-IV (d) A-IV, B-II, C-III, D-1

**13.** Given below are two statements: *[2025, 3 April Shift-II]*

**Statement I** Hyperconjugation is not a permanent effect.

**Statement II** In general, greater the number of alkyl groups attached to a positively charged C-atom, greater is the hyperconjugation interaction and stabilisation of the cation.

In the light of the above statements, choose the correct answer from the options given below.

(a) Statement I is false but Statement II is true.
(b) Both Statement I and Statement II are false.
(c) Statement I is true but Statement II is false.
(d) Both Statement I and Statement II are true.

**14.** Consider the following compound (*X*)

$$\overset{\text{I}}{\text{H}}—\text{C}\equiv\overset{\text{II}}{\text{C}}—\overset{\text{III}}{\text{CH}_2}—\underset{\substack{|\\ \text{CH}_3}}{\text{CH}}—\overset{\text{IV}}{\text{CH}_3}$$

(*X*)

The most stable and least stable carbon radicals, respectively, produced by homolytic cleavage of corresponding C—H bond are: *[2025, 2 April Shift-I]*

(a) I, IV (b) III, II (c) II, I (d) II, IV

**15** Which one of the carbocations from the following is most stable? *[2025, 24 Jan Shift-I]*

(a) 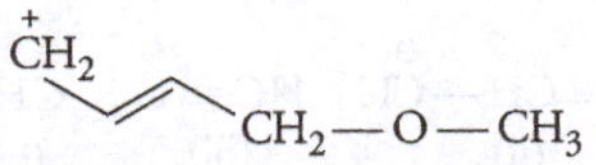

(b) 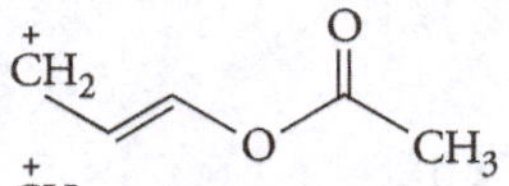

(c) 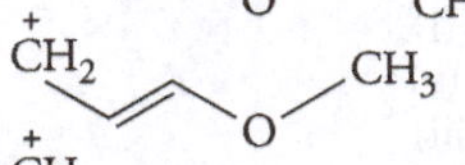

(d) 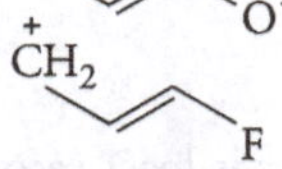

**16** Relative stability of the contributing structures is

$$\text{CH}_2{=}\text{CH}—\underset{\text{(I)}}{\overset{\overset{\ddot{\text{O}}:}{\|}}{\text{C}}}—\text{H} \longleftrightarrow \overset{+}{\text{C}}\text{H}_2\,\text{CH}{=}\underset{\text{(II)}}{\overset{\overset{:\ddot{\text{O}}:^{-}}{|}}{\text{C}}}—\text{H} \longleftrightarrow$$

$$:\overset{-}{\text{C}}\text{H}_2—\text{CH}{=}\underset{\text{(III)}}{\overset{\overset{:\ddot{\text{O}}:^{+}}{|}}{\text{C}}}—\text{H}$$

*[2024, 9 April Shift-I]*

(a) (II) > (I) > (III) (b) (III) > (II) > (I)
(c) (I) > (II) > (III) (d) (I) > (III) > (II)

**Imp.** **Stability of Carbanion**

**17** Correct order of stability of carbanion is

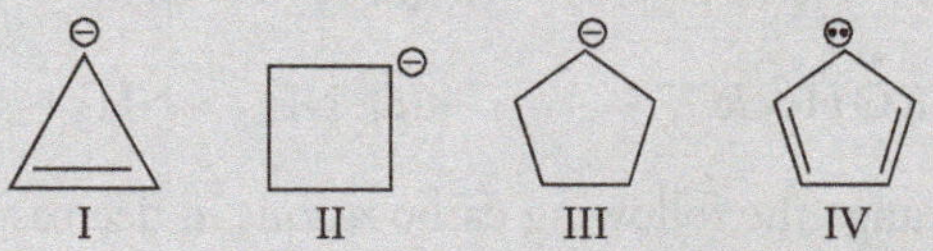

I II III IV

*[2024, 4 April Shift-II]*

(a) III > II > IV > I (b) I > II > III > IV
(c) IV > III > II > I (d) IV > I > III > II

**18** Which among the following is incorrect statement? *[2024, 4 April Shift-I]*

(a) Hydrogen ion ($H^+$) shows negative electromeric effect.
(b) Electromeric effect dominates over inductive effect.
(c) The electromeric effect is, temporary effect.
(d) The organic compound shows electromeric effect in the presence of the reagent only.

**19** The functional group that shows negative resonance effect is *[2024, 1 Feb Shift-II]*

(a) —*OR* (b) —COOH (c) —$NH_2$ (d) —OH

**20** A species having carbon with sextet of electrons and can act as electrophile is called *[2024, 31 Jan Shift-I]*

(a) pentavalent carbon (b) carbon free radical
(c) carbanion (d) carbocation

**Imp.** **Order of Hydride Affinity**

**21** The decreasing order of hydride affinity for following carbocations is

A. $\text{CH}_2{=}\text{CH}—\underset{\substack{|\\ \text{CH}_3}}{\overset{+}{\text{C}}}—\text{CH}_3$ B. $\text{C}_6\text{H}_5—\underset{\substack{|\\ \text{C}_6\text{H}_5}}{\overset{+}{\text{C}}}—\text{C}_6\text{H}_5$

C. $\text{CH}_3—\underset{\substack{|\\ \text{CH}_3}}{\overset{+}{\text{C}}}—\text{CH}_3$ D. 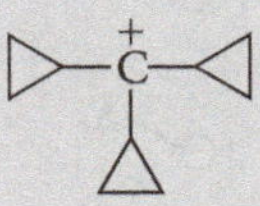

Choose the correct answer from the options given below *[2023, 10 April Shift-II]*

(a) C, A, B, D (b) C, A, D, B
(c) A, C, D, B (d) A, C, B, D

**22** The most stable carbocation for the following is *[2023, 30 Jan Shift-II]*

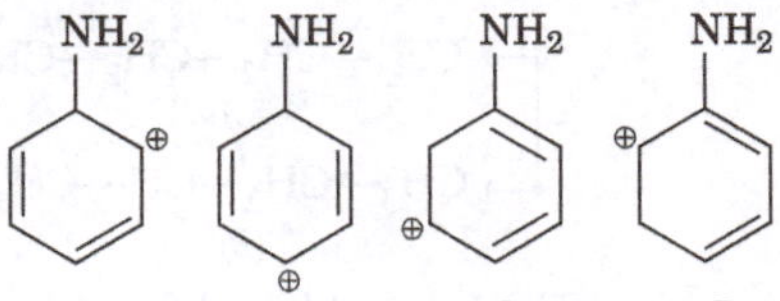

A B C D

(a) *D* (b) *C* (c) *B* (d) *A*

**23** The correct order of nucleophilicity is *[2022, 26 June Shift-II]*

(a) $F^- > OH^-$ (b) $H_2\ddot{\underset{\cdot\cdot}{O}} > OH^-$

(c) $R\ddot{\underset{\cdot\cdot}{O}}H > RO^-$ (d) $NH_2^- > NH_3$

**24** Arrange the following carbocations in decreasing order of stability. *[2022, 24 June Shift-II]*

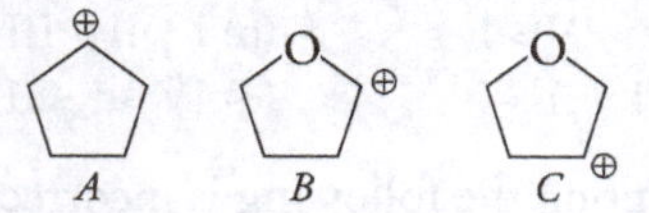

(a) $A > C > B$ (b) $A > B > C$
(c) $C > B > A$ (d) $C > A > B$

**25** 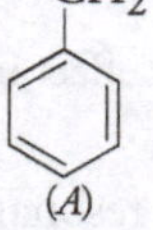 $\overset{\oplus}{CH_2}$ (A) $CH_2{=}\overset{\oplus}{C}H$ (B) $CH_3{-}\overset{\oplus}{C}H_2$ (C) $H_2C{=}\overset{\oplus}{C}$ (D)

The correct order of stability of given carbocation is *[2021, 27 July Shift-I]*

(a) $A > C > B > D$ (b) $D > B > C > A$
(c) $D > B > A > C$ (d) $C > A > D > B$

**26** Which of the following compounds does not exhibit resonance ? *[2021, 22 July Shift-II]*

(a) $CH_3CH_2OCH{=}CH_2$

(b) 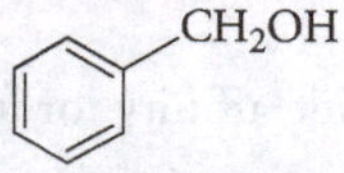

(c) $CH_3CH_2CH_2CONH_2$
(d) $CH_3CH_2CH{=}CHCH_2NH_2$

**27** (A) 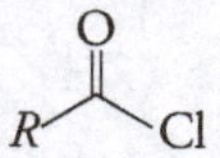 (B) 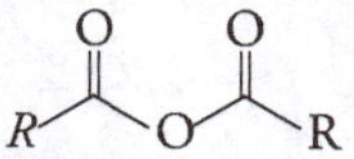

(C) 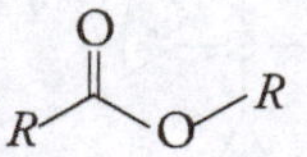 (D) 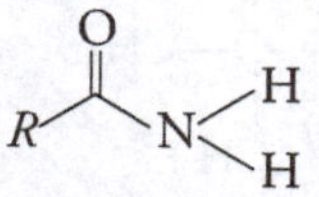

The correct order of their reactivity towards hydrolysis at room temperature is *[2021, 20 July Shift-II]*

(a) (A) > (B) > (C) > (D) (b) (D) > (A) > (B) > (C)
(c) (D) > (B) > (A) > (C) (d) (A) > (C) > (B) > (D)

**28** Choose the correct statement regarding the formation of carbocations *A* and *B*. *[2021, 17 March Shift-II]*

$CH_3CH_2CH{=}CH_2 + HBr$
$\rightarrow CH_3{-}CH_2{-}CH_2{-}\overset{+}{C}H_2 + Br^-$ (*A*)
$\rightarrow CH_3{-}CH_2{-}\overset{+}{C}H{-}CH_3 + Br^-$ (*B*)

(a) Carbocation *B* is more stable and formed relatively at faster rate.
(b) Carbocation *A* is more stable and formed relatively at slow rate.
(c) Carbocation *B* is more stable and formed relatively at slow rate.
(d) Carbocation *A* is more stable and formed relatively at faster rate.

**29** Which one of the following compounds possesses the most acidic hydrogen? *[2020, 3 Sep Shift-I]*

(a) 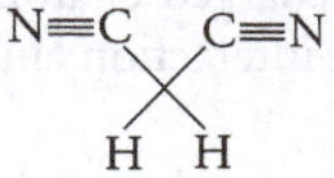 (b) $H_3C{-}C{\equiv}C{-}H$

(c) 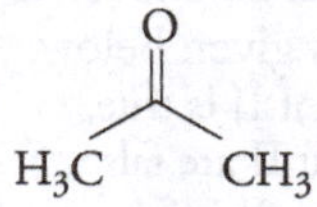 (d) 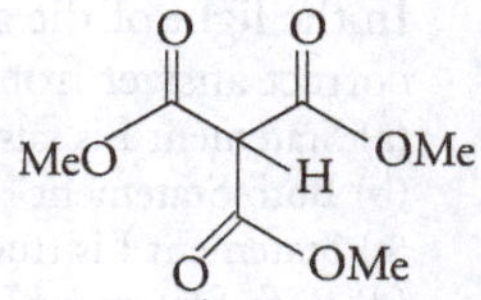

**30** Arrange the following labelled hydrogens in decreasing order of acidity *[2020, 2 Sep Shift-II]*

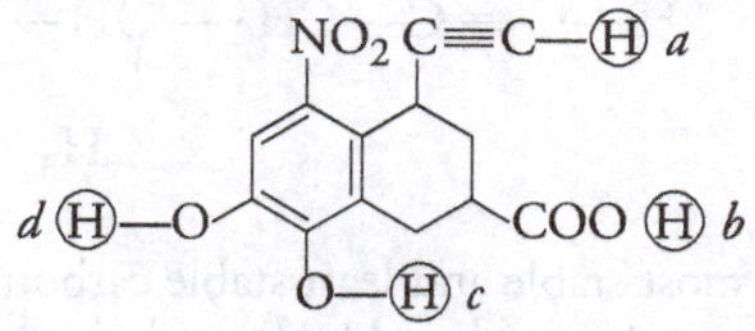

(a) $b > a > c > d$ (b) $c > b > d > a$
(c) $b > c > d > a$ (d) $c > b > a > d$

**31** The increasing order of basicity for the following intermediates is (from weak to strong) *[2020, 8 Jan Shift-II]*

$(CH_3)_3C^{\ominus}$ (i) $H_2C{=}CH{-}\overset{\ominus}{C}H_2$ (ii) $HC{\equiv}\overset{\ominus}{C}$ (iii) $\overset{\ominus}{C}H_3$ (iv) $\overset{\ominus}{C}N$ (v)

(a) (v) < (iii) < (ii) < (iv) < (i)
(b) (iii) < (i) < (ii) < (iv) < (v)
(c) (v) < (i) < (iv) < (ii) < (iii)
(d) (iii) < (iv) < (ii) < (i) < (v)

**32** Which of the following molecules is least resonance stabilised? *[JEE Main 2017]*

(a)  (b)  (c) 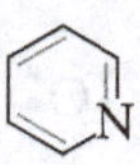 (d) 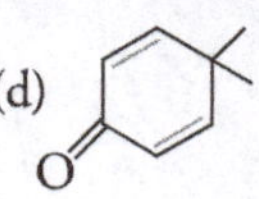

## Level 3

**33.** In which pairs, the first ion is more stable than the second? *[2025, 4 April Shift-II]*

A. 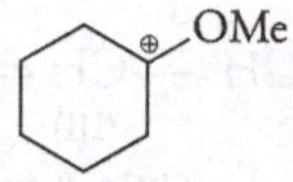 and 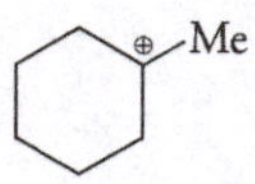

B. 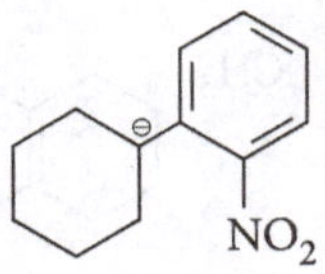 and 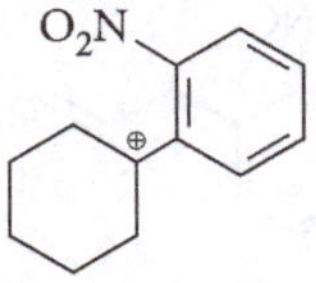

C. 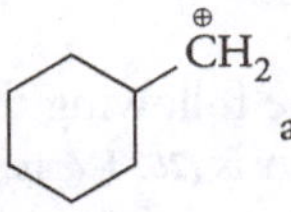 and 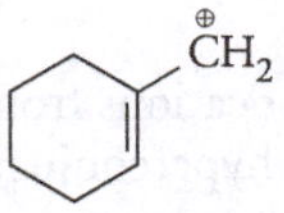

D. 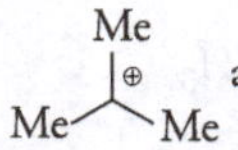 and 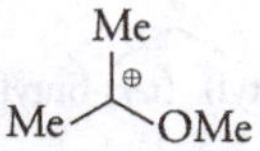

(a) (B) and (C) only (b) (A) and (C) only
(c) (B) and (D) only (d) (A) and (B) only

**Imp.** **Stability of Carbocation**

**34** The correct order of stability of following carbocations is **[2025, 28 Jan Shift-I]**

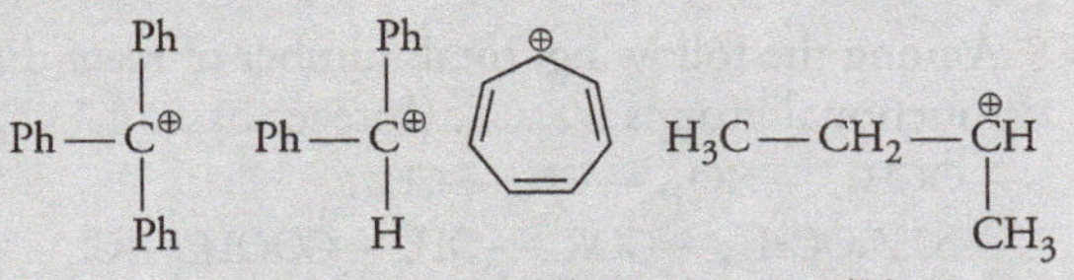

(A) (B) (C) (D)

(a) B > C > A > D (b) C > B > A > D
(c) C > A > B > D (d) A > B > C > D

**35** The most stable carbocation from the following is **[2025, 22 Jan Shift-II]**

(a) 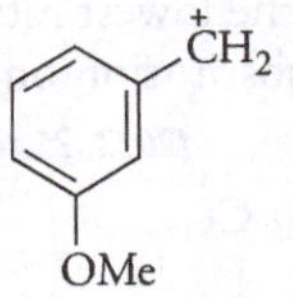 (b) 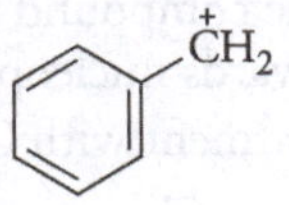

(c) 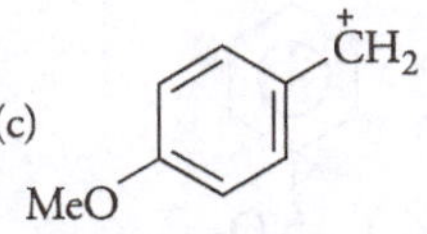 (d) 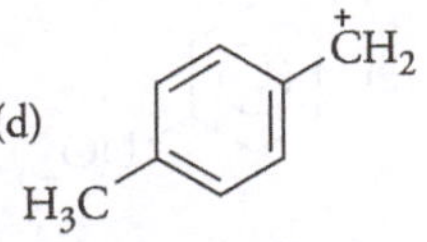

**36** Total number of nucleophiles from the following is :
$NH_3$, PhSH, $(H_3C)_2S$, $H_2C=CH_2$
$\overset{\ominus}{O}H^{\ominus}$, $H_3O^{\oplus}$, $(CH_3)_2CO$, 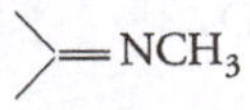 **[2025, 29 Jan Shift-I]**

(a) 5 (b) 6
(c) 7 (d) 4

**37** Match List I with List II.

| List I (Mechanism steps) | List II (Effect) |
|---|---|
| (A) 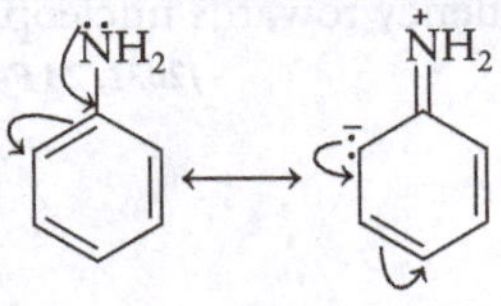 | I. $-E$ effect |
| (B) 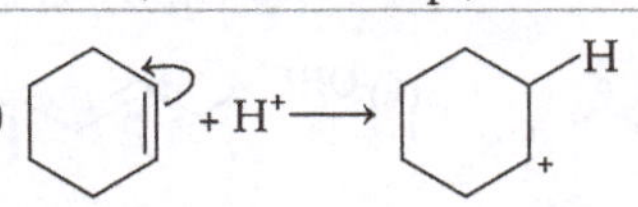 | II. $-R$ effect |
| (C) 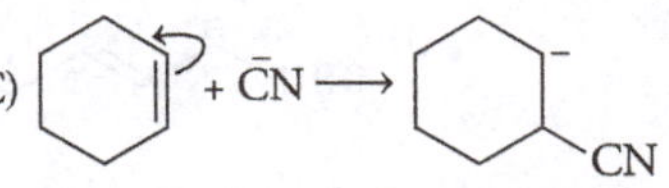 | III. $+E$ effect |
| (D) 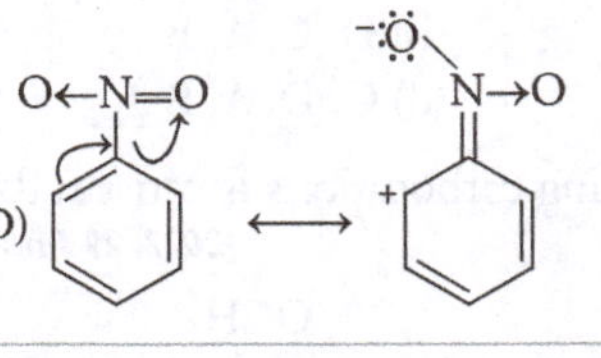 | IV. $+R$ effect |

Choose the correct answer from the options given below **[2024, 4 April Shift-I]**

(a) A-I, B-II, C-IV, D-III (b) A-II, B-IV, C-III, D-I
(c) A-III, B-I, C-II, D-IV (d) A-IV, B-III, C-I, D-II

**38** Ionic reactions with organic compounds proceed through
(A) homolytic bond cleavage
(B) heterolytic bond cleavage
(C) free radical formation
(D) primary free radical
(E) secondary free radical

Choose the **correct answer** from the options given below **[2024, 1 Feb Shift-I]**

(a) (C) only (b) (A) only
(c) (B) only (d) (D) and (E) only

**39** Which of the following compound will most easily be attacked by an electrophile? **[2024, 1 Feb Shift-I]**

(a) 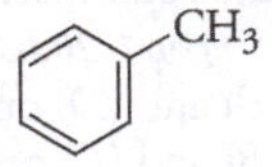 (b) 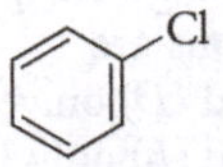

(c)  (d) 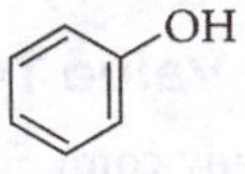

**40** The correct stability order of carbocations is **[2024, 30 Jan Shift-II]**

(a) $(CH_3)_3C^+ > CH_3-\overset{+}{C}H_2 > (CH_3)_2\overset{+}{C}H > \overset{+}{C}H_3$

(b) $\overset{+}{C}H_3 > (CH_3)_2\overset{+}{C}H > CH_3-\overset{+}{C}H_2 > (CH_3)_3\overset{+}{C}$

(c) $\overset{+}{C}H_3 > CH_3-\overset{+}{C}H_2 > CH_3-\overset{+}{C}H(CH_3) > (CH_3)_3\overset{+}{C}$

(d) $(CH_3)_3\overset{+}{C} > (CH_3)_2\overset{+}{C}H > CH_3-\overset{+}{C}H_2 > \overset{+}{C}H_3$

**41** Increasing order of stability of the resonance structures is **[2023, 24 Jan Shift-I]**

(A) 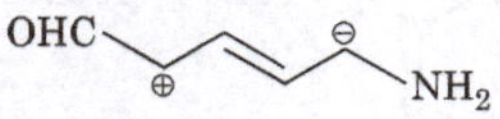
(B) 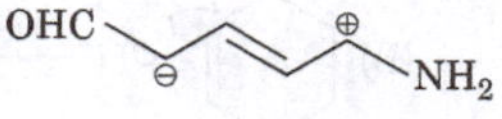

(C) 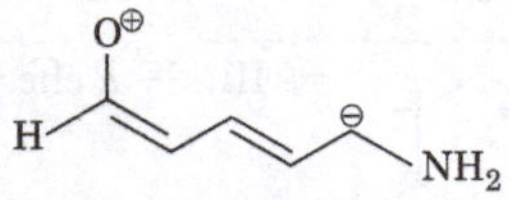
(D) 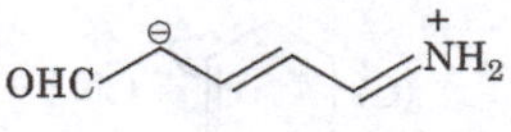

Choose the correct answer from the options given below.

(a) D, C, B, A
(b) D, C, A, B
(c) C, D, B, A
(d) C, D, A, B

**42** Which of the following carbocations is most stable? **[2022, 29 June Shift-II]**

(a) 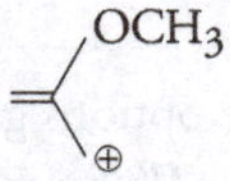
(b) 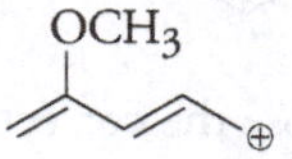

(c) 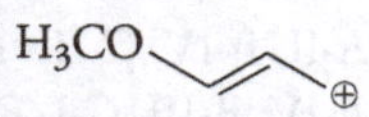
(d) 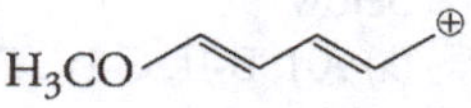

**43** Which of the following is most stable? **[2022, 27 June Shift-II]**

(a) 
(b) 

(c) 
(d) 

**44** 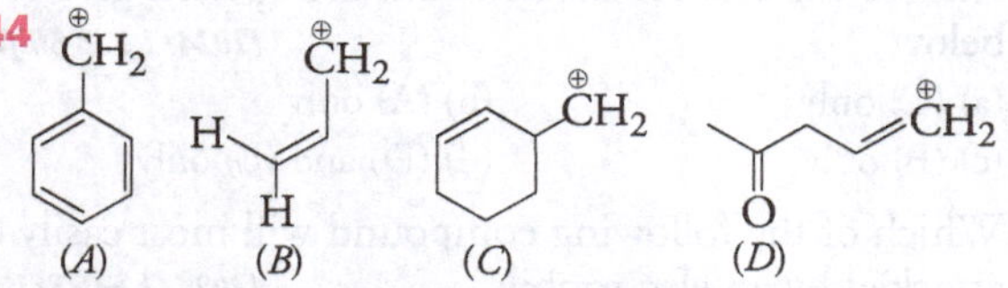

(A) (B) (C) (D)

Among the given species the resonance stabilised carbocations are **[2021, 20 July Shift-I]**

(a) (C) and (D) only
(b) (A), (B) and (D) only
(c) (A) and (B) only
(d) (A), (B) and (C) only

## Numerical Value Type Questions

**45** How many compounds among the following compounds show inductive, mesomeric as well as hyperconjugation effects? **[2024, 9 April Shift-I]**

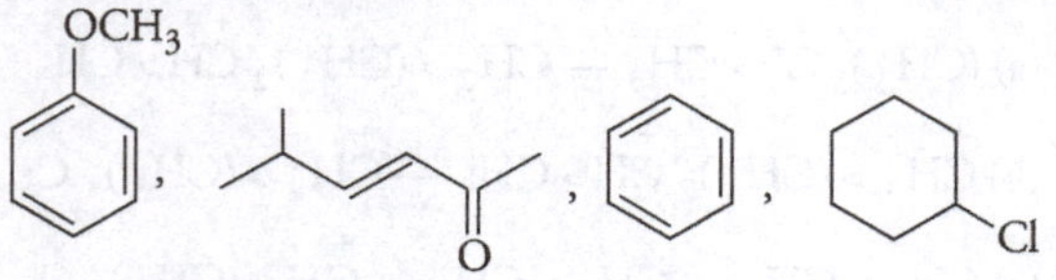

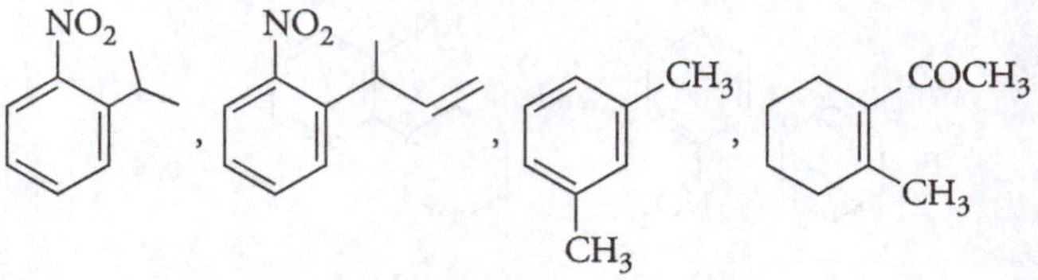

**46** Number of carbocations from the following that are not stablised by hyperconjugation is **[2024, 6 April Shift-II]**

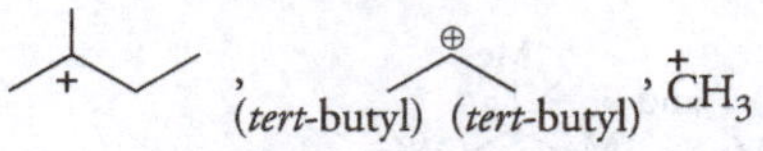

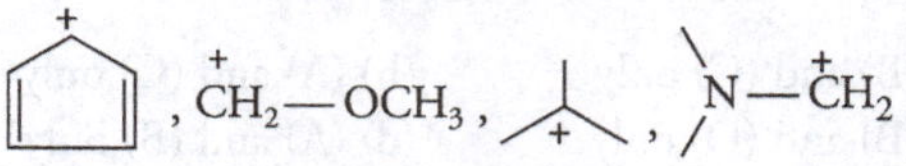

**47** Number of compounds among the following which contain sulphur as heteroatom is .......... .

Furan, thiophene, pyridine, pyrrole, cysteine, tyrosine **[2024, 29 Jan Shift-I]**

**48** Among the following, total number of *meta*-directing functional groups is ........ (Integer based)

$—OCH_3$, $—NO_2$, $—CN$, $—CH_3$, $—NHCOCH_3$, $—COR$, $—OH$, $—COOH$, $—Cl$ **[2024, 27 Jan Shift-I]**

## Topic 4 Types of Organic Reactions

### Level 1

**01** The compound which will have the lowest rate towards nucleophilic aromatic substitution on treatment with $OH^-$ is **[2023, 25 Jan Shift-I]**

(a) 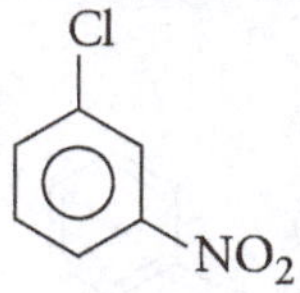
(b) 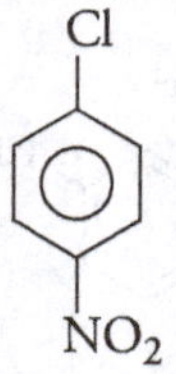

(c) 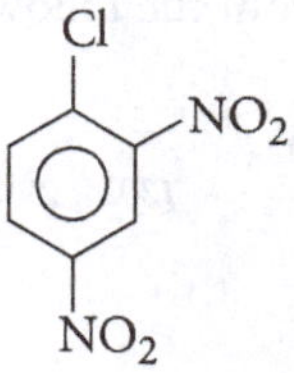
(d) 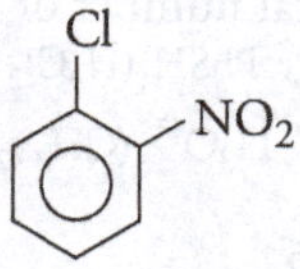

### Level 2

**02** The correct order of the following compounds showing increasing tendency towards nucleophilic substitution reaction is **[2021, 24 Feb Shift-III]**

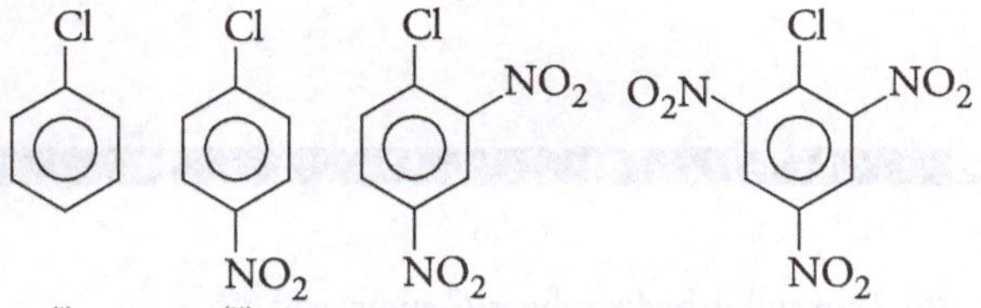

(i) (ii) (iii) (iv)

(a) (iv) < (iii) < (ii) < (i)
(b) (iv) < (i) < (ii) < (iii)
(c) (iv) < (i) < (iii) < (ii)
(d) (i) < (ii) < (iii) < (iv)

## Level 3

**03** In the following given reaction, 'A' is *[2023, 24 Jan Shift-I]*

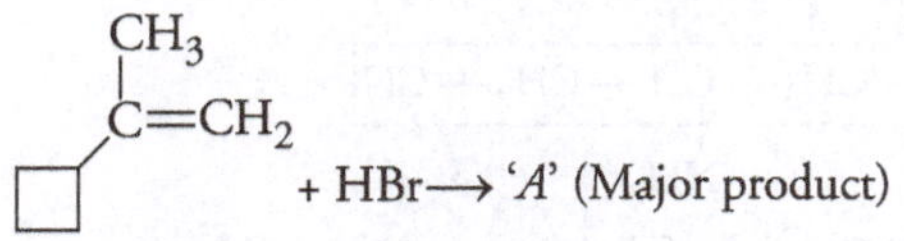

(a) 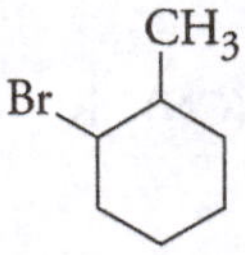

(b) 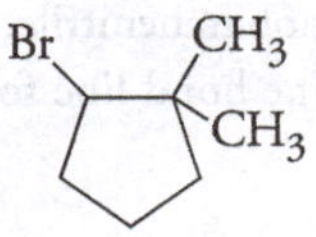

(c) 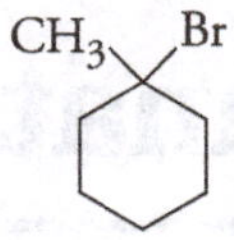

(d) 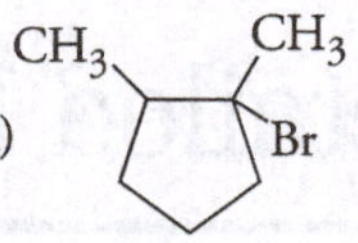

**04** Arrange the following in increasing order of reactivity towards nitration.

A. *p*-xylene
B. Bromobenzene
C. Mesitylene
D. Nitrobenzene
E. Benzene

*[2022, 28 July Shift-II]*

(a) C < D < E < A < B
(b) D < B < E < A < C
(c) D < C < E < A < B
(d) C < D < E < B < A

## Numerical Value Type Questions

**Imp.** Number of Hyperconjugative Structure

**05** 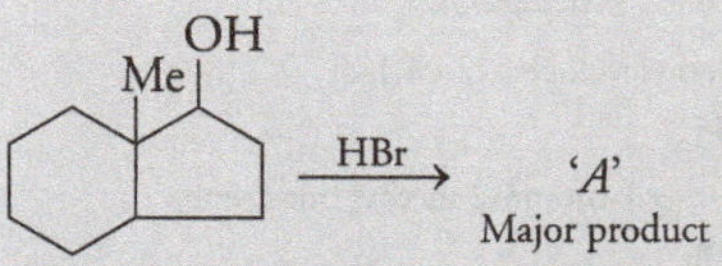

The number of hyperconjugation structure involved to stablise carbocation formed in the above reaction is ............ . *[2023, 11 April Shift-I]*

## Key Index

### Topic 1

| | | | | | | | | | |
|---|---|---|---|---|---|---|---|---|---|
| 1 (d) | 2 (a) | 3 (a) | 4 (b) | 5 (a) | 6 (d) | 7 (a) | 8 (d) | 9 (b) | 10 (a) |
| 11 (d) | 12 (a) | 13 (b) | 14 (b) | 15 (b) | 16 (c) | 17 (c) | 18 (d) | 19 (d) | 20 (a) |
| 21 (b) | 22 (a) | 23 (d) | 24 (d) | 25 (c) | 26 (c) | 27 (c) | 28 (c) | 29 (c) | 30 (15) |
| 31 (9) | 32 (5) | 33 (3) | 34 (18) | 35 (1) | 36 (10) | | | | |

### Topic 2

| | | | | | | | | | |
|---|---|---|---|---|---|---|---|---|---|
| 1 (b) | 2 (c) | 3 (d) | 4 (d) | 5 (d) | 6 (b) | 7 (d) | 8 (c) | 9 (a) | 10 (c) |
| 11 (d) | 12 (b) | 13 (b) | 14 (b) | 15 (8) | 16 (4) | 17 (2) | 18 (3) | 19 (42) | 20 (3) |
| 21 (5) | 22 (6) | 23 (5) | | | | | | | |

### Topic 3

| | | | | | | | | | |
|---|---|---|---|---|---|---|---|---|---|
| 1 (a) | 2 (b) | 3 (d) | 4 (d) | 5 (c) | 6 (a) | 7 (d) | 8 (b) | 9 (d) | 10 (c) |
| 11 (b) | 12 (b) | 13 (a) | 14 (c) | 15 (c) | 16 (c) | 17 (c) | 18 (a) | 19 (b) | 20 (d) |
| 21 (a) | 22 (b) | 23 (d) | 24 (a) | 25 (a) | 26 (d) | 27 (a) | 28 (a) | 29 (d) | 30 (c) |
| 31 (a) | 32 (d) | 33 (d) | 34 (c) | 35 (c) | 36 (a) | 37 (d) | 38 (c) | 39 (d) | 40 (d) |
| 41 (d) | 42 (d) | 43 (a) | 44 (c) | 45 (4) | 46 (5) | 47 (2) | 48 (4) | | |

### Topic 4

| | | | | |
|---|---|---|---|---|
| 1 (a) | 2 (d) | 3 (d) | 4 (b) | 5 (7) |

# Detailed **Explanations**

## Topic 1

**01.** *(d)* $\overset{7}{HC}\equiv\overset{6}{C}-\overset{5}{CH_2}-\overset{4}{\underset{}{CH}}(OH)-\overset{3}{CH_2}-\overset{2}{CH}=\overset{1}{CH_2}$

Hept-1-en-6-yn-4-ol

**02.** *(a)*

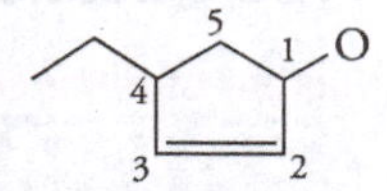

4-ethylcyclopent-2-en-1-ol

**03.** *(a)*

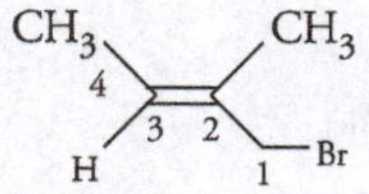

1-bromo-2 methyl but-2-ene

**04.** *(b)*

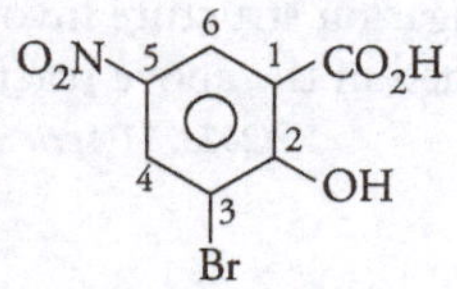

3-bromo-2 hydroxy-5-nitrobenzoic acid

**05.** *(a)* The structure of 3, 3-dimethylhex-1-ene-4-yne is

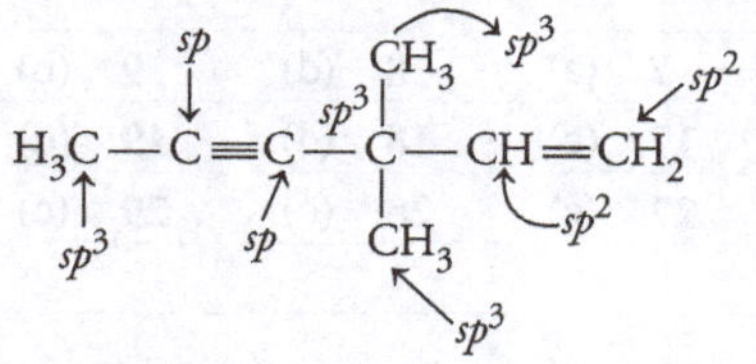


There are 4-$sp^3$, 2-$sp^2$, 2-$sp$ hybridised carbon atoms respectively.

**06.** *(d)* $H_3C-\underset{2}{CH}(\overset{1}{COOH})-\underset{3}{CH_2}-\underset{4}{CH_2}-\underset{5}{CH}(\overset{6}{COOCH_3})-CH_3$

So, the IUPAC name of the given compound is 6-methoxycarbonyl -2, 5- dimethylhexanoic acid.

**07.** *(a)* The hybridisation of each carbon in the given compound is

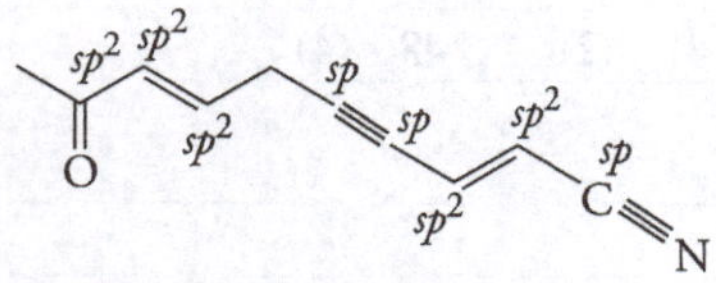


Hence, number of $sp$ and $sp^3$ hybridised carbon atom are 3 and 5.

**08.** *(d)* The IUPAC name of given hydrocarbon ($X$) is 2, 5, 6-trimethyloctane.

**09.** *(b)* 2-formyl-4-hydroxyhept-6-enoic acid.

Hence, option (b) is correct.

**10.** *(a)* The structure of 2, 3-dibromo-1-phenylpentane is

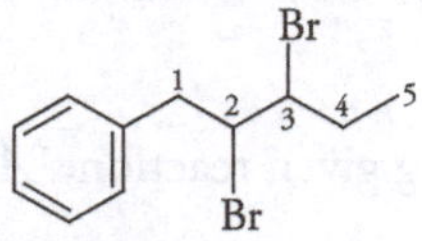

**11.** *(d)* $\overset{4}{CH_3}-\overset{3}{CH}(NH_2)-\overset{2}{CH_2}-\overset{1}{CN}$

IUPAC name of the given compound is 3-aminobutanenitrile.

**12.** *(a)* The bond line formula for $HOCH(CN)_2$ is

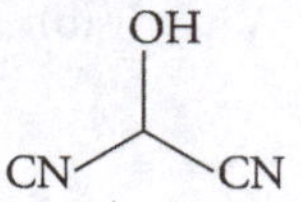

**13.** *(b)*

Cyclohexene is an alicyclic compound.

An alicyclic compound are the organic compounds that are both aliphatic and cyclic. These are either saturated or unsaturated but do not have aromatic character.

**14.** *(b)*

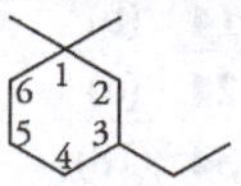

3-ethyl-1,1-dimethylcyclohexane.

**15.** *(b)* For naming the polyfunctional organic compounds, IUPAC has given an order of decreasing priority for the functional groups as follows

$-COOH > -SO_3H > -COOR > -COCl > -CONH_2 > -CN > -CHO > \rangle C=O > -OH, -NH_2 > \rangle C=C\langle > -C\equiv C-$

**16.** *(c)* Priority order for functional groups present in the given compound is

aldehyde > ketone > alkene > nitro.

The carbon atom of aldehyde group get lowest possible number in longest carbon chain of seven carbon atoms.

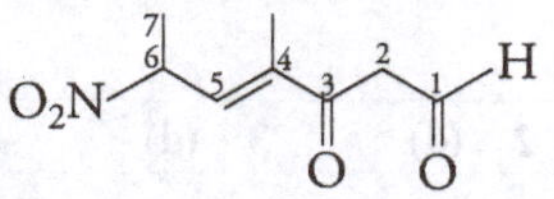

So, correct IUPAC name is 4-methyl-6-nitro-3-oxohept-4-enal.

Thus, option (c) is the correct answer.

**17.** *(c)* Hybridisation of carbon *a*, *b* and *c* in the given molecule respectively are $sp^3$, $sp^2$ and $sp^2$.

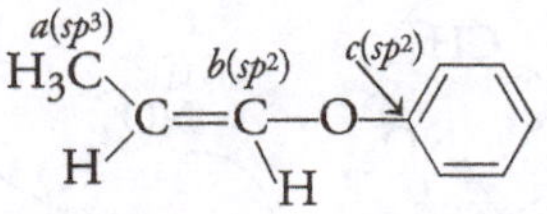


**18.** *(d)* The hybridisation of carbon 1, 2, 3 and 4 respectively are

$$\underset{sp^2}{\overset{1}{CH_2}}=\underset{sp}{\overset{2}{C}}=\underset{sp^2}{\overset{3}{CH}}-\underset{sp^3}{\overset{4}{CH_3}}$$

**19.** *(d)* The IUPAC name of the given compound,

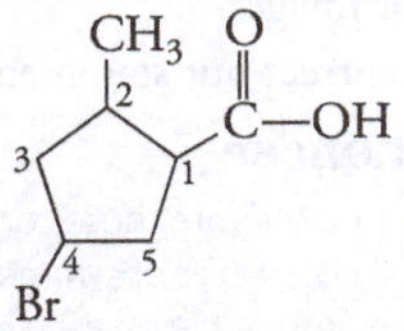

is 4-bromo-2-methylcyclopentane carboxylic acid.

**20.** *(a)* Statement I is incorrect but Statement II is correct. The correct form of statement I is, IUPAC name of compound (*A*) 1-chloro-2,4-dinitrobenzene.

**21.** *(b)* Statement I is correct but Statement II is incorrect. It is because the IUPAC name of the compound is

$$\overset{7}{HOCH_2}\overset{6}{CH_2}\overset{5}{CH_2}\overset{4}{CH_2}\overset{3}{CH_2}\underset{2}{\overset{O}{\overset{\|}{C}}}\overset{1}{CH_3}$$

7-hydroxyheptan-2-one

**22.** *(a)*

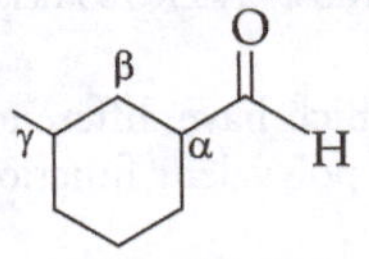

γ-methylcyclohexane carbaldehyde

Here, cyclohexane is considered as parent chain, aldehyde is functional group but since it is not included in principal chain, it is written as carbaldehyde.

Third position to the aldehyde group is called γ-position.

**23.** *(d)* The structure of ethylidene chloride is

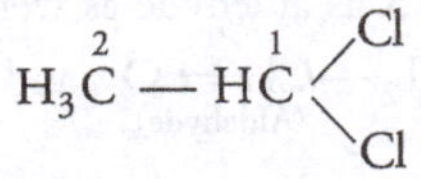

The IUPAC name is 1,1-dichloroethane.

**24.** *(d)* Mesityl oxide is a common name of 4-methyl pent-3-en-2-one.

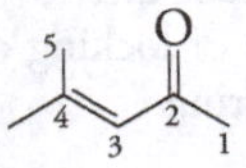 (4-methylpent-3-en-2-one)

**25.** *(c)* The IUPAC name of the given compound is 2-chloro-1-methyl-4-nitrobenzene .

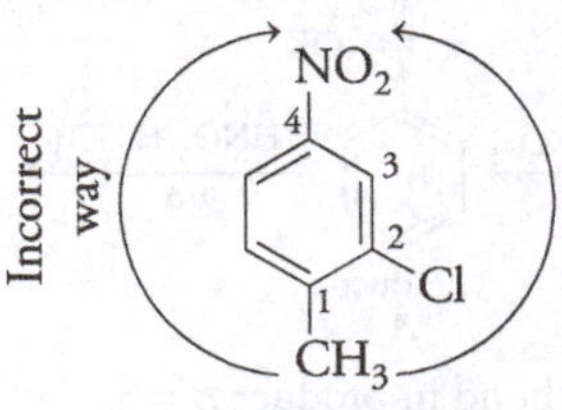


2- chloro-1-methyl-4-nitrobenzene

**26.** *(c)* The correct match is A-II, B-III, C-IV, D-I

**27.** *(c)* The correct match is A-(III), B-(IV), C-(I), D-(II)

**28.** *(c)*

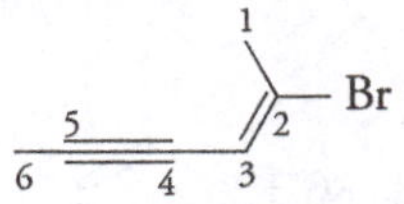

The IUPAC name of above compound is 2-bromohex-2-en-4-yne.

Now, to assign *E/Z* first assign priority to groups along the double bond.

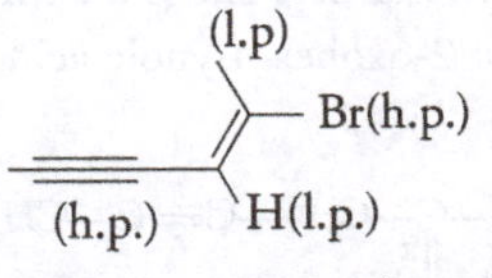

h.p.=high priority
l.p.=low priority

Since, h.p. groups are in opposite direction. So, *E* configuration will be assigned to double bond at second position.

Hence, full IUPAC name of given compound is (2*E*)-2-bromohex-2-en-4-yne.

Hence, correct option is (c).

**29.** *(c)*

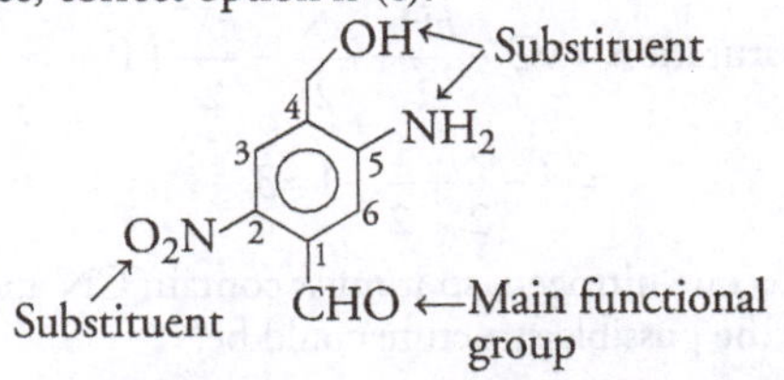


5-amino-4-hydroxymethyl-2-nitrobenzaldehyde

**30.** *(15)* The structure of hex-1, 3-dien-5-yne is

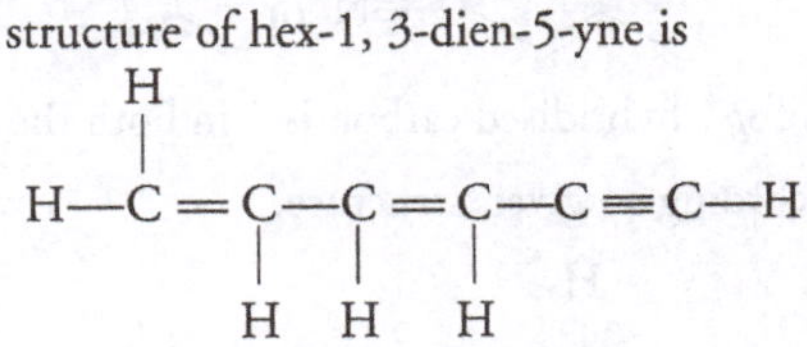

So, σ bond $=11$, π bond $=4$

Thus, total $\sigma+\pi=11+4=15$

**31.** *(9)* The reaction mentioned in question is as follows,

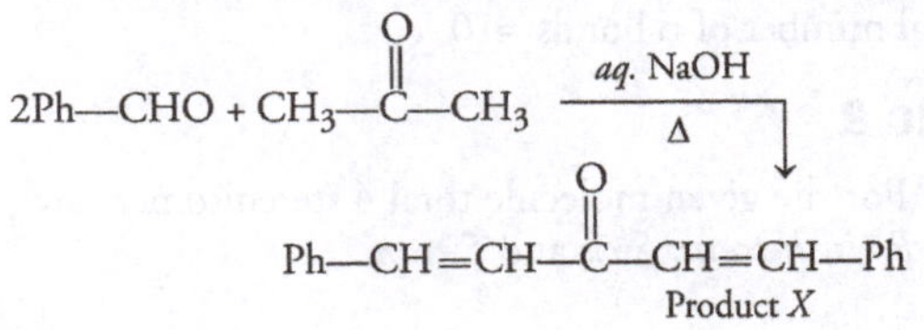

Total number of π bonds in $X=9$.

**32.** *(5)* The complete reaction is as follow,

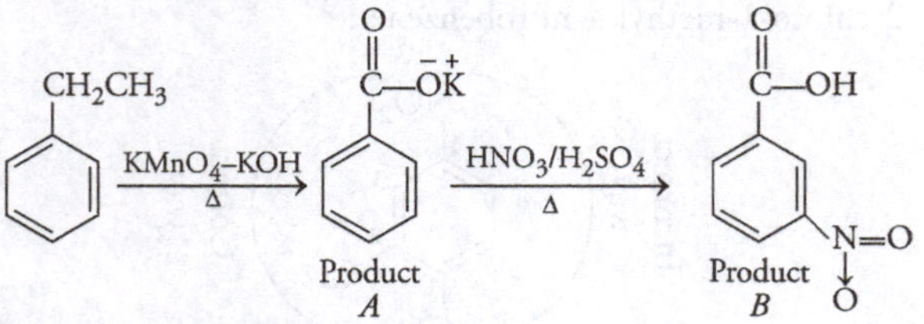

Total number of $\pi$ bond in product $B = 5$.

**33.** *(3)* The complete reaction is as follows,

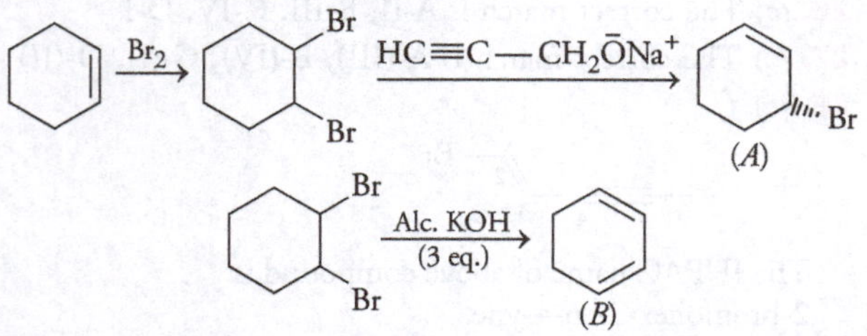

So, compound $A$ contains 1 $\pi$ bond while compound $B$ contains $2\pi$ bond. Thus total $\pi$ bond in $A$ and $B$ are three.

**34.** *(18)* The structure of 2-oxohex-4-ynoic acid is

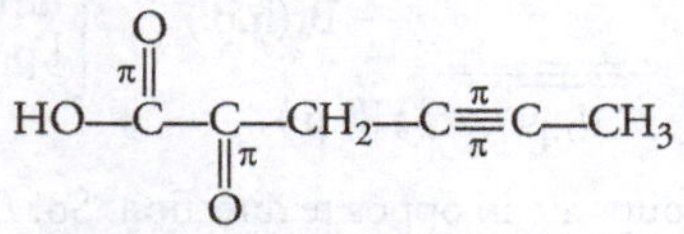

Thus, the given compound has $4\pi$ bonds and 14 sigma bonds. Hence, sum of sigma and pi bonds is 18.

**35.** *(1)* Molecular formula $= C_4H_5N$

$$\text{Degree of unsaturation} = nC - \frac{nH}{2} + \frac{nN}{2} - \frac{nX}{2} + 1$$

$$= 4 - \frac{5}{2} + \frac{1}{2} + 1 = 3$$

Compound has one nitrogen, so it must contain CN and it is acyclic. Thus, the possible structure could be.

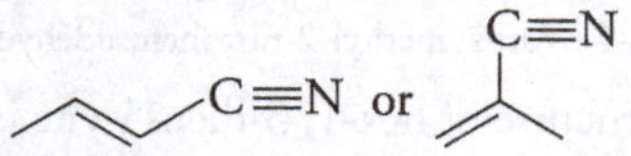

Number of $sp^3$ hybridised carbon is 1 in both the cases.

**36.** *(10)* According to given structure,

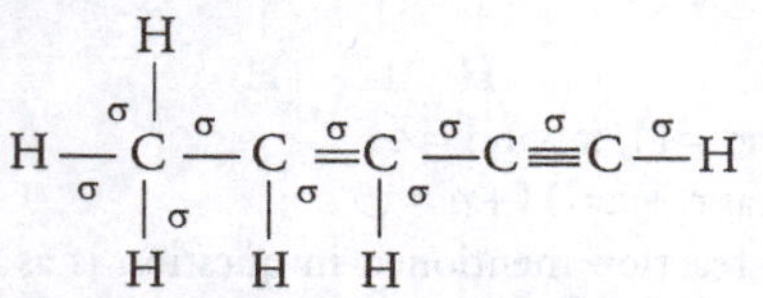

Total number of $\sigma$ bonds $= 10$

## Topic 2

**01.** *(b)* For the given molecule total 4 stereoisomers are possible i.e., *R cis*, *R trans*, *S cis* and *S trans*.

**02.** *(c)* Compound $A$ does not contain axes of symmetry and hence, it is optically active.

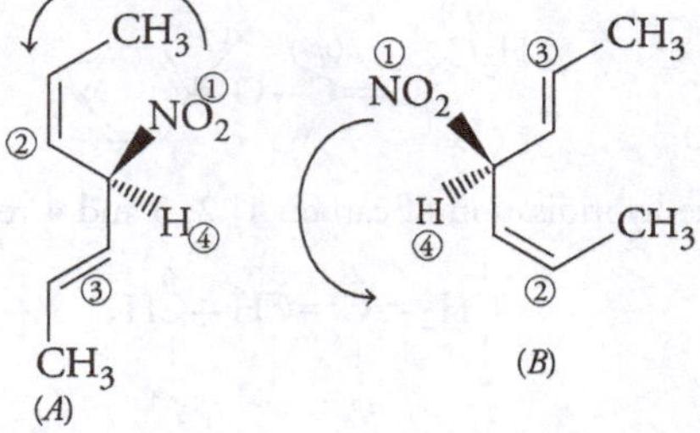

Both compounds ($A$) and ($B$) have same $S$-configurations, so they are not mirror images.

So, statement I is correct but statement II is incorrect.

> **Concept Enhancer**
>
> Optically active compounds are those compounds which have chiral carbon or absence of axes of symmetry or whose mirror images are non super imposable on each other.

**03.** *(d)* 1,2-dimethylcyclopropane is

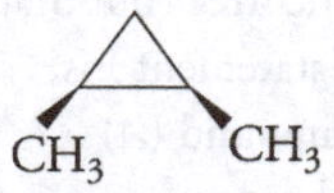

Hence, stereoisomers are as follows

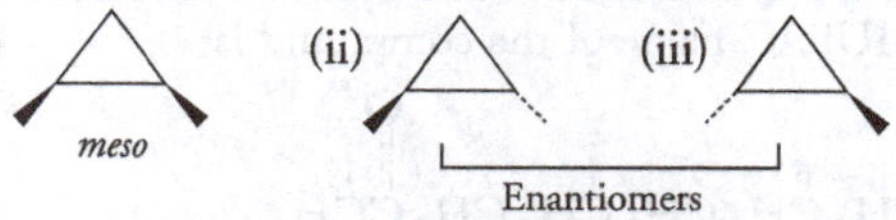

*meso*-form in optically inactive, whereas enantiomeric pairs are optically active. Therefore, total number of stereoisomers are three (3).

**04.** *(d)* Metamers are compounds which have different alkyl groups present along both side of polyvalent functional group.

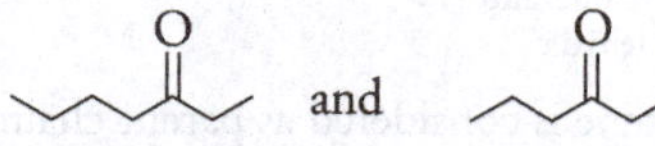

3-heptanone and 4-heptanone

(Metamers)

**05.** *(d)* Functional isomers are those which have same molecular formula but different functional groups. $C_3H_6O$ molecule can exists as aldehyde as well as ketone as follows

$$\underset{\text{(Aldehyde)}}{CH_3—CH_2—CH{=}O} \text{ and } \underset{\text{(Ketone)}}{CH_3—\overset{\|}{\underset{O}{C}}—CH_3}$$

So, they are functional isomers of each other.

**06.** *(b)* Geometrical isomers are molecule that are locked into their spatial positions with respect to one another due to double bond or a ring structure. The presence of locking or restricted rotation is due to double bond or a ring structure.

Two different groups must be attached to any two carbon atoms of restricted rotation.

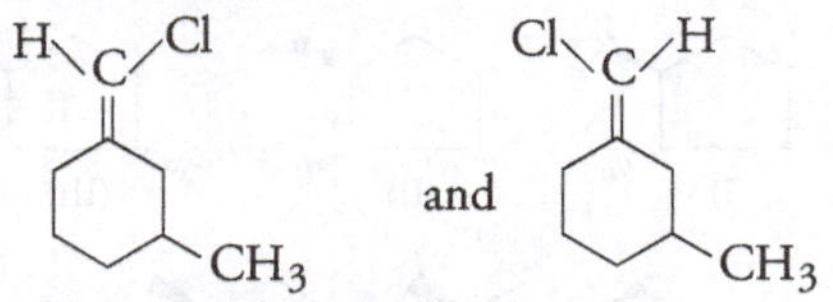

show geometrical (possible *cis* and *trans* isomer)

**07.** *(d)* For the structure.

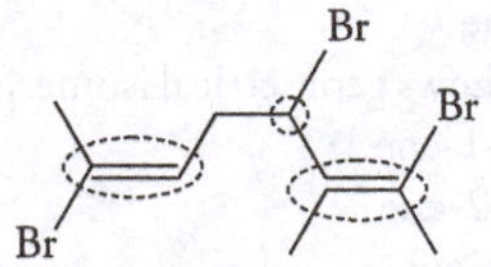

Total 3 stereocentres are there

So, total stereoisomers possible are $2^3 = 8$.

**08.** *(c)* Metamer are those isomers having same molecular formula, same functional group, but different alkyl/aryl groups on either side of functional group.

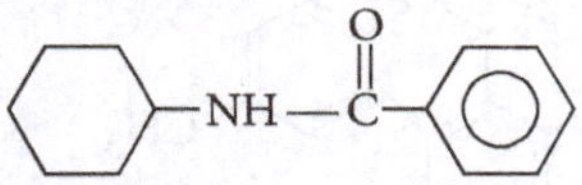

Above drawn compound is metamer of given compound ($X$).

**09.** *(a)* The isomeric deuterated bromide with molecular formula $C_4H_8DBr$ having two chiral atoms is 2-bromo-3-deuterobutane

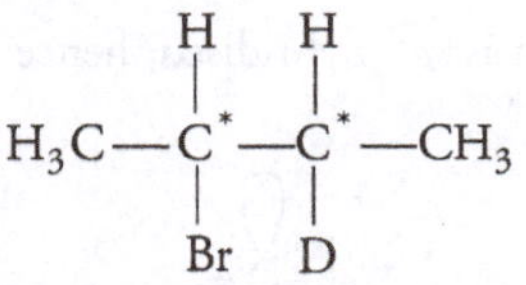

**10.** *(c)* 3-ethylhex-3-ene does not show stereoisomerism as it optically inactive due to absence of chiral carbon centre.

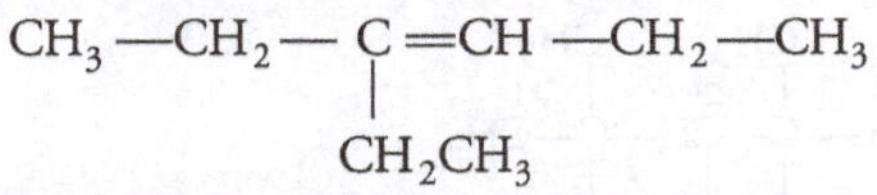

> **• Concept Enhancer**
>
> Compounds having the same molecular formula (constitution) and the same connectivity of atoms in their molecules but different-3-dimensional orientation of their atoms in space are called stereoisomers.

**11.** *(d)* Reaction of all options are as follows

(a) 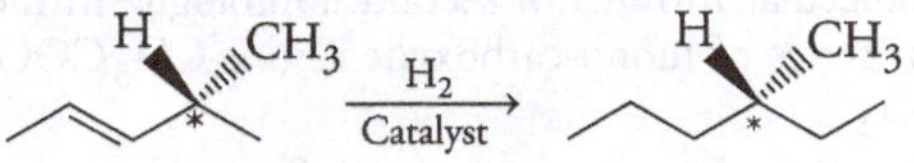

(b) 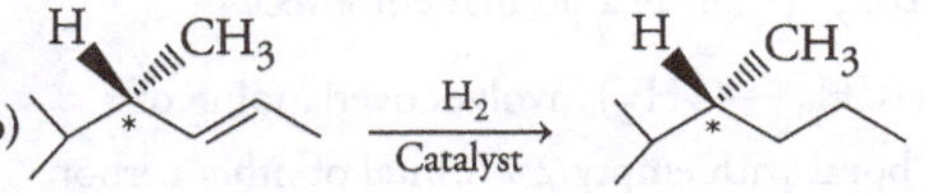

(c) 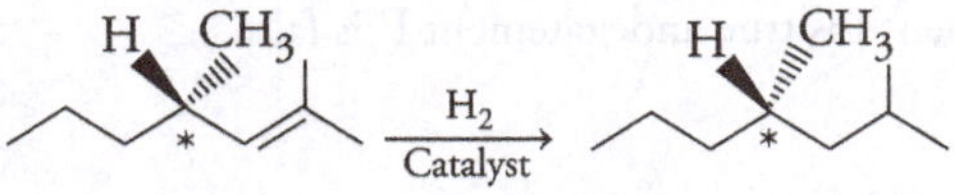

(d) 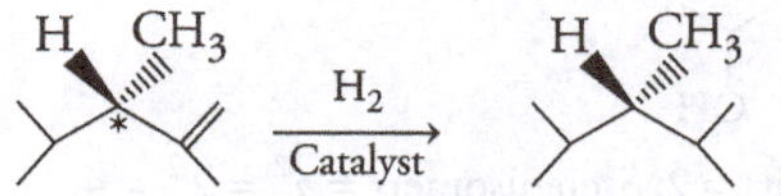

No change in chirality or optical activity due to hydrogenation of (a), (b) and (c) but in (d) the substrate (alkene) on hydrogenation becomes optically inactive as the product obtained does not have any chiral atom.

**12.** *(b)*

(a) 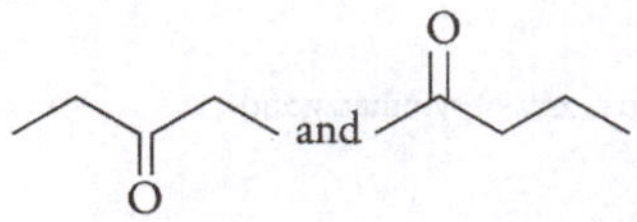 are metamers

(b) 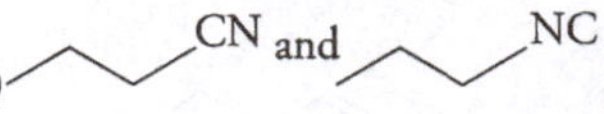 are functional isomers

(c) 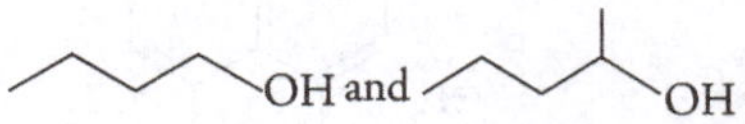 are homologous to each other

(d) 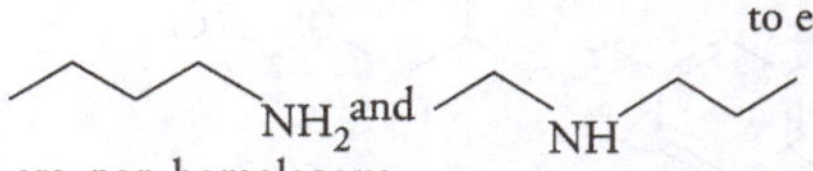

are non-homologous

So, A and B, i.e. option (b) is correct.

**13.** *(b)* The correct match is A-III, B-I, C-IV, D-II.

- *n*-propanol and *iso*-propanol

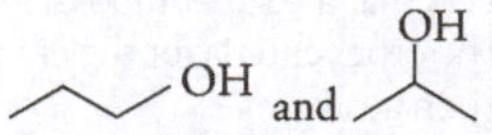 = Position isomers

- Methoxypropane and ethoxyethane

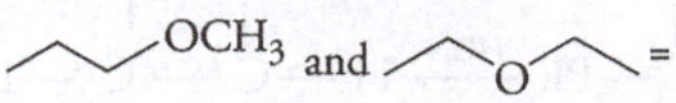 = Metamers

- Propanone and propanal

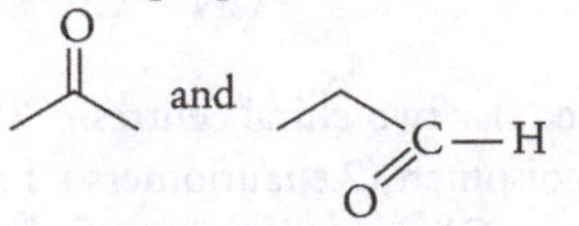 = Functiornal isomers

- Neopentane and *iso*-pentane

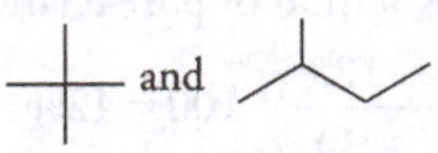 = Chain isomers

**14.** *(b)* The correct match is A-III, B-I, C-IV, D-II.

**15.** *(8)* There are total eight structural isomers possible for $C_9H_{12}$.

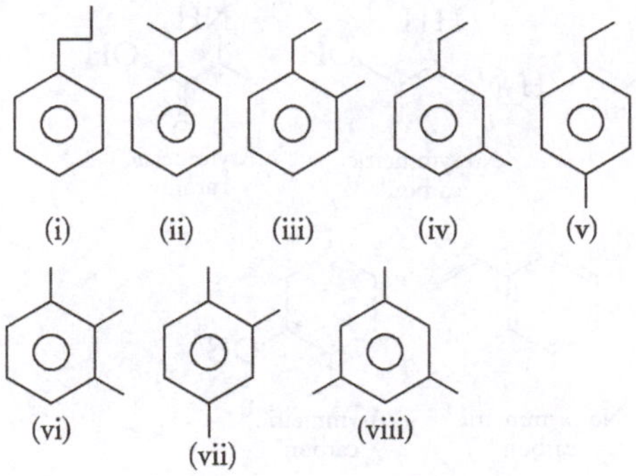

**16.** *(4)*

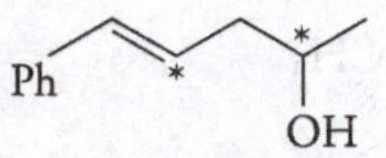

$n$ (stereogenic unit) = 2, Stereoisomers $= n^2 = 2^2 = 4$

Thus, 4 stereoisomers are possible.

**17.** *(2)* There are two chiral compounds [(i) and (ii)]

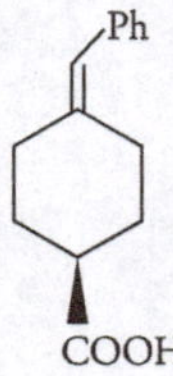

[No, plane of symmetry, axes of symmetry and central of symmetry]

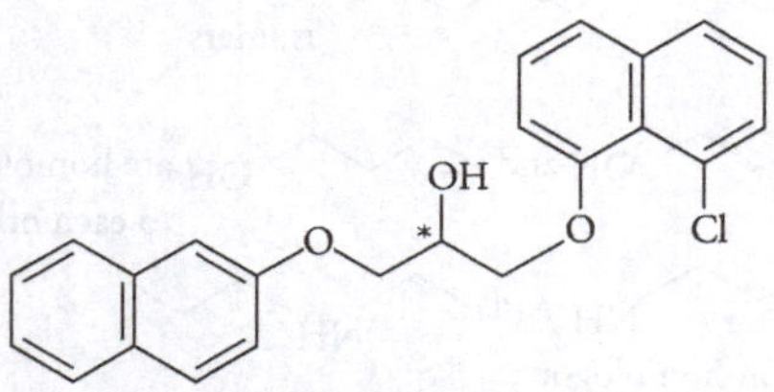

[Presence of asymmetric carbon]

> • Concept **Enhancer**
> The objects which are non-superimposable on their mirror image are said be chiral while the objects that are superimposable on each other are called as achiral. The other criteria for the chirality is the presence of asymmetric carbon atom.

**18.** *(3)*

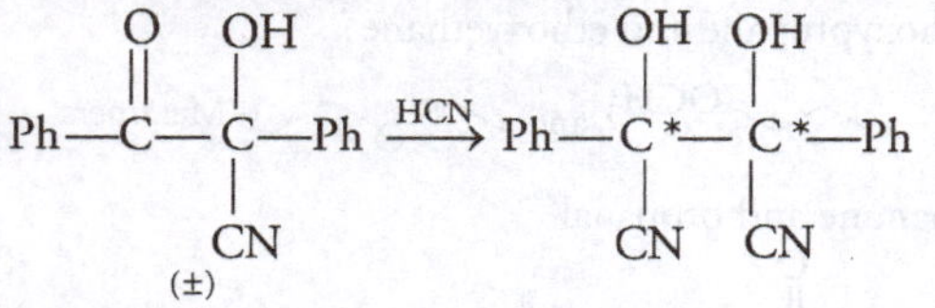

The compound formed has two chiral centres.

So, it forms three stereoisomers (2 enantiomers + 1 *meso* isomer)

**19.** *(42)* % Optical purity $= \dfrac{\text{Observed rotation of mixture}}{\text{Rotation of pure enantiomer}} \times 100$

$$= \frac{+12.6}{+30} \times 100 = 42\%$$

**20.** *(3)* An asymmetric carbon atom is a carbon atom which is attached to four different atoms or group of atoms. There are three compounds which possess asymmetric carbon atoms.

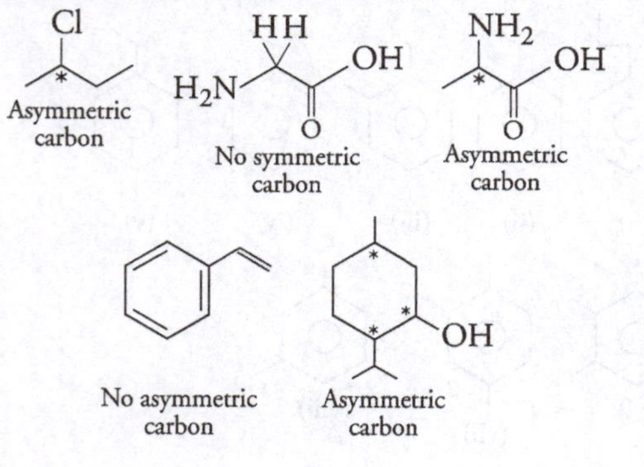


**21.** *(5)* Total number of possible stereoisomers of dimethyl cyclopentane is 5 which are as follows

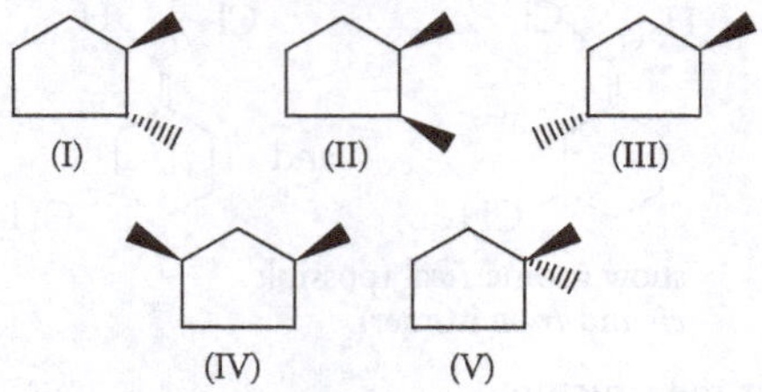


**22.** *(6)* (i) Pent-1-ene
(ii) Pent-2-ene it shows geometrical isomerism, i.e. *cis* and *trans*
(iii) 2-methyl but-1-ene
(iv) 2-methyl but-2-ene
(v) 3-methyl but-1-ene

Total number of acyclic structural isomers (including geometrical isomers) for pentene is 6.

**23.** *(5)*

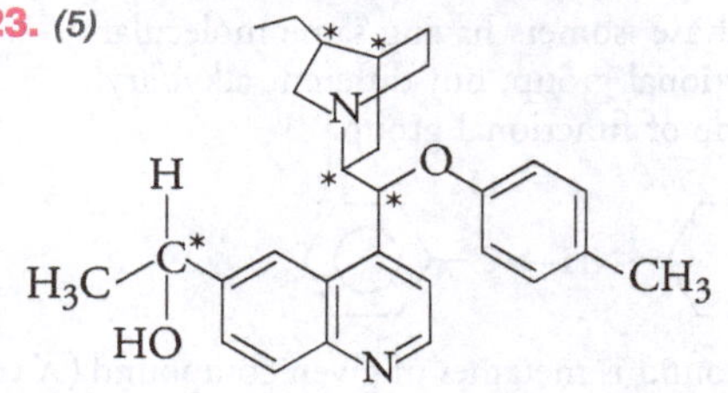

The molecule contains five chiral carbon (C*) atoms.

## Topic 3

**01.** *(a)* Carbocation is $sp^2$ hybridised, hence its shape is trigonal planar.

**02.** *(b)*

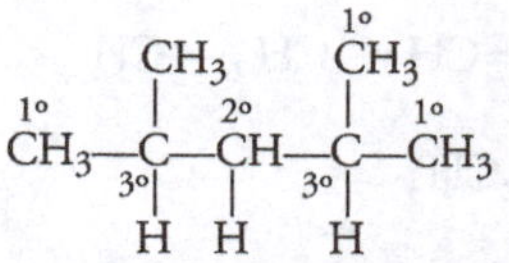

Only one 2° carbon is present in this compound.
Hence, option (b) is correct.

**03.** *(d)* The difference in energy between the actual structure and the lowest energy resonance structure for the given compound is called as resonance energy. Hence, option (d) is correct.

**04.** *(d)* The molecular formula of second homologue in the homologous series of mono-carboxylic acid is $CH_3COOH$ i.e., $C_2H_4O_2$.

**05.** *(c)* Hyperconjugation is a permanent effect.

Ethyl cation ($CH_3—\overset{+}{C}H_2$) involves overlapping of $C_{sp^3}—H_{1s}$ bond with empty 2$p$-orbital of other carbon.

∴ Statement I is true and statement II is false.

**06.** *(a)* Stability of a resonating structure depends on

(i) Charge separation between opposite charges should be minimum.

(ii) Charge separation between same charges should be maximum.

Thus, in option (a) two cationic charges are very close to each other. So, these charge repel each other. Therefore, structure (a) is unstable.

**07.** *(d)* The acid whose conjugate base is most stable will be strongest acid.

Conjugate base of molecule (d) is most stable as an aromatic compound is formed after donating a proton (conjugative base).

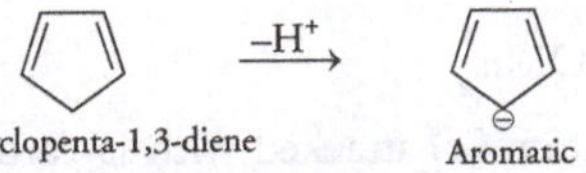


Cyclopenta-1,3-diene Aromatic

**08.** *(b)* Due to $-R$ (or $-M$) effect of '$NO_2$', there is stronger double bond character for the 'C—Cl' bond in Cl— CH =CH—$NO_2$. (option-b)

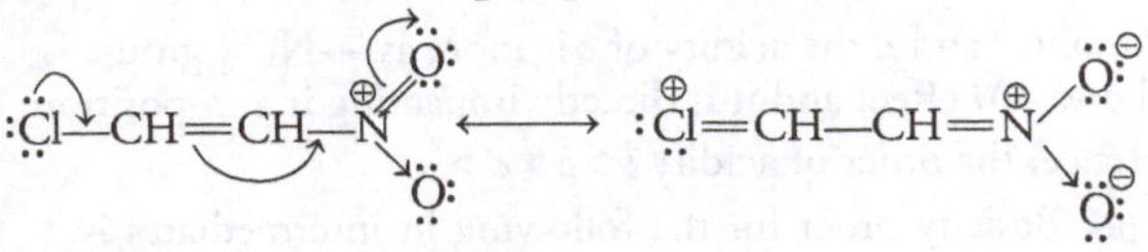

∴ The bond length is shortest in (b).

**09.** *(d)* We know, a stronger acid produces its stable or weaker conjugate base. Here, $CH(CN)_3$ produces the most stable conjugate base $(NC)_3C^-$. Stronger $-R$ and $-I$ effects of the $CN^-$ group, make the carbanion (conjugate base) very stable.

**10.** *(c)* Ethene ($H_2C{=}CH_2$) is $sp^2$-hybridised and ethyne ($HC \equiv CH$) is sp-hybridised. In ethyne, the sp-hybridised carbon atom possesses maximum s-character and hence, maximum electronegativity. Due to which, it attracts the shared electron pair of C—H bond to a greater extent and makes the removal of proton easier. Hence, alkyne is much more acidic than alkene. Further, among all the acidic alkynes $CH{\equiv}CH$ is most acidic.

Among $CH_2{=}CH_2$ and $CH_3-C{\equiv}CH$, $CH_2{=}CH_2$ is more acidic as in $CH_3-C{\equiv}CH$ the ERG i.e., $CH_3-$ is attached. Thus, the order for acidic strength of the compounds is $CH{\equiv}CH > CH_2{=}CH_2 > CH_3-C{\equiv}CH$

**11.** *(b)*

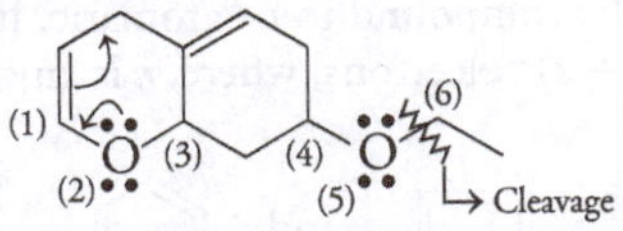


When above compound is treated with a strong acid, it will undergo hydrolysis involving the cleavage of $[O_5-C_6]$ bond which results into the formation of a stable carbocation.

**12.** *(b)* The correct match is A-III, B-IV, C-I, D-II

**13.** *(a)* Statement I is false but Statement II is true.

The correct form of Statement I is

Hyperconjugation is permanent effect because external reagent is not required.

**14.** *(c)* The given compound is

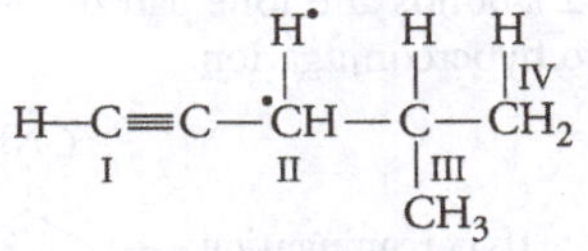


II carbon is most stable carbon radical due to resonance stabilising factor.

I carbon is least stable carbon radical due to no stabilising factor.

**15.** *(c)* Carbocation intermediate is stabilised by $+I$, $+M$ and hyperconjugation effect. Since in

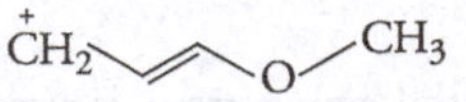

Carbocation is in conjugation with stronger $+M$ with group $-OCH_3$, hence it is most stable.

**16.** *(c)* Generally, neutral structures are more stable than charged ones. Therefore, I is more stable than II and III. While positive charge on less electronegative atom is more stable i.e. $C^+$ is more stable than $O^+$.

Thus, the correct order is I > II > III.

**17.** *(c)* The compound I is antiaromatic hence it is least stable. Among compound II and III, due to angle strain theory, III is more stable than II while compound IV is aromatic hence most stable. So, the correct order of stability of given carbanion i.e.

IV > III > II > I

**18.** *(a)* Among the given statements, incorrect statement is given in option (a). Its correct form is hydrogen ion ($H^+$) shows positive electromeric effect.

**19.** *(b)* Among the given options —COOH group shows negative *R*-effect, while other three functional groups shows positive *R*-effect.

e.g.

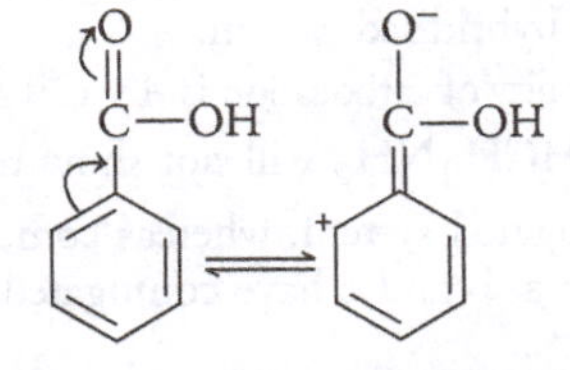

**20.** *(d)* Carbocations are the species that have sextet of electrons and can act as an electrophile i.e.,

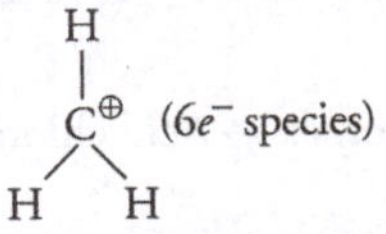


**21.** *(a)* The decreasing order of hydride affinity for the given carbocations is C > A > B > D.

The correct hydride affinity order of carbocations will be decided by the stability of carbocation. Higher the stability of carbocation, lower will be hydride affinity.

**22.** *(b)*

**key idea** Use the concept of stability of carbocation to solve this problem. This stability can be judged on the basis of resonance and hyperconjugation.

Carbocation (*C*) is most stable because positive charge delocalises with 2 $\pi$ bonds and lone pair of N-atom in $NH_2$ group and due to hyperconjugation.

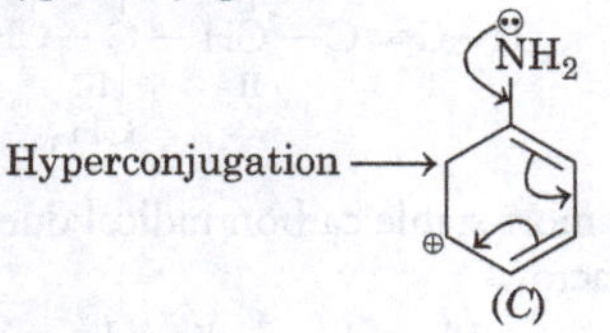

(*C*)

Thus, option (b) is correct.

**23.** *(d)* $NH_2^-$ is more nucleophilic than $NH_3$ due to the presence of the presence of more electron density of nitrogen atom.

**24.** *(a)*

(*A*) No electron withdrawing group is attached to it so it is the most stable structure among them.

(*B*) Oxygen is electron withdrawing group and directly attached to the carbon containing the positive charge which makes it least stable.

(*C*) In this structure oxygen is indirectly attached to carbon containing the positive charge. So, the $-I$ effect gets lesser with increase in position.

So, decreasing order of stability $A > C > B$.

**25.** *(a)* Correct order of stability of carbocations will be

(A) 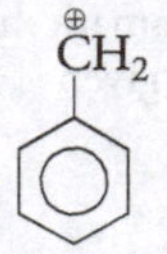 Benzyl carbocation get stabilise by resonance.

(C) $CH_3-\overset{\oplus}{C}H_2$ gets stabilised due to hyperconjugation.

Electronegativity of carbon increases with increase in % *s*-character.

For $sp^3$ -hybridised carbon, % *s*-character is 33%.

For $sp^3$ -hybridised carbon, % *s*-character is 50%. So, positive charge is less stable on *sp*-hybridised carbon.

So, the correct order of stability of carbocation is $A > C > B > D$.

**26.** *(d)* $CH_3CH_2CH=CHCH_2NH_2$ will not show resonance as it does not have conjugated system, whereas compounds (or their parts) in option a, b and c have conjugated system. So, they exhibit resonance.

**27.** *(a)* Rate of hydrolysis is directly proportional to positive charge present on carbon of $>C=O$ group.

Therefore,

Order of rate of hydrolysis $\rightarrow$ Acid chloride > Acid anhydride > Ester > Amide.

**28.** *(a)* Carbocation *B* (2°) is more stable due to more hyperconjugation and formed relatively at faster rate compared to *A* (1°).

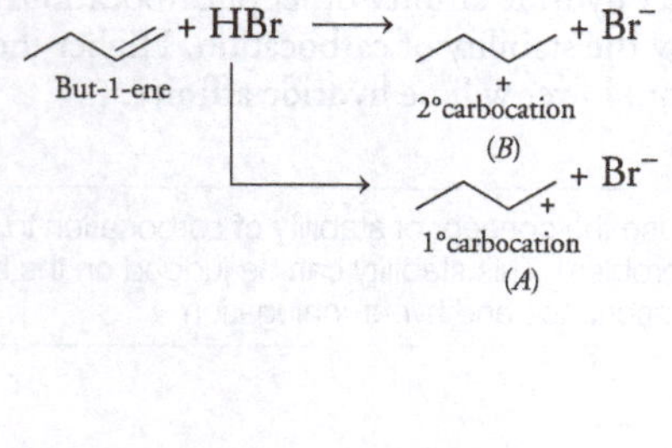


**29.** *(d)* Acidity of H-atom $\propto$ Attachment with more and stronger electron withdrawing groups.

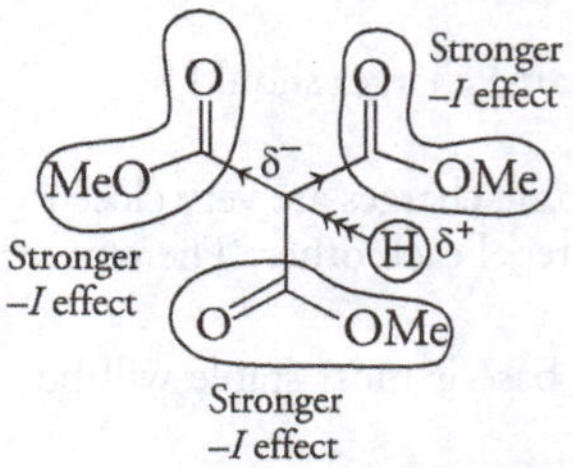


$\longrightarrow$ It is most acidic, because the carbon atom bearing this H-atom has stronger $-I$-effects by three ester groups $(-\overset{\overset{O}{\|}}{C}-OMe.)$

**30.** *(c)* The H-atom (*a*) attacked with *sp*-carbon ($s\%=50$) is also acidic but lesser acidic than H-atoms (*b*, *c* and *d*) who are attached with more electronegative O-atoms.

$\therefore$ Electronegativity : $O(sp^3) > C(sp)$

Among *c* and *d* the acidity of *c* is more as $-NO_2$ group shows $-M$ effect and it is directly impacting it at *p*-position Hence, the order of acidity $b > c > d > a$.

**31.** *(a)* Basicity order for the following in intermediates is

$$\underset{(v)}{CN^-} < \underset{(iii)}{CH\equiv C^{\ominus}} < \underset{(ii)}{H_2C=CH-\overset{\ominus}{C}H_2} < \underset{(iv)}{\overset{\ominus}{C}H_3} < \underset{(i)}{CH_3-\overset{\ominus}{C}(CH_3)-CH_3}$$

which is also the opposite of the order of stability of carbanions.

$$:\overset{\ominus}{C}\equiv N: > \underset{(sp\text{-carbon})}{CH\equiv\overset{\ominus}{C}} > \underset{\text{(resonance) stabilised}}{CH_2=CH-\overset{\ominus}{C}H_2} > \overset{\ominus}{C}H_3 > \underset{\text{(less stable due to }+I\text{ effect of }CH_3\text{ groups)}}{\overset{\ominus}{C}(CH_3)_3}$$

**32.** *(d)*

> **key idea** Use the concept of Huckel's rule. This rule helps to detect aromaticity in a compound.

Aromatic compounds are stable due to resonance while non-aromatics are not. According to Huckel's rule (or $4n+2$ rule), "For a planar, cyclic compound to be aromatic, its $\pi$ cloud must contain $(4n+2)\pi$ electrons, where, *n* is any whole number." Thus,

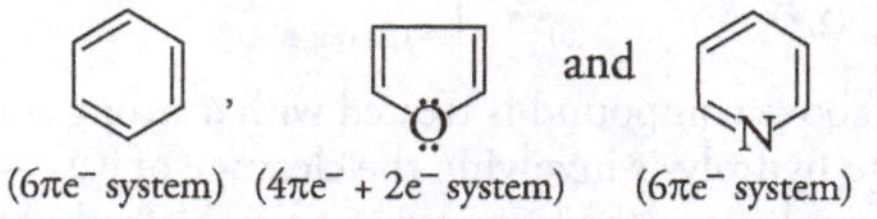


are aromatic and stabilised by resonance.

They follow Huckel's rule.

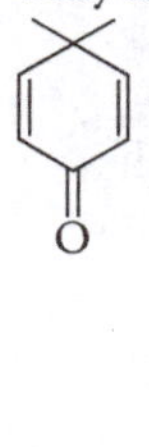
Compound (d) is non-aromatic, hence, least stabilised by resonance.

**33.** *(d)* The order of stability for all the given options are as follows

(A) 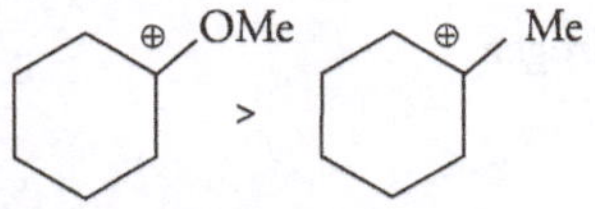


Stable by backbonding

(B) 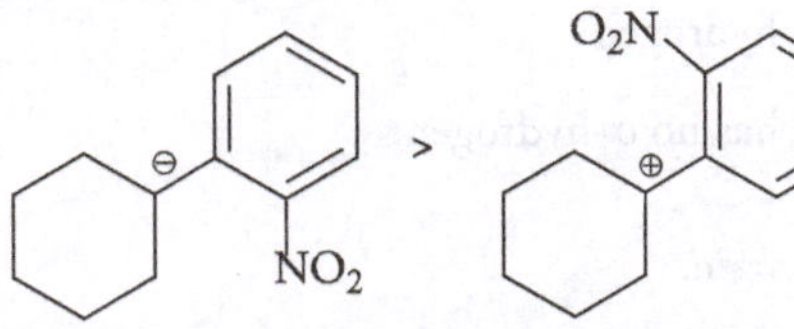


Stable due to –*M*, –*I* effect of $NO_2$ group

(C) 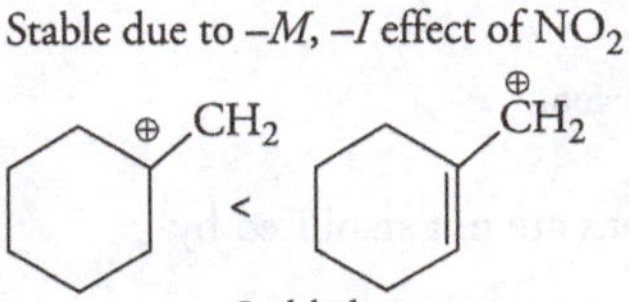


Stable by resonance

(D) 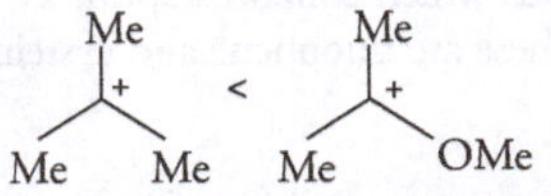


Stable by backbonding

**34.** *(c)*

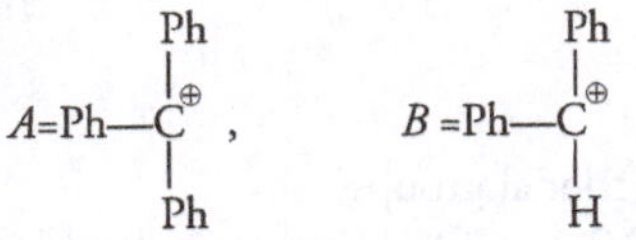


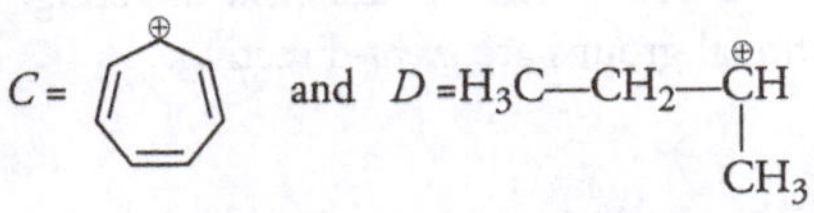


Now, *C* is aromatic due to positive charge and hence it is most stable.

*A* has more resonance structure.

*B* has less resonance structure

*D* has only hyperconjugation.

∴Stability is in order,

Aromaticity > Resonance > Hyper conjugation

Hence, the correct order of stability is *C* > *A* > *B* > *D*.

**35.** *(c)* The order of stability of given carbocations are as follows.

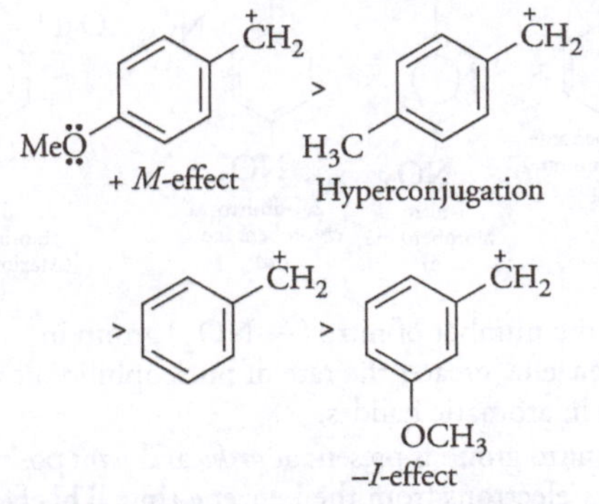


So, carbocation given in option (c) is most stable.

**36.** *(a)* Among the given options, there are five nucleophiles i.e,

$$NH_3,\ PhSH,\ (H_3C)_2S,\ \overset{\ominus}{O}H \text{ and } CH_2=CH_2$$

**37.** *(d)* The correct match is A-IV, B-III, C-I, D-II.

(A) Here, +*R* effect is shown due to presence of electron donating group. $\ddot{N}H_2$.

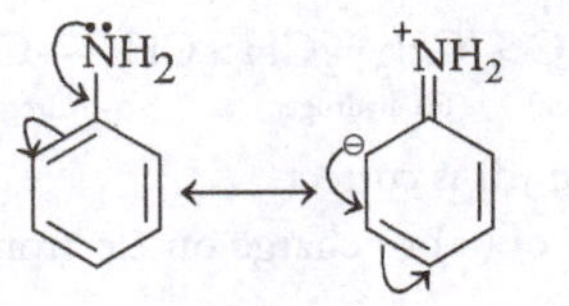


(B) 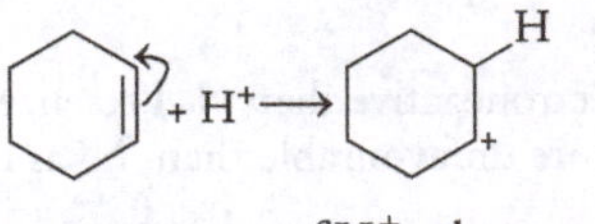


Here, due to presence of $H^+$, electron deficient species, +*E* effect is observed in the given mechanism.

(C) 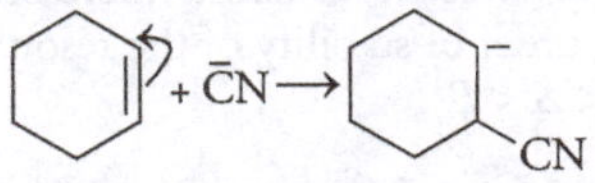


Here, –*E* effect is observed as the shared pair of electron is transferred away from the attacking reagent.

(D) 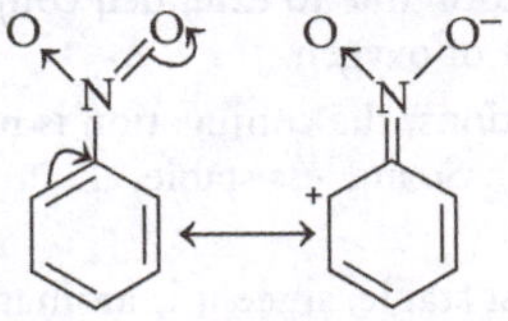


Here –*R* effect is shown due to presence of electron withdrawing group $NO_2$.

**Time Saver Tech**

Use elimination technique as time saver. As all the given option are different for *A*, *B*, *C*, *D*. So, you can solve for any one reaction and can choose your answer. The presence of electron donating group will show +*R* effect so, A-IV is given in option (d) only.

**38.** *(c)* Ionic reaction with organic compound proceed through heterolytic bond cleavage.

Hence, option (c) is correct.

**39.** *(d)* Greater the electron density in the benzene ring more easily it will be attacked by an electrophile. Phenol has highest electron density among the given options, therefore it will be most easily attacked by electrophile.

Hence, option (d) is correct.

**• Concept Enhancer**

$$\text{Electrophilic substitution} \propto \text{Electron releasing group} \propto \frac{1}{\text{Electron withdrawing group}}$$

—OH and—$CH_3$ are electron releasing while —Cl is $e^-$ withdrawing in nature.

**40.** *(d)* The stability order of carbocation depends on the number of hyperconjugable hydrogens.
$(CH_3)_3C^+$ has highest hyperconjugable hydrogen then $(CH_3)_2\overset{+}{C}H$, than $CH_3—\overset{+}{C}H_2$ and $\overset{+}{C}H_3$ has lowest among the given option.
Therefore, the correct order is

$$\underset{(9\alpha\text{-hydrogen})}{(CH_3)_3\overset{+}{C}} > \underset{(6\alpha\text{-hydrogen})}{(CH_3)_2\overset{+}{C}H} > \underset{3\alpha\text{-hydrogen}}{CH_3—\overset{+}{C}H_2} > \underset{(\text{zero }\alpha—H)}{\overset{+}{C}H_3}$$

Hence, option (d) is correct.

**41.** *(d)* Presence of (+) ve charge on electronegative atom (O) is unfavourable.
∴(C) is least stable.
'O' being more electronegative than N. Presence of '+ ve' charge on 'O' is more unfavourable than 'N' as in the case of (D).
Between A and B in (B) '– ve' charge is able to delocalise on both C=O and C=C. Hence, stability increases of (B) than (A). The increasing order of stability of the resonance structure is $C < D < A < B$

**42.** *(d)*

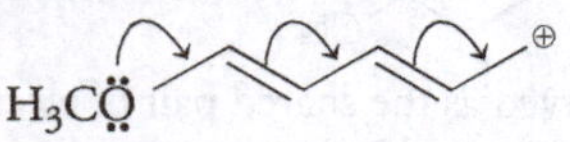

is most stable carbocation due to extended conjugation which also involves lone pair of oxygen.
In rest of the carbocations, the conjugation is not that extended as it is in (d). So are less stable.

**43.** *(a)*  is most stable, since it is aromatic compound, follows Huckel rule $(4n+2)$

$$4n+2=2$$
$$4n=0$$
$$n=0$$

**44.** *(c)* Resonance is shown by compound *(A)* and *(B)* only, hence they form stable carbocations.
Compound *(C)* and compound *(D)* do not show resonance because double bond and positive charge are not in conjugation.

**45.** *(4)* Among the given compounds, total four compounds shows inductive, mesomeric as well as hyperconjugation effects i.e.

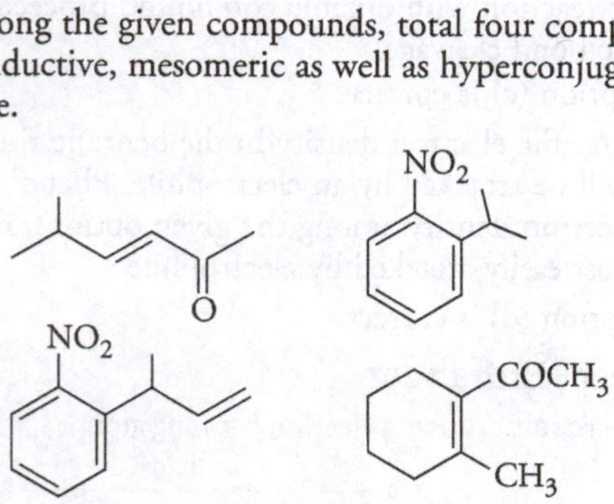


**46.** *(5)*

> **key idea** If carbocation has no α-hydrogen, it will not be stabilised by hyperconjugation.

The carbocation 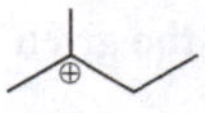 has 8α-hydrogen

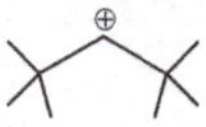 has no α-hydrogen

$\overset{+}{C}H_3$ has no α-hydrogen.

 has no α-hydrogen

$\overset{+}{C}H_2—OCH_3$ has no α-hydrogen.

 has 9 α-hydrogen.

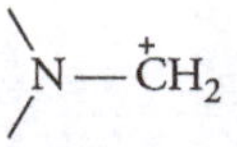 has no α-hydrogen.

Hence, total 5 carbocations are not stabilised by hyperconjugation.

**47.** *(2)* The number of compounds which contain sulphur as heteroatom in question is 2. These are thiophene and cysteine.

Thiophene ⟶ 

Cysteine ⟶ 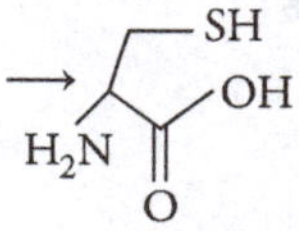


**48.** *(4)* Among the given functional groups, $—NO_2$, $—C≡N$, $—COR$, $—COOH$ are *meta*-directing.
Hence, total 4 functional groups are *meta*-directing.

## Topic 4

**01.** *(a)* Nitrochlorobenzene with strong E.W.G. (i.e. $—NO_2$) undergoes nucleophilic substitution at faster rate with $OH^-$ as nucleophile.
3-chloronitrobenzene has weaker E.W.G. nature of $NO_2$ (*–I* only) as compared to other compounds. So, it gives lowest rate towards nucleophilic aromatic substitution reaction ($ArS_N2$).
Thus, option (a) is correct.

**02.** *(d)* Rate of nucleophilic substitution in aromatic halide are as follows

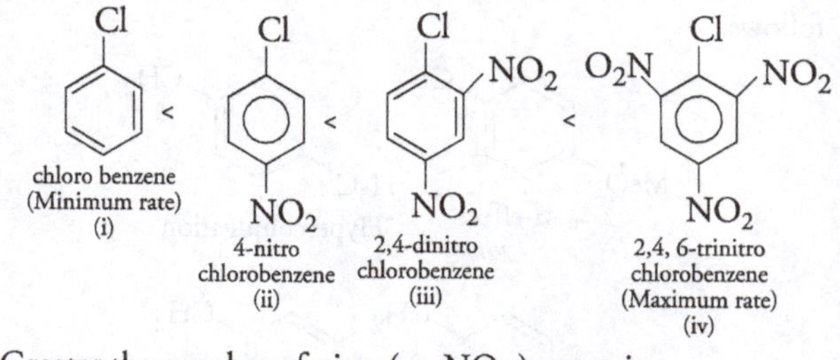


Greater the number of nitro ($—NO_2$) group in chlorobenzene, greater the rate of nucleophilic substitution reaction in aromatic halides.
When a nitro group is present at *ortho* and *para* position, it can withdraw electrons from the benzene ring. This facilitates, the attack of nucleophiles on haloarenes.

**03.** *(d)*

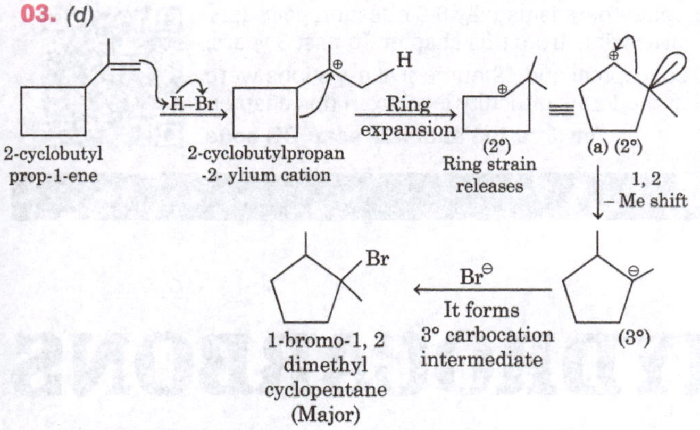


> • Concept **Enhancer**
>
> 2-cyclobutylprop-1-ene reacts with HBr to give a 3° carbocation, which undergoes ring expansion to release ring strain then again it rearranges to 3° carbocation *via* 1,2-methyl shift attack of $Br^-$ to give major product.

**04.** *(b)* Nitration of aromatic compounds proceeds *via* electrophilic substitution reaction. It means, higher the electron density on benzene, more will be the reactivity. The order of electron density on the benzene ring is

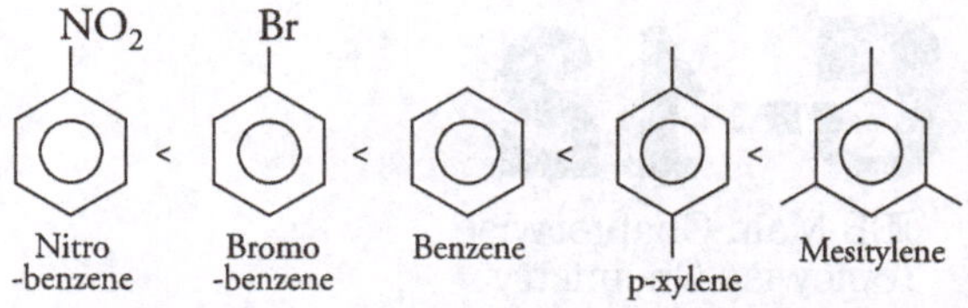


Hence, the increasing order of reactivity towards nitration is

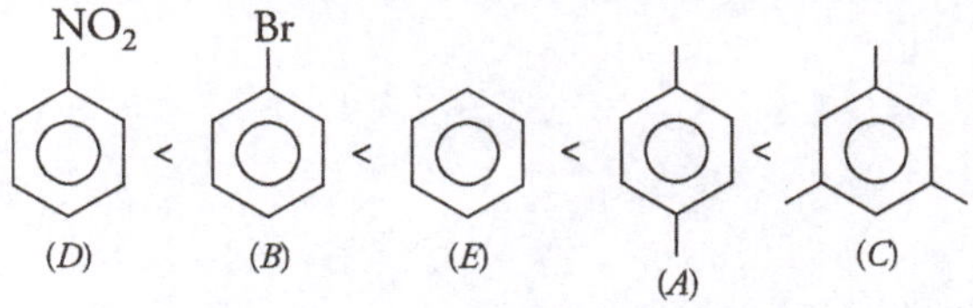


**05.** *(7)* To stabilise the given carbocation 7 hyperconjugation (including final product) structures are involved.

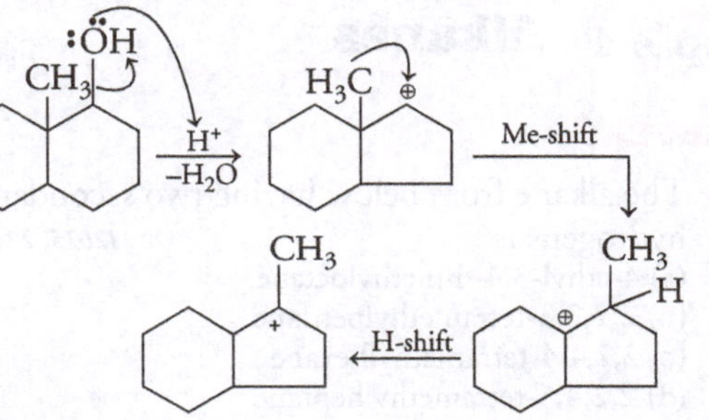

# C-12

JEE Main Chapterwise
Topicwise Chemistry

In JEE Main there is usually 0-2 question per paper are visible from this chapter. In past 3 years, 35 conceptual and 12 numerical questions were asked of moderate difficulty level from this chapter.

**For detailed analysis scan QR code**

# HYDROCARBONS

## Topic 1 Alkanes

### Level 1

**01** The alkane from below having two secondary hydrogens is *[2025, 22 Jan Shift-II]*

(a) 4-ethyl-3,4-dimethyloctane
(b) 2,2,3,3-tetramethylpentane
(c) 2,2,4,4-tetramethylhexane
(d) 2,2,4,5-tetramethylheptane

**02** Which of the following conformations will be the most stable? *[2023, 25 Jan Shift-I]*

(a) 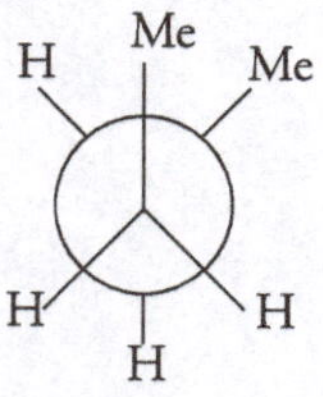


(b) 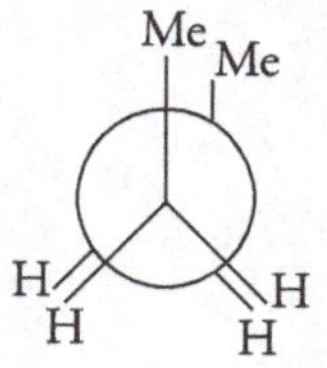


(c) 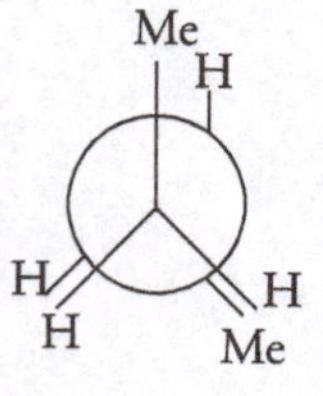


(d) 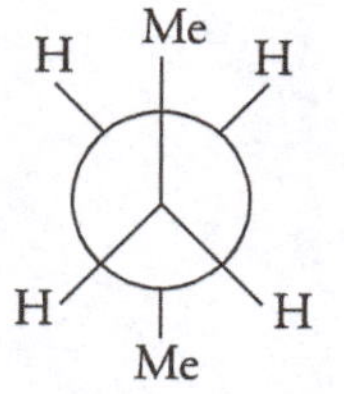


**03** In the following structures, which one is having staggered conformation with maximum dihedral angle? *[2022, 25 June Shift-I]*

(a) 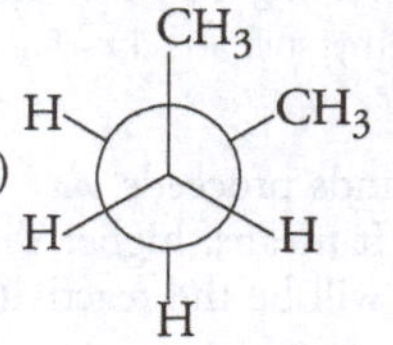


(b) 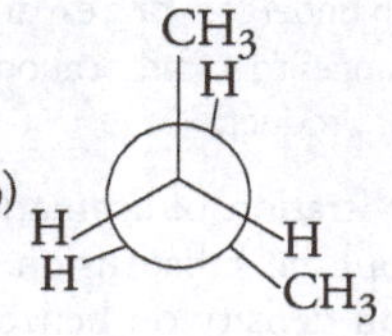


(c) 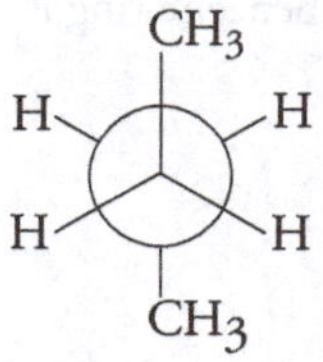


(d) 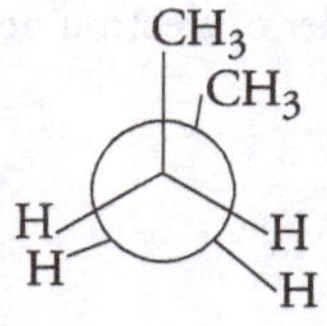


**04** Excess of isobutane on reaction with $Br_2$ in presence of light at 125°C gives which one of the following as the major product? *[2021, 26 Aug Shift-I]*

(a) $CH_3-\underset{\underset{CH_2Br}{|}}{CH}-CH_2Br$

(b) $CH_3-\underset{\underset{CH_3}{|}}{CH}-CH_2Br$

(c) $CH_3-\overset{\overset{Br}{|}}{\underset{\underset{CH_3}{|}}{C}}-CH_2Br$

(d) $CH_3-\overset{\overset{CH_3}{|}}{\underset{\underset{CH_3}{|}}{C}}-Br$

**05** Staggered and eclipsed conformers of ethane are *[2021, 27 July Shift-I]*

(a) polymers
(b) mirror images
(c) rotamers
(d) enantiomers

**06** For the given reaction

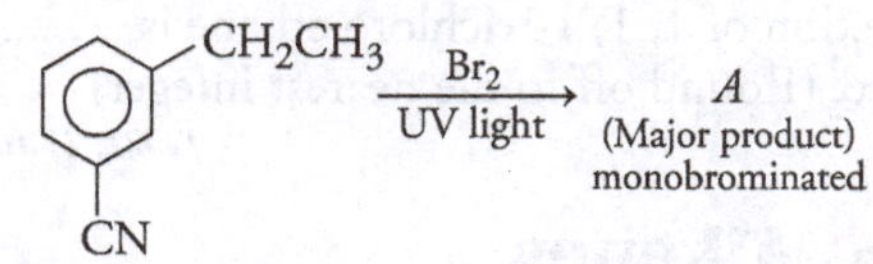


What is '*A*'? **[2021, 26 Feb Shift-I]**

(a) 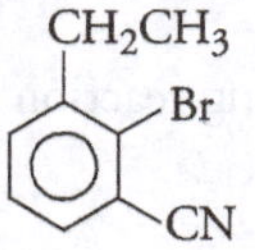
(b) 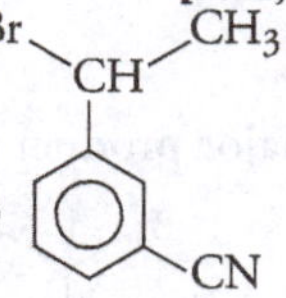
(c) 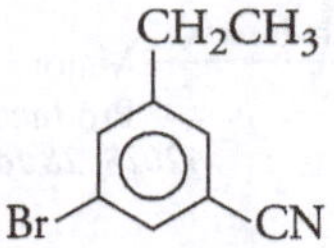
(d) 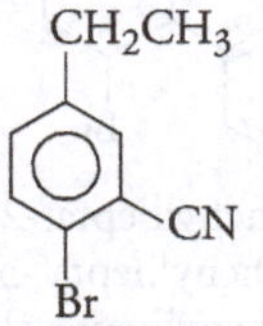

**07** Identify *A* in the given chemical reaction. **[2021, 25 Feb Shift-I]**

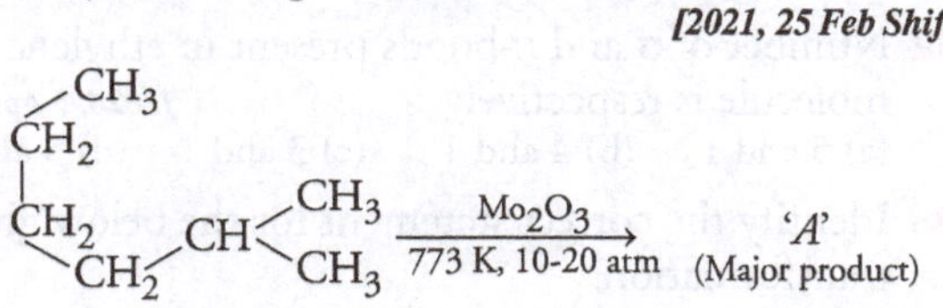


(a)  (b) 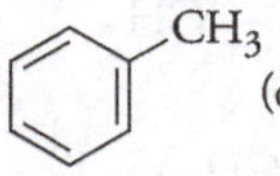 (c)  (d) 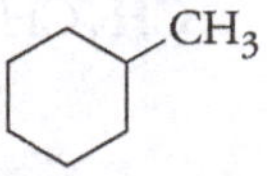

**Imp.** Order of Conformational Isomers

**08** Arrange the following conformational isomers of *n*-butane in order of their increasing potential energy **[2021, 31 Aug Shift-II]**

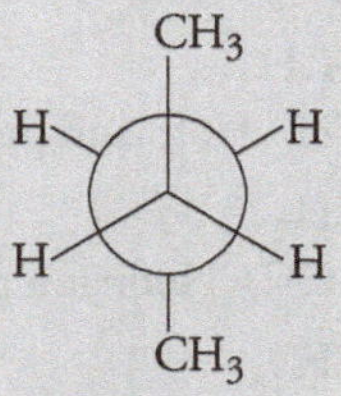
I

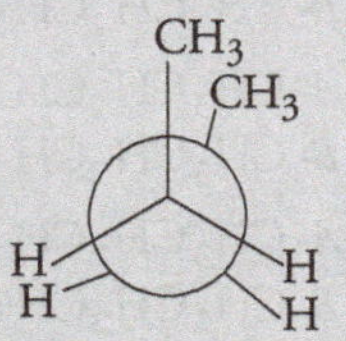
II

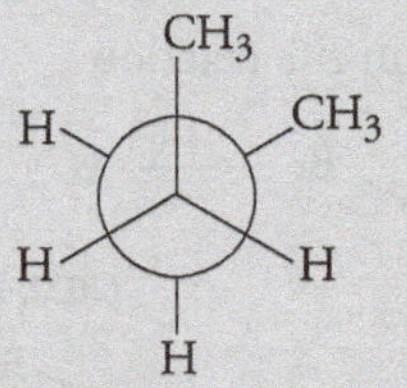
III

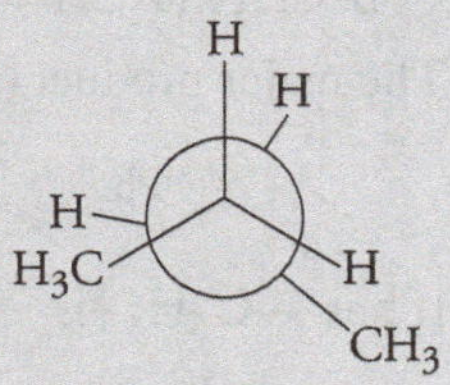
IV

(a) I < III < IV < II (b) I < IV < III < II
(c) II < IV < III < I (d) II < III < IV < I

**09** Which of the following reagent is used for the following reaction? **[2021, 24 Feb Shift-I]**

$$CH_3CH_2CH_3 \xrightarrow{?} CH_3CH_2CHO$$

(a) Copper at high temperature and pressure
(b) Molybdenum oxide
(c) Potassium permanganate
(d) Manganese acetate

**10** In the following skew conformation of ethane, H′—C—C—H″ dihedral angle is

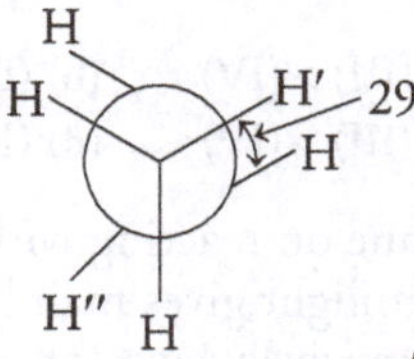


**[2019, 12 April Shift-II]**

(a) 58°
(b) 149°
(c) 151°
(d) 120°

## Level 2

**11** Given below are two statements. **[2025, 2 April Shift-II]**

**Statement I** Neopentane forms only one monosubstituted derivative.

**Statement II** Melting point of neopentane is higher than *n*-pentane.

In the light of the above statements, choose the most appropriate answer from the options given below.

(a) Statement I is correct but Statement II is incorrect.
(b) Both Statement I and Statement II are incorrect.
(c) Both Statement I and Statement II are correct.
(d) Statement I is incorrect but Statement II is correct.

**12** Given below are two statements.

**Statement-I** 2-methylbutane on oxidation with $KMnO_4$ gives 2-methylbutan-2-ol.

**Statement-II** *n*-alkanes can be easily oxidised to corresponding alcohols with $KMnO_4$.

Choose the correct option. **[2021, 17 March Shift-II]**

(a) Both statements I and II are correct.
(b) Both statements I and II are incorrect.
(c) Statement I is correct but statement II is incorrect.
(d) Statement I is incorrect but statement II is correct.

**13** Arrange in the correct order of stability (decreasing order) for the following molecules : *[JEE Main 2013]*

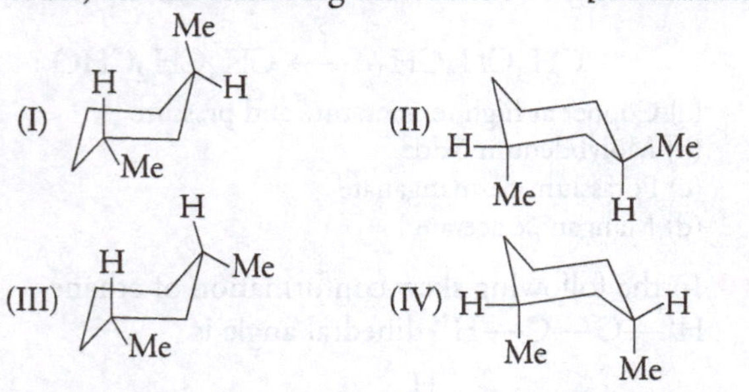


(a) (I) > (II) > (III) > (IV) (b) (IV) > (III) > (II) ≈ (I)
(c) (I) > (II) ≈ (III) > (IV) (d) (III) > (I) ≈ (II) > (IV)

**14** 2-methylbutane on reacting with bromine in the presence of sunlight gives mainly *[AIEEE 2005]*
(a) 1-bromo-3-methylbutane (b) 2-bromo-3-methylbutane
(c) 2-bromo-2-methylbutane (d) 1-bromo-2-methylbutane

**15** Which one of the following is reduced with zinc and hydrochloric acid to give the corresponding hydrocarbon? *[AIEEE 2004]*
(a) Ethyl acetate (b) Acetic acid
(c) Acetamide (d) Butan-2-one

**16** On mixing a certain alkane with chlorine and irradiating it with ultraviolet light, it forms only one monochloroalkane. This alkane could be *[AIEEE 2003]*
(a) propane (b) pentane
(c) *iso*-pentane (d) *neo*-pentane

## Level 3

**17** The incorrect statement regarding conformations of ethane is *[2024, 27 Jan Shift-II]*
(a) the conformations of ethane are inter-convertible to one-another.
(b) the dihedral angle in staggered conformation is 60°.
(c) eclipsed conformation is the most stable conformation.
(d) ethane has infinite number of conformations.

### Numerical Value Type Questions

**18** The number of different chain isomers for $C_7H_{16}$ is .... *[2024, 4 April Shift-I]*

**19** Total number of isomeric compounds (including stereoisomers) formed by monochlorination of 2-methylbutane is ......... . *[2024, 31 Jan, 1st Feb Shift-I]*

**20** Number of alkanes obtained on electrolysis of a mixture of $CH_3COONa$ and $C_2H_5COONa$ is ....... *[2024, 31 Jan Shift-I]*

**21** The total number of monobromo derivatives formed by the alkanes with molecular formula $C_5H_{12}$ is (excluding stereo isomers) *[2022, 25 July Shift-II]*

**22** Total number of possible stereoisomers of dimethyl cyclopentane is ......... . *[2022, 27 june Shift-I]*

**23** The dihedral angle in staggered form of Newman projection of 1, 1, 1-trichloro ethane is ........... degree. (Round off to the nearest integer) *[2021, 27 July Shift-II]*

# Topic 2 Alkenes

## Level 1

**01** The major product of the following reaction is

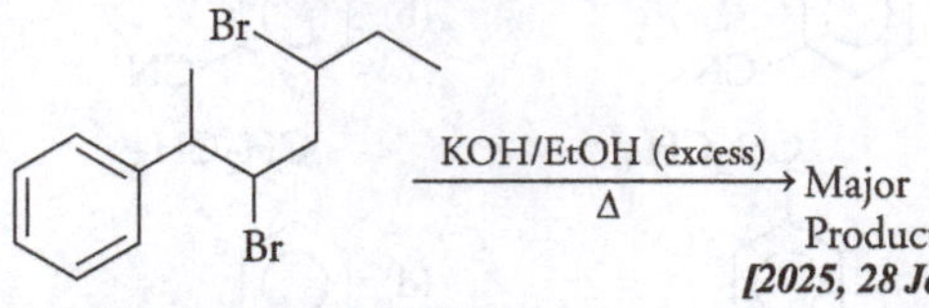


*[2025, 28 Jan Shift-II]*

(a) 6-phenylhepta -2, 4-diene
(b) 6-phenylhepta -3, 5-diene
(c) 2-phenylhepta -2, 4-diene
(d) 2-phenylhepta -2, 5-diene

**02** Number of σ and π-bonds present in ethylene molecule is respectively *[2024, 5 April Shift-I]*
(a) 5 and 1 (b) 4 and 1 (c) 3 and 1 (d) 5 and 2

**03** Identify the correct statement for the below given transformation.

$$CH_3CH_2CH_2—\underset{\overset{|}{\overset{+}{N}(CH_3)_3}}{CH}—CH_3 \xrightarrow[C_2H_5OH]{C_2H_5ONa} \underset{(Major)}{A} + \underset{(Minor)}{B}$$

*[2022, 28 July Shift-I]*

(a) $A$- $CH_3CH_2CH{=}CH—CH_3$, $B$- $CH_3CH_2CH_2CH{=}CH_2$ Saytzeff products
(b) $A$- $CH_3CH_2CH{=}CH—CH_3$, $B$- $CH_3CH_2CH_2CH{=}CH_2$ Hofmann products
(c) $A$- $CH_3CH_2CH_2CH{=}CH_2$, $B$- $CH_3CH_2CH{=}CH—CH_2$ Hofmann poducts
(d) $A$- $CH_3CH_2CH_2CH{=}CH_2$, $B$- $CH_3CH_2CH{=}CH—CH_3$ Saytzeff products

**04** The major product ($P$) in the reaction

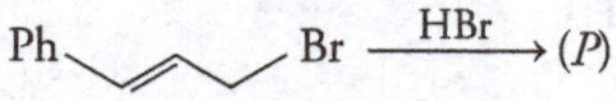


[Ph is $—C_6H_5$ ] is *[2022, 28 June Shift-I]*

(a) 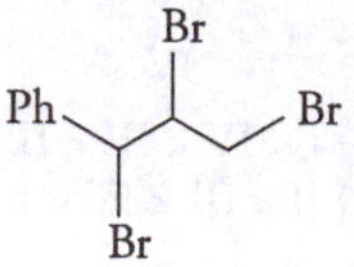


(b) 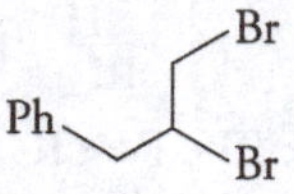


(c) 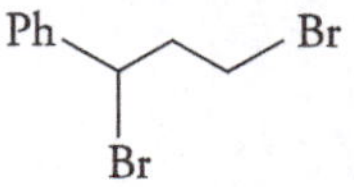


(d) 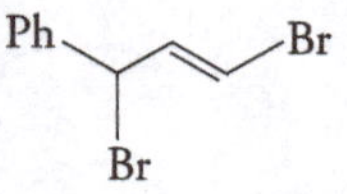

**05** Given below are two statements.

**Statement I** The presence of weaker π-bonds make alkenes less stable than alkanes.

**Statement II** The strength of the double bond is greater than that of carbon-carbon single bond.

In the light of the above statements, choose the correct answer from the options given below. *[2022, 24 June Shift-II]*

(a) Both Statement I and Statement II are correct.
(b) Both Statement I and Statement II are incorrect.
(c) Statement I is correct but Statement II is incorrect.
(d) Statement I is incorrect but Statement II is correct.

**06** Given below are two statements: One is labelled as Assertion (A) and the other is labelled as Reason (R). *[2021, 31 Aug Shift-I]*

**Assertion (A)** Treatment of bromine water with propene yields 1-bromopropan-2-ol.

**Reason (R)** Attack of water on bromonium ion follows Markownikoff rule and results in 1-bromopropan-2-ol.

In the light of the above statements, choose the most appropriate answer from the options given below.

(a) Both assertion and reason are true but reason is not the correct explanation of assertion.
(b) Assertion is false but reason is true.
(c) Both assertion and reason are true and reason is the correct explanation of assertion.
(d) Assertion is true but reason is false.

**07** 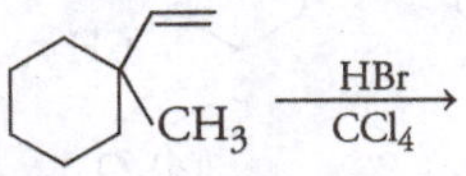 A (Major product)

Product "A" in the above chemical reaction is *[2021, 17 March Shift-I]*

(a) 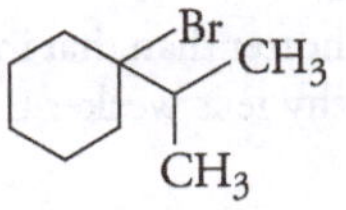
(b) 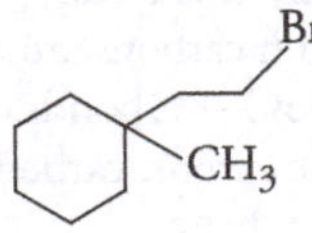
(c) 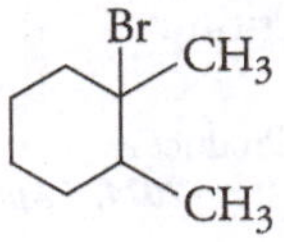
(b) 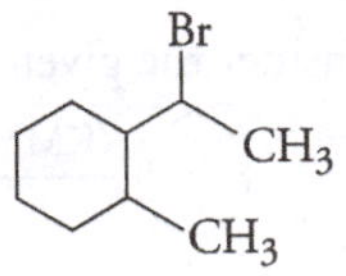

**08** 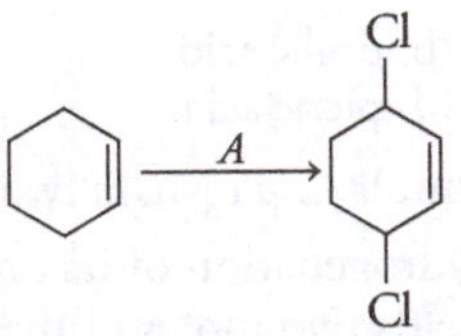

Identify the reagent(s) 'A' and condition(s) for the reaction. *[2021, 16 March Shift-II]*

(a) $A = HCl, ZnCl_2$
(b) $A = HCl$ ; anhydrous $AlCl_3$
(c) $A = Cl_2$ ; dark, anhydrous $AlCl_3$
(d) $A = Cl_2$ ; UV light

**09** Which of the following is Lindlar catalyst? *[2021, 16 March Shift-I]*

(a) Cold dilute solution of $KMnO_4$
(b) Partially deactivated palladised charcoal
(c) Zinc chloride and HCl
(d) Sodium and liquid $NH_3$

**10** The major product formed in the following reaction is *[2020, 05 Sept Shift-II]*

$$CH_3CH{=}CHCH(CH_3)_2 \xrightarrow{HBr}$$

(a) $CH_3CH(Br)CH_2CH(CH_3)_2$
(b) $CH_3CH_2CH(Br)CH(CH_3)_2$
(c) $Br(CH_2)_3CH(CH_3)_2$
(d) $CH_3CH_2CH_2C(Br)(CH_3)_2$

**11** The major product in the following reaction is

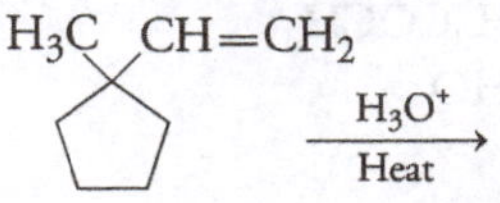

*[2020, 2 Sep Shift-I]*

(a) 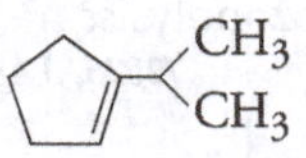
(b) 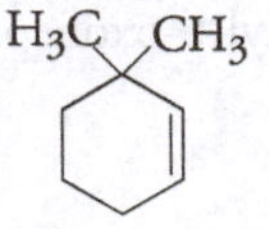
(c) 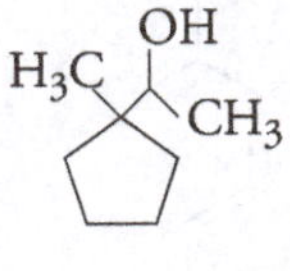
(d) 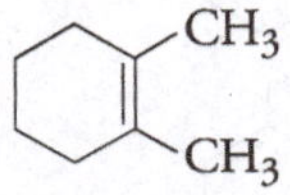

**Imp.** Order of Heat of Combustion

**12** The correct order of heat of combustion for following alkadienes is *[2020, 9 Jan Shift-I]*

(A) 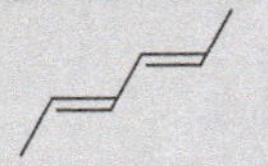
(B) 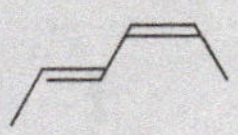
(C) 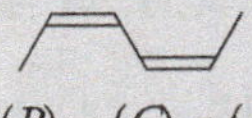

(a) (B) < (C) < (A)
(b) (A) < (B) < (C)
(c) (A) < (C) < (B)
(d) (C) < (B) < (A)

**13** But-2-ene on reaction with alkaline $KMnO_4$ at elevated temperature followed by acidification will give *[2019, 12 April Shift-I]*

(a) $CH_3—CH(OH)—CH(OH)—CH_3$
(b) one molecule of $CH_3CHO$ and one molecule of $CH_3COOH$
(c) 2 molecules of $CH_3COOH$
(d) 2 molecules of $CH_3CHO$

**14** The reaction of propene with HOCl ($Cl_2 + H_2O$) proceeds through the intermediate *[JEE Main 2016]*

(a) $CH_3\overset{+}{C}HCH_2Cl$
(b) $CH_3CH(OH)\overset{+}{C}H_2$
(c) $CH_3CHCl\overset{+}{C}H_2$
(d) $CH_3\overset{+}{C}HCH_2OH$

**15** Ozonolysis of an organic compound gives formaldehyde as one of the products. This confirms the presence of **[AIEEE 2011]**
(a) two ethylenic double bonds
(b) a vinyl group
(c) an *iso* propyl group
(d) an acetylenic triple bond

**16** In the following sequence of reactions, the alkene affords the compound '*B*'

$$CH_3CH{=}CHCH_3 \xrightarrow{O_3} A \xrightarrow[Zn]{H_2O} B$$

The compound *B* is **[AIEEE 2008]**
(a) $CH_3CH_2CHO$
(b) $CH_3COCH_3$
(c) $CH_3CH_2COCH_3$
(d) $CH_3CHO$

## Level 2

**17** Which compound would give 3-methyl-6-oxoheptanal upon ozonolysis? **[2025, 3 April Shift-I]**

(a) 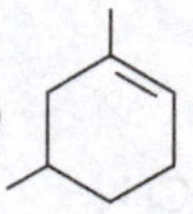 (b) 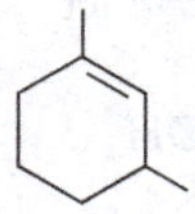

(c) 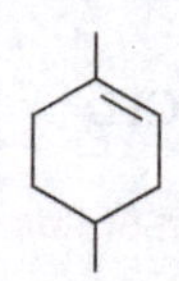 (d) 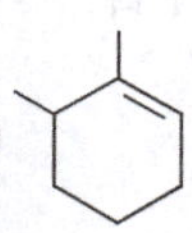

**18** The number of optically active products obtained from the complete ozonolysis of the given compound is:

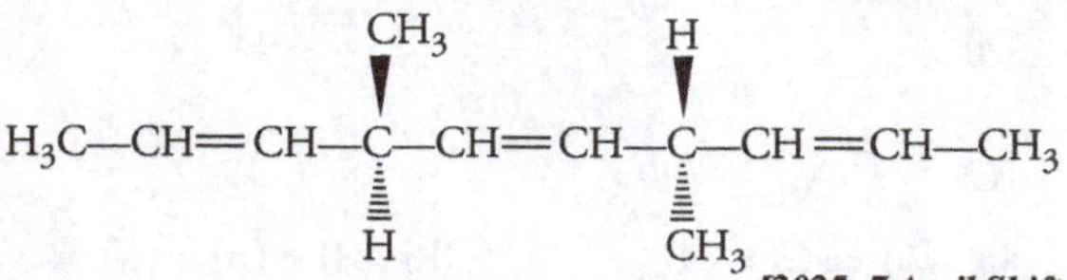

**[2025, 7 April Shift-II]**
(a) 2 (b) 4
(c) 0 (d) 1

**19** Given below are two statements: **[2025, 7 April Shift-I]**

**Statement I** Ozonolysis followed by treatment with Zn, $H_2O$ of *cis*-2-butene gives ethanal.

**Statement II** The product obtained by ozonolysis followed by treatment with Zn, $H_2O$ of 3, 6-dimethyloct-4-ene has no chiral carbon atom.

In the light of the above statements, choose the correct answer from the options given below.
(a) Statement I is false but Statement II is true.
(b) Statement I is true but Statement II is false.
(c) Both Statement I and Statement II are true.
(d) Both Statement I and Statement II are False.

**Imp.** Aldol Condensation in Blend with Ozonolysis

**20** Aman has been asked to synthesise the molecule

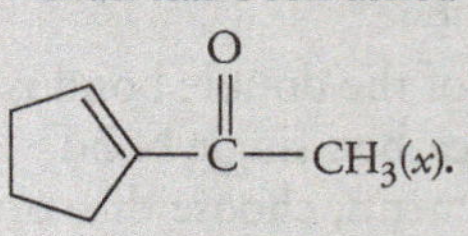

(*x*). He thought of preparing the molecule using an aldol condensation reaction. He found a few cyclic alkenes in his laboratory. He thought of performing ozonolysis reaction on alkene to produce a dicarbonyl compound followed by aldol reaction to prepare "*x*". Predict the suitable alkene that can lead to the formation of "*x*". **[2025, 24 Jan Shift-I]**

(a) 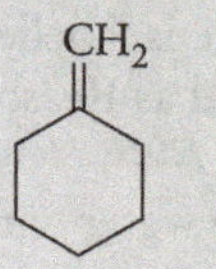 (b) 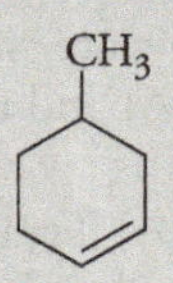

(c) 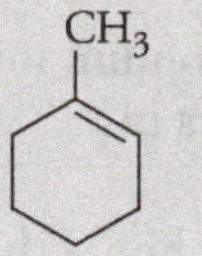 (d) 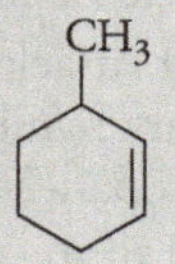

**21** Following are the four molecules "*P*", "*Q*", "*R*" and "*S*". **[2025, 24 Jan shift-I]**

Which one among the four molecules will react with H – Br(*aq*) at the fastest rate?

 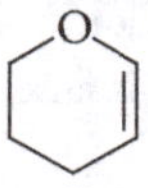 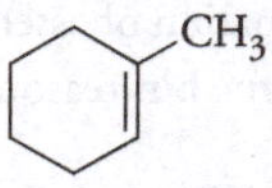 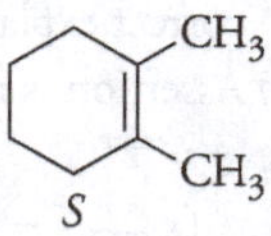

*P* *Q* *R* *S*

(a) *R* (b) *S* (c) *P* (d) *Q*

**22** The incorrect statement regarding ethyne is **[2024, 9 April Shift-II]**
(a) ethyne is linear
(b) both carbons are *sp*-hybridised
(c) the C—C bonds in ethyne is shorter than that in ethene
(d) the carbon-carbon bonds in ethyne is weaker than that in ethene

**23** Consider the given chemical reaction :

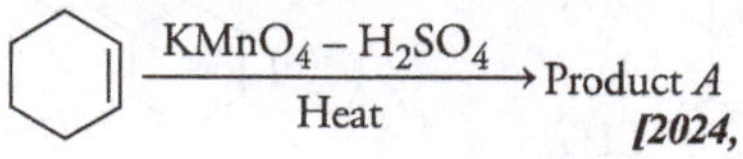

**[2024, 5 April Shift-II]**

Product *A* is
(a) adipic acid (b) oxalic acid
(c) acetic acid (d) picric acid

**24** A hydrocarbon '*X*' with formula $C_6H_8$ uses two moles of $H_2$ on catalytic hydrogenation of its one mole. On ozonolysis, '*X*' yields two moles of methane dicarbaldehyde. The hydrocarbon '*X*' is **[2023, 31 Jan Shift-II]**
(a) 1-methylcyclopenta-l, 4-diene
(b) cyclohexa - 1, 3 - diene
(c) hexa-1, 3, 5-triene
(d) cyclohexa-1, 4-diene

**25** In the following given reaction, '*A*' is *[2023, 24 Jan Shift-I]*

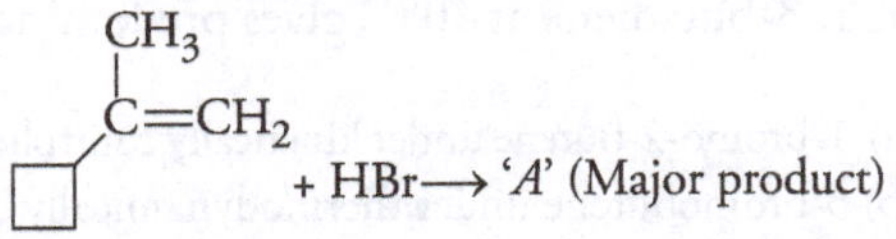

(a) 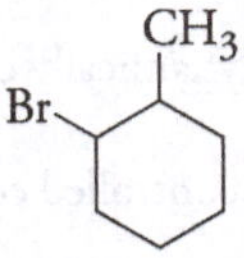
(b) 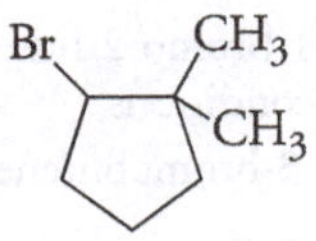

(c) 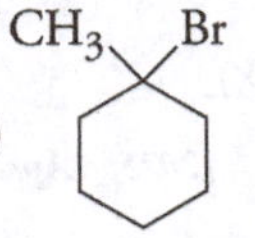
(d) 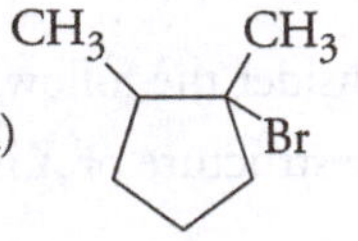

**26** When ethanol is heated with conc. $H_2SO_4$, a gas is produced. The compound formed, when this gas is treated with cold dilute aqueous solution of Baeyer's reagent, is *[2022, 29 July Shift-II]*

(a) formaldehyde
(b) formic acid
(c) glycol
(d) ethanoic acid

**27** The major product (*P*) of the given reaction is (where, Me is $—CH_3$) *[2022, 28 June Shift-II]*

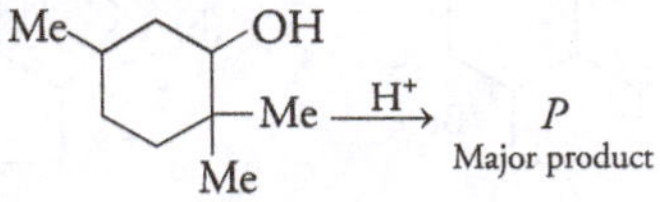

(a) 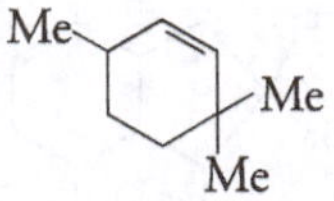
(b) 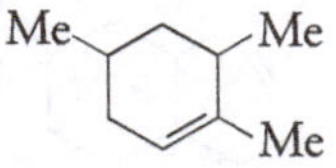

(c) 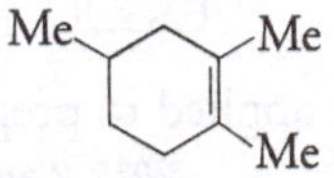
(d) 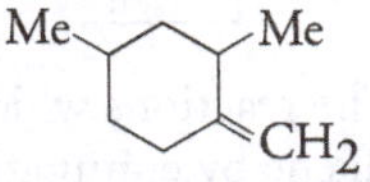

**28** Amongst the following, the major product of the given chemical reaction is *[2022, 25 June Shift-II]*

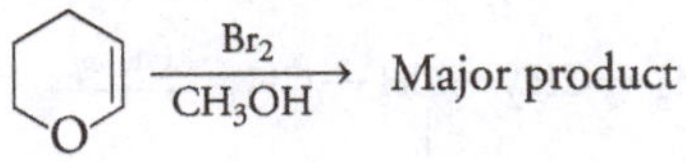

(a) 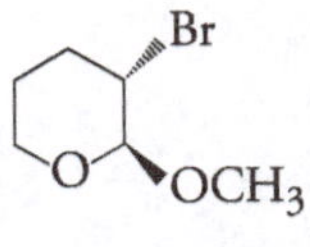
(b) 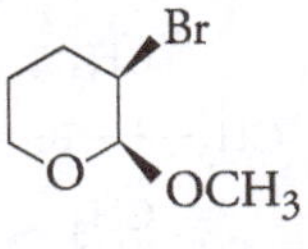

(c) 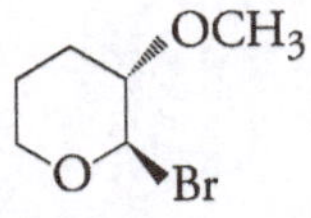
(d) 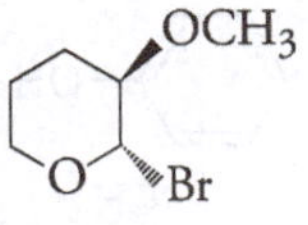

**Imp.** Electrophilic Addition of Bromine

**29** The stereoisomers that are formed by electrophilic addition of bromine to *trans*-but-2-ene is/are *[2021, 01 Sept Shift-II]*

(a) 2 identical mesomers
(b) 2 enantiomers
(c) 2 enantiomers and 2 mesomers
(d) 1 racemic and 2 enantiomers

**30** The major product formed in the following reaction is *[2021, 31 Aug Shift-I]*

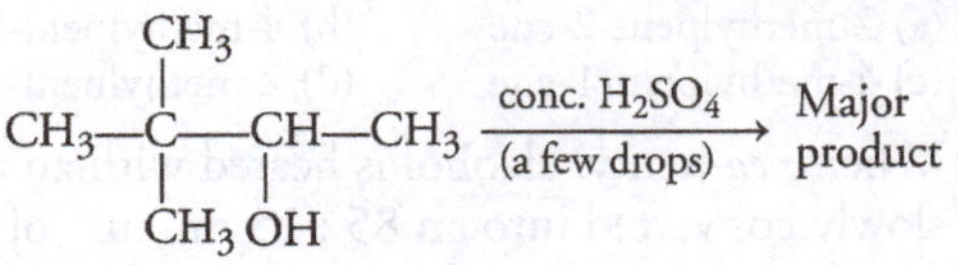

(b) 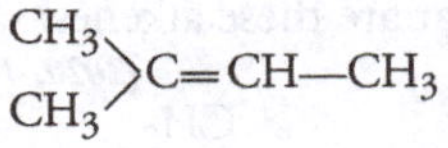
(b) 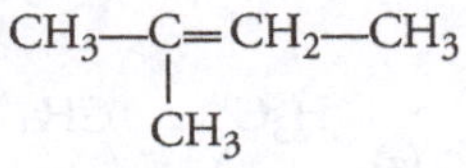

(c) 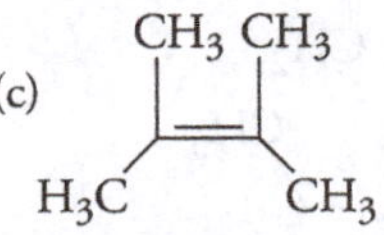
(d) 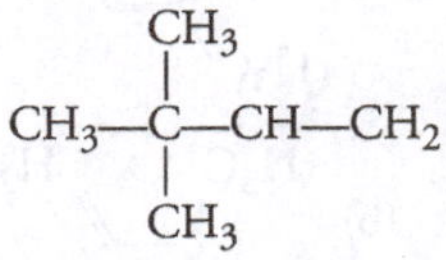

**31** Which of the following molecules does not show stereo isomerism? *[2021, 22 July Shift-II]*

(a) 3,4-dimethylhex-3-ene
(b) 3-methylhex-1-ene
(c) 3-ethylhex-3-ene
(d) 4-methylhex-1-ene

**32** 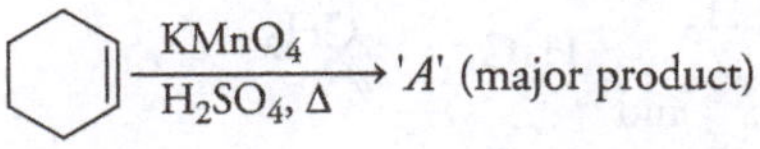

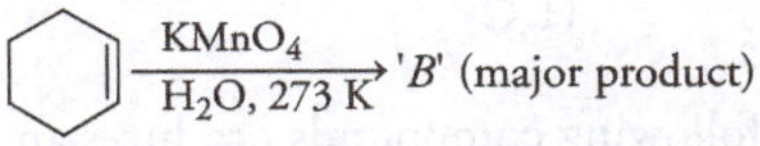

For above chemical reactions, identify the correct statement from the following. *[2021, 20 July Shift-I]*

(a) Compound '*A*' is diol and compound '*B*' is dicarboxylic acid.
(b) Both compound '*A*' and compound '*B*' are dicarboxylic acids.
(c) Compound '*A*' is dicarboxylic acid and compound '*B*' is diol.
(d) Both compound '*A*' and compound '*B*' are diols.

**33** Identify products *A* and *B*.

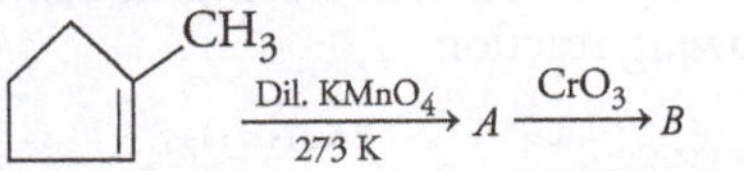

*[2021, 24 Feb Shift-I]*

(a) A : 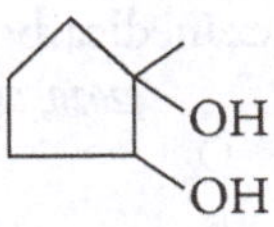 *B* : 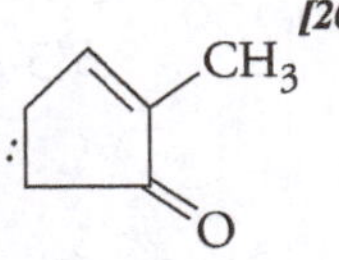

(b) *A* : 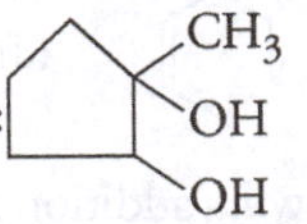 *B* : 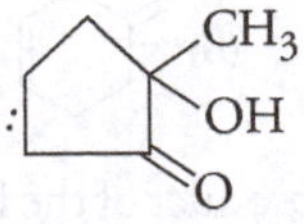

(c) *A* : 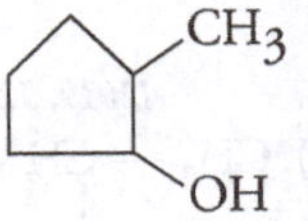 *B* : 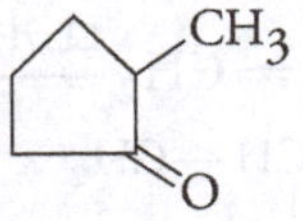

(d) *A* : $OHC—CH_2CH_2CH_2—\overset{O}{\overset{\|}{C}}—CH_3$

*B* : $HOOC—CH_2CH_2CH_2—\overset{O}{\overset{\|}{C}}—CH_3$

**34** Which of the following compounds shows geometrical isomerism? *[2020, 06 Sept Shift-I]*
(a) 2-methylpent-2-ene (b) 4-methylpent-2-ene
(c) 4-methylpent-1-ene (d) 2-methylpent-1-ene

**35** When *neo*-pentyl alcohol is heated with an acid, it slowly converted into an 85 : 15 mixture of alkenes *A* and *B*, respectively. What are these alkenes? *[2020, 4 Sep Shift-I]*

(a) 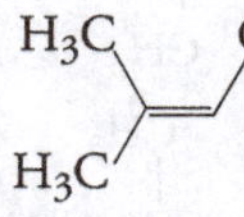 and 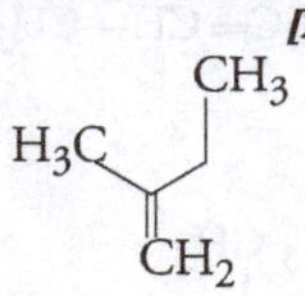

(b) 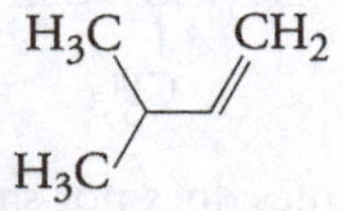 and 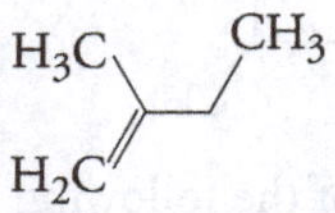

(c) 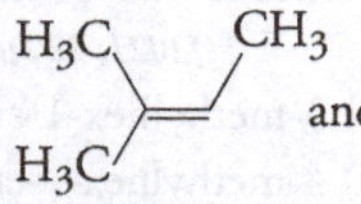 and 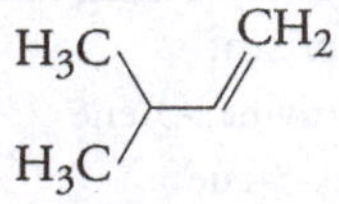

(d) 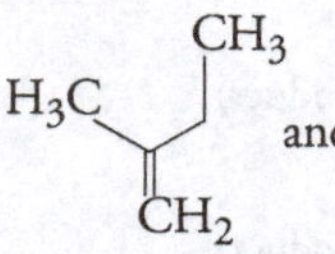 and 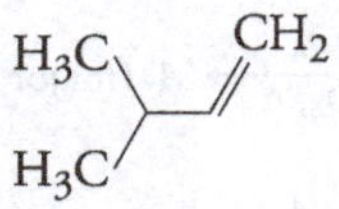

**36** Which of the following compounds produces an optically inactive compound on hydrogenation? *[2020, 03 Sept Shift-I]*

(a) 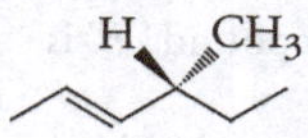 (b) 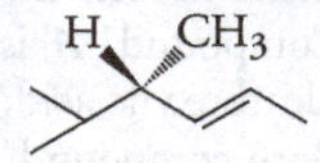

(c)  (d) 

**37** An unsaturated hydrocabon *X* absorbs two hydrogen molecules on catalytic hydrogenation and also gives following reaction

$$X \xrightarrow[\text{Zn}/H_2O]{O_3} A \xrightarrow{[Ag(NH_3)_2]^+} B\text{(3-oxo-hexanedicarboxylic acid)}$$

*X* will be *[2020, 08 Jan Shift-II]*

(a) 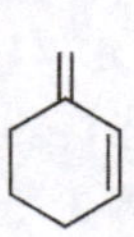 (b) 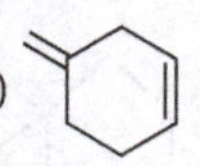 (c) 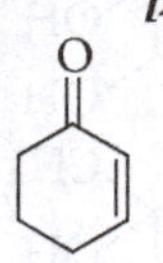 (d) 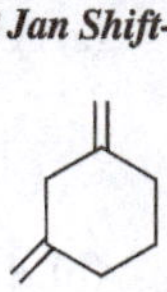

**38** The major product of the following addition reaction is

$$H_3C—CH=CH_2 \xrightarrow{Cl_2/H_2O}$$ *[2019, 12 April Shift-I]*

(a) $CH_3—\underset{\text{Cl}}{\underset{|}{CH}}—\underset{\text{OH}}{\underset{|}{CH_2}}$ (b) $CH_3—\underset{\text{OH}}{\underset{|}{CH}}—\underset{\text{Cl}}{\underset{|}{CH_2}}$

(c) 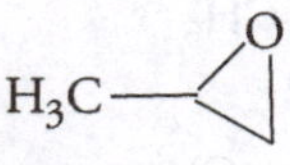 (d) 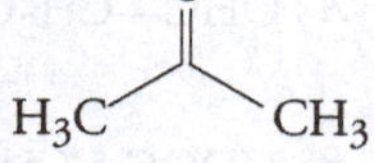

**39** Reaction of one molecule of HBr with one molecule of 1, 3-butadiene at 40°C gives predominantly *[AIEEE 2005]*
(a) 1-bromo-2-butene under kinetically controlled conditions
(b) 3-bromobutene under thermodynamically controlled conditions
(c) 1-bromo-2-butene under thermodynamically controlled conditions
(d) 3-bromobutene under kinetically controlled conditions

## Level 3

**40** Consider the following molecule (*X*).
The structure of *X* is *[2025, 4 April Shift-II]*

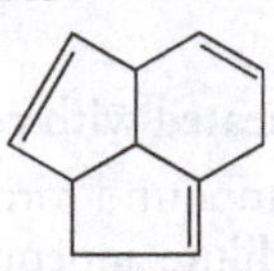

(*X*)

The major product formed when the given molecule (*X*) is treated with HBr (1 eq) is:

(a) 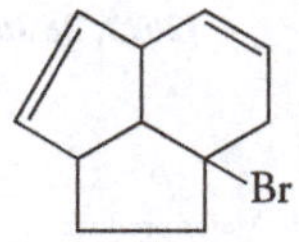 (b) 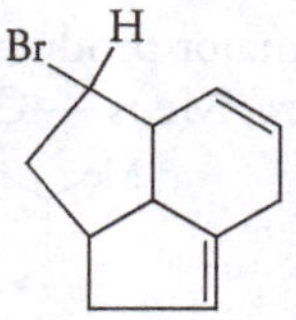

(c) 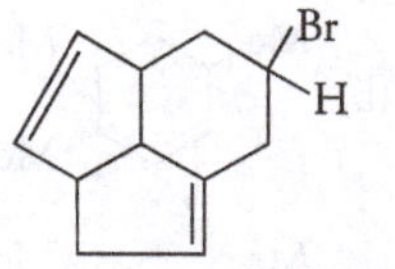 (d) 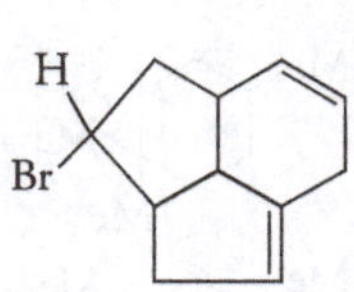

**41** The reactions which cannot be applied to prepare an alkene by elimination, are *[2025, 7 April Shift-I]*

(A) 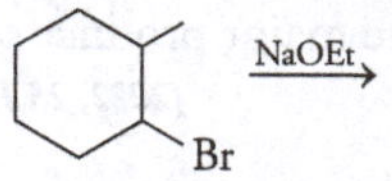

(B) $CH_3—CH_2—\underset{\text{Br}}{\underset{|}{CH}}—CH_3 \xrightarrow{\text{KOH}(aq)}$

(C) $CH_3—\underset{CH_3}{\overset{CH_3}{C}}—Br \xrightarrow{\text{NaOMe}}$

(D) 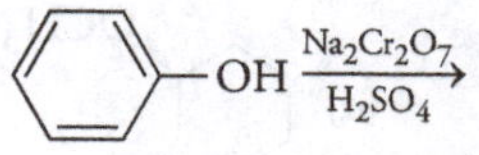

(E) $CH_3—\underset{CH_3}{\overset{CH_3}{C}}—OH \xrightarrow[573\text{ K}]{\text{Cu}}$

Choose the correct answer from the options given below.
(a) A, C and D only (b) B and D only
(c) B, C and D only (d) B and E only

**42** Match List-I with List-II. *[2025, 23 Jan Shift-II]*

| | List-I (Isomers of $C_{10}H_{14}$) | | List-II (Ozonolysis product) |
|---|---|---|---|
| (a) | 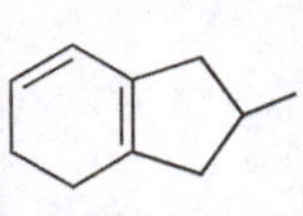 | I. | 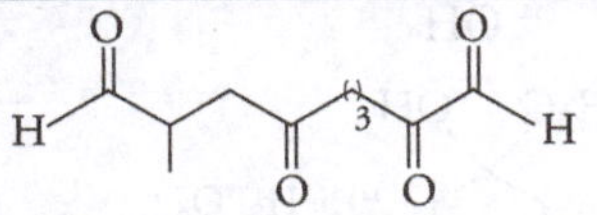 |
| (b) | 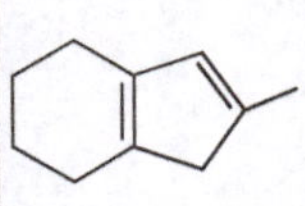 | II. | 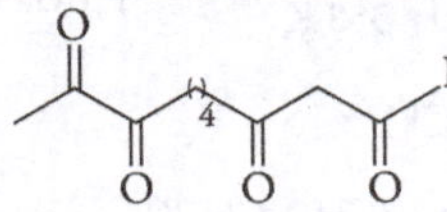 |
| (c) | 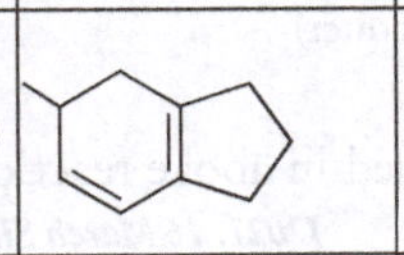 | III. | 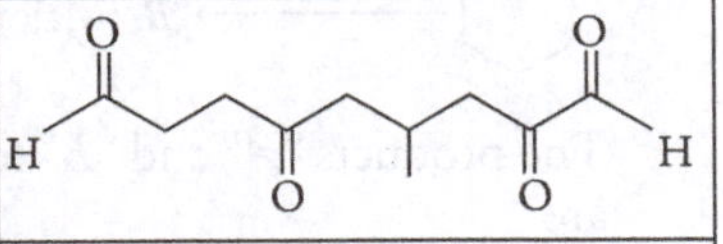 |
| (d) | 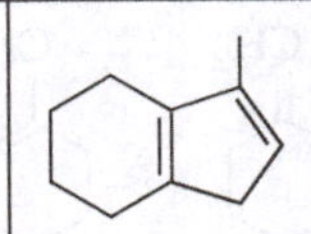 | IV. | 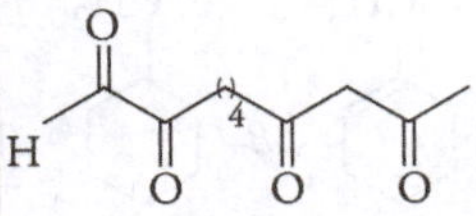 |

Choose the **correct** answer from the options given below

(a) (A)-(II), (B)-(III), (C)-(I), (D)-(IV)
(b) (A)-(I), (B)-(IV), (C)-(III), (D)-(II)
(c) (A)-(III), (B)-(II), (C)-(I), (D)-(IV)
(d) (A)-(III), (B)-(IV), (C)-(I), (D)-(II)

**43** The incorrect statements regarding geometrical isomerism are *[2025, 22 Jan Shift-I]*

(A) Propene shows geometrical isomerism.
(B) *trans* isomer has identical atoms/groups on the opposite sides of the double bond.
(C) *cis*-but-2-ene has higher dipole moment than *trans*-but-2-ene.
(D) 2-methylbut-2-ene shows two geometrical isomers.
(E) *trans*-isomer has lower melting point than *cis* isomer.

Choose the correct answer from the options given below.

(a) (A), (D) and (E) Only
(b) (B) and (C) Only
(c) (A) and (E) Only
(d) (C), (D) and (E) Only

**44** The incorrect statement regarding the geometrical isomers of 2-butene is *[2024, 6 April Shift-II]*

(a) *trans*-2-butene is more stable than *cis*-2-butene.
(b) *cis*-2-butene and *trans*-2 butene are not interconvertible at room temperature.
(c) *cis*-2-butene has less dipole moment than *trans*-2-butene.
(d) *cis*-2-butene and *trans*-2-butene are stereoisomers.

**45**

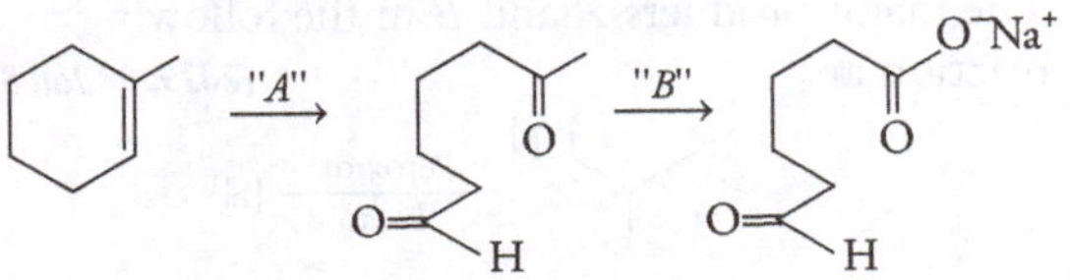


In the above chemical reaction sequence "*A*" and "*B*" respectively are *[2024, 4 April Shift-II]*

(a) $H_2O$, $H^+$ and $KMnO_4$
(b) $O_3$, $Zn/H_2O$ and $KMnO_4$
(c) $O_3$, $Zn/H_2O$ and $NaOH(alc)/I_2$
(d) $H_2O$, $H^+$ and $NaOH(alc)/I_2$

**46** The product formed in the following multi-step reaction is

$$CH_3CH{=}CH_2 \xrightarrow[\text{(iii) PCC} \quad \text{(iv) } CH_3MgBr]{\text{(i) } B_2H_6 \quad \text{(ii) } H_2O_2,\ NaOH}$$

*[2023, 15 April Shift-I]*

(a) $CH_3CH_2CH_2CH_2OH$
(b) $CH_3-\overset{OH}{\underset{CH_3}{C}}-CH_3$
(c) $CH_3CH_2\overset{O}{\overset{\|}{C}}OCH_3$
(d) $CH_3CH_2\overset{OH}{C}HCH_3$

**47** The major products '*A*' and '*B*' respectively, are *[2023, 30 Jan Shift-I]*

$$\text{'}A\text{'} \xleftarrow[H_2SO_4]{\text{Cold}} H_3C-\overset{CH_3}{C}{=}CH_2 \xrightarrow[80°C]{H_2SO_4} \text{'}B\text{'}$$

(a) $H_3C-\overset{CH_3}{\underset{OSO_3H}{C}}-CH_3$ and $CH_3-\overset{CH_3}{CH}-CH_2CH_2-\overset{CH_3}{HC}-CH_3$

(b) $CH_3-\overset{CH_3}{C}{=}CH-\overset{CH_3}{\underset{CH_3}{C}}-CH_3$ and $CH_3-\overset{CH_3}{\underset{OSO_3H}{C}}-CH_3$

(c) $CH_3-\overset{CH_3}{CH}-CH_2CH_2-\overset{CH_3}{CH}-CH_3$ and $H_3C-\overset{CH_3}{\underset{OSO_3H}{C}}-CH_3$

(d) $H_3C-\overset{CH_3}{\underset{OSO_3H}{C}}-CH_3$ and $CH_3-\overset{CH_3}{C}{=}CH-\overset{CH_3}{\underset{CH_3}{C}}-CH_3$

**48** The one giving maximum number of isomeric alkenes on dehydrohalogenation reaction is (excluding rearrangement) *[2023, 29 Jan Shift-II]*

(a) l-bromo-2-methylbutane
(b) 2-bromopentane
(c) 2-bromo-3 ,3-dimethylpentane
(d) 2-bromopropane

**49** A compound *A* on reaction with *X* and *Y* produces the same major product but different by-product *a* and *b*. Oxidation of *a*, gives a substance produced by ants,

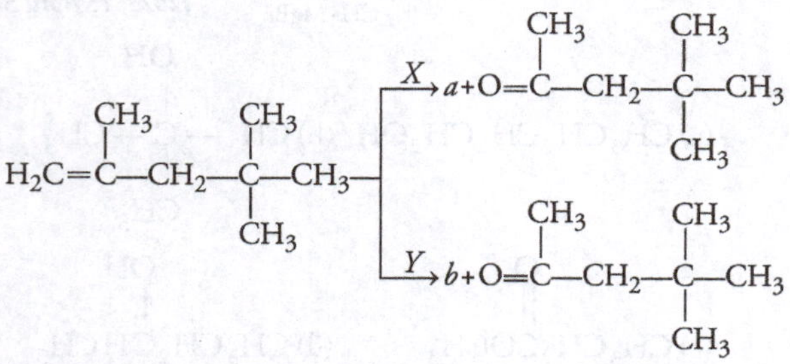

*X* and *Y* respectively are *[2022, 25 July Shift-I]*

(a) $KMnO_4/H^+$ and dil. $KMnO_4$, 273 K
(b) $KMnO_4$ (dilute), 273 K and $KMnO_4/H^+$
(c) $KMnO_4/H^+$ and $O_3$, $H_2O/Zn$
(d) $O_3$, $H_2O/Zn$ and $KMnO_4/H^+$

**50** Two isomers '*A*' and '*B*' with molecular formula $C_4H_8$ give different products on oxidation with $KMnO_4$ in acidic medium. Isomer '*A*' on reaction with $KMnO_4/H^+$ results in effervescence of a gas and gives ketone. The compound '*A*' is *[2022, 29 June Shift-II]*

(a) but-1-ene (b) *cis*-but-2-ene
(c) *trans*-but-2-ene (d) 2-methylpropene

**51** *A* and *B* respectively are

$$A \xrightarrow[(2)\ Zn\text{-}H_2O]{(1)\ O_3} \text{ethane-1,2-dicarbaldehyde} + \text{glyoxal/oxaldehyde}$$

$$B \xrightarrow[(2)\ Zn\text{-}H_2O]{(1)\ O_3} \text{5-oxohexanal}$$

*[2022, 27 June Shift-II]*

(a) 1-methylcyclohex-1,3-diene and cyclopentene
(b) cyclohex-1,3-diene and cyclopentene
(c) 1-methylcyclohex-1,4-diene and 1-methylcyclopent-1-ene
(d) cyclohex-1,3-diene and 1-methylcyclopent-1-ene

**52** The major product in the following reaction *[2022, 26 June Shift-II]*

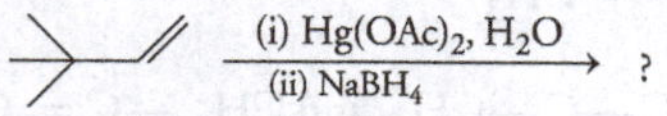

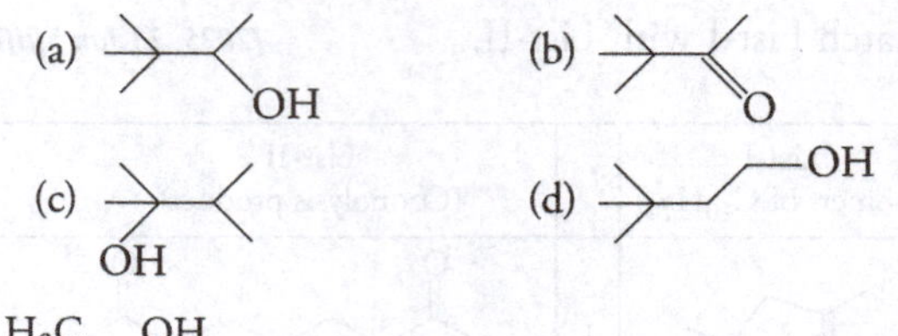

**53**

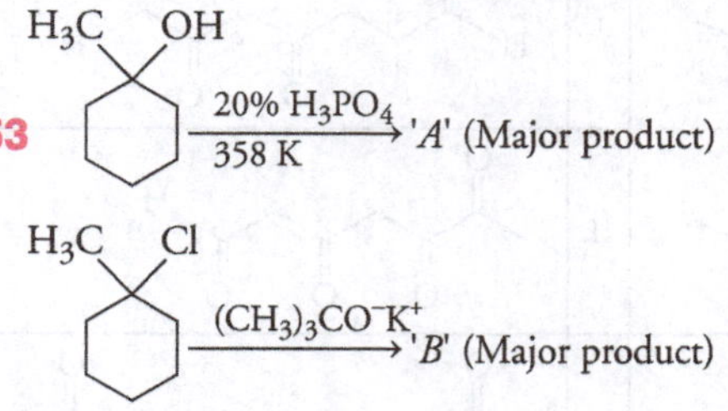

The products "*A*" and "*B*" formed in above reactions are *[2021, 16 March Shift-I]*

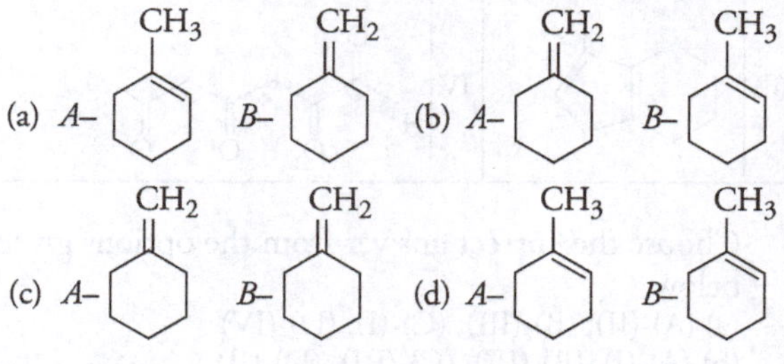

**54** The major product [*B*] in the following sequence of reactions is

$$CH_3-\underset{\underset{CH(CH_3)_2}{|}}{C}=CH-CH_2CH_3 \xrightarrow[(ii)\ H_2O_2,\ \overset{\ominus}{O}H]{(i)\ B_2H_6} [A] \xrightarrow[\Delta]{\text{Dil. } H_2SO_4} [B]$$

*[2020, 8 Jan Shift-II]*

(a) $CH_3-\underset{\underset{CH(CH_3)_2}{|}}{C}-CH=CH-CH_3$

(b) $CH_3-\underset{\underset{CH(CH_3)_2}{|}}{C}=CH-CH_2CH_3$

(c) $CH_3-\underset{\underset{\underset{H_3C\ \ CH_3}{C}}{||}}{C}-CH_2CH_2CH_3$

(d) $CH_2=\underset{\underset{CH(CH_3)_2}{|}}{C}-CH_2CH_2CH_3$

**55** The major products *A* and *B* in the following reaction are *[2020, 08 Jan Shift-I]*

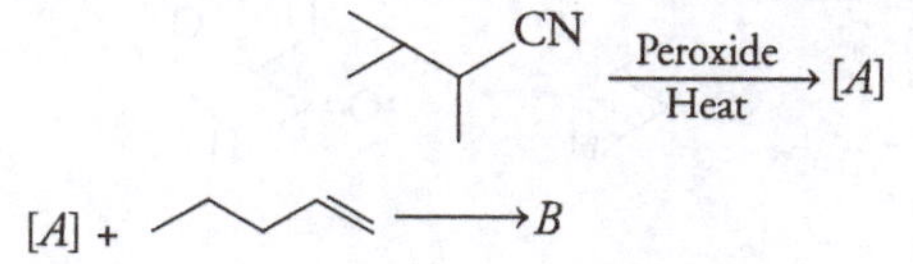

(a) 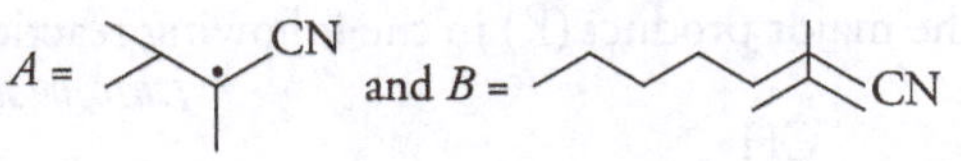

(b) 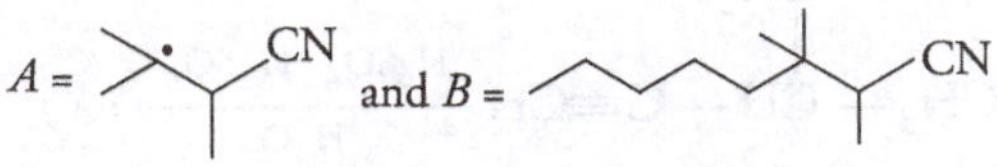

(c) 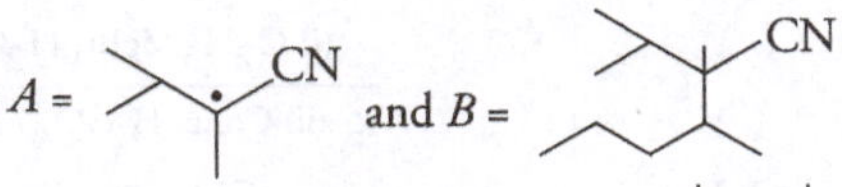

(d) 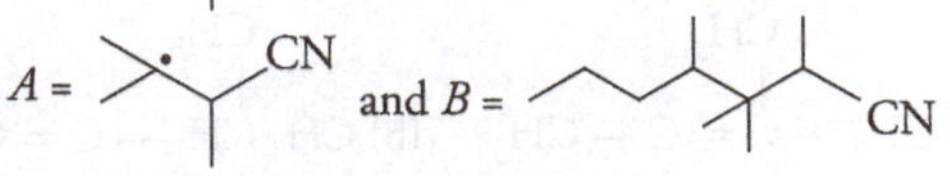

## Numerical Value Type Questions

**56** The hydrocarbon ($X$) with molar mass 80 g $mol^{-1}$ and 90% carbon has .............. degree of unsaturation. *[2025, 24 Jan Shift-II]*

**57** A compound '$X$' absorbs 2 moles of hydrogen and '$X$' upon oxidation with $KMnO_4|H^+$ gives

$$CH_3-\underset{\underset{O}{\|}}{C}-CH_3,\ CH_3-\underset{\underset{O}{\|}}{C}-OH$$

$$\text{and } CH_3-\underset{\underset{O}{\|}}{C}-CH_2CH_2-\underset{\underset{O}{\|}}{C}-OH$$

The total number of σ-bonds present in the compound '$X$' is ......... . *[2025, 23 Jan Shift-II]*

**58** 3-methylhex-2-ene on reaction with HBr in presence of peroxide forms and addition product ($A$). The number of possible stereoisomers for '$A$' is ........ . *[2024, 27 Jan Shift-I]*

**59** A hydrocarbon '$X$' is found to have molar mass of 80. A 10.0 mg of compound '$X$' on hydrogenation consumed 8.40 mL of $H_2$ gas (measured at STP). Ozonolysis compound '$X$' yields only formaldehyde and dialdehyde. The total number of fragments/molecules produced from the ozonolysis of compound '$X$' is ......... . *[2022, 30 June Shift-I]*

**60** The major product of the following reaction contains ......... bromine atom(s).

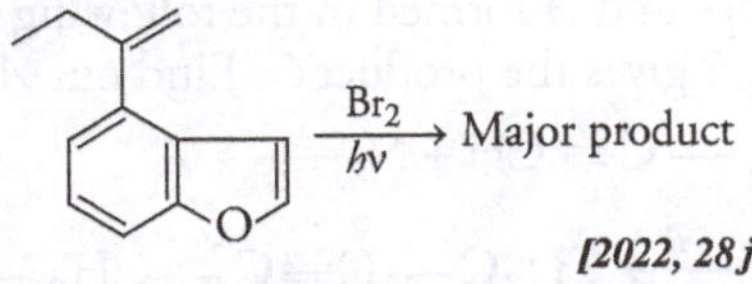

*[2022, 28 june Shift-II]*

**61** The major product '$A$' of the following given reaction has ......... $sp^2$ hybridised carbon atoms.

$$\text{2,7-dimethyl-2,6-octadiene} \xrightarrow{H^+} \underset{\text{(Major product)}}{A}$$

*[2022, 24 June Shift-I]*

**62** 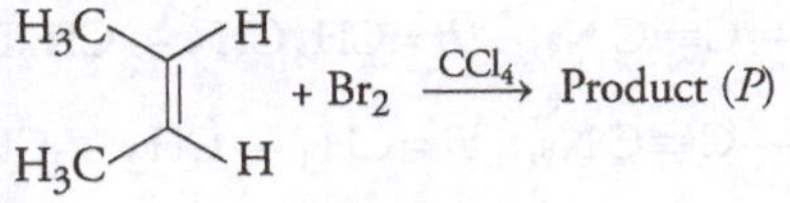

Consider the above chemical reaction. The total number of stereoisomers possible for product $P$ is ...... *[2021, 25 July Shift-II]*

**63** The number of acyclic structural isomers (including, geometrical isomers) for pentene are ......... . *[2021, 22 July Shift-II]*

## Topic 3 Alkynes

### Level 1

**01** Metallic sodium does not react normally with *[2021, 20 July Shift-II]*

(a) but-2-yne
(b) *tert*-butyl alcohol
(c) gaseous ammonia
(d) ethyne

**02** The major product obtained from the following reaction is *[2020, 6 Sep Shift-I]*

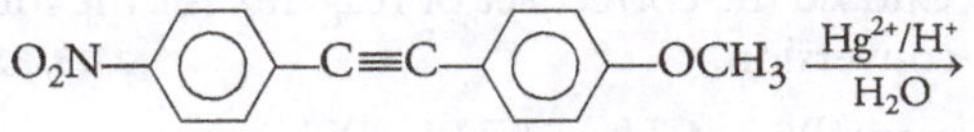

(a) 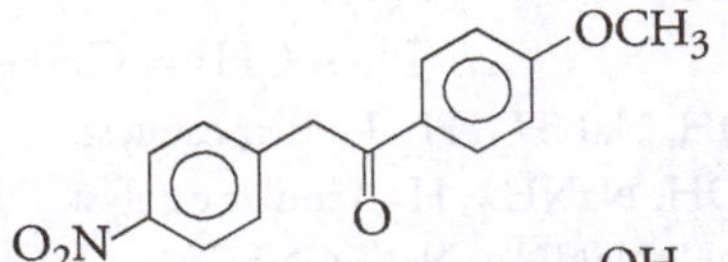

(b) 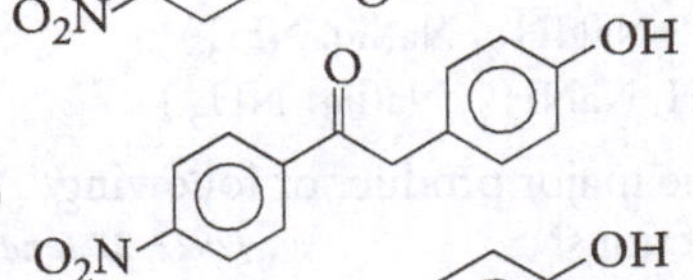

(c) 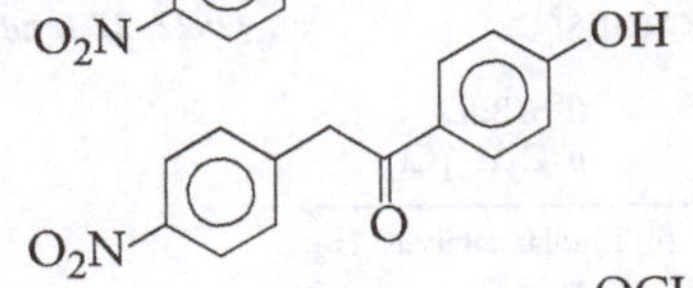

(d) 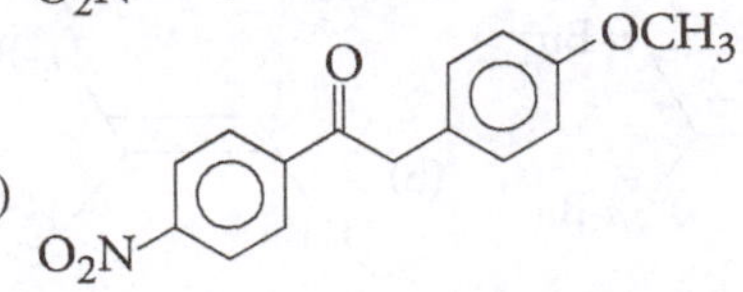

**03** The major product of the following reaction is

$$CH_3CH_2\underset{\underset{Br}{|}}{CH}-\underset{\underset{Br}{|}}{CH_2} \xrightarrow[\text{(ii) } NaNH_2 \text{ in liq. } NH_3]{\text{(i) alc. KOH}}$$

*[2019, 12 Jan Shift-II]*

(a) $CH_3CH_2\underset{\underset{NH_2}{|}}{CH}-\underset{\underset{NH_2}{|}}{CH_2}$
(b) $CH_3CH{=}CHCH_2NH_2$
(c) $CH_3CH{=}C{=}CH_2$
(d) $CH_3CH_2C{\equiv}CH$

**04** The *trans*-alkenes are formed by the reduction of alkynes with *[JEE Main 2018]*

(a) $H_2$-Pd/C, $BaSO_4$
(b) $NaBH_4$
(c) Na/liq. $NH_3$
(d) Sn-HCl

**05** The major organic compound formed by the reaction of 1, 1, 1–trichloroethane with silver powder is **[JEE Main 2014]**

(a) acetylene (b) ethene
(c) 2–butyne (d) 2–butene

**06** 2-hexyne gives *trans*-2-hexene on treatment with **[AIEEE 2012]**

(a) $Pt/H_2$ (b) $Li/NH_3$
(c) $Pd/BaSO_4$ (d) $LiAlH_4$

**07** The treatment of $CH_3MgX$ with $CH_3C\equiv C-H$ produces **[AIEEE 2008]**

(a) $CH_3-CH=CH_2$ (b) $CH_3C\equiv C-CH_3$
(c) $CH_3-\overset{H}{C}=\overset{H}{C}-CH_3$ (d) $CH_4$

**08** Acetylene does not react with **[AIEEE 2002]**

(a) Na (b) ammoniacal $AgNO_3$
(c) HCl (d) NaOH

## Level 2

**09** Choose the correct set of reagents for the following conversion. **[2023, 31 Jan Shift-I]**

$$trans\,(Ph-CH=CH-CH_3) \longrightarrow cis\,(Ph-CH=CH-CH_3)$$

(a) $Br_2$, aq. KOH, $NaNH_2$, $H_2$ Lindlar catalyst
(b) $Br_2$, alc. KOH, $NaNH_2$, $H_2$ Lindlar catalyst
(c) $Br_2$, aq. KOH, $NaNH_2$, Na(liq. $NH_3$)
(d) $Br_2$, alc. KOH, $NaNH_2$, Na(liq. $NH_3$)

**10** What will be the major product of following sequence of reactions? **[2022, 27 June Shift-II]**

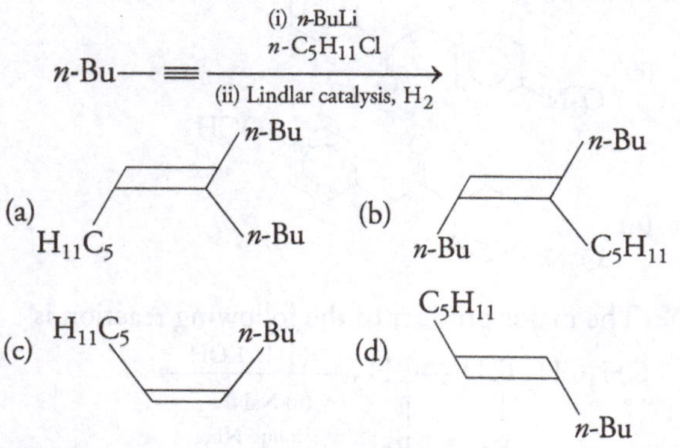


**11** An unsaturated hydrocarbon $X$ on ozonolysis gives $A$. Compound $A$ when warmed with ammoniacal silver nitrate forms a bright silver mirror along the sides of the test tube. The unsaturated hydrocarbon $X$ is **[2021, 16 March Shift-II]**

(a) $HC\equiv C-CH_2-CH_3$ (b) 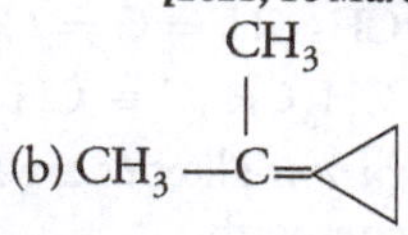
(c) $CH_3-C\equiv C-CH_3$ (d) $CH_3-C=C(CH_2CH_3)-CH_3$

**12** The major product ($Y$) in the following reactions is **[2020, 09 Jan Shift-I]**

$$CH_3-\underset{}{CH}(CH_3)-C\equiv CH \xrightarrow[H_2O]{HgSO_4,\ H_2SO_4} (X) \xrightarrow[\text{(ii) Conc. } H_2SO_4/\Delta]{\text{(i) } C_2H_5MgBr,\ H_2O} (Y)$$

(a) $CH_3-C(CH_3)=C(CH_2CH_3)-CH_3$ (b) $CH_3CH-C(CH_3)(CH_3)=CHCH_3$
(c) $H_3C-C(=CH_2)-CH(C_2H_5)-CH_3$ (d) $CH_3-CH(CH_3)-C(CH_2CH_3)=CH_2$

**13** The major product of the following reaction is **[2019, 9 April Shift-I]**

$$CH_3C\equiv CH \xrightarrow[\text{(ii) DI}]{\text{(i) DCl (1 equiv.)}}$$

(a) $CH_3CD(Cl)CHD(I)$
(b) $CH_3CD_2CH(Cl)(I)$
(c) $CH_3CD(I)CHD(Cl)$
(d) $CH_3C(I)(Cl)CHD_2$

## Level 3

**14** In the given reactions identify $A$ and $B$.

$$H_2 + A \xrightarrow{Pd/C} \underset{H}{\overset{CH_3}{}}C=C\underset{H}{\overset{C_2H_5}{}}$$

$$CH_3-C\equiv C-CH_3 + H_2 \xrightarrow{Na/Liquid\ NH_3} B$$

**[2024, 01 Feb Shift-II]**

(a) $A$ : 2-pentyne $B$ : *trans*-2-butene
(b) $A$ : *n*-pentane $B$ : *cis*-2-butene
(c) $A$ : *n*-pentane $B$ : *trans*-2-butene
(d) $A$ : 2-pentyne $B$ : *cis*-2-butene

**15** Compound $A$ formed in the following reaction reacts with $B$ gives the product $C$. Find out $A$ and $B$.

$$CH_3-C\equiv CH + Na \longrightarrow A \xrightarrow{B} \underset{(C)}{CH_3C-C\equiv C-CH_2-CH_2(CH_3)} + NaBr$$

**[2024, 30 Jan Shift-I]**

(a) $A = CH_3-CH_2CH_3$, $B = CH_3-C\equiv CH$
(b) $A = CH_3-CH=CH_2$, $B = CH_3CH_2-CH_2Br$
(c) $A = CH_3-C\equiv \overset{-}{C}\overset{+}{Na}$, $B = CH_3CH_2-CH_2Br$
(d) $A = CH_3-C\equiv C\overset{+}{Na}$, $B = CH_3-CH_2-CH_3$

**Imp.** Chemical Reactions of Ethyne

**16** The major product [$R$] in the following sequence of reactions is *[2020, 04 Sept Shift-II]*

$$HC\equiv CH \xrightarrow[\text{(ii) } (H_3C)((CH_3)_2CH)CH-Br]{\text{(i) } LiNH_2/\text{ether}} [P] \xrightarrow[\text{(ii) } NaBH_4]{\text{(i) } HgSO_4/H_2SO_4} [Q] \xrightarrow[\Delta]{\text{Conc. } H_2SO_4} [R]$$

(a) $(H_3C)((CH_3)_2CH)C{=}C{-}CH_2{-}CH_3$

(b) $(H_3C)((CH_3)_2CH)CH{-}CH{=}CH_2$

(c) $(H_3C)(H_3CCH_2)C{=}C(CH_3)_2$

(d) $(H_3C)((CH_3)_2CH)C{=}CH_2{-}CH_3$

**17** Consider the following reactions,

$$A \xrightarrow[\Delta]{Ag_2O} \text{ppt}$$

$$A \xrightarrow{Hg^{2+}/H^+} B \xrightarrow{NaBH_4} C \xrightarrow[\text{Conc. HCl}]{ZnCl_2} \text{Turbidity within 5 minutes}$$

$A$ is *[2019, 12 April Shift-II]*

(a) $CH\equiv CH$ (b) $CH_3C\equiv CCH_3$
(c) $CH_3-C\equiv CH$ (d) $CH_2=CH_2$

## Numerical Value Type Question

**18** The compound with molecular formula $C_6H_6$, which gives only one monobromo derivative and takes up four moles of hydrogen per mole for complete hydrogenation has ......... π-electrons. *[2025, 22 Jan shift-II]*

**19** The major product of the following reaction is $P$.

$$CH_3C\equiv C-CH_3 \xrightarrow[\text{(ii) dil. } KMnO_4,\ 273\text{ K}]{\text{(i) Na/liq. } NH_3} 'P'$$

Number of oxygen atoms present in product '$P$' is ............ (Nearest integer) *[2024, 6 April Shift-I]*

# Topic 4 Aromatic Hydrocarbons

## Level 1

**01** Which of the following is not an example of benzenoid compound? *[2022, 26 July shift-II]*

(a) 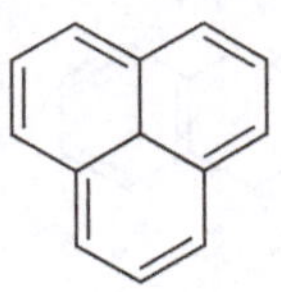

(b) 

(c) 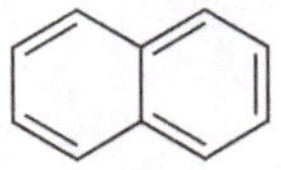

(d) 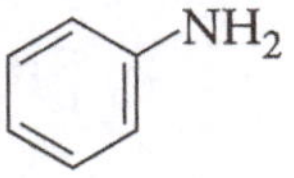


**02**  $+ CH_3CH_2CH_2Cl \xrightarrow{\text{Anhydrous, } AlCl_3} A$ (Major product)

The stable carbocation formed in the above reaction is *[2022, 29 June Shift-II]*

(a) $CH_3CH_2\overset{\oplus}{C}H_2$ (b) $CH_3\overset{\oplus}{C}H_2$

(c) $CH_3-\overset{\oplus}{C}H-CH_3$ (d) 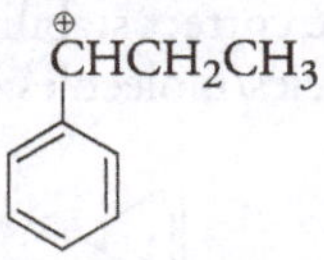


**03** Benzene on nitration gives nitrobenzene in presence of $HNO_3$ and $H_2SO_4$ mixutre, where *[2021, 20 July Shift-II]*

(a) both $H_2SO_4$ and $HNO_3$ act as a bases
(b) $HNO_3$ acts as an acid and $H_2SO_4$ acts as a base
(c) both $H_2SO_4$ and $HNO_3$ act as an acids
(d) $HNO_3$ acts as a base and $H_2SO_4$ acts as an acid

**04** Which of the following is an aromatic compound? *[2021, 17 March Shift-I]*

(a) 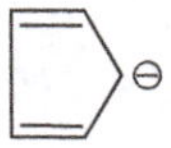

(b) 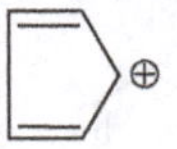

(c) 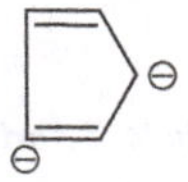

(d) 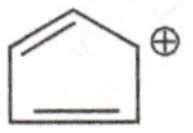

**05** The increasing order of the reactivity of the following compounds towards electrophilic aromatic substitution reaction is *[2019, 10 April Shift-I]*

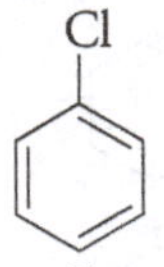

(I)

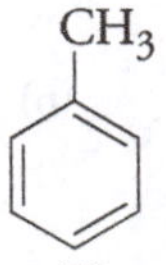

(II)

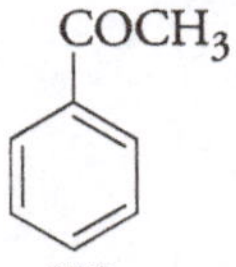

(III)

(a) III < I < II (b) II < I < III
(c) III < II < I (d) I < III < II

**06** Which compound (s) out of the following is/are not aromatic? *[2019, 11 Jan Shift-I]*

(A)

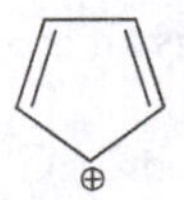
(B)

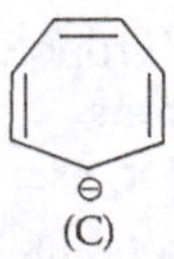
(C)

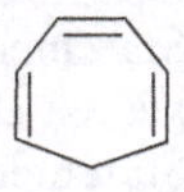
(D)

(a) (B), (C) and (D) (b) (C) and (D)
(c) (B) (d) (A) and (C)

**07** Which of the following compounds is not aromatic? *[2019, 09 Jan Shift-II]*

(a) 

(b) 

(c) 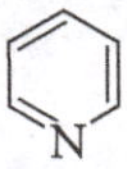


(d) 

**08** Presence of a nitro group in a benzene ring *[AIEEE 2007]*
(a) activates the ring towards electrophilic substitution
(b) renders the ring basic
(c) deactivates the ring towards nucleophilic substitution
(d) deactivates the ring towards electrophilic substitution

## Level 2

**09** The correct stability order of the following species/molecules is *[2025, 23 Jan Shift-I]*

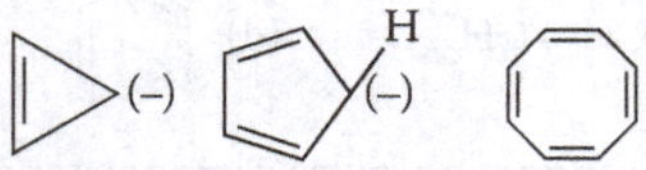

p q r

(a) $q > p > r$ (b) $p > q > r$
(c) $q > r > p$ (d) $r > q > p$

**10** Which of the following are aromatic?

A. 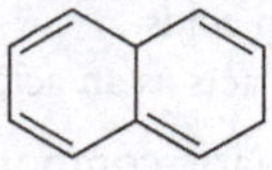 B. 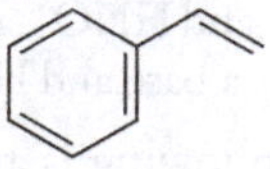

C. 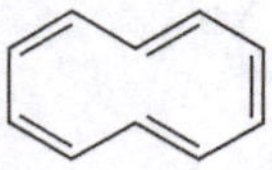 D. 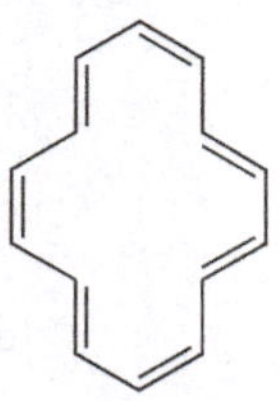

*[2024, 8 April Shift-I]*

(a) C and D only (b) B and D only
(c) A and C only (d) A and B only

**11** Which of the following molecule/species is most stable? *[2024, 30 Jan Shift-I]*

(a)  (b) 

(c) 

(d) 

**Imp.** Aromaticity of Tropolone

**12** Given below are two statements.
**Statement I** Tropolone is an aromatic compound and has 8 $\pi$ electrons.
**Statement II** $\pi$ electrons of >C=O group in tropolone are involved in aromaticity.
In the light of the above statements, choose the correct answer from the options given below. *[2023, 13 April Shift-II]*
(a) Both Statement I and Statement II are true.
(b) Statement I is true but Statement II is false.
(c) Statement I is false but Statement II is true.
(d) Both Statement I and Statement II are false.

**13** 'X' is : 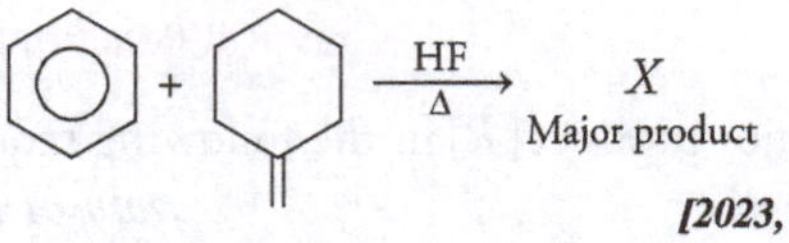


*[2023, 01 Feb Shift-II]*

(a) 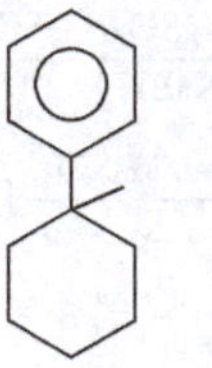 (b) 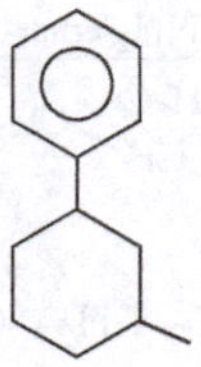

(c) 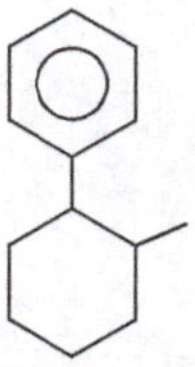 (d) 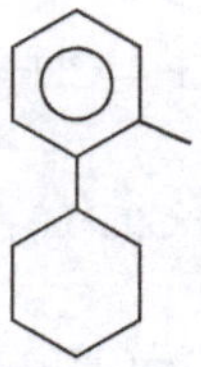

**14** Given below are two statements, one is labelled as Assertion (A) and, the other is labelled as Reason (R).
**Assertion (A)** [6] Annulene, [8] Annulene and *cis*-[10] Annulene, are respectively aromatic, non-aromatic and aromatic. *[2022, 27 July shift-I]*

[6] Annulene 

[8] Annulene 

*cis*-[10] Annulene 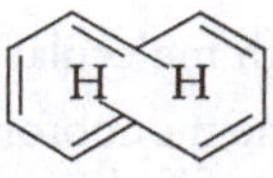


**Reason (R)** Planarity is one of the requirements of aromatic systems.
In the light of the above statements, choose the most appropriate answer from the options given below.
(a) Both A and R are correct and R is the correct explanation of A.
(b) Both A and R are correct but R is not the correct explanation of A.
(c) A is correct but R is not correct.
(d) A is not correct but R is correct.

**15** Which of the following compounds is non-aromatic? *[2022, 26 July shift-I]*

(a) 

(b) 

(c) 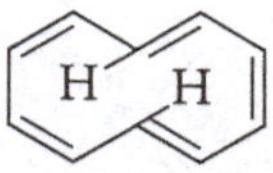

(d) 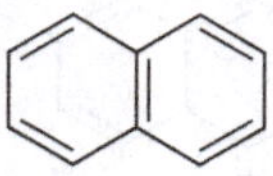

**16** The major product of the following reaction is *[2022, 26 July shift-I]*

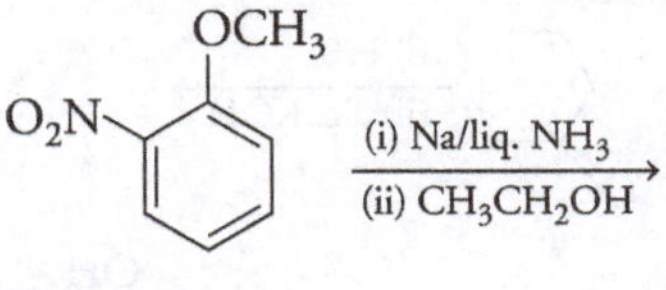

(a) 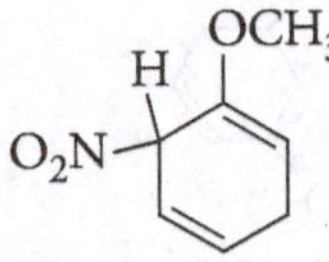 (b) 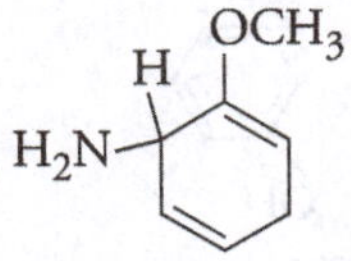

(c) 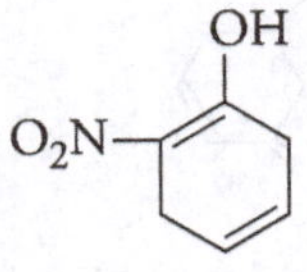 (d) 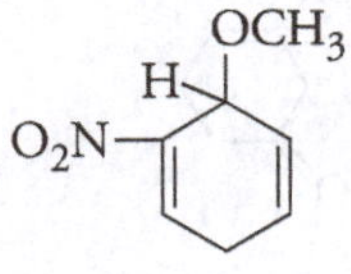

**17** Which one of the following compound(s) is/are aromatic in nature ? *[2021, 1 Sep Shift-II]*

(a) 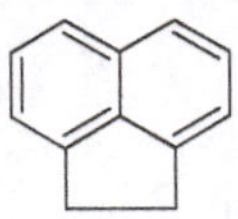 (b) 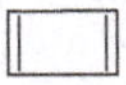

(c) 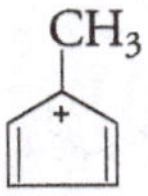 (d) 

**18** For the following

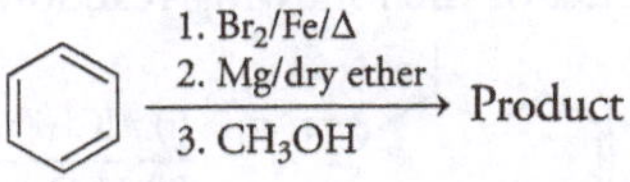

*[2021, 31 Aug Shift-II]*

(a) 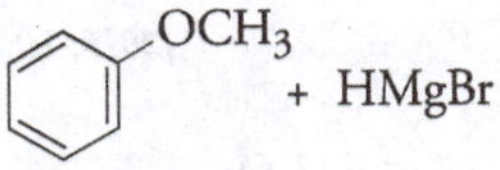

(b) 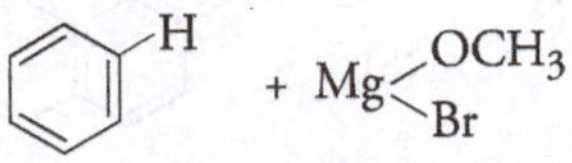

(c) 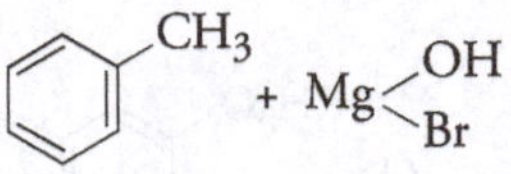

(d) 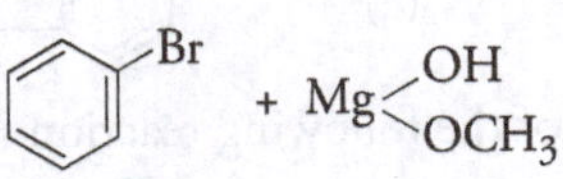

**19** Which one of the following compounds is not aromatic ? *[2021, 26 Aug Shift-II]*

(a) 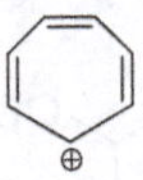 (b) 

(c)  (d) 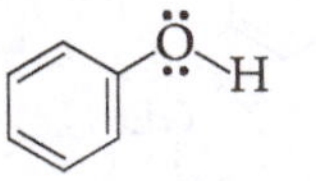

**Imp.** Decreasing Order of Density

**20** The correct decreasing order of densities of the following compounds is *[2021, 25 July Shift-II]*

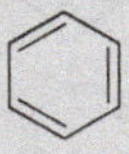 (A)

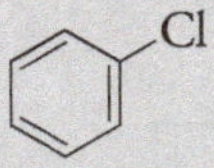 (B)

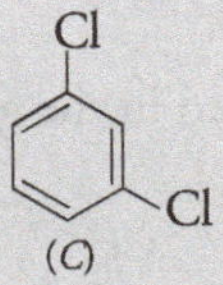 (C)

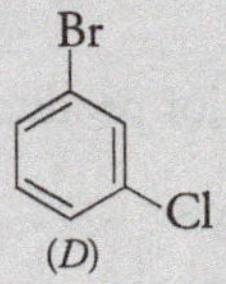 (D)

(a) (D) > (C) > (B) > (A)
(b) (C) > (D) > (A) > (B)
(c) (C) > (B) > (A) > (D)
(d) (A) > (B) > (C) > (D)

**21** Which among the following is the strongest acid? *[2021, 25 July Shift-II]*

(a) 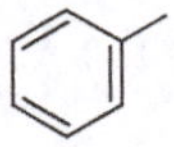 (b) 

(c)  (d) $CH_3CH_2CH_2CH_3$

**22** Among the following, the aromatic compounds are *[2021, 16 March Shift-I]*

(A) 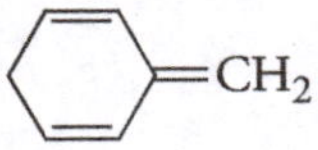 (B) 

(C)  (D) 

Choose the correct answer from the following options.
(a) Only (A) and (B) (b) Only (B) and (C)
(c) (B), (C) and (D) (d) (A), (B) and (C)

**23** The major product of the following reaction is *[2021, 25 Feb Shift-II]*

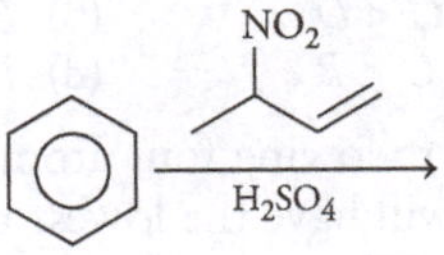

(a) 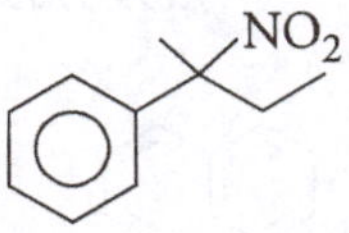 (b) 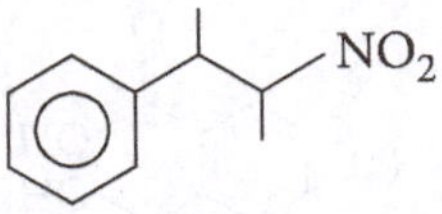

(c) 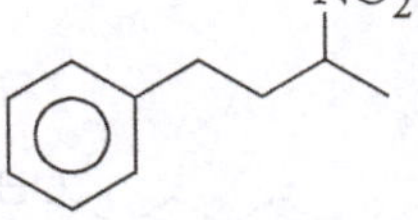 (d) 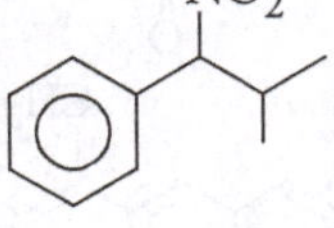

**24** Which one of the following compounds is non-aromatic ? *[2021, 24 Feb Shift-II]*

(a) 

(b) 

(c) 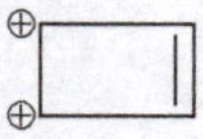


(d) 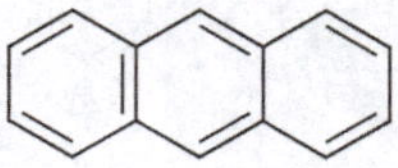

**25** The major product obtained in the given reaction is

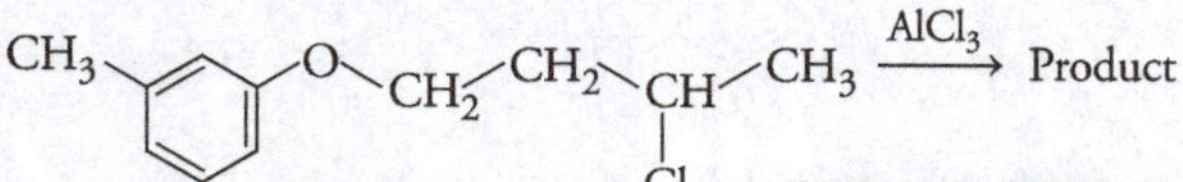


*[2019, 10 April Shift-II]*

(a) 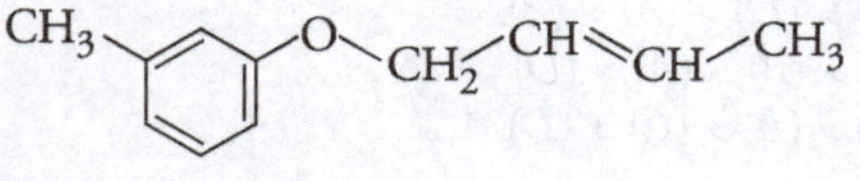


(b) 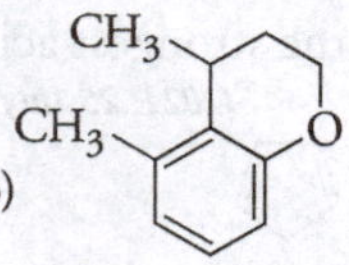


(c) 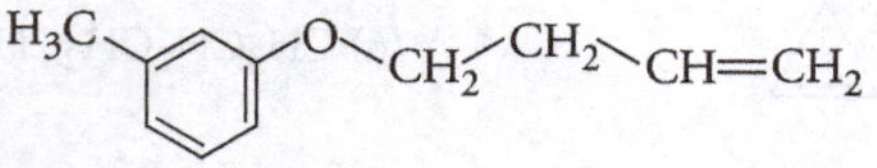


(d) 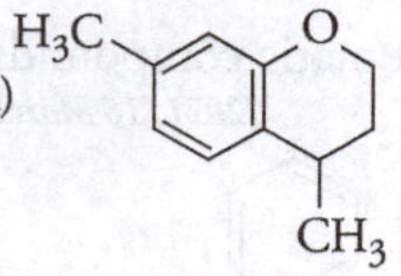


**26** The increasing order of reactivity of the following compounds towards aromatic electrophilic substitution reaction is *[2019, 9 April Shift-I]*

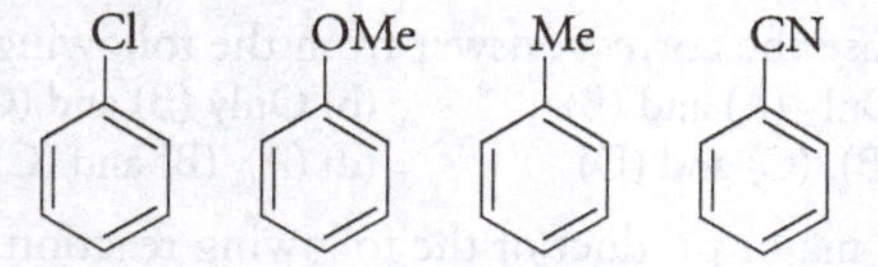


*A* *B* *C* *D*

(a) $A < B < C < D$
(b) $B < C < A < D$
(c) $D < A < C < B$
(d) $D < B < A < C$

**27** Among the following four aromatic compounds, which one will have the lowest melting point? *[2019, 12 Jan Shift-II]*

(a) 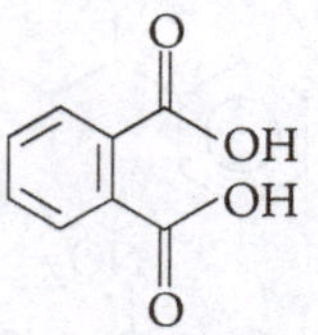


(b) 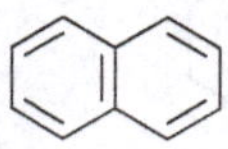

(c) 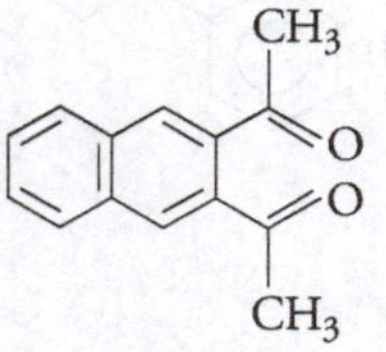


(d) 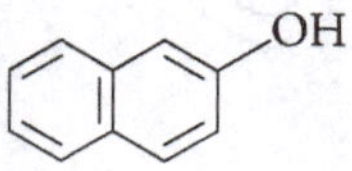


**28** The major product of the following reaction is

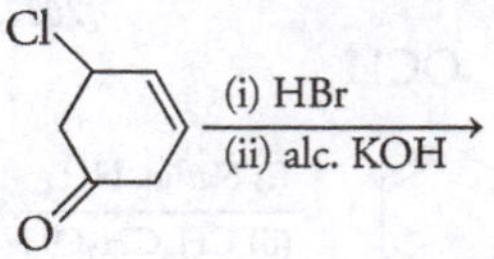


*[2019, 11 Jan Shift-I]*

(a) 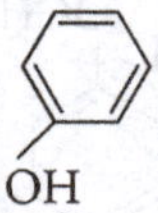


(b) 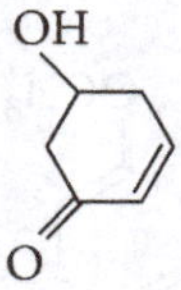


(c) 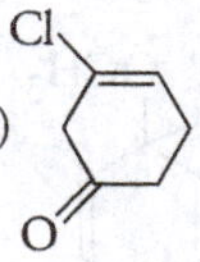


(d) 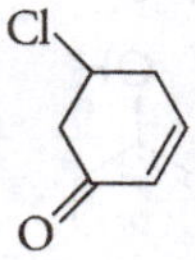


**29** The major product of the following reaction is

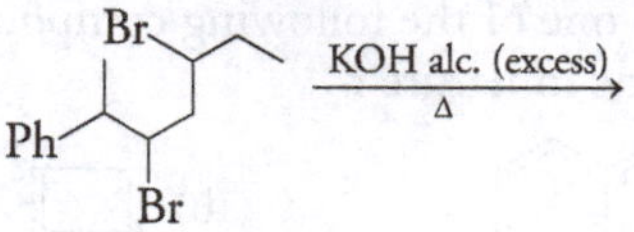


(a) 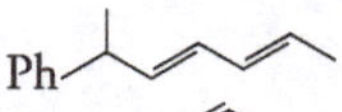


(b) 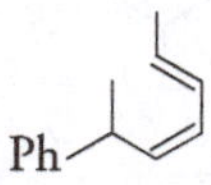


(c) 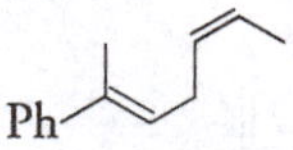


(d) 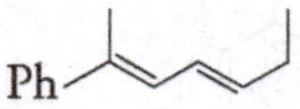


*[2019, 10 Jan Shift-I]*

**30** The major product of the following reaction is

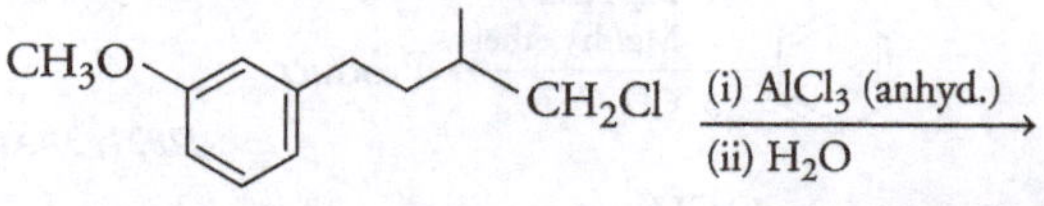


*[2019, 10 Jan Shift-I]*

(a) 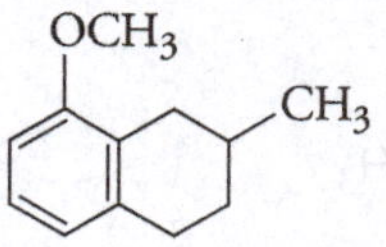


(b) 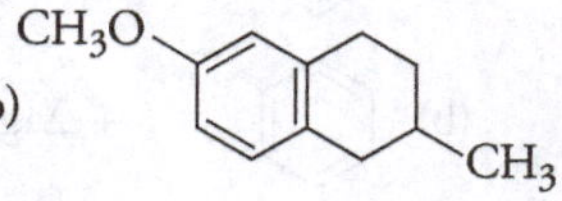


(c) 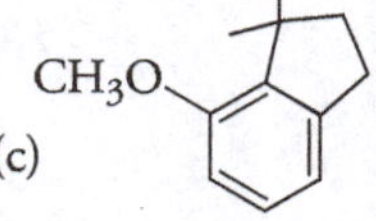


(d) 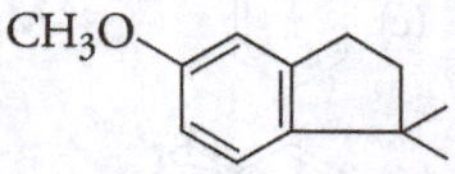


**31** The major product of the following reaction is

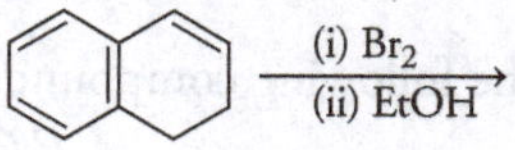


*[2019, 09 Jan Shift-I]*

(a) 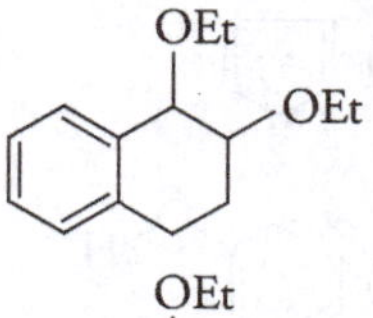


(b) 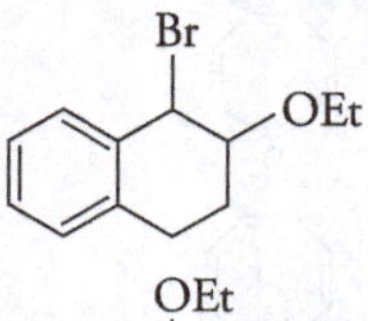


(c) 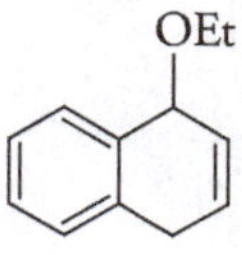


(d) 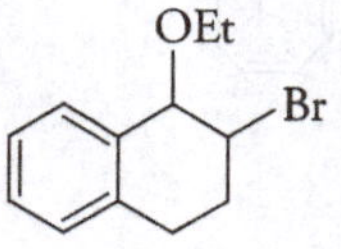

**32** The electrophile, $E^+$ attacks the benzene ring to generate the intermediate σ-complex. Which of the following σ-complex is of lowest energy? *[AIEEE 2008]*

(a) 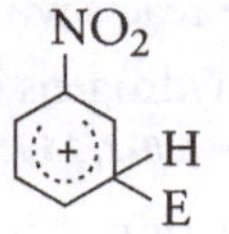
(b) 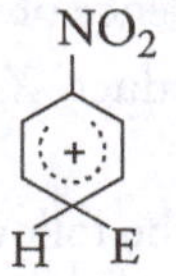
(c) 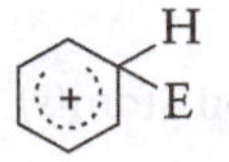
(d) 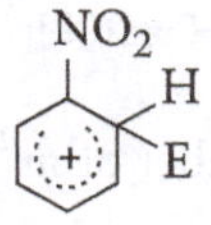

## Level 3

**33** Designate whether each of the following compounds is aromatic or not aromatic.

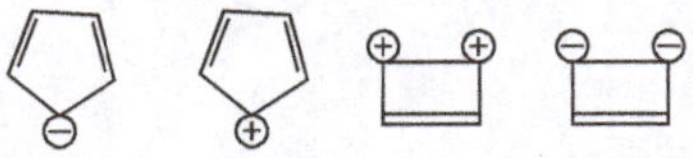

(A) (B) (C) (D)

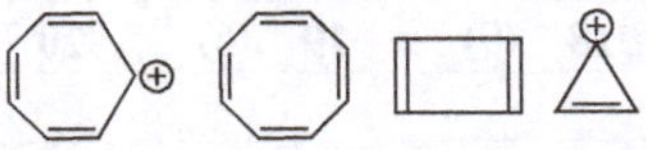

(E) (F) (G) (H)

(a) A, C, D, E, H aromatic and B, F, G not aromatic
(b) E, G aromatic and A, B, C, D, F, H not aromatic
(c) B, E, F, G aromatic and A, C, D, H not aromatic
(d) A, B, C, D aromatic and E, F, G, H not aromatic

**34** Given below are two statements. *[2025, 29 Jan Shift-II]*

**Statement I** On nitration of *m*-xylene with $HNO_3, H_2SO_4$ followed by oxidation, 4-nitrobenzene-1,3-dicarboxylic acid is obtained as the major product.

**Statement II** —$CH_3$ group is *o/p*-directing while —$NO_2$ group is *m*-directing group.

In the light of the above statements, choose the correct answer from the options given below:
(a) Both Statement I and Statement II are false.
(b) Statement I is false but Statement II is true.
(c) Statement I is true but Statement II is false.
(d) Both Statement I and Statement II are true.

**35** Identify correct statement/s *[2025, 24 Jan Shift-II]*
(A) — $OCH_3$ and — $NHCOCH_3$ are activating groups.
(B) — CN and — OH are *meta* directing groups.
(C) — CN and — $SO_3H$ are *meta* directing groups.
(D) Activating groups act as *ortho* and *para* directing groups.
(E) Halides are activating groups.

Choose the correct answer from the options given below
(a) (A), (B) and (E) only (b) (A), (C) and (D) only
(c) (A) and (C) only (d) (A) only

**36** Match List-I and List-II.

| List-I (Chemicals) | List-II (Use/Preparation/ Constituent) |
|---|---|
| (A) Alcoholic potassium hydroxide | (i) Electrodes in batteries |
| (B) Pd/$BaSO_4$ | (ii) Obtained by addition reaction |
| (C) BHC (Benzene hexachloride) | (iii) Used for β-elimination reaction |
| (D) Polyacetylene | (iv) Lindlar's catalyst |

Choose the most appropriate match *[2021, 18 March Shift-I]*
(a) (A)-(ii), (B)-(iv), (C)-(i), (D)-(iii)
(b) (A)-(iii), (B)-(iv), (C)-(ii), (D)-(i)
(c) (A)-(iii), (B)-(i), (C)-(iv), (D)-(ii)
(d) (A)-(ii), (B)-(i), (C)-(iv), (D)-(iii)

**37** The correct sequence of reagents used in the preparation of 4-bromo-2-nitroethyl benzene from benzene is *[2021, 25 Feb Shift-II]*
(a) $CH_3COCl/AlCl_3, Br_2/AlBr_3, HNO_3/H_2SO_4$, Zn/HCl
(b) $CH_3COCl / AlCl_3$, Zn-Hg/HCl, $Br_2/AlBr_3$, $HNO_3/H_2SO_4$
(c) $Br_2/AlBr_3, CH_3COCl/AlCl_3, HNO_3/H_2SO_4$, Zn/HCl
(d) $HNO_3/H_2SO_4, Br_2/AlCl_3, CH_3COCl/AlCl_3$, Zn-Hg/HCl

## Numerical Value Type Questions

**38** Total number of aromatic compounds among the following compounds is ........ *[2024, 8 April Shift-II]*

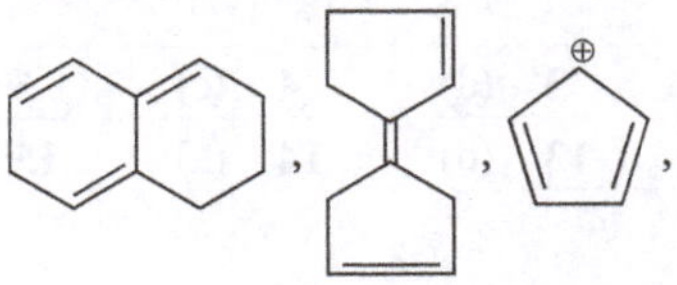

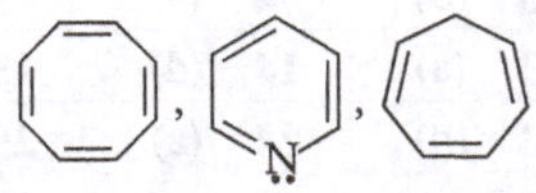

**39** Total number of deactivating groups in aromatic electrophilic substitution reaction among the following is .......... *[2024, 1st Feb Shift-I]*

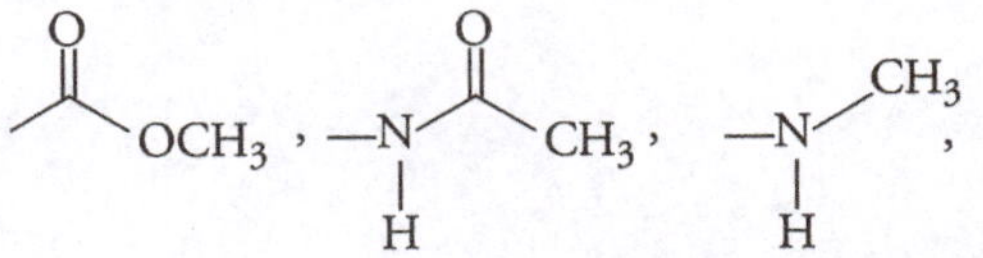

—C≡N, —$OCH_3$

**40** Among the given organic compounds, the total number of aromatic compounds is ........ **[2024, 27 Jan Shift-I]**

A. 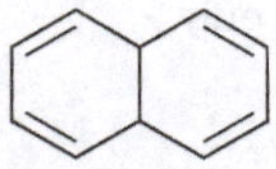

B. 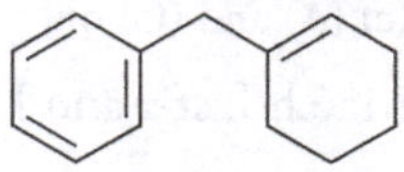

C. 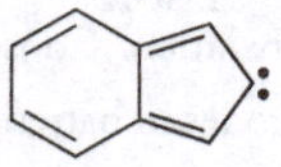

D. 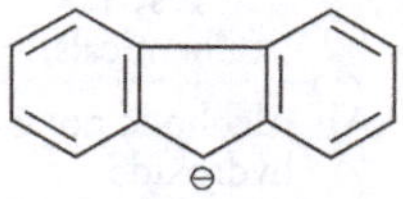

**41** A trisubstituted compound '*A*', $C_{10}H_{12}O_2$ gives neutral $FeCl_3$ test positive. Treatment of compound '*A*' with NaOH and $CH_3Br$ gives $C_{11}H_{14}O_2$, with hydroiodic acid gives methyl iodide and with hot conc. NaOH gives a compound *B*, $C_{10}H_{12}O_2$. Compound '*A*' also decolourises alkaline $KMnO_4$. The number of $\pi$ bonds present in the compound '*A*' is ........ **[2023, 30 Jan Shift-I]**

**42** In the presence of sunlight, benzene reacts with $Cl_2$ to give product, *X*. the number of hydrogens in *X* is ......... . **[2022, 26 July Shift-I]**

**43** Consider the following chemical reaction.

$$CH\equiv CH \xrightarrow[\text{(2) CO, HCl, AlCl}_3]{\text{(1) Red hot Fe tube, 873 K}} \text{Product}$$

The number of $sp^2$ hybridised carbon atom(s) present in the product is ... . **[2021, 25 Feb Shift-I]**

## • Key Index

### Topic 1

| | | | | | | | | | |
|---|---|---|---|---|---|---|---|---|---|
| 1 (b) | 2 (d) | 3 (c) | 4 (d) | 5 (c) | 6 (b) | 7 (b) | 8 (a) | 9 (b) | 10 (b) |
| 11 (c) | 12 (c) | 13 (d) | 14 (c) | 15 (d) | 16 (d) | 17 (c) | 18 (9) | 19 (6) | 20 (3) |
| 21 (8) | 22 (6) | 23 (60) | | | | | | | |

### Topic 2

| | | | | | | | | | |
|---|---|---|---|---|---|---|---|---|---|
| 1 (c) | 2 (a) | 3 (c) | 4 (c) | 5 (a) | 6 (c) | 7 (a) | 8 (d) | 9 (b) | 10 (d) |
| 11 (d) | 12 (b) | 13 (c) | 14 (a) | 15 (b) | 16 (d) | 17 (c) | 18 (c) | 19 (b) | 20 (c) |
| 21 (d) | 22 (d) | 23 (a) | 24 (d) | 25 (d) | 26 (c) | 27 (c) | 28 (a) | 29 (a) | 30 (c) |
| 31 (c) | 32 (c) | 33 (d) | 34 (b) | 35 (a) | 36 (d) | 37 (b) | 38 (b) | 39 (c) | 40 (a) |
| 41 (b) | 42 (d) | 43 (a) | 44 (c) | 45 (c) | 46 (d) | 47 (d) | 48 (b) | 49 (d) | 50 (d) |
| 51 (d) | 52 (a) | 53 (a) | 54 (c) | 55 (a) | 56 (3) | 57 (27) | 58 (4) | 59 (4) | 60 (1) |
| 61 (2) | 62 (2) | 63 (6) | | | | | | | |

### Topic 3

| | | | | | | | | | |
|---|---|---|---|---|---|---|---|---|---|
| 1 (a) | 2 (a) | 3 (d) | 4 (c) | 5 (c) | 6 (b) | 7 (d) | 8 (d) | 9 (b) | 10 (c) |
| 11 (a) | 12 (a) | 13 (d) | 14 (a) | 15 (c) | 16 (c) | 17 (c) | 18 (8) | 19 (2) | |

### Topic 4

| | | | | | | | | | |
|---|---|---|---|---|---|---|---|---|---|
| 1 (b) | 2 (c) | 3 (d) | 4 (a) | 5 (a) | 6 (a) | 7 (a) | 8 (d) | 9 (c) | 10 (b) |
| 11 (c) | 12 (d) | 13 (a) | 14 (d) | 15 (c) | 16 (a) | 17 (a,d) | 18 (b) | 19 (c) | 20 (a) |
| 21 (b) | 22 (b) | 23 (b) | 24 (a) | 25 (d) | 26 (c) | 27 (b) | 28 (a) | 29 (d) | 30 (d) |
| 31 (d) | 32 (c) | 33 (a) | 34 (d) | 35 (b) | 36 (b) | 37 (b) | 38 (1) | 39 (2) | 40 (3) |
| 41 (4) | 42 (6) | 43 (7) | | | | | | | |

# Detailed **Explanations**

## Topic 1

**01.** *(b)* The structure of given alkanes are as follows.

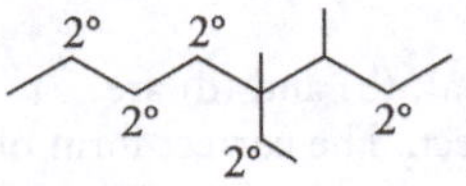


⇒ 10 secondary hydrogens are present.

4-ethyl-3, 4-dimethyloctane

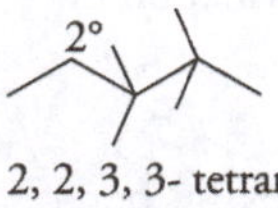


⇒ 2 secondary hydrogens are present.

2, 2, 3, 3- tetramethylpentane

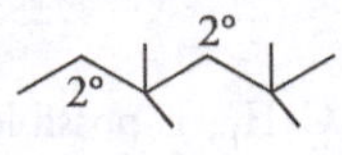


⇒ 4 secondary hydrogens are present.

2, 2, 4, 4- tetramethylhexane

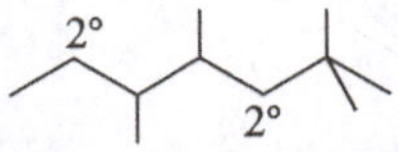


⇒ 4 secondary hydrogens are present.

2, 2, 4, 5- tetramethylheptane

So, 2, 2, 3, 3-tetramethylpentane has 2 secondary hydrogen.

**02.** *(d)* Following are the conformations of *n*-butane with their stability

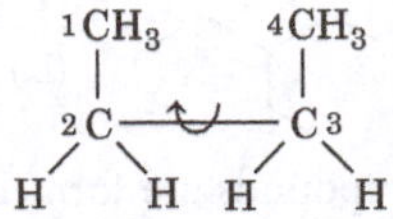


*n*-butane

(a)

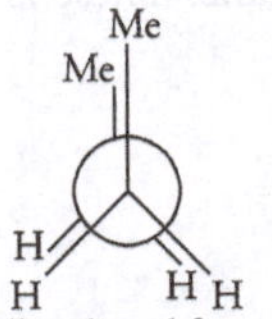


Fully eclipsed form

Fully eclipsed form has maximum steric repulsion between alkyl group ($CH_3$). So, it is least stable.

(b)

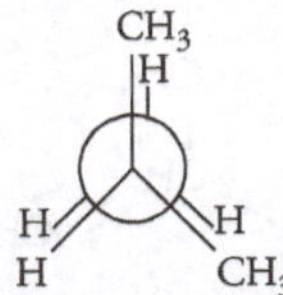


Partially eclipsed form

Partially eclipsed form has lesser steric repulsion than the fully eclipsed form. So, it is more stable than the fully eclipsed form.

(c)

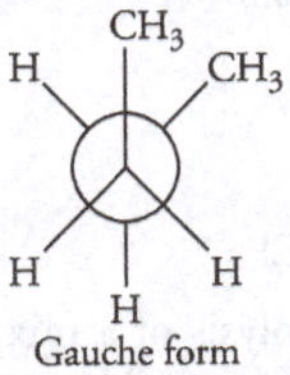


Gauche form

Gauche form is more stable than the partially eclipsed form.

(d)

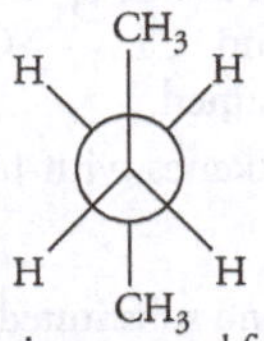


Anti or staggered form

Antistaggered or staggered form is most stable due to lowest steric and torsional strains.

**03.** *(c)* Among the given conformation(a) is gauche and (c) is staggered configurations (d) is eclipsed configurations. Whereas (b) is partially eclipsed form. In the staggered or anti-conformation (c), the dihedral angle is the largest (180°) between the two methyl groups. Thus, structure (c) has staggered conformation with maximum dihedral angle.

**04.** *(d)*

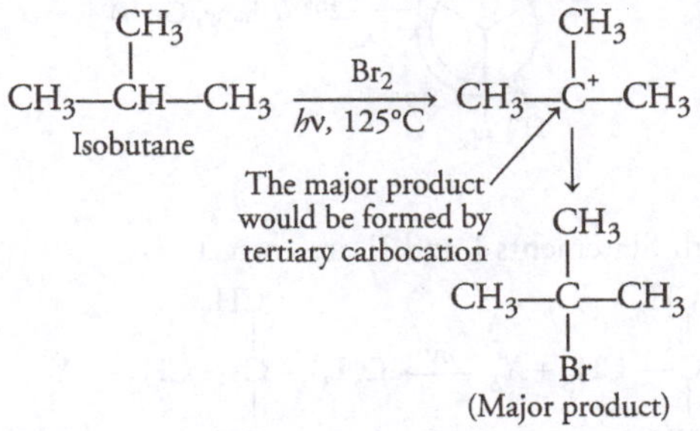


**05.** *(c)* Staggered and eclipsed conformers of ethane are also called as rotamers.

> **• Concept Enhancer**
>
> Staggered conformation is that conformation in which the substituents attached to the carbon atoms are far apart. While Eclipsed conformation is that in which the substituents are closely attached as possible.

**06.** *(b)*

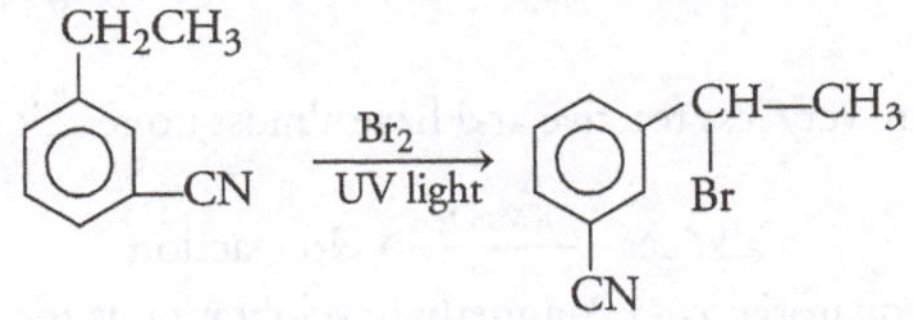


Secondary hydrogen → $CH_2$ of ethyl group is replaced by bromine atom.

**07.** *(b)* The reaction is called aromatisation.

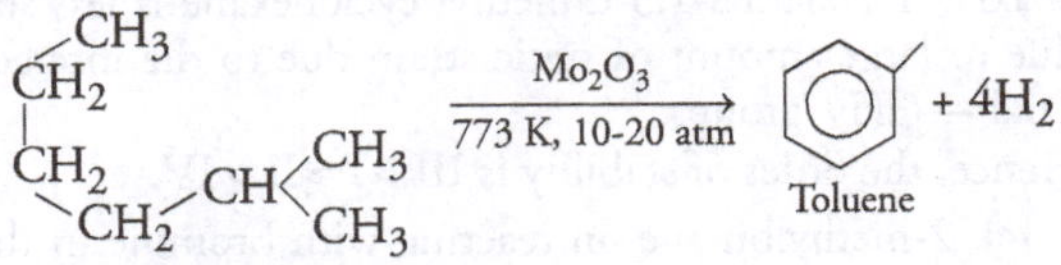


2-methylhexane

Oxides of molybdenum, vanadium and chromium, i.e. $Mo_2O_3$, $V_2O_5$ and $Cr_2O_3$ work as catalyst. When they are operated under certain temperature and pressure with 2-methyl hexane then they form toluene which is aromatic compound. $Mo_2O_3$ behave as aromatising agent.

**8.** *(a)*

> **key idea** Stability of any compound is inversely proportional to its energy. To achieve stability energy should be minimised.

Higher the stability, lower will be its potential energy.

Order of stability : I > III > IV > II

Order of potential energy I < III < IV < II.

**09.** *(b)* $CH_3CH_2CH_3 \xrightarrow{Mo_2O_3} CH_3CH_2CHO$

Hence, correct option is (b).

**10.** *(b)* A dihedral angle is the angle between two C—H bonds projected on a plane orthogonal to the C—C bond. In the given skew conformation, having Newman's projection the dihedral angle is

$H'—C—C—H'' = (H'—C—C—H_a) + (H_aC—C—H'')$
$= 29° + 120° = 149°$

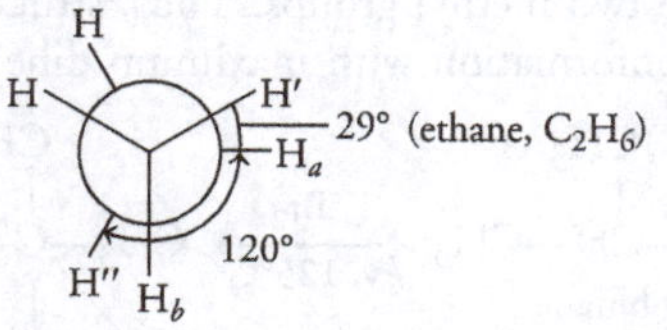


**11.** *(c)* Both Statements I and II are correct.

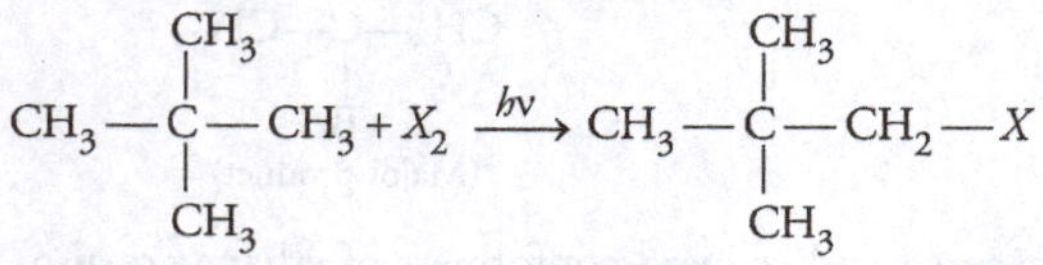

($X$ = Cl, Br)

Melting point  > 

*neo*-pentane *n*-pentane

**12.** *(c)* Statement I is correct, whereas Statement II is incorrect. Corrected statement is as follows:

Tertiary alkanes can be oxidised to alcohol by treating with $KMnO_4$.

*n*-alkane are very less reactive and have almost no reaction with $KMnO_4$.

$n\text{-alkanes} \xrightarrow{KMnO_4} \text{No reaction}$

**13.** *(d)* Diequatorial *cis*-1,3-dimethyl cyclohexane is more stable as both methyl groups are equatorial and are away from each other. It is followed by *trans*-1,3-dimethyl cyclohexane where one $—CH_3$ group is equatorial and other is axial. Diaxial *cis*-1,3-dimethyl cyclohexane is less stable due to large amount of steric strain due to the interaction of axial $—CH_3$ groups.

Hence, the order of stability is III > I ≈ II > IV.

**14.** *(c)* 2-methylbutane on reacting with bromine in the presence of sunlight gives 2-bromo-2-methyl butane.

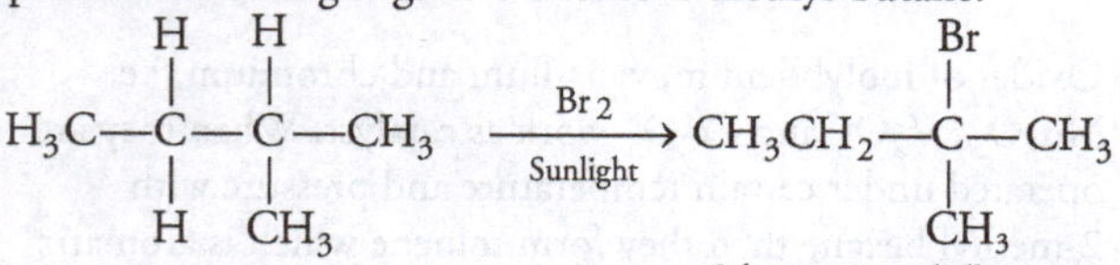

2-bromo-2-methylbutane

**15.** *(d)* Carbonyl compounds are reduced to corresponding alkanes with (Zn + conc. HCl). It is called Clemmensen reduction.

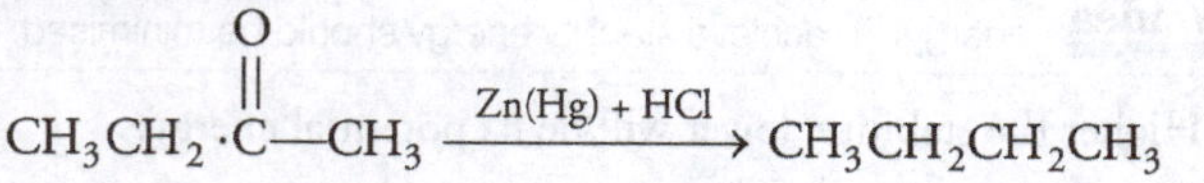

**16.** *(d)* In *neo*-pentane, all H are equivalents thus it forms only one monochloroalkane.

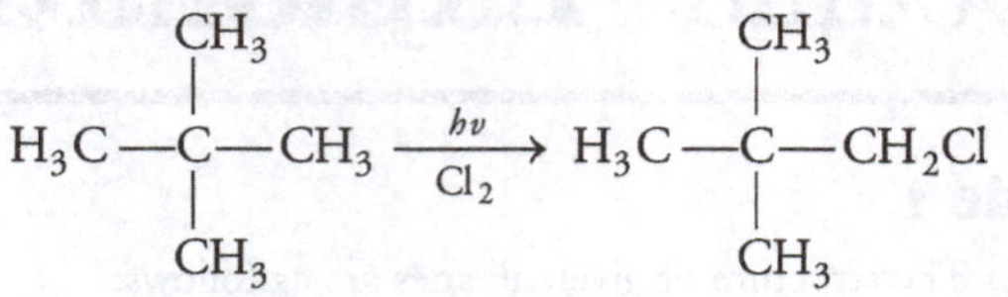

*neo*-pentane

**17.** *(c)* The statement given in option (a), (b) and (d) are correct, while statement (c) is incorrect. The correct form of statement (c) is :

Eclipsed conformation is the least stable conformation of ethane.

**• Concept Enhancer**

It is interesting to note that in all the conformations, the bond angles and the bond lengths remain the same.

**18.** *(9)* Total nine different chain isomers for $C_7H_{16}$ is possible. These are as follows.

(i)  (ii) 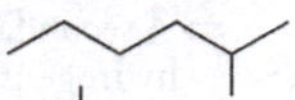

(iii) 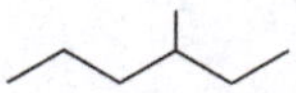 (iv) 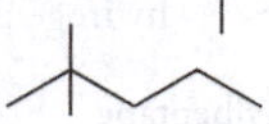

(v) 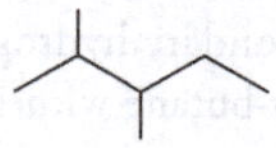 (vi) 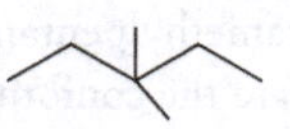

(vii)  (viii) 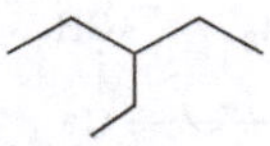 (ix) 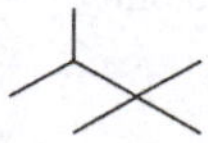

**19.** *(6)* Total 6 isomeric compounds are formed by monochlorination of 2-methylbutane. These are as follows

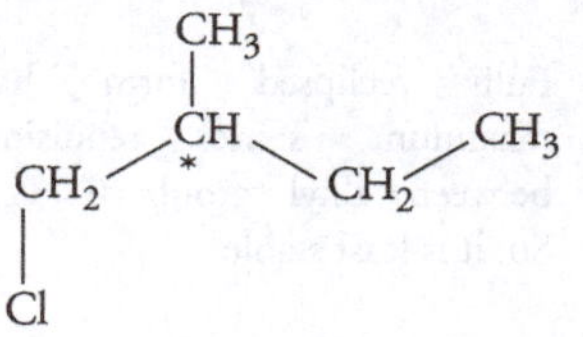

(Two isomers i.e., *R/S*)

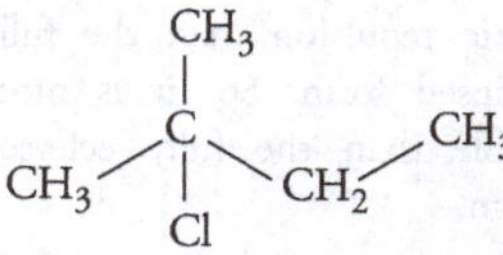 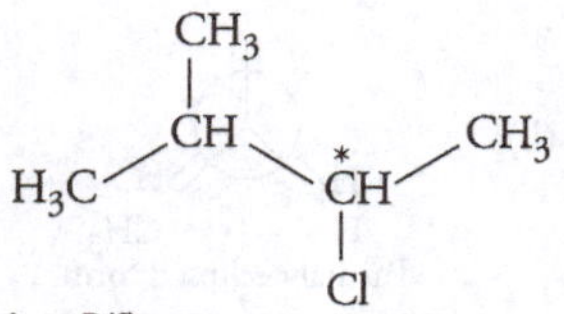

(Two isomers i.e., *R/S*)

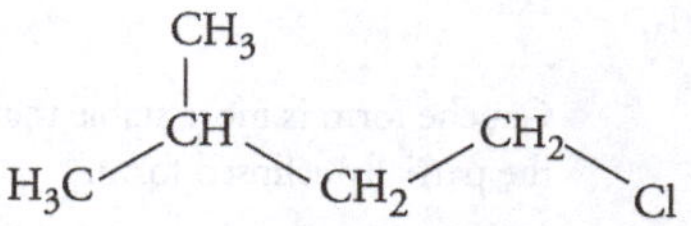

**20.** *(3)* Alkanes obtained on electrolysis of a mixture of $CH_3COONa$ and $C_2H_5COONa$ are $CH_3—CH_3$, $CH_3—CH_2—CH_2—CH_3$ and $CH_3—CH_2—CH_3$. Thus, total three alkanes are obtained.

**21.** *(8)* The possible structures of alkanes with formula $C_5H_{12}$ are

 = 3 mono substituted products

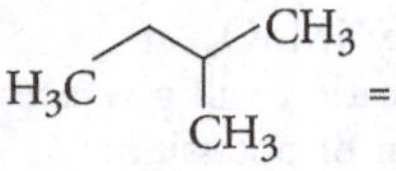

= 4 mono substituted product

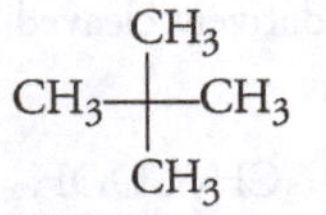

= 1 mono substituted product

Total number of monobromo derivatives formed by alkane with formula $C_5H_{12}$ are $= 3 + 4 + 1 = 8$

**22.** *(6)*

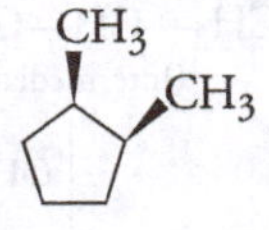

It has 2 stereoisomers.

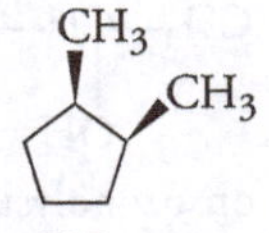

It has only one form due to plane of symmetry.

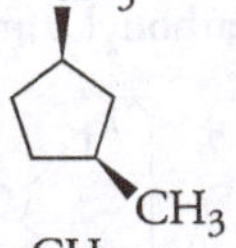

It has only one stereo form due to plane of symmetry.

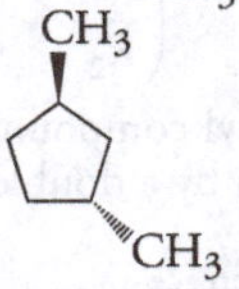

It has 2 stereoisomers.

Total stereoisomers of dimethyl cyclopentane $= 2 + 1 + 1 + 2 = 6$

**• Concept Enhancer**

A structure of a compound has plane of symmetry if it can be divided into two halves by a plane and have each half be a reflection of the other across the plane.

**23.** *(60)* Newman staggered form of 1,1,1-trichloro ethane ($CCl_3—CH_3$) is as follows.

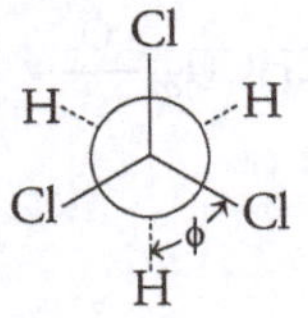

It's dihedral angles ($\phi$) is 60°.

## Topic 2

**01.** *(c)* The complete reactions is as follow

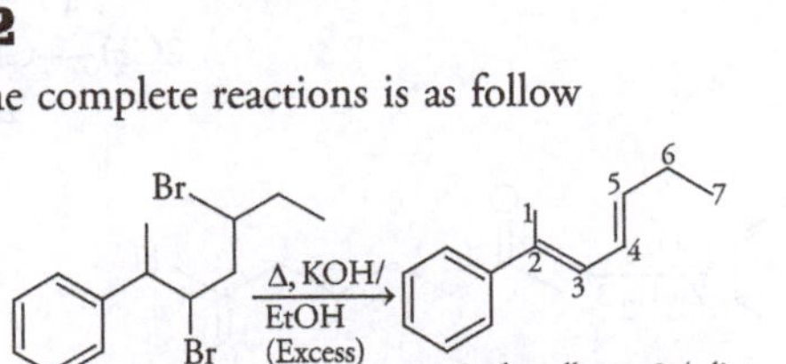

2-phenylhepta-2,4-diene
(Major product)

**02.** *(a)* The bond structure of ethylene is

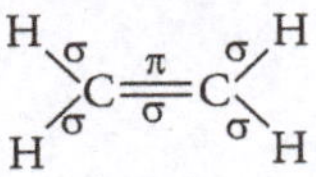

It has $5\sigma$ bonds and $1\pi$ bond.

**03.** *(c)* This is β-elimination reaction of quaternary ammonium salts. It is also known as Hofmann elimination as the major alkene formed will be a Hofmann product i.e., less substituted alkene.

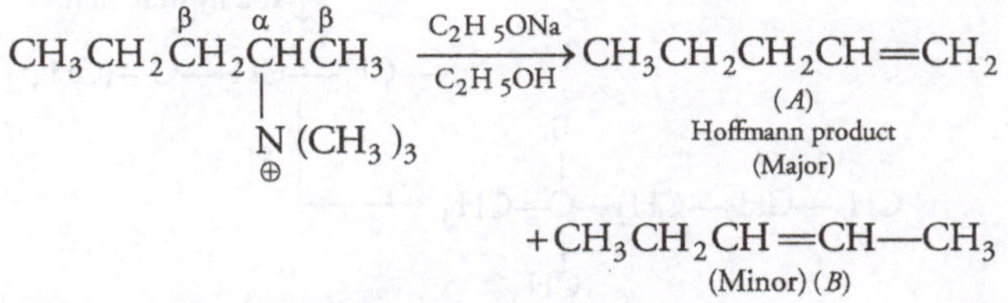

**04.** *(c)* The major product is formed according to Markownikoff's rule.

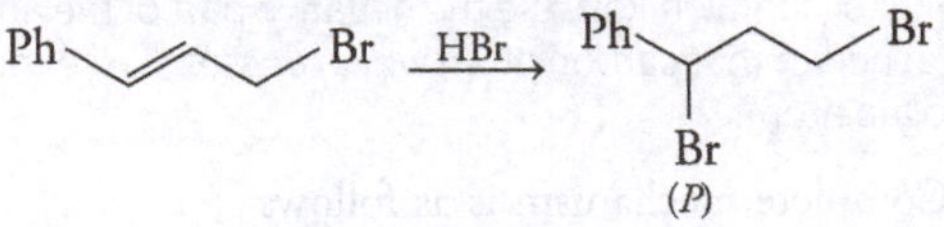

**05.** *(a)* Due to presence of π-bonds, alkenes are less stable than alkanes.
Stability order of hydrocarbons is $C\equiv C < C=C < C—C$
The strength of hydrocarbons is $C\equiv C > C=C > C—C$
This is because bond length is minimum in alkyne and maximum in alkane.
So, both Statement I and Statement II are correct.

**06.** *(c)* Both Assertion and Reason are true and Reason is correct explanation of Assertion.

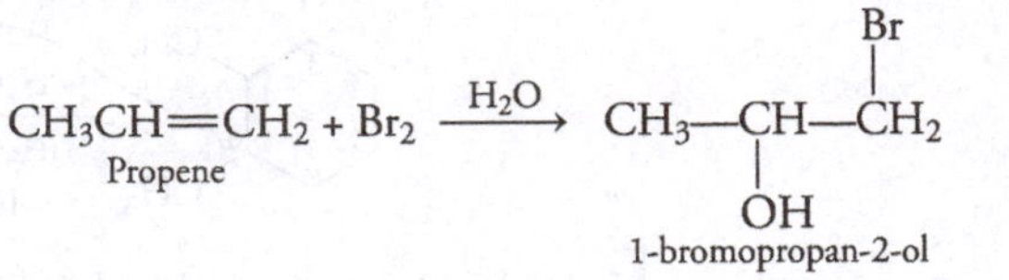

**Mechanism**

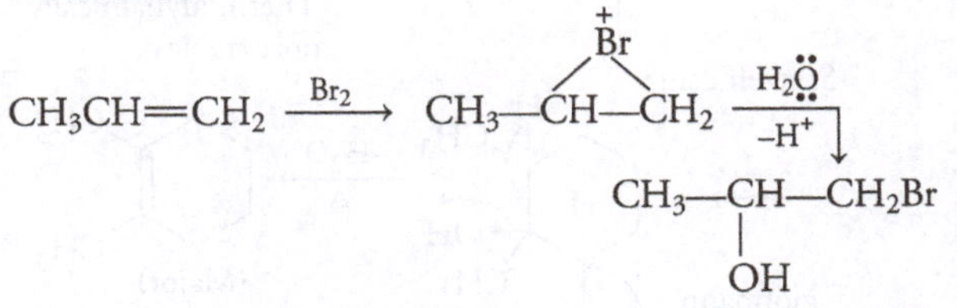

**07.** *(a)*

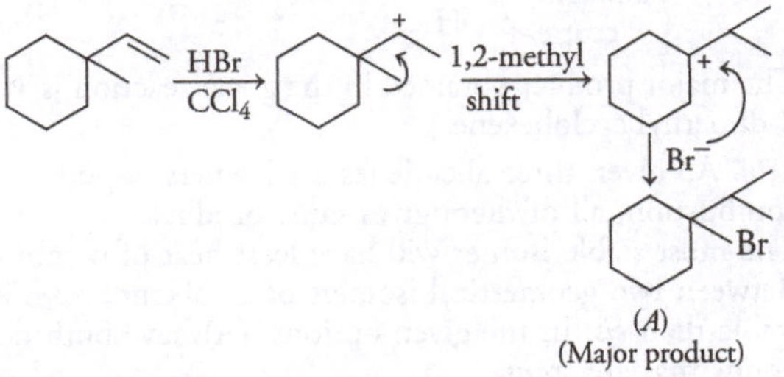

It shows Markownikoff's addition.

**08.** *(d)*

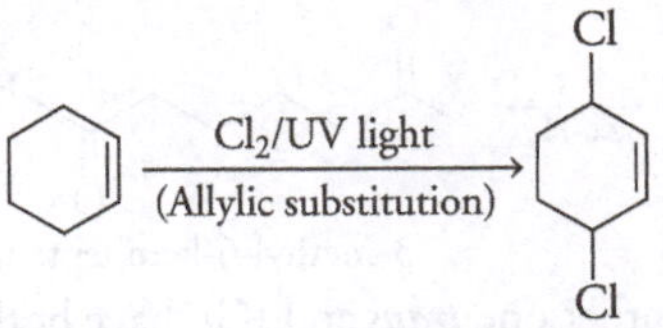

(Cl group attach to the allylic position.)

**09.** *(b)* Lindlar's catalyst consist of palladium catalyst poisoned with traces of lead and quinoline.
*i.e.*, $Pd/CaCO_3 + (CH_3COO)_2Pb$ + Quinoline.

**10.** *(d)* The major product formed in the given reaction is an alkane.

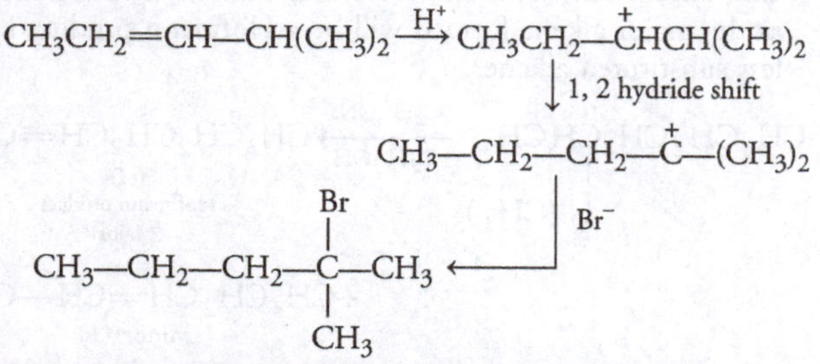


> • Concept **Enhancer**
> According to Markownikoff rule the negative part of the addition gets attached to that carbon atom which possess lesser number of hydrogen atom.

**11.** *(d)* Complete mechanism is as follows

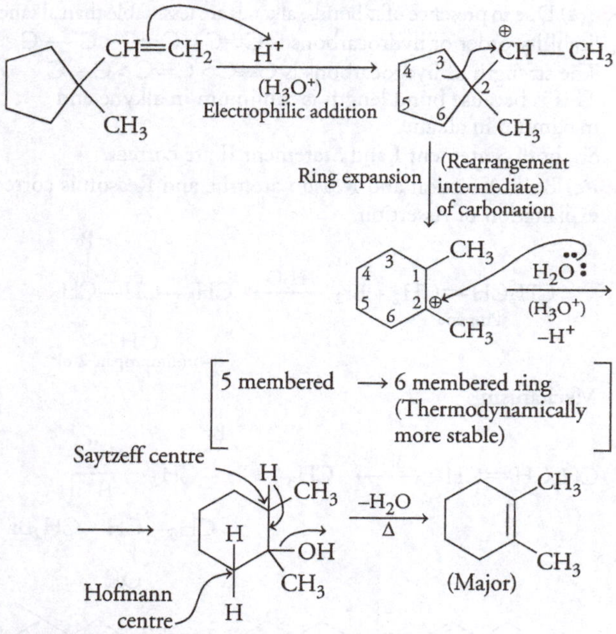


The major product obtained in the given reaction is 1, 2-dimethyl cyclohexene.

**12.** *(b)* All given three alkadienes are isomers. Upon combustion, all of them gives same product.
The most stable isomer will have least heat of combustion.
Between two geometrical isomers of an alkene, *trans* is more stable than *cis*. In the given options, (A) have both double bonds that are *trans,*

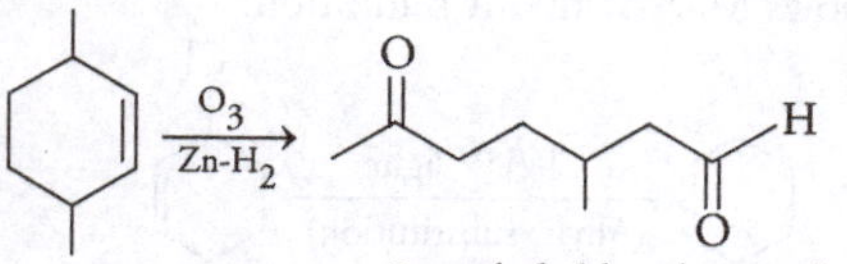

3-methyl-6-ketoheptanal

(B) have one *cis* and other one *trans* and (C) have both *cis*. So, the correct stability order of heat of combustion is *A* < *B* < *C*.

**13.** *(c)* But-2-ene on reaction with alkaline $KMnO_4$ at elevated temperature followed by acidification will give acetic acid ($CH_3COOH$). Hot alkaline solution of potassium permanganate followed by acidification oxidatively cleaved alkenes. The reaction proceed as follows :

$$\underset{\text{But-2-ene}}{CH_3-CH{=}CH-CH_3} \xrightarrow[H_3O^+]{\text{Alk. } KMnO_4\text{, heat}} \underset{\text{Acetic acid}}{2CH_3COOH}$$

**14.** *(a)*

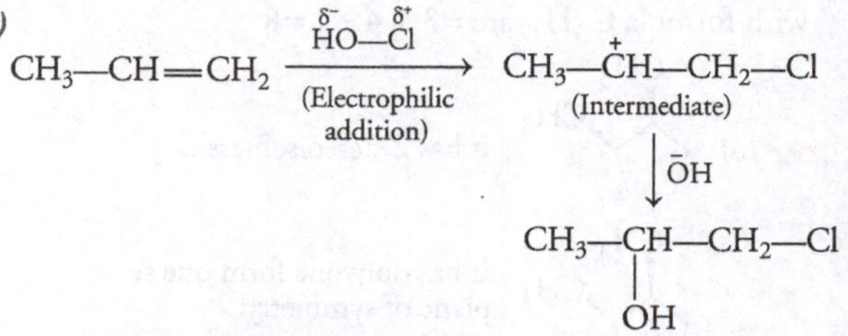


**15.** *(b)* Alkenes give carbonyl compounds on ozonolysis.

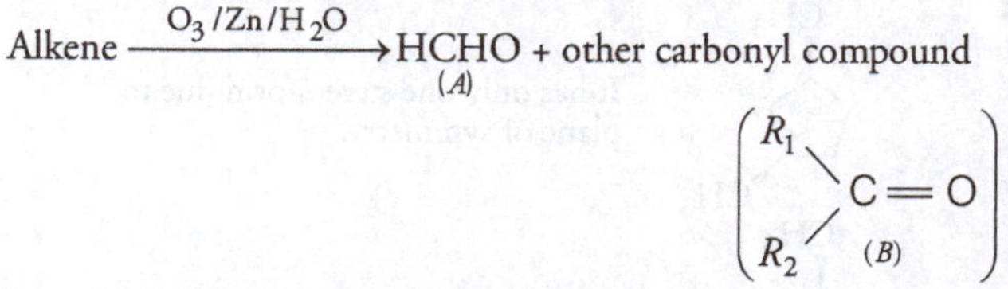

In order to determine alkene, place carbonyl compounds with their O-atom face to face. Replace O-atom by a double bond.

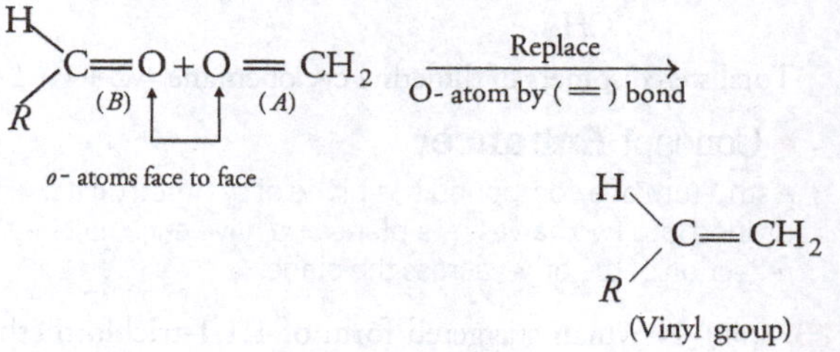


**16.** *(d)*

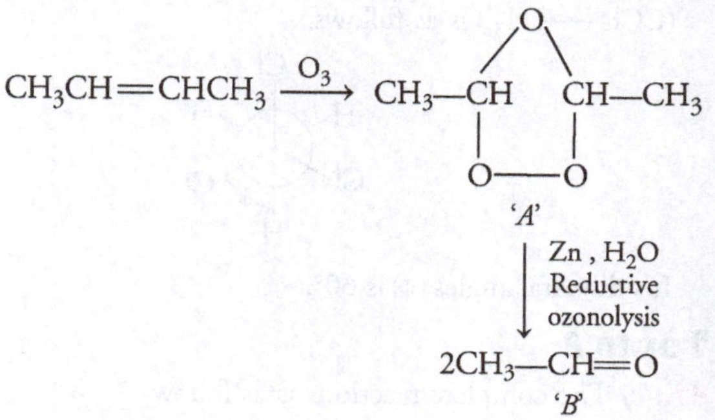


**17.** *(c)*

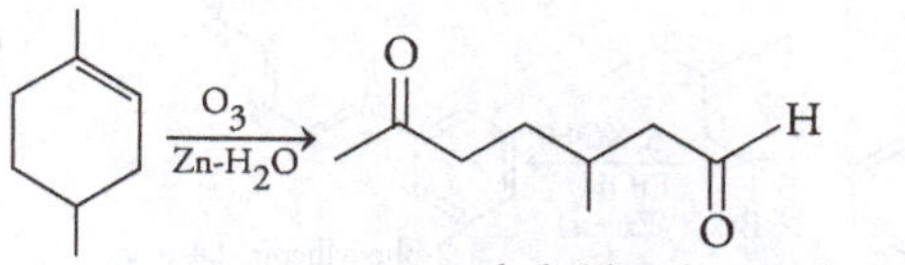

3-methyl-6-ketoheptanal

**18.** *(c)*

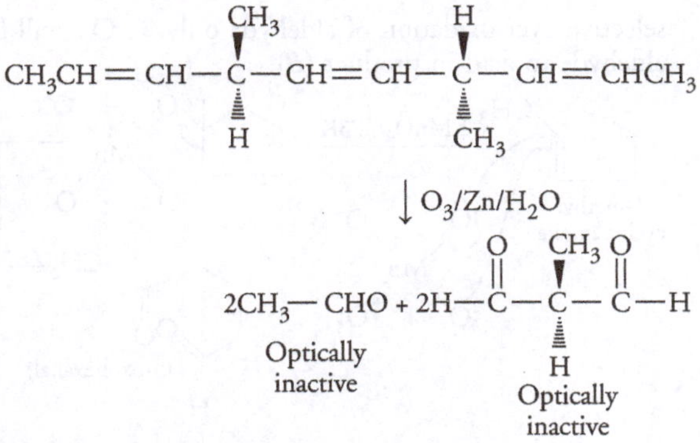


**19.** *(b)* Statement I is true but Statement II is false:

The reaction involved to obtain product by ozonolysis followed by treatment with Zn, $H_2O$.

$$H_3C-CH_2-\underset{|}{\underset{CH_3}{CH}}-HC=CH-\underset{|}{\underset{CH_3}{C}}-CH_2CH_3$$

$$\downarrow \text{(i) } O_3/Zn, \text{ (ii) } H_2O$$

$$2[CH_3-CH_2-\overset{*}{\underset{CH_3}{CH}}-\underset{O}{\overset{\|}{C}}-]$$

It has chiral centre.

**20.** *(c)* The complete reaction is as follows

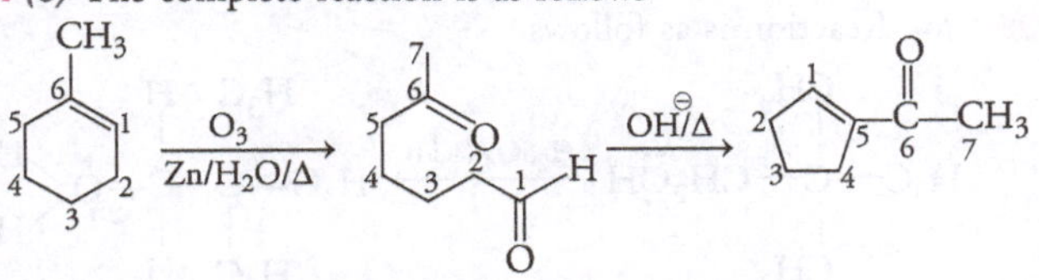


So, the suitable alkene is given in option (c).

**21.** *(d)* Addition of H—Br *(aq)* to an alkene follows electrophilic addition mechanism. In the rate determining step a carbocation intermediate is formed. Among *P*, *Q*, *R*, and *S* compound, *Q* will form most stable carbocation intermediate since it is resonance stabilised.

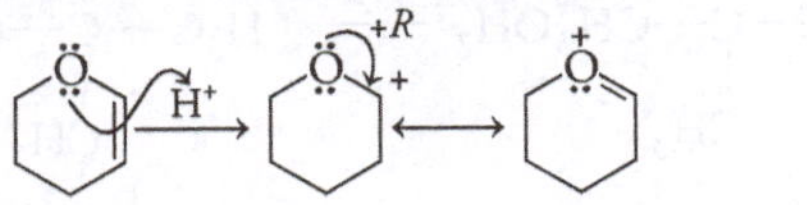


**22.** *(d)*

**key idea** Use Bond strength $\propto \dfrac{1}{\text{Bond length}}$

Statement given in option *(d)* is incorrect. Its correct form is, the carbon-carbon bonds in ethyne is stronger than that in ethene.

**23.** *(a)* The complete reaction is as follows,

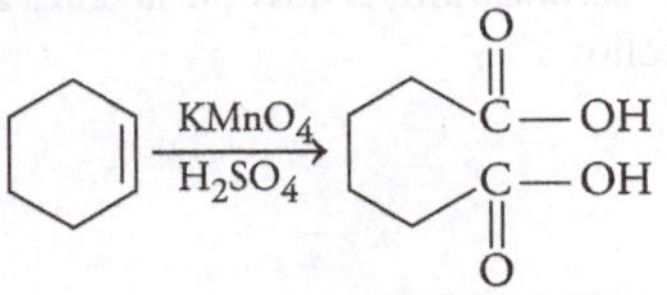


Adipic acid
*(A)*

Hence, option (a) is correct.

**24.** *(d)*

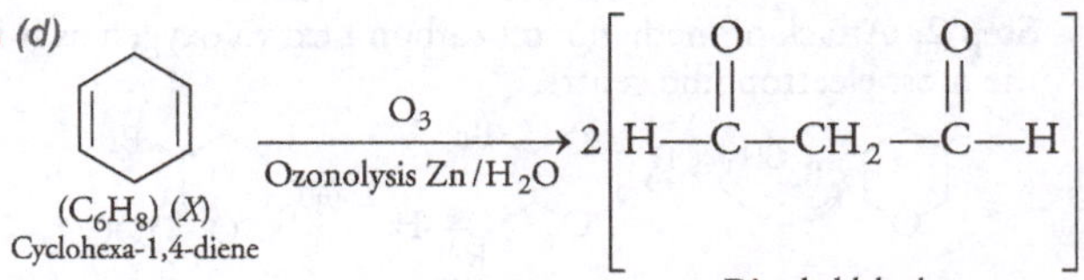


**25.** *(d)*

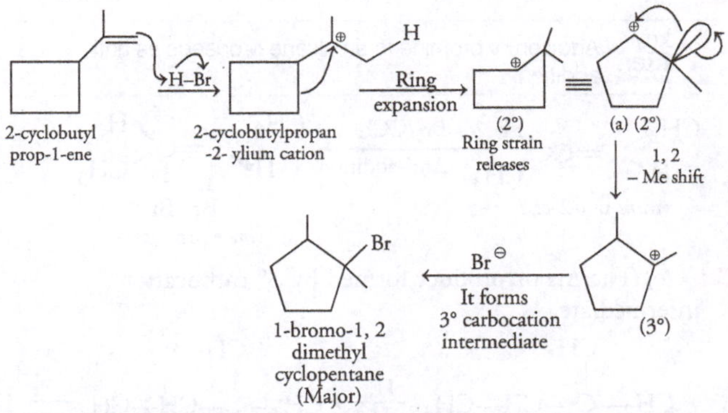


**• Concept Enhancer**

2-cyclobutylprop-1-ene reacts with HBr to give a 3° carbocation, which undergoes ring expansion to release ring strain then again it rearranges to 3° carbocation *via* 1,2-methyl shift attack of $Br^-$ to give major product.

**26.** *(c)*

$$\underset{\text{Ethanol}}{CH_3CH_2OH} \xrightarrow{\text{Conc. } H_2SO_4} \underset{\text{Ethene}}{CH_2=CH_2}$$

$$\xrightarrow[\text{(Baeyer's reagent)}]{\text{Cold/dil. } KMnO_4} \underset{\text{Ethan-1,2-diol}}{\underset{OH\quad\ \ OH}{CH_2-CH_2}}$$

In presence of conc. $H_2SO_4$ and Baeyer's reagent ethanol gets converted into diol.

**27.** *(c)* The alcoholic (—OH) group attacks / takes the proton ($H^+$) and water leaves to form carbocation.

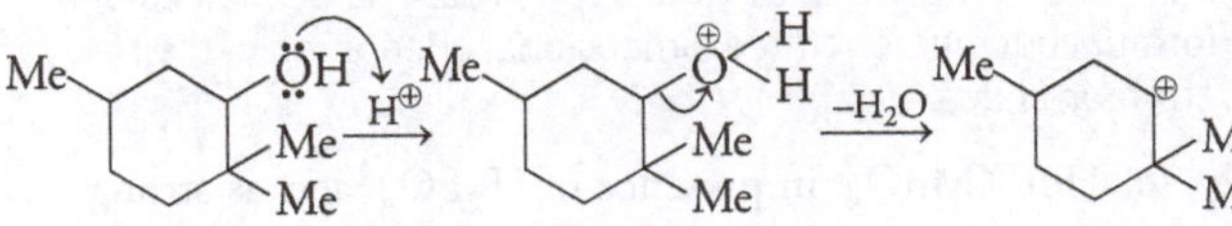


Now, rearrangement of carbocation takes place by 1,2-methyl shift to form more stable carbocation (3°-carbocation) followed by removal of $H^+$ ion to given more substituted alkene (major product).

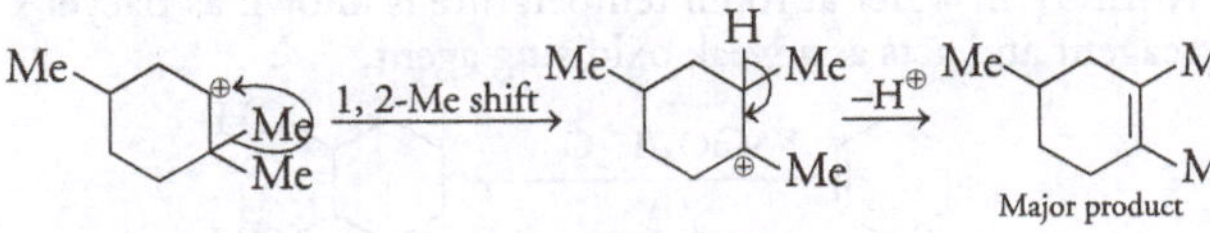


**28.** *(a)* This reaction proceeds in two steps :

**Step 1.** *syn*-addition of bromine or formation of bromonium ion.

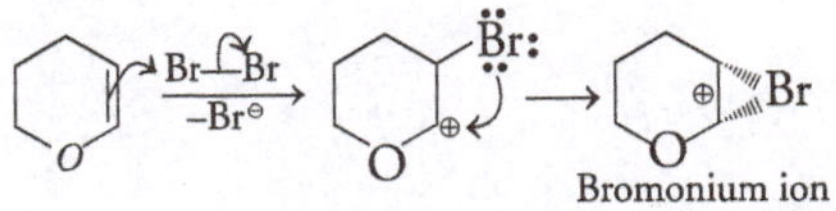


Bromonium ion

**Step 2.** Attack of methanol on carbon next to oxygen as it is the most electrophilic centre.

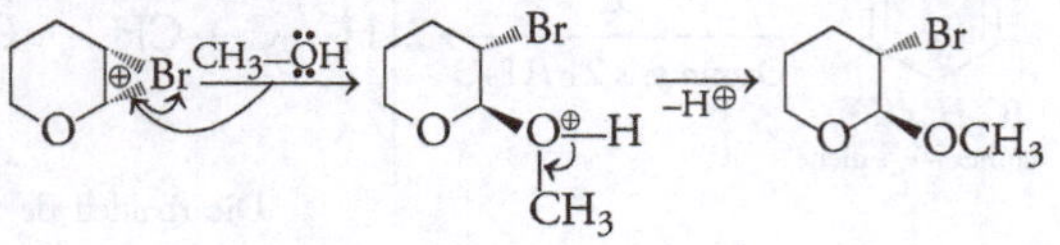

**29.** *(a)*

**key idea** Addition of bromine to an alkene proceeds as anti addition.

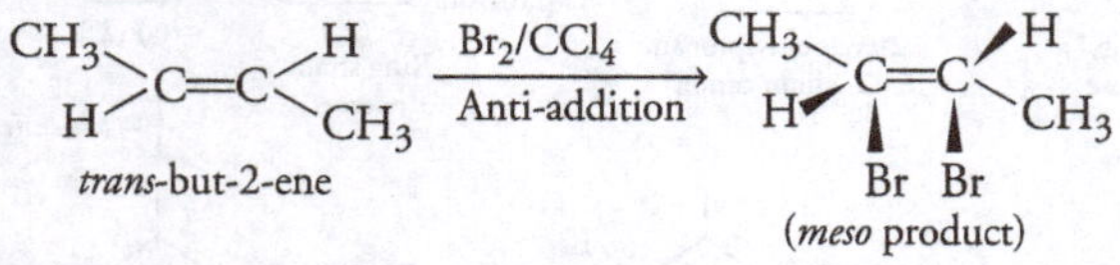

**30.** *(c)* The major product formed by 3° carbocation intermediate.

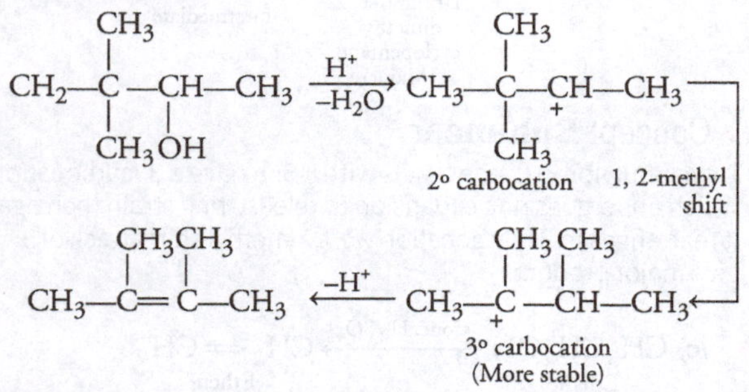

**31.** *(c)* Stereoisomers will be exhibited by those alkenes which have two different subtituent groups attached to each carbon of double bond.

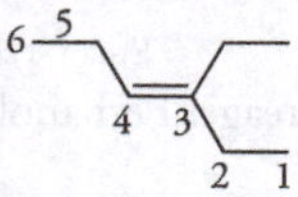

Thus, 3 ethylhex-3-ene do not show stereoisomerism.

**• Concept Enhancer**
Stereoisomers are those isomers which have same molecular formula but different three dimensional orientation of their atoms in space.

**32.** *(c)* Hot $KMnO_4$ in presence of $H_2SO_4$ acts as strong oxidising agent.

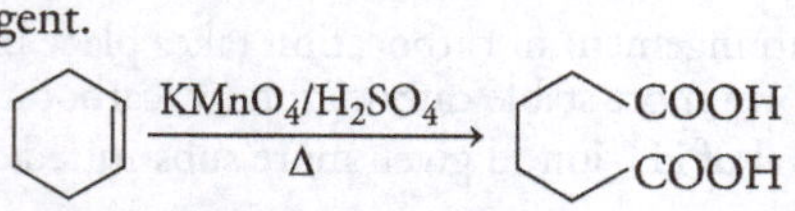

$KMnO_4$ in water at room temperature is known as Baeyer's reagent and acts as a weak oxidising agent.

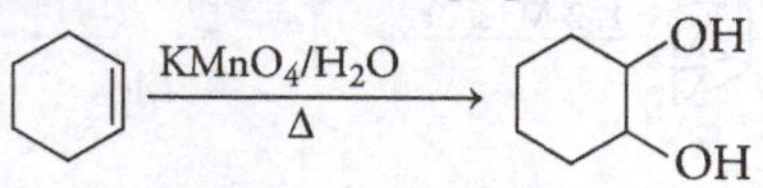

**33.** *(d)* In first step, $KMnO_4$ (dil.) help to break alkene in ketone and aldehyde and in second step, $CrO_3$ is used for selective over oxidation of aldehyde only. $CrO_3$ will form aldehyde to acid in product (*B*).

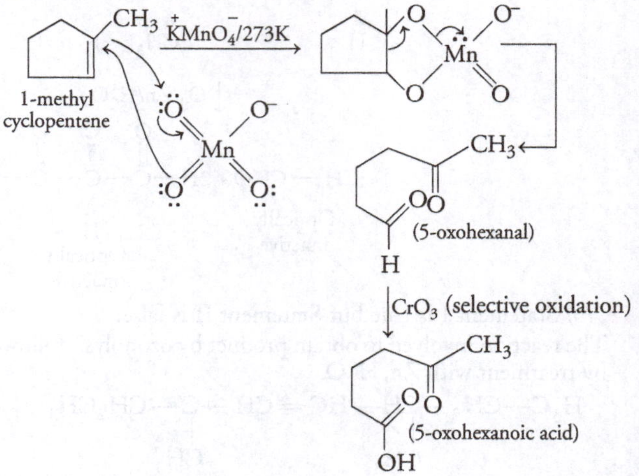

**34.** *(b)* Geometrical isomerisms is shown by those alkenes which have different type of groups attached to the carbon atoms around the double bond.

e.g., 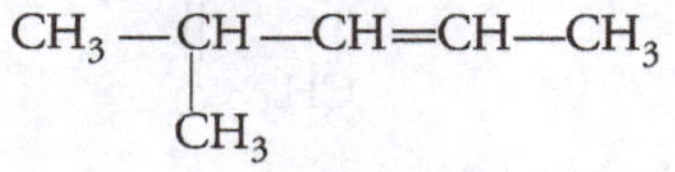

4-methylpent-2-ene

**35.** *(a)* Reaction is as follows

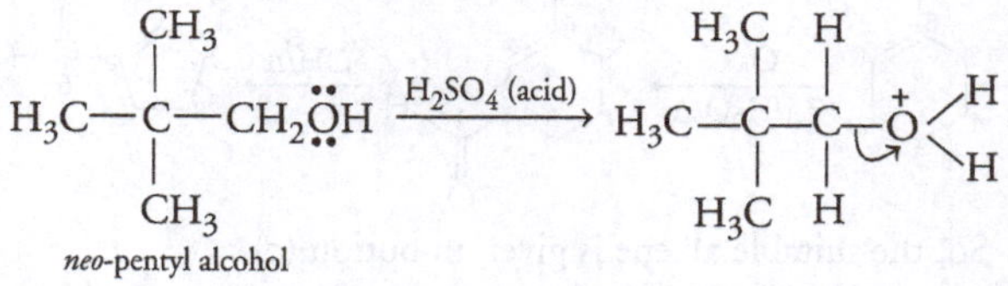

Here O being eletronegative contains positive charge thus it is unstable. Then $H_2O$ molecule will remove from here.

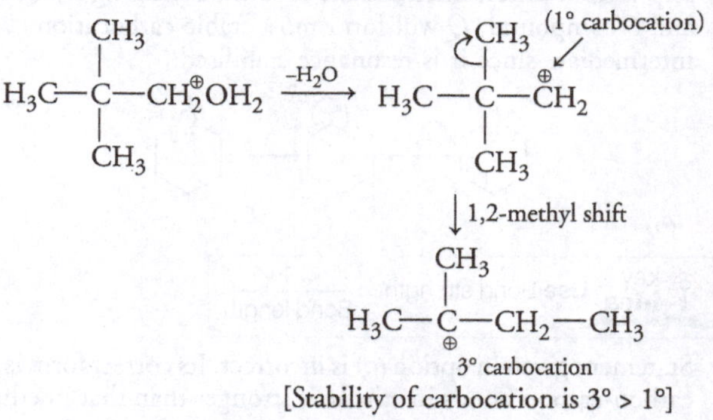

On heating 3° carbocation, 2 possible alkenes are possible. These are as follows

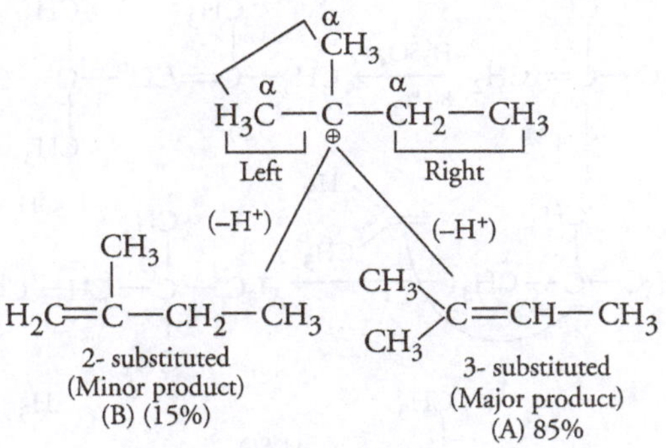

So, correct option is (a).

**• Concept Enhancer**

The major and minor product (alkene) formed is governed by Saytzeff rule according to which a more substituted alkene will form a major product.

**36.** *(d)*

**key idea** In this question we will use the concept of optical activity. Compound which contain a chiral carbon or doesn't contain axes of symmetry is an optically active compound.

Hydrogenation involve addition of hydrogen atom.

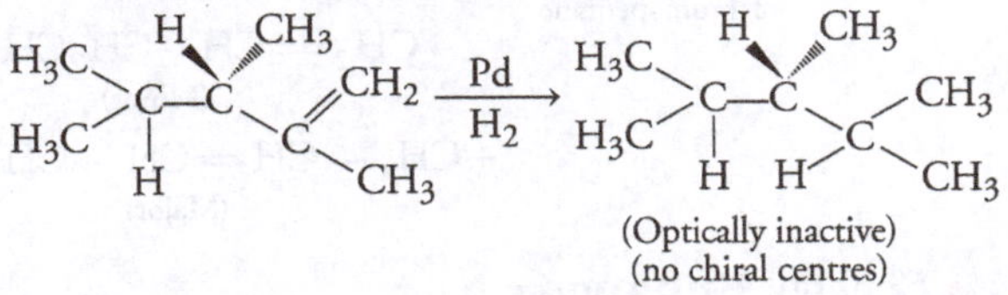

**37.** *(b)*

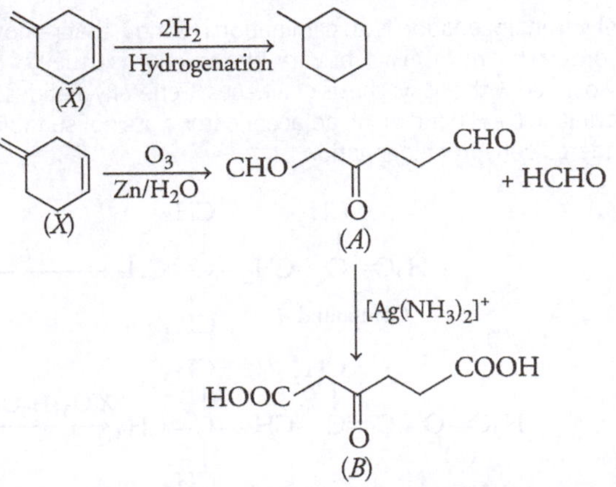

3-oxo-hexanedicarboxylic acid

The correct option is (b).

**38.** *(b)* The major product of the given addition reaction is

$$H_3C-\underset{OH}{\underset{|}{CH}}-\underset{Cl}{\underset{|}{CH_2}}.$$

The reaction proceeds through following mechanism

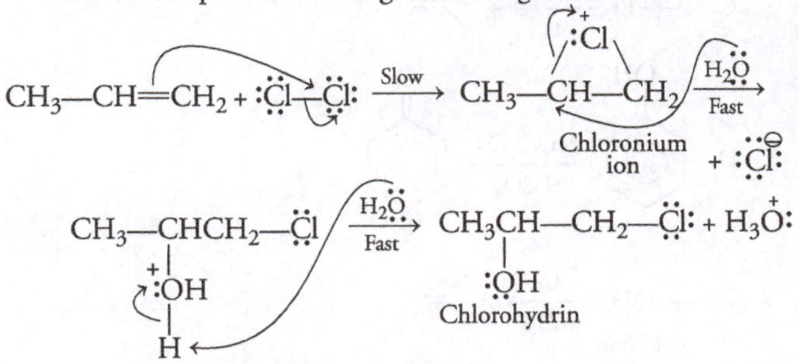

**39.** *(c)* $CH_2{=}CHCH{=}CH_2 + HBr \longrightarrow$

$$\underset{\substack{Br\\ \text{1, 2-addition product}}}{CH_3\underset{|}{C}HCH{=}CH_2} + \underset{\text{1, 4-addition product}}{CH_3CH{=}CHCH_2Br}$$

Addition is through the formation of allylic carbocation.

$$\underset{\substack{(2^\circ \text{ allylic})\\ \text{(more stable)}}}{CH_2{=}CH\overset{\oplus}{C}HCH_3} \xleftrightarrow{\text{Hydride shift}} \underset{\substack{(1^\circ \text{ allylic})\\ \text{(less stable)}}}{CH_3CH{=}CH\overset{\oplus}{C}H_2}$$

Under mild conditions (temperature $\approx -80°C$) kinetic product is the 1, 2-addition product and under vigorous conditions, (temp. $\approx 40°C$) thermodynamic product is the 1, 4-addition product.

Thus, 1-bromo-2-butene is the major product under given condition.

**40.** *(a)* Complete reaction sequence involved when molecule (*X*) reacts with 1 eq of HBr is as follows.

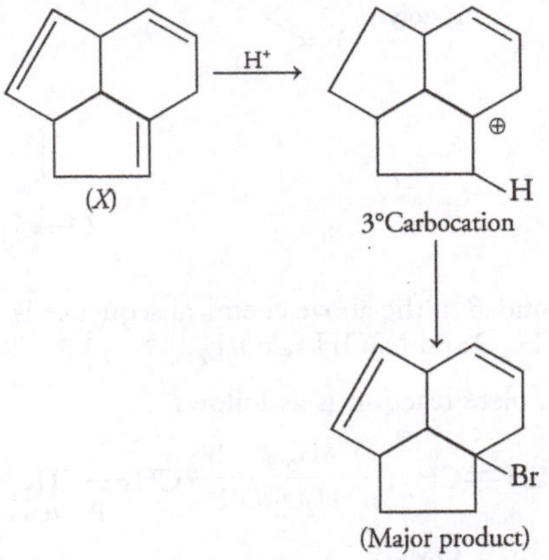

**41.** *(b)* Reaction given in B and D will not form alkene by elimination. The product formed in each reactions are as follows.

A. 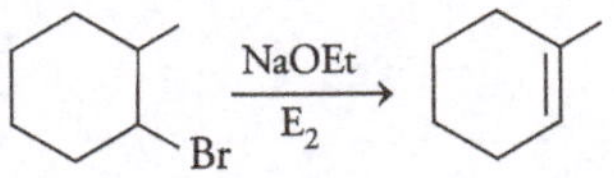

B. 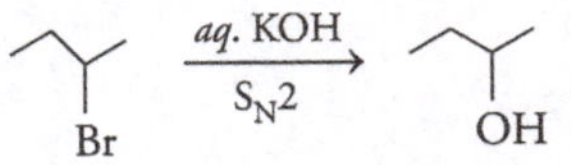

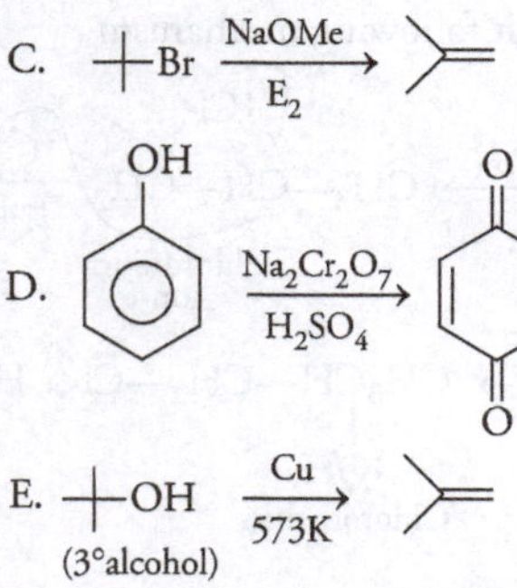

**42.** *(d)* The correct match is A-III, B-IV, C-I, D-II.

**43.** *(a)* Statements given in A, D and E are incorrect while B and C are correct. The correct form of incorrect statements are

(A) Propene does not show geometrical isomerism.
(D) 2-methylbut-2-ene does not show geometrical isomerism.
(E) *trans* isomer has higher melting point than *cis* isomer.

**44.** *(c)* Statement given in option (c) is incorrect. Its correct form is,

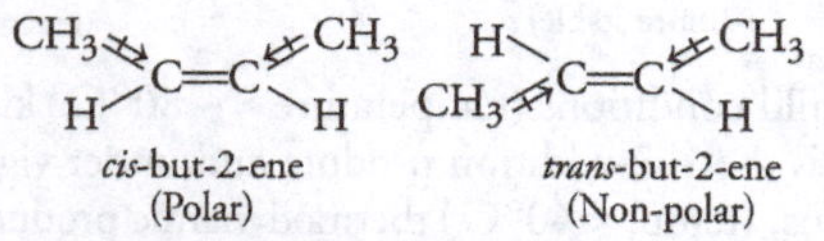

*cis* but-2-ene has higher dipole moment that *trans* but-2-ene.

**45.** *(c)*

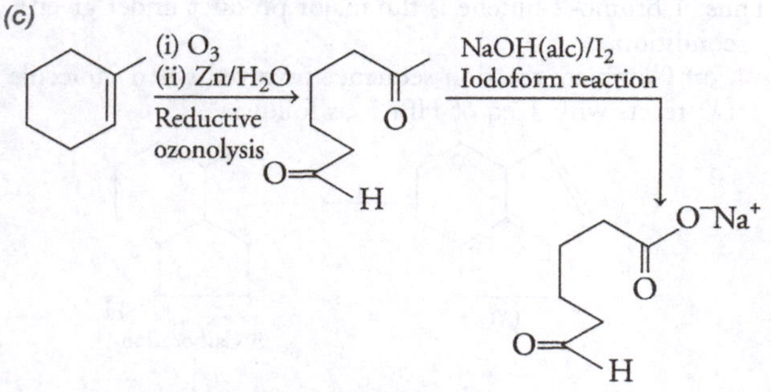

Thus, *A* and *B* in the above chemical sequence is $O_3$, $Zn/H_2O$ and NaOH (alc.)/$I_2$.

**46.** *(d)* Complete reaction is as follows

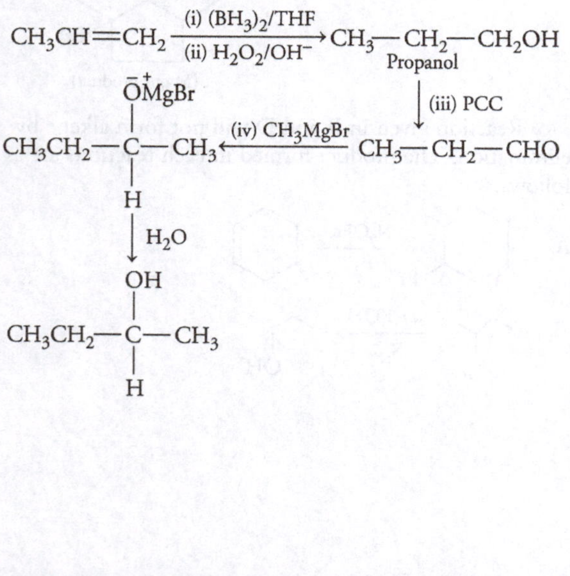

**47.** *(d)*

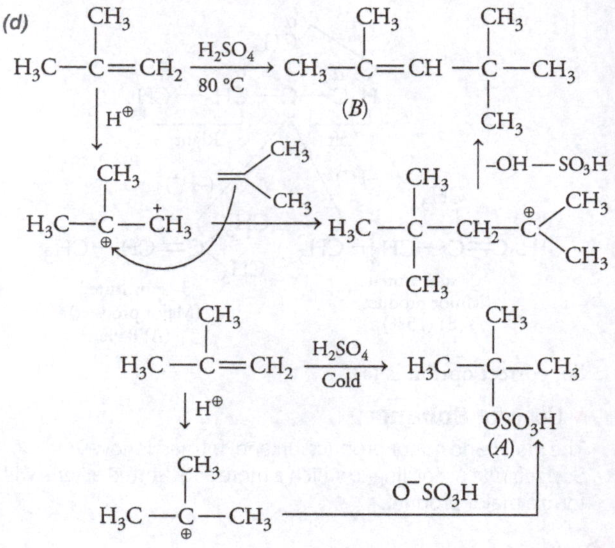

**48.** *(b)* The one giving maximum number of isomeric alkane on dehydrohalogenation will be

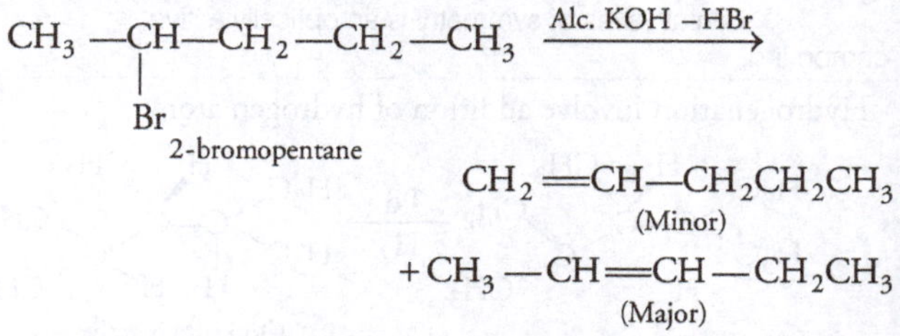

### • Concept **Enhancer**

Dehydrohalogenation is an elimination reaction, that removes hydrogen halide from a substrate. The reaction is usually associated with the synthesis of alkenes. Generally, halides having no C—H bond on an adjacent carbon are not suitable to undergo dehydrohalogenation.

**49.** *(d)*

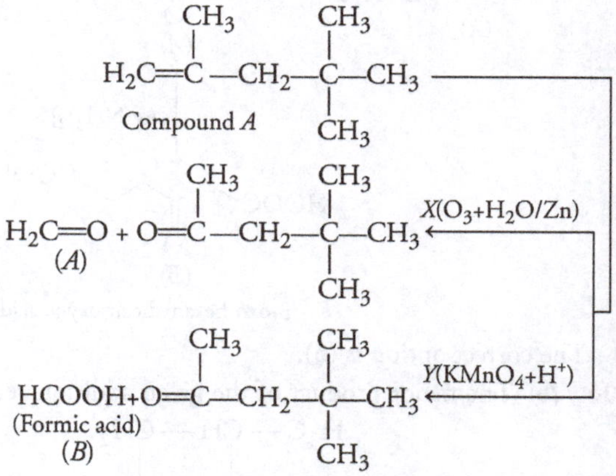

Oxidation of '*a*', i.e., HCHO produces formic acid which is produced by ants.

**50.** *(d)* $CH_3-\underset{(C_4H_8)}{\overset{CH_3}{\overset{|}{C}}}=CH_2 \xrightarrow{\text{Acidic KMnO}_4} CH_3-\overset{CH_3}{\overset{|}{C}}=O + HCOOH$

Acetone (ketone group) $\xrightarrow{\Delta}$ $H_2O + CO_2\uparrow$ (Effervescence)

$CH_3-\overset{H}{\overset{|}{C}}=\overset{H}{\overset{|}{C}}-CH_3 \xrightarrow{\text{Acidic KMnO}_4} 2CH_3CHO$ Ethanal (Aldehyde)

$CH_3-CH_2-CH=CH_2 \xrightarrow{\text{Acidic KMnO}_4} CH_3-CH_2-CHO + HCOOH$

The compound (*A*) is $CH_3-\underset{CH_3}{\underset{|}{C}}=CH_2$ (2-methylpropene)

So, correct option is (d).

**51.** *(d)*

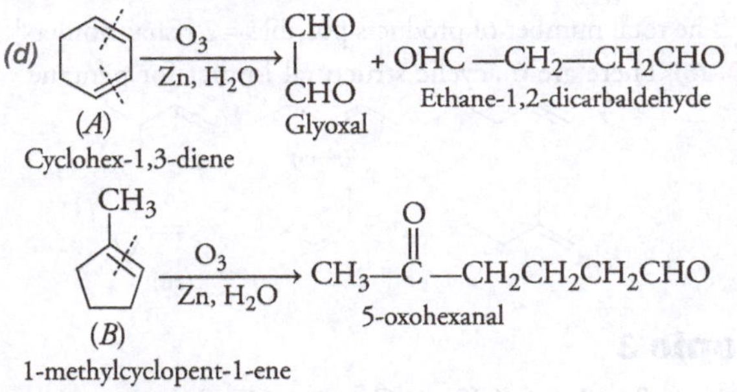

*A* is six membered ring and *B* is five membered ring.

**52.** *(a)* This is oxymercuration- demercuration reaction.

**Step 1** Alkene reacts with mercury (II) acetate and forms cyclic mercurinium ion.

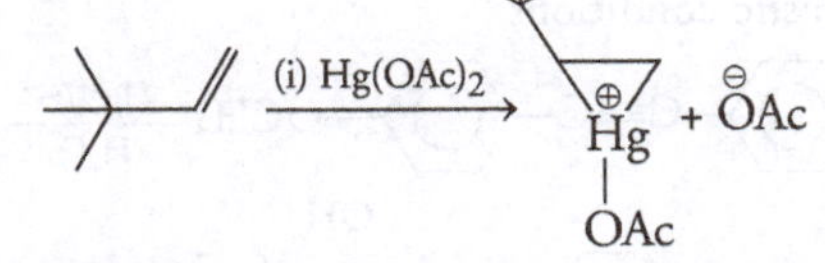

**Step 2** Attack of nucleophile occurs and this mechanism follows Markownikoff's rule of regioselectivity.

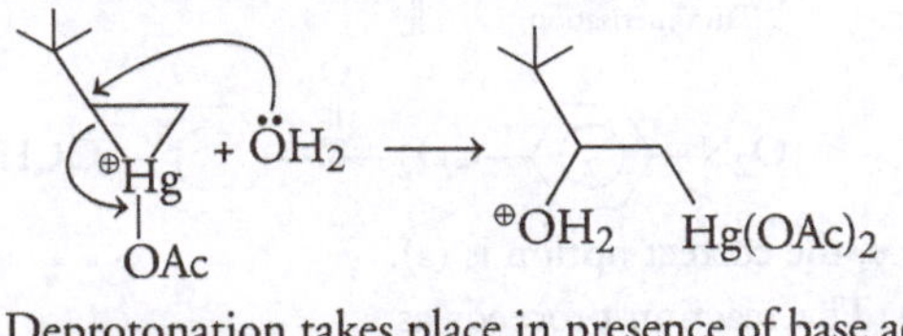

**Step 3** Deprotonation takes place in presence of base acetate ion which gives alcohol.

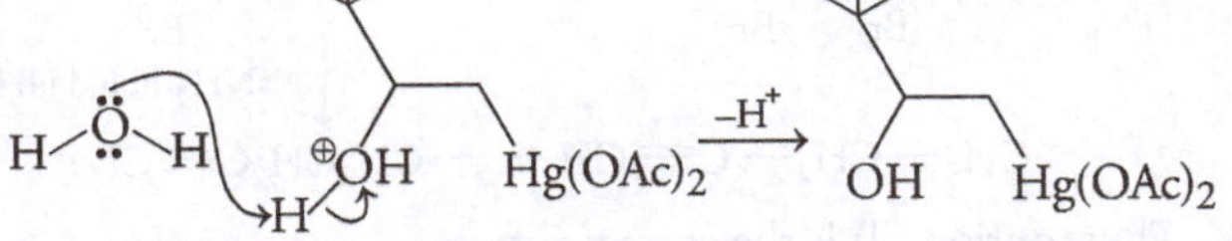

**Step 4** Addition of $NaBH_4$ reduces the acetyl mercury by a hydrogen atom. $NaBH_4$ acts as a reducing agent.

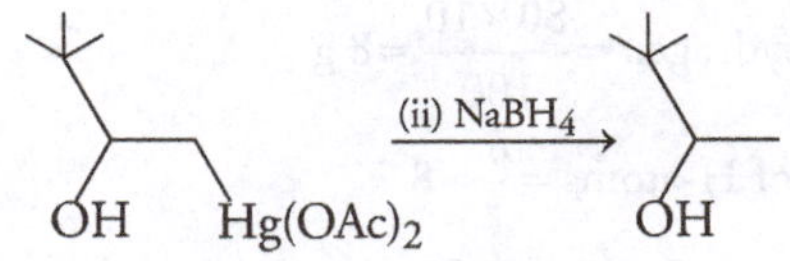

**53.** *(a)*

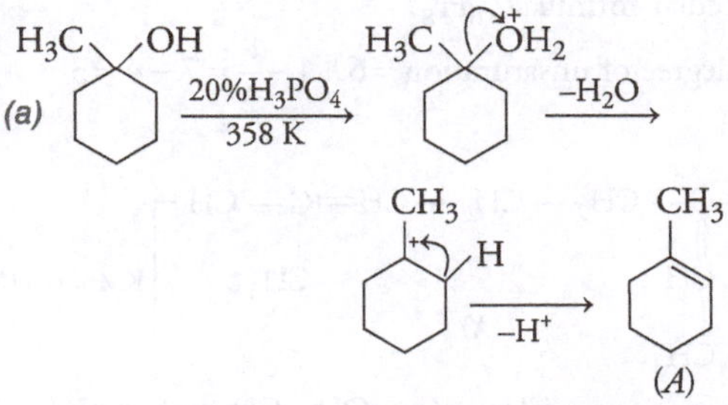

Alcohols gives E1 elimination with conc. $H_2SO_4$ and $H_3PO_4$ on heating. Thus, OH group is removed and alkene get formed.

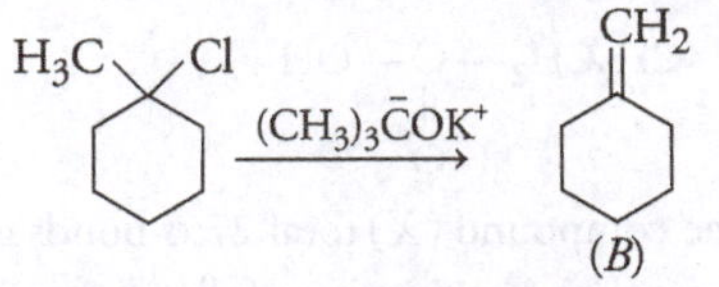

*t*-$BuO^-$ is a bulky base and will form Hoffmann alkene via elimination.

**54.** *(c)*

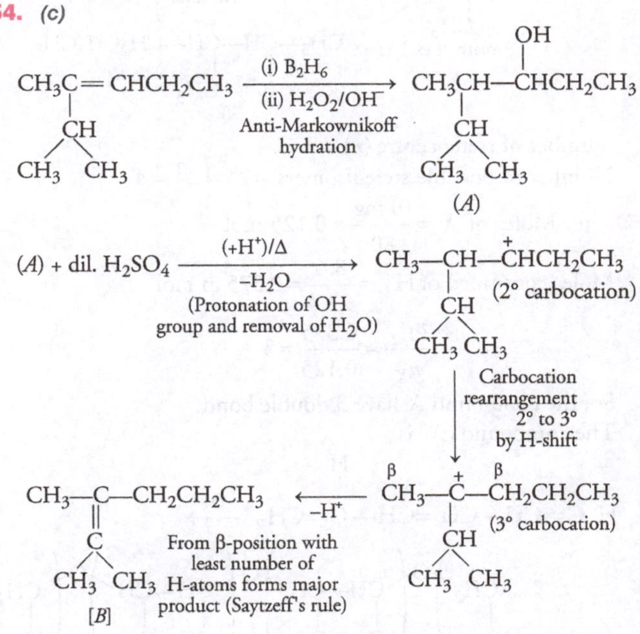

**55.** *(a)*

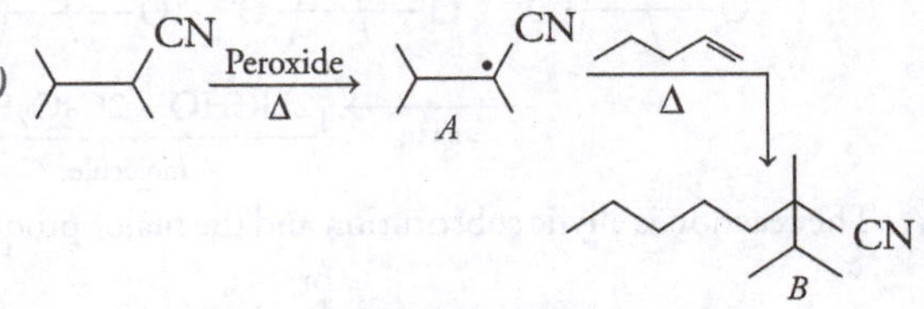

Reaction proceed *via* free radical formation.

**56.** *(3)* Mass of carbon $= \frac{80 \times 90}{100} = 72$ g

Number of C-atoms $= \frac{72}{12} = 6$

Mass of hydrogen $= \frac{80 \times 10}{100} = 8$ g

Number of H-atoms $= \frac{8}{1} = 8$

So, molecular formula $C_6H_8$.

Hence, degree of unsaturation $= 6 + 1 - \frac{8}{2} = 7 - 4 = 3$

**57.** *(27)*

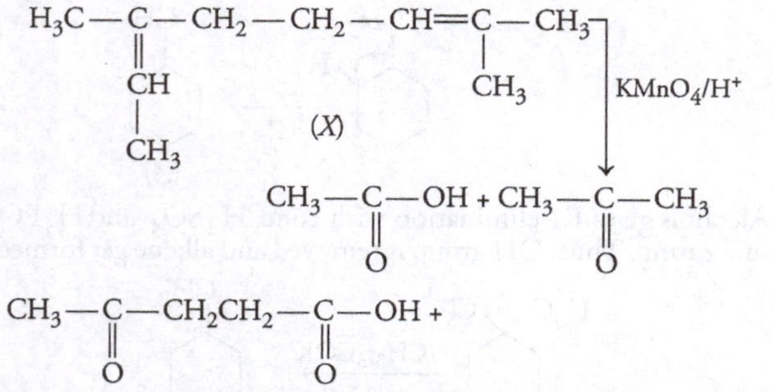

Thus, in the compound ($X$) total 27 σ-bonds are present.

**58.** *(4)* The complete reaction is as follow

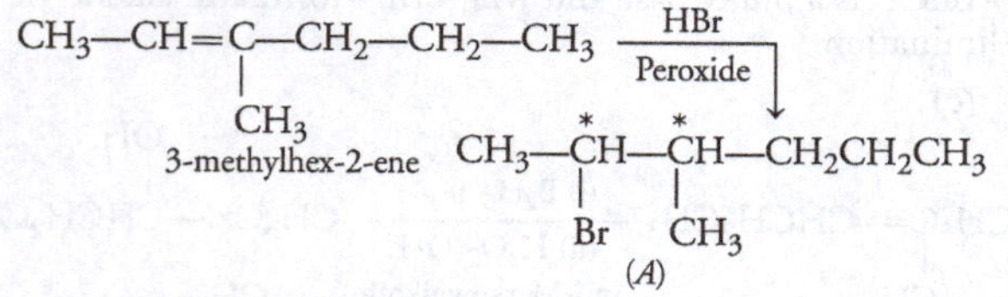

Number of stereocentre ($n$) = 2

Number of possible stereoisomers $= 2^n = 2^2 = 4$

**59.** *(4)* Moles of $X = \frac{10 \text{ mg}}{80} = 0.125$ mol

Moles consumed of $H_2 = \frac{8.4}{22.4} = 0.375$ m mol

$\frac{n_{H_2}}{n_X} = \frac{0.375}{0.125} = 3$

So, the compound $X$ have 3 double bond.

The compound ($X$) is

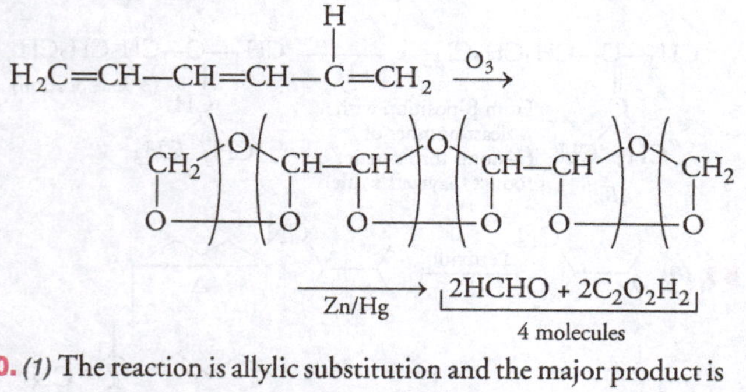

**60.** *(1)* The reaction is allylic substitution and the major product is

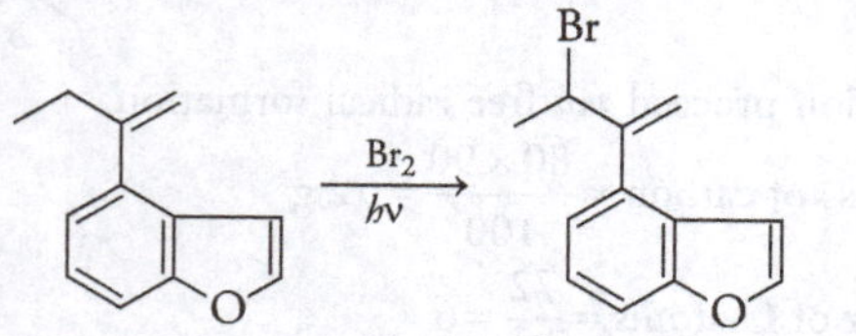

Number of Br atoms = 1

**61.** *(2)* The major product formed has two $sp^2$ hybridised carbon.

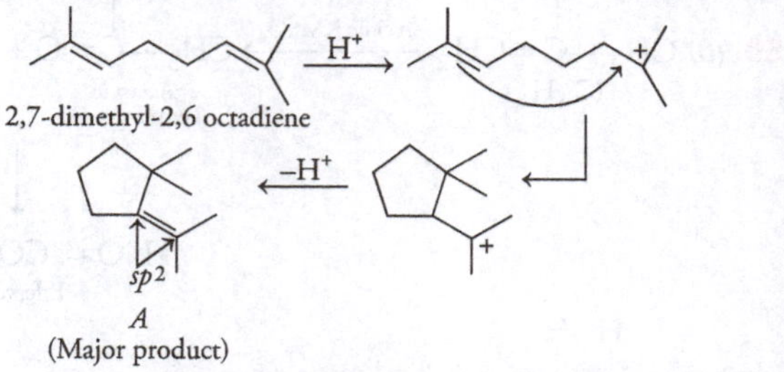

**62.** *(2)* $Br_2/CCl_4$ gives anti-addition of bromine with alkene but product formed in the reaction depends upon the stereo of alkene ($E$ or $Z$ alkene).

Here, *cis* alkene /Z-alkene is given, so we get racemic mixture of two isomers.

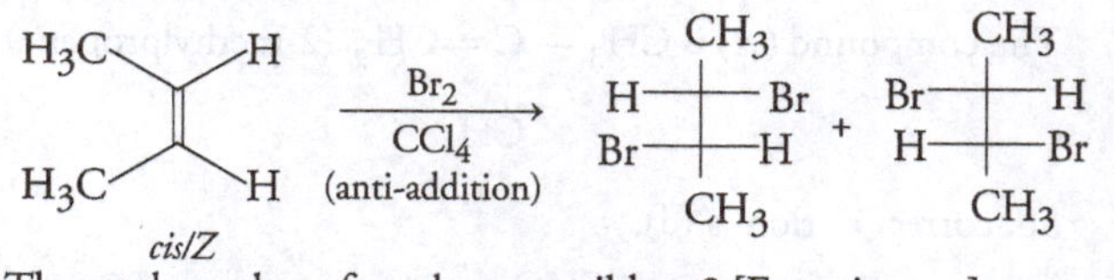

The total number of products possible = 2 [Enantiomers]

**63.** *(6)* There are 6 acyclic structural isomer for pentane

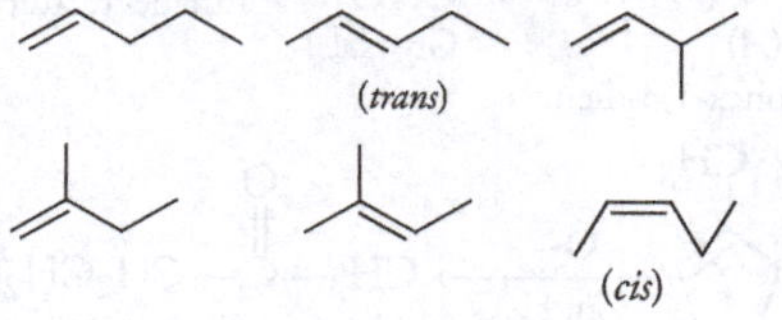

## Topic 3

**01.** *(a)* But-2-yne ($CH_3-C \equiv C-CH_3$) does not have any acidic hydrogen and hence does not react with metallic sodium under normal conditions.

**02.** *(a)* Alkynes undergo hydration in presence of $HgSO_4$ and ether are least reactive and their cleavage take place only under drastic conditions.

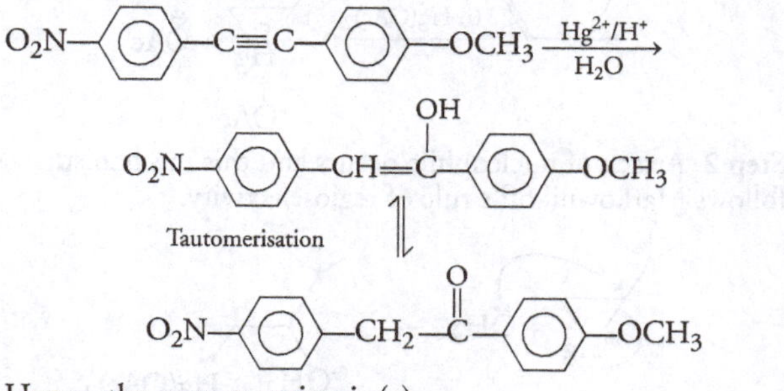

Hence, the correct option is (a).

**03.** *(d)* The reaction proceeds as

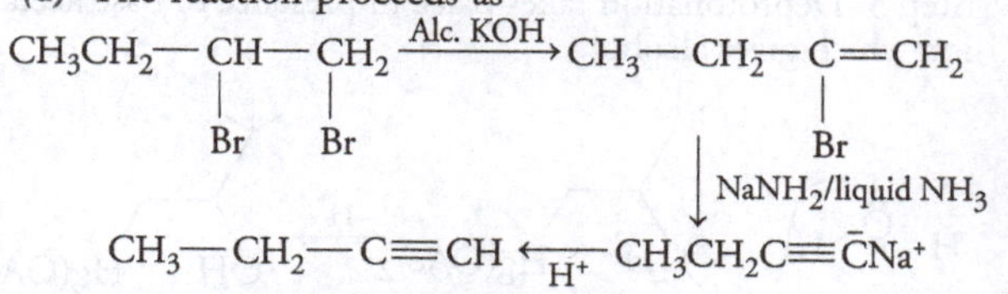

Thus, option (d) is the correct answer.

**04.** *(c)* Sodium metal in liquid ammonia reduces alkynes with anti stereochemistry to give *trans* alkenes. The reduction is selectively *anti* since the vinyl radical formed during reduction is more stable in *trans* configuration.

**05.** *(c)*

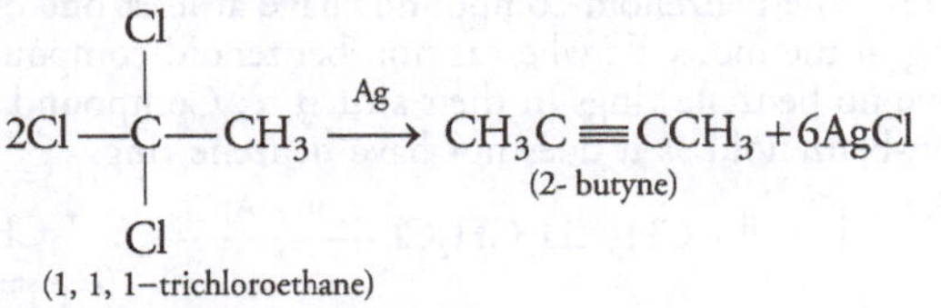

**• Concept Enhancer**

Different type of alkenes are formed by differnt catalyst but with same kind of alkynes. For e.g.

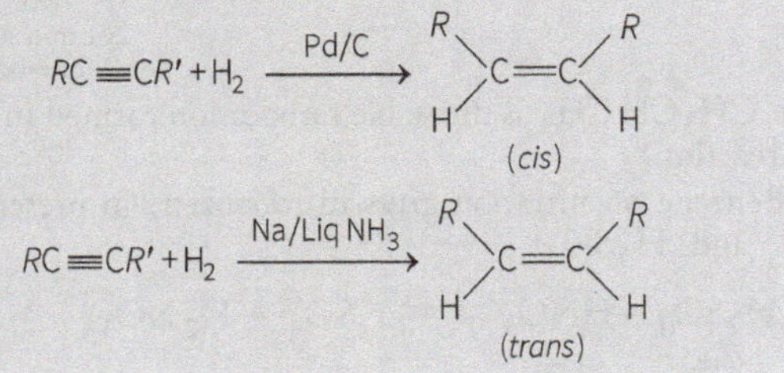

**06.** *(b)*

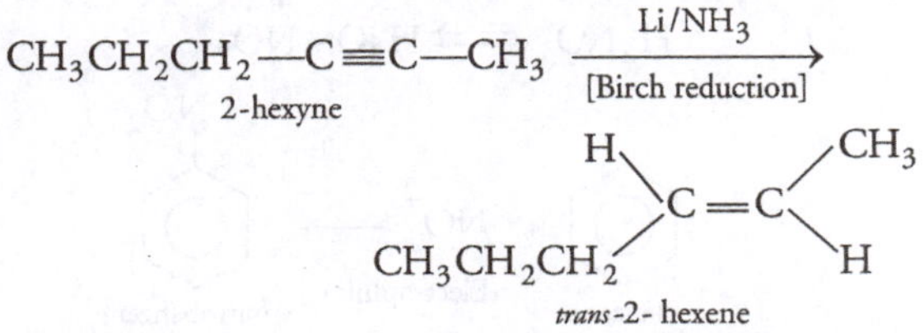

**07.** *(d)* Terminal alkyne has acidic hydrogen which is enough to protonate the Grignard reagent.

$$CH_3MgX + CH_3C\equiv CH \longrightarrow CH_4 + CH_3C\equiv CMgX$$

**08.** *(d)* $CH\equiv CH + NaOH \longrightarrow$ no reaction
Acetylene

It is clear from the hard acid, hard base and soft acid, soft base (HASB) theory.

**09.** *(b)* Correct option is (b) $Br_2$, alc. KOH, $NaNH_2$, Lindlar catalyst $H_2$. Lindlar's catalyst is a palladium catalyst poisoned with traces of lead and quinoline. It is used for formation of *cis*-product.

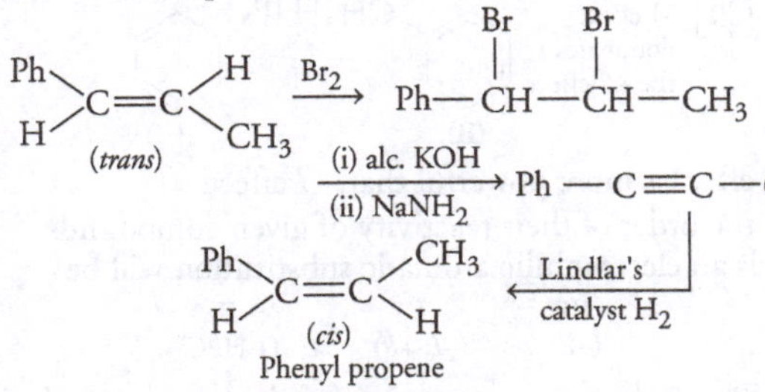

**10.** *(c)* Alkyne undergoes hydrogenation reaction to give alkene in presence of Lindlar's catalyst.

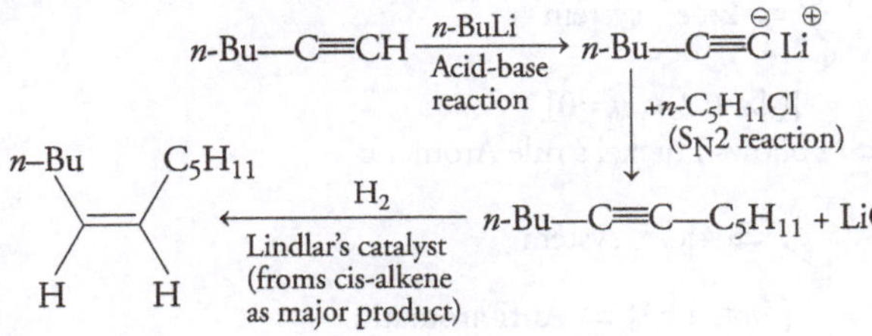

**11.** *(a)* $HC\equiv C-CH_2-CH_3 \xrightarrow[H_2O]{O_3}$ $CH_3CH_2COOH + HCOOH$
(X) (A)

$$HCOOH \xrightarrow[\text{[Ag(NH}_3)_2]^+]{\text{Tollen's test}} CO_2 + H_2O + 2Ag\downarrow$$
(Bright silver mirror)

**12.** *(a)*

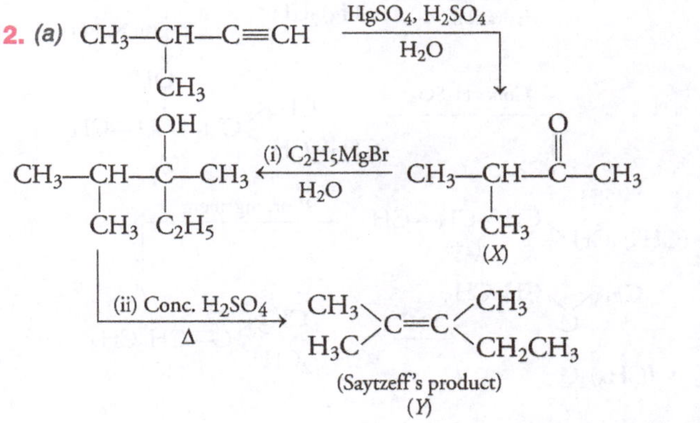

**13.** *(d)* The major product obtained in the given reaction is $CH_3C(I)(Cl)CHD_2$.
Reaction proceed as follows

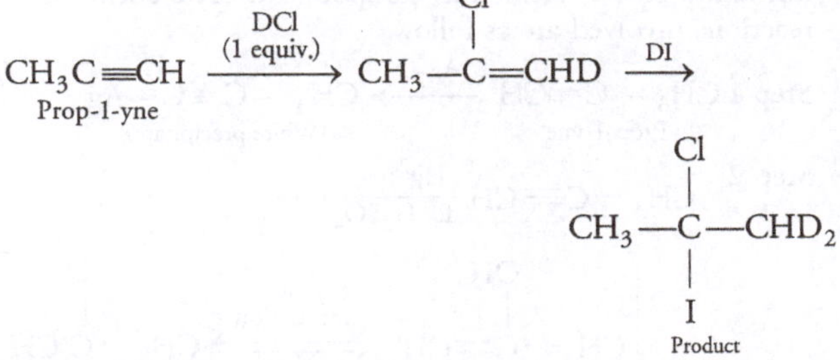

**14.** *(a)* *A* will be 2-pentyne i.e.,
$H_3C-C\equiv C-CH_2-CH_3$
as *cis*-2-pentene is formed on reacting with Pd/C.
The complete reaction is given as,

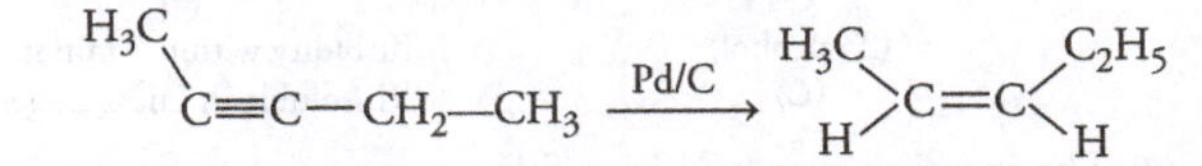

Also, *B* is *trans*-2-butene as Na/Liq. $NH_3$ is used as reducing agent.
The complete reaction is

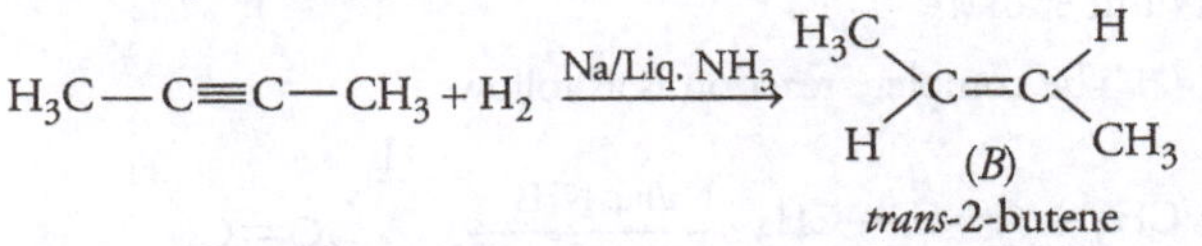

**15.** *(c)* The complete reaction is as follows,

$$CH_3-C\equiv CH \xrightarrow{Na} \underset{(A)}{CH_3-C\equiv C^-Na^+} \xrightarrow[(B)]{CH_3CH_2CH_2Br} CH_3-C\equiv C-CH_2CH_2CH_3 + NaBr$$

Therefore, $A = CH_3-C\equiv C^-Na^+$
$B = CH_3CH_2CH_2Br$

Hence, option (c) is correct.

**16.** *(c)*

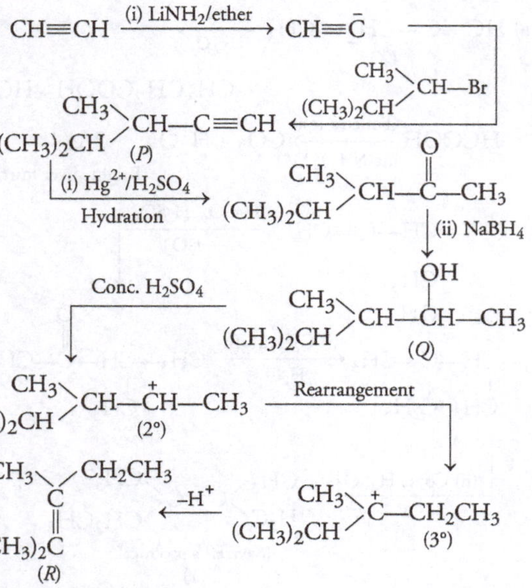

**17.** *(c)* According to the given conditions, the compound should be alkyne with triple bond present at the terminal. Secondly, appearance of turbidity in 5 minutes indicates the formation of a 2° alcohol as compound *C*. The chemical reactions involved are as follows

Step 1 $\underset{\text{Prop-1-yne}}{CH_3-C\equiv CH} \xrightarrow{Ag_2O} \underset{\text{(White precipitates)}}{CH_3-C\equiv C-Ag}$

Step 2

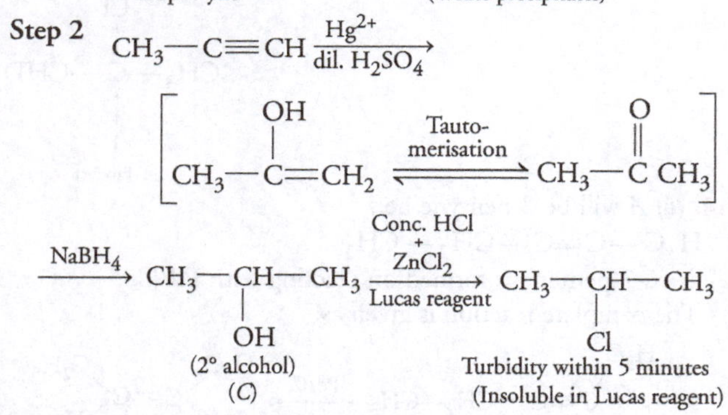

**18.** *(8)* The reaction involved is as follows.

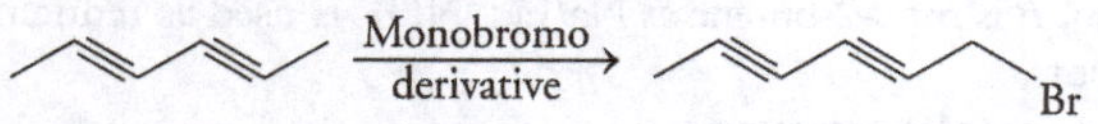

Number of $\pi e^- = 8$

**19.** *(2)* The complete reaction is as follows,

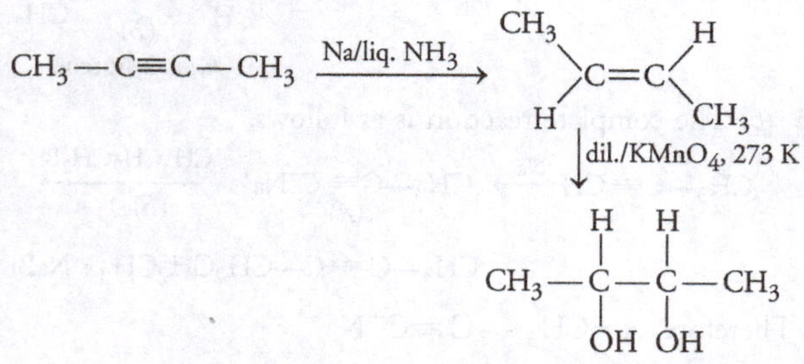

Final product (*P*)

Hence, the number of oxygen atom present in product *P* is 2.

## Topic 4

**01.** *(b)* The benzenoid compounds have at least one benzene ring in the molecule whereas non-benzenoid compounds have no benzene rings in their structure. Compound (b) is non-benzenoid as it does not have benzene ring.

**02.** *(c)*

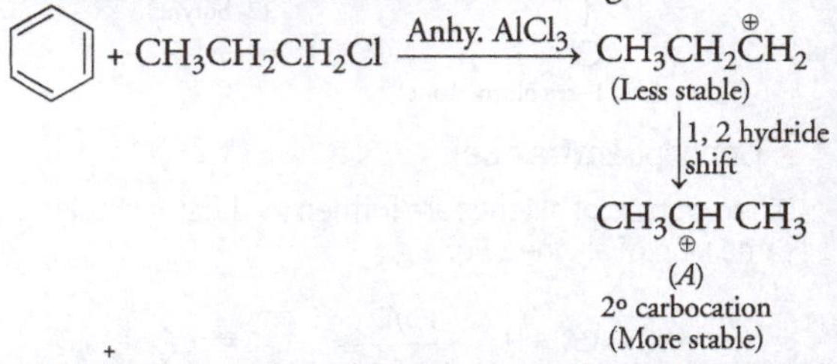

Hence, $CH_3\overset{+}{C}HCH_3$ is the stable carbocation formed in the given reaction.

**03.** *(d)* Benzene on nitration gives nitrobenzene in presence of $HNO_3$ and $H_2SO_4$.

$$\underset{\text{(Acid)}}{H_2SO_4} + \underset{\text{(Base)}}{HNO_3} \rightleftharpoons HSO_4^{\ominus} + H_2\overset{\oplus}{N}O_3$$

$$H_2\overset{\oplus}{N}O_3 \rightleftharpoons H_2O + \overset{\oplus}{N}O_2$$

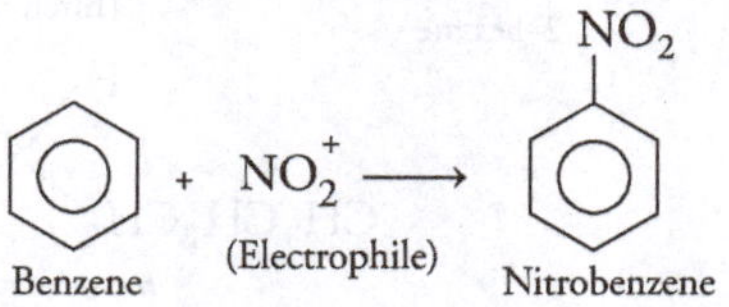

Benzene + $\overset{+}{N}O_2$ (Electrophile) ⟶ Nitrobenzene

**04.** *(a)* 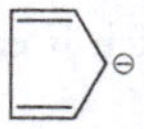 is an aromatic compound ($6\pi e^-$).

As it follows huckel's rule, cyclic and planar and threat is delocalisation of electrons.

**05.** *(a)* Higher is electron density in benzene nucleus faster will the rate of an electrophilic aromatic substitute reaction. Electron-donating groups (EDG) will increase electron density in benzene nucleus by any or both of their $+R/+M$ and hyperconjugative (HPC) effects.

Now, let us consider the given substrate.

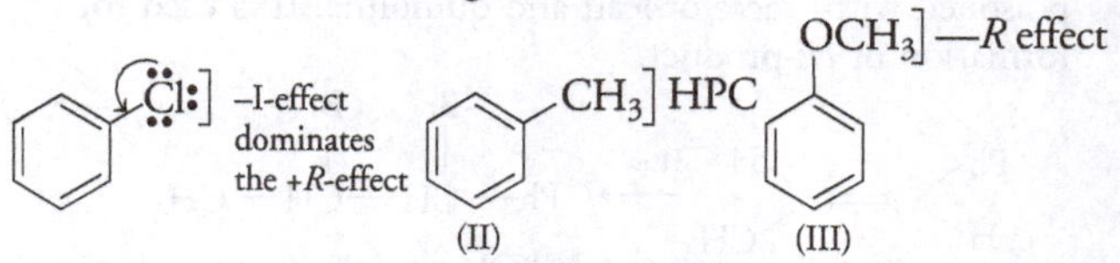

As $-R$ effect is more powerful than $-I$ effect.

Thus, the order of their reactivity of given compounds towards an electrophilic aromatic substitution will be

$$\underset{(-R)}{\text{III}} < \underset{(-I>+R)}{\text{I}} < \underset{(+\text{HPC})}{\text{II}}$$

**06.** *(a)* The $\pi$ electrons present in the given compounds are as follows

(A) 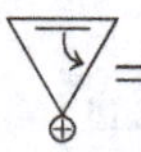 $\Rightarrow 2\pi\text{-}e^-$ system

$[(4n+2)\pi, n=0]$

$\Rightarrow$ Follows Huckel's rule Aromatic

(B)  $\Rightarrow 4\pi\text{-}e^-$ system

$[4n\pi, n=1] \Rightarrow$ Anti-aromatic

(C) 

$\Rightarrow$ $8\pi$-$e^-$ system $[4n\pi, n=2]$ $\Rightarrow$ Anti-aromatic

(D) 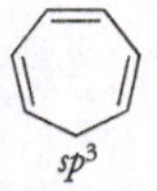

$\Rightarrow$The $sp^3$-carbon does not support conjugation and also devoids planarity.

$\Rightarrow$ Non-aromatic

**07.** *(a)* Aromaticity of a compound can be decided by Huckel's rule.

Hence, 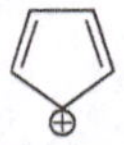 is anti-aromatic. It does not follow Huckel's rule as it has conjugated $4\pi$-electron $(4n\,\pi, n=1)$ system. Rest of the species are aromatic as each of them belongs to $6\pi$-electrons $[(4n+2)\,\pi, n=1]$ system and thus follow Huckel's rule.

**08.** *(d)* $NO_2$ group withdraw electron from the ring, thus it shows $-M$ effect and makes ring more electron deficient. Hence, $NO_2$ deactivates the ring for electrophilic substitution.

**09.** *(c)* 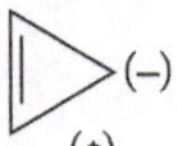 (p) is an anti-aromatic compound, 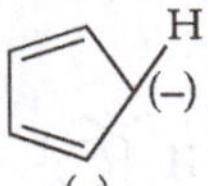 (q) is an aromatic compound, while  (r) is non-aromatic compound. As we know more is the aromatic character higher is the stability. Thus, the correct order of stability is $q > r > p$.

**10.** *(b)*

> **key idea** Any compound can be aromatic if it follows the following conditions
> - The compound should be cyclic and planar.
> - The compound should follow Huckle rule i.e., have $(4n+2)\pi$ electrons.

Structure in B and D are aromatic compound. Whereas structure in A and C are non-aromatic compound.

**11.** *(c)*  is most stable due to aromatic character. It has $2\pi e^-$ and follow $(4n+2)\,\pi e^-$ (Huckel's rule).

**12.** *(d)* 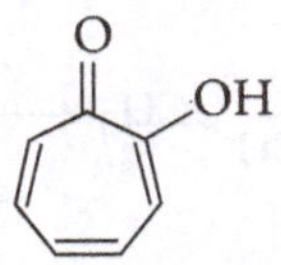

(Tropolone)

- Non-benzoid aromatic compound.
- Follows $(4n+2)\pi$ rule.
- Only $6\pi\, e^-$ of the ring are involved in aromaticity while $\pi$-electrons of $>C{=}O$ group are not involved.

Thus, both statement I and statement II are false.

**13.** *(a)* The correct option is (a).

Mechanism involve in the reaction of cycloalkene with acid hydrogen fluoride so that a proton is abstracted from hydrogen fluoride to form carbocation of cyclohexene is as follows

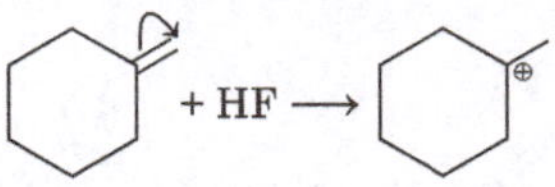

The benzene attacks on the cyclohexyl cation to form the product with two rings joined to each other.

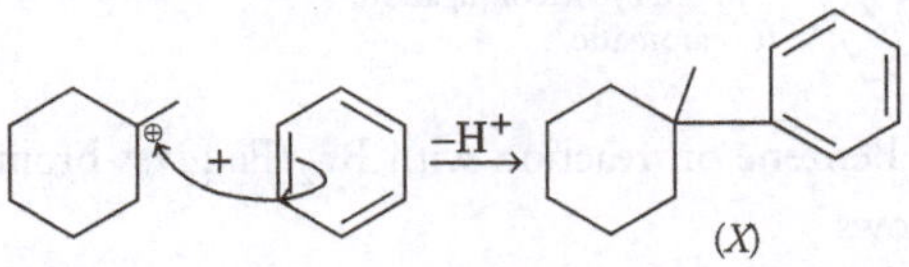

(X)

The positive charge on cyclohexyl ring is stabilised by the delocalisation in benzene ring. The abstraction of hydrogen from benzene ring gives final product as shown above.

**14.** *(d)*

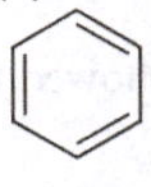 [6] annulene (aromatic)

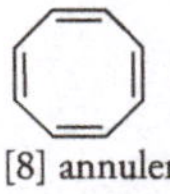 (Non aromatic)

[8] annulene

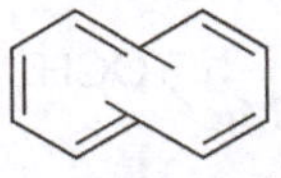 (Non aromatic)

cis-[10] annulene

In cis [10] annulene $\rightarrow$ the H-atoms at the 1 and 6 position interfere with each other and force the molecule out of planarity. Thus, it is ion aromatic planarity is one of the requirements of aromatic system.

**15.** *(c)*

> **key idea** In this question, we use concept of aromaticity. There are three requirements for a compound to be aromatic which are as follows
> (i) It should be cyclic and planar.
> (ii) Follows's Huckel's rule i.e., $(4n+2)\,\pi$ electrons.

Among the given options only (c) is non-aromatic while other are aromatic. This is because it doesn't follows Huckel's rule and is also non-planar.

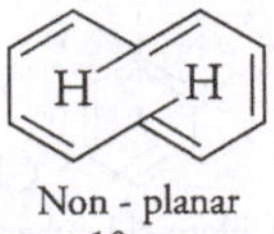

Non - planar
$10\,\pi e^-$
(Non - aromatic)

**16.** *(a)* The given reaction is Birch reduction.

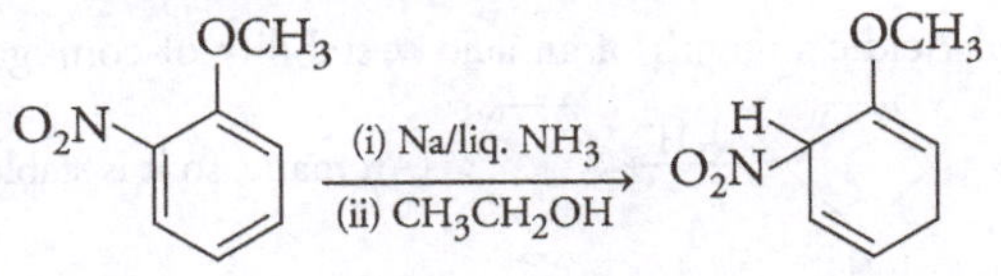

**17.** *(a,d)* Aromatic compounds are conjugated planar ring system having delocalised $\pi$-electron clouds and follows $(4n + 2)\pi$-electron rule, i.e. Huckel's rule. $\{n = 0, 1, 2, \ldots\}$

(a) 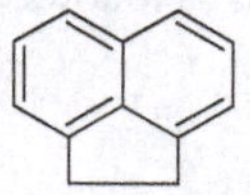

$\longrightarrow$ $10\pi e^-$ in cyclic conjugation place.
It is aromatic.

(d)  $6\pi e^-$ in cyclic conjugation.
It is aromatic.

**18.** *(b)* Benzene on reaction with $Br_2$/Fe gives bromo benzene as follows

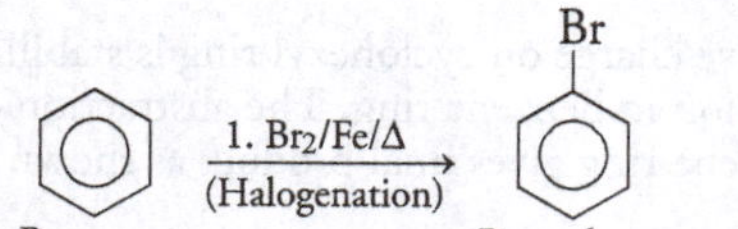


Bromobenzene on reaction with Mg / dry ether followed by $CH_3OH$ gives benzene as follows

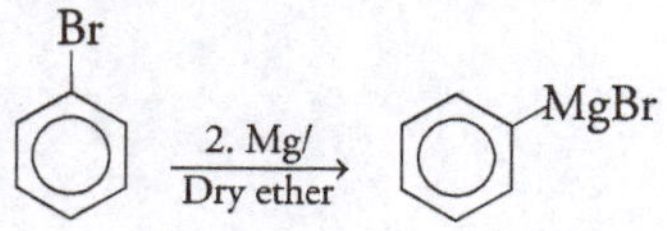


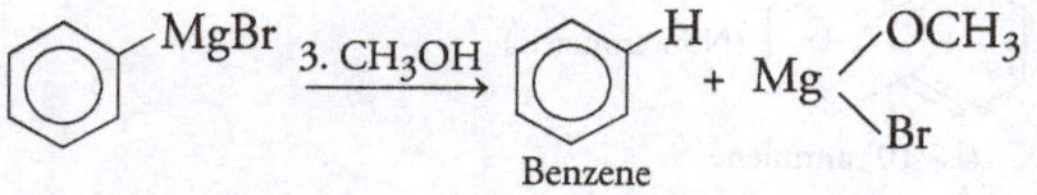


**19.** *(c)* The compound which follows Huckel's rule, has planar structure and has delocalisation of $\pi e^-$s is aromatic in nature.

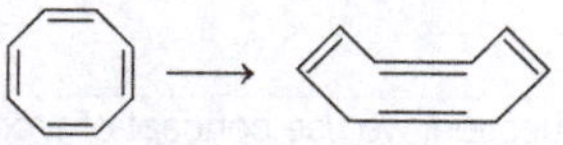

It exist in a non-planar tube shape and does not follows Huckle rule, so it is non-aromatic.

**20.** *(a)* Density is directly proportional to mass of the halogen. On moving down the group, mass increases. Order of decreasing density in benzene ring by halogen is I > Br > Cl > F.
Thus, density order of the given compounds is

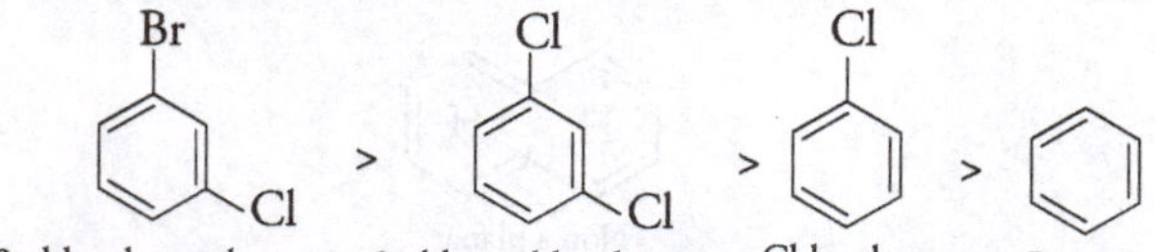


3-chlorobromobenzene (*D*)   3-chloro-chlorobenzene or 1, 3 dichlorobenzene (*C*)   Chlorobenzene (*B*)   Benzene (*A*)

**21.** *(b)* Acidic strength of an acid $\propto$ stability of conjugate base.

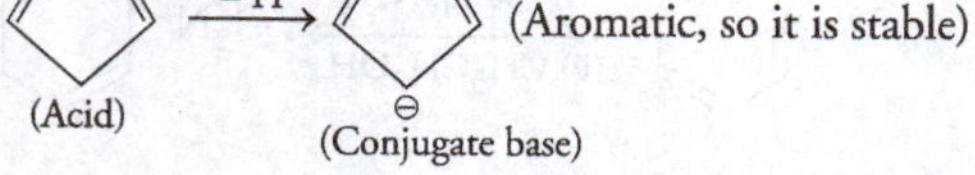


The negative charge on ring makes stable conjugate base and thus, compound (b) is the strongest acid.

**22.** *(b)*

(B) 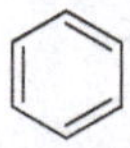

(Planar) $6\pi e^-$
(Huckel's rule followed, i.e. aromatic)

(C) 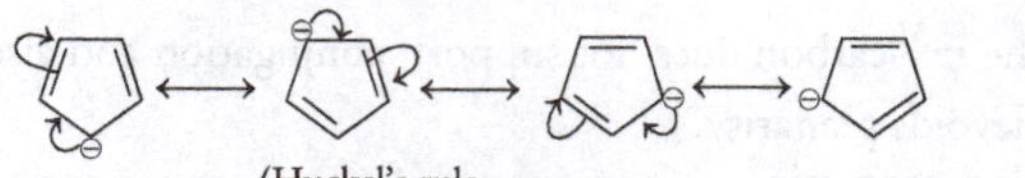

$6\pi e^-$ planar   (Huckel's rule followed, i.e. aromatic)

**23.** *(b)* The reaction take place as follows

Step I 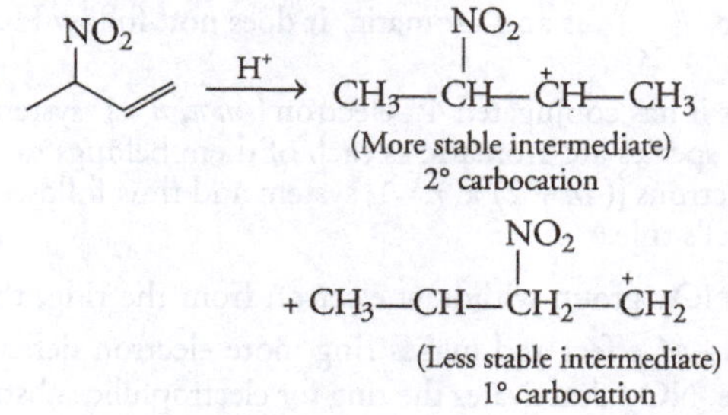


Step II 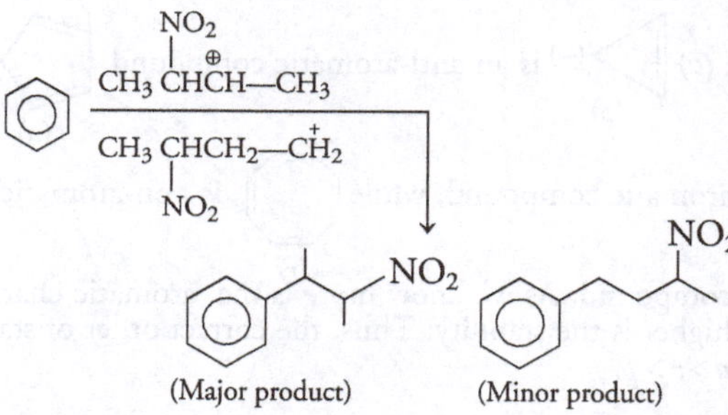


**24.** *(a)* Compound which contain one or more $sp^3$ carbon in its structure will be non-aromatic compound.

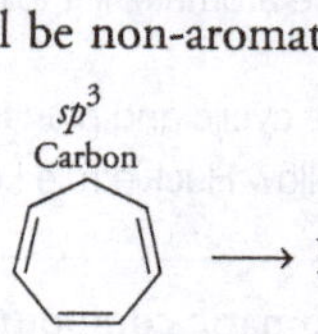


$\longrightarrow$ Non-aromatic

Only (a) is non-aromatic and rest all are aromatic.

**25.** *(d)* 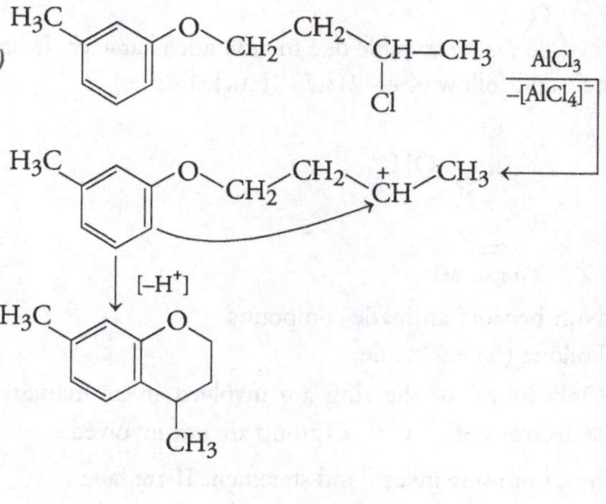

**26.** *(c)* Electron withdrawing groups (EWGs) reduces the electron density in the benzene ring due to its strong $-I$ effect and makes further substitution difficult. Hence, they are called as deactivating groups (like —CN, —Cl etc). While electron releasing groups (ERGs) increases the electron density in the benzene due to resonance. Therefore, activates the benzene ring for the attack by an electrophile. Hence, they are called as activating groups (like —Me, $—OCH_3$, etc.)

So, the correct order is

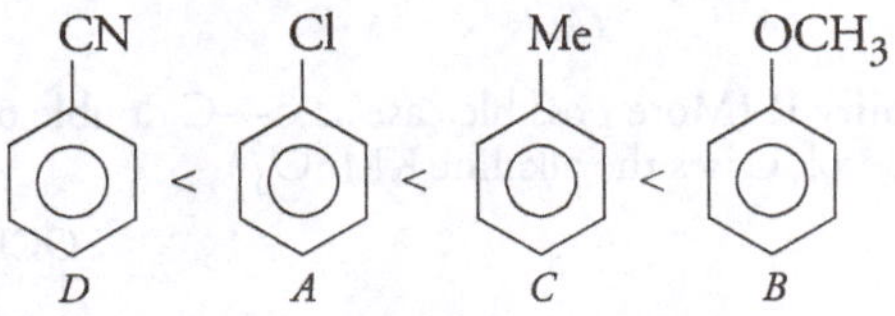


**27.** *(b)* Melting point of a compound depends on size and force of attraction between the molecules. Compound (c) has largest size and also possess weak intermolecular association due to dipole-dipole interactions.

Compound (a) is a dicarboxylic acid and possess high melting point due to intermolecular hydrogen bonding.

As a result, it exist as associated molecule. Compound (d) is an alcohol and possess intermolecular H-bonding. No interactions are present in hydrocarbon (naphthalene) compound (b).

Hence, melting point is lowest for naphthalene ($\approx 80°C$), i.e. compound (b).

**28.** *(a)*

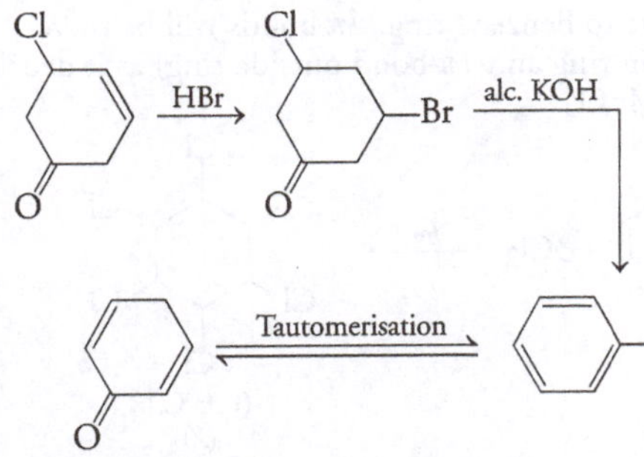


**29.** *(d)* The reaction follows α, β-elimination mechanism to give a more substituted stable alkene as a major product. As the substrate is a α, γ -dibromo (1, 3-) compound it gives a conjugated diene.

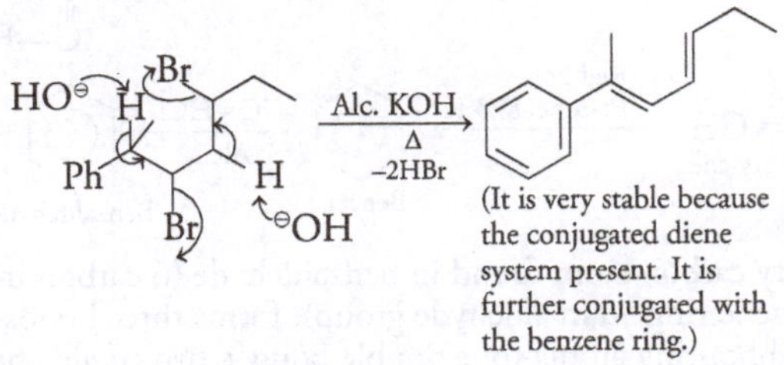


**30.** *(d)*

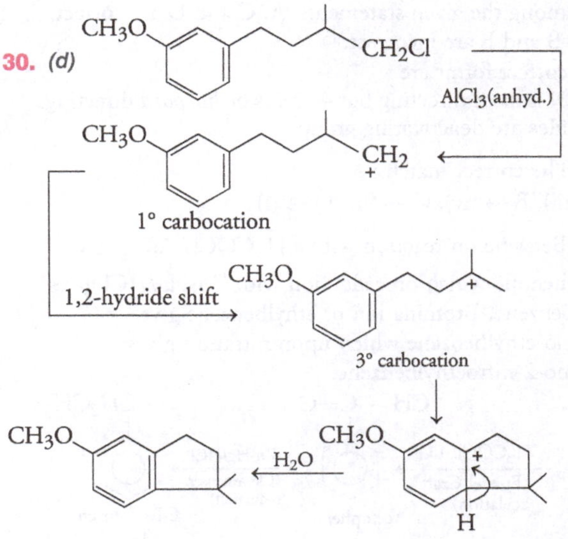


**31.** *(d)*

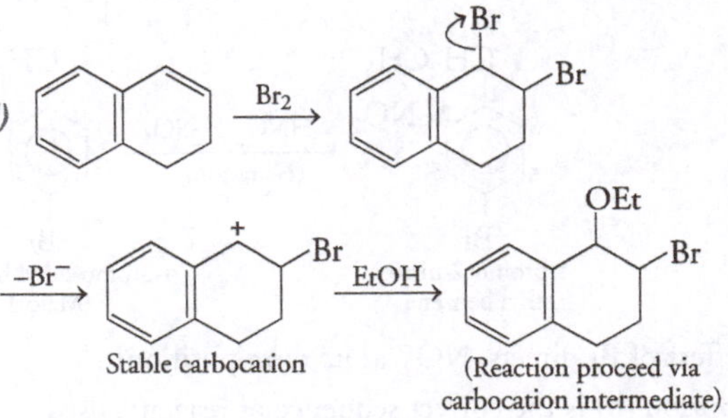


**32.** *(c)* $—NO_2$ being an electron withdrawing group destabilises the carbocation by decreasing electron density to form σ– complex. Thus, the presence of $NO_2$ group on a benzene ring of σ– complex decreases their stability. However, the carbocation (c) does not contain any $—NO_2$ group.

Thus, it would be highly stable than other σ-complexes. Hence, σ-complex (c) is of lowest energy.

**33.** (a) According to Huckel's rule, planar, cyclic fully conjugated molecules is aromatic if it contains $(4n+2)\pi$ electrons.

A, C, D, E and H follows Huckel's rule hence they are aromatic compounds. While B, F, G are not aromatic compounds. As they do not follow Huckel's rule.

**34.** *(d)* Both Statement I and II are true. Reaction for statement I is

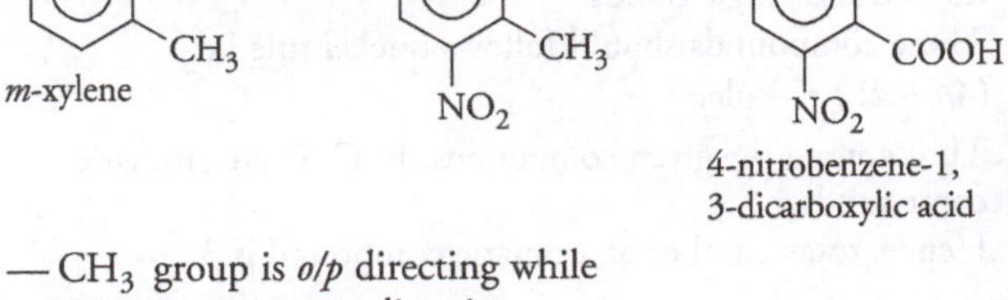


$— CH_3$ group is *o*/*p* directing while
$—NO_2$ group is *meta* directing.

**35.** *(b)* Among the given statements, A, C and D are correct. While B and E are incorrect.
Their correct forms are
—CN is *meta* directing but —OH is *ortho/para* directing.
Halides are deactivating group.

**36.** *(b)* The correct match is
A → (iii), B → (iv), C → (ii), D → (i)

**37.** *(b)* Benzene on reaction with $CH_3COCl/AlCl_3$ gives acetophenone which on reduction with Zn-Hg/HCl gives ethyl benzene. Bromination of ethylbenzene give 4-bromo-ethylbenzene which upon nitration gives 4-bromo-2-nitroethylbenzene.

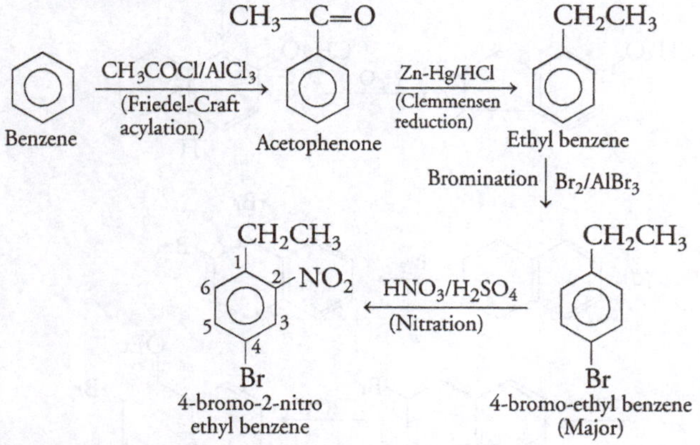


[$-I$-effect of Br directs $NO_2^-$ at its *meta*-position]

So, option (b) is the correct sequence of reagents used.

**38.** *(1)* There is only one aromatic compound among the given compounds.

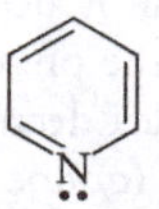

**39.** *(2)* Total number of deactivating groups in aromatic electrophilic substitution reaction is 2.

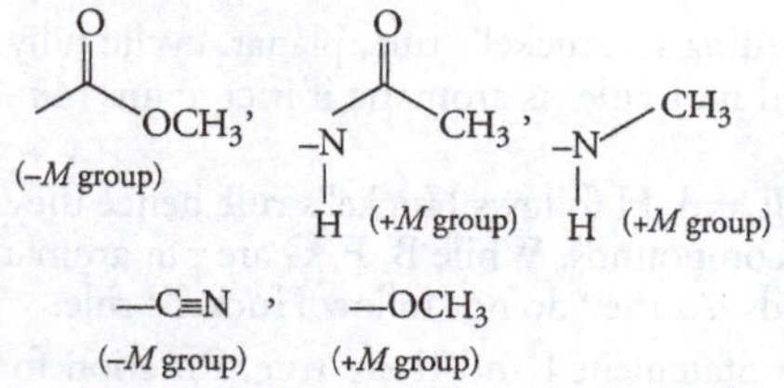


**40.** *(3)* Aromatic compounds are chemical compounds that consist of conjugated planar ring systems accompanied by delocalised π-electron clouds in place of individual alternating double and single bonds.
These compounds should follows huckel rule i.e., $(4n+2)\,\pi e^-$ rule.

Thus among the given compounds. B, C, D are aromatic compounds.
Hence, total number of aromatic compound is 3.

**41.** *(4)* It is given that the given compound ($A$) gives neutral $FeCl_3$ test that means it is a phenol secondly, it decolourises the alkaline $KMnO_4$ that means an unsaturated carbon is present (may be double bond or triple bond). By using these information there are two following possibility 1.

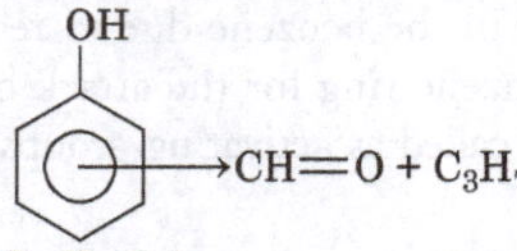


$(C_{10}H_{12}O_2)$
($A$)

**Possibility II** (More possible case as C—C double or triple bond decolourises the alkaline $KMnO_4$)

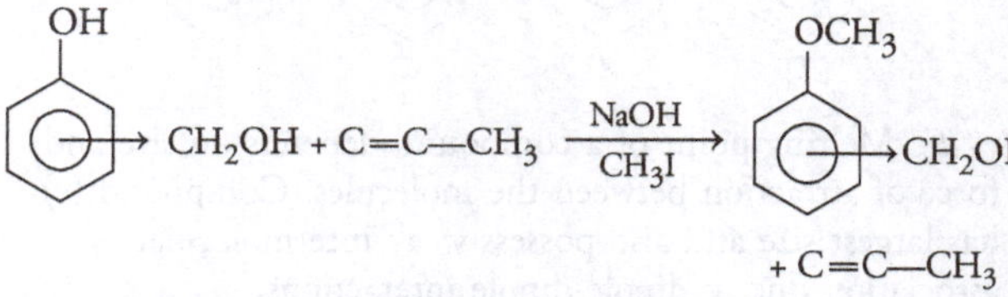


However, from both the possibility the answer will be same i.e., 4π bond.

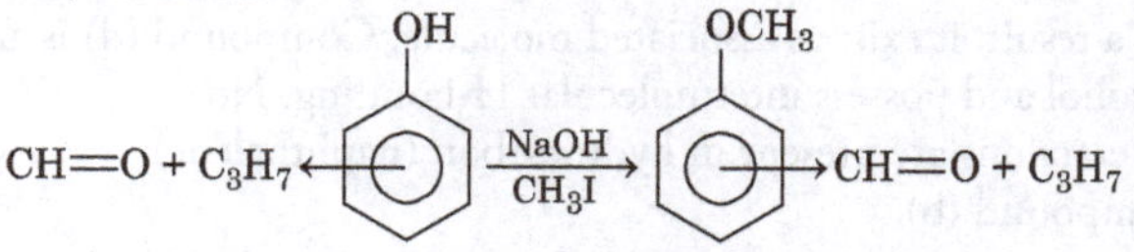


Degree of unsaturation (DU) of $A = \frac{22-12}{2} = 5$

1 DU is due to benzene ring. 4π-bonds will be there (3π-bonds in ring and 1π-bond outside ring) as it decolourises alkaline $KMnO_4$.

**42.** *(6)*

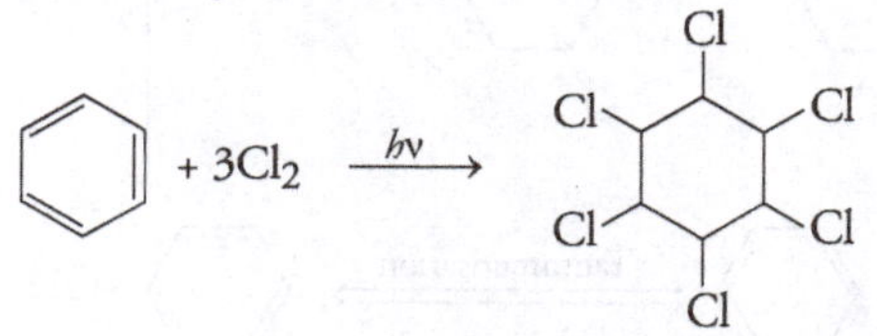


(B.H.C.)
($X$)

Product '$X$' contains 6 hydrogen atom at all the six position of hexane ring.

**43.** *(7)* The reaction take place as follows

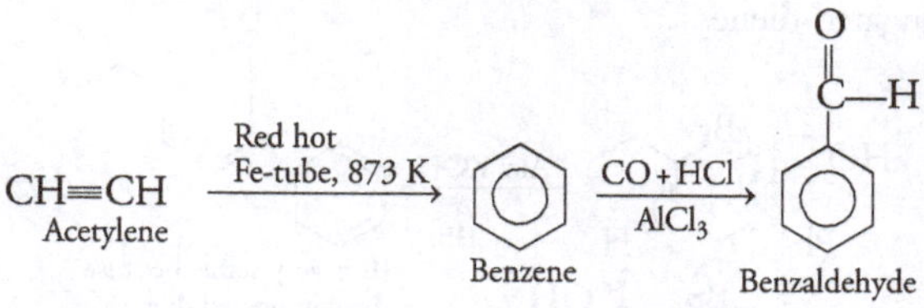


Every carbon atom found in benzaldehyde (6 carbon in benzene ring +1 in aldehyde group), forms three bonds with neighbouring atoms-(one double bond + two single bond). Therefore, in benzaldehyde total number of $sp^2$ 'C' are 7.

# C-13

JEE Main Chapterwise Topicwise Chemistry

In JEE Main there is usually 0-1 question per paper are visible from this chapter. In past 3 years, 32 conceptual and 39 numerical questions were asked of moderate difficulty level from this chapter.

**For detailed analysis scan QR code**

# SOLUTIONS

## Topic 1 Expression of Concentrations of Solutions

### Level 1

**01** The molarity ($M$) of an aqueous solution containing 5.85 g of NaCl in 500 mL water is (Given, Molar mass Na: 23 and Cl: 35.5 $gmol^{-1}$) *[2024, 4 April Shift-I]*

(a) 4 (b) 0.2 (c) 2 (d) 20

**Imp.** Relation of Mole Fraction

**02** If a substance '$A$' dissolves in solution of a mixture of '$B$' and '$C$' with their respective number of moles as $n_A, n_B$ and $n_C$. Mole fraction of $C$ in the solution is *[2024, 30 Jan Shift-I]*

(a) $\frac{n_C}{n_A \times n_B \times n_C}$ (b) $\frac{n_C}{n_A - n_B - n_C}$

(c) $\frac{n_C}{n_A + n_B + n_C}$ (d) $\frac{n_B}{n_A + n_B}$

**03** The quantity which changes with temperature is *[2024, 27 Jan Shift-II]*

(a) molality (b) mole fraction

(c) mass percentage (d) molarity

**04** What would be the molality of 20% (mass/mass) aqueous solution of KI? (Molar mass of KI = 166 g $mol^{-1}$) *[2019, 9 April Shift-II]*

(a) 1.48 (b) 1.51 (c) 1.35 (d) 1.08

**05** A solution of sodium sulphate contains 92 g of $Na^+$ ions per kilogram of water. The molality of $Na^+$ ions in that solution in mol $kg^{-1}$ is *[2019, 9 Jan Shift-I]*

(a) 16 (b) 4 (c) 132 (d) 8

**06** The molality of a urea solution in which 0.0100 g of urea, $[(NH_2)_2CO]$ is added to 0.3000 $dm^3$ of water at STP is *[AIEEE 2011]*

(a) 0.555 m (b) $5.55 \times 10^{-4}$ m

(c) 33.3 m (d) $3.33 \times 10^{-2}$ m

**07** Two solutions of a substance (non-electrolyte) are mixed in the following manner. 480 mL of 1.5 M first solution + 520 mL of 1.2 M second solution. What is the molarity of the final mixture ? *[AIEEE 2005]*

(a) 2.70 M (b) 1.34 M

(c) 1.50 M (d) 1.20 M

### Level 2

**08** '$x$' g of NaCl is added to water in a beaker with a lid. The temperature of the system is raised from 1°C to 25°C. Which out of the following plots, is best suited for the change in the molarity ($M$) of the solution with respect to temperature?

[Consider the solubility of NaCl remains unchanged over the temperature range] *[2025, 2 April Shift-II]*

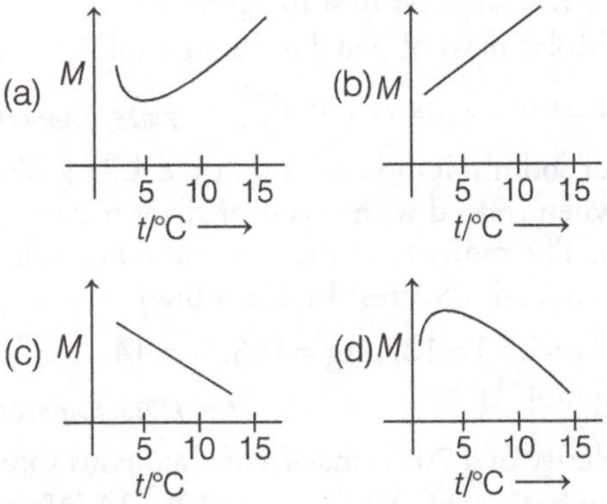


**09** 250 g of solution of $D$-glucose in water contains 10.8% of carbon by weight. The molality of the solution is nearest to (Given : Atomic weights are H : 1u, C : 12 u, O : 16 u) *[2022, 27 July Shift-I]*

(a) 1.03 (b) 2.06 (c) 3.09 (d) 5.40

**10** The mole fraction of a solvent in aqueous solution of a solute is 0.8. The molality (in mol $kg^{-1}$) of the aqueous solution is *[2019, 12 April Shift-I]*

(a) $13.88 \times 10^{-2}$ (b) $13.88 \times 10^{-1}$

(c) 13.88 (d) $13.88 \times 10^{-3}$

**11** A 5.2 molal aqueous solution of methyl alcohol, $CH_3OH$, is supplied. What is the mole fraction of methyl alcohol in the solution ? *[AIEEE 2011]*
(a) 0.100 (b) 0.190 (c) 0.086 (d) 0.050

## Level 3

**12** Concentrated nitric acid is labelled as 75% by mass. The volume in mL of the solution which contains 30g of nitric acid is......... . *[2025, 28 Jan, Shift-II]*
(Given, density of nitric acid solution is 1.25 g/mL.)
(a) 32 (b) 55 (c) 45 (d) 40

**13** A commercially sold conc. HCl is 35% HCl by mass. If the density of this commercial acid is 1.46 g/mL, the molarity of this solution is
(Atomic mass : Cl = 35.5 amu, H = 1 amu) *[2022, 26 June, Shift-I]*
(a) 10.2 M (b) 12.5 M (c) 14.0 M (d) 18.2 M

**14** 8 g of NaOH is dissolved in 18 g of $H_2O$. Mole fraction of NaOH in solution and molality (in mol $kg^{-1}$) of the solution respectively are *[2019, 12 Jan Shift-II]*
(a) 0.2, 11.11 (b) 0.167, 22.20
(c) 0.2, 22.20 (d) 0.167, 11.11

### Numerical Value Type Questions

**15** Sea water, which can be considered as a 6 molar (6 M) solution of NaCl, has a density of 2 g $mL^{-1}$. The concentration of dissolved oxygen ($O_2$) in sea water is 5.8 ppm. Then the concentration of dissolved oxygen ($O_2$) in sea water is $x \times 10^{-4}$ m.
$x$ = ............... . (Nearest integer)
Given: Molar mass of NaCl is 58.5 g $mol^{-1}$
Molar mass of $O_2$ is 32 g $mol^{-1}$ *[2025, 4 April shift-II]*

**16** 20 mL of sodium iodide solution gave 4.74 g silver iodide when treated with excess of silver nitrate solution. The molarity of the sodium iodide solution is ........... M. (Nearest Integer value)
Given Na = 23, I = 127, Ag = 108, N = 14, O = 16 g $mol^{-1}$) *[2025, 8 April Shift-II]*

**17** The molarity of a 70% (mass/mass) aqueous solution of a monobasic acid ($X$) is ..... $\times 10^{-1}$ M (Nearest integer) *[2025, 28 Jan, Shift-I]*
[Given, density of aqueous solution of ($X$) is 1.25 g $mL^{-1}$
Molar mass of the acid is 70 g $mol^{-1}$]

**18** Molarity ($M$) of an aqueous solution containing $x$ g of anhyd. $CuSO_4$ in 500 mL solution at 32°C is $2 \times 10^{-1}$ M. Its molality will be ........... $\times 10^{-3}$ m. (Nearest integer).
[Given density of the solution = 1.25 g/mL.] *[2024, 9 April Shift-I]*

**19** Molality of an aqueous solution of urea is 4.44 m. Mole fraction of urea in solution is $x \times 10^{-3}$. Value of $x$ is .......... (Integer answer) *[2024, 8 April Shift-II]*

**20** Molality of 0.8 M $H_2SO_4$ solution (density 1.06 g $cm^{-3}$) is ......... $\times 10^{-3}$ m. *[2024, 29 Jan Shift-II]*

**21** A solution of $H_2SO_4$ is 31.4% $H_2SO_4$ by mass and has a density of 1.25 g/mL. The molarity of the $H_2SO_4$ solution is .......... M (nearest integer).
[Given, molar mass of $H_2SO_4$ = 98 g $mol^{-1}$] *[2024, 29 Jan Shift-I]*

**22** Some amount of dichloromethane ($CH_2Cl_2$) is added to 671.141 mL of chloroform ($CHCl_3$) to prepare $2.6 \times 10^{-3}$ M solution of $CH_2Cl_2$(DCM). The concentration of DCM is ........ ppm (by mass).
Given, atomic mass C = 12, H = 1, Cl = 35.5,
density of $CHCl_3$ = 1.49 g $cm^{-3}$ *[2023, 30 Jan Shift-I]*

**23** The number of units, which are used to express concentration of solutions from the following is ........
mass per cent, mole, mole-fraction, molarity, ppm, molality *[2023, 24 Jan Shift-II]*

**24** When 800 mL of 0.5 M nitric acid is heated in a beaker, its volume is reduced to half and 11.5 g of nitric acid is evaporated. The molarity of the remaining nitric acid solution is $x \times 10^{-2}$ M. (Nearest integer) *[2022, 26 July Shift-I]*

**25** If 80 g of copper sulphate $CuSO_4 \cdot 5H_2O$. is dissolved in deionised water to make 5 L of solution. The concentration of the copper sulphate solution is $x \times 10^{-3}$ mol $L^{-1}$. The value of $x$ is ......... .
[Atomic masses Cu = 63.54 u,
S = 32 u, O = 16 u, H = 1 u] *[2021, 1 Sep Shift-II]*

**26** An aqueous KCl solution of density 1.20 g $mL^{-1}$ has a molality of 3.30 mol $kg^{-1}$. The molarity of the solution in mol $L^{-1}$ is ......... .
[Molar mass of KCl = 74.5] *[2021, 26 Aug Shift-I]*

**27** The density of NaOH solution is 1.2 g $cm^{-3}$. The molality of this solution is ......... m. (Round off to the nearest integer)
[Use : Atomic masses : Na = 23.0 u, O = 16.0 u, H = 1.0 u. Density of $H_2O$ = 1.0 g $cm^{-3}$] *[2021, 27 July Shift-I]*

**28** The mole fraction of a solute in a 100 molal aqueous solution is ......... $\times 10^{-2}$.
(Round off to the Nearest Integer)
[Given : Atomic masses : H : 1.0 u, O : 16.0 u] *[2021, 17 March Shift-I]*

**29** A 6.50 molal solution of KOH ($aq$) has a density of 1.89 g $cm^{-3}$. The molarity of the solution is ...... mol $dm^{-3}$ (Round off to the nearest integer).
[Atomic masses: K : 39.0 u, O : 16.0u, H : 1.0 u] *[2021, 16 March Shift-I]*

**30** The hardness of a water sample containing $10^{-3}$ M $MgSO_4$ expressed as $CaCO_3$ equivalents (in ppm) is...... (molar mass of $MgSO_4$ is 120.37 g/mol) *[2020, 9 Jan Shift-II]*

**31** 10.30 mg of $O_2$ is dissolved into a litre of sea water of density 1.03 g/mL. The concentration of $O_2$ in ppm is *[2020, 9 Jan Shift-II]*

## Topic 2 Vapour Pressure, Henry's Law and Raoult's Law

### Level 1

**01** A solution is made by mixing one mole of volatile liquid $A$ with 3 moles of volatile liquid $B$. The vapour pressure of pure $A$ is 200 mm Hg and that of the solution is 500 mm Hg. The vapour pressure of pure $B$ and the least volatile component of the solution, respectively, are: *[2025, 2 April Shift-I]*
(a) 600 mm Hg, $A$ (b) 1400 mm Hg, $B$
(c) 1400 mm Hg, $A$ (d) 600 mm Hg, $B$

**02** On mixing, heptane and octane form an ideal solution. At 373 K, the vapour pressures of the two liquid components (heptane and octane) are 105 kPa and 45 kPa, respectively. Vapour pressure of the solution obtained by mixing 25 g of heptane and 35 g of octane will be (molar mass of heptane= 100 g $mol^{-1}$ and of octane= 114 g $mol^{-1}$). *[AIEEE 2010]*
(a) 72.0 kPa (b) 36.1 kPa
(c) 96.2 kPa (d) 144.5 kPa

**03** A mixture of ethyl alcohol and propyl alcohol has a vapour pressure of 290 mm at 300 K. The vapour pressure of propyl alcohol is 200 mm. If the mole fraction of ethyl alcohol is 0.6, its vapour pressure (in mm) at the same temperature will be *[AIEEE 2007]*
(a) 350 (b) 300 (c) 700 (d) 360

### Level 2

**04** When a non-volatile solute is added to the solvent, the vapour pressure of the solvent decreases by 10 mm of Hg. The mole fraction of the solute in the solution is 0.2. What would be the mole fraction of the solvent if decrease in vapour pressure is 20 mm of Hg ? *[2025, 23 Jan Shift-II]*
(a) 0.4 (b) 0.2 (c) 0.8 (d) 0.6

**05** Liquid $A$ and $B$ form an ideal solution. The vapour pressures of pure liquids $A$ and $B$ are 350 and 750 mm Hg respectively at the same temperature. If $x_A$ and $x_B$ are the mole fraction of $A$ and $B$ in solution while $y_A$ and $y_B$ are the mole fraction of $A$ and $B$ in vapour phase then, *[2024, 7 April Shift-II]*
(a) $\frac{x_A}{x_B} = \frac{y_A}{y_B}$ (b) $(x_A - y_A) < (x_B - y_B)$
(c) $\frac{x_A}{x_B} > \frac{y_A}{y_B}$ (d) $\frac{x_A}{x_B} < \frac{y_A}{y_B}$

**Imp.** Graphical Interpretation of Vapour Pressure vs Temperature Graph

**06** A graph of vapour pressure and temperature for three different liquids $X, Y$, and $Z$ is shown below : *[2020, 8 Jan Shift-I]*

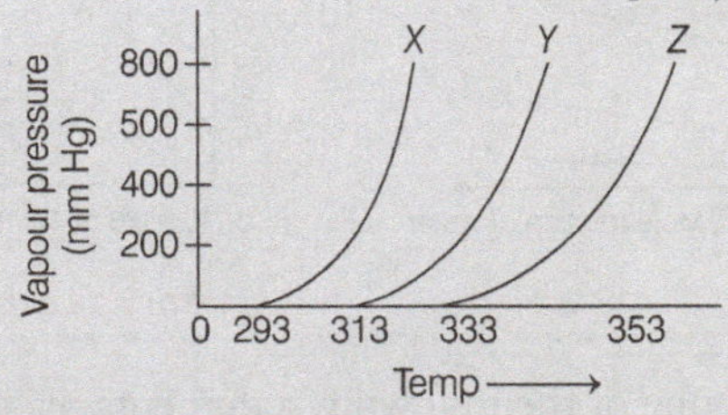


The following inference are made :
A. $X$ has higher intermolecular interactions compared to $Y$.
B. $X$ has lower intermolecular interactions compared to $Y$.
C. $Z$ has lower intermolecular interactions compared to $Y$.
The correct inference(s) is/are :
(a) (C) (b) (B)
(c) (A) (d) (A) and (C)

**07** At 35°C, the vapour pressure of $CS_2$ is 512 mm Hg and that of acetone is 344 mm Hg. A solution of $CS_2$ in acetone has a total vapour pressure of 600 mm Hg. The false statement amongst the following is *[2020, 7 Jan Shift-I]*
(a) Raoult's law is not obeyed by this system
(b) $CS_2$ and acetone are less attracted to each other than to themselves
(c) a mixture of 100 mL $CS_2$ and 100 mL acetone has a volume < 200 mL
(d) heat must be absorbed in order to produce the solution at 35°C

**08** Liquid $M$ and liquid $N$ form an ideal solution. The vapour pressures of pure liquids $M$ and $N$ are 450 and 700 mmHg, respectively, at the same temperature. Then correct statement is *[2019, 9 April Shift-I]*
$x_M$ = mole fraction of $M$ in solution;
$x_N$ = mole fraction of $N$ in solution;
$y_M$ = mole fraction of $M$ in vapour phase;
$y_N$ = mole fraction of $N$ in vapour phase
(a) $\frac{x_M}{x_N} > \frac{y_M}{y_N}$ (b) $\frac{x_M}{x_N} = \frac{y_M}{y_N}$
(c) $\frac{x_M}{x_N} < \frac{y_M}{y_N}$ (d) $(x_M - y_M) < (x_N - y_N)$

**09** For the solution of the gases $w, x, y$ and $z$ in water at 298 K, the Henry's law constants ($K_H$) are 0.5, 2, 35 and 40 K bar, respectively. The correct plot for the given data is *[2019, 8 April Shift-II]*

(a)
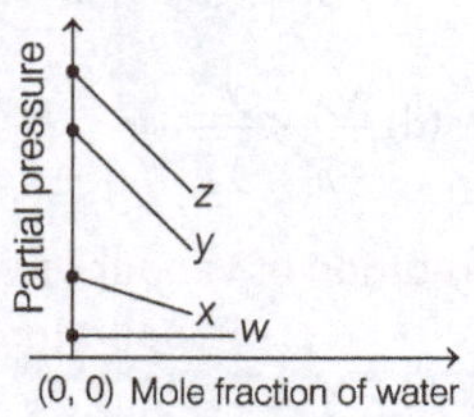


(b)
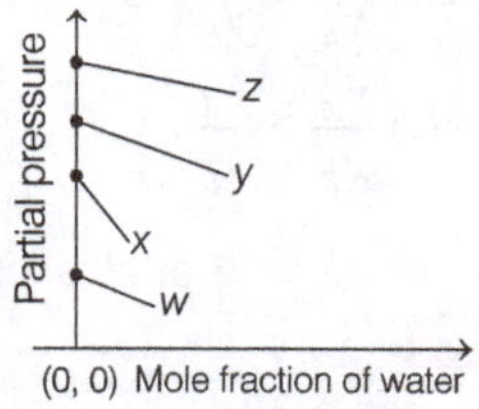


(c)
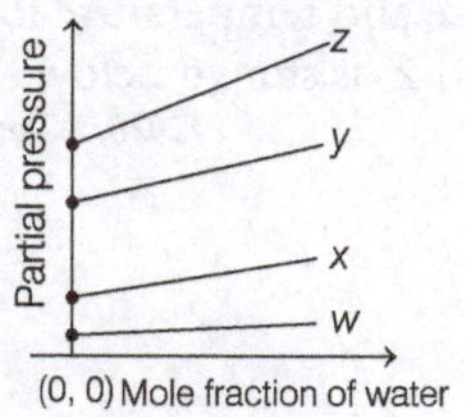


(d)
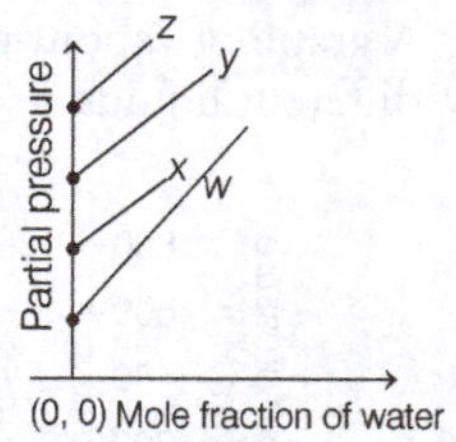


**Imp.** Henry's Law Statement

**10** Which one of the following statements regarding Henry's law is not correct?
(a) Different gases have different $K_H$ (Henry's law constant) values at the same temperature
(b) Higher the value of $K_H$ at a given pressure, higher is the solubility of the gas in the liquids
(c) The value of $K_H$ increases with increase of temperature and $K_H$ is function of the nature of the gas
(d) The partial pressure of the gas in vapour phase is proportional to the mole fraction of the gas in the solution *[2019, 9 Jan Shift-I]*

**11** A binary liquid solution is prepared by mixing *n*-heptane and ethanol. Which one of the following statements is correct regarding the behaviour of the solution? *[AIEEE 2009]*
(a) The solution formed is an ideal solution
(b) The solution is non-ideal, showing positive deviation from Raoult's law
(c) The solution is non-ideal, showing negative deviation from Raoult's law
(d) *n*-heptane shows positive deviation while ethanol show negative deviation from Raoult's law

**12** Benzene and toluene form nearly ideal solutions. At 20°C, the vapour pressure of benzene is 75 torr and that of toluene is 22 torr. The partial vapour pressure of benzene at 20°C for a solution containing 78 g of benzene and 46 g of toluene in torr is *[AIEEE 2005]*
(a) 53.5 (b) 37.5 (c) 25 (d) 50

## Level 3

**13** Which of the following binary mixture does not show the behaviour of minimum boiling azeotropes? *[2025, 8 April Shift-II]*
(a) $CS_2 + CH_3COCH_3$ (b) $CH_3OH + CHCl_3$
(c) $H_2O + CH_3COC_2H_5$ (d) $C_6H_5OH + C_6H_5NH_2$

**14** Match List - I with List - II. *[2025, 7 April Shift-II]*

| List-I | List-I |
|---|---|
| (A) Solution of chloroform and acetone | (I) Minimum boiling azeotrope |
| (B) Solution of ethanol and water | (II) Dimerizes |
| (C) Solution of benzene and toluene | (III) Maximum boiling azeotrope |
| (D) Solution of acetic acid in benzene | (IV) $\Delta V_{mix} = 0$ |

Choose the correct answer from the options given below:
(a) (A)-(III), (B)-(I), (C)-(IV), (D)-(II)
(b) (A)-(II), (B)-(IV), (C)-(I), (D)-(III)
(c) (A)-(II), (B)-(I), (C)-(IV), (D)-(III)
(d) (A)-(III), (B)-(IV), (C)-(I), (D)-(II)

**15** At the sea level, the dry air mass percentage composition is given as nitrogen gas: 70.0, oxygen gas: 27.0 and argon gas: 3.0. If total pressure is 1.15 atm, then calculate the ratio of following respectively:
(i) partial pressure of nitrogen gas to partial pressure of oxygen gas.
(ii) partial pressure of oxygen gas to partial pressure of argon gas.
(Given: Molar mass of N, O and Ar are 14, 16 and 40 g mol$^{-1}$ respectively) *[2025, 7 April Shift-I]*
(a) 5.46, 17.8 (b) 2.96, 11.2
(c) 2.59, 11.85 (d) 4.26. 19.3

**16** Which of the following graph correctly represents the plots of $K_H$ at 1 bar for gases in water *versus* temperature? *[2025, 2 April Shift-I]*

(a)
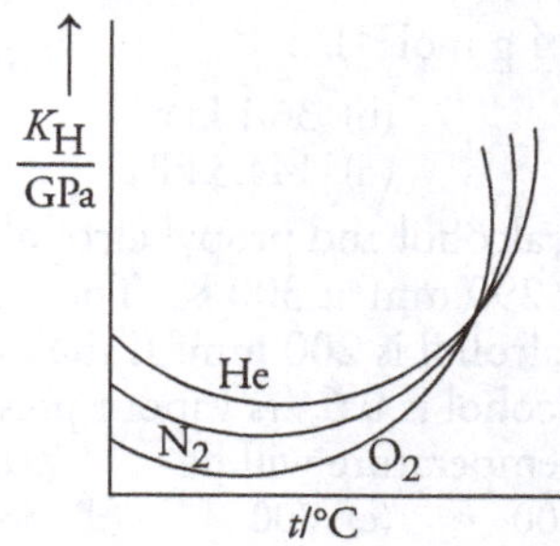


(b)
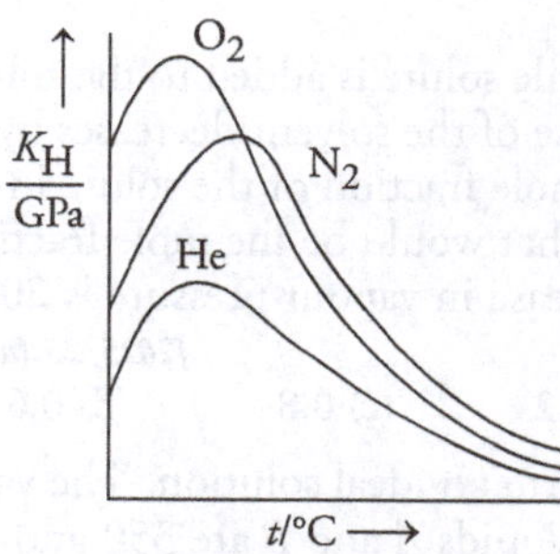

(c) 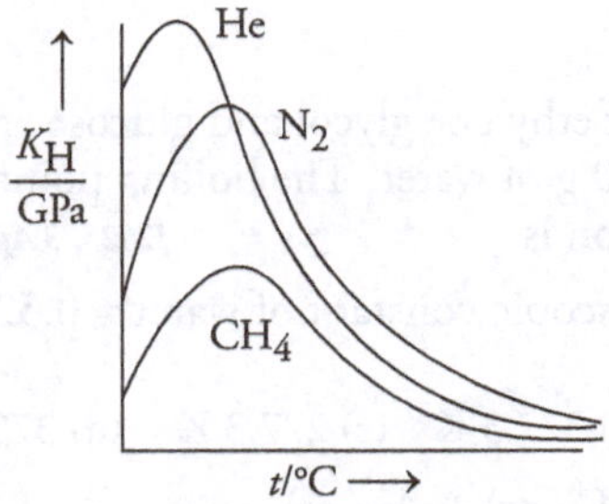


(d) 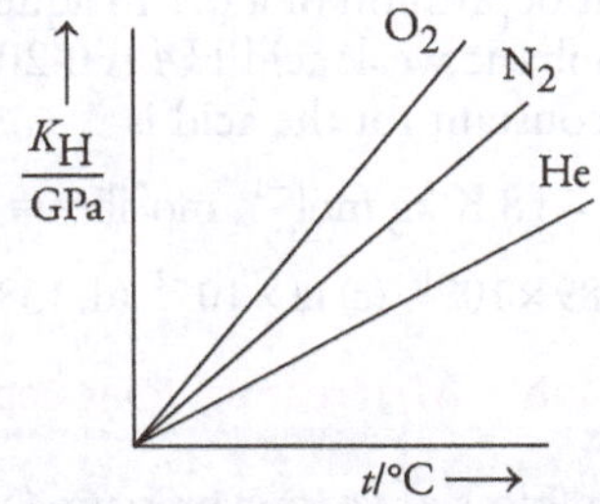


**17** Consider a binary solution of two volatile liquid components 1 and 2. $x_1$ and $y_1$ are the mole fractions of component 1 in liquid and vapour phase, respectively. The slope and intercept of the linear plot of $1/x_1$ *vs* $1/y_1$ are given respectively as *[2025, 23 Jan Shift-II]*

(a) $\frac{p_2^\circ}{p_1^\circ}, \frac{p_2^\circ - p_1^\circ}{p_2^\circ}$ (b) $\frac{p_1^\circ}{p_2^\circ}, \frac{p_1^\circ - p_2^\circ}{p_2^\circ}$

(c) $\frac{p_2^\circ}{p_1^\circ}, \frac{p_1^\circ - p_2^\circ}{p_2^\circ}$ (d) $\frac{p_1^\circ}{p_2^\circ}, \frac{p_2^\circ - p_1^\circ}{p_2^\circ}$

**18** Identify the mixture that shows positive deviations from Raoult's law. *[2024, 31 Jan Shift-I]*

(a) $(CH_3)_2CO + CS_2$
(b) $CHCl_3 + C_6H_6$
(c) $(CH_3)_2CO + C_6H_5NH_2$
(d) $CHCl_3 + (CH_3)_2CO$

**19** A solution of two miscible liquids showing negative deviation from Raoult's law will have *[2024, 27 Jan Shift-I]*

(a) increased vapour pressure, increased boiling point
(b) decreased vapour pressure, increased boiling point
(c) increased vapour pressure, decreased boiling point
(d) decreased vapour pressure, decreased boiling point

**20** The vapour pressures of pure liquids *A* and *B* are 400 and 600 mmHg, respectively at 298 K. On mixing the two liquids, the sum of their initial volumes is equal to the volume of the final mixture. The mole fraction of liquid *B* is 0.5 in the mixture. The vapour pressure of the final solution, the mole fractions of components *A* and *B* in vapour phase, respectively are *[2019, 8 April Shift-I]*

(a) 450 mmHg, 0.4, 0.6 (b) 500 mmHg, 0.5, 0.5
(c) 450 mmHg, 0.5,0.5 (d) 500 mmHg, 0.4,0.6

**21** Liquids *A* and *B* form an ideal solution in the entire composition range. At 350 K, the vapour pressures of pure *A* and pure *B* are $7 \times 10^3$ Pa and $12 \times 10^3$ Pa, respectively. The composition of the vapour in equilibrium with a solution containing 40 mole percent of *A* at this temperature is *[2019, 10 Jan Shift-I]*

(a) $x_A = 0.76; x_B = 0.24$ (b) $x_A = 0.28; x_B = 0.72$
(c) $x_A = 0.4; x_B = 0.6$ (d) $x_A = 0.37; x_B = 0.63$

## Numerical Value Type Questions

**22** The vapour pressure of pure benzene and methyl benzene at 27°C is given as 80 torr and 24 torr, respectively. The mole fraction of methyl benzene in vapour phase, in equilibrium with an equimolar mixture of those two liquids (ideal solution) at the same temperature is ........... $\times 10^{-2}$. (Nearest integer) *[2024, 9 April Shift-II]*

**23** The vapour pressure *vs*, temperature curve for a solution solvent system is shown below.

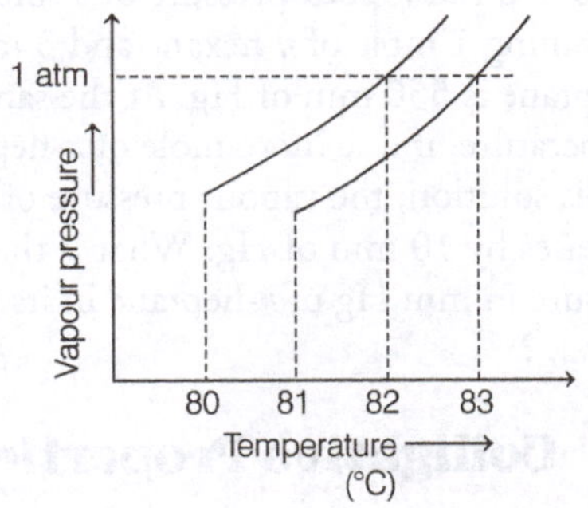


The boiling point of the solvent is .......... °C. *[2023, 8 April Shift-I]*

**24** The total pressure observed by mixing two liquids *A* and *B* is 350 mm Hg when their mole fractions are 0.7 and 0.3 respectively.

The total pressure becomes 410 mm Hg if the mole fractions are changed to 0.2 and 0.8 respectively for *A* and *B*. The vapour pressure of pure *A* is ........ mm Hg. (Nearest integer)

Consider the liquids and solutions behave ideally. *[2023, 24 Jan Shift-II]*

**25** '*x*' g of molecular oxygen ($O_2$) is mixed with 200 g of neon (Ne). The total pressure of the non-reactive mixture of $O_2$ and Ne in the cylinder is 25 bar. The partial pressure of Ne is 20 bar at the same temperature and volume. The value of '*x*' is ......... .

[Given : Molar mass of $O_2 = 32$ g mol$^{-1}$.

Molar mass of Ne $= 20$ g mol$^{-1}$] *[2022, 29 July Shift-II]*

**26** If $O_2$ gas is bubbled through water at 303 K, the number of millimoles of $O_2$ gas that dissolve in 1 litre of water is ....... (Nearest integer).

(Given, Henry's law constant for $O_2$ at 303 K is 46.82 k bar and partial pressure of $O_2 = 0.920$ bar)
(Assume, solubility of $O_2$ in water is too small, nearly negligible.) *[2022, 29 July Shift-I]*

**27** $CO_2$ gas is bubbled through water during a soft drink manufacturing process at 298 K. If $CO_2$ exerts a partial pressure of 0.835 bar then $x$ m mol of $CO_2$ would dissolve in 0.9 L of water. The value of $x$ is ......... . (Nearest integer)
(Henry's law constant for $CO_2$ at 298 K is $1.67 \times 10^3$ bar) *[2021, 25 July Shift-I]*

**28** At 363 K the vapour pressure of $A$ is 21 kPa and that of $B$ is 18 kPa. One mole of $A$ and 2 moles of $B$ are mixed. Assuming that this solution is ideal, the vapour pressure of the mixture is ...... kPa.
[Round off to the Nearest Integer]. *[2021, 16 March Shift-II]*

**29** At 300 K, the vapour pressure of a solution containing 1 mole of $n$-hexane and 3 moles of $n$-heptane is 550 mm of Hg. At the same temperature, if one more mole of $n$-heptane is added to this solution, the vapour pressure of the solution increases by 10 mm of Hg. What is the vapour pressure in mm Hg of $n$-heptane in its pure state ........... ? *[2020, 4 Sep Shift-I]*

## Topic 3 Colligative Properties

### Level 1

**01** What is the freezing point depression constant of a solvent, 50 g of which contain 1 g non-volatile solute (molar mass 256 g $mol^{-1}$) and the decrease in freezing point is 0.40 K? *[2025, 28 Jan Shift-I]*
(a) 5.12 K kg $mol^{-1}$ (b) 3.72 K kg $mol^{-1}$
(c) 4.43 K kg $mol^{-1}$ (d) 1.86 K kg $mol^{-1}$

**02** 1 g of a non-volatile, non-electrolyte solute is dissolved in 100 g of two different solvents $A$ and $B$, whose ebullioscopic constants are in the ratio of 1 : 5. The ratio of the elevation in their boiling points, $\frac{\Delta T_b(A)}{\Delta T_b(B)}$, is *[2019, 10 April Shift-II]*
(a) 5 : 1 (b) 10 : 1
(c) 1 : 5 (d) 1 : 0.2

**03** $K_f$ for water is 1.86 K kg $mol^{-1}$. If your automobile radiator holds 1.0 kg of water, then how many grams of ethylene glycol ($C_2H_6O_2$) must you add to get the freezing point of the solution lowered to – 2.8°C ? *[AIEEE 2012]*
(a) 72 g (b) 93 g (c) 39 g (d) 27 g

### Level 2

**04** 2 moles each of ethylene glycol and glucose are dissolved in 500 g of water. The boiling point of the resulting solution is *[2025, 3 April Shift-I]*
(Given, ebullioscopic constant of water = 0.52 K kg $mol^{-1}$)
(a) 375.3 K (b) 377.3 K (c) 277.3 K (d) 379.2 K

**05** $HA(aq) \rightleftharpoons H^+(aq) + A^-(aq)$ *[2025, 8 April Shift-II]*
The freezing point depression of a 0.1 in aqueous solution of a monobasic weak acid $HA$ is 0.20 °C. The dissociation constant for the acid is
Given, $K_f(H_2O) = 1.8$ K kg $mol^{-1}$, molality ≡ molarity
(a) $1.90 \times 10^{-3}$ (b) $1.89 \times 10^{-1}$ (c) $1.1 \times 10^{-2}$ (d) $1.38 \times 10^{-3}$

**Imp.** Graph of $\Delta T_f$ (Freezing Point Depression)

**06** Consider the given plots of vapour pressure (VP) *vs* temperature ($T$ / K). Which amongst the following options is correct graphical representation showing $\Delta T_f$ depression in the freezing point of a solvent in a solution? *[2025, 24 Jan Shift-I]*

(a)
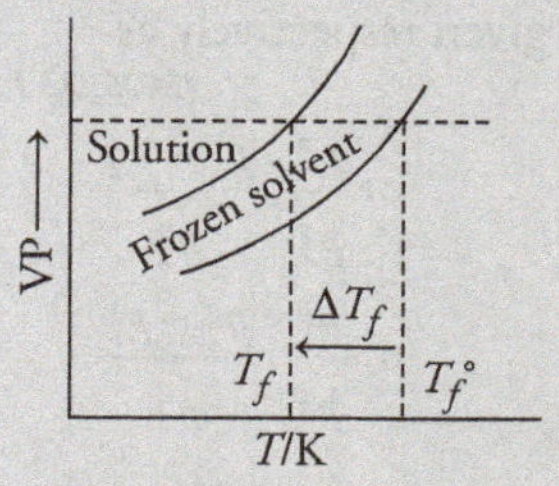


(b)
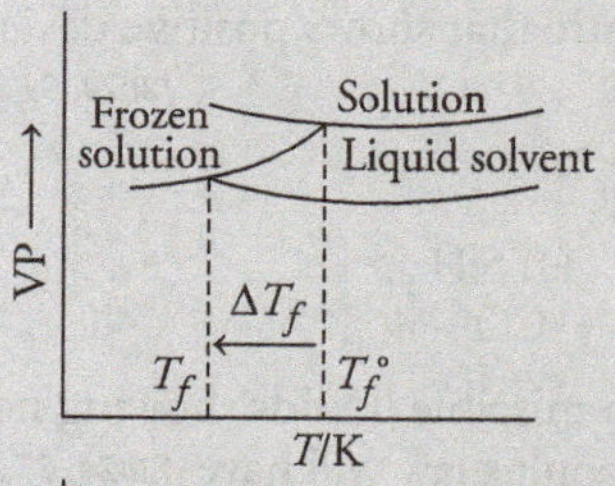


(c)
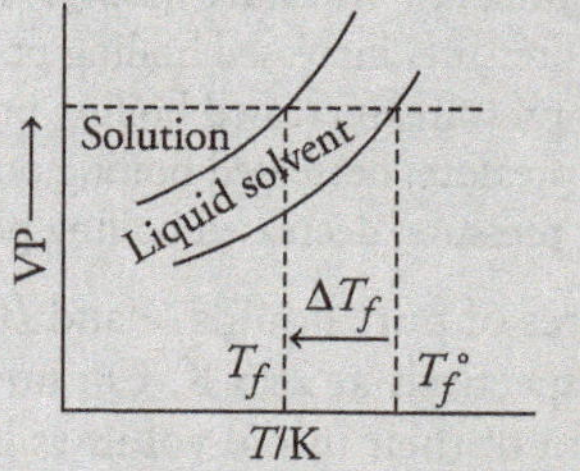


(d)
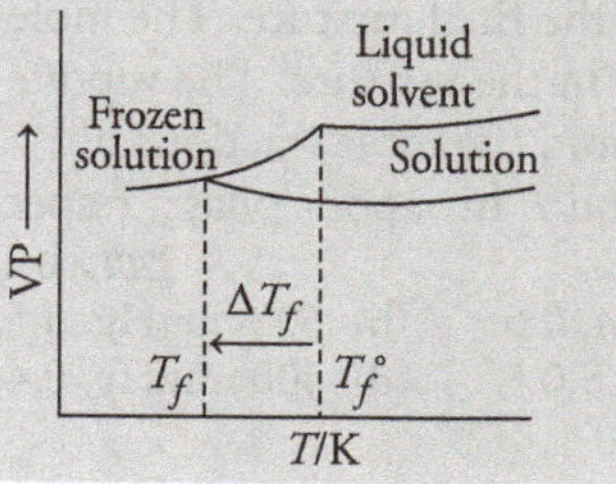

**07** Given below are two statements. *[2025, 29 Jan Shift-II]*

**Statement I** NaCl is added to the ice at 0°C, present in the ice cream box to prevent the melting of ice cream.

**Statement II** On addition of NaCl to ice at 0°C, there is a depression in freezing point.

In the light of the above statements, choose the correct answer from the options given below:

(a) Both Statement I and Statement II are false.
(b) Statement I is false but Statement II is true.
(c) Statement I is true but Statement II is false.
(d) Both Statement I and Statement II are true.

**08** What happens to freezing point of benzene when small quantity of naphthalene is added to benzene?

(a) First decreases and then increases *[2024, 30 Jan Shift-I]*
(b) Remains unchanged
(c) Increases
(d) Decreases

**09** What weight of glucose must be dissolved in 100 g of water to lower the vapour pressure by 0.20 mm Hg ? (Assume, dilute solution is being formed)

Given, vapour pressure of pure water is 54.2 mm Hg at room temperature. Molar mass of glucose is 180 g $mol^{-1}$ *[2023, 11 April Shift-II]*

(a) 3.69 g (b) 3.59 g
(c) 2.59 g (d) 4.69 g

**10** In the depression of freezing point experiment.

A. Vapour pressure of the solution is less than that of pure solvent
B. Vapour pressure of the solution is more than that of pure solvent
C. Only solute molecules solidify at the freezing point
D. Only solvent molecules solidify at the freezing point

Choose the most appropriate answer from the options given below. *[2023, 24 Jan Shift-I]*

(a) B and C only (b) A and D only
(c) A and C only (d) A only

**11** Two solutions *A* and *B* are prepared by dissolving 1 g of non-volatile solutes *X* and *Y*, respectively in 1 kg of water. The ratio of depression in freezing points for *A* and *B* is found to be 1 : 4. The ratio of molar masses of *X* and *Y* is *[2022, 25 July Shift-II]*

(a) 1 : 4 (b) 1 : 0.25
(c) 1 : 0.20 (d) 1 : 5

**12** A solution is prepared by dissolving 0.6 g of urea (molar mass = 60 g $mol^{-1}$) and 1.8 g of glucose (molar mass = 180 g $mol^{-1}$) in 100 mL of water at 27°C. The osmotic pressure of the solution is ($R = 0.08206$ L atm $K^{-1}$ $mol^{-1}$) *[2019, 12 April Shift-II]*

(a) 8.2 atm (b) 2.46 atm
(c) 4.92 atm (d) 1.64 atm

**13** At room temperature, a dilute solution of urea is prepared by dissolving 0.60 g of urea in 360 g of water. If the vapour pressure of pure water at this temperature is 35 mm Hg, lowering of vapour pressure will be (Molar mass of urea = 60 g $mol^{-1}$) *[2019, 10 April Shift-I]*

(a) 0.027 mmHg (b) 0.031 mmHg
(c) 0.017 mmHg (d) 0.028 mmHg

**14** Elevation in the boiling point for 1 molal solution of glucose is 2 K. The depression in the freezing point for 2 molal solution of glucose in the same solvent is 2 K. The relation between $K_b$ and $K_f$ is *[2019, 10 Jan Shift-II]*

(a) $K_b = 1.5\, K_f$ (b) $K_b = 0.5\, K_f$
(c) $K_b = K_f$ (d) $K_b = 2K_f$

**15** A 5.25% solution of a substance is isotonic with a 1.5% solution of urea (molar mass = 60 g $mol^{-1}$) in the same solvent. If the densities of both the solutions are assumed to be equal to 1.0 g $cm^{-3}$, molar mass of the substance will be *[AIEEE 2007]*

(a) 90.0 g $mol^{-1}$ (b) 115.0 g $mol^{-1}$
(c) 105.0 g $mol^{-1}$ (d) 210.0 g $mol^{-1}$

**16** Equimolar solutions in the same solvent have *[AIEEE 2005]*

(a) different boiling point and different freezing point
(b) same boiling point and same freezing point
(c) same freezing point but different boiling point
(d) same boiling point but different freezing point

## Level 3

**17** Given below are two statements: *[2025, 4 April Shift-II]*

**Statement I** Molal depression constant $K_f$ is given by $\frac{M_1 RT_f}{\Delta S_{fus}}$, where symbols have their usual meaning.

**Statement II** $K_f$ for benzene is less than the $K_f$ for water.

In the light of the above statements, choose the most appropriate answer from the options given below:

(a) Both Statement I and Statement II are correct
(b) Statement I is incorrect but Statement II is correct
(c) Statement I is correct but Statement II is incorrect
(d) Both Statement I and Statement II are incorrect

**18** $CrCl_3 \cdot xNH_3$ can exist as a complex. 0.1 molal aqueous solution of this complex shows a depression in freezing point of 0.558°C. Assuming 100% ionisation of this complex and coordination number of Cr is 6, the complex will be (Given, $K_f = 1.86$ K kg $mol^{-1}$) *[2025, 23 Jan Shift-I]*

(a) $[Cr(NH_3)_3Cl_3]$ (b) $[Cr(NH_3)_4Cl_2]Cl$
(c) $[Cr(NH_3)_6]Cl_3$ (d) $[Cr(NH_3)_5Cl]Cl_2$

**19** 0.05 M $CuSO_4$ when treated with 0.01 M $K_2Cr_2O_7$ gives colour solution of $Cu_2Cr_2O_7$. The two solutions are separated as shown below.
[SPM : Semi-permeable membrane] ***[2024, 9 April Shift-I]***

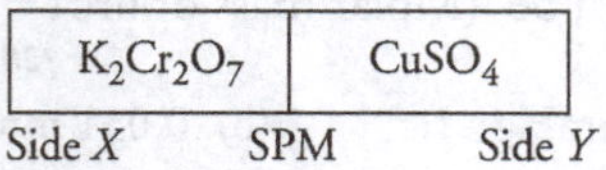


Due to osmosis
(a) molarity of $K_2Cr_2O_7$ solution is lowered.
(b) green colour formation observed on side $X$.
(c) molarity of $CuSO_4$ solution is lowered.
(d) green colour formation observed on side $Y$.

**20** Boiling point of a 2% aqueous solution of a non-volatile solute $A$ is equal to the boiling point of 8% aqueous solution of a non-volatile solute $B$. The relation between molecular weights of $A$ and $B$ is ***[2022, 27 July Shift-I]***
(a) $M_A = 4M_B$ (b) $M_B = 4M_A$
(c) $M_A = 8M_B$ (d) $M_B = 8M_A$

**21** 18 g of glucose ($C_6H_{12}O_6$) is added to 178.2 g water. The vapour pressure of water (in torr) for this aqueous solution is ***[JEE Main 2016; AIEEE 2006]***
(a) 76.0 (b) 752.4 (c) 759.0 (d) 7.6

**22** The vapour pressure of water at 20°C is 17.5 mmHg. If 18 g of glucose ($C_6H_{12}O_6$) is added to 178.2 g of water at 20°C, the vapour pressure of the resulting solution will be ***[AIEEE 2008]***
(a) 17.675 mmHg (b) 15.750 mmHg
(c) 16.500 mmHg (d) 17.325 mmHg

## Numerical Value Type Questions

**23** When $x \times 10^{-2}$ mL methanol (molar mass $= 32$ g mol$^{-1}$ : density $= 0.792$ g/cm$^2$) is added to 100 mL water (density $= 1$ g/cm$^3$), the following diagram is obtained.

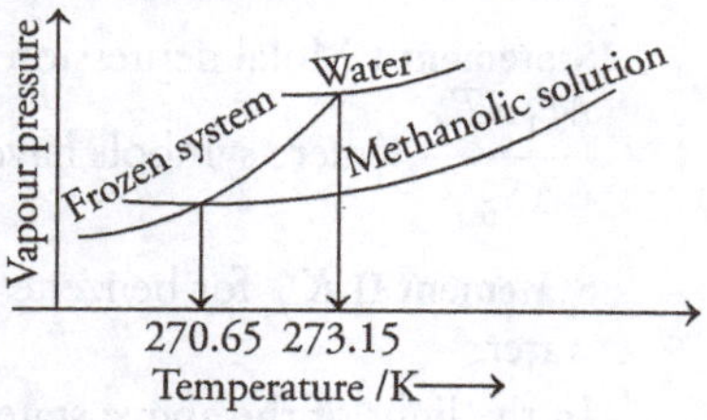


***[2024, 6 April Shift-II]***

$x = $......... (nearest integer)
[Given : Molal freezing point depression constant of water at 273.15 K is 1.86 K kg mol$^{-1}$]

**24** An artificial cell is made by encapsulating 0.2 M glucose solution within a semipermeable membrane. The osmotic pressure developed when the artificial cell is placed within a 0.05 M solution of NaCl at 300 K is ....... $\times 10^{-1}$ bar. (Nearest integer).
[Given : $R = 0.083$ L bar mol$^{-1}$ K$^{-1}$]
Assume complete dissociation of NaCl. ***[2024, 5 April Shift-I]***

**25** 2.7 kg of each of water and acetic acid are mixed. The freezing point of the solution will be $-x$°C. Consider the acetic acid does not dimerise in water, nor dissociates in water, $x = $........ (Nearest integer)
[given: Molar mass of water $= 18$ g mol$^{-1}$, acetic acid $= 60$ g mol$^{-1}$
$K_{f\,H_2O}$ : 1.86 K kg mol$^{-1}$
$K_{f\,\text{acetic acid}}$ : 3.90 K kg mol$^{-1}$
Freezing point: $H_2O = 273$ K, acetic acid $= 290$ K] ***[2024, 4 April Shift-II]***

**26** 2.5 g of a non-volatile, non-electrolyte is dissolved in 100g of water at 25°C. The solution showed a boiling point elevation by 2°C. Assuming the solute concentration is negligible with respect to the solvent concentration, the vapor pressure of the resulting aqueous solution is...... mm of Hg. (Nearest integer)
[Given: Molal boiling point elevation constant of water ($K_b$) $= 0.52$ K kg mol$^{-1}$, 1 atm pressure $= 760$ mm of Hg, molar mass of water $= 18$g mol$^{-1}$] ***[2024, 4 April Shift-I]***

**27** The osmotic pressure of a dilute solution is $7 \times 10^5$ Pa at 273 K. Osmotic pressure of the same solution at 283 K is ......... $\times 10^4$ Nm$^{-2}$. ***[2024, 29 Jan Shift-I]***

**28** The vapour pressure of 30% ($w/v$) aqueous solution of glucose is ........... mm Hg at 25°C.
[Given, the density of 30% ($w/v$), aqueous solution of glucose is 1.2 g cm$^{-3}$ and vapour pressure of pure water is 24 mm Hg.] (Molar mass of glucose is 180 g mol$^{-1}$) ***[2023, 15 April Shift-I]***

**29** If the boiling points of two solvents $X$ and $Y$ (having same molecular weights) are in the ratio 2:1 and their enthalpy of vaporisations are in the ratio 1:2, then the boiling point elevation constant of $X$ is $m$ times the boiling point elevation constant of $Y$. The value of $m$ is ......... (nearest integer). ***[2023, 8 April Shift-II]***

**30** Mass of urea ($NH_2CONH_2$) required to be dissolved in 1000 g of water in order to reduce the vapour pressure of water by 25% is ......... g. (Nearest integer)
Given, molar mass of N, C, O and H are 14, 12, 16 and 1g mol$^{-1}$ respectively. ***[2023, 6 April Shift-I]***

**31** At 27°C, a solution containing 2.5 g of solute in 250.0 mL of solution exerts an osmotic pressure of 400 Pa. The molar mass of the solute is ....... g mol$^{-1}$. (Nearest integer) ***[2023, 31 Jan Shift-I]***
(Given : $R = 0.083$ L bar K$^{-1}$ mol$^{-1}$)

**32** A solution containing 2 g of a non-volatile solute in 20 g of water boils at 373.52 K. The molecular mass of the solute is .......... g $mol^{-1}$. (Nearest integer) [Given : water boils at 373 K, $K_b$ for water $= 0.52$ K kg $mol^{-1}$] ***[2023, 30 Jan Shift-I]***

**33** The osmotic pressure of solutions of PVC in cyclohexanone at 300K are plotted on the graph. The molar mass of PVC is ........ g $mol^{-1}$. (Nearest integer) ***[2023, 25 Jan Shift-I]***

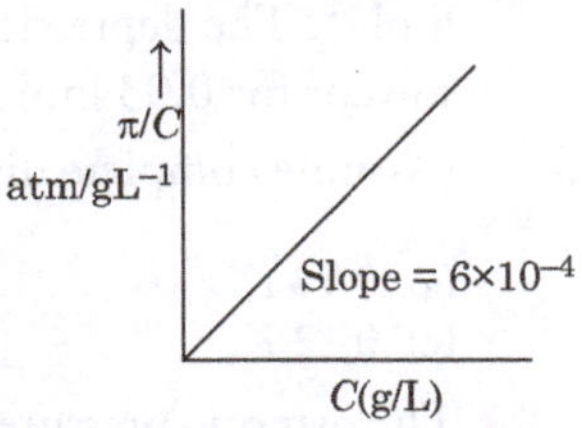


(Given : $R = 0.083$ L atm $K^{-1}$ $mol^{-1}$).

**34** 150 g of acetic acid was contaminated with 10.2 g ascorbic acid ($C_6H_8O_6$) to lower down its freezing point by $(x \times 10^{-1})$°C. The value of $x$ is ......... (Nearest integer) [Given $K_f = 3.9$ K kg $mol^{-1}$; molar mass of ascorbic acid = 176 g $mol^{-1}$] ***[2022, 28 July Shift-I]***

**35** When a certain amount of solid $A$ is dissolved in 100 g of water at 25°C to make a dilute solution, the vapour pressure of the solution is reduced to one-half of that of pure water. The vapour pressure of pure water is 23.76 mmHg. The number of moles of solute $A$ added is ............... (nearest integer) ***[2022, 27 July Shift-II]***

**36** The elevation in boiling point for 1 molal solution of non-volatile solute $A$ is 3 K. The depression in freezing point for 2 molal solution of $A$ in the same solvent is 6 K. The ratio of $K_b$ and $K_f$ i.e., $K_b / K_f$ is 1 : $x$. The value of $x$ is......... [nearest integer] ***[2022, 26 July Shift-II]***

**37** A solution containing $2.5 \times 10^{-3}$ kg of a solute dissolved in $75 \times 10^{-3}$ kg of water boils at 373.535 K. The molar mass of the solute is ........ g $mol^{-1}$. [Nearest integer] (Given, $K_b(H_2O) = 0.52$ K kg $mol^{-1}$ and boiling point of water = 373.15 K). ***[2022, 27 June Shift-II]***

**38** 2 g of a non-volatile non-electrolyte solute is dissolved in 200 g of two different solvents $A$ and $B$ whose ebullioscopic constants are in the ratio of 1:8. The elevation in boiling points of $A$ and $B$ are in the ratio $\frac{x}{y}$ $(x : y)$. The value of $y$ is ......... (Nearest integer) ***[2022, 27 June Shift-I]***

**39** The osmotic pressure exerted by a solution prepared by dissolving 2.0 g of protein of molar mass 60 kg $mol^{-1}$ in 200 mL of water at 27°C is ...... Pa. [Integer value] (Use, $R = 0.083$ L bar $mol^{-1}$ $K^{-1}$) ***[2022, 26 June Shift-II]***

**40** The osmotic pressure of blood is 7.47 bar at 300 K. To inject glucose to a patient intravenously, it has to be isotonic with blood. The concentration of glucose solution in $gL^{-1}$ is ...... . (Molar mass of glucose = 180 g $mol^{-1}$ $R = 0.083$ L bar $K^{-1}$ $mol^{-1}$) (Nearest integer) ***[2022, 24 June, Shift-I]***

**41** 40 g of glucose (Molar mass = 180) is mixed with 200 mL of water. The freezing point of solution is ........... K. (Nearest integer) [Given, $K_f = 1.86$ K kg $mol^{-1}$, density of water = 1.00 g $cm^{-3}$, freezing point of water = 273.15 K] ***[2021, 27 Aug Shift-II]***

**42** 1 kg of 0.75 molal aqueous solution of sucrose can be cooled up to −4°C before freezing. The amount of ice (in g) that will be separated out is ......... . (Nearest integer) [Given, $K_f(H_2O) = 1.86$ K kg $mol^{-1}$] ***[2021, 27 Aug Shift-I]***

**43** 1.46 g of a biopolymer dissolved in a 100 mL water at 300 K exerted an osmotic pressure of $2.42 \times 10^{-3}$ bar. The molar mass of the biopolymer is ... $\times 10^4$ g $mol^{-1}$. (Round off to the nearest integer) [Use : $R = 0.083$ L bar $mol^{-1}$ $K^{-1}$] ***[2021, 27 July Shift-I]***

**44** When 3.00 g of a substance '$X$' is dissolved in 100 g of $CCl_4$, it raises the boiling point by 0.60 K. The molar mass of the substance '$X$' is ........ g $mol^{-1}$. (Nearest integer) [Given, $K_b$ for $CCl_4$ is 5.0 K kg $mol^{-1}$] ***[2021, 25 July Shift-II]***

**45** A 1 molal $K_4Fe(CN)_6$ solution has a degree of dissociation of 0.4. Its boiling point is equal to that of another solution which contains 18.1 weight per cent of a non- electrolytic solute $A$. The molar mass of $A$ is ....... u. (Round off to the nearest integer). [Density of water = 1.0 g $cm^{-3}$] ***[2021, 17 March Shift-II]***

**46** A cylinder containing an ideal gas (0.1 mol of 1.0 $dm^3$) is in thermal equilibrium with a large volume of 0.5 molal aqueous solution of ethylene glycol at its freezing point. If the stoppers $S_1$ and $S_2$ (as shown in the figure) are suddenly withdrawn, the volume of the gas in litres after equilibrium is achieved will be ........ (Given, $K_f$ (water) = 2.0 K kg $mol^{-1}$, $R = 0.08$ $dm^3$ atm $K^{-1}mol^{-1}$) ***[2020, 9 Jan Shift-II]***

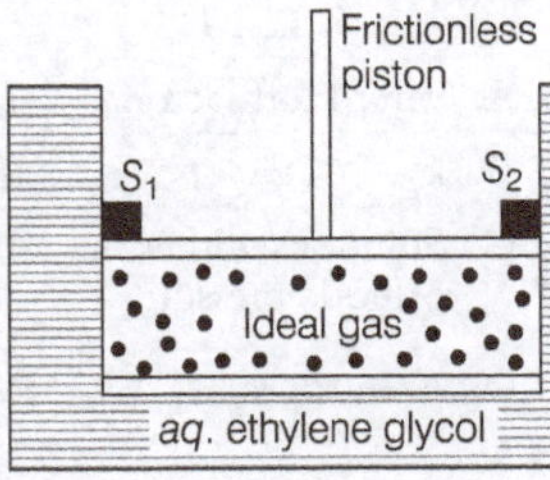

## Topic 4 Abnormal Molecular Masses and van't Hoff Factor

### Level 1

**01** Which one of the following 0.10 M aqueous solutions will exhibit the largest freezing point depression ? **[2021, 31 Aug Shift-I]**
(a) Hydrazine (b) Glucose (c) Glycine (d) $KHSO_4$

**02** If sodium sulphate is considered to be completely dissociated into cations and anions in aqueous solution, the change in freezing point of water ($\Delta T_f$), when 0.01 mole of sodium sulphate is dissolved in 1 kg of water, is ($K_f = 1.86$ K kg mol$^{-1}$) **[AIEEE 2010]**
(a) 0.0372 K (b) 0.0558 K (c) 0.0744 K (d) 0.0186 K

**03** If $\alpha$ is the degree of dissociation of $Na_2SO_4$, the van't Hoff factor ($i$) used for calculating the molecular mass is **[AIEEE 2005]**
(a) $1-2\alpha$ (b) $1+2\alpha$ (c) $1-\alpha$ (d) $1+\alpha$

### Level 2

**04** The solution from the following with highest depression in freezing point/lowest freezing point is **[2024, 30 Jan Shift-II]**
(a) 180 g of acetic acid dissolved in water
(b) 180 g of glucose dissolved in water
(c) 180 g of acetic acid dissolved in benzene
(d) 180 g of benzoic acid dissolved in benzene

**05** We have three aqueous solutions of NaCl labelled as '$A$', '$B$' and '$C$' with concentration 0.1 M, 0.01 M and 0.001 M, respectively. The value of van't Hoff factor ($i$) for these solutions will be in the order **[2024, 1 Feb Shift-I]**
(a) $i_A = i_B = i_C$ (b) $i_A > i_B > i_C$
(c) $i_A < i_C < i_B$ (d) $i_A < i_B < i_C$

**Imp.** Colligative Properties and Azeotropes

**06** Match List-I with List-II

| List-I | List-II |
|---|---|
| A. van't Hoff factor, $i$ | I. Cryoscopic constant |
| B. $k_f$ | II. Isotonic solutions |
| C. Solutions with same osmotic pressure | III. $\dfrac{\text{Normal molar mass}}{\text{Abnormal molar mass}}$ |
| D. Azeotropes | IV. Solutions with same composition of vapour above it |

Choose the correct answer from the options given below : **[2023, 29 Jan Shift-II]**
(a) A-III, B-I, C-IV, D-II (b) A-I, B-III, C-II, D-IV
(c) A-III, B-I, C-II, D-IV (d) A-III, B-II, C-I, D-IV

**07** Which one of the following 0.06 M aqueous solutions has lowest freezing point? **[2021, 22 July Shift-II]**
(a) $Al_2(SO_4)_3$ (b) $C_6H_{12}O_6$
(c) KI (d) $K_2SO_4$

**08** Molal depression constant for a solvent is 4.0 K kg mol$^{-1}$. The depression in the freezing point of the solvent for 0.03 mol kg$^{-1}$ solution of $K_2SO_4$ is (Assume complete dissociation of the electrolyte) **[2019, 9 April Shift-II]**
(a) 0.18 K (b) 0.36 K
(c) 0.12 K (d) 0.24 K

**09** The osmotic pressure of a dilute solution of an ionic compound $XY$ in water is four times that of a solution of 0.01 M $BaCl_2$ in water. Assuming complete dissociation of the given ionic compounds in water, the concentration of $XY$ (in mol L$^{-1}$) in solution is **[2019, 9 April Shift-I]**
(a) $4\times10^{-2}$ (b) $16\times10^{-4}$
(c) $4\times10^{-4}$ (d) $6\times10^{-2}$

**10** $K_2HgI_4$ is 40% ionised in aqueous solution. The value of its van't Hoff factor ($i$) is **[2019, 11 Jan Shift-II]**
(a) 1.6 (b) 1.8 (c) 2.2 (d) 2.0

**11** For 1 molal aqueous solution of the following compounds, which one will show the highest freezing point? **[JEE Main 2018]**
(a) $[Co(H_2O)_6]Cl_3$
(b) $[Co(H_2O)_5Cl]Cl_2\cdot H_2O$
(c) $[Co(H_2O)_4Cl_2]Cl\cdot 2H_2O$
(d) $[Co(H_2O)_3Cl_3]\cdot 3H_2O$

**12** The degree of dissociation ($\alpha$) of a weak electrolyte, $A_xB_y$ is related to van't Hoff factor ($i$) by the expression **[AIEEE 2011]**
(a) $\alpha = \dfrac{i-1}{(x+y-1)}$ (b) $\alpha = \dfrac{i-1}{x+y+1}$
(c) $\alpha = \dfrac{x+y-1}{i-1}$ (d) $\alpha = \dfrac{x+y+1}{i-1}$

**13** Which one of the following aqueous solutions will exhibit highest boiling point? **[AIEEE 2004]**
(a) 0.01 M $Na_2SO_4$ (b) 0.01 M $KNO_3$
(c) 0.015 M urea (d) 0.015 M glucose

### Level 3

**14** $XY$ is the membrane/partition between two chambers 1 and 2 containing sugar solutions of concentration $c_1$ and $c_2$ ($c_1 > c_2$) mol L$^{-1}$. For the reverse osmosis to take place identify the correct condition. (Here $p_1$ and $p_2$ are pressures applied on chamber 1 and 2). **[2025, 4 April Shift-I]**

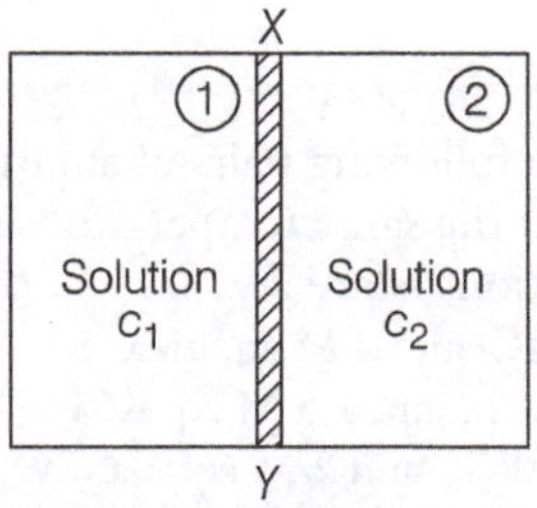


A. Membrane/Partition: Cellophane, $p_1 > \pi$
B. Membrane/Partition: Porous, $p_2 > \pi$
C. Membrane/Partition: Parchment paper, $p_1 > \pi$
D. Membrane/Partition: Cellophane, $p_2 > \pi$

Choose the correct answer from the option given below:
(a) A and C Only (b) B and D Only
(c) C Only (d) A and D Only

**Imp.** Order of Boiling Point

**15** Arrange the following solutions in order of their increasing boiling points. *[2025, 22 Jan Shift-I]*
(i) $10^{-4}$ M NaCl (ii) $10^{-4}$ M Urea
(iii) $10^{-3}$ M NaCl (iv) $10^{-2}$ M NaCl
(a) (ii) < (i) < (iii) < (iv) (b) (ii) < (i) < (iii) < (iv)
(c) (iv) < (iii) < (i) < (ii) (d) (i) < (ii) < (iii) < (iv)

**16** 1.24 g of $AX_2$ (molar mass 124 g mol$^{-1}$) is dissolved in 1 kg of water to form a solution with boiling point of 100.0156°C, while 25.4 g of $AY_2$ (molar mass 250 g mol$^{-1}$) in 2 kg of water constitutes a solution with a boiling point of 100.0260°C. *[2025, 29 Jan Shift-I]*
[$K_b(H_2O) = 0.52$ K kg mol$^{-1}$]
Which of the following is correct?
(a) $AX_2$ and $AY_2$ (both) are fully ionised.
(b) $AX_2$ is fully ionised while $AY_2$ is completely unionised.
(c) $AX_2$ is completely unionised while $AY_2$ is fully ionised.
(d) $AX_2$ and $AY_2$ (both) are completely unionised.

**17** Evaluate the following statements for their correctness.
A. The elevation in boiling point temperature of water will be same for 0.1 M NaCl and 0.1 M urea.
B. Azeotropic mixtures boil without change in their composition.
C. Osmosis always takes place from hypertonic to hypotonic solution.
D. The density of 32% $H_2SO_4$ solution having molarity 4.09 M is approximately 1.26 g mL$^{-1}$.
E. A negatively charged sol is obtained when KI solution is added to silver nitrate solution.

Choose the correct answer from the options given below. *[2023, 31 Jan Shift-II]*
(a) B and D (b) A and C
(c) A, B and D (d) B, D and E

**18** The depression in freezing point observed for a formic acid solution of concentration 0.5 mol L$^{-1}$ is 0.0405°C. Density of formic acid is 1.05 g mol L$^{-1}$. The van't Hoff factor of the formic acid solution is nearly (Given, for water $K_f = 1.86$ K kg mol$^{-1}$) *[2022, 25 July Shift-I]*
(a) 0.8 (b) 1.1 (c) 1.9 (d) 2.4

**19** Solute *A* associates in water. When 0.7 g of solute *A* is dissolved in 42.0 g of water, it depresses the freezing point by 0.2°C. The percentage association of solute *A* in water, is
[Given, molar mass of $A = 93$ g mol$^{-1}$. Molal depression constant of water is 1.86 K kg mol$^{-1}$] *[2022, 25 June Shift-II]*
(a) 50% (b) 60%
(c) 70% (d) 80%

**20** The freezing point of benzene decreases by 0.45°C when 0.2 g of acetic acid is added to 20 g of benzene. If acetic acid associates to form a dimer in benzene, percentage association of acetic acid in benzene will be ($K_f$ for benzene = 5.12 K kg mol$^{-1}$) *[JEE Main 2017]*
(a) 64.6 % (b) 80.4 % (c) 74.6 % (d) 94.6 %

**21** Consider the separate solution of 0.500 M $C_2H_5OH$ (*aq*), 0.100 M $Mg_3(PO_4)_2$(*aq*), 0.250 M KBr (*aq*) and 0.125 M $Na_3PO_4$(*aq*) at 25°C. Which statement is true about these solutions, assuming all salts to be strong electrolytes? *[JEE Main 2014]*
(a) They all have same osmotic pressure
(b) 0.100 M $Mg_3(PO_4)_2$ (*aq*) has the highest osmotic pressure
(c) 0.125 M $Na_3PO_4$ (*aq*) has the highest osmotic pressure
(d) 0.5000 M $C_2H_5OH$ (*aq*) has the highest osmotic pressure

## Numerical Value Type Questions

**22** The percentage dissociation of a salt ($MX_3$) solution at given temperature (van't Hoff factor $i = 2$) is .............. % (Nearest integer) *[2025, 7 April Shift-I]*

**23** When 1 g each of compounds *AB* and $AB_2$ are dissolved in 15 g of water separately, they increased the boiling point of water by 2.7 K and 1.5 K respectively. The atomic mass of *A* (in amu) is ........ × $10^{-1}$ (Nearest integer)
(Given : Molal boiling point elevation constant is 0.5 K kg mol$^{-1}$) *[2025, 2 April Shift-II]*

**24** The observed and normal molar masses of compound $MX_2$ are 65.6 and 164 respectively. The percent degree of ionisation of $MX_2$ is ...........%. (Nearest integer) *[2025, 24 Jan Shift-II]*

**25** If $A_2B$ is 30% ionised in an aqueous solution, then the value of van't Hoff factor ($i$) is ........... $\times 10^{-1}$ **[2025, 29 Jan Shift-I]**

**26** A solution containing 10 g of an electrolyte $AB_2$ in 100 g of water boils at 100.52°C. The degree of ionisation of the electrolyte ($\alpha$) is ........ $\times 10^{-1}$. (Nearest integer)

[Given : Molar mass of $AB_2 = 200$ g mol$^{-1}$, $K_b$ (Molal boiling point elevation const. of water) = 0.52 K kg mol$^{-1}$, boiling point of water = 100°C, $AB_2$ ionises as $AB_2 \longrightarrow A^{2+} + 2B^-$] **[2024, 8 April Shift-II]**

**27** Considering acetic acid dissociates in water, its dissociation constant is $6.25 \times 10^{-5}$. If 5 mL of acetic acid is dissolved in 1 L water, the solution will freeze at $-x \times 10^{-2}$ °C, provided pure water freezes at 0 °C.

$x$ = .......... (Nearest integer)

Given: $(K_f)_{\text{water}} = 1.86$ K kg mol$^{-1}$.

density of acetic acid is 1.2 g mL$^{-1}$

molar mass of water = 18 g mol$^{-1}$.

molar mass of acetic acid = 60 g mol$^{-1}$

density of water = 1 g cm$^{-3}$

Acetic acid dissociates as

$CH_3COOH \rightleftharpoons CH_3COO^{\ominus} + H^{\oplus}$ **[2024, 5 April Shift-II]**

**28** Mass of ethylene glycol (antifreeze) to be the added to 18.6 kg of water to protect the freezing point at –24°C is ........ kg (Molar mass in g mol$^{-1}$ for ethylene glycol is 62, $K_f$ of water = 1.86 K kg mol$^{-1}$) **[2024, 1 Feb Shift-II]**

**29** 80 mole per cent of $MgCl_2$ is dissociated in aqueous solution. The vapour pressure of 1.0 molal aqueous solution of $MgCl_2$ at 38°C is ........... mm Hg. (Nearest integer)

(Given, vapour pressure of water at 38°C is 50 mm Hg) **[2023, 12 April Shift-I]**

**30** Sea water contains 29.25% NaCl and 19% $MgCl_2$ by weight of solution. The normal boiling point of the sea water is ........ °C (Nearest integer)

Assume 100% ionisation for both NaCl and $MgCl_2$.

(Given, $K_b(H_2O) = 0.52$ K kg mol$^{-1}$, Molar mass of NaCl and $MgCl_2$ is 58.5 and 95 g mol$^{-1}$ respectively.) **[2023, 10 April Shift-II]**

**31** If the degree of dissociation of aqueous solution of weak monobasic acid is determined to be 0.3, then the observed freezing point will be ..........% higher than the expected/theoretical freezing point. (Nearest integer) **[2023, 10 April Shift-I]**

**Imp.** **Isotonic Solutions**

**32** Consider the following pairs of solution which will be isotonic at the same temperature. The number of pairs of solutions is/are .......... **[2023, 6 April Shift-II]**

A. 1 M aq. NaCl and 2 M aq. urea

B. 1 M aq. $CaCl_2$ and 1.5 M aq. KCl

C 1.5 M aq. $AlCl_3$ and 2 M aq. $Na_2SO_4$

D. 2.5 M aq. KCl and 1 M aq. $Al_2(SO_4)_3$

**33** 20% of acetic acid is dissociated when its 5 g is added to 500 mL of water. The depression in freezing point of such water is ........... $\times 10^{-3}$°C. Atomic mass of C, H and O are 12, 1 and 16 amu, respectively. [Given, molal depression constant and density of water are 1.86 K kg mol$^{-1}$ and 1 g cm$^{-3}$ respectively. **[2023, 1 Feb Shift-II]**

**34** 25 mL of an aqueous solution of KCl was found to require 20 mL of 1 M $AgNO_3$ solution when titrated using $K_2CrO_4$ as an indicator. What is the depression in freezing point of KCl solution of the given concentration? .......... (Nearest integer)

(Given, $K_f = 2.0$ K kg mol$^{-1}$)

**Assume**

1. 100% ionisation and
2. density of the aqueous solution as 1 g mL$^{-1}$

**[2023, 1 Feb Shift-I]**

**35** Lead storage battery contains 38% by weight solution of $H_2SO_4$. The van't Hoff factor is 2.67 at this concentration. The temperature in Kelvin at which the solution in the battery will freeze is ......... (Nearest integer). **[2023, 30 Jan Shift-II]**

[Given : $K_f = 1.8$ K kg mol$^{-1}$]

**36** The number of pairs of the solutions having the same value of the osmotic pressure from the following is ........ (Assume 100% ionisation)

A. 0.500 M $C_2H_5OH(aq)$ and 0.25 M $KBr(aq)$

B. 0.100 M $K_4[Fe(CN)_6](aq)$ and 0.100 M $FeSO_4(NH_4)_2SO_4(aq)$

C. 0.05 M $K_4[Fe(CN)_6](aq)$ and 0.25 M $NaCl(aq)$

D. 0.15 M $NaCl(aq)$ and 0.1 M $BaCl_2(aq)$

E. 0.02 M $KCl \cdot MgCl_2 \cdot 6H_2O(aq)$ and 0.05 M $KCl(aq)$ **[2023, 25 Jan Shift-II]**

**37** 1.2 mL of acetic acid is dissolved in water to make 2.0 L of solution. The depression in freezing point observed for this strength of acid is 0.0198°C. The percentage of dissociation of the acid is ...... . (Nearest integer)

[Given, density of acetic acid is 1.02 g $mL^{-1}$, molar mass of acetic acid is 60 g $mol^{-1}$ $K_f(H_2O) = 1.85$ K kg $mol^{-1}$] **[2022, 29 June Shift-I]**

**38** A 0.5 per cent solution of potassium chloride was found to freeze at – 0.24°C. The percentage dissociation of potassium chloride is ...... . (Nearest integer)
(Molal depression constant for water is 1.80 K kg $mol^{-1}$ and molar mass of KCl is 74.6 g $mol^{-1}$) **[2022, 26 June Shift-I]**

**39** 1.22 g of an organic acid is separately dissolved in 100 g of benzene ($K_b$ = 2.6 K kg $mol^{-1}$) and 100 g of acetone ($K_b$ = 1.7 K kg $mol^{-1}$). The acid is known to dimerise in benzene but remain as a monomer in acetone. The boiling point of the solution in acetone increases by 0.17°C. The increase in boiling point of solution in benzene in °C is $x \times 10^{-2}$. The value of $x$ is ........ (Nearest integer)
[Atomic mass : C = 12.0, H = 1.0, O = 16.0] **[2021, 31 Aug Shift-II]**

**40** 83 g of ethylene glycol dissolved in 625 g of water. The freezing point of the solution is .........K. (Nearest integer)
[Use, molal freezing point depression constant of water = 1.86 K kg $mol^{-1}$, Freezing point of water = 273 K and Atomic masses : C = 12.0 u, O = 16.0 u, H = 1.0 u] **[2021, 26 Aug Shift-II]**

**41** Of the following four aqueous solutions, total number of those solutions whose freezing point is lower than that of 0.10 M $C_2H_5OH$ is ............ . (Integer answer)
(i) 0.10 M $Ba_3(PO_4)_2$ (ii) 0.10 M $Na_2SO_4$
(iii) 0.10 M KCl (iv) 0.10 M $Li_3PO_4$ **[2021, 26 Aug Shift-I]**

**42** In a solvent 50% of an acid $HA$ dimerises and the rest dissociates. The van't Hoff factor of the acid is ........... $\times 10^{-2}$.
(Round off to the nearest integer) **[2021, 27 July Shift-II]**

**43** A solute $A$ dimerises in water. The boiling point of a 2 molar solution of $A$ is 100.52°C. The percentage association of $A$ is ......... . (Round off to the nearest integer)
[Use : $K_b$ for water = 0.52 K kg $mol^{-1}$, boiling point of water = 100°C] **[2021, 18 March Shift-II]**

**44** 2 molal solution of a weak acid $HA$ has a freezing point of 3.885°C. The degree of dissociation of this acid is ........ $\times 10^{-3}$. (Round off to the nearest integer).
[Given : Molal depression constant of water = 1.85 K kg $mol^{-1}$, freezing point of pure water = 0°C] **[2021, 18 March Shift-I]**

**45** When 12.2 g of benzoic acid is dissolved in 100 g of water, the freezing point of solution was found to be – 0.93°C ($K_f(H_2O)$ = 1.86 K kg $mol^{-1}$). The number ($n$) of benzoic acid molecules associated (assuming 100% association) is ......... . **[2021, 26 Feb Shift-II]**

**46** If a compound $AB$ dissociates to the extent of 75% in an aqueous solution, the molality of the solution which shows a 2.5 K rise in the boiling point of the solution is ........ molal. (Rounded off to the nearest integer) [$K_b$ = 0.52 K kg $mol^{-1}$] **[2021, 25 Feb Shift-II]**

**47** 1 molal aqueous solution of an electrolyte $A_2B_3$ is 60% ionised. The boiling point of the solution at 1 atm is ........... K (Rounded off to the nearest integer). [Given, $K_b$ for ($H_2O$) = 0.52 K kg $mol^{-1}$] **[2021, 25 Feb Shift-I]**

**48** $C_6H_6$ freezes at 5.5°C. The temperature at which a solution 10 g of $C_4H_{10}$ in 200 g of $C_6H_6$ freeze is .........°C. (The molal freezing point depression constant of $C_6H_6$ is 5.12°C/m.) **[2021, 24 Feb Shift-II]**

**49** The elevation of boiling point of 0.10 m aqueous $CrCl_3 \cdot x\,NH_3$ solution is two times that of 0.05 m aqueous $CaCl_2$ solution. The value of $x$ is ....... .
[Assume 100% ionisation of the complex and $CaCl_2$, coordination number of Cr as 6, and that all $NH_3$ molecules are present inside the coordination sphere] **[2020, 6 Sep Shift-I]**

**50** How much amount of NaCl should be added to 600 g of water (ρ = 1.00 g/mL) to decrease the freezing point of water to – 0.2°C? ...... (The freezing point depression constant for water = 2 K kg $mol^{-1}$) **[2020, 9 Jan Shift-I]**

# Key Index

## Topic 1

| | | | | | | | | | |
|---|---|---|---|---|---|---|---|---|---|
| **1** (b) | **2** (c) | **3** (d) | **4** (b) | **5** (b) | **6** (b) | **7** (b) | **8** (d) | **9** (b) | **10** (c) |
| **11** (c) | **12** (a) | **13** (c) | **14** (d) | **15** (2) | **16** (1) | **17** (125) | **18** (164) | **19** (74) | **20** (815) |
| **21** (4) | **22** (148) | **23** (5) | **24** (54) | **25** (64) | **26** (3.17) | **27** (5) | **28** (64) | **29** (9) | **30** (100) |
| **31** (10) | | | | | | | | | |

## Topic 2

| | | | | | | | | | |
|---|---|---|---|---|---|---|---|---|---|
| **1** (a) | **2** (a) | **3** (a) | **4** (d) | **5** (c) | **6** (b) | **7** (c) | **8** (a) | **9** (a) | **10** (b) |
| **11** (b) | **12** (d) | **13** (d) | **14** (a) | **15** (b) | **16** (c) | **17** (d) | **18** (a) | **19** (b) | **20** (d) |
| **21** (b) | **22** (23) | **23** (82) | **24** (314) | **25** (80) | **26** (1) | **27** (25) | **28** (19) | **29** (600) | |

## Topic 3

| | | | | | | | | | |
|---|---|---|---|---|---|---|---|---|---|
| **1** (a) | **2** (c) | **3** (b) | **4** (b) | **5** (d) | **6** (d) | **7** (d) | **8** (d) | **9** (a) | **10** (b) |
| **11** (b) | **12** (c) | **13** (c) | **14** (d) | **15** (d) | **16** (b) | **17** (c) | **18** (d) | **19** (c) | **20** (b) |
| **21** (b) | **22** (d) | **23** (543) | **24** (25) | **25** (31) | **26** (707) | **27** (73) | **28** (23) | **29** (8) | **30** (833) |
| **31** (62250) | **32** (100) | **33** (41500) | **34** (15) | **35** (6) | **36** (1) | **37** (45) | **38.** (8) | **39** (415) | **40** (54) |
| **41** (271) | **42** (518) | **43** (15) | **44** (250) | **45** (85) | **46** (2.18) | | | | |

## Topic 4

| | | | | | | | | | |
|---|---|---|---|---|---|---|---|---|---|
| **1** (d) | **2** (b) | **3** (b) | **4** (a) | **5** (d) | **6** (c) | **7** (a) | **8** (b) | **9** (d) | **10** (b) |
| **11** (d) | **12** (a) | **13** (a) | **14** (a) | **15** (b) | **16** (b) | **17** (a) | **18** (c) | **19** (d) | **20** (d) |
| **21** (a) | **22** (33) | **23** (25) | **24** (75) | **25** (16) | **26** (5) | **27** (19) | **28** (15) | **29** (48) | **30** (116) |
| **31** (30) | **32** (4) | **33** (372) | **34** (3) | **35** (243) | **36** (4) | **37** (5) | **38** (98) | **39** (13) | **40** (269) |
| **41** (4) | **42** (125) | **43** (100) | **44** (50) | **45** (2) | **46** (3) | **47** (375) | **48** (1) | **49** (5) | **50** (1.76) |

# Detailed Explanations

## Topic 1

**01.** *(b)* Given,

Mass of NaCl = 5.85 g

Volume of water = 500 mL = 0.5 L

Using, Molarity

$$= \frac{\text{No. of moles of NaCl}}{\text{Volume of solution (in L)}}$$

$$M = \frac{\frac{5.85}{58.5}}{0.5} \Rightarrow M = 0.2\text{ M}$$

**02.** *(c)* Mole fraction

$$= \frac{\text{Number of moles of component}}{\text{Total number of moles of all the component}}$$

From above definition, mole fraction of *C* will be

$$= \frac{n_C}{n_A + n_B + n_C}$$

where $n_A, n_B, n_C$ are number of moles of substance *A*, *B* and *C*.

Hence, option (c) is correct.

**03.** *(d)* Concentration terms such as mass percentage, mole fraction and molality are independent of temperature, However, molarity is a function of temperature. Since it depend on the total volume of the solution.

$$\text{Molarity} = \frac{\text{Mole of solute}}{\text{Volume of solution}}$$

**• Concept Enhancer**

On increasing the temperature, volume of solution decreases and molarity increases.

**04.** *(b)* This question involves the concept of molality.

**key idea** Molality is defined as number of moles of solute per kg of solvent.

$$m = \frac{w_2}{Mw_2} \times \frac{1000}{w_1}$$

$w_2$ = mass of solute;
$Mw_2$ = molecular mass of solute
$w_1$ = mass of solvent

20% aqueous solution of KI means that 20 g of KI is present in 80 g solvent.

$$m = \frac{20}{166} \times \frac{1000}{80} = 1.506 \approx 1.51\text{ mol/kg}$$

**05.** *(b)* Molality $(m) = \dfrac{w_{\text{Na}^+} \times 1000}{M_{\text{Na}^+} \times w_{H_2O}}$

$$= \frac{92 \times 1000}{23 \times 1000} = 4\text{ mol kg}^{-1}$$

**06.** *(b)* Moles of urea $= \dfrac{0.010}{60}$ mol

Water at STP

$(d = 1\text{g/cm}^3 = 1\text{ kg/dm}^3)$

$$= 0.3\text{ dm}^3 = 0.3\text{ kg}$$

$$\therefore \text{ Molality} = \frac{0.010}{60 \times 0.3} = 5.55 \times 10^{-4}\text{ m}$$

**07.** *(b)* The molarity of the resulting solution is given by

$$\text{Total molarity} = \frac{M_1V_1 + M_2V_2}{V_1 + V_2}$$

$$= \frac{1.5 \times 480 + 1.2 \times 520}{480 + 520} = 1.34\text{ M}$$

**08.** *(d)* The best suited graph is given in option (d). The reason is explained through the given flow chart.

$$\text{Water}_{1°C} \xrightarrow[\text{decrease}]{\text{Volume}} \text{Water}_{4°C} \xrightarrow[\text{Thermal expansion}]{\text{Volume increases}} \text{Water}_{25°C}$$

$$\text{Molarity} = \frac{\text{No. of moles of solute}}{\text{Volume of solution in litre}}$$

Hence,

$$1°\text{C} \xrightarrow[\text{Increases}]{\text{Molarity}} 4°\text{C} \xrightarrow[\text{decreases}]{\text{Molarity}} 25°\text{C}$$

**09.** *(b)* The molecular mass of glucose is 180 g/mol

Mass of carbon in 250 gram solution is

$$= \frac{250 \times 10.8}{100} = 27\text{ g}$$

We know that, 72 g of C is present in 180 g glucose

so, 27 g of carbon is present in

$$\frac{180}{72} \times 27\text{ g glucose}$$

$\Rightarrow$ Weight of glucose

$$= \frac{180}{72} \times 27 = 67.5\text{ g}$$

So, mass of solvent

$$= 250 - 67.5 = 182.5\text{ g}$$

$$\Rightarrow \text{Molality} = \frac{67.5 \times 1000}{180 \times 182.5}$$

$$= 2.055 \approx 2.06\text{ m}$$

**10.** *(c)* $X_{\text{solvent}} = 0.8$ (Given)

It means that $n_{\text{solvent}}(n_1) = 0.8$

and $n_{\text{solute}}(n_2) = 0.2$

Using formula, $m = n_2 \times \dfrac{1000}{n_1 \times M_1}$

$$= 0.2 \times \frac{1000}{0.8 \times 18} = 13.88\text{ mol kg}^{-1}$$

**11.** *(c)* 5.2 molal aqueous solution of $CH_3OH$ suggests that 5.2 moles of $CH_3OH$ are dissolved in 1000 g of water.

$$n_1(CH_3OH) = 5.2$$

$$n_2\ (H_2O) = \frac{1000}{18} = 55.56$$

$$\therefore \quad n_1 + n_2 = 5.20 + 55.56 = 60.76\text{ mol}$$

$$\therefore X_{CH_3OH} = \text{Mole fraction of } CH_3OH$$

$$= \frac{n_1}{n_1 + n_2} = \frac{5.2}{60.76} = 0.086$$

**12.** *(a)* % $w/w$ of $HNO_3$ = 75%,

Means that 100 g, of solution contains 75 g of $HNO_3$

and $\left(\dfrac{g}{mL}\right)_{\text{Solution}} = 1.25 = \dfrac{100}{V}$ g

$\because$ 75 g of $HNO_3$ present in $\dfrac{100}{1.25}$ mL solution.

$\therefore$ 30 g of $HNO_3$ present in

$$\frac{100}{1.25 \times 75} \times 30 = 32\text{ mL solution}$$

**13.** *(c)* Given, density of commercial acid = 1.46 g/mL

So, mass of commercial acid

$$= \text{Density} \times \text{Volume}$$

$$= 1.46 \times 1000 = 1460\text{ g}$$

$\therefore$ Mass of given 35% HCl

$$= \frac{35}{100} \times 1460 = 511\text{ g}$$

Molarity of solution

$$= \frac{\text{Number of moles of acid}}{\text{Volume of acid (in litre)}}$$

$= \frac{\text{Weight of acid}}{\text{M.M of acid} \times \text{Volume of acid (in litre)}}$

$= \frac{511}{36.5 \times 1\text{ L}} = 14\text{ M}$

**14.** *(d)* $\chi_{\text{Solute}} = \frac{\frac{w_{\text{Solute}}}{Mw_{\text{Solute}}}}{\frac{w_{\text{Solute}}}{Mw_{\text{Solute}}} + \frac{w_{\text{Solvent}}}{Mw_{\text{Solvent}}}}$

Given, $w_{\text{Solute}} = w_{\text{NaOH}} = 8\text{ g}$

$Mw_{\text{Solute}} = Mw_{\text{NaOH}} = 40\text{ g mol}^{-1}$

$w_{\text{Solvent}} = w_{H_2O} = 18\text{ g}$

$Mw_{\text{Solvent}} = 18\text{ g mol}^{-1}$

$\therefore \quad \chi_{\text{Solute}} = \chi_{\text{NaOH}} = \frac{\frac{8}{40}}{\frac{8}{40} + \frac{18}{18}}$

$= \frac{0.2}{0.2 + 1} = \frac{0.2}{1.2} = 0.167$

$m = \frac{\frac{w_{\text{Solute}}}{Mw_{\text{Solute}}}}{w_{\text{Solvent}}\text{ (in g)}} \times 1000$

$= \frac{\frac{8}{40}}{18} \times 1000 = \frac{0.2}{18} \times 1000$

$= 11.11\text{ mol kg}^{-1}$

Thus, mole fraction of NaOH in solution and molality of the solution respectively are 0.167 and 11.11 mol kg$^{-1}$

**15.** *(2)* Given that, sea water is 6M solution of NaCl

So, 1000 ml of sea water contains 6mol of NaCl

Mass of solution = Volume × density

$= 1000 \times 2 = 2000\text{ g}$

$ppm = \frac{\text{mass of } O_2}{2000} \times 10^6$

Mass of $O_2 = 5.8 \times 2 \times 10^{-3}$

$= 1.16 \times 10^{-2}\text{ g}$

Molality for $O_2 = \frac{1.16 \times 10^{-2} / 32 \times 1000}{(2000 - 6 \times 58.5)}$

$= \frac{1.16 \times 10}{32 \times 1649} = 2.19 \times 10^{-4}$

$\approx 2$

**16.** *(1)* The reaction involved is as follows

$\underset{\text{(M) 20mL}}{NaI(aq)} + \underset{\text{Excess}}{AgNO_3(aq)} \longrightarrow \underset{4.74\text{ g}}{AgI(s)} + NaNO_3(aq)$

Moles of $I^-$ in NaI = moles of $I^-$ in AgI $= \frac{4.74}{235}$

Moles of NaI $= \frac{4.74}{235}$

Molarity [NaI] $= \frac{4.74}{235 \times 0.02} = 1.008 = 1$

**17.** *(125)* Given, $\frac{\text{Mass}}{\text{Mass}}\% = 70\%$

Molarity = 2

Let 100 g solution contain 70 g solute.

Volume of 100 g solution will be $\frac{100}{1.25}$ mL

Molarity $= \frac{70/70}{100/1.25} \times 1000$

$= 12.5 \text{ or } 125 \times 10^{-1}$

**18.** *(164)* $M_{\text{soln}} = V_{\text{soln}} \times d_{\text{soln}}$

$= 500 \times 1.25 = 625\text{ g}$

Mass of solute $(x) = 0.2 \times 0.5 \times 159.5$

$= 15.95\text{ g}$

$n_{\text{solute}} = \frac{15.95\text{ g}}{159.5\text{ g/mol}} = 0.1$

Mass of solvent = Mass of solution − Mass of solute

$= 625 - 15.95 = 609.05\text{ g}$

Now using formula of molality

Molality $= \frac{\text{No. of moles of solute} \times 1000}{\text{Mass of solvent (in kg)}}$

$m = \frac{0.1 \times 1000}{609.05}$

$m = 0.164\text{ m}$

$m = 164 \times 10^{-3}\text{ m}$

Answer is different from NTA i.e., 81.

**19.** *(74)* $\frac{x_B}{x_A} = \frac{m.M_A}{1000}$

Where $x_B, x_A$ = Mole fraction,

$m$ = molality

$M_A$ = Molar mass of solvent

$\frac{x_B}{1 - x_B} = \frac{4.44 \times 18}{1000} \quad [x_A + x_B = 1]$

$\frac{x_B}{1 - x_B} = 0.08$

$x_B = 0.074 = 74 \times 10^{-3}$

Compare it with given value, $x = 74$.

**20.** *(815)* Molality is given by formula

$m = \frac{M \times 1000}{d_{\text{sol}} \times 1000 - M \times \text{Molar mass solute}} \quad \text{...(i)}$

$M = 0.8\text{ M [given]}$

$d_{\text{sol}} = 1.06\text{ g/cm}^3$

Molar mass of $H_2SO_4 = 98\text{ g/mol}$

Put all the value in Eq. (i),

$= \frac{0.8 \times 1000}{1.06 \times 1000 - 0.8 \times 98}$

Molality $= 815 \times 10^{-3}\text{ m}$

**21.** *(4)* Given, let mass of solution = 100 g

Mass of $H_2SO_4$ in solution = 31.4%

density = 1.25 g/mL

Molar mass of $H_2SO_4$ = 98 g/mol

Molarity, $M = \frac{n_{\text{solute}}}{V} \times 1000$

$= \frac{(31.4/98)}{(100/1.25)} \times 1000 = 4.005 \approx 4\text{ M}$

Hence, nearest integer is 4.

**22.** *(148)* Molarity

$= \frac{\text{No. of moles of solute}}{\text{Volume of solution }(L)}$

$2.6 \times 10^{-3} = \frac{x/85}{0.67141}$

$x = 0.148\text{ g}$

Conc. of dichloromethane in ppm

$= \frac{0.148}{1.49 \times 671.141} \times 10^6$

$= 148\text{ ppm}$

**23.** *(5)* Mass per cent, mole fraction, molarity, ppm and molality are the units which are used to express concentration of solution.

So, total number of units = 5.

**24.** *(54)* Moles of nitric acid

$= 800 \times 10^{-3} \times 0.5 = 0.4$

Moles of niric acid evaporated

$= \frac{11.5}{63} = 0.182$

Moles left = 0.4 - 0.182 = 0.218

Molarity $= \frac{0.218 \times 10^3}{400}$

$= 0.545\text{ M} = 54.5 \times 10^{-2}\text{ M}$

**25.** *(64)* Given, mass of $CuSO_4 \cdot 5H_2O = 80\text{ g}$

The concentration of copper sulphate solution is $x \times 10^{-3}$ mol/L.

Molar mass of

$CuSO_4 \cdot 5H_2O = 63.54 + 32 + 16 \times 4 + 5 \times 18$

$= 249.54\text{ g / mol}$

Number of moles of solute

$$= \frac{\text{Weight of solute}}{\text{Molecular mass of solute}}$$

$$= \frac{80 \text{ g}}{249.54 \text{ g / mol}}$$

$$= 0.32 \text{ mol}$$

Volume of solution = 5 L

$$\text{Molarity} = \frac{\text{Number of moles of solute}}{\text{Volume of solution (in L)}}$$

$$\therefore \text{Molarity} = \frac{0.32}{5}$$

$$= 64.11 \times 10^{-3} \text{ mol / L}$$

$$\therefore \quad x = 64.11 \text{ or } x \approx 64$$

Hence, answer is 64.

**26.** *(3.17)* Using the formula of molality

$$m = \frac{M \times 1000}{(1000\, d) - (M \times M_{\text{solute}})}$$

$$3.3 = \frac{M \times 1000}{1000 \times 1.2 - 74.5 \times M}$$

$$3960 - 245.8\, M = 1000\, M$$

$$1245.85\, \text{M} = 3960$$

$$\Rightarrow \quad M = 3.17 \text{ M}$$

**27.** *(5)* Given density $= 1.2 \text{ g/cm}^3$

So, mass of solvent

$$= 1 \text{ cm}^3 \times 1 \text{ g/cm}^3 = 1 \text{ g}$$

Mass of solute $= 1.2 - 1 = 0.2$ g

$$m = \frac{0.2}{40} \times 1000 = 5$$

**28.** *(64)* 100 molal aqueous solution means 100 mole solute in 1000 g water.

$$\therefore \text{mole fraction of solute} = \frac{100}{100 + \frac{1000}{18}}$$

$$= \frac{1800}{2800} = 64.28 \times 10^{-2}$$

**29.** *(9)* Molality

$$(m) = \frac{M \times 1000}{\{(1000\, d) - M \times M_{\text{solute}}\}}$$

$M_{\text{solute}}$ = molar mass of solute

Putting the values,

$$6.50 = \frac{M \times 1000}{(1000 \times 1.89) - (M \times 56)}$$

$$\text{On solving, } M = \frac{12285}{1364}$$

$$M = 9 \text{ mol dm}^{-3}$$

**30.** *(100)* Hardness in ppm

$$= \frac{\text{Hardness causing substances} \times 10^6}{\text{Mass of water sample}}$$

Here, $[\text{Mg}^{2+}] = 10^{-3}$ M

$n_{\text{Mg}^{2+}} = 10^{-3}$ per 1 L water ≅ 1000 g

Mass of $10^{-3}$ moles of $CaCO_3$

$$= 10^{-3} \times 100 \text{ g} = 0.1 \text{ g}$$

$$\therefore \text{ppm} = \frac{0.1 \times 10^6}{1000} = 100 \text{ ppm}$$

**31.** *(10)* Mass of $O_2$ = 10.30 mg (Given)

Mass of sea water solution

$$= \text{volume} \times \text{density}$$

$$= 1 \text{ L} \times 1.03 \text{ g/mL}$$

$$= 1000 \text{ mL} \times 1.03 \text{ g mL}^{-1} = 1030 \text{ g}$$

$$\text{ppm} = \frac{\text{mass of } O_2}{\text{mass of solution}} \times 10^6$$

$$= \frac{10.30 \times 10^{-3} \text{ g} \times 10^6}{1030 \text{ g}} = 10 \text{ ppm}$$

## Topic 2

**01.** *(a)* Using Dalton's Law of partial pressure

$$p_S = p_A^{o} X_A + p_B^{o} X_B$$

$$500 = 200 \times \frac{1}{4} + p_B^{o} \times \frac{3}{4}$$

$$p_B^{o} = 600 \text{ mmHg}$$

As $p_A^{o} < p_B^{o} \Rightarrow A$ is least volatile.

**02.** *(a)*

$$p_T = X_{\text{H}} \cdot p_{\text{H}}^{o} + X_{\text{O}} \cdot p_{\text{O}}^{o}$$

$$X_{\text{H}} = \frac{25/100}{25/100 + 35/114} = 0.45$$

$$\therefore \quad X_{\text{O}} = 0.55 \quad [\because X_{\text{H}} + X_{\text{O}} = 1]$$

$$p_T = 0.45 \times 105 + 0.55 \times 45$$

$$= 72.0 \text{ kPa}$$

**03.** *(a)* According to Raoult's law,

$$p = p^{\circ}_A X_A + p^{\circ}_B X_B$$

$$290 = 200 \times 0.4 + p^{\circ}_B \times 0.6$$

$$p^{\circ}_B = 350 \text{ mm}$$

**04.** *(d)* Given, $\Delta p_1 = 10$ mm Hg

Initial mole fraction of solute $\chi_{B_1} = 0.2$

$$\Delta p_2 = 20 \text{ mm Hg}$$

Final mole fraction of solute, $\chi_{B_2} = ?$

Mole fraction of solvent when vapour pressure is 20 mm Hg

$$\chi_{A_2} = ?$$

$$\frac{\Delta p_1}{p^{\circ}} = \chi_{B_1} \Rightarrow \frac{\Delta p_2}{p^{\circ}} = \chi_{B_2}$$

$$\frac{\Delta p_1}{\Delta p_2} = \frac{\chi_{B_1}}{\chi_{B_2}}$$

$$\Rightarrow \quad \frac{10}{20} = \frac{0.2}{\chi_{B_2}}$$

$$\chi_{B_2} = 0.2 \times 2 = 0.4$$

$$\chi_{A_2} = 1 - \chi_{B_2} = 1 - 0.4 = 0.6$$

**05.** *(c)* Given, $p_A^{\circ} < p_B^{\circ}$

$$\frac{p_A^{\circ}}{p_B^{\circ}} < 1$$

$$\frac{y_A}{y_B} = \frac{p_A^{\circ}}{p_B^{\circ}} \frac{x_A}{x_B}$$

$$\frac{\frac{y_A}{y_B}}{\frac{x_A}{x_B}} < 1 \Rightarrow \frac{y_A}{y_B} < \frac{x_A}{x_B}$$

**06.** *(b)* From given graph it can be seen that, $X$ has higher vapour pressure as compared to $Y$ at lower temperature (or to attain same vapour pressure, $Y$ need higher temperature than $X$). It means that intermolecular interactions must be weaker in $X$ as compared to $Y$.

**07.** *(c)* According to Raoult's law, vapour pressure of the solution $p_{\text{ideal}}$ (ideal vapour pressure) $= p_A^{\circ} + (p_B^{\circ} - p_A^{\circ}) x_B$

($A$ = solvent : acetone, $B$ = solute : $CS_2$)

Given; $p_A^{\circ} = 344$ mm Hg.

$$p_B^{\circ} = 512 \text{ mm Hg.}$$

$$\therefore \quad p_{\text{ideal}} = 344 + (512 - 344) x_B$$

$$\text{or} \quad p_{\text{ideal}} = 344 + 168\, x_B$$

So, $\quad 0 < x_B < 1$.

Also, total vapour pressure ($p$)

$$= 600 \text{ mmHg}$$

∴ For any value of $x_B$, $p > p_{\text{ideal}}$.

Option (c) is incorrect. It is because, mixture of pure $CS_2$ and $CH_3COCH_3$ shows positive deviation from Raoult's law. For such solution $\Delta V_{\text{min}} > 0$ and $\Delta H_{\text{mix}} > 0$.

**08.** *(a)*

> **key idea** For a solution of volatile liquids the partial vapour pressure of each component of the solution is directly proportional to its mole fraction present in solution. This is known as Raoult's law.

Liquid $M$ and $N$ form an ideal solution. Vapour pressures of pure liquids $M$ and $N$ are 450 and 700 mm Hg, respectively.

$$\therefore \quad p^{\circ}_N > p^{\circ}_M$$

So, by using Raoult's law

$$y_N > x_N \quad \ldots(i)$$

and

$$x_M > y_M \quad \ldots(ii)$$

Multiplying (i) and (ii) we get

$$y_N x_M > y_M x_N$$

$$\therefore \quad \frac{x_M}{x_N} > \frac{y_M}{y_N}$$

Thus, correct relation is (a).

**09.** *(a)* According to Henry's law (at constant temperature)

$$p_{gas} = K_H \times \chi_{gas\,(solute)}$$
$$= K_H \times [1 - \chi_{H_2O\,(solvent)}]$$
$$p_{gas} = K_H - K_H \chi_{H_2O}$$

Comparing the above equation with the equation of straight line $y = mx + c$

Slope $= -K_H$, intercept $= K_H$

So, higher the value of $K_H$, more (–ve) will be the slope.

Therefore, the intercept on *Y*-axis is increases from *w* to *z*.

**10.** *(b)* Henry's law constant ($K_H$)

$$K_H = \frac{\text{Pressure}}{\text{Solubility of a gas in a liquid}} = \frac{p}{S}$$

Thus, higher the value of $K_H$ at a given pressure, the lower is the solubility of the gas in the liquids.

**11.** *(b)* *n*-heptane and ethanol forms non-ideal solution. As a result, solution shows positive deviation from Raoult's law.

**12.** *(d)* Mixture contains 78 g benzene = 1 mol benzene and 46 g toluene = 0.5 mol toluene

Total moles of benzene and toluene

= 1.5 mol

Mole fraction of benzene in mixture

$$= \frac{1}{1.5} = \frac{2}{3}$$

Vapour pressure of benzene

$$p_b^\circ = 75 \text{ torr}$$

∴Partial vapour pressure of benzene

$$= p^\circ_b X_b = 75 \times \frac{2}{3} = 50 \text{ torr}$$

**13.** *(d)* Binary mixture of $C_6H_5OH$ and $C_6H_5NH_2$ shows negative deviation from Raoult's law.

So, vapour pressure of solution is less than vapour pressure of pure $C_6H_5OH$ and $C_6H_5NH_2$.

Thus, boiling point of solution is greater than boiling point of pure $C_6H_5OH$ and $C_6H_5NH_2$.

So, it shows maximum boiling azeotrope.

**14.** *(a)* The correct match is A-III, B-I, C-IV, D-II

- Solution of chloroform and acetone shows –ve deviation, so maximum boiling azeotrope.
- Solution of ethanol and water shows +ve deviation, so minimum boiling azeotrope.
- Solution of benzene and toluene from ideal solution. $\Delta V_{mix} = 0$.
- Acetic acid in benzene form dimer.

**15.** *(b)* Using Dalton's Law of partial pressure,

$$\frac{p_{N_2}}{p_{O_2}} = \frac{x_{N_2} \cdot p_T}{x_{O_2} \cdot p_T} = \frac{n_{N_2}}{n_{O_2}}$$

$$= \frac{\frac{70}{28}}{\frac{27}{32}} = 2.96$$

$$\frac{p_{O_2}}{p_{Ar}} = \frac{n_{O_2}}{n_{Ar}} = \frac{\frac{27}{32}}{\frac{3}{40}} = 11.25 \approx 11.2$$

**16.** *(c)* When we increase the temperature of solutions, first solubility decreases than increases.

Hence $K_H$ first increase than decrease.

Also at moderate temperature $K_H$ value is

$$He > N_2 > CH_4.$$

**17.** *(d)* For the given binary solution of two volatile liquid components 1 and 2.

$$p_1 = p_T y_1 = p_1^\circ x_1$$

$$\frac{p_T}{x_1} = \frac{p_1^\circ}{y_1}$$

$$\text{So,} \quad \frac{p_2^\circ + x_1(p_1^\circ - p_2^\circ)}{x_1} = \frac{p_1^\circ}{y_1}$$

$$\therefore \quad \frac{p_2^\circ}{x_1} + (p_1^\circ - p_2^\circ) = \frac{p_1^\circ}{y_1}$$

$$\therefore \quad \frac{1}{x_1} = \left(\frac{p_1^\circ}{p_2^\circ}\right)\left(\frac{1}{y_1}\right) + \left(\frac{p_2^\circ - p_1^\circ}{p_2^\circ}\right)$$

$$\therefore \quad \text{Slope} = \left(\frac{p_1^\circ}{p_2^\circ}\right)$$

$$\text{So, Intercept} = \left(\frac{p_2^\circ - p_1^\circ}{p_2^\circ}\right)$$

Thus, option (d) is the correct answer.

**18.** *(a)* Among the given options $(CH_3)_2CO + CS_2$ i.e., shows positive deviation from Raoult's law. Since acetone-$CS_2$ attractions are weaker than acetone-acetone attractions and $CS_2 - CS_2$ attractions.

> **• Concept Enhancer**
>
> The liquid solutions which shows positive deviation from the Raoult's law form minimum boiling azeotropes.

**19.** *(b)* The $A-B$ interactions in negative deviation solutions are stronger than the $A-A$ and $B-B$ interactions in the two liquids.

Due to greater interactions, *A* and *B* types of molecules have a lower tendency to escape from solutions than pure liquids.

As a result based on Raoult's law, each component of solution has a partial pressure lower than expected.

If vapour pressure decreases so boiling point increases.

**20.** *(d)* According to Dalton's law of partial pressure, $p_{total} = p_A + p_B$

$$= p_A^\circ \chi_A + p_B^\circ \chi_B \quad \ldots(i)$$

Given, $p_A^\circ = 400$ mm Hg,

$$p_B^\circ = 600 \text{ mm Hg}$$
$$\chi_B = 0.5, \ \chi_A + \chi_B = 1$$
$$\therefore \quad \chi_A = 0.5$$

On substituting the given values in Eq. (i), We get,

$$p_{total} = 400 \times 0.5 + 600 \times 0.5$$
$$= 500 \text{ mm Hg}$$

Mole fraction of *A* in vapour phase,

$$= \frac{0.5 \times 400}{500} = 0.4$$

Mole fraction of *B* in vapour phase,

$$Y_A + Y_B = 1$$
$$Y_B = 1 - 0.4 = 0.6$$

**21.** *(b)* For ideal solution,

$$p = x'_A p^\circ_A + x'_B p^\circ_B \quad \ldots(i)$$

$$\because \quad x'_A = 0.4, \ x'_B = 0.6$$
$$p^\circ_A = 7 \times 10^3 \text{ Pa},$$
$$p^\circ_B = 12 \times 10^3 \text{ Pa}$$

On substituting the given values in Eq. (i), we get

$$p = 0.4 \times 7 \times 10^3 + 0.6 \times 12 \times 10^3$$
$$= 10 \times 10^3 \text{ Pa}$$
$$= 1 \times 10^4 \text{ Pa}$$

In vapour phase,

$$x_A = \frac{0.4 \times 7 \times 10^3}{1 \times 10^4} = 0.28$$

$\therefore \quad x_B = 1 - 0.28 = 0.72 \; [\because x_A + x_B = 1]$

**22.** *(23)* $y_A = \frac{p_{\text{methyl benzene}}}{p_{\text{total}}}$

$$y_A = \frac{24}{80 + 24} = 0.2307$$

$$y_A = 23.07 \times 10^{-2}$$

$$y_A \approx 23 \times 10^{-2}$$

**23.** *(82)*

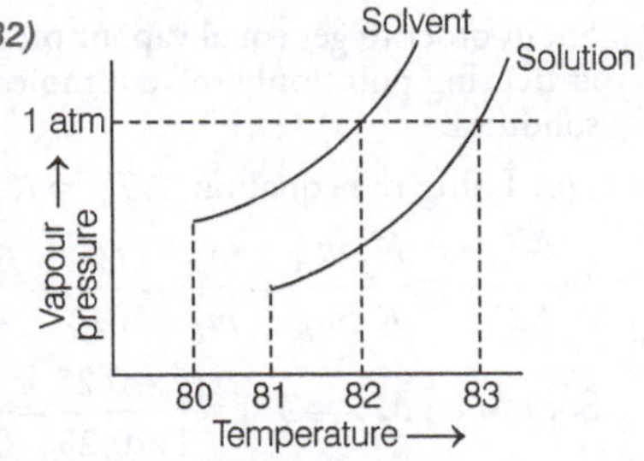


From the graph at 1 atm pressure, boiling point of solvent = 82°C

**24.** *(314)* Given;

Mole fraction of $A_1 = 0.7$

Mole fraction of $B_1 = 0.3$

Total pressure, $p_1 = 350$ mm Hg

Mole fraction of $A_2 = 0.2$

Mole fraction of $B_2 = 0.8$

Total pressure, $p_2 = 410$ mm Hg

Let vapour pressure of pure $A$ be $p_A^\circ$ and $B$ be $p_B^\circ$.

When, mole fraction of

$$A_1 = \chi_{A_1} = 0.7$$

and mole fraction of $B_1$,

$$= \chi_{B_1} = 0.3$$

and total pressure $p_1 = 350$.

$\Rightarrow (p_A^\circ \times 0.7) + (p_B^\circ \times 0.3) = 350 \quad \ldots$(i)

When, mole fraction of $A_2$,

$$= \chi_{A_2} = 0.2$$

mole fraction of $B_2$, $= \chi_{B_2} = 0.8$

and, total pressure $p_2 = 410$

$(p_A^\circ \times 0.2) + (p_B^\circ \times 0.8) = 410 \quad \ldots$(ii)

Now, solving Eq. (i) and (ii), we get

$$0.7p_A^\circ + 0.3p_B^\circ = 350 \quad \ldots\text{(i)}$$

$$0.2p_A^\circ + 0.8p_B^\circ = 410 \quad \ldots\text{(ii)}$$

Multiply Eq. (i) by 0.2 and Eq. (ii) by 0.7, we get

$$0.14p_A^\circ + 0.06p_B^\circ = 70 \quad \ldots\text{(iii)}$$

$$0.14p_A^\circ + 0.56p_B^\circ = 287 \quad \ldots\text{(iv)}$$

$$0 - 0.5p_B^\circ = -217$$

$\Rightarrow \quad -0.5p_B^\circ = -217$

$\Rightarrow \quad p_B^\circ = \frac{217}{0.5} = 434$ mm Hg

Now put the value of $p_B^\circ$ in Eq. (i),

$$0.7p_A^\circ + 0.3(434) = 350$$

$$0.7p_A^\circ = 350 - 130.2$$

$$0.7p_A^\circ = 219.8$$

$$p_A^\circ = \frac{219.8}{0.7}$$

$$= 314$$

Therefore, the vapour pressure of pure '$A$' is 314 mm Hg.

**25.** *(80)* Total pressure of cylinder ,

$$p_{\text{total}} = 25 \text{bar}$$

Partial pressure of Ne, $p_{\text{Ne}} = 20$ bar

$\therefore$ Partial pressure of $O_2$

$$p_{O_2} = p_{\text{total}} - p_{\text{Ne}}$$
$$= 25 - 20 = 5 \text{ bar}$$

$$\text{Moles of } O_2 = \frac{x}{32} \text{ mol}$$

Mass of Ne = 200 g

$$\text{Moles of Ne} = \frac{200}{20} \text{ mol}$$

Now, putting value of $p_{O_2}$

$\Rightarrow \quad 5 = \dfrac{\frac{x}{32}}{\frac{x}{32} + \frac{200}{20}} \times 25$

$\Rightarrow \quad \dfrac{1}{5} = \dfrac{\frac{x}{32}}{\frac{x}{32} + 10}$

$\Rightarrow \quad 5x = x + 320$

$\Rightarrow \quad 4x = 320 \Rightarrow x = 80$ g

So, the value of '$x$' is 80.

**26.** *(1)* Given, Henry's constant,

$$K_H = 46.82 \text{ k bar}$$
$$= 46.82 \times 10^3 \text{ bar}$$

Partial pressure of $O_2$, $p_{O_2} = 0.92$ bar

Using the Henry's equation,

$p_{O_2} = K_H \cdot \chi_{O_2} \quad [\chi = \text{mole fraction}]$

So, $\chi_{O_2} = \dfrac{p_{O_2}}{K_H}$

$$\chi_{O_2} = \frac{0.92 \times 10^{-3}}{46.82} = 0.02 \times 10^{-3}$$

Mole fraction of water,

$$\chi_{H_2O} = 1 - \chi_{O_2} = 1 - 0.02 \times 10^{-3} = 0.99$$

1 L water = 1000 gm ($\rho_{H_2O} - 1$ gmL$^{-1}$)

$$\text{Moles of water} = \frac{1000}{18} = 55.55 \text{ mol}$$

So, moles of $O_2$,

$$n_{O_2} = \frac{55.55 \times 0.02 \times 10^{-3}}{0.99}$$
$$= 1.13 \times 10^{-3} \text{ mol}$$
$$= 1.13 \text{ m mol} \approx 1 \text{ m mol}$$

**27.** *(25)* According to Henry's law,

$$p_{CO_2} = K_H \chi_{CO_2}$$

$p_{CO_2}$, $K_H$ and $\chi_{CO_2}$ are pressure of $CO_2$ gas, Henry's constant and mole fraction of $CO_2$ respectively.

$\Rightarrow \quad \chi_{CO_2} = \dfrac{p_{CO_2}}{K_H} \quad \ldots$(i)

Given, $K_H = 0.835 \times 2 \times 10^3$ bar

$$= 1.67 \times 10^3 \text{ bar}$$

$$p_{CO_2} = 0.835 \text{ bar}$$

Put all values in equation (i), we get

$$\chi_{CO_2} = \frac{0.835}{0.835 \times 10^3 \times 2} = 0.5 \times 10^{-3}$$

Volume of water = 0.9 L = 900 mL

Density of water = 1 g/mL

Mass of water = density × volume

$$= 900 \text{ g}$$

So, number of mole of water

$$= \frac{900}{18} = 50 \text{ moles}$$

So, $\quad \dfrac{n_{CO_2}}{n_{CO_2} + n_{H_2O}} = 0.5 \times 10^{-3}$

$$n_{CO_2} << n_{H_2O} = 50 \text{ moles}$$

So, $\quad n_{CO_2} + n_{H_2O} = n_{H_2O}$

$$n_{CO_2} = 0.5 \times 10^{-3} \times 50$$
$$= 25 \times 10^{-3} \text{ mol}$$

Millimoles of $CO_2$ = $0.025 \times 1000 = 25$

**28.** *(19)* Given, $p_A^\circ = 21$ kPa, $p_B^\circ = 18$ kPa

$$n_A = 1 \text{ mole}, \; n_B = 2 \text{ mole}$$

$$x_A = \frac{1}{3} \quad x_B = \frac{2}{3}$$

$$P_T = 21 \times \frac{1}{3} \text{ kPa} + 18 \times \frac{2}{3} \text{ kPa}$$
$$= 7 + 12 = 19 \text{ kPa}$$

**29.** (600) $p_{\text{Total}} = x_{\text{hexane}} p^\circ_{\text{hexane}} + x_{\text{heptane}} p^\circ_{\text{heptane}}$

Given that, $p_{\text{total}} = 550$ mm

$$x_{\text{hexane}} = \frac{1}{1+3} = \frac{1}{4},\ x_{\text{heptane}} = \frac{3}{3+1} = \frac{3}{4}$$

Substituting the values in above formula

$$550 = \frac{1}{4} \times p^\circ_{\text{hexane}} + \frac{3}{4} \times p^\circ_{\text{heptane}} \quad \ldots\text{(i)}$$

After mixing 1 more mole of *n*-heptane

$$560 = \frac{1}{5} \times p^\circ_{\text{hexane}} + \frac{4}{5} \times p^\circ_{\text{heptane}} \quad \ldots\text{(ii)}$$

From Eq. (i) and (ii)

$$p^\circ_{\text{heptane}} = 600 \text{ mm Hg.}$$

## Topic 3

**01.** (a) Given molar mass of non-volatile, Solute = 256 g/mol

Mass of solvent = 50 g

Mass of solute = 1 g

$\Delta T_f = 0.40$ g

The depression in freezing point is given by the formula,

$$\Delta T_f = K_F m$$

On substituting the values, we get

$$0.4 = \frac{K_F \times \frac{1}{256}}{50 \times 10^{-3}}$$

$$\Rightarrow \quad K_F = 5.12 \text{ K kg/mol}$$

**02.** (c) The expression of elevation of boiling point,

$$\Delta T_b = K_b \times m$$

So, the expression becomes,

$$\Delta T_b \propto K_b$$

$$\Rightarrow \frac{\Delta T_b(A)}{\Delta T_b(B)} = \frac{K_b(A)}{K_b(B)} = \frac{1}{5}$$

$$\left[\text{Given } \frac{K_b(A)}{K_b(B)} = \frac{1}{5}\right]$$

**03.** (b) Coolant is glycol ($C_2H_6O_2$) and it is non-electrolyte.

$$\Delta T_f = 2.8°\text{C} \Rightarrow \Delta T_f = \frac{1000 K_f w_1}{m_1 w_2}$$

$$2.8 = \frac{1000 \times 1.86 \times w_1}{62 \times 1000}$$

$$\Rightarrow \quad w_1 = 93.33 \text{ g}$$

**04.** (b) Given $K_b = 0.52$ K kg mol$^{-1}$

Mass of water = 500 g = 0.5 kg

Number of moles of ethylene glycol = 2

Number of moles of water = 2

Using $\Delta T_b = i_1 m_1 k_b + i_2 m_2 k_b$

$$= 1 \times \frac{2}{0.5} \times 0.52 + \frac{1 \times 2}{0.5} \times 0.52 = 4.16$$

$$(T_b)_{\text{Solution}} = 373.16 + 4.16 = 377.3 \text{ K}$$

**05.** (d) Depression in freezing point is given by

$$\Delta T_f = iK_f m$$

$$0.2 = i \times 1.8 \times 0.1$$

$$i = \frac{20}{18} = \frac{10}{9}$$

For, $HA(aq) \rightleftharpoons H^+(aq) + A^-(aq)$

| | $HA(aq)$ | $H^+(aq)$ | $A^-(aq)$ |
|---|---|---|---|
| $t = 0$ | 1 | | |
| $t = t_{eq}$ | $1-\alpha$ | $\alpha$ | $\alpha$ |

$i = 1 + \alpha$

$$\frac{10}{9} = 1 + \alpha \Rightarrow \alpha = \frac{1}{9}$$

$$K_{eq} = \frac{[H^+][A^-]}{[HA]} = \frac{C\alpha^2}{1-\alpha} = \frac{0.1\left(\frac{1}{9}\right)^2}{1 - \frac{1}{9}}$$

$$K_{eq} = 1.38 \times 10^{-3}$$

**06.** (d) When a non-volatile solute is added to solvent, freezing point of solution decreases.

$$T_f < T_f^\circ \text{ i.e.,}$$

Freezing point of solution < freezing point of pure solvent.

Also vapour pressure of solution decreases on adding a non-volatile solute in a solvent.

Hence graph (d) is correct.

**07.** (d) Both Statement I and Statement II are true.

**08.** (d) When small amount of naphthalene is added to benzene, depression in freezing point takes place and freezing point of benzene decreases.

**09.** (a)

> **key idea** This question involves the concept of Raoult's law and mole concept. Use Raoult's law to calculate the number of moles of glucose. Then use this value to find out the weight by multiplying it with molar mass of glucose.

According to Raoult's law, for dilute solution

$$\frac{p^\circ - p}{p^\circ} = \frac{n}{N}$$

[where $n$ = number of moles of glucose, $N$ = number of moles of $H_2O$]

On substituting the values

$$\frac{0.2}{54.2} = \frac{n \times 18}{100} \Rightarrow n = \frac{100}{271 \times 18}$$

Now weight of glucose = number of moles × molar mass

$$w = \frac{100}{271 \times 18} \times 180 \Rightarrow w = 3.69 \text{ g}$$

**10.** (b) Upon addition of non-volatile solute, less number of solvent particle can evaporate. Hence, vapour pressure of solution molecule decreases than pure solvent and hence, vapour pressure of pure solvent is more than that of solution. So, in order to get equal vapour pressure at freezing point only solvent molecule solidifies.

**11.** (b) Using the equation, $\Delta T_f = K_f m$

$$\therefore \frac{\Delta T_{fA}}{\Delta T_{fB}} = \frac{K_f m_A}{K_f m_B} = \frac{m_A}{m_B} = \frac{M_Y}{M_X} = \frac{1}{4}$$

$$\text{So, } M_X : M_Y = 4:1 = \frac{4 \times 0.25}{1 \times 0.25} = \frac{1}{0.25}$$

$$M_X : M_Y = 1 : 0.25$$

So, option (b) is correct.

**12.** (c)

> **key idea** Osmotic pressure is proportional to the molarity ($C$) of the solution at a given temperature ($T$). So, use the formula $\pi = CRT$ to calculate the value of $\pi$.

For the relation,

$$\pi = CRT = \frac{n}{V} RT$$

Given, mass of urea = 0.6 g

Molar mass of urea = 60 g mol$^{-1}$

Mass of glucose = 1.8 g

Molar mass of glucose = 180 g mol$^{-1}$

$$\pi = \frac{\left(\frac{0.6}{60} + \frac{1.8}{180}\right)}{100} \times 1000 \times 0.0821 \times 300$$

$$= (0.01 + 0.01) \times 10 \times 0.0821 \times 300$$

$$\pi = 4.92 \text{ atm}$$

**13.** (c) $\chi_B$ = mole fraction of solute (urea)

$$= \frac{n_B}{n_A + n_B} = \frac{0.60/60}{\frac{360}{18} + \frac{0.60}{60}} = \frac{0.01}{20 + 0.01}$$

$$= \frac{0.01}{20.01} = 0.0005$$

Now, according to Raoult's law,

$$\Delta p = \chi_B \times p^\circ$$

On substituting the above given values, we get

$$\Delta p = 0.0005 \times 35 = 0.0175 \text{ mm Hg}$$

**14.** *(d)* Elevation in boiling point

$(\Delta T_b) = K_b \times m$

Depression in freezing point

$(\Delta T_f) = K_f \times m$

where, $m$ = molality

$\Delta T_b^{1m} = \Delta T_f^{2m} = 2K$

So, $K_b \times 1 = K_f \times 2$

$\Rightarrow K_b = 2K_f$

**15.** *(d)* Osmotic pressure of isotonic solutions are equal.

Therefore, $\pi_1 = \pi_2$

$\Rightarrow C_1RT = C_2RT$

$\Rightarrow C_1 = C_2$

If densities of both the solutions are equal, then 5.25 g of a substance and 1.5 g of urea are dissolved in the same volume of the solvent, thus,

$$C_1 = \frac{5.25/M}{V}; \quad C_2 = \frac{1.5/60}{V}$$

$$\therefore \frac{5.25/M}{V} = \frac{1.5/60}{V}$$

$\Rightarrow M = 210.0 \text{ g mol}^{-1}$

**16.** *(b)* Boiling point and freezing point depend on $K_b$ (molal elevation constant) and $K_f$ (molal depression constant) of the solvent. Thus, equimolar solution (of the non-electrolyte), will have same boiling point and also same freezing point.

$\Delta T_f = K_f \times$ molality;

$\Delta T_b = K_b \times$ molality

**Note** In this question, nature of solute has not been mentioned. Hence, we have assumed that solute is non-electrolyte.

**17.** *(c)* Statement I is correct but Statement II is incorrect.

The correct form of Statement II is $k_f$ for benzene is more than $k_f$ for water.

**18.** *(d)* Given, $K_f = 1.86 \text{ K kg mol}^{-1}$

$\Delta T_f = 0.558$ °C

Molality = 0.1 molal

Using formula,

$\Delta T_f = iK_fm$

$0.558 = i \times 1.86 \times 0.1$

On solving, we get

$i = 3$

So, among the given options, compound given in option (d) have the value of $i = 3$. So, the complex is $[Cr(NH_3)_5Cl]Cl_2$.

**19.** *(c)* Due to osmosis, net movement of water molecules takes place from low osmotic pressure to high osmotic pressure. Since the value of concentration of 0.05 M $CuSO_4$ solution is higher hence, it has higher osmotic pressure. So, movement of water molecule will take towards $CuSO_4$ solution. Due to which molarity of $CuSO_4$ is lowered.

**20.** *(b)* $\Delta T_b = K_b \times m$

$(\Delta T_b)_A = (\Delta T_b)_B$ (given)

$\therefore m_A = m_B$

Molality

$$= \frac{\text{Mass of solute} \times 1000}{\text{M.M. of solute} \times \text{Weight of solvent}}$$

$$m_A = \frac{2 \times 1000}{M_A \times 98}; \quad m_B = \frac{8 \times 1000}{M_B \times 92}$$

$$\frac{2 \times 1000}{M_A \times 98} = \frac{8 \times 1000}{M_B \times 92}$$

$\therefore M_B = 4M_A$

**21.** *(b)* Vapour pressure of water $(p°) = 760$ torr

$$n_g = \frac{18 \text{ g}}{180 \text{ gmol}^{-1}} = 0.1 \text{ mol}$$

Molar mass of water = 18 g/mol

Mass of water (given) = 178.2 g

Number of moles of water

$$n_w = \frac{178.2 \text{g}}{18 \text{ g/mol}} = 9.9 \text{ mol}$$

Total number of moles = (0.1 + 9.9) moles = 10 moles

Now, mole fraction of glucose in solution = Change in pressure with respect to initial pressure

i.e. $\frac{\Delta p}{p°} = \frac{0.1}{10}$

or $\Delta p = 0.01p°$

$= 0.01 \times 760 = 7.6$ torr

$\therefore$ Vapour pressure of solution

$= (760 - 7.6)$ torr $= 752.4$ torr

**22.** *(d)* This problem is related to the relative lowering of vapour pressure as glucose is a non-volatile solute. Thus, according to Raoult's law,

$$\frac{p^0 - p_{\text{solute}}}{p^0} = X_{\text{solute}}$$

Moles of glucose $= \frac{18}{180} = 0.1$

Moles of $H_2O = \frac{178.2}{18} = 9.9$

$$X_{\text{glucose}} = \frac{0.1}{9.9 + 0.1} = \frac{0.1}{10}$$

Thus, $\frac{17.5 - p_s}{17.5} = \frac{0.1}{10}$

So, $p_s = 17.325$ mm Hg

**23.** *(543)* Given,

$\Delta T_f = 2.5$°C (from diagram)

$K_f = 1.86 \text{ K kg mol}^{-1}$

$\Delta T_f = K_f \times m$

$2.5 = 1 \times 1.86 \times \frac{n_B}{0.1}$

$n_B = 0.1344$ mole

Mass of methanol $= 0.1344 \times 32 = 4.3$ g

$d = \frac{m}{V} \Rightarrow V = \frac{m}{d}$

$V = \frac{4.3}{0.792} = 5.43$ mL

$= 543 \times 10^{-2}$ mL

Thus, the value of $x$ is 543.

**24.** *(25)* $NaCl \rightarrow Na^+ + Cl^-$

0.05 M 0.05 M 0.05 M

Total $C_1 = 0.05 + 0.05 = 0.1$ M (NaCl)

$C_2 = 0.2$ M (glucose)

Osmotic pressure is given by,

$\pi = (C_2 - C_1)RT$

$= (0.2 - 0.1) \times 0.083 \times 300 = 2.49$ bar

$= 24.9 \times 10^{-1}$ bar

$\approx 25 \times 10^{-1}$ bar

**25.** *(31)*

> **key idea** To solve this question, first of all see that, the component which is present in excess will act as solvent. Then using this data calculate the molality, after that use the formula $\Delta T_f = K_fm$ to calculate the freezing point of the solution.

As the number of moles of water is greater than the moles of acetic acid. Thus, water acts as a solvent.

Now, calculating the molality of acetic acid $= \frac{2700}{60} \times \frac{1}{2.7}$ mol/kg

$= 16.667$ mol/kg

Now, using $\Delta T_f = K_fm$

$= 1.86 \times 16.667 = 31$ K

**26.** *(707)*

> **key idea** To solve this question, use the formula $\Delta T_b = k_b m$ to calculate the value of molality ($m$). After that use this value of molality in the equation $\frac{\Delta p}{p^\circ} = \frac{m}{100} \times M_{solvent}$, as the solution is diluted, to calculate $\Delta p$. Then using this $\Delta p$ calculate the required value of $p_s$.

Given that,

$\Delta T_b = 2°C$

$K_b = 0.52 \text{ K kg mol}^{-1}$

$p° = 760 \text{ mm Hg}$

$M_{solvent} = 18 \text{ g mol}^{-1}$

Now using $\Delta T_b = K_b m$

$$2 = 0.52 \times m$$

$$m = \frac{2}{0.52}$$

Now, according to question, since the solution is diluted much. Hence,

$$\frac{\Delta p}{p^\circ} = \frac{n_{solute}}{n_{solvent}} \Rightarrow \frac{\Delta p}{p^\circ} = \frac{m}{1000} \times M_{solvent}$$

$$\Rightarrow \Delta p = 760 \times \frac{\left(\frac{2}{0.52}\right)}{1000} \times 18$$

$\Delta p = 52.615$

So, $p_s = 760 - 52.615$

$p_s = 707.385$ mm Hg

Hence, nearest integer is 707.

**27.** *(73)* Given, $T_1 = 273$ K

Osmotic pressure at $T_1 = 7 \times 10^5$ Pa

$T_2 = 283$ K

Osmotic pressure, $\pi = CRT$

$\pi_1 = CRT_1$ and $\pi_2 = CRT_2$

$$\frac{\pi_1}{\pi_2} = \frac{T_1}{T_2} \Rightarrow \pi_2 = \frac{\pi_1 T_2}{T_1}$$

$$\Rightarrow \pi_2 = \frac{7 \times 10^5 \times 283}{273}$$

$$\pi_2 = 72.56 \times 10^4 \text{ N/m}^2$$

$$\approx 73 \times 10^4 \text{ N/m}^2$$

Hence, nearest integer value is 73.

**28.** *(23)* Given,

weight of solute = 30 g

volume of solution = 100 mL

weight of solution = 1.2 × 100 = 120 g

weight of solvent = 120 − 30 = 90 g

$$m = \frac{30 \times 1000}{180 \times 90} = 1.85$$

$$\frac{24 - p_S}{p_S} = \frac{1.85 \times 18}{1000}$$

$$\Rightarrow 24 - p_S = 0.0333 p_S$$

$$p_S (1.033) = 24$$

$$\Rightarrow p_S = 23.22$$

**29.** *(8)* The boiling point elevation constant ($K_b$) of a volatile liquid is given by the following expression

$$K_b = \frac{RT_b^2 M}{\Delta Hv}$$

where, $T_b$ = boiling point of the liquid

$M$ = molar mass of the liquid

$\Delta Hv$ = enthalpy of vaporisation

Given, $M_x = M_y$

$(Tb)_x = 2(Tb)_y$

$2(\Delta Hv)_x = (\Delta Hv)_y$

$$\therefore \frac{(K_b)_x}{(K_b)_y} = \left(\frac{2}{1}\right)^2 \times \frac{2}{1}$$

$$= \frac{8}{1} = m$$

$\therefore \quad m = 8$

**30.** *(833)* Given, mass of water = 1000 g

Molar mass of mass of urea = 60 g mol$^{-1}$

Mass of urea = ?

We know that, vapour pressure decreases by 25%.

$$\text{So, } \frac{p^\circ - p}{p^\circ} = \frac{\text{Moles of urea}}{\text{Moles of water}}$$

$$\frac{100 - 75}{100} = \frac{\frac{x}{60}}{\frac{1000}{18}}$$

$$x = 833.3 \text{ g} \approx 833$$

**31.** *(62250)* Osmotic pressure ($\pi$) = $CRT$

Given : solute mass = 2.5 g

Volume of solution = 250 ml

osmotic pressure = 400 pa

$$\text{So, } \frac{400\text{Pa}}{10^5} = \frac{2.5\text{g}/M}{250/1000} \times 0.83 \frac{\text{L-bar}}{\text{K mol}} \times 300\text{K}$$

$\therefore \quad M = 62250 \text{ gmol}^{-1}$

**32.** *(100)* Given, mass of solute

$= 2\text{g} = 2 \times 10^{-3}$ kg

Mass of solvent = 20g = $20 \times 10^{-3}$ kg

Temperature = 373.52 K

Elevation in boiling point,

$\Delta T_b = 373.52 \text{ K} - 373 \text{ K}$

$= 0.52$ K

Use the formula,

$$\Delta T_b = K_b \times m$$

Ebullioscopic constant of water

$(K_b) = 0.52 \text{ K kg mol}^{-1}$

Mass of solute = 2g = 0.002 kg

So,

$$0.52 = 0.52 \times \frac{2}{\text{molar mass}} \times \frac{1}{20 \times 10^{-3}}$$

$$\text{Molar mass} = \frac{0.52 \times 2 \times 1}{0.52 \times 20 \times 10^{-3}}$$

$$= 100 \text{ g/mol}$$

**33.** *(41500)*

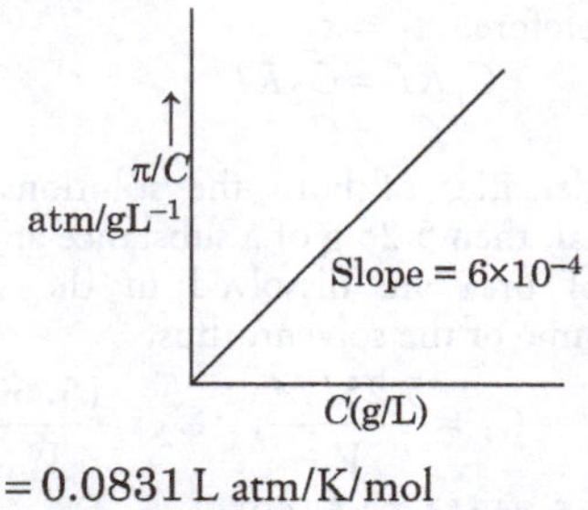


$R = 0.0831$ L atm/K/mol

Osmotic pressure, $\pi$ = Molar concentration × $RT$

$$\pi = \frac{C}{M} RT \Rightarrow \frac{\pi}{C} = \frac{RT}{M}$$

Assuming $\pi/C$ *vs* $C$ graph

$$\frac{RT}{\text{Slope}} = \frac{RT}{M} = \frac{0.083 \times 300}{6 \times 10^{-4}}$$

$$M = 41500 \text{ g/mol}$$

**34.** *(15)* We know that,

$$\Delta T_f = K_f \times m$$

$$= 3.9 \times \frac{10.2 \times 1000}{176 \times 150} = 1.5°C$$

or $15 \times 10^{-1}$ °C

$\therefore x = 15$

**35.** *(6)* Vapour pressure of pure water,

$p^\circ_{H_2O} = 23.76$ mmHg

Vapour pressure of solution,

$$p_s = \frac{p^\circ_{H_2O}}{2} = 11.88 \text{ mmHg}$$

According to the expression of relative lowering of vapour pressure.

$$\frac{p^\circ_{H_2O} - p_s}{p^\circ_{H_2O}} = \chi_A$$

$$\frac{23.76 - 11.88}{23.76} = \frac{n_A}{n_A + n_{H_2O}} \quad \left[n_{H_2O} = \frac{100}{18} = 5.55\right]$$

$$\frac{1}{2} = \frac{n_A}{n_A + 5.55} \text{ or } n_A \approx 6$$

**36.** *(1)* Given,

$$\Delta T_b = K_b \times m = 3$$

or, $\Delta T_b = K_b \times 1 = 3$

$$\Delta T_f = K_f \times m = 6$$

or, $\Delta T_f = K_f \times 2 = 6$

$$\frac{\Delta T_b}{\Delta T_f} = \frac{K_b \times 1}{K_f \times 2}$$

$$\frac{3}{6} = \frac{K_b}{K_f \times 2} \Rightarrow \frac{6}{6} = \frac{K_b}{K_f}$$

According to data given,

$$\frac{K_b}{K_f} = \frac{1}{1} = \frac{1}{x} \Rightarrow x = 1$$

**37.** *(45)* Mass of solute $= 2.5 \times 10^{-3}$ kg

Mass of solvent $= 75 \times 10^{-3}$ kg

Boiling point of solution $= 373.535$ K

Boiling point of water $= 373.15$ K

$\therefore$ Elevation in boiling point $\Delta T_b$

= Boiling point of solution − Boiling point of water

$= 373.535 - 373.15 = 0.385$ K

Also, $\Delta T_b = K_b \times$ molality

$$0.385\text{ K} = 0.52 \times \left(\frac{2.5 \times 10^{-3}}{M \times 75 \times 10^{-3}}\right)$$

$M = 0.045$ kg/mol or $M = 45$ g/mol

Hence, molar mass of solute is 45 g/mol.

**38.** *(8)* We know that $\Delta T_b \propto K_b$, as

$$\Delta T_b = K_b m$$

$$\Rightarrow \frac{(\Delta T_b)_A}{(\Delta T_b)_B} = \frac{(K_b)_A}{(K_b)_B} \Rightarrow \frac{x}{y} = \frac{1}{8}$$

$\therefore$ $y = 8$

**39.** *(415)* Osmotic pressure, $\pi = CRT$

Given, mass of protein = 2 g

Molar mass of protein $= 60$ kg mol$^{-1}$ $= 60{,}000$ g mol$^{-1}$

Volume of water = 200 mL = 0.2 L

Concentration,

$$C = \frac{\text{Given mass}}{\text{Molar mass}} \times \frac{1}{\text{Volume solution}}$$

$$\pi = \frac{2 \times 0.083 \times 300}{60000\text{ g/mol} \times 0.2\text{ L}}$$

$\pi = 0.00415$ bar (1 bar $= 10^5$ Pascal)

So, $\pi = 0.00415 \times 10^5$ Pa

$\pi = 415$ Pa

**40.** *(54)* Molar mass of glucose $= 180$ g mol$^{-1}$.

We know, $\pi = CRT$

$$7.47 = C \times 0.083 \times 300$$

$$\Rightarrow C = \frac{7.47}{0.083 \times 300} = 0.3\text{ M}$$

Concentration of glucose solution $= 0.3$ M

Now, concentration in gL$^{-1}$ $= 0.3 \times 180 = 54$ gL$^{-1}$

**41.** *(271)* 40 g of glucose mixed with 200 mL of water

180 g of glucose = 1 moles of glucose

40 g of glucose = 0.22 mol of glucose

1 mL of water = 1 g of water

$[d = 1\text{ g/cm}^2,\ 1\text{ mL} = 1\text{ cm}^3]$

200 mL of water = 200 g of water

$$\Delta T_f = K_f m$$

where,

$\Delta T_f$ = depression in freezing point,

$K_f$ = molal elevation constant $= 1.86$ K kg mol$^{-1}$ and

$m$ = molality of solution

$$\Delta T_f = \frac{1.86 \times 0.22}{200} \times 1000 \quad (\because 1000\text{ g} = 1\text{ kg})$$

$$\Delta T_f = 2\text{ K}$$

$\therefore (T_f - T_f') = 2$

[$T_f$ = freezing point of water (273 K), $T_f'$ = freezing point of solution]

$$273\text{ K} - T_f' = 2\text{ K}$$

$$T_f' = (273 - 2)\text{K} = 271\text{ K}$$

**42.** *(518)* Let mass of water (initially present) $= x$ g

Mass of sucrose $= (1000 - x)$ g

$$\text{Moles of sucrose} = \frac{1000 - x}{342}$$

$$0.75 = \frac{\left(\frac{1000 - x}{342}\right)}{\left(\frac{x}{1000}\right)}$$

$$\frac{x}{1000} = \frac{1000 - x}{342 \times 0.75}$$

$$256.5x = 10^6 - 1000x$$

$$\Rightarrow x = 795.86\text{ g}$$

$$\text{Moles of sucrose} = \frac{1000 - 795.86}{342} = 0.5969$$

New mass of $H_2O = a$ kg

Depression in freezing point

$$\Delta T_f = K_f \times m$$

$$4 = \frac{0.5969}{a} \times 1.86 \Rightarrow a = 0.2775\text{ kg}$$

Ice separated $= 795.86 - 277.5 = 518.3\text{ g} \approx 518$ g.

**43.** *(15)* Using formula; $\pi = CRT$

$$\text{Molarity} = \frac{1.46 \times 1000}{M \times 100\text{ (mL)}} = \frac{1460}{100 \times M}$$

$$\therefore \pi = \frac{1460}{100 \times M} \times 0.083 \times 300$$

Given, $\pi = 2.42 \times 10^{-3}$ bar

Therefore, $M = 15.02 \times 10^4$ g/mol

**44.** *(250)* Weight of substance 'X' = 3.00 g

Elevation in boiling point, $\Delta T_b = 0.60$

$K_b$ (boiling constant) of $CCl_4$ $= 5$ K kg mol$^{-1}$

$$m = \frac{\frac{3}{M}}{\frac{100}{1000}} \quad \text{...(i)}$$

Now using the formula,

$$\Delta T_b = K_b \times \text{molality}$$

Put value of (i) we get,

$$0.60 = 5 \times \frac{\frac{3}{M}}{\frac{100}{1000}}$$

$$\Rightarrow 0.60 = 5 \times \frac{3 \times 1000}{M \times 100}$$

$$0.60\,M = 5 \times 30 \Rightarrow M = \frac{150}{0.60} = 250$$

$\therefore$ Molar mass of substance 'X' is 250 g mol$^{-1}$.

**45.** *(85)*

$$K_4Fe(CN)_6 \rightleftharpoons 4K^+ + [Fe(CN)_6]^{4-}$$

| | $K_4Fe(CN)_6$ | $4K^+$ | $[Fe(CN)_6]^{4-}$ |
|---|---|---|---|
| Initial conc. | 1 m | 0 | 0 |
| Final conc. | (1 − 0.4) m = 0.6 m | 4 × 0.4 = 1.6 m | 0.4 m |

Effective molality $= 0.6 + 1.6 + 0.4 = 2.6$ m

Molality of non-electrolyte solution = molality of $K_4[Fe(CN)_6] = 2.6$ m

Now, 18.1 weight per cent solution means 18.1 g solute is present in 100 g solution and hence, (100 − 18.1 =) 81.9 g water.

Molality

$$\text{Now, } 2.6 = \frac{18.1/M}{81.9/1000}$$

Molar mass of solute, $M = 85$

**46.** (2.18) $K_f = 2.0$ (Given) $m = 0.5$ m (Given)

$$\Delta T_f = K_f \times m = 2 \times 0.5 = 1$$

$\therefore T_f$ = Actual freezing point of water $- \Delta T_f$.

$= 273 - 1 = 272$ K.

Pressure of gas after the stoppers are withdrawn = 1 atm (atmospheric pressure)

$$\therefore \quad V = \frac{nRT}{p} = \frac{0.1 \times 0.08 \times 272}{1 \text{ atm}}$$

$= 2.176 \text{ L} \approx 2.18 \text{ L}$

## Topic 4

**01.** (d) The van't Hoff factor ($i$) is highest for $KHSO_4$ as it dissociate into its ions. Hydrazine, glucose and glycine will not dissociate to produce ions.

Hence, correct option is (d).

**02.** (b) $Na_2SO_4 \longrightarrow 2Na^+ + SO_4^{2-}$

van't Hoff factor ($i$) for $Na_2SO_4 = 3$

$$\Delta T_f = i \times K_f \times m$$
$$= 3 \times 1.86 \times 0.01$$

$$\left[\because m = \frac{0.01 \text{ mol}}{1 \text{ kg}} = 0.01 \text{ mol kg}^{-1}\right]$$

$= 0.0558$ K

**03.** (b) $Na_2SO_4 \rightleftharpoons 2Na^+ + SO_4^{2-}$

van't Hoff factor, $i = [1 + (y - 1)\alpha]$

where $y$ is the number of ions from one mole solute, (in this case = 3), $\alpha$ is the degree of dissociation, $i = (1 + 2\alpha)$

**04.** (a) Highest depression in freezing point (colligative properties) is inversely proportional to experimental molar mass of non-volatile solute.

- On dissociation, experimental molar mass decreases.
- On association, experimental molar mass increases.

**05.** (d) As dilution increases concentration of solution decreases and degree of dissociation increases, resulting in increase of van't Hoff factor.

So, the correct order will be $i_A < i_B < i_C$.

Hence, option (d) is correct.

**06.** (c) The correct match is

A-III, B-I, C-II, D-IV

**07.** (a) van't Hoff factor of $Al_2(SO_4)_3$ is 5 (maximum), so this will have lowest freezing point.

Depression in freezing point,

$$\Delta T_f = iK_f m$$

For freezing point to be lowest, $\Delta T_f$ should be highest and the $i$ values are as follows

| Solute | $i$ |
|---|---|
| $C_6H_{12}O_6$ | 1 |
| $Al_2(SO_4)_3$ | 5 $\rightarrow i$ highest, $\Delta T_f$ maximum, $T_f$ minimum |
| KI | 2 |
| $K_2SO_4$ | 3 |

Hence, for $Al_2(SO_4)_3$ freezing point is lowest.

**08.** (b) $K_f = 4.0$ K kg mol$^{-1}$ (Given)

$m = 0.03$ mol kg$^{-1}$ (Given)

$\Delta T_f = ?$

For $K_2SO_4$, $i = 3$

It can be verified by the following equation :

$$K_2SO_4 \rightleftharpoons 2K^+ + SO_4^{2-}$$

Using formula

$$\Delta T_f = iK_f \times m$$
$$= 3 \times 4 \times 0.03 = 0.36 \text{ K}$$

**09.** (d)

**key idea** Osmotic pressure is proportional to the molarity ($C$) of the solution at a given temperature, $\pi = CRT$.

Concentration of

$BaCl_2 = 0.01$ M (Given)

$\pi_{XY} = 4\pi_{BaCl_2}$ (Given)

$i \times CRT = 4 \times i \times CRT$ ...(i)

For the calculation of $i$,

$XY \longrightarrow X^+ + Y^-$ (Here, $i = 2$)

$BaCl_2 \longrightarrow Ba^{2+} + 2Cl^-$ (Here, $i = 3$)

Putting the values of $i$ in (i)

$$2 \times [XY] = 4 \times 3 \times [BaCl_2]$$
$$2 \times [XY] = 12 \times 0.01$$
$$[XY] = \frac{12 \times 0.01}{2}$$

So, the concentration of $XY$

$= 0.06$ mol L$^{-1}$

$= 6 \times 10^{-2}$ mol L$^{-1}$

**10.** (b) The ionisation of $K_2HgI_4$ in aqueous solution is as follows:

$$K_2[HgI_4] \rightleftharpoons 2K^+ + [HgI_4]^{2-}$$

van't Hoff factor ($i$) for ionisation reaction is given as, $i = 1 + \alpha(n - 1)$

where, $n$ = number of ions,

$\alpha$ = degree of ionisation or dissociation

From above equation, it is clear that $n = 3$

$i = 1 + 0.4(3 - 1) = 1.8$

[Given, % $\alpha$ = 40% or $\alpha = 0.4$]

**11.** (d)

**key idea** "Addition of solute particles to a pure solvent results to depression in its freezing point." All the compounds given in question are ionic in nature so, consider their van't Hoff factor (i) to reach at final conclusion.

Number of solute particles $\propto$ van't Hoff factor ($i$)

Thus, we can say directly

Solution with maximum freezing point will be the one in which solute with minimum van't Hoff factor is present

Now, for

$$Co(H_2O)_6]Cl_3 \rightleftharpoons [Co(H_2O)_6]^{3+} + 3Cl^-$$

van't Hoff factor ($i$) is 4. Similarly for,

$[Co(H_2O)_5Cl]Cl_2 \cdot H_2O \rightleftharpoons [Co(H_2O)_5Cl]^{2+} + 2Cl^-$ '$i$' is 3

$[Co(H_2O)_4Cl_2]Cl \cdot 2H_2O \rightleftharpoons [Co(H_2O)_4Cl_2]^+ + Cl^-$ '$i$' is 2

and for $[Co(H_2O)_3Cl_3] \cdot 3H_2O$, '$i$' is 1 as it does not show ionisation.

Hence, $[Co(H_2O)_3Cl_3] \cdot 3H_2O$ have minimum number of particles in the solution.

So, freezing point of its solution will be maximum.

**12.** (a) van't Hoff factor ($i$) is related to degree of dissociation ($\alpha$) as

$$\alpha = \frac{i - 1}{n - 1}$$

Here, $n$ are the moles of an electrolyte ($A - B$) dissolved in a solvent.

For $A_x B_y$,

$$A_x B_y \rightleftharpoons xA^{+y} + yB^{-x}$$

Therefore, $\alpha = \dfrac{i - 1}{(x + y - 1)}$

**13.** (a) The colligative property will be given as

$$\Delta T_b = i \times k_b \times m$$

From the above equation, it is clear that,

$$\Delta T_b \propto i$$

This suggests that higher the value of $i$, larger is the boiling point of the solution.

For 100% dissociation, $i$ = number of ions produced by one mole of solute

Thus, for

0.01 M, $Na_2SO_4$

$$Na_2SO_4 \rightleftharpoons \underset{[3 \text{ ions}]}{2Na^+ + SO_4^{2-}}$$

$\therefore \quad i = 3$

**14.** *(a)* Given that, $C_1 > C_2$

So, normal osmosis will take place from 2 to 1 while reverse osmosis will take place from 1 to 2.

Pressure, $p_1 > \pi$

So, statement A and C are correct.

**15.** *(b)* We know that,

$$\Delta T_b = i \times m \times k_b$$

$$\Rightarrow \quad \Delta T_b \propto iM$$

(Where, $M$ = concentration)

For NaCl, the value of $i$ is 2 while for urea the value of $i$ is 1.

(i) For $10^{-4}$ M NaCl, the value of

$$\Delta T_b \propto 10^{-4} \times 2$$

(ii) For $10^{-4}$ M urea, the value of

$$\Delta T_b \propto 10^{-4} \times 1$$

(iii) For $10^{-3}$ M NaCl, the value of

$$\Delta T_b \propto 10^{-3} \times 2$$

(iv) For $10^{-2}$ M NaCl, the value of

$$\Delta T_b \propto 10^{-2} \times 2$$

So, the correct increasing order of boiling points is

(ii) < (i) < (iii) < (iv)

**16.** *(b)* The correct statement is $AX_2$ is fully ionised while $AY_2$ is completly unionised.

For $AX_2, \Delta T_b = K_b \times m \times i$

Given, $T_b = 100.0156°C = 373.0156$ K

$$\therefore \Delta T_b = T_b - T_b^\circ$$

$$= (373.0156 - 373) \text{ K}$$

$$= 0.0156 \text{ K}$$

$K_b = 0.52$ K kg mol$^{-1}$,

$$\text{Molality } (m) = \frac{1.24 \text{ g} / 124 \text{ g mol}^{-1}}{1 \text{ kg}}$$

$$0.0156 = 0.52 \times \frac{0.01}{1} \times i_{AX_2}$$

$$\Rightarrow \quad i_{AX_3} = 3$$

$\Rightarrow$ Completely ionised

For $AY_2$

$$\therefore \quad \Delta T_b = K_b \times m \times i$$

Given, $T_b = 100.0260°C = 373.0260$ K

$$\therefore \Delta T_b = T_b - T_b^\circ$$

$$= (373.0260 - 373) \text{ K} = 0.0260 \text{ K}$$

$$\text{Molality } (m) = \frac{25.4 \text{ g} / 250 \text{ g mol}^{-1}}{2 \text{ kg}}$$

$$= 0.0508 \text{ mol kg}^{-1}$$

$$0.026 = 0.52 \times 0.0508 \times i_{AY_2}$$

$$= i_{AY_2} \approx 1$$

$\Rightarrow$ Completely unionised.

**17.** *(a)* Statement B and D are correct. The corrected form of Statements A, C and E are respectively

(A) $\Delta T_b \propto i \times C$.

$NaCl \rightleftharpoons Na^+ + Cl^- \quad (i = 2)$

$NH_2CONH_2 \longrightarrow$ Non-ionic solution

(C) Osmosis always takes place from hypotonic to hypertonic solution.

(E) When KI solutions is added to $AgNO_3$ solution, positively charged solution results due to adsorption of $Ag^+$ ions from dispersion medium.

$$\underset{\text{Positively charged}}{AgI/Ag^+}$$

**18.** *(c)* Given, concentration of formic acid = 0.5 mol L$^{-1}$

Density of formic acid = 1.05 g mL$^{-1}$.

$$K_f = 1.86 \text{ K kg mol}^{-1}$$

$$\Delta T_f = 0.0405°C$$

Now, $\quad d = \frac{M}{V} \Rightarrow M = d \cdot V$

$$= 1.05 \times 0.5 = 0.525 \text{ g}$$

According to formula,

$$\Delta T_f = iK_f m \quad (m = \text{molality})$$

Let us take, mass of solution = 1 kg

$$= 1000 \text{ g}$$

Mass of solvent = 1000 − 0.525

$$= 999.475 \text{ g (formic acid)}$$

So $\quad m = \frac{0.525 \times 1000}{46 \times 999.475} = 0.0114$

$$\Delta T_f = iK_f \; (0.0114)$$

$$\Rightarrow \quad i = \frac{0.0405}{1.86 \times 0.0114} = 1.92$$

**19.** *(d)* Given, Mass of $A$ = 0.7 g

Molar mass of $A$ = 93 g mol$^{-1}$

$$\text{Moles of } A = \frac{0.7}{93} \text{ mol}$$

Mass of solvent = 42 g

$$\text{Molality} = \frac{0.7 \times 1000}{93 \times 42}$$

Depression in freezing point,

$$\Delta T_f = 0.2°C$$

Molal depression constant,

$$K_f = 1.86 \text{ K kg mol}^{-1}$$

Using the equation,

$$\Delta T_f = iK_f m$$

$$0.2 = i \times 1.86 \times \frac{0.7 \times 1000}{93 \times 42}$$

$$i = \frac{0.2 \times 93 \times 42}{1.86 \times 0.7 \times 1000} = 0.6$$

For association,

$$\beta = \frac{i - 1}{\left(\frac{1}{n} - 1\right)}$$

So, $\quad \beta = \frac{0.6 - 1}{\frac{1}{2} - 1} = \frac{-0.4}{-0.5}$

$\beta = 0.8 \quad \Rightarrow \%\beta = 0.8 \times 100 = 80\%$

**20.** *(d)* Let the degree of association of acetic acid ($CH_3COOH$) in benzene is α, then

| | $2CH_3COOH \rightleftharpoons$ | $(CH_3COOH)_2$ |
|---|---|---|
| Initial moles | 1 | 0 |
| Moles at equilibrium | $1 - \alpha$ | $\frac{\alpha}{2}$ |

$\therefore$ Total moles $= 1 - \alpha + \frac{\alpha}{2} = 1 - \frac{\alpha}{2}$

or $\quad i = 1 - \frac{\alpha}{2}$

Now, depression in freezing point ($\Delta T_f$) is given as

$$\Delta T_f = i K_f m \quad \ldots \text{(i)}$$

$$m = \frac{0.2}{60} \times \frac{1000}{20}$$

Putting the values in Eq. (i)

$$\therefore \; 0.45 = \left[1 - \frac{\alpha}{2}\right](5.12)\left[\frac{0.2}{60} \times \frac{1000}{20}\right]$$

$$1 - \frac{\alpha}{2} = \frac{0.45 \times 60 \times 20}{5.12 \times 0.2 \times 1000}$$

$$\Rightarrow \quad 1 - \frac{\alpha}{2} = 0.527 \Rightarrow \frac{\alpha}{2} = 1 - 0.527$$

$\therefore \quad \alpha = 0.946$

Thus, percentage of association

$$= 94.6\%$$

**21.** *(a)* Osmotic pressure, $\pi = i \times MRT$

Where, $i$ = van't Hoff factor,

$M$ = Molarity

$R$ = Gas constant,

$T$ = Temperature

For 0.500 M $C_2H_5OH$ $(aq)$, $i = 1$, thus,

$$\pi_{C_2H_5OH} (aq) = 0.5 \text{ RT}$$

For 0.100 M $Mg_3(PO_4)_2 (aq), i = 5$

$[\because Mg_3(PO_4)_2 \rightleftharpoons 3Mg^{2+} + 2PO_4^{3-}]$

Thus, $\pi_{Mg_3(PO_4)_2}(aq) = 5 \times 0.1 \times RT$
$= 0.5\ RT$

For 0.250 M KBr (*aq*), $i = 2$

$[\because KBr \rightleftharpoons K^+ + Br^-]$

Thus, $\pi_{KBr}(aq) = 2 \times 0.250 \times RT$
$= 0.5\ RT$

For 0.125 M $Na_3PO_4(aq)$, $i = 4$

$[\because Na_3PO_4 \rightleftharpoons 3Na^+ + PO_4^{3-}]$

Thus, $\pi_{Na_3PO_4}(aq) = 4 \times 0.125 \times RT$
$= 0.5\ RT$

Hence, all solutions have same osmotic pressure.

**22.** *(33)* The dissociation reaction is

$$MX_3 \longrightarrow M^{3+} + 3X^-$$

$$i = 1 + (n-1)\ \alpha$$
$$i = 1 + (4-1)\ \alpha = 2$$
$$\alpha = \frac{1}{3} = 33.33 \text{ or } 33\%$$

**23.** *(25)* For *AB*

Change in boiling point, $\Delta T_b = 2.7$ K

Using formula, $\Delta T_b = iK_b m$

$$2.7 = 1 \times 0.5 \times m \Rightarrow m = \frac{27}{5}$$

Let molar mass of *AB* = *X*

So, $\dfrac{1/X}{15} \times 1000 = \dfrac{27}{5}$

$$X = 12.34$$

For $AB_2$

$$\Delta T_b = 1.5 \text{ K}$$
$$1.5 = 1 \times 0.5 \times m$$
$$m = 3$$

Let molar mass of $AB_2 = Y$

$$\frac{\frac{1}{Y}}{15} \times 1000 = 3 \Rightarrow Y = 22.22$$

Now, $A + b = 12.34$ ...(i)

$A + 2b = 22.27$ ...(ii)

Where, *a* and *b* be atomic mass of *A* and *B*

$$b = 9.88$$
$$a = 24.6 \times 10^{-1}$$
$$\simeq 25 \times 10^{-1}$$

**24.** *(75)* The dissociation takes place as follows

$$MX_2 \longrightarrow M^{2+} + 2X^-$$

$$i = \frac{\text{Normal molar mass}}{\text{Observed molar mass}}$$

$$i = \frac{164}{65.6}$$

$$1 + (3-1)\ \alpha = \frac{164}{65.6}$$

$$2\alpha = \frac{98.4}{65.6}$$

$$\alpha = 0.75$$

Per cent degree of ionisation = 75%

**25.** *(16)* The reaction involved is as follows

$$A_2B \longrightarrow 2A^+ + B^{2-},\ y = 3$$
$$\alpha = 0.3$$
$$i = 1 + (y-1)\alpha$$

On substituting the value of *y* and α in above equation, we get

$$i = 1 + (3-1)\ (0.3) = 1.6 \text{ or } 16 \times 10^{-1}$$

**26.** *(5)* $\Delta T_b = 0.52°C$

$$\Delta T_b = i \times K_b \times m$$

$$0.52 = i \times 0.52 \times \frac{10}{200 \times 0.1}$$

$$i = 2 \quad \text{...(i)}$$

$$\alpha = \frac{i-1}{n-1}$$

$$AB_2 \longrightarrow A^{2+} + 2B^-$$
$$n = 3$$

$$\alpha = \frac{2-1}{3-1} \Rightarrow \alpha = \frac{1}{2} = 0.5$$

$$\alpha = 5 \times 10^{-1}$$

Hence, degree of ionisation is $5 \times 10^{-1}$.

**27.** *(19)* Mass of $CH_3COOH = V \times d = 6$ g

$\because V = 5$ mL, $d = 1.2$ g/mL (given)

$$n_{CH_3COOH} = \frac{\text{Mass}}{\text{Molar mass}} = \frac{6}{60} = 0.1 \text{ mol}$$

$$m_{CH_3COOH} \simeq M_{CH_3COOH} = \frac{0.1}{1} = 0.1 \text{ M}$$

$$CH_3COOH \rightleftharpoons CH_3COO^- + H^+$$

| | | |
|---|---|---|
| $C$ | 0 | 0 |
| $C - C\alpha$ | $C\alpha$ | $C\alpha$ |

$$K_\alpha = \frac{C\alpha^2}{1-\alpha}$$

$$1 - \alpha \simeq 1 \Rightarrow K_\alpha = C\alpha^2$$

$$\alpha = \sqrt{\frac{K_\alpha}{C}} = \sqrt{\frac{6.25 \times 10^{-5}}{0.1}} = 25 \times 10^{-3}$$

Using,

$$i = 1 + \alpha\ (n-1)$$
$$= 1 + \alpha\ (2-1) = 1 + \alpha$$
$$= 1 + 25 \times 10^{-3} = 1.025$$

$$\Delta T_f = iK_f m = (1.025)\ (1.86)\ (0.1)$$
$$= 0.19 = 19 \times 10^{-2}$$

Now, $\Delta T_f = T_f^\circ - T_f$

$\therefore \quad T_f = -19 \times 10^{-2}$

$= -x \times 10^{-2}$

$\therefore \quad x = 19$

**28.** *(15)*

**key idea** This problem is based on the concept of depression in freezing point. To solve this, one should know about temperature difference ($\Delta T_f$) and van't Hoff factor ($i$) as $\Delta T_f = iK_f m$

In this case, the value of *i* is 1 as ethylene glycol is a non-electrolyte.

Using, $\Delta T_f = iK_f \times m$

$i = 1$ (for ethylene glycol)

$$\Delta T_f = 0 - (-24)$$
$$= 24°C \text{ or } 24 \text{ K}$$

On substituting the given values in above formula we get

$$24 = \frac{1.86 \times x \times 1000}{62} \times \frac{1}{18.6}$$

$$x = \frac{24 \times 62 \times 18.6}{1860}$$

$$= 14.88 \approx 15 \text{ kg}$$

**29.** *(48)* Molality = *m*, means that 1 mol of $MgCl_2$ is present in 1 kg of solvent ($H_2O$)

$p_S = ?$

$$MgCl_2 \longrightarrow Mg^{2+} + 2Cl^-$$
$$n = 3$$

$$i = 1 - \alpha + n\alpha$$
$$i = 1 - 0.8 + 3\ (0.8) = 2.6$$

$$\frac{p_A^\circ - p_S}{p_S} = i \times \frac{n_B \text{ (moles of solute)}}{n_A \text{ (moles of solvent)}}$$

$$\frac{50 - p_S}{p_S} = 2.6 \times \frac{18}{1000}$$

$[\because p^\circ_A = 50 \text{ (given)}]$

$\Rightarrow \quad p_S = 47.76 \approx 48$ mm Hg

**30.** *(116)* Elevation in boiling point

$= [\Delta T_b = i \times K_b \times \text{molality}]$

Since two solute are present which dissociates completely, hence for NaCl ($i = 2$),

$MgCl_2$, ($i = 3$)

$\Rightarrow$ Mass of NaCl = 29.25% (by weight)
= 29.25 g

$\Rightarrow$ Mass of $MgCl_2$ = 19 g

So, total mass of solute = 29.25 + 19
= 48.25 g ...(i)

So, mass of solvent = 100 − 48.25
= 51.75 g ...(ii)

$$\Delta T_b = K_b \left[\frac{i \times n_1 + i \times n_2}{\text{Mass of solvent (kg)}}\right]$$

$$= 0.52 \left[\frac{2 \times \frac{29.25}{58.5} + 3 \times \frac{19}{95}}{\frac{51.75}{1000}}\right]$$

$$\Rightarrow \frac{0.52\,(1 + 0.6)}{0.05175}$$

$\Rightarrow [16.07]$

So, boiling point of solution

$= 100 + 16.07$

$= 116°C$

**31.** *(30)* Given degree of dissociation of aqueous solution of weak monobasic acid $= 0.3$

$$HA \rightleftharpoons H^+ + A^-$$
$$(1-a) \quad (a) \quad (a)$$

$i = 1 + a$

$a = 0.3$ (given)

So, $i = 1 + 0.3 = 1.3$

$(\Delta T_f)_{obs} = 1.3 \times K_f \times m$

$(\Delta T_f)_{cal} = 1 \times K_f \times m$

$$\Rightarrow \frac{(\Delta T_f)_{obs} - (\Delta T_f)_{cal}}{(\Delta T_f)_{cal}} \times 100$$

$$= \frac{0.3}{1} \times 100 = 30\%$$

**32.** *(4)* For isotonic solutions, osmotic pressure should be same or equal,

(A) 1 M aq. NaCl and 2 M aq. urea

0.1 M aq NaCl, $i = 2$

Osmotic pressure $\pi = iCRT$

$= 2 \times 1 \times RT$

$= 2RT$

2 M aq. urea, $i = 1$

$\pi = iCRT$

$= 1 \times 2\,RT = 2RT$

It is pair of isotonic solution.

(B) 1 M aq. $CaCl_2$ and 1.5 M aq KCl

1 M aq. $CaCl_2$, $i = 3$

$\pi = iCRT = 3RT$

1.5 M aq. KCl, $i = 2$

$\pi = iCRT = 2 \times 1.5\,RT = 3RT$

$\therefore$ It is pair of isotonic solution.

(C) 1.5 M aq. $AlCl_3$ and 2 M aq. $Na_2SO_4$

1.5 M aq. $AlCl_3$, $i = 4$

$\pi = 6RT$

2 M aq. $Na_2SO_4$, $i = 3$

$\pi = 6RT$

$\therefore$ It is a pair of isotonic solution.

(D) 2.5 M aq. KCl and 1 M aq. $Al_2(SO_4)_3$

2.5 M aq. KCl, $i = 2$

$\pi = 5RT$

1 M aq. $Al_2(SO_4)_3$, $i = 5$

$\pi = 5RT$

$\therefore$ It is pair of isotonic solution.

**33.** *(372)* $i = 1 + (n - 1)\alpha$

$[i = 1 + 0.2(2 - 1) = 1.2]$

$\Delta T_f = iK_b m$

$$= 1.2 \times 1.86 \times \frac{5 \times 1000}{60 \times 500}$$

$= 0.372 = 372 \times 10^{-3}$

Thus, the depression in freezing point of water is $372 \times 10^{-3}$ °C.

**34.** *(3)*

> **key idea** To calculate the value of depression in freezing point, we need to know the value of molality and van't Hoff factor. So, first of all calculate the above two values. Then, substitute the obtained value in the equation
>
> $\Delta T_f = i \times K_f \times m$

Reaction equation of titration is

$$\underset{\substack{V = 25\text{ mL}\\ 1\text{ M}}}{KCl} + \underset{\substack{20\text{ mL}\\ 1\text{ M}}}{AgNO_3} \longrightarrow AgCl + KNO_3$$

At equivalent point

millimole of KCl = millimole of $AgNO_3$ = 20 millimole

(i) Volume of KCl solution = 25 mL

(ii) Mass of KCl solution = 25 (g) (assuming, $d = 1$ g/mL)

(iii) Mass of KCl (solute)

$= (20 \times 10^{-3} \times 74.5)$

[$\because$ molar mass of KCl = 39 + 35.5 = 74.5 g/mol]

(iv) Mass of solvent = mass of solution − mass of solute

$= 25 - [20 \times 10^{-3} \times 74.5]$

$= 23.51$ g

$= 23.51 \times 10^{-3}$ kg

$$m_{KCl} = \frac{20 \times 10^{-3}}{23.51 \times 10^{-3}} = 0.85\text{ m}$$

Depression of freezing point of the KCl solution,

$\Delta T_f = i \times K_f \times m$

[$\because$ $i$ for KCl = 2 assuming it gets 100% dissociated in water]

$= 2 \times 2 \times 0.85$

$= 3.4 \approx 3$

**35.** *(243)*

> **key idea** Consider the percentage concentration in grams and use the formula $\Delta T = i_{max} \times K_f \times$ molality.

Given, $W_{H_2SO_4} = 38$ g, (van't Hoff factor, $i = 2.67$)

$K_f = 1.8$ K Kg/mol

Let us consider 100 g solution, then the mass of solvent will be

$= 100 - 38 = 62$ g

From equation $\Delta T = i_{max} \times K_f \times$ molality

$$= 2.67 \times 1.8 \times \frac{38 \times 1000}{98 \times 62} = 30°C$$

Freezing point of liquid in battery is − 30°C.

Thus, the temperature in Kelvin at which the solution in the battery will freeze is 243 K.

**36.** *(4)* 'A', 'B', 'D' and 'E' have same value of osmotic pressure.

Osmotic pressure can be calculated with the help of following formula :

$\pi = iCRT$, also $\pi \propto iC$

A. $(iC)_{C_2H_5OH} = 0.5$

$\therefore$ $(iC)_{KBr} = 2 \times 0.25 = 0.5$

Osmotic pressure will be same.

B. $(iC)_{K_4[Fe(CN)_6]} = 0.1 \times 5 = 0.5$

$(iC)_{FeSO_4(NH_4)_2SO_4} = 0.1 \times 5 = 0.5$

Osmotic pressure will be same.

C. $(iC)_{K_4[Fe(CN)_6]} = 5 \times 0.05 = 0.25$

$(iC)_{NaCl} = 0.25 \times 2 = 0.5$

Osmotic pressure will not be same.

D. $(iC)_{NaCl} = 0.15 \times 2 = 0.3$

$(iC)_{BaCl_2} = 0.1 \times 3 = 0.3$

Osmotic pressure will be same.

E. $(iC)_{KCl \cdot MgCl_2 \cdot 6H_2O} = 0.02 \times 5 = 0.1$

$(iC)_{KCl} = 0.05 \times 2 = 0.1$

Osmotic pressure will be same.

**37.** *(5)* Molar mass of acetic acid

$= 60$ g mol$^{-1}$

Number of moles of acetic acid

$$= \frac{V \times \rho}{\text{Molar mass}}$$

$$= \frac{(1.2)\text{mL} \times (1.02)\text{g/mL}}{(60)\text{g/mol}}$$

$= 0.0204$ mol

Molality

$$(m) = \frac{\text{Moles of solute}}{\text{Mass of solvent }(g)} \times 1000$$

$$= \frac{0.0204}{2000} \times 1000$$

(As 2 L solution in water is prepared so mass of solvent becomes 2000 g).

$$= 0.0102$$

$$\Delta T_f = i \times K_f \times m$$

$\Delta T_f$ = Depression in freezing point

$$= 0.0198°\text{C}$$

$$i = \frac{\Delta T_f}{K_f \times m} = \frac{0.0198}{1.85 \times 0.0102}$$

$$= 1.049$$

$$CH_3COOH \rightleftharpoons CH_3COO^- + H^+$$

| | | | |
|---|---|---|---|
| Initial | $n$ | 0 | 0 |
| Final | $n(1-x)$ | $nx$ | $nx$ |

[$n$ = number of moles, $x$ = degree of dissociation]

Total moles $= n - nx + 2nx$

$$= n - nx$$

$$i = \frac{n(1-x) + nx + nx}{n}$$

$$= \frac{n[1 - x + x + x]}{n}$$

$\Rightarrow \quad i = 1 + x$

$\therefore \quad 1.049 = 1 + x$

$x$ (degree of dissociation) $= 1.049 - 1$

$$= 0.049$$

$$\approx 0.05$$

% degree of dissociation $= 0.05 \times 100$

$$= 5\%$$

**38.** *(98)* We are given with 0.5% of KCl solution.

$\therefore$ Molality of KCl solution,

$$m = \frac{0.5}{74.6} \times \frac{100}{99.5}$$

$$0.24 = i \times 1.80 \times \frac{0.5}{74.6} \times \frac{1000}{99.5}$$

$$\Rightarrow \quad i = \frac{0.24 \times 74.6 \times 99.5}{0.5 \times 1.80 \times 1000} = 1.979$$

$$KCl \rightleftharpoons K^+ + Cl^-$$

| | | | |
|---|---|---|---|
| Initial | 1 | 0 | 0 |
| At equilibrium | $1-x$ | $x$ | $x$ |

{$x$ is the amount of KCl dissociated}

$$i = x + x + (1 - x)$$

$$1.979 = 1 + x$$

$$x = 0.979 \Rightarrow \%x = 97.9\% \approx 98\%$$

$\therefore$ Percentage of dissociation of KCl is 98%

**39.** *(13)* With benzene as solvent, $R$COOH dimerises.

$$2RCOOH \rightleftharpoons (RCOOH)_2$$

$$\therefore \quad i = \frac{1}{2}$$

$$\Delta T_b = i \times K_b \times m$$

On substituting the values, we get

$$\Delta T_b = \frac{1}{2} \times 2.6 \times \frac{1.22/M_w}{100/1000} \quad \ldots\text{(i)}$$

With acetone as solvent, no dimerisation.

$$\therefore \quad i = 1$$

$$\Delta T_b = i \times K_b \times m$$

$$0.17 = 1 \times 1.7 \times \frac{1.22/M_w}{100/1000} \quad \ldots\text{(ii)}$$

Eq. (i) divide by Eq. (ii),

$$\frac{\Delta T_b}{0.17} = \frac{\frac{1}{2} \times 2.6 \times \frac{1.22/M_w}{100/1000}}{1 \times 1.7 \times \frac{1.22/M_w}{100/1000}}$$

$$\Rightarrow \quad \Delta T_b = \frac{0.26}{2} = 13 \times 10^{-2}$$

$$x = 13$$

**40.** *(269)* $\Delta T_f = iK_f \times m$

For ethylene glycol, $i = 1$

$$\text{Molality} = \frac{\text{Number of moles of solute}}{\text{Weight of solvent (kg)}}$$

$$\Delta T_f = 1 \times 1.86 \times \frac{83 \times 1000}{62 \times 625} \simeq 3.98$$

Freezing point of solution

= Freezing point of water $- \Delta T_f$

$= 273 - 3.98$

$\approx 269$ K.

**41.** *(4)* As 0.10 M $C_2H_5OH$ is non-dissociative in nature and all salts [0.10 M $Ba_3(PO_4)_2$, 0.10 M, $Na_2SO_4$, 0.10 M KCl, 0.10 M $Li_3PO_4$] are electrolyte and dissociative in nature.

$$\Delta T_f = iK_f m$$

Where $i$ is van't Hoff factor.

$i = 1$ for $C_2H_5OH$ (no dissociation)

$i > 1$ for all other electrolytes where dissociation takes place.

So, for all these electrolytes effective molarity is more than 0.10. So, all the elecrolytes given will have low freezing point.

**42.** *(125)*

Dimerisation $2HA \rightleftharpoons H_2A_2$

| | | |
|---|---|---|
| Initial | 0.5 | 0 |
| Final | 0 | 0.25 |

Dissociation

$$HA \rightleftharpoons H^+ + A^-$$

| | | | |
|---|---|---|---|
| Initial | 0.5 | 0 | 0 |
| Final | 0 | 0.5 | 0.5 |

$$i = \frac{0.25 + 0.5 + 0.5}{1}$$

$$= 1.25 = 125 \times 10^{-2}$$

**43.** *(100)* Given : $T_b = 100.52°\text{C}$

$$T_b^\circ = 100\ °\text{C}$$

$\Rightarrow \quad K_b = 0.52 \text{ K kg mol}^{-1}$

Elevation in boiling point,

$$\Delta T_b = T_b - T_b^\circ$$

$$\Delta T_b = 100.52 - 100$$

$$\Delta T_b = 0.52°\text{C}$$

For dimerisation (van't Hoff factor)

$$i = \left(1 - \frac{\alpha}{2}\right)$$

$\because \quad \Delta T_b = i \times K_b \times m$

$$0.52 = \left(1 - \frac{\alpha}{2}\right) \times 0.52 \times 2$$

$$\alpha = 1$$

So, percentage association = 100%

**44.** *(50)* Given,

$$K_f = 1.85 \text{ K kg mol}^{-1}$$

$$T_f^\circ = 0°\text{C}$$

$$T_f = 3.885°\text{C}$$

$$\Delta T_f = T_f - T_f^\circ$$

$$= (3.885 - 0)°\text{C}$$

$$= 3.885°\text{C}$$

$$\Delta T_f = 3.885°\text{C}$$

Depression in freezing point,

$$\Delta T_f = i \times K_f \times m$$

$$3.885 = i \times 1.85 \times 2$$

but, $\quad i = (1 + \alpha)$

$$3.885 = (1 + \alpha) \times 1.85 \times 2$$

$$\Rightarrow \quad \alpha = \frac{0.185}{3.7}$$

$$\Rightarrow \quad \alpha = 0.05 \Rightarrow \alpha = 50 \times 10^{-3}$$

**45.** *(2)*

$$n \underset{\text{Benzoic acid}}{\text{PhCOOH}} \xrightarrow{\text{Association}} (\text{PhCOOH})_n$$

Assuming 100% association ($\alpha = 1$),

$$\Rightarrow \quad i = 1 - \alpha\left(1 - \frac{1}{n}\right) = \frac{1}{n} \quad [\because \alpha = 1]$$

Now, $\Delta T_f = K_f \times m \times i$

$$0-(0.93)=1.86\times\frac{w_B\times1000}{w_A\times M_B}\times\frac{1}{n}$$

[$\because w_B$ = mass of PhCOOH =12.2 g

$w_A$ = mass of $H_2O$=100 g

$M_B$ = molar mass of PhCOOH =122g $mol^{-1}$ ]

$$0.93=1.86\times\frac{12.2\times1000}{100\times122}\times\frac{1}{n}$$

$$\Rightarrow\ n=\frac{1.86\times12.2\times1000}{0.93\times100\times122}=2$$

$\therefore$ Number of benzoic acid molecules associated, $n = 2$

**46.** *(3)* As $AB$ is a binary electrolyte,

$\therefore\quad AB \rightleftharpoons A^+ + B^-, n=2$

$$i=1+\alpha(n-1)=1+\frac{75}{100}(2-1)=1.75$$

Given, $\Delta T_b = 2.5$ K

$K_b = 0.52$ K kg $mol^{-1}$

$\therefore\quad \Delta T_b = K_b \times m \times i$

$$\Rightarrow\ m=\frac{\Delta T_b}{K_b\times i}=\frac{2.5}{0.52\times1.75}$$

$=2.74 \simeq 3$ mol / kg

**47.** *(375)* Given,

$K_b(H_2O) = 0.52$ K kg $mol^{-1}$

$$A_2B_3 \longrightarrow 2A^{+3} + 3B^{-2}$$

No. of ions, $n=5$, concentration, $m=1$ molal (Given)

$$\alpha=\frac{60}{100}=0.6$$

$$\Delta T_b = i\cdot K_f\cdot m$$
$$=[(1+(n-1)\alpha]\times K_f\times m$$
$$\Delta T_b=[(1+(5-1)0.6]\times0.52\times1$$
$$=(1+2.4)\times0.52$$
$$\Delta T_b = 1.768$$
$$T_b = 1.768 + 373.15$$
$$= 375.268 = 375 \text{ K}$$

**48.** *(1)* Pure solvent : $C_6H_6(l)$

Given, $T_f^{\circ} = 5.5°C$

$K_f = 5.12°$ C/m, $m_{solvent} = 200$ g

$m_{solute} = 10$ g

Molar mass of solute

$C_4H_{10} = 12\times 4 + 10 = 58$

Solute ($C_4H_{10}$) is non-dissociative;

$\therefore\quad i = 1$

$\therefore\quad \Delta T_f = iK_f\ m$

$$\Rightarrow (T_f^{\circ}-T_f^1)=1\times5.12\times\frac{(10/58)}{(200/1000)}$$

$$5.5-T_f^1=\frac{5.12\times5\times10}{58}$$

$\Rightarrow\quad T_f^1 = 1.086$℃ or $T_f^1 \approx 1$℃

**49.** *(5)* $\Delta T_b = i\times m\times K_b$

For $CaCl_2$, $i=3$, $m_{(CaCl_2)} = 0.05$ m

$\therefore\quad \Delta T_b = 3\times0.05\times K_b = 0.15K_b$

Molality of $CrCl_3 \cdot x\, NH_3$ = 0.1 m

$\therefore\quad \Delta T_b' = i\times m\times K_b = 0.1\, iK_b$

Given that : $0.1 iK_b = 2\times0.15\times K_b$

$\therefore i=3$ and coordination number is 6.

Thus, complex will be $[Cr(NH_3)_5Cl]Cl_2$

$\therefore\quad x = 5$

Hence, the correct answer is 5.

**50.** *(1.76)*

$$\Delta T_f=\frac{i\times K_f\times w_{NaCl}\times1000}{M_{NaCl}\ (= 58.5)\times w_{H_2O}}$$

$$(m = \text{molality}) = \frac{w_B\times1000}{M_B\times w_A}$$

$\Delta T_f = 0.2$ because freezing point $=-0.2$℃

$K_f$ of water 2 K kg $mol^{-1}$ (Given)

$i$ = van't Hoff factor, for NaCl = 2

$$\therefore w_{NaCl}=\frac{\Delta T_f\times M_{NaCl}\times w_{H_2O}}{i\times K_f\times1000}$$

$$=\frac{0.2\times58.5\times600}{2\times2\times1000}$$

$=1.755$ g $\approx 1.76$ g

# C-14

JEE Main Chapterwise
Topicwise Chemistry

In JEE Main there is usually 0-1 question per paper are visible from this chapter. In past 3 years, 34 conceptual and 32 numerical questions were asked of moderate difficulty level from this chapter.
**For detailed analysis scan QR code**

# ELECTROCHEMISTRY

## Topic 1 Conductance and Electrolysis

### Level 1

**01** A solution of aluminium chloride is electrolysed for 30 minutes using a current of 2A. The amount of the aluminium deposited at the cathode is .......
[Given, molar mass of aluminium and chlorine are 27 g $mol^{-1}$ and 35.5 g $mol^{-1}$ respectively. Faraday constant = 96500 C $mol^{-1}$] *[2025, 22 Jan Shift-I]*
(a) 1.007 g (b) 0.336 g (c) 0.441 g (d) 1.660 g

**02** Which of the following electrolyte can be used to obtain $H_2S_2O_8$ by the process of electrolysis? *[2025, 22 Jan Shift-I]*
(a) Acidified dilute solution of sodium sulphate.
(b) Dilute solution of sulphuric acid.
(c) Concentrated solution of sulphuric acid.
(d) Dilute solution of sodium sulphate.

**03** The variation of molar conductivity with concentration of an electrolyte ($X$) in aqueous solution is shown in the given figure.

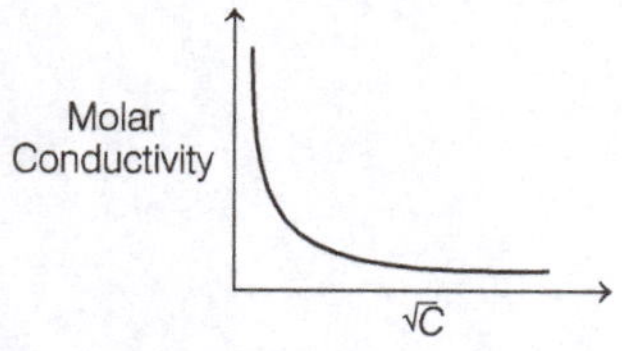


The electrolyte $X$ is *[2020, 5 Sep Shift-II]*
(a) HCl (b) NaCl
(c) $KNO_3$ (d) $CH_3COOH$

**04** Two Faraday of electricity is passed through a solution of $CuSO_4$. The mass of copper deposited at the cathode is (at. mass of Cu = 63.5 u) *[JEE Main 2015]*
(a) 0g (b) 63.5g (c) 2g (d) 127g

**05** The equivalent conductance of NaCl at concentration $C$ and at infinite dilution are $\lambda_c$ and $\lambda_\infty$, respectively. The correct relationship between $\lambda_c$ and $\lambda_\infty$ is given as *[JEE Main 2014]*
(a) $\lambda_c = \lambda_\infty + (B)C$ (b) $\lambda_c = \lambda_\infty - (B)C$
(c) $\lambda_c = \lambda_\infty - (B)\sqrt{C}$ (d) $\lambda_c = \lambda_\infty + (B)\sqrt{C}$

**06** The equivalent conductances of two strong electrolytes at infinite dilution in $H_2O$ (where ions move freely through a solution) at 25°C are given below $\Lambda^\circ_{CH_3COONa} = 91.0$ S $cm^2$ / equiv
$\Lambda^\circ_{HCl} = 426.2$ S $cm^2$/equiv
What additional information/ quantity one needs to calculate $\Lambda^\circ$ of an aqueous solution of acetic acid ? *[AIEEE 2007]*
(a) $\lambda^\circ$ of NaCl
(b) $\lambda^\circ$ of $CH_3COOH$
(c) The limiting equivalent conductance of $H^+$ ($\lambda^\circ_{H^+}$)
(d) $\lambda^\circ$ of chloroacetic acid ($ClCH_2COOH$)

**07** The molar conductivities $\lambda^\circ_{NaOAc}$ and $\lambda^\circ_{HCl}$ at infinite dilution in water at 25°C are 91.0 and 426.2S $cm^2$/mol, respectively. To calculate $\Lambda^\circ_{HOAc}$, the additional value required is *[AIEEE 2006]*
(a) $\lambda^\circ_{H_2O}$ (b) $\lambda^\circ_{KCl}$ (c) $\lambda^\circ_{NaOH}$ (d) $\lambda^\circ_{NaCl}$

**08**

| Electrolyte | KCl | $KNO_3$ | HCl | NaOAc | NaCl |
|---|---|---|---|---|---|
| $\Lambda^\infty$ (S $cm^2$ $mol^{-1}$) | 149.9 | 145.0 | 426.2 | 91.0 | 126.5 |

Calculate $\Lambda^\infty_{HOAc}$ using appropriate molar conductances of the electrolytes listed above at infinite dilution in $H_2O$ at 25°C. *[AIEEE 2005]*
(a) 217.5 (b) 390.7 (c) 552.7 (d) 517.2

**09** The limiting molar conductivities $\Lambda^\circ$ for NaCl, KBr and KCl are 126, 152 and 150 S $cm^2$ $mol^{-1}$, respectively. The $\Lambda^\circ$ for NaBr is *[AIEEE 2004]*
(a) 128 S $cm^2 mol^{-1}$ (b) 176 S $cm^2 mol^{-1}$
(c) 278 S $cm^2 mol^{-1}$ (d) 302 S $cm^2 mol^{-1}$

**10** Conductivity (Seimen's S) is directly proportional to area of the vessel and the concentration of the solution in it and is inversely proportional to the length of the vessel, then constant of proportionality is expressed in *[AIEEE 2002]*
(a) S m $mol^{-1}$ (b) $S^2$ $m^2$ $mol^{-2}$
(c) S $m^2$ $mol^{-1}$ (d) $S^2$ $m^2$ mol

## Level 2

**11** Correct order of limiting molar conductivity for cations in water at 298 K is: *[2025, 3 April Shift-I]*

(a) $Mg^{2+} > H^{+} > Ca^{2+} > K^{+} > Na^{+}$

(b) $H^{+} > Ca^{2+} > Mg^{2+} > K^{+} > Na^{+}$

(c) $H^{+} > Na^{+} > K^{+} > Ca^{2+} > Mg^{2+}$

(d) $H^{+} > Na^{+} > Ca^{2+} > Mg^{2+} > K^{+}$

**12** The molar conductivity of a weak electrolyte when plotted against the square root of its concentration, which of the following is expected to be observed? *[2025, 29 Jan Shift-I]*

(a) Molar conductivity decreases sharply with increase in concentration.

(b) A small decrease in molar conductivity is observed at infinite dilution.

(c) A small increase in molar conductivity is observed at infinite dilution.

(d) Molar conductivity increases sharply with increase in concentration.

**Imp.** Graphical Representation of Strong & Weak Electrolyte

**13** The molar conductivity for electrolytes *A* and *B* are plotted against $C^{1/2}$ as shown below. Electrolytes *A* and *B* respectively are *[2024, 9 April Shift-I]*

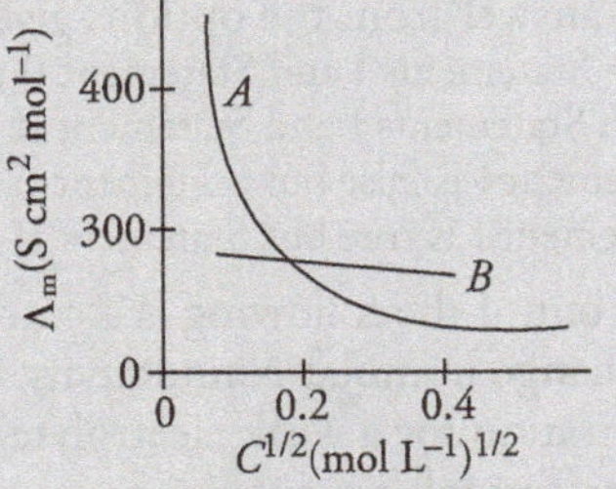


| | A | B |
|---|---|---|
| (a) | Weak electrolyte | Strong electrolyte |
| (b) | Weak electrolyte | Weak electrolyte |
| (c) | Strong electrolyte | Weak electrolyte |
| (d) | Strong electrolyte | Strong electrolyte |

**14** A conductivity cell with two electrodes (dark side) are half-filled with infinitely dilute aqueous solution of a weak electrolyte. If volume is doubled by adding more water at constant temperature, the molar conductivity of the cell will *[2024, 6 April Shift-I]*

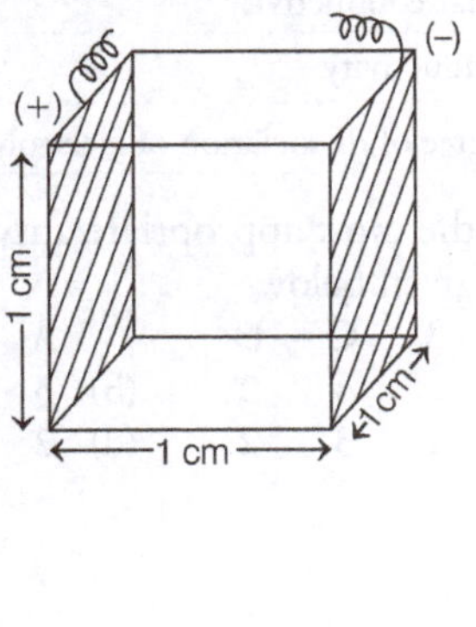


(a) depend upon type of electrolyte

(b) decrease sharply

(c) increase sharply

(d) remain same or can not be measured accurately

**15** The quantity of silver deposited when one coulomb charge is passed through $AgNO_3$ solution *[2024, 5 April Shift-II]*

(a) 1 electrochemical equivalent of silver

(b) 1 chemical equivalent of silver

(c) 1 g of silver

(d) 0.1 g atom of silver

**16** Molar ionic conductivities of divalent cation and anion are 57 S cm$^2$ mol$^{-1}$ and 73 S cm$^2$ mol$^{-1}$ respectively. The molar conductivity of solution of an electrolyte with the above cation and anion will be *[2024, 5 April Shift-I]*

(a) 130 S cm$^2$ mol$^{-1}$

(b) 187 S cm$^2$ mol$^{-1}$

(c) 260 S cm$^2$ mol$^{-1}$

(d) 65 S cm$^2$ mol$^{-1}$

**Imp.** Unit of Molar Conductivity

**17** For a strong electrolyte, a plot of molar conductivity against (concentration)$^{1/2}$ is a straight line, with a negative slope, the correct unit for the slope is *[2024, 4 April Shift-II]*

(a) S cm$^2$ mol$^{\frac{-3}{2}}$ L$^{\frac{-1}{2}}$

(b) S cm$^2$ mol$^{-1}$ L$^{\frac{1}{2}}$

(c) S cm$^2$ mol$^{\frac{-3}{2}}$ L$^{\frac{1}{2}}$

(d) S cm$^2$ mol$^{\frac{-3}{2}}$ L

**18** One of the commonly used electrode is calomel electrode. Under which of the following categories, calomel electrode comes? *[2024, 4 April Shift-I]*

(a) Gas- ion electrodes

(b) Oxidation-reduction electrodes

(c) Metal- insoluble salt-anion electrodes

(d) Metal ion-metal electrodes

**19** Identify the factor from the following that does not affect electrolytic conductance of a solution. *[2024, 31 Jan Shift-I]*

(a) The nature of the electrode used.

(b) Concentration of the electrolyte.

(c) The nature of the electrolyte added.

(d) The nature of solvent used.

**20** Which one of the following statements is correct for electrolysis of brine solution? *[2023, 31 Jan Shift-I]*

(a) $Cl_2$ is formed at cathode.

(b) $OH^-$ is formed at cathode.

(c) $O_2$ is formed at cathode.

(d) $H_2$ is formed at anode.

**21** Choose the correct representation of conductometric titration of benzoic acid *vs* sodium hydroxide. ***[2023, 24 Jan Shift-II]***

(a)
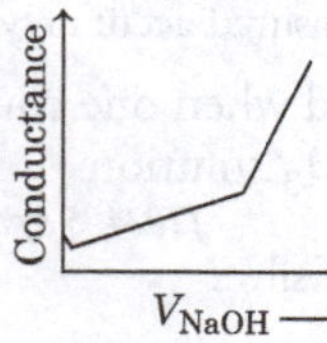


(b)
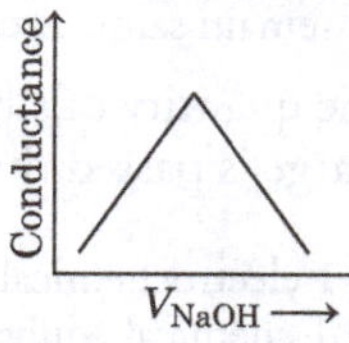


(c)
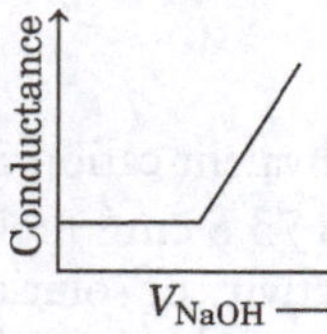


(d)
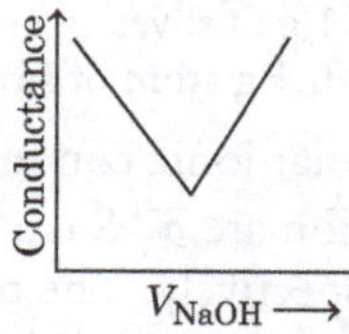


**22** Given below are two statements.
**Statement I** For KI, molar conductivity increases steeply with dilution.
**Statement II** For carbonic acid, molar conductivity increases slowly with dilution.
In the light of the above statements, choose the correct answer from the options given below. ***[2022, 27 July Shift-II]***
(a) Both Statement I and Statement II are true.
(b) Both Statement I and Statement II are false.
(c) Statement I is true but Statement II is false.
(d) Statement I is false but Statement II is true.

**23** The molar conductivity of a conductivity cell filled with 10 moles of 20 mL NaCl solution is $\Lambda_{M_1}$ and that of 20 moles another identical cell having 80 mL NaCl solution is $\Lambda_{M_2}$. The conductivities exhibited by these two cells are same. The relationship between $\Lambda_{M_2}$ and $\Lambda_{M_1}$ is ***[2022, 25 July Shift-II]***
(a) $\Lambda_{M_2} = 2\Lambda_{M_1}$ (b) $\Lambda_{M_2} = \Lambda_{M_1} / 2$
(c) $\Lambda_{M_2} = \Lambda_{M_1}$ (d) $\Lambda_{M_2} = 4\Lambda_{M_1}$

**24** Given below are two statements.
**Statement I** The limiting molar conductivity of KCl (strong electrolyte) is higher compared to that of $CH_3COOH$ (weak electrolyte).
**Statement II** Molar conductivity decreases with decrease in concentration of electrolyte.
In the light of the above statements, choose the most appropriate answer from the options given below ***[2021, 26 Aug Shift-I]***
(a) Statement I is true but statement II is false.
(b) Statement I is false but statement II is true.
(c) Both statement I and statement II are true.
(d) Both statement I and statement II are false.

**25** A solution of $Ni(NO_3)_2$ is electrolysed between platinum electrodes using 0.1 Faraday electricity. How many mole of Ni will be deposited at the cathode? ***[2019, 9 April Shift-II]***
(a) 0.20 (b) 0.10
(c) 0.15 (d) 0.05

**26** The anodic half-cell of lead-acid battery is recharged using electricity of 0.05 Faraday. The amount of $PbSO_4$ electrolysed in g during the process is (Molar mass of $PbSO_4 = 303\ g\ mol^{-1}$) ***[2019, 9 Jan Shift-I]***
(a) 11.4 (b) 7.6 (c) 15.2 (d) 22.8

**27** The highest electrical conductivity from the following aqueous solutions is of ***[AIEEE 2005]***
(a) 0.1 M difluoroacetic acid (b) 0.1 M fluoroacetic acid
(c) 0.1 M chloroacetic acid (d) 0.1 M acetic acid

**28** When during electrolysis of a solution of $AgNO_3$, 9650 C of charge pass through the electroplating bath, the mass of silver deposited on the cathode will be ***[AIEEE 2003]***
(a) 1.08 g (b) 10.8 g (c) 21.6 g (d) 108 g

## Level 3

**29** Given below are two statements: ***[2025, 7 April Shift-I]***
**Statement I** Mohr's salt is composed of only three types of ions-ferrous, ammonium and sulphate.
**Statement II** If the molar conductance at infinite dilution for ferrous, ammonium and sulphate ions are $x_1, x_2$ and $x_3$ S $cm^2\ mol^{-1}$, respectively then the molar conductance for Mohr's salt solution at infinite dilution would be given by $x_1 + x_2 + 2x_3$
In the light of the given statements, choose the correct answer from the options given below:
(a) Both Statements I and Statement II are true.
(b) Both Statements I and Statement II are false.
(c) Statement I is false but Statement II is true.
(d) Statement I is true but Statement II are false.

**30** Which out of the following is a correct equation to show change in molar conductivity with respect to concentration for a weak electrolyte, if the symbols carry their usual meaning : ***[2024, 9 April Shift-II]***
(a) $\Lambda_m - \Lambda_m^\circ - AC^{1/2} = 0$
(b) $\Lambda_m^2 C - K_a \Lambda_m^{\circ 2} + K_a \Lambda_m \Lambda_m^\circ = 0$
(c) $\Lambda_m - \Lambda_m^\circ + AC^{1/2} = 0$
(d) $\Lambda_m^2 C + K_a \Lambda_m^{\circ 2} - K_a \Lambda_m \Lambda_m^\circ = 0$

**31** Match List-I with List-II.

| List-I (Parameter) | List-II (Unit) |
|---|---|
| A. Cell constant | 1. S $cm^2\ mol^{-1}$ |
| B. Molar conductivity | 2. Dimensionless |
| C. Conductivity | 3. $m^{-1}$ |
| D. Degree of dissociation of electrolyte | 4. $\Omega^{-1} m^{-1}$ |

Choose the most appropriate answer from the options given below ***[2021, 31 Aug Shift-II]***

| | A | B | C | D | | A | B | C | D |
|---|---|---|---|---|---|---|---|---|---|
| (a) | 3 | 1 | 4 | 2 | (b) | 3 | 1 | 2 | 4 |
| (c) | 1 | 4 | 3 | 2 | (d) | 2 | 1 | 3 | 4 |

**32** $\Lambda^{\circ}_m$ for NaCl, HCl and Na*A* are 126.4, 425.9 and 100.5 S $cm^2$ $mol^{-1}$, respectively. If the conductivity of 0.001 M H*A* is $5 \times 10^{-5}$ S $cm^{-1}$, degree of dissociation of H*A* is *[2019, 12 Jan Shift-II]*
(a) 0.25 (b) 0.50
(c) 0.75 (d) 0.125

**33** How long (approximate) should water be electrolysed by passing through 100 amperes current so that the oxygen released can completely burn 27.66 g of diborane?
(Atomic weight of $B = 10.8$ u) *[JEE Main 2018]*
(a) 6.4 hours (b) 0.8 hours
(c) 3.2 hours (d) 1.6 hours

**34** Resistance of 0.2 M solution of an electrolyte is 50 Ω. The specific conductance of the solution is 1.4 $Sm^{-1}$. The resistance of 0.5 M solution of the same electrolyte is 280 Ω. The molar conductivity of 0.5 M solution of the electrolyte in S $mol^{-1}$ is *[JEE Main 2014]*
(a) $5 \times 10^{-4}$ (b) $5 \times 10^{-3}$
(c) $5 \times 10^{3}$ (d) $5 \times 10^{2}$

## Numerical Value Type Questions

**35** The molar conductance of an infinitely dilute solution of ammonium chloride was found to be 185 S $cm^2$ $mol^{-1}$ and the ionic conductance of hydroxyl and chloride ions are 170 and 70 S $cm^2$ $mol^{-1}$, respectively. If molar conductance of 0.02 M solution of ammonium hydroxide is 85.5 S $cm^2 mol^{-1}$, its degree of dissociation is given by $x \times 10^{-1}$.

The value of *x* is ............ .(Nearest integer) *[2025, 4 April Shift-II]*

**36** Given below is the plot of the molar conductivity *vs* $\sqrt{\text{concentration}}$ for KCl in aqueous solution. *[2025, 28 Jan Shift-I]*

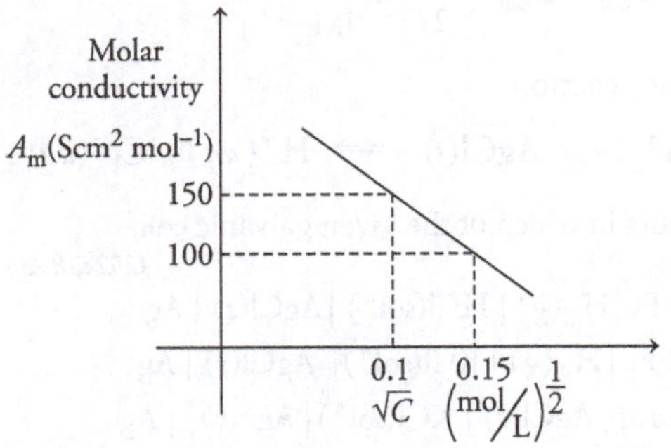


If, for the higher concentration of KCl solution, the resistance of the conductivity cell is 100 Ω, then the resistance of the same cell with the dilute solution is '*x*' Ω
The value of *x* is ........ (Nearest integer)

**37** Electrolysis of 600 mL aqueous solution of NaCl for 5 min changes the pH of the solution to 12.
The current in Amperes used for the given electrolysis is.......(Nearest integer). *[2025, 28 Jan Shift-II]*

**38** The amount of electricity in coulomb required for the oxidation of 1 mol of $H_2O$ to $O_2$ is ......... $\times 10^5$ C. *[2024, 1 Feb, Shift-I]*

**39** The value of conductivity of some materials at 298.15 K in $Sm^{-1}$ are $2.1 \times 10^3$, $1.0 \times 10^{-16}$, $1.2 \times 10$, 3.91, $1.5 \times 10^{-2}$, $1 \times 10^{-7}$, $1.0 \times 10^3$. The number of conductors among the materials is ... . *[2024, 31 Jan, Shift-II]*

**40** One Faraday of electricity liberates $x \times 10^{-1}$ gram atom of copper from copper sulphate, *x* is ......... *[2024, 31 Jan, Shift-I]*

**41** A constant current was passed through a solution of $AuCl_4^-$ ion between gold electrodes. After a period of 10.0 minutes, the increase in mass of cathode was 1.314 g. The total charge passed through the solution is ......... $\times 10^{-2}$ F.
(Given, atomic mass of Au = 197) *[2024, 29 Jan, Shift-II]*

**42** The mass of zinc produced by the electrolysis of zinc sulphate solution with a steady current of 0.015 A for 15 minutes is ......... $\times 10^{-4}$ g.
(Atomic mass of zinc = 65.4 amu) *[2024, 29 Jan, Shift-I]*

**43** The number of correct statement from the following is ........... . *[2023, 15 April, Shift-I]*
A. Conductivity always decreases with decrease in concentration for both strong and weak electrolytes.
B. The number of ions per unit volume that carry in a solution increases on dilution.
C. Molar conductivity increases with decrease in concentration.
D. The variation in molar conductivity is different for strong and weak electrolytes.
E. For weak electrolytes, the change in molar conductivity with dilution is due to decrease in degree of dissociation.

**44** A metal surface of 100 $cm^2$ area has to be coated with nickel layer of thickness 0.001 mm. A current of 2A was passed through a solution of $Ni(NO_3)_2$ for '*x*' seconds to coat the desired layer. The value of *x* is ........... (Nearest integer) ($\rho_{Ni}$ (density of nickel) is 10 g $mL^{-1}$. Molar mass of nickel is 60 g $mol^{-1}$ $F = 96500$ C $mol^{-1}$) *[2023, 13 April Shift-I]*

**45** The specific conductance of 0.0025 M acetic acid is $5 \times 10^{-5}$ S $cm^{-1}$ at a certain temperature. The dissociation constant of acetic acid is ........ $\times 10^{-7}$. (Nearest integer)
Consider limiting molar conductivity of $CH_3COOH$ as 400 S $cm^2$ $mol^{-1}$. *[2023, 10 April Shift-II]*

**46** $1\times10^{-5}$ M $AgNO_3$ is added to 1 L of saturated solution of AgBr. The conductivity of this solution at 298 K is .............. $\times10^{-8}\,S\,m^{-1}$.

[Given, $K_{sp}(AgBr)=4.9\times10^{-13}$ at 298 K

$\lambda^{\circ}_{Ag^+}=6\times10^{-3}\,S\,m^2\,mol^{-1}$

$\lambda^{\circ}_{Br^-}=8\times10^{-3}\,S\,m^2\,mol^{-1}$

$\lambda^{\circ}_{NO_3^-}=7\times10^{-3}\,S\,m^2\,mol^{-1}$ *[2023, 1 Feb Shift-II]*

**47** The resistivity of a 0.8 M solution of an electrolyte is $5\times10^{-3}\Omega$ cm. Its molar conductivity is .......... $\times10^4$ $\Omega^{-1}cm^2\,mol^{-1}$. (Nearest integer) *[2023, 31 Jan Shift-II]*

**48** Resistance of a conductivity cell (cell constant $129\,m^{-1}$) filled with 74.5 ppm solution of KCl is 100 Ω (labelled as solution 1). When the same cell is filled with KCl solution of 149 ppm, the resistance is 50 Ω (labelled as solution 2). The ratio of molar conductivity of solution 1 and solution 2 is i.e. $\frac{\Lambda_1}{\Lambda_2}=x\times10^{-3}$. The value of $x$ is ......... .

(Nearest integer) *[2022, 29 July Shift-I]*

**49** The quantity of electricity in Faraday needed to reduce 1 mole of $Cr_2O_7^{2-}$ to $Cr^{3+}$ is ........ *[2022, 28 June Shift-I]*

**50** The limiting molar conductivities of NaI, $NaNO_3$ and $AgNO_3$ are 12.7, 12.0 and 13.3 mS $m^2\,mol^{-1}$, respectively (all at 25°C). The limiting molar conductivity of AgI at this temperature is ...... mS $m^2\,mol^{-1}$. *[2022, 27 June Shift-I]*

**51** A solution of $Fe_2(SO_4)_3$ is electrolysed for '$x$' min with a current of 1.5 A to deposit 0.3482 g of Fe. The value of $x$ is ............ . [Nearest integer]

Given, 1 F = 96500 C $mol^{-1}$,
atomic mass of Fe = 56 g $mol^{-1}$ *[2022, 25 June Shift-II]*

**52** The molar conductivities at infinite dilution of barium chloride, sulphuric acid and hydrochloric acid are 280, 860 and 426 $Scm^2\,mol^{-1}$ respectively. The molar conductivity at infinite dilution of barium sulphate is ... S $cm^2\,mol^{-1}$ (Round off to the nearest Integer). *[2021, 18 March Shift-II]*

**53** A KCl solution of conductivity 0.14 S $m^{-1}$ shows a resistance of 4.19 Ω in a conductivity cell. If the same cell is filled with an HCl solution, the resistance drops to 1.03 Ω. The conductivity of the HCl solution is ........ $\times10^{-2}$ S $m^{-1}$ (Round off to the nearest integer) *[2021, 17 March Shift-II]*

**54** A 5.0 m mol $dm^{-3}$ aqueous solution of KCl has a conductance of 0.55 mS when measured in a cell constant 1.3 $cm^{-1}$. The molar conductivity of this solution is ...... $mSm^2\,mol^{-1}$.
(Round off to the nearest integer) *[2021, 16 March Shift-II]*

**55** Consider the following reaction,
$MnO_4^- + 8H^+ + 5e^- \rightarrow Mn^{2+} + 4H_2O,\ E^{\circ}=1.51V.$
The quantity of electricity required in Faraday to reduce five moles of $MnO_4^-$ is ......... . *[2021, 26 Feb Shift-I]*

**56** Potassium chlorate is prepared by the electrolysis of KCl in basic solution
$6OH^- + Cl^- \longrightarrow ClO_3^- + 3H_2O + 6e^-$. If only 60% of the current is utilised in the reaction, the time (rounded to the nearest hour) required to produce 10 g of $KClO_3$ using a current of 2 A is ........ .
(Given : $F$ = 96,500 C $mol^{-1}$; molar mass of $KClO_3$ = 122 g $mol^{-1}$) *[2020, 6 Sep Shift-I]*

## Topic 2 Electrochemical Series, Cells and Their EMF

### Level 1

**01** For a Mg |$Mg^{2+}(aq)$||$Ag^+(aq)$| Ag the correct Nernst equation is *[2025, 29 Jan Shift-I]*

(a) $E_{cell}=E^{\circ}_{cell}-\frac{RT}{2F}\ln\frac{[Mg^{2+}]}{[Ag^+]}$

(b) $E_{cell}=E^{\circ}_{cell}+\frac{RT}{2F}\ln\frac{[Ag^+]^2}{[Mg^{2+}]}$

(c) $E_{cell}=E^{\circ}_{cell}-\frac{RT}{2F}\ln\frac{[Ag^+]}{[Mg^{2+}]}$

(d) $E_{cell}=E^{\circ}_{cell}-\frac{RT}{2F}\ln\frac{[Ag^+]^2}{[Mg^{2+}]}$

**02** The reaction

$\frac{1}{2}H_2(g)+AgCl(s) \rightleftharpoons H^+(aq)+Cl^-(aq)+Ag(s)$

occurs in which of the given galvanic cell. *[2024, 8 April Shift-II]*

(a) Pt |$H_2(g)$| HCl(sol") |AgCl(s)| Ag
(b) Pt |$H_2(g)$| KCl(sol")| AgCl(s) | Ag
(c) Ag | AgCl(s) | KCl(sol") | $AgNO_3$ | Ag
(d) Pt |$H_2(g)$| HCl(sol") | $AgNO_3$ (sol)| Ag

**03** The standard electrode potential of $M^+/M$ in aqueous solution does not depend on *[2023, 6 April Shift-I]*
(a) ionisation of a solid metal atom
(b) ionisation of a gaseous metal atom
(c) hydration of a gaseous metal ion
(d) sublimation of a solid metal

**Imp.** Calculation of Gibbs Energy

**04** The standard Gibbs energy for the given cell reaction in kJ $mol^{-1}$ at 298 K is

$$Zn(s) + Cu^{2+}(aq) \longrightarrow Zn^{2+}(aq) + Cu(s),\ E° = 2\ V \text{ at } 298\ K$$

(Faraday's constant, $F$ = 96000 C $mol^{-1}$) **[2019, 9 April Shift-I]**

(a) 384 (b) 192 (c) −384 (d) −192

**05** Given, $E^{\circ}_{Cr^{3+}/Cr} = -0.74$ V; $E^{\circ}_{MnO_4^-/Mn^{2+}} = 1.51$ V

$E^{\circ}_{Cr_2O_7^{2-}/Cr^{3+}} = 1.33$ V; $E^{\circ}_{Cl/Cl^-} = 1.36$ V

Based on the data given above, strongest oxidising agent will be **[JEE Main 2013]**

(a) Cl (b) $Cr^{3+}$
(c) $Mn^{2+}$ (d) $MnO_4^-$

**Imp.** Order of Reducing Power

**06** Standard reduction electrode potentials of three metals $A$, $B$ and $C$ are + 0.5 V, − 3.0 V and − 1.2 V, respectively. The reducing power of these metals are **[AIEEE 2003]**

(a) $B > C > A$ (b) $A > B > C$
(c) $C > B > A$ (d) $A > C > B$

**07** For the redox reaction

$$Zn(s) + Cu^{2+}(0.1\ M) \longrightarrow Zn^{2+}(1\ M) + Cu(s)$$

taking place in a cell, $E°_{cell}$ is 1.10 V. $E_{cell}$ for the cell will be $\left(2.303\dfrac{RT}{F} = 0.0591\right)$ **[AIEEE 2003]**

(a) 2.14 V (b) 1.80 V (c) 1.07 V (d) 0.82 V

**08** Which of the following reaction is possible at anode ? **[AIEEE 2002]**

(a) $Fe_2 + 2e^- \longrightarrow 2F^-$

(b) $2H^+ + \frac{1}{2}O_2 + 2e^- \longrightarrow H_2O$

(c) $2Cr^{3+} + 7H_2O \rightarrow Cr_2O_7^{2-} + 14H^+ + 6e^-$

(d) None of the above

## Level 2

**09** $FeO_4^{2-} \xrightarrow{+2.0\ V} Fe^{3+} \xrightarrow{0.8\ V} Fe^{2+} \xrightarrow{-0.5\ V} Fe^0$

In the above diagram, the standard electrode potentials are given in volts (over the arrow). The value of $E^{\circ}_{FeO_4^{2-}/Fe^2}$ is **[2025, 23 Jan Shift-I]**

(a) 2.1 V (b) 1.2 V
(c) 1.7 V (d) 1.4 V

**10** How can an electrochemical cell be converted into an electrolytic cell? **[2024, 6 April Shift-II]**

(a) Reversing the flow of ions in salt bridge.
(b) Exchanging the electrodes at anode and cathode.
(c) Applying an external opposite potential greater than $E^{\circ}_{cell}$
(d) Applying an external opposite potential lower than $E^0_{cell}$

**11** For the electrochemical cell

$$M|M^{2+}||X|X^{2-}$$

If, $E^0_{(M^{2+}/M)} = 0.46$ V

and $E^0_{(X/X^{2-})} = 0.34$ V.

Which of the following is correct? **[2024, 5 April Shift-II]**

(a) $E_{cell} = 0.80$ V
(b) $E_{cell} = -0.80$ V
(c) $M + X \rightarrow M^{2+} + X^{2-}$ is a spontaneous reaction
(d) $M^{2+} + X^{2-} \longrightarrow M + X$ is a spontaneous reaction

**12** What pressure (bar) of $H_2$ would be required to make emf of hydrogen electrode zero in pure water at 25°C? **[2024, 4 April Shift-I]**

(a) 0.5 (b) $10^{-14}$
(c) 1 (d) $10^{-7}$

**13** The $\left(\dfrac{\partial E}{\partial T}\right)_p$ of different types of half-cells are as follows

| $A$ | $B$ | $C$ | $D$ |
|---|---|---|---|
| $1\times10^{-4}$ | $2\times10^{-4}$ | $0.1\times10^{-4}$ | $0.2\times10^{-4}$ |

(where, $E$ is the electromotive force.)
Which of the above half-cells would be preferred to be used as reference electrode ? **[2022, 26 June Shift-I]**

(a) $A$ (b) $B$ (c) $C$ (d) $D$

**14**

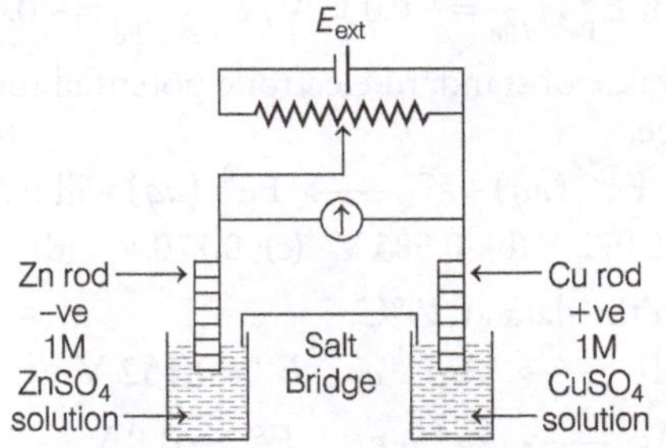


$E^{\circ}_{Cu^{2+}|Cu} = +0.34$ V

$E^{\circ}_{Zn^{2+}|Zn} = -0.76$ V

Identify the incorrect statement from the options below for the above cell. **[2020, 4 Sep Shift-I]**

(a) If $E_{ext} > 1.1$ V, Zn dissolves at Zn electrode and Cu deposits at Cu electrode.
(b) If $E_{ext} = 1.1$ V, no flow of electrons or current occurs.
(c) If $E_{ext} > 1.1$ V, electrons flows from Cu to Zn.
(d) If $E_{ext} < 1.1$ V, Zn dissolves at anode and Cu deposits at cathode.

**15** Given that the standard potentials ($E^\circ$) of $Cu^{2+}/Cu$ and $Cu^+/Cu$ are 0.34 V and 0.522 V respectively, the $E^\circ$ of $Cu^{2+}/Cu^+$ is *[2020, 7 Jan Shift-I]*
(a) $-0.158$ V (b) $+0.158$ V
(c) $-0.182$ V (d) 0.182 V

**16** If the standard electrode potential for a cell is 2V at 300 K, the equilibrium constant ($K$) for the reaction, $Zn(s) + Cu^{2+}(aq) \rightleftharpoons Zn^{2+}(aq) + Cu(s)$ at 300 K is approximately ($R = 8\ JK^{-1}mol^{-1}$, $F = 96000\ C\ mol^{-1}$) *[2019, 9 Jan Shift-II]*
(a) $e^{-160}$ (b) $e^{160}$ (c) $e^{-80}$ (d) $e^{320}$

**17** Given below are the half-cell reactions
$$Mn^{2+} + 2e^- \longrightarrow Mn,\ E^\circ = -1.18\ V$$
$$2(Mn^{3+} + e^- \longrightarrow Mn^{2+}),\ E^\circ = +1.51\ V$$
The $E^\circ$ for $3Mn^{2+} \longrightarrow Mn + 2Mn^{3+}$ will be *[JEE Main 2014]*
(a) $-2.69$ V, the reaction will not occur
(b) $-2.69$ V the reaction will occur
(c) $-0.33$ V, the reaction will not occur
(d) $-0.33$ V, the reaction will occur

**18** The Gibbs energy for the decomposition of $Al_2O_3$ at 500°C is as follows *[AIEEE 2010]*
$$\frac{2}{3}Al_2O_3 \longrightarrow \frac{4}{3}Al + O_2,\ \Delta_rG = +966\ kJ\ mol^{-1}$$
The potential difference needed for electrolytic reduction of $Al_2O_3$ at 500°C is at least
(a) 4.5 V (b) 3.0 V (c) 2.5 V (d) 5.0 V

**19** Given, $E^\circ_{Fe^{3+}/Fe} = -0.036\ V$, $E^\circ_{Fe^{2+}/Fe} = -0.439\ V$
The value of standard electrode potential for the change, *[AIEEE 2009]*
$$Fe^{3+}(aq) + e^- \longrightarrow Fe^{2+}(aq)$$ will be
(a) $-0.072$ V (b) 0.385 V (c) 0.770 V (d) $-0.270$ V

**20** Given the data at 25°C,
$$Ag + I^- \longrightarrow AgI + e^-;\ E^\circ = 0.152\ V$$
$$Ag \longrightarrow Ag^+ + e^-;\ E^\circ = -0.8\ V$$
What is the value of $\log K_{sp}$ for AgI ?
$\left(2.303\frac{RT}{F} = 0.059\ V\right)$ *[AIEEE 2006]*
(a) $-8.12$ (b) $+8.612$
(c) $-37.83$ (d) $-16.13$

**21** The $E^\circ_{M^{3+}/M^{2+}}$ values for Cr, Mn, Fe and Co are $-0.41$, $+1.57$, $+0.77$ and $+1.97$ V, respectively. For which one of these metals, the change in oxidation state from +2 to +3 is easiest ? *[AIEEE 2004]*
(a) Cr (b) Mn
(c) Fe (d) Co

**22** The standard emf of a cell, involving one electron change is found to be 0.591 V at 25° C. The equilibrium constant of the reaction is ($F = 96500\ C\ mol^{-1}$) *[AIEEE 2004]*
(a) $1.0 \times 10^1$ (b) $1.0 \times 10^5$
(c) $1.0 \times 10^{10}$ (d) $1.0 \times 10^{30}$

**23** For a cell reaction involving a two electron change, the standard emf of the cell is found to be 0.295 V at 25°C. The equilibrium constant of the reaction at 25°C will be *[AIEEE 2003]*
(a) $1 \times 10^{-10}$ (b) $29.5 \times 10^{-2}$
(c) 10 (d) $1 \times 10^{10}$

**24** For a cell given below
$$\underset{-}{Ag} \mid Ag^+ \parallel Cu^{2+} \mid \underset{+}{Cu}$$
$$Ag^+ + e^- \longrightarrow Ag;\ E^\circ = x$$
$$Cu^{2+} + 2e^- \longrightarrow Cu;\ E^\circ = y$$
$E^\circ$ cell is *[AIEEE 2002]*
(a) $x + 2y$ (b) $2x + y$
(c) $y - x$ (d) $y - 2x$

## Level 3

**25** Given below are two statements: *[2025, 7 April Shift-II]*
1 M aqueous solutions of each of $Cu(NO_3)_2$, $AgNO_3$, $Hg_2(NO_3)_2$, $Mg(NO_3)_2$ are electrolysed using inert electrodes. Given: $E^\theta_{Ag^+/Ag} = 0.80$ V, $E^\theta_{Hg_2^{2+}/Hg} = 0.79$ V, $E^\theta_{Cu^{2+}/Cu} = 0.24$ V and $E^\theta_{Mg^{2+}/Mg} = -2.37$ V.

**Statement I** With increasing voltage, the sequence of deposition of metals on the cathode will be Ag, Hg and Cu.

**Statement II** Magnesium will not be deposited at cathode instead oxygen gas will be evolved at the cathode.

In the light of the above statements, choose the most appropriate answer from the options given below:
(a) Statement I is incorrect but Statement II is correct.
(b) Both Statement I and Statement II are incorrect.
(c) Both Statement I and Statement II are correct.
(d) Statement I is correct but Statement II is incorrect.

**26** The standard cell potential ($E^\ominus_{cell}$) of a fuel cell based on the oxidation of methanol in air that has been used to power television relay station is measured as 1.21 V. The standard half cell reduction potential for $O_2$ ($E^\circ_{O_2/H_2O}$) is 1.229 V. *[2025, 3 April Shift-II]*

Choose the correct statement.
(a) The standard half cell reduction potential for the reduction of $CO_2$ ($E^\circ_{CO_2/CH_3OH}$) is 19 mV.
(b) Oxygen is formed at the anode.
(c) Reactants are fed at one go to each electrode.
(d) Reduction of methanol takes place at the cathode.

**27** For the given cell
$Fe^{2+}(aq) + Ag^+(aq) \rightarrow Fe^{3+}(aq) + Ag(s)$
The standard cell potential of the above reaction is (Given :
$Ag^+ + e^- \rightarrow Ag$ $E^\circ = x$V
$Fe^{2+} + 2e^- \rightarrow Fe$ $E^\circ = y$V
$Fe^{3+} + 3e^- \rightarrow Fe$ $E^\circ = z$V) *[2025, 24 Jan Shift-I]*
(a) $y - 2x$ (b) $x + 2y - 3z$
(c) $x + y - z$ (d) $x + 2y$

**28** Standard electrode potentials for a few half cells are mentioned below.
$E^\circ_{Cu^{2+}/Cu} = 0.34$ V, $E^\circ_{Zn^{2+}/Zn} = -0.76$ V
$E^\circ_{Ag^+/Ag} = 0.80$ V, $E^\circ_{Mg^{2+}/Mg} = -2.37$ V
Which one of the following cells given the most negative value of $\Delta G^\circ$? *[2025, 23 Jan Shift-II]*
(a) $Ag|Ag^+(1M)||Mg^{2+}(1M)|Mg$
(b) $Cu|Cu^{2+}(1M)||Ag^+(1M)|Ag$
(c) $Zn|Zn^{2+}(1M)||Ag^+(1M)|Ag$
(d) $Zn|Zn^{2+}(1M)||Mg^{2+}(1M)|Mg$

**29** The emf of cell $Tl|\underset{(0.001\text{ M})}{Tl^+}||\underset{(0.01\text{ M})}{Cu^{2+}}|Cu$ is 0.83 V at 298 K. It could be increased by *[2024, 8 April Shift-II]*
(a) decreasing concentration of both $Tl^+$ and $Cu^{2+}$ ions
(b) increasing concentration of $Tl^+$ ions
(c) increasing concentration of both $Tl^+$ and $Cu^{2+}$ ions
(d) increasing concentration of $Cu^{2+}$ ions

**30** The correct order of reduction potentials of the following pairs is
A. $Cl_2/Cl^-$ B. $I_2/I^-$ C. $Ag^+/Ag$
D. $Na^+/Na$ E. $Li^+/Li$
Choose the correct answer from the options given below. *[2022, 25 June Shift-II]*
(a) A > C > B > D > E (b) A > B > C > D > E
(c) A > C > B > E > D (d) A > B > C > E > D

**31** The standard reduction potentials for $Zn^{2+}/Zn$, $Ni^{2+}/Ni$ and $Fe^{2+}/Fe$ are – 0.76, – 0.23 and – 0.44 V, respectively. The reaction $X + Y^{2+} \longrightarrow X^{2+} + Y$ will be spontaneous when *[AIEEE 2012]*
(a) $X = Ni, Y = Fe$ (b) $X = Ni, Y = Zn$
(c) $X = Fe, Y = Zn$ (d) $X = Zn, Y = Ni$

**32** Given, $E^\circ_{Cr^{3+}/Cr} = -0.72$ V, $E^\circ_{Fe^{2+}/Fe} = -0.42$ V.
The potential for the cell
$Cr | Cr^{3+}$ (0.1M) || $Fe^{2+}$ (0.01M) | Fe is *[AIEEE 2008]*
(a) 0.26 V (b) 0.399 V (c) – 0.339 V (d) – 0.26 V

**33** The cell, $Zn|Zn^{2+}(1M)||Cu^{2+}(1M)|Cu$ ($E^\circ_{cell} = 1.10$ V), was allowed to be completely discharged at 298 K. The relative concentration of $Zn^{2+}$ to $Cu^{2+}$ $\left(\frac{[Zn^{2+}]}{[Cu^{2+}]}\right)$ is *[AIEEE 2007]*
(a) antilog (24.08) (b) 37.3
(c) $10^{37.3}$ (d) $9.65 \times 10^4$

**34** In a cell that utilises the reaction
$Zn(s) + 2H^+(aq) \longrightarrow Zn^{2+}(aq) + H_2(g)$
addition of $H_2SO_4$ to cathode compartment will *[AIEEE 2004]*
(a) lower the $E$ and shift equilibrium to the left
(b) lower the $E$ and shift the equilibrium to the right
(c) increase the $E$ and shift the equilibrium to the right
(d) increase the $E$ and shift the equilibrium to the left

## Numerical Value Type Questions

**35** Consider the following half cell reaction
$Cr_2O_7^{2-}(aq) + 6e^- + 14H^+(aq) \longrightarrow 2Cr^{3+}(aq) + 7H_2O(l)$
The reaction was conducted with the ratio of $\frac{[Cr^{3+}]^2}{[Cr_2O_7^{2-}]} = 10^{-6}$. The pH value at which the EMF of the half cell will become zero is ........... (nearest integer value)
[Given : standard half cell reduction potential $E^\circ_{Cr_2O_7^{2-}, H^+/Cr^{3+}} = 1.33$ V, $\frac{2.303RT}{F} = 0.059$ V.] *[2025, 8 April Shift-II]*

**36** 1 Faraday electricity was passed through $Cu^{2+}$ (1.5 M, 1 L)/ Cu and 0.1 Faraday was passed through $Ag^+$(0.2 M, 1 L)/Ag electrolytic cells. After this the two cells were connected as shown below to make an electrochemical cell. The emf of the cell thus formed at 298 K is ........... mV (Nearest integer)

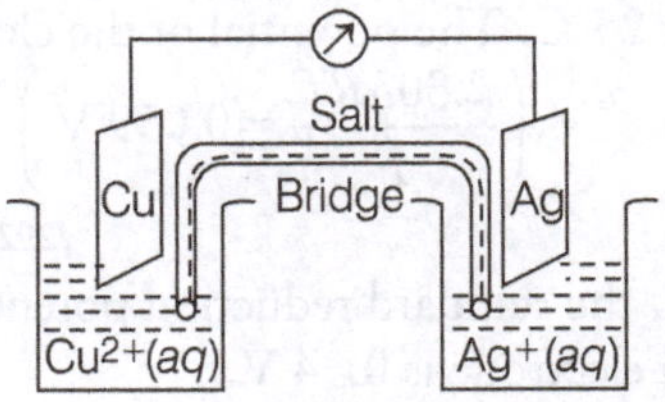


Given, $E^\circ_{Cu^{2+}/Cu} = 0.34$ V $\Rightarrow E^\circ_{Ag^+/Ag} = 0.8$ V
$\frac{2.303RT}{F} = 0.06$ V *[2025, 7 April Shift-I]*

**37** 0.2% (*w/V*) solution of NaOH is measured to have resistivity 870.0 m$\Omega$ m. The molar conductivity of the solution will be ........ $\times 10^2$ mS dm$^2$ mol$^{-1}$. (Nearest integer) **[2025, 2 April Shift-II]**

**38** Consider the following electrochemical cell at standard condition. **[2025, 2 April Shift-I]**

Au(*s*) | QH$_2$, Q | NH$_4$*X* (0.01 M) || Ag$^+$ (1 M) | Ag(*s*) $E_{cell} = +0.4$ V

The couple QH$_2$/Q represents quinhydrone electrode, the half cell reaction is given below:

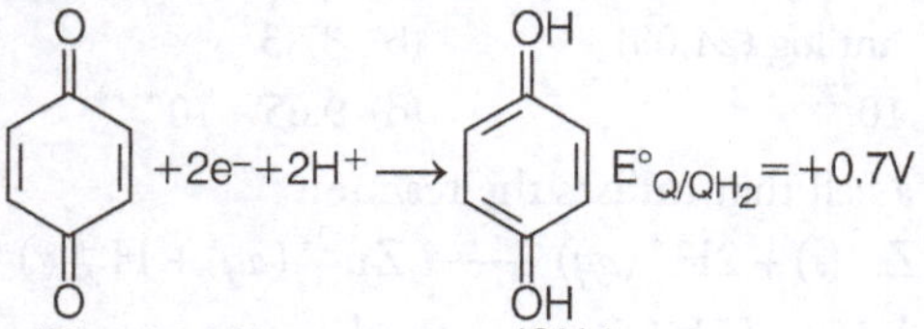


(Q) (QH$_2$)

[Given: $E^\circ_{Ag^+/Ag} = +0.8$ V and $\frac{2.303RT}{F} = 0.06$ V]

The p$K_b$ value of the ammonium halide salt (NH$_4$*X*) used here is ............ (Nearest integer)

**39** Consider the following redox reaction,

$MnO_4^- + H^+ + H_2C_2O_4 \longrightarrow Mn^{2+} + H_2O + CO_2$

The standard reduction potentials are given as below ($E^\circ_{red}$).

$$E^\circ_{MnO_4^-/Mn^{2+}} = +1.51 \text{ V}$$

$$E^\circ_{CO_2/H_2C_2O_4} = -0.49 \text{ V}$$

If the equilibrium constant of the above reaction is given as $K_{eq} = 10^x$, then the value of $x$ = ........ (Nearest integer) **[2024, 1 Feb Shift-II]**

**40** The potential for the given half-cell at 298 K is (–) ........ $\times 10^{-2}$ V.

$2H^+(aq) + 2e^- \longrightarrow H_2(g)$

$[H^+] = 1M$, $p_{H_2} = 2$ atm

(Given, $2.303\,RT/F = 0.06$ V, $\log 2 = 0.3$) **[2024, 1 Feb Shift-I]**

**41** The hydrogen electrode is dipped in a solution of pH = 3 at 25°C. The potential of the electrode will be ....... $\times 10^{-2}$ V. $\left(\frac{2.303RT}{F} = 0.059 \text{ V}\right)$ **[2024, 27 Jan Shift-II]**

**42** At 298 K, the standard reduction potential for Cu$^{2+}$/Cu electrode is 0.34 V.

Given : $K_{sp}$ Cu(OH)$_2$ = $1 \times 10^{-20}$

Take $\frac{2.303\,RT}{F} = 0.059$ V

The reduction potential at pH = 14 for the above couple is (–)$x \times 10^{-2}$ V.

The value of $x$ is ........ **[2023, 13 April Shift-II]**

**43** The number of correct statements from the following is .....

A. $E_{cell}$ is an intensive parameter.

B. A negative $E^\ominus$ means that the redox couple is a stronger reducing agent than the H$^+$ / H$_2$ couple.

C. The amount of electricity required for oxidation or reduction depends on the stoichiometry of the electrode reaction.

D. The amount of chemical reaction which occurs at any electrode during electrolysis by a current is proportional to the quantity of electricity passed through the electrolyte. **[2023, 11 April Shift-II]**

**44** In an electrochemical reaction of lead, at standard temperature, if $E^\circ_{(Pb^{2+}/Pb)} = m$ volt and $E^\circ_{(Pb^{4+}/Pb)} = n$ volt, then the value of $E^\circ_{(Pb^{2+}/Pb^{4+})}$ is given by $m - xn$. Then value of $x$ is ........... (Nearest integer). **[2023, 11 April Shift-I]**

**45** $FeO_4^{2-} \xrightarrow{+2.2\text{ V}} Fe^{3+} \xrightarrow{+0.70\text{ V}} Fe^{2+} \xrightarrow{-0.45\text{ V}} Fe^0$

$E^\circ_{FeO_4^{2-}/Fe^{2+}}$ is $x \times 10^{-3}$ V. The value of $x$ is ...... **[2023, 10 April Shift-I]**

**46** The number of incorrect statements from the following is ...........

A. the electrical work that a reaction can perform at constant pressure and temperature is equal to the reaction Gibbs energy.

B. $E^\circ_{cell}$ is dependent on the pressure.

C. $\frac{dE^\circ_{cell}}{dT} = \frac{\Delta_r S^\circ}{nF}$

D. a cell is operating reversibly if the cell potential is exactly balanced by an opposing source of potential difference. **[2023, 8 April Shift-II]**

**47** The standard reduction potentials at 298 K for the following half cells are given below

$NO_3^- + 4H^+ + 3e^- \longrightarrow NO(g) + 2H_2O$ $\quad E^\theta = 0.97$ V

$V^{2+}(aq) + 2e^- \longrightarrow V$ $\quad E^\theta = -1.19$ V

$Fe^{3+}(aq) + 3e^- \longrightarrow Fe$ $\quad E^\theta = -0.04$ V

$Ag^+(aq) + e^- \longrightarrow Ag(s)$ $\quad E^\theta = 0.80$ V

$Au^{3+}(aq) + 3e^- \longrightarrow Au(s)$ $\quad E^\theta = 1.40$ V

The number of metal(s) which will be oxidised by $NO_3^-$ in aqueous solution is ....... **[2023, 6 April Shift-II]**

**48** At what pH, given half-cell $MnO_4^-(0.1\text{ M}) \mid Mn^{2+}(0.001\text{ M})$ will have electrode potential of 1.282 V? ........... (Nearest integer)

Given, $E^\circ_{MnO_4/Mn^{2+}} = 1.54\text{ V}, \frac{2.303RT}{F} = 0.059\text{ V}$

*[2023, 1 Feb Shift-I]*

**49** The logarithm of equilibrium constant for the reaction $Pd^{2+} + 4Cl^- \rightleftharpoons PdCl_4^{2-}$ is ....... (Nearest integer)

Given : $\frac{2.303RT}{F} = 0.06\text{ V}$

$Pd^{2+}(aq) + 2e^- \rightleftharpoons Pd(s), E^\ominus = 0.83\text{ V}$

$PdCl_4^{2-}(aq) + 2e^- \rightleftharpoons Pd(s) + 4Cl^-(aq),$
$E^\ominus = 0.65\text{ V}$

*[2023, 31 Jan Shift-I]*

**50** The electrode potential of the following half cell at 298 K $X \mid X^{2+}(0.001\text{ M}) \parallel Y^{2+}(0.01\text{ M}) \mid Y$ is ....... $\times 10^{-2}$ V. (Nearest integer).

[Given : $E^\circ_{X^{2+}/X} = -2.36\text{ V}$

$E^\circ_{Y^{2+}/Y} = +0.36\text{ V}$

$\frac{2.303RT}{F} = 0.06\text{ V}$]

*[2023, 30 Jan Shift-II]*

**51** Consider the cell

$Pt(s) \mid H_2(g, 1\text{ atm}) \mid H^+(aq, 1\text{ M}) \parallel Fe^{3+}(aq), Fe^{2+}(aq) \parallel Pt(s)$

When the potential of the cell is 0.712 V at 298 K, the ratio $[Fe^{2+}]/[Fe^{3+}]$ is ......... (Nearest integer)

[Given : $Fe^{3+} + e^- \rightleftharpoons Fe^{2+}$,

$E^\circ_{Fe^{3+}, Fe^{2+}/Pt} = 0.771\text{ V}$

$\frac{2.303RT}{F} = 0.06\text{ V}$

*[2023, 30 Jan Shift-I]*

**52** The equilibrium constant for the reaction

$Zn(s) + Sn^{2+}(aq) \rightleftharpoons Zn^{2+}(aq) + Sn(s)$ is $1 \times 10^{20}$ at 298 K. The magnitude of standard electrode potential of $Sn/Sn^{2+}$ if $E^\circ_{Zn^{2+}/Zn} = -0.76\text{ V}$ is ......... $\times 10^{-2}$ V. (Nearest integer).

Given, $\frac{2.303RT}{F} = -0.059\text{ V}$

*[2023, 29 Jan Shift-II]*

**53** $Pt(s) \mid H_2(g)\,(1\text{ bar}) \mid H^+(aq)\,(1\text{ M}) \parallel M^{3+}(aq) \mid M^+(aq) \mid Pt(s)$

The $E_{cell}$ for the given cell is 0.1115 V at 298 K when $\frac{[M^+(aq)]}{[M^{3+}(aq)]} = 10^a$

The value of $a$ is .........

[Given, $E^\circ_{M^{3+}/M^+} = 0.2\text{ V}$

$\frac{2.303RT}{F} = 0.059\text{ V}$]

*[2023, 25 Jan Shift-II]*

**54** At 298 K, a 1 L solution containing 10 mmol of $Cr_2O_7^{2-}$ and 100 mmol of $Cr^{3+}$ shows a pH of 3.0.

Given : $Cr_2O_7^{2-} \longrightarrow Cr^{3+}$; $E^\circ = 1.330\text{ V}$ and

$\frac{2.303RT}{F} = 0.059\text{ V}$

The potential for the half-cell reaction is $x \times 10^{-3}$ V. The value of $x$ is ......... .

*[2023, 24 Jan Shift-I]*

**55** For a cell,

$Cu(s) \mid Cu^{2+}(0.001\text{ M}) \parallel Ag^+(0.01\text{ M} \mid Ag(s)$

the cell potential is found to be 0.43 V at 298 K. The magnitude of standard electrode potential for $Cu^{2+}/Cu$ is ........... $\times 10^{-2}$ V.

[Given : $E^\circ_{Ag^+/Ag} = 0.80\text{ V}$ and $\frac{2.303\,RT}{F} = 0.06\text{ V}$]

*[2022, 29 July Shift-II]*

**56** The cell potential for $Zn \mid Zn^{2+}(aq) \parallel Sn^{x+} \mid Sn$ is 0.801 V at 298 K. The reaction quotient for the above reaction is $10^{-2}$. The number of electrons involved in the given electrochemical cell reaction is ........... .

(Given, $E^\circ_{Zn^{2+}/Zn} = -0.763\text{ V}, E^\circ_{Sn^{x+}/Sn} = +0.008\text{ V}$ and $\frac{2.303\,RT}{F} = 0.06\text{ V}$)

*[2022, 25 July Shift-I]*

**57** The cell potential for the given cell at 298 K $Pt \mid H_2(g, 1\text{ bar}) \mid H^+(aq) \parallel Cu^{2+}(aq) \mid Cu(s)$ is 0.31 V. The pH of the acidic solution is found to be 3, whereas the concentration of $Cu^{2+}$ is $10^{-x}$ M. The value of $x$ is ...... .

(Given, $E^\ominus_{Cu^{2+}/Cu} = 0.34\text{ V}$ and $\frac{2.303\,RT}{F} = 0.06\text{ V}$)

*[2022, 29 June Shift-II]*

**58** For the given reactions, $Sn^{2+} + 2e^- \longrightarrow Sn$

$Sn^{4+} + 4e^- \longrightarrow Sn$

The electrode potentials are $E^\circ_{Sn^{2+}/Sn} = -0.140\text{ V}$ and $E^\circ_{Sn^{4+}/Sn} = 0.010\text{ V}$. The magnitude of standard electrode potential for $Sn^{4+}/Sn^{2+}$ i.e., $E^\circ_{Sn^{4+}/Sn^{2+}}$ is ........ $\times 10^{-2}$ V. (Nearest integer) *[2022, 28 June shift-II]*

**59** For the reaction taking place in the cell:
$Pt(s) \mid H_2(g) \mid H^+(aq) \| Ag^+(aq) \mid Ag(s)$
$E^\circ_{cell} = +0.5332$ V.
The value of $\Delta_f G^\ominus$ is – ......... kJ $mol^{-1}$.
[Nearest integer] *[2022, 27 June Shift-II]*

**60** $Cu(s) + Sn^{2+}(0.001M) \longrightarrow Cu^{2+}(0.01\text{ M}) + Sn(s)$
The Gibb's free energy change for the above reaction at 298 K is $x \times 10^{-1}$ kJ $mol^{-1}$.
The value of $x$ is ...... . [Nearest integer]
[Given, $E^\ominus_{Cu^{2+}/Cu} = 0.34$ V; $E^\ominus_{Sn^{2+}/Sn} = -0.14$ V; $F = 96500$ C $mol^{-1}$]. *[2022, 26 June Shift-II]*

**61** In a cell, the following reactions takes place *[2022, 25 June Shift-I]*
$$Fe^{2+} \longrightarrow Fe^{3} + e^-;\ E^\circ_{Fe^{3+}/Fe^{2+}} = 0.77\text{ V}$$
$$2I^- \longrightarrow I_2 + 2e^-;\ E^\circ_{I_2/I^-} = 0.54\text{ V}$$
The standard electrode potential for the spontaneous reaction in the cell is $x \times 10^{-2}$ V 298 K. The value of $x$ is ... [Nearest integer]

**62** Consider the following cell reaction,
$$Cd(s) + Hg_2SO_4(s) + \frac{9}{5}H_2O(l) \rightleftharpoons CdSO_4 \cdot \frac{9}{5}H_2O(s) + 2Hg(l)$$
The value of $E^\circ_{cell}$ is 4.315 V at 25°C. If $\Delta H^\circ = -825.2$ kJ $mol^{-1}$, the standard entropy change $\Delta S^\circ$ in J $K^{-1}$ is ....... (Nearest integer)
[Given, Faraday constant = 96487 C $mol^{-1}$] *[2021, 31 Aug Shift-I]*

**63** For the galvanic cell,
$Zn(s) + Cu^{2+}(0.02\text{ M}) \longrightarrow Zn^{2+}(0.04\text{M}) + Cu(s)$,
$E_{Cell}$ = ......... $\times 10^{-2}$ V.
(Nearest integer)
[Use $E^\circ_{Cu/Cu^{2+}} = -0.34$ V, $E^\circ_{Zn/Zn^{2+}} = +0.76$ V, $\frac{2.303\,RT}{F} = 0.059$ V] *[2021, 26 Aug Shift-II]*

**64** These are physical properties of an element.
A. Sublimation enthalpy
B. Ionisation enthalpy
C. Hydration enthalpy
D. Electron gain enthalpy
The total number of above properties that affect the reduction potential is ......... (Integer answer) *[2021, 26 Aug Shift-I]*

**65** For the cell
$Cu(s) \mid Cu^{2+}(aq)\ (0.1)\text{ M}) \| Ag^+(aq)\ (0.01\text{M}) \mid Ag(s)$
the cell potential, $E_1 = 0.3095$ V
For the cell,
$Cu(s) \mid Cu^{2+}(aq)\ (0.01\text{ M}) \| Ag^+(aq)\ (0.001\text{M}) \mid Ag(s)$
the cell potential = ......$\times 10^{-2}$ V.
(Round off the nearest integer).
$\left[\text{Use: } \frac{2.303\,RT}{F} = 0.059\right]$ *[2021, 27 July Shift-II]*

**66** Consider the cell at 25° C
$Zn \mid Zn^{2+}(aq), (1\text{M}) \| Fe^{3+}(aq), Fe^{2+}(aq) \mid Pt(s)$
The fraction of total iron present as $Fe^{3+}$ ion at the cell potential of 1.500 V is $x \times 10^{-2}$. The value of $x$ is ......... . (Nearest integer)
(Given, $E^\circ_{Fe^{3+}/Fe^{2+}} = 0.77$ V, $E^\circ_{Zn^{2+}/Zn} = -0.76$ V) *[2021, 25 July Shift-I]*

**67** Assume a cell with the following reaction
$$Cu(s) + 2Ag^+(1 \times 10^{-3}\text{ M}) \longrightarrow Cu^{2+}(0.250\text{M}) + 2Ag(s),\ E^\ominus_{cell} = 2.97\text{ V}$$
$E_{cell}$ for the above reaction is ......... V.
(Nearest integer)
[Given : log 2.5 = 0.3979, $T$ = 298 K] *[2021, 22 July Shift -II]*

**68** For the reaction,
$2Fe^{3+}(aq) + 2I^-(aq) \longrightarrow 2Fe^{2+}(aq) + I_2(s)$
The magnitude of the standard molar Gibbs free energy change, $\Delta_r G^\circ_m$ = – .......... kJ (Round off to the nearest integer).
$\left[E^\circ_{Fe^{2+}/Fe(s)} = -0.440\text{ V};\ E^\circ_{Fe^{3+}/Fe(s)} = -0.036\text{ V};\ E^\circ_{I_2/2I^-} = 0.539\text{ V};\ F = 96500\text{ C}\right]$ *[2021, 18 March Shift-I]*

**69** Emf of the following cell at 298 K in V is $x \times 10^{-2}$,
$Zn \mid Zn^{2+}(0.1\text{ M}) \| Ag^+(0.01\text{ M}) \mid Ag$
The value of $x$ is ......... .
(Rounded off to the nearest integer).
[Given, $E^\circ_{Zn^{2+}/Zn} = -0.76$ V, $E^\circ_{Ag^+/Ag} = +0.80$ V, $\frac{2.303RT}{F} = 0.059$] *[2021, 26 Feb Shift-II]*

**70** The magnitude of the change in oxidising power of the $MnO_4^- / Mn^{2+}$ couple is $x \times 10^{-4}$ V, if the $H^+$ concentration is decreased from 1 M to $10^{-4}$ M at 25°C. (Assume concentration of $MnO_4^-$ and $Mn^{2+}$ to

be same on change in $H^+$ concentration). The value of $x$ is ......... (Rounded off to the nearest integer).
$\left[\text{Given, } \frac{2.303\,RT}{F} = 0.059\right]$ **[2021, 24 Feb Shift-II]**

**71** An oxidation-reduction reaction in which 3 electrons are transferred has a $\Delta G^\circ$ of 17.37 kJ $mol^{-1}$ at 25° C. The value of $E^\circ_{cell}$ (in V) is – ...... $\times 10^{-2}$.
($1\text{ F} = 96{,}500\text{ C mol}^{-1}$) **[2020, 5 Sep Shift-I]**

**72** The Gibbs energy change (inJ) for the given reaction at $[Cu^{2+}] = [Sn^{2+}] = 1\text{M}$ and 298 K is
$Cu(s) + Sn^{2+}(aq) \longrightarrow Cu^{2+}(aq) + Sn(s)$;
($E^\circ_{Sn^{2+}/Sn} = -0.16\text{ V}$, $E^\circ_{Cu^{2+}/Cu} = 0.34\text{ V}$,
Take, $F = 96500\text{ C mol}^{-1}$) **[2020, 2 Sep Shift-I]**

**73** For the disproportionation reaction
$2Cu^+(aq) \rightleftharpoons Cu(s) + Cu^{2+}(aq)$ at 298 K, In $K$ (where $K$ is the equilibrium constant) is ...... $\times 10^{-1}$.
Given : ($E^\circ_{Cu^{2+}/Cu^+} = 0.16\text{ V}$,
$E^\circ_{Cu^+/Cu} = 0.52\text{ V}$ and $\frac{RT}{F} = 0.025$) **[2020, 2 Sep Shift-II]**

## Topic 3 Batteries, Fuel Cells and Corrosion

### Level 1

**01** Galvanisation is applying a coating of **[JEE Main 2016]**
(a) Cr (b) Cu (c) Zn (d) Pb

### Level 2

**02** On charging the lead storage battery, the oxidation state of lead changes from $x_1$ to $y_1$ at the anode and from $x_2$ to $y_2$ at the cathode. The values of $x_1$, $y_1$, $x_2$, $y_2$ are respectively: **[2025, 4 April Shift-I]**
(a) + 4, + 2, 0, + 2
(b) 0, + 2, + 4, + 2
(c) + 2, 0, 0, + 4
(d) + 2, 0, + 2, + 4

**03** Given below are two statements:
**Statement I** Corrosion is an electrochemical phenomenon in which pure metal acts as an anode and impure metal as a cathode.
**Statement II** The rate of corrosion is more in alkaline medium than in acidic medium.
In the light of the above statements, choose the correct answer from the options given below: **[2025, 22 Jan Shift-II]**
(a) Statement I is true but Statement II is false.
(b) Both Statement I and Statement II are true.
(c) Statement I is false but Statement II is true.
(d) Both Statement I and Statement II are false.

**04** Compound $A$ used as a strong oxidising agent is amphoteric in nature. It is the part of lead storage batteries. Compound $A$ is **[2021, 26 Feb Shift-I]**
(a) $PbO_2$
(b) PbO
(c) $PbSO_4$
(d) $Pb_3O_4$

### Level 3

**Imp.** Applications of Cells and Batteries

**05** Match List-I with List-II.

| List-I (Applications) | List-II (Batteries/Cell) |
|---|---|
| A. Transistors | I. Anode – Zn/Hg; Cathode –HgO+C |
| B. Hearing aids | II. Hydrogen fuel cell |
| C. Invertors | III. Anode-Zn;Cathode-Carbon |
| D. Apollo space ship | IV. Anode-Pb;Cathode-Pb\|$PbO_2$ |

Choose the correct answer from the options given below **[2025, 29 Jan Shift-II]**
(a) (A)-(III), (B)-(II), (C)-(IV), (D)-(I)
(b) (A)-(II), (B)-(III), (C)-(IV), (D)-(I)
(c) (A)-(III), (B)-(I), (C)-(IV), (D)-(II)
(d) (A)-(IV), (B)-(III), (C)-(II), (D)-(I)

**06** Match List-I with List-II.

| List-I (Cell) | List-II (Use/Property/Reaction) |
|---|---|
| A. Leclanche cell | I. Converts energy of combustion into electrical energy |
| B. Ni – Cd cell | II. Does not involve any ion in solution and is used in hearing aids |
| C. Fuel cell | III. Rechargeable |
| D. Mercury cell | IV. Reaction at anode $Zn \longrightarrow Zn^{2+} + 2e^-$ |

Choose the correct answer from the options given below **[2024, 9 April Shift-II]**
(a) A-III, B-I, C-IV, D-II
(b) A-I, B-II, C-III, D-IV
(c) A-II, B-III, C-IV, D-I
(d) A-IV, B-III, C-I, D-II

**07** Fuel cell, using hydrogen and oxygen as fuels,
A. has been used in spaceship
B. has as efficiency of 40% to produce electricity
C. uses aluminum as catalysts
D. is eco-friendly
E. is actually a type of Galvanic cell only.
Choose the correct answer from the options given below: **[2024, 4 April Shift-II]**
(a) A, B, C only
(b) A, B, D only
(c) A, B, D, E only
(d) A, D, E only

**Imp.** Rusting of Iron

**08** Which of the following statements is not correct about rusting of iron? *[2024, 27 Jan Shift-II]*

(a) Dissolved acidic oxides $SO_2$, $NO_2$ in water act as catalyst in the process of rusting.
(b) Rusting of iron is envisaged as setting up of electrochemical cell on the surface of iron object.
(c) When pH lies above 9 or 10, rusting of iron does not take place.
(d) Coating of iron surface by tin prevents rusting, even if the tin coating is peeling off.

**08** For lead storage battery pick the correct statements.

A. During charging of battery, $PbSO_4$ on anode is converted into $PbO_2$.
B. During charging of battery, $PbSO_4$ on cathode is converted into $PbO_2$.
C. Lead storage battery consists of grid of lead packed with $PbO_2$ as anode.
D. Lead storage battery ~ 38% solution of sulphuric acid as an electrolyte. *[2023, 12 April Shift-I]*

Choose the correct answer from the options given below.

(a) B, D only (b) B, C only
(c) B, C, D only (d) A, B, D only

**09** Match List I with List II.

| | List I | | List II |
|---|---|---|---|
| (A) | $Cd(s) + 2Ni(OH)_3(s) \longrightarrow CdO(s) + 2Ni(OH)_2(s) + H_2O(l)$ | (I) | Primary battery |
| (B) | $Zn(Hg) + HgO(s) \longrightarrow ZnO(s) + Hg(l)$ | (II) | Discharging of secondary battery |
| (C) | $2PbSO_4(s) + 2H_2O(l) \longrightarrow Pb(s) + PbO_2(s) + 2H_2SO_4(aq)$ | (III) | Fuel cell |
| (D) | $2H_2(g) + O_2(g) \longrightarrow 2H_2O(l)$ | (IV) | Charging of secondary battery |

Choose the correct answer from the options given below. *[2022, 28 July Shift-I]*

(a) (A) → (I), (B) → (II), (C) → (III), (D) → (IV)
(b) (A) → (IV), (B) → (I), (C) → (II), (D) → (III)
(c) (A) → (II), (B) → (I), (C) → (IV), (D) → (III)
(d) (A) → (II), (B) → (I), (C) → (III), (D) → (IV)

## Key Index

### Topic 1

| | | | | | | | | | | | | | | | | | | | |
|---|---|---|---|---|---|---|---|---|---|---|---|---|---|---|---|---|---|---|---|
| 1 | (b) | 2 | (c) | 3 | (d) | 4 | (b) | 5 | (c) | 6 | (a) | 7 | (d) | 8 | (b) | 9 | (a) | 10 | (c) |
| 11 | (b) | 12 | (a) | 13 | (a) | 14 | (d) | 15 | (a) | 16 | (a) | 17 | (c) | 18 | (c) | 19 | (a) | 20 | (b) |
| 21 | (a) | 22 | (b) | 23 | (a) | 24 | (d) | 25 | (d) | 26 | (b) | 27 | (a) | 28 | (b) | 29 | (d) | 30 | (b) |
| 31 | (a) | 32 | (d) | 33 | (c) | 34 | (a) | 35 | (3) | 36 | (150) | 37 | (2) | 38 | (2) | 39 | (4) | 40 | (5) |
| 41 | (2) | 42 | (46) | 43 | (3) | 44 | (161) | 45 | (66) | 46 | (14) | 47 | (25) | 48 | (1000) | 49 | (6) | 50 | (14) |
| 51 | (20) | 52 | (288) | 53 | (57) | 54 | (14) | 55 | (25) | 56 | (11) | | | | | | | | |

### Topic 2

| | | | | | | | | | | | | | | | | | | | |
|---|---|---|---|---|---|---|---|---|---|---|---|---|---|---|---|---|---|---|---|
| 1 | (b) | 2 | (a) | 3 | (a) | 4 | (c) | 5 | (d) | 6 | (a) | 7 | (c) | 8 | (c) | 9 | (c) | 10 | (c) |
| 11 | (d) | 12 | (b) | 13 | (c) | 14 | (a) | 15 | (b) | 16 | (b) | 17 | (a) | 18 | (c) | 19 | (c) | 20 | (d) |
| 21 | (a) | 22 | (c) | 23 | (d) | 24 | (c) | 25 | (d) | 26 | (a) | 27 | (b) | 28 | (c) | 29 | (d) | 30 | (a) |
| 31 | (d) | 32 | (a) | 33 | (c) | 34 | (c) | 35 | (10) | 36 | (400) | 37 | (23) | 38 | (6) | 39 | (338) | 40 | (1) |
| 41 | (18) | 42 | (25) | 43 | (4) | 44 | (2) | 45 | (1825) | 46 | (1) | 47 | (3) | 48 | (3) | 49 | (6) | 50 | (275) |
| 51 | (10) | 52 | (17) | 53 | (3) | 54 | (917) | 55 | (34) | 56 | (4) | 57 | (7) | 58 | (16) | 59 | (51) | 60 | (983) |
| 61 | (23) | 62 | (25) | 63 | (109) | 64 | (3) | 65 | (28) | 66 | (24) | 67 | (3) | 68 | (45) | 69 | (147) | 70 | (3776) |
| 71 | (6) | 72 | (96500) | 73 | (144) | | | | | | | | | | | | | | |

### Topic 3

| | | | | | | | | | | | | | | | | | |
|---|---|---|---|---|---|---|---|---|---|---|---|---|---|---|---|---|---|
| 1 | (c) | 2 | (d) | 3 | (a) | 4 | (a) | 5 | (a) | 6 | (d) | 7 | (d) | 8 | (a) | 9 | (c) |

# Detailed **Explanations**

## Topic 1

**01.** *(b)* From Faraday's law,
gram equivalent of Al deposited

$$=\frac{It}{96500}$$

$$\frac{w}{27}\times 3=\frac{2\times 30\times 60}{96500}$$

so $w=0.336$ g

Thus, the amount of aluminium deposited at cathode is 0.336 g.

**02.** *(c)* Concentrated solution of sulphuric acid is used as an electrolyte to obtain $H_2S_2O_8$ by the process of electrolysis.

**03.** *(d)* The given graph of molar conductivity *vs* $\sqrt{C}$ is for weak electrolyte. Hence among the given options $CH_3COOH$ is the correct answer, as it is a weak electrolyte $(X)$.

**04.** *(b)* Given, $Q=2F$

Atomic mass of Cu = 63.5 u

Valency of the metal $Z=2$

We have, $CuSO_4 \longrightarrow Cu^{2+}+SO_4^{2-}$

$$\underset{1\,\text{mol}}{Cu^{2+}}+\underset{\substack{2\text{ mol}\\2F}}{2e^-}\longrightarrow \underset{1\,\text{mol}\,=\,63.5\,\text{g}}{Cu}$$

Alternatively, $W=ZQ=\frac{E}{F}\cdot 2F=2E$

$$=\frac{2\times 63.5}{2}=63.5$$

**05.** *(c)* Debye-Huckel Onsager equation can be written as

$$\lambda_c=\lambda_\infty-(B)\sqrt{C}$$

> • Concept **Enhancer**
>
> Debye-Huckel Onsager equation is used to explain the behaviour of molar conductivity wrt to dilution of strong electrolytes, It is not applicable for weak electrolyte. It's graph can plotted as
>
> 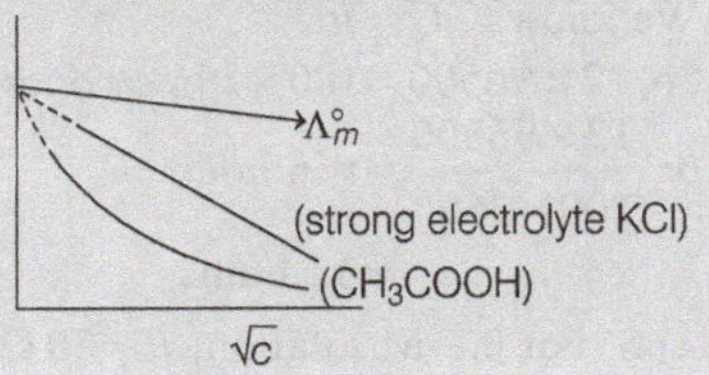
> 

**06.** *(a)* Molar conductivity of aqueous $CH_3COOH$ solution can be calculated from Kohlrausch's law as

$$\lambda^\circ_{CH_3COOH}=\lambda^\circ_{CH_3COONa}+\lambda^\circ_{HCl}-\lambda^\circ_{NaCl}$$

Hence, for determining the molar conductivity of $CH_3COOH$ solution, $\lambda^\circ_{NaCl}$ is required.

**07.** *(d)* According to Kohlrausch's law

$$\lambda^\circ_{CH_3COOH}=\lambda^\circ_{CH_3COO^-}+\lambda^\circ_{H^+} \quad ...(i)$$

$$\lambda^\circ_{HCl}=\lambda^\circ_{H^+}+\lambda^\circ_{Cl^-} \quad ...(ii)$$

$$\lambda^\circ_{CH_3COONa}=\lambda^\circ_{CH_3COO^-}+\lambda^\circ_{Na^+} \quad ..(iii)$$

Thus, on adding Eqs. (ii) and (iii), if $\lambda^\circ_{Na^+}$ and $\lambda^\circ_{Cl^-}$ are subtracted, we can obtain the value of $\lambda^\circ_{HOAc}$. Thus, additional value required is $\lambda^\circ_{NaCl}$.

**08.** *(b)* $\Lambda^\infty_{AcOH}=\Lambda^\infty_{AcONa}+\Lambda^\infty_{HCl}-\Lambda^\infty_{NaCl}$
$=91.0+426.2-126.5=390.7$

**09.** *(a)* By Kohlrausch's law.

$$\Lambda^\circ_{NaBr}=\Lambda^\circ_{NaCl}+\Lambda^\circ_{KBr}-\Lambda^\circ_{KCl}$$
$$=126+152-150$$
$$=128\text{ S cm}^2\text{ mol}^{-1}$$

**10.** *(c)* $S\propto$ area $(m^2)\propto$ conc $(mol/m^3)$

$$\propto\frac{1}{\text{length}}(m^{-1})$$

$$S=\kappa\, m^2\,\frac{mol}{m^3}\,m^{-1}\Rightarrow \kappa=S\,mol^{-1}\,m^2$$

**11.** *(b)* The value of limiting molar conductivities for given cations in water are

| | |
|---|---|
| $H^+$ | 349.6 S $cm^2$ $mol^{-1}$ |
| $Na^+$ | 50.1 S $cm^2$ $mol^{-1}$ |
| $K^+$ | 73.5 S$cm^2$$mol^{-1}$ |
| $Ca^{2+}$ | 119 S$cm^2$$mol^{-1}$ |
| $Mg^{2+}$ | 106 S$cm^2$$mol^{-1}$ |

**12.** *(a)* The relation between molar conductivity and concentration is as follows

$$\text{Molar conductivity}\propto\frac{1}{\sqrt{C}}$$

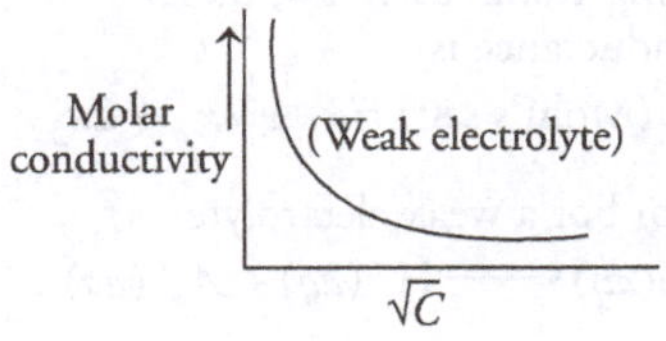


Thus, for weak electrolyte molar conductivity decreases sharply as there is an increases in concentration. Thus, option (a) is the correct answer.

**13.** *(a)* The electrolytes *A* and *B* are weak electrolyte and strong electrolyte respectively.

**14.** *(d)* As the given solution is already infinitely dilute hence no change in molar conductivity is observed upon addition of water.

**15.** *(a)* Amount of substance deposited is given by Faraday's law,

$W=Zit$

$W=ZQ$ $\quad [Q=it]$

$Q=\frac{W}{Z}$

$W=QZ$ when $Q=1$ then $W=Z$ where, $Z$ is electrochemical equivalent

Hence, option (a) is correct.

**16.** *(a)* Given,

Ionic conductivities,

$\lambda_{C^{2+}}=57$ Scm$^2$ / mol

$\lambda_{A^{2-}}=73$ Scm$^2$ / mol

Molar conductivity of solution is given by

$$\Lambda_{solution}=\lambda_{C^{2+}}+\lambda_{A^{2-}}$$
$$=57+73=130\text{ Scm}^2/\text{mol}$$

Hence, option (a) is correct.

**17.** *(c)* Units of molar conductivity $=$ Scm$^2$mol$^{-1}$.

Unit of (concentration)$^{1/2}$ $=mol^{1/2}L^{-1/2}$

Units for slope

$$=\frac{\text{Unit of molar conductivity}}{\text{Unit of (concentration)}^{1/2}}$$

$$=\text{Scm}^2\text{mol}^{-3/2}L^{1/2}$$

**18.** *(c)* Calomel electrodes comes in the category of metal-insoluble salt-anion electrodes.

**19.** *(a)* Among the given options only nature of electrode does not affect electrolytic conductance of a solution.

**20.** *(b)* Reaction involved in electrolysis of brine solution is as follows

$$NaCl(aq)\longrightarrow Na^+(aq)+Cl^-(aq)$$

At anode $2Cl^-(aq)\longrightarrow Cl_2(g)+2e^-$

$$2H_2O(l)\longrightarrow O_2(g)+4H^++4e^-$$

At cathode

$$2H_2O(l) + 2e^- \longrightarrow H_2(g)\uparrow + 2OH^-(aq)$$

$$2Na^+ + 2OH^- \longrightarrow 2NaOH$$

Thus, $OH^-$ is formed at cathode.

**21.** *(a)* The correct representation of conductometric titration of benzoic acid *vs* sodium hydroxide is (a).

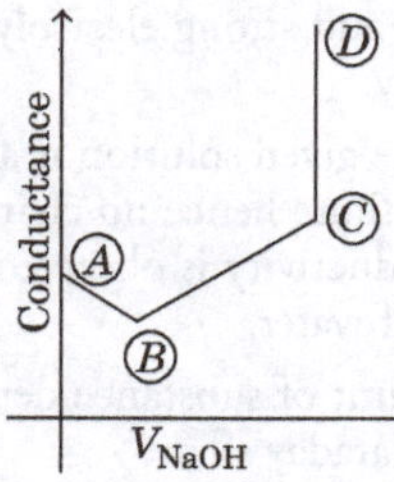


In *A* to *B*, free $H^+$ ions are replaced by $Na^+$ which decreases conductance.

In *B* to *C*, un-dissociated benzoic acid reacts with NaOH and forms salt which increases number of ions and thus conductance increases.

In *C* to *D*, NaOH added further increases $Na^+$ and $OH^-$ ions which further increases the conductance.

**22.** *(b)* Molar conductivity of strong electrolyte (i.e., KI) increases slowly on dilution, while molar conductivity of weak electrolyte (i.e., $H_2CO_3$) increases steeply with dilution.

**23.** *(a)* Molar conductivity,

$$\Lambda_M = \frac{\kappa \times 1000}{M}$$

$$\Lambda_{M_1} = \frac{\kappa_1 \times 1000}{\frac{10}{20} \times 1000} = 2\kappa_1$$

$$\Lambda_{M_2} = \frac{\kappa_2 \times 1000}{\frac{20}{80} \times 1000} = 4\kappa_2$$

Given, $\kappa_1 = \kappa_2 = \kappa$

$\therefore \quad \Lambda_{M_1} = 2\kappa; \; \Lambda_{M_2} = 4\kappa$

$$\frac{\Lambda_{M_2}}{\Lambda_{M_1}} = \frac{4\kappa}{2\kappa} = \frac{4}{2}$$

$\Rightarrow \quad \Lambda_{M_2} = 2\Lambda_{M_1}$

**24.** *(d)* Limiting molar conductivity of $CH_3COOH$ is more than KCl. Hence, statement I is false.

With decrease in concentration of electrolyte, molar conductivity increases.

Hence, statement II is also false.

**25.** *(d)* A solution of $Ni(NO_3)_2$ is electrolysed between platinum electrodes using 0.1 Faraday electricity. It means that 0.1 equivalent of $Ni^{2+}$ will be discharged.

Electrolysis of $Ni(NO_3)_2$ gives

$$Ni^{2+} + 2e^- \longrightarrow Ni$$

(Atomic mass of Ni = 58.7)

Number of equivalents = Number of moles × number of electrons transferred

0.1 = Number of moles of Ni × 2

$\therefore$ Number of moles of Ni $= \frac{0.1}{2} = 0.05$

**26.** *(b)*

**key idea** This question is based upon Faraday's first law which states that "Mass of any substance deposited or liberated at any electrode is directly proportional to the quantity of electricity passed."

During charging :

$$Pb + SO_4^{2-} \longrightarrow PbSO_4 + 2e^-$$

$\Rightarrow \quad Q = nF$

$Q = 2F$

2F gives 303 g/mol

Amount of $PbSO_4$ electrolysed by 0.05 F is

$\therefore \quad = \frac{303}{2} \times 0.05$ g of $PbSO_4$

$= 7.575 \approx 7.6$ g of $PbSO_4$

**27.** *(a)* Difluoroacetic acid being strongest acid will furnish maximum number of ions and thus will show highest electrical conductivity.

**28.** *(b)* $Ag^+ + e^- \longrightarrow Ag$

Number of moles of Ag $= \frac{9650}{96500}$

$= \frac{1}{10}$ moles

$\therefore$ Mass of Ag produced $= \frac{1}{10} \times 108$

$= 10.8$ g

**29.** *(d)* Statement I is true but Statement II is false.

Mohr's salt molecular formula is

$$FeSO_4 \cdot (NH_4)_2SO_4 \cdot 6H_2O$$

Using Kohlrausch law, molar conductance is

$\lambda_m^\infty$ (Mohr's salt) $= x_1 + 2x_2 + 2x_3$.

**30.** *(b)* For a weak electrolyte

$$HA(aq) \rightleftharpoons H^+(aq) + A^-(aq)$$

$$K_a = \frac{\alpha^2 C}{1-\alpha}$$

$$\alpha^2 C = K_a - K_a\alpha$$

$$\alpha^2 C + K_a\alpha - K_a = 0$$

$$\left(\frac{\Lambda_m}{\Lambda_m^\circ}\right)^2 C + K_a\left(\frac{\Lambda_m}{\Lambda_m^\circ}\right) - K_a = 0$$

$$\Lambda_m^2 C + K_a\Lambda_m\Lambda_m^\circ - K_a\Lambda_m^{\circ 2} = 0$$

**31.** *(a)* The correct match is

A → 3, B → 1, C → 4, D → 2.

**32.** *(d)* According to Kohlrausch's law,

$$\Lambda_m^\circ(HA) = [\Lambda_m^\circ(H^+) + \Lambda_m^\circ(Cl^-)] + [\Lambda_m^\circ(Na^+) + \Lambda_m^\circ(A^-)] - [\Lambda_m^\circ(Na^+) + \Lambda_m^\circ(Cl^-)]$$

$= 425.9 + 100.5 - 126.4$

$= 400 \text{ S cm}^2 \text{ mol}^{-1}$

Also, molar conductivity at given concentration is given as,

$$\Lambda_m = \frac{1000 \times \kappa}{M}$$

$$\therefore \quad \Lambda_m = \frac{1000 \times 5 \times 10^{-5} \text{ S cm}^{-1}}{10^{-3} \text{ M}}$$

$= 50 \text{ S cm}^2 \text{ mol}^{-1}$

Therefore, degree of dissociation ($\alpha$), of *HA* is,

$$\alpha = \frac{\Lambda_m}{\Lambda_m^\circ} = \frac{50 \text{ S cm}^2\text{mol}^{-1}}{400 \text{ S cm}^2\text{mol}^{-1}}$$

$= 0.125$

**33.** *(c)* Consider the equation

$$B_2H_6 + 3O_2 \longrightarrow B_2O_3 + 3H_2O$$

Now, the oxygen is produced by the electrolysis of $H_2O$,

$$2H_2O \xrightarrow{4F} 2H_2 + O_2$$

1 mole of $O_2$ is produced by 4F charge.

So, for the production of 3 moles of $O_2$ 12(3 × 4) moles of electrons will be needed.

So, the total amount of charge required to produce 3 moles of oxygen will be

$12 \times F$ or $12 \times 96500$

We know $\quad Q = it$

So, $12 \times 96500 = 100 \times t$ in seconds

or $\quad \frac{12 \times 96500}{100 \times 3600} = t$ in hours

$= 3.2$ hours

**34.** *(a)* For 0.2 M solution, $R = 50\,\Omega$

$\kappa = 1.4 \text{ S m}^{-1} = 1.4 \times 10^{-2} \text{ Scm}^{-1}$

The resistivity of the solution is related to specific conductance by

$$\rho = \frac{1}{\kappa} = \frac{1}{1.4 \times 10^{-2}} \Omega \text{ cm}$$

It is known that, $R = \rho \frac{l}{a}$

$\Rightarrow$ cell constant $= \frac{l}{a} = R \times \frac{1}{\rho}$

$= 50 \times 1.4 \times 10^{-2}$ cm

For 0.5 M solution, $R = 280\,\Omega$; $\kappa = ?$

cell constant $= \frac{l}{a} = 50 \times 1.4 \times 10^{-2}$ cm

The specific conductance of the solution is given by

$$\kappa = \frac{1}{\rho} = \frac{1}{R} \times \frac{l}{a}$$

$$\kappa = \frac{1}{280} \times 50 \times 1.4 \times 10^{-2}$$

$= 2.5 \times 10^{-3}$ S cm$^{-1}$

Now, the molar conductivity of the solution is given by

$$\Lambda_m = \frac{\kappa \times 1000}{M} = \frac{2.5 \times 10^{-3} \times 1000}{0.5}$$

$= 5 \text{ S cm}^2 \text{mol}^{-1} = 5 \times 10^{-4} \text{ S m}^2 \text{ mol}^{-1}$

**35.** *(3)* Given,

$\lambda_m$ of $NH_4Cl = 185$

$(\lambda^\circ_m)_{NH_4^+} + (\lambda^\circ_m)_{Cl^-} = 185$

$(\lambda^\circ_m)_{NH_4^+} = 185 - 70 = 115 \text{ Scm}^2 \text{mol}^{-1}$

$(\lambda^\circ_m)_{NH_4OH} = (\lambda^\circ_m)_{NH_4^+} + (\lambda^\circ_m)_{OH^-}$

$= 115 + 170$

$(\lambda^\circ_m)_{NH_4OH} = 285 \text{ Scm}^2 \text{mol}^{-1}$

Degree of dissociation $= \frac{(\lambda_m)_{NH_4OH}}{(\lambda^\circ_m)_{NH_4OH}}$

$= \frac{85.5}{285} = 0.3$

$= 3 \times 10^{-1}$

**36.** *(150)* Given, $R = 100\,\Omega$

Resistance of same cell with dilute solution $x = ?$

The resistance is given by the formule

$R = \rho \frac{l}{A}$

$\kappa = G \times G^*$ (where, $G = \frac{1}{R}$)

and $G^*$ (Cell constant) $= \frac{l}{A}$

$$\frac{\kappa_C}{\kappa_d} = \frac{R_d}{R_C};$$

$$\lambda_m = \frac{\kappa \times 1000}{C}$$

$$\frac{\kappa_C}{\kappa_d} = \frac{(\lambda_m . C)_c}{(\lambda_m . C)_d} = \frac{R_d}{R_C}$$

$C \rightarrow$ concentrated solution,

$d \rightarrow$ dilute solution.

$$\frac{100 \times (0.15)^2}{150 \times (0.1)^2} = \frac{R_d}{100}$$

$\Rightarrow \quad R_d = 150\Omega$

So, the value of $x$ is 150.

**37.** *(2)*

**key idea** To solve this problem, first of all find the value of $[OH^-]$ by using the pH value. After that use Faraday's law to calculate the value of current in ampere.

Electrolysis of NaCl is

$$NaCl + H_2O(aq) \longrightarrow NaOH(aq) + \frac{1}{2}Cl_2(g) + \frac{1}{2}H_2(g)$$

Since during electrolysis pH changes to 12.

So, $[OH^-] = 10^{-2}$ and $[H^+] = 10^{-12}$

So, according to Faraday law,

Gram amount of substance deposited = Amount of electricity passed

$$10^{-2} \times \frac{600}{1000} \times 96500 = I \times t$$

$$\frac{10^{-2} \times 600}{1000} \times 96500 = I \times 5 \times 60$$

$I = 1.93 \text{ A} \simeq 2$ ampere.

**38.** *(2)* $H_2O \longrightarrow 2H^+ + \frac{1}{2}O_2 + 2e^-$

So, we can say, oxidation of 1 mole of water required $96500 \times 2$C

$= 96500 \times 2 = 193000$

$= 1.93 \times 10^5 \text{ C} \approx 2 \times 10^5$ C

**39.** *(4)* Conductivity (Sm$^{-1}$)

$2.1 \times 10^3$, $1.2 \times 10$, $3.91$, $1 \times 10^3$ } Conductor at 298.15 K

$1 \times 10^{-16}$ Insulator at 298.15 K

**40.** *(5)* $Cu^{2+} + 2e^- \longrightarrow Cu$

From the reaction, it is clear that 2 Faraday liberates 1 mol Cu.

So, 1 Faraday will librate 0.5 mol Cu.

0.5 mol = 0.5 g atom $= 5 \times 10^{-1}$

Thus, the value of $x = 5$

**41.** *(2)* Given, increase in mass at cathode, $W = 1.314$ g

$Au^{3+} + 3e^- \longrightarrow Au$

$$\frac{W}{E} = \frac{\text{Charge}}{1}$$

$$\frac{1.314}{\frac{197}{3}} = \frac{Q}{1\text{ F}} \quad \text{[At. mass of Au = 197]}$$

$Q = 0.02 \text{ F} = 2 \times 10^{-2}$ F.

**42.** *(46)*

**key idea** This problem is based on the faraday's first law of electrolysis. To solve this question, first write the electrolysis reaction of zinc sulphate. Using this equation calculate the value of $n$. Then, put this value along with given values in the formula $W = Zit$, to calculate the value of $w$.

The electrolysis reaction of zinc sulphate is, $Zn^{2+} + 2e^- \longrightarrow Zn$

$SO_4^{2-} + H_2O \longrightarrow H_2SO_4$

$n = 2$

$\underset{\text{(mass)}}{W} = Z \times i \times t$

$$= \frac{\text{Atomic mass} \times i \times t}{nF}$$

Mass of zinc deposit is

$$= \frac{65.4}{2 \times 96500} \times 0.015 \times 15 \times 60$$

[Given, $i = 0.015$, $t = 15 \times 60$, Atomic mass of Zn = 65.4 amu]

$W = 45.75 \times 10^{-4} \text{ g} \approx 46 \times 10^{-4}$ g

Hence, the nearest integer is 46.

**43.** *(3)* Only three statements A, C and D are correct whereas all other statements B, E are incorrect.

The correct form of B and E respectively are :

(B) The number of ions per unit volume in a solution decreases on dilution, because the total number of ions in solution remains the same, but the volume increases.

(E) For weak electrolyte, the change in molar conductivity with dilution is due to an increase in the degree of dissociation.

**44.** *(161)* Volume of Ni deposited $= v \times d$

$= (10^{-4})(100) \text{ cm}^3 = 10^{-2} \text{ cm}^3$

$\therefore$ Weight of Ni deposited

$= (10^{-2}) \times (10)$

$= 0.1$ gram

No. of moles of (Ni) deposited

$$= \frac{\text{Given mass}}{\text{Molar mass}} = \frac{0.1}{60} = \left[\frac{1}{6} \times 10^{-2}\right]$$

$Ni^{2+} + 2e^{\ominus} \longrightarrow Ni$

$\Rightarrow$ Charge required $= \frac{1}{6} \times 10^{-2} \times 2F$

$= i \times t$

$t = \frac{1}{6} \times 10^{-2} \times \frac{2 \times 96500}{2}$

$t = 161\,s$

**45.** *(66)* Given,

$C = 0.0025$ M acetic acid

($\kappa$) specific conductance of 0.0025 M acetic acid $= 5 \times 10^{-5}$ S cm mol$^{-1}$

Limiting molar conductivity of $CH_3COOH$, $\lambda_m^\infty = 400$ S cm mol$^{-1}$

Dissociation constant, $\Lambda_m = \frac{\kappa}{C} \times 1000$

$= \frac{5 \times 10^{-5} \times 10^{-3}}{0.0025}$

$= \frac{5 \times 10^{-2}}{2.5 \times 10^{-3}}$

$= 20$ S cm$^2$ mol$^{-1}$

$\alpha = \lambda_m / \lambda_m^\infty = \frac{20}{400} = \frac{1}{20}$

$K_a = \frac{C\alpha^2}{1-\alpha} = \frac{0.0025 \times \frac{1}{20} \times \frac{1}{20}}{\frac{19}{20}}$

$= \frac{0.0025}{19 \times 20} = 6.6 \times 10^{-6} = 66 \times 10^{-7}$

**46.** *(14)* $\underset{10^{-5}}{AgNO_3} \longrightarrow \underset{10^{-5}}{Ag^+} + \underset{10^{-5}}{NO_3^-}$

$AgBr \rightleftharpoons \underset{S}{Ag^+} + \underset{S}{Br^-}$

$K_{sp} = [Ag^+][Br^-] = 4.9 \times 10^{-13}$

$\Rightarrow S^2 = 4.9 \times 10^{-13}$

$\Rightarrow S = 7 \times 10^{-7}$ M

$\therefore [Ag^+] = 7 \times 10^{-7} + 100 \times 10^{-7}$

$= 107 \times 10^{-7}$ M

$\therefore [Br^-] = \frac{K_{sp}}{[Ag^+]}$

$= \frac{4.9 \times 10^{-13}}{107 \times 10^{-7}} = 4.58 \times 10^{-8}$

Conductivity

$= [Ag^+]\lambda^\circ_{Ag^+} + [Br^-]\lambda^\circ_{Br^-} + [NC_3^-]\lambda^\circ_{NO_3^-}$

$= (107 \times 10^{-7} \times 6 \times 10^{-3}) + (4.58 \times 10^{-8} \times 8 \times 10^{-3}) + (10^{-5} \times 7 \times 10^{-3})$

$= (642 \times 10^{-10}) + (36.64 \times 10^{-11}) + (7 \times 10^{-8})$

$= (642 + 3.664 + 700) \times 10^{-10}$

$= 1346 \times 10^{-10}$ Sm$^{-1}$

or $13.46 \times 10^{-8}$ Sm$^{-1}$ $\approx 14 \times 10^{-8}$ Sm$^{-1}$

**47.** *(25)*

> **key idea** This question can be solved by using the concept of molar conductivity for this, use the formula $\Lambda_m = \frac{\kappa \times 1000}{M}$, where $\kappa$ can be expressed in terms of resistivity as $\frac{1}{\rho}$

$\Lambda_m = \frac{\kappa \times 1000}{M} = \frac{1}{\rho} \times \frac{1000}{M}$

$= \frac{1}{5 \times 10^{-3}} \times \frac{1000}{0.8}$

$= 25 \times 10^4\ \Omega^{-1}$ cm$^2$ mol$^{-1}$

**48.** *(1000)* In case of solution 1, cell constant

$\frac{l}{A} = 129$ m$^{-1}$

Resistance, $R = 100\,\Omega$

Specific conductivity,

$\kappa = \frac{1}{R} \cdot \frac{l}{A} = \frac{1}{100} \times 129$

$= 1.29\Omega^{-1}$ m$^{-1}$

Concentration, $C = 74.5$ ppm

Molarity, $M_1 = \frac{74.5}{\text{molar mass} \times 1000}$

$= \frac{74.5}{74.5 \times 1000} = 10^{-3}$ M

Molar conductivity $\Lambda_1 = \frac{\kappa}{M}$

$\Lambda_1 = \frac{1.29}{10^{-3}}$

$= 1290\Omega^{-1}$ m$^{-1}$ M$^{-1}$

or $\frac{129 \times 10^3}{100}$

In case of solution 2,

Resistance, $R = 50\Omega$

$\kappa = \frac{1}{50} \times 129 = \frac{129}{50}\Omega^{-1}$ m$^{-1}$

$M = \frac{149}{74.5 \times 1000}$

$= 2 \times 10^{-3}$ M

$\Lambda_2 = \frac{\kappa}{M} = \frac{129 \times 10^3}{50 \times 2}\Omega^{-1}$ m$^{-1}$M$^{-1}$

$\frac{\Lambda_1}{\Lambda_2} = \frac{129 \times 10^3 \times 50 \times 2}{100 \times 129 \times 10^3} = 1$

$\frac{\Lambda_1}{\Lambda_2} = 1 = 1000 \times 10^{-3}$.

Thus, the value of $x$ is 1000.

**49.** *(6)* Reaction involved is as follows

$Cr_2O_7^{2-} + 14H^{\oplus} + 6e^{\ominus} \longrightarrow 2Cr^{3+} + 7H_2O$

So, 1 mole $Cr_2O_7^{2-}$ requires 6 moles electrons

1 mole $= nF$

1 mole electron $= 1$ Faraday

6 moles electrons $= 6$ Faraday

**50.** *(14)* Given,

$\Lambda^\infty_{AgI} = \Lambda^\infty_{AgNO_3} + \Lambda^\infty_{NaI} - \Lambda^\infty_{NaNO_3}$

$= 13.3 + 12.7 - 12.0$

$= 14$ mS m$^2$ mol$^{-1}$.

**51.** *(20)* Given, time, $t = x$ min

Using Faraday's first law of electrolysis,

$w = ZIt = \frac{EIt}{96500}$

$\frac{0.3482 \times 96500 \times 3}{56 \times 1.5} = t$

$\Rightarrow t = 1200$ s.

So, time, $x = 20$ min.

**52.** *(288)* Molar conductivity of $BaCl_2 = 280$ Scm$^2$mol$^{-1}$

Molar conductivity of $H_2SO_4$ $= 860$ Scm$^2$mol$^{-1}$

Molar conductivity of HCl $= 426$ S cm$^2$mol$^{-1}$

Molar conductivity of $BaSO_4$ = ?

From Kohlrausch's law,

$\Lambda^\infty_m(BaSO_4) = \lambda^\infty_m(Ba^{2+}) + \lambda^\infty_m(SO_4^{2-})$

$\Lambda^\infty_m(BaSO_4) = \Lambda^\infty_m(BaCl_2) + \Lambda^\infty_m(H_2SO_4) - 2\Lambda^\infty_m(HCl)$

On putting given values we get,

$= 280 + 860 - 2(426)$

$= 288$ S cm$^2$ mol$^{-1}$

> **Concept Enhancer**
> According to Kohlrausch law of independent migration of ions, the limiting molar conductivity of an electrolyte can be represented as the sum of the individual contribution of the anion and cation of the electrolyte.

**53.** *(57)* We know that,

$\kappa = 1/R \cdot G^*$

$\because$ The conductivity cell is same, $G^*$ is constant and hence $\kappa \cdot R$ = constant

$\therefore \quad \kappa_{KCl} \cdot R_{KCl} = \kappa_{HCl} \cdot R_{HCl}$

Therefore, $0.14 \times 4.19 = \kappa \times 1.03$

or, $\kappa$ of HCl solution $= \dfrac{0.14 \times 4.19}{1.03}$

$= 0.5695 \text{ Sm}^{-1}$

$\approx 57 \times 10^{-2} \text{ Sm}^{-1}$

**54.** *(14)*

> **key idea** To find out molar conductivity, first of all we will calculate the conductivity by using the formula $\kappa = G\,l/A$ then we will use this value in the formula of molar conductivity
>
> $\Lambda_m = \dfrac{\kappa\,(\text{S cm}^{-1}) \times 1000\,(\text{cm}^3/\text{L})}{\text{Molarity (mol/L)}}$

Using the formula

$\kappa = G\dfrac{l}{A} = 0.55 \times 10^{-3} \times 1.3 \text{ S cm}^{-1}$

Now using,

$\Lambda_m = \dfrac{\text{Conductivity (S cm}^{-1}) \times 1000\,(\text{cm}^3/\text{L})}{\text{Molarity (mol/L)}}$

$= \dfrac{0.55 \times 10^{-3} \times 1.3 \times 10^3}{5 \times 10^{-3}}$

$= 143 \text{ S cm}^2 \text{ mol}^{-1}$

$\therefore 1 \text{ S cm}^2 \text{ mol}^{-1} = 10^{-4} \text{ S m}^2 \text{ mol}^{-1}$

$\Rightarrow 143 \text{ S cm}^2 \text{ mol}^{-1}$

$= 143 \times 10^{-4} \text{ S m}^2 \text{ mol}^{-1}$

$= 143 \times 10^{-4} \times 10^3 \text{ m Sm}^2 \text{ mol}^{-1}$

$\Rightarrow 14.3 \text{ m Sm}^2 \text{ mol}^{-1}$

$\approx 14 \text{ m Sm}^2 \text{ mol}^{-1}$

**55.** *(25)* In the reduction half-cell,

$MnO_4^- + 8H^+ + 5e^- \longrightarrow Mn^{2+} + 4H_2O$

As one mole of $MnO_4^-$ required 5F of charge,

$\therefore$ 5 moles of $MnO_4^-$ will require charge,

$Q = 5 \times 5 \text{ Faraday} = 25 \text{ Faraday}$

**56.** *(11)* For 1 mole of $ClO_3^-$, 6 F charge required but current efficiency is only 60%. So, to obtain 1 mol $ClO_3^-$, charge required is 10 F.

Now, moles of $KClO_3 = \dfrac{10 \text{ g}}{122}$

$\therefore$ Charge required $= \dfrac{10}{122} \times 10 \text{ F}$

$\therefore$ Current $= 2 \text{ A}$

Apply, $Q = i \cdot t$

$\therefore \quad 100 \times \dfrac{96500}{122} = 2t$

$\Rightarrow t = 39549.18 \text{ sec} = 10.98 \text{ h} \approx 11 \text{ h}$

Hence, the correct answer is (11).

## Topic 2

**01.** *(b)*

> **key idea** To solve this question, first of all write the balanced chemical equation, then using this find the value of change in number of $e^-$. After that use Nernst-equation to obtain the correct option.

According to Nernst equation,

$$E_{cell} = E^\circ_{cell} - \frac{RT}{nF} \ln Q$$

The cell reaction given in question is

$Mg(s) + 2Ag^+(aq) \rightleftharpoons 2Ag(s) + Mg^{2+}(aq)$

$$Q = \frac{[Mg^{2+}]}{[Ag^+]^2}$$

Also, as there is change in 2 number of electrons. So, $n = 2$

$$\therefore \quad E = E^\circ_{Cell} - \frac{RT}{2F} \ln \left[\frac{[Mg^{2+}]}{[Ag^+]^2}\right]$$

$$\text{or} \quad E = E^\circ_{Cell} + \frac{RT}{2F} \ln \left[\frac{[Ag^+]^2}{[Mg^{2+}]}\right]$$

**02.** *(a)* For the reaction,

$\frac{1}{2} H_2(g) + AgCl(s) \rightleftharpoons H^+(aq) + Cl^-(aq) + Ag(s)$

At anode : $\frac{1}{2} H_2(g) \rightleftharpoons H^+(aq) + e^-$

At cathode :

$AgCl(s) + e^- \longrightarrow Ag(s) + Cl^-(aq)$

$\therefore$ Cell representation.

$Pt|H_2(g)|HCl(sol^n) \mid AgCl(s)|Ag$

**03.** *(a)* The standard electrode potential of $M^+/M$ in aqueous solution does not depend on ionisation of a solid metal atom. So, option (a) is correct.

**04.** *(c)* To find the value of $\Delta G^\circ$ (kJ mol), we use the formula

$\Delta G^\circ = -nFE^\circ_{cell}$

$\Delta G^\circ = -2 \times 96000 \times 2$

$= -384000 \text{ J/mol}$

In terms of kJ/mol,

$$\Delta G^\circ = \frac{-384000}{1000} = -384 \text{ kJ/mol}$$

**05.** *(d)* Higher the SRP (Standard Reduction Potential), stronger is the oxidising agent. Among the given, electrode potentials, $E^\circ_{MnO_4^-/Mn^{2+}}$ is highest. Hence, $MnO_4^-$ is the strongest oxidising agent.

**06.** *(a)* More negative value of $E^\circ$ means higher reducing power. Thus, the correct order of reducing power of the metals is given by, $B > C > A$.

**07.** *(c)* By using Nernst equation,

$$E_{cell} = E^\circ_{cell} - \frac{0.0591}{n} \frac{[P]}{[R]}$$

$$\underset{0.1\text{ M}}{Cu^{2+}} + Zn \longrightarrow \underset{1\text{M}}{Zn^{2+}} + Cu$$

$$\frac{[P]}{[R]} = \frac{[Zn^{2+}]}{[Cu^{2+}]} = \frac{1}{0.1} = 10$$

$$E_{cell} = 1.10 - \frac{0.0591}{2} \log 10$$

$= 1.10 - 0.0295 = 1.0705 \text{ V}$

**08.** *(c)* Oxidation reaction takes place at anode. Hence, (a), (b) is not feasible as these are reduction reactions. But $Cr^{3+}$ is oxidised to $Cr^{6+}$ $(Cr_2O_7^{2-})$ under given condition.

**09.** *(c)*

$$FeO_4^{2-} \xrightarrow[n_1=3]{E_1^\circ = +2.0\text{ V}} Fe^{3+} \xrightarrow[n_2=1]{E_2^\circ = 0.8\text{ V}} Fe^{2+} \xrightarrow{E_3^\circ = -0.5\text{ V}} Fe^0$$

$$FeO_4^{2-} \xrightarrow[n_4=4]{E_4^\circ = ?} Fe^{2+}$$

$-n_4FE_4^\circ = -n_1FE_1^\circ - n_2FE_2^\circ$

On substituting the values, we get

$+4E_4^\circ = 3 \times 2 + (1 \times 0.8)$

$E_4^\circ = \dfrac{6.8}{4} = 1.7 \text{ V}$

So, the value of $E^\circ_{FeO_4^{2-}/Fe^{2+}} = 1.7 \text{ V}$

**10.** *(c)* An electrochemical cell can be converted into an electrolytic cell by applying an external potential greater than $E^\circ_{cell}$.

The entire cell reaction will be reversed. The oxidised species will be reduced and the reduced species will be oxidised.

**11.** *(d)* $M \mid M^{+2} \parallel X \mid X^{2-}$

The cell potential is given by

$E^\circ_{cell} = E^\circ_{M/M^{+2}} + E^\circ_{X/X^{-2}}$

$= -0.46 + 0.34 = -0.12 \text{ V}$

As $E^\circ_{cell}$ is negative, so anode becomes cathode and cathode becomes anode. Spontaneous reaction will be

$M^{+2} + X^{2-} \rightarrow M + X$

Hence, option (d) is correct.

**12.** (b)

**key idea** To solve this question, first of all write the balanced chemical equation to find the value of $n$, then use the equation $E = E^\circ - \frac{0.059}{n}\log\frac{p_{H_2}}{[H^+]^2}$ to calculate the pressure of $H_2$.

The required reaction is

$$2e^- + 2H^+(aq) \rightarrow H_2(g)$$

Using, $E = E^\circ - \frac{0.059}{n}\log\frac{p_{H_2}}{[H^+]^2}$

$$0 = 0 - \frac{0.059}{2}\log\frac{p_{H_2}}{(10^{-7})^2}$$

$$\log\frac{p_{H_2}}{(10^{-7})^2} = 0 \Rightarrow \frac{p_{H_2}}{(10^{-7})^2} = 1$$

$$p_{H_2} = 10^{-14} \text{ bar}$$

**13.** (c) A cell with less value of $\left(\frac{\partial E}{\partial T}\right)_P$ is preferred to be used as reference electrode. Because it can be used for wider range of temperature without much deviation from standard value. Thus, half cell $C$ would be preferred to be used as reference electrode.

**14.** (a) $E^\circ_{cell} = E^\circ_{Cu^{2+}/Cu} - E^\circ_{Zn^{2+}/Zn}$

$$= 0.34 - (-0.76) = 1.10 \text{ volt}$$

If $E_{ext} < E^\circ_{cell}$, then $e^-$s flow from Zn to Cu and Zn dissolves at anode and Cu deposites at cathode.

Then, statement (a) is incorrect whereas all other statements are correct.

**15.** (b) The half cell reactions are :

(1) $Cu^{2+} + 2e^- \rightleftharpoons Cu$ ; $E^\circ = 0.34$ V

(2) $Cu^+ + e^- \rightleftharpoons Cu$ ; $E^\circ = 0.522$ V

(3) $Cu^{2+} + e^- \rightleftharpoons Cu^+$ ; $E^\circ = x$

As we know, $\Delta G^\circ = -nFE$

∴ To obtain the $E^\circ$ of required equation, subtract $\Delta G^\circ$ of equation (2) from $\Delta G^\circ$ of (1) as

$\Delta G^\circ$ of (3) = $\Delta G^\circ$ of (1) $-\Delta G^\circ$ of (2)

or $-1 \times F \times x =$

$(-2 \times F \times 0.34) - (-1 \times F \times 0.522)$

$$\therefore \quad x = 2 \times 0.34 - 0.522$$

$$= 0.68 - 0.522 = 0.158 \text{ V}$$

**16.** (b) We know that,

$$E^\circ = \frac{RT}{nF}\ln K \Rightarrow K = e^{nFE^\circ/RT}$$

Given, $n = 2$, $F = 96000 \text{ C mol}^{-1}$

$E^\circ = 2$ V, $R = 8 \text{ JK}^{-1} \text{ mol}^{-1}$

$T = 300$ K

On substituting the values,

$$\therefore \quad K = e^{\frac{2 \times 96000 \times 2}{8 \times 300}} = e^{160}$$

**17.** (a) $E^\circ_{cell}$ is non-additive property while $\Delta G^\circ$ is an additive property. So, $E^\circ_{cell}$ is not directly added or subtracted.

$$Mn^{2+} + 2e^- \longrightarrow Mn, \quad E^\circ = -1.18 \text{ V} \quad \ldots(i)$$

$$\Delta G^\circ = -nFE^\circ$$

[here, $n$ = number of $e^-$ involved in the reaction]

$$\Delta G^\circ_1 = -2F(-1.18) = 2.36 \text{ F}$$

$$2Mn^{3+} + 2e^- \longrightarrow 2Mn^{2+}, \quad E^\circ = +1.51 \text{ V} \quad \ldots(ii)$$

$$\Delta G^\circ_2 = -2F[1.51] = -3.02 \text{ F}$$

Subtracting Eq. (ii) from Eq. (i), we get

$$3\,Mn^{2+} \longrightarrow Mn + 2Mn^{3+}, \quad [n = 2]$$

$$\Delta G^\circ_3 = -2FE^\circ$$

$$\Delta G^0_3 = \Delta G^0_1 - \Delta G^0_2 = 5.38 \text{ F}$$

$$\Rightarrow -2FE^\circ = 5.38 \text{ F} \Rightarrow E^\circ = -2.69 \text{ V}$$

∵ We get negative value EMF (i.e., $\Delta G$ = + ve). So, the reaction will be non-spontaneous.

**18.** (c) $\frac{2}{3}Al_2O_3 \longrightarrow \frac{4}{3}Al + O_2$

i.e., $\frac{2}{3} \times 3(O^{2-}) \longrightarrow O_2 + 4e^-$

$$[\because n = 4]$$

$$\Delta_r G = +966 \text{ kJ mol}^{-1}$$

$$= 966 \times 10^3 \text{ J mol}^{-1}$$

$$\Delta_r G = -nFE_{cell}$$

$$966 \times 10^3 = -4 \times 96500 \times E_{cell}$$

$$\Rightarrow \quad E_{cell} = -2.5 \text{ V}$$

Thus, the difference is 2.5 V.

**19.** (c) It is given that,

$$Fe^{3+} + 3e^- \longrightarrow Fe, \quad E^\circ_1 = -0.036 \text{ V} \quad \ldots(i)$$

$$Fe^{2+} + 2e^- \longrightarrow Fe, \quad E^\circ_2 = -0.439 \text{ V} \quad \ldots(ii)$$

We need to calculate

$$Fe^{3+} + e^- \longrightarrow Fe^{2+}, E^\circ_3 = ? \ldots(iii)$$

In order to calculate $E^\circ_3$, concept of $\Delta G^\circ$ is used.

$\Delta G^\circ_1$ for Eq. (i) is given by

$$\Delta G^\circ_1 = -3 \times F \times (-0.036)$$

$$\Rightarrow \quad \Delta G^\circ_1 = +0.108 \text{ F}$$

$\Delta G^\circ_2$ for Eq. (ii) is given by

$$\Delta G^\circ_2 = -2 \times F \times (-0.439)$$

$$\Rightarrow \quad \Delta G^\circ_2 = 0.878 \text{ F}$$

$\Delta G^\circ_3$ for Eq. (iii) is given by

$$\Delta G^\circ_3 = -1 \times F \times E^\circ_3$$

$$\Rightarrow \quad \Delta G^\circ_3 = \Delta G^\circ_1 - \Delta G^\circ_2$$

$$\Rightarrow \quad -1 \times F \times E^\circ_3 = 0.108 \text{ F} - 0.878 \text{ F}$$

$$\Rightarrow \quad E^\circ_3 = 0.770 \text{V}$$

**20.** (d) $AgI(s) + e^- \longrightarrow Ag(s) + I^-$; $\Delta G^\circ_1 = 0.152$ F

$Ag(s) \longrightarrow Ag^+ + e^-$; $\Delta G^\circ_2 = 0.8$ F

Net reaction is given by,

$$AgI(s) \longrightarrow Ag^+ + I^-;$$

$$\Delta G^\circ = \Delta G^\circ_1 + \Delta G^\circ_2$$

$$-FE^\circ_{cell} = 0.152 \text{ F} + 0.8 \text{ F}$$

$$= -0.952$$

$$E^\circ_{cell} = \frac{0.059}{n}\log K_{sp}$$

$$-0.952 = \frac{0.059}{1}\log K_{sp}$$

$$\log K_{sp} = \frac{-0.952}{0.059} = -16.135$$

**21.** (a) More negative value of $E^\circ_{red}$ indicates better reducing agent. Thus, easily oxidised. Hence, oxidation of $Cr^{2+}$ to $Cr^{3+}$ is the easiest.

**22.** (c) Relation between $K_{eq}$ and $E^\circ_{cell}$ is

$$E^\circ_{cell} = \frac{2.303\,RT}{nF}\log K_{eq}$$

$$E^\circ_{cell} = \frac{0.0591}{n}\log K_{eq} \text{ (at 298 K)}$$

$$0.591 = \frac{0.0591}{1}\log K_{eq}$$

$$\therefore \quad \log K_{eq} = 10 \Rightarrow K_{eq} = 1 \times 10^{10}$$

**23.** (d) The standard emf of the cell,

$$E^\circ_{cell} = \frac{2.303\,RT}{nF}\log K_{eq}$$

$$\Rightarrow \quad 0.295 = \frac{0.0591}{2}\log K_{eq}$$

$$\therefore \log K_{eq} = 10 \Rightarrow K_{eq} = 1 \times 10^{10}$$

**24.** (c) At LHS (oxidation)

$$2 \times (Ag \longrightarrow Ag^+ + e^-); \; E^\circ_{oxi} = -x$$

At RHS (reduction)

$$Cu^{2+} + 2e^- \longrightarrow Cu$$

$$E^\circ_{red} = +y$$

$$2Ag + Cu^{2+} \longrightarrow Cu + 2Ag^+,$$

$$E^\circ_{cell} = (y - x)$$

**25.** *(d)* Statement I is correct but Statement II is incorrect.

The correct form of Statement II is,

At cathode, instead of Mg, $H_2O(l)$ will reduce and evolve $H_2$ gas.

**26.** (a) Among the given statements only (a) is correct

Cell potential is given by

$$E^\circ_{cell} = E^\circ_{cathode} - E^\circ_{anode}$$

$$1.21 = 1.229 - E^\circ_{anode}$$

$$= 19 \text{ mV}$$

∴Fuel cell involves oxidation of methanol which will occur at anode and reduction of $O_2$ will occur at cathode.

**27.** *(b)* For given cell;

$$Fe^{2+}(aq) + Ag^+(aq) \longrightarrow Fe^{3+}(aq) + Ag(s)$$

$$Fe \xrightarrow{\Delta G_1^\circ} Fe^{2+} \xrightarrow{\Delta G_2^\circ} Fe^{3+}$$

$$\Delta G_3^\circ = \Delta G_1^\circ + \Delta G_2^\circ$$

$$-3F(-z) = -2F(-y) + \Delta G_2^\circ$$

$$\Delta G_2^\circ = 3F_z - 2F_y$$

Also $\Delta G_2^\circ = -nFE^\circ_{Fe^{2+}/Fe^{3+}}$

$$3F_z - 2F_y = -IF(E^0_{Fe^{2+}/Fe^{+3}})$$

$$E^\circ_{Fe^{2+}/Fe^{3+}} = 2y - 3z$$

$E^\circ_{cell}$ for reaction will be

$$E^\circ_{Ag^+/Ag} + E^\circ_{Fe^{2+}/Fe^{3+}} = x + 2y - 3z$$

**28.** *(c)*

> **key idea** To solve this problem first of all calculate the value of $E^\circ$ of corresponding option. After that use this value of $E^\circ$ into the equation, $\Delta G^\circ = -nFE^\circ$ to obtain the value of $\Delta G^\circ$.

Using the formula,

$$\Delta G^\circ = -nFE^\circ$$

For option (a),

$$E^\circ = -2.37 - 0.8 = -3.17 \text{ V}$$

$$\Delta G^\circ = -2 \times F \times (-3.17) = +6.34 \text{ FV}$$

For option (b),

$$E^\circ = 0.8 - 0.34 = 0.46 \text{ V}$$

$$\Delta G^\circ = -2 \times F \times 0.46$$

$$= -0.92 \text{ FV}$$

For option (c),

$$E^\circ = 0.8 + 0.76 = 1.56 \text{ V}$$

Now, $\Delta G^\circ = -2 \times F \times 1.56 = -3.12 \text{ FV}$

For option (d),

$$E^\circ = -2.37 + 0.76 = -1.61 \text{ V}$$

Now, $\Delta G^\circ = -2 \times F \times (-1.61)$

$$= +3.22 \text{ FV}$$

Thus, the most negative value of $\Delta G^\circ$ is obtained for option (c).

**29.** *(d)* Anode reaction

$$[Tl(s) \longrightarrow Tl^+(aq) + e^-] \times 2$$

Cathode reaction

$$Cu^{+2}(aq) + 2e^- \longrightarrow Cu(s)$$

Overall redox reaction

$$2Tl(s) + Cu^{+2}(aq) \longrightarrow 2Tl^+(aq) + Cu(s)$$

$$E_{cell} = E^\circ_{cell} - \frac{0.0591}{2} \log \frac{[Tl^+]^2}{[Cu^{+2}]}$$

$E_{cell}$ increases by increasing concentration of $[Cu^{+2}]$.

**30.** *(a)* The values of reduction potentials are as follows

A. $Cl_2 / Cl^-$; $E^\circ_{Cl_2/Cl^-} = +1.36$ V

B. $I_2 / I^-$; $E^\circ_{I_2/I^-} = +0.54$ V

C. $Ag^+ / Ag$; $E^\circ_{Ag^+/Ag} = +0.80$ V

D. $Na^+ / Na$; $E^\circ_{Na^+/Na} = -2.71$ V

E. $Li^+ / Li$; $E^\circ_{Li^+/Li} = -3.05$ V

Hence, the correct order of reduction potentials is

$$A > C > B > D > E.$$

> **• Concept Enhancer**
> Reduction potential represents the tendency of a substance to loose or gain electrons from an electrode. Standard reduction potential $= E^\circ_{Cathode} - E^\circ_{Anode}$.

**31.** *(d)* A cell reaction is spontaneous, if $\Delta G^\circ < 0$.

Since, $\Delta G^\circ = -nFE^\circ_{cell}$

Thus, $E^\circ_{cell} > 0$,

$$X + Y^{2+} \longrightarrow X^{2+} + Y \quad (\text{Red.: } Y^{2+} \to Y; \ \text{Oxi.: } X \to X^{2+})$$

$$X = Zn, \quad Y = Ni$$

$$Zn + Ni^{2+} \longrightarrow Zn^{2+} + Ni$$

$$E^\circ_{cell} = E^\circ_{Zn/Zn^{2+}} + E^\circ_{Ni^{2+}/Ni}$$

$$= 0.76 + (-0.23) = 0.53 \text{ V}$$

$$E^\circ_{cell} > 0$$

⇒ Spontaneous reaction.

While for other options $E^\circ_{cell}$ is negative. Hence, the reaction is non-spontaneous.

**32.** *(a)* The two half cell reactions are as follows

$$Cr \longrightarrow Cr^{3+} + 3e^-$$

$$Fe^{2+} + 2e^- \longrightarrow Fe$$

Net cell reaction is given by

$$2Cr + 3Fe^{2+} \longrightarrow 2Cr^{3+} + 3Fe; \ n = 6$$

$$E^\circ_{cell} = E^\circ_{Fe^{2+}/Fe} - E^\circ_{Cr^{3+}/Cr}$$

$$= -0.42 - (-0.72) = 0.30 \text{V}$$

$$E_{cell} = E^\circ_{cell} - \frac{0.0591}{n} \log \frac{[Cr^{3+}]^2}{[Fe^{2+}]^3}$$

$$= 0.30 - \frac{0.0591}{6} \log \frac{(0.1)^2}{(0.01)^3}$$

$$= 0.30 - \frac{0.0591}{6} \log \frac{10^{-2}}{10^{-6}}$$

$$= 0.30 - \frac{0.0591}{6} \log 10^4$$

$$E_{cell} = 0.2606 \text{ V}$$

**33.** *(c)* Cell is completely discharged. It means equilibrium gets established,

$E_{cell} = 0$

Net cell reaction is given by

$$Zn + Cu^{2+} \rightleftharpoons Zn^{2+} + Cu$$

We know,

$$E_{cell} = E^\circ_{cell} - \frac{0.0591}{n} \log \frac{[P]}{[R]}$$

$$E^\circ_{cell} = \frac{0.0591}{n} \log \left[\frac{Zn^{2+}}{Cu^{2+}}\right]$$

or $$1.10 = \frac{0.0591}{2} \log \left[\frac{Zn^{2+}}{Cu^{2+}}\right]$$

$$\frac{[Zn^{2+}]}{[Cu^{2+}]} = \text{antilog} \frac{2.20}{0.0591}$$

$$= \text{antilog } 37.3 = 10^{37.3}$$

**34.** *(c)*

$$Zn(s) + 2H^+(aq) \longrightarrow Zn^{2+}(aq) + H_2(g)$$

Reaction quotient, $Q = \dfrac{[Zn^{2+}]}{[H^+]^2}$

Corresponding cell is

$$Zn \mid Zn^{2+}(C_1) \parallel H^+(aq) \mid Pt(H_2)$$

Anode Cathode

$$= E^\circ_{cell} - \frac{0.0591}{2} \log \frac{[Zn^{2+}]}{[H^+]^2}$$

$$= E^\circ_{cell} + \frac{0.0591}{2} \log \frac{[H^+]^2}{[Zn^{2+}]}$$

If $H_2SO_4$ is added then $[H^+]$, (towards reactant side) increases.
Hence, equilibrium shift towards right due to Le-chatelier's principle and $E_{cell}$ increases.

**35.** *(10)* The balanced cell reaction is as follows,

$$Cr_2O_7^{2-}(aq)+14H^+(aq)+6e^- \longrightarrow 2Cr^{3+}(aq)+7H_2O(l)$$

Using nernst equation

$$E_R = E_R^o - \frac{0.059}{6}\log\frac{[Cr^{3+}]^2}{[Cr_2O_7^{2-}][H^+]^{14}}$$

$$0 = 1.33 - \frac{0.059}{6}\log\frac{10^{-6}}{[H^+]^{14}}$$

$141.254 = 14\,pH \Rightarrow pH = 10.08$

**36.** *(400)* Cell reactions are as follows,

$$Cu^{2+} \quad + \quad 2e^- \quad \longrightarrow \quad Cu$$

(1 Faraday = charge on 1 mole electrons)

| | | | |
|---|---|---|---|
| $t=0$ | 1.5 | 1 mole | |
| $t=t$ | 1 | – | 0.5 mole |

$[Cu^{2+}] = 1$ M after electrolysis.

$$Ag^+ \quad + \quad e^- \longrightarrow Ag$$

| | | | |
|---|---|---|---|
| $t=0$ | 0.2 | 0.1 mole | |
| $t=t$ | 0.1 | – | – |

$[Ag^+] = 0.1$ M after electrolysis.
Cell reaction:

$$Cu(s) + 2Ag^+(aq) \longrightarrow Cu^{2+}(aq) + 2Ag(s)$$

$$E = E° - \frac{0.06}{n}\log\frac{[Cu^{2+}]}{[Ag^+]^2}$$

$$E = (0.8 - 0.34) - \frac{0.06}{2}\log\frac{(1)}{(0.1)^2}$$

$= 0.4$ V = or 400 mV

**37.** *(23)* Concentration of NaOH $= 0.2\%\ (w/v)$

$\therefore 0.2$ g of NaOH in 100 mL of solution

Molarity of NaOH solution

$$= \frac{\text{Mole of solute}}{V_{mL}} \times 1000$$

$$\frac{\frac{0.2}{40}}{100} \times 1000 = \frac{2}{40} \text{ M}$$

Given, resistivity of solution

$= 870$ m ohm m $= 8.7$ ohm dm

Now conductivity,

$$\kappa = \frac{1}{\rho} = \frac{1}{8.7}\ \text{ohm}^{-1}\ \text{dm}^{-1}$$

Molar conductivity,

$$\lambda_m = \frac{\kappa}{M} = \frac{\frac{1}{8.7}}{\frac{2}{40}} = 2.29\ \text{S dm}^2\ \text{mol}^{-1}$$

or $23 \times 10^2$ m S dm$^2$ mol$^{-1}$

**38.** *(6)* Consider the given electrochemical cell reaction,

$$QH_2 + 2Ag^+ \longrightarrow 2Ag + Q + 2H^+$$

$$E = E° \frac{-0.06}{2}\log[H^+]^2$$

$$E = E° - 0.06 \times \log[H^+]$$

$$pH = -\log[H^+] = \frac{E - E°}{0.06}$$

$$= \frac{0.4 - 0.1}{0.06} = \frac{0.3}{0.06} = 5$$

$$pH + NH_4X = 7 - \frac{1}{2}pK_b - \frac{1}{2}\log C$$

$$5 = 7 - \frac{1}{2} \times pK_b - \frac{1}{2}\log(10^{-2})$$

$pK_b = 6$

**39.** *(338)* $[Mn^{+7} + 5e^- \rightarrow Mn^{+2}] \times 2$

$[C_2O_4^{-2} \rightarrow 2CO_2 + 2e^-] \times 5$

$$2Mn^{+7} + 5C_2O_4^{-2} \rightarrow 2Mn^{+2} + 10CO_2$$

$E°_{cell} = 0.49 + 1.51 = 2.00$ V

At equilibrium, $E_{cell} = 0$

So, $E°_{cell} = \dfrac{0.0591}{n}\log K_{eq}$

$$\log K_{eq} = \frac{2 \times 10}{0.0591}$$

$\log K_{eq} = 338.40$

$\log K_{eq} \approx 338$

$K_{eq} = 10^{338}$

Thus, $x = 338$

**40.** *(1)* $2H^+(aq) + 2e^- \longrightarrow H_2(g)$

$$E_{cell} = E° - \frac{RT}{nF}\log\frac{[\text{Product}]}{[\text{Reactant}]}$$

$$E_{cell} = E^{\circ}_{H^+/H_2} - \frac{0.06}{2}\log\frac{pH_2}{[H^+]^2}$$

$$= 0.00 - \frac{0.06}{2}\log\frac{2}{[1]^2}$$

$= -0.03 \times 0.3$

$= -0.9 \times 10^{-2}$ V

Nearest integer value, $-1 \times 10^{-2}$ V

**41.** *(18)* The reaction in cell is,

$$2H^+(aq) + 2e^- \longrightarrow H_2(g)$$

$$E_{cell} = E° - \frac{0.059}{2}\log\left[\frac{p_{H_2}}{[H^+]^2}\right]$$

$= 0 - 0.059 \times 3 = -0.177$ V

or $= -17.7 \times 10^{-2}$ V

$\approx -18 \times 10^{-2}$ V

Hence, the missing term will be 18.

**42.** *(25)* For the reaction,

$$Cu^{2+} + 2e^- \longrightarrow Cu$$

pH = 14

means, $p_{OH} = 0$ or $[OH^-] = 1$ M

Also, $K_{sp} = [Cu^{2+}][OH^-]^2$

Hence, $1 \times 10^{-20} = [Cu^{2+}][1]^2$

or, $[Cu^{2+}] = 1 \times 10^{-20}$ ...(1)

The reduction potential of copper can be written as

$$E_{cell} = E^{\circ}_{cell} - \frac{0.0591}{2} \times \log[1/Cu^{2+}]$$

$$E_{cell} = E^{\circ}_{cell} + \frac{0.059}{2} \times \log[Cu^{2+}]$$

$E_{cell} = 0.34 + 0.0295 \times \log[10^{-20}]$

$E_{cell} + 0.34 + [0.0295 \times (-20)]$

$= 0.34 + [-0.59] = -0.25$

or $-25 \times 10^{-2}$ V

Thus, the value of $x = 25$

**43.** *(4)* All the given statements i.e., *A*, *B*, *C* and *D* are correct.

**44.** *(2)* $Pb^{2+} + 2e^- \longrightarrow Pb;$ $\quad \Delta G°_1 = -2FE°_1$

$Pb^{4+} + 4e^- \longrightarrow Pb;$ $\quad \Delta G°_2 = -4FE°_2$

$Pb^{2+} \longrightarrow Pb^{4+} + 2e^-$ $\quad \Delta G°_3 = -2FE°_3$

$[\because \Delta G° = -nFE°]$

$\Rightarrow \Delta G°_3 = \Delta G°_1 - \Delta G°_2$

$-2FE°_3 = 2F(2n - m)$ $\quad \left[\because E°_{Pb^{4+}/Pb} = n,\ E°_{Pb^{2+}/Pb} = m\right]$

$\Rightarrow E°_3 = m - 2n = m - xn$

$\Rightarrow x = 2$

**45.** *(1825)* Given,

For the reaction,

$$FeO_4^{2-} + 4e^- \longrightarrow Fe^{2+}$$

$$E°_{FeO_4^{2-}/Fe^{2+}} = x \times 10^{-3}\ \text{V}$$

$$FeO_4^{2-} \xrightarrow[n=3]{2.2V} Fe^{3+} \xrightarrow[n=1]{0.7\,V} Fe^{2+} \xrightarrow{-0.45V} Fe$$

$$-nFE°_{FeO_4^{2-}/Fe^{2+}} = -nFE°_{FeO_4^{2-}/Fe^{3+}} + (-nFE°_{Fe^{3+}/Fe^{2+}})$$

$\Rightarrow \quad 4\times x = 3\times 2.2 + 1 + 0.7$

$\Rightarrow \quad x = \frac{7.3}{4} = 1.825 \text{ V}$

$= 1825\times 10^{-3} \text{ V}$

**46.** *(1)* Only statement B is incorrect. $E°_{cell}$ of an electrochemical cell depends on equilibrium constant of the cell reaction, nature of the reactant as well as temperature and does not depend on the pressure of gas involved in the cell.

**47.** *(3)* For the redox reaction to be feasible, $E°_{cell}$ must be positive. Therefore, $E°_{cell}$ will be positive in case of 3 metals i.e., V, Fe and Au.

(i) $E°_{cell} = E°_{reduction} - E°_{oxidation}$

$E°_{cell} = E°_{NO_3^-/NO} - E°_{V/V^{2+}}$

$= 0.97 - (-1.19) = 2.16 \text{V}$

(ii) $E°_{cell} = E°_{NO_3^-/NO} - E°_{Fe/Fe^{3+}}$

$= 0.97 - (-0.04) = 1.01 \text{V}$

(iii) $E°_{cell} = E°_{NO_3^-/NO} - E°_{Ag/Ag^+}$

$= 0.97 - (0.80) = 0.17 \text{ V}$

> ***Alternate Solution***
> Metals having lower standard reduction potential than that of given redox reaction i.e. 0.97 V will be oxidised by $NO_3^-$. So, among the given options metals with SRP value −1.19 V, −0.04 V and 0.80 V will be oxidised by $NO_3^-$.

**48.** *(3)* Reaction equation is

$MnO_4^- + 8H^+ + 5e^- \rightleftharpoons Mn^{2+} + 4H_2O$

$\Rightarrow E = E° - \frac{0.059}{5}\log\frac{[Mn^{2+}]}{[MnO_4^-][H^+]^8}$

$\Rightarrow \quad 1.282 = 1.54 - \frac{0.059}{5}\log\frac{10^{-3}}{10^{-1}\times[H^+]^8}$

$\Rightarrow 1.282 = 1.54 - 0.0118\log\frac{10^{-2}}{[H^+]^8}$

$\Rightarrow \quad -0.258 = -0.0118\log\frac{10^{-2}}{[H^+]^8}$

$\Rightarrow \quad \frac{0.258}{0.0118} = \log\frac{10^{-2}}{[H^+]^8}$

$\Rightarrow \quad 21.86 = -2 - 8\log[H^+]$

$\Rightarrow \quad -\log[H^+] = 2.9825 \simeq 3$

$\Rightarrow \quad pH = 3$

**49.** *(6)*

> **key idea** To find out the value of log *k*, first of all we will calculate the value of *n* and *E*° from the given equation. Then we will substitute these values in the equation $\Delta G° = -RT\ln K$.

$Pd^{+2}(aq) + 2e^- \rightleftharpoons Pd(s),$

$E°_{\text{Cathode, (reduction)}} = 0.83 \text{ V}$

$Pd(s) + 4Cl^-(aq) \rightleftharpoons PdCl_4^{2-}(aq) + 2e^-,$

$E°_{\text{Anode, (oxidation)}} = 0.65$

Net reaction is

$Pd^{2+}(aq) + 4Cl^-(aq) \rightleftharpoons PdCl_4^{2-}(aq)$

$E°_{cell} = E°_{\text{Cathode, (reduction)}} - E°_{\text{Anode, (oxidation)}}$

$E°_{cell} = 0.83 - 0.65 = 0.18 \quad \text{...(i)}$

Also, $n = 2 \quad \text{...(ii)}$

$\Delta G° = -RT\ln K_{eq}$

$-nFE°_{cell} = -RT\times 2.303(\log_{10} K_{eq})$

$\frac{E°_{cell}}{0.06}\times n = \log K_{eq} \quad \text{...(iii)}$

Using Eq. (i), (ii) and (iii), we get

$\frac{0.18\times 2}{0.06} = \log K_{eq} \Rightarrow \log K = 6$

**50.** *(275)* $E°_{cell} = E°_{Y^{2+}/Y} - E°_{X^{2+}/X}$

$= 0.36 - (-2.36) = 2.72$

From the Nernst equation,

$E_{cell} = E°_{cell} - \frac{0.06}{2}\log\frac{[X^{2+}]}{[Y^{2+}]}$

$= 2.72 - 0.03\log\left(\frac{0.001}{0.01}\right)$

$= 2.72 - 0.03\log(10^{-1})$

$= 2.72 - 0.03\times(-1)$

$= 2.72 + 0.03 = 2.75$

$= 275\times 10^{-2} \text{ V}$

**51.** *(10)* Reactions involved in the cell are

At anode : $H_2 \longrightarrow 2H^+ + 2e^-$

At cathode :

$(Fe^{3+} + e^- \longrightarrow Fe^{2+})\times 2$

Overall equation

$H_2 + 2Fe^{3+} \longrightarrow 2H^+ + 2Fe^{2+}$

$E_{cell} = E°_{cell} - \frac{0.059}{2}\log\left(\frac{Fe^{2+}}{Fe^{3+}}\right)^2$

$0.712 = 0.771 - 0.059\log\left(\frac{Fe^{2+}}{Fe^{3+}}\right)^2$

$0.712 - 0.771 = -0.059\log\left(\frac{Fe^{2+}}{Fe^{3+}}\right)^2$

$\log\left(\frac{Fe^{2+}}{Fe^{3+}}\right)^2 = \frac{0.059}{0.059} = 1$

$\Rightarrow \quad \frac{Fe^{2+}}{Fe^{3+}} = 10$

**52.** *(17)* $Zn(s) + Sn^{2+}(aq) \rightleftharpoons Zn^{2+}(aq) + Sn(s)$

$\Rightarrow \Delta G° = -2.303\, RT\log_{10} K_{eq}$

$-nE°_{cell} = -\frac{2.303\, RT}{F}\log_{10} K_{eq}$

$\Rightarrow E°_{Zn/Zn^{2+}} + E°_{Sn^{2+}/Sn} = \frac{0.059}{2}\log_{10} K_{eq} \quad [\because n = 2]$

$0.76 + E°_{Sn^{2+}/Sn} = \frac{0.059}{2}\log_{10} 10^{20}$

$[\because E°_{Zn^{2+}/Zn} = -0.76 \text{ V}]$

$0.76 + E°_{Sn^{2+}/Sn} = \frac{0.059\times 20}{2}$

$\Rightarrow \quad E°_{Sn^{2+}/Sn} = 0.59 - 0.76$

$E°_{Sn^{2+}/Sn} = -0.17$

$\Rightarrow \quad E°_{Sn/Sn^{2+}} = -17\times 10^{-2} \text{ V}$

**53.** *(3)*

> **key idea** To calculate the value of a, Nernst equation is used. First determine the value of $E°_{cell}$ from given $E°_{X^{2+}/X}$ and $E°_{Y^{2+}/Y}$ value. Then use this value in Nernst Equation.

Given, $E°_{M^{3+}/M^+} = 0.2 \text{ V}$

$\frac{2.303\, RT}{F} = 0.059 \text{ V}$

$\frac{[M^+(aq)]}{[M^{3+}(aq)]} = 10^a$

$E_{cell} = 0.1115 \text{ V at } 298 \text{ K}$

Overall reaction of the cell is given by

$H_2(g) + M^{3+}(aq) \longrightarrow M^+(aq) + 2H^+(aq)$

According to Nernst Equation

$E_{cell} = (E°_{Cathode} - E°_{Anode}) - \frac{0.059}{2}\log\frac{[M^+]}{[M^{+3}]}$

$0.1115 = (E°_{M^{3+}/M^+}) - \frac{0.059}{2}\log\frac{[M^+]}{[M^{+3}]}$

$0.1115 = 0.2 - \frac{0.059}{2}\log 10^a$

$\left(\therefore \frac{[M^+]}{[M^{+3}]} = 10^a\right)$

$$\Rightarrow \frac{0.059}{2}\log 10^a = 0.2 - 0.115$$

$$\Rightarrow \frac{0.059}{2}\log 10^a = 0.0885$$

$$\Rightarrow \log 10^a = \frac{0.0885 \times 2}{0.059}$$

$$\Rightarrow \log 10^a = 3 \Rightarrow a = 3$$

So, the value of '$a$' is 3.

**54.** *(917)*

> **key idea** To calculate the potential for the half-cell reaction. First of all, write the balanced chemical equation. Then calculate the value of $n$ from it. After that use Nernst equation to calculate potential for half-cell reaction.

$[Cr_2O_7]^{-2} = 10$ m mol; $[Cr^{+3}] = 100$ m mol $= 0.1$ mol $= 10^{-1}$ mol

$$\overset{+6}{Cr_2}O_7^{-2} \longrightarrow Cr^{+3};\ E° = 1.33\ V$$

$$\overset{+6}{Cr_2}O_7^{-2} + 14H^+ + 6e^- \xrightarrow{\text{Reduction}} 2Cr^{+3} + 7H_2O$$

$\therefore n = 6$

Using Nernst Equation

$$E_{cell} = 1.33 - \frac{0.059}{n}\log\frac{[Cr^{3+}]^2}{[Cr_2O_7^{2-}][H^+]^{14}}$$

$$= 1.33 - \frac{0.059}{6}\log\left[\frac{(0.1)^2}{(10^{-2})(10^{-3})^{14}}\right]$$

$$E = 1.33 - \frac{0.059}{6} \times 42 = 0.917$$

$$E = 917 \times 10^{-3} \Rightarrow x = 917$$

**55.** *(34)* $Cu(s)\,|Cu^{2+}(0.001)\ M\,||$ $Ag^+(0.01\ M)|Ag(s);\ E_{cell} = 0.43\ V$

**At anode**, $Cu \longrightarrow Cu^{2+} + 2e^-$

**At cathode**, $2Ag^+ + 2e^- \longrightarrow 2Ag$

Net, cell reaction is

$$Cu + 2Ag^+ \longrightarrow Cu^{2+} + 2Ag$$

According to Nernst equation,

$$E_{cell} = E°_{cell} - \frac{2.303\,RT}{nF}\log\frac{[Cu^+]}{[Ag^+]^2}$$

$$E_{cell} = E°_{cell} - \frac{0.06}{n}\log\frac{[Cu^{2+}]}{[Ag^+]^2}$$

[$n$ = number of $e^-$ involved]

$$\Rightarrow 0.43 = E°_{cell} - \frac{0.06}{2}\log\frac{(0.001)}{(0.01)^2}$$

$$\Rightarrow 0.43 = E°_{cell} - 0.03\log\frac{10^{-3}}{10^{-4}}$$

$$\Rightarrow 0.43 = E°_{cell} - 0.03\log 10$$

$$\Rightarrow E°_{cell} = 0.43 + 0.03 = 0.46\ V$$

Now, $E°_{cell} = E°_{cathode} - E°_{anode}$

$$\therefore E°_{cell} = E°_{Ag^+/Ag} - E°_{Cu^{2+}/Cu}$$

$$\Rightarrow 0.46 = 0.80 - E°_{Cu^{2+}/Cu}$$

$$\Rightarrow E°_{Cu^{2+}/Cu} = 0.34\ V = 34 \times 10^{-2}\ V$$

**56.** *(4)* Reactions of half-cell are given as

At anode $Zn \longrightarrow Zn^{2+} + 2e^-$;

**At cathode :** $Sn^{x+} + xe^- \longrightarrow Sn$

$$Zn + Sn^{x+} + xe^- \rightarrow Sn + Zn^{2+} + 2e^-;$$

$$E°_{cell} = E°_{Sn^{x+}/Sn} - E_{Zn/Zn^{2+}}$$

$$= 0.008 - (-0.763)$$

$$E°_{cell} = 0.771\ V$$

Applying Nernst equation,

$$E_{cell} = E°_{cell} - \frac{2.303\,RT}{nF}\log Q$$

$$0.801 = 0.771 - \frac{0.06}{n}\log 10^{-2}$$

$$0.801 - 0.771 = \frac{-0.06}{n}(-2\log 10)$$

$$0.03 = \frac{0.12}{n}$$

$$\therefore n = 4$$

**57.** *(7)* The two half-reactions are

$$H_2 \longrightarrow 2H^+ + 2e^- \quad E° = 0$$

$$Cu^{2+} + 2e^- \longrightarrow Cu \quad E° = 0.34$$

$$H_2 + Cu^{2+} \longrightarrow 2H^+ + Cu(s)\ E° = 0.34$$

Using Nernst Equation

$$E_{cell} = E°_{cell} - \frac{0.0591}{2}\log\frac{[H^+]^2}{[Cu^{2+}]\,p_{H_2}}$$

$$0.31 = 0.34 - 0.03\log\frac{[10^{-3}]^2}{Cu^{2+} \times 1}$$

$$0.03\log\frac{10^{-6}}{Cu^{2+}} = 0.03$$

$$\log\frac{10^{-6}}{Cu^{2+}} = 1$$

Taking antilog,

$$\frac{10^{-6}}{Cu^{2+}} = 10$$

$$\Rightarrow Cu^{2+} = 10^{-7} \quad \ldots(i)$$

Given $[Cu^{2+}] = 10^{-x}M$ ...(ii)

$\therefore$ $x = 7$ [from Eq. (i) and (ii)]

**58.** *(16)* Given,

$Sn^{2+}(aq) + 2e^- \longrightarrow Sn(s)$;

$$E°_{Sn^{2+}/Sn} = -0.140\ V$$

$$\Delta G_1° = -nFE°_{Sn^{2+}/Sn}$$

$$= -2 \times (-0.14)\ F = 0.28\ F \ldots(i)$$

$Sn^{4+}(aq) + 4e^- \longrightarrow Sn(s)$;

$$E°_{Sn^{4+}/Sn} = 0.010\ V$$

$$\Delta G_2° = -nFE°_{Sn^{4+}/Sn}$$

$$= -4 \times 0.01\ F = -0.04\ F \quad \ldots(ii)$$

$Sn^{4+}(aq) + 2e^- \longrightarrow Sn^{2+}(aq);\ \Delta G_3°$ ...(iii)

To get Eqs. (iii), subtract Eq. (i) from (ii)

So, we get

$$Sn^{4+} + 4e^- \longrightarrow Sn;\quad \Delta G_2° = -0.04\ F$$

$$Sn \rightarrow Sn^{2+} + 2e^-;\ -\Delta G_1° = -0.28F$$

$$Sn^{4+} + 2e^- \longrightarrow Sn^{2+};$$

$$\Delta G_3° = \Delta G_2° - \Delta G_1°$$

$$-nFE°_{Sn^{4+}/Sn^{2+}} = -0.04F - 0.28\ F$$

$$-2FE°_{Sn^{4+}/Sn^{2+}} = -0.32\ F$$

$$E°_{Sn^{4+}/Sn^{2+}} = \frac{0.32\ F}{2\ F} = 0.16\ V$$

$$E°_{Sn^{4+}/Sn^{2+}} = 16 \times 10^{-2}\ V$$

> **Alternate Solution**
>
> This question can be solved using Latimer diagram.
>
> $Sn^{4+} \xrightarrow{E°_{Sn^{4+}/Sn^{2+}}} Sn^{2+} \xrightarrow{-0.140\ V} Sn$ (overall $Sn^{4+} \to Sn$: 0.010 V)
>
> $$nE°_{Sn^{4+}/Sn} = nE°_{Sn^{4+}/Sn^{2+}} - nE°_{Sn^{2+}/Sn}$$
>
> So, $4 \times 0.01 = 2E°_{Sn^{4+}/Sn^{2+}} - 2 \times -0.14$
>
> $$2E°_{Sn^{4+}/Sn^{2+}} = 0.04 + 0.28 = 0.32$$
>
> $$E°_{Sn^{4+}/Sn^{2+}} = \frac{0.32}{2} = 16 \times 10^{-2}\ V$$

**59.** *(51)* The reaction taking place in cell is written as

$$\frac{1}{2}H_2 + Ag^+ \longrightarrow H^+ + Ag$$

Standard Gibb's free energy is given as

$$\Delta G^- = -nFE°_{cell}$$

$$= -1 \times 96500\ J \times 0.5332\ V$$

$$= -51.35\ kJ \approx -51\ kJ$$

**60.** *(983)* The given reaction is as follows

$$Cu(s) + Sn^{2+}(0.001\ M) \longrightarrow Cu^{2+}(0.01\ M) + Sn(s)$$

$$E°_{cell} = E°_{cathode} - E°_{anode}$$

$$= (-0.14) - (0.34) = -0.48\ V$$

$$E_{cell} = E°_{cell} - \frac{0.059}{2}\log\frac{[Cu^{2+}]}{[Sn^{2+}]}$$

$$= -0.48 - 0.0295\log\frac{0.01}{0.001}$$

$$= -0.5095\ V$$

$$\Delta G = -nFE_{cell}$$

$= -2 \times 96500 \times (-0.5095)$

$= 98333.5\ J/mol$

$= 98.3335\ kJ/mol$

or $983.335 \times 10^{-1}$ kJ/mol

$\Delta G \approx 983 \times 10^{-1}\ kJ\ mol^{-1}$

**61.** *(23)* From the following oxidation reactions, the spontaneous cell reaction will be in which $Fe^{3+}$ gets reduced and $I^-$ gets oxidised as the reduction potential of $Fe^{3+}$ is higher than $I_2$.

Reduction (cathode) half-cell reaction

$Fe^{3+} + e^- \longrightarrow Fe^{2+}$

$E^\circ_{Fe^{3+}/Fe^{2+}} = 0.77\ V$

Oxidation (anode) half-cell reaction

$2I^- \longrightarrow I_2 + 2e^-;\ E^\circ_{I_2/I^-} = 0.54\ V$

The cell reaction

$Fe^{3+} + 2I^- \longrightarrow Fe^{2+} + I_2$

$E^\circ_{cell} = E^\circ_{cathode} - E^\circ_{anode}$

$= 0.77\ V - 0.54\ V$

$= 0.23\ V = 23 \times 10^{-2}\ V$

**62.** *(25)* $\Delta G^\circ = \Delta H^\circ - T\Delta S^\circ$

Also, $\Delta G^\circ = -nFE^\circ$

$\therefore\ -nFE^\circ = \Delta H^\circ - T\Delta S^\circ$

$\Rightarrow\ \Delta S^\circ = \dfrac{nFE^\circ + \Delta H^\circ}{T}$

$= \dfrac{(2 \times 96487\ C\ mol^{-1} \times 4.315\ V) + (-825.2\ kJ/mol)}{298\ K}$

$\Delta S^\circ = 25.11\ J/K\ mol \approx 25\ J/K\ mol$

**63.** *(109)* Cell reaction,

$\underset{0.02\ M}{Cu^{2+}(aq)} + Zn(s) \rightleftharpoons Cu(s) + \underset{0.04\ M}{Zn^{2+}(aq)}$

$E^\circ_{cell} = E^\circ_{Cu^{2+}/Cu} - E^\circ_{Zn^{2+}/Zn}$

$= 0.34 - (-0.76) = 1.1\ V$

$E_{cell} = E^\circ_{cell} - \dfrac{0.059}{2}\log\dfrac{[Zn^{2+}]}{[Cu^{2+}]}$

$= 1.1 - \dfrac{0.059}{2}\log\dfrac{[0.04]}{[0.02]}$

$= 1.1 - 0.03\log 2$

$= 1.1 - 0.03 \times 0.30$

$= 1.09\ V = 109 \times 10^{-2}\ V$

**64.** *(3)* The physical properties of an element that affects the reduction potential is sublimation enthalpy, ionisation enthalpy and hydration enthalpy. So, there are total 3 properties that will affect the reduction potential.

**65.** *(28)* Cell reaction,

$Cu(s) + 2Ag^+(aq) \longrightarrow Cu^{2+}(aq) + 2Ag(s)$

$E_{cell} = E^\circ_{cell} - \dfrac{0.0591}{2}\log\dfrac{[Cu^{2+}]}{[Ag^+]^2}$

$E_1 = 0.3095$

$= E^\circ_{cell} - \dfrac{0.0591}{2}\log\dfrac{(0.1)}{(0.01)^2}$

$E^\circ_{cell} = 0.3095 + \dfrac{0.0591}{2}\log(1000)$

$= 0.3095 + \dfrac{0.0591}{2}\log 10^3$

$E^\circ_{cell} = 0.3095 + \dfrac{0.0591}{2} \times 3$ ...(i)

$E_2 = E^\circ_{cell} - \dfrac{0.0591}{2}\log\dfrac{(0.01)}{(0.001)^2}$

$= E^\circ_{cell} - \dfrac{0.0591}{2}\log 10^4$

$E_2 = E^\circ_{cell} - \dfrac{0.0591}{2} \times 4$

From Eq. (i), we substitute $E^\circ_{cell}$

$E_2 = 0.3095 + \dfrac{0.0591}{2} \times 3 - \dfrac{0.0591}{2} \times 4$

$= 0.3095 - 0.0295 = 0.28 = 28 \times 10^{-2}\ V$

**66.** *(24)* Reaction cell

$Zn \mid Zn^{2}(aq), (1M) \| Fe^{3+}(aq)|, Fe^{2+}(aq), Pt(s)$

(Anode) $Zn \longrightarrow Zn^{2+} + 2e^-$

(Cathode) $2Fe^{3+} + 2e^- \longrightarrow 2Fe^{2+}$

Overall $Zn + 2Fe^{3+} \longrightarrow Zn^{2+} + 2Fe^{2+}$

So, $E^\circ_{cell} = E^\circ_{Fe^{3+}/Fe^{2+}} - E^\circ_{Zn^{2+}/Zn}$

$= 0.77 - (-0.76) = 1.53\ V$

Using Nernst equation,

$E_{cell} = E^\circ_{cell} - \dfrac{0.0591}{n}\log\dfrac{[P]}{[R]}$

$1.50 = 1.53 - \dfrac{0.0591}{2}\log\left(\dfrac{Fe^{2+}}{Fe^{3+}}\right)^2$

$\log\left(\dfrac{Fe^{2+}}{Fe^{3+}}\right)^2 = \dfrac{0.06}{0.06}$

$\Rightarrow\ \dfrac{[Fe^{2+}]}{[Fe^{3+}]} = 10^{1/2} = \sqrt{10}$

$\dfrac{[Fe^{3+}]}{[Fe^{2+}]} = \dfrac{1}{\sqrt{10}}$

Fractin of total iron present as $Fe^{3+}$ ion.

$= \dfrac{[Fe^{3+}]}{[Fe^{2+}] + [Fe^{3+}]} = \dfrac{1}{1 + \sqrt{10}} = \dfrac{1}{4.16}$

$= 0.2402 = 24 \times 10^{-2}$

Thus, the value of $x = 24$

**67.** *(3)* Given $E^\circ_{cell} = 2.97\ V$

$n = 2$ (for 2 electrons)

Using nernst equation,

$E_{cell} = E^\circ_{cell} - \dfrac{0.0591}{n}\log\dfrac{[Cu^{2+}]}{[Ag^+]^2}$

$= 2.97 - \dfrac{0.0591}{2}\log\dfrac{0.250}{(10^{-3})^2}$

$= 2.81 \approx 3$ (approx.)

Hence, $E^\circ_{cell}$ of the reaction is 3V.

**68.** *(45)* Given reaction is

$2Fe^{3+}(aq) + 2I^-(aq) \longrightarrow 2Fe^{2+}(aq) + I_2(s)$

Now, $Fe^{3+} \xrightarrow{E_1^\circ} Fe^{2+} \xrightarrow{E_2^\circ} Fe$ (overall $Fe^{3+} \xrightarrow{E_3^\circ} Fe$)

$\Delta G_1^\circ + \Delta G_2^\circ = \Delta G_3^\circ$

$nE_1^\circ + nE_2^\circ = nE_3^\circ$

[$n$ = number of electron transferred]

$1E_1^\circ + 2E_2^\circ = 3E_3^\circ$

$E^\circ_{Fe^{3+}/Fe^{2+}} + 2E^\circ_{Fe^{2+}/Fe\ (s)} = 3E^\circ_{Fe^{3+}/Fe\ (s)}$

$E^\circ_{Fe^{3+}/Fe^{2+}} + (-0.440) \times 2 = (-0.036) \times 3$

$E^\circ_{Fe^{3+}/Fe^{2+}} = 0.772\ V$

The standard electrode potential for the above given reaction,

$E^\circ_{cell} = E^\circ_{Fe^{3+}|Fe^{2+}} - E^\circ_{I^-|I_2}$

$E^\circ_{cell} = E^\circ_{cathode} - E^\circ_{anode}$

$= 0.772 - 0.539 = 0.233\ V$

The magnitude of the standard molar Gibb's free energy,

$\Delta G^\circ = -nFE^\circ_{cell}$

$= -2 \times 96500 \times 0.233$

$\Delta G^\circ = -44969\ J = -44.9\ kJ \approx -45\ kJ$

**69.** *(147)*

$Zn | Zn^{2+}(0.1M) \ \|\ Ag^+(0.01M) | Ag$

Cell reaction

$Zn(s) + \underset{0.01M}{2Ag^+} \longrightarrow \underset{0.1M}{Zn^{2+}} + 2Ag(s)$

$\Rightarrow\ \dfrac{[P]}{[R]} = \dfrac{[Zn^{2+}]\ [Ag]^2}{[Zn]\ [Ag^+]^2}$ ([Ag] = [Zn] = 1)

$= \dfrac{0.1 \times 1^2}{1 \times (0.01)^2} = 10^3$

$n = 2$

and $E^\circ_{cell} = E^\circ_{Ag^+/Ag} - E^\circ_{Zn^{2+}/Zn}$

$= 0.8 - (-0.76) = 1.56\ V$

$\Rightarrow E_{cell} = E^\circ_{cell} - \dfrac{0.059}{n} \log \dfrac{[Zn^{2+}]}{[Ag^+]}$

$= 1.56 - \dfrac{0.059}{2} \log 10^3$

$= 1.4715\ V$

$= 147.15 \times 10^{-2}\ V$

$= 147 \times 10^{-2}\ V$

$x = 147$

**70.** *(3776)* Reaction,

$MnO_4^- + H^+ + 5e^- \longrightarrow Mn^{2+} + 4H_2O$

$n = 5$

Applying Nernst equation,

$E_{cell} = E^\circ_{cell} - \dfrac{0.0591}{n} \log \dfrac{[P]}{[R]}$

or $E_{cell} = E^\circ_{cell} - \dfrac{0.0591}{5} \log \dfrac{[Mn^{2+}]}{[MnO_4^-]} \left[\dfrac{1}{H^+}\right]^8$

Case (I) Given, $[H^+] = 1\ M$

$E_1 = E^\circ - \dfrac{0.0591}{5} \log \dfrac{[Mn^{2+}]}{[MnO_4^-]}$

Case (II) Now, $[H^+] = 10^{-4}\ M$

$E_2 = E_1 - \dfrac{0.0591}{5} \log \dfrac{[Mn^{2+}]}{[MnO_4^-]} \times \dfrac{1}{(10^{-4})^8}$

$\therefore$ Magnitude of change in oxidising power of $MnO_4^- | Mn^{2+}$

$|E_1 - E_2| = \dfrac{0.0591}{5} \times 32$

$= 0.3776\ V$

$= 3776 \times 10^{-4}$

$\Rightarrow x = 3776$

**71.** *(6)* We know that,

$\Delta G^\circ = -nFE^\circ_{cell}$

Here, $\Delta G^\circ = 17.37\ kJ\ mol^{-1}$

$17.37 \times 1000 = -3 \times 96500 \times E^\circ_{cell}$

$E^\circ_{cell} = \dfrac{-17.37 \times 1000}{3 \times 96500}$

$E^\circ_{cell} = -0.06 = -6 \times 10^{-2}$

**72.** *(96500)*

$E^\circ_{cell} = (E^\circ_{Sn^{2+}/Sn} - E^\circ_{Cu^{2+}/Cu}) - \dfrac{0.0591}{2} \log \dfrac{[Cu^{2+}]}{[Sn^{2+}]}$

$= (-0.16 - 0.34) - \dfrac{0.0591}{2} \log \dfrac{1}{1}$

$= -0.50\ V$

$\Delta G = -nFE^\circ_{cell}$

$= -2 \times 96500 \times (-0.50)\ J$

$= 96500.00\ J$

**73.** *(144)*

$2Cu^+(aq) \rightleftharpoons Cu(s) + Cu^{2+}(aq)$

For the cell reaction,

$E^\circ_{cell} = E^\circ_{Cu^+/Cu} - E^\circ_{Cu^{2+}/Cu^+}$

$= (0.52 - 0.16)\ V = 0.36\ V$

$\Rightarrow \Delta G^\circ = -nFE^\circ_{cell}$

$\Delta G = -RT \ln K$

$\Rightarrow -RT \ln K = -nFE^\circ_{cell}$

$\Rightarrow \ln K = n \times \dfrac{F}{RT} \times E^\circ_{cell}$

$= 1 \times \dfrac{1}{0.025} \times 0.36$

$\ln K = 14.4 = 144 \times 10^{-1}$

## Topic 3

**01.** *(c)* Zinc metal is the most stable metal to cover iron surfaces. The process of coating the iron surface by zinc is called galvanisation.

**02.** *(d)* On charging the lead storage battery, following reactions takes place.

$2PbSO_4(s) + 2H_2O(l) \longrightarrow Pb(s) + PbO_2(s) + 2H_2SO_4(aq)$

At anode $PbSO_4$ is reduced to Pb while at cathode $PbSO_4$ is oxidised to $PbO_2$.

Thus,. $x_1 = +2,\ y_1 = 0$

and $x_2 = +2,\ y_2 = 4$

**03.** *(a)* Statement I is true but Statement II is false. It is because corrosion is more favourable in acidic medium than alkaline medium. So, the rate of corrosion is high in acidic medium than alkaline.

**04.** *(a)* $PbO_2$ or $Pb_3O_4$ can be the compound *A*. But out of these two compounds only $PbO_2$ is used in lead storage batteries where a grid of lead packed with $PbO_2$ acts as cathode and also it is amphoteric in nature. It reacts with both acids and alkali.

(i) $\overset{+4}{PbO_2} + 4H\overset{-1}{Cl} \longrightarrow \overset{+2}{PbCl_2} + \overset{0}{Cl_2} + 2H_2O$

Here, $PbO_2$ acts as a basic oxide as well as an oxidising agent.

(ii) $PbO_2 + 2NaOH \longrightarrow Na_2PbO_2 + H_2O$

Here, $PbO_2$ acts as an acidic oxide.

So, the compound *A* is $PbO_2$ (option-a).

**05.** *(a)* The correct match is

A-III, B-II, C-IV, D-I

**06.** *(d)* The correct match is

A-IV, B-III, C-I, D-II

**07.** *(d)* Statements given in A, D and E are correct regarding fuel cells while B and C are incorrect.

It is because fuel cells produces electricity with an efficiency of about 70% and aluminium is not used a catalyst in it.

**08.** *(a)* Statements given in B, D only are correct.

On charging the lead storage battery, the reaction is reversed and $PbSO_4(s)$ on anode is converted into Pb instead of $PbO_2$.

Lead storage battery consists of lead anode and a grid of lead packed with lead oxide as cathode, a 38% solution of $H_2SO_4$ is used as an electrolyte.

**09.** *(c)* The correct match is

A-II, B-I, C-IV, D-III.

In JEE Main there is usually 0-1 question per paper are visible from this chapter. In past 3 years, 27 conceptual and 40 numerical questions were asked of moderate difficulty level from this chapter.
**For detailed analysis scan QR code**

# CHEMICAL KINETICS

## Topic 1 Rate of Chemical Reaction and Rate Expression

### Level 1

**01** Rate law for a reaction between $A$ and $B$ is given by

$$r = k[A]^n\,[B]^m$$

If concentration of $A$ is doubled and concentration of $B$ is halved from their initial value, the ratio of new rate of reaction to the initial rate of reaction $\left(\frac{r_2}{r_1}\right)$ is **[2025, 4 April Shift-I]**

(a) $2^{(n-m)}$ (b) $\frac{1}{2^{m+n}}$
(c) $(n-m)$ (d) $(m+n)$

**02** Consider an elementary reaction,

$$A(g) + B(g) \longrightarrow C(g) + D(g)$$ **[2025, 28 Jan Shift-II]**

If the volume of reaction mixture is suddenly reduced to $\frac{1}{3}$ of its initial volume, the reaction rate will become '$x$' times of the original reaction rate. The value of $x$ is

(a) 1/9 (b) 1/3 (c) 9 (d) 3

**03** For the reaction $2A + 3B + \frac{3}{2}C \rightarrow 3P$, which statement is correct? **[2020, 3 Sep Shift-II]**

(a) $\frac{dn_A}{dt} = \frac{2}{3}\frac{dn_B}{dt} = \frac{3}{4}\frac{dn_C}{dt}$ (b) $\frac{dn_A}{dt} = \frac{3}{2}\frac{dn_B}{dt} = \frac{3}{4}\frac{dn_C}{dt}$

(c) $\frac{dn_A}{dt} = \frac{dn_B}{dt} = \frac{dn_C}{dt}$ (d) $\frac{dn_A}{dt} = \frac{2}{3}\frac{dn_B}{dt} = \frac{4}{3}\frac{dn_C}{dt}$

**04** For the reaction

$$2H_2(g) + 2NO(g) \longrightarrow N_2(g) + 2H_2O(g)$$

the observed rate expression is, rate $= K_f[NO]^2[H_2]$.

The rate expression for the reverse reaction is **[2020, 7 Jan Shift-II]**

(a) $K_b[N_2][H_2O]^2/[H_2]$ (b) $K_b[N_2][H_2O]$
(c) $K_b[N_2][H_2O]^2/[NO]$ (d) $K_b[N_2][H_2O]^2$

**05** $A + 2B \longrightarrow C$, the rate equation for this reaction is given as

$$\text{Rate} = k[A][B].$$

If the concentration of $A$ is kept the same but that of $B$ is doubled what will happen to the rate itself? **[JEE Mains 2015]**

(a) Halved (b) The same (c) Doubled (d) Quadrupled

**06** For a reaction $\frac{1}{2}A \longrightarrow 2B$, rate of disappearance of '$A$' is related to the rate of appearance of '$B$' by the expression **[AIEEE 2008]**

(a) $-\frac{d[A]}{dt} = \frac{1}{2}\frac{d[B]}{dt}$ (b) $-\frac{d[A]}{dt} = \frac{1}{4}\frac{d[B]}{dt}$

(c) $-\frac{d[A]}{dt} = \frac{d[B]}{dt}$ (d) $-\frac{d[A]}{dt} = 4\frac{d[B]}{dt}$

### Level 2

**07** For a reaction $A \xrightarrow{K_1} B \xrightarrow{K_2} C$

If the rate of formation of $B$ is set to be zero then the concentration of $B$ is given by : **[2024, 8 April Shift-II]**

(a) $\left(\frac{K_1}{K_2}\right)[A]$ (b) $(K_1 + K_2)[A]$
(c) $(K_1 - K_2)[A]$ (d) $K_1K_2\,[A]$

**08** $NO_2$ required for a reaction is produced by the decomposition of $N_2O_5$ in $CCl_4$ as per the equation,

$$2N_2O_5(g) \longrightarrow 4NO_2(g) + O_2(g)$$

The initial concentration of $N_2O_5$ is 3.00 mol $L^{-1}$ and it is 2.75 mol $L^{-1}$ after 30 minutes. The rate of formation of $NO_2$ is **[2019, 12 April Shift-II]**

(a) $4.167 \times 10^{-3}$ mol $L^{-1}$ $min^{-1}$
(b) $1.667 \times 10^{-2}$ mol $L^{-1}$ $min^{-1}$
(c) $8.333 \times 10^{-3}$ mol $L^{-1}$ $min^{-1}$
(d) $2.083 \times 10^{-3}$ mol $L^{-1}$ $min^{-1}$

**09** Consider the reaction,

$$Cl_2(aq) + H_2S(aq) \longrightarrow S(s) + 2H^+(aq) + 2Cl^-(aq)$$

The rate equation for this reaction is,

$$\text{rate} = k[Cl_2][H_2S]$$

Which of these mechanisms is/are consistent with this rate equation? **[AIEEE 2010]**

I. $Cl_2 + H_2S \longrightarrow H^+ + Cl^- + Cl^+ + HS^-$ (slow)

$Cl^+ + HS^- \longrightarrow H^+ + Cl^- + S$ (fast)

II. $H_2S \rightleftharpoons H^+ + HS^-$ (fast)

$Cl_2 + HS^- \longrightarrow 2Cl^- + H^+ + S$ (slow)

(a) II only (b) Both I and II
(c) Neither I nor II (d) I only

## Level 3

**10** A person's wound was exposed to some bacteria and then bacterial growth started to happen at the same place. The wound was later treated with some antibacterial medicine and the rate of bacterial decay($r$) was found to be proportional with the square of the existing number of bacteria at any instance. Which of the following set of graphs correctly represents the 'before' and 'after situation of the application of the medicine? **[2025, 7 April Shift-I]**

[Given, $N$ = No. of bacteria, $t$ = time, bacterial growth follows 1st order kineties.]

(a) 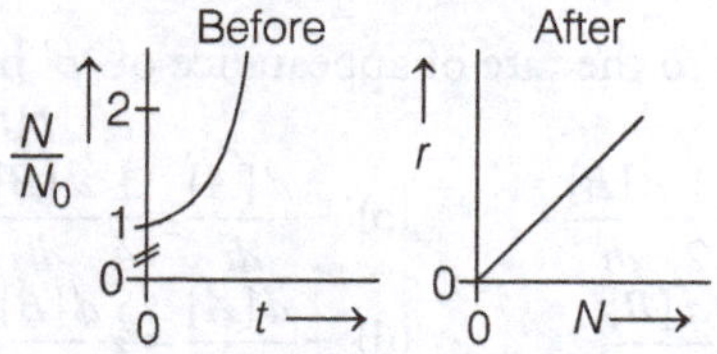


(b) 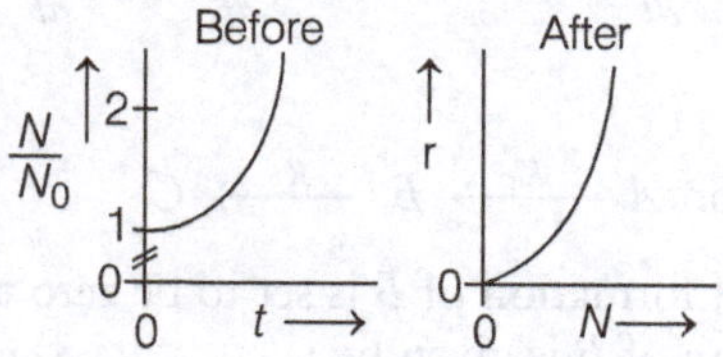


(c) 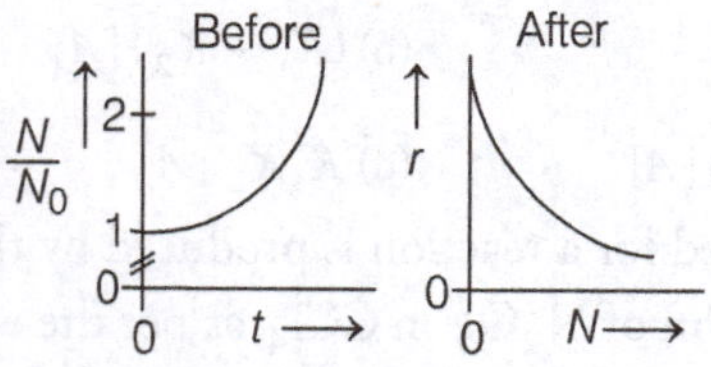


(d) 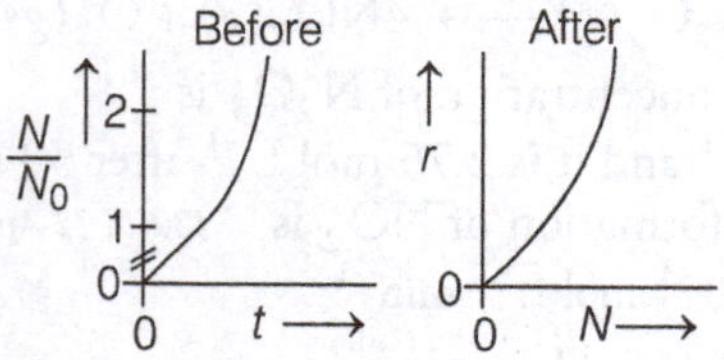


**11** For an elementary chemical reaction, $A_2 \underset{k_{-1}}{\overset{k_1}{\rightleftharpoons}} 2A$, the expression for $\frac{d[A]}{dt}$ is **[2019, 10 Jan Shift-II]**

(a) $2k_1[A_2] - k_{-1}[A]^2$ (b) $k_1[A_2] - k_{-1}[A]^2$
(c) $2k_1[A_2] - 2k_{-1}[A]^2$ (d) $k_1[A_2] + k_{-1}[A]^2$

**Imp.** **Graphical Representation of Concentration vs Time**

**12** A bacterial infection in an internal wound grows as $N'(t) = N_0 \exp(t)$, where the time $t$ is in hours. A dose of antibiotic, taken orally, needs 1 hour to reach the wound. Once it reaches there, the bacterial population goes down as $\frac{dN}{dt} = -5N^2$. What will be the plot of $\frac{N_0}{N}$ vs $t$ after 1 hour? **[2019, 10 April Shift-I]**

(a) 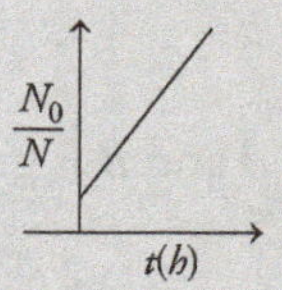


(b) 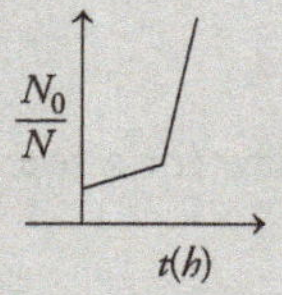


(c) 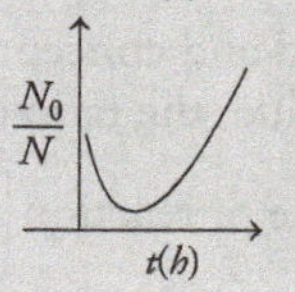


(d) 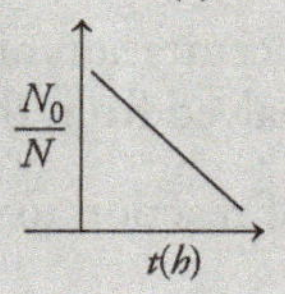


**13** For the non-stoichiometric reaction, $2A + B \longrightarrow C + D$, the following kinetic data were obtained in three separate experiments, all at 298 K. **[JEE Main 2014]**

| Initial concentration ($A$) | Initial concentration ($B$) | Initial rate of formation of $C$ ($mol\ L^{-1}\ s^{-1}$) |
|---|---|---|
| 0.1 M | 0.1 M | $1.2 \times 10^{-3}$ |
| 0.1 M | 0.2 M | $1.2 \times 10^{-3}$ |
| 0.2 M | 0.1 M | $2.4 \times 10^{-3}$ |

The rate law for the formation of $C$ is

(a) $\frac{dC}{dt} = k[A][B]$ (b) $\frac{dC}{dt} = k[A]^2[B]$
(c) $\frac{dC}{dt} = k[A][B]^2$ (d) $\frac{dC}{dt} = k[A]$

## Numerical Value Type Questions

**14** Consider the following first order gas phase reaction at constant temperature **[2024, 9 April Shift-II]**

$$A(g) \longrightarrow 2B(g) + C(g)$$

If the total pressure of the gases is found to be 200 torr after 23 sec, and 300 torr upon the complete decomposition of $A$ after a very long time, then the rate constant of the given reaction is ……….. $\times 10^{-2} s^{-1}$ (Nearest integer) [Given : $\log_{10}(2) = 0.301$]

**15** Given below are two statements: *[2024, 9 April Shift-I]*

**Statement I** The rate law for the reaction $A + B \longrightarrow C$ is rate $(r) = k[A]^2[B]$. When the concentration of both $A$ and $B$ is doubled, the reaction rate is increased "$x$" times.

**Statement II**

The figure is showing "the variation in concentration against time plot" for a "$y$" order reaction.

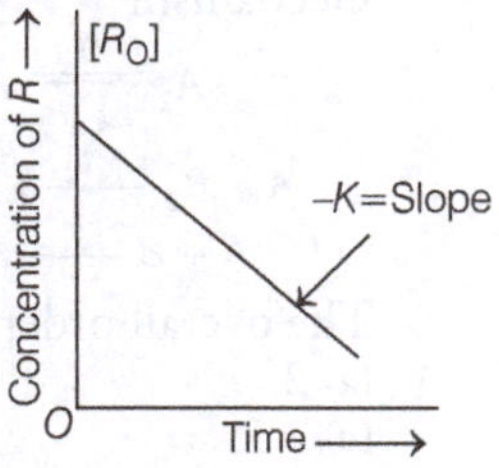


The value of $x + y$ is ..........

**16** $NO_2$ required for a reaction is produced by decomposition of $N_2O_5$ in $CCl_4$ as by equation

$$2N_2O_5(g) \longrightarrow 4NO_2(g) + O_2(g)$$

The initial concentration of $N_2O_5$ is 3 mol $L^{-1}$ and it is 2.75 mol $L^{-1}$ after 30 minutes.

The rate of formation of $NO_2$ is $x \times 10^{-3}$ mol $L^{-1}min^{-1}$, value of $x$ is ......... .

(nearest integer) *[2024, 30 Jan Shift-II]*

**17** The following data were obtained during the first order thermal decomposition of a gas $A$ at constant volume. *[2024, 1 Feb Shift-II]*

$$A(g) \longrightarrow 2B(g) + C(g)$$

| S. No. | Time/(s) | Total pressure/(atm) |
|---|---|---|
| 1. | 0 | 0.1 |
| 2. | 115 | 0.28 |

The rate constant of the reaction is ......... $10^{-2}\ s^{-1}$ (nearest integer)

**18** For a given chemical reaction,

$$\gamma_1 A + \gamma_2 B \longrightarrow \gamma_3 C + \gamma_4 D$$

concentration of $C$ changes from 10 m mol $dm^{-3}$ to 20 mmol $dm^{-3}$ in 10 seconds. Rate of appearance of $D$ is 1.5 times the rate of disappearance of $B$ which is twice the rate of disappearance $A$. The rate of appearance of $D$ has been experimentally determined to be 9 m mol $dm^{-3}\ s^{-1}$. Therefore, the rate of reaction is ... mmol $dm^{-3}\ s^{-1}$. [Nearest integer] *[2022, 25 June Shift-I]*

**19** The reaction that occurs in a breath analyser, a device used to determine the alcohol level in a person's blood stream is

$$2K_2Cr_2O_7 + 8H_2SO_4 + 3C_2H_6O \longrightarrow 2Cr_2(SO_4)_3 + 3C_2H_4O_2 + 2K_2SO_4 + 11H_2O$$

If the rate of appearance of $Cr_2(SO_4)_3$ is 2.67 mol $min^{-1}$ at a particular time, the rate of disappearance of $C_2H_6O$ at the same time is ........ mol $min^{-1}$. (Nearest integer) *[2021, 27 Aug Shift-I]*

**20** For a chemical reaction $A \longrightarrow B$, it was found that concentration of B is increased by 0.2 mol $L^{-1}$ in 30 min. The average rate of the reaction is ...... $\times 10^{-1}$ mol $L^{-1}\ h^{-1}$. (Nearest integer) *[2021, 25 July Shift-II]*

**21** The reaction $2A + B_2 \longrightarrow 2AB$ is an elementary reaction. For a certain quantity of reactants, if the volume of the reaction vessel is reduced by a factor of 3, the rate of the reaction increases by a factor of ......... (Round off to the nearest integer). *[2021, 17 March Shift-II]*

## Topic 2 Order, Molecularity and Half-life Period

### Level 1

**01** Which of the following graphs most appropriately represents a zero order reaction? *[2025, 23 Jan Shift-II]*

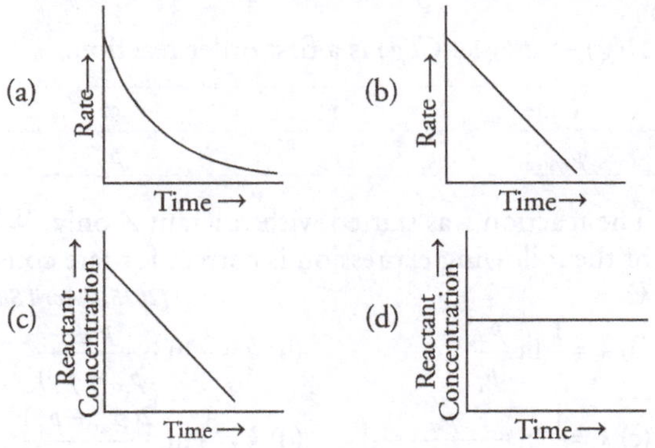


**Imp.** Unit of Rate Constant

**02** For a reaction of order $n$, the unit of the rate constant is *[2021, 27 July Shift-I]*

(a) $mol^{1-n}\ L^{1-n}\ s$
(b) $mol^{1-n}\ L^{2n}\ s^{-1}$
(c) $mol^{1-n}\ L^{n-1}\ s^{-1}$
(d) $mol^{1-n}\ L^{1-n}\ s^{-1}$

**03** For a reaction $A \rightarrow$ Products, a plot of $\log t_{1/2}$ *versus* $\log a_0$ as shown in figure. If the initial concentration of $A$ is represented by $a_0$, the order of the reaction is *[AIEEE 2012]*

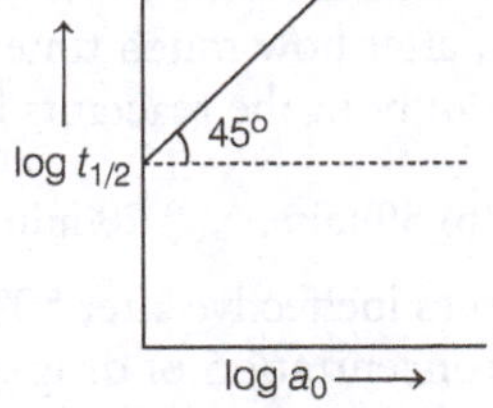


(a) one (b) zero (c) two (d) three

**04** The half-life period of a first order chemical reaction is 6.93 min. The time required for the completion of 99% of the chemical reaction will be ($\log 2 = 0.301$) ***[AIEEE 2009]***

(a) 230.3 min (b) 23.03 min
(c) 46.06 min (d) 460.6 min

**05** A reaction was found to be second order with respect to the concentration of carbon monoxide. If the concentration of carbon monoxide is doubled, with everything else kept the same, the rate of reaction will ***[AIEEE 2006]***

(a) remain unchanged (b) triple
(c) increase by a factor of 4 (d) double

## Level 2

**06** In a first order decomposition reaction, the time taken for the decomposition of reactant to one fourth and one eighth of its initial concentration are $t_1$ and $t_2$(s), respectively. The ratio $\frac{t_1}{t_2}$ will be ***[2025, 8 April Shift-II]***

(a) $\frac{3}{2}$ (b) $\frac{4}{3}$ (c) $\frac{3}{4}$ (d) $\frac{2}{3}$

**07** $A(g) \to B(g) + C(g)$ is a first order reaction.

| Time | $t$ | $\infty$ |
|---|---|---|
| $p_{system}$ | $p_t$ | $p_\infty$ |

The reaction was started with reactant $A$ only. Which of the following expression is correct for rate constant $k$? ***[2025, 7 April Shift-II]***

(a) $k = \frac{1}{t} \ln \frac{p_\infty}{p_t}$ (b) $k = \frac{1}{t} \ln \frac{p_\infty}{(p_\infty - p_t)}$

(c) $k = \frac{1}{t} \ln \frac{p_\infty}{2(p_\infty - p_t)}$ (d) $k = \frac{1}{t} \ln \frac{2(p_\infty - p_t)}{p_t}$

**08** Half-life of zero order reaction $A \to$ product is 1 hour, when initial concentration of reactant is 2.0 mol $L^{-1}$. The time required to decrease concentration of $A$ from 0.50 to 0.25 mol $L^{-1}$ is ***[2025, 4 April Shift-II]***

(a) 0.5 hour (b) 4 hour (c) 15 min (d) 60 min

**09** In a reaction $A + B \to C$, initial concentrations of $A$ and $B$ are related as $[A]_0 = 8[B]_0$. The half-lifes of $A$ and $B$ are 10 min and 40 min, respectively. If they start to disappear at the same time, both following first order kinetics, after how much time will the concentration of both the reactants be same? ***[2025, 3 April Shift-I]***

(a) 40 min (b) 80 min (c) 20 min (d) 60 min

**10** Drug $X$ becomes ineffective after 50% decomposition. The original concentration of drug in a bottle was 16 mg/ mL which becomes 4 mg/mL in 12 months. The expiry time of the drug in months is ........ ***[2025, 29 Jan Shift-II]***

Assume that the decomposition of the drug follows first order kinetics.

(a) 2 (b) 6
(c) 12 (d) 3

**11** The reaction, $A_2 + B_2 \longrightarrow 2AB$ follows the mechanism

$$A_2 \underset{k_{-1}}{\overset{k_1}{\rightleftharpoons}} A + A \text{ (Fast)}$$
$$A + B_2 \xrightarrow{k_2} AB + B \text{ (Slow)}$$
$$A + B \longrightarrow AB \text{ (Fast)}$$

The overall order of the reaction is ***[2025, 29 Jan Shift-I]***

(a) 2 (b) 1.5
(c) 2.5 (d) 3

**12** Which of the following statement is not true for radioactive decay? ***[2025, 22 Jan Shift-I]***

(a) Amount of radioactive substance remained after three Half-lifes is 1/8th of original amount.
(b) Half-life is ln2 times of $\frac{1}{\text{rate constant}}$.
(c) Decay constant does not depend upon temperature.
(d) Decay constant increases with increase in temperature.

**Imp.** **Graph of First Order Reactions**

**13** Given below are two statements. ***[2025, 24 Jan Shift-II]***

**Statement I**

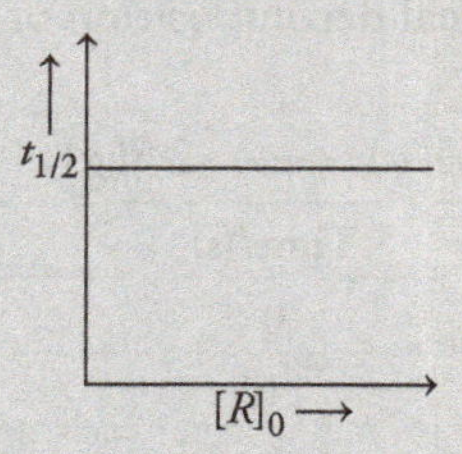


is valid for first order reaction.

**Statement II**

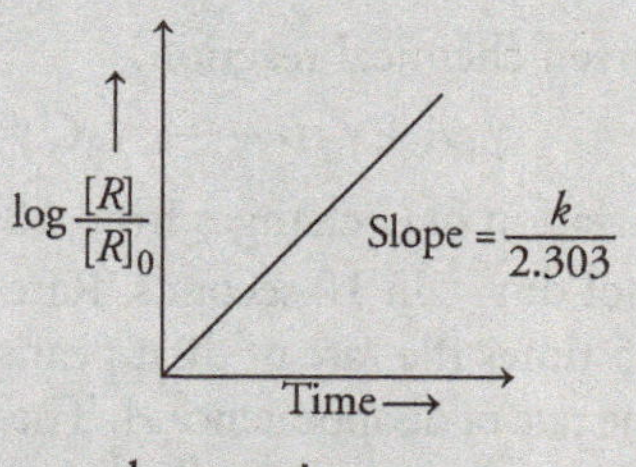


is valid for first order reaction.

In the light of the above statements, choose the correct answer from the options given below

(a) Statement I is true but Statement II is false.
(b) Both Statement I and Statement II are false.
(c) Both Statement I and Statement II are true.
(d) Statement I is false but Statement II is true.

**14** For bacterial growth in a cell culture, growth law is very similar to the law of radioactive decay. Which of the following graphs is most suitable to represent bacterial colony growth? ***[2025, 28 Jan Shift-II]***

Where N-number of bacteria at any time, $N_0$ –Initial number of bacteria.

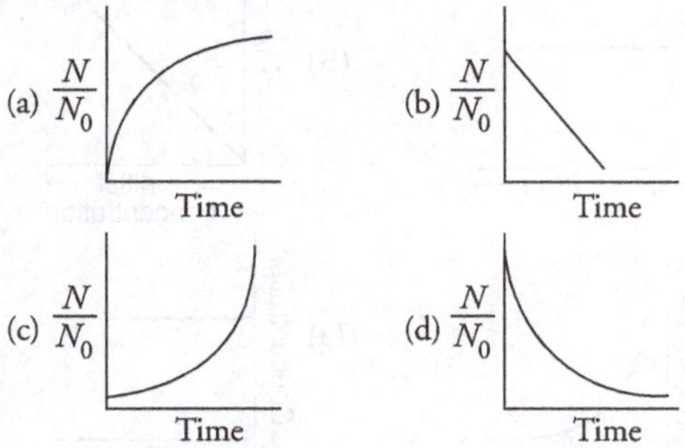


**15** Integrated rate law equation for a first order gas phase reaction is given by (where $p_i$ is initial pressure and $p_t$ is total pressure at time $t$) *[2024, 31 Jan Shift-I]*

(a) $k = \frac{2.303}{t} \times \frac{p_i}{(2p_i - p_t)}$

(b) $k = \frac{2.303}{t} \times \log \frac{p_i}{(2p_i - p_t)}$

(c) $k = \frac{2.303}{t} \times \log \frac{2p_i}{(2p_i - p_t)}$

(d) $k = \frac{2.303}{t} \times \log \frac{(2p_i - p_t)}{p_i}$

**16** A student has studied the decomposition of a gas $AB_3$ at 25°C. He obtained the following data.

| $p$ (mm Hg) | 50 | 100 | 200 | 400 |
|---|---|---|---|---|
| Relative $t_{1/2}$ (s) | 4 | 2 | 1 | 0.5 |

The order of the reaction is *[2023, 24 Jan Shift-II]*

(a) 1 (b) 0 (zero) (c) 0.5 (d) 2

**Imp.** Half-Life of Ist Order Reaction

**17** For a first order reaction, the time required for completion of 90% reaction is '$x$' times the half-life of the reaction. The value of '$x$' is (Given, ln10 = 2.303 and log 2 = 0.3010) *[2022, 24 June Shift-II]*

(a) 1.12 (b) 2.43
(c) 3.32 (d) 33.31

**18** Consider the following reactions

$A \rightarrow P1; B \rightarrow P2; C \rightarrow P3; D \rightarrow P4,$

The order of the above reactions are $a, b, c$ and $d$, respectively. The following graph is obtained when log[rate] vs log[conc.] are plotted :

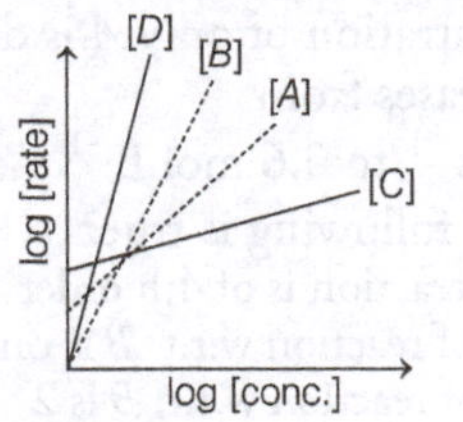


Among the following, the correct sequence for the order of the reactions is *[2020, 6 Sep Shift-I]*

(a) $D > A > B > C$
(b) $A > B > C > D$
(c) $C > A > B > D$
(d) $D > B > A > C$

**19** It is true that *[2020, 3 Sep Shift-I]*

(a) a second order reaction is always a multistep reaction.
(b) a zero order reaction is a multistep reaction.
(c) a first order reaction is always a single step reaction.
(d) a zero order reaction is a single step reaction.

**20** The reaction, $2X \rightarrow B$ is a zeroth order reaction. If the initial concentration of $X$ is 0.2 M, the half-life is 6 h. When the initial concentration of $X$ is 0.5 M, the time required to reach its final concentration of 0.2 M will be *[2019, 11 Jan Shift-II]*

(a) 7.2 h (b) 18.0 h
(c) 12.0 h (d) 9.0 h

**21** For a first order reaction, $(A) \longrightarrow$ products the concentration of $A$ changes from 0.1 M to 0.025 M in 40 min. The rate of reaction when the concentration of $A$ is 0.01 M is *[AIEEE 2012]*

(a) $1.73 \times 10^{-5}$ M / min
(b) $3.47 \times 10^{-4}$ M / min
(c) $3.47 \times 10^{-5}$ M / min
(d) $1.73 \times 10^{-4}$ M / min

**22** In a chemical reaction $A$ is converted into $B$. The rates of reaction, starting with initial concentrations of $A$ as $2 \times 10^{-3}$ M and $1 \times 10^{-3}$ M, are equal to $2.40 \times 10^{-4}$ Ms$^{-1}$ and $0.60 \times 10^{-4}$ Ms$^{-1}$ respectively. The order of reaction with respect to reactant $A$ will be *[AIEEE 2012]*

(a) 0 (b) 1.5
(c) 1 (d) 2

**23** Consider the reaction, $2A + B \rightarrow$ product. When concentration of $B$ alone was doubled, the half-life did not change. When the concentration of A alone was doubled, the rate increased by two times. The unit of rate constant for this reaction is *[AIEEE 2007]*

(a) L mol$^{-1}$ s$^{-1}$ (b) no unit
(c) mol L$^{-1}$ s$^{-1}$ (d) s$^{-1}$

**24** The rate equation for the reaction $2A + B \rightarrow C$ is found to be rate $= k\,[A]\,[B]$

The correct statement in relation to this reaction is that the *[AIEEE 2004]*

(a) unit of $k$ must be s$^{-1}$
(b) $t_{1/2}$ is a constant
(c) rate of formation of $C$ is twice the rate of disappearance of $A$
(d) value of $k$ is independent of the initial concentrations of $A$ and $B$

## Level 3

**25** Reaction $A(g) \rightarrow 2B(g) + C(g)$ is a first order reaction. It was started with pure $A$ *[2025, 7 April Shift-I]*

| *t*/min | Pressure of system at time *t*/time Hg |
|---|---|
| 10 | 160 |
| $\infty$ | 240 |

Which of the following option is incorrect?
(a) The reaction never goes to completion.
(b) Initial pressure of $A$ is 80 mm Hg
(c) Rate constant of the reaction is 1.693 $min^{-1}$
(d) Partial pressure of $A$ after 10 minute is 40 mm Hg

**26**

| $[A]_0$ / mol $L^{-1}$ | $t_{1/2}$ / min |
|---|---|
| 0.100 | 200 |
| 0.025 | 100 |

For a given reaction $R \rightarrow P$, $t_{1/2}$ is related to $[A]_0$ as given in table.
Given, $\log 2 = 0.30$
Which of the following is true?
A. The order of the reaction is 1/2
B. If $[A]_0$ is 1M, then $t_{1/2}$ is $200\sqrt{10}$ min.
C. The order of the reaction changes to 1 if the concentration of reactant changes from 0.100 M to 0.500 M.
D. $t_{1/2}$ is 800 min for $[A]_0 = 1.6$ M
Choose the correct answer from the options given below. *[2025, 28 Jan Shift-I]*
(a) C and D only (b) A and C only
(c) A, B and D only (d) A and B only

**27** For a chemical reaction, $A + B \longrightarrow$ Product, the order is 1 with respect to $A$ and $B$.

| Rate mol $L^{-1}$ $s^{-1}$ | $[A]$ mol $L^{-1}$ | $[B]$ mol $L^{-1}$ |
|---|---|---|
| 0.10 | 20 | 0.5 |
| 0.40 | $x$ | 0.5 |
| 0.80 | 40 | $y$ |

What is the value of $x$ and $y$? *[2023, 11 April Shift-II]*
(a) 80 and 4 (b) 80 and 2
(c) 40 and 4 (d) 160 and 4

**28** At 30°C, the half-life for the decomposition of $AB_2$ is 200 s and is independent of the initial concentration of $AB_2$. The time required for 80% of the $AB_2$ to decompose is
(Given, $\log 2 = 0.30$, $\log 3 = 0.48$) *[2022, 26 July Shift-II]*
(a) 200 s (b) 323 s (c) 467 s (d) 532 s

**29** For the following graphs

(A)
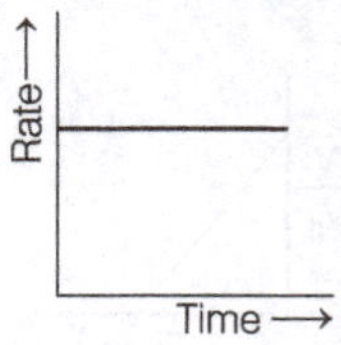


(B)
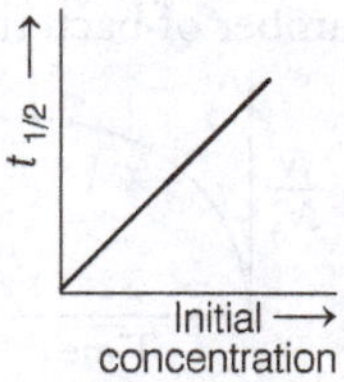


(C)
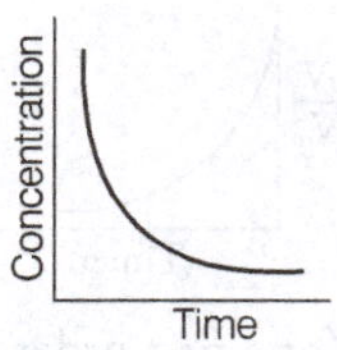


(D)
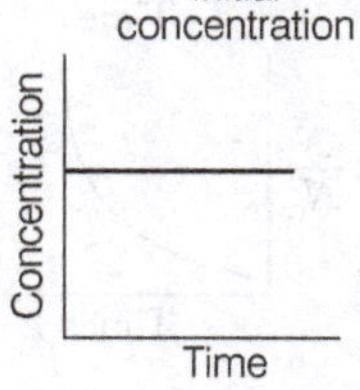


(E)
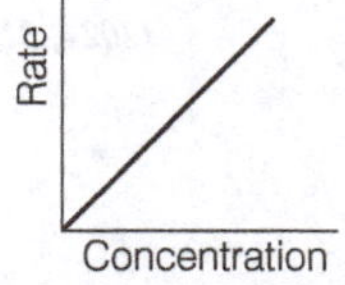


Choose from the options given below, the correct one regarding order of reaction is *[2021, 25 July Shift-I]*
(a) (B) zero order (C) and (E) first order
(b) (A) and (B) zero order (E) first order
(c) (B) and (D) zero order (E) first order
(d) (A) and (B) zero order (C) and (E) first order

**30** The results given in the below table were obtained during kinetic studies of the following reaction

$$2A + B \longrightarrow C + D$$

| Experiment | $[A]$/mol $L^{-1}$ | $[B]$/mol $L^{-1}$ | Initial rate/mol $L^{-1}$ $min^{-1}$ |
|---|---|---|---|
| I | 0.1 | 0.1 | $6.00 \times 10^{-3}$ |
| II | 0.1 | 0.2 | $2.40 \times 10^{-2}$ |
| III | 0.2 | 0.1 | $1.20 \times 10^{-2}$ |
| IV | $X$ | 0.2 | $7.20 \times 10^{-2}$ |
| V | 0.3 | $Y$ | $2.88 \times 10^{-1}$ |

$X$ and $Y$ in the given table are respectively *[2020, 2 Sep Shift-II]*
(a) 0.4, 0.4 (b) 0.4, 0.3 (c) 0.3, 0.4 (d) 0.3, 0.3

**31** For the following reaction, $2A + B \longrightarrow$ products
When concentration of both ($A$ and $B$) becomes double, then rate of reaction increases from 0.3 mol $L^{-1}$ $s^{-1}$ to 2.4 mol $L^{-1}$ $s^{-1}$.
When concentration of only $A$ is doubled, the rate of reaction increases from
0.3 mol $L^{-1}$ $s^{-1}$ to 0.6 mol $L^{-1}$ $s^{-1}$.
Which of the following is true? *[2019, 9 Jan Shift-II]*
(a) The whole reaction is of 4th order
(b) The order of reaction w.r.t. $B$ is one
(c) The order of reaction w.r.t. $B$ is 2
(d) The order of reaction w.r.t. $A$ is 2

**32** The following results were obtained during kinetic studies of the reaction;

$$2A + B \longrightarrow \text{Products}$$

| Experiment | [A] (mol $L^{-1}$) | [B] (mol $L^{-1}$) | Initial rate of reaction (in mol $L^{-1}$ $min^{-1}$) |
|---|---|---|---|
| I. | 0.10 | 0.20 | $6.93 \times 10^{-3}$ |
| II. | 0.10 | 0.25 | $6.93 \times 10^{-3}$ |
| III. | 0.20 | 0.30 | $1.386 \times 10^{-2}$ |

The time (in minutes) required to consume half of $A$ is **[2019, 9 Jan Shift-I]**
(a) 5 (b) 10 (c) 100 (d) 1

**33** At 518°C, the rate of decomposition of a sample of gaseous acetaldehyde, initially at a pressure of 363 Torr, was 1.00 Torr $s^{-1}$ when 5% had reacted and 0.5 Torr $s^{-1}$ when 33% had reacted. The order of the reaction is : **[JEE Main 2018]**
(a) 2 (b) 3 (c) 1 (d) 0

**34** Decomposition of $H_2O_2$ follows a first order reaction. In 50 min, the concentration of $H_2O_2$ decreases from 0.5 to 0.125 M in one such decomposition. When the concentration of $H_2O_2$ reaches 0.05 M, the rate of formation of $O_2$ will be **[JEE Main 2016]**
(a) $6.93 \times 10^{-4}$ mol $min^{-1}$
(b) 2.66 L $min^{-1}$ at STP
(c) $1.34 \times 10^{-2}$ mol $min^{-1}$
(d) $6.93 \times 10^{-2}$ mol $min^{-1}$

**35** Rate constant, $k$ of the first order reaction when initial concentration ($C_0$) and concentration ($C_t$) at time $t$, is given by equation $kt = \log C_0 - \log C_t$ Graph is a straight line if we plot **[AIEEE 2002]**
(a) $t$ *vs* $\log C_0$ (b) $t$ *vs* $\log C_t$
(c) $t^{-1}$ *vs* $\log C_t$ (d) $\log C_0$ *vs* $\log C_t$

## Numerical Value Type Questions

**36** For the reaction $A \rightarrow B$ the following graph was obtained. The time required (in seconds) for the concentration of $A$ to reduce to 2.5 g $L^{-1}$ (if the initial concentration of $A$ was 50 g $L^{-1}$) is ......... . (Nearest integer)
Given : $\log 2 = 0.3010$ **[2025, 2 April Shift-II]**

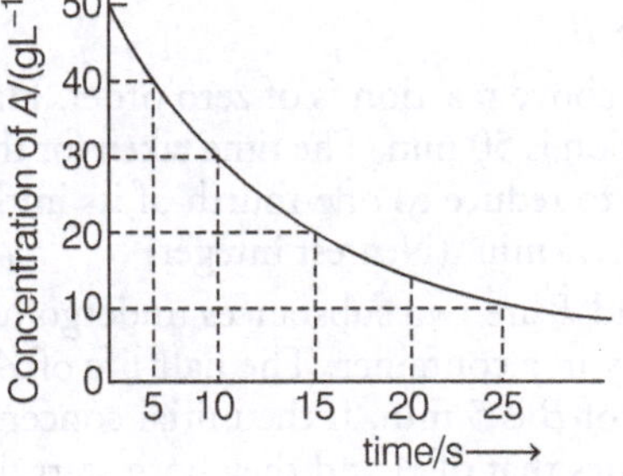


**37** For the reaction $A \rightarrow$ products

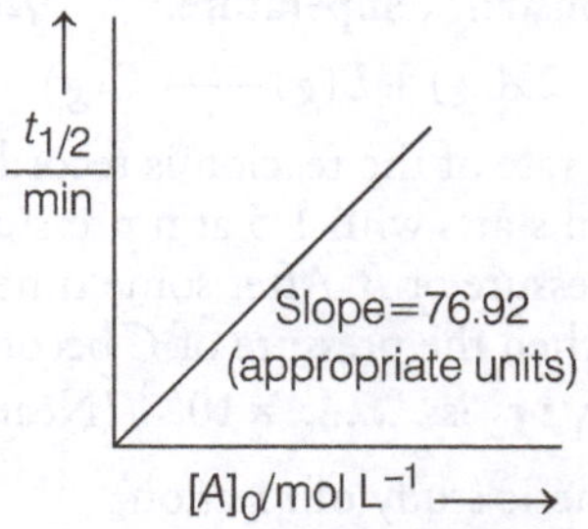


The concentration of $A$ at 10 minutes is ....... $\times 10^{-3}$ mol $L^{-1}$ (nearest integer). The reaction was started with 2.5 mol $L^{-1}$ of A. **[2025, 2 April Shift-I]**

**38** For the thermal decomposition of $N_2O_5$ (g) at constant volume, the following table can be formed. For the reaction mentioned below. **[2025, 23 April Shift-I]**

$$2N_2O_5(g) \longrightarrow 2N_2O_4(g) + O_2(g)$$

| S.No. | Time | Total pressure (atm) |
|---|---|---|
| 1. | 0 | 0.6 |
| 2. | 100 | 'x' |

$x =$ ........... $\times 10^{-3}$ atm [Nearest integer]
(Given, rate constant for the reaction is $4.606 \times 10^{-2} s^{-1}$.)

**39** Consider the following reaction **[2024, 8 April Shift-I]**

$$A + B \longrightarrow C$$

The time taken for $A$ to become $1/4^{th}$ of its initial concentration is twice the time taken to become 1/2 of the same. Also, when the change of concentration of $B$ is plotted against time, the resulting graph gives a straight line with a negative slope and a positive intercept on the concentration axis.
The overall order of the reaction is .........

**40** Consider the two different first order reactions given below. **[2024, 6 April Shift-II]**

$$A + B \longrightarrow C \text{ (Reaction 1)}$$
$$P \longrightarrow Q \text{ (Reaction 2)}$$

The ratio of the half-life of reaction 1 : reaction 2 is 5 : 2. If $t_1$ and $t_2$ represent the time taken to complete $\frac{2}{3}$rd and $\frac{4}{5}$th of reaction 1 and reaction 2, respectively, then the value of the ratio $t_1 : t_2$ is ...... $\times 10^{-1}$ (Nearest integer).

[Given, $\log_{10}(3) = 0.477$ and $\log_{10}(5) = 0.699$]

**41** Time required for 99.9% completion of a first order reaction is ........ times the time required for completion of 90% reaction. (Nearest integer) **[2024, 6 April Shift-I]**

**42** Consider the following single step reaction in gas phase at constant temperature. *[2024, 5 April Shift-II]*

$$2A(g) + B(g) \longrightarrow C(g)$$

The initial rate of the reacion is recorded as $r_1$, when the reaction starts with 1.5 atm pressure of $A$ and 0.7 atm pressure of $B$. After some time, the rate $r_2$, is recorded when the pressure of $C$ becomes 0.5 atm. The ratio $r_1 : r_2$ is .......... $\times 10^{-1}$ (Nearest integer)

**43** During kinetic study of reaction, $2A + B \longrightarrow C + D$, the following results were obtained. *[2024, 5 April Shift-II]*

| | $A$[M] | $B$[M] | Initial rate of formation of $D$ |
|---|---|---|---|
| I. | 0.1 | 0.1 | $6.0 \times 10^{-3}$ |
| II. | 0.3 | 0.2 | $7.2 \times 10^{-2}$ |
| III. | 0.3 | 0.4 | $2.88 \times 10^{-1}$ |
| IV. | 0.4 | 0.1 | $2.40 \times 10^{-2}$ |

Based on above data, overall order of the reaction is .........

**44** Consider the following reaction, the rate expression of which is given below *[2024, 4 April Shift-II]*

$$A + B \rightarrow C$$

$$\text{Rate} = k[A]^{1/2}[B]^{1/2}$$

The reaction is initiated by taking 1 M concentration of $A$ and $B$ each. If the rate constant ($k$) is $4.6 \times 10^{-2} s^{-1}$, then the time taken for $A$ to become 0.1 M is ...... sec. (Nearest integer)

**45** The ratio of $\frac{^{14}C}{^{12}C}$ in a piece of wood is $\frac{1}{8}$ part that of atmosphere. If half-life of $^{14}C$ is 5730 years, the age of wood sample is ......... years. *[2024, 1 Feb Shift-I]*

**46** $r = k[A]$ for a reaction, 50% of $A$ is decomposed in 120 minutes. The time taken for 90% decomposition of $A$ is ........... minutes. *[2024, 31 Jan Shift-II]*

**47** The rate of first order reaction is 0.04 mol $L^{-1}s^{-1}$ at 10 minutes and 0.03 mol $L^{-1}s^{-1}$ at 20 minutes after initiation. Half-life of the reaction is ......... minutes. (Given, $\log 2 = 0.3010$, $\log 3 = 0.4771$) *[2024, 30 Jan Shift-I]*

**48** The half-life of radioisotope bromine-82 is 36 hours. The fraction which remains after one day is ......... $\times 10^{-2}$.

(Given, antilog 0.2006 = 1.587) *[2024, 29 Jan Shift-II]*

**49** Time required for completion of 99.9% of a first order reaction is ........ times of half-life ($t_{1/2}$) of the reaction. *[2024, 27 Jan Shift-II]*

**50** Consider the following data for the given reaction

$$2HI(g) \longrightarrow H_2(g) + I_2(g)$$

| | 1 | 2 | 3 |
|---|---|---|---|
| HI (mol $L^{-1}$) | 0.05 | 0.01 | 0.02 |
| Rate (mol $L^{-1}s^{-1}$) | $7.5 \times 10^{-4}$ | $3.0 \times 10^{-3}$ | $1.2 \times 10^{-2}$ |

The order of the reaction is .......... *[2024, 27 Jan Shift-I]*

**51** $A(g) \longrightarrow 2B(g) + C(g)$ is a first order reaction. The initial pressure of the system was found to be 800 mm Hg which increased to 1600 mm Hg after 10 min. The total pressure of the system after 30 min will be ........ mm Hg. (Nearest integer) *[2023, 13 April Shift-II]*

**Imp.** **Relation between $t_{87.5}$ and $t_{50}$ of Ist Order Reaction**

**52** $t_{87.5}$ is the time required for the reaction to undergo 87.5% completion and $t_{50}$ is the time required for the reaction to undergo 50% completion. The relation between $t_{87.5}$ and $t_{50}$ for a first order reaction is $t_{87.5} = x \times t_{50}$

The value of $x$ is ........... . (Nearest integer) *[2023, 13 April Shift-I]*

**53** The reaction $2NO + Br_2 \longrightarrow 2NOBr$ takes places through the mechanism given below

$$NO + Br_2 \rightleftharpoons NOBr_2 \text{(fast)}$$

$$NOBr_2 + NO \longrightarrow 2NOBr \text{ (Slow)}$$

The overall order of the reaction is ........... . *[2023, 12 April Shift-I]*

**54** The number of incorrect statements from the following is *[2023, 10 April Shift-II]*

A. The successive half lives of zero order reactions decreases with time.
B. A substance appearing as reactant in the chemical equation may not affect the rate of reaction
C. Order and molecularity of a chemical reaction can be a fractional number
D. The rate constant units of zero and second order reaction are mol $L^{-1}s^{-1}$ and mol $L^{-1}s^{-1}$ respectively

**55** A molecule undergoes two independent first order reactions whose respective half lives are 12 min and 3 min. If both the reactions are occurring then the time taken for the 50% consumption of the reactant is .......... min. (Nearest integer) *[2023, 10 April Shift-I]*

**56** $A \rightarrow B$

The above reaction is of zero order. Half-life of this reaction is 50 min. The time taken for the concentration of $A$ to reduce to one-fourth of its initial value is ......... min. (Nearest integer) *[2023, 1 Feb Shift-II]*

**57** $A$ and $B$ are two substances undergoing radioactive decay in a container. The half-life of $A$ is 15 min and that of $B$ is 5 min. If the initial concentration of $B$ is 4 times that of $A$ and they both start decaying at the

same time, how much time will it take for the concentration of both of them to be same?
............ min. *[2023, 1 Feb Shift-I]*

**58** The rate constant for a first order reaction is 20 $min^{-1}$. The time required for the initial concentration of the reactant to reduce to its $\frac{1}{32}$ level is ......... $\times 10^{-2}$ min. (Nearest integer)
(Given, ln 10 = 2.303, log 2 = 0.3010)
*[2023, 31 Jan Shift-II]*

**59** An organic compound undergoes first order decomposition. If the time taken for the 60% decomposition is 540 s, then the time required for 90% decomposition will be is .......... s. (Nearest integer) [Given : ln 10 = 2.3; log 2 = 0.3]
*[2023, 30 Jan Shift-II]*

**60** If compound $A$ reacts with $B$ following first order kinetics with rate constant $2.011 \times 10^{-3}\ s^{-1}$. The time taken by $A$ (in seconds) to reduce from 7 g to 2 g will be ......... (Nearest integer) [log 5 = 0.698, log 7 = 0.845, log 2 = 0.301]
*[2023, 30 Jan Shift-I]*

**61** For conversion of compound $A \rightarrow B$, the rate constant of the reaction was found to be $4.6 \times 10^{-5}$ L $mol^{-1}\ s^{-1}$. The order of the reaction is ..........
*[2023, 29 Jan Shift-II]*

**62** For certain chemical reaction $X \rightarrow Y$, the rate of formation of product is plotted against the time as shown in the figure. The number of correct statements from the following is ....... *[2023, 29 Jan Shift-I]*

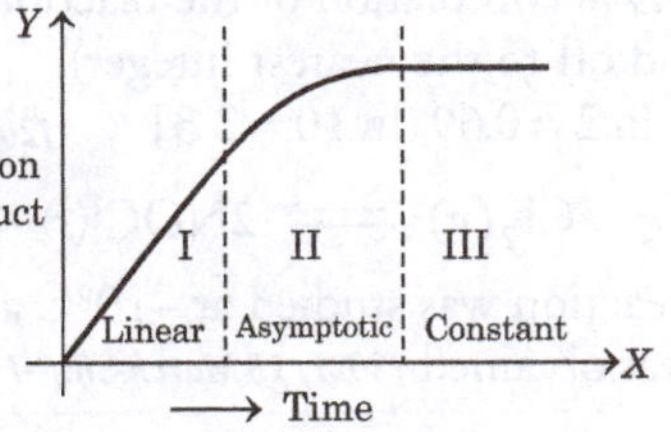


(A) over all order of this reaction is one.
(B) order of this reaction can't be determined.
(C) in region I and III, the reaction is of first and zero order respectively.
(D) in region-II, the reaction is of first order.
(E) in region-II, the order of reaction is in the range of 0.1 to 0.9.

**63** A first order reaction has the rate constant, $k = 4.6 \times 10^{-3}\ s^{-1}$. The number of correct statements from the following is/are ........
[Given, log 3 = 0.48] *[2023, 25 Jan Shift-II]*

A. reaction completes in 1000 s.
B. the reaction has a half-life of 500 s.
C. the time required for 10% completion is 25 times the time required for 90% completion.
D. the degree of dissociation is equal to $(1 - e^{-kt})$.
E. the rate and the rate constant have the same unit.

**64** For the first order reaction $A \rightarrow B$, the half-life is 30 min. The time taken for 75% completion of the reaction is ......... min. (Nearest integer).
[Given : log 2 = 0.3010
log 3 = 0.4771
log 5 = 0.6989] *[2023, 25 Jan Shift-I]*

**65** Assuming 1 μg of trace radioactive element $X$ with a half-life of 30 years is absorbed by a growing tree. The amount of $X$ remaining in the tree after 100 years is ........ $\times 10^{-1}$ μg.
[Given : ln 10 = 2.303; log 2 = 0.30] *[2022, 29 July Shift-II]*

**66** The reaction between $X$ and $Y$ is first order with respect to $X$ and zero order with respect to $Y$.

| Experiment | $[X]$ ($mol\ L^{-1}$) | $[Y]$ ($mol\ L^{-1}$) | Initial rate ($mol\ L^{-1}\ min^{-1}$) |
|---|---|---|---|
| I. | 0.1 | 0.1 | $2 \times 10^{-3}$ |
| II. | $L$ | 0.2 | $4 \times 10^{-3}$ |
| III. | 0.4 | 0.4 | $M \times 10^{-3}$ |
| IV. | 0.1 | 0.2 | $2 \times 10^{-3}$ |

Examine the data of table and calculate ratio of numerical values of $M$ and $L$. (Nearest integer)
*[2022, 29 July Shift-I]*

**67** For the given first order reaction $A \rightarrow B$, the half-life of the reaction is 0.3010 min. The ratio of the initial concentration of reactant to the concentration of reactant at time 2.0 min will be equal to ............ . (Nearest integer) *[2022, 28 July Shift-I]*

**68** $[A] \longrightarrow [B]$
Reactant Product
If formation of compound $[B]$ follows the first order of kinetics and after 70 minutes the concentration of $[A]$ was found to be half of its initial concentration. Then the rate constant of the reaction is $x \times 10^{-6} s^{-1}$. The value of $x$ is ........... . (nearest integer)
*[2022, 27 July Shift-II]*

**69** $2NO + 2H_2 \longrightarrow N_2 + 2H_2O$

The above reaction has been studied at 800°C. The related data are given in the table below.

| Reaction serial number | Initial pressure of $H_2$/kPa | Initial pressure of NO/kPa | Initial rate $\left(\frac{-dp}{dt}\right)$/ (kPa/s) |
|---|---|---|---|
| 1 | 65.6 | 40.0 | 0.135 |
| 2 | 65.6 | 20.1 | 0.033 |
| 3 | 38.6 | 65.6 | 0.214 |
| 4 | 19.2 | 65.6 | 0.106 |

The order of the reaction with respect to NO is...... . **[2022, 27 July Shift-I]**

**70** For a reaction, $A \longrightarrow 2B + C$ the half-lifes are 100 s and 50 s when the concentration of reactant $A$ is 0.5 and 1.0 mol $L^{-1}$ respectively. The order of the reaction is ........... . (Nearest integer) **[2022, 26 July Shift-I]**

**71** For the decomposition of azomethane, $CH_3N_2CH_3(g) \longrightarrow CH_3CH_3(g) + N_2(g)$, a first order reaction, the variation in partial pressure with time at 600 K is given as

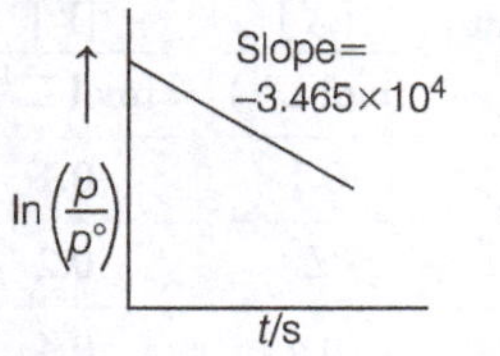


The half-life of the reaction is ...... $\times 10^{-5}$ s. **[2022, 25 July Shift-II]**

**72** For a first order reaction, $A \rightarrow B$, the rate constant, $k = 5.5 \times 10^{-14}\ s^{-1}$. The time required for 67% completion of reaction is $x \times 10^{-1}$ times the half-life of reaction. The value of $x$ is ...... . [Nearest integer] (Given, log 3 = 0.4771). **[2022, 28 June Shift-I]**

**73** A flask is filled with equal moles of $A$ and $B$. The half-lives of $A$ and $B$ are 100 s and 50 s respectively and are independent of the initial concentration. The time required for the concentration of $A$ to be four times that of $B$ is ...... s. (Given, ln 2 = 0.693) **[2022, 26 June Shift-I]**

**74** At 345 K, the half-life for the decomposition of a sample of a gaseous compound initially at 55.5 kPa was 340 s. When the pressure was 27.8 kPa, the half-life was found to be 170 s. The order of the reaction is ........... . [Nearest integer] **[2022, 25 June Shift-II]**

**75** For a first order reaction, the ratio of the time for 75% completion of a reaction to the time for 50% completion is ...... (Integer answer) **[2021, 31 Aug Shift-I]**

**76** The following data was obtained for chemical reaction given below at 975 K.

$2NO(g) + 2H_2(g) \longrightarrow N_2(g) + 2H_2O(g)$

| | [NO] mol $L^{-1}$ | [$H_2$] mol $L^{-1}$ | [Rate] mol $L^{-1}$ |
|---|---|---|---|
| A. | $8 \times 10^{-5}$ | $8 \times 10^{-5}$ | $7 \times 10^{-9}$ |
| B. | $24 \times 10^{-5}$ | $8 \times 10^{-5}$ | $21 \times 10^{-8}$ |
| C. | $24 \times 10^{-5}$ | $32 \times 10^{-5}$ | $8.4 \times 10^{-8}$ |

The order of the reaction with respect to NO is ........... [Integer answer] **[2021, 26 Aug Shift-I]**

**77** For the first order reaction, $A \longrightarrow 2B$, 1 mole of reactant $A$ gives 0.2 moles of $B$ after 100 minutes. The half-life of the reaction is ............. min. (Round off to the nearest integer). [Use : ln 2 = 0.69, ln 10 = 2.3; properties of logarithms : $\ln x^y = y \ln x$; $\ln\left(\frac{x}{y}\right) = \ln x - \ln y$]

(Round off to the nearest integer) **[2021, 27 July Shift-II]**

**78** The inactivation rate of a viral preparation is proportional to the amount of virus. In the first minute after preparation, 10% of the virus is inactivated. The rate constant for viral inactivation is ...... $\times 10^{-3}\ min^{-1}$. (Nearest integer) [Use : ln 10 = 2.303; $\log_{10} 3 = 0.477$; property of logarithm : $\log x^y = y \log x$] **[2021, 20 July Shift-I]**

**79** A reaction has a half-life of 1 min. The time required for 99.9% completion of the reaction is ........ min (Round off to the nearest integer). [Use : ln 2 = 0.69, ln 10 = 2.3] **[2021, 18 March Shift-II]**

**80** $2NO(g) + Cl_2(g) \rightleftharpoons 2NOCl(s)$

This reaction was studied at −10°C and the following data was obtained **[2021, 18 March Shift-I ; 2019, 8 April Shift-I]**

| Run | $[NO]_0$ | $[Cl_2]_0$ | $r_0$ |
|---|---|---|---|
| 1 | 0.10 | 0.10 | 0.18 |
| 2 | 0.10 | 0.20 | 0.35 |
| 3 | 0.20 | 0.20 | 1.40 |

$[NO]_0$ and $[Cl_2]_0$ are the initial concentrations and $r_0$ is the initial reaction rate.

The overall order of the reaction is ......... . (Rounded off to the nearest integer)

**81** For a certain first order reaction 32% of the reactant is left after 570 s. The rate constant of this reaction is ...... $\times 10^{-3}\ s^{-1}$. (Round off to the nearest integer). [Given, $\log_{10} 2 = 0.301$, ln 10 = 2.303] **[2021, 17 March Shift-I]**

**82** $A$ and $B$ decompose *via* first order kinetics with half-lives 54.0 min and 18.0 min respectively. Starting from an equimolar non-reactive mixture of $A$ and $B$, the time taken for the concentration of $A$ to become 16 times that of $B$ is ......... min. (Round off to the nearest integer). **[2021, 16 March Shift-II]**

**83** Sucrose hydrolyses in acid solution into glucose and fructose following first order rate law with a half-life of 3.33 h at 25°C. After 9 h, the fraction of sucrose remaining is $f$. The value of $\log_{10}\left(\frac{1}{f}\right)$ is ... $\times 10^{-2}$

(Rounded off to the nearest integer).
[Assume, $\ln 10 = 2.303$, $\ln 2 = 0.693$] **[2021, 24 Feb Shift-II]**

**84** Gaseous cyclobutene isomerises to butadiene in a first order process which has a '$k$' value of $3.3 \times 10^{-4}\ s^{-1}$ at 153°C. The time in minutes it takes for the isomerisation to proceed 40 % to completion at this temperature is ....... .
(Rounded off to the nearest integer) **[2021, 24 Feb Shift-I]**

**85** If 75% of a first order reaction was completed in 90 minutes, 60% of the same reaction would be completed in approximately (in minutes) ........... .
(Take : $\log 2 = 0.30$; $\log 2.5 = 0.40$) **[2020, 4 Sep Shift-I]**

**86** During the nuclear explosion, one of the products is $^{90}Sr$ with half-life of 6.93 years. If 1 μg of $^{90}Sr$ was absorbed in the bones of a newly born baby in place of Ca, how much time, in years, is required to reduce it by 90% if it is not lost metabolically ......... .
**[2020, 7 Jan Shift-I]**

## Topic 3 Arrhenius Theory, Activation Energy, Collision and Related Theories

### Level 1

**01** Consider the given figure and choose the correct option **[2025, 22 Jan Shift-II]**

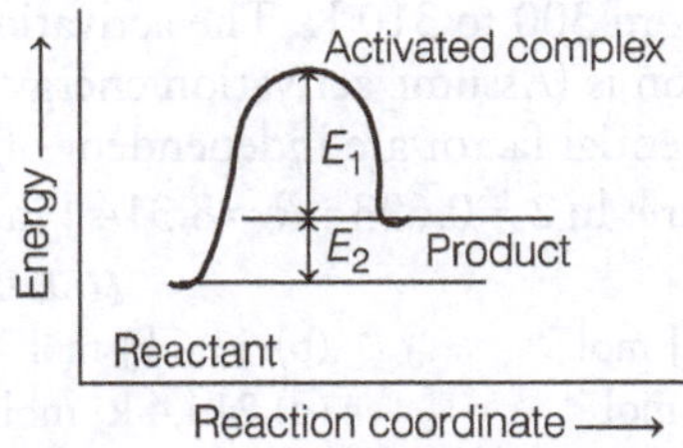


(a) Activation energy of forward reaction is $E_1 + E_2$ and product is less stable than reactant.
(b) Activation energy of both forward and backward reaction is $E_1 + E_2$ and reactant is more stable than product.
(c) Activation energy of forward reaction is $E_1 + E_2$ and product is more stable than reactant.
(d) Activation energy of backward reaction is $E_1$ and product is more stable than reactant.

**Imp.** **Energy Profile Graph of Positive Catalyst**

**02** The correct reaction profile diagram for a positive catalyst reaction is **[2023, 8 April Shift-II]**

(a) 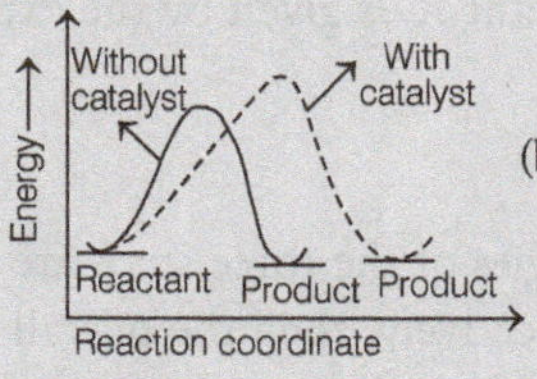


(b) 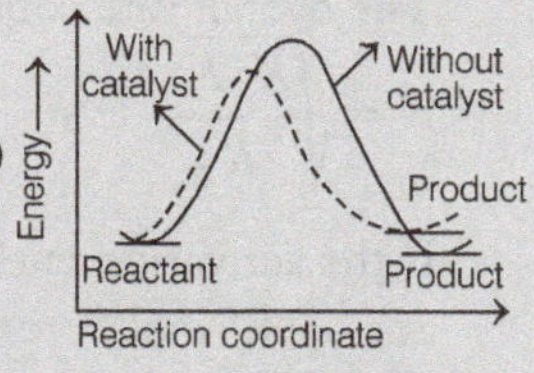


(c) 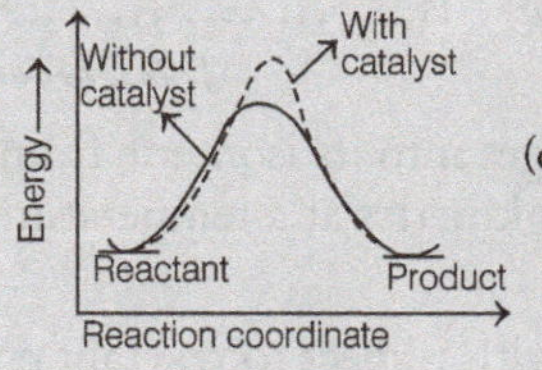


(d) 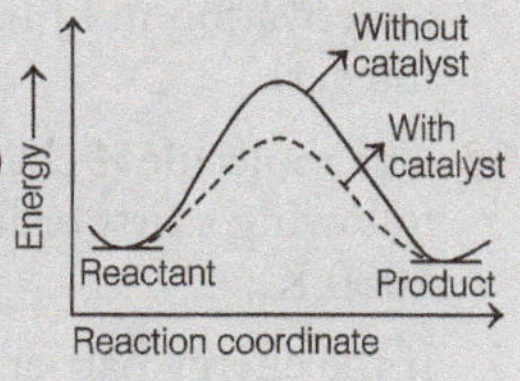


**03** Which one of the following given graphs represents the variation of rate constant ($k$) with temperature ($T$) for an endothermic reaction? **[2021, 1 Sep Shift-II]**

(a) 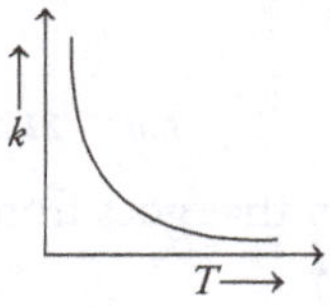


(b) 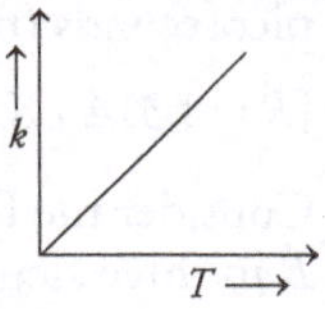


(c) 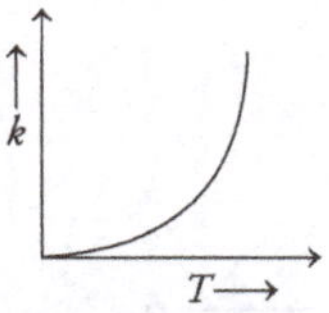


(d) 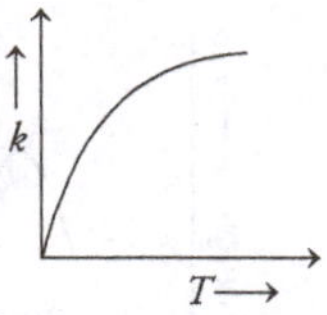


**04** Higher order (>3) reactions are rare due to **[JEE Main 2015]**
(a) low probability of simultaneous collision of all the reacting species
(b) increase in entropy and activation energy as more molecules are involved
(c) shifting of equilibrium towards reactants due to elastic collisions
(d) loss of active species on collision

**05** The reaction $X \rightarrow Y$ is an exothermic reaction. Activation energy of the reaction for $X$ into $Y$ is 150 kJ mol$^{-1}$. Enthalpy of reaction is 135 kJ mol$^{-1}$. The activation energy for the reverse reaction, $Y \rightarrow X$ will be : **[JEE Main 2013]**
(a) 280 kJ mol$^{-1}$ (b) 285 kJ mol$^{-1}$
(c) 270 kJ mol$^{-1}$ (d) 15 kJ mol$^{-1}$

**06** A reactant ($A$) forms two products

$$A \xrightarrow{k_1} B\text{, Activation energy } E_{a_1}$$
$$A \xrightarrow{k_2} C\text{, Activation energy } E_{a_2}$$

If $E_{a_2} = 2E_{a_1}$, then $k_1$ and $k_2$ are related as **[AIEEE 2011]**
(a) $k_1 = 2k_2 e^{E_{a_2}/RT}$ (b) $k_1 = k_2 e^{E_{a_1}/RT}$
(c) $k_2 = k_1 e^{E_{a_2}/RT}$ (d) $k_1 = A k_2 e^{E_{a_1}/RT}$

## Level 2

**07** Consider a complex reaction taking place in three steps with rate constants $K_1$, $K_2$ and $K_3$ respectively. The overall rate constant $K$ is given by the expression $K = \sqrt{\frac{K_1 K_3}{K_2}}$.

If the activation energies of the three steps are 60, 30 and 10 kJ mol$^{-1}$ respectively, then the overall energy of activation in kJ mol$^{-1}$ is ................ .(Nearest integer) ***[2025, 24 Jan Shift-II]***

**08** The molecule $A$ changes into its isomeric from $B$ by following a first order kinetics at a temperature of 1000 K.

If the energy barrier with respect to reactant energy for such isomeric transformation is 191.48 kJ mol$^{-1}$ and the frequency factor is $10^{20}$, the time required for 50% molecules of $A$ to become $B$ is ........ picoseconds (nearest integer).

$[R = 8.314\ \text{J K}^{-1}\text{mol}^{-1}]$ ***[2025, 22 Jan Shift-I]***

**09** Consider the following reaction that goes from $A$ to $B$ in three steps as shown below

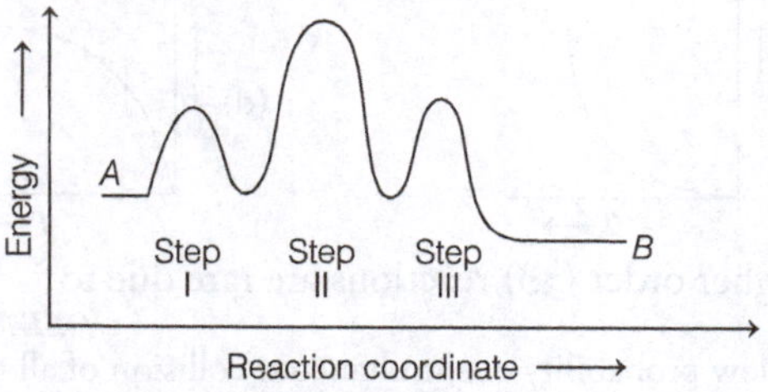


Choose the correct option ***[2023, 6 April Shift-II]***

| | Number of intermediates | Number of Activated complexes | Rate determining step |
|---|---|---|---|
| (a) | 2 | 3 | III |
| (b) | 3 | 2 | II |
| (c) | 2 | 3 | II |
| (d) | 2 | 3 | I |

**10** The rate constant ($k$) of a reaction is measured at different temperature ($T$), and the data are plotted in the given figure. The activation energy of the reaction in kJ mol$^{-1}$ is ($R$ is gas constant)

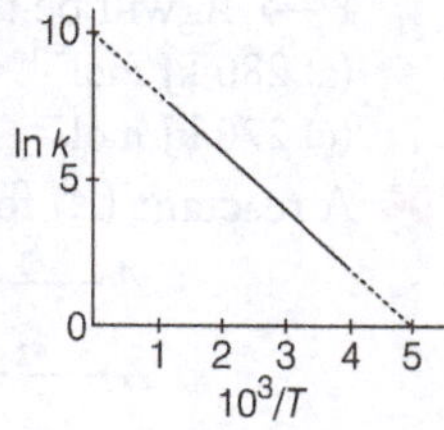


***[2020, 5 Sep Shift-II]***

(a) $\frac{2}{R}$ (b) $\frac{1}{R}$
(c) $R$ (d) $2R$

**11** Consider the following plots of rate constant *versus* $1/T$ for four different reactions. Which of the following orders is correct for the activation energies of these reactions?

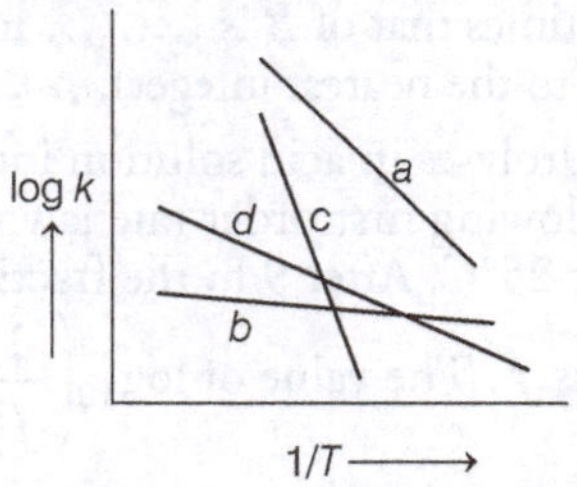


***[2020, 8 Jan Shift-II]***

(a) $E_b > E_a > E_d > E_c$ (b) $E_a > E_c > E_d > E_b$
(c) $E_b > E_d > E_c > E_a$ (d) $E_c > E_a > E_d > E_b$

**12** The rate of a certain biochemical reaction at physiological temperature ($T$) occurs $10^6$ times faster with enzyme than without. The change in the activation energy upon adding enzyme is ***[2020, 8 Jan Shift-I]***

(a) $+6RT$ (b) $-6(2.303)RT$
(c) $+6(2.303)RT$ (d) $-6RT$

**13** For the reaction of $H_2$ with $I_2$, the rate constant is $2.5 \times 10^{-4}\ \text{dm}^3\text{mol}^{-1}\text{s}^{-1}$ at 327°C and $1.0\ \text{dm}^3\ \text{mol}^{-1}\ \text{s}^{-1}$ at 527°C.

The activation energy for the reaction, in kJ mol$^{-1}$ is ($R = 8.314\ \text{JK}^{-1}\ \text{mol}^{-1}$) ***[2019, 10 April Shift-II]***

(a) 59 (b) 72 (c) 150 (d) 164

**14** For a reaction, consider the plot of $\ln k$ *versus* $1/T$ given in the figure. If the rate constant of this reaction at 400 K is $10^{-5}\ \text{s}^{-1}$, then the rate constant at 500 K is ***[2019, 12 Jan Shift-II]***

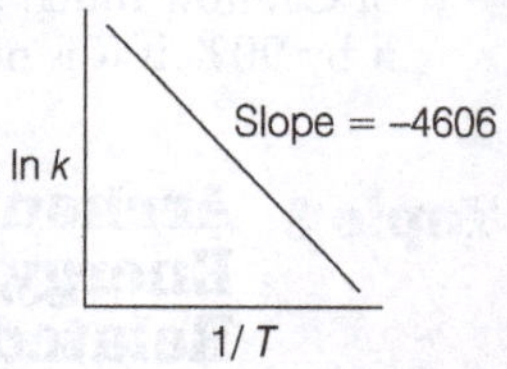


(a) $4 \times 10^{-4}\ \text{s}^{-1}$ (b) $10^{-6}\ \text{s}^{-1}$
(c) $10^{-4}\ \text{s}^{-1}$ (d) $2 \times 10^{-4}\ \text{s}^{-1}$

**15** The rate of a reaction quadruples when the temperature changes from 300 to 310 K. The activation energy of this reaction is (Assume activation energy and pre-exponential factor are independent of temperature; $\ln 2 = 0.693$; $R = 8.314\ \text{J mol}^{-1}\ \text{K}^{-1}$) ***[JEE Main 2013, 2017]***

(a) 107.2 kJ mol$^{-1}$ (b) 53.6 kJ mol$^{-1}$
(c) 26.8 kJ mol$^{-1}$ (d) 214.4 kJ mol$^{-1}$

**16** The rate of a reaction double when its temperature changes from 300 K to 310 K. Activation energy of such a reaction will be ($R = 8.314\ \text{JK}^{-1}\ \text{mol}^{-1}$ and $\log 2 = 0.301$) ***[JEE Main 2013]***

(a) 53.6 kJ mol$^{-1}$ (b) 48.6 kJ mol$^{-1}$
(c) 58.5 kJ mol$^{-1}$ (d) 60.5 kJ mol$^{-1}$

**17** The energies of activation for forward and reverse reactions for $A_2 + B_2 \rightleftharpoons 2AB$ are 180 kJ mol$^{-1}$ and 200 kJ mol$^{-1}$, respectively. The presence of a catalyst lowers the activation energy of both (forward and reverse) reactions by 100 kJ mol$^{-1}$. The enthalpy change of the reaction ($A_2 + B_2 \rightarrow 2AB$) in the presence of catalyst will be (in kJ mol$^{-1}$) **[AIEEE 2007]**

(a) 300 (b) 120 (c) 280 (d) 20

## Level 3

**18** Consider the following plots of log of rate constant $k$ ($\log k$) *vs* $\frac{1}{T}$ for three different reactions. The correct order of activation energies of these reactions is:

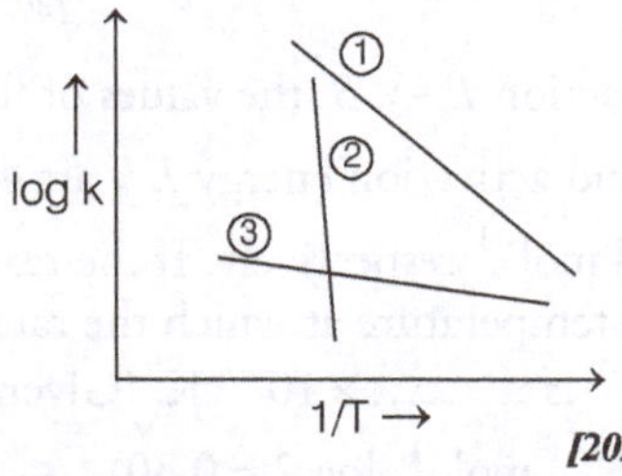


**[2025, 4 April Shift-II]**

(a) $Ea_3 > Ea_2 > Ea_1$ (b) $Ea_1 > Ea_3 > Ea_2$
(c) $Ea_2 > Ea_1 > Ea_3$ (d) $Ea_1 > Ea_2 > Ea_3$

**19** Consider the following statements related to temperature dependence of rate constants.
Identify the correct statements. **[2025, 3 April Shift-II]**

A. The Arrhenius equation holds true only for an elementary homogenous reaction.
B. The unit of $A$ is same as that of $k$ in Arrhenius equation.
C. At a given temperature, a low activation energy means a fast reaction.
D. $A$ and $E_a$ as used in Arrhenius equation depend on temperature.
E. When $E_a >> RT$, $A$ and $E_a$ become interdependent.

Choose the correct answer from the options given below.

(a) B and C only (b) B, D and E only
(c) A and B only (d) A, C and D only

**20** For $A_2 + B_2 \rightleftharpoons 2AB$
$E_a$ for forward and backward reaction are 180 and 200 kJ mol$^{-1}$ respectively. **[2025, 4 April Shift-I]**
If catalyst lowers $E_a$ for both reaction by 100 kJ mol$^{-1}$.
Which of the following statement is correct?

(a) Catalyst does not alter the Gibbs energy change of a reaction.
(b) Catalyst can cause non-spontaneous reactions to occur.
(c) The enthalpy change for the reaction is +20 kJ mol$^{-1}$.
(d) The enthalpy change for the catalysed reaction is different from that of uncatalysed reaction.

**21** Consider the given plot of enthalpy of the following reaction between $A$ and $B$.

$$A + B \longrightarrow C + D$$

Identify the incorrect statement. **[2019, 9 April Shift-II]**

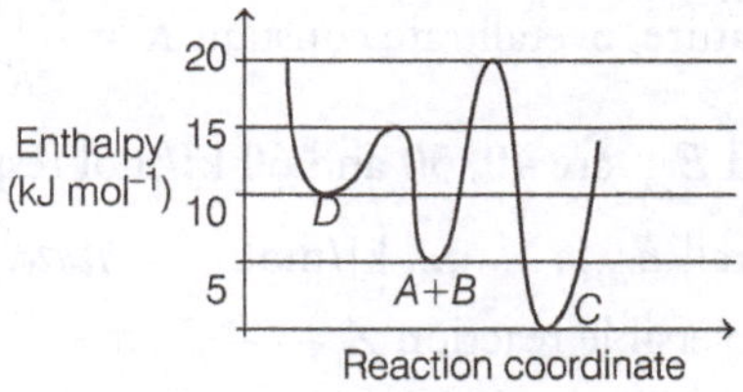


(a) $D$ is kinetically stable product.
(b) Formation of $A$ and $B$ from $C$ has highest enthalpy of activation.
(c) $C$ is the thermodynamically stable product.
(d) Activation enthalpy to form $C$ is 5 kJ mol$^{-1}$ less than that to form $D$.

**22** Two reactions $R_1$ and $R_2$ have identical preexponential factors. Activation energy of $R_1$ exceeds that of $R_2$ by 10 kJ mol$^{-1}$. If $k_1$ and $k_2$ are rate constants for reactions $R_1$ and $R_2$, respectively at 300 K, then $\ln (k_2 / k_1)$ is equal to ($R = 8.314$ J mol$^{-1}$K$^{-1}$) **[JEE Main 2017]**

(a) 8 (b) 12 (c) 6 (d) 4

**23** For the equilibrium, $A(g) \rightleftharpoons B(g)$, $\Delta H$ is –40 kJ/mol. If the ratio of the activation energies of the forward ($E_f$) and reverse ($E_b$) reactions is $\frac{2}{3}$ then: **[JEE Main 2015]**

(a) $E_f = 80$ kJ/mol ; $E_b = 120$ kJ/mol
(b) $E_f = 60$ kJ/mol ; $E_b = 100$ kJ/mol
(c) $E_f = 30$ kJ/mol ; $E_b = 70$ kJ/mol
(d) $E_f = 70$ kJ/mol ; $E_b = 30$ kJ/mol

## Numerical Value Type Questions

**24** Consider the following transformation involving first order elementary reaction in each step at constant temperature as shown below. **[2024, 4 April Shift-I]**

$$A + B \underset{\text{Step 3}}{\overset{\text{Step 1}}{\rightleftharpoons}} C \xrightarrow{\text{Step 2}} P$$

Some details of the above reactions are listed below.

| Step | Rate constant (sec$^{-1}$) | Activation energy (kJ mol$^{-1}$) |
|---|---|---|
| 1 | $K_1$ | 300 |
| 2 | $K_2$ | 200 |
| 3 | $K_3$ | $Ea_3$ |

If the overall rate constant of the above transformation ($K$) is given as $K = \frac{K_1 K_2}{K_3}$ and the overall activation energy ($E_a$) is 400 kJ $mol^{-1}$, then the value of $E_{a_3}$ is...... kJ $mol^{-1}$. (Nearest integer)

**25** For a reaction taking place in three steps at same temperature, overall rate constant $K = \frac{K_1 K_2}{K_3}$. If $E_{a_1}$, $E_{a_2}$ and $E_{a_3}$ are 40, 50 and 60 kJ/mol respectively, the overall $E_a$ is ......... kJ/mol. ***[2024, 29 Jan Shift-I]***

**26** For a reversible reaction $A \rightleftharpoons B$, the $\Delta H_{\text{forward reaction}} = 20$ kJ $mol^{-1}$. The activation energy of the uncatalysed forward reaction is 300 kJ $mol^{-1}$. When the reaction is catalysed keeping the reactant concentration same, the rate of the catalysed forward reaction at 27°C is found to be same as that of the uncatalysed reaction at 327°C is found to be same as that of the uncatalysed reaction at 327°C. The activation energy of the catalysed backward reaction is ........... kJ $mol^{-1}$. ***[2023, 15 April Shift-I]***

**27** The number of given statement/s which is/are correct is ***[2023, 8 April Shift-I]***

(A) The stronger the temperature dependence of the rate constant, the higher is the activation energy.
(B) If a reaction has zero activation energy, its rate is independent of temperature.
(C) The stronger the temperature dependence of the rate constant, the smaller is the activation energy.
(D) If there is no correlation between the temperature and the rate constant then it means that the reaction has negative activation energy.

**28** $A \rightarrow B$

The rate constants of the above reaction at 200 K and 300 K are 0.03 $min^{-1}$ and 0.05 $min^{-1}$ respectively. The activation energy for the reaction is ........ J. (Nearest integer)[Given : ln 10 = 2.3, $R = 8.3$ J $K^{-1}$ $mol^{-1}$, log 5 = 0.70, log 3 = 0.48, log 2 = 0.30] ***[2023, 31 Jan Shift-I]***

**Imp.** **Characteristic of Activation Energy**

**29** The number of correct statements from the following is .......... ***[2023, 24 Jan Shift-I]***

(A) larger the activation energy, smaller is the value of the rate constant.
(B) the higher is the activation energy, higher is the value of the temperature coefficient.
(C) at lower temperature, increase in temperature causes more change in the value of $k$ than at higher temperature.
(D) a plot of ln $k$ vs $\frac{1}{T}$ is a straight line with slope equal to $-\frac{E_a}{R}$

**30** For a reaction, given below is the graph of ln $k$ vs $1/T$. The activation energy for the reaction is equal to ........... cal $mol^{-1}$. (Nearest integer) (Given, $R = 2$ cal $K^{-1}$ $mol^{-1}$)

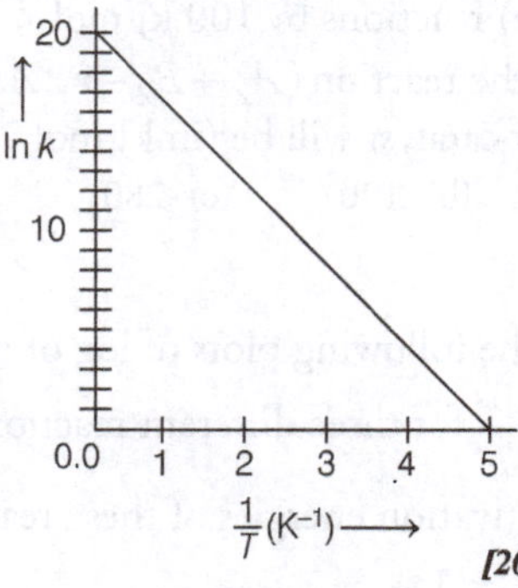


***[2022, 28 July Shift-II]***

**31** For the reaction $P \rightarrow B$, the values of frequency factor $A$ and activation energy $E_A$ are $4 \times 10^{11} s^{-1}$ and 8.3 kJ $mol^{-1}$ respectively. If the reaction is of first order, the temperature at which the rate constant is $2 \times 10^{-6} s^{-1}$ is ......... $\times 10^{-1}$ K. (Given, ln 10 = 2.3, $R = 8.3$ J $K^{-1}$ $mol^{-1}$, log 2 = 0.30) ***[2022, 30 June Shift-I]***

**32** It has been found that for a chemical reaction with rise in temperature by 9K the rate constant gets doubled. Assuming a reaction to be occurring at 300 K, the value of activation energy is found to be ......... kJ $mol^{-1}$. [Nearest integer] (Given, ln 10 = 2.3, $R = 8.3$ J$K^{-1}$$mol^{-1}$, log 2 = 0.30) ***[2022, 27 June Shift-II]***

**33** The rate constant for a first order reaction is given by the following equation

$$\ln k = 33.24 - \frac{2.0 \times 10^4 \text{ K}}{T}$$

The activation energy for the reaction is given by ...... kJ $mol^{-1}$. [In nearest integer] (Given, $R = 8.3$ J $K^{-1}$ $mol^{-1}$) ***[2022, 27 June Shift-I]***

**34** Catalyst $A$ reduces the activation energy for a reaction by 10 kJ $mol^{-1}$ at 300 K. The ratio of rate constants, $\frac{K', \text{catalysed}}{K, \text{uncatalysed}}$ is $e^x$. The value of $x$ is ...... [Nearest integer]
[Assume that, the pre-exponential factor is same in both the cases. Given, $R = 8.31$ J$K^{-1}$ $mol^{-1}$] ***[2022, 26 June Shift-II]***

**35** The rate constants for decomposition of acetaldehyde have been measured over the temperature range 700-1000 K. The data has been analysed by plotting In $k$ $vs \frac{10^3}{T}$ graph.

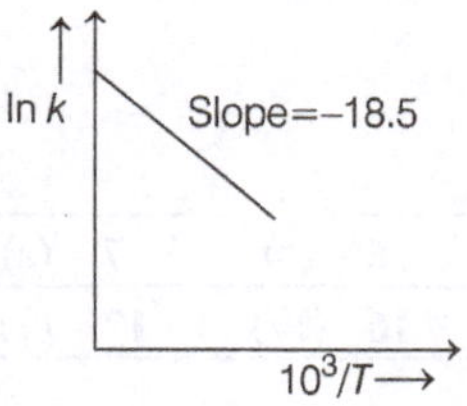


The value of activation energy for the reaction is ...... kJ $mol^{-1}$. [Nearest integer]
(Given, $R = 8.31$ J $K^{-1}$ $mol^{-1}$ ) *[2022, 24 June Shift-I]*

**36** For the reaction $A \rightarrow B$, the rate constant $k$ (in $s^{-1}$) is given by $\log_{10} k = 20.35 - \frac{(2.47 \times 10^3)}{T}$.
The energy of activation in kJ $mol^{-1}$ is ......... (Nearest integer)
[Given : $R = 8.314$ J $K^{-1}$ $mol^{-1}$] *[2021, 31 Aug Shift-II]*

**37** The first order rate constant for the decomposition of $CaCO_3$ at 700 K is $6.36 \times 10^{-3}$ $s^{-1}$ and activation energy is 209 kJ $mol^{-1}$. Its rate constant (in $s^{-1}$) at 600 K is $x \times 10^{-6}$. The value of $x$ is .......... . (Nearest integer)
[Given, $R = 8.31$ $JK^{-1}mol^{-1}$, $\log 6.36 \times 10^{-3} = -2.19$, $10^{-4.79} = 1.62 \times 10^{-5}$] *[2021, 27 Aug Shift-II]*

**38** The decomposition of formic acid on gold surface follows first order kinetics. If the rate constant at 300K is $1.0 \times 10^{-3}$ $s^{-1}$ and the activation energy, $E_a = 11.488$ kJ $mol^{-1}$, the rate constant at 200 K is ...... $\times 10^{-5} s^{-1}$ (Round off to the nearest integer).
[Given, $R = 8.314$ J $mol^{-1}$ $K^{-1}$] *[2021, 16 March Shift-I]*

**39** If the activation energy of a reaction is 80.9 kJ $mol^{-1}$, the fraction of molecules at 700 K, having enough energy to react to form products is $e^{-x}$. The value of $x$ is ......... . (Rounded off to the nearest integer)
[Use, $R = 8.31$ $JK^{-1}$ $mol^{-1}$]. *[2021, 26 Feb Shift-II]*

**40** An exothermic reaction $X \rightarrow Y$ has an activation energy 30 kJ $mol^{-1}$. If energy change $\Delta E$ during the reaction is –20 kJ $mol^{-1}$, then the activation energy for the reverse reaction in kJ is ............ . *[2021, 26 Feb Shift-I]*

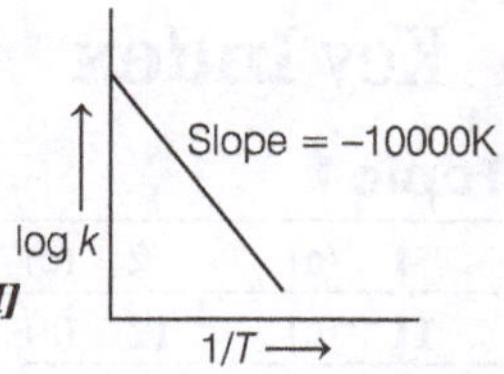


**41** The rate constant of a reaction increases by five times on increase in temperature from 27°C to 52°C. The value of activation energy in kJ $mol^{-1}$ is ........ (Rounded off to the nearest integer)
[$R = 8.314$ J $K^{-1}$ $mol^{-1}$] *[2021, 25 Feb Shift-II]*

**42** For the reaction, $aA + bB \rightarrow cC + dD$, the plot of $\log k$ $vs \frac{1}{T}$ is given below
The temperature at which the rate constant of the reaction is $10^{-4}$ $s^{-1}$ is ......... K
(Rounded off to the nearest integer).
[Given, The rate constant of the reaction is $10^{-5}$ $s^{-1}$at 500 K] *[2021, 25 Feb Shift-I]*

**43** The rate of a reaction decreased by 3.555 times when the temperature was changed from 40° C to 30° C. The activation energy (in kJ $mol^{-1}$) of the reaction is ......... .
(Take; $R = 8.314$ J $mol^{-1}K^{-1}$, $\ln 3.555 = 1.268$) *[2020, 6 Sep Shift-II]*

**44** The number of molecules with energy greater than the threshold energy for a reaction increases five fold by a rise of temperature from 27 °C to 42 °C. Its energy of activation in J/mol is ......... .
(Take $\ln 5 = 1.6094$; $R = 8.314$ J $mol^{-1}$ $K^{-1}$) *[2020, 4 Sep Shift-II]*

**45** A sample of milk splits after 60 min at 300 K and after 40 min. at 400 K when the population of *lactobacillus acidophilus* in it doubles. The activation energy (in kJ/mol) for this process is closest to
(Given, $R = 8.3$ J $mol^{-1}K^{-1}$
$\ln\left(\frac{2}{3}\right) = 0.4$, $e^{-3} = 4.0$) *[2020, 9 Jan Shift-II]*

## • Key Index

### Topic 1

| | | | | | | | | | |
|---|---|---|---|---|---|---|---|---|---|
| 1 (a) | 2 (c) | 3 (d) | 4 (a) | 5 (c) | 6 (b) | 7 (a) | 8 (b) | 9 (d) | 10 (b) |
| 11 (c) | 12 (a) | 13 (d) | 14 (3) | 15 (8) | 16 (17) | 17 (2) | 18 (1) | 19 (4) | 20 (4) |
| 21 (27) | | | | | | | | | |

### Topic 2

| | | | | | | | | | |
|---|---|---|---|---|---|---|---|---|---|
| 1 (c) | 2 (c) | 3 (b) | 4 (c) | 5 (c) | 6 (d) | 7 (c) | 8 (c) | 9 (a) | 10 (b) |
| 11 (b) | 12 (d) | 13 (a) | 14 (c) | 15 (b) | 16 (d) | 17 (c) | 18 (d) | 19 (b) | 20 (b) |
| 21 (b) | 22 (d) | 23 (a) | 24 (d) | 25 (c) | 26 (c) | 27 (b) | 28 (c) | 29 (b) | 30 (c) |
| 31 (c) | 32 (b) | 33 (a) | 34 (a) | 35 (b) | 36 (43) | 37 (2435) | 38 (897) | 39 (1) | 40 (17) |
| 41 (3) | 42 (315) | 43 (3) | 44 (50) | 45 (17190) | 46 (399) | 47 (24) | 48 (63) | 49 (10) | 50 (2) |
| 51 (2200) | 52 (3) | 53 (3) | 54 (1) | 55 (2) | 56 (75) | 57 (15) | 58 (17) | 59 (1350) | 60 (623) |
| 61 (2) | 62 (2) | 63 (2) | 64 (60) | 65 (1) | 66 (40) | 67 (100) | 68 (165) | 69 (2) | 70 (2) |
| 71 (2) | 72 (16) | 73 (200) | 74 (0) | 75 (2) | 76 (1) | 77 (656) | 78 (106) | 79 (10) | 80 (3) |
| 81 (2) | 82 (108) | 83 (81) | 84 (26) | 85 (60) | 86 (23.03) | | | | |

### Topic 3

| | | | | | | | | | |
|---|---|---|---|---|---|---|---|---|---|
| 1 (a) | 2 (d) | 3 (c) | 4 (a) | 5 (b) | 6 (b) | 7 (20) | 8 (69) | 9 (c) | 10 (d) |
| 11 (d) | 12 (b) | 13 (d) | 14 (c) | 15 (a) | 16 (a) | 17 (d) | 18 (c) | 19 (a) | 20 (a) |
| 21 (d) | 22 (d) | 23 (a) | 24 (100) | 25 (30) | 26 (130) | 27 (2) | 28 (2520) | 29 (3) | 30 (8) |
| 31 (225) | 32 (59) | 33 (166) | 34 (4) | 35 (154) | 36 (47) | 37 (16) | 38 (10) | 39 (14) | 40 (50) |
| 41 (52) | 42 (526) | 43 (100) | 44 (84297.47) | 45 (3.98) | | | | | |

# Detailed Explanations

### Topic 1

**01.** *(a)* $r_1 = k[A]^n[B]^m$

Given that, $A$ is doubled and $B$ is halved in concentration.

So, $r_2 = k2^n[A]^n \dfrac{[B]^m}{2^m} \Rightarrow \dfrac{r_2}{r_1} = 2^{n-m}$

**02.** *(c)* The rate of reaction is given as

$$R_1 = K[A]^1[B]^1$$

$$R_1 = K\left[\frac{n_A}{V}\right]^1\left[\frac{n_B}{V}\right]^1$$

Now, for $R_2$ rate is given as

$$R_2 = K\left[\frac{3n_A}{V}\right]^1\left[\frac{3n_B}{V}\right]^1 = 9R_1$$

**03.** *(d)* Rate of reaction

$$\frac{-1}{2}\frac{d[A]}{dt} = \frac{-1}{3}\frac{d[B]}{dt} = \frac{-2}{3}\frac{d[C]}{dt}$$

$$= \frac{1}{3}\frac{d[P]}{dt}$$

$$\frac{dn_A}{dt} = \frac{2}{3}\frac{dn_B}{dt} = \frac{4}{3}\frac{dn_C}{dt}$$

**04.** *(a)* For the reaction,

$$2H_2(g) + 2NO(g) \longrightarrow N_2(g) + 2H_2O(g)$$

Rate law for forward reaction is given as:

$$\text{rate} = K_f\,[NO]^2\,[H_2]^2$$

Rate of forward reaction is equal to rate of reverse reaction at equilibrium and also,

at equilibrium $(K) = \dfrac{K_f}{K_b}$

where, $K$ (equilibrium constant)

$$= \frac{[N_2]\,[H_2O]^2}{[H_2]^2\,[NO]^2}$$

$$\Rightarrow \frac{K_f}{K_b} = \frac{[N_2]\,[H_2O]^2}{[H_2]^2\,[NO]^2}$$

On rearranging :

$$K_f[H_2][NO]^2 = K_b[N_2][H_2O]^2/[H_2]$$

At Equilibrium, $r_f = r_b$

∴ Rate of reverse reaction

$$= K_b\,[N_2]\,[H_2O]^2/[H_2]$$

**05.** *(c)* Rate $(R_1) = k\,[A]\,[B]$ ...(i)

Rate $(R_2) = k\,[A]\,[2B]$ ...(ii)

On dividing Eq. (ii) by (i)

$$\frac{R_2}{R_1} = 2$$

$$\Rightarrow \quad R_2 = 2R_1$$

**06.** *(b)* For the reaction,

$$\frac{1}{2}A \longrightarrow 2B$$

Rate of reaction $= -\dfrac{2d[A]}{dt} = \dfrac{1}{2}\dfrac{d[B]}{dt}$

Rate of disappearance of $A = -\frac{d[A]}{dt}$

$$= \frac{1}{2\times 2}\frac{d[B]}{dt}$$

$$= \frac{1}{4}\frac{d[B]}{dt}$$

**07.** *(a)* Given reaction,

$$A \xrightarrow{K_1} B \xrightarrow{K_2} C$$

Rate of formation of $B$ is,

$$\frac{d[B]}{dt} = K_1[A] - K_2[B]$$

$$0 = K_1[A] - K_2[B]$$

$$\left[\frac{K_1}{K_2}\right][A] = [B]$$

**08.** *(b)*

**key idea** The rate of formation of product i.e. $NO_2$ can be calculated by equating the rate of disappearance of reactant i.e. $N_2O_5$ with rate of appearance of product, $NO_2$.

$$2N_2O_5(g) \longrightarrow 4NO_2(g) + O_2(g)$$

$t = 0$ 3.0 M

$t = 30$ 2.75 M

From the equation, rate of reaction can be concluded that

$$\frac{1}{2}\times\frac{-\Delta[N_2O_5]}{\Delta t} = \frac{1}{4}\times\frac{\Delta[NO_2]}{\Delta t}$$

and $\frac{\Delta[NO_2]}{\Delta t} = -2\frac{\Delta(N_2O_5)}{\Delta t}$

$$\Rightarrow \frac{\Delta[NO_2]}{\Delta t} = 2\times\frac{0.25}{30}$$

$$= 1.667\times 10^{-2}\ \text{mol L}^{-1}\ \text{min}^{-1}$$

**09.** *(d)* For more than one elemental step, the slowest step is the rate determining step. Thus, in case (I), rate law is given as rate $= k[Cl_2][H_2S]$

While for the reaction given in case (II), rate law is given as rate

$$= k[Cl_2][HS^-].$$

Hence, only mechanism (I) is consistent with the given rate equation.

**10.** *(b)* Before applying medicine,

Rate law for first order growth.

$$\frac{dA}{dt} = k[A]$$

$$\frac{A}{A_0} = \frac{N}{N_0} = e^{kt}$$

After applying medicine

Active bacteria $\longrightarrow$ Inactive bacteria

$(A)$ $(I)$

$$r = \frac{-dA}{dt} = k[A]^2 \text{ (Rate law)}$$

$$\gamma = kx^2$$

Hence, a parabolic curve will form.

**11.** *(c)*

(i) Rate of the reaction,

$$A_2 \underset{K_{-1}}{\overset{K_1}{\rightleftharpoons}} 2A$$

$$r = r_{\text{forward}} - r_{\text{backward}}$$
$$= k_1[A_2] - k_{-1}[A]^2 \quad \dots \text{(i)}$$

(ii) Also differential, rate of the reaction can be expressed as,

$$r = -\frac{d[A_2]}{dt} = +\frac{1}{2}\frac{d[A]}{dt}$$

So, the rate of appearance of $A$,

i.e. $\frac{d[A]}{dt} = 2r$

[From Eq. (i)]

$$\frac{d[-A]}{dt} = 2k_1[A_2] - 2k_{-1}[A]^2$$

**12.** *(a)* The expression for bacterial growth is

$$N = N_0 e^t$$

From 0 to 1 hour $N'(t) = N_0 e^t$

From 1 hour onwards, $\frac{dN}{dt} = -5N^2$

On differentiating the above equation from $N'$ to $N$ we get.

$$\int_{eN_0}^{N} N^{-2}dN = -5\int_1^t dt$$

$[\because$ At 1 hour, $N' = eN_0]$

$$\left[\frac{1}{N} - \frac{1}{eN_0}\right] = 5(t-1)$$

Multiply both sides by $N_0$, we get

$$\frac{N_0}{N} - \frac{1}{e} = 5N_0(t-1)$$

$$\frac{N_0}{N} = 5N_0 t + \left[\frac{1}{e} - 5N_0\right]$$

On comparing the above equation with equation of straight line, $y = mx + c$

We get $m = 5N_0 \Rightarrow c = \frac{1}{e} - 5N_0$

$\therefore$ Plot of $\frac{N_0}{N}$ *vs* $t$ is shown in option (a).

**13.** *(d)* Rate law equation is given by,

$$\frac{dC}{dt} = k[A]^x[B]^y$$

$\therefore \frac{dC}{dt}$ for $C$ at various concentrations are

$$1.2\times 10^{-3} = k(0.1)^x(0.1)^y \quad \dots \text{(i)}$$
$$1.2\times 10^{-3} = k(0.1)^x(0.2)^y \quad \dots \text{(ii)}$$
$$2.4\times 10^{-3} = k(0.2)^x(0.1)^y \quad \dots \text{(iii)}$$

Solving for $x$ [divide Eq. (iii) by Eq. (i)]

$$\frac{2.4\times 10^{-3}}{1.2\times 10^{-3}} = \frac{k(0.2)^x(0.1)^y}{k(0.1)^x(0.1)^y}$$

$$2 = (2)^x$$

$$\Rightarrow x = 1$$

Solving for y [divide Eq. (iii) by Eq. (ii)]

$$\frac{2.4\times 10^{-3}}{1.2\times 10^{-3}} = \frac{k(0.2)^x(0.1)^y}{k(0.1)^x(0.2)^y}$$

$$\Rightarrow 2 = (2)^x\left(\frac{0.1}{0.2}\right)^y$$

Put $x = 1$ in above Eq., we get

$$2 = (2)^1\left(\frac{0.1}{0.2}\right)^y$$

$$\Rightarrow 1 = (0.5)^y$$

$$(1)^0 = (0.5)^y \Rightarrow y = 0$$

Thus, $\frac{dC}{dt} = k[A]^1[B]^0$

$$\Rightarrow \frac{dC}{dt} = k[A]^1$$

**14.** *(3)* The given reaction is

$$A(g) \longrightarrow 2B(g) + C(g)$$

| | | | |
|---|---|---|---|
| At time $t = 0$ | $a - x$ | $2x$ | $x$ |
| On completion | 0 | $2a$ | $a$ |

Given that, at 23 seconds

$(p)_t = a - x + 2x + x = 200$ torr ...(i)

at completion $2a + a = 300$ torr ...(ii)

$$3a = 300 \text{ torr}$$

$$[p]_0 = a = 100 \text{ torr}$$

From Eq. (i),

$$a + 2x = 200$$

$$100 + 2x = 200$$

$$\Rightarrow x = 50 \text{ torr}$$

Now, $K = \frac{2.303}{t}\log\frac{a}{a-x}$

$$= \frac{2.303}{23}\log\frac{100}{100-50}$$

$$K = 3\times 10^{-2}\ s^{-1}$$

**15.** *(8)* Given $r = k[A]^2[B]$

If concentration is doubled

$$r' = k[2A]^2[2B]^1$$

$$r' = 8r \Rightarrow x = 8$$

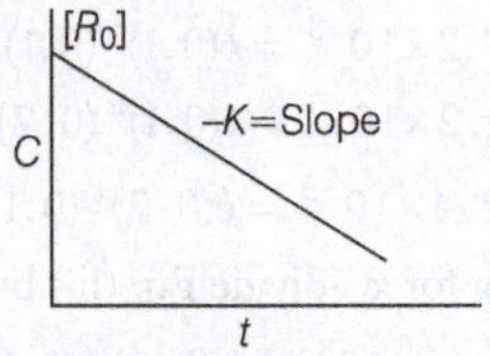


From the graph, the concentration of reactant $R$ decrease linearly with time, which indicates a zero order reaction thus, $y = 0$

So $x + y = 8$

**16.** *(17)* For reaction,

$$2N_2O_5(g) \rightleftharpoons 4NO_2(g) + O_2(g)$$

Rate of reaction (ROR)

$$= -\frac{1}{2}\frac{\Delta[N_2O_5]}{\Delta t}$$

$$= \frac{1}{4}\frac{\Delta[NO_2]}{\Delta t} = \frac{\Delta[O_2]}{\Delta t}$$

$$\text{ROR} = -\frac{1}{2}\frac{\Delta[N_2O_5]}{\Delta t}$$

$$= -\frac{1}{2}\frac{(2.75-3)}{30}\ \text{mol L}^{-1}\text{min}^{-1}$$

$$\text{ROR} = -\frac{1}{2}\frac{(-0.25)}{30}$$

$$= \frac{1}{240}\ \text{mol L}^{-1}\ \text{min}^{-1}$$

Rate of formation of $NO_2$

$$= \frac{\Delta[NO_2]}{\Delta t} = 4 \times \text{ROR}$$

$$= \frac{4}{240}$$

$$= 16.66 \times 10^{-3}\ \text{mol L}^{-1}\ \text{min}^{-1}$$

$$\approx 17 \times 10^{-3}\ \text{mol L}^{-1}\ \text{min}^{-1}$$

**17.** *(2)*

$$A(g) \longrightarrow 2B(g) + C(g)$$

| | | | |
|---|---|---|---|
| At $t = 0$ | $p_0$ | 0 | 0 |
| $t = t$ | $p_0 - x$ | $2x$ | $x$ |

Total pressure after time $t$,

$$p_t = p_0 - x + x + 2x$$
$$= p_0 + 2x$$
$$p_t = 0.1 + 2x$$
$$x = \frac{0.28 - 0.1}{2} = 0.09$$
$$k = \frac{2.303}{t}\log\frac{p_0}{p_0 - x}$$
$$= \frac{2.303}{115}\log\frac{0.1}{0.1 - 0.09}$$
$$= \frac{2.303}{115}\log 10$$
$$= 0.020 = 2 \times 10^{-2}\ \text{s}^{-1}$$

Thus, the nearest integer is 2.

**18.** *(1)* Rate of reaction,

$$= -\frac{1}{\gamma_1}\frac{dA}{dt} = -\frac{1}{\gamma_2}\frac{dB}{dt}$$

$$= \frac{1}{Y_3}\frac{dC}{dt} = \frac{1}{\gamma_4}\frac{dD}{dt}$$

$$= -\frac{1}{\gamma_1}\times 3 = -\frac{1}{\gamma_2}\times 6$$

$$= \frac{1}{\gamma_3}\times 1 = \frac{1}{\gamma_4}\times 9$$

Let us assume $\gamma_1 = 1$ then

$$\gamma_2 = 2;\ \gamma_3 = \frac{1}{3};\ \gamma_4 = 3$$

Substituting in the reaction,

$$1A + 2B \longrightarrow \frac{1}{3}C + 3D$$

Multiplying the coefficients by 3 to remove fraction 1/3 gives values of $\gamma_1 = 3;\ \gamma_2 = 6;\ \gamma_3 = 1$ and $\gamma_4 = 9$

$$3A + 6B \longrightarrow C + 9D$$

Rate of reaction

$$= \frac{1}{\gamma_4}\frac{dD}{dt} = \frac{1}{9}\times 9\ \text{m mol dm}^{-3}\ \text{s}^{-1}$$

$$= 1\ \text{m mol dm}^{-3}\ \text{s}^{-1}$$

**19.** *(4)* For the given reaction,

$$\text{rate of reaction} = -\frac{1}{2}\frac{d[K_2Cr_2O_7]}{dt}$$

$$= \frac{-1}{8}\frac{d[H_2SO_4]}{dt} = \frac{-1}{3}\frac{d[C_2H_6O]}{dt}$$

$$= \frac{1}{2}\frac{d[Cr_2(SO_4)_3]}{dt} = \frac{1}{3}\frac{d[C_2H_4O_2]}{dt}$$

$$= \frac{1}{2}\frac{d[K_2SO_4]}{dt} = \frac{1}{11}\frac{d[H_2O]}{dt}$$

Rate of appearence of 2 moles of $Cr_2(SO_4)_3$ is given as

$$\frac{d}{dt}[Cr_2(SO_4)_3] = 2.67\ \text{mol/min}$$

Rate of disappearence of $C_2H_6O$ can be calculated as

$$\frac{1}{3}\frac{d[C_2H_6O]}{dt} = \frac{+1}{2}\frac{d[Cr_2(SO_4)_3]}{dt}$$

$$= \frac{3}{2}\frac{d[Cr_2(SO_4)_3]}{dt}$$

$$= \frac{2.67}{2}\times 3$$

$$= 4.005 = 4\ \text{mol/min}$$

**20.** *(4)* Reaction, $A \longrightarrow B$

| | | | |
|---|---|---|---|
| At | $t = 0$ | | 0 |
| At | $t = 30$ min | | 0.2 M |

Average rate of reaction

$$= -\frac{\Delta[A]}{\Delta t} = \frac{\Delta[B]}{\Delta t} = \frac{(0.2-0)}{\frac{1}{2}}$$

$$= 0.4\ \text{mol L}^{-1}\ \text{h}^{-1}$$
$$= 4 \times 10^{-1}\ \text{mol L}^{-1}\ \text{h}^{-1}$$

**21.** *(27)* Given, the reaction $2A + B_2 \longrightarrow 2AB$ is an elementary reaction.

So, rate of reaction $= k[A]^2[B_2]$

Initial rate $= k(a/V)^2(b/V)^1$

where, $a$ and $b$ are the number of moles of $A$ and $B$ respectively and $V$ is the volume.

On reducing the volume by a factor of 3, the concentrations of A and $B_2$ will become 3 times

Final rate $= k(3a/V)^2(3b/V)^1$

$$3^2 \times 3k\left(\frac{a}{V}\right)^2\left(\frac{b}{V}\right)^1 = 27 \times \text{initial rate}$$

Hence, the rate becomes = 27 times of initial rate.

## Topic 2

**01.** *(c)* For zero order reaction,

$$A \longrightarrow P,\ \text{rate} = k$$
$$[A]_t = [A]_0 - kt$$

So, the most suitable graph is given in option (c).

i.e.,

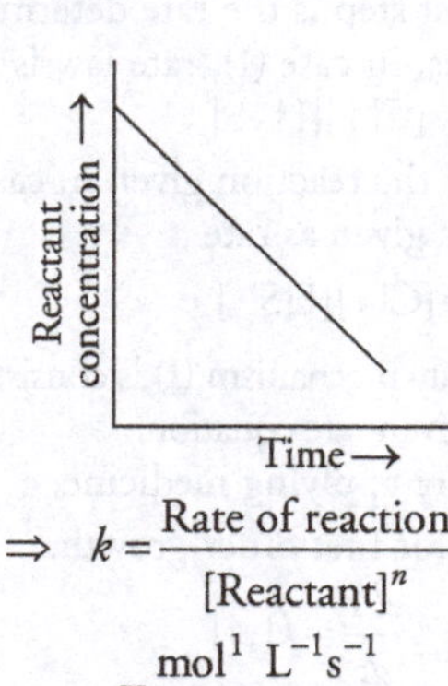


**02.** *(c)* $\Rightarrow k = \dfrac{\text{Rate of reaction}}{[\text{Reactant}]^n}$

$$= \frac{\text{mol}^1\ \text{L}^{-1}\text{s}^{-1}}{[\text{mol}^n/\text{L}^n]}$$

$$= \text{mol}^{1-n}\text{L}^{n-1}\text{s}^{-1}$$

Therefore, option (c) is correct.

**03.** *(b)* The given graph is of zero order reaction.

**04.** *(c)* Half-life $=6.93$ min,

Half-life of a Ist order reaction

$$k=\frac{0.693}{t_{1/2}}$$

$$k=\frac{0.693}{6.93}=0.1\ \text{min}^{-1}$$

Rate constant for the first order reaction is given by

$$k=\frac{2.303}{t}\log\frac{[A_0]}{[A]}$$

$$0.1=\frac{2.303}{t}\times\log\frac{100}{1}$$

$$0.1=\frac{2.303}{t}\log 10^2$$

$$t=\frac{2.303\times 2}{0.1}\times 2\log 10$$

$$=46.06\ \text{min}$$

**05.** *(c)* Rate of the reaction $=k[CO]^2$

If the concentration of carbon monoxide is double, the rate of reaction becomes

$$r'=k[2CO]^2$$

$$r'=4k\,[CO_2]^2 \Rightarrow r'=4r.$$

**06.** *(d)* For 1st order reaction,

When concentration is $C_t$ at given time $t$

$$C_t=\frac{C_0}{4}$$

$$t_1=2_{t_{50\%}}$$

When $\quad C_t=C_0/8$

$$t_2=3t_{50\%}$$

$$\therefore\quad \frac{t_1}{t_2}=\frac{2}{3}$$

**07.** *(c)* Given first order reaction is,

$$A(g)\rightarrow B(g)+C(g)$$

| | | | |
|---|---|---|---|
| $t=0$ | $p^\circ$ | 0 | 0 |
| $t=t$ | $p^\circ-x$ | $x$ | $x$ |
| $t=\infty$ | 0 | $p^\circ$ | $p^\circ$ |

$$p_t=p^\circ+x\Rightarrow x=p_t-p^\circ=p_t-\frac{p_\infty}{2}$$

$$p_\infty=2p^\circ\Rightarrow p^\circ=\frac{p_\infty}{2}$$

$$k=\frac{1}{t}\ln\frac{p^\circ}{p^\circ-x}$$

$$\Rightarrow\quad k=\frac{1}{t}\ln\frac{p_\infty}{2(p_\infty-p_t)}$$

**08.** *(c)* For zero order reaction, half-life is given by

$$\text{half-life}=\frac{A_0}{2k}$$

$$60=\frac{2}{2k}\Rightarrow k=\frac{1}{60}\text{M/ min}$$

Now, we know that $A_t=A_0-kt$

$$\Rightarrow\quad t=\frac{A_0-A_t}{k}=\frac{0.5-0.25}{1/60}=15\ \text{min}$$

$$\Rightarrow\quad t=15\ \text{min}$$

**09.** *(a)* Given, $\quad [A_0]=8[B_0]$

$$[t_{1/2}]_A=10\ \text{min}$$

$$[t_{1/2}]_B=40\ \text{min}$$

$A$ and $B$ follows first order kinetics

$$t=?$$

$$[A]_t=[B]_t$$

$$-k_At=-k_Bt$$

or $$\frac{[A_0]}{[B_0]}=e^{(k_A-k_B)t}$$

$$8=e^{(k_A-k_B)t}$$

$$\ln 8=(k_A-k_B)t$$

$$\ln 8=\ln 2\left(\frac{1}{(t_{1/2})_A}-\frac{1}{(t_{1/2})_B}\right)\times t$$

$$2.08=0.693\left(\frac{1}{10}-\frac{1}{40}\right)t$$

$$\Rightarrow\quad t=40\ \text{min}$$

**10.** *(b)* Given, $[A]=16$ mg/mL

$[A_0]=4$ mg/mL, $T=12$ months

Using the formula of rate constant for 1st order reaction.

$$k=\frac{2.303}{t}\log\frac{A}{[A_0]}\quad\ldots\text{(i)}$$

$$=\frac{2.303}{12}\log\frac{16}{4}$$

On solving, we get

$$k=0.1155$$

Now, since 50% decomposition takes place. So,

$$[A]=[A_0]-\frac{50}{100}[A_0]$$

$$[A]=0.5[A_0]$$

Now, substitute the obtained values in Eq. (i) to calculate the expiry time

$$0.1155=\frac{2.303}{t}\log\frac{[A_0]}{[0.5A_0]}$$

$$t=6\ \text{months}$$

**11.** *(b)* The rate of reaction is given by

$$\text{Rate}=k_2[A][B_2]\quad\ldots\text{(i)}$$

Also $$\left(\frac{k_1}{k_{-1}}\right)=\left(\frac{[A]^2}{[A_2]}\right)$$

Or $$[A]=\sqrt{\frac{k_1}{k_{-1}}}\sqrt{A_2}$$

Substituting the value of $[A]$ in Eq (i),

$$\text{rate}=k_2\sqrt{\frac{k_1}{k_{-1}}}.[A_2]^{1/2}[B_2]$$

So, the order will be given by $\frac{1}{2}+1=\frac{3}{2}$

$$\Rightarrow\quad \text{Order}=1.5$$

**12.** *(d)* Statement given in option (d) is not true for radioactive decay. Its correct form is decay constant is independent of temperature.

**13.** *(a)* Statement I is true but statement II is false.

The correct explanation for statement II is :

For first order reaction, $t_{1/2}=\frac{\ln 2}{k}$

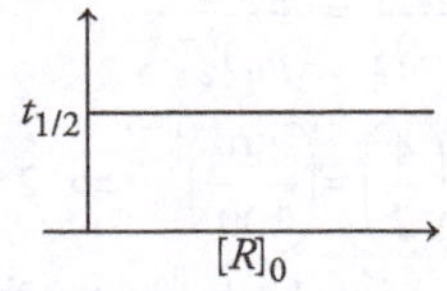


For first order reaction,

$$\log\frac{[R]_0}{[R]}=\frac{1}{2.303}kt$$

$$\log\frac{[R]_0}{[R]}=\left[\frac{k}{2.303}\right]t$$

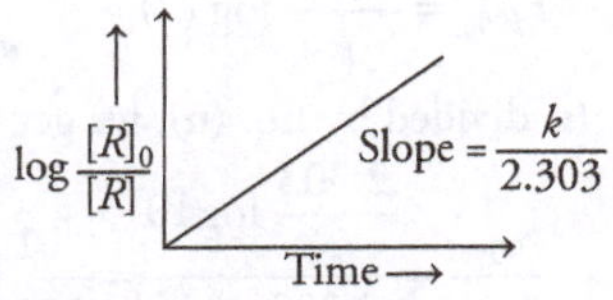


**14.** *(c)* Initial number of bacteria $=N_0$

Number of bacteria at any given time $=N$

Bacterial growth is given as

$$N=N_0e^{kt}$$

Where, $k=$ growth constant for bacterial growth.

The most suitable graph to show bacterial growth is

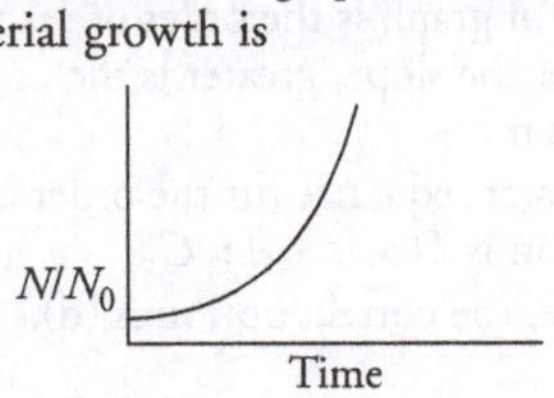

**15.** (b) $A(g) \longrightarrow B(g) + C(g)$

At $t = 0$ $\quad p_i \quad 0 \quad 0$

At $t = t$ $\quad p_i - x \quad x \quad x$

$$p_t = p_i - x + x + x$$
$$p_t = p_i + x$$
$$p_i - x = p_i - p_t + p_i$$
$$= 2p_i - p_t$$

So, $k = \dfrac{2.303}{t} \log \dfrac{p_i}{(2p_i - p_t)}$

**16.** (d)

**key idea** In this question, we will use the concept of half life. It is the time in which the concentration of a reactant is reduced to one half of its initial concentration.

$$AB_3(g) \longrightarrow A(g) + 3B(g)$$

As decomposition reaction of $AB_3(g)$ is not given, we assume that $p$ (mm Hg) is for $AB_3(g)$ only.

So, half-life $t_{1/2} \propto (p_0)^{1-n}$

$$\frac{(t_{1/2})_1}{(t_{1/2})_2} = \frac{(p_0)_1^{1-n}}{(p_0)_2^{1-n}}$$

$$\Rightarrow \left(\frac{4}{2}\right) = \left(\frac{50}{100}\right)^{1-n} \Rightarrow 2 = \left(\frac{1}{2}\right)^{1-n}$$

$$\Rightarrow \quad n = 1 + 1 \Rightarrow n = 2$$

Therefore, the order of reaction is 2.

**17.** (c) Given, $t_{90\%} = x\, t_{50\%}$

$$t_{90\%} = \frac{2.303}{k} \log\left(\frac{100}{100-90}\right)$$

$$t_{50\%} = \frac{2.303}{k} \log(2) \qquad \ldots\text{(ii)}$$

Eq. (i) divided by Eq. (ii), we get

$$\frac{t_{90\%}}{t_{50\%}} = \frac{\frac{2.303}{k}\log 10}{\frac{2.303}{k}\log 2} = \frac{1}{0.3010}$$

So, $\dfrac{t_{90\%}}{t_{50\%}} = 3.32 \Rightarrow x = 3.32$

**18.** (d) We know that,

Rate = $k$ (concentration)$^n$

Taking log both side,

log (Rate) = log $k$ + $n$ log [concentration]

Slope of graph is the order of reaction greater the slope, greater is the order of reaction.

∴Correct sequence for the order of reaction is $D > B > A > C$.

Hence, the correct option is (d).

**19.** (b) Only statement (b) is true whereas all other statements are incorrect.

**20.** (b) For zero order reaction,

$$[A_0] - [A_t] = kt \qquad \ldots\text{(i)}$$

Also, for zero order reaction, $t_{1/2} = \dfrac{[A_0]}{2k}$

$$\therefore \quad 6 = \frac{0.2}{2k}$$

or $\quad k = \dfrac{0.2}{2 \times 6} = \dfrac{1}{60}$

Now, from Eq. (i)

$$[A_0] - [A_t] = kt$$

Given, $\quad [A_0] = 0.5$ M, $[A_t] = 0.2$ M

$$\therefore \quad 0.5 - 0.2 = \frac{1}{60} \times t \qquad \left[\because k = \frac{1}{60}\right]$$

$$0.3 = \frac{1}{60} \times t$$

$$\Rightarrow \quad t = 0.3 \times 60 = 18\text{h}$$

**21.** (b) Rate constant for the first order reaction is given by

$$k = \frac{2.303}{t} \log\left(\frac{a}{a-x}\right)$$

$$a = 0.1 \text{ M}$$
$$(a - x) = 0.025 \text{ M}, \quad t = 40 \text{ min}$$

$$\therefore \quad k = \frac{2.303}{40} \log \frac{0.1}{0.025 \text{ M}}$$

$$= 0.0347 \text{ min}^{-1}$$

$$\text{Rate} = \left(\frac{dx}{dt}\right) = k[A]^1$$

$$= 0.0347 \times 0.01$$
$$= 3.47 \times 10^{-4} \text{ M min}^{-1}$$

**22.** (d) For the reaction, $A \rightarrow B$

| Initial concentration | Rate of reaction |
|---|---|
| $2 \times 10^{-3}$ M | $2.40 \times 10^{-4}$ Ms$^{-1}$ |
| $1 \times 10^{-3}$ M | $0.60 \times 10^{-4}$ Ms$^{-1}$ |

$r = k[A]^n$ [where, $n$ = order of reaction]

$$2.40 \times 10^{-4} = k[2 \times 10^{-3}]^n \quad \ldots\text{(i)}$$
$$0.60 \times 10^{-4} = k[1 \times 10^{-3}]^n \quad \ldots\text{(ii)}$$

On dividing Eq. (i) by (ii), we will get

$$4 = (2)^n \Rightarrow n = 2$$

∴ Order of reaction = 2

**• Concept Enhancer**

Order of the reaction is defined as the sum of powers of the concentration of the reactants in the rate law expression.

**23.** (a) $2A + B \longrightarrow$ Products

$[B]$ is doubled, half-life didn't change.

First order *wrt* to $B$

When $[A]$ is doubled, rate increased by two times.

⇒ The reaction is first order *wrt* $A$

Hence, net order of reaction = 1 + 1 = 2

Unit for the rate constant

$$= \text{conc}^{(1-n)}\, t^{-1}$$
$$= (\text{mol L}^{-1})^{1-2} \cdot \text{s}^{-1}$$
$$= \text{L mol}^{-1}\,\text{s}^{-1}$$

**24.** (d) It represents second order reaction.

Thus, unit of $k$ is L mol$^{-1}$ s$^{-1}$.

Hence, (a) is false.

$t_{1/2}$ is dependent on concentration but not constant.

The rate of formation of $C$ will be half of the rate of disappearance of $A$.

**25.** (c) Among the given statements, statement given in option C is incorrect. It is because

$$A(g) \longrightarrow 2B(g) + C(g)$$

at $t = 0$ $\quad p_0 \quad — \quad —$

$t = \infty$ $\quad 0 \quad 2p_0 \quad p_0$

$$p_\infty = 3p_0 = 240 \Rightarrow p_0 = 80 \text{ mm Hg}$$

$$kt = \ln\left[\frac{p_\infty - p_0}{p_\infty - p_t}\right]$$

$$\Rightarrow k \times 10 = \ln\left[\frac{240 - 80}{240 - 160}\right]$$

$$k = 0.0693 \text{ min}^{-1}$$

So, the rate constant of reaction is 0.0693 min$^{-1}$.

**26.** (c) Among the given statements, A, B and D are correct, while statement C is incorrect.

The correct form of statement C is : 1st order kinetics have half-life ($t_{1/2}$) independent of concentration. So, upon changing the concentration $t_{1/2}$ should not change for 1st order reaction.

**27.** (b) We know that

$$r = k[A]^1[B]^1$$

From the given data

$$0.1 = k[20]^1[0.5]^1 \qquad \ldots\text{(i)}$$
$$0.40 = k[x]^1[0.5]^1 \qquad \ldots\text{(ii)}$$
$$0.80 = k[40]^1[y]^1 \qquad \ldots\text{(iii)}$$

Dividing Eqs. (i) and (ii)

$$\frac{0.1}{0.40} = \frac{K[20]^1\,[0.5]^1}{K[x]^1\,[0.5]^1}$$

$$\frac{10}{40}=\frac{[20]^1}{[x]^1}$$

$\therefore \quad x=80$

Putting the value of $x$ in Eq. (ii), we get

$$0.40=K\,[80]^1\,[0.5]^1$$

Dividing the above Eq. by Eq. (iii) we get

$$\frac{0.40=K\,[80]^1\,[0.5]^1}{0.80=K\,[40]^1\,[y]^1}$$

$$\frac{1}{2}=\frac{2\,[0.5]^1}{[y]^1} \Rightarrow y=2$$

**28.** *(c)* It is a first order reaction because the half-life is independent of initial concentration of reactants.

For first order reaction,

$$k=\frac{2.303}{t}\log\frac{[A]_0}{[A]}$$

For 50% reaction,

$$k=\frac{2.303}{t}\log\frac{100}{(100-50)}$$

or, $\quad k=\frac{2.303}{200}\log 2$

$\Rightarrow \quad k=0.0034545$

For 80% reaction,

$$k=\frac{2.303}{t}\log\frac{100}{100-80}$$

or, $\quad 0.0034545=\frac{2.303}{t}\log\frac{100}{20}$

or, $\quad t=467\text{ s}$

**29.** *(b)* From graph (A) it can be interpreted that the rate of the reaction is independent of time. Thus ($A$) is a zero order as its

$$\text{Rate}=K\,[\text{Reactant}]^0=K.$$

From graph (B) it can be interpreted that the half-life is directly proportional to initial concentration of reactant. Thus, (B) is also a zero order. As half life of zero order is given by $t_{\frac{1}{2}}=\frac{[R_0]}{2K}$.

From graph (E) it can be interpreted that it is first order as the rate of reaction is increasing with increase in time.

$$\text{Rate}=K\,[\text{Reactant}]^1$$

**30.** *(c)* Let, the rate law is : $r\propto[A]^a[B]^b$

$$\frac{\text{Experiment II}}{\text{Experiment I}}=\left(\frac{0.1}{0.1}\right)^a\times\left(\frac{0.2}{0.1}\right)^b$$

$$\Rightarrow \quad \frac{2.4\times10^{-2}}{6\times10^{-3}}=1\times2^b$$

$$\Rightarrow \quad 2^b=4\Rightarrow b=2$$

$$\frac{r_{III}}{r_I}=\left(\frac{0.2}{0.1}\right)^a\times\left(\frac{0.1}{0.1}\right)^2$$

$$\Rightarrow \quad \frac{1.2\times10^{-2}}{6\times10^{-3}}=2^a\times1$$

$$\Rightarrow \quad 2^a=2\Rightarrow a=1$$

$$\Rightarrow \quad r_1=k\times(0.1)^1\,(0.1)^2$$

$$\Rightarrow 6\times10^{-3}=k\times10^{-3}$$

$$\Rightarrow \quad k=6\text{ L}^2\text{ mol}^{-2}\text{ min}^{-1}$$

Experiment IV : $r_{II}=6\times(X)^1\times(0.2)^2$

$$\Rightarrow \quad 7.2\times10^{-2}=6\times(X)\times(4\times10^{-2})$$

$$\Rightarrow \quad X=0.3\text{mol L}^{-1}$$

$\Rightarrow$ Experiment V: $r_V=6\times(0.3)^1\times(Y)^2$

$$\Rightarrow \quad 2.88\times10^{-1}=6\times0.3\times(Y)^2$$

$$\Rightarrow \quad Y=0.4\text{ mol L}^{-1}$$

**31.** *(c)* For the reaction,

$$2A+B\longrightarrow\text{products.}$$

Let, the rate expression is $r\propto[A]^a[B]^b$

Let the initial rate of the reaction

$$r_1=[A]^a[B]^b \qquad \ldots\text{(i)}$$

where $a$ and $b$ are the respective orders with respect to reactants $A$ and $B$.

Rate of reaction when conc. of both $A$ and $B$ gets doubled,

$$r_2=[2A]^a[2B]^b$$

$$\frac{r_2}{r_1}=\left(\frac{2A}{A}\right)^a\left(\frac{2B}{B}\right)^b$$

$$\Rightarrow \quad \frac{2.4}{0.3}=2^a\times2^b\Rightarrow2^3=2^{a+b}$$

$$\Rightarrow \quad 3=a+b \qquad \ldots\text{(i)}$$

Rate of reaction when only the concentration of $A$ gets doubled.

$$r_3=K\,[2A]^a[B]^b$$

$$\frac{r_3}{r_1}=\left(\frac{2A}{A}\right)^a\left(\frac{B}{B}\right)^b\Rightarrow\frac{0.6}{0.3}=2^a\times1$$

$$\Rightarrow \quad 2^1=2^a\Rightarrow a=1 \qquad \ldots\text{(ii)}$$

$\therefore$ From Eq. (i), $1+b=3\Rightarrow b=2$

$\Rightarrow$ Order of the reaction

$(n)=a+b=1+2=3$

$\Rightarrow$ Order of the reaction wrt $A=1$

$\Rightarrow$ Order of the reaction wrt $B=2$

**32.** *(b)* Let, the rate expression is $r\propto[A]^a\,[B]^b$.

From experiment I,

$$\frac{r_2}{r_1}=\left(\frac{0.1}{0.1}\right)^a\times\left(\frac{0.25}{0.20}\right)^b$$

$$\Rightarrow \quad \frac{6.93\times10^{-3}}{6.93\times10^{-3}}=1\times\left(\frac{5}{4}\right)^b$$

$$\Rightarrow \quad 1=\left(\frac{5}{4}\right)^b\Rightarrow\left(\frac{5}{4}\right)^0=\left(\frac{5}{4}\right)^b\Rightarrow b=0$$

From experiment II,

$$\frac{r_3}{r_1}=\left(\frac{0.2}{0.1}\right)^a\times\left(\frac{0.30}{0.20}\right)^b$$

$$\Rightarrow \quad \frac{1.386\times10^{-2}}{0.693\times10^{-2}}=(2)^a\times(1.5)^0$$

$$\Rightarrow \quad 2=2^a\times1\Rightarrow2^1=2^a\Rightarrow a=1$$

$\Rightarrow$ So, $\quad r\propto[A]^1[B]^0\Rightarrow r\propto[A]$

Order of the reaction $(n)=1$

$\Rightarrow$ Now, for the 1st experiment,

$$r_1=k\cdot[A]$$

$$\Rightarrow \quad k=\frac{r_1}{[A]}=\frac{6.93\times10^{-3}}{0.1}$$

$$=6.93\times10^{-2}\text{ s}^{-1}$$

$$\Rightarrow \quad t_{50}=\frac{0.693}{k}=\frac{0.693}{6.93\times10^{-2}}=10\text{ s}$$

**33.** *(a)* For the reaction,

$$CH_3CHO(g)\xrightarrow{\text{Decomposes}}CH_4+CO$$

Let order of reaction with respect to $CH_3CHO$ is $m$.

It is given that, $r_1=1$ torr/sec. when $CH_3CHO$ is 5% reacted i.e. 95% unreacted. Similarly, $r_2=0.5$ torr/sec when $CH_3CHO$ is 33% reacted i.e., 67% unreacted.

Use the formula,

$$r\propto(a-x)^m$$

where $(a-x)=$ amount unreacted

or $\quad \frac{r_1}{r_2}=\left[\frac{a-x_1}{a-x_2}\right]^m$

Now putting the given values

$$\frac{1}{0.5}=\left(\frac{0.95}{0.67}\right)^m$$

$$\Rightarrow \quad 2=(1.41)^m \quad\text{or } m=2$$

**34.** *(a)* For first order reaction,

$$k=\frac{2.303}{t}\log\frac{a}{a-x}$$

$$\therefore \quad k = \frac{2.303}{50} \log \frac{0.5}{0.125} = 0.0277 \text{ min}^{-1}$$

Now, as per reaction

$$2H_2O_2 \longrightarrow 2H_2O + O_2$$

$$-\frac{1}{2}\frac{d[H_2O_2]}{dt} = \frac{1}{2}\frac{d[H_2O]}{dt} = \frac{d[O_2]}{dt}$$

Rate of reaction,

$$-\frac{d[H_2O_2]}{dt} = k[H_2O_2]$$

$$\therefore \quad \frac{d[O_2]}{dt} = -\frac{1}{2}\frac{d[H_2O_2]}{dt}$$

$$= \frac{1}{2}k[H_2O_2] \quad \text{...(i)}$$

When the concentration of $[H_2O_2]$ reaches 0.05 M,

$$\frac{d[O_2]}{dt} = \frac{1}{2} \times 0.0277 \times 0.05 \quad \text{[from Eq. (i)]}$$

or $\frac{d[O_2]}{dt} = 6.93 \times 10^{-4}$ mol min$^{-1}$

**Alternate Method**

In fifty minutes, the concentration of $H_2O_2$ decreases from 0.5 to 0.125 M or in one half-life, concentration of $H_2O_2$ decreases from 0.5 to 0.25 M. In two half-lifes, concentration of $H_2O_2$ decreases from 0.5 to 0.125 M or $2t_{1/2} = 50$ min

$$t_{1/2} = 25 \text{ min}$$

$$\Rightarrow \quad k = \left(\frac{0.693}{25}\right) \text{min}^{-1}$$

$$\text{or} \quad \frac{d[O_2]}{dt} = -\frac{1}{2}\frac{d[H_2O_2]}{dt} = \frac{k[H_2O_2]}{2}$$

$$= 6.93 \times 10^{-4} \text{ mol min}^{-1}$$

**35.** *(b)* For Ist order reaction

$$kt = -\log_e C_t + \log_e C_0$$

$$t = -\frac{1}{k}\log_e C_t + \log_e C_0$$

$y = mx + c$, we get straight line with a negative slope when we plot $t$ *vs* $\log C_t$. Comparing the above with straight line equation.

**36.** *(43)* For specific time interval 0-5 sec, 5-10 sec and 10-15 sec, order comes to be zero, but graph is not straight line. Assuming 1st order kinetics,

$$k = \frac{1}{t}\ln\frac{A_0}{A_t} \Rightarrow k = \frac{1}{t}\ln\frac{50}{2.5}$$

$$\frac{1}{10}\ln 2 = \frac{1}{t}\ln 20$$

$$\Rightarrow \quad t = \frac{13010 \times 10}{0.3010} = 43.3 \text{ sec} \approx 43 \text{ sec}$$

**37.** *(2435)* Half-life is directly proportional to $[A]_0$

$\therefore t_{1/2} \propto [A]_0 \Rightarrow$ order = zero

$$t_{1/2}\ \frac{A_0}{2k} \Rightarrow \text{Slope} = \frac{1}{2k} = 76.92$$

$$k = \frac{1}{2 \times 76.92} \Rightarrow [A]_{10} = -kt + A_0$$

$$= -\frac{1}{2 \times 76.92} \times 10 + 2.5 = 2.435$$

$$= 2435 \times 10^{-3} \text{ mol/L}$$

**38.** *(897)*

$$2N_2O_5(g) \longrightarrow 2N_2O_4(g) + O_2(g)$$

| | | | |
|---|---|---|---|
| $p_i$ | 0.6 | 0 | 0 |
| $p_t$ | $0.6 - p$ | $p$ | $p/2$ |

Using $k = \frac{2.303}{t}\log\frac{p_i}{p_t}$

$$4.606 \times 10^{-2} = \frac{2.303}{100}\log\frac{0.6}{0.6-p}$$

$$10^2 = \frac{0.6}{0.6-p} \Rightarrow 10^4 = \frac{0.6}{0.6-p}$$

On solving, we get $p = 0.594$

$$p_{\text{total}} = 0.6 + \frac{p}{2}$$

$$= 0.6 + 0.297$$

$$p_{\text{Total}} = 897 \times 10^{-3}$$

So, the value of approx is 897.

**39.** *(1)* Given reaction,

$$A + B \longrightarrow C$$

$$A \xrightarrow{t=x} \frac{A}{2} \xrightarrow{t=2x} \frac{A}{4}$$

Order w.r.t. $A = 1$

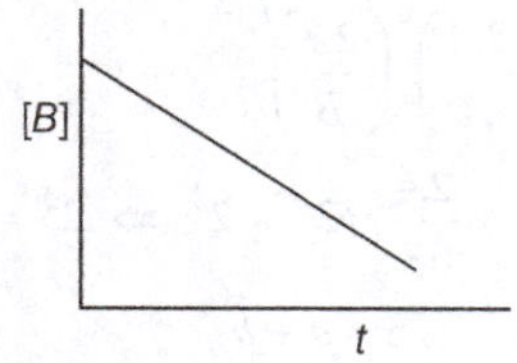


$$[B] = -kt + [B]_0$$

Order w.r.t. $B$ = zero order

Overall order = 1 + 0 = 1

Hence, the overall order is 1.

**40.** *(17)*

$$A + B \rightarrow C \quad \text{...(i)}$$

$$P \rightarrow Q \quad \text{...(ii)}$$

$$t_{1/2} \propto \frac{1}{K}$$

We know that

$$\frac{(t_{1/2})_1}{(t_{1/2})_2} = \frac{K_2}{K_1} = \frac{5}{2}$$

$$\frac{t_{2/3}}{t_{4/5}} = \frac{K_2 \log 3}{K_1 \log 5}$$

$$= \left(\frac{5}{2}\right)\frac{0.477}{0.699} = 1.7 = 17 \times 10^{-1}$$

Comparing it with given equation the value is 17.

**41.** *(3)* Reaction rate,

$$k = \frac{1}{\text{time required}} \ln\left[\frac{\text{initial concentration}}{\text{final concentration}}\right]$$

According to question,

$$k = \frac{100}{t_{99.9\%}}\ln\left[\frac{100}{0.1}\right] = \frac{1}{t_{90\%}}\ln\left(\frac{100}{10}\right)$$

$$t_{99.9\%} = t_{90\%}\frac{\ln[10^3]}{\ln[10]}$$

$$t_{99.9\%} = 3 \times t_{90\%}$$

Hence, time required for 99.9% completion of first order reaction is three times the time required for 90%.

**42.** *(315)* The reaction given to us is

| | | | |
|---|---|---|---|
| | $2A(g)$ + | $B(g)$ $\longrightarrow$ | $C(g)$ |
| $r_1$ | 1.5 atm | 0.7 atm | |
| $r_2$ | 0.5 atm | 0.2 atm | 0.5 atm |

We know that,

Reaction rate, $r = k[p_A]^2[p_B]$

$$r_1 = k[1.5]^2[0.7] \quad \text{...(i)}$$

$$r_2 = k[0.5]^2[0.2] \quad \text{...(ii)}$$

Divide equation (i) by (ii)

$$\frac{r_1}{r_2} = [3]^2\left[\frac{0.7}{0.2}\right] = 31.5 = 315 \times 10^{-1}$$

Hence, the value is 315.

**43.** *(3)* Reaction rate, $r = k[A]^x[B]^y$

for I, $r = 6 \times 10^{-3}$, $A = 0.1$, $B = 0.1$

$$6 \times 10^{-3} = k[0.1]^x[0.1]^y$$

Similarly for IV,

$$2.4 \times 10^{-2} = [0.4]^x[0.1]^y$$

Now divide IV by I

$$4 = 4^x \Rightarrow x = 1$$

For III,

$$2.88 \times 10^{-1} = k[0.3]^x[0.4]^y$$

For II,

$$7.2 \times 10^{-2} = k[0.3]^x[0.2]^y$$

III/II, $4 = 2^y \Rightarrow y = 2$

Overall order $= x + y = 1 + 2 = 3$

**44.** *(50)* Given $[A_0]=1$

$[A]=0.1$

$K=4.6\times10^{-2}\ s^{-1}$

Since the given reaction is of first order. Hence, using

$$t=\frac{2.303}{K}\log\frac{[A_0]}{[A]}$$

On substituting the values

$$t=\frac{2.303}{4.6\times10^{-2}}\log\frac{1}{0.1}\Rightarrow t=50\text{ sec}$$

**45.** *(17190)* $\lambda t=\ln\frac{(^{14}C/^{12}C)_{\text{atmosphere}}}{(^{14}C/^{12}C)_{\text{wood sample}}}$ ...(i)

According to data given in question,

$$\frac{(^{14}C/^{12}C)_{\text{wood}}}{(^{14}C/^{12}C)_{\text{atmosphere}}}=\frac{1}{8}$$

Then, Eq. (i) will become,

$$\lambda t=\ln 8\Rightarrow\frac{\ln 2}{t_{1/2}}t=\ln 8$$

$$t=3\times t_{1/2}=3\times5730$$

$$=17190\text{ years}$$

Hence, the age of wood sample is 17190 years.

**46.** *(399)* Rate of reaction $r=k[A]$

So, order of reaction $=1$

$t_{1/2}=120$ min

For 90% completion of reaction

$$k=\frac{2.303}{t}\log\left(\frac{a}{a-x}\right)$$

$$\frac{0.693}{t_{1/2}}=\frac{2.303}{t}\log\left(\frac{100}{10}\right)$$

$$t=398.7\approx399=399\text{ min}$$

Hence, 399 min is required for 90% decomposition.

**47.** *(24)* Rate of first order reaction is given by

$$R=k[A]e^{-kt}\quad\text{...(i)}$$

$=0.04$ mol $L^{-1}s^{-1}$ 10 minutes

substitute in Eq. (i)

$$0.04=k[A]_0e^{-k\times10\times60}\quad\text{...(ii)}$$

$R=0.03$ mol $L^{-1}s^{-1}$ for 20 min

$$0.03=k[A]_0e^{-k\times20\times60}\quad\text{...(iii)}$$

Divide Eq. (ii) by (iii),

$$\frac{4}{3}=e^{600(k)(2-1)}$$

$$\frac{4}{3}=e^{600k}\Rightarrow\ln\frac{4}{3}=600k$$

$$\ln\frac{4}{3}=600\times\frac{\ln 2}{t_{1/2}}$$

$$\Rightarrow t_{1/2}=\frac{600\times\log 2}{\log 4-\log 3}\text{ s}$$

$$=\frac{10\times0.3010}{0.6020-0.477}\text{ min}$$

$$=24.08\text{ min}\approx24\text{ min}$$

Hence, half-life of reaction is 24 min.

**48.** *(63)* Given, half-life of Br $=36$ hrs

$$36=\frac{0.693}{k}\Rightarrow k=0.01925\text{ hr}^{-1}$$

Ist order kinetic equation

$$t=\frac{2.303}{k}\log\frac{a}{a-x}$$

$$\log\frac{a}{a-x}=\frac{t\times k}{2.303}\quad(t=1\text{ day}=24\text{ hrs})$$

$$\log\frac{a}{a-x}=\frac{24\times0.01925}{2.303}$$

$$\Rightarrow\frac{a}{a-x}=1.587$$

$$\text{If }a=1\Rightarrow\frac{1}{1-x}=1.587$$

$$1-x=0.6301=63\times10^{-2}$$

(Fractional remain after 1 day)

Hence, missing term is 63.

**Time Saver Tech**

$n$ = Number of half-life $=\frac{24}{36}=\frac{2}{3}$

Amount left $=\frac{N_0}{2^{2/3}}=63\times10^{-2}N_0$

So, the fraction left $=63\times10^{-2}$

Thus, the missing term is 63.

**49.** *(10)*

**key idea** To solve this problem first put the given values in the formula, $t_{1/2}=\frac{2.303}{k}\log 2$ and $t_{99.9\%}=\frac{2.303}{k}\log\left(\frac{a}{a-x}\right)$ then compare these equation to get the required relation between $t_{1/2}$ and $t_{99.2}$.

$$t_{1/2}\text{ is given by}=\frac{2.303}{k}\log 2$$

$$t_{99.9\%}=\frac{2.303}{k}\log\left(\frac{a}{a-x}\right)$$

$$\frac{t_{99.9\%}}{t_{1/2}}=\frac{\log\left(\frac{a}{a-x}\right)}{\log 2}$$

$$\frac{t_{99.9}}{t_{1/2}}=\frac{\log\left(\frac{100}{100-99.9}\right)}{\log 2}$$

$$\frac{t_{99.9}}{t_{1/2}}=\frac{\log10^3}{\log 2}=\frac{3}{0.3}$$

$$t_{99.9}=10t_{1/2}$$

**50.** *(2)* The reaction is

$$2HI(g)\longrightarrow H_2(g)+I_2(g)$$

The rate of reaction $R$ is given by

$$R=k[HI]^n$$

By using data of (1) and (2) we get,

$$\frac{3\times10^{-3}}{7.5\times10^{-4}}=\left(\frac{0.01}{0.005}\right)^n$$

$$n=2$$

Hence, the order of reaction is 2.

**51.** *(2200)* The value of $x$ is 3.

| | $A(g)$ | $\longrightarrow 2B(g)$ | $+\,C(g)$ |
|---|---|---|---|
| Initially ($t=0$) | 800 | – | – |
| When ($t=10$ min) | $(800-p)$ | $2p$ | $p$ |

Total pressure $(p_t)=800-p+2p+p$

$=800+2p=1600$

or $[p=400\text{ mm}]$ ...(i)

Since pressure reduce to half of its initial pressure. Hence, 10 min will be the half life.

So, $[t_{1/2}=10\text{ min}]$

So,

$$p\xrightarrow{10\text{ min}}p/2\,p\xrightarrow{10\text{ min}}p/4\,p\xrightarrow{10\text{ min}}p/8$$

Total = 30 min

Hence, after 30 min pressure will left or $=(p-p/8)=(7p/8)$

| | $A\longrightarrow$ | $2B$ | $+\,C$ |
|---|---|---|---|
| $T=0$ | 800 | – | – |
| $T=10$ min | $(800-p)$ | $2p$ | $p$ |
| $T=30$ min | $\left(800-\frac{7p}{8}\right)$ | $(14p/8)$ | $(7p/8)$ |

$$p_T=p/8+1+14p/8+7p/8$$

$$p_T=22p/8=\frac{22\times800}{8}=2200\text{ mm Hg}$$

**52.** *(3)* $t_{87.5}=\frac{1}{K}\ln\left(\frac{100}{100-87.5}\right)$

$$=\frac{1}{K}\ln(8)=\frac{3\ln 2}{K}\quad\text{...(i)}$$

$$t_{50\%} = \frac{1}{K}\ln\left(\frac{100}{100-50}\right) = \frac{1}{K}\ln 2 \ldots\text{(ii)}$$

$$\therefore \frac{t_{87.5}}{t_{50}} = \frac{\frac{3\ln 2}{K}}{\frac{\ln 2}{K}} = [3]$$

or $[t_{87.5} = 3 \times t_{50}]$

The value of $x$ is 3.

**53.** *(3)* Slowest step is the rate determining step.

$$NOBr_2 + NO \longrightarrow 2NOBr$$

$$r = K\,[NOBr_2]\,[NO] \quad \ldots\text{(i)}$$

$$NO + Br_2 \rightleftharpoons NOBr_2$$

$$K_{eq} = \frac{[NOBr_2]}{[NO]\,[Br_2]}$$

$$[NOBr_2] = K_{eq}[NO][Br_2] \quad \ldots\text{(ii)}$$

Subtituting Eqs. (ii) in (i)

$$r = K \cdot K_{eq}\,[NO]\,[Br_2]\,[NO]$$

$$r = K^1[NO]^2[Br_2]$$

Thus, the overall order = 3

**54.** *(1)*

(A) For zero order, $t_{1/2} = \frac{[A]_0}{2K}$ as concentration decreases, half-life decreases. So, statement '*A*' is correct.

(B) If order w.r.t that reactant is zero then it will not affect rate of reaction. So statement '*B*' is correct.

(C) Order can be fractional but molecularity cannot be. So statement 'C' is incorrect.

(D) For zero order reaction unit is $mol\,L^{-1}s^{-1}$ and for second order reaction unit is $mol^{-1}Ls^{-1}$. So statement 'D' is correct.

**55.** *(2)* Given, $(t_{1/2})_A = 12$ minute, $(t_{1/2})_B = 3$ minute

$$\frac{1}{(t_{1/2})_C} = \frac{1}{(t_{1/2})_A} + \frac{1}{(t_{1/2})_B}$$

$$= \frac{1}{12} + \frac{1}{3} = \frac{1+4}{12} = \frac{5}{12}$$

$$\Rightarrow (t_{1/2})_C = \frac{12}{5}\text{ minutes} = 2.4\text{ minutes}$$

**56.** *(75)* Let initial conc. of $A = 1$ mol

For zero order reaction half-life period,

$$t_{1/2} = \frac{a}{2k}$$

$$\Rightarrow \quad k = \frac{a}{2t_{1/2}} = \frac{a}{2\times 50} = \frac{1}{2\times 50}$$

For 75% completion

$$\left(a - \frac{a}{4}\right) = kt \Rightarrow \frac{4a-a}{4} = kt \Rightarrow \frac{3a}{4} = kt$$

$$t = \frac{3}{4}\frac{a}{k} = \frac{3}{4} \times \frac{100}{1} = 75\text{ min}$$

**57.** *(15)* Given that $[B] = 4[A]$

For $A$ decay, $[A] \xrightarrow{15\text{ min}} \frac{[A]}{2}$

For $B$ decay, $[B]$

$$4[A] \xrightarrow{5\text{ min}} 2[A] \xrightarrow{5\text{ min}} [A] \xrightarrow{5\text{ min}} \frac{[A]}{2}$$

So, after $(3 \times 5\text{ min})$ concentration of $A$ and $B$ will be same.

**58.** *(17)* From first order

$$C = \frac{C_0}{2^n} = \frac{C_0}{32} = \frac{C_0}{2^5} \Rightarrow n = 5$$

$$t = 5t_{1/2} = \frac{5 \times 0.693}{20}$$

$$= \frac{0.693}{4} = 0.17325\text{ min}$$

$$= 17.325 \times 10^{-2}\text{ min}$$

**59.** *(1350)* $c_{t(60\%)} = 100 - 60 = 40$

$c_{t(90\%)} = 100 - 90 = 10$

From first order reaction,

$$t = \frac{1}{k}\ln\frac{c_0}{c_t}$$

So, in cases of 60% and 90% decomposition,

$$t_{60\%} = 540 = \frac{1}{k}\ln\frac{100}{40} \quad \ldots\text{(i)}$$

$$t_{90\%} = \frac{1}{k}\ln\frac{100}{10} \quad \ldots\text{(ii)}$$

From Eq. (i) and (ii), we get

$$\frac{540}{t_{90\%}} = \frac{\ln\frac{100}{40}}{\ln\frac{100}{10}}$$

On substituting the values, of ln gives in the question we obtained

$$\frac{540}{t_{90\%}} = \frac{2-1.6}{2-1} = 0.4$$

$$t_{90\%} = \frac{540}{0.4} = 1350\text{ seconds}$$

**60.** *(623)* $A + B \longrightarrow P$

$t = 0 \quad 7\text{g}$

$t = t \quad 2\text{g}$

For Ist order

$$t = \frac{2.303}{k}\log\frac{[A]_0}{[A]_t}$$

$$= \frac{2.303}{2.011 \times 10^{-3}}\log\frac{7}{2}$$

$$= \frac{2.303 \times 10^3}{2.011}(0.845 - 0.301)$$

$$= 622.99 \approx 623\text{ sec}$$

**61.** *(2)*

> **key idea** To approach this type of questions, simply solve the unit of rate constant by using the general expression $[mol\,L^{-1}]^{1-n}\ sec^{-1}$ for determining the order of a reaction.

The unit of rate constant can be the expressed as

$$[mol\,L^{-1}]^{1-n}s^{-1}$$

On putting ($n = 2$), unit of $K$ becomes $(mol\,L^{-1})^{1-2}s^{-1}$ or $(mol\,L^{-1})^{-1}s^{-1}$

or $\quad [mol^{-1}L\,s^{-1}]$

Hence, it is a second order reaction.

**62.** *(2)* Region (I) is the straight line for first order reaction and Region (III) is the constant line for zero order reaction, since for zero order, rate do not depend on the time but on the condition. (catalyst, $h\nu$) etc.

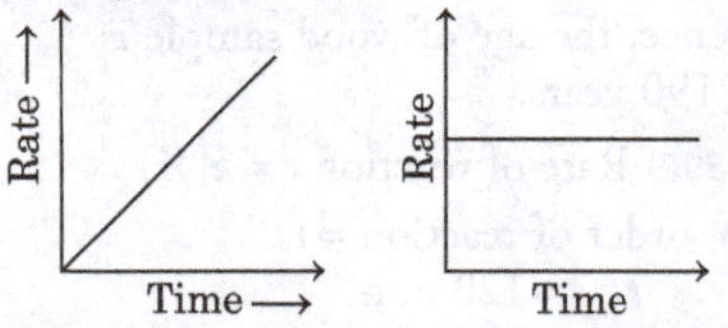


First order reaction Zero order reaction

Hence, statement (C) is correct.

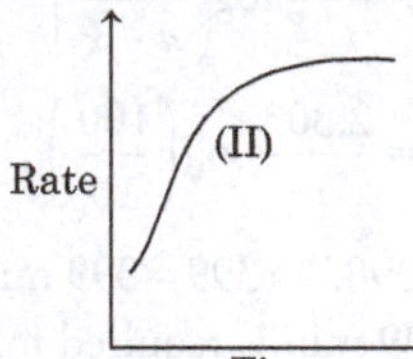


Since region (II) is asymptotic region it shows the running time, hence order cannot be determined.

Thus, statement $A$ and $C$ are correct.

**63.** *(2)* Time required for 10% completion,

$$t_{10\%} = \frac{1}{\text{Rate constant }(k)}\ln\left(\frac{a}{a-x}\right)$$

$$\Rightarrow \quad t_{10\%} = \frac{1}{k}\ln\left(\frac{100}{100-10}\right)$$

$$t_{10\%} = \frac{2.303}{k}\ln\left(\frac{100}{90}\right)$$

$$t_{10\%} = \frac{2.303}{k} \times (0.04)$$

Similarly, $t_{90\%} = \frac{1}{k} \ln\left(\frac{a}{a-x}\right)$

$$t_{90\%} = \frac{1}{k} \ln\left(\frac{100}{100-90}\right) = \frac{1}{k} \ln\left(\frac{100}{10}\right)$$

$$t_{90\%} = \frac{2.303}{k}$$

Now, $\frac{t_{90\%}}{t_{10\%}} = \frac{\frac{2.303}{k}}{\frac{2.303 \times 0.04}{k}} = \frac{1}{0.04} = 25$

So, statement C is correct.

$$e^{kt} = \frac{a}{a-x}$$

By reversing the above equation

$$\frac{a-x}{a} = e^{-kt} \text{ or } 1 - \frac{x}{a} = e^{-kt}$$

$\Rightarrow \quad x = a(1 - e^{-kt})$

$\Rightarrow \quad d = \frac{x}{a} = (1 - e^{-kt})$

Where, $d$ is degree of dissociation.

So, statement D is also correct.

**64.** *(60)* Given, $t_{1/2} = 30$ min

For first order $K = \frac{0.693}{t_{1/2}}$

for 75% completion

$$t = \frac{2.303}{K} \log \frac{100}{25}$$

$$= \frac{2.303}{K} \log 2^2 = 2t_{1/2}$$

$\Rightarrow t_{75\%} = 2t_{1/2}$ (for first order reaction)

$t_{75\%} = 2 \times 30 = 60$ min

**65.** *(1)* Given, $t_{1/2} = 30$ years

Initial amount of elements $x$, $a = 1$ μg

Now, for first order reaction

$$t_{1/2} = \frac{0.693}{k}$$

$$\Rightarrow k = \frac{0.693}{t_{1/2}} = \frac{0.693}{30} = 0.0231 \text{ years}^{-1}$$

$$t = \frac{2.303}{k} \log\left(\frac{a}{a-x}\right)$$

$$\Rightarrow 100 = \frac{2.303}{0.0231} \log\left(\frac{1}{w}\right)$$

[let weight remaining be '$w$']

$\Rightarrow \quad \log \frac{1}{w} = 1 \Rightarrow \log \frac{1}{w} = \log 10$

$\Rightarrow \quad \frac{1}{w} = 10$

$\Rightarrow \quad w = 0.1\ \mu g = 1 \times 10^{-1}\ \mu g.$

**66.** *(40)* Rate $= k[X][Y]^0 = k[X]$

Using data of experiment I and II,

$$\frac{\text{Rate}_I}{\text{Rate}_{II}} = \frac{k[X_I]}{k[X_{II}]}$$

$$\frac{2 \times 10^{-3}}{4 \times 10^{-3}} = \frac{0.1}{L}$$

$\Rightarrow \quad L = 0.1 \times 2 = 0.2$

Using experiment III and IV,

$$\frac{\text{Rate}_{III}}{\text{Rate}_{IV}} = \frac{K\ [X_{III}]}{K\ [X_{IV}]}$$

$$\frac{M \times 10^{-3}}{2 \times 10^{-3}} = \frac{0.4}{0.1}$$

$$M = 4 \times 2 = 8$$

So, $M : L = \frac{8}{0.2} = \frac{80}{2} = 40.$

**67.** *(100)* For a first order reaction,

$$\text{(half-life) } t_{1/2} = \frac{0.693}{k}$$

$$= \frac{0.693}{0.3010} \quad [\text{Given, } t_{1/2} = 0.3010]$$

$$= 2.303$$

The rate law for 1st order reaction is given as

$$k = \frac{2.303}{t} \log \frac{[A_0]}{[A_t]}$$

$$\therefore 2.303 = \frac{2.303}{2} \log \frac{[A_0]}{[A_t]}$$

$\Rightarrow \quad 2 = \log \frac{[A_0]}{[A_t]}$

$\Rightarrow \quad \frac{[A_0]}{[A_t]} = \text{antilog } (2)$

$\Rightarrow \quad \frac{[A_0]}{[A_t]} = 100$

**68.** *(165)* Since after 70 minutes, the concentration of reactant $[A]$ is halved.

So, $t_{1/2} = 70$ min

Half-life of first order reaction is given as

$$t_{1/2} = \frac{0.6932}{k}$$

[$k$ = rate constant for 1st order reaction]

$$k = \frac{0.6932}{t_{1/2}} = \frac{0.6932}{70}$$

$$= 0.0099 \text{ min}^{-1}$$

$$= \frac{0.0099}{60} \text{ s}^{-1} = 165 \times 10^{-6} \text{s}^{-1}$$

The value of $x$ is 165.

**69.** *(2)* Rate of reaction $(r) = k[\text{NO}]^x [H_2]^y$

From I data,

$$0.135 = k[40]^x [65.6]^y \quad \ldots\text{(i)}$$

From II data,

$$0.033 = k[20.1]^x [65.6]^y \quad \ldots\text{(ii)}$$

On dividing Eq. (i) by Eq. (ii),

$$\frac{0.135}{0.033} = \left[\frac{40}{20.1}\right]^x$$

$$4 = (2)^x \Rightarrow x = 2$$

$\therefore$ Order of reaction with respect to NO is 2.

**70.** *(2)* The half-life for $n$th order

$$t_{1/2} \propto \frac{1}{(a_0)^{n-1}}$$

when $t_{1/2}$ is 100 s then the equation is

$$100 \propto \frac{1}{(0.5)^{n-1}} \quad \ldots\text{(i)}$$

when $t_{1/2}$ is 50 s then the equation is

$$50 \propto \frac{1}{(1)^{n-1}} \quad \ldots\text{(ii)}$$

Dividing Eq. (i) by Eq. (ii), we get,

$$\frac{100}{50} = \left(\frac{1}{0.5}\right)^{n-1} \Rightarrow 2 = (2)^{n-1}$$

$\therefore \quad n - 1 = 1 \Rightarrow n = 1 + 1 = 2$

**71.** *(2)* For first order reaction,

$$\ln A = \ln A_0 - kT$$

As slope $-k = -3.465 \times 10^4 \quad \ldots$(i)

Also, $\quad k = \frac{0.69}{t_{1/2}} \quad \ldots$(ii)

From (i) and (ii) case, we get

$$3.465 \times 10^4 = \frac{0.69}{t_{1/2}}$$

$\Rightarrow \quad t_{1/2} = 2 \times 10^{-5}$ s.

**72.** *(16)* For first order reaction,

$$t = \frac{2.303}{k} \log \frac{A_0}{A_0 - A_t}$$

$$t_{67\%} = \frac{2.303}{k} \log\left(\frac{100}{33}\right) \quad \ldots\text{(i)}$$

$$t_{50\%} = \frac{0.693}{k} \quad \ldots\text{(ii)}$$

Dividing Eqs. (i) by (ii),

$$\frac{t_{67\%}}{t_{50\%}} = \frac{2.303 \times k}{k \times 2.303 \times 03} \times \log \frac{100}{33}$$

$$= \frac{0.4814}{03} = 1.604$$

So, $t_{67\%} = 16.04 \times 10^{-1} t_{50\%}$

$= x \times 10^{-1} t_{50\%}$

Hence, value of $x$ is 16.

**73.** *(200)* Since, half-lives are independent of initial concentration, that means reaction is Ist order.

So, $t_{1/2} = \dfrac{0.6932}{k}$

$k_A = \dfrac{0.6932}{100}, \; k_B = \dfrac{0.6932}{50}$

Rate law for Ist order reaction is $C_t = C_0 e^{-kt}$

For $A$, $A_t = A_0 e^{\left(\frac{-0.6932}{100} \times t\right)}$

For $B$, $B_t = B_0 e^{\left(\frac{-0.6932}{50} \times t\right)}$

$A_0 = B_0$ and $A_t = 4B_t$

$$e^{\left(\frac{-0.6932}{100} \times t\right)} = 4e^{\left(\frac{-0.6932}{50} \times t\right)}$$

Taking ln both sides,

$$\left(-\frac{0.6932}{100} \times t\right) = \ln 4 + \left(\frac{-0.6932}{50} \times t\right)$$

$$\frac{-0.6932\,t}{100} + \frac{0.6932\,t}{50} = 2 \ln 2$$

$$+\,0.6932t\left[\frac{-1}{100} + \frac{1}{50}\right] = 2 \times 0.6932$$

$$= 2 \times 100 = 200 \text{ s}$$

**74.** *(0)* Using the equation,

$$\frac{(t_{1/2})_1}{(t_{1/2})_2} = \left(\frac{p_1}{p_2}\right)^{1-n}$$

$$\frac{340}{170} = \left(\frac{55.5}{27.8}\right)^{1-n}$$

$$(2)^1 = (2)^{1-n}$$

So, $1 = 1 - n$

$\Rightarrow n = 1 - 1 = 0$

**75.** *(2)* For 50% completion,

$$k = \frac{2.303}{t_{50\%}} \log \frac{100}{100-50} \quad \ldots\text{(i)}$$

For 75% completion,

$$k = \frac{2.303}{t_{75\%}} \log \frac{100}{100-75} \quad \ldots\text{(ii)}$$

From Eqs. (i) and (ii),

$$\frac{2.303}{t_{50\%}} \log \frac{100}{50} = \frac{2.303}{t_{75\%}} \log \frac{100}{25}$$

$$\frac{0.3010}{t_{50\%}} = \frac{0.6020}{t_{75\%}}$$

$$t_{75\%} = 2t_{50\%}$$

$$\therefore \quad \frac{t_{75\%}}{t_{50\%}} = \frac{2}{1}$$

**76.** *(1)* Rate of the reaction

$$= K[\text{NO}]^x \, [H_2]^y$$

where $K$ is rate constant and $x$, $y$ are order of reaction w.r.t. NO and $H_2$ respectively.

From observations $A$ and $B$

$$7 \times 10^{-9} = K(8 \times 10^{-5})^x (8 \times 10^{-5})^y \ldots\text{(i)}$$

and

$$2.1 \times 10^{-8} = K\,(24 \times 10^{-5})^x \, (8 \times 10^{-5})^y \quad \ldots\text{(ii)}$$

Dividing Eqs. (i) by (ii)

$$\frac{7 \times 10^{-9}}{2.1 \times 10^{-8}} = \frac{K\,(8 \times 10^{-5})^x \, (8 \times 10^{-5})^y}{K(24 \times 10^{-5})^x \, (8 \times 10^{-5})^y}$$

$$\frac{1}{3} = \left(\frac{1}{3}\right)^x \Rightarrow x = 1$$

$\therefore$ Order of reaction w.r.t. NO is 1.

**77.** *(656)*

| | $A$ | $\longrightarrow$ | $2B$ |
|---|---|---|---|
| Initial | 1 mol | | 0 |
| Final | $1-x$ | | $2x$ |

Given, $2x = 0.2 \Rightarrow x = 0.1$

$\therefore$ Final 0.9 0.2

Now, according to first order kinetics

$$t = \frac{2.303}{k} \log_{10} \frac{[A_0]}{[A_t]} \quad \ldots\text{(i)}$$

$$t_{1/2} = \frac{0.691}{k} \quad \ldots\text{(ii)}$$

Eq. (i) divided by Eq. (ii).

$$\frac{t}{t_{1/2}} = \frac{2.303}{0.691} \log_{10} \frac{(1)}{(0.9)} = \frac{1}{0.3} \log_{10}\left(\frac{10}{9}\right)$$

$$\frac{t}{t_{1/2}} = \frac{0.0457}{0.3}$$

$$t_{1/2} = \frac{3 \times t}{0.457} = \frac{3 \times 100}{0.457} \approx 656$$

**78.** *(106)* Unit of rate constant is $\text{min}^{-1}$, so it must be a first order reaction.

For first order reaction,

$$k \times t = 2.303 \log \frac{A_0}{A_t}$$

Where,

$A_0 = 100$, $A_t = 90$ in 1 min

So, $k \times 1 = 2.303 \times \log \dfrac{100}{90}$

$$= 2.303 \times (\log 10 - 2 \log 3)$$

$$= 2.303 \times (1 - 2 \times 0.477)$$

$$= 0.10593$$

$$= 105.93 \times 10^{-3} \text{ min}^{-1}$$

$$\approx 106 \times 10^{-3} \text{ min}^{-1}$$

Hence, the rate constant for viral inactivation is $106 \times 10^{-3}$ $\text{min}^{-1}$.

**79.** *(10)* The rate constant for first order reaction is given by

$$k = \frac{1}{t} \ln\left(\frac{C_0}{C_t}\right)$$

As $k = \dfrac{\ln 2}{t_{1/2}}$

$t_{1/2} = 1 \text{ min}$ [Given]

For 99.9% completion the rate constant will be

$$k = \frac{1}{t} \ln \frac{100}{(100-99.9)}$$

$$\therefore \quad \frac{\ln 2}{1} = \frac{1}{t} \ln\left(\frac{100}{0.1}\right)$$

$$\Rightarrow \quad t = \frac{\ln 1000}{\ln 2} \Rightarrow t = \frac{3 \times 2.3}{0.69} = 10$$

**80.** *(3)* General form of rate law is given by $r = k\,[\text{NO}]^x [\text{Cl}_2]^y$

Using the values of concentrations and rate constants from all 3 experiments

$$(r_1)\, 0.18 = k\,[0.10]^x [0.10]^y \quad \ldots\text{(i)}$$

$$(r_2)\, 0.35 = k\,[0.10]^x [0.20]^y \quad \ldots\text{(ii)}$$

$$(r_3)\, 1.40 = k\,[0.20]^x [0.20]^y \quad \ldots\text{(iii)}$$

Dividing Eq. (i) and (ii),

$$\frac{0.18}{0.35} = \frac{[0.10]^x}{[0.10]^x} \frac{[0.10]^y}{[0.20]^y}$$

$$\Rightarrow \quad \frac{0.18}{0.35} = \left(\frac{1}{2}\right)^y \Rightarrow y = 1$$

So, order w.r.t $[\text{Cl}_2] = 1$

Dividing Eq. (ii) and (iii) we get

$$\frac{0.35}{1.40} = \frac{[0.10]^x [0.20]^y}{[0.20]^x [0.20]^y}$$

$$\Rightarrow \quad \frac{0.35}{1.40} = \left[\frac{1}{2}\right]^x \Rightarrow x = 2$$

Order w.r.t. [NO] = 2

$$\therefore \quad r = k\,[\text{NO}]^2 [\text{Cl}_2]^1$$

**81.** *(2)* For first order reaction,

$$k \times t = \ln \frac{A_0}{A_t} \quad \ldots\text{(i)}$$

$$k \times 570 = \ln \frac{100}{32}$$

$$k \times 570 = 2.303\ (\log 10^2 - \log 2^5)$$

$$\Rightarrow \quad k = \frac{2.303}{570}(2 - 5\log 2)$$

$$= \frac{2.303}{570} \times 0.5$$

$$k = 2 \times 10^{-3}\ s^{-1}$$

**82.** *(108)* Given, $(t_{1/2})_1 = 54$ min

$(t_{1/2})_2 = 18$ min

$$\begin{matrix} A & & B \\ t = 0, \text{'}x\text{' M} & & t = 0, \text{'}x\text{' M} \end{matrix}$$

For Ist order reaction, $[A_t] = \dfrac{A_0}{(2)^n}$

$n$ = number of half-lives

According to question,

$$[A_t] = 16 \times [B_t]$$

$$\frac{x}{(2)^{n_1}} = \frac{x}{(2)^{n_2}} \times 16$$

$$(2)^{n_2} = (2)^{n_1} \times (2)^4$$

$$\Rightarrow \quad n_2 = n_1 + 4$$

$$\frac{t}{(t_{1/2})_2} = \frac{t}{(t_{1/2})_1} + 4$$

$$\Rightarrow t\left(\frac{1}{18} - \frac{1}{54}\right) = 4$$

$$\Rightarrow \quad t = \frac{4 \times 18 \times 54}{36} = 108 \text{ min}$$

**83.** *(81)* Given,

$$\underset{\text{Sucrose}}{C_{12}H_{22}O_{11}} + H_2O \xrightarrow{\text{1 st order}} \underset{\text{Glucose}}{C_6H_{12}O_6} + \underset{\text{Fructose}}{C_6H_{12}O_6}$$

$$t_{1/2} = \frac{10}{3} \text{h}$$

At $t = 0$, $a = [A]_0$ (initial conc.)

$t = 9$ h $a - x = [A]_t$ [conc. at time $t$]

On using 1st order equation,

$$k = \frac{2.303}{t} \log \frac{[A]_0}{[A]_t}$$

$$\Rightarrow \quad \frac{k \times t}{2.303} = \log \frac{[A]_0}{[A]_t}$$

$$\frac{\ln 2 \times 9}{10/3 \times 2.303} = \log\left(\frac{1}{f}\right) \left(\because k = \frac{\ln 2}{t_{1/2}}\right)$$

$$\Rightarrow \quad \log\left(\frac{1}{f}\right) = 0.8124$$

$$\log\left(\frac{1}{f}\right) = 81.24 \times 10^{-2}$$

$$x = 81.24$$

or $\quad x \approx 81$

● Concept **Enhancer**

The sucrose hydrolysis is an example of pseudo first order reaction. These reactions are not Ist order reaction but under certain conditions becomes Ist order.

**84.** *(26)* Integrated rate law for 1st order reaction is

$$kt = \ln \frac{[A]_0}{[A]_t} \quad \ldots\text{(i)}$$

Given, $k = 3.3 \times 10^{-4}\ s^{-1}$

$[A]_0 = 100$

$\Rightarrow \quad [A]_t = 100 - 40 = 60$

Put values in Eq. (i), we get

$$3.3 \times 10^{-4} s^{-1} \times t = \ln \frac{100}{60}$$

$$t = 1547.95 \text{ s} = 25.79 \text{ min}$$

$$\left(1 \text{ min} = 60 \text{ s or } 1 \text{ s} = \frac{1}{60} \text{ min}\right)$$

$$= 26 \text{ minutes}$$

**85.** *(60)* For first order reaction,

$$t = \frac{2.303}{k} \log\left(\frac{a_0}{a}\right)$$

For 75%,

$$t_{75\%} = \frac{2.303}{k} \log\left(\frac{100}{25}\right) \quad \ldots\text{(i)}$$

For 60%,

$$t_{60\%} = \frac{2.303}{k} \log\left(\frac{100}{40}\right) \quad \ldots\text{(ii)}$$

On dividing Eqs. (ii) by (i),

$$\frac{t_{60\%}}{t_{75\%}} = \frac{\dfrac{2.303}{k} \log\left(\dfrac{100}{40}\right)}{\dfrac{2.303}{k} \log\left(\dfrac{100}{25}\right)}$$

$$t_{60\%} = t_{75\%} \times \left(\frac{0.3979}{0.602}\right)$$

$$\Rightarrow \quad t_{60\%} = 90 \times \frac{0.3979}{0.602}$$

$$t_{60\%} = 59.48 \text{ minute}$$

$$\approx 60 \text{ min}$$

**86.** *(23.03)* All the radioactive reactions follow Ist order kinetics. So, time taken for decay from $N_0$ to $N_t$ is ($N$ = number of nuclei)

$$t = \frac{1}{\lambda} \ln \frac{N_0}{N_t}$$

$$t = \frac{1}{\lambda} \times 2.303 \log \frac{N_0}{N_t} \quad \ldots\text{(i)}$$

Also, we know $\lambda$ = (decay constant)

$$= \frac{0.693}{t_{1/2}}$$

where, $t_{1/2} = 6.93$ yr (given)

Also, we know 90% nuclei are decayed

$$\therefore \quad \frac{N_0}{N_t} = \frac{100}{10} = 10$$

Put the values of $\lambda$ and $\dfrac{N_0}{N_t}$ in Eq. (i), we get

$$\therefore t = \frac{6.93}{0.693} \times 2.303 \times \log 10 = 23.03 \text{ (yr)}$$

## Topic 3

**01.** *(a)* For the given graph, activation energy of forward reaction is $E_1 + E_2$. Also energy of product > Energy of reactant.

So, order of stability is reactant > product.

**02.** *(d)* A catalyst provides an alternate pathway or reaction mechanism by reducing the activation energy between reactants and products and hence lowering the potential energy barrier. Hence, the correct reaction profile diagram is as follows.

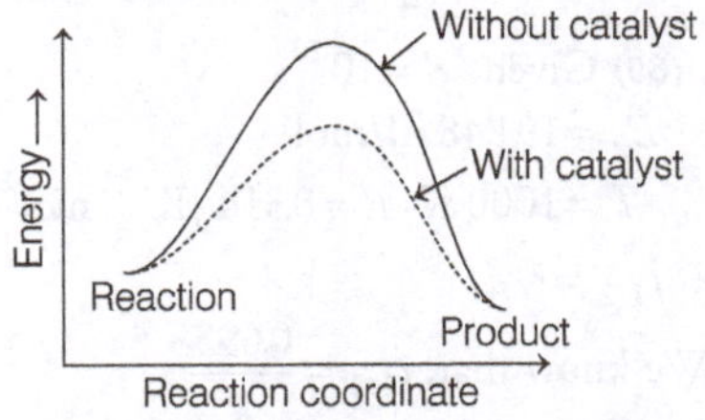


**03.** *(c)* According to Arrhenius equation,

$$k = Ae^{-E_a/RT}$$

As the temperature increases, rate constant $k$ also increases which will generally speed up the process (increase the rate). So, graph will exponentially increases as in option (c).

**04.** *(a)* The main conditions for the occurrence of a reaction is proper orientation and effective collision of the reactants. Since the chances of simultaneous collision with proper orientation between more than 3 species is very rare, so reaction with order greater than 3 are rare.

**05.** *(b)* We know that, for an exothermic reaction

$$\Delta H = E_{a(F.R)} - E'_{a(B.R)}$$

$$E'_{a(B.R)} = 150 - (-135) = 285 \text{ kJ/mol}$$

**06.** *(b)* By Arrhenius equation,

$$k_1 = A' e^{-E_{a_1}/RT}$$

and $$k_2 = A' e^{-E_{a_2}/RT}.$$

[$A'$ is Arrhenius constant]

$\because$ $E_{a_2} = 2E_{a_1}$

$\Rightarrow$ $k_2 = A' e^{-2E_{a_1}/RT}$

$$\frac{k_1}{k_2} = \frac{A' e^{-E_{a_1}/RT}}{A' e^{-2E_{a_1}/RT}} = e^{E_{a_1}/RT}$$

$\therefore$ $k_1 = k_2 e^{E_{a_1}/RT}$

**07.** *(20)* The resultant constant is given by

$$k = \sqrt{\frac{k_1 k_3}{k_2}}$$

$$Ae^{-E_a/RT} = \sqrt{\frac{A_1 e^{-E_{a_1}/RT} \times A_3 e^{-E_{a_3}/RT}}{A_2 e^{-E_{a_2}/RT}}}$$

By comparing exponential term

$$\frac{E_a}{RT} = \frac{1}{2} \times \left[\frac{E_{a_1}}{RT} + \frac{E_{a_3}}{RT} - \frac{E_{a_2}}{RT}\right]$$

$$E_a = \frac{(E_{a_1} + E_{a_3} - E_{a_2})}{2}$$

$$= \frac{(60+10-30)}{2} = 20 \text{ kJ/mol}$$

**08.** *(69)* Given, $A = 10^{20}$

$E_a = 191.48$ kJ/mol

$T = 1000$ K $R = 8.314$ JK$^{-1}$ mol$^{-1}$

$t_{1/2} = ?$

We know that, $t_{1/2} = \frac{0.693}{k}$

and $k$ can be calculated by using,

$$= Ae^{\frac{-E_a}{RT}}$$

$$= 10^{20} \times e^{\frac{191.48 \times 10^3}{8.314 \times 1000}}$$

$$= 10^{20} \times e^{-23.031}$$

$$= \frac{10^{20}}{10^{10}} = 10^{10} s^{-1}$$

Now, substitute this value in half-life equation

$$t_{1/2} = \frac{0.693}{10^{10}} = 6.93 \times 10^{-11} \text{ s}$$

or $t_{1/2} = 69.3 \times 10^{-12}$ s

So, time required for 50% molecule of $A$ to become molecule $B$ is 69 picoseconds.

**09.** *(c)*

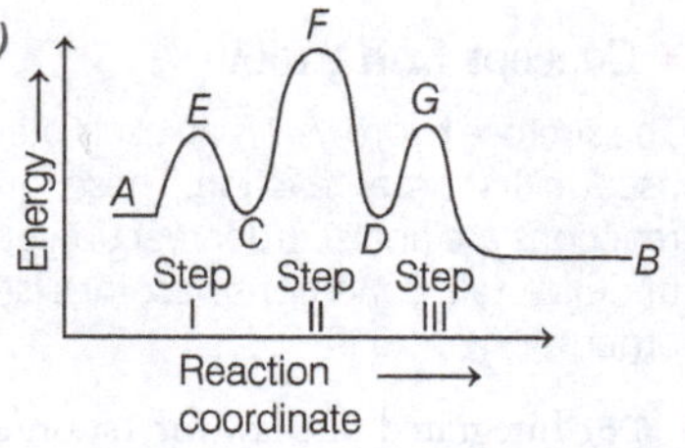


This is a 3-step reaction. Total number of intermediates are 2 ($C$ and $D$) and transition states are 3 ($E$, $F$ and $G$). The rate determining step is step II. Since it has highest activation energy and higher energy correspond to slowest step.

**10.** *(d)* $k = Ae^{-E_a/RT}$

Taking natural logarithm on both sides, the Arrhenius equation becomes.

$$\Rightarrow \quad \ln k = \ln A - \left(\frac{E_a}{R}\right)\frac{1}{T}$$

$$\Rightarrow \quad \ln k = \ln A - \left(\frac{E_a}{R \times 10^3}\right) \times \frac{10^3}{T}$$

$$\text{Slope of a graph} = \frac{-E_a}{R \times 10^3} = \frac{-10}{5}$$

$$E_a = 2R \times 10^3 \text{ J mol}^{-1}$$

$$\Rightarrow \quad E_a = 2R \text{ kJ}$$

**11.** *(d)* According to Arrhenius equation

$$k = Ae^{-E_a/RT}$$

$$\Rightarrow \quad \log k = \log A - \frac{E_{act}}{2.303RT}$$

Slope of the graph of 'log $k$' $vs \frac{1}{T}$ is $\frac{-E_{act}}{2.303R}$

Greater the value of $E_{act}$, more will be the negative slope.

From the graph,

most $-ve$ slope $= c$, followed by $a$, $d$ and $b$.

$\therefore$ Order of activation energies must be

$$E_c > E_a > E_d > E_b$$

**12.** *(b)* Rate constant of reaction without catalyst is given by Arrhenius equation.

$$k = Ae^{-E_a/RT}$$

Taking log on both sides, we get

$$\log k = \log A - \frac{E_a}{RT(2.303)} \quad \text{...(i)}$$

With catalyst, rate constant is $10^6$ times more

$$\therefore \quad 10^6 k = Ae^{\frac{-E_a}{RT}}$$

$$\log k + \log 10^6 = \log A - \frac{E_a^*}{RT(2.303)}$$

$$\therefore \quad \log k = \log A - \frac{E_a^*}{RT(2.303)} - 6 \quad \text{...(ii)}$$

Comparing Eqs. (i) and (ii)

$$\frac{E_a}{RT(2.303)} = \frac{E_a^*}{RT(2.303)} + 6$$

$$\therefore \quad E_a^* - E_a = -6(2.303)RT$$

**13.** *(d)*

**key idea** To calculate the activation energy, use Arrhenius equation for rate constant at two different temperatures (in K) i.e.

$$\log \frac{k_2}{k_1} = \frac{E_a}{2.303\, R}\left[\frac{T_2 - T_1}{T_1 T_2}\right]$$

For the reaction,

$$H_2 + I_2 \longrightarrow 2HI$$

Given $k_1 = 2.5 \times 10^{-4}$ dm$^3$ mol$^{-1}$s$^{-1}$

$T_1 = (273 + 327)$ K $= 600$ K

$k_2 = 1$ dm$^3$ mol$^{-1}$ s$^{-1}$

at $T_2 = (273 + 527)$ K $= 800$ K

$$\text{Now, } \log \frac{k_2}{k_1} = \frac{E_a}{2.303R}\left(\frac{T_2 - T_1}{T_1 T_2}\right)$$

$$\Rightarrow \log \frac{1}{2.5 \times 10^{-4}} = \frac{E_a}{2.303 \times 8.314 \times 10^{-3}}\left(\frac{800 - 600}{600 \times 800}\right)$$

$$\Rightarrow \log \frac{(10 \times 10^3)}{2.5} = \frac{E_a}{0.019} \times \frac{200}{48 \times 10^4}$$

$$\Rightarrow \quad \log 4 + 3 \log 10 = E_a \times 0.022$$

$$\Rightarrow \quad E_a = \frac{2 \times \log 2 + 3}{0.022}$$

$$= \frac{3.6}{0.022} = 163.6 \approx 164 \text{ kJ mol}^{-1}$$

**14.** *(c)* Using the Arrhenius equation,

$$k = Ae^{-E_a/RT} \quad \text{...(i)}$$

Taking natural logarithm on both sides, the Arrhenius equation becomes,

$$\ln k = \ln A - \frac{E_a}{RT}$$

On comparing the above equation with the straight line equation.

$y_{axis} = \ln K$

$x_{axis} = \frac{1}{T}$,

Slope $= \frac{-E_a}{R}$,

$$I = \ln A$$

⇒ From given graph,

$$\frac{-Ea}{R} = -4606$$

$$\therefore \quad \ln\frac{k_2}{k_1} = \frac{E_a}{R}\left(\frac{1}{T_1} - \frac{1}{T_2}\right) \quad \ldots\text{(ii)}$$

On substituting all the given values in Eq. (ii), we get

$$\ln\frac{k_2}{10^{-5}} = 4606\left(\frac{1}{400} - \frac{1}{500}\right)$$

$$\ln\frac{k_2}{10^{-5}} = 2.303$$

$$\frac{k_2}{10^{-5}} = 10 \Rightarrow k_2 = 10^{-4}\ s^{-1}$$

Therefore, rate constant for the reaction at 500 K is $10^{-4}\ s^{-1}$.

**15.** *(a)* Using the formula,

$$\ln\frac{k_1}{k_2} = \frac{E_a}{R}\left[\frac{1}{T_1} - \frac{1}{T_2}\right]$$

$$\therefore \ \ln 4 = \frac{E_a}{8.314} \times \left[\frac{310-300}{310\times 300}\right]$$

$$2\ln 2 = \frac{E_a}{8.314}\ \frac{10}{310\times 300}$$

$$\Rightarrow \ E_a = 107.2\ \text{kJ/mol}$$

**16.** *(a)* From Arrhenius equation,

$$\log\frac{k_2}{k_1} = \frac{-E_a}{2.303\,R}\left(\frac{1}{T_2} - \frac{1}{T_1}\right)$$

Given, $\frac{k_2}{k_1} = 2;\ T_2 = 310\ \text{K}$

⇒ $T_1 = 300$ K

On putting values,

$$\Rightarrow \log 2 = \frac{-E_a}{2.303\times 8.314}\left(\frac{1}{310} - \frac{1}{300}\right)$$

$$\Rightarrow \quad E_a = 53603.93\ \text{J/mol}$$
$$= 53.6\ \text{kJ / mol}$$

**17.** *(d)* $A_2 + B_2 \rightleftharpoons 2AB$

$E_a$ (forward) = 180 kJ $mol^{-1}$

$E_a$ (backward) = 200 kJ $mol^{-1}$

In the presence of catalyst,

$E_a$ (forward) = 180 − 100 = 80 kJ $mol^{-1}$

$E_a$ (backward) = 200 − 100 = 100 kJ $mol^{-1}$

$\Delta H = E_a$ (forward) − $E_a$ (backward) = 80 − 100 = − 20 kJ $mol^{-1}$

For the value of enthalpy change of the reaction, negative sign can be neglected.

**18.** *(c)* We know that,

$$k = Ae^{-E_a/RT}$$

On taking log on both side we get

$$\log k = \log A - \frac{E_a}{2.303\,RT}$$

For graph between $\log k$ with $\frac{1}{T}$

Slope of curve is given by $\frac{E_a}{2.303\,R}$

So, the magnitude of slope will be

$$E_{a_2} > E_{a_1} > E_{a_3}$$

**19.** *(a)* Among the given statements, only B and C are correct, while A, D and E are incorrect. Their correct form are as follows.

- Arrhenius equation hold true for elementry as well as complex reactions.
- Unit of $A$ is same as unit of $k$, rate of reaction is high if activation energy is low.
- $A$ and $E_a$ are temperature independent.

**20.** *(a)* For reaction, $A_2 + B_2 \rightleftharpoons 2AB$

Given, $E_F$ = 180kJ/mol

$E_B$ = 200 kJ/mol

$\Delta H = E_F - E_B = -20$ kJ/mol

In presence of catalyst,

$E_F$ = 180 − 100 = 80 kJ/mol

$E_B$ = 200 − 100 = 100 kJ/mol

Also, catalyst does not change $\Delta H$ and $\Delta G$ of a reaction. Thus, statement given in option (a) is correct.

**21.** *(d)* Only statement (d) is incorrect. Corrected statement is "Activation enthalpy to form $C$ is 15 kJ $mol^{-1}$ which is more than 5 kJ $mol^{-1}$ that is required to form $D$."

**22.** *(d)* From Arrhenius equation,

$$K = Ae^{-E_a/RT}$$

For reaction $R_1$, $k_1 = Ae^{-E_{a_1}/RT}$ ...(i)

For reaction $R_2$, $k_2 = Ae^{-E_{a_2}/RT}$ ...(ii)

On dividing Eq. (ii) by Eq. (i), we get

$$\frac{k_2}{k_1} = e^{-\frac{(E_{a_2} - E_{a_1})}{RT}} \quad \ldots\text{(iii)}$$

Taking ln on both the sides of Eq. (iii), we get

$$\ln\left(\frac{k_2}{k_1}\right) = \frac{E_{a_1} - E_{a_2}}{RT}$$

Given, $E_{a_1} = E_{a_2} + 10\ \text{kJ mol}^{-1}$
$= E_{a_2} + 10{,}000\ \text{J mol}^{-1}$

$$\therefore \ln\frac{k_2}{k_1} = \frac{10{,}000\ \text{J mol}^{-1}}{8.314\ \text{J mol}^{-1}\ \text{K}^{-1} \times 300\ \text{K}} = 4$$

**23.** *(a)* Let activation energy of the reaction be $x$,

Since, $\frac{E_f}{E_b} = \frac{2}{3}$, therefore, $E_f = \frac{2x}{5}$ and

$$E_b = \frac{3x}{5}$$

$$E_b - E_f = -\Delta H = +40$$

$$\Rightarrow \quad \frac{3x}{5} - \frac{2x}{5} = 40 \Rightarrow x = 200$$

Therefore,

$$E_b = \frac{3x}{5} = \frac{3\times 200}{5} = 120\ \text{kJ mol}^{-1}$$

$$E_f = \frac{2x}{5} = \frac{2\times 200}{5} = 80\ \text{kJ mol}^{-1}$$

**24.** *(100)* Given that, $K = \frac{K_1K_2}{K_3}$

$$\text{or } Ae^{-E_a/RT} = \frac{A_1e^{-\frac{E_{a_1}}{RT}}\ A_2e^{-\frac{E_{a_2}}{RT}}}{A_3\ e^{-E_{a_3}/RT}}$$

$$Ae^{-E_a/RT} = \frac{A_1A_2}{A_3}\ e^{-\frac{(E_{a_1} + E_{a_2} - E_{a_3})}{RT}}$$

$$E_a = E_{a_1} + E_{a_2} - E_{a_3}$$

$$400 = 300 + 200 - E_{a_3}$$

$$E_{a_3} = 100\ \text{kJ/mole}$$

**25.** *(30)* Given,

Overall rate constant, $k = \frac{K_1K_2}{K_3}$

$E_{a_1}$ = 40 kJ/mol, $E_{a_2}$ = 50 kJ/mol

$E_{a_3}$ = 60 kJ/mol

$$k = \frac{K_1 \cdot K_2}{K_3}$$

$$Ae^{-\frac{E_a}{RT}} = \frac{A_1A_2}{A_3}\ e^{-\left(\frac{E_{a_1} + E_{a_2} - E_{a_3}}{RT}\right)}$$

$$A = \frac{A_1A_2}{A_3}, \text{ and } E_a = E_{a_1} + E_{a_2} - E_{a_3}$$

$$E_a = 40 + 50 - 60 = 30\ \text{kJ/mol}$$

**26.** *(130)* Given, $E_a$ = 300 kJ/mol

$\frac{E_a}{T} = \frac{E_a'}{T'}$ (∵ Rate of catalysed and uncatalysed reaction is same)

$$\frac{300}{600} = \frac{E_{a\,\text{Forward}}}{300}$$

$\Rightarrow \quad E_{a\,\text{Forward}} = 150$

Now, $\quad 20 = 150 - E_{a\,\text{Backward}}$

$\Rightarrow \quad E_{a\,\text{Backward}} = 130$

**27.** *(2)* Among the given statements $A$ and $B$ are correct.

We know that

$$K = Ae^{-E_a/RT}$$

$$\text{or } \log K = \log A - \frac{E_a}{2.303R} \times \frac{1}{T}$$

Thus, higher is the $E_a$ values, stronger will be the temperature dependence of $K$.

Also, if activation energy $(E_a)$ of a reaction is zero, then the rate is independent of temperature.

**28.** *(2520)* From Arrhenius equation of kinetics

$$\ln\left(\frac{k_2}{k_1}\right) = \frac{E_a}{R}\left(\frac{1}{T_1} - \frac{1}{T_2}\right)$$

$$\log\left(\frac{0.05}{0.03}\right) = \frac{E_a}{2.3 \times 8.3}\left[\frac{1}{200} - \frac{1}{300}\right]$$

$$\log(1.66) = \frac{E_a}{2.3 \times 8.3}\left[\frac{300-200}{300 \times 200}\right]$$

$$\Rightarrow 0.22 = \frac{E_a}{2.3 \times 8.3}\left(\frac{1}{600}\right)$$

$$E_a = 0.22 \times 2.3 \times 8.3 \times 600$$
$$= 2519.88 \text{ J} \approx 2520 \text{ J}$$

**29.** *(3)* Among the given statements A, B and D are correct. However, statement C is incorrect.

The correct statement for Statement (C) is that the value of $k$ may be greater or lesser depending upon the temperature.

**30.** *(8)* Using Arrhenius equation,

$$k = Ae^{-E_a/RT}$$

So, $\quad \ln k = \ln A - \dfrac{E_a}{RT}$

Thus, slope of the graph $\ln k$ *vs* $\dfrac{1}{T}$ is $-\dfrac{E_a}{R}$

$$\Rightarrow \left(\frac{0-20}{5-0}\right) = \frac{-E_a}{R}$$

$$\Rightarrow \frac{20}{5} = \frac{E_a}{R}$$

$$\Rightarrow E_a = 4R = 4 \times 2 = 8 \text{ cal mol}^{-1}$$

**31.** *(225)* Using Arrhenius equation,

$$K = Ae^{-E_a/RT}$$

Taking ln on both sides,

$$\ln k = \ln A - \frac{E_a}{RT}$$

$$\Rightarrow \log(2 \times 10^{-6}) = \log(4 \times 10^{13}) - \frac{8.3 \times 10^3}{8.3 \times T \times 2.3}$$

$$\log 2 - 6 = 2\log 2 + 13 - \frac{8.3 \times 10^3}{8.3 \times T \times 2.3}$$

$$-19.3 = \frac{-10^3}{T \times 2.3}$$

$$\Rightarrow \quad T = 225 \times 10^{-1} \text{ K}$$

**32.** *(59)* According to Arrhenius equation,

$$k = Ae^{-E_a/RT}$$

Taking ln both sides

$$\ln k = \ln A - \frac{E_a}{RT}$$

Rate constant when reaction occur at 300 K :

$$\ln k_1 = \ln A - \frac{E_a}{R \times 300} \quad \ldots \text{(i)}$$

Rate constant when reaction occur at temperature 309 K

$$\ln k_2 = \ln A - \frac{E_a}{R \times 309} \quad \ldots \text{(ii)}$$

$\because \quad k_2 = 2k_1$ (given)

$$\ln(2k_1) = \ln A - \frac{E_a}{309R} \quad \ldots \text{(iii)}$$

Subtract Eqs. (i) from (iii),

$$\ln(2k_1) - \ln k_1 = \frac{-E_a}{309R} + \frac{E_a}{300R}$$

$$\log(2) = \frac{E_a}{2.303\,R}\left[\frac{9}{300 \times 309}\right]$$

$$0.3 = \frac{E_a}{2.303 \times 8.3} \times \frac{9}{300 \times 309}$$

$$\Rightarrow \quad E_a = 59065.04 \text{ J} = 59.06 \text{ kJ}$$

**33.** *(166)* Given,

$$\ln k = 33.24 - \frac{2.0 \times 10^4}{T}$$

Also, $\quad \ln k = \ln A - \dfrac{E_a}{RT}$

On comparing both above equations,

$$\frac{E_a}{R} = 2.0 \times 10^4$$

$$\therefore \quad E_a = 2.0 \times 10^4 \times R$$
$$= 2.0 \times 10^4 \times 8.3 \text{ J}$$
$$= 16.6 \times 10^4 \text{ J} = 166 \text{ kJ}$$

**34.** *(4)* For catalyst $A$,

$$K_{\text{catalysed}} = Ae^{-E'_a/RT}$$

(From Arrhenius equation)

For uncatalysed, $K_{\text{uncatalysed}} = Ae^{-E_a/RT}$

So,

$$\frac{K_{\text{catalysed}}}{K_{\text{uncatalysed}}} = e^{(E_a - E'_a)/RT}$$

$$= e^{10000\text{ J/mol}/(8.314 \times 300)}$$

$$= e^{4.009} = e^x$$

$\therefore \quad x = 4$

**35.** *(154)* According to given Arrhenius plot,

$$\ln k = \ln A - \frac{E_a}{10^3\,RT} \times 10^3$$

$$\ln k = \ln A + \frac{10^3}{T}\left[\frac{-E_a}{10^3\,R}\right]$$

where, $\dfrac{-E_a}{10^3\,R}$ represent slope.

$$\therefore \text{ Slope} = \frac{-E_a}{10^3\,R} = -18.5$$

$$\therefore \quad E_a = 18.5 \times 10^3 \times 8.31$$
$$= 153.735 \text{ kJ/mol}$$
$$\approx 154 \text{ kJ/mol}$$

**36.** *(47)* According to Arrhenius theory,

$$\log k = \log A - \frac{E_a}{2.303RT} \quad \ldots \text{(i)}$$

$$\log_{10} k = 20.35 - \frac{2.47 \times 10^3}{T} \quad \ldots \text{(ii)}$$

On comparing Eqs. (i) and (ii),

$$\Rightarrow \frac{-2.47 \times 10^3}{T} = -\frac{E_a}{2.303RT}$$

$$E_a = 2.47 \times 2.303 \times 8.314$$
$$= 47.29 \text{ kJ / mol}$$
$$\simeq 47 \text{ kJ / mol}$$

**37.** *(16)* According to Arrhenius theory,

$$\log\frac{k_1}{k_2} = \frac{E_a}{2.303\,R}\left[\frac{1}{T_2} - \frac{1}{T_1}\right]$$

At, $T_1 = 700$ K, $k_1 = 6.36 \times 10^{-3}$ s$^{-1}$,

$E_a = 209$ kJ mol$^{-1}$,

$R = 8.314$ J K$^{-1}$ mol$^{-1}$

At, $T_2 = 600$ K, $k_2 = x \times 10^{-6}$ s$^{-1}$

$$\log\frac{6.36 \times 10^{-3}}{x \times 10^{-6}} = \frac{209 \times 10^3}{2.303 \times 8.314}\left(\frac{1}{600} - \frac{1}{700}\right)$$

$$=\frac{209\times 10^3}{2.303\times 8.314}\left(\frac{100}{420000}\right)$$

$$=16\times 10^{-6}\ s^{-1}$$

$$x=16\,s^{-1}$$

**38.** *(10)* We know, from Arrhenius equation,

$$\log\frac{k_2}{k_1}=\frac{E}{2.303\,R}\left(\frac{1}{T_1}-\frac{1}{T_2}\right)$$

Putting the all given values,

$$\log\frac{10^{-3}}{k_1}=\frac{11.488\times 10^3}{2.303\times 8.314}\left(\frac{1}{200}-\frac{1}{300}\right)$$

$$\log\frac{10^{-3}}{k_1}=1$$

$$\Rightarrow\quad \log\frac{10^{-3}}{k_1}=\log 10$$

$$\Rightarrow\quad k_1=10^{-4}=10\times 10^{-5}\ s^{-1}$$

**39.** *(14)* According to Arrhenius equation, the factor $e^{-E_a/RT}$ corresponds to the fraction of molecules that have kinetic energy greater than $E_a$ (activation energy).

$$\Rightarrow\quad e^{-E_a/RT}=e^{\frac{-80.9\times 1000}{8.31\times 700}}=e^{-13.90}$$

$$\simeq e^{-14}=e^{-x}$$

$$x=14$$

**40.** *(50)* $X\longrightarrow Y$, it is an exothermic reaction whose $\Delta E$ or

$$\Delta H=-20\ \text{kJ mol}^{-1}$$

$$E_a^f=30\ \text{kJ mol}^{-1}\quad\text{(Given)}$$

Activation energy of the backward or reverse reaction $(Y\rightarrow X)$, $E_a^b$ can be calculated as,

$$\Delta H=E_a^f-E_a^b$$

$$\Rightarrow\quad E_a^b=E_a^f-\Delta H$$

$$=30-(-20)=50\ \text{kJ mol}^{-1}$$

**41.** *(52)* $T_1=(273+27)=300$ K,

$$T_2=(273+52)=325\ \text{K}$$

Given, temperature coefficient of the reaction,

$$\alpha_T=\frac{K_{325}}{K_{300}}=5$$

From Arrhenius equation,

$$\log\frac{K_{T_2}}{K_{T_1}}=\frac{E_a}{2.303R}\times\left(\frac{T_2-T_1}{T_1T_2}\right)$$

$$\log\frac{K_{325}}{K_{300}}=\frac{E_a}{2.303\times 8.314}\left(\frac{325-300}{300\times 325}\right)$$

$$\log 5=\frac{E_a}{2.303\times 8.314}\times\frac{25}{300\times 325}$$

$$E_a=52271\ \text{J}$$

$$=52.271\ \text{kJ mol}^{-1}$$

$$\simeq 52\ \text{kJ mol}^{-1}$$

**42.** *(526)* According to Arrhenius equation,

$$\log K=\log A-\frac{E_a}{2.303\,RT}$$

$$\text{Given, Slope}=\frac{E_a}{2.303\,R}=10{,}000$$

$$K_1=10^{-5},\ K_2=10^{-4}$$

$$T_1=500\ \text{K}$$

$$\log_{10}\left[\frac{K_1}{K_2}\right]=\frac{Ea}{2.303R}\left[\frac{1}{T_1}-\frac{1}{T_2}\right]$$

$$\log\left[\frac{10^{-4}}{10^{-5}}\right]=10000\left[\frac{1}{500}-\frac{1}{T_2}\right]$$

$$\log 10=10000\left[\frac{1}{500}-\frac{1}{T_2}\right]$$

$$T_2=526.31\simeq 526\ \text{K}$$

**43.** *(100)* The Arrhenius Equation is

$$\ln\left(\frac{k_2}{k_1}\right)=\frac{E_a}{R}\left[\frac{1}{T_1}-\frac{1}{T_2}\right]$$

$$\text{Given,}\quad\frac{k_2}{k_1}=3.555$$

$$T_1=30+273=303\ \text{K};$$

$$T_2=40+273=313\ \text{K}$$

Substituting the given values in the Arrhenius equation,

$$\ln(3.555)=\frac{E_a}{8.314}\left[\frac{1}{303}-\frac{1}{313}\right]$$

$$1.268=\frac{E_a}{8.314}\left[\frac{313-303}{303\times 313}\right]$$

$$E_a=\frac{1.268\times 8.314\times 303\times 313}{10}$$

$$=99980.7\ \text{J mol}=99.98\ \text{kJ/mol}$$

$$\simeq 100\ \text{kJ/mol}$$

**44.** *(84297.47)* $T_1=273+27=300$ K

$$T_2=273+42=315\ \text{K}$$

Now, using (Arrhenius equation)

$k=Ae^{-E_a/RT}$

Taking ln on both sides

$$\ln\left(\frac{k_2}{k_1}\right)=\frac{E_a}{R}\left(\frac{1}{T_1}-\frac{1}{T_2}\right)$$

$$\text{Given,}\ \frac{k_2}{k_1}=5$$

$$\ln 5=\frac{E_a}{R}\left(\frac{1}{300}-\frac{1}{315}\right)$$

$$1.6094=\frac{E_a}{8.314}\times\frac{15}{300\times 315}$$

$$E_a=\frac{1.6094\times 8.314\times 300\times 315}{15}$$

$$E_a=84297.47\ 50\ \text{J/mol}$$

**45.** *(3.98)* Time taken for milk to split $\propto\frac{1}{k}$

where, $k$ = rate constant of the process and, $k$ at 2 different temperatures and activation energy are related.

$$\ln\frac{k_2}{k_1}=\frac{-E_a}{R}\left(\frac{1}{T_1}-\frac{1}{T_2}\right)$$

$$\therefore\quad\ln\left(\frac{60}{40}\right)=\frac{-E_a}{8.3}\left(\frac{1}{400}-\frac{1}{300}\right)$$

$$E_a=0.4\times 8.3\times 1200$$

$$=3984\ \text{J mol}^{-1}$$

$$=3.984\ \text{kJ mol}^{-1}$$

**Note** Given, $\ln 2/3=0.4\Rightarrow\ln\frac{3}{2}=-0.4$

(These are wrong values, because $\ln 2/3$ is –ve and $\ln 3/2$ is +ve, but assume them to be correct).

# C-16

JEE Main Chapterwise Topicwise Chemistry

In JEE Main there is usually 0-1 question per paper are visible from this chapter. In past 3 years, 40 conceptual and 5 numerical questions were asked of moderate difficulty level from this chapter.

**For detailed analysis scan QR code**

# *p*-BLOCK ELEMENTS-2

## Topic 1 Group 15 Elements

### Level 1

**01** The maximum covalency of a non-metallic group 15 element '$E$' with weakest $E—E$ bond is *[2025, 22 Jan Shift-I]*

(a) 4 (b) 3
(c) 6 (d) 5

**02** Concentrated $HNO_3$ reacts with iodine to give *[2022, 28 July Shift-II]*

(a) HI, $NO_2$ and $H_2O$ (b) $HIO_2$, $N_2O$ and $H_2O$
(c) $HIO_3$, $NO_2$ and $H_2O$ (d) $HIO_4$, $N_2O$ and $H_2O$

**03** The gas produced by treating an aqueous solution of ammonium chloride with sodium nitrite is *[2022, 27 June Shift-II]*

(a) $NH_3$ (b) $N_2$
(c) $N_2O$ (d) $Cl_2$

**04** $PCl_5$ is well known but $NCl_5$ is not, because *[2022, 24 June Shift-II]*

(a) nitrogen is less reactive than phosphorus.
(b) nitrogen doesn't have *d*-orbitals in its valence shell.
(c) catenation tendency is weaker in nitrogen than phosphorus.
(d) size of phosphorus is larger than nitrogen.

**05** The most stable trihalide of nitrogen is *[2022, 24 June Shift-I]*

(a) $NF_3$ (b) $NCl_3$
(c) $NBr_3$ (d) $NI_3$

**06** Which one of the following group-15 hydride is the strongest reducing agent? *[2021, 22 July Shift-II]*

(a) $AsH_3$ (b) $BiH_3$
(c) $PH_3$ (d) $SbH_3$

**07** A group 15 element, which is a metal and forms a hydride with strongest reducing power among group 15 hydrides. The element is *[2021, 16 March Shift-I]*

(a) Sb (b) P
(c) As (d) Bi

**Imp.** Structure of Pyrophosphoric Acid

**08** In a molecule of pyrophosphoric acid, the number of P—OH, P=O and P—O—P bonds/moiety(ies) respectively are *[2020, 3 Sep Shift-I]*

(a) 2, 4 and 1 (b) 3, 3, and 3
(c) 4, 2, and 0 (d) 4, 2, and 1

### Level 2

**09** Given below are two statements: *[2025, 4 April Shift-I]*

**Statement I** Nitrogen forms oxides with +1 to +5 oxidation states due to the formation of $p\pi$-$p\pi$ bond with oxygen.

**Statement II** Nitrogen does not form halides with +5 oxidation state due to the absence of *d*-orbital in it.

In the light of given statements, choose the correct answer from the options given below.

(a) Statement I is true but Statement II is false.
(b) Both Statement I and Statement II are true.
(c) Both Statement I and Statement II are False.
(d) Statement I is false but Statement II are true.

**10** On reaction of lead sulphide with dilute nitric acid which of the following is not formed? *[2024, 9 April Shift-I]*

(a) Sulphur (b) Nitric oxide
(c) Lead nitrate (d) Nitrous oxide

**11** Given below are two statements. *[2024, 8 April Shift-I]*

**Statement I** $N(CH_3)_3$ and $P(CH_3)_3$ can act as ligands to form transition metal complexes.

**Statement II** As N and P are from same group, the nature of bonding of $N(CH_3)_3$ and $P(CH_3)_3$ is always same with transition metals.

In the light of the above statements, choose the most appropriate answer from the options given below:

(a) Statement I is correct but Statement II is incorrect.
(b) Statement I is incorrect but Statement II is correct.
(c) Both Statement I and Statement II are incorrect.
(d) Both Statement I and Statement II are correct.

**12** The strongest reducing agent among the following is *[2024, 1 Feb Shift-II]*
(a) $NH_3$ (b) $PH_3$ (c) $BiH_3$ (d) $SbH_3$

**13** Given below are two statements : one is labelled as **Assertion (A)** and the other is labelled as **Reason (R)**.
**Assertion (A)** $PH_3$ has lower boiling point than $NH_3$.
**Reason (R)** In liquid state $NH_3$ molecules are associated through van der Waals' forces, but $PH_3$ molecules are associated through hydrogen bonding. *[2024, 1 Feb Shift-I]*
In the light of the above statements, choose the **most appropriate** answer from the options given below
(a) A is false but R is true.
(b) Both A and R are true but R is not the correct explanation of A.
(c) Both A and R are true and R is the correct explanation of A.
(d) A is true but R is false.

**14** Given below are two statements : one is labelled as Assertion (A) and the other is labelled as Reason (R).
**Assertion (A)** There is a considerable increase in covalent radius from N to P. However from As to Bi only a small increase in covalent radius is observed.
**Reason (R)** Covalent and ionic radii in a particular oxidation state increases down the group. *[2024, 30 Jan Shift-I]*
In the light of the above statements, choose the **most appropriate** answer from the options given below
(a) Both A and R are true and R is the correct explanation of A.
(b) A is true but R is false.
(c) A is false but R is true.
(d) Both A and R are true but R is not the correct explanation of A.

**15** Which of the phosphorus oxoacid can create silver mirror from $AgNO_3$ solution? *[2023, 24 Jan, Shift-I]*
(a) $H_4P_2O_6$ (b) $H_4P_2O_7$
(c) $(HPO_3)_n$ (d) $H_4P_2O_5$

**16** Dinitrogen and dioxygen, the main constituents of air do not react with each other in atmosphere to form oxides of nitrogen because *[2022, 28 July Shift-II]*
(a) $N_2$ is unreactive in the condition of atmosphere
(b) oxides of nitrogen are unstable
(c) reaction between them can occur in the presence of a catalyst
(d) the reaction is endothermic and require very high temperature

**17** White phosphorus reacts with thionyl chloride to give *[2022, 28 July, Shift-II]*
(a) $PCl_5$, $SO_2$ and $S_2Cl_2$
(b) $PCl_3$, $SO_2$ and $S_2Cl_2$
(c) $PCl_3$, $SO_2$ and $Cl_2$
(d) $PCl_5$, $SO_2$ and $Cl_2$

**18** $A \xrightarrow{573\text{ K}}$ Red phosphorus $\xrightarrow[\text{Under pressure}]{\text{Heat ; 803 K}} B$
Red phosphorus is obtained by heating "$A$" at 573 K, and can be converted to "$B$" by heating at 803 K under pressure. $A$ and $B$, respectively, are *[2022, 30 June, Shift-I]*
(a) β-black phosphorus and white phosphorus
(b) White phosphorus and β-black phosphorus
(c) α-black phosphorus and white phosphorus
(d) White phosphorus and α-black phosphorus

**19** The oxoacid of phosphorus that is easily obtained from a reaction of alkali and white phosphorus and has two P—H bonds, is *[2022, 29 June Shift-I]*
(a) phosphonic acid (b) phosphinic acid
(c) pyrophosphorus acid (d) hypophosphoric acid

**20** Nitrogen gas is obtained by thermal decomposition of *[2022, 28 June, Shift-I]*
(a) $Ba(NO_3)_2$ (b) $Ba(N_3)_2$
(c) $NaNO_2$ (d) $NaNO_3$

**21** Heating white phosphorus with conc. NaOH solution gives mainly *[2022, 27 June Shift-I]*
(a) $Na_3P$ and $H_2O$ (b) $H_3PO$ and NaH
(c) $P(OH)_3$ and $NaH_2PO_4$ (d) $PH_3$ and $NaH_2PO_2$

**22** The oxide which contains an odd electron at the nitrogen atom is *[2022, 26 June Shift-II]*
(a) $N_2O$ (b) $NO_2$ (c) $N_2O_3$ (d) $N_2O_5$

**23** The oxide without nitrogen- nitrogen bond is *[2021, 1 Sep Shift-II]*
(a) $N_2O$ (b) $N_2O_4$ (c) $N_2O_3$ (d) $N_2O_5$

**24** Which one of the following is formed (mainly) when red phosphorus is heated in a sealed tube at 803 K ? *[2021, 27 Aug Shift-II]*
(a) White phosphorus (b) Yellow phosphorus
(c) β-black phosphorus (d) α-black phosphorus

**25** Chemical nature of the nitrogen oxide compound obtained from a reaction of concentrated nitric acid and $P_4O_{10}$ (in 4 : 1 ratio) is *[2021, 20 July Shift-I]*
(a) acidic (b) basic
(c) amphoteric (d) neutral

**Imp.** Order of Oxidation States of Nitrogen

**26** The oxidation states of nitrogen in NO, $NO_2$, $N_2O$ and $NO_3^-$ are in the order of *[2021, 18 March Shift-II]*
(a) $NO_3^- > NO_2 > NO > N_2O$
(b) $NO_2 > NO_3^- > NO > N_2O$
(c) $N_2O > NO_2 > NO > NO_3^-$
(d) $NO > NO_2 > N_2O > NO_3^-$

**27** The set that represents the pair of neutral oxides of nitrogen is *[2021, 17 March Shift-II]*
(a) NO and $N_2O$ (b) $N_2O$ and $N_2O_3$
(c) $N_2O$ and $NO_2$ (d) NO and $NO_2$

**28** Which of the following compound cannot act as a Lewis base? *[2021, 17 March Shift-I]*

(a) $NF_3$ (b) $PCl_5$
(c) $SF_4$ (d) $ClF_3$

**29** **Statement I** Sodium hydride can be used as an oxidising agent.
**Statement II** The lone pair of electrons on nitrogen in pyridine makes it basic.
Choose the correct answer from the options given below. *[2021, 16 March Shift-II]*

(a) Both Statement I and Statement II are true
(b) Both Statement I and Statement II are false
(c) Statement I is true but Statement II is false
(d) Statement I is false but Statement II is true

**30** The reaction of NO with $N_2O_4$ at 250 K gives *[2020, 6 Sep Shift-II]*

(a) $N_2O$ (b) $NO_2$
(c) $N_2O_3$ (d) $N_2O_5$

**31** Reaction of ammonia with excess $Cl_2$ gives *[2020, 5 Sep Shift-II]*

(a) $NH_4Cl$ and $N_2$ (b) $NH_4Cl$ and HCl
(c) $NCl_3$ and $NH_4Cl$ (d) $NCl_3$ and HCl

**32** *Aqua-regia* is used for dissolving noble metals (Au, Pt, etc.). The gas evolved in this processes is *[2020, 03 Sep, Shift-I]*

(a) NO (b) $N_2O_5$
(c) $N_2$ (d) $N_2O_3$

**33** The number of P—OH bonds and the oxidation state of phosphorus atom in pyrophosphoric acid ($H_4P_2O_7$) respectively are *[2020, 03 Sep, Shift-I, 2017]*

(a) four and four (b) five and four
(c) five and five (d) four and five

**34** An organic compound *A* is oxidised with $Na_2O_2$ followed by boiling with $HNO_3$. The resultant solution is then treated with ammonium molybdate to yield a yellow precipitate.
Based on above observation, the element present in the given compound is *[2019, 12 April Shift-I]*

(a) nitrogen (b) phosphorus
(c) fluorine (d) sulphur

**35** Good reducing nature of $H_3PO_2$ is attributed to the presence of *[2019, 9 Jan Shift-II]*

(a) two P— H bonds (b) one P— H bond
(c) two P— OH bonds (d) one P — OH bond

**36** The number of P—O bonds in $P_4O_6$ is *[JEE Main 2018]*

(a) 9 (b) 6 (c) 12 (d) 18

**37** The pair in which phosphorus atoms have a formal oxidation state of +3 is *[JEE Main 2016]*

(a) pyrophosphorus and hypophosphoric acids
(b) orthophosphorus and hypophosphoric acids
(c) pyrophosphorus and pyrophosphoric acids
(d) orthophosphorus and pyrophosphorous acids

**38** Which of the following statements is wrong?

(a) The stability of hydrides increases from $NH_3$ to $BiH_3$ in group 15 of the periodic table.
(b) Nitrogen can't form $d\pi$-$p\pi$ bond.
(c) Single N—N bond is weaker than the single P—P bond.
(d) $N_2O_4$ has two resonance structure. *[AIEEE 2011]*

**39** Regular use of which of the following fertilizers increases the acidity of soil ? *[AIEEE 2007]*

(a) Potassium nitrate (b) Urea
(c) Superphosphate of lime (d) Ammonium sulphate

**40** The number of hydrogen atom(s) attached to phosphorus atom in hypophosphorous acid is *[AIEEE 2005]*

(a) three (b) one (c) two (d) zero

## Level 3

**41** Given below are two statements:

**Statement I** The N—N single bond is weaker and longer than that of P—P single bond.
**Statement II** Compounds of group 15 elements in +3 oxidation states readily undergo disproportionation reactions. *[2025, 3 April Shift-II]*
In the light of the above statements, choose the correct answer from the options given below

(a) Statement I is true but Statement II is false.
(b) Both Statement I and Statement II are false.
(c) Both Statement I and Statement II are true.
(d) Statement I is false but Statement II is true.

**42** The incorrect statement among the following is *[2025, 23 Jan Shift-I]*

(a) $PH_3$ shows lower proton affinity than $NH_3$.
(b) $NO_2$ can dimerise easily.
(c) $SO_2$ can act as an oxidising agent, but not as a reducing agent.
(d) $PF_3$ exists but $NF_5$ does not.

**43** Identify the incorrect statements about group 15 elements:

(A) Dinitrogen is a diatomic gas which acts like an inert gas at room temperature.
(B) The common oxidation states of these elements are −3, +3 and +5.
(C) Nitrogen has unique ability to form $p\pi$-$p\pi$ multiple bonds.
(D) The stability of +5 oxidation states increases down the group.
(E) Nitrogen shows a maximum covalency of 6.

Choose the correct answer from the options given below: *[2024, 8 April Shift-II]*

(a) (A), (C), (E) only
(b) (D) and (E) only
(c) (B), (D), (E) only
(d) (A), (B), (D) only

**44** Choose the correct statements about the hydrides of group 15 elements. **[2024, 30 Jan Shift-II]**

A. The stability of the hydrides decreases in the order $NH_3 > PH_3 > AsH_3 > SbH_3 > BiH_3$.

B. The reducing ability of the hydride increases in the order $NH_3 < PH_3 < AsH_3 < SbH_3 < BiH_3$.

C. Among the hydride, $NH_3$ is strong reducing agent while $BiH_3$ is mild reducing agent.

D. The basicity of the hydrides increases in the order $NH_3 < PH_3 < AsH_3 < SbH_3 < BiH_3$.

Choose the **most appropriate** from the options given below:

(a) A and D only (b) B and C only
(c) C and D only (d) A and B only

**45** The number of P—O—P bonds in $H_4P_2O_7$, $(HPO_3)_3$ and $P_4O_{10}$ are respectively **[2023, 15 April Shift-II]**

(a) 1, 2, 4 (b) 0, 3, 4
(c) 0, 3, 6 (d) 1, 3, 6

**46** Reaction of thionyl chloride with white phosphorus forms a compound [*A*], which on hydrolysis gives [*B*], a dibasic acid. [*A*] and [*B*] are respectively **[2023, 25 Jan, Shift-I]**

(a) $PCl_3$ and $H_3PO_3$ (b) $PCl_5$ and $H_3PO_4$
(c) $POCl_3$ and $H_3PO_4$ (d) $P_4O_6$ and $H_3PO_3$

**47** Dinitrogen is a robust compound, but reacts at high altitude to form oxides. The oxide of nitrogen that can damage plant leaves and retard photosynthesis is **[2022, 29 July, Shift-II]**

(a) NO (b) $NO_3^-$ (c) $NO_2$ (d) $NO_2^-$

**48** Which oxo-acid of phosphorus has the highest number of oxygen atoms present in its chemical formula? **[2022, 27 July Shift-I]**

(a) Pyrophosphorus acid (b) Hypophosphoric acid
(c) Phosphoric acid (d) Pyrophosphoric acid

**49** Among the given oxides of nitrogen; $N_2O$, $N_2O_3$, $N_2O_4$ and $N_2O_5$, the number of compounds having N—N bond is **[2022, 28 June Shift-II]**

(a) 1 (b) 2 (c) 3 (d) 4

**50** Given below are two statements. **[2022, 28 June Shift-I]**

**Statement I** The pentavalent oxide of group-15 element, $E_2O_5$, is less acidic than trivalent oxide, $E_2O_3$ of the same element.

**Statement II** The acidic character of trivalent oxide of group 15 elements, $E_2O_3$, decreases down the group.

In light of the above statements, choose most appropriate answer from the options given below

(a) Both Statement I and Statement II are true.
(b) Both Statement I and Statement II are false.
(c) Statement I is true but Statement II is false.
(d) Statement I is false but Statement II is true.

**51** Consider the following reaction,

$A + \text{alkali} \rightarrow B$ (major product)

If *B* is an oxoacid of phosphorus with no P—H bond, then *A* is **[2022, 26 June Shift-I]**

(a) white $P_4$ (b) red $P_4$
(c) $P_2O_3$ (d) $H_3PO_3$

**52** The number of ionisable hydrogens present in the product obtained from a reaction of phosphorus trichloride and phosphonic acid is **[2021, 18 March Shift-I]**

(a) 3 (b) 0 (c) 2 (d) 1

**53** Match List-I with List-II : **[2021, 16 March, Shift-I]**

| | List-I (Name of oxo acid) | | List-II (Oxidation state of 'P') |
|---|---|---|---|
| (A) | Hypophosphorus acid | (I) | +5 |
| (B) | Orthophosphoric acid | (II) | +4 |
| (C) | Hypophosphoric acid | (III) | +3 |
| (D) | Orthophosphorus acid | (IV) | +2 |
| | | (V) | +1 |

Choose the correct answer from the options given below

(a) (A)-(IV), (B)-(V), (C)-(II), (D)-(III)
(b) (A)-(V), (B)-(IV), (C)-(II), (D)-(III)
(c) (A)-(V), (B)-(I), (C)-(II), (D)-(III)
(d) (A)-(IV), (B)-(I), (C)-(II), (D)-(III)

**54** Find *A*, *B* and *C* in the following reactions

$$NH_3 + A + CO_2 \longrightarrow (NH_4)_2CO_3$$
$$(NH_4)_2CO_3 + H_2O + B \longrightarrow NH_4HCO_3$$
$$NH_4HCO_3 + NaCl \longrightarrow NH_4Cl + C$$

**[2021, 26 Feb Shift-I]**

(a) $A - O_2$, $B - CO_2$, $C - Na_2CO_3$
(b) $A - H_2O$, $B - O_2$, $C - Na_2CO_3$
(c) $A - H_2O$, $B - O_2$, $C - NaHCO_3$
(d) $A - H_2O$, $B - CO_2$, $C - NaHCO_3$

**55** On heating, lead (II) nitrate gives a brown gas (*A*). The gas (*A*) on cooling changes to a colourless solid-liquid (*B*). (*B*) on heating with NO changes to a blue solid (*C*). The oxidation number of nitrogen in solid (*C*) is **[2020, 04 Sep, Shift-I]**

(a) +5 (b) +2 (c) +3 (d) +4

**56** The correct order of the oxidation states of nitrogen in NO, $NO_2$, $NO_2$ and $N_2O_3$ is **[2019, 9 April Shift-I]**

(a) $NO_2 < NO < N_2O_3 < N_2O$
(b) $N_2O < NO < N_2O_3 < NO_2$
(c) $O_2 < N_2O_3 < NO < N_2O$
(d) $N_2O < N_2O_3 < NO < NO_2$

**57** The pair that contains two P—H bonds in each of the oxoacids is **[2019, 10 Jan Shift-II]**

(a) $H_4P_2O_5$ and $H_4P_2O_6$
(b) $H_3PO_3$ and $H_3PO_2$
(c) $H_4P_2O_5$ and $H_3PO_3$
(d) $H_3PO_2$ and $H_4P_2O_5$

**58** Which one of the following properties is not shown by NO? *[JEE Main 2014]*
(a) It is diamagnetic in gaseous state
(b) It is a neutral oxide
(c) It combines with oxygen to form nitrogen dioxide
(d) Its bond order is 2.5

**59** In which of the following arrangements, the sequence is not strictly according to the property written against it? *[AIEEE 2009]*
(a) $CO_2 < SiO_2 < SnO_2 < PbO_2$ increasing oxidising power
(b) $HF < HCl < HBr < HI$ increasing acid strength
(c) $NH_3 > PH_3 < AsH_3 < SbH_3$ increasing basic strength
(d) $B < C < O < N$ increasing first ionisation enthalpy

## Numerical Value Type Questions

**60** A group 15 element forms $d\pi - d\pi$ bond with transition metals. It also forms hydride, which is a strongest base among the hydrides of other group members that form $d\pi - d\pi$ bond. The atomic number of the element is...... *[2025, 28 Jan Shift-II]*

**61** The ratio of σ and π bonds present in pyrophosphoric acid is ....... *[2023, 8 April Shift-II]*

**62** The number of non-ionisable protons present in the product $B$ obtained from the following reaction is ........

$$C_2H_5OH + PCl_3 \longrightarrow C_2H_5Cl + A$$
$$A + PCl_3 \longrightarrow B$$ *[2022, 26 July, Shift-II]*

**63** Consider the following reactions,

$$PCl_3 + H_2O \longrightarrow A + HCl$$
$$A + H_2O \longrightarrow B + HCl$$

The number of ionisable protons present in the product $B$ is ......... *[2022, 25 June Shift-II]*

**64** The reaction of white phosphorus on boiling with alkali in inert atmosphere resulted in the formation of product $A$. The reaction of 1 mol of $A$ with excess of $AgNO_3$ in aqueous medium gives ...... moles of Ag (Round off to the nearest integer). *[2021, 17 March Shift-I]*

## Topic 2 Group 16 Elements

### Level 1

**01** Anomalous behaviour of oxygen is due to its *[2024, 29 Jan Shift-II]*
(a) large size and high electronegativity
(b) small size and low electronegativity
(c) small size and high electronegativity
(d) large size and low electronegativity

**02** The correct order of melting point of hydrides of group 16 elements is *[2022, 26 June Shift-I]*
(a) $H_2S < H_2Se < H_2Te < H_2O$
(b) $H_2O < H_2S < H_2Se < H_2Te$
(c) $H_2S < H_2Te < H_2Se < H_2O$
(d) $H_2Se < H_2S < H_2Te < H_2O$

**03** In polythionic acid, $H_2S_xO_6 (x = 3 \text{ to } 5)$ the oxidation states of sulphur is/are *[2021, 27 Aug Shift-I]*
(a) only +5
(b) only +6
(c) +3 and +5
(d) 0 and +5

**04** Chalcogen group elements are *[2021, 26 Aug Shift-II]*
(a) Se, Tb and Pu
(b) Se, Te and Po
(c) S, Te and Pm
(d) O, Ti and Po

**05** The correct order of electron gain enthalpy is *[2021, 26 Feb Shift-II]*
(a) $S > Se > Te > O$
(b) $Te > Se > S > O$
(b) $O > S > Se > Te$
(d) $S > O > Se > Te$

### Level 2

**06** Given below are two statements: *[2025, 8 April Shift-II]*

**Statement I** $H_2Se$ is more acidic than $H_2Te$.

**Statement II** $H_2Se$ has higher bond enthalpy for dissociation than $H_2Te$.

In the light of the above statements, choose the correct answer from the options given below
(a) Both Statement I and Statement II are true.
(b) Statement I is false but Statement II is true.
(c) Both Statement I and Statement II are false.
(d) Statement I is true but Statement II is false.

**07** Which among the following molecules is (a) involved in $sp^3d$ hybridisation, (b) has different bond lengths and (c) has lone pair of electrons on the central atom? *[2025, 2 April Shift-II]*
(a) $PF_5$
(b) $XeF_4$
(c) $XeF_2$
(d) $SF_4$

**08** Given below are two statements. *[2025, 24 Jan Shift-II]*

**Statement I** Experimentally determined oxygen-oxygen bond lengths in the $O_3$ are found to be same and the bond length is greater than that of a O=O (double bond) but less than that of a single (O—O) bond.

**Statement II** The strong lone pair-lone pair repulsion between oxygen atoms is solely responsible for the fact that the bond length in ozone is smaller than that of a double bond (O=O) but more than that of a single bond (O—O).

In the light of the above statements, choose the correct answer from the options given below
(a) Both Statement I and Statement II are false.
(b) Statement I is false but Statement II is true.
(c) Statement I is true but Statement II is false.
(d) Both Statement I and Statement II are true.

**09** Given below are two statements : one is labelled as Assertion (A) and other is labelled as Reason (R). *[2024, 9 April Shift-I]*

**Assertion (A)** Both rhombic and monoclinic sulphur exist as $S_8$ while oxygen exists as $O_2$.

**Reason (R)** Oxygen forms $p\pi$-$p\pi$ multiple bonds with itself and other elements having small size and high electronegativity like C, N, which is not possible for sulphur.

In the light of the above statements, choose the **most appropriate** answer from the options given below.

(a) Both A and R are correct but R is not the correct explanation of A.
(b) A is correct but R is not correct.
(c) A is not correct but R is correct.
(d) Both A and R are correct and R is the correct explanation of A.

**10** Given below are two statements : one is labelled as **Assertion (A)** and the other is labelled as **Reason (R)**. *[2024, 30 Jan Shift-II]*

**Assertion (A)** $H_2Te$ is more acidic than $H_2S$.

**Reason (R)** Bond dissociation enthalpy of $H_2Te$ is lower than $H_2S$.

In the light of the above statements, choose the **most appropriate** answer from the options given below:

(a) A is false but R is true.
(b) Both A and R are true but R is not the correct explanation of A.
(c) A is true but R is not true.
(d) Both A and R are true and R is the correct explanation of A.

**11** Given below are two statements. *[2024, 27 Jan Shift-I]*

**Statement I** Oxygen being the first member of group 16 exhibits only −2 oxidation state.

**Statement II** Down the group 16 stability of +4 oxidation state decreases and +6 oxidation state increases.

In the light of the above statements, choose the **most appropriate** answer from the options given below.

(a) Statement I is correct but Statement II is incorrect.
(b) Both Statement I and Statement II are incorrect.
(c) Statement I is incorrect but Statement II is correct.
(d) Both Statement I and Statement II are correct.

**12** Sulphur dioxide is one of the components of polluted air. $SO_2$ is also a major contributor to acid rain. The correct and complete reaction to represent acid rain caused by $SO_2$ is *[2022, 29 June Shift-II]*

(a) $2SO_2 + O_2 \longrightarrow 2SO_3$
(b) $SO_2 + O_3 \longrightarrow SO_3 + O_2$
(c) $SO_2 + H_2O_2 \longrightarrow H_2SO_4$
(d) $2SO_2 + O_2 + 2H_2O \longrightarrow 2H_2SO_4$

**13** The depositions of *X* and *Y* on ground surfaces is referred to as wet and dry depositions, respectively. *X* and *Y* are *[2021, 31 Aug Shift-II]*

(a) *X* = Ammonium salts, *Y* = $CO_2$
(b) *X* = $SO_2$, *Y* = Ammonium salts
(c) *X* = Ammonium salts, *Y* = $SO_2$
(d) *X* = $CO_2$, *Y* = $SO_2$

**14** If the boiling point of $H_2O$ is 373 K, the boiling point of $H_2S$ will be *[2020, 03 Sep, Shift-I]*

(a) less than 300 K
(b) equal to 373 K
(c) more than 373 K
(d) greater than 300 K but less than 373 K

**15** Which of the following statements regarding sulphur is incorrect? *[AIEEE 2011]*

(a) $S_2$ molecule is paramagnetic.
(b) The vapour at 200°C consists mostly of $S_8$ rings.
(c) At 600°C, the gas mainly consists of $S_2$ molecules.
(d) The oxidation state of sulphur is never less than +4 in its compounds.

## Level 3

**16** The large difference between the melting and boiling points of oxygen and sulphur may be explained on the basis of *[2025, 24 Jan Shift-I]*

(a) electronegativity (b) atomic size
(c) electron gain enthalpy (d) atomicity

**17** Bond dissociation energy of "*E*—H" bond of the "$H_2E$" hydrides of group 16 elements (given below), follows order. *[2023, 30 Jan, Shift-II]*

A. O B. S
C. Se D. Te

Choose the correct from the options given below:

(a) A > B > C > D
(b) D > C > B > A
(c) B > A > C > D
(d) A > B > D > C

**18** Which of the following oxoacids of sulphur contains "S" in two different oxidation states? *[2022, 28 June Shift-II]*

(a) $H_2S_2O_3$ (b) $H_2S_2O_6$
(c) $H_2S_2O_7$ (d) $H_2S_2O_8$

**19** The number of S=O bonds present in sulphurous acid, peroxodisulphuric acid and pyrosulphuric acid, respectively are *[2021, 31 Aug Shift-II ; 2017]*

(a) 2, 3 and 4 (b) 1, 4 and 3
(c) 2, 4 and 3 (d) 1, 4 and 4

**20** The product obtained from the electrolytic oxidation of acidified sulphate solutions, is *[2021, 27 July Shift-I]*

(a) $HSO_4^-$
(b) $HO_3SOOSO_3H$
(c) $HO_2SOSO_2H$
(d) $HO_3SOSO_3H$

Imp. **Application of Industrial Process**

**21** Match List -I with List II.

| List-I (Industrial process) | List-II (Application) |
|---|---|
| A. Haber's process | (i) $HNO_3$ synthesis |
| B. Ostwald's process | (ii) Aluminium extraction |
| C. Contact process | (iii) $NH_3$ synthesis |
| D. Hall-Heroult process | (iv) $H_2SO_4$ synthesis |

Choose the correct answer from the options given below *[2021, 16 March Shift-I]*
(a) A-(ii), B-(iii), C-(iv), D-(i)
(b) A-(iii), B-(iv), C-(i), D-(ii)
(c) A-(iii), B-(i), C-(iv), D-(ii)
(d) A-(iv), B-(i), C-(ii), D-(iii)

**22** Given below are two statements.
**Statement I** α and β-forms of sulphur can change reversibly between themselves with slow heating or slow cooling.
**Statement II** At room temperature, the stable crystalline form of sulphur is monoclinic sulphur.
In the light of the above statements, choose the correct answer from the options given below.
(a) Both statements I and II are true. *[2021, 25 Feb Shift-II]*
(b) Both statements I and II are false.
(c) Statement I is true but statement II is false.
(d) Statement I is false but statement II is true.

**23** Choose the correct statements from the following.
A. All group 16 elements form oxides of general formula $EO_2$ and $EO_3$, where $E =$ S, Se, Te and Po. Both the types of oxides are acidic in nature.
B. $TeO_2$ is an oxidising agent while $SO_2$ is reducing in nature.
C. The reducing property decreases from $H_2S$ to $H_2Te$ down the group.
D. The ozone molecule contains five lone pairs of electrons.

Choose the correct answer from the options given below. *[2020, 31 Jan, Shift-I]*
(a) A and B only
(b) B and C only
(c) C and D only
(d) A and D only

**24** Reaction of an inorganic sulphite $X$ with dilute $H_2SO_4$ generates compound $Y$. Reaction of $Y$ with NaOH gives $X$. Further, the reaction of $X$ with $Y$ and water affords compound $Y$ and $Z$. respectively are *[2020, 06 Sep, Shift-II]*
(a) $SO_2$ and $Na_2SO_3$
(b) $SO_3$ and $NaHSO_3$
(c) $SO_2$ and $NaHSO_3$
(d) S and $Na_2SO_3$

**25** The number of bonds between sulphur and oxygen atoms in $S_2O_8^{2-}$ and the number of bonds between sulphur and sulphur atoms in rhombic sulphur, respectively, are *[2020, 8 Jan Shift-I]*
(a) 4 and 6 (b) 8 and 8
(c) 4 and 8 (d) 8 and 6

## Numerical Value Type Questions

Imp. **Structure of Oxoacids**

**26** Consider the following sulphur based oxoacids. $H_2SO_3$, $H_2SO_4$, $H_2S_2O_8$ and $H_2S_2O_7$. Amongst these oxo-acids, the number of those with peroxo (O—O) bond is ......... . *[2022, 29 July Shift-II]*

**27** Consider the sulphides HgS, PbS, CuS, $Sb_2S_3$, $As_2S_3$ and CdS. Number of these sulphides soluble in 50% $HNO_3$ is *[2021, 31 Aug Shift-I]*

**28** Among the following allotropic forms of sulphur, the number of allotropic forms, which will show paramagnetism is ......... . *[2021, 24 Feb Shift-II]*
A. α-sulphur B. β-sulphur
C. $S_2$-form

**29** Number of oxygen atoms present in chemical formula of fuming sulphuric acid is ........... . *[2020, 9 April Shift-II]*

# Topic 3 Group 17 Elements

## Level 1

**01** Which halogen is known to cause the reaction given below?

$$2Cu^{2+} + 4X^- \longrightarrow Cu_2X_2(s) + X_2$$

*[2023, 8 April Shift-I]*
(a) all halogens (b) only chlorine
(c) only iodine (d) only bromine

**02** Element not present in Nessler's reagent is *[2023, 6 April Shift-II]*
(a) N (b) Hg
(c) K (d) I

**03** Which element is not present in Nessler's reagent? *[2023, 1 Feb Shift-II]*
(a) Oxygen (b) Iodine
(c) Potassium (d) Mercury

**04** Formula for Nessler's reagent is *[2023, 30 Jan Shift-II]*
(a) $KHgI_3$ (b) $HgI_2$
(c) $K_2HgI_4$ (d) $KHg_2I_2$

**05** Given below are two statements. One is labelled as Assertion A and the other is labelled as Reason R.
**Assertion (A)** Fluorine forms one oxoacid.
**Reason (R)** Fluorine has smallest size amongst all halogens and is highly electronegative.

In the light of the above statements, choose the most appropriate answer from the options given below. *[2022, 27 June Shift-II]*

(a) Both A and R are correct and R is the correct explanation of A.
(b) Both A and R are correct but R is not the correct explanation of A.
(c) A is correct but R is not correct.
(d) A is not correct but R is correct.

**06** The absolute value of the electron gain enthalpy of halogens satisfies *[2021, 17 March Shift-I]*

(a) I > Br > Cl > F (b) Cl > Br > F > I
(c) Cl > F > Br > I (d) F > Cl > Br > I

**07** Which among the following is the most reactive? *[JEE Main 2015]*

(a) $Cl_2$ (b) $Br_2$
(c) $I_2$ (d) ICl

## Level 2

**08** Among the following halogens $F_2, Cl_2, Br_2$ and $I_2$ Which can undergo disproportionation reactions? *[2024, 8 April Shift-II]*

(a) $F_2, Cl_2$ and $Br_2$ (b) $F_2$ and $Cl_2$
(c) Only $I_2$ (d) $Cl_2$, $Br_2$ and $I_2$

**09** Given below are two statements: *[2024, 29 Jan Shift-II]*

**Statement I** Fluorine has most negative electron gain enthalpy in its group.

**Statement II** Oxygen has least negative electron gain enthalpy in its group.

In the light of the above statements, choose the **most appropriate** from the options given below

(a) Statement I is false but Statement II is true.
(b) Both Statement I and Statement II are false.
(c) Statement I is true but Statement II is false.
(d) Both Statement I and Statement II are true.

**10** Element not showing variable oxidation state is *[2024, 27 Jan Shift-II]*

(a) bromine (b) chlorine
(c) fluorine (d) iodine

**Imp.** Oxidation of Halide Ion in Acidic Medium

**11** The correct group of halide ions which can be oxidised by oxygen in acidic medium is *[2023, 13 April Shift-II]*

(a) $I^-$ only (b) $Br^-$ only
(c) $Cl^-$, $Br^-$ and $I^-$ only (d) $Br^-$ and $I^-$ only

**12** Given below are two statements.

**Statement I** $SbCl_5$ is more covalent than $SbCl_3$.

**Statement II** The higher oxides of halogens also tend to be more stable than the lower ones.

In the light of the above statements, choose the correct answer from the options given below. *[2023, 12 April Shift-I]*

(a) Statement I is incorrect but Statement II is correct.
(b) Both statement I and Statement II are correct.
(c) Statement I is correct but Statement II is incorrect.
(d) Both Statement I and Statement II are incorrect.

**13** The interhalogen compound formed from the reaction of bromine with excess of fluorine is a *[2022, 25 July Shift-I]*

(a) hypohalite (b) halate
(c) perhalate (d) halite

**14** Which one of the following correctly represents the order of stability of oxides, $X_2O$ ($X$ = halogen)? *[2021, 31 Aug Shift-II]*

(a) Br > Cl > I (b) Br > I > Cl
(c) Cl > I > Br (d) I > Cl > Br

**15** Which one of the following is used to remove most of plutonium from spent nuclear fuel? *[2021, 27 Aug Shift-II]*

(a) $ClF_3$ (b) $O_2F_2$
(c) $I_2O_5$ (d) $BrO_3$

**16** Choose the incorrect statement. *[2021, 26 Aug Shift-I]*

(a) $Cl_2$ is more reactive than ClF.
(b) $F_2$ is more reactive than ClF.
(c) On hydrolysis ClF froms HOCl and HF.
(d) $F_2$ is a stronger oxidising agent than $Cl_2$ in aqueous solution.

**17** Number of Cl=O bonds in chlorous acid, chloric acid and perchloric acid respectively are *[2021, 27 July Shift-II]*

(a) 3, 1 and 1 (b) 4, 1 and 0
(c) 1, 1 and 3 (d) 1, 2 and 3

**18** HF has highest boiling point among hydrogen halides, because it has *[2019, 10 April Shift-I]*

(a) lowest ionic character
(b) strongest van der Waals' interactions
(c) strongest hydrogen bonding
(d) lowest dissociation enthalpy

**19** What products are expected from the disproportionation reaction of hypochlorous acid? *[2019, 8 April Shift I]*

(a) $HClO_3$ and $Cl_2O$ (b) $HClO_2$ and $HClO_4$
(c) HCl and $Cl_2O$ (d) HCl and $HClO_3$

**20** Among the following reactions of hydrogen with halogens, the one that requires a catalyst is *[2019, 10 Jan Shift-II]*

(a) $H_2 + Cl_2 \longrightarrow 2HCl$ (b) $H_2 + I_2 \longrightarrow 2HI$
(c) $H_2 + F_2 \longrightarrow 2HF$ (d) $H_2 + Br_2 \longrightarrow 2HBr$

**21** The products obtained when chlorine gas reacts with cold and dilute aqueous NaOH are *[JEE Main 2017]*

(a) $ClO^-$ and $ClO_3^-$
(b) $ClO_2^-$ and $ClO_3^-$
(c) $Cl^-$ and $ClO^-$
(d) $Cl^-$ and $ClO_2^-$

## Level 3

**22** The property/properties that show irregularity in first four elements of group-17 is/are: *[2025, 2 April Shift-I]*

A. covalent radius
B. electron affinity
C. ionic radius
D. first ionisation energy

Choose the correct answer from the options given below:

(a) B only (b) A, B, C and D
(c) A and C only (c) B and D only

**23** Match List I with List II, match the gas evolved during each reaction.

| List I | List II |
|---|---|
| (A) $(NH_4)_2Cr_2O_7 \xrightarrow{\Delta}$ | (I) $H_2$ |
| (B) $KMnO_4 + HCl \longrightarrow$ | (II) $N_2$ |
| (C) $Al + NaOH + H_2O \longrightarrow$ | (III) $O_2$ |
| (D) $NaNO_3 \longrightarrow$ | (IV) $Cl_2$ |

Choose the correct answer from the options given below. *[2022, 28 July Shift-I]*

(a) (A) → (II), (B) → (III), (C) → (I), (D) → (IV)
(b) (A) → (III), (B) → (I), (C) → (IV), (D) → (II)
(c) (A) → (II), (B) → (IV), (C) → (I), (D) → (III)
(d) (A) → (III), (B) → (IV), (C) → (I), (D) → (II)

**24** The correct order of bond dissociation enthalpy of halogens is *[2021, 25 Feb Shift-II]*

(a) $F_2 > Cl_2 > Br_2 > I_2$
(b) $I_2 > Br_2 > Cl_2 > F_2$
(c) $Cl_2 > Br_2 > F_2 > I_2$
(d) $Cl_2 > F_2 > Br_2 > I_2$

**25** Among the following oxoacids, the correct decreasing order of acid strength is *[JEE Main 2014]*

(a) $HOCl > HClO_2 > HClO_3 > HClO_4$
(b) $HClO_4 > HOCl > HClO_2 > HClO_3$
(c) $HClO_4 > HClO_3 > HClO_2 > HOCl$
(d) $HClO_2 > HClO_4 > HClO_3 > HOCl$

### Numerical Values Type Questions

**26** Among the following oxides of *p*-block elements, number of oxides having amphoteric nature is $Cl_2O_7$, CO, $PbO_2$, $N_2O$, NO, $Al_2O_3$, $SiO_2$, $N_2O_5$, $SnO_2$ *[2024, 1 Feb, Shift-I]*

**27** In the following reactions, the total number of oxygen atoms in *X* and *Y* is ........

$$Na_2O + H_2O \longrightarrow 2X$$
$$Cl_2O_7 + H_2O \longrightarrow 2Y$$
*[2023, 10 April Shift-I]*

**28** The number of interhalogens from the following having square pyramidal structure is
$ClF_3$, $IF_7$, $BrF_5$, $BrF_3$, $I_2Cl_6$, $IF_5$, ClF, $ClF_5$ *[2022, 28 July Shift-I]*

**29** The number of halogens forming halic (V) acid is ......... . *[2021, 31 Aug Shift-I]*

**30** Among the following, the number of halides which is/are inert to hydrolysis is .............. . *[2021, 25 Feb Shift-I]*

(i) $BF_3$ (ii) $SiCl_4$
(iii) $PCl_5$ (iv) $SF_6$

**31** The number of Cl = O bonds in perchloric acid is "........". *[2020, 6 Sep Shift-I]*

**32** Chlorine reacts with hot and concentrated NaOH and produces compounds (*X*) and (*Y*). Compound (*X*) gives white precipitate with silver nitrate solution. The average bond order between Cl and O atoms in (*Y*) is ...... . *[2020, 7 Jan Shift-I]*

# Topic 4 Group 18 Elements

## Level 1

**01** The noble gas that does not occur in the atmosphere is *[2019, 10 April Shift-II]*

(a) Ra (b) Kr
(c) He (d) Ne

**02** Which one has the highest boiling point? *[JEE Main 2015]*

(a) He (b) Ne
(c) Kr (d) Xe

## Level 2

**03** Given below are two statements.

**Statement I** Noble gases have very high boiling points.

**Statement II** Noble gases are monoatomic gases. They are held together by strong dispersion forces. Because of this they are liquefied at very low temperature. Hence, they have very high boiling points.

In the light of the above statements, choose the correct answer from the options given below.

(a) Statement I is true but Statement II is false.
(b) Statement I is false but Statement II is true.
(c) Both Statement I and Statement II are true.
(d) Both Statement I and Statement II are false.

**04** The difference between electron gain enthalpies will be maximum between *[2023, 6 April Shift-I]*

(a) Ar and Cl (b) Ar and F
(c) Ne and Cl (d) Ne and F

**05** Match List I with List II.

| List I (Molecule) | | List II (Hybridisation shape) |
|---|---|---|
| A. | $XeO_3$ | I. $sp^3d$; linear |
| B. | $XeF_2$ | II. $sp^3$ ; pyramidal |
| C. | $XeOF_4$ | III. $sp^3d^3$; distorted octahedral |
| D. | $XeF_6$ | IV. $sp^3d^2$; square pyramidal |

Choose the correct answer from the options given below. *[2022, 25 July Shift-II]*

(a) A-II, B-I, C-IV, D-III
(b) A-II, B-IV, C-III, D-I
(c) A-IV, B-II, C-III, D-I
(d) A-IV, B-II, C-I, D-III

**06** The shape/structure of $[XeF_5]^-$ and $XeO_3F_2$, respectively, are *[2020, 2 Sep Shift-II]*

(a) pentagonal planar and trigonal bipyramidal
(b) octahedral and square pyramidal
(c) trigonal bipyramidal and pentagonal planar
(d) trigonal bipyramidal and trigonal bipyramidal

**07** The type of hybridisation and number of lone pair(s) of electrons of Xe in $XeOF_4$, respectively, are *[2019, 10 Jan Shift-I]*

(a) $sp^3d^2$ and 1
(b) $sp^3d$ and 2
(c) $sp^3d$ and 1
(d) $sp^3d^2$ and 2

## Level 3

**08** Match the List-I with List-II. *[2025, 7 April Shift-I]*

| List-I (Molecule/ion) | | List-II Bond pair : lone pair (on the central atom) | |
|---|---|---|---|
| A. | $ICl_2^-$ | I. | 4 : 2 |
| B. | $H_2O$ | II. | 4 : 1 |
| C. | $SO_2$ | III. | 2 : 3 |
| D. | $XeF_4$ | IV. | 2 : 2 |

Choose the correct answer from the options given below:

(a) A-III, B-IV, C-I, D-II
(b) A-III, B-IV, C-II, D-I
(c) A-IV, B-III, C-II, D-I
(d) A-II, B-I, C-IV, D-III

**09** A molecule with the formula $AX_4Y$ has all it's elements from p-block. Element A is rarest monoatomic, non-radioactive from its group and has the lowest ionisation enthalpy value among A, X and Y. Elements X and Y have first and second highest electronegativity values respectively among all the known elements. The shape of the molecule is: *[2025, 2 April Shift-I]*

(a) pentagonal planar
(b) trigonal bipyramidal
(c) square pyramidal
(d) octahedral

**10** Match List-I with List-II.

| List-I (Species) | | List-II (Number of lone pair of electrons on the central atom) | |
|---|---|---|---|
| A. | $XeF_2$ | 1. | 0 |
| B. | $XeO_2F_2$ | 2. | 1 |
| C. | $XeO_3F_2$ | 3. | 2 |
| D. | $XeF_4$ | 4. | 3 |

Choose the most appropriate answer from the options given below *[2021, 27 Aug Shift-I]*

| | A | B | C | D |
|---|---|---|---|---|
| (a) | 4 | 1 | 2 | 3 |
| (b) | 3 | 4 | 2 | 1 |
| (c) | 3 | 2 | 4 | 1 |
| (d) | 4 | 2 | 1 | 3 |

**11** The reaction in which the hybridisation of the underlined atom is affected is *[2020, 4 Sep Shift-II]*

(a) $\underline{Xe}F_4 + SbF_5 \longrightarrow$
(b) $H_2\underline{S}O_4 + NaCl \xrightarrow{420\text{ K}}$
(c) $H_3\underline{P}O_2 \xrightarrow{\text{Disproportionation}}$
(d) $\underline{N}H_3 \xrightarrow{H^+}$

## Numerical Value Type Questions

**12** The difference in the oxidation state of Xe between the oxidised product of Xe formed on complete hydrolysis of $XeF_4$ and $XeF_4$ is ........ *[2023, 10 April Shift-II]*

**13** $XeF_4$ reacts with $SbF_5$ to form $[XeFm]^n [SbF_y]^{z-}$

$m + n + y + z =$ ...... *[2023, 8 April Shift-I]*

**14** The sum of number of lone pairs of electrons present on the central atoms of $XeO_3$, $XeOF_4$ and $XeF_6$, is ............ *[2022, 25 July Shift-II]*

**15** A xenon compound 'A' upon partial hydrolysis gives $XeO_2F_2$. The number of lone pair of electrons present in compound A is ......... (Round off to the nearest integer). *[2021, 18 March Shift-II]*

## • Key **Index**

### Topic 1

| | | | | | | | | | |
|---|---|---|---|---|---|---|---|---|---|
| 1 (a) | 2 (c) | 3 (b) | 4 (b) | 5 (a) | 6 (b) | 7 (d) | 8 (d) | 9 (b) | 10 (d) |
| 11 (a) | 12 (c) | 13 (d) | 14 (d) | 15 (d) | 16 (d) | 17 (b) | 18 (d) | 19 (b) | 20 (b) |
| 21 (d) | 22 (b) | 23 (d) | 24 (d) | 25 (a) | 26 (a) | 27 (a) | 28 (b) | 29 (d) | 30 (c) |
| 31 (d) | 32 (a) | 33 (d) | 34 (b) | 35 (a) | 36 (c) | 37 (d) | 38 (a) | 39 (d) | 40 (c) |
| 41 (b) | 42 (c) | 43 (b) | 44 (d) | 45 (d) | 46 (a) | 47 (c) | 48 (d) | 49 (c) | 50 (d) |
| 51 (b) | 52 (c) | 53 (c) | 54 (d) | 55 (c) | 56 (b) | 57 (d) | 58 (a) | 59 (c) | 60 (15) |
| 61 (6) | 62 (2) | 63 (2) | 64 (4) | | | | | | |

### Topic 2

| | | | | | | | | | |
|---|---|---|---|---|---|---|---|---|---|
| 1 (c) | 2 (a) | 3 (d) | 4 (b) | 5 (a) | 6 (b) | 7 (d) | 8 (c) | 9 (d) | 10 (d) |
| 11 (b) | 12 (d) | 13 (c) | 14 (a) | 15 (d) | 16 (d) | 17 (a) | 18 (a) | 19 (d) | 20 (b) |
| 21 (c) | 22 (c) | 23 (a) | 24 (c) | 25 (b) | 26 (1) | 27 (4) | 28 (1) | 29 (7) | |

### Topic 3

| | | | | | | | | | |
|---|---|---|---|---|---|---|---|---|---|
| 1 (c) | 2 (a) | 3 (d) | 4 (c) | 5 (a) | 6 (c) | 7 (d) | 8 (d) | 9 (a) | 10 (c) |
| 11 (a) | 12 (b) | 13 (b) | 14 (d) | 15 (b) | 16 (a) | 17 (d) | 18 (c) | 19 (d) | 20 (b) |
| 21 (c) | 22 (a) | 23 (c) | 24 (c) | 25 (c) | 26 (3) | 27 (5) | 28 (3) | 29 (3) | 30 (1) |
| 31 (3) | 32 (1.67) | | | | | | | | |

### Topic 4

| | | | | | | | | | |
|---|---|---|---|---|---|---|---|---|---|
| 1 (a) | 2 (d) | 3 (d) | 4 (c) | 5 (a) | 6 (a) | 7 (a) | 8 (b) | 9 (c) | 10 (d) |
| 11 (a) | 12 (2) | 13 (11) | 14 (3) | 15 (19) | | | | | |

# Detailed **Explanations**

### Topic 1

**01.** *(a)* In group 15 elements, weakest $E—E$ bond is present in nitrogen. Its maximum possible covalency is 4.

**02.** *(c)* $HNO_3$ is a strong oxidising agent. It oxidises $I_2$ to $HIO_3$.

$$I_2 + 10\,\underset{(\text{Conc.})}{HNO_3} \longrightarrow 2HIO_3 + 10NO_2 + 4H_2O$$

**03.** *(b)* The gas produced by treating an aqueous solution of ammonium chloride ($NH_4Cl$) with sodium nitrite ($NaNO_2$) is $N_2$.

$$NH_4Cl + NaNO_2 \xrightarrow{\Delta} NH_4NO_2 \downarrow N_2\uparrow + 2H_2O + NaCl$$

**04.** *(b)* $PCl_5$ forms five bonds using the vacant *d*-orbitals to "expand the octet". $NCl_5$ has only *s* and *p*-orbitals and no *d*-orbitals in the second energy level. Therefore, $NCl_5$ does not exist.

**05.** *(a)* $NF_3$ is the most stable trihalide of nitrogen because both nitrogen and fluorine have comparable sizes and electronegativity.

**06.** *(b)* $BiH_3$ is a strongest reducing agent among group-15. It is because as we move down the group, the $M—H$ bond can break more easily to evolve $H_2$ gas which acts as the reducing agent. Thus, $BiH_3$ is least stable among the hydrides of group 15.

**07.** *(d)* $BiH_3$ is the strongest reducing agent among group 15 hyrdrides. The reducing character of the group 15 element hydrides increases down the group. So, correct answer is Bi (bismuth).

**08.** *(d)* The structure of pyrophosphoric acid ($H_4P_2O_7$) molecule is

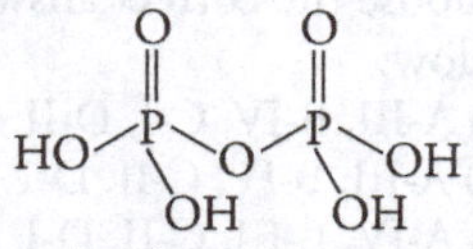


Pyrophosphoric acid ($H_4P_2O_7$)

Number of P—OH bonds = 4, P═O bonds = 2 and P—O—P bond = 1.

**09.** *(b)* Both Statement I and Statement II are true.

**10.** *(d)* Reaction of lead sulphide with dilute nitric acid takes place as follows

$$PbS + HNO_3 \longrightarrow Pb(NO_3)_2 + NO + S + H_2O$$

Thus, nitrous oxide ($N_2O$) is not formed during the reaction.

**11.** *(a)* Statement I is correct but Statement II is incorrect. The correct of Statement II is,
Bonding of $N(CH_3)_3$ and $P(CH_3)_3$ are different as $P(CH_3)_3$ has vacant *d*-orbital, hence acts as sigma donor as well as π-acceptor but $N(CH_3)_3$ can acts only as σ donor.
Hence, option (a) is correct.

**12.** *(c)* The reducing character of hydrides of group 15 elements increases from top to bottom because the strength of *M*—H (where, *M* = N, As, Sb or Bi) bond decreases down the group due to increase in the size of central atom.
Hence, among the given options, $BiH_3$ will be strongest reducing agent.

**13.** *(d)* A is correct but R is not correct.
The correct form of R is:
Unlike $NH_3$, $PH_3$ molecules are not associated through hydrogen bonding in liquid state.
Hence, option (d) is correct.

**14.** *(d)* Both (A) and (R) are true but (R) is not the correct explanation of (A).
The correct explanation of A is:
From As to Bi only small increase in covalent radius is observed. This is due presence of completely filled *d* and *f* orbitals of heavier members.
Hence, option (d) is correct.

**15.** *(d)* Due to presence of two P—H bond and +3 oxidation state of P in $H_4P_2O_5$, it acts as a reducing agent and hence reduces $AgNO_3$ to metallic Ag which forms silver mirror.

**16.** *(d)* The reaction of dinitrogen and dioxygen is highly endothermic. Thus, requires very high temperature.

**17.** *(b)* When white phosphorus reacts with thionyl chloride, it produces phosphorus trichloride and sulphur dioxide.

$$P_4 + 8SOCl_2 \longrightarrow 4PCl_3 + 4SO_2 + 2S_2Cl_2$$

**18.** *(d)* Red phosphrous is produced by heating white phosphorous at 573 K in inert atmosphere. When red phosphorus is heated in a sealed tube at 803 K, we get α-black phosphorus.

**19.** *(b)* White phosphrus reacts with alkali in presence of $CO_2$ (inert atmosphere) as

$$\underset{\text{White phosphorus}}{P_4} + NaOH \xrightarrow[CO_2]{\text{Inert atmosphere}} PH_3 + NaH_2PO_2 \xrightarrow{H^+} H_3PO_2$$

The structure of $H_3PO_2$ is

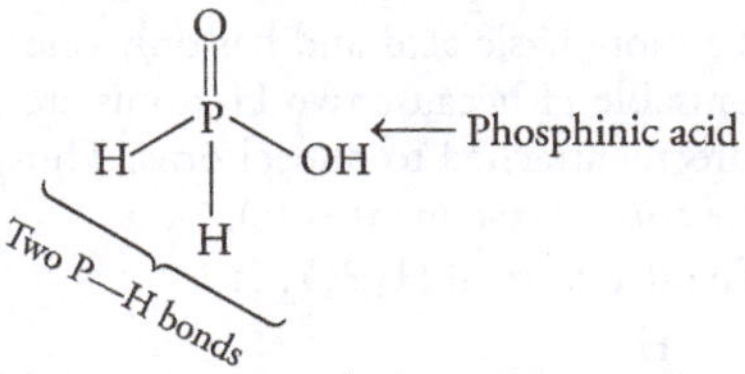


**20.** *(b)* $Ba(N_3)_2 \xrightarrow{\Delta} Ba + 3N_2$ (pure)

$$2NaNO_2 \xrightarrow{>800°C} Na_2O + N_2 + \frac{3}{2}O_2$$

$$2NaNO_3 \xrightarrow{<800°C} Na_2O + N_2\uparrow + \frac{5}{2}O_2\uparrow$$

**Note** Thermal decomposition means, usually $T < 300°C$.

**21.** *(d)* White phosphorus is $P_4$. Its reaction with conc. NaOH produces sodium hypophosphite and phosphine.

$$P_4 + 3NaOH + 3H_2O \longrightarrow 3NaH_2PO_2 + PH_3$$

**22.** *(b)* Electronic configuration of $N = 1s^2\ 2s^2\ 2p^3$.
∴ N has 5 valence electrons.

$NO_2$ 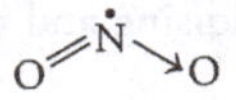 (odd electron species)

**23.** *(d)*

Structure of $N_2O_5$ ⟶ 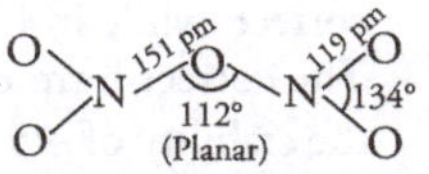


Thus, oxide without nitrogen-nitrogen bond is $N_2O_5$.

**24.** *(d)* When red phosphorus is heated in a sealed tube at 803 K, α-black phosphorus is formed. Therefore , option (d) is correct.

**25.** *(a)* Chemical nature of the nitrogen pentaoxide compound obtained from a reaction of concentrated nitric acid and $P_4O_{10}$ (in 4 : 1 ratio) is acidic.
Chemical reaction is as follows

$$\underset{\text{Phosphorus pentaoxide}}{P_4O_{10}} + \underset{\text{Nitric acid}}{4HNO_3} \longrightarrow \underset{\textit{Meta}\text{-phosphoric acid}}{4HPO_3} + \underset{\text{Nitrogen pentaoxide (Acidic)}}{2N_2O_5}$$

**26.** *(a)* Oxidation states of 'N' in the given compounds are as follows

$$NO \rightarrow x + 1(-2) = 0$$
$$x = +2$$
$$NO_2 \rightarrow x + 2 \times (-2) = 0$$
$$x = +4$$
$$N_2O \rightarrow 2x + (-2) = 0$$
$$x = +1$$
$$NO_3^- \rightarrow x + 3(-2) = -1$$
$$\Rightarrow \quad x = +5$$

Thus, $NO_3^- > NO_2 > NO > N_2O$.

**27.** *(a)* $N_2O$ and NO are neutral oxides of nitrogen, whereas $NO_2$ and $N_2O_3$ are acidic oxides.

> • Concept **Enhancer**
> Neutral oxides neither react with acids nor with bases.

**28.** *(b)* $PCl_5$ has no lone-pair on the central atom but has empty *d*-orbital in valence shell. So that it can accept a pair of electrons (from a Lewis base). Hence, $PCl_5$ act as a Lewis acid.

**29.** *(d)* NaH is not an oxidising agent as $Na^+$ and $H^-$ ions are formed. Na exists only in +1 oxidation state and hydrogen is present in minimum −1 oxidation state, so cannot be reduced further. Thus, NaH is used as reducing agent.
Lone pair on nitrogen in pyridine is localised and does not take part in resonance. So, it can accept $H^+$ acting as a base.

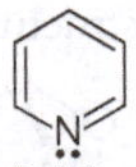

Pyridine

So, statement I is false but statement II is true.

**30.** *(c)* Nitrogen monoxide reacts with dinitrogen tetraoxide at 250 K temperature forms blue solid or $N_2O_3$.

$$2NO + N_2O_4 \xrightarrow{250\text{ K}} 2N_2O_3$$

**31.** *(d)* A yellow explosive liquid nitrogen trichloride is formed

$$NH_3 + \underset{\text{(Excess)}}{3Cl_2} \longrightarrow 3HCl + \underset{\text{Nitrogen trichloride}}{NCl_3}$$

> • Concept **Enhancer**
> During the reaction of ammonia with $Cl_2$, if ammonia is taken in excess then ammonium chloride is formed.
> $$\underset{\text{(Excess)}}{7NH_3} + 3Cl_2 \longrightarrow NH_4Cl + N_2$$

**32.** *(a)* *Aqua-regia* is conc.$HNO_3$ : conc.HCl
1 : 3 [*V/V*]

$$Au + 4H^+ + NO_3^- + 4Cl^- \longrightarrow [AuCl_4]^- + NO + 2H_2O$$

$$3Pt + 16H^+ + 4NO_3^- + 18Cl^- \longrightarrow 3[PtCl_6]^{2-} + 4NO + 8H_2O$$

∴ NO gas is evolved in this processes.

**33.** *(d)*

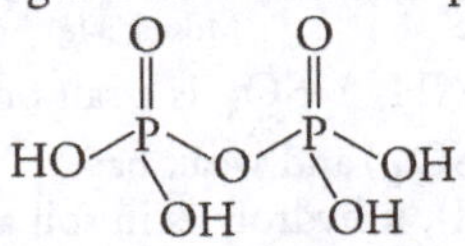

Pyrophosphoric acid ($H_4P_2O_7$)

Number of P—OH bonds = 4
Oxidation state of P in $H_4P_2O_7$ (Pyrophosphoric acid)

$\Rightarrow \quad 4 \times (+1) + 2x + 7 \times (-2) = 0$

$\Rightarrow \quad x = +5$

**34.** *(b)* Organic compound '*A*' contain phosphorus as it gives positive test with ammonium molybdate. Phosphorus present in organic compound '*A*' get oxidised with $Na_2O_2$ and form $Na_3PO_4$.

**35.** *(a)* The structure of $H_3PO_2$ (hypophosphorus) acid is

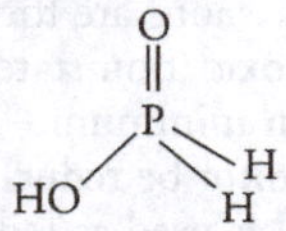

Due to the presence of two P— H bonds, $H_3PO_2$ acts a strong reducing agent.

**36.** *(c)* The no. of P—O bonds in $P_4O_6$ = 12. Its structure is shown below

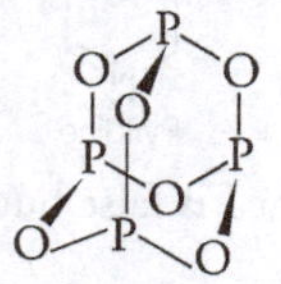

**37.** *(d)* Orthophosphorous acid, $H_3PO_3$

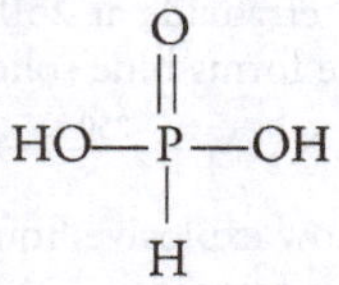

$H_3\overset{x}{P}O_3 = 3 + x + 3(-2) = 0$ or $x = +3$

Pyrophosphorous acid,

$H_4P_2O_5$ : 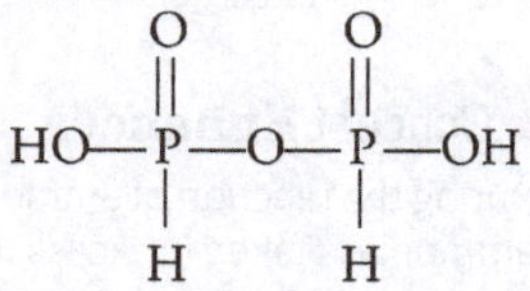

$H_4\overset{x}{P_2}O_5 = 4 + 2x + 5(-2) = 0$

$4 + 2x - 10 = 0 \Rightarrow x = +3$

**Time Saver Tech**

Phosphorus series contains P in +3 state while phosphoric series contains P in +5 state.

**38.** *(a)* Thermal stability of the hydrides decrease as we go down the group in periodic table for group 15 (N-family)

$BiH_3 < SbH_3 < AsH_3 < PH_3 < NH_3$
Least stable — Most stable

**39.** *(d)* An $(NH_4)_2SO_4$ is a salt of strong acid ($H_2SO_4$) and weak base ($NH_4OH$), it hydrolysis in soil and makes it acidic after some time.

**40.** *(c)* Hypophosphorous acid ($H_3PO_2$) is a monobasic acid and has only one ionisable H because two H-atoms are directly attached to phosphorus. Thus, the correct statement is (c).

The structure of $H_3PO_2$ is

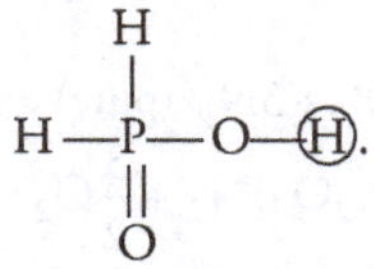

. Ionisable as directly attached to O.

**41.** *(b)* Both Statement I and Statement II are false. It is because N—N single bond is weaker than P—P single bond due to more $lp - lp$ repulsion.

In group 15 elements only N and P show disproportionation in +3 oxidation state, while as Sb and Bi are almost inert for disproportionation in +3 oxidation state.

**42.** *(c)* In correct statement is given in option (c). Its correct form is $SO_2$ can acts as both, oxidising and reducing agent.

**43.** *(b)* Statements given in A, B and C are correct, while in D and E are incorrect. The correct form of D and E are the stability of +5 oxidation states decreases down the group due to inert pair effect.

Nitrogen can show a maximum covalency of 4.

**44.** *(d)* Among the given statements A and B are correct while statement C and D are incorrect.

The correct form of statements C and D are :

**Statement C** On moving down the group, bond strength of *M*—H bond decreases, which reduces the thermal stability but increases reducing nature of hydrides. Therefore, $BiH_3$ is a strong reducing agent.

**Statement D** The basicity of hydrides decreases in order

$NH_3 > PH_3 > AsH_3 > SbH_3 > BiH_3$

Hence, option (d) is correct.

**45.** *(d)* The structures of $H_4P_2O_7$, $P_4O_{10}$ and $(HPO_3)_3$ are as follows:

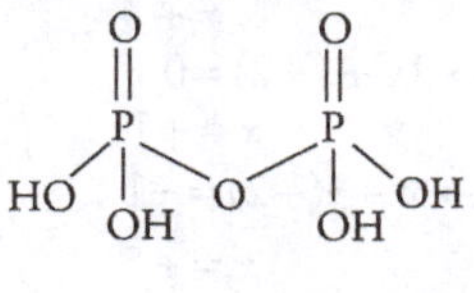

$H_4P_2O_7$
P—O—P bond = 1

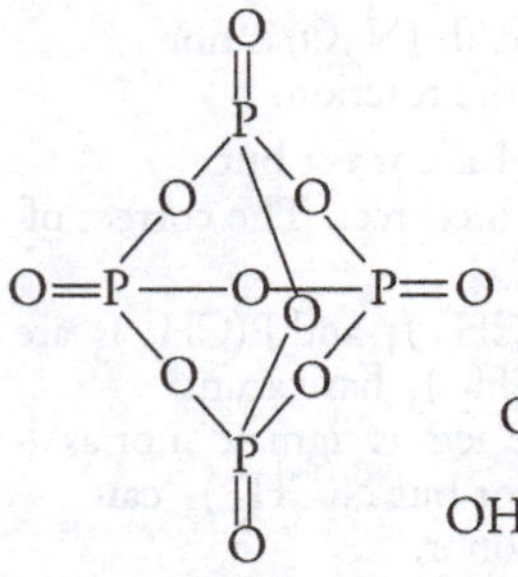

$P_4O_{10}$
P—O—P bonds = 6

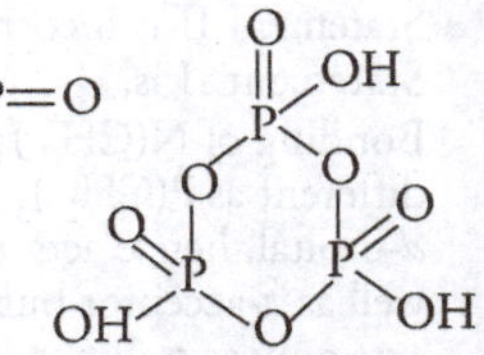

$(HPO_3)_3$
P—O—P bonds=3

**46.** *(a)* The reactions will takes place as follows:

$P_4 + 8SOCl_2 \longrightarrow \underset{(A)}{4PCl_3} + 4SO_2 + 2S_2Cl_2$

$\underset{(A)}{PCl_3} + 3H_2O \longrightarrow \underset{(B)}{H_3PO_3} + 3HCl$

$H_3PO_3$ or $HPO(OH)_2$ is a dibasic acid. Thus, option (a) is correct.

**Concept Enhancer**

Basicity of an acid is defined as the number of hydronium ions ($H_3O^+$) produced by the ionisation of one molecule of that acid in an aqueous solution. Depending upon their basicity i.e., 1, 2 or 3, they are classified as monobasic, dibasic and tribasic respectively.

**47.** *(c)* $N_2(g) + O_2(g) \longrightarrow 2NO(g)$

$2NO(g) + O_2(g) \longrightarrow 2NO_2(g)$

Higher concentration of $NO_2$ damage the leaves of the plant and retard the rate of photosynthesis.

**48.** *(d)* Pyrophosphorus acid $\longrightarrow H_4P_2O_5$

Hypophosphoric acid $\longrightarrow H_4P_2O_6$

Phosphoric acid $\longrightarrow H_3PO_4$

Pyrophosphoric acid $\longrightarrow H_4P_2O_7$

**49.** *(c)* Structure of $N_2O$

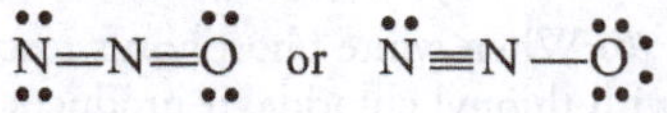

Structure of $N_2O_3$

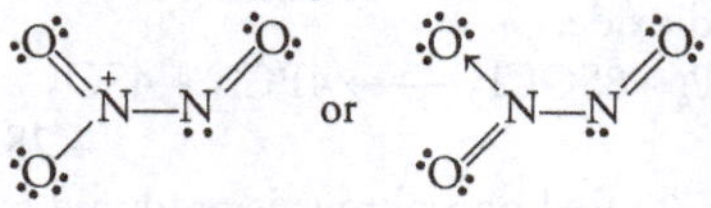

Structure of $N_2O_4$

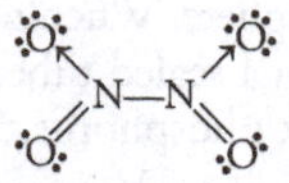

Structure of $N_2O_5$

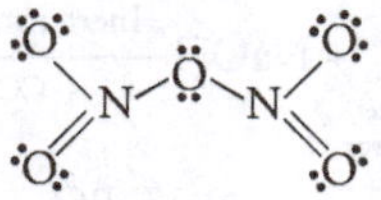

So, from the above structures, it is clear that $N_2O$, $N_2O_3$ and $N_2O_4$ contains N—N bond.

**50.** *(d)* Higher the oxidation state of the central atom means higher will be the acidic nature of their oxides. Thus, pentavalent oxides of group 15 element, $E_2O_5$ are more acidic than trivalent oxide, $E_2O_3$ of same element.
On moving down the group, metallic character increases that means the acidic nature of the oxides of group 15 elements decreases down the group.
Hence, given statement I is false but statement II is true.

**51.** *(b)* $\underset{(A)}{\text{Red } P_4} + \text{Alkali} \longrightarrow \underset{\substack{\text{(Hypophosphoric}\\ \text{acid)}\\ (B)}}{H_4P_2O_6}$

Hypophosphoric acid contains four P—OH bonds, two P=O bonds and one P—P bond. No P—H bond is present in $H_4P_2O_6$.

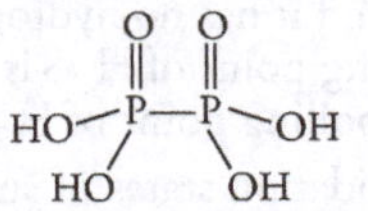

(Hypophosphoric acid)

**52.** *(c)* $PCl_3 + H_3PO_3 + H_2O \longrightarrow H_4P_2O_5 + HCl$

Product is pyrophosphoric acid, i.e. $H_4P_2O_5$.
The structure of pyrophosphoric acid shows that it has two acidic or ionisable hydrogen, i.e. *H.

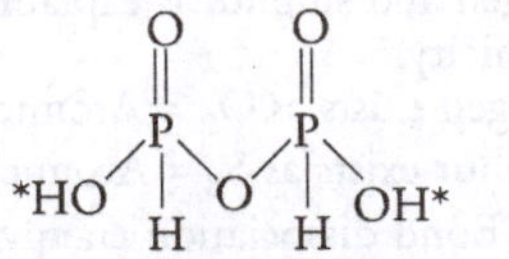

(Two ionisable H)

**53.** *(c)* The correct match is
A-V, B-I, C-II, D-III

**54.** *(d)* The given reaction takes place as follows

(i) $2NH_3 + \underset{(A)}{H_2O} + CO_2 \longrightarrow (NH_4)_2CO_3$

(ii) $(NH_4)_2CO_3 + H_2O + \underset{(B)}{CO_2} \longrightarrow 2NH_4HCO_3$

(iii) $NH_4HCO_3 + NaCl \longrightarrow NH_4Cl + \underset{(C)}{NaHCO_3}$

So, $A$ - $H_2O$, $B$ - $CO_2$ ; $C$ - $NaHCO_3$.

**55.** *(c)* $Pb(NO_3)_2 \xrightarrow{\Delta} PbO + \underset{\substack{(A)\\ \text{(Brown gas)}}}{2NO_2(g)}$

$$\underset{(A)}{2NO_2(g)} \underset{\text{Heating}}{\overset{\text{Cooling}}{\rightleftharpoons}} \underset{(B)}{N_2O_4(s)}$$

$$\underset{(B)}{N_2O_4(s)} + NO(g) \xrightarrow{\Delta} \underset{\substack{(C)\\ \text{(Blue solid)}}}{\overset{+3}{N_2O_3(s)}}$$

Solid $C$ is $N_2O_3$ and oxidation number of nitrogen in $N_2O_3$ is +3.

**56.** *(b)* The correct increasing order of oxidation state of nitrogen for nitrogen oxides is

$$\overset{+1}{N_2O} < \overset{+2}{NO} < \overset{+3}{N_2O_3} < \overset{+4}{NO_2}$$

**57.** *(d)* Let us consider the structure of the phosphorus oxoacids,

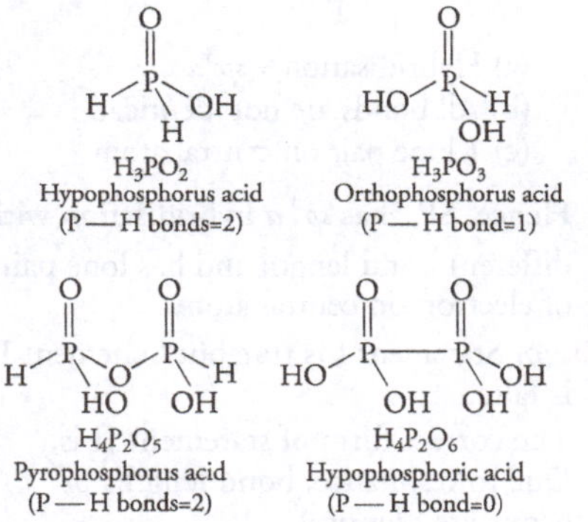


**58.** *(a)* The electronic configuration of $NO[15e^-]$ is given by

$$\sigma 1s^2, \sigma^* 1s^2, \sigma 2s^2, \sigma^* 2s^2, \sigma 2p_z^2, \pi 2p_x^2 = \pi 2p_y^2, \pi^* 2p_z^1$$

Thus, NO possess one unpaired electron, hence, it is paramagnetic in nature. However, it dimerises at low temperature to become diamagnetic.

$$2NO \rightleftharpoons N_2O_2$$

$$\text{Bond order of NO} = \frac{1}{2}[10-5] = 2.5$$

**59.** *(c)* The correct order of increasing basic strength

$$SbH_3 < AsH_3 < PH_3 < NH_3$$

$NH_3$ is most basic because of its small size, the electron density of electron pair is concentrated over small region.

**60.** *(15)* Phosphorus belong to 15th group and forms $d\pi - d\pi$ bond with transition metal and $PH_3$ is strongest base among the other group member except $NH_3$.
So, atomic number of element is 15.

**61.** *(6)* The structure of pyro-phosphoric acid is

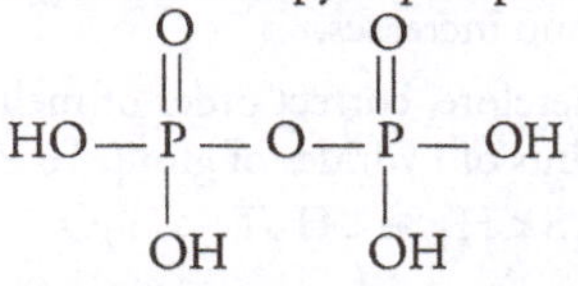


$$\frac{\text{Number of } \sigma\text{-bonds}}{\text{Number of } \pi\text{-bonds}} = \frac{12}{2} = 6$$

**62.** *(2)*

$$C_2H_5OH + PCl_3 \rightarrow C_2H_5Cl + \underset{(A)}{H_3PO_3}$$

$$\underset{(A)}{H_3PO_3} + PCl_3 \longrightarrow \underset{(B)}{H_4P_2O_5} + HCl$$

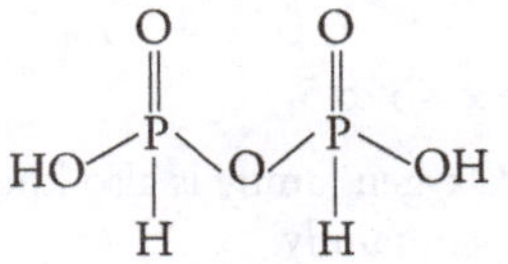

(B)

Number of non-ionisable protons in $B = 2$.

**63.** *(2)* $PCl_3$ reacts with $H_2O$ to give $POCl_3$ and HCl. $POCl_3$ on further hydrolysis for phosphorus acid ($H_3PO_3$) which has two ionisable protons.

$$PCl_3 + H_2O \longrightarrow \underset{(A)}{POCl_3} + HCl$$

$$\underset{(A)}{POCl_3} + H_2O \longrightarrow \underset{(B)}{H_3PO_3} + HCl$$

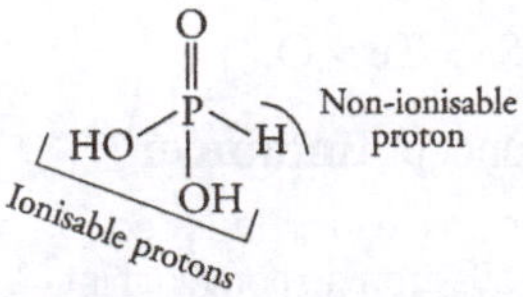


**64.** *(4)* Reaction of white phosphorus with alkali in inert atmosphere forms sodium hypophosphite ($NaH_2PO_2$).

$$P_4 + 3NaOH + 3H_2O \longrightarrow PH_3 + \underset{\substack{(A)\\ \text{Sodium}\\ \text{hypophosphite}}}{3NaH_2PO_2}$$

1 mole of $NaH_2PO_2$ reacts with excess of $AgNO_3$ in aqueous medium to give 4 moles of Ag.

$$NaH_2PO_2 + 4AgNO_3 + 2H_2O \longrightarrow 4Ag + H_3PO_4 + NaNO_3 + 3HNO_3$$

Hence, 1 mole produce 4 moles of Ag.

## Topic 2

**01.** *(c)* Anomalous behaviour of oxygen is due to its small size and high electronegativity.

**02.** *(a)* As we move down the group, molecular size and surface area increase and also van der Waals' forces increase.

Hence, the boiling point down the group increases.

Therefore, correct order of melting points of hydrides of group 16 element is $H_2S < H_2Se < H_2Te < H_2O$

**03.** *(d)* Polythionic acid with a small number of sulphur atoms (3 to 5) are most stable. The oxidation state of sulphur in polythionic acid are 0 and +5 only. They have straight chain of sulphur.

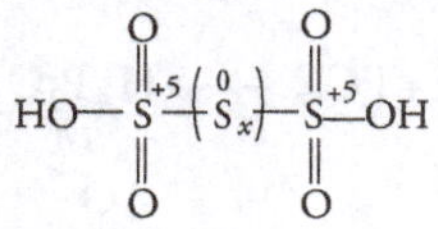

where $x = 3$ to 5.

**04.** *(b)* Oxygen family is also known as chalcogen family.

O, Se, S, Te, Po and Lv are chalcogen family member.

**05.** *(a)* Oxygen is the second most electronegative element in comparison to fluorine. O-atom is smallest in size. $|\Delta_{eg}H|$ of oxygen decreases to a higher extent in comparison to other elements of group -16 who have larger size and lower electronegativity.

So, the correct order of EA or $|\Delta_{eg}H|$ of group -16 elements will be

S > Se > Te > O.

**• Concept Enhancer**

When an electron is added to a neutral gaseous atom to convert it into a negative ion, the enthalpy change accompanying the process is defined as the electron gain enthalpy.

Electrons gain enthalpy provides a measure of the ease with which an atom adds an electron to form anion.

$$X(g) + e^- \longrightarrow X^-(g)$$

**06.** *(b)* Statement I is false, but Statement II is true.

The correct form of statement II is,

Acidic strength : $H_2Se < H_2Te$

$\Delta H_{\text{dissociation}}$ : $H_2Se$ [276 kJ/mol] > $H_2Te$ [238 kJ/mol]

**07.** *(d)*

• $PF_5$

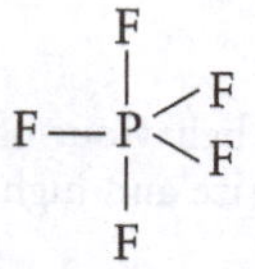

(a) Hybridisation $= sp^3d$

(b) All bonds are not identical

(c) No lone pair on central atom

•

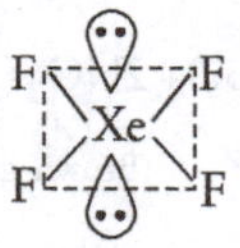

(a) Hybridisation $= sp^3d^2$

(b) All bonds are identical

(c) 2 lone pairs on central atom.

• $XeF_2$

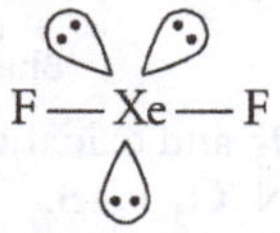

(a) Hybridisation $= sp^3d$

(b) All bonds are identical

(c) 3 lone pairs on central atom

• $SF_4$

(a) Hybridisation $= sp^3d$

(b) All bonds are not identical

(c) 1 lone pair on central atom

Hence, $SF_4$ has $sp^3d$ hybridisation with different bond length and has lone pair of electron on central atom.

**08.** *(c)* Statement I is true but Statement II is false.

The correct form of statement II is,

Due to resonance, bond lengths of ozone are identical.

**09.** *(d)* Both A and R are correct and R is the correct explanation of A.

It is because oxygen can form $2p\pi - 2p\pi$ multiple bond with itself due to its smaller size while sulphur can not form multiple bond with itself as $3p\pi - 3p\pi$ bond will be unstable due to large size of sulphur, but it can form multiple bond with small size atom like C and N.

**10.** *(d)* Both A and R are true and R is the correct explanation of A.

**• Concept Enhancer**

As the atomic size increases down the group the bond length increases and hence the bond strength decreases, thereby leading to the decrease in bond enthalpy down the group.

Consequently the cleavage of E—H bond become easier. As a result the tendency to release hydrogen as proton increases. Thus, as the conclusion acidic strength of metal hydrides increases down the group.

**11.** *(b)* Both Statement I and Statement II are incorrect.

Their correct forms are as follows.

**Statement I** Oxygen can have oxidation state from −2 to +2.

**Statement II** On moving down the group, stability of +4 oxidation state increases whereas stability of +6 oxidation state decreases due to inert pair effect.

**12.** *(d)* $SO_2$ on reaction with water and oxygen forms sulphuric acid which is responsible for acid rain.

$$2SO_2 + O_2 + 2H_2O \longrightarrow \underset{\text{(Acid rain)}}{2H_2SO_4}$$

**13.** *(c)* Oxides of nitrogen and sulphur are acidic and settle down on ground as dry deposition.

Ammonium salts in rain drops result in wet deposition.

∴ $X$ = Ammonium salts

$Y = SO_2$

**14.** *(a)* At room temperature, water is liquid and has boiling point 373 K because of hydrogen bonding. But $H_2S$ is a gas and it has no hydrogen bonding. So, boiling point of $H_2S$ is less than 300 K (boiling point of $H_2S$ is −60°C).

**15.** *(d)* Oxidation states of sulphur are

| | | |
|---|---|---|
| −2 | in | $H_2S$ |
| 0 | in | $S_8$ |
| +2 | in | $S_2O_3^{2-}$ |
| +4 | in | $SO_2$ |
| +6 | in | $SO_3$ |

Hence, correct option is (d).

**16.** *(d)* The large difference between the melting point and boiling point of oxygen and sulphur is explained by atomicity.

Oxygen exists as $O_2$ = Atomicity = 2

Sulphur exists as $S_8$ = Atomicity = 8

**17.** *(a)* Bond dissociation energy depends on bond length.

i.e. smaller the bond length, greater is the bond energy. In the given compound, order of bond length is

$$\underset{(D)}{H—Te} > \underset{(C)}{H—Se} > \underset{(B)}{H—S} > \underset{(A)}{H—O}$$

So, the order of bond energy will be

$H_2O > H_2S > H_2Se > H_2Te$

**• Concept Enhancer**

Bond dissociation energy is the energy required to break a bond and form two atomic or molecular fragments or species. It is a means of calculating the strength of a chemical bond. For example, dissociation enthalpy of $Cl_2$ is +59 kcal mol$^{-1}$.

:Cl—Cl: ⟶ 2 :Cl

**18.** *(a)* Structures of oxoacids of sulphur are

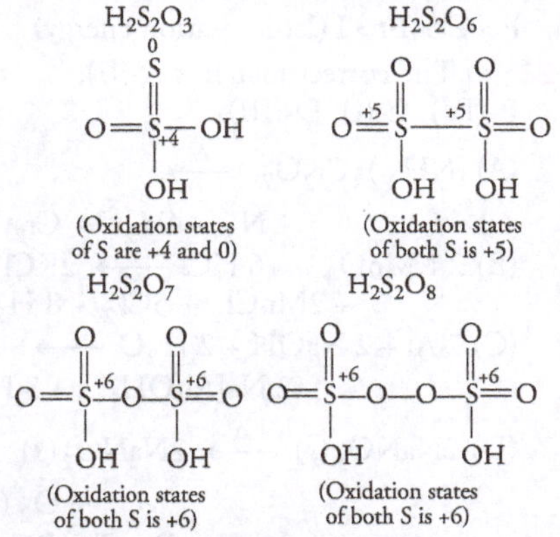


So, only (a) $H_2S_2O_3$ contains "S" in two different oxidation states (i.e. +4 and 0).

**19.** *(d)*

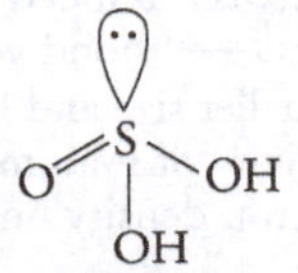

($H_2SO_3$)
Sulphurous acid
No. of S═O bond = 1

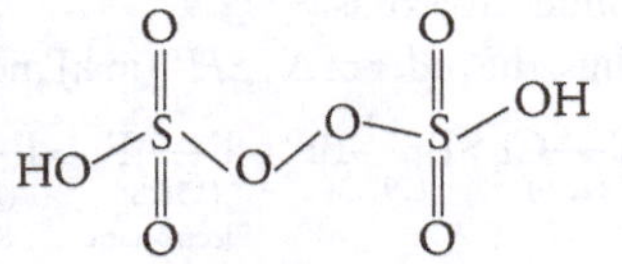

($H_2S_2O_8$)
Peroxidesulphuric acid
No. of S═O bonds = 4

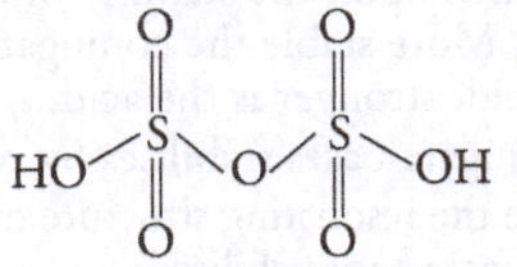

$H_2S_2O_7$
Pyrosulphuric acid
No. of S═O bonds = 4

**20.** *(b)* Electrolysis of concentrated solution of acidified sulphate solution yields peroxydisulphuric acid ($H_2S_2O_8$).

During electrolysis,

**At cathode**

$$2H_2O + 2e^- \longrightarrow 2OH^- + H_2\uparrow$$

**At anode**

$$2H_2SO_4 \longrightarrow H_2S_2O_8 + 2H^+ + 2e^-$$

$$2H_2O \longrightarrow 4H^+ + O_2 + 2e^-$$

**21.** *(c)* The correct match is,

A → (iii), B → (i), C → (iv), D → (ii)

**22.** *(c)* Two crystalline allotropic forms of sulphur ($S_\alpha$ and $S_\beta$) can change reversibly between themselves with slow heating (above 369 K) or cooling.

$$\underset{\text{(Rhombic)}}{S_\alpha} \underset{\text{Cooling}}{\overset{>369\text{ K}}{\rightleftharpoons}} \underset{\text{(Monoclinic)}}{S_\beta}$$

So, statement I is true.

At room temperature or at standard conditions of pressure and temperature, rhombic sulphur is the thermodynamically most stable crystalline allotrope of sulphur ($\Delta_f H^\circ = 0$).

So, statement II is false.

**23.** *(a)* Statements given in option (a) i.e., A and B are correct, while other two statements are incorrect.

Their correct form are :

C. The reducing property increases from $H_2S$ to $H_2Te$ down the group.

D. The ozone molecules contains 6 lone pairs of electrons.

**24.** *(c)* $X = Na_2SO_3$

$$\underset{(X)}{SO_3^{2-}} + H_2SO_4 \longrightarrow [H_2SO_3] \longrightarrow H_2O + \underset{(Y)}{SO_2}$$

$$SO_2 + 2NaOH \longrightarrow \underset{(X)}{Na_2SO_3} + H_2O$$

$$\underset{(X)}{Na_2SO_3} + \underset{(Y)}{SO_2} + H_2O \longrightarrow \underset{(Z)}{2NaHSO_3}$$

Hence, $Y = SO_2$, $Z = NaHSO_3$

**25.** *(b)* $S_2O_8^{2-}$ is

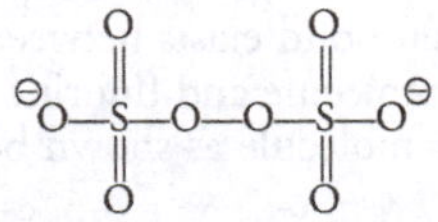

Total number of S—O or S═O bond = 8

Rhombic sulphur is

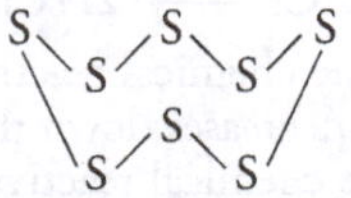

Total number of S—S bond = 8

Thus, the correct answer is (b).

**26.** *(1)* Structure of oxoacids of sulphur are

$H_2SO_3$ : 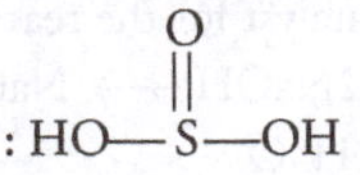

$H_2SO_4$ : 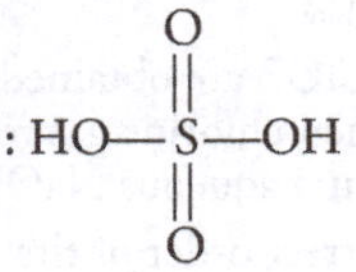

$H_2S_2O_8$ :

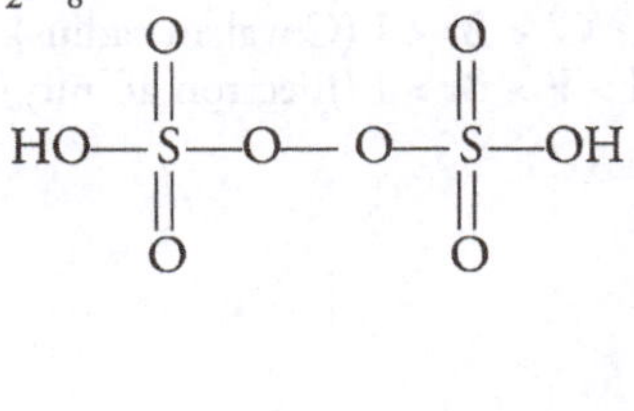

$H_2S_2O_7$ : 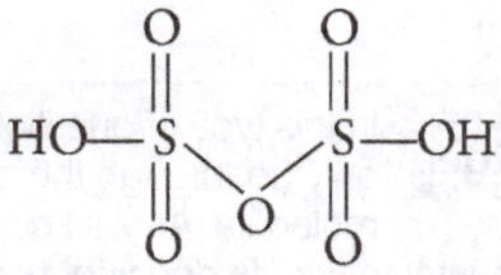

Therefore, number of oxoacids of sulphur with peroxo bond (O—O) is 1.

**27.** *(4)* PbS, CuS, $As_2S_3$, CdS, are soluble in 50% $HNO_3$ while HgS, $Sb_2S_3$ are insoluble in 50% $HNO_3$.

**28.** *(1)* Only $S_2$-form of sulphur is paramagnetic in nature. Because $S_2$ is like $O_2$ i.e., paramagnetic as per molecular orbital theory. It contains unpaired electron.

**29.** *(7)* Fuming sulphuric acids formula is $H_2S_2O_7$. So, total 7 oxygen atoms are present in it.

## Topic 3

**01.** *(c)* Iodine is the halogen which reacts with $Cu^{2+}$ ion to give a salt of copper.

$$2Cu^{2+} + 4I^- \longrightarrow Cu_2I_2 + I_2$$

**02.** *(a)* Among the given options nitrogen is not present in Nessler's reagent. It is because Nessler's reagent is potassium tetraiodomercurate $[K_2(HgI_4)]$ with KOH solution. Hence, it contains K (potassium), Hg (mercury) and I (iodine).

**03.** *(a)* Nessler's reagent is potassium tetraiodomercurate II $[K_2(HgI_4)]$. Thus oxygen element is not present in it. It is prepared by combining potassium iodide (KI), mercuric chloride ($HgCl_2$) and KOH.

**04.** *(c)* Nessler's reagent = $K_2HgI_4$ + KOH

It is used for confirmation test of cation ($NH_4^+$).

Thus, option (c) is correct.

**05.** *(a)* Fluorine forms only one oxoacid, HOF due to its small size and high electronegativity. Hence, both A and R are correct and R is the correct explanation of A.

**06.** *(c)* Order of electron gain enthalpy (absolute value)

Cl > F > Br > I

| Element | F | Cl | Br | I |
|---|---|---|---|---|
| $\Delta H_{eg}$ (kJ/mol) | −328 | −349 | −325 | −295 |

Absolute value of electron gain enthalpy decreases down the group in a periodic table due to increase in size, with an reception in group 17 where, $|\Delta_{eg} H|$ : Cl > F. It is because of smaller size than Cl and highest electronegativity of F.

**07.** *(d)*

**key idea** In this type of question of halogen, only go through the polarity of the molecule. As we know, diatomic elemental molecule does not have polarity but hetero-nuclear molecules with dissimilar sizes have polarity resulting in more reactivity.

$Cl_2$, $Br_2$ and $I_2$ are homonuclear diatomic molecule in which electronegativity of the combining atoms is same, so they are more stable and less reactive, whereas, I and Cl have different electronegativities and bond between them are polarised and hence, reactive. Therefore, interhalogen compounds are more reactive.

**08.** *(d)* Among the given halogen, $Cl_2$, $Br_2$ and $I_2$ undergoes disproportionation reaction.
$F_2$ does not disproportionate because fluorine do not exist in positive oxidation state.

**09.** *(a)* Statement I is false and Statement II is true.
The correct form of Statement I is :
Chlorine instead of fluorine has most negative electron gain enthalpy in its group.

**10.** *(c)* Fluorine has no *d*-orbitals (2*d*-orbitals) and there is no scope for any electron excitation.
Therefore, flourine can show only oxidation state of –1 in its compounds.
Other members of family have vacant *d*-orbitals. Hence, they can show variable oxidation state.

**11.** *(a)* Among the given options only $I^-$ ions can be oxidised by oxygen in acidic medium.

$$4I^- + 4H^+ + O_2 \longrightarrow 2I_2 + 2H_2O$$

**12.** *(b)* $SbCl_5$ is more covalent than $SbCl_3$ due to more polarising power of $Sb^{5+}$ than that of $Sb^{3+}$, as per Fajan's rule.
Stability of higher oxides of halogens is due to more electronegativity and higher oxidation state. Thus, both statement I and statement II are correct.

**13.** *(b)* $Br_2 + 5F_2(\text{Excess}) \longrightarrow 2BrF_5$

Due to highly reactive nature of $BrF_5$, it absorbs moisture from surrounding results into the formation of $HBrO_3$.

$$\overset{+5\ -1}{BrF_5} + 3H_2O \longrightarrow \underset{\text{[Bromate or Halate, } XO_3^-]}{\overset{+5}{HBrO_3}} + \underset{[X'^{(-)}]}{5H\overset{-1}{F}}$$

$(XX'_5)$

**14.** *(d)* Iodine oxygen bond is stable due to greater polarity of bond. Stability of chlorine oxygen bond is due to multiple bond formation of oxygen with *d*-orbital of chlorine atom.
Bromine lacks both these characteristics.
So, the stability order of oxide is

$$I > Cl > Br.$$

**15.** *(b)* Dioxygen difluoride is used to remove the plutonium from spent nuclear fuel.

$$3F\text{—}O\text{—}O\text{—}F + \underset{\text{(From nuclear fuel)}}{Pu} \longrightarrow PuF_6 + 3O_2\uparrow$$

**16.** *(a)* The $Cl_2$ is more reactive than ClF is incorrect statement as reactivity order can be written as follows
$F_2 >$ ClF (interhalogen compounds) $> Cl_2$.
$Cl_2$ is a non-polar bond with lower reactivity than Cl—F due to a difference in electronegativity between two bonds. As a result, $Cl_2$ is less reactive than ClF.

**17.** *(d)*

| Formula | $HClO_2$ | $HClO_3$ | $HClO_4$ |
|---|---|---|---|
| No. of Cl$=$O bond | 1 | 2 | 3 |

**18.** *(c)* HF has highest boiling point among hydrogen halides because it has strongest hydrogen bonding. Here, the hydrogen bond exists between hydrogen of one molecule and fluorine atom of another molecule as shown below.

$$\ldots\overset{\delta+}{H}\text{—}\overset{\delta-}{F}\ldots\overset{\delta+}{H}\text{—}\overset{\delta-}{F}\ldots\overset{\delta+}{H}\text{—}\overset{\delta-}{F}$$

**19.** *(d)* $3HO\overset{+1}{Cl} \longrightarrow 2H\overset{-1}{Cl} + H\overset{+5}{Cl}O_3$

**20.** *(b)* As the chemical reactivity of halogens decreases down the group. Thus, the chemical reactivity follows the order

$$F_2 > Cl_2 > Br_2 > I_2.$$

$I_2$ being the least reactive halogen, it requires a catalyst for the reaction.

**21.** *(c)* $Cl_2 + \underset{\text{Cold and dilute}}{2NaOH} \longrightarrow NaCl + NaClO + H_2O$

$\therefore$ $Cl^-$ and $ClO^-$ are obtained as products when chlorine gas reacts with cold and dilute aqueous NaOH.

**22.** *(a)* The correct order of first four elements of group 17 are as follows,
F < Cl < Br < I (Covalent radius)
Cl > F > Br > I (Electron affinity)
$F^- < Cl^- < Br^- < I^-$ (Ionic radius)
F > Cl > Br > I (Ist ionisation energy)

**23.** *(c)* The correct match is A-(II), B-(IV), C-(I), D-(III).

(A) $(NH_4)_2Cr_2O_7 \xrightarrow{\Delta} N_2 + 4H_2O + Cr_2O_3$

(B) $2KMnO_4 + 16HCl \longrightarrow 2KCl + 2MnCl_2 + 5Cl_2 + 8H_2O$

(C) $2Al + 2NaOH + 2H_2O \longrightarrow 2Na[Al(OH)_4] + 3H_2$

(D) $2NaNO_3(s) \xrightarrow{\Delta} 2NaNO_2(s) + O_2(g)$

**24.** *(c)* Among halogens ($F_2$, $Cl_2$, $Br_2$ and $I_2$), bond dissociation enthalpy ($\Delta_{diss}H°$) of $I_2$ is minimum because of larger size of I-atom there is a steric repulsion between bonded I-atoms, which makes I— I bond weakest.
Whereas, smaller size and highest electronegativity of F-atom cause highest electron density on F-atom of $F_2$ molecule.
As a result, F—F bond becomes weaker due to electrostatic repulsion between bonded F-atoms.
Thus, the order of $\Delta_{diss}H°$ (in kJ $mol^{-1}$) is

$$\underset{(242.6)}{Cl\text{—}Cl} > \underset{(192.3)}{Br\text{—}Br} > \underset{\substack{(158.8)\\ \text{Electrostatic repulsion}}}{F\text{—}F} > \underset{\substack{(151.1)\\ \text{Steric repulsion}}}{I\text{—}I}$$

**25.** *(c)* Acidic strength of oxoacids depends upon the stability of conjugate base. More stable the conjugate base of an acid, stronger is the acid.
Conjugate bases stabilises by resonance. More the resonating structure more the conjugate base stabilised.
$ClO^-$ does not show resonance. Thus, the stability order of conjugate base due to resonance is

$$ClO_4^- > ClO_3^- > ClO_2^- > ClO^-$$

Hence, the acidic strength of given oxoacids is as follows

$$HClO_4 > HClO_3 > HClO_2 > HClO$$

**26.** *(3)* The oxides of *p*-block elements are differentiated as below,
Acidic oxide : $Cl_2O_7$, $SiO_2$, $N_2O_5$
Neutral oxide : CO, NO, $N_2O$
Amphoteric oxide : $Al_2O_3$, $SnO_2$, $PbO_2$
Amphoteric oxides are compounds that react with both acids and bases to form salt and water.
Hence, total number of amphoteric oxide is 3.

**27.** *(5)* $Na_2O + H_2O \longrightarrow 2NaOH$ ...(i)

$Cl_2O_7 + H_2O \longrightarrow 2HClO_4$ ...(ii)

Adding number of oxygen atoms in eqs (i) and eqs (ii), we get

$$1 + 4 = 5$$

So, the total number of oxygen atoms in $X$ and $Y$ is 5.

**28.** *(3)*

| Interhalogen compounds | Bond pair | Lone pair | Structure | Shape |
|---|---|---|---|---|
| 1. $ClF_3$ | 3 | 2 | 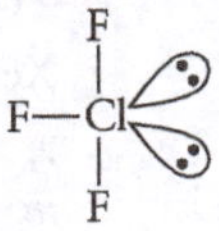 | T-shape |
| 2. $IF_7$ | 7 | 0 | 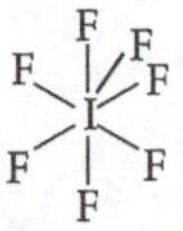 | Pentagonal bipyramidal |
| 3. $BrF_5$ | 5 | 1 | 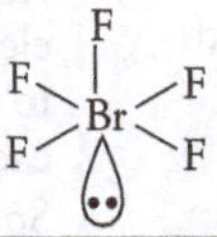 | Square pyramidal |
| 4. $I_2Cl_6$ | 4 For each I atom | 2 | 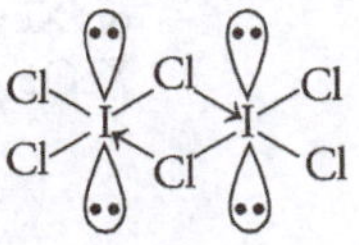 | Square planar |
| 5. $IF_5$ | 5 | 1 | 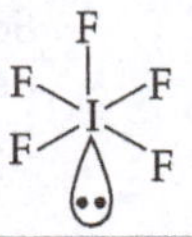 | Square pyramidal |
| 6. ClF | 1 | 3 | $\ddot{\underset{..}{:Cl}}—F$ | Linear |
| 7. $ClF_5$ | 5 | 1 | 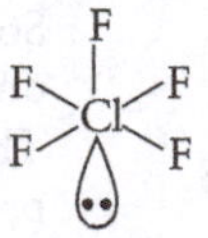 | Square pyramidal |

**29.** *(3)* The halic (V) acids formed by chlorine (Cl), bromine (Br) and iodine (I) are $HClO_3$, $HBrO_3$ and $HIO_3$ respectively.

**30.** *(1)* $BF_3$, $SiCl_4$ and $PCl_5$ are easily hydrolysed while $SF_6$ is inert to hydrolysis. In a polar covalent bond, the electrophilic centre ($\delta^+$) able to expand covalency by the nucleophilic attack of $H_2\ddot{O}:$ undergoes hydrolysis. In $SF_6$, S is the electrophilic centre ($S^{\delta+}$ —— $F^{\delta-}$) but here S is being $sp^3d^2$ hybridised, it cannot expand covalency beyond 6. Hence, $SF_6$ is non-hydrolysable.

Thus, only one halide is inert to hydrolysis.

**31.** *(3)*

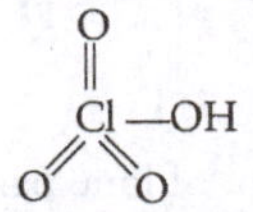

Thus, Cl=O bond is 3.

**32.** *(1.67)* The reaction is,

$$3Cl_2 + 6NaOH \longrightarrow \underset{(X)}{5NaCl} + \underset{(Y)}{NaClO_3} + 3H_2O$$

$X \xrightarrow{+AgNO_3}$ AgCl (white ppt.)

$Y = NaClO_3$ ($Na^+$ and $ClO_3^-$)

The structure of $ClO_3^-$ (chlorate ion) is,

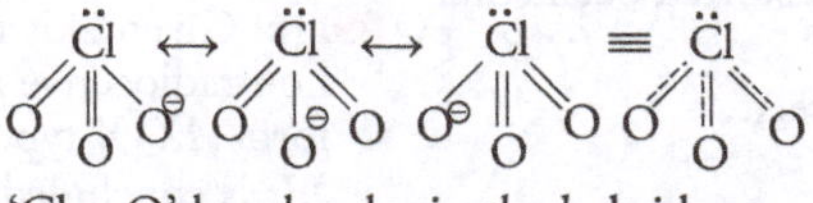

'Cl—O' bond order in the hybrid

$$= \frac{\text{Number of bonds between Cl and O}}{\text{Total number of O (surrounding atoms)}} = \frac{5}{3} = 1.66 \text{ or } 1.67$$

## Topic 4

**01.** *(a)* Radium (Ra) is a radioactive element. Ra belongs to group 2 (alkaline earth metals), it is not a noble gas.

**Note** In question noble gas which does not exist in the atmosphere is asked and answer is Ra. But Ra (radium) is an alkaline earth metal and not noble gas. It can be Rn (radon).

**02.** *(d)* As we move down the group of noble gases, molecular size increases by which dipole produced for a moment and hence London forces increases from He to Xe. Therefore, more amount of energy is required to break these forces, thus boiling point also increases from He to Xe.

**03.** *(d)* Both Statement I and Statement II are false. Their correct form are as follows

- Noble gases have very low boiling points.
- Noble gases are held together by weak dispersion forces.

**04.** *(c)* Among the given options Cl has the most negative electron gain enthalpy while neon has the most positive electron gain enthalpy. Thus, Ne and Cl has the maximum difference of $\Delta_{eg}H$.

**05.** *(a)* The correct match is A-II, B-I, C-IV, D-III

(A) $XeO_3$;

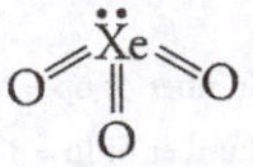

Hybridisation = lone pair + bond pair

$$= 1 + 3 = 4 \Rightarrow sp^3$$

**Note** Multiple bonds are treated as single bonds.

For $AB_3E$, shape = pyramidal

where, $A$ = Xe; $B$ = O and $E$ = lone pair.

(B) $XeF_2$

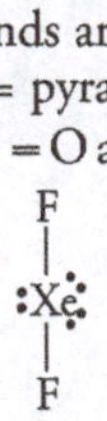

Hybridisation = 3 + 2 = 5; $sp^3d$

$AB_2E_3$, shape = linear

(C) $XeOF_4$

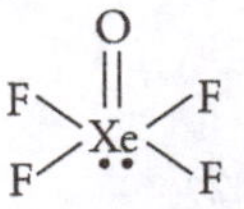

Hybridisation = 5 + 1 = 6 $\Rightarrow sp^3d^2$

For $AB_5E$, shape = square pyramidal

($\because B$ = F/O)

(D) $XeF_6$

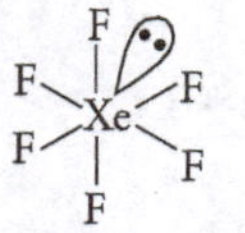

Hybridisation = 1 + 6 = 7

$\Rightarrow sp^3d^3$

For $AB_6E$, shape = distorted octahedral

($\because B$ = F)

So, option (a) is correct.

**06.** *(a)* $[XeF_5]^-$

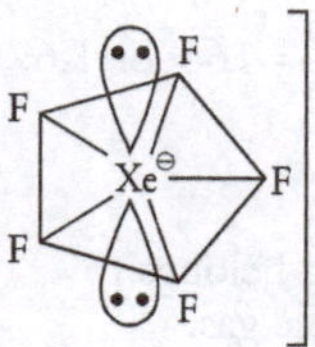

Xe : $sp^3d^3$-hybridised
Geometry = Pentagonal bipyramidal
Shape/Structure = Pentagonal planar

$XeO_3F_2$

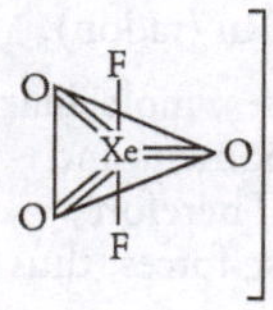

Xe: $sp^3d$-hybridised
Geometry = Shape/ structure = Trigonal bipyramidal
[because, Xe in $XeO_3F_2$ does not have lone pair of electrons]

**07.** *(a)* In $XeOF_4$, Xe is $sp^3d^2$-hybridised. Geometry of the molecule is octahedral, but shape of the molecule is square pyramidal.
According to VSEPR, theory it has one $\pi$-bond. Remaining six electron pairs form an octahedron with one position occupied by a lone pair.

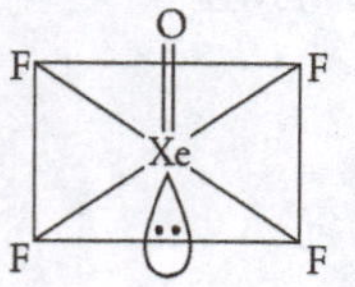

Number of bp = 5
Number of lp = 1

Here, Xe contains one lone pair of electrons.

**08.** *(b)* The correct match is A-III, B-IV, C-II, D-I

$ICl_2^-$

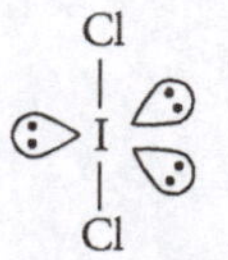

Bond pair : lone pair = 2:3

$H_2O$

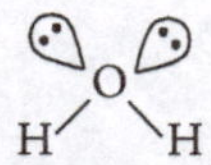

Bond pair : lone pair = 2:2

$SO_2$

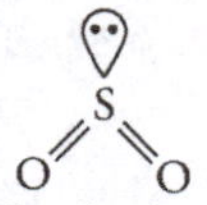

Bond pair : lone pair = 4:1

$XeF_4$

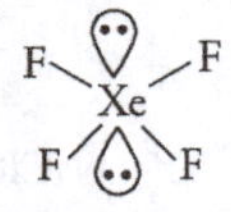

Bond pair : lone pair = 4:2

**09.** *(c)* Given *A* is rarest, monoatomic, non-radioactive *p*-block element and form $AX_4Y$ type of molecule.
$\therefore$ It is concluded that it is Xe.
It is given that electronegativity of *A* is less than *X* and *Y*.
It is given that electronegativity of *X* and *Y* is highest and second highest respectively among all element.
$\therefore$ *X* and *Y* are F and O.
Compound is consider as $XeOF_4$ with square pyramidal shape.

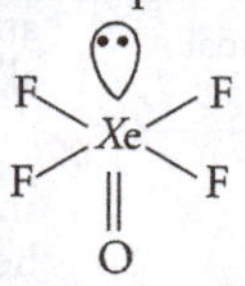

**10.** *(d)* The correct match is
A-4, B-2, C-1, D-3

A. $XeF_2$

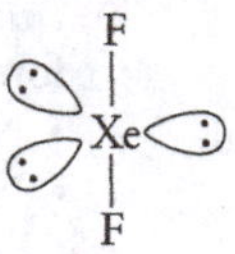

Three lone pairs

B. $XeO_2F_2$

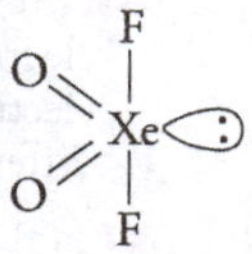

One lone pair

C. $XeO_3F_2$

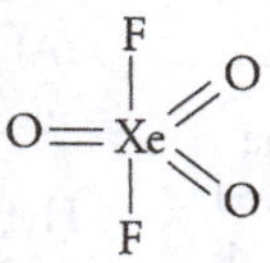

Zero lone pair

D. $XeF_4$

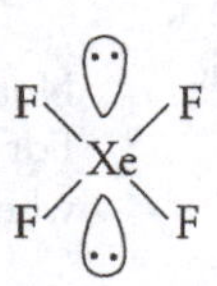

Two lone pairs

**11.** *(a)* Complete reaction with change in hybridisation is as follows :

$$\underline{XeF_4} + SbF_5 \longrightarrow [XeF_3]^+[SbF_6]^-$$

$sp^3d^2$ → $sp^3d$ (Change in hybridisation)

In the reaction (a), hybridisation of Xe in $XeF_4$ is $sp^3d^2$ and gets converted into $[XeF_3]^+$ and hybridisation of Xe in $[XeF_3]^+$ is $sp^3d$.

In this reaction, $sp^3d^2$-hybridisation changes into $sp^3d$.

**12.** *(2)* $6XeF_4 + 12H_2O \longrightarrow 2XeO_3 + 4Xe + 24HF + 3O_2$

In $XeO_3$, oxidation state of Xe = +6
In $XeF_4$, oxidation state of Xe = +4
So, difference in oxidation state = 6 − 4 = 2

**13.** *(11)* When $XeF_4$ reacts with $SbF_5$

$$XeF_4 + SbF_5 \longrightarrow [XeF_3]^+[SbF_6]^-$$

$\therefore \quad m = 3, \quad n = 1, \quad y = 6, \quad z = 1$

$m + n + y + z = 3 + 1 + 6 + 1 = 11$

**14.** *(3)* $XeO_3$

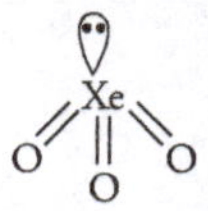

Xe is noble gas and therefore contain 8 electrons in valence shell. It gets bonded to oxygen by sharing its 2 electrons with each oxygen atom.
So, in this case the number of lone pair on Xe = 1

$XeOF_4$

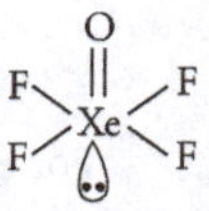

So, in this case the number of lone pair on Xe = 1

$XeF_6$

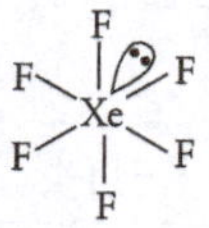

So, in this case the number of lone pair on Xe = 1
Thus the sum of lone pairs of electrons present on the Xe atoms of $XeO_3$, $XeOF_4$ and $XeF_6$ = 1 + 1 + 1 = 3.

**15.** *(19)* Xenon hexafluoride (*A*) on partial hydrolysis gives $XeO_2F_2$ as a product.

$$\underset{\text{(A) Xenon hexafluoride}}{XeF_6} + 2H_2O \xrightarrow{\text{Partially}} XeO_2F_2 + 4HF$$

Structure of $XeF_6$ is given as

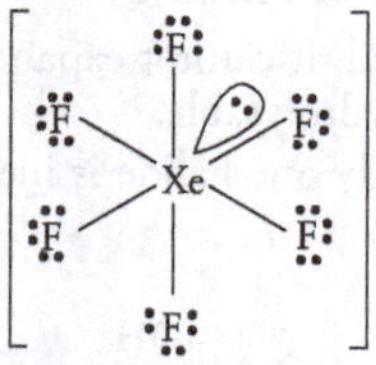

$\therefore$ Thus, total no. of lone pair
= lone pair on Xe + 6 × lone pair on F
= 1 + 6 × 3 = 19.

# C-17

JEE Main Chapterwise Topicwise Chemistry

In JEE Main there is usually 1-2 question per paper are visible from this chapter. In past 3 years, 41 conceptual and 6 numerical questions were asked of moderate difficulty level from this chapter.

**For detailed analysis scan QR code**

# *d*- and *f*-Block Elements

## Topic 1 Properties of Transition Elements

### Level 1

**01** The correct decreasing order of spin only magnetic moment values (BM) of $Cu^+$, $Cu^{2+}$, $Cr^{2+}$ and $Cr^{3+}$ ions is *[2025, 8 April Shift-II]*

(a) $Cr^{2+} > Cr^{3+} > Cu^{2+} > Cu^+$
(b) $Cr^{3+} > Cr^{2+} > Cu^+ > Cu^{2+}$
(c) $Cu^+ > Cu^{2+} > Cr^{3+} > Cr^{2+}$
(d) $Cu^{2+} > Cu^+ > Cr^{2+} > Cr^{3+}$

**02** The metal that shows highest and maximum number of oxidation state is *[2024, 5 April Shift-I]*

(a) Ti (b) Fe
(c) Co (d) Mn

**03** Which of the following $3d$-metal ion will give the lowest enthalpy of hydration ($\Delta_{hyd} H$), when dissolved in water ? *[2022, 29 July Shift-II]*

(a) $Cr^{2+}$ (b) $Mn^{2+}$ (c) $Fe^{2+}$ (d) $Co^{2+}$

**Imp.** Electronic Configuration of 5*d*-block Element

**04** The electronic configuration of Pt (atomic number 78) is *[2022, 29 June Shift-I]*

(a) $[Xe]4f^{14}5d^9 6s^1$ (b) $[Kr]4f^{14}5d^{10}$
(c) $[Xe]4f^{14}5d^{10}$ (d) $[Xe]4f^{14}5d^8 6s^2$

**05** In $3d$-series, the metal having the highest $M^{2+}/M$ standard electrode potential is *[2022, 27 June Shift-II]*

(a) Cr (b) Fe
(c) Cu (d) Zn

**06** Metals generally melt at very high temperature. Amongst the following, the metal with highest melting point will be *[2022, 24 June Shift-II]*

(a) Hg (b) Ag (c) Ga (d) Cs

**07** The common positive oxidation states for an element with atomic number 24, are *[2021, 17 March Shift-II]*

(a) + 2 to + 6 (b) + 1 and + 3 to + 6
(c) + 1 and + 3 (d) + 1 to + 6

**08** What is the spin only magnetic moment value (BM) of a divalent metal ion with atomic number 25, in its aqueous solution? *[2021, 17 March Shift-I]*

(a) 5.26 (b) 5.0
(c) 5.92 (d) Zero

**09** The major components of German silver are *[2021, 25 Feb Shift-II]*

(a) Cu, Zn and Ag (b) Cu, Zn and Ni
(c) Ge, Cu and Ag (d) Zn, Ni and Ag

**10** The major components in Gun metal are *[2021, 24 Feb Shift-I]*

(a) Cu, Sn and Zn (b) Cu, Zn and Ni
(c) Cu, Ni and Fe (d) Al, Cu, Mg and Mn

**11** The electrode potential of $M^{2+}/M$ of $3d$-series elements shows positive value for *[2021, 24 Feb Shift-I]*

(a) Fe (b) Co (c) Zn (d) Cu

**12** The atomic radius of Ag is closest to *[2020, 7 Jan Shift-I]*

(a) Hg (b) Au (c) Ni (d) Cu

**13** The statement that is incorrect about the interstitial compounds is *[2019, 8 April Shift-II]*

(a) they are very hard
(b) they have metallic conductivity
(c) they have high melting points
(d) they are chemically reactive

**14** The element that usually does not show variable oxidation states is *[2019, 1 Jan Shift-I]*

(a) Sc (b) Cu (c) Ti (d) V

**15** Which of the following ions does not liberate hydrogen gas on reaction with dilute acids? *[JEE Main 2017]*

(a) $Mn^{2+}$ (b) $Ti^{2+}$ (c) $V^{2+}$ (d) $Cr^{2+}$

**16** Match the catalysts to the correct processes. *[JEE Main 2015]*

| | Catalyst | | Processes |
|---|---|---|---|
| (A) | $TiCl_4$ | (i) | Wacker processes |
| (B) | $PdCl_2$ | (ii) | Ziegler-Natta polymerisation |
| (C) | $CuCl_2$ | (iii) | Contact processes |
| (D) | $V_2O_5$ | (iv) | Deacon's processes |

(a) (A)-(ii), (B)-(iii), (C)-(iv), (D)-(i)
(b) (A)-(iii), (B)-(i), (C)-(ii), (D)-(iv)
(c) (A)-(iii), (B)-(ii), (C)-(iv), (D)-(i)
(d) (A)-(ii), (B)-(i), (C)-(iv), (D)-(iii)

**17** Four successive members of the first row transition elements listed below with atomic numbers. Which one of them is expected to have the highest $E^{\circ}_{M^{3+}/M^{2+}}$ value? *[JEE Main 2013]*
(a) Cr $(Z = 24)$ (b) Mn $(Z = 25)$
(c) Fe $(Z = 26)$ (d) Co $(Z = 27)$

**18** Which pair of elements with the given atomic numbers is expected to have similar properties? *[JEE Main 2012]*
(a) 40, 72 (b) 20, 36 (c) 10, 28 (d) 11, 12

**19** The "spin-only" magnetic moment [in units of Bohr magneton, $(\mu_B)$] of $Ni^{2+}$ in aqueous solution would be (atomic number of Ni = 28) *[AIEEE 2006]*
(a) 2.84 (b) 4.90 (c) 0 (d) 1.73

**20** Of the following outer electronic configurations of atoms, the highest oxidation state is achieved by which one of them ? *[AIEEE 2004]*
(a) $(n-1)\,d^8 ns^2$ (b) $(n-1)\,d^5 ns^1$
(c) $(n-1)\,d^3 ns^2$ (d) $(n-1)\,d^5 ns^2$

**21** The number of $d$-electrons retained in $Fe^{2+}$ (Atomic number Fe = 26) ions is *[AIEEE 2003]*
(a) 3 (b) 4 (c) 5 (d) 6

**22** Which of the following groups of transition metals is called coinage metals? *[AIEEE 2003]*
(a) Cu, Ag, Au (b) Ru, Rh, Pd
(c) Fe, Co, Ni (d) Os, Ir, Pt

**23** For making good quality mirrors, plates of float glass are used. These are obtained by floating molten glass over a liquid metal which does not solidify before glass. The metal used can be *[AIEEE 2003]*
(a) mercury (b) tin
(c) sodium (d) magnesium

## Level 2

**24** The metal ions that have the calculated spin-only magnetic moment value of 4.9. BM are
A. $Cr^{2+}$ B. $Fe^{2+}$
C. $Fe^{3+}$ D. $Co^{2+}$
E. $Mn^{3+}$

Choose the correct answer from the options given below: *[2025, 3 April Shift-I]*
(a) A, B and E Only (b) A, D and E Only
(c) B and E Only (d) A, C and E Only

**25** The incorrect relationship in the following pairs in relation to ionisation enthalpies is *[2025, 4 April Shift-II]*
(a) $Fe^{2+} < Fe^{3+}$ (b) $Mn^{2+} < Fe^{2+}$
(c) $Mn^{+} < Cr^{+}$ (d) $Mn^{+} < Mn^{2+}$

**26** The first transition series metal '$M$' has the highest enthalpy of atomisation in its series. One of its aquated ion $(M^{n+})$ exists in green colour. The nature of the oxide formed by the above $(M^{n+})$ ion is *[2025, 7 April Shift-I]*
(a) neutral (b) amphoteric
(c) acidic (d) basic

**27** The number of valence electrons present in the metal among Cr, Co, Fe and Ni which has the lowest enthalpy of atomisation is *[2025, 7 April Shift-I]*
(a) 8 (b)10 (c) 6 (d) 9

**28** Pair of transition metal ions having the same number of unpaired electrons is *[2025, 4 April Shift-I]*
(a) $V^{2+}, Co^{2+}$ (b) $Ti^{3+}, Mn^{2+}$
(c) $Ti^{2+}, Co^{2+}$ (d) $Fe^{3+}, Cr^{2+}$

**29** Given below are two statements. *[2025, 3 April Shift-II]*

**Statement I** $CrO_3$ is a stronger oxidising agent than $MoO_3$.

**Statement II** Cr(VI) is more stable than Mo(VI).

In the light of the above statements, choose the correct answer from the options given below.
(a) Statement I is false but Statement II is true.
(b) Both Statement I and Statement II are false.
(c) Both Statement I and Statement II are true.
(d) Statement I is true but Statement II is false.

**30** The amphoteric oxide among $V_2O_3$, $V_2O_4$ and $V_2O_5$, upon reaction with alkali leads to formation of an oxide anion. The oxidation state of V in the oxide anion is *[2025, 28 Jan Shift-II]*
(a) +4 (b) +7 (c) +5 (d) +3

**31** Given below are two statements. *[2024, 9 April Shift-II]*

**Statement I** The higher oxidation states are more stable down the group among transition elements unlike $p$-block elements.

**Statement II** Copper cannot liberate hydrogen from weak acids.

In the light of the above statements, choose the correct answer from the options given below.
(a) Both Statement I and Statement II are false.
(b) Statement I is true but Statement II is false.
(c) Statement I is false but Statement II is true.
(d) Both Statement I and Statement II are true.

**32** The electronic configuration of Cu(II) is $3d^9$ whereas that of Cu(I) is $3d^{10}$. Which of the following is correct? *[2024, 9 April Shift-I]*
(a) Cu(II) is more stable
(b) Cu(II) is less stable
(c) Stability of Cu(I) and Cu(II) depends on nature of copper salts
(d) Cu(I) and Cu(II) are equally stable

**33** A first row transition metal in its +2 oxidation state has a spin-only magnetic moment value of 3.86 BM. The atomic number of the metal is *[2024, 4 April Shift-II]*
(a) 25 (b) 26 (c) 22 (d) 23

**34** The element which shows only one oxidation state other than its elemental form is *[2024, 4 April Shift-I]*
(a) Nickel (b) Cobalt (c) Titanium (d) Scandium

**35** Choose the correct option having all the elements with $d^{10}$ electronic configuration from the following. *[2024, 27 Jan Shift-II]*
(a) $^{27}Co$, $^{28}Ni$, $^{26}Fe$, $^{24}Cr$ (b) $^{28}Ni$, $^{24}Cr$, $^{26}Fe$, $^{29}Cu$
(c) $^{46}Pd$, $^{28}Ni$, $^{26}Fe$, $^{24}Cr$ (d) $^{29}Cu$, $^{30}Zn$, $^{48}Cd$, $^{47}Ag$

**36** Identify the incorrect pair from the following:
(a) Haber process—Iron *[2024, 27 Jan Shift-II]*
(b) Polythene preparation—$TiCl_4$, $Al(CH_3)_3$
(c) Wacker process—$PtCl_2$
(d) Photography—AgBr

**37** Given below are two statements.
One is labelled as Assertion (A) and the other is labelled as Reason (R).
**Assertion (A)** The first ionisation enthalpy of $3d$ series elements is more than that of group 2 metals.
**Reason (R)** In $3d$ series of elements successive filling of $d$-orbitals takes place.
In the light of the above statements, choose the correct answer from the options given below.
(a) (A) is true but (R) is false. *[2023, 31 Jan Shift-II]*
(b) (A) is false but (R) is true.
(c) Both (A) and (R) are true and (R) is the correct explanation of (A).
(d) Both (A) and (R) are true but (R) is not the correct explanation of (A).

**38** Which of the following has least tendency to liberate $H_2$ from mineral acids? *[2022, 28 July Shift-I]*
(a) Cu (b) Mn
(c) Ni (d) Zn

**39** Among the following, which is the strongest oxidising agent? *[2022, 25 June Shift-I]*
(a) $Mn^{3+}$ (b) $Fe^{3+}$ (c) $Ti^{3+}$ (d) $Cr^{3+}$

**40** Identify the elements for which electronic configuration in +3 oxidation state is $[Ar]3d^5$. *[2021, 1 Sep Shift-II]*
(a) Ru (b) Mn (c) Co (d) Fe

**41** Which one of the following when dissolved in water gives coloured solution in nitrogen atmosphere? *[2021, 26 Aug Shift-I]*
(a) $CuCl_2$ (b) AgCl (c) $ZnCl_2$ (d) $Cu_2Cl_2$

**42** The spin only magnetic moments (in BM) for free $Ti^{3+}$, $V^{2+}$ and $Sc^{3+}$ ions respectively are (Atomic number Sc : 21, Ti : 22, V : 23) *[2021, 25 July Shift-II]*
(a) 1.73, 0, 3.87 (b) 3.87, 1.73, 0
(c) 0, 3.87, 1.73 (d) 1.73, 3.87, 0

**43** In which of the following pairs, the outer most electronic configuration will be the same? *[2021, 25 Feb Shift-I]*
(a) $Ni^{2+}$ and $Cu^+$ (b) $Fe^{2+}$ and $Co^+$
(c) $Cr^+$ and $Mn^{2+}$ (d) $V^{2+}$ and $Cr^+$

**44** The set that contains atomic numbers of only transition elements, is *[2020, 6 Sep Shift-I]*
(a) 37, 42, 50, 64 (b) 21, 25, 42, 72
(c) 9, 17, 34, 38 (d) 21, 32, 53, 64

**45** The pair that has similar atomic radii is *[2019, 12 April Shift-II]*
(a) Mn and Re (b) Ti and Hf
(c) Sc and Ni (d) Mo and W

**46** Which of the following compounds is metallic and ferromagnetic? *[JEE Main 2016]*
(a) $CrO_2$ (b) $VO_2$ (c) $MnO_2$ (d) $TiO_2$

**47** Which one of the following species is stable in aqueous solution? *[JEE Main 2016]*
(a) $Cr^{2+}$ (b) $MnO_4^{2-}$
(c) $MnO_4^{3-}$ (d) $Cu^+$

**48** The element with which of the following outer electronic configuration may exhibit the largest number of oxidation states in its compounds *[JEE Main 2013]*
(a) $3d^5 4s^2$ (b) $3d^8 4s^2$
(c) $3d^7 4s^2$ (d) $3d^6 4s^2$

**49** The correct order of $E^\circ_{M^{2+}/M}$ values with negative sign for the four successive elements Cr, Mn, Fe and Co is *[AIEEE 2010]*
(a) Mn > Cr > Fe > Co (b) Cr > Fe > Mn > Co
(c) Fe > Mn > Cr > Co (d) Cr > Mn > Fe > Co

## Level 3

**50** The correct set of ions (aqueous solution) with same colour from the following is *[2025, 23 Jan Shift-I]*
(a) $V^{2+}$, $Cr^{3+}$, $Mn^{3+}$ (b) $Zn^{2+}$, $V^{3+}$, $Fe^{3+}$
(c) $Sc^{3+}$, $Ti^{3+}$, $Cr^{2+}$ (d) $Ti^{4+}$, $V^{4+}$, $Mn^{2+}$

**51** The correct option with order of melting points of the pairs (Mn, Fe), (Tc, Ru) and (Re, Os) is *[2025, 29 Jan Shift-I]*
(a) Mn < Fe, Tc < Ru and Os < Re
(b) Fe < Mn, Ru < Tc and Os < Re
(c) Mn < Fe, Tc < Ru and Re < Os
(d) Fe < Mn, Ru < Tc and Re < Os

**52** Match List-I with List-II. *[2025, 24 Jan Shift-II]*

| | List-I (Transition metal ion) | | List-II (Spin only magnetic moment (BM)) |
|---|---|---|---|
| (A) | $Ti^{3+}$ | (I) | 3.87 |
| (B) | $V^{2+}$ | (II) | 0.00 |
| (C) | $Ni^{2+}$ | (III) | 1.73 |
| (D) | $Sc^{3+}$ | (IV) | 2.84 |

Choose the correct answer from the options given below

(a) (A) - (IV), (B) - (II), (C)-(III), (D) - (I)
(b) (A) - (III), (B) - (I), (C) - (IV), (D) - (II)
(c) (A) - (III), (B) - (I), (C) - (II), (D) - (IV)
(d) (A) - (II), (B) - (IV), (C) - (I), (D)-(III)

**53** Consider '*n*' is the number of lone pair of electrons present in the equatorial position of the most stable structure of $ClF_3$. The ions from the following with '*n*' number of unpaired electrons are

A. $V^{3+}$ B. $Ti^{3+}$ C. $Cu^{2+}$ D. $Ni^{2+}$
E. $Ti^{2+}$

Choose the correct answer from the options given below *[2025, 28 Jan Shift-I]*

(a) B and C only (b) A, D and E only
(c) B and D only (d) A and C only

**Imp.** Calculation of Number of Unpaired Electrons

**54** Arrange the following elements in the increasing order of number of unpaired electrons in it.

A. Sc B. Cr C. V D. Ti
E. Mn

Choose the correct answer from the options given below *[2024, 6 April Shift-II]*

(a) C < E < B < A < D (b) A < D < C < B < E
(c) A < D < C < E < B (d) B < C < D < E < A

**55** Which of the following electronic configuration would be associated with the highest magnetic moment? *[2024, 27 Jan Shift-I]*

(a) $[Ar]\,3d^7$ (b) $[Ar]\,3d^3$
(c) $[Ar]\,3d^6$ (d) $[Ar]\,3d^8$

**56** In chromyl chloride, the number of *d*-electrons present on chromium is same as in (Given atomic number of Ti : 22, V : 23, Cr : 24, Mn : 25, Fe : 26) *[2023, 8 April Shift-I]*

(a) V (IV) (b) Ti (I)
(c) Mn (VII) (d) Fe (II)

**57** In following pairs, the one in which both transition metal ions are colourless is *[2022, 29 July Shift-I]*

(a) $Sc^{3+}, Zn^{2+}$ (b) $Ti^{4+}, Cu^{2+}$
(c) $V^{2+}, Ti^{3+}$ (d) $Zn^{2+}, Mn^{2+}$

**58** The metal ion (in gaseous state) with lowest spin only magnetic moment value is *[2022, 25 June Shift-II]*

(a) $V^{2+}$ (b) $Ni^{2+}$
(c) $Cr^{2+}$ (d) $Fe^{2+}$

**Imp.** Colour and Magnetic Properties of *d*-block Species

**59** The set having ions which are coloured and paramagnetic both is *[2021, 22 July Shift-II]*

(a) $Cu^{+}, Zn^{2+}, Mn^{4+}$ (b) $Ni^{2+}, Mn^{7+}, Hg^{2+}$
(c) $Cu^{2+}, Cr^{3+}, Sc^{+}$ (d) $Sc^{3+}, V^{5+}, Ti^{4+}$

**60** Match List-I and List-II.

| List-I | | List-II | |
|---|---|---|---|
| (A) | Chlorophyll | (i) | Ruthenium |
| (B) | Vitamin-$B_{12}$ | (ii) | Platinum |
| (C) | Anticancer drug | (iii) | Cobalt |
| (D) | Grubbs catalyst | (iv) | Magnesium |

Choose the most appropriate answer from the options given below. *[2021, 18 March Shift-I]*

(a) (A)-(iv), (B)-(ii), (C)-(iii), (D)-(i)
(b) (A)-(iv), (B)-(iii) (C)-(i), (D)-(ii)
(c) (A)-(iii), (B)-(ii), (C)-(iv), (D)-(i)
(d) (A)-(iv), (B)-(iii), (C)-(ii), (D)-(i)

**61** Consider the hydrated ions of $Ti^{2+}$, $V^{2+}$, $Ti^{3+}$ and $Sc^{3+}$. The correct order of their spin-only magnetic moment is *[2019, 10 April 12 Shift-I]*

(a) $Sc^{3+} < Ti^{3+} < Ti^{2+} < V^{2+}$
(b) $Sc^{3+} < Ti^{3+} < V^{2+} < Ti^{2+}$
(c) $Ti^{3+} < Ti^{2+} < Sc^{3+} < V^{2+}$
(d) $V^{2+} < Ti^{2+} < Ti^{3+} < Sc^{3+}$

**62** Which of the following arrangements does not represent the correct order of the property stated against it? *[JEE Main 2013]*

(a) $V^{2+} < Cr^{2+} < Mn^{2+} < Fe^{2+}$ : paramagnetic behaviour
(b) $Ni^{2+} < Co^{2+} < Fe^{2+} < Mn^{2+}$ : ionic size
(c) $Co^{3+} < Fe^{3+} < Cr^{3+} < Sc^{3+}$ : stability in aqueous solution
(d) Sc < Ti < Cr < Mn : number of oxidation states

**63** In context with the transition elements, which of the following statements is incorrect? *[AIEEE 2009]*

(a) In addition to the normal oxidation state, the zero oxidation state is also shown by these elements in complexes
(b) In the highest oxidation states, the transition metal shows basic character and form cationic complexes
(c) In the highest oxidation states of the first five transition elements (Sc to Mn), all the 4*s* and 3*d* electrons are used for bonding
(d) Once the $d^5$ configuration is exceeded, the tendency to involve all the 3*d* electrons in bonding decreases

**64** The atomic numbers of vanadium (V), chromium, (Cr), manganese (Mn) and iron (Fe) are respectively 23, 24, 25 and 26. Which one of these may be expected to have the highest second ionisation enthalpy? *[AIEEE 2003]*

(a) V (b) Cr (c) Mn (d) Fe

**65** Which of the following ions has the maximum magnetic moment? *[AIEEE 2002]*

(a) $Mn^{2+}$ (b) $Fe^{2+}$
(c) $Ti^{2+}$ (d) $Cr^{2+}$

## Numerical Value Type Questions

**66** Among Sc, Mn, Co and Cu. identify the element with highest enthalpy of atomisation. The spin only magnetic moment value of that element in its +2 oxidation state is ........ BM (in Nearest integer) *[2025, 3 April Shift-II]*

**67** Niobium (Nb) and ruthenium (Ru) have '$x$' and '$y$' number of electrons in their respective $4d$-orbitals. The value of $x + y$ is ........ . *[2025, 22 Jan Shift-II]*

**68** The spin only magnetic moment ($\mu$) value (BM) of the compound with strongest oxidising power among $Mn_2O_3$, TiO and VO is ........ BM (Nearest integer). *[2025, 28 Jan Shift-II]*

**69** A transition metal '$M$' among Sc, Ti, V, Cr, Mn and Fe has the highest second ionisation enthalpy. The spin-only magnetic moment value of $M^+$ ion is ........... BM. (Nearest integer)

(Given, atomic number Sc : 21, Ti : 22, V : 23, Cr : 24, Mn : 25, Fe : 26) *[2024, 9 April Shift-II]*

**70** The 'spin only' magnetic moment value of $MO_4^{2-}$ is ........ BM. (Where $M$ is a metal having least metallic radii among Sc, Ti, V, Cr, Mn and Zn).

(Given atomic number : Sc = 21, Ti = 22, V = 23, Cr = 24, Mn = 25 and Zn = 30) *[2024, 8 April Shift-II]*

**71** The spin-only magnetic moment value of the ion among $Ti^{2+}$, $V^{2+}$, $Co^{3+}$ and $Cr^{2+}$, that acts as strong oxidising agent in aqueous solution is ........ BM (Nearest integer).

(Given, atomic numbers : Ti : 22, V : 23, Cr : 24, Co : 27) *[2024, 5 April Shift-I]*

**72** A first row transition metal with highest enthalpy of atomisation, upon reaction with oxygen at high temperature forms oxides of formula $M_2O_n$ (where $n = 3, 4, 4$). The 'spin-only' magnetic moment value of the amphoteric oxide from the above oxides is...... BM. (Nearest integer)

(given, atomic number: Sc : 21, Ti : 22, V : 23, Cr : 24, Mn : 25, Fe : 26, Co : 27, Ni : 28, Cu : 29, Zn : 30) *[2024, 4 April Shift-II]*

**73** Consider the following reaction

$$MnO_2 + KOH + O_2 \longrightarrow A + H_2O$$

Product '$A$' in neutral or acidic medium disproportionate to give products '$B$' and '$C$' along with water. The sum of spin-only magnetic moment values of $B$ and $C$ is...... BM. (Nearest integer). (Given atomic number of Mn is 25) *[2024, 4 April Shift-I]*

**74** The spin - only magnetic moment value of $M^{3+}$ ion (in gaseous state) from the pairs $Cr^{3+}/Cr^{2+}$, $Mn^{3+}/Mn^{2+}$, $Fe^{3+}/Fe^{2+}$ and $Co^{3+}/Co^{2+}$ that has negative standard electrode potential, is ........... BM. [nearest integer] *[2022, 25 July Shift-II]*

**75** Among $Co^{3+}$, $Ti^{2+}$, $V^{2+}$ and $Cr^{2+}$ ions, one if used as a reagent cannot liberate $H_2$ from dilute mineral acid solution, its spin-only magnetic moment in gaseous state is ........ BM (Nearest integer) *[2022, 25 JulyShift-I]*

**76** Number of electrons that vanadium ($Z = 23$) has in $p$-orbitals is equal to ......... . *[2021, 22 July Shift-II]*

**77** In the ground state of atomic Fe($Z = 26$), the spin-only magnetic moment is ........... $\times 10^{-1}$ BM. (Round off to the nearest integer).
[Given : $\sqrt{3} = 1.73$, $\sqrt{2} = 1.41$] *[2021, 17 March Shift-II]*

## Topic 2 Important Compounds of Transition Elements

### Level 1

**01** Dihydrogen reacts with CuO to give *[2022, 28 June Shift-I]*

(a) $CuH_2$ (b) Cu
(c) $Cu_2O$ (d) $Cu(OH)_2$

**02** The nature of oxides $V_2O_3$ and CrO is indexed as '$X$' and '$Y$' type respectively. The correct set of $X$ and $Y$ is *[2021, 27 Aug Shift-I]*

(a) $X$ = acidic $Y$ = acidic
(b) $X$ = basic $Y$ = amphoteric
(c) $X$ = basic $Y$ = basic
(d) $X$ = amphoteric $Y$ = basic

**03** In the given chemical reaction, colours of the $Fe^{2+}$ and $Fe^{3+}$ ions, are respectively

$$5Fe^{2+} + MnO_4^- + 8H^+ \longrightarrow Mn^{2+} + 4H_2O + 5Fe^{3+}$$

*[2021, 1 Sep Shift-II]*

(a) yellow, orange (b) yellow, green
(c) green, orange (d) green, yellow

**04** In the following reactions, ZnO is respectively acting as a/an *[JEE Main 2017]*

(i) $ZnO + Na_2O \longrightarrow Na_2ZnO_2$
(ii) $ZnO + CO_2 \longrightarrow ZnCO_3$

(a) base and acid (b) base and base
(c) acid and acid (d) acid and base

**05** Calomel ($Hg_2Cl_2$) on reaction with ammonium hydroxide gives *[AIEEE 2005]*

(a) HgO (b) $Hg_2O$
(c) $NH_2—Hg—HgCl$ (d) $HgNH_2Cl$

## Level 2

**06** Find the compound '$A$' from the following reaction sequences. *[2025, 24 Jan Shift-II]*

$$A \xrightarrow{\text{Aqua-regia}} B \xrightarrow[\text{(ii) AcOH}]{\text{(i) } KNO_2/[NH_4OH]} \text{yellow ppt.}$$

(a) ZnS (b) MnS
(c) CoS (d) NiS

**07** Consider the following reactions, *[2025, 23 Jan Shift-II]*

$$K_2Cr_2O_7 \xrightarrow[-H_2O]{KOH} [A] \xrightarrow[-H_2O]{H_2SO_4} [B] + K_2SO_4$$

The products [$A$] and [$B$], respectively are
(a) $K_2CrO_4$ and $K_2Cr_2O_7$
(b) $K_2Cr(OH)_6$ and $Cr_2O_3$
(c) $K_2CrO_4$ and CrO
(d) $K_2CrO_4$ and $Cr_2O_3$

**08** The orange colour of $K_2Cr_2O_7$ and purple colour of $KMnO_4$ is due to *[2024, 30 Jan Shift-II]*
(a) $d \rightarrow d$ transitions in $KMnO_4$ and charge transfer transitions in $K_2Cr_2O_7$
(b) $d \rightarrow d$ transitions in $K_2Cr_2O_7$ and charge transfer transitions in $KMnO_4$
(c) Charge transfer transition in both
(d) $d \rightarrow d$ transitions in both

**09** Alkaline oxidative fusion of $MnO_2$ gives "$A$" which on electrolytic oxidation in alkaline solution produces $B$. $A$ and $B$ respectively are *[2024, 30 Jan Shift-II]*
(a) $MnO_4^{2-}$ and $MnO_4^-$ (b) $Mn_2O_3$ and $MnO_4^{2-}$
(c) $Mn_2O_7$ and $MnO_4^-$ (d) $MnO_4^{2-}$ and $Mn_2O_7$

**10** Highest oxidation state of Mn is exhibited in $Mn_2O_7$. The correct statements about $Mn_2O_7$ are
A. Mn is tetrahedrally surrounded by oxygen atoms.
B. Mn is octahedrally surrounded by oxygen atoms.
C. contains Mn—O—Mn bridge.
D. contains Mn—Mn bond.
Choose the correct answer from the options given below. *[2023, 1 Feb Shift-I]*
(a) B and D only (b) B and C only
(c) A and C only (d) A and D only

**11** When $Cu^{2+}$ ion is treated with KI, a white precipitate, $X$ appears in solution. The solution is titrated with sodium thiosulphate, the compound $Y$ is formed. $X$ and $Y$ respectively are *[2023, 31 Jan Shift-I 2023]*
(a) $X = Cu_2I_2$ $Y = Na_2S_4O_5$
(b) $X = CuI_2$ $Y = Na_2S_2O_3$
(c) $X = CuI_2$ $Y = Na_2S_4O_6$
(d) $X = Cu_2I_2$ $Y = Na_2S_4O_6$

**Imp.** Structure of $Mn_2O_7$

**12** The total number of Mn ═O bonds in $Mn_2O_7$ is *[2022, 27 July Shift-I]*
(a) 4 (b) 5 (c) 6 (d) 3

**13** The products obtained from a reaction of hydrogen peroxide and acidified potassium permanganate are *[2022, 26 July Shift-II]*
(a) $Mn^{4+}$, $H_2O$ only (b) $Mn^{2+}$, $H_2O$ only
(c) $Mn^{4+}$, $H_2O$, $O_2$ only (d) $Mn^{2+}$, $H_2O$, $O_2$ only

**14** In the structure of the dichromate ion, there is a *[2021, 31 Aug Shift-I]*
(a) linear symmetrical Cr—O—Cr bond
(b) non-linear unsymmetrical Cr—O—Cr bond
(c) non-linear symmetrical Cr—O—Cr bond
(d) linear unsymmetrical Cr—O—Cr bond

**15** The oxide that shows magnetic property is *[2021, 18 March Shift-II]*
(a) $SiO_2$ (b) $Mn_3O_4$
(c) $Na_2O$ (d) MgO

**16** $Fex_2$ and $Fey_3$ are known when $x$ and $y$ are *[2021, 16 March Shift-II]*
(a) $x$ = F, Cl, Br, I and $y$ = F, Cl and Br
(b) $x$ = F, Cl, Br and $y$ = F, Cl, Br and I
(c) $x$ = Cl, Br, I and $y$ = F, Cl, Br and I
(d) $x$ = F, Cl, Br, I and $y$ = F, Cl, Br and I

**17** The colour of $KMnO_4$ is due to *[JEE Main 2015]*
(a) $M \rightarrow L$ charge transfer transition
(b) $d - d$ transition
(c) $L \rightarrow M$ charge transfer transition
(d) $\sigma - \overset{*}{\sigma}$ transition

**18** Excess of KI reacts with $CuSO_4$ solution and then $Na_2S_2O_3$ solution is added to it. Which of the following statements is incorrect for this reaction? *[AIEEE 2004]*
(a) $Cu_2I_2$ is formed
(b) $CuI_2$ is formed
(c) $Na_2S_2O_3$ is oxidised
(d) Evolved $I_2$ is reduced

## Level 3

**Imp.** Oxidation Reaction of $K_2Cr_2O_7$ and $KMnO_4$

**19** Which of the following oxidation reactions are carried out by both $K_2Cr_2O_7$ and $KMnO_4$ in acidic medium? *[2025, 28 Jan Shift-I]*
A. $I^- \rightarrow I_2$ B. $S^{2-} \rightarrow S$ C. $Fe^{2+} \rightarrow Fe^{3+}$
D. $I^- \rightarrow IO_3^-$ E. $S_2O_3^{2-} \rightarrow SO_4^{2-}$
Choose the correct answer from the options given below
(a) C, D and E only (b) A, B and C only
(c) B, C and D only (d) A, D and E only

**20** Preparation of potassium permanganate from $MnO_2$, involves two step process in which the 1st step is a reaction with KOH and $KNO_3$ to produce *[2025, 24 Jan Shift-I]*
(a) $K_3MnO_4$ (b) $K_4[Mn(OH)_6]$
(c) $K_2MnO_4$ (d) $KMnO_4$

**Imp.** Composition of Alloys

**21** Match List -I with List -II. *[2025, 23 Jan Shift-II]*

| | List-I | | List-II |
|---|---|---|---|
| (A) | Bronze | (I) | Cu, Ni |
| (B) | Brass | (II) | Fe, Cr, Ni, C |
| (C) | UK silver coin | (III) | Cu, Zn |
| (D) | Stainless steel | (IV) | Cu, Sn |

Choose the correct answer from the options given below.
(a) (A)-(IV), (B)-(III), (C)-(I), (D)-(II)
(b) (A)-(III), (B)-(I), (C)-(IV), (D)-(II)
(c) (A)-(III), (B)-(IV), (C)-(II), (D)-(I)
(d) (A)-(IV), (B)-(II), (C)-(III), (D)-(I)

**Imp.** Reactions of $KMnO_4$

**22** During the reaction of permanganate with thiosulphate, the change in oxidation of manganese occurs by value of 3. Identify which of the below medium will favour the reaction. *[2023, 6 April Shift-II]*
(a) Both aqueous acidic and faintly alkaline
(b) Aqueous acidic
(c) Aqueous neutral
(d) Both aqueous acidic and neutral

**23** In neutral or faintly alkaline medium, $KMnO_4$ being a powerful oxidant can oxidise, thiosulphate almost quantitatively, to sulphate. In this reaction overall change in oxidation state of manganese will be *[2022, 29 July Shift-I]*
(a) 5 (b) 1
(c) 0 (d) 3

**24** Given below are two statements.
**Statement I** Iron (III) catalyst, acidified $K_2Cr_2O_7$ and neutral $KMnO_4$ have the ability to oxidise $I^-$ to $I_2$ independently.
**Statement II** Manganate ion is paramagnetic in nature and involves $p\pi - p\pi$ bonding.
In the light of the above statements, choose the correct answer from the options given below. *[2022, 27 July Shift-I]*
(a) Both Statement I and Statement II are true.
(b) Both Statement I and Statement II are false.
(c) Statement I is true but Statement II is false.
(d) Statement I is false but Statement II is true.

**25** The dark purple colour of $KMnO_4$ disappears in the titration with oxalic acid in acidic medium. The overall change in the oxidation number of manganese in the reaction is *[2022, 26 July Shift-I]*
(a) 5 (b) 1
(c) 7 (d) 2

**26** In which one of the following sets all species show disproportionation reaction *[2021, 31 Aug Shift-II]*
(a) $ClO_2^-$, $F_2$, $MnO_4^{2-}$ and $Cr_2O_7^{2-}$
(b) $Cr_2O_7^{2-}$, $MnO_4^-$, $ClO_2^-$ and $Cl_2$
(c) $ClO_2^-$, $Cl_2$ and $Mn^{3+}$
(d) $ClO_4^-$, $MnO_4^{2-}$, $ClO_2^-$ and $F_2$

**27** Potassium permanganate on heating at 513 K gives a product which is *[2021, 27 Aug Shift-II]*
(a) paramagnetic and colourless
(b) diamagnetic and green
(c) diamagnetic and colourless
(d) paramagnetic and green

**28** The correct order of following $3d$-metal oxides, according to their oxidation numbers is
(A) $CrO_3$ (B) $Fe_2O_3$
(C) $MnO_2$ (D) $V_2O_5$
(E) $Cu_2O$ *[2021, 25 July Shift-I]*
(a) (D) > (A) > (B) > (C) > (E)
(b) (A) > (C) > (D) > (B) > (E)
(c) (A) > (D) > (C) > (B) > (E)
(d) (C) > (A) > (D) > (E) > (B)

**29** Given below are two statements
**Statement I** Potassium permanganate, on heating at 573 K forms potassium manganate.
**Statement II** Both potassium permanganate and potassium manganate are tetrahedral and paramagnetic in nature.
In the light of the above statements, choose the most appropriate answer from the options given below *[2021, 17 March Shift-I]*
(a) Both statement I and statement II are true.
(b) Both statement I and statement II are false.
(c) Statement I is false but statement II is true.
(d) Statement I is true but statement II is false.

**30** The incorrect statement among the following is *[2021, 24 Feb Shift-II]*
(a) red colour of ruby is due to the presence of $Co^{3+}$
(b) $RuO_4$ is an oxidizing agent
(c) $Cr_2O_3$ is an amphoteric oxide
(d) $VOSO_4$ is a reducing agent

**31** The incorrect statement is : *[2020, 3 Sep Shift-II]*
(a) Manganate ion is green in colour and permanganate ion is purple in colour.
(b) Manganate and permanganate ions are paramagnetic.
(c) In manganate and permanganate ions, the $\pi$-bonding takes place by overlap of $p$-orbitals of oxygen and $d$-orbitals of manganese.
(d) Manganate and permanganate ions are tetrahedral.

**32** $\underline{A} \xrightarrow{4\,KOH,\,O_2} 2\underline{B} + 2H_2O$ (Green)

$3\underline{B} \xrightarrow{4\,HCl} 2\underline{C} + MnO_2 + 2H_2O$ (Purple)

$2\underline{C} \xrightarrow{H_2O,\,KI} 2\underline{A} + 2KOH + \underline{D}$

In the above sequence of reactions, $\underline{A}$ and $\underline{D}$, respectively, are *[2019, 11 Jan Shift-II]*

(a) KI and $KMnO_4$ (b) $MnO_2$ and $KIO_3$
(c) KI and $K_2MnO_4$ (d) $KIO_3$ and $MnO_2$

**33** Iron exhibits +2 and +3 oxidation states. Which of the following statements about iron is incorrect? *[AIEEE 2012]*

(a) Ferrous oxide is more basic in nature than the ferric oxide
(b) Ferrous compounds are relatively more ionic than the corresponding ferric compounds
(c) Ferrous compounds are less volatile than the corresponding ferric compounds
(d) Ferrous compounds are more easily hydrolysed than the corresponding ferric compounds

**34** When $KMnO_4$ acts as oxidising agent and ultimately forms $[MnO_4]^{-2}$, $MnO_2$, $Mn_2O_3$, $Mn^{+2}$ then the number of electrons transferred in each case respectively is *[AIEEE 2002]*

(a) 4, 3, 1, 5 (b) 1, 5, 3, 7 (c) 1, 3, 4, 5 (d) 3, 5, 7, 1

## Numerical Value Type Questions

**35.** Consider the following reactions. *[2025, 3 April Shift-I]*

$A + \underset{\text{Little amount}}{NaCl} + H_2SO_4 \longrightarrow CrO_2Cl_2 + \text{side products}$

$CrO_2Cl_2(\text{Vapour}) + NaOH \longrightarrow B + NaCl + H_2O$

$B + H^+ \longrightarrow C + H_2O$

The number of terminal 'O' present in the compound 'C' is ........

**36.** $KMnO_4$ acts as an oxidising agent in acidic medium. '$X$' is the difference between the oxidation states of Mn in reactant and product. '$Y$' is the number of '$d$' electrons present in the brown red precipitate formed at the end of the acetate ion test with neutral ferric chloride. The value of $X + Y$ is *[2025, 4 April Shift-I]*

**Reaction of $KMnO_4$ in Blend with Spin only Magnetic Moment**

**Imp.**

**37** The difference in the 'spin-only' magnetic moment values of $KMnO_4$ and the manganese product formed during titration of $KMnO_4$, against oxalic acid in acidic medium is ........ BM. (Nearest integer) *[2024, 6 April Shift-I]*

**38** Among CrO, $Cr_2O_3$ and $CrO_3$ the sum of spin-only magnetic moment values of basic and amphoteric oxides is ......... $10^{-2}$ BM (Nearest integer).
(Given, atomic number of Cr is 24) *[2024, 6 April Shift-I]*

**39** The fusion of chromite ore with sodium carbonate in the presence of air leads to the formation of products $A$ and $B$ along with the evolution of $CO_2$. The sum of spin-only magnetic moment values of $A$ and $B$ is ......... B.M. (Nearest integer)
[Given, atomic number : C : 6, Na : 11, O : 8, Fe : 26, Cr : 24] *[2024, 5 April Shift-II]*

**40** The total change in the oxidation state of manganese involved in the reaction of $KMnO_4$ and potassium iodide in the acidic medium is ........... . *[2023, 15 April Shift-I]*

**41** The spin-only magnetic moment value of the compound with strongest oxidising ability among $MnF_4$, $MnF_3$ and $MnF_2$ is .............. BM. [nearest integer] *[2022, 26 July Shift-II]*

**42** In chromyl chloride, the oxidation state of chromium is (+) ........... . *[2023, 15 April Shift-I]*

**43** The oxidation states of transition metal atoms in $K_2Cr_2O_7$, $KMnO_4$ and $K_2FeO_4$, respectively, are $x$, $y$ and $z$. The sum of $x$, $y$ and $z$ is ......... *[2022, 2 Sept Shift-II]*

**44** The disproportionation of $MnO_4^{2-}$ in acidic medium resulted in the formation of two manganese compounds $A$ and $B$. If the oxidation state of Mn in $B$ is smaller than that of $A$, then the spin-only magnetic moment ($\mu$) value of $B$ in BM is ........... . (Nearest integer) *[2022, 28 July Shift-I]*

**45** The number of terminal oxygen atoms present in the product $B$ obtained from the following reaction is ...... . *[2022, 29 June Shift-I]*

$FeCr_2O_4 + Na_2CO_3 + O_2 \longrightarrow A + Fe_2O_3 + CO_2$

$A + H^+ \longrightarrow B + H_2O + Na^+$

**46** An acidified manganate solution undergoes disproportionation reaction. The spin-only magnetic moment value of the product having manganese in higher oxidation state is ...... BM. (Nearest integer) *[2022, 29 June Shift-I]*

**47** The number of statement(s) correct from the following for copper (Atomic number 29) is/are ...... *[2022, 27 June Shift-I]*

(A) Cu(II) complexes are always paramagnetic.
(B) Cu(I) complexes are generally colourless.
(C) Cu(I) is easily oxidised.
(D) In Fehling solution, the active reagent has Cu(I).

**48** The spin–only magnetic moment value of the most basic oxide of vanadium among $V_2O_3$, $V_2O_4$ and $V_2O_5$ is ...... BM. (Nearest integer) *[2022, 26 June Shift-I]*

**49** Manganese (VI) has ability to disproportionate in acidic solution. The difference in oxidation states of two ions it forms in acidic solution is ......... .

*[2022, 24 June Shift-II]*

**50** The difference in oxidation state of chromium in chromate and dichromate salt is ......... .

*[2022, 24 June Shift-I]*

**51** In mild alkaline medium, thiosulphate ion is oxidised by $MnO_4^-$ to "A". The oxidation state of sulphur in "A" is ......... . *[2021, 26 Feb Shift-II]*

**52** Dichromate ion is treated with base, the oxidation number of Cr in the product formed is ......... .

*[2021, 26 Feb Shift-I]*

**53** The sum of the total number of bonds between chromium and oxygen atoms in chromate and dichromate ions is ........ *[2020, 9 Jan Shift-II]*

## Topic 3 Inner Transition Elements (Lanthanoids and Actinoids)

### Level 1

**01** Which of the following ions is the strongest oxidising agent? (Atomic Number of Ce = 58, Eu = 63, Tb = 65, Lu = 71) *[2025, 24 Jan Shift-I]*

(a) $Lu^{3+}$ (b) $Ce^{3+}$
(c) $Tb^{4+}$ (d) $Eu^{2+}$

**02** The electronic configuration of Einsteinium is (Given atomic number of Einsteinium = 99)

*[2024, 9 April Shift-II]*

(a) $[Rn]\,5f^{12}\,6d^0\,7s^2$ (b) $[Rn]\,5f^{11}\,6d^0\,7s^2$
(c) $[Rn]\,5f^{13}\,6d^0\,7s^2$ (d) $[Rn]\,5f^{10}\,6d^0\,7s^2$

**03** The number of element from the following that do not belong to lanthanoids is
Eu, Cm, Er, Tb, Yb and Lu *[2024, 6 April Shift-I]*
(a) 5 (b) 4 (c) 3 (d) 1

**Imp.** Electronic Configuration of Lanthanoids

**04** The electronic configuration for neodymium is [Atomic number for neodymium 60]

*[2024, 27 Jan Shift-I]*

(a) $[Xe]\,5f^7\,7s^2$ (b) $[Xe]\,4f^4\,6s^2$
(c) $[Xe]\,4f^1\,5d^1\,6s^2$ (d) $[Xe]\,4f^6\,6s^2$

**05** Given below are two statements, one is labelled as Assertion (A) and the other is labelled as Reason (R).
**Assertion** (A) $5f$-electrons can participate in bonding to a far greater extent than $4f$-electrons.
**Reason** (R) $5f$-orbitals are not as buried as $4f$-orbitals.

In the light of the above statements, choose the correct answer from the options given below.

*[2023, 12 April Shift-I]*

(a) A is true but R is false
(b) Both A and R are true and R is the correct explanation of A
(c) Both A and R are true but R is not the correct explanation of A
(d) A is false but R is true

**06** $Nd^{2+}$ = ........ *[2023, 31 Jan Shift-I]*

(a) $4f^3$ (b) $4f^4$
(c) $4f^4 6s^2$ (d) $4f^2 6s^2$

**07** Which one amongst the following are good oxidising agents?

A. $Sm^{2+}$ B. $Ce^{2+}$
C. $Ce^{4+}$ D. $Tb^{4+}$

Choose the most appropriate answer from the options given below. *[2023, 24 Jan Shift-II]*
(a) C and D (b) A and B (c) Only C (d) Only D

**08** The most common oxidation state of lanthanoid elements is +3. Which of the following is likely to deviate easily from +3 oxidation state?

*[2022, 26 June Shift-II]*

(a) Ce (58) (b) La (57)
(c) Lu (71) (d) Gd (64)

**09** The $Eu^{2+}$ ion is a strong reducing agent in spite of its ground state electronic configuration (outermost) [Atomic number of Eu = 63] *[2021, 31 Aug Shift-II]*

(a) $4f^7 6s^2$ (b) $4f^6$
(c) $4f^7$ (d) $4f^6 6s^2$

**10** Misch metal is an alloy consisting mainly of
(a) lanthanoid metals *[2020, 6 Sep Shift-II]*
(b) actinoid and transition metals
(c) lanthanoid and actinoid metals
(d) actinoid metals

**11** The lanthanoid that does not show +4 oxidation state is *[2020, 6 Sep Shift-I]*

(a) Dy (b) Ce
(c) Eu (d) Tb

**12** The elements with atomic numbers 101 and 104 belong to, respectively, *[2020, 4 Sep Shift-I]*
(a) actinoids and group 4 (b) actinoids and group 6
(c) group 6 and actinoids (d) group 11 and group 4

**13** The highest possible oxidation states of uranium and plutonium, respectively, are *[2019, 10 April Shift-II]*
(a) 7 and 6 (b) 6 and 7 (c) 6 and 4 (d) 4 and 6

**14** The correct order of atomic radii is

*[2019, 12 Jan Shift-II]*

(a) Ho > N > Eu > Ce (b) N > Ce > Eu > Ho
(c) Eu > Ce > Ho > N (d) Ce > Eu > Ho > N

**15** The effect of lanthanoid contraction in the lanthanoid series of elements by and large means *[2019, 10 Jan Shift-I]*
(a) increase in atomic radii and decrease in ionic radii
(b) decrease in both atomic and ionic radii
(c) increase in both atomic and ionic radii
(d) decrease in atomic radii and increase in ionic radii

**16** The outer electron configuration of Gd (At. no. 64) is *[AIEEE 2011]*
(a) $4f^3, 5d^5, 6s^2$ (b) $4f^8, d^0, 6s^2$
(c) $4f^4, 5d^4, 6s^2$ (d) $4f^7, 5d^1, 6s^2$

**17** Larger number of oxidation states are exhibited by the actinoides than those by the lanthanoides, the main reason being *[AIEEE 2008]*
(a) $4f$ orbitals more diffused than the $5f$ orbitals
(b) lesser energy difference between $5f$ and $6d$ than between $4f$ and $5d$ orbitals
(c) more energy difference between $5f$ and $6d$ than between $4f$ and $5d$ orbitals
(d) more reactive nature of the actinoides than the lanthanoides

**Imp.** Lanthanoid Contraction

**18** Lanthanoid contraction is caused due to *[AIEEE 2006]*
(a) the appreciable shielding on outer electrons by $4f$ electrons from the nuclear charge
(b) the appreciable shielding on outer electrons by $5d$ electrons from the nuclear charge
(c) the same effective nuclear charge from Ce to Lu
(d) the imperfect shielding on outer electrons by $4f$ electrons from the nuclear charge

**19** Most common oxidation states of Ce (Cerium) are *[AIEEE 2002]*
(a) +3, +4 (b) +2, +3
(c) +2, +1 (d) +3, +5

## Level 2

**20** Lanthanoid ions with $4f^7$ configuration are
(A) $Eu^{2+}$ (B) $Gd^{3+}$ (C) $Eu^{3+}$ (D) $Tb^{3+}$ (E) $Sm^{2+}$
Choose the correct answer from the options given below *[2025, 22 Jan Shift-I]*
(a) (*A*) and (*D*) only (b) (*A*) and (*B*) only
(c) (*B*) and (*E*) only (d) (*B*) and (*C*) only

**21** Given below are two statements. *[2024, 27 Jan Shift-II]*
**Statement I** In the lanthanoids, the formation of $Ce^{4+}$ is favoured by its noble gas configuration.
**Statement II** $Ce^{4+}$ is a strong oxidant reverting to the common +3 state.
In the light of the above statements, choose the **most appropriate** answer from the options given below.
(a) Statement I is false but Statement II is true.
(b) Statement I is true but Statement II is false.
(c) Both Statement I and Statement II are true.
(d) Both Statement I and Statement II are false.

**Imp.** Electronic Configuration of Lanthanoids

**22** The '*f*' orbitals are half and completely filled, respectively in lanthanide ions
(Given, atomic number : Eu 63; Sm 62; Tm 69; Tb 65; Yb 70; Dy 66] *[2022, 27 June Shift-II]*
(a) $Eu^{2+}$ and $Tm^{2+}$ (b) $Sm^{2+}$ and $Tm^{3+}$
(c) $Tb^{4+}$ and $Yb^{2+}$ (d) $Dy^{3+}$ and $Yb^{3+}$

**23** Cerium (IV) has a noble gas configuration. Which of the following is correct statement about it? *[2022, 25 June Shift-I]*
(a) It will not prefer to undergo redox reactions.
(b) It will prefer to gain electron and act as an oxidising agent.
(c) It will prefer to give away an electron and behaves as reducing agent.
(d) It acts as both, oxidising and reducing agent.

**24** Given below are two statements.
**Statement I** The $E°$ value of $Ce^{4+}/Ce^{3+}$ is +1.74 V.
**Statement II** Ce is more stable in $Ce^{4+}$ state than $Ce^{3+}$ state.
In the light of the above statements, choose the most appropriate answer from the options given below. *[2021, 16 March Shift-I]*
(a) Both statement I and statement II are correct.
(b) Statement I is incorrect but statement II is correct.
(c) Both statement I and statement II are incorrect.
(d) Statement I is correct but statement II is incorrect.

**25** Given below are two statements. One is labelled as Assertion (A) and the other is labelled as Reason (R).
**Assertion (A)** Size of $Bk^{3+}$ ion is less than $Np^{3+}$ ion.
**Reason (R)** The above is a consequence of the lanthanoid contraction. In the light of the above statements, choose the correct answer from the options given below *[2021, 16 March Shift-I]*
(a) A is false but R is true.
(b) Both A and R are true but *R* is not the correct explanation of (a)
(c) Both A and R are true and R is the correct explanation of (a)
(d) A is true but R is false.

**26** Which one of the following lanthanoids does not form $MO_2$?
[*M* is lanthanoid metal] *[2021, 26 Feb Shift-I]*
(a) Pr (b) Dy
(c) Nd (d) Yb

**27** The correct electronic configuration and spin-only magnetic moment (BM) of $Gd^{3+}$ ($Z = 64$), respectively, are *[2020, 5 Sep Shift-I]*
(a) [Xe] $4f^7$ and 8.9 (b) [Xe] $4f^7$ and 7.9
(c) [Xe] $5f^7$ and 8.9 (d) [Xe] $5f^7$ and 7.9

**28** The maximum number of possible oxidation states of actinides are shown by *[2019, 9 April Shift-II]*
(a) berkelium (Bk) and californium (Cf)
(b) neptunium (Np) and plutonium (Pu)
(c) nobelium (No) and lawrencium (Lr)
(d) actinium (Ac) and thorium (Th)

**29** In context of the lanthanoids, which of the following statements is not correct? *[AIEEE 2011]*
(a) There is a gradual decrease in the radii of the members with increasing atomic number in the series.
(b) All the member exhibit +3 oxidation state.
(c) Because of similar properties the separation of lanthanoids is not easy.
(d) Availability of $4f$ electrons results in the formation of compounds in +4 state for all the members of the series.

**30** Cerium ($Z = 58$) is an important member of the lanthanides. Which of the following statements about cerium is incorrect? *[AIEEE 2004]*
(a) The common oxidation states of cerium are +3 and +4
(b) The +3 oxidation state of cerium is more stable than the +4 oxidation state
(c) The +4 oxidation state of cerium is not known in solutions
(d) Cerium (IV) acts as an oxidising agent

**31** The radius of $La^{3+}$ (atomic number : La = 57) is 1.06 Å. Which one of the following given values will be closest to the radius of $Lu^{3+}$ (atomic number : Lu = 71) ? *[AIEEE 2003]*
(a) 1.60Å (b) 1.40 Å (c) 1.06 Å (d) 0.85 Å

## Level 3

**32** The pair of lanthanides in which both elements have high third - ionisation energy *[2023, 13 April Shift-I]*
(a) Lu, Yb (b) Dy, Gd
(c) Eu, Yb (d) Eu, Gd

**33** Which of the following pair is not isoelectronic species?
(Atomic number : Sm 62; Er 68; Yb 70; Lu 71; Eu 63; Tb 65; Tm 69) *[2022, 28 July Shift-II]*
(a) $Sm^{2+}$ and $Er^{3+}$ (b) $Yb^{2+}$ and $Lu^{3+}$
(c) $Eu^{2+}$ and $Tb^{4+}$ (d) $Tb^{2+}$ and $Tm^{4+}$

**34** Which one of the lanthanoids given below is the most stable in divalent form? *[2022, 28 June Shift-I]*
(a) Ce (atomic number 58)
(b) Sm (atomic number 62)
(c) Eu (atomic number 63)
(d) Yb (atomic number 70)

**35** Which one of the following lanthanides exhibits +2 oxidation state with diamagnetic nature ?
(Given, $Z$ for Nd = 60, Yb = 70, La = 57, Ce = 58) *[2021, 31 Aug Shift-I]*
(a) Nd (b) Yb
(c) La (d) Ce

**36** Given below are two statements
**Statement I** $CeO_2$ can be used for oxidation of aldehydes and ketones.
**Statement II** Aqueous solution of $EuSO_4$ is a strong reducing agent.
In the light of the above statements, choose the correct answer from the options given below. *[2021, 25 Feb Shift-I]*
(a) Both statement I and statement II are true.
(b) Both statement I and statement II are false.
(c) Statement I is true but statement II is false.
(d) Statement I is false but statement II is true.

**37** The incorrect statement(s) among (1)-(3) is (are)
(1) W(VI) is more stable than Cr(VI).
(2) In the presence of HCl, permanganate titrations provide satisfactory results.
(3) Some lanthanoid oxides can be used as phosphorus. *[2020, 4 Sep Shift-II]*
(a) (2) and (3) only (b) (1) and (2) only
(c) (2) only (d) (1) only

**38** Knowing that the chemistry of lanthanoids (Ln) is dominated by its +3 oxidation state, which of the following statements is incorrect? *[AIEEE 2009]*
(a) Because of the large size of the Ln (III) ions the bonding in its its compounds is predominantly ionic in character
(b) The ionic sizes of Ln (III) decrease in general with increasing atomic number
(c) Ln (III) compounds are generally colourless
(d) Ln (III) hydroxide are mainly basic in character

### Numerical Value Type Questions

**39** Number of colourless lanthanoid ions among the following is ........
$Eu^{3+}$, $Lu^{3+}$, $Nd^{3+}$, $La^{3+}$, $Sm^{3+}$ *[2024, 9 April Shift-I]*

**40** The number of $f$-electrons in the ground state electronic configuration of Np ($Z = 93$) is ...... . (Nearest integer) *[2021, 27 Aug Shift-I]*

**41** The number of $4f$ electrons in the ground state electronic configuration of $Gd^{2+}$ is ......... . [Atomic number of Gd is 64.] *[2021, 26 Aug Shift-I]*

**42** Number of electrons present in $4f$ orbital of $Ho^{3+}$ ion is ...... . (Given, atomic number of Ho = 67) *[2021, 25 July Shift-II]*

# Key Index

## Topic 1

| | | | | | | | | | |
|---|---|---|---|---|---|---|---|---|---|
| 1 (a) | 2 (d) | 3 (b) | 4 (a) | 5 (c) | 6 (b) | 7 (a) | 8 (c) | 9 (b) | 10 (a) |
| 11 (d) | 12 (b) | 13 (d) | 14 (a) | 15 (a) | 16 (d) | 17 (d) | 18 (a) | 19 (a) | 20 (d) |
| 21 (d) | 22 (a) | 23 (a) | 24 (a) | 25 (b) | 26 (d) | 27 (c) | 28 (a) | 29 (d) | 30 (c) |
| 31 (d) | 32 (a) | 33 (d) | 34 (d) | 35 (d) | 36 (c) | 37 (c) | 38 (a) | 39 (a) | 40 (d) |
| 41 (a) | 42 (d) | 43 (c) | 44 (b) | 45 (d) | 46 (a) | 47 (b) | 48 (a) | 49 (a) | 50 (a) |
| 51 (a) | 52 (b) | 53 (b) | 54 (c) | 55 (c) | 56 (c) | 57 (a) | 58 (b) | 59 (c) | 60 (d) |
| 61 (a) | 62 (a) | 63 (b) | 64 (b) | 65 (a) | 66 (4) | 67 (11) | 68 (5) | 69 (6) | 70 (0) |
| 71 (5) | 72 (0) | 73 (4) | 74 (4) | 75 (5) | 76 (12) | 77 (49) | | | |

## Topic 2

| | | | | | | | | | |
|---|---|---|---|---|---|---|---|---|---|
| 1 (b) | 2 (c) | 3 (d) | 4 (d) | 5 (d) | 6 (c) | 7 (a) | 8 (c) | 9 (a) | 10 (c) |
| 11 (d) | 12 (c) | 13 (d) | 14 (c) | 15 (b) | 16 (a) | 17 (c) | 18 (b) | 19 (b) | 20 (c) |
| 21 (a) | 22 (c) | 23 (d) | 24 (b) | 25 (a) | 26 (c) | 27 (d) | 28 (c) | 29 (d) | 30 (a) |
| 31 (b) | 32 (b) | 33 (d) | 34 (c) | 35 (6) | 36 (10) | 37 (6) | 38 (877) | 39 (6) | 40 (5) |
| 41 (5) | 42 (6) | 43 (19) | 44 (4) | 45 (6) | 46 (0) | 47 (3) | 48 (3) | 49 (3) | 50 (0) |
| 51 (6) | 52 (6) | 53 (12) | | | | | | | |

## Topic 3

| | | | | | | | | | |
|---|---|---|---|---|---|---|---|---|---|
| 1 (c) | 2 (b) | 3 (d) | 4 (b) | 5 (b) | 6 (b) | 7 (a) | 8 (a) | 9 (c) | 10 (a) |
| 11 (c) | 12 (a) | 13 (b) | 14 (c) | 15 (b) | 16 (d) | 17 (b) | 18 (d) | 19 (a) | 20 (b) |
| 21 (c) | 22 (c) | 23 (b) | 24 (d) | 25 (d) | 26 (d) | 27 (b) | 28 (b) | 29 (d) | 30 (c) |
| 31 (d) | 32 (c) | 33 (a,d) | 34 (c) | 35 (b) | 36 (a) | 37 (c) | 38 (c) | 39 (2) | 40 (18) |
| 41 (7) | 42 (10) | | | | | | | | |

# Detailed **Explanations**

## Topic 1

**01.** (a) The correct decreasing order of spin only magnetic moment is ($\mu_s$),
$Cr^{2+} > Cr^{3+} > Cu^{2+} > Cu^+$
$Cu^+$ : $[Ar]3d^{10}$, unpaired electron = 0, $\mu_s = 0$ BM
$Cu^{2+}$ : $[Ar]3d^9$, unpaired electrons = 1, $\mu_s = \sqrt{3}$ BM
$Cr^{2+}$ : $[Ar]3d^4$, unpaired electrons = 4, $\mu_s = \sqrt{24}$ BM
$Cr^{3+}$ : $[Ar]3d^3$, unpaired electrons = 3, $\mu_s = \sqrt{15}$ BM

**02.** (d) The metal which shows highest and maximum number of oxidation state is Mn.
Oxidation number (Highest) = +7

**03.** (b) In $3d$ metal ions, the enthalpy of hydration generally increases as the charge density of the ion increases. Also the charge density is inversly proportional to the ionic radius. Thus, among the given options metal ion ($Mn^{2+}$) with largest ionic radius should have the lowest charge density and weakest ion-water interactions (i.e. hydration enthalpy)

**04.** (a) The atomic number of Pt is 78. Its exceptional electronic configuration can be written as $[Xe]\,4f^{14}5d^9 6s^1$.
According to Aufbau's principle of filling electrons, the electronic configuration of Pt should be $[Xe]4f^{14}5d^8 6s^2$. But as the number of electrons increases, interelectronic repulsion also increases. In higher atomic number elements energy difference between shells decreases. Hence, $6s$ and $5d$ come closer due to which electronic transition from $6s$ to $5d$ is easily done. Therefore, we have the electronic configuration as [Xe] $4f^{14}5d^9 6s^1$.

**05.** (c) The standard electrode potential of given elements are

$$Cr^{2+}/Cr \longrightarrow -0.91\text{ V}$$
$$Fe^{2+}/Fe \longrightarrow -0.44\text{ V}$$
$$Cu^{2+}/Cu \longrightarrow +0.34\text{ V}$$
$$Zn^{2+}/Zn \longrightarrow -0.76\text{ V}$$

So, it is clearly seen from above values that copper (Cu) has highest $M^{2+}/M$ standard electrode potential.

**06.** (b)

| Metal | Melting point (°C) |
|---|---|
| Hg | 38.83 |
| Ag | 961.8 |
| Ga | 29.76 |
| Cs | 28.44 |

The metal with highest melting point is Ag.

**07.** (a) Common positive oxidation states for an element with atomic number 24 (Cr), are + 2 to + 6.

**08.** (c) The element with atomic number 25 is Mn. $Mn^{2+}$ is divalent ion and has the outer electronic configuration $3d^5$ giving 5 unpaired electrons.

So, $\mu = \sqrt{5(5+2)}$
$= \sqrt{35} = 5.916$
$\approx 5.92$ BM

**09.** (b) The major components of German silver are Cu : 25-30%, Zn : 25-30% and Ni : 40-30%.

**10.** (a) 'Gun metal' is also called as red brass (in USA). It is greyish in colour and originally used for making guns. The major components of 'Gun metal' are

Cu = 87%
Zn = 3%
Sn = 10%

**11.** (d) In the electrode potential series, only copper have positive value for electrode potential because copper has lower tendency than hydrogen to form ions.

∴ Electrode potential of Cu ($E^\circ_{Cu^{2+}/Cu}$) show positive value.

**12.** (b) The atomic radii of Ag is almost equal to Au due to lanthanide contraction.

**13.** (d) Interstitial compounds are non-stoichiometric (Berthollide) in composition. They are very hard with very high melting points. The electrical conductivity of interstitial compounds are comparable to that of the pure metal. These are chemically unreactive in nature.

**14.** (a) The most stable oxidation states in the compounds of the given transition metals of $3d$-series are,
Sc: +3; Ti: +3, +4; V: +2, +3, + 4, +5;
Cu: +1, +2
Due to the presence of only one $3d$-electron (no pairing energy) and two $4s$-electrons, they easily ionise to achieve most stable +3 oxidation state.

**15.** (a) The third ionisation energy of $Mn^{2+}$ is too high due to stable half filled $3d$-orbital. Hence, it cannot further get oxidised to liberate hydrogen. So, it unable to release electron.

**16.** (d) The correct match is
A-(ii), B-(i), C-(iv), D-(iii)

**17.** (d) Standard reduction potential value normally increases from left to right in the period of $d$-block elements. Some SRP value are exceptionally higher due to stability of product ion. *e.g.*,

$$E^\circ_{Fe^{3+}/Fe^{2+}} = +0.77\text{ V};$$
$$E^\circ_{Mn^{3+}/Mn^{2+}} = +1.57\text{ V};$$
$$E^\circ_{Cr^{3+}/Cr^{2+}} = -0.411\text{ V};$$
$$E^\circ_{Co^{3+}/Co^{2+}} = +1.97\text{ V}$$

Thus, $E^\circ_{M^{3+}/M^{2+}}$ is highest for Co.

**18.** (a) Zr (40), Hf (72) have similar properties as they belong to same group (group 4) and have almost similar atomic radii (~160 pm) due to lanthanoid contraction.

**19.** (a) $_{28}Ni \rightarrow [Ar]\,3d^8 4s^2$
⇒ Number of unpaired electrons ($n$) = 2 ($3d^8$) in $Ni^{2+}$

$$\mu = \sqrt{n(n+2)} = \sqrt{2(2+2)}$$
$$= \sqrt{8} \approx 2.84$$

So, option (a) is correct.

**20.** *(d)* The highest oxidation state is achieved by Mn with outer electronic configuration as $(n-1)d^5ns^2$.

**21.** *(d)* $Fe^{2+} = 1s^2 2s^2 2p^6 3s^2 3p^6 3d^6$

The number of *d*-electrons retained in $Fe^{2+} = 6$.

**22.** *(a)* Cu, Ag, Au (Group 11) are called coinage metals because in the early times, they were used in making coins.

**23.** *(a)* Mercury (Hg) solidifies after glass, so used mercury for making good quality mirrors.

**24.** *(a)* Given that, magnetic moment value is 4.9 BM.

So, $\mu = \sqrt{n(n+2)}$ BM

$4.9 = \sqrt{n(n+2)}$ BM

On solving, we get $n = 4$

Now,

(A) $Cr^{2+} = [Ar]\,3d^4$ (4 unpaired $e^-$)

(B) $Fe^{2+} = [Ar]3d^6$ [4 unpaired $e^-$]

(C) $Fe^{3+} = [Ar]\,3d^5$ [5 unpaired $e^-$]

(D) $Co^{2+} = [Ar]\,3d^7$ (3 unpaired $e^-$)

(E) $Mn^{3+} = [Ar]\,3d^4$ [4 unpaired $e^-$]

So, among the given species $Cr^{2+}$, $Fe^{2+}$ and $Mn^{3+}$ has 4.9 BM value.

**25.** *(b)* Incorrect relationship is given in option (b) i.e., $Mn^{2+} < Fe^{2+}$.

It is because due to half-filled orbital of $Mn^{2+}$, it has more I.E. than $Fe^{2+}$.

Thus, the correct order is $Mn^{2+} > Fe^{2+}$.

**26.** *(d)* In 3*d* series, vanadium has highest enthalpy of atomisation and colour of $V^{3+}$ is green.

Oxide formed by $V^{3+}$ is $V_2O_3$ which is basic in nature.

**27.** *(c)* Among the given elements i.e. Cr, Co, Fe, Ni.

Cr has lowest heat of atomisation.

Electronic configuration of Cr is given by

$Cr = [Ar]\,3d^5\,4s^1$

∴Total six valence $e^-$ in Cr.

**28.** *(a)*

| S.No. | Species | Configuration | No. of electrons |
|---|---|---|---|
| 1 | $V^{2+}$ | $[Ar]\,3d^3 4s^0$ | 3 |
| 2 | $Co^{2+}$ | $[Ar]\,3d^7 4s^0$ | 3 |
| 3 | $Ti^{3+}$ | $[Ar]\,3d^1 4s^0$ | 1 |
| 4 | $Mn^{2+}$ | $[Ar]\,3d^5 4s^0$ | 5 |
| 5 | $Ti^{2+}$ | $[Ar]\,3d^2 4s^0$ | 2 |
| 6 | $Co^{2+}$ | $[Ar]\,3d^7 4s^0$ | 3 |
| 7 | $Fe^{3+}$ | $[Ar]\,3d^5 4s^0$ | 5 |
| 8 | $Cr^{2+}$ | $[Ar]\,3d^4 4s^0$ | 4 |

So, $V^{2+}$ and $Co^{2+}$ have same number of unpaired electrons.

**29.** *(d)* Statement I is true but Statement II is false.

The correct form of statement II is Cr(VI) is less stable than Mo(VI).

Hence, $CrO_3$ easily reduce into $Cr^{3+}$ as compare to $MoO_3$ and shows stronger oxidising nature.

**30.** *(c)* Among the given oxides, $V_2O_3$, $V_2O_4$ and $V_2O_5$. The amphoteric oxide is $V_2O_5$. The reaction of $V_2O_5$ with alkali is given by

$$V_2O_5 + \text{Alkali} \longrightarrow VO_4^{3-}$$

In $VO_4^{3-}$ ion, vanadium is in +5 oxidation state.

**31.** *(d)* Both Statement I and Statement II are true.

**32.** *(a)* Among the given options correct statement is given in option (a) i.e. Cu(II) is more stable than Cu(I) because hydration energy of $Cu^{2+}$ ion compensates second ionisation energy of copper.

**33.** *(d)* Given $\mu = 3.86$ B.M.

So, using $\mu = \sqrt{n(n+2)}$

we get $n = 3$

Thus, electronic configuration of ion can be $[Ar]\,3d^3$.

So, the electronic configuration of atom will be $[Ar]\,3d^3\,4s^2$.

Hence, atomic number is 23.

**34.** *(d)* Among the given elements, scandium shows only one oxidation state in its compound i.e. +3 while cobalt, titanium and nickel can show +2, +3 and +4 oxidation state.

**35.** *(d)* All the elements of option (d) have $d^{10}$ electronic configuration in its outermost shell. The electronic configuration of given elements are

$[Cd] = [Kr]\,4d^{10}\,5s^2$

$[Cu] = [Ar]\,3\,d^{10}\,4s^1$

$[Ag] = [Kr]\,4d^{10}\,5s^1$

$[Zn] = [Ar]\,3d^{10}\,4s^2$

**36.** *(c)* The pairs given in option (a), (b) and (d) are correctly match, while option (c) is incorrectly matched.

The correct form is :

The catalyst used in Wacker's process is $PdCl_2$.

Hence, option (c) is final answer.

**37.** *(c)* Correct option is (c).

Since, in *d*-block elements the successive filling of *d*-orbital takes place due to which the outer shell is poorly shielded as compared to group 2 elements thus *d* -block elements have more ionisation enthalpy than group 2 elements.

**38.** *(a)* The metal atom whose oxidation potential is greater than that of hydrogen can release $H_2$ from mineral acid. ($E^\circ_{H/H^+} = 0$)

| Metal | Oxidation potential |
|---|---|
| $E^\circ_{Zn/Zn^{2+}}$ | 0.76 |
| $E^\circ_{Ni/Ni^{2+}}$ | 0.25 |
| $E^\circ_{Mn/Mn^{2+}}$ | 1.18 |
| $E^\circ_{Cu/Cu^{2+}}$ | − 0.34 |

Therefore, Cu has least tendency to liberate hydrogen from mineral acids.

**39.** *(a)* The electronic configuration of ions are

$Mn^{3+} = 1s^2 2s^2 2p^6 3s^2 3p^6 3d^4 4s^0$

$Fe^{3+} = 1s^2 2s^2 2p^6 3s^2 3p^6 3d^5 4s^0$

$Ti^{3+} = 1s^2 2s^2 2p^6 3s^2 3p^6 3d^1 4s^0$

$Cr^{3+} = 1s^2 2s^2 2p^6 3s^2 3p^6 3d^3 4s^0$

Strongest oxidising ion is that which has a strong tendency to gain the electrons. Among the given option $Mn^{3+}$ is strongest oxidising agent as it needs just 1 electron to attain half-filled $3d^5$ subshell which is comparatively more stable.

**40.** *(d)*

| Metal atomic number | E.C. | +3 (Oxidation state) |
|---|---|---|
| Ru(44) | $[Kr]\,4d^7 5s^1$ | $[Kr]\,4d^5$ |
| Mn(25) | $[Ar]\,4s^2 3d^5$ | $[Ar]\,3d^4$ |
| Co(27) | $[Ar]\,4s^2 3d^7$ | $[Ar]\,3d^6$ |
| Fe(26) | $[Ar]\,4s^2 3d^6$ | $[Ar]\,3d^5$ |

Thus, Fe has configuration of $3d^5$ in +3 oxidation state.

**41.** *(a)* $CuCl_2$ dissolve in water to give blue coloured $Cu^{2+}$ ions. The colour is due to the presence of unpaired $e^-$ $(3d^9)$ in $Cu^{2+}$ ions.

$$CuCl_2 + nH_2O \longrightarrow \underset{\text{(Blue colour)}}{Cu^{2+}(aq)}$$

AgCl and $Cu_2Cl_2$ are insoluble in water while $ZnCl_2$ on dissolving in water forms colourless solution due to formation of $Zn^{2+}$ ions, which has no unpaired $e^-$ $(3d^{10})$.

**42.** *(d)* $Ti^{3+}$ : [Ar] $3d^1$

(Unpaired electrons = 1)

$$\mu = \sqrt{1(1+2)} = 1.73 \text{ B.M.}$$

$V^{2+}$ : [Ar] $3d^3$

(Unpaired electron = 3)

$$\mu = \sqrt{3(3+2)} = \sqrt{15} = 3.87 \text{ B.M.}$$

$Sc^{3+}$ : [Ar] $3d^0$

$$\mu = 0 \text{ B.M.}$$

**43.** *(c)* $Cr^+ - 4s^0 3d^5$

$Mn^{2+} - 4s^0 3d^5$

So, the outermost electronic configuration will be same in case of $Cr^+$ and $Mn^{2+}$ metal ions.

**44.** *(b)* *d*-block elements in the periodic table are called transition elements. The set that contains atomic number of only transition elements is 21, 25, 42, 72.

**45.** *(d)* The pair that has similar atomic radii is Mo and W. It is due to lanthanoid contraction.

**46.** *(a)* $CrO_2$ is also a metallic and ferromagnetic compound which is used to make magnetic tapes for cassette recorders.

**47.** *(b)* $MnO_4^{2-}$ is stable in aqueous solution as +6 oxidation state of Mn is stable.

**48.** *(a)* The element with outer electron configuration $3d^5 4s^2$ is Mn which exhibits oxidation states from +2 to +7 due to presence of electrons in outermost orbital $3d^5$ and $4s^2$.

**49.** *(a)* Usually across the first transition series, the negative values for standard electrode potential decrease except for Mn due to stable $d^5$-configuration.

So, correct order : Mn > Cr > Fe > Co

**50.** *(a)* The colour of the given species are as follows

(a) $V^{2+}$ (Violet), $Cr^{3+}$ (Violet), $Mn^{3+}$ (Violet)

(b) $Zn^{2+}$ (Colourless), $V^{3+}$ (Green), $Fe^{3+}$ (Yellow)

(c) $Ti^{4+}$ (Colourless), $V^{4+}$ (Blue), $Mn^{2+}$ (Pink)

(d) $Sc^{3+}$ (Colourless), $Ti^{3+}$ (Purple), $Cr^{2+}$ (Blue)

So, option (a) contains all the species with same colour i.e violet.

**51.** *(a)* Melting point of the given pair is in order (increasing)

Mn < Fe, Tc < Ru, Os < Re

So, the order given in option (a) is correct.

**52.** *(b)*

**Key idea**

To solve this question, simply find out the number of unpaired electrons.

Then, substitute this value in formula $\mu = \sqrt{n(n+2)}$ BM for required value.

The correct match is A-III, B-I, C-IV, D-II.

| Ions | Electronic configuration | $\mu_s$ |
|---|---|---|
| $Sc^{3+}$ | $3d^0$ | 0 |
| $V^{2+}$ | $3d^3$ | 3.87 BM |
| $Ni^{2+}$ | $3d^8$ | 2.84 BM |
| $Ti^{3+}$ | $3d^1$ | 1.73 BM |

**53.** *(b)* The most stable structure of $ClF_3$ is

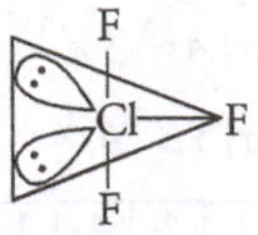


Number of lone pair present in equitorial plane,

$$n = 2$$

| Ions | Configuration | Unpaired electrons |
|---|---|---|
| $V^{3+}$ | [Ar] $3d^2$ | 2 |
| $Ti^{3+}$ | [Ar] $3d^1$ | 1 |
| $Cu^{2+}$ | [Ar] $3d^9$ | 1 |
| $Ni^{2+}$ | [Ar] $3d^8$ | 2 |
| $Ti^{2+}$ | [Ar] $3d^2$ | 2 |

Hence, $V^{3+}$, $Ni^{2+}$ and $Ti^{2+}$ have same *n* number of unpaired electrons i.e. 2.

**54.** *(c)* The electronic configuration of given elements are

| | Elements | Electronic configuration | Unpaired electron |
|---|---|---|---|
| A. | Sc | [Ar] $4s^2 3d^1$ | 1 |
| B. | Cr | [Ar] $4s^1 3d^5$ | 6 |
| C. | V | [Ar] $4s^2 3d^3$ | 3 |
| D. | Ti | [Ar] $4s^2 3d^2$ | 2 |
| E. | Mn | [Ar] $4s^2 3d^5$ | 5 |

Thus, the correct order (increasing) of number of unpaired electron is

A < D < C < E < B

**55.** *(c)*

**Time Saver Tech**

To save time while solving this question, one should look for option containing maximum number of unpaired electron. More the number of unpaired electron more will be its magnetic moment.

Magnetic moment is given by

$$\mu = \sqrt{n(n+2)} \text{ BM}$$

where, *n* is number of unpaired electron.

[Ar] $3d^7$ has 3 unpaired electrons.

$$\mu = \sqrt{3(3+2)}$$
$$= \sqrt{15} \text{ BM}$$

[Ar] $3d^3$ has 3 unpaired electrons.

$$\mu = \sqrt{3(3+2)}$$
$$= \sqrt{15} \text{ BM}$$

[Ar] $3d^6$ has 4 unpaired electrons.

$$\mu = \sqrt{4(4+2)}$$
$$= \sqrt{24} \text{ BM}$$

[Ar] $3d^8$ has 2 unpaired electrons.

$$\mu = \sqrt{2(2+2)}$$
$$= \sqrt{8} \text{ BM}$$

Hence, [Ar] $3d^6$ has highest value of magnetic moment.

**56.** *(c)* In $CrO_2Cl_2$ (chromyl chloride), the oxidation state of Cr is + 6

So, Cr (VI) = [Ar], $3d^0$

Mn(VII) = [Ar], $3d^0$

Fe(III) = [Ar], $3d^5$

Ti(III) = [Ar], $3d^1$

V(IV) = [Ar], $3d^1$

Hence, Cr (VI) and Mn(VII) have same $d^0$ configuration.

**57.** *(a)* Ions with $d^0$ and $d^{10}$ electronic configurations are colourless in aqueous solution.

In $Sc^{3+} \longrightarrow 3d^0 4s^0$, (colourless)

$Zn^{2+} \longrightarrow 3d^{10} 4s^0$, (colourless)

Whereas rest of the metal ions are coloured in nature.

**58.** *(b)*

| Ion | Electronic configuration | No. of unpaired electron | $\mu = (\sqrt{n(n+2)})$ |
|---|---|---|---|
| $V^{2+}$ | [↑ \| ↑ \| ↑ \|   \|  ] $3d^3$ | 3 | 3.87 |
| $Ni^{2+}$ | [↑↓ \| ↑↓ \| ↑↓ \| ↑ \| ↑] $3d^8$ | 2 | 2.83 |
| $Cr^{2+}$ | [↑ \| ↑ \| ↑ \| ↑ \|  ] $3d^4$ | 4 | 4.90 |
| $Fe^{2+}$ | [↑↓ \| ↑ \| ↑ \| ↑ \| ↑] $3d^6$ | 4 | 4.90 |

So, $Ni^{2+}$ ion (in gaseous state) has lowest spin-only magnetic moment value.

**59.** *(c)*

> **key idea** First write the electronic configuration of all the given metal ion and check the presence of unpaired electron. Compounds containing unpaired electron(s) are generally coloured and paramagnetic in nature.

| Ion | Number if unpaired $e^-$ |
|---|---|
| $Cu^{2+}$ | 1 |
| $Sc^+$ | 2 |
| $Cr^{3+}$ | 3 |

Due to presence of unpaired electrons these ions show colour and also are paramagnetic in nature.

**60.** *(d)* The correct match is

A → (iv), B → (iii), C → (ii), D → (i)

So, correct option is (d).

**61.** *(a)* $\mu = \sqrt{n(n+2)}$ BM

[∵ $n$ = No. of unpaired electron(s)]

| Metal ion | Z | n (for metal ion) | μ (BM) | Nature |
|---|---|---|---|---|
| $Ti^{2+}$ | 22 | 2 $(3d^2)$ | $\sqrt{8}$ | Paramagnetic |
| $V^{2+}$ | 23 | 3 $(3d^3)$ | $\sqrt{15}$ | Paramagnetic |
| $Ti^{3+}$ | 22 | 1 $(3d^1)$ | $\sqrt{3}$ | Paramagnetic |
| $Sc^{3+}$ | 21 | 0 $(3d^0)$ | 0 | Diamagnetic |

Thus, the correct order of spin only magnetic moment of given hydrated ions will be

$$Sc^{3+} < Ti^{3+} < Ti^{2+} < V^{2+}$$

**62.** *(a)* $V^{2+}$ = 3 unpaired electrons

$Cr^{2+}$ = 4 unpaired electrons

$Mn^{2+}$ = 5 unpaired electrons

$Fe^{2+}$ = 4 unpaired electrons

Hence, the order of paramagnetic behaviour should be

$$V^{2+} < Cr^{2+} = Fe^{2+} < Mn^{2+}$$

**63.** *(b)* As oxidation state increases, electronegativity increases thus acidic characteristic increases not basic

**64.** *(b)* The second ionisation potential values of Cu and Cr are sufficiently higher than those of neighbouring elements. This is because of the electronic configuration of $Cu^+$ which is $3d^{10}$ (completely filled) and of $Cr^+$ which is $3d^5$ (half-filled), i.e., for the second ionisation potentials, the electron is to be removed from very stable configurations.

**65.** *(a)* Larger the number of unpaired electrons, greater is the magnetic moment, $\mu = \sqrt{n(n+2)}$ BM

where, $n$ = Number of unpaired electrons

$Mn_{25} = [Ar]\, 4s^2\, 3d^5$

$Mn^{2+} = [Ar]\, 3d^5 4s^0$

[↑ | ↑ | ↑ | ↑ | ↑] $3d^5$

Number of unpaired electrons = 5

Hence, $Mn^{2+}$ : [Ar]$3d^5$ with five unpaired electrons have maximum magnetic moment.

**66.** *(4)*

| Element | Enthaply of atomisation (kJ/mol) |
|---|---|
| Sc | 326 |
| Mn | 281 |
| Co | 425 |
| Cu | 339 |

Among the given options, Co has highest enthalpy of atomisation.

$Co^{2+}$ = [Ar]$3d^7 \longrightarrow$ [↑↓ | ↑↓ | ↑ | ↑ | ↑]

∴ $n = 3$

Spin only magnetic moment

$$u = \sqrt{n(n+2)} = \sqrt{15} = 3.87 \text{ BM}$$

Nearest integer = 4.

**67.** *(11)* The electronic configuration of niobium (Nb) with atomic number 41 = [Kr]$4d^4 5s^1$

Number of electrons in $4d$ orbitals = 4 = $x$

Similarly the electronic configuration of ruthenium (Ru) with atomic number 44 = [Kr] $4d^7 5s^1$

Number of electrons in $4d$ orbitals is = 7 = $y$

So, $x + y = 11$

**68.** *(5)* $Mn_2O_3$ has strongest oxidising power among the given compounds. This is due to value of $E°$.

$$E°_{Mn^{3+}/Mn^{2+}} = +1.57 \text{ V}$$

$Mn^{3+} \longrightarrow d^4$ configuration

(So, 4 unpaired $e^-$)

$$\mu = \sqrt{24} \text{ BM} = 4.89 \text{ BM} \approx 5 \text{ BM}$$

**69.** *(6)* Among the given metals, chromium has maximum second ionisation enthalpy because 2nd electron is removed from stable configuration $3d^5$.

$Cr^+ = [Ar]3d^5 4s^0$

∴ Number of unpaired $e^-$ in $Cr^+$ is 5 i.e. $n = 5$

So, $\mu = \sqrt{n(n+2)}$ BM

$= \sqrt{5(5+2)}$

$= 5.92 \text{ BM} \approx 6 \text{ BM}$

**70.** *(0)* Metal having least metallic radii among given is Cr.

Spin only magnetic moment of $CrO_4^{2-} = \sqrt{n(n+2)}$ B.M.

Here $Cr^{+6}$ is in $d^0$ configuration, means $n = 0$

This spin only magnetic moment is 0.

**71.** *(5)* Among the given ions, $Co^{3+}$ is strong oxidising agent.

Number of unpaired electron in $Co^{3+} = 4$

Spin only magnetic moment

$$\mu = \sqrt{n(n+2)}\ \text{BM}$$

$$\mu = \sqrt{4(4+2)}\ \text{BM} = \sqrt{24} \approx 5\ \text{BM}$$

Hence, nearest integer value is 5.

**72.** *(0)* In first transition series elements, vanadium, 'V' has highest enthalpy of atomisation (515 kJ/mol). Its oxide i.e. $V_2O_5$ has V in +5 oxidation state.

So the configuration of $V^{+5}$ is $1s^2 2s^2 2p^6 3s^2 3p^6$. Hence it has no unpaired electrons.

**73.** *(4)* The given reaction proceeds as follows

$$MnO_2 + KOH + O_2 \longrightarrow \underset{(A)}{K_2MnO_4} + H_2O$$

$$K_2MnO_4 \xrightarrow[\text{medium}]{\text{Neutral / Acidic}} \underset{(B)}{KMnO_4} + \underset{(C)}{MnO_2}$$

Here Mn is in +4 oxidation state, in $MnO_2$

$$Mn^{+4} = [Ar]\,3d^3$$

$$\Rightarrow n = 3,\ \mu = \sqrt{3(3+2)} = 3.87\ \text{BM}$$

$\mu = 0$ for $KMnO_4$

sum of spin only magnetic moment of $B + C$ is 3.87.

So, the Nearest integer is 4.

**74.** *(4)* $Cr^{3+}$ has lowest number of *d*-electrons, so it has lowest reduction potential and is actually negative as shown in electrochemical series.

$Cr^{3+}$ has 3 unpaired electrons.

So, spin-only magnetic moment is

$$\mu = \sqrt{n(n+2)} = \sqrt{3(3+2)}$$
$$= \sqrt{15} = 3.87\ \text{BM}$$
$$\approx 4\ \text{BM}$$

**75.** *(5)* On the basis of electrochemical series, $Co^{3+}$ cannot liberate $H_2$ from dilute mineral acid solution. In gaseous state, its electronic configuration is

$3d^6$ [⥮ | ↑ | ↑ | ↑ | ↑]  $4s^0$ [ ]  i.e. $n = 4$

$$\mu = \sqrt{n(n+2)}$$
$$= \sqrt{4(4+2)}$$
$$= 4.89 \approx 5\ \text{BM}$$

**76.** *(12)*

$${}_{23}V = 1s^2 2s^2 2p^6 3s^2 3p^6 3d^3 4s^2$$

**77.** *(49)* Electronic configuration of Fe : $1s^2\ 2s^2\ 2p^6\ 3s^2 3p^6\ 4s^2\ 3d^6$

In $3d^6$, number of unpaired electrons $(n) = 4$

Spin only magnetic moment

$$= \sqrt{n(n+2)}\ \text{BM}$$
$$= \sqrt{4(4+2)}$$
$$= \sqrt{24} = 4.898$$
$$= 48.98 \times 10^{-1}$$
$$\approx 49 \times 10^{-1}\ \text{BM}$$

Nearest integer value will be 49.

## Topic 2

**01.** *(b)* Copper get reduced and hydrogen get oxidised. The reaction is as follows.

$CuO + H_2 \longrightarrow Cu + H_2O$ (under hot conditions). Cu and $H_2O$ are formed.

**02.** *(c)* $V_2O_3$ is a basic oxide as on dissolving in acids, give solution of vanadium (III) complexes. In case of transition metal oxide, the metal with lower oxidation states are basic. So, CrO is also basic as oxidation state of Cr is +2.

**03.** *(d)* The $Fe^{2+}$ ion has green colour while $Fe^{3+}$ is yellow in colour.

**04.** *(d)* Zinc oxide (ZnO) when react with $Na_2O$ it act as an acid while with $CO_2$ it act as a base. Therefore, it is an amphoteric oxide.

**05.** *(d)*

$$Hg_2Cl_2 + 2NH_4OH \longrightarrow \underset{\text{white}}{Hg(NH_2)Cl} + \underset{\text{black}}{Hg\downarrow} + NH_4Cl + 2H_2O$$

**06.** *(c)* The complete reaction is as follows

$$\underset{(A)}{CoS} \xrightarrow{\textit{Aqua-regia}} \underset{(B)}{CoCl_2} \xrightarrow[\text{(ii) AcOH}]{\text{(i) } KNO_2/NH_4OH} \underset{\text{Yellow ppt.}}{K_3[Co(NO_2)_6]\downarrow}$$

**07.** *(a)* The complete reaction sequence is as follows

$$K_2Cr_2O_7 \xrightarrow[-H_2O]{KOH} \underset{[A]}{K_2CrO_4} \xrightarrow[-H_2O]{H_2SO_4} \underset{[B]}{K_2Cr_2O_7} + K_2SO_4$$

Thus, [*A*] and [*B*] respectively are $K_2CrO_4$ and $K_2Cr_2O_7$.

**08.** *(c)* Potassium dichromate ($K_2Cr_2O_7$) and $KMnO_4$ both are coloured due to ligand to metal charge transfer in both.

**09.** *(a)* Alkaline oxidative fusion of $MnO_2$ gives, $MnO_4^{2-}$

$$2MnO_2 + 4OH^- + O_2 \longrightarrow \underset{(A)}{2MnO_4^{2-}} + 2H_2O$$

On electrolysis of $MnO_4^{2-}$ in alkaline medium, the reaction is

$$MnO_4^{2-} \longrightarrow \underset{(B)}{MnO_4^-} + e^-$$

Hence, option (a) is correct.

**10.** *(c)* The correct statements regarding $Mn_2O_7$, are A and C only. In the structure of $Mn_2O_7$, each Mn atom is tetrahedrally surrounded by four oxygen atoms. Also there is a bridging of one coordinated oxygen atom between two tetrahedral units.

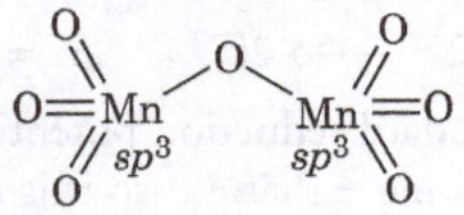

**11.** *(d)* $X = Cu_2I_2$, $Y = Na_2S_4O_6$

$$CuSO_4 + KI \longrightarrow \underset{\substack{\text{White ppt.}\\ [X]}}{Cu_2I_2} + \underset{\text{Violet}}{I_2} + K_2SO_4 \xrightarrow{Na_2S_2O_3} NaI + \underset{[Y]}{Na_2S_4O_6}$$

$I^-$ being strong reducing agent reduces $Cu^{2+}$ to $Cu^+$.

**12.** *(c)* Structure of $Mn_2O_7$ is as follows :

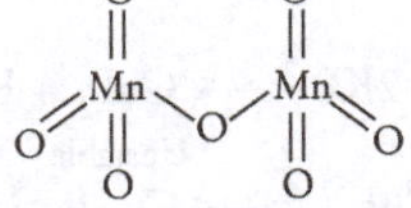

∴ Total 6 Mn =O bonds are present in $Mn_2O_7$ compound.

**13.** *(d)* In acidic medium,

$$2KMnO_4 + 3H_2SO_4 + 5H_2O_2 \longrightarrow K_2SO_4 + 2MnSO_4 + 8H_2O + 5O_2$$

**14.** *(c)* Non linear symmetrical Cr—O—Cr bond.

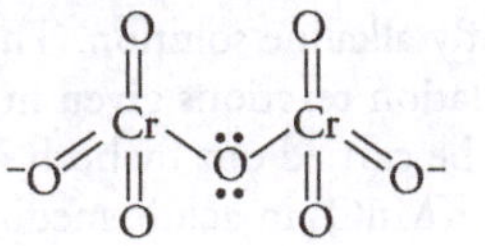

**15.** (b) Mn in $Mn_3O_4$, $(MnO \cdot Mn_2O_5)$ i.e. manganese shows + 2 and +3 oxidation state.

Both having unpaired electrons as shown below

$$_{25}Mn = [Ar]3d^5 4s^2$$

$Mn^{2+}(3d^5 4s^0)$ having 5 unpaired electrons in $3d$ orbital.

$Mn^{3+}$, $(3d^4 4s^0)$ having 4 unpaired electrons in $3d$ orbital and hence paramagnetic.

$\Rightarrow$ MgO and $Na_2O$ have no unpaired electrons and hence are diamagnetic.

Hence, $Mn_3O_4$ shows magnetic property.

**16.** (a) $FeCl_3$ does not exist because of $I^-$. Being a very good reducing agent, $I^-$ reduces $Fe^{3+}$ to $Fe^{2+}$.

$$2Fe^{3+} + 2I^- \longrightarrow 2Fe^{2+} + I_2$$

Reducing nature order I > Br > Cl > F

$$Fe^{3+} + e^- \longrightarrow Fe^{2+} \quad E° = 0.77 \text{ V}$$

$$I_2 + 2e^- \longrightarrow 2I^- \quad E° = -0.54 \text{ V}$$

As standard reduction potential value of $Fe^{3+}$ is more than $I_2$, so it is reduced to $Fe^{2+}$. Hence, except $FeI_3$, all other halogens can form iron halide compounds.

**17.** (c) In $MnO_4^-$, Mn has +7 oxidation state having no electron in $d$-orbitals. It is considered that higher the oxidation state of metal, greater is the tendency to occur $L \rightarrow M$ charge transfer, because ligand is able to donate the electrons into the vacant $d$-orbital of metal.

**18.** (b)

$$CuSO_4 + 2KI \longrightarrow \underset{\text{Unstable}}{CuI_2} + K_2SO_4$$

$$2CuI_2 \longrightarrow Cu_2I_2 + I_2$$

Thus, $CuI_2$ is not formed.

**19.** (b) The conversions given in $I^- \longrightarrow I_2$, $S^{2-} \longrightarrow S$ and $Fe^{2+} \longrightarrow Fe^{3+}$ occurs in acidic medium, while $I^- \longrightarrow IO_3^-$, $S_2O_3^{2-} \longrightarrow SO_4^{2-}$ occurs in neutral or faintly alkaline solution. Thus, oxidation reactions given in $A$, $B$ and $C$ can be carried out by both $K_2Cr_2O_7$ and $KMnO_4$ in acidic medium.

**20.** (c) The first step of the reaction for preparation of potassium permanganate from $MnO_2$ is

$$MnO_2 \xrightarrow[KNO_3,\ \Delta]{KOH} K_2MnO_4$$

Thus, the product formed is $K_2MnO_4$.

**21.** (a) The correct match is A-IV, B-III, C-I, D-II.

**22.** (c) In aqueous neutral medium, thiosulphate is oxidised by $KMnO_4$ to $SO_4^{2-}$. In this manganese changes its oxidation state from +7 to +4 i.e. by value of 3

$$8\overset{+7}{Mn}O_4^- + 3S_2O_3^{2-} + H_2O \longrightarrow 8\overset{+4}{Mn}O_2 + 6SO_4^{2-} + 2OH^-$$

Oxidation state of Mn is changes by value of 3.

**23.** (d) The reaction involved in neutral or faintly alkaline medium is

$$8\overset{+7}{Mn}O_4^- + 3S_2O_3^{2-} + H_2O \longrightarrow 8\overset{+4}{Mn}O_2 + 6SO_4^{2-} + 2\bar{O}H$$

So, change in oxidation state of Mn is $7 - 4 = 3$

**24.** (b) Neutral $KMnO_4$ oxidises $I^-$ to $IO_3^-$. Acidified potassium dichromate reacts with potassium iodide and oxidises it to $I_2$.

Manganate ion $(MnO_4^{2-})$ is paramagnetic $[Mn(VI): 3d^1]$ with tetrahedral structural and has $d\pi$ - $p\pi$ bonding, i.e. $p$-orbital of oxygen and $d$-orbital of Mn.

Both statements I and II are false.

**25.** (a) $KMnO_4$ on reaction with oxalic acid in acidic medium undergoes following reaction.

$$2K\overset{+7}{Mn}O_4 + 5H_2C_2O_4 + 3H_2SO_4 \longrightarrow K_2SO_4 + 2\overset{+2}{Mn}SO_4 + 10CO_2 + 8H_2O$$

The oxidation state of Mn changes from +7 (in $KMnO_4$) to +2 (in $MnSO_4$). Thus, the overall change in the oxidation number of Mn is of 5.

**26.** (c) Disproportionation reactions are redox reaction in which a compound undergoes oxidation as well as reduction.

$$3\overset{+3}{Cl}O_2^- \longrightarrow \overset{-1}{Cl}^- + 2\overset{+5}{Cl}O_3^-$$

$$\overset{0}{Cl}_2 + OH^- \longrightarrow \overset{-1}{Cl}^- + \overset{+5}{Cl}O_3^- + H_2O$$

$$2Mn^{3+} + 2H_2O \longrightarrow Mn^{2+} + \overset{+4}{Mn}O_2 + 4H^+$$

But $MnO_4^-$ does not get disproportonated.

Hence, option (c) is the correct answer.

**27.** (d) When $KMnO_4$ is heated at 513 K, it forms $K_2MnO_4$ and $MnO_2$ and $O_2$.

$$2K\overset{+7}{Mn}O_4 \xrightarrow[513\text{ K}]{\Delta} \underset{\text{(Green)}}{K_2\overset{+6}{Mn}O_4} + \underset{\text{(Black residue)}}{\overset{+4}{Mn}O_2} + O_2$$

$Mn_{25} = 1s^2, 2s^2, 2p^6, 3s^2, 3p^6, 4s^2, 3d^5$

$Mn^{+6} = 1s^2, 2s^2, 2p^6, 3s^2, 3p^6, 3d^1$; 1 unpaired electron, so paramagnetic

So, option (d) is correct.

**28.** (c) Oxidation number of $3d$-metal oxides are

A. $CrO_3$ here metal atom is Cr $\Rightarrow$ +6

B. $Fe_2O_3$ here metal atom is Fe $\Rightarrow$ +3

C. $MnO_2$ here metal atom is Mn $\Rightarrow$ +4

D. $V_2O_5$ here metal atom is V $\Rightarrow$ +5

E. $Cu_2O$ here metal atom is Cu $\Rightarrow$ +1

So, oxidation state order is

$$CrO_3 > V_2O_5 > MnO_2 > Fe_2O_3 > Cu_2O$$

or A > D > C > B > E.

**29.** (d)
$$2KMnO_4 \xrightarrow{\Delta} \underset{\text{Potassium manganate}}{K_2MnO_4} + MnO_2 + O_2$$

Potassium permanganate is diamagnetic while potassium manganate is paramagnetic. Both of these are tetrahedral in structure.

**30.** (a) Ruby is the mineral also called as red corundum which has the traces of chromium element which causes the red colour in the ruby. It contain $Cr^{3+}$ ions in $Al_2O_3$ lattice.

**31.** *(b)*

| Manganate | Permanganate |
|---|---|
| $\overset{+6}{MnO_4^{2-}}$ | $\overset{+7}{MnO_4^{\ominus}}$ |

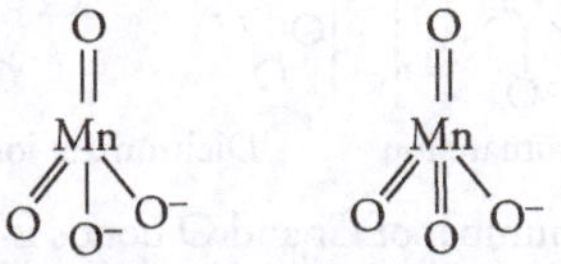

| | |
|---|---|
| Hybridisation-$sp^3$ | Hybridisation-$sp^3$ |
| Colour - Green | Colour - Purple |
| Unpair $e^-$ = 1 $e^-$ | Unpair $e^-$ = No unpair $e^-$ |
| Magnetic nature = Paramagnetic | Magnetic nature = Diamagnetic |

Statement (b) is incorrect while all other statements are correct.

**32.** *(b)* The reaction can be shown as

(i) $\underset{(A)}{2\overset{+4}{Mn}O_2} \xrightarrow[\Delta]{4\,KOH,\ \overset{0}{O_2}} \underset{(B)}{2K_2\overset{+6}{Mn}\overset{-2}{O_4}}$
Potassium manganate (Green)

(ii) $\underset{(B)}{3K_2\overset{+6}{Mn}O_4} \xrightarrow{4HCl} \underset{(C)}{2K\overset{+7}{Mn}O_4} + \overset{+4}{MnO_2} + 2H_2O + 4KCl + 2H_2\overset{-2}{O}$
Potassium permanganate (Purple)

(iii) $\underset{(C)}{2K\overset{+7}{Mn}O_4} \xrightarrow{K\overset{-1}{I},\ H_2O} \underset{(A)}{2\overset{+4}{Mn}O_2} + 2KOH + \underset{(D)}{K\overset{+5}{I}O_3}$

Thus, $A$ and $D$ are $MnO_2$ and $KIO_3$ respectively.

**33.** *(d)* Greater the covalent nature, more easily they are hydrolysed Thus, $FeCl_3$ is more hydrolysed than $FeCl_2$.
Whereas all other statements are correct
Thus, statement (d) is incorrect.

**34.** *(c)*

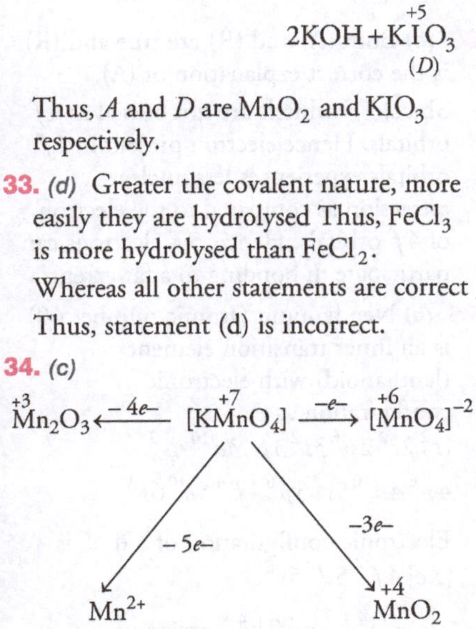


**35.** *(6)* The complete reaction sequence is as follows

$\underset{(A)}{Cr_2O_7^{2-}} + NaCl + H_2SO_4 \rightarrow CrO_2Cl_2$

$\underset{(Vapour)}{CrO_2Cl_2} + NaOH \longrightarrow \underset{(B)}{Na_2CrO_4} + NaCl + H_2O$

$\underset{(B)}{Na_2CrO_4} + H^+ \longrightarrow \underset{(C)}{Na_2Cr_2O_7} + H_2O$

$Na_2Cr_2O_7 \longrightarrow 2Na^+ + Cr_2O_7^{2-}$

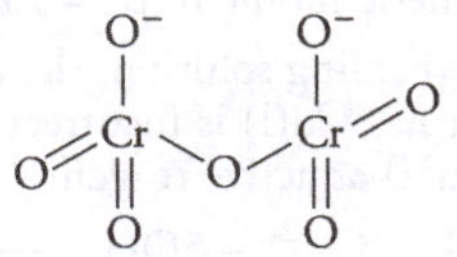

So, number of terminal oxygen is 6.

**36.** *(10)* $K\overset{+7}{Mn}O_4 \xrightarrow[\text{medium}]{\text{Acidic}} Mn^{2+}$

$X$ is difference in oxidation state.
So, $x = 5$

$6CH_3COO^- + 3Fe^{3+} + 2H_2O \longrightarrow [Fe_3(OH)_2(CH_3COO)_6]^+ + 2H^+$

$[Fe_3(OH)_2(CH_3COO)_6]^+ + 4H_2O \longrightarrow \underset{\substack{\text{Brown red ppt.} \\ \text{Iron (III) dihydroxy acetate}}}{3[Fe(OH)_2CH_3COO]} + 3CH_3COOH + H^+$

Electronic configuration of $Fe^{3+}$ is given by
$[Ar]3d^5$ i.e. $5d$ electrons.
Thus, $Y = 5$
$X + Y = 10$

**37.** *(6)* Spin only magnetic moment of Mn in $KMnO_4 = 0$ as no unpaired electron is present.
$(\mu = \sqrt{n(n+2)}$ B.M.)
Spin only value of manganese product formed during titration of $KMnO_4$. against oxalic acid in acidic medium is,
$n = 5$ (number of unpaired electron)
$\mu_B = \sqrt{5(5+2)}$
$= \sqrt{35} = 5.2$ B.M. $\approx 6$ B.M.
Nearest integer value of spin only magnetic moment is 6.

**38.** *(877)* Among the given compound, CrO is basic oxide.
In CrO, Cr exist as $Cr^{+2}$ and have 4 unpaired electron.
Spin only magnetic moment
$\mu = \sqrt{n(n+2)}$
$\mu = \sqrt{4(4+2)}$
$= 4.90$ B.M.
$Cr_2O_3$ is amphoteric oxide.
$Cr^{+3}$ has 3 unpaired electrons.
Spin only magnetic moment
$\mu = \sqrt{3(3+2)} = 3.87$ B.M.
Sum $= 4.90 + 3.87$
$= 8.77$
$\mu_{total} = 877 \times 10^{-2}$
Comparing with given equation, $X$ will be 877.

**39.** *(6)* The reaction that is mentioned in question is given below

$4FeCr_2O_4 + 8Na_2CO_3 + 7O_2 \longrightarrow \underset{A}{8Na_2CrO_4} + \underset{B}{2Fe_2O_3} + 8CO_2$

Spinly only magnetic moment
$\mu_B = \sqrt{n(n+2)}$ BM
For product $A$,
$Na_2CrO_4$, $\mu_B = 0$,
unpaired electron = 0
For product $B$,
$Fe_2O_3$, $\mu_B = 5.92$ as unpaired electron, $n = 5$
Sum of magnetic moment = 5.92 BM.
Hence, nearest integer value is 6.

**40.** *(5)* Complete reaction is as follows:
$2KMnO_4 + 10KI + 3H_2SO_4 \longrightarrow 2MnSO_4 + K_2SO_4 + 5I_2 + 8H_2O$
The oxidation Mn in $KMnO_4$ is
$+1 + x + 4(-2) = 0$
$x - 7 = 0$
$x = +7$
Oxidation state of Mn in $MnSO_4$ is +2.
So, the change in oxidation state of Mn is 5.

**41.** *(5)*

| Compound | Oxidation state of central metal | Configuration of central metal |
|---|---|---|
| $MnF_4$ | $Mn^{4+}$ | $[Ar]\,3d^3 4s^0$ |
| $MnF_3$ | $Mn^{3+}$ | $[Ar]\,3d^4 4s^0$ |
| $MnF_2$ | $Mn^{2+}$ | $[Ar]\,3d^5 4s^0$ |

⇒ Here, $Mn^{3+}[d^4]$ (of $MnF_3$) in the strongest oxidising agent.
⇒ Spin only magnetic moment
$= \sqrt{n(n+2)}$ BM
$n$ = Number of unpaired electrons = 4
So, spin only magnetic moment
$= \sqrt{4(4+2)}$
$= \sqrt{24} \approx 5$

**42.** *(6)* Chromyl chloride chemical formula is $CrO_2Cl_2$. In $CrO_2Cl_2$ oxidation state of Cr is +6.

**43.** *(19)* $K_2Cr_2O_7 : +2+2x+7(-2)=0$

$2x = 14-2 = 12, x = +6$

$KMnO_4 : +1+y+4(-2)=0$
$+1+y-8=0 \Rightarrow y=+7$

$K_2FeO_4 : 2(+1)+z+4(-2)=0$
$+2+z-8=0$

$\Rightarrow \quad z=+6$

Thus, sum of oxidation state $=+6+7+6=19$.

**44.** *(4)* The disproportionation reaction of $MnO_4^{2-}$ in acidic medium is as follows

$$3MnO_4^{2-} + 4H^+ \longrightarrow \underset{(B)}{MnO_2} + \underset{(A)}{MnO_4^-}$$

Oxidation state of Mn in $MnO_2$ is +4
Oxidation state of Mn in $MnO_4^-$ is +7

Hence, $A = MnO_4^-$
$B = MnO_2$

$Mn = [Ar]3d^5\ 4s^2$

$Mn^{4+} = [Ar], 3d^3$

$Mn^{4+} = [Ar]$ | 1 | 1 | 1 | | | (3d)

Total number of unpaired electrons $(n) = 3$

Magnetic moment (μ) spin only of $B$ =

$\therefore \quad \mu = \sqrt{n(n+2)} = \sqrt{3(3+2)} = \sqrt{15}$

$= 3.87\text{ BM} \approx 4\text{ BM}$

**45.** *(6)* The balanced equation will be

$$4FeCr_2O_4 + 7O_2 + 8Na_2CO_3 \longrightarrow \underset{(A)}{8Na_2CrO_4} + 2Fe_2O_3 + 8CO_2$$

$$2Na_2CrO_4 + H^+ \longrightarrow \underset{(B)}{Na_2Cr_2O_7} + 2Na^+ + H_2O$$

$Na_2Cr_2O_7$

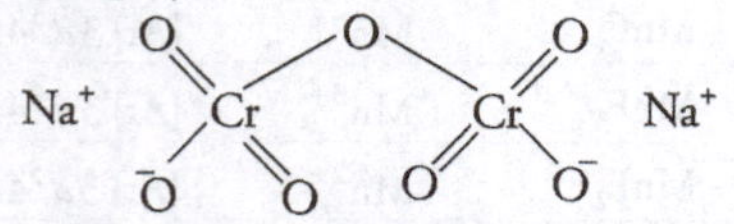

Number of terminal oxygen atoms is 6.

**46.** *(0)* Disproportionation reaction of acidified manganate solution is

$$3MnO_4^{2-} + 4H^+ \longrightarrow 2\overset{+7}{Mn}O_4^- + \overset{+4}{Mn}O_2 + 2H_2O$$

Higher oxidation state of manganese is +7 (in $MnO_4^-$).

Electronic configuration of Mn in +7 oxidation state will be

Mn (25) : $[Ar]\ 4s^2 3d^5$

$Mn^{(+7)}$ : [Ar]

i.e. It does not have any unpaired electron.

$\therefore$ Magnetic moment (μ) = 0 as $n=0$.

**47.** *(3)* In Fehling solution, the active reagent has Cu(I) is incorrect because it has Cu(II) as active reagent.

$$RCHO + 2Cu^{2+} + 5OH^- \longrightarrow RCOO^- + Cu_2O + 3H_2O$$

Therefore, rest of the three statements are correct.

**48.** *(3)* Most basic oxide of vanadium is $V_2O_3$.
(An oxide with less positive charge and less number of oxygen atom will be more basic.)

$V = [Ar]\,3d^3\ 4s^2$

$V^{3+} = [Ar]\,3d^2$

$\mu = \sqrt{n(n+2)}$

($n$ = number of unpaired electrons)

$= \sqrt{2(2+2)} = 2.84\text{ BM} \approx 3$

**49.** *(3)* $3MnO_4^{2-} + 3H^+ \longrightarrow 2MnO_4^- + MnO_2 + 2H_2O$

O.S. of Mn in $MnO_4^- = +7$
O.S. of Mn in $MnO_2 = +4$
Difference $= 7-4 = 3$

**50.** *(0)* Chromate ion, $CrO_4^{2-}$;

Let the oxidation state of Cr be $x$.
Then, $x+4(-2)=-2 \Rightarrow x=+6$
Dichromate ion, $Cr_2O_7^{2-}$;
Then, $2x+7(-2)=-2 \Rightarrow x=+6$
Difference $=6-6=0$
So, difference in their oxidation state is zero.

**51.** *(6)*

$$\underset{\substack{\text{Thiosulphate ion}\\ \text{[Reducing agent]}}}{S_2O_3^{2-}} + \underset{\substack{\text{Permanganate ion}\\ \text{[ Oxidising agent]}}}{MnO_4^-} \xrightarrow[\text{Alkaline medium}]{\text{Mild}} \underset{\substack{\text{Sulphate ion}\\ (A)}}{SO_4^{2-}} + \underset{\text{Manganate ion}}{MnO_4^{2-}}$$

Oxidation number of sulphur in $SO_4^{2-}$ (A) is +6.

$[\because\ x+4(-2)=-2 \Rightarrow x=+6]$

**52.** *(6)* $K_2Cr_2O_7 + 2KOH \longrightarrow 2K_2CrO_4 + H_2O$

Oxidation state of Cr in $K_2CrO_4$ is

$2\times1+x+4(-2)=0$
$2+x-8=0 \Rightarrow x=+6$

**53.** *(12)*

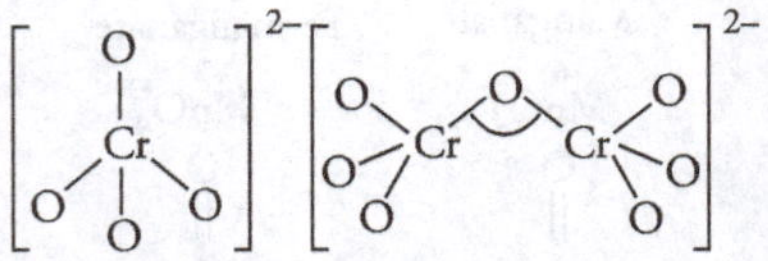

Chromate ion Dichromate ion

Total number of Cr and O bonds is 12, four in chromate ions and eight in dichromate ions.

## Topic 3

**01.** *(c)* $Tb^{4+}$ is the strongest oxidising agent. As it can readily gain an electron to reach its more stable +3 oxidation state.

**02.** *(b)* Given, atomic number of Es = 99

So, its electronic configuration
$=[Rn]\,5f^{11}6d^0 7s^2$

**03.** *(d)* Lanthanoids = Eu, Er, Tb, Yr and Lu

Actinoides = Cm

Hence, Cm does not belongs to lanthanoid.

**04.** *(b)* Atomic number for Nd = 60

Electronic configuration is
$1s^2 2s^2 2p^6 3s^2 3p^6 3d^{10} 4s^2$
$4p^6 4d^{10} 4f^4 5s^2 5p^6 6s^2$

$=[Xe]\,4f^4 6s^2$

**05.** *(b)* Both (A) and (R) are true and (R) is the correct explanation of (A).

Since, $5f$ orbitals are not buried as $4f$ orbitals. Hence electron present in $5f$ orbitals experiences less nuclear attraction as compared to the electron of $4f$ orbitals. Hence, $5f$ electrons can participate in bonding to a far extent.

**06.** *(b)* Neodymium (atomic number 60) is an inner transition element (lanthanoid) with electronic configuration as
$1s^2 2s^2 2p^6 3s^2 3p^6 3d^{10} 4s^2$
$4p^6 4d^{10} 5s^2 5p^6 4f^4 5d^0 6s^2$

Electronic configuration of $Nd^{2+}$ is $[Xe]\,4f^4 5d^0 5s^0$.

**07.** *(a)* $Ce^{4+}$ and $Tb^{4+}$ are good oxidising agents.

Ce is most stable in +3 state. In +4 state it is most oxidising because in +4 state it can take an electron to become more stable at +3 state.

$Tb^{4+}$ can be easily converted into stable $Tb^{3+}$ by gain of an electron. Due to this reason, $Tb^{4+}$ is also a good oxidising agent.

> • Concept **Enhancer**
>
> In general, all inner transition elements that have an oxidation state of +2, changes to +3 and acts as reducing agents. As the oxidation number is increased. While those elements that have an oxidation state of +4 tends to change to +3 oxidation state and acts as an oxidising agent, as the oxidation number is decreased.

**08.** (a) Ce will show +3 oxidation state but it is likely to deviate easily to +4 oxidation state to attain a noble gas configuration.

$$Ce = [Xe]\,4f^1\,5d^1\,6s^2$$

$$Ce^{3+} = [Xe]\,4f^1\,5d^0$$

$$Ce^{4+} = [Xe]\,4f^0$$

(Noble gas configuration)

**09.** (c) The electronic configuration of Eu and $Eu^{2+}$ ion is as follows

$$Eu\ (Z=63) = [Xe]\,4f^7 6s^2$$

$$Eu^{2+}\ (Z=63) = [Xe]\,4f^7$$

**10.** (a) A well known alloy is mischmetal which has the following composition.

Mischmetal – lanthanoid metals = 95% (About 25% La, 50% Ce and the rest other lanthanoids) with and traces of iron = 5% S, C, Ca and Al.

**11.** (c) Europium (Eu = atomic number 63)

Electronic configuration of $Eu = [Xe]4f^7, 6s^2$

It can show +2 and +3 oxidation state in +2 oxidation:

$$Eu^{2+} = [Xe]4f^7, 6s^0$$

in +3 oxidation :

$$Eu^{3+} = [Xe]4f^6, 6s^0$$

Being a typical member of the lanthanide, Eu usually assumes the oxidation state +3 and +2 but it doesn't show +4 oxidation state due to very high ionisation energy (4140 kJ/mol). Hence, the correct option is (c).

**12.** (a) Actinoid series have atomic number from 90 to 103 (5f-block element).

So, element with atomic number (Z) 101 belongs to actinoid series and elements with atomic number (Z) 104 belongs to 4th group ($6d^2$).

**13.** (b) In the actinoids family (5f-block), uranium (U) neptunium (Np), plutonium (Pu) and americium (Am) have highest possible oxidation states of +6, +7, +7 and +6 respectively.

**14.** (c) The correct order of atomic radii is

Europium (Eu) [199 pm] > Cerium (Ce) [183 pm] > Holmium (Ho) [176 pm] > Nitrogen (N) [65 pm]

**15.** (b) Lanthanoid contraction in the lanthanoid series takes place due to the presence of electron(s) in the 4f-orbitals. f-orbitals have poor shielding effect. As a result, the effective nuclear charge will be more experienced by the 5d and 6s- electrons and it will cause contraction or decrease in both atomic and ionic radii.

**16.** (d) All the electrons in 4f-orbital are unpaired, hence stable. Thus, Gd (64) has electronic configuration as $[Xe]_{54}\ 4f^7, 5d^1, 6s^2$

Instead of $[Xe]_{54}\ 4f^8, 6s^2$.

**17.** (b) Lanthanoides : $[Xe]\,4f^{1-14}\,5d^{0-1}\,6s^2$

Actinoides : $[Rn]5f^{1-14}\,6d^{0-1}\,7s^2$

Lanthanoides and actinoides use core d and f orbitals also to show higher oxidation state. As actinoides have comparatively low energy difference between f and d orbitals, thus, they show more number of oxidation states.

**18.** (d) Imperfect shielding on outer electrons by 4f electrons from the nuclear charge is due to the diffused large shape of 4f-orbitals.

This imperfect shielding is unable to counterbalance the effect of the increased nuclear charge. Hence, the net result shows contraction in size.

**19.** (a) Cerium, $Ce_{58} = [Xe]\,4f^1 5d^1 6s^2$.

Its most stable oxidation state is +3 but +4 is also existing.

**20.** (b) The electronic configuration of given species are

(A) $Eu^{2+} : [Xe]\,4f^7 6s^0$

(B) $Gd^{3+} : [Xe]\ 4f^7 5d^0 6s^0$

(C) $Eu^{3+} : [Xe]\,4f^6 6s^0$

(D) $Tb^{3+} : [Xe]\,4f^8 6s^0$

(E) $Sm^{2+} : [Xe]\,4f^6 6s^0$

Thus, $Eu^{2+}$ and $Gd^{3+}$ have $4f^7$ configuration i.e. A and B only.

**21.** (c) Both Statement I and II are true. $Ce^{4+}$ has noble gas elecronic configuration. Also due to high reduction potential of $Ce^{4+}/Ce^{3+}$ and stability of $Ce^{3+}$ is greater than $Ce^{4+}$, thus $Ce^{4+}$ acts as strong oxidising agent.

**22.** (c) $Tb \longrightarrow [Xe]4f^9, 6s^2$

$Tb^{4+} \longrightarrow [Xe], 4f^7$

(Half-filled configuration)

$Yb \longrightarrow [Xe]\,4f^{14}, 6s^2$

$Yb^{2+} \longrightarrow [Xe]\,4f^{14}$

(Completely filled configuration)

According to the above given configurations, $Tb^{4+}$ and $Yb^{2+}$ have half-filled and completely filled f-orbitals respectively.

**23.** (b) All the lanthanides show +3 oxidation state as most stable oxidation state hence, Ce(IV) act as oxidising agent so that it can get converted to Ce(III).

> • Concept **Enhancer**
>
> Cerium is the primary lanthanide (50%) utilised for heat resistance material along with 25% trace amount of lanthanum.

**24.** (d) $E^\circ_{Ce^{4+}/Ce^{+3}} = 1.74$ (data given)

$Ce^{4+}$ can be reduced to $Ce^{3+}$ and can oxidise water.

$$Ce^{4+} + e^- \longrightarrow Ce^{3+}$$

$Ce^{4+}$ is having noble gas configuration $[Xe]4f^0$ but it is strong oxidant and gets converted to $Ce^{3+}$.

Statement I is correct but statement II is incorrect.

**25.** (d) $_{97}Bk$ (Berkelium) and $_{93}Np$ (Neptunium) are present in actinide series. Due to actinoid contraction, there is gradual decrease in size of $M^{3+}$ ions across the period due to poor shielding of 5f-electrons.

So, $Bk^{3+}$ is smaller than that of $Np^{3+}$ due to actinoid contraction.

So, A is true and R is false.

**26.** *(d)* In oxides, $MO_2$ ($M$ is lanthanoid metal) only four lanthanoids exhibit +4 oxidation state. These lanthanoids are praseodymium (Pr, $Z = 59$), neodymium (Nd, $Z = 60$) terbium (Tb, $Z = 65$) dysprosium (Dy, $Z = 66$). So, Yb (ytterbium) option (d) does not form $MO_2$ type of oxide.

**27.** *(b)* Electronic configuration of $Gd^{3+}$: $[Xe]\ 4f^7$

$\therefore$ Number of unpaired electrons = 7

Magnetic moment, $\mu = \sqrt{n(n+2)}$ B.M.

$= \sqrt{7(7+2)}$ B.M. = 7.93 B.M.

So, option (b) is correct.

**28.** *(b)* Neptunium and plutonium show highest number of oxidation states (+3, +4, +5, +6, +7).

**29.** *(d)* Lanthanoids generally shows the oxidation state of +3. Formation of +4 state requires very high energy, thus incorrect and other statements are correct.

**30.** *(c)* +3 and +4 states are shown by Ce in aqueous solution. Thus, statement (c) is incorrect.

**31.** *(d)* Due to lanthanide contraction, there occurs net decrease in size. Only, 0.85 Å is smaller one.

**32.** *(c)* $Eu = [Xe] 4f^7\ 5d^0\ 6s^2$

$Yb = [Xe]\ 4f^{14}\ 5d^0\ 6s^2$

Due to its stable half-filled ($4f^7$ of $Eu^{2+}$) and fully filled ($4f^{14}$ of $Yb^{2+}$) electronic configuration, the removal of 3rd electron requires more energy from the elements.

**33.** *(a, d)* Isoelectronic species are species which have same number of electrons.

| Ions | Z | No. of electron in neutral atom | No. of electron in ion | Isoelectronic nature |
|---|---|---|---|---|
| $Sm^{2+}$ | 62 | 62 | 60 | Not isoelectronic |
| $Er^{3+}$ | 68 | 68 | 65 | |
| $Tb^{2+}$ | 65 | 65 | 63 | Not isoelectronic |
| $Tm^{4+}$ | 69 | 69 | 65 | |

Hence, $Sm^{2+}$ and $Er^{3+}$ and $Tb^{2+}$ and $Tm^{4+}$ are not isoelectronic species.

**34.** *(c)* Electronic configuration of $Eu^{2+}$

$$[Xe]\ 4f^7 5d^0 6s^0$$

Electronic configuration of $Yb^{2+}$

$$[Xe]\ 4f^{14}\ 5d^0 6s^0$$

After loosing two electrons, Eu and Yb form stable half-filled $f$-orbitals and fully-filled $f$-orbitals.

$$E^\circ_{Eu^{3+}/Eu^{2+}} = -0.35$$

$$E^\circ_{Yb^{3+}/Yb^{2+}} = -1.05$$

Due to more reduction potential in Eu than Yb, $Eu^{2+}$ is more stable than $Yb^{2+}$.

**35.** *(b)* Ytterbium shows +2 oxidation state and it is diamagnetic in nature.

$$Yb = [Xe],\ 4f^{14} 6s^2$$

$$Yb^{2+} = [Xe],\ 4f^{14}$$

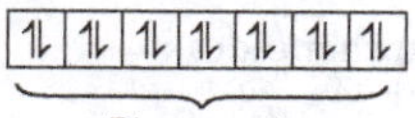

Diamagnetic

Since, there is no unpaired electron, it is diamagnetic in nature.

**36.** *(a)* Both statement I and statement II are true.

**37.** *(c)* Permanganate titrations in presence of HCl are unsatisfactory as HCl is oxidised to $Cl_2$.

Whereas other statements are correct. Therefore, only statement (2) is incorrect.

**38.** *(c)* The most common oxidation state of lanthanoid is +3. Lanthanoids in +3 oxidation state usually have unpaired electron in $f$-subshell and impart characteristic colour in solid as well as in solution state due to $f$-$f$ transition. (Except lanthanum and lutetium)

**39.** *(2)* Colour of lanthanoid ions are due to presence of unpaired electron.

The electronic configuration of given lanthanoid ions are

$La^{3+}$ — $[Xe]\ 4f^0$

$Nd^{3+}$ — $[Xe]\ 4f^3$

$Sm^{3+}$ — $[Xe]\ 4f^5$

$Eu^{3+}$ — $[Xe]\ 4f^6$

$Lu^{3+}$ — $[Xe]\ 4f^{14}$

$La^{+3}$ and $Lu^{+3}$ do not show any colour due to absence of unpaired electrons.

**40.** *(18)* The electronic configuration of $Np = 93$ is

$1s^2\ 2s^2 2p^6 3s^2 3p^6 3d^{10}\ 4s^2 4p^6 4d^{10}$
$5s^2 5p^6\ 4f^{14}\ 5d^{10}\ 6s^2\ 6p^6\ 5f^4$
$6d^1 7s^2$

In ground state, total number of $f$ electrons = 14 + 4 = 18

**41.** *(7)* The electronic configuration of Gd is as follows

$$Gd = [Xe]\ 4f^7, 5d^1, 6s^2$$

So, electronic configuration of $Gd^{2+}$ will be

$$Gd^{2+} = [Xe],\ 4f^7, 5d^1, 6s^0$$

$\therefore$ The number of $4f$ electrons in ground state configuration of $Gd^{2+}$ is 7.

**42.** *(10)* Holmium is $5f$-block element (actinide) having atomic number 67.

Ho(Holmium) = $[Xe]\ 4f^{11} 6s^2$

Electronic configuration of

$Ho^{3+} = [Xe]\ 4f^{10} 6s^0$

So, number of electrons present in $4f$ is 10.

# C-18

JEE Main Chapterwise Topicwise Chemistry

In JEE Main there is usually 0-3 question per paper are visible from this chapter. In past 3 years, 85 conceptual and 17 numerical questions were asked of moderate difficulty level from this chapter.

**For detailed analysis scan QR code**

# COORDINATION COMPOUNDS

## Topic 1 Werner's Theory, Coordination Number, Nomenclature and Isomerism of Coordination Compounds

### Level 1

**01** The correct IUPAC name of $[PtBr_2(PMe_3)_2]$ is *[2024, 6 April Shift-II]*

(a) *bis*(trimethylphosphine)dibrompolatinum (II)
(b) dibromo*bis*(trimethylphosphine) platinum(II)
(c) *bis*[bromo(trimethylphosphine)] platinum (II)
(d) dibromodi(trimethylphosphine) platinum (II)

**02** The correct IUPAC name of $K_2MnO_4$ is *[2024, 29 Jan Shift-II]*

(a) potassium tetraoxopermanganate (VI)
(b) dipotassium tetraoxidomanganate (VII)
(c) potassium tetraoxidomanganate (VI)
(d) potassium tetraoxidomanganese (VI)

**03** The IUPAC name of $K_3[Co(C_2O_4)_3]$ is *[2023, 6 April Shift-II]*

(a) potassium trioxalatocobaltate (III)
(b) potassium tris(oxalato)cobalt (III)
(c) potassium tris(oxalato)cobaltate (III)
(d) potassium trioxalatocobalt (III)

**04** The correct structures of *trans*-$[NiBr_2(PPh_3)_2]$ and meridional-$[Co(NH_3)_3(NO_2)_3]$, respectively are *[2021, 18 March Shift-I]*

(a) 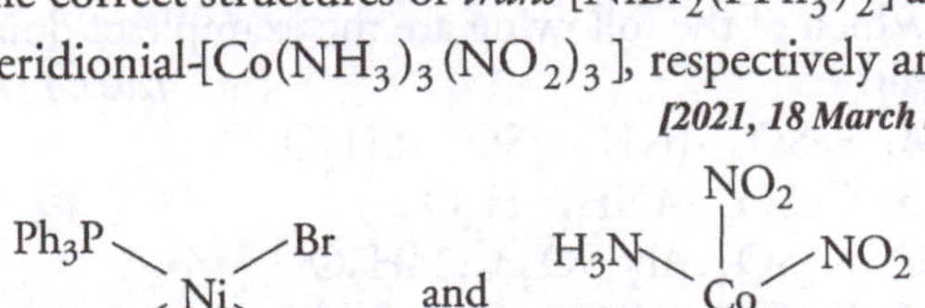


(b) 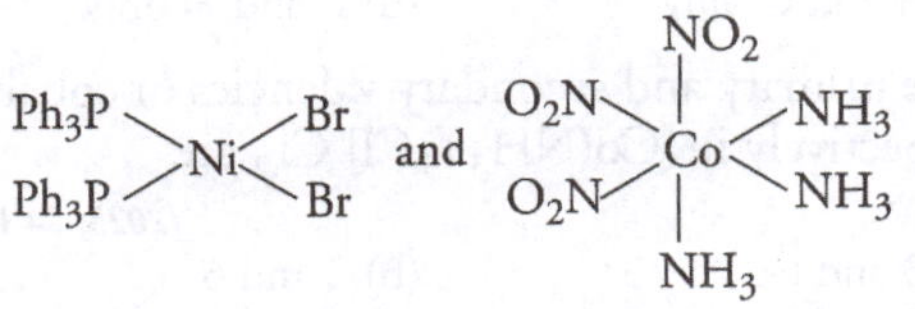


(c) 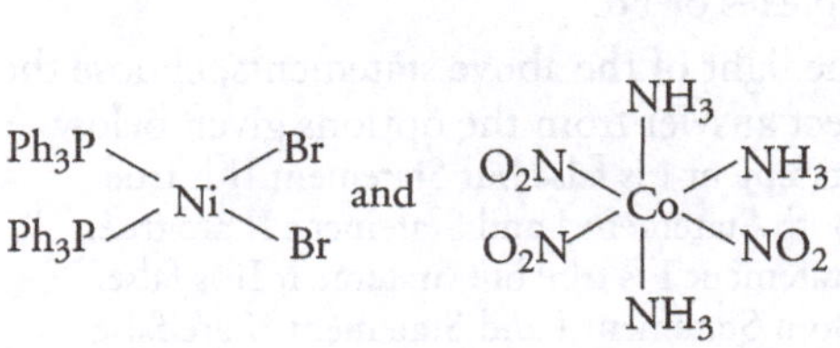


(d) 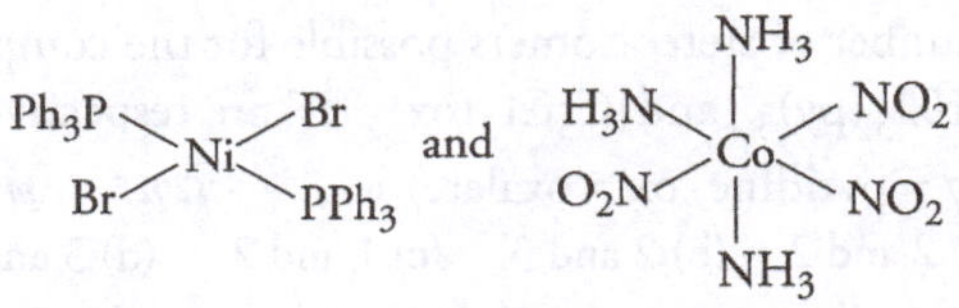


**05** Which of the following is an example of homoleptic complex? *[JEE Main 2016]*

(a) $[Co(NH_3)_6]Cl_3$ (b) $[Pt(NH_3)_2Cl_2]$
(c) $[Co(NH_3)_4Cl_2]$ (d) $[Co(NH_3)_5Cl]Cl_2$

**06** The equation which is balanced and represents the correct product(s) is *[JEE Main 2013]*

(a) $Li_2O + 2KCl \longrightarrow 2LiCl + K_2O$
(b) $[CoCl(NH_3)_5]^+ + 5H^+ \longrightarrow Co^{2+} + 5NH_4^+ + Cl^-$
(c) $[Mg(H_2O)_6]^{2+} + (EDTA)^{4-} \xrightarrow{\text{excess NaOH}} [Mg(EDTA)]^{2+} + 6H_2O$
(d) $CuSO_4 + 4KCN \longrightarrow K_2[Cu(CN)_4] + K_2SO_4$

**07** Which one of the following complex ions has geometrical isomers? *[AIEEE 2011]*

(a) $[Co(en)_3]^{3+}$ (b) $[Ni(NH_3)_5Br]^+$
(c) $[Co(NH_3)_2(en)_2]^{3+}$ (d) $[Cr(NH_3)_4(en)]^{3+}$

**08** How many EDTA (ethylenediamine tetraacetic acid) molecules are required to make an octahedral complex with a $Ca^{2+}$ ion? *[AIEEE 2006]*

(a) Six (b) Three
(c) One (d) Two

**09** In the coordination compound, $K_4[Ni(CN)_4]$, the oxidation state of nickel is *[AIEEE 2003]*

(a) –1 (b) 0
(c) +1 (d) +2

## Level 2

**10** '$X$' is the number of acidic oxides among $VO_2$, $V_2O_3$, $CrO_3$, $V_2O_5$ and $Mn_2O_7$. The primary valency of cobalt in $[Co(H_2NCH_2CH_2NH_2)_3]_2(SO_4)_3$ is $Y$. The value of $X+Y$ is ............... . *[2025, 7 April Shift-II]*

(a) 2 (b) 4 (c) 3 (d) 5

**11** Given below are two statements: *[2025, 8 April Shift-II]*

**Statement I :** A homoleptic octahedral complex, formed using monodentate ligands, will not show stereoisomerism.

**Statement II :** *cis* - and *trans* - platin are heteroleptic complexes of Pd.

In the light of the above statements, choose the correct answer from the options given below

(a) Statement I is false but Statement II is true.
(b) Both Statement I and Statement II are true.
(c) Statement I is true but Statement II is false.
(d) Both Statement I and Statement II are false.

**12** Number of stereoisomers possible for the complexes, $[CrCl_3(py)_3]$ and $[CrCl_2(ox)_2]^{3-}$ are respectively (py = pyridine, ox = oxalate) *[2025, 4 April Shift-I]*

(a) 2 and 2 (b) 2 and 3 (c) 1 and 2 (d) 3 and 3

**13** The complex that shows *facial - meridional* isomerism is *[2025, 23 Jan Shift-I]*

(a) $[Co(NH_3)_4Cl_2]^+$ (b) $[Co(NH_3)_3Cl_3]$
(c) $[Co(en)_2Cl_2]^+$ (d) $[Co(en)_3]^{3+}$

**14** One mole of the octahedral complex compound $Co(NH_3)_5Cl_3$ gives 3 moles of ions on dissolution in water. One mole of the same complex reacts with excess of $AgNO_3$ solution to yield two moles of $AgCl(s)$. The structure of the complex is *[2025, 24 Jan Shift-I]*

(a) $[Co(NH_3)_4Cl_2]Cl \cdot NH_3$
(b) $[Co(NH_3)_3Cl_3] \cdot 2NH_3$
(c) $[Co(NH_3)_5Cl]Cl_2$
(d) $[Co(NH_3)_4Cl] \cdot Cl_2 \cdot NH_3$

**15** Given below are two statements : one is labelled as Assertion (A) and other is labelled as Reason (R). *[2024, 9 April Shift-II]*

**Assertion (A)** The total number of geometrical isomers shown by $[Co(en)_2Cl_2]^+$ complex ion is three.

**Reason (R)** $[Co(en)_2Cl_2]^+$ complex ion has an octahedral geometry.

In the light of the above statements, choose the most appropriate answer from the options given below.

(a) A is not correct but R is correct.
(b) Both A and R are correct but R is not the correct explanation of A.
(c) A is correct but R is not correct.
(d) Both A and R are correct but R is the correct explanation of A.

**16** An octahedral complex with the formula $CoCl_3.nNH_3$ upon reaction with excess of $AgNO_3$ solution gives 2 moles of AgCl. Consider the oxidation state of Co in the complex is '$x$'. The value of "$x+n$" is .......... . *[2024, 8 April Shift-I]*

(a) 3 (b) 6 (c) 5 (d) 8

Imp. **Geometrical Isomers**

**17** The metal atom present in the complex $MABXL$ (where $A$, $B$, $X$ and $L$ are unidentate ligands and $M$ is metal) involves $sp^3$-hybridisation. The number of geometrical isomers exhibited by the complex is *[2024, 5 April Shift-II]*

(a) 0 (b) 4 (c) 2 (d) 3

**18** Which of the following complex is homoleptic? *[2024, 1 Feb Shift-I]*

(a) $[Ni(NH_3)_2Cl_2]$ (b) $[Fe(NH_3)_4Cl_2]^+$
(c) $[Co(NH_3)_4Cl_2]^+$ (d) $[Ni(CN)_4]^{2-}$

**19** The total number of stereoisomers for the complex $[Cr(ox)_2ClBr]^{3-}$ (where ox = oxalate) is *[2023, 13 April Shift-II]*

(a) 1 (b) 3 (c) 2 (d) 4

**20** Which of the following complex has a possibility to exist as meridional isomer? *[2023, 11 April Shift-II]*

(a) $[Pt(NH_3)_2Cl_2]$ (b) $[Co(NH_3)_3(NO_2)_3]$
(c) $[Co(en)_2Cl_2]$ (d) $[Co(en)_3]$

**21** The set which does not have ambidentate ligand(s) is *[2023, 11 April Shift-I]*

(a) $NO_2^-$, $C_2O_4^{2-}$, $EDTA^{4-}$
(b) $EDTA^{4-}$, $NCS^-$, $C_2O_4^{2-}$
(c) $C_2O_4^{2-}$, $NO_2^-$, $NCS^-$
(d) $C_2O_4^{2-}$, ethylene diammine, $H_2O$

**22** The complex cation which has two isomers is *[2023, 1 Feb Shift-II]*

(a) $[Co(NH_3)_5Cl]^+$ (b) $[Co(H_2O)_6]^{3+}$
(c) $[Co(NH_3)_5NO_2]^{2+}$ (d) $[Co(NH_3)_5Cl]^{2+}$

**23** Which of the following are the example of double salt? *[2023, 1 Feb Shift-I]*

A. $FeSO_4 \cdot (NH_4)_2SO_4 \cdot 6H_2O$
B. $CuSO_4 \cdot 4NH_3 \cdot H_2O$
C. $K_2SO_4.Al_2(SO_4)_3.24H_2O$
D. $Fe(CN)_2.4KCN$

Choose the correct answer given below.

(a) A and B only (b) A, B and D only
(c) A and C only (d) B and D only

**24** The primary and secondary valencies of cobalt respectively in $[Co(NH_3)_5Cl]Cl_2$ are *[2023, 24 Jan Shift-I]*

(a) 3 and 6 (b) 2 and 6
(c) 3 and 5 (d) 2 and 8

**25** White precipitate of AgCl dissolves in aqueous ammonia solution due to formation of: *[2022, 25 June 25 Shift-I]*

(a) $[Ag(NH_3)_4]Cl_2$ (b) $[Ag(Cl)_2(NH_3)_2]$
(c) $[Ag(NH_3)_2]Cl$ (d) $[Ag(NH_3)Cl]Cl$

**26** Indicate the complex/complex ion which did not show any geometrical isomerism. *[2021, 26 Aug Shift-II]*

(a) $[CoCl_2(en)_2]$ (b) $[Co(CN)_5(NC)]^{3-}$
(c) $[Co(NH_3)_3(NO_2)_3]$ (d) $[Co(NH_3)_4Cl_2]^+$

**27** Which one of the following metal complexes is most stable? *[2021, 25 July Shift-II]*

(a) $[Co(en)(NH_3)_4]Cl_2$ (b) $[Co(en)_3]Cl_2$
(c) $[Co(en)_2(NH_3)_2]Cl_2$ (d) $[Co(NH_3)_6]Cl_2$

**28** Match List-I and List-II.

| List-I | List-II |
|---|---|
| A. $[Co(NH_3)_6][Cr(CN)_6]$ | 1. Linkage isomerism |
| B. $[Co(NH_3)_3(NO_2)_3]$ | 2. Solvate isomerism |
| C. $[Cr(H_2O)_6]Cl_3$ | 3. Co-ordination isomerism |
| D. *cis*-$[CrCl_2(ox)_2]^{3-}$ | 4. Optical isomerism |

Choose the correct answer from the options given below. *[2021, 17 March Shift-II]*

| | A | B | C | D | | A | B | C | D |
|---|---|---|---|---|---|---|---|---|---|
| (a) | 3 | 1 | 2 | 4 | (b) | 4 | 2 | 3 | 1 |
| (c) | 2 | 1 | 3 | 4 | (d) | 1 | 2 | 3 | 4 |

**29** The number of isomers possible for $[Pt(en)(NO_2)_2]$ is *[2020, 4 Sep Shift-I]*

(a) 3 (b) 2
(c) 4 (d) 1

**30** The one that is not expected to show isomerism is *[2020, 2 Sep Shift-II]*

(a) $[Ni(NH_3)_4(H_2O)_2]^{2+}$ (b) $[Ni(en)_3]^{2+}$
(c) $[Ni(NH_3)_2Cl_2]$ (d) $[Pt(NH_3)_2Cl_2]$

**Imp.** Optical Isomers

**31** The number of possible optical isomers for the complexes $MA_2B_2$ with $sp^3$ and $dsp^2$ hybridised metal atom, respectively, is

*Note* : *A* and *B* are unidentate neutral and unidentate monoanionic ligands, respectively. *[2020, 7 Jan Shift-II]*

(a) 0 and 0 (b) 0 and 1
(c) 2 and 2 (d) 0 and 2

**32** The complex that can show optical activity is *[2020, 3 Sep Shift-I]*

(a) *trans*-$[Cr(Cl_2)(ox)_2]^{3-}$
(b) *trans*-$[Fe(NH_3)_2(CN)_4]^-$
(c) *cis*-$[Fe(NH_3)_2(CN)_4]^-$
(d) *cis*-$[CrCl_2(ox)_2]^{3-}$ (ox = oxalate)

**33** The complex that can show *fac*- and *mer*-isomers is : *[2020, 8 Jan Shift-I]*

(a) $[Pt(NH_3)_2Cl_2]$ (b) $[Co(NH_3)_3(NO_2)_3]$
(c) $[Co(NH_3)_4Cl_2]^+$ (d) $[CoCl_2(en)_2]$

**34** The species that can have a *trans*-isomer is (en = ethane -1, 2-diamine, ox = oxalate) *[2019, 10 April Shift-I]*

(a) $[Pt(en)Cl_2]$ (b) $[Cr(en)_2(ox)]^+$
(c) $[Pt(en)_2Cl_2]^{2+}$ (d) $[Zn(en)Cl_2]$

**35** Which one of the following complexes shows optical isomerism? *[JEE Main 2016]*

(a) *cis* $[Co(en)_2Cl_2]Cl$ (b) *trans* $[Co(en)_2Cl_2]Cl$
(c) $[Co(NH_3)_4Cl_2]Cl$ (d) $[Co(NH_3)_3Cl_3]$

**36** The correct statement on the isomerism associated with the following complex ions,

(A) $[Ni(H_2O)_5NH_3]^{2+}$
(B) $[Ni(H_2O)_4(NH_3)_2]^{2+}$ and
(C) $[Ni(H_2O)_3(NH_3)_3]^{2+}$ is *[JEE Main 2015]*

(a) (A) and (B) show only geometrical isomerim
(b) (A) and (B) show geometrical and optical isomerism
(c) (B) and (C) show geometrical and optical isomerism
(d) (B) and (C) show only geometrical isomerism

**37** The number of geometric isomers that can exist for square planar $[Pt(Cl)(py)(NH_3)(NH_2OH)]^+$ is (py = pyridine) *[JEE Main 2015]*

(a) 2 (b) 3
(c) 4 (d) 6

**Imp.** Concept of Lewis Acid

**38** Consider the coordination compound, $[Co(NH_3)_6]Cl_3$. In the formation of this complex, the species which acts as the Lewis acid is *[JEE Main April 11, 2014]*

(a) $[Co(NH_3)_6]^{3+}$ (b) $Cl^-$
(c) $Co^{3+}$ (d) $NH_3$

**39** Which of the following complex species is not expected to exhibit optical isomerism? *[JEE Main 2013]*

(a) $[Co(en)_3]^{3+}$ (b) $[Co(en)_2Cl_2]^+$
(c) $[Co(NH_3)_3Cl_3]$ (d) $[Co(en)(NH_3)Cl_2]^+$

**40** Which among the following will be named as dibromidobis- (ethylenediamine) chromium (III) bromide? *[AIEEE 2012]*

(a) $[Cr(en)_3]Br_3$ (b) $[Cr(en)_2Br_2]Br$
(c) $[Cr(en)Br_4]^-$ (d) $[Cr(en)Br_2]Br$

**41** Which one of the following has an optical isomer? (en = ethylenediamine) *[AIEEE 2010]*

(a) $[Zn(en)(NH_3)_2]^{2+}$ (b) $[Co(en)_3]^{3+}$
(c) $[Co(H_2O)_4(en)]^{3+}$ (d) $[Zn(en)_2]^{2+}$

**42** Which of the following has an optical isomer? **[AIEEE 2009]**

(a) $[Co(NH_3)_3Cl]^+$ (b) $[Co(en)(NH_3)_2]^{2+}$

(c) $[Co(H_2O)_4(en)]^{3+}$ (d) $[Co(en)_2(NH_3)_2]^{3+}$

**43** Which of the following pairs represents linkage isomers? **[AIEEE 2009]**

(a) $[Cu(NH_3)_4][PtCl_4]$ and $[Pt(NH_3)_4][CuCl_4]$

(b) $[Pd(PPh_3)_2(NCS)_2]$ and $[Pd(PPh_3)_2(SCN)_2]$

(c) $[Co(NH_3)_5]NO_3SO_4$ and $[Co(NH_3)_5SO_4]NO_3$

(d) $[PtCl_2(NH_3)_4]Br_2$ and $[PtBr_2(NH_3)_4]Cl_2$

**44** The coordination number and the oxidation state of the element '$E$' in the complex $[E(en)_2(C_2O_4)]NO_2$ (where (en) is ethylene diamine) are, respectively **[AIEEE 2008]**

(a) 6 and 2 (b) 4 and 2 (c) 4 and 3 (d) 6 and 3

**45** Which of the following compounds shows optical isomerism? **[AIEEE 2005]**

(a) $[Co(CN)_6]^{3-}$ (b) $[Cr(C_2O_4)_3]^{3-}$

(c) $[ZnCl_4]^{2-}$ (d) $[Cu(NH_3)_4]^{2+}$

**46** The coordination number of a central metal atom in a complex is determined by **[AIEEE 2004]**

(a) the number of ligands around a metal ion bonded by sigma and pi-bonds both

(b) the number of ligands around a metal ion bonded by pi-bonds

(c) the number of ligands around a metal ion bonded by sigma bonds

(d) the number of only anionic ligands bonded to the metal ion.

**47** One mole of the complex compound $Co(NH_3)_5Cl_3$, gives 3 moles of ions on dissolution in water. One mole of the same complex reacts with two moles of $AgNO_3$ solution to yield two moles of $AgCl(s)$. The structure of the complex is **[AIEEE 2003]**

(a) $[Co(NH_3)_5Cl]Cl_2$

(b) $[Co(NH_3)_3Cl_3]\cdot 2NH_3$

(c) $[Co(NH_3)_4Cl_2]Cl\cdot NH_3$

(d) $[Co(NH_3)_4Cl]Cl_2\cdot NH_3$

**48** The type of isomerisation present in nitropentamine chromium (III) chloride is **[AIEEE 2002]**

(a) optical (b) linkage

(c) ionisation (d) polymerisation

## Level 3

**49.** '$X$' is the number of electrons in $t_{2g}$ orbitals of the most stable complex ion among $[Fe(NH_3)_6]^{3+}$, $[FeCl_6]^{3-}$, $[Fe(C_2O_4)_3]^{3-}$ and $[Fe(H_2O)_6]^{3+}$. The nature of oxide of vanadium of the type $V_2O_X$ is **[2025, 4 April Shift-II]**

(a) amphoteric (b) neutral

(c) basic (d) acidic

**50.** An octahedral complex having molecular composition $Co\cdot 5NH_3\cdot Cl\cdot SO_4$, has two isomers $A$ and $B$. The solution of $A$ gives a white precipitate with $AgNO_3$ solution and the solution of $B$ gives white precipitate with $BaCl_2$, solution. The type of isomerism exhibited by the complex is, **[2025, 7 April Shift-I]**

(a) ionisation isomerism (b) linkage isomerism

(c) geometrical isomerism (d) co-ordinate isomerism

**51.** Match List -I with List - II. **[2025, 7 April Shift-II]**

| List -I (Complex) | | List -II (Primary valency and Secondary valency) | |
|---|---|---|---|
| (A) $[Co(en)_2Cl_2]Cl$ | (I) | 3 | 6 |
| (B) $[Pt(NH_3)_2Cl(NO_2)]$ | (II) | 3 | 4 |
| (C) $Hg[Co(SCN)_4]$ | (III) | 2 | 6 |
| (D) $[Mg(EDTA)]^{2-}$ | (IV) | 2 | 4 |

Choose the correct answer from the options given below:

(a) (A)-(III), (B)-(I), (C)-(II), (D)-(IV)

(b) (A)-(II), (B)-(III), (C)-(IV), (D)-(I)

(c) (A)-(I), (B)-(III), (C)-(II), (D)-(IV)

(d) (A)-(I), (B)-(IV), (C)-(II), (D)-(III)

**52** Identify the coordination complexes in which the central metal ion has $d^4$ configuration.

(A) $[FeO_4]^{2-}$ (B) $[Mn(CN)_6]^{3-}$

(C) $[Fe(CN)_6]^{3-}$

(D) $Cr_2(O{-}\overset{\overset{O}{\|}}{C}{-}Me)_4 = (H_2O)_2$

(E) $[NiF_6]^{2-}$

Choose the correct answer from the option given below. **[2025, 23 Jan Shift-II]**

(a) (A), (B) and (E) only (b) (C) and (E) only

(c) (B) and (D) only (d) (B), (C) and (D) only

**53** Match List-I with List-II.

| | List-I (Complex) | | List-II (Hybridisation of central metal ion) |
|---|---|---|---|
| (A) | $[CoF_6]^{3-}$ | I. | $d^2sp^3$ |
| (B) | $[NiCl_4]^{2-}$ | II. | $sp^3$ |
| (C) | $[Co(NH_3)_6]^{3+}$ | III. | $sp^3d^2$ |
| (D) | $[Ni(CN)_4]^{2-}$ | IV. | $dsp^2$ |

Choose the correct answer from the options given below. **[2025, 28 Jan Shift-II]**

(a) (A)-(III), (B)-(II), (C)-(I), (D)-(IV)

(b) (A)-(I), (B)-(II), (C)-(III), (D)-(IV)

(c) (A)-(I), (B)-(IV), (C)-(III), (D)-(II)

(d) (A)-(III), (B)-(IV), (C)-(I), (D)-(II)

**54** Identify the homoleptic complexes with odd number of $d$-electrons in the central metal *[2025, 29 Jan Shift-II]*

(A) $[FeO_4]^{2-}$ (B) $[Fe(CN)_6]^{3-}$

(C) $[Fe(CN)_5NO]^{2-}$ (D) $[CoCl_4]^{2-}$

(E) $[Co(H_2O)_3F_3]$

Choose the correct answer from the options given below

(a) (A), (B) and (D) only (b) (C) and (E) only
(c) (B) and (D) only (d) (A), (C) and (E) only

**55** The $F^-$ ions make the enamel on teeth much harder by converting hydroxyapatite (the enamel on the surface of teeth) into much harder fluoroapatite having the formula. *[2024, 9 April Shift-I]*

(a) $[3(Ca_3(PO_4)_2 \cdot CaF_2]$
(b) $[3(Ca_3(PO_4)_2) \cdot Ca(OH)_2]$
(c) $[3(Ca_2(PO_4)_2) \cdot Ca(OH)_2]$
(d) $[3(Ca_3)(PO_4)_3) \cdot CaF_2]$

**56** Correct formula of the compound which gives a white precipitate with $BaCl_2$ solution, but not with $AgNO_3$ solution is *[2022, 30 June Shift-I]*

(a) $[Co(NH_3)_5Br]SO_4$ (b) $[Co(NH_3)_5SO_4]Br$
(c) $[Pt(NH_3)_4Cl_2]Br_2$ (d) $[Pt(NH_3)_4Br_2]Cl_2$

**57** The number of geometrical isomers found in the metal complexes $[PtCl_2(NH_3)_2]$, $[Ni(CO)_4]$, $[Ru(H_2O)_3Cl_3]$ and $[CoCl_2(NH_3)_4]^+$ respectively, are *[2021, 27 July Shift-I]*

(a) 1, 1, 1, 1 (b) 2, 1, 2, 2
(c) 2, 0, 2, 2 (d) 2, 1, 2, 1

**Imp.** Mole Concept in Blend with Formula of Coordination Compounds

**58** Complex ($A$) has a composition of $H_{12}O_6Cl_3Cr$. If the complex on treatment with conc. $H_2SO_4$ loses 13.5% of its original mass, the correct molecular formula of ($A$) is[Given : atomic mass of Cr = 52 amu and Cl = 35 amu] *[2020, 3 Sep Shift-II]*

(a) $[Cr(H_2O)_4Cl_2]Cl \cdot 2H_2O$
(b) $[Cr(H_2O)_6]Cl_3$
(c) $[Cr(H_2O)_5Cl]Cl_2 \cdot H_2O$
(d) $[Cr(H_2O)_3Cl_3] \cdot 3H_2O$

**59** Among (A) - (D), the complexes that can display geometrical isomerism are

(A) $[Pt(NH_3)_3Cl]^+$ (B) $[Pt(NH_3)Cl_5]^-$

(C) $[Pt(NH_3)_2Cl(NO_2)]$ (D) $[Pt(NH_3)_4ClBr]^{2+}$ *[2020, 8 Jan Shift-II]*

(a) (D) and (A) (b) (C) and (D)
(c) (A) and (B) (d) (B) and (C)

**60** Which of the following complexes will show geometrical isomerism? *[JEE Main 2018]*

(a) Potassium tris(oxalato) chromate (III)
(b) Pentaaquachlorochromium (III) chloride
(c) Aquachlorobis (ethylenediamine) cobalt (II) chloride
(d) Potassium amminetrichloroplatinate (II)

**61** The total number of possible isomers for square-planar $[Pt(Cl)(NO_2)(NO_3)(SCN)]^{2-}$ is: *[JEE Main 2018]*

(a) 16 (b) 12 (c) 8 (d) 24

**62** The oxidation states of Cr, in $[Cr(H_2O)_6]Cl_3$, $[Cr(C_6H_6)_2]$, and $K_2[Cr(CN)_2(O)_2(O_2)(NH_3)]$ respectively are *[JEE Main 2018]*

(a) +3, +4 and +6 (b) +3, +2 and +4
(c) +3, 0 and +6 (d) +3, 0 and +4

**Imp.** *cis* and *trans* Isomers

**63** Consider the following reaction and statements *[JEE Main 2018]*

$$[Co(NH_3)_4Br_2]^+ + Br^- \rightarrow [Co(NH_3)_3Br_3] + NH_3$$

I. Two isomers are produces if the reactant complex ion is a *cis*-isomer.
II. Two isomers are produced if the reactant complex ion is a *trans*-isomer.
III. Only one isomer is produced if the reactant complex ion is a *trans*-isomer.
IV. Only one isomer is produced if the reactant complex ion is a *cis*-isomer.

The correct statements are

(a) (I) and (II) (b) (I) and (III)
(c) (III) and (IV) (d) (II) and (IV)

**64** On treatment of 100 mL of 0.1 M solution of $CoCl_3 \cdot 6H_2O$ with excess of $AgNO_3$; $1.2 \times 10^{22}$ ions are precipitated. The complex is *[JEE Main 2017]*

(a) $[Co(H_2O)_4Cl_2]Cl \cdot 2H_2O$
(b) $[Co(H_2O)_3Cl_3] \cdot 3H_2O$
(c) $[Co(H_2O)_6]Cl_3$
(d) $[Co(H_2O)_5Cl]Cl_2 \cdot H_2O$

**65** Which of the following name formula combinations is not correct? *[JEE Main 2014]*

| | Formula | Name |
|---|---|---|
| (a) | $K_2[Pt(CN)_4]$ | Potassium tetracyanoplatinate (II) |
| (b) | $[Mn(CN)_5]^{2-}$ | Pentacyanomangnate ion (II) |
| (c) | $K[Cr(NH_3)_2Cl_4]$ | Potassium diammine tetrachlorochromate (III) |
| (d) | $[Co(NH_3)_4(H_2O)I]SO_4$ | Tetraammineaquaiodido cobalt (III) sulphate |

**66** A solution containing 2.675 g of $CoCl_3 \cdot 6NH_3$ (molar mass = 267.5 g $mol^{-1}$) is passed through a cation exchanger. The chloride ions obtained in solution were treated with excess of $AgNO_3$ to give

4.78 g of AgCl (molar mass = 143.5 g $mol^{-1}$). The formula of the complex is (Atomic mass of Ag = 108 u) *[AIEEE 2010]*
(a) $[Co(NH_3)_6]Cl_3$ (b) $[CoCl_2(NH_3)_4]Cl$
(c) $[CoCl_3(NH_3)_3]$ (d) $[CoCl(NH_3)_5]Cl_2$

**67** Which one of the following has largest number of iosmers? *[AIEEE 2004]*
(a) $[Ir(PR_3)_2H(CO)]^{2+}$ (b) $[Co(NH_3)_5Cl]^{2+}$
(c) $[Ru(NH_3)_4Cl_2]^+$ (d) $[Co(en)_2Cl_2]^+$
($R$ = alkyl group, en = ethylene diamine)

## Numerical Value Type Questions

**68** The number of optical isomers exhibited by the iron complex ($A$) obtained from the following reaction is ........
$FeCl_3 + KOH + H_2C_2O_4 \rightarrow A$ *[2025, 3 April Shift-I]*

**69** A metal complex with a formula $MCl_4.3NH_3$ is involved in $sp^3d^2$ hybridisation. It upon reaction with excess of $AgNO_3$ solution gives '$x$' moles of AgCl. Consider '$x$' is equal to the number of lone pairs of electron present in central atom of $BrF_5$. Then the number of geometrical isomers exhibited by the complex is ........... . *[2025, 4 April Shift-II]*

**70** Number of ambidentate ligands among the following is ........
$NO_2^-, SCN^-, C_2O_4^{2-}, NH_3, CN^-, SO_4^{2-}, H_2O$ *[2024, 9 April Shift-I]*

**71** Number of complexes which show optical isomerism among the following is ......... .
*cis*-$[Cr(ox)_2Cl_2]^{3-}$, $[Co(en)_3]^{3+}$, *cis*-$[Pt(en)_2Cl_2]^{2+}$, *cis*-$[Co(en)_2Cl_2]^+$, *trans*-$[Pt(en)_2Cl_2]^{2+}$, *trans* -$[Cr(ox)_2Cl_2]^{3-}$ *[2024, 30 Jan Shift-II]*

**72** Total number of moles of AgCl precipitated on addition of excess of $AgNO_3$ to one mole each of the following complexes $[Co(NH_3)_4Cl_2]Cl$, $[Ni(H_2O_6]Cl_2$, $[Pt(NH_3)_2Cl_2]$ and $[Pd(NH_3)_4]Cl_2$ is ...... . *[2023, 25 Jan Shift-II]*

**73** Sum of oxidation state (magnitude) and coordination number of cobalt in $Na[Co(bpy)Cl_4]$ is ........... .
(Given, bpy = 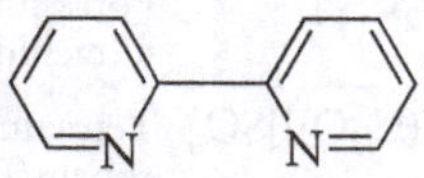 ) *[2022, 29 July Shift-II]*

**74** For the reaction given below
$CoCl_3 \cdot xNH_3 + AgNO_3(aq) \longrightarrow$
If two equivalents of AgCl precipitate out, then the value of $x$ will be ...... . *[2022, 29 June Shift-II]*

**75** (i) $CoCl_3 \cdot 4NH_3$, (ii) $CoCl_3 \cdot 5NH_3$, (iii) $CoCl_3 \cdot 6NH_3$ and (iv) $CoCl(NO_3)_2 \cdot 5NH_3$
Number of complex(es) which will exist in *cis-trans* form is/are ........... . *[2022, 28 June Shift-II]*

**76** The sum of oxidation states of two silver ions in $[Ag(NH_3)_2]$, $[Ag(CN)_2]$ complex is ........ . *[2021, 1 Sep Shift-II]*

**Imp.** Werner's Theory

**77** 3 moles of metal complex with formula $Co(en)_2Cl_3$ gives 3 moles of silver chloride on treatment with excess of silver nitrate. The secondary valency of Co in the complex is ........... .
(Round off to the nearest integer) *[2021, 27 July Shift-II]*

**78** The number of geometrical isomers possible in triamminetrinitrocobalt (III) is $X$ and in trioxalatochromate (III) is $Y$. Then, the value of $X + Y$ is ........... *[2021, 27 July Shift-I]*

**79** The equivalents of ethylene diamine required to replace the neutral ligands from the coordination sphere of the *trans*-complex of $CoCl_3 \cdot 4NH_3$ is ...... (Round off to the nearest integer). *[2021, 16 March Shift-I]*

**80** The total number of coordination sites in ethylenediaminetetraacetate $[EDTA^{4-}]$ is .... . *[2020, 5 Sep Shift-I]*

## Topic 2 Valence Bond Theory and CFT

### Level 1

**01** In which of the following complexes the CFSE, $\Delta_0$ will be equal to zero? *[2025, 22 Jan Shift-I]*
(a) $[Fe(NH_3)_6]Br_2$ (b) $K_3[Fe(SCN)_6]$
(c) $[Fe(en)_3]Cl_3$ (d) $K_4[Fe(CN)_6]$

**02** Which of the following is correct order of ligand field strength? *[2023, 31 Jan Shift-I]*
(a) $NH_3 < en < CO < S^{2-} < C_2O_4^{2-}$
(b) $CO < en < NH_3 < C_2O_4^{2-} < S^{2-}$
(c) $S^{2-} < C_2O_4^{2-} < NH_3 < en < CO$
(d) $S^{2-} < NH_3 < en < CO < C_2O_4^{2-}$

**03** Low oxidation state of metals in their complexes are common when ligands *[2022, 27 July Shift-II]*
(a) have good $\pi$-accepting character
(b) have good $\sigma$-donor character
(c) are having good $\pi$-donating ability
(d) are having poor $\sigma$-donating ability

**04** Spin only magnetic moment of an octahedral complex of $Fe^{2+}$ in the presence of a strong field ligand in BM is *[2021, 20 July Shift-II]*
(a) 0 (b) 4.89 (c) 2.82 (d) 3.46

**05** According to the valence bond theory the hybridisation of central metal atom is $dsp^2$ for which one of the following compounds? *[2021, 20 July Shift-I]*
(a) $NiCl_2 \cdot 6H_2O$ (b) $K_2[Ni(CN)_4]$
(c) $[Ni(CO)_4]$ (d) $Na_2[NiCl_4]$

**06** For a $d^4$ metal ion in an octahedral field, the correct electronic configuration is *[2020, 6 Sep Shift-II]*
(a) $t_{2g}^3 e_g^1$ when $\Delta_0 < P$ (b) $t_{2g}^3 e_g^1$ when $\Delta_0 > P$
(c) $t_{2g}^4 e_g^0$ when $\Delta_0 < P$ (d) $t_g^2 e_{2g}^2$ when $\Delta_0 < P$

**07** The species that has a spin-only magnetic moment of 5.9 BM, is ($T_d$ = tetrahedral) *[2020, 6 Sep Shift-I]*
(a) $[Ni(CN)_4]^{2-}$ (b) $[NiCl_4]^{2-}(T_d)$
(c) $[Ni(CO)_4](T_d)$ (d) $[MnBr_4]^{2-}(T_d)$

**08** The crystal field stabilisation energy (CFSE) of $[CoF_3(H_2O)_3]$ $(\Delta_0 < P)$ is *[2020, 4 Sep Shift-II]*
(a) $-0.8\,\Delta_o$ (b) $-0.8\Delta_o + 2P$
(c) $-0.4\Delta_o$ (d) $-0.4\Delta_o + P$

**09** The $d$-electron configuration of $[Ru(en)_3]Cl_2$ and $[Fe(H_2O)_6]Cl_2$, respectively are *[2020, 3 Sep Shift-II]*
(a) $t_{2g}^6 e_g^0$ and $t_{2g}^6 e_g^0$ (b) $t_{2g}^4 e_g^2$ and $t_{2g}^6 e_g^0$
(c) $t_{2g}^6 e_g^0$ and $t_{2g}^4 e_g^2$ (d) $t_{2g}^4 e_g^2$ and $t_{2g}^4 e_g^2$

**10** The correct statements among I to III are:
(I) Valence bond theory cannot explain the colour exhibited by transition metal complexes.
(II) Valence bond theory can predict quantitatively the magnetic properties of transition metal complexes.
(III) Valence bond theory cannot distinguish ligands as weak and strong field one. *[2019, 9 April Shift-II]*
(a) (II) and (III) only (b) (I), (II) and (III)
(c) (I) and (III) only (d) (I) and (II) only

**11** In Wilkinson's catalyst, the hybridisation of central metal ion and its shape are respectively: *[JEE Main 2019, 2018, 10 Jan Shift-I]*
(a) $sp^3d$, trigonal bipyramidal
(b) $d^2sp^3$, octahedral
(c) $dsp^2$, square planar
(d) $sp^3$, tetrahedral

**12** The color of $KMnO_4$ is due to *[JEE Main 2015]*
(a) $L \rightarrow M$ charge transfer transition
(b) $\sigma - \sigma^*$ transition
(c) $M \rightarrow L$ charge transfer
(d) $d-d$ transition

**13** The correct statement about the magnetic properties of $[Fe(CN)_6]^{3-}$ and $[FeF_6]^{3-}$ is (Z = 26)
(a) both are paramagnetic. *[JEE Main April 9, 2014]*
(b) both are diamagnetic.
(c) $[Fe(CN)_6]^{3-}$ is diamagnetic, $[FeF_6]^{3-}$ is paramagnetic.
(d) $[Fe(CN)_6]^{3-}$ is paramagnetic, $[FeF_6]^{3-}$ is diamagnetic.

**14** The octahedral complex of a metal ion $M^{3+}$ with four monodentate ligands $L_1$, $L_2$, $L_3$ and $L_4$ absorb wavelengths in the region of red, green, yellow and blue, respectively. The increasing order of ligand strength of the four ligands is *[JEE Main 2014]*
(a) $L_4 < L_3 < L_2 < L_1$ (b) $L_1 < L_3 < L_2 < L_4$
(c) $L_3 < L_2 < L_4 < L_1$ (d) $L_1 < L_2 < L_4 < L_3$

**15** Among the following species, the one which causes the highest CFSE, $\Delta_0$ as a ligand is *[JEE Main 2014, 13, 08]*
(a) $CN^-$ (b) $NH_3$ (c) F (d) CO

**16** Homoleptic octahedral complexes of a metal ion '$M^{3+}$' with three monodentate ligands $L_1$, $L_2$ and $L_3$ absorb wavelengths in the region of green, blue and red respectively. The increasing order of the ligand strength is: *[2014, 9 Jan Shift-II]*
(a) $L_3 < L_1 < L_2$ (b) $L_3 < L_2 < L_1$
(c) $L_1 < L_2 < L_3$ (d) $L_2 < L_1 < L_3$

**17** The magnetic moment of the complex anion $[Cr(NO)(NH_3)(CN)_4]^{2-}$ is *[JEE Main 2013]*
(a) 5.91 BM (b) 3.87 BM (c) 1.73 BM (d) 2.82 BM

**18** The correct order of ligands in the spectrochemical series is *[JEE Main 2012]*
(a) $Cl^- > en > CN^- > NCS^-$ (b) $CN^- > en > NCS^- > Cl^-$
(c) $NCS^- > CN^- > Cl^- > en$ (d) $en > CN^- > Cl^- > NCS^-$

**19** Which of the following facts about the complex $[Cr(NH_3)_6]Cl_3$ is wrong? *[AIEEE 2011]*
(a) The complex involves $d^2sp^3$ hybridization and is octahedral in shape
(b) The complex is paramagnetic
(c) The complex is an outer orbital complex
(d) The complex gives white precipitate with silver nitrate solution

**20** In which of the following octahedral complexes of Co (At. no. 27), will the magnitude of $\Delta_o$ be the highest? *[AIEEE 2008]*
(a) $[Co(CN)_6]^{3-}$ (b) $[Co(C_2O_4)_3]^{3-}$
(c) $[Co(H_2O)_6]^{3+}$ (d) $[Co(NH_3)_6]^{3+}$

**21** Which one of the following has a square planar geometry? (At. no. Co = 27, Ni = 28, Fe = 26, Pt = 78) *[AIEEE 2007]*
(a) $[CoCl_4]^{2-}$ (b) $[FeCl_4]^{2-}$
(c) $[NiCl_4]^{2-}$ (d) $[PtCl_4]^{2-}$

## Level 2

**22** The correct order of $[FeF_6]^{3-}$, $[CoF_6]^{3-}$, $[Ni(CO)_4]$ and $[Ni(CN)_4]^{2-}$ complex species based on the number of unpaired electrons present is: *[2025, 4 April Shift-II]*
(a) $[FeF_6]^{3-} > [CoF_6]^{3-} > [Ni(CN)_4]^{2-} > [Ni(CO)_4]$
(b) $[CoF_6]^{3-} > [FeF_6]^{3-} > [Ni(CO)_4] > [Ni(CN)_4]^{2-}$
(c) $[Ni(CN)_4]^{2-} > [FeF_6]^{3-} > [CoF_6]^{3-} > [Ni(CO)_4]$
(d) $[FeF_6]^{3-} > [CoF_6]^{3-} > [Ni(CN)_4]^{2-} = [Ni(CO)_4]$

**23** The number of species from the following that are involved in $sp^3d^2$ hybridisation is $[Co(NH_3)_6]^{3+}$, $SF_6$, $[CrF_6]^{3-}$, $[CoF_6]^{3-}$, $[Mn(CN)_6]^{3-}$ and $[MnCl_6]^{3-}$ **[2025, 8 April Shift-II]**
(a) 6 (b) 3
(c) 5 (d) 4

**24** Which one of the following complexes will have $\Delta_0 = 0$ and $\mu = 5.96$ B.M? **[2025, 4 April Shift-I]**
(a) $[Co(NH_3)_6]^{3+}$ (b) $[Fe(CN)_6]^{4-}$
(c) $[FeF_6]^{4-}$ (d) $[Mn(SCN)_6]^{4-}$

**25** The type of hybridisation and the magnetic property of $[MnCl_6]^{3-}$ are **[2025, 2 April Shift-II]**
(a) $sp^3d^2$, paramagnetic with four unpaired electrons.
(b) $d^2sp^3$, paramagnetic with four unpaired electrons.
(c) $sp^3d^2$, paramagnetic with two unpaired electrons.
(d) $d^2sp^3$, paramagnetic with two unpaired electrons.

**26** Given below are two statements: **[2025, 2 April Shift-I]**
**Statement I** In octahedral complexes, when $\Delta_0 < P$ high spin complexes are formed. When $\Delta_0 > P$ low spin complexes are formed.
**Statement II** In tetrahedral complexes because of $\Delta_t < P$, low spin complexes are rarely formed.
In the light of the above statements, choose the most appropriate answer from the options given below:
(a) Statement I is correct but Statement II is incorrect.
(b) Both Statement I and Statement II are incorrect.
(c) Both Statement I and Statement II are correct.
(d) Statement I is incorrect but Statement II is correct.

**27** The $d$-electronic configuration of an octahedral Co(II) complex having magnetic moment of 3.95 BM is **[2025, 23 Jan Shift-I]**
(a) $e^4t_2^3$ (b) $t_{2g}^6e_g^1$
(c) $t_{2g}^3e_g^0$ (d) $t_{2g}^5e_g^2$

**28** The conditions and consequence that favours the $t_{2g}^3e_g^1$ configuration in a metal complex are **[2025, 24 Jan Shift-II]**
(a) weak field ligand, low spin complex
(b) strong field ligand, high spin complex
(c) weak field ligand, high spin complex
(d) strong field ligand, low spin complex

**29** The calculated spin-only magnetic moments of $K_3[Fe(OH)_6]$ and $K_4[Fe(OH)_6]$ respectively are **[2025, 29 Jan Shift-II]**
(a) 3.87 and 4.90 BM
(b) 4.90 and 5.92 BM
(c) 4.90 and 4.90 BM
(d) 5.92 and 4.90 BM

**30** When ethane-1,2-diamine is added progressively to an aqueous solution of nickel (II) chloride, the sequence of colour change observed will be **[2025, 24 Jan Shift-II]**
(a) Violet → Blue → Pale blue → Green
(b) Green → Pale blue → Blue → Violet
(c) Pale blue → Blue → Green → Violet
(d) Pale blue → Blue → Violet → Green

**31** The coordination environment of $Ca^{2+}$ ion in its complex with $EDTA^{4-}$ is **[2024, 9 April Shift-II]**
(a) tetrahedral (b) trigonal prismatic
(c) square planar (d) octahedral

**32** If an iron (III) complex with the formula $[Fe(NH_3)_x(CN)_y]^-$ has no electron in its $e_g$ orbital, then the value of $x + y$ is **[2024, 4 April Shift-II]**
(a) 5 (b) 3
(c) 4 (d) 6

**33** The correct sequence of ligands in the order of decreasing field strength is: **[2024, 4 April Shift-I]**
(a) $NCS^- > EDTA^{4-} > CN^- > CO$
(b) $CO > H_2O > F^- > S^{2-}$
(c) $^-OH > F^- > NH_3 > CN^-$
(d) $S^{2-} > ^-OH > EDTA^{4-} > CO$

**34** $[Co(NH_3)_6]^{3+}$ and $[CoF_6]^{3-}$ are respectively known as **[2024, 1 Feb Shift-II]**
(a) spin paired complex, spin free complex
(b) inner orbital complex, spin paired complex
(c) outer orbital complex, inner orbital complex
(d) spin free complex, spin paired complex

**35** Given below are two statements. **[2024, 1 Feb Shift-I]**
**Statement I** A solution of $[Ni(H_2O)_6]^{2+}$ is green in colour.
**Statement II** A solution of $[Ni(CN)_4]^{2-}$ is colourless.
In the light of the above statements, choose the **most appropriate** answer from the options given below.
(a) Statement I is correct but Statement II is incorrect.
(b) Both Statement I and Statement II are incorrect.
(c) Both Statement I and Statement II are correct.
(d) Statement I is incorrect but Statement II is correct.

**36** Select the option with correct property. **[2024, 31 Jan Shift-II]**
(a) $[Ni(CO)_4]$ and $[NiCl_4]^{2-}$ both paramagnetic.
(b) $[Ni(CO)_4]$ and $[NiCl_4]^{2-}$ both diamagnetic.
(c) $[Ni(CO)_4]$ diamagnetic, $[NiCl_4]^{2-}$ paramagnetic.
(d) $[NiCl_4]^{2-}$ diamagnetic, $[Ni(CO)_4]$ paramagnetic.

**37** The coordination geometry around the manganese in decarbonyldimanganese (0) is **[2024, 30 Jan Shift-II]**
(a) square planar (b) octahedral
(c) trigonal bipyramidal (d) square pyramidal

**38** The octahedral diamagnetic low spin complex among the following is *[2023, 10 April Shift-I]*
(a) $[CoF_6]^{3-}$ (b) $[Co(NH_3)_6]^{3+}$
(c) $[NiCl_4]^{2-}$ (d) $[CoCl_6]^{3-}$

**39** The correct order of spin only magnetic moments for the following complex ions is *[2023, 8 April Shift-I]*
(a) $[Fe(CN)_6]^{3-} < [CoF_6]^{3-} < [MnBr_4]^{2-} < [Mn(CN)_6]^{3-}$
(b) $[Fe(CN)_6]^{3-} < [Mn(CN)_6]^{3-} < [CoF_6]^{3-} < [MnBr_4]^{2-}$
(c) $[MnBr_4]^{2-} < [CoF_6]^{3-} < [Fe(CN)_6]^{3-} < [Mn(CN)_6]^{3-}$
(d) $[CoF_6]^{3-} < [MnBr_4]^{2-} < [Fe(CN)_6]^{3-} < [Mn(CN)_6]^{3-}$

**40** A solution of $FeCl_3$ when treated with $K_4[Fe(CN)_6]$ gives a prussian blue precipitate due to the formation of *[2023, 1 Feb Shift-I]*
(a) $K[Fe_2(CN)_6]$ (b) $Fe_4[Fe(CN)_6]_3$
(c) $Fe_3[Fe(CN)_6]_2$ (d) $Fe[Fe(CN)_6]$

**Imp.** Colour and Geometry of the Coordination Compounds

**41** Cobalt chloride when dissolved in water forms pink coloured complex $\underline{X}$ which has octahedral geometry. This solution on treating with conc. HCl forms deep blue complex, $\underline{Y}$ which has a $\underline{Z}$ geometry. $X, Y$ and $Z$, respectively, are *[2023, 31 Jan Shift-I]*
(a) $X = [Co(H_2O)_4Cl_2]^+, Y = [CoCl_4]^{2-}, Z =$ tetrahedral
(b) $X = [Co(H_2O)_6]^{2+}, Y = [CoCl_6]^{3-}, Z =$ octahedral
(c) $X = [Co(H_2O)_6]^{2+}, Y = [CoCl_4]^{4-}, Z =$ tetrahedral
(d) $X = [Co(H_2O)_6]^{3+}, Y = [CoCl_6]^{3+}, Z =$ octahedral

**42** Correct order of spin only magnetic moment of the following complex ions is
(Given : Atomic number Fe = 26, Co = 27) *[2023, 29 Jan Shift-II]*
(a) $[CoF_6]^{3-} > [FeF_6]^{3-} > [Co(C_2O_4)_3]^{3-}$
(b) $[Co(C_2O_4)_3]^{3-} > [CoF_6]^{3-} > [FeF_6]^{3-}$
(c) $[FeF_6]^{3-} > [CoF_6]^{3-} > [Co(C_2O_4)_3]^{3-}$
(d) $[FeF_6]^{3-} > [Co(C_2O_4)_3]^{3-} > [CoF_6]^{3-}$

**43** The hybridisation and magnetic behaviour of cobalt ion in $[Co(NH_3)_6]^{3+}$ complex, respectively is *[2023, 24 Jan Shift-II]*
(a) $d^2sp^3$ and diamagnetic (b) $d^2sp^3$ and paramagnetic
(c) $sp^3d^2$ and diamagnetic (d) $sp^3d^2$ and paramagnetic

**44** If $Ni^{2+}$ is replaced by $Pt^{2+}$ in the complex $[NiCl_2Br_2]^{2-}$, which of the following properties are expected to get changed?
A. Geometry
B. Geometrical isomerism
C. Optical isomerism
D. Magnetic properties *[2023, 11 April Shift-II]*
(a) A, B and D (b) A and D
(c) A, B and C (d) B and C

**45** Given below are two statements, one is labelled as Assertion (A) and the other is labelled as Reason (R).
**Assertion** (A) $[CoCl(NH_3)_5]^{2-}$ absorbs at lower wavelength of light with respect to $[Co(NH_3)_5(H_2O)]^{3+}$.
**Reason** (R) It is because the wavelength of the light absorbed depends on the oxidation state of the metal ion.
In the light of the above statements, choose the most appropriate answer from the options given below. *[2023, 11 April Shift-II]*
(a) Both A and R are true and R is the correct explanation of A
(b) A is true but R is false
(c) Both A and R are true but R is not the correct explanation of A
(d) A is false but R is true

**Imp.** Jahn-Teller Distortion

**46** Octahedral complexes of copper (II) undergo structural distortion (Jahn-Teller). Which one of the given copper (II) complexes will show the maximum structural distortion? (en–ethylenediamine; $H_2N—CH_2—CH_2—NH_2$) *[2022, 29 July Shift-II]*
(a) $[Cu(H_2O)_6]SO_4$ (b) $[Cu(en)(H_2O)_4]SO_4$
(c) *cis*-$[Cu(en)_2Cl_2]$ (d) *trans*-$[Cu(en)_2Cl_2]$

**47** The correct order of energy of absorption for the following metal complexes is
A. $[Ni(en)_3]^{2+}$ B. $[Ni(NH_3)_6]^{2+}$
C. $[Ni(H_2O)_6]^{2+}$ *[2022, 25 July Shift-II]*
(a) C < B < A (b) B < C < A
(c) C < A < B (d) A < C < B

**48** Given below are two statements.
**Statement I** In $CuSO_4 \cdot 5H_2O$, Cu—O bonds are present.
**Statement II** In $CuSO_4 \cdot 5H_2O$, ligands coordinating with Cu(II) ion are O-and S-based ligands.
In the light of the above statements, choose the correct answer from the options given below *[2022, 29 June Shift-II]*
(a) Both Statement I and Statement II are correct.
(b) Both Statement I and Statement II are incorrect.
(c) Statement I is correct but Statement II is incorrect.
(d) Statement I is incorrect but Statement II is correct.

**49** Given below are two statements :

**Statement I :** $[Ni(CN)_4]^{2-}$ is square planar and diamagnetic complex with $dsp^2$ hybridisation for Ni but $[Ni(CO)_4]$ is tetrahedral paramagnetic and with $sp^3$-hybridisation for Ni.

**Statement II :** $[NiCl_4]^{2-}$ and $[Ni(CO)_4]$ both have same *d*-electron configuration have same geometry and are paramagnetic.

In light of the above statements, choose the correct answer from the options given below: *[2022, 28 June Shift-I]*

(a) Both Statement I and Statement II are correct.
(b) Both Statement I and Statement II are incorrect.
(c) Statement I is correct but Statement II is incorrect.
(d) Statement I is incorrect but Statement II is correct.

**50** Arrange the following coordination compounds in the increasing order of magnetic moments. (Atomic numbers : Mn = 25 ; Fe = 26)

(A) $[FeF_6]^{3-}$ (B) $[Fe(CN)_6]^{3-}$
(C) $[MnCl_6]^{3-}$ (high spin) (D) $[Mn(CN)_6]^{3-}$

Choose the correct answer from the options given below : *[2022, 27 June Shift-II]*

(a) A < B < D < C (b) B < D < C < A
(c) A < C < D < B (d) B < D < A < C

**Imp.** Crystal Field splitting

**51** Transition metal complex with highest value of crystal field splitting ($\Delta_o$) will be *[2022, 24 June Shift-II]*

(a) $[Cr(H_2O)_6]^{3+}$ (b) $[Mo(H_2O)_6]^{3+}$
(c) $[Fe(H_2O)_6]^{3+}$ (d) $[Os(H_2O)_6]^{3+}$

**52** Arrange the following cobalt complexes in the order of increasing crystal field stabilisation energy (CFSE) value.

$[CoF_6]^{3-}$ (A), $[Co(H_2O)_6]^{3+}$ (B), $[Co(NH_3)_6]^{3+}$ (C) and $[Co(en)_3]^{3+}$ (D)

Choose the correct option. *[2021, 26 Aug Shift-II]*

(a) $A < B < C < D$ (b) $B < A < C < D$
(c) $B < C < D < A$ (d) $C < D < B < A$

**53** Which one of the following complexes is violet in colour? *[2021, 26 Aug Shift-I]*

(a) $[Fe(CN)_6]^{4-}$ (b) $[Fe(SCN)_6]^{4-}$
(c) $Fe_4[Fe(CN_6)]_3 \cdot H_2O$ (d) $[Fe(CN)_5NOS]^{4-}$

**54** The type of hybridisation and magnetic property of the complex $[MnCl_6]^{3-}$, respectively, are *[2021, 27 July Shift-I]*

(a) $sp^3d^2$ and diamagnetic
(b) $d^2sp^3$ and diamagnetic
(c) $d^2sp^3$ and paramagnetic
(d) $sp^3d^2$ and paramagnetic

**55** Which one of the following species responds to an external magnetic field *[2021, 25 July Shift-I]*

(a) $[Fe(H_2O)_6]^{3+}$ (b) $[Ni(CN)_4]^{2-}$
(b) $[Co(CN)_6]^{3-}$ (d) $[Ni(CO)_4]$

**56** The correct order of intensity of colors of the compound is *[2021, 20 July 20 Shift-I]*

(a) $[Ni(CN)_4]^{2-} > [NiCl_4]^{2-} > [Ni(H_2O)_6]^{2+}$
(b) $[Ni(H_2O)_6]^{2+} > [NiCl_4]^{2-} > [Ni(CN)_4]^{2-}$
(c) $[NiCl_4]^{2-} > [Ni(H_2O)_6]^{2+} > [Ni(CN)_4]^{2-}$
(d) $[NiCl_4]^{2-} > [Ni(CN)_4]^{2-} > [Ni(H_2O)_6]^{2+}$

**57** The hybridisation and magnetic nature of $[Mn(CN)_6]^{4-}$ and $[Fe(CN)_6]^{3-}$, respectively are *[2021, 25 Feb Shift-I]*

(a) $d^2sp^3$ and paramagnetic (b) $sp^3d^2$ and diamagnetic
(c) $d^2sp^3$ and diamagnetic (d) $sp^3d^2$ and paramagnetic

**58** The calculated magnetic moments (spin only value) for species $[FeCl_4]^{2-}$, $[Co(C_2O_4)_3]^{3-}$ and $MnO_4^{2-}$ respectively are *[2021, 24 Feb Shift-II]*

(a) 5.82, 0 and 0 BM (b) 4.90, 0 and 1.73 BM
(c) 5.92, 4.90 and 0 BM (d) 4.90, 0 and 2.83 BM

**59** The molecule in which hybrid MOs involve only one *d*-orbital of the central atom is *[2020, 4 Sep Shift-II]*

(a) $XeF_4$ (b) $[CrF_6]^{3-}$ (c) $BrF_5$ (d) $[Ni(CN)_4]^{2-}$

**60** The pair in which both the species have the same magnetic moment (spin only) is *[2020, 4 Sep Shift-I]*

(a) $[Mn(H_2O)_6]^{2+}$ and $[Cr(H_2O)]^{2+}$
(b) $[Cr(H_2O)_6]^{2+}$ and $[Fe(H_2O)_6]^{2+}$
(c) $[Co(OH)_4]^{2-}$ and $[Fe(NH_3)_6]^{2+}$
(d) $[Cr(H_2O)_6]^{2+}$ and $[CoCl_4]^{2-}$

**61** The electronic spectrum of $[Ti(H_2O)_6]^{3+}$ shows a single broad peak with a maximum at 20,300 $cm^{-1}$. The crystal field stabilisation energy (CFSE) of the complex ion, in kJ $mol^{-1}$, is (1 kJ $mol^{-1}$ = 83.7 $cm^{-1}$) *[2020, 3 Sep Shift-I]*

(a) 145.5 (b) 242.5
(c) 83.7 (d) 97

**62** Consider that a $d^6$ metal ion ($M^{2+}$) forms a complex with *aqua* ligands and the spin only magnetic moment of the complex is 4.90 BM. The geometry and the crystal field stabilisation energy of the complex is *[2020, 2 Sep Shift-I]*

(a) tetrahedral and $-1.6\Delta_t + 1P$
(b) octahedral and $-2.4\,\Delta_0 + 2P$
(c) octahedral and $-1.6\,\Delta_0$
(d) tetrahedral and $-0.6\,\Delta_t$

**63** The correct order of the calculated spin only magnetic moments of complexes (A) to (D) is *[2020, 8 Jan Shift-II]*

(A) $Ni(CO)_4$ (B) $[Ni(H_2O)_6]Cl_2$
(C) $Na_2[Ni(CN)_4]$ (D) $PdCl_2(PPh_3)_2$

(a) (A) ≈ (C) < (B) ≈ (D) (b) (C) ≈ (D) < (B) < (A)
(c) (C) < (D) < (B) < (A) (d) (A) ≈ (C) ≈ (D) < (B)

**Imp.** Splitting Pattern of Octachedral Complex

**64** Complete removal of both the axial ligands (along the z-axis) from an octahedral complex leads to which of the following splitting patterns? (relative orbital energies not on scale). *[2019, 12 April Shift-I]*

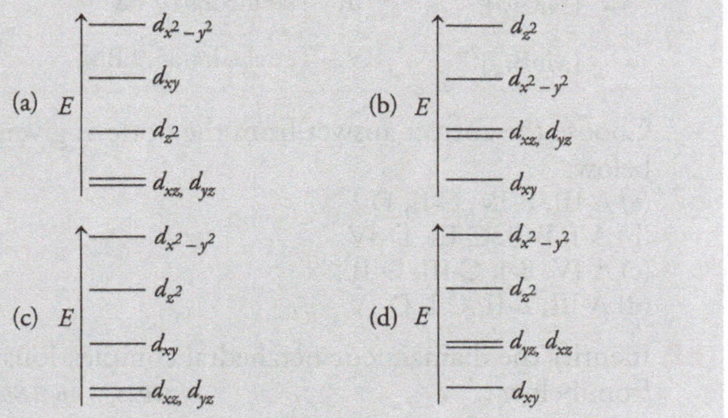


**65** The crystal field stabilisation energy (CFSE) of $[Fe(H_2O)_6]Cl_2$ and $K_2[NiCl_4]$, respectively, are *[2019, 10 April Shift-II]*

(a) $-0.4\,\Delta_o$ and $-1.2\,\Delta_t$ (b) $-0.4\,\Delta_o$ and $-0.8\,\Delta_t$
(c) $-2.4\,\Delta_o$ and $-1.2\,\Delta_t$ (d) $-0.6\,\Delta_o$ and $-0.8\,\Delta_t$

**66** The correct order of the spin only magnetic moment of metal ions in the following low spin complexes, $[V(CN)_6]^{4-}$, $[Fe(CN)_6]^{4-}$, $[Ru(NH_3)_6]^{3+}$, and $[Cr(NH_3)_6]^{2+}$, is *[2019, 8 April Shift-I]*

(a) $Cr^{2+} > Ru^{3+} > Fe^{2+} > V^{2+}$
(b) $V^{2+} > Cr^{2+} > Ru^{3+} > Fe^{2+}$
(c) $V^{2+} > Ru^{3+} > Cr^{2+} > Fe^{2+}$
(d) $Cr^{2+} > V^{2+} > Ru^{3+} > Fe^{2+}$

**Imp.** Colour of Coordination Compounds

**67** Two complexes $[Cr(H_2O)_6]Cl_3$ (*A*) and $[Cr(NH_3)_6]Cl_3$ (*B*) are violet and yellow coloured, respectively. The incorrect statement regarding them is *[2019, 9 Jan Shift-I]*

(a) $\Delta_o$ value for (*A*) is less than that of (*B*)
(b) both absorb energies corresponding to their complementary colours
(c) $\Delta_o$ values of (*A*) and (*B*) are calculated from the energies of violet and yellow light, respectively
(d) both are paramagnetic with three unpaired electrons

**68** The correct combination is *[JEE Main 2018]*

(a) $[Ni(CN)_4]^{2-}$ - tetrahedral; $[Ni(CO)_4]$ - paramagnetic
(b) $[NiCl_4]^{2-}$ - paramagnetic; $[Ni(CO)_4]$ - tetrahedral
(c) $[NiCl_4]^{2-}$ - diamagnetic; $[Ni(CO)_4]$ - square-planar
(d) $[NiCl_4]^{2-}$ -square-planar; $[Ni(CN)_4]^{2-}$ -paramagnetic

**69** The correct order of spin-only magnetic moments among the following is : (Atomic number : Mn = 25, Co = 27, Ni = 28, Zn = 30) *[JEE Main 2018]*

(a) $[ZnCl_4]^{2-} > [NiCl_4]^{2-} > [CoCl_4]^{2-} > [MnCl_4]^{2-}$
(b) $[CoCl_4]^{2-} > [MnCl_4]^{2-} > [NiCl_4]^{2-} > [ZnCl_4]^{2-}$
(c) $[NiCl_4]^{2-} > [CoCl_4]^{2-} > [MnCl_4]^{2-} > [ZnCl_4]^{2-}$
(d) $[MnCl_4]^{2-} > [CoCl_4]^{2-} > [NiCl_4]^{2-} > [ZnCl_4]^{2-}$

**70** When concentrated HCl is added to an aqueous solution of $CoCl_2$, its colour changes from reddish pink to deep blue. Which complex ion gives blue colour in this reaction? *[JEE Main 2018]*

(a) $[CoCl_6]^{4-}$ (b) $[CoCl_6]^{3-}$
(c) $[CoCl_4]^{2-}$ (d) $[Co(H_2O)_6]^{2+}$

**71** Identify the correct trend given below
(Atomic number = Ti : 22, Cr : 24 and Mo : 42) *[JEE Main 2016]*

(a) $\Delta_0$ of $[Cr(H_2O)_6]^{2+} > [Mo(H_2O)_6]^{2+}$ and $\Delta_0$ of $[Ti(H_2O)_6]^{3+} > [Ti(H_2O)_6]^{2+}$
(b) $\Delta_0$ of $[Cr(H_2O)_6]^{2+} > [Mo(H_2O)_6]^{2+}$ and $\Delta_0$ of $[Ti(H_2O)_6]^{3+} < [Ti(H_2O)_6]^{2+}$
(c) $\Delta_0$ of $[Cr(H_2O)_6]^{2+} < [Mo(H_2O)_6]^{2+}$ and $\Delta_0$ of $[Ti(H_2O)_6]^{3+} > [Ti(H_2O)_6]^{2+}$
(d) $\Delta_0$ of $[Cr(H_2O)_6]^{2+} < [Mo(H_2O)_6]^{2+}$ and $\Delta_0$ of $[Ti(H_2O)_6]^{3+} < [Ti(H_2O)_6]^{2+}$

**72** Which of the following complex ions has electrons that are symmetrically filled in both $t_{2g}$ and $e_g$ orbitals? *[JEE Main 2015]*

(a) $[FeF_6]^{3-}$ (b) $[Mn(CN)_6]^{4-}$
(c) $[CoF_6]^{3-}$ (d) $[CO(NH_3)_6]^{2+}$

**73** Which of the following compounds is not coloured yellow? *[JEE Main 2015]*

(a) $Zn_2[Fe(CN)_6]$ (b) $K_3[Co(NO_2)_6]$
(c) $(NH_4)_3[As(Mo_3O_{10})_4]$ (d) $BaCrO_4$

**74** Nickel (Z = 28) combines with a uninegative monodentate ligand to form a diamagnetic complex $[NiL_4]^{2-}$. The hybridisation ivolved and the number of unpaired electrons present in the complex are respectively : *[Online April 19, 2014]*

(a) $sp^3$, two (b) $dsp^2$, zero
(c) $dsp^2$, one (d) $sp^3$, zero

**75** Which one of the following complexes will most likely absorb visible light?
(Atomic number Sc = 21, Ti = 22, V = 23, Zn = 30) *[Online April 12, 2014]*

(a) $[Sc(H_2O)_6]^{3+}$ (b) $[Ti(NH_3)_6]^{4+}$
(c) $[V(NH_3)_6]^{3+}$ (d) $[Zn(NH_3)_6]^{2+}$

**76** An octahedral complex of $Co^{3+}$ is diamagnetic. The hybridisation involved in the formation of the complex is *[JEE Main April 9, 2014]*
(a) $sp^3d^2$ (b) $dsp^2$
(c) $d^2sp^3$ (d) $sp^3d$

**77** Which of the following is diamagnetic? *[JEE Main 2013]*
(a) $[Fe(CN)_6]^{3-}$ (b) $[Co(ox)_3]^{3-}$
(c) $[FeF_6]^{3-}$ (d) $[Co(F_6)]^{3-}$

**78** The structure of which of the following chloro species can be explained on the basis of $dsp^2$ hybridisation? *[JEE Main 2013, 12, 07, 06]*
(a) $[PdCl_4]^{2-}$ (b) $[FeCl_4]^{2-}$
(c) $[CoCl_4]^{2-}$ (d) $[NiCl_4]^{2-}$

**79** The $d$-electron configurations of $Cr^{2+}$, $Mn^{2+}$, $Fe^{2+}$ and $Co^{2+}$ are $d^4$, $d^5$, $d^6$ and $d^7$ respectively. Which one of the following will exhibit the lowest paramagnetic behaviour? (Atomic number Cr = 24, Mn = 25, Fe = 26, Co = 27). *[JEE Main 2012]*
(a) $[Co(H_2O)_6]^{2+}$ (b) $[Cr(H_2O)_6]^{2+}$
(c) $[Mn(H_2O)_6]^{2+}$ (d) $[Fe(H_2O)_6]^{2+}$

**80** Which one of the following cyano complexes would exhibit the lowest value of paramagnetic behaviour ? (At. no. of Cr = 24, Mn =25, Fe = 26, Co = 27) *[AIEEE 2005]*
(a) $[Co(CN)_6]^{3-}$ (b) $[Fe(CN)_6]^{3-}$
(c) $[Mn(CN)_6]^{3-}$ (d) $[Cr(CN)_6]^{3-}$

**81** The correct order of magnetic moments (spin only values in BM) among the following is (At. no of Mn = 25, Fe = 26, Co = 27) *[AIEEE 2004]*
(a) $[MnCl_4]^{2-} > [CoCl_4]^{2-} > [Fe(CN)_6]^{4-}$
(b) $[MnCl_4]^{2-} > [Fe(CN)_6]^{4-} > [CoCl_4]^{2-}$
(c) $[Fe(CN)_6]^{4-} > [MnCl_4]^{2-} > [CoCl_4]^{2-}$
(d) $[Fe(CN)_6]^{4-} > [CoCl_4]^{2-} > [MnCl_4]^{2-}$

**82** The most stable ion is *[AIEEE 2002]*
(a) $[Fe(OH)_5]^{3-}$ (b) $[FeCl_6]^{3-}$
(c) $[Fe(CN)_6]^{3-}$ (d) $[Fe(H_2O)_6]^{3+}$

## Level 3

**83** The correct order of the complexes $[Co(NH_3)_5(H_2O)]^{3+}$ $(A)$, $[Co(NH_3)_6]^{3+}$ $(B)$, $[Co(CN)_6]^{3-}$ $(C)$ and $[CoCl(NH_3)_5]^{2+}$ $(D)$ in terms of wavelength of light absorbed is *[2025, 3 April Shift-I]*
(a) $D > C > B > A$ (b) $C > B > D > A$
(c) $D > A > B > C$ (d) $C > B > A > D$

**84** The number of unpaired electrons responsible for the paramagnetic nature of the following complex species are respectively:
$[Fe(CN)_6]^{3-}, [FeF_6]^{3-}, [CoF_6]^{3-}, [Mn(CN)_6]^{3-}$ *[2025, 7 April Shift-II]*
(a) 1, 5, 4, 2 (b) 1, 5, 5, 2
(c) 1, 1, 4, 2 (d) 1, 4, 4, 2

**85** Match the List-I with List-II. *[2025, 8 April Shift-II]*

| List-I (Complex/ Species) | List-II (Shape and magnetic moment) |
|---|---|
| A. $[Ni(CO)_4]$ | I. Tetrahedral, 2.8 BM |
| B. $[Ni(CN)_4]^{2-}$ | II. Square planar, 0 BM |
| C. $[NiCl_4]^{2-}$ | III. Tetrahedral, 0 BM |
| D. $[MnBr_4]^{2-}$ | IV. Tetrahedral, 5.9 BM |

Choose the correct answer from the options given below:
(a) A-III, B-IV, C-II, D-I
(b) A-I, B-II, C-III, D-IV
(c) A-IV, B-I, C-III, D-II
(d) A-III, B-II, C-I, D-IV

**86** Identify the diamagnetic octahedral complex ions from below : *[2025, 3 April Shift-II]*
A. $[Mn(CN)_6]^{3-}$
B. $[Co(NH_3)_6]^{3+}$
C. $[Fe(CN)_6]^{4-}$
D. $[Co(H_2O)_3F_3]$
Choose the correct answer from the options given below.
(a) A and D only (b) B and C only
(c) A and C only (d) B and D only

**87** The $d$-orbital electronic configuration of the complex among $[Co(en)_3]^{3+}$, $[CoF_6]^{3-}$, $[Mn(H_2O)_6]^{2+}$ and $[Zn(H_2O)_6]^{2+}$ that has the highest CFSE is *[2025, 2 April Shift-II]*
(a) $t_{2g}^6e_g^0$ (b) $t_{2g}^3e_g^2$
(c) $t_{2g}^4e_g^2$ (d) $t_{2g}^6e_g^4$

**88** From the magnetic behaviour of $[NiCl_4]^{2-}$ (paramagnetic) and $[Ni(CO)_4]$ (diamagnetic), choose the correct geometry and oxidation state. *[2025, 22 Jan Shift-I]*
(a) $[NiCl_4]^{2-}$ : $Ni^{II}$, square planar
$[Ni(CO)_4]$ : Ni(0), square planar
(b) $[NiCl_4]^{2-}$ : $Ni^{II}$, tetrahedral
$[Ni(CO)_4]$ : $Ni^{II}$, square planar
(c) $[NiCl_4]^{2-}$ : Ni(0), tetrahedral
$[Ni(CO)_4]$ : Ni(0), square planar
(d) $[NiCl_4]^{2-}$ : $Ni^{II}$, tetrahedral
$[Ni(CO)_4]$ : Ni(0), tetrahedral

**89** Identify the homoleptic complex(es) that is/are low spin. *[2025, 22 Jan Shift-II]*

(A) $[Fe(CN)_5NO]^{2-}$ (B) $[CoF_6]^{3-}$

(C) $[Fe(CN)_6]^{4-}$ (D) $[Co(NH_3)_6]^{3+}$

(E) $[Cr(H_2O)_6]^{2+}$

Choose the correct answer from the options given below:

(a) (C) and (D) only (b) (A) and (C) only
(c) (C) only (d) (B) and (E) only

**Imp.** Order of CFSE of Coordination Compounds

**90** The correct order of the following complexes in terms of their crystal field stabilisation energies is *[2025, 22 Jan Shift-II]*

(a) $[Co(NH_3)_4]^{2+} < [Co(NH_3)_6]^{2+} < [Co(NH_3)_6]^{3+} < [Co(en)_3]^{3+}$

(b) $[Co(en)_3]^{3+} < [Co(NH_3)_6]^{3+} < [Co(NH_3)_6]^{2+} < [Co(NH_3)_4]^{2+}$

(c) $[Co(NH_3)_6]^{2+} < [Co(NH_3)_6]^{3+} < [Co(NH_3)_4]^{2+} < [Co(en)_3]^{3+}$

(d) $[Co(NH_3)_4]^{2+} < [Co(NH_3)_6]^{2+} < [Co(en)_3]^{3+} < [Co(NH_3)_6]^{3+}$

**91** The correct increasing order of stability of the complexes based on $\Delta_o$ value is *[2025, 29 Jan Shift-I]*

I. $[Mn(CN)_6]^{3-}$ II. $[Co(CN)_6]^{4-}$

III. $[Fe(CN)_6]^{4-}$ IV. $[Fe(CN)_6]^{3-}$

(a) I < II < IV < III
(b) II < III < I < IV
(c) IV < III < II < I
(d) III < II < IV < I

**92** Match List-I with List-II. *[2025, 29 Jan Shift-I]*

| | List-I (Complex) | | List-II (Hybridisation and Magnetic characters) |
|---|---|---|---|
| (A) | $[MnBr_4]^{2-}$ | I. | $d^2sp^3$ and diamagnetic |
| (B) | $[FeF_6]^{3-}$ | II. | $sp^3d^2$ and paramagnetic |
| (C) | $[Co(C_2O_4)_3]^{3-}$ | III. | $sp^3$ and diamagnetic |
| (D) | $[Ni(CO)_4]$ | IV. | $sp^3$ and paramagnetic |

Choose the correct answer from the options given below

(a) (A)-(IV), (B)-(II), (C)-(I), (D)-(III)
(b) (A)-(III), (B)-(II), (C)-(I), (D)-(IV)
(c) (A)-(IV), (B)-(I), (C)-(II), (D)-(III)
(d) (A)-(III), (B)-(I), (C)-(II), (D)-(IV)

**93** Match List-I with List-II.

| | List-I (Complex ion) | | List-II (Spin only magnetic moment in B.M.) |
|---|---|---|---|
| A. | $[Cr(NH_3)_6]^{3+}$ | I. | 4.90 |
| B. | $[NiCl_4]^{2-}$ | II. | 3.87 |
| C. | $[CoF_6]^{3-}$ | III. | 0.0 |
| D. | $[Ni(CN_4)]^{2-}$ | IV. | 2.83 |

Choose the correct answer from the options given below: *[2024, 8 April Shift-II]*

(a) A-I, B-IV, C-II, D-III
(b) A-II, B-III, C-I, D-IV
(c) A-II, B-IV, C-I, D-III
(d) A-IV, B-III, C-I, D-II

**94** Match List-I with List-II

| | List-I | | List-II |
|---|---|---|---|
| A. | $K_2[Ni(CN)_4]$ | I. | $sp^3$ |
| B. | $[Ni(CO)_4]$ | II. | $sp^3d^2$ |
| C. | $[Co(NH_3)_6]Cl_3$ | III. | $dsp^2$ |
| D. | $Na_3[CoF_6]$ | IV. | $d^2sp^3$ |

Choose the correct answer from the options given below: *[2024, 9 April Shift-II]*

(a) A-III, B-I, C-II, D-IV (b) A-III, B-I, C-IV, D-II
(c) A-I, B-III, C-II, D-IV (d) A-III, B-II, C-IV, D-I

**95** Number of complexes with even number of electrons in $t_{2g}$ orbitals is $[Fe(H_2O)_6]^{2+}$, $[Co(H_2O)_6]^{2+}$, $[Co(H_2O)_6]^{3+}$, $[Cu(H_2O)_6]^{2+}$, $[Cr(H_2O)_6]^{2+}$ *[2024, 8 April Shift-I]*

(a) 5 (b) 3
(c) 1 (d) 2

**96** Match List-I with List-II *[2024, 6 April Shift-II]*

| | List-I (Tetrahedral complex) | | List-II (Electronic configuration) |
|---|---|---|---|
| A. | $TiCl_4$ | I. | $e^2, t_2^0$ |
| B. | $[FeO_4]^{2-}$ | II. | $e^4, t_2^3$ |
| C. | $[FeCl_4]^-$ | III. | $e^0, t_2^0$ |
| D. | $[CoCl_4]^{2-}$ | IV. | $e^2, t_2^3$ |

Choose the correct answer from the options given below.

| | A | B | C | D |
|---|---|---|---|---|
| (a) | I | III | IV | II |
| (b) | III | IV | II | I |
| (c) | III | I | IV | II |
| (d) | IV | III | I | II |

**97** Consider the following complexes. *[2024, 6 April Shift-I]*

$[CoCl(NH_3)_5]^{2+}$ (*A*), $[Co(CN)_6]^{3-}$ (*B*),

$[Co(NH_3)_5(H_2O)]^{3+}$ (*C*), $[Cu(H_2O)_4]^{2+}$ (*D*)

The correct order of *A, B, C* and *D* in terms of wave number of light absorbed is :

(a) $B < C < A < D$ (b) $D < A < C < B$
(c) $C < D < A < B$ (d) $A < C < B < D$

**98** The number of complexes from the following with no electrons in the $t_2$ orbital is .........

$TiCl_4$, $[MnO_4]^-$, $[FeO_4]^{2-}$, $[FeCl_4]^-$, $[CoCl_4]^{2-}$

*[2024, 5 April Shift-II]*

(a) 3 (b) 1 (c) 4 (d) 2

**99** Which one of the following complexes will exhibit the least paramagnetic behaviour?

[Atomic number, Cr = 24, Mn = 25, Fe = 26, Co = 27]

*[2024, 5 April Shift-I]*

(a) $[Mn(H_2O)_6]^{2+}$ (b) $[Fe(H_2O)_6]^{2+}$
(c) $[Co(H_2O)_6]^{2+}$ (d) $[Cr(H_2O)_6]^{2+}$

**100** Match List-I with List-II. *[2024, 31 Jan Shift-II]*

| List-I (Complex ion) | | List-II (Electronic configuration) | |
|---|---|---|---|
| A. | $[Cr(H_2O)_6]^{3+}$ | I. | $t_{2g}^2 e_g^0$ |
| B. | $[Fe(H_2O)_6]^{3+}$ | II. | $t_{2g}^3 e_g^0$ |
| C. | $[Ni(H_2O)_6]^{2+}$ | III. | $t_{2g}^3 e_g^2$ |
| D. | $[V(H_2O)_6]^{3+}$ | IV. | $t_{2g}^6 e_g^2$ |

Choose the correct answer from the options given below.

(a) A-IV, B-III, C-I, D-II (b) A-III, B-II, C-IV, D-I
(c) A-IV, B-I, C-II, D-III (d) A-II, B-III, C-IV, D-I

**101** The correct statements from following are: *[2024, 31 Jan Shift-I]*

A. The strength of anionic ligands can be explained by crystal field theory.
B. Valence bond theory does not give a quantitative interpretation of kinetic stability of coordination compounds.
C. The hybridisation involved in formation of $[Ni(CN)_4]^{2-}$ complex is $dsp^2$.
D. The number of possible isomer(s) of *cis*-$[PtCl_2(en)_2]^{2+}$ is one

Choose the correct answer from the options given below:

(a) B, D only (b) B, C only
(c) A, D only (d) A, C only

**102** Consider the following complex ions

$P = [FeF_6]^{3-}$, $Q = [V(H_2O)_6]^{2+}$, $R = [Fe(H_2O)_6]^{2+}$

The correct order of the complex ions, according to their spin only magnetic moment values (in BM) is *[2024, 27 Jan Shift-I]*

(a) $R < P < Q$ (b) $R < Q < P$
(c) $Q < R < P$ (d) $Q < P < R$

**103** Match List-I with List-II.

| List-I (Complex) | | List-II (Colour) | |
|---|---|---|---|
| A. | $Mg(NH_4)PO_4$ | I. | brown |
| B. | $K_3[Co(NO_2)_6]$ | II. | white |
| C. | $MnO(OH)_2$ | III. | yellow |
| D. | $Fe_4[Fe(CN)_6]_3$ | IV. | blue |

Choose the correct answer from the options given below. *[2023, 11 April Shift-II]*

(a) A-III, B-IV, C-II, D-I (b) A-II, B-III, C-IV, D-I
(c) A-II, B-IV, C-I, D-III (d) A-II, B-III, C-I, D-IV

**104** The complex with highest magnitude of crystal field splitting energy ($\Delta_0$) is *[2023, 15 April Shift-I]*

(a) $[Ti(OH_2)_6]^{3+}$ (b) $[Cr(OH_2)_6]^{3+}$
(c) $[Fe(OH_2)_6]^{3+}$ (d) $[Mn(OH_2)_6]^{3+}$

**105** Which of the following complexes will exhibit maximum attraction to an applied magnetic field? *[2023, 13 April Shift-II]*

(a) $[Co(H_2O)_6]^{2+}$ (b) $[Co(en)_3]^{3+}$
(c) $[Ni(H_2O)_6]^{2+}$ (d) $[Zn(H_2O)_6]^{2+}$

**106** Match List-I with List-II.

| List-I (Complex) | | List-II (CFSE ($\Delta_0$)) | |
|---|---|---|---|
| A. | $[Cu(NH_3)_6]^{2+}$ | I. | −0.6 |
| B. | $[Ti(H_2O)_6]^{3+}$ | II. | −2.0 |
| C. | $[Fe(CN)_6]^{3-}$ | III. | −1.2 |
| D. | $[NiF_6]^{4-}$ | IV. | −0.4 |

Choose the correct answer from the options given below. *[2023, 12 April Shift-I]*

(a) A-I, B-II, C-IV, D-III (b) A-II, B-III, C-I, D-IV
(c) A-I, B-IV, C-II, D-III (d) A-III, B-IV, C-I, D-II

**107** Match List-I with List-II.

| List-I (Complex) | | List-II (Crystal field splitting energy ($\Delta_0$)) | |
|---|---|---|---|
| A. | $[Ti(H_2O)_6]^{2+}$ | I. | −1.2 |
| B. | $[V(H_2O)_6]^{2+}$ | II. | −0.6 |
| C. | $[Mn(H_2O)_6]^{3+}$ | III. | 0 |
| D. | $[Fe(H_2O)_6]^{3+}$ | IV. | −0.8 |

Choose the correct answer from the options given below *[2023, 10 April Shift-II]*

(a) A-II, B-IV, C-I, D-III
(b) A-IV, B-I, C-III, D-II
(c) A-IV, B-I, C-II, D-III
(d) A-II, B-IV, C-III, D-I

**108** The correct order of the number of unpaired electrons in the given complexes is

A. $[Fe(CN)_6]^{3-}$ B. $[FeF_6]^{3-}$
C. $[CoF_6]^{3-}$ D. $[Cr(oxalate)_3]^{3-}$
E. $[Ni(CO)_4]$

Choose the correct answer from the options given below : *[2023, 10 April Shift-II]*

(a) E < A < B < D < C
(b) A < E < C < B < D
(c) A < E < D < C < B
(d) E < A < D < C < B

**109** Match List-I with List-II.

| List-I (Coordination complex) | | List-II (Number of unpaired electrons) |
|---|---|---|
| A. $[Cr(CN)_6]^{3-}$ | I. | 0 |
| B. $[Fe(H_2O)_6]^{2+}$ | II. | 3 |
| C. $[Co(NH_3)_6]^{3+}$ | III. | 2 |
| D. $[Ni(NH_3)_6]^{2+}$ | IV. | 4 |

Choose the correct answer from the options given below: *[2023, 8 April Shift-II]*

(a) A-II, B-IV, C-I, D-III
(b) A-III, B-IV, C-I, D-II
(c) A-II, B-I, C-IV, D-III
(d) A-IV, B-III, C-II, D-I

**110** Match List-I with List-II. *[2023, 30 Jan, Shift-II]*

| List-I (Complexes) | | List-II (Hybridisation) |
|---|---|---|
| A. $[Ni(CO)_4]$ | I. | $sp^3$ |
| B. $[Cu(NH_3)_4]^{2+}$ | II. | $dsp^2$ |
| C. $[Fe(NH_3)_6]^{2+}$ | III. | $sp^3d^2$ |
| D. $[Fe(H_2O)_6]^{2+}$ | IV. | $d^2sp^3$ |

(a) A-II, B-I, C-III, D-IV
(b) A-II, B-I, C-IV, D-III
(c) A-I, B-II, C-IV, D-III
(d) A-I, B-II, C-III, D-IV

**111** Match List-I with List-II.

| List-I (Coordination entity) | | List-II (Wavelength of light absorbed in nm) |
|---|---|---|
| A. $[CoCl(NH_3)_5]^{2+}$ | I. | 310 |
| B. $[Co(NH_3)_6]^{3+}$ | II. | 475 |
| C. $[Co(CN)_6]^{3-}$ | III. | 535 |
| D. $[Cu(H_2O)_4]^{2+}$ | IV. | 600 |

Choose the correct answer from the options given below. *[2023, 25 Jan Shift-II]*

(a) A-II, B-III, C-IV, D-I
(b) A-III, B-II, C-I, D-IV
(c) A-IV, B-I, C-III, D-II
(d) A-III, B-I, C-II, D-IV

**112** Match List-I with List-II.

| List-I (Complex) | | List-II (Hybridisation) |
|---|---|---|
| A. $Ni(CO)_4$ | I. | $sp^3$ |
| B. $[Ni(CN)_4]^{2-}$ | II. | $sp^3d^2$ |
| C. $[Co(CN)_6]^{3-}$ | III. | $d^2sp^3$ |
| D. $[CoF_6]^{3-}$ | IV. | $dsp^2$ |

Choose the correct answer from the options given below *[2022, 28 July Shift-II]*

(a) A-IV, B-I, C-III, D-II
(b) A-I, B-IV, C-III, D-II
(c) A-I, B-IV, C-II, D-III
(d) A-IV, B-I, C-II, D-III

**113** The metal complex that is diamagnetic is (Atomic number Fe, 26; Cu, 29) *[2022, 26 July Shift-II]*

(a) $K_3[Cu(CN)_4]$ (b) $K_2[Cu(CN)_4]$
(c) $K_3[Fe(CN)_6]$ (d) $K_4[FeCl_6]$

**114** Arrange the following metal complex/compounds in the increasing order of spin only magnetic moment. Presume all the three, high spin system. (Atomic numbers, Ce = 58, Gd = 64 and Eu = 63.)

A. $(NH_4)_2[Ce(NO_3)_6]$
B. $Gd(NO_3)_3$ and
C. $Eu(NO_3)_3$ *[2021, 16 March Shift-II]*

(a) (B) < (A) < (C) (b) (C) < (A) < (B)
(c) (A) < (B) < (C) (d) (A) < (C) < (B)

**115** In which of the following order the given complex ions are arranged correctly with respect to their decreasing spin only magnetic moment? *[2021, 25 Feb Shift-II]*

(i) $[FeF_6]^{3-}$ (ii) $[Co(NH_3)_6]^{3+}$
(iii) $[NiCl_4]^{2-}$ (iv) $[Cu(NH_3)_4]^{2+}$

(a) (i) > (iii) > (iv) > (ii) (b) (ii) > (iii) > (i) > (iv)
(c) (iii) > (iv) > (ii) > (i) (d) (ii) > (i) > (iii) > (iv)

**Imp.** Absorption Spectra of Complex in Relation with Wavelength

**116** Simplified absorption spectra of three complexes [(i), (ii) and (iii)] of $M^{n+}$ ion are provided below; their $\lambda_{max}$ values are marked as $A$, $B$ and $C$ respectively. The correct match between the complexes and their $\lambda_{max}$ values is

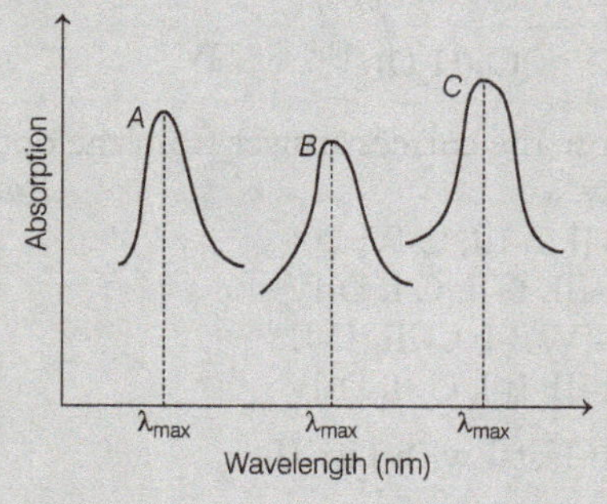


(i) $[M(NCS)_6]^{(-6+n)}$ (ii) $[MF_6]^{(-6+n)}$
(iii) $[M(NH_3)_6]^{n+}$ **[2020, 2 Sep Shift-II]**
(a) $A$-(iii), $B$-(i), $C$-(ii) (b) $A$-(ii), $B$-(i), $C$-(iii)
(c) $A$-(ii), $B$-(iii), $C$-(i) (d) $A$-(i), $B$-(ii), $C$-(iii)

**117** $[Pd(F)(Cl)(Br)(I)]^{2-}$ has $n$ number of geometrical isomers. Then, the spin-only magnetic moment and crystal field stabilisation energy [CFSE] of $[Fe(CN)_6]^{n-6}$ respectively, are **[2020, 9 Jan Shift-II]**
[**Note** Ignore the pairing energy]
(a) 1.73 BM and $-2.0\Delta_o$ (b) 2.84 BM and $-1.6\Delta_o$
(c) 0 BM and $-2.4\Delta_o$ (d) 5.92 BM and 0

**118** The correct order of the spin only magnetic moments of the following complexes is **[2020, 9 Jan Shift-II]**
(I) $[Cr(H_2O)_6]Br_2$ (II) $Na_4[Fe(CN)_6]$
(III) $Na_3[Fe(C_2O_4)_3](\Delta_0 > P)$
(IV) $(Et_4N)_2[CoCl_4]$
(a) (II) ≈ (I) > (IV) > (III) (b) (I) > (IV) > (III) > (II)
(c) (III) > (I) > (IV) > (II) (d) (III) > (I) > (II) > (IV)

**119** The complex ion that will lose its crystal field stabilisation energy upon oxidation of its metal to +3 state is **[2019, 12 April Shift-I]**

(Phen = 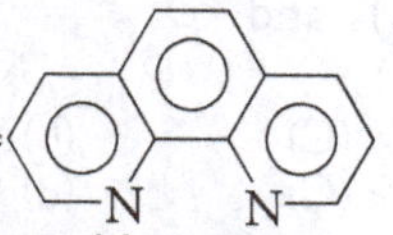


Ignore pairing energy

(a) $[Co(phen)_3]^{2+}$ (b) $[Ni(phen)_3]^{2+}$
(c) $[Zn(phen)_3]^{2+}$ (d) $[Fe(phen)_3]^{2+}$

**120** The difference in the number of unpaired electrons of a metal ion in its high-spin and low-spin octahedral complexes is two. The metal ion is: **[2019, 10 Jan Shift-II]**
(a) $Ni^{2+}$ (b) $Fe^{2+}$ (c) $Co^{2+}$ (d) $Mn^{2+}$

## Numerical Value Type Questions

**Imp.** Calculation of Spin Only Magnetic Moment

**121** Consider the following low-spin complexes $K_3[Co(NO_2)_6]$, $K_4[Fe(CN)_6]$, $K_3[Fe(CN)_6]$, $Cu_2[Fe(CN)_6]$ and $Zn_2[Fe(CN)_6]$.
The sum of the spin-only magnetic moment values of complexes having yellow colour is ...... BM (answer in nearest integer). **[2025, 29 Jan Shift-II]**

**122** The number of paramagnetic complex species among $[Co(NH_3)_6]^{3+}$, $[Co(C_2O_4)_3]^{3-}$, $[MnCl_6]^{3-}$, $[Mn(CN)_6]^{3-}$, $[CoF_6]^{3-}$, $[Fe(CN)_6]^{3-}$ and $[FeF_6]^{3-}$ with same number of unpaired electrons is ............... . **[2025, 7 April Shift-II]**

**123** The number of paramagnetic complexes among $[FeF_6]^{3-}$, $[Fe(CN)_6]^{3-}$, $[Mn(CN)_6]^{3-}$, $[Co(C_2O_4)_3]^{3-}$, $[MnCl_6]^{3-}$, and $[CoF_6]^{3-}$, which involved $d^2sp^3$ hybridisation is .............. **[2025, 7 April Shift-I]**

**124** The spin-only magnetic moment value of $M^{n+}$ ion formed among Ni, Zn, Mn and Cu that has the least enthalpy of atomisation is ......... . (in nearest integer)
Here $n$ is equal to the number of diamagnetic complexes among $K_2[NiCl_4]$, $[Zn(H_2O)_6]Cl_2$, $K_3[Mn(CN)_6]$ and $[Cu(PPh_3)_3I]$. **[2025, 2 April Shift-II]**

**125** A transition metal ($M$) among Mn, Cr, Co and Fe has the highest standard electrode potential ($M^{3+}/M^{2+}$). It forms a metal complex of the type $[M(CN)_6]^{4-}$. The number of electrons present in the $e_g$ orbital of the complex is ........ **[2025, 2 April Shift-I]**

**126** The complex of $Ni^{2+}$ ion and dimethyl glyoxime contains ........ number of hydrogen (H) atoms. **[2025, 22 Jan Shift-II]**

**127** The spin only magnetic moment for $[Ni(NH_3)_6]^{2+}$ is ..........$\times 10^{-1}$ BM.
(Given, atomic number of Ni = 28) **[2024, 31 Jan Shift-I]**

**128** The spin only magnetic moment value of square planar complex $[Pt(NH_3)_2Cl(NH_2CH_3)]Cl$ is ........ B.M. (Nearest integer)
(Given, atomic number for Pt = 78) **[2024, 27 Jan Shift-II]**

**129** The homoleptic and octahedral complex of $Co^{2+}$ and $H_2O$ has ........... unpaired electron(s) in the $t_{2g}$ set of orbitals. **[2023, 15 April Shift-I]**

**130** In potassium ferrocyanide, there are ......... pairs of electrons in the $t_{2g}$ set of orbitals. **[2023, 10 April Shift-I]**

**131** The observed magnetic moment of the complex $[Mn(NCS)_6]^{x-}$ is 6.06 BM. The numerical value of $x$ is ........ **[2023, 8 April Shift-II]**

**132** If the CFSE of $[Ti(H_2O)_6]^{3+}$ is $-96.0$ kJ/mol, this complex will absorb maximum at wavelength ......... nm. (Nearest integer) **[2023, 31 Jan Shift-II]**
[Assume Planck's constant $(h) = 6.4 \times 10^{-34}$ Js, speed of light $(c) = 3.0 \times 10^{8}$ m/s and Avogadro's constant $(N_A) = 6 \times 10^{23}$/mol]

**133** The number of paramagnetic species from the following is ........ **[2023, 25 Jan Shift-I]**
$[Ni(CN)_4]^{2-}$, $[Ni(CO)_4]$, $[NiCl_4]^{2-}$,
$[Fe(CN)_6]^{4-}$, $[Cu(NH_3)_4]^{2+}$
$[Fe(CN)_6]^{3-}$ and $[Fe(H_2O)_6]^{2+}$

**134** $[Fe(CN)_6]^{3-}$ should be an inner orbital complex. Ignoring the pairing energy, the value of crystal field stabilisation energy for this complex is (–) ......... $\Delta_o$. (Nearest integer) **[2022, 29 July Shift-I]**

**135** The difference between spin only magnetic moment value of $[Co(H_2O)_6]Cl_2$ and $[Cr(H_2O)_6]Cl_3$ is ............ . **[2022, 26 July Shift-I]**

**136** Consider the following metal complexes
$[Co(NH_3)_6]^{3+}$; $[CoCl(NH_3)_5]^{2+}$
$[Co(CN)_6]^{3-}$; $[Co(NH_3)_5(H_2O)]^{3+}$
The spin-only magnetic moment value of the complex that absorbs light with shortest wavelength is ........... BM. (Nearest integer) **[2022, 25 July Shift-I]**

**137** Reaction of $[Co(H_2O)_6]^{2+}$ with excess ammonia and in the presence of oxygen results into a diamagnetic product. Number of electrons present in $t_{2g}$-orbitals of the product is ...... . **[2022, 26 June Shift-II]**

**138** Amongst $FeCl_3 \cdot 3H_2O$, $K_3[Fe(CN)_6]$ and $[Co(NH_3)_6]Cl_3$, the spin-only magnetic moment value of the inner-orbital complex that absorbs light at shortest wavelength is ........... BM. [Nearest integer] **[2022, 25 June Shift-II]**

**139** If $[Cu(H_2O)_4]^{2+}$ absorbs a light of wavelength 600 nm for $d$-$d$ transition, then the value of octahedral crystal field splitting energy for $[Cu(H_2O)_6]^{2+}$ will be ... $\times 10^{-21}$ J. [Nearest integer]
(Given, $h = 6.63 \times 10^{-34}$ Js and $c = 3.08 \times 10^{8}$ $ms^{-1}$) **[2022, 25 June Shift-I]**

**140** The total number of unpaired electrons present in $[Co(NH_3)_6]Cl_2$ and $[Co(NH_3)_6]Cl_3$ is ...... . **[2021, 22 July Shift-II]**

**141** An aqueous solution of $NiCl_2$ was heated with excess sodium cyanide in presence of strong oxidising agent to form $[Ni(CN)_6]^{2-}$. The total change in number of unpaired electrons on metal centre is ...... . **[2021, 20 July Shift-II]**

**142** The spin only magnetic moment value for the complex $[Co(CN)_6]^{4-}$ is ...... BM.
[Atomic number of Co = 27] **[2021, 20 July Shift-I]**

**143** The total number of unpaired electrons present in the complex $K_3[Cr(oxalate)_3]$ is .......... . **[2021, 18 March Shift-I]**

**144** $[Ti(H_2O)_6]^{3+}$ absorbs light of wavelength 498 nm during a $d$-$d$ transition. The octahedral splitting energy for the above complex is ........ $\times 10^{-1}$ J (Round off to the nearest intger).
$[h = 6.626 \times 10^{-34}$ Js, $c = 3 \times 10^{8}$ $ms^{-1}]$ **[2021, 16 March Shift-II]**

**145** Considering that $\Delta_0 > P$, the magnetic moment (in BM) of $[Ru(H_2O)_6]^{2+}$ would be ........ . **[2020, 5 Sep Shift-II]**

## Topic 3 Application of Coordination Compounds and Organometallic Compounds

### Level 1

**01** The compound used in the treatment of lead poisoning is **[2019, 12 April Shift-II]**
(a) D-penicillamine (b) desferrioxime-B
(c) *cis*-platin (d) EDTA

**02** The compound that inhibits the growth of tumors is **[2019, 8 April Shift-II]**
(a) *trans*-$[Pt(Cl)_2(NH_3)_2]$
(b) *cis*-$[Pd(Cl)_2(NH_3)_2]$
(c) *cis*-$[Pt(Cl)_2(NH_3)_2]$
(d) *trans*-$[Pd(Cl)_2(NH_3)_2]$

**03** $Mn_2(CO)_{10}$ is an organometallic compound due to the presence of **[2019, 12 Jan Shift-I]**
(a) Mn—C bond
(b) Mn—Mn bond
(c) Mn—O bond
(d) C—O bond

**04** $CH_3$ — Mg — Br is an organo metallic compound due to **[AIEEE 2004]**
(a) Mg—Br bond (b) C—Mg bond
(c) C—Br bond (d) C—H bond

## Level 2

**05** The mismatched combinations are
A. chlorophyll - Co
B. water hardness - EDTA
C. photography -$[Ag(CN)_2]^-$
D. Wilkinson catalyst - $[(Ph_3P)_3RhCl]$
E. chelating ligand -D-penicillamine
Choose the correct answer from the options given below *[2023, 13 April Shift-I]*
(a) A and C only (b) D and E only
(c) A, C, and E only (d) A and E only

**06** The complex that dissolves in water is *[2023, 11 April Shift-I]*
(a) $[Fe_3(OH)_2(OAc)_6]Cl$ (b) $Fe_4[Fe(CN)_6]_3$
(c) $K_3[Co(NO_2)_6]$ (d) $(NH_4)_3[As(Mo_3O_{10})_4]$

## Level 3

**07** Choose the correct statements from the following. *[2024, 30 Jan Shift-I]*
(A) Ethane-1, 2-diamine is a chelating ligand.
(B) Metallic aluminium is produced by electrolysis of aluminium oxide in presence of cryolite.
(C) Cyanide ion is used as ligand for leaching of silver.
(D) Phosphine act as a ligand in Wilkinson catalyst.
(E) The stability constants of $Ca^{2+}$ and $Mg^{2+}$ are similar with EDTA complexes.

Choose the correct answer from the options given below.
(a) (A), (B), (C) only (b) (B), (C), (E) only
(c) (C), (D), (E) only (d) (A), (D), (E) only

**08** In which one of the following metal carbonyls, CO forms a bridge between metal atoms? *[2024, 29 Jan Shift-I]*
(a) $[Mn_2(CO)_{10}]$ (b) $[Ru_3(CO)_{12}]$
(c) $[Os_3(CO)_{12}]$ (d) $[Co_2(CO)_8]$

**09** The number of bridging CO ligand(s) and Co—Co bond(s) in $Co_2(CO)_8$, respectively are *[2019, 11 Jan Shift-II]*
(a) 2 and 0 (b) 0 and 2
(c) 4 and 0 (d) 2 and 1

**10** $[Co_2(CO)_8]$ displays *[JEE Main 2017]*
(a) one Co—Co bond, four terminal CO and four bridging CO
(b) one Co—Co bond, six terminal CO and two bridging CO
(c) no Co—Co bond, four terminal CO and four bridging CO
(d) no Co—Co bond, six terminal CO and two bridging CO

**11** Coordination compounds have great importance in biological systems. In this context which of the following statements is incorrect? *[AIEEE 2004]*
(a) Cyanocobalamin is $B_{12}$ and contains cobalt.
(b) Haemoglobin is the red pigment of blood and contains iron.
(c) Chlorophylls are green pigments in plants and contain calcium.
(d) Carboxpeptidase-A is an enzyme and contains zinc.

## Numerical Value Type Questions

**12** In the cobalt-carbonyl complex $[Co_2(CO)_8]$, number of Co — Co bonds is "$X$" and terminal CO ligands is "$Y$". $X + Y =$ ...... . *[2022, 24 June Shift-I]*

**13** Number of complexes which will exhibit synergic bonding amongst, $[Cr(CO)_6]$, $[Mn(CO)_5]$ and $[Mn_2(CO)_{10}]$ is ...... *[2022, 28 June Shift-I]*

**14** The spin only magnetic moment of the complex present in Fehling's reagent is ........... BM. (nearest integer). *[2022, 27 July Shift-II]*

**15** Number of bridging CO ligands in $[Mn_2(CO)_{10}]$ is ............ . *[2021, 26 Feb Shift-I]*

**Imp.** Stability Constant of the Coordination Compounds

**16** The stepwise formation of $[Cu(NH_3)_4]^{2+}$ is given below

$$Cu^{2+} + NH_3 \overset{K_1}{\rightleftharpoons} [Cu(NH_3)]^{2+}$$
$$[Cu(NH_3)]^{2+} + NH_3 \overset{K_2}{\rightleftharpoons} [Cu(NH_3)_2]^{2+}$$
$$[Cu(NH_3)_2]^{2+} + NH_3 \overset{K_3}{\rightleftharpoons} [Cu(NH_3)]^{2+}$$
$$[Cu(NH_3)_3]^{2+} + NH_3 \overset{K_4}{\rightleftharpoons} [Cu(NH_3)_4]^{2+}$$

The value of stability constants $K_1$, $K_2$, $K_3$ and $K_4$ are $10^4$, $1.58 \times 10^3$, $5 \times 10^2$ and $10^2$ respectively. The overall equilibrium constants for dissociation of $[Cu(NH_3)_4]^{2+}]$ is $x \times 10^{-12}$. The value of $x$ is ........ . (Rounded off to the nearest integer) *[2021, 24 Feb Shift-I]*

# Key Index

## Topic 1

| | | | | | | | | | | | | | | | | | | | |
|---|---|---|---|---|---|---|---|---|---|---|---|---|---|---|---|---|---|---|---|
| **1** | (b) | **2** | (c) | **3** | (a) | **4** | (d) | **5** | (a) | **6** | (b) | **7** | (c) | **8** | (c) | **9** | (b) | **10** | (d) |
| **11** | (c) | **12** | (b) | **13** | (b) | **14** | (c) | **15** | (a) | **16** | (d) | **17** | (a) | **18** | (d) | **19** | (b) | **20** | (b) |
| **21** | (d) | **22** | (c) | **23** | (c) | **24** | (a) | **25** | (c) | **26** | (b) | **27** | (b) | **28** | (a) | **29** | (a) | **30** | (c) |
| **31** | (a) | **32** | (d) | **33** | (b) | **34** | (c) | **35** | (a) | **36** | (d) | **37** | (b) | **38** | (c) | **39** | (c) | **40** | (b) |
| **41** | (b) | **42** | (d) | **43** | (b) | **44** | (d) | **45** | (b) | **46** | (c) | **47** | (a) | **48** | (b) | **49** | (a) | **50** | (a) |
| **51** | (d) | **52** | (c) | **53** | (a) | **54** | (c) | **55** | (a) | **56** | (a) | **57** | (c) | **58** | (a) | **59** | (b) | **60** | (c) |
| **61** | (b) | **62** | (c) | **63** | (b) | **64** | (d) | **65** | (b) | **66** | (a) | **67** | (d) | **68** | (2) | **69** | (2) | **70** | (3) |
| **71** | (4) | **72** | (5) | **73** | (9) | **74** | (5) | **75** | (1) | **76** | (2) | **77** | (6) | **78** | (2) | **79** | (2) | **80** | (6) |

## Topic 2

| | | | | | | | | | | | | | | | | | | | |
|---|---|---|---|---|---|---|---|---|---|---|---|---|---|---|---|---|---|---|---|
| **1** | (b) | **2** | (c) | **3** | (a) | **4** | (a) | **5** | (b) | **6** | (a) | **7** | (d) | **8** | (c) | **9** | (c) | **10** | (c) |
| **11** | (c) | **12** | (a) | **13** | (a) | **14** | (b) | **15** | (d) | **16** | (a) | **17** | (d) | **18** | (b) | **19** | (c) | **20** | (a) |
| **21** | (d) | **22** | (d) | **23** | (b) | **24** | (d) | **25** | (a) | **26** | (c) | **27** | (d) | **28** | (c) | **29** | (d) | **30** | (b) |
| **31** | (d) | **32** | (d) | **33** | (b) | **34** | (a) | **35** | (c) | **36** | (c) | **37** | (b) | **38** | (b) | **39** | (b) | **40** | (b) |
| **41** | (c) | **42** | (c) | **43** | (a) | **44** | (a) | **45** | (d) | **46** | (a) | **47** | (a) | **48** | (c) | **49** | (b) | **50** | (b) |
| **51** | (d) | **52** | (a) | **53** | (d) | **54** | (d) | **55** | (a) | **56** | (c) | **57** | (a) | **58** | (b) | **59** | (d) | **60** | (b) |
| **61** | (d) | **62** | (d) | **63** | (d) | **64** | (a) | **65** | (b) | **66** | (b) | **67** | (c) | **68** | (b) | **69** | (d) | **70** | (c) |
| **71** | (c) | **72** | (a) | **73** | (a) | **74** | (b) | **75** | (c) | **76** | (c) | **77** | (b) | **78** | (a) | **79** | (a) | **80** | (a) |
| **81** | (a) | **82** | (c) | **83** | (c) | **84** | (a) | **85** | (d) | **86** | (b) | **87** | (a) | **88** | (d) | **89** | (a) | **90** | (a) |
| **91** | (a) | **92** | (a) | **93** | (c) | **94** | (b) | **95** | (b) | **96** | (c) | **97** | (b) | **98** | (a) | **99** | (c) | **100** | (d) |
| **101** | (b) | **102** | (c) | **103** | (d) | **104** | (b) | **105** | (a) | **106** | (c) | **107** | (c) | **108** | (d) | **109** | (a) | **110** | (c) |
| **111** | (b) | **112** | (b) | **113** | (a) | **114** | (d) | **115** | (a) | **116** | (a) | **117** | (a) | **118** | (b) | **119** | (d) | **120** | (c) |
| **121** | (0) | **122** | (2) | **123** | (2) | **124** | (0) | **125** | (1) | **126** | (14) | **127** | (28) | **128** | (0) | **129** | (1) | **130** | (3) |
| **131** | (4) | **132** | (480) | **133** | (4) | **134** | (2) | **135** | (0) | **136** | (0) | **137** | (6) | **138** | (2) | **139** | (766) | **140** | (1) |
| **141** | (2) | **142** | (2) | **143** | (3) | **144** | (4) | **145** | (0) | | | | | | | | | | |

## Topic 3

| | | | | | | | | | | | | | | | | | | | |
|---|---|---|---|---|---|---|---|---|---|---|---|---|---|---|---|---|---|---|---|
| **1** | (d) | **2** | (c) | **3** | (a) | **4** | (b) | **5** | (a) | **6** | (a) | **7** | (a) | **8** | (d) | **9** | (d) | **10** | (b) |
| **11** | (c) | **12** | (7) | **13** | (3) | **14** | (2) | **15** | (0) | **16** | (1) | | | | | | | | |

# Detailed **Explanations**

## Topic 1

**01.** *(b)* The correct IUPAC name of $[PtBr_2(PMe_3)_2]$ is dibromo*bis* (trimethylphosphine) platinum (II).

**02.** *(c)* Compound $= K_2MnO_4$

Oxidation state of Mn,

$0 = 2\times1 + x - 4\times2$

$x = +6$

Therefore, IUPAC name

= Potassium tetraoxidomanganate (VI)

**03.** *(a)* $C_2O_4^{2-}$ is an oxalato ion which is a bidentate ligand. In this complex, let the oxidation state of cobalt is $\begin{bmatrix} +1\times3 + x + (-2)\times3 = 0 \\ x = +3 \end{bmatrix}$. So, the IUPAC name of $K_3[Co(C_2O_4)_3]$ is potassium trioxalatocobaltate (III).

**04.** *(d)* The correct structure *trans* $[Ni\,Br_2(PPh_3)_2]$ is

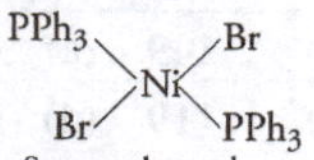


Square planar shape

and of meridional $[Co(NH_3)_3(NO_2)_3]$ is

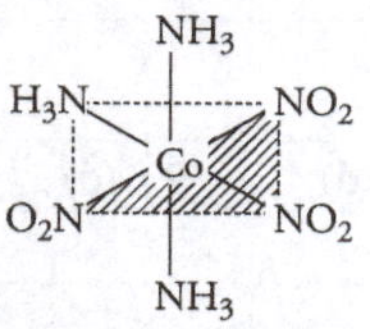


Meridional (mer)
[*trans*]

> **Concept Enhancer**
>
> Meridional isomer exist for those complexes having $[Ma_3b_3]$ type possessing octahedral geometry. Its other isomer is facial. The facial isomer of $[Co(NH_3)_3(NO_2)_3]$ can be drawn as
>
> 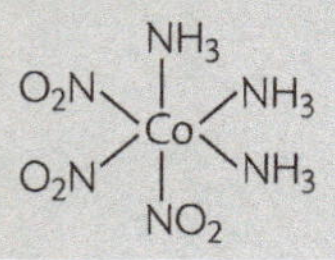
> 

**05.** *(a)* Homoleptic complex are those complexes which contain only one type of ligand. Thus, $[Co(NH_3)_6]Cl_3$ is an examples of it.

**06.** *(b)* $[CoCl(NH_3)_5]^+ + 5H^+ \longrightarrow Co^{2+} + 5NH_4^+ + Cl^-$

The above equation is correct because amine complexes decomposes under acidic medium. Thus, the complex $[CoCl(NH_3)_5]$ decomposes under acidic conditions to give ammonium ions.

**07.** *(c)*

| Complex | Isomerism shown |
|---|---|
| (a) $[Co(en)_3]^{3+}$ | Optical only |
| (b) $[Ni(NH_3)_5Br]^+$ | No geometrical isomer |
| (c) $[Co(NH_3)_2(en)_2]^{3+}$ | *cis* and *trans* |
| (d) $[Cr(NH_3)_4(en)]^{3+}$ | No geometrical isomer |

**08.** *(c)* EDTA (Ethylenediamine tetraacetic acid)

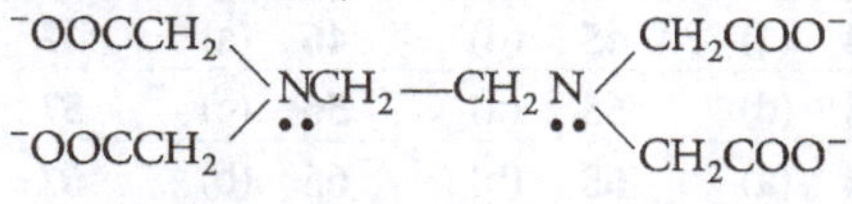


It is hexadentate (6 electron pairs). That's why for octahedral complex, only one EDTA is required. To form complex with $Ca^{2+}$ ion.

**09.** *(b)* Let oxidation no. of Ni $= x$

So, $1\times4 + x + (-1)\times4 = 0$

$4 + x - 4 = 0 \Rightarrow x = 0$

**10.** *(d)* $CrO_3$ is acidic nature.

$Mn_2O_7$ is acidic in nature.

$\therefore \quad X = 2$

The reaction is,

$[Co(H_2NCH_2CH_2NH_2)_3]_2(SO_4)_3 \rightleftharpoons 2[Co(H_2NCH_2CH_2NH_2)_3]^{3+} + 3SO_4^{2-}$

$\therefore$ Primary valency $= 3$

$\therefore \quad X + Y = 5$

**11.** *(c)* Statement I is true but Statement II is false.

Homoleptic complex of type $[Ma_6]$ (where *a* is monodentate ligand) cannot show geometrical as well as optical isomerism.

*cis*-platin and *trans*-platin has formula $[Pt(NH_3)_2Cl_2]$ which is a heteroleptic complex of platinum.

**12.** *(b)* For complex $[CrCl_3(py)_3]$ the stereoisomers are

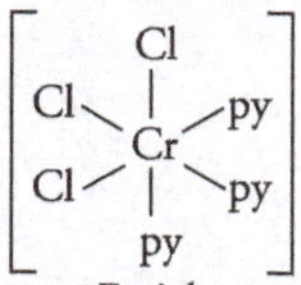


Facial

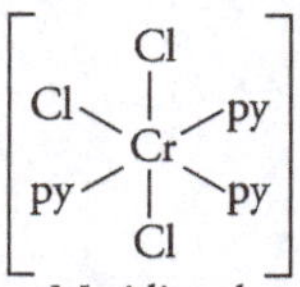


Meridional

So, total 2 stereoisomers.

$[CrCl_3(ox)_2]^{3-}$

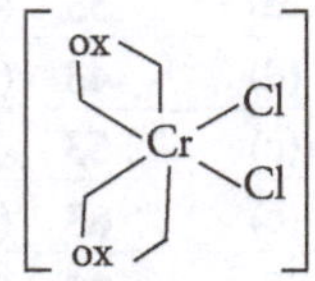


*cis*-isomer

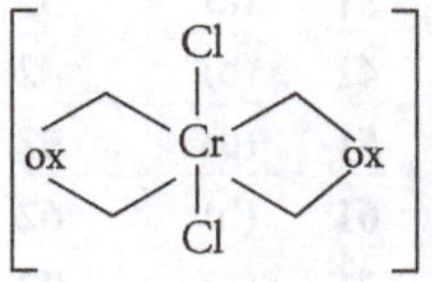


*trans*-isomer

So, geometricaly isomer = 2 (1 *cis* + 1 *trans*)

Optical isomer = 3 (2 optically active + 1 optically inactive). So, total 3 stereoisomer are there.

**13.** *(b)* Complexes of the type $Ma_3b_3$ shows facial-meridional isomerism. So, among the given options, compound given in option (b) i.e., $[Co(NH_3)_3Cl_3]$ is of type $Ma_3b_3$. Thus, it will show facial-meridional isomerism.

> **Time Saver Tech**
>
> Coordination compound of the type $[Ma_3b_3]$ shows facial-meridional isomerism.
>
> So, among the given options only option (b) is of type $[Ma_3b_3]$.

**14.** *(c)* The complete reaction is,

$[Co(NH_3)_5Cl]Cl_2$

$\downarrow$

$\underbrace{[Co(NH_3)_5Cl]^{2+}(aq) + 2Cl^-(aq)}_{\text{3 ions in water}}$

$[Co(NH_3)_5Cl]Cl_2(aq) + 2AgNO_3)(aq) \longrightarrow [Co(NH_3)_5Cl](NO_3)_2(aq) + 2AgCl(s)$

Thus, the structure of the complex is $[Co(NH_3)_5Cl]Cl_2$.

**15.** *(a)* A is not correct but R is correct. The correct form of A is the total number of geometrical isomers shown by $[Co(en)_2Cl_2]^+$ complex ion is two.

> **Concept Enhancer**
>
> Geometrical isomerism arises in heteroleptic complexes with coordination number 4 and 6 like $[MX_2L_2]$, where X and L are unidentate, MABXL (where A, B, X and L are unidentate), $MX_2L_4$ and $[MX_2(L-L)_2]$ where L-L are didentate ligand.
>
> Note that tetrahedral complexes do not show geometrical isomerism. The complex with coordination number 4 are square planar complexes.

**16.** *(d)* $CoCl_2 \cdot nNH_3 \xrightarrow[\text{Excess}]{AgNO_3} 2$ mol of AgCl

It means $2Cl^-$ ions will be present in ionisation sphere. So coordination compound should be $[Co(NH_3)_5Cl]Cl_2$.
Oxidation state of cobalt is +3

$$x = +3, n = 5 \Rightarrow x + n = 8$$

**17.** *(a)* The metal atom present, have $sp^3$ hybridisation. The complex will have tetrahedral structure. Tetrahedral complex does not show geometrical isomerism.
Hence, option (a) is correct.

**18.** *(d)* Homoleptic complex are those complex in which all the ligand attached with central atom are same. Among the given options, only $[Ni(CN)_4]^{2-}$ is a homoleptic complex.
Hence, option (d) is correct.

**19.** *(b)* $[Cr(ox_2)ClBr)]^{3-} = [M(AA)_2bc]$ type gives 2-Geometrical isomers - *cis* and *trans* where the *cis*- form is optically active

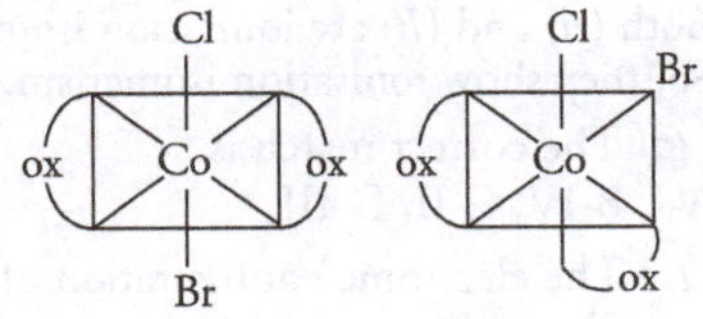


(*trans*) [(+) *cis*]

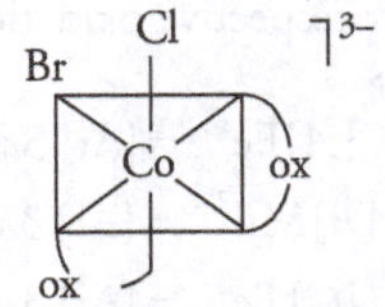


[(–) *cis*]

Thus, the total three stereoisomers for the $[Cr(ox)_2ClBr]^{3-}$ are possible.

**20.** *(b)* $[MA_3B_3]$ type of compound exists as facial and meriodional isomer. Thus $[Co(NH_3)_3(NO_2)_3]$ will exist as meriodional isomer.

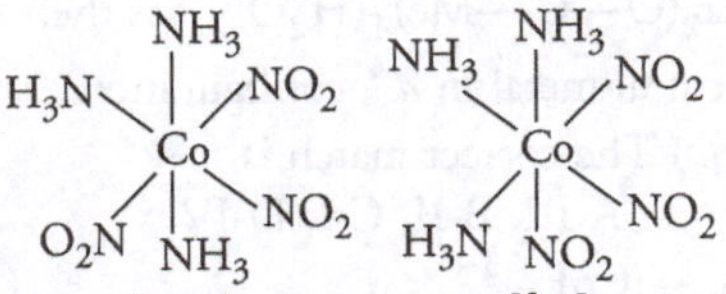


[mer-] [fac-]

**21.** *(d)* Ambidentate ligand has two different donor atoms and either of the two donor atom is attached to the metal during complex formation.

Among the given ligands, only $NCS^-$ and $NO_2^-$ are ambidentate ligand while $C_2O_4^{2-}$, ethylene diamine, $H_2O$ and EDTA are not ambidentate ligand.

**22.** *(c)* The correct option is (c).
The complex $[Co(NH_3)_5NO_2]^{2+}$ shows linkage isomerism. $NO_2$ group is an ambidentate ligand which can bind to a metal atom either through N or O-atom and hence two different isomers are formed.

**23.** *(c)* The examples of double salt are given in A and C only.
$FeSO_4 \cdot (NH_4)_2SO_4 \cdot 6H_2O$ is hydrated ferrous ammonium sulphate known as Mohr's salt.
$K_2SO_4 \cdot Al_2(SO_4)_3 \cdot 24H_2O$, potash alum is a double salt of $K_2SO_4$ and $Al_2(SO_4)_3$.

**24.** *(a)* Let, oxidation state of Co is *x*.
$[Co(NH_3)_5Cl]Cl_2$
$\therefore x + 0 - 1 = +2$ ($NH_3$ = Neutral ligand)
$x = +3$ (Primary valency)
$\therefore$ Coordination number of Co = 6 (Secondary valency)
Hence, primary valency is 3 and secondary valency is 6.

**25.** *(c)* $AgCl + 2NH_3 \longrightarrow \underset{\text{soluble}}{[Ag(NH_3)_2]^+Cl^-}$

Thus, white ppt of AgCl is soluble in $NH_3$ due to the formation of $[Ag(NH_3)_2]Cl$.

**26.** *(b)* $[Co(CN)_5(NC)]^{3-}$ is of $MA_5B$ type complex, which will not show any isomerism and the rest complexes will show geometrical isomerism as compounds of $MA_4B_2$, $M(AA)_2B_2$ and $MA_3B_3$ shows geometrical isomerism.
$\therefore$ Option (b) is correct.

**27.** *(b)* Stability of a complex depends on the chelating effect. Complex $[Co(en)_3]Cl_2$ is most stable complex among the given complex compounds because more number of chelate rings are present in this complex as compare to others.

**28.** *(a)* The correct match is (A)-3, (B)-1, (C)-2, (D)-4.

**29.** *(a)* Total number of possible isomers for $[Pt(en)(NO_2)_2]$ are 3.
These are $[Pt(en)(NO_2)_2]$, $[Pt(en)(ONO)(NO_2)]$ and $[Pt(en)(ONO)_2]$.

**30.** *(c)* In $[Ni(NH_3)_2Cl_2]$, Ni(II) is $sp^3$-hybridised. So, the complex has tetrahedral geometry. So, stereoisomers (geometrical and optical) as well as structural isomerism will not be shown by the complex.
On the other hand, rest of them shows geometrical isomerism.

**31.** *(a)* There are clearly many planes of symmetry in the tetrahedral (*M* is $sp^3$ in $MA_2B_2$) structure and the structure is superimposable on its mirror image.
$\therefore$ It has no optical isomers.
An $MA_2B_2$ complex of $dsp^2$-hybridised metal (*M*) has square planar geometry.

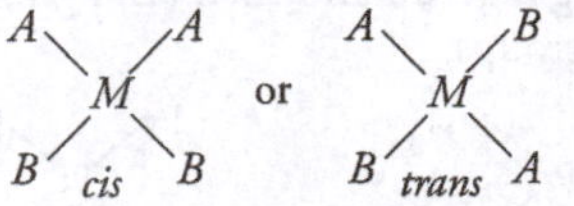


There are two geometrical isomers *cis* and *trans*.
But, since they are planar and therefore they always have a plane of symmetry.
$\therefore$ No optical isomerism also.

**32.** *(d)* The *cis*-$[CrCl_2(ox)_2]^{3-}$ will give non-super imposable mirror image.
So, it is optical active.
Whereas other complexes will give super imposable mirror image hence they are optically inactive.
So, option (d) is correct.

**33.** *(b)* Octahedral coordination entities of the type $[Ma_3b_3]$ shows geometrical isomers: *fac* and *mer* isomers. Among the given complexes, the complex with general formula $[Ma_3b_3]$ is $[Co(NH_3)_3(NO_2)_3]$. Thus, it can show *fac* and *mer* isomers.

**34.** *(c)* $[Pt(en)_2(Cl_2)]^{2+}$ with formula $[M(AA)_2b_2]$ will show geometrical isomerism as follows :

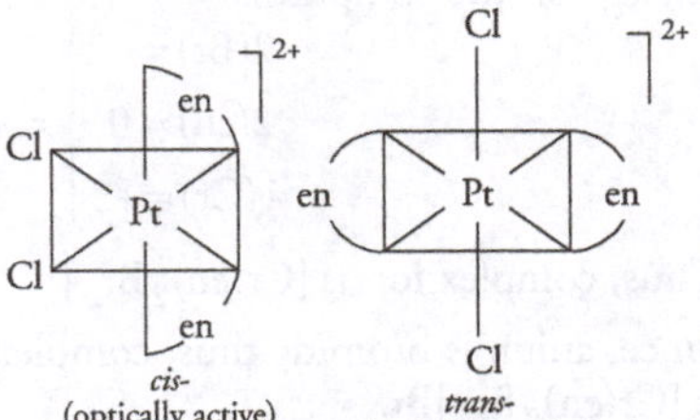


*cis*- (optically active) (I) *trans*- (optically inactive)

**35.** *(a)* Between *cis* and *trans* isomers of $[Co(en)_2Cl_2]Cl$, only *cis* isomer will exhibit optical isomer. Octahedral complexes of type $[Ma_4b_2]$ and $[Ma_3b_3]$ will not exhibit this type of isomerism. Thus *cis* $[Co(en)_2Cl_2]Cl$ will show optical isomerism.

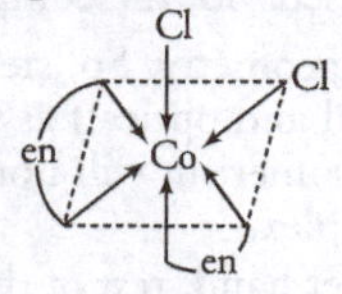


*cis*-$[Co(en)_2Cl_2]Cl$
(optically active)

**36.** *(d)* $[Ni(H_2O)_4(NH_3)_2]^{2+}$ and $[Ni(H_2O)_3(NH_3)_3]^{2+}$ show geometrical isomerism.

**37.** *(b)* $[Pt(Cl)(py)(NH_3)(NH_2OH)]^+$ is square planar complex of type [*Mabcd*]. The following structures are formed by fixing a group and then arranging all the groups.

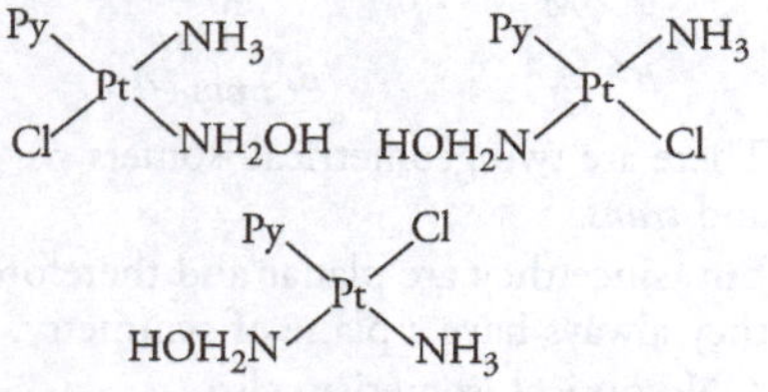


Hence, this complex shows three geometrical isomers.

**38.** *(c)* $Co^{3+} + 6NH_3 \rightarrow [Co(NH_3)_6]^{3+}$
Lewis acid, Lewis base, adduct

$Co^{3+}$ acts as Lewis acid.

**39.** *(c)* Optical isomerism is exhibited by only those complexes which lacks elements of symmetry.
$[Co(NH_3)_3Cl_3]$ shows geometrical isomerism i.e., facial as well as meridional isomerism. But both the forms of compound contain plane of symmetry. Thus, this complex does not exhibit optical isomerism.

**40.** *(b)* Two Br, two (en) and one Cr are parts of complex.
Charge on the complex is

$$\left.\begin{array}{r} 2(Br) = -2 \\ 2(en) = 0 \\ 1(Cr) = +3 \end{array}\right\} = +1$$

Thus, complex ion is $[Cr(en)_2Br_2]^+$.
Since, anion is bromide thus, complex is $[Cr(en)_2Br_2]Br$.

**41.** *(b)* Complex $[Co(en)_3]^{3+}$ has no plane of symmetry and centre of symmetry. That's why, it is optically active.

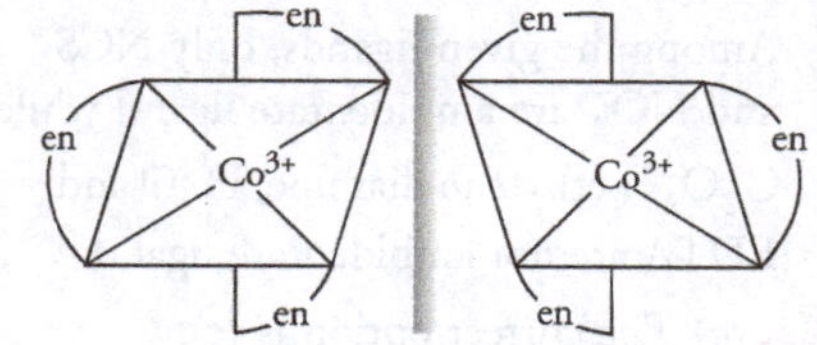


**42.** *(d)* $[Co(en)_2(NH_3)_2]^{3+}$ (*cis*-form) has an optical isomer, where en $= NH_2CH_2CH_2NH_2$ i.e.,

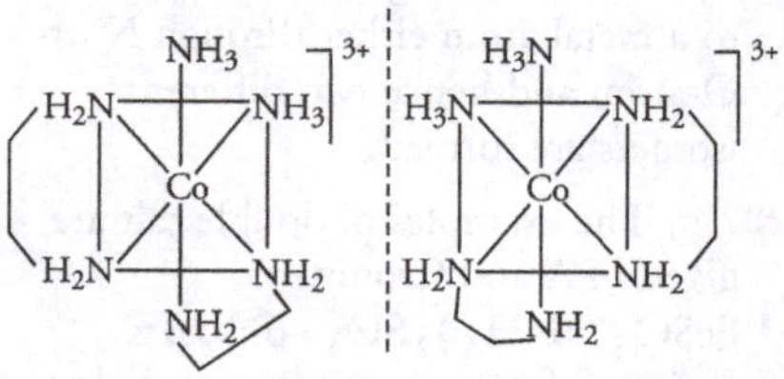


**43.** *(b)* Linkage isomers are caused due to the presence of ambidentate ligands $[Pd(PPh_3)_2(NCS)_2]$ and $[Pd(PPh_3)_2(SCN)_2]$ are linkage isomers due to SCN which is an ambidentate ligand.

**44.** *(d)* en, $\begin{array}{l} CH_2—NH_2 \\ | \\ CH_2—NH_2 \end{array}$ is a neutral bidentate ligand and $C_2O_4^{2-}$ is a binegative bidentate ligand.
Hence, coordination number of the element *E* in the given complex comes out to be 6.
The complex can be ionised as

$$[E(en)_2(C_2O_4)]NO_2 \longrightarrow [E(en)_2(C_2O_4)]^+ + NO_2^-$$

Oxidation number $x + 0 + (-2) = +1$
$\Rightarrow$ Oxidation number, $x = 3$

**45.** *(b)* Among the given only octahedral complex having bidentate i.e., $[Cr(C_2O_4)_3]^{3-}$ will show optical isomer. Thus, it can exist in the form of *dextro* and *leavo* form

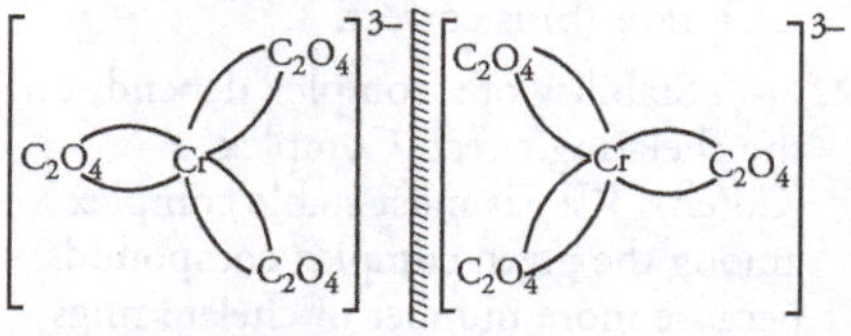


**46.** *(c)* The coordination number of central metal atom in a complex is equal to number around the metal ion bonded by coordinate bonds.
Hence, coordination number is equal to number of sigma bonds formed by the metal.

**47.** *(a)* $[Co(NH_3)_5Cl]Cl_2 \rightleftharpoons \underbrace{[Co(NH_3)_5Cl]^{2+} + 2Cl^-}_{\text{three ions}}$

$$[Co(NH_3)_5Cl]Cl_2 + 2AgNO_3 \longrightarrow [Co(NH_3)_5Cl](NO_3)_2 + 2AgCl$$

Thus, the correct strcuture of the complex is $[Co(NH_3)_5Cl]Cl_2$

**48.** *(b)* The linkage isomerism is present in nitropentamine chromium (III) chloride due to the presence of $NO_2$ group which is an ambidenate ligand.

**49.** *(a)* Due to Chelation effect most stable complex is $[Fe(C_2O_4)_3]^{3-}$
$[Fe(C_2O_4)_3]^{3-}$ electronic configuration is given by

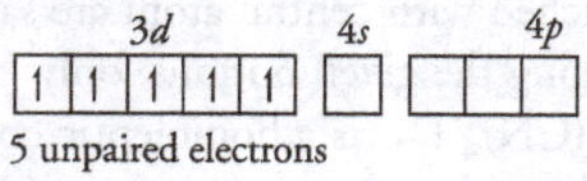


5 unpaired electrons

So, $x = 5$
thus, $V_2O_5$ is amphoteric.

**50.** *(a)*
(*A*) Complex is $[Co(NH_3)_5(SO_4)]Cl$
(*B*) Complex is $[Co(NH_3)_5Cl]SO_4$
Both (*A*) and (*B*) are ionisation isomers. So, they show ionisation isomerism.

**51.** *(d)* The correct match is
A-I, B-IV, C-II, D-III

**52.** *(c)* The electronic configuration of the central metal ion in the given complex with their respective oxidation state are as follows

$$[A]Fe^{6+} = [Ar]\,3d^2$$
$$[B]Mn^{3+} = [Ar]\,3d^4$$
$$[C]Fe^{3+} = [Ar]\,3d^5$$
$$[D]Cr^{2+} = [Ar]\,3d^4$$
$$[E]\,Ni^{4+} = [Ar]\,3d^6$$

So, among the given complexes $[Mn(CN)_6]^{3-}$ and $Cr_2(O—\overset{\overset{O}{||}}{C}—Me)_4(H_2O)_2$ has their central metal in $d^4$ configuration.

**53.** *(a)* The correct match is
A-III, B-II, C-I, D-IV
$A \rightarrow [CoF_6]^{3-}$
Here, Co is in +3 oxidation state. Thus, outer electronic configuration is

$$Co^{3+} \longrightarrow 3d^6$$

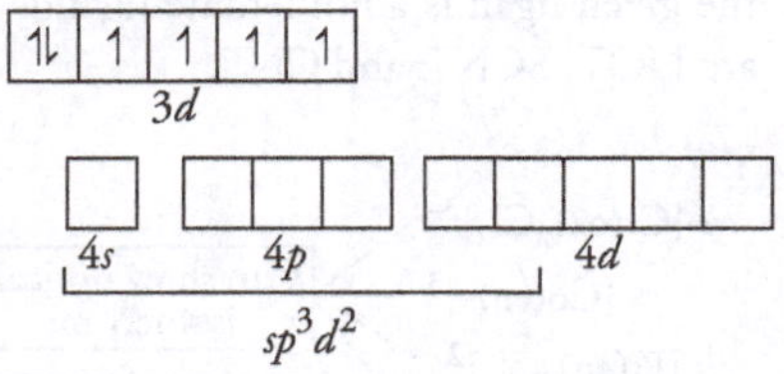


( No pairing will take place as $F^-$ is a weak field ligand. )

$B \longrightarrow [NiCl_4]^{2-}$

Here, Ni is in +2 oxidation state. Thus, outer electronic configuration is

$Ni^{2+} \longrightarrow 3d^8$

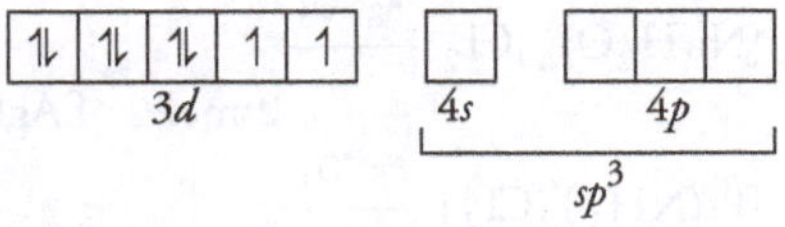


(No pairing will take place as $Cl^-$ is a weak field ligand.)

$C \longrightarrow [Co(NH_3)_6]^{3+}$

Here, Co is in +3 oxidation state. Thus, outer electronic configuration is

$Co^{3+} \longrightarrow 3d^6$

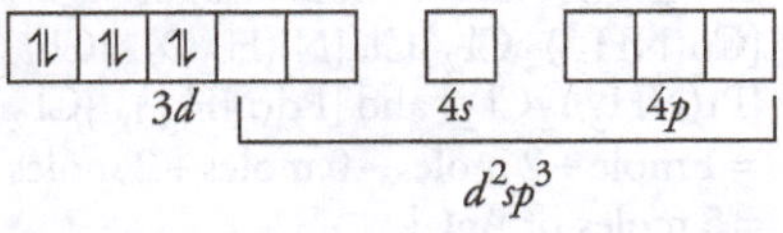


(Pairing will take place as $NH_3$ will act as strong field ligand as Co is in +3 oxidation state.)

$D \longrightarrow [Ni(CN)_4]^{2-}$

Here, Ni is in +2 oxidation state, thus outer electronic configuration is

$Ni^{2+} \longrightarrow 3d^8$

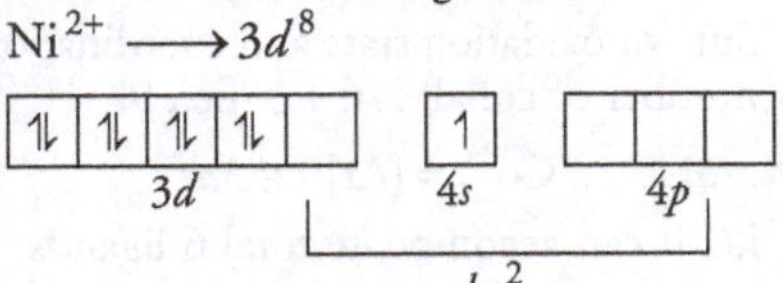


(Pairing will take place as CN is a strong field ligand.)

**54.** *(c)*

**Key idea**

Homoleptic complexes are those in which all the ligand which are connected with central atom are same.

So, look for the complex in which same ligands are present in central atom.

Among the given complexes, the one which is homoleptic complex with odd number of *d*-electrons in the central metal are $[Fe(CN)_6]^{3-}$ with $d^5$ and $[CoCl_4]^{2-}$ with $d^7$ configuration i.e., B and D only.

**55.** *(a)* The formula of fluoroapatite is $[3(Ca_3(PO_4)_2 \cdot CaF_2)]$.

**56.** *(a)* Only the counter ions ate exchanged by other reagents.

$$[Co(NH_3)_5Br]SO_4 + BaCl_2 \longrightarrow [Co(NH_3)_5Br]Cl_2 + \underset{\text{White ppt.}}{BaSO_4 \downarrow}$$

It does not have chloride ion as counter ion so it will not give white ppt with $AgNO_3$.

**57.** *(c)* Number of geometrical isomers of $[PtCl_2(NH_3)_2]$ = two geometrical isomers.

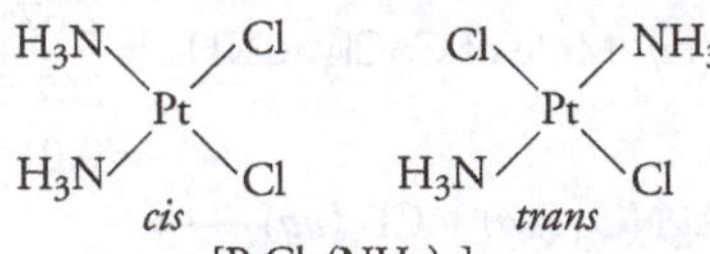


$[PtCl_2(NH_3)_2]$

In $[Ni(CO)_4]$ complex all ligands are same. Therefore, it does not show geometrical isomerism.

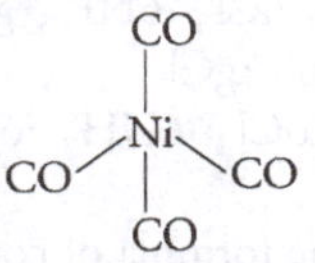

$[Ni(CO)_4]$
(no isomerism)

$[Ru(H_2O)_3Cl_3]$ : two geometrical isomers.

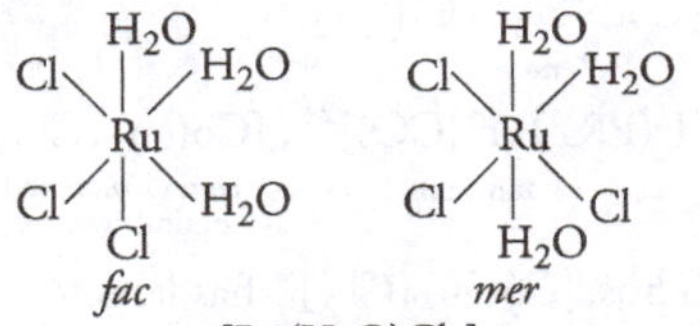


$[Ru(H_2O)Cl_3]$

$[CoCl_2(NH_3)_4]^+$: two geometrical isomerism.

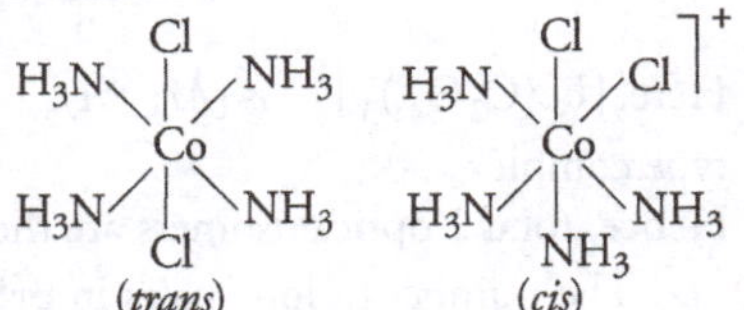


$[CoCl_2(NH_3)_4]^+$

**58.** *(a)* Complex

$(A) = [Cr(H_2O)_4Cl_2]Cl \cdot 2H_2O$

Molar mass of complex = 266.5 g/mol

On reaction with conc. $H_2SO_4$ the

$$\text{mass lost by complex} = \frac{13.5}{100} \times 265 = 35.78 \approx 36 \text{ g}$$

It means that 2 moles of $H_2O$ are present in complex.

[$\because$ Molar mass of $H_2O$ = 18 g/mol]

Then, formula of complex is

$[Cr(H_2O)_4Cl_2]Cl \cdot 2H_2O$

**59.** *(b)* $Pt^{2+}$ complexes with coordination number equals to 4 are square planar. In $[Pt(NH_3)_2Cl(NO_2)]$ two $NH_3$ ligands could arrange in *cis* or *trans* positions.

$\therefore$ There are two geometrical isomers.

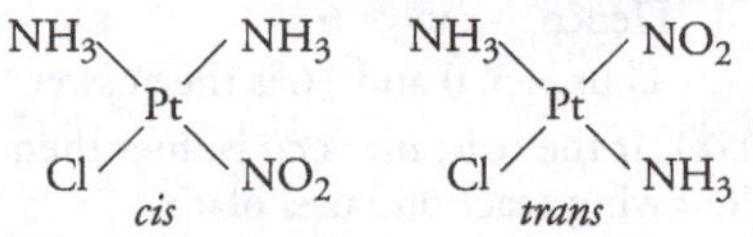


$[Pt(NH_3)_4ClBr]$ : Two geometrical isomers are possible coordination number = 6 and Octahedral geometry Cl and Br can be arranged *cis* or *trans*

$\therefore$ 2 geometrical isomers

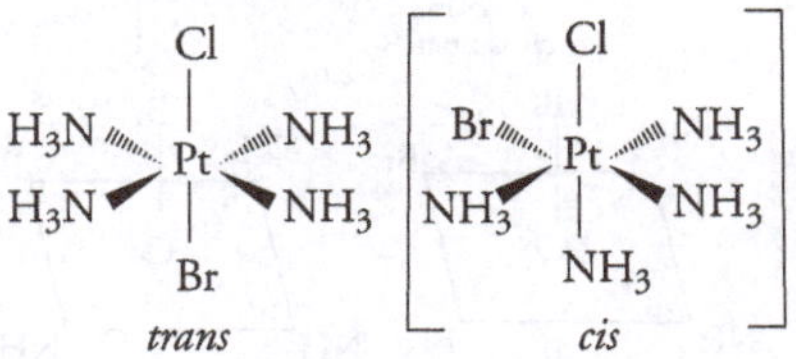


Hence, (C) and (D) will show geometrical isomerism.

**60.** *(c)* Aquachlorobis (ethylene diamine cobalt(II) chloride) i.e., $[Co(H_2O)Cl(en)_2]Cl$ show geometrical isomerism

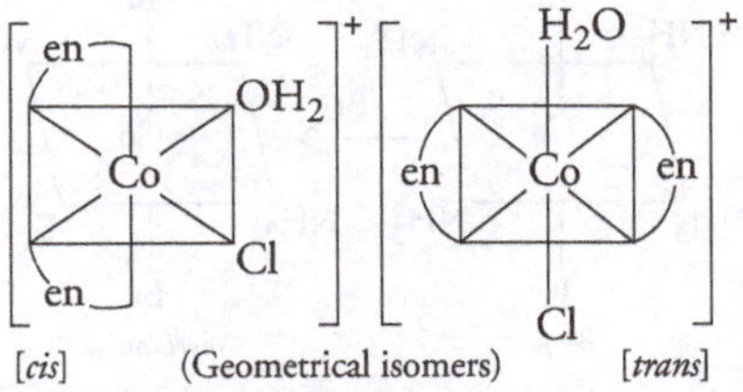


(Geometrical isomers)

**61.** *(b)* The square planar complex of the type $[Mabcd]^{n\pm}$, where all four ligands are different forms 3 geometrical isomers. But if one of the ligands is ambidentate, then $2 \times 3 = 6$ geometrical isomers are possible and if two of them are ambidentate, then $2 \times 2 \times 3 = 12$ geometrical isomers are possible. In the given example, $NO_2^-$ and $SCN^-$ are ambidentate ligands thus it will have 12 possible isomers.

**62.** *(c)* Let, the oxidation state of Cr in all cases is '*x*'

(i) Oxidation state of Cr in $[Cr(H_2O)_6]Cl_3$

$x + (0 \times 6) + (-1 \times 3) = 0$

or $x + 0 - 3 = 0$ or $x = +3$

(ii) Oxidation state of Cr in $[Cr(C_6H_6)_2]$

$x + (2 \times 0) = 0$ or $x = 0$

(iii) Oxidation state of Cr in

$K_2[Cr(CN)_2(O)_2(O_2)(NH_3)]$

$1\times 2 + x + (-1\times 2) + (-2\times 2) + (-2) + 0 = 0$

or $2 + x - 2 - 4 - 2 = 0$

or $x - 6 = 0$

Hence $x = +6$

Thus, +3, 0 and +6 is the answer.

**63.** *(b)* If the reactant is *cis* isomer then following reaction takes place.

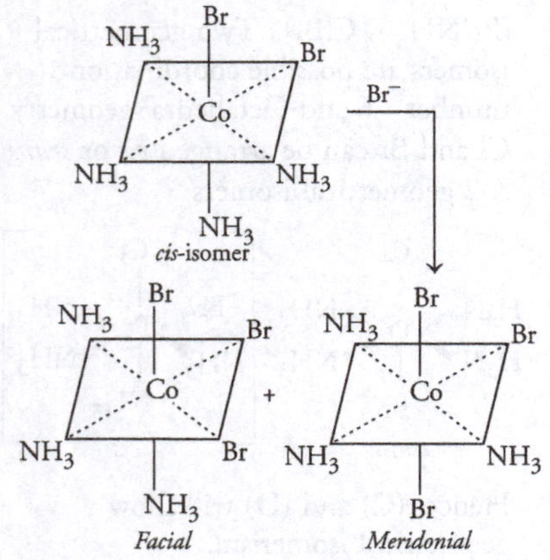


i.e. two isomers are produced.

If the reactant is *trans* isomer then following reaction takes place.

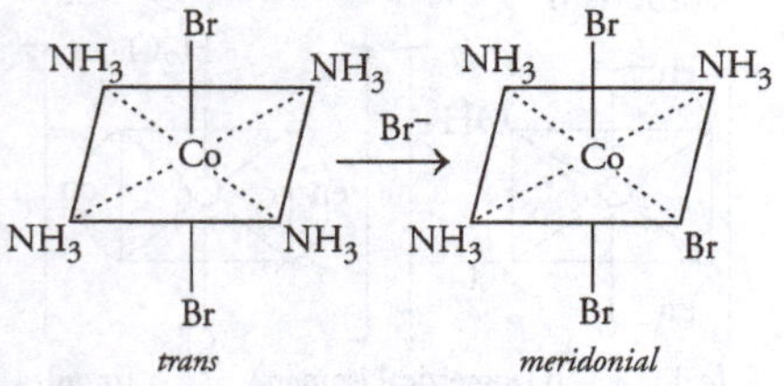


i.e. only 1 isomer is produced. Thus, statement (I) and (III) are correct resulting to option (b) as the correct answer.

**Note** In the given reaction, "Trans effect" is observed, i.e., $NH_3$ (leaving group) *trans* to Br gets substituted by the nucleophile (ligand), $Br^-$.

**64.** *(d)* Molarity *(M)*

$\therefore$ Number of moles of complex

$= \dfrac{\text{Molarity} \times \text{volume (in mL)}}{1000}$

$= \dfrac{0.1\times 100}{1000} = 0.01$ mole

Number of moles of ions precipitate

$= \dfrac{1.2\times 10^{22}}{6.02\times 10^{23}} \approx 0.02$ moles

$\therefore$ Number of $Cl^-$ present in ionisation sphere

$= \dfrac{\text{Number of moles of ions precipitated}}{\text{Number of moles of complex}}$

$= \dfrac{0.02}{0.01} = 2$

$\therefore$ 2 $Cl^-$ are present outside the square brackets, i.e. in ionisation sphere. Thus, the formula of complex is $[Co(H_2O)_5Cl]\,Cl_2\cdot H_2O$.

**65.** *(b)* Pentacyanomangnate (III) ion is the IUPAC name of complex $[Mn(CN)_5]^{2-}$.

**66.** *(a)* Mole of $CoCl_3\cdot 6NH_3 = \dfrac{2.675}{267.5} = 0.01$ mol

$AgNO_3(aq) + Cl^-(aq) \longrightarrow AgCl\downarrow(\text{white}) + NO_3^-(aq)$

Moles of $AgCl = \dfrac{4.78}{143.5} = 0.03$ mol

0.01 mol $CoCl_3\cdot 6NH_3$ gives = 0.03 mol AgCl

$\therefore$ 1 mol $CoCl_3\cdot 6NH_3$ ionises to give = 3 mol $Cl^-$

Hence, the formula of compound is $[Co(NH_3)_6]Cl_3$.

**67.** *(d)* $[Ru(NH_3)_4Cl_2]^+$, *cis* and *trans*

$[Co(NH_3)_5Cl]^{2+}$ none

$[Ir(PR_3)_2H(CO)]^{2+}$, *cis* and *trans*

$[Co(en)_2Cl_2]^+$ *cis* and *trans* and optical isomers

Thus, $[Co(en)_2Cl_2]^+$ has largest number of isomers.

**68.** *(2)* $FeCl_3 + KOH + H_2C_2O_4 \rightarrow \underset{(A)}{K_3(Fe(C_2O_4)_3]}$

Here, $[Fe(C_2O_4)_3]^{3-}$ is $[M(AA)_3]$ type complex

Hence, total 2 optical isomers are there.

**69.** *(2)* The number of lone pairs in $BrF_5$ are one.

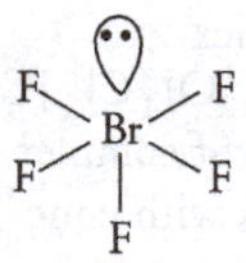


Thus, 1 mole of AgCl complex is $[M(NH_3)_3Cl_3]Cl$.

It shows 2 geometrical isomers ($Ma_3b_3$ type) i.e., facial and meridional.

**70.** *(3)* Ligands which have two different donor sites but at a time connects with only one donor site to central metal are called as ambidentate ligands. Among the given ligands ambidentate ligands are $NO_2^-$, $SCN^-$ and $CN^-$.

**71.** *(4)*

*cis*-$[Cr(ox)_2Cl_2]^{3-}$

$[Co(en)_3]^{3+}$

*cis*-$[Pt(en)_2Cl_2]^{2+}$

*cis*-$[Co(en)_2Cl_2]^+$

→ Can show optical isomerism (no plane of symmetry and no centre of symmetry)

Hence, total 4 complex shows optical isomerism.

**72.** *(5)* $[Co(NH_3)_4Cl_2]Cl \xrightarrow{AgNO_3}$ 1 moles of AgCl

$[Ni(H_2O)_6]Cl_2 \xrightarrow{AgNO_3}$ 2 moles of AgCl

$[Pt(NH_3)_2Cl_2] \xrightarrow{AgNO_3}$ zero mole of AgCl

$[Pd(NH_3)_4]Cl_2 \xrightarrow{AgNO_3}$ 2 moles of AgCl

Therefore, total number of moles of AgCl precipitated on addition of excess of $AgNO_3$ to one mole each of $[Co(NH_3)_4Cl_2]Cl$, $[Ni(H_2O)_6]Cl_2$, $[Pt(NH_3)_2Cl_2]$ and $[Pd(NH_3)_4]Cl_2$ is

= 1 mole + 2 moles + 0 moles + 2 moles

= 5 moles of AgCl.

**73.** *(9)* Here, bpy is bidentate ligand. So, coordination number of cobalt is 6.

Let the oxidation state of Co in $[Co(bpy)Cl_4]^-$ be $x$.

$\therefore \quad x + 0 + 4(-1) = -1 \Rightarrow x = 3$

$\therefore$ Oxidation state of Co = 3

Sum of oxidation state and coordination number of cobalt is 6 + 3 i.e., 9.

**74.** *(5)* $\therefore \quad Co^{3+} = [Ar]\,4s^1 3d^5$

i.e. it can accomodate total 6 ligands.

As

$CoCl_3\cdot x\,NH_3 + Ag\,NO_3\,(aq) \longrightarrow$ 2 moles of AgCl

$\therefore$ 2 chloride ions in $CoCl_3\cdot x\,NH_3$ are primary/ionisable valency (present outside the square bracket)

$[CoCl(NH_3)_x]Cl_2$

Now, as total number of ligands that can bond to cobalt are 6, out of which one is Cl.

$\therefore \quad x = 5$

$[Co(Cl)(NH_3)_5]Cl_2$. might be the complex.

**75.** *(1)* For the complex $CoCl_3\cdot 4NH_3$ with formula $[Co(NH_3)_4Cl_2]Cl$ geometrical isomers i.e., *cis* and *trans* are possible.

Their structure are as follows

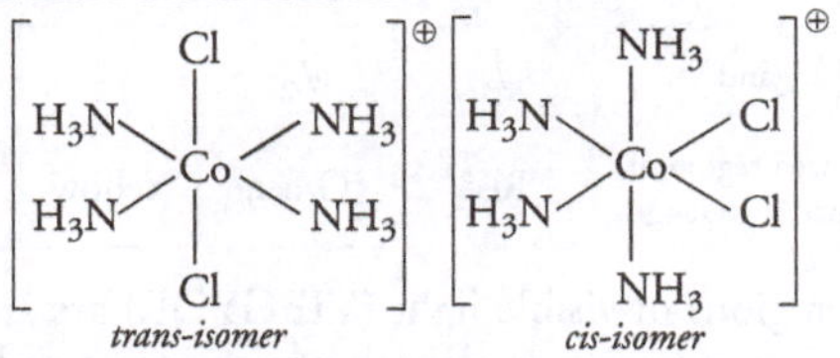


*trans-isomer* *cis-isomer*

**76.** *(2)* $\overset{+1}{[Ag(NH_3)_2]^+}\overset{+1}{[Ag(CN)_2]^-}$

Total sum of oxidation states of two silver ions $=1+1=2$

**77.** *(6)* 1 mole of $[Co(en)_2Cl_3]$ produces 1 mole of AgCl.
3 moles of $[Co(en)_2Cl_3]$ produces 3 moles of AgCl.
$\therefore$ Given compound is $[Co(en)_2Cl_2]Cl$.

$$[Co(en)_2Cl_2]Cl\,(aq) + \underset{\text{(excess)}}{AgNO_3\,(aq)} \rightarrow [Co(en)_2Cl_2]\,NO_3\,(aq) + \underset{\text{(white)}}{AgCl\,(s)}\downarrow$$

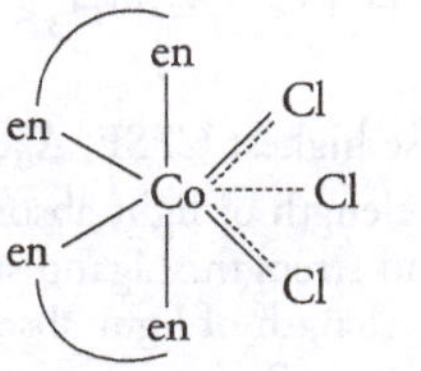


Dotted line represents primary valency = 3.
Solid line represents secondary valency (coordinated ions) of Co = 6.

> • Concept **Enhancer**
> Primary valences are normally ionisable and are satisfied by negative ions.

**78.** *(2)* Formula of triamminetrinitrocobalt (III) is $[Co(NO_2)_3(NH_3)_3]$. Its geometrical isomers are as follows :
Thus $X=2$

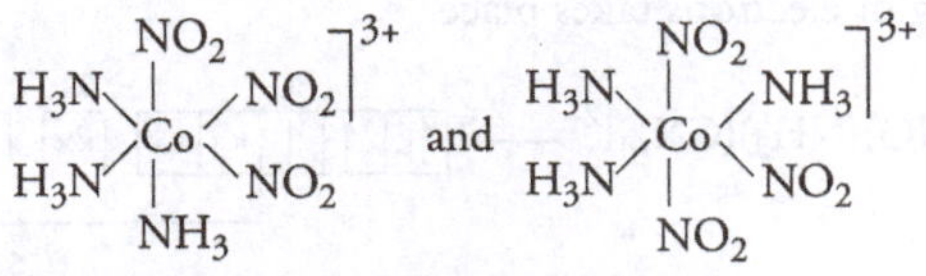


The formula of trioxalatochromate (III) ion is $[Cr(C_2O_4)_3]^{3-}$. It has no geometrical isomers. Thus, $Y=0$

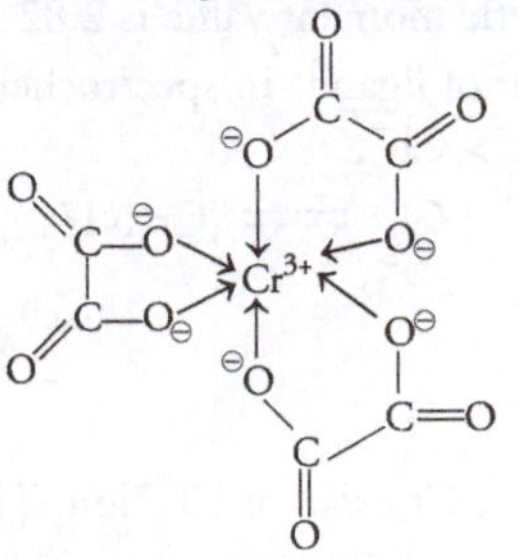

Zero geometrical isomer

Thus, the value of $X+Y=2+0=2$

**79.** *(2)* $[Co(NH_3)_4Cl_2]\,Cl + 2en \longrightarrow [Co(en)_2Cl_2] + 4NH_3$

$NH_3$ is the neutral monodentate ligand. Ethylene diamine is a neutral didentate ligand.

$$H_2\ddot{N}—CH_2—CH_2—\ddot{N}H_2\,(en)$$

So, two ethylene diamine equivalent are required to replace four '$NH_3$' ligands.

**80.** *(6)*

> **key idea** $EDTA^{4-}$ is ethylenediaminetetraacetate anion. It is a polyprotic acid containing four carboxylic acid groups and two amine groups with lone-pair electrons that chelate calcium and several other metal ions, it has six coordination sites.

The structure of $EDTA^{-4}$ is

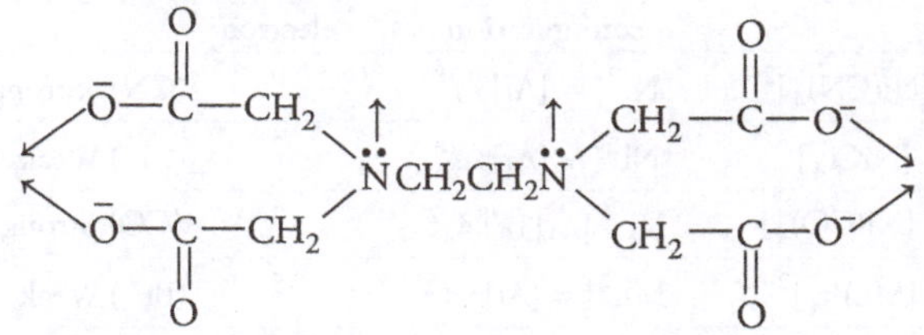

It is a hexadentate ligand. Thus it has six coordination sites.

## Topic 2

**01.** *(b)* Electronic configuration of $Fe^{3+}$ ion in complex $K_3[Fe(SCN)_6]$ is given by

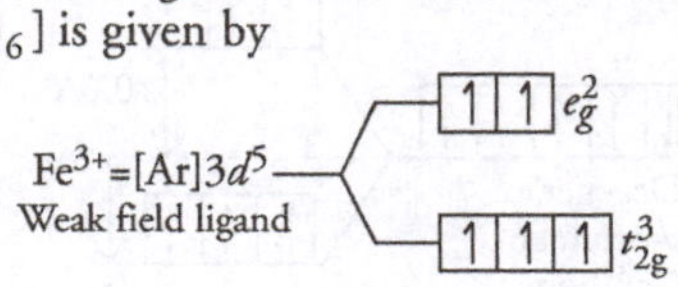


So, using the formula,

$$=(-0.4\times t_{2g}e^- + 0.6\times e_g e^-)\Delta_0$$
$$=(-0.4\times 3 + 0.6\times 2)\Delta_0 = 0\Delta_0$$

**02.** *(c)* In general, ligands can be ordered in the following sequence in terms of increasing field strength and in accordance with the spectrochemical series.

$I^- < Br^- < SCN^- < Cl^- < S^{2-} < F^- < OH^- < C_2O_4^{2-} < H_2O < NCS^- < EDTA^{4-} < NH_3 < en < CN^- < CO.$

According to this series the ligand is correctly arranged in option (c).

> • Concept **Enhancer**
> Spectrochemical series is a series in which ligands are arranged in the order of increasing field strength. It is an experimentally determined series based on the absorption of light by complexes with different ligands.

**03.** *(a)* Low oxidation state of metals in their complexes is favoured when ligands have good $\pi$-accepting ability in addition to sigma bonding (Synergic effect).

**04.** *(a)* $Fe^{2+} = [Ar]\,3d^6$

In presence of strong field ligand, low spin complex is formed. Hence, all electrons get paired with the configuration. The electronic configuration of $Fe^{+2} = t_{2g}^6\,e_g^0$

Number of unpaired electrons = 0

$\mu = \sqrt{n\,(n+2)} = 0$ BM

**05.** *(b)* The complex $K_2[Ni(CN)_4]$ has $dsp^2$ hybridisation of Ni(II).

$$Ni^{2+} = [Ar]\,3d^8$$

**06.** *(a)* In octahedral complexes
If $\Delta_0 < P$ (weak field, high spin situation).
The fourth electron enters one of the $e_g$ orbitals giving the configuration $t_{2g}^3 e_g^1$ (metal ion : $sp^3d^2$).

**07.** *(d)* Given that $\sqrt{n(n+2)} = 5.91$ BM.
$\therefore \quad n = 5$

| | Complex | Electronic configuration | Unpaired electron | Ligand |
|---|---|---|---|---|
| (a) | $[Ni(CN)_4]^{2-}$ | $Ni^{2+} = [Ar]3d^8$ | 0 | $(CN)^-$ Strong |
| (b) | $[NiCl_4]^{2-}$ | $Ni^{2+} = [Ar]3d^8$ | 2 | $(Cl^-)$ Weak |
| (c) | $[Ni(CO)_4]$ | $Ni = [Ar]3d^8 4s^2$ | 2 | (CO) Strong |
| (d) | $[MnBr_4]^{2-}$ | $Mn^{2+} = [Ar]3d^5$ | 5 | $(Br^-)$ Weak |

Thus, 5 unpaired electrons are present in Mn.

**08.** *(c)* So the electronic configuration of Co (III) will be $t_{2g}^4 e_g^2$, i.e., it has four unpaired electrons.

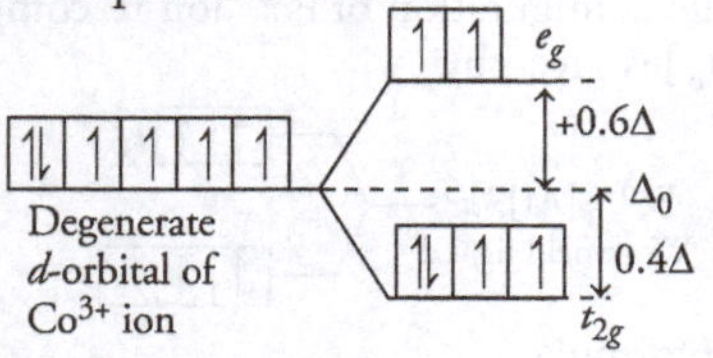


Here, CFSE can be calculated as

$$CFSE = [-0.4x + 0.6y]\,\Delta_0 = [-0.4 \times 4 + 0.6 \times 2]\,\Delta_0 = [-1.6 + 1.2]\,\Delta_0 = [-0.4\,\Delta_0]$$

Hence, the crystal field stabilisation energy (CFSE) of $[CoF_3(H_2O)_3]$ will be $-0.4\,\Delta_0$.

**09.** *(c)* The electronic configuration is

$[Ru(en)_3]^{2+} \Rightarrow Ru^{2+} = 4d^6 = t_{2g}^6, e_g^0$

$[Fe(H_2O)_6]^{2+} \Rightarrow Fe^{2+} = 3d^6 = t_{2g}^4, e_g^2$

As, en is strong field whereas $H_2O$ is a weak field ligand.
So, correct answer is (c).

**10.** *(c)* Valence bond theory cannot distinguish between weak field ligands and strong field ligands. It also can't explain the colour exhibits by transition metal complex.

**11.** *(c)* Wilkinson catalyst is $[RhCl(PPh_3)_3]$.

$$[RhCl(PPh_3)_3] \equiv [\overset{}{Rh}\ \overset{(-1)(-1)}{Cl}\ (\overset{(0)}{P}Ph_3)_3]$$

$Rh^+ \rightarrow [Kr]$
The hybridisation of central metal ion is $dsp^2$ and its shape is square planar.

**12.** *(a)* In permanganate ion the source of color is due to ligand to metal ($L \rightarrow M$) charge transfer which takes place between *p*-orbital of oxygen and the empty *d*-orbitals on the metal.

**13.** *(a)* Both are paramagnetic, but $CN^-$ will result in one unpaired $e^-$ in $[Fe(CN)_6]^{3-}$ and $F^-$ will result in 5 unpaired electrons in $[FeF_6]^{3-}$ because $CN^-$ is strong field ligand.

**14.** *(b)*

| Ligand | $L_1$ | $L_2$ | $L_3$ | $L_4$ |
|---|---|---|---|---|
| Coloured region of absorbed wavelength | Red | Green | Yellow | Blue |

Coloured regions of visible light (VIBGYOR) are arranged in the increasing order of wavelength or decreasing order of energy [$\because E \propto 1/\lambda$]. Therefore,

$$\underset{(L_1)}{E_{Red}} < \underset{(L_2)}{E_{yellow}} < \underset{(L_3)}{E_{green}} < \underset{(L_4)}{E_{blue}}$$

Since, higher absorbed energy corresponds to greater extent of crystal field splitting and thus, higher field strength of the ligand. Hence, the increasing order of ligand strength of the four ligands is $L_1 < L_3 < L_2 < L_4$

**15.** *(d)* Following is the order of the ligands in increasing value of crystal field splitting

$I^- < Br^- < Cl^- < F < H_2O < Py{=}NH_3 < en < NO_2^- << CN^- <$ CO

Thus, CO will cause highest CFSE, $\Delta_0$.

**16.** *(a)* Lesser the wavelength of light absorbed (more energy) greater will be ligand strength. Ligand strength is inversely proportional to wavelength of light absorbed.
Energy : Blue > Green > Red
Ligand strength : $L_3 < L_1 < L_2$

**17.** *(d)* $[Cr(NO)(NH_3)(CN)_4]^{2-}$

But Cr does not exist in +1 O.S. Now, let O.S. of NO be zero.

$\therefore x + (0) + (0) + 4(-1) = -2 \Rightarrow x = +2$

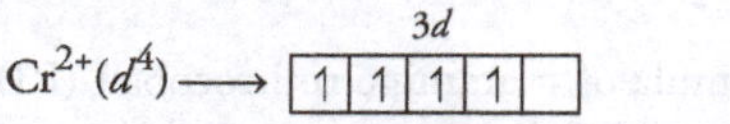


In presence of strong field ligands ($CN^-$, $NH_3$ and NO), the pairing of electrons takes place.

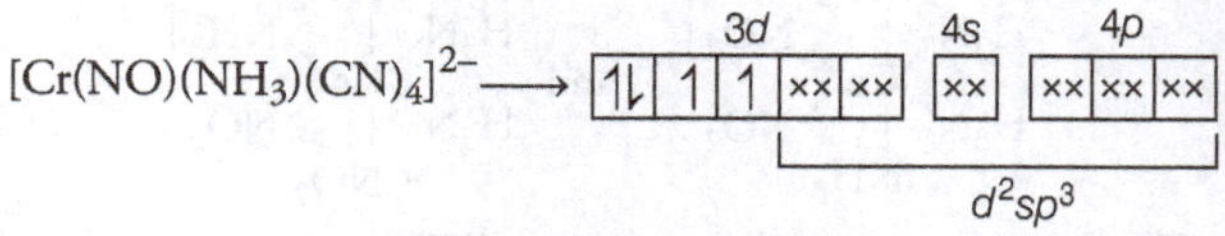


$$\mu = \sqrt{2(2+2)} = \sqrt{2 \times 4} = \sqrt{8} = 2.82 \text{ BM}$$

Therefore, magnetic moment value is 2.82 BM.

**18.** *(b)* Correct order of ligands in spectrochemical series is :
$CN^- >$ en $> NCS^- > Cl^-$.

**19.** *(c)* $\underbrace{[Cr(NH_3)_6]}_{\text{Coordinate sphere}}\ \underbrace{Cl_3}_{\text{Ionisable}} \rightleftharpoons [Cr(NH_3)_6]^{3+} + 3Cl^-$

$$\xrightarrow{AgNO_3} \underset{\text{White precipitate}}{AgCl\downarrow}$$

In $[Cr(NH_3)_6]^{3+}$, Cr exists as $Cr^{3+}$ ion. The electronic configuration of Cr is $3d^3\ 4s^0$. The orbital diagram is represented by

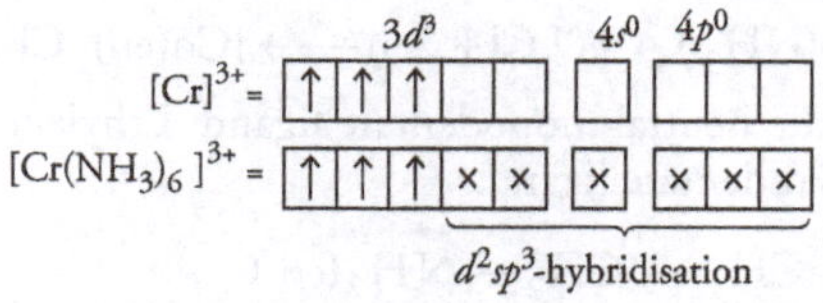

Here, [×] indicates lone-pair of $NH_3$ donated to Cr. The given complex has $d^2sp^3$ hybridisation and is octahedral in shape. It is paramagnetic and is an inner orbital complex.
Therefore, (c) is wrong statement while other statements are correct.

**20.** *(a)* CFSE (crystal field splitting energy) for octahedral complex, $\Delta_o$ depends on the strength of negative ligand.
Spectrochemically, it has been found that the strength of splitting is as follows
$CO > CN^- > NO_2^- > en > NH_3 > py >$
$NCS^- > H_2O > O^{2-} > Ox^{2-}$
$> OH^- > F^- > Cl^- > SCN^-$
$> S^{2-} > Br^- > I^-$

**21.** *(d)* $Cl^-$ is a weak ligand but it cause the pairing of electrons with large $Pt^{2+}$ and consequently give $dsp^2$ hybridisation and square planar geometry.

**22.** *(d)* Compounds with their electronic configuration and number of unpaired electrons are as follows.

| | | |
|---|---|---|
| $[FeF_6]^{3-}$ | $[Ar]3d^5$ | 5 unpaired $e^-$ |
| $[CoF_6]^{3-}$ | $[Ar]3d^6$ | 4 unpaired $e^-$ |
| $[Ni(CN)_4]^{2-}$ | $[Ar]3d^8$ | zero unpaired $e^-$ |
| $[Ni(CO)_4]$ | $[Ar]3d^{10}$ | zero unpaired $e^-$ |

So, the correct order is
$[FeF_6]^{3-} > [CoF_6]^{3-} > [Ni(CN)_4]^{2-}$
$= [Ni(CO)_4]$

**23.** *(b)* In $[Co(NH_3)_6]^{3+}$, $Co^{3+}$: $[Ar]3d^6$
$NH_3$ is strong field ligand,
$Co^{3+} = d^2sp^3$

- $[CrF_6]^{3-}$, $Cr^{3+}$: $[Ar]3d^3$
Hybridisation of $Cr^{3+} = d^2sp^3$.
- $SF_6 = sp^3d^2$ of sulphur
- $[CoF_6]^{3-}$, $Co^{3+}$ : $[Ar]3d^6$, $F^-$ is strong field ligand.
Hybridisation of $Co^{3+} = sp^3d^2$
- $[Mn(CN)_6]^{3-}$, $Mn^{3+}$ :
$[Ar]3d^4$ $CN^-$ is strong field ligand.
Hybridisation of $Mn^{3+} = d^2sp^3$
- $[MnCl_6]^{3-}$, $Mn^{3+} = [Ar]3d^4$, $Cl^-$ is weak field ligand
Hybridisation of $Cl^-$ is $sp^3d^2$.

Hence, total three $sp^3d^2$ hybridised molecules.

**24.** *(d)* $[Mn(SCN)_6]^{4-}$
(a) $Co^{3+}$ $[Ar]3d^6 4s^0$
Also, $NH_3$ is a SFL thus

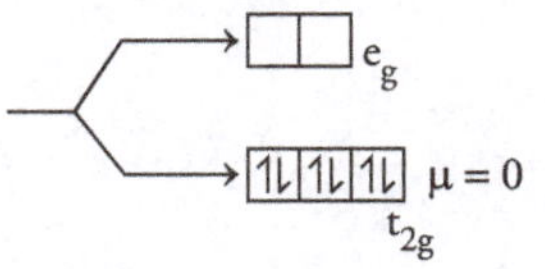


Now, $CFSE = [-0.4 \times 6 + 0.6 \times 0]\Delta_o$
$= -2.4\Delta_0$
(b) $[Fe(CN)_6]^{4-}$
$Fe^{2+} = [Ar]3d^6 4s^0$
Also, $\bar{C}N$ is a SFL

thus

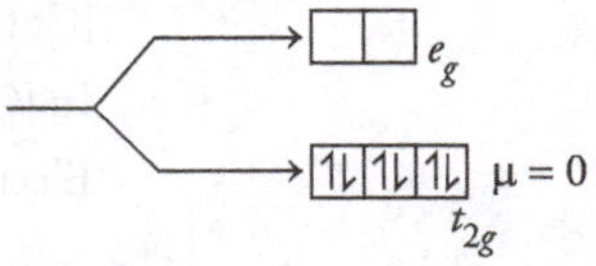


$\mu = 0$
$CFSE = -2.4\Delta_o$
(c) $[FeF_6]^{4-}$
$Fe^{2+} = [Ar]3d^6 4s^0$
Also, $F^-$ is a weak field ligand

thus

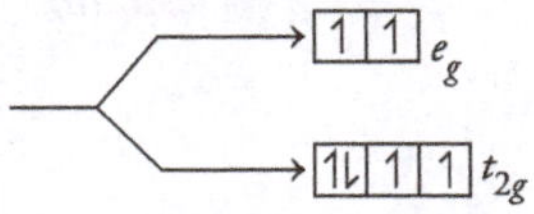


$\mu = \sqrt{24}$ BM $= 4.89$ BM
$CFSE = (-0.4 \times 4 + 0.6 \times 2)\Delta_0$
$= -1.2\Delta_o$
(d) $[Mn(SCN)_6]^{4-}$
$Mn^{2+} = [Ar]3d^5 4s^0$
Also, $\bar{S}CN$ is a weak field ligand.
Then,

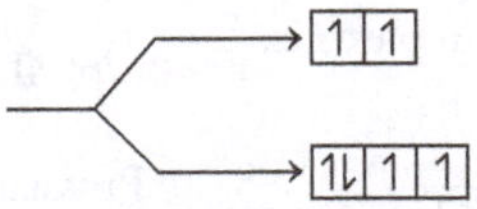

$\mu = \sqrt{35} = 5.96$ BM
$CFSE = (-0.4 \times 3 + 0.6 \times 2)\Delta_o$
So, $\Delta_o = 0$

**25.** *(a)* $[MnCl_6]^{3-}$ contains $Mn^{3+}$
$[Mn^{3+}] = [Ar]3d^4$
Ligand $= Cl^- =$ weak field ligand

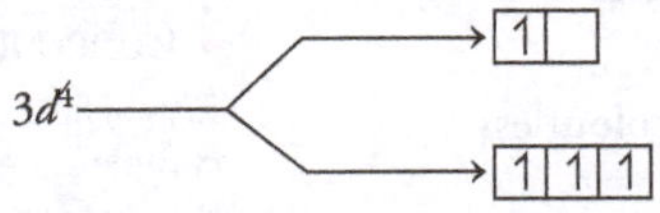


Hybridisation $= sp^3d^2$ and 4 unpaired electrons.

**26.** *(c)* Both Statement I and II are correct.
In octahedral complex, coordination number is 6.
If $\Delta_0 <$ P.E., then high spin complexes are formed.
If $\Delta_0 >$ P.E., then low spin complexes are formed.
But in tetrahedral complex (CN = 4) $\Delta_t <$ P.E., then mainly high spin complexes are formed and rarely low spin complexes are formed.

**27.** *(d)* The electronic configuration of $Co^{2+}$ is given by $[Ar]3d^7 4s^0$. Also, since it is given that magnetic moment of the octahedral complex of Co(II) is 3.95. So, it has 3 unpaired electrons.
Thus, its configuration is $t_{2g}^5 e_g^2$.

**28.** *(c)* For $3d^4$,
if ligand is strong field ligand :
$t_{2g}^4 e_g^0$ (low spin)
If ligand is weak field ligand :
$t_{2g}^3 e_g^1$ (high spin)

> **Concept Enhancer**
> Ligands for which $\Delta_o < P$ are called as weak field ligands, they form high spin complexes.
> Ligands for which $\Delta_o > P$ are called as strong filed ligands, they form low spin complexes.

**29.** *(d)* In $K_3[Fe(OH)_6]$, Fe is in +3 oxidation state. So, it has $d^5$ electronic configuration. Thus, the number of unpaired electrons = 5
So, $\mu = \sqrt{5(5+2)} = 5.92$ BM
Similarly, in $K_4[Fe(OH)_6]$, Fe is +2 oxidation state. So, it has $d^4$ electronic configuration. Thus, the number of unpaired electrons = 4
So, $\mu = \sqrt{4(4+2)} = 4.90$ BM
Thus, the calculated spin-only magnetic moment of $K_3[Fe(OH)_6]$ and $K_4[Fe(OH)_6]$, respectively are 5.92 and 4.90 BM.

**30.** *(b)* When ethane -1, 2-diamine is added progressively to an aqueous solution of nickel (II) chloride.

- $[Ni(H_2O)_6]^{2+}(aq) + en\,(aq)$
(Green)
↓
$[Ni(H_2O)_4(en)]^{2+}(aq) + 2H_2O$

- $[Ni(H_2O)_4\,en]^{2+}(aq) + en\,(aq)$
  (Pale blue)
  ↓
  $[Ni(H_2O)_2(en)_2]^{2+}(aq) + 2H_2O$
- $[Ni(H_2O)_2\,en]^{+2}(aq) + en(aq)$
  (Blue/purple)
  ↓
  $[Ni(en)_3]^{2+}(aq) + 2H_2O$
  Violet

**31.** *(d)* The coordination environment of $Ca^{2+}$ ion in its complex with $EDTA^{4-}$ is octahedral. As $EDTA^{4-}$ is hexadentate ligand.

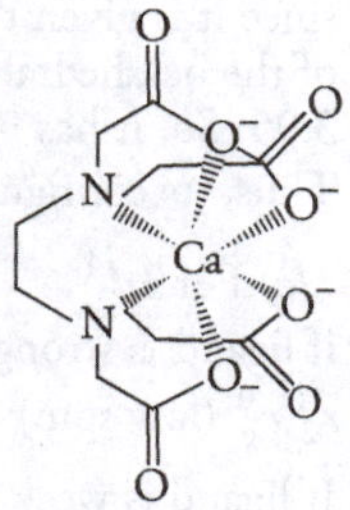

**32.** *(d)* For the given complex
$[Fe(NH_3)_x(CN)_y]^-$ if it has no electron in its $e_g$ orbital then, it will become
$[Fe^{III}(NH_3)_2(CN)_4]^-$
So, $x = 2$, $y = 4$
Hence, $x + y = 6$.

**33.** *(b)* According to spectrochemical series the correct sequence of ligands in the order of decreasing field strength is
$CO > H_2O > F^- > S^{2-}$.

**34.** *(a)* In complex $[Co(NH_3)_6]^{3+}$, Co is present in +3 oxidation state. So, configuration of $Co^{3+}$ in presence of strong field ligand (i.e., $NH_3$) is $3d^6(t_{2g}^6 e_g^0)$.
Its hybridisation is $d^2sp^3$ and is an inner orbital complex *i.e.,* (spin paired complex).
In $[CoF_6]^{3-}$, Co is present in +3 oxidation state. So, configuration of $Co^{3+}$ in presence of weak field ligand (i.e. $F^-$) is $3d^6(t_{2g}^4 e_g^2)$.
Its hybridisation is $sp^3d^2$ and is an outer orbital complex *i.e.,* (spin free complex).

**35.** *(c)* Both Statement I and Statement II are correct.
$[Ni(H_2O)_6]^{2+}$ is green colour solution because of *d-d* transition.
$[Ni(CN_4)]^{2-}$ is diamagnetic in nature and is colourless.
Hence, option (c) is correct.

**36.** *(c)* $[Ni(CO)_4] \rightarrow$ Diamagnetic,
$sp^3$-hybridisation, number of unpaired electrons $= 0$
$[NiCl_4]^{2-} \rightarrow$ Paramagnetic,
$sp^3$-hybridisation number of unpaired electrons $= 2$

**37.** *(b)* The coordination geometry around the manganese in decacarbonyldimanganese (0) is octahedral. The structure of $Mn_2(CO)_{10}$ is as follows.

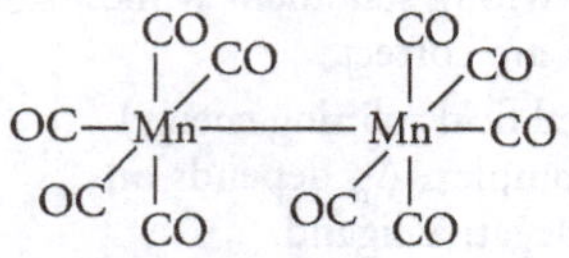

Hence, option (b) is correct.

**38.** *(b)* $[Co(NH_3)_6]^{3+}$ is octahedral diamagnetic low spin complex.
$NH_3$ is a strong field ligand which causes larger splitting of *d*-orbital and pairing of electrons is favoured. Hence, $[Co(NH_3)_6]^{3+}$ is a low spin complex as well as diamagnetic.
In $[Co(NH_3)_6]^{3+}$
Electronic configuration after complex formation is

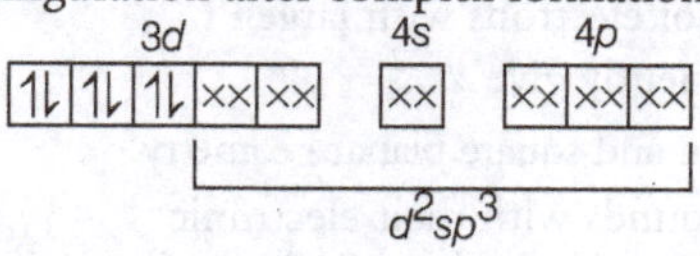


Number of unpaired electrons $= 0$
Geometry of complex ion $=$ Octahedral.

**39.** *(b)*

| Complex | Configuration of *d*-orbital | Unpaired electron | Spin moment, $\mu = \sqrt{n(n+2)}$ |
|---|---|---|---|
| $[Fe(CN)_6]^{3-}$ | $Fe^{3+} \Rightarrow t_{2g}^5 e_g^0$ | 1 | 1.732 |
| $[Mn(CN)_6]^{3-}$ | $Mn^{3+} \Rightarrow t_{2g}^4 e_g^0$ | 2 | 2.828 |
| $[CoF_6]^{3-}$ | $Co^{3+} \Rightarrow t_{2g}^4 e_g^2$ | 4 | 4.899 |
| $[MnBr_4]^{2-}$ | $Mn^{2+} \Rightarrow e^2 t_2^3$ | 5 | 5.916 |

Hence, correct order of spin only magnetic moment is
$$[Fe(CN)_6]^{3-} < [Mn(CN)_6]^{3-} < [CoF_6]^{3-} < [MnBr_4]^{2-}.$$

**40.** *(b)* $4FeCl_3 + 3K_4[Fe(CN)_6] \longrightarrow 12KCl + \overset{III}{Fe_4}[\overset{II}{Fe}(CN)_6]_3$
Prussian blue

Prussian blue colour is due to formation of coordination compound of iron $Fe_4[Fe(CN)_6]_3$ (ferri-ferrocyanide).

**41.** *(c)* $CoCl_2 + 6H_2O \longrightarrow [Co(H_2O)_6]Cl_2$
Pink (*X*) octahedral
$\xrightarrow{\text{(conc.) HCl}} [CoCl_4]^{2-}$
(*Y*) Blue solution
(*Z*) Tetrahedral

Thus, $X = [Co(H_2O)_6]^{2+}$, $Y = [CoCl_4]^{2-}$, $Z =$ tetrahedral

**• Concept Enhancer**

Crystal field theory attributes to the colour of the coordination complex due to *d-d* transition of electron. It should be noted that in the absence of ligand, crystal field splitting does not takes place. Hence, the substance is colourless. Although in some complexes the colour is due to charge transfer spectra.

**42.** *(c)* Spin only magnetic moment can be calculated by using the formula

$$\mu = \sqrt{n(n+2)}\ \text{BM}$$

For $[FeF_6]^{3-}$

$Fe^{3+}$ : $_{18}[Ar]3d^5$ so, 5 unpaired $e^-$.

$\Rightarrow \quad \mu = \sqrt{5(5+2)} = 5.91$ BM

For $[CoF_6]^{3-}$ :

$Co^{3+}$ : $_{18}[Ar]3d^6$ So, 4 unpaired $e^-$

$\Rightarrow \quad \mu = \sqrt{4(4+2)} = 4.89$ BM

For $[Co(C_2O_4)_3]^{3-}$

$Co^{3+}$ : $_{18}[Ar]3d^6$ but due to pairing no unpaired $e^-$ are present

$\Rightarrow \quad \mu = \sqrt{0(0+2)} = 0$ BM

$\therefore$ Correct order of spin only magnetic moment's

$$[FeF_6]^{3-} > [CoF_6]^{3-} > [Co(C_2O_4)_3]^{3-}$$

**43.** *(a)* In $[Co(NH_3)_6]^{3+}$, the oxidation state of cobalt is +3.

Ammonia is a strong field ligand so it will pair up four unpaired electron present in $3d$-orbitals.

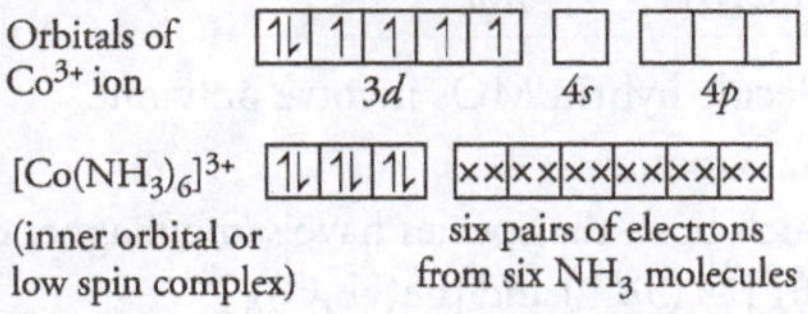


Since, it has no unpaired electrons, so, $[Co(NH_3)_6]^{3+}$ is diamagnetic and its hybridisation is $d^2sp^3$.

**44.** *(a)* In the complex $[NiCl_2Br_2]^{2-}$, the geometry is tetrahedral due to presence of weak field ligands and do not show geometrical isomerism.

- However for complex $[PtCl_2Br_2]^{2-}$ the geometry is square planar and it shows two geometrical isomers.
- Both the complex species are optically inactive.
- Also they contain different number of unpaired electron. Hence, their magnetic properties varies.

**45.** *(d)* A is false but R is true.

Correct statement for Assertion is that, $[CoCl(NH_3)_5]^{2+}$ absorbs at higher wavelength of light with respect to $[Co(NH_3)_5(H_2O)]^{3+}$.

Presence of weak field ligand like $Cl^-$ as compared to $H_2O$, decreases the CFSE for $[CoCl(NH_3)_5]^{2+}$. Thus it absorbs at higher wavelength.

**46.** *(a)* Jahn-Teller distortion is shown in those octahedral complexes which have unsymmetrically filled $t_{2g}$ or $e_g$ orbital.

In case of $[Cu(H_2O)_6]SO_4$, uneven filling of electron in $e_g$ orbital ($t_{2g}^6\, e_g^3$) leads to structural distortion of octahedral geometry.

**47.** *(a)* As strength of ligand increases, crystal field splitting energy increases. In same way energy of absorption for the metal complexes also increases.

Thus, the correct order is

$$\underset{(C)}{[Ni(H_2O)_6]^{2+}} < \underset{(B)}{[Ni(NH_3)_6]^{2+}} < \underset{(A)}{[Ni(en)_3]^{2+}}$$

Because the order of field strength of the ligands is $H_2O < NH_3 <$ (en).

**48.** *(c)*

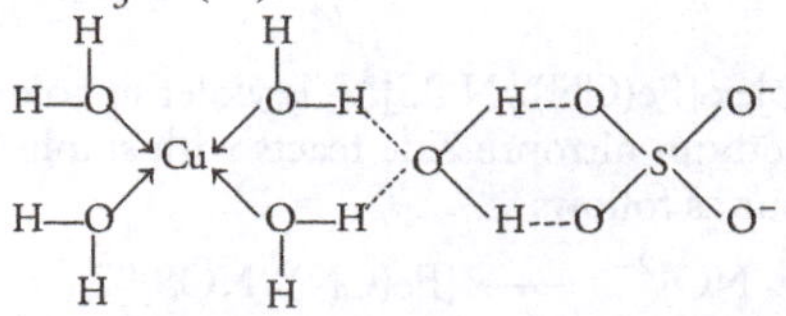

In $CuSO_4 \cdot 5H_2O$, Cu—O bonds are present. It is a correct statement.

In $CuSO_4 \cdot 5H_2O$, ligands coordinating with Cu(II) ion are O - and S-based ligands. The statement is incorrect. The ligands coordinating with Cu(II) ion is O-based and not S-based. It is a incorrect statement.

**49.** *(b)* $[Ni(CN)_4]^{2-}$ : $Ni^{2+}$ has $d^8$ configuration, Due to strong field ligand it will have sq. planar geometry ($dsp^2$). It has all paired electrons. Therefore is diamagnetic.

$[Ni(CO)_4]$ : Ni has $d^{10}$ configuration also it has, SFL, tetrahedral geometry ($sp^3$) and diamagnetic nature.

$[NiCl_4]^{2-}$ : has $d^8$ configuration, due to weak field ligand, tetrahedral geometry ($sp^3$), paramagnetic (2 unpaired $e^-$)

**50.** *(b)* As the number of unpaired electrons ($n$) increases magnetic moment of complex also increases.

(A) $[FeF_6]^{3-}$ is a high-spin complex. So, pairing of electrons will not occur.

$Fe^{3+} \Rightarrow t_{2g}^3\, e_g^2\ (n=5)$

Where $n$ = number of unpairs electrons

(B) $[Fe(CN)_6]^{3-}$ is a low-spin complex. So, pairing of electrons occur is $d$-orbital.

$Fe^{3+} \Rightarrow t_{2g}^5\, e_g^0\ (n=1)$

(C) $[MnCl_6]^{3-}$ is a high spin complex. So, pairing of electrons will not occur.

$Mn^{3+} \Rightarrow t_{2g}^3\, e_g^1\ (n=4)$

(D) $[Mn(CN)_6]^{3-}$ is a low-spin complex. So, pairing of electrons take place.

$Mn^{2+} \Rightarrow t_{2g}^3\, e_g\ (n=2)$

So, increasing order of magnetic moment : B < D < C < A.

**51.** *(d)* All the transition metals have same oxidation states.

$[Cr(H_2O)_6]^{3+} \longrightarrow 3d$ series

$[Mo(H_2O)_6]^{3+} \longrightarrow 4d$ series

$[Fe(H_2O)_6]^{3+} \longrightarrow 3d$ series

$[Os(H_2O)_6]^{3+} \longrightarrow 5d$ series

For the same oxidation state, CFSE increasing order will be $3d < 4d < 5d$.

So, $[Os(H_2O)_6]^{3+}$ has maximum CFSE ($\Delta_o$).

**52.** *(a)* For the same central metal ion, the CFSE increases with increase in strength of the ligand. The order of strength of ligand is given by the spectrochemical series :

$I^- < Br^- < SCN^- < Cl^- < S^{2-} < F^- < OH^- < C_2O_4^{2-}$
$< H_2O < NCS^- < EDTA^{4-} < NH_3 < en < CN^- < CO$

Thus, the order of CFSE of given complexes will be :

$\underset{(A)}{[CoF_6]^{3-}} < \underset{(B)}{[Co(H_2O)_6]^{3+}} < \underset{(C)}{[Co(NH_3)_6]^{3+}} < \underset{(D)}{[Co(en)_3]^{3+}}$.

**53.** *(d)* The complex $[Fe(CN)_5NOS]^{4-}$ is violet in colour. It is formed when sodium nitroprusside reacts with sulphide ion to give violet colour as follows

$$S^{2-} + [Fe(CN)_5NO]^{2-} \longrightarrow \underset{\text{(Violet colour)}}{[Fe(CN)_5NOS]^{4-}}$$

**54.** *(d)* $[MnCl_6]^{3-}$

$Mn^{3+}$ in $[MnCl_6]^{3-} \Rightarrow$

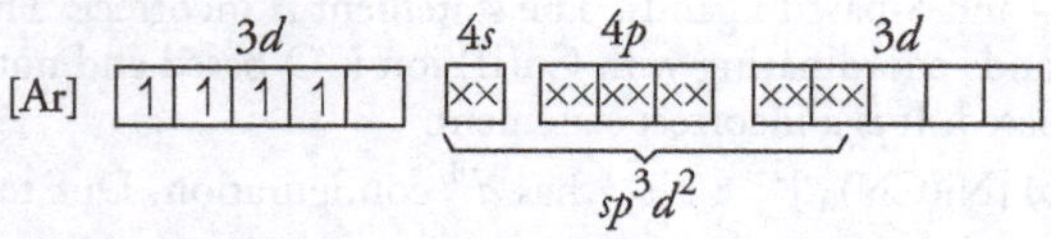


$Cl^-$ is a weak field ligand, so $Mn^{3+}$ forms high spin complex or outer orbital complex. Hence, it will be $sp^3d^2$ hybridised and is paramagentic.

**55.** *(a)* Only paramagnetic substance or complex response to external magnetic field.

For $[Fe(H_2O)_6]^{3+}$; Electronic configuration of

$Fe^{3+} = [Ar]3d^5 4s^0$

All electrons are unpaired as $H_2O$ is a weak field ligand

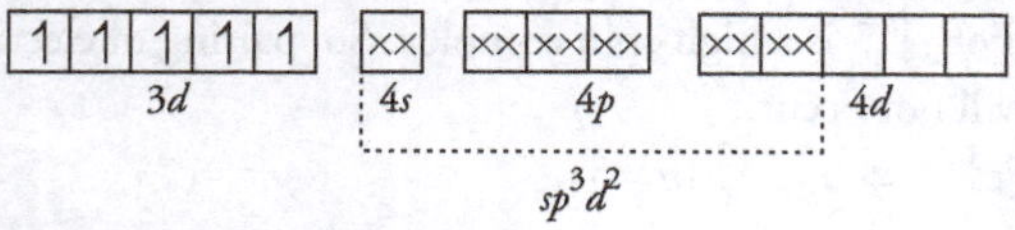


Unpaired electrons = 5

Magnetic nature = Paramagnetic (so this complex response to external magnetic field)

**56.** *(c)* The CFSE order of different geometric is

$\Delta_t < \Delta_0 < \Delta_{sq.}$

$[NiCl_4]^{2-} < [Ni(H_2O)_6]^{2+} < [Ni(CN)_4]^2$

CFSE $\propto$ Energy of absorption $\propto$ Intensity of color

$\Rightarrow$ Intensity of (complementary) colour of compounds will be $[NiCl_4]^{2-} > [Ni(H_2O)_6]^{2+} > [Ni(CN)_4]^2$

**57.** *(a)* $[Mn(CN)_6]^{4-}$ In this complex, oxidation state of Mn is + 2.

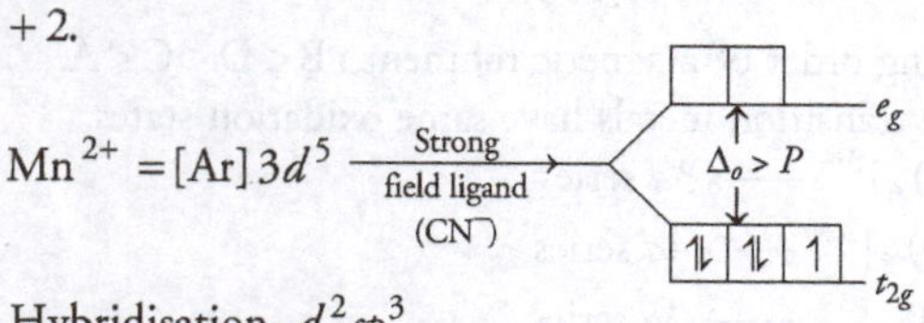


Hybridisation- $d^2sp^3$

Magnetic nature-Paramagnetic

$[Fe(CN)_6]^{-3}$

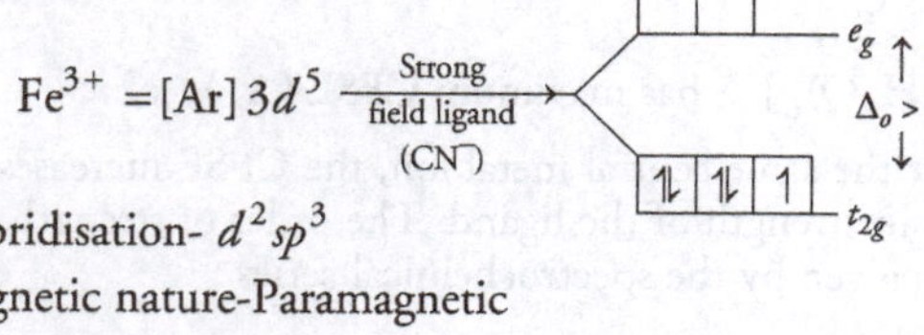


Hybridisation- $d^2sp^3$

Magnetic nature-Paramagnetic

**58.** *(b)* So, magnetic moment

$$(\mu) = \sqrt{n(n+2)}\text{ BM}$$

$n = 4$

(i) $[FeCl_4]^{2-} \longrightarrow Fe^{2+} \longrightarrow [Ar]\,3d^6$

$= \sqrt{4(4+2)}\text{ BM} = \sqrt{24}\text{ BM} = 4.90\text{ BM}$

(ii) $[Co(C_2O_4)_3]^{3-} \longrightarrow Co^{3+} \longrightarrow [Ar]3d^6$

All electrons are paired, $n = 0$

hence, $\mu = 0$

(iii) $MnO_4^{2-} \rightarrow Mn^{+6} \rightarrow [Ar]3d^1$

$n = 1$

$= \sqrt{n(n+2)}\text{ BM}$

$= \sqrt{1(1+2)}\text{ BM} = \sqrt{3}\text{BM} = 1.73\text{ BM}$

**59.** *(d)*

| | Molecule | Hybridisation |
|---|---|---|
| (a) | $XeF_4$ | $sp^3d^2$ |
| (b) | $[CrF_6]^{3-}$ | $d^2sp^3$ |
| (c) | $BrF_5$ | $sp^3d^2$ |
| (d) | $[Ni(CN)_4]^{2-}$ | $dsp^2$ |

In $[Ni(CN)_4]^{2-}$ molecule hybrid MOs involve only one *d*-orbital of the central atom.

**60.** *(b)* The pairs in which both the species have same magnetic moment is $[Cr(H_2O)_6]^{2+}$ $(3d^4)$ and $[Fe(H_2O)_6]^{2+}$ $3d^6$ due to presence of same number of unpaired electron. i.e., $n = 4$

**61.** *(d)* Under octahedral crystal field, the configuration of $Ti^{3+}$ will be, $t_{2g}^1 e_g^0$.

So, CFSE $= [(-0.4) \times 1 + (+0.6) \times 0]\,\Delta_0$

$= -0.4 \times \Delta_0 = -0.4 \times 20{,}300\text{ cm}^{-1}$ $\quad [\because \Delta_0 = 20{,}300\text{ cm}^{-1}]$

$= -0.4 \times \dfrac{20{,}300}{83.7}\text{ kJ mol}^{-1}$ $\quad [\because 1\text{ kJ mol}^{-1} = 83.7\text{ cm}^{-1}]$

$= -97.013\text{ kJ mol}^{-1} \Rightarrow \text{CFSE} = 97\text{ kJ mol}^{-1}$

**62.** *(d)* $\underset{(d^6)}{M^{2+}} + mH_2O \longrightarrow [M(H_2O)_m]^{2+}$

Here, $H_2O$ is a weak field ligand and it should give a high spin, paramagnetic ($\mu = 4.90$ BM) complex ion.

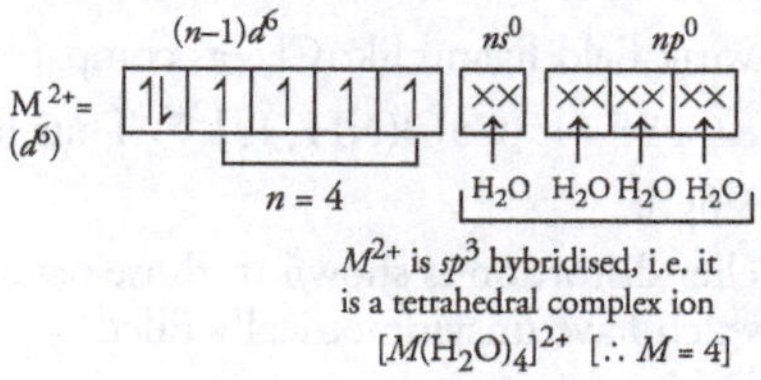


So, the complex will experience splitting of $(n-1)\,d$-orbitals in tetrahedral field. $(e_g^3\, t_g^3)$

CFSE $= [(-0.6) \times 3 + (0.4) \times 3]\Delta_t = -0.6\,\Delta_t$

So, the geometry and the crystal field stabilisation energy of the complex is tetrahedral and $-0.6\,\Delta_t$ respectively.

**63.** *(d)* (A) In $Ni(CO)_4$ Ni has $3d^{10}4s^2$ configuration ($sp^3d^2$) and tetrahedral (strong field ligand).

$\therefore$ Unpaired electrons = 0, $\mu = 0$ BM

(B) In $[Ni(H_2O)_6]Cl_2$ $Ni^{2+}$ has ($3d^8$) configuration ($sp^3d^2$) and has two unpaired electrons.

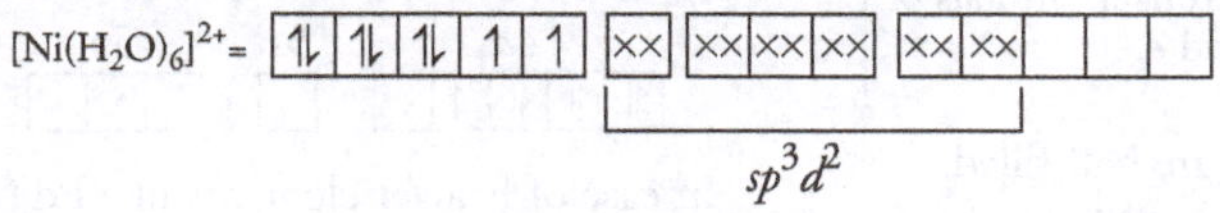


$$\mu = \sqrt{n(n+2)} = \sqrt{2\times 4}$$
$$= \sqrt{8}\text{ B.M} = 2.82\text{ BM}$$

(C) In $Na_2[Ni(CN)_4]$ $Ni^{2+}$ has ($d^8$) configuration and square planar (strong field ligand).

$\therefore$ $dsp^2$ hybridisation and no unpaired electron

$\therefore$ $\mu = 0$ BM

(D) In $PdCl_2(PPh_3)_2$ $Pd^{2+}$ has ($3d^8$) configuration and coordination number = 4

Complexes of $Pd^{2+}$ are square planar.

$\therefore$ $dsp^2$ hybridised.

No unpaired electron, $\mu = 0$ BM

The correct order of the calculated spin only magnetic moments of complexes is

$$(A) \approx (C) \approx (D) < (B)$$

**64.** *(a)* Complete removal of both the axial ligands (along the *z*-axis) from an octahedral complex leads to the following splitting pattern.

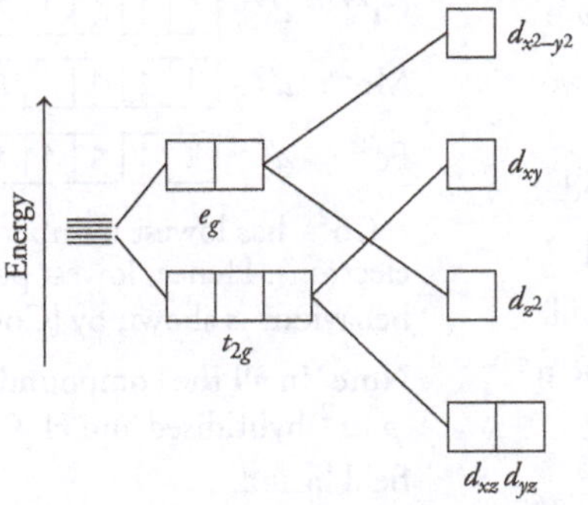


A square planar complex is formed, the attempt to form an octahedral complex being unsuccessful.

**65.** *(b)*

> **key idea** Crystal field stabilisation energy (CFSE) for octahedral complexes = $(-0.4x + 0.6y)\Delta_o$
> where,
> $x$ = number of electrons occupying $t_{2g}$ orbital.
> $y$ = number of electrons occupying $e_g$ orbital.

In $[Fe(H_2O)_6]Cl_2$, $H_2O$ is a weak field ligand, so it is a high spin (outer orbital) octahedral complex of $Fe^{2+}$.

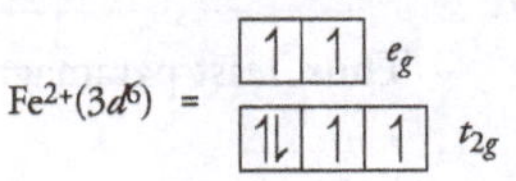


$$\therefore \text{CFSE} = (-0.4x + 0.6y)\Delta_o$$
$$= [-0.4\times 4 + 0.6\times 2]\Delta_o = -0.4\Delta_o$$

In $K_2[NiCl_4]$, $Cl^-$ is a weak field ligand, so it is a high spin tetrahedral complex of $Ni^{2+}$.

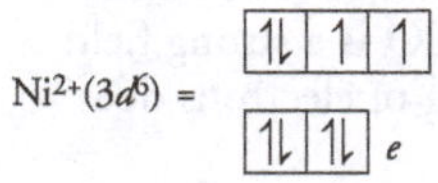


$$\therefore \text{CFSE} = (-0.6\times 4 + 0.4\times 4)\Delta_t$$
$$= -0.8\Delta_t$$

**66.** *(b)*

> **key idea** In presence of strong field ligands, $\Delta_o > P$. for fourth electron it is more energetically favourable to occupy $t_{2g}$ orbital with configuration $t_{2g}^4 e_g^0$ and form low spin complexes.

The correct order of the spin only magnetic moment of metal ions in the given low-spin complexes is $V^{2+} > Cr^{2+} > Ru^{3+} > Fe^{2+}$.

| Complex | Oxidation state | Configuration | Orbital splitting | No. of unpaired electrons |
|---|---|---|---|---|
| $[V(CN)_6]^{4-}$ | $V^{2+}$ | $t_{2g}^3 e_g^0$ | 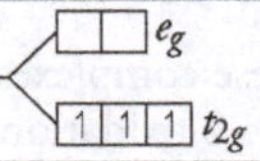 | 3 |
| $[Cr(NH_3)_6]^{2+}$ | $Cr^{2+}$ | $t_{2g}^4 e_g^0$ | 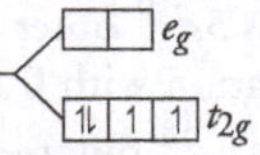 | 2 |
| $[Ru(NH_3)_6]^{3+}$ | $Ru^{3+}$ | $t_{2g}^5 e_g^0$ | 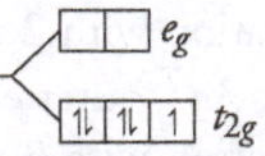 | 1 |
| $[Fe(CN)_6]^{4-}$ | $Fe^{2+}$ | $t_{2g}^6 e_g^0$ | 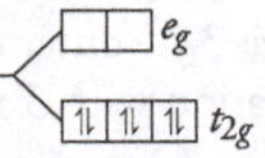 | 0 |

**67.** *(c)* '*A*' absorbs yellow light of less energy and emits violet light of high energy (complementary colour) because $H_2O$ is a weak field ligand. But in case of '*B*', due to presence of strong field ligand ($NH_3$), it absorbs high energy violet light and emits low energy complementary yellow colour.

$\Rightarrow$ Both the complexes contain three unpaired electrons. Therefore, both are paramagnetic.

$\Rightarrow$ $A(t_{2g}^3 e_g^0)$; CFSE $= (-0.4\times 3)\,\Delta_0 = -1.2\Delta_0$

$B(t_{2g}^3 e_g^0)$; CFSE $= [(-0.4\times 3) + P]\,\Delta_0 = [-1.2 + P]\,\Delta_0$

$\therefore |\text{CFSE}|_A < |\text{CFSE}|_B$

(absorbs, yellow light) (absorbs violet light)

**68.** *(b)* $[NiCl_4]^{2-}$ : Oxidation state of Ni in $[NiCl_4]^{2-} = +2$

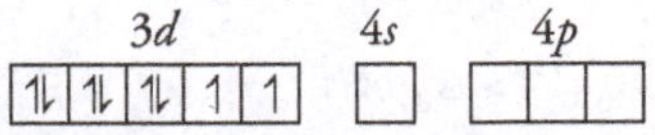


$Cl^-$ is a weak field ligand and pairing does not occur.

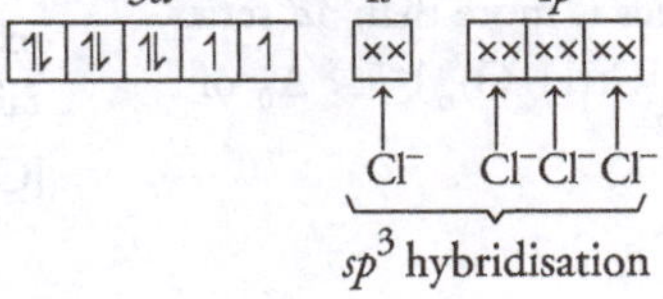

Hence, the complex is tetrahedral and paramagnetic with two unpaired electrons.

$[Ni(CO)_4]$ : Oxidation state of Ni in $[Ni(CO)_4]$ is zero. CO is a strong field ligand. Thus, pairing of electrons does takes place.

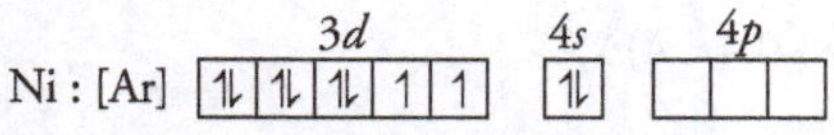


Hence, the complex is tetrahedral and diamagnetic.

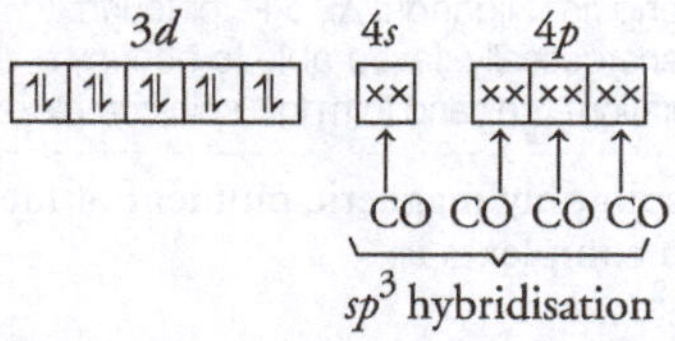


**69.** *(d)* The complex having higher number of unpaired elecrtons will have higher value of spin only magnetic moment.

In all these complexes, the central metal ion is in +2 oxidation state.

$Zn^{2+}$ has $3d^{10}$ outer electronic configuration with 0 unpaired electron.

$Ni^{2+}$ has $3d^8$ outer electronic configuration with 2 unpaired electron.

$Co^{2+}$ has $3d^7$ outer electronic configuration with 3 unpaired electron.

$Mn^{2+}$ has $3d^5$ outer electronic configuration with 5 unpaired electron.

The order of unpaired electron $Zn^{2+} < Ni^{2+} < Co^{2+} < Mn^{2+}$

Hence, the correct order of spin-only magnetic moments is $[MnCl_4]^{2-} > [CoCl_4]^{2-} > [NiCl_4]^{2-} > [ZnCl_4]^{2-}$.

**70.** *(c)* $\underset{\text{Reddish pink}}{[Co(H_2O)_6]^{2+}} + 4Cl^- \longrightarrow \underset{\text{Deep blue}}{[CoCl_4]^{2-}} + 6H_2O$

$[CoCl_4]^{2-}$ give blue colour in the reaction.

**71.** *(c)* Higher the oxidation state, higher is the splitting.

$\Delta_0$ of $[Ti(H_2O)_6]^{3+} > \Delta_0$ of $[Ti(H_2O)_6]^{2+}$

Further $\Delta_0$ also depends of $Z_{eff}$ and $Z_{eff}$ of 4*d* series is more than 3*d* series.

Hence, $\Delta_0$ of $[Cr(H_2O)_6]^{2+} < \Delta_0$ of $[Mo(H_2O)_6]^{2+}$

**72.** *(a)* Symmetrically filled $t_{2g}$ and $e_g$ are those, which has equal distribution of electrons.

$[FeF_6]^{3-}$

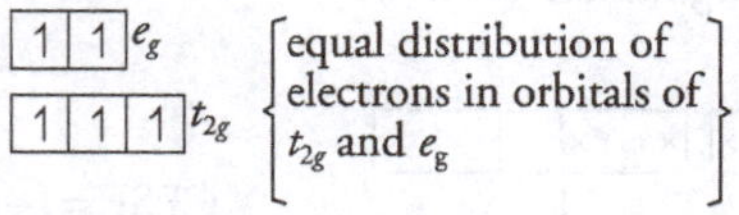


As, both $t_{2g}$ and $e_g$ are half filled.

**73.** *(a)* $Zn_2[Fe(CN)_6]$, will not show *d-d* transition due to absence of unpaired electrons. Thus, $K_3[Co(NO_2)_6]$ and $[(NH_4)_3As(Mo_3O_{10})_4]$ show colour due to *d-d* transition while $BaCrO_4$ is coloured due to charge transfer phenomenon.

Therefore, complexes containing $NO_2$, $NH_4^+$, $O^{2-}$ etc ligands may show yellow colour while $CN^-$ forces the complex to impart white colour.

**74.** *(b)* $[NiL_4]^{2-} \Rightarrow$ Diamagnetic in nature (given)

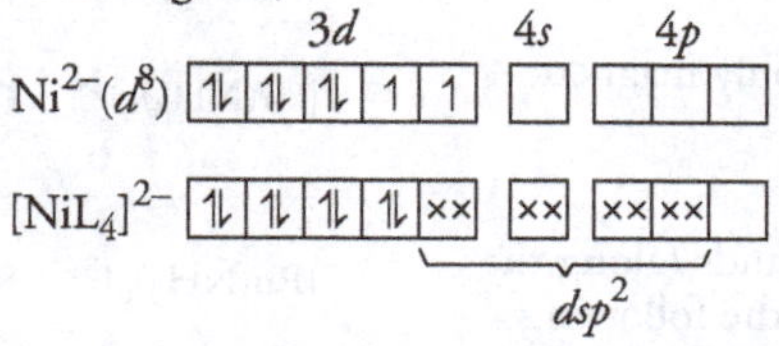


So, number of unpaired electron = 0.

Therefore, the hybridisation is $dsp^2$.

**75.** *(c)* $V^{3+}$ in $[V(NH_3)_6]^{3+}$ have two unpaired electrons whereas $[Sc(H_2O)_6]^{3+}$, $[Ti(NH_3)_6]^{4+}$ and $[Zn(NH_3)_6]^{2+}$ have no unpaired electron. Hence, $[V(NH_3)_6]^{3+}$ will most likely absorb visible light due to presence of unpaired electron.

**76.** *(c)*

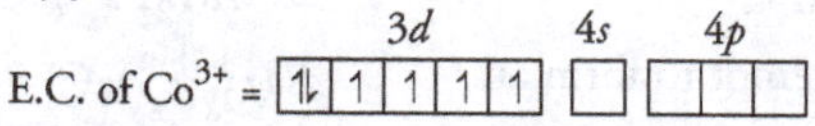


As octahedral compound of $Co^{3+}$ is diamagnetic thus pairing of electrons takes place.

∴ For the complex

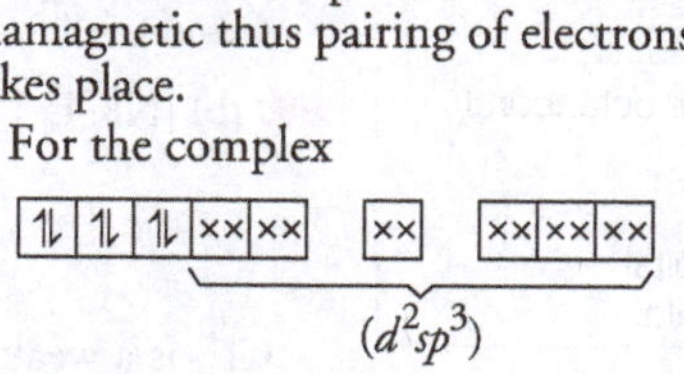


**77.** *(b)* Among four ligands, $CN^-$ and $ox^{2-}$ are strong field ligands. So pairing takes places in $[Fe(CN)_6]^{3-}$ and $[Co(ox)_3]^{3-}$ complexes. $Fe^{3+}$ having one unpaired electron in it complex while $Co^{3+}$ does not have unpaired elecrton in its complex. Thus, $[Co(ox)_3]^{3-}$ is diamagnetic.

**78.** *(a)* $[PdCl_4]^{2-}$ : $Pd^{2+}(d^8)$

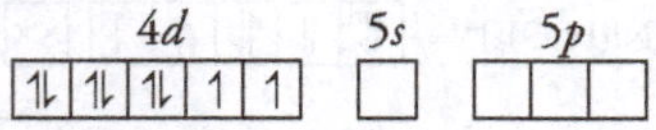


In case of heavier elements like Pd (4*d* series), the CFSE for square planar complex ($\Delta_s$) increases such that a weak field ligand can also force the $e^-$ from highest energy orbital ($d_{x^2-y^2}$) will pair with $e^-$ of lower energy orbital ($d_{xy}$).

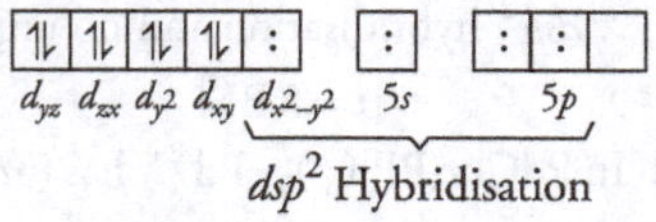


Whereas, other complexes shows $sp^3$ hybridisation of the metal (II) ion. Therefore, option (a) is correct.

**79.** *(a)* Paramagnetism is directly proportional to the number of unpaired electron.

| | Electronic configuration | | No. of unpaired electrons |
|---|---|---|---|
| $Co^{2+}$ | $d^7$ | ↑↓ ↑↓ ↑ ↑ ↑ | 3 |
| $Cr^{2+}$ | $d^4$ | ↑ ↑ ↑ ↑ | 4 |
| $Mn^{2+}$ | $d^5$ | ↑ ↑ ↑ ↑ ↑ | 5 |
| $Fe^{2+}$ | $d^6$ | ↑↓ ↑ ↑ ↑ ↑ | 4 |

∴ $Co^{2+}$ has lowest number of unpaired electrons. Hence, lowest paramagnetic behaviour is shown by $[Co(H_2O)_6]^{2+}$.

**Note** In all the compounds, $M^{2+}$ is $sp^3d^2$ hybridised and $H_2O$ is a weak field ligand.

**80.** *(a)*

| | Hybridisation Unpaired electrons | Magnetic moment |
|---|---|---|
| (a) $[Co(CN)_6]^{3-}$ | $d^2sp^3$ (0) | 0 |
| (b) $[Fe(CN)_6]^{3-}$ | $d^2sp^3$ (1) | $\sqrt{3}$ BM |
| (c) $[Mn(CN)_6]^{3-}$ | $d^2sp^3$ (2) | $\sqrt{8}$ BM |
| (d) $[Cr(CN)_6]^{3-}$ | $d^2sp^3$ (3) | $\sqrt{15}$ BM |

Thus, least paramagnetism is (a).

**81.** *(a)* Number of unpaired electrons in $[Fe(CN)_6]^{4-}$ is zero.

Thus, magnetic moment

$$=\sqrt{n(n+2)}=0 \text{ BM}$$

($n$ = unpaired electrons)

$n$ in $[MnCl_4]^{2-}=5, \mu=\sqrt{35}$ BM

$n$ in $[CoCl_4]^{2-}=3, \mu=\sqrt{15}$ BM

**82.** *(c)* Stability of complex increases with increase in charge on the central metal ion and with increase in basic strength of ligand i.e., with strong field ligand. In $[Fe(CN)_6]^{3-}$, oxidation state of Fe is +3 and basicity of $CN^-$ is higher than $OH^-$, $Cl^-$ and $H_2O$.

**83.** *(c)* As we know that $E=\frac{hc}{\lambda}$

$$E \propto \frac{1}{\lambda}$$

Here in all the species Co is in +3 oxidation state. So, as the ligand field strength increases, CFSE also increases. So, the order of field strength of ligand is $CN^- > NH_3 > H_2O > Cl^-$

(C) (B) (A) (D)

So, the order of wavelength is $D > A > B > C$

**84.** *(a)*

- $[Fe(CN)_6]^{3-}$ $Fe^{3+}$ $3d^5$ $t_{2g}^{2,2,1}$ $e_g^{0,0}$ unpaired $e^-=1$
- $[FeF_6]^{3-}$ $Fe^{3+}$ $3d^5$ $t_{2g}^{1,1,1}$ $e_g^{1,1}$ unpaired $e^-=5$
- $[CoF_6]^{3-}$ $Co^{3+}$ $3d^6$ $t_{2g}^{2,1,1}e_g^1$ unpaired $e^-=4$
- $[Mn(CN)_6]^{3-}$ $Mn^{3+}$ $3d^4$ $t_{2g}^{2,1,1}$ $e_g^{0,0}$ unpaired $e^-=2$

**85.** *(d)* The correct match is A-III, B-II, C-I, D-IV

- $[Ni(CO)_4]$, $Ni^0:[Ar]3d^8 4s^2$

Valence orbitals of $Ni^0$ in pre-hydridisation state :

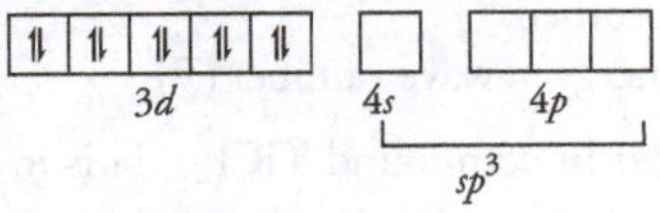


Tetrahedral, diamagnetic, $\mu_s=0$ BM.

- $[Ni(CN)_4]^{2-}$, $Ni^{2+}:[Ar]3d^8 4s^0$

Valence orbitals of $Ni^{2+}$ in pre-hydridisation state.

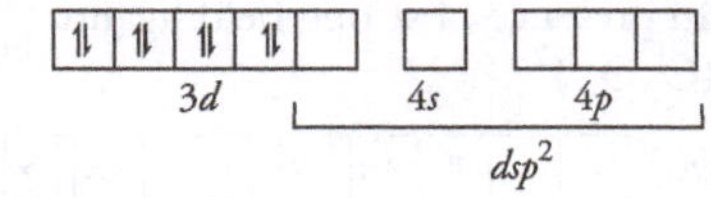


Square planar, diamagnetic, $\mu_s=0$ B.M.

- $[NiCl_4]^{2-}$, $Ni^{2+}$ ; $[Ar]3d^8 4s^0$

Valence Orbitals of $Ni^{2+}$ in ground state:

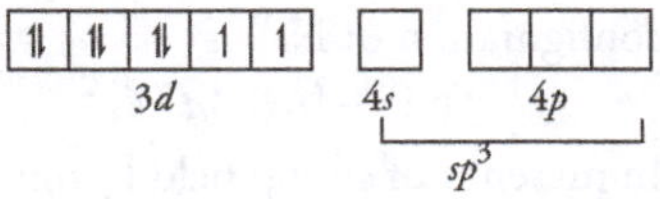


Tetrahedral, paramagnetic, $\mu_s=\sqrt{8}=2.8$ B.M.

- $[MnBr_4]^{2-}$, $Mn^{2+}:[Ar]3d^5$

Valence orbitals of $Mn^{2+}$ in ground state

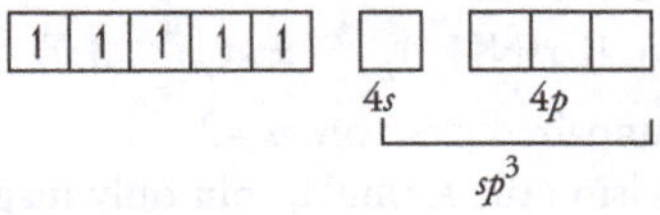


Tetrahedral, paramagnetic, $\mu_s=\sqrt{35}=5.9$ B.M.

**86.** *(b)*

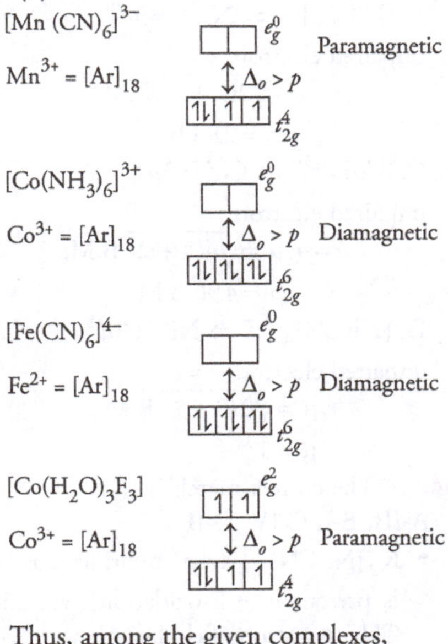


Thus, among the given complexes, diamagnetic octahedral complex ions are $[Co(NH_3)_6]^{3+}$ and $[Fe(CN)_6]^{4-}$.

**87.** *(a)* In $[Co(en)_3]^{3+}$, strong field ligand is present and hence highest value of CFSE.

In rest complexes, weak field ligand are present and hence CFSE is low.

$[Co(en)_3]^{3+} \Rightarrow Co^{3+}$ and en(SFL)

$Co^{3+} \Rightarrow [Ar]3d^6$

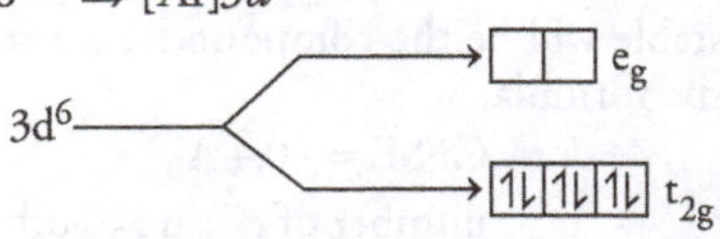


$\therefore$ CFSE is $t_{2g}^6 e_g^0$.

**88.** *(d)* In complex $[NiCl_4]^{2-}$, Ni is in + 2 oxidation state. So, configuration of $Ni^{2+}=[Ar]3d^8$. Also, since Cl is a weak field ligand. So, the geometry of compound is tetrahedral with $sp^3$ hybridisation.

Similarly in complex $[Ni(CO)_4]$, Ni is in zero oxidation state. So, configuration of $Ni^0=[Ar]3d^8 4s^2$. Also, since CO is a strong field ligand. So, after rearrangement the configuration of Ni $=[Ar]3d^{10} 4s^0$ So, the geometry of the compound is tetrahedral with $sp^3$ hybridisation.

So, option (d) is the correct answer.

**89.** *(a)*

(A) $[Fe(CN)_5NO]^{2-}$ is a heteroleptic complex with $d^2sp^3$ hybridisation, low spin (as strong field ligand).

(B) $[CoF_6]^{3-}$ is a homoleptic complex with $sp^3d^2$ hybridisation. It is high spin (as weak field ligand).

(C) $[Fe(CN)_6]^{4-}$ is a homoleptic complex, with $d^2sp^3$ hybridisation low spin (as strong field ligand).

(D) $[Co(NH_3)_6]^{3+}$ is a homoleptic complex with, $d^2sp^3$ hybridisation low spin (as Co is in + 3 oxidation state. Thus, $NH_3$ will act (as strong field ligand).

(E) $[Cr(H_2O)_6]^{2+}$ is a homoleptic complex with $d^2sp^3$ hybridisation high spin (as $H_2O$ is a weak field ligand).

Thus, option a is correct.

**90.** *(a)* As the charge on complex increases, the value of CFSE increases. Also, as the strength of ligand increases the value of CFSE increases.

Thus, the correct order of CFSE is $[Co(NH_3)_4]^{2+} < [Co(NH_3)_6]^{2+} < [Co(NH_3)_6]^{3+} < [Co(en)_3]^{3+}$

**91.** *(a)* The less the value of CFSE the less stable will be the compound, so using the formula,

$$CFSE = -0.4\,\Delta_0 \text{ (number of } e^- \text{ in } t_{2g} \text{ orbital)}$$
$$+0.6\,\Delta_0 \text{ (number of } e^- \text{ in } e_g \text{ orbital)}$$

For $[Mn(CN)_6]^{3-}$ the electronic configuration is $t_{2g}^4 e_g^0$.

On substituting the values in above formula, we get

$$CFSE = -1.6\,\Delta_0$$

Similarly for $[Co(CN)_6]^{4-}$, $[Fe(CN)_6]^{4-}$, $[Fe(CN)_6]^{3-}$ the value of CFSE are $-1.8\Delta_0, -2.4\Delta_0, -2.0\Delta_0$ respectively

Thus, the correct increasing order of stability is

$[Mn(CN)_6]^{3-} < [Co(CN)_6]^{4-} < [Fe(CN)_6]^{3-} < [Fe(CN)_6]^{4-}$

i.e., I < II < IV < III

**92.** *(a)* The correct match is A-IV, B-II, C-I, D-III.

A. $[MnBr_4]^{2-}$

In the given complex Mn is in +2 oxidation state. So, electronic configuration of $Mn^{2+}$ is

$$Mn^{2+} \rightarrow [Ar]\,3d^5$$

In presence of weak field ligand, Br

[Ar] | 1 | 1 | 1 | 1 | 1 | (3d) | ×× | (4s) | ×× | ×× | ×× | (4p)

$= sp^3$ hybridisation, paramagnetic

B. $[FeF_6]^{3-}$

In the complex Fe is in +3 oxidation state. So, electronic configuration of $Fe^{3+}$ is

$$Fe^{3+} = [Ar]\,3d^5$$

In presence of weak field ligand, F

[Ar] | 1 | 1 | 1 | 1 | 1 | (3d) | ×× | (4s) | ×× | ×× | ×× | (4p) | ×× | ×× | | | | (4d)

$sp^3d^2$ hybridisation, paramagnetic

C. $[Co(C_2O_4)_3]^{3-}$

In the complex Co is in +3 oxidation state. So, electronic configuration of $Co^{3+}$ is

$$Co^{3+} = [Ar]3d^6$$

In presence of strong field ligand, $(C_2O_4)^{2-}$

[Ar] | ⥮ | ⥮ | ⥮ | ×× | ×× | (3d) | ×× | (4s) | ×× | ×× | ×× | (4p)

$d^2sp^3$ hybridisation, diamagnetic

D. $[Ni(CO)_4]$

In this complex Ni is present in zero oxidation state. So, electronic configuration of $Ni^0$ is

$$Ni^0 = [Ar]\,3d^8 4s^2$$

In presence of strong field ligand, CO

| ⥮ | ⥮ | ⥮ | ⥮ | ⥮ | (3d) | ×× | (4s) | ×× | ×× | ×× | (4p)

$sp^3$ hybridisation, diamagnetic

**93.** *(c)* The correct match is A-II, B-IV, C-I, D-III

A. $[Cr(NH_3)_6]^{3+} \Rightarrow Cr^{3+}; 3d^3$

unpaired electron, $n = 3$

Using the formula, spin only magnetic moment

$$= \sqrt{n(n+2)} \text{ B.M.}$$
$$\mu = 3.87 \text{ B.M.}$$

B. $[NiCl_4]^{2-} \Rightarrow Ni^{2+} = 3d^8$

unpaired electron, $n = 2$,

$$\mu = \sqrt{2(2+2)} \text{ B.M.}$$
$$\mu = 2.83 \text{ B.M.}$$

C. $[CoF_6]^{3-} \Rightarrow Co^{3+}; 3d^6$

unpaired electron,

$$n = 4, \mu = \sqrt{4(4+2)} \text{ B.M.}$$
$$\mu = 4.90 \text{ B.M.}$$

D. $[Ni(CN)_4]^{2-} \Rightarrow Ni^{2+}; 3d^8$

unpaired electron, $n = 0$,

$$\mu = \sqrt{0(0+2)} \text{ B.M.}$$
$$\mu = 0$$

**94.** *(b)* The correct match is A-III, B-I, C-IV, D-II

- $K_2[Ni(CN)_4]$ is $dsp^2$ hybridised as Ni is present in +2 oxidation state and $CN^-$ is a strong field ligand.
- $[Ni(CO)_4]$ is $sp^3$ hybridised as Ni is present in zero oxidation state.
- $[Co(NH_3)_6]Cl_3$, Co is $d^2sp^3$ hybridised as $NH_3$ acts as strong field ligand for $Co^{3+}$.
- $Na_3[CoF_6]$ is $sp^3d^2$ hybridised, F act as weak field ligand.

**95.** *(b)* $[Fe(H_2O)_6]^{2+}$

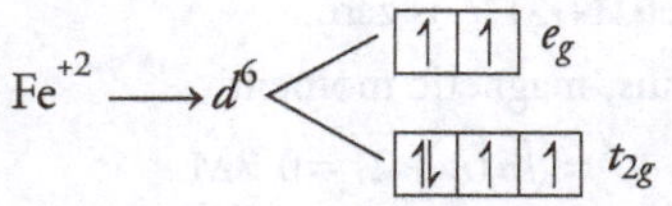


Electron in $t_{2g} = 4$ (even)

$[Co(H_2O)_6]^{2+}$

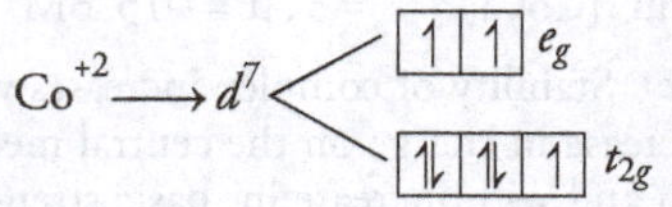


Electron in $t_{2g} = 5$ (odd)

$[Co(H_2O)_6]^{3+}$

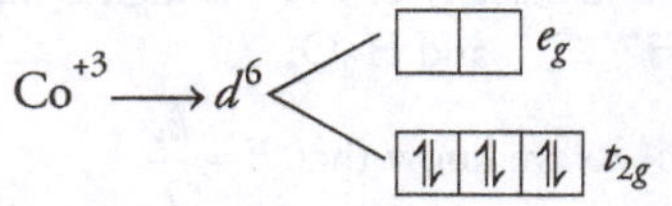


Electron in $t_{2g} = 6$ (even)

$[Cu(H_2O)_6]^{2+}$

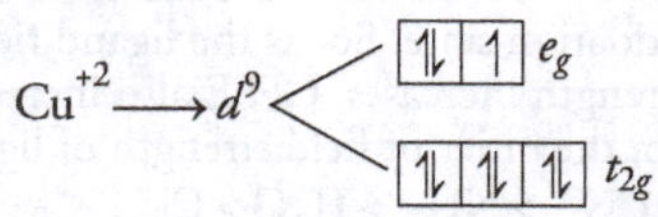


Electron in $t_{2g} = 6$ (even)

$[Cr(H_2O)_6]^{2+}$

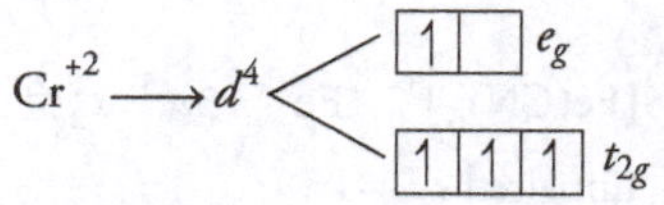


Electron in $t_{2g} = 3$ (odd)

Hence, total 3 complexes have even no. of electrons.

**96.** *(c)* The correct match is, A-III, B-I, C-IV, D-II

**97.** *(b)* The correct order of wave number of light absorbed is

$[Cu(H_2O)_4]^{2+} < [CoCl(NH_3)_5]^{2+} < [Co(NH_3)_5(H_2O)]^{3+} < [Co(CN)_6]^{3-}$

The reason is, as ligand field strength increases, light of more energy is absorbed.

Energy $\propto$ wave number $(\bar{\nu})$

**98.** *(a)* In compound $TiCl_4$, Ti is in +4 oxidation state. Hence its configuration will be $e^0 t_2^0$.

Similarly, for other complexes their configuration is given as

$[MnO_4]^- = Mn^{7+} = e^0 t_2^0$

$[FeO_4]^{2-} = Fe^{6+} = e^2 t_2^0$

$[FeCl_4]^{2-} = Fe^{2+} = e^3 t_2^3$

$[CoCl_4]^{2-} = Co^{2+} = e^4 t_2^3$

Hence, in $[MnO_4]^-$, $[FeO_4]^{2-}$ and $TiCl_4$ does not have any electron in $t_2$ orbital. Hence, option (a) is correct.

**99.** *(c)* $[Mn(H_2O)_6]^{2+} = 5$

(number of unpaired electrons)

$\mu = \sqrt{n(n+2)}$ B.M

$\mu = \sqrt{5(5+2)}$ B.M $= 5.92$ B.M

$[Fe(H_2O)_6]^{2+} = 4$

(number of unpaired electrons)

$[Co(H_2O)_6]^{2+} = 3$

(number of unpaired electrons)

$\mu = \sqrt{n(n+2)}$ B.M

$= \sqrt{3(3+2)}$ B.M $= 3.87$ B.M

$[Cr(H_2O)_6]^{2+} = 4$

(number of unpaired electrons)

$\mu = \sqrt{n(n+2)}$ B.M

$= \sqrt{4(4+2)}$ B.M

$= 4.89$ B.M

Hence, least paramagnetic behaviour is shown by $= [Co(H_2O)_6]^{2+}$

**Time Saver Tech**

Connect paramagnetic behaviour with number of unpaired electrons directly. The complex with least number of unpaired electrons will show least paramagnetic behaviour.

**100.** *(d)* The correct match is A-II, B-III, C-IV, D-I

$[Cr(H_2O)_6]^{3+}$ contains $Cr^{3+}$ :

$[Ar]3d^3 : t_{2g}^3 e_g^0$

$[Fe(H_2O)_6]^{3+}$ contains $Fe^{3+}$ :

$[Ar]3d^5 : t_{2g}^3 e_g^2$

$[Ni(H_2O)_6]^{2+}$ contains $Ni^{2+}$ :

$[Ar]3d^8 : t_{2g}^6 e_g^2$

$[V(H_2O)_6]^{3+}$ contains $V^{3+}$ :

$[Ar]3d^2 : t_{2g}^2 e_g^0$

**101.** *(b)* Statement given in B and C are correct while statements A and D are incorrect. Their correct form are as follows

- The strength of anionic ligands cannot be explained by crystal field theory.
- The number of possible isomers of *cis* $[PtCl_2(en)_2]^{2+}$ is more than one.

**102.** *(c)* Magnetic moment is given by

$\mu_{\text{spin only}} = \sqrt{n(n+2)}$ BM

where, $n$ is number of unpaired electrons.

(*P*) $[FeF_6]^{3-}$ has Fe in +3 oxidation state.

$Fe^{3+} : 3d^5$

So, it has 5 unpaired electrons as $F^-$ is a weak field ligand.

$\therefore\ \mu = \sqrt{5(5+2)}$

$= \sqrt{35}$ BM (*P*)

(*Q*) $[V(H_2O_6)]^{2+}$ has V in + 2 oxidation state.

$V^{2+} : 3d^3$

So, it has 3 unpaired electrons also, $H_2O$ is a weak field ligand.

$\mu = \sqrt{3(3+2)}$

$= \sqrt{15}$ BM (*Q*)

(R) $[Fe(H_2O)_6]^{2+}$ has Fe in +2 oxidation state.

$Fe^{2+} : 3d^6$

So, it has 4 unpaired electrons,

$\mu = \sqrt{4(4+2)}$

$= \sqrt{24}$ BM (*R*)

Hence, the correct sequence of spin only magnetic moment is $Q < R < P$.

**103.** *(d)* The correct match is A-II, B-III, C-I, D-IV.

| Complex | Colour |
|---|---|
| (A) $Mg(NH_4)PO_4$ | II. White |
| (B) $K_3[Co(NO_2)_6]$ | III. Yellow |
| (C) $MnO(OH)_2$ | I. Brown |
| (D) $Fe_4[Fe(CN)_6]_3$ | IV. Blue |

**104.** *(b)* (a) $[Ti(OH_2)_6]^{3+} \Rightarrow t_{2g}^1\ e_g^0$

$\Rightarrow$ CFSE

$= [-0.4 \times 1 + 0.6 \times 0]\ \Delta_0 = -0.4\Delta_0$

(b) $[Cr(OH_2)_6]^{3+} \Rightarrow t_{2g}^3 e_g^0$

$\Rightarrow$ CFSE $= [-0.4 \times 3 + 0.6 \times 0]\ \Delta_0$

$= -1.2\ \Delta_0$

(c) $[Fe(OH_2)_6]^{3+} \Rightarrow t_{2g}^3 e_g^2$

$\Rightarrow$ CFSE $= [-0.4 \times 3 + 0.6 \times 2]\ \Delta_0 = 0$

(d) $[Mn(OH_2)_6]^{3+} \Rightarrow t_{2g}^3 e_g^1$

$\Rightarrow$ CFSE $= [-0.4 \times 3 + 0.6 \times 1]\ \Delta_0$

$= -0.6\ \Delta_0$

$\Rightarrow$ So, $|\text{CFSE}|_{max}$ is for option (b).

**105.** *(a)* More the number of unpaired electron in the *d*-orbital more it will be affected by magnetic field.

$[Co(H_2O)_6]^{2+} \longrightarrow Co^{2+} = 3d^7$

$\Rightarrow t_{2g}^5 e_g^2$

$\Rightarrow$ unpaired electron $= 3$

$[Ni(H_2O)_6]^{2+} \longrightarrow Ni^{2+}$

$\Rightarrow 3d^8 \Rightarrow t_{2g}^6 e_g^2$

$\Rightarrow$ unpaired electron $= 2$

$[Zn(H_2O)_6]^{2+} \longrightarrow Zn^{2+} \Rightarrow 3d^{10}$

$\Rightarrow t2_g^6 e_g^4$

$\Rightarrow$ unpaired electron $= 0$

$[Co(en)_3]^{3+} \longrightarrow Co^{3+} \Rightarrow 3d^6$

$\Rightarrow t_{2g}^6 e_g^0$

$\Rightarrow$ unpaired electron $= 0$

**106.** *(c)* The correct match is A-I, B-IV, C-II, D-III.

Use formula CFSE

$= (-0.4\ nt_{2g} + 0.6ne_g)\ \Delta_0$

where,

$nt_{2g}$ = number of electrons $t_{2g}$ orbitals

$ne_g$ = number of electrons in $ne_g$ orbitals

(A) $[Cu(NH_3)_6]^2$ has $9e^-$ in *d*-orbital

i.e. $t_{2g}^6 e_g^3$

CFSE $= (-0.4 \times 6 + 0.6 \times 3)\ \Delta_0$

$= -0.6\Delta_0$

(B) $[Ti(H_2O)_6]^{3-}$ has $1e^-$ in *d*-orbital

i.e. $t_{2g}^1 e_g^0$

CFSE $= (-0.4 \times 1 + 0.6 \times 0)\ \Delta_0$

$= -0.4\ \Delta_0$

(C) $[Fe(CN)_6]^{3-}$ has $5e^-$ in *d*-orbital,

i.e. $t_{2g}^5 e_g^0$

CFSE $= (-0.4 \times 5 + 0.6 \times 0) = -2\Delta_0$

(D) $[NiF_6]^{4-}$ has $8e^-$ in *d*-orbital, i.e.

$t_{2g}^6 e_g^2$

CFSE $= (-0.4 \times 6 + 0.6 \times 2)$

$= -1.2\Delta_0$.

**107.** *(c)* The correct match is A-IV, B-I, C-II, D-III.

(A) $[Ti(H_2O)_6]^{2+}$

$Ti^{2+} \Rightarrow 3d^2 4s^0$

$\Rightarrow\ t_{2g}e^- = 2,\ e_g e^- = 0$

CFSE $= [-0.4 \times 2 + 0.6 \times 0]\ \Delta_0$

$= -0.8\ \Delta$

(B) $[V(H_2O)_6]^{2+}$

$V^{2+} \Rightarrow 3d^3 4s^0$

$\Rightarrow\ t_{2g}e^- = 3$

$\Rightarrow\ e_g e^- = 0$

CFSE $= [-0.4 \times 3 + 0.6 \times 0]\ \Delta_0 = -1.2\ \Delta_0$

(C) $[Mn(H_2O)_6]^{3+}$

$Mn^{3+} \Rightarrow 3d^4\,4s^0$

$\Rightarrow\ t_{2g}e^- = 3,\ e_g e^- = 1$

CFSE $= [-0.4 \times 3 + 0.6 \times 1]\ \Delta_0$

$= -0.6\ \Delta_0$

(D) $[Fe(H_2O)_6]^{3+}$

$Fe^{3+} \Rightarrow 3d^5\,4s^0$

$\Rightarrow\ t_{2g} = 3, e = 2$

CFSE $= [-0.4 \times 3 + 0.6 \times 2]\ \Delta_0 = 0$

**108.** *(d)* The correct order of the number of unpaired electrons in the given complexes is E < A < D < C < B.

(A) $[Fe(CN)_6]^{3-}$ $n = 1$

(B) $[FeF_6]^{3-}$ $n = 5$

(C) $[CoF_6]^{3-}$ $n = 4$

(D) $[Cr(oxalate)_3]^{3-}$ $n = 3$

(E) $[Ni(CO)_4]$ $n = 0$

**109.** *(a)*

| Complex | Configuration | Number of unpaired electrons |
|---|---|---|
| (A) $[Cr(CN)_6]^{3-}$ | $3d^3$: [1 \| 1 \| 1 \| xx \| xx] $4s^0$: [xx] $4p$: [xx \| xx \| xx] | 3 (II) |
| (B) $[Fe(H_2O)_6]^{2+}$ Weak field ligand | $3d^6$: [1l \| 1 \| 1 \| 1 \| 1] $4s^0$: [xx] $4p$: [xx \| xx \| xx] $4d$: [xx \| xx \| \| \| ] | 4 (IV) |
| (C) $[Co(NH_3)_6]^{3+}$ Strong field ligand | $3d^6$: [1l \| 1l \| 1l \| xx \| xx] $4s^0$: [xx] $4p$: [xx \| xx \| xx] | 0 (I) |
| (D) $[Ni(NH_3)_6]^{3-}$ Strong field ligand | $3d^8$: [1l \| 1l \| 1l \| 1 \| 1] $4s^0$: [xx] $4p$: [xx \| xx \| xx] $4d$: [xx \| xx \| \| \| ] | 2 (III) |

**110.** *(c)* The correct match is A-I, B-II, C-IV, D-III.

**111.** *(b)* The correct match is A-III, B-II, C-I, D-IV.

Ligand field strength : ($CN^- > NH_3 > H_2O > Cl^-$) $\propto E_{abs} \propto \frac{1}{\lambda_{abs}}$

**112.** *(b)* The correct match is A-I, B-IV, C-III, D-II.

**113.** *(a)*

| Metal complex | Configuration of central metal atom | Magnetic behaviour |
|---|---|---|
| $K_3[Cu(CN)_4]$ | $Cu^+(3d^{10}4s^0)$ | Diamagnetic (All electrons paired) |
| $K_2[Cu(CN)_4]$ | $Cu^{2+}(3d^9 4s^0)$ | Paramagnetic (1 unpaired electron) |
| $K_3[Fe(CN)_6]$ | $Fe^{3+}(3d^5 4s^0)$ | Paramagnetic (1 unpaired electron) |
| $K_4[FeCl_6]$ | $Fe^{2+}(3d^6 4s^0)$ $(t_{2g}^4\, e^2)$ | Diamagnetic (4 unpaired electrons) |

**114.** *(d)* Magnetic moment $\propto$ No. of unpaired electrons

(A) ${}_{58}Ce \longrightarrow [Xe]\,4f^2 5d^0 6s^2$

In this complex $Ce^{+4} \longrightarrow [Xe]\,4f^0 5d^0 6s^0$

There is no unpaired electron, so $\mu_m = 0$ BM

(B) ${}_{64}Gd \longrightarrow [Xe]4f^7 5d^1 6s^2$

In this complex, ${}_{64}Gd^{2+} \longrightarrow [Xe]4f^7 5d^1$

There are 8 unpaired electrons.

(C) ${}_{63}Eu \longrightarrow [Xe]4f^9 5d^0 6s^0$

${}_{63}Eu^{+3} \longrightarrow [Xe]\,4f^6 5d^0 6s^0$

In this complex six unpaired electrons are present.

Hence, order of spin only magnetic moment $A < C < B$.

**115.** *(a)* $\mu \propto n$.

(i) $[FeF_6]^{3-}$

$\Rightarrow Fe^{3+} = (3d^5)$

(weak field ligand). $(n = 5)$

(ii) $[Co(NH_3)_6]^{3+}$

$\Rightarrow Co^{3+} = (3d^6)$ (strong field ligand). Thus, pairing of electron takes place. $(n = 0)$

(iii) $[NiCl_4]^{2-} \Rightarrow Ni^{2+} = (3d^8)$

(weak field ligand). Thus, pairing of electron will not takes place. $(n = 2)$

(iv) $[Cu(NH_3)_4]^{2+} \Rightarrow$

$Cu^{2+}(3d^9)$ (strong field ligand). Thus, pairing of electron takes place. $(n = 1)$

So, the decreasing order of $\mu$ is

$$\underset{(n=5)}{\mu_i} > \underset{(n=2)}{\mu_{iii}} > \underset{(n=1)}{\mu_{iv}} > \underset{(n=0)}{\mu_{ii}}$$

**116.** *(a)* We have to compare their CFSE $(\Delta_0)$ as well as wavelength $(\lambda)$ values, where

$\Delta E(\text{CFSE}) = \frac{hc}{\lambda}$ or, $\Delta_0 \propto \frac{1}{\lambda}$

Power of ligand $\propto \Delta_0 \propto \frac{1}{\lambda}$,

$F^- < NCS^- < NH_3; \bar{F} > NC\bar{S} > NH_3$

$\longrightarrow$ Power of Ligand; $\longleftarrow \lambda \longrightarrow \Delta_o$

So, for the given complexes, it is evidented from the plot (Absorption vs. λ)

$$\lambda_{max}^{(ii)} > \lambda_{max}^{(i)} > \lambda_{max}^{(iii)}$$
$$(C) \quad (B) \quad (A)$$
$$[F^-] \quad [NCS^-] \quad [NH_3]$$

**117.** *(a)* $Pd^{2+}$ complexes with coordination number = 4 are of square planar geometry, irrespective of the field strength of ligands.

For square planar complex with four different ligands, three, geometrical isomers are possible.

$[Fe(CN)_6]^{n-6} = [Fe(CN)_6]^{-3}$. So, oxidation state of iron in $[Fe(CN)_6]^{-3}$ is +3. $Fe^{3+} = d^5$.

It forms a low spin complex ($CN^-$ is a strong field ligand).

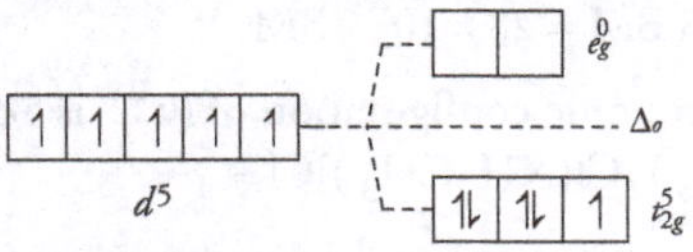

∴ Number of unpaired electrons $(n) = 1$

Spin only magnetic moment $\mu_s = \sqrt{n(n+2)}$ BM

$= \sqrt{3} = 1.73$ BM

CFSE $= (t_{2g}e^- \times -0.4\Delta_o) + (e_g e^- \times 0.6\Delta_o)$

$= 5 \times -0.4\Delta_o = -2.0\Delta_o$.

**118.** *(b)*

| Complex | Electronic configuration | Number of unpaired electron | Magnetic moment (BM) |
|---|---|---|---|
| $[Cr(H_2O)_6]Br_2$ | $t_{2g}^3 e_g^1$ | 4 | $\sqrt{24}$ |
| $Na_4[Fe(CN)_6]$ | $t_{2g}^6 e_g^0$ | 0 | 0 |
| $Na_3[Fe(C_2O_4)_3]$ $(\Delta_0 > P)$ | $t_{25g}^5 e_g^0$ | 1 | 1.73 |
| $[Et_4N)_2[CoCl_4]$ | $e^4 t_2^3$ | 3 | $\sqrt{15}$ |

Order is I > IV > III > II.

**119.** *(d)* The complex ion that will lose its crystal field stabilisation energy upon oxidation of its metal to +3 state is $[Fe(phen)_3]^{2+}$.

$$[Fe(phen)_3]^{2+} \xrightarrow{-e^-} [Fe(phen)_3]^{3+}$$

In $[Fe(phen)_3]^{2+}$, electronic configuration of $Fe^{2+}$ is $3d^6 4s^0$.

Phenanthrene is a strong field symmetrical bidentate ligand.

The splitting of orbital in $Fe^{2+}$ is as follows

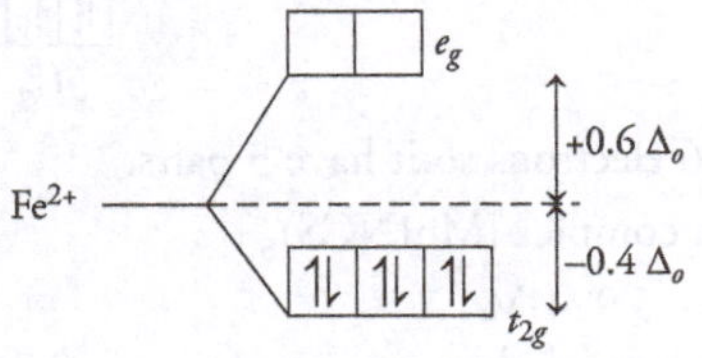

CFSE $= 6 \times -0.4\Delta_o = -2.4\Delta_o$.

The splitting of orbital and arrangement of electrons in $Fe^{3+}$ is as follows

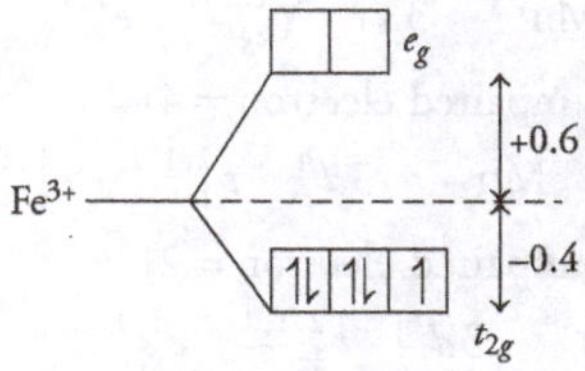

CFSE $= 5 \times -0.4\,\Delta_o = -2.0\,\Delta_o$

$Fe^{2+}$ upon oxidation of its metal to +3 state lose its CFSE from $-2.4\,\Delta_o$ to $-2.0\Delta_o$.

**120.** *(c)*

| Metal ion | No. of unpaired elec. High spin | No. of unpaired elec. Low spin | Difference in the unpaired electrons |
|---|---|---|---|
| $Ni^{2+}(3d^8)$ | 2 | 2 | 0 |
| $Mn^{2+}(3d^5)$ | 5 | 1 | 4 |
| $Fe^{2+}(3d^6)$ | 4 | 0 | 4 |
| $Co^{2+}(3d^7)$ | 3 | 1 | 2 |

Thus, $Co^{2+}$ has difference of 2.

**121.** *(0)* $K_3[Co(NO_2)_6] \rightarrow$ Zero unpaired electron.

$K_4[Fe(CN)_6] \rightarrow$ Zero unpaired electron.

$K_3[Fe(CN)_6] \rightarrow$ One unpaired electron.

$Cu_2[Fe(CN)_6] \rightarrow$ Zero unpaired electron.

$Zn_2[Fe(CN)_6] \rightarrow$ Zero unpaired electron.

Yellow colour compounds are

$K_3[Co(NO_2)_6]$ and $K_4[Fe(CN)_6]$

Now, in $K_3[Co(NO_2)_6]$, the outer electronic configuration of $Co^{3+}$ is $[Ar]3d^6$

In presence of strong field ligand, $NO_2$

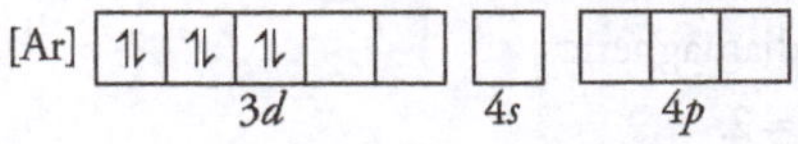

Magnetic moment $(\mu) = \sqrt{n(n+2)}$ BM

$= 0$ BM

In $K_4[Fe(CN)_6]$, $Fe^{2+}$ is $[Ar]3d^6$

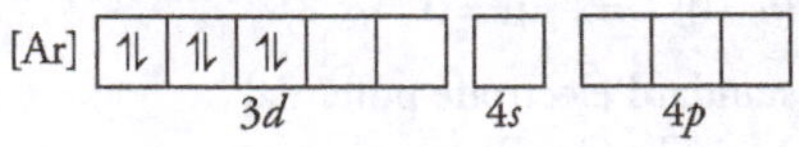

$\mu = \sqrt{n(n+2)}$ BM

$= 0$ BM

So, total spin only magnetic moment value of complex with yellow colour is zero.

**122.** *(2)*

- $[Co(NH_3)_6]^{3+}$ $Co^{3+}$ $3d^6$ $t_{2g}^{2,2,2}$ $e_g^{0,0}$

  Diamagnetic [as unpaired electron = 0]

- $[Co(C_2O_4)_3]^{3-}]$ $Co^{3+}$ $3d^6$ $t_{2g}^{2,2,2}$ $e_g^{0,0}$
Diamagnetic [unpaired electron = 0]
- $[MnCl_6]^{3-}$ $Mn^{3+}$ $3d^4$ $t_{2g}^{1,1,1}$ $e_g^{1,0}$
Paramagnetic [unpaired electron = 4]
- $[Mn(CN)_6]^{3-}$ $Mn^{3+}$ $3d^4$ $t_{2g}^{2,1,1}$ $e_g^{0,0}$
Paramagnetic [unpaired electron = 2]
- $[CoF_6]^{3-}$ $Co^{3+}$ $3d^6$ $t_{2g}^{2,1,1}$ $e_g^{1,1}$
Paramagnetic [unpaired electron = 4]
- $[Fe(CN)_6]^{3-}$ $Fe^{3+}$ $3d^5$ $t_{2g}^{2,2,1}$ $e_g^{0,0}$
Paramagnetic [unpaired electron = 1]
- $[FeF_6]^{3-}$ $Fe^{3+}$ $3d^5$ $t_{2g}^{1,1,1}$ $e_g^{1,1}$
Paramagnetic [unpaired electron = 5]

Hence, total two complex have same number of unpaired electron.

**123.** *(2)*

| Complex | Magnetic nature | Hybridisation |
|---|---|---|
| $[FeF_6]^{3-}$ | Paramagnetic | $sp^3d^2$ |
| $[Fe(CN_6)]^{3-}$ | Paramagnetic | $d^2sp^3$ |
| $[Mn(CN)_6]^{3-}$ | Paramagnetic | $d^2sp^3$ |
| $[Co(C_2O_4)_3]^{3-}$ | Diamagnetic | $d^2sp^3$ |
| $[MnCl_6]^{3-}$ | Paramagnetic | $sp^3d^2$ |
| $[CoF_6]^{3-}$ | Paramagnetic | $sp^3d^2$ |

Only $[Fe(CN)_6]^{3-}$ and $[Mn(CN)_6]^{3-}$ are paramagnetic with $d^2sp^3$ hybridisation.

**124.** *(0)* $K_2[NiCl_4] = sp^3$, paramagnetic
$[Zn(H_2O)_6]Cl_2 = sp^3d^2$, diamagnetic
$[K_3[Mn(CN)_6] = d^2sp^3$, paramagnetic
$[Cu(PPh_3)_3I] = sp^3$, diamagnetic
Hence, the value of $n = 2$.
Least value of enthalpy of atomisation among Ni, Zn, Mn and Cu is Zn.
$Zn^{2+}$ : $[Ar]3d^{10}$
$\mu = \sqrt{n(n+2)}$ $(n = 0)$ $\Rightarrow$ $\mu = 0$

**125.** *(1)* Co has highest standard electrode potential ($M^{3+}/M^{2+}$) among Mn, Cr, Co, Fe
∴ Complex is $[Co(CN)_6]^{4-}$ and its splitting is as follows

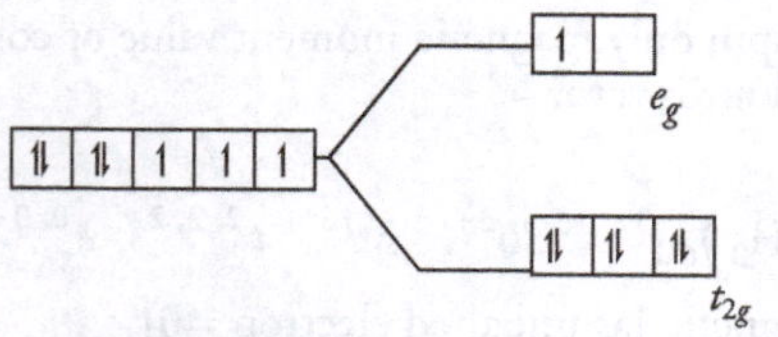

∴ Electron in $e_g$ orbital is one.

**126.** *(14)* The structure of complex $Ni^{2+}$ ion and dimethyl glyoxime is

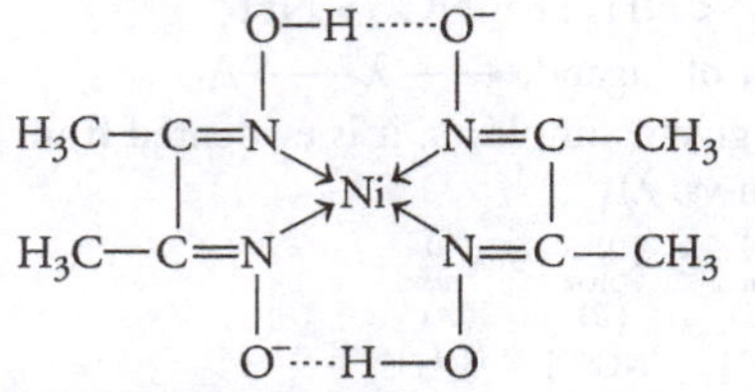

So, the number of H-atom = 14

**127.** *(28)* The electronic configuration of $Ni^{2+}$ is $[Ar]3d^8 4s^0$.

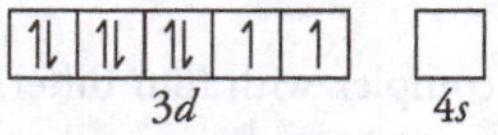

So, no. of unpaired electrons = 2
using $\mu = \sqrt{n(n+2)}$ BM
$= \sqrt{2(2+2)}$ BM $= \sqrt{8}$ BM
$= 2.8$ BM $= 28 \times 10^{-1}$ BM

**128.** *(0)* The electronic configuration of $Pt^{2+}$ is $5d^8 6s^0$.
Pt in $[Pt(NH_3)_2Cl(NH_2CH_3)]Cl$ =

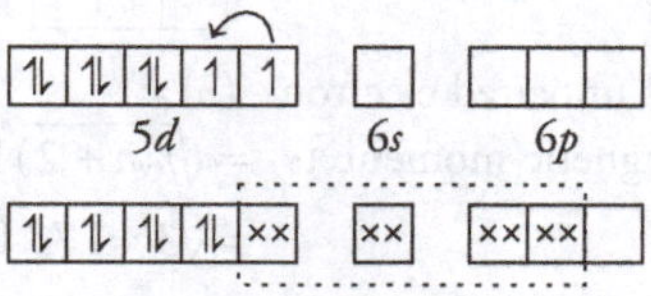

Thus, the complex is $dsp^2$ hybridisation and have no unpaired electrons $(n = 0)$
Magnetic moment, $\mu = \sqrt{n(n+2)} = 0$
Hence, magnetic moment is zero.

**129.** *(1)* $Co^{2+} = [Ar]3d^7 4s^0$

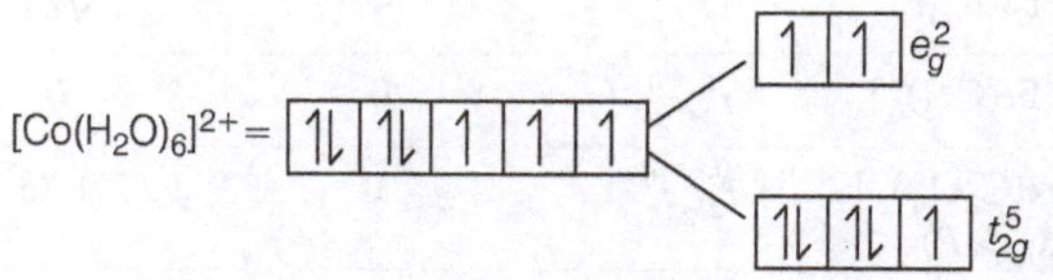

Only one unpaired electron in $t_{2g}$ set of orbital is present.

**130.** *(3)* In potassium ferrocyanide $K_4[Fe(CN)_6]$, there are three pairs of electrons in the $t_{2g}$ set of orbitals.
In $K_4[Fe(CN)_6]$, oxidation state of iron is +2.
In case of strong field ligand i.e., $CN^-$, all electron are paired up then the configuration of this complex will be $t_{2g}^6$.

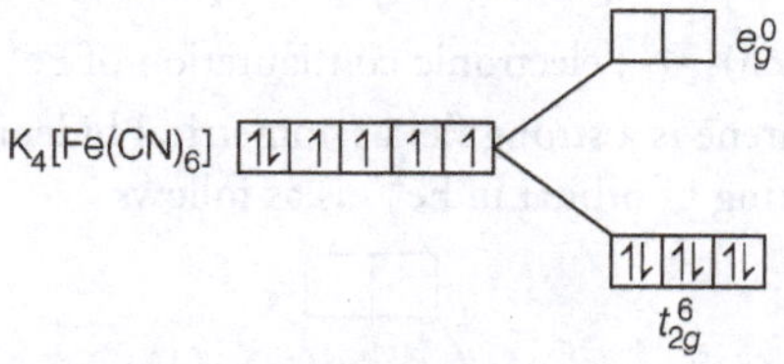

$t_{2g}$ contains 6 electrons so it have 3 pairs.

**131.** *(4)* For the complex $[Mn(NCS)_6]^{x-}$
$\mu = 6.06$ BM $\simeq 5.92$ BM
Number of unpaired electrons = 5

Here, ligand is $NCS^{\ominus}$ and $Mn^{2+}$ has electronic configuration of $3d^5$. So, $x = 4$.

**132.** *(480)* In $[Ti(H_2O)_6]^{3+}$, $Ti^{3+}$ has outer electronic configuration $3d^1$

CFSE for $[Ti^{3+}] = -0.4 \times \Delta_o$

$$= -\frac{96 \times 10^3}{N_A} \text{ J}$$

$$\Rightarrow \quad \Delta_o = \frac{96 \times 10^3}{0.4 \times 6 \times 10^{23}} = \frac{hc}{\lambda}$$

$\therefore$

$$\lambda = \frac{0.4 \times 6 \times 10^{23} \times 6.4 \times 10^{-34} \times 3 \times 10^8}{96 \times 10^3}$$

$$= 0.48 \times 10^{-6} \text{ m}$$

$$= 480 \times 10^{-9} = 480 \text{ nm}$$

**133.** *(4)*

(i) $[NiCl_4]^{2-}$; $Ni^{2+} = 3d^8$, $sp^3$ (Paramagnetic)

(ii) $[Fe(CN)_6]^{3-}$; $Fe^{3+} = 3d^5$, $d^2sp^3$ (Paramagnetic)

(iii) $[Fe(H_2O)_6]^{2+}$; $3d^6$, $sp^3d^2$ (Paramagnetic)

(iv) $[Cu(NH_3)_4]^{2+}$; $3d^9$, $dsp^2$ (Paramagnetic)

**134.** *(2)* In $[Fe(CN)_6]^{3-}$

Oxidation state of Fe is +3.

As complex is inner sphere, so electrons are paired.

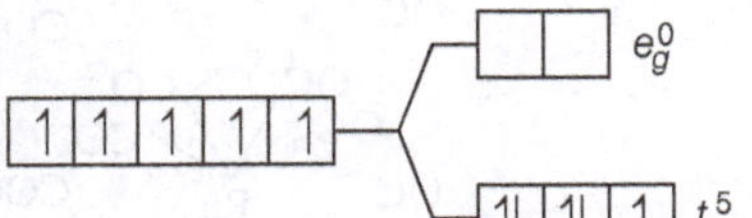


$$\text{CFSE} = -0.4 \times 5\Delta_o = -2\Delta_o$$

So, answer is 2.

**135.** *(0)* $[\overset{+2}{Co}(H_2O)_6]Cl_2$

Electronic configuration of

$$Co^{2+} = [Ar]\,4s^0\,3d^7$$

= [Ar]

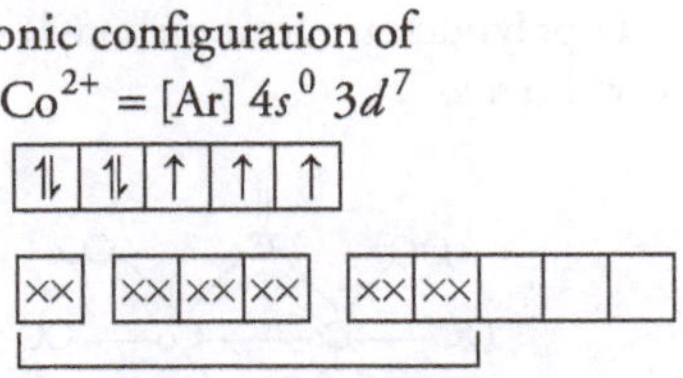


i.e., there are 3 unpaired electrons.

Magnetic moment, $\mu = \sqrt{n(n+2)}$

$$\therefore \quad \mu_1 = \sqrt{3(3+2)} = \sqrt{3 \times 5} = \sqrt{15} \text{ BM}$$

In complex $[\overset{+3}{Cr}(H_2O)_6]Cl_3$

Electronic configuration of

$Cr^{3+} \rightarrow [Ar]\,4s^0\,3d^3$

= [Ar]

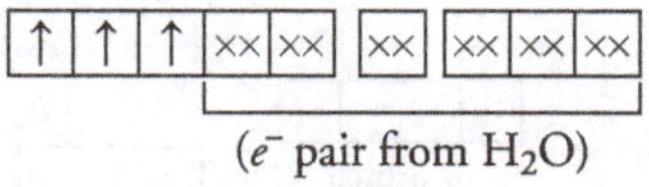


$$n = 3 \Rightarrow \mu_2 = \sqrt{15} \text{ BM}$$

Hence, $\mu_1 - \mu_2 = 0$

**136.** *(0)* Oxidation state of Co in all the given complexes = +3.

The inner orbital and outer orbital nature of complex depends on the chemical nature of ligands and also crystal field splitting depends on strength of ligands.

Ligand strength order is

$$\overset{\ominus}{C}N > NH_3 > H_2O > Cl^-$$

$$\Delta_0 \propto \frac{1}{\lambda}$$

So, shortest wavelength is corresponds to $[Co(CN)_6]^{3-}$

The electronic configuration of $Co^{3+}$ in the given complex is $3d^6\,4s^0$

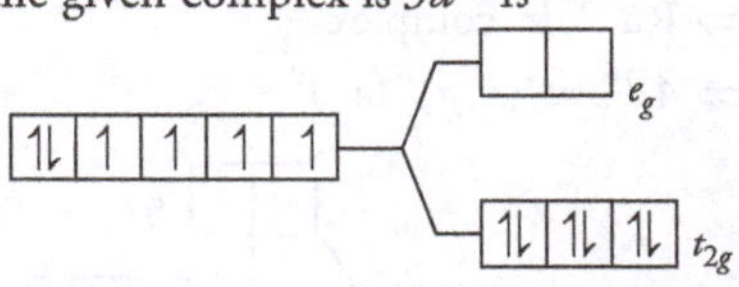


Hence, number of unpaired electrons $(n) = 0$

So, spin only magnetic moment

$$= \sqrt{n(n+2)}$$

$$\mu_{spin} = 0$$

**137.** *(6)*

$$[Co(H_2O)_6]^{2+} + NH_3 \text{ (excess)} \xrightarrow{O_2} \underset{\substack{\text{Diamagnetic} \\ \text{(means no unpaired electrons)}}}{[Co(NH_3)_6]^{3+}} + 6H_2O$$

$[Co(NH_3)_6]^{3+}$ is low spin complex and diamagnetic in nature. So, configuration of in this complex

$$Co^{3+} = [Ar]\,3d^6\,4s^0$$

$$= [Ar]\,(t_{2g}^6 e_g^0)\,4s^0$$

$\therefore$ Total number of electrons in $t_{2g}$ orbital of $[Co(NH_3)_6]^{3+}$ is 6.

**138.** *(2)* The order of strength of ligand is

$$\overset{\ominus}{C}N > NH_3 > H_2O > Cl$$

So, splitting in $K_3[Fe(CN)_6]$ is highest and thus, it absorbs light of more energy and shortest wavelength.

$$Fe^{3+} \longrightarrow 3d^5$$

$K_3[Fe(CN)_6] =$

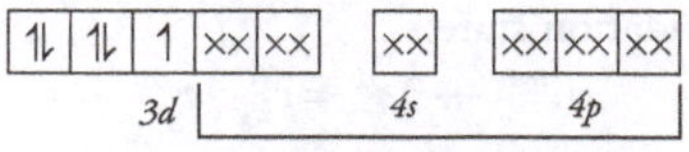


So, number of unpaired electrons $n$ is 1.

$\mu$ (spin only) $= \sqrt{n(n+2)}$

$$= \sqrt{1(1+2)} = \sqrt{3}$$

$$\mu_{spin} = 1.73 \text{ BM} \approx 2 \text{ BM}$$

**139.** *(766)* $\Delta_t = \frac{4}{9}\Delta_0$ where $\Delta_t$ and $\Delta_0$ are crystal field splitting energy for tetrahedral and octahedral complexes respectively.

$$\Delta_t \propto \frac{1}{\lambda_t};\ \Delta_0 \propto \frac{1}{\lambda_0}$$

$$\frac{1}{\lambda_t} = \frac{4}{9}\frac{1}{\lambda_0}$$

$$\lambda_0 = \frac{4}{9} \times 600 \text{ nm} = 266.6 \text{ nm}$$

$$\Delta_0 = \frac{hc}{\lambda_0} = \frac{6.63 \times 10^{-34} \times 3.08 \times 10^8}{266.6}$$

$$= 765.95 \times 10^{-21} \text{ J} \approx 766 \times 10^{-21} \text{ J}$$

**140.** *(1)* (i) $[Co(NH_3)_6]Cl_2$

$\Rightarrow$ Oxidation state of cobalt $Co^{2+}$ is +2.

$Co^{2+} = [Ar]\,4s^0\,3d^7$ i.e.

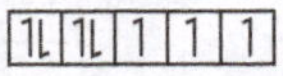

$NH_3$ acts as SFL (strong field ligand) due to small size of cobalt with +2 charge, so pairing will took place.

$\Rightarrow \quad t_{2g}^6\,e_g^1$

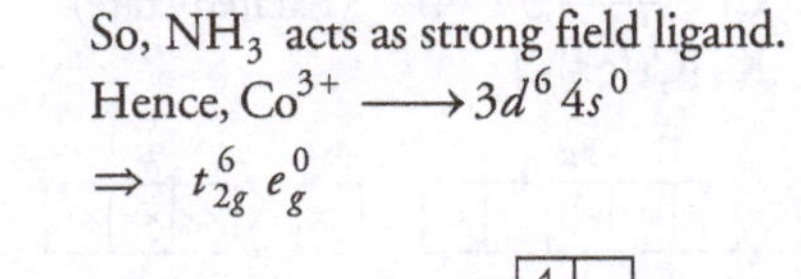


$\therefore \quad n = 1$

(ii) $[Co(NH_3)_6]Cl_3 \Rightarrow$ Oxidation state of cobalt $(Co^{2+})$ is +3.

So, $NH_3$ acts as strong field ligand.

Hence, $Co^{3+} \longrightarrow 3d^6\,4s^0$

$\Rightarrow t_{2g}^6\,e_g^0$

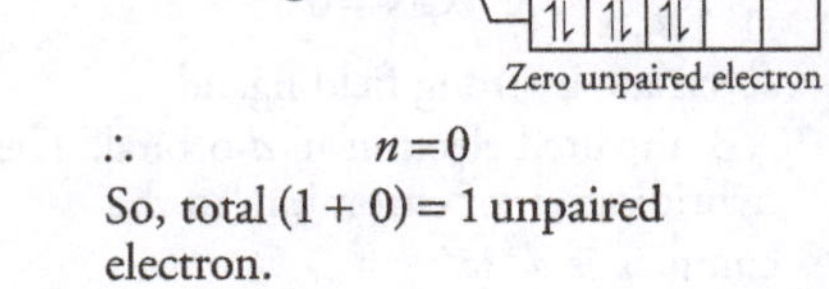


$\therefore \quad n = 0$

So, total $(1 + 0) = 1$ unpaired electron.

**141.** *(2)* In $[Ni(CN)_6]^{2-}$, Ni is in +4 oxidation state.

$Ni^{+4} \rightarrow 3d^6 = t_{2g}^6, e_g^0$

(Presence of strong field)

Unpaired electron = 0

Similarly, $NiCl_2 \longrightarrow Ni^{2+} = t_{2g}^6, e_g^2$

(Two unpaired electrons)

Number of unpaired electron in $NiCl_2$ and $[Ni(CN)_6]^{2-}$ are two and zero respectively.

Change in number of unpaired electrons ⇒ 2 to 0.

Change = 2

**142.** *(2)* Electronic configuration of $Co^{2+}$: [Ar] $3d^7$ and $CN^-$ is a strong field ligand which can pair electron of central atom.

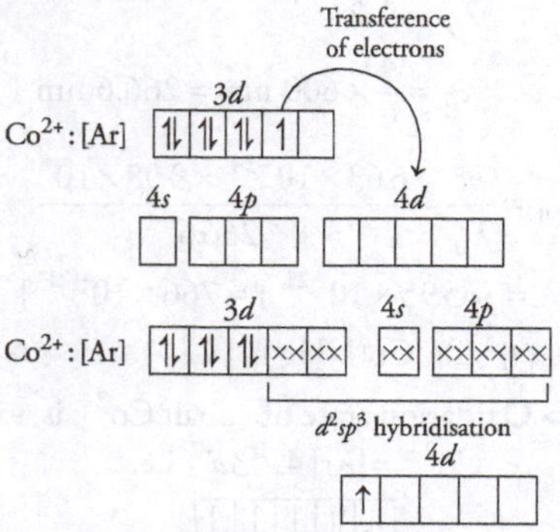


So, spin only magnetic moment $(\mu) = \sqrt{n(n+2)}$ BM

where, $n$ = number of unpaired electrons

$= \sqrt{1(1+2)}$ BM $= \sqrt{3}$ BM

$\mu = \sqrt{3}$ BM $= 1.73$ BM

Nearest integer = 2

**143.** *(3)* $K_3[Cr(ox)_3]$ the oxidation state of Cr in $K_3[Cr(C_2O_4)_3]$ is + 3.

$Cr \longrightarrow 3d^5 4s^1$ (ground state electronic configuration)

$Cr^{3+} \longrightarrow 3d^3 4s^0$ (Excited state)

$K_3[Cr(ox)_3] =$

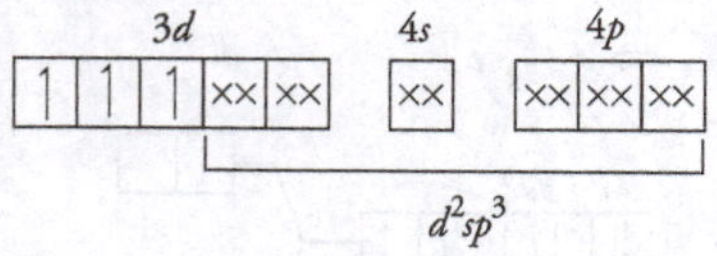


$\Rightarrow t_{2g}^3 e_g^0$ CN = 6

∴ oxalate is strong field ligand

∴ 3 unpaired electron in $d$-orbital. The hybridisation of chromium in the complex is $d^2sp^3$.

**144.** *(4)* $[Ti(H_2O)_6]^{3+}$ is an octahedral complex.

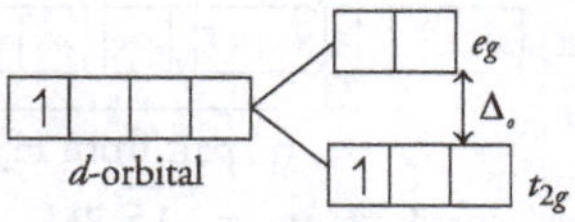


For transition of electron from $t_{2g}$ to $e_g$ orbital, $\Delta_o$ amount of energy is required.

Given, $\lambda = 498$ nm $= 498 \times 10^{-9}$ m

$\Delta_o = hc/\lambda$

where, $h$ = Planck's constant

$c$ = speed of light

$\lambda$ = wavelength of radiation

So, $\Delta_o = \dfrac{6.626 \times 10^{-34} \times 3 \times 10^8}{498 \times 10^{-9}}$

$= 3.99 \times 10^{-19}$ J $= 4 \times 10^{-19}$ J

**145.** *(0)* Magnetic moment (in BM) of $[Ru(H_2O)_6]^2$ would be; while considering that $\Delta_0 > P$,

$Ru_{(44)}$; [Kr]$4d^7 5s^1$ (in ground state)

⇒ $Ru^{2+}$ in complex

⇒ $4d^6 \Rightarrow (t_2g)^6 (e_g)^0$

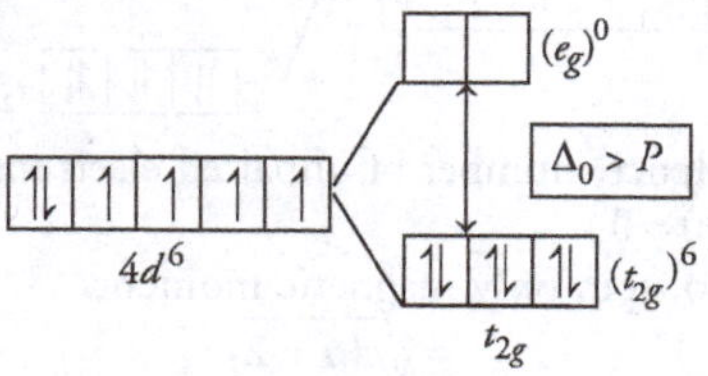


Here, number of unpaired electrons is zero and hence

$\mu = \sqrt{n(n+2)}$ BM = 0 BM

## Topic 3

**01.** *(d)* The compound used in the treatment of lead poisoning is EDTA. Medication occurs through chelation therapy.

**02.** *(c)* *cis*-platin is used as an anti-cancer drug which helps in treatment of cancer.

**03.** *(a)* Compounds having atleast one carbon metal (Mn—C) bond are known as organometallic compounds.

**04.** *(b)* Compounds that contain carbon-metal bond are known as organometallic compounds.

In $CH_3$—Mg—Br (Grignard's reagent), one bond is present between carbon and Mg (metal), hence it acts as organometallic compound.

**05.** *(a)* Chlorophyll is a coordination compound of magnesium (Mg). $[Ag(S_2O_3)_2]^3$ is used in photography. Hence, (A) and (C) are mismatched combinations.

**06.** *(a)* Among the given options only $[Fe_3(OH)_2(OAc)_6]Cl$ is water soluble, while all other options are water insoluble.

**07.** *(a)* Statement given in A, B and C are correct, while in D and E are incorrect. The correct form of D and E are: In (D) Wilkinson's catalyst $[Rh(PPh_3)_3Cl]$, phosphine does not act as a ligand.

(E) $Ca^{2+}$ ion form more stable complex with EDTA.

**08.** *(d)*

$[Co_2(CO)_8]$

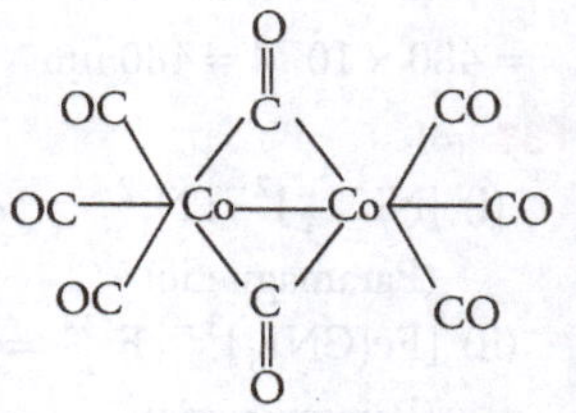


$[Mn_2(CO)_{10}]$

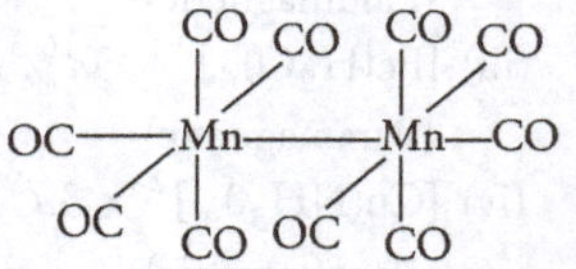


$[Os_3(CO)_{12}]$

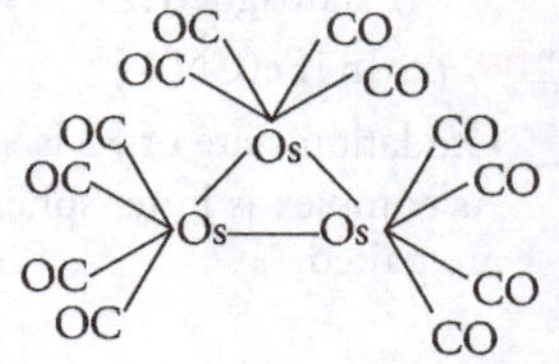


$[Ru_3(CO)_{12}]$

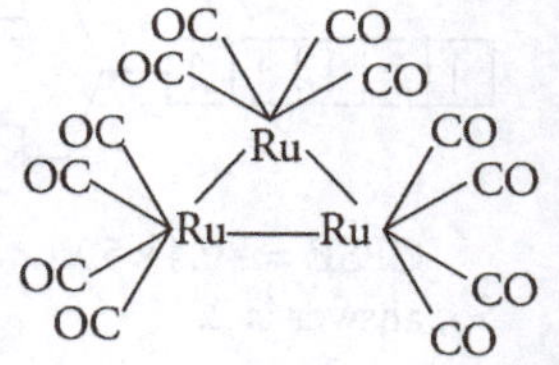


**09.** *(d)* The structure of $Co_2(CO)_8$ (a polynuclear metal carbonyl) can be written as

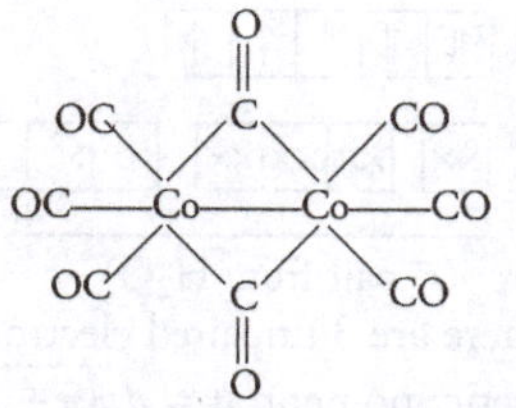


Total number of bridging CO ligands = 2 and the Co—Co bond = 1

**10.** *(b)* $Co_2(CO)_8$ show one Co—Co bond, 6-terminal CO ligands and 2 bridging CO-ligands.

**11.** *(c)* The chlorophyll molecule plays an important role in processes of photosynthesis. It has porphyrin rings and the central metal Mg not Ca.

**12.** *(7)* The structure of $[Co_2(CO)_8]$ is

Number of Co—Co bond $(X) = 1$
Number of terminal CO ligands $(Y) = 6$
$\therefore \quad X + Y = 1 + 6 = 7$

**13.** *(3)* The $(M - C)\pi$-bond is formed by the donation of a pair of electrons from a filled $d$-orbital of metal to the vacant antibonding, $\overset{*}{\pi}$-orbital of carbon of carbon ligand. Thus, CO acts as a $\sigma$-donor $(OC \longrightarrow M)$ and a $\pi$-acceptor $(M \longrightarrow CO)$, with the two interaction creating a synergic effect which strengthens the bond between CO and the metal.

So, all the given three complexes have CO ligands and all will exhibit synergic bonding.

**14.** *(2)* Fehling's solution contains $Cu^{2+} \rightarrow [Ar]\, 3d^9$ configuration

= | ⇅ | ⇅ | ⇅ | ⇅ | ↿ |

$\therefore$ Number of unpaired $e^-$ $(n) = 1$

So, spin magnetic moment ($\mu$).

$$= \sqrt{n(n+2)} \text{ BM}$$
$$= \sqrt{1(3)} = 1.73 \text{ BM} \approx 2 \text{ BM}$$

**15.** *(0)* $[Mn_2(CO)_{10}]$ is a polynuclear metal carbonyl compound. It is decarboxyldimanganese (0) which is made up of two square pyramidal $Mn(CO)_5$ units joined by a Mn—Mn bond. The structure of $[Mn_2(CO)_{10}]$ is

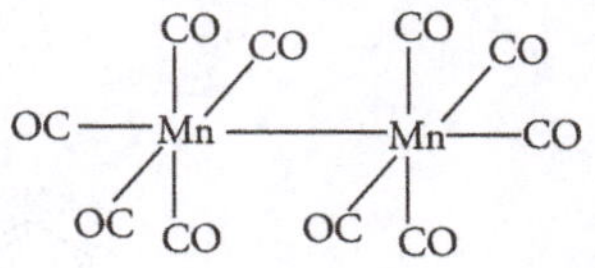

Here, no bridging CO ligand is present. So, number of bridging CO ligands in $[Mn_2(CO)_{10}]$ is 0 (zero).

**16.** *(1)* Given, stability constant value,

$K_1 = 10^4$; $K_2 = 1.58 \times 10^3$

$K_3 = 5 \times 10^2 \Rightarrow K_4 = 10^2$

On adding all equations,

$$Cu^{2+} + 4NH_3 \overset{K}{\rightleftharpoons} [Cu(NH_3)_4]^{2+}$$

$\therefore$ The overall reaction constant $(K)$ or equilibrium constant for formation of $[Cu(NH_3)_4]^{2+}$ is

$$K = K_1 \times K_2 \times K_3 \times K_4$$
$$K = 10^4 \times 1.58 \times 10^3 \times 5 \times 10^2 \times 10^2$$
$$K = 7.9 \times 10^{11}$$

where, $K$ = equilibrium constant for formation of $[Cu(NH_3)_4]^{2+}$

So, equilibrium constant $K'$ for dissociation of $[Cu(NH_3)_4]^{2+}$ is $\frac{1}{K}$.

$$K' = \frac{1}{K} = \frac{1}{7.9 \times 10^{11}} = 1.26 \times 10^{-12}$$

Hence, $K' = x \times 10^{-12} \Rightarrow x = 1.26 \approx 1$

# C-19

JEE Main Chapterwise Topicwise Chemistry

In JEE Main there is usually 0-1 question per paper are visible from this chapter. In past 3 years, 56 conceptual and 9 numerical questions were asked of moderate difficulty level from this chapter.
**For detailed analysis scan QR code**

# ORGANIC COMPOUNDS CONTAINING HALOGENS

## Topic 1 Haloalkanes

### Level 1

**01** When *sec*-butylcyclohexane reacts with bromine in the presence of sunlight, the major product is *[2025, 22 Jan Shift-II]*

(a) 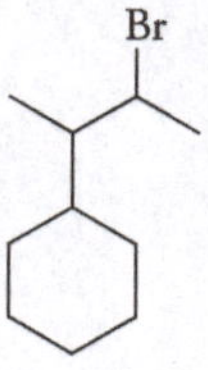
(b) 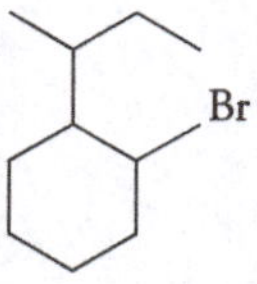
(c) 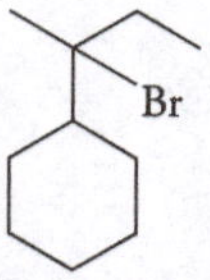
(d) 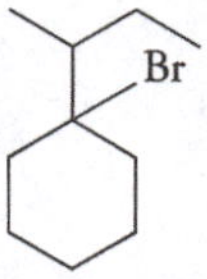

**02** Total number of stereoisomers possible for the given structure *[2024, 9 April Shift-II]*

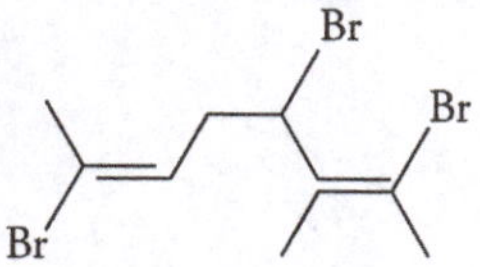

(a) 4 (b) 3
(c) 2 (d) 8

**03** 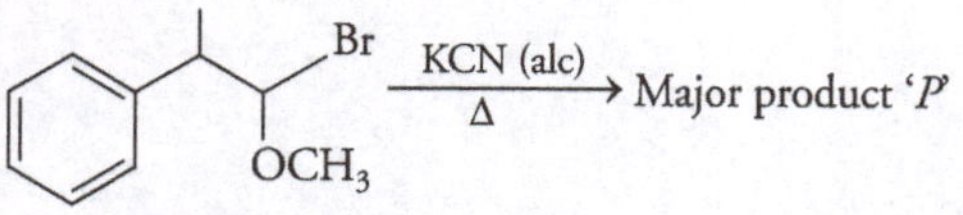


In the above reaction product '*P*' is *[2024, 9 April Shift-II]*

(a) 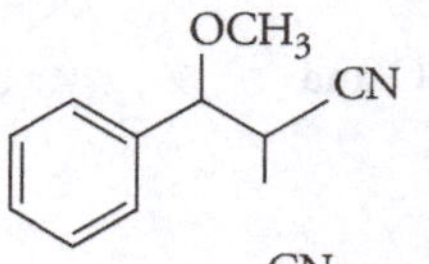
(b) 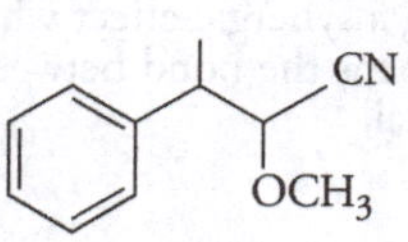
(c) 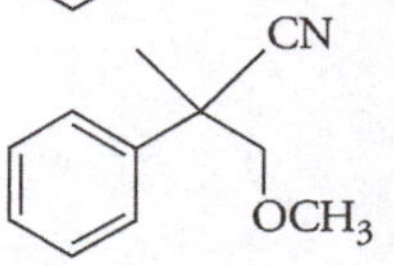
(d) 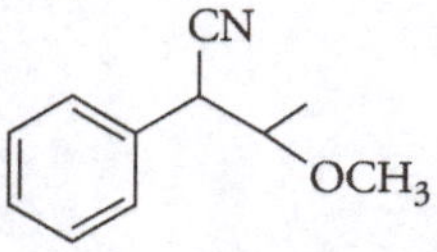

**04** Identify the major product in the following reaction. *[2024, 5 April Shift-II]*

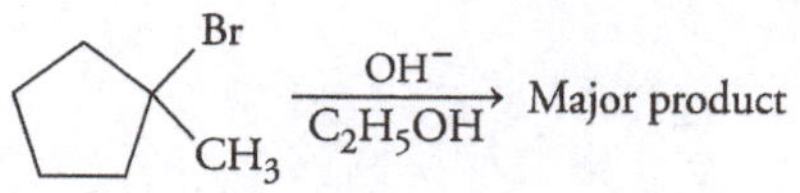


(a) 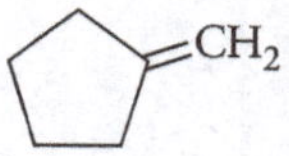
(b) 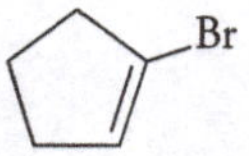
(c) 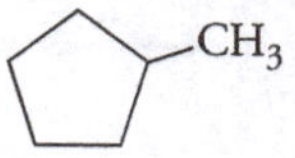
(d) 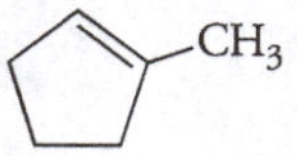

**05** 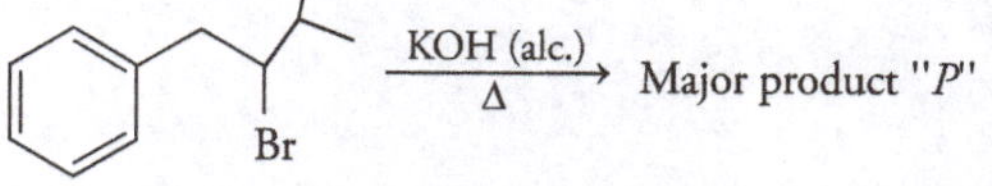


Product *P* is *[2024, 4 April Shift-II]*

(a) 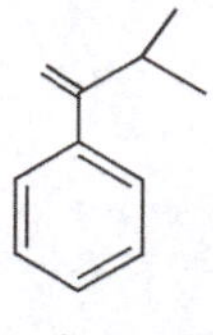
(b) 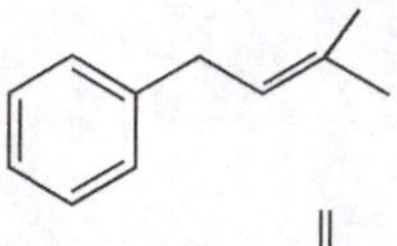
(c) 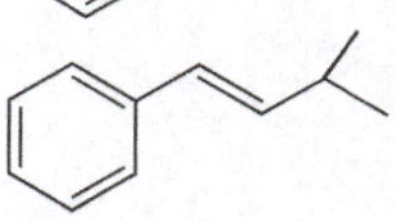
(d) 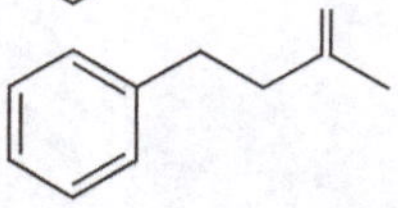

**06** Identify structure of 2, 3-dibromo-1-phenylpentane. *[2024, 31 Jan Shift-II]*

(a) 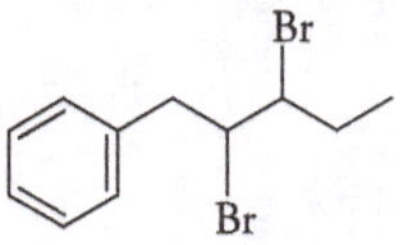
(b) 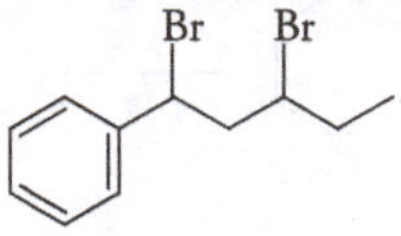
(c) 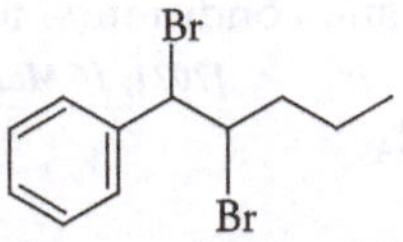
(d) 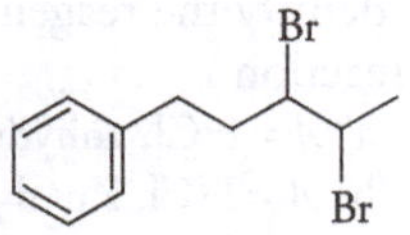

**07** Example of vinylic halide is *[2024, 30 Jan Shift-I]*

(a) 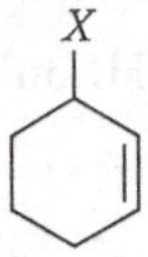
(b) 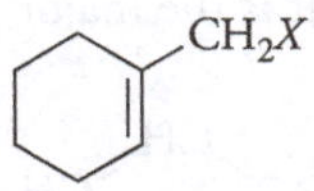
(c) 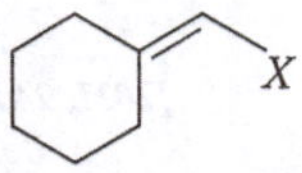
(d) 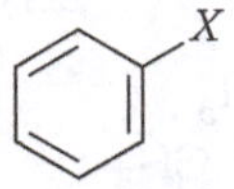

**08** Alkyl halide is converted into alkyl isocyanide by reaction with *[2024, 29 Jan Shift-II]*

(a) KCN (b) NaCN (c) $NH_4CN$ (d) AgCN

**Imp.** Reactivity of Halogens towards $S_N1$ Reaction

**09** Choose the halogen which is most reactive towards $S_N1$ reaction in the given compounds (*A*, *B*, *C* and *D*) *[2023, 8 April Shift-I]*

(A) 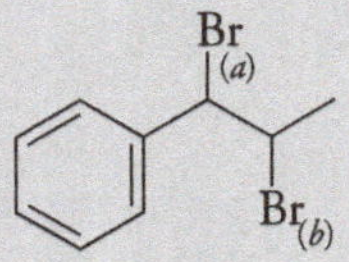
(B) 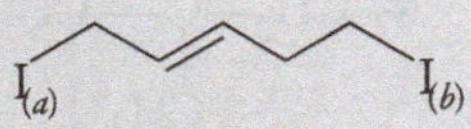
(C) 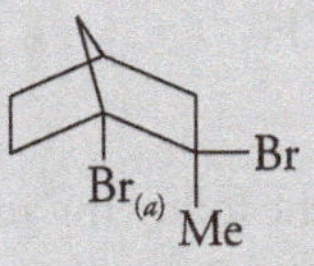
(D) 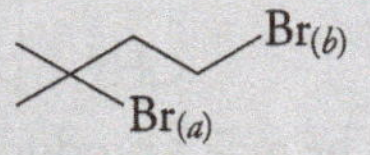

(a) *A*—Br(*a*); *B*—I(*a*); *C*—Br(*b*); *D*—Br(*a*)
(b) *A*—Br(*b*); *B*—I(*b*); *C*—Br(*b*); *D*—Br(*b*)
(c) *A*—Br(*b*); *B*—I(*a*); *C*—Br(*a*); *D*—Br(*a*)
(d) *A*—Br(*a*); *B*—I(*a*); *C*—Br(*a*); *D*—Br(*a*)

**10** **Assertion (A)** Hydrolysis of an alkyl chloride is a slow reaction but in the presence of NaI, the rate of the hydrolysis increases.

**Reason (R)** $I^-$ is a good nucleophile as well as a good leaving group. *[2023, 24 Jan Shift-I]*

In the light of the above statements, choose the correct answer from the options given below.

(a) Both A and R are true and R is the correct explanation of A.
(b) Both A and R are true but R is not the correct explanation of A.
(c) A is false but R is true.
(d) A is true but R is false.

**11** The major product (*P*) in the reaction. *[2022, 28 June Shift-I]*

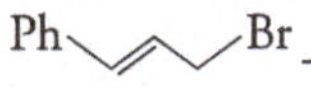 $\xrightarrow{HBr}$ (*P*) [Ph is $-C_6H_5$] is

(a) 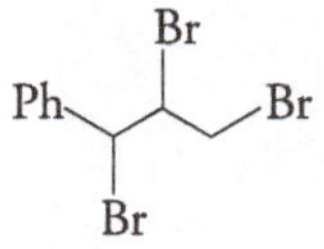
(b) 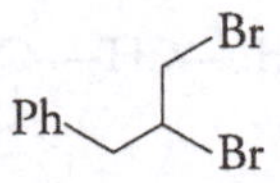
(c) 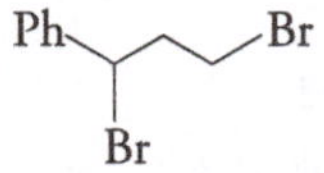
(d) 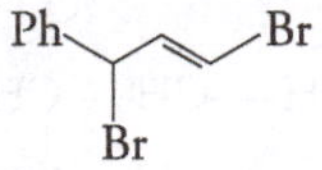

**12** Which one of the following compounds is inactive towards $S_N1$ reaction? *[2022, 28 June Shift-I]*

(a) 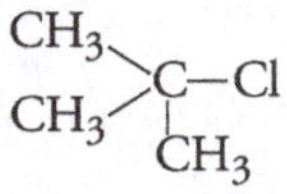
(b) 
(c) 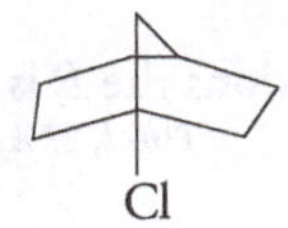
(d) 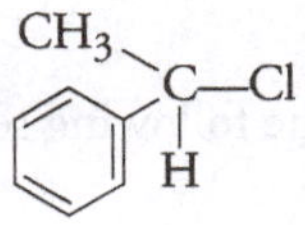

**13** The major product in the reaction *[2022, 25 June Shift-I]*

$$CH_3-\underset{CH_3}{\overset{CH_3}{C}}-Cl + \overset{+}{K}\,\overset{-}{O}-\underset{CH_3}{\overset{CH_3}{C}}-CH_3 \longrightarrow \text{? is}$$

(a) *t*-butyl ethyl ether
(b) 2,2 dimethyl butane
(c) 2-methyl pent-1-ene
(d) 2-methylprop-1-ene

**14** The major product of the following reaction is *[2021, 31 Aug Shift-II]*

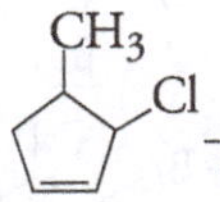 $\xrightarrow[C_2H_5OH]{NaOH}$ Major product

(a) 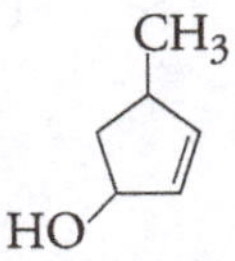
(b) 
(c) 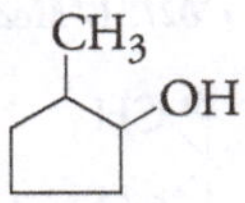
(d) 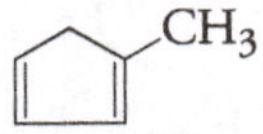

**15** The major product (*A*) formed in the reaction given below is *[2021, 27 Aug Shift-II]*

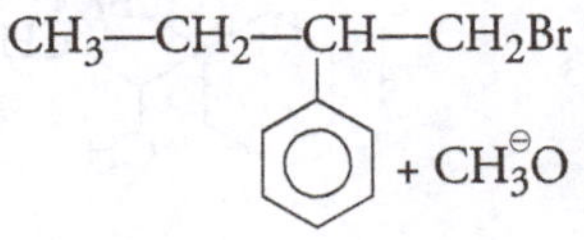 + $CH_3\overset{\ominus}{O}$ $\xrightarrow{CH_3OH}$ *A*
Major product

(a) 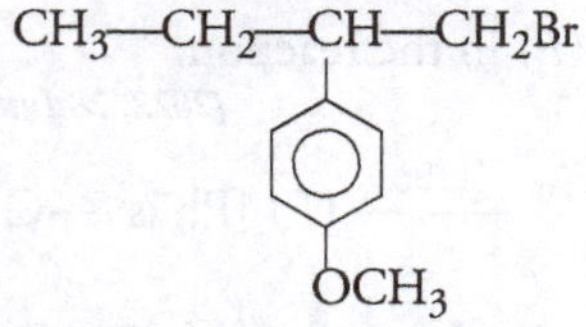


(b) 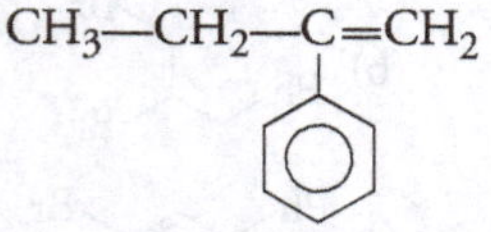


(c) 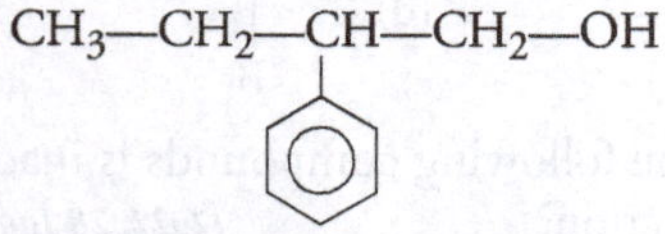


(d) 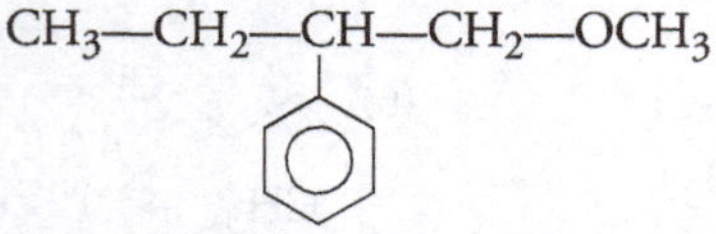


**16** In the following sequence of reactions the *P* is *[2021, 27 Aug Shift-I]*

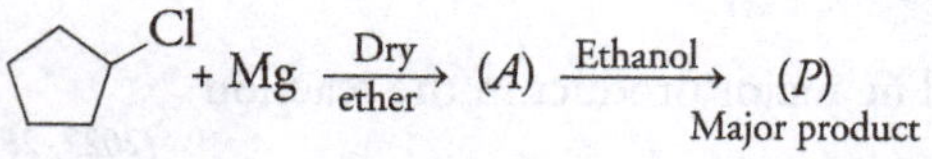


(a) 
(b) 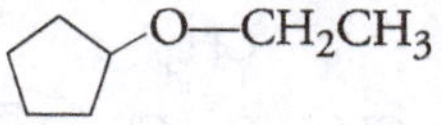


(c) 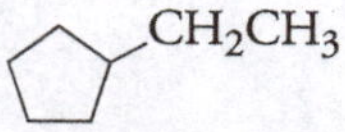


(d) 

**17** The major product formed in the following reaction is *[2021, 26 Aug Shift-I]*

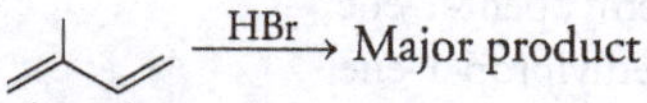


(a) 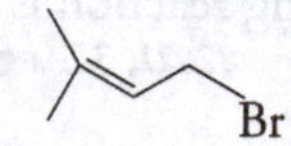


(b) 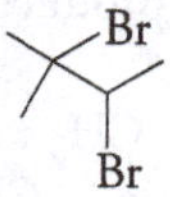


(c) 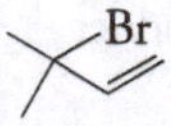


(d) 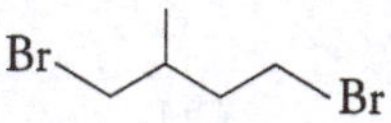


**18** 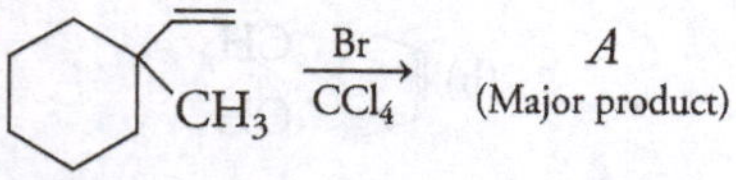


Product *A* is *[2021, 17 March Shift-I]*

(a) 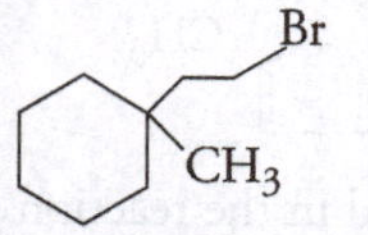


(b) 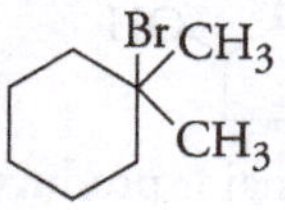


(c) 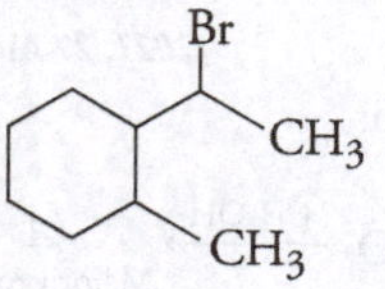


(d) 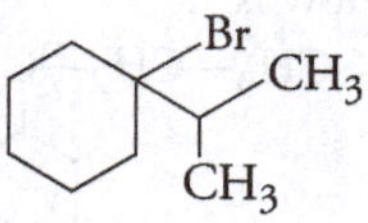


**19** 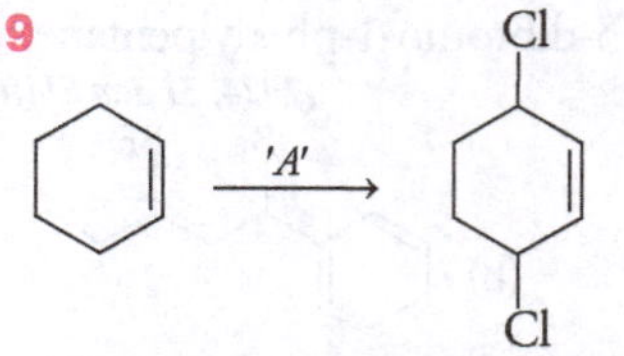


Identify the reagent(s) '*A*' and condition(s) for the reaction *[2021, 16 March Shift-II]*

(a) $A$ = HCl, anhydrous $AlCl_3$
(b) $A$ = HCl, $ZnCl_2$
(c) $A$ = $Cl_2$, UV light
(d) $A$ = $Cl_2$, dark, anhydrous $AlCl_3$

**20** What is the major product formed by HI on reaction with

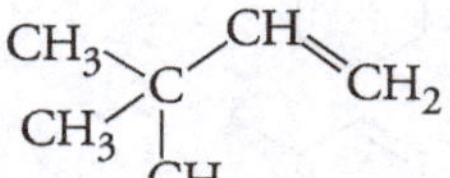


*[2021, 24 Feb Shift-I]*

(a) 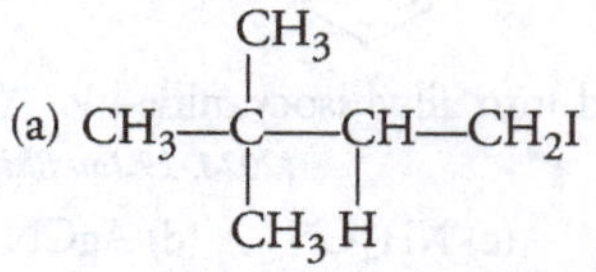


(b) 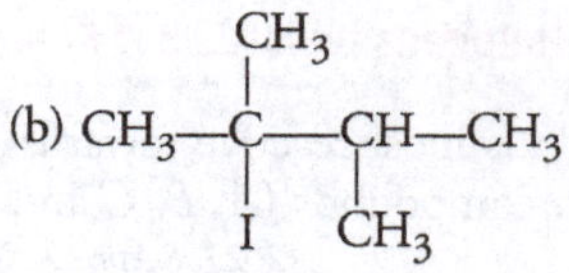


(c) 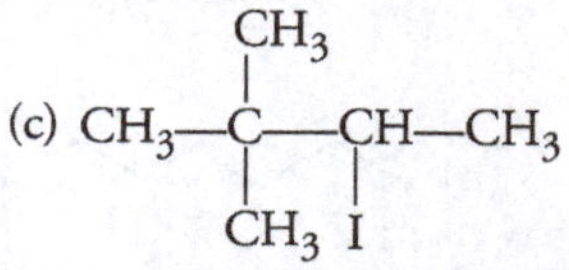


(d) 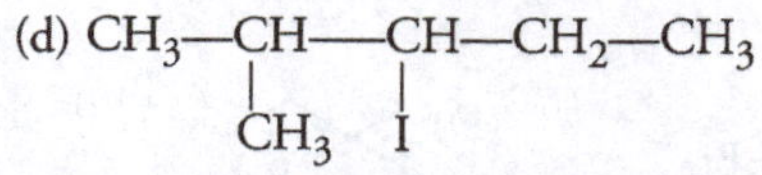


**21** The major product formed in the following reaction is

$$CH_3CH{=}CHCH(CH_3)_2 \xrightarrow{HBr}$$ *[2020, 5 Sep Shift-II]*

(a) $CH_3CH(Br)CH_2CH(CH_3)_2$
(b) $CH_3CH_2CH(Br)CH(CH_3)_2$
(c) $Br(CH_2)_3CH(CH_3)_2$
(d) $CH_3CH_2CH_2C(Br)(CH_3)_2$

**22** Among the following compounds, which one has the shortest C—Cl bond? *[2020, 4 Sep Shift-II]*

(a) $H_3C—Cl$

(b) 

(c) 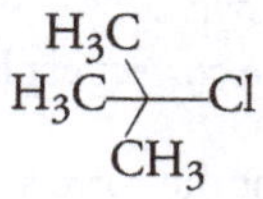


(d) 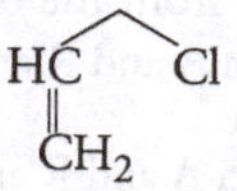

**23** The major product in the following reaction is

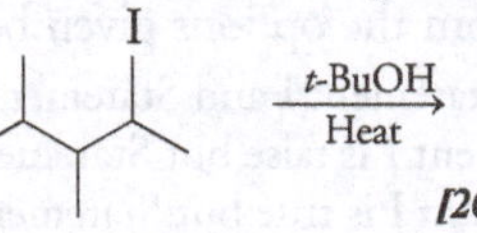

[2020, 3 Sep Shift-II]

(a) 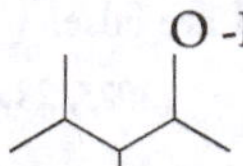 (b) 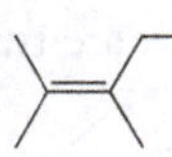

(c) 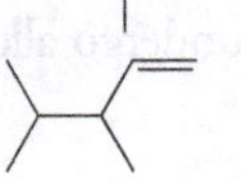 (d) 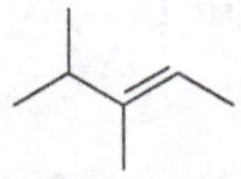

**24** The major product obtained in the given reaction is

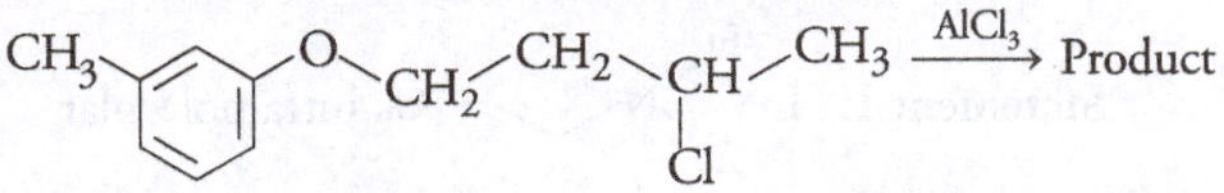

[2019, 10 April Shift II]

(a) 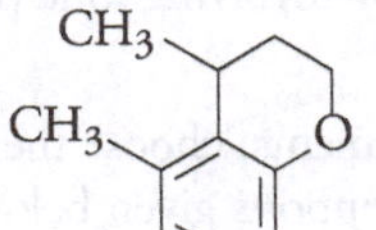

(b) 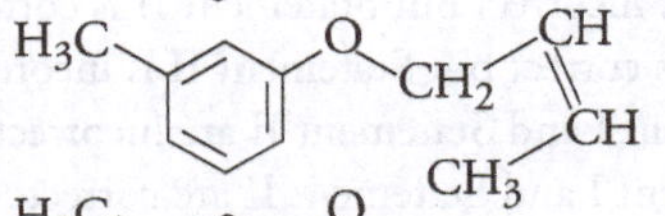

(c) 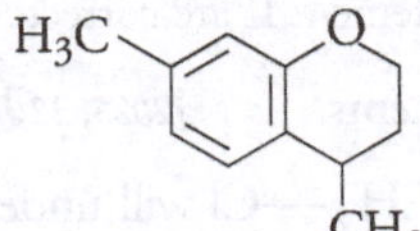

(d) 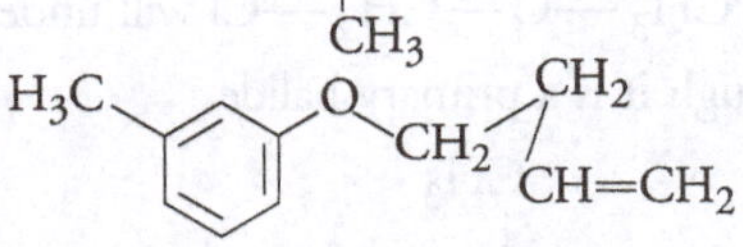

**25** The major product of the following reaction is

$$CH_3-\underset{H}{\overset{CH_3}{\overset{|}{\underset{|}{C}}}}-\underset{Br}{\underset{|}{C}}HCH_3 \xrightarrow{CH_3OH}$$

[2019, 10 April Shift-I]

(a) $CH_3-\overset{CH_3}{\overset{|}{C}}=CHCH_3$ (b) $CH_3-\underset{H}{\underset{|}{\overset{CH_3}{\overset{|}{C}}}}-CH=CH_2$

(c) $CH_3-\underset{OCH_3}{\underset{|}{\overset{CH_3}{\overset{|}{C}}}}-CH_2CH_3$ (d) $CH_3-\underset{H}{\underset{|}{\overset{CH_3}{\overset{|}{C}}}}-\underset{OCH_3}{\underset{|}{C}}HCH_3$

**26** The major product in the following conversion is

[2019, 12 Jan Shift-II]

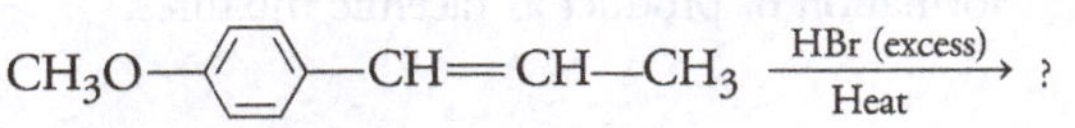

(a) 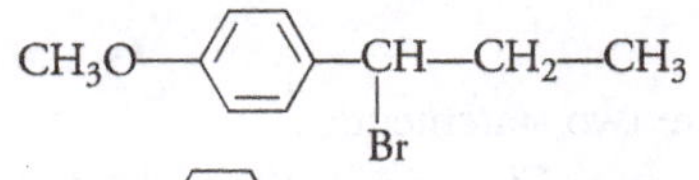

(b) 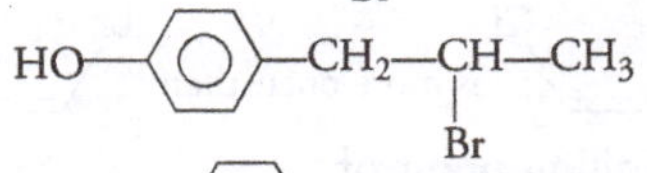

(c) 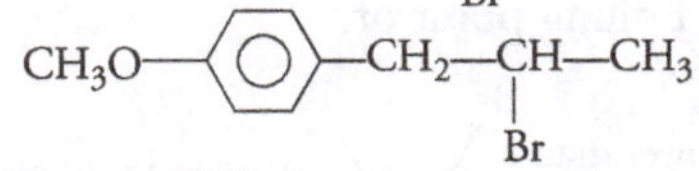

(d) 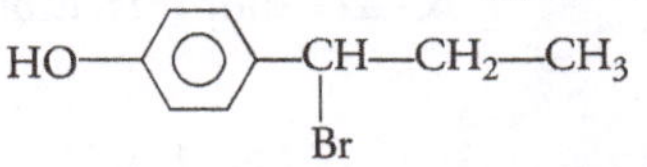

**27** Which hydrogen in compound ($E$) is easily replaceable during bromination reaction in presence of light?

$$\overset{\delta}{CH_3}-\overset{\gamma}{CH_2}-\overset{\beta}{CH}=\overset{\alpha}{CH_2}$$
($E$)

[2019, 10 Jan Shift-I]

(a) β-hydrogen (b) δ-hydrogen
(c) γ-hydrogen (d) α-hydrogen

**28** The major product of the following reaction is

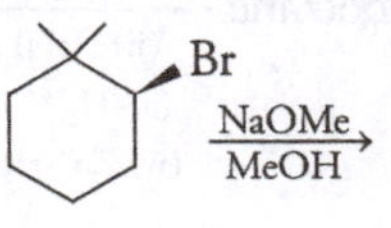

[JEE Main 2018]

(a) 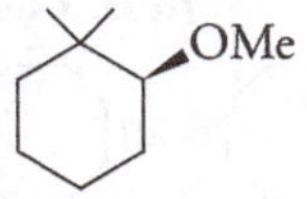 (b) 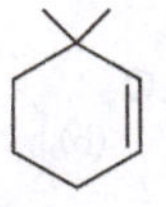

(c) 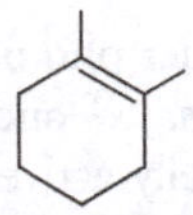 (d) 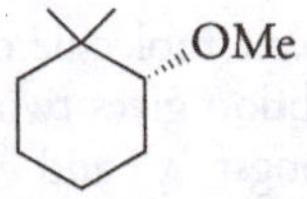

**29** The major product obtained in the following reaction is [JEE Main 2017]

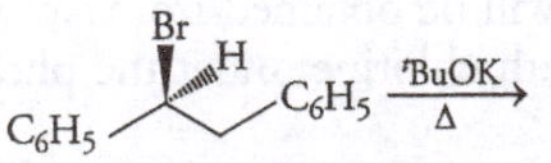

(a) (±) $C_6H_5CH(O^tBu)CH_2C_6H_5$
(b) $C_6H_5CH=CHC_6H_5$
(c) (+) $C_6H_5CH(O^tBu)CH_2C_6H_5$
(d) (−)$C_6H_5CH(O^tBu)CH_2C_6H_5$

**30** The synthesis of alkyl fluorides is best accomplished by [JEE Main 2015]

(a) free radical fluorination (b) Sandmeyer's reaction
(c) Finkelstein reaction (d) Swarts reaction

**31** A solution of (−)-1-chloro-1- phenylethane in toluene racemises slowly in the presence of a small amount of $SbCl_5$, due to the formation of [JEE Main 2013]

(a) carbanion (b) carbene
(c) carbocation (d) free radical

**32** How many chiral compounds are possible on monochlorination of 2-methyl butane? [AIEEE 2012]

(a) 8 (b) 2 (c) 4 (d) 6

## Level 2

**33** Given below are two statements:

**Statement I**  is more polar than 

**Statement (II):** Boiling point of

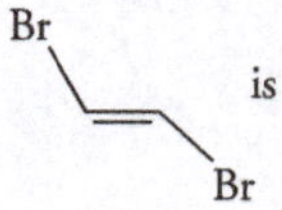 is lower than 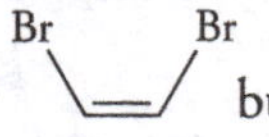 but it is more polar than 

In the light of the above statements, choose the most appropriate answer from the options given below: *[2025, 7 April Shift-II]*

(a) Statement I is incorrect but Statement II is correct.
(b) Both Statement I and Statement II are correct.
(c) Statement I is correct but Statement II is incorrect.
(d) Both Statement I and Statement II are incorrect.

**34** 1,2-dibromocyclooctane $\xrightarrow[\text{(ii) } NaNH_2 \text{ (iii) } Hg^{2+}/H^+ \text{ (iv) Zn–Hg}/H^+]{\text{(i) KOH (alc.)}}$ $P$ Major product

'$P$' is *[2025, 8 April Shift-II]*

(a) 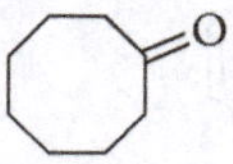 (b) 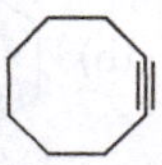 (c) 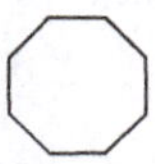 (d) 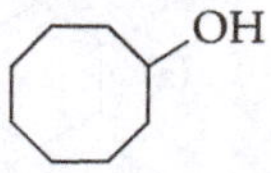

**35** Propane molecule on chlorination under photochemical condition gives two di-chloro products. "$x$" and "$y$". Amongst "$x$" and "$y$", "$x$" is an optically active molecule. *[2025, 23 Jan Shift-I]*

How many tri-chloro products (consider only structural isomers) will be obtained from "$x$" when it is further treated with chlorine under the photochemical condition?

(a) 4 (b) 2 (c) 3 (d) 5

**36** Given below are two statements:

**Statement I** The conversion proceeds well in the less polar medium.

$$CH_3—CH_2—CH_2—CH_2—Cl \xrightarrow{HO^-} CH_3—CH_2—CH_2—CH_2—OH + Cl^{\ominus}$$

**Statement II** The conversion proceeds well in the more polar medium.

$$CH_3—CH_2—CH_2—CH_2—Cl \xrightarrow{R_3\ddot{N}} CH_3—CH_2—CH_2—CH_2—\overset{R}{\underset{R}{\overset{|(+)}{\underset{|}{N}}}}—RCl^{\ominus}$$

In the light of the above statements, choose the correct answer from the options given below *[2025, 24 Jan Shift-I]*

(a) Both Statement I and Statement II are true.
(b) Statement I is false but Statement II is true.
(c) Statement I is true but Statement II is false.
(d) Both Statement I and Statement II are false.

**37** Given below are two statements: *[2025, 28 Jan Shift-I]*

**Statement I** 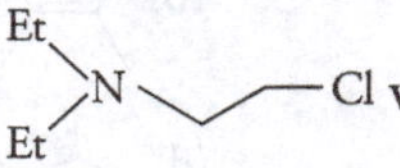 will undergo alkaline hydrolysis at a faster rate than 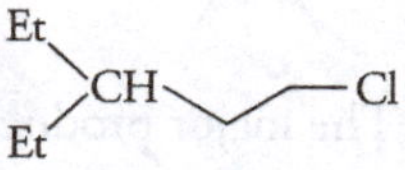

**Statement II** In 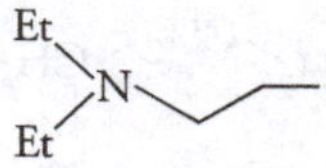, intramolecular substitution takes place first by involving lone pair of electrons on nitrogen.

In the light of the above statements, choose the most appropriate answer from the options given below

(a) Statement I is incorrect but Statement II is correct.
(b) Statement I is correct but Statement II is incorrect.
(c) Both Statement I and Statement II are incorrect.
(d) Both Statement I and Statement II are correct.

**38** Given below are two statements. *[2025, 22 Jan Shift-I]*

**Statement I** $CH_3—O—CH_2—Cl$ will undergo $S_N1$ reaction though it is a primary halide.

**Statement II** $CH_3—\overset{CH_3}{\underset{CH_3}{\overset{|}{\underset{|}{C}}}}—CH_2—Cl$ will not undergo $S_N2$ reaction very easily though it is a primary halide.

In the light of the above statements, choose the most appropriate answer from the options given below

(a) Both Statement I and Statement II are correct.
(b) Statement I is incorrect but Statement II is correct.
(c) Statement I is correct but Statement II is incorrect.
(d) Both Statement I and Statement II are incorrect.

**39** Given below are two statements : *[2024, 8 April Shift-II]*

**Statement I** $S_N2$ reactions are 'stereospecific', indicating that they result in the formation of only one *stereo*-isomer as the product.

**Statement II** $S_N1$ reactions generally result in formation of product as racemic mixtures.

In the light of the above statements, choose the correct answer from the options given below:
(a) Statement I is true but statement II is false.
(b) Statement I is false but statement II is true.
(c) Both statement I and statement II are true.
(d) Both statement I and statement II are false.

**40** Which among the following compounds will undergo fastest $S_N2$ reaction. *[2024, 8 April Shift-I]*

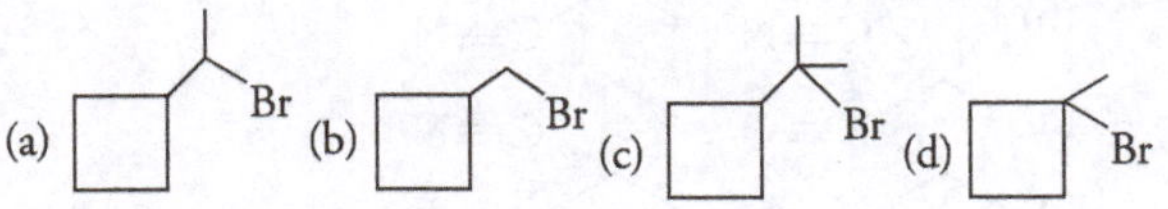

**41**

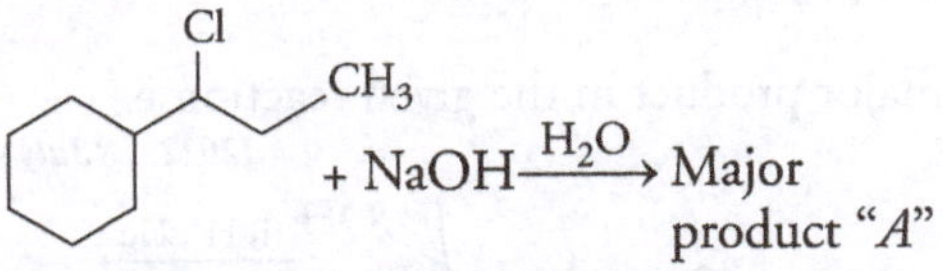

Consider the above chemical reaction. Product '*A*' is *[2024, 6 April Shift-II]*

(a) 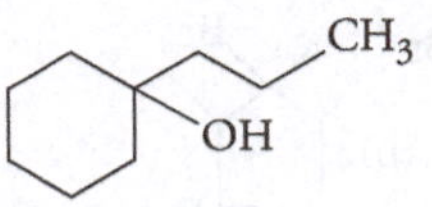 (b) 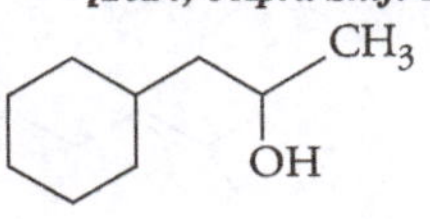

(c) 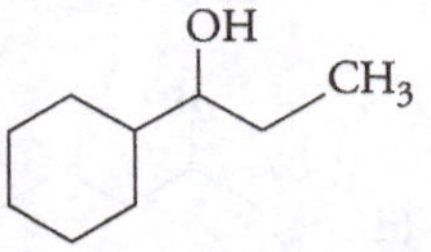 (d) 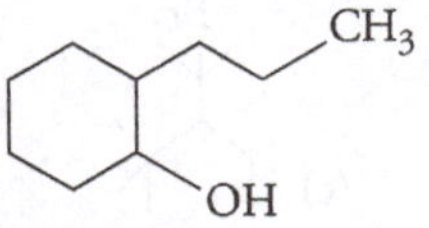

**42** Identify the correct set of reagents or reaction conditions '*X*' and '*Y*' in the following set of transformation. *[2024, 4 April Shift-I]*

$$CH_3—CH_2—CH_2—Br \xrightarrow{'X'} \text{Product}$$

$$\xrightarrow{'Y'} CH_3—\underset{\underset{Br}{|}}{CH}—CH_3$$

(a) $X$ = conc. alc. NaOH, 80°C, $Y$ = HBr/Acetic acid
(b) $X$ = dil. aq. NaOH, 20°C, $Y$ = HBr/Acetic acid
(c) $X$ = dil. aq. NaOH, 20°C, $Y = Br_2/CHCl_3$
(d) $X$ = conc. alc. NaOH, 80°C, $Y = Br_2/CHCl_3$

**43** Given below are two statements : one is labelled as **Assertion (A)** and other is labelled as **Reason (R).** *[2024, 1 Feb Shift-I]*

**Assertion (A)** Haloalkanes react with KCN to form alkyl cyanides as a main product while with AgCN form isocyanide as the main product.
**Reason (R)** KCN and AgCN both are highly ionic compounds.
In the light of the above statements, choose the **most appropriate** answer from the options given below.
(a) Both (A) and (R) are correct but (R) is not the correct explanation of (A).
(b) (A) is correct but (R) is not correct.
(c) Both (A) and (R) are correct and (R) is the correct explanation of (A).
(d) (A) is not correct but (R) is correct.

**44** Given below are two statements. *[2024, 30 Jan Shift-II]*

**Statement I** High concentration of strong nucleophilic reagent with secondary alkyl halides which do not have bulky substituents will follow $S_N2$ mechanism.
**Statement II** A secondary alkyl halide when treated with a large excess of ethanol follows $S_N1$ mechanism.
In the light of the above statements, choose the **most appropriate** from the options given below :
(a) Both Statement I and Statement II are true.
(b) Statement I is false but Statement II is true.
(c) Statement I is true but Statement II is false.
(d) Both Statement I and Statement II are false.

**45** Given below are two statements : one is labelled as **Assertion (A)** and the other is labelled as **Reason (R).**
**Assertion (A)** $CH_2{=}CH—CH_2—Cl$ is an example of allyl halide. *[2024, 30 Jan Shift-I]*
**Reason (R)** Allyl halides are the compounds in which the halogen atom is attached to $sp^2$-hybridised carbon atom.
In the light of the above statements, choose the **most appropriate** answer from the options given below:
(a) A is false but R is true.
(b) Both A and R are true and R is the correct explanation of A.
(c) Both A and R are true but R is not the correct explanation of A.
(d) A is true but R is false.

**46** Identify product *A* and product *B*. *[2024, 29 Jan Shift-I]*

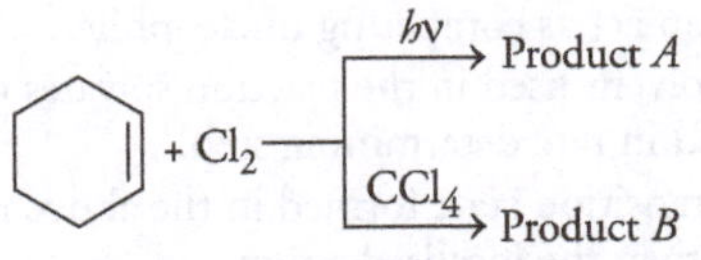

(a) *A* : 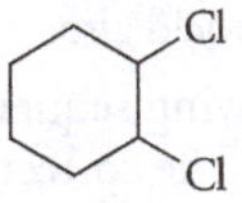 *B* : 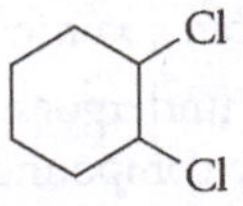

(b) *A* : 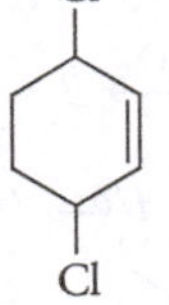 *B* : 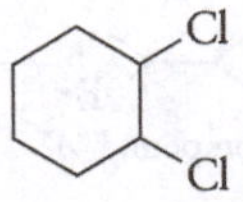

(c) *A* : 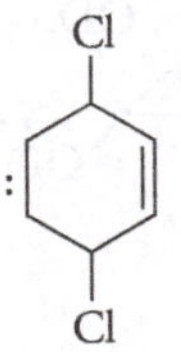 *B* : 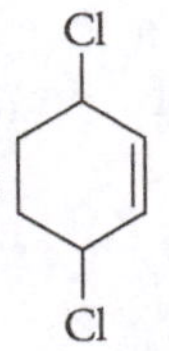

(d) $A$: 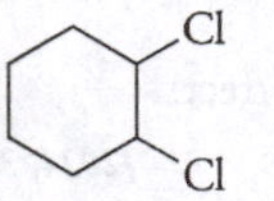 $B$: 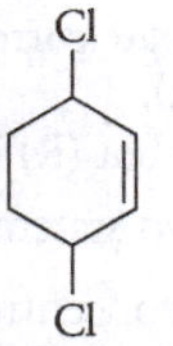

**47** Compound from the following that will not produce precipitate on reaction with $AgNO_3$ is *[2023, 11 April Shift-II]*

(a) 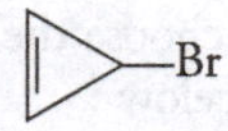 (b) 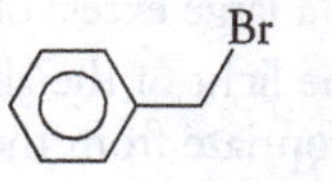

(c) 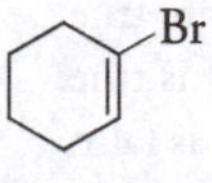 (d) 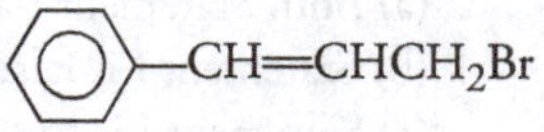

**48** Major product '$P$' formed in the following reaction is

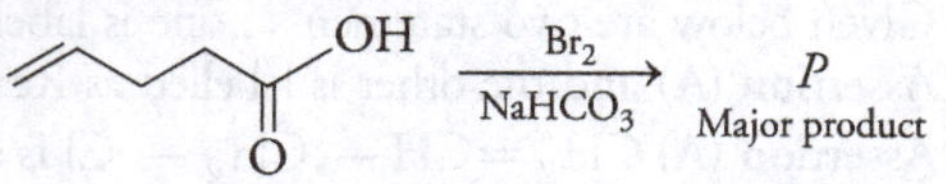

*[2023, 8 April Shift-II]*

(a) 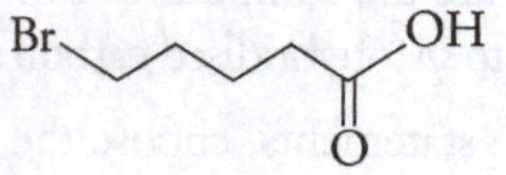 (b) 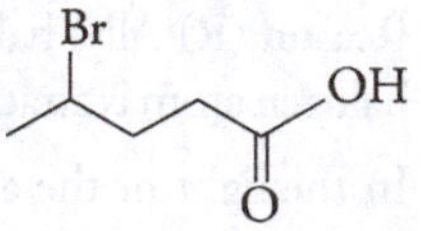

(c) 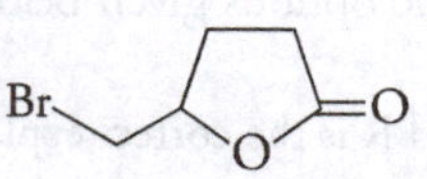 (d) 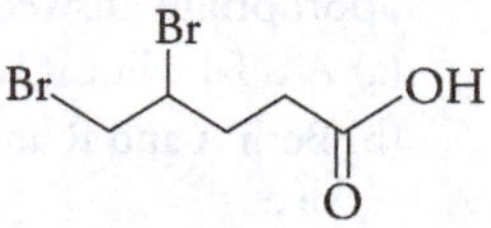

**49** For the reaction,

$$RCH_2Br + I^- \xrightarrow{\text{Acetone}} \underset{\text{Major}}{RCH_2I} + Br^-$$

The correct statement is *[2023, 6 April Shift-I]*

(a) $Br^-$ can act as competing nucleophile.
(b) The solvent used in the reaction solvates the ions formed in rate determining step.
(c) The transition state formed in the above reaction is less polar than the localised anion.
(d) The reaction can occur in acetic acid also.

**50** Compound '$A$' undergoes following sequence of reactions to give compound '$B$'. The correct structure and chirality of compound '$B$' is [where, Et is $-C_2H_5$]

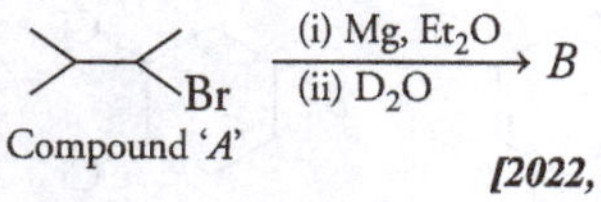

*[2022, 29 July Shift-II]*

(a) 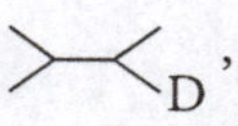, Achiral (b) 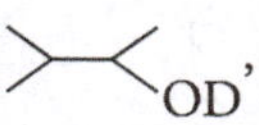, Chiral

(c) 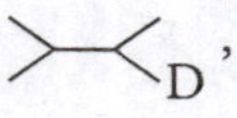, Chiral (d) 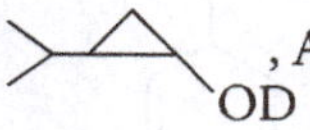, Achiral

**Imp.** **Reaction of Haloalkanes with AgCN and NaCN**

**51** 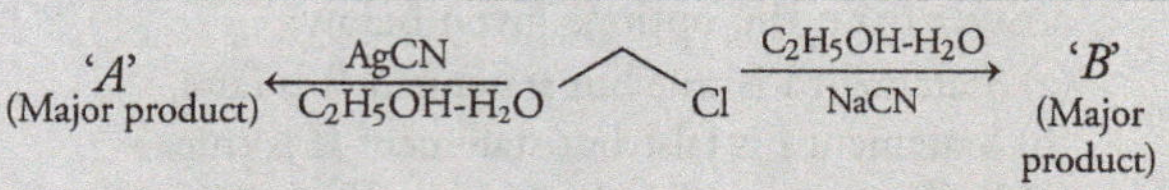

Considering the above reactions, the compound '$A$' and compound '$B$' respectively are *[2022, 29 July shift-I]*

(a) 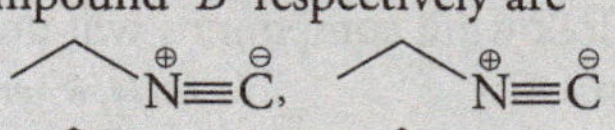

(b) 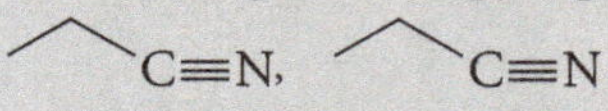

(c) 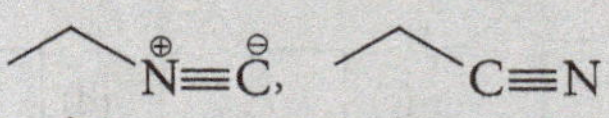

(d) 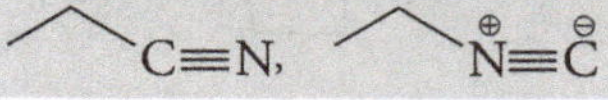

**52** The major product in the given reaction is *[2022, 28 July shift-II]*

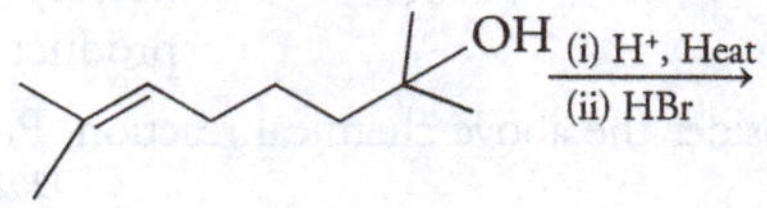

(a) 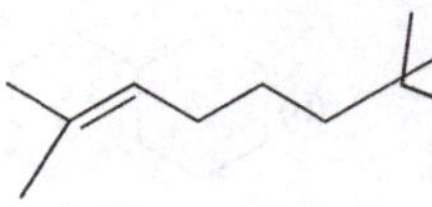 (b) 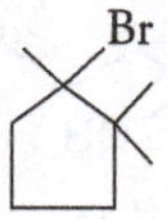

(c) 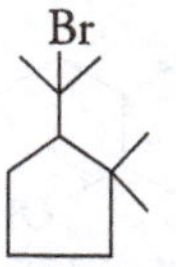 (d) 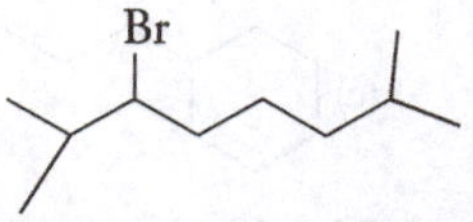

**53** Major product '$B$' of the following reaction sequence is *[2022, 27 July shift-II]*

$$CH_3-\underset{\underset{CH_3}{|}}{C}=CH-CH_3 \xrightarrow[CH_3OH]{Br_2} A \xrightarrow{HI} \underset{\text{(Major product)}}{B}$$

(a) $CH_3-\overset{HO}{\underset{CH_3}{C}}-\overset{Br}{CH}-CH_3$ (b) $CH_3-\overset{I}{\underset{CH_3}{C}}-\overset{Br}{CH}-CH_3$

(c) $CH_3-\overset{Br}{\underset{CH_3}{C}}-\overset{OH}{CH}-CH_3$ (d) $CH_3-\overset{Br}{\underset{CH_3}{C}}-\overset{I}{CH}-CH_3$

**54** The correct order of reactivity of the given chlorides with acetate in acetic acid is *[2021, 31 Aug Shift-I]*

(a) 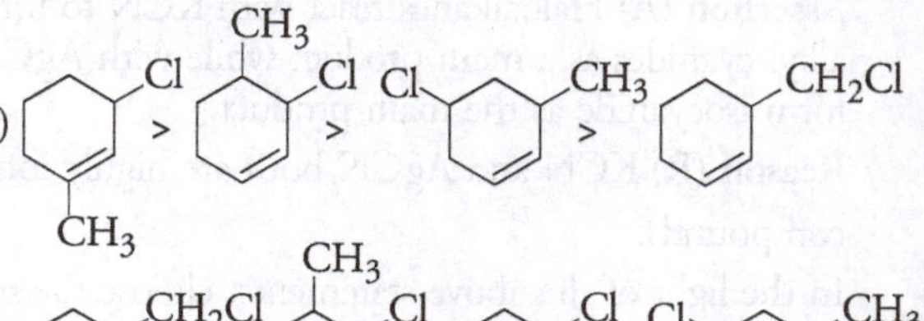

(b) 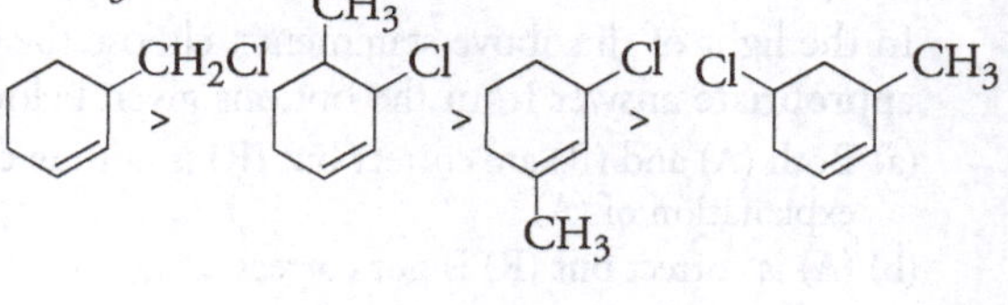

(c) 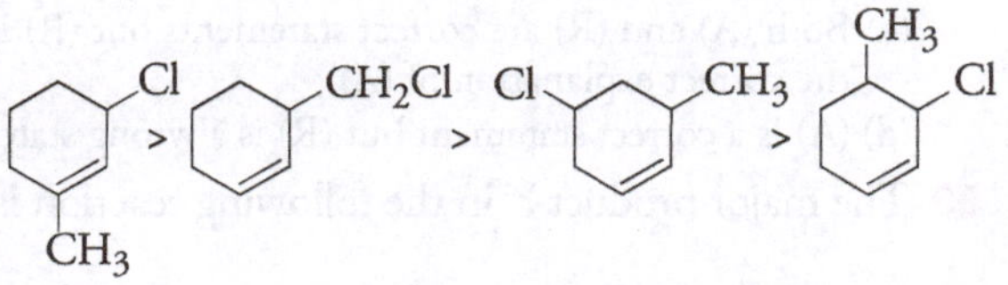

(d) 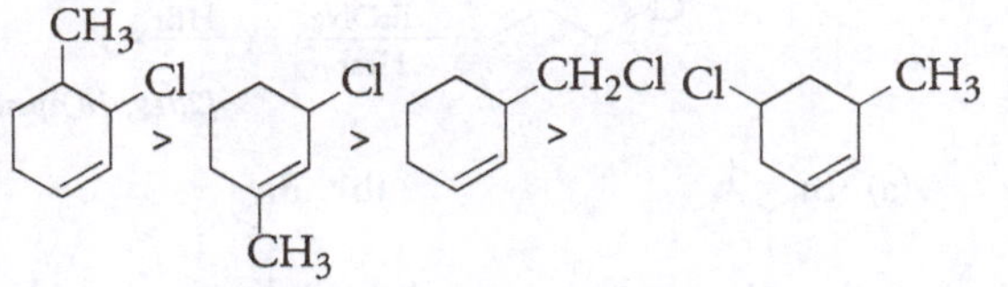

**55** Excess of isobutane on reaction with $Br_2$ in presence of light at 125°C gives which one of the following, as the major product? *[2021, 26 Aug Shift-I]*

(a) $CH_3-C(Br)(CH_3)-CH_2Br$ (b) $CH_3-CH(CH_2Br)-CH_2Br$

(c) $CH_3-CH(CH_3)-CH_2Br$ (d) $CH_3-C(CH_3)(CH_3)-Br$

**56** 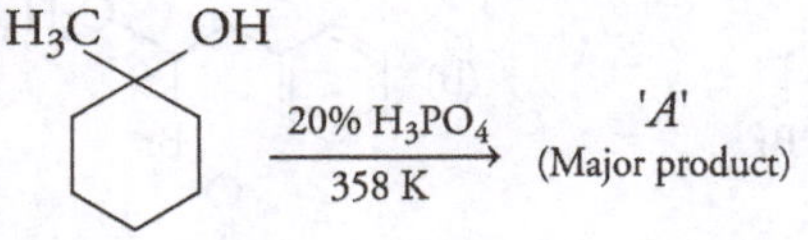

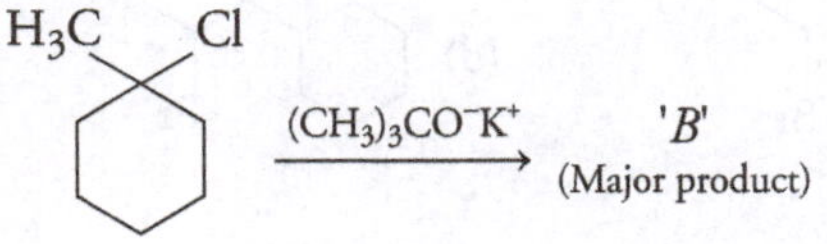

The product *A* and *B* formed in above reactions are *[2021, 16 March Shift-I]*

(a) *A*- 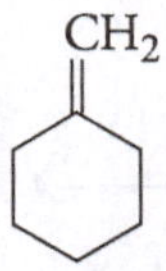 *B*- 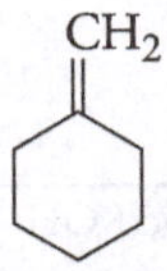 (b) *A*- 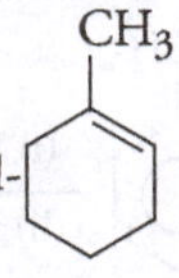 *B*- 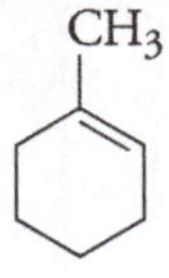

(c) *A*- 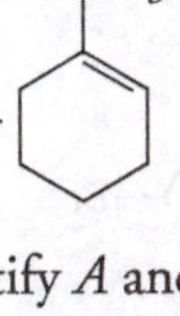 *B*- 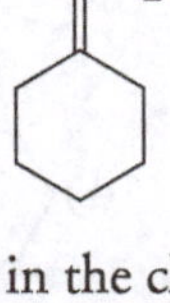 (d) *A*- 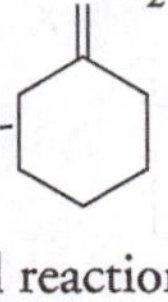 *B*- 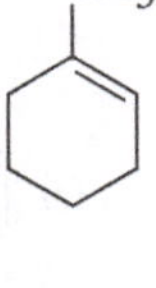

**57** Identify *A* and *B* in the chemical reaction.

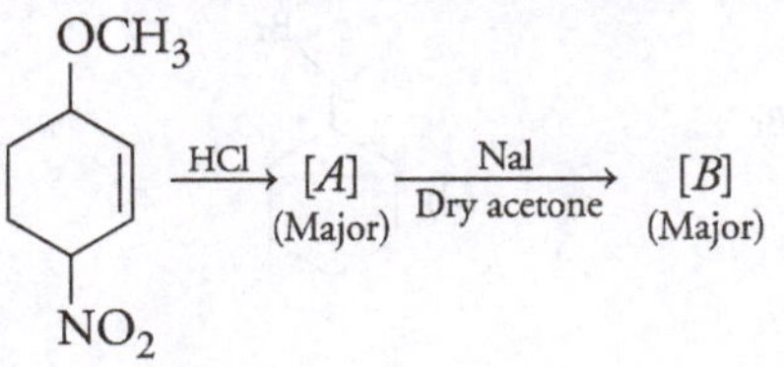

*[2021, 25 Feb Shift-I]*

(a) *A* = 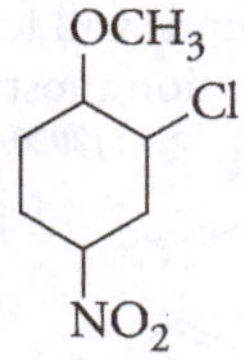 *B* = 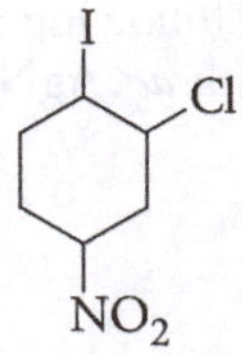

(b) *A* = 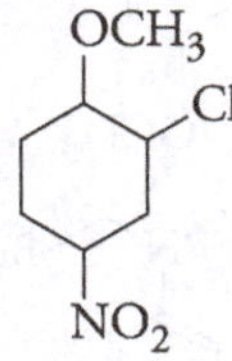 *B* = 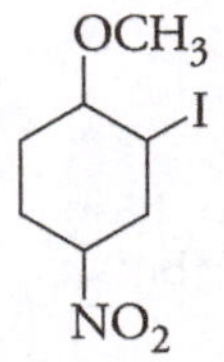

(c) *A* = 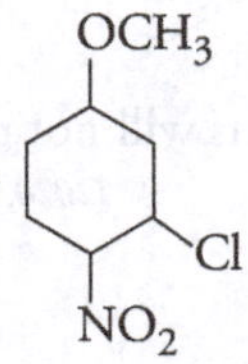 *B* = 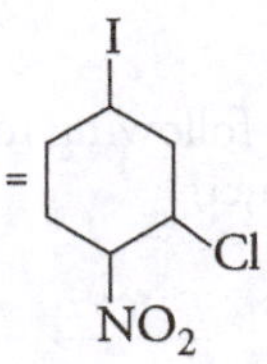

(d) *A* = 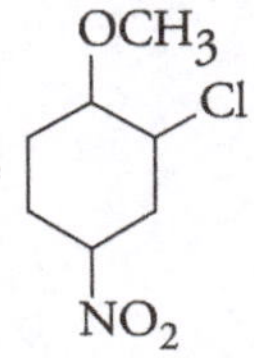 *B* = 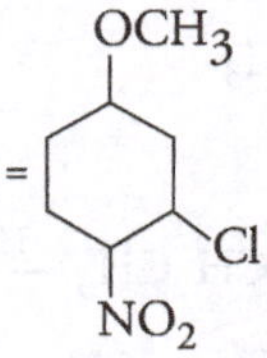

**58** The product formed in the first step of the reaction of $CH_3-CH_2-CH(Br)-CH_2-CH(Br)-CH_3$ with excess $Mg/Et_2O$ ($Et = C_2H_5$) is *[2021, 24 Feb Shift-I]*

(a) 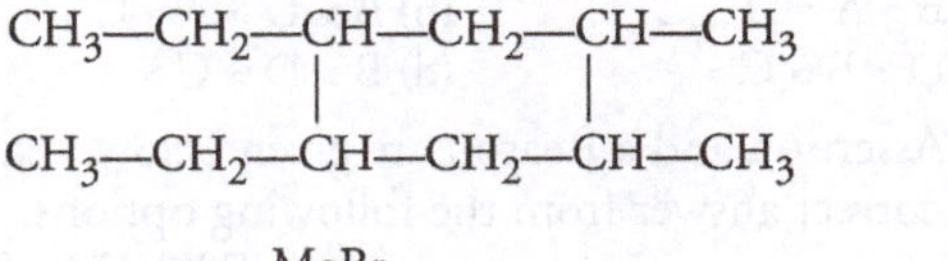

(b) 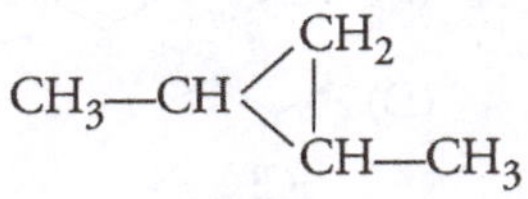

(c) 

(d) $CH_3CH_2-CH(MgBr)-CH_2-CH(MgBr)-CH_3$

**59** Which of the following compounds will form the precipitate with *aq*. $AgNO_3$ solution most readily? *[2020, 4 Sep Shift-II]*

(a) 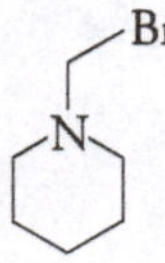
(b) 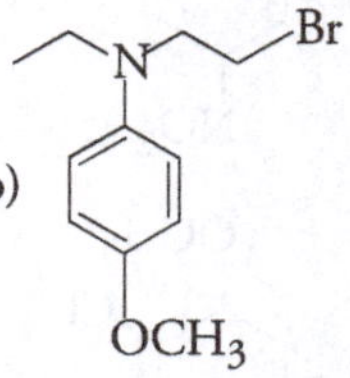
(c) 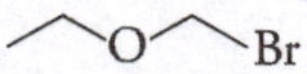
(d) 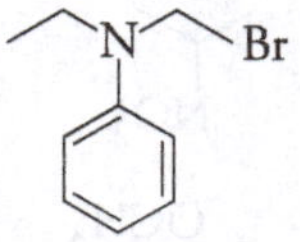

**60** Which of the following reactions will not produce a racemic product? *[2020, 9 Jan Shift-II]*

(a) 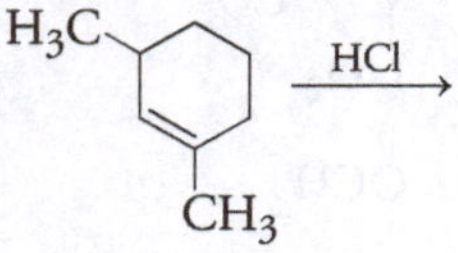

(b) $CH_3-\overset{\overset{O}{\|}}{C}CH_2CH_3 \xrightarrow{HCN}$

(c) $CH_3CH_2CH=CH_2 \xrightarrow{HBr}$

(d) $CH_3-\underset{\underset{H}{|}}{\overset{\overset{CH_3}{|}}{C}}-CH=CH_2 \xrightarrow{HCl}$

**61** The decreasing order of reactivity towards dehydrohalogenation ($E_1$) reaction of the following compounds is *[2020, 8 Jan Shift-I]*

(A) 
(B) 
(C) 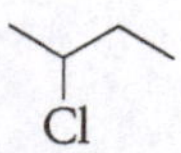
(D) 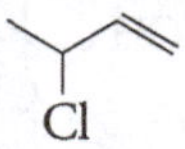

(a) B > A > D > C
(b) B > D > A > C
(c) D > B > C > A
(d) B > D > C > A

**62** An Assertion and a Reason are given below. Choose the correct answer from the following options. *[2019, 12 April Shift-II]*

**Assertion** (A) Vinyl halides do not undergo nucleophilic substitution easily.

**Reason** (R) Even though the intermediate carbocation is stabilised by loosely held $\pi$-electrons, the cleavage is difficult because of strong bonding.

(a) Both (A) and (R) are wrong statements.
(b) Both (A) and (R) are correct statements and (R) is correct explanation of (A).
(c) Both (A) and (R) are correct statements but (R) is not the correct explanation of (A).
(d) (A) is a correct statement but (R) is a wrong statement.

**63** The major product *Y* in the following reaction is

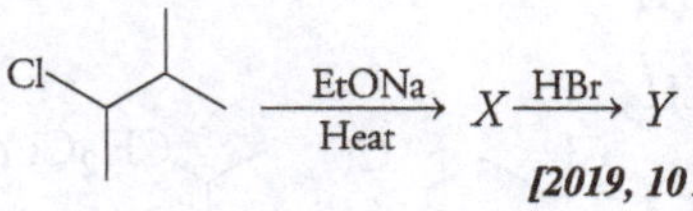

*[2019, 10 April Shift-II]*

(a) 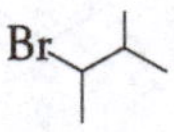
(b) 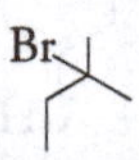
(c) 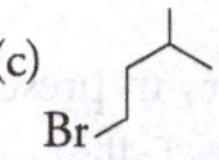
(d) 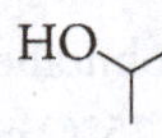

**64** The increasing order of reactivity of the following halides for the $S_N1$ reaction is

I. $CH_3CH(Cl)CH_2CH_3$
II. $CH_3CH_2CH_2Cl$
III. $p$-$H_3CO-C_6H_4-CH_2Cl$ *[JEE Main 2017]*

(a) (III) < (II) < (I)
(b) (II) < (I) < (III)
(c) (I) < (III) < (II)
(d) (II) < (III) < (I)

**65** Which of the following, upon treatment with *tert*-$BuO^-Na^+$ followed by addition of bromine water, fails to decolourise the colour of bromine? *[JEE Main 2017]*

(a) 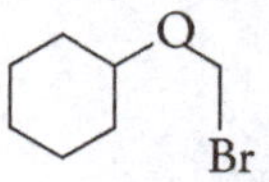
(b) 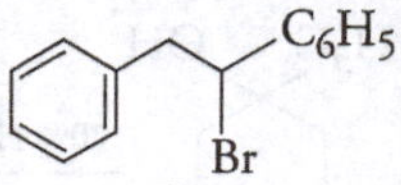
(c) 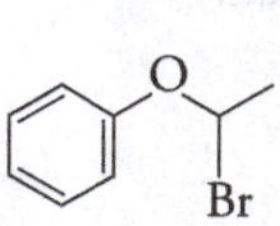
(d) 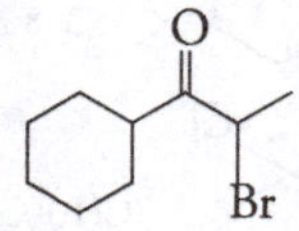

## Level 3

**66** Predict the major product of the following reaction sequence: *[2025, 4 April Shift-I]*

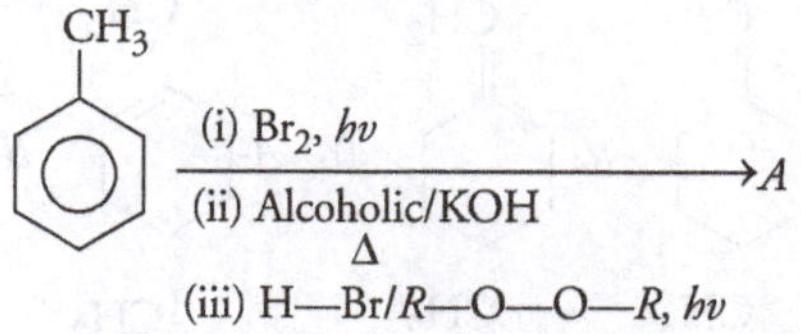

(a) 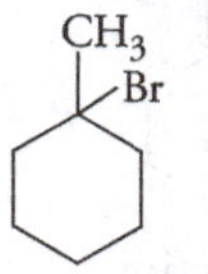
(b) 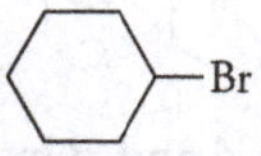
(c) 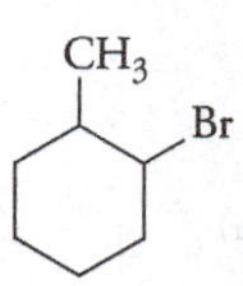
(d) 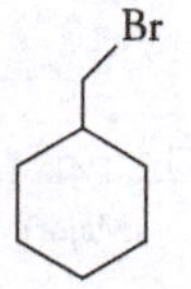

**67** Match the List-I with List-II *[2025, 23 Jan Shift-I]*

| List-I (Name reaction) | List-II (Product obtainable) |
|---|---|
| A. Swarts reaction | I. Ethyl benzene |
| B. Sandmeyer's reaction | II. Ethyl iodide |
| C. Wurtz Fittig reaction | III. Cyanobenzene |
| D. Finkelstein reaction | IV. Ethyl fluoride |

Choose the correct answer from the options given below:

(a) A-IV, B-I, C-III, D-II (b) A-IV, B-III, C-I, D-II
(c) A-II, B-III, C-I, D-IV (d) A-II, B-I, C-III, D-IV

**68** The ascending order of relative rate of solvolysis of following compounds is *[2025, 23 Jan Shift-II]*

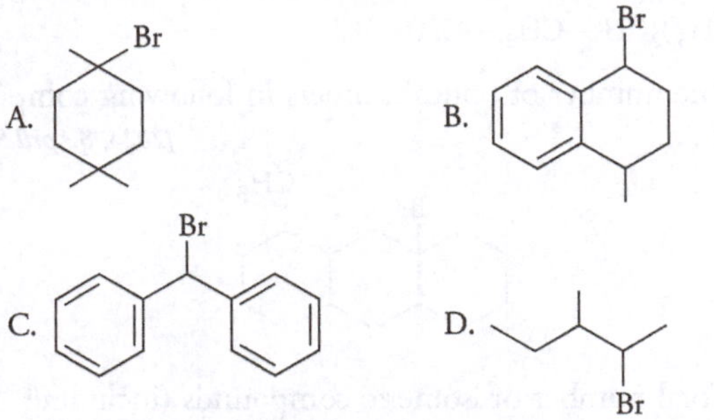

(a) (C) < (B) < (A) < (D) (b) (D) < (B) < (A) < (C)
(c) (C) < (D) < (B) < (A) (d) (D) < (A) < (B) < (C)

**69** $C_2H_5Br \xrightarrow{\text{Alc. KOH}} A \xrightarrow[CCl_4]{Br_2} B \xrightarrow[\text{Excess}]{KCN} C \xrightarrow[\text{Excess}]{H_3O^+} D$

Acid $D$ formed in above reaction is *[2024, 1 Feb Shift-II]*

(a) oxalic acid (b) malonic acid
(c) gluconic acid (d) Succinic acid

**Imp.** **Reaction of Haloalkanes with KOH (alc.) and (aq.) KOH in the Presence of Heat**

**70** The product ($C$) in the below mentioned reaction is

$$CH_3-CH_2-CH_2-Br \xrightarrow[\Delta]{\text{KOH (alc.)}} A \xrightarrow{HBr} B \xrightarrow[\text{KOH}(aq)]{\Delta} C$$

*[2024, 31 Jan Shift-I]*

(a) Propan-1-ol (b) Propan-2-ol
(c) Propene (d) Propyne

**71** Which of the following reaction is correct? *[2024, 29 Jan Shift-II]*

(a) 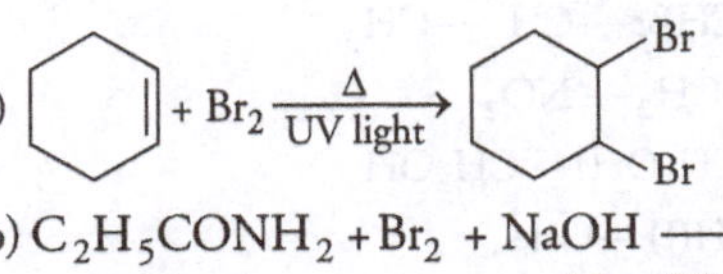

(b) $C_2H_5CONH_2 + Br_2 + NaOH \longrightarrow C_2H_5CH_2NH_2 + Na_2CO_3 + NaBr + H_2O$

(c) 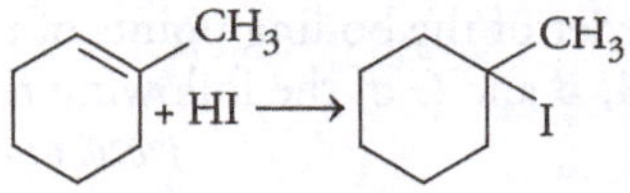

(d) $CH_3CH_2CH_2NH_2 \xrightarrow[H_2O]{HNO_2,\ 0°C} CH_3CH_2OH + N_2 + HCl$

**72** 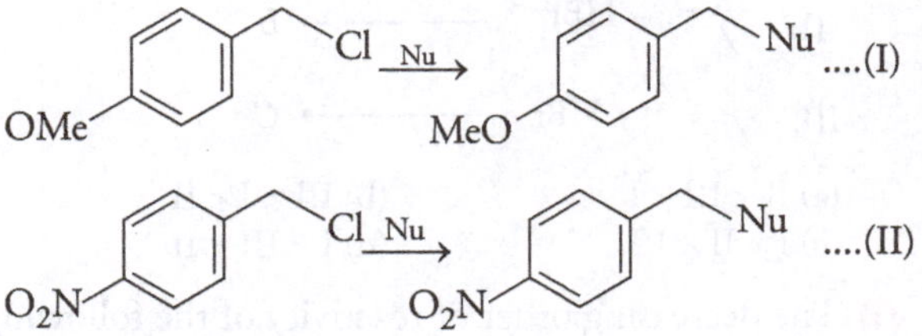

where, Nu = Nucleophile *[2023, 11 April Shift-I]*

Find out the correct statement from the options given below for the above 2 reactions.

(a) Reaction (I) and (II) both are of 1st order
(b) Reactions (I) and (II) both are of 2nd order
(c) Reactions (I) is of 2nd order and reaction (II) is of 1st order
(d) Reaction (I) is of 1st order and reaction (II) is of 2nd order

**73** The pair from the following pairs having both compounds with net non-zero dipole moment is *[2023, 10 April Shift-I]*

(a) 1, 4-dichlorobenzene, 1,3-dichlorobenzene
(b) $CH_2Cl_2$, $CHCl_3$
(c) Benzene, anisidine
(d) *cis*-butene, *trans*-butene

**74** In the given reaction sequence, the major product $C$ is *[2022, 24 June Shift-I]*

$$C_8H_{10} \xrightarrow[H_2SO_4]{HNO_3} A \xrightarrow[\Delta]{Br_2} B \xrightarrow[\text{KOH}]{\text{Alcoholic}} C$$

(a) 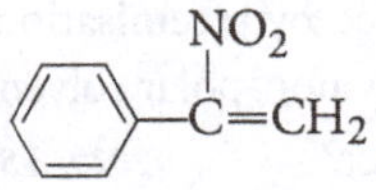

(b) 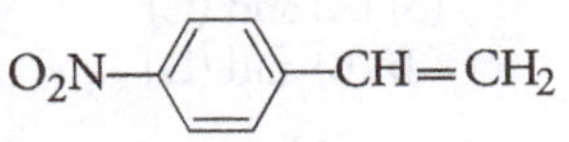

(c) 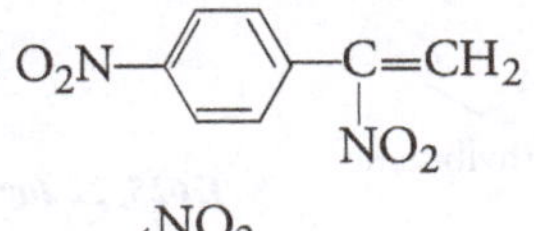

(d) 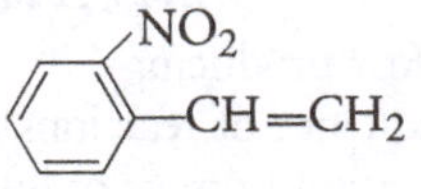

**75** The increasing order of the boiling points of the major products *A*, *B* and *C* of the following reactions will be *[2020, 6 Sep Shift-II]*

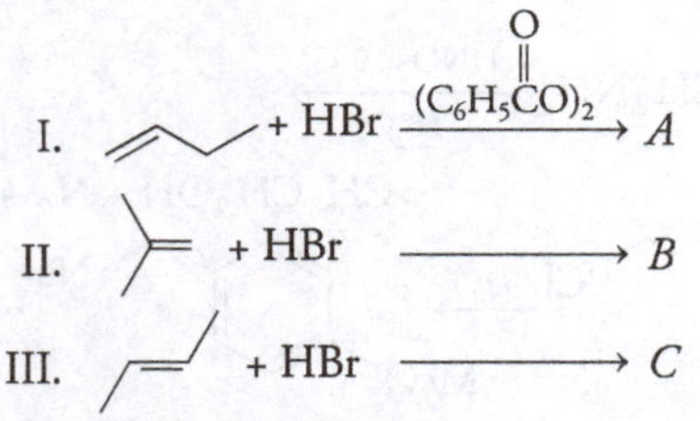


(a) II < III < I (b) III < I < II
(c) I < II < III (d) I < III < II

**76** The decreasing order of reactivity of the following organic molecules towards $AgNO_3$ solution is *[2020, 4 Sep Shift-I]*

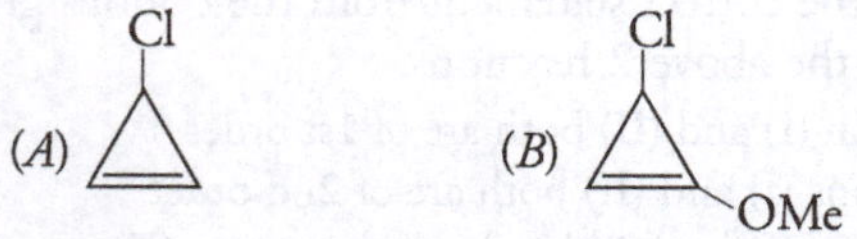


(C) $CH_3CH(Cl)CH_3$ (D) $CH_3CH(Cl)CH_2NO_2$

(a) (*A*) > (*B*) > (*C*) > (*D*) (b) (*C*) > (*D*) > (*A*) > (*B*)
(c) (*B*) > (*A*) > (*C*) > (*D*) (d) (*A*) > (*B*) > (*D*) > (*C*)

**77** The mechanism of $S_N1$ reaction is given as

$$RX \longrightarrow \underset{\text{Ion pair}}{R^{\oplus}X^{\ominus}} \longrightarrow \underset{\text{Solvent Separated ion pair}}{R^{\oplus}||X^{\ominus}} \xrightarrow{Y^{\ominus}} RY + X^{\ominus}$$

A student writes general characteristics based on the given mechanism as :

(A) The reaction is favoured by weak nucleophiles.
(B) $R^{\oplus}$ would be easily formed if the substituents are bulky.
(C) The reaction is accompanied by recemisation.
(D) The reaction is favoured by non-polar solvents.

Which observations are correct? *[2020, 3 Sep Shift-I]*

(a) (A) and (B) (b) (A) and (C)
(c) (A), (B) and (C) (d) (B) and (D)

## Numerical Value Type Questions

**78**

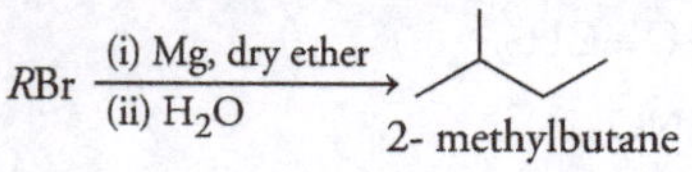


*[2025, 22 Jan Shift-II]*

The maximum number of *R*Br producing 2-methylbutane by above sequence of reactions is ......... . (Consider the structural isomers only)

(a) 5 (b) 3
(c) 4 (d) 1

**79** Total number of optically active compounds from the following is ........ *[2024, 8 April Shift-II]*

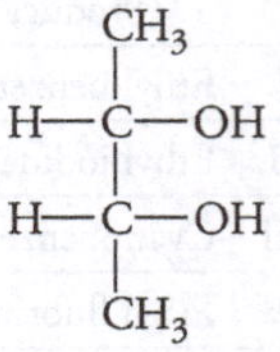

,

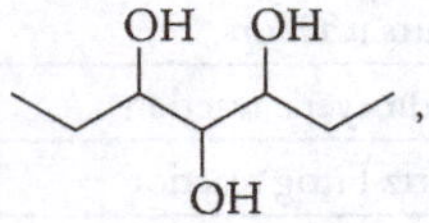

,

$CH_3—CH_2—CH_2—CH_2—OH$,

$CH_3—CH_2—CH(Cl)—CH_3$

$CH_3—CH_2—CH_2—CH_2—Cl$

$(CH_3)_2CH—CH_2—CH_2—Cl$

**80** The number of optical isomers in following compound is ...... *[2024, 8 April Shift-I]*

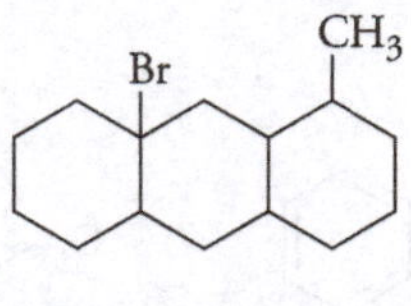


**81** Total number of isomeric compounds (including stereoisomers) formed by monochlorination of 2-methylbutane is ......... . *[2024, 1 Feb Shift-II]*

**82** Number of optical isomers possible for 2-chlorobutane ......... . *[2024, 1 Feb Shift-I]*

**83**

$$CH_3CH_2Br + NaOH \begin{cases} \xrightarrow{C_2H_5OH} \text{Product } A \\ \xrightarrow{H_2O} \text{Product } B \end{cases}$$

The total number of hydrogen atoms in product *A* and product *B* is ........ *[2024, 31 Jan Shift-I]*

**84** Total number of compounds with chiral carbon atoms from following is ....... *[2024, 27 Jan Shift-II]*

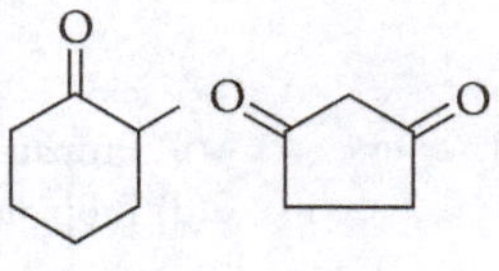


$CH_3—CH_2—CH(NO_2)—COOH$

$CH_3—CH_2—CHBr—CH_2—CH_3$

$CH_3—CH(I)—CH_2—NO_2$

$CH_3—CH_2—CH(OH)—CH_2OH$

$CH_3—CH(I)—CH(I)—C_2H_5$

## Topic 2 Haloarenes

### Level 1

**01** The structure of the major product formed in the following reaction is *[2025, 24 Jan Shift-II]*

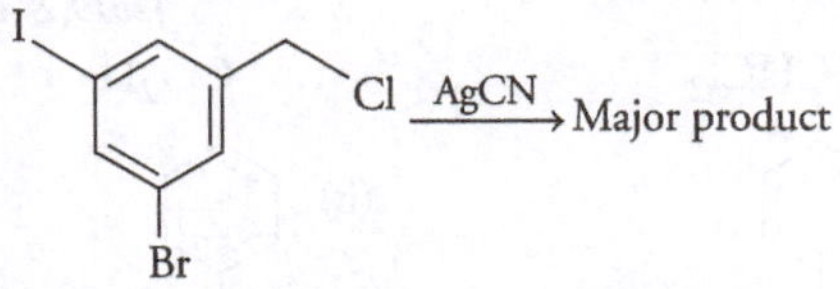

(a) 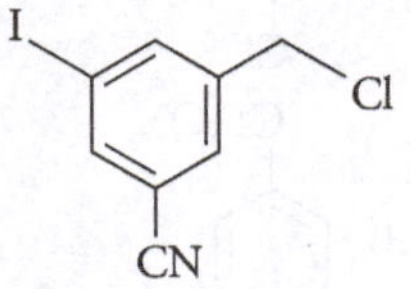 (b) 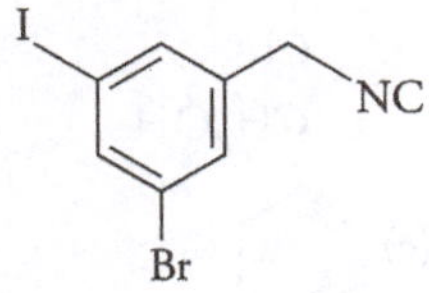

(c) 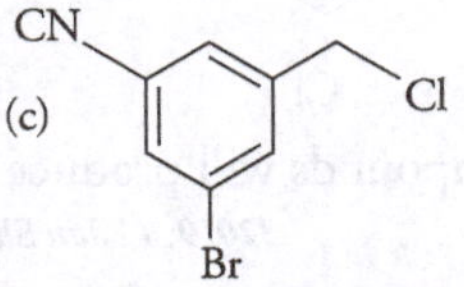 (d) 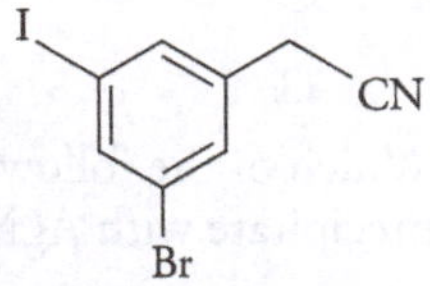

**02** Which among the following halides will generate the most stable carbocation in the nucleophilic substitution reaction? *[2025, 29 Jan Shift-II]*

(a)  (b) 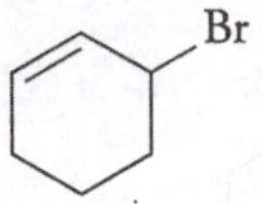

(c) 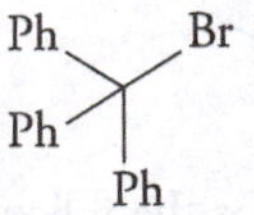 (d) 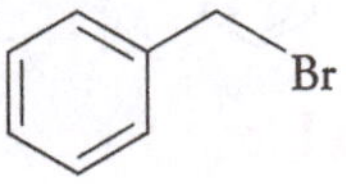

**03** In the following substitution reaction,

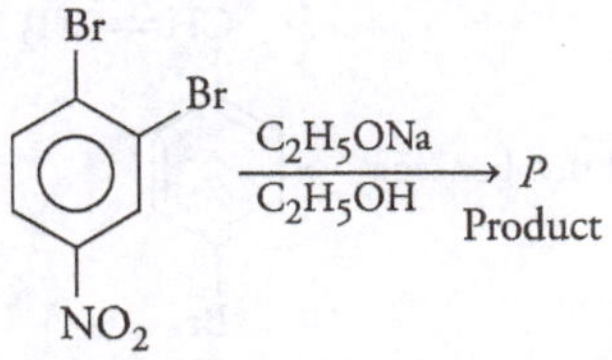

Product $P$ formed is *[2025, 29 Jan Shift-I]*

(a) 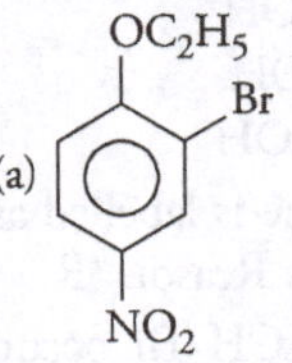 (b) 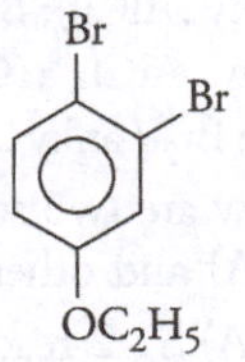

(c) 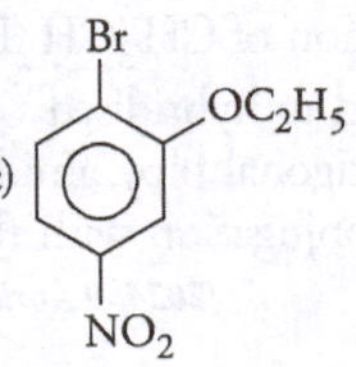 (d) 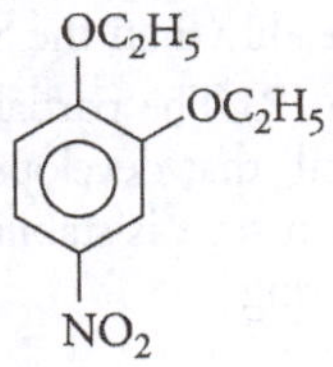

**04** The major product formed in the Friedel-Craft acylation of chlorobenzene is *[2023, 15 April Shift-I]*

(a) 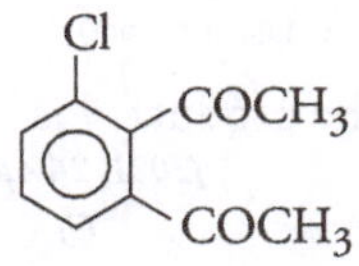 (b) 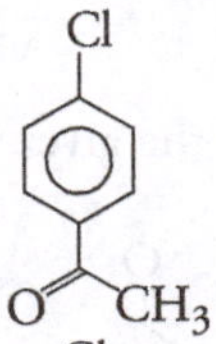

(c) 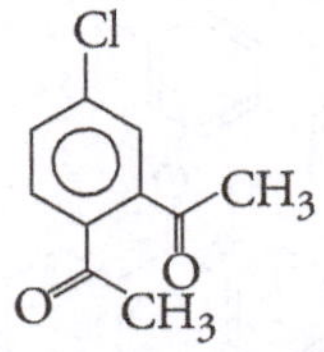 (d) 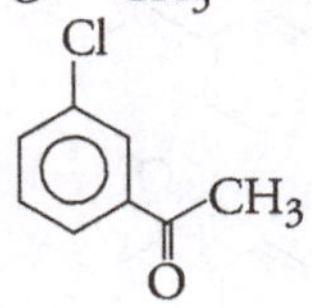

**05** The compound which will have the lowest rate towards nucleophilic aromatic substitution on treatment with $OH^-$ is *[2023, 25 Jan Shift-I]*

(a) 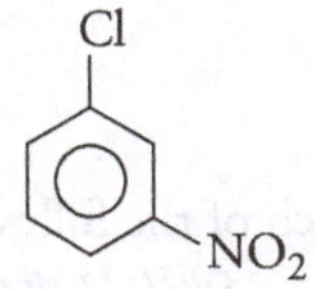 (b) 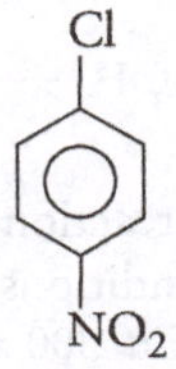

(c) 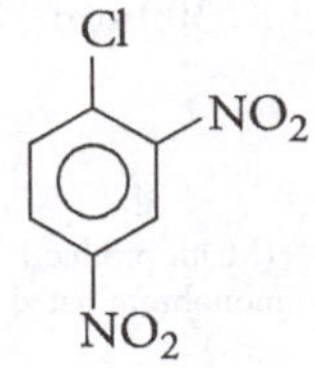 (d) 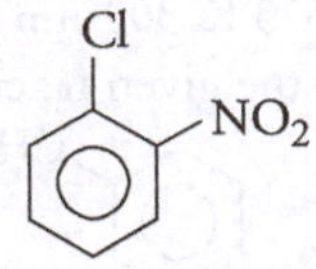

**06** 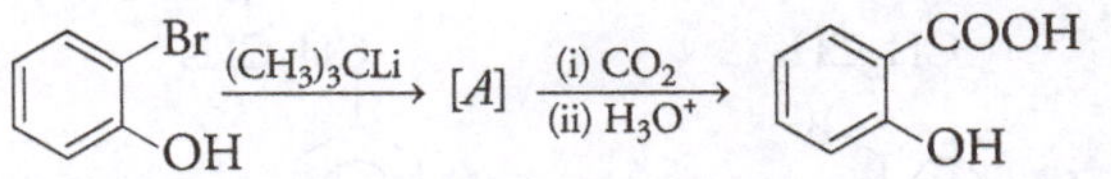

In the given conversion the compound $A$ is *[2022, 29 June Shift-I]*

(a) 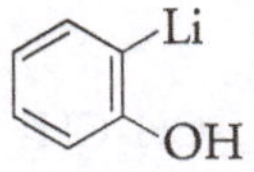 (b) 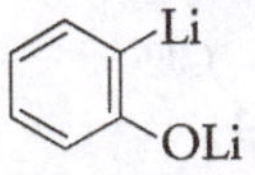

(c) 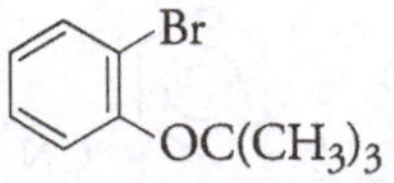 (d) 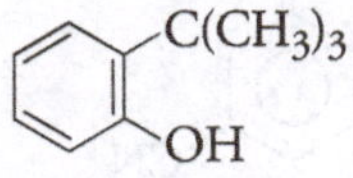

**07** Halogenation of which one of the following will yield *m*-substituted product with respect to methyl group as a major product? *[2022, 26 June Shift-II]*

(a) 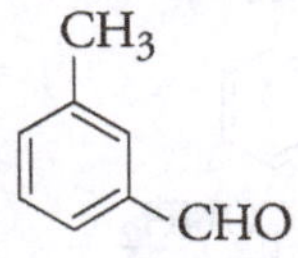 (b) 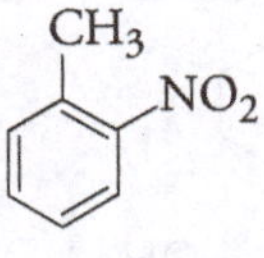

(c) 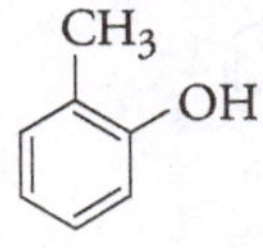 (d) 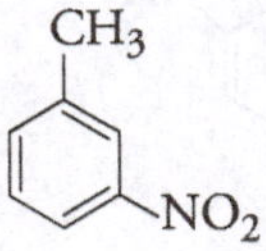

**08** 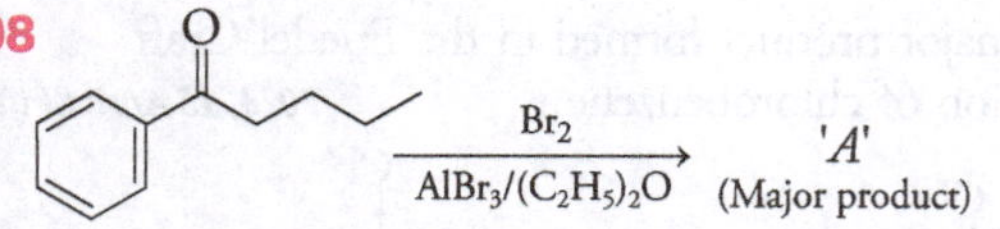

Consider the given reaction, the product *A* is **[2021, 26 Aug Shift-II]**

(a) 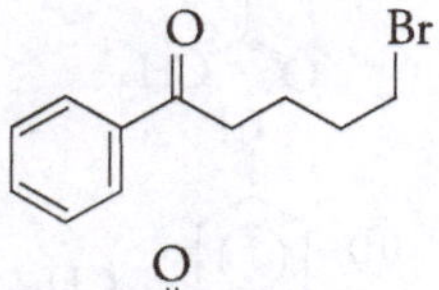 (b) 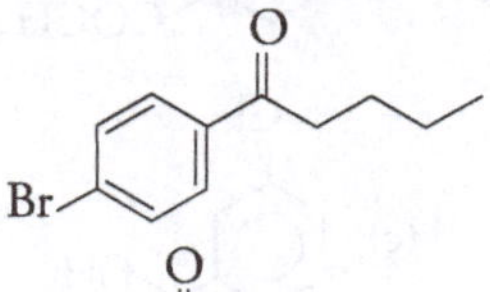

(c) 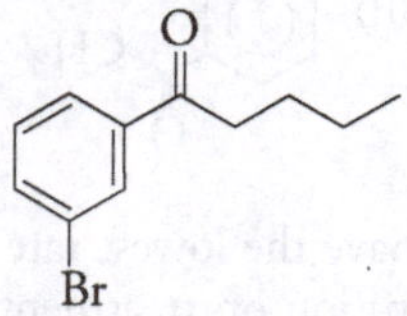 (d) 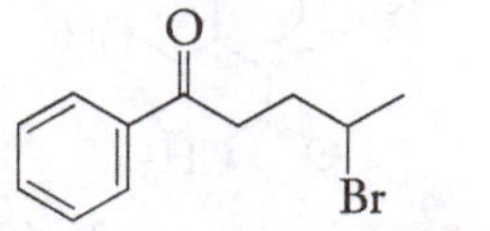

**09** 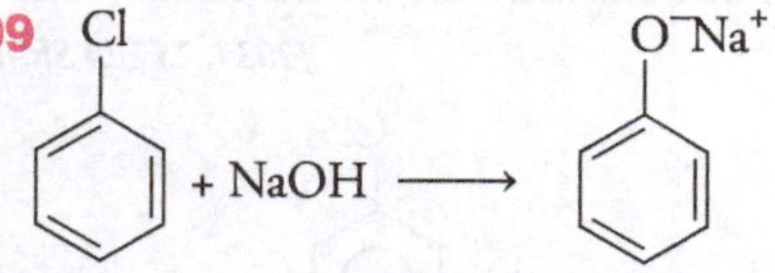

The above reaction requires which of the following reaction conditions? **[2021, 17 March Shift-I]**

(a) 573 K, Cu, 300 atm (b) 623 K, Cu, 300 atm
(c) 573 K, 300 atm (d) 623 K, 300 atm

**10** For the given reaction,

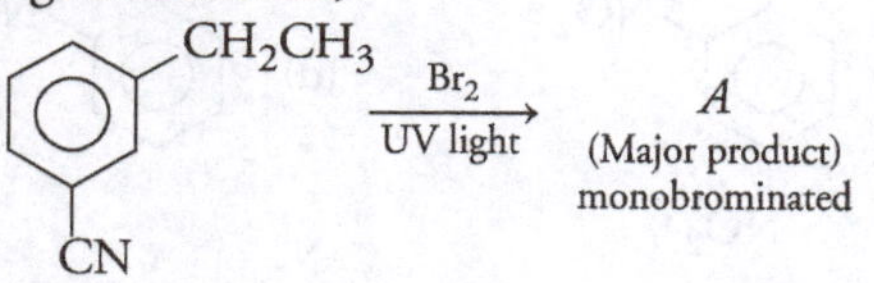

What is *A*? **[2021, 26 Feb Shift-I]**

(a) 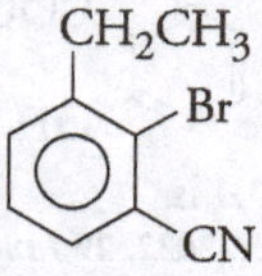 (b) 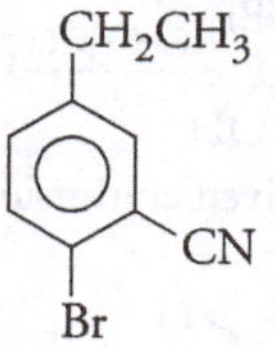

(c) 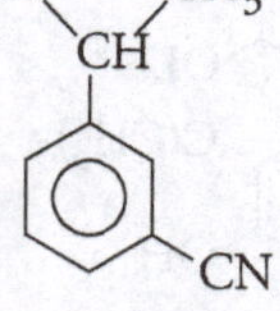 (d) 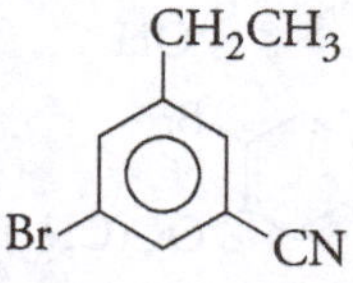

**11** Which of these will produce the highest yield in Friedel-Crafts reaction? **[2020, 9 Jan Shift-I]**

(a) 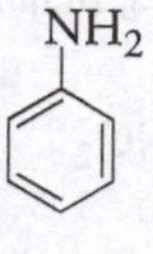 (b) 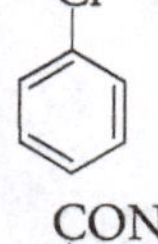

(c) 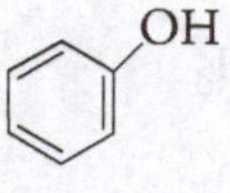 (d) 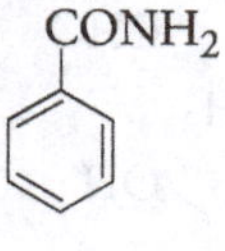

**12** The major product of the following reaction is

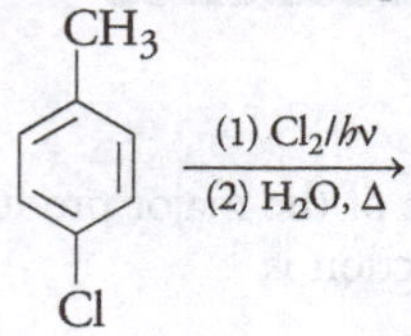

**[2019, 8 April Shift-II]**

(a) 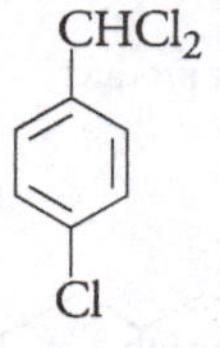 (b) 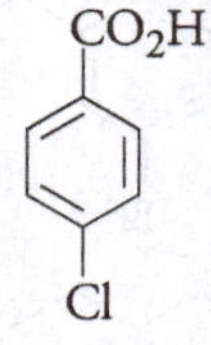

(c) 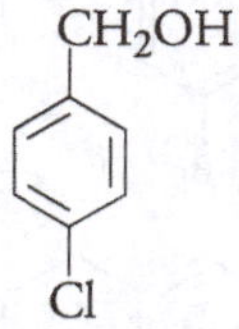 (d) 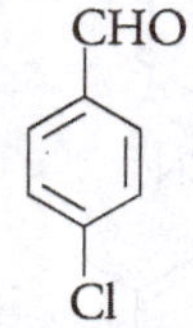

**13** Which of the following compounds will produce a precipitate with $AgNO_3$? **[2019, 11 Jan Shift-II]**

(a) 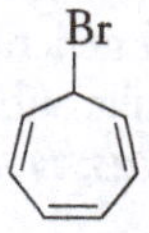 (b) 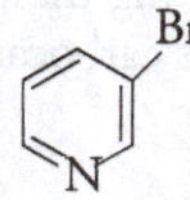

(c) 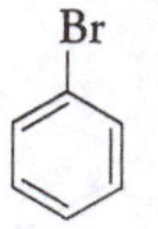 (d) 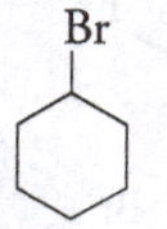

## Level 2

**14.** Choose the correct set of reagents for the following conversion. **[2025, 8 April Shift-II]**

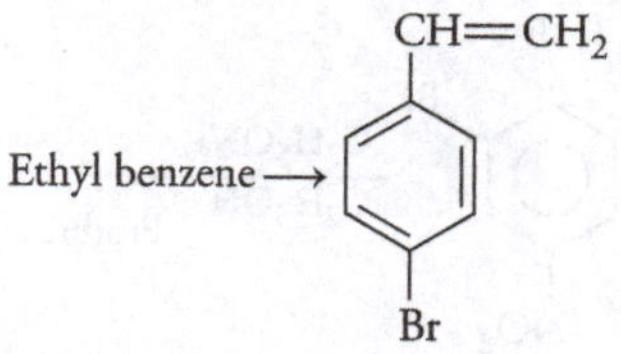

(a) $Br_2$ / Fe; $Cl_2, \Delta$; alc. KOH
(b) $Cl_2$ / anhy. $AlCl_3$; $Br_2$ / Fe; alc. KOH
(c) $Br_2$ / anhy. $AlCl_3$; $Cl_2, \Delta$; *aq*. KOH
(d) $Cl_2$ / Fe; $Br_2$ / anhy. $AlCl_3$; *aq*. KOH

**15** Given below are two statements : one is labelled as Assertion (A) and other is labelled as Reason (R).

**Assertion (A)** $S_N2$ reaction of $C_6H_5CH_2Br$ occurs more readily than the $S_N2$ reaction of $CH_3CH_2Br$.

**Reason (R)** The partially bonded unhybridised *p*-orbital that develops in the trigonal bipyramidal transition state is stabilised by conjugation with the phenyl ring. **[2024, 9 April Shift-I]**

In the light of the above statements, choose the most appropriate answer from the options given below.

(a) Both A and R are correct but R is not the correct explanation of A.
(b) A is not correct, but R is correct.
(c) Both A and R are correct but R is the correct explanation of A.
(d) A is correct but R is not correct.

**16** Given below are two statements one is labelled as **Assertion (A)** and the other is labelled as **Reason (R).** *[2024, 29 Jan Shift-II]*

**Assertion (A)** Aryl halides cannot be prepared by replacement of hydroxyl group of phenol by halogen atom.

**Reason (R)** Phenols react with halogen acids violently.

In the light of the above statements, choose the **most appropriate** from the options given below

(a) Both A and R are true but R is not the correct explanation of A.
(b) Both A and R are true and R is the correct explanation of A.
(c) A is false but R is true.
(d) A is true but R is false.

**17** The correct order of reactivity of following haloarenes towards nucleophilic substitution with aqueous NaOH is *[2023, 8 April Shift-II]*

(A) 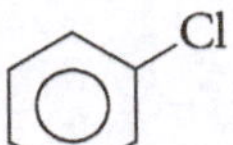


(B) 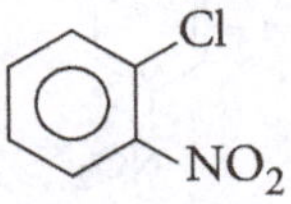


(C) 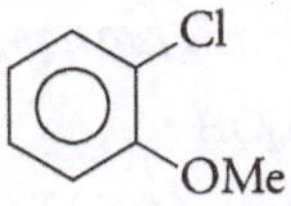


(D) 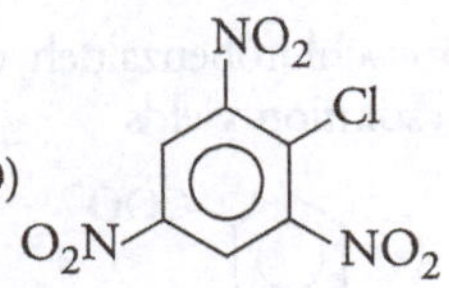


Choose the correct answer from the options given below:

(a) C > A > D > B (b) A > B > D > C
(c) D > B > A > C (d) D > C > B > A

**18** The correct order of melting points of dichlorobenzenes is *[2023, 31 Jan Shift-I]*

(a) 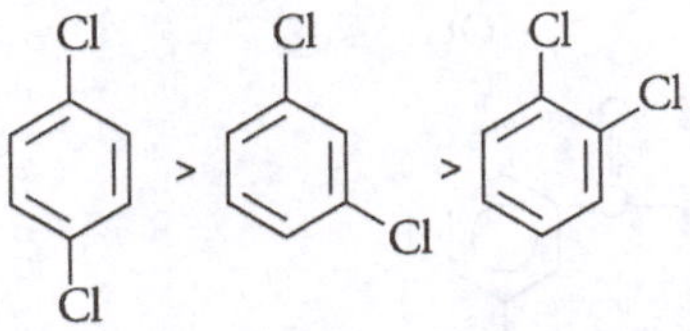


(b) 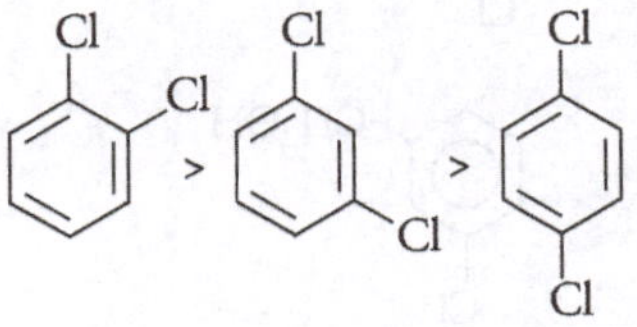


(c) 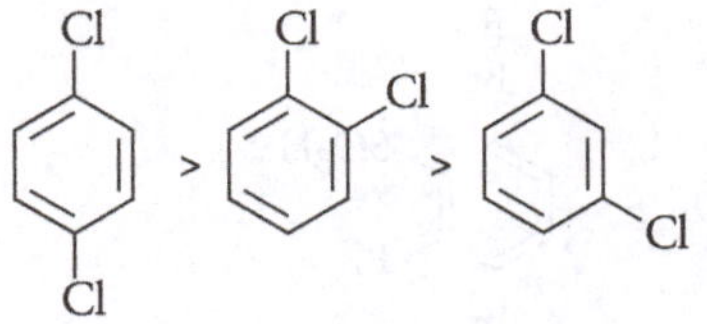


(d) 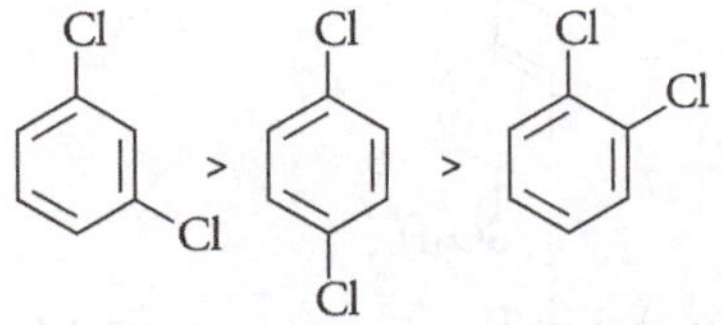


**19** 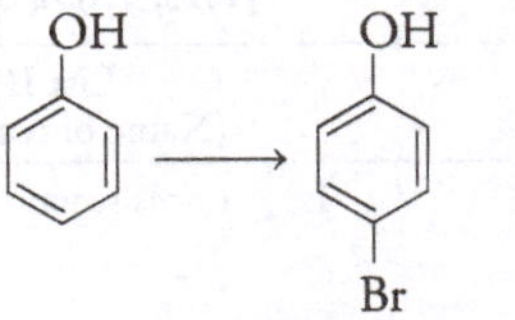


(Major product)

The given reaction can occur in the presence of

(A) bromine water (B) $Br_2$ in $CS_2$, 273 K
(C) $Br_2$ / $FeBr_3$ (D) $Br_2$ in $CHCl_3$, 273 K

Choose the correct answer from the options given below *[2021, 22 July Shift-I]*

(a) (B) and (D) only (b) (A) and (C) only
(c) (B), (C) and (D) only (d) (A), (B) and (D) only

**20** The compounds *A* and *B* in the following reaction are, respectively

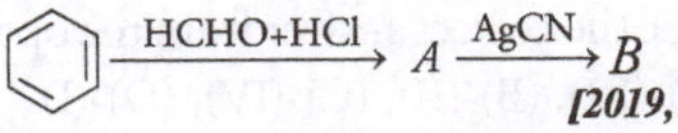


*[2019, 9 Jan Shift-I]*

(a) *A* = Benzyl alcohol, *B* = Benzyl isocyanide
(b) *A* = Benzyl alcohol, *B* = Benzyl cyanide
(c) *A* = Benzyl chloride, *B* = Benzyl isocyanide
(d) *A* = Benzyl chloride, *B* = Benzyl cyanide

## Level 3

**21** In the following series of reactions identify the major products *A* and *B* respectively *[2025, 3 April Shift-II]*

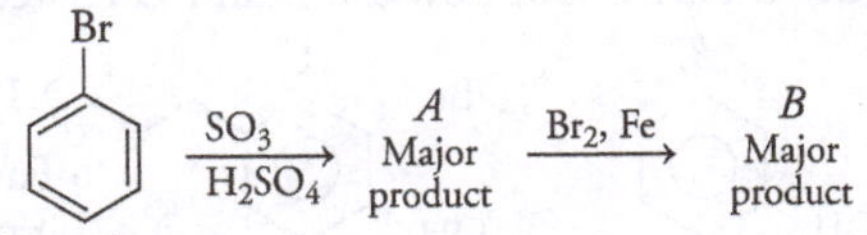


Bromobenzene

A B

(a) 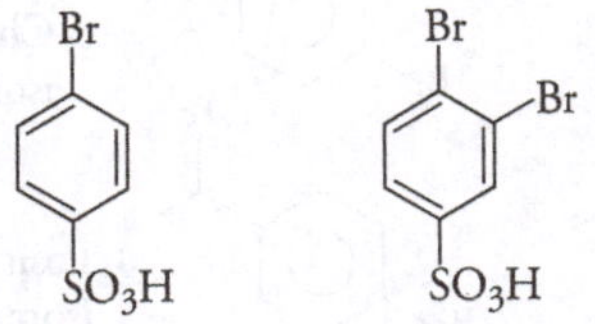


(b) 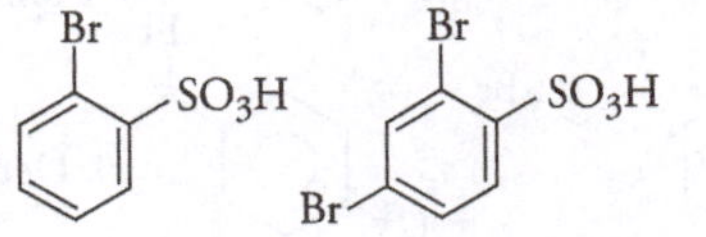

(c) 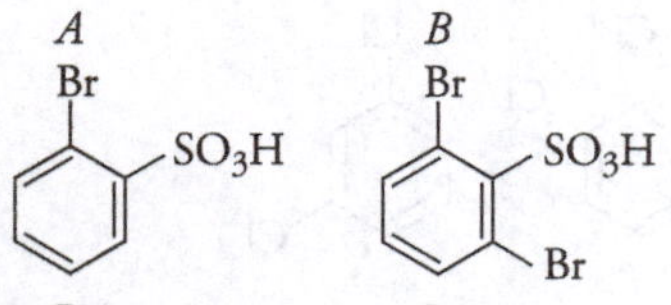

(d) 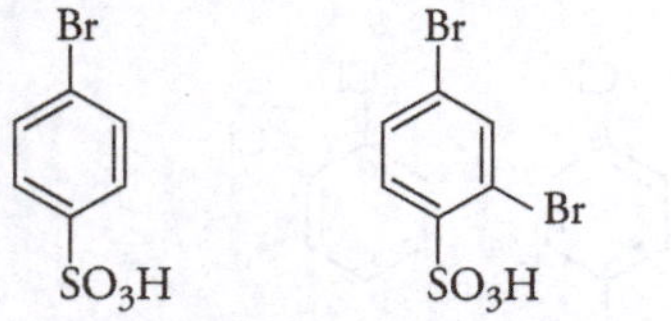

**22** Match List I with List II **[2025, 2 April Shift-II]**

| List I (Reaction) | | List II (Name of reaction) |
|---|---|---|
| A. | 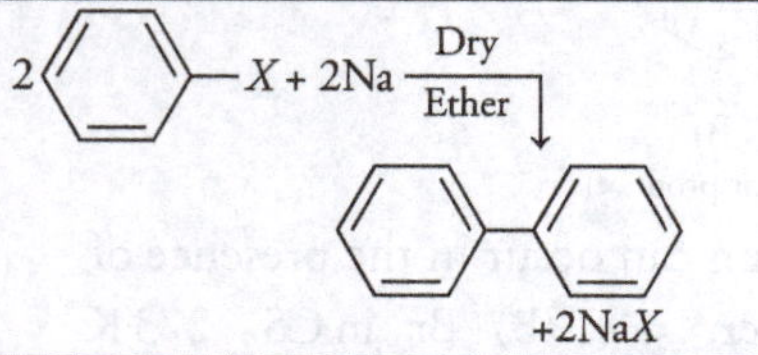 | I. Lucas reaction |
| B. | $ArN_2^+X^- \xrightarrow[HCl]{Cu} ArCl + N_2\uparrow + CuX$ | II. Finkelstein reaction |
| C. | $C_2H_5Br + NaI \xrightarrow[Acetone]{Dry} C_2H_5I + NaBr$ | III. Fittig reaction |
| D. | $CH_3C(OH)(CH_3)CH_3 \xrightarrow[ZnCl_2]{HCl} CH_3C(Cl)(CH_3)CH_3$ | IV. Gattermann reaction |

Choose the correct answer from the options given below.

(a) (A)-(III), (B)-(II), (C)-(IV), (D)-(I)
(b) (A)-(IV), (B)-(I), (C)-(II), (D)-(III)
(c) (A)-(IV), (B)-(III), (C)-(I), (D)-(II)
(d) (A)-(III), (B)-(IV), (C)-(II), (D)-(I)

**23** 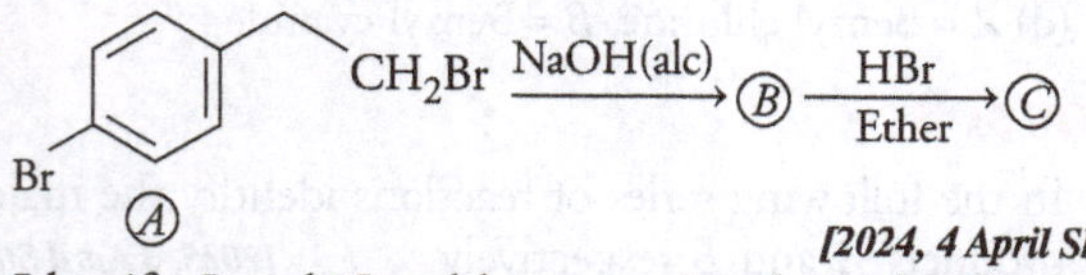

**[2024, 4 April Shift-I]**

Identify *B* and *C* and how are *A* and *C* related?

(a) 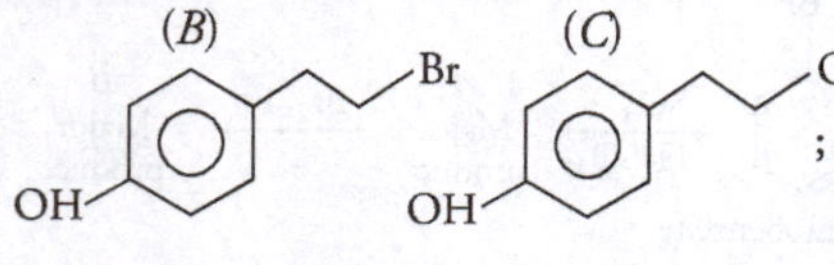 ; Functional group isomers

(b) 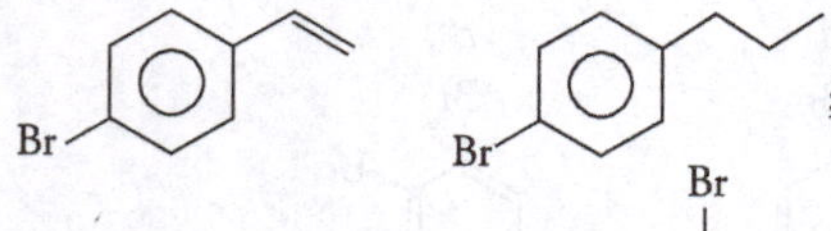 ; Chain isomers

(c) 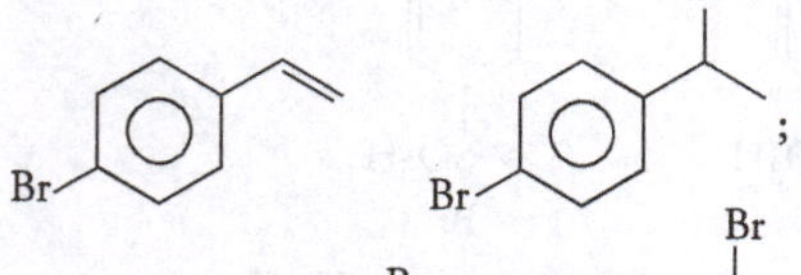 ; Position isomers

(d) 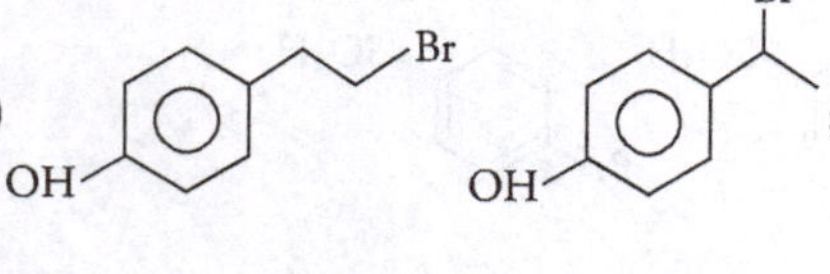 ; Derivative

**24** Identify *A* and *B* in the following reaction sequence.

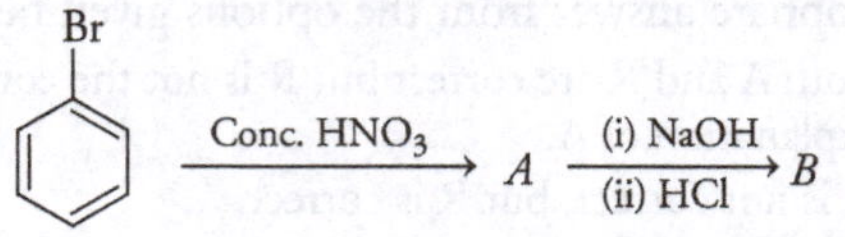

**[2024, 31 Jan Shift-II]**

(a) 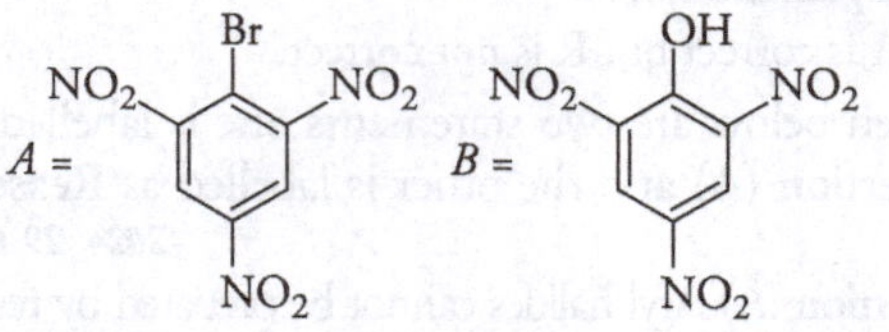

(b) 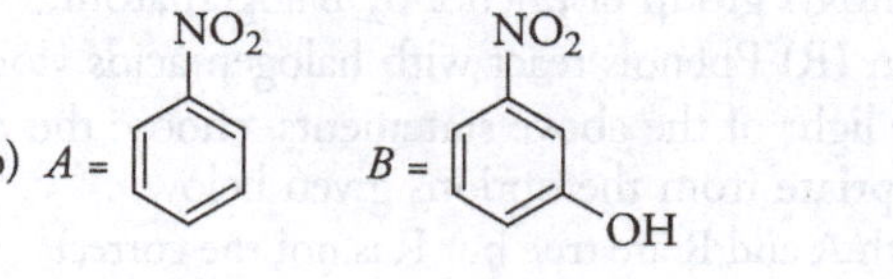

(c) 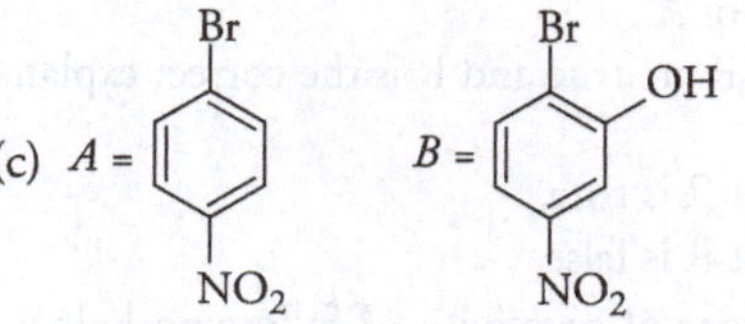

(d) 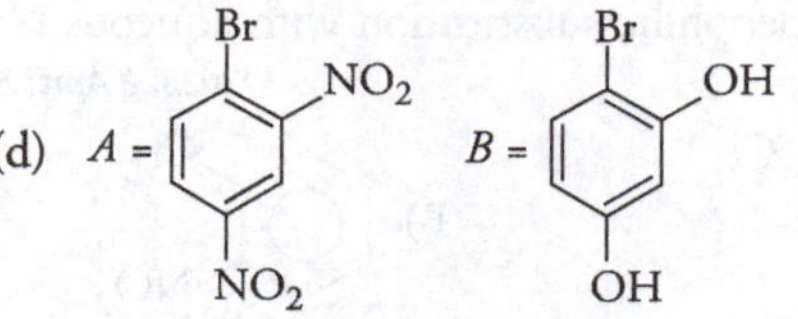

**25** *m*-chlorobenzaldehyde on treatment with 50% KOH solution yields **[2024, 30 Jan Shift-II]**

(a) 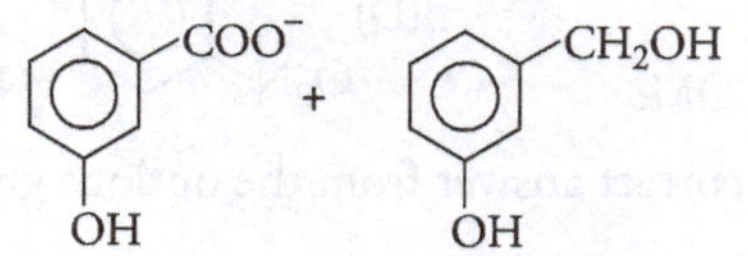

(b) 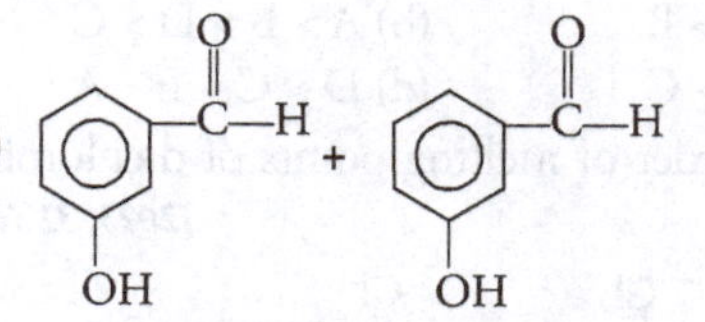

(c) 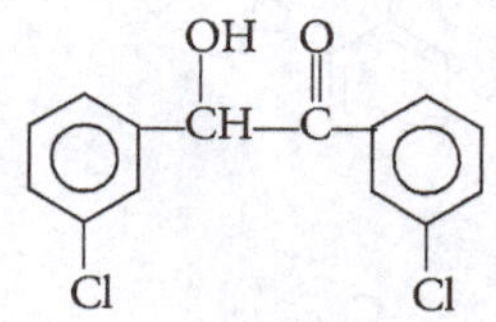

(d) 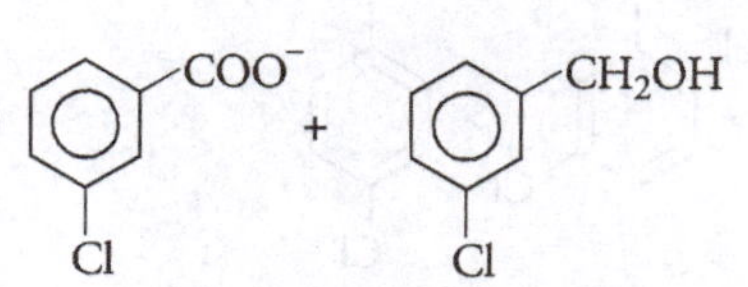

**26** Identify the correct order of reactivity for the following pairs towards the respective mechanism *[2023, 10 April Shift-I]*

(A) $S_N2$ 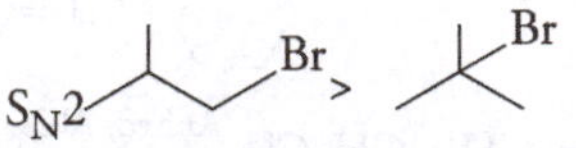

(B) $S_N1$ 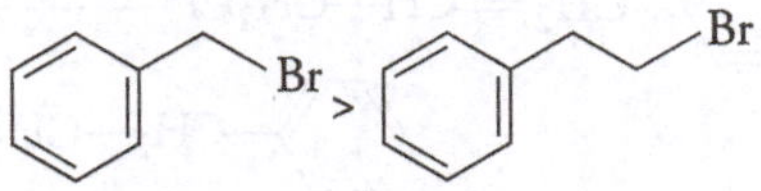

(C) Electrophilic substitution 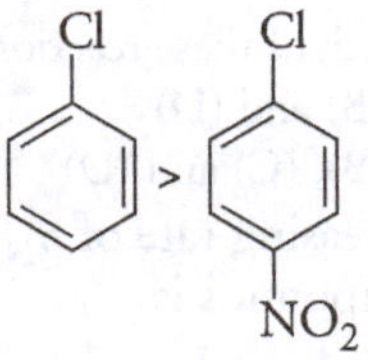

(D) Nucleophilic substitution 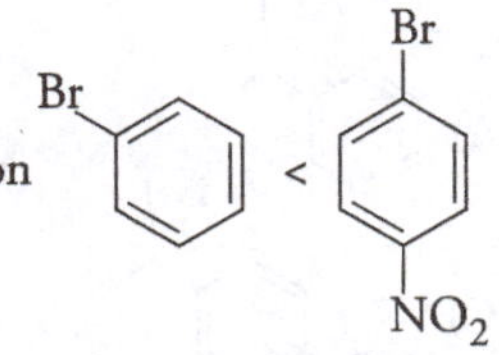

Choose the correct answer from the options given below:

(a) (A), (C) and (D) only (b) (A), (B), (C) and (D) only
(c) (A), (B) and (D) only (d) (B), (C) and (D) only

**27** Match List-I with List-II

| | List-I | | List-II |
|---|---|---|---|
| A | 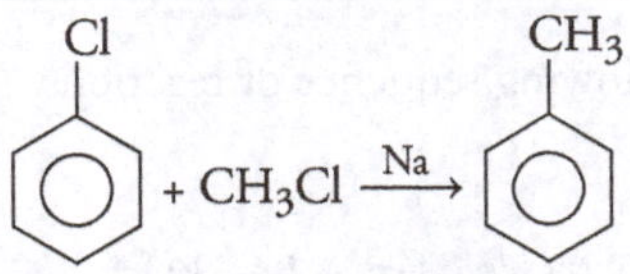 | I. | Fittig reaction |
| B | 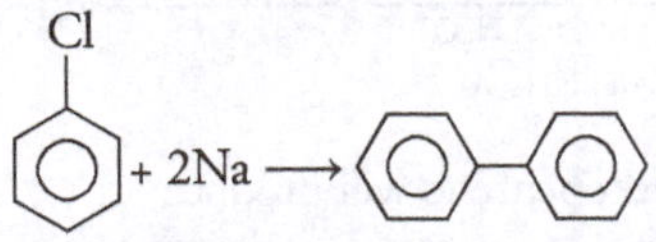 | II. | Wurtz Fittig reaction |
| C | 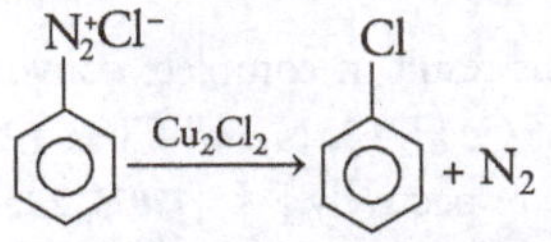 | III. | Finkelstein reaction |
| D | $C_2H_5Cl + NaI \longrightarrow C_2H_5I + NaCl$ | IV. | Sandmeyer reaction |

Choose the correct answer from the options given below. *[2023, 30 Jan Shift-I]*

(a) A-II, B-I, C-IV, D-III (b) A-III, B-II, C-IV, D-I
(c) A -II, B-I, C-III, D-IV (d) A-IV, B-II, C-III, D-I

**28** Two isomers (*A*) and (*B*) with molar mass 184 g/mol and elemental composition *C*, 52.2%; H, 4.9% and Br, 42.9% gave benzoic acid and *p*-bromobenzoic acid, respectively on oxidation with $KMnO_4$. Isomer *A* is optically active and gives a pale yellow precipitate when warmed with alcoholic $AgNO_3$. Isomers *A* and *B* are, respectively *[2022, 29 June Shift-II]*

(a) $H_3C$—CHBr—$C_6H_5$ and 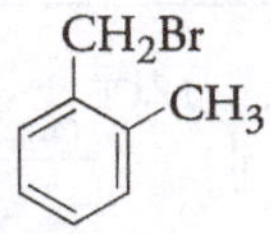

(b) 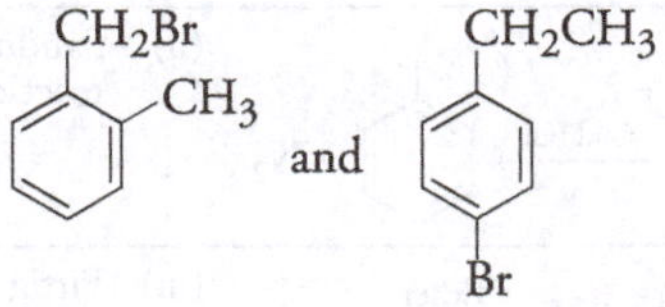

(c) $H_3C$—CHBr—$C_6H_5$ and 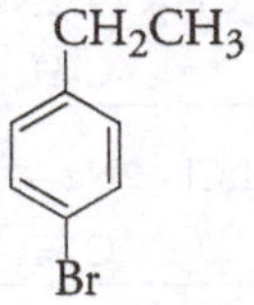

(d) 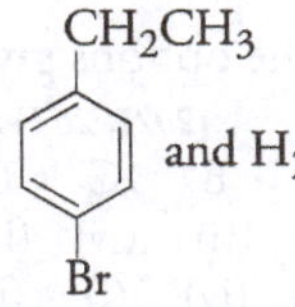 and $H_2C$—CHBr—$C_6H_5$

**29** $A \xrightarrow[\text{(ii) } CN^- \text{ (iii) } H_2O/H^+]{\text{(i) } Cl_2, \Delta}$ 4-bromophenyl acetic acid.

In the above reaction '*A*' is *[2022, 28 June Shift-II]*

(a) 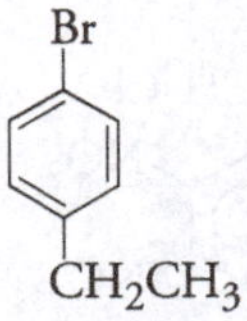 (b) 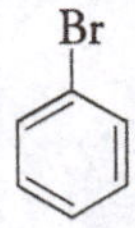

(c) 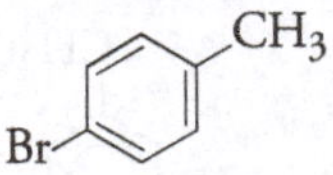 (d) 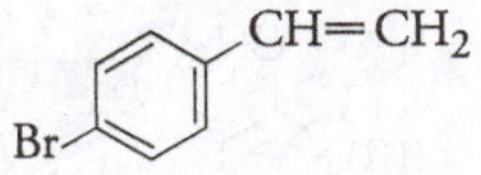

**Imp.** Reagents that forms Yellow Precipitates with $AgNO_3$

**30** Among the following compounds I-IV, which one forms a yellow precipitate on reacting sequentially with
(i) NaOH (ii) dil. $HNO_3$
(iii) $AgNO_3$? *[2021, 26 Aug Shift-I]*

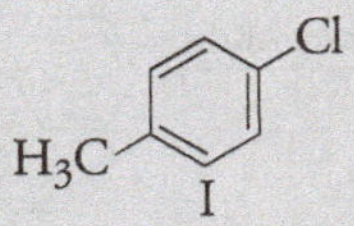
I

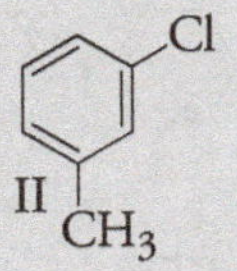
II

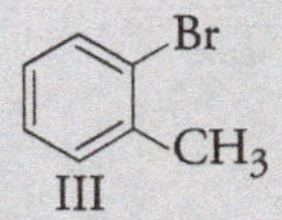
III

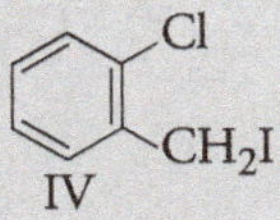
IV

(a) II (b) IV (c) I (d) III

**31** Match List-I with List-II.

| List-I | List-II |
|---|---|
| A. 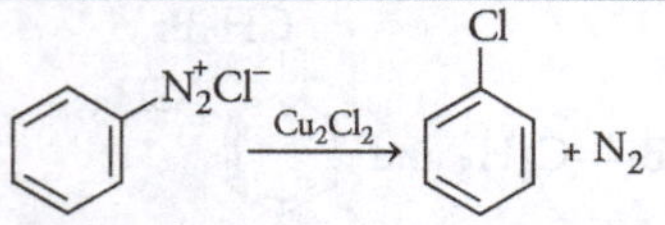 | (i) Wurtz reaction |
| B. 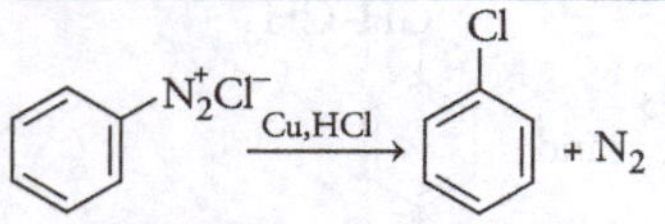 | (ii) Sandmeyer reaction |
| C. $2CH_3CH_2Cl + 2Na \xrightarrow{\text{Ether}} C_2H_5—C_2H_5 + 2NaCl$ | (iii) Fittig reaction |
| D. $2C_6H_5Cl + 2Na \xrightarrow{\text{Ether}} C_6H_5—C_6H_5 + 2NaCl$ | (iv) Gattermann reaction |

Choose the correct answer from the options given below. **[2021, 26 Feb Shift-II]**

| | A | B | C | D |
|---|---|---|---|---|
| (a) | (iii) | (i) | (iv) | (ii) |
| (b) | (ii) | (i) | (iv) | (iii) |
| (c) | (ii) | (iv) | (i) | (iii) |
| (d) | (iii) | (iv) | (i) | (ii) |

**Imp.** Reactivity of Haloarenes towards $S_N2$ Reaction

**32** The decreasing order of reactivity of the following compounds towards nucleophilic substitution ($S_N2$) reaction is

(I) 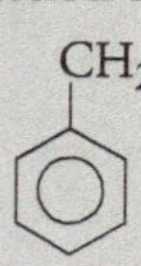 (II) 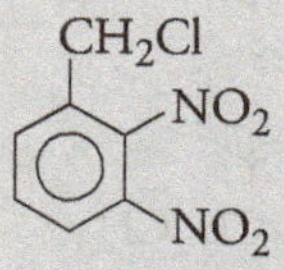

(III) 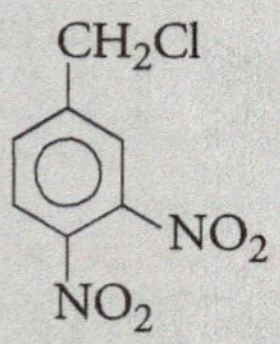 (IV) 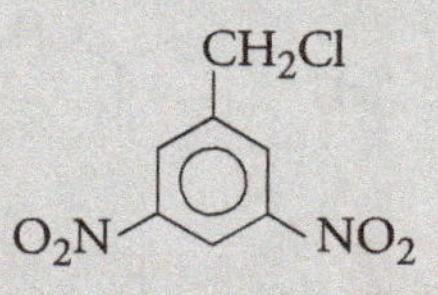

**[2020, 3 Sep Shift-II]**

(a) (II) > (III) > (IV) > (I) (b) (IV) > (II) > (III) > (I)
(c) (III) > (II) > (IV) > (I) (d) (II) > (III) > (I) > (IV)

**33** Consider the following reactions :

(A) 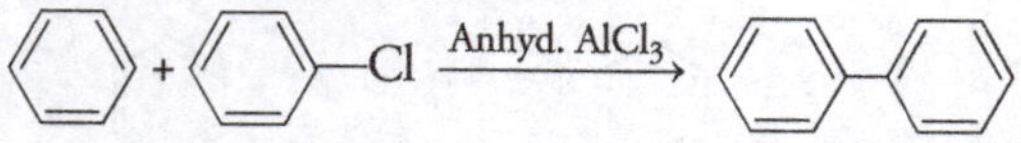

(B) 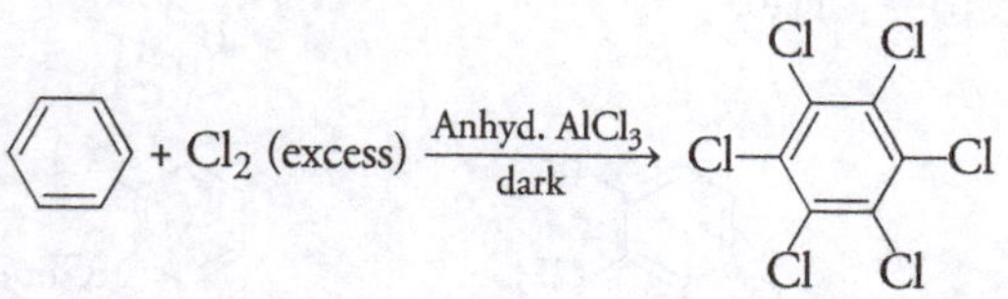

(C) 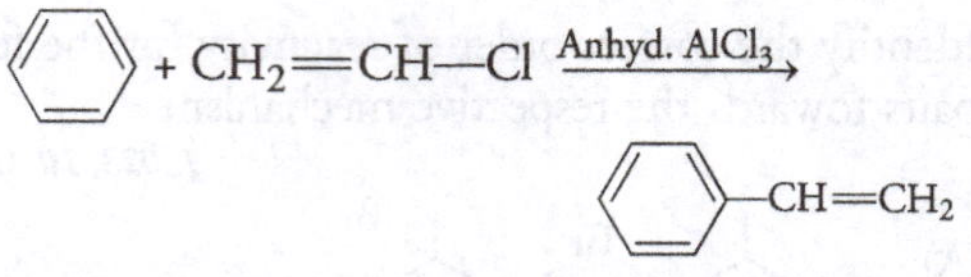

(D) 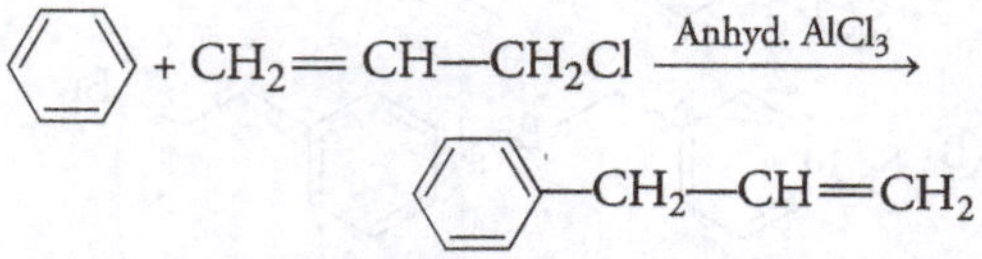

Which of these reactions are possible? **[2020, 7 Jan Shift-II]**

(a) (B) and (D) (b) (A) and (D)
(c) (B), (C) and (D) (d) (A) and (B)

**34** Increasing rate of $S_N1$ reaction in the following compounds is **[2019, 10 April Shift-I]**

(A) 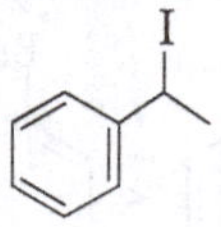 (B) 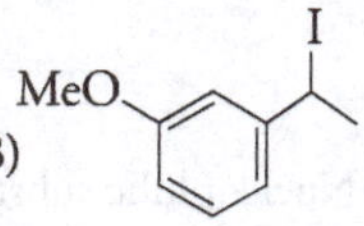

(C) 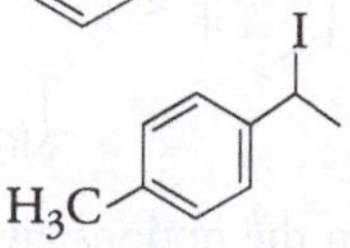 (D) 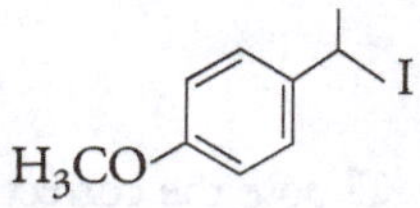

(a) (A) < (B) < (C) < (D) (b) (B) < (A) < (C) < (D)
(c) (A) < (B) < (D) < (C) (d) (B) < (A) < (D) < (C)

## Numerical Value Type Questions

**Imp.** Reaction of Chlorobenzene in Blend with Stoichiometry

**35** Consider the following sequence of reactions

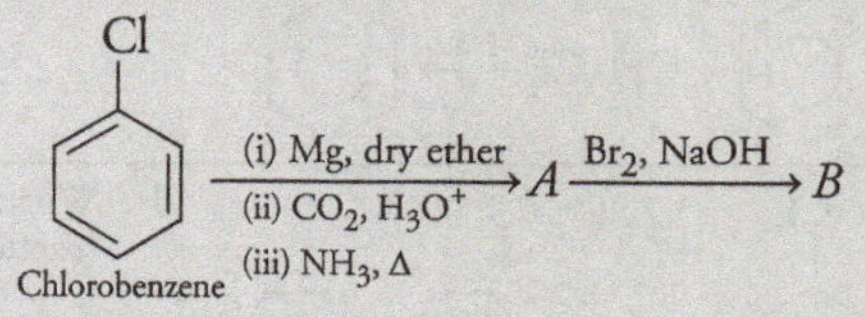

11.25 mg of chlorobenzene will produce ......... $\times 10^{-1}$ mg of product *B*.
(Consider the reactions result in complete conversion.)
[Given, molar mass of C, H, O, N and Cl as 12, 1, 16, 14 and 35.5 g mol$^{-1}$ respectively.] **[2025, 28 Jan Shift-I]**

**36** The number of halobenzenes from the following that can be prepared by Sandmeyer's reaction is ..........

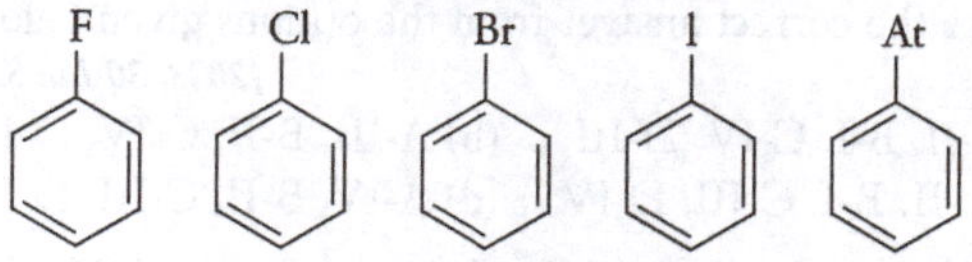

I II III IV V

**[2024, 5 April Shift-I]**

**37** In the presence of sunlight, benzene reacts with $Cl_2$ to give product $X$. The number of hydrogens in $X$ is ............ . *[2022, 26 July Shift-I]*

**38** The major product of the following reaction contains ............ bromine atom (s). *[2022, 28 June Shift-II]*

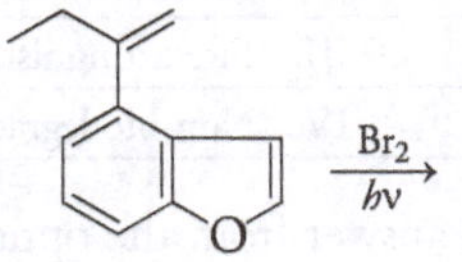
Major product

## Topic 3 Some Important Polyhalogen Compounds

### Level 1

**01** In the following halogenated organic compounds the one with maximum number of chlorine atoms in its structure is *[2023, 31 Jan Shift-II]*

(a) gammaxene
(b) chloropicrin
(c) freon-12
(d) chloral

**02** The major organic compound formed by the reaction of 1,1,1-trichloroethane with silver powder is: *[JEE Mains 2014]*

(a) acetylene (b) ethene
(c) 2-butyne (d) 2-butene

**03** Chlorobenzene reacts with trichloro acetaldehyde in the presence of $H_2SO_4$.

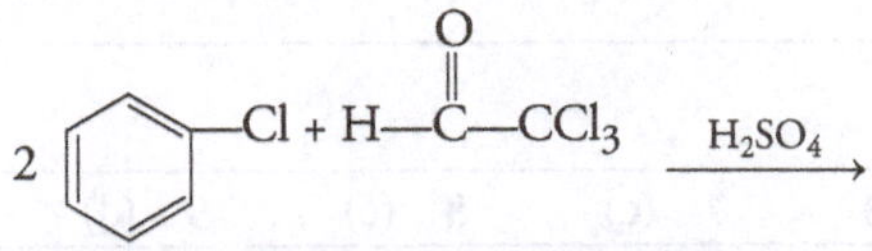

The major product formed is: *[JEE Main 2014 ; AIEEE 2004]*

(a) 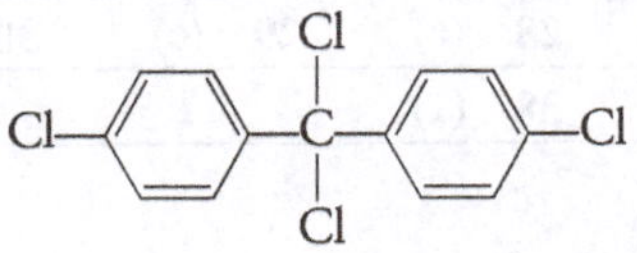

(b) 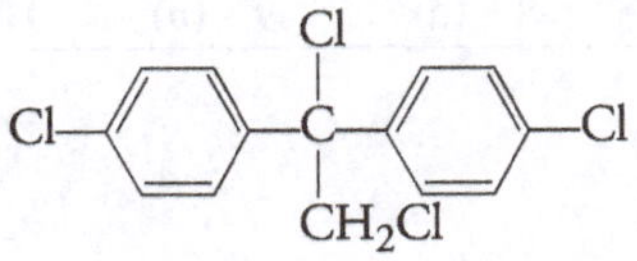

(c) 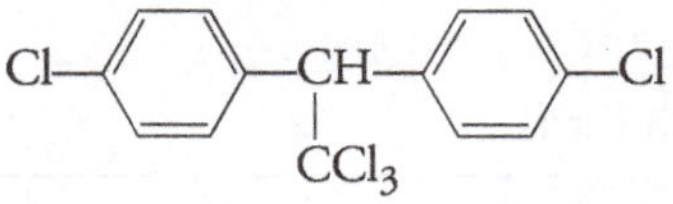

(d) 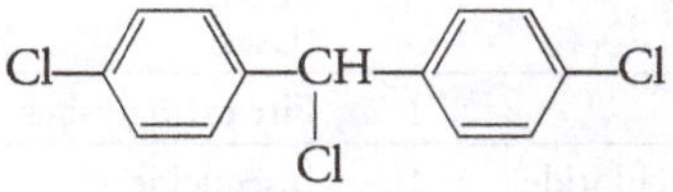

**Imp.** Reaction of Polyhalogen Compound with Aqueous KOH

**04** The major product formed when 1,1,1-trichloropropane is treated with aqueous potassium hydroxide, is *[JEE Main 2014]*

(a) propyne (b) 1-propanol
(c) 2-propanol (d) propionic acid

**05** What is DDT among the following? *[AIEEE 2012]*

(a) A fertilizer
(b) Biodegradable pollutant
(c) Non-biodegradable pollutant
(d) Greenhouse gas

### Level 2

**06** In the following sequence of reaction, the major products $B$ and $C$ respectively are

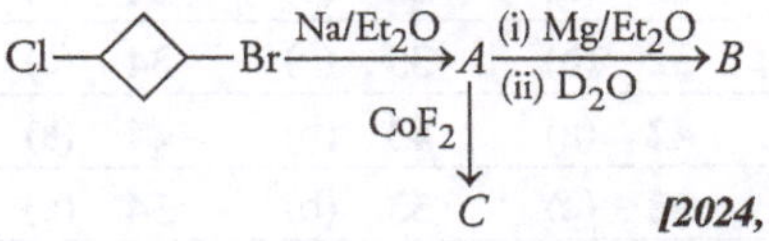

*[2024, 9 April Shift-I]*

(a) 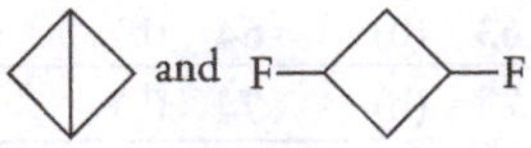

(b) 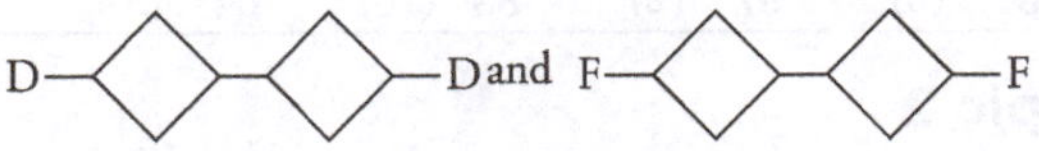

(c) 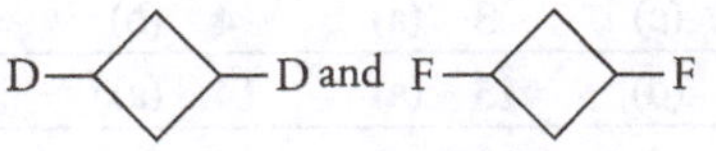

(d) 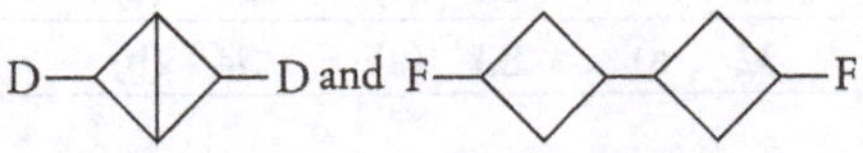

**07** Iodoform can be prepared from all except *[AIEEE 2012]*

(a) isopropyl alcohol (b) 3-methyl-2-butanone
(c) isobutyl alcohol (d) ethyl methyl ketone

**08** Which of the following on heating with aqueous KOH produces acetaldehyde? *[AIEEE 2009]*

(a) $CH_3COCl$ (b) $CH_3CH_2Cl$
(c) $CH_2ClCH_2Cl$ (d) $CH_3CHCl_2$

## Level 3

**09** Match list I with list II.

| | List-I (Compound) | | List-II (Uses) |
|---|---|---|---|
| A. | Iodoform | I. | Fire extinguisher |
| B. | Carbon tetrachloride | II. | Insecticide |
| C. | CFC | III. | Antiseptic |
| D. | DDT | IV. | Refrigerants |

Choose the correct answer from the options given below. *[2024, 6 April Shift-I]*

(a) A-II, B-IV, C-I, D-III (b) A-III, B-I, C-IV, D-II
(c) A-III, B-II, C-IV, D-I (d) A-I, B-II, C-III, D-IV

**10** Match List-I with List-II

| | List-I (Compound) | | List-II (Use) |
|---|---|---|---|
| A. | Carbon tetrachloride | I. | Paint remover |
| B. | Methylene chloride | II. | Refrigerators and air conditioners |
| C. | DDT | III. | Fire extinguisher |
| D. | Freons | IV. | Non-biodegradable insecticide |

Choose the correct answer from the options given below *[2024, 1 Feb Shift-II]*

(a) A-IV, B-III, C-II, D-I (b) A-III, B-I, C-IV, D-II
(c) A-II, B-III, C-I, D-IV (d) A-I, B-II, C-III, D-IV

## Key Index

### Topic 1

| | | | | | | | | | | | | | | | | | | | |
|---|---|---|---|---|---|---|---|---|---|---|---|---|---|---|---|---|---|---|---|
| 1 | (a) | 2 | (d) | 3 | (b) | 4 | (d) | 5 | (c) | 6 | (a) | 7 | (c) | 8 | (d) | 9 | (a) | 10 | (a) |
| 11 | (c) | 12 | (c) | 13 | (d) | 14 | (d) | 15 | (b) | 16 | (a) | 17 | (a) | 18 | (d) | 19 | (c) | 20 | (b) |
| 21 | (d) | 22 | (b) | 23 | (b) | 24 | (c) | 25 | (c) | 26 | (d) | 27 | (c) | 28 | (b) | 29 | (b) | 30 | (d) |
| 31 | (c) | 32 | (b) | 33 | (c) | 34 | (c) | 35 | (c) | 36 | (a) | 37 | (d) | 38 | (a) | 39 | (c) | 40 | (b) |
| 41 | (a) | 42 | (a) | 43 | (b) | 44 | (a) | 45 | (d) | 46 | (b) | 47 | (c) | 48 | (c) | 49 | (c) | 50 | (c) |
| 51 | (c) | 52 | (c) | 53 | (b) | 54 | (a) | 55 | (d) | 56 | (c) | 57 | (b) | 58 | (d) | 59 | (a) | 60 | (d) |
| 61 | (c) | 62 | (c) | 63 | (b) | 64 | (b) | 65 | (a) | 66 | (c) | 67 | (b) | 68 | (d) | 69 | (d) | 70 | (b) |
| 71 | (c) | 72 | (d) | 73 | (b) | 74 | (b) | 75 | (a) | 76 | (c) | 77 | (c) | 78 | (c) | 79 | (1) | 80 | (32) |
| 81 | (6) | 82 | (2) | 83 | (10) | 84 | (5) | | | | | | | | | | | | |

### Topic 2

| | | | | | | | | | | | | | | | | | | | |
|---|---|---|---|---|---|---|---|---|---|---|---|---|---|---|---|---|---|---|---|
| 1 | (b) | 2 | (c) | 3 | (a) | 4 | (b) | 5 | (a) | 6 | (b) | 7 | (c) | 8 | (c) | 9 | (d) | 10 | (c) |
| 11 | (b) | 12 | (d) | 13 | (a) | 14 | (a) | 15 | (c) | 16 | (d) | 17 | (c) | 18 | (c) | 19 | (c) | 20 | (c) |
| 21 | (a) | 22 | (d) | 23 | (c) | 24 | (a) | 25 | (d) | 26 | (b) | 27 | (a) | 28 | (c) | 29 | (c) | 30 | (b) |
| 31 | (c) | 32 | (a) | 33 | (a) | 34 | (b) | 35 | (93) | 36 | (2) | 37 | (6) | 38 | (1) | | | | |

### Topic 3

| | | | | | | | | | | | | | | | | | | | |
|---|---|---|---|---|---|---|---|---|---|---|---|---|---|---|---|---|---|---|---|
| 1 | (a) | 2 | (c) | 3 | (c) | 4 | (d) | 5 | (c) | 6 | (b) | 7 | (c) | 8 | (d) | 9 | (b) | 10 | (b) |

# Detailed **Explanations**

## Topic 1

**01.** *(a)* The reaction involved is as follows.

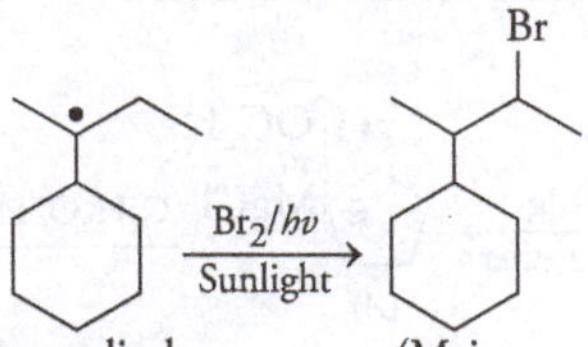

Free radical intermediate (Major product)

**02.** *(d)* For the structure.

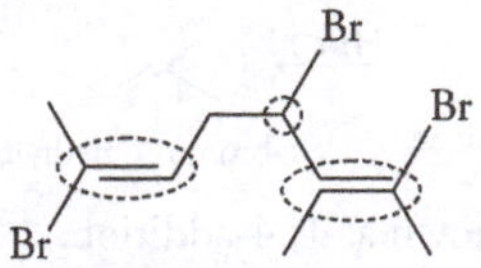

Total 3 stereocentres are there

So, total stereoisomers possible are $2^3 = 8$.

**03.** *(b)* The complete reaction takes place as follows

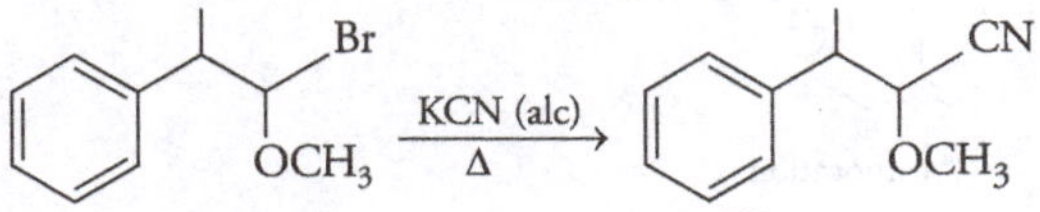

(*P*)

Nucleophilic substitution reaction will occur.

**04.** *(d)* The complete reaction is as follows

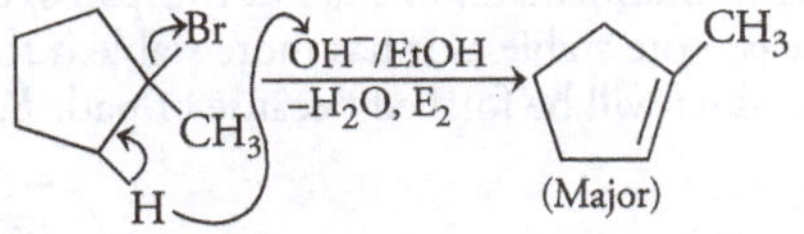

Hence, option (d) is correct.

**05.** *(c)* The complete reaction proceeds as follows

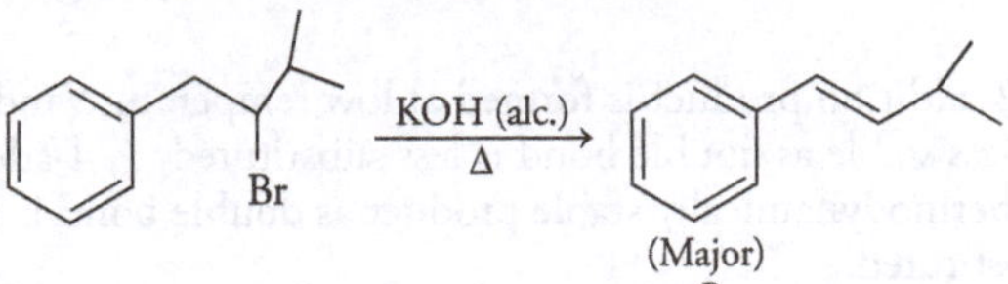

Here, more stable double bond will be formed, as double bond is in the resonance with the ring.

**06.** *(a)* The structure of 2, 3-dibromo-1-phenylpentane is

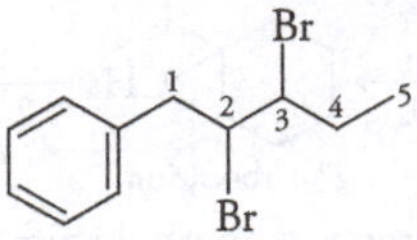

**07.** *(c)* Vinyl carbon is $sp^2$ hybridised aliphatic carbon.

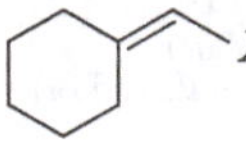 is vinyl halide.

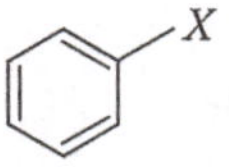 is aryl halide.

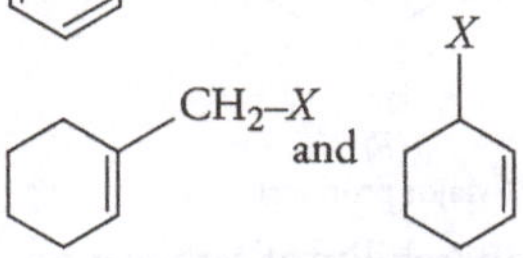 are allyl halide.

> **• Concept Enhancer**
>
> Vinylic halides are the compounds in which the halogen atom is bonded to a $sp^2$ hybridised carbon atom of C ═ C. In such compounds, if lone pair is present, then they are always in conjugation with π-bond.

**08.** *(d)* Alkyl halide is converted into alkyl isocyanide by reaction with AgCN.

$$\underset{\text{Alkyl halide}}{R—X} + AgCN \xrightarrow[\text{reaction}]{\text{Substitution}} \underset{\text{Alkyl isocyanide}}{R—NC} + AgX$$

**09.** *(a)* Substrate which will form most stable carbocation will be highly reactive towards $S_N1$ reaction, their stability orders.

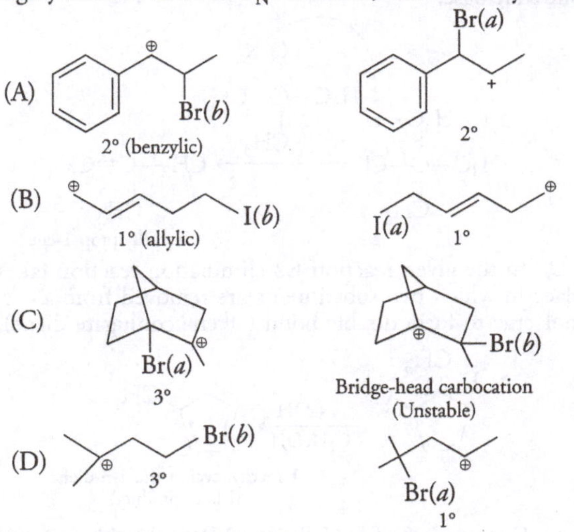

So, the carbocation resulted from in option (a)

**10.** *(a)* Both A and R are true and R is the correct explanation of A.

> **• Concept Enhancer**
>
> When alkyl chloride undergoes hydrolysis, the reaction is slow because $Cl^-$ is a poor leaving group, hence the rate of $S_N2$ reaction is slow. But in presence of NaI, $I^\ominus$ is a good leaving group that facilitates the hydrolysis. Hence, rate of hydrolysis increases.

**11.** *(c)* The given reaction proceeds in two steps:

**Step 1.** Addition of $H^+$ or attack of alkene on $H^+$ (electrophilic addition)

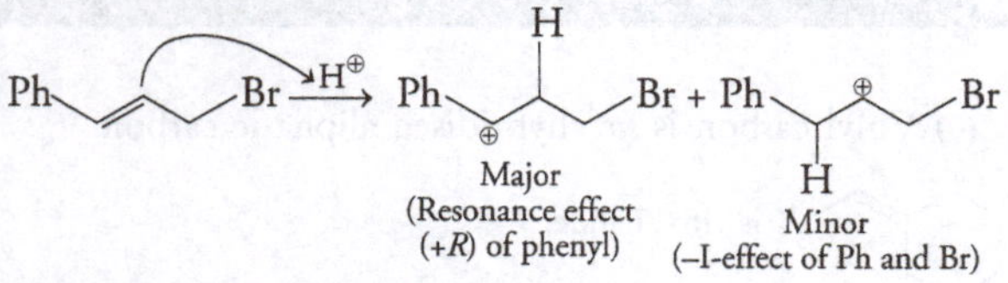


**Step 2.** Addition of $Br^-$ to carbocation.

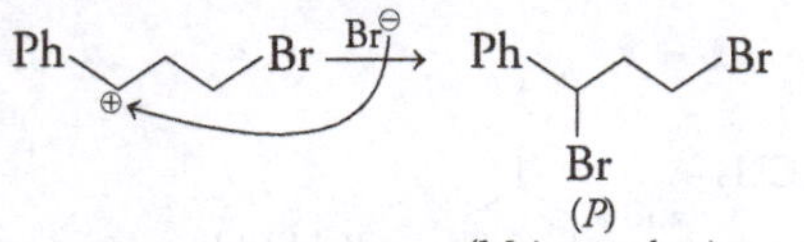


**12.** *(c)* Rate of $S_N1$ reaction depends on stability of carbocation.

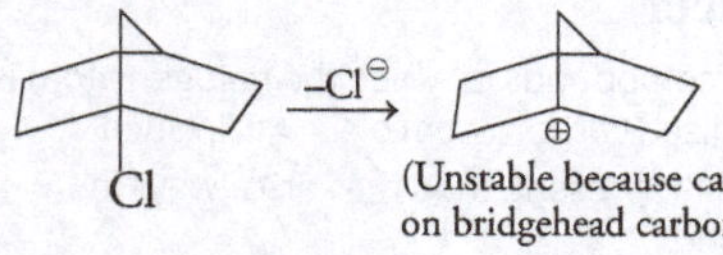


Hence, compound of option (c) is inactive towards $S_N1$ reaction. However, carbocation resultant from other options are stabilised by + *R*-effect and hyperconjugation.

**13.** *(d)* In 3° alkyl halides E2 elimination will dominate with most nucleophiles (even if they are weak bases). No $S_N2$ substitution take place due to sterically hindered butoxide base.

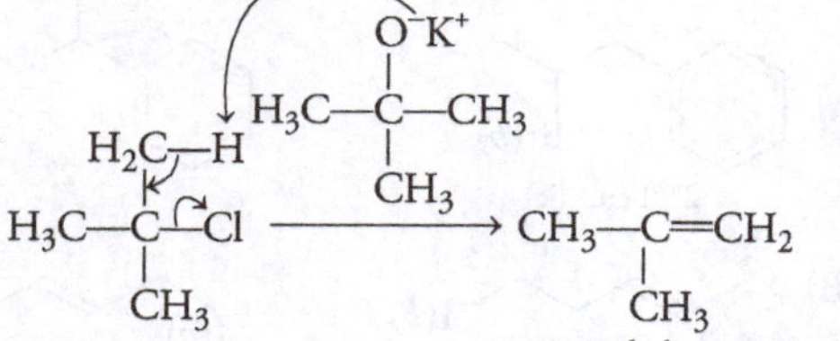


**14.** *(d)* In the given reaction E2 elimination reaction takes place in which two substituents are removed from a molecule to form double bond (alkene conjugate diene).

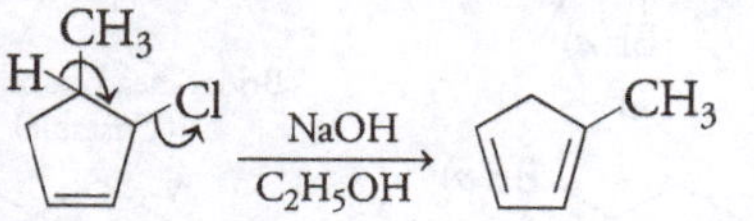


**15.** *(b)* On reaction of haloalkane with methoxide ion, alkene is formed.

**Mechanism**

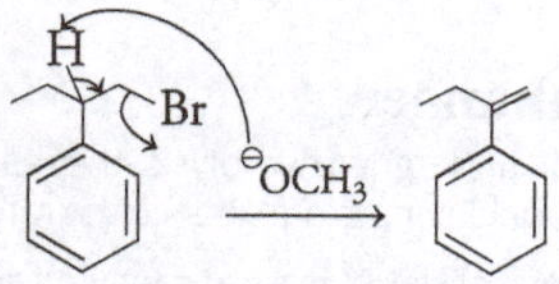


- $\overset{\ominus}{O}CH_3$ will act as a base due to its small size and high electron density and therefore, abstracts proton to form double bond which is in conjugation with aromatic ring.

**16.** *(a)* In the first step, Grignard reagent is obtained. In the second step, acid-base reaction occurs. Due to presence of an acidic hydrogen in alcohol, neutralisation reaction takes place that produces alkane and water.

The reaction is

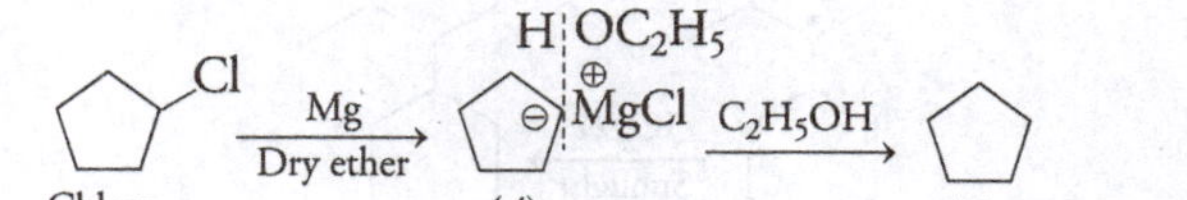


**17.** *(a)*

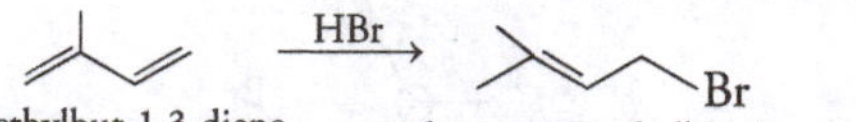


This addition is known as 1, 4-addition.

**Mechanism** This reaction can undergoes two pathways

Path *A*

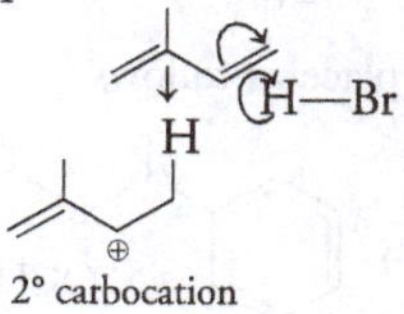


Path *B*

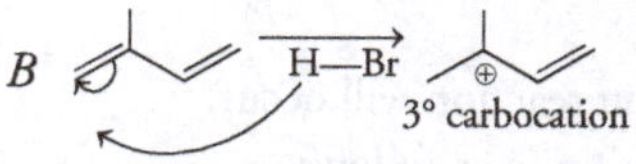


Out of the two intermediates formed in two paths, the path *B* intermediate is more stable as it has more stable carbocation. So, major product will be formed because of path *B*.

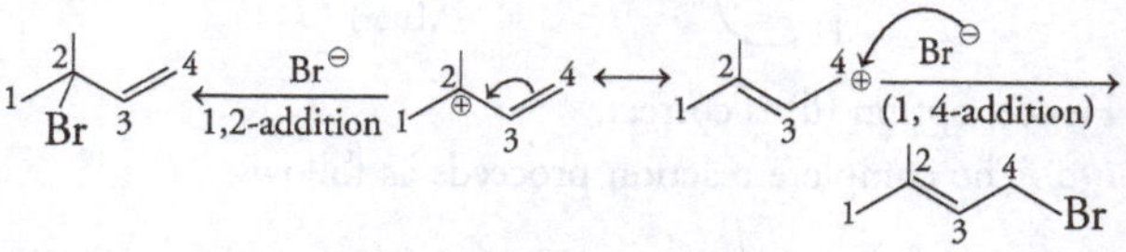


1, 2-addition product is formed at low temperature and will be less stable as double bond is less substituted. 1, 4-addition is thermodynamically stable product as double bond is more substituted.

**18.** *(d)* Alkene undergoes electrophilic addition with HBr to give a major product.

**Step 1.** Addition of electrophile to form carbocation.

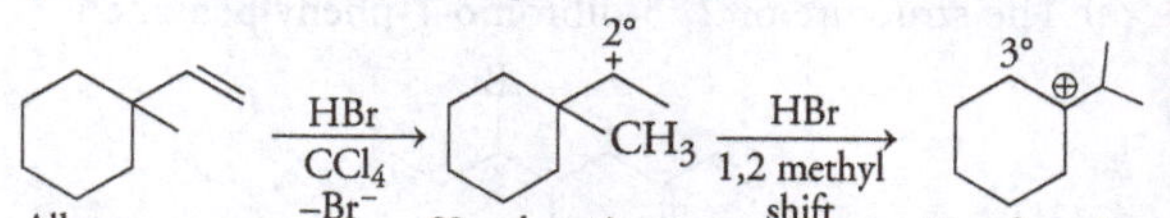


2° carbocation undergoes α-methyl shift to form more stable 3° carbocation.

**Step 2.** Attack of nucleophile on carbocation.

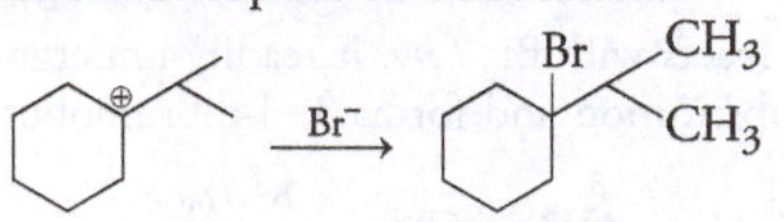

Product (*A*)

**19.** *(c)* The reaction can be achieved by $Cl_2$/UV light (*A*). Given, reaction follows free radical substitution mechanism. (allylic)

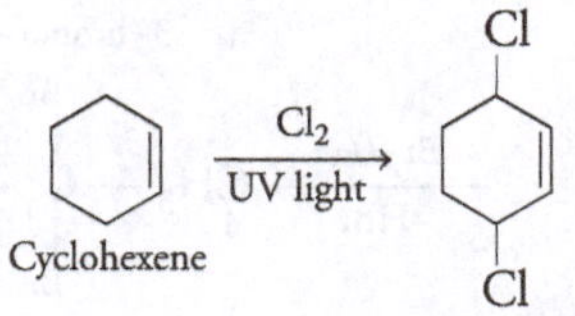


Cyclohexene

1, 4-dichloro cyclohexene

**Mechanism**

(I) Initiation This involves formation of radicals.

$$Cl_2 \xrightarrow[\text{light}]{\text{UV}} 2Cl^{\bullet}$$

(II) Propagation : Reactive free radicals react with stable molecules to form new free radicals.

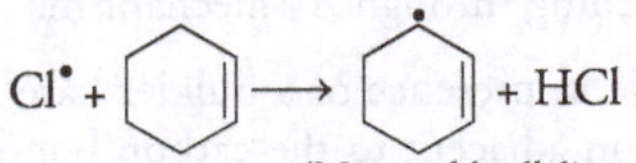


(More stable allylic free radical is formed)

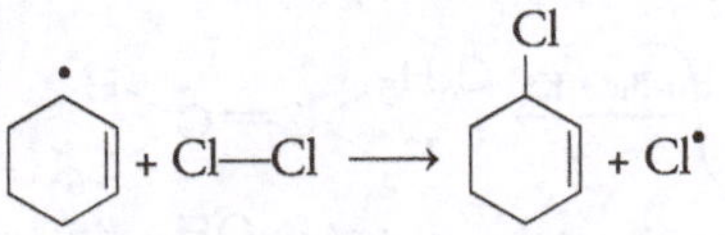


Similarly second Cl atom will also be attached at another allylic position.

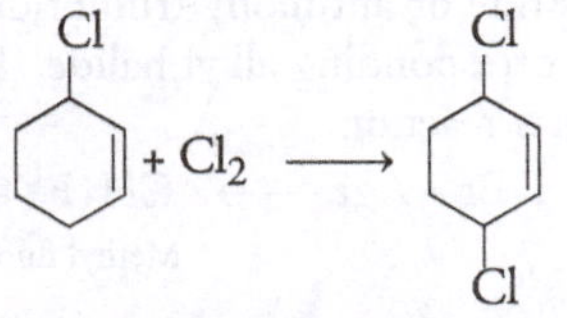


**20.** *(b)* The major product formed is 2-iodo-2, 3-dimethylbutane.

Mechanism Involved in the reaction is as follows

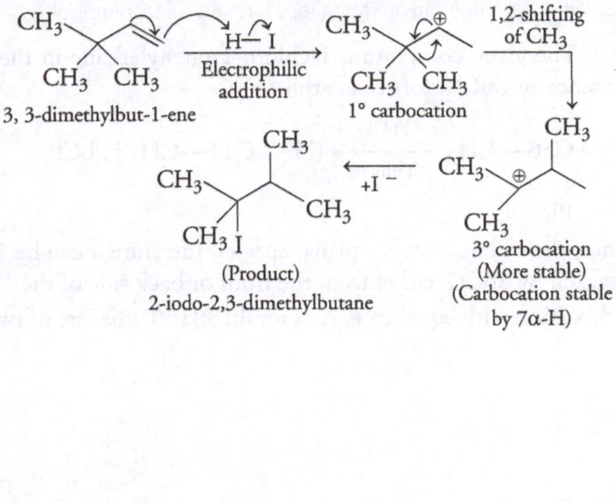


**21.** *(d)*

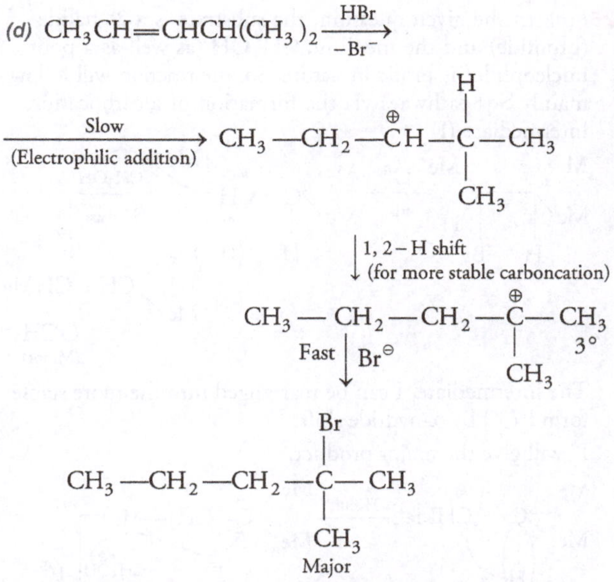


**22.** *(b)* In option (b) vinyl chloride, C—Cl bond is shortest due to resonance of lone pair of — Cl (chlorine atom) (+*R*). Due to resonance C—Cl bond aquire partial double bond character.

Hence, vinyl chloride have shortest C—Cl bond length.

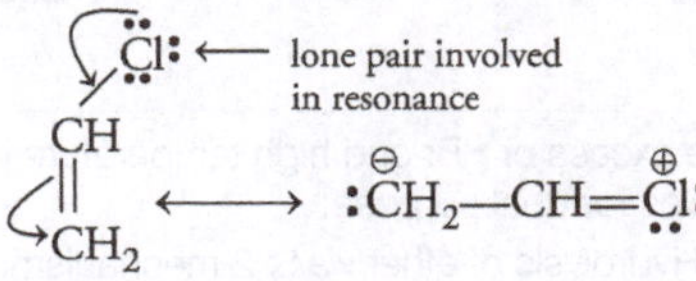


**23.** *(b)* Complete reaction is as follows

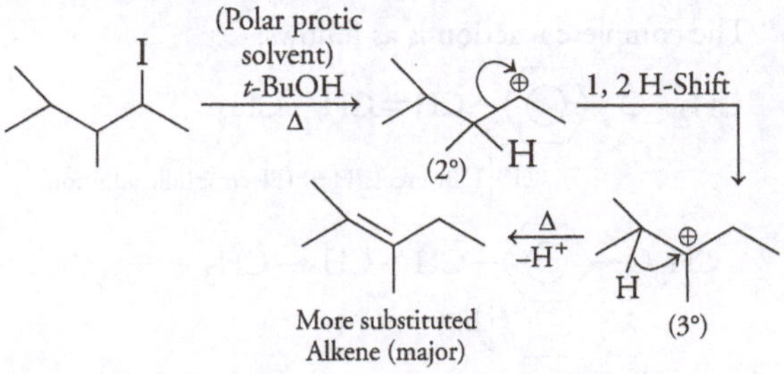


The alkene formed is stable according to Saytzeff's rule.

**24.** *(c)* The given reaction takes place as follows

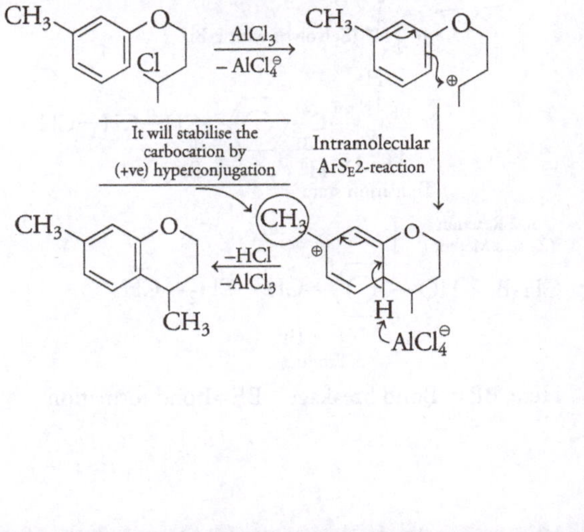

**25.** *(c)* In the given question, the substrate is a 2°-halide (bromide) and the medium, $CH_3OH$ (as well as a poor nucleophile) is protic in nature, So, the reaction will follow mainly $S_N1$ pathways via the formation of a carbocation intermediate (I).

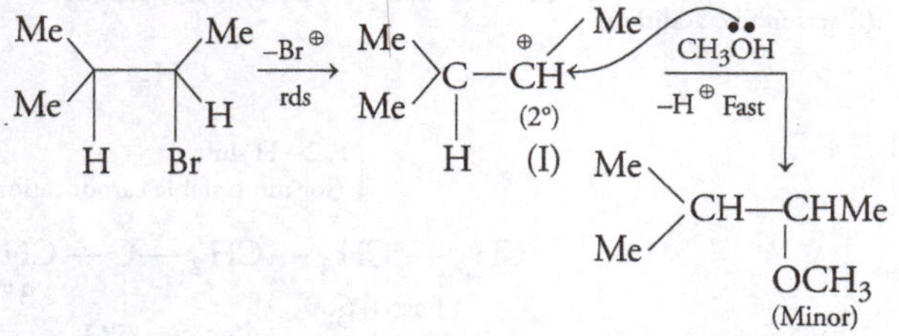

The intermediate, I can be rearranged into the more stable form I′ (3°) by α-hydride shift.

I′ will give the major product.

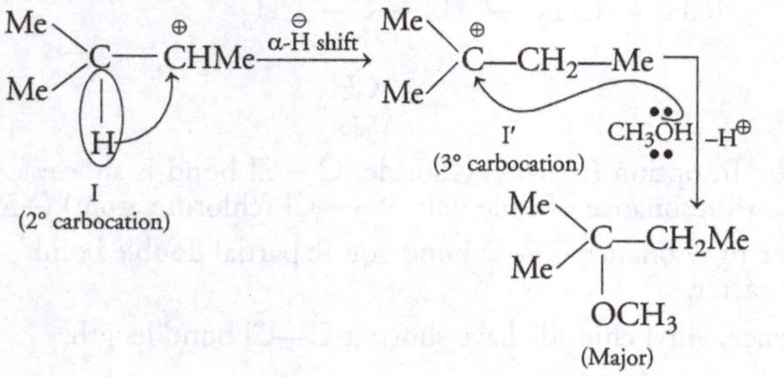

**26.** *(d)*

> **key idea** The excess of HBr and high temperature in given reaction serves for dual purpose:
> (i) Hydrolysis of ether *via* $S_N2$ mechanism.
> (ii) Markownikoff addition at double bond of the branch.

The complete reaction is as follows:

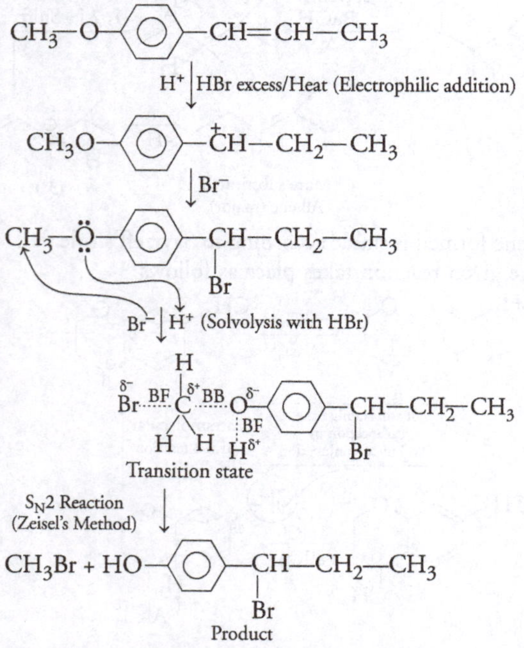

Here, BB = Bond breakage BF = Bond formation

**27.** *(c)* The compound (*E*) has two allyl-hydrogen atoms (γ). When *E* reacts with $Br_2 / h\nu$, it readily undergoes allylic free radical substitution and forms 3, 3-dibromobut-1-ene

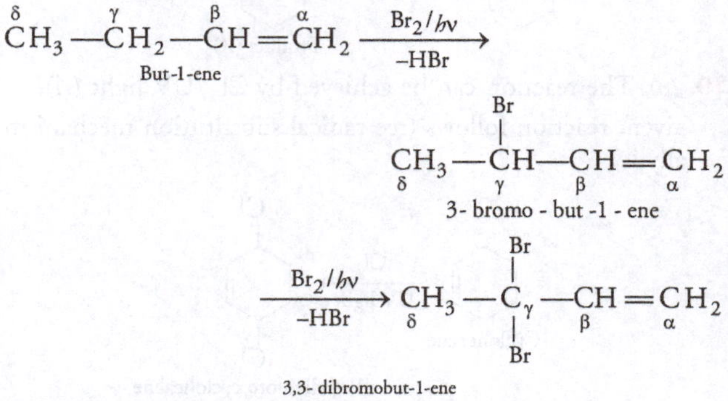

**28.** *(b)* Complete reaction can be represented as

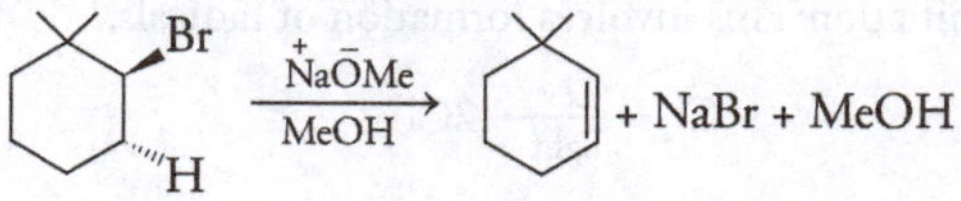

Thus, the given reaction is dehydrohalogenation which is a β-elimination proceeding through $E_2$ mechanism.

**29.** *(b)* An alkyl halide in presence of a bulkier base removes a proton from a carbon adjacent to the carbon bonded to the halogen. This reaction is called E2 (β-elimination reaction).

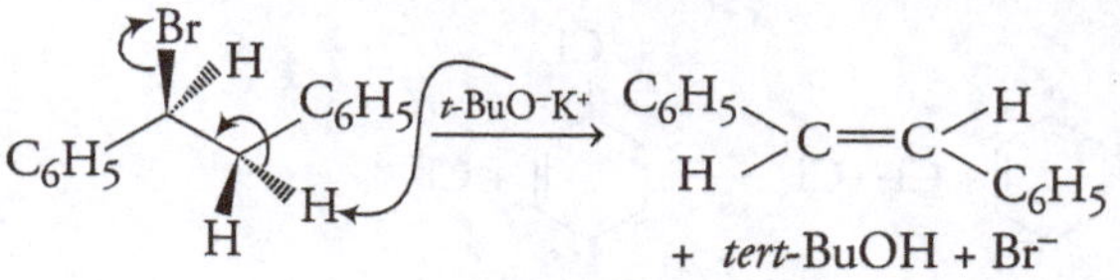

**30.** *(d)* Alkyl fluorides can be prepared by the action of mercurous fluoride or antimony trifluorides (inorganic fluorides) on corresponding alkyl halide. This reaction is known as Swarts reaction.

$$CH_3Br + AgF \longrightarrow \underset{\text{Methyl fluoride}}{CH_3F} + AgBr$$

> **Concept Enhancer**
> Free Radical fluorination involves the reaction of an independently generated carbon-centered radical with an atomic fluorine source and yields an organo fluorine compound. Sandmeyer's reaction is widely used in the production of aryl halides form aryl diazonium salts.

**31.** *(c)* The given compound, 1-chloro-1-phenylethane in the presence of $SbCl_5$ forms a carbocation.

$$Cl-\underset{\substack{|\\ Ph}}{CH}-CH_3 \xrightarrow[\text{Toluene}]{SbCl_5} [Ph-\overset{+}{C}H-CH_3]\,SbCl_6^-$$

Since, the carbocation is a planar species, therefore it can be attacked by $SbCl_6^-$ either from the front or back side of the carbocation with equal ease. As a result, 50 : 50 mixture of two enantiomers of 1-chloro-1-phenylethane undergoes racemisation due to the formation of a carbocation intermediate.

$$[\mathrm{Ph{-}\overset{+}{C}H{-}CH_3}]\,\mathrm{SbCl_6^-} \longrightarrow \mathrm{Ph{-}CH(Cl){-}CH_3} + \mathrm{SbCl_5}$$

[ *d* and *l* forms]

**32.** *(b)* 2-methylbutane on monochlorination gives 4 isomers (structure), among which I and III are chiral in nature.

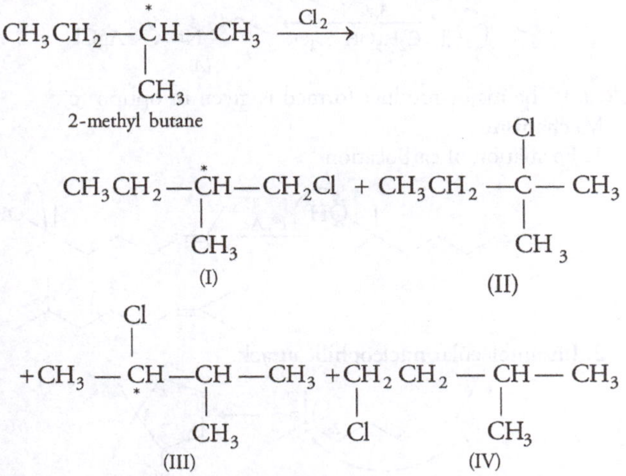


Hence, 2 chiral compounds are formed in the above reaction.

**33.** *(c)* Statement I is correct but Statement II is incorrect.

(i)

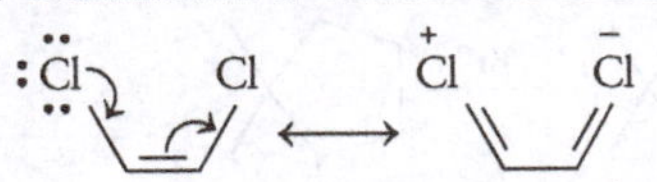


(ii)

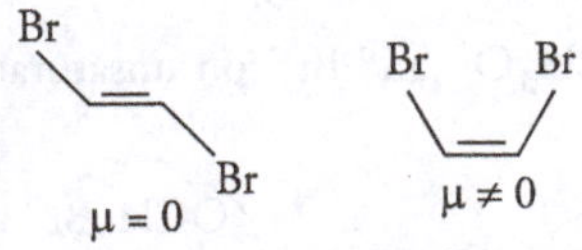


**34.** *(c)* The complete reaction is as follows

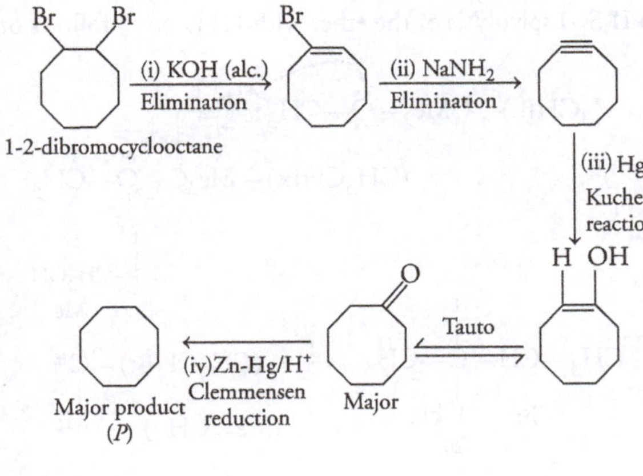


**35.** *(c)* According to question, the product '*x*' formed is

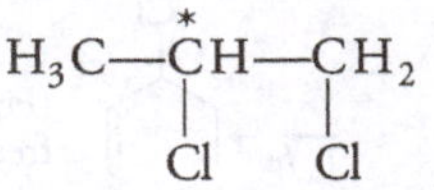

So, reaction of '*x*' with chlorine under the photochemical condition is as follows

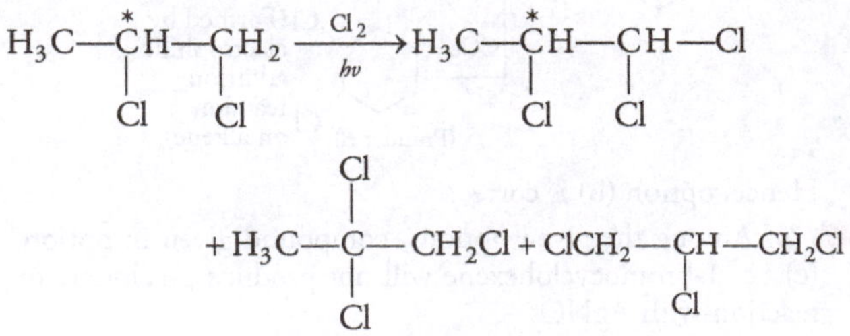

So, total '3' tri-chloro products are obtained.

**36.** *(a)* Both Statement I and Statement II are true.

**37.** *(d)* Both Statements I and II are correct.

**38.** *(a)* Both Statement I and Statement II are correct.

**39.** *(c)* Both Statement I and Statement II are correct.

**40.** *(b)* Rate of $S_N2$ reaction depends on steric hindrance, less the steric hindrance more will be rate of reaction

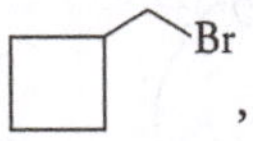

, it will undergo $S_N2$ reaction fastest.

Rate of $S_N2$ reaction is

$Me-X > 1^\circ-X > 2^\circ-X > 3^\circ-X$

**41.** *(a)* The complete reaction is as follow,

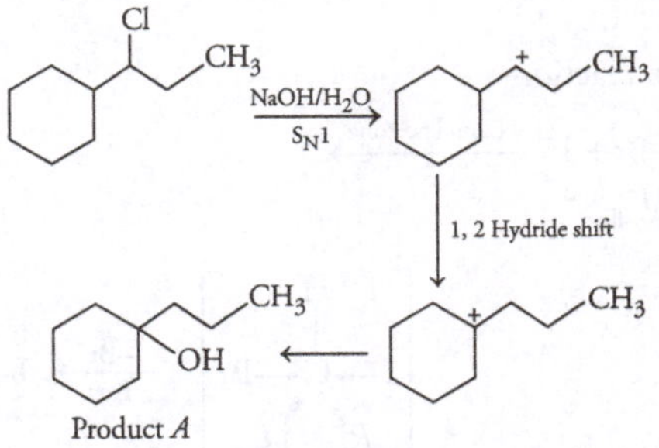


**42.** *(a)* The complete reaction takes place as follows

$$CH_3-CH_2-CH_2-Br \xrightarrow[80^\circ C\,(X)]{\text{Conc. alc. NaOH}}$$

$$CH_3-CH=CH_2 \xrightarrow{\text{HBr/ Acetic acid }(Y)} CH_3-CHBr-CH_3$$

**43.** *(b)* A is correct but R is not correct.

The correct form of R is :

AgCN is mainly covalent in nature.

**44.** *(a)* Both Statement I and Statement II are true.

**45.** *(d)* A is true but R is false.

The correct form of R is :

Allyl halides are the compounds in which the halogen atom is bonded to an $sp^3$ hybridised carbon atom adjacent to C=C.

**46.** *(b)* The complete reaction is as follow

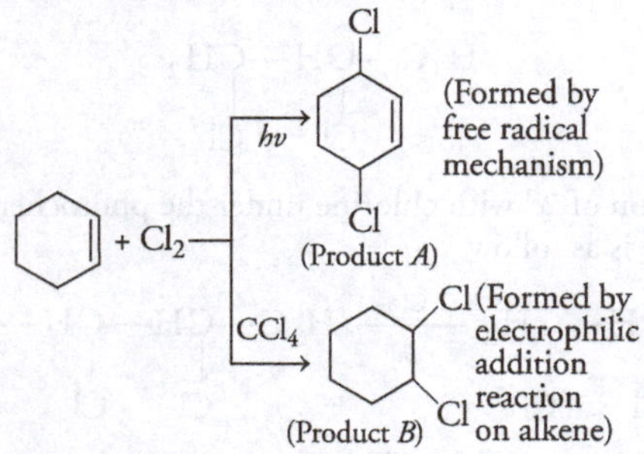

Hence, option (b) is correct.

**47.** *(c)* Among the given options, compound given in option (c) i.e. 1-bromocyclohexene will not produce precipitate on reaction with $AgNO_3$.

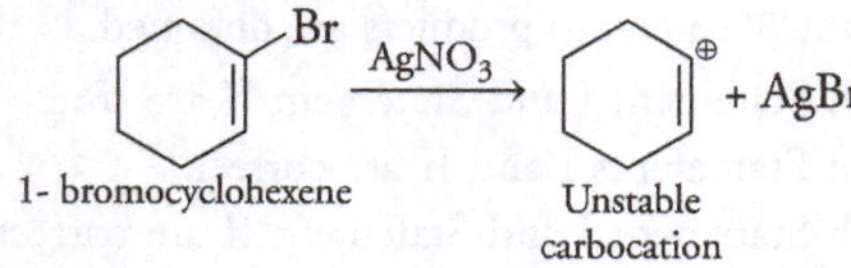

**48.** *(c)* The complete reaction is as follows.

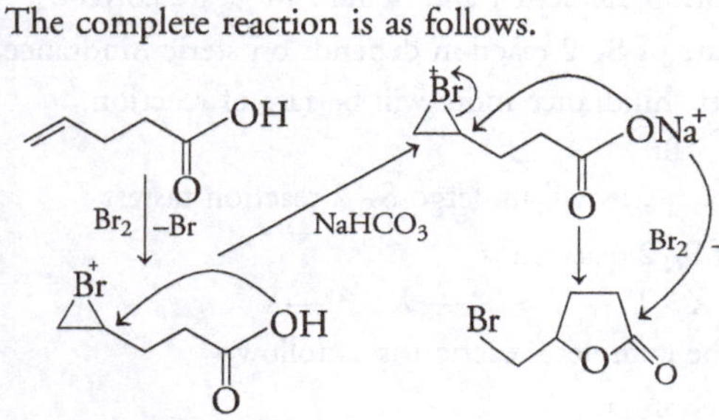

**49.** *(c)* For the reaction

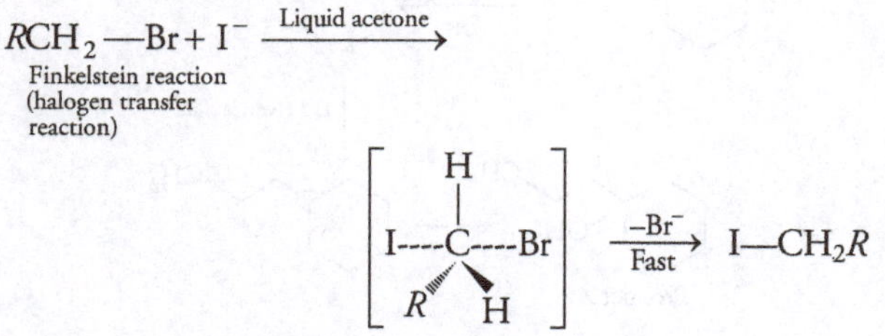

Penta-coordinated TS

During the reaction, transition state (TS) formed in the above reaction is less polar than the localised anion $Br^-$ or $I^-$.

**50.** *(c)*

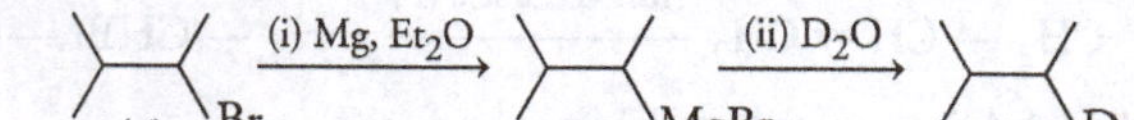

Compound (*A*) reacts with magnesium to form Grignard reagent which undergoes acid-base reaction with $D_2O$ to form compound *B*.

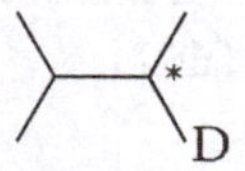

Compound *B* has one chiral centre (*) therefore, it is chiral in nature.

**51.** *(c)* NaCN is ionic and provides cyanide ion ($CN^-$) in solution. The attack (nucleophilic) takes place mainly through carbon atom and not through nitrogen atom.

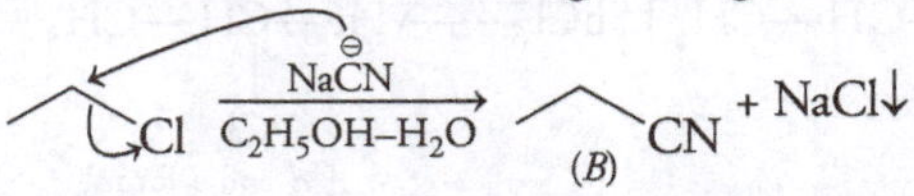

AgCN is covalent in nature and nitrogen is free to donate electron pair forming isocyanide as the main product.

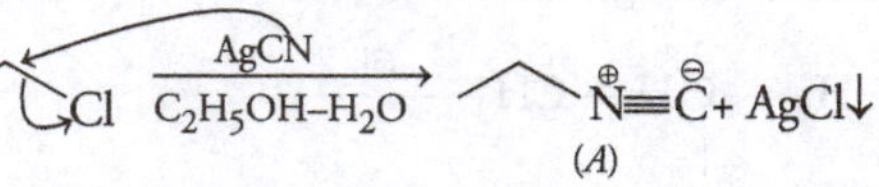

**52.** *(c)* The major product formed is given in option (c).

**Mechanism**

1. Formation of carbocation

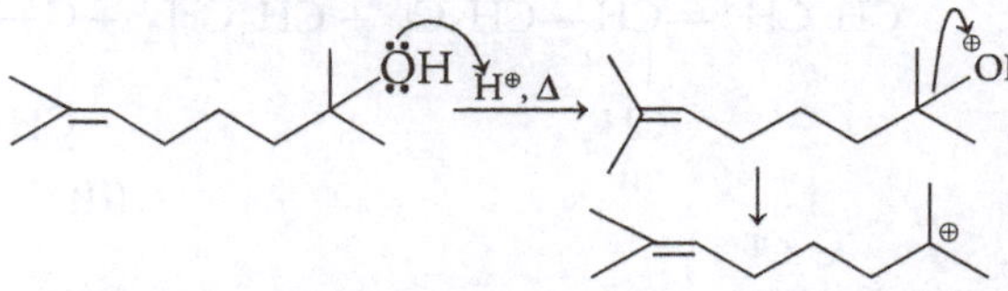

2. Intramolecular nucleophilic attack

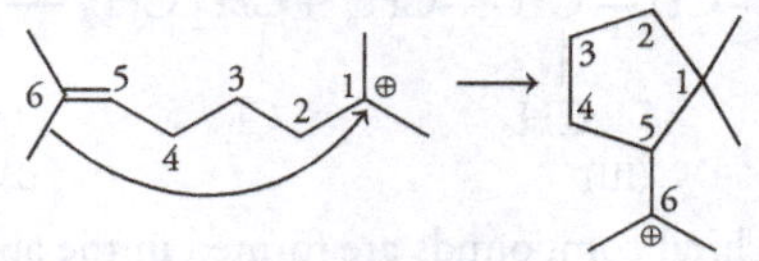

3. Nucleophilic attack of $Br^{\ominus}$

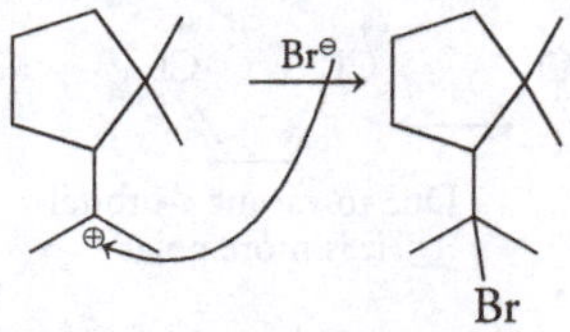

**53.** *(b)* **Step I** Addition of $CH_3O^-$ and $Br^-$ on unsaturated part of alkene.

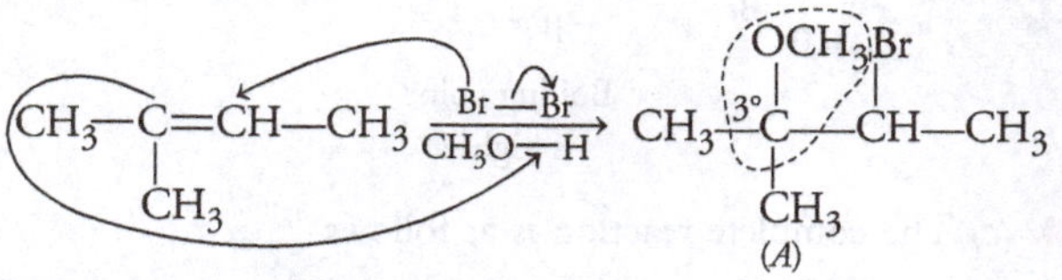

**Step II** $S_N1$ solvolysis of the ether with HI (1 eqv.) follows on

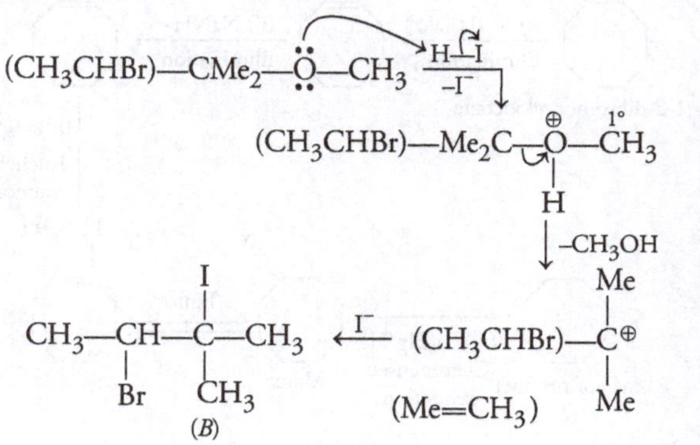

**54.** *(a)* The correct order of reactivity of chlorides with acetate in acetic acid is

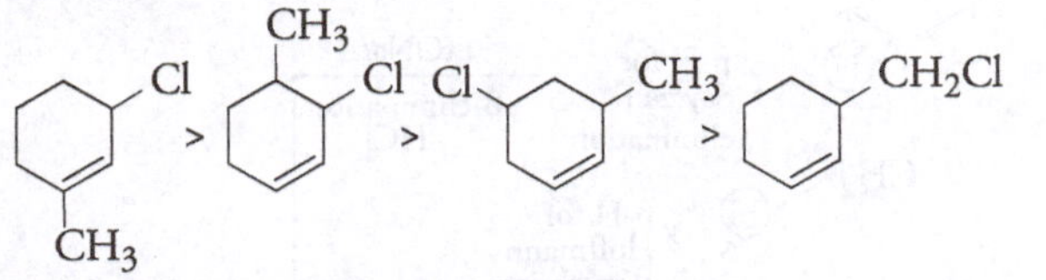

The given chlorides undergoes $S_N1$ reaction. So, as the stability of carbocation formed increases, rate of reaction increases. Stability of carbocation is as follows

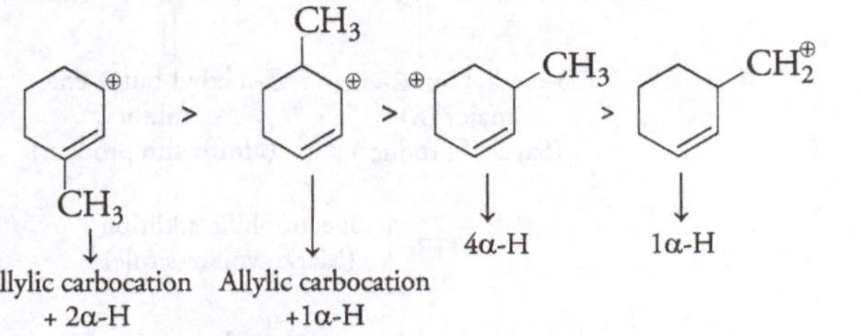


Hence, correct option is (a).

**55.** *(d)* Excess of isobutane reacts with $Br_2$ in presence of light at 125°C gives 2-bromo-2-methyl propane as major product.

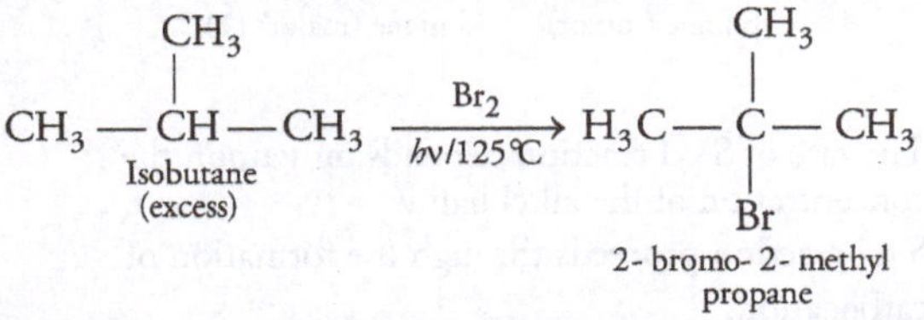


**Mechanism** The above reaction is halogenation of alkane *via* free radical substitution reaction.

**Initiation step**

$$Br_2 \xrightarrow{h\nu} Br^\bullet + Br^\bullet$$

**Propagation step**

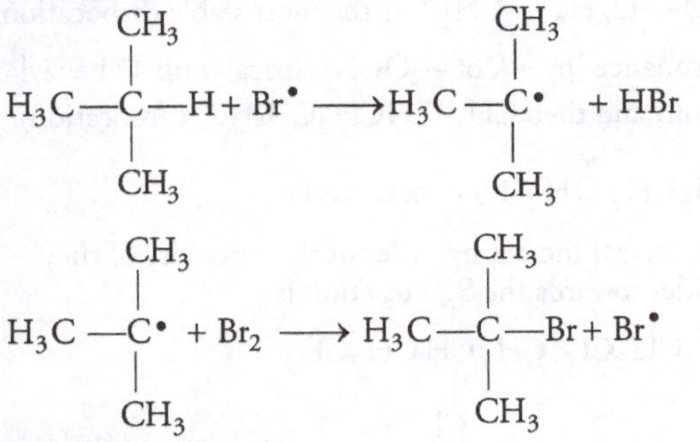

**Termination step**

$$Br^\bullet + Br^\bullet \longrightarrow Br_2$$

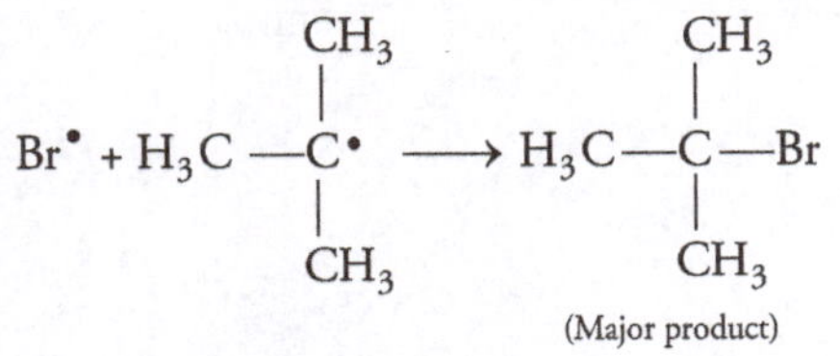


Since, isobutane is in excess, so dibromination of single isobutane is not favourable reaction. This make (a) and (b) incorrect options.

**56.** *(c)* Alcohols undergo dehydration (removal of water molecule) in the presence of protic acids (conc. $H_2SO_4, H_3PO_4$) at high temperature to give Saytzeff alkene (more alkylated) as major product.

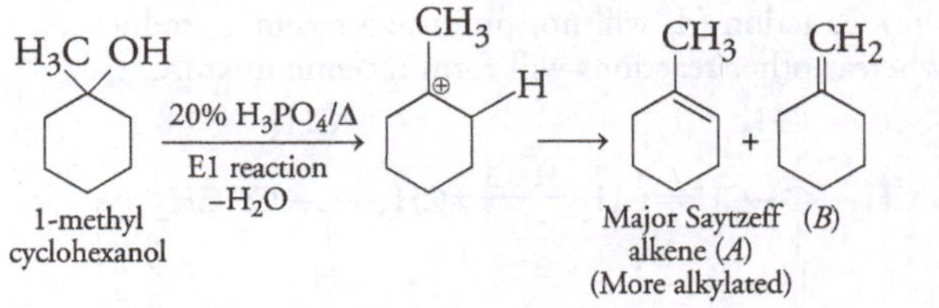


Bulky bases like *tert*-butoxide, due to steric hindrance, a abstracts a less hindered β-proton, that leads to formation of the less substituted product as, the major one, called the Hofmann product.

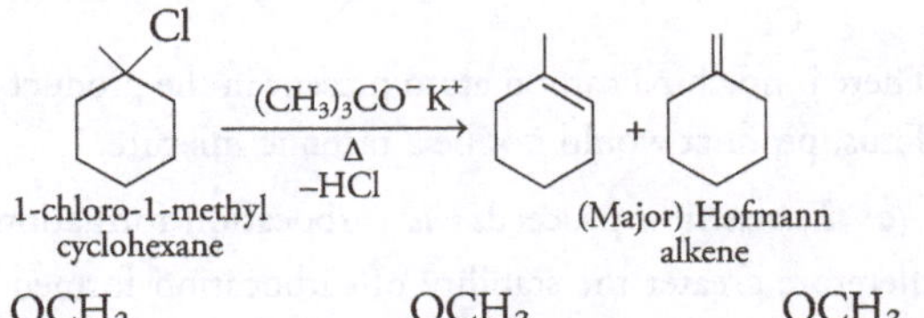


**57.** *(b)*

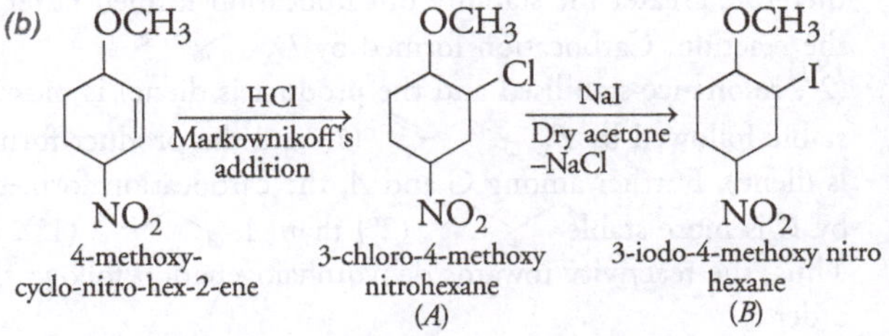


Ist reaction is electrophilic addition of HCl on double bond while 2nd reaction is halide substitution by Finkelstein reaction in which chlorine get displaced by iodine.

**58.** *(d)* Here, only one equiv. of Mg/ $Et_2O$ attacks on one bromine.

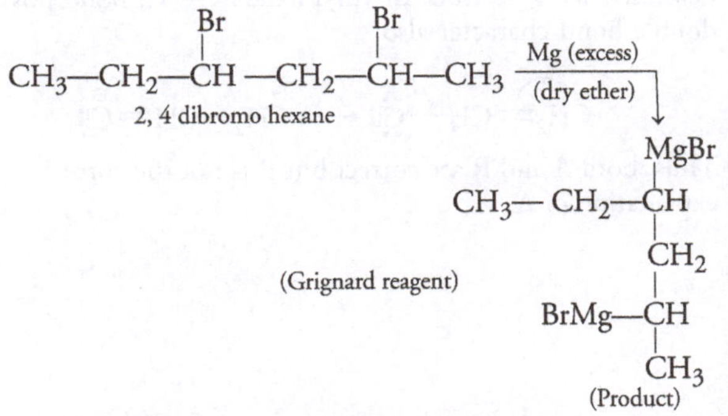


**59.** *(a)* All given species in the question are alkyl halide. They react with aqueous $AgNO_3$ solution with different rate. Hence, rate of precipitate formation of $AgX$ depends on stability of carbocation ($R^+$), obtained by heterolysis of the most polar (ionic in nature) of the C—Br bond

Among the given option(s) carbocation of the bromide (a) is most stable due to stronger + $M$ effect, so 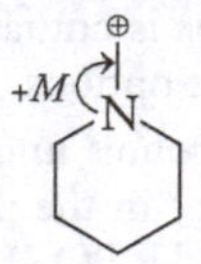 give precipitate with aqueous $AgNO_3$ solution most readily.

**60.** *(d)* Reaction (d) will not produce a racemic product whereas other reactions will form racemic mixture.

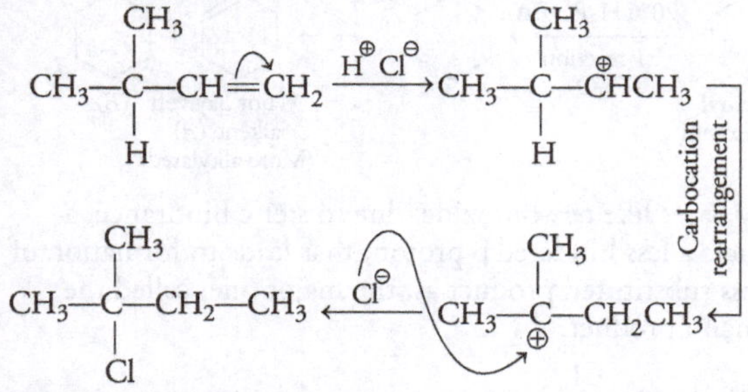


There is no chiral carbon atom present in the product (major). Thus, product would not be a racemic mixture.

**61.** *(c)* $E_1$ reaction proceeds *via* carbocation formation, therefore greater the stability of carbocation formed faster is the reaction. Carbocation formed by $D$, 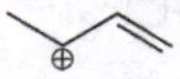 (2°, resonance stabilised and the product is diene) is most stable followed by $B$, 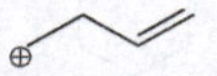 (1° and the product formed is diene). Further among C and $A$, the carbocation formed by $C$ is more stable 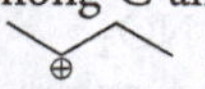 (2°) than $A$, 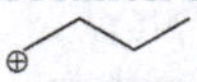 (1°). Thus, the reactivity towards dehydrohalogenation follows the order :

$$D > B > C > A$$

**62.** *(c)* Vinyl halide ($CH_2=CH-Cl$) do not undergo nucleophilic substitution reactions. This is because it forms highly unstable carbocation ($CH_2=\overset{\oplus}{C}H$). It cannot delocalise its $\pi$-electron. In vinyl halide C—Cl bond possess double bond character also.

$$CH_2=CH-\ddot{\underset{..}{Cl}}: \longleftrightarrow \overset{\ominus}{C}H_2-CH=\overset{\oplus}{\underset{..}{Cl}}:$$

Thus, both A and R are correct but R is not the correct explanation of A.

**63.** *(b)* The given reaction takes place as follows

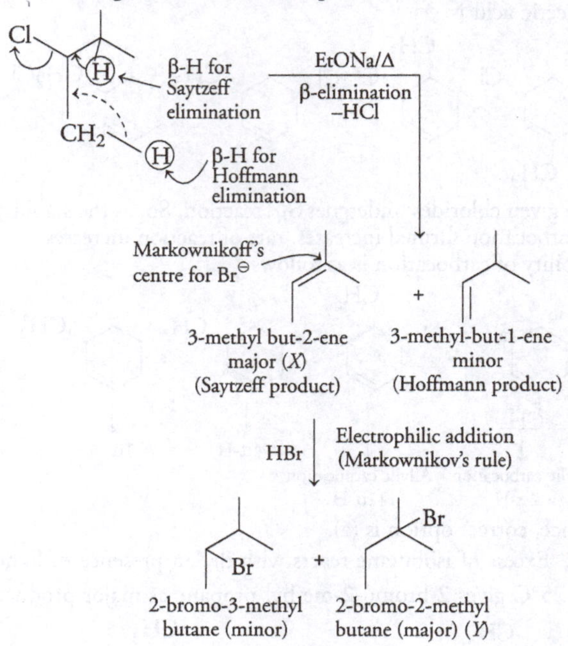


**64.** *(b)*

(i) The rate of $S_N1$ reaction depends only upon the concentration of the alkyl halide.

(ii) $S_N1$ reaction proceeds through the formation of carbocation.

The reactivity is decided by ease of dissociation of alkyl halide.

$$R-X \rightleftharpoons R^{\oplus} + X^{\ominus}$$

Higher the stability of $R^+$ (carbocation), higher would be the reactivity towards $S_N1$ reaction.

$p$-$H_3CO-C_6H_4-\overset{\oplus}{CH_2}$ is the most stable carbocation due to resonance [by $+R$ of $-OCH_3$ (para) with 1° benzyl carbocation] and then $CH_3\underset{\oplus}{C}HCH_2CH_3$ (2° carbocation) while $CH_3CH_2\overset{\oplus}{C}H_2$ (1°) is least stable.

Thus, the correct increasing order of the reactivity of the given halides towards the $S_N1$ reaction is

$$\underset{(II)}{CH_3CH_2CH_2Cl} < \underset{(I)}{CH_3\underset{Cl}{\underset{|}{C}H}CH_2CH_3} < \underset{(III)}{p\text{-}CH_3CO-C_6H_4-CH_2Cl}$$

**65.** *(a)* To show decolourisation, compound must be unsaturated.

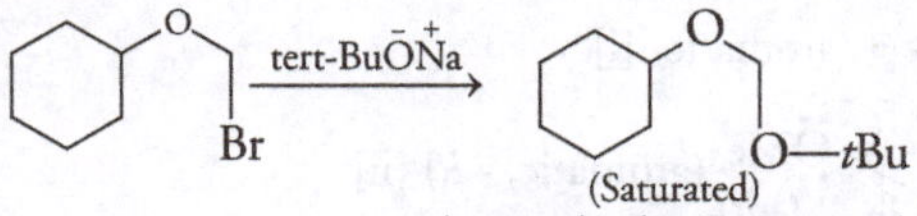


(Saturated)
(cannot decolourise $Br_2$ water)

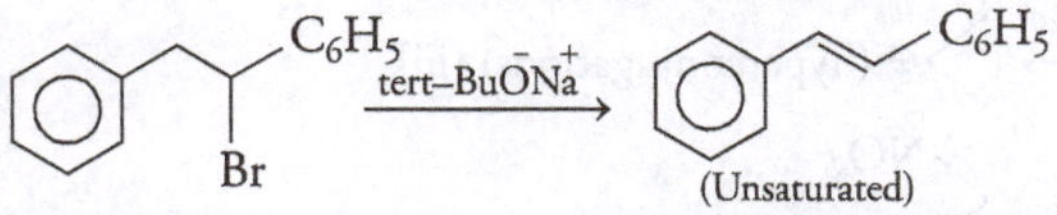


(Unsaturated)
(decolourise $Br_2$ water)

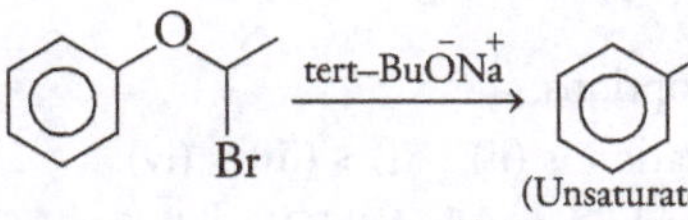


(Unsaturated)
(decolourise $Br_2$ water)

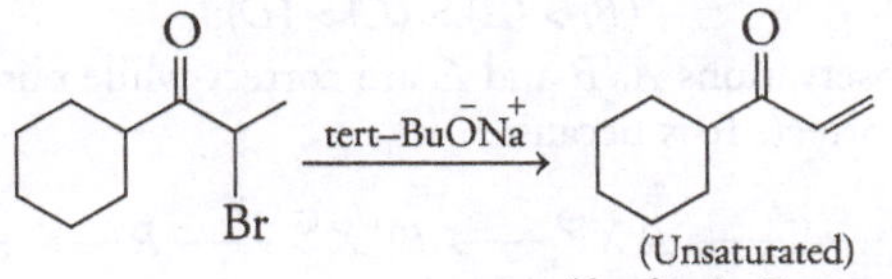


(Unsaturated)
(decolourise $Br_2$ water)

**66.** *(c)* Complete reaction sequence is as follows

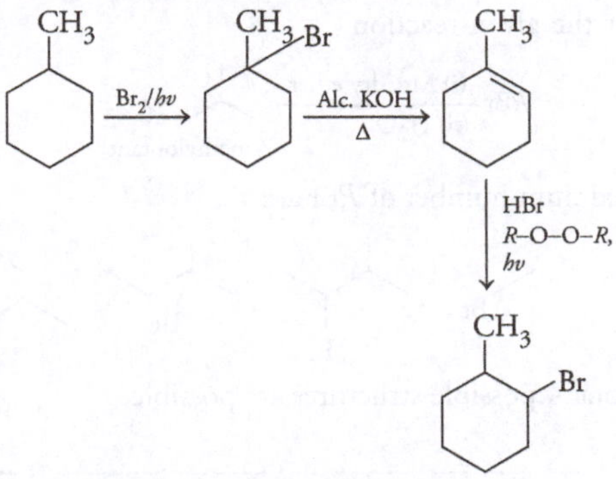


**67.** *(b)* The correct match is A-IV, B-III, C-I, D-II.

- Swarts reaction →

$$C_2H_5—Br + AgF \longrightarrow \underset{\text{Ethyl fluoride}}{C_2H_5—F} + AgBr$$

- Sandmeyer's reaction →

$$PhN_2^+Cl^- \xrightarrow{\text{CuCN/KCN}} \underset{\text{Cyanobenzene}}{PhCN} + N_2$$

- Wurtz-Fittig reaction →

$$Ph—Cl + EtCl \xrightarrow[\text{Ether}]{\text{Na}} \underset{\text{Ethyl benzene}}{Ph\text{-}Et}$$

- Finkelstein reaction →

$$Et\text{-}Cl \xrightarrow[\text{Acetone}]{\text{NaI}} \underset{\text{Ethyl iodide}}{Et\text{-}I} + NaCl$$

**68.** *(d)* Rate of solvolysis depends on stability of carbocation. As the stability of carbocation increases the rate of solvolysis also increases. The order of stability of the carbocations of the given compounds is

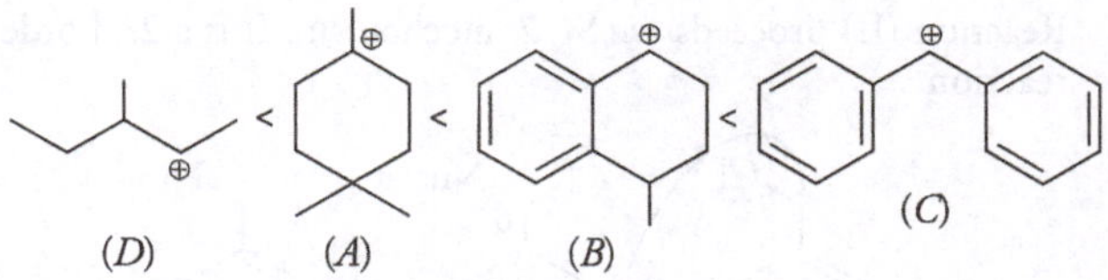


So, the order of solvolysis is also $D < A < B < C$.

**69.** *(d)* The complete reaction sequence is as follows

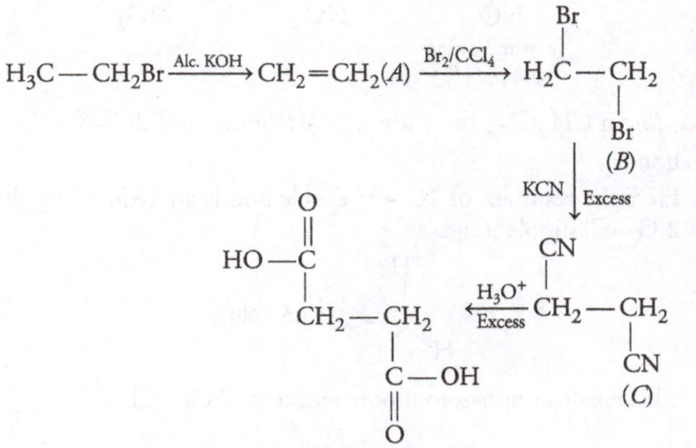


(*D*) Succinic acid

**70.** *(b)* $CH_3—CH_2—CH_2Br \xrightarrow[\Delta]{\text{alc. KOH}}$

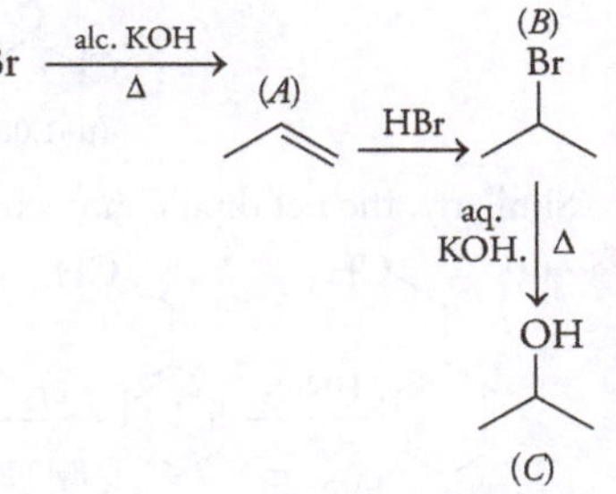


Propan 2- ol

**71.** *(c)* Reaction given in option (c) is correct while reactions in (a), (b) and (d) are incorrect. According to Markownikoff's rule when reagent adds to an unsymmetrical alkene, the more positive part of the reagent goes to the carbon that have more number of hydrogen atoms.
Reaction in option (c) is an example of Markownikoff addition reaction.

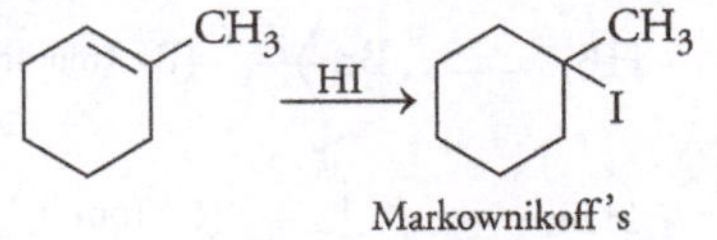


Markownikoff's addition

**72.** *(d)* Reaction (I) *via* $S_N1$ mechanism which involves the formation of an intermediate. It is a 1st order reaction.

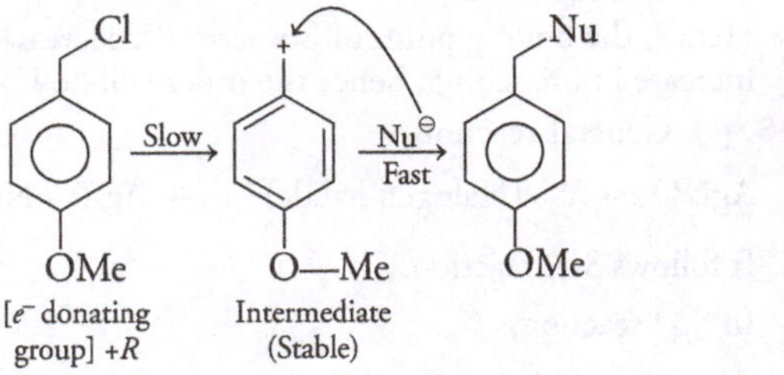

Reaction (II) proceeds *via* $S_N2$ mechanism. It is a 2nd order reaction.

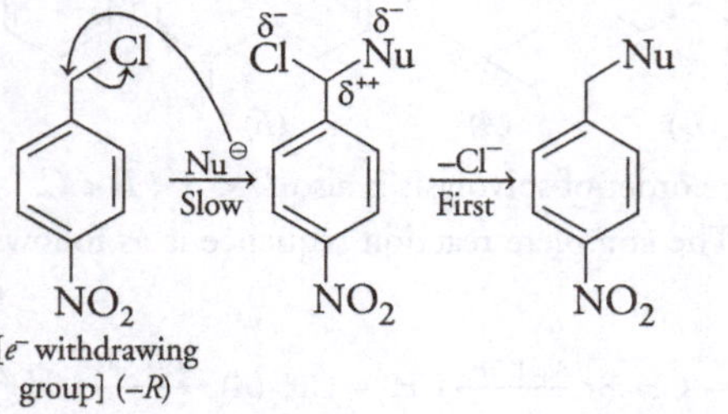


**73.** *(b)* In $CH_2Cl_2$, there are 2C—H bonds and 2C—Cl bonds.

Here the resultant of 2C—H dipole bonds are reduced by the 2 C—Cl dipole bond

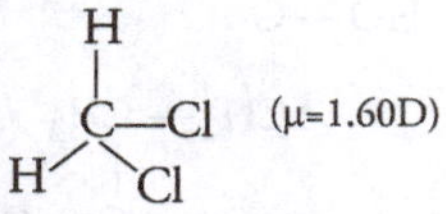


This result in non-zero dipole moment of $CH_2Cl_2$.

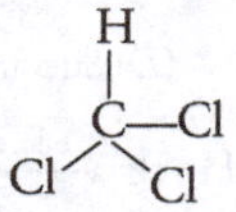


(μ=1.08D)

Similarly, the net dipole moment of $CHCl_3$ is non-zero.

**74.** *(b)*

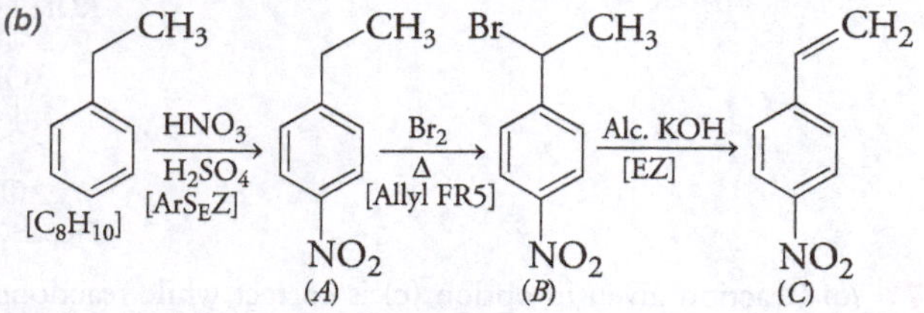


**75.** *(a)* Complete reaction (peroxide effect, free radical addition) of I, II and III are as follow:

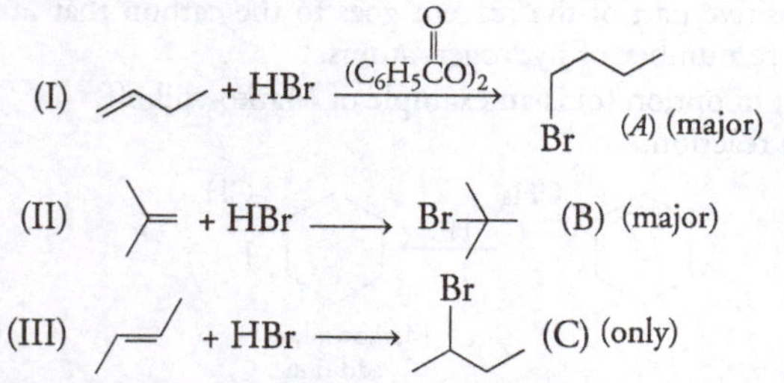


Branching decreases the surface area as in alkane because the branched *C* chains are more spherical -like, which results in lower boiling point.

Hence, the boiling point of isomeric *RX* decreases with increase in branching, hence the order will be II < III< I

**76.** *(c)* General reaction :

$AgNO_3 + X^-$ (Halogen halide) $\longrightarrow AgX + NO_3^{2-}$

It follows $S_N1$ reaction.

In $S_N1$ reaction,

Rate of reaction ∝ stability of carbocation formed

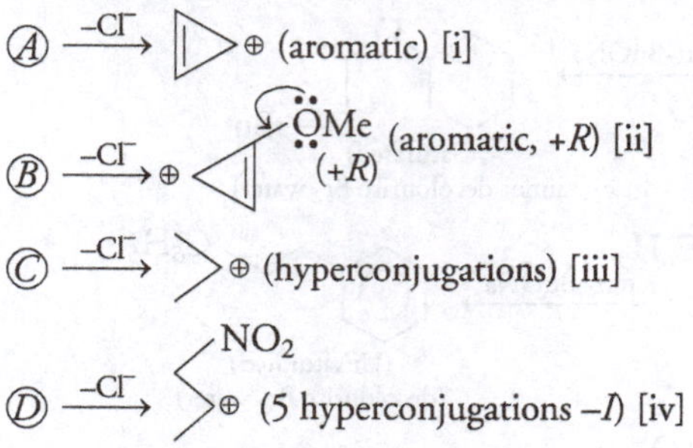


Among the given options.

Stability of carbocation is (ii) > (i) > (iii) > (iv).

So, the reactivity order towards $AgNO_3$ solution is

$$(B) > (A) > (C) > (D).$$

**77.** *(c)* Observations *A*, *B* and *C* are correct while observation *D* is incorrect. It is because

$$R—X \longrightarrow \overset{\oplus}{R}\, X^{\ominus} \longrightarrow \overset{\oplus}{R}||X^{\ominus} \xrightarrow{Y^{\ominus}} R—Y + X^{\ominus}$$

$R^{\oplus}$ (carbocation) formation takes place and $R^{\oplus}$ got stabilised by electronic factors and polar solvent molecules.

**78.** *(c)* For the given reaction

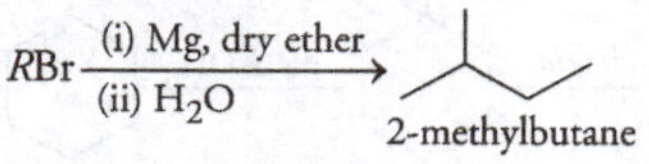


The maximum number of *R*Br are

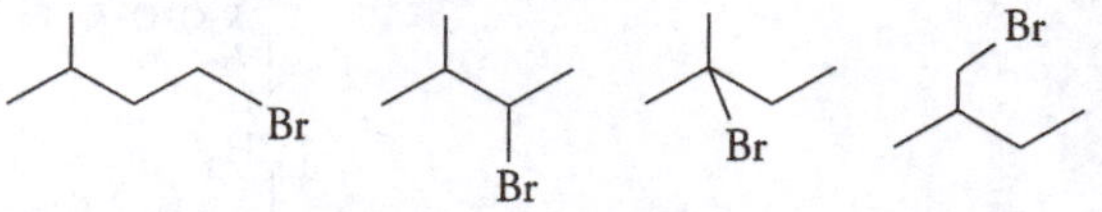


Thus, total 4 possible structures are possible.

**79.** *(1)*

**key idea** All optically active compounds are chiral asymmetrical.

Total number of optically active compounds among the given compounds is one.

$$CH_3—CH_2—\underset{\displaystyle Cl}{\underset{|}{C}}H—CH_3$$

**80.** *(32)*

**key idea** To calculate the total number of optical isomers, first of all look for number of chiral centres, then use the formula $2^n$, where $n$ = no. of chiral centre.

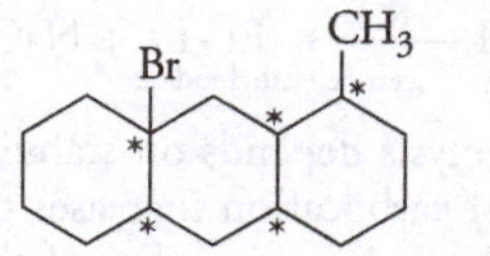


Total chiral centre = 5

Number of optical isomers = $2^5$ = 32

Hence, number of optical isomers is 32.

**81.** *(6)* Total 6 isomeric compounds are formed by monochlorination of 2-methylbutane. These are as follows

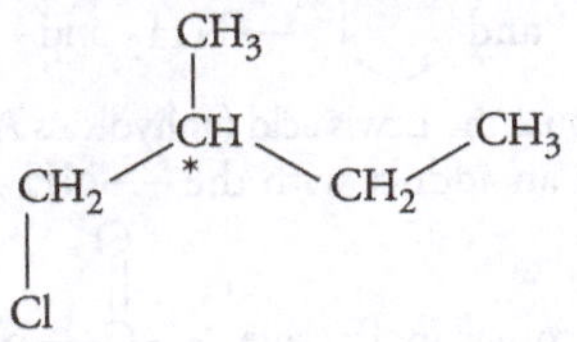

(Two isomers i.e., *R/S*)

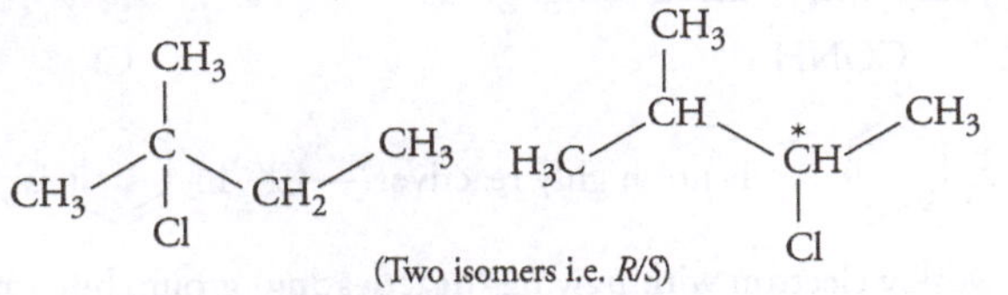

(Two isomers i.e. *R/S*)

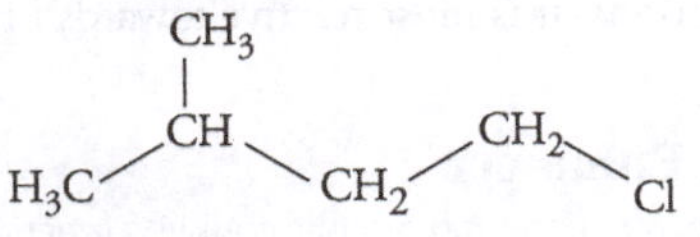

**82.** *(2)* 2-chlorobutane 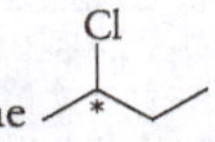

There is one chiral centre present in given compound.

So, total optical isomer = 2

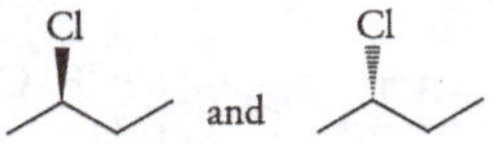

Hence, total isomer is 2.

**83.** *(10)* $CH_3CH_2Br + NaOH \xrightarrow{C_2H_5OH} \underset{(A)}{CH_2{=}CH_2}$

$CH_3CH_2Br + NaOH \xrightarrow{H_2O} \underset{(B)}{CH_3{-}CH_2OH}$

Thus total number of hydrogen atom in *A* and *B* is 10.

**84.** *(5)* The total number of chiral carbon compounds are 5.

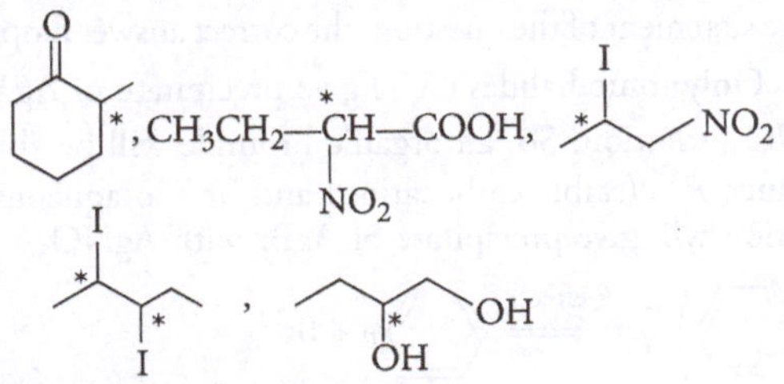

Chiral carbons are marked by (*).

## Topic 2

**01.** *(b)* The complete reaction is as follows

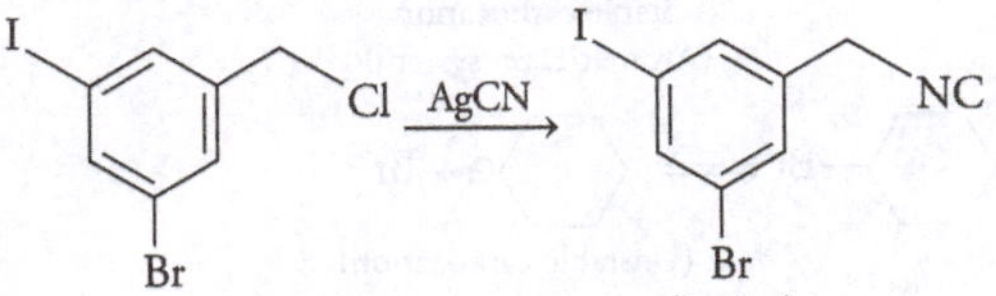

(Major)

**02.** *(c)* Stability order of carbocation is

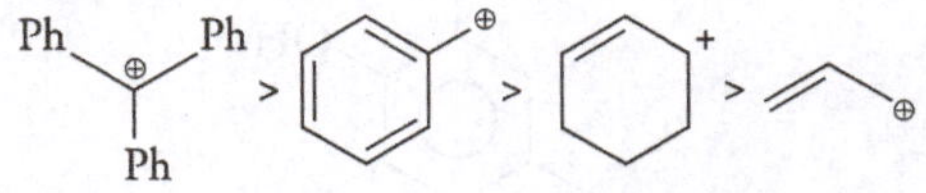

So, the most stable carbocation is given in option (c).

**03.** *(a)* The given reaction is an example of nucleophilic aromatic substitution reaction and gives the following product '*P*'.

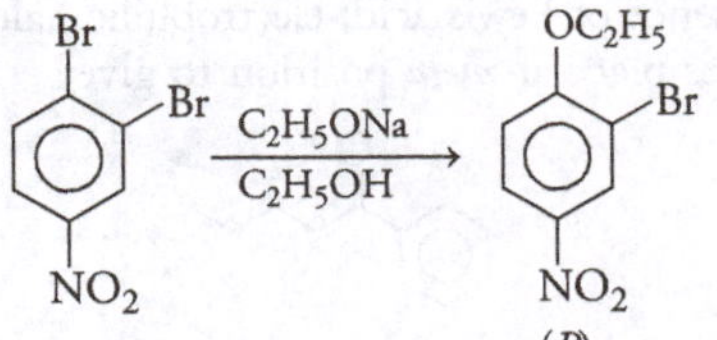

(*P*)

**04.** *(b)* Complete reaction is as follows.

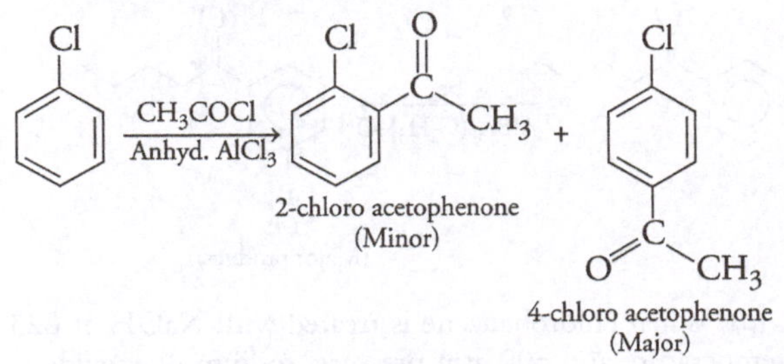

This reaction is called Friedel-Crafts reaction.

**05.** *(a)*

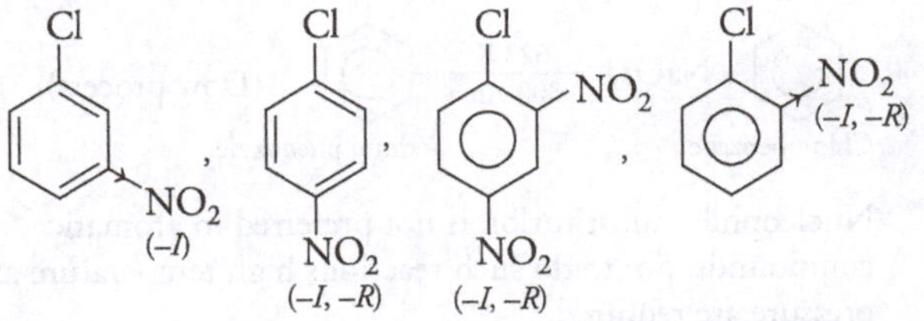

3-chloronitrobenzene has weaker electron withdrawing nature of $NO_2$ (−*I* only)as compared to other compounds. So, it gives lowest rate towards nucleophilic aromatic substitution reaction ($ArS_N2$).

Thus, option (a) is correct.

**06.** *(b)*

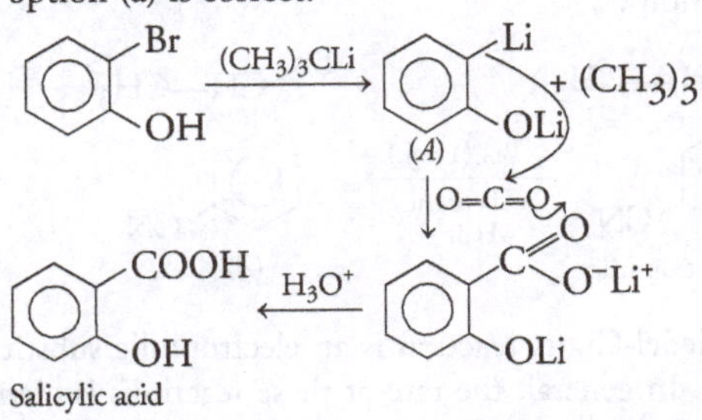

**07.** *(c)* Since —OH is a better electron releasing group than $—CH_3$.

So, electrophile will attack at *ortho* and *para*-position with respect to —OH group.

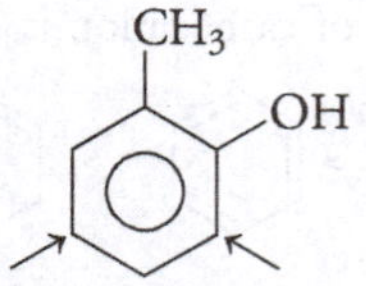


*para*-position with respect to —OH group is the *meta*-position with respect to —$CH_3$ group.

—CHO and —$NO_2$ are electron withdrawing groups which lead to electrophilic attack at *meta*-position.

**08.** *(c)* In presence of Lewis acid, electrophilic halogenation reaction takes place at *meta* position to give

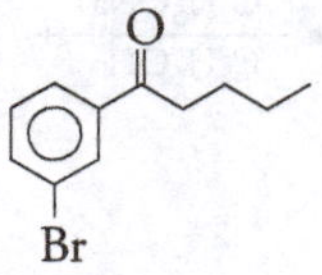


as follows

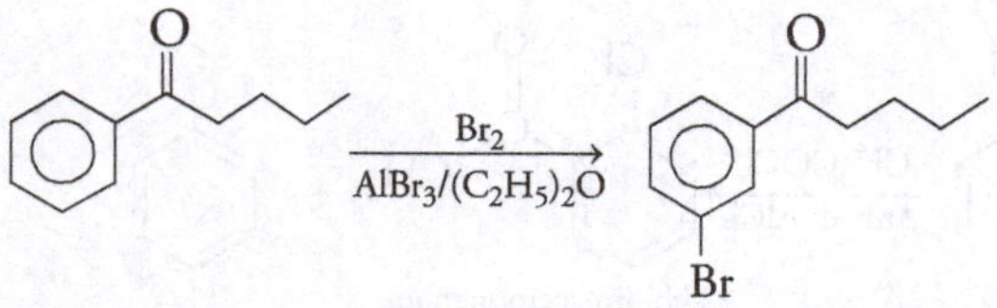


*(A)*
(Major product)

**09.** *(d)* When chlorobenzene is treated with NaOH at 623 K temperature and 300 atm pressure, sodium phenoxide is formed by nucleophilic substitution.

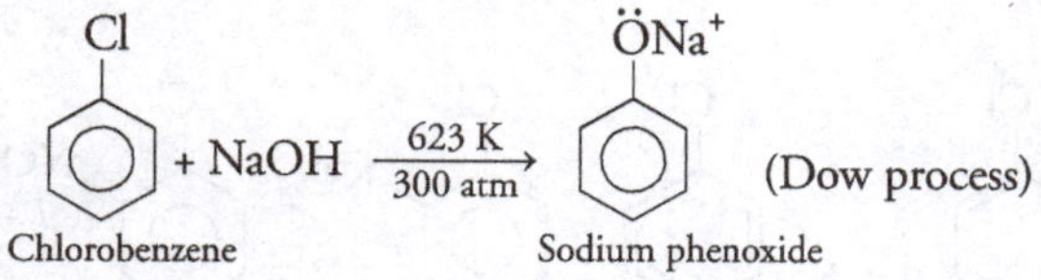


Nucleophilic substitution is not preferred in aromatic compounds. So, to do such reactions high temperature and pressure are required.

**10.** *(c)* When 3-ethyl benzonitrile undergoes photochemical reaction (UV) with bromine (1 equivalent), we get a monobrominated product 1 bromo -1 (3-cyanophenyl) ethane *(A)* as the major product (benzyl free radical substitution).

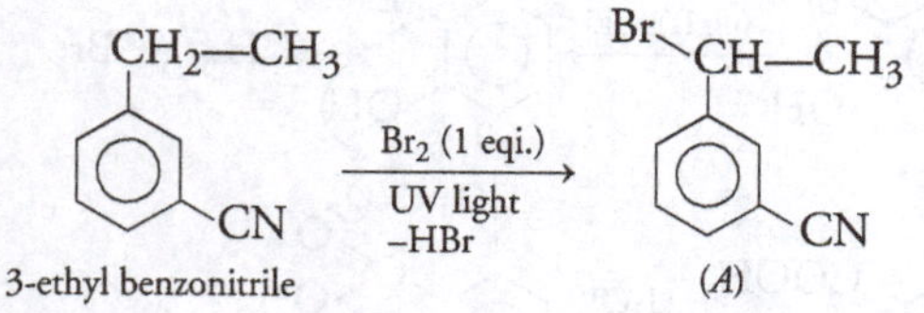


**11.** *(b)* Friedel-Crafts reaction is an electrophilic substitution reaction. In general, the rate of these reactions is highest when benzene have some electron releasing groups, (i.e. —$NH_2$, —OH) attached to the nucleus.

But in 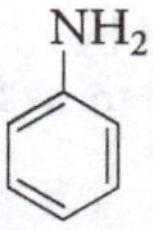

and 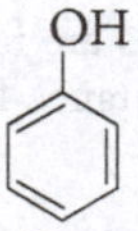

, —$\ddot{N}H_2$ and —$\ddot{O}H$ groups are deactivated because the Lewis acid (anhydrous $AlCl_3$) used in these reactions form an adduct with the '—$NH_2$' or —OH' groups and becomes ineffective. '—C(=O)—$NH_2$' is an electron withdrawing (deactivating) group.

∴ 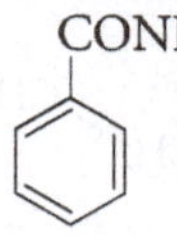

is not highly reactive. '—Cl' in 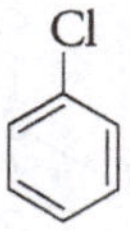

is weakly electron withdrawing (deactivating) group, but among the given options, it is most reactive towards Friedel-Crafts reaction.

> **• Concept Enhancer**
>
> Friedel-Crafts reaction is an organic coupling reaction involving an electrophilic aromatic substitution that is used for the attachment of substitutents to aromatic rings. The two primary types of friedel-crafts reactions are the alkylation and acylation reactions.

**12.** *(d)*

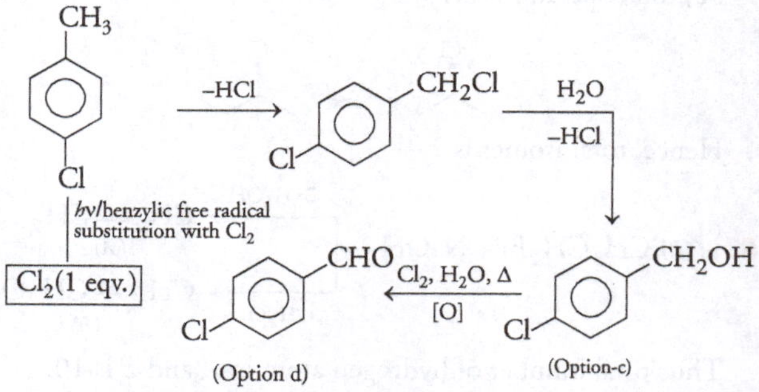


So, considering the second reaction condition, $Cl^-$, $H_2O$ and $\Delta$ in the statement of the question, the correct answer is option (d).

**13.** *(a)* Only ionic halides ($X^-$) give precipitate of Ag*X* with $AgNO_3$ solution. So, an organic bromide will be able to produce $R^{\oplus}$ (stable carbocation) and $Br^-$ in aqueous solution will give precipitate of AgBr with $AgNO_3$.

(c) 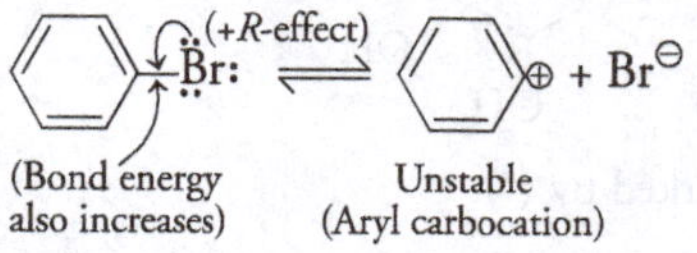


(a) 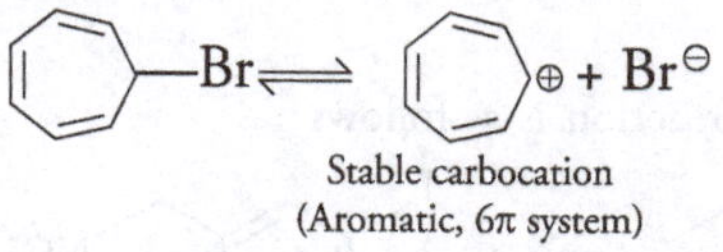


(d) 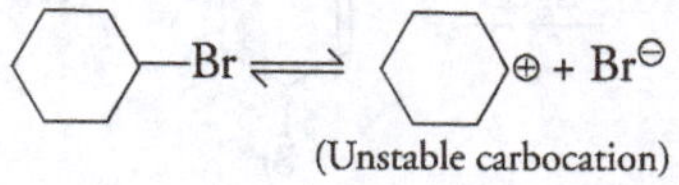

(b) 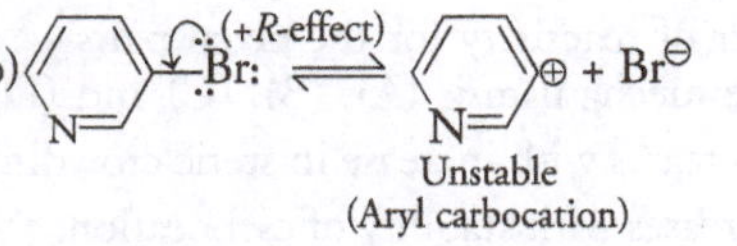


So, only 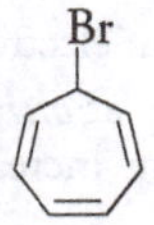 produces a precipitate of AgBr with $AgNO_3$ solution.

**14.** *(a)* The complete reaction is as follows,

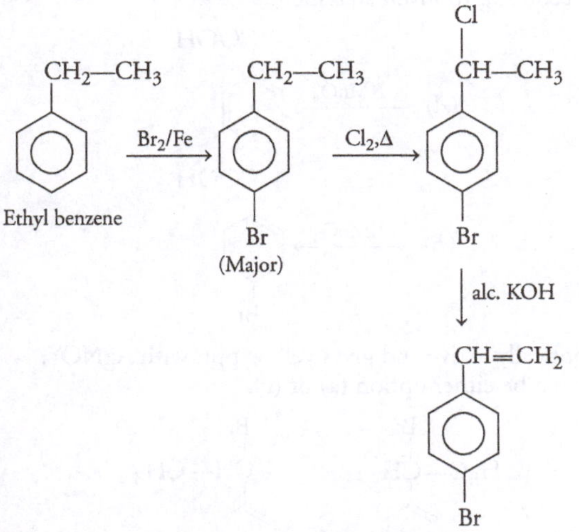


**15.** *(c)* Both A and R are correct and R is the correct explanation of A.

**16.** *(d)* A is true but R is false.

The correct form of R is : No reaction occurs between phenol and halogen acid. It is due to the presence of the double bond character in C—O bond.

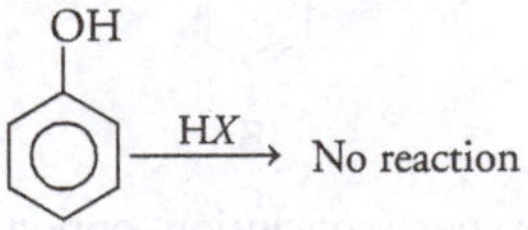


Hence, option (d) is correct.

**17.** *(c)* The correct reactivity of haloarenes towards nucleophilic substitutions with *aq.* NaOH is

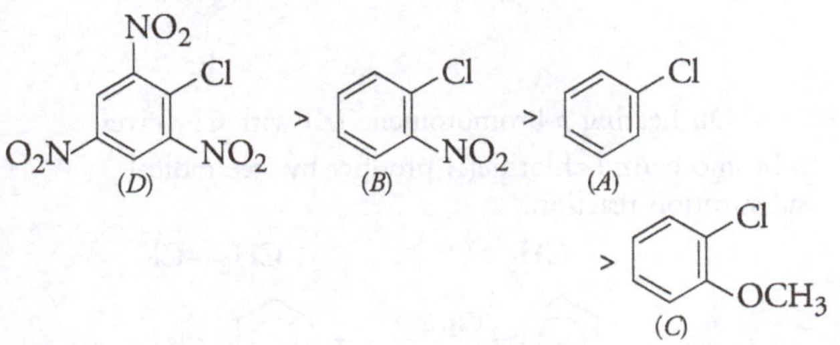


Generally, aryl halides are less reactive towards nucleophilic substitutions with aq. NaOH. Due to presence of electron releasing group i.e., (—$OCH_3$) further decreases its reactivity in (c) towards nucleophilic substitution whereas the presence of an electron withdrawing group (–*R*) i.e., (—$NO_2$) at *ortho* and *para* position increases its reactivity towards nucleophilic substitution. As the number of electron withdrawing groups increases, reactivity increases.

**18.** *(c)* correct option is (c)

Melting point of molecule depends on packing efficiency. If a molecule is symmetrical then close packing is possible due to better crystal fitting, which results in strong intermolecular attractive force. Thus, more energy is required to break the bond, which results in increased melting point.

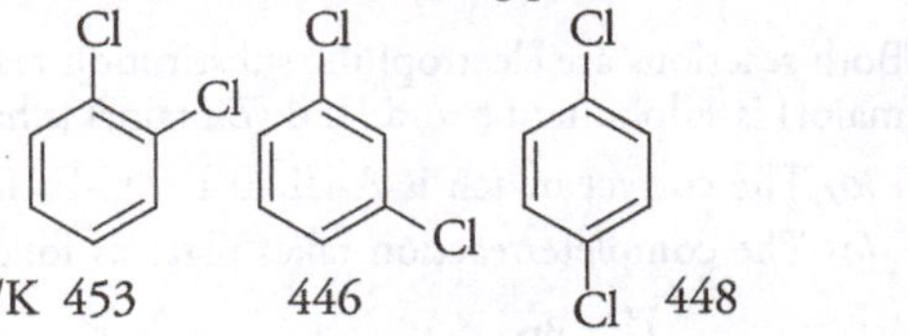


| | | | |
|---|---|---|---|
| Boiling point/K | 453 | 446 | 448 |
| Meting point/K | 256 | 249 | 323 |

**19.** *(c)* Only bromine water gives tribromo products, while all other gives monobromo products in which *para* bromo is major product.

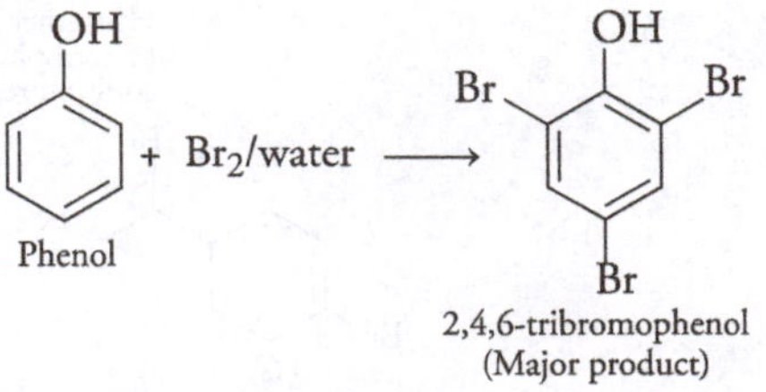


When phenol reacts with bromine in the presence of $CS_2$ or $FeBr_3$ or $CHCl_3$, it results in the formation of two products.

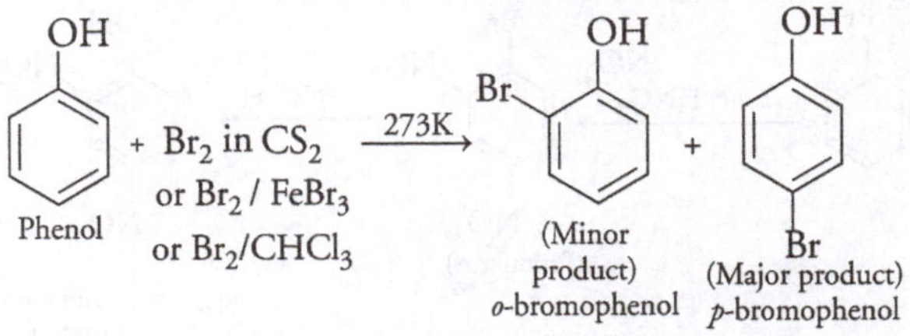


**20.** *(c)* The mechanism of the given reaction is as follows

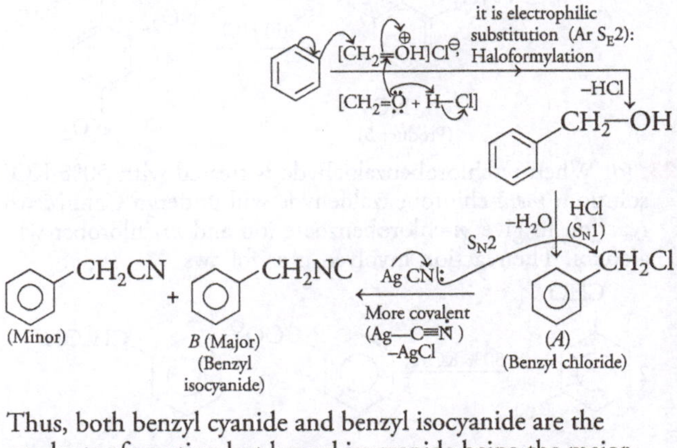


Thus, both benzyl cyanide and benzyl isocyanide are the products of reaction but benzyl isocyanide being the major product gives the correct option as (c).

**21.** *(a)* The complete reaction sequence is as follows,

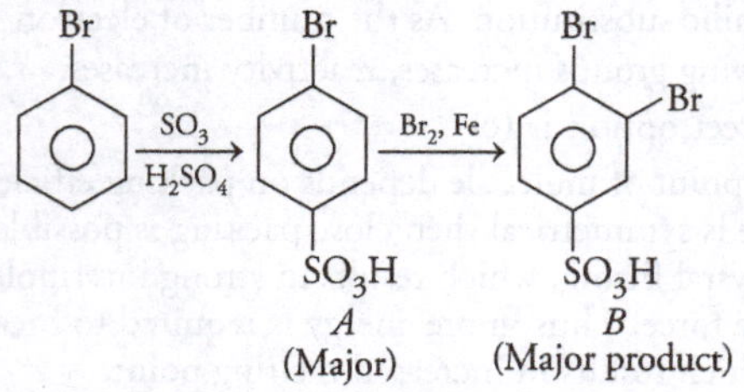


Both reactions are electrophilic substitution reaction, Ist (*A* major) is sulphonation and IInd (*B* major) is halogenation.

**22.** *(d)* The correct match is A-III, B-IV, C-II, D-I

**23.** *(c)* The complete reaction takes place as follows

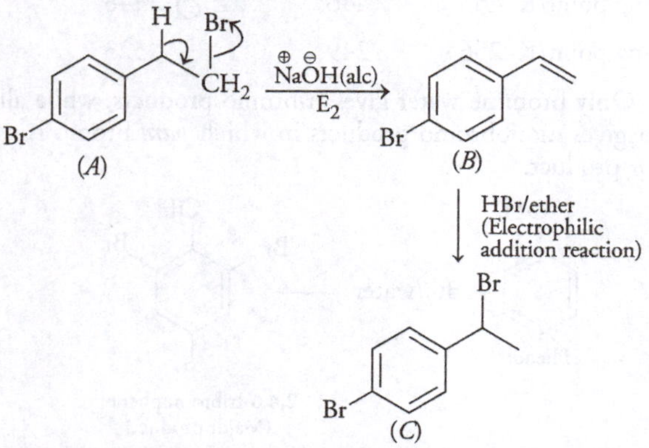


Here A and C are position isomers.

**24.** *(a)* The complete reaction is as follow,

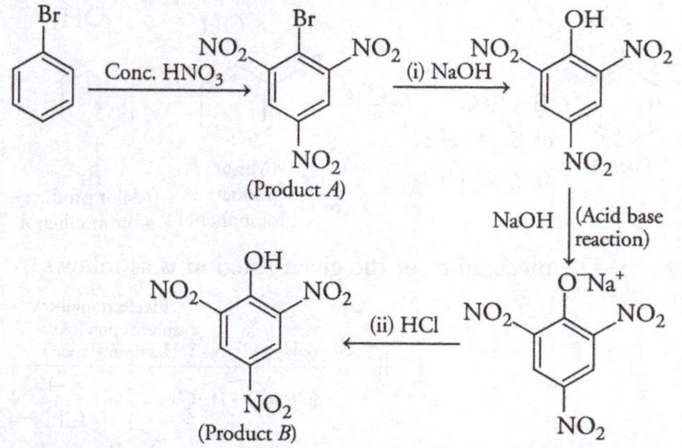


**25.** *(d)* When *m*-chlorobenzaldehyde is treated with 50% KOH solution, *meta*-chlorobenzaldehyde will undergo Cannizzaro reaction to give *m*-chlorobenzoate ion and *m*-chlorobenzyl alcohol. The reaction involved is as follows

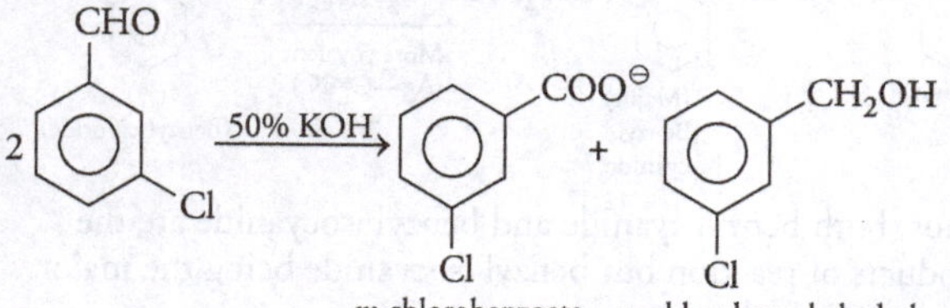


*m*-chlorobenzoate *m*-chlorobenzyl alcohol

Hence, option (d) is correct.

**26.** *(b)* The correct order of reactivity for the given pairs towards the respective mechanism is (A), (B), (C) and (D).

(A) $S_N2$ reaction decreases with increase in steric crowding.

(B) $S_N1$ reaction increases with stability of carbocation.

(C) EAS reaction decreases with decreases in electron density.

(D) Presence of electron withdrawing group at *ortho* and *para*-position w.r.t. halogen in haloarene increases nucleophilic aryl substitution.

**27.** *(a)* The correct match is A-II, B-I, C-IV, D-III.

**28.** *(c)* According to given statement,

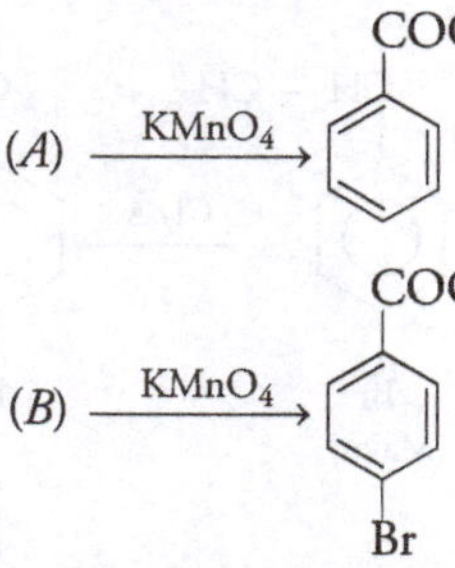


(*A*) is optically active and gives yellow ppt. with $AgNO_3$. So, (*A*) can be either option (a) or (c).

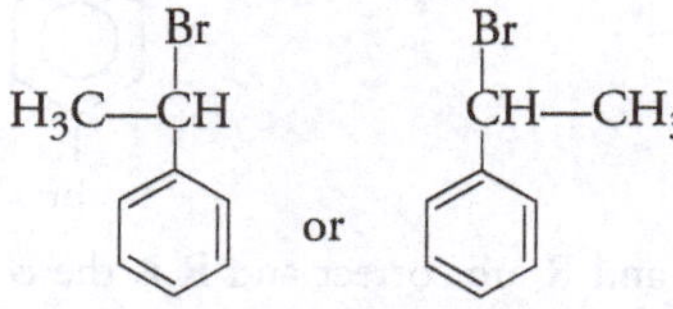


But as *B* gives *p*-bromobenzene on oxidation, it will be option (b) or option (c).

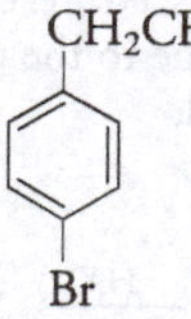


Hence, from the above conclusion, option (c) is correct, i.e.,

*A* is $CH_3$ — CH(Br) — $C_6H_5$ and *B* is 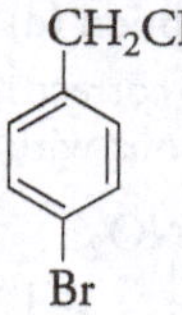


**29.** *(c)* On heating *p*-bromotoluene (*A*) with $Cl_2$ gives *p*-bromo benzyl chloride(I) product by free radical substitution reaction.

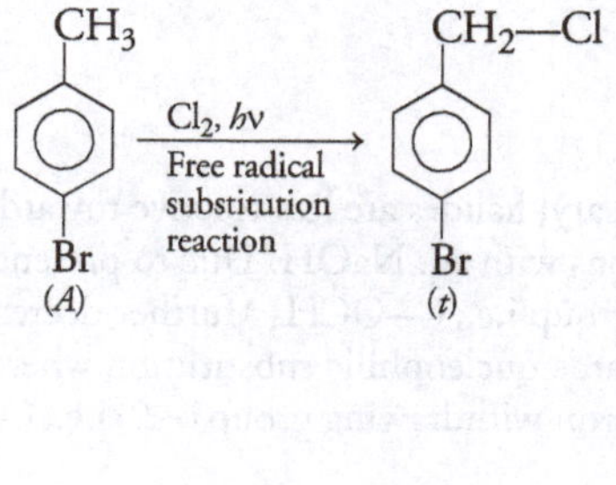

Product (I) on treating with $CN^-$ shows $S_N2$ reaction at benzylic carbon. $CN^-$ replace $Cl^-$ to form *p*-bromo benzylic nitrile (II).

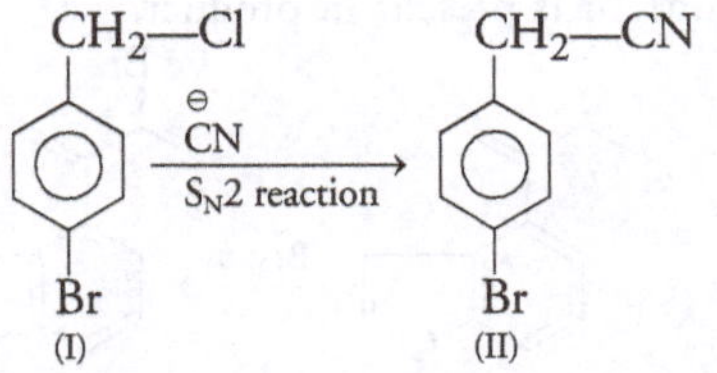


Benzyl nitrile (II) on complete hydrolysis with acidic water gives 4-bromophenyl acetic acid.

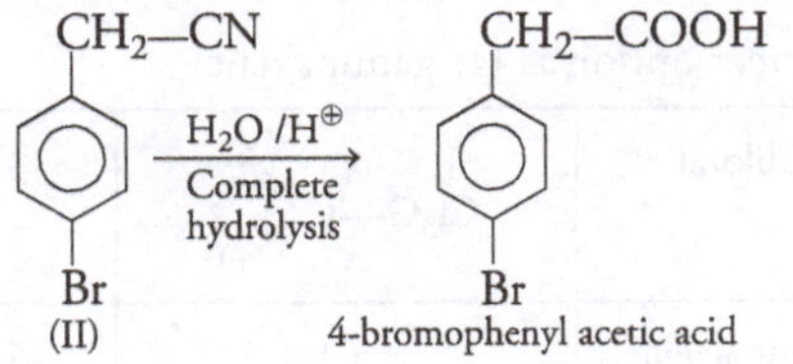


**30.** *(b)*

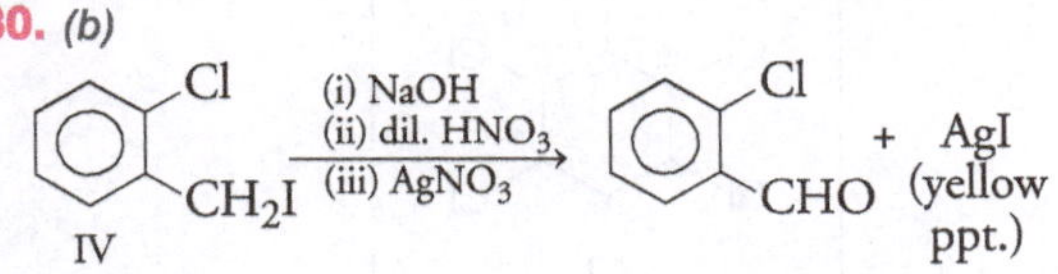


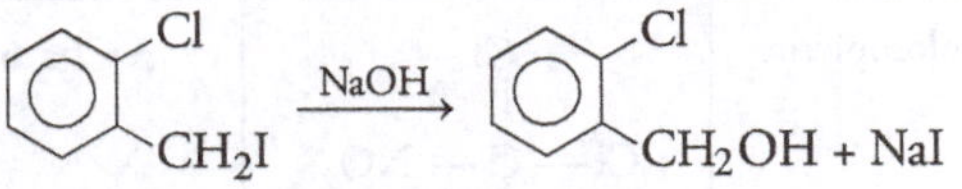


This compound halide will only give yellow ppt. as benzyl carbocation formed shown below is highly stable by conjugation.

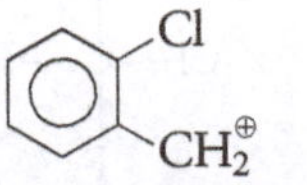


Benzyl carbocation

While other compounds halide cannot be removed because their corresponding carbocation is highly unstable.

**31.** *(c)* Benzene diazonium chloride undergoes Sandmeyer reaction in presence of $Cu_2Cl_2$ while in presence of Cu/HCl it undergo. Gattermann reaction to give chlorobenzene along with nitrogen gas ($N_2$).

While in Wurtz reaction ethyl chloride reacts with sodium in presence of dry ether to give *n*-butane and the reaction of chlorobenzene in presence of dry ether to give biphenyl is Fittig reaction.

So, correct match is A-(ii), B-(iv), C-(i) and D-(iii).

**32.** *(a)* Rate of $S_N2$ reaction $\propto \dfrac{1}{\text{Stability of carbocation}}$

Stability of carbocation $\propto \dfrac{1}{-I\,/\,-M}$ (effective group)

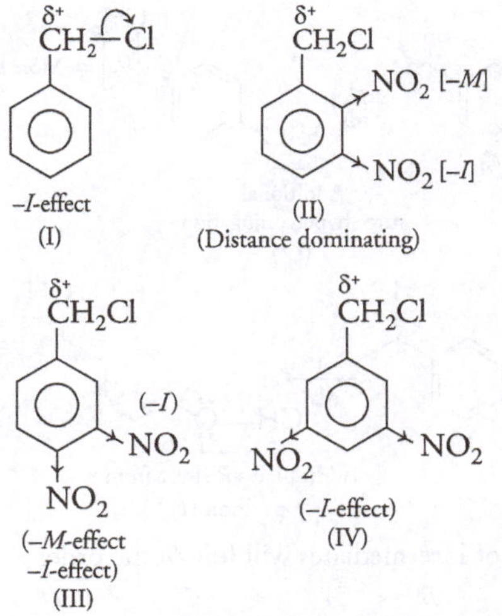


So, the decreasing order of reactivity of the given compounds toward $S_N2$ reaction is

(II) > (III) > (IV) > (I).

**33.** *(a)* B and D are possible while (A) and (C) are not possible. $AlCl_3$ do not act as electrophiles to give Friedel-Crafts reaction with benzene. Therefore, A and C reactions are not possible.

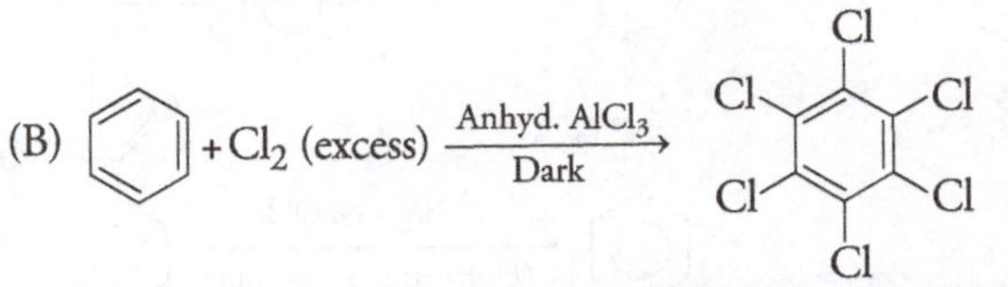


(This reaction is possible)

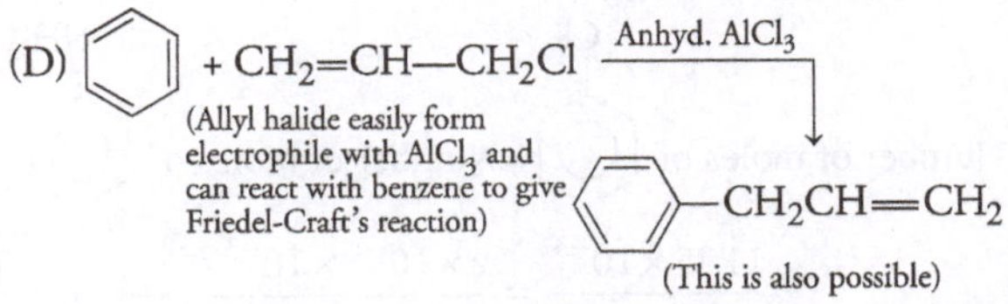


(This is also possible)

**34.** *(b)* More stable the carbocation intermediate, higher will be the rate of $S_N1$ reaction.

The reaction involving carbocation intermediate formation for the given compounds are as follows

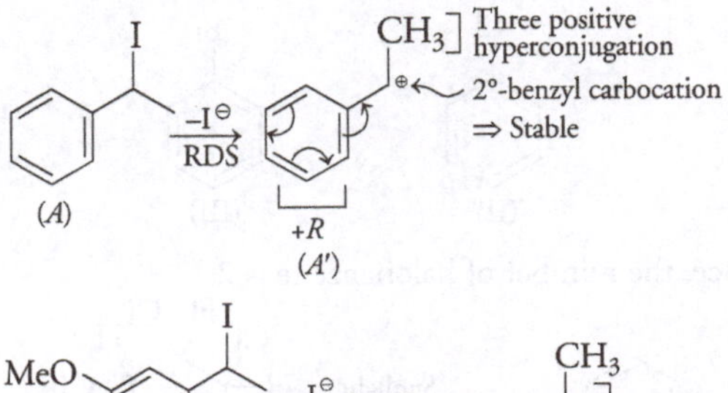


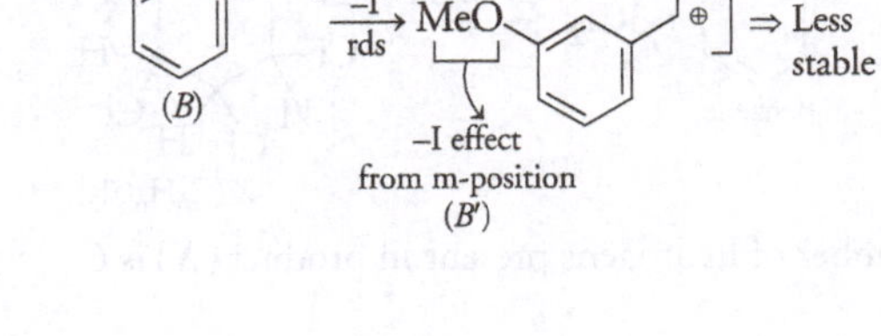

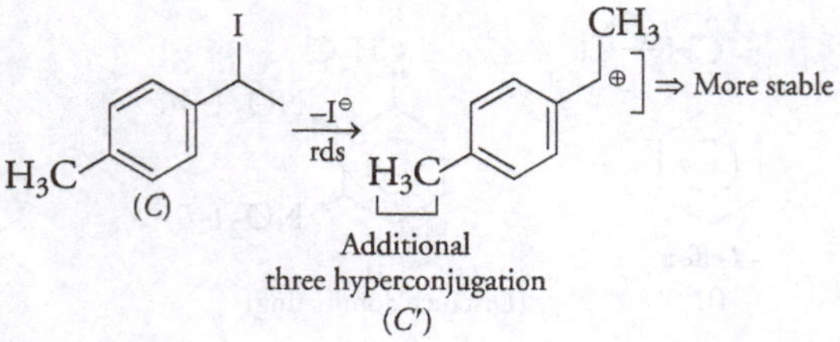


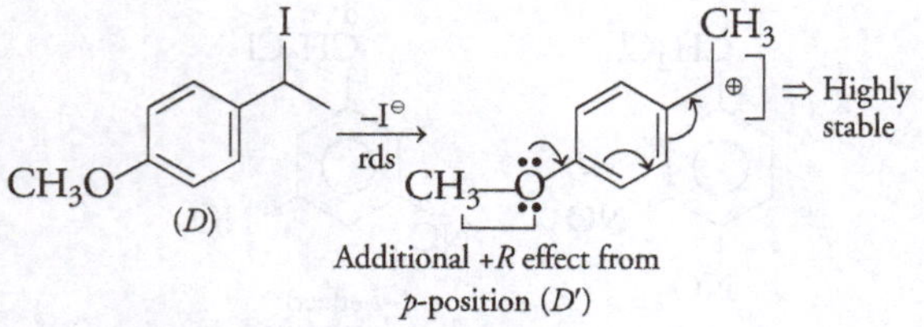


Stability of intermediates will follow the order

$$\underset{(-I)}{B'} < \underset{(-\text{HPC}=3)}{A'} < \underset{(+\text{HPC}=6)}{C'} < \underset{\left(\substack{+\text{HPC}=3\\ +R=1}\right)}{D'}$$

So, the rate of $S_N1$ reaction with the given compounds will be, $B < A < C < D$

**35.** *(93)* The complete reaction is as follows

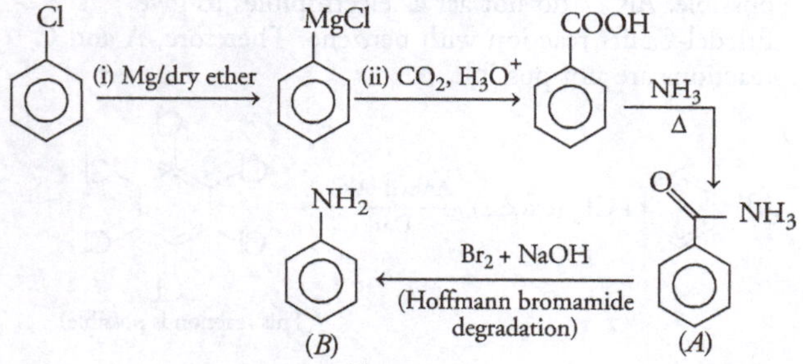


Number of moles of 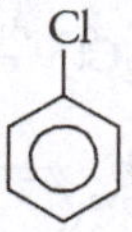 = Number of moles of 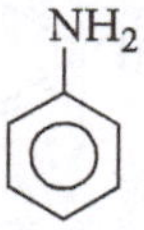

$$\frac{11.25 \times 10^{-3}}{112.5} = \frac{x \times 10^{-1} \times 10^{-3}}{93}$$

$$x = 93 \text{ mg}$$

**36.** *(2)* In Sandmeyer's reaction only bromobenzene and chlorobenzene are prepared.

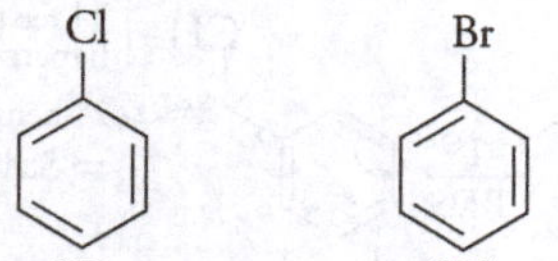


Hence, the number of halobenzene is 2.

**37.** *(6)*

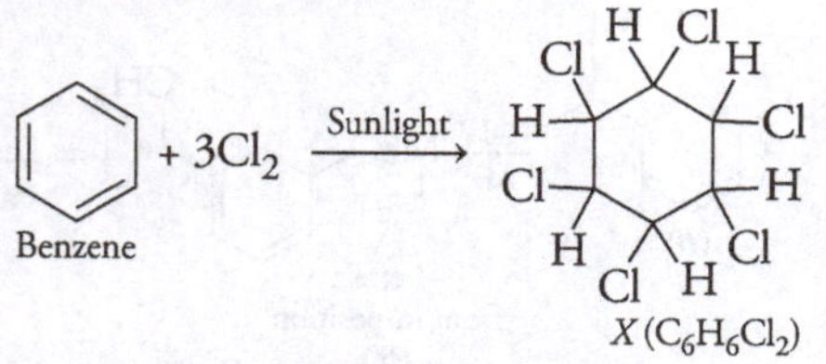


Number of hydrogens present in product ($X$) is 6.

**38.** *(1)* Bromine in presence of sunlight is added to the allylic $sp^3$ carbon atom. This reaction proceeds *via* free radical substitution reaction. As there is only one allylic position(*). So, only one Br is present in product.

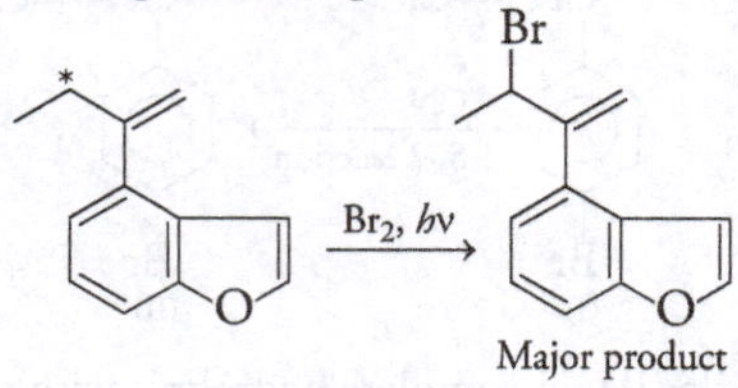


## Topic 3

**01.** *(a)* Correct option is (a) gammaxene

| | | |
|---|---|---|
| Chloral | 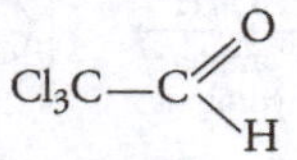 | 3 Cl-atoms |
| Gammaxene | 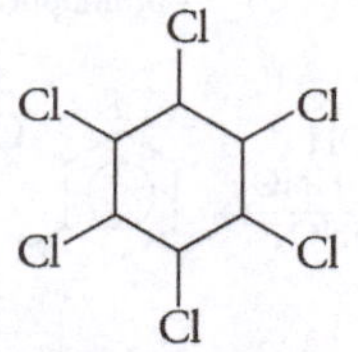 | 6 Cl-atoms |
| Chloropicrin | 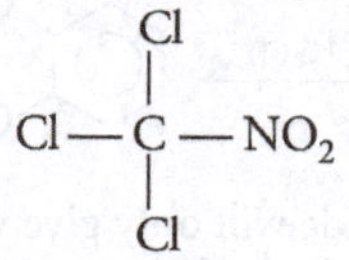 | 3 Cl-atoms |
| Freon-12 | 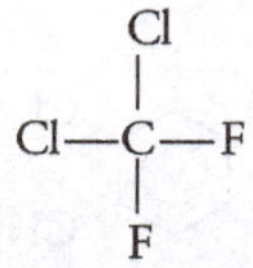 | 2 Cl-atoms |

**02.** *(c)*

$$\underset{\text{1,1,1-trichloroethane}}{H_3C-CCl_3} + 6Ag + Cl_3C-CH_3 \longrightarrow \underset{\text{But-2-yne}}{H_3C-C\equiv C-CH_3} + 6AgCl$$

Product formed is 2-butyne.

**03.** *(c)* Trichloro acetaldehyde reacts with chlorobenzene in the presence of sulphuric acid and gives DDT (dichloro-diphenyl trichloro ethane) which is a powerful insecticide.

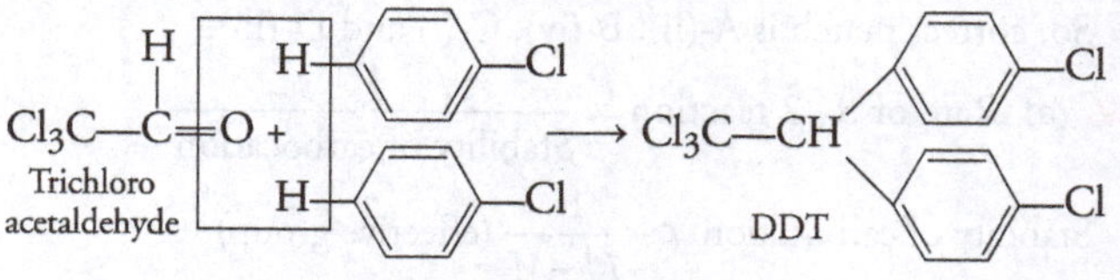


**04.** *(d)* Propionic acid is the main product of the reaction of 1,1,1-trichloropropane with aqueous potassium hydroxide. Alcoholic KOH leads to substitution while aqueous KOH causes elimination in the haloalkene molecule generally.

$$CH_3CH_2CCl_3 + 3KOH(aq) \longrightarrow CH_3CH_2C(OH)_3 + 3KCl$$

$$\underset{\text{(Unstable)}}{[CH_3CH_2C(OH)_3]} \xrightarrow{-H_2O} \underset{\text{Propionic acid}}{CH_3CH_2COOH}$$

$$(CH_3CH_2COOK)$$

**05.** *(c)* DDT stands for dichlorodiphenyltrichloro ethane. It is non-biodegradable pollutant, it gets saturated in the environment.

**06.** *(b)* The complete reaction sequence is as follows

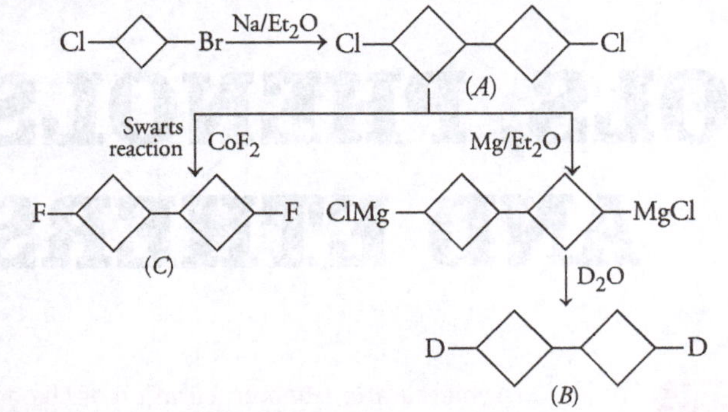


**07.** *(c)* Iodoform test should be given by methyl ketones and alcohols of the type $R$—CH(OH)CH$_3$ where $R$ can be hydrogen also. So, isobutyl alcohol [$(CH_3)_2CHCH_2OH$] does not give the iodoform test.

**08.** *(d)* When *geminal* dihalides are heated in presence of KOH, they form unstable diols.

$$\underset{\text{1,1-Dichloroethane}}{CH_3CH\begin{matrix}Cl\\Cl\end{matrix}} + 2KOH(aq) \xrightarrow{\Delta} \underset{\text{Unstable diol}}{\left[CH_3CH\begin{matrix}OH\\OH\end{matrix}\right]} + 2KCl$$

Diols are highly stable, so undergoes self-rearrangement to remove water molecules.

$$\underset{\text{Unstable diol}}{\left[CH_3CH\begin{matrix}OH\\OH\end{matrix}\right]} \longrightarrow \underset{\text{Acetaldehyde}}{CH_3-CHO} + H_2O$$

**09.** *(b)*

> **Time Saver Tech**
> Use elimination technique as time saver. DDT with insectide (D-II) is given in option (b) only.

The correct match is A-III, B-I, C-IV, D-II.

- Iodoform is used as antiseptic.
- Carbon tetrachloride is used in fire extinguisher.
- CFC are used in refrigerants.
- DDT is used as insecticide in farming.

**10.** *(b)* The correct match is A-III, B-I, C-IV, D-II.

| Compound | Use |
|---|---|
| (A) Carbon tetrachloride | (III) Fire extinguisher |
| (B) Methylene chloride | (I) Paint remover |
| (C) DDT | (IV) Non-biodegradable insecticide |
| (D) Freons | (II) Refrigerators and air conditioners |

# C-20

JEE Main Chapterwise Topicwise Chemistry

In JEE Main there is usually 0-2 questions per paper are visible from this chapter. In past 3 years, 47 conceptual and 2 numerical questions were asked of moderate difficulty level from this chapter.
**For detailed analysis scan QR code**

# ALCOHOLS, PHENOLS AND ETHERS

## Topic 1 Preparation, Properties and Uses of Alcohols

### Level 1

**01** Incorrect method of preparation of alcohols from the following is *[2023, 10 April Shift-II]*
(a) reaction of ketone with *R*MgBr followed by hydrolysis.
(b) hydroboration-oxidation of alkene.
(c) ozonolysis of alkene.
(d) reaction of alkyl halide with aqueous NaOH.

**02** The water gas on reacting with cobalt as a catalyst forms *[2023, 8 April Shift-I]*
(a) methanoic acid (b) ethanol
(c) methanal (d) methanol

**03** Given below are two statements: One is labelled as Assertion A and the other is labelled as Reason R.
**Assertion (A)** Butan-1-ol has higher boiling point than ethoxyethane.
**Reason (R)** Extensive hydrogen bonding leads to stronger association of molecules. *[2023, 8 April, Shift-I]*
In the light of the above statements, choose the correct answer from the options given below
(a) A is true but R is false
(b) Both A and R are true but R is not the correct explanation of A
(c) Both A and R are true and R is the correct explanation of A
(d) A is false but R is true

**04** A solution of $CrO_5$ in amyl alcohol has a ........ colour. *[2023, 29 Jan Shift-I]*
(a) orange-red (b) yellow
(c) green (d) blue

**Imp.** Oxymercuration-Demerucuration of Alkenes

**05** The major product in the following reaction

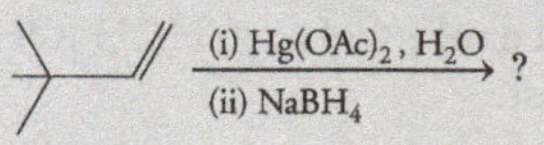

*[2022, 26 June Shift-II]*

(a) 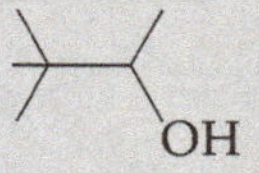 (b) 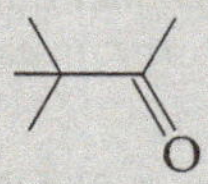

(c) 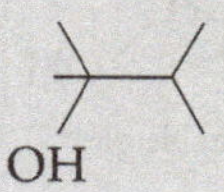 (d) 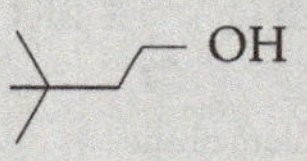

**06** What is '*X*' in the given reaction?

$$\begin{matrix} CH_2OH \\ | \\ CH_2OH \end{matrix} + \text{oxalic acid} \xrightarrow{210^\circ C} \underset{\text{Major product}}{X}$$

*[2021, 25 Feb Shift-II]*

(a) $\begin{matrix} CHO \\ | \\ CHO \end{matrix}$ (b) $\begin{matrix} CH_2 \\ \| \\ CH_2 \end{matrix}$ (c) $\begin{matrix} CHOH \\ \| \\ CH_2 \end{matrix}$ (d) $\begin{matrix} CH_2OH \\ \| \\ CHO \end{matrix}$

**07** The major product of the following reaction is

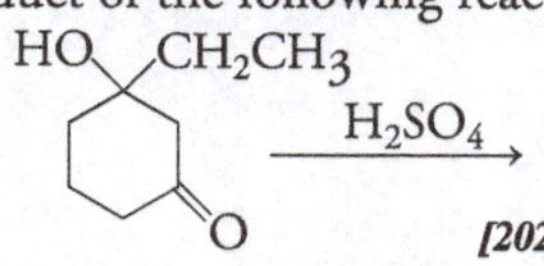

*[2020, 5 Sep Shift-II]*

(a) 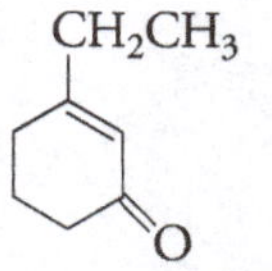 (b) 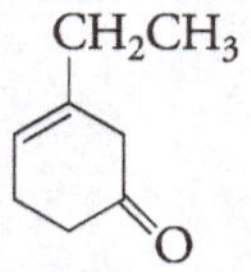

(c) 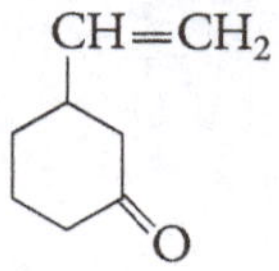 (d) 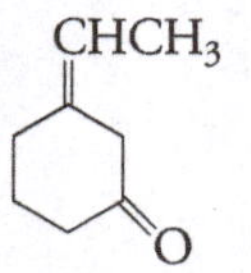

**08** When neopentyl alcohol is heated with an acid, it is slowly converted into an 85 : 15 mixture of alkenes *A* and B, respectively. What are these alkenes? **[2020, 04 Sep Shift-I]**

(a) 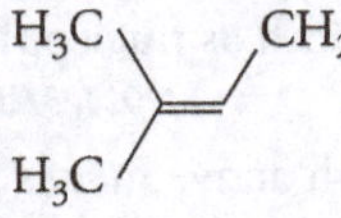 and 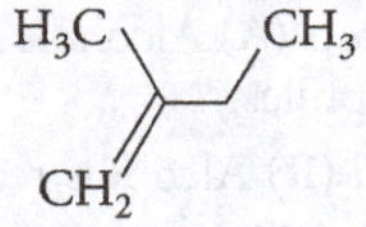

(b) 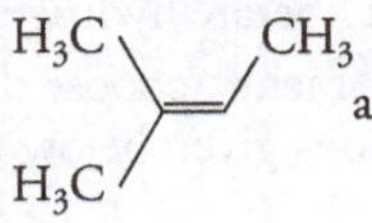 and 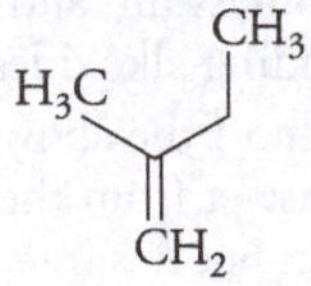

(c) 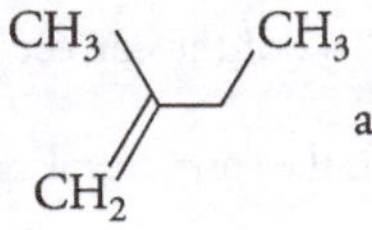 and 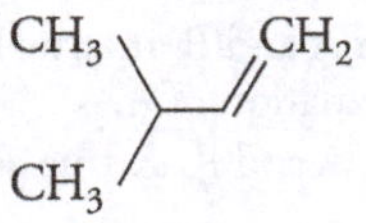

(d) 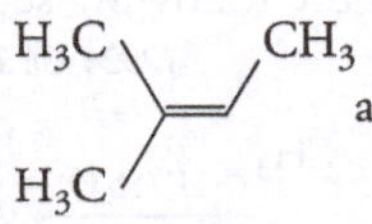 and 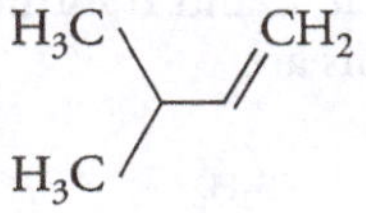

**09** Consider the following reaction

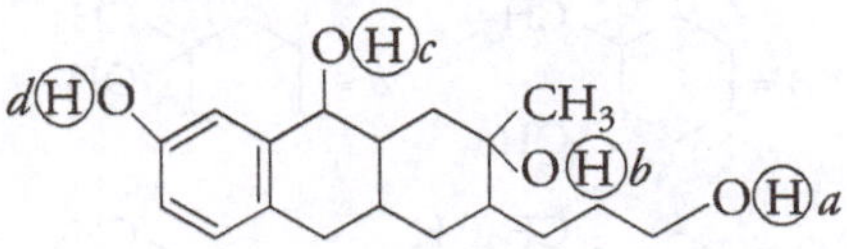


$\xrightarrow[\text{anhydride}]{\text{Chromic}}$ '*P*' The product '*P*' gives positive ceric ammonium nitrate test. This is because of the presence of which of these —OH group(s)? **[2020, 3 Sep Shift-II]**

(a) (*b*) and (*d*) (b) (*d*) only
(c) (*b*) only (d) (*c*) and (*d*)

**10** $CH_3CH_2—\underset{Ph}{\overset{OH}{C}}— CH_3$ cannot be prepared by **[2019, 12 Jan Shift-I]**

(a) $CH_3CH_2COCH_3 + PhMgX$
(b) $PhCOCH_3 + CH_3CH_2MgX$
(c) $PhCOCH_2CH_3 + CH_3MgX$
(d) $HCHO + PhCH(CH_3)CH_2MgX$

**11** The most suitable reagent for the conversion of $R—CH_2—OH \rightarrow R—CHO$ is **[JEE Main 2014]**

(a) $KMnO_4$
(b) $K_2Cr_2O_7$
(c) $CrO_3$
(d) PCC [Pyridinium chlorochromate]

**12** An unknown alcohol is treated with the 'Lucas reagent' to determine whether the alcohol is primary, secondary or tertiary. Which alcohol reacts fastest and by what mechanism? **[JEE Main 2013]**

(a) Secondary alcohol by $S_N1$
(b) Tertiary alcohol by $S_N1$
(c) Secondary alcohol by $S_N2$
(d) Tertiary alcohol by $S_N2$

**13** From amongst the following alcohols the one that would react fastest with conc. HCl and anhydrous $ZnCl_2$ is **[AIEEE 2010]**

(a) 2-butanol (b) 2-methylpropan-2-ol
(c) 2-methylpropanol (d) 1-butanol

**14** Among the following the one that gives positive iodoform test upon reaction with $I_2$ and NaOH is **[AIEEE 2006]**

(a) $CH_3CH_2CH(OH)CH_2CH_3$
(b) $C_6H_5CH_2CH_2OH$
(c) 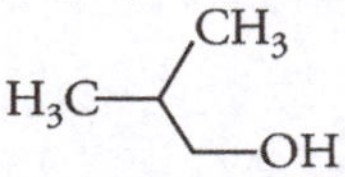

(d) $PhCHOHCH_3$

**15** Phenyl magnesium bromide reacts with methanol to give **[AIEEE 2006]**

(a) a mixture of anisole and Mg(OH)Br
(b) a mixture of benzene and Mg(OMe)Br
(c) a mixture of toluene and Mg(OH)Br
(d) a mixture of phenol and Mg(Me)Br

**16** Acid catalysed hydration of alkenes except ethene leads to the formation of **[AIEEE 2005]**

(a) mixture of secondary and tertiary alcohols
(b) mixture of primary and secondary alcohols
(c) secondary or tertiary alcohol
(d) primary alcohol

**17** Maximum dehydration takes place in that of **[AIEEE 2002]**

(a) 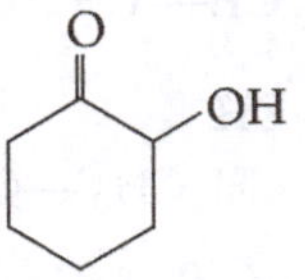

(b) 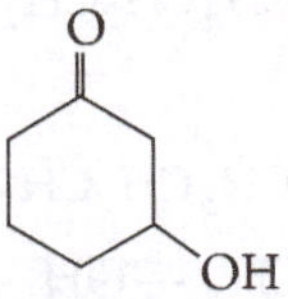

(c) 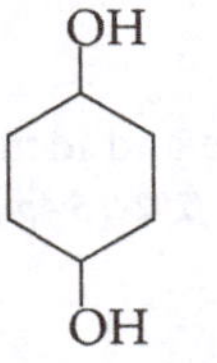

(d) 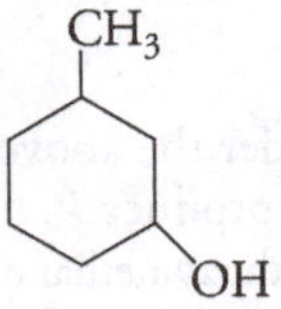

## Level 2

**18** Identify the products [A] and [B], respectively in the following reaction

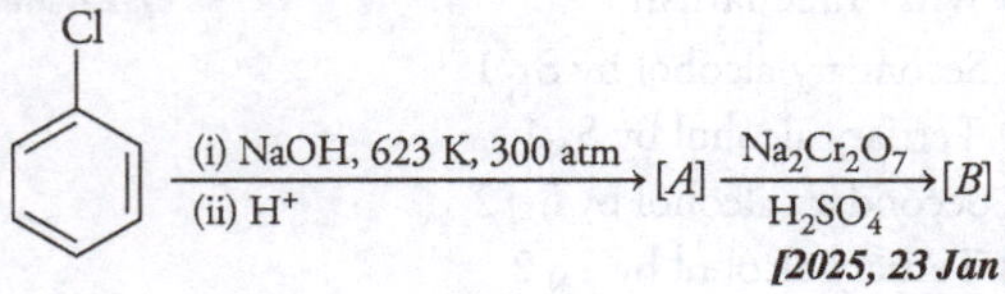

[2025, 23 Jan Shift-II]

(a) (A) 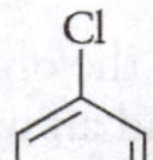, (B) 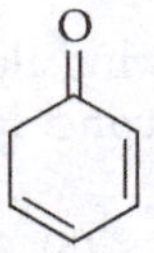

(b) (A) 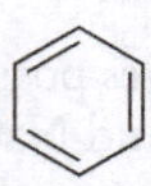, (B) 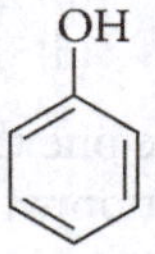

(c) (A) 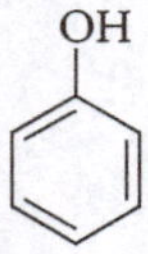, (B) 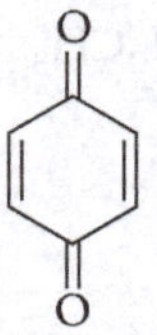

(d) (A) 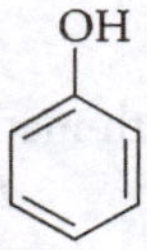, (B) 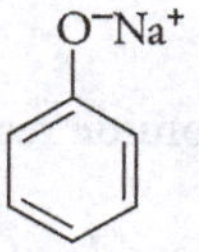

**19** Identify the product *A* and product *B* in the following set of reactions. [2024, 9 April Shift-I]

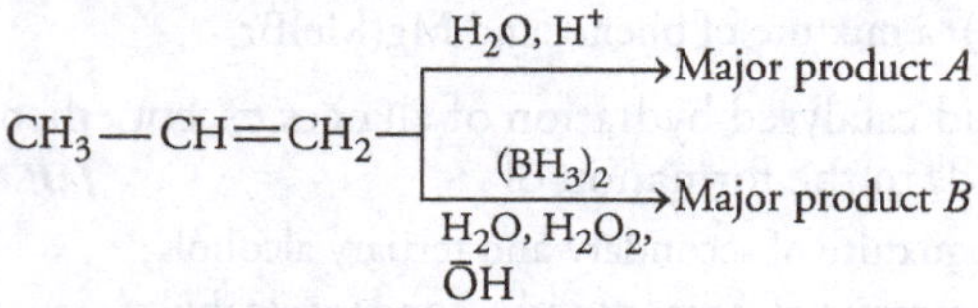

(a) $A$-$CH_3CH_2CH_3$, $B$-$CH_3CH_2CH_3$

(b) $A$-$CH_3-\underset{\displaystyle OH}{\underset{|}{C}H}-CH_3$, $B$-$CH_3CH_2CH_2-OH$

(c) $A$-$CH_3CH_2CH_2-OH$, $B$-$CH_3\underset{\displaystyle OH}{\underset{|}{C}H}-CH_3$

(d) $A$-$CH_3CH_2CH_2-OH$, $B$-$CH_3CH_2CH_2-OH$

**20** $CH_3CH_2-OH \xrightarrow[\text{(ii) } KMnO_4 \text{ (iii) NaOH, CaO, } \Delta]{\text{(i) Jone's reagent}} P$

Consider the above reaction sequence and identify the major product *P*. [2024, 5 April Shift-II]

(a) Methoxymethane
(b) Methanoic acid
(c) Methane
(d) Methanal

**Imp.** Properties of Alcohols

**21** Given below are two statements : one is labelled as **Assertion** (A) and the other is labelled as **Reason** (R).

**Assertion** (A) Alcohols react both as nucleophiles and electrophiles. [2024, 31 Jan Shift-I]

**Reason** (R) Alcohols react with active metals such as sodium, potassium and aluminium to yield corresponding alkoxides and liberate hydrogen.

In the light of the above statements, choose the correct answer from the options given below.

(a) A is false but R is true.
(b) A is true but R is false.
(c) Both A and R are true but R is not the correct explanation of A.
(d) Both A and R are true and R is the correct explanation of A.

**22** Products *A* and *B* formed in the following set of reactions are [2024, 30 Jan Shift-II]

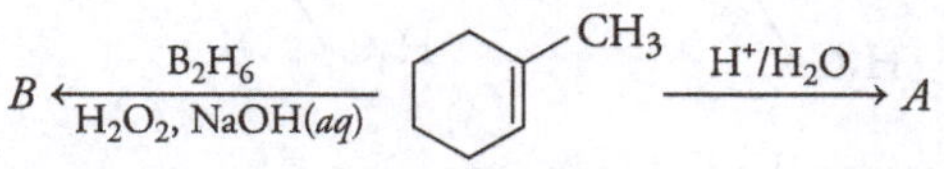

(a) $A$ = 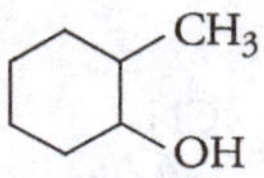 $B$ = 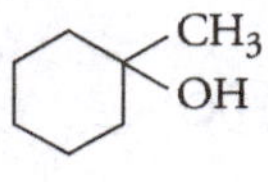

(b) $A$ = 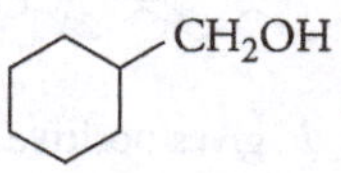 $B$ = 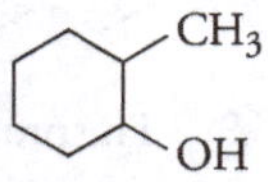

(c) $A$ = 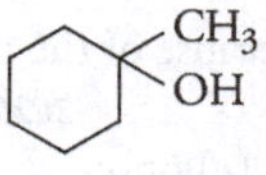 $B$ = 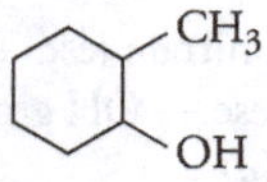

(d) $A$ = 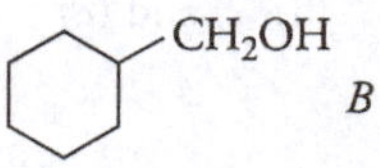 $B$ = 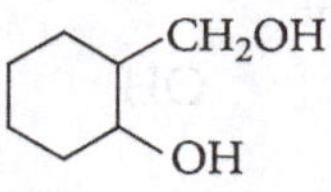

**23** The major product for the following reaction is

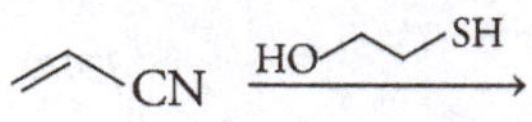

[2023, 13 April Shift-II]

(a) 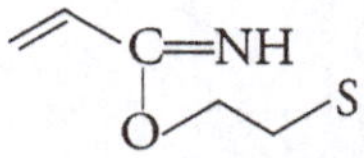 (b) 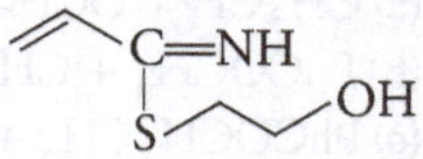

(c) 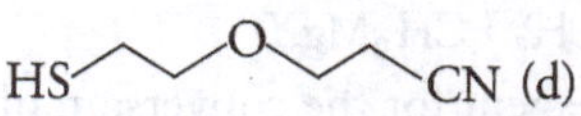 (d) 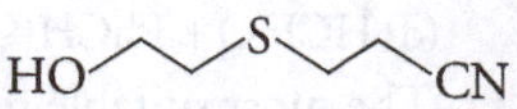

**24** Given below are two statements, one is labelled as Assertion (A) and the other is labelled as Reason (R).

**Assertion** (A) Order of acidic nature of the following compounds is A > B > C.

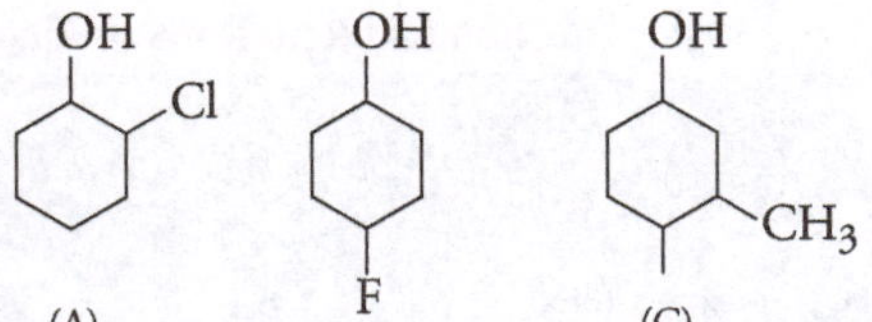

(A) (B) (C) *[2023, 13 April Shift-II]*

**Reason** (R) Fluoro is a stronger electron withdrawing group than chloro group.

In the light of the above statements, choose the most appropriate answer from the options given below.

(a) A is true but R is false
(b) Both A and R are correct and R is the correct explanation of A
(c) Both A and R are correct but R is not the correct explanation of A
(d) A is false but R is true

**25** $H_3C—CH_2—CH(OH)—CH_3 \xrightarrow[\text{(iii) } D_2O]{\text{(i) NaI, } H_3PO_4 \text{ (ii) Mg, Dry ether}} [X]$ Product

Product $[X]$ formed in the above reaction is *[2023, 11 April Shift-II]*

(a) $H_3C—CH=CH—CH_3$
(b) $H_3C—CH_2—CH=CH_2$
(c) $H_3C—CH_2—CH(D)—CH_3$
(d) $H_3C—CH_2—C(H)(OH)—CH_3$

**26** Decreasing order of dehydration of the following alcohols is *[2023, 1 Feb Shift-II]*

(i) 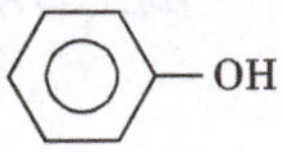 (ii) 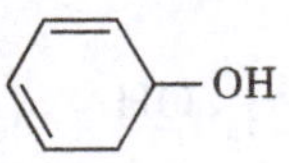

(iii) 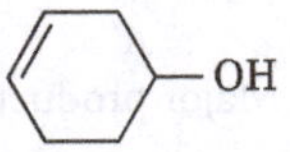 (iv) 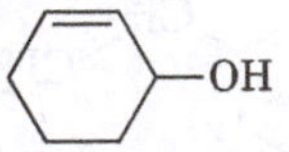

(a) (iv) > (ii) > (iii) > (i)
(b) (i) > (iv) > (ii) > (iii)
(c) (ii) > (i) > (iv) > (iii)
(d) (ii) > (iv) > (iii) > (i)

**27** Find out the major product from the following reaction.

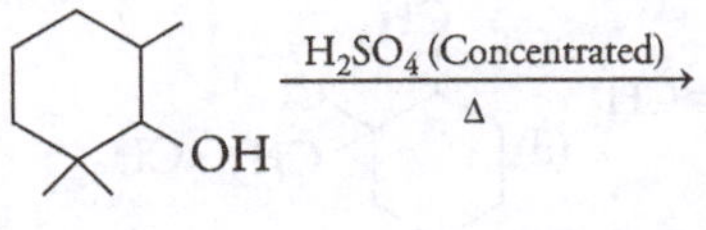

*[2023, 25 Jan Shift-II]*

(a) 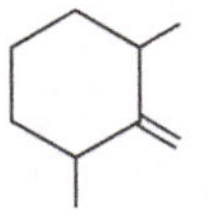 (b) 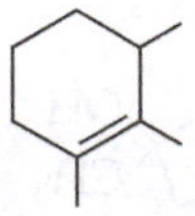

(c) 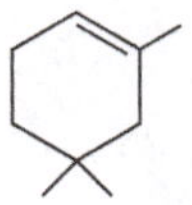 (d) 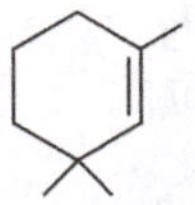

**28** When ethanol is heated with conc. $H_2SO_4$, a gas is produced. The compound formed, when this gas is treated with cold dilute aqueous solution of Baeyer's reagent, is *[2022, 29 July Shift-II]*

(a) formaldehyde
(b) formic acid
(c) glycol
(d) ethanoic acid

**29** Which reactant will give the following alcohol on reaction with one mole of phenyl magnesium bromide (PhMgBr) followed by acidic hydrolysis? *[2022, 26 July Shift-I]*

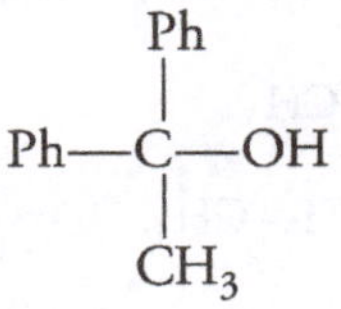

(a) $CH_3—C≡N$ (b) $Ph—C≡N$
(c) $CH_3—C(=O)—O—Ph$ (d) $Ph—C(=O)—CH_3$

**30** The major product ($P$) of the given reaction is (where, Me is $—CH_3$) *[2022, 28 June Shift-II]*

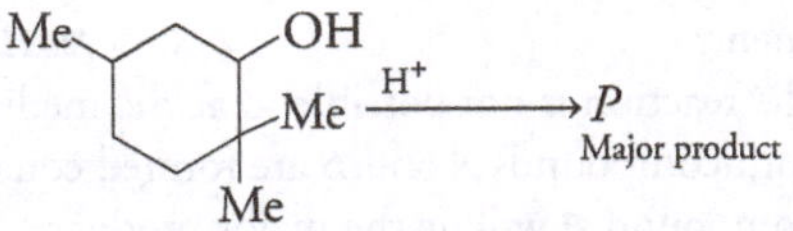

(a) 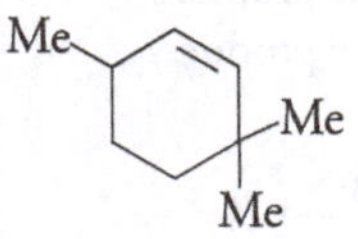 (b) 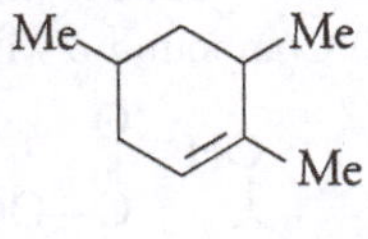

(c) 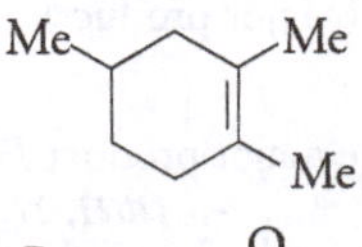 (d) 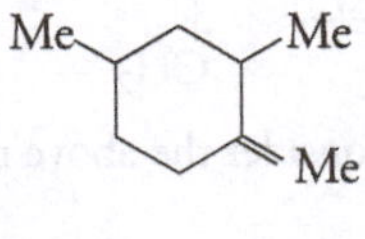

**31**

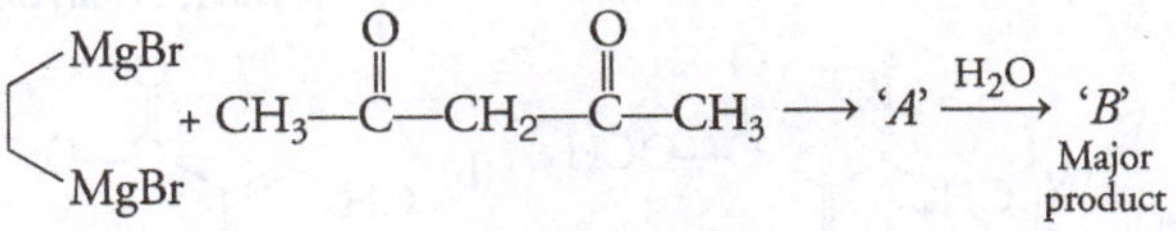

Consider the above reaction sequence and identify the product $B$. *[2022, 26 June Shift-I]*

(a) 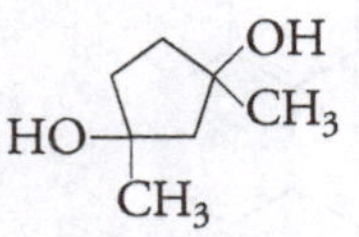 (b) 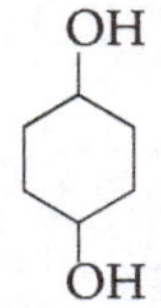

(c) 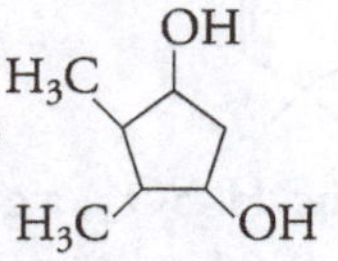 (d) 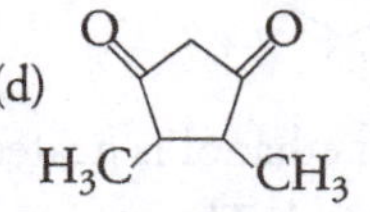

**32** The major product formed in the following reaction is

$$CH_3-\underset{CH_3}{\overset{CH_3}{C}}-\underset{OH}{CH}-CH_3 \xrightarrow[\text{a few drops}]{\text{Conc.}H_2SO_4} \text{Major product}$$

[2021, 31 Aug Shift-I]

(a) $CH_3-\underset{CH_3}{C}=CH-CH_2CH_3$

(b) 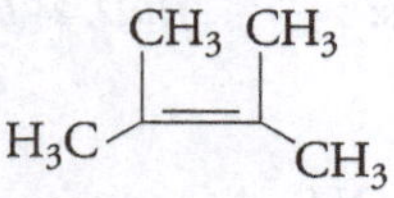

(c) $\overset{CH_3}{\underset{CH_3}{>}}C=CH-CH_3$

(d) $CH_3-\underset{CH_3}{\overset{CH_3}{C}}-CH=CH_2$

**33** 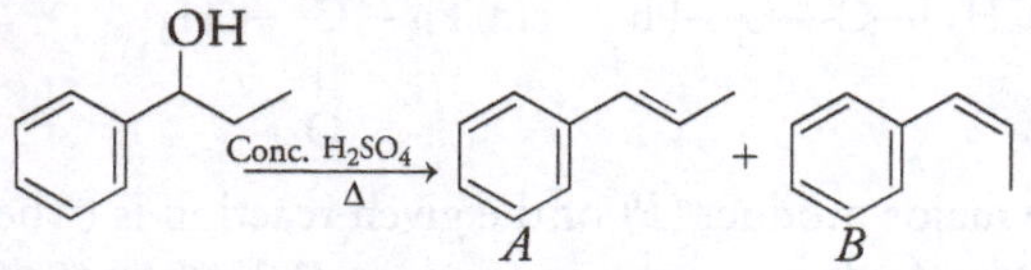

Consider the above reaction, and choose the correct statement. [2021, 27 July Shift-II]

(a) The reaction is not possible in acidic medium.

(b) Both compounds *A* and *B* are formed equally.

(c) Compound *A* will be the major product.

(d) Compound *B* will be the major product.

**34** 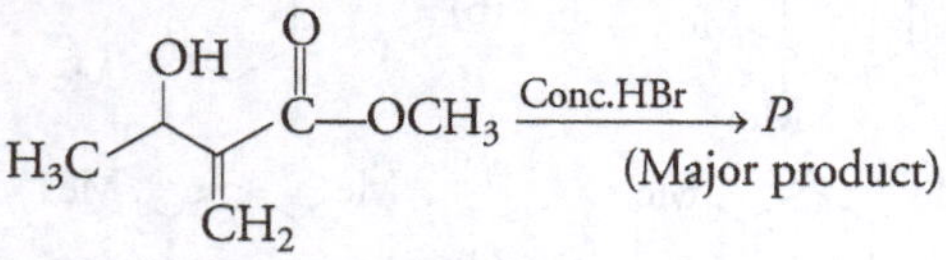

Consider the above reaction, the major product *P* formed is [2021, 27 July Shift-II]

(a) 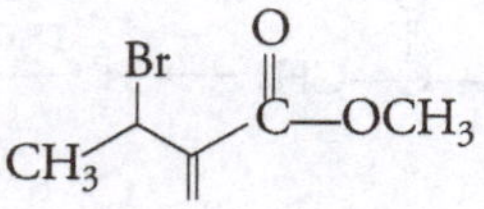 (b) 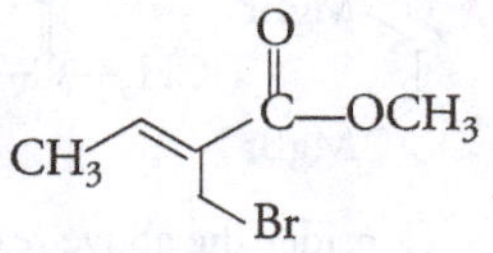

(c) 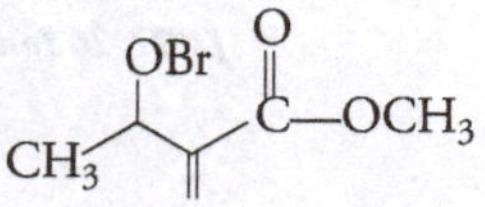 (d) 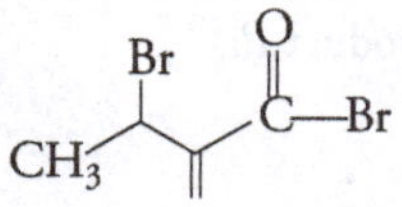

Imp.

## Chemical Reactions of Alcohols

**35** 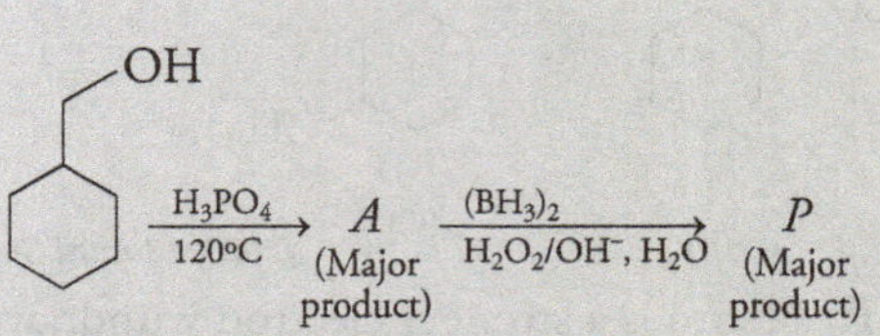

Consider the above reaction and identify the product (*P*). [2021, 27 July Shift-I]

(a) 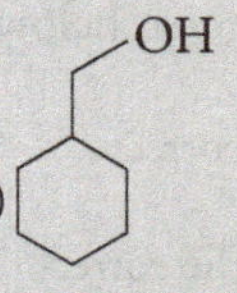 (b) 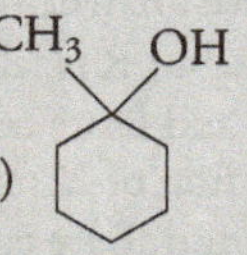

(c) 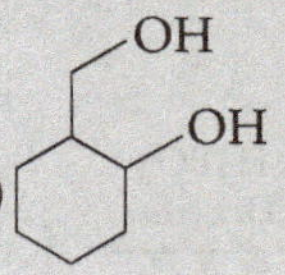 (d) 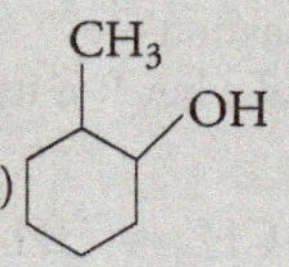

**36** 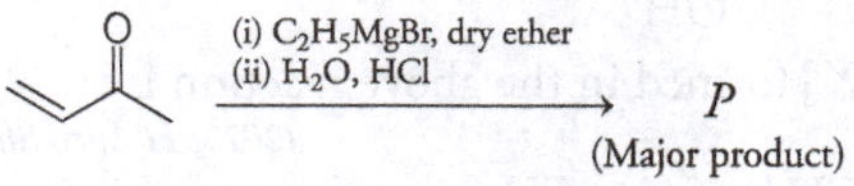

Consider the above reaction, the major product *P* is [2021, 25 July Shift-I]

(a) 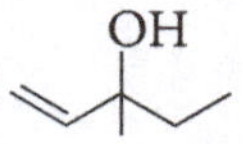 (b) 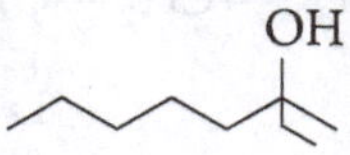

(c) 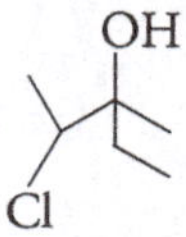 (d) 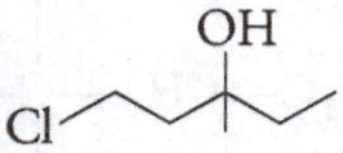

**37** What is the final product (major) '*A*' in the given reaction? [2021, 24 Feb Shift-I]

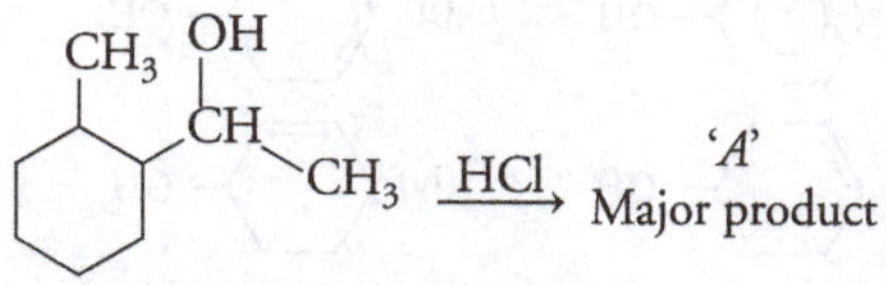

(a) 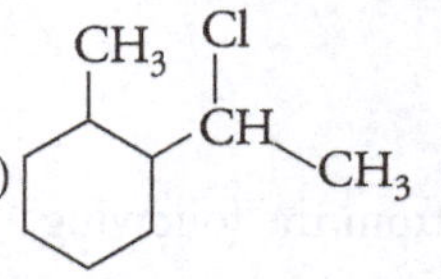 (b) 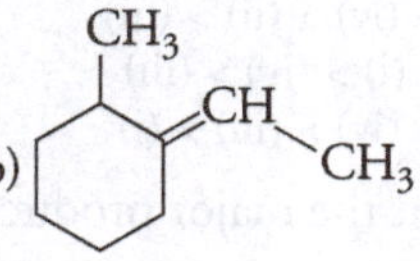

(c) 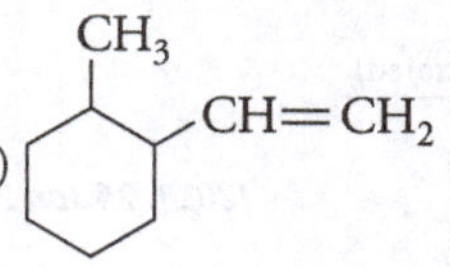 (d) 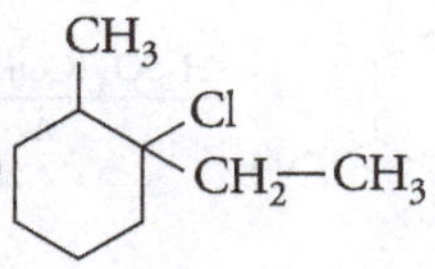

**38** The major product of the following reaction is

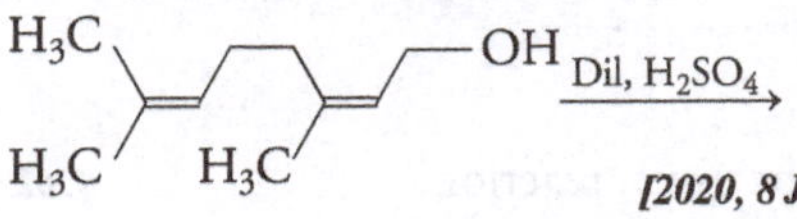


*[2020, 8 Jan Shift-I]*

(a) 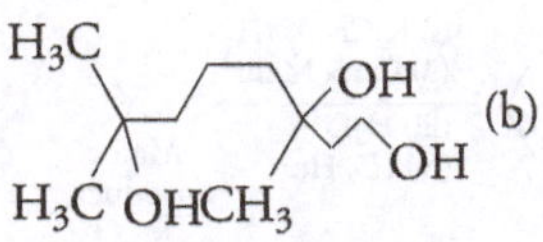


(b) 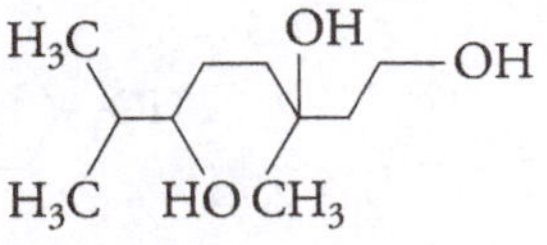


(c) 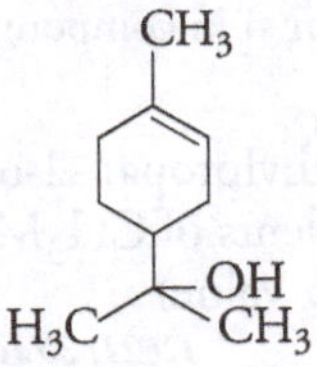


(d) 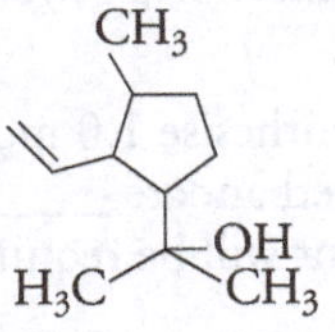


**39** The product of the reaction given below is *[JEE Main 2016]*

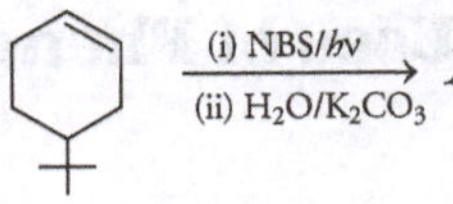


(a) 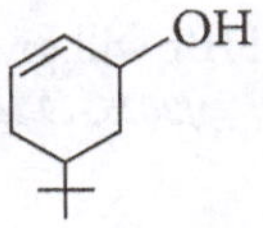


(b) 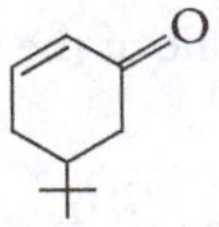


(c) 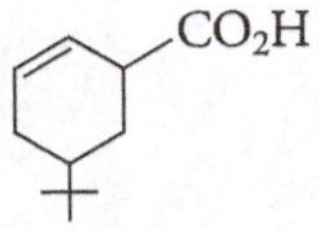


(d) 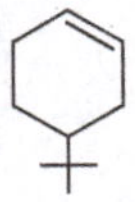

**40** Acetyl bromide reacts with excess of $CH_3MgI$ followed by treatment with a saturated solution of $NH_4Cl$ gives *[AIEEE 2004]*

(a) acetone
(b) acetamide
(c) 2-methyl-2-propanol
(d) acetyl iodide

## Level 3

**41.** Given below are two statements: *[2025, 4 April Shift-II]*

**Statement I** Alcohols are formed when alkyl chlorides are treated with aqueous potassium hydroxide by elimination reaction.

**Statement II** In alcoholic potassium hydroxide, alkyl chlorides form alkenes by abstracting the hydrogen from the β-carbon.

In the light of the above statements, choose the most appropriate answer from the options given below:

(a) Both Statement I and Statement II are incorrect
(b) Both Statement I and Statement II are correct
(c) Statement I is incorrect but Statement II is correct
(d) Statement I is correct but Statement II is incorrect

**42** A molecule ("*P*") on treatment with acid undergoes rearrangement and gives ("*Q*"). ("*Q*") on ozonolysis followed by reflux under alkaline condition gives ("*R*"). The structure of ("*R*") is given below.

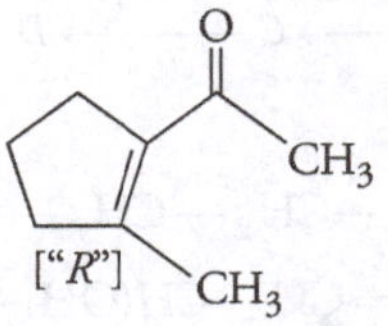


The structure of ("*P*") is *[2025, 28 Jan Shift-I]*

(a) 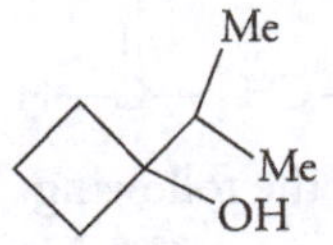


(b) 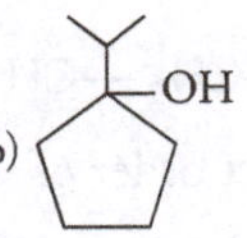


(c) 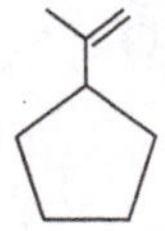

(d) 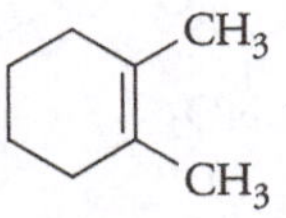


**43** $CH_3—CH_2—CH_2—Br + NaOH \xrightarrow{C_2H_5OH}$ Product '*A*'

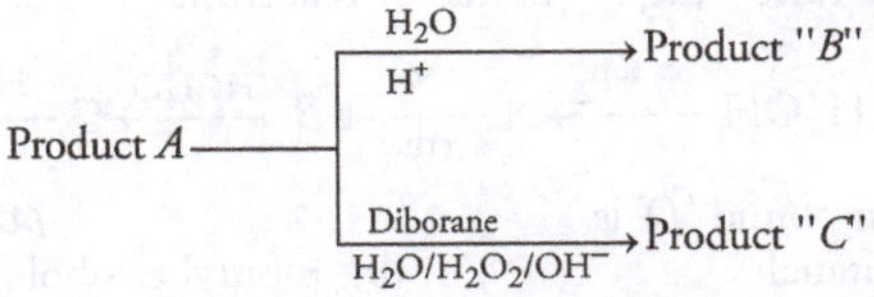


Consider the above reactions, identify product *B* and product *C*. *[2024, 4 April Shift-II]*

(a) $B = C =$ 2- Propanol
(b) $B = C =$1- Propanol
(c) $B =$ 2- Propanol $C =$1- Propanol
(d) $B =$1-Propanol $C =$ 2- Propanol

**44** Most stable product of the following reaction is *[2022, 25 July Shift-I]*

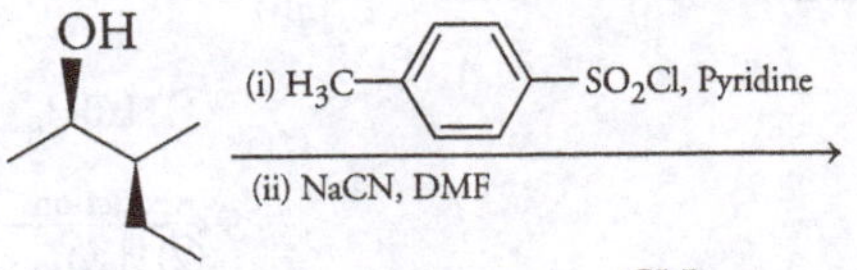


(a) 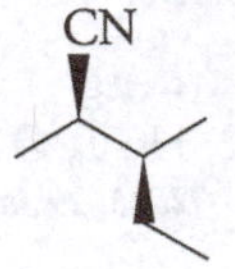


(b) 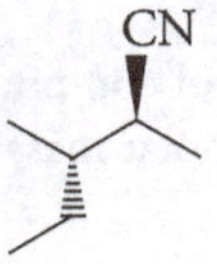


(c) 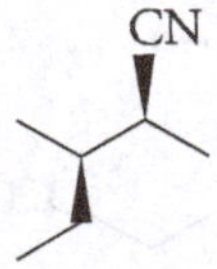


(d) 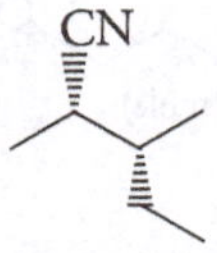

**45** In the following sequence of reactions, the final product $D$ is.

$CH_3—C{\equiv}C—H + NaNH_2 \longrightarrow A$

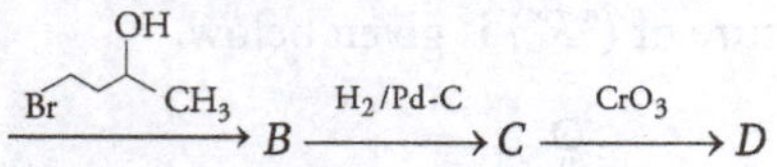

**[2021, 27 Aug Shift-I]**

(a) $CH_3—CH_2—CH_2—CH_2—CH_2—C(=O)—CH_3$

(b) $H_3C—CH=CH—CH(OH)—CH_2—CH_2CH_3$

(c) $CH_3—CH=CH—CH_2—CH_2—CH_2—COOH$

(d) $H_3C—CH_2—CH_2—CH_2—CH_2—C(=O)—H$

**46** The correct order of stability for the following alkoxides is **[2020, 7 Jan Shift-II]**

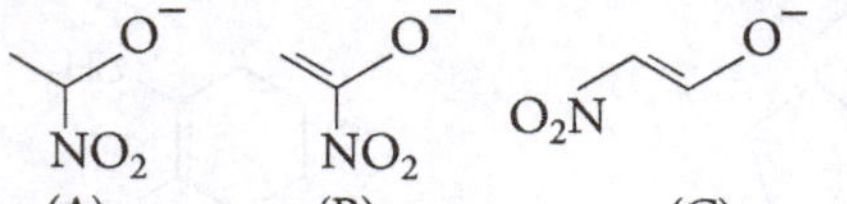

(a) (C) > (B) > (A)  (b) (B) > (C) > (A)
(c) (B) > (A) > (C)  (d) (C) > (A) > (B)

**47** In the following sequence of reactions,

$$CH_3CH_2OH \xrightarrow{P + I_2} A \xrightarrow[\text{ether}]{Mg} B \xrightarrow{HCHO} C \xrightarrow{H_2O} D$$

the compound '$D$' is **[AIEEE 2007]**

(a) butanal  (b) $n$-butyl alcohol
(c) $n$-propyl alcohol  (d) propanal

## Numerical Value Type Questions

**Imp.** Chemical Reactions of Alcohols in Blend with Stoichiometry

**48**

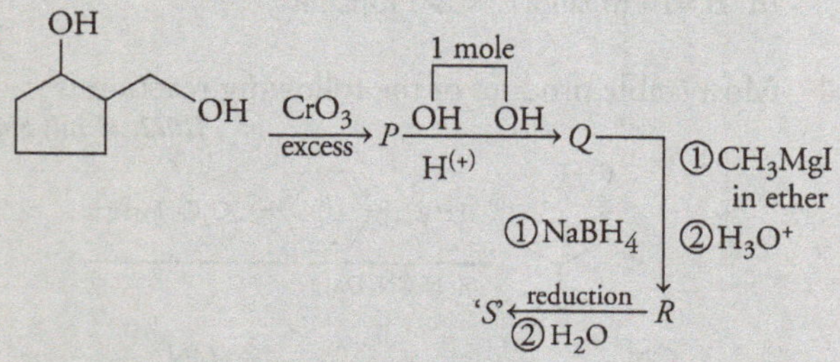

0.1 mole of compound '$S$' will weigh ...... g.
(Given, molar mass in g $mol^{-1}$ C : 12, H : 1, O : 16)
**[2025, 29 Jan Shift-I]**

**49**

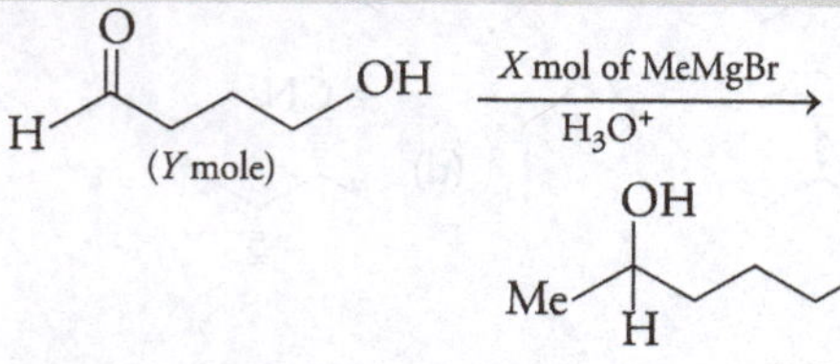

**[2023, 11 April Shift-I]**

The ratio $X / Y$ on completion of the above reaction is ............ .

**50** In the given reaction, **[2022, 29 June Shift-II]**

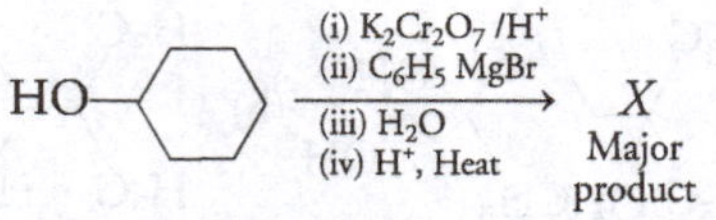

the number of $sp^2$ hybridised carbon(s) in compound '$X$' is ....... .

**51** To synthesise 1.0 mole of 2-methylpropan-2-ol from Ethylethanoate ________ equivalents of $CH_3MgBr$ reagent will be required. (Integer value)
**[2021, 20 July Shift-I]**

# Topic 2 Preparation, Properties and Uses of Phenols

## Level 1

**01** Which one of the following, with HBr will give a phenol? **[2025, 29 Jan Shift-II]**

(a) 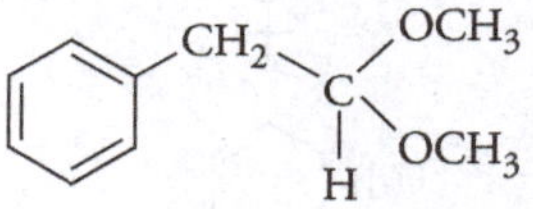

(b) 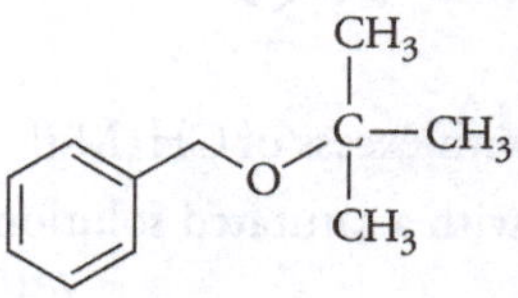

(c) 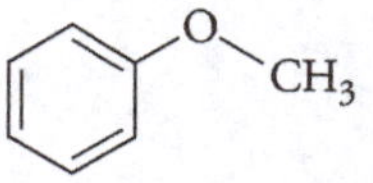

(d) 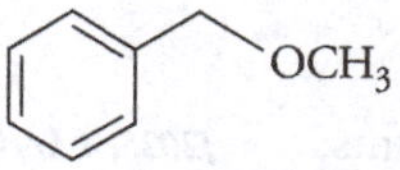

**Imp.** Intermediate Involved in Reimer Tiemann Reaction

**02** In Reimer-Tiemann reaction, phenol is converted into salicylaldehyde through an intermediate. The structure of intermediate is **[2024, 6 April Shift-I]**

(a) 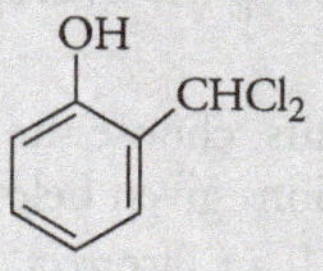
(b) 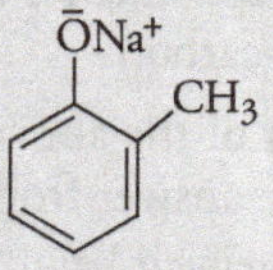
(c) 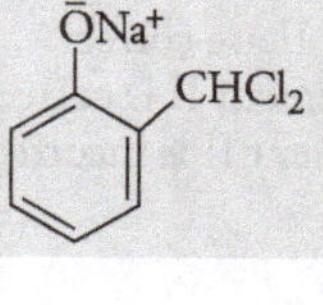
(d) 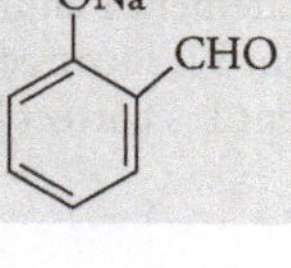

**03** Salicylaldehyde is synthesised from phenol, when reacted with *[2024, 30 Jan Shift-II]*

(a) $CCl_4$, NaOH (b) $HCCl_3$, NaOH

(c) $CO_2$, NaOH (d) 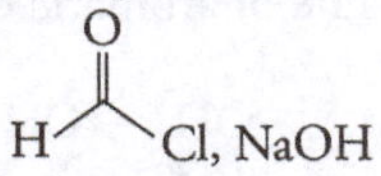

**04** The strongest acid from the following is *[2023, 6 April Shift-II]*

(a) 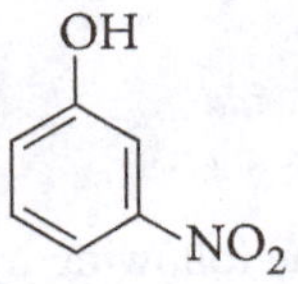 (b) 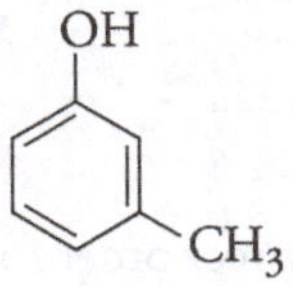

(c) 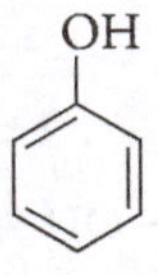 (d) 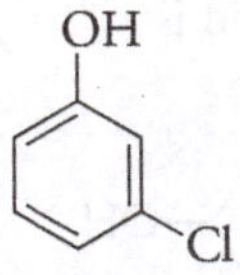

**05** The difference in the reaction of phenol with bromine in chloroform and bromine in water medium is due to *[2022, 26 July Shift-I]*

(a) hyperconjugation in substrate
(b) polarity of solvent
(c) free radical formation
(d) electromeric effect of substrate

**06** Given below are two statements. *[2022, 29 June Shift-I]*

**Statement I** Phenols are weakly acidic.

**Statement II** Therefore they are freely soluble in NaOH solution and are weaker acids than alcohols and water.

Choose the most appropriate option.

(a) Both Statement I and Statement II are correct.
(b) Both Statement I and Statement II are incorrect.
(c) Statement I is correct but Statement II is incorrect.
(d) Statement I is incorrect but Statement II is correct.

**07** The intermediate *X*, in the reaction *[2022, 25 June Shift-I]*

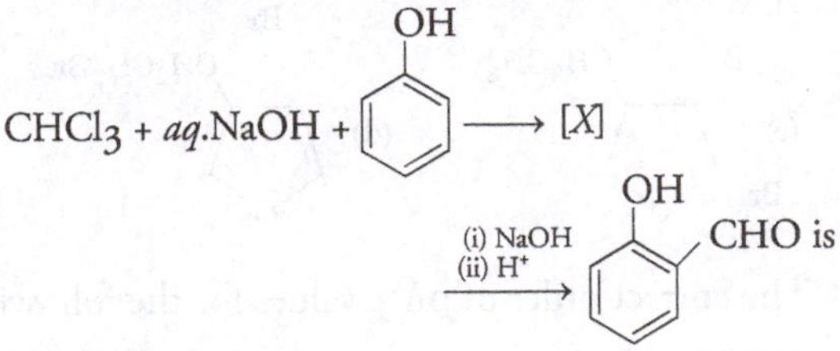

(a) 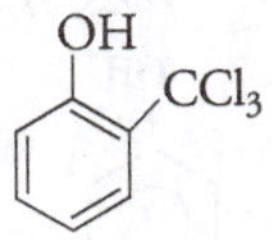 (b) 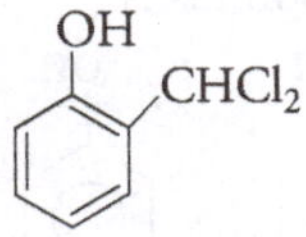

(c) 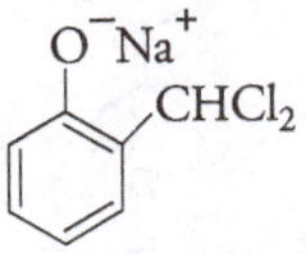 (d) 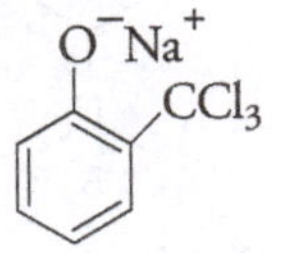

**08** Which one of the following phenols does not give colour when condensed with phthalic anhydride in presence of conc. $H_2SO_4$ ? *[2021, 26 Aug Shift-II]*

(a) 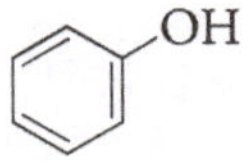 (b) 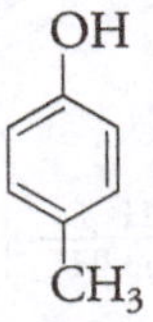

(c) 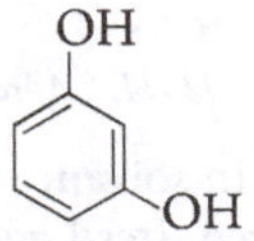 (d) 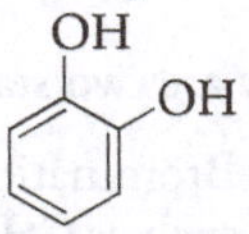

**09** *ortho*-nitrophenol is less soluble in water than *p*- and *m*-nitrophenols because *[AIEEE 2012]*

(a) *o*-nitrophenol is more volatile steam than those of m- and p-isomers.
(b) *o*-nitrophenol shows intramolecular H-bonding
(c) *o*-nitrophenol shows intermolecular H-bonding
(d) melting point of *o*-nitrophenol is lower than those of *m*- and *p*-isomers

**10** Phenol is heated with a solution of mixture of KBr and $KBrO_3$. The major product obtained in the above reaction is *[AIEEE 2011]*

(a) 2-bromophenol (b) 3-bromophenol
(c) 4-bromophenol (d) 2, 4, 6-tribromophenol

**11** Phenol, when it first reacts with concentrated sulphuric acid and then with concentrated nitric acid gives *[AIEEE 2008]*

(a) 2, 4, 6-trinitrobenzene (b) *o*-nitrophenol
(c) *p*-nitrophenol (d) nitrobenzene

## Level 2

**12** Given below are two statetments. *[2024, 6 April Shift-I]*

**Statement I** Picric acid is 2, 4, 6-trinitrotoluene

**Statement II** Phenol-2, 4-disulphonic acid is treated with conc. $HNO_3$, to get picric acid.

In the light of the above statements, choose the most appropriate answer from the options given below.

(a) Both statement I and Statement II are incorrect.
(b) Statement I is incorrect but Statement II is correct.
(c) Both Statement I and Statement II are correct.
(d) Statement I is correct but Statement II is incorrect.

**13** Which one of the following reaction is not possible ? *[2024, 5 April Shift-II]*

(a) 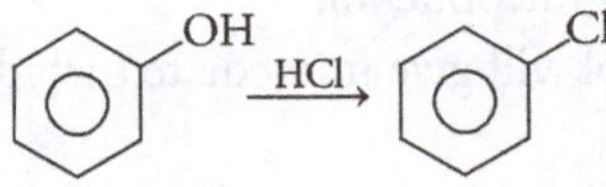

(b) 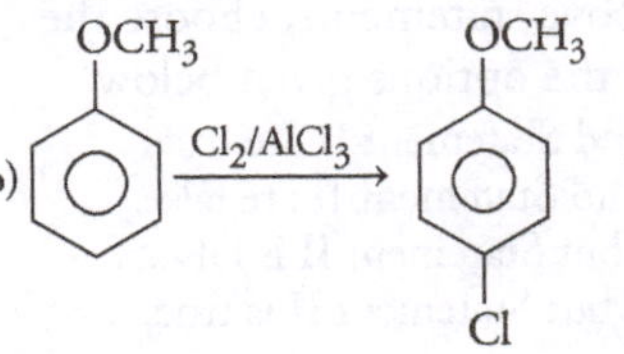

(c) 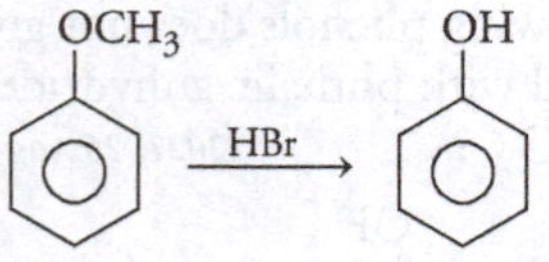

(d) 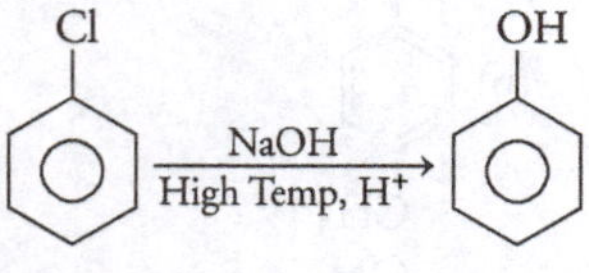

**14** Given below are two statement. ***[2024, 5 April Shift-I]***

**Statement I** Bromination of phenol in solvent with low polarity such as $CHCl_3$ or $CS_2$ requires Lewis acid catalyst.

**Statement II** The lewis acid catalyst polarises the bromine to generate $Br^+$.

In the light of the above statements, choose the correct appropriate answer from the options given below:

(a) Both Statement I and Statement II are false.
(b) Statement I is false but Statement II is true.
(c) Statement I is true but Statement II is false.
(d) Both Statement I and Statement II are true.

**15** The major product (*P*) in the following reaction is ***[2024, 29 April Shift-I]***

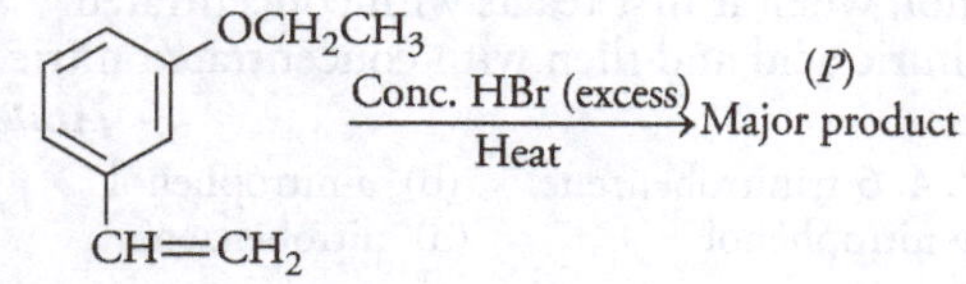

(a) 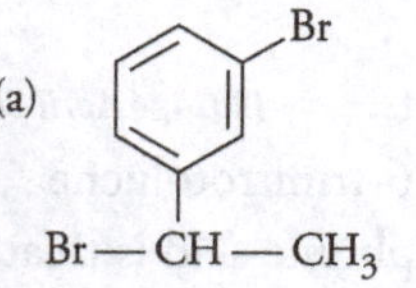 (b) 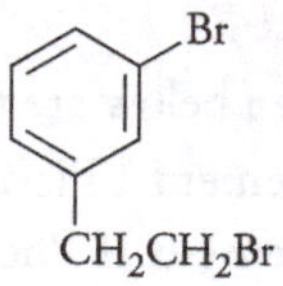

(c) 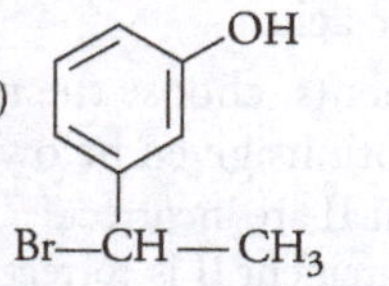 (d) 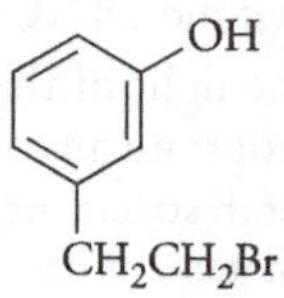

**16** Given below are two statements. ***[2024, 27 Jan Shift-I]***

**Statement I** *p*-nitrophenol is more acidic than *m*-nitrophenol and *o*-nitrophenol.

**Statement II** Ethanol will give immediate turbidity with Lucas reagent.

In the light of the above statements, choose the correct answer from the options given below

(a) Both Statement I and Statement II are true.
(b) Both Statement I and Statement II are false.
(c) Statement I is true but Statement II is false.
(d) Statement I is false but Statement II is true.

**Imp.** Order of Electrophilic Substitution Reaction of Alcohols

**17** Arrange the following compounds in increasing order of rate of aromatic electrophilic substitution reaction ***[2023, 11 April Shift-I]***

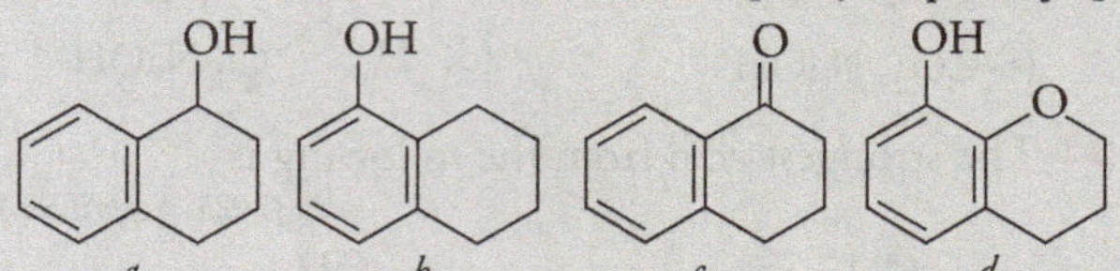

*a* *b* *c* *d*

(a) *d, b, c, a* (b) *b, c, a, d*
(c) *c, a, b, d* (d) *d, b, a, c*

**18** The correct order for acidity of the following hydroxyl compound is

*A.* $CH_3OH$ *B.* $(CH_3)_3COH$

*C.* 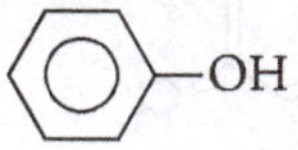 *D.* 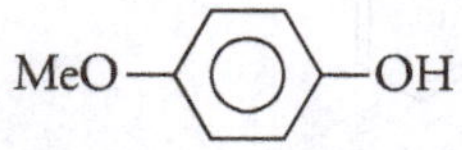

*E.* 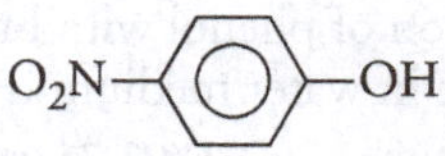

Choose the correct answer from the options given below ***[2023, 10 April Shift-II]***

(a) $C > E > D > B > A$ (b) $E > D > C > B > A$
(c) $E > C > D > A > B$ (d) $D > E > C > A > B$

**19** An organic compound '*A*' with empirical formula $C_6H_6O$ gives sooty flame on burning. Its reaction with bromine solution in low polarity solvent results in high yield of *B*. *B* is ***[2023, 31 Jan Shift-I]***

(a) 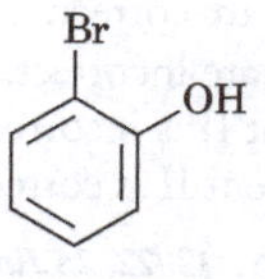 (b) 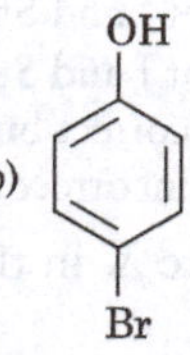

(c) 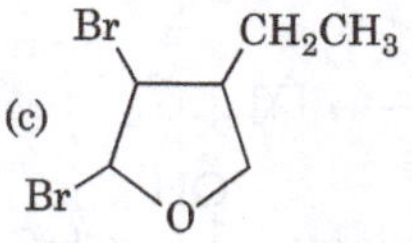 (d) 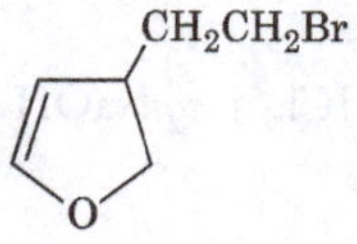

**20** The correct order of p$K_a$ values for the following compounds is ***[2023, 31 Jan Shift-I]***

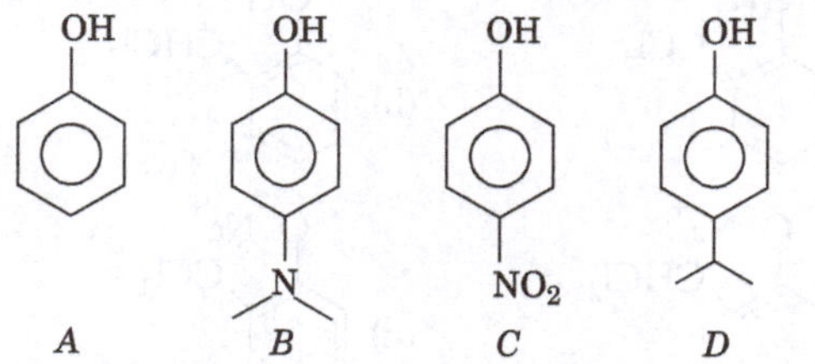

*A* *B* *C* *D*

(a) $B > D > A > C$ (b) $B > A > D > C$
(c) $C > A > D > B$ (d) $A > B > C > D$

**Imp.** Order of p$K_a$ of Phenols

**21** The increasing order of p$K_a$ for the following phenols is *[2023, 29 Jan Shift-I]*

(A) 2, 4 - dinitrophenol (B) 4 - nitrophenol
(C) 2, 4, 5 - trimethylphenol (D) phenol
(E) 3-chlorophenol

Choose the correct answer from the option given below.

(a) A, E, B, D, C (b) A, B, E, D, C
(c) C, D, E, B, A (d) C, E, D, B, A

**22**

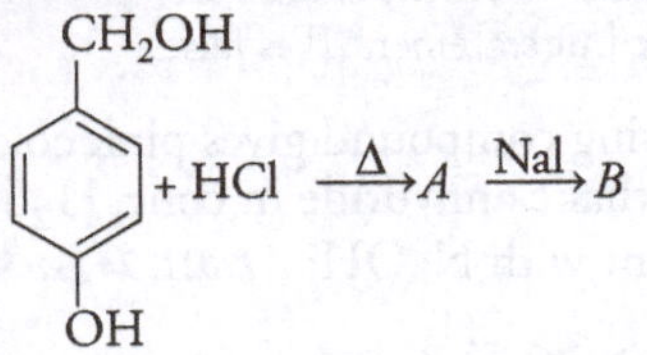

*[2022, 27 July Shift-I]*

(a) 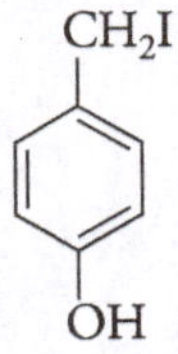 (b) 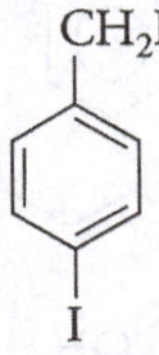

(c) 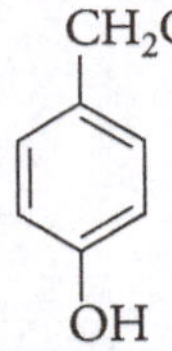 (d) 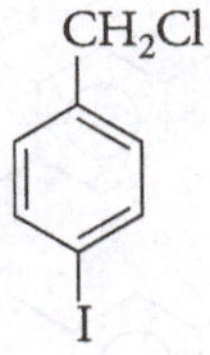

**23** In the given reaction

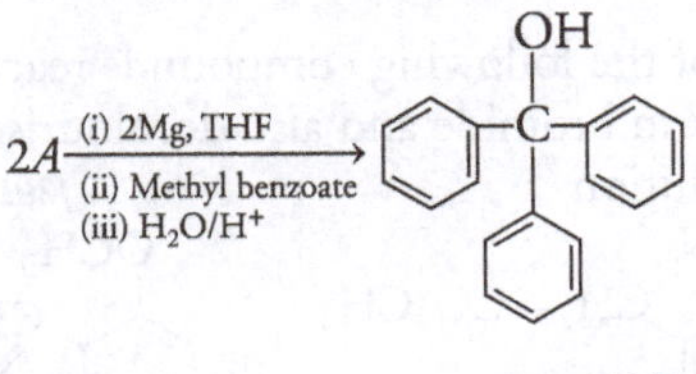

*A* can be *[2022, 25 June Shift-II]*

(a) benzyl bromide
(b) bromobenzene
(c) cyclohexyl bromide
(d) methyl bromide

**24** In the following reaction, *[2022, 25 June Shift-I]*

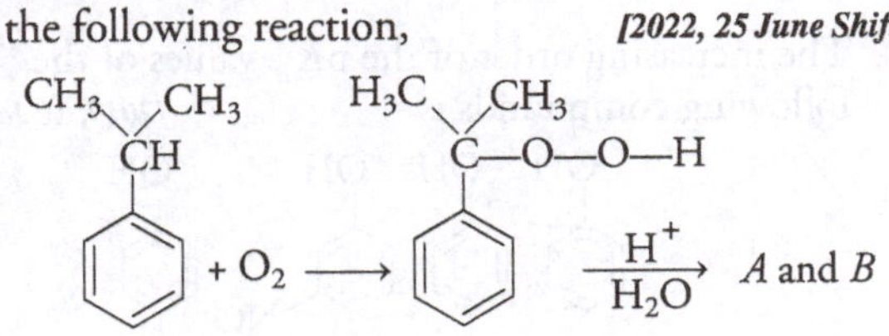

The compounds *A* and *B* respectively are

(a) 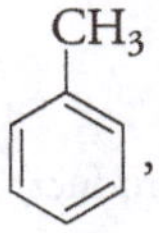, $CH_3COOH$ (b) 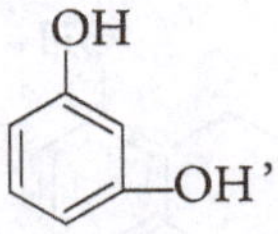, $CH_3COOH$

(c) 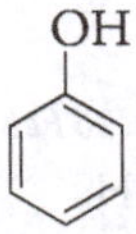, $CH_3COCH_3$ (d) 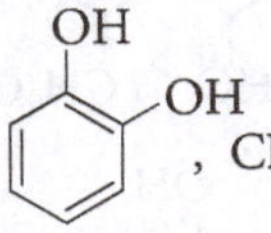, $CH_3COCH_3$

**25** The major product of the following reaction, if it occurs by $S_N2$ mechanism is *[2021, 27 Aug Shift-II]*

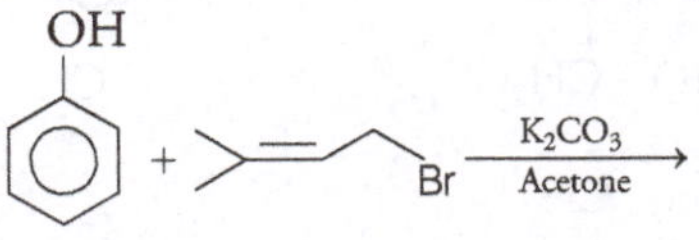

(a) 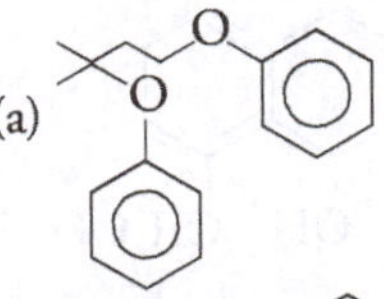 (b) 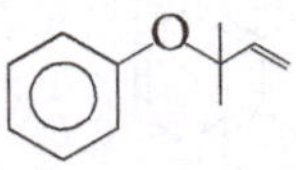

(c) 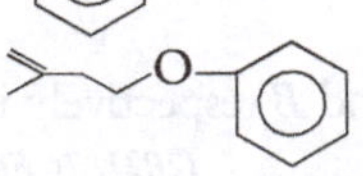 (d) 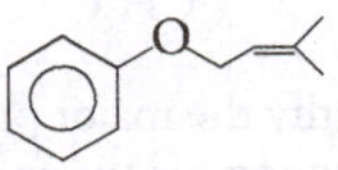

**26** The correct options for the products *A* and *B* of the following reactions are *[2021, 26 Aug Shift-I]*

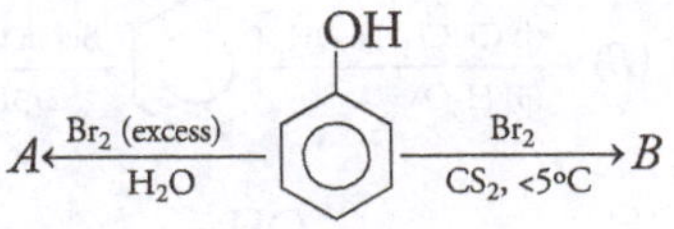

(a) *A* = 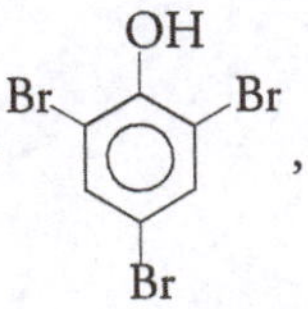, *B* = 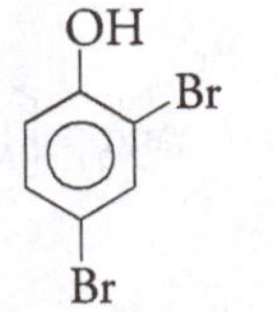

(b) *A* = 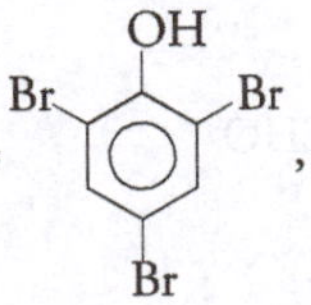, *B* = 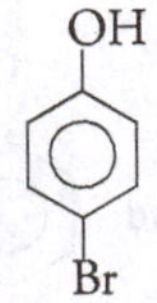

(c) *A* = 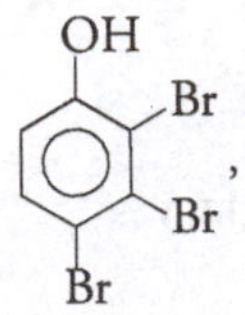, *B* = 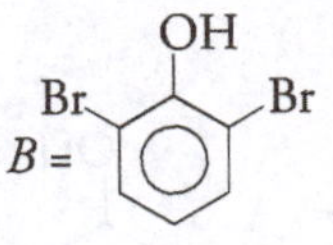

(d) *A* = 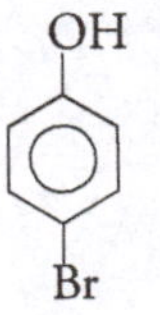, *B* = 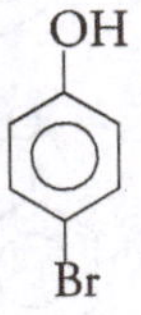

**27** Identify $A$ in the given reaction.

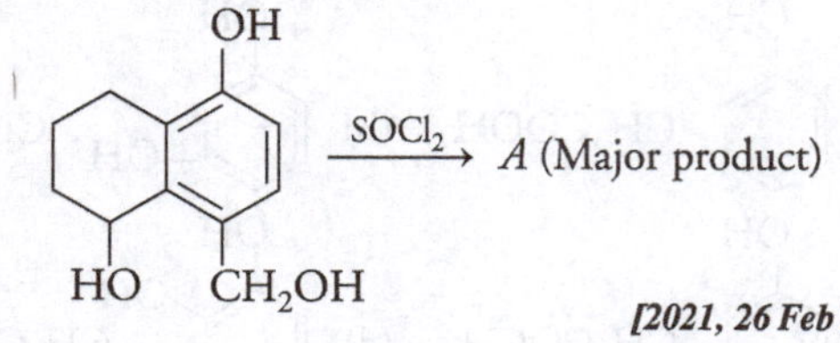


[2021, 26 Feb Shift-II]

(a) 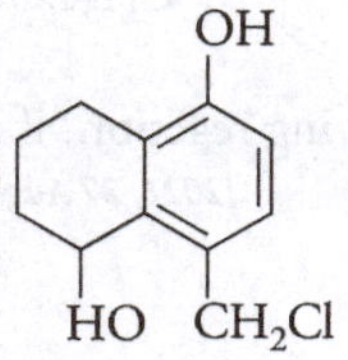


(b) 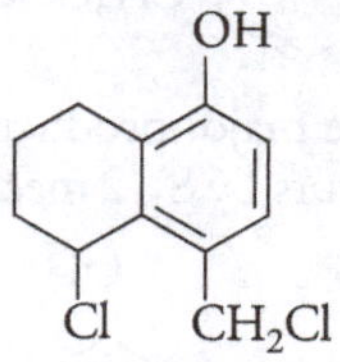


(c) 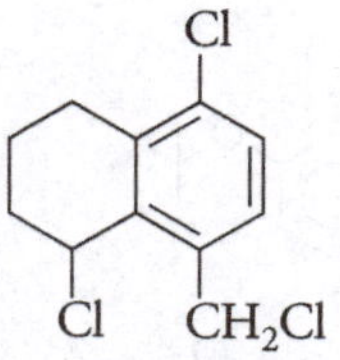


(d) 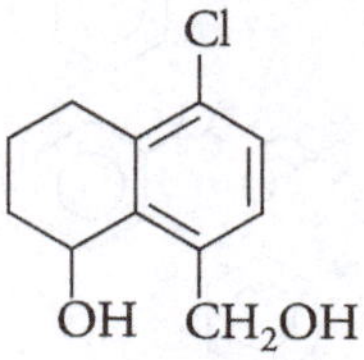


**28** Identify the major products $A$ and $B$ respectively in the following reactions of phenol. [2021, 26 Feb Shift-I]

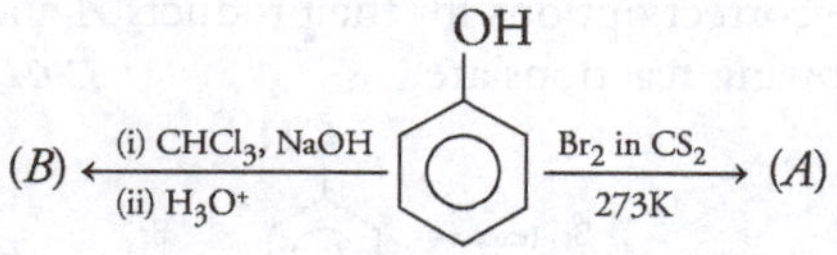


(a) 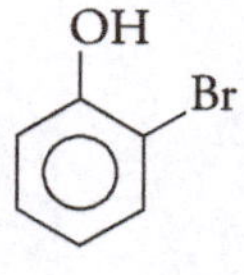 and 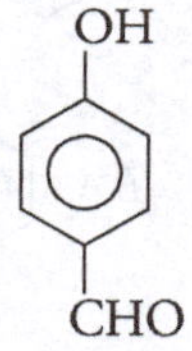

(b)  and 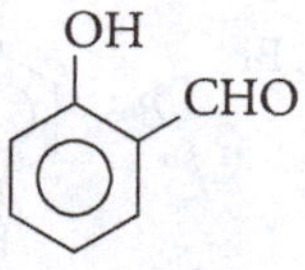

(c) 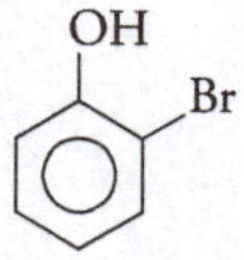 and 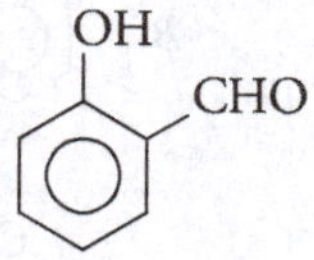

(d) 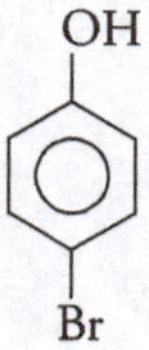 and 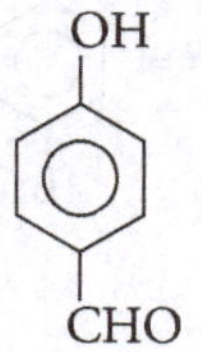

**Imp.** Types of H-bonding in Nitrophenols

**29** Given below are two statements

**Statement I** *o*-nitrophenol is steam volatile due to intramolecular hydrogen bonding.

**Statement II** *o*-nitrophenol has high melting due to hydrogen bonding.

In the light of the above statements, choose the most appropriate answer from the options given below. [2021, 26 Feb Shift-I]

(a) Statement I is false but statement II is true
(b) Both statement I and statement II are true
(c) Both statement I and statement II are false
(d) Statement I is true but statement II is false

**30** Which of the following compound gives pink colour on reaction with phthalic anhydride in conc. $H_2SO_4$ followed by treatment with NaOH? [2021, 24 Feb Shift-I]

(a) 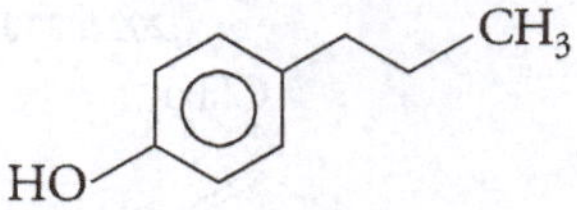

(b) 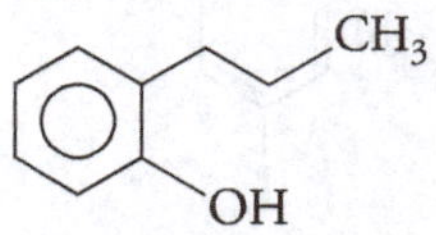

(c) 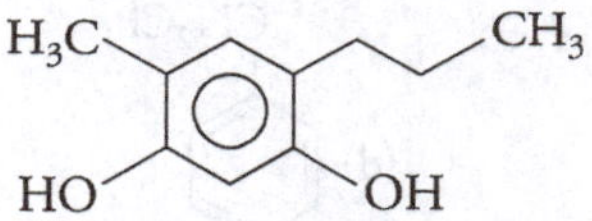

(d) 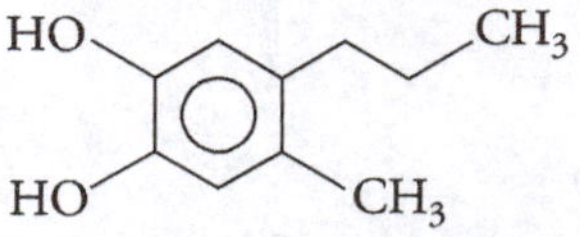

**31** Which of the following compounds reacts with ethyl magnesium bromide and also decolourises bromine water solution [2019, 11 Jan Shift-II]

(a) 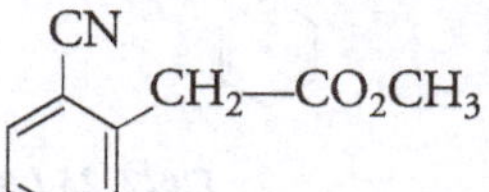

(b) 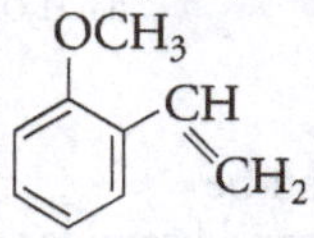

(c) 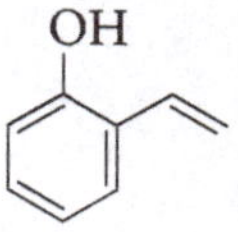

(d) 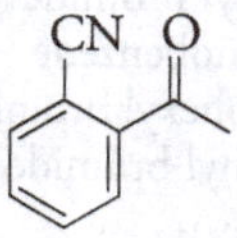

**32** The increasing order of the p$K_a$ values of the following compounds is [2019, 10 Jan Shift-I]

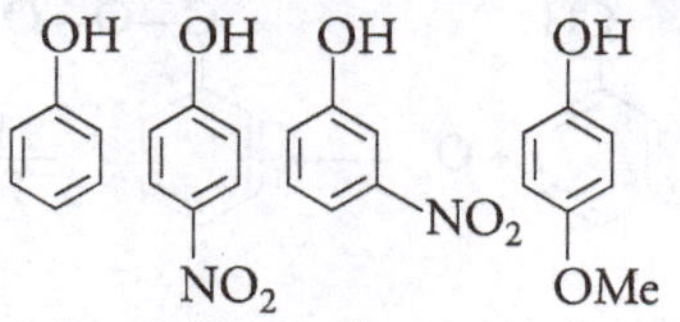


$A$ $B$ $C$ $D$

(a) $D < A < C < B$
(b) $B < C < A < D$
(c) $C < B < A < D$
(d) $B < C < D < A$

**33** Phenol on treatment with $CO_2$ in the presence of NaOH followed by acidification produces compound $X$ as the major product. $X$ on treatment with $(CH_3CO)_2O$ in the presence of catalytic amount of $H_2SO_4$ produces *[JEE Main 2018]*

(a) 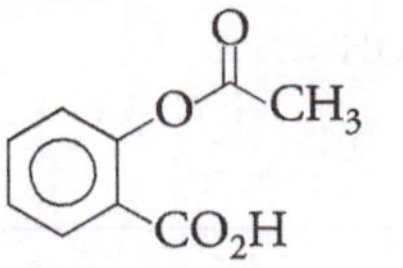

(b) 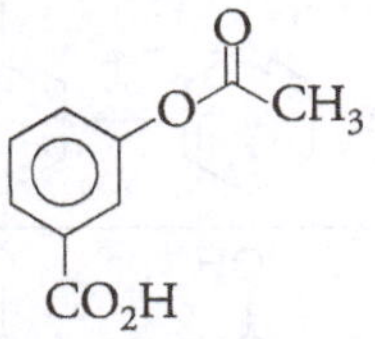

(c) 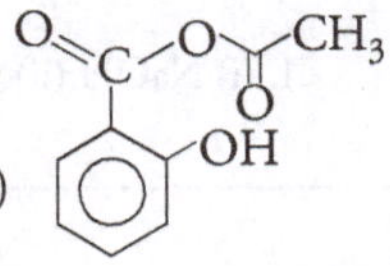

(d) 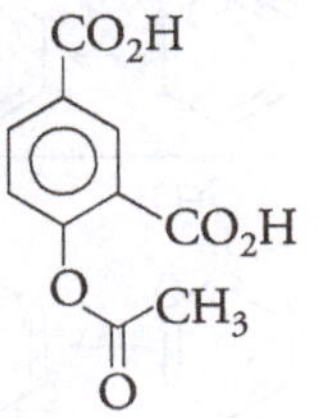

**Imp.** Order of Acidity of Phenols

**34** Arrange the following compounds in the order of decreasing acidity *[JEE Main 2013]*

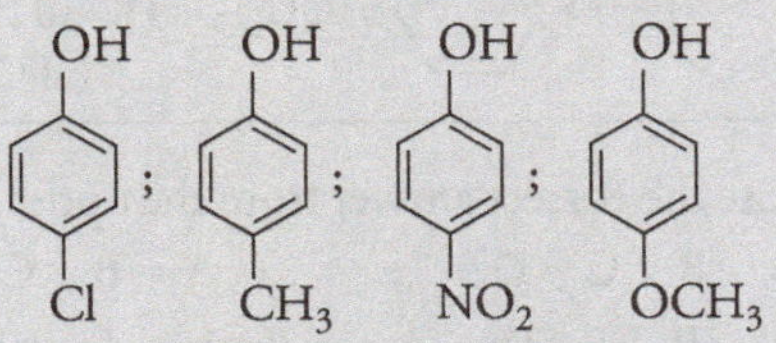

I II III IV

(a) II > IV > I > III (b) I > II > III > IV
(c) III > I > II > IV (d) IV > III > I > II

## Level 3

**35.** Match List - I with List - II. *[2025, 7 April Shift-II]*

| | List - I (Conversion) | | List - II (Reagents, conditions used) |
|---|---|---|---|
| A. | 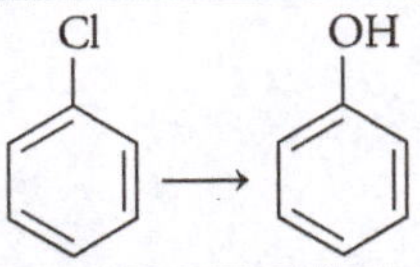 | I. | Warm, $H_2O$ |
| B. | 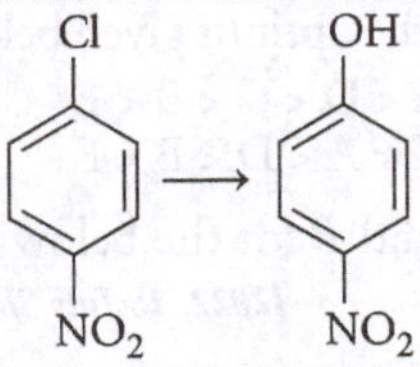 | II. | (a) NaOH, 368 K; (b) $H_3O^+$ |
| C. | 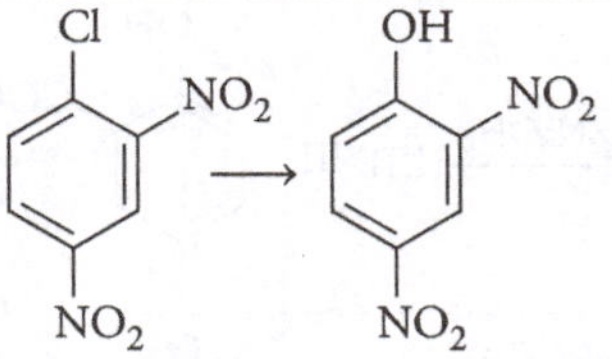 | III. | (a) NaOH, 443 K; (b) $H_3O^+$ |
| D. | 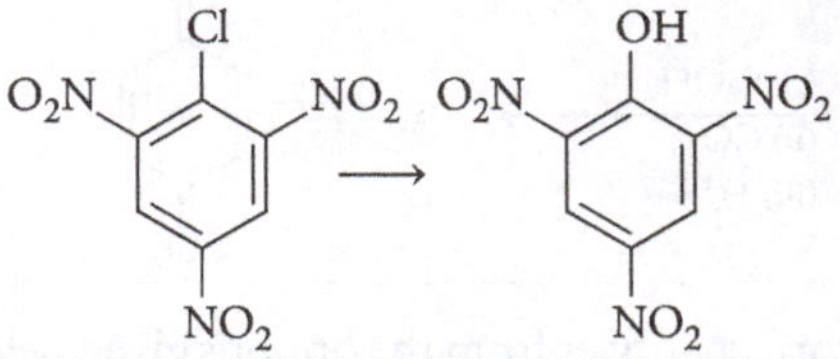 | IV. | (a) NaOH, 623 K, 300 atm; (b) $H_3O^+$ |

Choose the correct answer from the options given below:

(a) (A)-(IV), (B)-(III), (C)-(I), (D)-(II)
(b) (A)-(II), (B)-(III), (C)-(I), (D)-(IV)
(c) (A)-(IV), (B)-(III), (C)-(II), (D)-(I)
(d) (A)-(III), (B)-(IV), (C)-(II), (D)-(I)

**36.** Benzene is treated with oleum to produce compound ($X$) which when further heated with molten sodium hydroxide followed by acidification produces compound ($Y$). The compound $Y$ is treated with zinc metal to produce compound ($Z$). Identify the structure of compound ($Z$) from the following option. *[2025, 4 April Shift-I]*

(a) 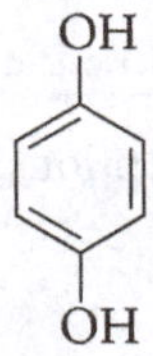

(b) 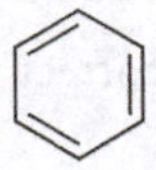

(c) 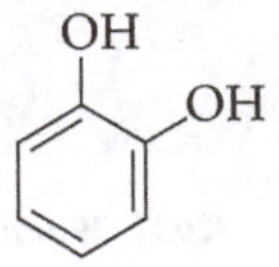

(d) 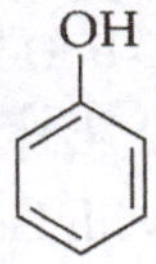

**37** Match List-I with List-II. *[2024, 8 July Shift-II]*

| | List-I (Reactions) | | List-II (Products) |
|---|---|---|---|
| A. | 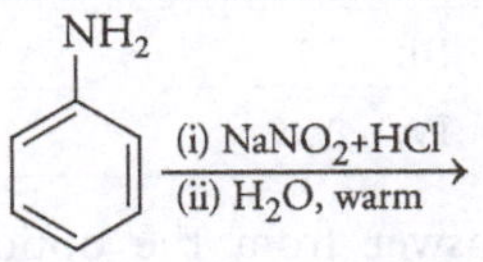 | I. | 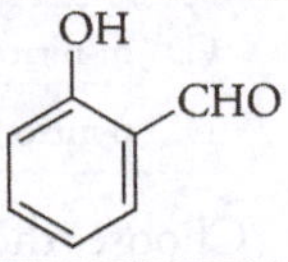 |
| B. | 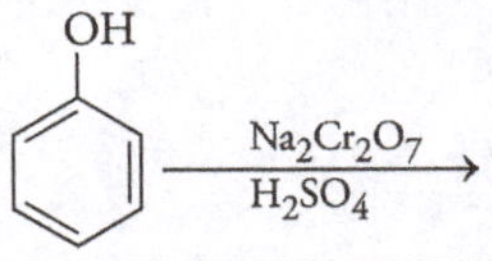 | II. | 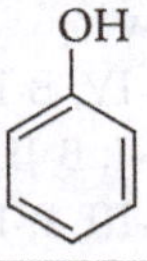 |

C. 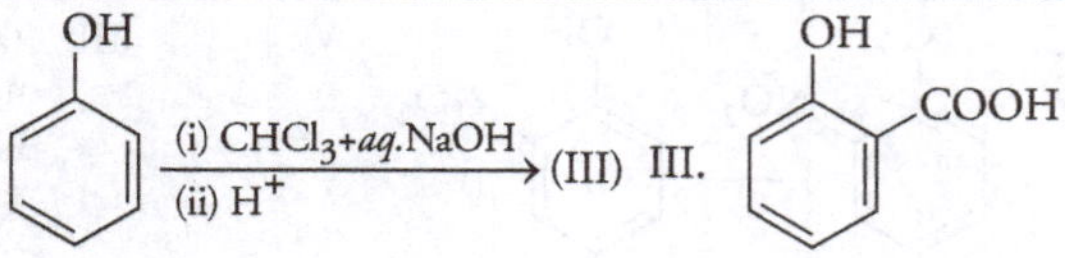

D. 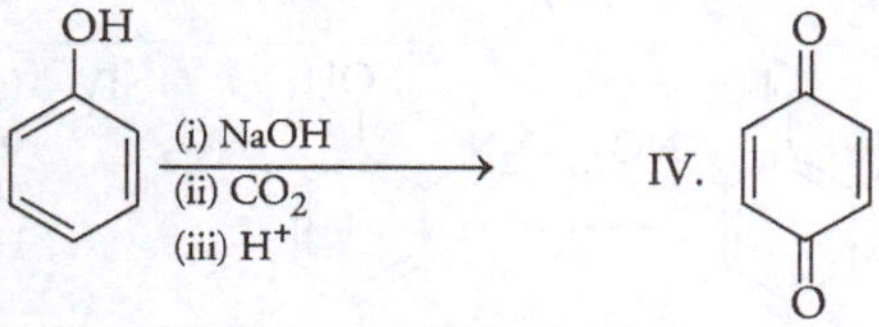

Choose the correct answer from the options given below
(a) A-I, B-IV, C-II, D-III
(b) A-III, B-II, C-I, D-IV
(c) A-IV, B-II, C-III, D-I
(d) A-II, B-IV, C-I, D-III

**Imp.** Chemical Reactions of Phenol

**38** Match List-I with List-II. *[2024, 1 Feb Shift-II]*

| List-I (Reactants) | List-II (Product) |
|---|---|
| A. Phenol, Zn /Δ | I. Salicylaldehyde |
| B. Phenol, $CHCl_3$, NaOH, HCl | II. Salicylic acid |
| C. Phenol, $CO_2$, NaOH, HCl | III. Benzene |
| D. Phenol, Conc. $HNO_3$ | IV. Picric acid |

Choose the correct answer from the options given below
(a) A-IV, B-I, C-II, D-III
(b) A-III, B-IV, C-I, D-II
(c) A-III, B-I, C-II, D-IV
(d) A-IV, B-II, C-I, D-III

**39** Match List-I with List-II. *[2024, 29 Jan Shift-II]*

| List I (Compound) | List II ($pK_a$ value) |
|---|---|
| A. Ethanol | I. 10.0 |
| B. Phenol | II. 15.9 |
| C. m-nitrophenol | III. 7.1 |
| D. p-nitrophenol | IV. 8.3 |

Choose the correct answer from the options given below
(a) A-IV, B-I, C-II, D-III
(b) A-I, B-II, C-III, D-IV
(c) A-III, B-IV, C-I, D-II
(d) A-II, B-I, C-IV, D-III

**40** Match List-I with List-II. *[2024, 27 Jan Shift-II]*

| List-I (Reaction) | List-II (Reagent(s)) |
|---|---|
| A. 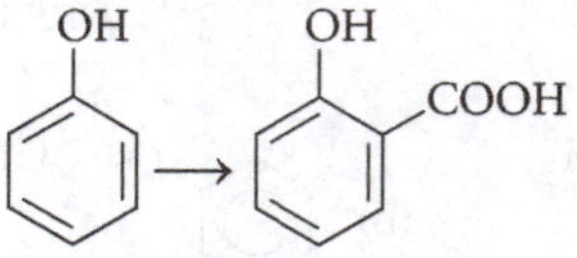 | I. $Na_2Cr_2O_7$, $H_2SO_4$ |
| B. 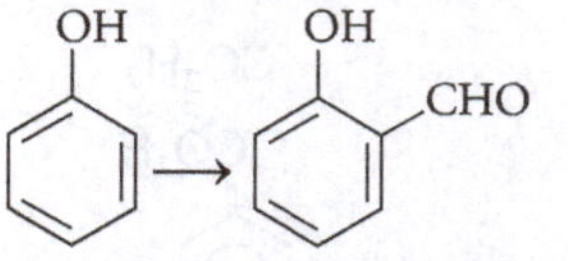 | II. (i) NaOH (ii) $CH_3Cl$ |
| C. 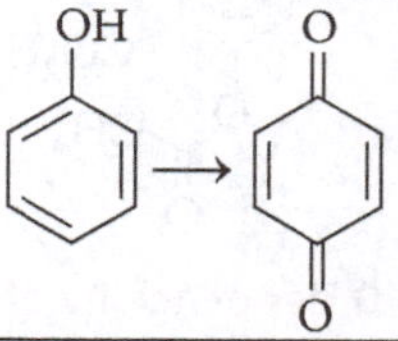 | III. (i) NaOH, $CHCl_3$ (ii) NaOH (iii) HCl |
| D. 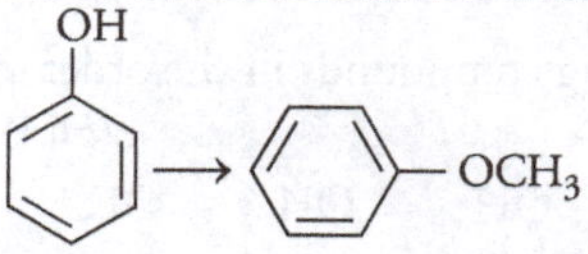 | IV. (i) NaOH (ii) $CO_2$ (iii) HCl |

Choose the correct answer from the options given below

| | A | B | C | D |
|---|---|---|---|---|
| (a) | II | III | I | IV |
| (b) | II | I | III | IV |
| (c) | IV | I | III | II |
| (d) | IV | III | I | II |

**41** The ascending order of acidity of —OH group in the following compounds is *[2024, 27 Jan Shift-I]*

A. Bu—OH

B. 

C. 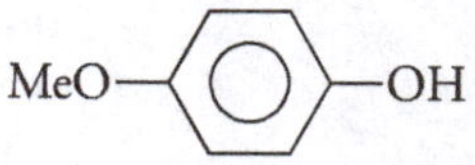

D. 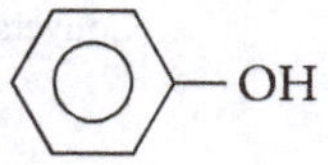

E. 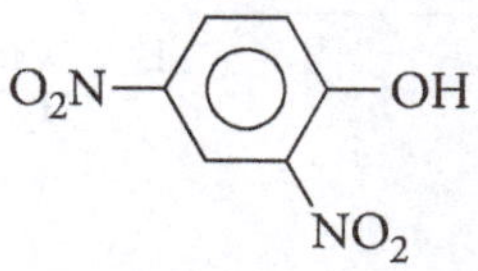

Choose the correct answer from the options given below:
(a) A < C < D < B < E
(b) A < D < C < B < E
(c) C < D < B < A < E
(d) C < A < D < B < E

**42** Identify the major products *A* and *B* for the below given reaction sequence. *[2022, 28 July Shift-I]*

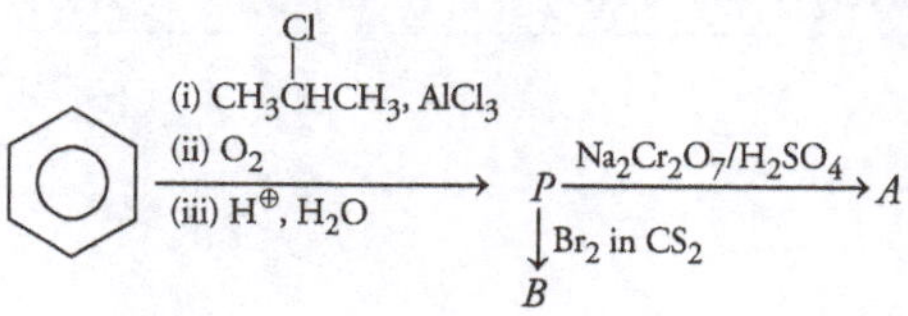

(a) 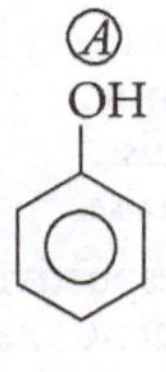 and 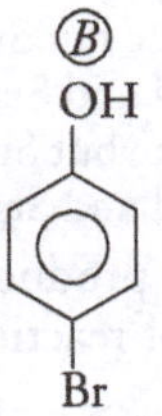

(b) 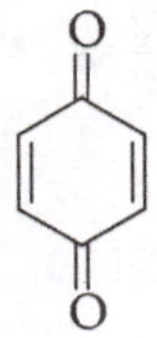 and 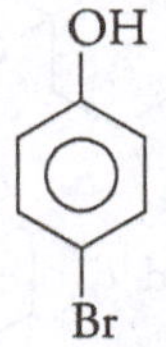

(c) 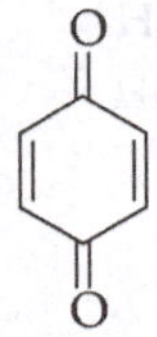 and 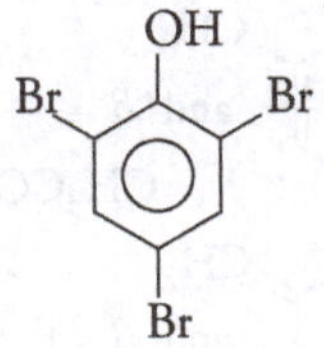

(d)  and 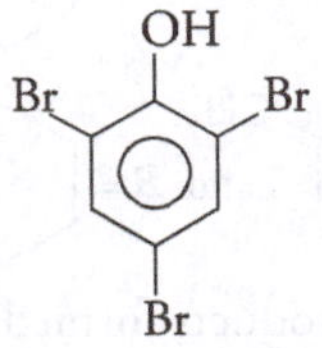

**43** Match List I with List II. *[2022, 27 June Shift-II]*

| | List-I | | List-II |
|---|---|---|---|
| A. | 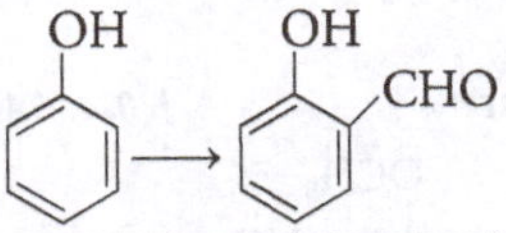 | I. | $Br_2$ in $CS_2$ |
| B. | 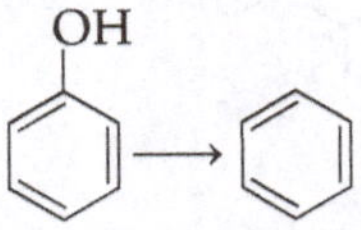 | II. | $Na_2Cr_2O_7/H_2SO_4$ |
| C. | 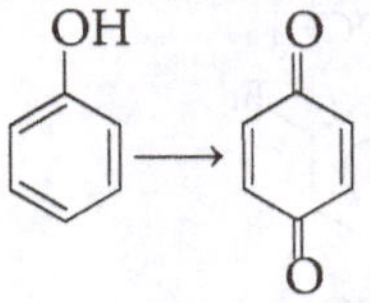 | III. | Zn |
| D. | 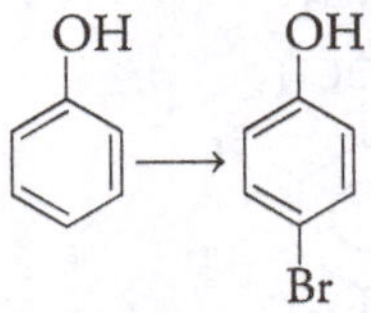 | IV. | $CHCl_3$ / NaOH |

Choose the correct answer from the options given below.

| | A | B | C | D | | A | B | C | D |
|---|---|---|---|---|---|---|---|---|---|
| (a) | IV | III | II | I | (b) | IV | III | I | II |
| (c) | II | III | I | IV | (d) | IV | II | III | I |

## Numerical Value Type Questions

**44** In the following reaction, *[2022, 27 July Shift-I]*

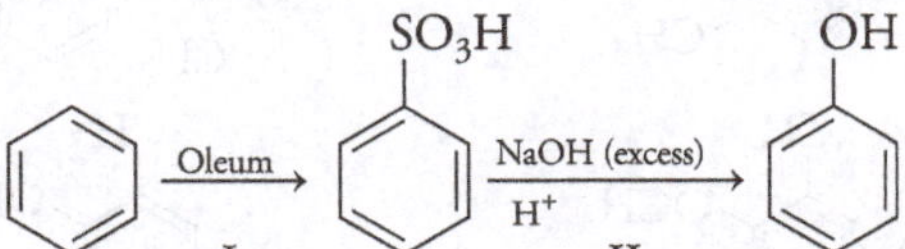

I II

the % yield for reaction I is 60% and that of reaction II is 50%. The overall yield of the complete reaction is ............%. [Nearest integer]

**45** Compound (*P*) on nitration with dil. $HNO_3$ yields two isomers (*A*) and (*B*). These isomers can be separated by steam distillation. Isomers (*A*) and (*B*) show the intramolecular and intermolecular hydrogen bonding respectively. Compound (*P*) on reaction with conc. $HNO_3$ yields a yellow compound (*C*), a strong acid. The number of oxygen atoms present in compound (*C*) is. *[2022, 26 June Shift-I]*

# Topic 3 Preparation, Properties and Uses of Ethers

## Level 1

**01** Suitable reaction condition for preparation of methyl phenyl ether is *[2023, 10 April Shift-I]*

(a) $Ph\overset{\ominus}{O}Na^{\oplus}$, MeOH

(b) Ph—Br, $Me\overset{\ominus}{O}Na^{\oplus}$

(c) Benzene, MeBr

(d) $Ph\overset{\ominus}{O}Na^{\oplus}$, MeBr

**02** Given below are two statements, one is labelled as Assertion (A) and the other is labelled as Reason (R).

**Assertion (A)** Acetal/ketal is stable in basic medium.

**Reason (R)** The high leaving tendency of alkoxide ion gives the stability to acetal/ketal in basic medium.

In the light of the above statements, choose the correct answer from the options given below : *[2023, 25 Jan Shift-I]*

(a) A is true but R is false.

(b) A is false but R is true.

(c) Both A and R are true but R is not the correct explanation of A.

(d) Both A and R are true and R is the correct explanation of A.

**03** The major product of the following reaction is

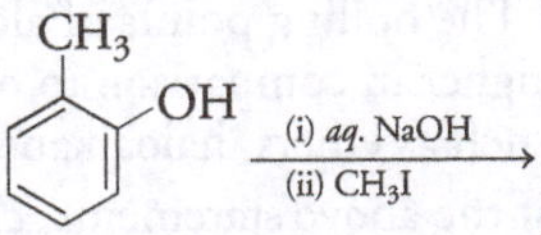

*[2019, 10 Jan Shift-II]*

(a) 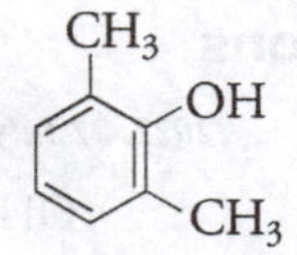
(b) 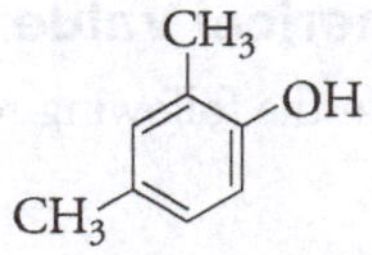

(c) 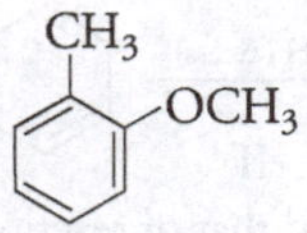
(d) 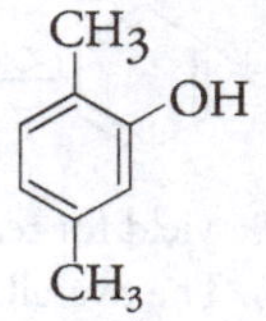

**04** An ether is more volatile than an alcohol having the same molecular formula. This is due to *[AIEEE 2003]*

(a) dipolar character of ethers
(b) alcohols having resonance structures
(c) intermolecular hydrogen bonding in ethers
(d) intermolecular hydrogen bonding in alcohols

## Level 2

**05** Which one of the following reactions will not lead to the desired ether formation in major proportion?
(*iso*-Bu ⇒ *iso*butyl, *sec*-Bu ⇒ *sec*-butyl, *n*Pr ⇒ *n*-propyl, $^t$Bu ⇒ *tert*-butyl, Et ⇒ ethyl) *[2025, 8 April Shift-II]*

(a) *iso*-BuO$^-$Na$^+$ + *sec*-BuBr ⟶ *sec*-Bu—O—*iso*-Bu

(b) 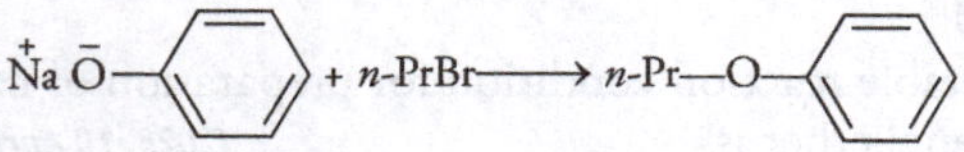

(c) 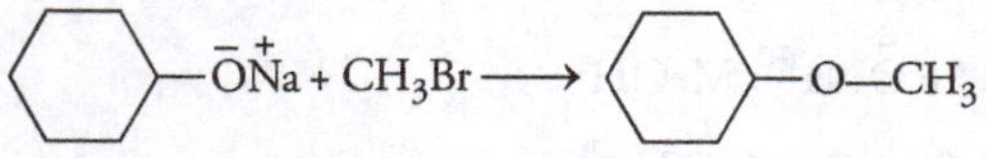

(d) $^t$BuO$^-$Na$^+$ + EtBr ⟶ $^t$Bu—O—Et

**06** Given below are two statements: *[2025, 7 April Shift-I]*

**Statement I** Dimethyl ether is completely soluble in water. However, diethyl ether is soluble in water to a very small extent.

**Statement II** Sodium metal can be used to dry diethyl ether and not ethyl alcohol.

In the light of given statements, choose the correct answer from the options given below

(a) Statement I is true but Statement II is false.
(b) Both Statement I and Statement II are true.
(c) Both Statement I and Statement II are false.
(d) Statement I is false but Statement II are true.

**07** Given below are two statements. *[2025, 23 Jan Shift-II]*

**Statement I** The boiling points of alcohols and phenols increase with increase in the number of C-atoms.

**Statement II** The boiling points of alcohols and phenols are higher in comparison to other class of compounds such as ethers, haloalkanes.

In the light of the above statements, choose the correct answer from the options given below

(a) Statement I is false but Statement II is true.
(b) Both Statement I and Statement II are true.
(c) Statement I is true but Statement II is false.
(d) Both Statement I and Statement II are false.

**08** Identify the major products *A* and *B* respectively in the following set of reactions. *[2024, 8 April Shift-I]*

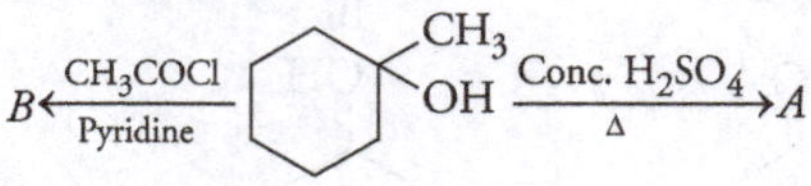

(a) $A$ = 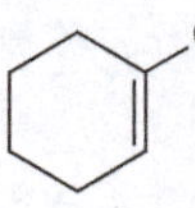 and $B$ = 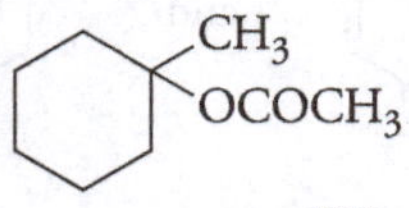

(b) $A$ = 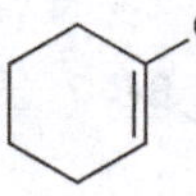 and $B$ = 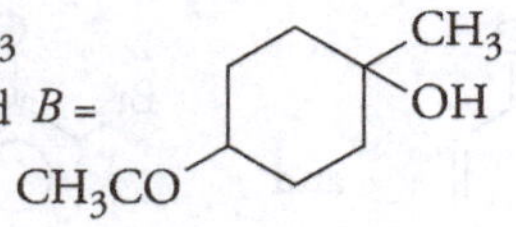

(c) $A$ = 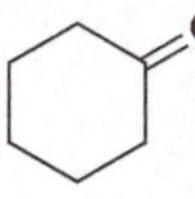 and $B$ = 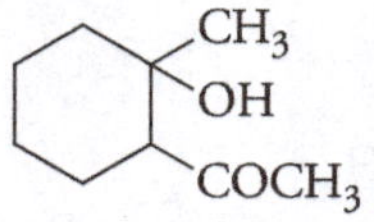

(d) $A$ = 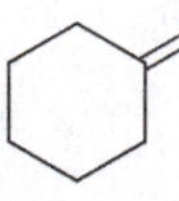 and $B$ = 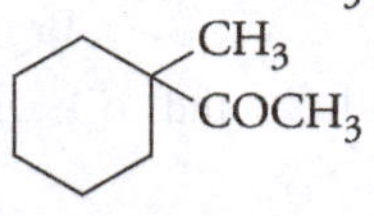

**09** The major products formed

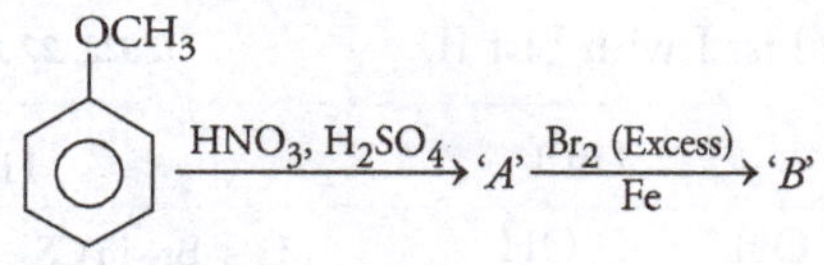

*A* and *B* respectively are *[2024, 6 April Shift-II]*

(a) 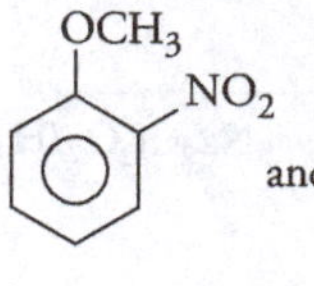 and 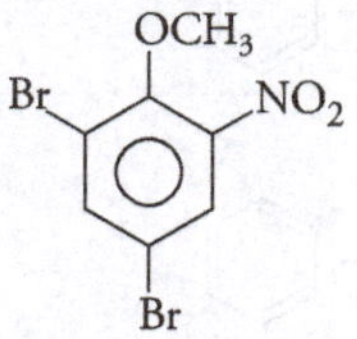

(b) 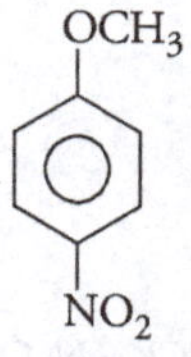 and 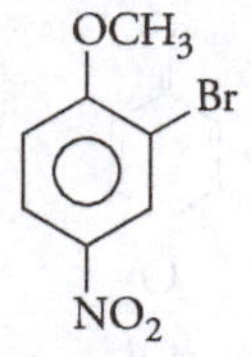

(c) 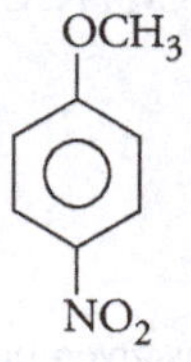 and 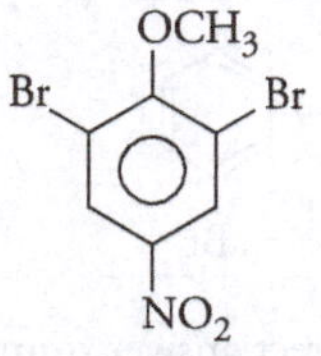

(d) 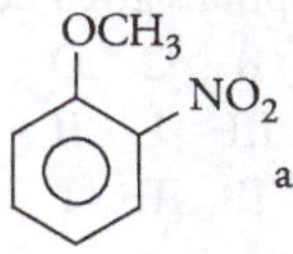 and 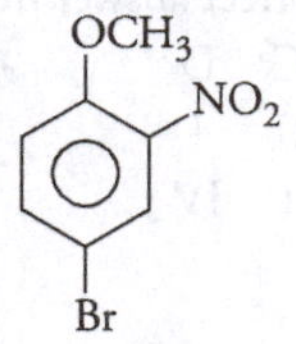

**10** Compound I is heated with conc. HI to give a hydroxy compound *A* which is further heated with Zn dust to give compound *B*. Identify *A* and *B*. *[2022, 28 July Shift-II]*

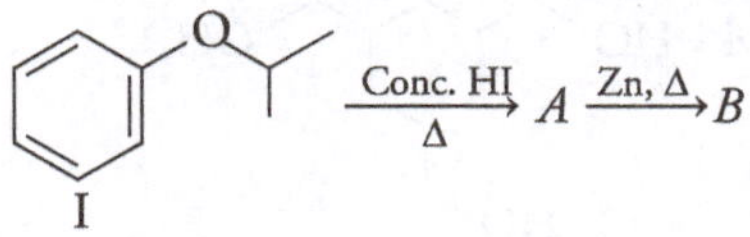

(a) *A* = 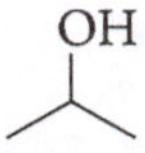 *B* = 

(b) *A* = 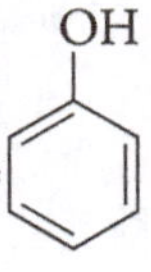 *B* = 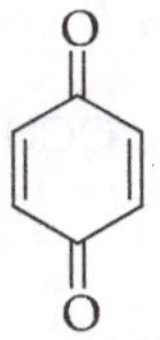

(c) *A* = 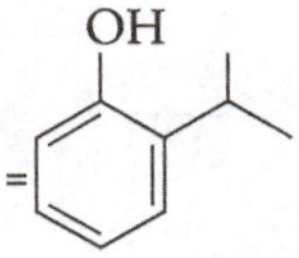 *B* = 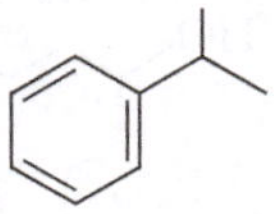

(d) *A* = 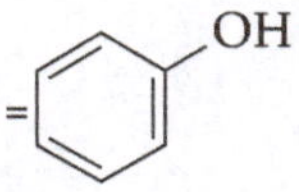 *B* = 

**11** The major product formed in the following reaction, is *[2022, 25 June Shift-II]*

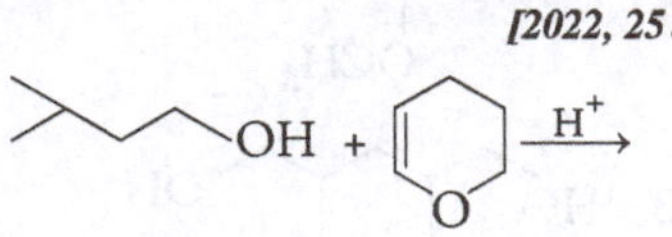

(a) 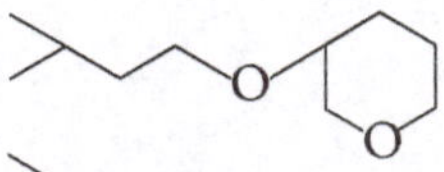

(b) 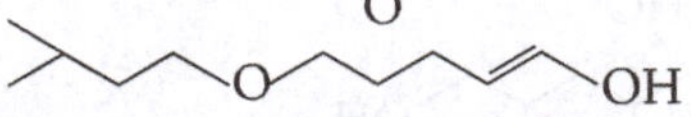

(c) 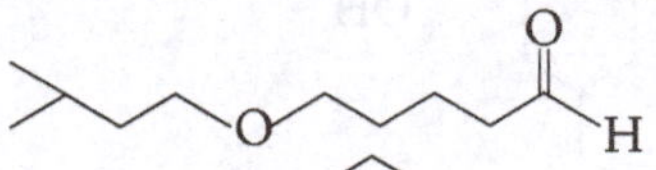

(d) 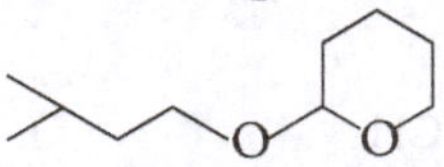

**12** Given below are two statements one is labelled as Assertion (A) and the other is labelled as Reason (R).

**Assertion** (A) Synthesis of ethyl phenyl ether may be achieved by Williamson synthesis.

**Reason** (R) Reaction of bromobenzene with sodium ethoxide yields ethyl phenyl ether.

In the light of the above statements, choose the most appropriate answer from the options given below *[2021, 27 Aug Shift-I]*

(a) Both (A) and (R) are correct and (R) is the correct explanation of (A)
(b) (A) is correct but (R) is incorrect
(c) (A) is incorrect but (R) is correct
(d) Both (A) and (R) are correct but (R) is not the correct explanation of (A)

**13** In the given reaction, 3-bromo-2, 2-dimethyl butane $\xrightarrow{C_2H_5OH}$ $\underset{\text{Major product}}{(A)}$.

Product *A* is *[2021, 20 July Shift-I]*

(a) 2-ethoxy-3, 3-dimethyl butane
(b) 1-ethoxy-3, 3-dimethyl butane
(c) 2-ethoxy-2, 3-dimethyl butane
(d) 2-hydroxy-3, 3-dimethyl butane

**14** Main products formed during a reaction of 1-methoxy naphthalene with hydroiodic acid are *[2021, 18 March Shift-II]*

(a) 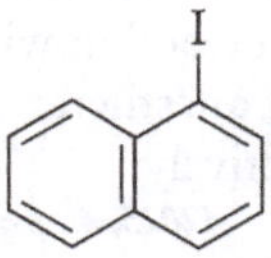 and $CH_3I$

(b) 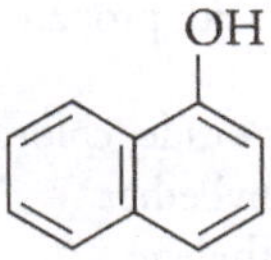 and $CH_3I$

(c) 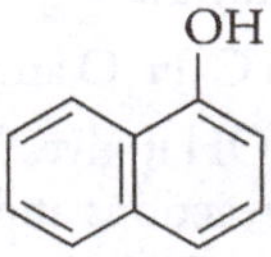 and $CH_3OH$

(d) 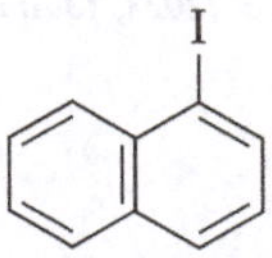 and $CH_3OH$

**15** Arrange the following compounds in increasing order of C—OH bond length methanol, phenol, *p*-ethoxyphenol *[2020, 8 Jan Shift-I]*

(a) phenol < methanol < *p*-ethoxyphenol
(b) phenol < *p*-ethoxyphenol < methanol
(c) methanol < *p*-ethoxyphenol < phenol
(d) methanol < phenol < *p*-ethoxyphenol

**16** The major product of the following reaction is

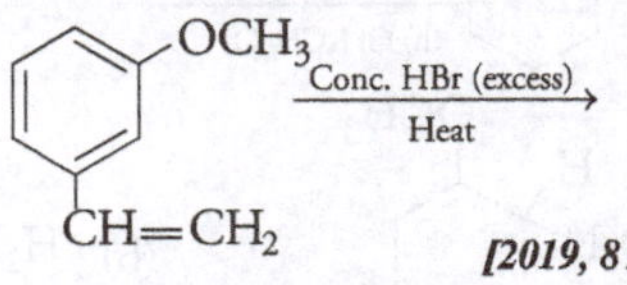

*[2019, 8 April Shift-I]*

(a) 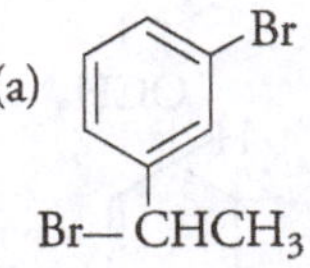

(b) 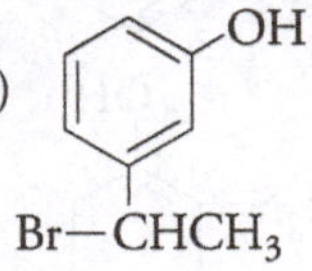

(c) 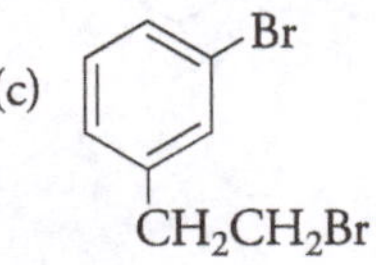

(d) 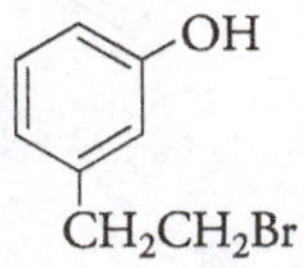

**17** Consider the following reaction,

$$C_2H_5OH + H_2SO_4 \longrightarrow \text{Product}$$

Among the following, which one cannot be formed as a product under any conditions? *[AIEEE 2011]*

(a) Ethyl hydrogen sulphate (b) Ethylene
(c) Acetylene (d) Diethyl ether

## Level 3

**18** A toxic compound '*A*' when reacted with NaCN in aqueous acidic medium yields an edible cooking component and food preservative 'B'. 'B' is converted to '*C*' by diborane and can be used as an additive to petrol to reduce emission. '*C*' upon reaction with oleum at 140°C yields an inhalable anesthetic '*D*'. Identify '*A*', '*B*', '*C*' and '*D*', respectively: *[2025, 4 April Shift-II]*

(a) Acetaldehyde; 2-hydroxypropanoic acid; propanoic acid; dipropyl ether
(b) Methanol; formaldehyde; methyl chloride; chloroform
(c) Methanol; acetic acid; ethanol; diethyl ether
(d) Ethanol; acetonitrile; ethylamine; ethylene

**19** 2-methyl propyl bromide reacts with $C_2H_5O$ and gives '*A*' whereas on reaction with $C_2H_5OH$ it gives '*B*'. The mechanism followed in these reactions and the products '*A*' and '*B*' respectively are *[2023, 13 April Shift-I]*

(a) $S_N1$, $A =$ *tert*-butyl ethyl ether; $S_N2$, $B =$ *iso*-butyl ethyl ether
(b) $S_N2$, $A =$ 2-butyl ethyl ether; $S_N2$, $B =$ iso-butyl ethyl ether
(c) $S_N2$, $A =$ iso-butyl ethyl ether; $S_N1$, $B =$ tert-butyl ethyl ether
(d) $S_N1$, $A =$ iso-butyl ethyl ether; $S_N1$, $B =$ 2-butyl ethyl ether

**20** The major product of the following reaction is

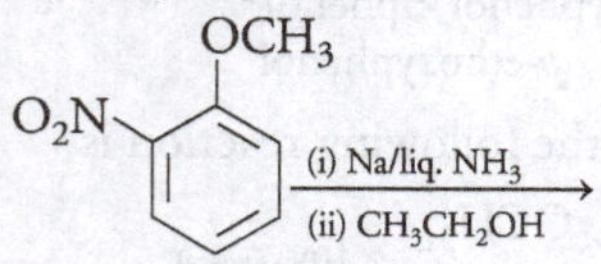


*[2022, 26 July Shift-I]*

(a) 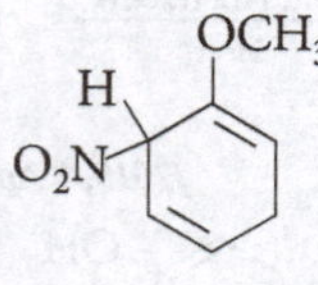
(b) 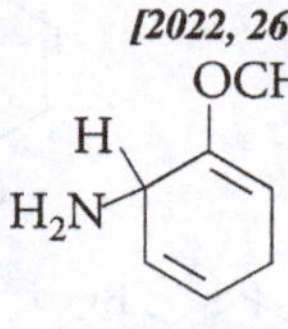
(c) 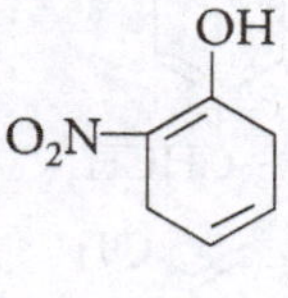
(d) 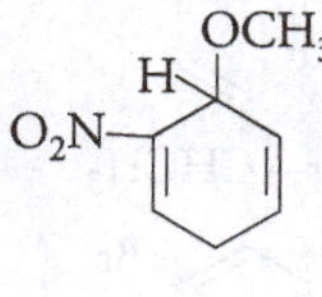

**21** Among the compounds *A* and *B* with molecular formula $C_9H_{18}O_3$, *A* is having higher boiling point the *B*. The possible structures of *A* and *B* are *[2020, 8 Jan Shift-II]*

(a) 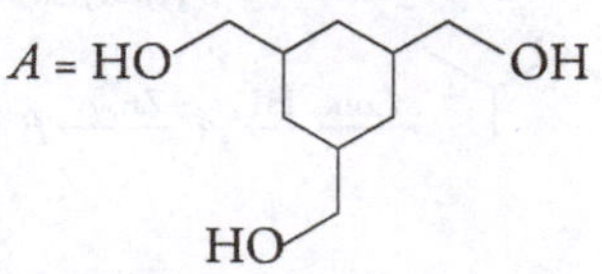

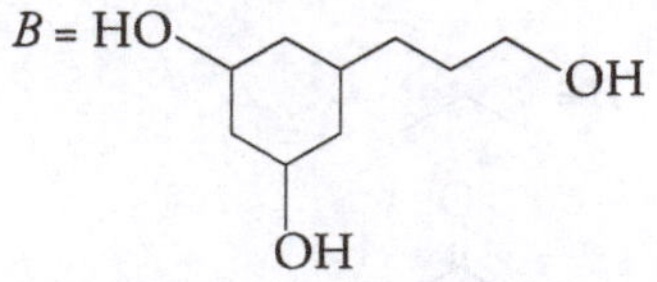

(b) 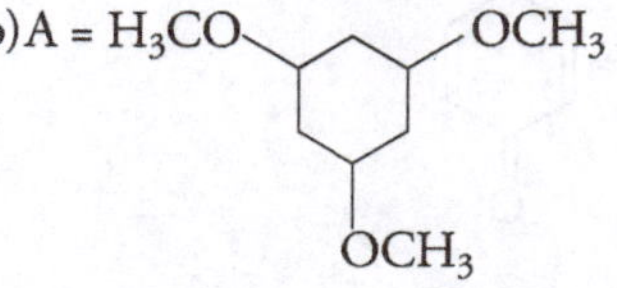

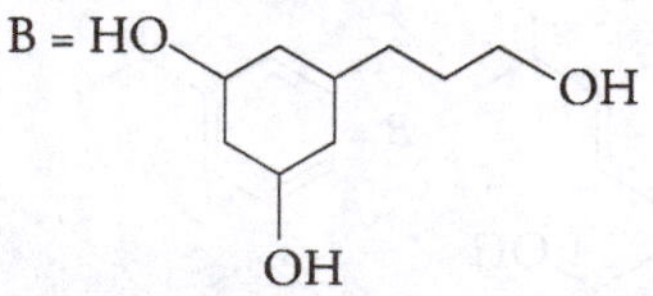

(c) 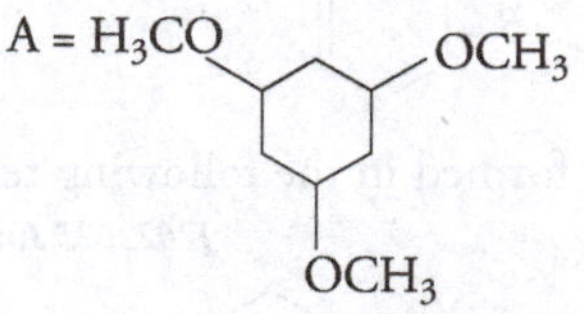

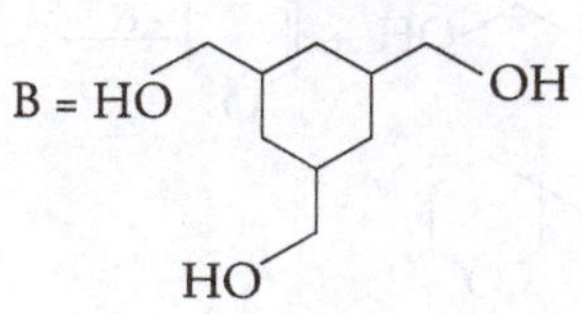

(d) 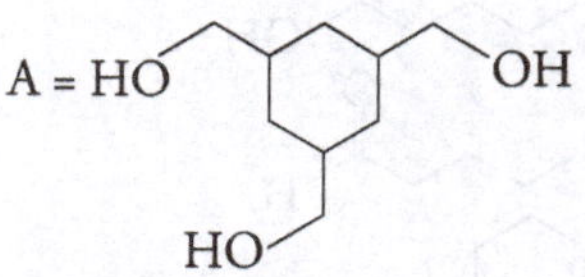

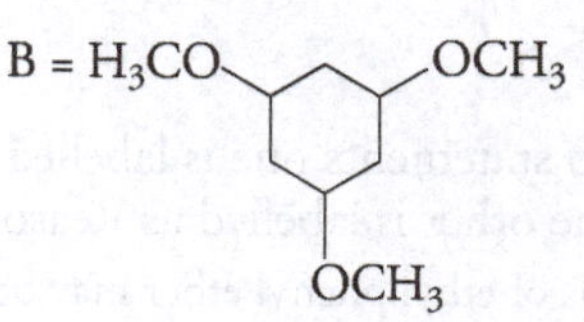

# • Key Index

## Topic 1

| | | | | | | | | | |
|---|---|---|---|---|---|---|---|---|---|
| 1 (c) | 2 (d) | 3 (c) | 4 (d) | 5 (a) | 6 (b) | 7 (a) | 8 (b) | 9 (c) | 10 (d) |
| 11 (d) | 12 (b) | 13 (b) | 14 (d) | 15 (b) | 16 (c) | 17 (b) | 18 (c) | 19 (b) | 20 (c) |
| 21 (c) | 22 (c) | 23 (d) | 24 (c) | 25 (c) | 26 (d) | 27 (b) | 28 (c) | 29 (d) | 30 (c) |
| 31 (a) | 32 (b) | 33 (c) | 34 (b) | 35 (d) | 36 (c) | 37 (d) | 38 (c) | 39 (a) | 40 (c) |
| 41 (c) | 42 (b) | 43 (c) | 44 (b) | 45 (a) | 46 (a) | 47 (c) | 48 (13) | 49 (2) | 50 (8) |
| 51 (2) | | | | | | | | | |

## Topic 2

| | | | | | | | | | |
|---|---|---|---|---|---|---|---|---|---|
| 1 (c) | 2 (c) | 3 (b) | 4 (a) | 5 (b) | 6 (c) | 7 (c) | 8 (b) | 9 (b) | 10 (d) |
| 11 (*) | 12 (b) | 13 (a) | 14 (b) | 15 (c) | 16 (c) | 17 (c) | 18 (c) | 19 (b) | 20 (a) |
| 21 (b) | 22 (a) | 23 (b) | 24 (c) | 25 (d) | 26 (b) | 27 (b) | 28 (b) | 29 (d) | 30 (b) |
| 31 (c) | 32 (b) | 33 (a) | 34 (c) | 35 (c) | 36 (b) | 37 (d) | 38 (c) | 39 (d) | 40 (d) |
| 41 (a) | 42 (b) | 43 (a) | 44 (30) | 45 (7) | | | | | |

## Topic 3

| | | | | | | | | | |
|---|---|---|---|---|---|---|---|---|---|
| 1 (d) | 2 (a) | 3 (c) | 4 (d) | 5 (a) | 6 (b) | 7 (b) | 8 (a) | 9 (c) | 10 (d) |
| 11 (d) | 12 (b) | 13 (c) | 14 (b) | 15 (b) | 16 (b) | 17 (c) | 18 (c) | 19 (c) | 20 (a) |
| 21 (d) | | | | | | | | | |

# Detailed **Explanations**

## Topic 1

**01.** *(c)* Among the given options, the incorrect method of preparation of alcohols is from ozonolysis of alkene. It is because on ozonolysis, alkenes gives aldehyde, ketones and carboxylic acids instead of alcohol.

**02.** *(d)* The water gas on reaction with cobalt as a catalyst will form methanol ($CH_3OH$).

$$CO + 2H_2 \xrightarrow{Co} \underset{\text{Methanol}}{CH_3OH}$$

**03.** *(c)* Butan-1-ol ($CH_3CH_2CH_2CH_2OH$) can undergo H—bonding while ethoxyethane ($CH_3CH_2OCH_2CH_3$) has no hydrogen attached to oxygen atom and it doesn't undergo H—bonding.

More is the extent of intermolecular H—bonding, more will be the association of molecules. Thus, butan-1-ol has higher boiling point.

Hence, both A and R are true and R is the correct explanation of A.

**04.** *(d)* A solution of $CrO_5$ in amyl alcohol has a blue colour. When an acidified cold solution of potassium chromate and $H_2O_2$ is layered with an equal volume of amyl alcohol, a blue alcohol layer is obtained. The blue colour is due to the formation of a (higher top layer of amyl alcohol) chromium (VI) compound, chromium peroxide ($CrO_5$).

**05.** *(a)* The given reaction involves Markownikov addition. It is known as oxymercuration-demercuration reaction

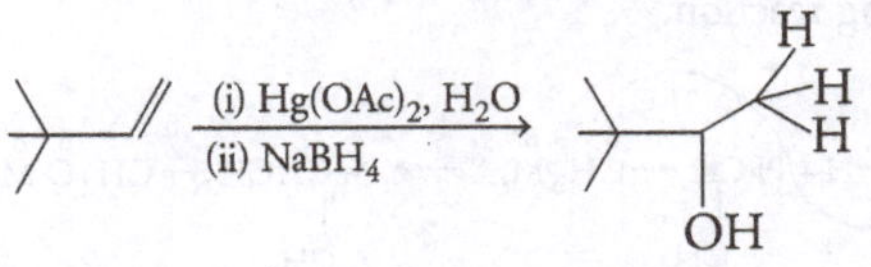


**06.** *(b)* Ethylene glycol reacts with oxalic acid at 210°C to form ethene.

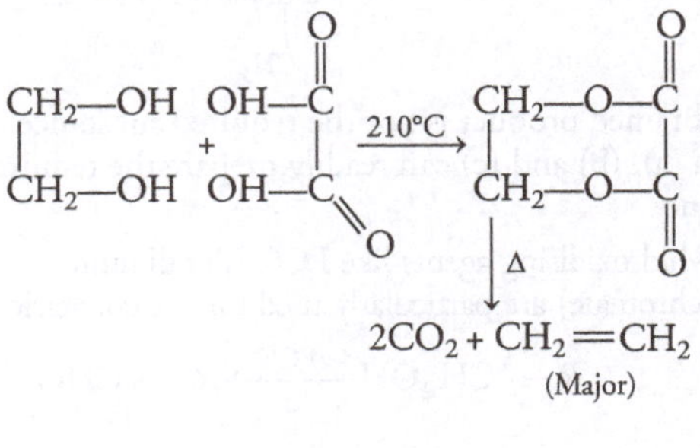

**07.** *(a)*

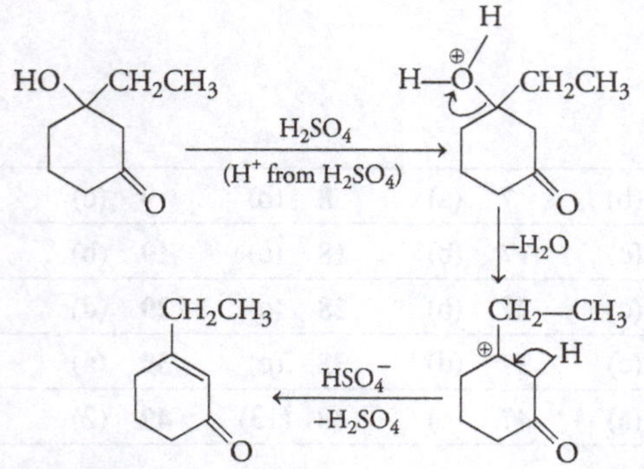


**08.** *(b)* As β-hydrogen is not present therefore E-2 elimination is not possible.

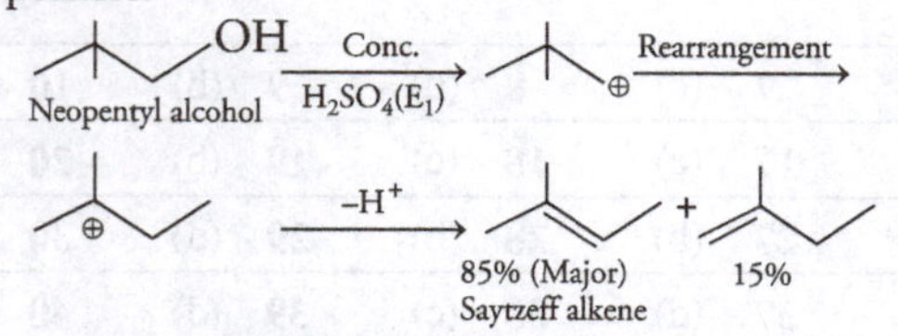


**09.** *(c)* Only 3° alcohol give positive ceric ammonium nitrate test.

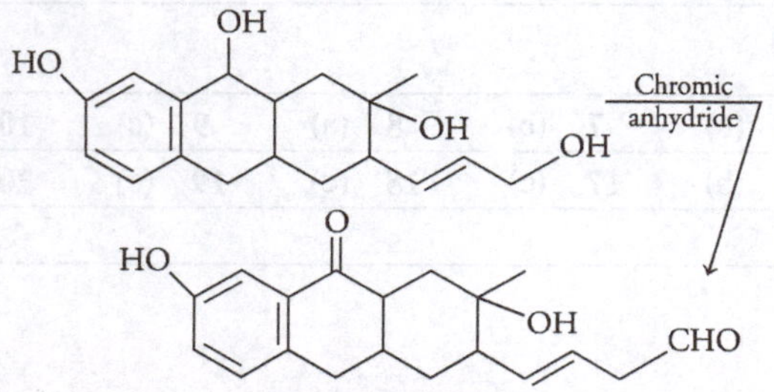


Product '*P*'
(3° alcohol gives red colour with ceric ammonium nitrate)

**10.** *(d)* $CH_3CH_2-\underset{\underset{Ph}{|}}{\overset{\overset{OH}{|}}{C}}-CH_3$ cannot be prepared by HCHO and $PhCH(CH_3)CH_2MgX$. This can be easily illustrated by following reaction.

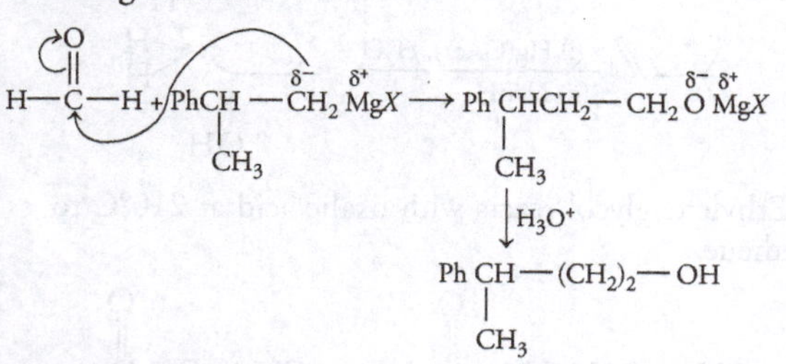


The obtained product is not the required substance. While option (a), (b) and (c) can readily prepare the required substance.

**11.** *(d)* Mild oxidising agents like PCC [Pyridinium chlorochromate] are particularly used for the conversion of

$$R-CH_2OH \xrightarrow{PCC} R-CHO.$$

**12.** *(b)* The reaction of alcohol with Lucas reagent is mostly $S_N1$ reaction and the rate of reaction is directly proportional to the stability of carbocation formed in the reaction.

Since 3° *R*—OH forms 3° carbocation (most stable) hence it will react fastest by $S_N1$ reaction.

**13.** *(b)* The reaction of alcohol with conc. HCl and anhydrous $ZnCl_2$ follows $S_N1$ pathway. So, greater the stability of carbocation formed, faster is the reaction. So, among the given options 2-methylpropan-2-ol gives 3° carbocation.

Hence, it reacts rapidly with conc. HCl and anhydrous $ZnCl_2$ (Lucas reagent).

**14.** *(d)* For positive iodoform test, alcohol molecule must have $CH_3-\underset{\underset{OH}{|}}{CH}-$ group.

Thus, iodoform test is given by only (d) $Ph-\underset{\underset{OH}{|}}{CH}-CH_3$

while others will not give this test.

$$Ph-\underset{\underset{OH}{|}}{CH}-CH_3 \xrightarrow{I_2+NaOH} CHI_3 + Ph-COO^-$$

**15.** *(b)* PhMgBr can be protonated by any of the protic solvent *e.g.*, $CH_3OH$.

$$Ph\,MgBr + CH_3O\,H \longrightarrow \underset{\text{Benzene}}{Ph-H} + \underset{\text{Methoxy magnesium bromide}}{Mg\begin{matrix}Br\\ OCH_3\end{matrix}}$$

**16.** *(c)* Hydration of ethene gives 1° alcohol (ethanol) while all other alkenes give either 2° or 3° alcohols.

**17.** *(b)* Dehydration takes place with the formation of more stable carbocation. Among the given compounds, only compound (b) forms conjugated carbocation. Thus, maximum amount of dehydration takes place in compound (*b*).

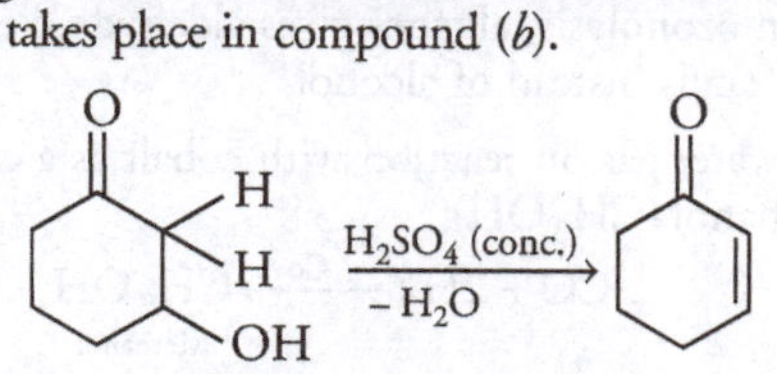


**18.** *(c)* The complete reaction sequence is as follows.

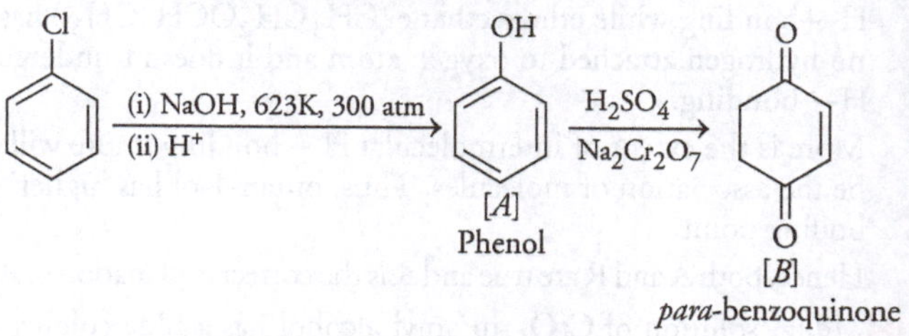


So, option (c) is the correct answer.

**19.** *(b)* The complete sequence of the given reaction is as follows

$$CH_3-CH=CH_2 \xrightarrow{H_2O,\ H^+} CH_3-CH(OH)-CH_3 \text{ (Markownikoff product) } (A)$$

$$CH_3-CH=CH_2 \xrightarrow{(BH_3)_2} (CH_3CH_2CH_2)_3B \xrightarrow{H_2O,\ H_2O_2,\ OH^-} CH_3CH_2CH_2OH \text{ (Hydroboration-oxidation reaction) } (B)$$

**20.** *(c)* The complete reaction is as follow,

$$CH_3-CH_2OH \xrightarrow[\text{(ii) } KMnO_4]{\text{(i) Jones reagent } (CrO_3+H^+)} CH_3-\overset{O}{\overset{\|}{C}}-OH \xrightarrow[\text{(iii) NaOH, CaO, } \Delta]{\text{Soda lime process}} CH_4 + Na_2CO_3$$

The major product is methane.
Hence, option (c) is correct.

**21.** *(c)* Both A and R are true but R is not the correct explanation of A.

> • Concept **Enhancer**
> Alcohols behave as electrophiles and nucleophiles both due to the presence of oxygen in them. Their nucleophilic nature is due to the ability of oxygen to donate its lone pair. On the other hand, its electrophilic activity is due to the polarity in the C—O bond.

**22.** *(c)* The complete reaction is as follows,

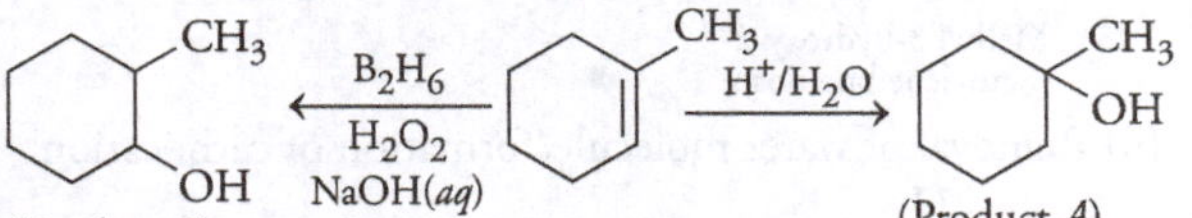


(Product *B*) (Product *A*)

Hence, option (c) is correct.

**23.** *(d)*

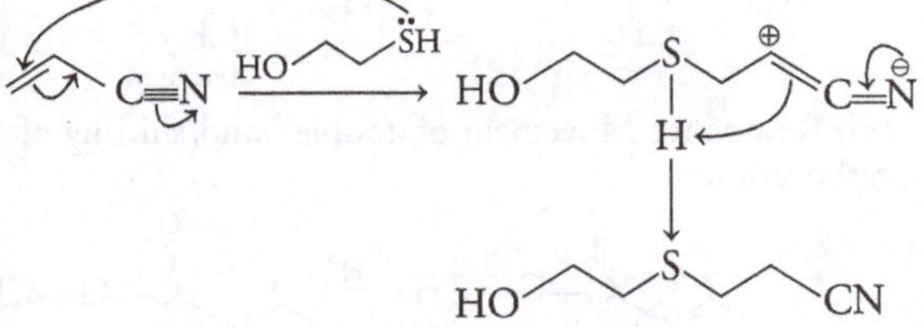


**24.** *(c)*

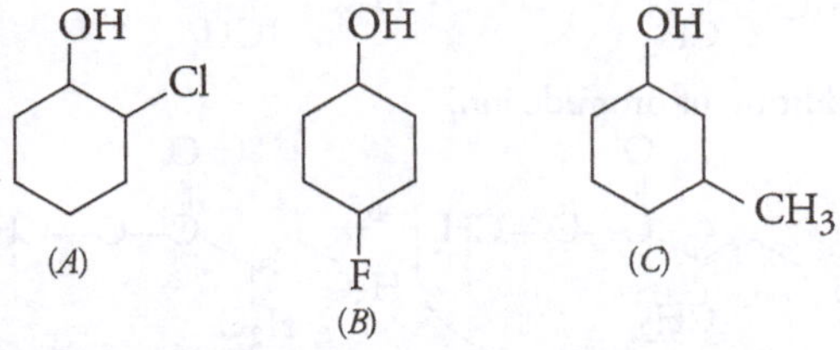


Acidic character depends upon the presence of electron withdrawing group which shows (–*I*) effect with less distance.
Hence, (*A*) is more acidic than (*B*).
$CH_3$ shows (+*I*) effect hence least acidic.
Hence both Assertion and Reason are correct but Reason is not correct explanation of Assertion.

**25.** *(c)* The product [*X*] formed in the given reaction is given in option (c).

$$H_3C-CH_2-\underset{OH}{\underset{|}{CH}}-CH_3 \xrightarrow{\text{(i) NaI, } H_3PO_4} H_3C-CH_2-\underset{I}{\underset{|}{CH}}-CH_3$$

$$\xrightarrow{\text{(ii) Mg, Dry ether}} H_3C-CH_2-\underset{MgI}{\underset{|}{CH}}-CH_3 \xrightarrow{\text{(iii) } D_2O} H_3C-CH_2-\underset{D}{\underset{|}{CH}}-CH_3 \ [X]$$

**26.** *(d)* The correct order is (ii) > (iv) > (iii) > (i).
Dehydration of an alcohol in the presence of an acid involves the loss of —OH functional group as $H_2O$ after protonation of —OH group and formation of carbocation. Its reactivity is directly proportional to the stability of the carbocation. So, according to stability of a carbocation the order of dehydration will be

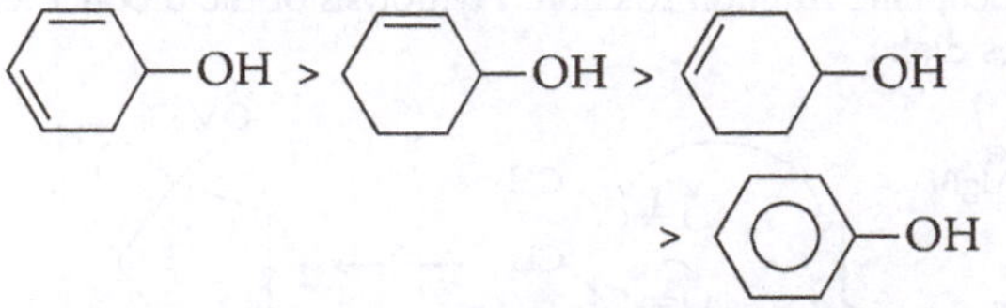


**27.** *(b)* The major product from the following reaction is given in option (b).

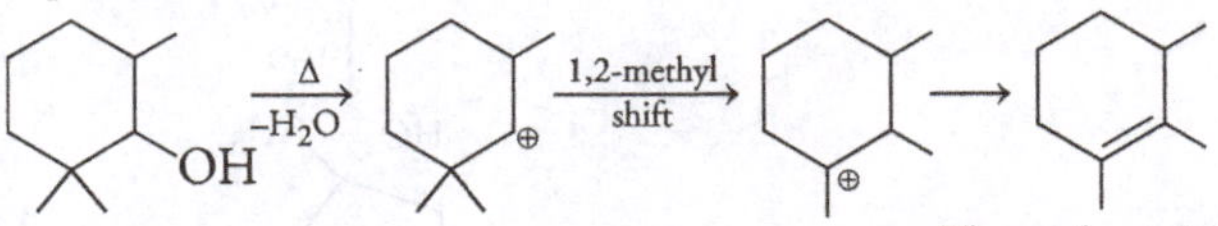


Thermodynamically controlled product

**28.** *(c)* When ethanol is heated with conc. $H_2SO_4$ dehydration reaction occurs and acetylene gas is formed which when treated with cold dilute aqueous solution of Baeyer's reagent glycol is formed.

$$\underset{\text{Ethanol}}{CH_3CH_2OH} \xrightarrow{\text{Conc.} H_2SO_4} \underset{\text{Ethene}}{CH_2=CH_2} \xrightarrow[\text{(Baeyer's reagent)}]{\text{alk. } KMnO_4} \underset{\text{Glycol}}{\underset{OH\quad\ OH}{\underset{|\qquad\ \ |}{CH_2-CH_2}}}$$

> • Concept **Enhancer**
> Baeyer's reagent is a cold potassium permanganate alkaline solution ($KMnO_4$). It is a strong oxidising agent that is used in qualitative organic analysis to check for unsaturation.

**29.** *(d)* Diphenylethanol can be obtained from acetophenone and phenyl magnesium bromide as shown below.

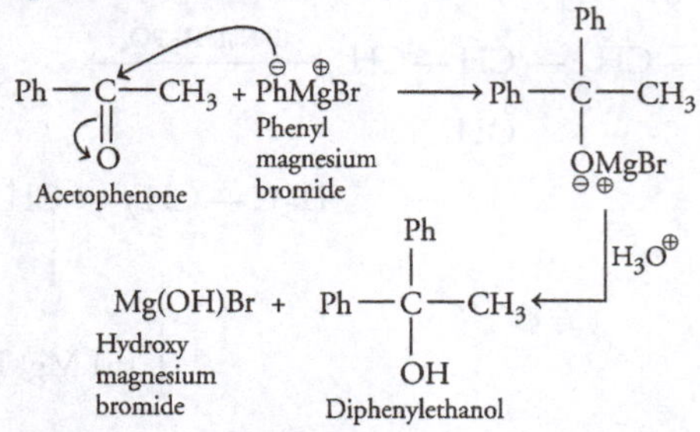


**30.** *(c)* The following reaction sequence are involved

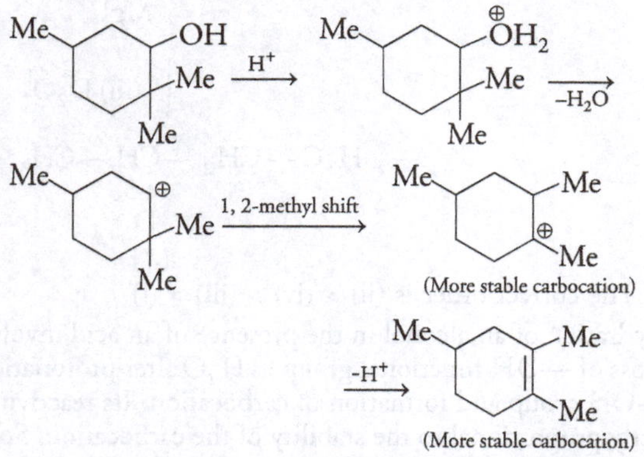


**31.** *(a)* According to the given options, the reaction would be nucleophilic addition reaction. Hydrolysis of the product formed gives diol.

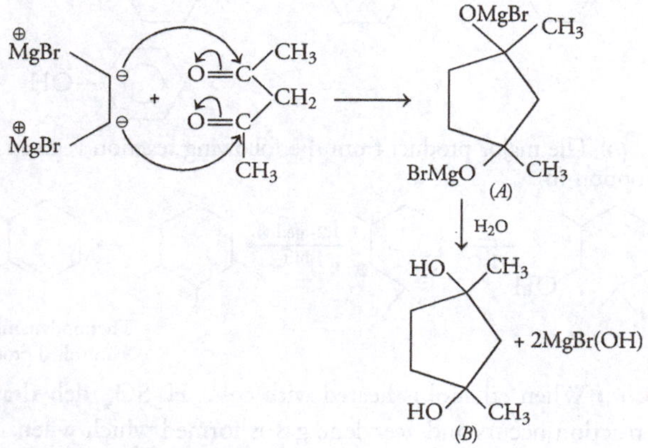

**32.** *(b)*

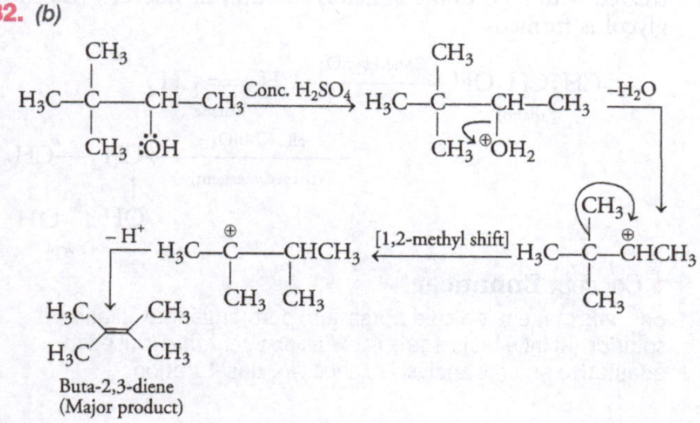


**33.** *(c)* Complete mechanism of given reaction is as follows
**Hydrolysis process.** Acid catalysed hydration of alcohol group.

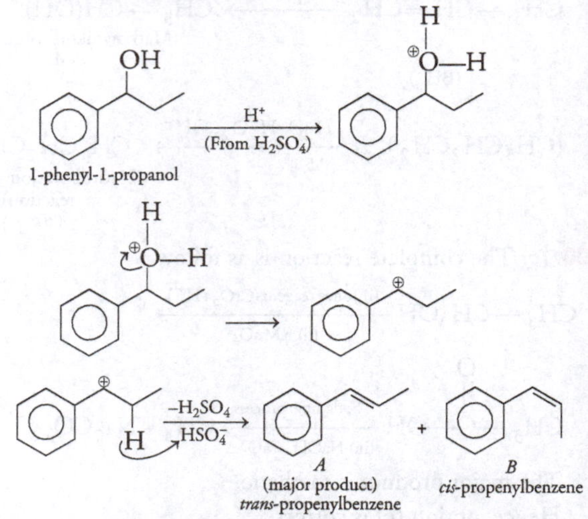


*trans* alkenes are more stable than *cis* alkene, so product *A* is major product formed in the reaction.

**34.** *(b)* Complete reaction with mechanism is as follows

(i) **Hydrolysis process** Attack of $H^+$ on $OH^-$ ion.

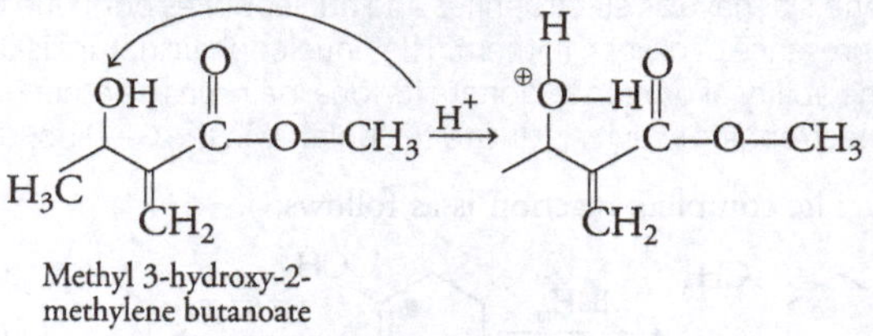


(ii) Removal of water molecule/formation of carbocation.

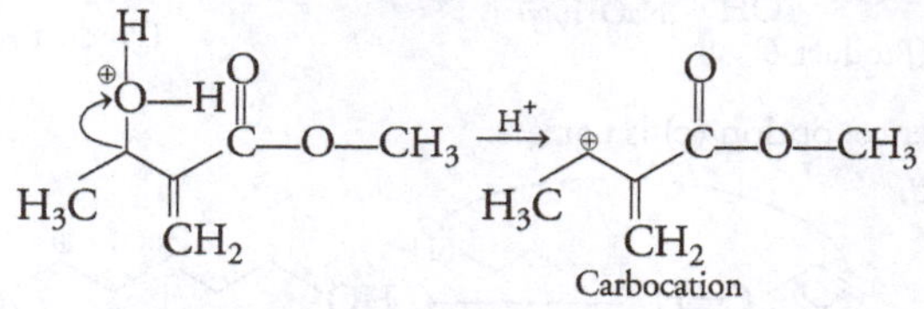


(iii) **Resonance** Movement of double bond/shifting of carbocation.

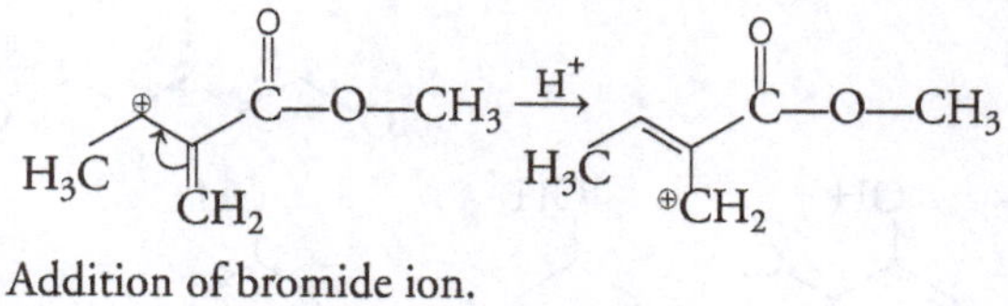

(iv) Addition of bromide ion.

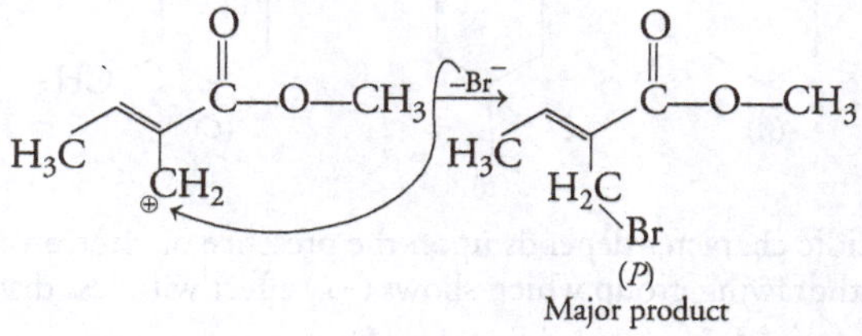

**35.** *(d)*

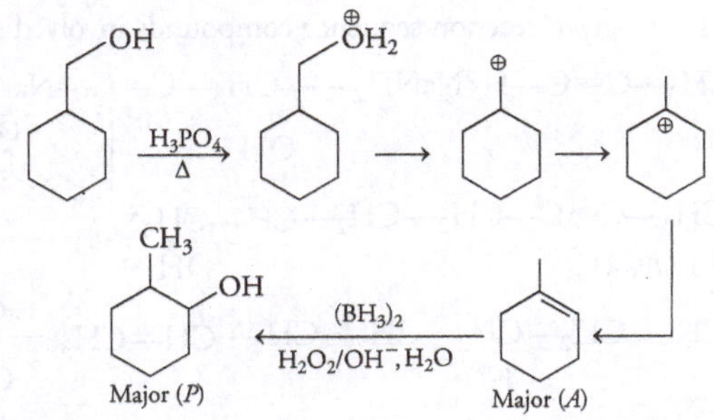


Conversion of compound *A* to compound *P* in undergoes through hydroboration-oxidation reaction that converts an alkene into an alcohol.

**36.** *(c)* In the given reaction, Grignard reagent is used for 1,2-addition in the carbonyl carbon and form alcohol as a major product.

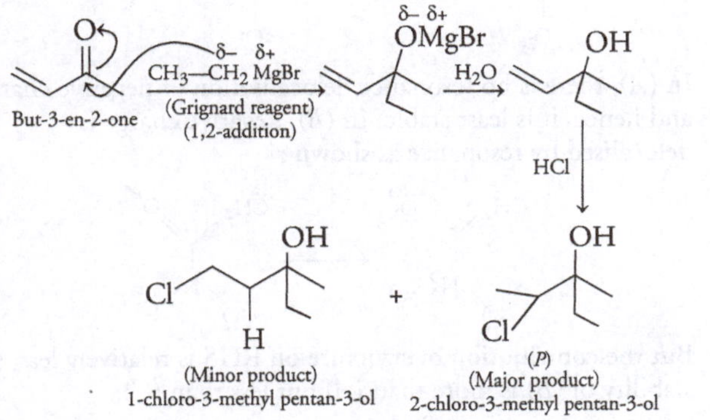


Addition of HCl at the alkene gives 2-chloro-3-methylpentan-3-ol as major product.

**37.** *(d)* Steps involved in this reaction are as follows

**Step 1** $HCl \longrightarrow H^+ + Cl^-$

$H^+$ attacks on —OH (lone pair) and formed $\overset{\oplus}{O}H_2$ ion. Here, $\overset{\oplus}{O}H_2$ is the good leaving group.

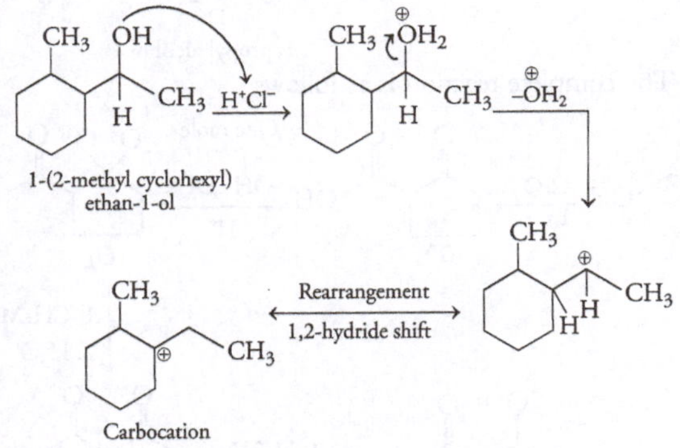


**Step 2** Elimination reaction always give minor product than substitution reaction.

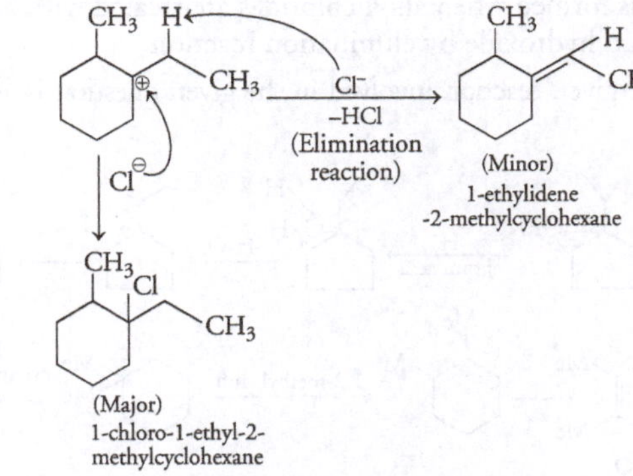


**38.** *(c)* The reaction proceeds as follows:

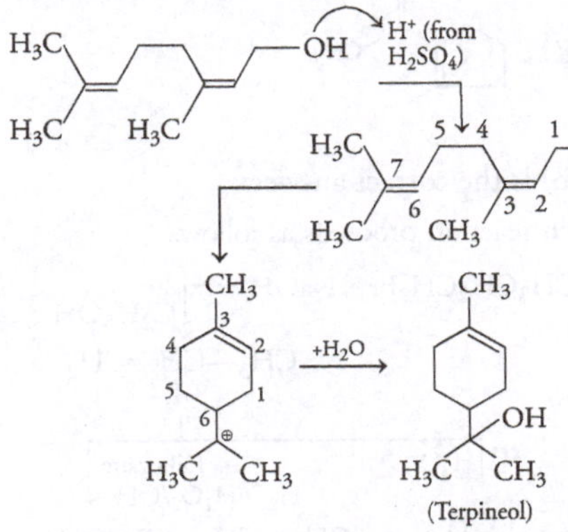


**39.** *(a)*

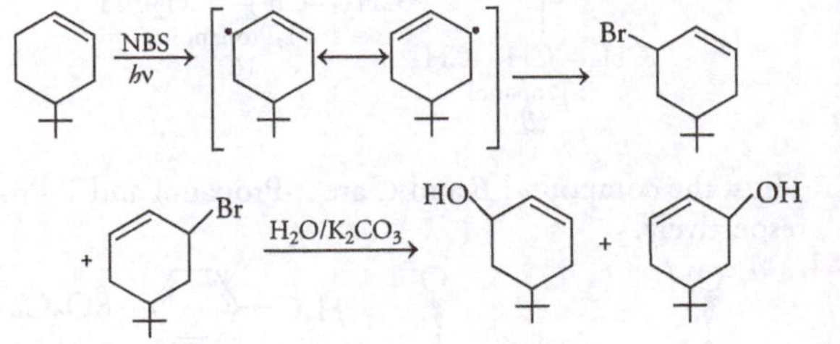


**40.** *(c)*

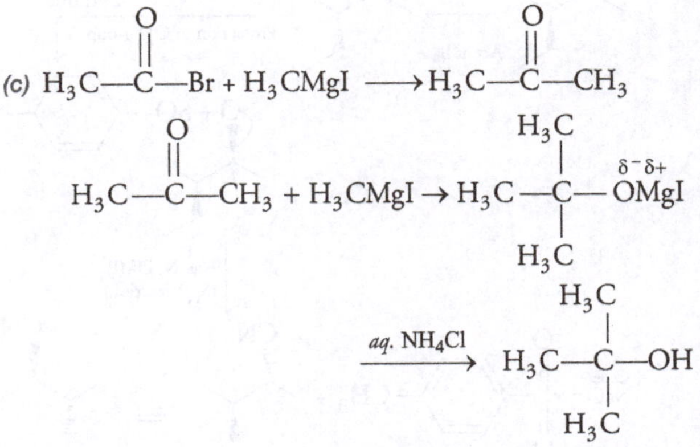


2-methyl-2-propanol

**41.** *(c)* Statement I is incorrect but Statement II is correct. The corrected form of statement I is

Alkene is formed when alkyl chlorides are treated with aqueous potassium hydroxide by elimination reaction.

**42.** *(b)* The given reaction involved in the given question is as follows

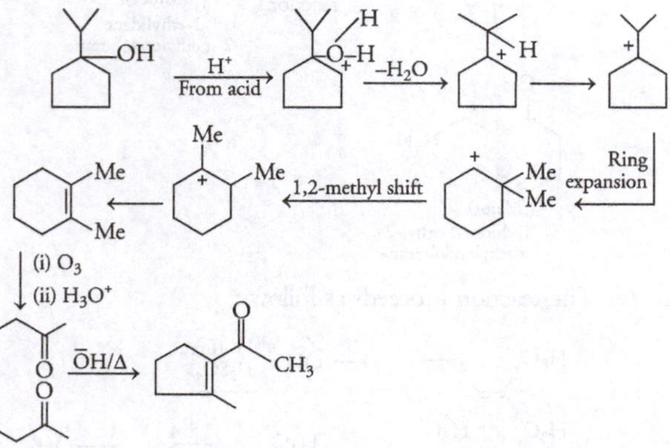


So, option (b) is the correct answer.

**43.** *(c)* The given reaction proceeds as follows

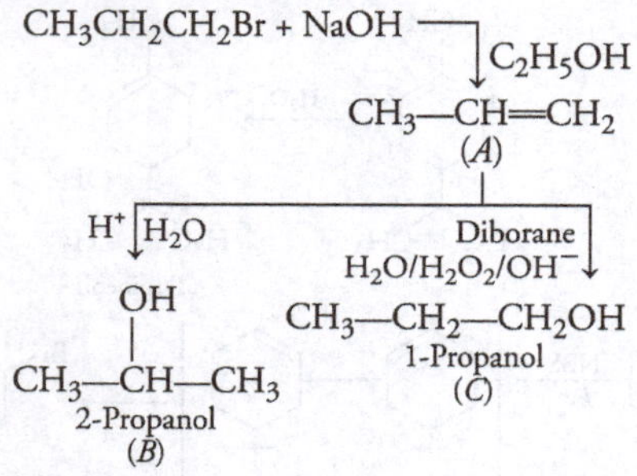


Thus, the compound *B* and *C* are 2-Propanol and 1-Propanol respectively.

**44.** *(b)*

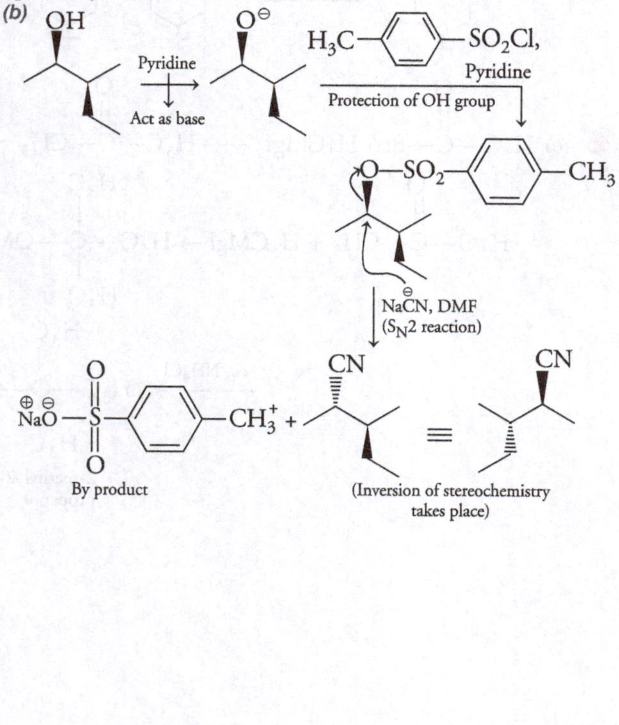


**45.** *(a)* In the given reaction sequence compounds involved are :

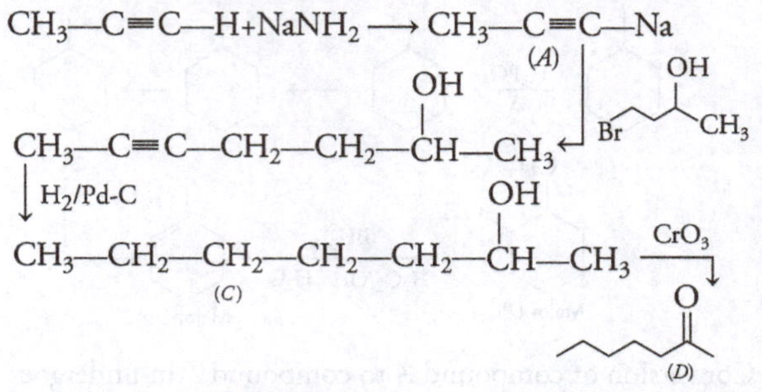


**46.** *(a)* (*C*) is most stable due to strong $-I$ effect as well as $-R$ (or $-M$) effect of '$—NO_2$' group. The negative charge is delocalised as represented by the following resonance structure

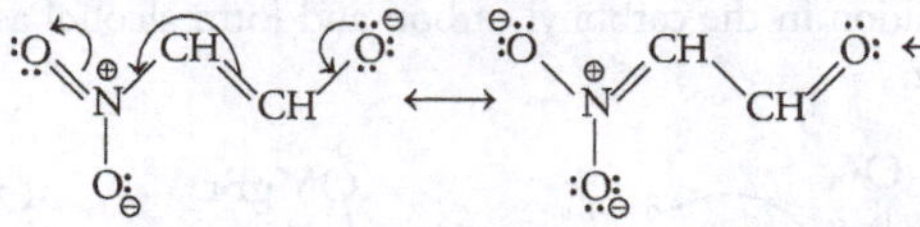

In (*A*), there is no resonance delocalisation of negative charge, and hence it is least stable. In (*B*), negative charge is delocalised by resonance as shown :

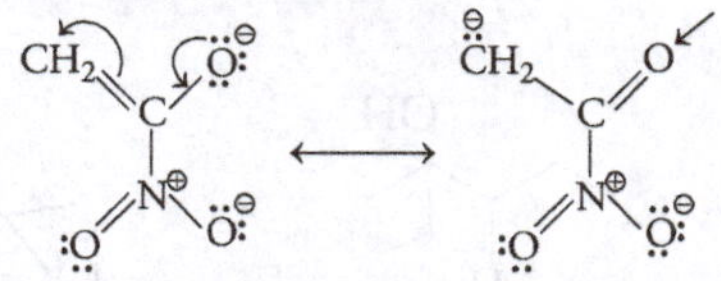

But the contribution of structure on RHS is relatively less, so stability of (*B*) is more than (*A*) but less than (*C*).

**47.** *(c)*

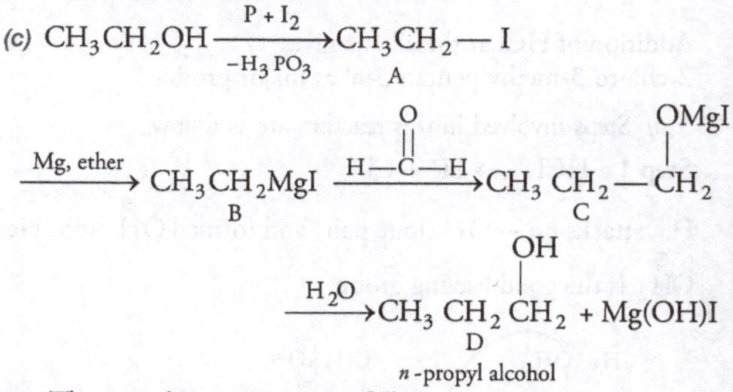


**48.** *(13)* The complete reaction is as follows

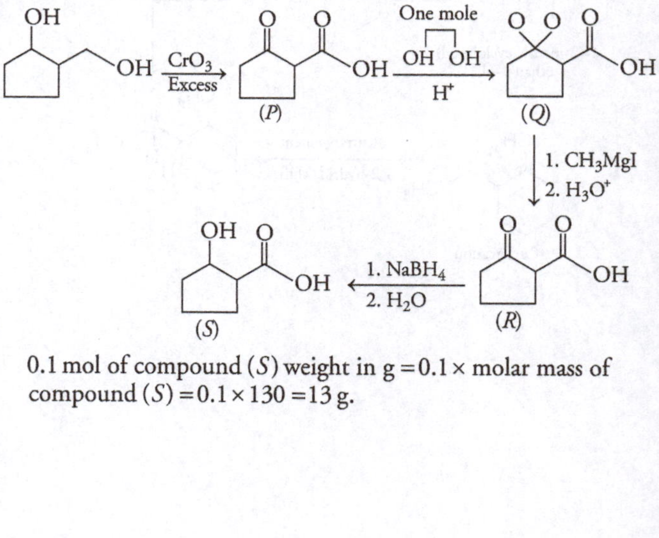


0.1 mol of compound (*S*) weight in g $= 0.1 \times$ molar mass of compound $(S) = 0.1 \times 130 = 13$ g.

**49.** *(2)*

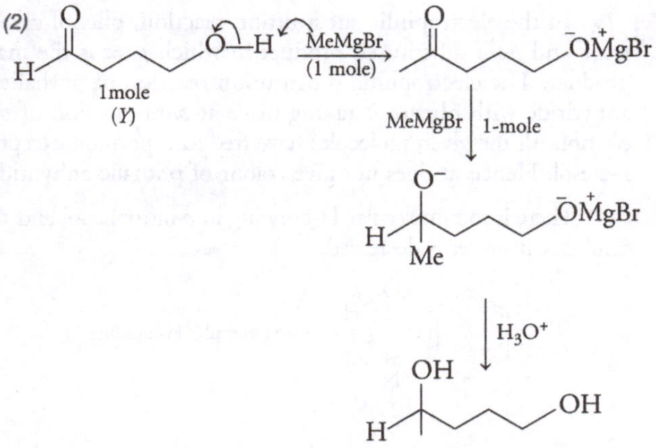

∴ 2 moles of MeMgBr are involved in this reaction and 1 mole of reactant. Thus $X = 2$ and $Y = 1$

$$\Rightarrow \frac{X}{Y} = \frac{2}{1} = 2$$

**50.** *(8)* The complete reaction is as follows

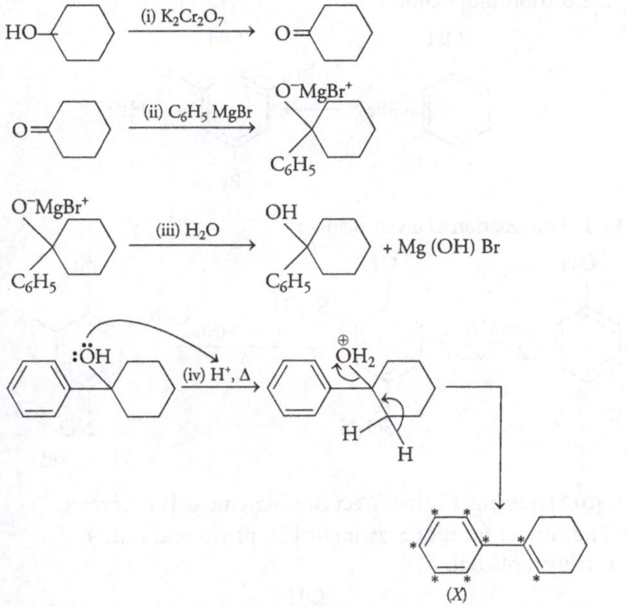

(X)

Number of $sp^2$ hybridised carbons in compound $X$ is 8.

**51.** *(2)* Two moles of $CH_3MgBr$ is required as follows.

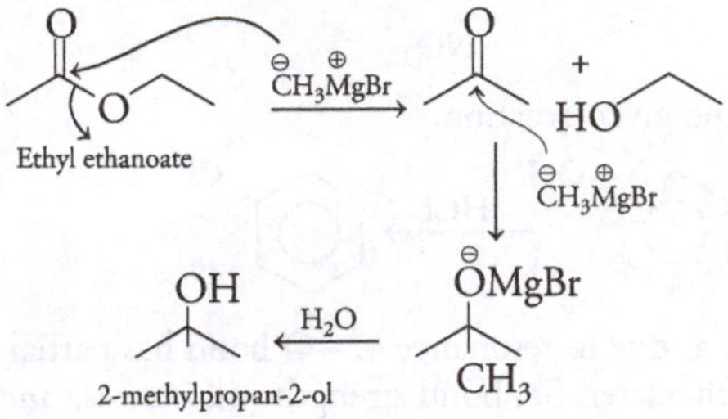

## Topic 2

**01.** *(c)* Among the given compounds, compound given in option (c) will react with HBr to give a phenol.

The complete reaction is as follows,

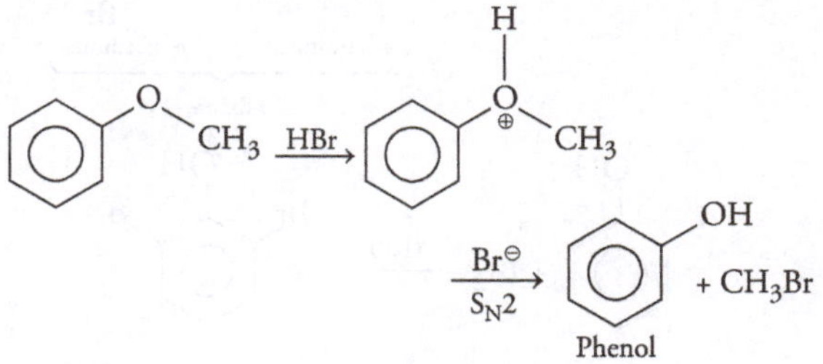

**02.** *(c)* The complete Reimer-Tiemann reaction is

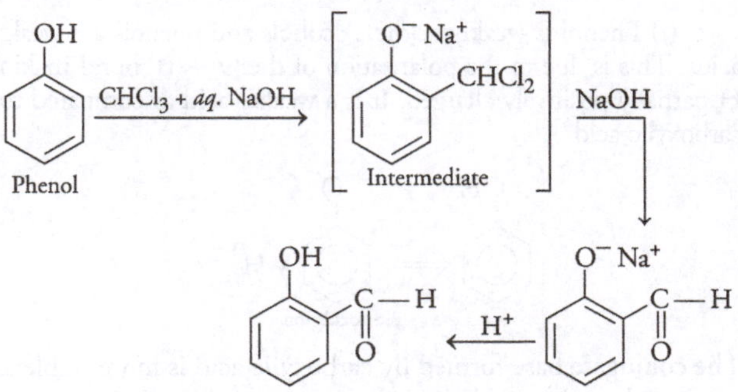

Hence, option (c) is correct.

> • Concept **Enhancer**
>
> Riemer-Tiemann reaction is an electrophilic substitution reaction. In this reaction, dichlorocarbene acts as an electrophile this dichlorocarbene is formed by abstraction of hydrogen from chloroform that creates a trichlorocarbanion, which loses a chloride ion to form dichlorocarbene.

**03.** *(b)* Salicylaldehyde is synthesised from phenol, when reacted with NaOH, $CHCl_3$. The complete reaction is as follows,

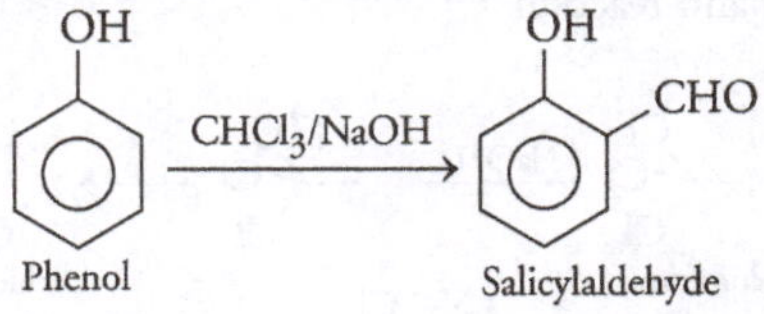

Hence, option (b) is correct.

**04.** *(a)* Among the given options *m*-nitrophenol i.e.,

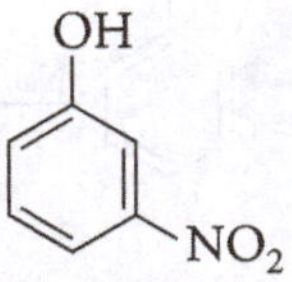

is most acidic due to presence of —$NO_2$ group which has more electron withdrawing nature. Hence, increases the acidic character.

**05.** *(b)* Phenol reacts differently with bromine in chloroform and bromine in water forming/yielding different products. The difference in the products obtained is due to the difference in polarity of the solvents.

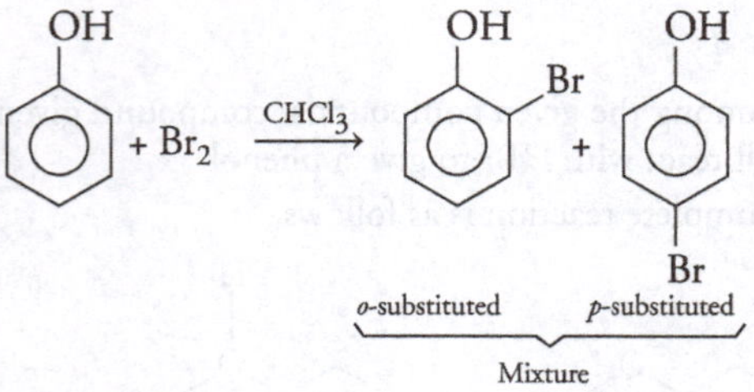


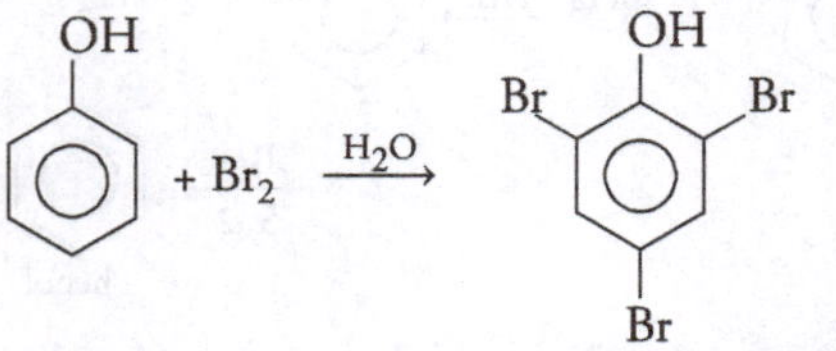


*tri*-substituted product

**06.** *(c)* (I) Phenol is weakly acidic. Alcohols and phenols are weak acids. This is due to the polarisation of the O—H bond making H partially positively charged. It is a weaker acid as compared to carboxylic acid.

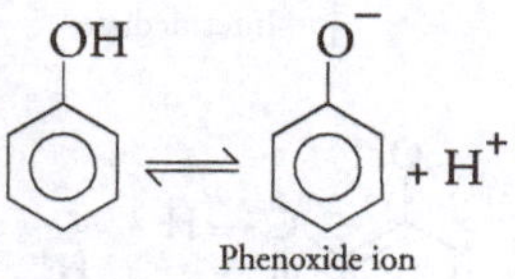


Phenoxide ion

The conjugate base formed by carboxylic acid is more stable as compared to phenoxide ion.

(II) Phenols are freely soluble in NaOH solution and are stronger acids than alcohols and water. This is due to more stability of phenoxide ion (conjugate base formed) as negative charge on the oxygen atom is delocalised on the ring (resonance) as compared to stability of alkoxide (conjugate base of alcohol) and hydroxide ion (conjugate base of water)

Hence, Statement I is correct but Statement II is incorrect.

**07.** *(c)* Phenol in presence of sodium hydroxide reacts with chloroform to form salicylaldehyde. The reaction is known as Reimer-Tiemann reaction.

**Mechanism**

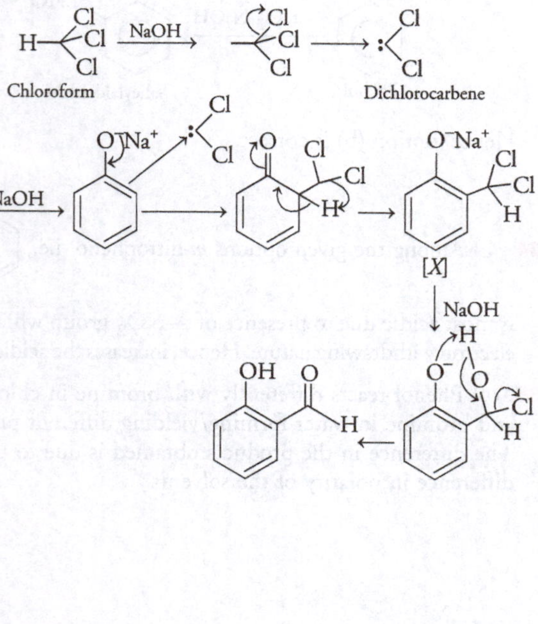


**08.** *(b)* In the electrophilic substitution reaction, phenol gives *ortho* and *para* substituted product in which *para* is the major product. The electrophilic substitution reaction of phthalic anhydride with phenol is taking place at *para* position of the phenol. All the given molecules have free *para* position except in *p*-cresol. Hence, it does not give colour of phthalic anhydride.

**09.** *(b)* There is intramolecular H-bonding in *o*-nitrophenol and thus solubility in water is decreased.

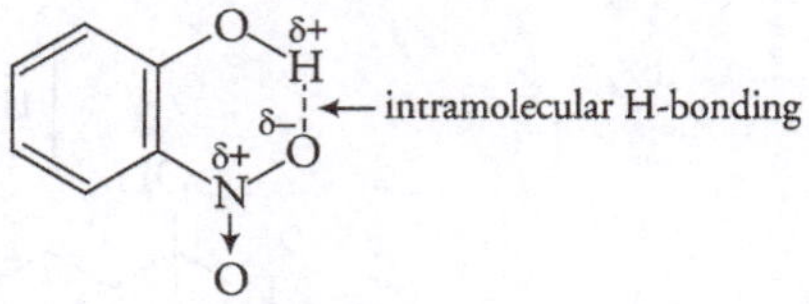


**10.** *(d)* $Br_2$ is formed by a redox reaction.

$$5Br^- + BrO_3^- + 6H^+ \longrightarrow 3Br_2 + 3H_2O$$

—OH group is the activating group and thus activates the benzene towards electrophilic substitution reaction at *o*-and *p*-positions giving yellowish white precipitate of 2,4,6-tribromophenol.

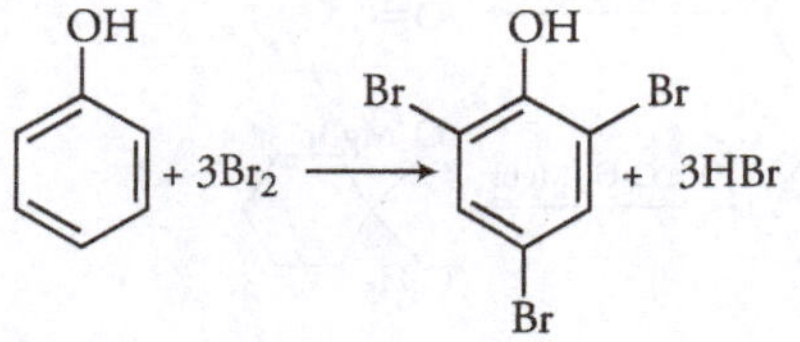


**11.** (*) The reaction is given below :

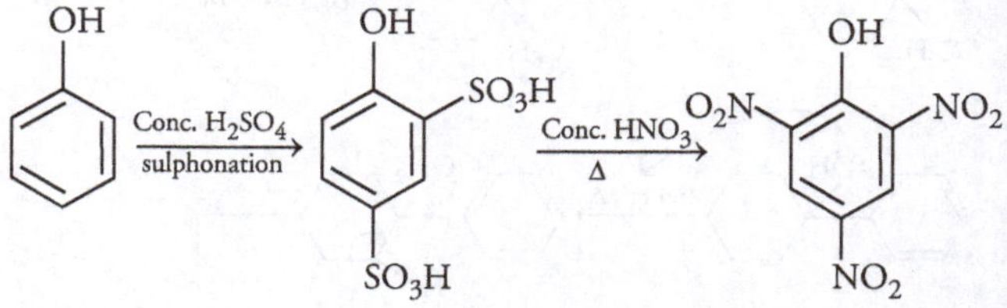


Picric acid

**12.** *(b)* Statement I is incorrect but Statement II is correct. The correct form of Statement I is, picric acid is 2, 4, 6-trinitrophenol.

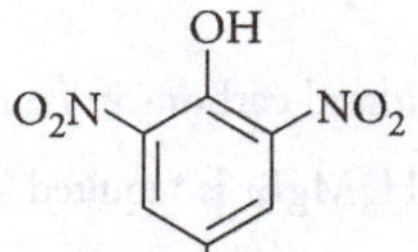


**13.** *(a)* Among the given reaction,

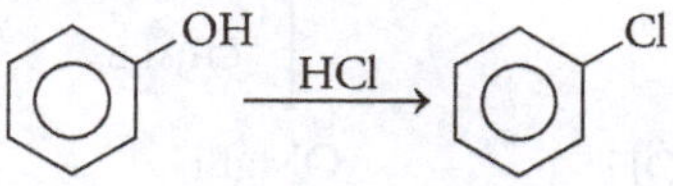


is not possible as due to resonance C—O bond has partial double bond character. So, bond strength will increase and substitution will be less likely.

**14.** *(b)* Statement I is false but Statement II is true.

The correct form of Statement I is,

Phenol is a highly activated compound which can undergo bromination directly with bromine without any Lewis acid.

**15.** *(c)* The complete reaction is as follows

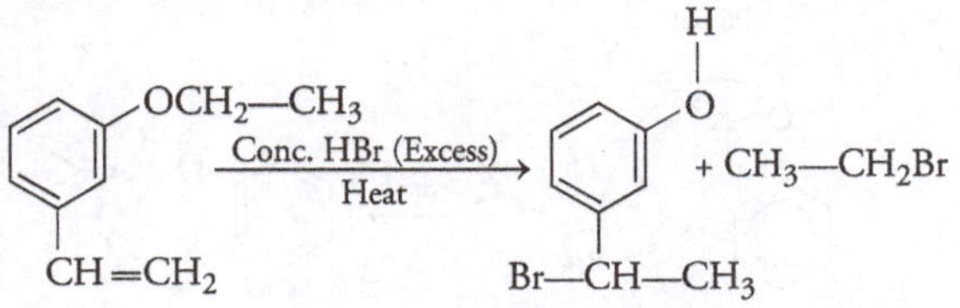


(*P*)

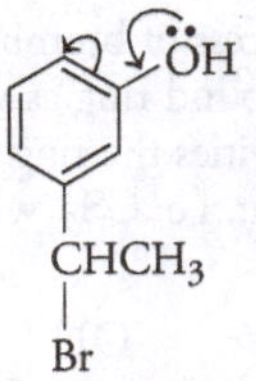


does not react further with HBr due to double bond character between C—O bond.

Hence, option (c) is correct.

**16.** *(c)* Statement I is true but Statement II is false.

The correct form of Statement II is :

Ethanol will not give immediate turbidity with Lucas reagent. The reason is that ethanol is a primary alcohol.

Tertiary alcohol gives immediate turbidity while secondary alcohol give turbidity within five minutes.

**17.** *(c)* The reactivity of benzene increases towards electrophilic substitution reaction when any substituent increases the electron density.

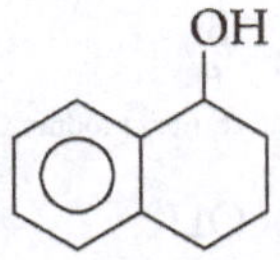


(*a*)

–$CH_2$ has +H effect

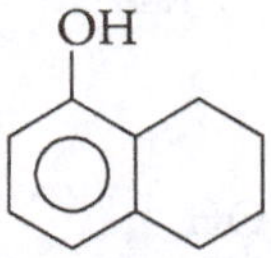


(*b*)

–OH has +*R* effect while –$CH_2$– group has +H effect

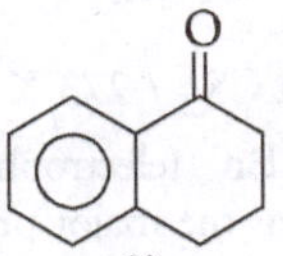


(*c*)

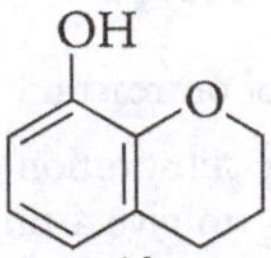


(*d*)

–OH and –O– both shows +*R* effect

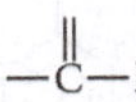

has, – *R* effect

Thus, the correct order is $c < a < b < d$.

**18.** *(c)* The correct order for acidity of the given hydroxyl compound is $E > C > D > A > B$.

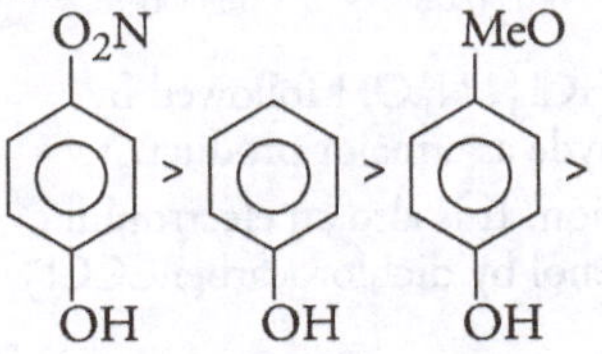


**19.** *(b)* Correct option is (b).

The organic compound that burns with sooty flame is phenol. Saturated hydrocarbon undergo complete combustion and give clean blue flame whereas unsaturated hydrocarbons burn with sooty flame.

Thus,

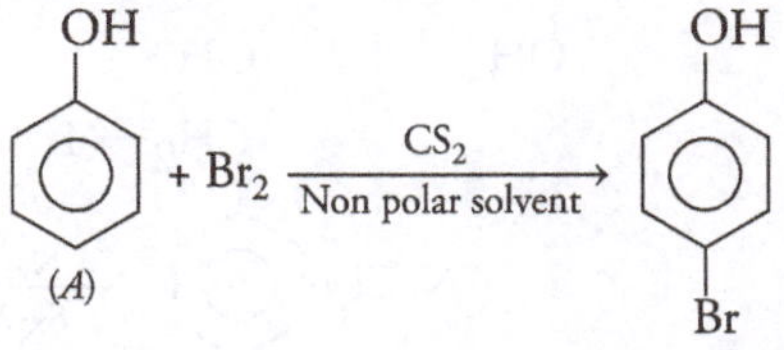


(*A*)

(*B*) (Major)
4-bromophenol

**• Concept Enhancer**

Unsaturated carbon compounds do not burn completely. Hence, they give yellow coloured flame that tends to pollute, such flames are called sooty flames. While the saturated compounds, burns with blue coloured non-sooty flame.

**20.** *(a)*

(*A*) 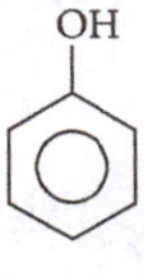


Phenol (Original solution)

(*B*) 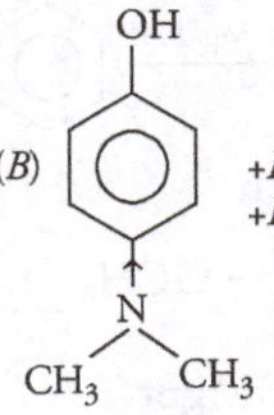


+*I*-effect, +*R*-effect

(*C*) 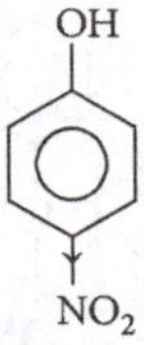


–*I*-effect, –*R*-effect

(*D*) 

+*I*-effect, +*R*-effect

$K_a$ value increases by electron withdrawing group i.e., $NO_2$.

Hence, the order of $K_a$ value will be $C > A > D > B$.

Also, we know that, $pK_a \propto \frac{1}{K_a}$

So the order of, $pK_a$ value is $B > D > A > C$.

**21.** *(b)* The increasing order of $pK_a$ for the given phenol is A, B, E, D, C. Presence of electron withdrawing group increases the acidic nature of the molecule and hence decreases the $pK_a$ value for an acid.

$$\left[\text{Acidic nature} \propto \text{electron withdrawing group} \propto \frac{1}{pK_a}\right]$$

Hence, correct order for increasing $pK_a$ value

[2, 4-dinitrophenol, 4-nitrophenol, 3-chlorophenol, phenol, 2, 4, 5-trimethylphenol.]

**22.** *(a)*

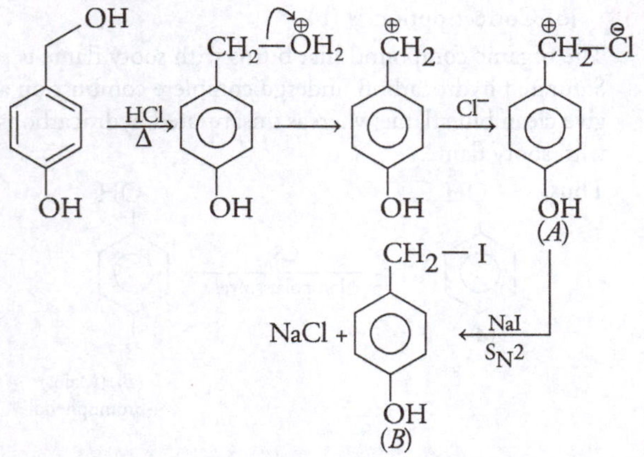

The final product formed in the given reaction is 4-iodomethylphenol.

**23.** *(b)* In the given reaction, starting compound is bromobenzene.

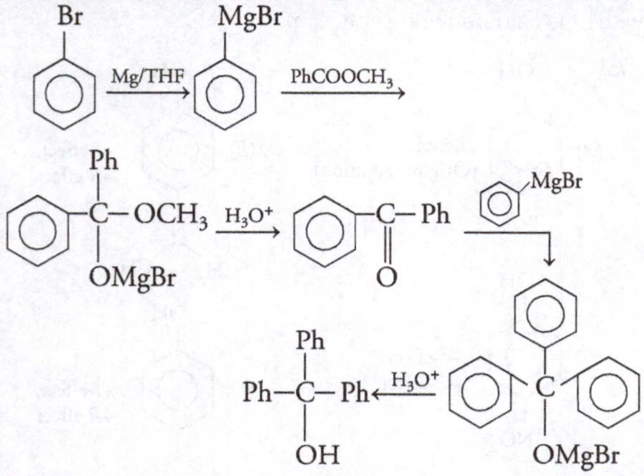

**24.** *(c)* Preparation of phenol from cumene hydroperoxide through air oxidation.

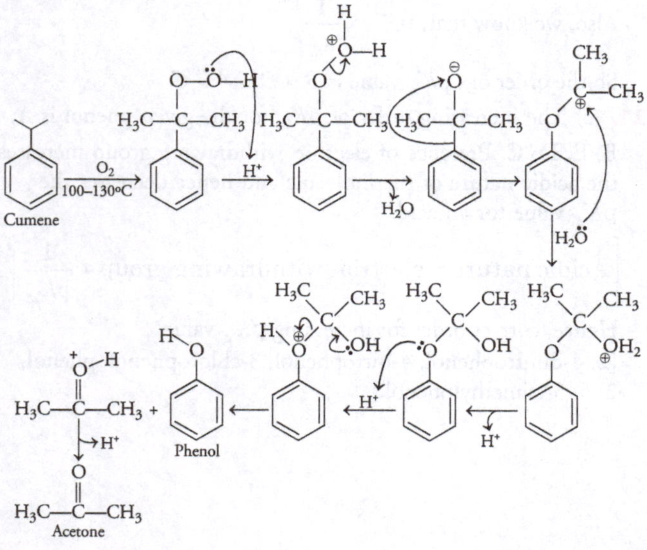

**25.** *(d)* Here, in the question

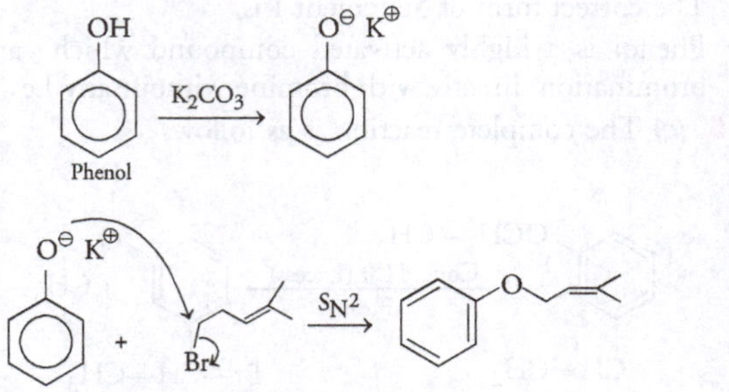

So, here option (d) is correct.

**26.** *(b)* Bromination of phenol with excess of bromine and water will result in multiple substitution around ring as water ionises phenol to phenoxide ion, which activities the ring for trisubstitution, while less polar solvent, i.e. $CS_2$ will give mono-substituted compound.

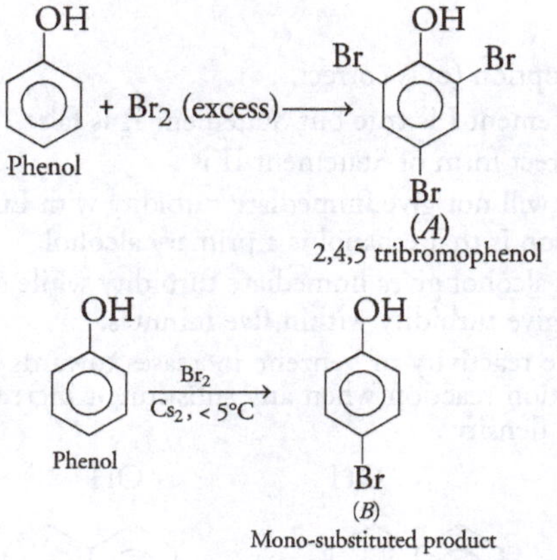

**27.** *(b)*

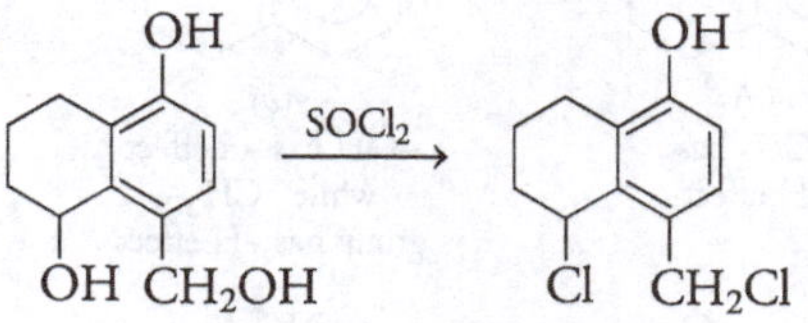

**28.** *(b)* Phenol on reaction with $Br_2$ in $CS_2$ / 273 K undergoes an electrophilic substitution reaction by $Br^+$ (electrophile) in aprotic solvent $CS_2$ to give 4-bromophenol as the major product.

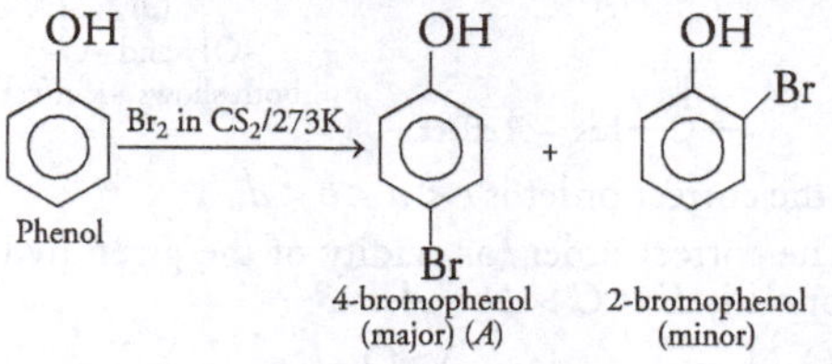

Phenol on reaction with $CHCl_3$, NaOH followed by hydrolysis gives salicylaldehyde as a major product.

It is Reimer-Tiemann reaction. It is also an electrophilic substitution reaction of phenol by dichlorocarbene $CCl_2$ (electrophilic).

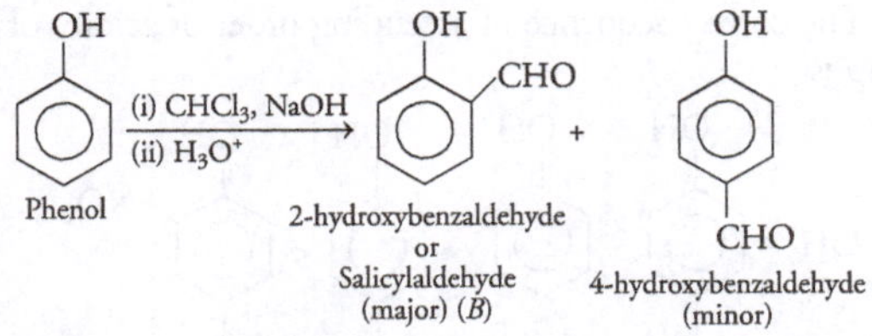


**29.** *(d)* Statement I is true. While statement II is false, because, due to the presence of intramolecular hydrogen bonding, boiling point and melting point of *o*-nitrophenol will be lower.

**30.** *(b)* Firstly phthalic anhydride in presence of conc. $H_2SO_4$ undergoes protonation to give an intermediate carbocation. This carbocation reacts with 2-propylphenol in presence of NaOH to give pink colour compound.

**Mechanism**

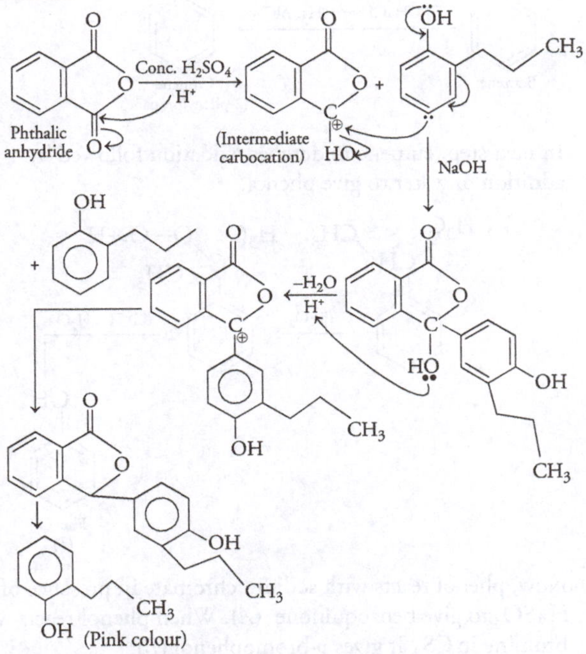


**31.** *(c)* Ethyl magnesium bromide is a Grignard reagent (GR), it constitutes $C_2H_5^-$ [$C_2H_5^- \overset{\oplus}{M}gBr$ in ether/aprotic medium] which can act as nucleophile as well as strong base. Bromine water ($Br_2/H_2O$) gets decolourised with phenol derivatives.

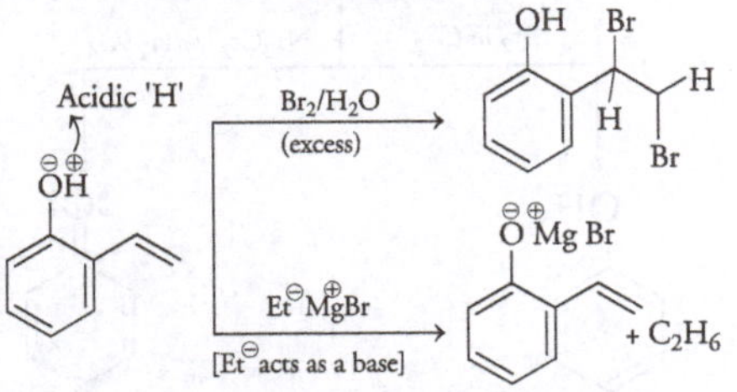


**32.** *(b)* Acidic strength is inversely proportional to $pK_a$ value. The acidity of phenols is due to greater resonance stabilisation of phenoxide ion relative to phenol.

Thus, order of acidic strength of nitrophenol is *p*-nitrophenol > *m*-nitrophenol and the correct order of the $pK_a$ values of give option is

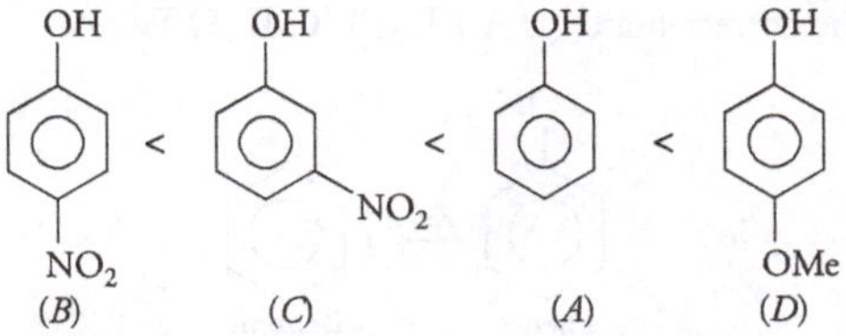


**33.** *(a)* Aspirin formation takes place as follows :

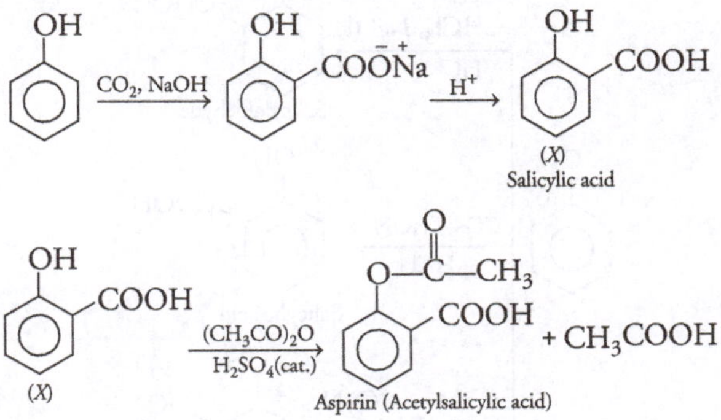


**34.** *(c)*

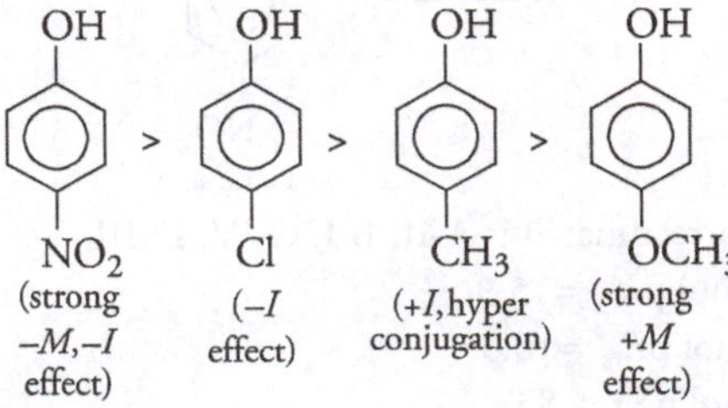


Electron releasing group decreases while electron withdrawing group increases acidic strength by destabilising and stabilising the phenoxide ion formed respectively.

**35.** *(c)* The correct match is A-IV, B-III, C-II, D-I

**36.** *(b)* Complete reaction sequence is as follows

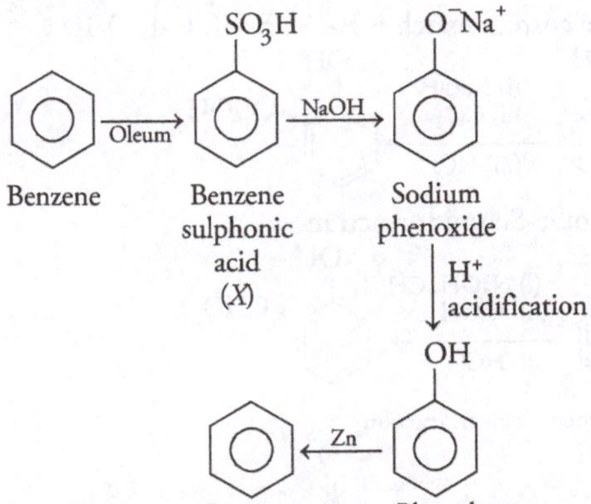


**37.** *(d)*

**Time Saver Tech**

Use elimination technique as time saver. As preparation of phenol from aniline i.e. A-II is given is option (d) only.

The correct match is A-II, B-IV, C-I, D-III.

**38.** *(c)* The correct match is A-III, B-I, C-II, D-IV.

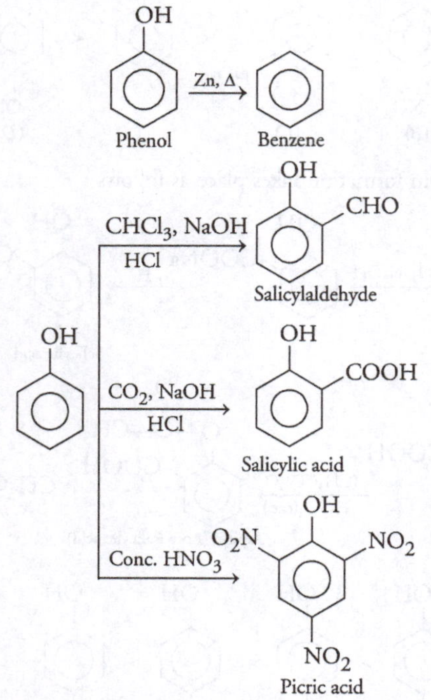


**39.** *(d)* The correct match is A-II, B-I, C-IV, D-III.

Ethanol $pK_a = 15.9$

Phenol $pK_a = 10.0$

*m*nirophenol $pK_a = 83$

*p*-nitrophenol $pK_a = 7.1$

• Concept **Enhancer**

The acidic strength of an acid can be determined with the help of $pK_a$ value it is the long inverse of dissociation constant of an acid. Hence, smaller is the value of $pK_a$ more strong is the acid.

**40.** *(d)* The correct match is A-IV, B-III, C-I, D-II.

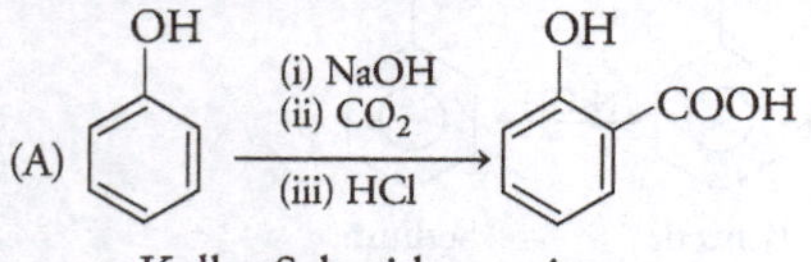


Kolbe-Schmidt reaction

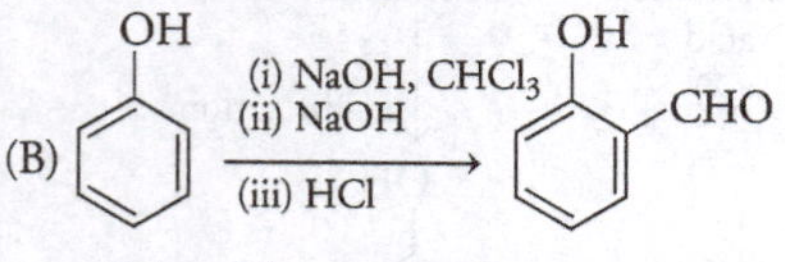


Reimer-Tiemann reaction

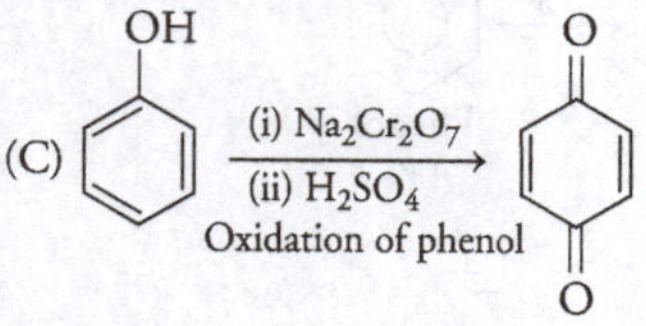


Oxidation of phenol

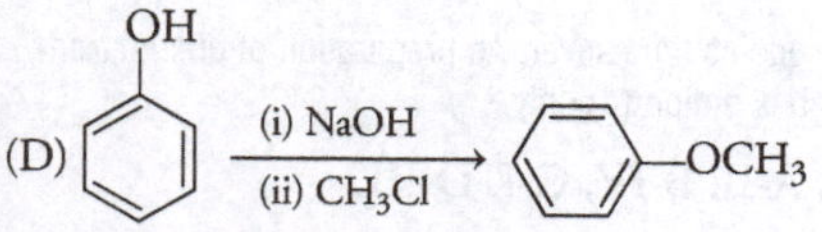


Hence, option (d) is correct.

**41.** *(a)* The correct sequence of ascending order of acidity of —OH group is

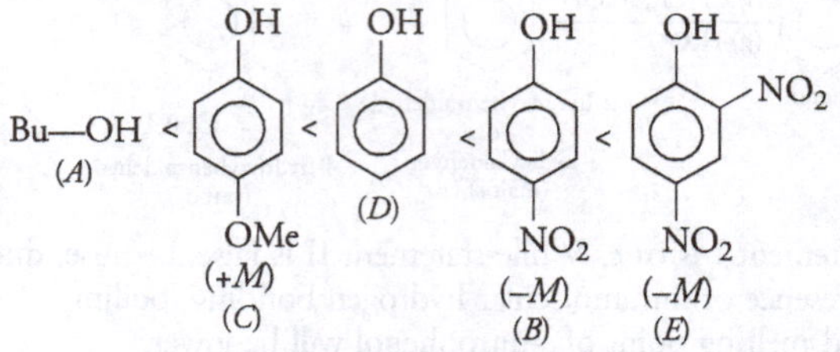


**42.** *(b)* Benzene undergoes Friedel-Craft alkylation reaction in presence of $AlCl_3$ and reacts with alkyl halide.

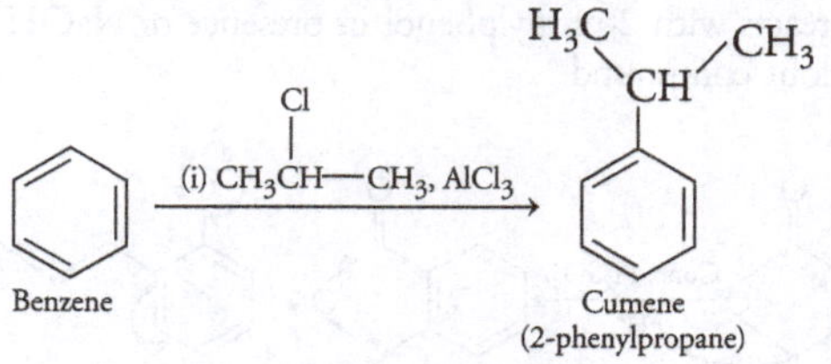


In next step, cumene undergoes oxidation followed by addition of water to give phenol.

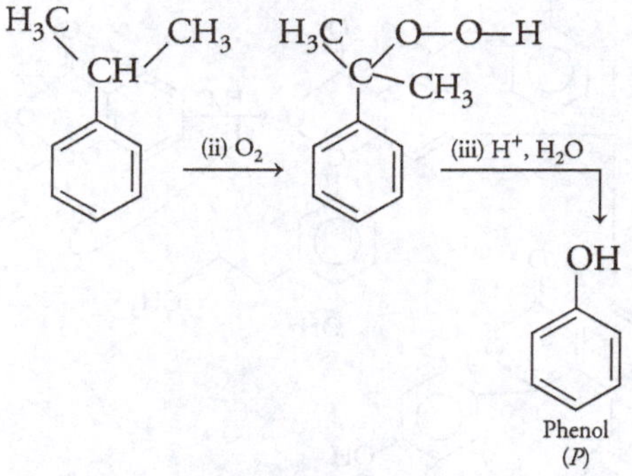


Now, phenol reacts with sodium chromate in presence of $H_2SO_4$ to give benzoquinone (*A*). When phenol reacts with bromine in $CS_2$ it gives *p*-bromophenol (*B*).

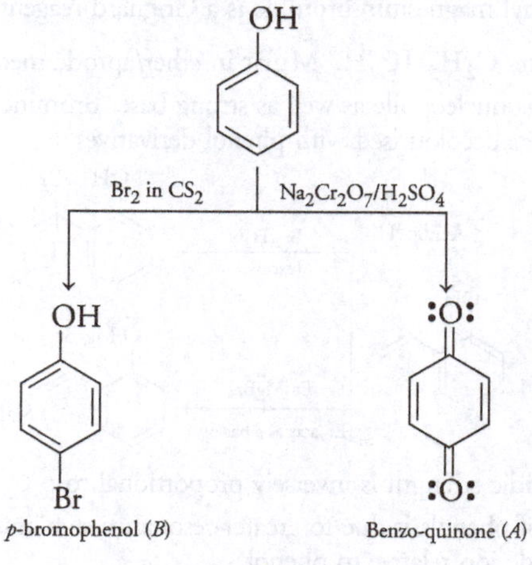


**43.** *(a)* Correct matching is A-IV, B-III, C-II, D-I.

**44.** *(30)*

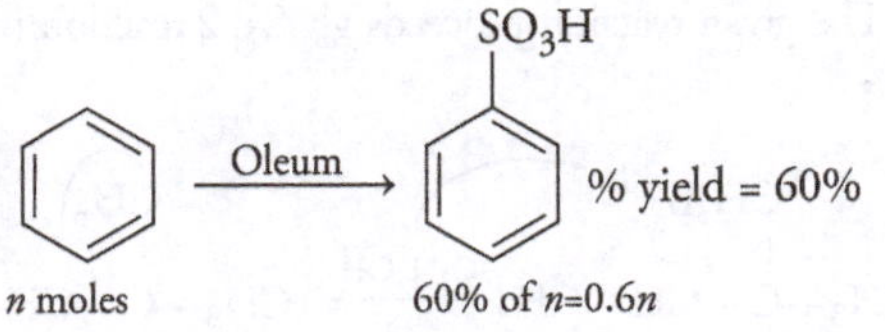

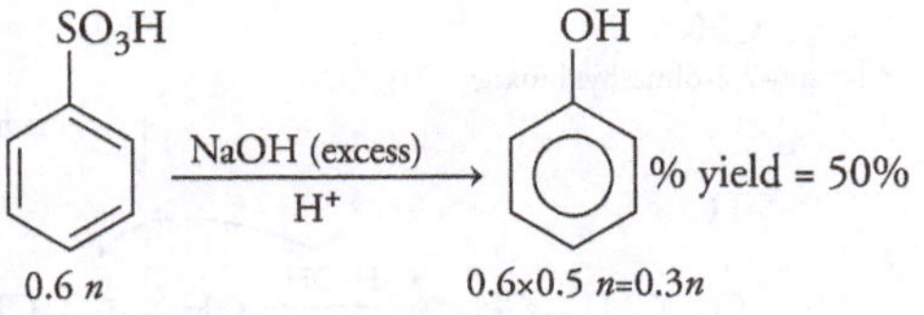

Overall % yield of reactions $= \frac{0.3n}{n} \times 100 = 30$

**45.** *(7)* Phenol on reaction with mixed acid forms picric acid. Picric acid contains seven oxygen atoms.

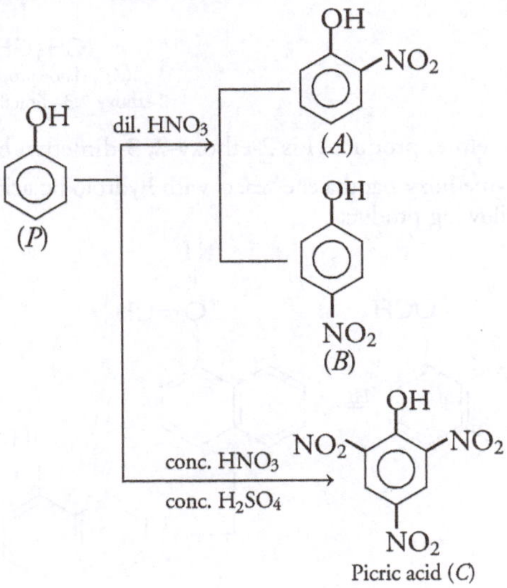

## Topic 3

**01.** *(d)* Suitable reaction condition for preparation of methyl phenyl ether is $PhO^-Na^+$ and MeBr.

$$PhO^{\ominus}Na^{\oplus} + Me—Br \xrightarrow{S_N2} Ph—O—Me$$

**02.** *(a)* A is true but R is false.

It is because that acetals and ketals are *gem*-di ethers (Lewis bases). Hence, they should be stable in basic medium. However, it will break in acidic medium. Hence, A is true, while R is false because alkoxide ion [$RO^-$] is not a good leaving group. Hence, it has poor leaving tendency.

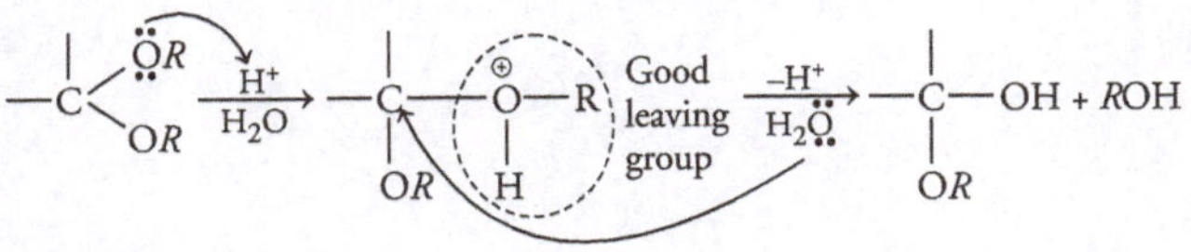

**03.** *(c)*

> **key idea** Ethers are least reactive functional groups. The cleavage of C—O bond in ethers take place under drastic conditions with excess of HX.

Substituted phenols react with aq. NaOH to form sodium phenoxides which on reaction with $CH_3I$ undergoes $S_N2$ reaction to give 2-methoxy-1-methyl benzene.

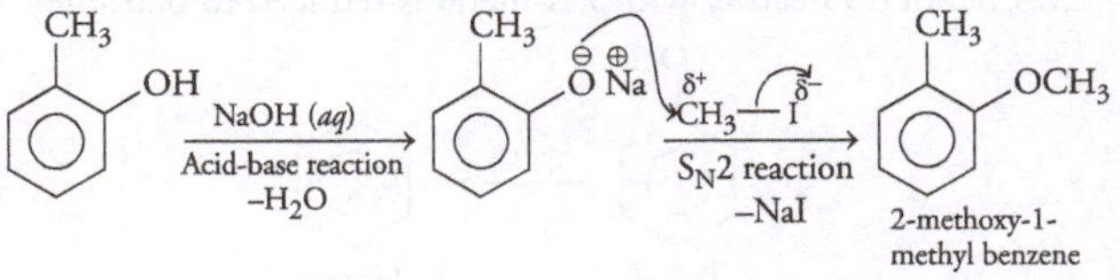

**04.** *(d)* Alcohol has polar H which makes intermolecular H-bonding possible. Ether is non-polar, hence has no H-bonding. Lack of H-bonding in ether makes it more volatile than alcohol.

**05.** *(a)* Reaction mention in (a) will not lead to desired ether formation in major proportion.

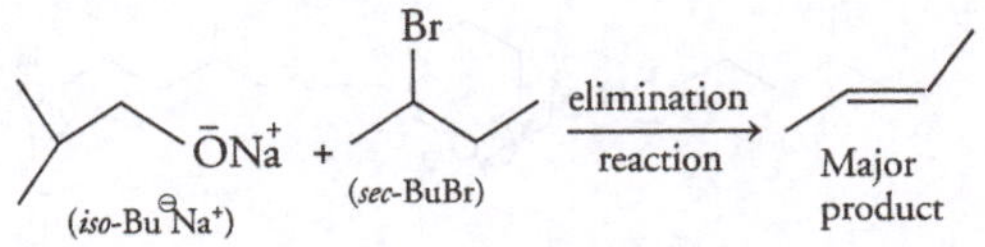

**06.** *(b)* Both Statement I and Statement II are true.

**07.** *(b)* Both statement I and statement II are true.

**08.** *(a)* The complete reaction is as follows,

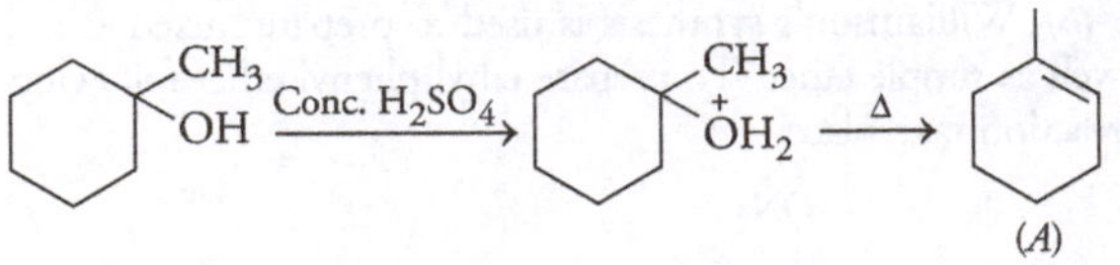

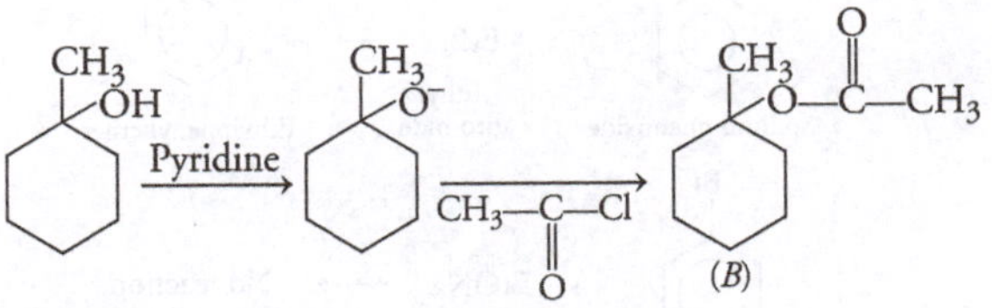

Hence, option (a) is correct.

**09.** *(c)* The complete reaction is as follows,

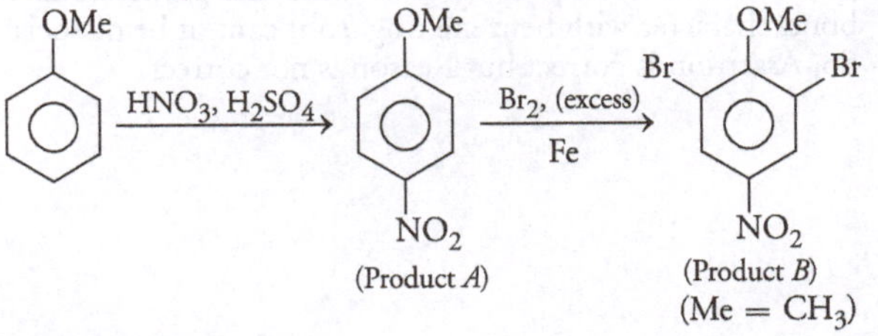

**10.** *(d)* **Reaction mechanism**

1. Proton capture

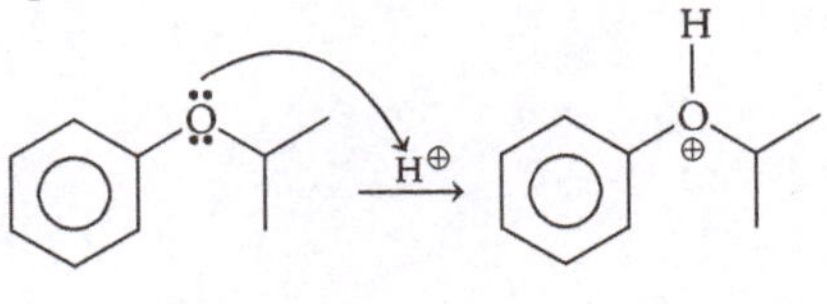

2. Nucleophilic attack of $I^{\ominus}$

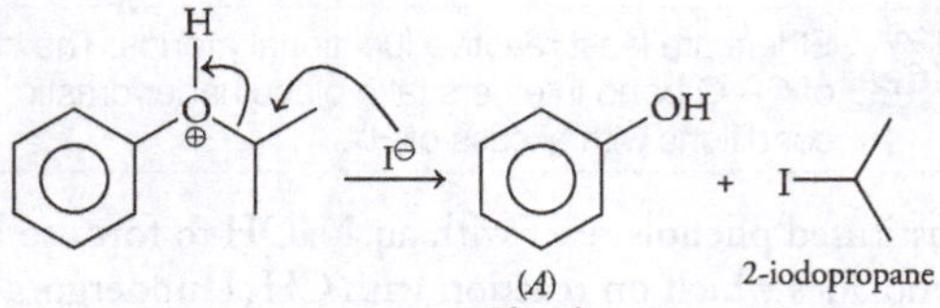

(A) Phenol   2-iodopropane

3. Phenol on heating with Zn-metal is reduced to benzene.

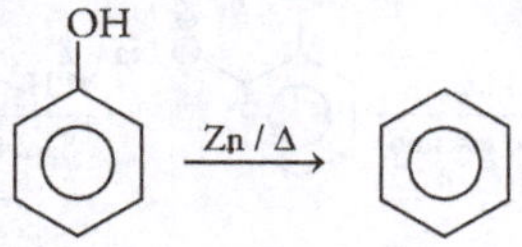

Phenol (A)   Benzene (B)

**11.** *(d)* The mechanism of the given reaction is as follows :

**Step 1.**

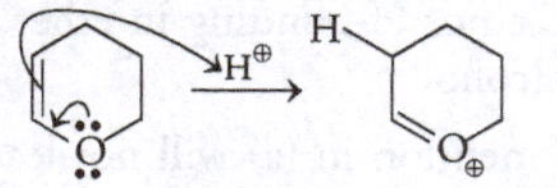

**Step 2.**

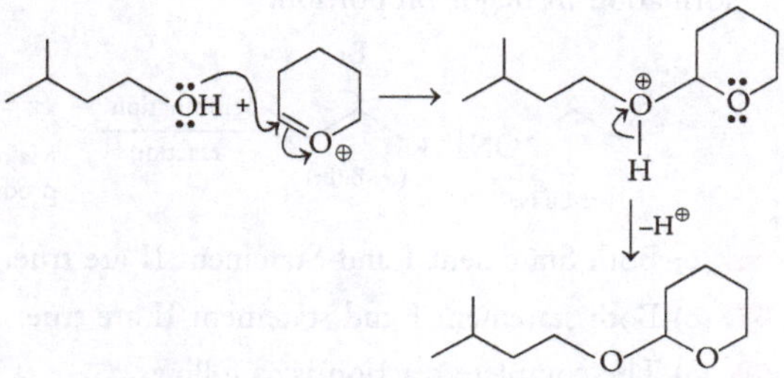

**12.** *(b)* Williamson's synthesis is used to prepare mixed ethers as well as simple ether. To prepare ethyl phenyl ether following reaction take place.

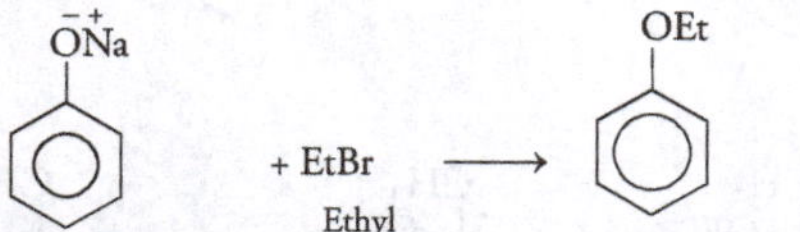

Sodium phenoxide   Ethyl bromide   Ethylphenyl ether

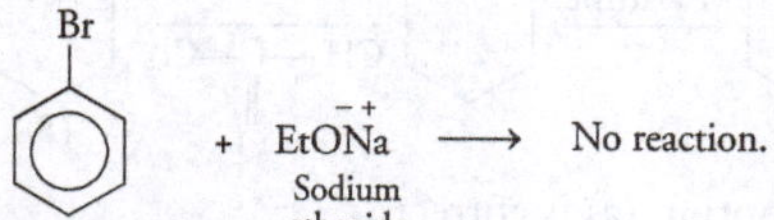

Bromo benzene   Sodium ethoxide   No reaction.

This reaction is not possible as bromine has partial double bond character with benzene ring, so it cannot be dissociated. So, Assertion is correct but Reason is not correct.

**13.** *(c)* The given reaction proceeds via $S_N2$ reaction mechanism as follows

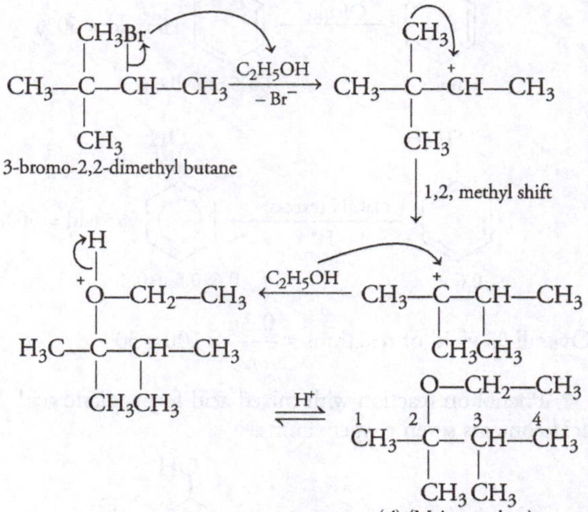

3-bromo-2,2-dimethyl butane

(A) (Major product) 2-ethoxy-2,3-dimethyl butane

∴ Therefore, product *A* is 2-ethoxy-2, 3-dimethyl butane.

**14.** *(b)* 1-methoxy napthalene reacts with hydroiodic acid to form the following products.

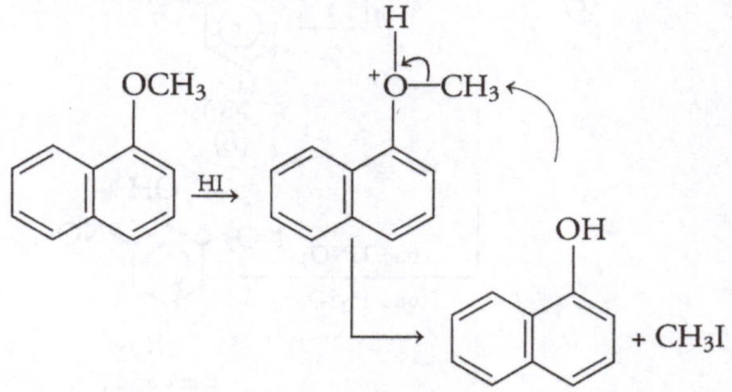

**15.** *(b)* The increasing order of C—OH bond length in the given compounds is : phenol < *p*-ethoxy phenol < methanol. It is explained as follows

Methanol has no resonance and no partial double bond character in C—OH bond.

∴It has maximum C—OH bond length.

In both phenol and *p*-ethoxyphenol, resonance is involved.

∴ C—OH bond has partial double bond character and hence C—OH bond length is less than methanol.

But in *p*-ethoxyphenol, there is +*R*-effect of '$OC_2H_5$' group also.

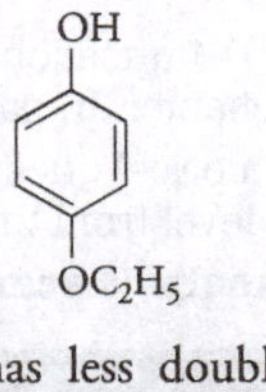


So the 'C—OH' bond has less double bond character and slightly more bond length than C—OH' bond length in phenol.

**16.** *(b)* The major product obtained in the reaction is as follows

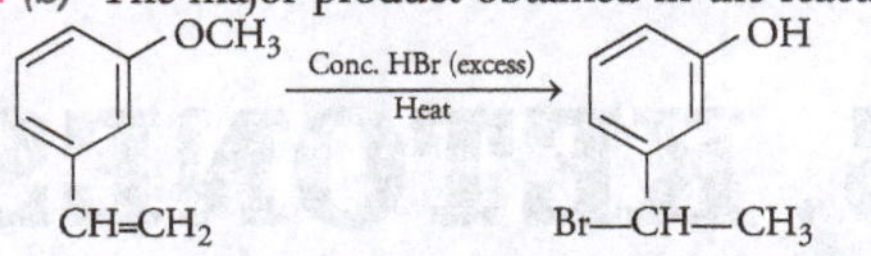


As conc. HBr is in excess. So, reaction will take place at both the substituents.

**17.** *(c)* (a), (b), (d) may be formed but (c) is never formed.

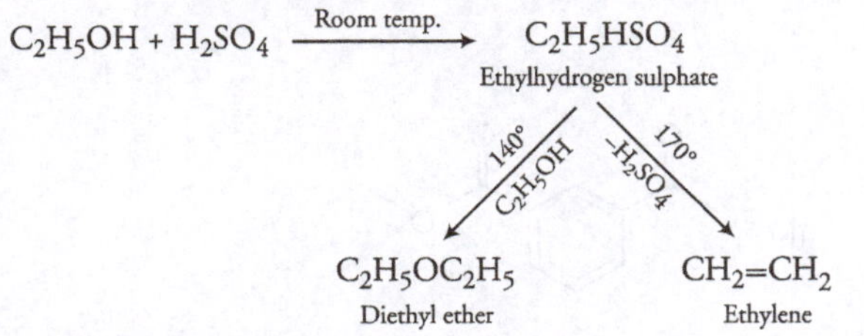


**18.** *(c)* Complete reaction sequence is as follows.

Methanol → Acetic acid → Ethanol → Diethylether
(*A*) (*B*) (*C*) (*D*)

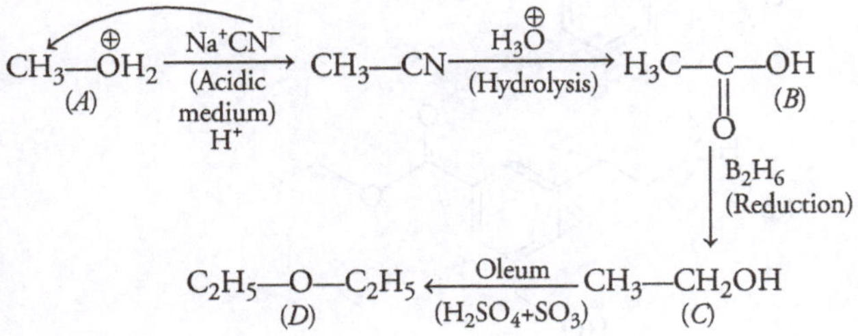


**19.** *(c)*

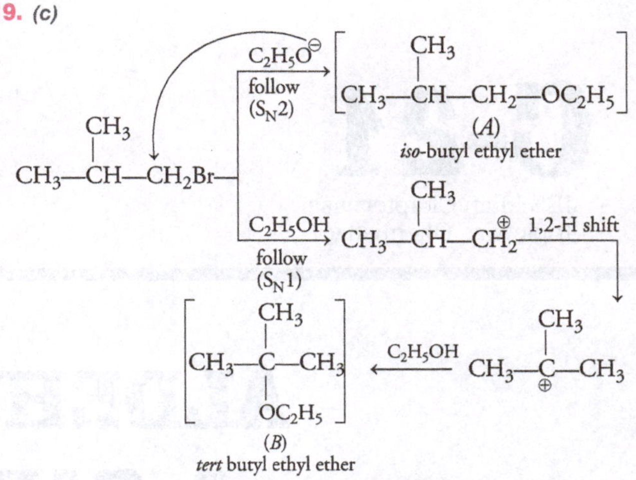


**20.** *(a)*

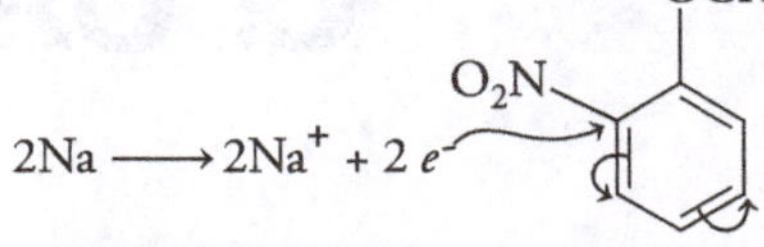


Sodium will donate electrons to the more electron deficient position.

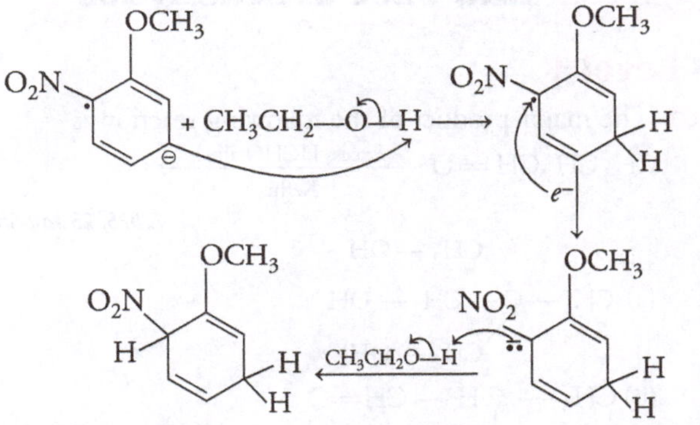


**21.** *(d)* Alcohols have higher boiling point than ethers of same molecular mass due to possibility of intermolecular hydrogen bonds in alcohols and not in ethers. So, *A* is alcohol derivative and *B* is ether derivative.

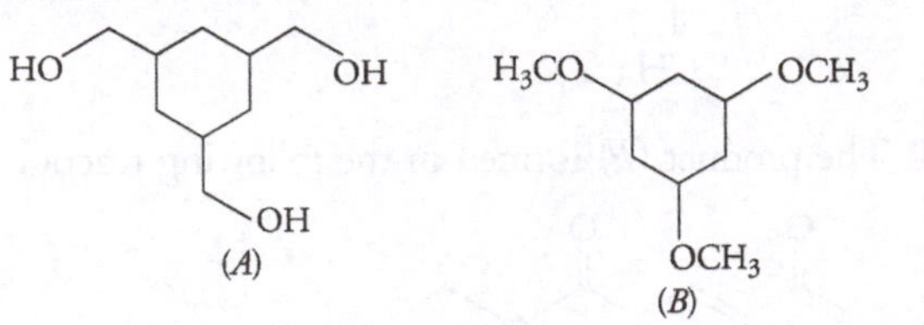

# C-21

JEE Main Chapterwise Topicwise Chemistry

In JEE Main there is usually 0-4 question per paper are visible from this chapter. In past 3 years, 52 conceptual and 3 numerical questions were asked of hard to moderate level from this chapter.
**For detailed analysis scan QR code**

# ALDEHYDES, KETONES & CARBOXYLIC ACIDS

## Topic 1 Preparation, Properties and Uses of Aldehydes

### Level 1

**01** The major product of the following reaction is

$$CH_3CH_2CH{=}O \xrightarrow[\text{Reflux}]{\text{Excess HCHO alkali}} ?$$

*[2025, 23 Jan Shift-I]*

(a) $CH_3-C(CH_2-OH)(CH_2-OH)-CH_2-OH$

(b) $CH_3-CH(CH_2-OH)-CH{=}O$

(c) $CH_3-CH_2-CH_2-OH$

(d) $CH_3-C(=CH_2)-CH{=}O$

**02** The product (*P*) formed in the following reaction is

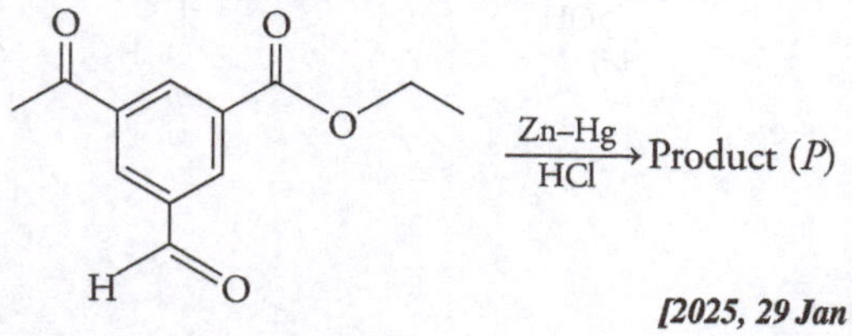


*[2025, 29 Jan Shift-I]*

(a) 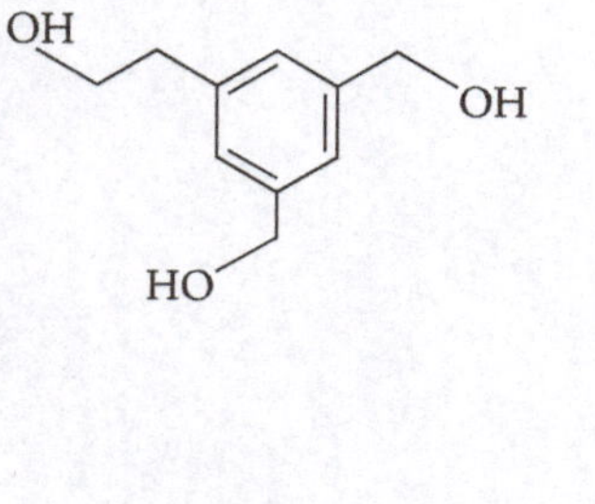

(b) 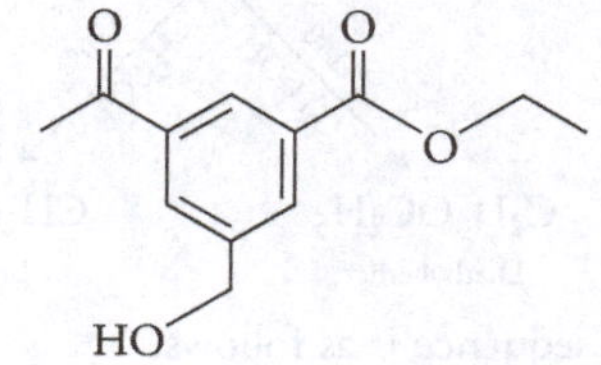

(c) 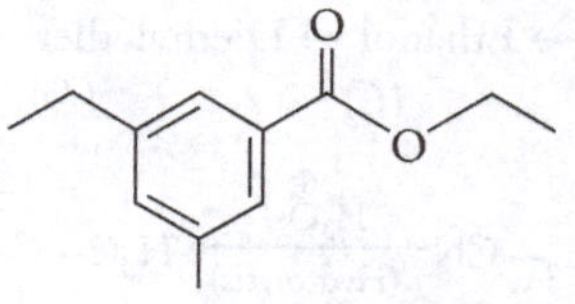

(d) 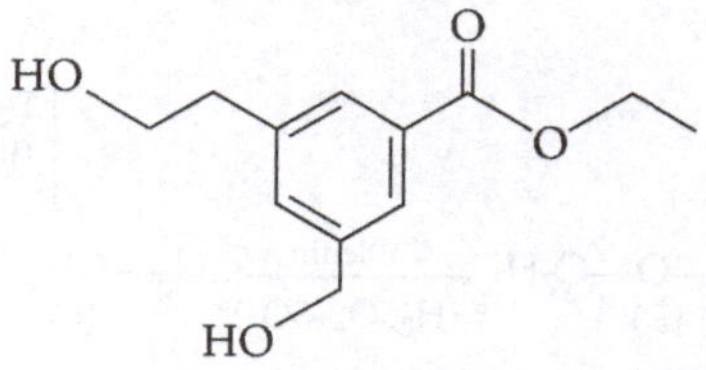

**03** Identify major product "*X*" formed in the following reaction: *[2024, 9 April Shift-I]*

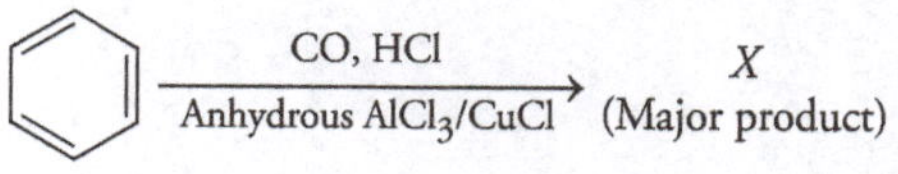


(a) 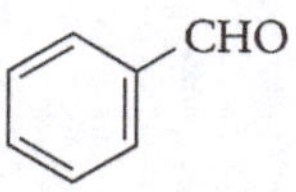

(b) 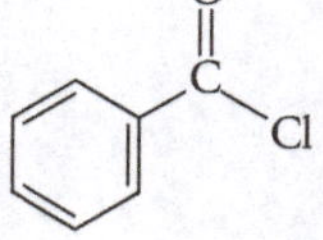

(c) 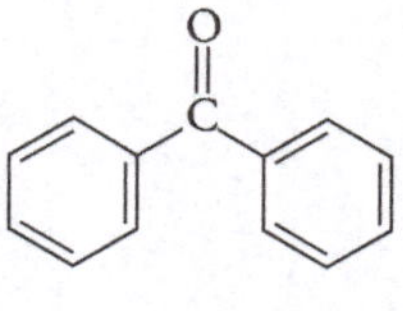

(d) 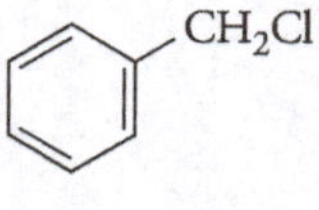

**04** Which among the following aldehydes is most reactive towards nucleophilic addition reactions? *[2024, 6 April Shift-I]*

(a) $C_2H_5-\overset{O}{\overset{\|}{C}}-H$ (b) $H-\overset{O}{\overset{\|}{C}}-H$

(c) $C_3H_7-\overset{O}{\overset{\|}{C}}-H$ (d) $CH_3-\overset{O}{\overset{\|}{C}}-H$

**05** Identify the name reaction. *[2024, 31 Jan Shift-II]*

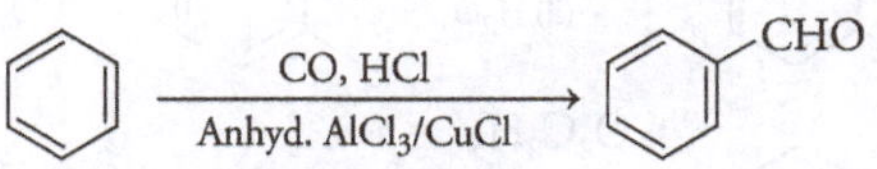

(a) Etard reaction
(b) Gatterman-Koch reaction
(c) Stephen reaction
(d) Rosenmund reduction

**06** Structure of 4-methylpent-2-enal is *[2024, 30 Jan Shift-I]*

(a) $CH_3-CH_2-\underset{CH_3}{\underset{|}{C}}=CH-\overset{O}{\overset{\|}{C}}-H$

(b) $H_2C=\underset{H}{\underset{|}{C}}-\overset{CH_3}{\overset{|}{\underset{H}{\underset{|}{C}}}}-CH_2-\overset{O}{\overset{\|}{C}}-H$

(c) $CH_3-\underset{CH_3}{\underset{|}{CH}}-CH=CH-\overset{O}{\overset{\|}{C}}-H$

(d) $CH_3-CH_2-CH=\underset{CH_3}{\underset{|}{C}}-\overset{O}{\overset{\|}{C}}-H$

**07** Correct structure of γ-methylcyclohexane carbaldehyde is *[2022, 29 July Shift-II]*

(a) 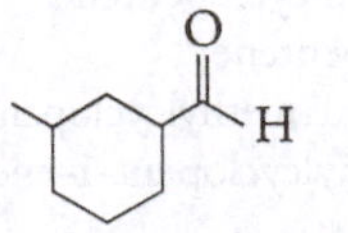 (b) 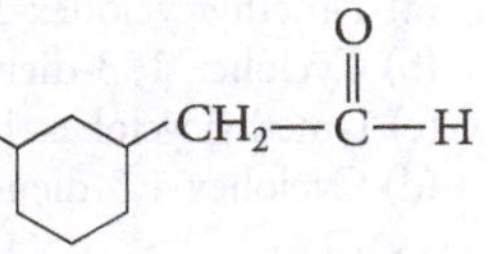

(c) 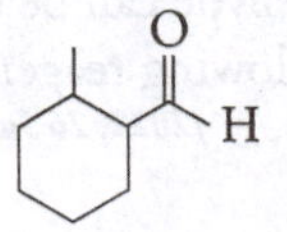 (d) 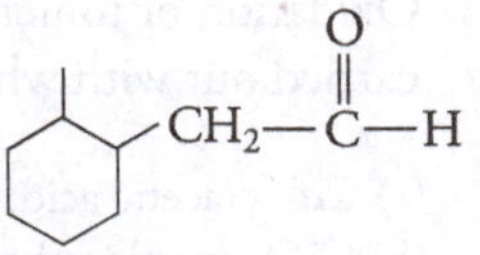

## Level 2

**08** Given below are two statements. *[2025, 2 April Shift-I]*

**Statement I** Vanillin 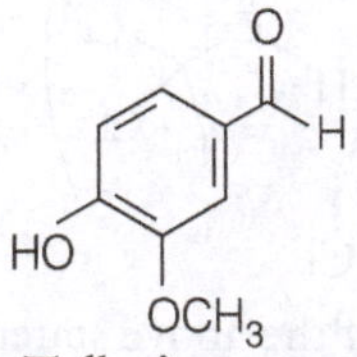 will react with NaOH and also with Tollen's reagent.

**Statement II** Vanillin 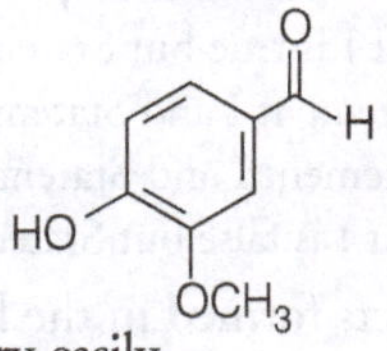 will undergo self aldol condensation very easily.

In the light of the above statements, choose the most appropriate answer from the options given below

(a) Statement I is correct but Statement II is incorrect.
(b) Statement I is incorrect but Statement II is correct.
(c) Both Statement I and Statement II are correct.
(d) Both Statement I and Statement II are incorrect.

**09** Given below are two statements. *[2025, 4 April Shift-I]*

**Statement I** The dipole moment of $\overset{4}{C}H_3-\overset{3}{C}H=\overset{2}{C}H-\overset{1}{C}H=O$ is greater than $\overset{4}{C}H_3-\overset{3}{C}H_2-\overset{2}{C}H_2-\overset{1}{C}H=O$

**Statement II** $C_1-C_2$ bond length of $\underset{4}{C}H_3\underset{3}{C}H=\underset{2}{C}H-\underset{1}{C}H=O$ is greater than $C_1-C_2$ bond length of $\underset{4}{C}H_3-\underset{3}{C}H_2-\underset{2}{C}H_2-\underset{1}{C}H=O$

In the light of the above statements, choose the correct answer from the options given below.

(a) Both Statement I and Statement II are false.
(b) Both Statement I and Statement II are true.
(c) Statement I is false but Statement II is true.
(d) Statement I is true but Statement II is false.

**10** Given below are two statements
Consider the following reaction, *[2025, 23 Jan Shift-II]*

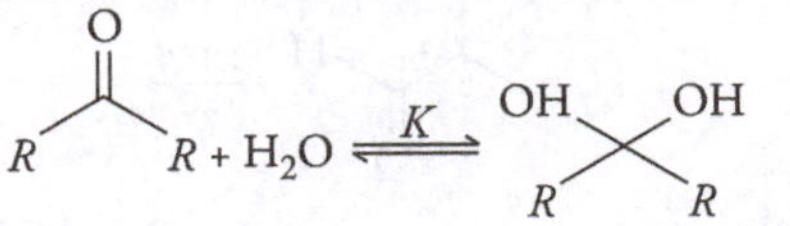

**Statement I** In the case of formaldehyde 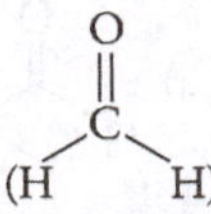, *K* is about 2280, due to small substituents hydration is faster.

**Statement II** In the case of trichloro acetaldehyde 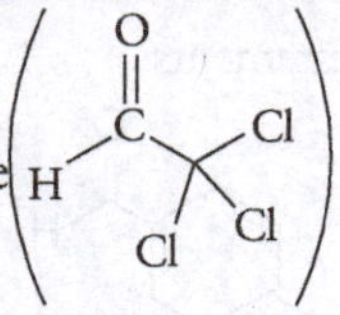, $K$ is about 2000 due to $-I$ effect of –Cl.

In the light of the above statements, choose the correct answer from the options given below

(a) Statement I is true but Statement II is false.
(b) Both Statement I and Statement II are false.
(c) Both Statement I and Statement II are true.
(d) Statement I is false but Statement II is true.

**11** The products formed in the following reaction, $A$ and $B$ are **[2022, 26 July Shift-I]**

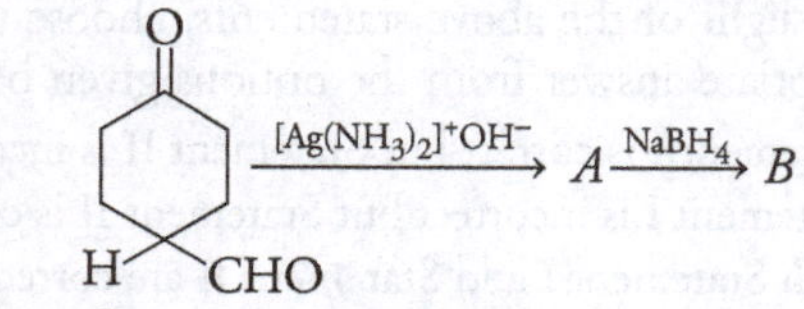

(a) $A =$ 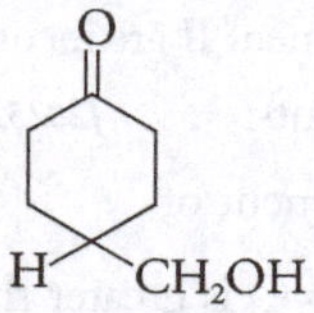 $B =$ 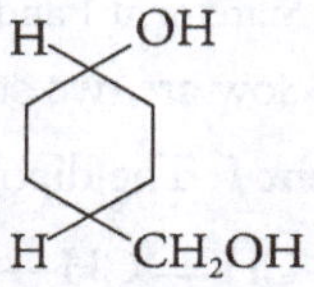

(b) $A =$ 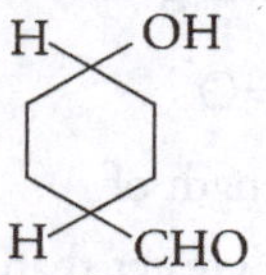 $B =$ 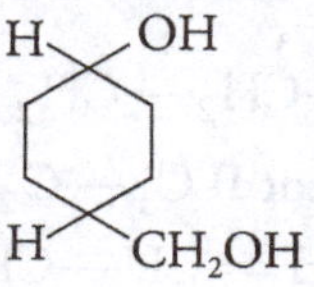

(c) $A =$ 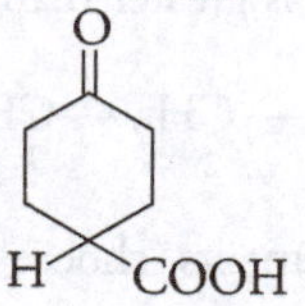 $B =$ 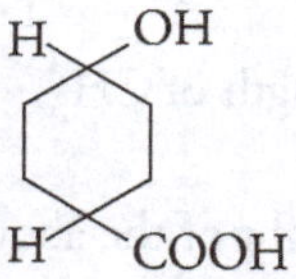

(d) $A =$ 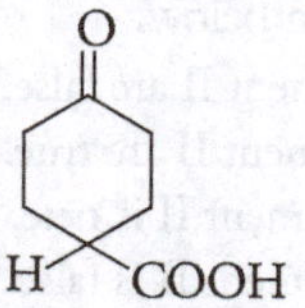 $B =$ 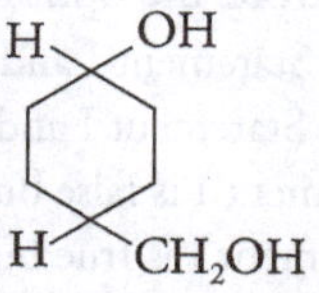

**12** What is the major product of the following reaction? **[2022, 25 July Shift-II]**

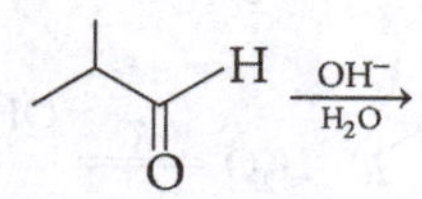

(a) 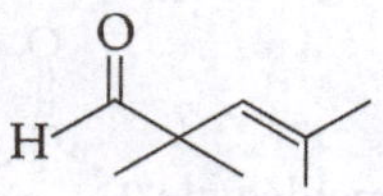 (b) 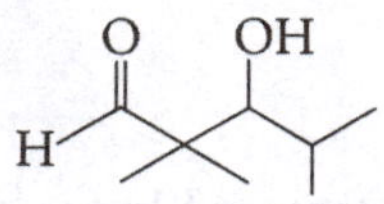

(c) 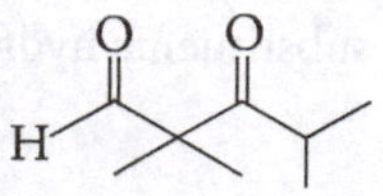 (d) 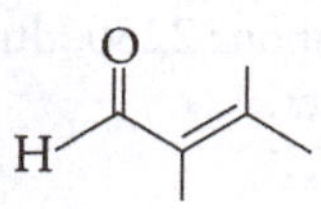

**13** Which one of the following reactions does not represent correct combination of substrate and product under the given conditions? **[2022, 25 July Shift-I]**

(a) 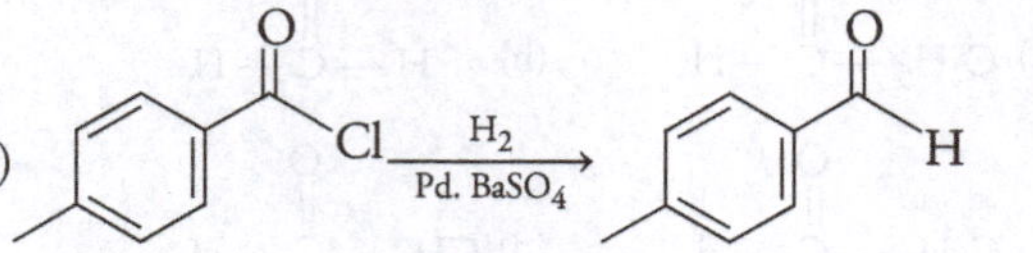

(b) 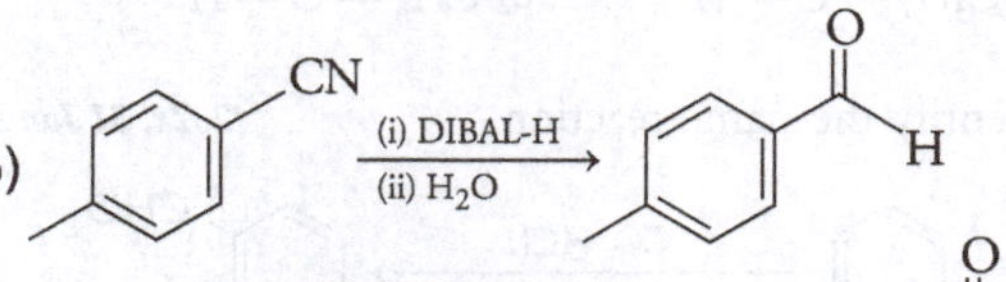

(c) 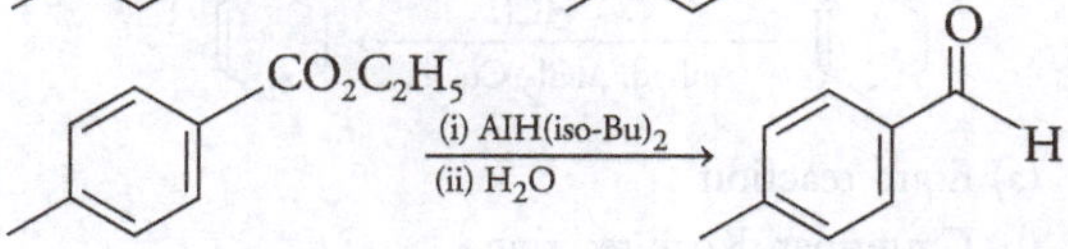

(d) 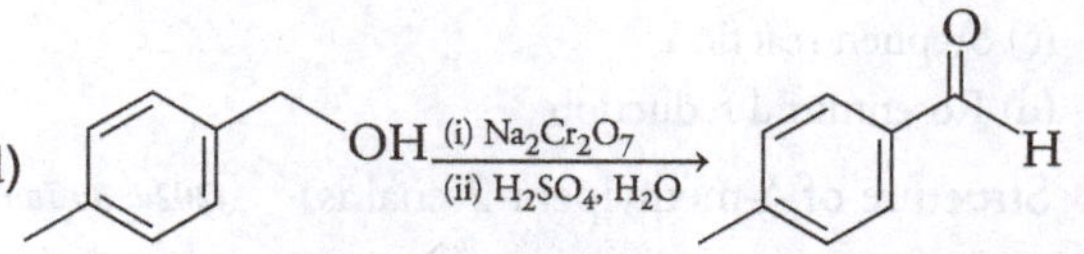

**14** The correct structure of product '$A$' formed in the following reaction is **[2022, 28 June Shift-I]**

$$PhCHO + PhCHO \xrightarrow[\text{in } D_2O]{NaOD} A + Ph—\overset{O}{\overset{\|}{C}}—O^-$$

(Ph is $—C_6H_5$)

(a) 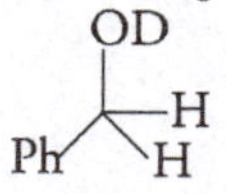 (b) 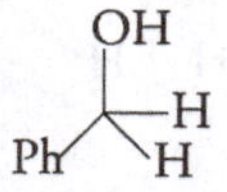

(c) 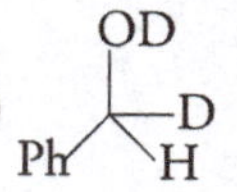 (d) 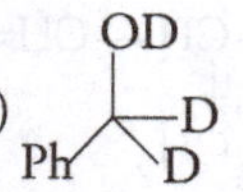

**15** '$A$' and '$B$' respectively are **[2022, 27 June Shift-I]**

$$A \xrightarrow[\text{(ii) Zn-}H_2O]{\text{(i) }O_3} \text{Ethane-1,2-dicarbaldehyde} + \text{Glyoxal/Oxaldehyde}$$

$$B \xrightarrow[\text{(ii) Zn-}H_2O]{\text{(i) }O_3} \text{5-oxohexanal}$$

(a) 1-methylcyclohex-1, 3-diene and cyclopentene
(b) Cyclohex-1, 3-diene and cyclopentene
(c) 1-methylcyclohex-1,4-diene and 1-methylcyclopent-1-ene
(d) Cyclohex-1,3-diene and 1-methylcyclopent-1-ene

**16** Oxidation of toluene to benzaldehyde can be easily carried out with which of the following reagents? **[2022, 26 June Shift-II]**

(a) $CrO_3$/acetic acid, $H_3O^+$
(b) $CrO_3$/acetic anhydride, $H_3O^+$
(c) $KMnO_4$/HCl, $H_3O^+$
(d) CO/HCl, anhydrous $AlCl_3$

**17** Which of the following reagents/reactions will convert '*A*' to '*B*'? *[2022, 24 June Shift-II]*

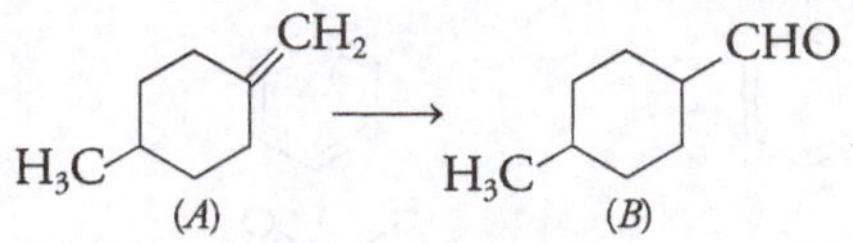

(a) PCC oxidation (b) Ozonolysis

(c) $BH_3, H_2O_2/OH^-$ followed by PCC oxidation

(d) HBr, hydrolysis followed by oxidation by $K_2Cr_2O_7$

**18** $R—CN \xrightarrow[\text{(ii) } H_2O]{\text{(i) DIBAL–H}} R—Y$

Consider the above reaction and identify *Y*. *[2021, 27 July Shift-II]*

(a) $—CH_2NH_2$ (b) $—CONH_2$
(c) —CHO (d) —COOH

**19** An unsaturated hydrocarbon *X* on ozonolysis gives *A*. Compound *A* when warmed with ammoniacal silver nitrate forms a bright silver mirror along the sides of the test tube. The unsaturated hydrocarbon *X* is *[2021, 17 March Shift-II]*

(a) $CH_3—C(CH_3)=C(CH_3)—CH_3$ (b)

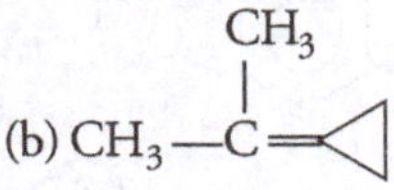

(c) $HC≡C—CH_2—CH_3$ (d) $CH_3—C≡C—CH_3$

**20**

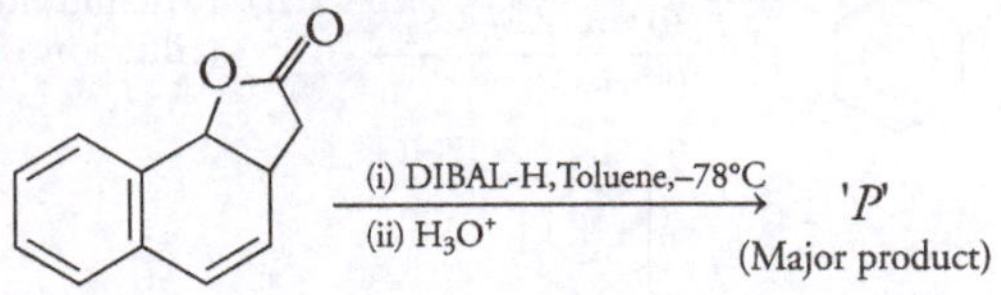

The product *P* in the above reaction is *[2021, 16 March Shift-I]*

(a) 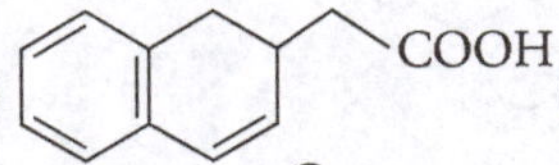 (b) 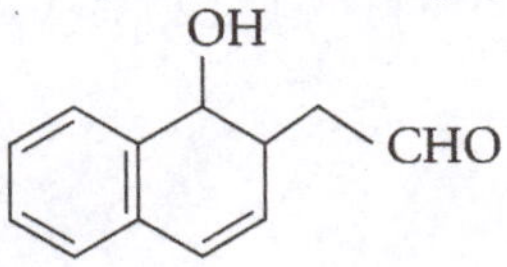

(c) 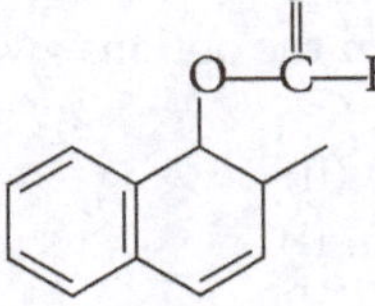 (d) 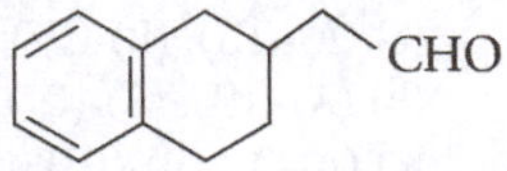

**21** The major product of the following reaction is

$CH_3CH_2CH=CH_2 \xrightarrow[\text{Rh catalyst}]{H_2/CO}$ *[2021, 25 Feb Shift-II]*

(a) $CH_3CH_2CH_2CH_2CHO$ (b) $CH_3CH_2CH_2CHO$

(c) $CH_3CH_2CH=CHCHO$ (d) $CH_3CH_2\underset{|}{\overset{}{C}}(CHO)=CH_2$

**22** Which one of the following reactions will not form acetaldehyde? *[2021, 25 Feb Shift-I]*

(a) $CH_2=CH_2 + O_2 \xrightarrow[H_2O]{\text{Pd (II)/Cu (II)}}$

(b) $CH_3CH_2OH \xrightarrow{CrO_3\text{-}H_2SO_4}$

(c) $CH_3CN \xrightarrow[\text{(ii) } H_2O]{\text{(i) DIBAL -H}}$

(d) $CH_3CH_2OH \xrightarrow[573\text{ K}]{Cu}$

**23** Which of the following reagent is used for the following reaction ? *[2021, 24 Feb Shift-I]*

$$CH_3CH_2CH_3 \xrightarrow{?} CH_3CH_2CHO$$

(a) Copper at high temperature and pressure
(b) Molybdenum oxide
(c) Manganese acetate
(d) Potassium permanganate

**24** The major product(s) obtained in the following reaction is/are *[2019, 12 April Shift-I]*

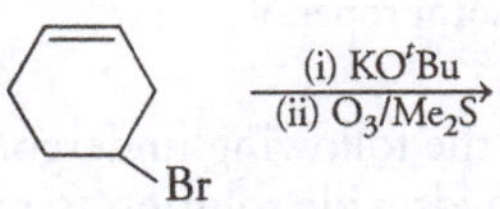

(a) 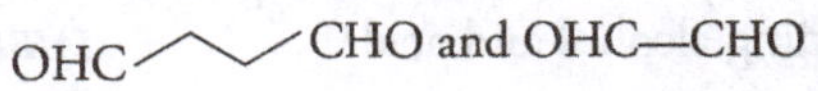

(b) 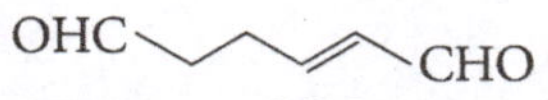

(c) 

(d) 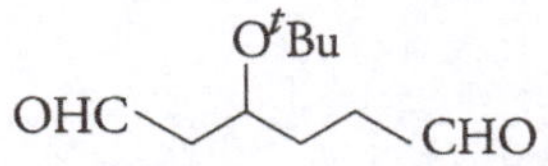

**25** In the following reaction,

$$\text{Carbonyl compound} + MeOH \overset{HCl}{\rightleftharpoons} \text{Acetal}$$

Rate of the reaction is the highest for *[2019, 9 April Shift-II]*

(a) Acetone as substrate and methanol in excess
(b) Propanal as substrate and methanol in stoichiometric amount
(c) Acetone as substrate and methanol in stoichiometric amount
(d) Propanal as substrate and methanol in excess

**26** The aldehydes which will not form Grignard product with one equivalent Grignard reagents are

(A) 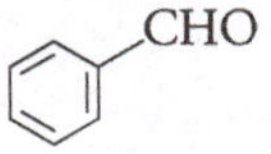 (B) 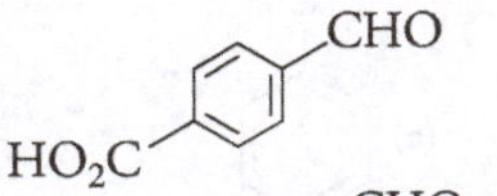

(C) 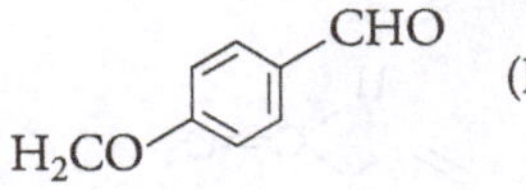 (D) 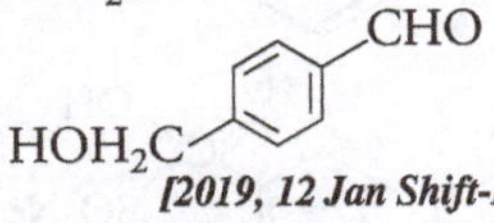

*[2019, 12 Jan Shift-II]*

(a) (C), (D) (b) (B), (D)
(c) (B), (C), (D) (d) (B), (C)

**27** The major product obtained in the following reaction is

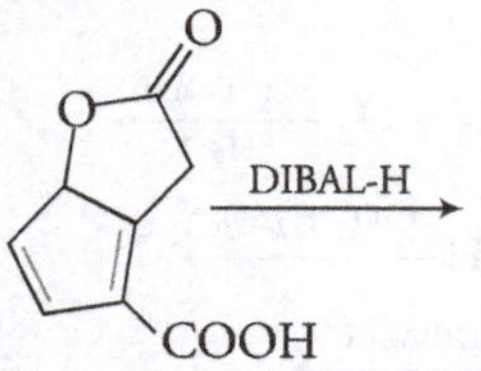

*[JEE Main 2017]*

(a) 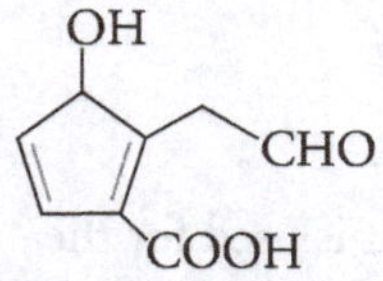

(b) 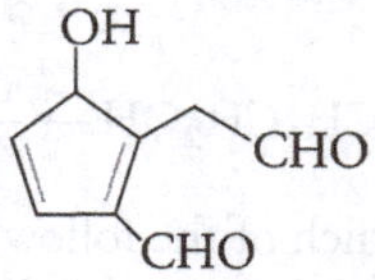

(c) 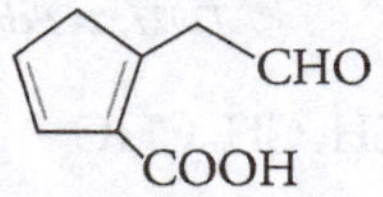

(d) 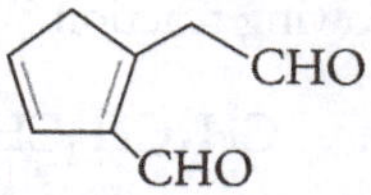

**28** Trichloroacetaldehyde was subjected to Cannizzaro's reaction by using NaOH. The mixture of the products contains sodium trichloroacetate ion and another compound. The other compound is *[AIEEE 2011]*

(a) 2, 2, 2-trichloroethanol
(b) trichloromethanol
(c) 2, 2, 2-trichloropropanol
(d) chloroform

**29** Which one of the following undergoes reaction with 50% sodium hydroxide solution to give the corresponding alcohol and acid? *[AIEEE 2004]*

(a) Phenol
(b) Benzaldehyde
(c) Butanal
(d) Benzoic acid

## Level 3

**30**

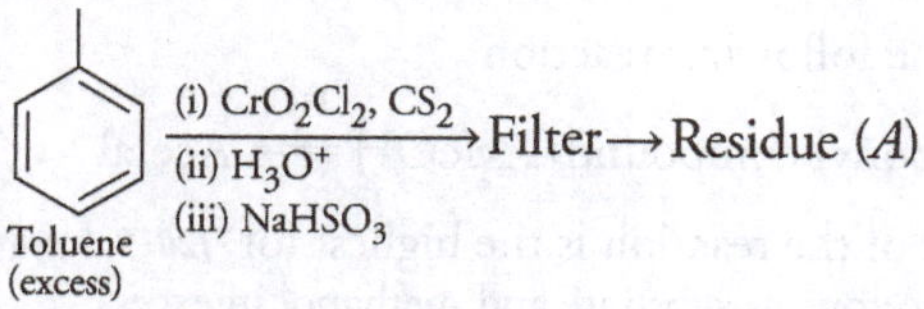

Residue $(A)$ + HCl(dil.) $\longrightarrow$ Compound $(B)$

Structure of residue $(A)$ and compound $(B)$ formed respectively is *[2025, 22 Jan Shift-II]*

| | [A] | [B] |
|---|---|---|
| (a) | 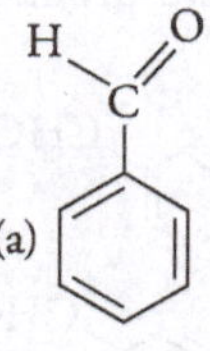 | 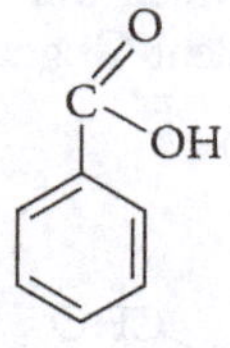 |
| (b) | 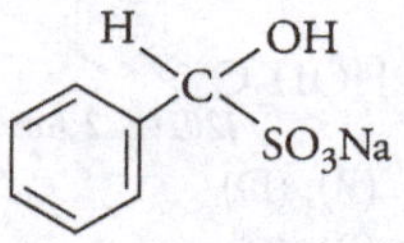 | 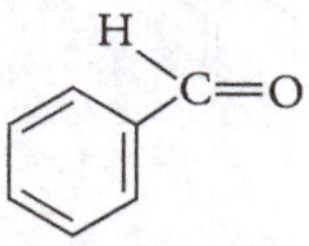 |
| (c) | 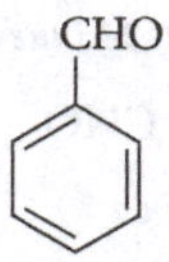 | 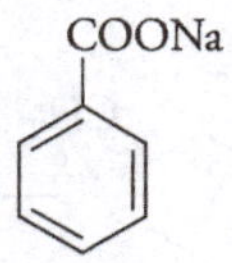 |
| (d) | 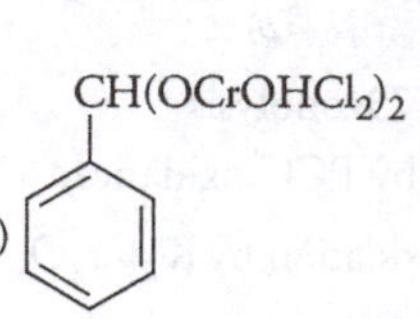 | 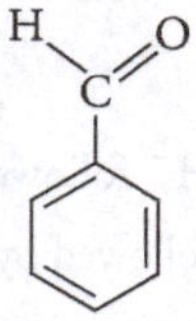 |

**Imp.** **Preparation of Benzaldehyde**

**31** Match List-I with List-II. *[2025, 24 Jan Shift-II]*

| | List-I | | List-II |
|---|---|---|---|
| (A) | $RCN \xrightarrow[\text{(ii) } H_3O^+]{\text{(i) } SnCl_2,\ HCl} R\,CHO$ | (I) | Etard reaction |
| (B) | 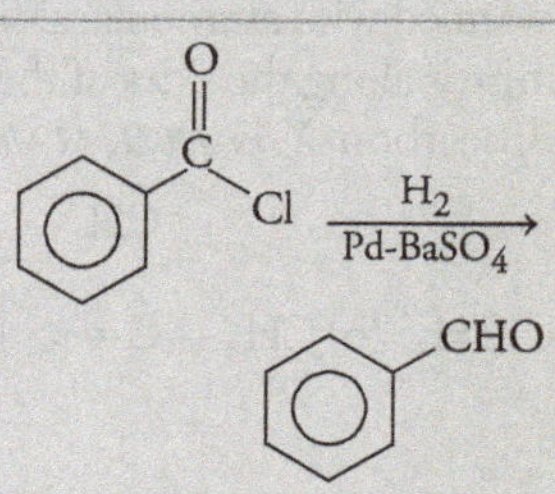 | (II) | Gattermann-Koch reaction |
| (C) | 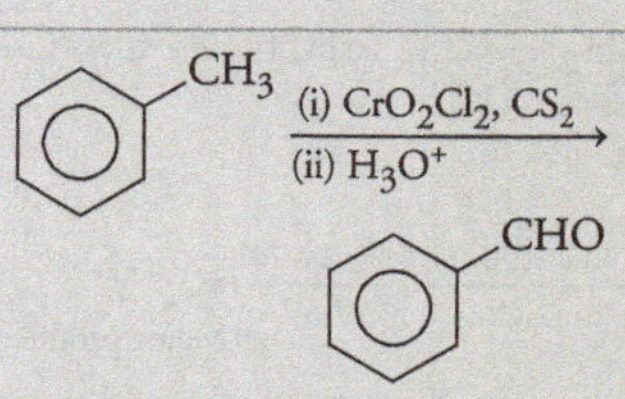 | (III) | Rosenmund reduction |
| (D) | 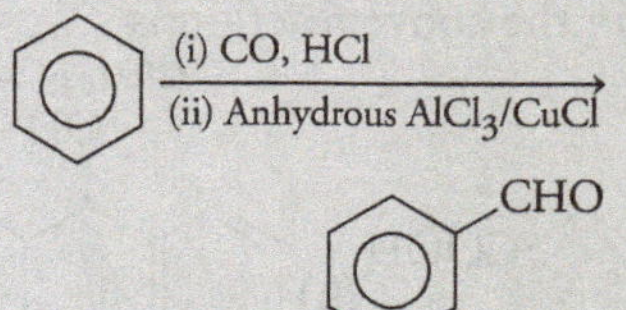 | (IV) | Stephen reaction |

Choose the correct answer from the options given below.

(a) (A)-(III), (B)-(IV), (C)-(I), (D)-(II)
(b) (A)-(III), (B)-(IV), (C)-(II), (D)-(I)
(c) (A)-(I), (B)-(III), (C)-(II), (D)-(IV)
(d) (A)-(IV), (B)-(III), (C)-(I), (D)-(II)

**32** Identify product $[A]$, $[B]$ and $[C]$ in the following reaction sequence *[2025, 28 Jan Shift-II]*

$$CH_3—C\equiv CH \xrightarrow[H_2]{Pd/C} [A] \xrightarrow[\text{(ii) Zn, } H_2O]{\text{(i) } O_3} [B]+[C]$$

(a) $[A]: CH_3—CH=CH_2$,
$[B]: CH_3CHO$, $[C]: HCHO$

(b) $[A]: CH_2=CH_2$, $[B]: H_3C-\overset{O}{\overset{\|}{C}}-CH_3$,
$[C]: HCHO$
(c) $[A]: CH_3CH_2CH_3$, $[B]: CH_3CHO$, $[C]: HCHO$
(d) $[A]: CH_3-CH=CH_2$, $[B]\ CH_3CHO$,
$[C]: CH_3CH_2OH$

**33** The structures of major products $A$, $B$ and $C$ in the following reaction sequence are

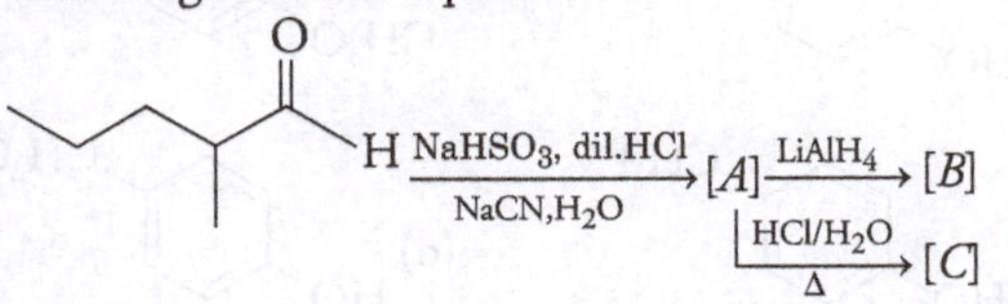


*[2023, 1 Feb Shift-II]*

(a)
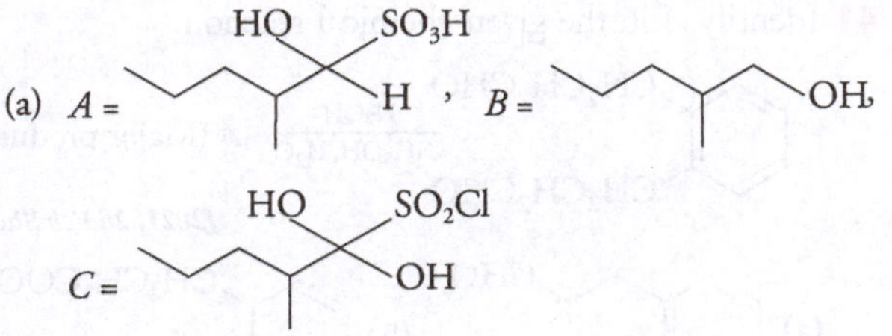


(b)
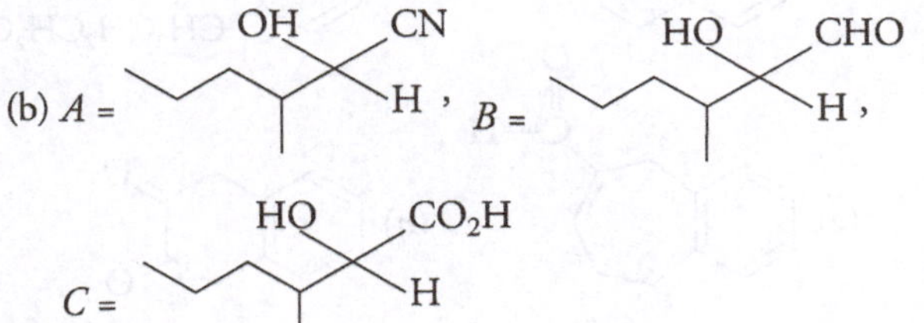


(c)
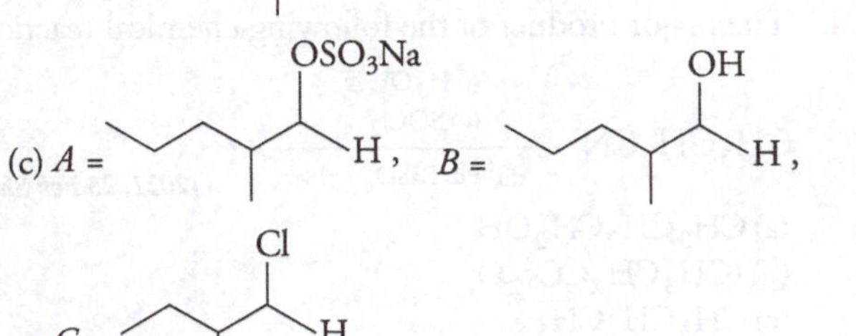


(d)
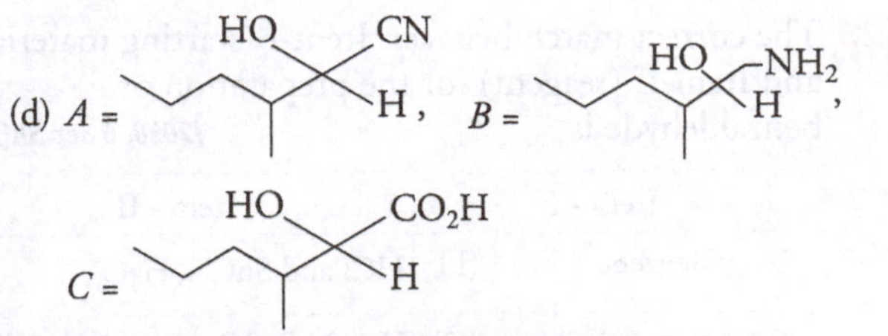


**34** Compound $(X)$ undergoes following sequence of reactions to give the lactone $(Y)$.

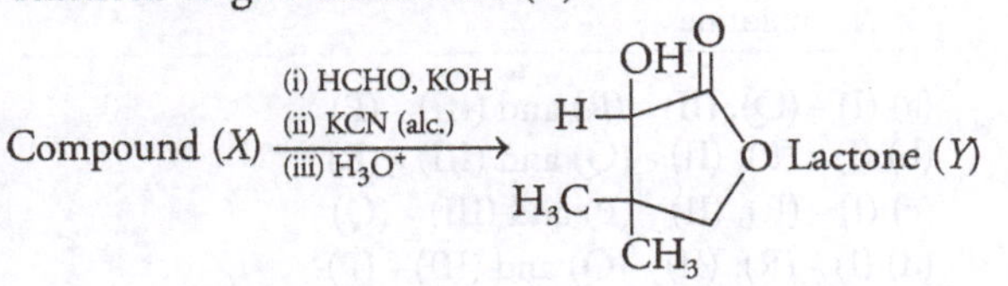


Compound $(X)$ is *[2023, 24 Jan Shift-I]*

(a) $HOH_2C-C(CH_3)_2-CHO$ (b) $H_2C(OH)CH_2CH_2CH_2CHO$

(c) $H-C(CH_3)_2-CHO$ (d) $H_2C(CH_3)-CH_2-CHO$

**35**

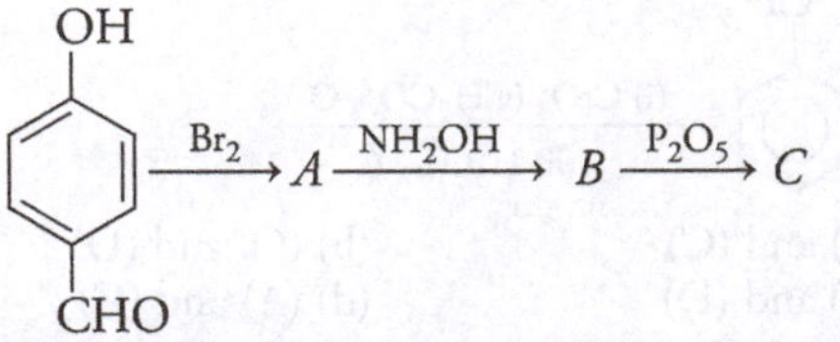


Consider the above reaction sequence, the product '$C$' is *[2022, 29 July Shift-I]*

(a)
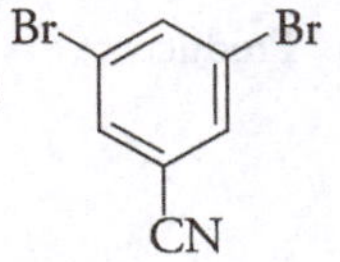


(b)
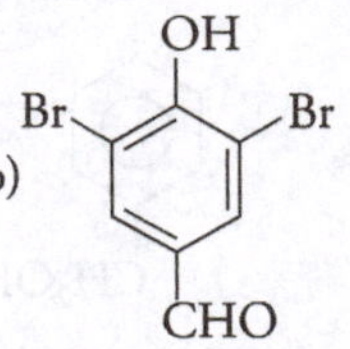


(c)
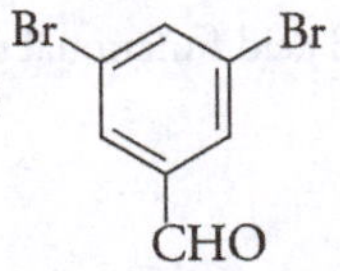


(d)
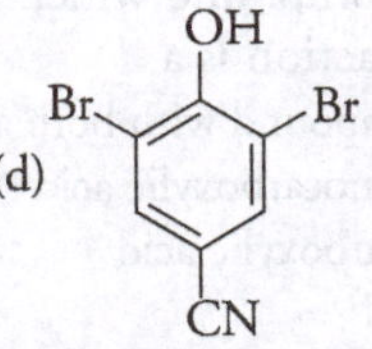


**36** Isobutyraldehyde on reaction with formaldehyde and $K_2CO_3$ gives compound $A$. Compound $A$ reacts with KCN and yields compound $B$, which on hydrolysis gives a stable compound $C$. The compound $C$ is *[2022, 28 June Shift-II]*

(a) $HO-CH_2-C(CH_3)_2-CH(OH)-COOH$

(b) $HO-CH_2-CH_2-CH(CH_3)-CH(OH)-COOH$

(c)
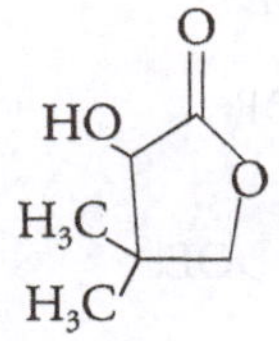


(d)
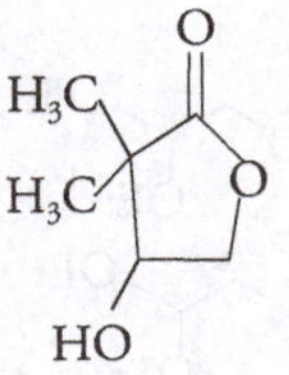

**37** Which of the following reactions will yield benzaldehyde as a product? *[2022, 27 June Shift-I]*

(A) 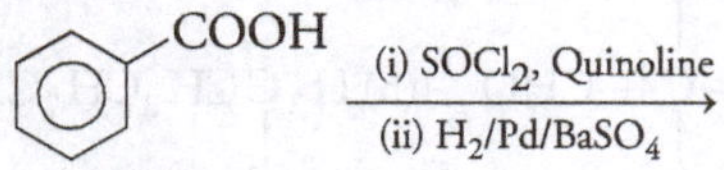


(B) 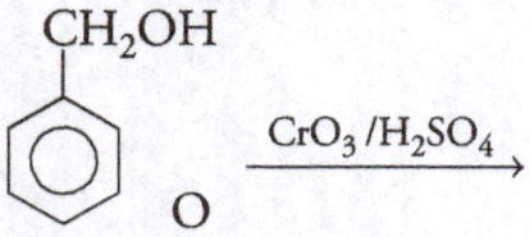


(C) 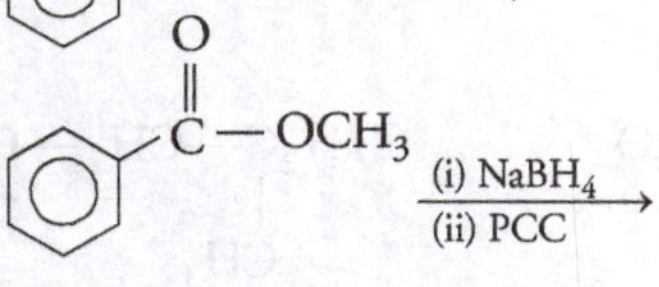


(D) 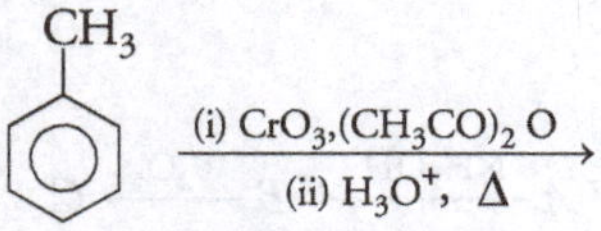


(a) (B) and (C)
(b) (C) and (D)
(c) (A) and (D)
(d) (A) and (C)

**38** For the reaction given below *[2021, 31 Aug Shift-II]*

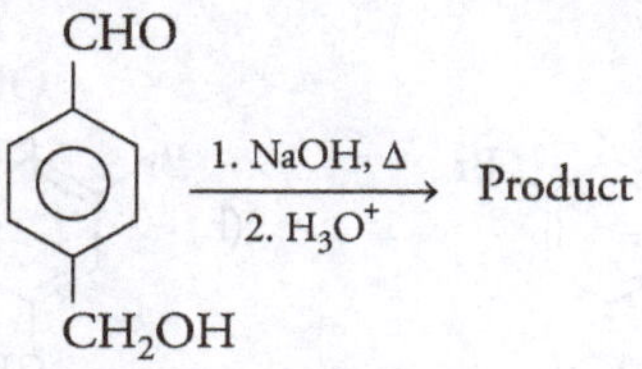


The compound which is not formed as a product in the reaction is a

(a) compound with both alcohol and acid functional groups
(b) monocarboxylic acid
(c) dicarboxylic acid
(d) diol

**39** 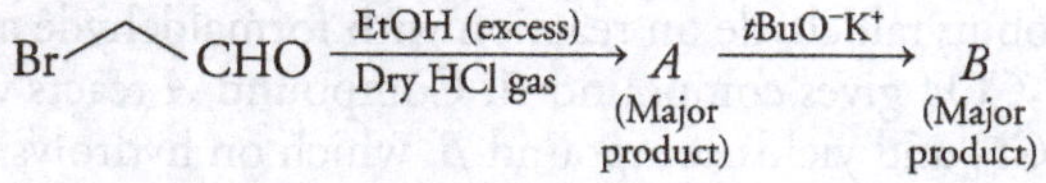


[where, Et $=C_2H_5$, $t$Bu $=(CH_3)_3C$—]

Consider the above reaction sequence, product $A$ and product $B$ formed respectively are *[2021, 25 July Shift-II]*

(a) 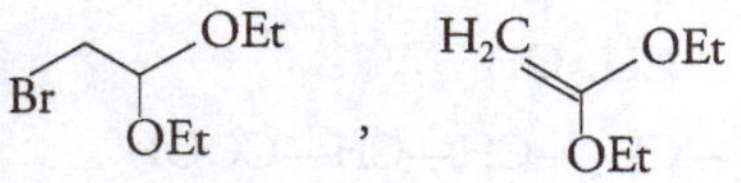


(b) 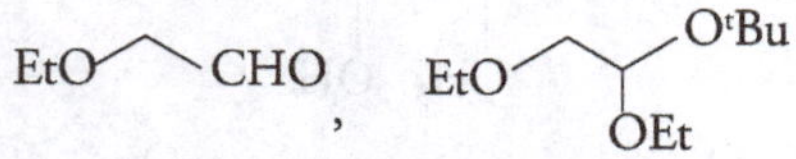


(c) 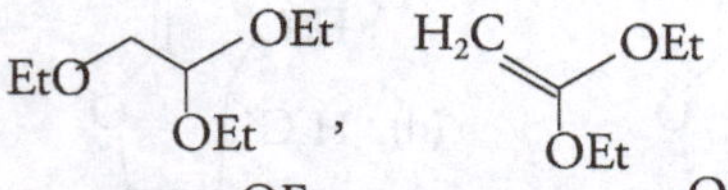


(d) 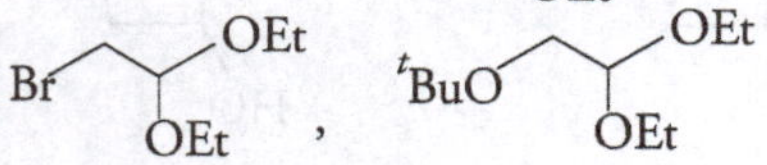


**40** Identify $A$ in the following chemical reaction.

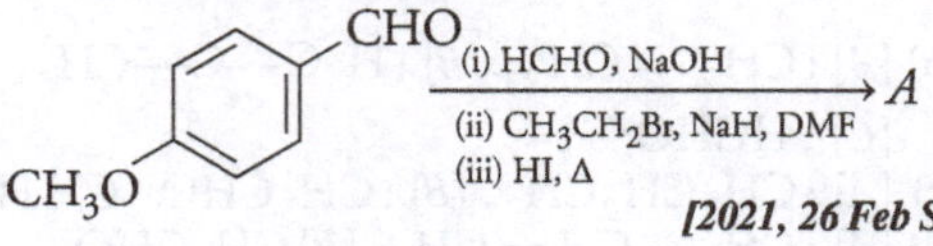


*[2021, 26 Feb Shift-II]*

(a) 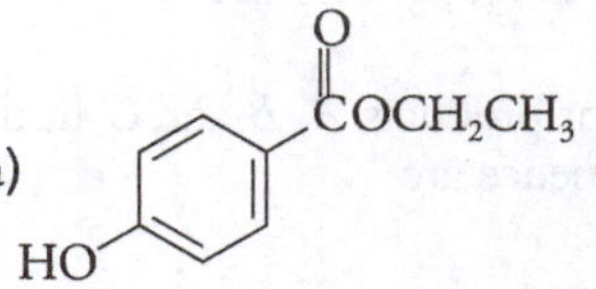


(b) 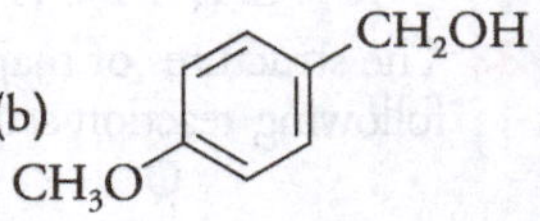


(c) 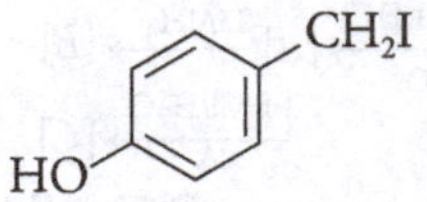


(d) 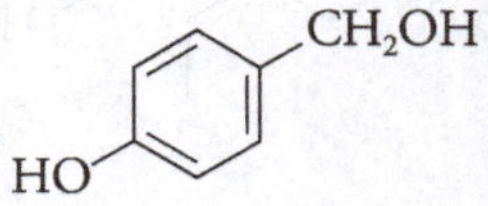


**41** Identify $A$ in the given chemical reaction,

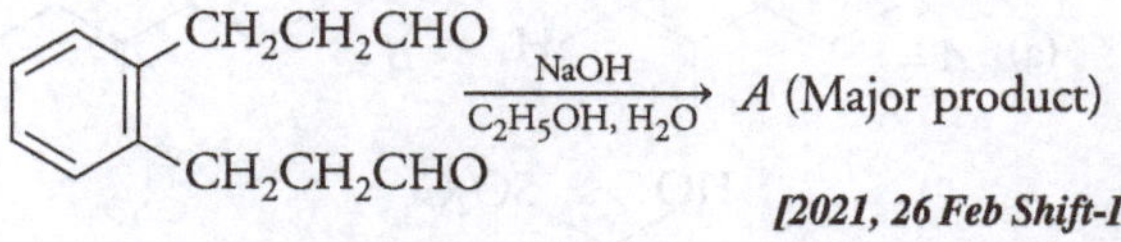


*[2021, 26 Feb Shift-II]*

(a) 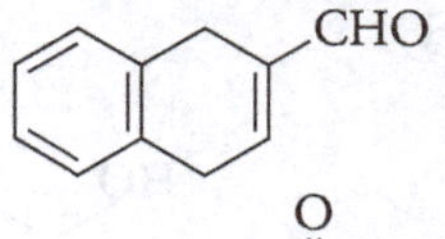


(b) 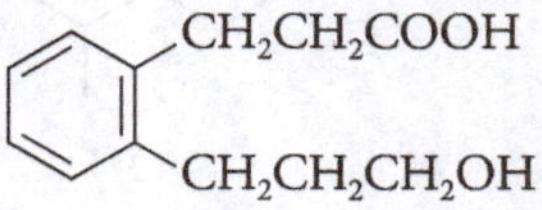


(c) 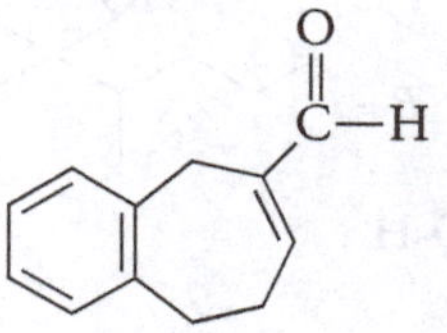


(d) 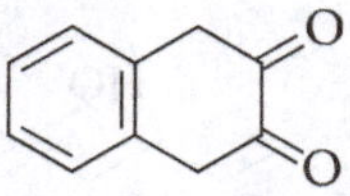


**42** The major product of the following chemical reaction is

$$CH_3CH_2CN \xrightarrow[\text{(iii) } Pd/BaSO_4,\ H_2]{\text{(i) } H_3O^+,\ \Delta \quad \text{(ii) } SOCl_2} ?$$

*[2021, 25 Feb Shift-I]*

(a) $CH_3CH_2CH_2OH$
(b) $(CH_3CH_2CO)_2O$
(c) $CH_3CH_2CH_3$
(d) $CH_3CH_2CHO$

**43** The correct match between Item-I (starting material) and Item-II (reagent) for the preparation of benzaldehyde is *[2020, 6 Sep Shift-II]*

| | Item - I | | Item - II |
|---|---|---|---|
| I | Benzene | (P) | HCl and $SnCl_2$, $H_3O^+$ |
| II | Benzonitrile | (Q) | $H_2$, Pd-$BaSO_4$, S and quinoline |
| III | Benzoyl chloride | (R) | CO, HCl and $AlCl_3$ |

(a) (I) - (Q), (II) - (R) and (III) - (P)
(b) (I) - (P), (II) - (Q) and (III) - (R)
(c) (I) - (R), (II) - (P) and (III) - (Q)
(d) (I) - (R), (II) - (Q) and (III) - (P)

**44** The increasing order of the following compounds towards HCN addition is *[2020, 2 Sep Shift-I]*

(i) 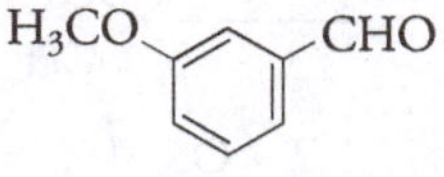 (ii) 

(iii) 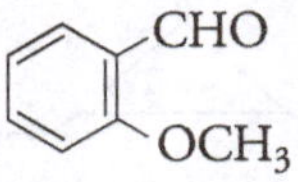 (iv) 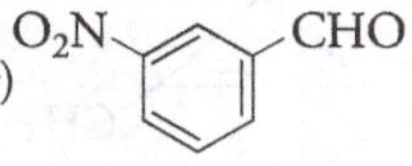

(a) (i) < (iii) < (iv) < (ii)
(b) (iii) < (iv) < (i) < (ii)
(c) (iii) < (iv) < (ii) < (i)
(d) (iii) < (i) < (iv) < (ii)

**45** In the following reaction *A* is *[2020, 9 Jan Shift-II]*

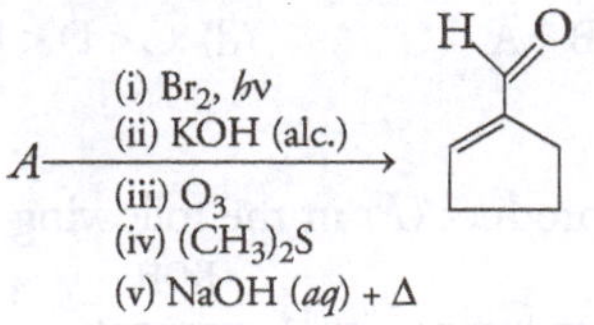

(a) 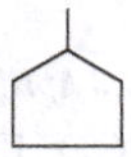 (b) 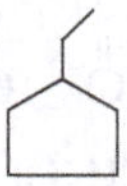

(c) 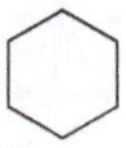 (d) 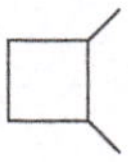

**Imp.** Oxidation of Toluene

**46** In the following sequence of reaction,

$$\text{Toluene} \xrightarrow{KMnO_4} A \xrightarrow{SOCl_2} B \xrightarrow[BaSO_4]{H_2/Pd} C$$

The product C is *[JEE Main 2015]*
(a) $C_6H_5COOH$ (b) $C_6H_5CH_3$
(c) $C_6H_5CH_2OH$ (d) $C_6H_5CHO$

**47** Which of the following on heating with aqueous KOH, produces acetaldehyde? *[AIEEE 2009]*
(a) $CH_3COCl$ (b) $CH_3CH_2Cl$
(c) $CH_2ClCH_2Cl$ (d) $CH_3CHCl_2$

### Numerical Value Type Questions

**48** The product of the following reaction is *P*. *[2024, 31 Jan Shift-I]*

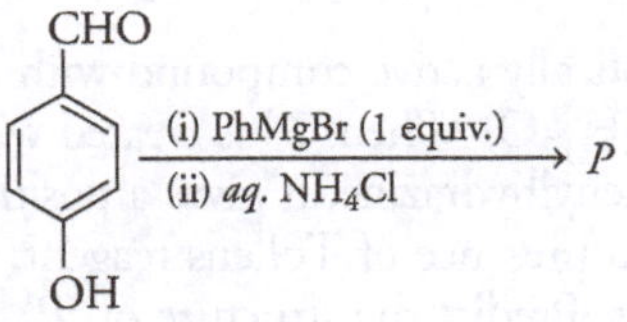

The number of hydroxyl groups present in the product *P* is .......... .

## Topic 2 Preparation, Properties and Uses of Ketones

### Level 1

**01** Identify the product in the following reaction:

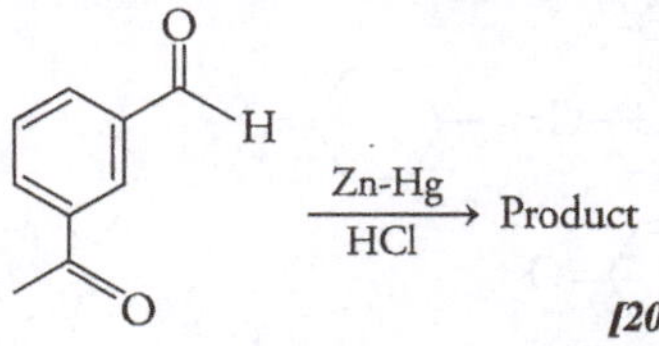

*[2024, 4 April Shift-I]*

(a) 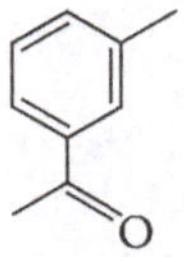 (b) 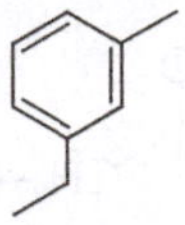

(c) 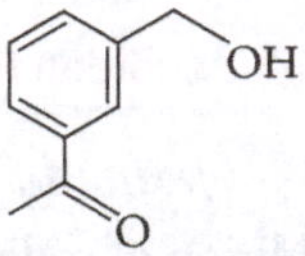 (d) 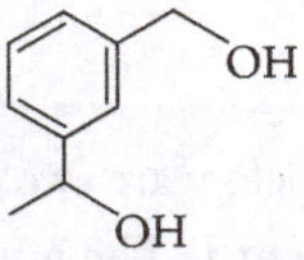

**02** Given below are two statements, one is labelled as Assertion (A) and the other is labelled as Reason (R).

**Assertion (A)** 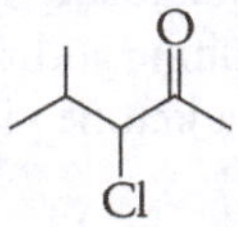 can be subjected to Wolff-Kishner reduction to give 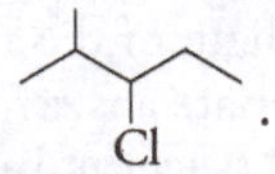.

**Reason (R)** Wolff-Kishner reduction is used to convert 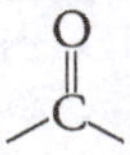 into $\diagup CH_2 \diagdown$.

In the light of the above statements, choose the most appropriate answer from the options given below. *[2023, 11 April Shift-II]*

(a) Both A and R are true and R is the correct explanation of A.
(b) A is true but R is false.
(c) A is false but R is true.
(d) Both A and R are true but R is not the correct explanation of A.

**Imp.** Basicity of Ketones

**03** Which will undergo deprotonation most readily in basic medium? *[2023, 24 Jan Shift-II]*

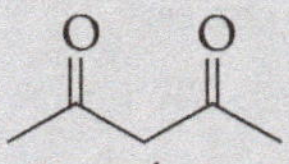
*A*

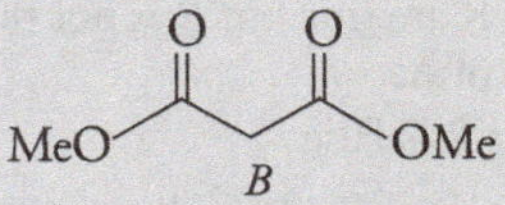
*B*

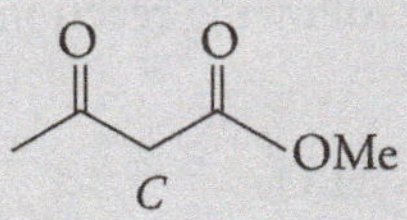
*C*

(a) *C* only (b) *B* only
(c) Both *A* and *C* (d) *A* only

**04** Which of the following is an example of conjugated diketone? *[2022, 24 June Shift-I]*

(a) $CH_3-\overset{O}{\overset{\|}{C}}-CH_2-CH_2-\overset{O}{\overset{\|}{C}}-CH_2-CH_3$

(b) 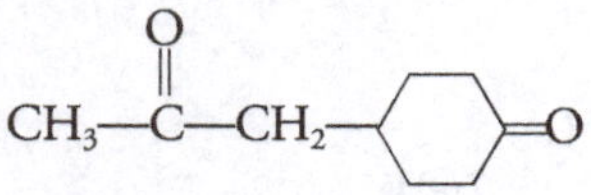

(c) 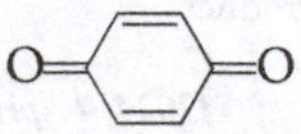

(d) 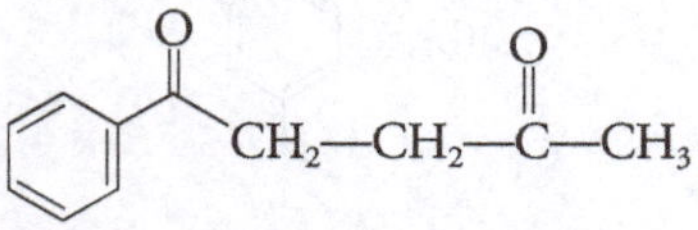

**Imp.** Nucleophilic Addition Reaction

**05** Given below are two statements. *[2021, 1 Sep Shift-II]*

**Statement I** The nucleophilic addition of sodium hydrogen sulphite to an aldehyde or a ketone involves proton transfer to form a stable ion.

**Statement II** The nucleophilic addition of hydrogen cyanide to an aldehyde or a ketone yields amine as final product.

In the light of the above statements, choose the most appropriate answer from the options given below.

(a) Both statement I and statement II are true.
(b) Statement I is false but statement II is true.
(c) Statement I is true but statement II is false.
(d) Both statement I and statement II are false.

**06** **Assertion (A)** Enol form of acetone $[CH_3COCH_3]$ exists in < 0.1% quantity. However, the enol form of acetyl acetone $[CH_3COCH_2OCCH_3]$ exists in approximately 15% quantity.

**Reason (R)** Enol form of acetyl acetone is stabilised by intramolecular hydrogen bonding, which is not possible in enol form of acetone.

Choose the correct statement. *[2021, 16 March Shift-I]*

(a) A is false but R is true.
(b) Both A and R are true and R is the correct explanation of A.
(c) Both A and R are true but R is not the correct explanation of A.
(d) A is true but R is false.

**07** The major product in the following reaction is

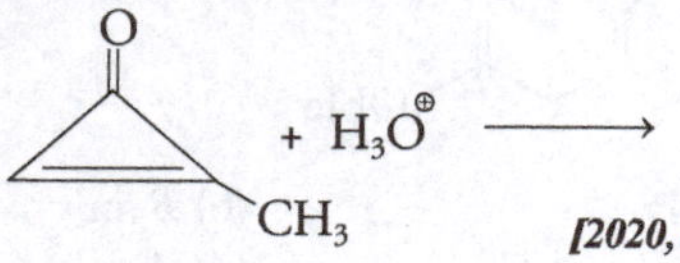

*[2020, 8 Jan Shift-II]*

(a) 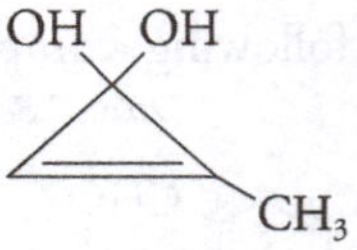

(b) 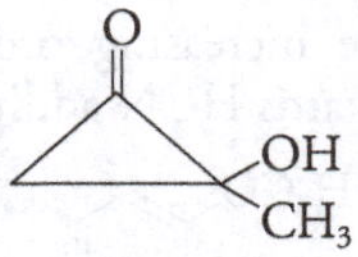

(c) 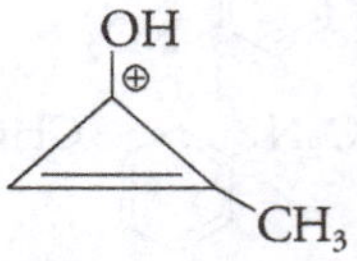

(d) 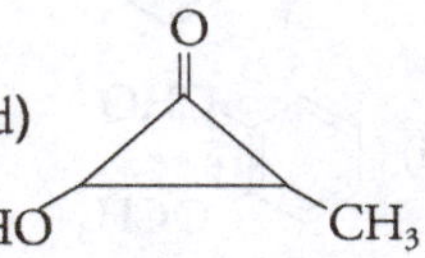

**08** The increasing order of the rate of HCN addition to compounds A-D is *[AIEEE 2006]*

(A) HCHO
(B) $CH_3COCH_3$
(C) $PhCOCH_3$
(D) PhCOPh

(a) A < B < C < D
(b) D < B < C < A
(c) D < C < B < A
(d) C < D < B < A

## Level 2

**09** The major product ($P$) in the following reaction is

$$Ph-\underset{O}{\underset{\|}{C}}-\underset{O}{\underset{\|}{C}}-H \xrightarrow[\Delta]{KOH} \underset{\text{Major product}}{P}$$

*[2025, 3 April Shift-II]*

(a) $Ph-\underset{O}{\underset{\|}{C}}-CH_2OH$
(b) $Ph-\underset{O}{\underset{\|}{C}}-COO^-K^+$
(c) $Ph-\overset{OH}{\overset{|}{CH}}-CH_2OH$
(d) $Ph-\underset{OH}{\underset{|}{CH}}-COO^-K^+$

**10** Aldol condensation is a popular and classical method to prepare $\alpha, \beta$-unsaturated carbonyl compounds. This reaction can be both intermolecular and intramolecular. Predict which one of the following is not a product of intramolecular aldol condensation? *[2025, 4 April Shift-I]*

(a) 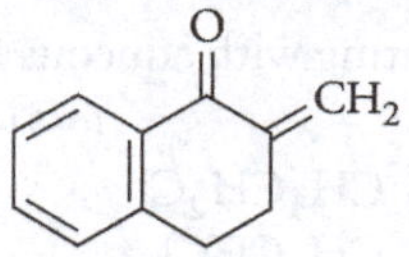

(b) 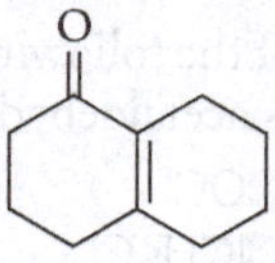

(c) 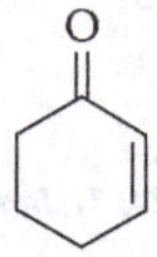

(d) 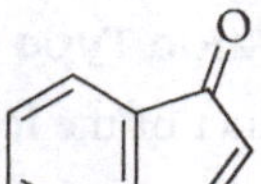

**11** '$P$' is an optically active compound with molecular formula $C_6H_{12}O$. When '$P$' is treated with 2, 4-dinitrophenylhydrazine, it gives a positive test. However, in presence of Tollens reagent, '$P$' gives a negative test. Predict the structure of '$P$'.

*[2025, 7 April Shift-II]*

(a) 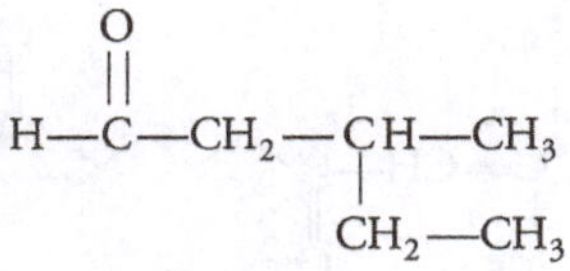

(b) 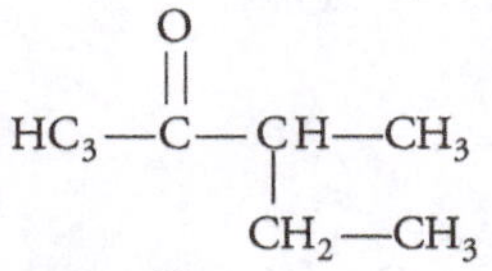

(c) 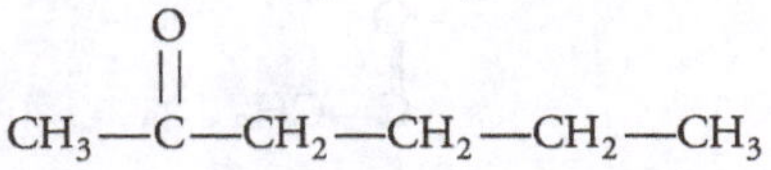

(d) 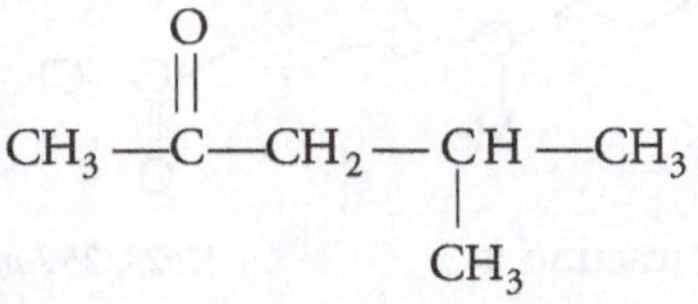

**12** Given below are two statements *[2024, 4 April Shift-I]*

**Statements I** Acidity of α-hydrogens of aldehydes and ketones is responsible for aldol reaction.

**Statements II** Reaction between benzaldehyde and ethanal will not give cross-aldol product.

In the light of the above statements, choose the most appropriate answer from the options given below:

(a) Both Statement I and Statement II are incorrect.
(b) Both Statement I and Statement II are correct.
(c) Statement I is incorrect but Statement II is correct.
(d) Statement I is correct but Statement II is incorrect.

**13** Which of the following compound can give positive idoform test when treated with aqueous KOH solution followed by potassium hypoiodite. *[2024, 9 April Shift-II]*

(a) 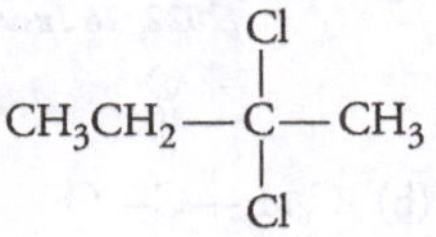

(b) 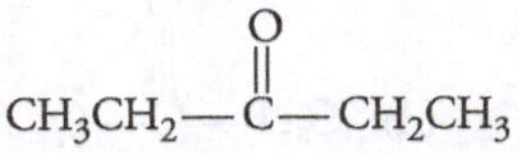

(c) $CH_3CH_2CH_2CHO$

(d) 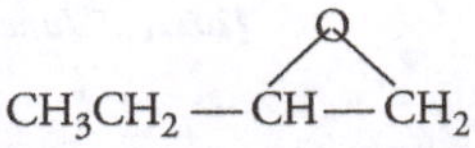

**14** Major product of the following reaction is

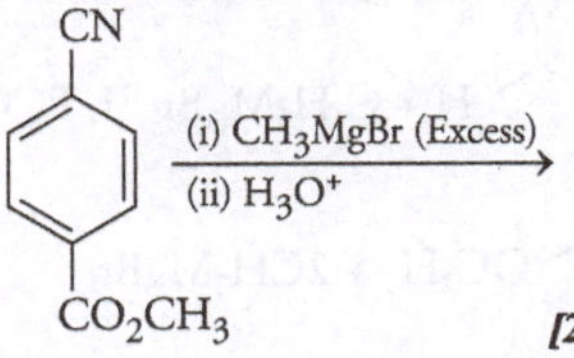


*[2024, 9 April Shift-II]*

(a) 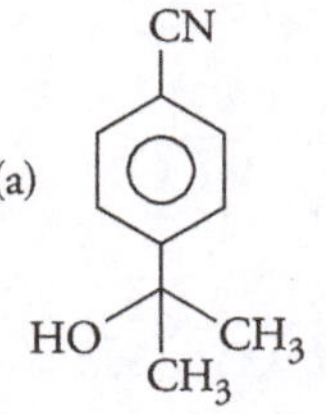

(b) 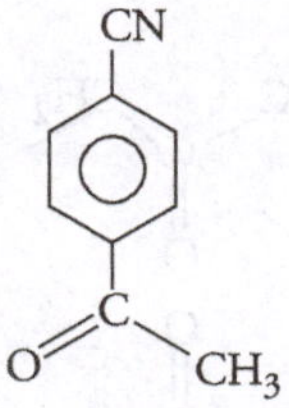

(c) 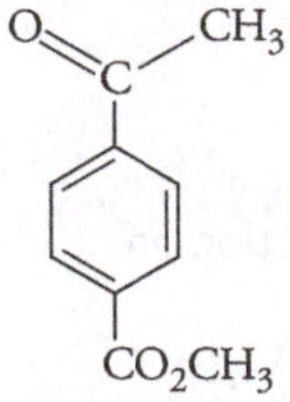

(d) 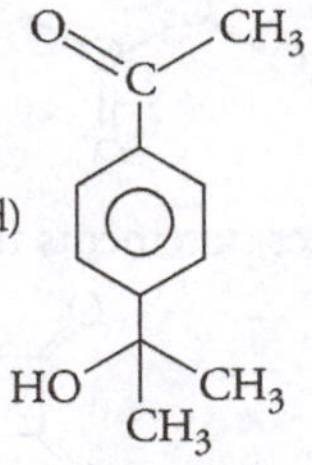

**15** Identify major product '*P*' formed in the following reaction. *[2024, 31 Jan Shift-II]*

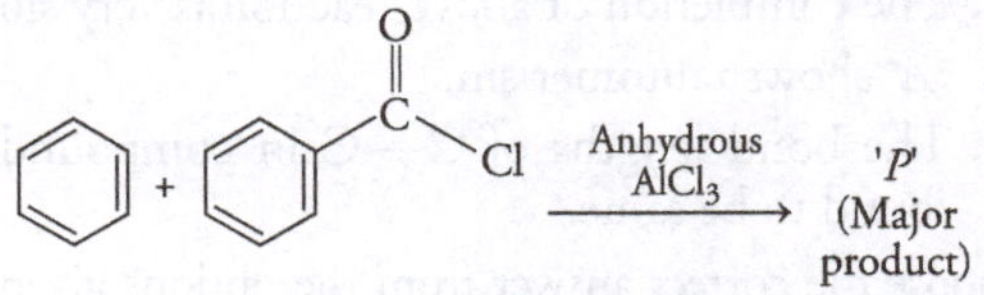


(a) 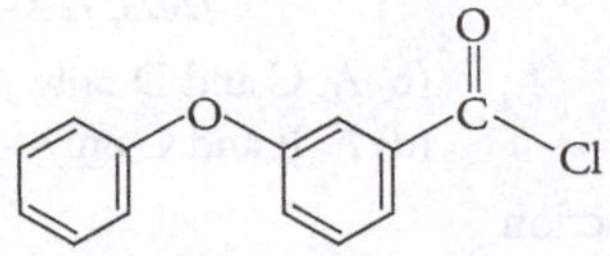

(b) 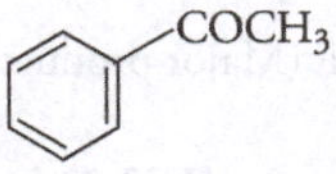

(c) 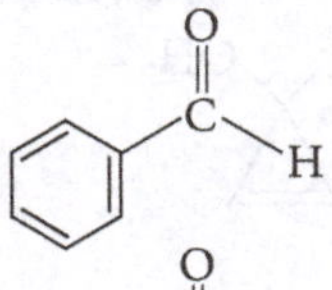

(d) 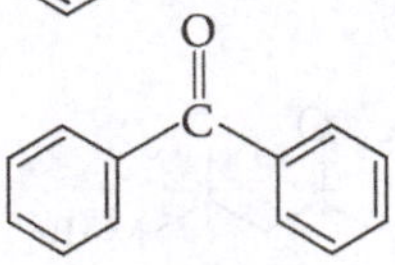

**16** Which of the following has highly acidic hydrogen? *[2024, 27 Jan Shift-I]*

(a) 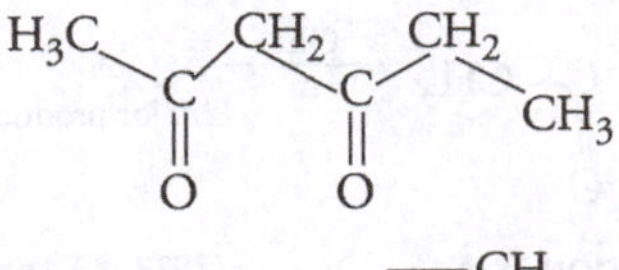

(b) 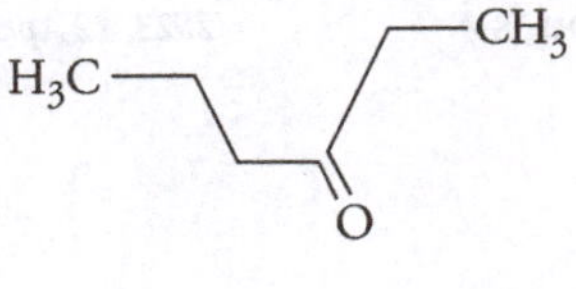

(c) 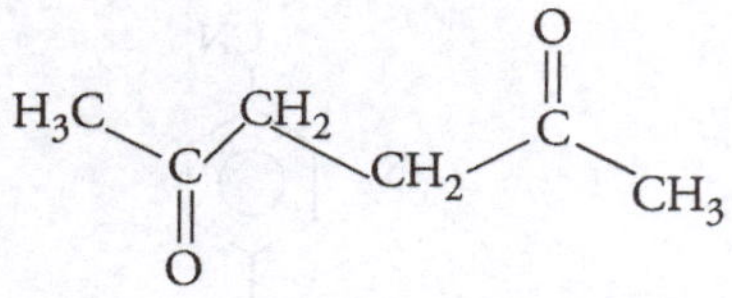

(d) 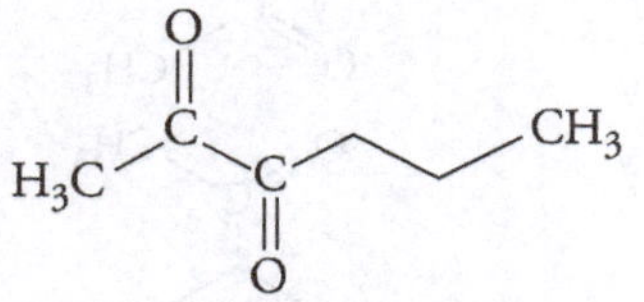

**17** Correct statements for the given reaction are

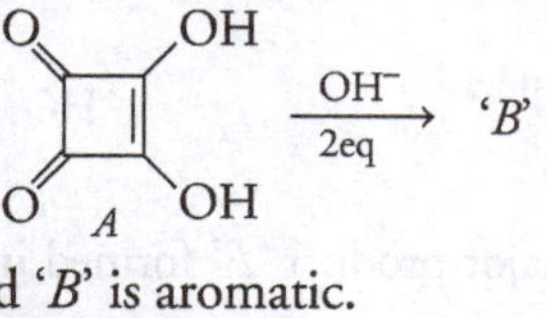

A. Compound '*B*' is aromatic.
B. The completion of above reaction is very slow.
C. '*A*' shows tautomerism.
D. The bond lengths of C—C in compound *B* are found to be same.

Choose the correct answer from the options given below. **[2023, 12 April Shift-I]**

(a) B, C and D only (b) A, C and D only
(c) A, B and D only (d) A, B and C only

**18** In the following reaction

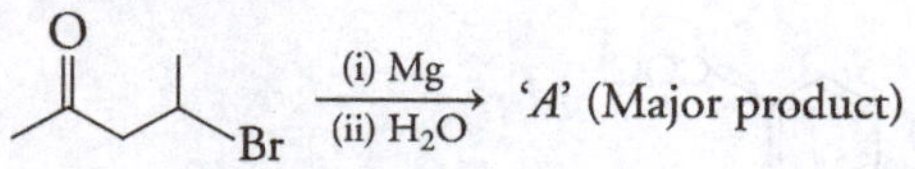

*A* is **[2023, 12 April Shift-I]**

(a) 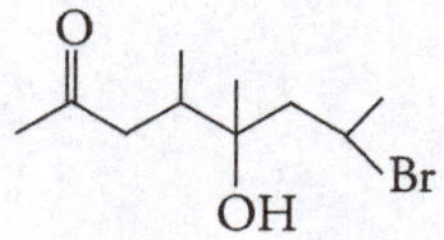 (b) 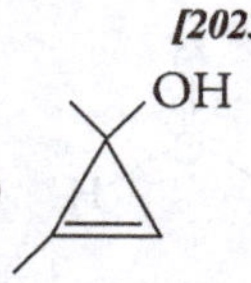

(c) 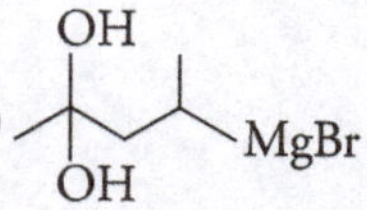 (d) 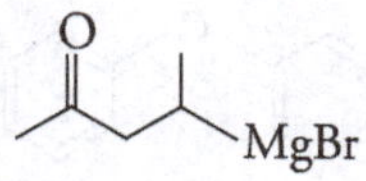

**19** 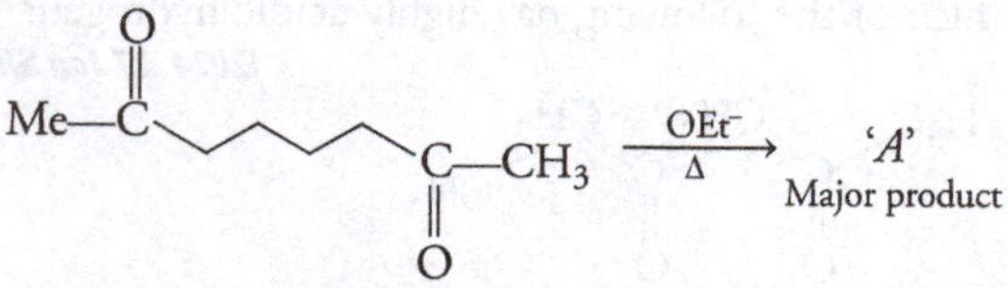

*A* in the above reaction is **[2023, 12 April Shift-I]**

(a) 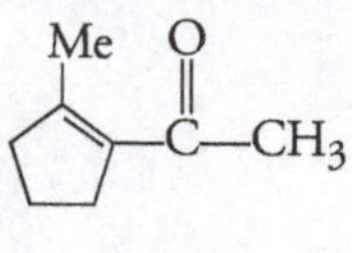

(b) 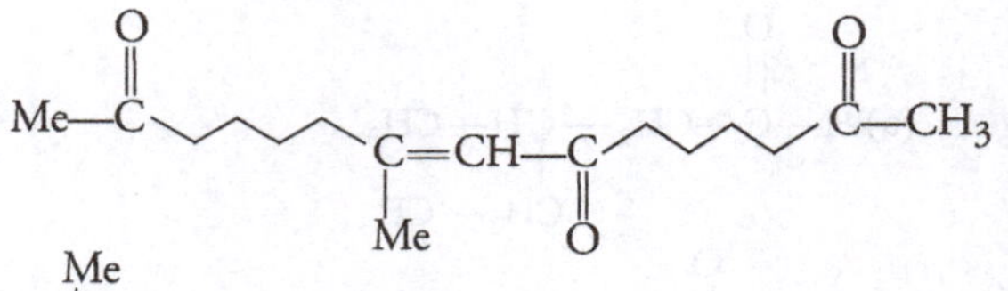

(c) 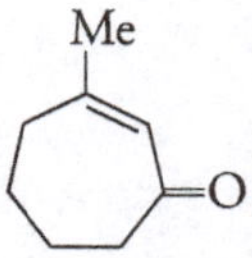

(d) 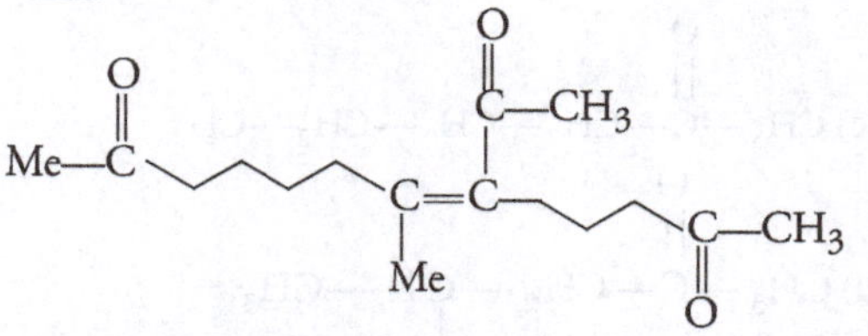

**20** '*A*' in the given reaction is **[2023, 25 Jan Shift-II]**

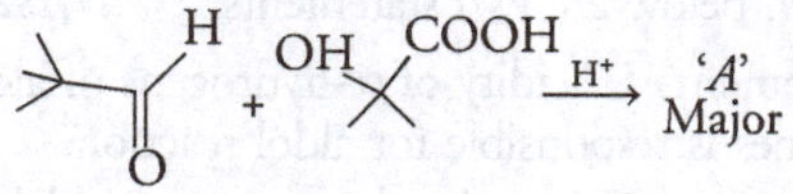

(a) 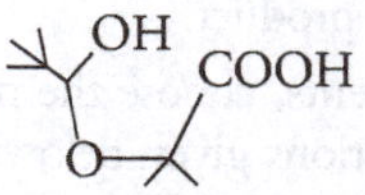 (b) 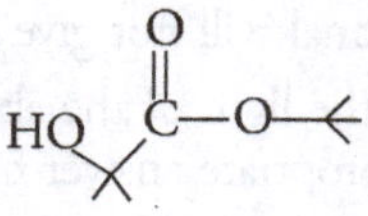

(c) 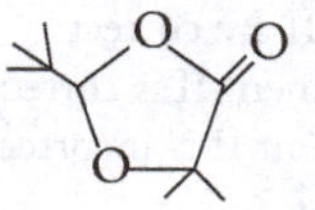 (d) 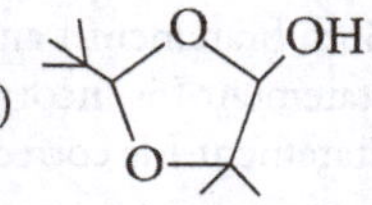

**21** $(C_7H_5O_2)_2 \xrightarrow{h\nu} [X] \longrightarrow 2C_6H_5 + 2CO_2$

Consider the above reaction and identify the intermediate *X*. **[2022, 26 June, Shift-I]**

(a) $C_6H_5-\overset{\overset{O}{\|}}{C^+}$ (b) $C_6H_5-\overset{\overset{O}{\|}}{C}-O^-$

(c) $C_6H_5-\overset{\overset{O}{\|}}{C^+}-\ddot{\underset{\cdot\cdot}{O}}$ (d) $C_6H_5-\overset{\overset{O}{\|}}{C}-\ddot{\underset{\cdot\cdot}{O}}\cdot$

**22** Which of the following conditions or reaction sequence will not give acetophenone as the major product ? **[2022, 25 June Shift-II]**

(a) I. 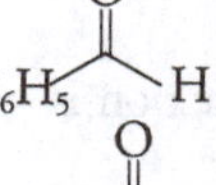 + $CH_3MgBr$ II. $Na_2Cr_2O_7$, $H^+$

(b) I. 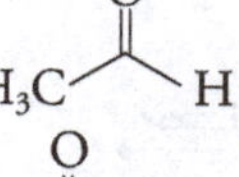 + $C_6H_5MgBr$ II. PCC, DCM

(c) 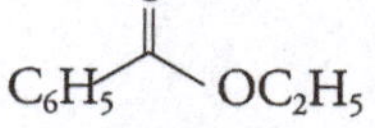 + $2CH_3MgBr$

(d) 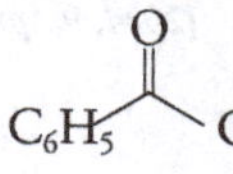 + $CH_3MgBr$ + $CdCl_2$

**23** Which of the following ketone will not give enamine on treatment with secondary amines? [where *t*-Bu is —$C(CH_3)_3$] *[2022, 25 June Shift-II]*

(a) 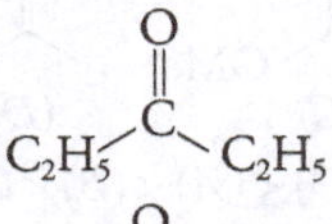
(b) 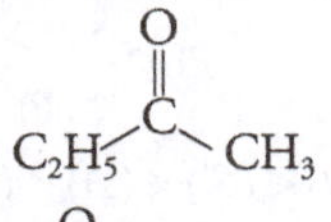
(c) 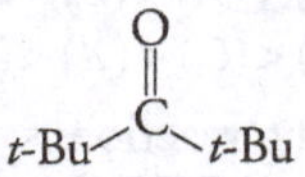
(d) 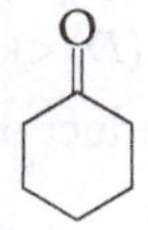

**24** Match List-I with List-II.

| List-I (Chemical reaction) | List-II (Reagent used) |
|---|---|
| A. $CH_3COOC_2H_5 \longrightarrow C_2H_5OH$ | 1. $CH_3MgBr/H_3O^+$ (1.equivalent) |
| B. $CH_3COOCH_3 \longrightarrow CH_3CHO$ | 2. $H_2SO_4/H_2O$ |
| C. $CH_3C\equiv N \longrightarrow CH_3CHO$ | 3. DIBAL-H/$H_2O$ |
| D. $CH_3C\equiv N \longrightarrow$ 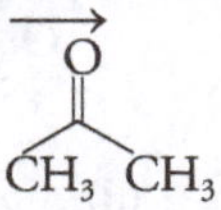 | 4. $SnCl_2$, HCl/$H_2O$ |

Choose the most appropriate answer from the options given below. *[2021, 26 Aug Shift-II]*

| | A | B | C | D | | A | B | C | D |
|---|---|---|---|---|---|---|---|---|---|
| (a) | 2 | 4 | 3 | 1 | (b) | 4 | 2 | 3 | 1 |
| (c) | 2 | 3 | 4 | 1 | (d) | 3 | 2 | 1 | 4 |

**25** The major products formed in the following reaction sequence *A* and *B* are *[2021, 26 Aug Shift-I]*

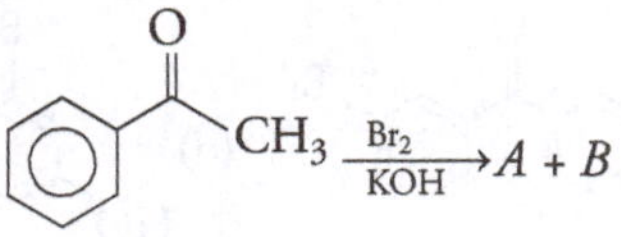

(a) *A* = 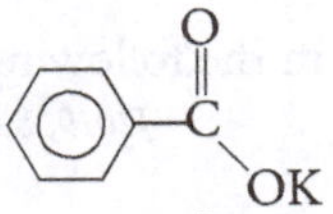 , *B* = $CHBr_3$

(b) *A* = 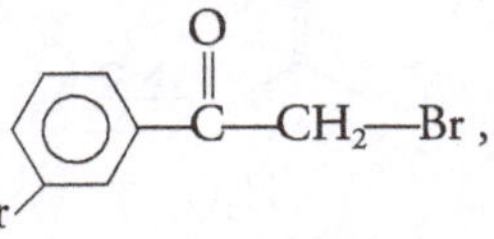 , *B* = 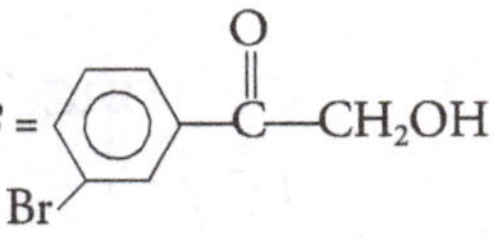

(c) *A* = 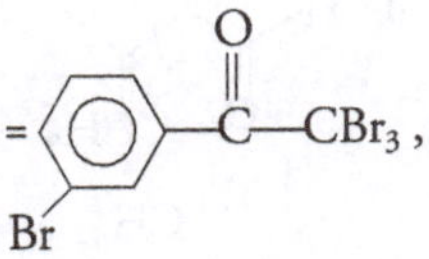 , *B* = 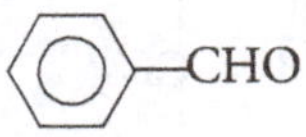

(d) *A* = 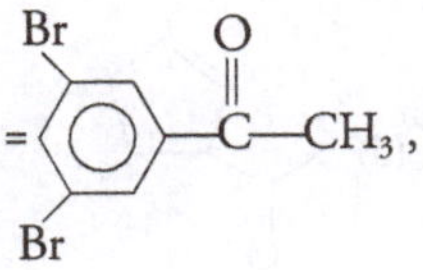 , *B* = 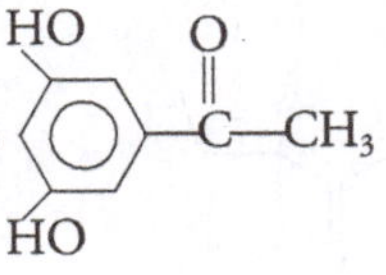

**26** Which one of the following compounds will give orange precipitate when treated with 2, 4 dinitrophenyl hydrazine? *[2021, 27 July Shift-I]*

(a) 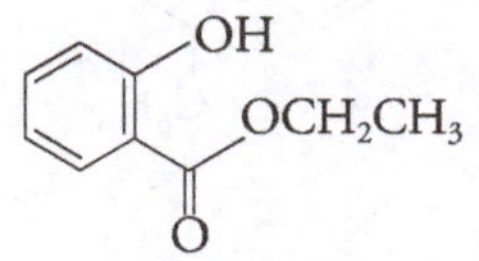
(b) 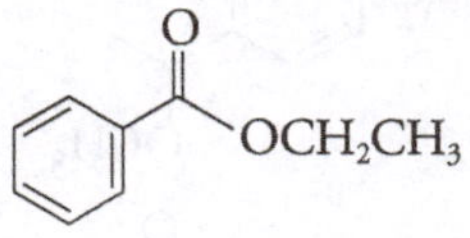
(c) 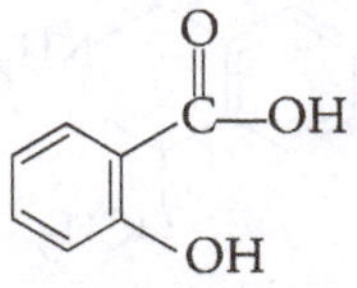
(d) 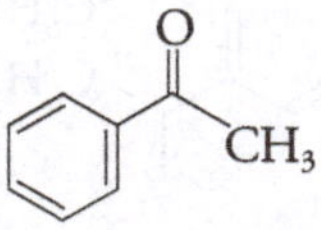

**27** A reaction of benzonitrile with one equivalent $CH_3MgBr$ followed by hydrolysis produces a yellow liquid *P*. The compound *P* will give positive *[2021 25 July Shift-II]*

(a) iodoform test
(b) Schiff's test
(c) ninhydrin's test
(d) Tollen's test

**28** 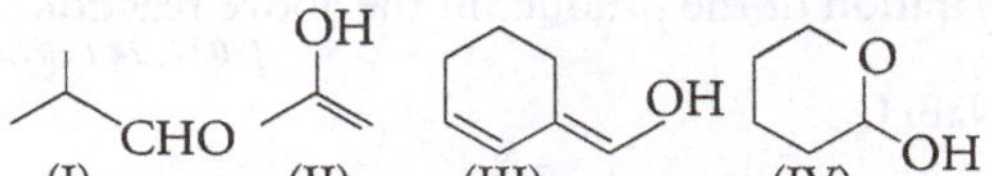

(I) (II) (III) (IV)

Which among the above compound/s does/do not form silver mirror when treated with Tollen's reagent? *[2021, 20 July Shift-I]*

(a) (I), (III) and (IV) only
(b) Only (IV)
(c) Only (II)
(d) (III) and (IV) only

**29** 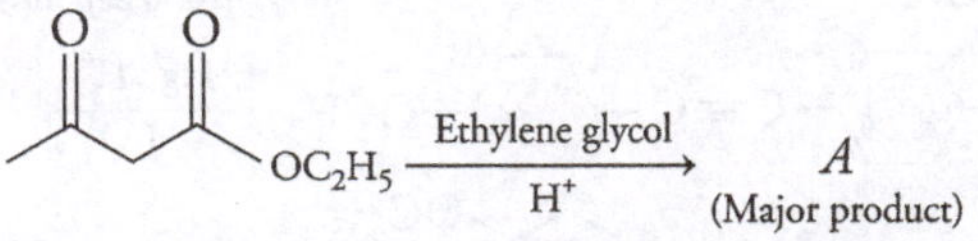


The product in the above reaction is *[2021, 17 March Shift-I]*

(a) 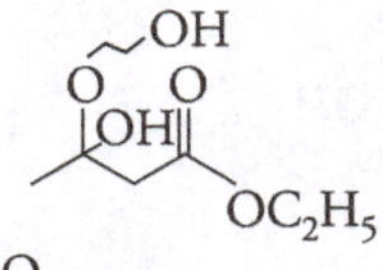
(b) 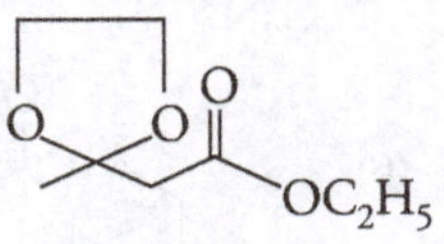
(c) 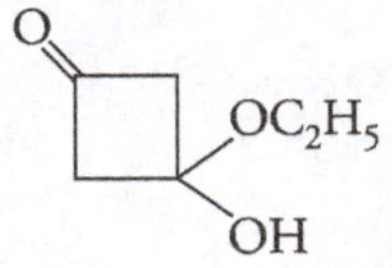
(d) 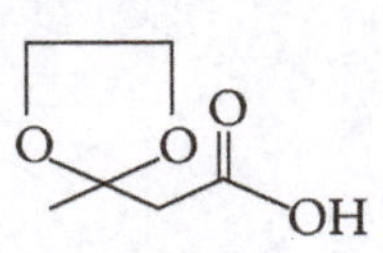

**30.** 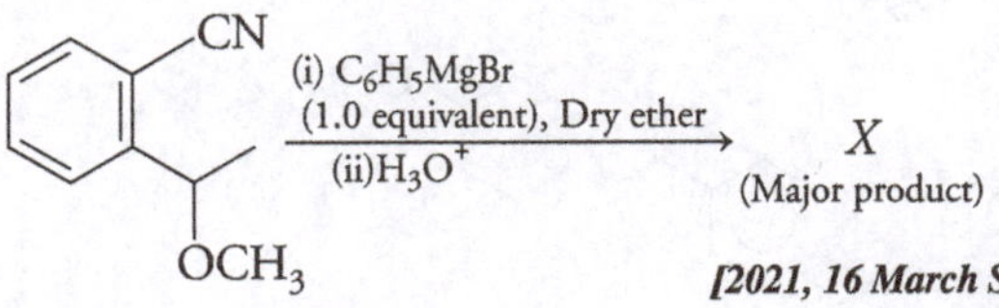


*[2021, 16 March Shift-II]*

(a) 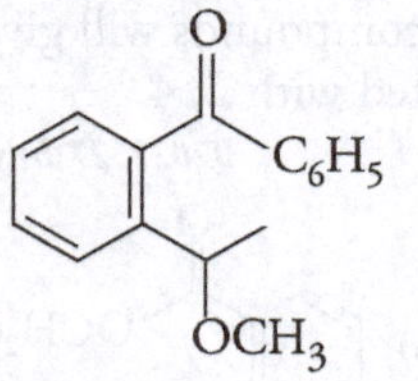 (b) 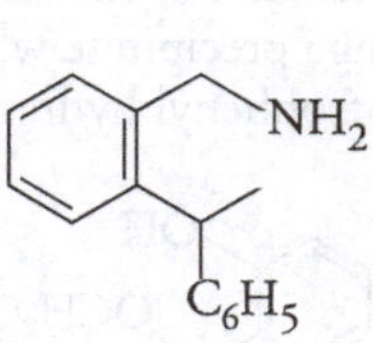

(c) 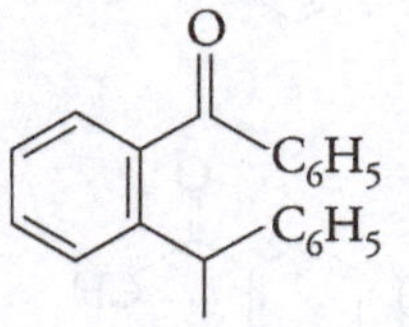 (d) 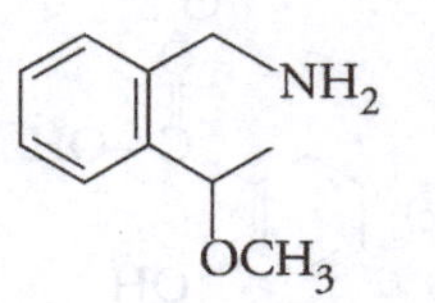

**31** 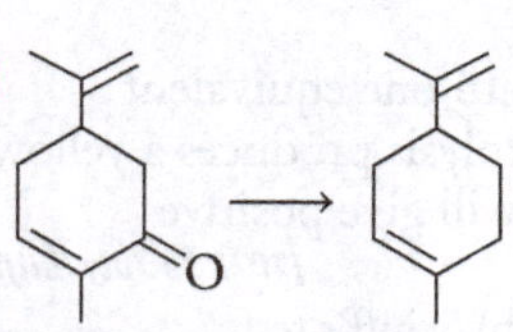

Which of the following reagent is suitable for the preparation of the product in the above reaction? *[2021, 24 Feb Shift-II]*

(a) $NaBH_4$

(b) $NH_2—NH_2/C_2H_5\overset{\ominus}{O}\overset{\oplus}{N}a$

(c) $Ni/H_2$

(d) Red P + $Cl_2$

**Imp.** Preparation of Ketone

**32** The major product obtained from the following reaction is *[2020, 6 Sep Shift-I]*

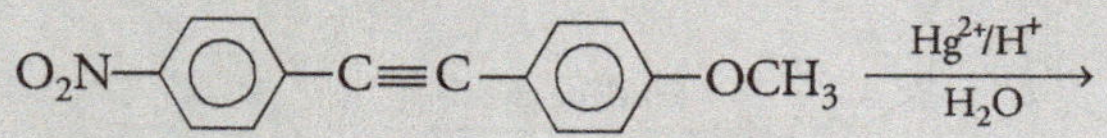

(a) 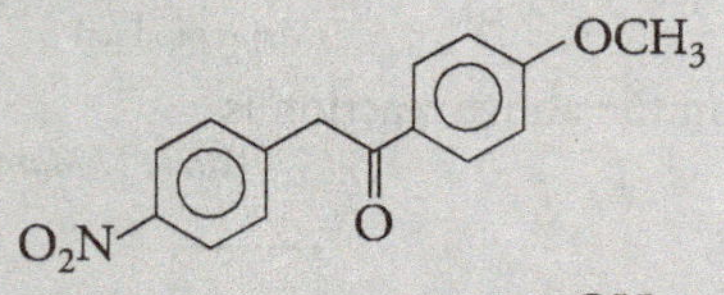

(b) 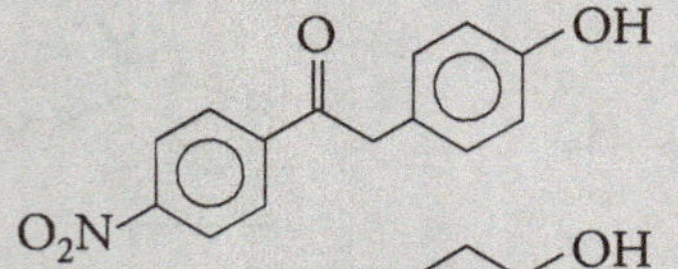

(c) 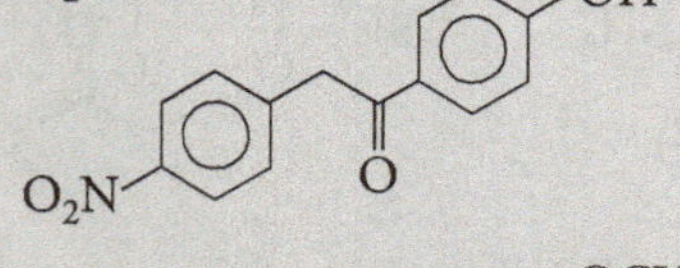

(d) 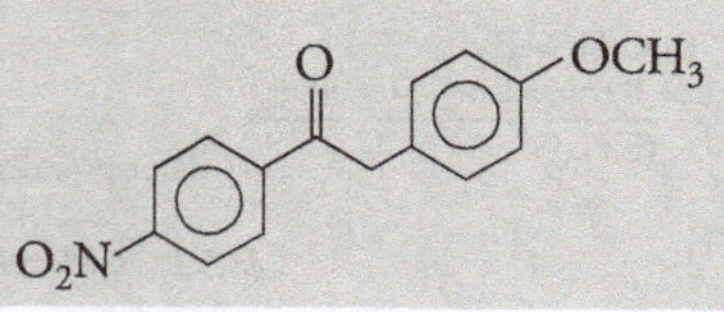

**33** The increasing order of the acidity of the α-hydrogen of the following compounds is *[2020, 5 Sep Shift-I]*

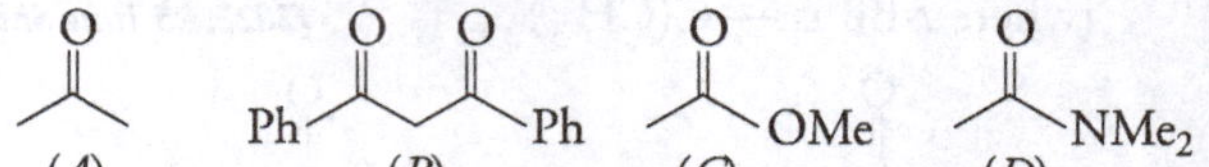

(A) (B) (C) (D)

(a) (D) < (C) < (A) < (B)
(b) (C) < (A) < (B) < (D)
(c) (A) < (C) < (D) < (B)
(d) (B) < (C) < (A) < (D)

**34** The major product of the following reaction is *[2019, 12 April Shift-I]*

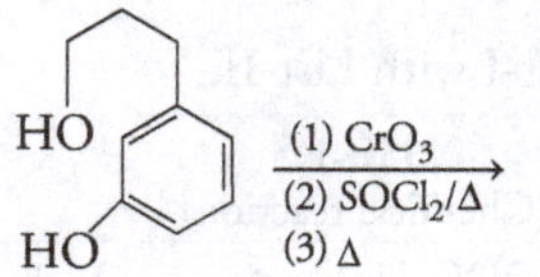

(a) 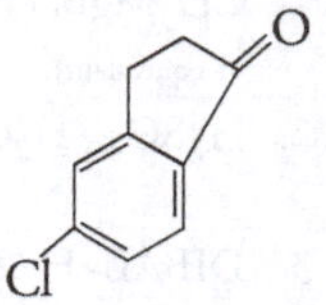 (b) 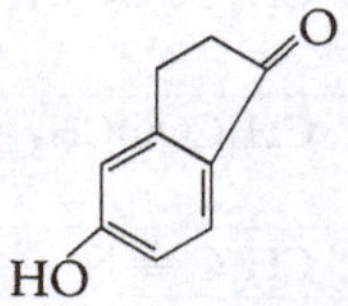

(c) 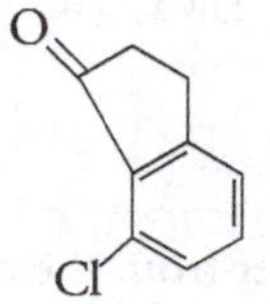 (d) 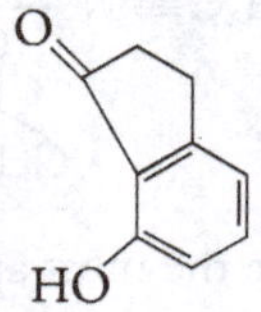

**35** *p*-hydroxybenzophenone upon reaction with bromine in carbon tetrachloride gives *[2019, 9 April Shift-II]*

(a) 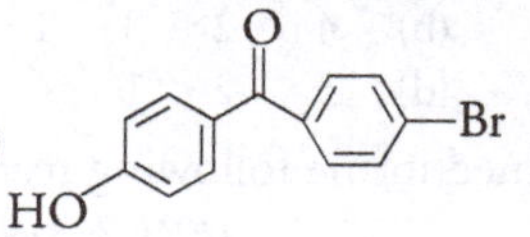 (b) 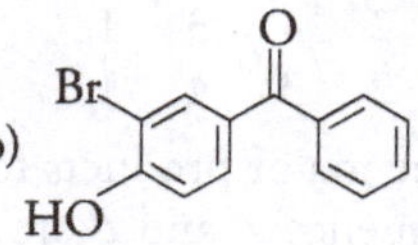

(c) 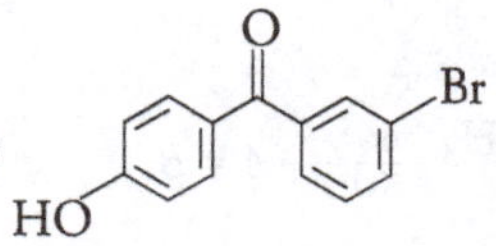 (d) 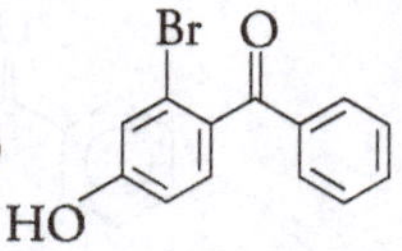

**36** The major product obtained in the following reaction is *[2019, 8 April Shift-II]*

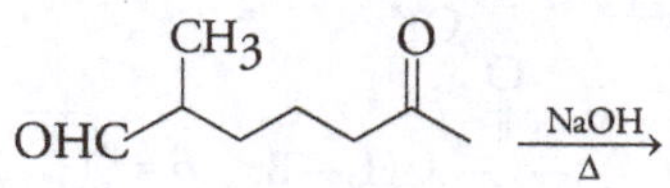

(a) 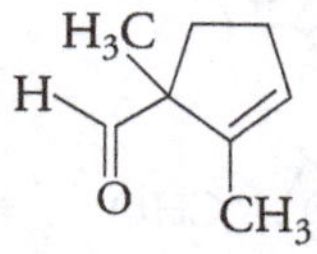 (b) 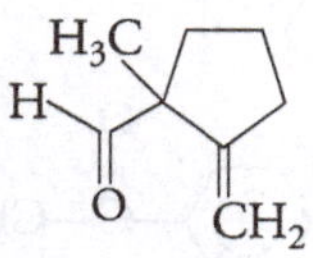

(c) 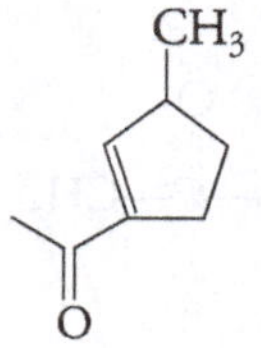 (d) 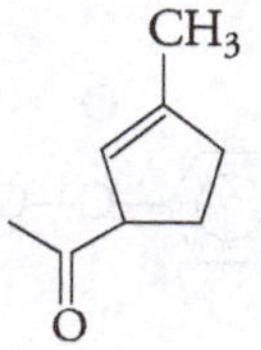

**37** The major product of the following reaction is

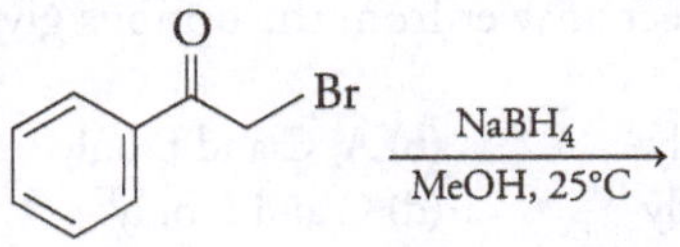

*[2019, 8 April Shift-I]*

(a) 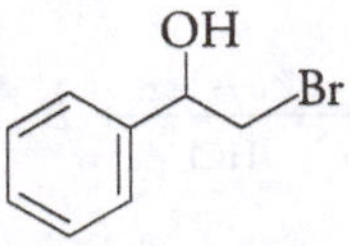 (b) 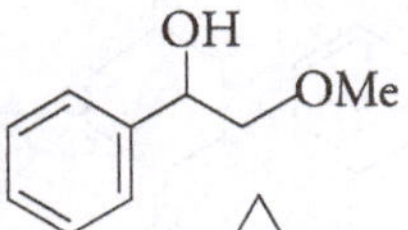

(c) 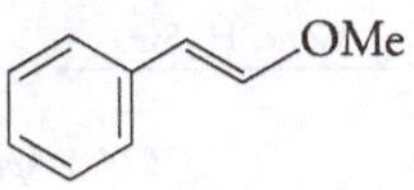 (d) 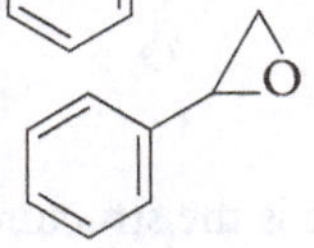

**38** Major product of the following reaction is

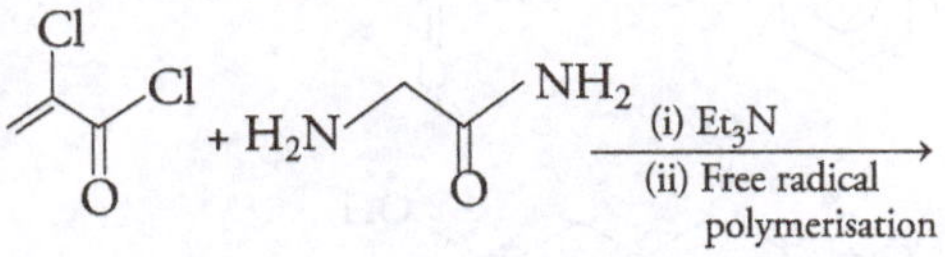

*[2019, 9 Jan Shift-I]*

(a) 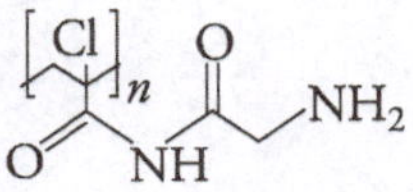 (b) 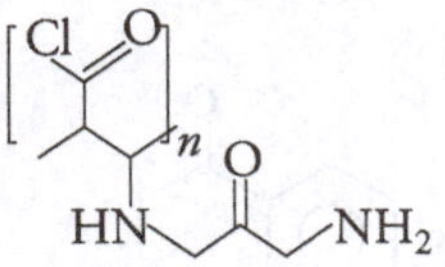

(c) 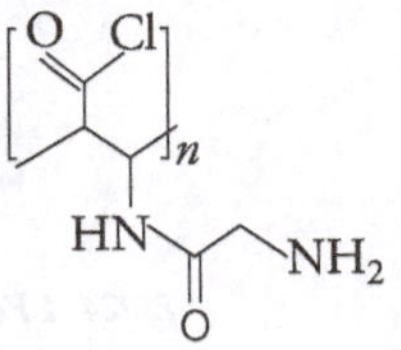 (d) 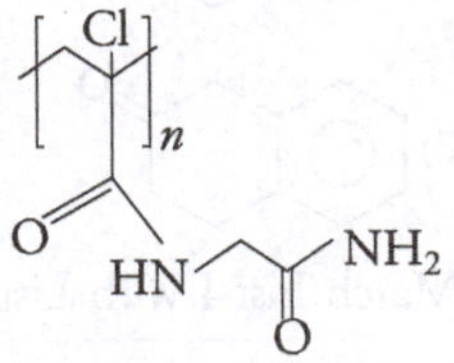

**39** The products formed in the reaction of cumene with $O_2$ followed by treatment with dil. HCl are *[2019, 9 Jan Shift-I]*

(a) 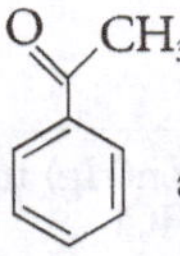 and $CH_3—OH$

(b) 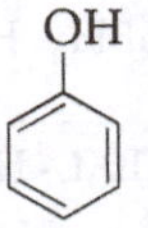 and 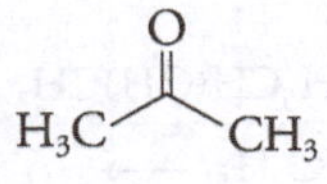

(c) 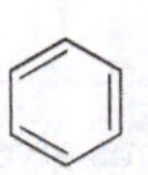 and 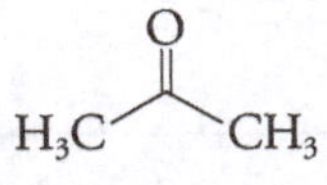

(d) 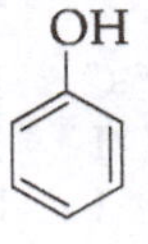 and 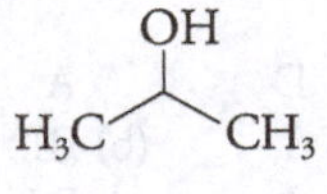

## Level 3

**40** An optically active alkyl halide $C_4H_9Br[A]$ reacts with hot KOH dissolved in ethanol and forms alkene [*B*] as major product which reacts with bromine to give dibromide [*C*]. The compound [*C*] is converted into a gas [*D*] upon reacting with alcoholic $NaNH_2$. During hydration 18 gram of water is added to 1 mole of gas [*D*]. On warming with mercuric sulphate and dilute acid at 333 K to form compound [*E*]. The IUPAC name of compound [*E*] is *[2025, 2 April Shift-I]*

(a) But-2-yne
(b) Butan-2-one
(c) Butan-1-al
(d) Butan-2-ol

**41** An organic compound (*X*) with molecular formula $C_3H_6O$ is not readily oxidised. On reduction it gives $C_3H_8O(Y)$ which reacts with HBr to give a bromide (*Z*) which is converted to Grignard reagent. This Grignard reagent on reaction with (*X*) followed by hydrolysis give 2, 3-dimethylbutan-2-ol. Compounds (*X*), (*Y*) and (*Z*) respectively are: *[2025, 4 April Shift-I]*

(a) $CH_3COCH_3$, $CH_3CH_2CH_2OH$, $CH_3CH(Br)CH_3$
(b) $CH_3COCH_3$, $CH_3CH(OH)CH_3$, $CH_3CH(Br)CH_3$
(c) $CH_3CH_2CHO$, $CH_3CH=CH_2$, $CH_3CH(Br)CH_3$
(d) $CH_3CH_2CHO$, $CH_3CH_2CH_2OH$, $CH_3CH_2CH_2Br$

**42** When 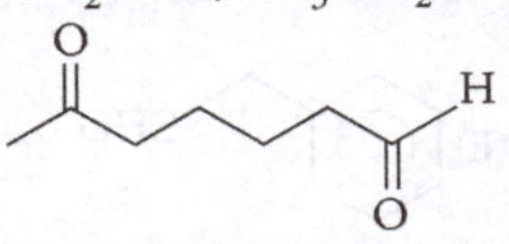 undergoes intramolecular aldol condensation, the major product formed is *[2025, 8 April Shift-II]*

(a) 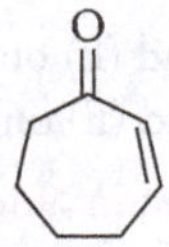 (b) 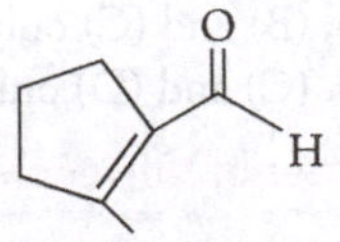

(c) 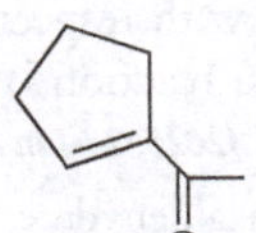 (d) 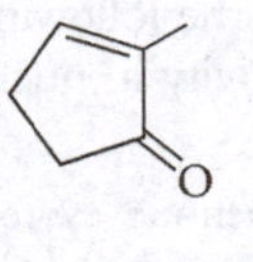

**43** The correct order of basicity for the following molecules is *[2025, 4 April Shift-II]*

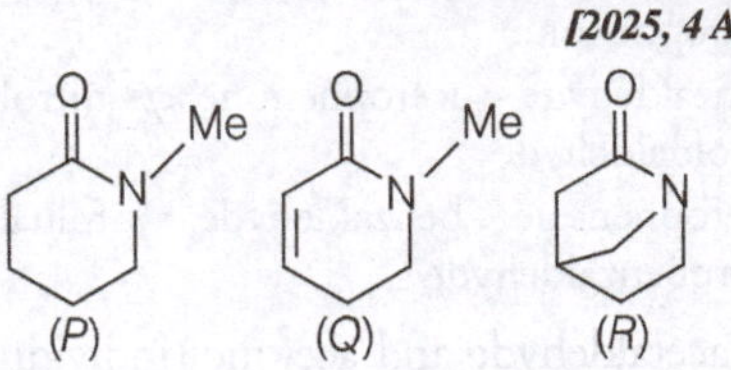

(*P*) (*Q*) (*R*)

(a) $R > Q > P$ (b) $P > Q > R$
(c) $Q > P > R$ (d) $R > P > Q$

**44** Consider the following molecules: *[2025, 2 April Shift-I]*

*p*. $CH_3-CH_2-\overset{\overset{O}{||}}{C}-Cl$

*q*. $CH_3-CH_2-\overset{\overset{O}{||}}{C}-O-\overset{\overset{O}{||}}{C}-CH_3$

*r*. $CH_3-CH_2-\overset{\overset{O}{||}}{C}-O-CH_2-CH_3$

*s*. $CH_3-CH_2-\overset{\overset{O}{||}}{C}-NH_2$

The correct order of rate of hydrolysis is

(a) $p > q > r > s$ (b) $q > p > r > s$
(c) $r > q > p > s$ (d) $p > r > q > s$

**45** The compounds which give positive Fehling's test are

(A) 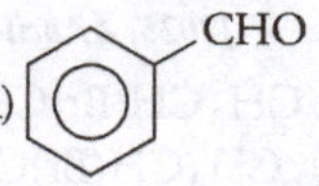 (B) 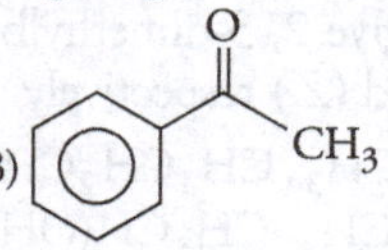

(C) $HOCH_2-CO-(CHOH)_3-CH_2-OH$

(D) $CH_3-\overset{\overset{O}{||}}{C}-H$ (E) 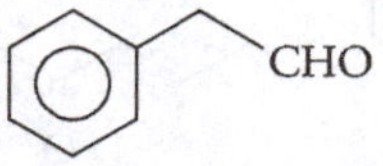

Choose the correct answer from the options given below. *[2025, 22 Jan Shift-I]*

(a) (A), (B) and (C) only (b) (C), (D) and (E) only
(c) (A), (C) and (D) only (d) (A), (D) and (E) only

**Imp.** **Order of Nucleophilic Addition Reaction in Aldehydes**

**46** Which of the following arrangements with respect to their reactivity in nucleophilic addition reaction is correct? *[2025, 24 Jan Shift-I]*

(a) acetophenone < *p*-tolualdehyde < benzaldehyde < *p*-nitrobenzaldehyde
(b) *p*-nitrobenzaldehyde < benzaldehyde < *p*-tolualdehyde < acetophenone
(c) benzaldehyde < acetophenone <*p*- nitrobenzaldehyde < *p*-tolualdehyde
(d) acetophenone < benzaldehyde < *p*-tolualdehyde < *p*-nitrobenzaldehyde

**47** Both acetaldehyde and acetone (individually) undergo which of the following reactions? *[2025, 28 Jan Shift-I]*

A. Iodoform reaction B. Cannizzaro reaction
C. Aldol condensation D. Tollen's test
E. Clemmensen reduction

Choose the correct answer from the options given below.

(a) A, B and D only (b) A, C and E only
(c) B, C and D only (d) C and E only

**48** 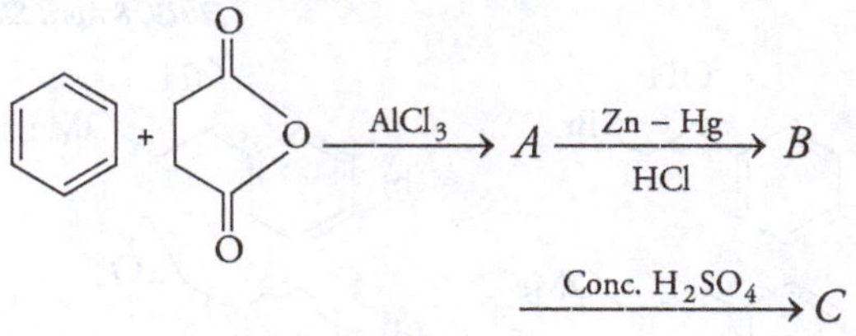

What is the structure of C? *[2024, 9 April Shift-I]*

(a) 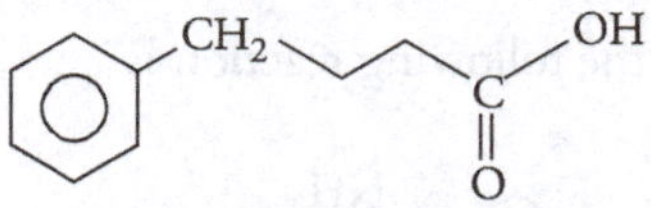

(b) 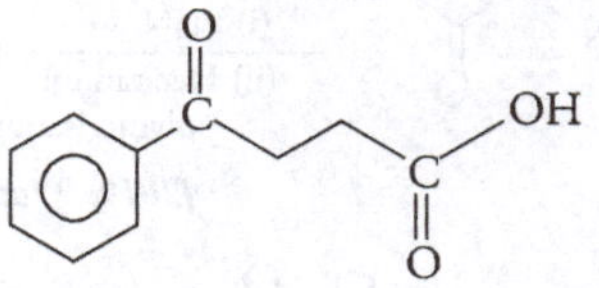

(c) 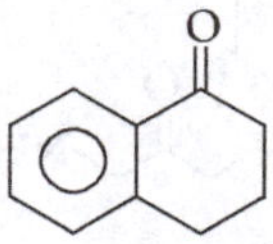

(d) 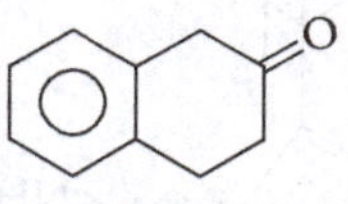

**49** Match List-I with List-II. *[2024, 1 Feb Shift-I]*

| List-I (Reactions) | List-II (Reagents) |
|---|---|
| A. $CH_3(CH_2)_5-\underset{\underset{O}{||}}{C}-OC_2H_5 \longrightarrow CH_3(CH_2)_5CHO$ | I. $CH_3MgBr, H_2O$ |
| B. $C_6H_5COC_6H_5 \longrightarrow C_6H_5CH_2C_6H_5$ | II. Zn(Hg) and conc. HCl |
| C. $C_6H_5CHO \longrightarrow C_6H_5CH(OH)CH_3$ | III. $NaBH_4, H^+$ |
| D. $CH_3COCH_2COOC_2H_5 \longrightarrow CH_3\underset{\underset{H}{|}}{C}(OH)CH_2COOC_2H_5$ | IV. DIBAL - H, $H_2O$ |

Choose the correct answer from the options given below.

| | A | B | C | D |
|---|---|---|---|---|
| (a) | III | IV | II | I |
| (b) | III | IV | I | II |
| (c) | IV | II | III | I |
| (d) | IV | II | I | III |

**50** Find out the major product from the following reaction.

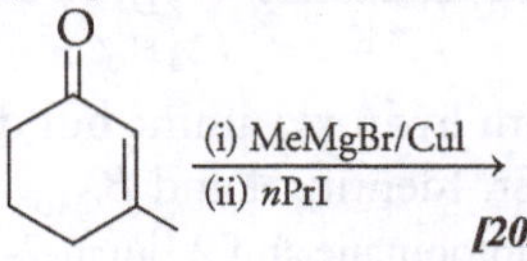

[2023, 6 April Shift-II]

(a) 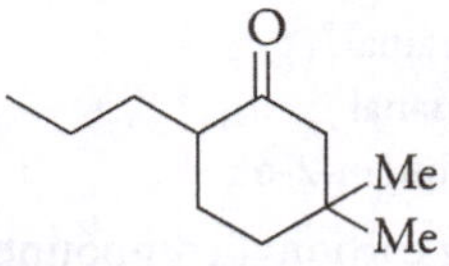

(b) 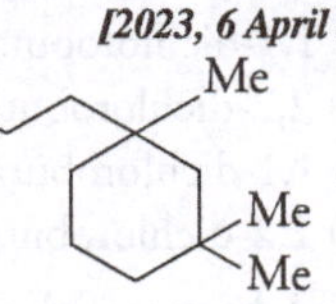

(c) 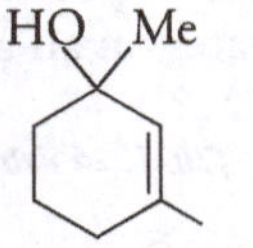

(d) 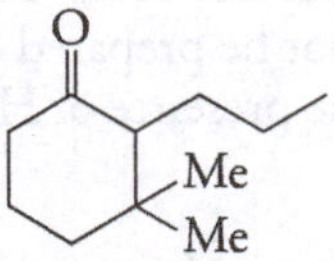

**51** Cyclohexylamine when treated with nitrous acid yields (*P*). On treating (*P*) with PCC results in (*Q*). When (*Q*) is heated with dil. NaOH we get (*R*). The final product (*R*) is [2023, 31 Jan Shift-II]

(a) 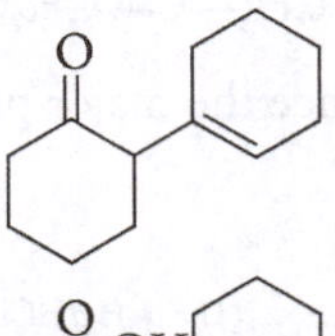

(b) 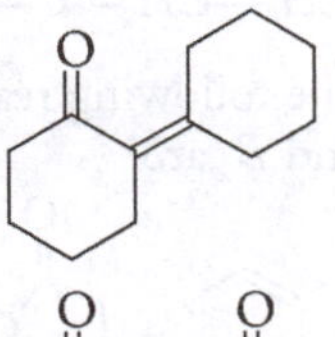

(c) 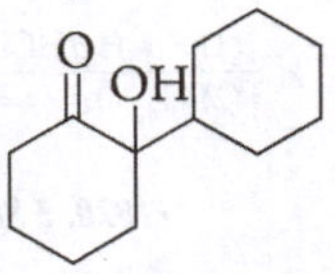

(d) 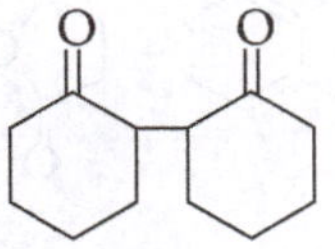

**52** Find out the major products from the following reaction sequence. [2023, 29 Jan Shift-II]

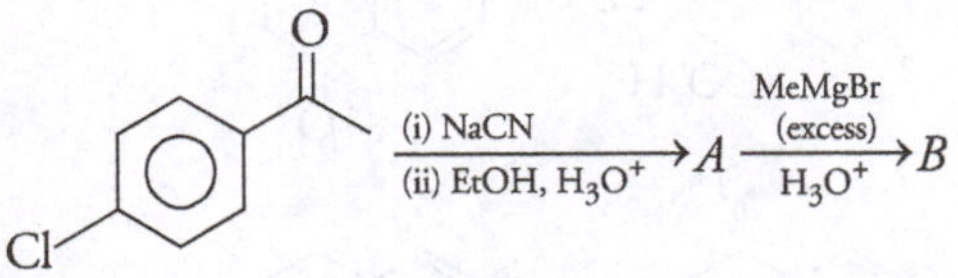

(a) $A$ = 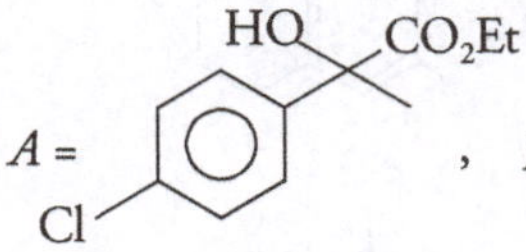, $B$ = 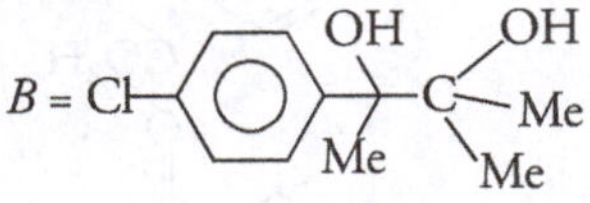

(b) $A$ = 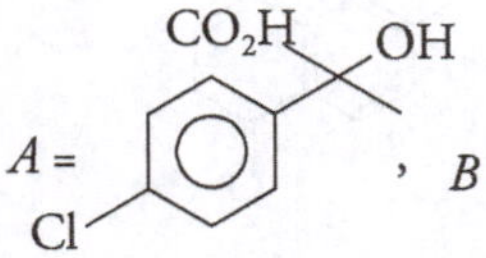, $B$ = 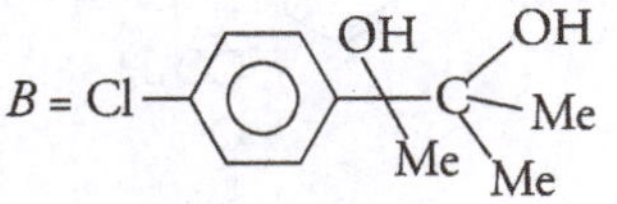

(c) $A$ = 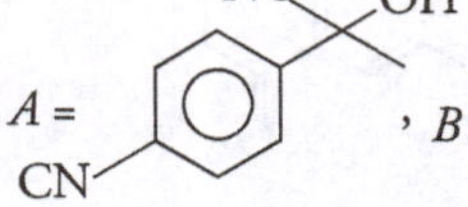, $B$ = 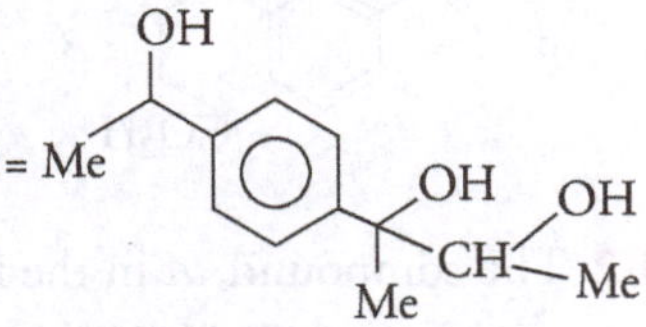

(d) $A$ = 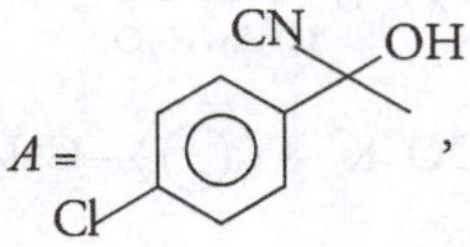, $B$ = 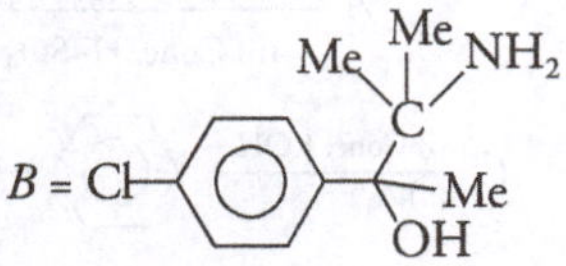

**53** '*R*' formed in the following sequence of reaction is

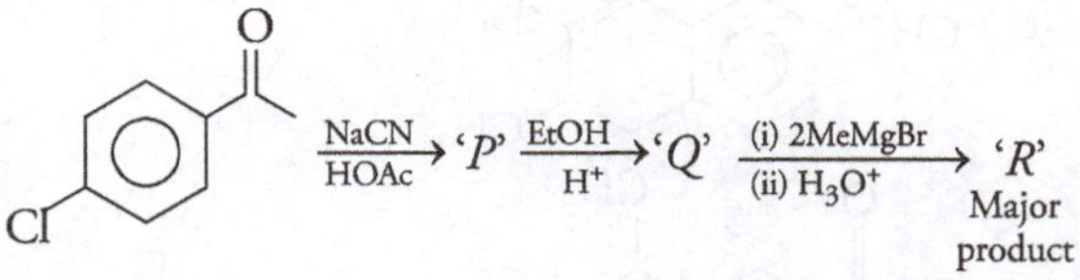

[2023, 24 Jan Shift-I]

(a) 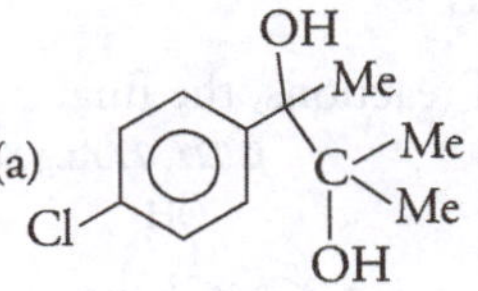

(b) 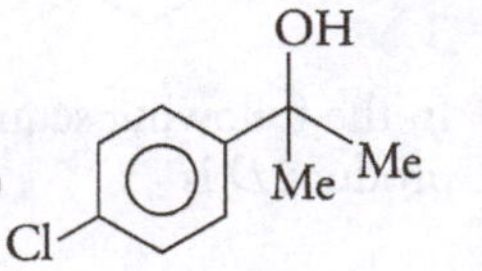

(c) 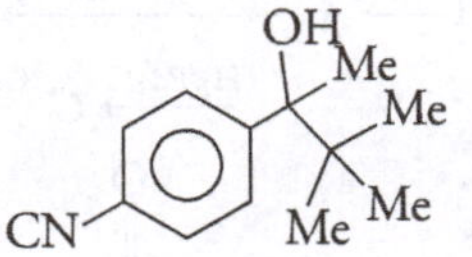

(d) 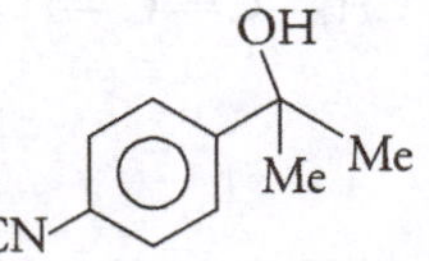

**54** $CH_3-CH_2C\equiv CN \xrightarrow[\text{Ether}]{CH_3MgBr} A \xrightarrow{H_3O^+} B \xrightarrow[\text{HCl}]{\text{Zn}-\text{Hg}} C$

The correct structure of *C* is [2022, 25 July Shift-II]

(a) $CH_3CH_2CH_2CH_3$

(b) $CH_3CH_2-\overset{O}{\overset{\|}{C}}-CH_3$

(c) $CH_3CH_2\overset{OH}{\overset{|}{C}H}CH_3$

(d) $CH_3CH_2CH=CH_2$

**55** Which one of the following is the major product of the given reaction? [2021, 27 Aug Shift-II]

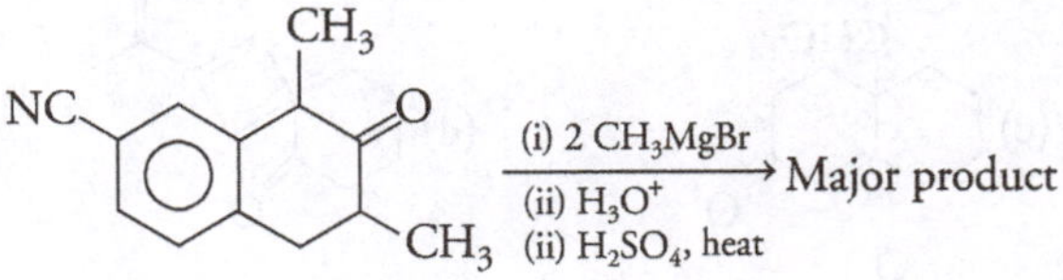

(a) 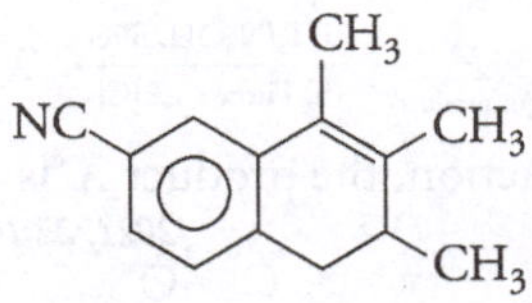

(b) 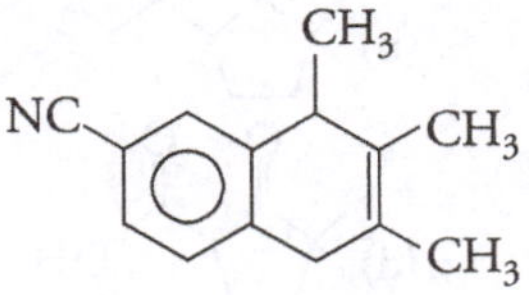

(c) 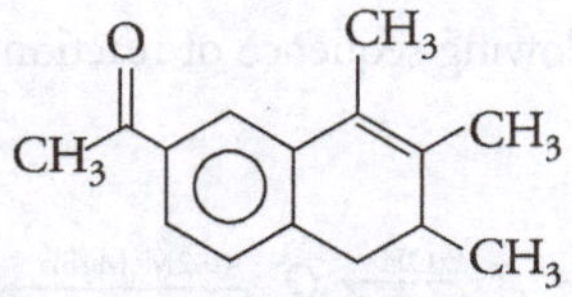

(d) 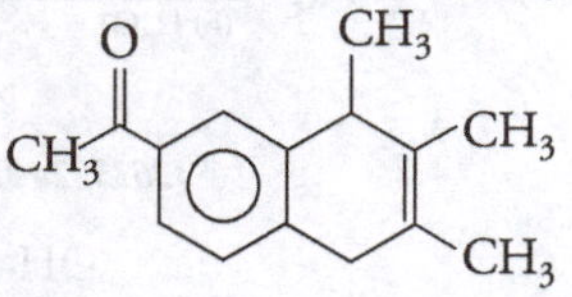

**56** In the following sequence of reactions, the final product $D$ is **[2021, 27 Aug Shift-I]**

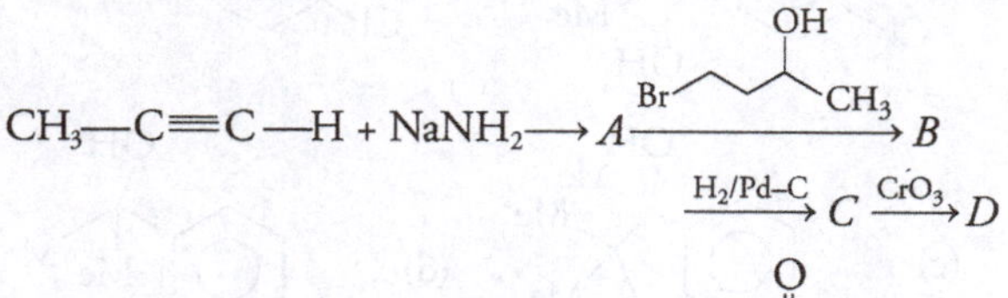

(a) $H_3C-CH_2-CH_2-CH_2-CH_2-\overset{O}{\overset{\|}{C}}-H$

(b) $CH_3-CH=CH-CH_2-CH_2-CH_2-COOH$

(c) $H_3C-CH=CH-CH(OH)-CH_2-CH_2CH_3$

(d) $CH_3-CH_2-CH_2-CH_2-CH_2-\overset{O}{\overset{\|}{C}}-CH_3$

**57** The major product $P$ in the following reaction is

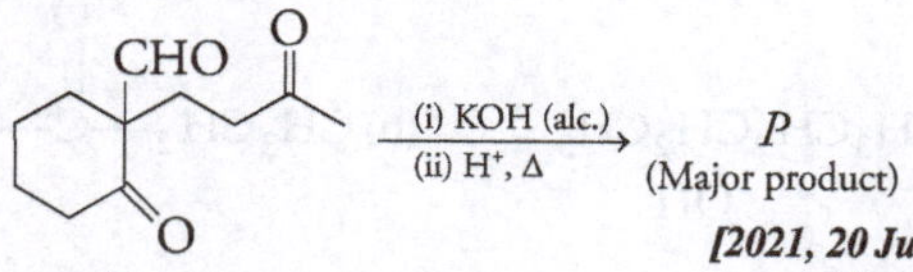

**[2021, 20 July Shift-II]**

(a) 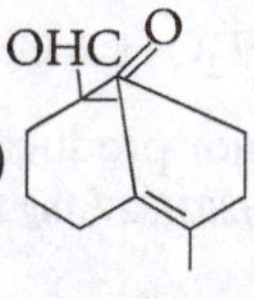

(b) 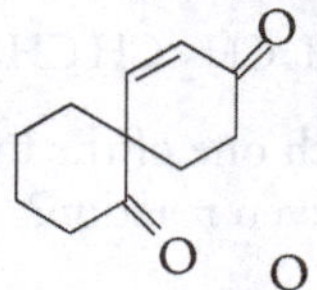

(c) 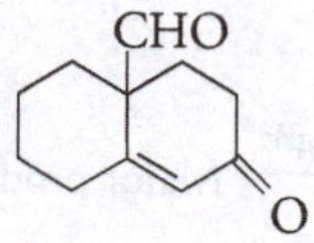

(d) 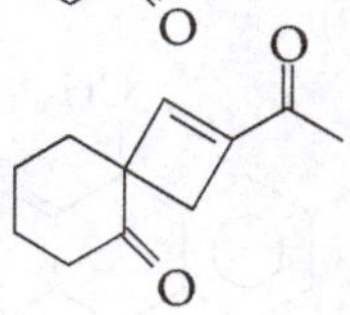

**58**

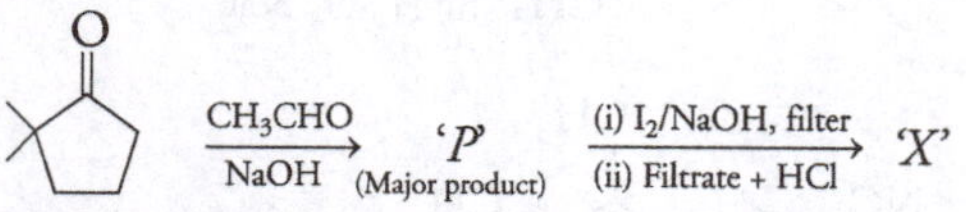

Consider the given reaction, the product $X$ is **[2021, 22 July Shift-I]**

(a) 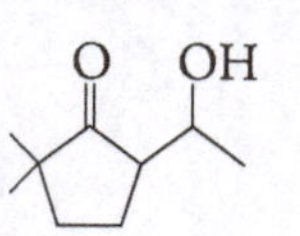

(b) 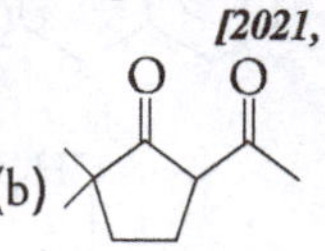

(c) 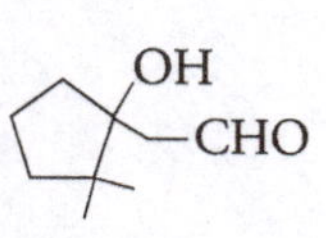

(d) 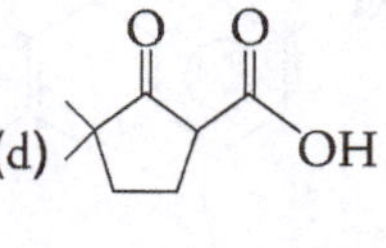

**59** $\underset{(C_4H_8Cl_2)}{A} \xrightarrow[373\text{ K}]{\text{Hydrolysis}} \underset{(C_4H_8O)}{B}$

$B$ reacts with hydroxyl amine but does not give Tollen's test. Identify $A$ and $B$. **[2021, 26 Feb Shift-I]**

(a) 1,1-dichlorobutane and 2-butan-2-one
(b) 2,2-dichlorobutane and butanal
(c) 1,1-dichlorobutane and butanal
(d) 2,2-dichlorobutane and 2-butan-2-one

**60** Which one of the following carbonyl compounds cannot be prepared by addition of water on an alkyne in the presence of $HgSO_4$ and $H_2SO_4$? **[2021, 24 Feb Shift-II]**

(a) $CH_3-\overset{O}{\overset{\|}{C}}-H$

(b) 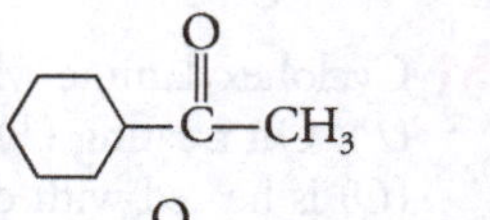

(c) $CH_3-CH_2-\overset{O}{\overset{\|}{C}}-H$

(d) $CH_3-\overset{O}{\overset{\|}{C}}-CH_2CH_3$

**61** In the following reaction sequence the major products $A$ and $B$ are

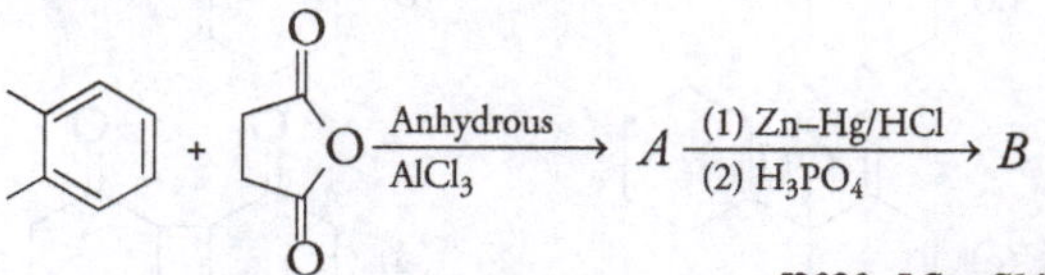

**[2020, 5 Sep Shift-I]**

(a) 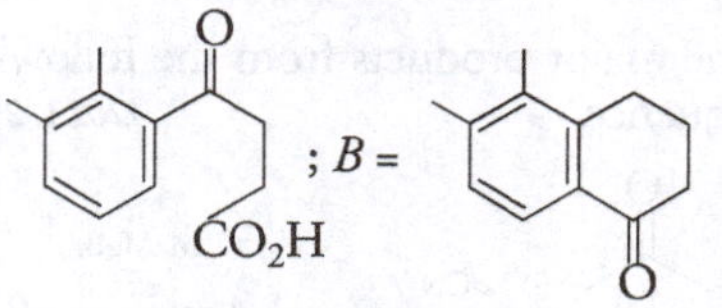

(b) 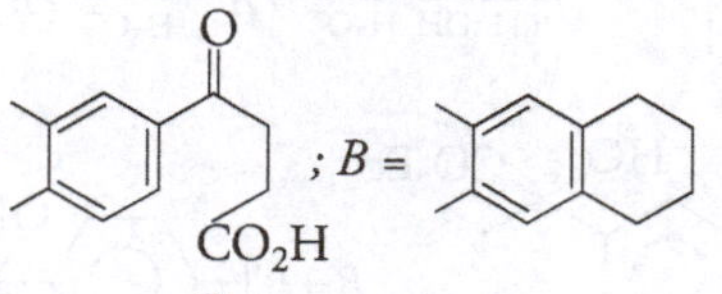

(c) 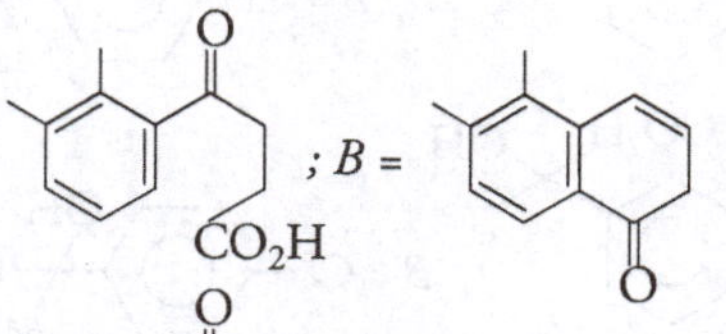

(d) 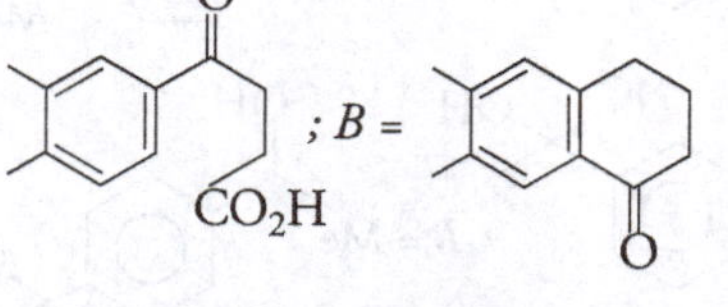

**62** The compound, $A$ in the following

$A \xrightarrow[\text{(ii) Conc. } H_2SO_4/\Delta]{\text{(i) } CH_3MgBr/H_2O} B \xrightarrow[\text{(ii) Zn}/H_2O]{\text{(i) } O_3} C + D$

$$D \xrightarrow[\Delta]{Ba(OH)_2} H_3C-\overset{\overset{CH_3}{|}}{C}=CH-\overset{\overset{O}{\|}}{C}-CH_3$$

**[2020, 3 Sep Shift-II]**

(a) $C_6H_5-CH_2-\overset{\overset{O}{\|}}{C}-CH_3$

(b) $C_6H_5-\overset{\overset{O}{\|}}{C}-CH_3$

(c) $C_6H_5-\overset{\overset{O}{\|}}{C}-CH_2CH_3$

(d) $C_6H_5-\overset{\overset{O}{\|}}{C}-CH(CH_3)_2$

**63** Identify (A) in the following reaction sequence :

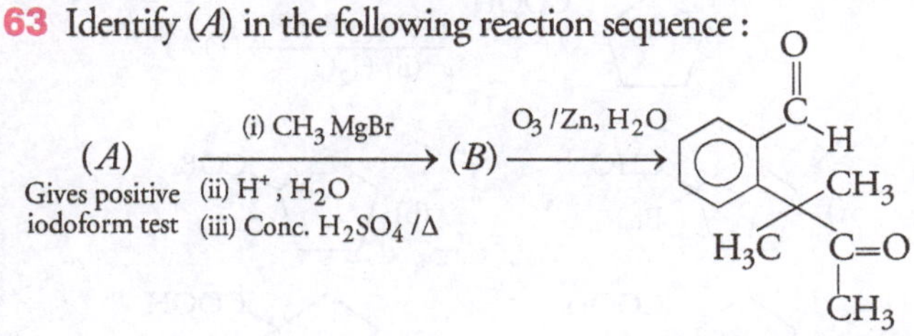

**[2020, 9 Jan Shift-I]**

(a) 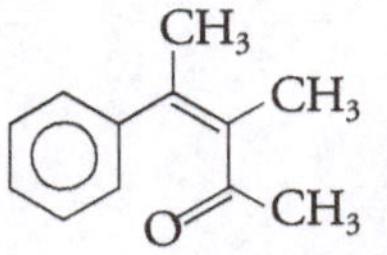 (b) 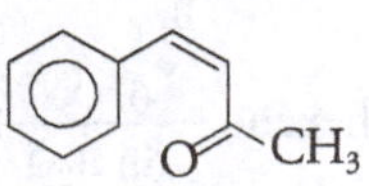

(c) 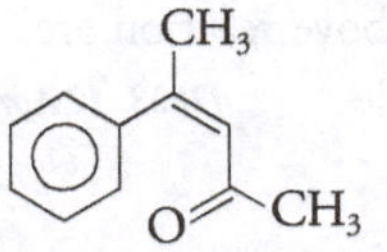 (d) 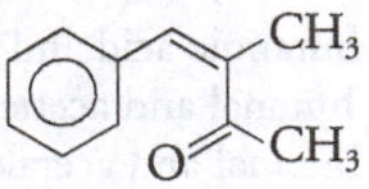

**64** The major product Y in the following reaction is

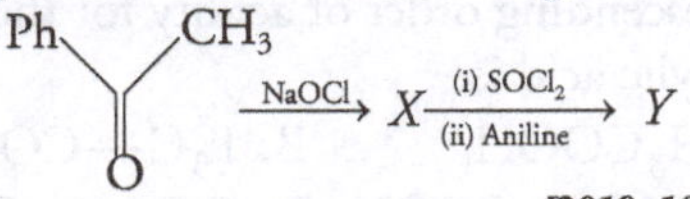

**[2019, 10 April Shift-II]**

(a) 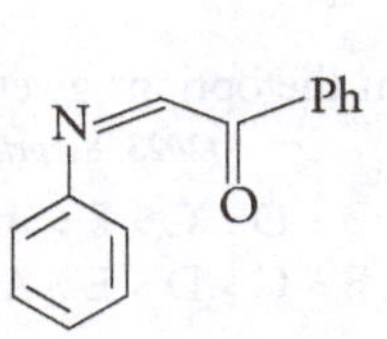 (b) 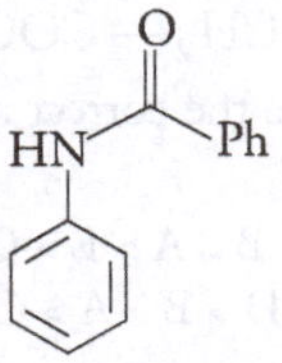

(c) 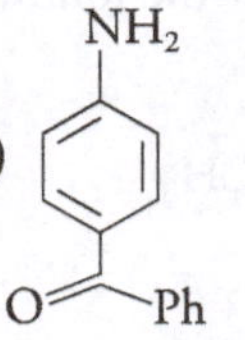 (d) 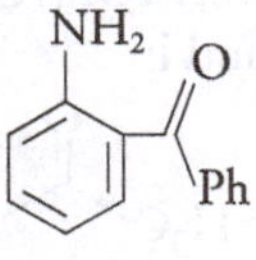

**65** In the following reactions, products A and B are

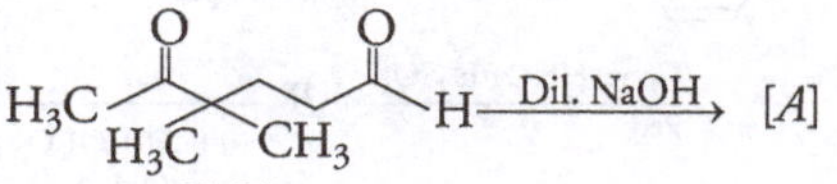

$[A] \xrightarrow[\Delta]{H_3O^+} [B]$

**[2019, 12 Jan Shift-I]**

(a) A = 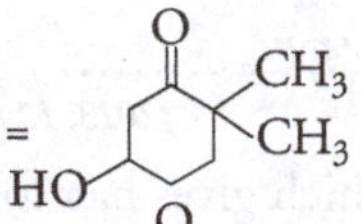 B = 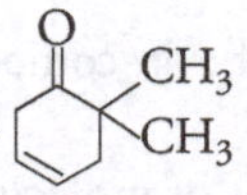

(b) A = 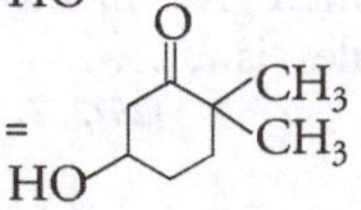 B = 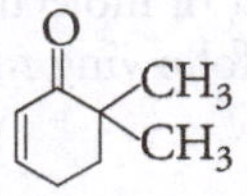

(c) A = 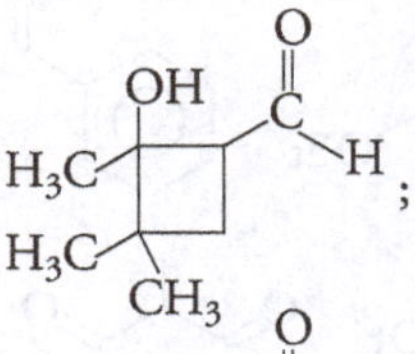 ; B = 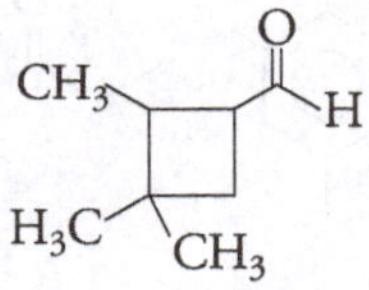

(d) A = 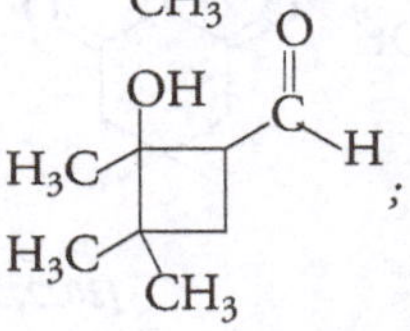 ; B = 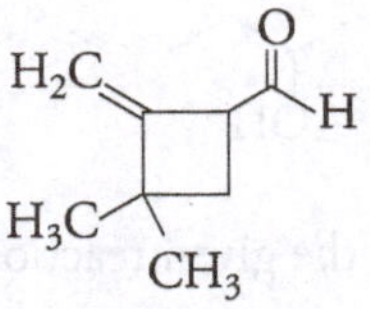

**66** The major product obtained in the following reaction is

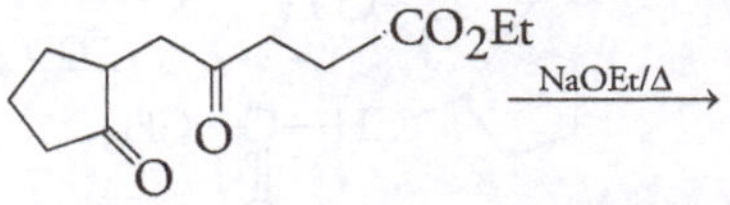

**[2019, 10 Jan Shift-II]**

(a) 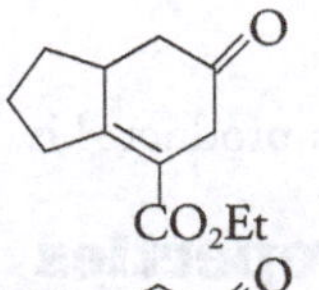 (b) 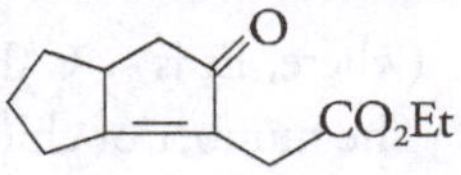

(c) 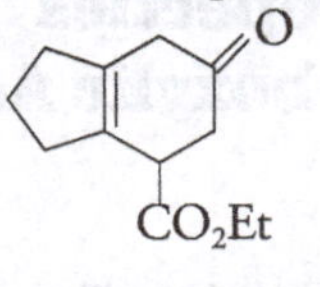 (d) 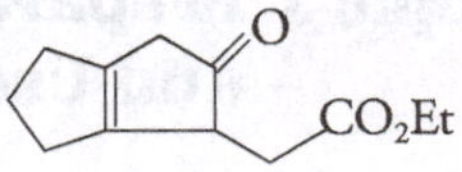

## Numerical Value Type Questions

**67** Identify the structure of the final product (D) in the following sequence of the reactions:

$$Ph-\overset{\overset{O}{\|}}{C}-CH_3 \xrightarrow[\Delta]{PCl_5} A \xrightarrow{3\text{ eq. }NaNH_2/NH_3}$$

$$B \xrightarrow{\text{Acidify}} C \xrightarrow[2.\ H_2O_2/OH^-]{1.\ B_2H_6} D$$

Total number of $sp^2$ hybridised carbon atoms in product D is ............ .

**[2025, 7 April Shift-II]**

**68** 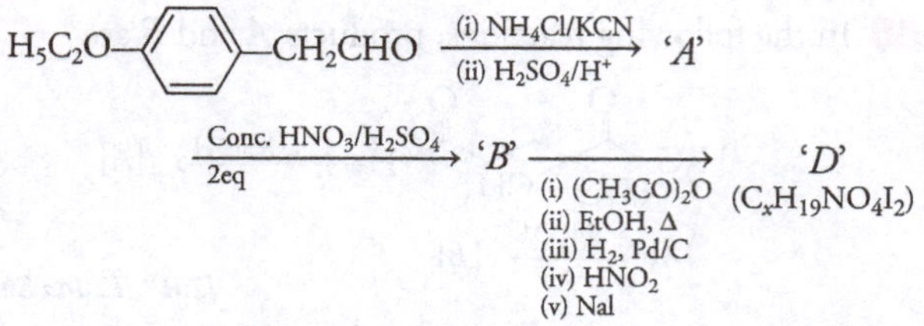


The value of $x$ in compound 'D' is ......... .

*[2023, 12 April Shift-I]*

**69** The number of molecules which gives haloform test among the following molecules is ...... .

*[2023, 31 Jan Shift-II]*

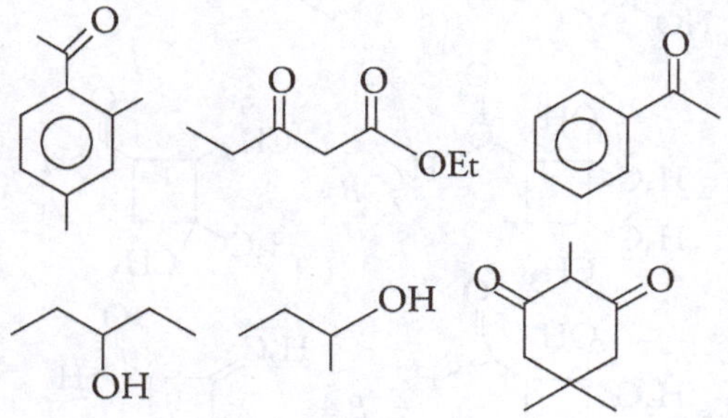


**70** In the given reaction *[2022, 25 July Shift-I]*

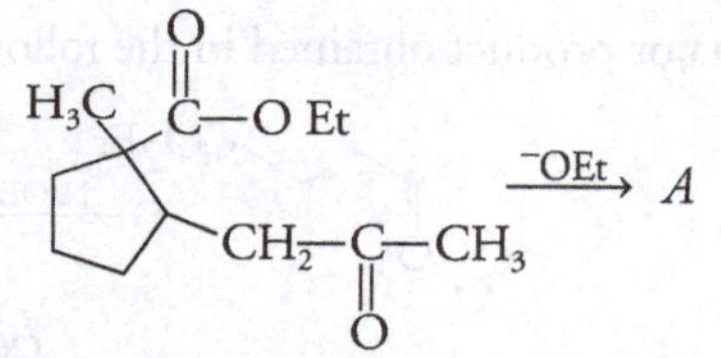


(where, Et is $—C_2H_5$)

The number of chiral carbon/s in product $A$ is ......

## Topic 3 Preparation, Properties and Uses of Carboxylic Acids

### Level 1

**01.** Which of the following compounds is least likely to give effervescence of $CO_2$ in presence of *aq.* $NaHCO_3$?

*[2024, 7 April shift-I]*

(a) $Ph—\overset{\oplus}{N}H_3\ \overset{\ominus}{Cl}$

(b) 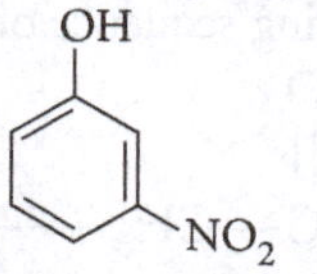


(c) 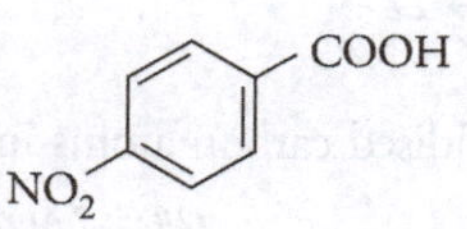


(d) 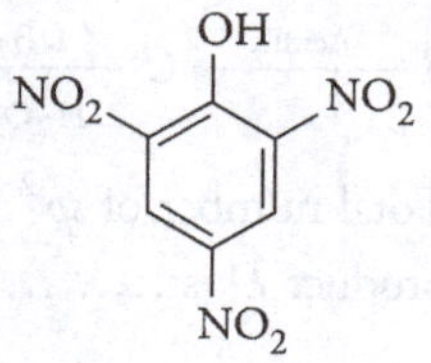


**Imp.** **Acidic Strength of Aliphatic Acid**

**02** The correct sequence of acidic strength of the following aliphatic acids in their decreasing order is

*[2024, 8 April Shift-II]*

$CH_3CH_2COOH, CH_3COOH, CH_3CH_2CH_2COOH, HCOOH$

(a) $CH_3COOH > CH_3CH_2COOH > CH_3CH_2CH_2COOH > HCOOH$

(b) $HCOOH > CH_3COOH > CH_3CH_2COOH > CH_3CH_2CH_2COOH$

(c) $HCOOH > CH_3CH_2CH_2COOH > CH_3CH_2COOH > CH_3COOH$

(d) $CH_3CH_2CH_2COOH > CH_3CH_2COOH > CH_3COOH > HCOOH$

**03** Identify the product $(P)$ in the following reaction :

*[2024, 8 April Shift-I]*

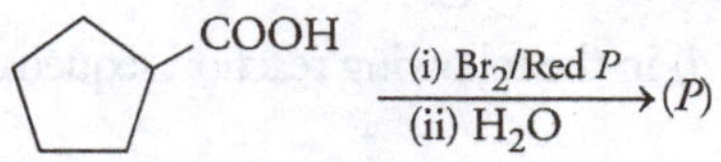


(a) 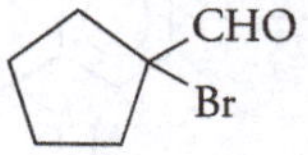


(b) 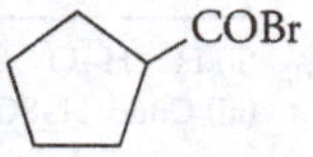


(c) 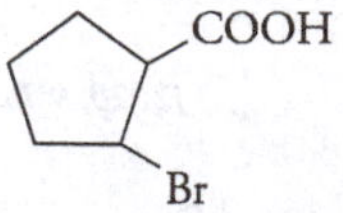


(d) 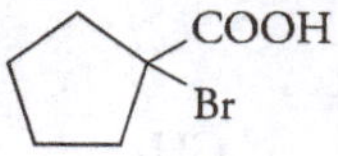


**04** 2-hexene $\xrightarrow[\text{(ii) } H_2O]{\text{(i) } O_3}$ Products

The two products formed in above reaction are

(a) butanoic acid and acetaldehyde *[2023, 12 April Shift-I]*

(b) butanal and acetic acid

(c) butanal and acetaldehyde

(d) butanoic acid and acetic acid

**05** The descending order of acidity for the following carboxylic acid is

A. $CH_3COOH$ B. $F_3C—COOH$

C. $ClCH_2—COOH$ D. $FCH_2—COOH$

E. $BrCH_2—COOH$

Choose the correct answer from the options given below:

*[2023, 8 April Shift-II]*

(a) D > B > A > E > C (b) B > D > C > E > A

(c) E > D > B > A > C (d) B > C > D > E > A

**06** The correct IUPAC nomenclature for the following compound is

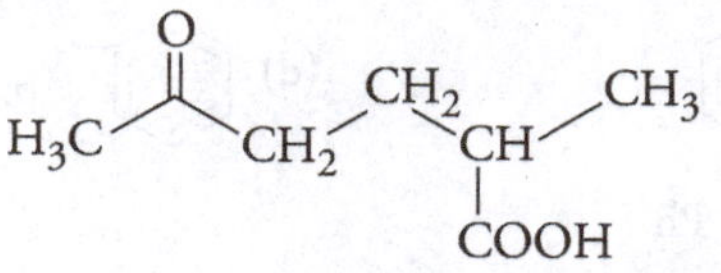


*[2023, 8 April Shift-II]*

(a) 5-formyl -2-methylhexanoic acid
(b) 5-methyl-2-oxohexan-6-oic acid
(c) 2-methyl-5-oxohexanoic acid
(d) 2-formyl -5-methylhexan-6-oic acid

**07** Given below are two statements.
One is labelled as Assertion (A) and the other is labelled as Reason (R).

**Assertion (A)** 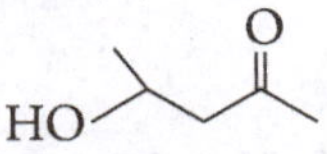 can be easily reduced using Zn-Hg / HCl to 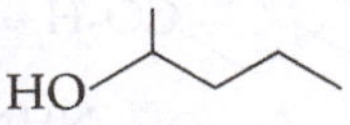.

**Reason (R)** Zn-Hg/HCl is used to reduce carbonyl group to —$CH_2$— group. *[2023, 30 Jan Shift-II]*

In the light of the above statements, choose the correct answer from the options given below.
(a) A is false but R is true
(b) Both A and R are true but R is not the correct explanation of A
(c) Both A and R are true and R is the correct explanation of A
(d) A is true but R is false

**08** Given below are two statements : *[2022, 29 July Shift-I]*

**Statement I** The esterification of carboxylic acid with an alcohol is a nucleophilic acyl substitution.

**Statement II** Electron withdrawing groups in the carboxylic acid will increase the rate of esterification reaction.

Choose the most appropriate answer from the options given below :
(a) Both Statement I and Statement II are correct.
(b) Both Statement I and Statement II are incorrect.
(c) Statement I is correct but Statement II is incorrect.
(d) Statement I is incorrect but Statement II is correct.

**09** Consider the given chemical reaction. *[2022, 30 June Shift-I]*

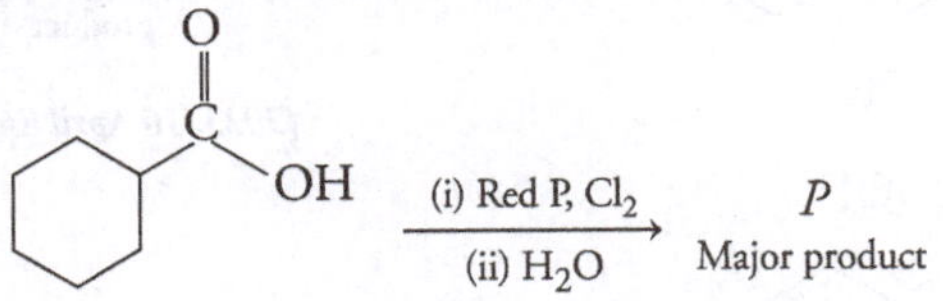


Identify the product *P*.

(a) 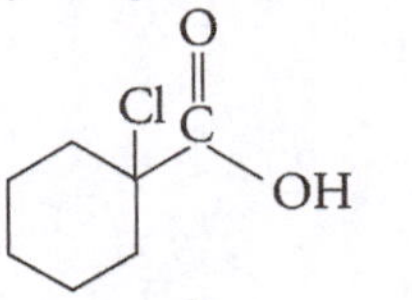
(b) 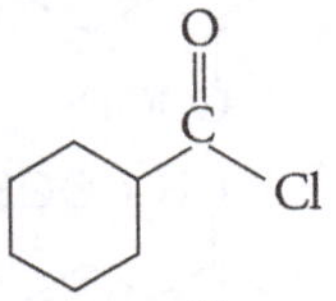
(c) 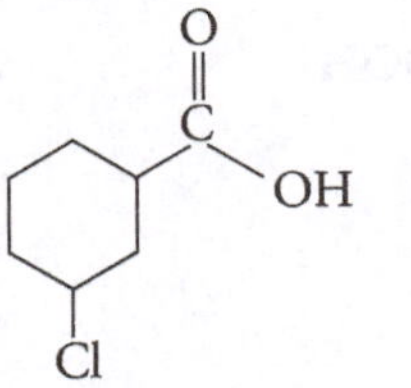
(d) 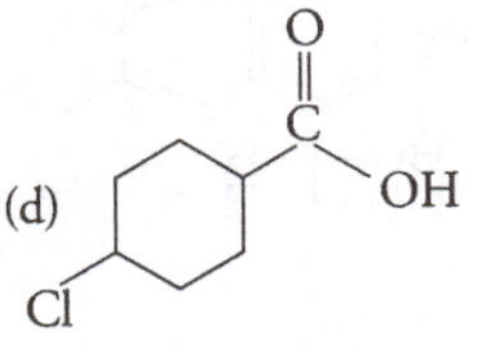

**10** The reagent, from the following, which converts benzoic acid to benzaldehyde in one step is *[2022, 26 June Shift-II]*

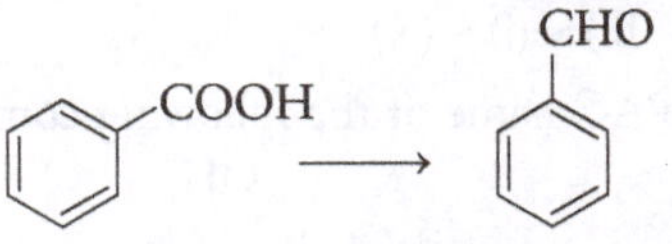


(a) $LiAlH_4$ (b) $KMnO_4$ (c) MnO (d) $NaBH_4$

**11** Two statements are given below:

**Statement I** The melting point of monocarboxylic acid with even number of carbon atoms is higher than that of with odd number of carbon atoms acid immediately below and above it in the series.

**Statement II** The solubility of monocarboxylic acids in water decreases with increase in molar mass.

Choose the most appropriate option. *[2022, 24 June Shift-I]*
(a) Both Statement I and Statement II are true.
(b) Both Statement I and Statement II are false.
(c) Statement I is true but Statement II is false.
(d) Statement I is false but Statement II is true.

**12** The structure of the starting compound *P* used in the reaction given below is *[2021, 27 Aug Shift-I]*

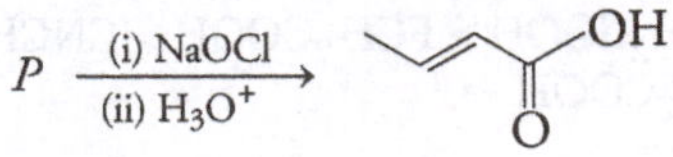


(a) 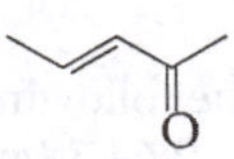
(b) 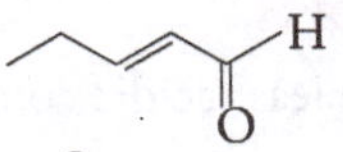
(c) 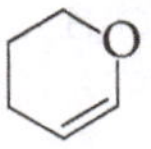
(d) 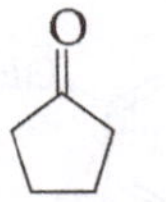

**13** (A) 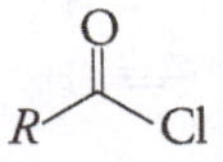 (B) 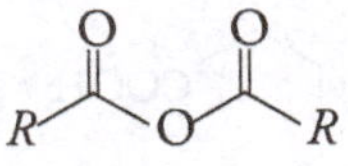

(C) 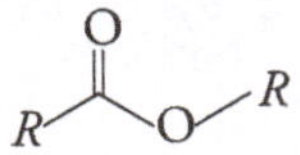 (D) 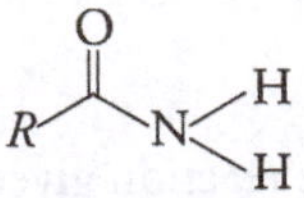

The correct order of their reactivity towards hydrolysis at room temperature is *[2021, 20 July Shift-II]*
(a) (D) > (A) > (B) > (C) (b) (A) > (B) > (C) > (D)
(c) (A) > (C) > (B) > (D) (d) (D) > (B) > (A) > (C)

**14** Arrange the following labelled hydrogens in decreasing order of acidity: *[2020, 2 Sep Shift-II]*

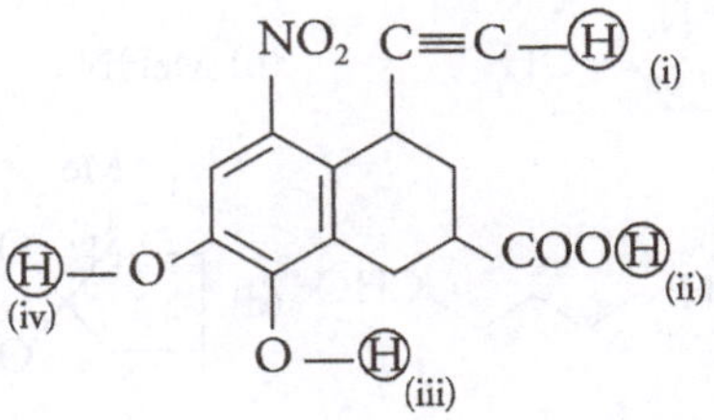

(a) (ii) > (i) > (iii) > (iv)
(b) (iii) > (ii) > (iv) > (i)
(c) (ii) > (iii) > (iv) > (i)
(d) (ii) > (iii) > (i) > (iv)

**15** The IUPAC name of the following compound is

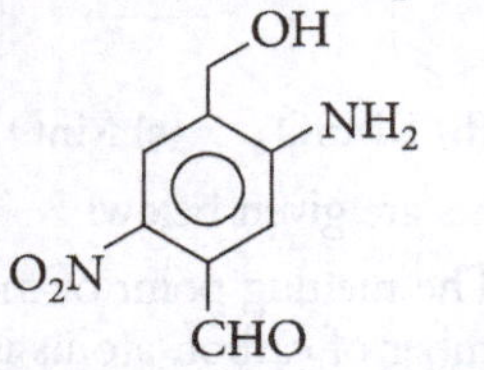

*[2020, 2 Sep Shift-II]*

(a) 2-nitro-4-hydroxymethyl-5-aminobenzaldehyde
(b) 3-amino-4-hydroxymethyl-5-nitrobenzaldehyde
(c) 5-amino-4-hydroxymethyl-2-nitrobenzaldehyde
(d) 2-amino-2-formyl-5-hydroxymethylnitrobenzene

**16** The correct decreasing order for acid strength is *[2019, 9 Jan Shift-I]*

(a) $FCH_2COOH > NCCH_2COOH > NO_2CH_2COOH > ClCH_2COOH$
(b) $CNCH_2COOH > O_2NCH_2COOH > FCH_2COOH > ClCH_2COOH$
(c) $NO_2CH_2COOH > NCCH_2COOH > FCH_2COOH > ClCH_2COOH$
(d) $NO_2CH_2COOH > FCH_2COOH > CNCH_2COOH > ClCH_2COOH$

## Level 2

**17** The least acidic compound, among the following is: *[2025, 3 April shift-I]*

(A) 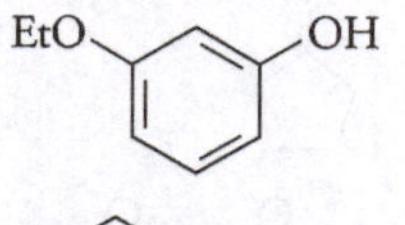 (B) 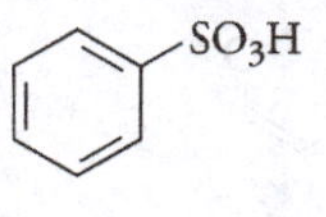

(C) 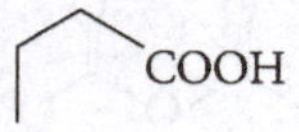 (D) $EtO_2C-\equiv-H$

(a) D (b) C
(c) A (d) B

**18** In the reaction given below

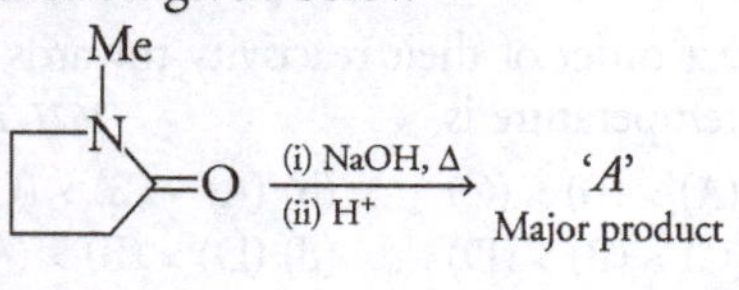

'A' is *[2023, 13 April Shift-I]*

(a) 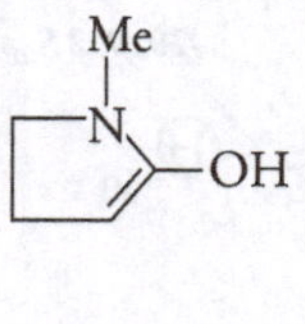 (b) 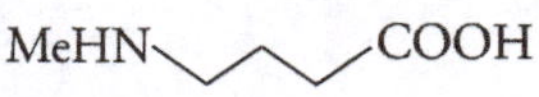

(c) 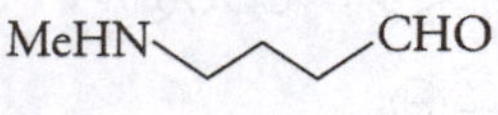 (d) 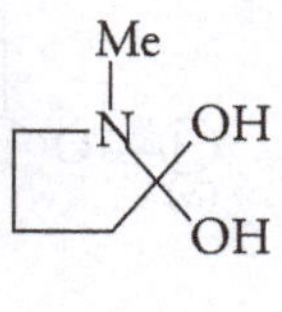

**19**

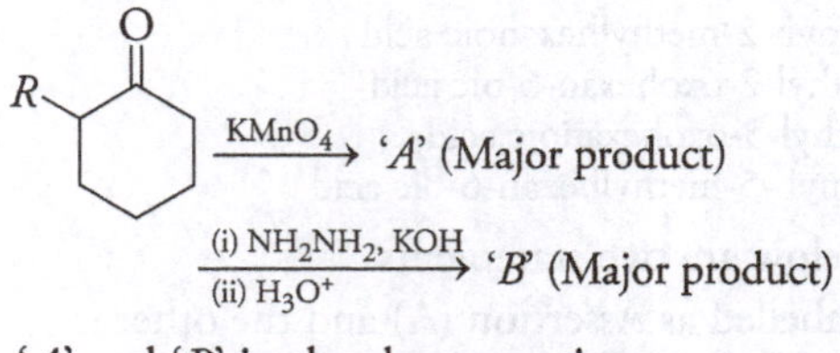

'A' and 'B' in the above reaction are *[2023, 11 April Shift-I]*

(a) 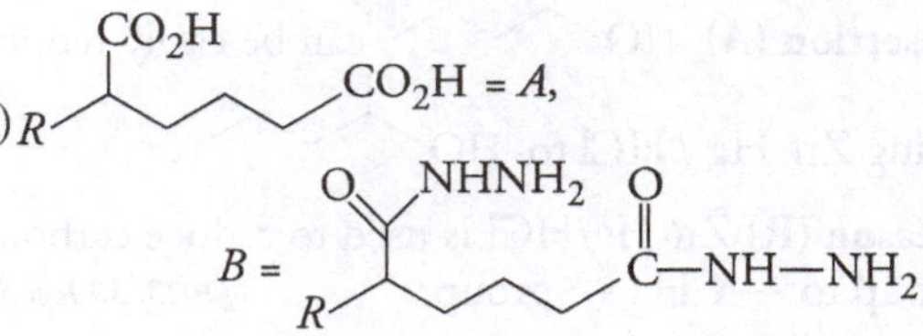

(b) 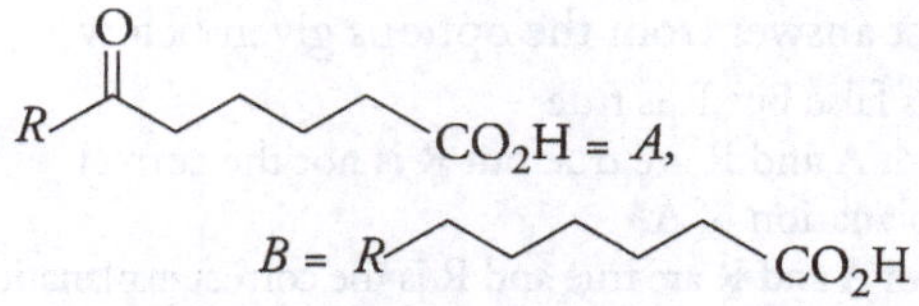

(c) 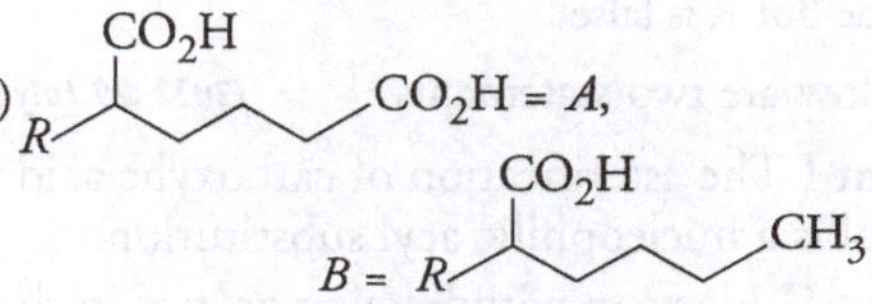

(d) 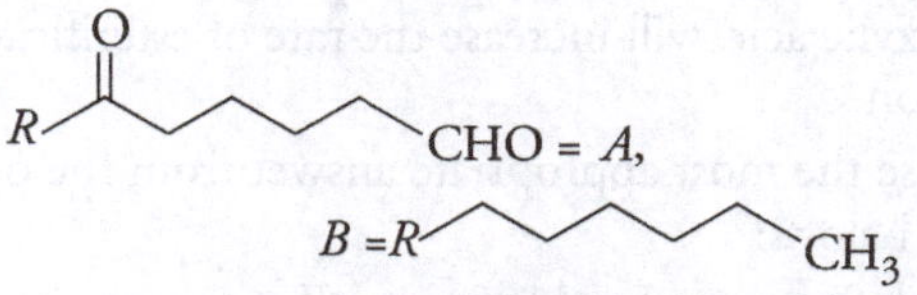

**20** The major product 'P' formed in the given reaction is

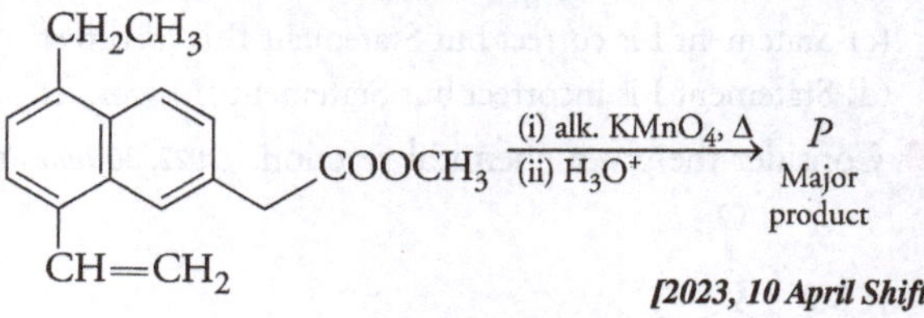

*[2023, 10 April Shift-I]*

(a) 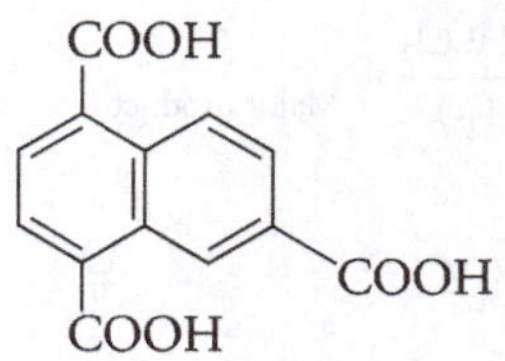

(b) 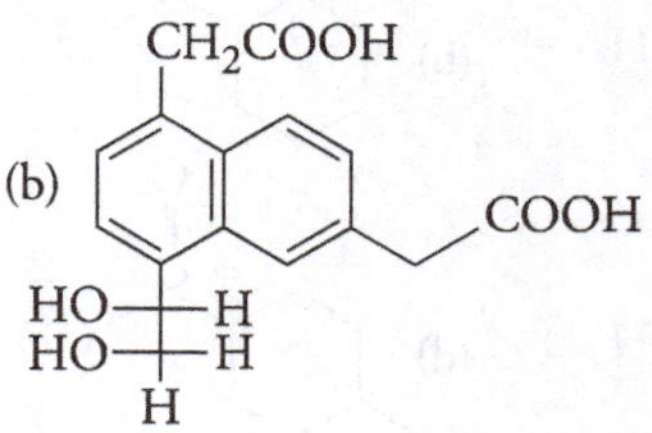

(c) 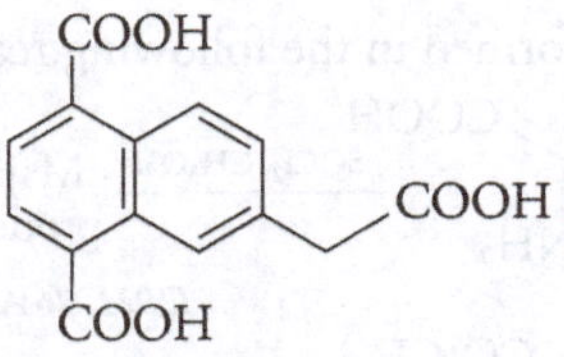

(d) 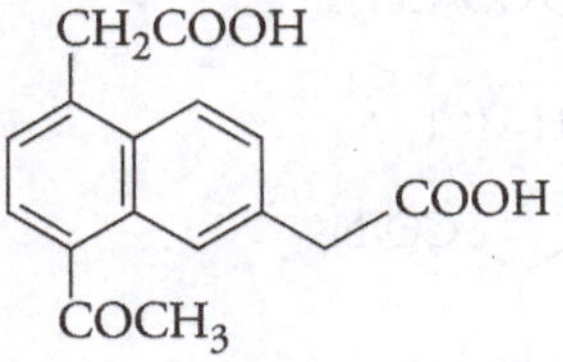

**21** The major product formed in the following reaction is

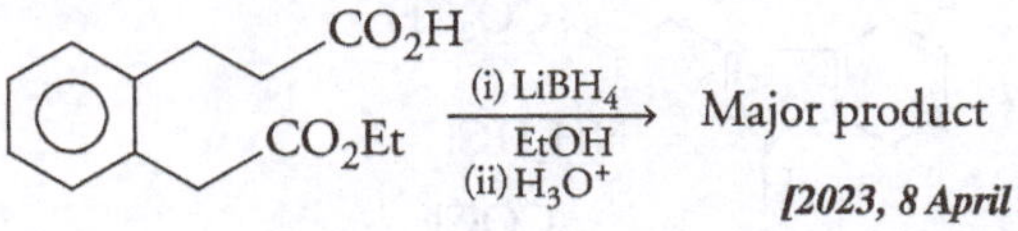

[2023, 8 April Shift-I]

(a) 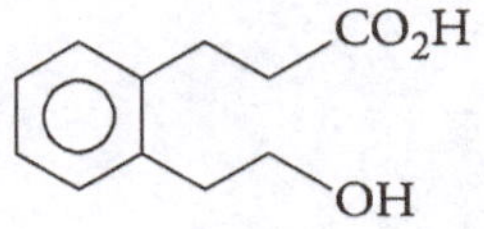 (b) 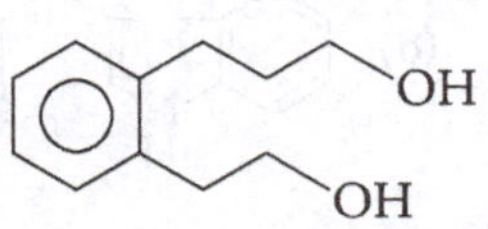

(c) 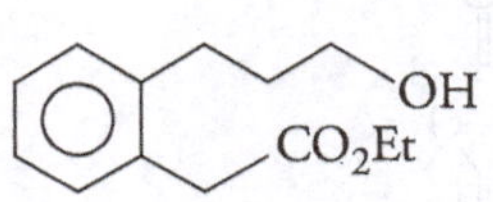 (d) 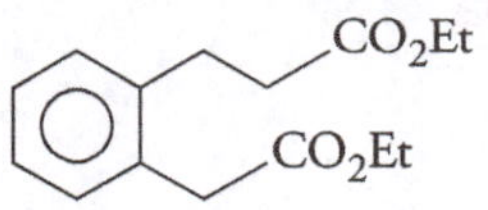

**Imp.** Name Reactions in Blend with Reagent Used

**22** Match List-I with List-II.

| List-I (Name of reaction) | List-II (Reagent used) |
|---|---|
| A. Hell-Volhard-Zelinsky reaction | I. $NaOH + I_2$ |
| B. Iodoform reaction | II. (i) $CrO_2Cl_2, CS_2$ (ii) $H_2O$ |
| C. Etard reaction | III. $Br_2$/red phosphorus |
| D. Gattermann-Koch reaction | IV. CO, HCl, anhyd. $AlCl_3$ |

Choose the correct answer from the options given below [2023, 6 April Shift-I]

(a) A-I, B-II, C-III, D-IV (b) A-III, B-I, C-II, D-IV
(c) A-III, B-II, C-I, D-IV (d) A-III, B-I, C-IV, D-II

**23** In a reaction,

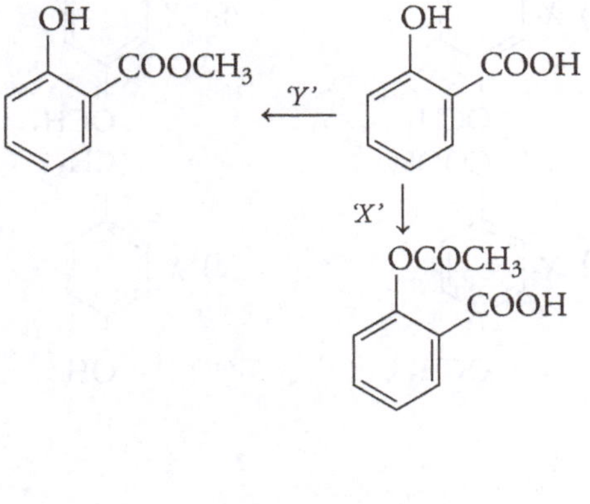

reagents 'X' and 'Y' respectively are [2023, 1 Feb Shift-II]

(a) $CH_3OH/H^+, \Delta$ and $(CH_3CO)_2O/H^+$
(b) $CH_3OH/H^+, \Delta$ and $CH_3OH/H^+, \Delta$
(c) $(CH_3CO)_2O/H^+$ and $CH_3OH/H^+, \Delta$
(d) $(CH_3CO)_2O/H^+$ and $(CH_3CO)_2O/H^+$

**24** 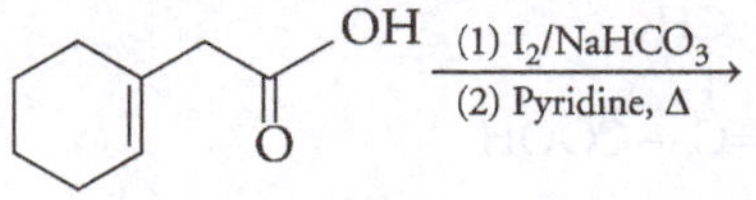

Find out the major product for the above reaction. [2022, 28 July Shift-II]

(a) 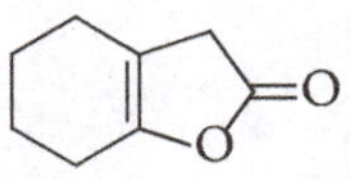 (b) 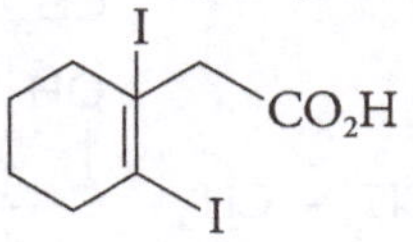

(c) 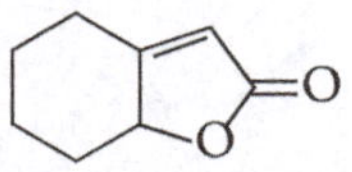 (d) 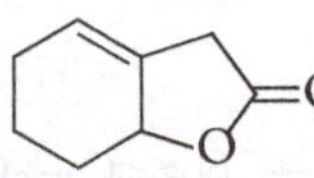

**25** Among the following marked proton which compound shows lowest $pK_a$ value? [2022, 28 July Shift-I]

(a) 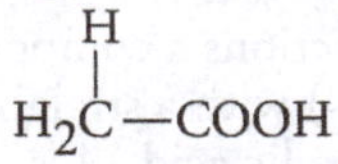 (b) 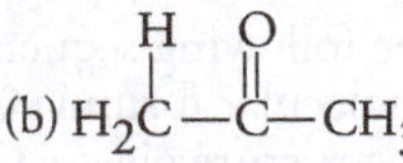

(c) 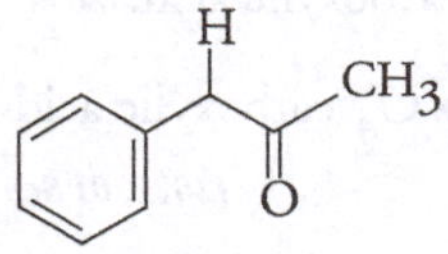 (d) 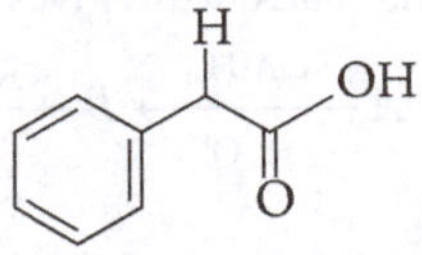

**26** 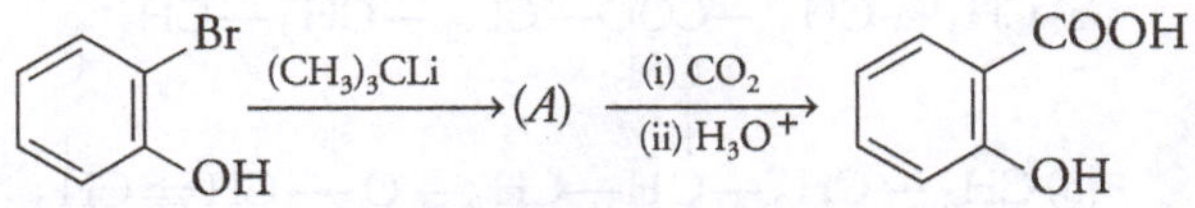

In the given conversion the compound A is: [2022, 29 June Shift-I]

(a) 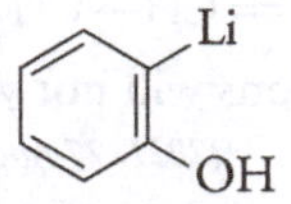 (b) 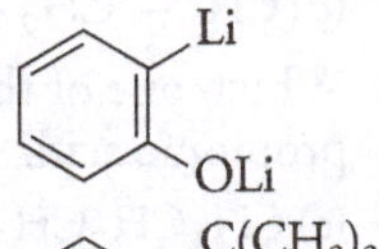

(c) 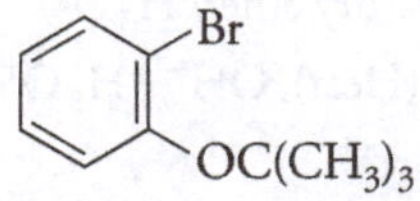 (d) 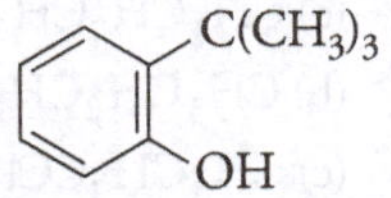

**27** $(A) \xrightarrow[\text{(ii) } CN^- \text{ (iii) } H_2O/H^+]{\text{(i) } Cl_2, \Delta}$ 4-Bromophenyl acetic acid.

In the above reaction 'A' is [2022, 28 June Shift-II]

(a) 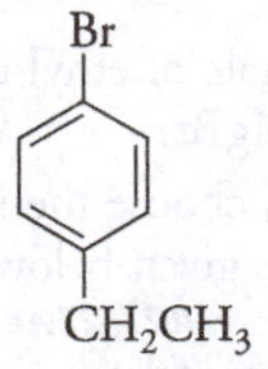 (b) 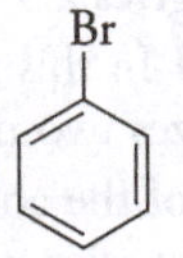

(c) 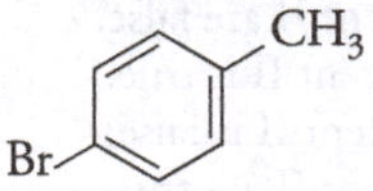 (d) 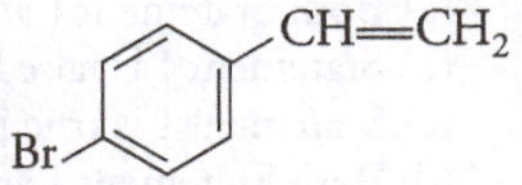

**28** The final product '*A*' in the following reaction sequence *[2022, 26 June Shift-II]*

$$CH_3CH_2-\underset{\underset{O}{\|}}{C}-CH_3 \xrightarrow{HCN} ? \xrightarrow[\text{Heat}]{95\%\ H_2SO_4} A$$

(a) $CH_3-CH=\overset{\overset{CH_3}{|}}{C}-COOH$

(b) $CH_3-CH=\underset{\underset{CH_3}{|}}{C}-CN$

(c) $CH_3-CH_2-\overset{\overset{OH}{|}}{\underset{\underset{CH_3}{|}}{C}}-COOH$

(d) $CH_3-CH=\underset{\underset{CH_3}{|}}{C}-CONH_2$

**29** In the following sequence of reactions a compound *A*, (molecular formula $C_6H_{12}O_2$) with a straight chain structure gives a $C_4$ carboxylic acid. *A* is

$$A \xrightarrow[H_3O^+]{LiAlH_4} B \xrightarrow{\text{Oxidation}} C_4 \text{ carboxylic acid}$$

*[2021, 01 Sep Shift-II]*

(a) $CH_3-CH_2-COO-CH_2-CH_2-CH_3$

(b) $CH_3-CH_2-\overset{\overset{OH}{|}}{CH}-CH_2-O-CH=CH_2$

(c) $CH_3-CH_2-CH_2-COO-CH_2-CH_3$

(d) $CH_3-CH_2-CH_2-O-CH=CH-CH_2-OH$

**30** Which one of the following reactions will not yield propionic acid? *[2021, 27 Aug Shift-II]*

(a) $CH_3CH_2CH_2Br + Mg, CO_2$ dry ether/ $H_3O^+$

(b) $CH_3CH_2CH_3 + KMnO_4$ (Heat), $OH^- / H_3O^+$

(c) $CH_3CH_2CCl_3 + OH^- / H_3O^+$

(d) $CH_3CH_2COCH_3 + OI^- / H_3O^+$

**31** Given below are two statements.

**Statement I** Ethyl pent-4-yn-oate on reaction with $CH_3MgBr$ gives a 3° alcohol.

**Statement II** In this reaction one mole of ethyl pent-4-yn-oate utilizes two moles of $CH_3MgBr$.

In the light of the above statements, choose the most appropriate answer from the options given below: *[2021, 27 Aug Shift-II]*

(a) Both Statement I and Statement II are false.
(b) Statement I is false but Statement II is true.
(c) Statement I is true but Statement II is false.
(d) Both Statement I and Statement II are true.

**32** The major product formed in the following reaction is

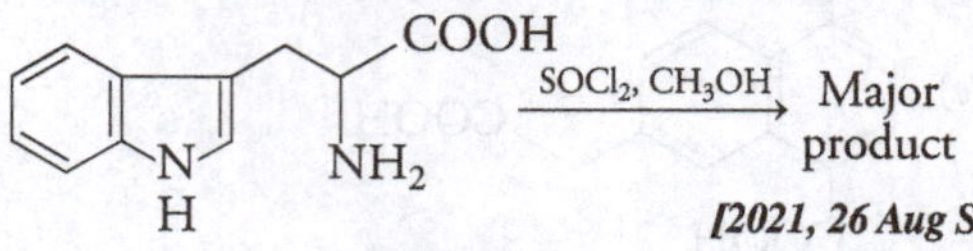

*[2021, 26 Aug Shift-I]*

(a) 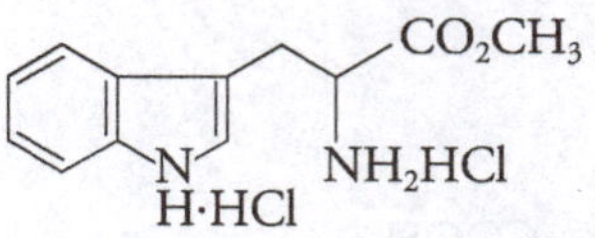

(b) 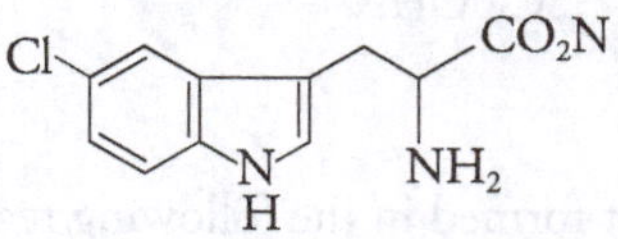

(c) 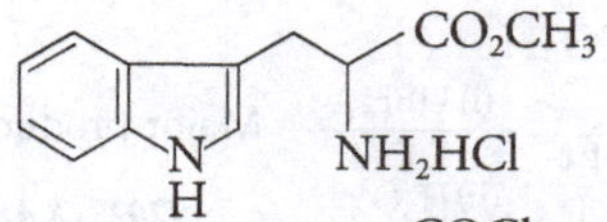

(d) 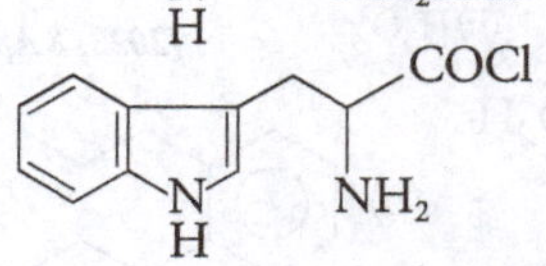

**33**

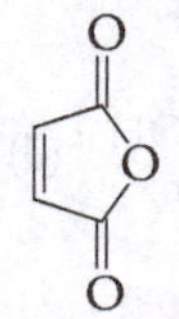

(Maleic anhydride)

Maleic anhydride can be prepared by *[2021, 25 July Shift-II]*

(a) heating *trans*-but-2-enedioic acid
(b) heating *cis*-but-2-enedioic acid
(c) treating *cis*-but-2-enedioic acid with alcohol and acid
(d) treating *trans*-but-2-enedioic acid with alcohol and acid

**34**

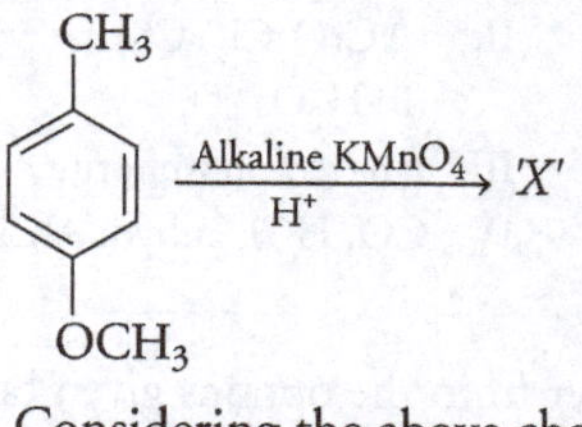

Considering the above chemical reaction, identify the product '*X*'. *[2021, 18 March Shift-I]*

(a) *X*- 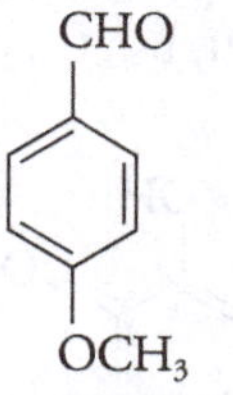

(b) *X*- 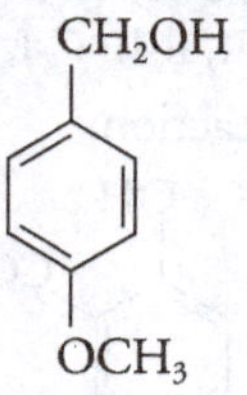

(c) *X*- 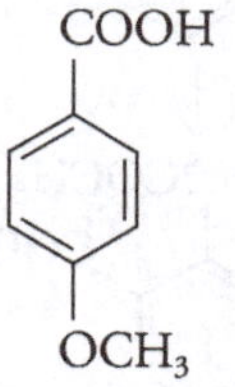

(d) *X*- 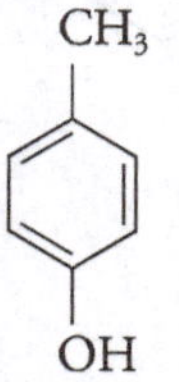

**35** 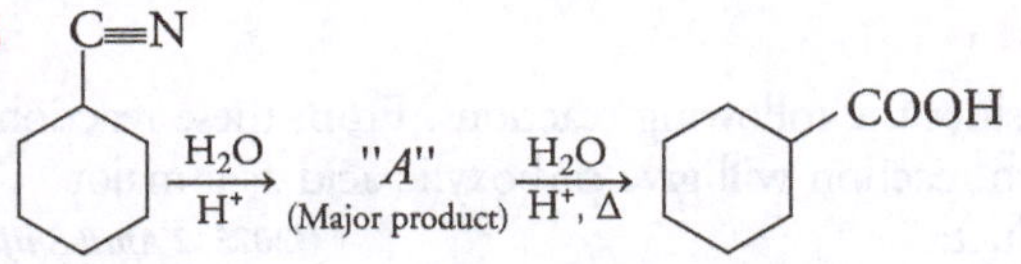

Consider the above chemical reaction and identify product "$A$". *[2021, 18 March Shift-I]*

(a) 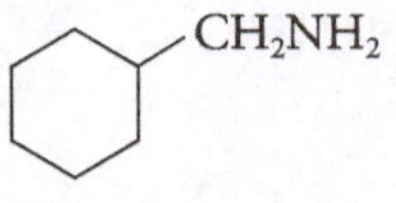 (b) 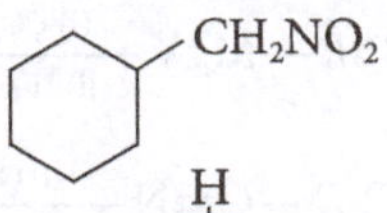

(c) 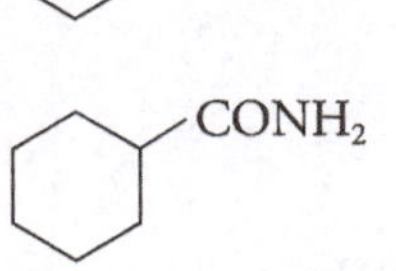 (d) 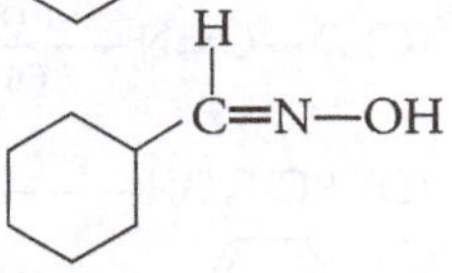

**36** 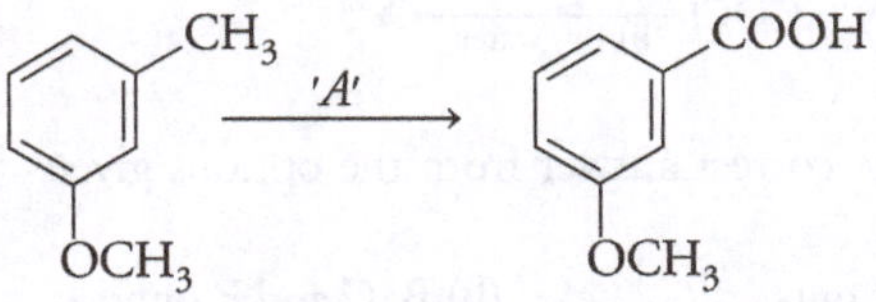

In the above reaction, the reagent '$A$' is *[2021, 16 March Shift-II]*

(a) $NaBH_4$, $H_3O^+$ (b) $LiAlH_4$

(c) Alkaline $KMnO_4$, $H^+$ (d) HCl, Zn - Hg

**37** The correct order of acid character of the following compounds is *[2021, 25 Feb Shift-II]*

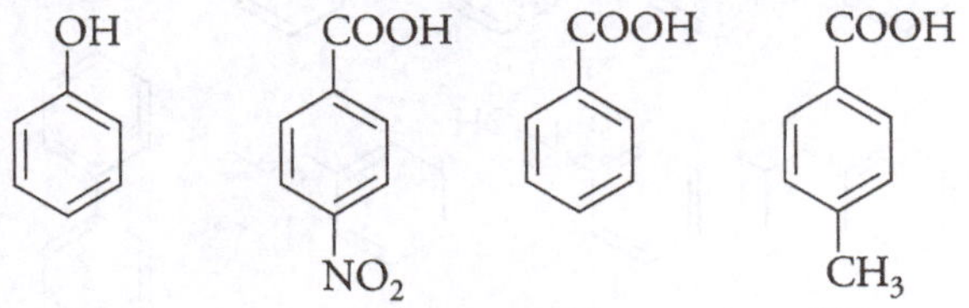

I II III IV

**Codes**

(a) I > II > III > IV (b) III > II > I > IV

(c) II > III > IV > I (d) IV > III > II > I

**38** Match List-I with List-II.

| | List-I | List-II |
|---|---|---|
| A. | $R—CO—Cl \longrightarrow R—CHO$ | 1. $Br_2$ /NaOH |
| B. | $R—CH_2—COOH \longrightarrow R—CH(Cl)—COOH$ | 2. $H_2$ /Pd - $BaSO_4$ |
| C. | $R—CO—NH_2 \longrightarrow R—NH_2$ | 3. Zn(Hg)/Conc.HCl |
| D. | $R—CO—CH_3 \longrightarrow RCH_2CH_3$ | 4. $Cl_2$ / Red P, $H_2O$ |

Choose the correct answer from the options given below. *[2021, 24 Feb Shift-II]*

| | A | B | C | D | | A | B | C | D |
|---|---|---|---|---|---|---|---|---|---|
| (a) | 2 | 1 | 4 | 3 | (b) | 3 | 4 | 1 | 2 |
| (c) | 2 | 4 | 1 | 3 | (d) | 3 | 1 | 4 | 2 |

**39** Consider the following molecules and statements related to them.

(A) 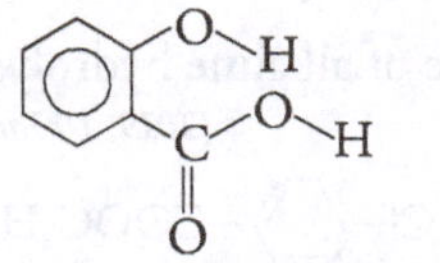 (B) 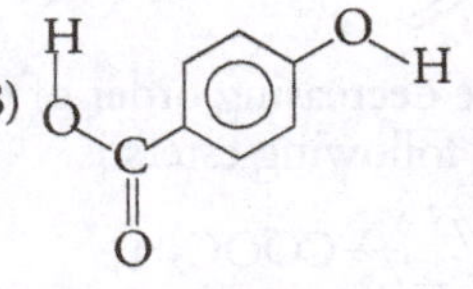

(I) ($B$) is more likely to be crystalline than ($A$).

(II) ($B$) has higher boiling point than ($A$).

(III) ($B$) dissolves more readily than ($A$) in water.

Identify the correct option from below : *[2020, 3 Sep Shift-II]*

(a) (I) and (III) are true. (b) Only (I) is true.

(c) (II) and (III) are true. (d) (I) and (II) are true.

**40** An unsaturated hydrocarbon $X$ absorbs two hydrogen molecules on catalytic hydrogenation, and also gives following reaction:

$$X \xrightarrow[Zn/H_2O]{O_3} A \xrightarrow{[Ag(NH_3)_2]^+} B \text{ (3-oxo-hexanedicarboxylic acid)}$$

$X$ will be *[2020, 08 Jan Shift-II]*

(a) 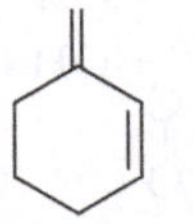 (b) 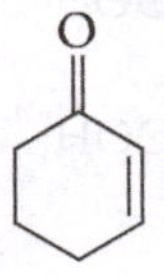

(c) 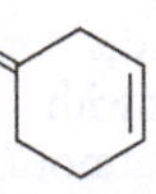 (d) 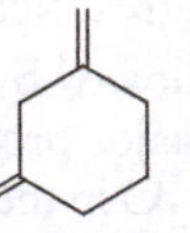

**41** The most suitable reagent for the given conversion is

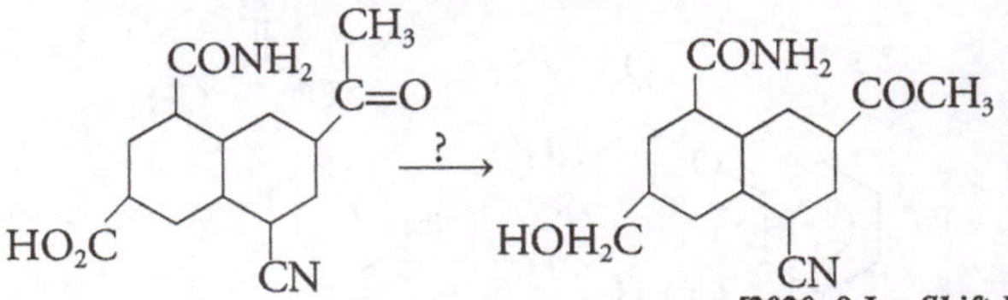

*[2020, 8 Jan Shift-I]*

(a) $B_2H_6$ (b) $LiAlH_4$

(c) $NaBH_4$ (d) $H_2$ / Pd

**42** The major product of the following reaction is

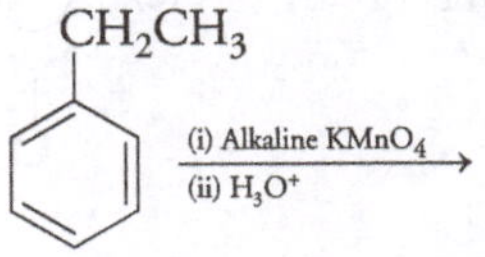

*[2019, 9 April Shift-I]*

(a) 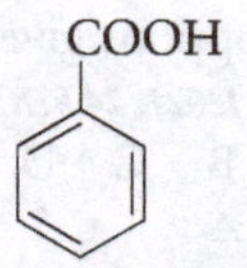
(b) 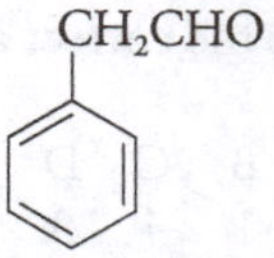
(c) 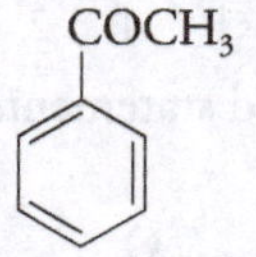
(d) 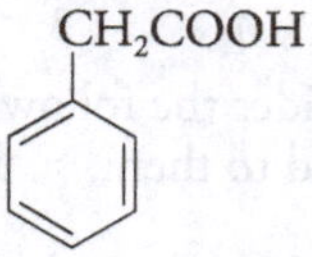

**43** The decreasing order of ease of alkaline hydrolysis for the following esters is **[2019, 10 Jan Shift-I]**

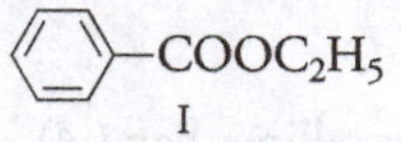
I
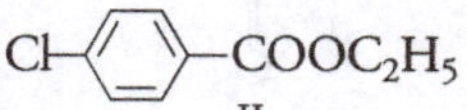
II
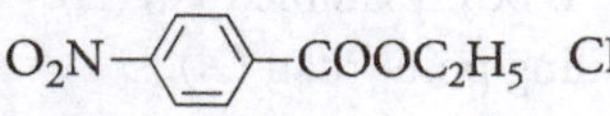
III
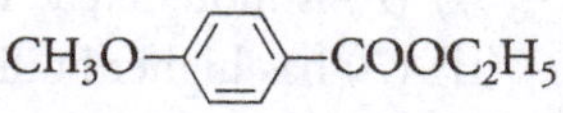
IV

(a) III > II > IV > I (b) III > II > I > IV
(c) II > III > I > IV (d) IV > II > III > I

**44** Which dicarboxylic acid in presence of a dehydrating agent is least reactive to give an anhydride? **[2019, 10 Jan Shift-I]**

(a) 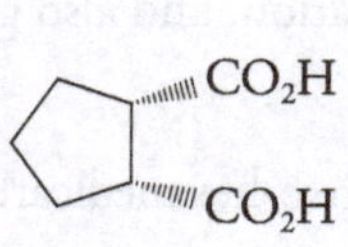
(b) 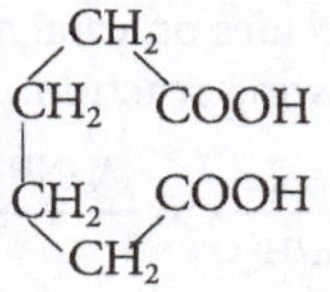
(c) 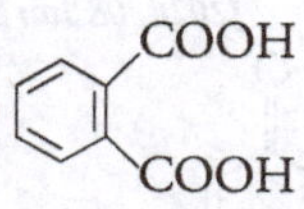
(d) 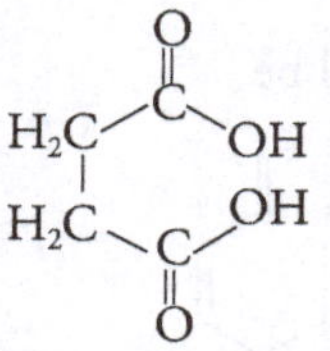

**45** Phenol on treatment with $CO_2$ in the presence of NaOH followed by acidification produces compound $X$ as the major product. $X$ on treatment with $(CH_3CO)_2O$ in the presence of catalytic amount of $H_2SO_4$ produces: **[JEE Main 2018]**

(a) 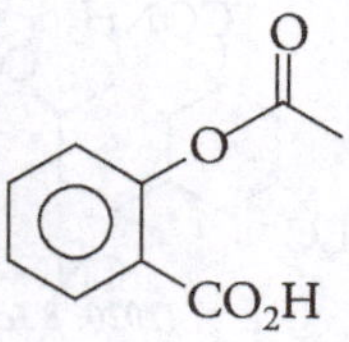
(b) 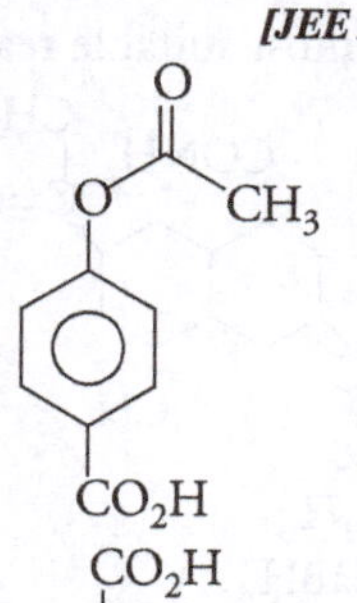
(c) 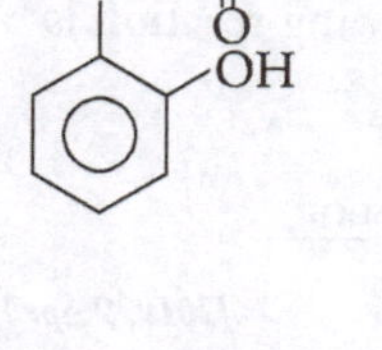
(d) 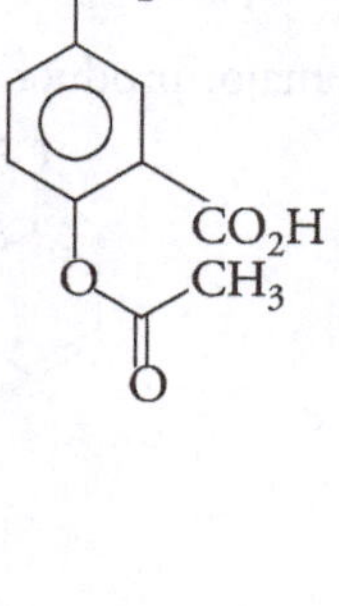

## Level 3

**46** Consider the following reactions. From these reactions which reaction will give carboxylic acid as a major product? **[2025, 2 April Shift-II]**

(A) $R—C{\equiv}N \xrightarrow[\text{Mild condition}]{\text{(i) } H^+/H_2O}$

(B) $R—MgX \xrightarrow[\text{(ii) } H_3O^+]{\text{(i) } CO_2}$

(C) $R—C{\equiv}N \xrightarrow[\text{(ii) } H_3O^+]{\text{(i) } SnCl_2/HCl}$

(D) $RCH_2OH \xrightarrow{\text{PCC}}$

(E) 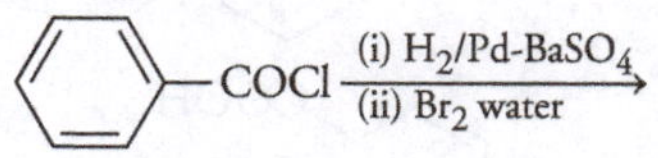

Choose the correct answer from the options given below.

(a) B and E only (b) B, C and E only
(c) A and D only (d) A, B and E only

**47** The total number of compounds from below when treated with hot $KMnO_4$ giving benzoic acid is **[2025, 28 Jan Shift-II]**

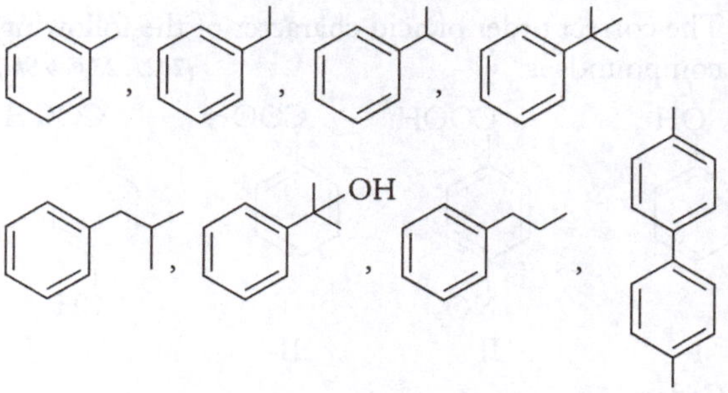

(a) 5 (b) 6 (c) 3 (d) 4

**48** The compounds that produce $CO_2$ with aqueous $NaHCO_3$ solution are **[2025, 28 Jan Shift-I]**

(A) 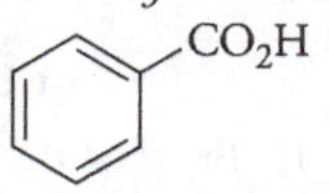
(B) 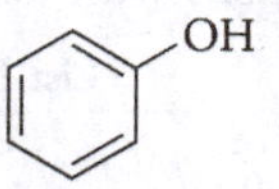
(C) 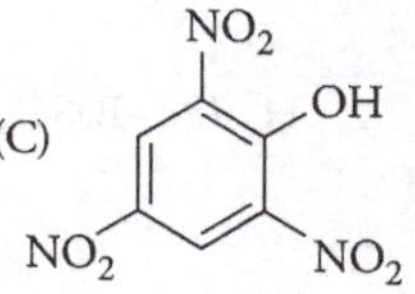
(D) 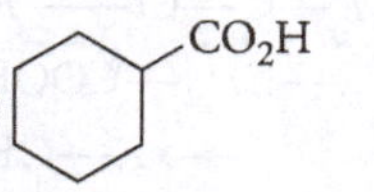
(E) 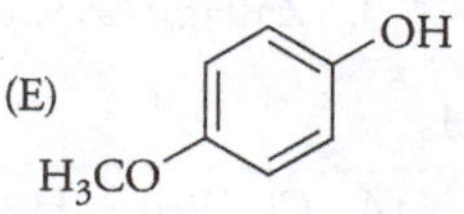

Choose the correct answer from the options given below.

(a) A and B only (b) A, B and E only
(c) A, C and D only (d) A and C only

**49** Consider the given reaction, identify the major product *P*. **[2024, 6 April Shift-II]**

$$CH_3-COOH \xrightarrow[\text{(iv) } H_2O/\bar{O}H, \Delta]{\text{(i) LiAlH}_4 \text{ (ii) PCC (iii) HCN}/\bar{O}H} \text{'}P\text{'}$$

(a) $CH_3-CH_2-CH_2-OH$

(b) $CH_3-CH_2-\overset{O}{\overset{\|}{C}}-NH_2$

(c) $CH_3-\overset{OH}{\overset{|}{CH}}-COOH$

(d) $CH_3-\overset{O}{\overset{\|}{C}}-CH_2CH_3$

**50** Identify *A* and *B* in the given chemical reaction sequence : **[2024, 5 April Shift-II]**

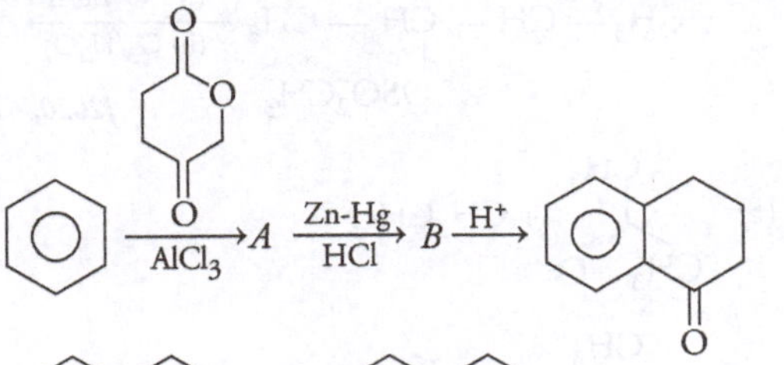

(a) *A*- 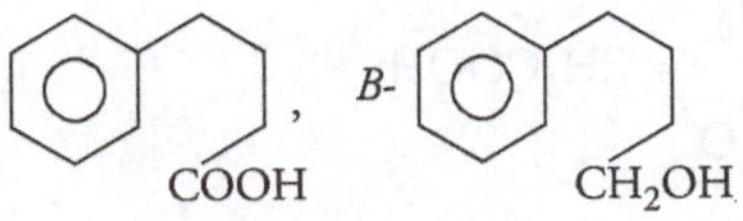

(b) *A*- 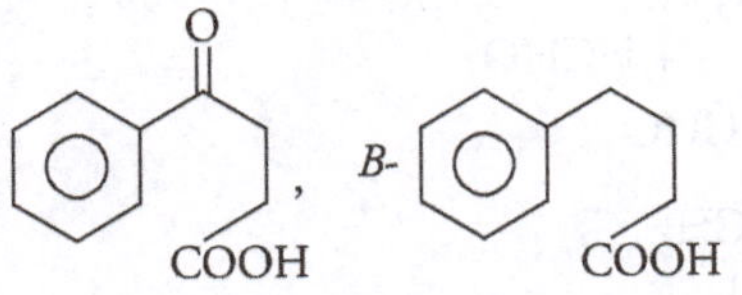

(c) *A*- 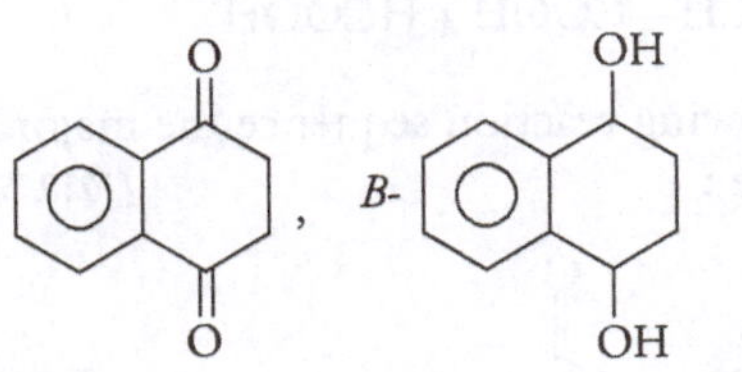

(d) *A*- 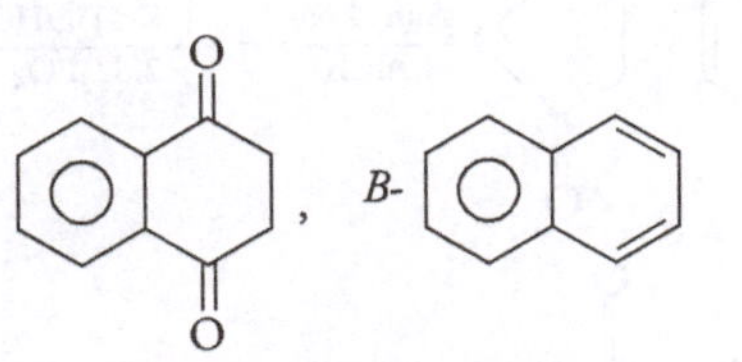

**51** Identify *A* and *B* in the following sequence of reaction.

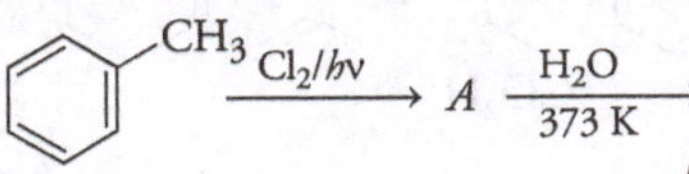

**[2024, 1 Feb Shift-I]**

(a) $A =$ 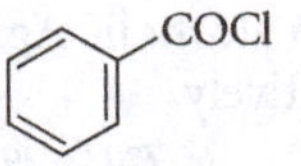 $B =$ 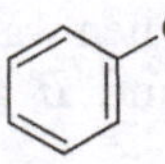

(b) $A =$ 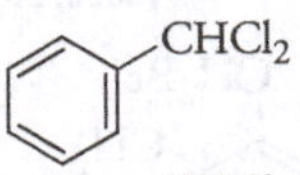 $B =$ 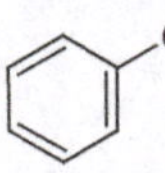

(c) $A =$ 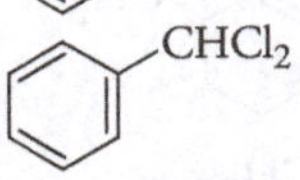 $B =$ 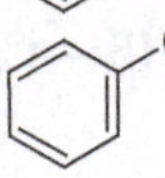

(d) $A =$ 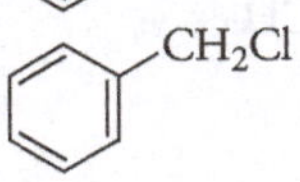 $B =$ 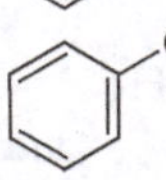

**52** 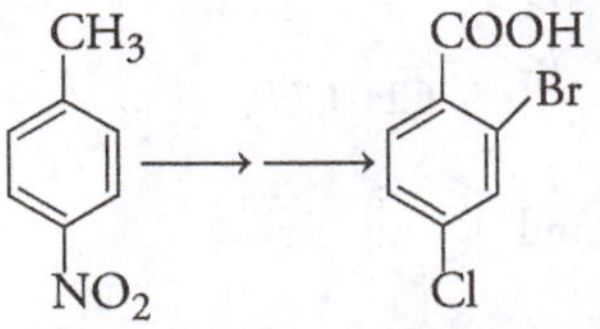

In the above conversion the correct sequence of reagents to be added is **[2023, 15 April Shift-I]**

(a) (i) $Br_2/Fe$, (ii) $Fe/H^+$, (iii) HONO, CuCl, (iv) $KMnO_4$
(b) (i) $Br_2/Fe$, (ii) $Fe/H^+$, (iii) $KMnO_4$, (iv) $Cl_2$
(c) (i) $KMnO_4$, (ii) $Br_2/Fe$, (iii) $Fe/H^+$, (iv) $Cl_2$
(d) (i) $Fe/H^+$, (ii) HONO, (iii) CuCl, (iv) $KMnO_4$, (v) $Br_2$

**53** The major product '*P*' formed in the following sequence of reactions is

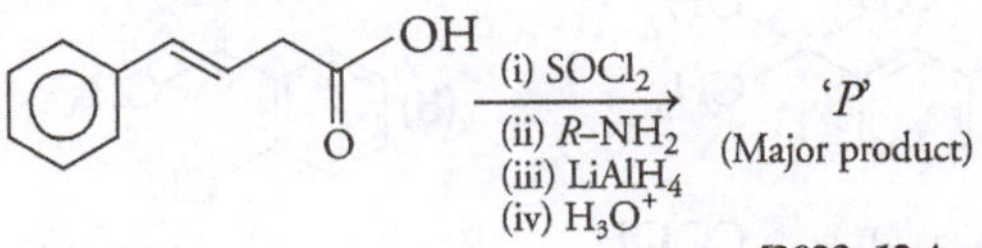

**[2023, 12 April Shift-I]**

(a) 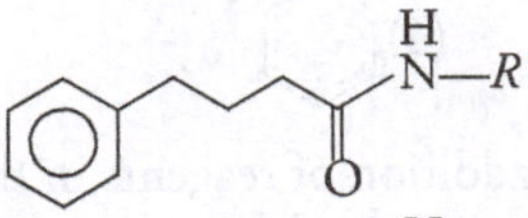

(b) 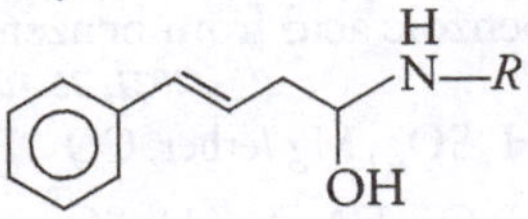

(c) 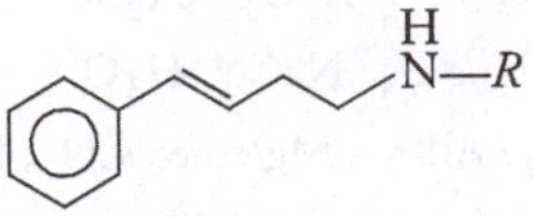

(d) 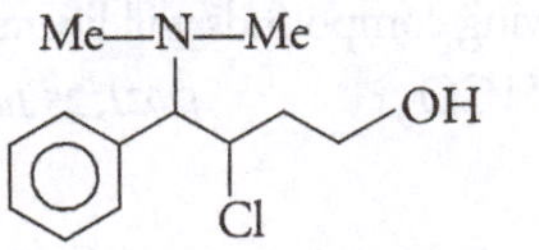

**54** Two isomers (*A*) and (*B*) with molar mass 184 g/mol and elemental composition '*C*', 52.2%; H, 4.9% and Br, 42.9% gave benzoic acid and *p*-bromobenzoic acid, respectively on oxidation with $KMnO_4$. Isomer '*A*' is optically active and gives a pale yellow precipitate when warmed with alcoholic $AgNO_3$. Isomers '*A*' and '*B*' are, respectively

**[2022, 29 June Shift-II]**

(a) $H_3C-CHBr-C_6H_5$ and 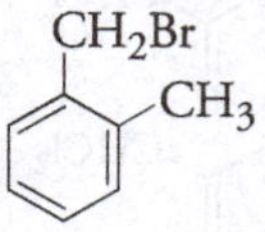

(b) 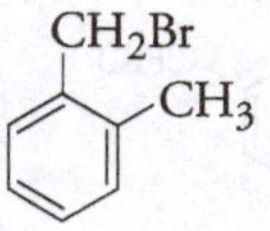 and 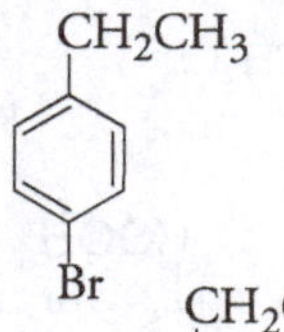

(c) $H_3C-CHBr-C_6H_5$ and 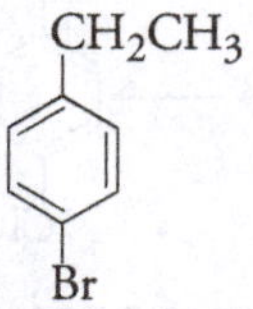

(d) 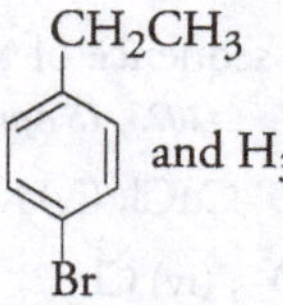 and $H_3C-CHBr-C_6H_5$

**55** The structure of product *C*, formed by the following sequence of reactions is

$$CH_3COOH + SOCl_2 \longrightarrow A \xrightarrow[AlCl_3]{Benzene} B \xrightarrow[-OH]{KCN} C$$

**[2021, 31 Aug Shift-I]**

(a) 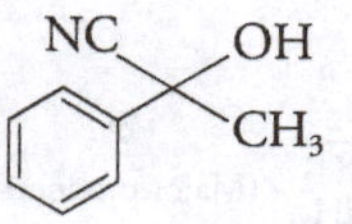 (b) 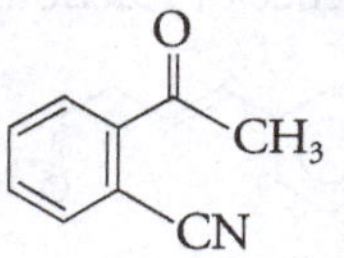

(c) 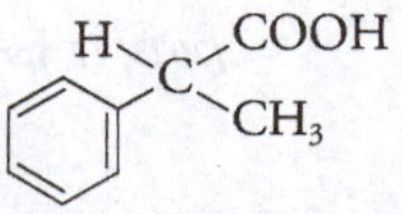 (d) 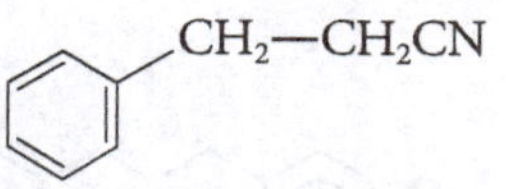

**56** The correct sequential addition of reagents in the preparation of 3-nitrobenzoic acid from benzene is

**[2021, 26 Aug Shift-I]**

(a) $Br_2$ /$AlBr_3$, $HNO_3$ /$H_2SO_4$, Mg /ether, $CO_2$, $H_3O^+$

(b) $Br_2$ / $AlBr_3$, NaCN, $H_3O^+$, $HNO_3$ / $H_2SO_4$

(c) $Br_2$ /$AlBr_3$, $HNO_3$ /$H_2SO_4$, NaCN, $H_3O^+$

(d) $HNO_3$ /$H_2SO_4$, $Br_2$ / $AlBr_3$, Mg/ether, $CO_2$, $H_3O^+$

**57** Which one of the following compounds will liberate $CO_2$, when treated with $NaHCO_3$?

**[2021, 25 July Shift-I]**

(a) $(CH_3)_3\overset{\oplus}{N}H\overset{\ominus}{Cl}$

(b) $(CH_3)_3\overset{\oplus\ominus}{N}OH$

(c) $CH_3-\overset{O}{\overset{\|}{C}}-NH_2$

(d) $CH_3NH_2$

**58** Compound(s) which will liberate carbon dioxide with sodium bicarbonate solution is/are **[2021, 25 Feb Shift-I]**

*A* = 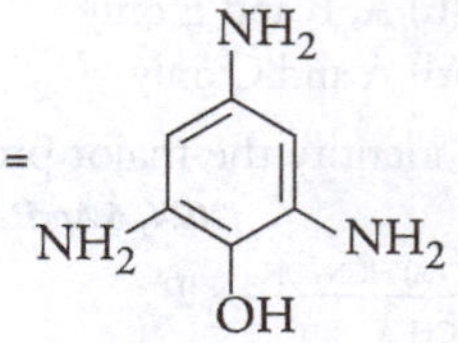 *B* = 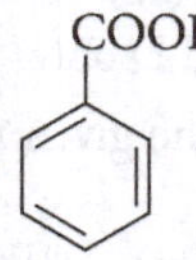

*C* = 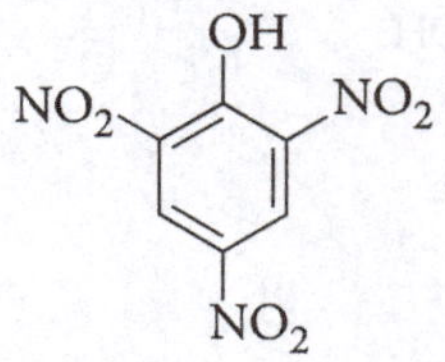

(a) *A* and *B* (b) *C* only

(c) *B* and *C* (d) *B* only

**59** The major products of the following reactions are

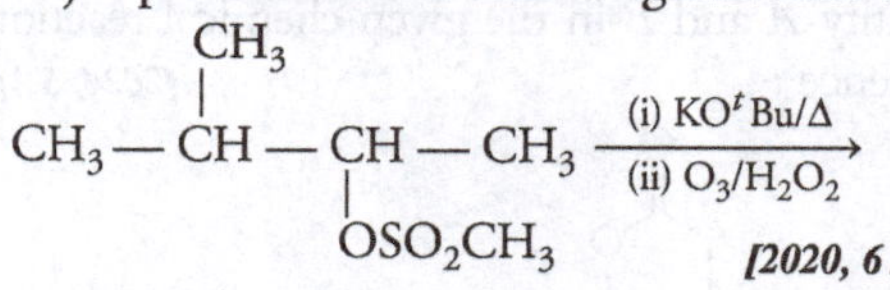

**[2020, 6 Sep Shift-I]**

(a) 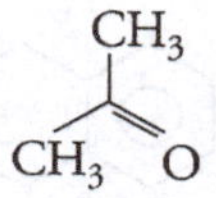 + $CH_3CHO$

(b) 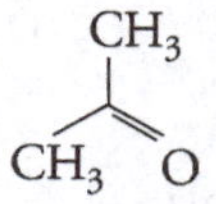 + $CH_3COOH$

(c) 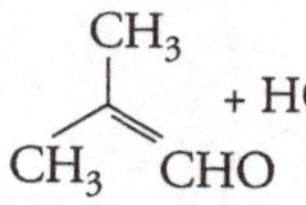 + HCHO

(d) $H_3C-\underset{}{\overset{CH_3}{\overset{|}{C}H}}-COOH + HCOOH$

**60** In the following reaction sequence the major products *A* and *B* are : **[2020, 5 Sep Shift-I]**

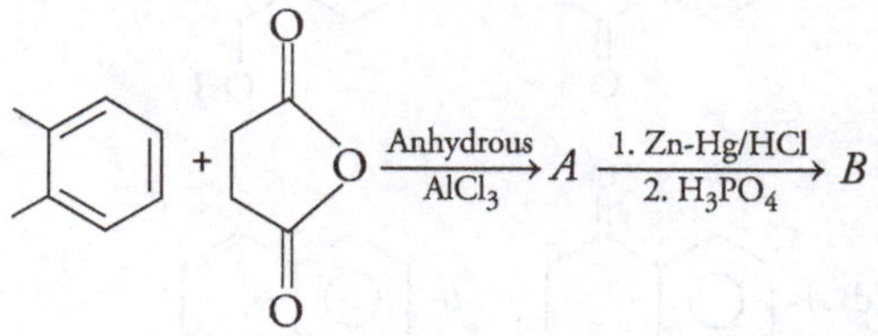

(a) *A* = 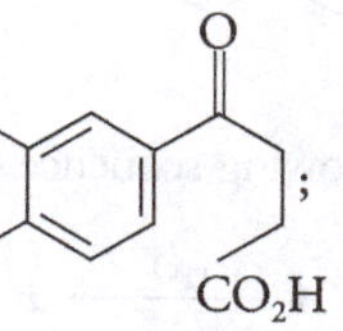 ; *B* = 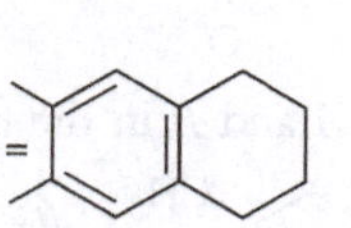

(b) $A$ = 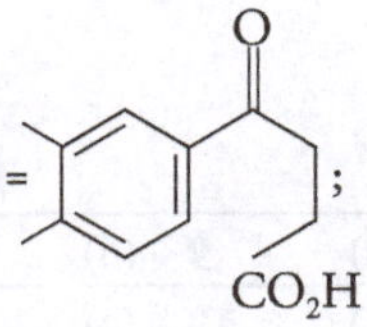 ; $B$ = 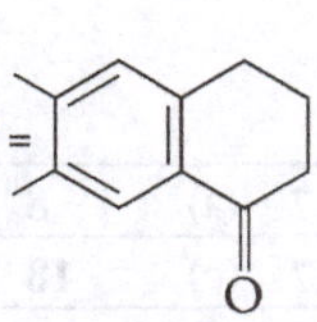

(c) $A$ = 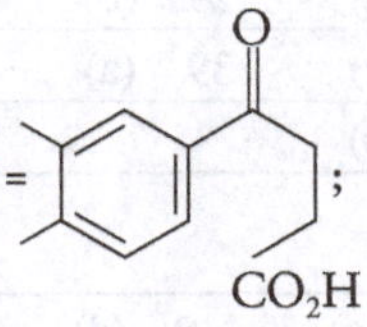 ; $B$ = 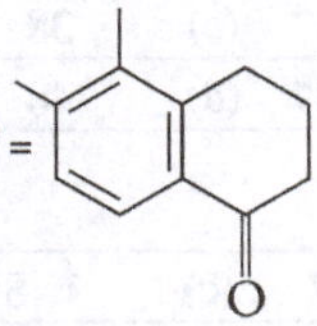

(d) $A$ = 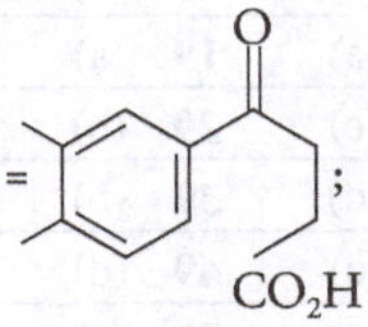 ; $B$ = 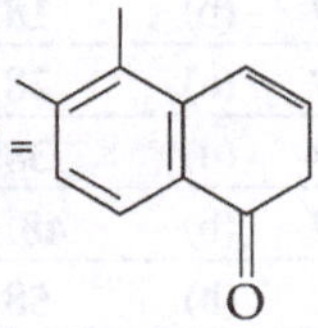

**61** The major product of the following reaction is

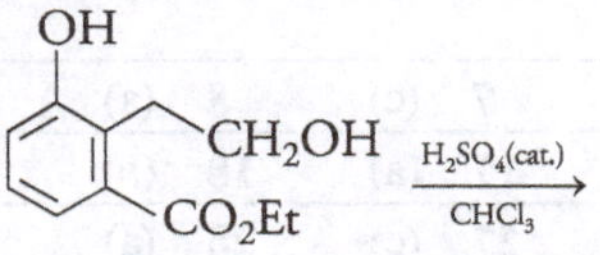

*[2019, 9 April Shift-II]*

(a) 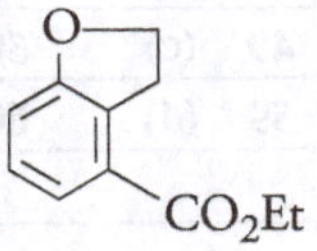

(b) 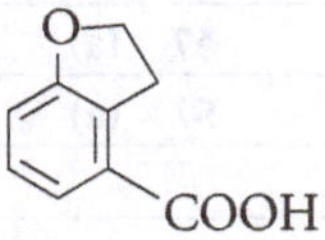

(c) 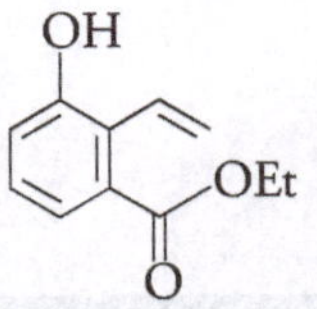

(d) 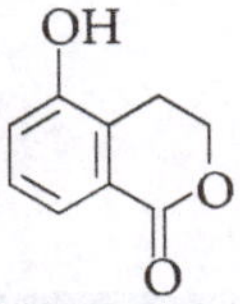

**62** Sodium phenoxide when heated with $CO_2$ under pressure at 125°C yields a product which on acetylation produces $C$. *[JEE Main 2014]*

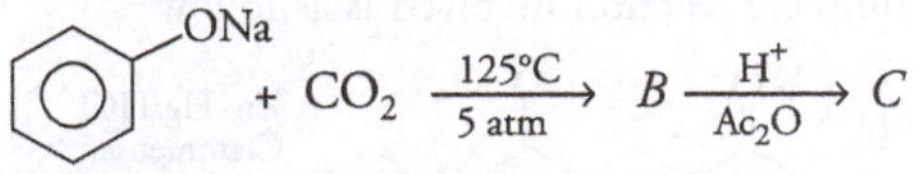

The major product $C$ would be

(a) 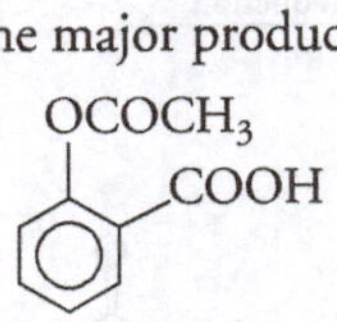

(b) 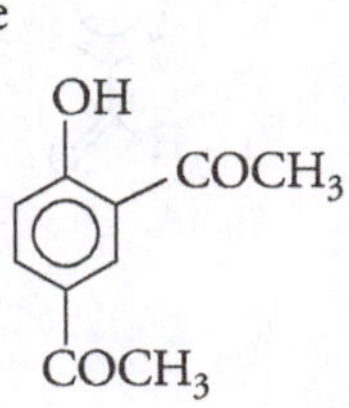

(c) 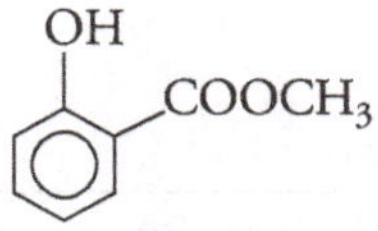

(d) 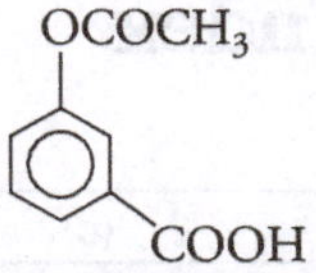

**63** *p*-cresol reacts with chloroform in alkaline medium to give the compound $A$ which adds hydrogen cyanide to form the compound $B$. The latter, on acidic hydrolysis gives chiral carboxylic acid. The structure of the carboxylic acid is *[AIEEE 2005]*

(a) 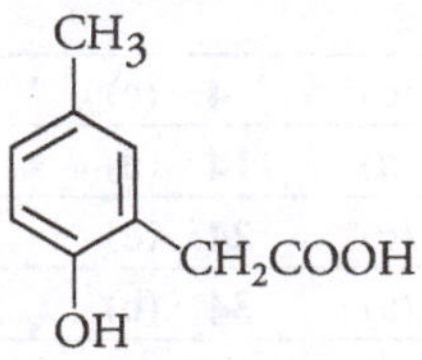

(b) 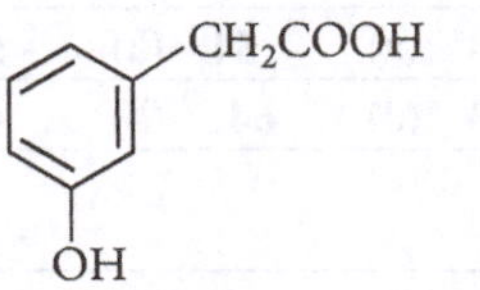

(c) 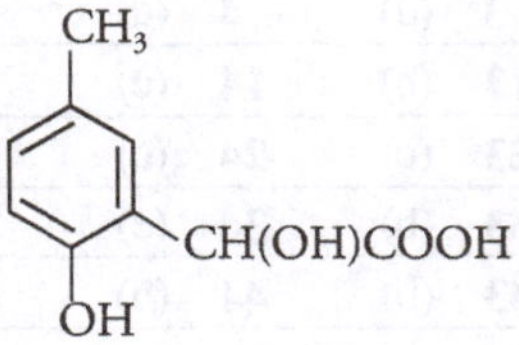

(d) 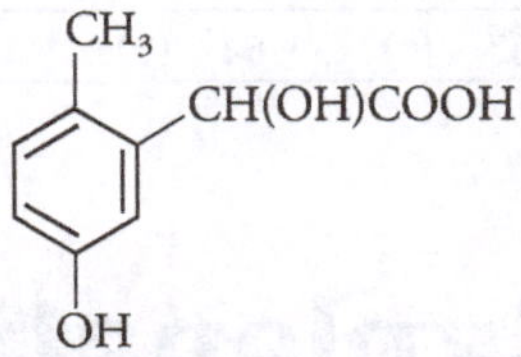

## Numerical Value Type Questions

**64** Number of compounds from the following which will not dissolve in cold $NaHCO_3$ and NaOH solutions but will dissolve in hot NaOH solution is ........ *[2023, 30 Jan Shift-II]*

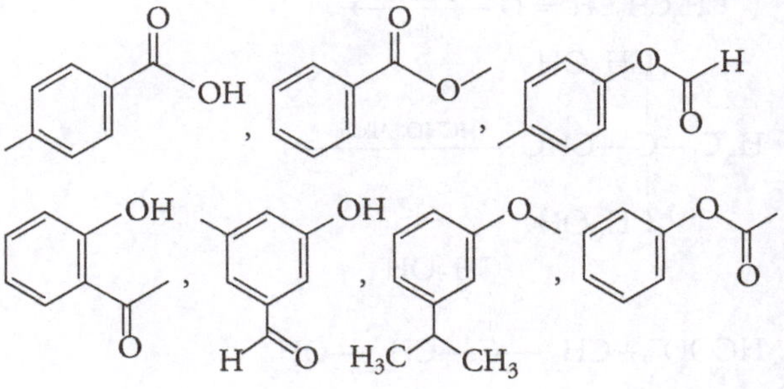

# • Key **Index**

## Topic 1

| | | | | | | | | | |
|---|---|---|---|---|---|---|---|---|---|
| **1** (a) | **2** (c) | **3** (a) | **4** (b) | **5** (b) | **6** (c) | **7** (a) | **8** (a) | **9** (d) | **10** (c) |
| **11** (c) | **12** (b) | **13** (d) | **14** (a) | **15** (d) | **16** (b) | **17** (c) | **18** (c) | **19** (c) | **20** (b) |
| **21** (a) | **22** (b) | **23** (b) | **24** (a) | **25** (d) | **26** (b) | **27** (a) | **28** (a) | **29** (b) | **30** (b) |
| **31** (d) | **32** (a) | **33** (d) | **34** (c) | **35** (d) | **36** (c) | **37** (c) | **38** (c) | **39** (a) | **40** (c) |
| **41** (c) | **42** (d) | **43** (c) | **44** (d) | **45** (c) | **46** (d) | **47** (d) | **48** (0) | | |

## Topic 2

| | | | | | | | | | |
|---|---|---|---|---|---|---|---|---|---|
| **1** (b) | **2** (a) | **3** (d) | **4** (c) | **5** (c) | **6** (b) | **7** (c) | **8** (c) | **9** (d) | **10** (a) |
| **11** (b) | **12** (d) | **13** (a) | **14** (d) | **15** (d) | **16** (a) | **17** (b) | **18** (a) | **19** (a) | **20** (c) |
| **21** (d) | **22** (c) | **23** (c) | **24** (c) | **25** (a) | **26** (d) | **27** (a) | **28** (c) | **29** (b) | **30** (a) |
| **31** (b) | **32** (a) | **33** (a) | **34** (b) | **35** (b) | **36** (c) | **37** (d) | **38** (d) | **39** (b) | **40** (b) |
| **41** (b) | **42** (c) | **43** (a) | **44** (a) | **45** (b) | **46** (a) | **47** (b) | **48.** (c) | **49** (d) | **50** (d) |
| **51** (b) | **52** (a) | **53** (a) | **54** (a) | **55** (c) | **56** (d) | **57** (b) | **58** (d) | **59** (d) | **60** (c) |
| **61** (d) | **62** (a) | **63** (d) | **64.** (b) | **65** (b) | **66** (b) | **67** (7) | **68** (15) | **69** (3) | **70** (2) |

## Topic 3

| | | | | | | | | | |
|---|---|---|---|---|---|---|---|---|---|
| **1** (b) | **2** (b) | **3** (d) | **4** (d) | **5** (b) | **6** (c) | **7** (c) | **8** (a) | **9** (a) | **10** (c) |
| **11** (a) | **12** (a) | **13** (b) | **14** (c) | **15** (c) | **16** (c) | **17** (a) | **18** (b) | **19** (b) | **20** (a) |
| **21** (a) | **22** (b) | **23** (c) | **24** (c) | **25** (c) | **26** (b) | **27** (c) | **28** (a) | **29** (c) | **30** (a) |
| **31** (c) | **32** (c) | **33** (b) | **34** (c) | **35** (c) | **36** (c) | **37** (c) | **38** (c) | **39** (c) | **40** (c) |
| **41** (a) | **42** (a) | **43** (b) | **44** (b) | **45** (a) | **46** (a) | **47** (a) | **48.** (c) | **49** (c) | **50** (b) |
| **51** (c) | **52** (a) | **53** (c) | **54** (c) | **55** (a) | **56** (d) | **57** (a) | **58** (c) | **59** (d) | **60** (b) |
| **61** (d) | **62** (a) | **63** (c) | **64** (3) | | | | | | |

# Detailed **Explanations**

## Topic 1

**01.** *(a)* Complete reaction sequence is as follows

$$CH_3CH_2CH{=}O \xrightarrow[\text{Alkali}]{\text{2HCHO}} H_3C{-}\underset{CH_2OH}{\overset{CH_2OH}{C}}{-}CHO \xrightarrow{\text{HCHO, Alkali}}$$

$$HCOO^- + CH_3{-}\underset{CH_2OH}{\overset{CH_2OH}{C}}{-}CH_2{-}OH$$

(Major product)

So, the major product obtained is given in option (a).

**02.** *(c)* The given reaction proceeds via Clemmensen reduction in which carbonyl group is reduced to methylene group. The complete reaction involved is as follow

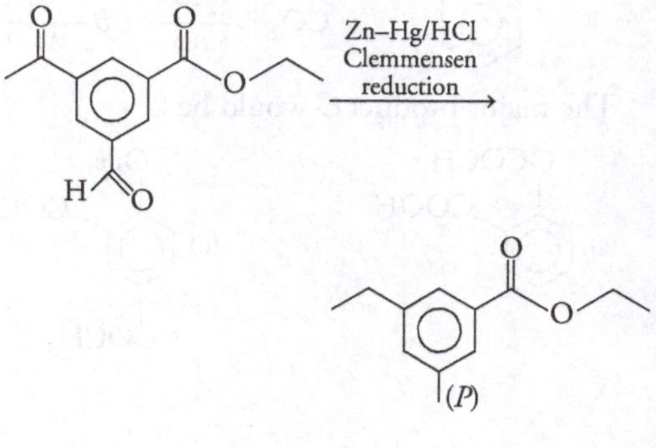

**03.** *(a)* The given reaction proceeds as follows

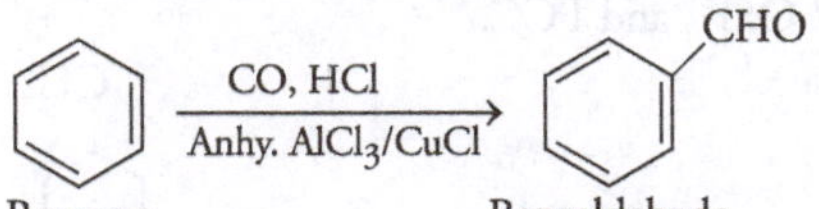


Benzene Benzaldehyde (*X*)

This reaction is called as Gatterman-Koch reaction.

**04.** *(b)* Among the given aldehydes, $H-\overset{O}{\overset{\|}{C}}-H$ is most reactive towards nucleophilic addition reaction.
The reason is that it has low steric hinderance at carbonyl carbon and high partial positive charge at carbonyl carbon.

**05.** *(b)* The given reaction is Gatterman-Koch reaction. In this reaction, benzene is treated with carbon monoxide in the acidic medium in presence of anhydrous aluminium chloride or cuprous chloride to give benzaldehyde.
The complete reaction is as follow,

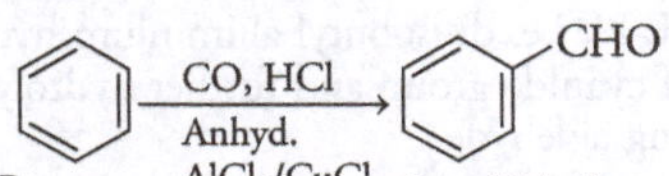


Benzene Benzaldehyde

**06.** *(c)* Structure of 4-methylpent-2-enal is

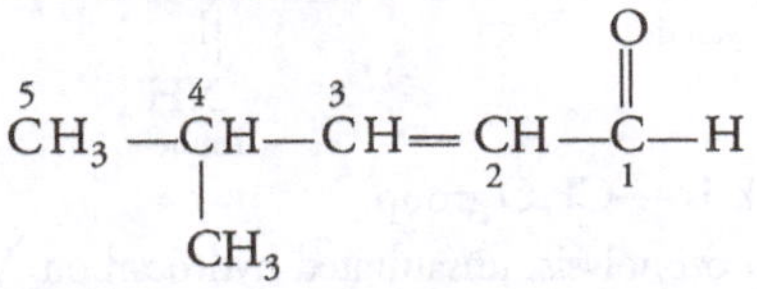


Hence, option (c) is correct.

**07.** *(a)* The structure of γ-methylcyclohexane carbaldehyde is given as

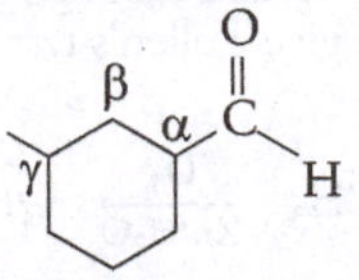


**08.** *(a)* Statement I is correct, whereas Statement II is incorrect.
The correct form of statement II is,
Vanillin does not give self aldol reaction due to lack of acidic H for condensation.

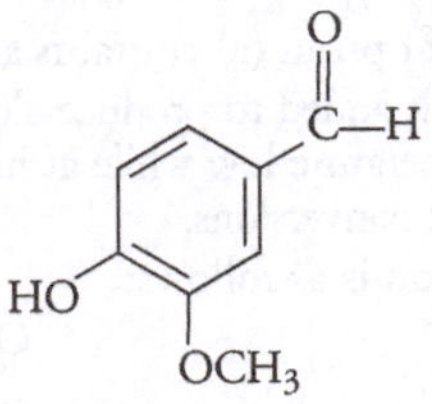


Phenolic group soluble in NaOH benzaldehyde derivative reacts with Tollen's reagent.

**09.** *(d)* Statement I is true but Statement II is false.
For statement II,

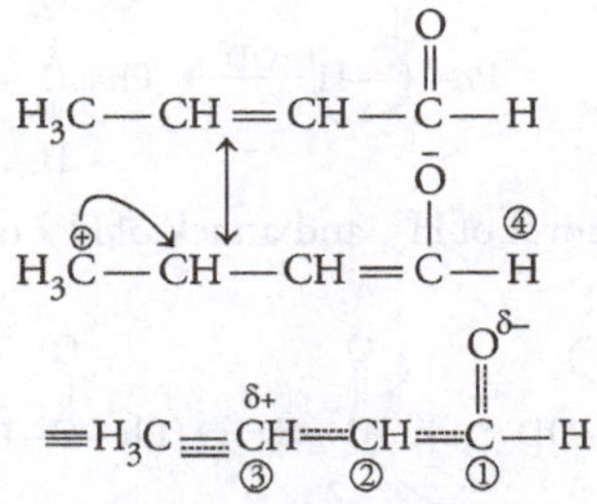


$C_1-C_2$ bond has partial double bond character that means lesser bond than $C_1-C_2$ bond of other compounds.

**10.** *(c)* Both Statement I and Statement II are true.

**11.** *(c)*

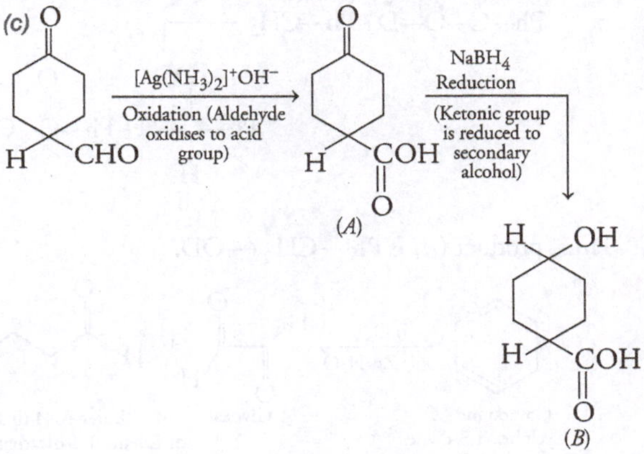


**12.** *(b)*

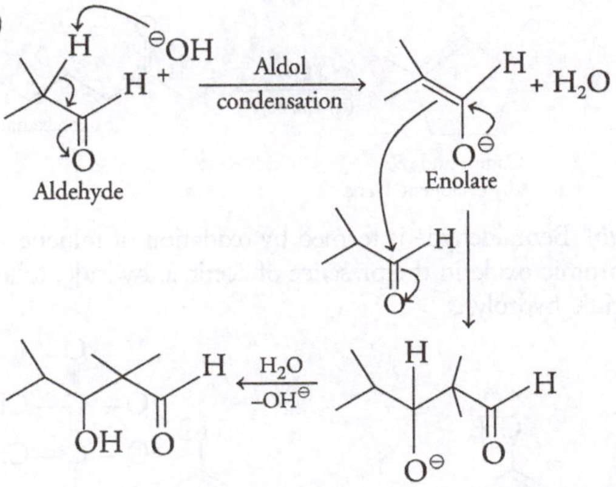


**13.** *(d)* $Na_2Cr_2O_7$, $H^+$ / $H_2O$ being strongest oxidising agent convert primary alcohols to carboxylic acids.

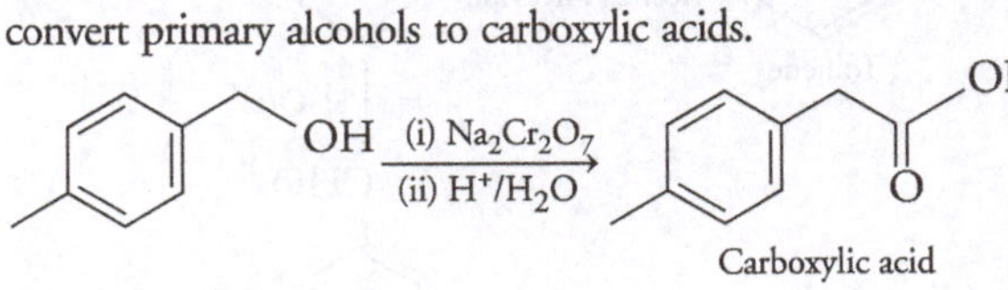


Carboxylic acid

**14.** *(a)* The given reaction is Cannizzaro reaction.

**Mechanism**

**Step 1.** Attack of nucleophile on carbonyl group.

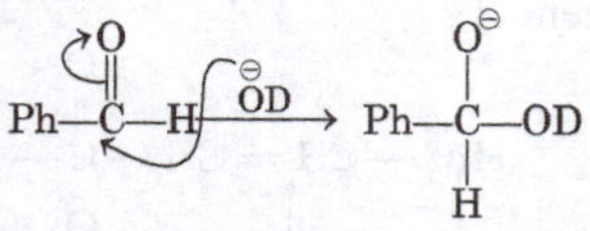

**Step 2.** Removal of $H^-$ and attack of $H^-$ on other carbonyl group.

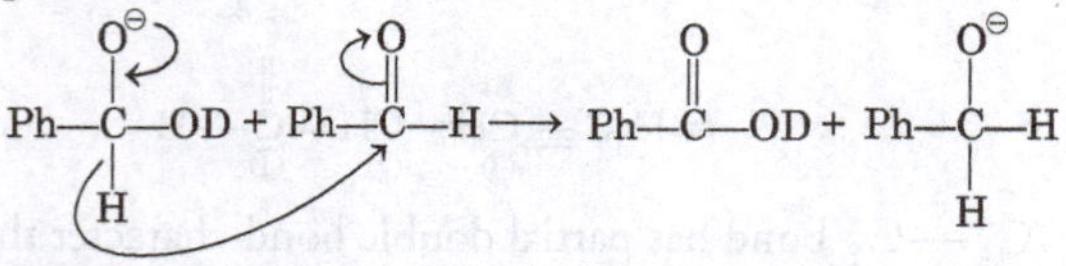

**Step 3.** Abstraction of $D^+$ from acid.

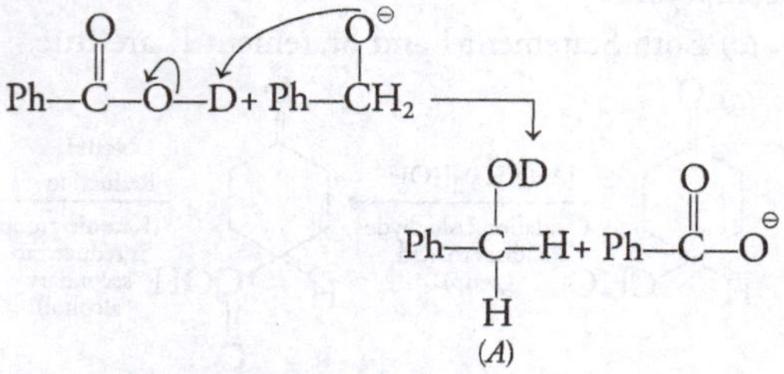

Thus, product (*A*) is Ph—$CH_2$—OD.

**15.** *(d)*

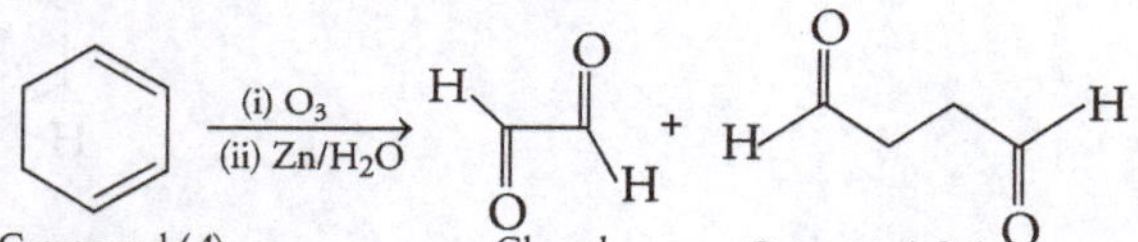

Compound (*A*) Cyclohex-1,3-diene; Glyoxal; Butene-1,4-dial or Ethane-1,2-dicarbaldehyde

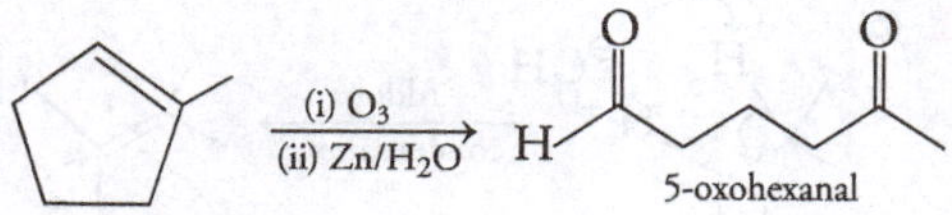

Compound (*B*) 1-methylcyclopent-1-ene; 5-oxohexanal

**16.** *(b)* Benzaldehyde is formed by oxidation of toluene with chromic oxide in the presence of acetic anhydride, followed by acidic hydrolysis.

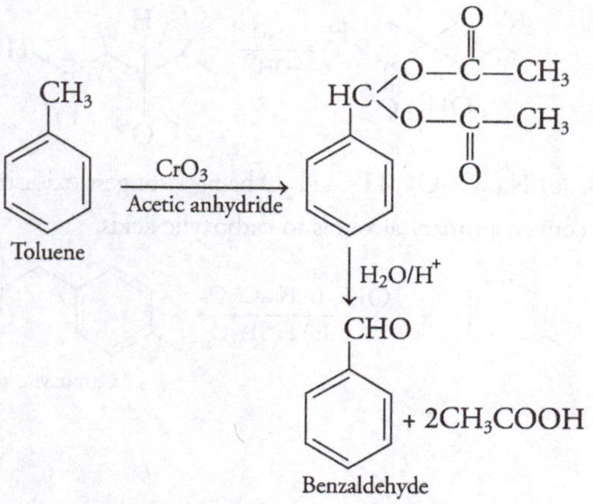


**17.** *(c)* Reagents used in the given reaction are $BH_3$, $H_2O_2$ / $OH^-$ and PCC.

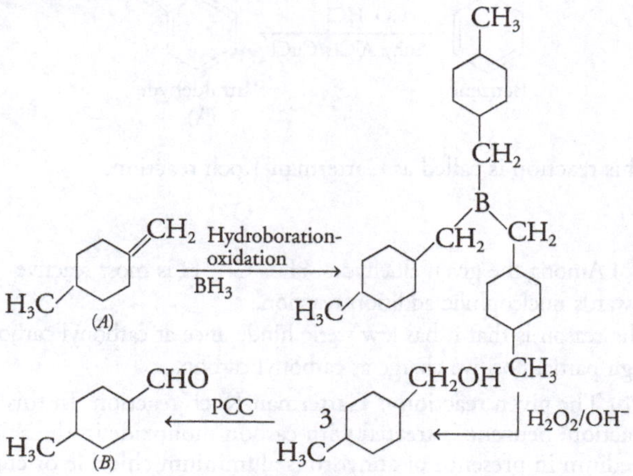

Hydroboration-oxidation is done by $BH_3$, $H_2O_2$ / $OH^-$ and further oxidation is done by PCC reagent.

**18.** *(c)* DIBAL-H, i.e. di-isobutyl aluminium hydride causes partial reduction of cyanide group and further hydrolysis forms corresponding aldehyde.

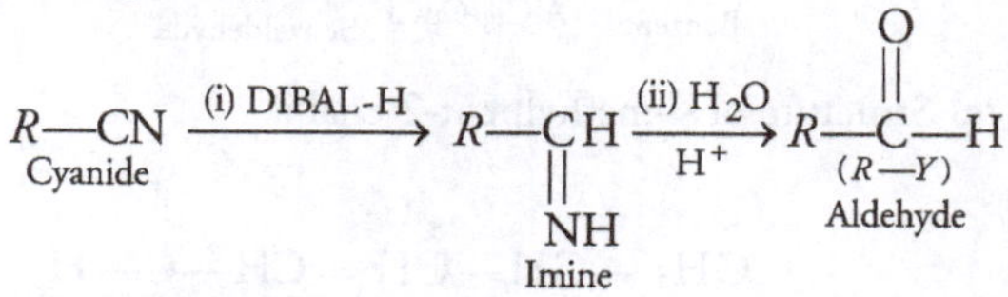

Hence, *Y* is —CHO group.

**19.** *(c)* On ozonolysis, unsaturated hydrocarbon *X* will form carbonyl compound, which gives positive Tollen's test. So, it should be aldehyde.

Butyne reacts with $O_3$ and $Zn/H_2O$ to give 2-oxobutyraldehyde. The aldehyde will give Tollen's test to give silver mirror.

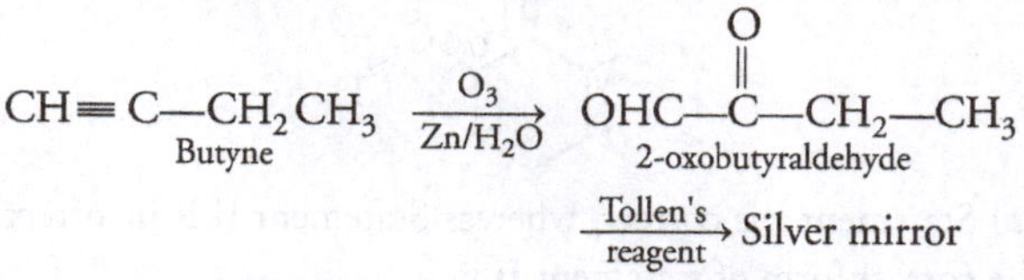

**Note** Tollen's reagent contain silver nitrate ($AgNO_3$), ammonia and sodium hydroxide (NaOH) which react with aldehydes to give Ag (silver mirror).

Only product of option (c) contains aldehyde group.

**20.** *(b)* DIBAL-H is added to produce aldehyde from ester by keeping the temperature low while at higher temperature it can perform different conversions.

Complete reaction is as follows:

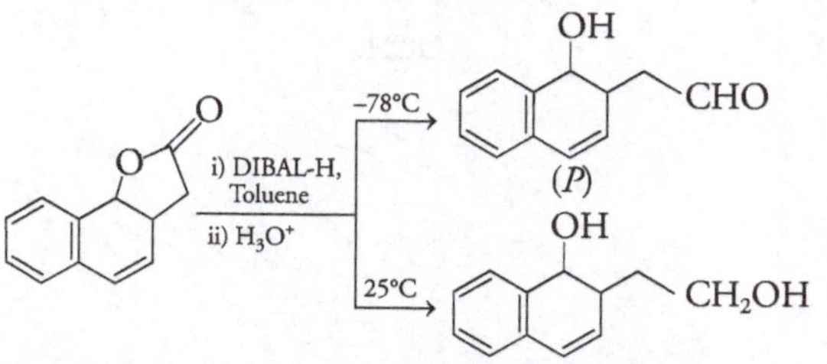

**Note** DIBAL-H does not reduce double bond.

**21.** *(a)* The major product of the reaction is $CH_3CH_2CH_2CH_2CHO$. Here, electrophilic addition of $H-\overset{+}{C}=O$ (formylation) take place to the alkene through Markownikoff addition.

(i) $H_2 + :C=\ddot{O}: \xrightarrow[\text{(Catalyst)}]{Rh} [H-\overset{\oplus}{C}=O]\,H^{\ominus}$
Water gas

(ii) $CH_3CH_2-CH=CH_2 \xrightarrow[\text{Slow}]{H-\overset{\oplus}{C}=O}$
But-1-ene

$CH_3CH_2-\overset{\oplus}{C}H-CH_2-CHO \xrightarrow[\text{Fast}]{H^{\ominus}}$ $CH_3CH_2CH_2CH_2CHO$ (major)
Pentanal

**22.** *(b)* $CH_3CH_2OH \xrightarrow{CrO_3-H_2SO_4} CH_3-\overset{O}{\overset{\|}{C}}-OH$
Ethanol — Ethanoic acid

Since, $CrO_3 \cdot H_2SO_4$ behave as strong oxidising agent and it converts alcohol directly to carboxylic acid. Thus, reaction (b) will not form acetaldehyde.

**23.** *(b)* Molybdenum oxide ($Mo_2O_3$) is used for oxidising alkanes to aldehyde. It is used to manufacture molybdenum metal, which serves as an additive to steel and corrosive resistant alloys.

$CH_3-CH_2-CH_3 \xrightarrow[\text{Oxidising agent}]{Mo_2O_3} CH_3-CH_2-CHO$
Propane — Propanal

**24.** *(a)* The reaction involved and the major product is given below

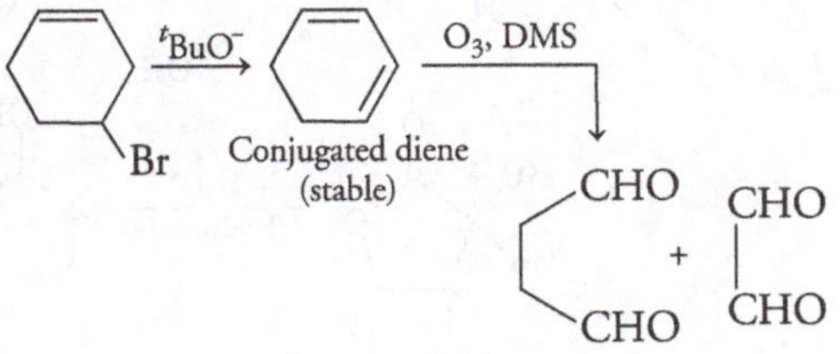


**25.** *(d)* For the reaction,

$$\text{Carbonyl compound} + MeOH \underset{}{\overset{HCl}{\rightleftharpoons}} \text{Acetal}$$

Rate of reaction is highest for propanal as substrate and methanol in excess. Propanal is an aldehyde and more reactive than ketones. When MeOH is taken in excess then reaction moves in the forward direction that results in the formation of acetal. Reaction involved is as follows :

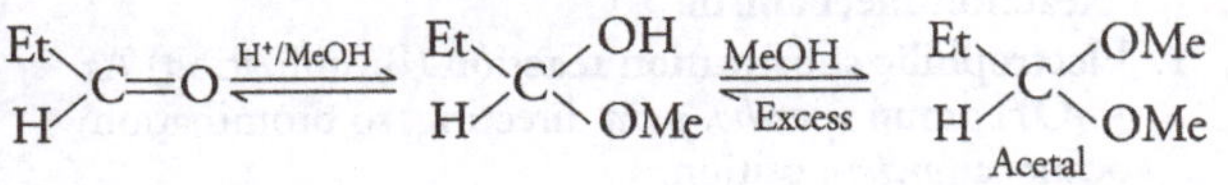


**26.** *(b)* The question is related to above reaction only with the condition that the consumption of *R*MgX will be more than 1 equivalent in some of the given cases.

Among the given compounds *B*, i.e.

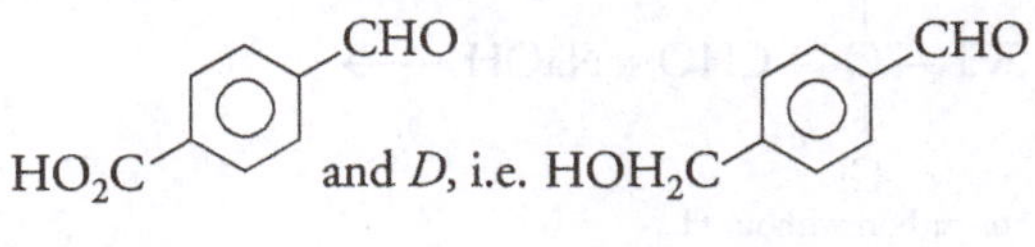


contain additional groups which can give active hydrogens.

**Concept Enhancer**

Grignard reagents produce alkanes whenever come in contact with any group or compound which can give active hydrogen as

$$ROH + R'MgX \xrightarrow[\text{ether}]{\text{Dry}} R'H + Mg\langle^{OR}_{X}$$

These reactions are equivalent to acid-base reactions. So, in both of these compounds, more than one equivalent will be required to form Grignard products. These compounds will give 2 type of products as:

(i) from the >C=O group
(ii) from the group which release active hydrogen

The additional reactions involved are:

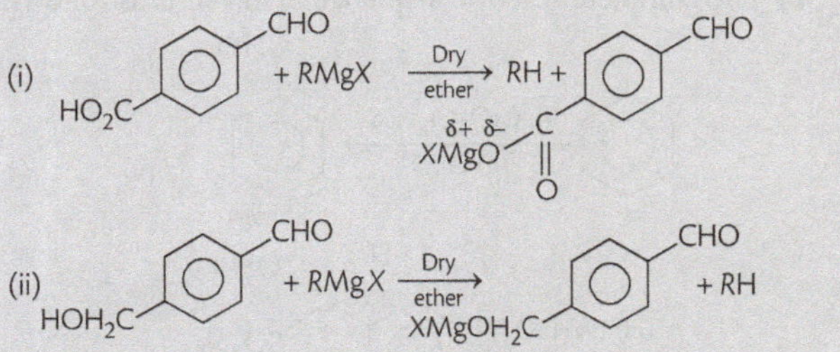


Grignard reagent usually attacks on >C=O group as

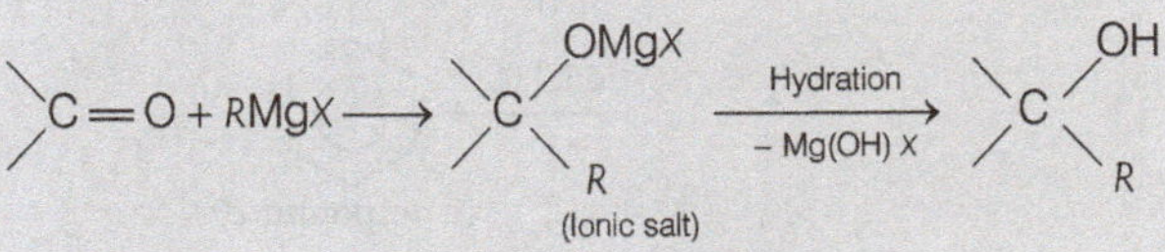


**27.** *(a)* DIBAL-H (Di-isobutyl aluminium hydride) is a reducing agent with formula

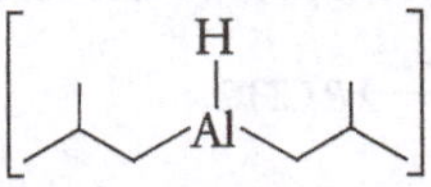


. This is generally used for the preparation of aldehydes. Using DIBAL —H, lactones are reduced directly to aldehydes.

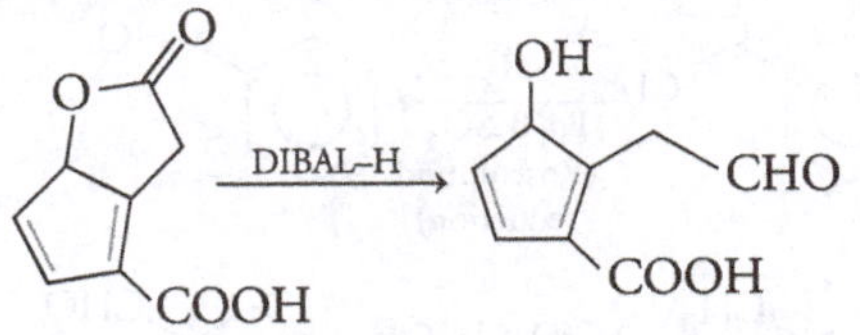


**28.** *(a)* Cannizzaro's reaction is given by aldehydes (*R*CHO) lacking H at α-carbon or lacking α-carbon (as in HCHO). With NaOH, there is formation of acid salt ($RCOO^-$) by oxidation and alcohol ($RCH_2OH$) by reduction.

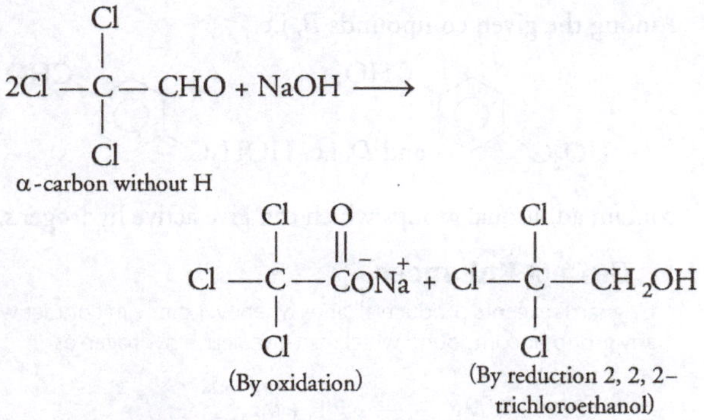


**29.** *(b)* $A + NaOH \longrightarrow$ Alcohol + Acid

Thus, it is Cannizzaro reaction.

*A* is thus aldehyde without α-H at α-carbon.

(as $C_6H_5CHO$, HCHO)

$$2C_6H_5CHO + NaOH \longrightarrow C_6H_5CH_2OH + C_6H_5COONa$$

**30.** *(b)* The complete reaction sequence involved is as follows.

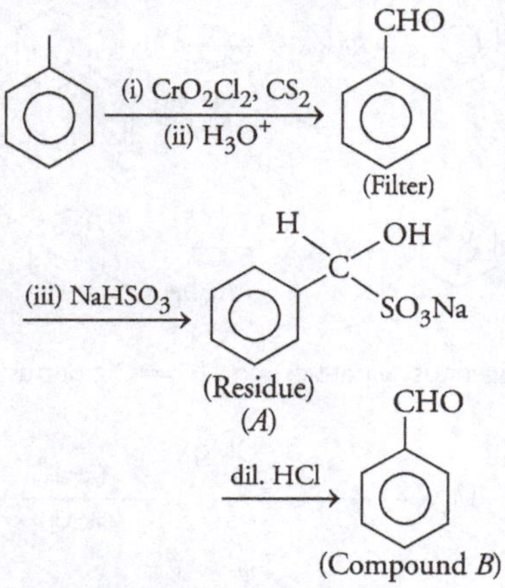


Thus option (b) is the correct answer.

**31.** *(d)* The correct match is A-IV, B-III, C-I, D-II

(A) $RCN \xrightarrow[\text{(ii) } H_3O^+]{\text{(i) } SnCl_2, HCl} RCHO$

(Stephen reaction)

(B) 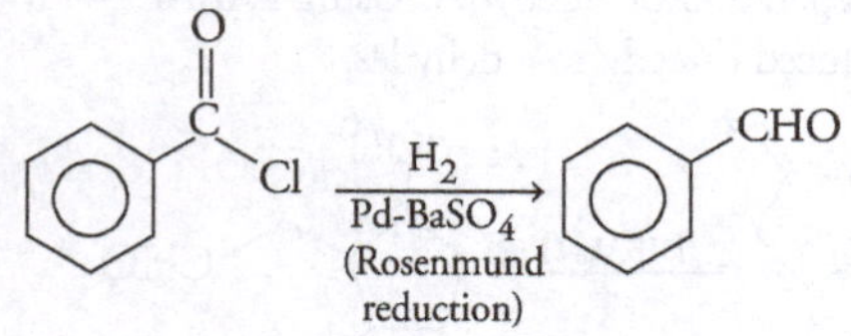


(C) 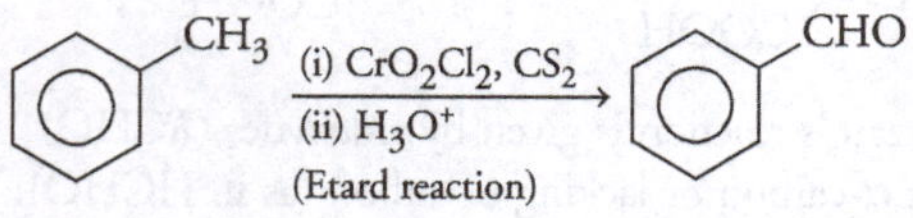


(D) 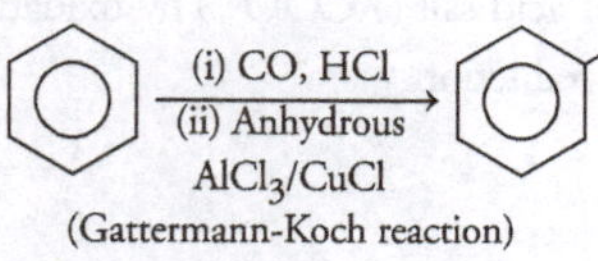


**32.** *(a)* The complete reaction is as follows.

$$CH_3-C\equiv CH \xrightarrow[H_2]{Pd/C} \underset{(A)}{CH_3-CH=CH_2} \xrightarrow[\text{(ii) Zn, } H_2O]{\text{(i) } O_3} \underset{(B)}{CH_3-CHO} + \underset{(C)}{HCHO}$$

**33.** *(d)*

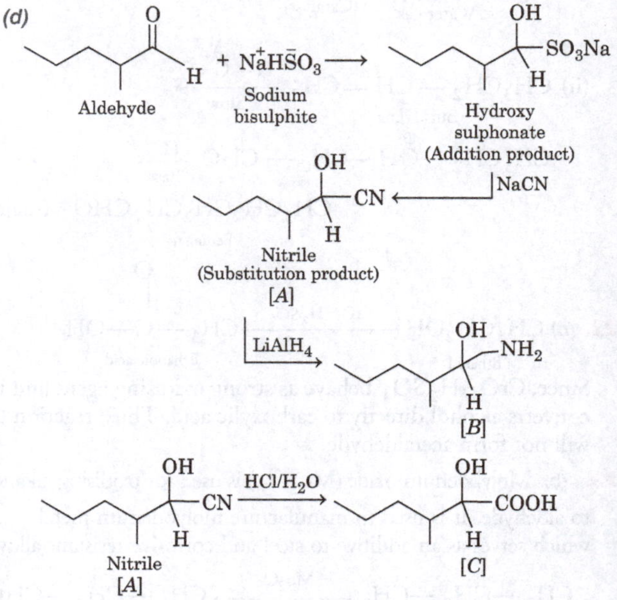


**34.** *(c)*

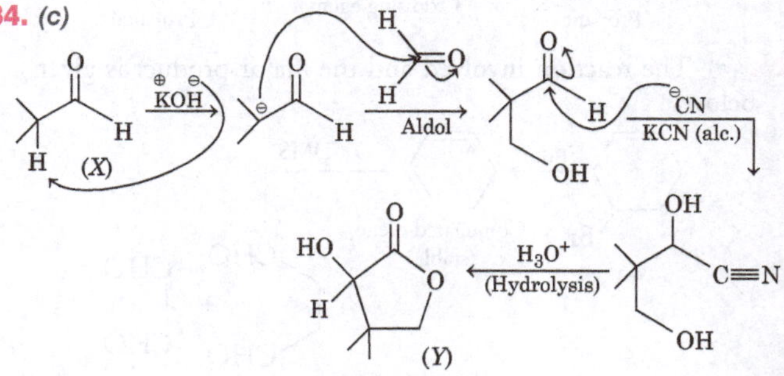


Compound [*X*] on reaction with —HCHO, KOH, gives aldol reaction. The product of aldol reaction upon reaction with cyanohydrin, gives nucleophilic addition reaction on >C=O of —CHO group, upon reaction with $H_3O^+$; —$CN^-$ hydrolysis to —$CO_2H$ and a ring closing reaction takes place to give product [*Y*].

**35.** *(d)* **Reaction mechanism**

1. **Electrophilic substitution reaction** (Bromination) As —OH group is *ortho, para* directing, so bromination occurs at *ortho*-position.

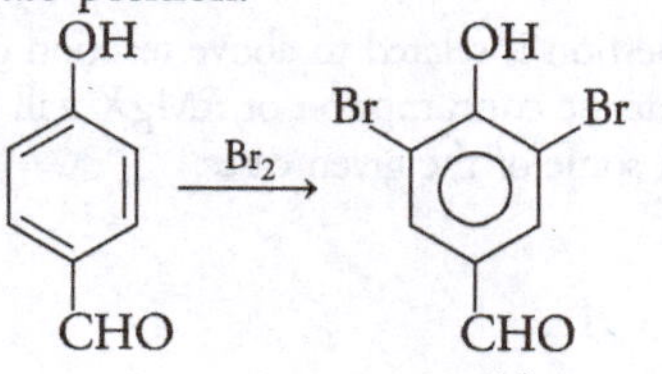

2. Hydroxylamine ($NH_2OH$) reacts with carbonyl group to form oximes.

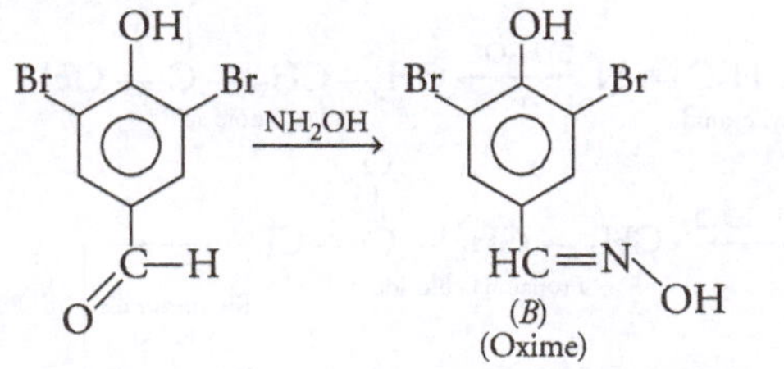

3. $P_2O_5$ is a dehydrating agent and upon heating, it removes water from —CH= NOH to form cyanide group.

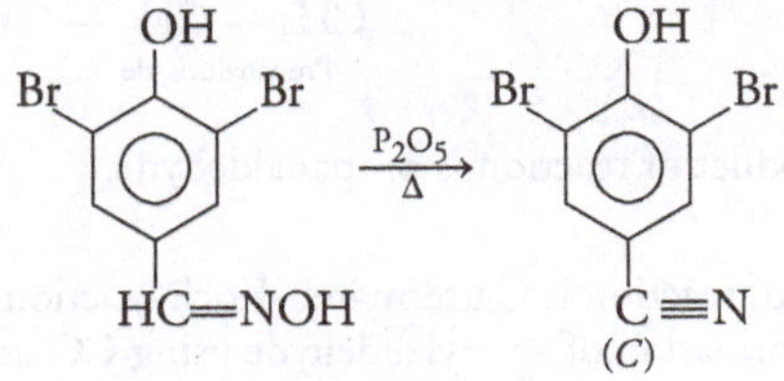

Hence, the final product *C* is

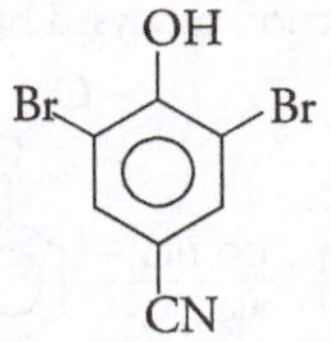

**36.** *(c)* The reaction sequence given is as follow :

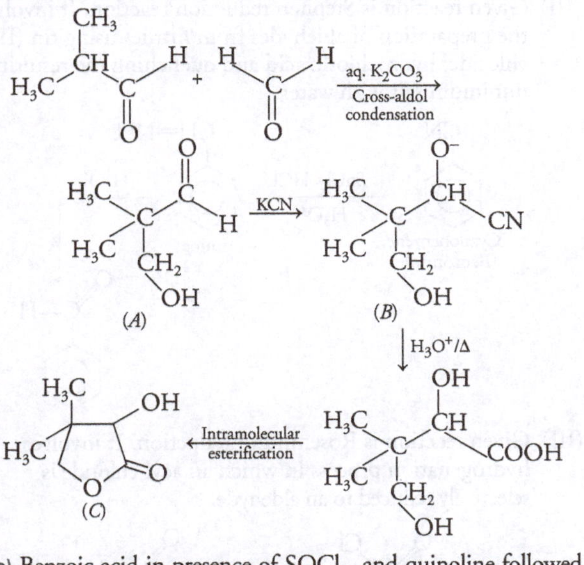

**37.** *(c)* Benzoic acid in presence of $SOCl_2$ and quinoline followed by reduction gives benzaldehyde.

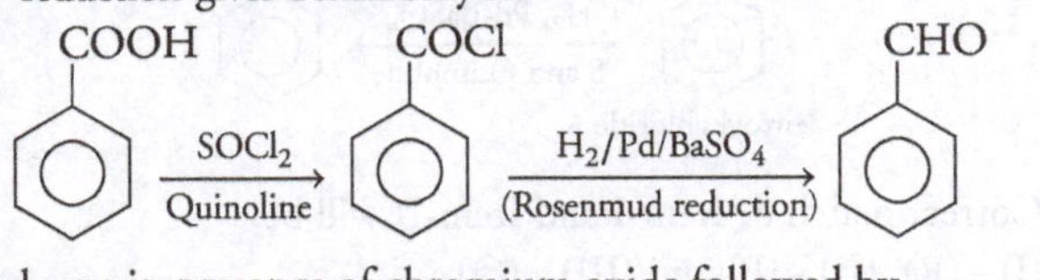

Toluene in presence of chromium oxide followed by hydrolysis gives benzaldehyde

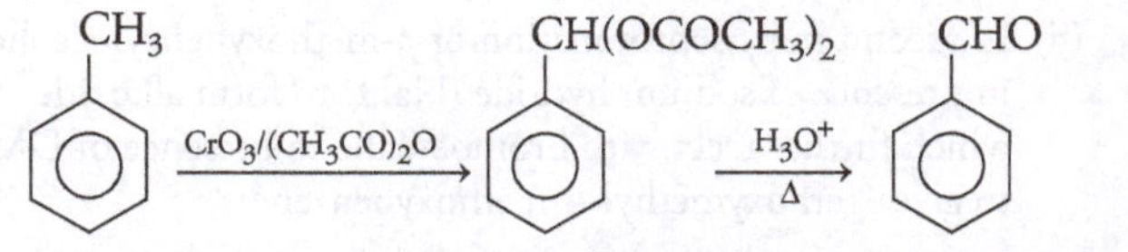

**38.** *(c)* The given compound does not form dicarboxylic acid on reaction with sodium hydroxide followed by hydrolysis in any case.

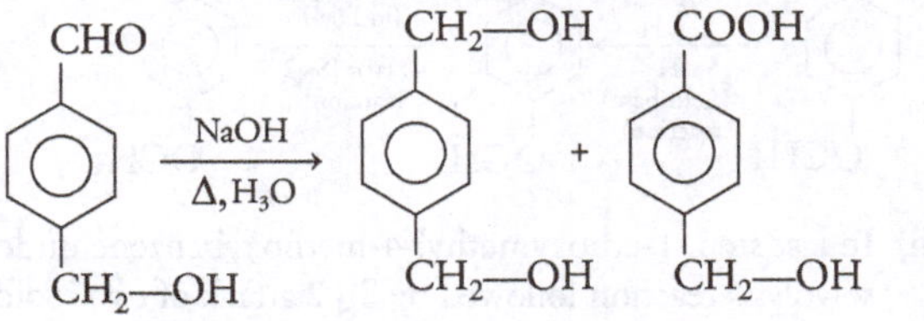

**39.** *(a)* In reaction $t$-$BuO^-H^+$ is used as a base to remove HBr and form alkene as a major product (*B*) *via* E2 elimination mechanism.

**Mechanism**

Firstly, 1-bromoacetaldehyde undergoes protonation which is then react with $t$-$BuO^-H^+$. ($t$-$BuO^-$ act as nucleophile and attack on carbonyl carbon). In next step, $t$-$BuO^{\ominus}K^+$ act as base and accepts proton and at last alkene is formed as major product.

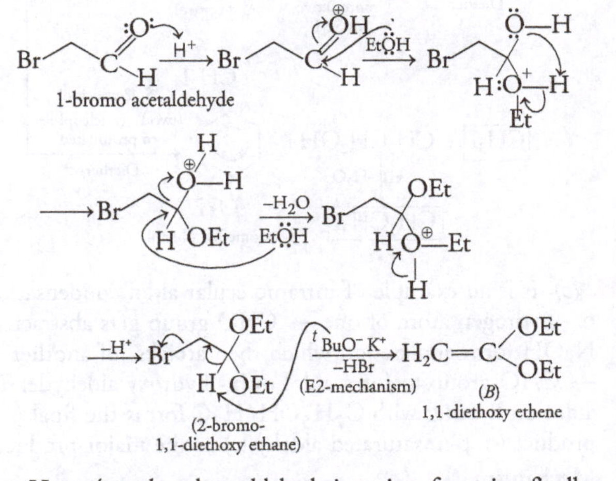

**40.** *(c)* Here, 4-methoxybenzaldehyde in series of reaction finally forms 4-iodomethyl phenol (*A*). Let us compute the reaction step by step with mechanism as follows

(i) In first step, oxidising agent (*p*-methoxybenzaldehyde) reacts with formaldehyde in presence of strong base NaOH to give *p*-methoxybenzyl alcohol along with sodium salt of methanoic acid. It is known as Cannizzaro's reaction.

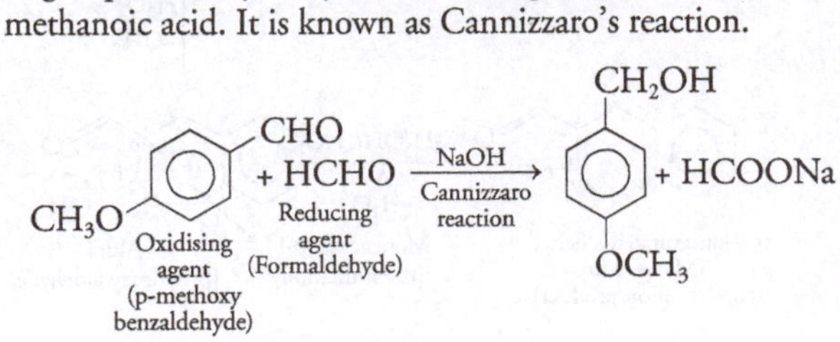

(ii) In second step, deprotonation of *p*-methoxybenzyl alcohol in presence of sodium hydride (NaH) to form alkoxide which further react with bromoethane in presence of DMF to give 1-ethoxymethyl-4-methoxybenzene.

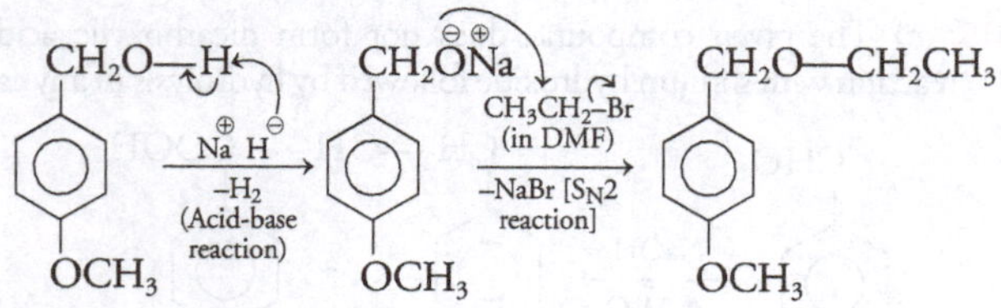


(iii) In last step, 1-ethoxymethyl-4-methoxybenzene undergoes solvolysis reaction followed by $S_N2$ attack of two iodide ion ($I^-$) to give 4-iodomethyl phenol, iodoform and iodoethane.

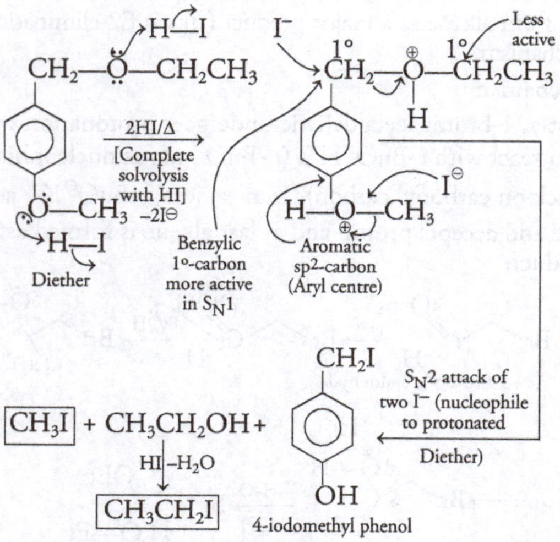


**41.** *(c)* It is an example of intramolecular aldol condensation. α - hydrogen atom of one — CHO group gets abstracted by NaOH form enolate ion which then attacks of another — CHO group to form aldol or β - hydroxy aldehyde. The aldol on heating with $C_2H_5OH/H_2O$ forms the final product (α, β-unsaturated aldehyde) as the major product.

**Mechanism**

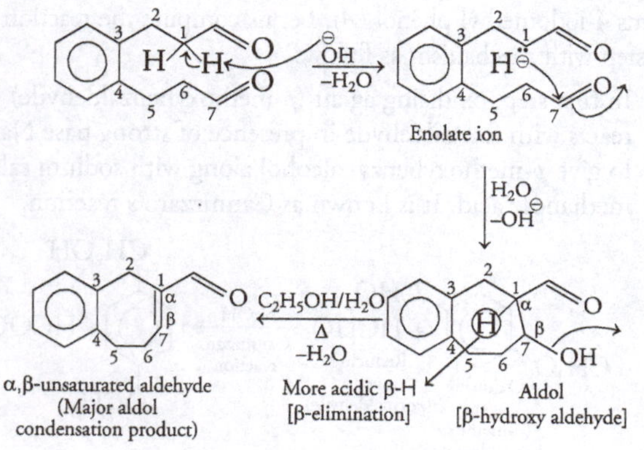


**42.** *(d)* The complete reaction take place as follows

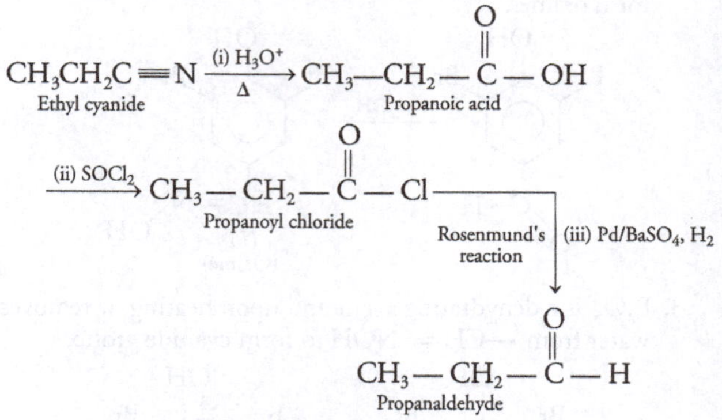


Final product of reaction is propanaldehyde.

**43.** *(c)*

(I) Given, reaction is Gattermann -Koch reaction. It involves the formation of an aryl aldehyde using CO and HCl as reactants. The reaction catalysed by $AlCl_3$.

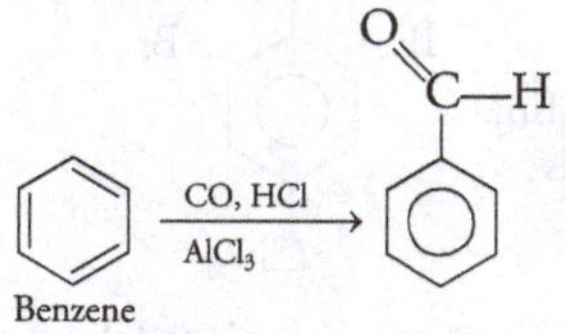


(II) Given reaction is Stephen reduction reaction. It involves the preparation of aldehydes from nitriles using tin (II) chloride, hydrochloric acid and quenching the resulting minimum salt with water.

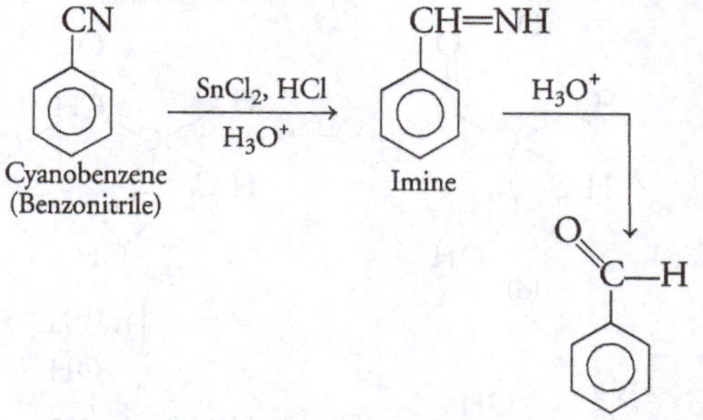


(III) Given reaction is Rosenmund reduction. It involves hydrogenation process in which an acyl chloride is selectively reduced to an aldehyde.

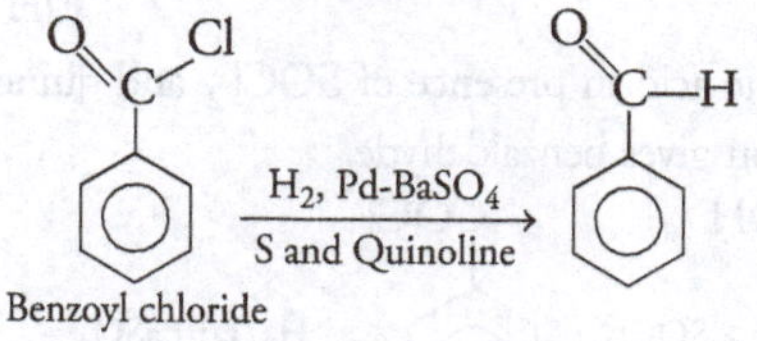


Correct match of Item-I and Item-II will be (I) - (R), (II) - (P) and (III) - (Q).

**44.** *(d)* Rate of nucleophilic addition ∝ Polarity of the $\overset{\delta+}{>C}=\overset{\delta-}{O}$ group ∝ Attachment of electron withdrawing group (EWG), i.e., $-R$ and $-I$ groups, (where $-R > -I$)

$$\propto \frac{1}{\text{Attachment of electron donating group (EDG) (i.e., } +R \text{ and } +I \text{ groups)}}$$

In substituted benzaldehyde, $—NO_2$ (EWG) shows $-R$ effect from *ortho*-and *para*-positions and $-I$ effect (only) from *meta*-positions. The group $—OCH_3$ (EDG) shows $+R$ effect from *ortho*-and *para*-position but $-I$ effect from *meta*-positions.

EWG ($-R > -I$) increases 'δ+' character of $sp^2$ carbon of the $>C=O$ group and hence attack of the nucleophile ($CN^-$) will be easier.

So, the increasing order of HCN addition is :

(iii) < (i) < (iv) < (ii)

(+R) (−I) (−I) (−R)

− I effect : $NO_2 > OCH_3$

**45.** *(c)*

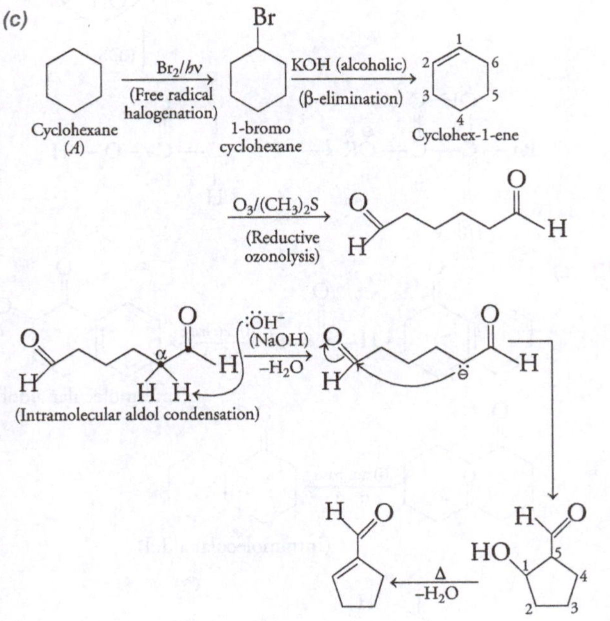


**46.** *(d)* The reaction mechanism is

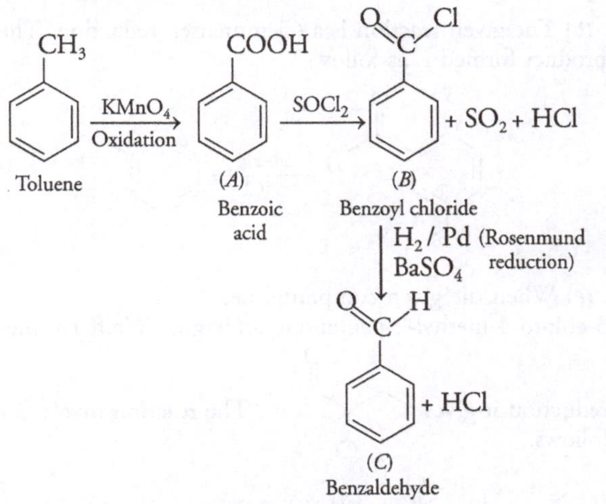


**47.** *(d)*

$$CH_3\overset{O}{\overset{||}{C}}—Cl \xrightarrow{KOH\,(aq)} CH_3—\overset{O}{\overset{||}{C}}—OH + KCl$$

[SN reaction]

$$CH_3—CH_2—Cl \xrightarrow{KOH\,(aq)} CH_3—CH_2—OH$$

[$S_N2$ reaction]

$$Cl—CH_2—CH_2—Cl \xrightarrow{2\,KOH\,(aq)} \underset{OH}{CH_2}—\underset{OH}{CH_2}$$

[$S_N2$ reaction]

$$CH_3—\underset{Cl}{CH}—Cl \xrightarrow{2\,KOH\,(aq)} CH_3—\underset{OH}{CH}—OH \text{ (unstable)}$$

$$\xrightarrow{-H_2O} CH_3—\overset{O}{\overset{||}{C}}—H + 2KCl$$

(acetaldehyde)

**48.** *(0)*

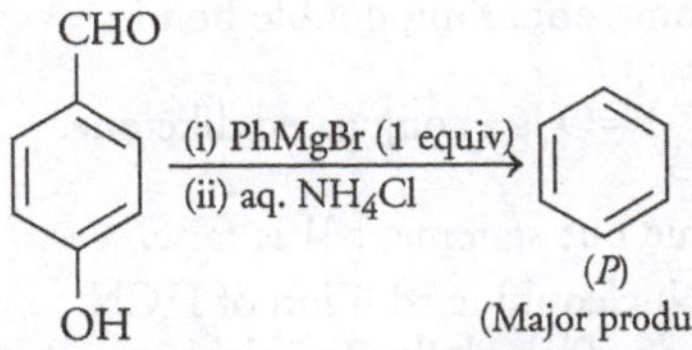


Since only 1 equivalent of Grignard reagent is used it will attack OH group and benzene is obtained as product *P*.

So, the number of hydroxyl group in product *P* is zero.

## Topic 2

**01.** *(b)* The given reaction is a Clemmensen reduction. The product formed is as follows

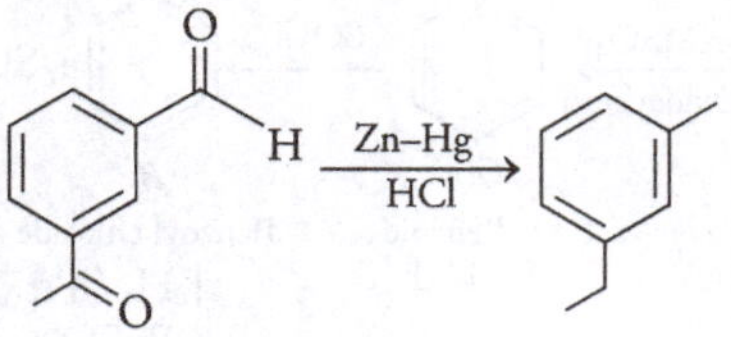

**02.** *(a)* When the given compound i.e. 3-chloro-4-methyl-2-pentanone undergoes Wolff Kishner reduction it gives 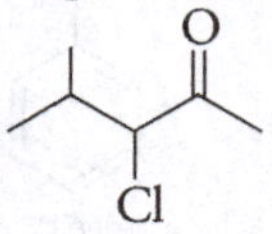 . The reaction involved is as follows.

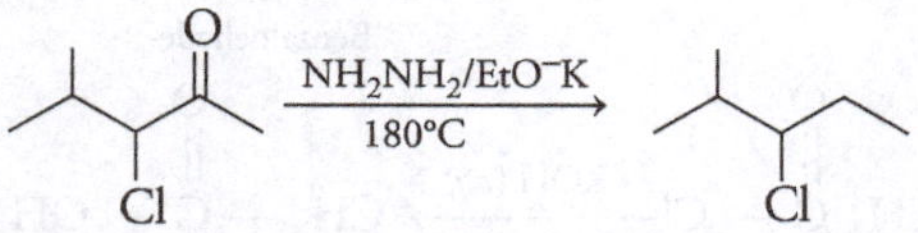

> • Concept **Enhancer**
>
> In Wolff-Kishner reaction the carbonyl group of aldehydes and ketones is reduced to $>CH_2$ group on treatment with hydrazine followed by heating with sodium or potassium hydroxide in high boiling solvent like ethylene glycol.

**03.** *(d)* Deprotonation will be easily possible for (*A*) because it is more resonance stabilised.

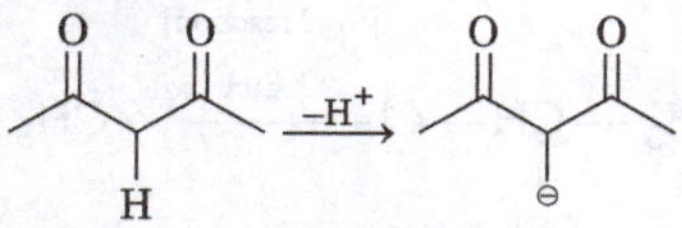

(More resonance stabilised)

(*B*) and (*C*) shows less tendency for deprotonation due to less conjugation.

Hence, option (d) is the correct answer.

**04.** *(c)* Conjugated diketone has two ketone group directly attached to carbon atom containing double bond.

Hence, 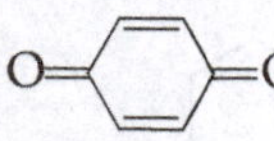 is a conjugated diketone.

**05.** *(c)* Statement I is true but statement II is false.

The correct form is : Nucleophilic addition of HCN (hydrogen cyanide) to an aldehyde/ketone yield cyanohydrin as final product.

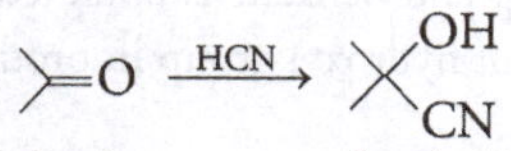

Carbonyl compound — Cyanohydrin

**06.** *(b)* Both A and R are true and R is the correct explanation of A.

**07.** *(c)* Option (c) is the major product formed.

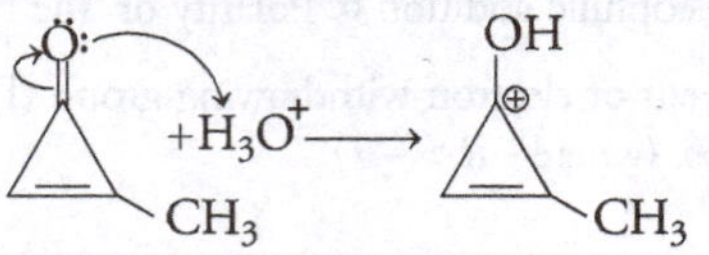

Aromatic product

**08.** *(c)* Addition of HCN is a nucleophilic reaction. Greater the electron deficiency of C-atom of carbonyl group, higher the rate of reaction.

Hence,

$$\underset{(D)}{Ph-\overset{O}{\overset{\|}{C}}-Ph} < \underset{(C)}{Ph-\overset{O}{\overset{\|}{C}}-CH_3} < \underset{(B)}{CH_3-\overset{O}{\overset{\|}{C}}-CH_3} < \underset{(A)}{H-\overset{O}{\overset{\|}{C}}-H}$$

**09.** *(d)* The complete reaction is as follows,

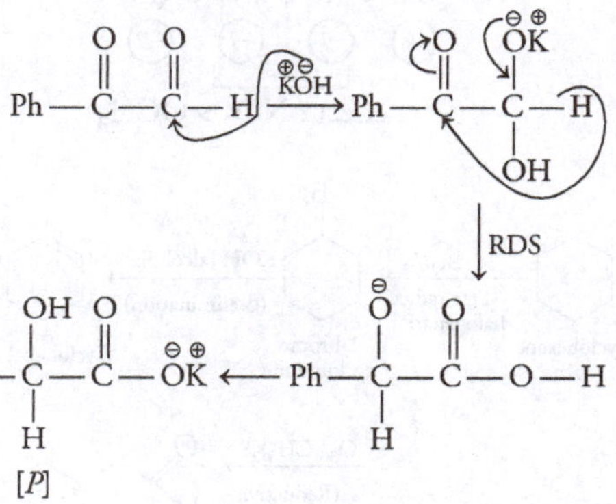

**10.** *(a)*

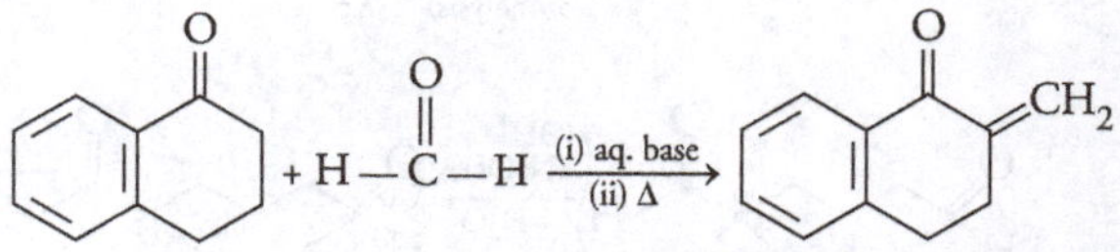

(Intermolecular aldol)

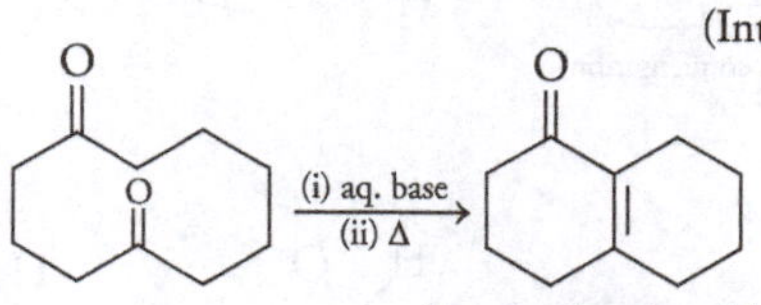

(Intramolecular aldol)

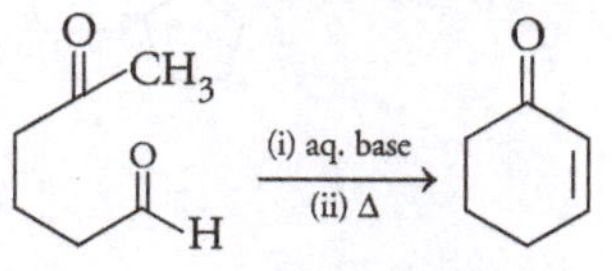

(Intermolecular aldol)

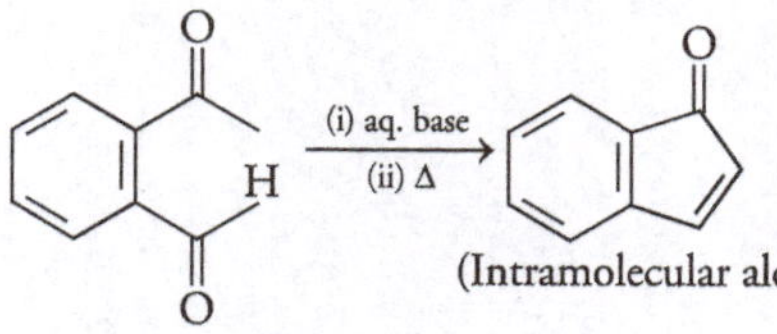

(Intramolecular aldol)

**11.** *(b)* The complete reaction is as follows,

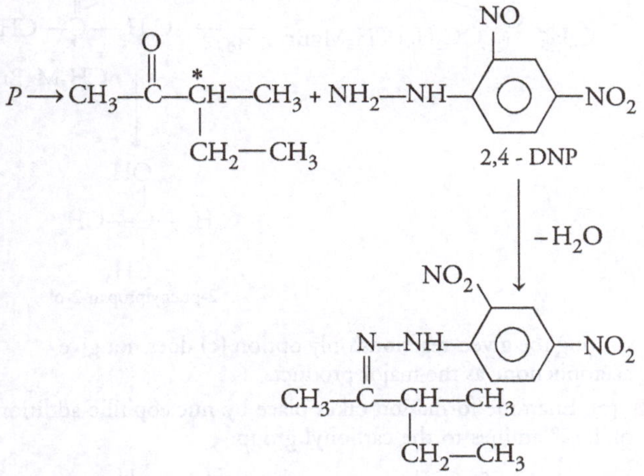

All ketones except α-hydroxy ketones chiral does not Tollen's test.

Gives test with 2,4-DNP.

**12.** *(d)* Statement I is correct but Statement II is incorrect. The correct form of Statement II is reaction between benzaldehyde and ethanal will give cross-aldol product.

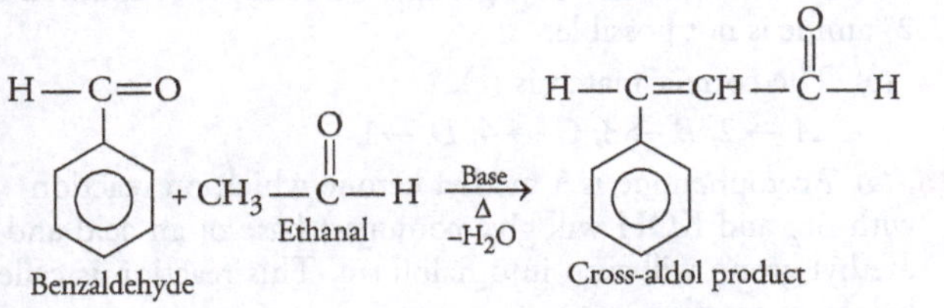

**13.** *(a)* The reaction sequence involved is as follows

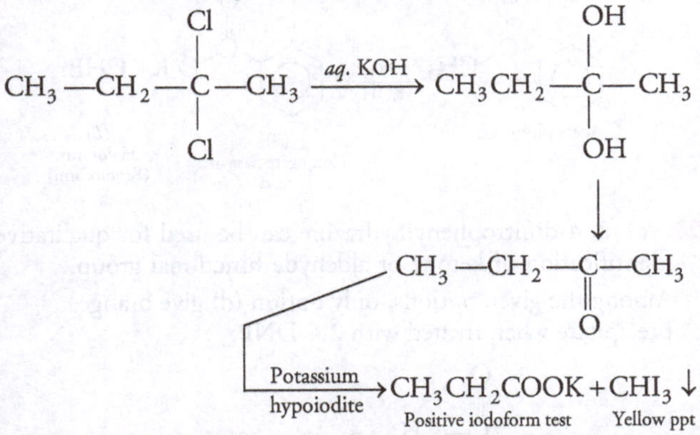

> • **Concept Enhancer**
>
> Positive iodoform test can be given by those compounds which have $CH_3-\overset{O}{\overset{\|}{C}}-$ and $CH_3-\overset{OH}{\overset{|}{C}}H$ group or their halogen derivatives in which halogen is substituted from $-CH_3$ group.

**14.** *(d)* The complete reaction takes place as follows

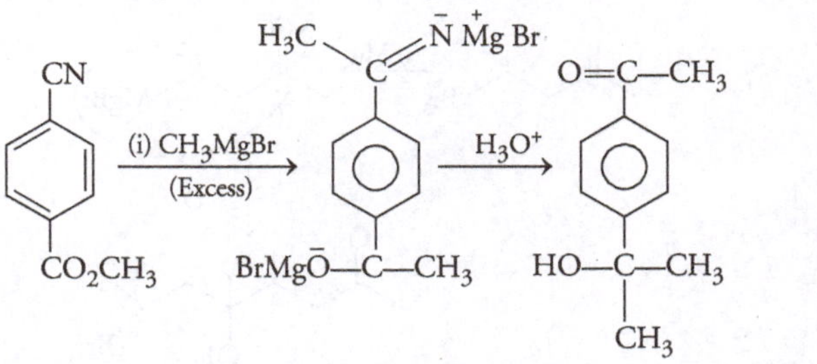

**15.** *(d)* The complete reaction is as follow,

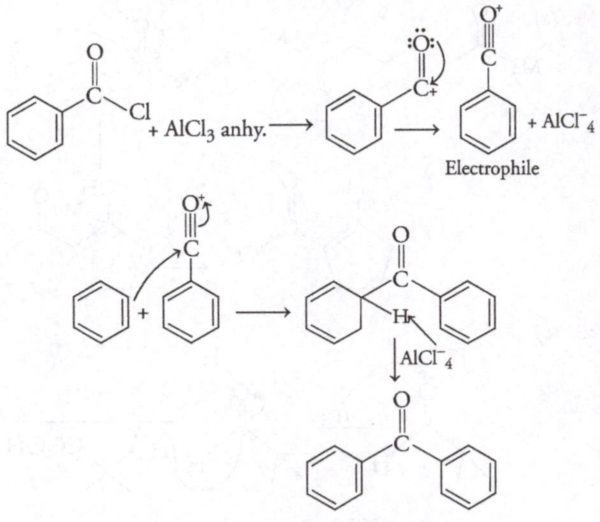

(Major product *P*)

**16.** *(a)* Among the given options, option (a) has most acidic hydrogen.

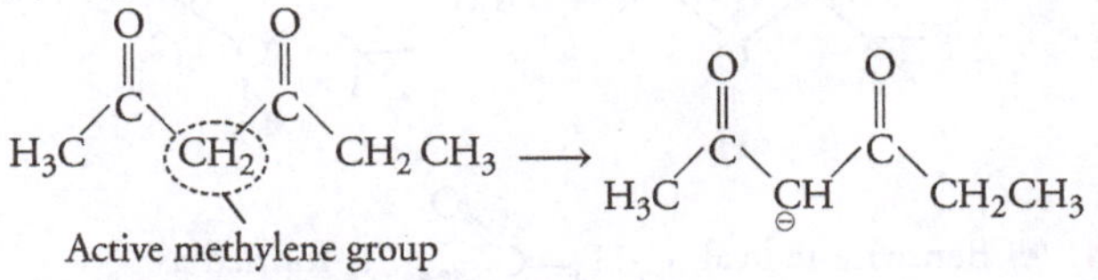

Conjugate base is more stable due to resonance.

**17.** *(b)* Statements given in A, C and D are correct.

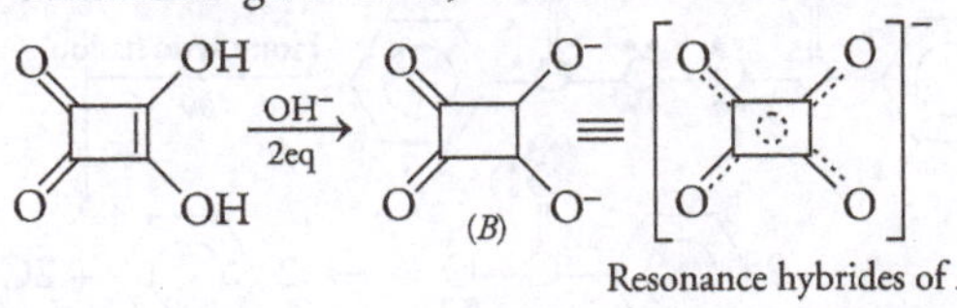

The above structures shows that *B* is an aromatic compound. Since it follows Huckel rule. Also, the resonance hybrid of *B* shows that all C—C bond length are same. Thus, A and D are correct.

Now,

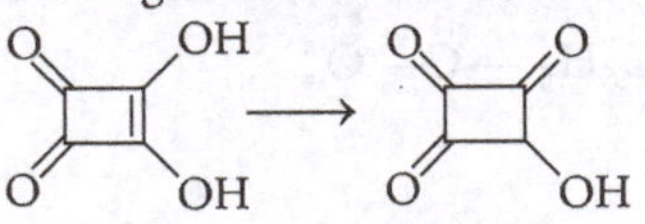

Thus, compound *A* shows tautomerism. Hence, C is also correct.

**18.** *(a)* The structure of product (A) is given in option (a).

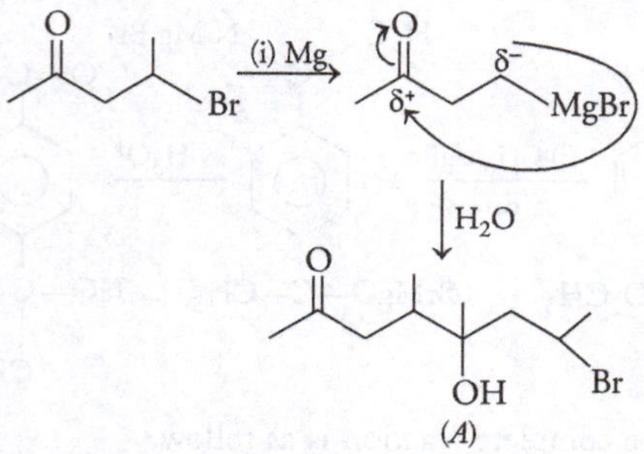

**19.** *(a)*

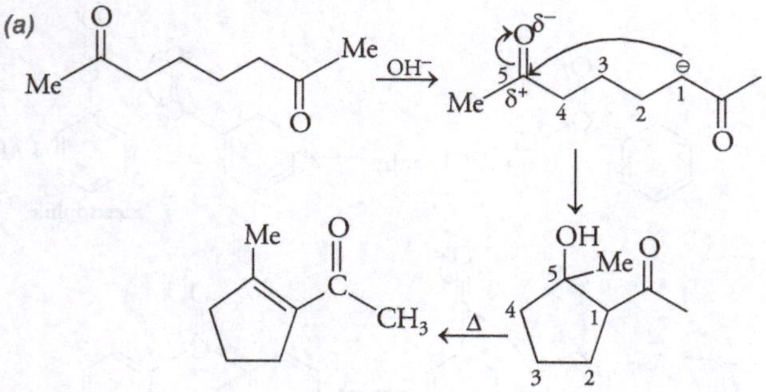

**20.** *(c)*

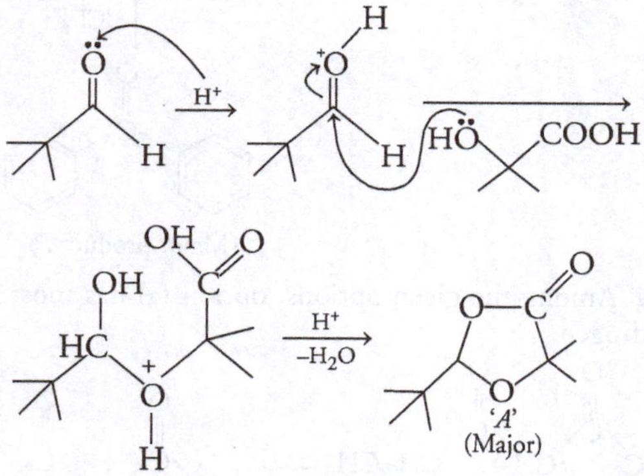

**21.** *(d)* Benzoate radical, $C_6H_5—\overset{O}{\overset{\|}{C}}—\ddot{\underset{\cdot\cdot}{O}}\cdot$ is formed as intermediate.

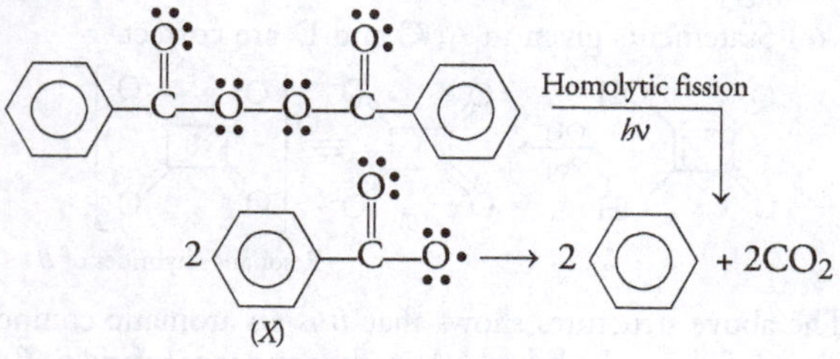

Hence, $X$ is $C_6H_5—\overset{O}{\overset{\|}{C}}—\ddot{O}\cdot$

**22.** *(c)*

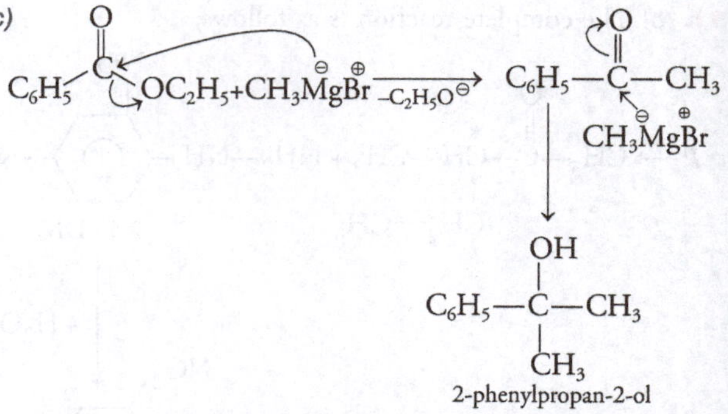

Among the given reactions only option (c) does not give acetophenone as the major product.

**23.** *(c)* Enamine formation takes place by nucleophilic addition of 1°/2° amines to the carbonyl group.

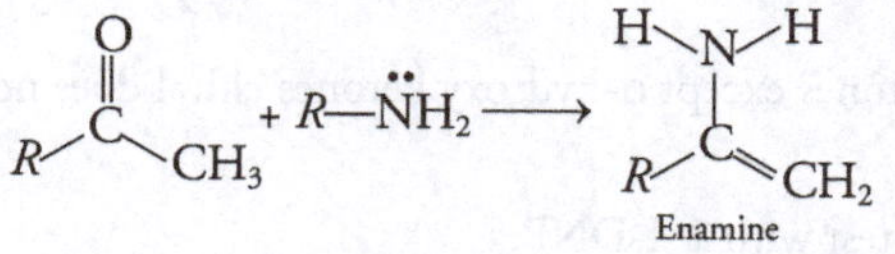

As $t$-Bu$—\overset{O}{\overset{\|}{C}}—t$-Bu has no α—H and also it has more steric crowd around the carbonyl group, so nucleophilic addition by 2° amine is not possible.

**24.** *(c)* The correct match is (c).

$A \rightarrow 2, B \rightarrow 3, C \rightarrow 4, D \rightarrow 1.$

**25.** *(a)* Acetophenone is a methyl ketone which on reaction with $Br_2$ and KOH will give conjugate base of an acid and methyl group will turn into haloform. This reaction is called haloform reaction.

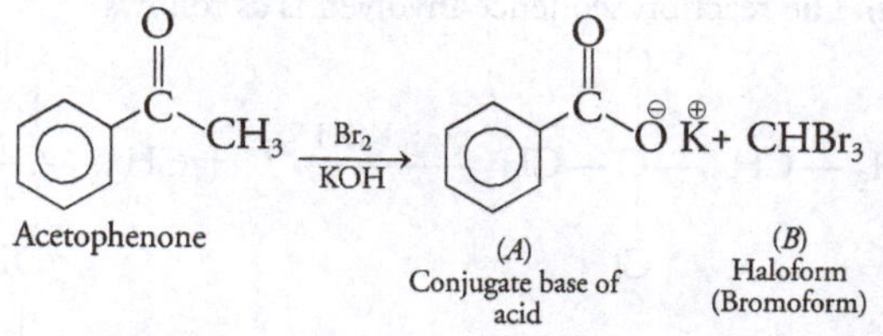

**26.** *(d)* 2, 4-dinitrophenylhydrazine can be used for qualitative identification of ketone or aldehyde functional group. Among the given options, only option (d) give orange precipitate when treated with 2,4-DNP.

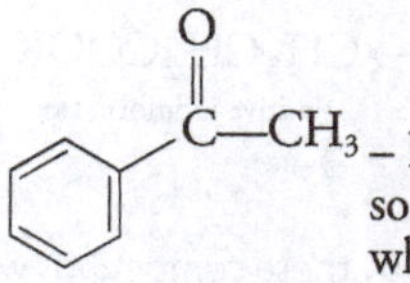 – Presence of ketonic group, so it will give orange ppt. when treated with 2, 4-DNP.

**27.** *(a)* Treatment of benzonitrile with Grignard reagent followed by acid hydrolysis gives acetophenone product.

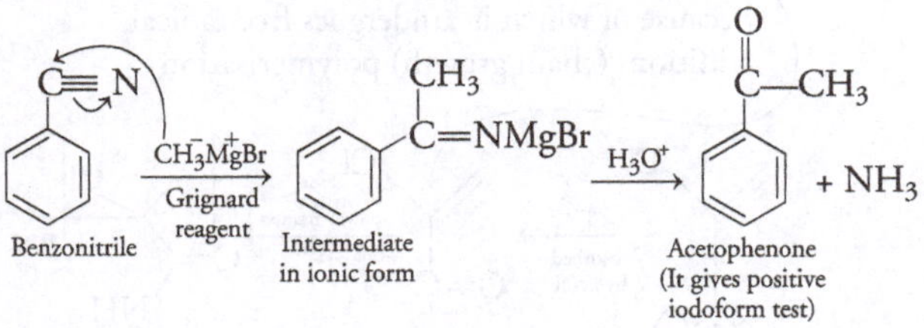


Iodoform test is used to check the presence of carbonyl compound with the structure $R—CO—CH_3$ or alcohols with the structure $R—CH(OH)—CH_3$ in a given unknown substance.

This conclude that acetophenone give iodoform test.

**28.** *(c)* Only compound (II) which has ketonic group does not form silver mirror when treated with Tollen's reagent. Because Tollen's reagent oxidises an aldehyde into the corresponding carboxylic acid. Ketones are not oxidised by Tollen's reagent, so the treatment of a ketone with Tollen's reagent in a glass test tube does not result in a silver mirror. Aldehydes give positive Tollen's test (silver mirror test).

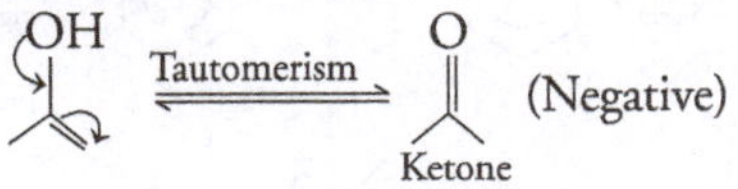


**29.** *(b)* Nucleophilic addition of ethylene glycol in ketone to give ketal.

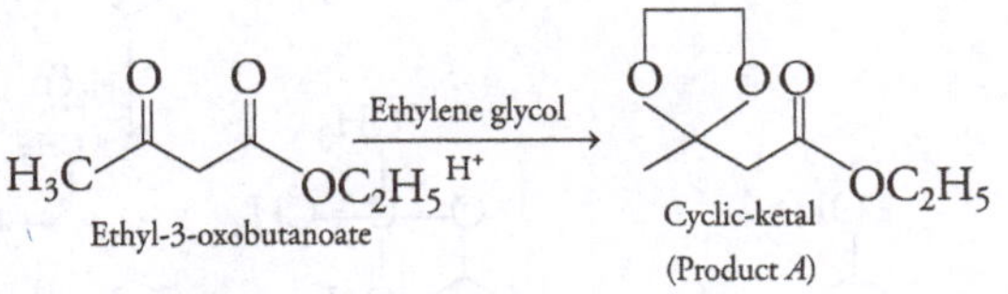


Addition of one molecule of alcohol on ketone give hemiketal and addition of two molecules give ketal.

**30.** *(a)* The reaction is represented as

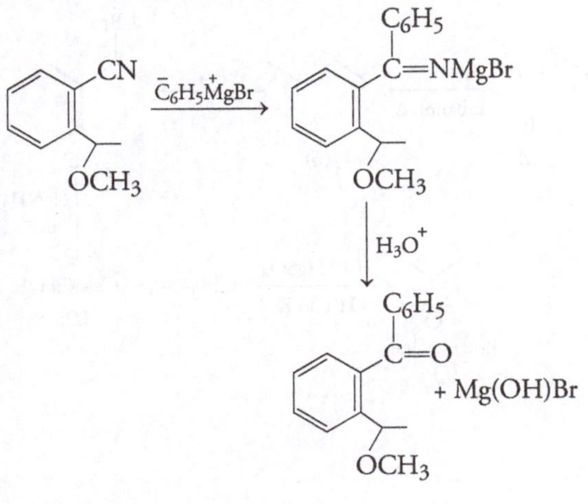


**31.** *(b)* To reduce the carbonyl groups into alkane, wolff-Kishner reduction is used, without affecting the double bond.

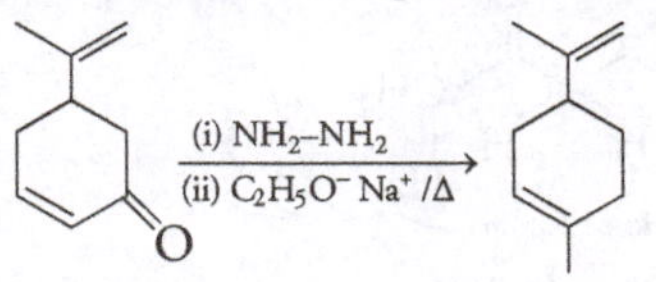


**32.** *(a)* Compound given in option (a) is formed. The steps involved are as follows:

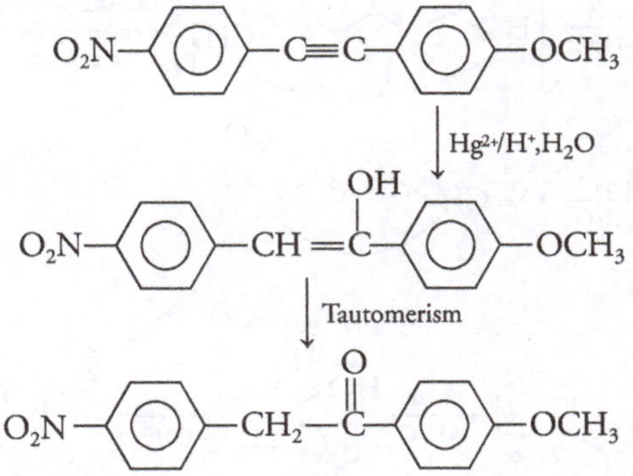


**33.** *(a)* Stability of conjugate determines the acidity of the given compounds

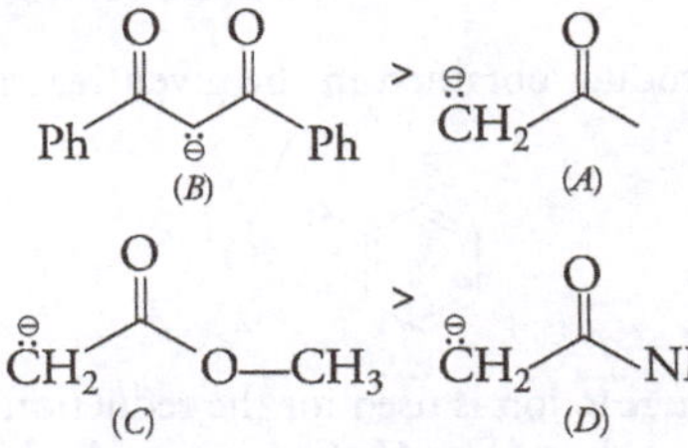


Thus increasing order of acidity is $D < C < A < B$.

**34.** *(b)* In the given reaction cyclisation occurs and the aliphatic alcoholic groups get converted to keto group as follows:

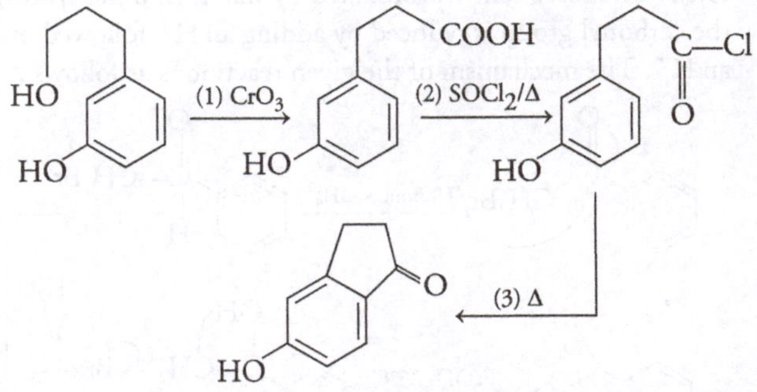


**35.** *(b)* *p*-hydroxy benzophenone upon reaction with bromine in carbon tetrachloride gives 3-bromo-4-hydroxy benzophenone.

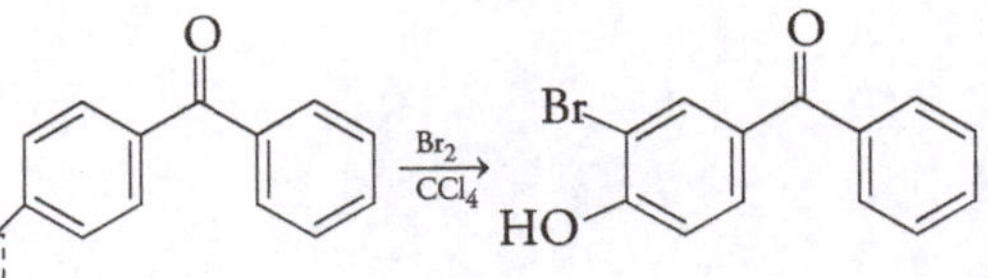

**36.** *(c)* It is an intermolecular aldol condensation reaction.

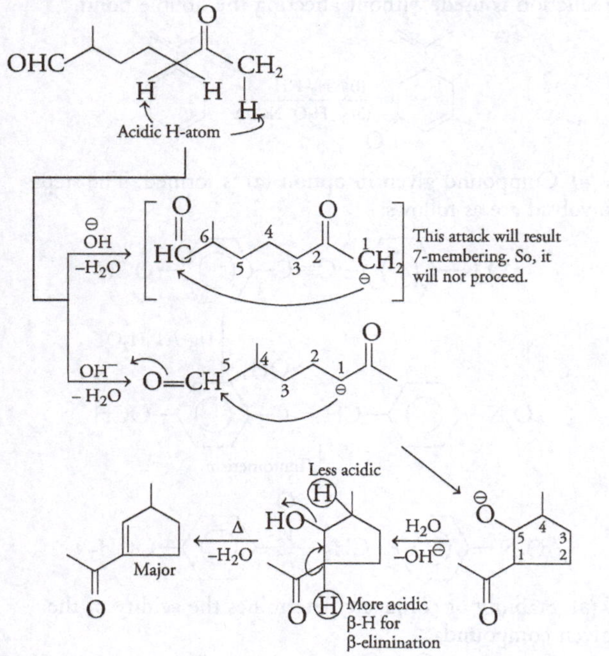


**37.** *(d)* Major product obtained in the given reaction is

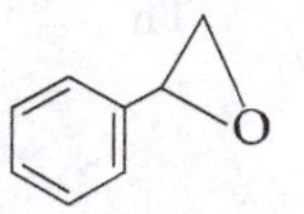

$NaBH_4$ in the reaction is used for the reduction by addition of a hydride ion and a proton. Carbon-oxygen double bonds are easily reduced by sodium borohydride. The actual reducing agent in these reductions is hydride ion ($H^-$). Hydride ion adds to the carbonyl carbon and the alkoxide ion that is formed is subsequently protonated by water. In other words, the carbonyl group is reduced by adding an $H^-$ followed by an $H^+$. The mechanism of the given reaction is as follows :

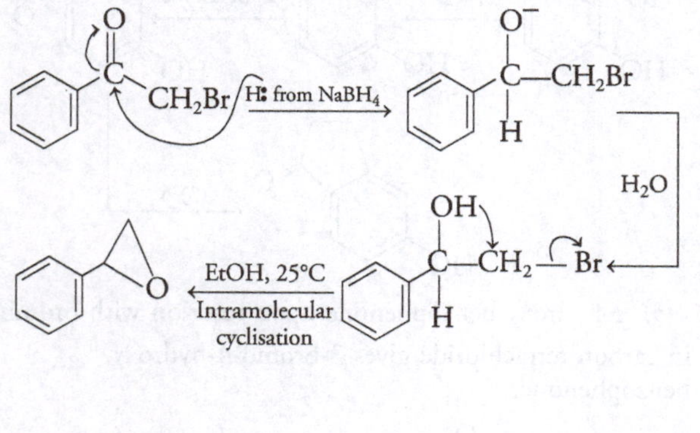


**38.** *(d)* The reaction can take place as follows

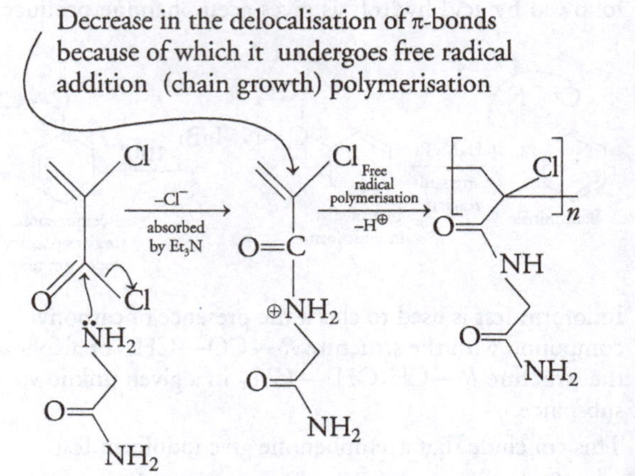


**39.** *(b)* The given process is cumene process (Hock process) to synthesise phenol and acetone industrially. In Hock process, Ph— group migrate and release $H_2O$.

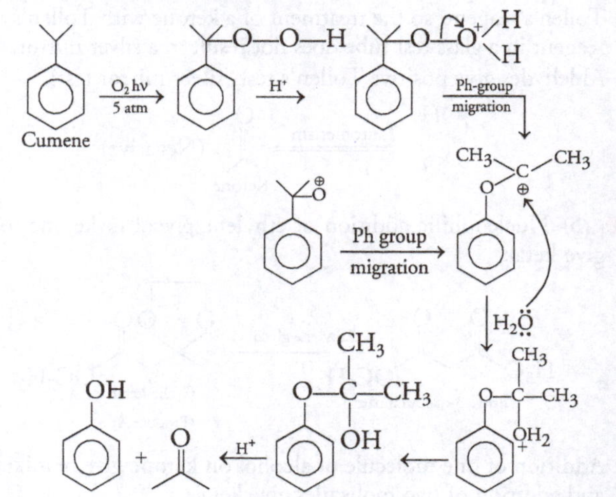


**40.** *(b)* The complete reaction is as follows,

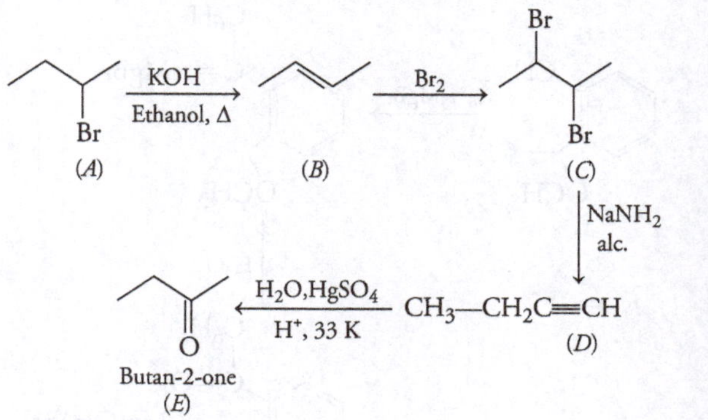

**41.** *(b)* Complete reaction sequence is as follows

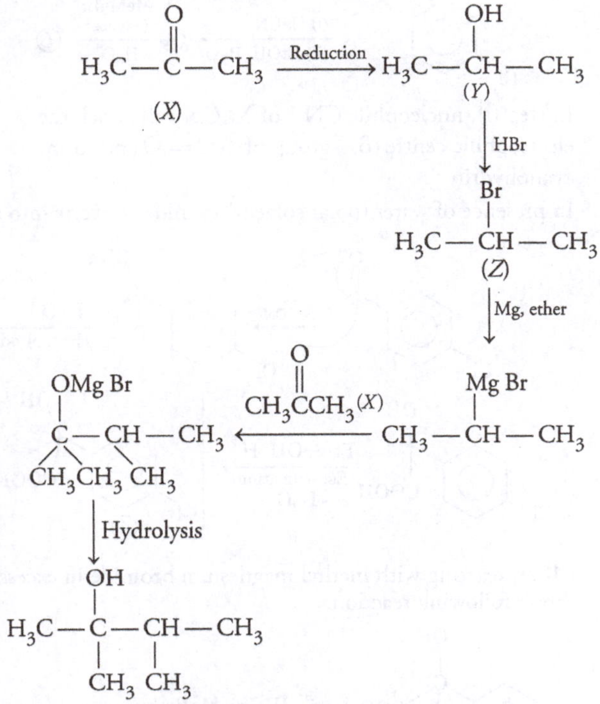

2,3 - dimethyl butan -2-ol

**42.** *(c)* The complete aldol condensation reaction is as follows,

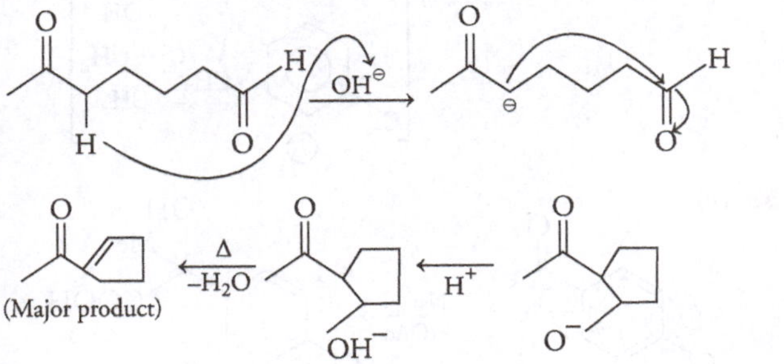

**43.** *(a)*

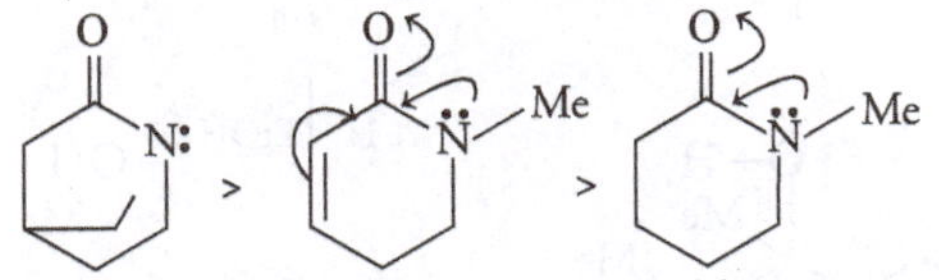

(R) (Q) (P)

It has delocalised lone pair according to Bredt's rule. Cross conjugation

Thus, the correct order of basicity is $R > Q > P$.

**44.** *(a)* The correct order of hydrolysis is

$$p > q > r > s$$

As we known that rate of hydrolysis is directly proportional to leaving group ability.

**45.** *(b)* Positive Fehling's test is given by aldehydes having alpha hydrogen.

Also, ketones generally do not react until they are α-hydroxy ketones.

So, among the given options, compounds given in C, D and E i.e.,

$HOCH_2—CO—(CHOH)_3—CH_2—OH$,

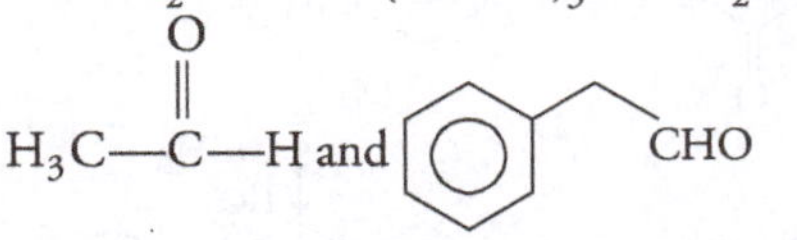

will give positive Fehling's test.

**46.** *(a)* The rate of nucleophilic addition reaction decreases due to steric crowding around carbonyl carbon and increases by electron withdrawing group if the steric crowding is same then the reactivity towards nucleophilic addition will be

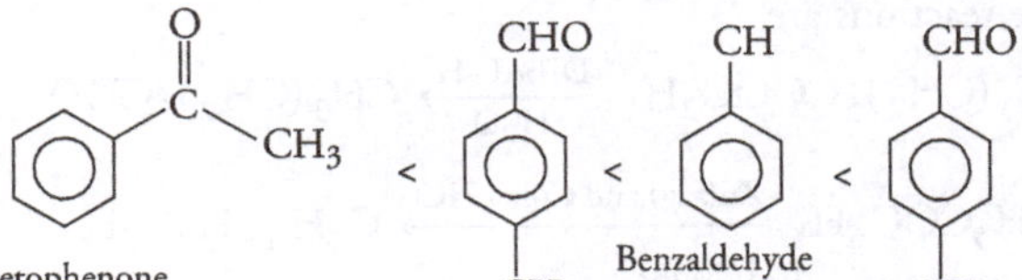

Acetophenone < *p*-tolualdehyde < Benzaldehyde < *p*-nitrobenzaldehyde

**47.** *(b)* Both acetaldehyde and acetone (individually) undergoes

(A) Iodoform reaction (C) Aldol condensation

(E) Clemmensen reduction

| Name of reaction | Acetaldehyde | Acetone |
|---|---|---|
| Iodoform | (+ ve) | (+ ve) |
| Cannizzaro | (− ve) | (− ve) |
| Aldol condensation | (+ ve) | (+ ve) |
| Tollen's test | (+ ve) | (− ve) |
| Clemmensen reduction | (+ ve) | (+ ve) |

## • Concept **Enhancer**

- Positive **Iodoform test** can be given by those compounds that have

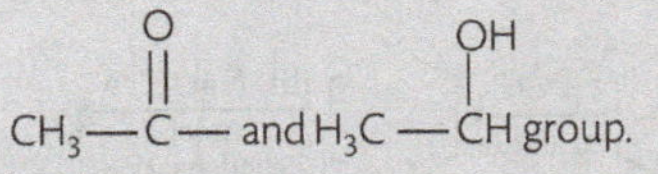

- **Cannizzaro reaction** Aldehydes that do not have an α-hydrogen undergoes self oxidation and reduction reaction of heating with concentrated alkali.
- **Aldol condensation** Aldehydes and ketones with at least one α-hydrogen undergo a reaction in the presence of dilute alkali as catalyst to form β-hydroxy aldehyde.
- **Tollen's test** It is a mild oxidising agent that converts aldehydes into corresponding carboxylate anion.

**Clemmensen reduction** In this reaction, carbonyl group of aldehydes and ketones is reduced to $CH_2$ group on treatment with zinc amalgam and concentrated hydrochloric acid

**48.** *(c)* The complete reaction sequence is as follows

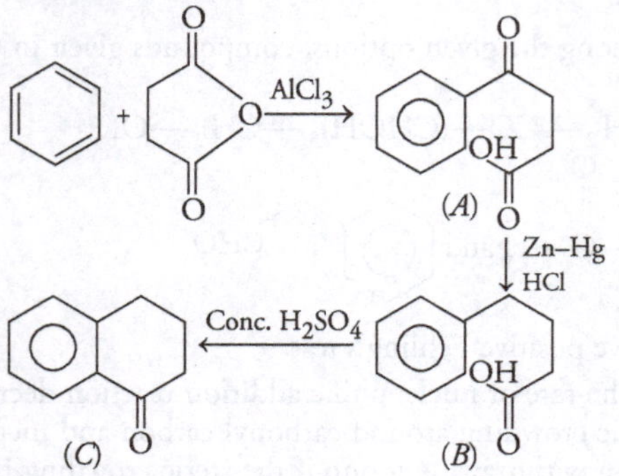

**49.** *(d)* The correct match is, A-IV, B-II, C-I, D-III.

The reactions are

$$CH_3(CH_2)_5COOC_2H_5 \xrightarrow[H_2O]{DIBAL\text{-}H} CH_3(CH_2)_5CHO$$

$$C_6H_5COC_6H_5 \xrightarrow{Zn(Hg)\text{ and Conc. HCl}} C_6H_5CH_2C_6H_5$$

$$C_6H_5CHO \xrightarrow[H_2O]{CH_3MgBr} C_6H_5CH(OH)CH_3$$

$$CH_3COCH_2COOC_2H_5 \xrightarrow{NaBH_4,\ H^+} CH_3CH(OH)CH_2COOC_2H_5$$

Hence, option (d) is correct.

**50.** *(d)* 3-methylcyclohex-2-enone reacts with Grignard's reagent in presence of copper iodide (CuI) to give '*A*'. '*A*' further reacts with *n*-propyliodide to give 3,3-dimethyl-2-propylcyclohexanone as final product.

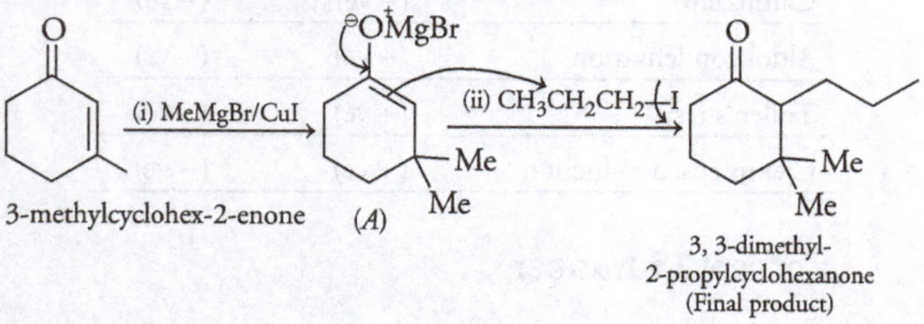

**51.** *(b)*

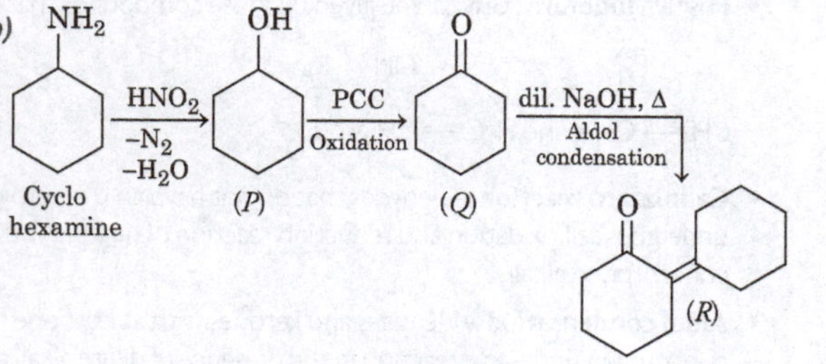

**52.** *(a)*

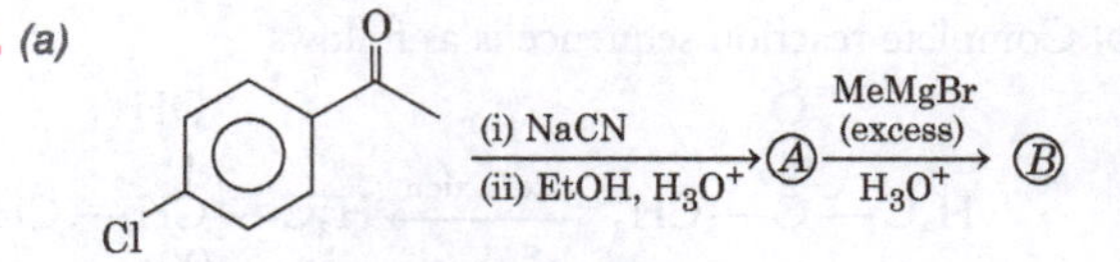

In step (i), nucleophile $CN^-$ of NaCN will attack the electrophilic centre ($\delta^+$) group of $>C=O$ and form cyanohydrin.

In presence of water (polar solvent) cyanide converts into acid.

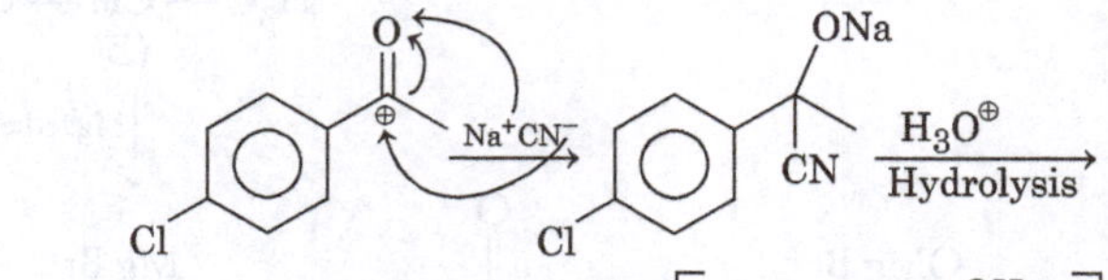

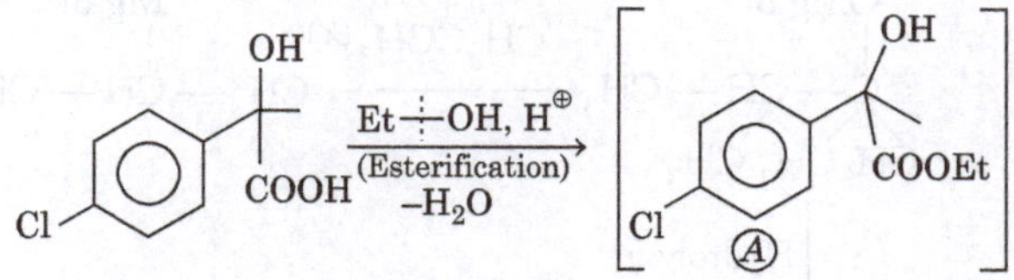

(*A*) on reacting with methyl magnesium bromide in excess shows following reaction.

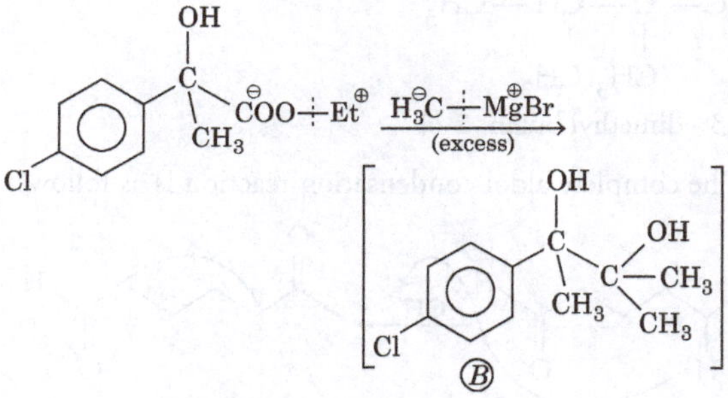

**53.** *(a)*

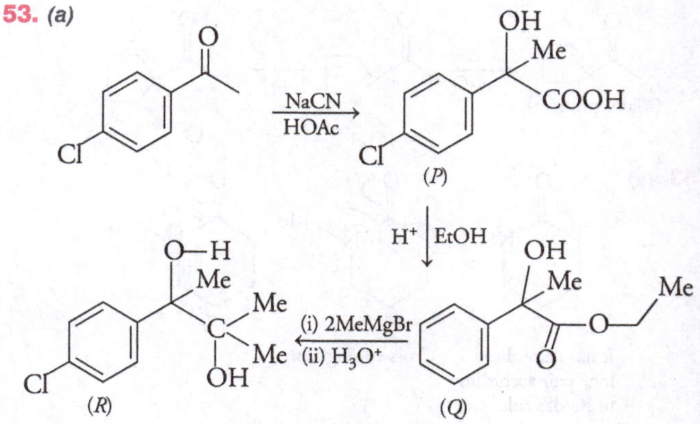

**54.** *(a)*

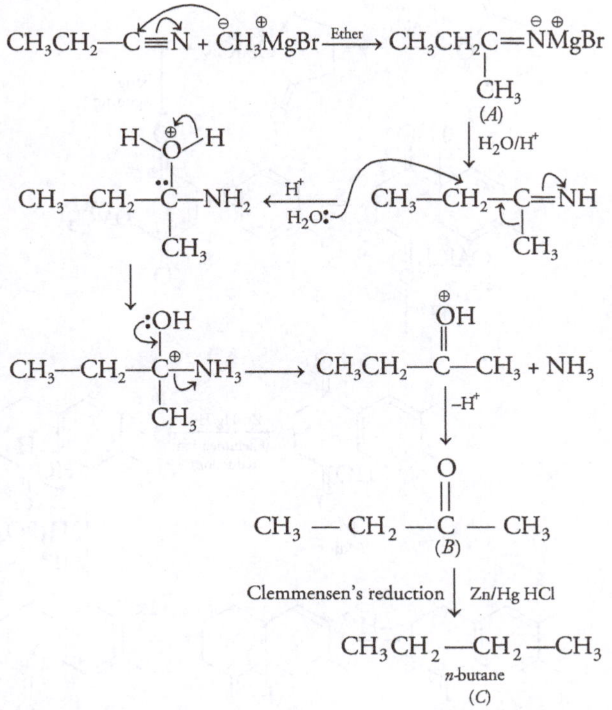

**55.** *(c)* The major product is given in option (c) which involves the following pathway.

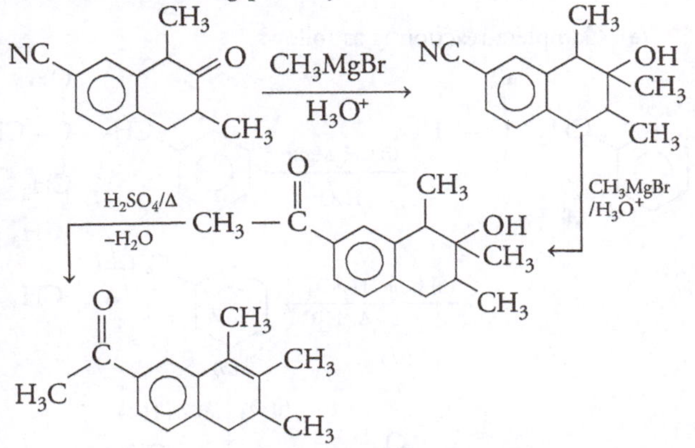

**56.** *(d)* The complete reaction take place as follows.

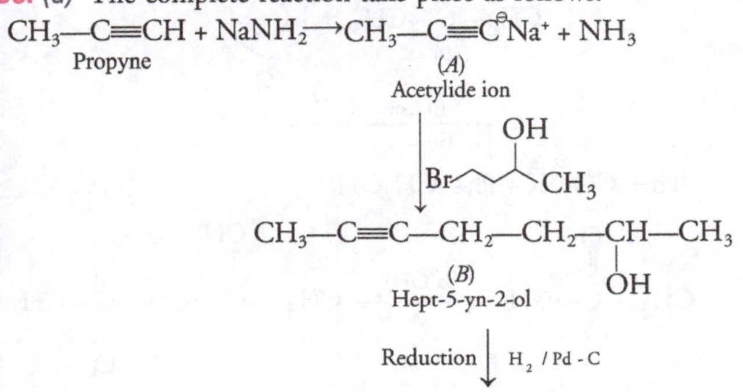

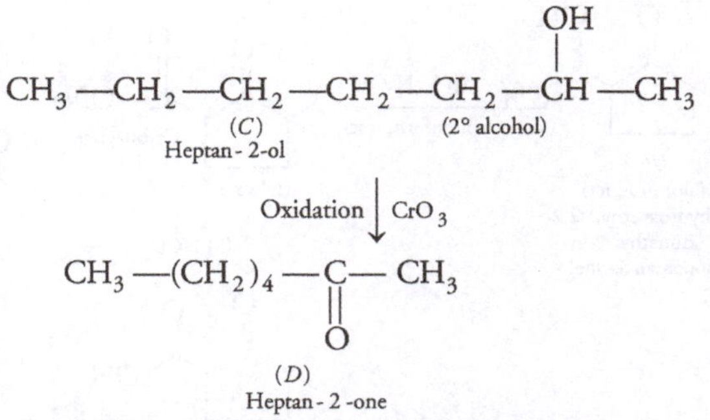

**57.** *(b)* In this reaction, substrate undergoes intramolecular aldol condensation in which condensation reaction of two aldehyde groups or ketone groups occur in same molecule.

**Mechanism**

**Step I** Abstraction of proton by base KOH to form enolate.

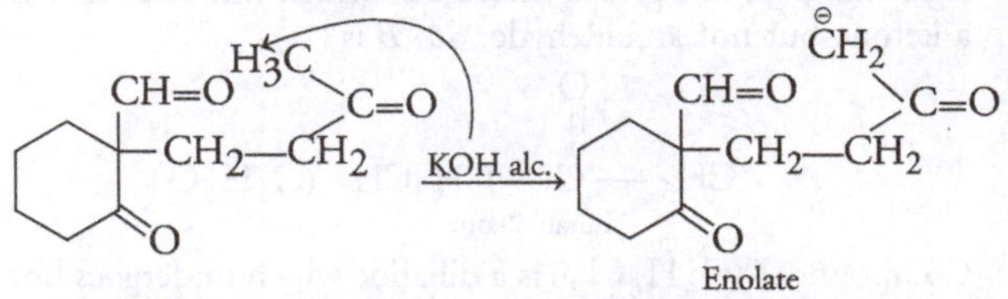

**Step II** Reaction of enolate and aldehyde group on substrate to form β-hydroxy ketone.

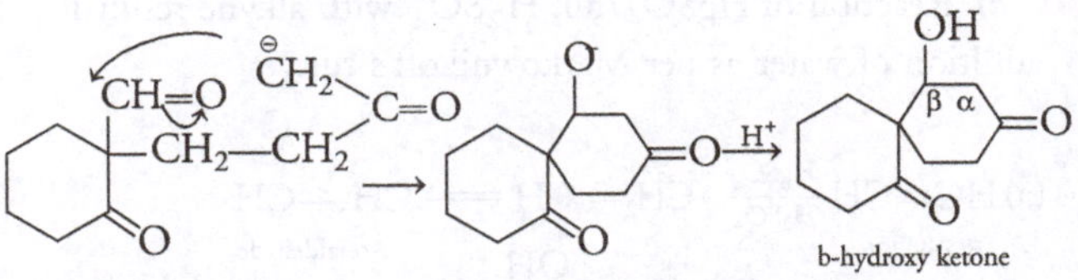

**Step III** β-hydroxy ketone undergoes dehydrate to form a conjugated enone.

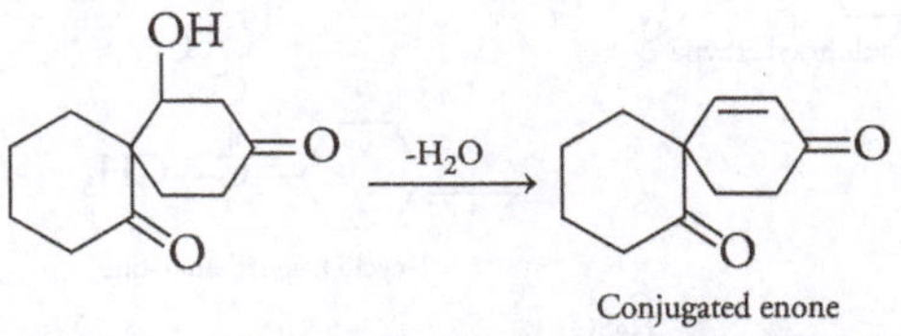

**58.** *(d)* In first step, aldol reaction is involved in reaction mechanism and in second step, ($I_2$, NaOH) is used for iodoform test, which gives $CHI_3$ ( yellow precipitate) as by product.

**Reaction Mechanism**

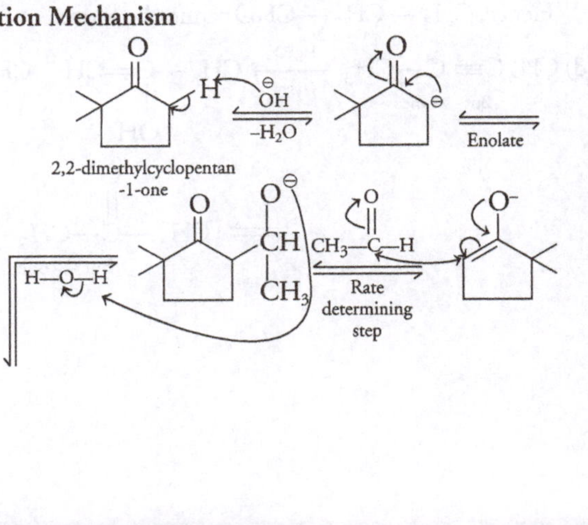

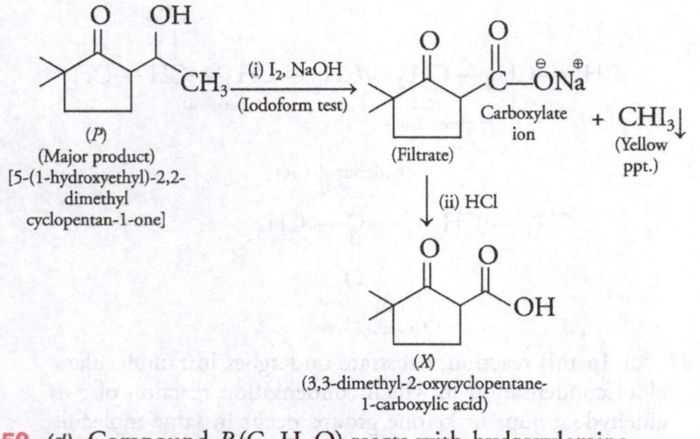


**59.** *(d)* Compound $B\,(C_4H_8O)$ reacts with hydroxylamine $(NH_2OH)$. So, compound $B$ is an aldehyde or a ketone. Again, $B$ does not give Tollen's test which indicates that $B$ is a ketone but not an aldehyde. So, $B$ is

$$CH_3-\overset{O}{\overset{\|}{C}}-CH_2CH_3\quad (C_4H_8O)$$

Butan-2-one

Compound $A\,(C_4H_8Cl_2)$ is a dihalide which undergoes hot hydrolysis ($H_2O$/ 373 K) to give $B$, a ketone. So, $A$ is a non-terminal *geminal* or *gem* dichloride and $B$ is 2-butan-2-one.

**60.** *(c)* Reaction of $HgSO_4$/dil. $H_2SO_4$ with alkyne result in addition of water as per Markownikoff's rule.

(a) $HC\equiv CH \xrightarrow[H_2SO_4]{HgSO_4} CH_2=\underset{OH}{\underset{|}{CH}} \rightleftharpoons CH_3-\overset{O}{\overset{\|}{C}}H$

Acetylene — Acetaldehyde

(b)

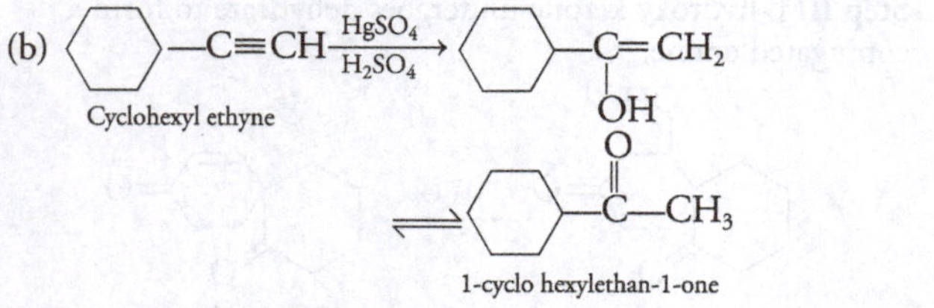


(c) $CH_3-C\equiv CH \xrightarrow[H_2SO_4]{HgSO_4} CH_3-\underset{OH}{\underset{|}{C}}=CH_2 \rightleftharpoons CH_3-\overset{O}{\overset{\|}{C}}-CH_3$

Prop-1-yne — Acetone

Hence, $CH_3-CH_2-CHO$ cannot be form.

(d) $CH_3C\equiv C-CH_3 \xrightarrow[H_2SO_4]{HgSO_4} CH_3-\underset{OH}{\underset{|}{C}}=CH-CH_3$

But - 2- yne

$\rightleftharpoons CH_3-\overset{O}{\overset{\|}{C}}-CH_2-CH_3$

But-2- one

**61.** *(d)*

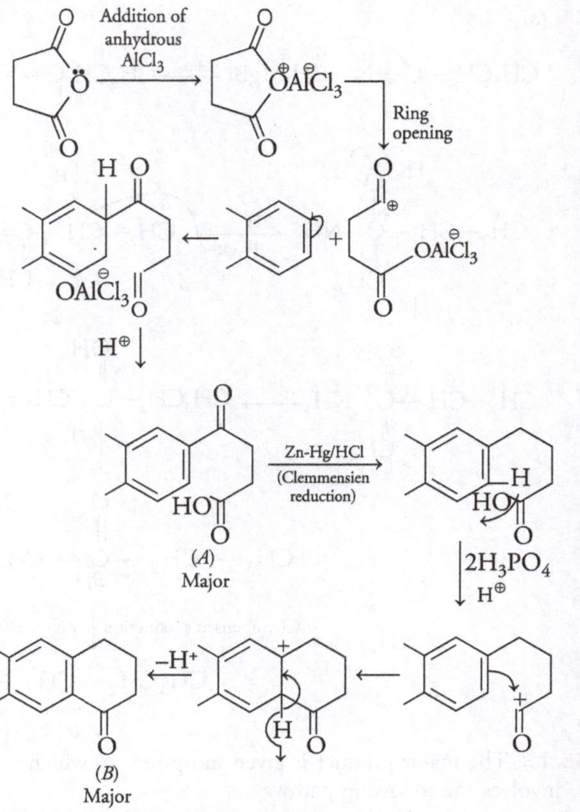


**62.** *(a)* Complete reaction is as follows

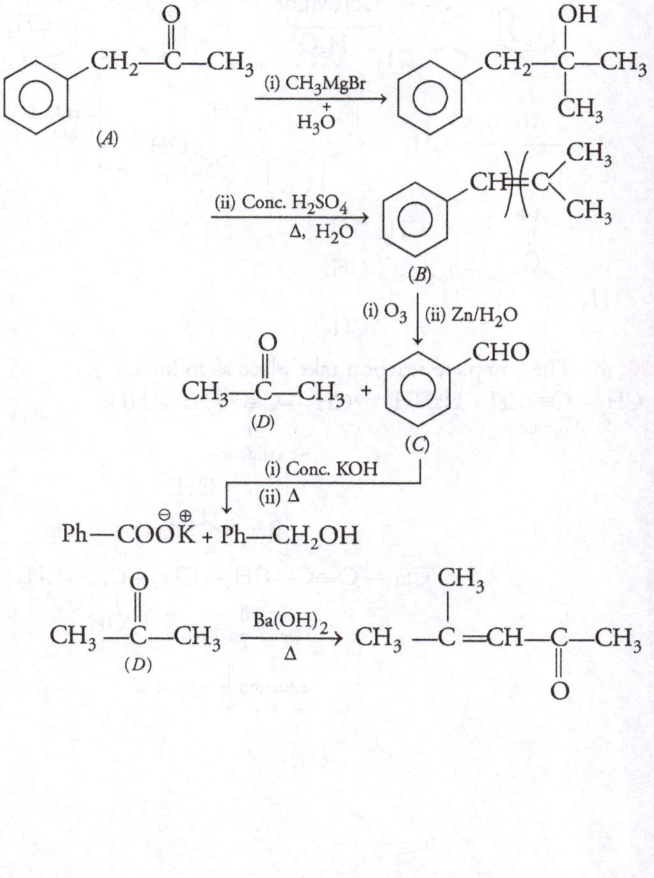

**63.** *(d)*

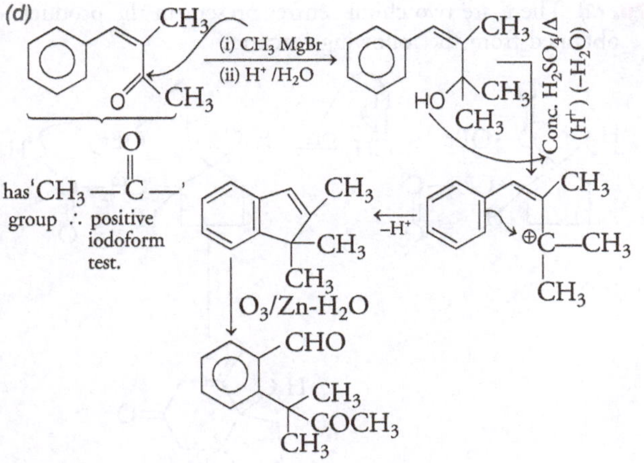

**64.** *(b)* NaOCl (sodium hypochlorite) is the reagent of haloform (chloroform formation) reaction.

The given reaction takes place as follows

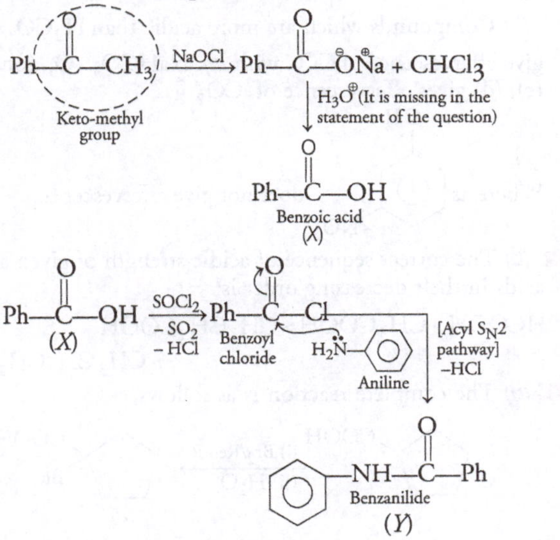

**65.** *(b)* The reactant in presence of dil. NaOH undergoes intramolecular aldol condensation reaction.

As a result of this, β-hydroxyketone (*A*) is obtained which on hydrolysis followed by heating produces α, β-unsaturated ketone (*B*).

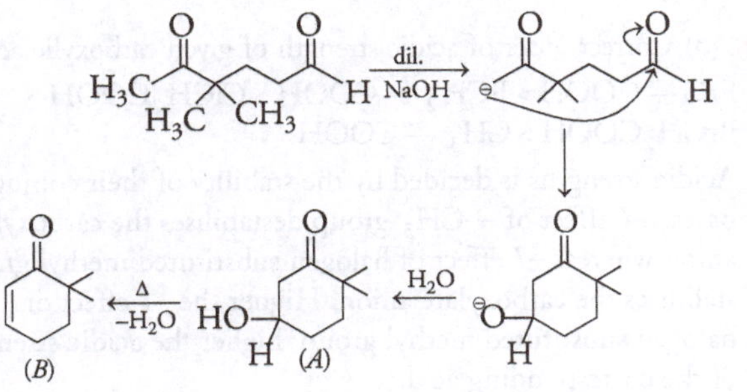

**66.** *(b)* In presence of strong base, acidic H can easily be removed that result in formation of anion. The resulting anion undergoes intramolecular nucleophilic addition which on hydrolysis followed by heating gives the required product.

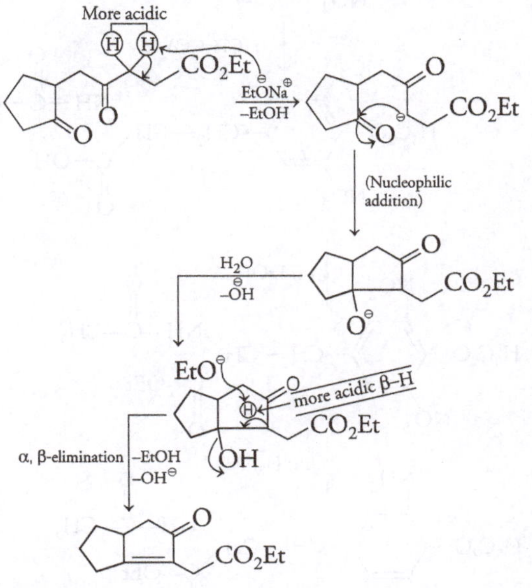

**67.** *(7)* The complete reaction is as follows,

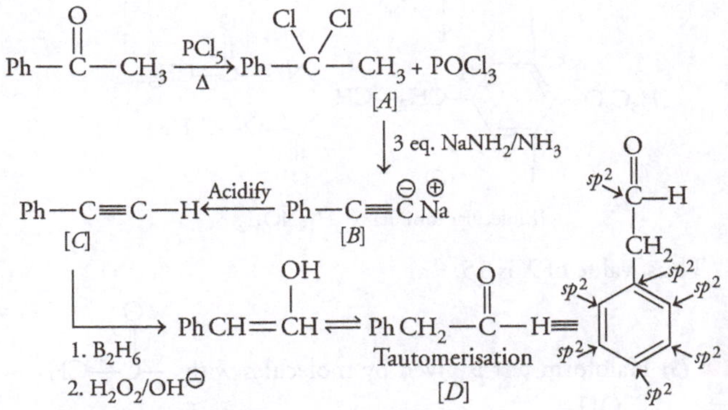

Thus, total number of $sp^2$-hybridised carbon atom in product *D* is 7.

**68.** *(15)*

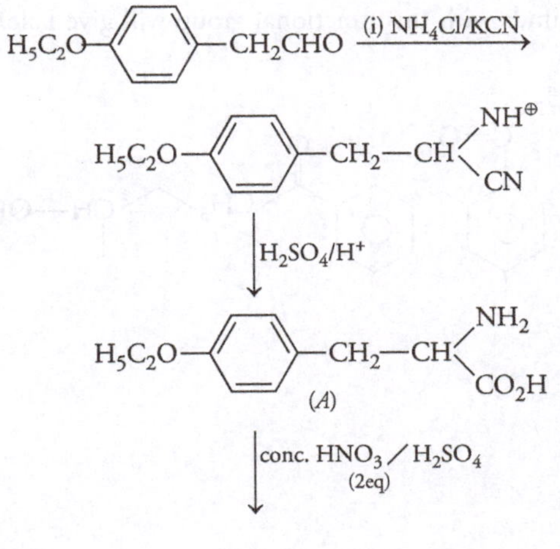

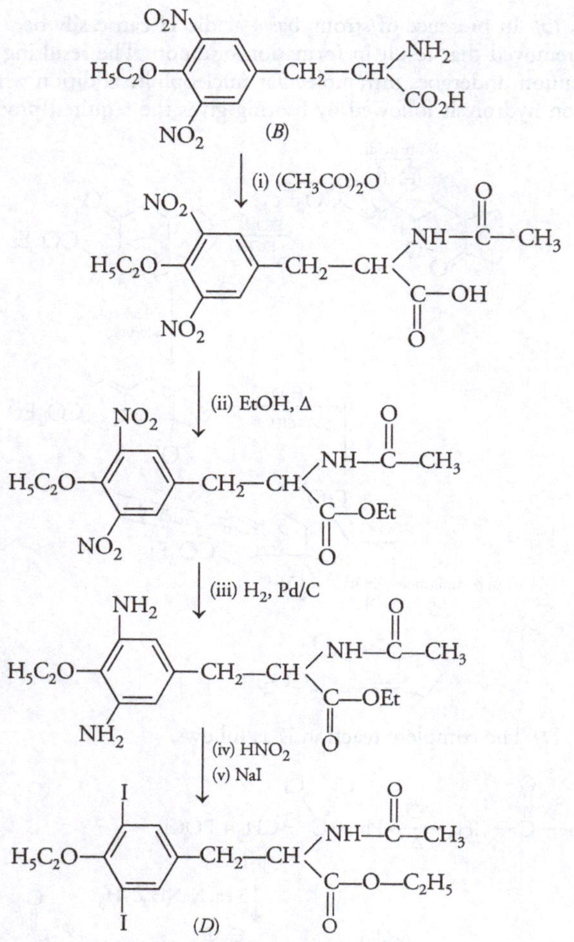

[Molecular formula $C_{15}H_{19}NO_4I_2$]

Thus, value of $X$ is 15.

**69.** *(3)* Haloform test is given by molecules with $-\overset{O}{\overset{\|}{C}}-CH_3$ and $-\overset{OH}{\overset{|}{C}H}-CH_3$ group present in them. Thus total three compounds with this functional group will give haloform test i.e.,

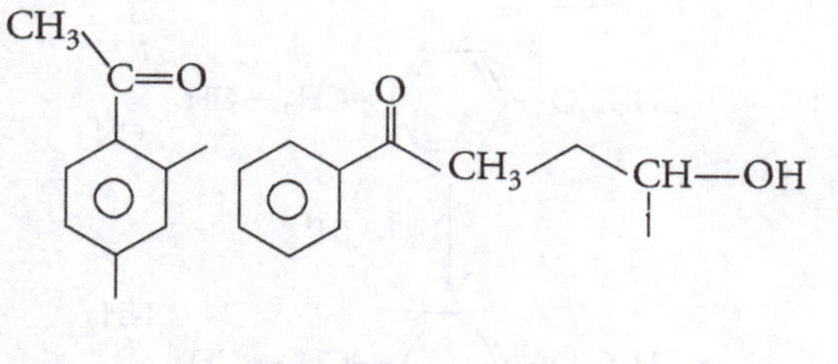

**70.** *(2)* There are two chiral centres present in the product obtained from the following reaction.

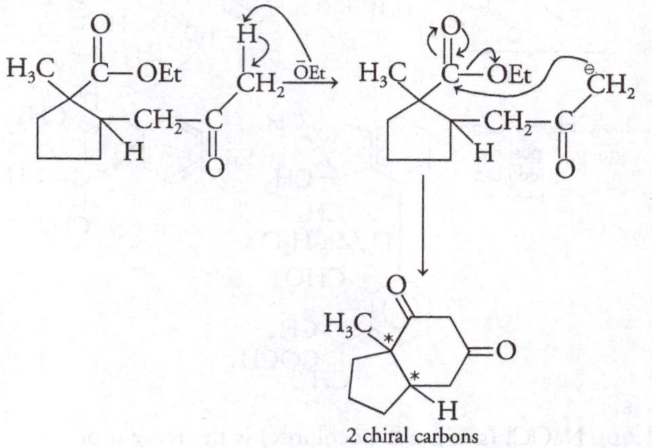

2 chiral carbons

## Topic 3

**01.** *(b)* Compounds which are more acidic than $H_2CO_3$ can give effervescence of $CO_2$ with *aq.* $NaHCO_3$. Options (a), (c), (d) gives effervescence of $CO_2$ gas.

Where as 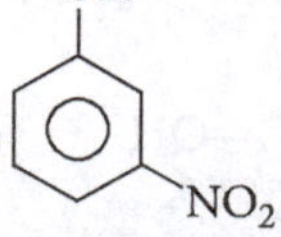 does not give effervescence.

**02.** *(b)* The correct sequence of acidic strength of given aliphatic acids in their decreasing order is;

$$HCOOH > CH_3COOH > CH_3CH_2COOH > CH_3CH_2CH_2COOH$$

**03.** *(d)* The complete reaction is as follows,

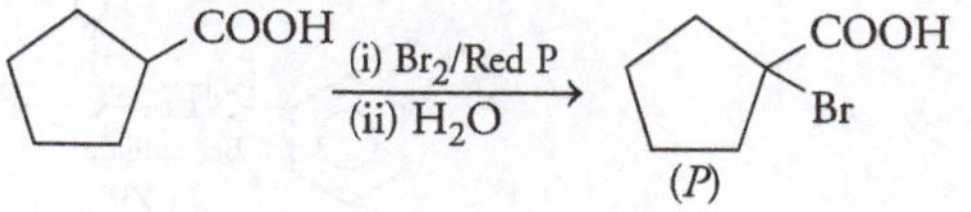

This is HVZ reaction.

**04.** *(d)* In the given reaction, product is formed by oxidative ozonolysis.

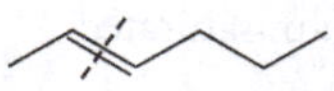 $\xrightarrow[\text{(ii) } H_2O]{\text{(i) } O_3}$ $CH_3COOH$ + $CH_3CH_2CH_2COOH$
Acetic acid  Butanoic acid

**05.** *(b)* Correct order of acidic strength of given carboxylic acid is:

$F_3C-COOH > FCH_2-COOH > ClCH_2COOH > BrCH_2COOH > CH_3-COOH$

Acidic strengths is decided by the stability of their conjugate bases. $+I$ effect of $-CH_3$ group destabilises the carboxylate anion whereas $-I$ effect of halogen substituted methyl group stabilises the carboxylate anion. Higher the $-I$ effect of halogen substituted methyl group, higher the acidic strength of the corresponding acid.

**06.** *(c)* IUPAC name of given compound is 2-methyl-5-oxohexanoic acid.

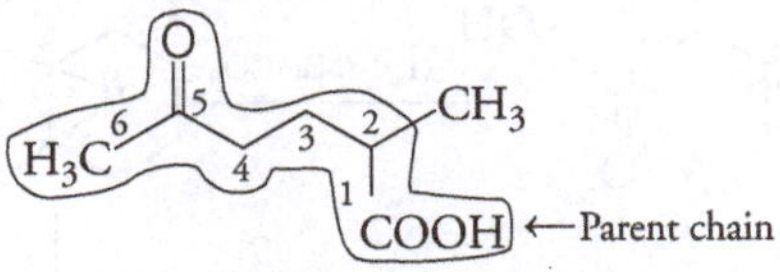


**07.** *(c)* Both A and R are true and R is the correct explanation of the A.

> **• Concept Enhancer**
>
> Clemmensen's reduction is most commonly used to convert acyl benzene to alkyl benzene in the presence of zinc amalgam and hydrochloric acid.
>
> 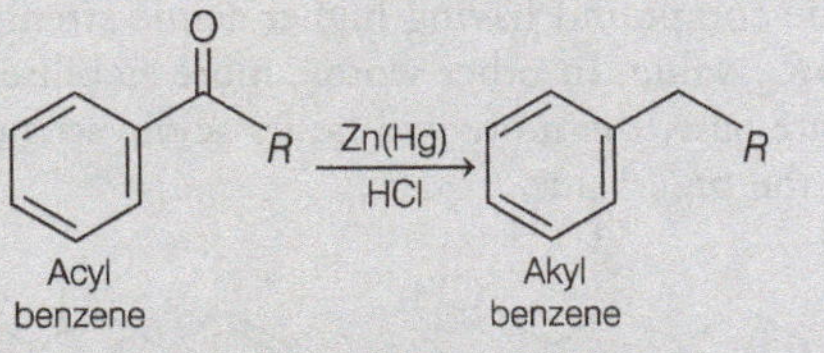
> 
>
> It involves the formation of saturated hydrocarbons from carbonyl compounds by the removal of oxygen.

**08.** *(a)* The esterification of carboxylic acid with an alcohol is a nucleophilic acyl substitution reaction.

Electron withdrawing groups in the carboxylic acid will increase the rate of esterification reaction by making carbonyl carbon of acyl group more electron deficient, thereby increasing the rate of nucleophilic attack by alcohol.

Hence, both the statements are correct.

**09.** *(a)* Benzoic acid in presence of red phosphorus undergoes halogenation at α-position. This reaction is known as Hell-volhard Zelinsky reaction.

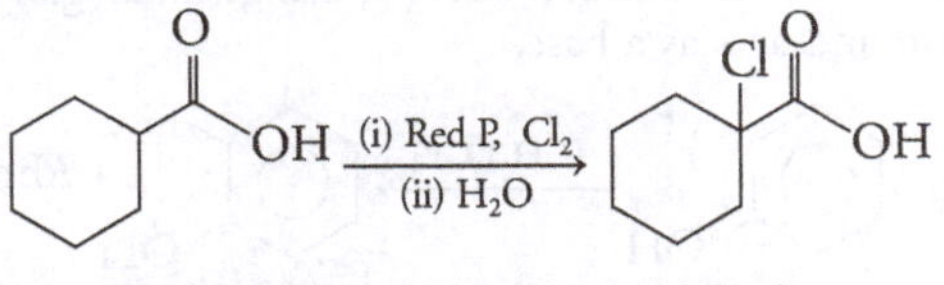


**10.** *(c)* Benzoic acid is converted to benzaldehyde in one step by the use of reagent MnO.

$$C_6H_5-\overset{O}{\overset{\|}{C}}-OH + HO-\overset{O}{\overset{\|}{C}}-H \xrightarrow[\Delta]{MnO} C_6H_5-\overset{O}{\overset{\|}{C}}-H + CO_2 + H_2O$$

**11.** *(a)* Both Statements I and II are correct.

**12.** *(a)* Sodium hypochlorite (NaOCl) is a strong oxidising agent that will convert ketone to carboxylic acid.

Since, product is carboxylic acid, therefore reactant (*P*) would be ketone.

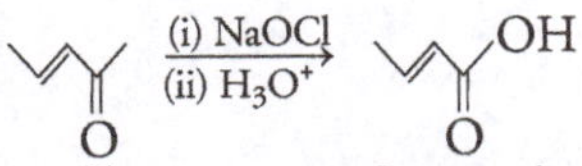


(*P*) Pent-3-en-2-one — But-2-enoic acid

**13.** *(b)* Rate of hydrolysis is directly proportional to positive charge present on carbon of $>C=O$ group.

Rate of hydrolysis : Acid chloride > Acid anhydride > ester > amide.

Hence, the order is $A > B > C > D$.

**14.** *(c)* Greater is stability of conjugate base more will be the acidic strength.

General order of acidic strength is

$R-COOH > Ph-OH > R-C\equiv CH$

In between (iii) and (iv), (iii) is more acidic due to -M effect of $-NO_2$.

Thus, decreasing order of acidity is (ii) > (iii) > (iv) > (i).

**15.** *(c)* The IUPAC name of the given compound is

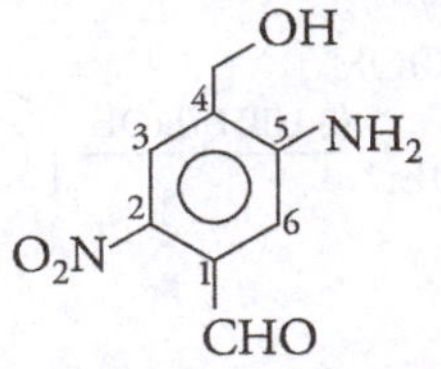


5-amino-4-hydroxymethyl-2-nitrobenzaldehyde.

**16.** *(c)* All the given compounds are α-monosubstituted acetic acid derivatives and the α-substitutions have been made by strong–*I* groups/atoms. More powerful the – *I* group, stronger will be the acid. –*I* power of different groups is as follows

$$-NO_2 > -CN > -F > -Cl.$$

Thus, the correct decreasing order for acid strength is:

$NO_2CH_2COOH > NCCH_2COOH > FCH_2COOH > ClCH_2COOH$.

**17.** *(a)* Among the given compounds $EtO_2C-\equiv-H$ is least acidic. It is because conjugate base of terminal alkyne will be *sp*-hybridised and localised, while in other compounds conjugate base will be resonance stabilised.

**18.** *(b)*

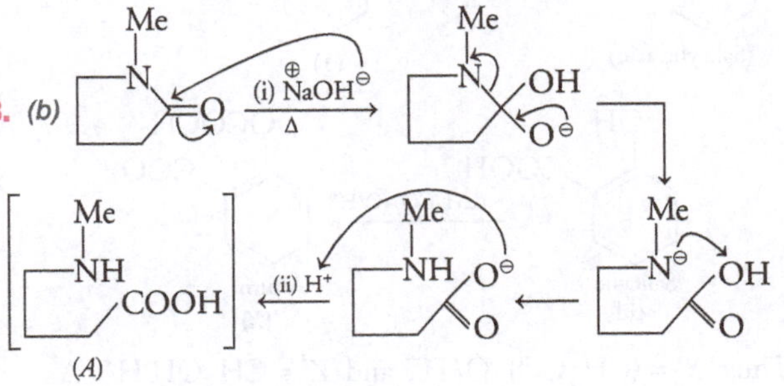


**19.** *(b)*

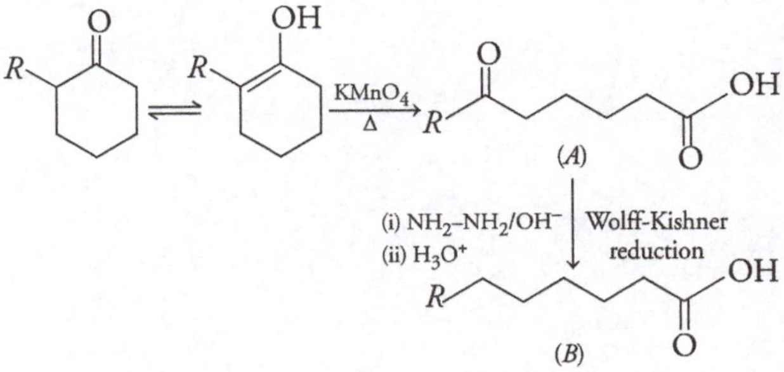

**20.** *(a)*

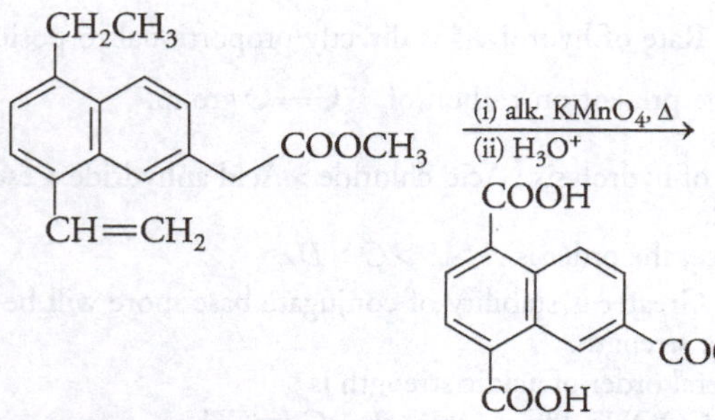

In the above reaction $KMnO_4$ oxidises benzylic carbon containing at least one α-hydrogen atom to —COOH.

**21.** *(a)* In first step, $LiBH_4$ can reduce ester selectively but not carboxylic acid. In second step, protonation of acid occur to give the final product.

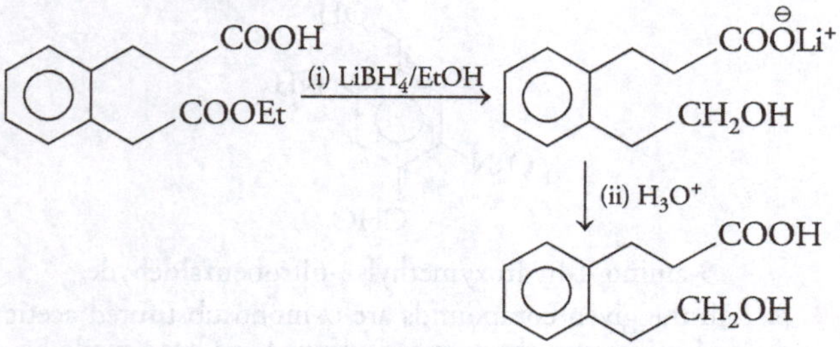

**22.** *(b)* The correct match is A-III, B-I, C-II, D-IV.

A. Hell-Volhard-Zelinsky reaction (III) reagent-$Br_2$/Red P
B. Iodoform reaction (I) $NaOH + I_2$
C. Etard reaction (II) $CrO_2Cl_2$, $CS_2$ and $H_2O$
D. Gattermann-Koch reaction (IV) CO, HCl, Anh. $AlCl_3$

**23.** *(c)* Correct option is (c).

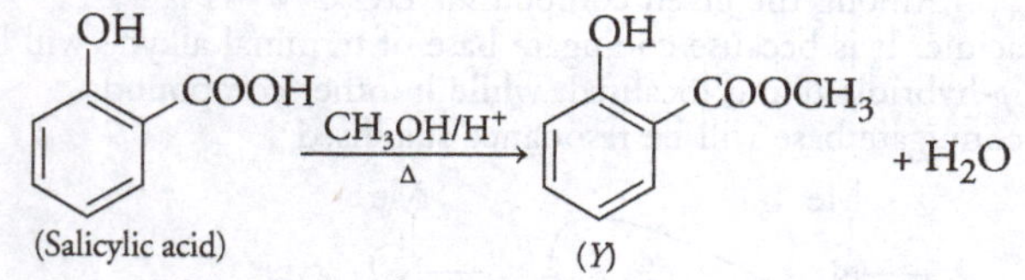

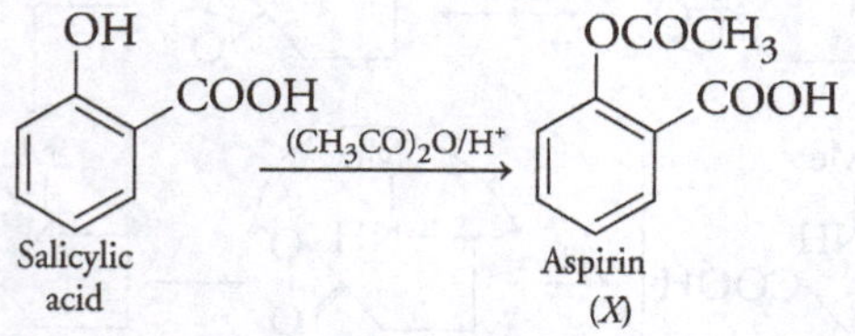

Thus '*X*' = $(CH_3CO)_2O/H^+$ and '*Y*' = $CH_3OH/H^+, \Delta$

**24.** *(c)* The given compound in presence of iodine sodium bicarbonate and pyridine undergoes the following reactions

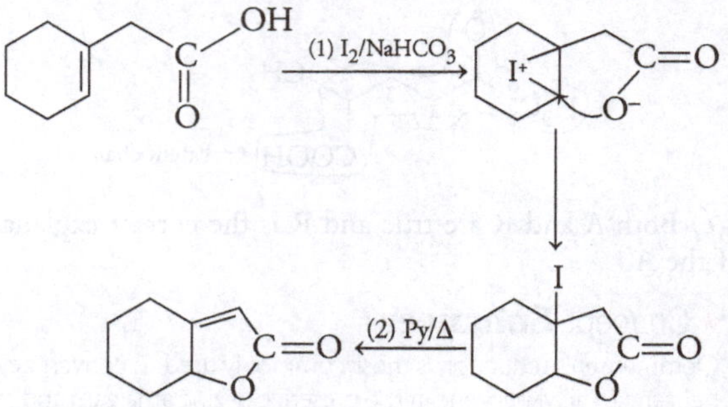

**25.** *(c)* The compound having higher acidic strength will have lower $pK_a$ value. In other words, more stabilised the conjugate base, the more will be its acidic strength and lower will be the $pK_a$ value.

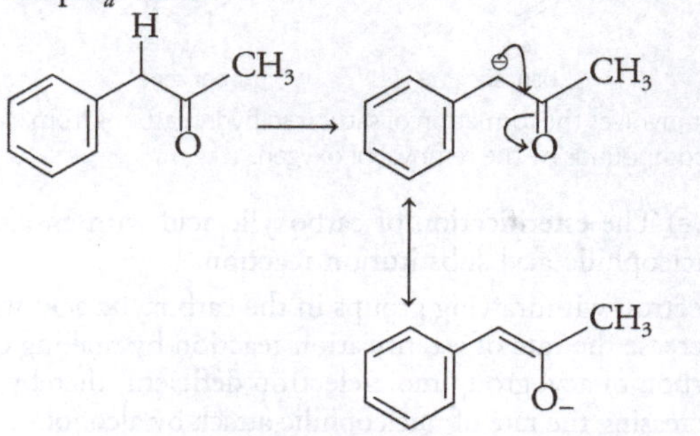

Extended conjugation

The conjugate base of (c) is stabilised by extended conjugation. Hence, the indicated proton in (c) has highest acidic strength and therefore lowest $pK_a$ value.

**26.** *(b)* In the given reaction $(CH_3)_3CLi$ give halogen exchange reaction and acts as a base.

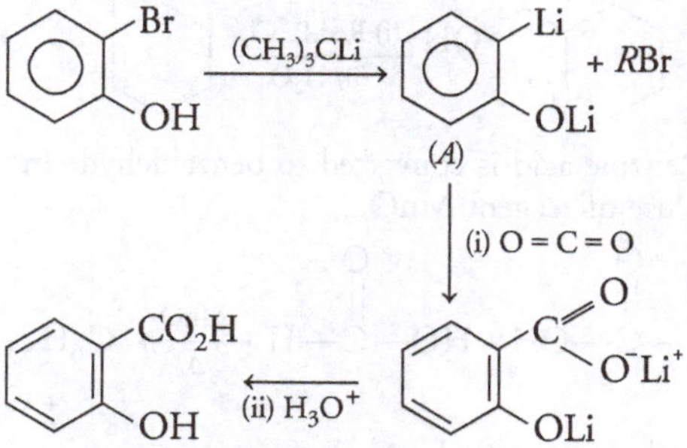

**27.** *(c)* In the given reaction the starting compound (*A*) is *p*-bromotoluene.

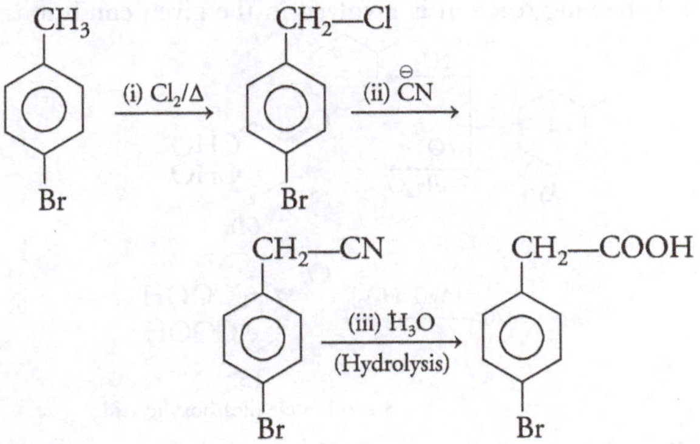

**28.** *(a)*

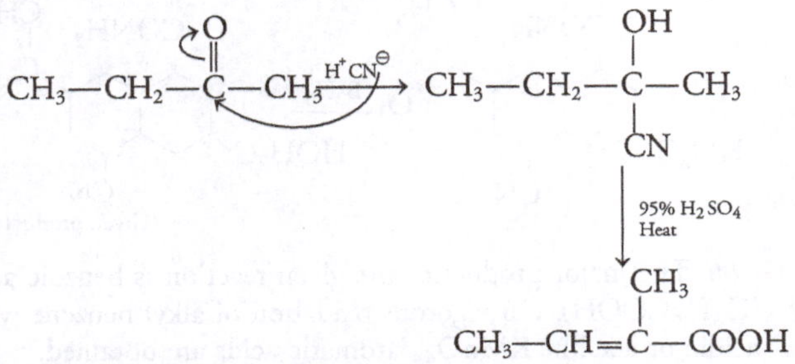

**29.** *(c)* The complete reaction is as follows

$CH_3-CH_2-CH_2-\underset{\underset{O}{\|}}{C}-O-CH_2-CH_3$

Ethylbutanoate (A)

$\downarrow LiAlH_4, H_3O^+$

$CH_3-CH_2-CH_2-CH_2-OH + CH_3-CH_2-OH$

Butanol (*B*) Ethanol

$\downarrow$ Oxidation

$CH_3-CH_2-CH_2-\underset{\underset{O}{\|}}{C}-OH$

**30.** *(a)* Bromopropane with Mg in presence of $CO_2$ and dry ether form butanoic acid as follows :

$CH_3CH_2CH_2Br \xrightarrow[\text{dry ether}]{Mg} CH_3CH_2CH_2MgBr$

$\downarrow CO_2$

$CH_3-CH_2CH_2-\underset{\underset{O}{\|}}{C}-OMgBr$

$\downarrow H_3O^+$

$CH_3CH_2CH_2COOH + Mg\begin{matrix} Br \\ OH \end{matrix}$

Butanoic acid

**31.** *(c)* Statement I is true but Statement II is false.

In this reaction one mole of Ethyl pent-4-yn-oate utilizes three moles of $CH_3MgBr$ as follows :

$HC\equiv C-CH_2-CH_2-\overset{\overset{O}{\|}}{C}-OC_2H_5 \xrightarrow[-C_2H_5OMgBr]{CH_3MgBr\ (3\ mole)}$

$Br\overset{\oplus}{Mg}\overset{\ominus}{C}\equiv C-CH_2-CH_2-\overset{\overset{\ominus}{O}MgBr}{\underset{CH_3}{\overset{|}{\underset{|}{C}}}}-CH_3$

$\downarrow HOH$

$HC\equiv C-CH_2-CH_2-\overset{OH}{\underset{CH_3}{\overset{|}{\underset{|}{C}}}}-CH_3$

**32.** *(c)* Esterification of carboxylic group (—COOH) take place in presence of $SOCl_2$ and $CH_3OH$ as follows.

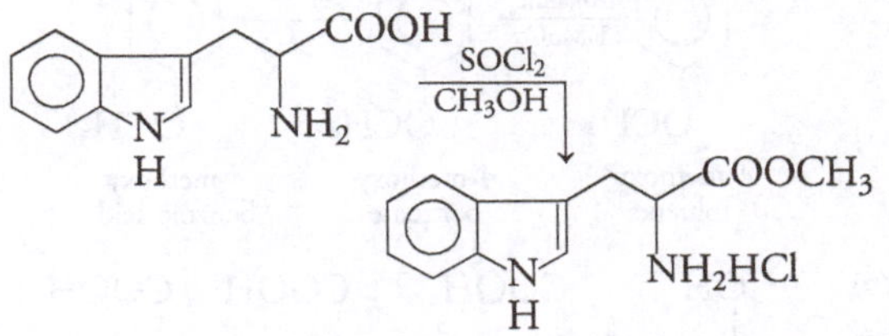

**33.** *(b)* Maleic anhydride can be prepared by heating *cis*-but-2-enedioic acid.

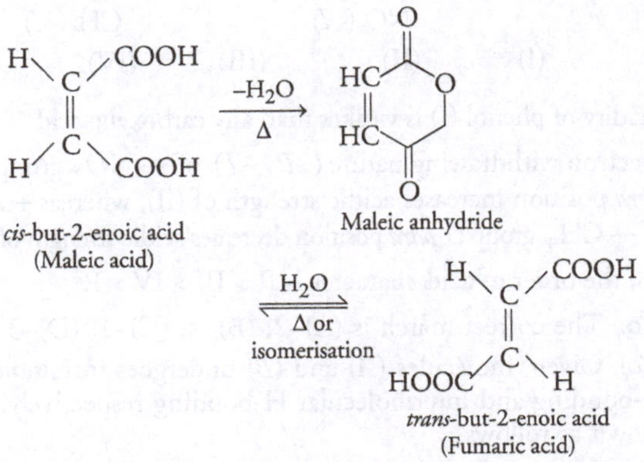

**34.** *(c)* 4-methoxy toluene in the presence of alkaline $KMnO_4$ to give 4-methoxy benzoic acid.

Chemical reaction is as follows :

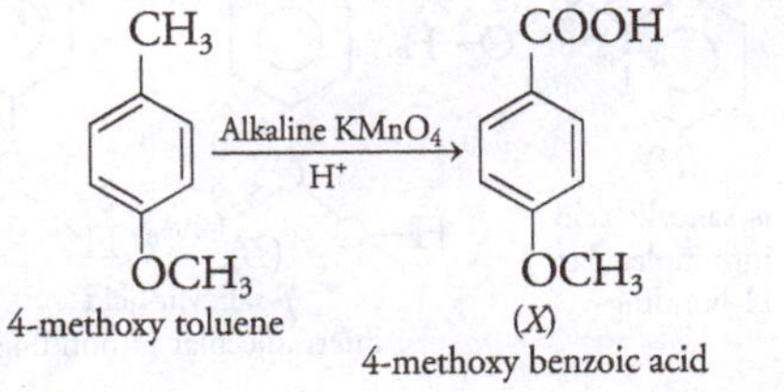

**35.** *(c)* 1-cyano cyclo hexane on complete hydrolysis gives 1-cyclo hexanoic acid. Intermediate 1- methanamide cyclohexane (*A*) is formed by partial hydrolysis.
Complete reaction is as follows

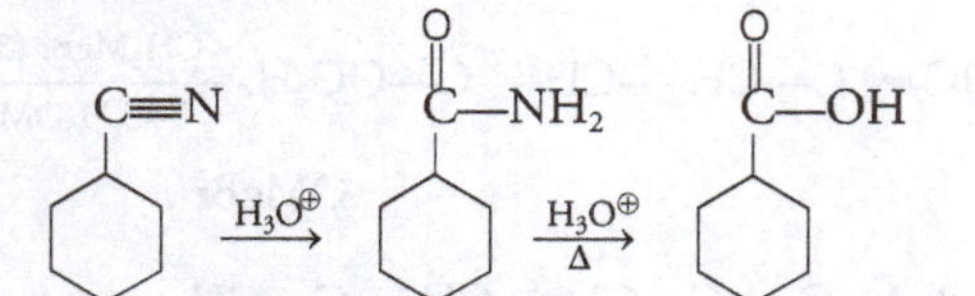

1-cyano cyclohexane — 1-methanamide cyclohexane 'A' Major product (Partial hydrolysis) — 1-cyclohexanoic acid (Complete hydrolysis)

**36.** *(c)* In the given reaction, *A* is alkaline $KMnO_4$, $H^+$. Alkaline $KMnO_4$ / $H^+$ is a strong oxidising agent and oxidises alkyl benzene to benzoic acid. While $NaBH_4$, $LiAlH_4$ and Zn-Hg/HCl are reducing agent.

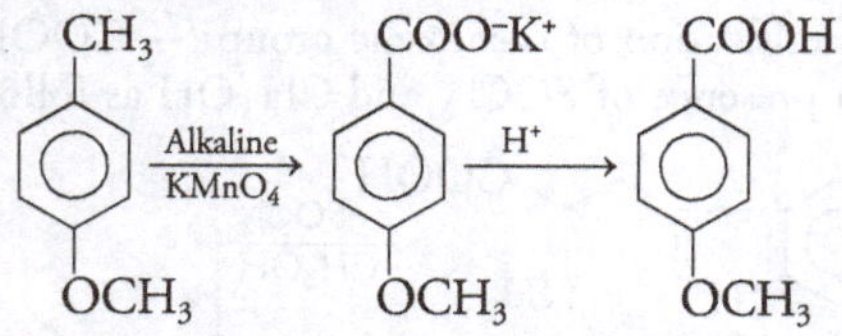

4-methoxy toluene — 4-methoxy benzoate — 4-methoxy benzoic acid

**37.** *(c)*

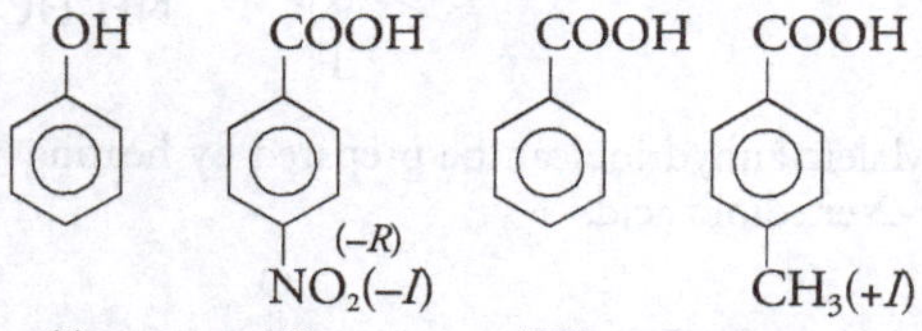

(I) (II) (III) (IV)

Acidity of phenol (I) is weaker than any carboxylic acid. Electron withdrawing nature (−*R*, −*I*), of —$NO_2$ group at *para* position increases acidic strength of (II), whereas + *I* effect of —$CH_3$ group at *para* position decreases acidic strength of (IV).
So, the order of acid character is II > III > IV > I.

**38.** *(c)* The correct match is (A)–2, (B)–4, (C)–1, (D)–3.

**39.** *(c)* Given, molecules (*A*) and (*B*) undergoes intramolecular H-bonding and intermolecular H-bonding respectively. It is shown as follows :

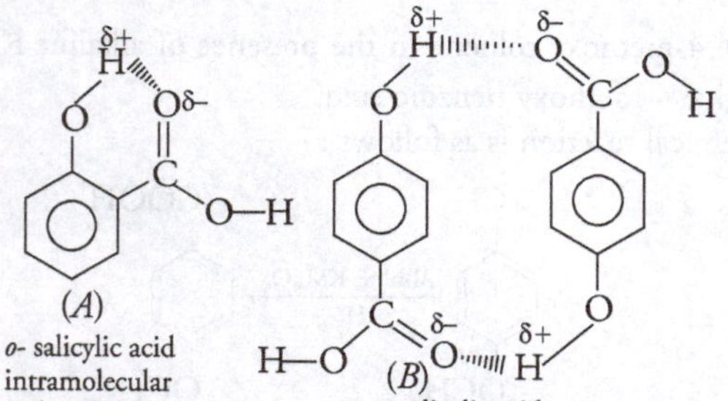

(*A*) *o*- salicylic acid intramolecular H-bonding

(*B*) *p*-salicylic acid intermolecular H-bonding

Among the given statements, II and III are true whereas I is false. Corrected statement I, (*A*) is more likely to crystalline than (*B*).

**40.** *(c)* Following reaction is involved in the given conditions

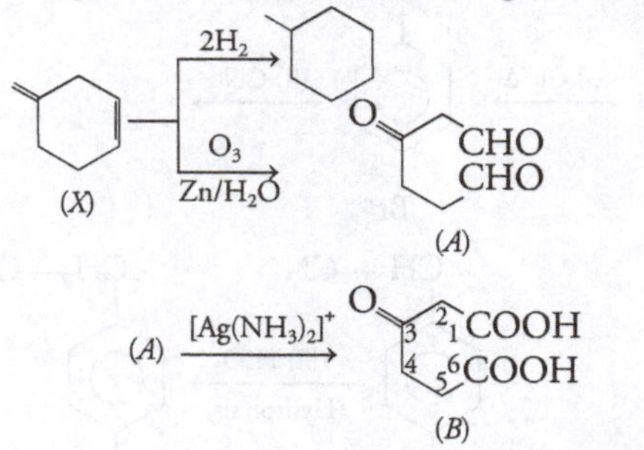

3-oxo-hexane dicarboxylic acid

**41.** *(a)* The most suitable reagent for the given conversion is $B_2H_6$.

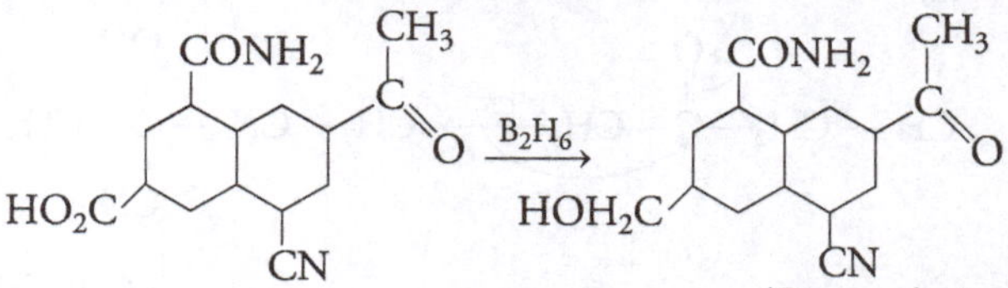

(Given product)

**42.** *(a)* The major product of the given reaction is benzoic acid ($C_6H_5COOH$). On vigorous oxidation of alkyl benzene with acidic or alkaline $KMnO_4$, aromatic acids are obtained. During oxidation of alkyl benzene, the aromatic nucleus remains intact and the entire chain is oxidised to —COOH group irrespective of the length of carbon chain.

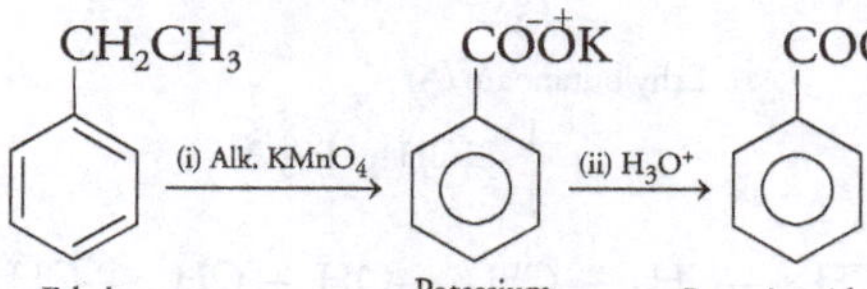

Ethyl benzene — Potassium benzoate — Benzoic acid

**43.** *(b)* Alkaline hydrolysis of an ester (carboxylic acid derivative) follows acyl $S_N2$ mechanism.
Rate of $S_N2$ mechanism depends on the polarity of group of —COO*R* group. Electron withdrawing group (−*R* > − *I*) increases the rate of $S_N2$ reaction whereas electron donating group (+ *R* > + *I*) decreases the rate of $S_N2$ reaction.
Here, the nature of functional groups attached *para* to the benzene ring are —$NO_2$ (−*R*) > —Cl (−*I*) > — $OCH_3$ (+*R*)

So, the order of hydrolysis will be, III (−*R*) > II (−*I*) > I > IV (+*R*)

**44.** *(b)* Heating of $(CH_2)_4 \langle {COOH \atop COOH}$ (adipic acid) with a dehydrating agent, decarboxylates (—$CO_2$) to give a ketone (cyclopentanone), not an anhydride.

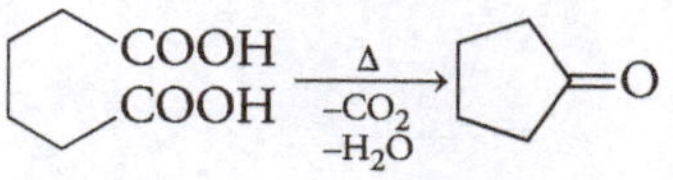

**45.** *(a)* The salicylic acid with acetic anhydride [$(CH_3CO)_2O$] in the presence of catalytic amount of conc. $H_2SO_4$ undergoes acylation to produce aspirin as

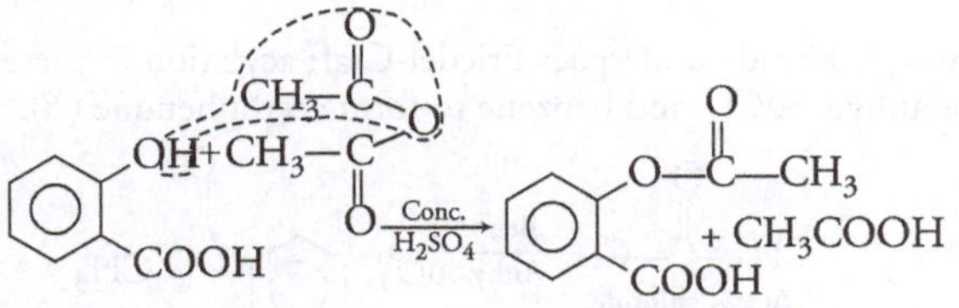


Acetyl salicylic acid (Aspirin)

**46.** *(a)* Reaction A and E gives carboxylic acid as a major product. The complete reaction are as follows.

(A) $R—C \equiv N \xrightarrow[\text{Mild condition}]{\text{(i) } H^+/H_2O} R—CONH_2$

Under mild condition amide is formed because this reaction is typically slow if further more heat will supplied then it converted into COOH.

(B) $R—MgX \xrightarrow[\text{(ii) } H_3O^+]{\text{(i) } CO_2} R—COOH$

(C) $R—C \equiv N \xrightarrow[\text{(ii) } H_3O^+]{\text{(i) } SnCl_2/HCl} R—CHO$

(D) $RCH_2OH \xrightarrow{PCC} R—CHO$

(E)

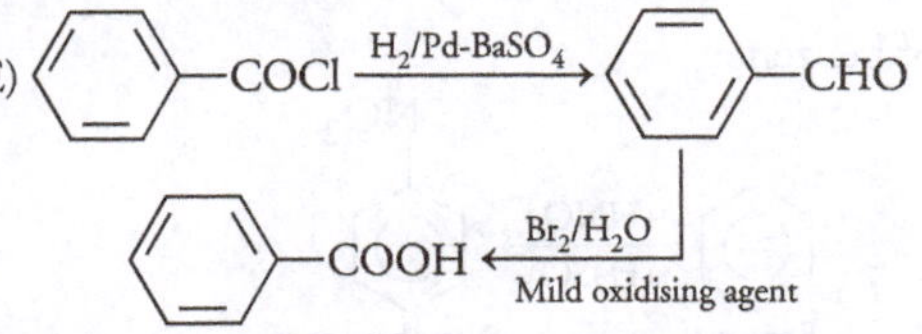


**47.** *(a)* Compounds which have at least one α-hydrogen will react with $KMnO_4$ and give benzoic acid.

There are five compounds with at least one α-hydrogen, these are

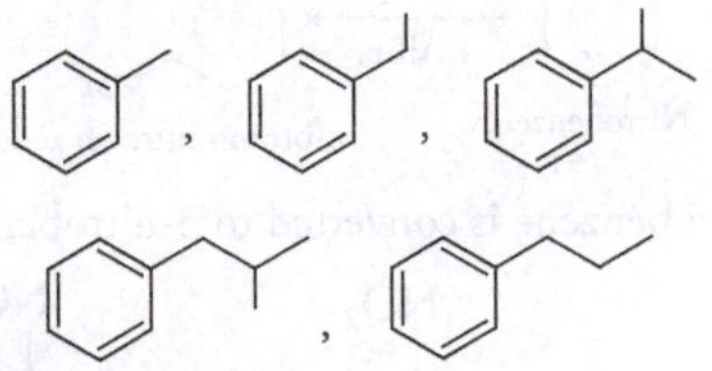

**48.** *(c)* Among the given options,

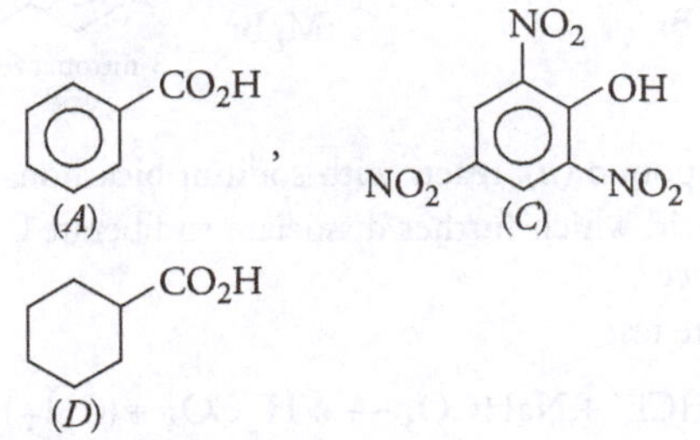


produces $CO_2$ with aqueous $NaHCO_3$ solution.

**49.** *(c)* The complete reaction is as follows:

$$CH_3COOH \xrightarrow{LiAlH_4} CH_3—CH_2OH \xrightarrow{PCC} CH_3—\overset{O}{\overset{\|}{C}}—H$$

$$\xrightarrow{HCN/OH^-} CH_3—\underset{H}{\overset{OH}{C}}—CN \xrightarrow{H_2O/\bar{O}H,\ \Delta} CH_3—\underset{OH}{CH}—COOH$$

Hence, final product is $CH_3—\underset{OH}{CH}—COOH$

**50.** *(b)* The complete reaction is as follows,

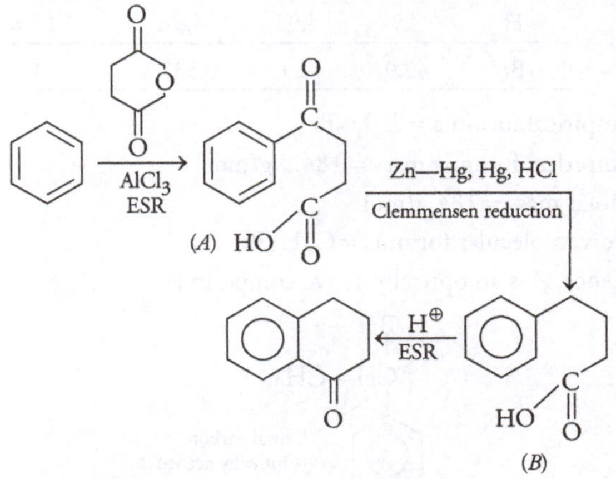


Hence, option (b) is correct.

**51.** *(c)* The complete reaction is as follow,

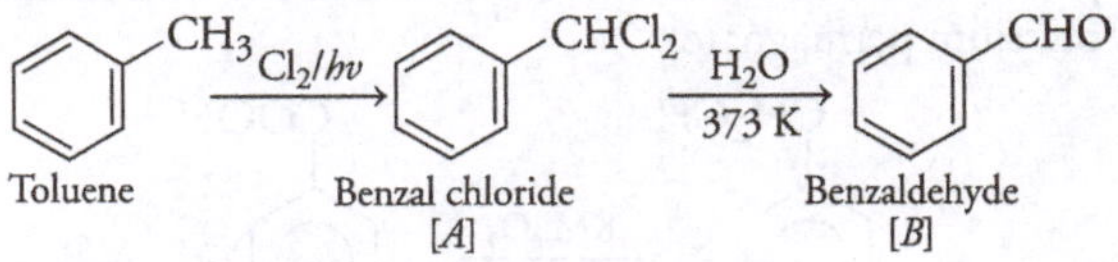


**52.** *(a)* Complete reaction is as follows:

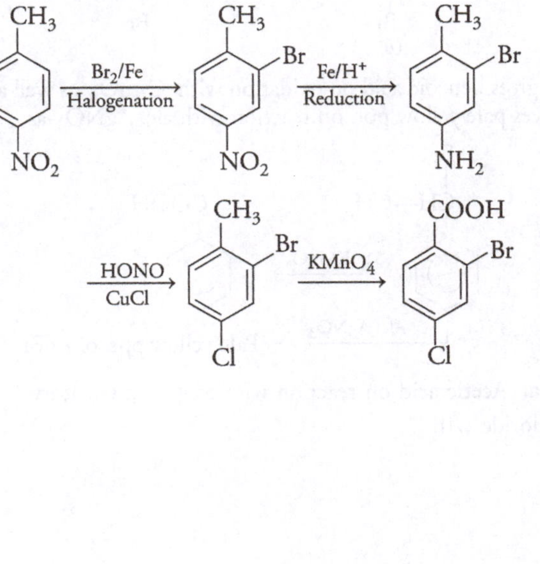

**53.** *(c)* The major product '*P*' is given in option (c).

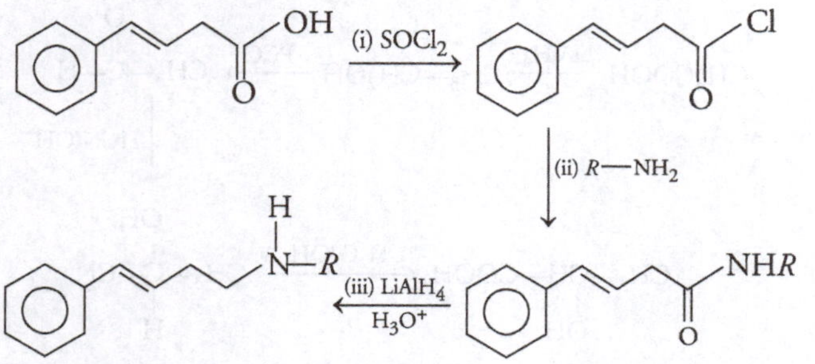


**54.** *(c)*

| Element | % | Mass | Mole | Simplest Ratio |
|---|---|---|---|---|
| C | 52.2 | 52.2 | 4.35 | 8.1 ≈ 8 |
| H | 4.9 | 4.9 | 4.9 | 9.12 ≈ 9 |
| Br | 42.9 | 42.9 | 0.537 | 1 |

Empirical formula $= C_8H_9Br$

Empirical formula mass = 184.9 g/mol

Molar mass = 184 g/mol

Now, molecular formula $= C_8H_9Br$

Hence, *A* is an optically active compound.

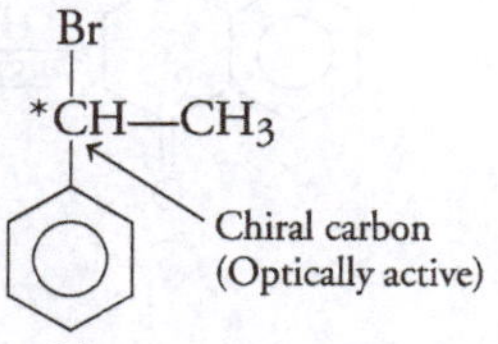


(*A*)

*B* produces *p*-bromobenzoic acid on oxidation with $KMnO_4$ (Potassium permagnate)

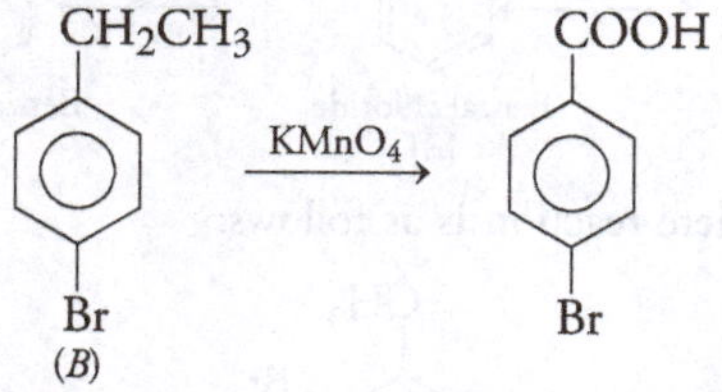


*A* gives benzoic acid on oxidation with $KMnO_4$ as well as it gives pale yellow ppt. on reaction with alc. $AgNO_3$ as

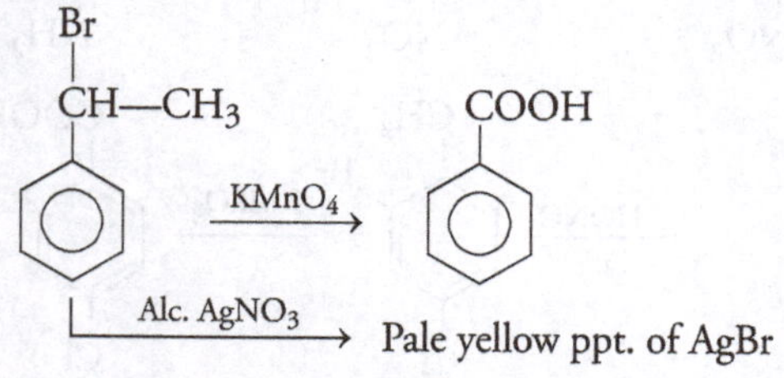


**55.** *(a)* Acetic acid on reaction with $SOCl_2$ gives acetyl chloride (*A*).

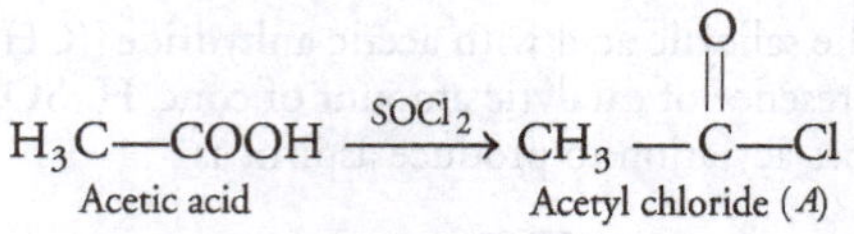


Acetyl chloride undergoes Friedel-Craft acylation in presence of anhyd. $AlCl_3$ and benzene to form acetophenone (*B*).

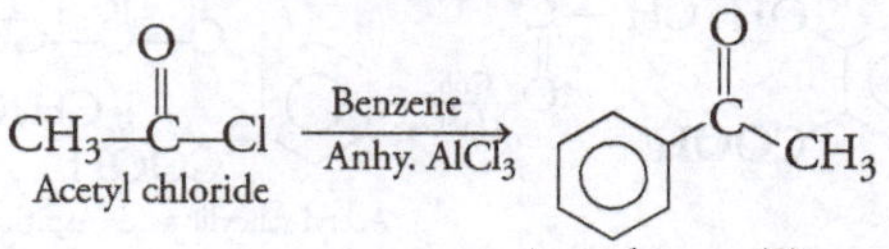


Acetophenone reacts with KCN to give (*C*) the final product, i.e. 2-hydroxy-2-phenolpropionitrile.

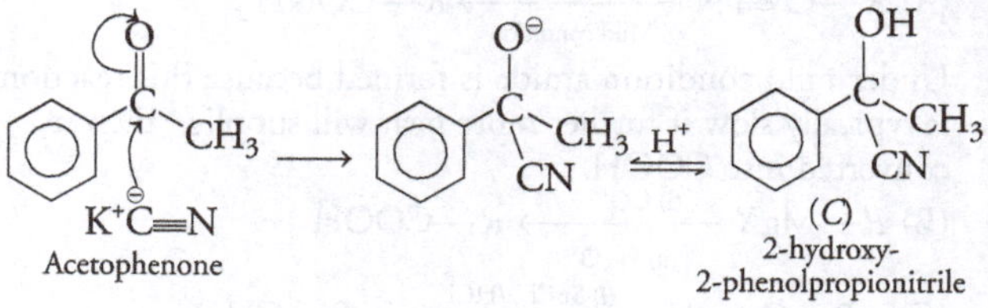


Hence, correct option is (a).

**56.** *(d)* The preparation of 3-nitrobenzoic acid from benzene is as follows

- Nitration of benzene

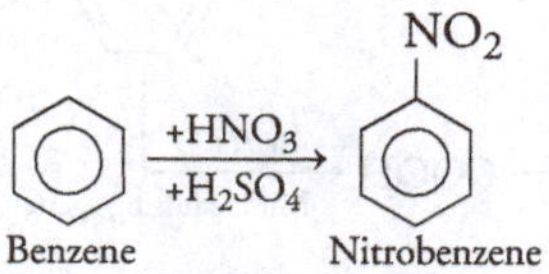


Electrophilic substitution of nitrobenzene

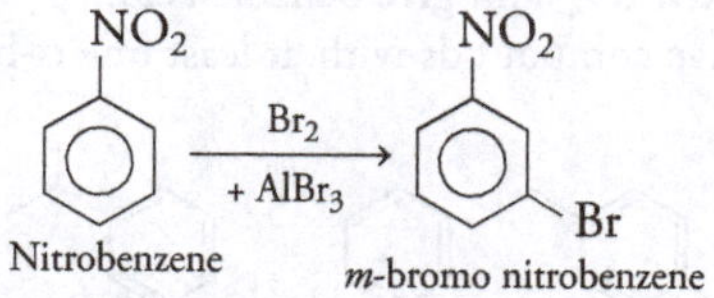


*m*-bromo nitrobenzene is converted to 3-nitrobenzoic acid.

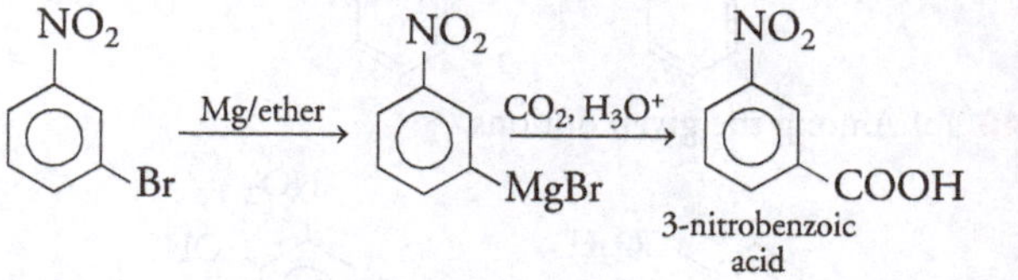


**57.** *(a)* Compound (*a*), reacts with sodium bicarbonate to form carbonic acid which further dissociate to liberate $CO_2$ effervescence.

Bicarbonate test

$$(CH_3)_3\overset{\oplus}{N}HCl^- + NaHCO_3 \longrightarrow H_2CO_3 + (CH_3)_3N + NaCl$$

(Sodium bicarbonate) (Carbonic acid) (Trimethyl-amine)

$$H_2CO_3 \longrightarrow CO_2 + H_2O$$

(Carbon dioxide (effervescence))

**58.** *(c)* The reactions of given compound with sodium bicarbonate solution are as follows

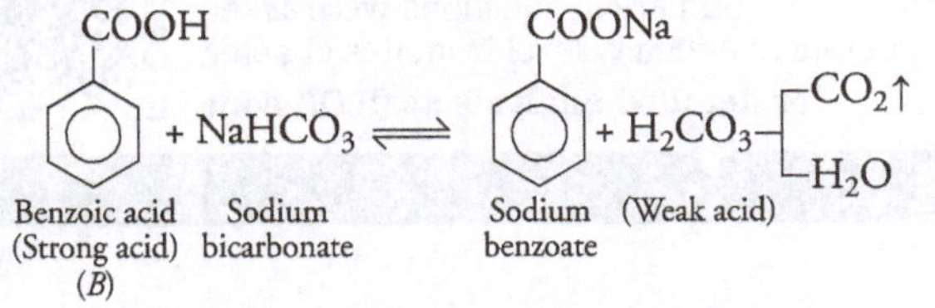

As $H_2CO_3$ is weak acid it dissociate to liberate $CO_2$ gas and $H_2O$.

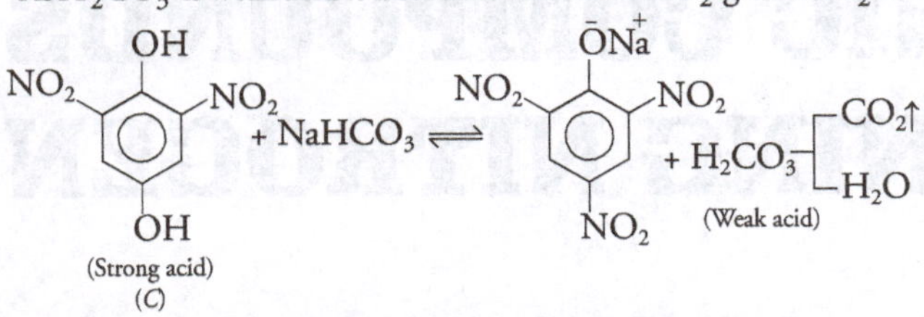

Equilibrium favours forward direction and $CO_2$ is liberated. In the above two reactions, $H_2CO_3$ is comparatively weak acid.
∴ $CO_2$ gas is liberated
Equilibrium favours backward direction and $CO_2$ is not liberated.
Thus, only *B* and *C* will liberate carbon dioxide with sodium bicarbonate solution.

**59.** *(d)* Tertiary butoxide is a bulky base, which favour elimination to form less substituted alkenes (Hofmann-product).

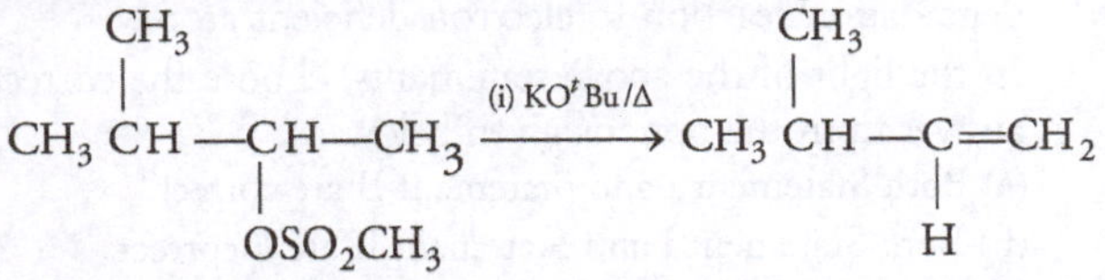

Now, alkene in presence of $O_3/H_2O_2$ undergoes oxidative ozonolysis to produce 2-methyl propionic acid and formic acid.

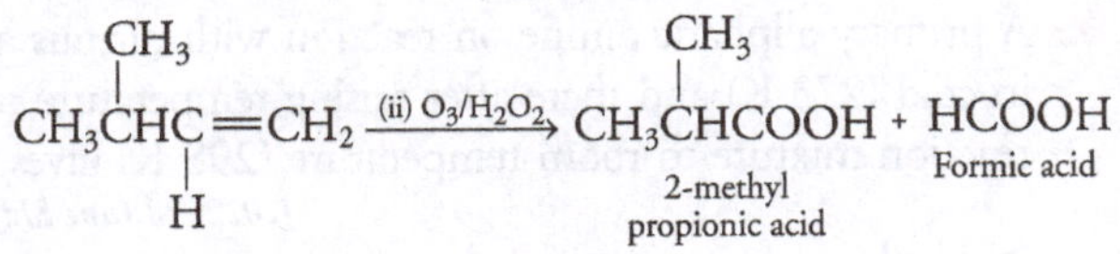

**60.** *(b)* The reaction sequence is given as :

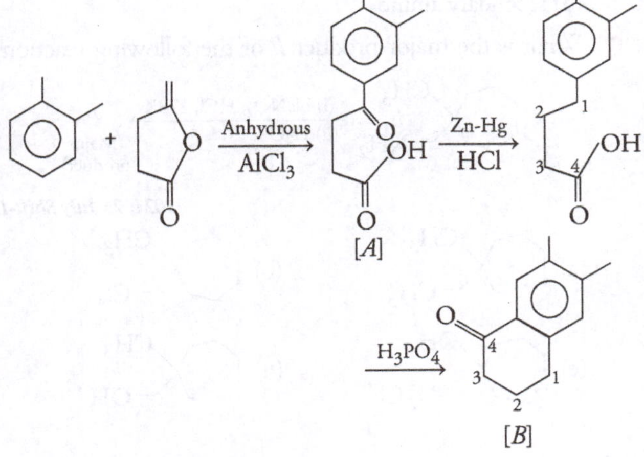

**61.** *(d)* Given reaction involves acidic hydrolysis of esters followed by the intramolecular cyclisation. The chemical equation is as follows:

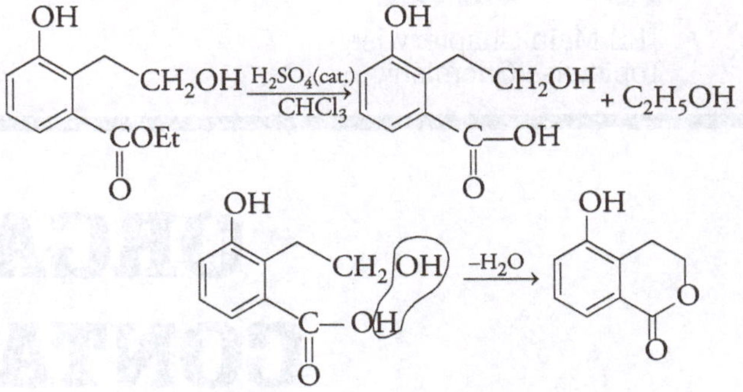

**62.** *(a)*

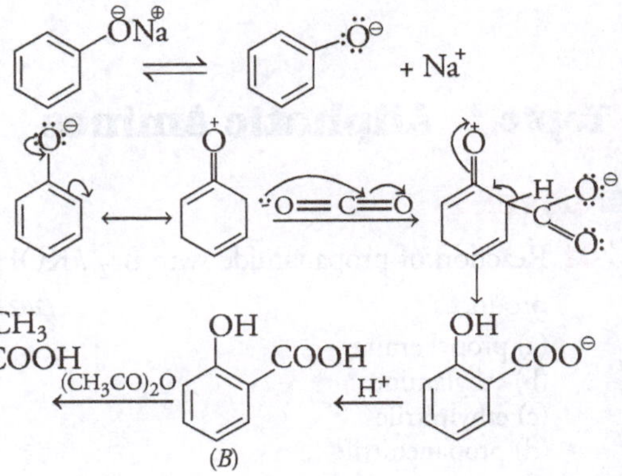

**63.** *(c)*

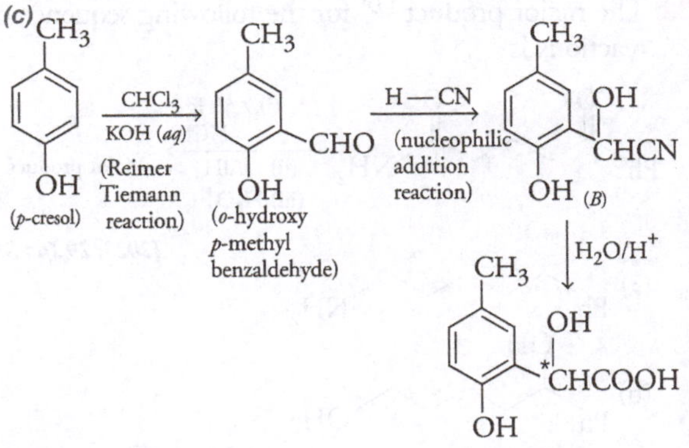

**64.** *(3)*

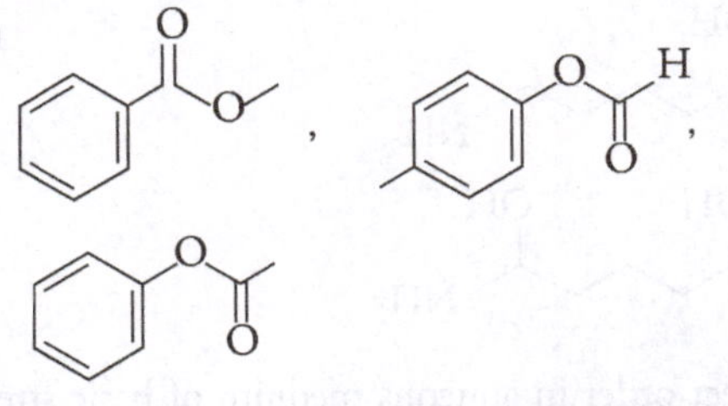

Those compounds dissolved in cold $NaHCO_3$ and NaOH solution that contain carboxylic acid and phenol group. But the compounds containing an ester functional group will only dissolve in hot NaOH solution.

# C-22

JEE Main Chapterwise Topicwise Chemistry

In JEE Main there is usually 0-2 question per paper are visible from this chapter. In past 3 years, 52 conceptual and 7 numerical questions were asked of moderate to difficulty level from this chapter.

**For detailed analysis scan QR code**

# ORGANIC COMPOUNDS CONTAINING NITROGEN

## Topic 1 Aliphatic Amines

### Level 1

**01** Reaction of propanamide with $Br_2$ / $KOH(aq)$ produces *[2023, 29 Jan Shift-II]*

(a) propyl amine
(b) ethylamine
(c) ethylnitrile
(d) propanenitrile

**02** The major product '*P*' for the following sequence of reactions is

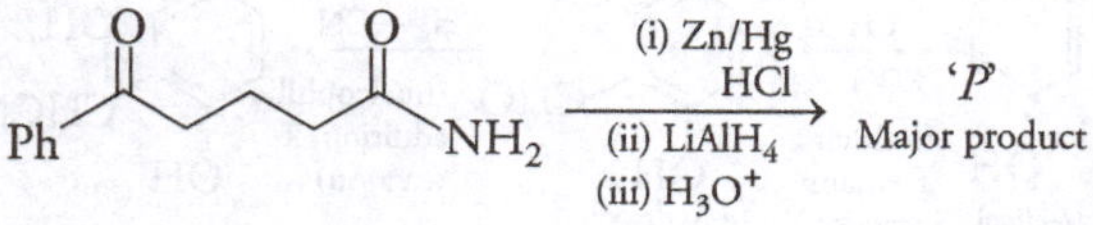

*[2023, 29 Jan Shift-I]*

(a)
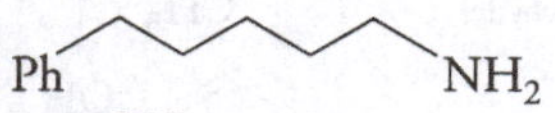

(b)
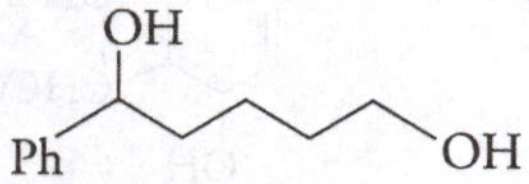

(c)
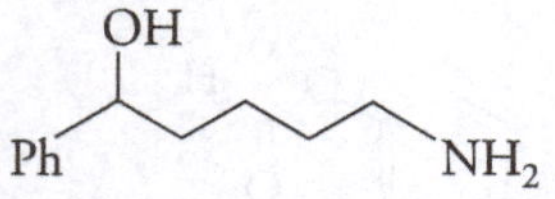

(d)
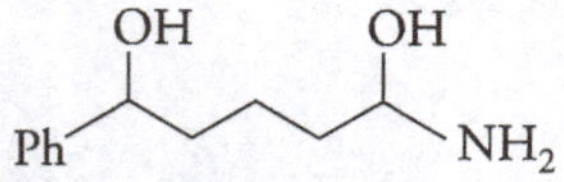

**03** The correct order in aqueous medium of basic strength in case of methyl substituted amines is *[2023, 25 Jan Shift-I]*

(a) $Me_2NH > Me_3N > MeNH_2 > NH_3$
(b) $NH_3 > Me_3N > MeNH_2 > Me_2NH$
(c) $Me_2NH > MeNH_2 > Me_3N > NH_3$
(d) $Me_3N > Me_2NH > MeNH_2 > NH_3$

**Imp.** Hofmann Degradation Reaction

**04** Given below are two statements. *[2022, 27 June Shift-I]*

**Statement I** In Hofmann degradation reaction, the migration of only an alkyl group takes place from carbonyl carbon of the amide to the nitrogen atom.

**Statement II** The group is migrated in Hofmann degradation reaction to electron deficient atom.

In the light of the above statements, choose the correct answer from the options given below.

(a) Both Statement I and Statement II are correct.
(b) Both Statement I and Statement II are incorrect.
(c) Statement I is correct but Statement II is incorrect.
(d) Statement I is incorrect but Statement II is correct.

**05** A primary aliphatic amine on reaction with nitrous acid in cold (273 K) and there after raising temperature of reaction mixture to room temperature (298 K) gives *[2022, 28 June Shift-I]*

(a) nitrile
(b) alcohol
(c) diazonium salt
(d) secondary amine

**06** What is the major product *P* of the following reaction ?

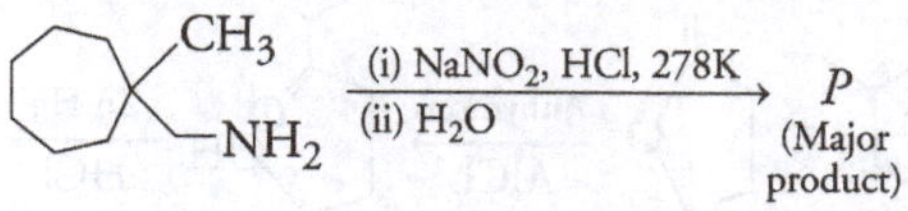

*[2021, 25 July Shift-II]*

(a)
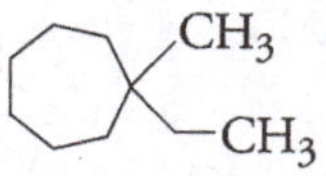

(b)
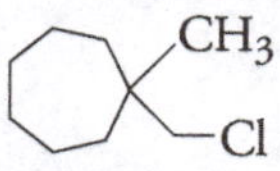

(c)
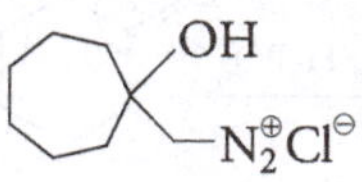

(d)
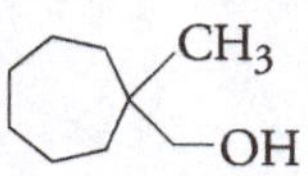

**07** Compound *A* is converted to *B* on reaction with $CHCl_3$ and KOH. The compound *B* is toxic and can be decomposed by *C*. *A*, *B* and *C* respectively are *[2021, 20 July Shift-I]*

(a) primary amine, nitrile compound, conc. HCl
(b) secondary amine, isonitrile compound, conc. NaOH
(c) primary amine, isonitrile compound, conc. HCl
(d) secondary amine, nitrile compound, conc. NaOH

**08** In the reaction of hypobromite with amide, the carbonyl carbon is lost as *[2021, 18 March Shift-II]*

(a) $CO_3^{2-}$ (b) $HCO_3^-$
(c) $CO_2$ (d) CO

**09** An organic compound "*A*" on treatment with benzene sulphonyl chloride gives compound *B*. *B* is soluble in dil. NaOH solution. Compound *A* is *[2021, 18 March Shift-II]*

(a) $C_6H_5—N—(CH_3)_2$ (b) $C_6H_5—NHCH_2CH_3$
(c) $C_6H_5—CH_2NHCH_3$ (d) $C_6H_5—\underset{\displaystyle CH_3}{\underset{|}{C}H}—NH_2$

**10** Which of the following is least basic? *[2021, 16 March Shift-II]*

(a) $(CH_3CO)\ddot{N}HC_2H_5$ (b) $(C_2H_5)_3\ddot{N}$
(c) $(CH_3CO)_2\ddot{N}H$ (d) $(C_2H_5)_2\ddot{N}H$

**11** The most appropriate reagent for conversion of $C_2H_5CN$ into $CH_3CH_2CH_2NH_2$ is *[2020, 5 Sep Shift-I]*

(a) $Na(CN)BH_3$ (b) $LiAlH_4$
(c) $NaBH_4$ (d) $CaH_2$

**Imp.** Order of Basic Strength

**12** In the following compounds, the decreasing order of basic strength will be *[2019, 8 April Shift-I]*

(a) $C_2H_5NH_2 > NH_3 > (C_2H_5)_2NH$
(b) $(C_2H_5)_2NH > NH_3 > C_2H_5NH_2$
(c) $(C_2H_5)_2NH > C_2H_5NH_2 > NH_3$
(d) $NH_3 > C_2H_5NH_2 > (C_2H_5)_2NH$

**13** A compound '*X*' on treatment with $Br_2/NaOH$,, provided $C_3H_9N$, which gives positive carbylamine test. Compound '*X*' is *[2019, 11 Jan Shift-II]*

(a) $CH_3COCH_2NHCH_3$
(b) $CH_3CH_2CH_2CONH_2$
(c) $CH_3CON(CH_3)_2$
(d) $CH_3CH_2COCH_2NH_2$

**14** The major product of following reaction is

$$R—C\equiv N \xrightarrow[\text{(ii) } H_2O]{\text{(i) AlH(i-Bu)}_2} ?$$

*[2019, 9 Jan Shift-I]*

(a) *R*CHO (b) $RCONH_2$
(c) *R*COOH (d) $RCH_2NH_2$

**15** The major product obtained in the following reaction is

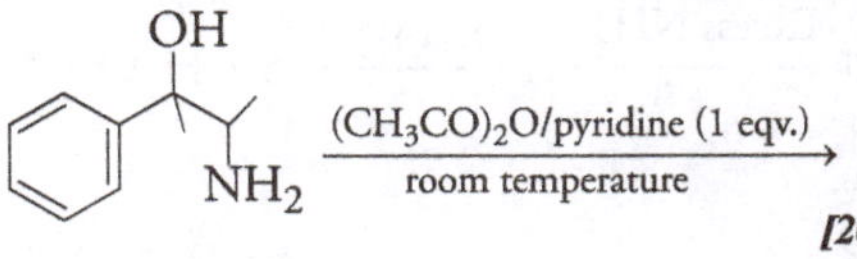


*[2019, 9 Jan Shift-II]*

(a) 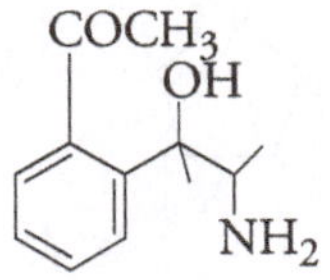


(b) 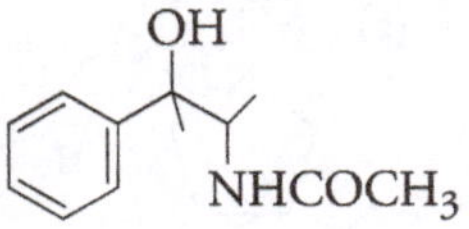


(c) 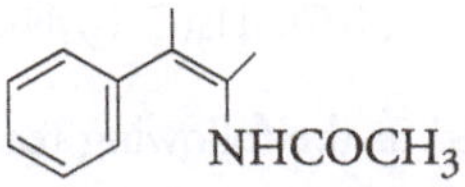


(d) 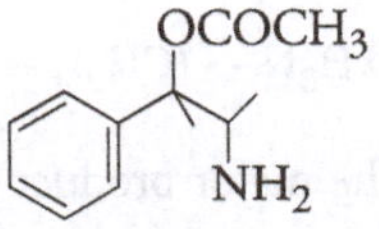


**16** In the Hofmann-bromamide degradation reaction, the number of moles of NaOH and $Br_2$ used per mole of amine produced are *[JEE Main 2016]*

(a) four moles of NaOH and two moles of $Br_2$
(b) two moles of NaOH and two moles of $Br_2$
(c) four moles of NaOH and one mole of $Br_2$
(d) one mole of NaOH and one mole of $Br_2$

**17** On heating an aliphatic primary amine with chloroform and ethanolic potassium hydroxide, the organic compound formed is *[JEE Main 2014]*

(a) an alkanol (b) an alkanediol
(c) an alkyl cyanide (d) an alkyl isocyanide

**18** Considering the basic strength of amines in aqueous solution, which one has the smallest $pK_b$ value? *[JEE Main 2014]*

(a) $(CH_3)_2NH$ (b) $CH_3NH_2$
(c) $(CH_3)_3N$ (d) $C_6H_5NH_2$

**19** Which one of the following is the strongest base in aqueous solution ? *[AIEEE 2007]*

(a) Trimethylamine (b) Aniline
(c) Dimethylamine (d) Methylamine

**20** Which one of the following methods is neither meant for the synthesis nor for separation of amines ? *[AIEEE 2005]*

(a) Curtius reaction (b) Wurtz reaction
(c) Hofmann method (d) Hinsberg method

## Level 2

**21** Find out the major product formed from the following reaction. [Me: $—CH_3$] *[2024, 4 April Shift-II]*

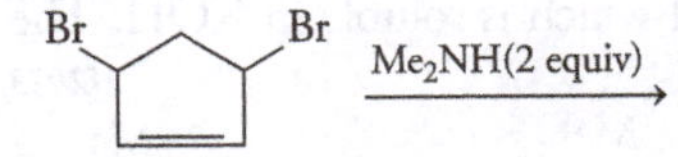


(a) 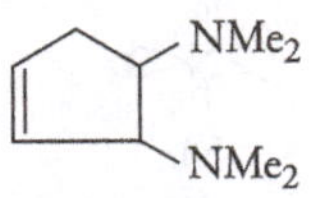


(b) 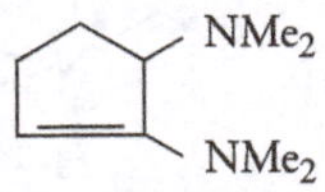


(c) 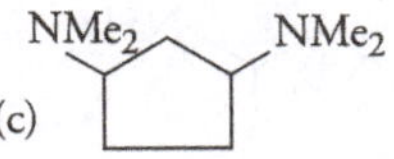


(d) 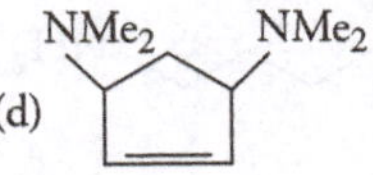

**22** Identify *B* formed in the reaction. *[2024, 27 Jan Shift-II]*

$$Cl(CH_2)_4Cl \xrightarrow{\text{Excess } NH_3} A \xrightarrow{NaOH} B + H_2O + NaCl$$

(a) 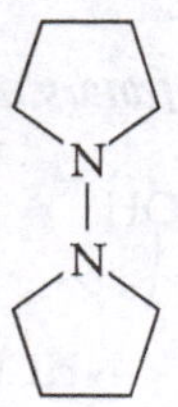

(b) 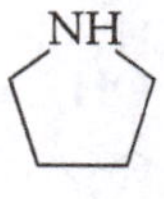

(c) $H_2N-(CH_2)_4-NH_2$

(d) $Cl\overset{+}{N}H_3(CH_2)_4\overset{+}{N}H_3Cl^-$

**23** The major product formed in the following reaction is

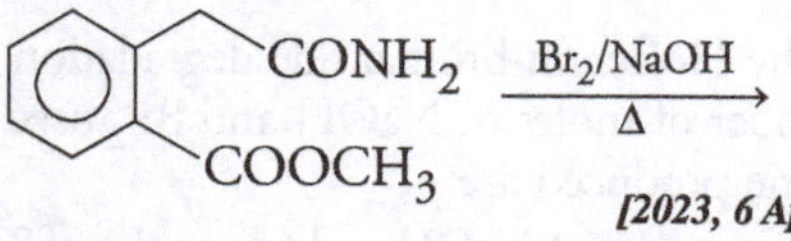

*[2023, 6 April Shift-I]*

(a) 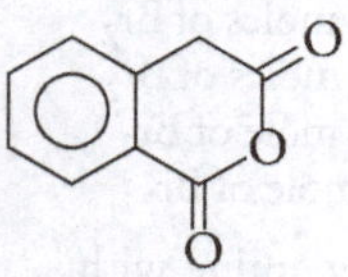

(b) 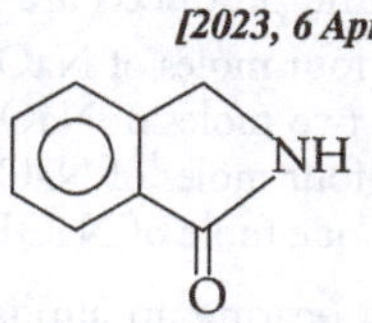

(c) 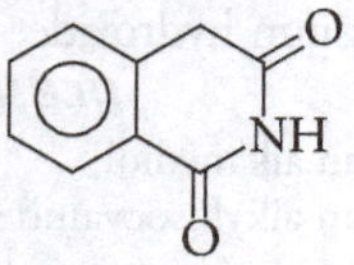

(d) 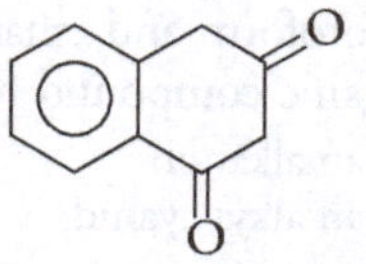

**24** In the following reaction, '*A*' is

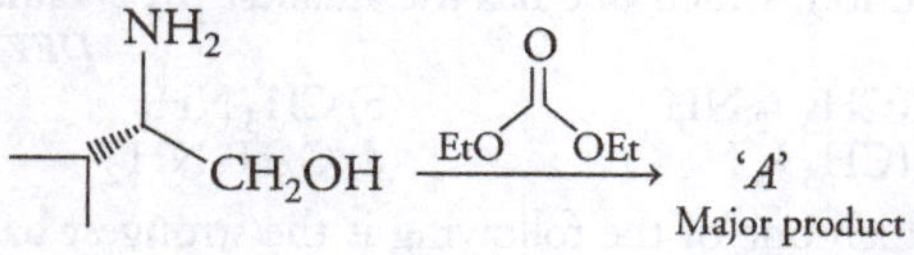

*[2023, 1 Feb Shift-I]*

(a) 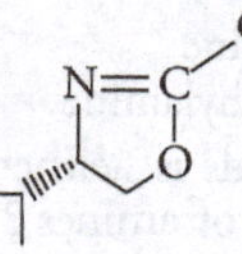

(b) 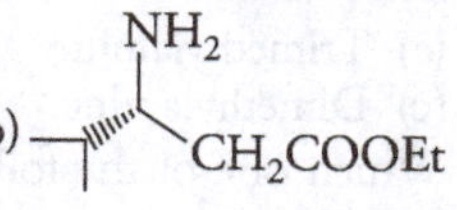

(c) 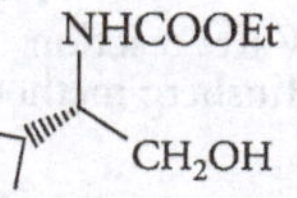

(d) 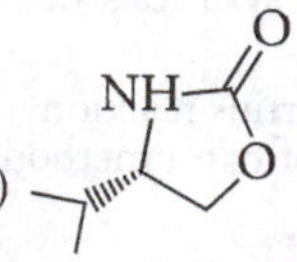

**25** An organic compound [*A*] ($C_4H_{11}N$), shows optical activity and gives $N_2$ gas on treatment with $HNO_2$. The compound [*A*] reacts with $PhSO_2Cl$ producing a compound which is soluble in KOH. The structure of *A* is *[2023, 31 Jan Shift-II]*

(a) 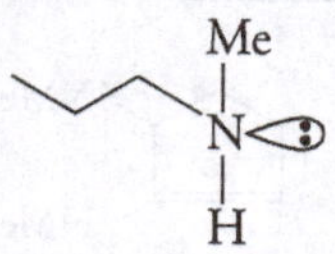

(b) 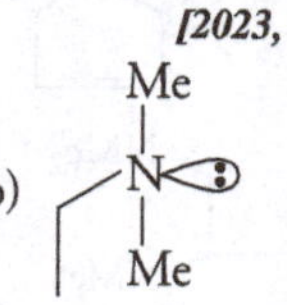

(c) 

(d) 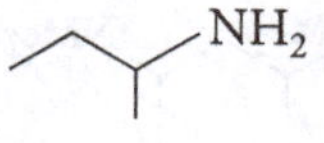

**26** Match List I with List II.

| | List I | | List II |
|---|---|---|---|
| A. | Benzene sulphonyl chloride | I. | Test for primary amines |
| B. | Hofmann bromamide reaction | II. | Anti Saytzeff |
| C. | Carbylamine reaction | III. | Hinsberg reagent |
| D. | Hofmann orientation | IV. | Known reaction of isocyanates |

Choose the correct answer from the options given below. *[2022, 27 July Shift-I]*

(a) A-IV, B-III, C-II, D-I (b) A-IV, B-II, C-I, D-II
(c) A-III, B-IV, C-I, D-II (d) A-IV, B-III, C-I, D-II

**27.** Among the following structures, which will show the most stable enamine formation? (where, Me is $-CH_3$) *[2022, 26 June Shift-I]*

(a) 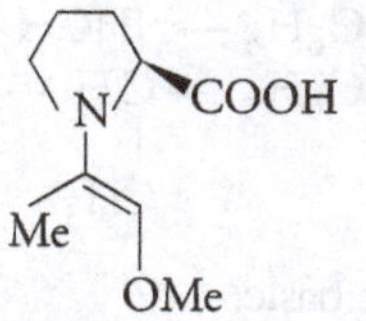

(b) 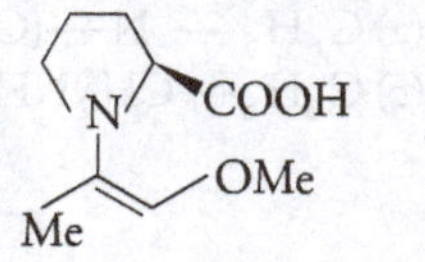

(c) 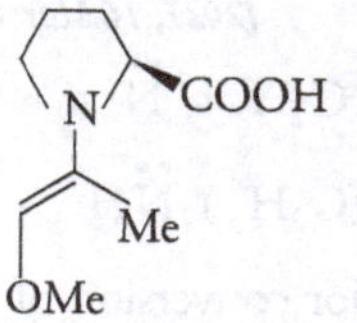

(d) 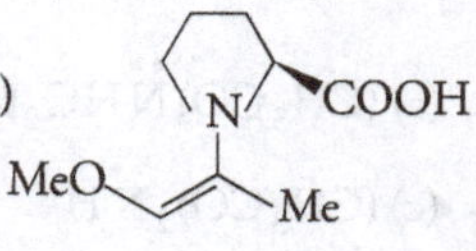

**28** The reaction of $R-\overset{\overset{O}{\|}}{C}-NH_2$ with bromine and KOH gives $RNH_2$ as the end product. Which one of the following is the intermediate product formed in this reaction? *[2022, 25 June Shift-I]*

(a) $R-\underset{\underset{O}{\|}}{C}-NH-Br$ (b) $R-NH-Br$

(c) $R-N=C=O$ (d) $R-\underset{\underset{O}{\|}}{C}-NBr_2$

**29** The major products *A* and *B* in the following set of reactions are *[2021, 31 Aug Shift-I]*

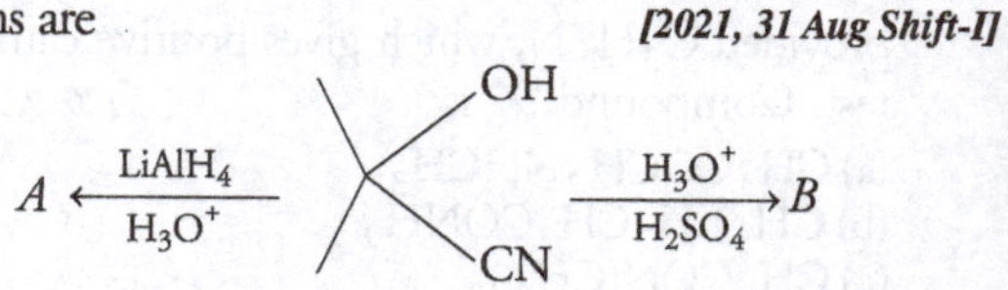

(a) *A* = 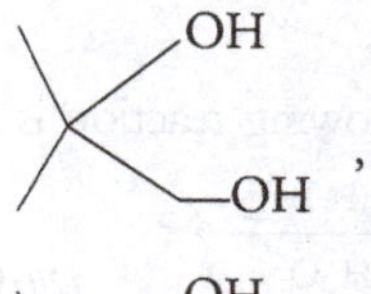, *B* = 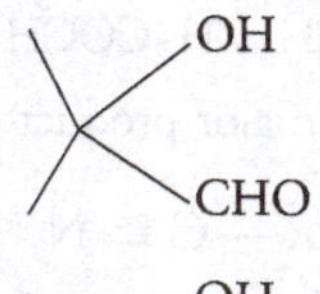

(b) *A* = 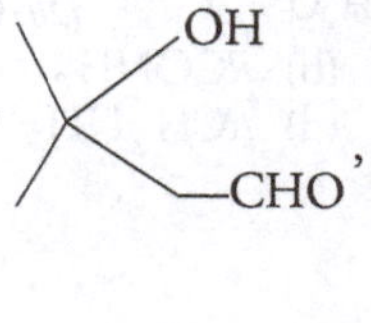, *B* = 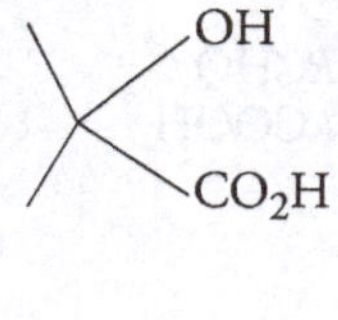

(c) $A$ = 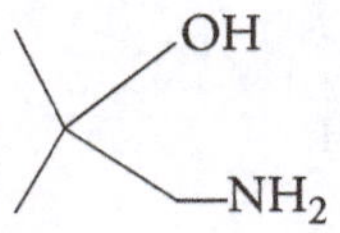 , $B$ = 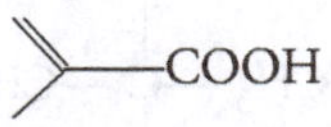

(d) $A$ = 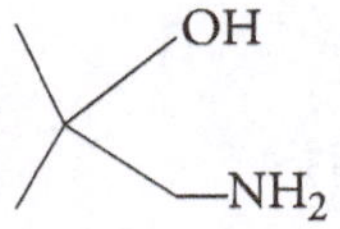 , $B$ = 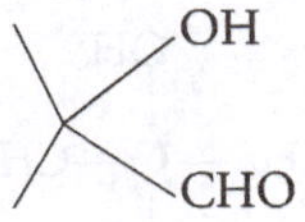

**30** Which of the following is not a correct statement for primary aliphatic amines? **[2021, 27 Aug Shift-I]**

(a) The intermolecular association in primary amines is less than the intermolecular association in secondary amines.
(b) Primary amines on treating with nitrous acid solution form corresponding alcohols except methyl amine.
(c) Primary amines are less basic than the secondary amines.
(d) Primary amines can be prepared by the gabriel phthalimide synthesis.

**31** 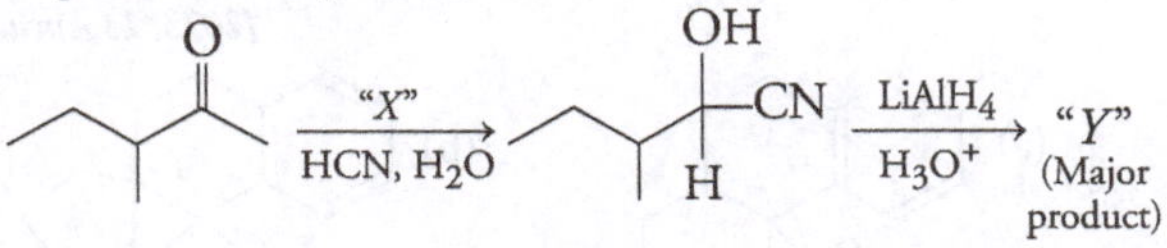

Consider the given reaction, identify '$X$' and '$Y$'. **[2021, 26 Aug Shift-II]**

(a) $X$ – NaOH $Y$ – 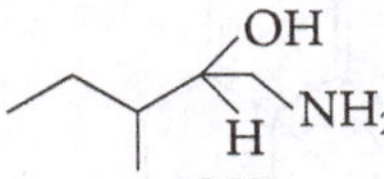

(b) $X$ – NaOH $Y$ – 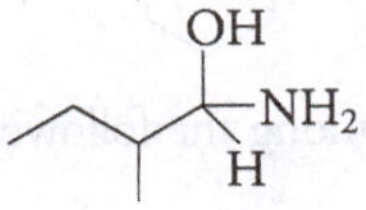

(c) $X$ – $HNO_3$ $Y$ – 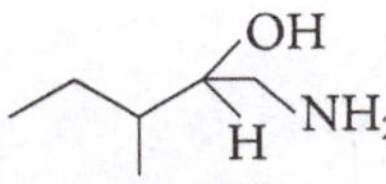

(d) $X$ – $HNO_3$ $Y$ – 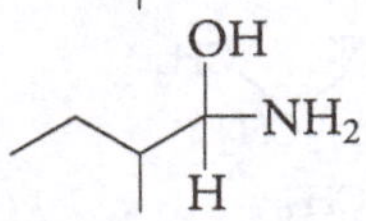

**32** What is $A$ in the following reaction?

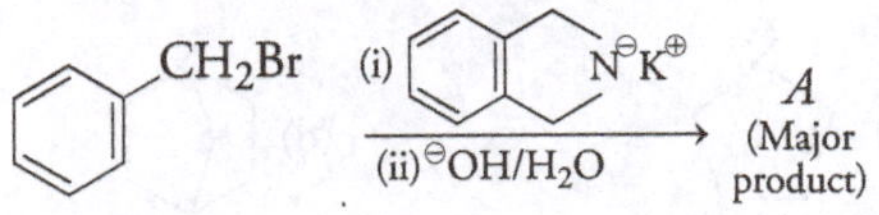

**[2021, 27 July Shift-II]**

(a) 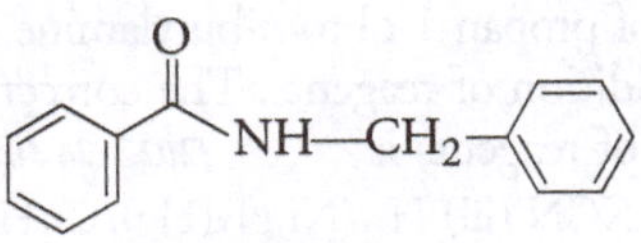

(b) 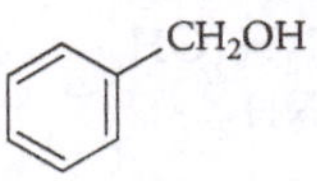

(c) 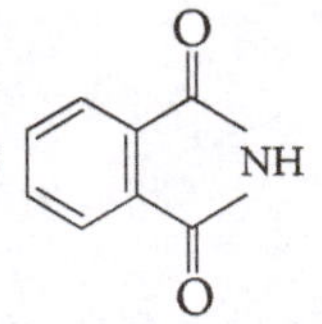

(d) 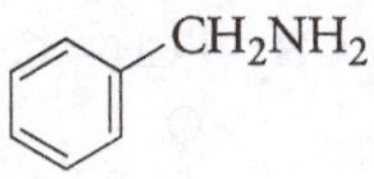

**33** Ammonolysis of alkyl halides followed by the treatment with NaOH solution can be used to prepare primary, secondary and tertiary amines. The purpose of NaOH in the reaction is **[2021, 16 March Shift-II]**

(a) to remove basic impurities
(b) to activate $NH_3$ used in the reaction
(c) to remove acidic impurities
(d) to increase the reactivity of alkyl halide

**34** The increasing order of p$K_b$ for the following compounds will be

$NH_2—CH{=}NH$, (A) 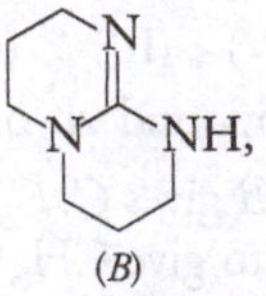 (B) $CH_3—NH—CH_3$ (C)

**[2020, 7 Jan Shift-I]**

(a) $(B) < (C) < (A)$ (b) $(C) < (A) < (B)$
(c) $(A) < (B) < (C)$ (d) $(B) < (A) < (C)$

**35** The major products $A$ and $B$ for the following reactions are, respectively **[2019, 9 April Shift-II]**

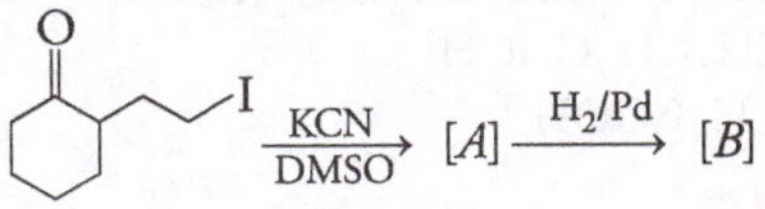

(a) 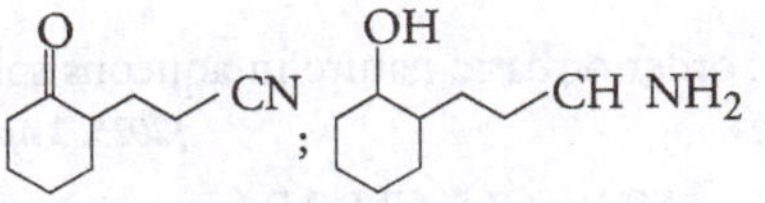

(b) 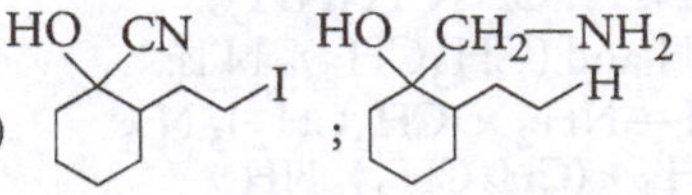

(c) 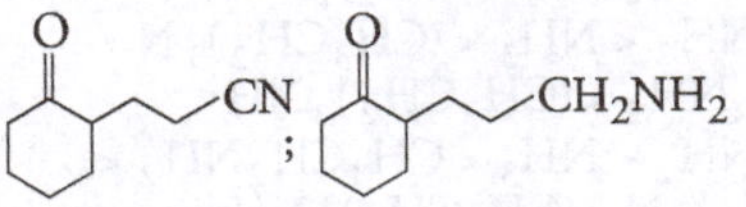

(d) 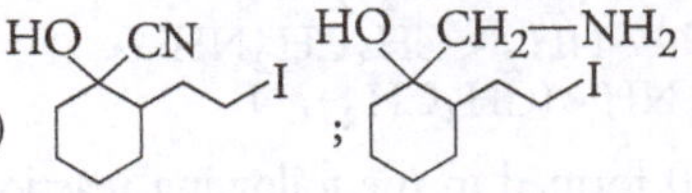

**36** The major product of the following reaction is

$$CH_3\overset{OH}{\overset{|}{C}}HCH_2CH_2NH_2 \xrightarrow[\text{Triethylamine}]{\text{Ethyl formate (1 equiv.)}}$$

**[2019, 10 April Shift-I]**

(a) $CH_3\overset{OH}{\overset{|}{C}}HCH_2CH_2NHCHO$
(b) $CH_3CH{=}CH—CH_2NH_2$

(c) $CH_3-\underset{|}{\overset{OH}{CH}}-CH=CH_2$

(d) 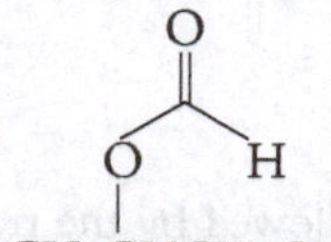

$CH_3CHCH_2CH_2NH_2$

**37** The increasing order of basicity of the following compounds is **[JEE Main 2018]**

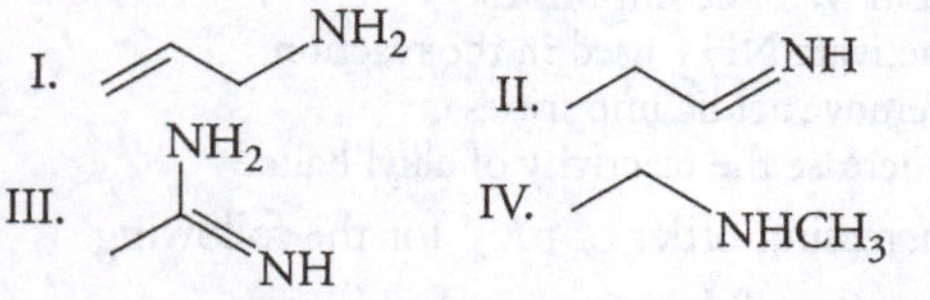


(a) (I) < (II) < (III) < (IV)
(b) (II) < (I) < (III) < (IV)
(c) (II) < (I) < (IV) < (III)
(d) (IV) < (II) < (I) < (III)

**38** An organic compound *A* on reacting with $NH_3$ gives *B*. On heating *B* gives *C*. *C* in the presence of KOH reacts with $Br_2$ to give $CH_3CH_2NH_2$. *A* is **[JEE Main 2013]**

(a) $CH_3-CH(CH_3)-COOH$
(b) $CH_3CH_2CH_2CH_2COOH$
(c) $CH_3CH_2COOH$
(d) $CH_3COOH$

## Level 3

**39** The correct order of basic nature in aqueous solution for the bases **[2025, 2 April Shift-I]**

$NH_3$, $H_2N-NH_2$, $CH_3CH_2NH_2$, $(CH_3CH_2)_2NH$ and $(CH_3CH_2)_3N$ is:

(a) $NH_3 < H_2N-NH_2 < (CH_3CH_2)_3N < CH_3CH_2NH_2 < (CH_3CH_2)_2NH$
(b) $H_2N-NH_2 < NH_3 < (CH_3CH_2)_3N < CH_3CH_2NH_2 < (CH_3CH_2)_2NH$
(c) $NH_2-NH_2 < NH_3 < CH_3CH_2NH_2 < (CH_3CH_2)_3N < (CH_3CH_2)_2NH$
(d) $NH_3 < H_2N-NH_2 < CH_3CH_2NH_2 < (CH_3CH_2)_2NH < (CH_3CH_2)_3N$

**40** The product (*A*) formed in the following reaction sequence is **[2025, 24 Jan Shift-I]**

$$CH_3-C\equiv CH \xrightarrow[\text{(iii) } H_2/Ni]{\text{(i) } Hg^{2+}, H_2SO_4 \text{ (ii) HCN}} \underset{\text{Product}}{(A)}$$

(a) $CH_3-CH_2-\underset{NH_2}{CH}-CH_2-OH$

(b) $CH_3-\underset{CH_3}{\overset{NH_2}{C}}-CH_2-OH$

(c) $CH_3-\underset{CH_3}{\overset{OH}{C}}-CH_2-NH_2$

(d) $CH_3-CH_2-\underset{}{\overset{OH}{CH}}-CH_2-NH_2$

**41** In the reaction given below

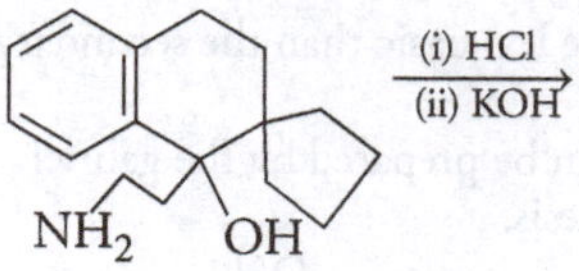


'*B*' Major product

**[2023, 13 April Shift-I]**

(a) 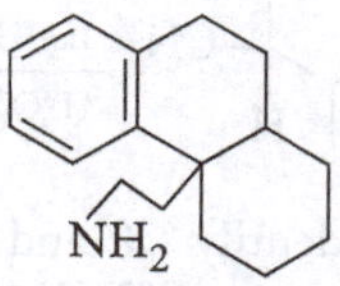


(b) 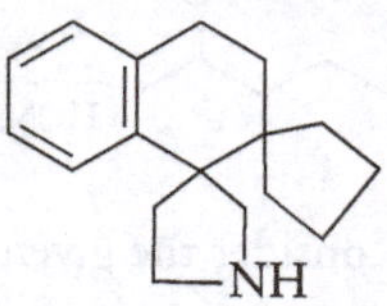


(c) 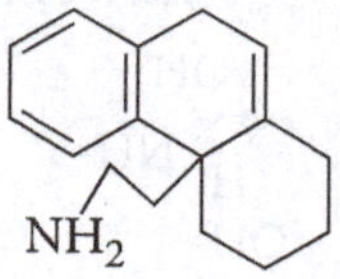


(d) 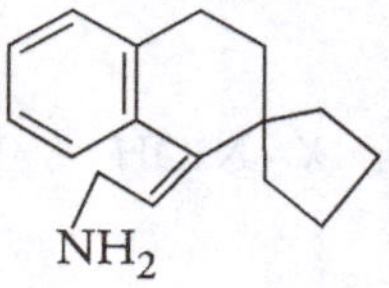


**42** Which among the following is the strongest Bronsted base? **[2022, 29 July Shift-I]**

(a) 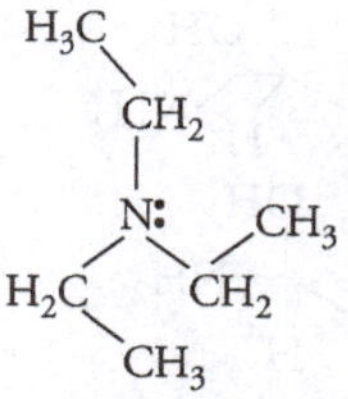


(b) 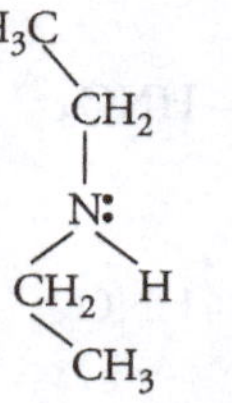


(c) 

(d) 

**43** The conversion of propan-1-ol to *n*-butylamine involves the sequential addition of reagents. The correct sequential order of reagents is **[2022, 24 June Shift-II]**

(a) (i) $SOCl_2$ (ii) KCN (iii) $H_2/Ni, Na(Hg)/C_2H_5OH$
(b) (i) HCl (ii) $H_2/Ni$, $Na(Hg)/C_2H_5OH$
(c) (i) $SOCl_2$ (ii) KCN (iii) $CH_3NH_2$
(d) (i) HCl (ii) $CH_3NH_2$

**44** 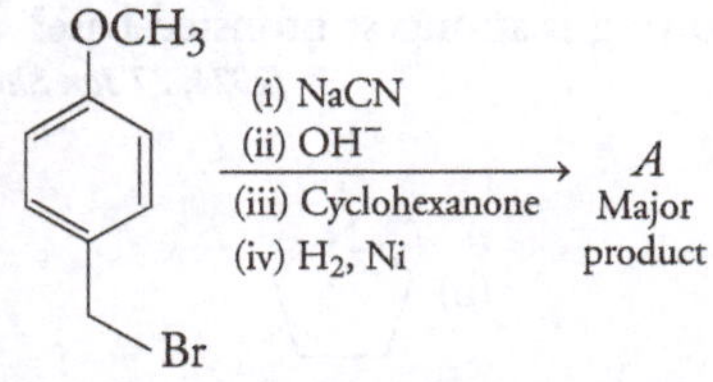


The major product of the above reaction is

*[2022, 24 June Shift-I]*

(a) 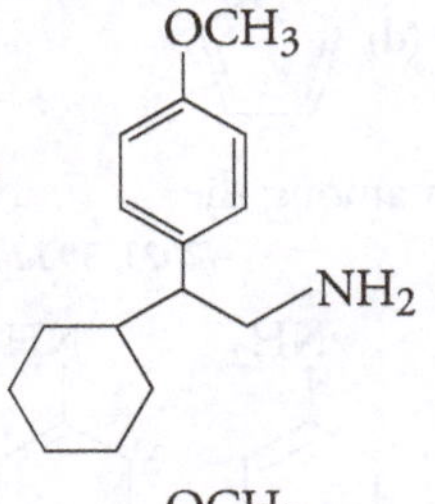


(b) 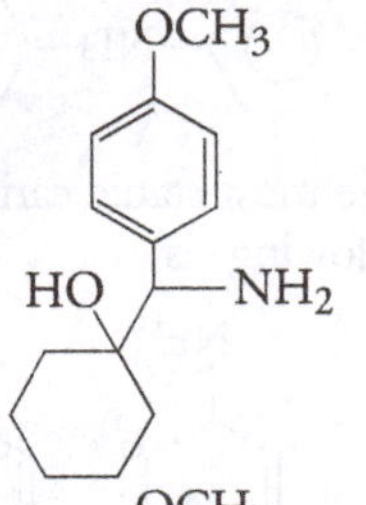


(c) 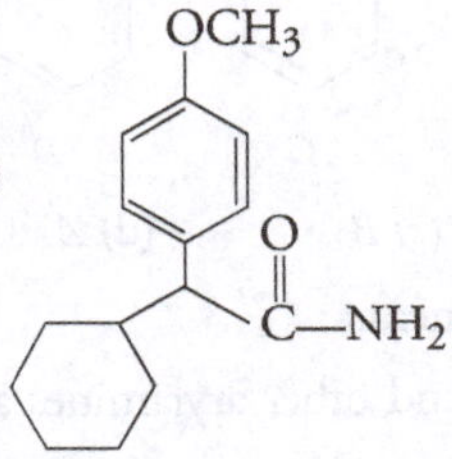


(d) 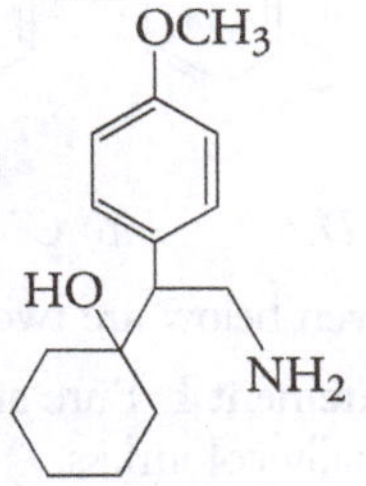


**45** The major product of the following reaction is

$$CH_3-\underset{\displaystyle CH_3}{\overset{}{C}}H-CH_2CH_2-\overset{O}{\overset{\|}{C}}-Cl \xrightarrow[\text{(iii) } NaNO_2,HCl \text{ (iv) } H_2O]{\text{(i) Alcoholic } NH_3 \text{ (ii) } NaOH,\, Br_2}$$

Major product

*[2021, 27 Aug Shift-I]*

(a) $CH_3-\underset{\displaystyle CH_3}{CH}-\overset{\displaystyle Br}{CH}-CH_2OH$

(b) $CH_3-\underset{\displaystyle CH_3}{CH}-CH_2-CH_2-CH_2OH$

(c) $CH_3-\underset{\displaystyle CH_3}{CH}-CH_2-CH_2OH$

(d) $CH_3-\underset{\displaystyle CH_3}{CH}-CH_2-CH_2-Cl$

**46** Which of the following reaction does not involve Hoffmann bromamide degradation? *[2021, 16 March Shift-I]*

(a) 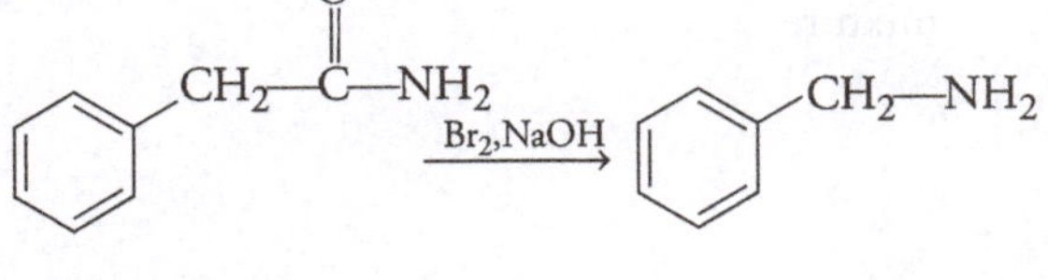


(b) 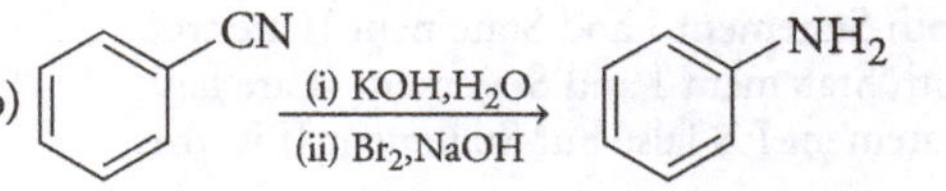


(c) 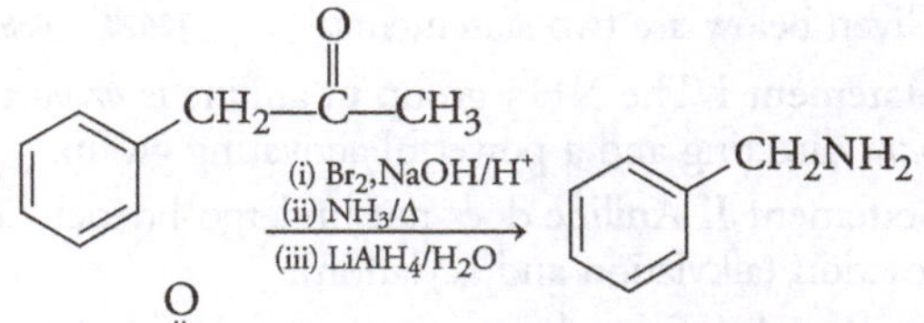


(d) 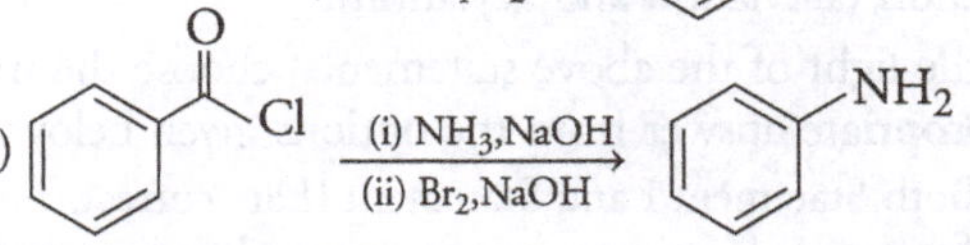


**47** Which of the following is not a correct method of the preparation of benzylamine from cyanobenzene?

(a) $H_2/Ni$ *[2019, 10 April Shift-II]*
(b) (i) $HCl/H_2O$ (ii) $NaBH_4$
(c) (i) $LiAlH_4$ (ii) $H_3O^+$
(d) (i) $SnCl_2 + HCl$ (gas) (ii) $NaBH_4$

## Numerical Value Type Questions

**48** A compound $(X)$ with molar mass 108 g $mol^{-1}$ undergoes acetylation to give product with molar mass 192 g $mol^{-1}$. The number of amino groups in the compound $(X)$ is ......... . *[2024, 31 Jan Shift-II]*

**49** In gaseous triethyl amine, the "—C—N—C—" bond angle is ........ degree. *[2021, 27 July Shift-I]*

## Topic 2 Aromatic Amines

### Level 1

**01** Correct order of basic strength of pyrrole , pyridine 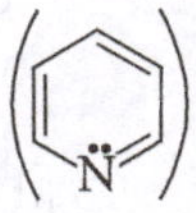 and piperidine 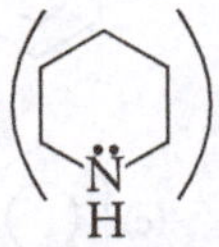 is

*[2024, 9 April Shift-I]*

(a) pyridine > piperidine > pyrrole
(b) piperidine > pyridine > pyrrole
(c) pyrrole > piperidine > pyridine
(d) pyrrole > pyridine > piperidine

**02** Given below are two statements : *[2024, 8 April Shift-II]*

**Statement I** All the following compounds react with *p*-toluenesulphonyl chloride.

$C_6H_5NH_2 \quad (C_6H_5)_2NH \quad (C_6H_5)_3N$

**Statement II** Their products in the above reaction are soluble is aqueous NaOH.

In the light of the above statements, choose the correct answer from the options given below:

(a) Statement I is true but Statement II is false.

(b) Both Statement I and Statement II are true
(c) Both Statement I and Statement II are false.
(d) Statement I is false but Statement II is true.

**03** Given below are two statements. *[2024, 1 Feb Shift-I]*

**Statement I** The $NH_2$ group in aniline is *ortho* and *para* directing and a powerful activating group.

**Statement II** Aniline does not undergo Friedel-Craft's reaction (alkylation and acylation).

In the light of the above statements, choose the **most appropriate** answer from the options given below.

(a) Both Statement I and Statement II are correct.
(b) Statement I is correct but Statement II is incorrect.
(c) Statement I is incorrect but Statement II is correct.
(d) Both Statement I and Statement II are incorrect.

**04** Given below are two statements. *[2024, 1 Feb Shift-I]*

**Statement I** Aminobenzene and aniline are same organic compounds.

**Statement II** Aminobenzene and aniline are different organic compounds.

In the light of the above statements, choose the **most appropriate** answer from the options given below.

(a) Statement I is incorrect but Statement II is correct.
(b) Both Statement I and Statement II are correct.
(c) Both Statement I and Statement II are incorrect.
(d) Statement I is correct but Statement II is incorrect.

**05** The product *A* formed in the following reaction is

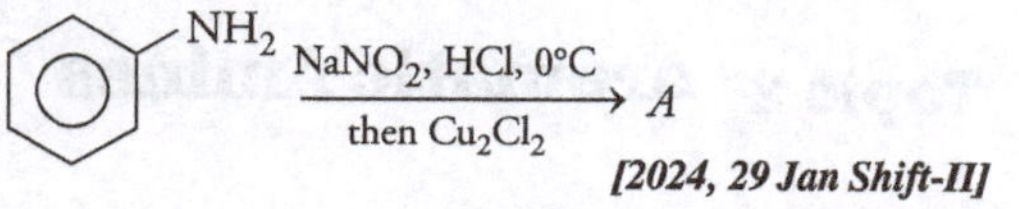

*[2024, 29 Jan Shift-II]*

(a) 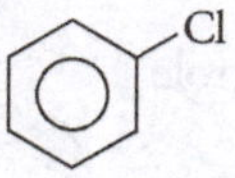 (b) 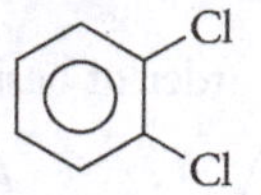

(c) 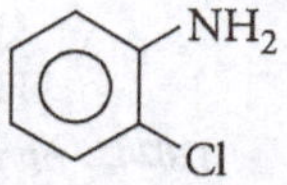 (d) 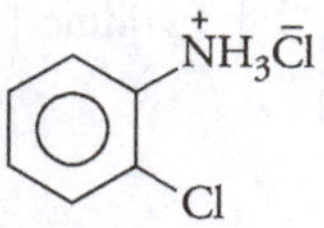

**06** The arenium ion which is not involved in the bromination of aniline is .......... *[2024, 29 Jan Shift-I]*

(a) 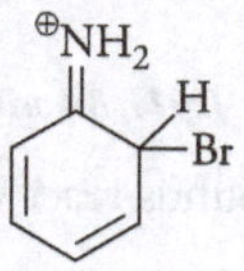 (b) 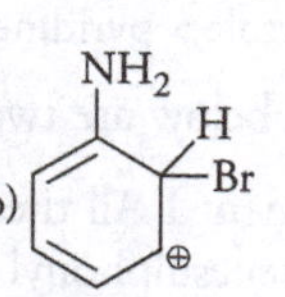

(c) 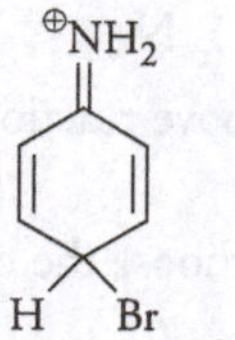 (d) 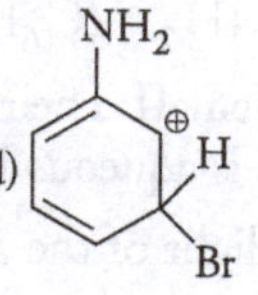

**07** Which of the following is strongest bronsted base? *[2024, 27 Jan Shift-II]*

(a) 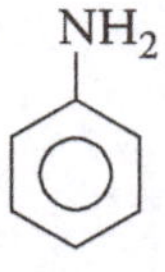 (b) 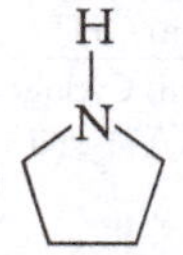

(c) 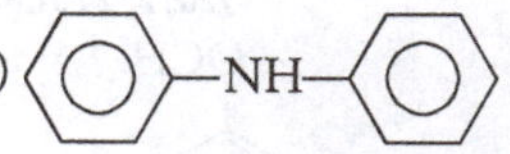 (d) 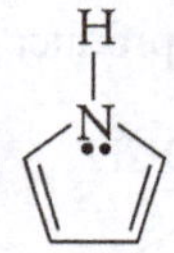

**08** The most stable carbocation among the following is *[2023, 30 Jan Shift-II]*

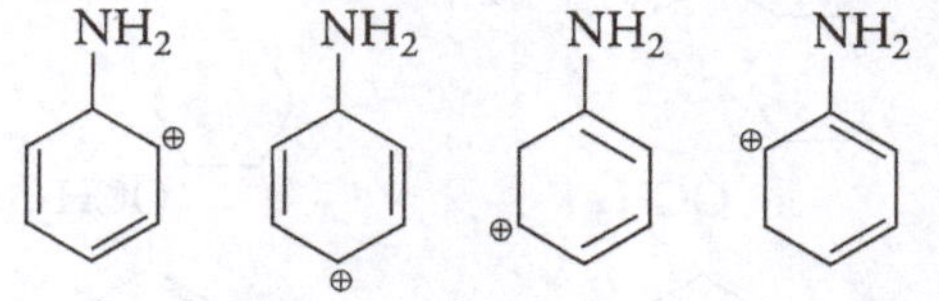

*A* *B* *C* *D*

(a) *D* (b) *C* (c) *B* (d) *A*

**09** Given below are two statements :

**Statement I** Pure aniline and other arylamines are usually colourless.

**Statement II** Arylamines get coloured on storage due to atmospheric reduction.

In the light of the above statements, choose the most appropriate answer form the options given below. *[2023, 24 Jan Shift-II]*

(a) Both Statement I and Statement II are incorrect.
(b) Both Statement I and Statement II are correct.
(c) Statement I is correct but Statement II is incorrect.
(d) Statement I is incorrect but Statement II is correct.

**10** The Hinsberg reagent is *[2022, 29 July Shift-II]*

(a) 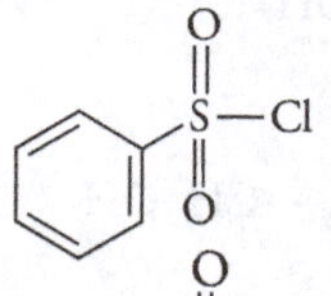 (b) 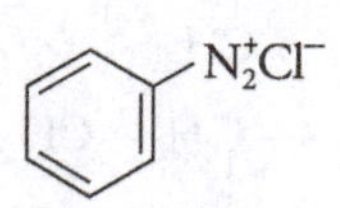

(c) 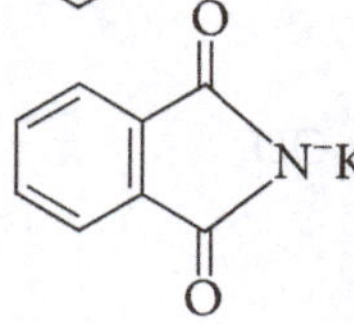 (d) 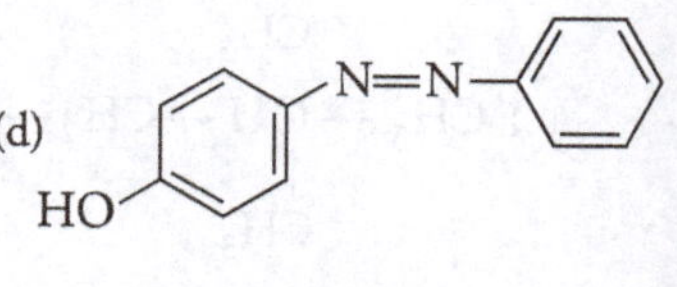

**11** Given below are two statements, one is labelled as Assertion (A) and the other is labelled as Reason (R).

**Assertion (A)** Aniline on nitration yields *ortho*, *meta* and *para* nitro derivatives of aniline.

**Reason (R)** Nitrating mixture is a strong acidic mixture.

In the light of the above statements, choose the correct answer from the options given below. *[2022, 28 July Shift-II]*

(a) Both A and R are true and R is the correct explanation of A
(b) Both A and R are true but R is not the correct explanation of A
(c) A is true but R is false
(d) A is false but R is true

**12** Which statement is not correct for *p*-toluene sulphonyl chloride? *[2022, 26 June Shift-II]*

(a) It is known as Hinsberg's reagent.
(b) It is used to distinguish primary and secondary amines.
(c) On treatment with secondary amine, it leads to a product, that is soluble in alkali.
(d) It doesn't react with tertiary amines.

**13** In Friedel-Craft's alkylation of aniline, one gets *[2022, 29 June Shift-II]*

(a) alkylated product with *ortho* and *para* substitution.
(b) secondary amine after acidic treatment.
(c) an amide product.
(d) positively charged nitrogen at benzene ring.

**14** Given below are two statements.

**Statement I** Aniline is less basic than acetamide.
**Statement II** In aniline, the lone pair of electrons on nitrogen atom is delocalised over benzene ring due to resonance and hence less available to a proton.

Choose the most appropriate option *[2021, 27 July Shift-I]*

(a) Statement I is true but Statement II is false.
(b) Statement I is false but Statement II is true.
(c) Both Statement I and Statement II are true.
(d) Both Statement I and Statement II are false.

**15**

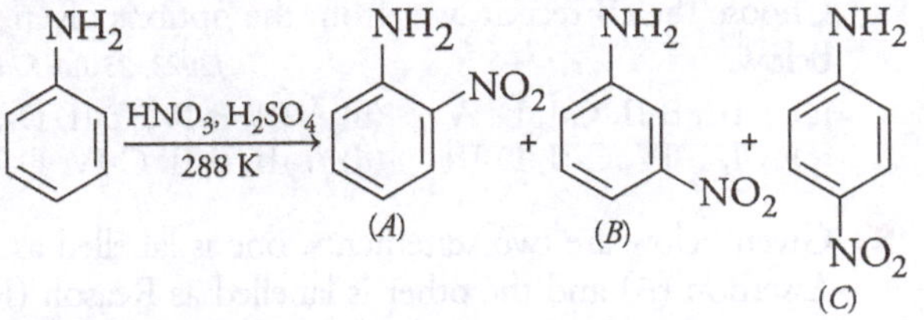


Consider the given reaction, percentage yield of *[2021, 18 March Shift-II]*

(a) (*C*) > (*A*) > (*B*) (b) (*B*) > (*C*) > (*A*)
(c) (*A*) > (*C*) > (*B*) (d) (*C*) > (*B*) > (*A*)

**16** Primary, secondary and tertiary amines can be separated using *[2021, 17 March Shift-II]*

(a) *para*-toluene sulphonyl chloride
(b) chloroform and KOH
(c) benzene sulphonic acid
(d) acetyl amide

**17** An amine on reaction with benzene sulphonyl chloride produces a compound insoluble in alkaline solution. This amine can be prepared by ammonolysis of ethyl chloride. The correct structure of amine is *[2021, 26 Feb Shift-I]*

(a) 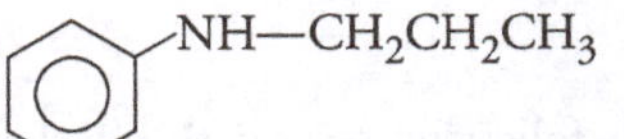


(b) $CH_3CH_2NH_2$
(c) $CH_3CH_2CH_2NHCH_3$
(d) $CH_3CH_2CH_2NH—CH_2CH_3$

**18** In the following reaction the reason why *meta*-nitro product also formed is *[2021, 24 Feb Shift-I]*

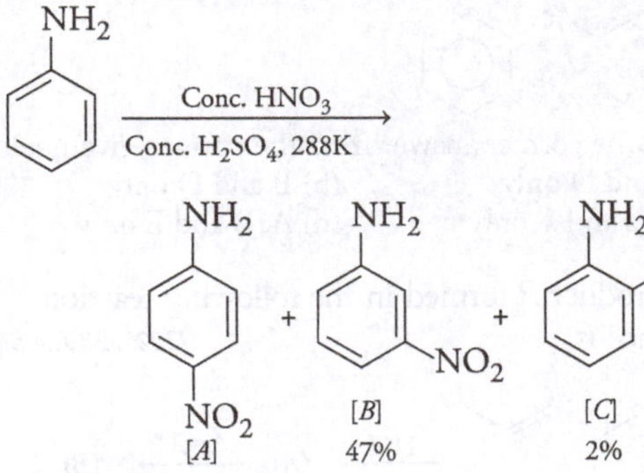


(a) —$NH_2$ group is highly *meta*-directive
(b) —$NO_2$ substitution always takes place at *meta*-position
(c) Formation of anilinium ion
(d) Low temperature

**19** Arrange the following amines in the decreasing order of basicity : *[2019, 9 Jan Shift-I]*

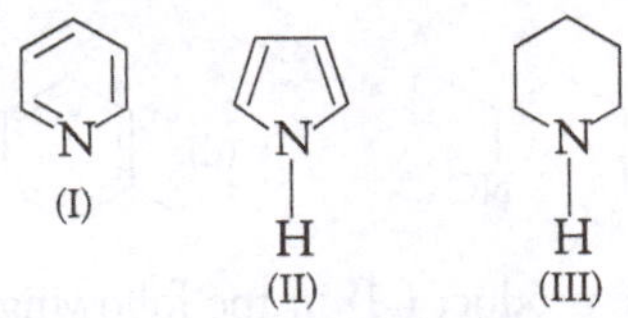


(a) I > II > III (b) III > II > I
(c) I > III > II (d) III > I > II

## Level 2

**20** Which of the following amine (s) show (s) positive carbylamine test?

(A) 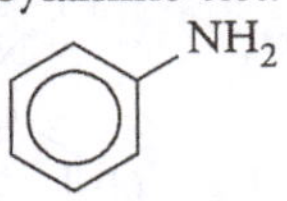


(b) $(CH_3)_2NH$
(C) $CH_3NH_2$
(D) $(CH_3)_3N$

(E) 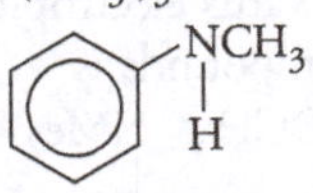


Choose the correct answer from the options given below: *[2025, 7 April Shift-I]*

(a) B, C and D only (b) C only
(c) A and E only (d) A and C only

**Imp.** Test of Amines

**21** Which among the following react with Hinsberg's reagent? *[2025, 23 Jan Shift-I]*

(A) 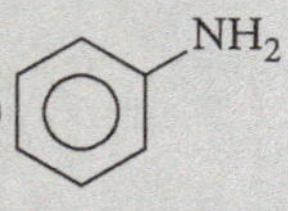

(B) 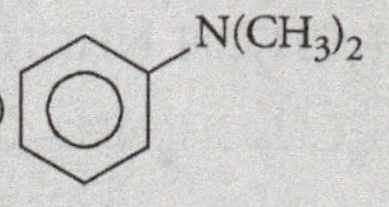


(C) $CH_3—NH_2$ (D) $N(CH_3)_3$

(E) 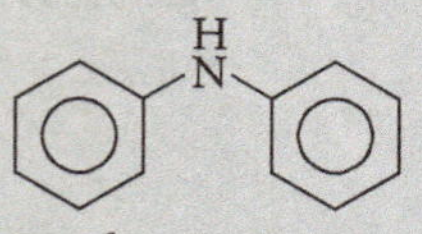


Choose the correct answer from the options given below

(a) C and D only (b) B and D only
(c) A, C and E only (d) A, B and E only

**22** The product *B* formed in the following reaction sequence is *[2025, 28 Jan Shift-II]*

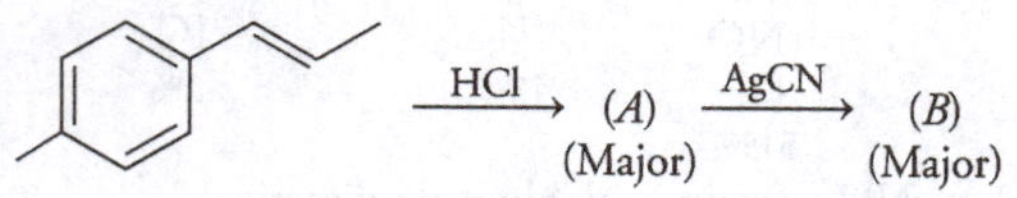


(a) 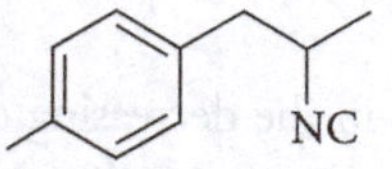

(b) 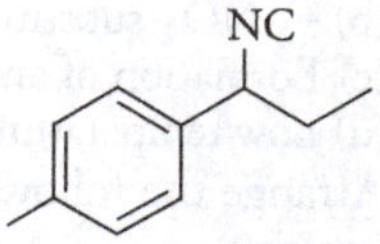


(c) 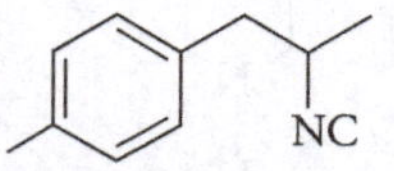

(d) 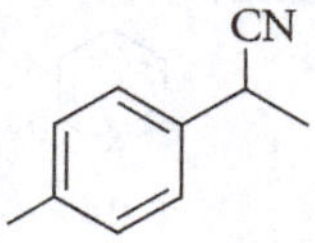


**23** Identify the product (*A*) in the following reaction. *[2024, 6 April Shift-II]*

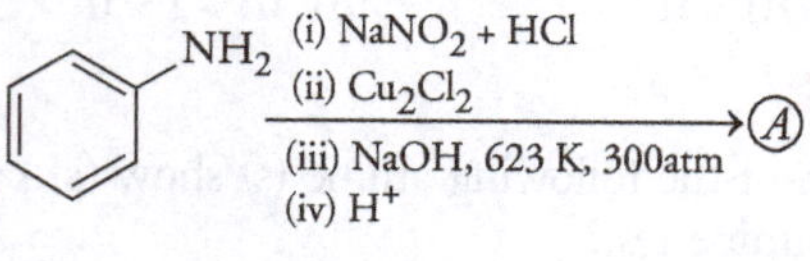


(a) 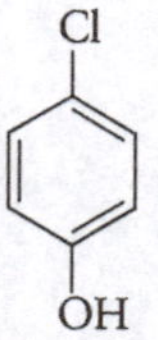

(b) 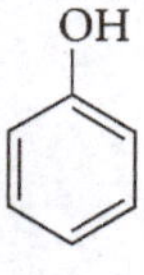

(c) 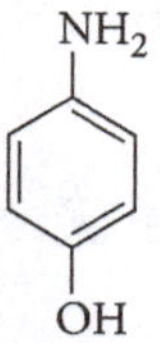

(d) 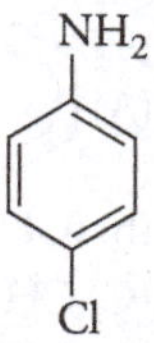


**24** Decreasing order of reactivity towards electrophilic substitution for the following compounds is

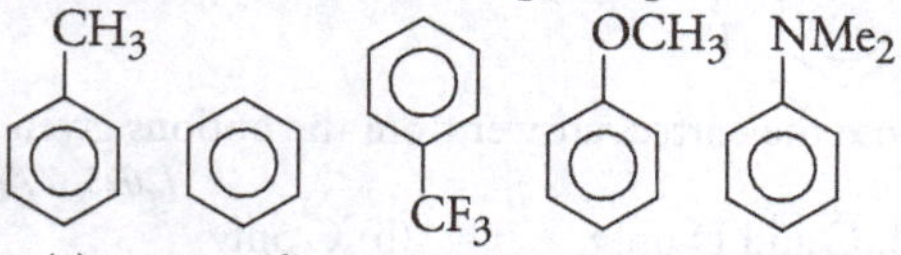


(*a*) (*b*) (*c*) (*d*) (*e*)

*[2023, 15 April Shift-I]*

(a) $a > d > e > b > c$ (b) $e > d > a > b > c$
(c) $c > b > a > d > e$ (d) $d > a > e > c > b$

**25** The major products *A* and *B* from the following reactions are *[2023, 6 April Shift-I]*

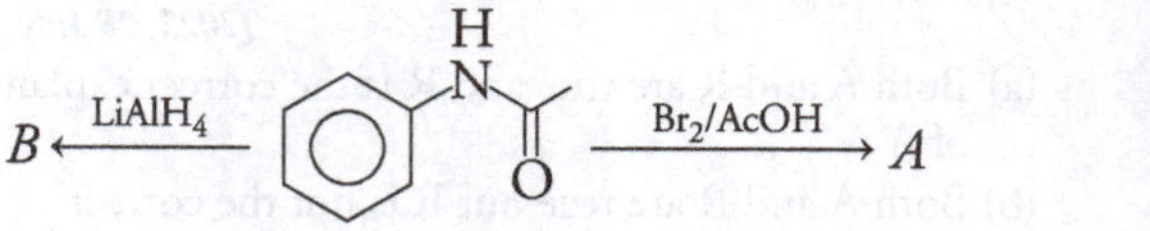


(a) *A* = 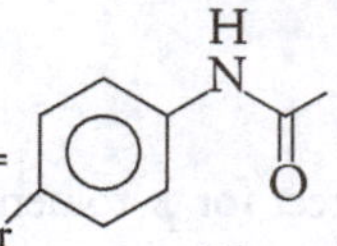 , *B* = 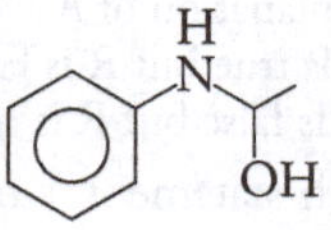

(b) *A* = 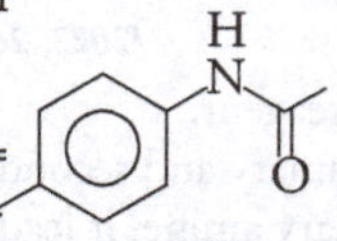 , *B* = 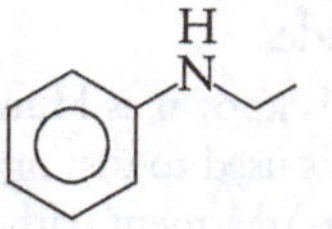

(c) *A* = 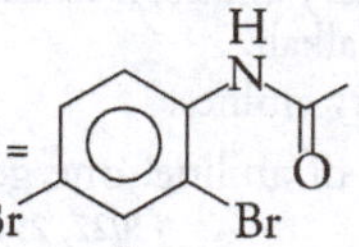 , *B* = 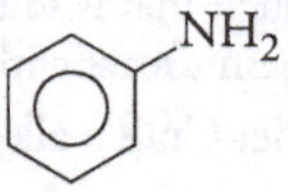


(d) *A* = 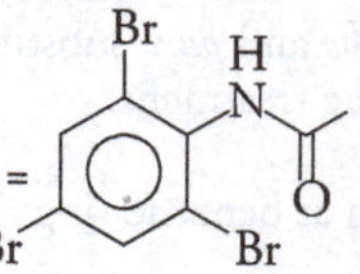 , *B* = 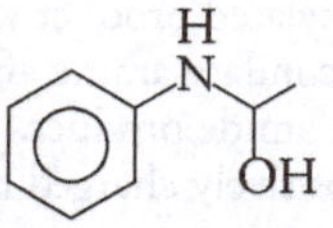

**Imp.** $pK_b$ Value of Amines

**26** Match List-I with List-II

| List-I (Amines) | | List-II ($pK_b$) | |
|---|---|---|---|
| A. | Aniline | I. | 3.25 |
| B. | Ethanamine | II. | 3.00 |
| C. | N-ethylethanamine | III. | 9.38 |
| D. | N, N-diethylethanamine | IV. | 3.29 |

Choose the correct answer from the options given below. *[2023, 25 Jan Shift-II]*

(a) A-III, B-II, C-I, D-IV (b) A-III, B-IV, C-II, D-I
(c) A-I, B-IV, C-II, D-III (d) A-III, B-II, C-IV, D-I

**27** Given below are two statements, one is labelled as Assertion (A) and the other is labelled as Reason (R).

**Assertion (A)** Experimental reaction of $CH_3Cl$ with aniline and anhydrous $AlCl_3$ does not give *o* and *p*-methylaniline. *[2022, 26 July Shift-I]*

**Reason (R)** The $—NH_2$ group of aniline becomes deactivating because of salt formation with anhydrous $AlCl_3$ and hence yields *m*-methyl aniline as the product.

In the light of the above statements, choose the most appropriate answer from the options given below

(a) Both (A) and (R) are true and (R) is the correct explanation of (A).
(b) Both (A) and (R) are true but (R) is not the correct explanation of (A).
(c) (A) is true, but (R) is false.
(d) (A) is false, but (R) is true.

**28** 'A' ($C_8H_6Cl_2O$) $\xrightarrow{NH_3}$ $C_8H_8ClNO$ $\xrightarrow[NaOH]{Br_2}$ 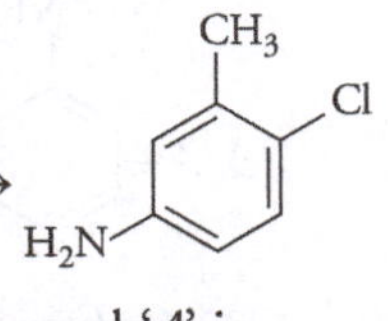

Consider the above reaction, the compound 'A' is *[2022, 29 July Shift-I]*

(a) 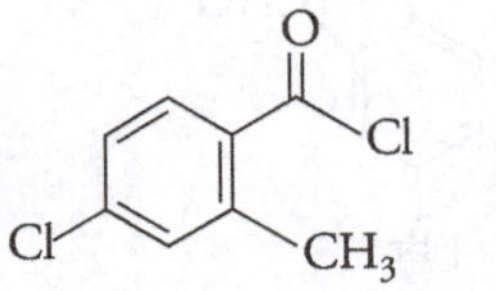 (b) 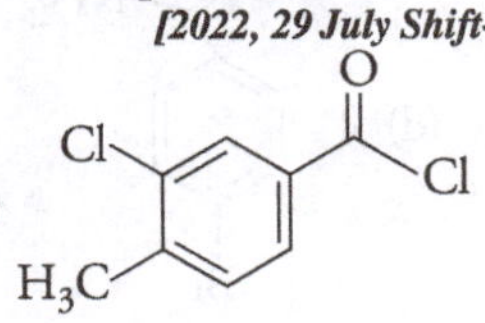

(c) 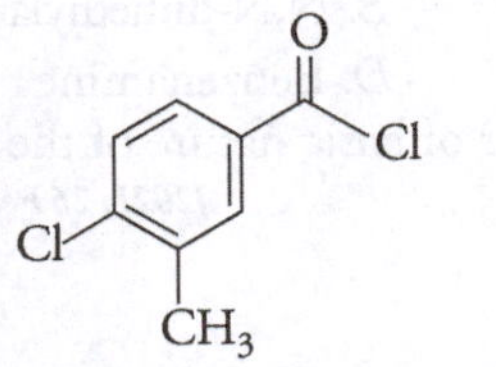 (d) 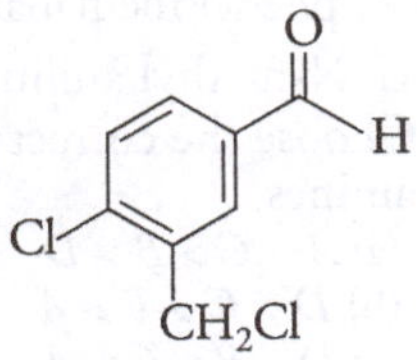

**29** With respect to the following reaction, consider the given statements. *[2022, 28 June Shift-II]*

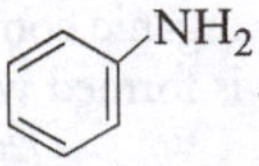 $\xrightarrow[H_2SO_4,\ 288\ K]{HNO_3}$ Products

(A) *o*-nitroaniline and *p*-nitroaniline are the predominant products.
(B) *p*-nitroaniline and *m*-nitroaniline are the predominant products.
(C) $HNO_3$ acts as an acid.
(D) $H_2SO_4$ acts as an acid.

Choose the correct option.
(a) (A) and (C) are correct statements.
(b) (A) and (D) are correct statements.
(c) (B) and (D) are correct statements.
(d) (B) and (C) are correct statements.

**30** Which one of the products of the following reactions does not react with Hinsberg reagent to form sulphonamide? *[2021, 25 July Shift-I]*

(a) 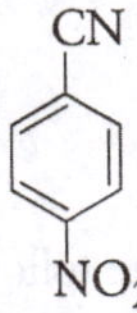 + Na/Hg $\xrightarrow{C_2H_5OH}$

(b) 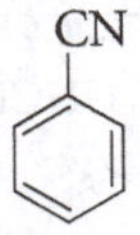 + $SnCl_2$ + HCl $\longrightarrow$

(c) 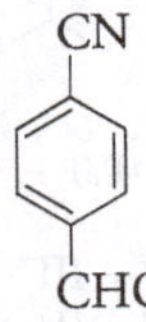 + $LiAlH_4$ $\xrightarrow{H_3O^+}$

(d) 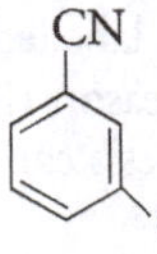 + $H_2$/Ni $\longrightarrow$

**31** The major products *A* and *B* formed in the following reaction sequence are *[2021, 31 Aug Shift-II]*

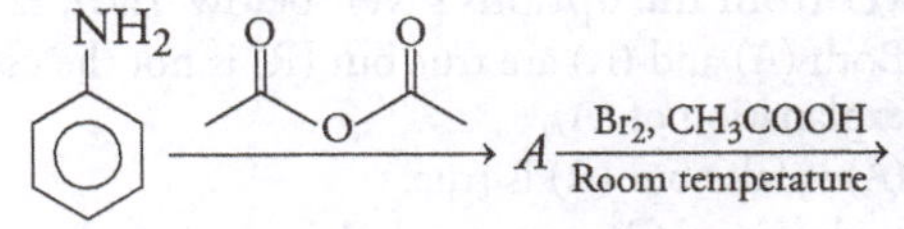

(a) *A* = 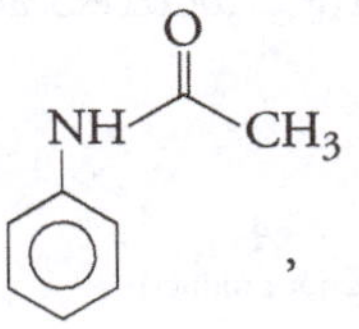, *B* = 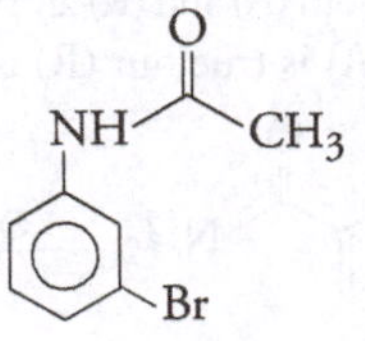

(b) *A* = 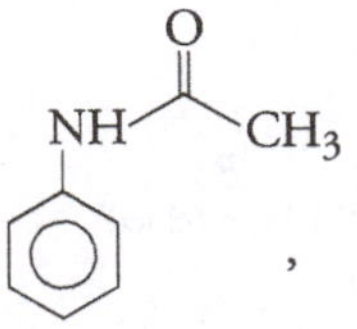, *B* = 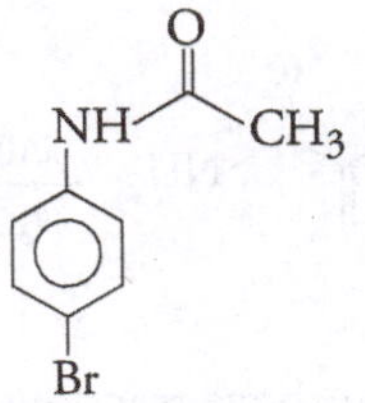

(c) *A* = 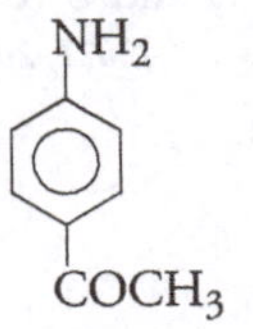, *B* = 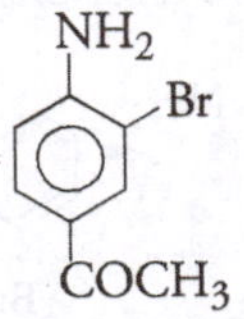

(d) *A* = 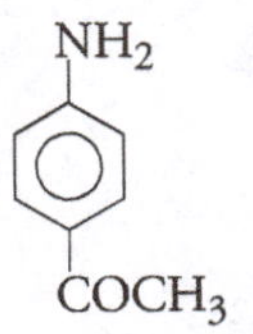, *B* = 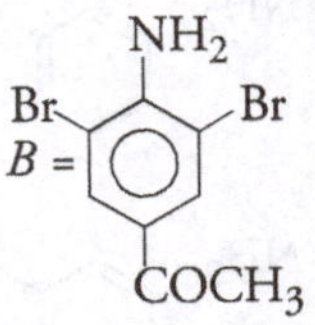

**32** The major product formed in the following reaction is *[2021, 26 Aug Shift-I]*

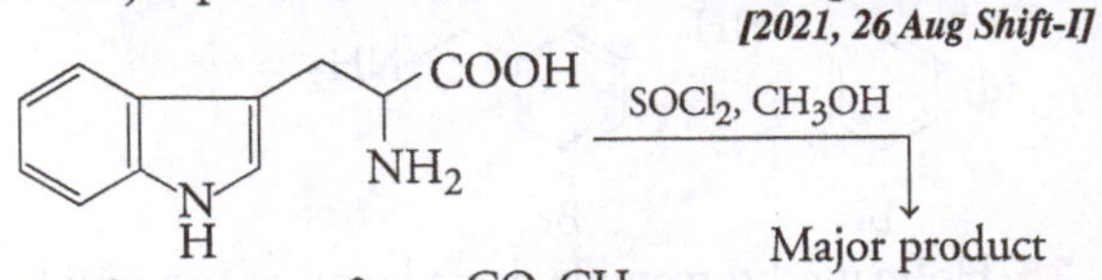

(a) 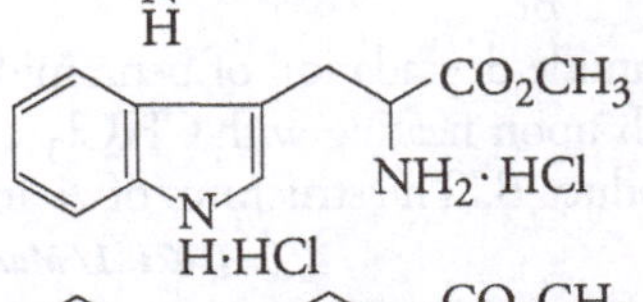

(b) 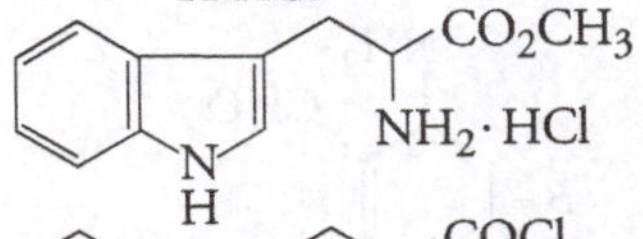

(c) 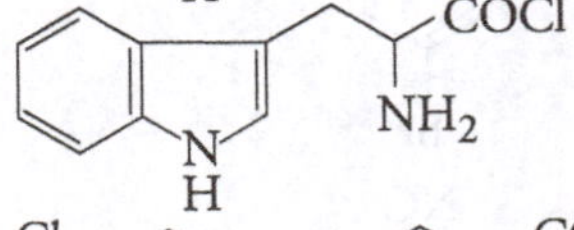

(d) 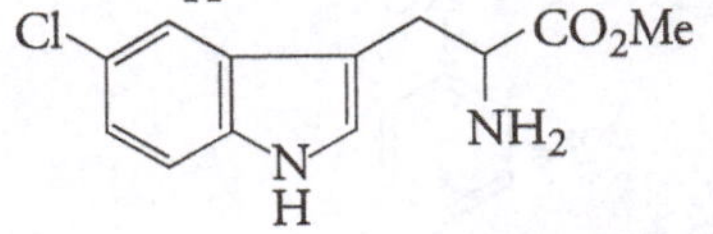

**33** Given below are two statements, one is labelled as Assertion (A) and other is labelled as Reason (R).
**Assertion (A)** Gabriel phthalimide synthesis cannot be used to prepare aromatic primary amines.
**Reason (R)** Aryl halides do not undergo nucleophilic substitution reaction.
In the light of the above statements, choose the correct answer from the options given below ***[2021, 22 July Shift-I]***
(a) Both (A) and (R) are true but (R) is not the correct explanation of (A).
(b) (A) is false but (R) is true.
(c) Both (A) and (R) are true and (R) is correct explanation of (A).
(d) (A) is true but (R) is false.

**34**

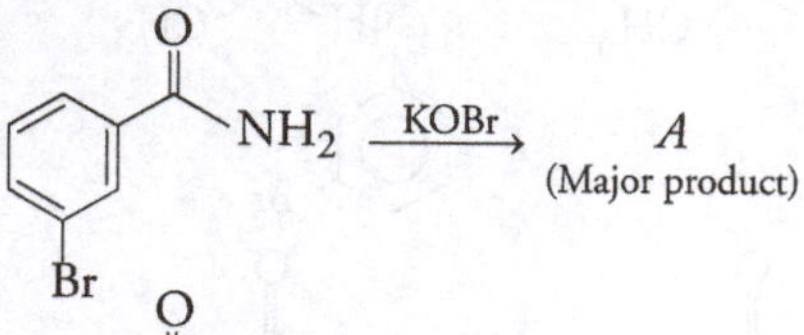


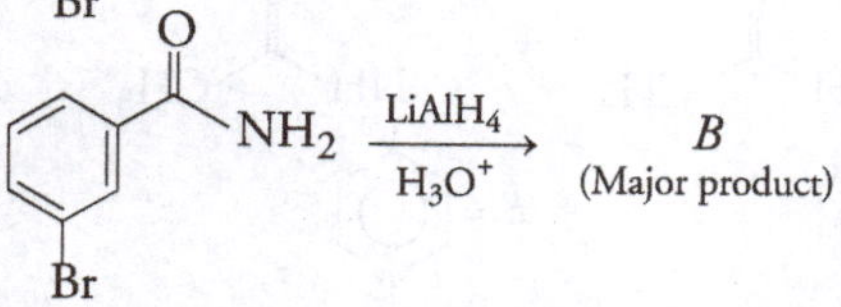


In the above reactions, products *A* and *B* respectively are ***[2021, 20 July Shift-II]***

(a) 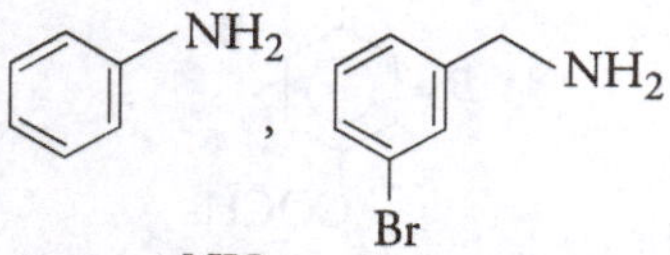

(b) 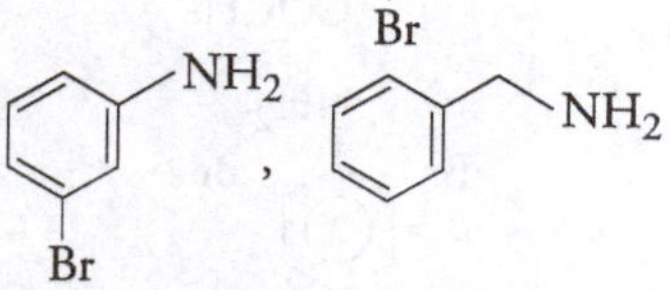

(c) 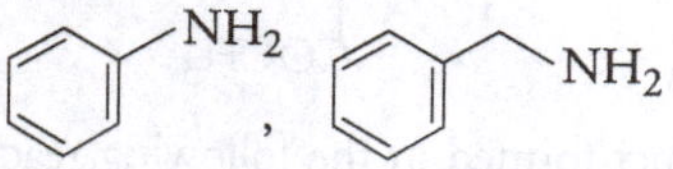

(d) 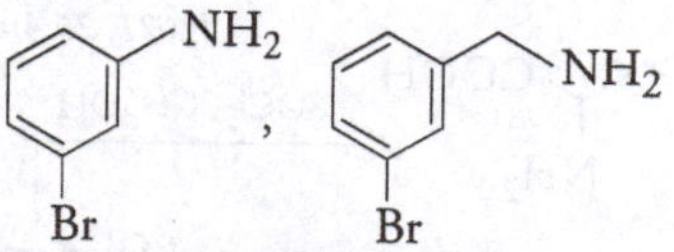

**35** Hofmann bromamide degradation of benzamide gives product *A*, which upon heating with $CHCl_3$ and NaOH gives product *B*. The structures of *A* and *B* are ***[2021, 17 March Shift-I]***

(a) *A* = 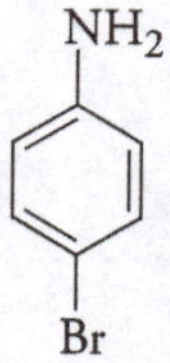 *B* = 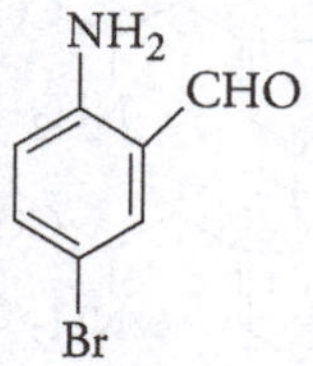

(b) *A* = 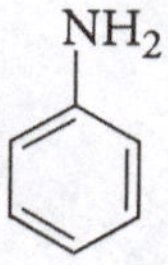 *B* = 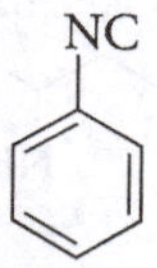

(c) *A* = 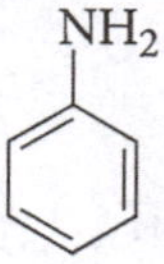 *B* = 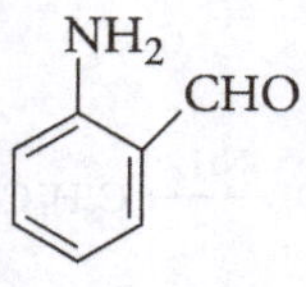

(d) *A* = 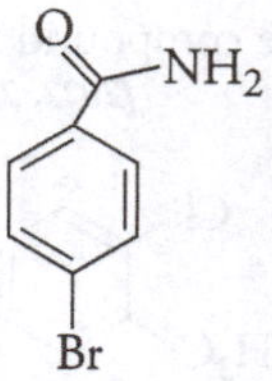 *B* = 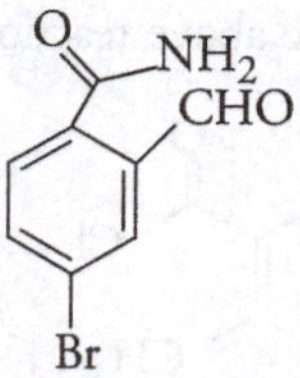

**36** *A*. phenyl methanamine *B*. N,N-dimethylaniline
*C*. N-methyl aniline *D*. Benzenamine
Choose the correct order of basic nature of the above amines. ***[2021, 26 Feb Shift-II]***
(a) $A > C > B > D$
(b) $D > C > B > A$
(c) $D > B > C > A$
(d) $A > B > C > D$

**37** Carbylamine test is used to detect the presence of primary amino group in an organic compound. Which of the following compound is formed when this test is performed with aniline? ***[2021, 25 Feb Shift-II]***

(a) 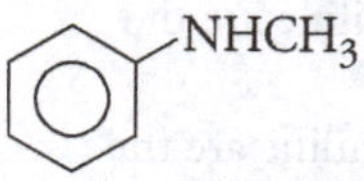 (b) 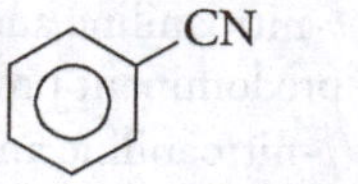

(c) 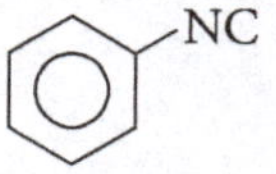 (d) 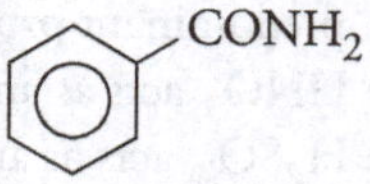

**38** The increasing order of basicity of the following compounds is ***[2020, 5 Sep Shift-I]***

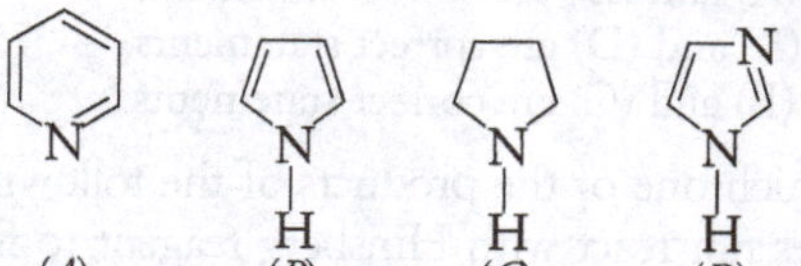

(A) (B) (C) (D)

(a) (*A*) < (*B*) < (*C*) < (*D*)
(b) (*D*) < (*A*) < (*B*) < (*C*)
(c) (*B*) < (*A*) < (*D*) < (*C*)
(d) (*B*) < (*A*) < (*C*) < (*D*)

**39** The increasing order of $pK_b$ values of the following compounds is ***[2020, 6 Sep Shift-I]***

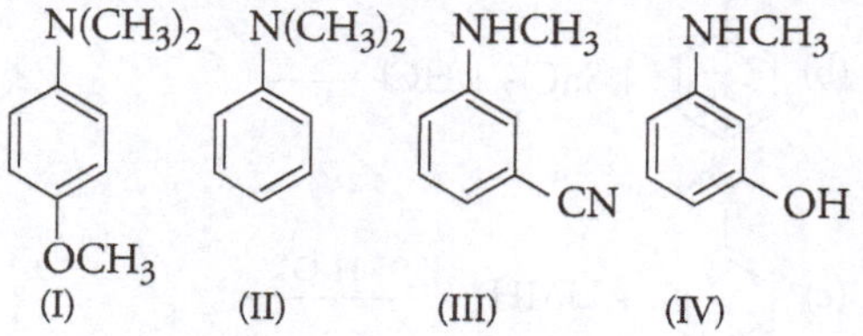

(I) (II) (III) (IV)

(a) II < IV < III < I (b) I < II < IV < III
(c) II < I < III < IV (d) I < II < III < IV

**40** The decreasing order of basicity of the following amines is *[2020, 9 Jan Shift-II]*

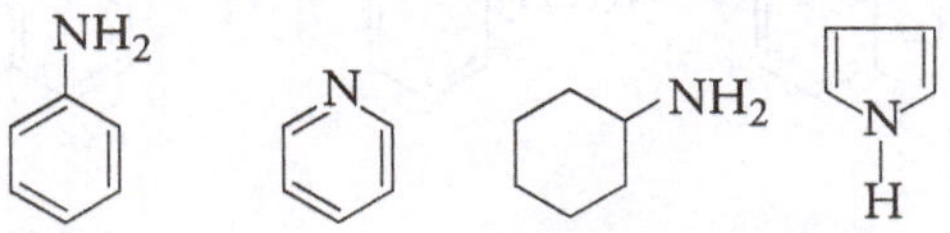

(I) (II) (III) (IV)

(a) (III) > (I) > (II) > (IV)
(b) (III) > (II) > (I) > (IV)
(c) (I) > (III) > (IV) > (II)
(d) (II) > (III) > (IV) > (I)

**41** The increasing order of reactivity of the following compounds towards reaction with alkyl halides directly is *[2019, 12 Jan Shift-I]*

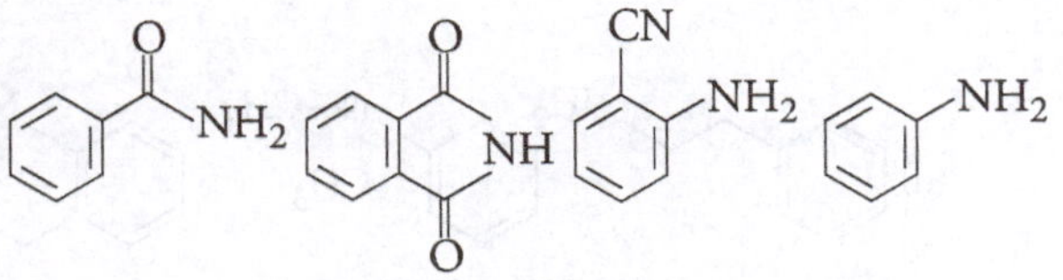

(*A*) (*B*) (*C*) (*D*)

(a) (*A*) < (*C*) < (*D*) < (*B*)
(b) (*B*) < (*A*) < (*C*) < (*D*)
(c) (*B*) < (*A*) < (*D*) < (*C*)
(d) (*A*) < (*B*) < (*C*) < (*D*)

**42** The major product of the following reaction is

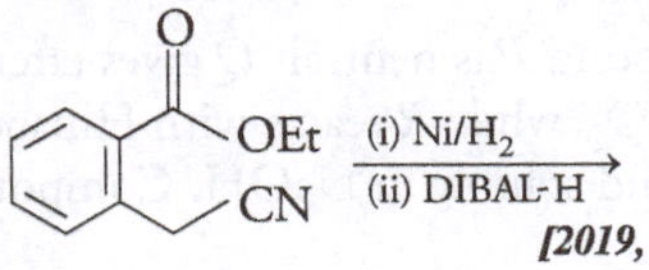


*[2019, 11 Jan Shift-I]*

(a) 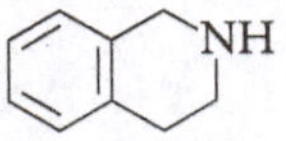 (b) 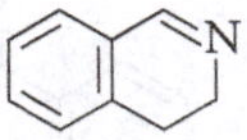

(c) 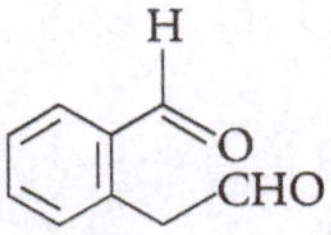 (d) 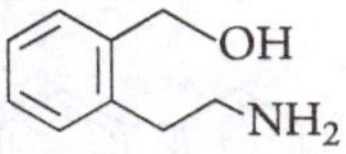

**43** An aromatic compound '*A*' having molecular formula $C_7H_6O_2$ on treating with aqueous ammonia and heating forms compound '*B*'. The compound '*B*' on reaction with molecular bromine and potassium hydroxide provides compound '*C*' having molecular formula $C_6H_7N$. The structure of '*A*' is *[2019, 10 Jan Shift-II]*

(a) 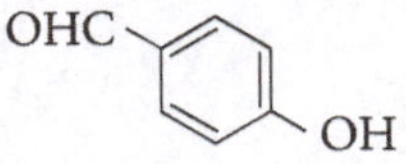 (b) 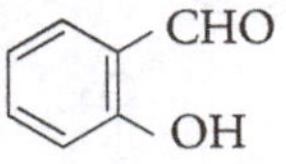

(c) 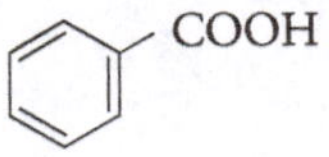 (d) 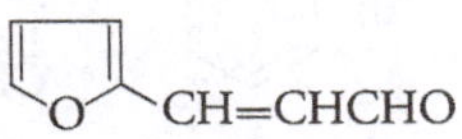

## Level 3

**44** The descending order of basicity of following amines is:

(A) 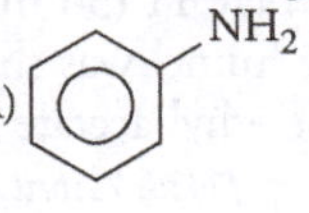 (B) 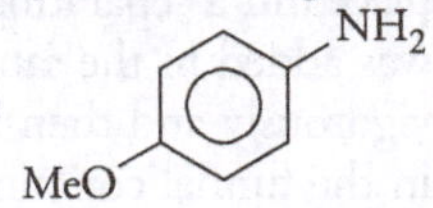

(C) 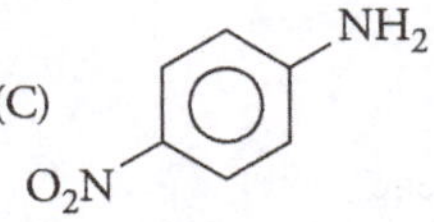 (D) $CH_3NH_2$

(E) $(CH_3)_2NH$

Choose the correct answer from the options given below: *[2025, 7 April Shift-II]*

(a) E > D > B > A > C (b) E > D > A > B > C
(c) E > A > D > C > B (d) B > E > D > A > C

**45** The major product (*A*) formed in the following reaction sequence is *[2025, 4 April Shift-I]*

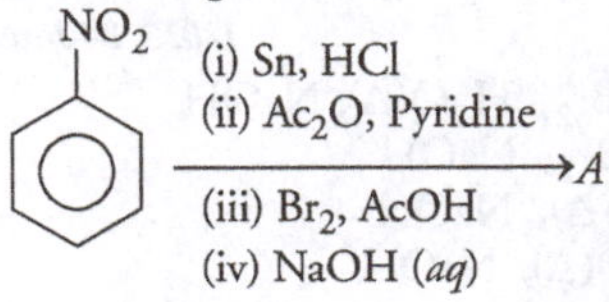


(a) 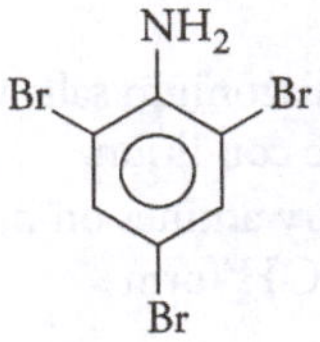 (b) 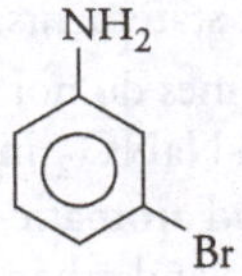

(c) 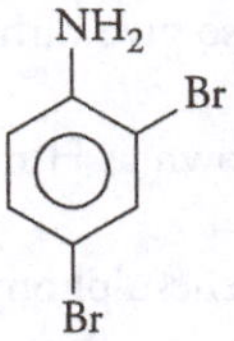 (d) 

**46** The sequence from the following that would result in giving predominantly 3, 4, 5-tribromoaniline is *[2025, 3 April Shift-II]*

(a) 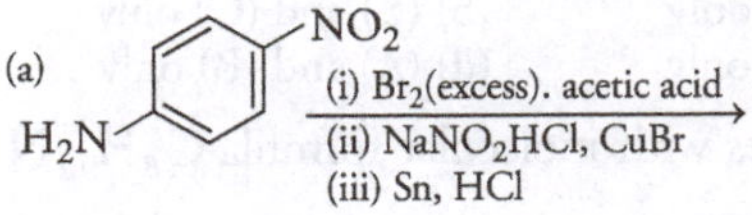


(b) 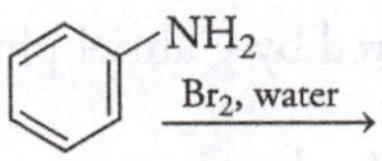


(c) 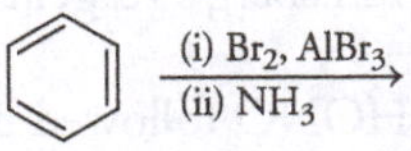


(d) 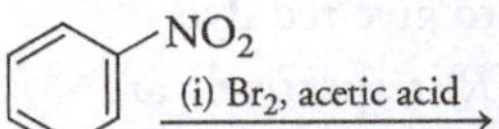

**47** Mixture of 1 g each of chlorobenzene, aniline and benzoic acid is dissolved in 50 mL ethyl acetate and placed in a separating funnel. 5 M NaOH (30 mL) was added in the same funnel. The funnel was shaken vigorously and then kept aside. The ethyl acetate layer in the funnel contains: ***[2025, 7 April Shift-II]***

(a) benzoic acid
(b) benzoic acid and aniline
(c) chlorobenzene and aniline
(d) benzoic acid and chlorobenzene

**48** For reaction,

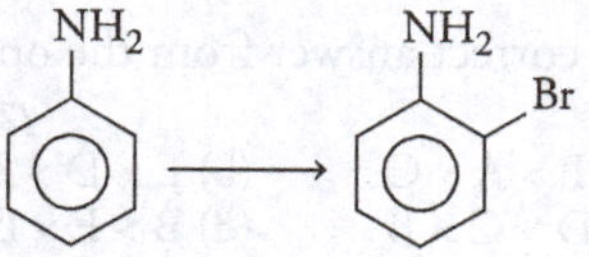

(Major)

The correct order of set of reagents for the above conversion is ***[2025, 24 Jan Shift-II]***

(a) $H_2SO_4$, $Ac_2O$, $Br_2$, $H_2O(\Delta)$, NaOH
(b) $Ac_2O$, $H_2SO_4$, $Br_2$, NaOH
(c) $Br_2$ |$FeBr_3$, $H_2O(\Delta)$, NaOH
(d) $Ac_2O$, $Br_2$, $H_2O(\Delta)$, NaOH

**49** Identify correct statements.

A. Primary amines do not give diazonium salts when treated with $NaNO_2$ in acidic condition.
B. Aliphatic and aromatic primary amines on heating with $CHCl_3$ and ethanolic KOH form carbylamines.
C. Secondary and tertiary amines also give carbylamine test.
D. Benzenesulphonyl chloride is known as Hinsberg's reagent.
E. Tertiary amines reacts with benzenesulphonyl chloride very easily.

Choose the correct answer from the options given below. ***[2025, 28 Jan Shift-II]***

(a) (D) and (E) only
(b) (B) and (C) only
(c) (B) and (D) only
(d) (A) and (B) only

**50** Isomeric amines with molecular formula $C_8H_{11}N$ give the following tests

Isomer (*P*) ⇒ can be prepared by Gabriel phthalimide synthesis.

Isomer (*Q*) ⇒ reacts with Hinsberg's reagent to give solid insoluble in NaOH.

Isomer (R) ⇒ reacts with HONO followed by β-naphthol in NaOH to give red dye.

Isomers (*P*), (*Q*) and (*R*) respectively are ***[2023, 10 April Shift-I]***

*P* *Q* *R*

(a) 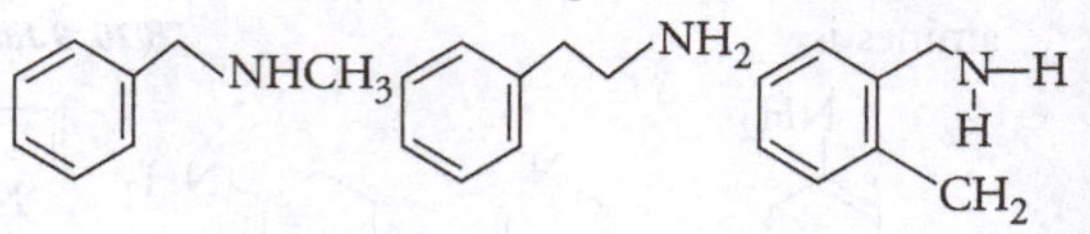

(b) 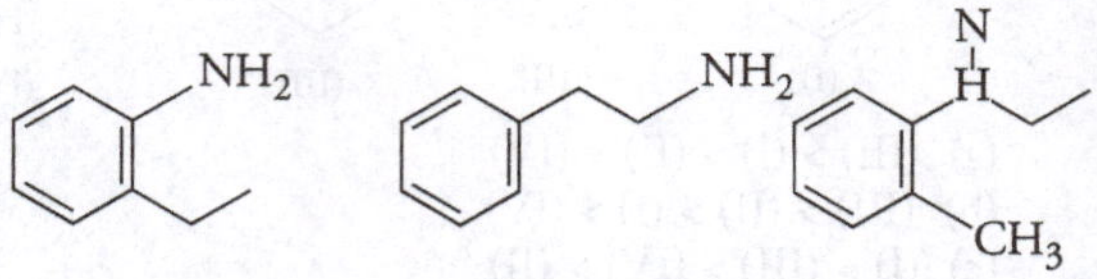

*P* *Q* *R*

(c) 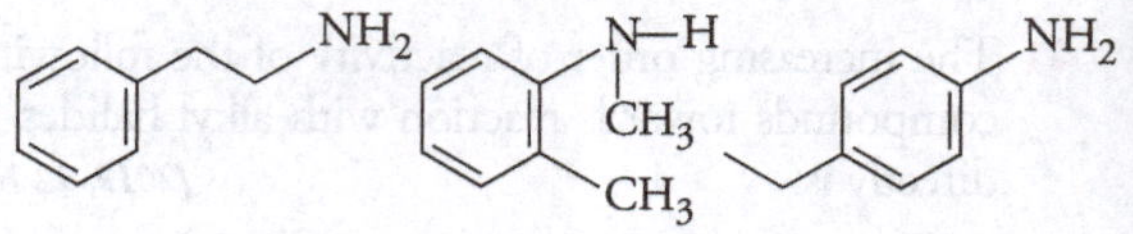

(d) 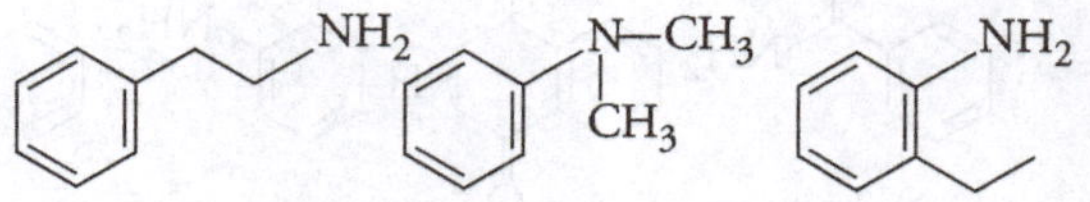

**51** Compound *P* $\xrightarrow{HCl, \Delta}$ Filter → Residue *Q*; → Filtrate $\xrightarrow{NaOH}$ Oily liquid *R*

(M.F. $C_{14}H_{13}ON$; M.F = Molecular formula)

Compound *P* is neutral, *Q* gives effervescence with $NaHCO_3$ while *R* reacts with Hinsbergs reagent to give solid soluble in NaOH. Compound *P* is ***[2023 6 April Shift-I]***

(a) 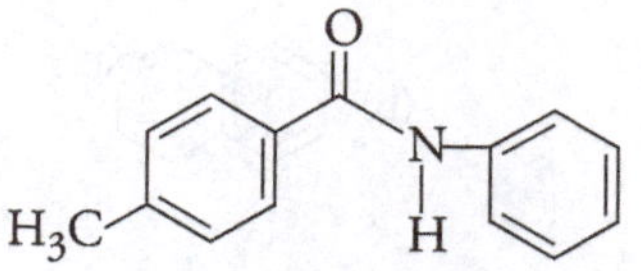

(b) 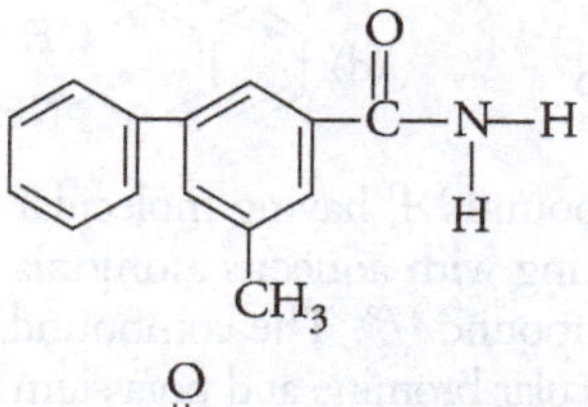

(c) 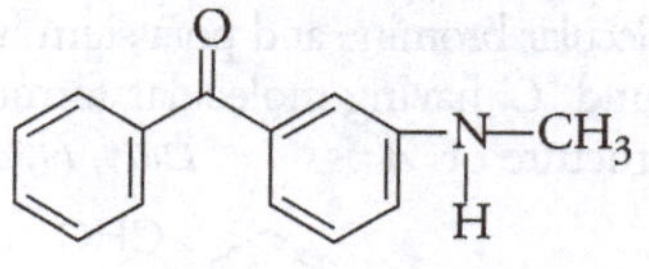

(d) 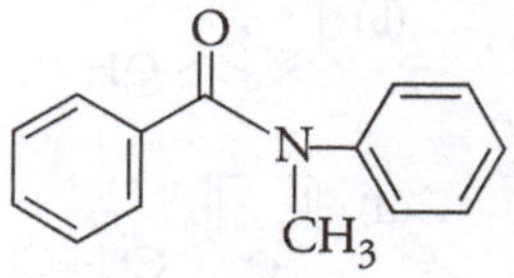

**52** 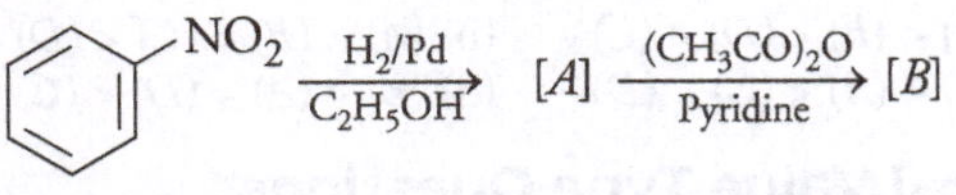

Consider the above reaction and identify the product *B*. *[2023, 31 Jan Shift-I]*

(a) 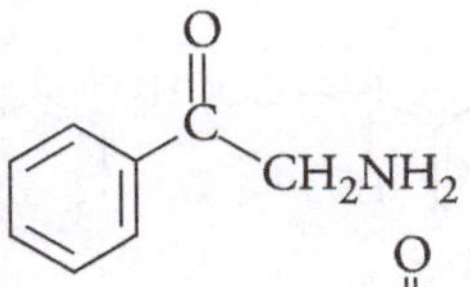

(b) 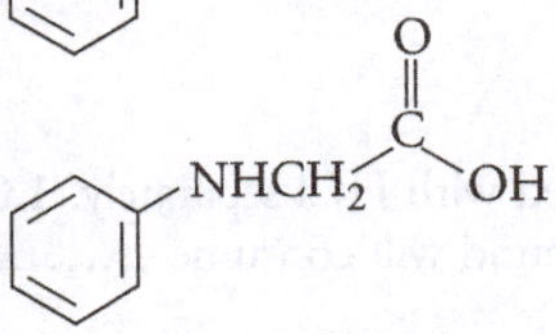

(c) 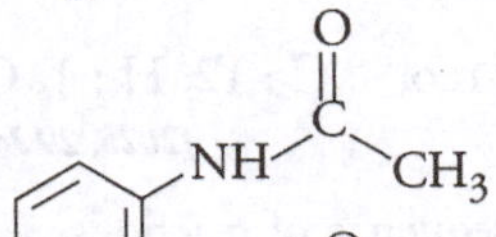

(d) 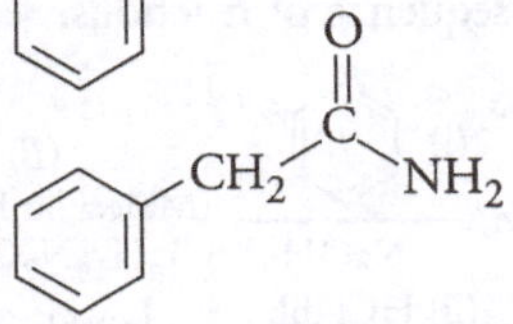

**53** Match List-I with List-II.

| List-I (Reaction) | List-II (Reagents) |
|---|---|
| A. Hoffmann degradation | I. Conc. KOH, Δ |
| B. Clemmensen reduction | II. $CHCl_3$, $NaOH/H_3O^+$ |
| C. Cannizzaro reaction | III. $Br_2$, NaOH |
| D. Reimer-Tiemann reaction | IV. Zn-Hg/HCl |

Choose the correct answer from the options given below. *[2023, 29 Jan Shift-I]*

(a) A-II, B-I, C-III, D-IV (b) A-II, B-IV, C-I, D-III
(c) A-III, B-IV, C-I, D-II (d) A-III, B-IV, C-II, D-I

**54** Consider the following reaction sequence

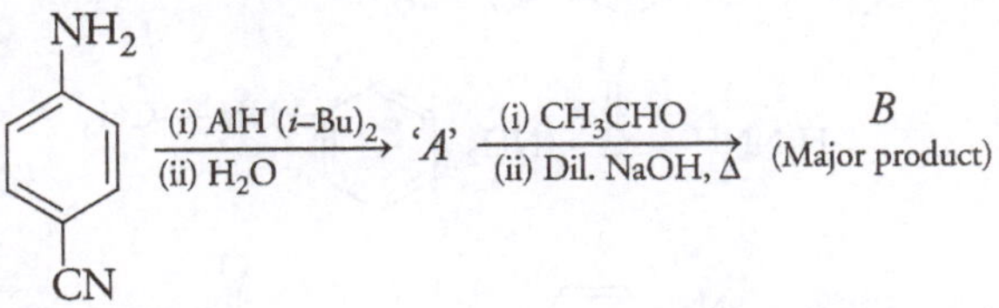

The product '*B*' is *[2022, 29 July Shift-I]*

(a) 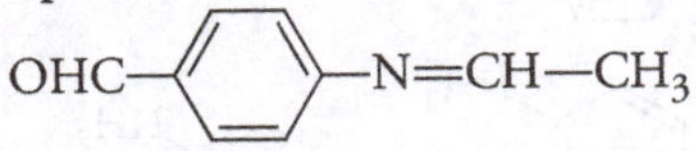

(b) 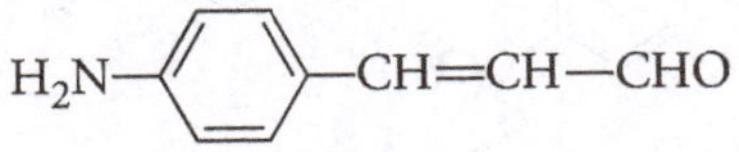

(c) 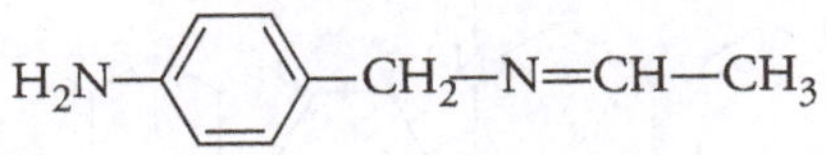

(d) 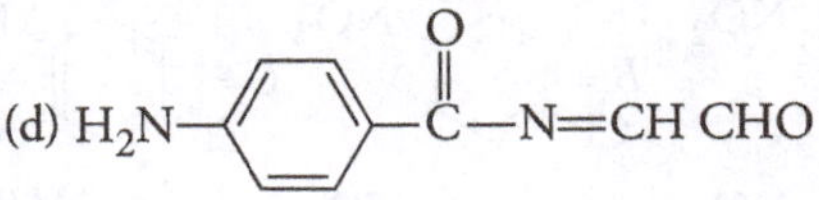

**55** 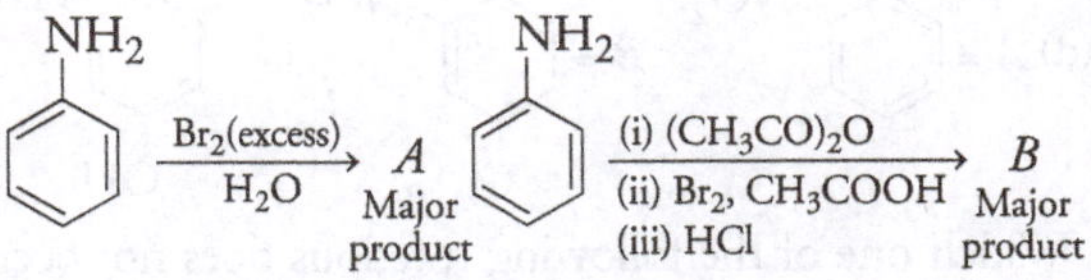

Consider the above reactions, the product *A* and product *B* respectively are *[2022, 29 June Shift-I]*

(a) 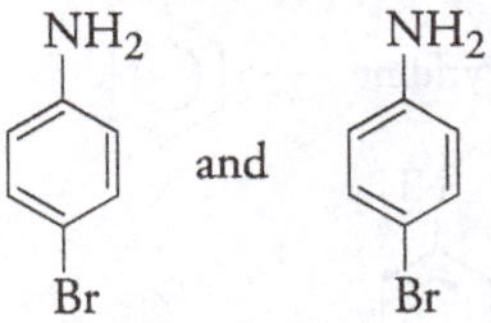

(b) 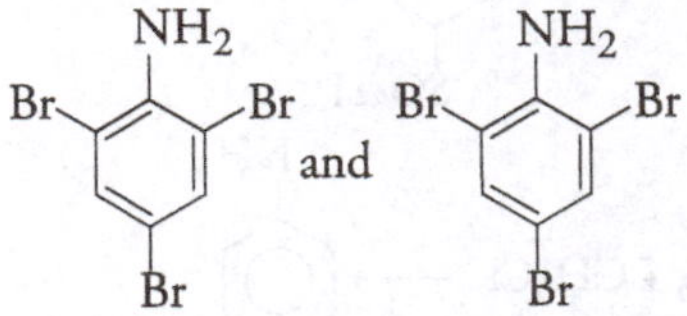

(c) 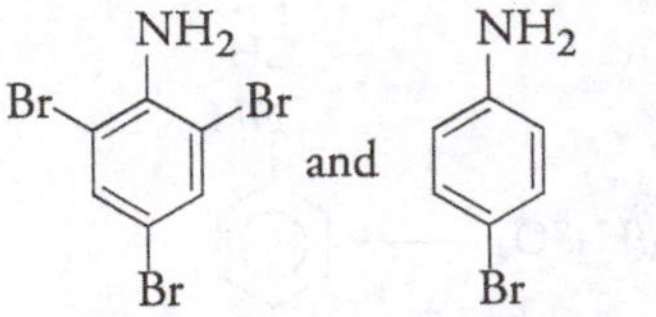

(d) 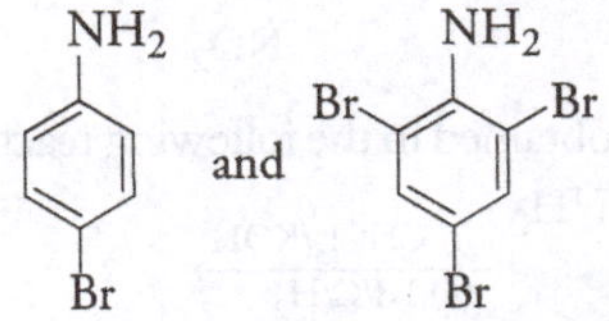

**56** Decarboxylation of all six possible forms of diaminobenzoic acid $C_6H_3(NH_2)_2COOH$ yields three products *A*, *B* and *C*. Three acids give a product *A*, two acids give a product '*B*' and one acid gives a product *C*. The melting point of product *C* is *[2022, 27 June Shift-II]*

(a) 63°C (b) 90°C (c) 104°C (d) 142°C

**57** Identify correct *A*, *B* and *C* in the reaction sequence given below. *[2021, 31 Aug Shift-II]*

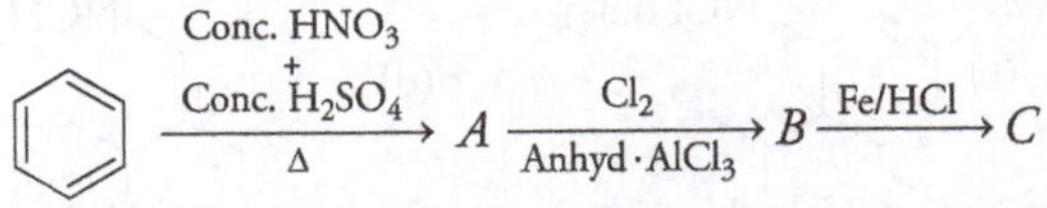

(a) 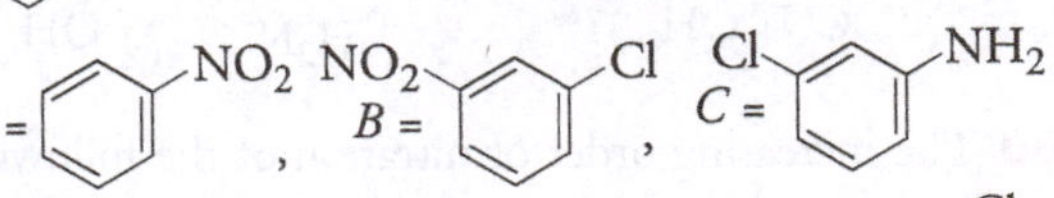

(b) 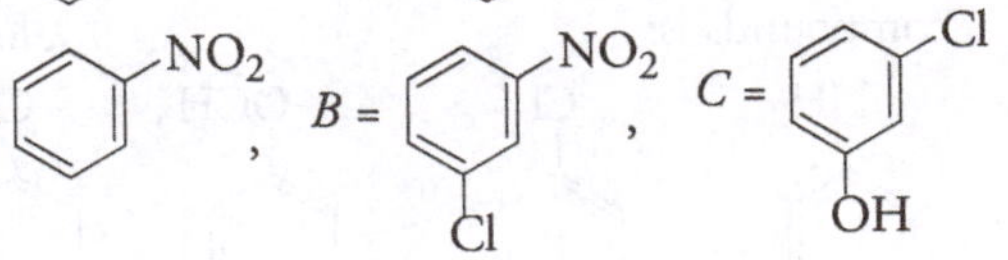

(c) 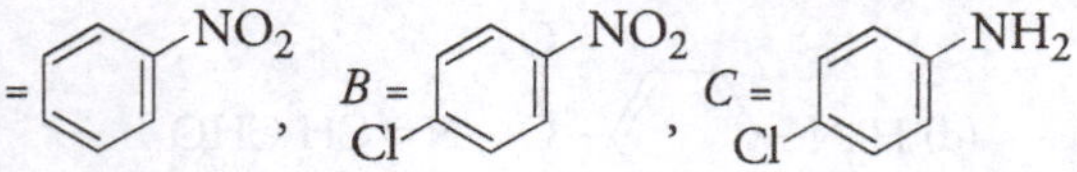

(d) 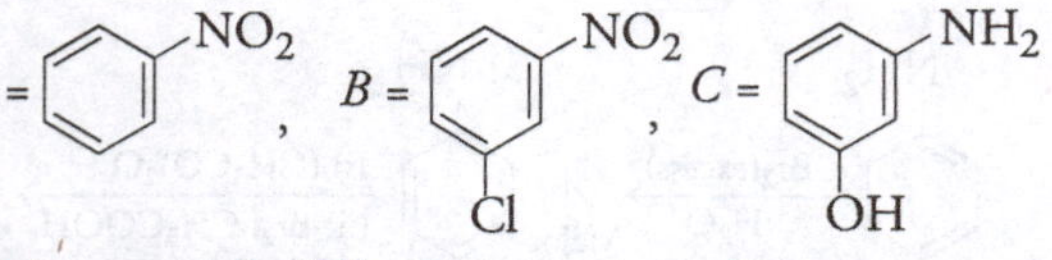

**58** Which one of the following reactions does not occur? *[2021, 22 July Shift-II]*

(a) 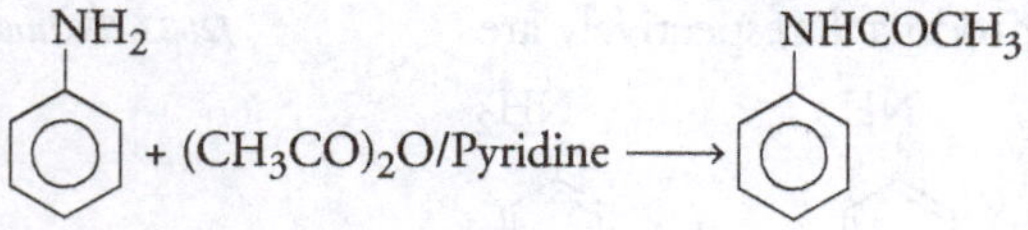

(b) 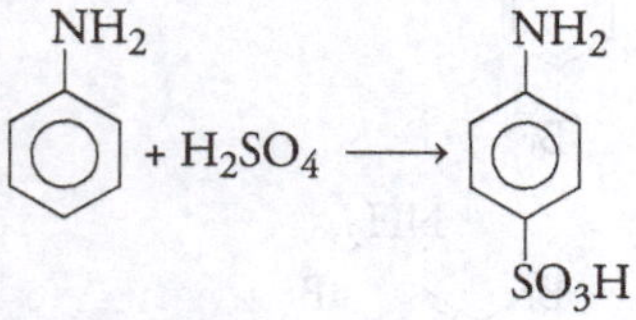

(c) 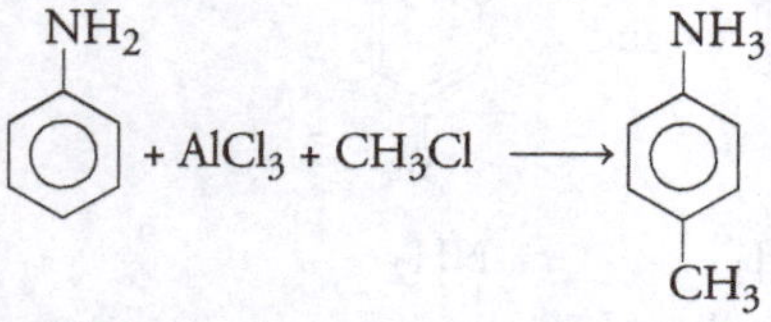

(d) 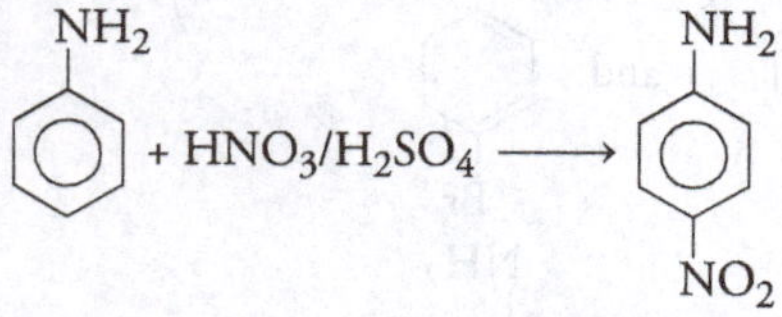

**59** The major product obtained in the following reaction is

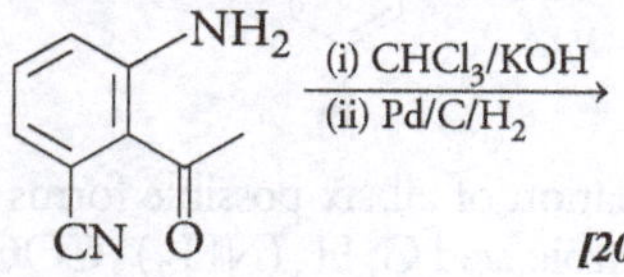

*[2019, 8 April Shift-II]*

(a) 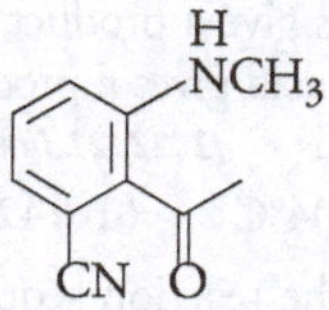

(b) 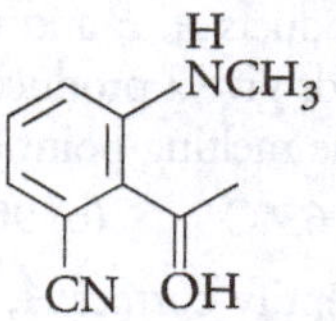

(c) 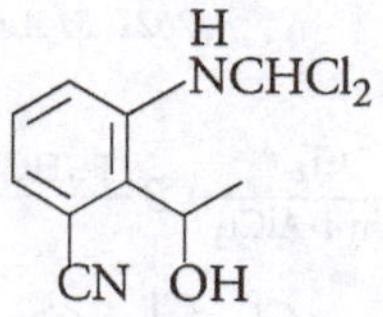

(d) 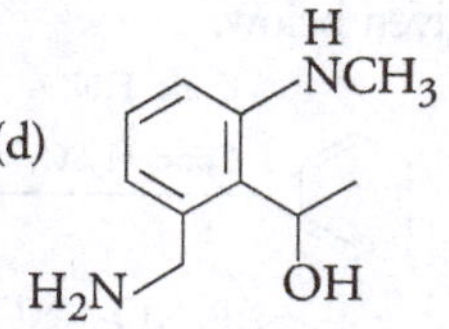

**60** The increasing order of nitration of the following compounds is: *[JEE Main 2018]*

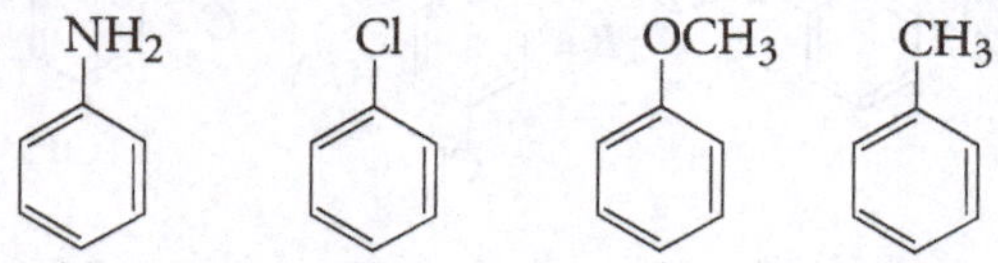

(A) (B) (C) (D)

(a) (A) < (B) < (D) < (C) (b) (A) < (B) < (C) < (D)
(c) (B) < (A) < (C) < (D) (d) (B) < (A) < (D) < (C)

## Numerical Value Type Questions

**61** Given below are some nitrogen containing compounds

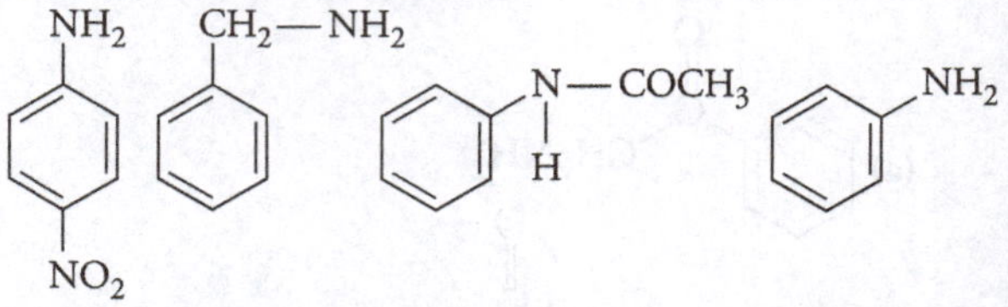

Each of them is treated with HCl separately. 1.0 g of the most basic compound will consume ........... mg of HCl.

(Given, molar mass in g $mol^{-1}$ C : 12, H : 1, O : 16, Cl : 35.5) *[2025, 29 Jan Shift-I]*

**62** Consider the following sequence of reactions.

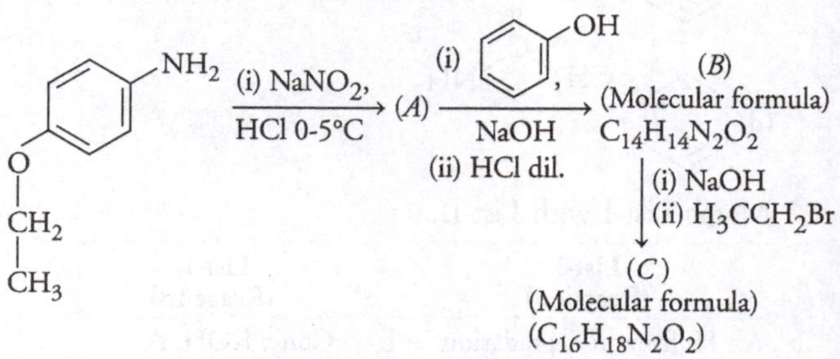

Total number of $sp^3$ hybridised carbon atoms in the major product $C$ formed is ......... *[2025, 23 Jan Shift-II]*

**63** Number of amine compounds from the following giving solids which are soluble in NaOH upon reaction with Hinsberg's reagent is ........ *[2024, 8 April Shift-I]*

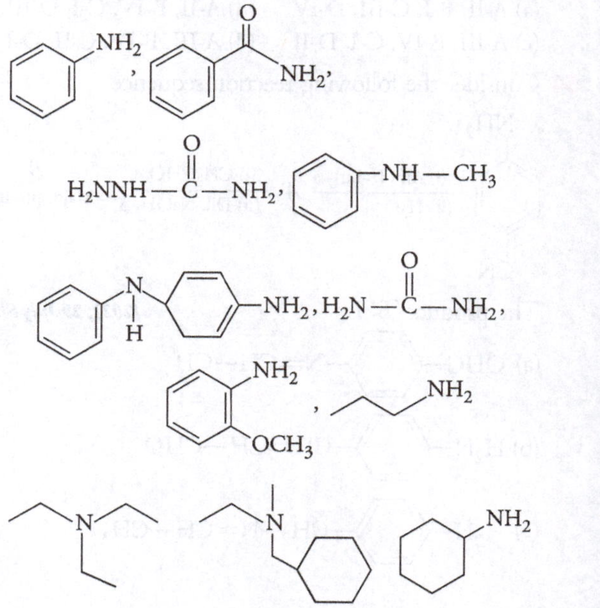

**64** Number of compounds which give reaction with Hinsberg's reagent is ......... . *[2024, 1 Feb Shift-II]*

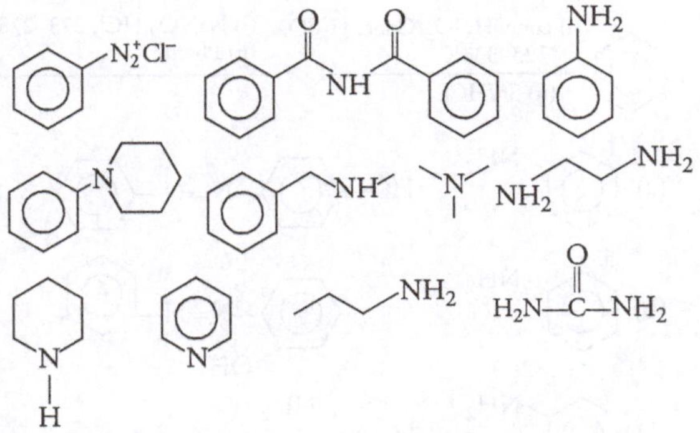

**65** How many of the transformations given below would result in aromatic amines? *[2023, 31 Jan Shift-I]*

(a) 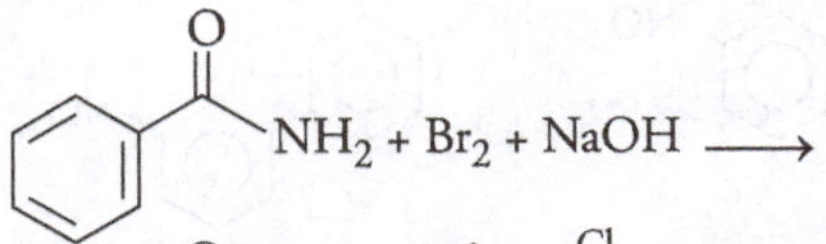

(b) 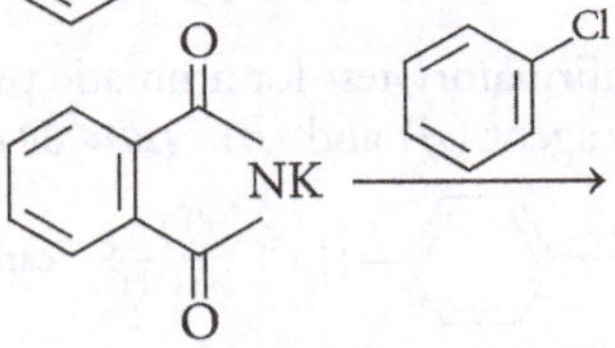

(c) 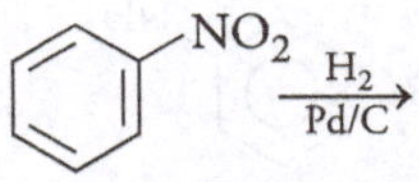

(d) 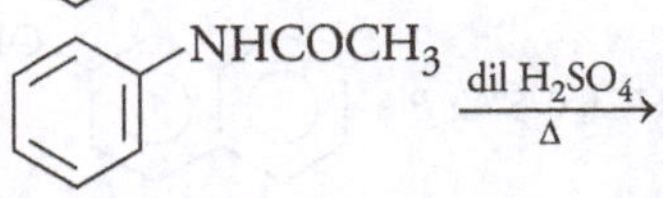

**66** The total number of reagents from those given below, that can convert nitrobenzene into aniline is ........ (Integer answer) *[2021, 31 Aug Shift-I]*

I. Sn – HCl II. Sn – $NH_4OH$
III. Fe – HCl IV. Zn – HCl
V. $H_2$ – Pd
VI. $H_2$-Raney nickel

**67** The total number of amines among the following which can be synthesised by Gabriel synthesis is ......... . *[2021, 24 Feb Shift-II]*

(a) $(CH_3)_2$CHCH$_2$NH$_2$

(b) $CH_3CH_2NH_2$

(c) 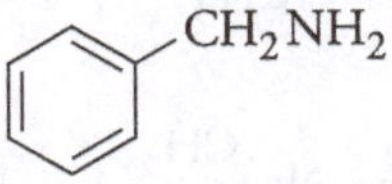

(d) 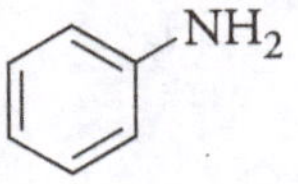

## Topic 3 Diazonium Salts and Other Nitrogen Containing Compounds

### Level 1

**01** *o*-phenylenediamine $\xrightarrow{HNO_2}$ '*X*' (Major)
Product '*X*' is *[2023, 11 April Shift-I]*

(a) 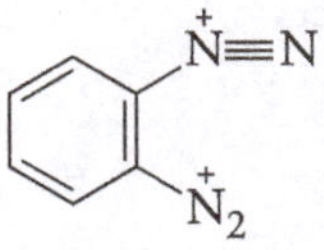 (b) 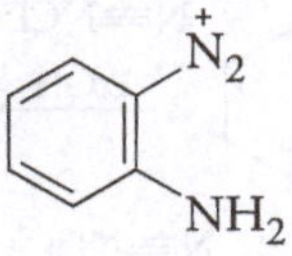

(c) 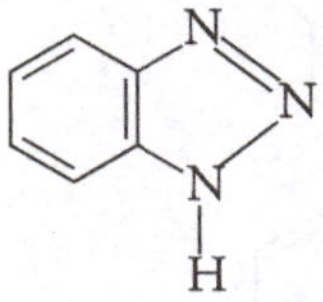 (d) 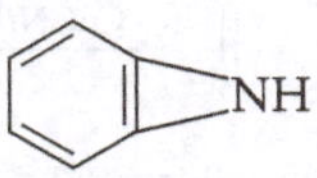

**02** 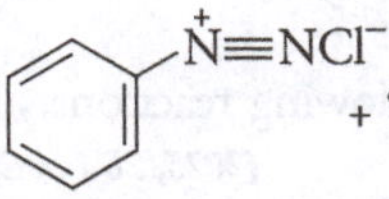 + '*A*' $\xrightarrow{NaOH}$ Product (Orange-red dye)

Which among the following represent reagent '*A*'? *[2022, 29 July Shift-I]*

(a) 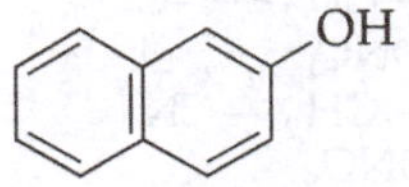 (b) 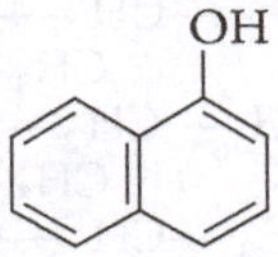

(c) 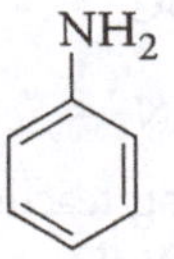 (d) 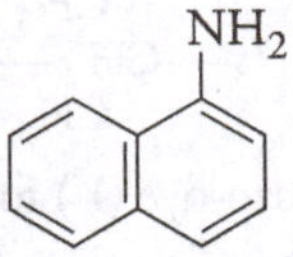

**03** Hydrolysis of which compound will give carbolic acid? *[2022, 26 July Shift-II]*

(a) Cumene
(b) Benzenediazonium chloride
(c) Benzal chloride
(d) Ethyene glycol ketal

**04** The correct stability order of the following diazonium salt is *[2022, 26 July Shift-I]*

(*A*) 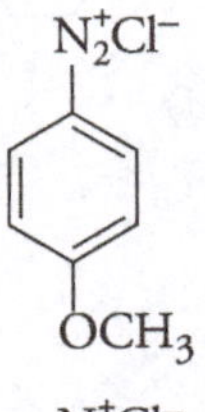 (*B*) 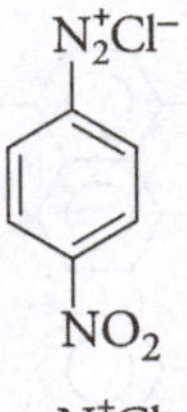

(*C*) 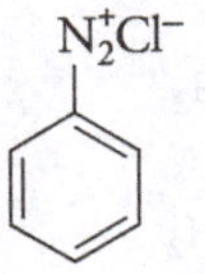 (*D*) 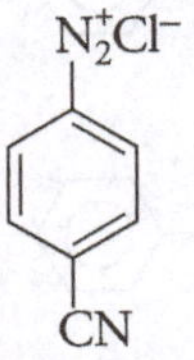

(a) (*A*) > (*B*) > (*C*) > (*D*) (b) (*A*) > (*C*) > (*D*) > (*B*)
(c) (*C*) > (*A*) > (*D*) > (*B*) (d) (*C*) > (*D*) > (*B*) > (*A*)

## Level 2

**05** In the following reactions, which one is NOT correct? **[2025, 3 April Shift-I]**

(a) 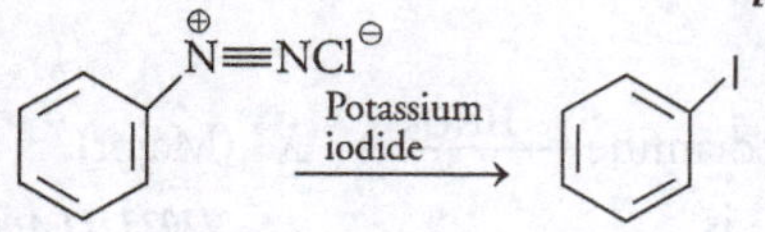

(b) 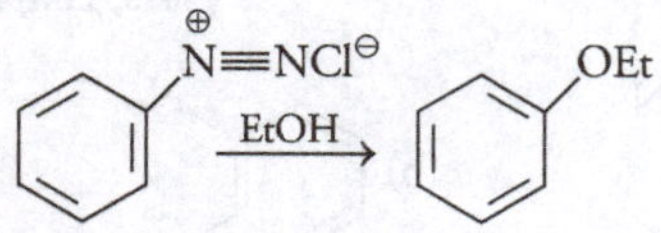

(c) 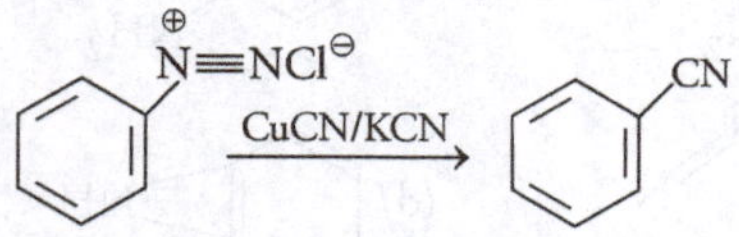

(d) 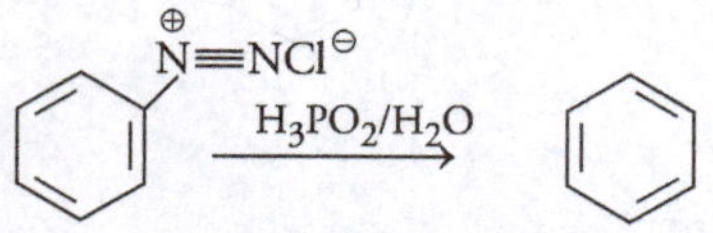

**06** The products $A$ and $B$ in the following reactions, respectively are **[2025, 28 Jan Shift-I]**

$$A \xleftarrow{Ag-NO_2} CH_3-CH_2-CH_2-Br \xrightarrow{AgCN} B$$

(a) $CH_3-CH_2-CH_2-NO_2$, $CH_3-CH_2-CH_2-NC$

(b) $CH_3-CH_2-CH_2-ONO$, $CH_3-CH_2-CH_2-CN$

(c) $CH_3-CH_2-CH_2-ONO$, $CH_3-CH_2-CH_2-NC$

(d) $CH_3-CH_2-CH_2-NO_2$, $CH_3-CH_2-CH_2-CN$

**07** The azo-dye ($Y$) formed in the following reactions is:

Sulphanilic acid + $NaNO_2$ + $CH_3COOH \longrightarrow X$

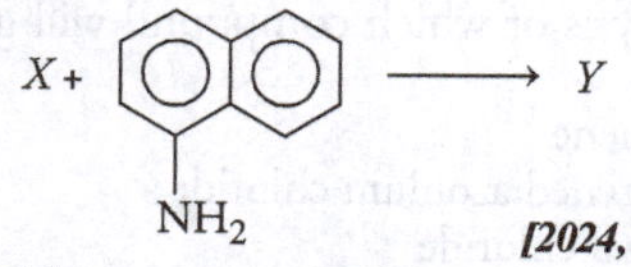

**[2024, 31 Jan Shift-II]**

(a) 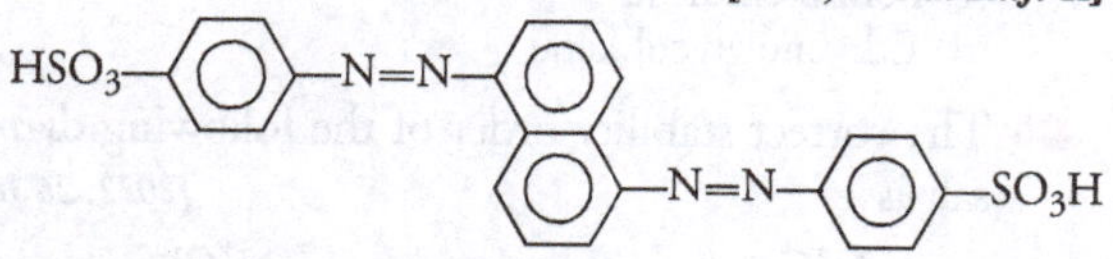

(b) 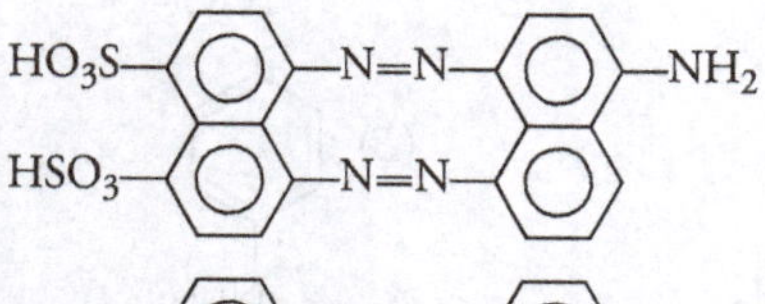

(c) 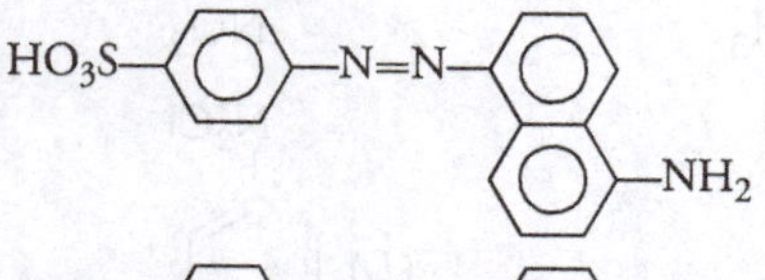

(d) 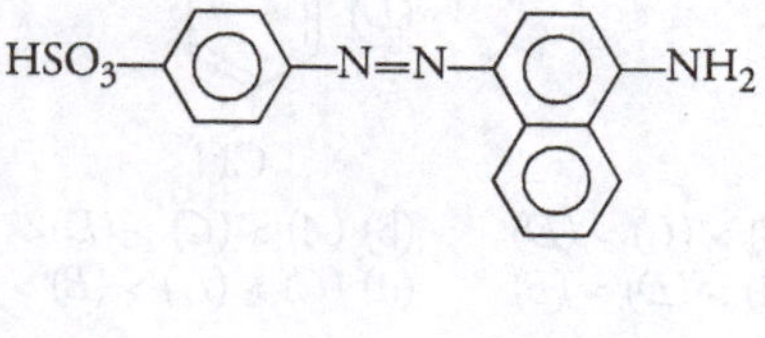

**08** The products $A$ and $B$ formed in the following reaction scheme are respectively **[2024, 30 Jan Shift-II]**

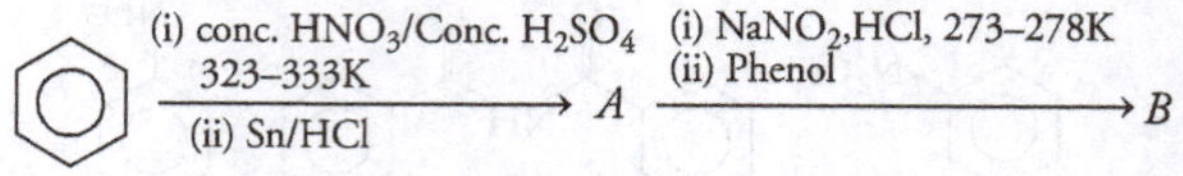

(a) 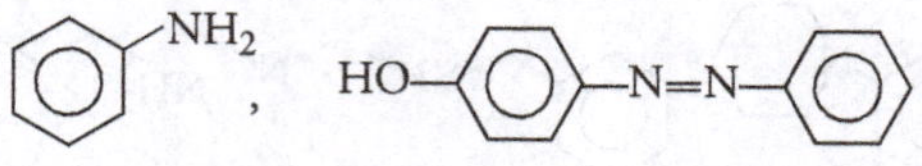

(b) 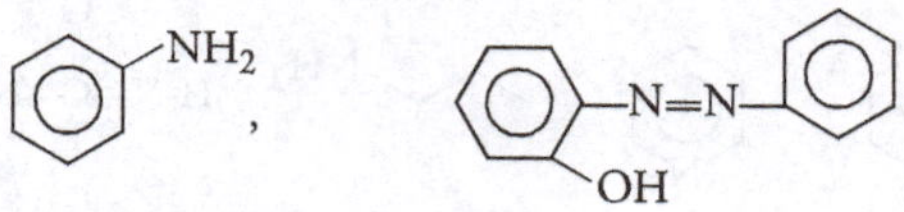

(c) 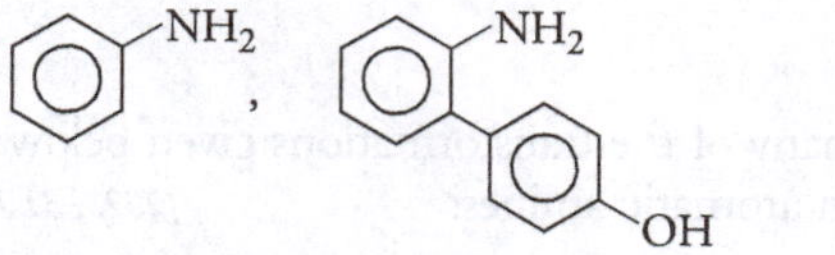

(d) 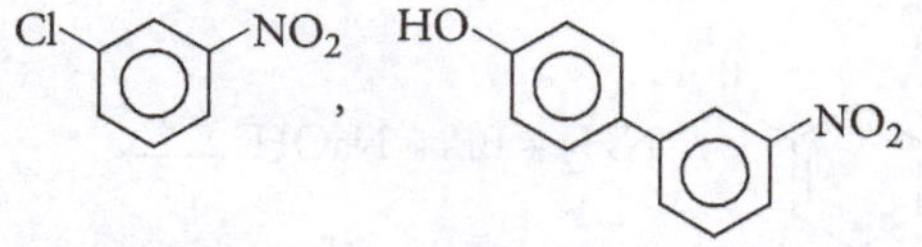

**09** Following is a confirmatory test for aromatic primary amines. Identify reagent ($A$) and ($B$). **[2024, 30 Jan Shift-I]**

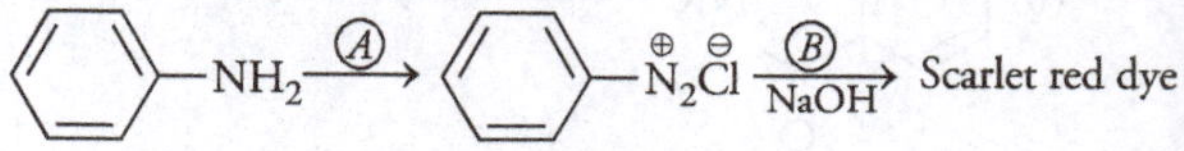

(a) $A$ = $NaNO_2$+HCl, 0-5°C; $B$ = 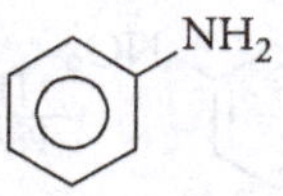

(b) $A$ = $NaNO_2$+HCl, 0-5°C; $B$ = 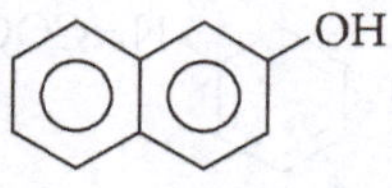

(c) $A$ = $HNO_3/H_2SO_4$ $B$ = 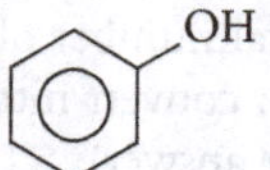

(d) $A$ = $NaNO_2$+HCl, 0-5°C; $B$ = 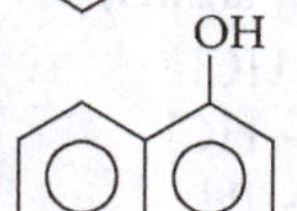

**10** Consider the following sequence of reactions

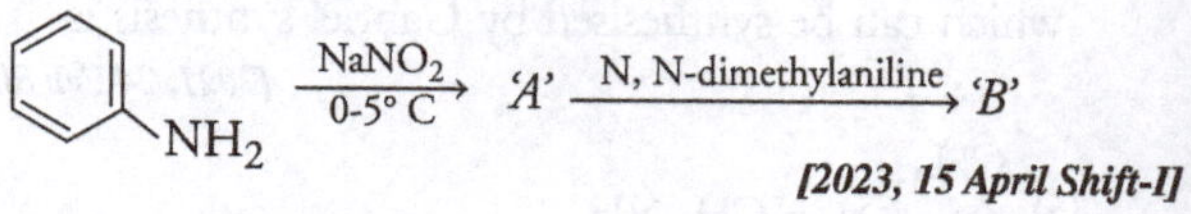

**[2023, 15 April Shift-I]**

The product 'B' is

(a) 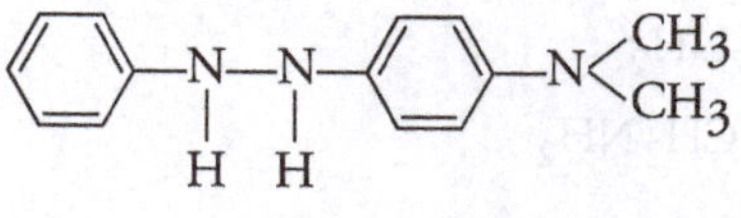

(b) 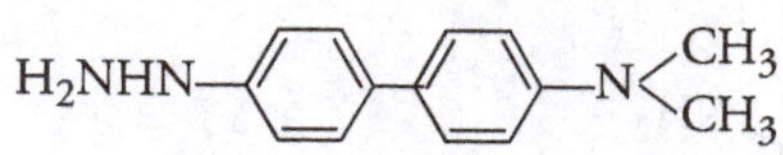

(c) 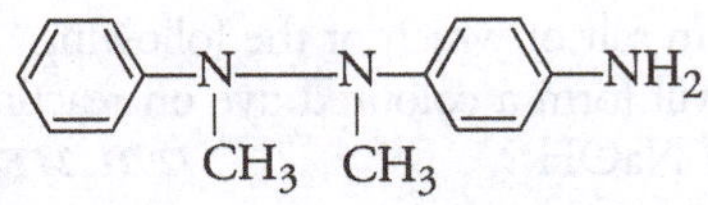

(d) 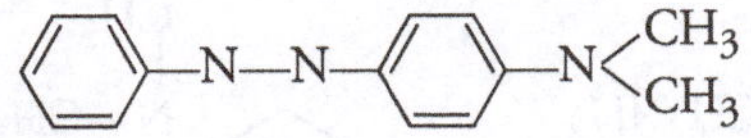

**11** Which one of the following gives the most stable diazonium salt ? **[2021, 01 Sep Shift-II]**

(a) $CH_3—CH_2—CH_2NH_2$ (b) 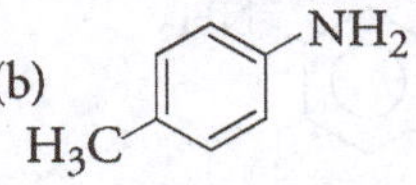

(c) 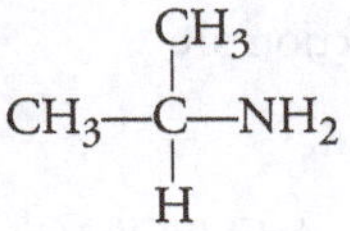 (d) 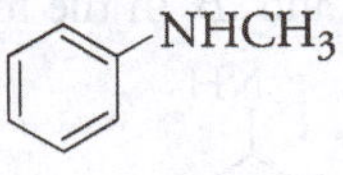

**12** $C_6H_5NO_2 \xrightarrow{Sn + HCl} A \xrightarrow[H^+]{C_6H_5N_2^+Cl^-} P$ (Yellow coloured compound)

Consider the above reaction, the product $P$ is **[2021, 25 July Shift-II]**

(a) 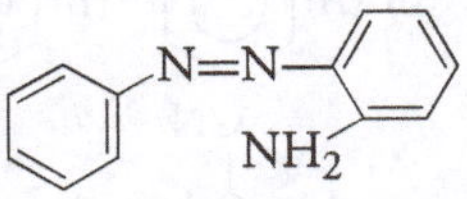

(b) 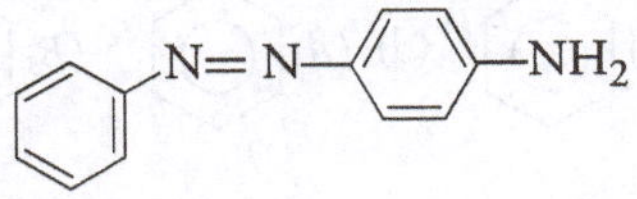

(c) 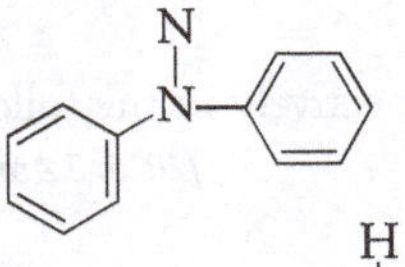

(d) 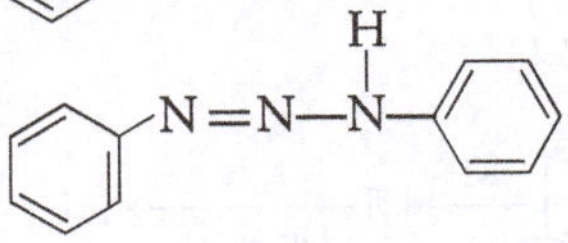

**Imp.** Reaction of Diazonium Salts

**13** 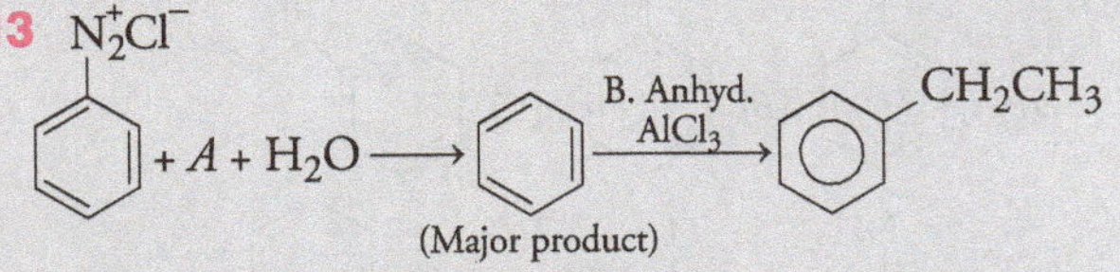


In the chemical reactions given above $A$ and $B$ respectively are **[2021, 22 July Shift-II]**

(a) $H_3PO_2$ and $CH_3CH_2Cl$
(b) $CH_3CH_2OH$ and $H_3PO_2$
(c) $H_3PO_2$ and $CH_3CH_2OH$
(d) $CH_3CH_2Cl$ and $H_3PO_2$

**14** 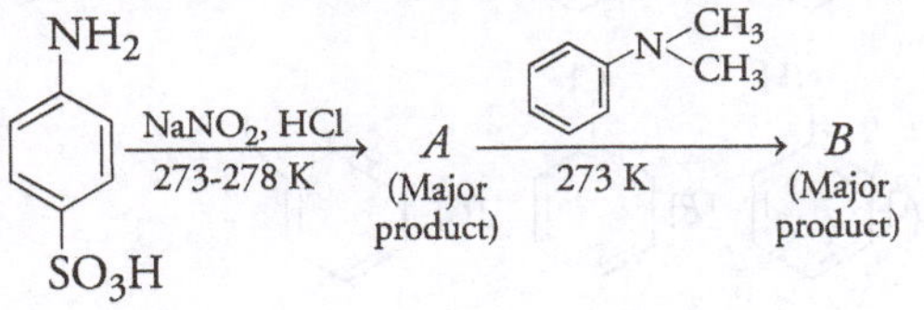


Consider the above reaction, compound $B$ is **[2021, 20 July Shift-II]**

(a) 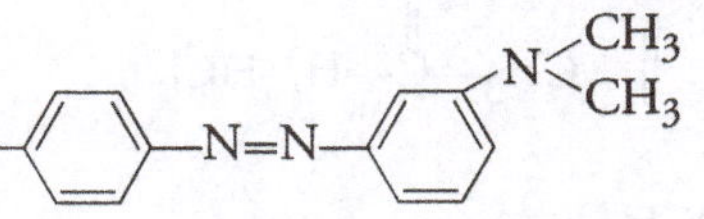

(b) 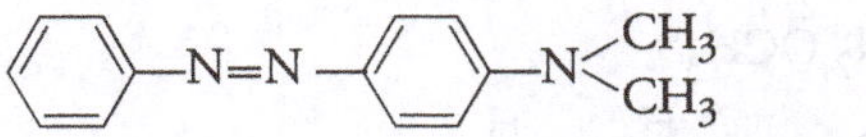

(c) 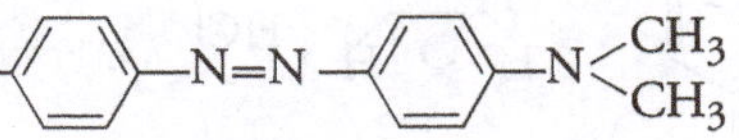

(d) 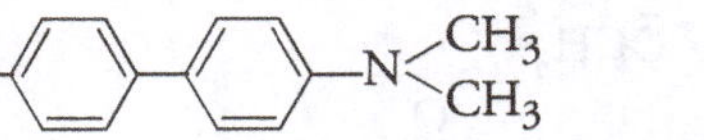

**15** 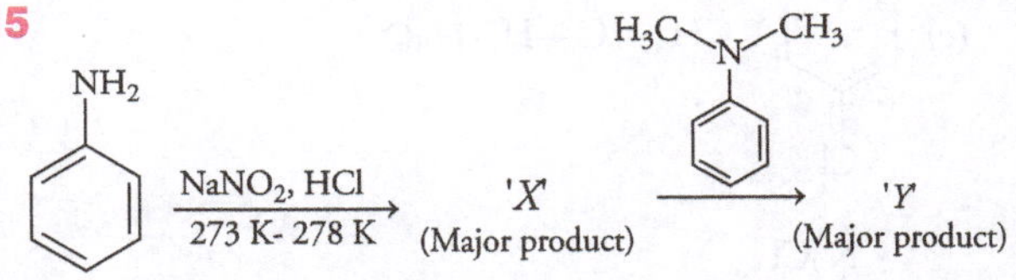


The product $X$ and $Y$ are **[2021, 18 March Shift-I]**

(a) 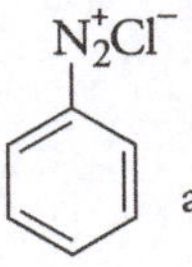 and 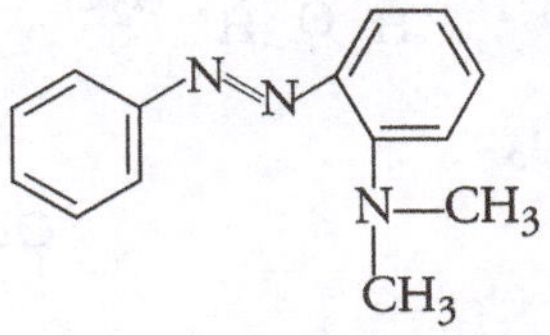

(b) 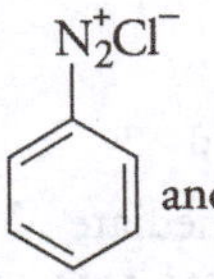 and 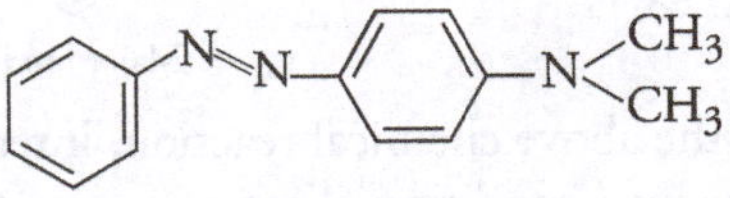

(c) 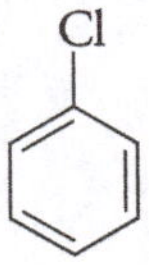 and 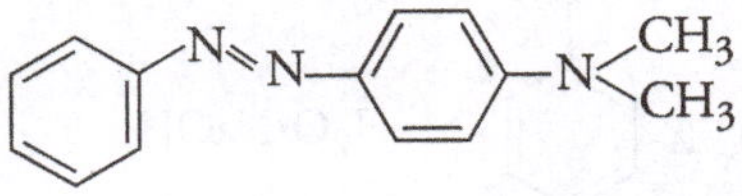

(d) 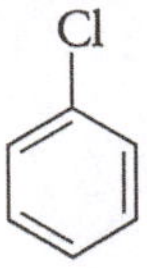 and 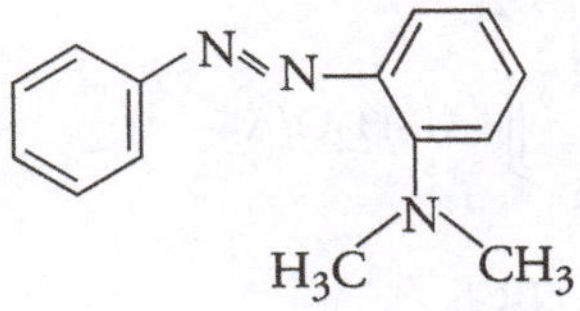

**16** $C_7H_7N_2OCl + C_2H_5OH \longrightarrow$ 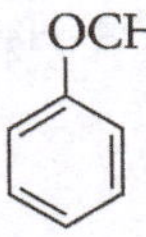 $+ N_2 + X + Y$
(A)

In the above reaction, the structural formula of $A$, $X$ and $Y$ respectively are **[2021, 17 March Shift-II]**

(a) 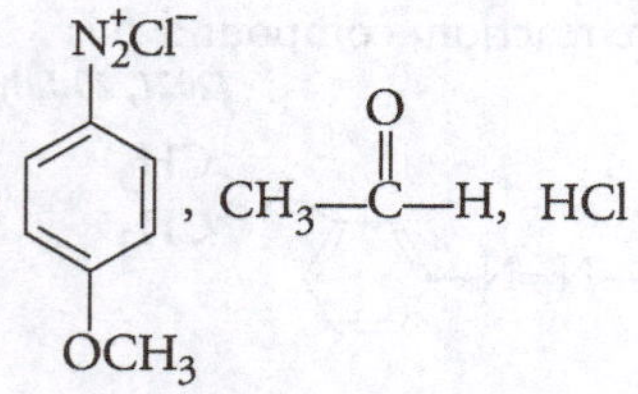

(b) 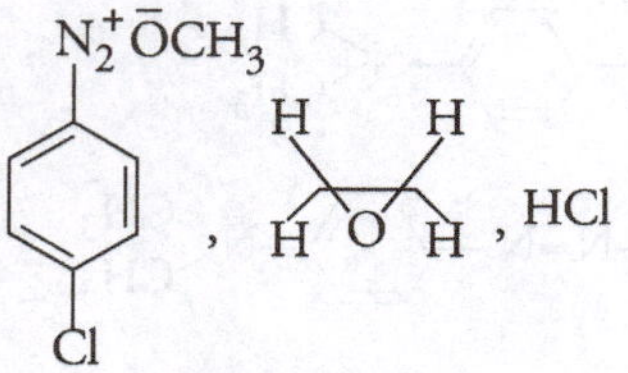

(c) 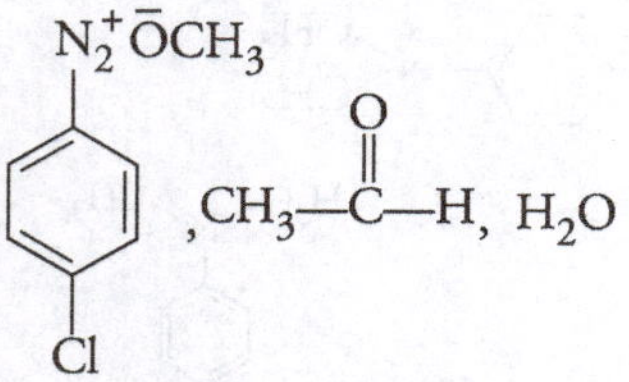

(d) 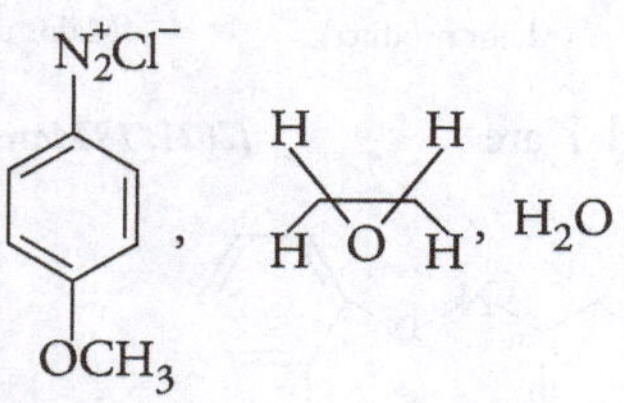

**17** 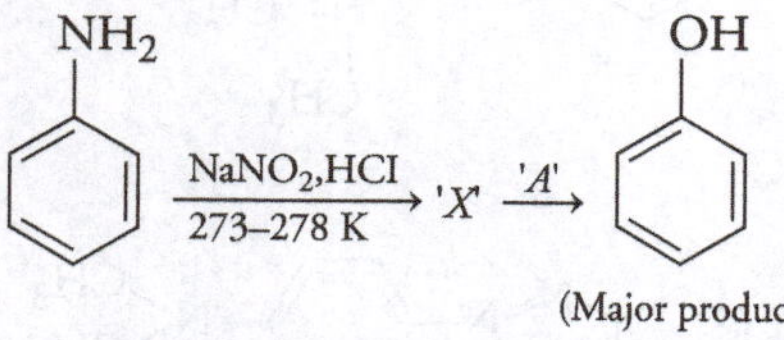

(Major product)

In the above chemical reaction, intermediate *X* and reagent / condition *A* are **[2021, 16 March Shift-I]**

(a) *X*- 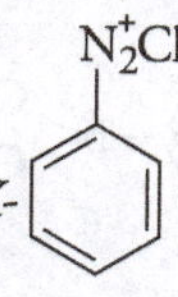 ; *A* -$H_2O$/NaOH

(b) *X*- 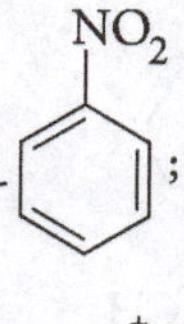 ; *A* -$H_2O/\Delta$

(c) *X*- 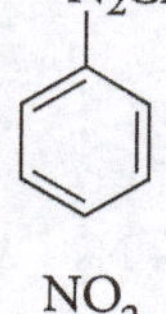 ; *A* -$H_2O/\Delta$

(d) *X*- 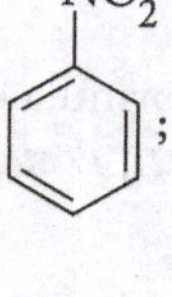 ; *A* -$H_2O$/NaOH

**18** The diazonium salt of which of the following compounds will form a coloured dye on reaction with β-naphthol in NaOH ? **[2021, 24 Feb Shift-II]**

(a) 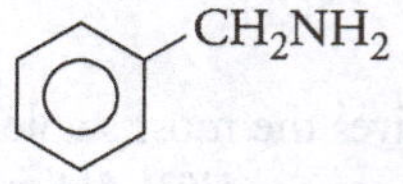

(b) 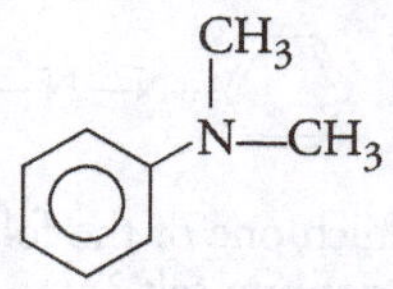

(c) 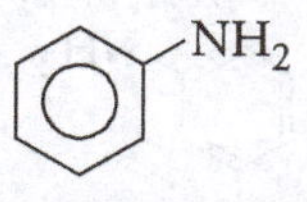

(d) 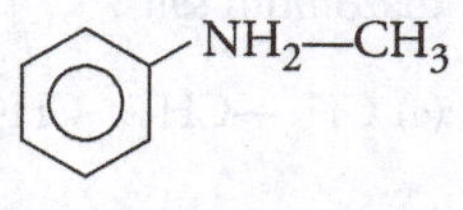

**19** '*A*' and '*B*' in the following reaction are

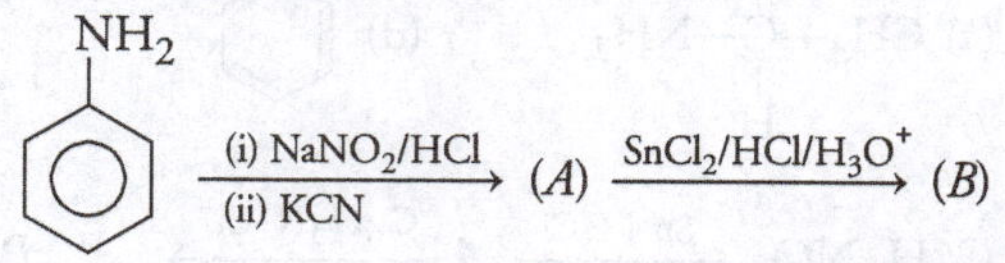

**[2021, 24 Feb Shift-I]**

(a) 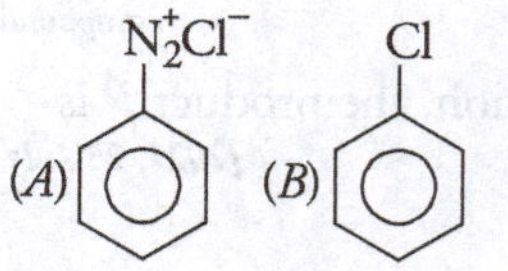

(b) 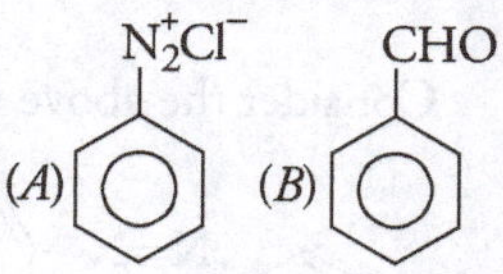

(c) 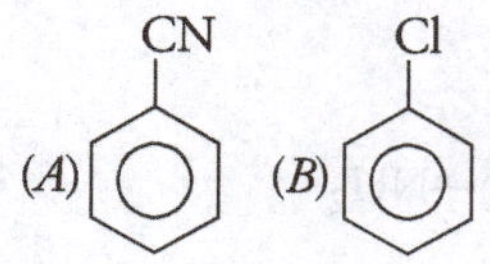

(d) 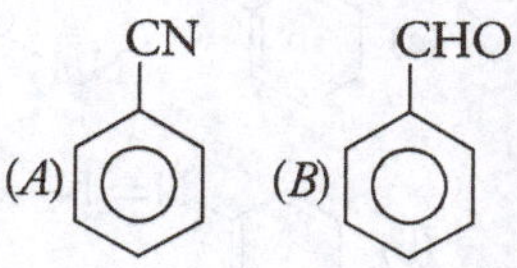

## Level 3

**20** Identify [*A*] [*B*] and [*C*], respectively in the following reaction sequence: **[2025, 3 April Shift-I]**

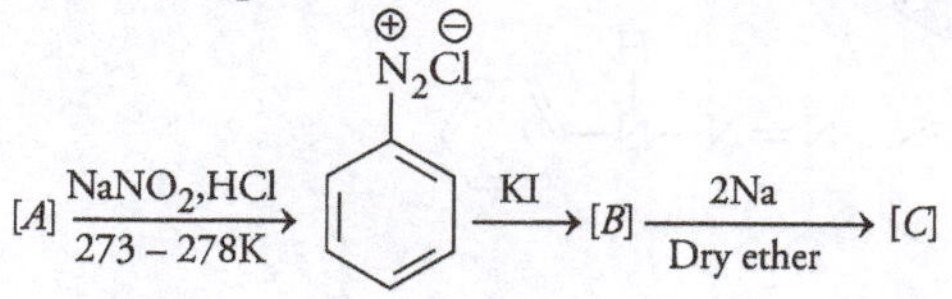

(a) 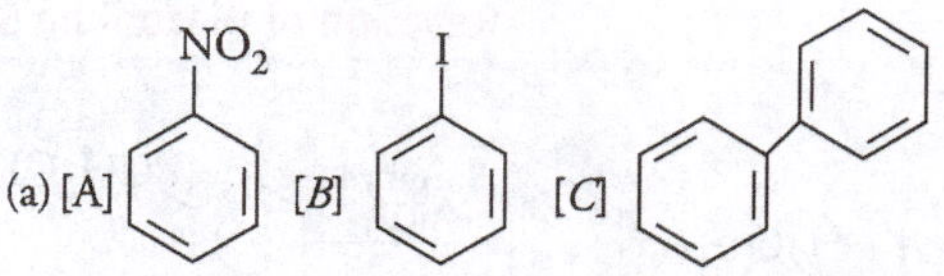

(b) 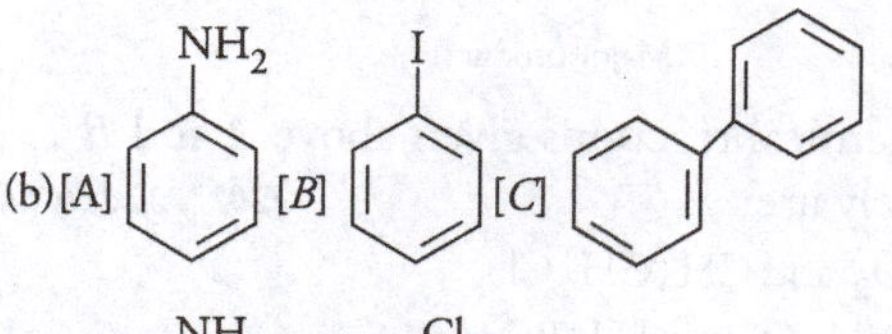

(c) 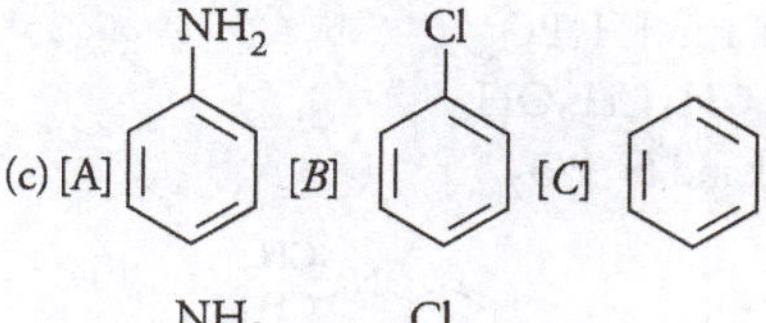

(d) 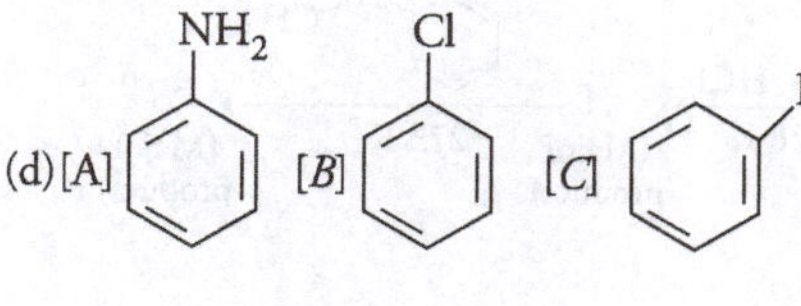

**21** The products formed in the following reaction sequence are *[2025, 22 Jan Shift-I]*

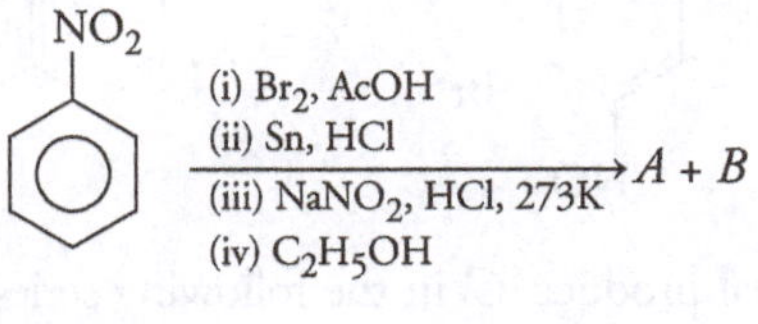

(a) 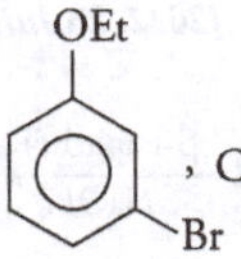 , $CH_3—COOH$

(b) 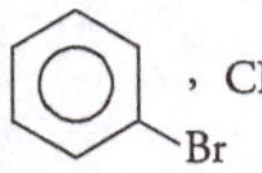 , $CH_3—CHO$

(c) 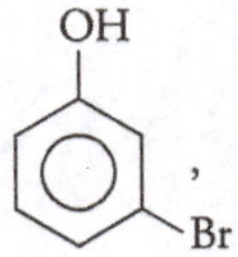 , 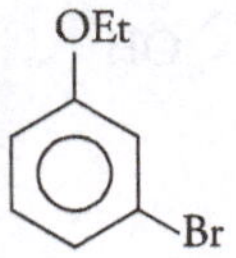

(d) 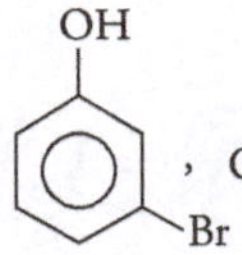 , $CH_3—CHO$

**22** Match the compounds (List-I) with the appropriate Catalyst/Reagents (List-II) for their reduction into corresponding amines. *[2025, 22 Jan Shift-II]*

| List-I (Compounds) | | List-II (Catalyst/Reagents) | |
|---|---|---|---|
| (A) | $R-\overset{\overset{O}{\|}}{C}-NH_2$ | (I) | NaOH (aqueous) |
| (B) | 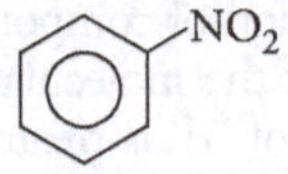 | (II) | $H_2$ / Ni |
| (C) | $R—C\equiv N$ | (III) | $LiAlH_4$, $H_2O$ |
| (D) | 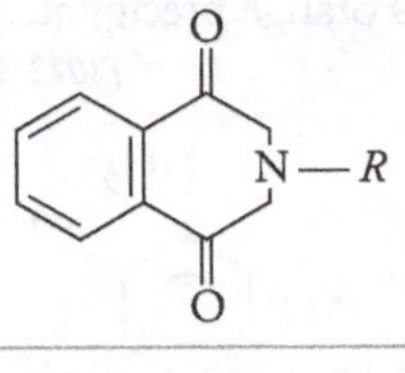 | (IV) | Sn, HCl |

Choose the correct answer from the options given below:

(a) (A)-(II), (B)-(I), (C)-(III), (D)-(IV)
(b) (A)-(III), (B)-(II), (C)-(IV), (D)-(I)
(c) (A)-(III), (B)-(IV), (C)-(II), (D)-(I)
(d) (A)-(II), (B)-(IV), (C)-(III), (D)-(I)

**Imp.** Preparation of Azo dye

**23** Which one of the following reaction sequences will give an azo dye? *[2025, 29 Jan Shift-II]*

(a) 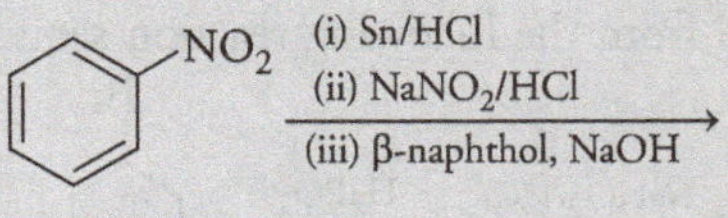

(b) 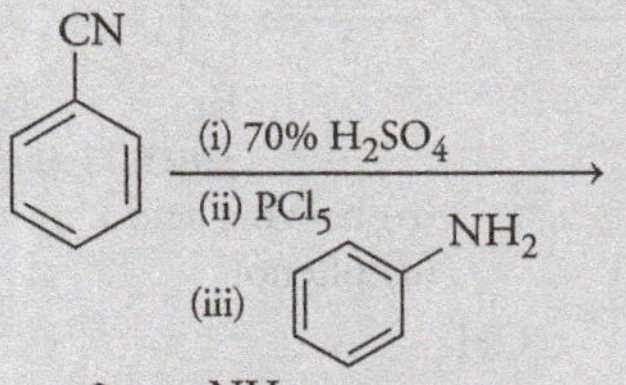

(c) 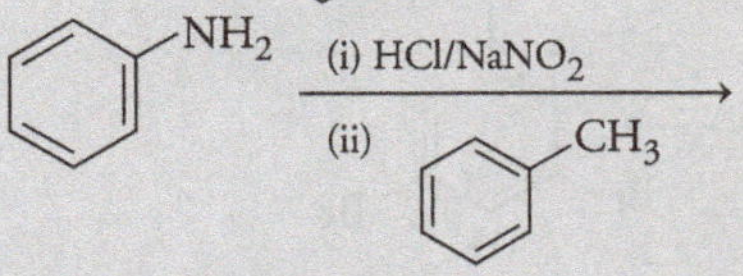

(d) 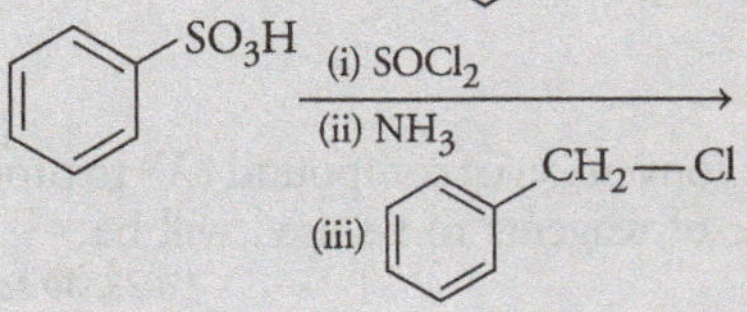

**24** The incorrect statement regarding the reaction given below is

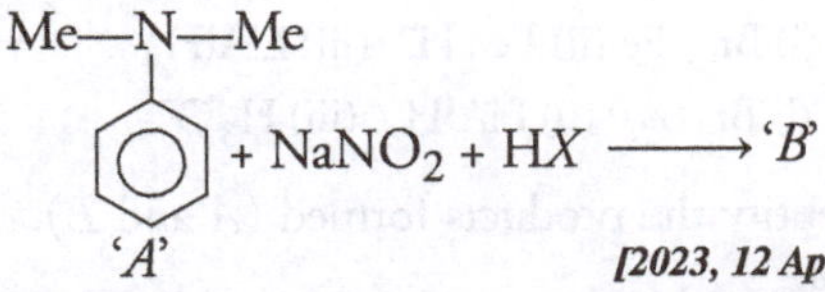

*[2023, 12 April Shift-I]*

(a) The electrophile involved in the reaction is $NO^+$.
(b) '*B*' is *N*-nitroso ammonium compound.
(c) The product '*B*' formed in the above reaction is *p*-nitroso compound at low temperature.
(d) The reaction occurs at low temperature.

**25** Match List-I with List-II.

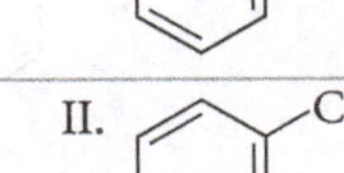

is reacted with reagents in List-I to form products in List-II

| List-I (Reagent) | | List-II (Product) | |
|---|---|---|---|
| A. | 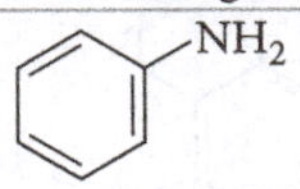 | I. | 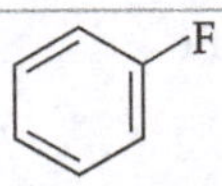 |
| B. | $HBF_4$, Δ | II. | 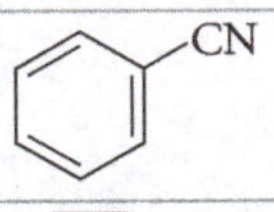 |
| C. | Cu, HCl | III. | 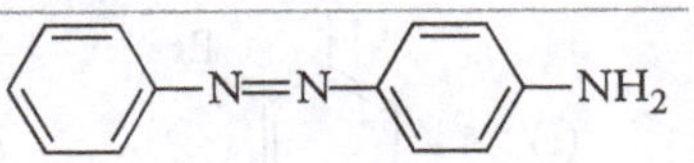 |
| D. | CuCN/KCN | IV. | 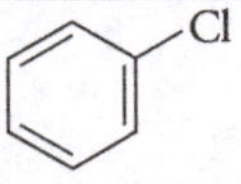 |

Choose the correct answer from the options given below **[2023, 8 April Shift-I]**

(a) A-IV, B-III, C-II, D-I (b) A-III, B-I, C-IV, D-II
(c) A-III, B-I, C-II, D-IV (d) A-I, B-III, C-IV, D-II

**26** Compound $A$ from the following reaction sequence is

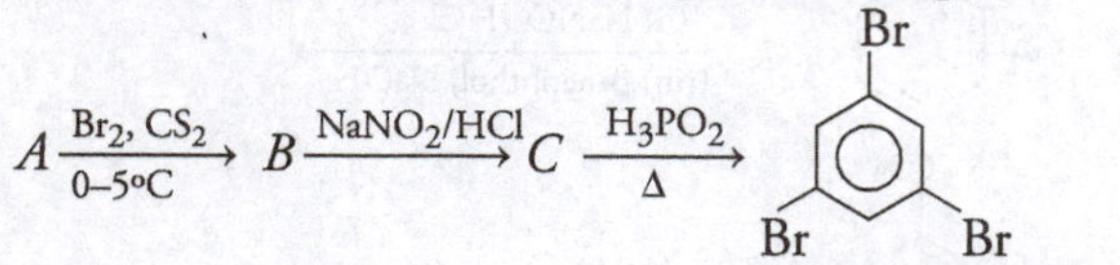

**[2023, 13 April Shift-II]**

(a) aniline (b) benzoic acid
(c) salicylic acid (d) phenol

**27**

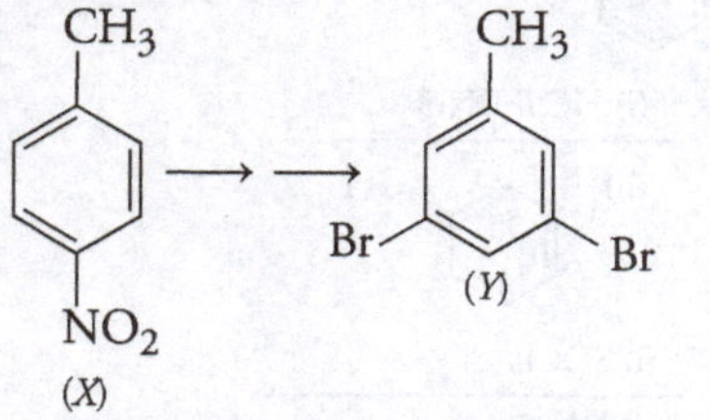

In the above conversion of compound ($X$) to product ($Y$), the sequence of reagents to be used will be **[2023, 30 Jan Shift-II]**

(a) (i) Fe, $H^+$ (ii) $Br_2(aq)$ (iii) $HNO_2$ (iv) CuBr
(b) (i) Fe, $H^+$ (ii) $Br_2(aq)$ (iii) $HNO_2$ (iv) $H_3PO_2$
(c) (i) $Br_2$, Fe (ii) Fe, $H^+$ (iii) $LiAlH_4$
(d) (i) $Br_2(aq)$ (ii) $LiAlH_4$ (iii) $H_3O^+$

**28** Identify the products formed ($A$ and $E$).

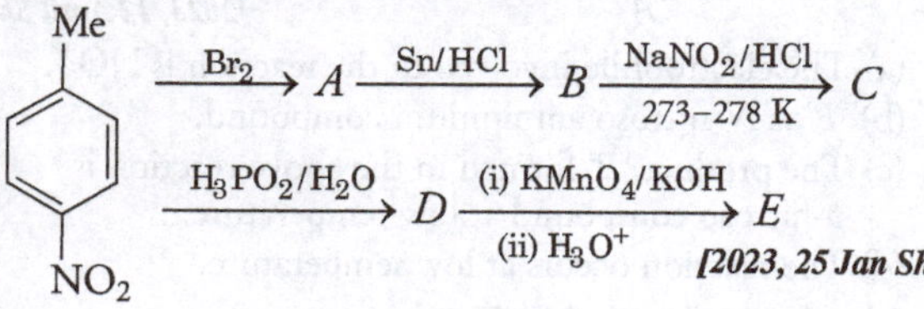

**[2023, 25 Jan Shift-I]**

(a) $A$ = 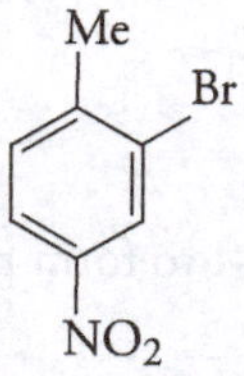 , $E$ = 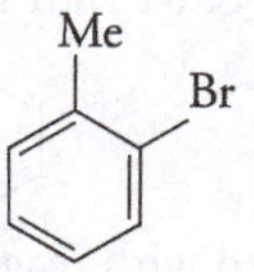

(b) $A$ = 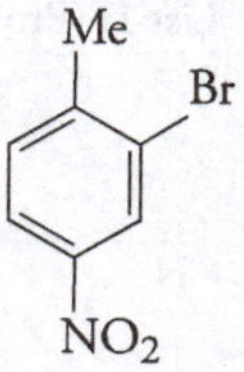 , $E$ = 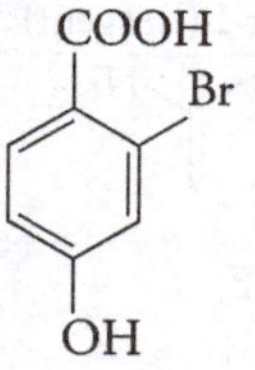

(c) $A$ = 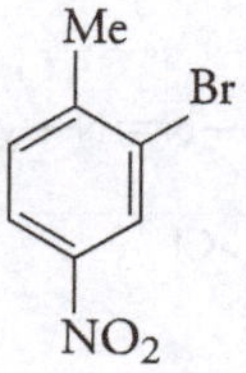 , $E$ = 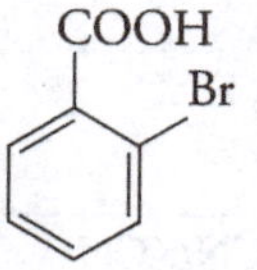

(d) $A$ = 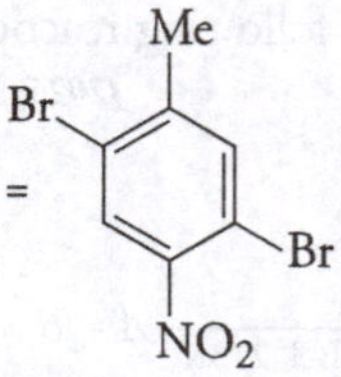 , $E$ = 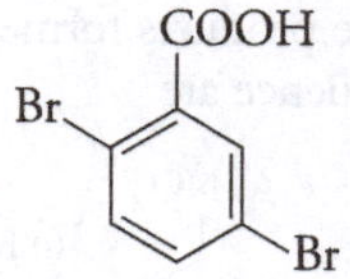

**29** The final product '$C$' in the following series of reactions. **[2022, 26 June Shift-II]**

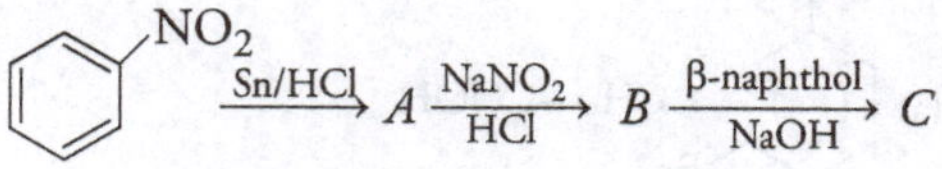

(a) 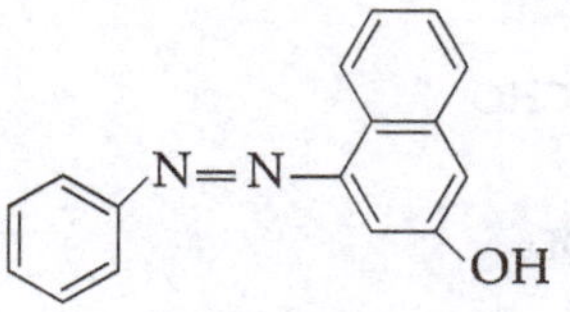

(b) 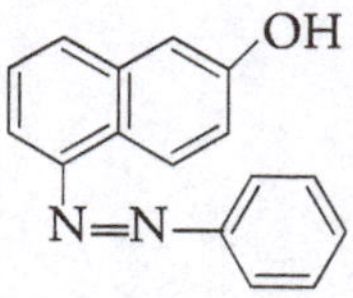

(c) 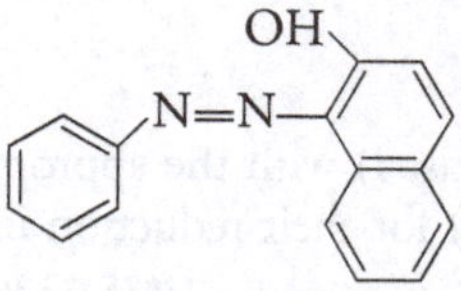

(d) 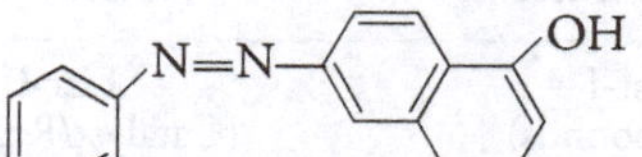

**30** An organic compound '$A$' contains nitrogen and chlorine. It dissolves readily in water to give a solution that turns litmus red. Titration of compound '$A$' with standard base indicates that the molecular weight of '$A$' is $131 \pm 2$. When a sample of '$A$' is treated with aq. NaOH, a liquid separates which contains N but not Cl. Treatment of the obtained liquid with nitrous acid followed by phenol gives orange precipitate. The compound $A$ is **[2022, 27 July Shift-II]**

(a) 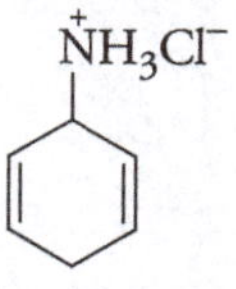 (b) 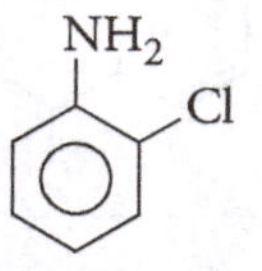

(c) 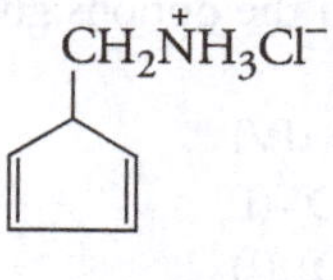 (d) 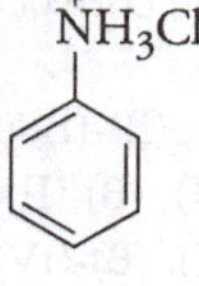

**31** Identify the major product formed in the following sequence of reactions. *[2022, 28 June Shift-I]*

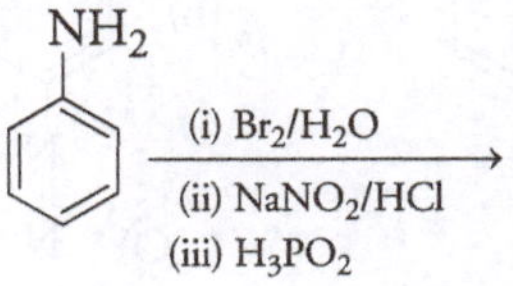


(a) 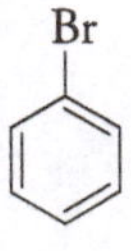
(b) 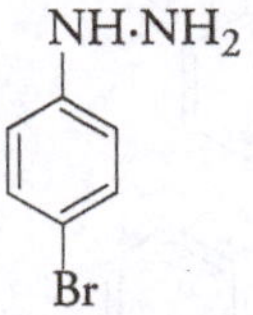
(c) 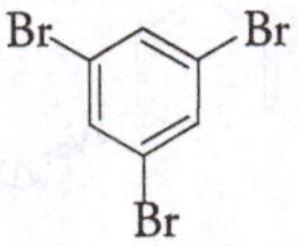
(d) 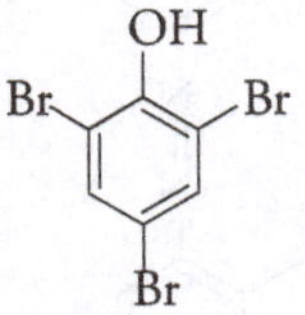

**32** The major product of the following reaction is *[2021, 27 Aug Shift-I]*

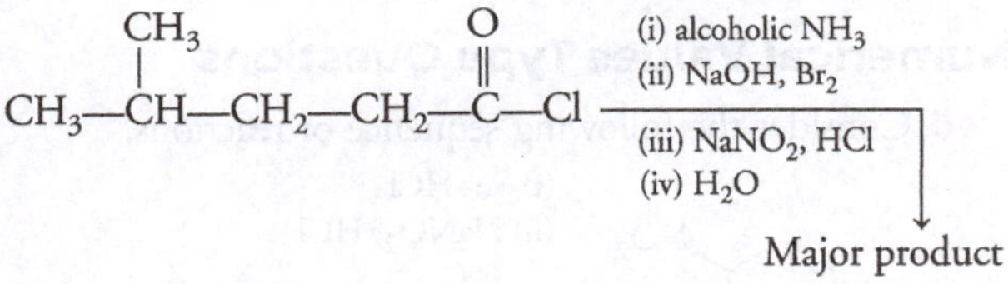


(a) 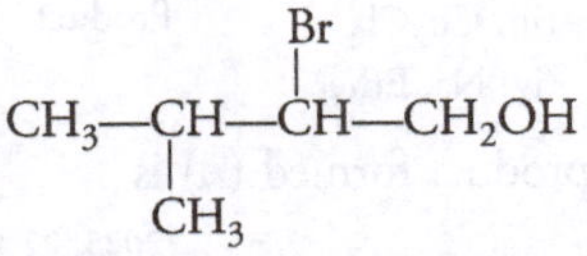
(b) 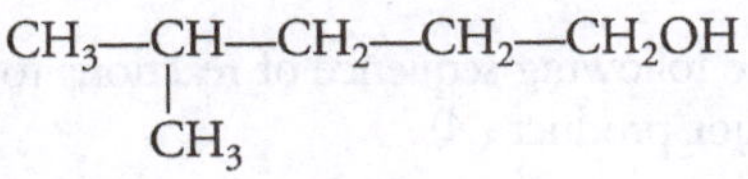
(c) 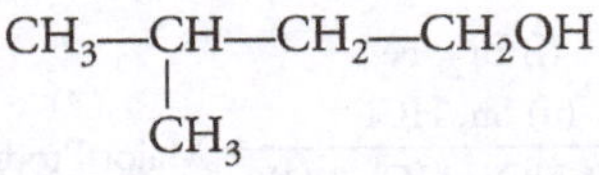
(d) 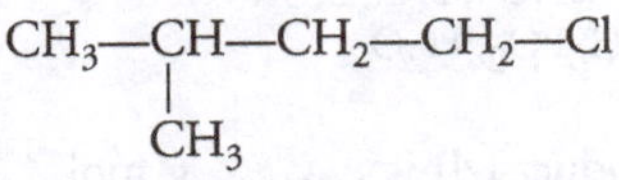

**33** Which of the following reaction(s) will not give *p*-aminoazobenzene? *[2021, 25 Feb Shift-I]*

$A =$ 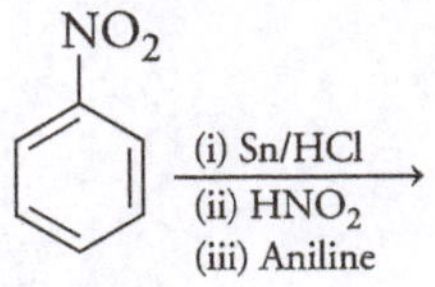


$B =$ 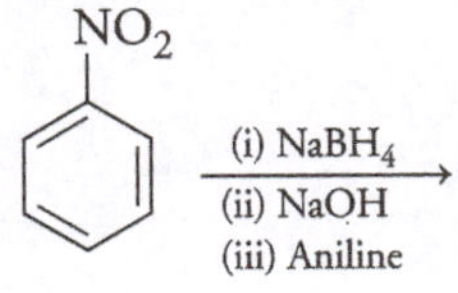


$C =$ 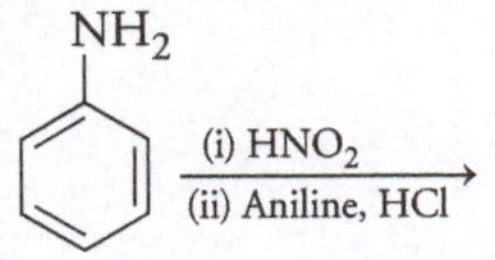


(a) *A* only (b) *C* only
(c) *B* only (d) *A* and *B*

**34** What is the correct sequence of reagents used for converting nitrobenzene into *m*-dibromobenzene?

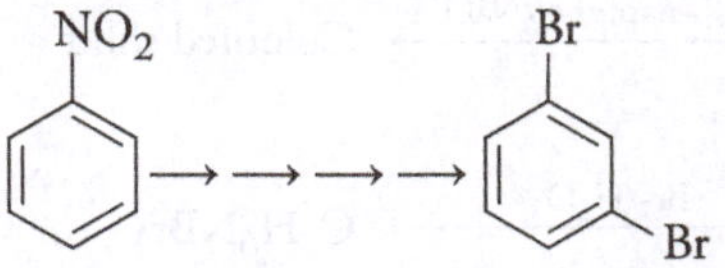


*[2021, 24 Feb Shift-II]*

(a) $\xrightarrow{NaNO_2} / \xrightarrow{HCl} / \xrightarrow{KBr} / \xrightarrow{H^+}$

(b) $\xrightarrow{Br_2/Fe} / \xrightarrow{Sn/HCl} / \xrightarrow{NaNO_2/HCl} / \xrightarrow{CuBr/HBr}$

(c) $\xrightarrow{Sn/HCl} / \xrightarrow{KBr} / \xrightarrow{Br_2} / \xrightarrow{H^+}$

(d) $\xrightarrow{Sn/HCl} / \xrightarrow{Br_2} / \xrightarrow{NaNO_2} / \xrightarrow{NaBr}$

**35** In the following reaction sequence, [*C*] is *[2020, 04 Sept Shift-II]*

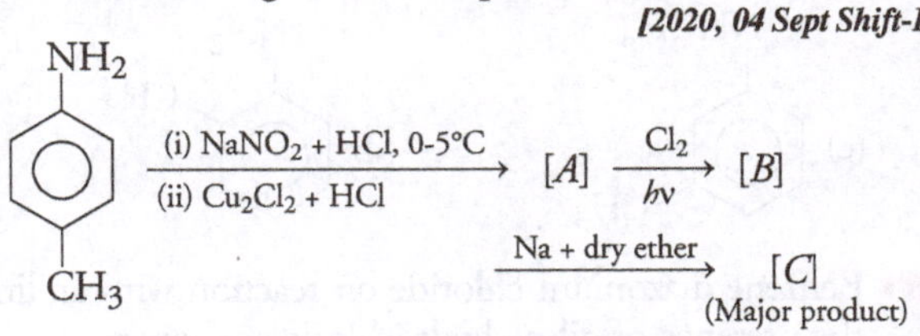


(a) 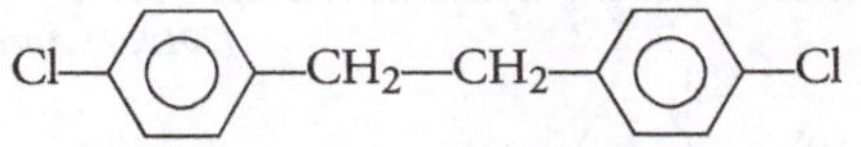
(b) 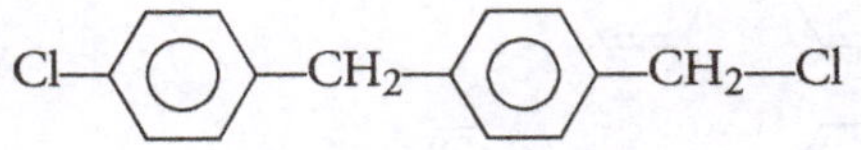
(c) 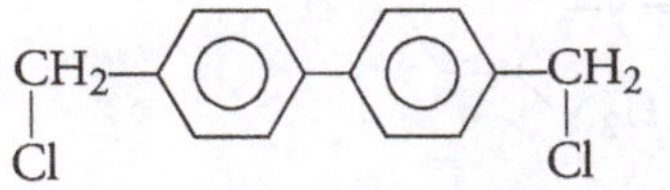
(d) 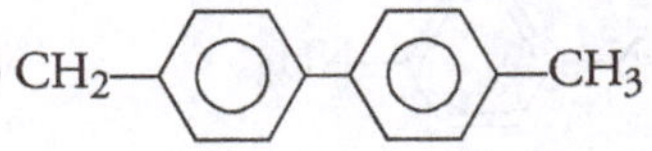

**36** The major product *Z* obtained in the following reaction scheme is *[2020, 9 Jan Shift-I]*

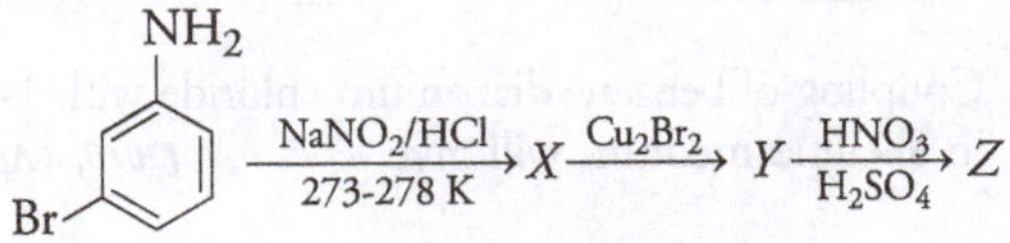


(a) 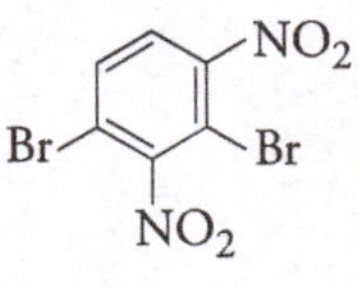
(b) 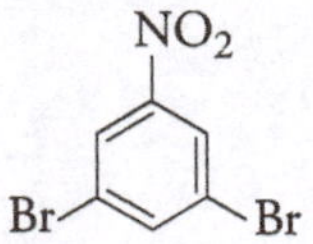
(c) 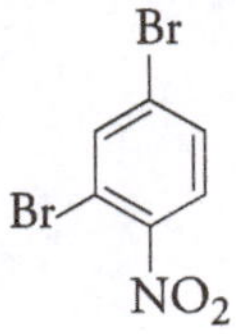
(d) 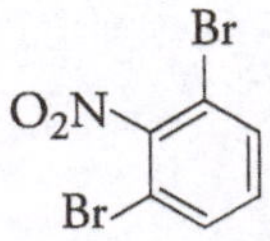

**37** Consider the following reactions

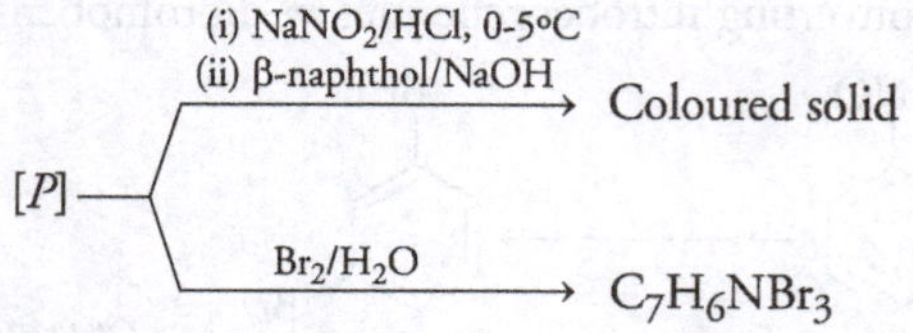


The compound [*P*] is **[2020, 9 Jan Shift-II]**

(a) 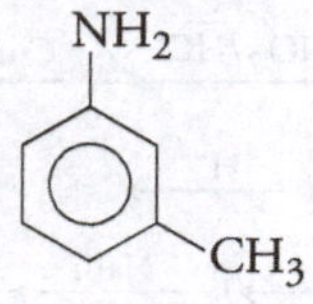


(b) 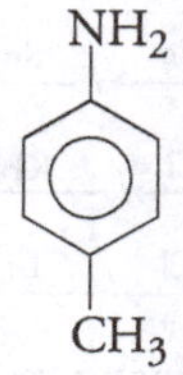


(c) 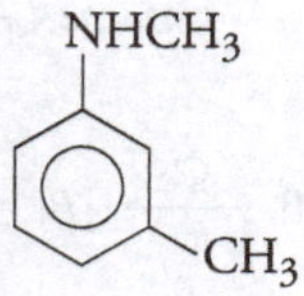


(d) 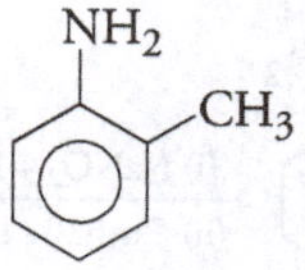


**38** Benzene diazonium chloride on reaction with aniline in the presence of dilute hydrochloric acid gives **[2019, 12 April Shift-II]**

(a) 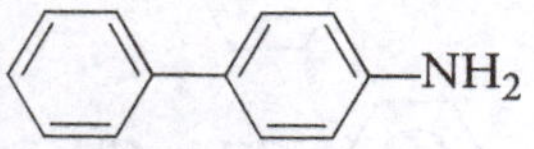


(b) 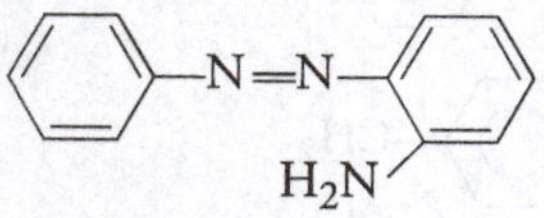


(c) 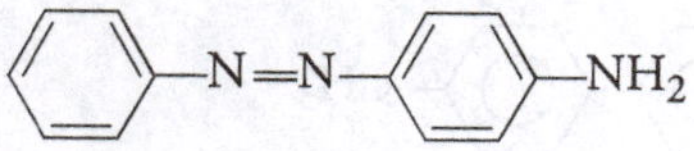


(d) 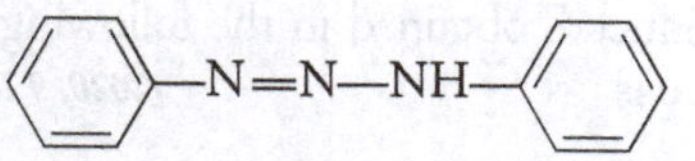


**39** Coupling of benzene diazonium chloride with 1-naphthol in alkaline medium will give **[2019, 8 April Shift-I]**

(a) 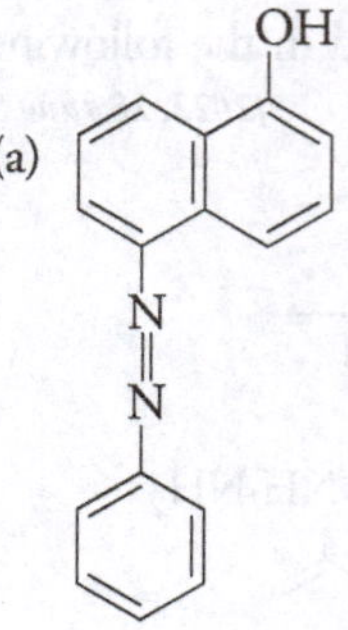


(b) 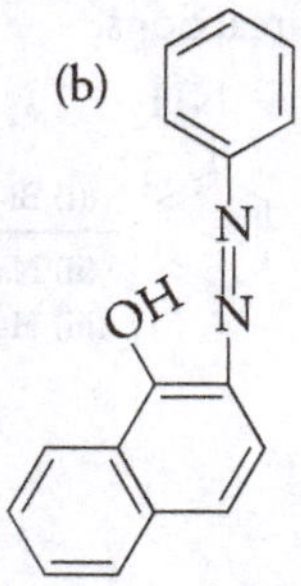


(c) 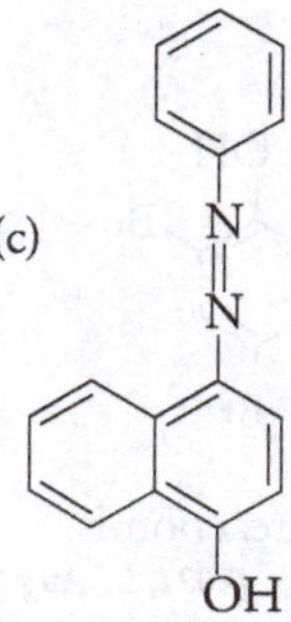


(d) 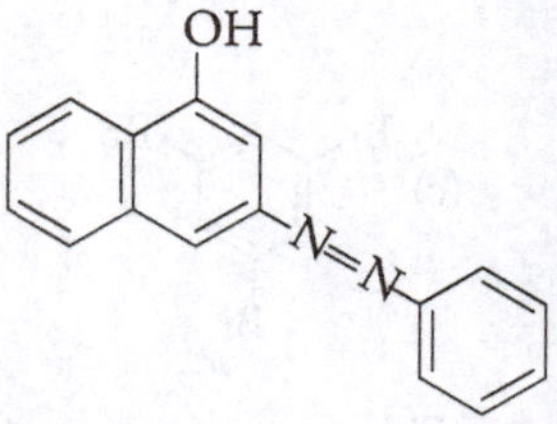


## Numerical Values Type Questions

**40** Consider the following sequence of reactions,

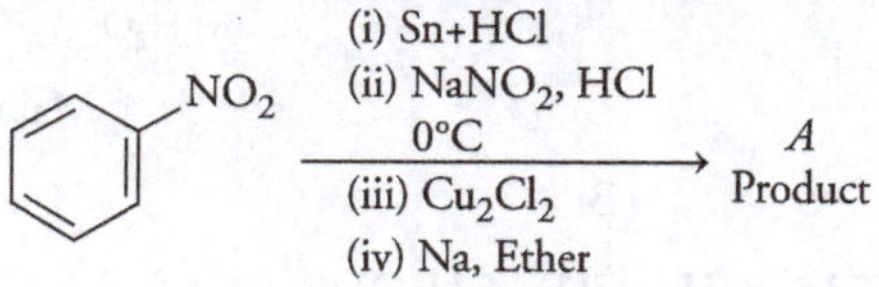


Molar mass of the product formed (*A*) is ...... g $mol^{-1}$. **[2025, 22 Jan Shift-I]**

**41** Consider the following sequence of reactions to produce major product (*A*)

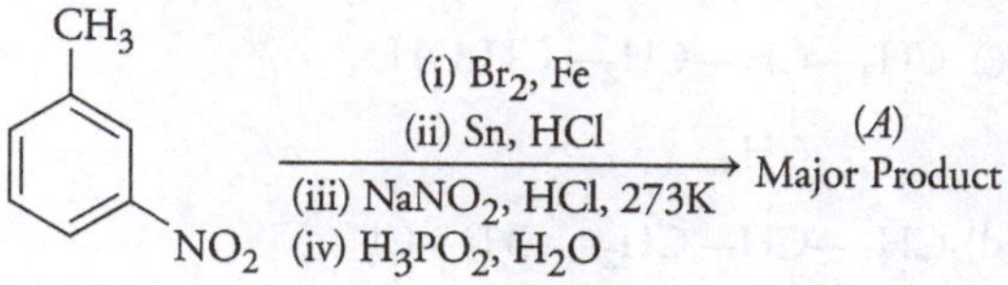


Molar mass of product (*A*) is ........ g $mol^{-1}$.
(Given, molar mass in g $mol^{-1}$ of C : 12, H : 1, O : 16, Br : 80, N : 14, P : 31)

# Key Index

## Topic 1

| | | | | | | | | | | | | | | | | | | | |
|---|---|---|---|---|---|---|---|---|---|---|---|---|---|---|---|---|---|---|---|
| **1** | (b) | **2** | (a) | **3** | (c) | **4** | (d) | **5** | (b) | **6** | (d) | **7** | (c) | **8** | (a) | **9** | (d) | **10** | (c) |
| **11** | (b) | **12** | (c) | **13** | (b) | **14** | (a) | **15** | (b) | **16** | (c) | **17** | (d) | **18** | (a) | **19** | (c) | **20** | (b) |
| **21** | (a) | **22** | (c) | **23** | (b) | **24** | (d) | **25** | (d) | **26** | (c) | **27** | (c) | **28** | (c) | **29** | (a) | **30** | (d) |
| **31** | (a) | **32** | (d) | **33** | (c) | **34** | (d) | **35** | (c) | **36** | (a) | **37** | (c) | **38** | (c) | **39** | (c) | **40** | (c) |
| **41** | (c) | **42** | (d) | **43** | (a) | **44** | (d) | **45** | (c) | **46** | (c) | **47** | (b) | **48** | (2) | **49** | (108°) | | |

## Topic 2

| | | | | | | | | | | | | | | | | | | | |
|---|---|---|---|---|---|---|---|---|---|---|---|---|---|---|---|---|---|---|---|
| **1** | (b) | **2** | (c) | **3** | (a) | **4** | (d) | **5** | (a) | **6** | (d) | **7** | (b) | **8** | (b) | **9** | (c) | **10** | (a) |
| **11** | (a) | **12** | (c) | **13** | (d) | **14** | (b) | **15** | (d) | **16** | (a) | **17** | (d) | **18** | (d) | **19** | (d) | **20** | (d) |
| **21** | (c) | **22** | (b) | **23** | (b) | **24** | (b) | **25** | (b) | **26** | (b) | **27** | (c) | **28** | (c) | **29** | (c) | **30** | (b) |
| **31** | (b) | **32** | (b) | **33** | (c) | **34** | (d) | **35** | (b) | **36** | (d) | **37** | (c) | **38** | (c) | **39** | (b) | **40** | (b) |
| **41** | (b) | **42** | (b) | **43** | (c) | **44** | (a) | **45** | (d) | **46** | (a) | **47** | (c) | **48** | (a) | **49** | (c) | **50** | (c) |
| **51** | (a) | **52** | (c) | **53** | (c) | **54** | (b) | **55** | (c) | **56** | (d) | **57** | (a) | **58** | (c) | **59** | (d) | **60** | (a) |
| **61** | (341) | **62** | (4) | **63** | (5) | **64** | (5) | **65** | (3) | **66** | (5) | **67** | (3) | | | | | | |

## Topic 3

| | | | | | | | | | | | | | | | | | | | |
|---|---|---|---|---|---|---|---|---|---|---|---|---|---|---|---|---|---|---|---|
| **1** | (c) | **2** | (a) | **3** | (b) | **4** | (b) | **5** | (b) | **6** | (a) | **7** | (d) | **8** | (a) | **9** | (b) | **10** | (d) |
| **11** | (b) | **12** | (b) | **13** | (a) | **14** | (c) | **15** | (b) | **16** | (a) | **17** | (c) | **18** | (c) | **19** | (d) | **20** | (b) |
| **21** | (b) | **22** | (c) | **23** | (a) | **24** | (b) | **25** | (b) | **26** | (a) | **27** | (b) | **28** | (c) | **29** | (c) | **30** | (d) |
| **31** | (c) | **32** | (c) | **33** | (c) | **34** | (b) | **35** | (a) | **36** | (c) | **37** | (a) | **38** | (c) | **39** | (c) | **40** | (154) |
| **41** | (171) | | | | | | | | | | | | | | | | | | |

# Detailed **Explanations**

## Topic 1

**01.** *(b)* $CH_3—CH_2—\overset{O}{\overset{||}{C}}—NH_2 \xrightarrow[-K_2CO_3,\ -KBr,\ -H_2O]{Br_2/KOH} CH_3CH_2NH_2$

Propanamide → Ethylamine

The above reaction is an example of Hoffmann bromamide reaction.

**02.** *(a)* The Clemmensen reduction is used to reduce aldehydes/ketones into alkanes using zinc amalgam and HCl. Also, $LiAlH_4$ is a good reducing agent, which reduces amides into amines.

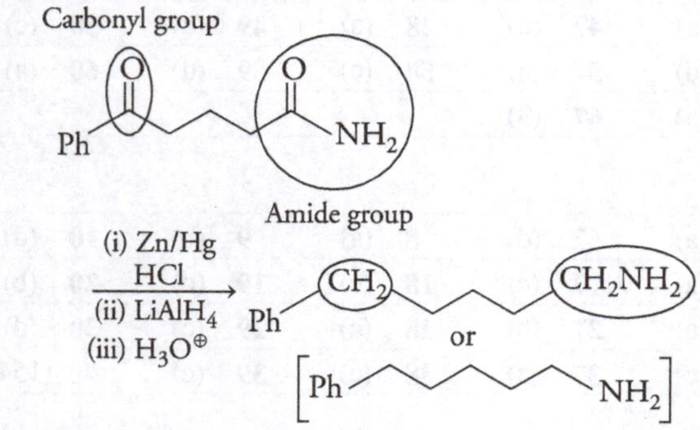


**03.** *(c)* In aqueous medium basic strength of methylsubstituted amines is dependent on inductive effect, solvation effect (H-bonding) and steric hinderance.

So, correct order is

$(CH_3)_2NH > CH_3—NH_2 > (CH_3)_3N > NH_3$

or $Me_2NH > MeNH_2 > Me_3N > NH_3$.

**04.** *(d)* Hofmann degradation reaction involves the migration of an alkyl as well as aryl group from the carbonyl carbon atom of the amide to the nitrogen atom.

Therefore, statement I is incorrect but statement II is correct.

**05.** *(b)* Primary amines react with nitrous acid at 273 K to form diazonium salt which is very unstable. It breaks to give nitrogen gas and alcohol on raising temperature to 298 K.

$$R—NH_2 \xrightarrow{HNO_2,\ 273K} R—\overset{+}{N}_2Cl^- \xrightarrow[H_2O]{298\ K} R—OH$$

Diazonium salt → Alcohol

**06.** *(d)*

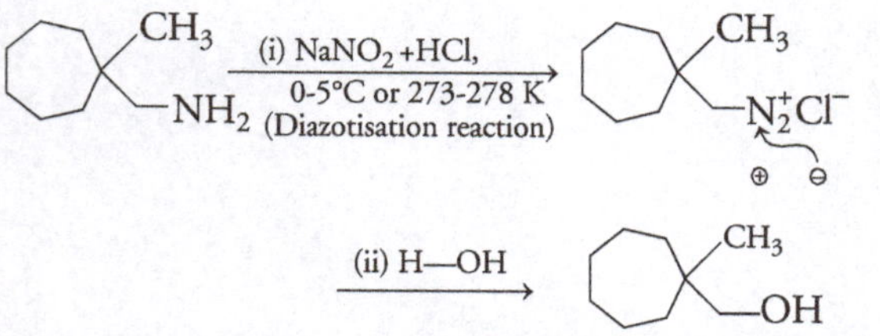


(1-methylcyclohexyl) methanol

(*P*)

(Major product)

**07.** *(c)* Chemical reaction is as follows

$$\underset{\substack{1°\ \text{amine} \\ (A)}}{RNH_2} \xrightarrow{CHCl_3} \underset{\substack{\text{Isonitrile} \\ (B)}}{R—\overset{\oplus}{N}\equiv\overset{\ominus}{C}} \xrightarrow[HCl\ (C)]{H_3O^+} RNH_2 + HCOOH$$

(*B*) ↓ Toxic

Hence, *A*-primary amine, *B*-isonitrile compound and *C*-conc. HCl.

**08.** *(a)* Hypobromite transforms the primary amide to carbamic acid which further loses carbonyl carbon as $CO_3^{2-}$, yielding the amine product.

$$R—\overset{O}{\overset{||}{C}}—NH_2 + BrO^- \longrightarrow R—NH_2 + CO_3^{2-} + Br^-$$

**09.** *(d)*

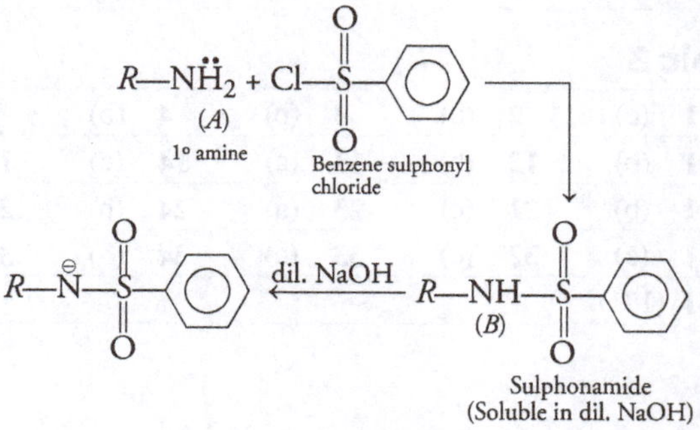


Hence, *A* is 1° amine i.e. $C_6H_5—\underset{CH_3}{\underset{|}{CH}}—NH_2$

**10.** *(c)* There is delocalisation of electron in both options (a) and (c). In option (c) there is delocalisation of electron with both carbonyl group and hence the lone pair is less available for donation. Hence, it is weakest base.

**11.** *(b)* For the conversion of $C_2H_5CN$ into $CH_3CH_2CH_2NH_2$, a strong reducing agent is required among the given reducing agent. Strong reducing agent is $LiAlH_4$. It is the most appropriate reagent for this reduction.

$$C_2H_5CN \xrightarrow{LiAlH_4} CH_3CH_2CH_2NH_2$$

**12.** *(c)* In first case, on increasing the size of alkyl group + *I* effect increases and the positive charge of ammonium cation gets dispersed more easily, Therefore, order of basicity is 2° amine > 1° amine > ammonia.

$(C_2H_5)_2NH > C_2H_5NH_2 > NH_3$

- Concept **Enhancer**

Basic strength can be compared by the reaction of an alkanamine and ammonia with proton.

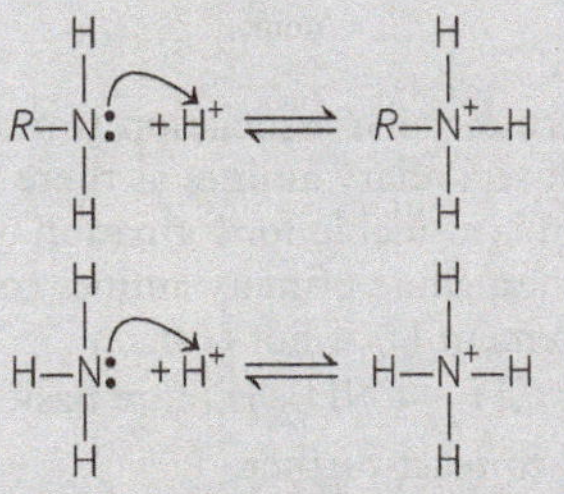

Basicity of an amine in aqueous solution depends upon the solubility of ammonium cation formed by accepting proton from water. The stability of ammonium cation in aqueous medium depends upon the following factors.

(a) +*I*-effect (b) Steric effect (c) Solvation effect

**13.** *(b)* $X \xrightarrow{Br_2/NaOH} C_3H_9N \xrightarrow[KOH]{CHCl_3} CH_3CH_2CH_2NC$

The first reaction is Hoffmann bromamide degradation reaction and the second reaction is carbylamine reaction. Since the compound $C_3H_9N$ gives positive carbylamine test, this means it is 1° amine. Thus, the compound *X* is $CH_3CH_2CH_2CONH_2$.

$$\underset{(X)}{CH_3CH_2CH_2CONH_2} \xrightarrow[NaOH]{Br_2} CH_3CH_2CH_2NH_2$$

$$\xrightarrow{CHCl_3 \;+\; KOH} CH_3CH_2CH_2NC + 3KCl + 3H_2O$$

**14.** *(a)*

**key idea** $AlH(i\text{-}Bu)_2$ also known DIBAL-H is di isobutyl aluminium hydride, $[(CH_3)_2CHCH_2]_2AlH$. It is a selective reducing agent. It reduces carboxylic acids, carboxylic acid derivatives and nitriles into aldehydes. It is an electrophilic reducing agent.

The mechanism of the reaction is as follows

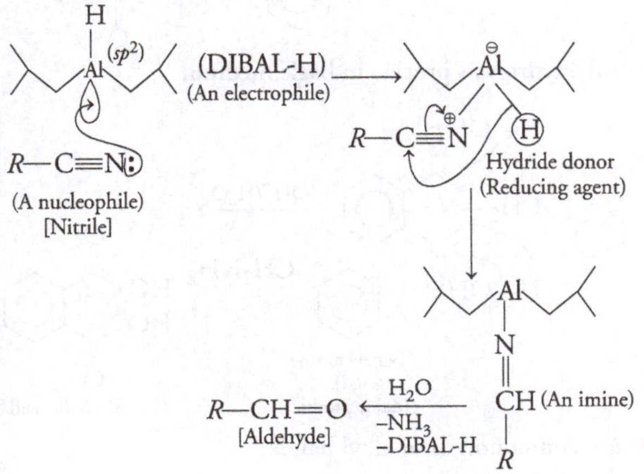

So, $R\text{—}\underset{\underset{O}{\|}}{C}H$ is the correct answer.

**15.** *(b)* Rate of acetylation $—NH_2 > —OH$ because N-bases are stronger than O-bases.

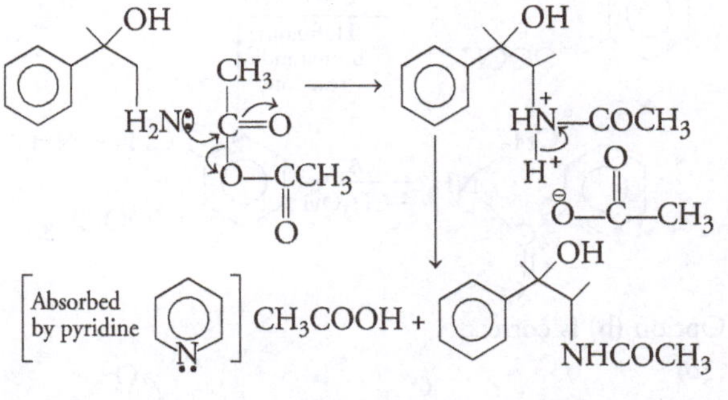

**16.** *(c)* Hofmann-bromamide degradation reaction is given as

$$RCONH_2 + 4NaOH + Br_2 \longrightarrow \underset{(1°\ \text{amine})}{RNH_2} + Na_2CO_3 + 2NaBr + 2H_2O$$

Hence, four moles of NaOH and one mole of $Br_2$ are used.

**17.** *(d)* Reaction of aliphatic primary amine with chloroform and ethanolic potassium hydroxide, leads to the formation of an alkyl isocyanides/carbylamines. Thus, this reaction is called **carbylamine reaction.**

$$RNH_2 + CHCl_3 + KOH \xrightarrow{\Delta} \underset{\text{Alkyl isocyanide}}{R\text{—}NC} + 3KCl + 3H_2O$$

**18.** *(a)* It is known that, $pK_b = -\log K_b$

Thus, larger the value of $K_b$, smaller is the value of $pK_b$ and hence, stronger is the base.

The order of basic strength of amines is as follows

$$(CH_3)_2\ddot{N}H > CH_3\ddot{N}H_2 > (CH_3)_3\ddot{N} > C_6H_5\ddot{N}H_2$$

Thus, $(CH_3)_2NH$ has the smallest $pK_b$ value.

**19.** *(c)* In aqueous solution, basicity order of amines is

$$\underset{2°}{\text{Dimethylamine}} > \underset{1°}{\text{methylamine}} > \underset{3°}{\text{trimethylamine}} > \text{aniline}$$

This order depends upon the inductive effect solvation energy and steric hinderance of alkyl groups.

**20.** *(b)* Wurtz reaction is used to prepare alkanes from alkyl halides.

$$2R\text{—}X + 2Na \xrightarrow{\text{Dry ether}} R\text{—}R + 2NaX$$

**21.** *(a)* The complete reaction, takes place as follows

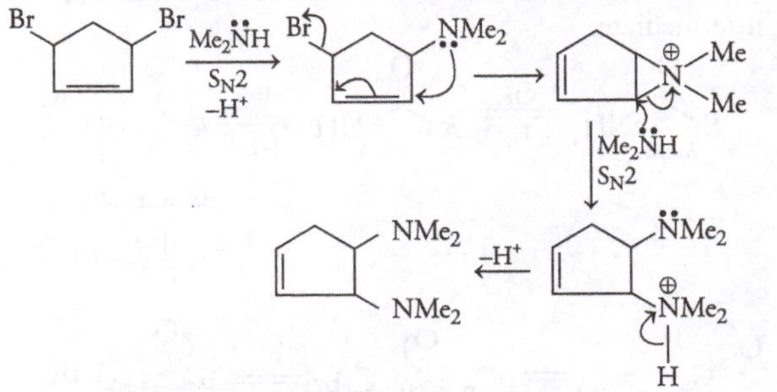

**22.** *(c)* The complete reaction is as follows

$$Cl(CH_2)_4Cl \xrightarrow{\text{Excess } NH_3} \underset{(A)}{Cl^-NH_3^+\text{—}(CH_2)_4\text{—}NH_3^+Cl^-}$$

$$\xrightarrow{NaOH} \underset{(B)}{H_2N\text{—}(CH_2)_4\text{—}NH_2} + 2NaCl + 2H_2O$$

**23.** *(b)* The reaction will takes place as follows

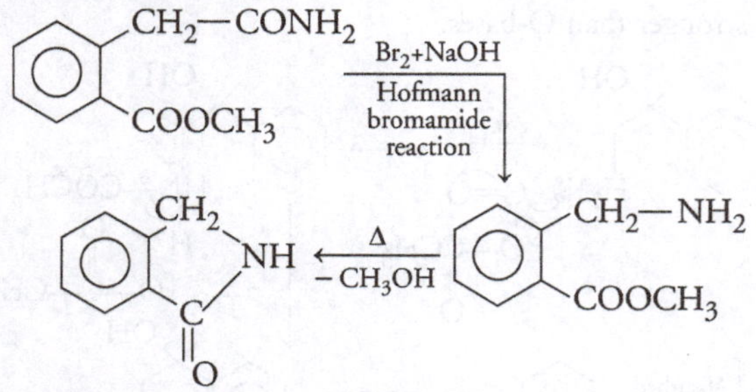

Option (b) is correct.

**24.** *(d)*

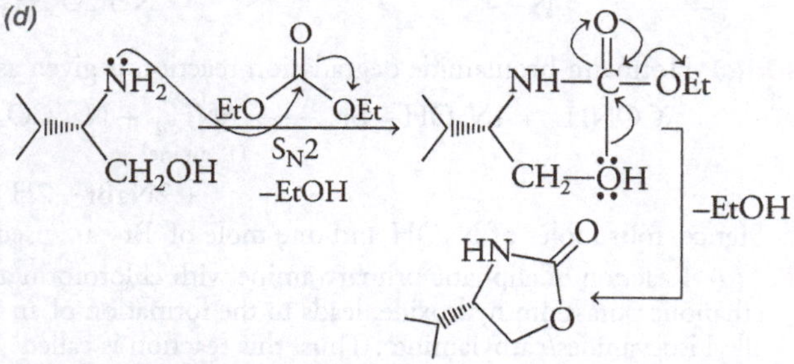

Initially, the lone pair electrons of $—NH_2$ will attack on electrophilic carbon of $O{=}C(OEt)_2$, after that lone pair electrons of oxygen of —OH group, will attack intramolecularly, leading to formation of the cyclic compound.

**25.** *(d)*

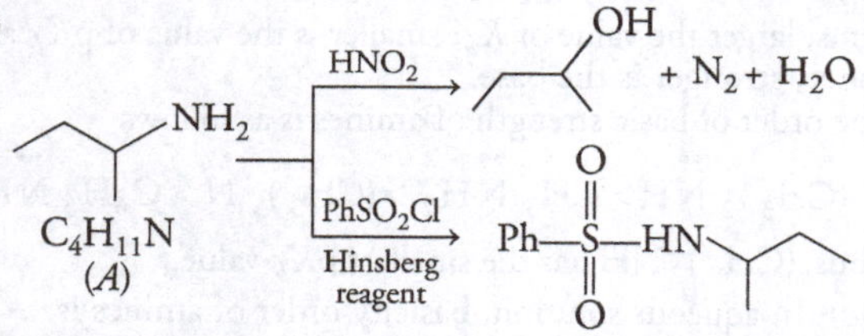

$C_4H_{11}N$ since it releases $N_2$ with $HNO_2$, it is a primary amine. After reacting with Hinsberg reagent it forms a compound which is soluble in KOH.

**26.** *(c)* The correct match is A-(III), B-(IV), C-(I), D-(II).

**27.** *(c)* All the given enamines are interconvertible. So, most stable form of enamine is '*c*' due to least steric hinderance between —COOH group and alkene substituents.

**28.** *(c)* Both bromoamide and isocyanate are isolated as intermediate.

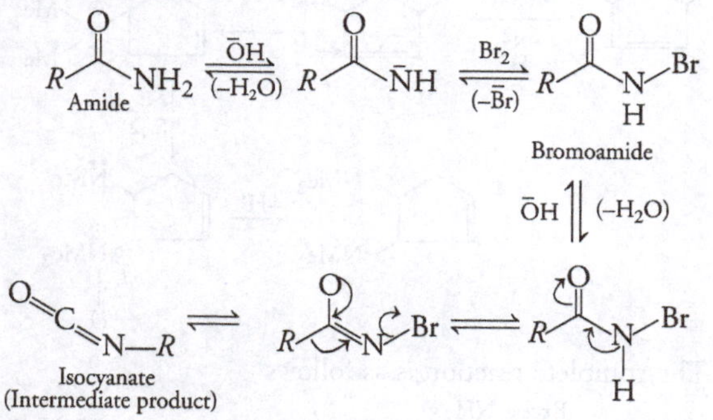

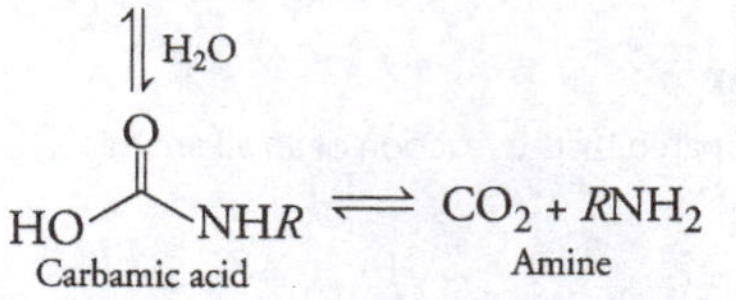

**29.** *(a)* The intermolecular association is more in primary amines than in secondary amines as there is only one hydrogen atom is available for formation of hydrogen bond in secondary amines, while primary amines contain two hydrogen atoms. So, statement (a) is not correct.

**30.** *(d)* Since $—CH_2—NH_2$ is more basic, the resulting amide will fail to react further.

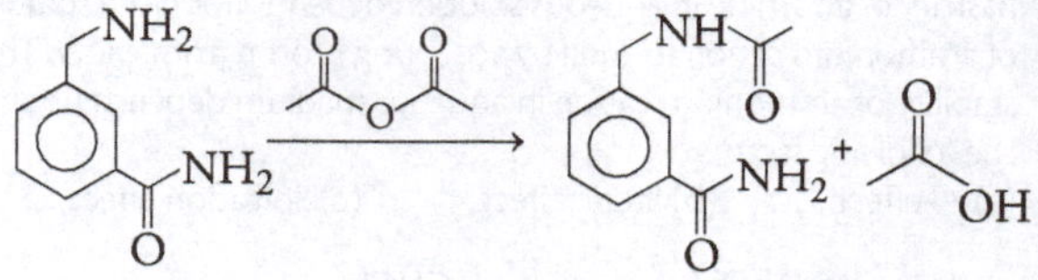

**31.** *(a)*

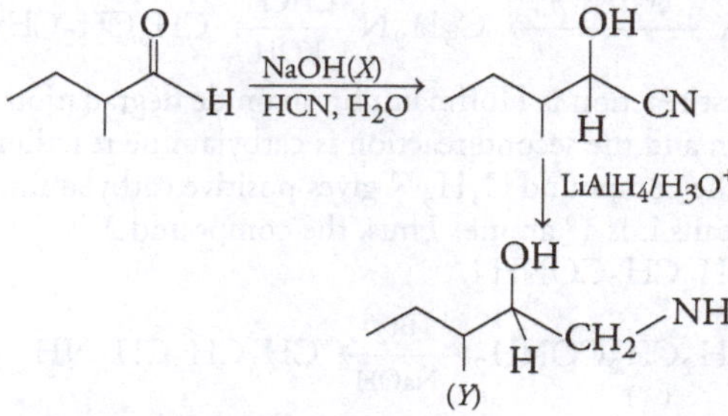

**32.** *(d)* Complete reaction is as follows

(i) **Removal of $Br^-$**

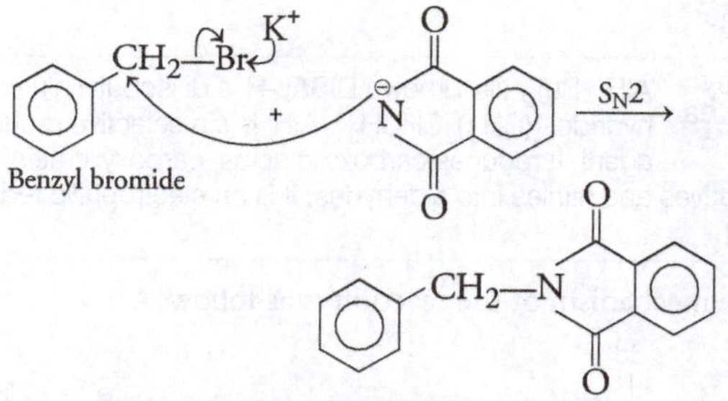

(ii) **Hydrolysis process in basic medium**

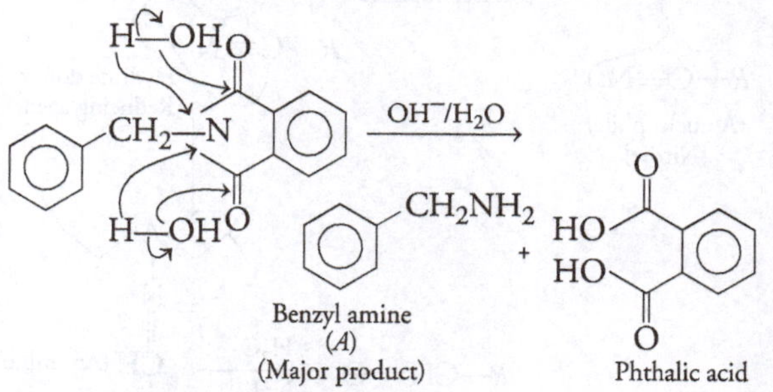

**33.** *(c)* Ammonolysis of alkyl halides

$$R\text{–}X \xrightarrow[-HX]{NH_3} R\text{—}NH_2 \xrightarrow[-HX]{R\text{—}X} R_2NH \xrightarrow[-HX]{R\text{—}X} R_3N \xrightarrow{RX} R_4\overset{\oplus\ominus}{NX}$$

During the reaction *HX* (acid) is formed. Hence, we use NaOH to remove this acidic impurities.

**34.** *(d)* (*B*) is most basic because of high resonance stabilisation of its conjugate acid.

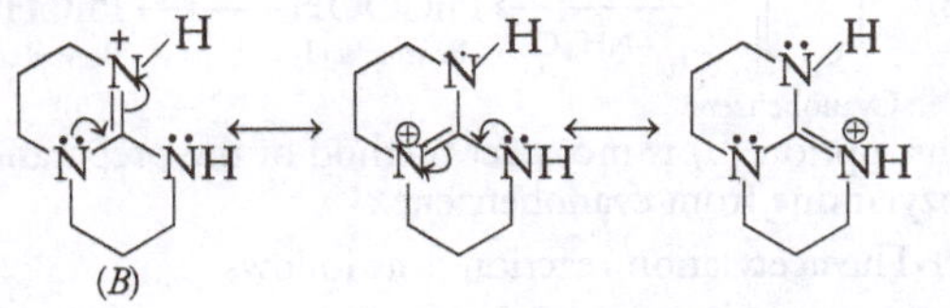


Among rest two compounds, $NH_2—CH=\ddot{N}H$ (*A*) has higher basicity than $CH_3—NH—CH_3$ (*C*) because conjugate acid of (*A*) is resonance stabilised but conjugate acid of (*C*) is not resonance stabilised at all.

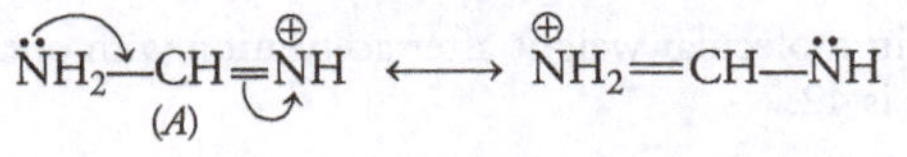


Hence, the correct order of basicity of the given compounds is, $B > A > C$. Stronger is the base, smaller will be its $pK_b$ value. Thus, the increasing order of $pK_b$ values is $B < A < C$.

**35.** *(c)*

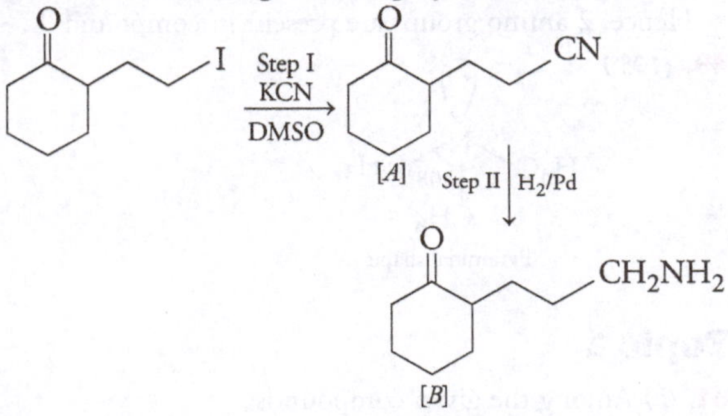


**36.** *(a)* The mechanism of the given reaction is as follows :

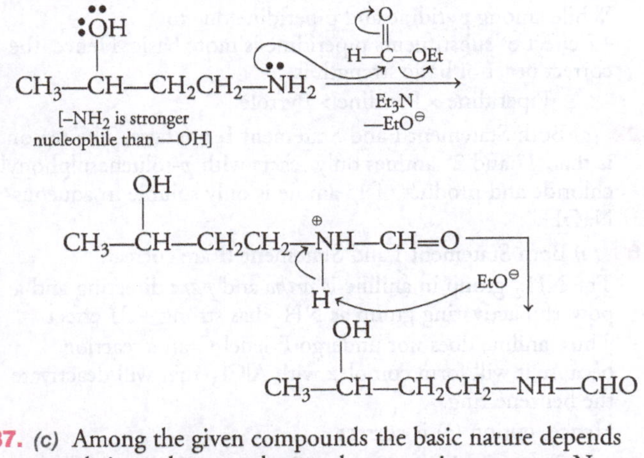


**37.** *(c)* Among the given compounds the basic nature depends upon their tendency to donate electron pair present on N atom

Hence, the correct order is (II) < (I) < (IV) < (III)

i.e. option (c) is correct.

**38.** *(c)* The complete series of reaction can be represented as

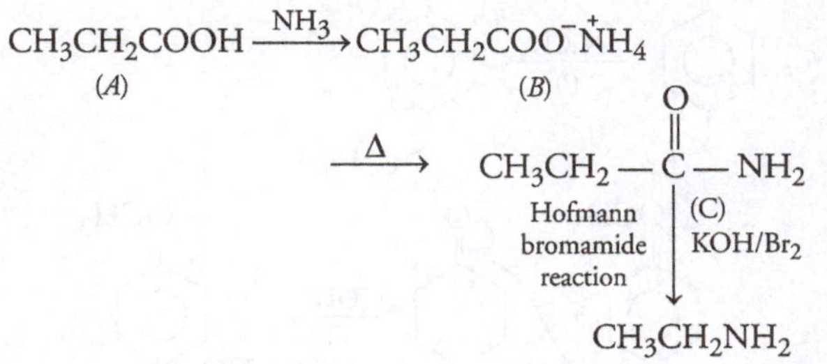


**39.** *(c)* Basic strength of amine depends on hydrogen bonding and electronic inductive effect.

Hence, the correct order is

$$NH_2-NH_2 < NH_3 < CH_3CH_2NH_2 < (CH_3CH_2)_3N < (CH_3CH_2)_2NH$$

**40.** *(c)* The complete reaction involved is as follows

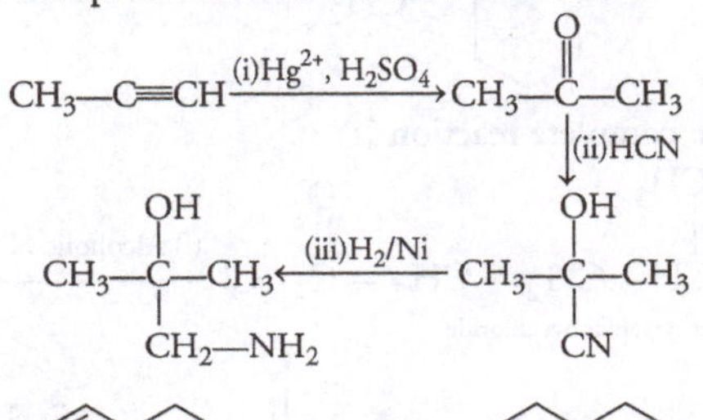


**41.** *(c)*

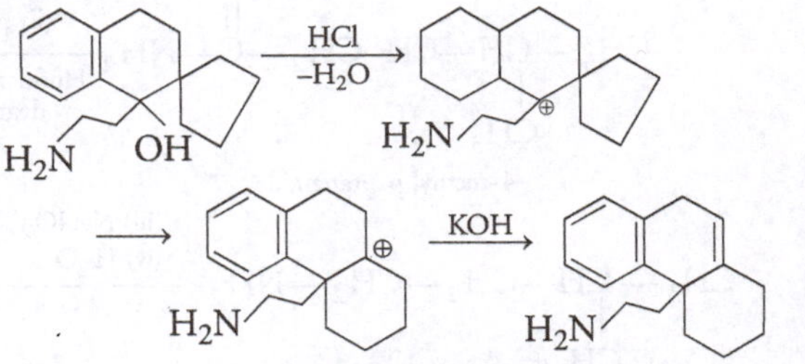


**42.** *(d)* Out of compounds a, b, c and d, d is the strongest Bronsted base as its lone pair of electrons can easily be donated.

**43.** *(a)*

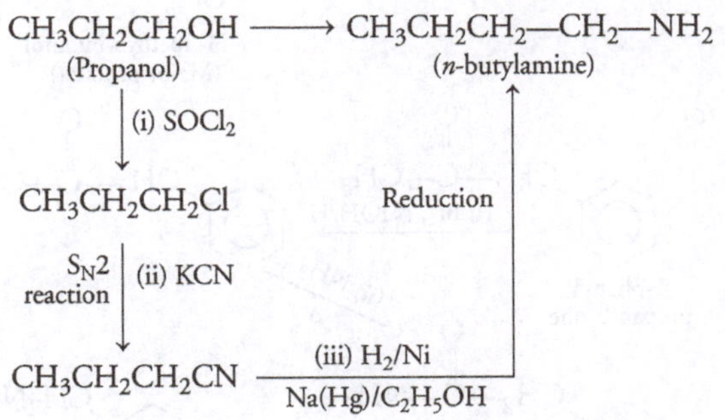

**44.** *(d)*

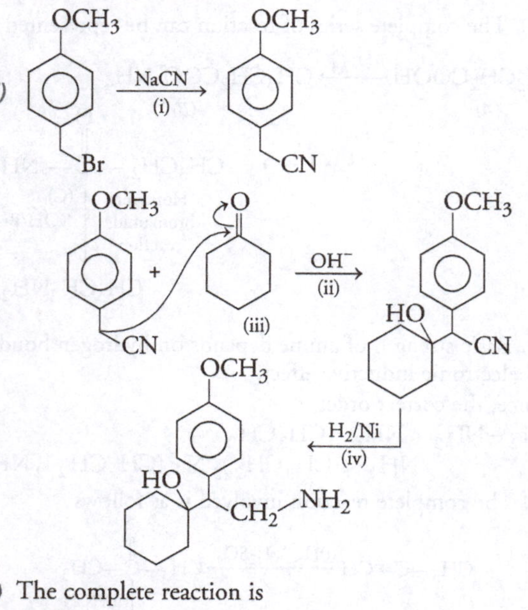


**45.** *(c)* The complete reaction is

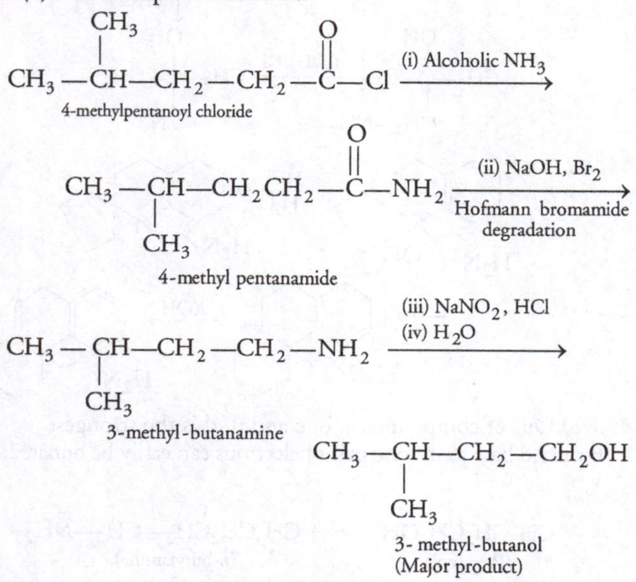


**46.** *(c)*

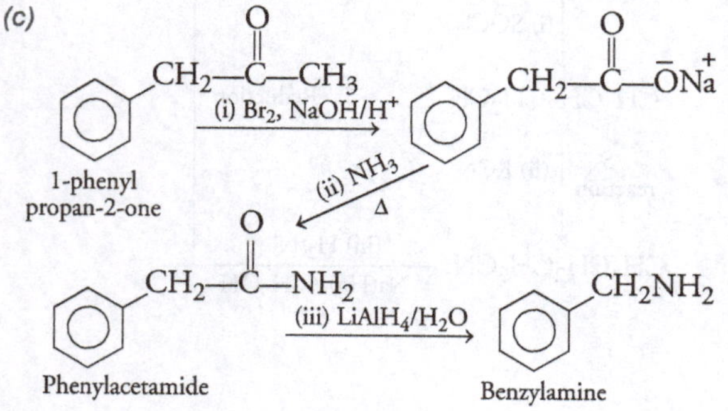


**47.** *(b)*

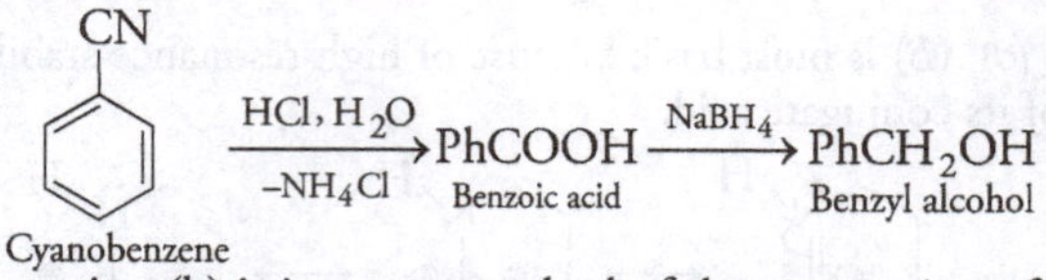


Thus, option (b) is incorrect method of the preparation of benzylamine from cyanobenzene..

**48.** *(2)* The acetylation reaction is as follows

$$R\text{—}NH_2 + CH_3\text{—}\overset{\overset{\large O}{\|}}{C}\text{—}Cl \longrightarrow R\text{—}NH\text{—}\overset{\overset{\large O}{\|}}{C}\text{—}CH_3$$

Compound $(X)$ molar mass $= 108$ g/mol

Product molar mass $= 192$ g $mol^{-1}$

Gain in molecular weight after acetylation with one —$NH_2$ group is 42.

Total increase in molecular weight $= 84$

$\therefore$ Number of amino group in $X = \frac{84}{42} = 2$

Hence, 2 amino groups are present in compound $X$.

**49.** *(108°)*

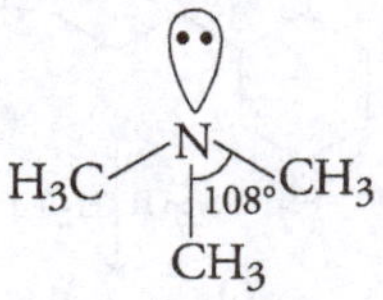

Pyramidal shape

## Topic 2

**01.** *(b)* Among the given compounds is weakest as in this the lone pair is completely delocalised *via* cyclic resonance to show aromaticity.

While among pyridine and piperidine due to $+I$ effect of substituents piperidine is more basic. Hence, the correct order of basic strength is

Piperidine > Pyridine > Pyrrole

**02.** *(c)* Both Statement I and Statement II are false. The reason is that, 1° and 2° amines only reacts with *p*-toluenesulphonyl chloride and product of 1° amine is only soluble in aqueous NaOH.

**03.** *(a)* Both Statement I and Statement II are correct.

The $NH_2$ group in aniline is *ortho* and *para* directing and a powerful activating group as $NH_2$ has strong $+M$ effect.

Thus, aniline does not undergo Friedel-Craft's reaction because it will form complex with $AlCl_3$ that will deactivate the benzene ring.

Hence, option (a) is correct.

**04.** *(d)* Statement I is correct but Statement II is incorrect.

Correct form of Statement II is

Aniline is also known as amino benzene. It is a liquid organic compound formed by bond formation between benzene and ammonia. Chemical formula $= C_6H_5NH_2$

Hence, option (d) is correct.

**05.** *(a)* The complete reaction takes place as follows,

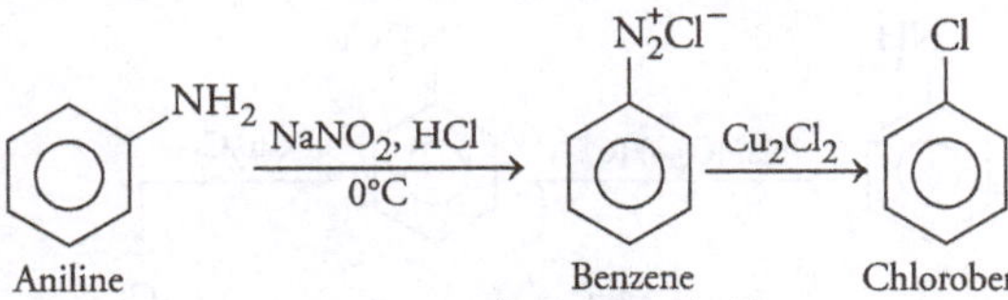


Aniline Benzene diazonium chloride Chlorobenzene

**06.** *(d)* As $-\ddot{N}H_2$ group is *o*/*p*-directing, hence arenium ion will be formed by attack at meta position i.e.

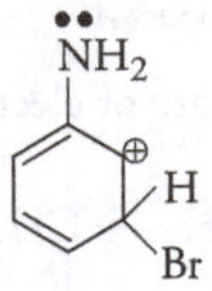


Hence, option (d) is correct.

**07.** *(b)* In option (a), (c) and (d) lone pairs is delocalised due to resonance.

While option (b) has highest availability of lone pair due to absence of resonance. Thus, it act as best proton acceptor or strongest bronsted base.

Localised lone pair

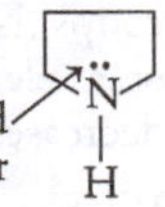


**08.** *(b)*

**key idea** Use the concept of stability of carbocation to solve this problem. This stability can be judged on the basis of resonance and hyperconjugation.

Carbocation (*C*) is most stable because positive charge delocalises with 2 $\pi$ bonds and lone pair of N-atom in $NH_2$ group and due to hyperconjugation.

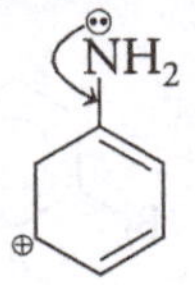


(*C*)

Thus, option (b) is correct.

**09.** *(c)* Statement I is correct but Statement II is incorrect. Aniline and other aryl amines are generally colourless, but they get coloured if stored outdoors due to atmospheric oxidation. Aniline becomes dark brown due to action of air and light [oxidation].

**10.** *(a)* Hinsberg reagent is another name used for benzene sulphonyl chloride.

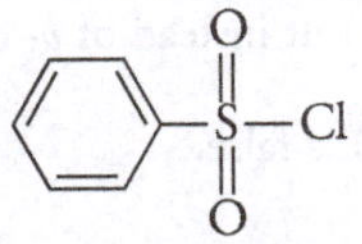


**• Concept Enhancer**

Hinsberg reagent name is given due to its use in the hinsberg test for the detection & distinction of primary, secondary and tertiary amines in a given sample. It is an organo-sulphur compound with chemical formula $C_6H_5SO_2Cl$.

**11.** *(a)* Nitration of aniline takes place in highly acidic medium. The electrophile formed in this reaction is $NO_2^+$. Moreover, aniline is also protonated to form anilinium ion which is *meta*-directing. That is why besides the *ortho* and *para* products, significant amounts of *meta* product is also formed.

Thus, both A and R are true and R is the correct explanation of A.

**12.** *(c)* Hinsberg's reagent is (*p*-toluene) sulphonyl chloride.

It reacts with primary amine to give product which is soluble in alkali while reacts with secondary amine to give product insoluble in alkali.

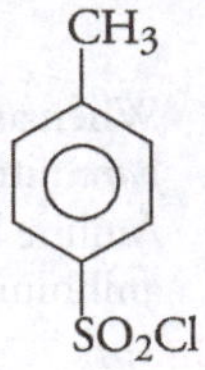


**13.** *(d)* Aniline when subjected to Friedel-Crafts alkylation reaction forms

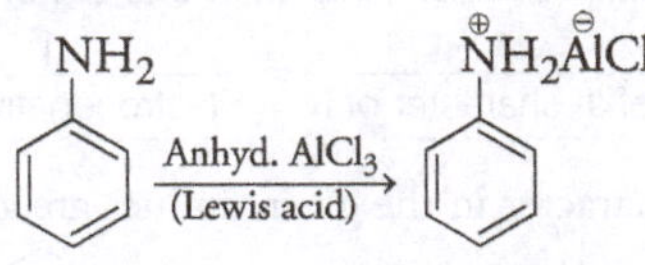


Weak base Forms salt

The positive charge on N makes it a deactivating group.

**14.** *(b)* Statement I is false but Statement II is true.

Aniline ($C_6H_5NH_2$) is more basic than acetamide ($CH_3CONH_2$) because in acetamide lone pair of electron present on nitrogen is delocalised to more electronegative element, i.e. oxygen.

As lone pair of N in acetamide is in conjugation with C=O that is strong electron withdrawing than phenyl group, so lone pair of N is unavailable for proton, so it is less basic than aniline.

**15.** *(d)* Reaction is as follows

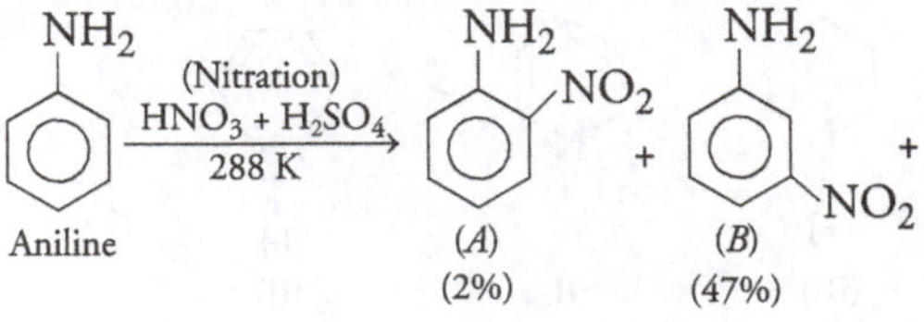


Aniline (*A*) (2%) (*B*) (47%) (*C*) (51%)

% yield order is $C > B > A$.

**16.** *(a)* Primary, secondary and tertiary amines can be separated using benzene sulphonyl chloride. Here it is present in its derivative form i.e., *p*-toluene sulphonyl chloride. This test is known as Hinsberg test.

**17.** *(d)* The amine on reaction with benzene sulphonyl chloride (Heisenberg reagent) produces a compound insoluble in alkali. It indicates the amine is a 2° amine. i.e. all options are possible except option (b) which is a 1° amine ($CH_3CH_2NH_2$).

As this 2° amine can be prepared by ammonolysis of ethyl chloride, the 2°-amine should have at least one ethyl ($C_2H_5$) group.

(a) $Ph-NH-CH_2CH_2CH_3$ does not have ethyl group.

(c) $CH_3CH_2CH_2-NH-CH_3$ does not have ethyl group.

(d) $CH_3CH_2CH_2-NH-CH_2CH_3$ has one ethyl group.

So, option (d) is the correct answer.

**18.** *(c)* Aniline on protonation forms anilinium ion, which is *meta*-directing. So, a considerable amount of *meta* product is formed alongwith *o*-nitroaniline and *p*-nitroaniline.

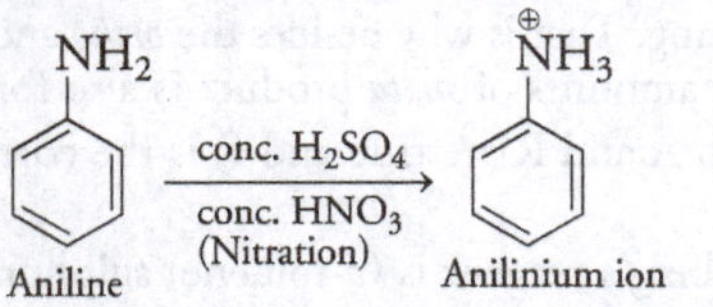


When aniline is treated with nitrating mixture *ortho*, *meta* and *para* nitroanilines are obtained.
Aniline being basic in nature, reacts with acids to form anilinium ion which is *meta* directing.

**19.** *(d)*

> **key idea** Basicity ∝ Ease of donation of lone pair of electrons
> $$\propto \frac{1}{\%\text{ of } s\text{-character of N}} \propto \frac{1}{\text{Electronegativity of N}}$$

The % of *s*-character in the given amines are as follows:

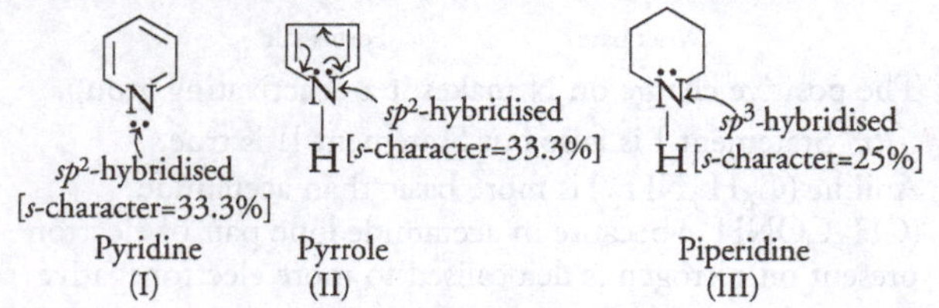


Therefore, piperidine (III) having minimum % s-character is most basic. Between, pyridine (I) and pyrrole (II) the lone pair of electrons of N in pyrrole (II) is involved in delocalisation and follows $(4n+2)$ π aromatic $(n=1)$ system. So, the N-atom of pyrrole (II) will show least basicity.
Thus, the order of basicity is as follows:

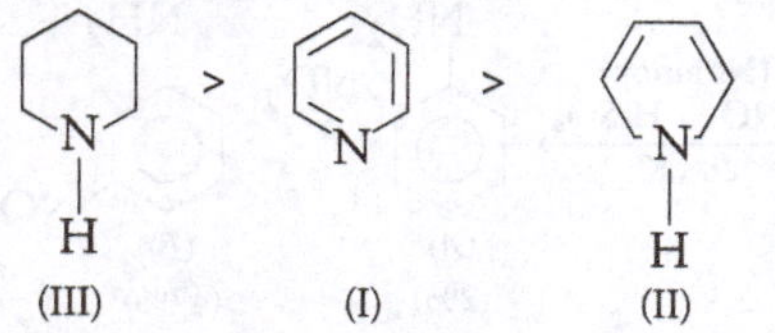


**20.** *(d)* Only primary amines (1°) gives carbylamine test. The reaction is as follows,

$$R—NH_2 \xrightarrow{CHCl_3 + KOH} R—N \overset{\rightarrow}{\equiv} C$$

Alkyl isocyanide (Pungent)

Option (A) and (C) are primary amine and therefore will give positive carbylamine test.

**21.** *(c)* Hinsberg reagent reacts with primary and secondary amines to form sulphonamides. But Hinsberg reagent do not react with tertiary amines. So, among the given options compound given in A, C and E will react Hinsberg's reagent while compound given in B and D will not react.

**22.** *(b)* The complete reaction is as follows.

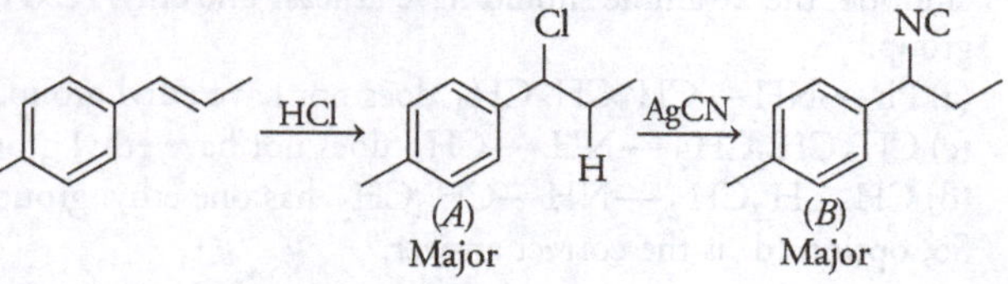


**23.** *(b)* The complete reaction is as follows:

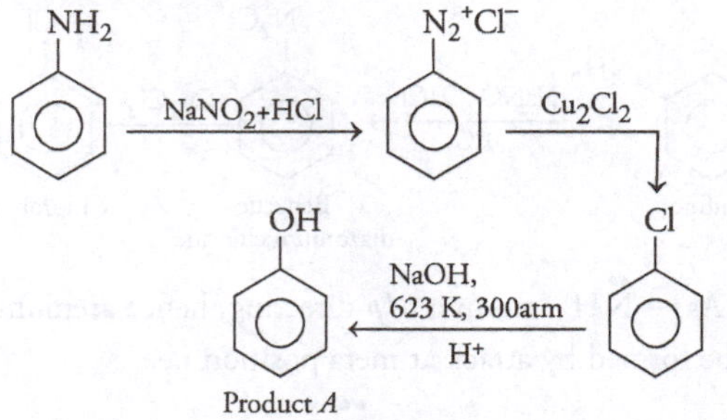


**24.** *(b)* Correct order of rate of electrophilic aromatic substitution is

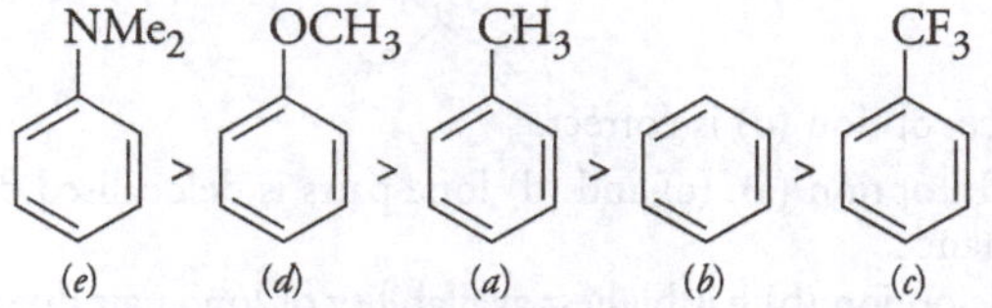


Rate of electrophilic aromatic substitution is directly proportional to the electron density on benzene ring.
Electron releasing groups ($NMe_2$ > $—OCH_3$) increases electron density whereas electron withdrawing groups ($—CF_3$ > $—NO_2$) decreases electron density.

**25.** *(b)* Option (b) is correct.

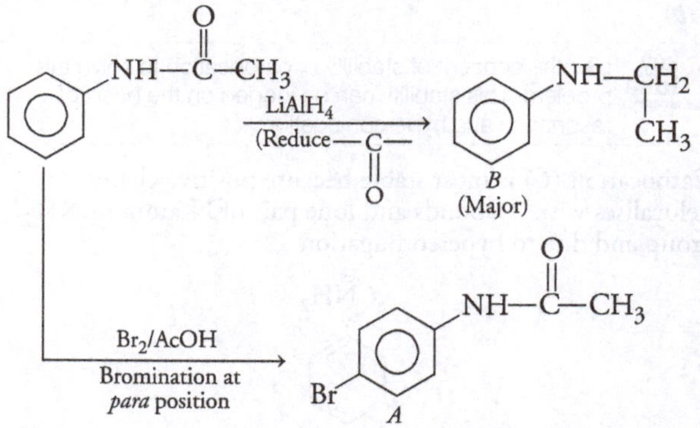


**26.** *(b)* The correct match is A-III, B-IV, C-II, D-I.

**27.** *(c)*

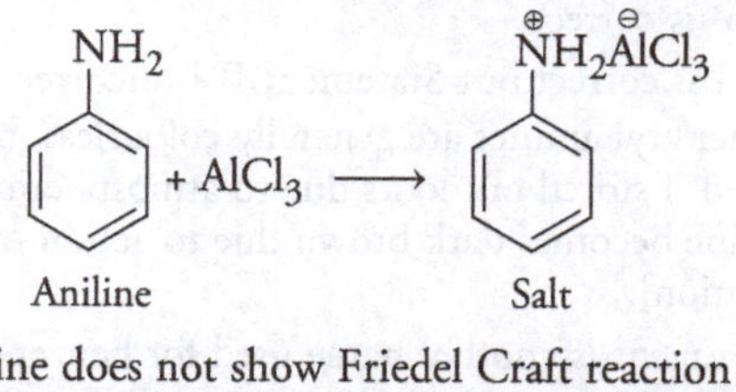


Aniline does not show Friedel Craft reaction as it is a weak base and reacts with $AlCl_3$ (electron deficient) which act as Lewis acid, forming salt instead of *o*- or *p*-substituted product.
∴ (A) is true but (R) is false.

**28.** *(c)*

1. **Addition-elimination reaction $\ddot{N}H_3$ attacks on carbonyl carbon and eliminates Cl to form amide.**

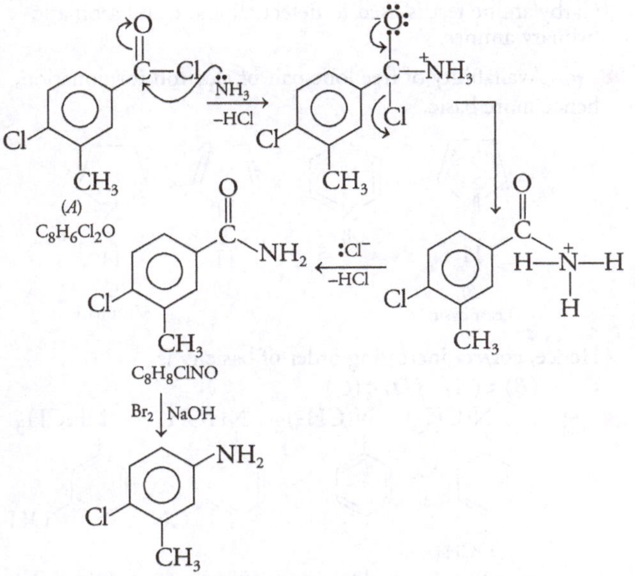

2. **Hoffmann bromamide degradation** Amide with $X_2$ in presence of NaOH undergoes Hoffmann bromamide degradation to form amine.

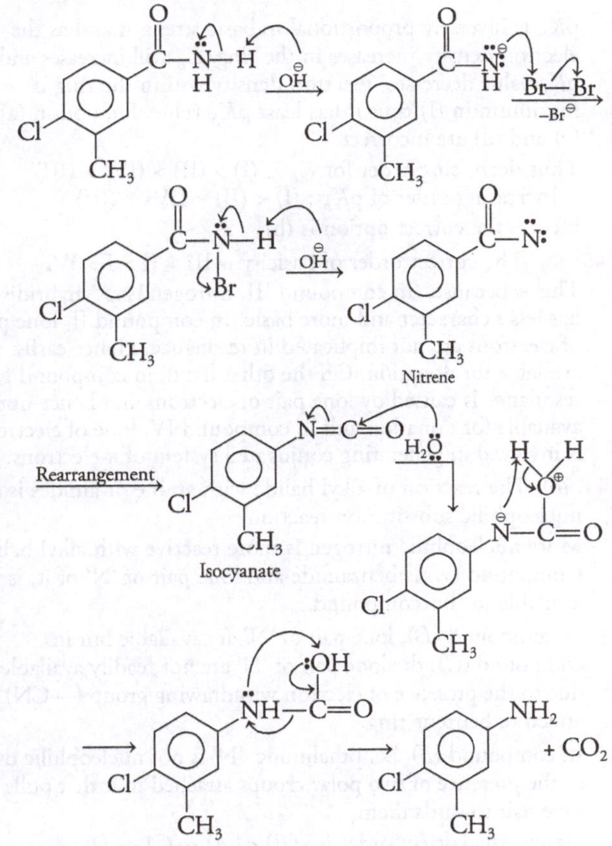

**29.** *(c)* Nitration of aniline gives three products as shown below in presence of $HNO_3$ and $H_2SO_4$ at 288 K.

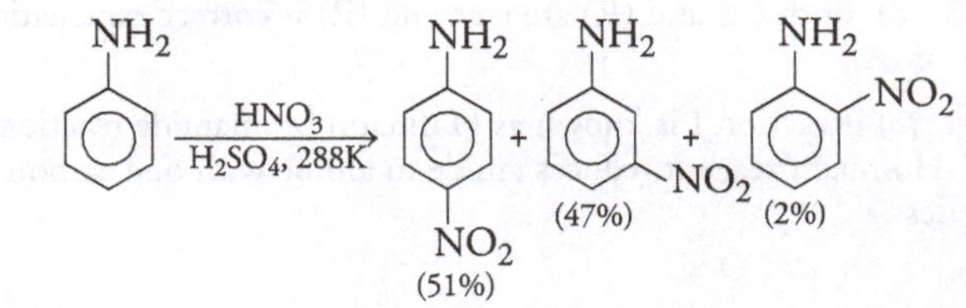

Here, $H_2SO_4$ acts as an acid.

$$H_2SO_4 + HNO_3 \longrightarrow NO_2^{\oplus} + HSO_4^{\ominus} + H_2O$$

Hence, statements (B) and (D) are correct.

**30.** *(b)* In reaction (b), imino group formed does not react with Hinsberg reagent.

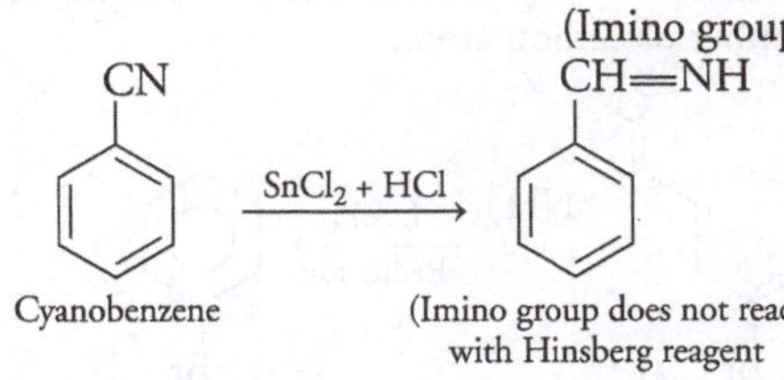

Products formed in all other reactions are a primary amine so, they react with Hinsberg reagent.

**31.** *(b)*

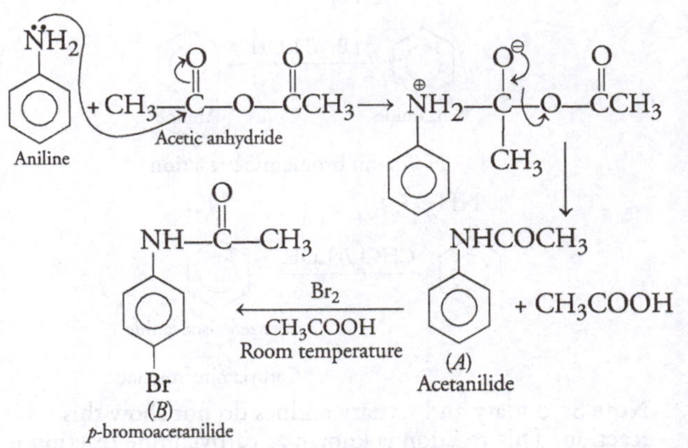

**32.** *(b)*

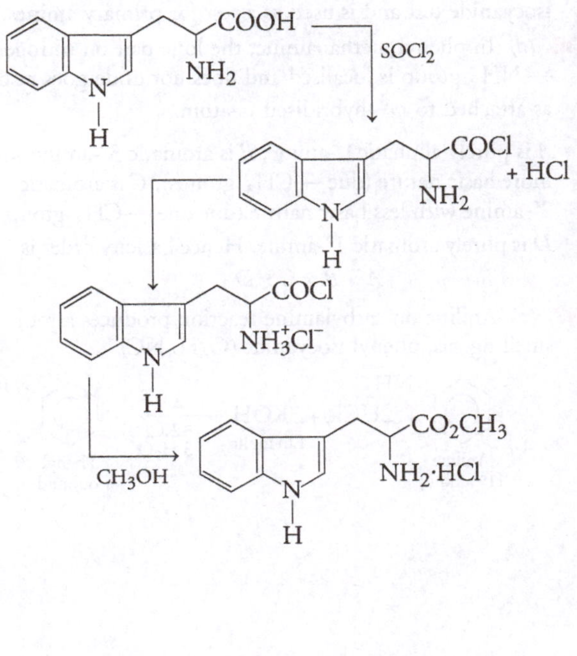

**33.** *(c)* Both (A) and (R) are true and (R) is correct explanation of (A).

**34.** *(d)* Reaction I is known as Hofmann bromamide reaction. Hofmann reagent reduces amide to amine with one carbon less.

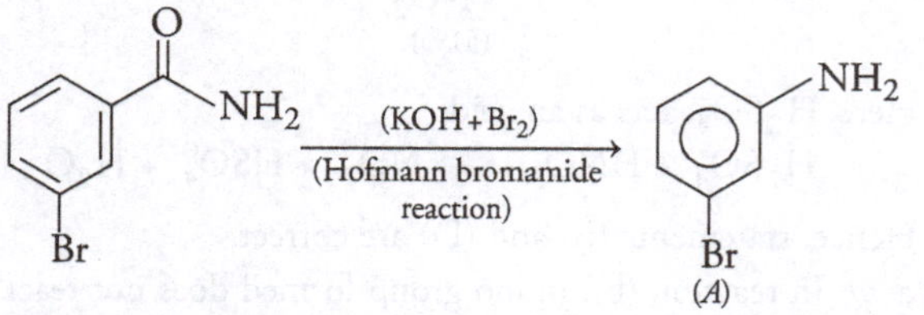


In reaction II, $LiAlH_4$ reduces amide to primary amine with same number of carbon atom.

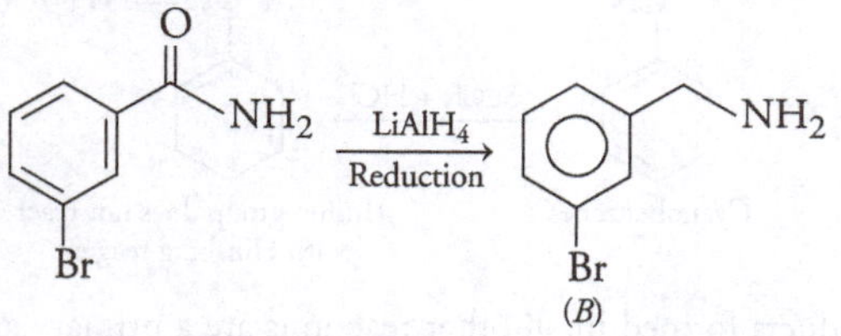


**35.** *(b)*

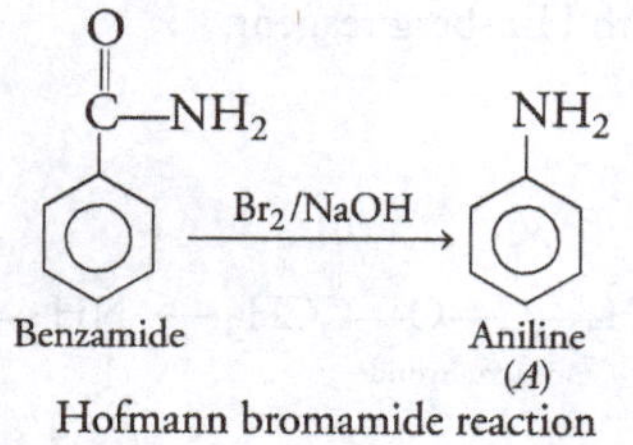


Hofmann bromamide reaction

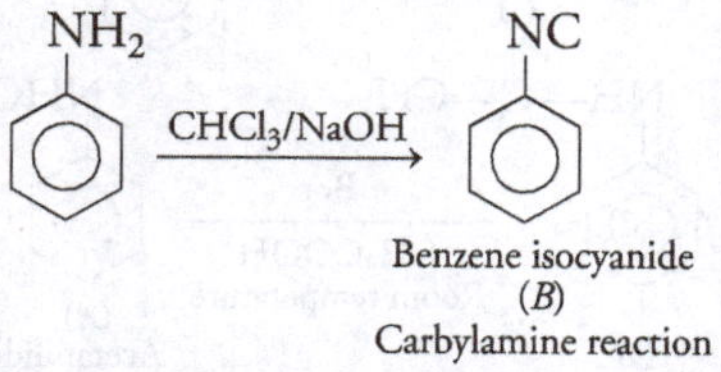


Carbylamine reaction

**Note** Secondary and tertiary amines do not show this reaction. This reaction is known as carbylamine reaction or isocyanide test and is used as a test for primary amines.

**36.** *(d)* In phenyl methanamine, the lone pair on nitrogen of $—NH_2$ group is localised and does not undergoes resonance as attached to $sp^3$-hybridised C-atom.

*A* is purely aliphatic 1°-amine. *B* is aromatic 3°-amine with more basic nature (due $—CH_3$ groups). *C* is aromatic 2°-amine with less basic nature (due one $—CH_3$ group). *D* is purely aromatic 1°-amine. Hence basicity order is

$$A > B > C > D.$$

**37.** *(c)* Aniline on carbylamine reaction produces a foul smelling gas, phenyl isocyanide ($C_6H_5NC$).

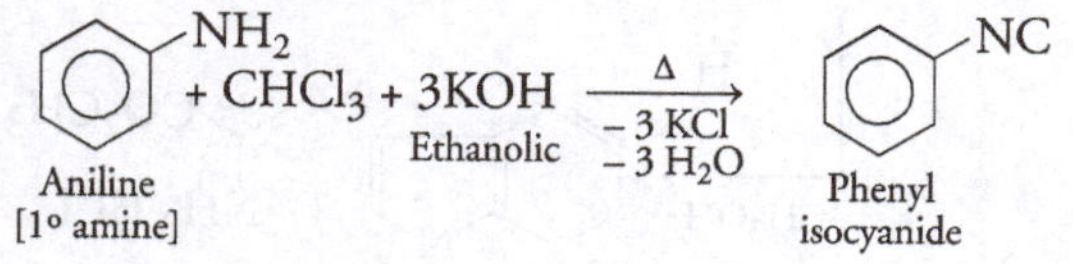


Carbylamine test is used to detect aliphatic and aromatic primary amines.

**38.** *(c)* Availability of free lone pair of electron for donation, hence more basic.

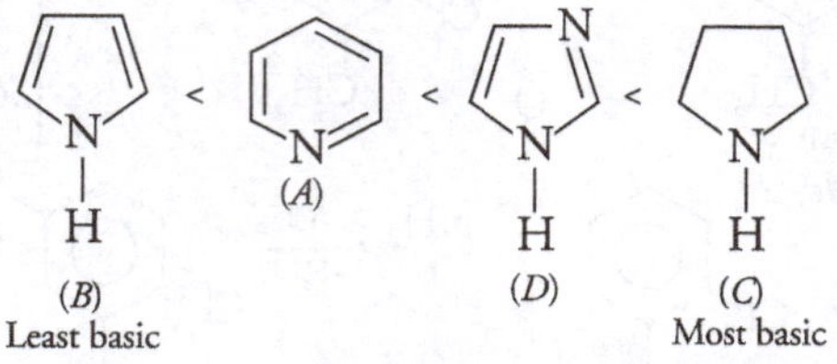


Hence, correct increasing order of basicity is

$$(B) < (A) < (D) < (C)$$

**39.** *(b)*

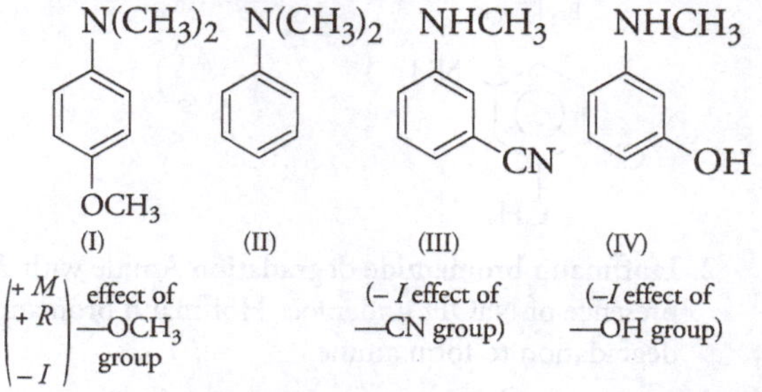


$pK_b$ is inversely proportional to basic strength and as the electron density increases in the ring, $K_b$ will increases and $pK_b$ value decreases. Electron density within the ring is maximum in (I) thus it has least $pK_b$ value. So, option (a), (c) and (d) are incorrect.

Thus decreasing order for $K_b$ is (I) > (II) > (IV) > (III)

$\therefore$ Increasing order of $pK_b$ : (I) < (II) < (IV) < (III).

Hence, the correct option is (b).

**40.** *(b)* The correct order of basicity is III > II > I > IV. This is because, in compound III, nitrogen is $sp^3$ hybridised has less *s* character and more basic. In compound II, lone pair of electrons are not implicated in resonance. Hence easily available for donation. On the other hand, in compound I, resonance is caused by lone pair of electrons and hence not available for donation and in compound IV, lone of electrons is involved in generating conjugated system of $\pi$-electrons.

**41.** *(b)* The reaction of alkyl halide with amine or amides is a nucleophilic substitution reaction.

More nucleophilic nitrogen is more reactive with alkyl halide. Compound (*A*) is benzamide and lone pair of 'N' of it, is not available in this compound.

In compound (*D*), lone pair of 'N' are available but in compound (*C*), the lone pair of 'N' are not readily available due to the presence of electron withdrawing group (—CN) attach to benzene ring.

In compound (*B*), i.e. pthalimide, 'N' is not nucleophilic due to the presence of two polar groups attached to it that pulls lone pair towards them.

Hence, the correct order is : (*B*) < (*A*) < (*C*) < (*D*).

**42.** *(b)* $Ni/H_2$ can reduce $—C\equiv N$ into $—CH_2—NH_2$ (1°-amine) but cannot reduce an ester group ($—CO_2Et$) whereas DIBAL-H, di-isobutylaluminium hydride,

$[(CH_3)_2CH]_2AlH$ reduces the ester group ($—CO_2Et$) into —CHO (an aldehyde).

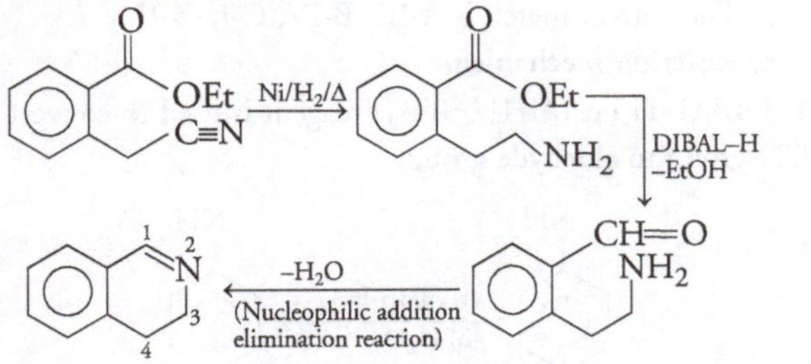


**43.** *(c)* Using retro-synthesis, to get the required compounds, *A*, *B* and *C*.

$C_6H_7N$ (*C*) [1°-amine] (No. of C=6, 1 carbon less) $\xleftarrow[\text{KOH}]{Br_2}$ Hoffmann bromamide reaction (step down) *B* [An amide] (No.of C=7) $\xleftarrow{NH_3/\Delta}$ *A* [Carboxylic acid $C_7H_6O_2$]

So, the reactions can be shown as:

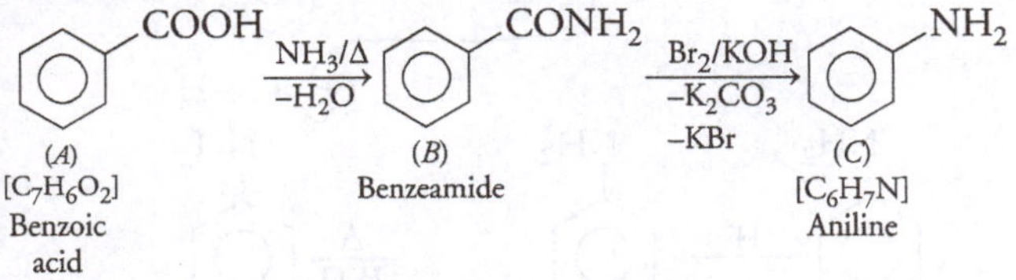


**44.** *(a)* The correct sequence is

E > D > B > A > C

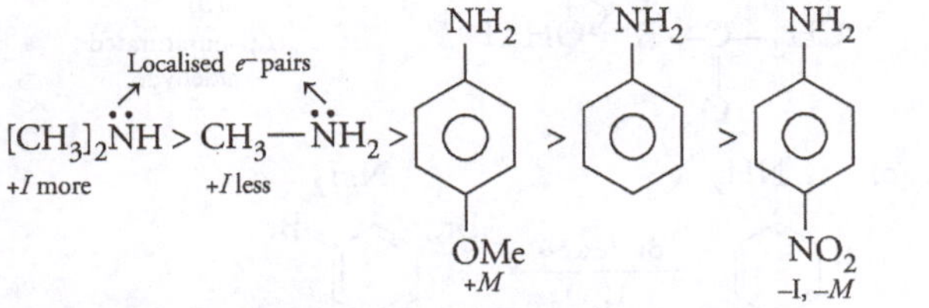


**45.** *(d)* Complete reaction sequence takes place as follows,

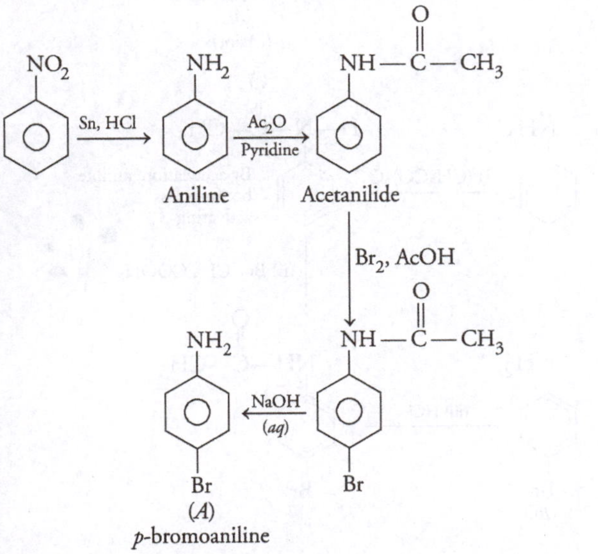


**46.** *(a)* Among the given options, reaction given in option (a) will give 3,4,5-tribromoaniline.

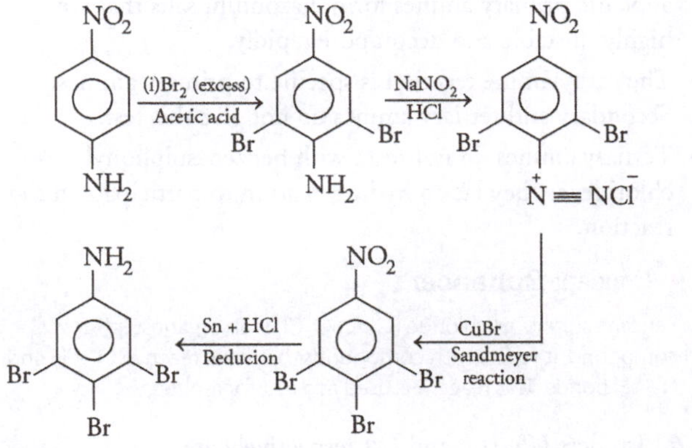


3, 4, 5-tribromoaniline

**47.** *(c)*

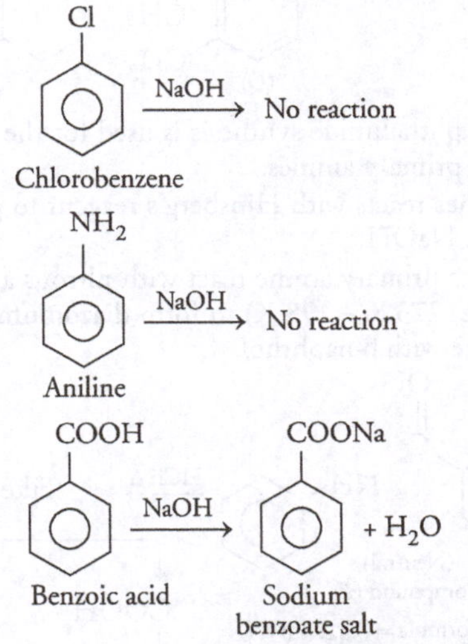


Organic layer in funnel are mixture of chlorobenzene and aniline.

**48.** *(a)* The complete reaction is as follows

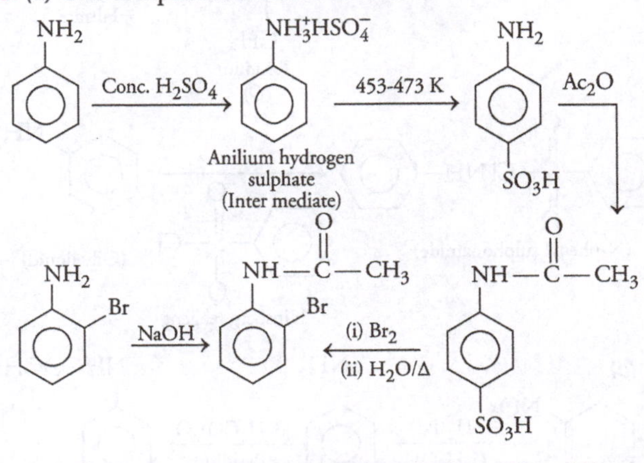


**49.** *(c)* The correct statements are given in B and D, while statements given is A , C and E are incorrect.

The correct form of A, C and E are

- Aromatic primary amines (like aniline) do form stable diazonium salts under the given conditions. However, aliphatic primary amines form diazonium salts that are highly unstable and decompose rapidly.
- The carbylamine reaction is specific to primary amines. Secondary and tertiary amines do not give this test.
- Tertiary amines do not react with benzenesulphonyl chloride, as they lack a hydrogen atom to participate in the reaction.

**Concept Enhancer**

Benzene sulphonyl chloride ($C_6H_5SO_2Cl$) is an organosulphur compound. It reacts with compounds that have reactive O—H and N—H bonds. This reagent is used as a test for amines.

**50.** *(c)* Isomers (*P*), (*Q*) and (*R*) respectively are

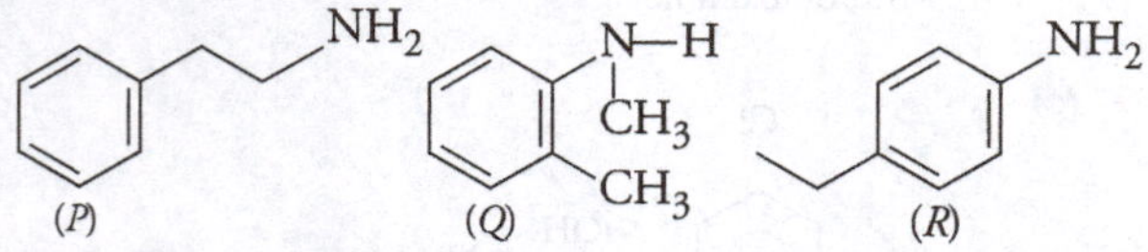


(*P*) Gabriel phthalimide synthesis is used for the preparation of aliphatic primary amines.
(*Q*) 2°-amines reacts with Hinsberg's reagent to give solid insoluble in NaOH.
(*R*) Aromatic primary amine react with nitrous acid at low temperature (273 K – 298 K) to form diazonium salts, which form red dye with β-naphthol.

**51.** *(a)*

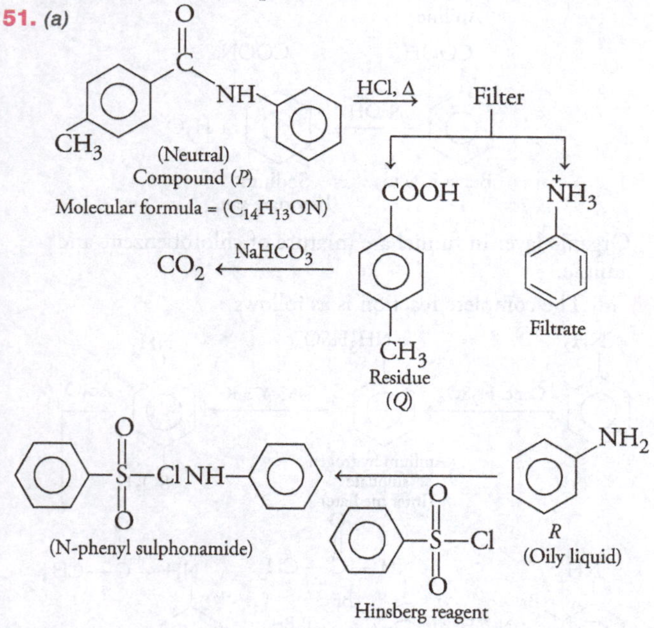


**52.** *(c)*

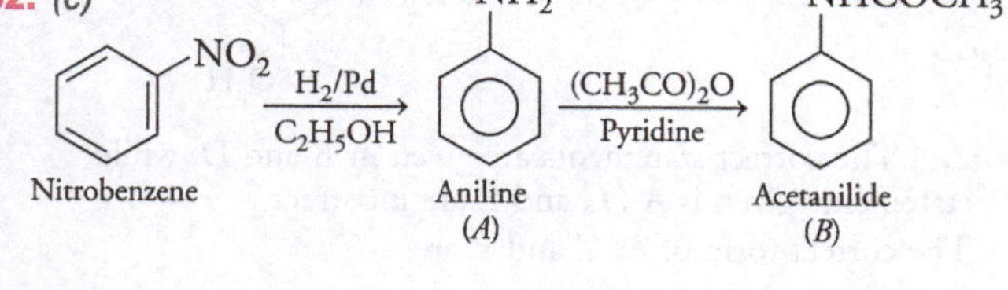


Amino group of aniline on acetylation with acetic anhydride/pyridine gives acetanilide.

**53.** *(c)* The correct match is A-III, B-IV, C-I, D-II.

**54.** *(b)* **Reaction mechanism**

1. DIBAL-H i.e. ($AlH(i\text{-}Bu)_2$) reagent is used to convert CN group to aldehyde group.

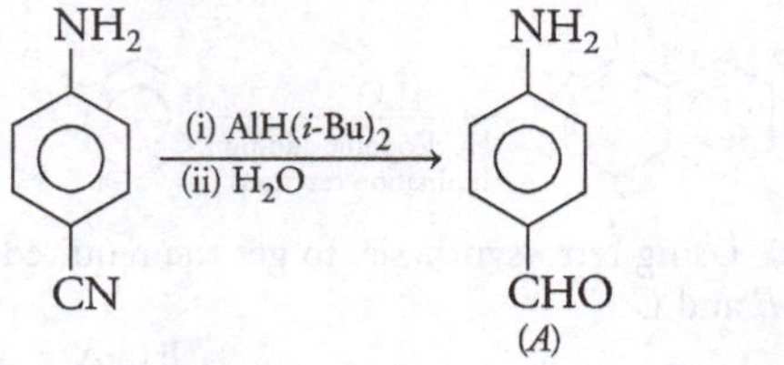


2. **Cross aldol condensation**

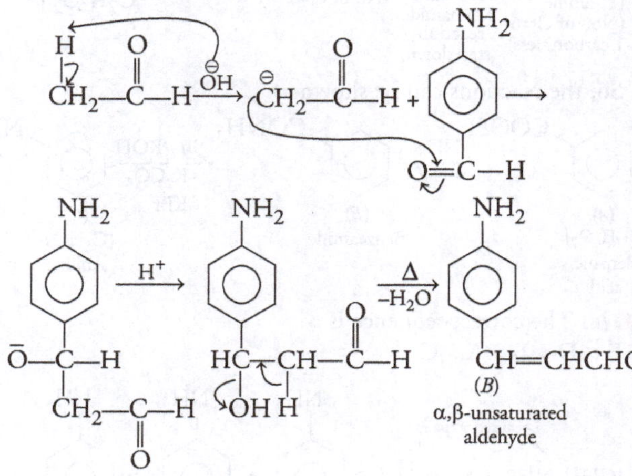


**55.** *(c)*

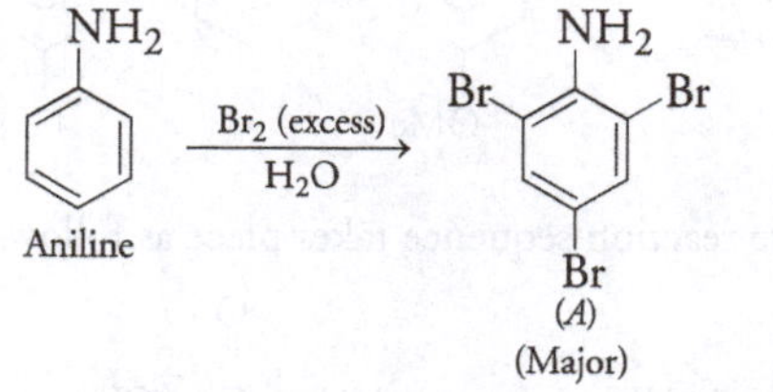


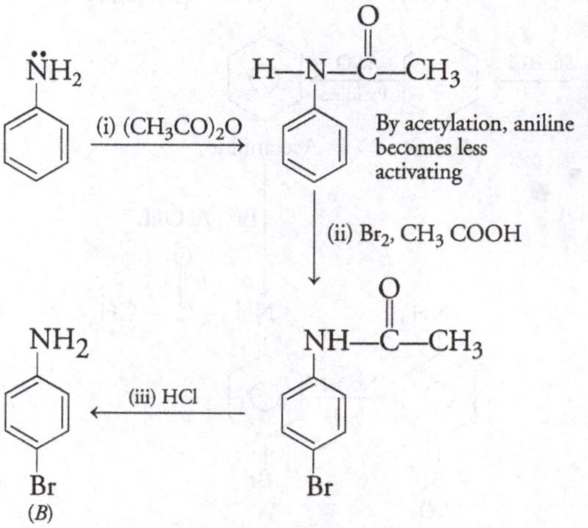

**56.** *(d)* 6 possible isomers of $C_6H_3(NH_2)_2COOH$ are

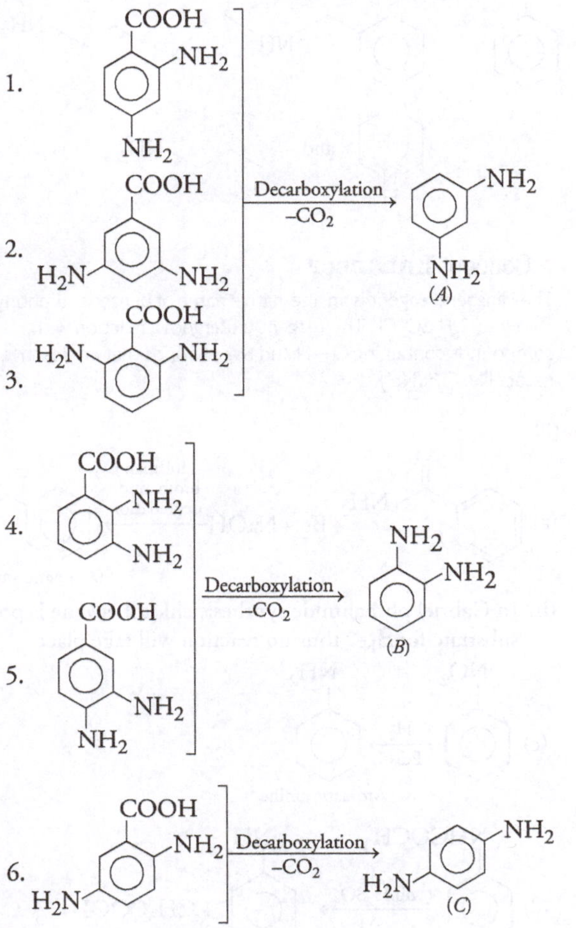


Product (*C*) have high melting point, i.e. 142°C more than products (*A*) and (*B*) because of the symmetry. It is packed more tightly in the crystal lattice. Hence, more heat is required to break the crystal lattice of product (*C*).

**57.** *(a)*

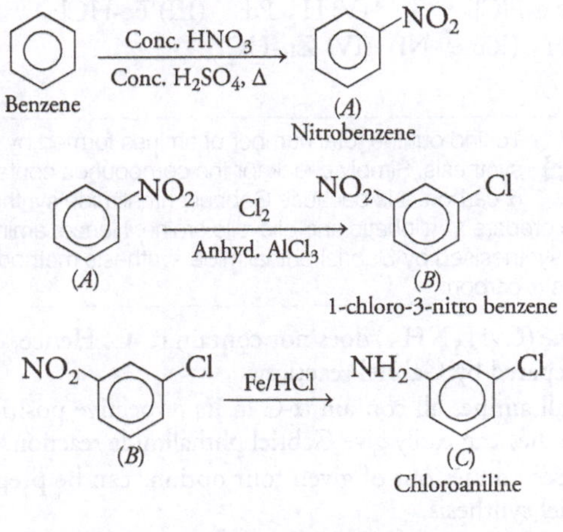


**58.** *(c)* Friedel-Craft reaction is not possible in highly activated ring, i.e. aniline because $NH_2$ group forms complex with Lewis acid and deactivates the ring.

Chemical reaction is as follows :

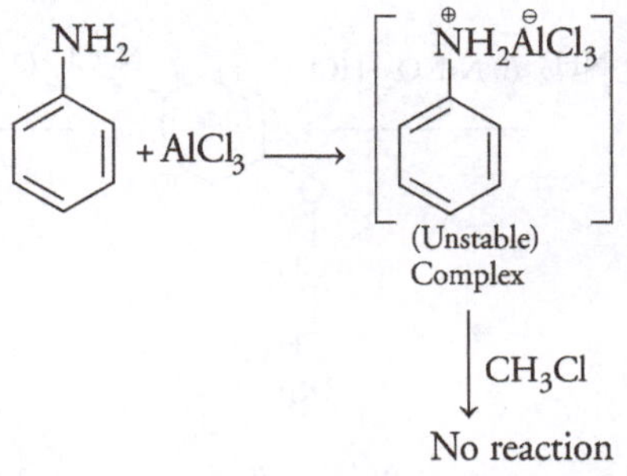


Due to lone pair of electrons on N, aniline is a strong Lewis base. So, an acid base reaction occur between aniline and $AlCl_3$ leading to anilinium salt formation (complex).

**59.** *(d)*

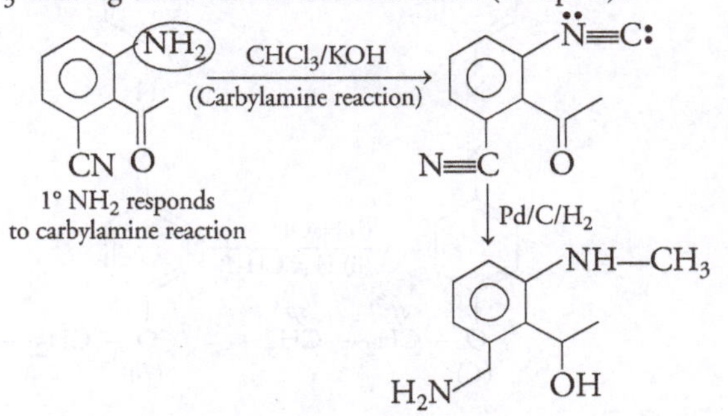


**60.** *(a)* In the given substituted benzene rings, the substitutents methoxy ($-OCH_3$) and amino ($-NH_2$) are strongly activating groups, while methyl ($-CH_3$) is weakly activating and chloro ($-Cl$) is deactivating group towards electrophilic aromatic substitution reaction.

Although amino group is strongly activating group, it gets protonated in presence of acid to form anilinium ion ($-\overset{+}{N}H_3$) which is strongly deactivating. Hence, (*A*) is less reactive than (*C*) and (*D*).

Chloro group is also deactivating group but less deactivating than ($-\overset{+}{N}H_3$). Thus, order is (*A*) < (*B*) < (*D*) < (*C*).

**61.** *(341)* Among the given compounds benzyl amine is most basic due to localised lone pair. The reaction involved as per question is as follow.

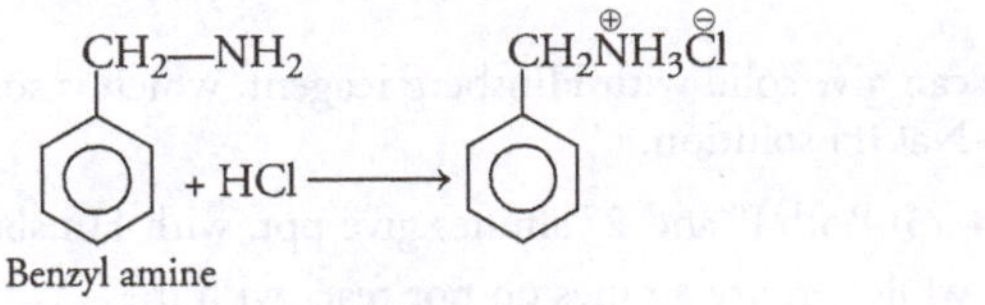


$$\text{Mole of benzyl amine} = \frac{1}{107} = 0.00934 \text{ mol}$$

1 mol of benzyl amine consumed 1 mole of HCl.

So, mole of HCl consumed = 0.00934 mol.

mass of HCl consumed = 0.00934 × molar mass of HCl

$$= 0.00934 \times 36.5 = 0.341 \text{ g} = 341 \text{ mg}$$

**62.** *(4)* The complete reaction involved is as follows

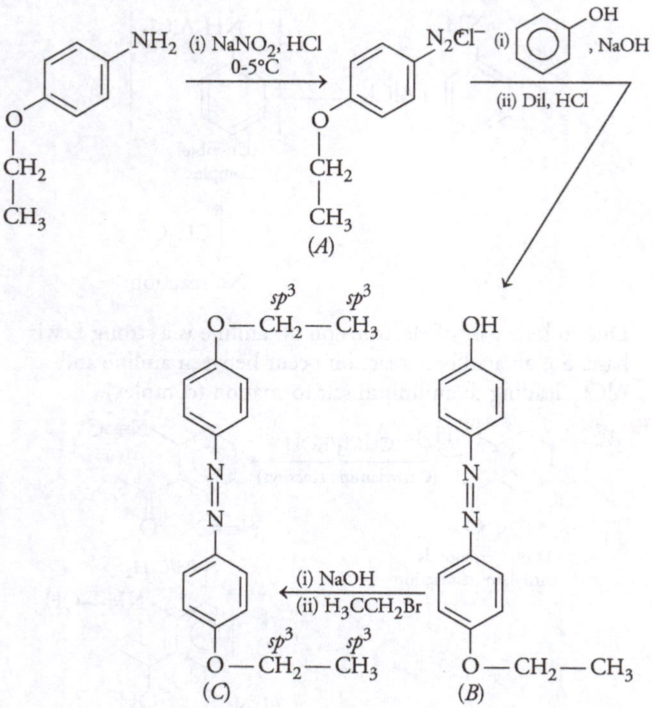

So, total four $sp^3$ hybridised carbon are present in product *C*.

**63.** *(5)*

> **key idea** The compound that contains one hydrogen atom attached to nitrogen will give alkali insoluble product on reaction with Hinsberg's reagent.

—$NH_2$ group containing compound can give solid with Hinsberg's reagent, which is soluble in NaOH solution.

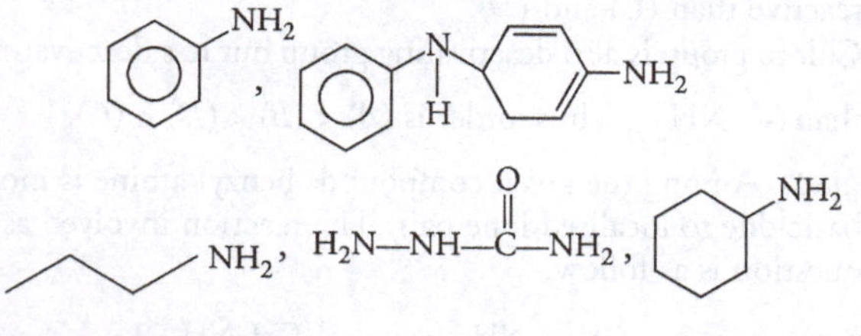

can give solid with Hinsberg reagent, which is soluble in NaOH solution.

**64.** *(5)* Both 1° and 2° amines give ppt. with Hinsberg's reagent while tertiary amines do not react with it.

Thus, among the given compounds only 5 gives Hinsberg's reagent test. These compounds are

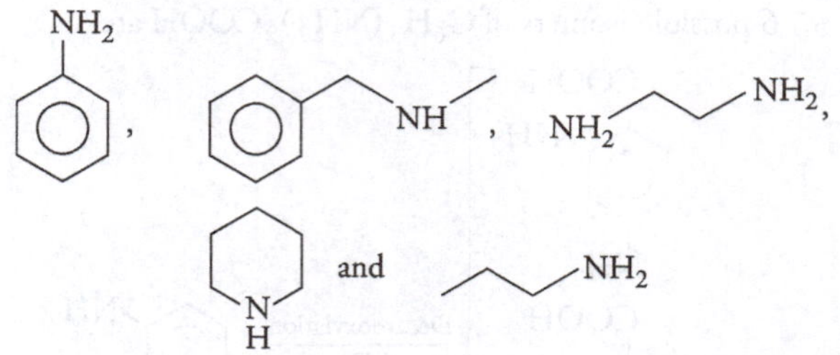

> **Concept Enhancer**
> The Hinsberg reagent is an alternative name of benzene sulphonyl chloride ($C_6H_5SO_2Cl$). This reagent undergoes a reaction with compounds containing O—H and N—H bonds that are reactive in nature like $C_2H_5NH_2$.

**65.** *(3)*

(a) 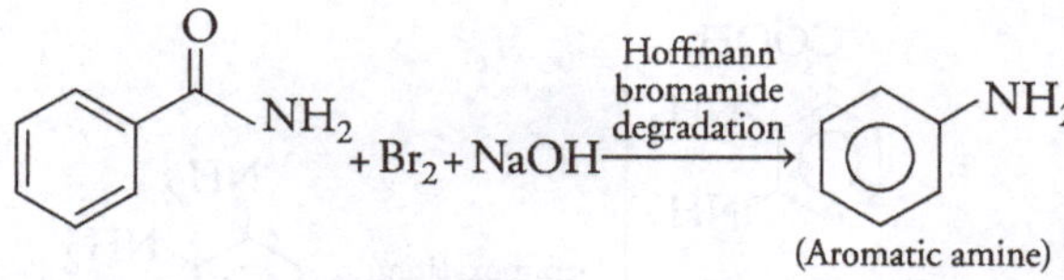

(b) In Gabriel phthalimide synthesis chlorobenzene is poor substrate for $S_N2$ thus no reaction will take place.

(c) 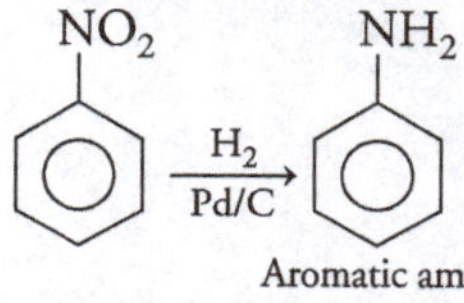

(d) 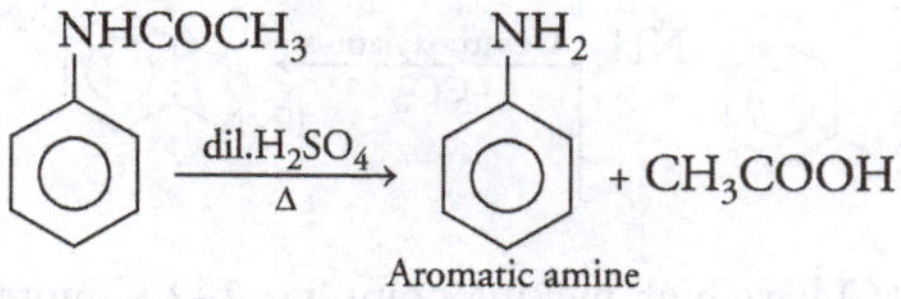

Aromatic amines will be formed in (a), (c) and (d).

**66.** *(5)* The nitrobenzene can be reduced to aniline *via* following reagents.

(I) Sn-HCl (V) $H_2$-Pd (III) Fe-HCl
(VI) $H_2$ (Raney Ni) (IV) Zn-HCl

**67.** *(3)*

> **key idea** To find out the total number of amines formed by Gabriel synthesis. Simply check for the compounds containing α-carbon. It is because Gabriel phthalimide synthesis is used to prepare 1° aliphatic and alicyclic amine. Hence, amine which can be synthesised by Gabriel phthalimide synthesis method contains α-carbon.

Aniline ($C_6H_5NH_2$) does not contain α -C. Hence, cannot be prepared by Gabriel reaction.

Rest all amines all contain α-C in its respective position, hence they can easily give Gabriel phthalimide reaction.

∴ Three amines out of given four options can be prepared by Gabriel synthesis.

## Topic 3

**01.** *(c)*

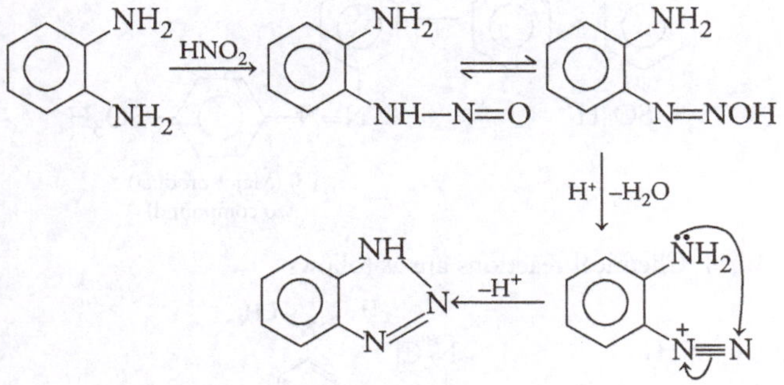

**02.** *(a)* Diazonium salts react with β- naphthol to form azo-compounds which are orange red dyes.

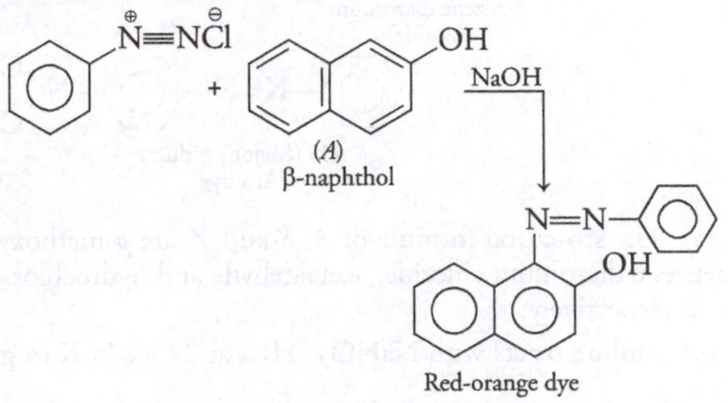

**03.** *(b)* Hydrolysis of benzene diazonium chloride will give carbolic acid/phenol.

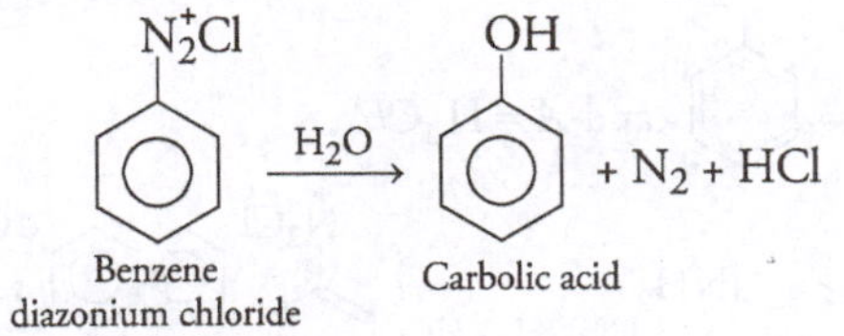

**04.** *(b)* Presence of electron donating groups in conjugation to diazonium salt containing aryl group stabilises the positive charge on —N due to resonance. Electron withdrawing groups when linked to aryl system containing diazonium group have a destabilising effect as it intensifies the positive charge.

Thus, the correct order of stability will be

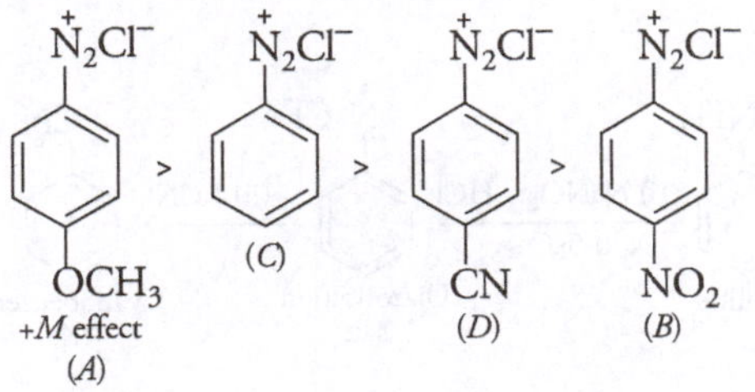

**05.** *(b)* Incorrect reaction is given in option (b). The correct reaction is as follows,

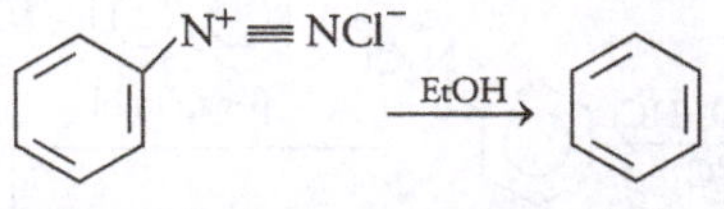

**06.** *(a)* The complete reaction involved is as follows.

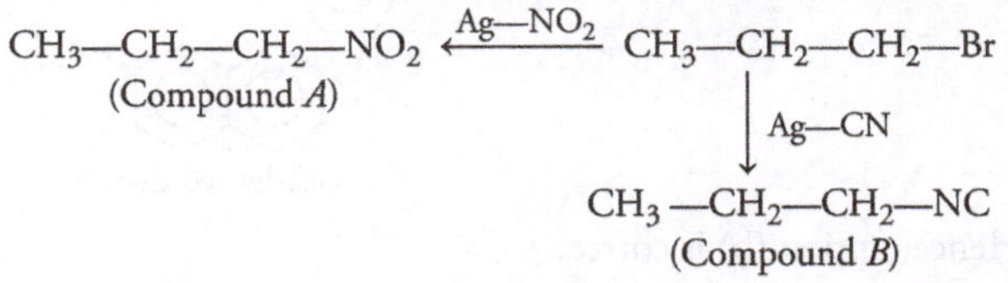

**07.** *(d)* The complete reaction is as follow,

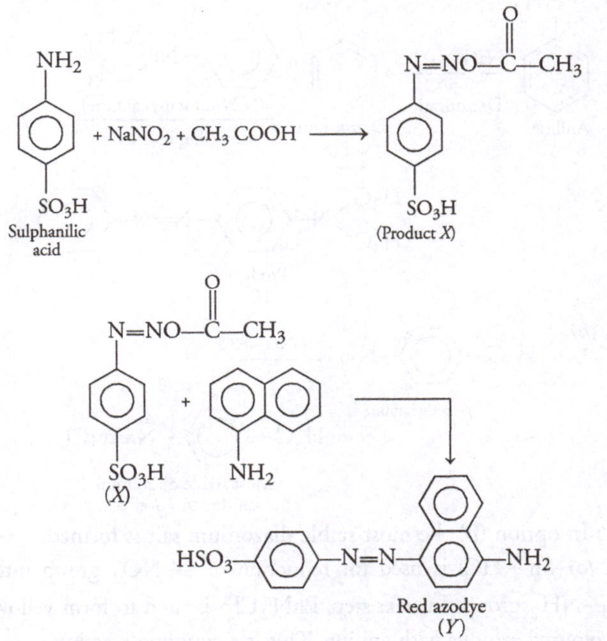

**08.** *(a)* The complete reaction is as follows:

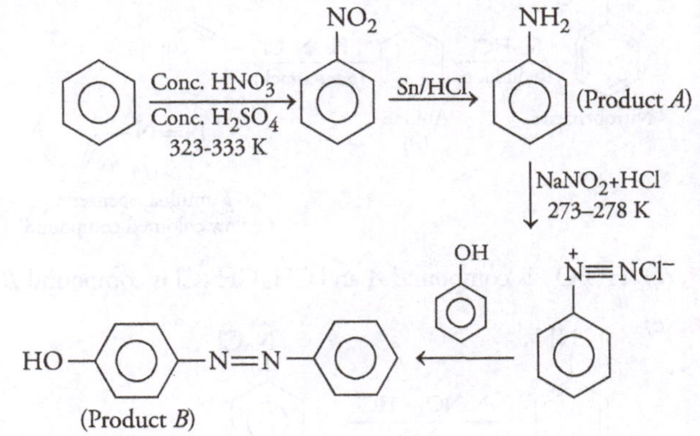

Hence, option (a) is correct.

**09.** *(b)* The complete reaction is as follows,

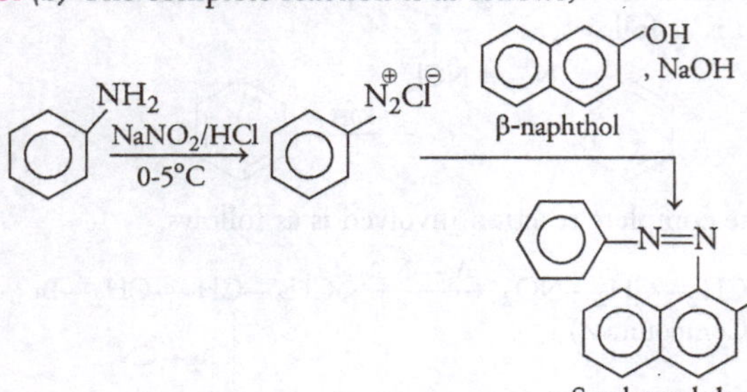


Scarlet red dye

Hence, option (b) is correct.

**10.** *(d)* Complete reaction is as follows.

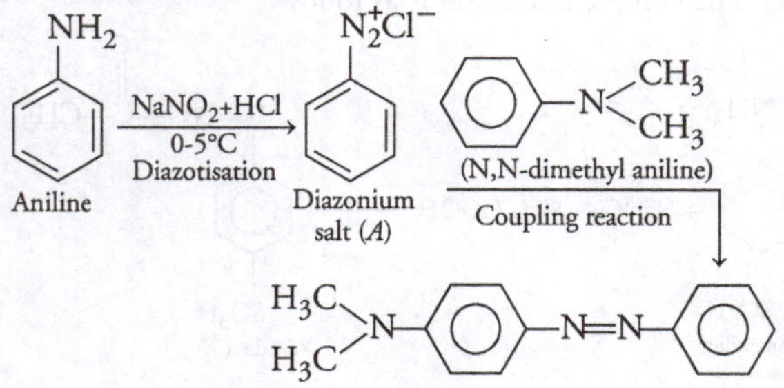


Product
(*B*)

**11.** *(b)*

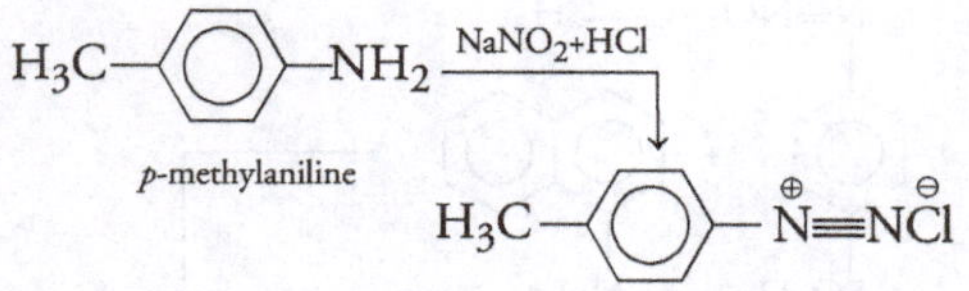


Most stable diazonium salt due to + *I*-effect

∴ In option (b) the most stable diazonium salt is formed.

**12.** *(b)* Sn + HCl is used for reduction of — $NO_2$ group into —$NH_2$ group. In next step, $PhN_2^+Cl^-$ is used to form yellow coloured azo dye with aniline. This is a coupling reaction.

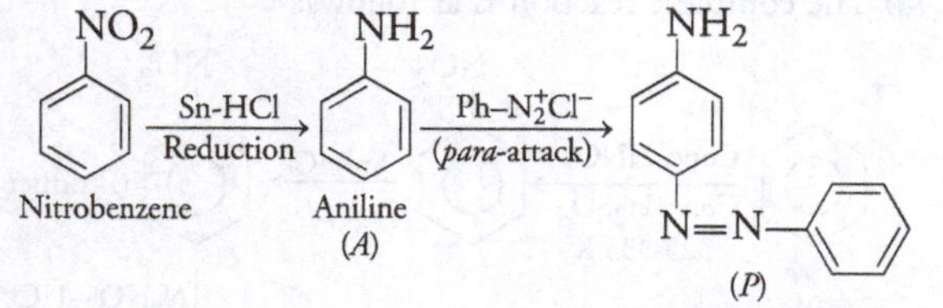


*p*-aminoazobenzene
(Yellow coloured compound)

**13.** *(a)* $H_3PO_2$ is compound *A* and $CH_3CH_2Cl$ is compound *B*.

**14.** *(c)*

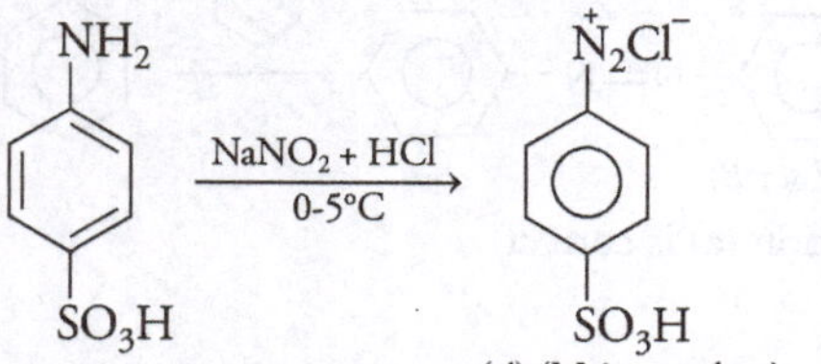


(*A*) (Major product)

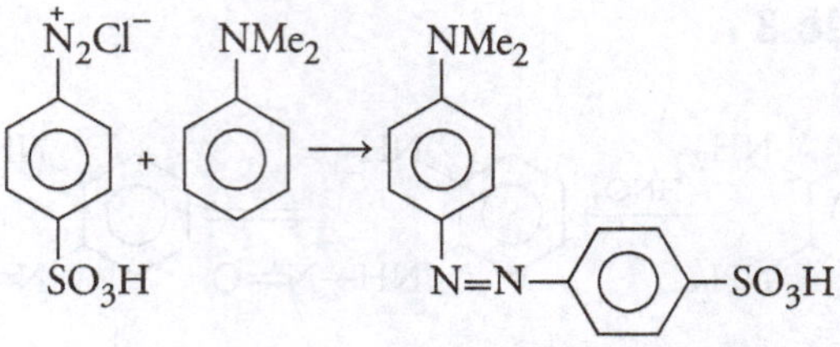


(*B*) (Major product)
(Azo compound)

**15.** *(b)* Chemical reactions are as follows

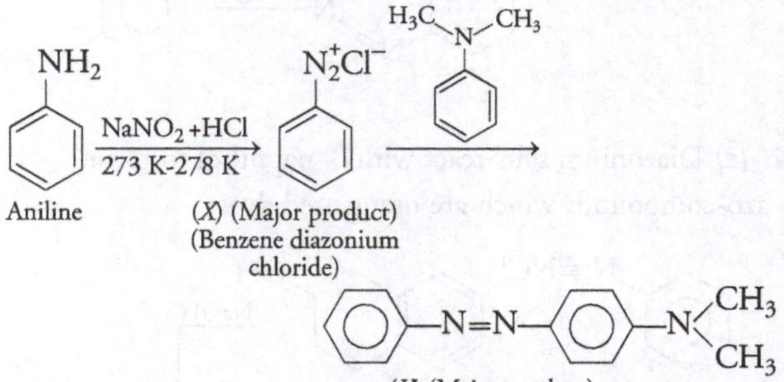


(*Y*) (Major product)
Azo dye

**16.** *(a)* The structural formula of *A*, *X* and *Y* are *p*-methoxy benzene diazonium chloride, acetaldehyde and hydrochloric acid respectively.

**17.** *(c)* Aniline reacts with $NaNO_2$, HCl at 273-278 K to give benzene diazonium salt ($C_6H_5\overset{+}{N}_2\overset{-}{Cl} = X$) which on hydrolysis ($H_2O/\Delta$) gives phenol.

So, $X =$ 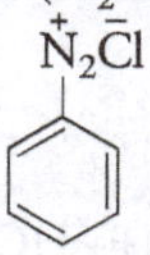

and $A = H_2O/\Delta$

**18.** *(c)*

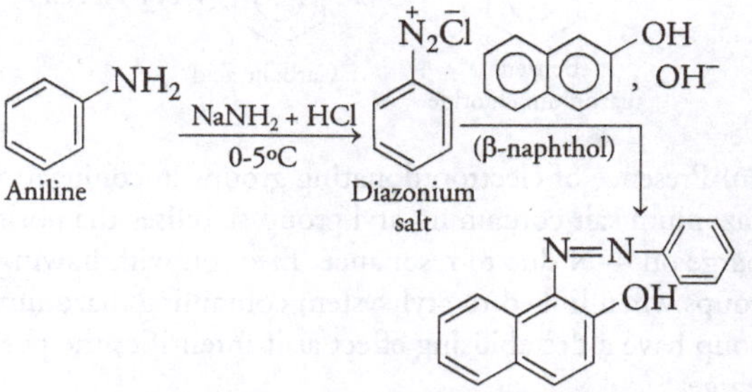


2-naphthol aniline
Orange-red dye

**19.** *(d)*

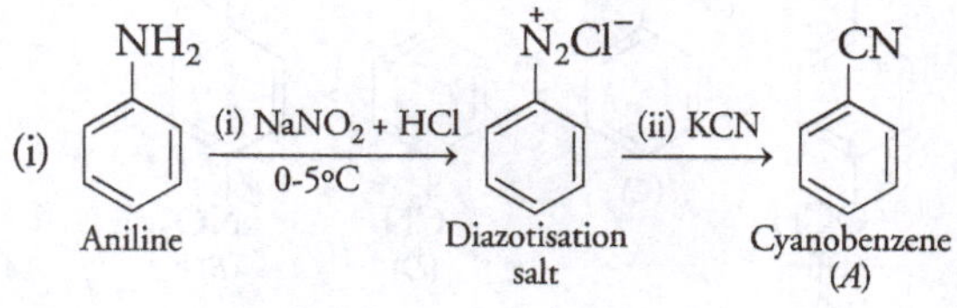

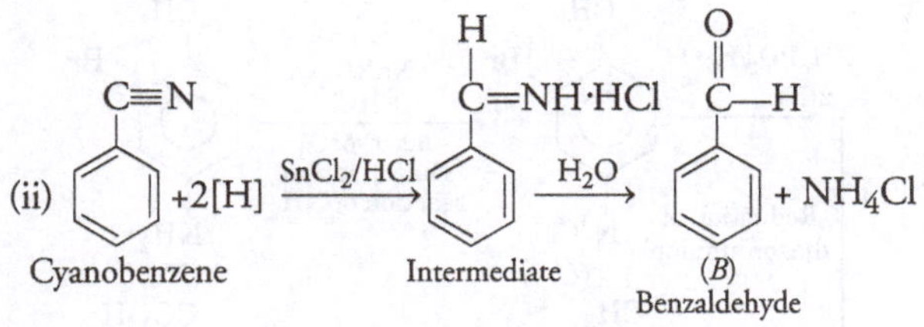


**20.** *(b)* Complete reaction sequence is as follows

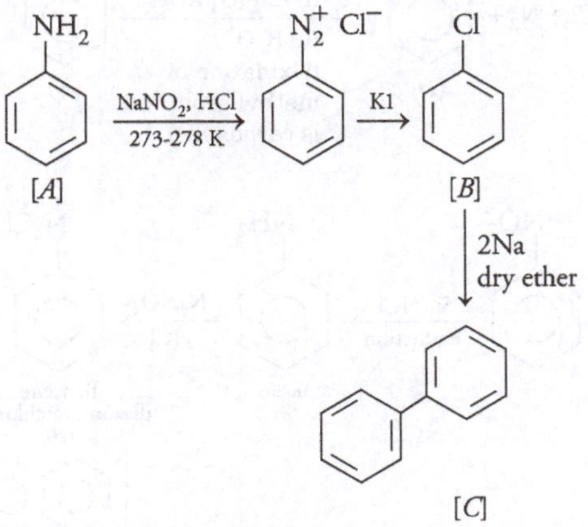


**21.** *(b)* The complete reaction sequence involved is as follows.

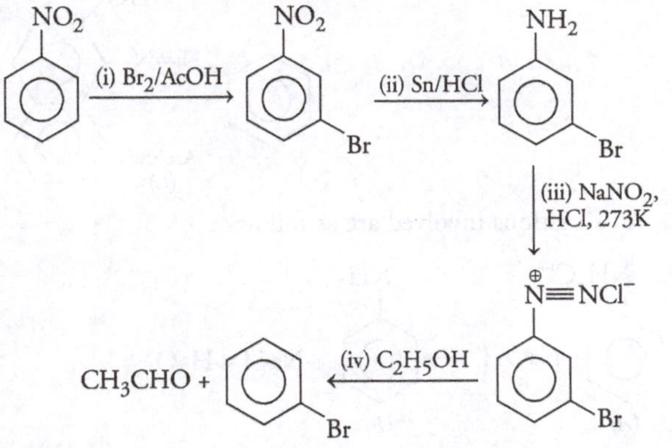


**22.** *(c)* The correct match is A-III, B-IV, C-II, D-I.

(A) $R-\overset{O}{\overset{\|}{C}}-NH_2 \xrightarrow[H_2O]{LiAlH_4} R-CH_2-NH_2$

(B) 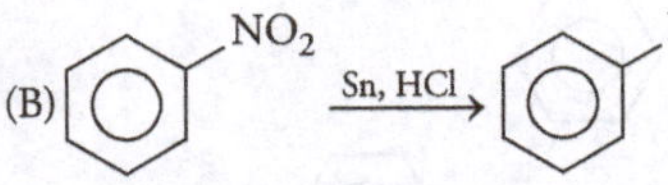


(C) $R-C\equiv N \xrightarrow{H_2/Ni} R-CH_2NH_2$

(D) 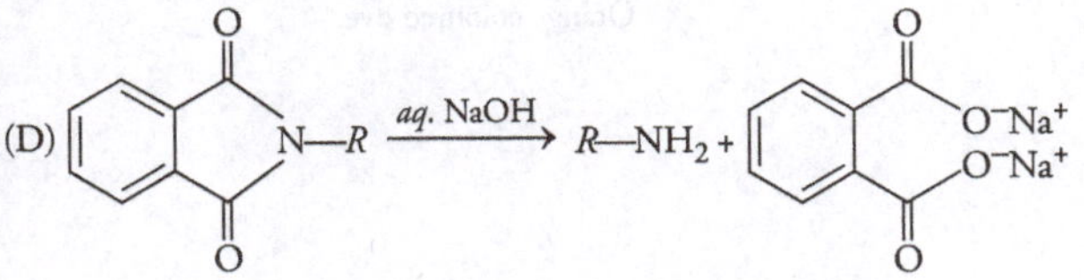


**23.** *(a)* Among the given reactions, nitrobenzene will give an azo dye.

The complete reaction is as follows :

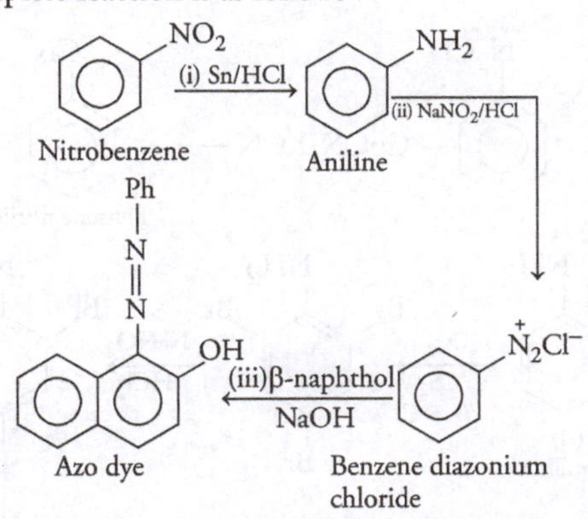


**24.** *(b)* Among the given options, incorrect statement is given in option (b). It is because '*B*' is *p*-nitroso ammonium compound. The product formation will takes place as follows

$$NaNO_2 + HX \longrightarrow HNO_2 + NaX$$

$$H-O-N=O \xrightarrow[-H_2O]{H^{\oplus}} NO^+ \text{ (nitrosonium ion)}$$

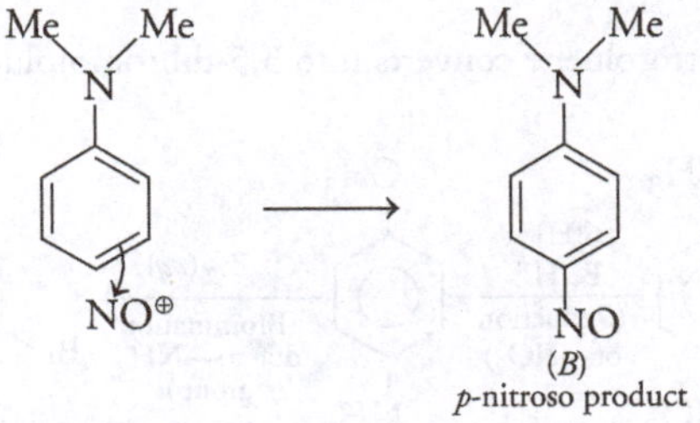


**25.** *(b)* The correct match is A-III, B-I, C-IV, D-II.

A. Aniline reacts with benzene diazonium chloride to give *p*-aminoazobenzene.

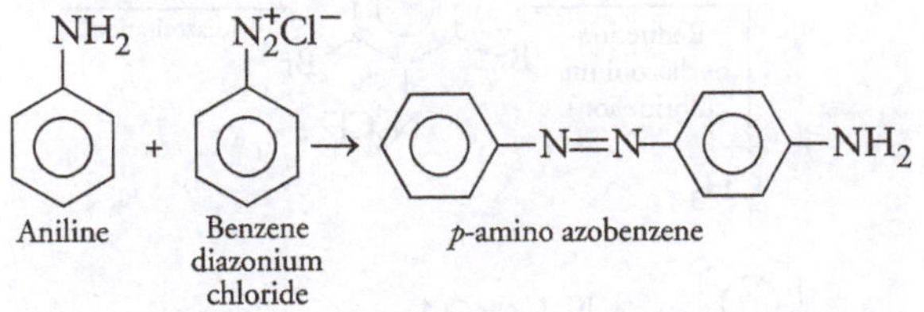


B. Benzene diazonium chloride reacts with $HBF_4$ to given fluorobenzene.

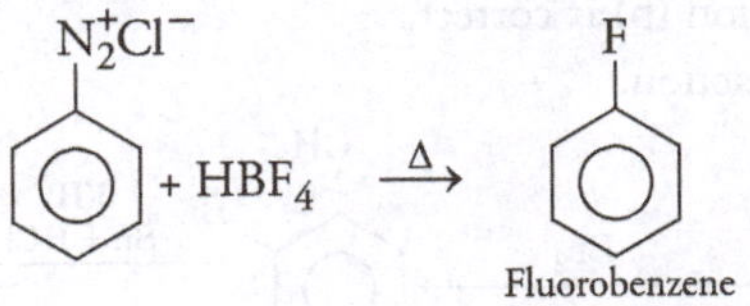


C. When benzene diazonium chloride is treated with copper in HCl, chlorobenzene is formed.

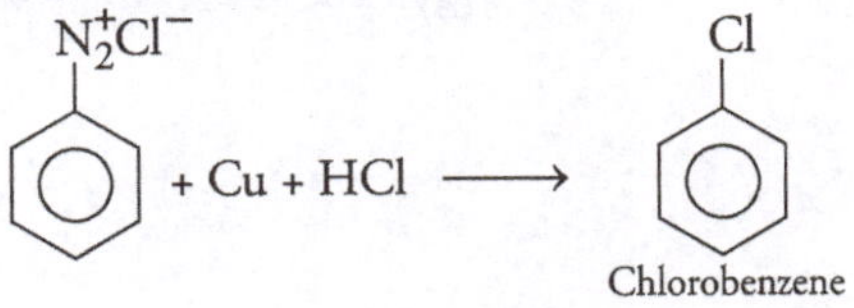

D. Benzene diazonium chloride reacts with CuCN/KCN to give benzene nitrile.

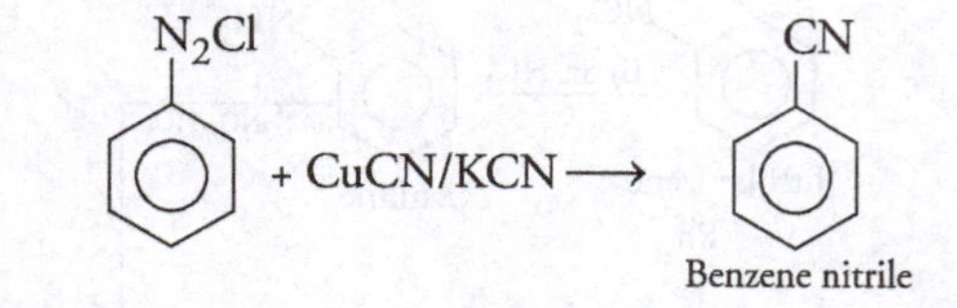


**26.** *(a)*

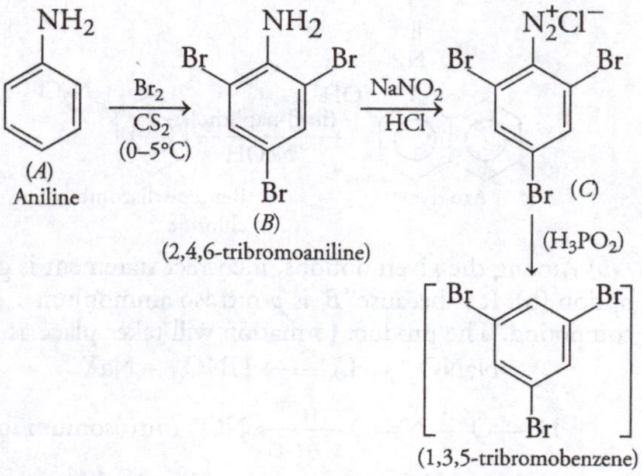


**27.** *(b)* *p*-nitrotoluene converts into 3,5-dibromotoluene as follows :

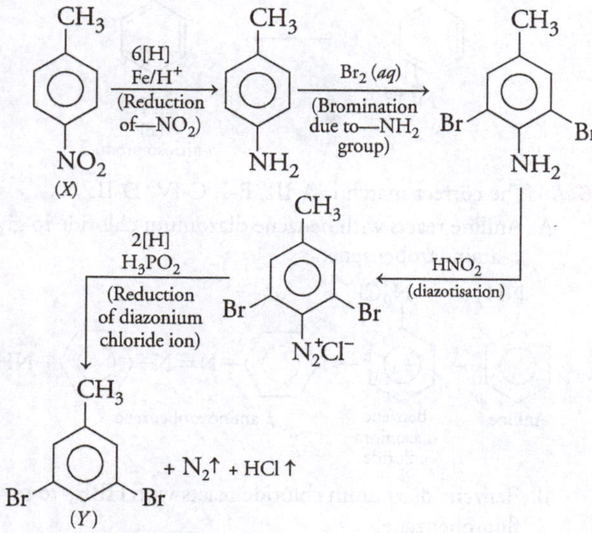


Thus, option (b) is correct.

**28.** *(c)* In reaction,

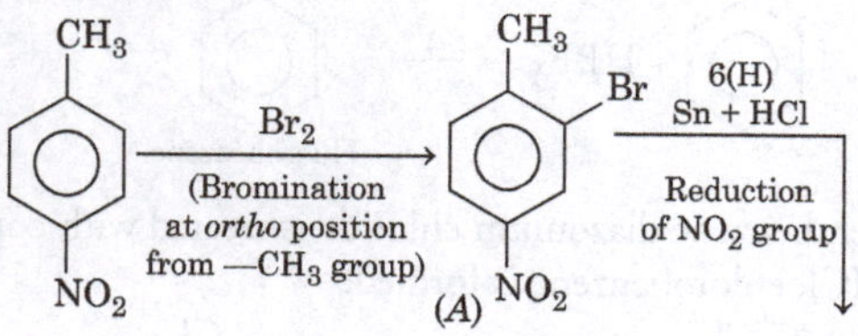


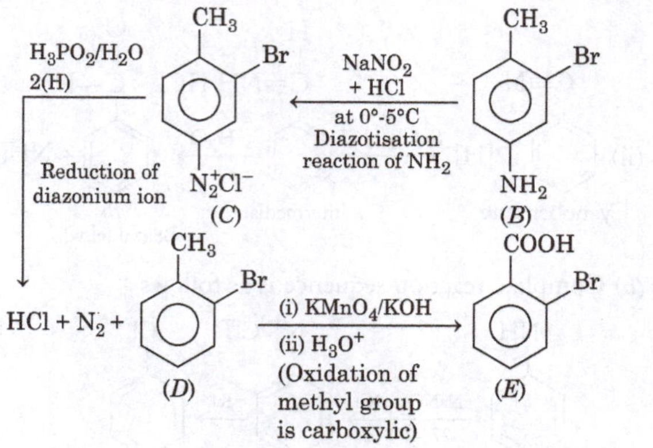


**29.** *(c)*

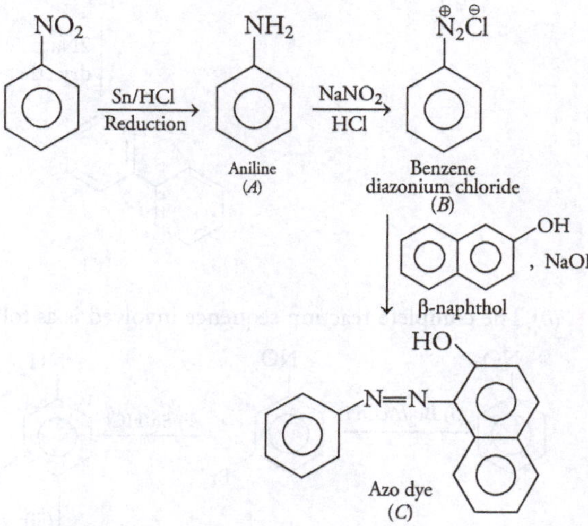


**30.** *(d)* Reactions involved are as follows :

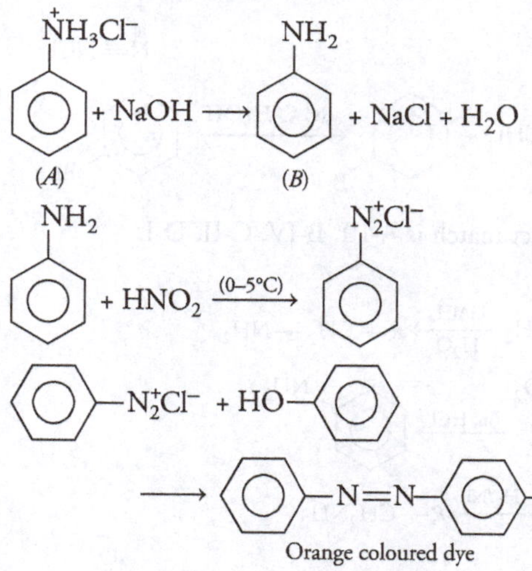

**31.** *(c)* The given reaction proceeds in three steps;

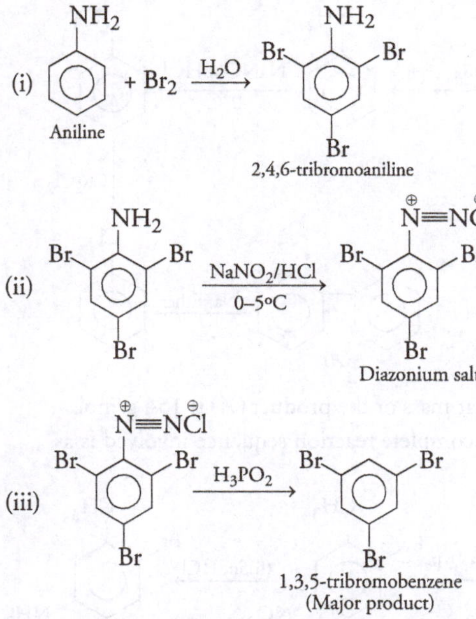

**32.** *(c)* The major product formed is 3-methyl propanol as follows :

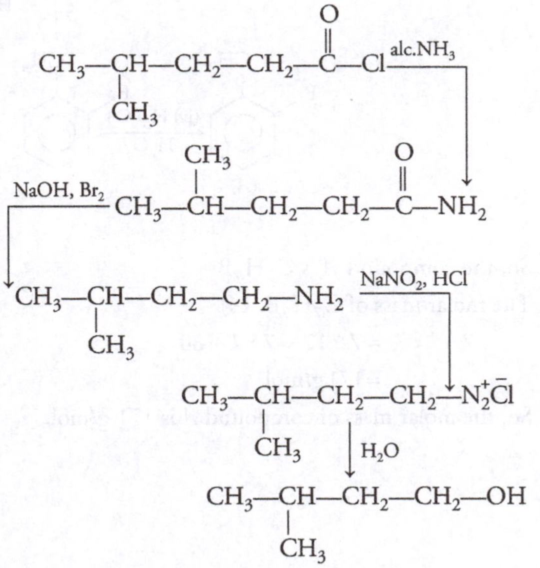

**33.** *(c)* Nitrobenzene in presence of $NaBH_4$, NaOH and aniline will not give *p*-amino azobenzene.

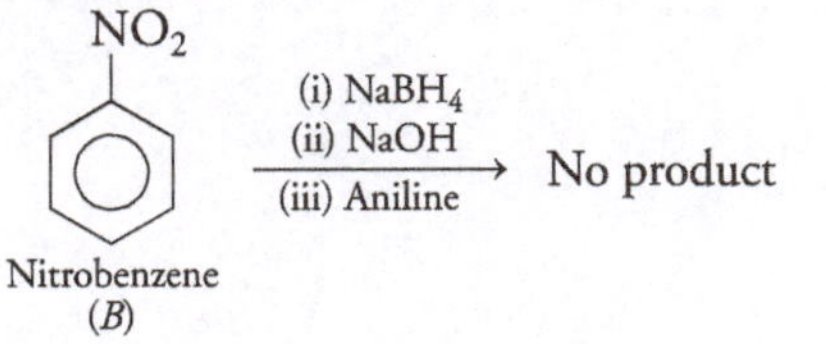

**34.** *(b)* Complete reaction is as follows

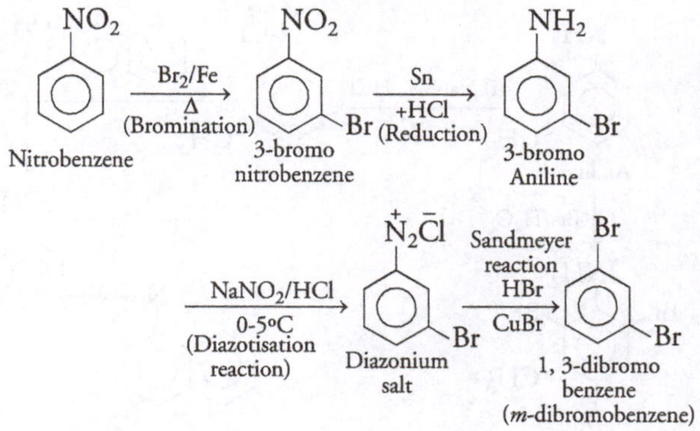

**35.** *(a)*

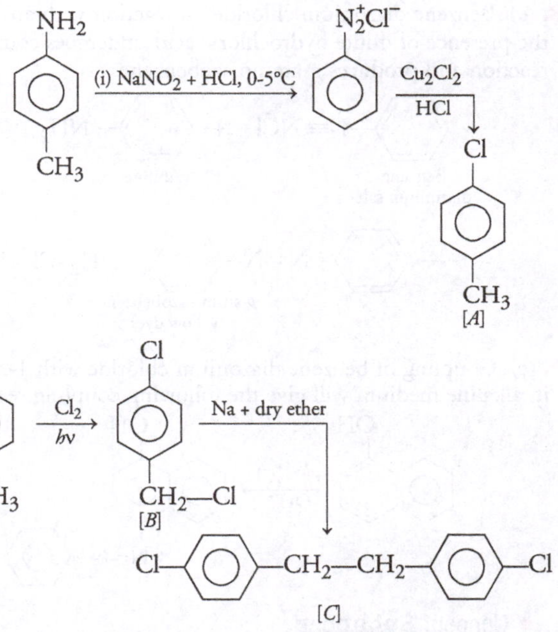

**36.** *(c)*

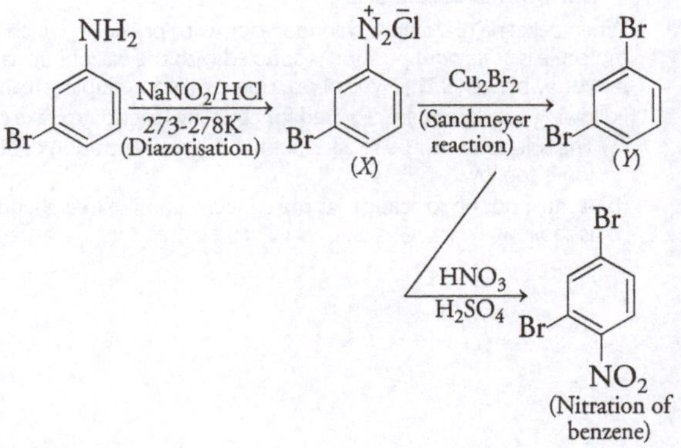

Br is *o*, *p*-directing group.

$\therefore$ $NO_2$ is attached at the *ortho* position of one of the Br atoms which is also *para* position with respect to the other Br atom.

**37.** *(a)* Compound [$P$] is 3-methylaniline.

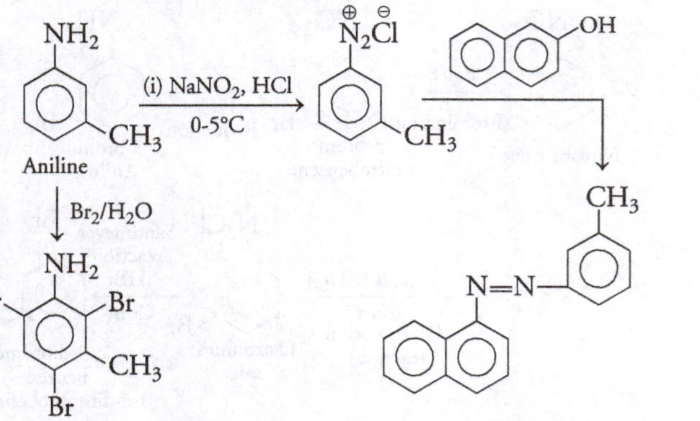

**38.** *(c)* Benzene diazonium chloride on reaction with aniline in the presence of dilute hydrochloric acid undergoes coupling reaction and produces *p*-amino azobenzene.

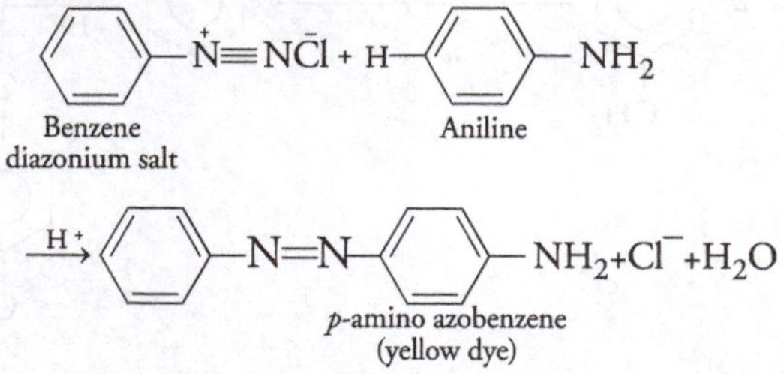

**39.** *(c)* Coupling of benzene diazonium chloride with 1-naphthol in alkaline medium will give the following coupling reaction.

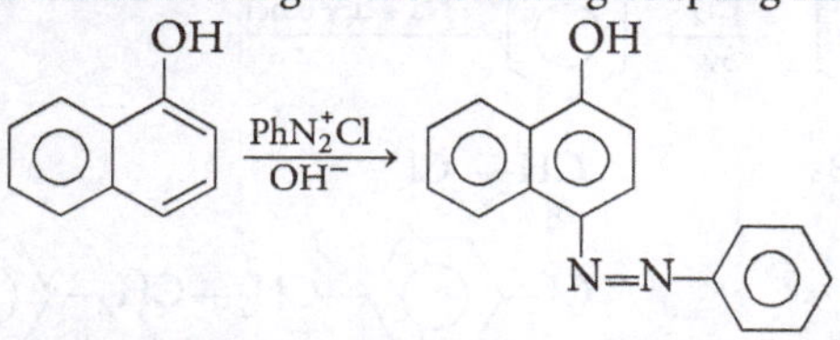

> • Concept **Enhancer**
>
> When benzene diazonium chloride reacts with phenol in which the phenol molecule is at its *para* position is coupled with the diazonium salt to form *p*-hydroxybenzene. This type of reaction is called a coupling reaction. Following conditions are required for coupling reaction to take place.
>
> (a) The solution must be so alkaline that the concentration of diazonium ion is too low.
>
> (b) It must not be so acidic that the concentration of free phenoxide ion is too low.

**40.** *(154)* Complete reaction sequence involved is as follows

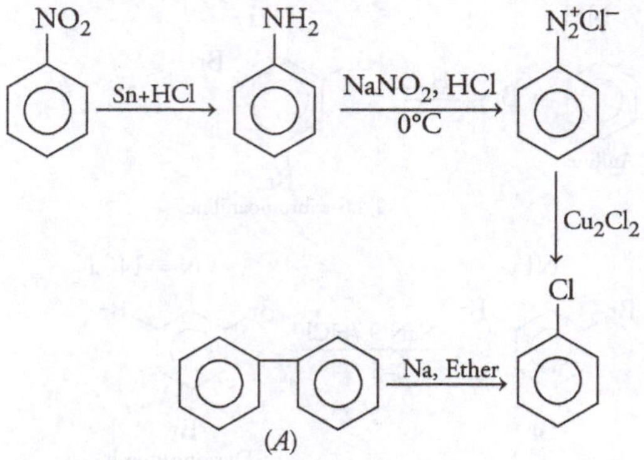

So, the molar mass of the product ($A$) is 154 g/mol.

**41.** *(171)* The complete reaction sequence involved is as follows

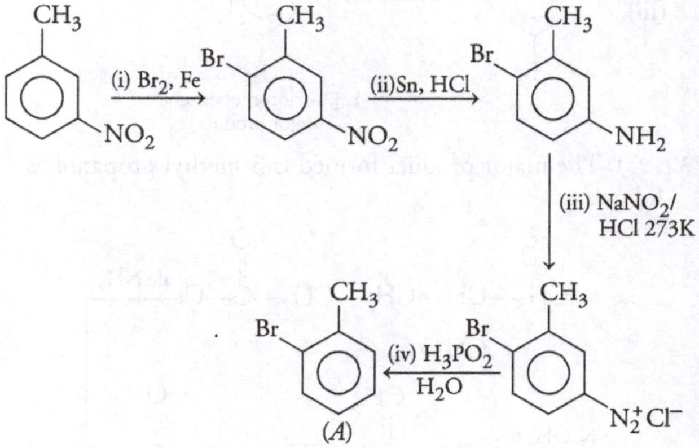

So, the compound $A$ is $C_7H_7Br$

The molar mass of $C_7H_7Br$ ($A$)

$$= 7 \times 12 + 7 \times 1 + 80$$

$$= 171 \text{ g/mol}$$

So, the molar mass of compound $A$ is 171 g/mol.

# C-23

JEE Main Chapterwise Topicwise Chemistry

In JEE Main there is usually 0-1 question per paper are visible from this chapter. In past 3 years, 50 conceptual and 11 numerical questions were asked of moderate difficulty level from this chapter.
**For detailed analysis scan QR code**

# BIOMOLECULES

## Topic 1 Carbohydrates

### Level 1

**01** Which of the following is the correct structure of L-glucose? *[2024, 4 April Shift-I]*

(a) 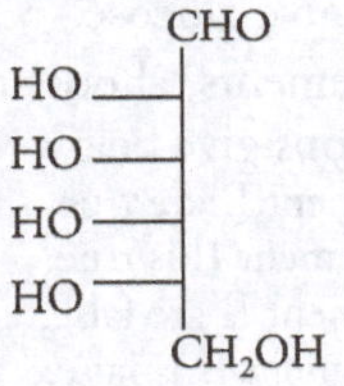
(b) 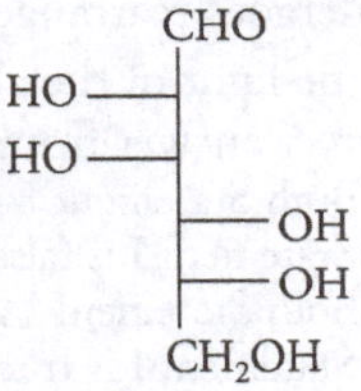
(c) 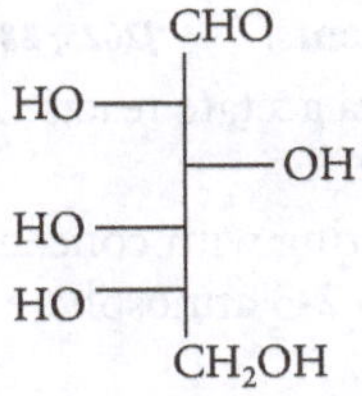
(d) 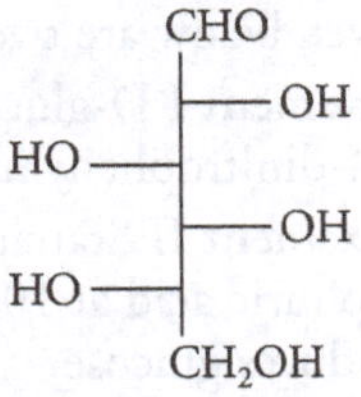

**02** Which of the following is the correct structure of L-Fructose? *[2023, 03 April Shift-I]*

(a) 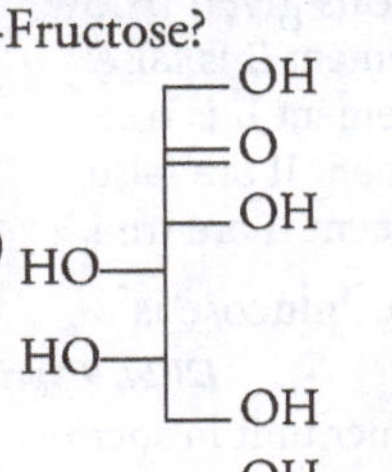
(b) 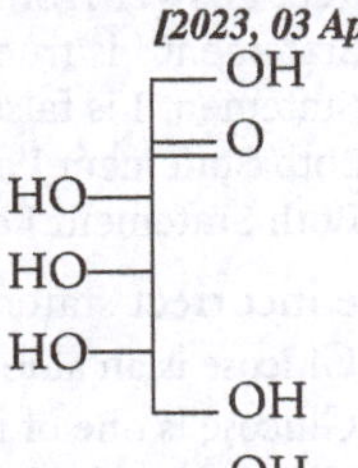
(c) 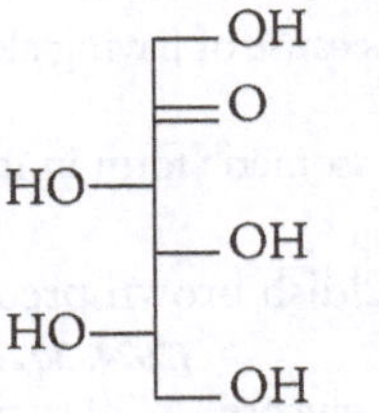
(d) 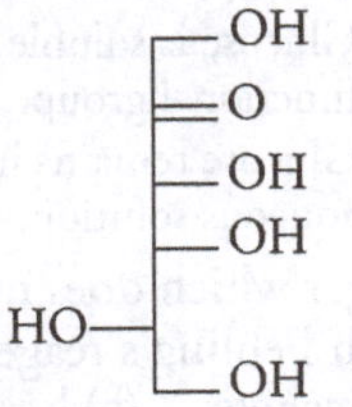

**03** The correct representation in six membered pyranose form for the following sugar $[X]$ is *[2023, 01 Feb Shift-I]*

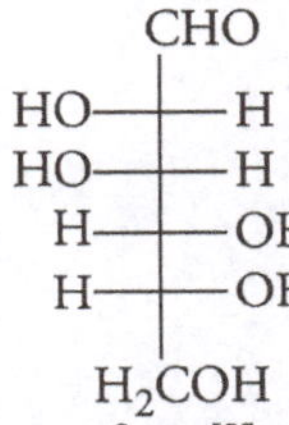
Sugar $[X]$

(a) 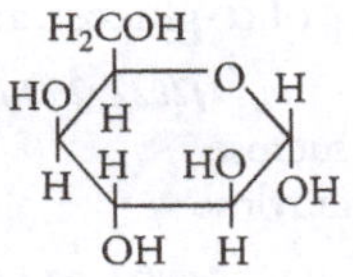
(b) 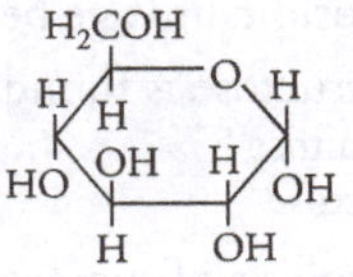
(c) 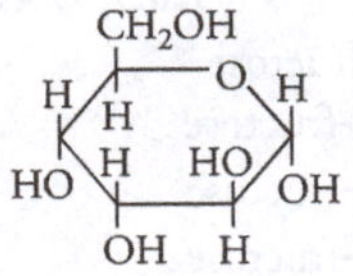
(d) 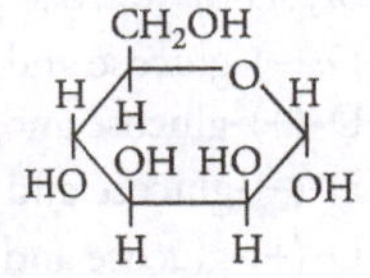

**04** Given below are two statements. One is labelled as Assertion (A) and the other is labelled as Reason (R).
**Assertion (A)** Amylose is insoluble in water.
**Reason (R)** Amylose is a long linear molecule with more than 200 glucose units.
In the light of the above statements, choose the correct answer from the options given below. *[2022, 29 July Shift-II]*

(a) Both A and R are correct and R is the correct explanation of A.
(b) Both A and R are correct and R is not the correct explanation of A.
(c) A is correct but R is not correct.
(d) A is incorrect but R is correct.

**05** Which one of the following is a reducing sugar? *[2022, 26 July Shift-I]*

(a) 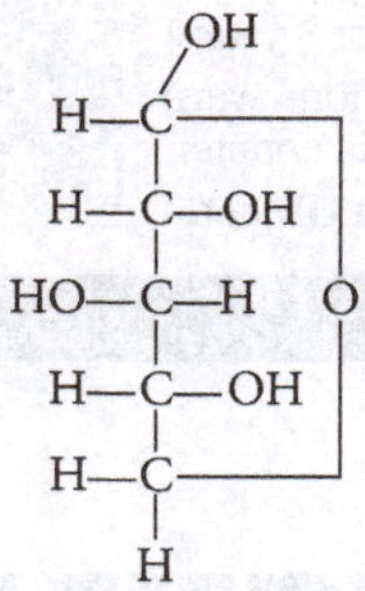 (b) 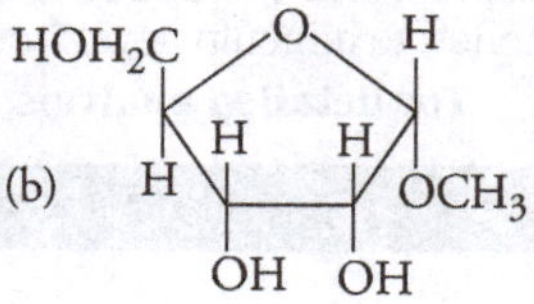

(c) 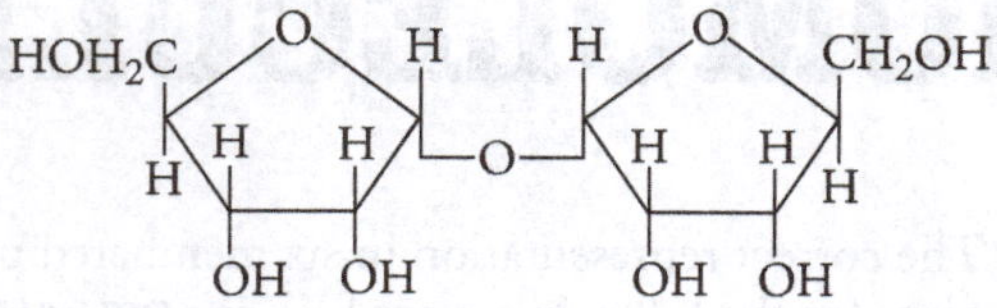

(d) 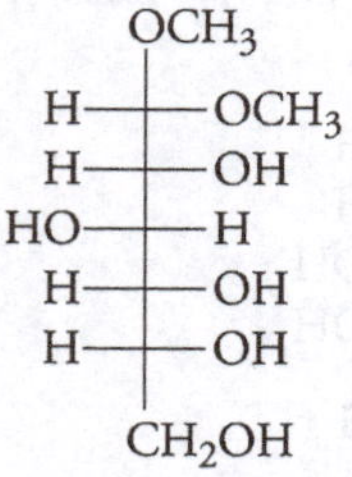

**06** Glycosidic linkage between $C_1$ of α-glucose and $C_2$ of β-fructose is found in *[2022, 25 July Shift-II]*
(a) maltose (b) sucrose
(c) lactose (d) amylose

**07** Hydrolysis of sucrose gives *[2021, 27 Aug Shift-II]*
(a) α-D-(–)-glucose and β-D(–)-fructose
(b) α-D-(+)-glucose and β-D-(+)-fructose
(c) α-D-(–)-glucose and β-D-(+)-fructose
(d) α-D-(+)-glucose and β-D-(–)-fructose

**08** Which one among the following chemical tests is used to distinguish monosaccharide from disaccharide? *[2021, 27 July Shift-I]*
(a) Seliwanoff's test (b) Iodine test
(c) Barfoed test (d) Tollen's test

**09** A non-reducing sugar "*A*" hydrolysis to give two reducing monosaccharides. Sugar *A* is *[2021, 18 March Shift-I]*
(a) fructose (b) galactose (c) glucose (d) sucrose

**10** Fructose is an example of *[2021, 17 March Shift-II]*
(a) pyranose (b) ketohexose
(c) aldohexose (d) heptose

**11** Match List-I with List-II.

| List-I | | List-II | |
|---|---|---|---|
| A. | Sucrose | (i) | β-D-galactose and β-D-glucose |
| B. | Lactose | (ii) | α-D-glucose and β-D-fructose |
| C. | Maltose | (iii) | α-D-glucose and α-D-glucose |

Choose the correct answer from the options given below. *[2021, 26 Feb Shift-II]*
(a) A → (i), B → (iii), C → (ii)
(b) A → (iii), B → (i), C → (iii)
(c) A → (ii), B → (i), C → (iii)
(d) A → (iii), B → (ii), C → (i)

**12** Two monomers in maltose are *[2020, 8 Jan Shift-II]*
(a) α-D-glucose and α-D-glucose
(b) α-D-glucose and α-D-galactose
(c) α-D-glucose and β-D-glucose
(d) α-D-glucose and α-D-fructose

**13** Glucose and galactose are having identical configuration in all the positions except position *[2019, 12 April Shift-II]*
(a) C-3 (b) C-4 (c) C-2 (d) C-5

## Level 2

**Imp.** Properties of Fructose

**14** Given below are two statements: *[2025, 23 Jan Shift-I]*
**Statement I** Fructose does not contain an aldehydic group but still reduces Tollen's reagent.
**Statement II** In the presence of base, fructose undergoes rearrangement to give glucose.
In the light of the above statements, choose the correct answer from the options give below
(a) Both Statement I and Statement II are true.
(b) Statement I is false but Statement II is true.
(c) Both Statement I and Statement II are false.
(d) Statement I is true but Statement II is false.

**15** Given below are two statements. *[2025, 28 Jan Shift-I]*
**Statement I** D-glucose penta acetate reacts with 2, 4-dinitrophenylhydrazine.
**Statement II** Starch, on heating with concentrated sulphuric acid at 100°C and 2-3 atmosphere pressure produces glucose.
In the light of the above statements, choose the correct answer from the options given below
(a) Statement I is true but Statement II is false.
(b) Statement I is false but Statement II is true.
(c) Both Statement I and Statement II are false.
(d) Both Statement I and Statement II are true.

**16** The incorrect statement about glucose is *[2024, 9 April Shift-II]*
(a) Glucose is an aldohexose.
(b) Glucose is one of the monomer unit in sucrose.
(c) Glucose is soluble in water because of having aldehyde functional group.
(d) Glucose remains in multiple isomeric form in its aqueous solution.

**17** Sugar which does not give reddish brown precipitate with Fehling's reagent, is *[2024, 30 Jan Shift-I]*
(a) maltose (b) lactose (c) glucose (d) sucrose

**18** L-isomer of tetrose $X(C_4H_8O_4)$ gives positive Schiff's test and has two chiral carbons. On acetylation, '*X*' yields triacetate. '*X*' also undergoes following reactions

$'A' \xleftarrow{HNO_3} 'X' \xrightarrow{NaBH_4} 'B'$ 'X' is

Chiral compound

*[2023, 11 April Shift-I]*

(a) 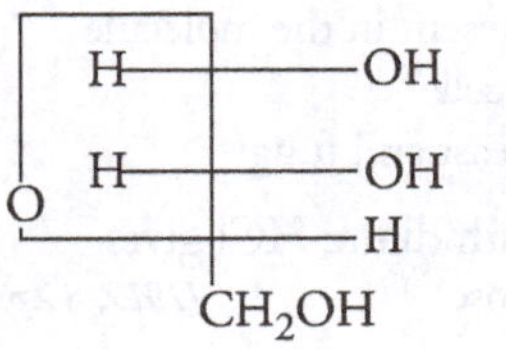
(b) 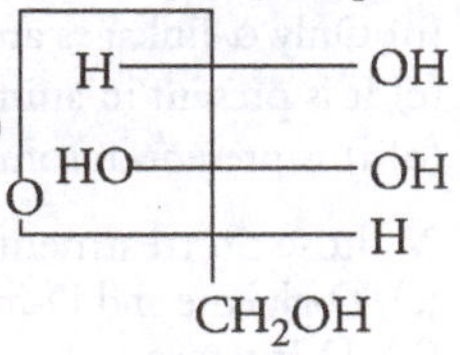

(c) 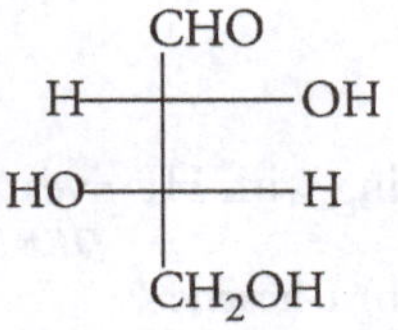
(d) 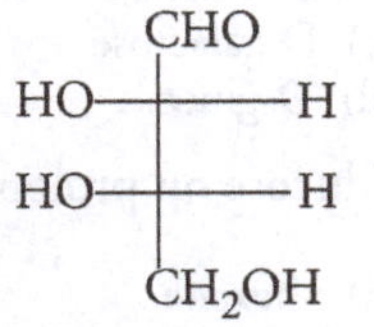

**19** Match items of Row-I with those of Row-II.

**Row I**

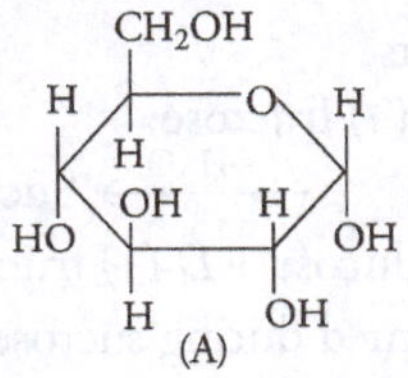
(A)

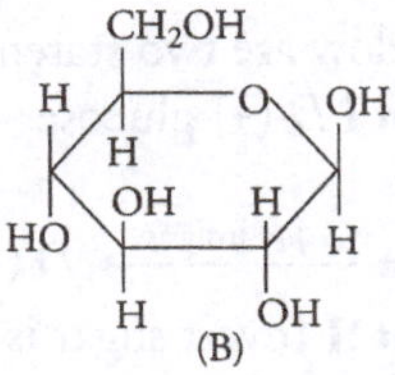
(B)

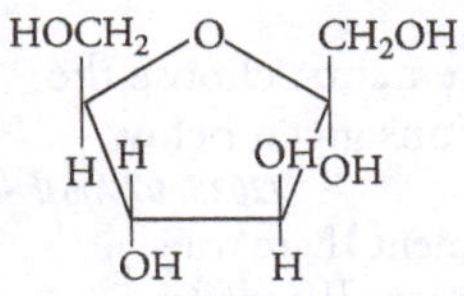
(C)

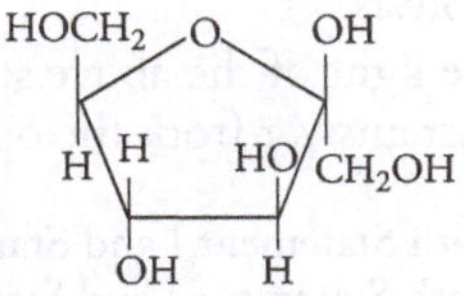
(D)

**Row II**

(i) α-D-(–)-fructofuranose,
(ii) β-D-(–)-fructofuranose
(iii) α-D-(–) glucopyranose,
(iv) β-D-(–)-glucopyranose

Correct match is *[2023, 25 Jan Shift-I]*

(a) A-(iv), B-(iii), C-(i), D-(ii)
(b) A-(iii), B-(iv), C-(i), D-(ii)
(c) A-(i), B-(ii), C-(iii), D-(iv)
(d) A-(iii), B-(iv), C-(ii), D-(i)

**20** The formulas of $A$ and $B$ for the following reaction sequence *[2022, 28 July Shift-II]*

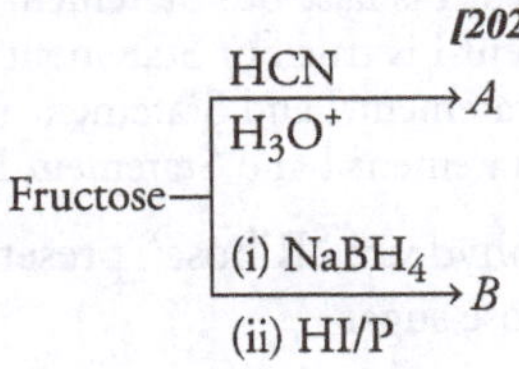


(a) $A = C_7H_{14}O_8$, $B = C_6H_{14}$
(b) $A = C_7H_{13}O_7$, $B = C_7H_{14}O$
(c) $A = C_7H_{12}O_8$, $B = C_6H_{14}$
(d) $A = C_7H_{14}O_8$, $B = C_6H_{14}O_6$

**21** For the below given cyclic hemiacetal ($X$), the correct pyranose structure is *[2022, 28 July Shift-I]*

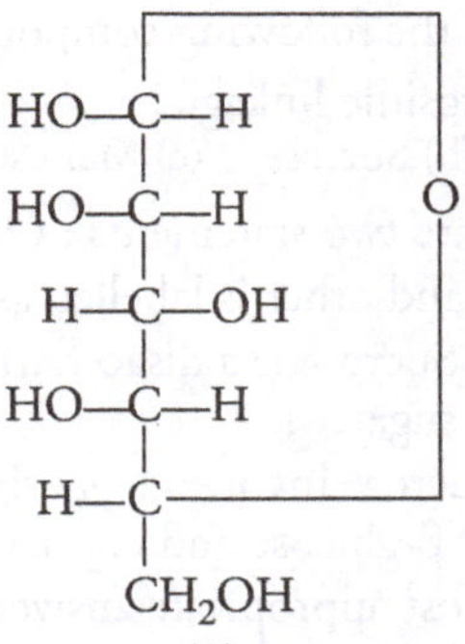
($X$)

(a) 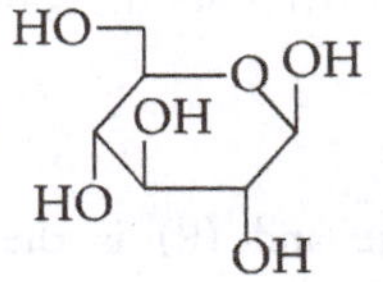
(b) 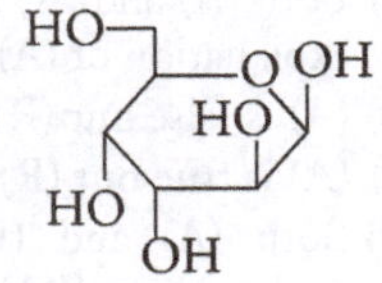

(c) 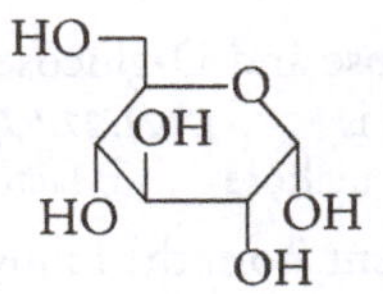
(d) 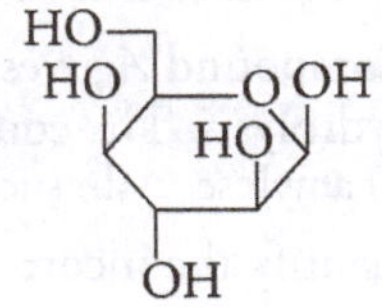

**Imp.** Chemical reactions of Glucose

**22** Match List-I with List-II.

| List-I | List-II |
|---|---|
| (A) Glucose + HI | (I) Gluconic acid |
| (B) Glucose + $Br_2$ water | (II) Glucose pentacetate |
| (C) Glucose + acetic anhydride | (III) Saccharic acid |
| (D) Glucose + $HNO_3$ | (IV) Hexane |

Choose the correct answer from the options given below: *[2022, 27 July Shift-II]*

(a) (A)-(IV), (B)-(I), (C)-(II), (D)-(III)
(b) (A)-(IV), (B)-(III), (C)-(II), (D)-(I)
(c) (A)-(III), (B)-(I), (C)-(IV), (D)-(II)
(d) (A)-(I), (B)-(III), (C)-(IV), (D)-(II)

**23** Given below are two statements.

**Statement I** Maltose has two α-D-glucose units linked at $C_1$ and $C_4$ and is a reducing sugar.

**Statement II** Maltose has two monosaccharides: α-D-glucose and β-D-glucose linked at $C_1$ and $C_6$ and it is a non-reducing sugar.

In the light of the above statements, choose the correct answer from the options given below. *[2022, 27 June Shift-II]*

(a) Both Statement I and Statement II are true
(b) Both Statement I and Statement II are false
(c) Statement I is true but Statement II are false
(d) Statement I is false but Statement II is true

**24** Which one of the following compounds contain β-$C_1$—$C_4$ glycosidic linkage ? *[2021, 31 Aug Shift-I]*
(a) Lactose (b) Sucrose (c) Maltose (d) Amylose

**25** Given below are two statements : One is labelled as Assertion (A) and other is labelled as Reason (R).
**Assertion** (A) Sucrose is a disaccharide and a non-reducing sugar.
**Reason** (R) Sucrose involves glycosidic linkage between $C_1$ of β-glucose and $C_2$ of α-fructose.
Choose the most appropriate answer from the options given below. *[2021, 26 Aug Shift-II]*
(a) Both (A) and (R) are true but (R) is not the correct explanation of (A).
(b) (A) is false but (R) is true.
(c) (A) is true but (R) is false.
(d) Both (A) and (R) are true and (R) is the correct explanation of (A).

**26** Compound *A* gives D-galactose and D-glucose on hydrolysis. The compound *A* is *[2021, 27 July Shift-II]*
(a) amylose (b) sucrose (c) maltose (d) lactose

**27** Identify the incorrect statement from the following. *[2021, 20 July Shift-I]*
(a) Amylose is a branched chain polymer of glucose.
(b) Starch is a polymer of α-D-glucose .
(c) β-glycosidic linkage makes cellulose polymer.
(d) Glycogen is called as animal starch.

**28** Which of the glycosidic linkage between galactose and glucose is present in lactose? *[2021, 25 Feb Shift-I]*
(a) C-1 of galactose and C-4 of glucose
(b) C-1 of galactose and C-6 of glucose
(c) C-1 of glucose and C-4 of galactose
(d) C-1 of glucose and C-6 of galactose

**29** What are the functional groups present in the structure of maltose? *[2020, 4 Sep Shift-I]*
(a) One ketal and one hemiketal
(b) One acetal and one ketal
(c) One acetal and one hemiacetal
(d) Two acetals

**30** Which of the following statement is not true for glucose? *[2020, 8 Jan Shift-I]*
(a) The pentaacetate of glucose does not react with hydroxylamine to give oxime
(b) Glucose exists in two crystalline forms α and β
(c) Glucose gives Schiff's test for aldehyde
(d) Glucose reacts with hydroxylamine to form oxime

**31** Which of the following statements is correct? *[2020, 7 Jan Shift-II]*
(a) Gluconic acid is obtained by oxidation of glucose with $HNO_3$
(b) Gluconic acid is a dicarboxylic acid
(c) Gluconic acid can form cyclic (acetal/hemiacetal) structure
(d) Gluconic acid is a partial oxidation product of glucose

**32** Which of the given statements is incorrect about glycogen? *[2019, 12 April Shift-II]*
(a) It is straight chain polymer similar to amylose
(b) Only α-linkages are present in the molecule
(c) It is present in animal cells
(d) It is present in some yeast and fungi

**33** Maltose on treatment with dilute HCl gives *[2019, 8 April Shift-I]*
(a) D-glucose and D-fructose
(b) D-fructose
(c) D-galactose
(d) D-glucose

**34** Glucose on prolonged heating with HI gives *[JEE Main 2018]*
(a) *n*-hexane (b) 1-hexene
(c) Hexanoic acid (d) 6-iodohexanal

## Level 3

**35** Given below are two statements
**Statement I** *D*-(+)-glucose + *D*-(+) fructose $\xrightarrow{-H_2O}$ Sucrose
Sucrose $\xrightarrow{\text{Hydrolysis}}$ *D*-(+) glucose + *D*-(+) fructose
**Statement II** Invert sugar is formed during sucrose hydrolysis.
In the light of the above statements, choose the correct answer from the options given below *[2025, 07 April Shift-I]*
(a) Both Statement I and Statement II are true.
(b) Both Statement I and Statement II are false.
(c) Statement I is true but Statement II is false.
(d) Statement I is false but Statement II are true.

**36** Given below are two statements.
**Statement I** Wet cotton clothes made of cellulose based carbohydrate takes comparatively longer time to get dried than wet nylon polymer based clothes.
**Statement II** Intermolecular hydrogen bonding with water molecule is more in nylon-based clothes than in the case of cotton clothes.
In the light of above statements, choose the correct answer from the options given below. *[2025, 03 April Shift-II]*
(a) Statement I is false but Statement II is true.
(b) Statement I is true but Statement II is false.
(c) Both Statement I and Statement II are true.
(d) Both Statement I and Statement II are false.

**37** The carbohydrate "Ribose" present in DNA, is
A. a pentose sugar
B. present in pyranose form
C. in "D" configuration
D. a reducing sugar, when free
E. in α-anomeric form

Choose the correct answer from the options given below: *[2025, 24 Jan Shift-I]*
(a) A, C and D only (b) A, B and E only
(c) B, D and E only (d) A, D and E only

**38** Match List-I with List-II. *[2025, 29 Jan Shift-I]*

| | List-I (Carbohydrate) | | List-II (Linkage source) |
|---|---|---|---|
| (A) | Amylose | I. | $\beta-C_1-C_4$, plant |
| (B) | Cellulose | II. | $\alpha-C_1-C_4$, animal |
| (C) | Glycogen | III. | $\alpha-C_1-C_4, \alpha-C_1-C_6$, plant |
| (D) | Amylopectin | IV. | $\alpha-C_1-C_4$, plant |

Choose the correct answer from the options given below :
(a) (A)-(IV), (B)-(I), (C)-(III), (D)-(II)
(b) (A)-(IV), (B)-(I), (C)-(II), (D)-(III)
(c) (A)-(III), (B)-(II), (C)-(I), (D)-(IV)
(d) (A)-(II), (B)-(III), (C)-(I), (D)-(IV)

**39** Identify the number of structure/s from the following which can be correlated to D-glyceraldehyde. *[2025, 22 Jan Shift-II]*

(A) 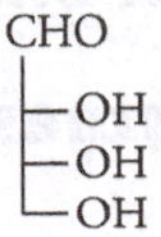
(B) 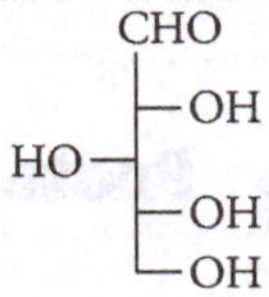
(C) 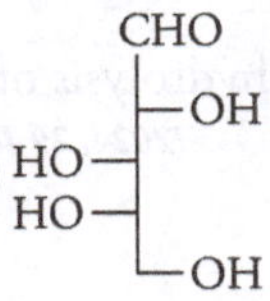
(D) 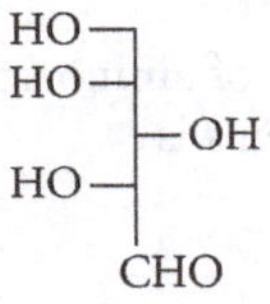

(a) four (b) one (c) two (d) three

**Imp.** Acidic Hydrolysis of Carbohydrates

**40** Identify correct conversion during acidic hydrolysis from the following. *[2025, 28 Jan Shift-II]*
A. Starch gives galactose.
B. Cane sugar gives equal amount of glucose and fructose.
C. Milk sugar gives glucose and galactose.
D. Amylopectin gives glucose and fructose.
E. Amylose gives only glucose.
Choose the correct answer from the options given below
(a) (A), (B) and (C) only (b) (B), (C) and (D) only
(c) (C), (D) and (E) only (d) (B), (C) and (E) only

**41** Match List-I with List-II. *[2025, 28 Jan Shift-II]*

| | List-I (Saccharides) | | List-II (Glycosidic-linkages found) |
|---|---|---|---|
| (A) | Sucrose | I. | $\alpha 1-4$ |
| (B) | Maltose | II. | $\alpha 1-4$ and $\alpha 1-6$ |
| (C) | Lactose | III. | $\alpha 1-\beta 2$ |
| (D) | Amylopectin | IV. | $\beta 1-4$ |

Choose the correct answer from the options given below :
(a) (A)-(III), (B)-(I), (C)-(IV), (D)-(II)
(b) (A)-(II), (B)-(IV), (C)-(III), (D)-(I)
(c) (A)-(IV), (B)-(II), (C)-(I), (D)-(III)
(d) (A)-(I), (B)-(II), (C)-(III), (D)-(IV)

**42** 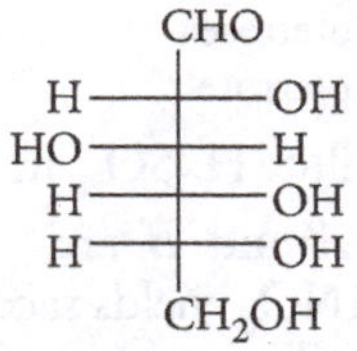

The incorrect statement regarding the given structure is *[2024, 8 April Shift-I]*
(a) will coexist in equilibrium with 2 other cyclic structure
(b) has 4 asymmetric carbon atom
(c) despite the presence of —CHO does not give Schiff's test
(d) can be oxidised to a dicarboxylic acid with $Br_2$ water

**43** Match List I with List II. *[2024, 4 April Shift-II]*

| List I | | List II |
|---|---|---|
| A. α- Glucose and α- Galactose | I. | Functional isomers |
| B. α- Glucose and β- Glucose | II. | Homologous |
| C. α- Glucose and α- Fructose | III. | Anomers |
| D. α- Glucose and α- Ribose | IV. | Epimers |

Choose the correct answer from the options given below:
(a) A- III, B- IV, C- II, D- I
(b) A- III, B- IV, C- I, D- II
(c) A- IV, B- III, C- II, D- I
(d) A- IV, B-III, C- I, D- II

**44** Match List-I with List-II. *[2024, 31 Jan Shift-I]*

| List-I | | List-II |
|---|---|---|
| A. Glucose/ $NaHCO_3$ / Δ | I. | Gluconic acid |
| B. Glucose/$HNO_3$ | II. | No reaction |
| C. Glucose HI / Δ | III. | *n*-hexane |
| D. Glucose/bromine water | IV. | Saccharic acid |

Choose the correct answer from the options given below:

| | A | B | C | D | | A | B | C | D |
|---|---|---|---|---|---|---|---|---|---|
| (a) | II | IV | III | I | (b) | IV | I | III | II |
| (c) | III | II | I | VI | (d) | I | IV | III | II |

**Imp.** Optical Activity of Carbohydrate

**45** D-(+)-Glyceraldehyde $\xrightarrow[\text{(ii) } H_2O\,/\,H^+ \text{ (iii) } HNO_3]{\text{(i) HCN}}$

The products formed in the above reaction are *[2023, 13 April Shift-I]*
(a) one optically inactive products
(b) two optically active products
(c) two optically inactive products
(d) one optically active and one meso product

**46** A sugar '$X$' dehydrates very slowly under acidic condition to give furfural which on further reaction with resorcinol gives the coloured product after sometime. Sugar '$X$' is *[2022, 27 July Shift-I]*
(a) aldopentose (b) aldotetrose
(c) oxalic acid (d) ketotetrose

**47** When sugar '$X$' is boiled with dilute $H_2SO_4$ in alcoholic solution, two isomers '$A$' and '$B$' are formed. '$A$' on oxidation with $HNO_3$ yields saccharic acid whereas '$B$' is laevorotatory. The compound '$X$' is *[2022, 28 June Shift-II]*
(a) maltose (b) sucrose (c) lactose (d) starch

**Imp.** **Preparation Reactions and Test of Carbohydrates**

**48** L-isomer of a compound ($A$) ($C_4H_8O_4$) gives a positive test with $[Ag(NH_3)_2]^+$. Treatment of $A$ with acetic anhydride yields triacetate derivative. Compound ($A$) produces an optically active compound ($B$) and an optically inactive compound ($C$) on treatment with bromine water and $HNO_3$ respectively. Compound ($A$) is *[2022, 27 June Shift-I]*

(a) 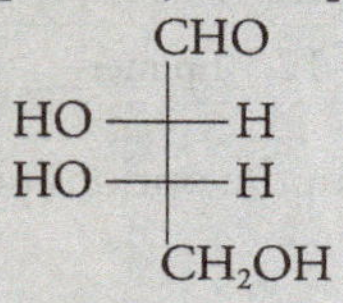
(b) 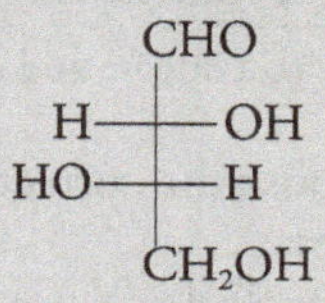
(c) 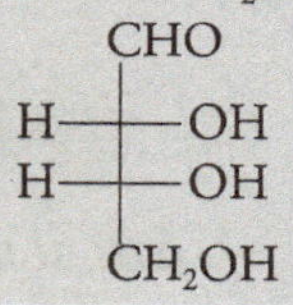
(d) 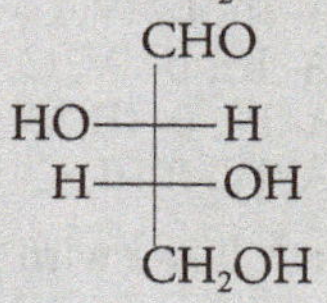

**49** A polysaccharide '$X$' on boiling with dil. $H_2SO_4$ at 393 K under 2-3 atm pressure yields '$Y$'. '$Y$' on treatment with bromine water gives gluconic acid. '$X$' contains β-glycosidic linkages only. Compound '$X$' is *[2022, 24 June Shift-I]*
(a) starch (b) cellulose
(c) amylose (d) amylopectin

**50** The correct observation in the following reactions, is

$$\text{Sucrose} \xrightarrow[\text{Cleavage (Hydrolysis)}]{\text{Glycosidic bond}} A + B \xrightarrow[\text{reagent}]{\text{Seliwanoff's}}$$

*[2020, 2 Sep Shift-II]*
(a) Formation of blue colour (b) Gives no colour
(c) Formation of red colour (d) Formation of violet colour

**51** Which of the following compounds will behave as a reducing sugar in an aqueous KOH solution? *[JEE Main 2017]*

(a) 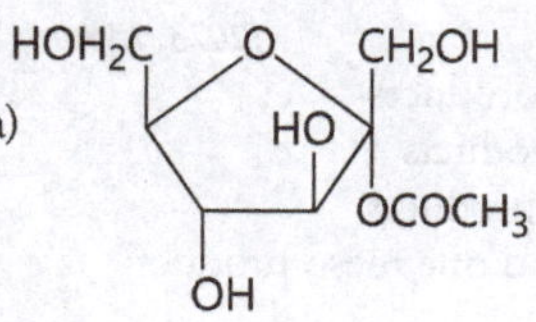
(b) 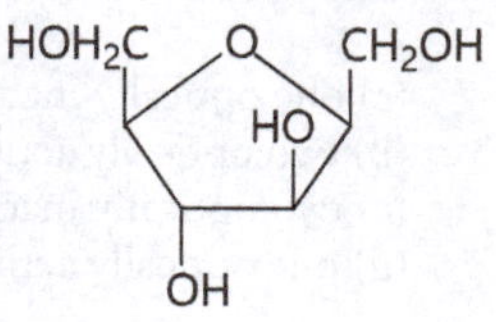
(c) 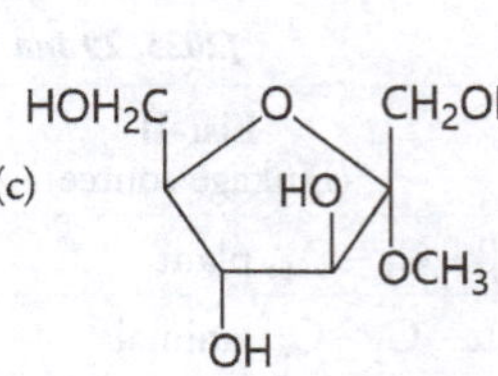
(d) 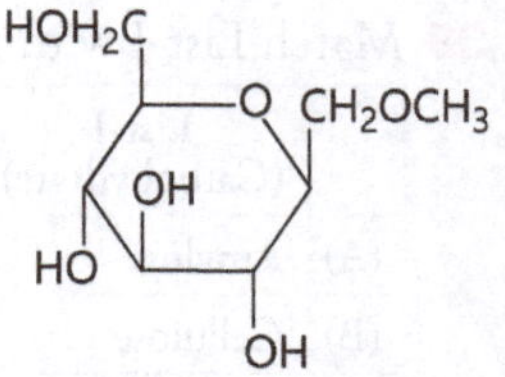

### Numerical Value Type Questions

**52** Number of compounds from the following which will not produce orange red precipitate with Benedict solution is ..........
Glucose, maltose, sucrose, ribose, 2-deoxyribose, amylose, lactose. *[2023, 11 April Shift-II]*

**53** $C_6H_{12}O_6 \xrightarrow{\text{Zymase}} A \xrightarrow[\Delta]{\text{NaOI}} B + CHI_3$
The number of carbon atoms present in the product $B$ is ......... . *[2022, 29 June Shift-I]*

## Topic 2 Proteins and Enzymes

### Level 1

**01** Type of amino acids obtained by hydrolysis of proteins is *[2024, 29 Jan Shift-I]*
(a) γ (b) α
(c) β (d) δ

**02** During the denaturation of proteins, which of these structures will remain intact? *[2022, 25 July Shift-I]*
(a) Primary
(b) Secondary
(c) Tertiary
(d) Quaternary

**03** The water soluble protein is *[2021, 22 July Shift-I]*
(a) fibrin (b) albumin
(c) myosin (d) collagen

**04** Which of the following is correct structure of tyrosine? *[2021, 17 March Shift-I]*

(a) 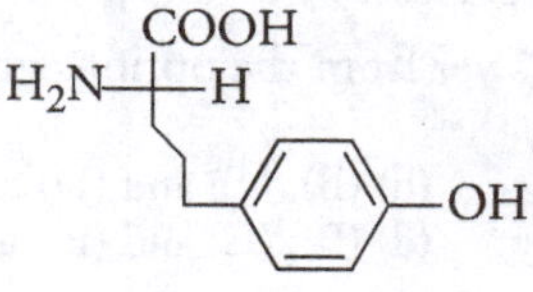
(b) 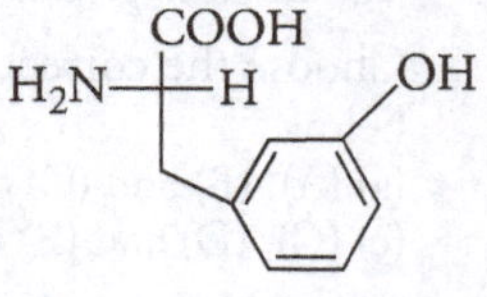
(c) 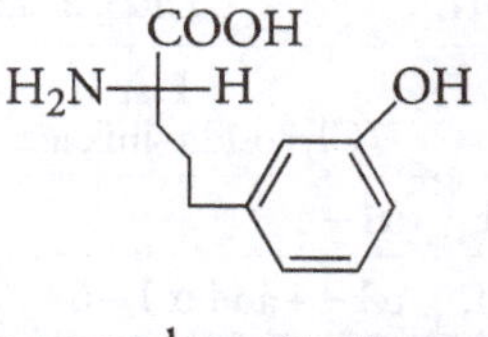
(d) 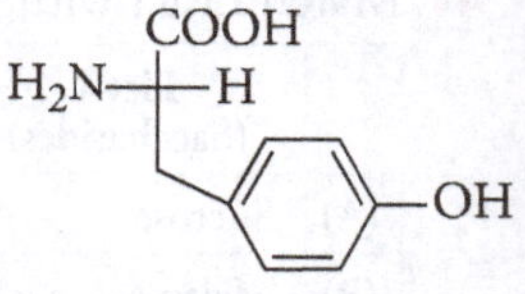

**05** The secondary structure of protein is stabilised by *[2021, 16 March Shift-II]*
(a) peptide bond
(b) glycosidic bond
(c) hydrogen bonding
(d) van der Waals' forces

## Level 2

**06** A dipeptide, '*x*' on complete hydrolysis gives '*y*' and '*z*'. '*y*' on treatment with aq. $HNO_2$, produces lactic acid. On the other hand '*z*' on heating gives the following cyclic molecule.

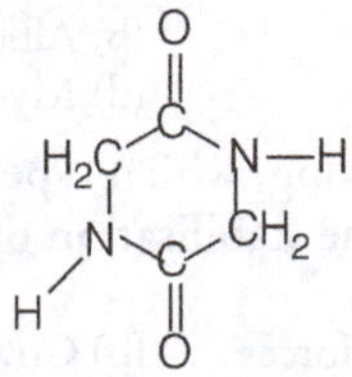


Based on the information given, the dipeptide *X* is *[2025, 04 April Shift-II]*

(a) alanine-alanine (b) valine-leucine
(c) valine-glycine (d) alanine-glycine

**07** Given below are two statements:

**Statement (I)** On hydrolysis, oligo peptides give rise to fewer number of α-amino acids while proteins give rise to a large number of β-amino acids.

**Statement (II)** Natural proteins are denatured by acids which convert the water soluble form of fibrous proteins to their water insoluble form.

In the light of the above statements, choose the most appropriate answer from the options given below *[2025, 07 April Shift-II]*

(a) Statement I is incorrect but Statement II is correct.
(b) Both Statement I and Statement II are correct.
(c) Statement I is correct but Statement II is incorrect.
(d) Both Statement I and Statement II are incorrect.

**08** A tetrapeptide, '*x*' on complete hydrolysis produced glycine (Gly), alanine (Ala), valine (Val), leucine (Leu) in equimolar proportion each. The number of tetrapeptides (sequences) possible involving each of these amino acids is *[2025, 02 April Shift-II]*

(a) 8 (b) 32
(c) 24 (d) 16

**09** Identify the pair of reactants that upon reaction, with elimination of HCl will give rise to the dipeptide Gly-Ala. *[2025, 04 April Shift-I]*

(a) $NH_2—CH_2—COOH$ and $NH_2—\overset{\overset{CH_3}{|}}{C}H—COOH$

(b) $NH_2—CH_2—COCl$ and $\overset{\oplus}{N}H_3—\overset{\overset{CH_3}{|}}{C}H—COCl$

(c) $NH_2—CH_2—COOH$ and $NH_2—\overset{\overset{CH_3}{|}}{C}H—COCl$

(d) $NH_2—CH_2—COCl$ and $NH_2—\overset{\overset{CH_3}{|}}{C}H—COOH$

**10** Coagulation of egg, on heating is because of *[2024, 5 April Shift-II]*

(a) biological property of protein remains unchanged
(b) the secondary structure of protein remains unchanged
(c) breaking of the peptide linkage in the primary structure of protein occurs
(d) denaturation of protein occurs.

**11** Which of the following gives a positive test with ninhydrin? *[2024, 5 April Shift-I]*

(a) Cellulose (b) Polyvinyl chloride
(c) Egg albumin (d) Starch

**12** Which structure of protein remains intact after coagulation of egg white on boiling? *[2024, 27 Jan Shift-II]*

(a) Secondary (b) Primary
(c) Quaternary (d) Tertiary

**13** Which is not true for arginine? *[2023, 15 April Shift-I]*

(a) It is associated with more than one $pK_a$ values.
(b) It is a crystalline solid.
(c) It has high solubility in benzene
(d) It has a fairly high melting point

**14** The naturally occurring amino acid that contains only one basic functional group in its chemical structure is *[2023, 13 April Shift-II]*

(a) histidine (b) asparagine
(c) lysine (d) arginine

**Imp.** Symbols of Amino Acids

**15** Match List I with List II.

| | List I (Natural amino acid) | | List II (One letter code) |
|---|---|---|---|
| A. | Glutamic acid | I. | Q |
| B. | Glutamine | II. | W |
| C. | Tyrosine | III. | E |
| D. | Tryptophan | IV. | Y |

Choose the correct answer from the options given below : *[2023, 8 April Shift-II]*

(a) A-III, B-IV, C-I, D-II (b) A-III, B-I, C-IV, D-II
(c) A-II, B-I, C-IV, D-III (d) A-IV, B-III. C-I.D-I

**16** Match List I with List II.

| | List I (Natural Amino acid) | | List II (One Letter Code) |
|---|---|---|---|
| A. | Arginine | I. | D |
| B. | Aspartic acid | II. | N |
| C. | Asparagine | III. | A |
| D. | Alanine | IV. | R |

Choose the correct answer from the options given below *[2023, 6 April Shift-II]*

(a) A-III, B-I, C-II, D-IV (b) A-IV, B-I, C-III, D-II
(c) A-I, B-III, C-IV, D-II (d) A-IV, B-I, C-II, D-III

**17** Match List I with List II

| List I (Enzymatic reaction) | | List II (Enzyme) |
|---|---|---|
| A. Sucrose → Glucose and fructose | I. | Zymase |
| B. Glucose → ethyl alcohol and $CO_2$ | II. | Pepsin |
| C. Starch → Maltose | III. | Invertase |
| D. Proteins → Amino acids | IV. | Diastase |

Choose the correct answer from the options given below *[2023, 6 April Shift-I]*
(a) A-I, B-IV, C-III, D-II (b) A-I, B-II, C-IV, D-III
(c) A-III, B-I, C-II, D-IV (d) A-III, B-I, C-IV, D-II

**18** The one that does not stabilise 2° and 3° structures of proteins is *[2023, 4 April Shift-I]*
(a) H-bonding (b) —S—S— linkage
(c) van der Waals forces (d) —O—O— linkage

**19** Given below are two statements, one is labelled as Assertion (A) and the other is labelled as Reason (R).
**Assertion (A)** $\alpha$-halocarboxylic acid on reaction with dil. $NH_3$ gives a good yield of $\alpha$-amino carboxylic acid whereas the yield of amines is very low when prepared from alkyl halides.
**Reason (R)** Amino acids exist in Zwitter ion form in aqueous medium.
In the light of the above statements, choose the correct answer from the options given below. *[2023, 01 Feb Shift-II]*
(a) A is incorrect but R is correct
(b) Both A and R are correct and R is the correct explanation of A
(c) A is correct but R is incorrect
(d) Both A and R are correct but R is not the correct explanation of A

**20** Number of cyclic tripeptides formed with 2 amino acids *A* and *B* is *[2023, 29 Jan Shift-I]*
(a) 3 (b) 4 (c) 2 (d) 5

**21** Two statements in respect of drug-enzyme interaction are given below
**Statement I** Action of an enzyme can be blocked only when an inhibitor blocks the active site of the enzyme.
**Statement II** An inhibitor can form a strong covalent bond with the enzyme.
In the light of the above statements, choose the correct answer from the options given below
*[2022, 28 July Shift-II]*
(a) Both Statement I and Statement II are true
(b) Both Statement I and Statement II are false
(c) Statement I is true but Statement II is false
(d) Statement I is false but Statement II is true

**22** Stability of $\alpha$-helix structure of proteins depends upon *[2022, 28 June Shift-I]*
(a) dipolar interaction (b) H-bonding interaction
(c) van der Waals' forces (d) $\pi$-stacking interaction

**23** Which of the following is not an example of fibrous protein ? *[2021, 31 Aug Shift-II]*
(a) Keratin (b) Albumin
(c) Collagen (d) Myosin

**24** Out of the following, which type of interaction is responsible for the stabilisation of $\alpha$-helix structure of proteins ? *[2021, 24 Feb Shift-I]*
(a) van der Waals' forces (b) Covalent bonding
(c) Ionic bonding (d) Hydrogen bonding

**25** Which of the following is not an essential amino acid? *[2020, 5 Sep Shift-I]*
(a) Leucine (b) Valine
(c) Lysine (d) Tyrosine

**26** Among the following compounds, most basic amino acid is *[2019, 12 Jan Shift-I]*
(a) serine (b) asparagine
(c) lysine (d) histidine

**Imp.** **p$K_a$ Order of Amino Acids**

**27** The increasing order of p$K_a$ of the following amino acids in aqueous solution is Gly, Asp, Lys, Arg *[2019, 9 Jan Shift-II]*
(a) Asp < Gly < Arg < Lys (b) Arg < Lys < Gly < Asp
(c) Gly < Asp < Arg < Lys (d) Asp < Gly < Lys < Arg

**28** The predominant form of histamine present in human blood is (p$K_a$, Histidine = 6.0) *[JEE Main 2018]*

(a) 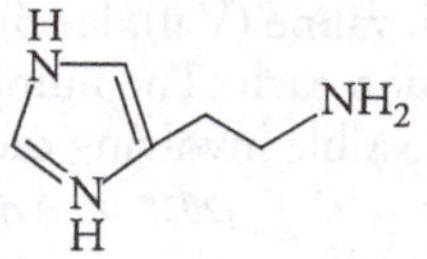
(b) 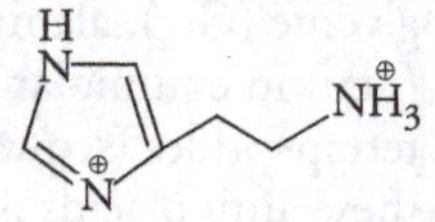
(c) 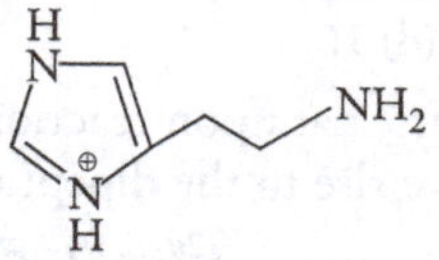
(d) 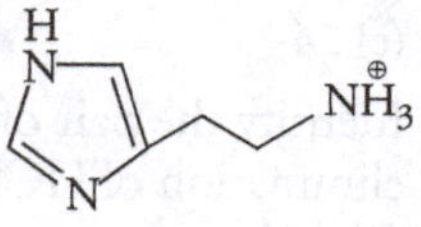

**29** Consider the following sequence for aspartic acid

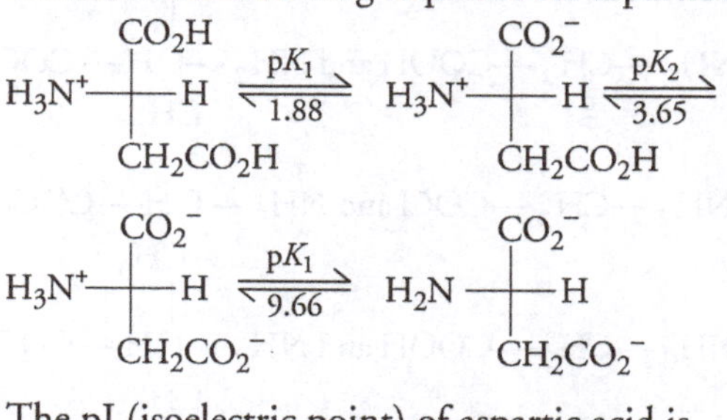

The pI (isoelectric point) of aspartic acid is *[JEE Main 2016]*
(a) 3.65 (b) 2.77 (c) 5.74 (d) 1.88

**30** Observation of "Ruhemann's purple" is a confirmatory test for the presence of **[JEE Main 2016]**
(a) starch (b) reducing sugar
(c) protein (d) cupric ion

**31** Thiol group is present in **[JEE Main 2016]**
(a) cystine (b) cysteine
(c) methionine (d) cytosine

**32** Biuret test is not given by **[AIEEE 2010]**
(a) carbohydrates (b) polypeptides
(c) urea (d) proteins

## Level 3

**33**
$$\underset{\text{CH(CH}_3)_2}{\overset{}{\text{H}_2\text{N—CH—COOH}}}\quad \xrightarrow{\text{pH}=2} A \qquad \xrightarrow{\text{pH}=10} B$$

(Side chain $CH(CH_3)_2$ on the CH carbon.)

Choose the correct option for structures of *A* and *B*, respectively **[2025, 08 April Shift-II]**

(a) $H_2N—CH(CH(CH_3)_2)—CO\bar{O}$ and $H_3\overset{+}{N}—CH(CH(CH_3)_2)—CO\bar{O}$

(b) $H_2N—CH(CH(CH_3)_2)—CO\bar{O}$ and $H_3\overset{+}{N}—CH(CH(CH_3)_2)—COOH$

(c) $H_3\overset{+}{N}—CH(CH(CH_3)_2)—COOH$ and $H_2N—CH(CH(CH_3)_2)—CO\bar{O}$

(d) $H_3\overset{+}{N}—CH(CH(CH_3)_2)—CO\bar{O}$ and $H_3\overset{+}{N}—CH(CH(CH_3)_2)—COOH$

**34** Identify the correct statement among the following **[2025, 02 April Shift-I]**
(a) All naturally occurring amino acids except glycine contain one chiral centre.
(b) All naturally occurring amino acids are optically active.
(c) Glutamic acid is the only amino acid that contains a —COOH group at the side chain.
(d) Amino acid, cysteine can easily undergo dimerisation due to the presence of free SH group.

**35** Identify the essential amino acids from below
(A) Valine (B) Proline
(C) Lysine (D) Threonine
(E) Tyrosine

Choose the correct answer from the options given below **[2025, 29 Jan Shift-II]**
(a) (A), (C) and (E) only
(b) (A), (C) and (D) only
(c) (C), (D) and (E) only
(d) (B), (C) and (E) only

**36** The incorrect statements regarding enzymes are
A. Enzymes are biocatalysts. **[2024, 6 April Shift-II]**
B. Enzymes are non-specific and can catalyse different kinds of reactions.
C. Most enzymes are globular proteins.
D. Enzyme-oxidase catalyses the hydrolysis of maltose into glucose.

Choose the correct answer from the options given below
(a) B and D (b) B and C
(c) A, B and C (d) B, C and D

**37** Sulphur (S) containing amino acids from the following are :
(A) isoleucine (B) cysteine
(C) lysine (D) methionine
(E) glutamic acid **[2023, 8 April Shift-I]**
(a) A, B, C (b) B, C (c) B, D (d) A, D

**38** A protein '*X*' with molecular weight of 70,000 u, on hydrolysis gives amino acids. One of these amino acid is **[2023, 31 Jan Shift-I]**
(a) $NH_2—CH_2—CH(CH_3)—CH_2CH_2COOH$
(b) $CH_3—CH(CH_3)—CH_2—CH(NH_2)—COOH$
(c) $CH_3—C(CH_3)(NH_2)—CH_2—CH_2COOH$
(d) $CH_3—CH(CH_3)—CH(NH_2)—CH_2—COOH$

**39** The peptide that gives positive ceric ammonium nitrate and carbylamine tests is **[2019, 9 April Shift-II]**
(a) Lys-Asp (b) Ser-Lys (c) Gln-Asp (d) Asp-Gln

**40** The correct structure of product '*P*' in the following reaction is **[2019, 10 Jan Shift-I]**

$$\text{Asn}-\text{Ser}+\underset{\text{(Excess)}}{(CH_3CO)_2O}\xrightarrow{NEt_3} P$$

(a)
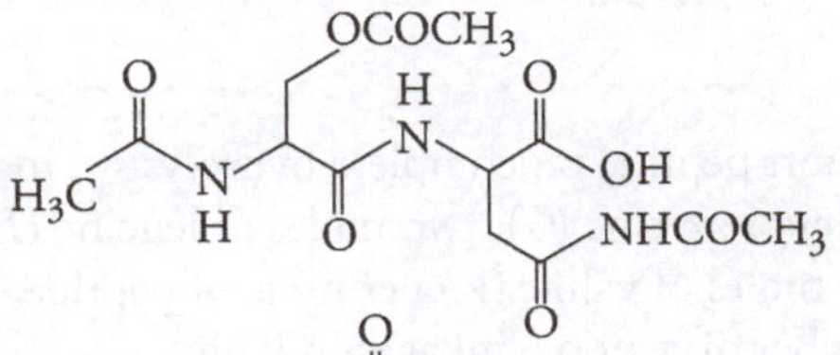


(b)
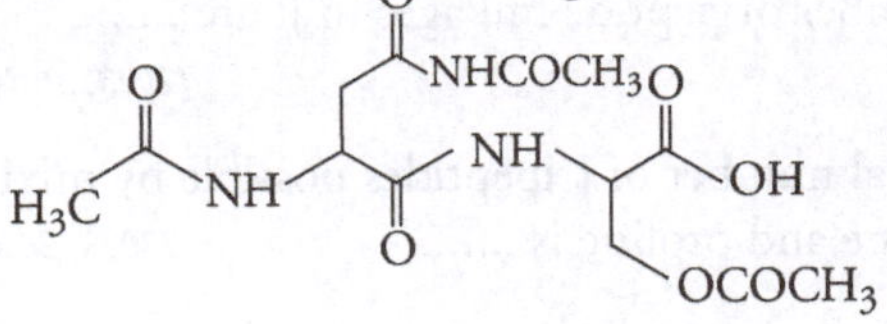

(c) 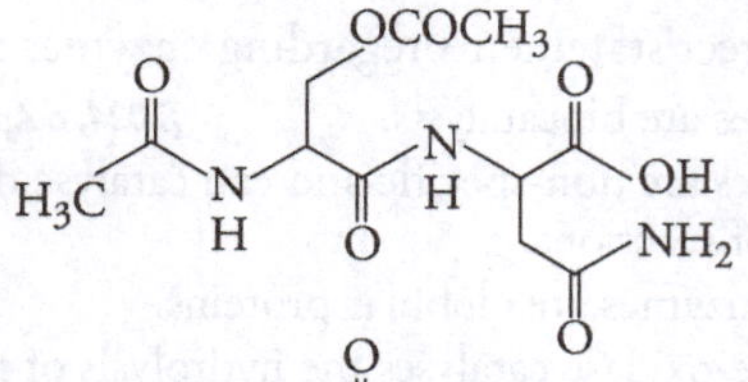

(d) 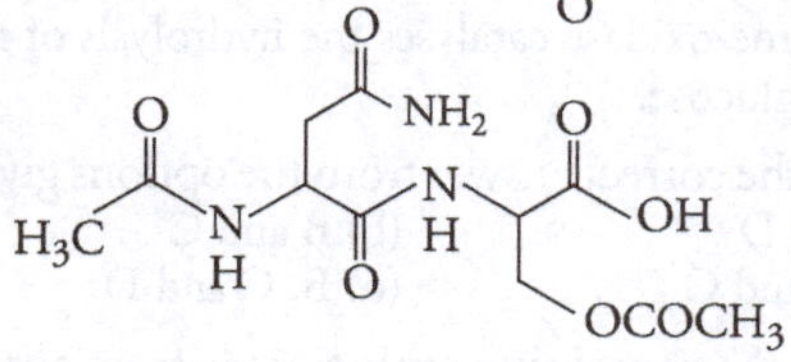

**Imp.** **Structures of Amino Acids**

**41** The correct sequence of amino acids present in the tripeptide given below is **[2019, 9 Jan Shift-II]**

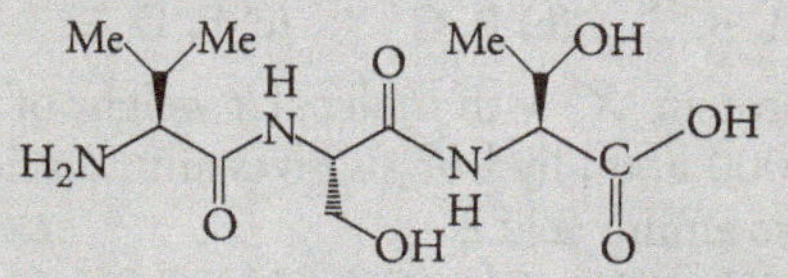

(a) Thr - Ser - Leu (b) Leu - Ser - Thr
(c) Val - Ser - Thr (d) Thr - Ser - Val

**42** Which one of the following statements is correct? **[AIEEE 2012]**

(a) All amino acids except lysine are optically active.
(b) All amino acids are optically active.
(c) All amino acids except glycine are optically active.
(d) All amino acids except glutamic acids are optically active.

## Numerical Value Type Questions

**43** Total number of essential amino acid among the given list of amino acids is .........

Arginine, Phenylalanine, Aspartic Acid, Cysteine, Histidine, Valine, Proline **[2024, 9 April Shift-I]**

**44** The total number of carbon atoms present in tyrosine, an amino acid is ...... **[2024, 8 April Shift-II]**

**45** The number of tripeptides formed by three different amino acids using each amino acid once is ......... . **[2024, 1 Feb Shift-I]**

**Imp.** **Calculation of Number of Peptide Linkages in a Amino Acids**

**46** A short peptide on complete hydrolysis produces 3 moles of glycine (*G*), two moles of leucine (*L*) and two moles of valine (*V*) per mole of peptide. The number of peptide linkages in it are ........ **[2023, 30 Jan Shift-II]**

**47** Total number of tripeptides possible by mixing of valine and proline is ....... **[2023, 24 Jan Shift-II]**

**48** In a linear tetrapeptide (Constituted with different amino acids), (number of amino acids) - (number of peptide bonds) is .......... . **[2022, 29 July Shift-I]**

**49** In alanylglycylleucylalanylvaline, the number of peptide linkages is ...... . **[2022, 24 June Shift-II]**

**50** How many of the given compounds will give a positive biuret test ...... ? **[2022, 25 June Shift-II]**
Glycine, Glycylalanine, Tripeptide, Biuret

**51** A peptide synthesised by the reactions of one molecule each of glycine, leucine, aspartic acid and histidine will have ......... peptide linkages. **[2021, 01 Sep Shift-II]**

**52** The total number of negative charge in the tetrapeptide Gly-Glu-Asp-Tyr, at pH 12.5 will be ......... . **[2021, 26 Aug Shift-I]**

**53** The number of chiral carbon(s) present in peptide, Ile-Arg-Pro, is ...... . **[2020, 5 Sep Shift-I]**

**54** The number of chiral centres present in threonine is ......... . **[2020, 04 Sept Shift-II]**

**55** The number of $\rangle C{=}O$ groups present in a tripeptide Asp-Glu-Lys is .......... . **[2020, 3 Sep Shift-II]**

## Topic 3 Vitamins, Nucleic Acids and Hormones

### Level 1

**01** Which of the following acids is a vitamin? **[2025, 22 Jan Shift-I]**

(a) Aspartic acid
(b) Ascorbic acid
(c) Saccharic acid
(d) Adipic acid

**02** All structures given below are of vitamin C. Most stable of them is **[2023, 01 Feb Shift-II]**

(a) 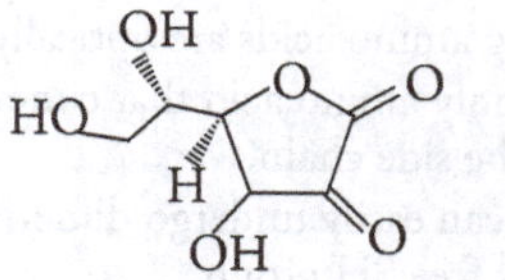

(b) 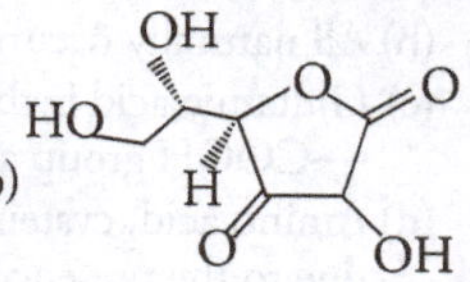

(c) 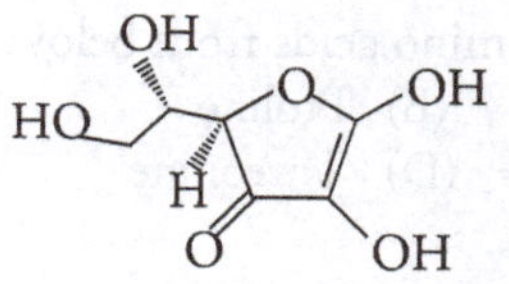

(d) 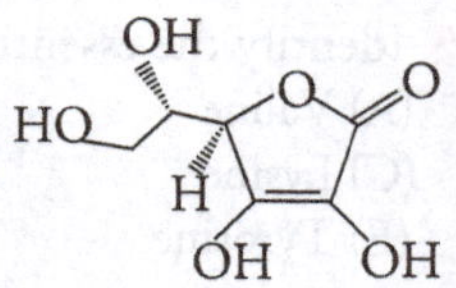

**03** Which one of the following is a water soluble vitamin, that is not excreted easily? **[2022, 26 June Shift-II]**

(a) Vitamin $B_2$ (b) Vitamin $B_1$
(c) Vitamin $B_6$ (d) Vitamin $B_{12}$

**04** Thiamine and pyridoxine are also known respectively as *[2021, 22 July Shift-II]*
(a) vitamin $B_2$ and vitamin E
(b) vitamin E and vitamin $B_2$
(c) vitamin $B_6$ and vitamin $B_2$
(d) vitamin $B_1$ and vitamin $B_6$

**05** Deficiency of vitamin K causes *[2021, 18 March Shift-II]*
(a) increase in blood clotting time
(b) increase in fragility of RBC's
(c) cheilosis
(d) decrease in blood clotting time

**06** Which of the following vitamin is helpful in delaying the blood clotting ? *[2021, 26 Feb Shift-I]*
(a) Vitamin C (b) Vitamin B (c) Vitamin E (d) Vitamin K

**07** Which of the vitamins given below is water soluble? *[JEE Main 2015]*
(a) Vitamin C (b) Vitamin D (c) Vitamin E (d) Vitamin K

**08** Which one of the following bases is not present in DNA? *[JEE Main 2014]*
(a) Quinoline (b) Adenine
(c) Cytosine (d) Thymine

**09** The pyrimidine bases present in DNA are *[AIEEE 2006]*
(a) cytosine and adenine (b) cytosine and guanine
(c) cytosine and thymine (d) cytosine and uracil

## Level 2

**Imp.** Structure of Nitrogenous Bases

**10** DNA molecule contains 4 bases whose structures are shown below. One of the structures is not correct. identify the incorrect base structure. *[2024, 6 April Shift-I]*

(a)
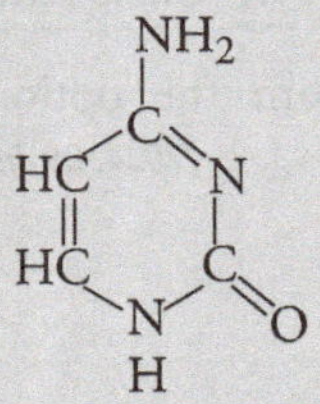

(b)
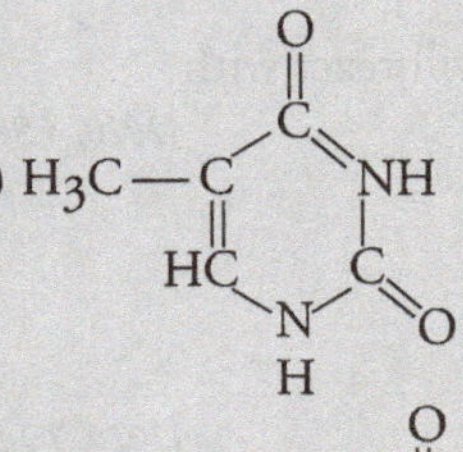

(c)
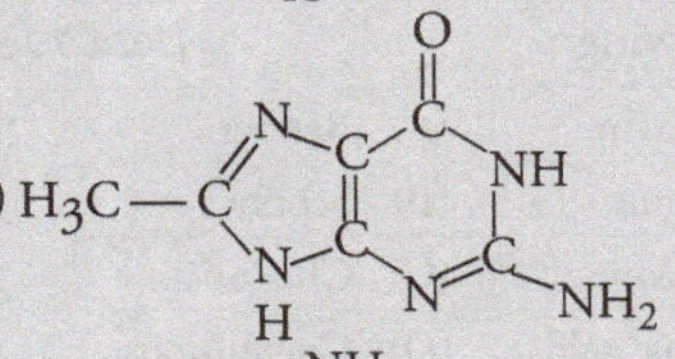

(d)
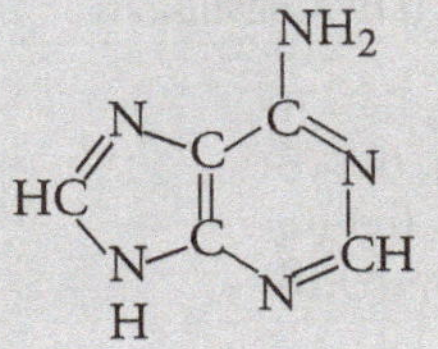

**11** If one strand of a DNA has the sequence ATGCTTCA, sequence of the bases in complementary strand is *[2024, 1 Feb Shift-II]*
(a) ATGCGACT (b) CATTAGCT
(c) TACGAAGT (d) GTACTTAC

**12** Two nucleotides are joined together by a linkage known as *[2024, 27 Jan Shift-I]*
(a) peptide linkage (b) phosphodiester linkage
(c) disulphide linkage (d) glycosidic linkage

**13** Sugar moiety in DNA and RNA molecules respectively are *[2022, 29 June Shift-I]*
(a) β-D-2-deoxyribose, β-D-deoxyribose
(b) β-D-2-deoxyribose, β-D-ribose
(c) β-D-ribose, β-D-2-deoxyribose
(d) β-D-deoxyribose, β-D-2-deoxyribose

**14**
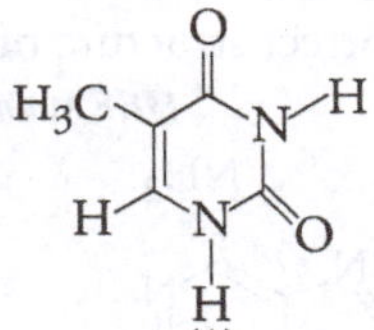
(A)

The compound '*A*' is a complementary base of .............. in DNA strands. *[2021, 27 July Shift-I]*
(a) uracil (b) guanine (c) adenine (d) cytosine

**15** Which one of the following is correct structure for cytosine? *[2021, 25 July Shift-II]*

(a)
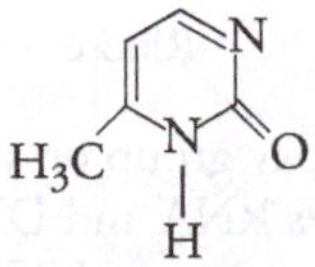
(b)
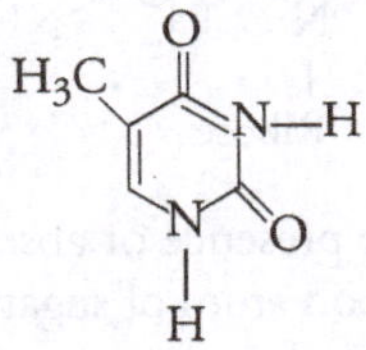
(c)
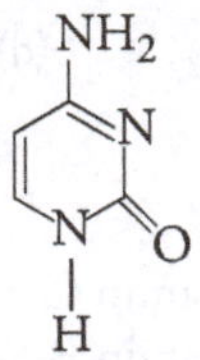
(d)
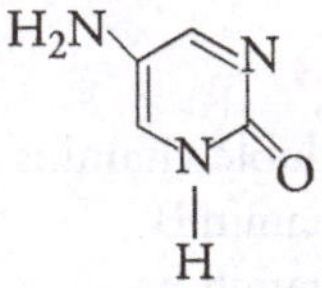

**16** Which among the following pair of vitamins is stored in our body relatively for longer duration ? *[2021, 16 March Shift-I]*
(a) Thiamine and vitamin A
(b) Vitamin A and vitamin D
(c) Thiamine and ascorbic acid
(d) Ascorbic acid and vitamin D

**17** Which of the following statement is not true about RNA? *[2019, 12 April Shift-I]*
(a) It controls the synthesis of protein
(b) It has always double stranded α-helix structure
(c) It usually does not replicate
(d) It is present in the nucleus of the cell

**18** Among the following compounds, which one is found in RNA? *[2019, 11 Jan Shift-I]*

(a) 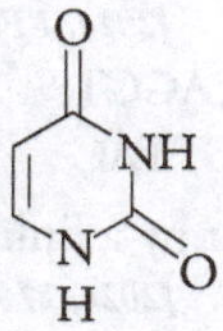
(b) 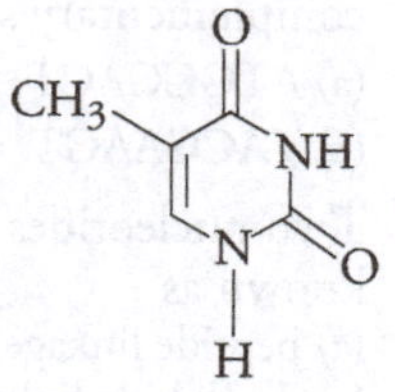
(c) 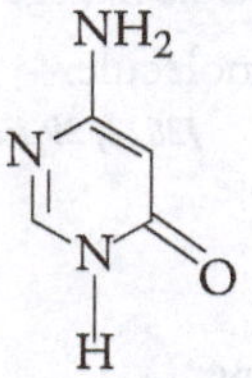
(d) 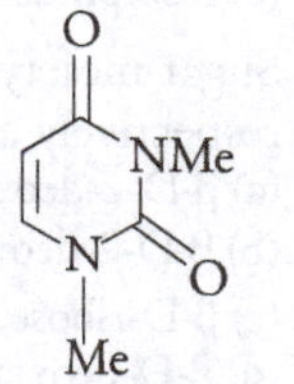

**19** Which of the following is the correct structure of adenosine? *[JEE Main 2018]*

(a) 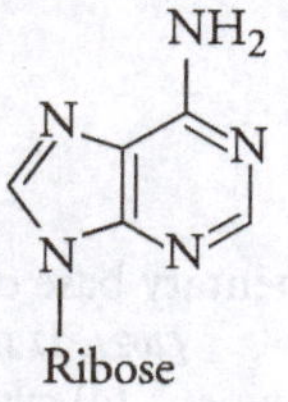
(b) 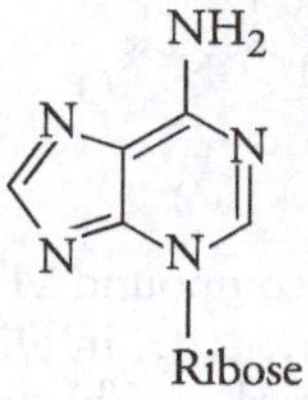
(c) 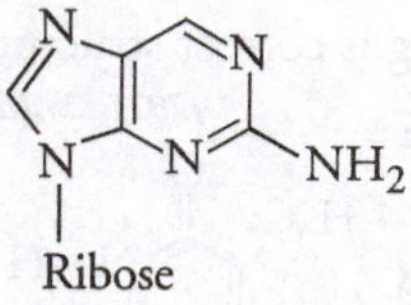
(d) 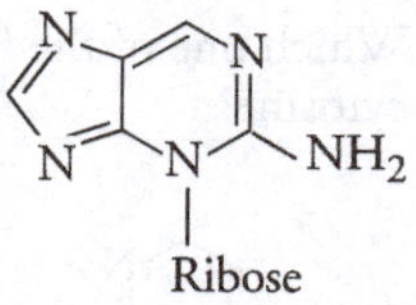

**20** The presence or absence of hydroxy group on which carbon atom of sugar differentiates RNA and DNA? *[AIEEE 2011]*

(a) 1st (b) 2nd (c) 3rd (d) 4th

## Level 3

**21** Fat soluble vitamins are

A. Vitamin $B_1$ B. Vitamin C
C. Vitamin E D. Vitamin $B_{12}$
E. Vitamin K

Choose the correct answer from the options given below. *[2025, 03 April Shift-II]*

(a) B and C only (b) C and D only
(c) C and E only (d) A and B only

**22** Match List - I with List - II. *[2025, 24 Jan Shift-II]*

| | List - I | | List - II |
|---|---|---|---|
| (A) | Adenine | (I) | 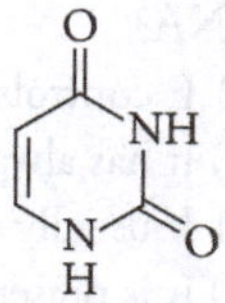 |
| (B) | Cytosine | (II) | 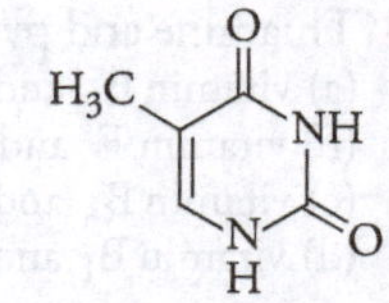 |
| (C) | Thymine | (III) | 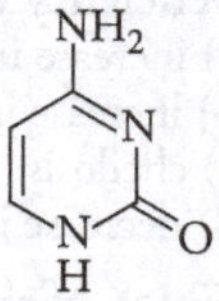 |
| (D) | Uracil | (IV) | 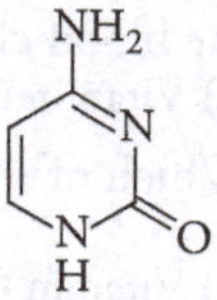 |

Choose the correct answer from the options given below

(a) (A) - (III), (B) - (I), (C) - (IV), (D) - (II)
(b) (A) - (IV), (B)-(III), (C) - (II), (D) - (I)
(c) (A) - (III), (B) - (IV), (C) - (II), (D) - (I)
(d) (A) - (III), (B)-(IV), (C) - (I), (D) - (II)

**23** Match List-I with List-II.

| | List-I (Biopolymer) | | List-II (Monomer) |
|---|---|---|---|
| A. | Starch | I. | Nucleotide |
| B. | Cellulose | II. | α-glucose |
| C. | Nucleic acid | III. | β-glucose |
| D. | Protein | IV. | α-amino acid |

Choose the correct answer from the options given below. *[2024, 29 Jan Shift-II]*

(a) A-IV, B-II, C-I, D-III
(b) A-II, B-I, C-III, D-IV
(c) A-I, B-III, C-IV, D-II
(d) A-II, B-III, C-I, D-IV

**24** Which of the following will react with $CHCl_3$ + alc. KOH ? *[2020, 4 Sep Shift-I]*

(a) Adenine and lysine
(b) Thymine and proline
(c) Adenine and thymine
(d) Adenine and proline

**25** Match the following : *[2020, 7 Jan Shift-I]*

| | | | |
|---|---|---|---|
| (i) | Riboflavin | (A) | Beriberi |
| (ii) | Thiamine | (B) | Scurvy |
| (iii) | Pyridoxine | (C) | Cheilosis |
| (iv) | Ascorbic acid | (D) | Convulsions |

(a) (i)-(C), (ii)-(D), (iii)-(A), (iv)-(B)
(b) (i)-(C), (ii)-(A), (iii)-(D), (iv)-(B)
(c) (i)-(D), (ii)-(B), (iii)-(A), (iv)-(C)
(d) (i)-(A), (ii)-(D), (iii)-(C), (iv)-(B)

**Imp.** **Linkages in DNA & RNA**

**26** In both DNA and RNA, heterocyclic base and phosphate ester linkages are at *[AIEEE 2005]*

(a) $C'_5$ and $C'_1$ respectively of the sugar molecule
(b) $C'_1$ and $C'_5$ respectively of the sugar molecule
(c) $C'_2$ and $C'_5$ respectively of the sugar molecule
(d) $C'_5$ and $C'_2$ respectively of the sugar molecule

## Numerical Value Type Questions

**27** Thyroxine, the hormone has given below structure *[2025, 7 April Shift-I]*

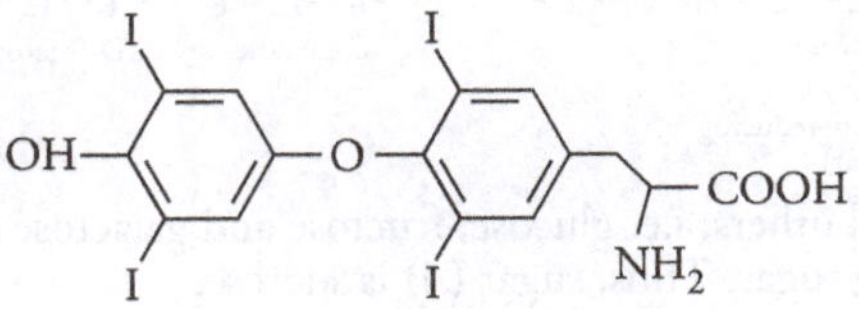


The percentage of iodine in thyroxine is ..........%. (Nearest integer)
(Given molar mass in g $mol^{-1}$ C : 12, H : 1, O : 16, N : 14, I : 127)

**28** The total number of hydrogen bonds of a DNA-double Helix strand whose one strand has the following sequence of bases is

5′-G-G-C-A-A-A-T-C-G-G-C-T-A-3′

*[2025, 4 April Shift-I]*

**29** From the vitamins A, $B_1$, $B_6$, $B_{12}$, C, D, E and K, the number of vitamins that can be stored in our body is ......... *[2024, 31 Jan Shift-II]*

**30** The total number of correct statements, regarding the nucleic acids is ......... . *[2024, 30 Jan Shift-II]*

A. RNA is regarded as the reserve of genetic information
B. DNA molecule self-duplicates during cell division
C. DNA synthesises proteins in the cell.
D. The message for the synthesis of particular proteins is present in DNA
E. Identical DNA strands are transferred to daughter cells

**31** Uracil is a base present in RNA with the following structure, % of N in uracil is ......... .

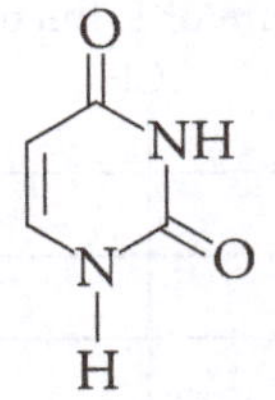


*[2023, 24 Jan Shift-I]*

[Given : Molar mass N = $14 g\ mol^{-1}$
$O = 16\ g\ mol^{-1}$, $C = 12\ g\ mol^{-1}$, $H = 1\ g\ mol^{-1}$]

**32** The number of oxygens present in a nucleotide formed from a base, that is present only in RNA is ......... . *[2022, 26 June Shift-I]*

## • Key Index

### Topic 1

| | | | | | | | | | | | | | | | | | | | |
|---|---|---|---|---|---|---|---|---|---|---|---|---|---|---|---|---|---|---|---|
| 1 | (c) | 2 | (a) | 3 | (d) | 4 | (d) | 5 | (a) | 6 | (b) | 7 | (d) | 8 | (c) | 9 | (d) | 10 | (b) |
| 11 | (c) | 12 | (a) | 13 | (b) | 14 | (a) | 15 | (b) | 16 | (c) | 17 | (d) | 18 | (c) | 19 | (b) | 20 | (a) |
| 21 | (d) | 22 | (a) | 23 | (c) | 24 | (a) | 25 | (c) | 26 | (d) | 27 | (a) | 28 | (a) | 29 | (c) | 30 | (c) |
| 31 | (d) | 32 | (a) | 33 | (d) | 34 | (a) | 35 | (d) | 36 | (b) | 37 | (a) | 38 | (b) | 39 | (d) | 40 | (d) |
| 41 | (a) | 42 | (d) | 43 | (d) | 44 | (a) | 45 | (d) | 46 | (a) | 47 | (b) | 48 | (a) | 49 | (b) | 50 | (c) |
| 51 | (a) | 52 | (2) | 53 | (1) | | | | | | | | | | | | | | |

### Topic 2

| | | | | | | | | | | | | | | | | | | | |
|---|---|---|---|---|---|---|---|---|---|---|---|---|---|---|---|---|---|---|---|
| 1 | (b) | 2 | (a) | 3 | (b) | 4 | (d) | 5 | (c) | 6 | (d) | 7 | (d) | 8 | (c) | 9 | (d) | 10 | (d) |
| 11 | (c) | 12 | (b) | 13 | (c) | 14 | (b) | 15 | (b) | 16 | (d) | 17 | (d) | 18 | (d) | 19 | (d) | 20 | (b) |
| 21 | (d) | 22 | (b) | 23 | (b) | 24 | (d) | 25 | (d) | 26 | (d) | 27 | (d) | 28 | (d) | 29 | (b) | 30 | (c) |
| 31 | (b) | 32 | (a) | 33 | (c) | 34 | (d) | 35 | (b) | 36 | (a) | 37 | (c) | 38 | (b) | 39 | (b) | 40 | (d) |
| 41 | (c) | 42 | (c) | 43 | (4) | 44 | (9) | 45 | (6) | 46 | (6) | 47 | (8) | 48 | (1) | 49 | (4) | 50 | (2) |
| 51 | (3) | 52 | (4) | 53 | (4) | 54 | (2) | 55 | (5) | | | | | | | | | | |

### Topic 3

| | | | | | | | | | | | | | | | | | | | |
|---|---|---|---|---|---|---|---|---|---|---|---|---|---|---|---|---|---|---|---|
| 1 | (b) | 2 | (d) | 3 | (d) | 4 | (d) | 5 | (a) | 6 | (d) | 7 | (a) | 8 | (a) | 9 | (c) | 10 | (c) |
| 11 | (c) | 12 | (b) | 13 | (b) | 14 | (c) | 15 | (c) | 16 | (b) | 17 | (b) | 18 | (a) | 19 | (a) | 20 | (b) |
| 21 | (c) | 22 | (c) | 23 | (d) | 24 | (a) | 25 | (b) | 26 | (b) | 27 | (65) | 28 | (33) | 29 | (5) | 30 | (3) |
| 31 | (25) | 32 | (9) | | | | | | | | | | | | | | | | |

# Detailed **Explanations**

## Topic 1

**01.** *(c)* The correct structure of L-glucose is given in option (c). i.e.,

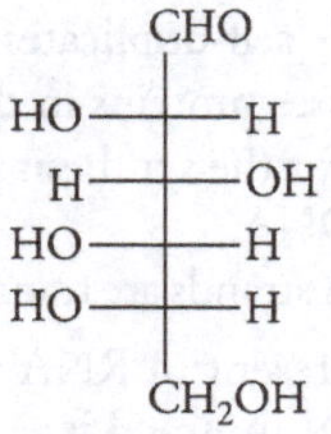

**02.** *(a)* The correct structure of L-fructose is

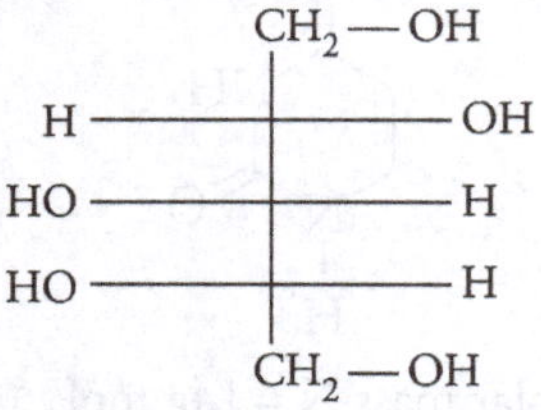

**03.** *(d)* The Haworth projection is cyclic representation of monosaccharides with simple three dimensional perspective. D-mannose Haworth structure is in the $\alpha$-pyranose configuration.

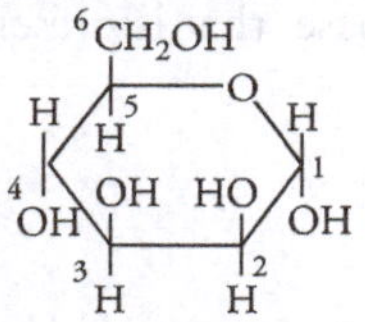

**04.** *(d)* Amylose is a linear polymer with 200 to 2000 $\alpha$-linked glucose units. It is soluble in water.
Therefore, (A) is incorrect but (R) is correct.

**05.** *(a)* Monosaccharides that contains hemiacetal group are reducing sugars (as their open chain structure contains an aldehyde group). While option (b), (c) and (d) compounds does not contain aldehydic or ketonic group in their open chain structure thus, they are non-reducing sugars.

**06.** *(b)* Sucrose is disaccharide of $\alpha$-glucose and $\beta$-fructose.

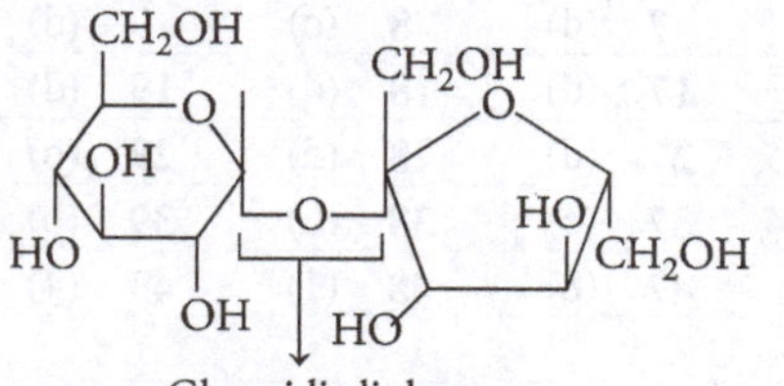

Glycosidic linkage

**07.** *(d)* Sucrose + $H_2O \rightleftharpoons \alpha$-D-(+)-glucose + $\beta$-D-(−)-fructose

**08.** *(c)* Barfoed test is used for distinguishing monosaccharides from disaccharides. Monosaccharides usually react in about 1-2 minutes while disaccharide take 7-12 minutes to react with reagent.

> • Concept **Enhancer**
> Barfoed's reagent is cupric acetate in 1% acetic acid. Brick red colour is obtained in this test due to formation of cuprous oxide.

**09.** *(d)* Sucrose is a non-reducing sugar because the two reducing monosaccharide units are held together by a glucosidic linkage between $C_1$ of $\alpha$-glucose and $C_2$ of $\beta$-fructose.

$$\underset{\substack{\text{Sucrose}\\(A)\\\text{(Non-reducing sugar)}}}{C_{12}H_{22}O_{11}} + H_2O \longrightarrow \underset{\text{D-glucose}}{C_6H_{12}O_6} + \underset{\text{D-fructose}}{C_6H_{12}O_6}$$

While all others, i.e. glucose, fructose and galactose are reducing sugar. Thus, sugar (*A*) is sucrose.

**10.** *(b)* Fructose is an example of ketohexose. It is a class of sugars that contains six-carbon atoms and a ketonic group. Fructose has a ketonic group with six carbon atoms.

**11.** *(c)* The correct match is A–(ii), B–(i), C–(iii).

**12.** *(a)* Maltose is a disaccharide formed from monosaccharides of 2 units of $\alpha$-D- glucose. These units are connected *via* $C_1 - C_4$ glycosidic linkage.

**13.** *(b)* Glucose and galactose differ at C-4 carbon positions.

**14.** *(a)* Both Statement I and Statement II are true.

**15.** *(b)* Statement I is false but Statement II is true.
The correct form of statement I is
Glucose does not respond to 2, 4-DNP test, as it does not contain any free aldehydic group in its cyclic form.

**16.** *(c)* The incorrect statement about glucose is given in option (c). It is because glucose readily dissolves in water, as it has lots of polar hydroxyl groups which can form hydrogen bond with water and not because of aldehyde functional group.

**17.** *(d)* Sucrose do not contain hemiacetal group. Hence, it does not give test with Fehling solution. All other given solutions give positive test with Fehling solution.
Hence, option (d) is correct.

**18.** *(c)* $X(C_4H_8O_4)$ gives Schiff's test due to presence of CHO group.

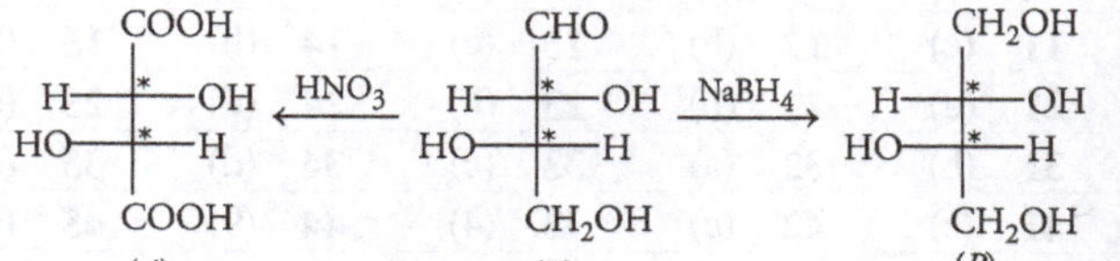

(*A*) (*X*) (*B*) Optically inactive (Chiral compound)

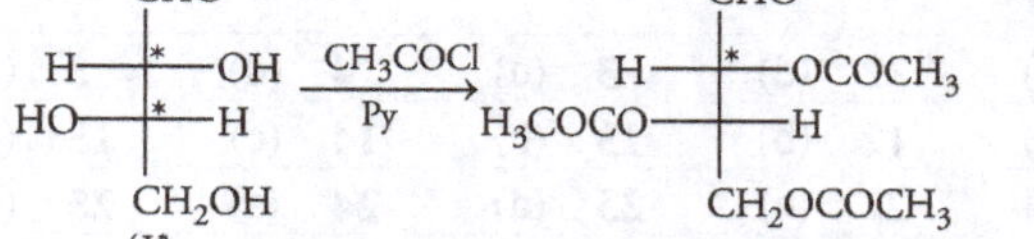

(*X*) (Triacetate)

**19.** *(b)* The correct match is A-(iii), B-(iv), C-(i), D-(ii).

**20.** *(a)* The product *A* and *B* formed are $C_7H_{14}O_8$ and $C_6H_{14}$.

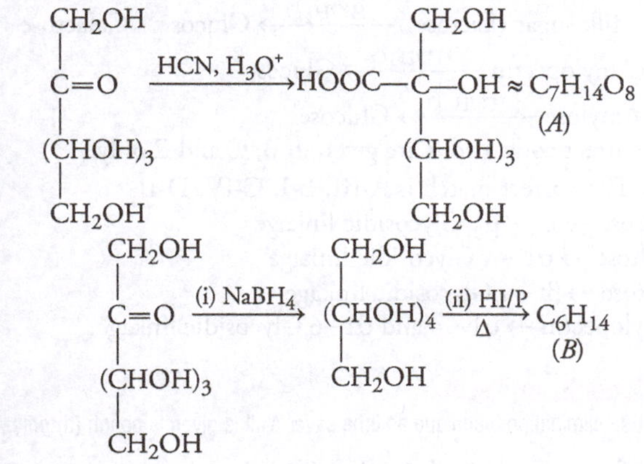


**21.** *(d)*

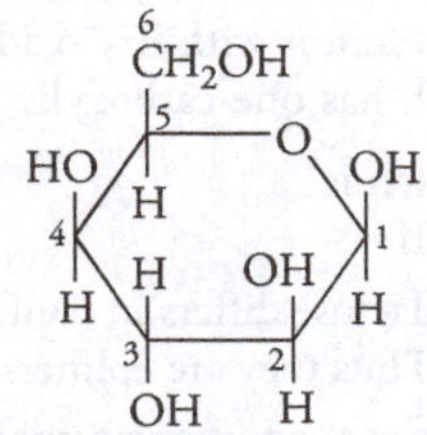

**22.** *(a)* The correct match is A-IV, B-I, C-II, D-III i.e.

$$\text{Glucose} + \text{HI} \longrightarrow n\text{-hexane}$$
$$\text{Glucose} + Br_2 \text{ water} \longrightarrow \text{Gluconic acid}$$
$$\text{Glucose} + \text{acetic anhydride} \longrightarrow \text{Glucose pentacetate}$$
$$\text{Glucose} + HNO_3 \longrightarrow \text{Saccharic acid}$$

**23.** *(c)* Maltose has two α-D-glucose units linked at $C_1$, and $C_4$ and is a reducing sugar as free aldehyde is formed by opening of ring and can react with fehling's solution.

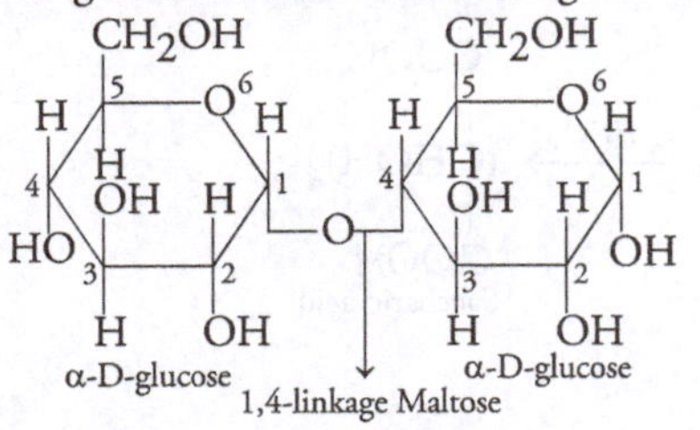


Thus, statement I is true while statement II is false.

**24.** *(a)* Lactose contains $\beta$-$C_1$—$C_4$ glycosidic linkage.

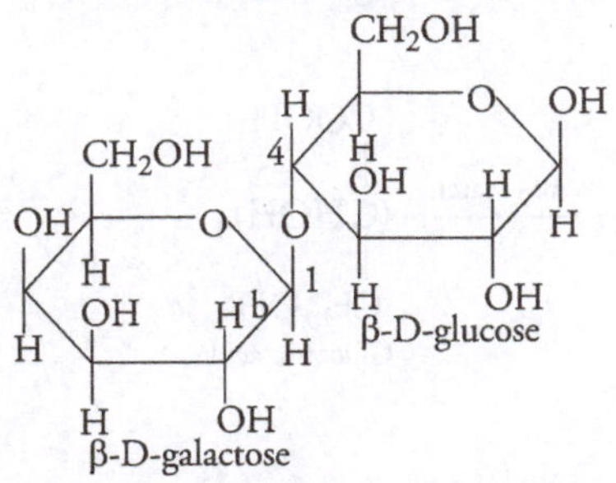


**25.** *(c)* Assertion is true but Reason is false. Sucrose is formed by linkage of $C_1$ of α-D glucose and $C_2$ of β-D fructose. The glycosidic bond in sucrose is formed between reducing ends of both glucose and fructose.

Sucrose is a disaccharide and a non-reducing sugar as it contains no free —CHO or >C=O group (anomeric carbon of both monosaccharides are involved in formation of glycosidic bond).

**26.** *(d)* Hydrolysis product of given compound are as follows

(a) $\text{Amylose} + H_2O \xrightarrow{H^+} \text{D-glucose}$

(b) $\text{Sucrose} + H_2O \xrightarrow{H^+} \text{D-glucose} + \text{D-fructose}$

(c) $\text{Maltose} + H_2O \xrightarrow{H^+} \text{D-glucose}$

(d) $\text{Lactose} + H_2O \xrightarrow{H^+} \text{D-glucose} + \text{D-galactose}$

Hence, option (d) is correct.

**27.** *(a)* Amylose consists of linear, helical chains of roughly 200 – 1,000 α-D-glucose monomers linked together through α -1,4 glycosidic bonds. In contrast to amylose, amylopectin is branched. Therefore, option (a) is incorrect statement.

**28.** *(a)* Glycosidic linkage is a type of covalent bond that joins carbohydrate molecules to another group.

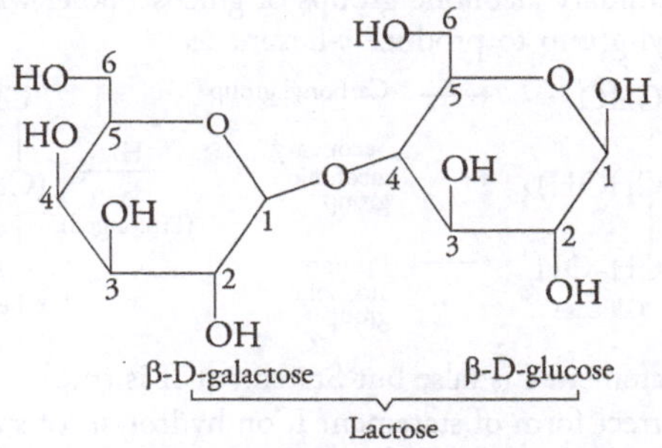


In lactose, glycosidic linkage is formed between C-1 of galactose and C-4, glycosidic of glucose.

**29.** *(c)* The functional groups present in the structure of maltose is one acetal and one hemiacetal. It is illustrated in following structure.

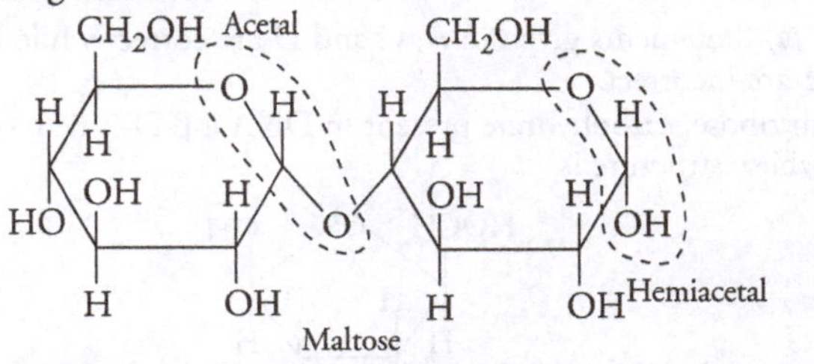


**30.** *(c)* The incorrect statement is (c). It can be corrected as follows : Glucose does not give Schiff's test due to absence of free —CHO in cyclic form.

**31.** *(d)* The correct chemical equations are given as follows :

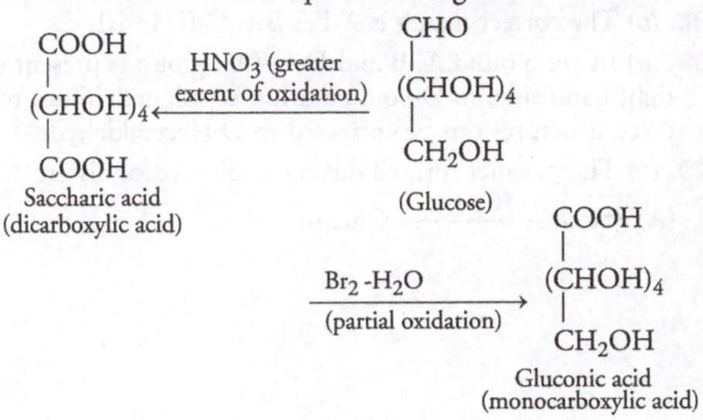

**32.** *(a)* Statement (a) is incorrect. Glycogen is highly branched structure similar to amylopectin. It is known to be the storage material of animals. It is found in liver, muscles and brain. It breaks down to glucose by the action of enzymes when body needs a glucose. It is also found in yeast and fungi.

**33.** *(d)*

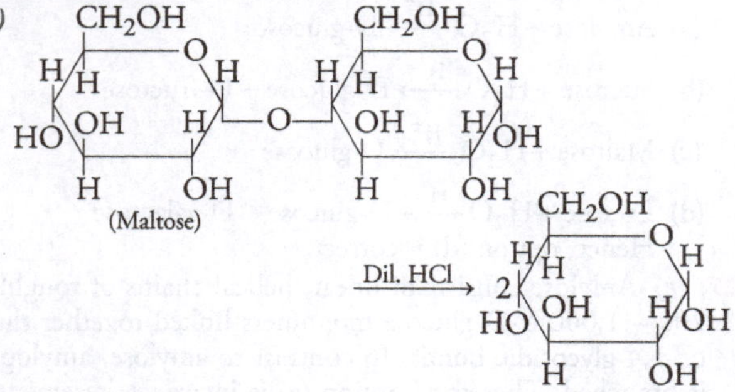


(Maltose) α-D-glucose

**34.** *(a)* HI is a strong reducing agent. It reduces both primary and secondary alcoholic groups of glucose along with the carbonyl group to produce *n*-hexane as

$$\underset{\text{Glucose}}{CHO\text{—}(CHOH)_4\text{—}CH_2OH} \xrightarrow[\Delta\ (\text{Prolonged})]{HI} \underset{n\text{-hexane}}{CH_3\text{—}(CH_2)_4\text{—}CH_3}$$

CHO ← Carbonyl group; $(CHOH)_4$ ← Secondary alcoholic group; $CH_2OH$ ← Primary alcoholic group

**35.** *(d)* Statement I is false but Statement II is true.
The correct form of statement I, on hydrolysis of sucrose gives *D*-(+)-glucose and *D*-(–)- fructose.

**36.** *(b)* Statement I is true but Statement II is false.
The correct form of statement II is, cellulose derivative has more number of hydroxy groups, so more H-bonding is present with water in cellulose derivatives cotton clothes.

**37.** *(a)* Statements given in A, C and D are correct while B and E are incorrect.
In ribose, carbohydrate present in DNA is β-D-2-deoxyribose whose structure is

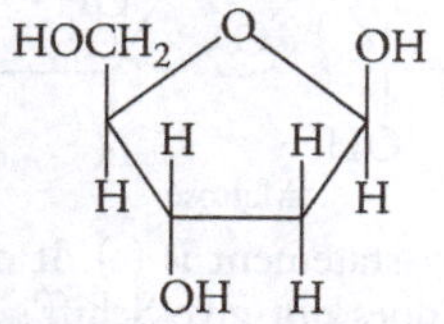

Which is a reducing D-sugar in β anomeric form (at C1) and it is a pentose sugar.

**38.** *(b)* The correct match is A-IV, B-I, C-II, D-III,

**39.** *(d)* In compound A, B and D —OH group is present on right hand side, so D-configuration is assigned. Thus, total three structures can be corrected to D-glyceraldehyde.

**40.** *(d)* The product formed during acidic hydrolysis are

(A) Starch $\xrightarrow{H^+/H_2O}$ Glucose

(B) Cane sugar (Sucrose) $\xrightarrow{H^+/H_2O}$ Glucose + Fructose (50%) (50%)

(C) Milk sugar (Lactose) $\xrightarrow{H^+/H_2O}$ Glucose + Galactose

(D) Amylopectin $\xrightarrow{H^+/H_2O}$ Glucose

(E) Amylose $\xrightarrow{H^+/H_2O}$ Glucose

So, correct conversions are given in B, C and E only.

**41.** *(a)* The correct match is A-III, B-I, C-IV, D-II
Sucrose → $\alpha_1 - \beta_2$ Glycosidic linkage
Maltose → α1 – 4 Glycosidic linkage
Lactose → β1 – 4 Glycosidic linkage
Amylopectin → α1 – 4 and α1 –6 Glycosidic linkage.

**Time Saver Tech**
Use elimination technique as time saver A-III is given is option (a) only.

**42.** *(d)* Statement given in option (d) is incorrect, its correct form is glucose on reaction with $Br_2$ oxidised to give gluconic acid. Which has one carboxylic acid group.

**43.** *(d)* The correct match is
A-IV, B-III, C-I, D-II
- α-Glucose and α-Galactose differs in configuration of one asymmetric carbon. Thus they are epimers.
- α and β-Glucose differs at anomeric carbon. Thus, they are anomers.
- α-Glucose and α-Fructose have different functional groups. Thus they are functional isomers.
- α-Glucose and α-Ribose are homologous.

**44.** *(a)* The correct match is A-II, B-IV, C-III, D-I

- Glucose $\xrightarrow[\Delta]{NaHCO_3}$ No reaction
- $CHO\text{—}(CHOH)_4\text{—}CH_2OH \xrightarrow[\Delta]{HNO_3} \underset{\text{Saccharic acid}}{COOH\text{—}(CHOH)_4\text{—}COOH}$
- $\underset{\text{Glucose}}{CHO\text{—}(CHOH)_4\text{—}CH_2OH} \xrightarrow{HI,\ \Delta} \underset{n\text{-hexane}}{(CH_3CH_2CH_2CH_2CH_2CH_3)}$
- $\underset{\text{Glucose}}{CHO\text{—}(CHOH)_4\text{—}CH_2OH} \xrightarrow{Br_2\ \text{water}} \underset{\text{Gluconic acid}}{COOH\text{—}(CHOH)_4\text{—}CH_2OH}$

**45.** *(d)*

Glyceraldelyde =

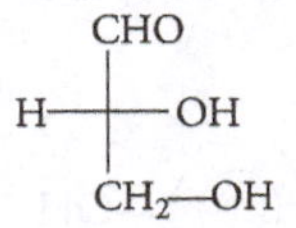


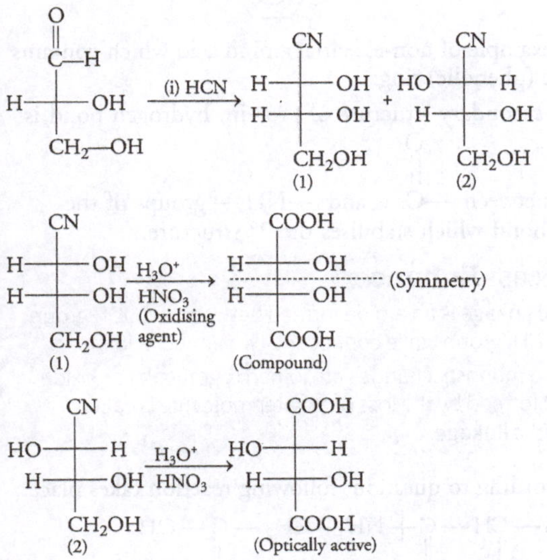


**46.** *(a)*

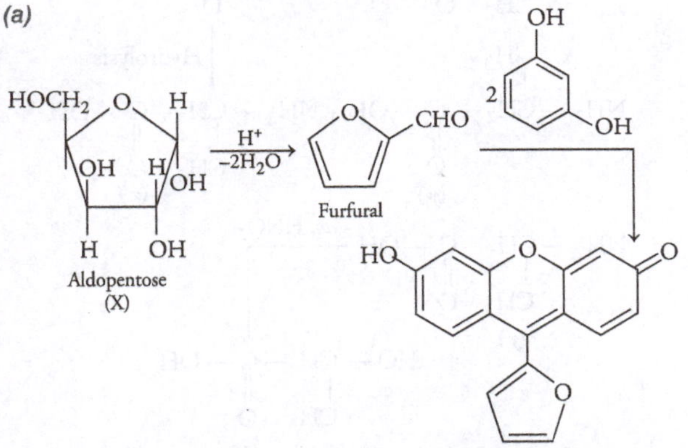


Cherry red product
(Seliwanoff's test)

**47.** *(b)* Compound '*X*' is sucrose.

$$\underset{\text{Sucrose }(X)}{C_{12}H_{22}O_{11}} \xrightarrow[C_2H_5OH]{\text{Dil. }H_2SO_4} \underset{\text{Glucose }(A)}{C_6H_{12}O_6} + \underset{\substack{\text{Fructose }(B)\\ \text{(Laevorotatory)}}}{C_6H_{12}O_6}$$

$$\underset{\substack{\text{Glucose}\\(A)}}{CHO-(CHOH)_4-CH_2OH} \xrightarrow{HNO_3} \underset{\text{Saccharic acid}}{COOH-(CHOH)_4-COOH}$$

**48.** *(a)*

- A positive test with Tollen's reagent confirms the presence of —CHO group.
- Formation of triacetate derivative confirms the presence of 3 —OH groups.
- L-isomer confirms that from bottom, at C-2 carbon OH is on left and H is on right side.
- $Br_2$ water will convert the aldehyde group to carboxylic acid group which is optically active.
- $HNO_3$ will oxidise both aldehyde and alcohol to carboxylic group which is optically inactive.

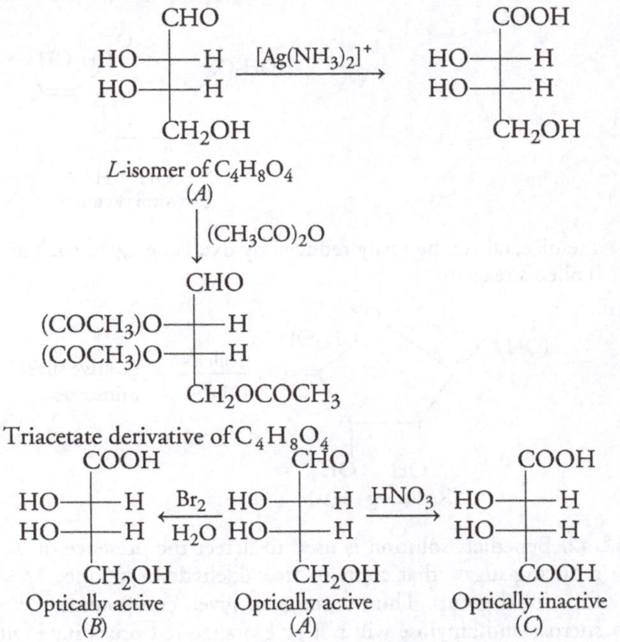


**49.** *(b)* Compound *X* is cellulose, as β-D-glucose unit in cellulose are joined by β-glycosidic linkage. Other given reactions are as follows

$$\underset{\substack{\text{Cellulose}\\(X)}}{(C_6H_{10}O_5)_n} \xrightarrow[\substack{393\text{ K},\\ 2\text{-}4\text{ atm}}]{\text{Dil. }H_2SO_4} n\underset{\substack{\text{Glucose}\\(Y)}}{(C_6H_{12}O_6)} \xrightarrow[Br_2/\text{water}]{[O]} \underset{\text{Gluconic acid}}{(C_6H_{12}O_7)_n}$$

**50.** *(c)* $\underset{\text{Sucrose}}{C_{12}H_{22}O_{11}} + H_2O \xrightarrow[\text{Cleavage (Hydrolysis)}]{\text{Glycosidic bond}}$

$$\underset{\substack{\text{Glucose}\\(A)}}{C_6H_{12}O_6} + \underset{\substack{\text{Fructose}\\(B)}}{C_6H_{12}O_6}$$

$\xrightarrow{\text{Seliwanoff's reagent}}$ Formation of red colouration

Seliwanoff's reagent test (dilute resorcinol + dil HCl) is given by ketohexoses (here, *B*) and sucrose only but not by aldohexose (here, *A*), glucose.

**51.** *(a)* Sugars that have an aldehyde, a ketone, a hemiacetal or a hemiketal group is able to reduce an oxidising agent are known as reducing sugars.

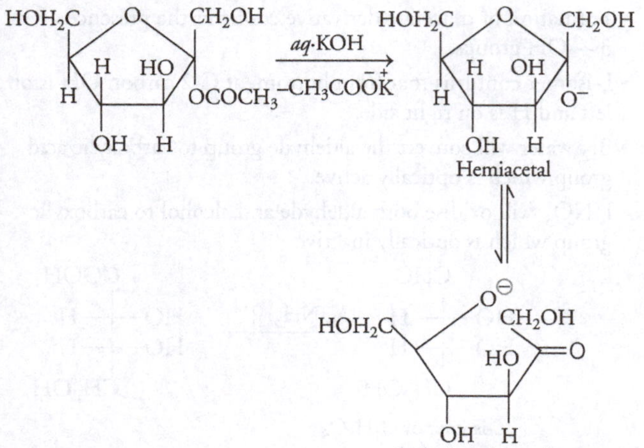

(α-hydroxyketone)

Hemiacetal can be easily reduced by oxidising agent such as Tollen's reagent.

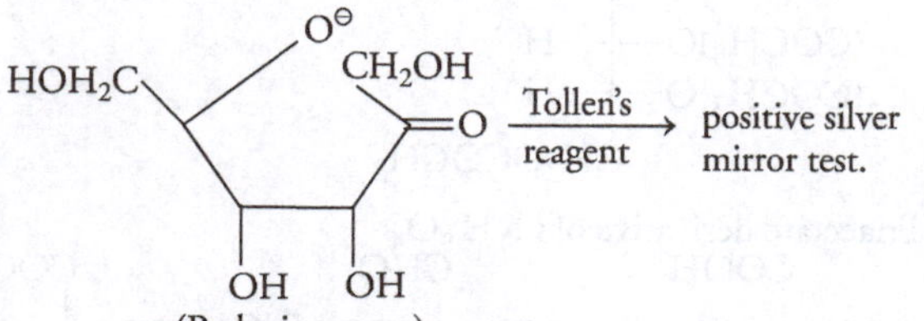

(Reducing sugar)

**52.** *(2)* Benedicts solution is used to detect the presence of reducing sugars that contains free aldehyde or ketone functional group. Thus, among the given compounds sucrose and amylose will not give orange-red precipitate with benedict solution as these sugars do not contain free aldehyde or ketone functional group.

**53.** *(1)* Only one carbon atom is present in the product *B*.

$$C_6H_{12}O_6 \xrightarrow{\text{Zymase}} \underset{(A)}{C_2H_5OH} \xrightarrow[H_3O^+]{I_2/OH^-} HCO\overset{-}{O}\overset{+}{N}a + CHI_3\downarrow$$

$$\downarrow HCl\,(aq)$$

$$\underset{(B)}{HCOOH} + NaCl$$

## Topic 2

**01.** *(b)* Proteins are natural polymers composed of α-amino acids which are connected by peptide linkage. Hence, proteins upon acidic hydrolysis produces α-amino acids.

**02.** *(a)* After denaturation, hydrogen bonds are broken hence, secondary and tertiary structures are destroyed. Primary structure remain intact.

**03.** *(b)* Albumin, a globular protein is soluble in water. It is half saturated with a salt such as ammonium sulphate. Rest all others i.e. fibrin, collagen and myosin are insoluble in water and all are fibrous proteins.

**04.** *(d)* The structure of tyrosine amino acid is

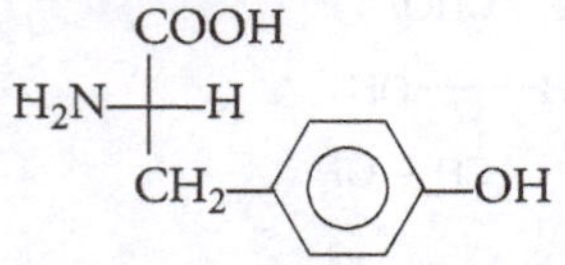

It is an example of non-essential amino acid which contains aromatic (phenolic) ring.

**05.** *(c)* In secondary structure of protein, hydrogen bond is formed between $-\overset{O}{\overset{\|}{C}}-$ and —NH— groups of the peptide bond which stabilises the 2° structure.

> **• Concept Enhancer**
>
> Peptide linkage is an amide formed between —COOH group and —$NH_2$ group while connecting two amino acids.
>
> The two monosaccharides are joined together by an oxide linkage formed by the loss of a water molecule is called glycosidic linkage.

**06.** *(d)* According to question, following reaction takes place.

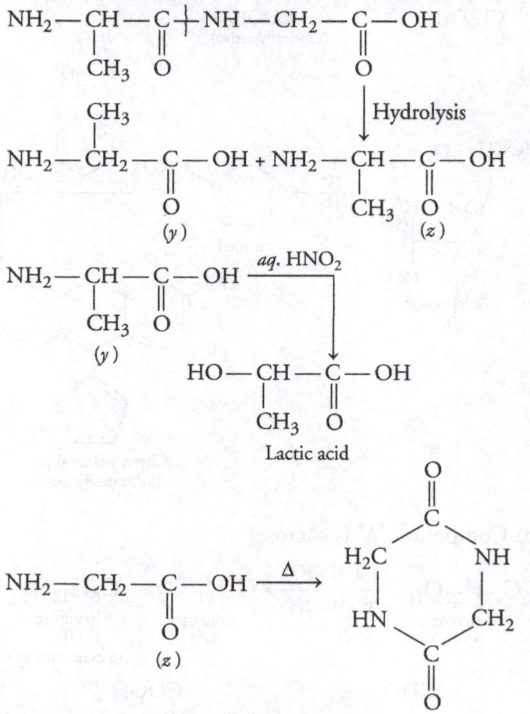

So, the dipeptide *X* is alanine-glycine.

**07.** *(d)* Both Statement I and II are incorrect. Their correct forms are

(i) Protein does not gives β-amino acid on hydrolysis.
(ii) Fibrous protein are not water soluble.

**08.** *(c)* The number of tetrapeptides (sequences) possible involving each of these amino acids.
(Glycine, Alanine, Valine, Leucine); It has three peptides linkage the number of permutation in which they can be arranged.

$$= 4 \times 3 \times 2 \times 1 = 24 \text{ ways.}$$

**09.** *(d)* Complete reaction sequence involved is as follows

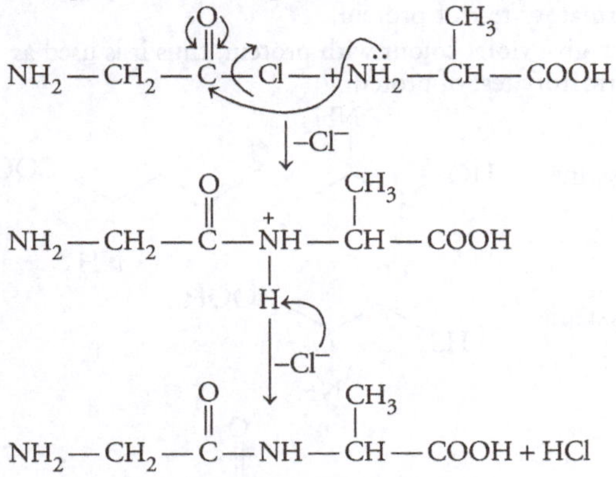

(gly-ala)

**10.** *(d)* Denaturation of proteins is a condition when the unique three-dimensional structure of a protein is exposed to changes. Coagulation of egg gives primary structure of protein, which changes on heating and is known as denaturation of protein. Hence, option (d) is correct.

**11.** *(c)* Ninhydrin test is a test use to detect amino acids. Egg albumin contain protein which is a natural polymer of amino acids. Hence, this will show positive ninhydrin test. Hence, option (c) is correct.

> **• Concept Enhancer**
> Ninhydrin test is used to check whether a given analyte contains amines or α-amino acids. In this test ninhydrin is added to a test solution of the analyte. The formation of deep blue colour indicates the presence of ammonia, primary or secondary amines or amino acids in the analyte.

**12.** *(b)* The coagulation of egg white on boiling is a common example of denaturation.
Due to denaturation, the secondary and tertiary structure are destroyed but primary structure remain intact.
Hence, option (b) is correct.

**13.** *(c)* Statement (c) is not true for arginine. It is a crystalline solid. It has a fairly high melting point and is associated with more than one $pK_a$ value. It has low solubility in benzene.

**14.** *(b)* Lysine, arginine and histidine are basic amino acids with one extra basic functional group, where asparagine has additional amide group, which is not classified as basic functional group.

**15.** *(b)* The correct match is A-III, B-I, C-IV, D-II.
One letter codes of the given natural amino acids are as follows
Glutamic acid → *E*; Glutamine → *Q*
Tyrosine → *Y*; Tryptophan → *W*

**16.** *(d)*

| Natural amino acid | One-letter code |
|---|---|
| (A) Arginine | R |
| (B) Aspartic acid | D |
| (C) Asparagine | N |
| (D) Alanine | A |

The correct match is (A)-IV, (B)-I, (C)-II, (D)-III.

**17.** *(d)* The correct match is A-III, B-I, C-IV, D-II

(A) Sucrose $\xrightarrow[\text{Enzyme}]{\text{(III) Invertase}}$ glucose + fructose

(B) Glucose $\xrightarrow[\text{Enzyme}]{\text{(I) Zymase}}$ $CO_2 + C_2H_5OH$

(C) Starch $\xrightarrow[\text{Enzyme}]{\text{(IV) Diastase}}$ maltose

(D) Protein $\xrightarrow[\text{Enzyme}]{\text{(III) Pepsin}}$ amino acid

**18.** *(d)* Hydrogen bonds, disulphide connections, van der Waal's forces and electrostatic forces of attraction are the primary factors that stabilize the secondary and tertiary structures of projections.

**19.** *(d)* Both A and R are correct but R is not the correct explanation of A. In α-halocarboxylic acid, the halogen group can be easily replaced by amino group but in case of alkylhalides replacement is not easy. Also amino acids act as Zwitter ion in aqueous solution but this has no relation with replacement reactions.

**20.** *(b)* Let's take an example of two amino acids

$A = H_2N—\underset{R_1}{\underset{|}{CH}}—COOH$ ; $B = H_2N—\underset{R_2}{\underset{|}{CH}}—COOH$

Tripeptides are three amino acid molecules joined together with the elimination of water molecule, leads to the formation of $\left[ H—N—\overset{\overset{O}{\|}}{C}— \right]$ bonds.

With two amino acids, four combinations of cyclic tripeptides are possible i.e., *AAB*, *ABB*, *AAA* and *BBB*. Hence, the answer is four.

**21.** *(d)* Action of enzymes can be altered by temperature, pH, presence of activator, coenzymes presence of inhibitors, poisons etc. So, statement I is false. Inhibitors interact with the active sites of enzymes and form a strong covalent bond with the enzyme. So, statement II is true.

**22.** *(b)* α-helix secondary structure is stabilised by hydrogen bonding exists between carbonyl oxygen (CO) and amino group (NH) of every third residue in the adjacent turn of the helix.

**23.** *(b)* Insulin and albumin are globular protein. Fibrous proteins are generally composed of long and narrow strands and have structural role while globular proteins generally have more compact and rounded shape and involved in metabolic functions.

**24.** *(d)* α-helix structure of proteins is stabilised by H-bonds between hydrogen of —NH and oxygen of —C=O group in amino acids.

**25.** *(d)* Tyrosine is not an essential amino acid and remaining all are essential amino acid.
Amino acids which are synthesised by the body, are called non-essential.

**26.** *(d)* The most basic amino acid is histidine. There are three basic amino acids (amino acids with more than one basic nitrogen containing group). These are lysine, arginine and histidine. Lysine have ∝ ε-amino group. On the other hand, histidine is an α-amino acid with an imidazole functional group.

**27.** *(d)* Amino acid molecules can be represented as,

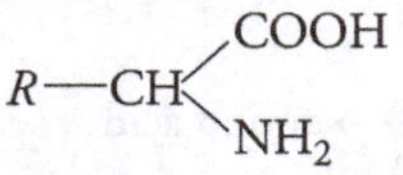

Nature of the '*R*' group will determine the basicity (hence, p$K_a$) of an amino acid.

| '*R*' in the amino group | Nature of *R* | Nature of the amino acids |
|---|---|---|
| 1. $—(CH_2)_3—NH—C(=\overset{+}{N}H_2)NH_2$ (Arginine : Arg) | Basic | More basic (due to the presence of acetamidine group) |
| 2. $—CH_2—C(=O)O^-$ (Aspartic acid: Asp) | Acidic | Acidic |
| 3. —H (Glycine) : Gly | Neutral | Neutral |
| 4. $—(CH_2)_4—\overset{+}{N}H_3$ (Lysine : Lys) | Basic | Basic |

The basicity order of given amino acids in aqueous solution is as follows:

Asp < Gly < Lys < Arg ⟶

Thus, order of p$K_a$ values is also,

| | Asp | < Gly | < Lys | < Arg |
|---|---|---|---|---|
| p$K_a$ value : | (3) | (6) | (9.8) | (10.8) |

**28.** *(d)* The structure of histamine is given below :

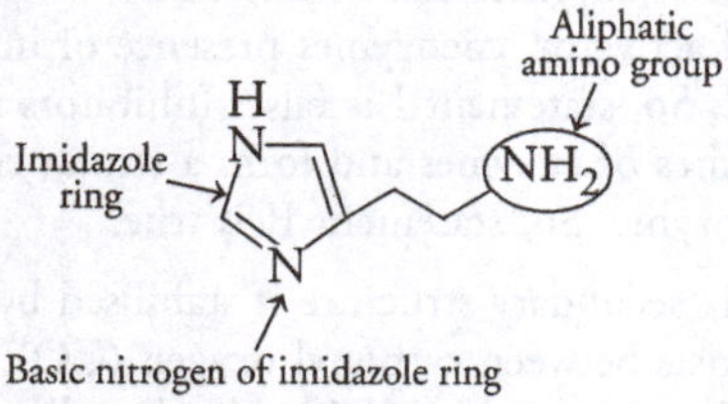


In blood with pH around 7.4 the aliphatic amino group of histamine become protonated to give a single charged cation as shown below

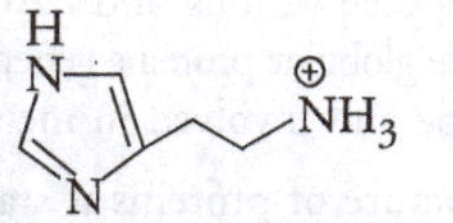

**29.** *(b)* The isoelectric point of aspartic acid can be calculated by using the formula

$$pI = \frac{pK_1 + pK_2}{2} = \frac{1.88 + 3.65}{2} \approx 2.77$$

**30.** *(c)* 'Ruhemann's Purple' is ninhydrin, which is used for the confirmatory test of protein.
Which give violet colour with protein, thus it is used as confirmatory test of protein.

**31.** *(b)*

Cystine

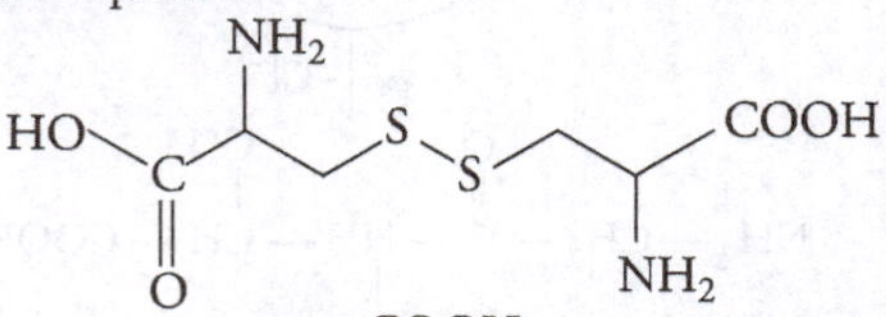

Cysteine

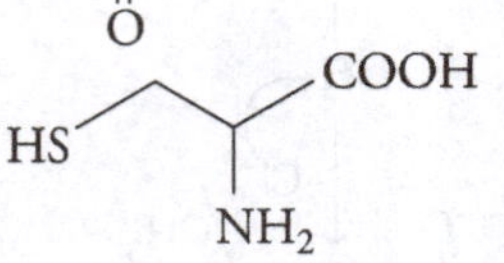

Methionine

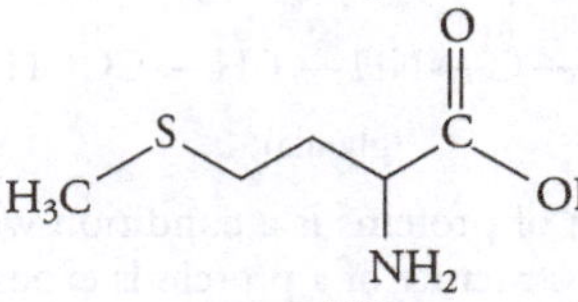

Cytosine

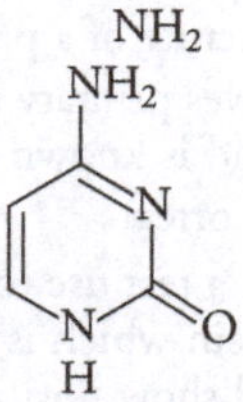

**32.** *(a)* Biuret test is characteristically given by the compound having $—\overset{O}{\overset{\|}{C}}—NH—$ functional group. Thus, it is not given by carbohydrates.

**33.** *(c)* The complete reaction is as follows,

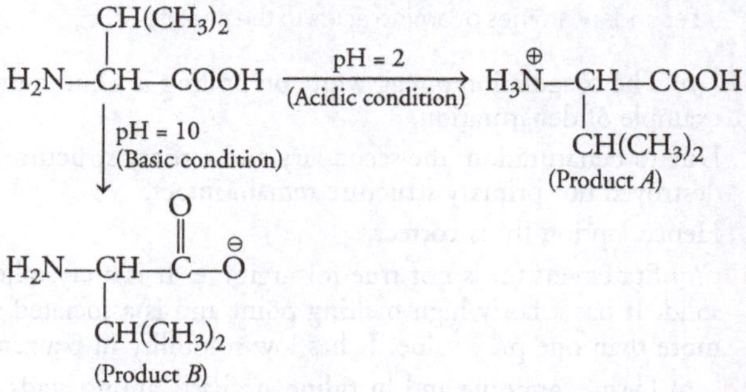

**34.** *(d)* Among the given statements, option 'd' is correct. While a, b, and c are incorrect. Their correct forms are
Isoleucine has 2 chiral centre.
Glycine is optically inactive.
Aspartic acid also contain COOH group at the side chain.
Cysteine easily dimmerise due to free SH group.

**35.** *(b)* The essential amino acids from the given options are valine, lysine and threonine. i.e., A, C and D only.

**36.** *(a)* Statements given is A and C are correct, while B and D are incorrect. It is because
Enzymes are biocatalysts and are very specific for a particular reaction and for a particular substrate. Almost all enzymes are globular proteins. Enzyme maltase hydrolyses maltose glucose.

$$\text{Maltose} \xrightarrow{\text{Maltase}} \text{Glucose}$$

**37.** *(c)* Sulphur containing amino acids are cysteine (B) and methionine (D).

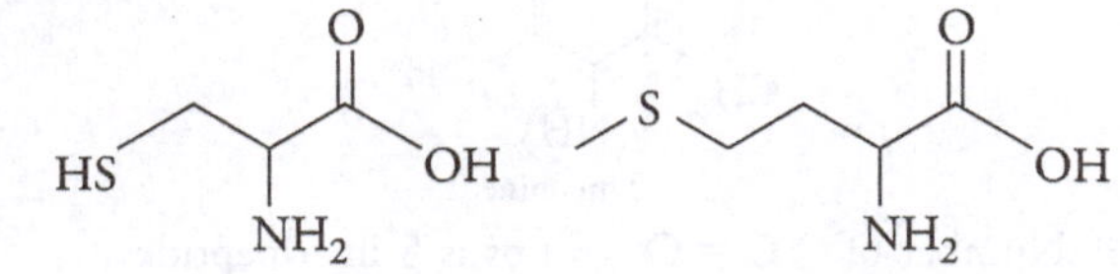


Cysteine  Methionine

**38.** *(b)* $CH_3-\underset{\overset{|}{CH_3}}{\overset{}{CH}}-CH_2-\underset{\underset{NH_2}{|}}{CH}-COOH$

Proteins on hydrolysis gives α-amino acid because amino acids are the building block of proteins. Among the given options, only in option (b), contains α-amino acids.

**39.** *(b)* The peptide that gives positive ceric ammonium nitrate and carbylamine tests is ser-lys. The structures of serine and lysine are,

$$HOCH_2-\underset{\underset{NH_2}{|}}{CH}-COOH;\quad H_2N(CH_2)_4-\underset{\underset{NH_2}{|}}{CH}-COOH$$

Serine  Lysine

**40.** *(d)* The product '*P*' formed from the given reaction is given in option (d).

**41.** *(c)* Formation of the tripeptide (Val-Ser-Thr) can be shown as:

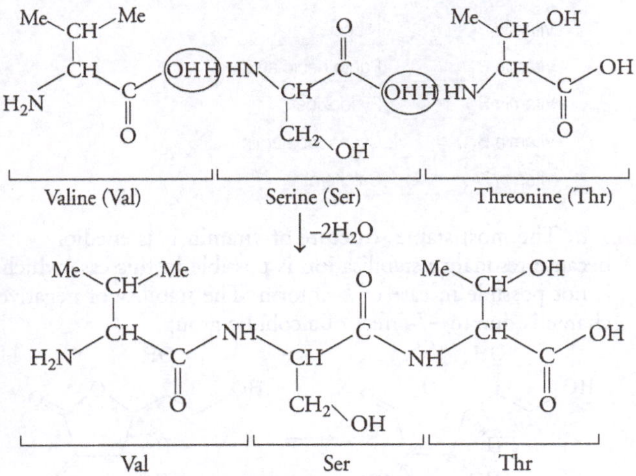


**42.** *(c)* Glycine $H_2C\begin{matrix}NH_2\\COOH\end{matrix}$ is the only α-amino acetic acid with no chiral carbon, thus it is optically inactive.

**43.** *(4)* Total number of essential amino acids are four i.e. arginine, phenylalanine, histidine and valine.

**Time Saver Tech**

You can remember the name of essential amino acid by memorising one sentence and keep the first letter in mind.

Any Help in Learning These Little Molecules Proves Truly Valuable

Histidine, Leucine, Lysine, Phenylalanine, Valine
Arginine, Isoleucine, Threonine, Methionine, Tryptophan

**44.** *(9)* The structure of tyrosine is,

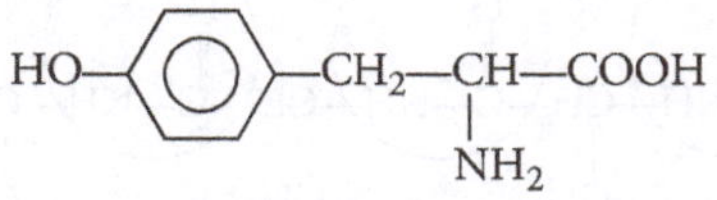


Number of carbon atom = 9

**45.** *(6)* If 1 amino acid gives 2 tripeptide
2 amino acid gives 4 tripeptide
3 amino acid gives 6 tripeptide.
If different amino acids are *X*, *Y* and *Z*, then their possible combination are as follows
*XYZ, XZY, YXZ, YZX, ZXY, ZYX*

**46.** *(6)* 3 moles of glycine (*G*), two moles of leucine (*L*) and two moles of valine (*V*) per moles of peptide show that peptide chain has 7 amino acid that is a heptapeptide chain. The number of peptide linkage is always one less than total amino acids. Hence, the number of peptide linkage in it are 6.

**47.** *(8)* The total number of tripeptides possible by mixing of valine and proline is $2^3$ i.e., 8.

(i) Val-val-val  (ii) Pro-pro-pro
(iii) Val-pro-pro  (iv) Pro-val-pro
(v) Val-val-pro  (vi) Val-pro-val
(vii) Pro-pro-val  (viii) Pro-val-val

**48.** *(1)* In a tetrapeptide, there are 4 amino acids linked with 3 peptide bonds. So, the number of amino acids − number of peptide bond will be $(4-3)=1$.

**49.** *(4)* Alanylglycylleucylalanylvaline

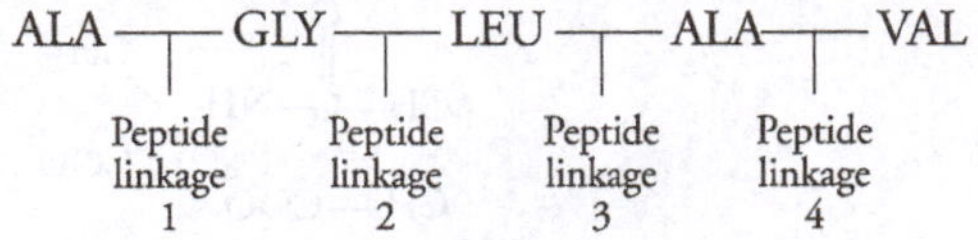


So, total four peptide linkages are present.

**50.** *(2)* Biuret test is used for the identification of peptide bond. Dipeptides and free amino acids do not give biuret test. Thus, glycine and glycylalanine do not give this test.

**51.** *(3)* Peptide synthesised by reaction of glycine, leucine, aspartic acid and histidine is as follows :

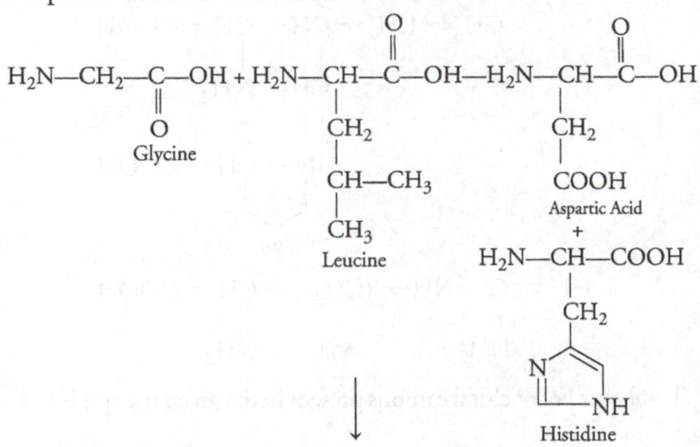

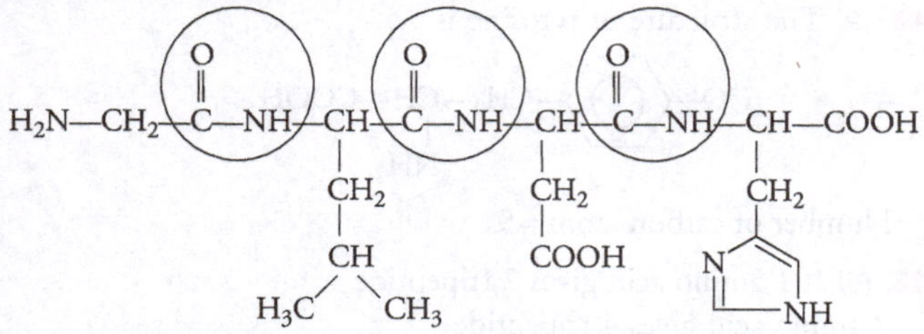

3 peptide linkage $(-\overset{O}{\overset{\|}{C}}-NH-)$ are present.
Hence, answer is 3.

**52.** *(4)* The structures of tetrapeptide **Gly-Glu-Asp-Tyr** is as follows

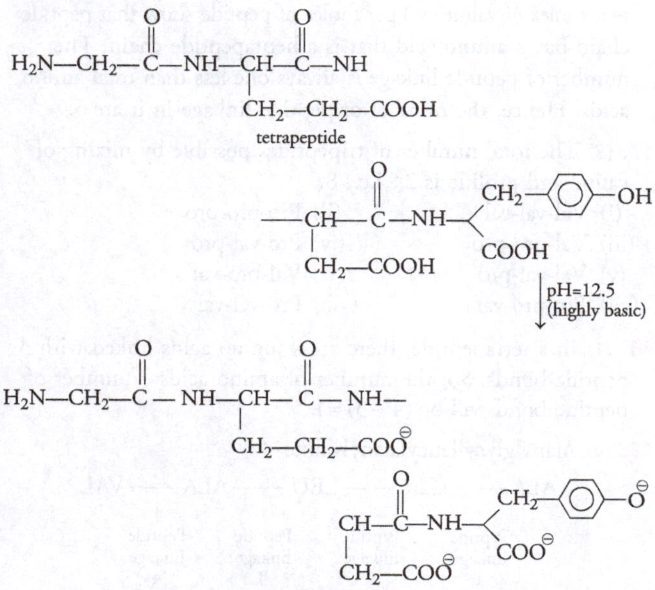


All protons ($H^+$) from carboxylic group and phenol are lost neutralised in basic medium)
Total negative charge = 4.

**53.** *(4)* The amino acids present in the given tripeptide Ile-Arg-pro are isoleucine, arginine and proline.

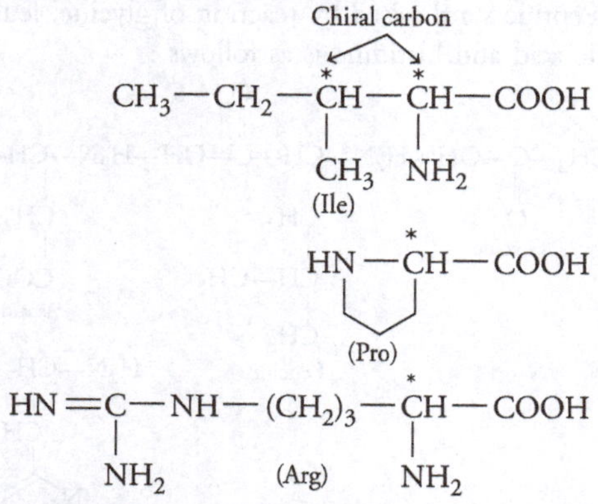


Total number of chiral carbons present in the given tripeptide is 4.

**54.** *(2)* Two chiral centres are present in threonine.

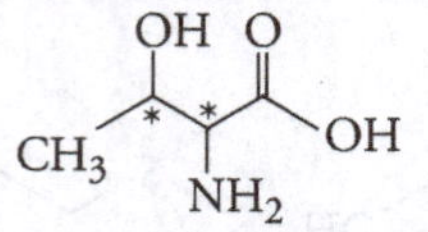

Threonine

**55.** *(5)* Number of $>C{=}O$ groups is 5 in tripeptide Asp - Glu - Lys.

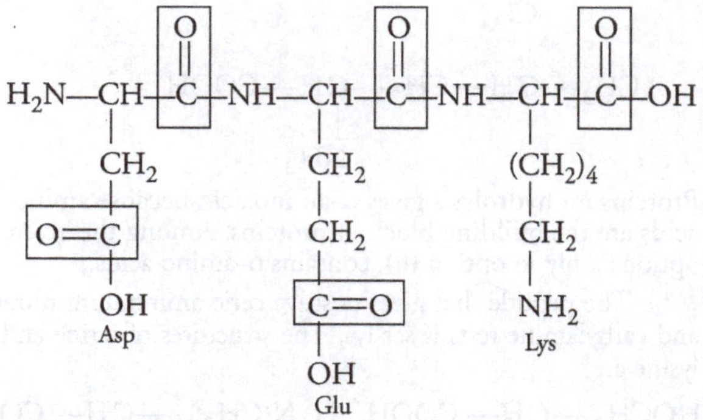


## Topic 3

**01.** *(b)* Ascorbic acid is vitamin C.

**Concept Enhancer**

| | |
|---|---|
| Vitamin A | Retinol |
| Vitamin $B_1$ | Thiamine |
| Vitamin $B_2$ | Niacin |
| Vitamin $B_5$ | Pantothenic acid |
| Vitamin $B_6$ | Pyridoxine |
| Vitamin $B_{12}$ | Cyanocobalamin |
| Vitamin D | Calciferol |

**02.** *(d)* The most stable structure of vitamin *C* is enediol because resonance stabilisation is possible in this case which is not possible in case of *keto* form. The stability of negative charge is due to –*I*-effect of alcoholic group.

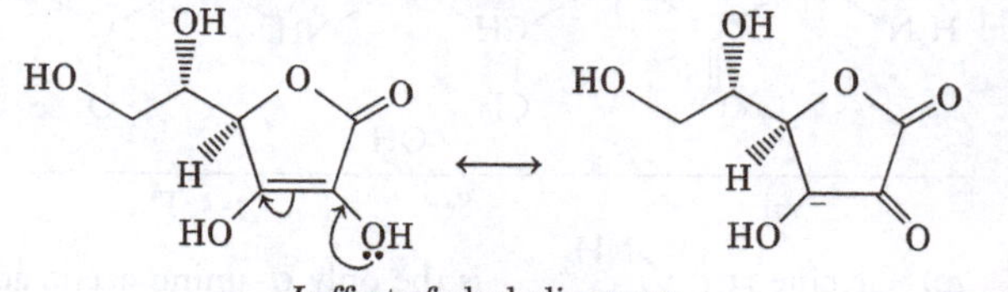

–*I*-effect of alcoholic group

**03.** *(d)* Vitamin $B_{12}$ is the water soluble vitamin that is not excreted in urine easily as excess of it is stored in liver.

**04.** *(d)* Thiamine is also called vitamin $B_1$ and pyridoxine is also known as vitamin $B_6$.

**05.** *(a)* Vitamin K deficiency causes increase in blood clotting time because their blood does not have enough vitamin K to form a clot. The bleeding can occur anywhere i.e. inside or outside the body.

**06.** *(d)* Deficiency of vitamin K increases blood clotting time. So, vitamin K is helpful in blood clotting. It is a fat-soluble vitamin.

**07.** *(a)* Vitamin B and C are water soluble while vitamin A, D, E and K are fat soluble or water insoluble.

**08.** *(a)* DNA contains four nitrogenous pyrimidine bases, adenine, guanine, cytosine, thymine. While quinoline is an alkaloid, hence, it is not present in DNA.

**09.** *(c)* In DNA, cytosine and thymine are pyrimidine bases.

**10.** *(c)* The four bases of DNA are A, T, C, G

A = Adenine C = Cytosine

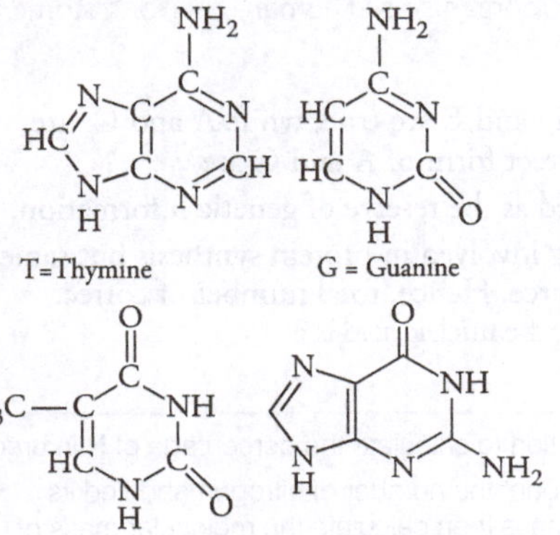

T=Thymine G = Guanine

**11.** *(c)*

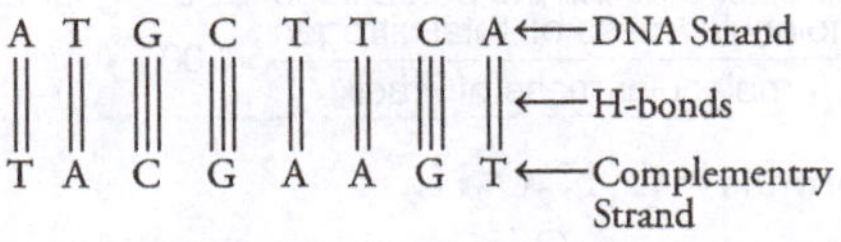


Adenine base pairs with thymine to form 2 hydrogen bond and cytosine base pair with guanine to form three H-bond. The complementary strand is TACGAAGT.
Hence, option (c) is correct.

**12.** *(b)* Nucleotides are joined together by phosphodiester linkage between 5′ and 3′ carbon atom of pentose sugar.

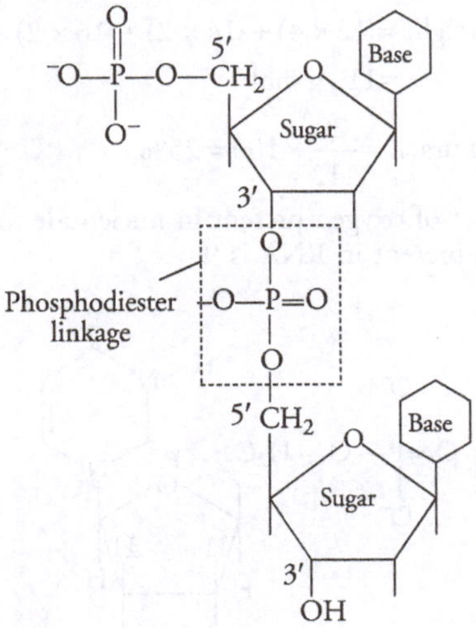


> • **Concept Enhancer**
> Usually phosphodiester bond is known to be present in DNA and RNA but it occur in other biomolecules also. Acyl carrier proteins, phospholipid and cyclic form of GMP and AMP.

**13.** *(b)*

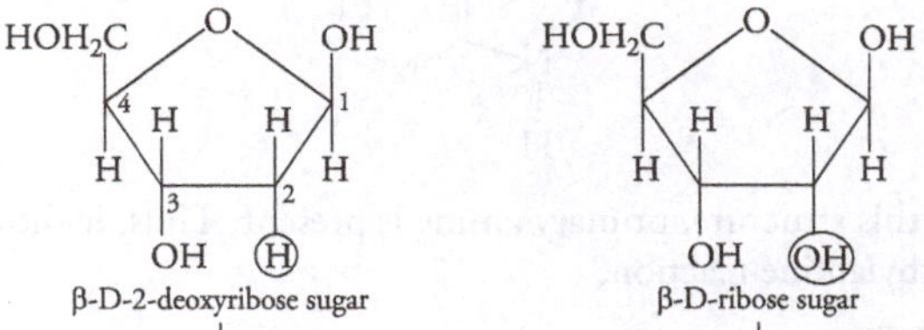

β-D-2-deoxyribose sugar → Sugar moiety in DNA
β-D-ribose sugar → Sugar moiety in RNA

**14.** *(c)* Given structure of compound *A* is thymine, so its complementary base will be adenine.

**15.** *(c)* The correct structure of cytosine is,

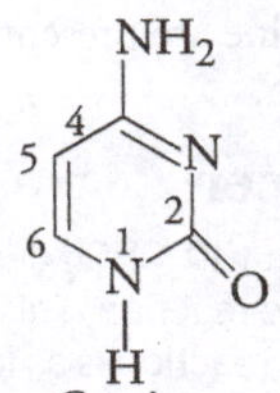

Cytosine

**16.** *(b)* Vitamin A and vitamin D are stored in our body relatively for longer duration. These are insoluble in water but are soluble in fats and oils. Hence, these are called fat soluble vitamins.

**17.** *(b)* RNA does not have double stranded α-helix structure. Helixes present in RNA are single-stranded but sometimes they fold back on themselves to form a double helix structure. RNA usually does not replicate.

**18.** *(a)* RNA contains, adenine (A), guanine (G), cytosine (C) and uracil (U). In the given options,

(a) 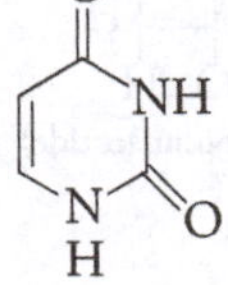 is uracil (present in RNA only)

**19.** *(a)* The correct structure of adenosine is given in option (a).

**20.** *(b)* In RNA molecule, sugar moeity is β-D-ribose in which —OH group is present at 2nd carbon whereas in DNA, it is β-D-2-deoxyribose in which —OH group is absent at 2nd carbon.

**21.** *(c)* Vitamin D, E, K, A are fat soluble. While Vitamin $B_1$, $B_{12}$ and C are water soluble.

**22.** *(c)* The correct match is, A-III, B-IV, C-II, D-I.

**23.** *(d)* The correct match is A-II, B-III- C-I, D-IV.

**24.** *(a)* Adenine and lysine will react with $CHCl_3$ + alc. KOH due to presence of primary amine.

**Adenine**

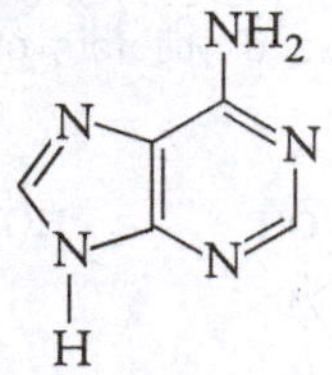

In this structure, primary amine is present. Thus, it shows carbylamine reaction.

**Lysine**

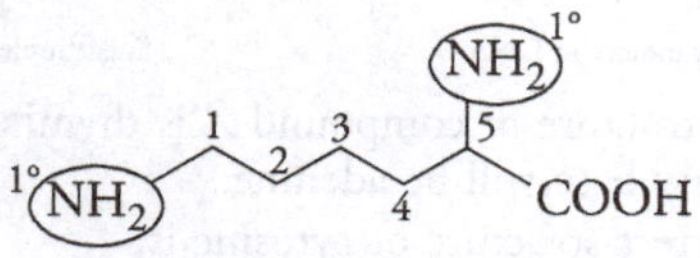

Here, 2 primary amine are present, thus it show carbylamine reaction.

> **• Concept Enhancer**
>
> Adenine and lysine react with $CHCl_3$ and Alc. KOH. This reaction is known as carbylamine reaction. This reaction is given by only primary amine. General reaction is as follows.
>
> $$\underset{1^\circ \text{amine}}{R\text{—}NH_2} + CHCl_3 + 3KOH \longrightarrow \underset{\text{isocyanide}}{R\text{—}N{\equiv}C} + 3KCl + 3H_2O$$

**25.** *(b)* The correct match is (i)-C, (ii)-A, (iii)-D, (iv)-B

**26.** *(b)* In both DNA and RNA, the heterocyclic base and phosphodiester linkage are at $C_1'$ and $C_5'$ respectively of the sugar molecule.

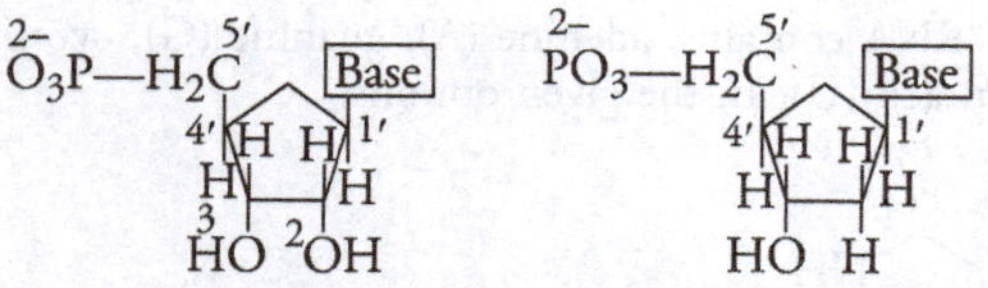

(Ribonucleotide) (Deoxyribonucleotide)

**27.** *(65)* Molecular formula of thyroxine is $[C_{15}H_{11}O_4NI_4]$

Molecular mass =

$C \rightarrow 15 \times 12 = 180,$

$O \rightarrow 16 \times 4 = 64$

$H \rightarrow 1 \times 11 = 11,$

$N \rightarrow 14 \times 1 = 14$

$I \rightarrow 127 \times 4 = 508$

Molecular mass of thyroxine = 777

$$\% \text{ of iodine} = \frac{508}{777} \times 100 = 65.38\%$$

Nearest integer = 65%.

**28.** *(33)* Adenine and thymine forms two hydrogen bond while guanine and cytosine form three hydrogen bond.

$5'$-G-G-C-A-A-A-T-C-G-G-C-T-A-$3'$

In given DNA strand total seven guanine and cytosine base which form 21 + 1 bond and six adenine and thymine base form total 12H-bond. Thus, total 33 hydrogen bonds are there.

**29.** *(5)* Vitamins A, D, E, K and $B_{12}$ are stored in liver and adipose tissue.

Hence, total number of vitamin stored in body is 5.

> **• Concept Enhancer**
>
> Usually, vitamin B and C cannot be stored in our body, but vitamin $B_{12}$ is an exception among vitamin B complex. It is synthesised by microorganisms in human colon and stored in liver.

**30.** *(3)* Statement B, D and E are true, while A and C are incorrect. The corrrect form of A and C are :

- RNA is not regarded as the reserve of genetic information.
- DNA is not directly involved in protein synthesis but same as the information source. Hence, total number of correct statements, regarding the nucleic acid is 3.

**31.** *(25)*

> **key idea** In this question to calculate the percentage of N in uracil. First of all count the number of nitrogen and find its molecular mass then calculate the molecular mass of uracil. After that find the percentage, by using formula
>
> $$\frac{\text{molecular mass of total nitrogen}}{\text{molecular mass of uracil}} \times 100$$

Molecular formula = $C_4N_2O_2H_4$

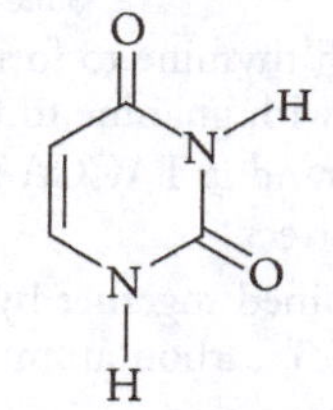

Molecular weight $= (12 \times 4) + (14 \times 2) + (16 \times 2) + (4 \times 1)$

$= 112 \text{ g mol}^{-1}$

$$\therefore \% \text{ of N in uracil} = \frac{28}{112} \times 100 = 25\%$$

**32.** *(9)* Number of oxygen present in nucleotide formed from a base, that is present in RNA is 9.

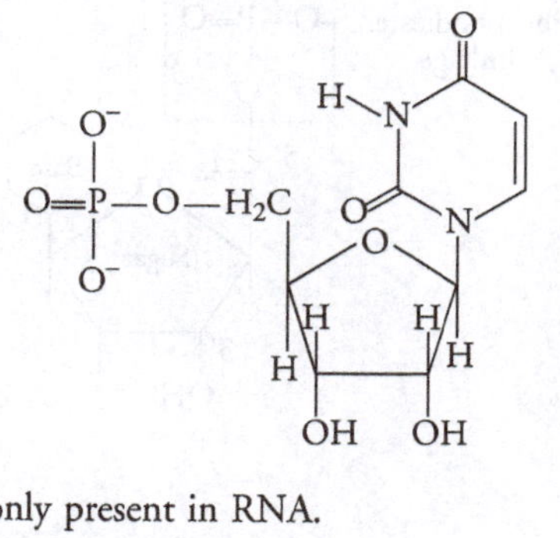

Uracil is only present in RNA.

# C-24

JEE Main Chapterwise Topicwise Chemistry

In JEE Main there is usually 0-1 question per paper are visible from this chapter. In past 3 years, 39 conceptual and 7 numerical questions were asked of moderate difficulty level from this chapter.

**For detailed analysis scan QR code**

# PRINCIPLES RELATED TO PRACTICAL CHEMISTRY

## Topic 1 Detection of Extra Elements and Functional Groups

### Level 1

**01** Lassaigne's test is used for detection of *[2024, 1 Feb Shift-II]*

(a) nitrogen and sulphur only
(b) phosphorus and halogens only
(c) nitrogen, sulphur and phosphorus only
(d) nitrogen, sulphur, phosphorus and halogens

**02** Phenolic group can be identified by a positive *[2024, 27 Jan Shift-II]*

(a) Tollen's test (b) carbylamine test
(c) phthalein dye test (d) Lucas test

**03** The formula of the purple colour formed in Lassaigne's test for sulphur using sodium nitroprusside is *[2022, 28 June Shift-I]*

(a) $NaFe[Fe(CN)_6]$ (b) $Na[Cr(NH_3)_2(NCS)_4]$
(c) $Na_2[Fe(CN)_5(NO)]$ (d) $Na_4[Fe(CN)_5(NOS)]$

**04** Which one of the following test used for the identification of functional groups in organic compounds does not use copper reagent ? *[2021, 27 Aug Shift-II]*

(a) Barfoed's test (b) Seliwanoff's test
(c) Benedict's test (d) Biuret test for peptide bond

**Imp.** Test for Detection of Alcohol & Amine

**05** Ceric ammonium nitrate and $CHCl_3$ /alc. KOH are used for the identification of functional groups present in ......... and ......... respectively. *[2021, 26 Feb Shift-II]*

(a) alcohol, phenol (b) amine, alcohol
(c) alcohol, amine (d) amine, phenol

**06** Which one of the following complexes is violet in colour? *[2021, 26 Aug Shift-I]*

(a) $Fe_4[Fe(CN)_6]_3 \cdot H_2O$ (b) $[Fe(CN)_5NOS]^{4-}$
(c) $[Fe(SCN)_6]^{4-}$ (d) $[Fe(CN)_6]^{4-}$

**07** Acidic ferric chloride solution on treatment with excess of potassium ferrocyanide gives a prussian blue coloured colloidal species. It is *[2021, 27 Aug Shift-I]*

(a) $HFe[Fe(CN)_6]$ (b) $K_5Fe[Fe(CN)_6]_2$
(c) $Fe_4[Fe(CN)_6]_3$ (d) $KFe[Fe(CN)_6]$

**Imp.** Silver Nitrate Test for Halogen Detection

**08** Which of the following compound is added to the sodium extract before addition of silver nitrate for testing of halogens? *[2021, 25 Feb Shift-II]*

(a) Hydrochloric acid (b) Sodium hydroxide
(c) Nitric acid (d) Ammonia

**09** Which of the following test cannot be used for identifying amino acids? *[2019, 10 Jan Shift-II]*

(a) Biuret test (b) Barfoed test
(c) Ninhydrin test (d) Xanthoproteic test

**10** Sodium extract is heated with concentrated $HNO_3$ before testing for halogens because *[JEE Main 2016]*

(a) $Ag_2S$ and AgCN are soluble in acidic medium
(b) silver halides are totally insoluble in nitric acid
(c) $S^{2-}$ and $CN^-$, if present, are decomposed by conc. $HNO_3$ and hence do not interfere in the test
(d) Ag reacts faster with halides in acidic medium

**11** Beilstein test is used for the estimation of which one of the following elements? *[JEE Main 2012]*

(a) S (b) Cl (c) C and H (d) N

**12** Copper wire test for halogens is known as *[AIEEE 2012]*
(a) Duma's Test (b) Beilstein's Test
(c) Liebig's Test (d) Lassaigne's Test

**13** Biuret test is not given by *[AIEEE 2010]*
(a) proteins (b) carbohydrates
(c) polypeptide (d) urea

**14** The compound formed in the positive test for nitrogen with the Lassaigne solution of an organic compound is *[AIEEE 2004]*
(a) $Fe_4[Fe(CN)_6]_3$ (b) $Na_3[Fe(CN)_6]$
(c) $Fe(CN)_3$ (d) $Na_4[Fe(CN)_5NOS]$

## Level 2

**15** Number of molecules from below which cannot give iodoform reaction is

Ethanol, Isopropyl alcohol, Bromoacetone, 2-Butanol, 2-Butanone, Butanal, 2-Pentanone, 3-Pentanone, Pentanal and 3-Pentanol.
*[2025, 3 April, Shift-I]*

(a) 2 (b) 5 (c) 4 (d) 3

**16** Which of the following compounds will give silver mirror with ammoniacal silver nitrate?
A. Formic acid B. Formaldehyde
C. Benzaldehyde D. Acetone

Choose the correct answer from the options given below. *[2024, 9 April Shift-II]*
(a) A only (b) A, B and C only
(c) C and D only (d) B and C only

**17** Appearance of blood red colour, on treatment of the sodium fusion extract of an organic compound with $FeSO_4$ in presence of concentrated $H_2SO_4$ indicates the presence of elements *[2024, 29 Jan Shift-I]*
(a) N and S (b) N
(c) Br (d) S

**Imp.** Detection of Extra Elements (P, N, S, Halogen)

**18** Match List-I with List-II.

| List-I (Element detected) | List-II (Reagent used/Product formed) |
|---|---|
| A. Nitrogen | I. $Na_2[Fe(CN)_5NO]$ |
| B. Sulphur | II. $AgNO_3$ |
| C. Phosphorus | III. $Fe_4[Fe(CN)_6]_3$ |
| D. Halogen | IV. $(NH_4)_2MoO_4$ |

Choose the correct answer from the options given below *[2023, 6 April Shift-I]*
(a) A-II, B-IV, C-I, D-III (b) A-III, B-I, C-IV, D-II
(c) A-IV, B-II, C-I, D-III (d) A-II, B-I, C-IV, D-III

**19** During halogen test, sodium fusion extract is boiled with concentrated $HNO_3$ to *[2022, 25 June Shift-II]*
(a) remove unreacted sodium
(b) decompose cyanide or sulphide of sodium
(c) extract halogen from organic compound
(d) maintain the pH of extract

## Level 3

**20** $A \xrightarrow[\text{(ii) } H_3O^+]{\text{(i) NaOH}} B \xrightarrow[\text{(ii) } H_2SO_4,\ \Delta]{\text{(i) EtOH}} C$

'*A*' shows positive Lassaigne's test for N and its molar mass is 121.
'*B*' gives effervescence with aq. $NaHCO_3$.
'*C*' gives fruity smell.
Identify *A*, *B* and *C* from the following.
*[2025, 8 April Shift-II]*

(a) *A* = 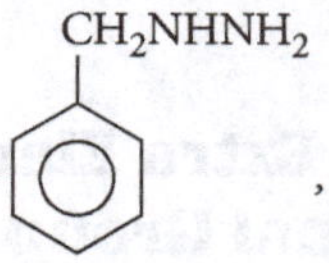 , *B* = 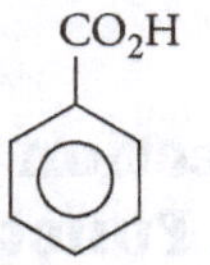 , *C* = 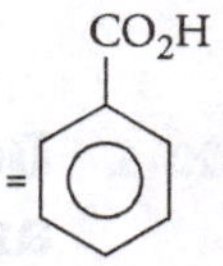

(b) *A* = 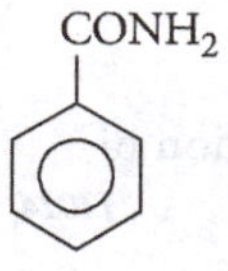 , *B* = 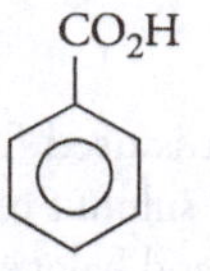 , *C* = 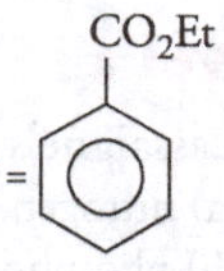

(c) *A* = 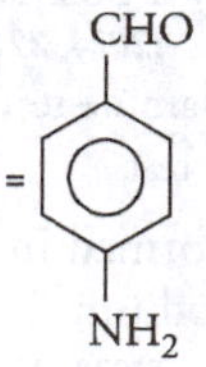 , *B* = 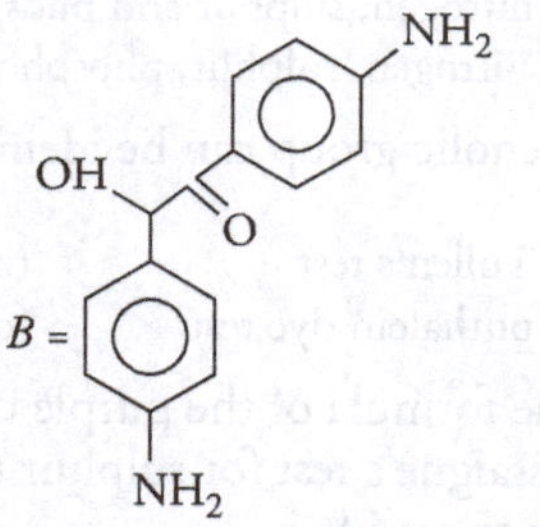 , *C* = 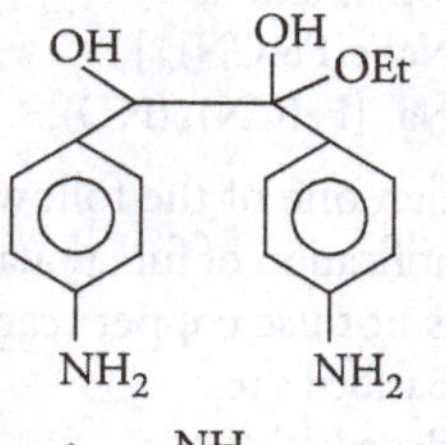

(d) *A* = 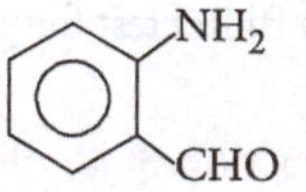 , *B* = 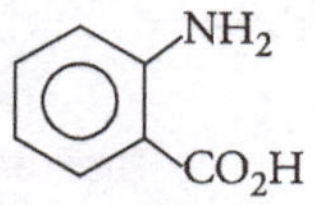 , *C* = 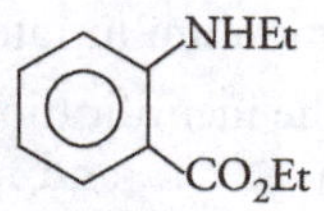

**21** Which among the following compounds give yellow solid when reacted with NaOI/NaOH? *[2025, 4 April, Shift-II]*

(A) 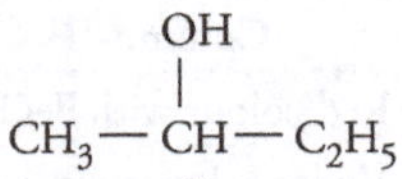 (B) $CH_3-CH_2-CH_2-OH$

(C) 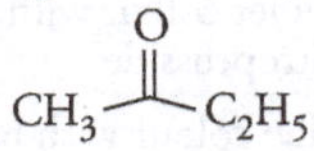 (D) 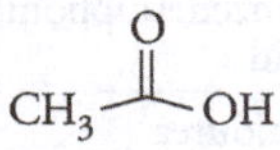

(E) 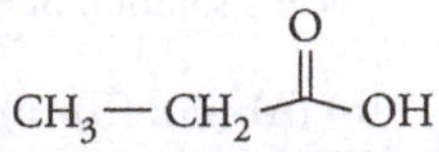

Choose the correct answer from the options given below

(a) (A), (C) and (D) only (b) (B), (C) and (E) only
(c) (C) and (D) only (d) (A) and (C) only

**22** Match the List-I with List-II *[2025, 8 April Shift-II]*

| List-I (Reagent) | List-II (Functional group detected) |
|---|---|
| A. Sodium bicarbonate solution | I. Double bond unsaturation |
| B. Neutral ferric chloride | II. Carboxylic acid |
| C. Ceric ammonium nitrate | III. Phenolic - OH |
| D. Alkaline $KMnO_4$ | IV. Alcoholic - OH |

Choose the correct answer from the options given below

(a) A-II, B-III, C-IV, D-I (b) A-III, B-II. C-IV, D-I
(c) A-II, B-III, C-I, D-IV (d) A-II, B-IV, C-III, D-I

**23** Match List-I with List-II. *[2024, 9 April Shift-II]*

| List-I (Test) | List-II (Observation) |
|---|---|
| A. $Br_2$ water test | I. Yellow orange or orange red precipitate formed |
| B. Ceric ammonium nitrate test | II. Reddish orange colour disappears |
| C. Ferric chloride test | III. Red colour appears |
| D. 2, 4-DNP test | IV. Blue, green, violet or red colour appear |

Choose the correct answer from the options given below.

(a) A-III, B-IV, C-I, D-II (b) A-I, B-II, C-III, D-IV
(c) A-II, B-III, C-IV, D-I (d) A-IV, B-I, C-II, D-III

**24** Match List-I with List-II. *[2024, 8 April Shift-II]*

| List-I (Test) | List-II (Identification) |
|---|---|
| A. Bayer's test | I. Phenol |
| B. Ceric ammonium nitrate test | II. Aldehyde |
| C. Phthalein dye test | III. Alcoholic —OH group |
| D. Schiff's test | IV. Unsaturation |

Choose the correct answer from the options given below:

(a) A-IV, B-I, C-II, D-III
(b) A-III, B-I, C-IV, D-II
(c) A-II, B-III, C-IV, D-I
(d) A-IV, B-III, C-I, D-II

**25** Match List-I with List-II. *[2024, 8 April Shift-I]*

| List-I (Compound) | List-II (Colour) |
|---|---|
| A. $Fe_4[Fe(CN)_6]_3 \cdot xH_2O$ | I. Violet |
| B. $[Fe(CN)_5NOS]^{4-}$ | II. Blood red |
| C. $[Fe(SCN)]^{2+}$ | III. Prussian blue |
| D. $(NH_4)_3PO_4 \cdot 12MoO_3$ | IV. Yellow |

Choose the correct answer from the options given below:

(a) A-II, B-III, C-IV, D-I (b) A-IV, B-I, C-II, D-III
(c) A-I, B-II, C-III, D-IV (d) A-III, B-I, C-II, D-IV

**Imp.** **Reagents Used for Functional Group Detection**

**26** Match List-I with List-II

| List-I (Reagents used) | List-II (Compound with Functional group detected) |
|---|---|
| A. Alkaline solution of copper sulphate and sodium citrate | I. 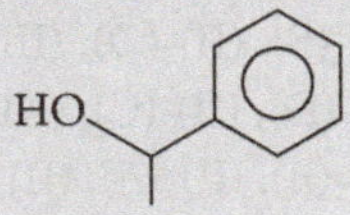 |
| B. Neutral $FeCl_3$, solution | II. 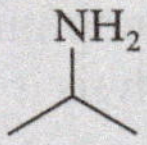 |
| C. Alkaline chloroform solution | III. 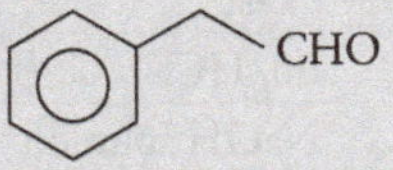 |
| D. Potassium iodide and sodium hypochlorite | IV. 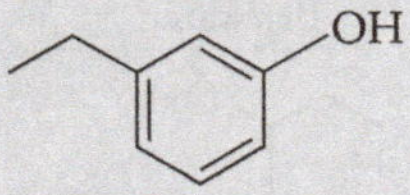 |

Choose the correct answer from the options given below: *[2023, 8 April Shift-I]*

(a) A-III, B-IV, C-I, D-II (b) A-II, B-IV, C-III, D-I
(c) A-IV, B-I, C-II, D-III (d) A-III, B-IV, C-II, D-I

**27** A compound '*X*' is acidic and it is soluble in NaOH solution, but insoluble in $NaHCO_3$ solution. Compound '*X*' also gives violet colour with neutral $FeCl_3$ solution. The compound '*X*' is *[2022, 29 July Shift-I]*

(a) 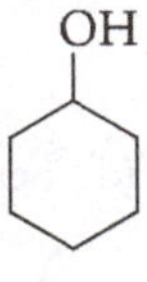 (b) 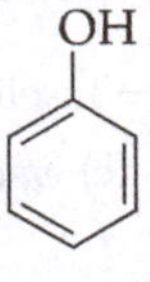 (c) 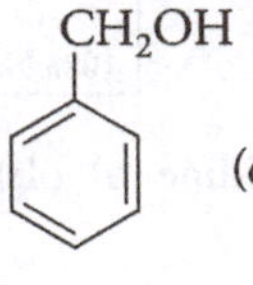 (d) 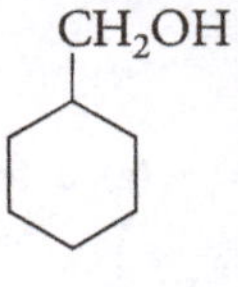

**28** Given below are two statements.

**Statement I :** In Lassaigne's test, when both nitrogen and sulphur are present in an organic compound, sodium thiocyanate is formed.

**Statement II :** If both nitrogen and sulphur are present in an organic compound, then the excess of sodium used in sodium fusion will decompose the sodium thiocyanate formed to give NaCN and $Na_2S$.

In the light of the above statements, choose the most appropriate answer from the options given below. *[2022, 26 June Shift-I]*

(a) Both Statement I and Statement II are correct.
(b) Both Statement I and Statement II are incorrect.
(c) Statement I is correct but Statement II is incorrect.
(d) Statement I is incorrect but Statement II is correct.

**29** Match the following. *[2020, 6 Sep Shift-II]*

| Column I (Test/Method) | | Column II (Reagent) | |
|---|---|---|---|
| (I) | Lucas test | (A) | $C_6H_5SO_2Cl$/ aq. KOH |
| (II) | Dumas method | (B) | $HNO_3/AgNO_3$ |
| (III) | Kjeldahl's method | (C) | $CuO/CO_2$ |
| (IV) | Hinsberg test | (D) | Conc. HCl and $ZnCl_2$ |
| | | (E) | $H_2SO_4$ |

(a) (I)-(D), (II)-(C), (III)-(B), (IV)-(A)
(b) (I)-(B), (II)-(D), (III)-(E), (IV)-(A)
(c) (I)-(D), (II)-(C), (III)-(E), (IV)-(A)
(d) (I)-(B), (II)-(A), (III)-(C), (IV)-(D)

**30** The organic compound that gives following qualitative analysis is *[2019, 9 April Shift-I]*

| | Test | Inference |
|---|---|---|
| (A) | dil. HCl | Insoluble |
| (B) | NaOH solution | Soluble |
| (C) | $Br_2$/water | Decolourisation |

(a) 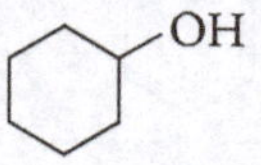

(b) 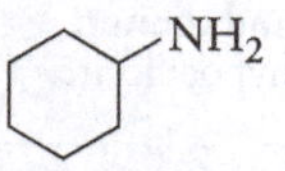

(c) 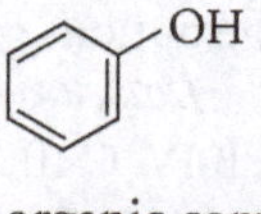

(d) 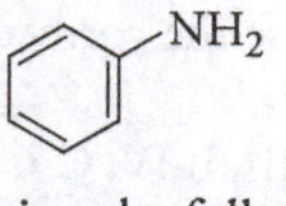


**31** An organic compound $X$ showing the following solubility profile is *[2019, 8 April Shift-I]*

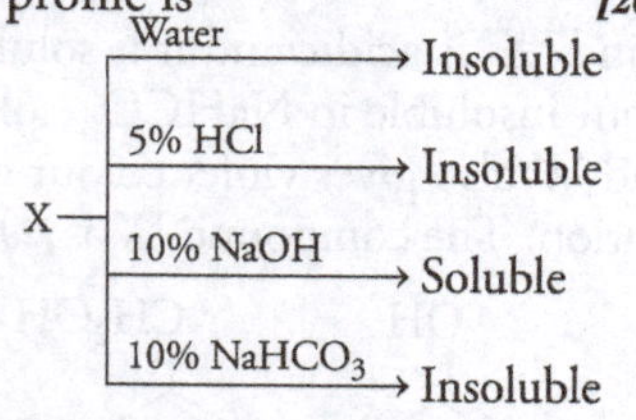


(a) *o*-toluidine (b) oleic acid (c) *m*-cresol (d) benzamide

**32** Match the organic compounds in column-I with the Lassaigne's test results in column-II appropriately. *[JEE Main 2015]*

| | Column-I | | Column-II |
|---|---|---|---|
| (A) | Aniline | (i) | Red colour with $FeCl_3$ |
| (B) | Benzenesulphonic acid | (ii) | Violet colour with sodium nitroprusside |
| (C) | Thiourea | (iii) | Blue colour with hot and acidic solution of $FeSO_4$ |

(a) (A)-(ii); (B)-(i); (C)-(iii) (b) (A)-(iii); (B)-(ii); (C)-(i)
(c) (A)-(ii); (B)-(iii); (C)-(i) (d) (A)-(iii); (B)-(i); (C)-(ii)

## Numerical Value Type Questions

**33** Isomeric hydrocarbons $\longrightarrow$ negative Baeyer's test (Molecular formula $C_9H_{12}$)
The total number of isomers from above with four different non-aliphatic substitution sites is *[2025, 29 Jan Shift-II]*

## Topic 2 Chemistry involved in the Preparation of Organic and Inorganic Compounds

**01** While preparing crystals of Mohr's salt dil. $H_2SO_4$. is added to a mixture of ferrous sulphate and ammonium sulphate, before dissolving this mixture in water, dil. $H_2SO_4$, is added here to *[2024, 5 April Shift-II]*

(a) make the medium strongly acidic
(b) increase the rate of formation of crystals
(c) prevent the hydrolysis of ammonium sulphate
(d) prevent the hydrolysis of ferrous sulphate

**02** The test performed on compound $X$ and their inferences are

| | Test | Inference |
|---|---|---|
| (i) | 2,4-DNP test | Coloured precipitate |
| (ii) | Iodoform test | Yellow precipitate |
| (iii) | Azo-dye test | No dye formation |

Compound '$X$' is *[2019, 9 Jan Shift-II]*

(a) 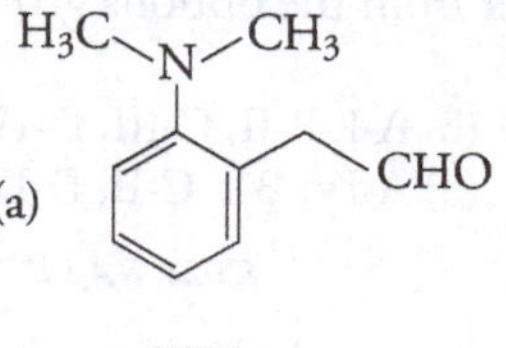

(b) 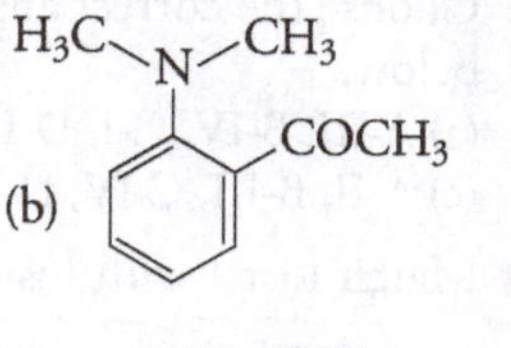

(c) 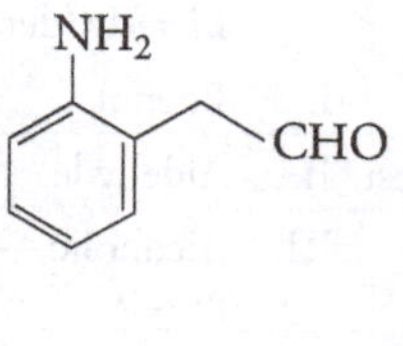

(d) 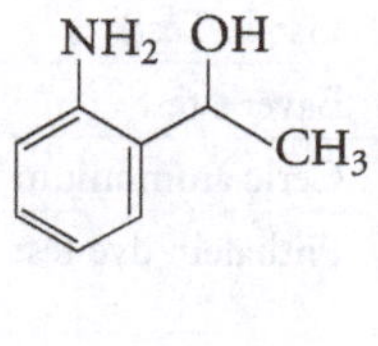

## Topic 3 Chemistry involved in Titrimetric Experiments

### Level 1

**01** Given below are two statements, one is labelled as Assertion A and the other is labelled as Reason R.

**Assertion (A)** Permanganate titrations are not performed in presence of hydrochloric acid.

**Reason (R)** Chlorine is formed as a consequence of oxidation of hydrochloric acid.

In the light of the above statements, choose the correct answer from the options given below. *[2022, 28 July Shift-II]*

(a) Both A and R are true and R is the correct explanation of A

(b) Both A and R are true but R is not the correct explanation of A

(c) A is true but R is false

(d) A is false but R is true

**02** In base *vs* acid titration, at the end point methyl orange is present as *[2022, 25 July Shift-II]*

(a) quinonoid form (b) heterocyclic form

(c) phenolic form (d) benzenoid form

**03** For standardising NaOH solution, which of the following is used as a primary standard? *[JEE Main 2018]*

(a) Oxalic acid

(b) Ferrous ammonium sulphate

(c) Sodium tetraborate

(d) Dil. HCl

### Level 2

**04** 10 mL of 2 M NaOH solution is added to 20 mL of 1 M HCl solution kept in a beaker. Now, 10 mL of this mixture is poured into a volumetric flask of 100 mL containing 2 moles of HCl and made the volume upto the mark with distilled water. The solution in this flask is *[2025, 3 April Shift-II]*

(a) Neutral solution

(b) 10 M HCl solution

(c) 0.2 M NaCl solution

(d) 20 M HCl solution

**05** A compound *X* when treated with phthalic anhydride in presence of concentrated $H_2SO_4$ yields '*Y*'. '*Y*' is used as an acid-base indicator. '*X*' and '*Y*' are respectively *[2023, 8 April Shift-II]*

(a) carbolic acid, phenolphthalein

(b) salicylaldehyde, phenolphthalein

(c) toludine, phenolphthalein

(d) anisole, methyl orange

**06** Given below are two statements. *[2023, 8 April Shift-II]*

**Statement I** Methyl orange is a weak acid.

**Statement II** The benzenoid form of methyl orange is more intense/deeply coloured than the quinonoid form.

In the light of the above statement, choose the most appropriate answer from the options given below

(a) Both Statement I and Statement II are correct

(b) Statement I is incorrect but Statement II is correct

(c) Both Statement I and Statement II are incorrect

(d) Statement I is correct but Statement II is incorrect

**Imp.** Properties of Phenolpthalein

**07** Given below are two statements.
One is labelled as Assertion A and the other is labelled as Reason R.

**Assertion (A)** Phenolphthalein is a pH dependent indicator, remains colourless in acidic solution and gives pink colour in basic medium.

**Reason (R)** Phenolphthalein is a weak acid. It doesn't dissociate in basic medium.

In the light of the above statements, choose the most appropriate answer from the options given below. *[2022, 26 July shift-II]*

(a) Both A and R are true and R is the correct explanation of A.

(b) Both A and R are true but R is not the correct explanation of A.

(c) A is true but R is false.

(d) A is false but R is true.

**08** The strength of an aqueous NaOH solution is most accurately determined by titrating

(Note : consider that an appropriate indicator is used.) *[2020, 8 Jan Shift-I]*

(a) *aq.* NaOH in a pipette and aqueous oxalic acid in a burette.

(b) *aq.* NaOH in a volumetric flask and concentrated $H_2SO_4$ in a conical flask.

(c) *aq.* NaOH in a burette and concentrated $H_2SO_4$ in a conical flask.

(d) *aq.* NaOH in a burette and aqueous oxalic acid in a conical flask.

**09** In an acid-base titration, 0.1 M HCl solution was added to the NaOH solution of unknown strength. Which of the following correctly shows the change of pH of the titration mixture in this experiment? *[2019, 9 April Shift-II]*

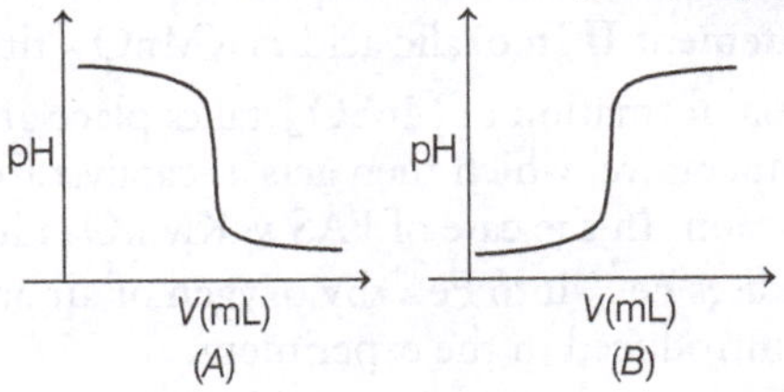

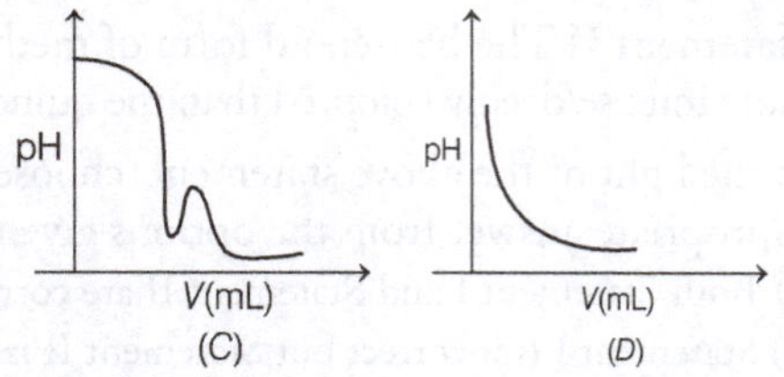


(a) (*D*) (b) (*A*) (c) (*B*) (d) (*C*)

## Level 3

**10** 40 mL of a mixture of $CH_3COOH$ and HCl (aqueous solution) is titrated against 0.1 M NaOH solution conductometrically. Which of the following statement is correct? *[2025, 3 April Shift-II]*

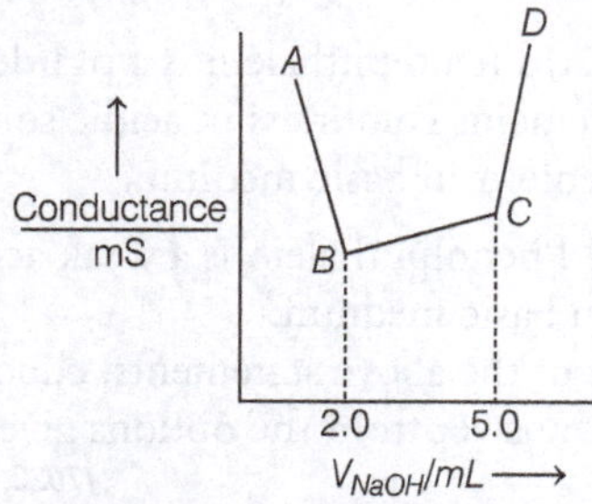


(a) The concentration of $CH_3COOH$ in the original mixture is 0.005 M.
(b) Point '*C*' indicates the complete neutralisation of HCl.
(c) The concentration of HCl in the original mixture is 0.005 M.
(d) $CH_3COOH$ is neutralised first followed by neutralisation of HCl.

**11** Compounds that should not be used as primary standards in titrimetric analysis are

A. $Na_2Cr_2O_7$ B. Oxalic acid
C. NaOH D. $FeSO_4 \cdot 6H_2O$
E. Sodium tetraborate

Choose the most appropriate answer from the options given below. *[2025, 3 April Shift-II]*

(a) B and D only (b) D and E only
(c) C, D and E only (d) A, C and D only

**12** Given below are two statements:

**Statement I** In the oxalic acid *vs* $KMnO_4$ (in the presence of dil. $H_2SO_4$) titration the solution needs to be heated initially to 60°C, but no heating is required in ferrous ammonium sulphate (FAS) *vs* $KMnO_4$ titration (in the presence of dil. $H_2SO_4$).

**Statement II** In oxalic acid *vs* $KMnO_4$ titration, the initial formation of $MnSO_4$ takes place at high temperature, which then acts as catalyst for further reaction. In the case of FAS *vs* $KMnO_4$, heating oxidises $Fe^{2+}$ into $Fe^{3+}$ by oxygen of air and error may be introduced in the experiment.

In the light of the above statements, choose the correct answer from the options given below *[2025, 28 Jan Shift-I]*

(a) Statement I is true but Statement II is false.
(b) Statement I is false but Statement II is true.
(c) Both Statement I and Statement II are false.
(d) Both Statement I and Statement II are true.

**13** Identify the incorrect statements regarding primary standard of titrimetric analysis. *[2024, 9 April Shift-I]*

(A) It should be purely available in dry form.
(B) It should not undergo chemical change in air.
(C) It should be hygroscopic and should react with another chemical instantaneously and stoichiometrically.
(D) It should be readily soluble in water.
(E) $KMnO_4$ and NaOH can be used as primary standard.

Choose the correct answer from the options below.

(a) (A) and (B) only (b) (B) and (E) only
(c) (C) and (D) only (d) (C) and (E) only

**14** A compound '*X*' is a weak acid and it exhibits colour change at pH close to the equivalence point during neutralisation of NaOH with $CH_3COOH$. Compound '*X*' exists in ionised form in basic medium. The compound '*X*' is *[2022, 29 July Shift-II]*

(a) methyl orange (b) methyl red
(c) phenolphthalein (d) erichrome black T

## Numerical Value Type Questions

***15** Only 2 mL of $KMnO_4$ solution of unknown molarity is required to reach the end point of a titration of 20 mL of oxalic acid (2 M) in acidic medium. The molarity of $KMnO_4$ solution should be...... M. *[2024, 4 April Shift-I]*

***16** 20 mL of calcium hydroxide was consumed when it was reacted with 10 mL of unknown solution of $H_2SO_4$. Also 20 mL standard solution of 0.5 M HCl containing 2 drops of phenolphthalein was titrated with calcium hydroxide, the mixture showed pink colour when burette displayed the value of 35.5 mL whereas the burette showed 25.5 mL initially. The concentration of $H_2SO_4$ is ........... M. (Nearest integer) *[2023, 13 April Shift-I]*

**17** $KMnO_4$ is titrated with ferrous ammonium sulphate hexahydrate in presence of dilute $H_2SO_4$. Number of water molecules produced for 2 molecules of $KMnO_4$ is ........... . *[2023, 13 April Shift-I]*

**18** When 10 mL of an aqueous solution of $KMnO_4$ was titrated in acidic medium, equal volume of 0.1 M of an aqueous solution of ferrous sulphate was required

for complete discharge of colour. The strength of $KMnO_4$ in g/L is ........ $\times 10^{-2}$. (Nearest integer)
[Atomic mass of K = 39, Mn = 55, O = 16]
*[2021, 27 Aug Shift-I]*

**19** 10.0 mL of 0.05 M $KMnO_4$ solution was consumed in a titration with 10.0 mL of given oxalic acid dihydrate solution. The strength of given oxalic acid solution is ........ $\times 10^{-2}$ g/L. (Round off to the Nearest Integer). *[2021, 27 July Shift-II]*

**20** When 10 mL of an aqueous solution of $Fe^{2+}$ ions was titrated in the presence of dil. $H_2SO_4$ using diphenylamine indicator, 15 mL of 0.02 M solution of $K_2Cr_2O_7$ was required to get the end point. The molarity of the solution containing $Fe^{2+}$ ions is $x \times 10^{-2}$ M. The value of $x$ is ......... .
(Nearest integer) *[2021, 25 July Shift-I]*

**21** A 10 mL of $Na_2CO_3$ solution is titrated against 0.2 M HCl solution. The following titre values were obtained in 5 readings
4.8 mL, 4.9 mL, 5.0 mL and 5.0 mL and 5.0 mL
Based on these readings, and convention of titrimetric estimation, the concentration of $Na_2CO_3$ solution is ........ mM. (Round off to the Nearest Integer). *[2021, 18 March Shift-II]*

**22** Consider titration of NaOH solution *versus* 1.25 M oxalic acid solution. At the end point following burette readings were obtained.
(i) 4.5 mL (ii) 4.5 mL
(iii) 4.4 mL (iv) 4.4 mL
(v) 4.4 mL
If the volume of oxalic acid taken was 10.0 mL, then the molarity of the NaOH solution is ........ M. (Rounded off to the nearest integer) *[2021, 25 Feb Shift-II]*

## Topic 4 Qualitative Salt Analysis

**01** The compound that is white in colour is *[2024, 31 Jan Shift-I]*
(a) lead sulphate
(b) ammonium arsinomolybdate
(c) ammonium sulphide
(d) lead iodide

**02** A reagent which gives brilliant red precipitate with nickel ions in basic medium is *[2024, 29 Jan Shift-II]*
(a) *meta*-dinitrobenzene
(b) dimethyl glyoxime
(c) sodium nitroprusside
(d) neutral $FeCl_3$

**03** In the wet tests for detection of various cations by precipitation, $Ba^{2+}$ cations are detected by obtaining precipitate of *[2023, 13 April Shift-II]*
(a) Ba(*ox*) : Barium oxalate
(b) $BaSO_4$
(c) $Ba(OAc)_2$
(d) $BaCO_3$

**Imp.** Confirmatory Test of Lead

**04** Formation of which complex, among the following, is not a confirmatory test of $Pb^{2+}$ ions *[2023, 6 April Shift-II]*
(a) lead chromate (b) lead nitrate
(c) lead sulphate (d) lead iodide

**05** In the wet tests for identification of various cations by precipitation. Which transition element cation doesn't belong to group IV in qualitative inorganic analysis? *[2023, 30 Jan Shift-I]*
(a) $Ni^{2+}$ (b) $Co^{2+}$
(c) $Fe^{3+}$ (d) $Zn^{2+}$

**06** $Fe^{3+}$ cation gives a prussian blue precipitate on addition of potassium ferrocyanide solution due to the formation of *[2022, 27 July Shift-II]*
(a) $[Fe(H_2O)_6]_2[Fe(CN)_6]$ (b) $Fe_2[Fe(CN)_6]_2$
(c) $Fe_3[Fe(OH)_2(CN)_4]_2$ (d) $Fe_4[Fe(CN)_6]_3$

**07** Which statement is not true with respect to nitrate ion test? *[2022, 26 June Shift-I]*
(a) A dark brown ring is formed at the junction of two solutions.
(b) Ring is formed due to nitroferrous sulphate complex.
(c) The brown complex is $[Fe(H_2O)_5(NO)SO_4]$.
(d) Heating the nitrate salt with conc. $H_2SO_4$, light brown fumes are evolved.

**08** Which one of the following when dissolved in water gives coloured solution in nitrogen atmosphere? *[2021, 26 Aug Shift-I]*
(a) $Cu_2Cl_2$ (b) $ZnCl_2$ (c) $CuCl_2$ (d) AgCl

**09** $Cu^{2+}$ salt reacts with potassium iodide to give *[2021, 20 July Shift-II]*
(a) $Cu_2I_2$ (b) CuI (c) $Cu(I_3)_2$ (d) $Cu_2I_3$

### Level 2

**10** When a salt is treated with sodium hydroxide solution it gives gas $X$. On passing gas $X$ through reagent $Y$ a brown coloured precipitate is formed. $X$ and $Y$ respectively, are
(a) $X = NH_3$ and $Y = HgO$
(b) $X = NH_4Cl$ and $Y = KOH$
(c) $X = HCl$ and $Y = NH_4Cl$
(d) $X = NH_3$ and $Y = K_2HgI_4 + KOH$

**Imp.** Colour of Salt (Salt Analysis)

**11** Identify the inorganic sulphides that are yellow in colour

(A) $(NH_4)_2S$ (B) PbS
(C) CuS (D) $As_2S_3$
(E) $As_2S_5$

Choose the correct answer from the options given below : *[2025, 28 Jan Shift-II]*

(a) (A) and (C) only (b) (D) and (E) only
(c) (A) and (B) only (d) (A), (D) and (E) only

**12** The metal ion whose electronic configuration is not affected by the nature of the ligand and which gives a violet colour in non-luminous flame under hot condition in borax bead test is *[2025, 28 Jan Shift-I]*

(a) $Ti^{3+}$ (b) $Ni^{2+}$ (c) $Mn^{2+}$ (d) $Cr^{3+}$

**13** During the detection of acidic radical present in a salt, a student gets a pale yellow precipitate soluble with difficulty in $NH_4OH$ solution when sodium carbonate extract was first acidified with dil. $HNO_3$ and then $AgNO_3$ solution was added. This indicates presence of *[2024, 6 April Shift-II]*

(a) $Cl^-$ (b) $CO_3^{2-}$ (c) $I^-$ (d) $Br^-$

**14** Given below are two statements. *[2024, 30 Jan Shift-I]*

**Statement I** The gas liberated on warming a salt with dil. $H_2SO_4$, turns a piece of paper dipped in lead acetate into black, it is a confirmatory test for sulphide ion.

**Statement II** In statement-I the colour of paper turns black because of formation of lead sulphite.

In the light of the above statements, choose the **most appropriate** answer from the options given below

(a) Statement I is true but Statement II is false.
(b) Both Statement I and Statement II are true.
(c) Both Statement I and Statement II are false.
(d) Statement I is false but Statement II is true.

**15** A white precipitate was formed when $BaCl_2$ was added to water extract of an inorganic salt. Further, a gas $X$ with characteristic odour was released when the formed white precipitate was dissolved in dilute HCl. The anion present in the inorganic salt is *[2022, 29 June Shift-II]*

(a) $I^-$ (b) $SO_3^{2-}$ (c) $S^{2-}$ (d) $NO_2^-$

**Imp.** Flame Test

**16** In the flame test of a mixture of salts, a green flame with blue centre was observe(d) Which one of the following cations may be present? *[2022, 24 June Shift-II]*

(a) $Cu^{2+}$ (b) $Sr^{2+}$
(c) $Ba^{2+}$ (d) $Ca^{2+}$

**17** The addition of dilute NaOH to $Cr^{3+}$ salt solution will give *[2021, 27 Aug Shift-II]*

(a) a solution of $[Cr(OH)_4]^-$
(b) precipitate of $[Cr(OH)_6]^{3-}$
(c) precipitate of $Cr_2O_3(H_2O)_n$
(d) precipitate of $Cr(OH)_3$

**18** To an aqueous solution containing ions such as $Al^{3+}$, $Zn^{2+}$, $Ca^{2+}$, $Fe^{3+}$, $Ni^{2+}$, $Ba^{2+}$ and $Cu^{2+}$ conc. HCl, was added followed by $H_2S$. The total number of cations precipitated during this reaction is/are *[2021, 27 July Shift-II]*

(a) 1 (b) 3
(b) 4 (d) 2

**19** Reagent 1-napthylamine and sulphanilic acid in acetic acid is used for the detection of *[2021, 18 March Shift-I]*

(a) $N_2O$ (b) $NO_2^-$
(c) $NO_3^-$ (d) NO

## Level 3

**20** Choose the correct tests with respective observations.

(A) $CuSO_4$ (acidified with acetic acid) $+ K_4[Fe(CN)_6] \rightarrow$ Chocolate brown precipitate.
(B) $FeCl_3 + K_4[Fe(CN)_6] \rightarrow$ Prussian blue precipitate.
(C) $ZnCl_2 + K_4[Fe(CN)_6]$, neutralised with $NH_4OH \rightarrow$ White or bluish white precipitate.
(D) $MgCl_2 + K_4[Fe(CN)_6] \rightarrow$ Blue precipitate.
(E) $BaCl_2 + K_4[Fe(CN)_6]$, neutralised with NaOH $\rightarrow$ White precipitate.

Choose the correct answer from the options given below *[2025, 2 April Shift-I]*

(a) A, D and E only
(b) A, B and C only
(c) C, D and E only
(d) B, D and E only

**21** Identify $A$, $B$ and $C$ in the given below reaction sequence

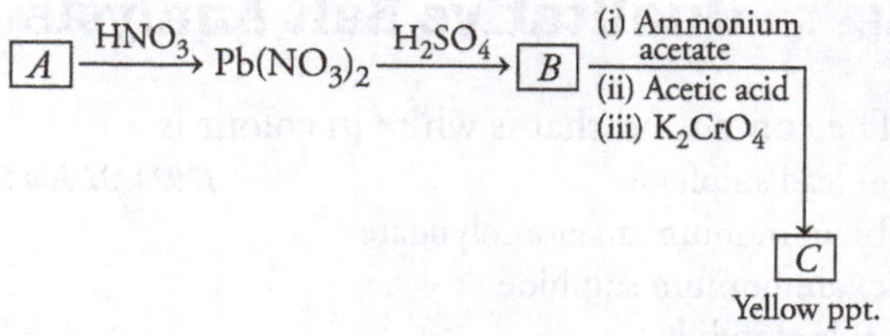


*[2025, 23 Jan Shift-II]*

(a) $PbCl_2$, $PbSO_4$, $PbCrO_4$
(b) PbS, $PbSO_4$, $PbCrO_4$
(c) PbS, $PbSO_4$, $Pb(CH_3COO)_2$
(d) $PbCl_2$, $Pb(SO_4)_2$, $PbCrO_4$

**22** Match List-I with List-II. *[2024, 8 April Shift-I]*

| List-I (Name of the test) | | List-II (Reaction sequence involved) [$M$ is metal] | |
|---|---|---|---|
| A. | Borax bead test | I. | $MCO_3 \rightarrow MO \xrightarrow[+\Delta]{Co(NO_3)_2} CoO \cdot MO$ |
| B. | Charcoal cavity test | II. | $MCO_3 \rightarrow MCl_2 \rightarrow M^{2+}$ |
| C. | Cobalt nitrate test | III. | $MSO_4 \xrightarrow[D]{Na_2B_4O_7} M(BO_2)_2 \rightarrow MBO_2 \rightarrow M$ |
| D. | Flame test | IV. | $MSO_4 \xrightarrow[\Delta]{Na_2CO_3} MCO_3 \rightarrow MO \rightarrow M$ |

Choose the correct answer from the options given below

(a) A-III, B-II, C-IV, D-I (b) A-III, B-IV, C-I, D-II
(c) A-III, B-I, C-II, D-IV (d) A-III, B-I, C-IV, D-II

**23** Match List-I with List-II.

| List-I (Precipitating reagent and conditions) | List-II (Cation) |
|---|---|
| A. $NH_4Cl + NH_4OH$ | I. $Mn^{2+}$ |
| B. $NH_4OH + Na_2CO_3$ | II. $Pb^{2+}$ |
| C. $NH_4OH + NH_4Cl + H_2S$ gas | III. $Al^{3+}$ |
| D. Dilute HCl | IV. $Sr^{2+}$ |

Choose the correct answer from the options given below *[2024, 6 April Shift-I]*

(a) A-III, B-IV, C-I, D-II (b) A-III, B-IV, C-II, D-I
(c) A-IV, B-III, C-I, D-II (d) A-IV, B-III, C-II, D-I

**24** When a solution of mixture having two inorganic salts was treated with freshly prepared ferrous sulphate in acidic medium, a dark brown ring was formed whereas on treatment with neutral $FeCl_3$, it gave deep red colour which disppeared on boiling and a brown red ppt was formed. The mixture contains *[2023, 11 April Shift-I]*

(a) $CH_3COO^-$ and $NO_3^-$ (b) $C_2O_4^{2-}$ and $NO_3^-$
(c) $SO_3^{2-}$ and $C_2O_4^{2-}$ (d) $SO_3^{2-}$ and $CH_3COO^-$

**25** Match List-I with List-II.

| List-I (Element) | | List-II (Colour imparted to the flame) | |
|---|---|---|---|
| A. | K | I. | Brick red |
| B. | Ca | II. | Violet |
| C. | Sr | III. | Apple green |
| D. | Ba | IV. | Crimson red |

Choose the correct answer from the options given below *[2023, 25 Jan Shift-I]*

(a) A-II, B-I, C-IV, D-III (b) A-II, B-IV, C-I, D-III
(c) A-IV, B-III, C-II, D-I (d) A-II, B-I, C-III, D-IV

**26** Match List-I with List-II.

| List-I (Cations) | | List-II (Group reagents) | |
|---|---|---|---|
| A. | $Pb^{2+}, Cu^{2+}$ | I. | $H_2S$ gas in presence of dilute HCl |
| B. | $Al^{3+}, Fe^{3+}$ | II. | $(NH_4)_2CO_3$ in presence of $NH_4OH$ |
| C. | $Co^{2+}, Ni^{2+}$ | III. | $NH_4OH$ in presence of $NH_4Cl$ |
| D. | $Ba^{2+}, Ca^{2+}$ | IV. | $H_2S$ in presence of $NH_4OH$ |

Correct match is *[2023, 25 Jan Shift-I]*

(a) A-III, B-I, C-IV, D-II (b) A-I, B-III, C-IV, D-II
(c) A -I, B-III, C-II, D-IV (d) A-IV, B-II, C-III, D-I

**27** During the qualitative analysis of salt with cation $Y^{2+}$, addition of a reagent ($X$) to alkaline solution of the salt gives a bright red precipitate. The reagent ($X$) and the cation ($Y^{2+}$) present respectively are

(a) dimethylglyoxime and $Ni^{2+}$ *[2022, 24 June Shift-I]*
(b) dimethylglyoxime and $Co^{2+}$
(c) Nessler's reagent and $Hg^{2+}$
(d) Nessler's reagent and $Ni^{2+}$

**28** Match List-I with List-II.

| List-I (Metal iron) | | List-II (Group in qualitative analysis) | |
|---|---|---|---|
| (A) | $Mn^{2+}$ | (i) | Group-III |
| (B) | $As^{3+}$ | (ii) | Group-IIA |
| (C) | $Cu^{2+}$ | (iii) | Group-IV |
| (D) | $Al^{3+}$ | (iv) | Group-IIB |

Choose the most appropriate answer from the options given below *[2021, 31 Aug Shift-II]*

(a) (A)-(iii), (B)-(iv), (C)-(ii), (D)-(i)
(b) (A)-(iv), (B)-(ii), (C)-(iii), (D)-(i)
(c) (A)-(i), (B)-(iv), (C)-(ii), (D)-(iii)
(d) (A)-(i), (B)-(ii), (C)-(iii), (D)-(iv)

**29** An inorganic compound '$X$' on treatment with concentrated $H_2SO_4$ produces brown fumes and gives dark brown ring with $FeSO_4$ in presence of concentrated $H_2SO_4$. Also compound '$X$' give precipitate '$Y$', when its solution in dilute HCl is treated with $H_2S$ gas. The precipitate '$Y$' on treatment with concentrated $HNO_3$ followed by excess of $NH_4OH$ further gives deep blue coloured solution, compound '$X$' is *[2021, 20 July Shift-II]*

(a) $Pb(NO_3)_2$ (b) $Pb(NO_2)_2$
(c) $Cu(NO_3)_2$ (d) $Co(NO_3)_2$

## Numerical Value Type Questions

**30** Among the following cations, the number of cations which will give characteristic precipitate in their identification tests with $K_4[Fe(CN)_6]$ is……

$Cu^{2+}, Fe^{3+}, Ba^{2+}, Ca^{2+}, NH_4^+, Mg^{2+}, Zn^{2+}$

***[2025, 24 Jan Shift-I]***

**31** The number of white coloured salts, among the following is …….. . ***[2024, 1 Feb Shift-I]***

(a) $SrSO_4$ (b) $Mg(NH_4)PO_4$

(b) $BaCrO_4$ (d) $Mn(OH)_2$

(e) $PbSO_4$ (f) $PbCrO_4$

(g) AgBr (g) $PbI_2$

(i) $CaC_2O_4$ (j) $[Fe(OH)_2(CH_3COO)]$

**32** Number of metal ions characterised by flame test among the following is ……… .

$Sr^{2+}, Ba^{2+}, Ca^{2+}, Cu^{2+}, Zn^{2+}, Co^{2+}, Fe^{2+}$

***[2024, 30 Jan Shift-II]***

# Key Index

### Topic 1

| | | | | | | | | | |
|---|---|---|---|---|---|---|---|---|---|
| 1 (d) | 2 (c) | 3 (d) | 4 (b) | 5 (c) | 6 (b) | 7 (d) | 8 (c) | 9 (b) | 10 (c) |
| 11 (b) | 12 (b) | 13 (b) | 14 (a) | 15 (c) | 16 (b) | 17 (a) | 18 (b) | 19 (b) | 20 (b) |
| 21 (d) | 22 (a) | 23 (c) | 24 (d) | 25 (d) | 26 (d) | 27 (b) | 28 (a) | 29 (c) | 30 (c) |
| 31 (c) | 32 (b) | 33 (2) | | | | | | | |

### Topic 2

| | |
|---|---|
| 1 (d) | 2 (b) |

### Topic 3

| | | | | | | | | | |
|---|---|---|---|---|---|---|---|---|---|
| 1 (a) | 2 (a) | 3 (a) | 4 (d) | 5 (a) | 6 (d) | 7 (c) | 8 (d) | 9 (b) | 10 (c) |
| 11 (d) | 12 (d) | 13 (d) | 14 (c) | 15 (8) | 16 (1) | 17 (68) | 18 (316) | 19 (1575) | 20 (18) |
| 21 (50) | 22 (6) | | | | | | | | |

### Topic 4

| | | | | | | | | | |
|---|---|---|---|---|---|---|---|---|---|
| 1 (a) | 2 (b) | 3 (d) | 4 (b) | 5 (c) | 6 (d) | 7 (b) | 8 (c) | 9 (a) | 10 (d) |
| 11 (d) | 12 (b) | 13 (d) | 14 (a) | 15 (b) | 16 (a) | 17 (c) | 18 (a) | 19 (b) | 20 (b) |
| 21 (b) | 22 (b) | 23 (a) | 24 (a) | 25 (a) | 26 (b) | 27 (a) | 28 (a) | 29 (c) | 30 (4) |
| 31 (5) | 32 (4) | | | | | | | | |

# Detailed **Explanations**

## Topic 1

**01.** *(d)* Lassaigne's test is used for detection of nitrogen, sulphur, phosphorus and halogens.

**02.** *(c)* Phenolic group can be identified by a positive phthalein dye test by giving pink colour solution. The complete reaction is as follows.

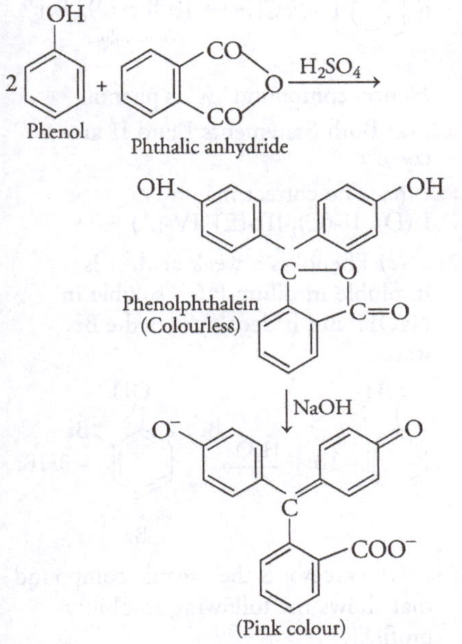


**03.** *(d)* $S^{2-} + [Fe(CN)_5NO]^{2-} \longrightarrow [Fe(CN)_5NOS]^{4-}$ Violet/purple

The formula of purple colour formed in Lassaigne's test for sulphur is $Na_4[Fe(CN)_5(NOS)]$.

**04.** *(b)* Seliwanoff's test which is used to distinguish between ketoses and aldoses. The Seliwanoff's reagent contains resorcinol and HCl.

**05.** *(c)* Ceric ammonium nitrate and $CHCl_3$ / alc. KOH are used for the identification of functional groups present in **alcohol** and **amine** respectively.

**06.** *(b)* $Na_2S + \underset{\text{Sodium nitroprusside}}{Na_2[Fe(CN)_5NO]} \longrightarrow Na_4[Fe(CN)_5(NOS)]$

$$\downarrow$$

$$4Na^+ + [Fe(CN)_5(NOS)]^{4-}$$ Violet in colour

**07.** *(d)* If $FeCl_3$ is added to excess of potassium ferrocyanide, it gives white precipitate that will become blue.

$$FeCl_3 + K_4[Fe(CN)_6] \longrightarrow KFe[Fe(CN)_6] + 3KCl$$

$FeCl_3$ is limiting reagent. So, all potassium is not replaced.

**08.** *(c)* Nitric acid is added to sodium extract before addition of silver nitrate for testing halogens. Because it decomposes NaCN and $Na_2S$ or else they interfere in the test.

The reaction are as follows :

$$NaCN + HNO_3 \longrightarrow NaNO_3 + HCN\uparrow$$

$$\underset{\text{Sodium extract}}{Na_2S} + \underset{\text{Nitric acid}}{HNO_3} \longrightarrow NaNO_3 + H_2S\uparrow$$

**09.** *(b)* To test the reducing nature of sugar Barfoed test is used.

**10.** *(c)* $S^{2-}$ and $CN^-$ anions, if present may interfere by giving white ppt. of AgCN and black ppt. of $Ag_2S$ with $AgNO_3$. Hence, before testing for halogens they are decomposed by concentrated $HNO_3$.

**11.** *(b)* Beilstein's test is used for chlorine.

**12.** *(b)* Copper wire, test for halogens is known as Beilsteins test.

**13.** *(b)* Biuret test is given by an amide linkage. It is present in urea, polypeptides and proteins. Carbohydrates do not have any amide linkages.

**14.** *(a)* If nitrogen is present in organic compound, then sodium extract contains NaCN.

$$Na + C + N \xrightarrow{\text{Fuse}} NaCN$$

$$FeSO_4 + 6NaCN \longrightarrow \underset{(A)}{Na_4[Fe(CN)_6]} + Na_2SO_4$$

*A* changes to Prussian blue $Fe_4[Fe(CN)_6]_3$ on reaction with $FeCl_3$.

$$4FeCl_3 + 3Na_4[Fe(CN)_6] \longrightarrow Fe_4[Fe(CN)_6]_3 + 12NaCl$$

**15.** *(c)* Among the given molecules butanal, 3-pentanone, pentanal and 3 pentanol will not give iodoform reactions while, other molecules will give iodoform reaction.

**16.** *(b)* Compounds containing

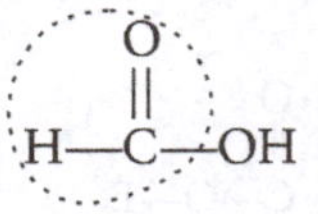


formic acid and aldehydes gives silver mirror test with ammoniacal silver nitrate.

**17.** *(a)* According to given condition the appearance of red colour indicates the presence of both nitrogen and sulphur. The reaction involve is as follow,

$$Fe^{2+} \xrightarrow[\text{Conc. } H_2SO_4]{H^+} Fe^{+3}$$

$$Fe^{3+} + SCN^- \xrightarrow{-SCN} \underset{\text{Blood red colour}}{[Fe(SCN)]^{2+}}$$

Hence, option (a) is correct.

**18.** *(b)* The correct match is A-III, B-I, C-IV, D-II.

Elements are detected by qualitative analysis in this processes some reagents are required. These are as follows

A. Nitrogen is detected by $Fe_4[Fe(CN)_6]_3$

B. Sulphur is detected by $Na_2[Fe(CN)_5 \cdot NO]$

C. Phosphorous is detected by $(NH_4)_2 \cdot MoO_4$

D. Halogens are detected by $AgNO_3$

**19.** *(b)* The sodium fusion extract is boiled with conc. $HNO_3$ to decompose any impurities of sodium cyanide or sulphide formed during Lassaigne's test.

> • Concept **Enhancer**
>
> The Lassaigne's Test is used in elemental analysis for the qualitative determination for the presence of extra elements namely halogens, nitrogen and sulphur. The halogen is covalently bonded to the organic compounds, are converted to sodium salt formed during the fusion. The aqueous extract is called sodium fusion extract or Lassaigne's extract. This extract is boiled with conc. $HNO_3$ followed by the addition of $AgNO_3$ solution which yields a white (AgCl) or yellow (AgBr or AgI) precipitate if halogen is present.
>
> $$NaX + AgNO_3 \longrightarrow AgX + NaNO_3$$

**20.** *(b)* The complete reaction is as follows,

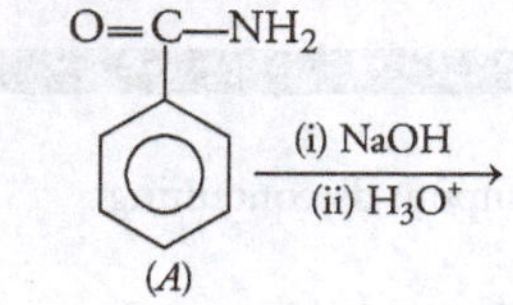


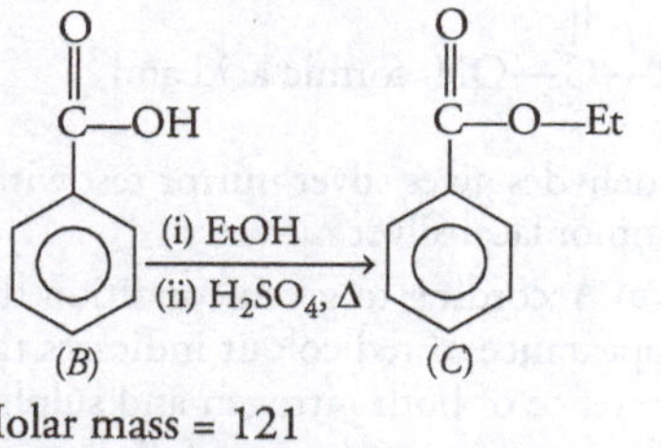


Molar mass = 121

$A$ = Benzamide show positive Lassaigne's test.

$B$ = Benzoic acid gives effervescence with aqueous $NaHCO_3$.

$C$ = Ester gives fruity smell.

**21.** *(d)* Compounds given in $A$ and $C$ will give yellow solid when reacted with NaOI / NaOH.

A. $CH_3-\underset{OH}{\underset{|}{CH}}-CH_2-CH_3 \xrightarrow[NaOH]{NaOI} \underset{\text{Yellow solid}}{CHI_3} + CH_3-CH_2-\overset{O}{\overset{||}{C}}-\bar{O}Na^+$

C. $CH_3-\overset{O}{\overset{||}{C}}-CH_2-CH_3 \xrightarrow[NaOH]{NaOI} \underset{\text{Yellow solid}}{CHI_3} + CH_3-CH_2-\overset{O}{\overset{||}{C}}-\bar{O}Na^+$

**22.** *(a)* The correct match is

A-II, B-III, C-IV, D-I

- Carboxylic acid gives effervescence with sodium bicarbonate solution.
- Phenolic —OH gives violet coloured complex with neutral $FeCl_3$.
- Alcoholic —OH gives red colour with ceric ammonium nitrate.
- When alkaline $KMnO_4$ react with an unsaturated compound (alkene or alkyne) the purple colour of $KMnO_4$ solution disappears indicating positive test for unsaturation.

**23.** *(c)* The correct match is A-II, B-III, C-IV, D-I

- $Br_2$ water test is a test which is used for unsaturation, in which reddish orange colour of bromine water disappears.
- Alcohol give reddish colour with ceric ammonium nitrate.
- Phenol gives violet colour with natural ferric chloride.
- Aldehyde and ketone give yellow/orange/red colour compounds with 2, 4-DNP i.e. 2, 4-dinitrophenyl hydrazine.

**24.** *(d)* The correct match is A-IV, B-III, C-I, D-II.

- Bayer's test–Used to test unsaturation.
- Ceric ammonium nitrate test–give red precipitate with alcohols.
- Schiff's test–gives pink colour with aldehydes.
- Phthalein dye test–used to test for phenol.

**25.** *(d)* The correct match is A-III, B-I, C-II, D-IV

**26.** *(d)* A. Alkaline solution of copper sulphate ($CuSO_4$) and sodium citrate is Benedict's solution which is used to test aliphatic aldehydes. Hence, it is used to test compound (III) i.e.,

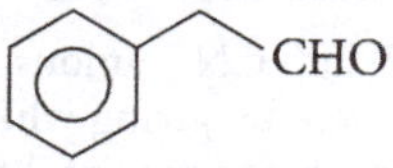


B. Neutral $FeCl_3$ solution is used to test phenolic compound (IV) i.e.,

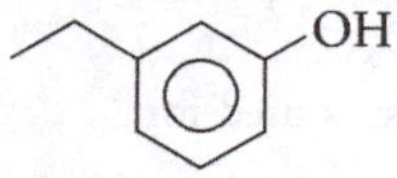


red, green, purple or blue colouration.

C. Alkaline chloroform is used to detect primary amines (II)

i.e., 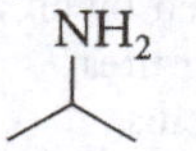


It is also called isocyanide test.

D. Potassium iodide (KI) and sodium hypochlorite (NaOCl) gives ($I_2$ + KOH) which is used to test compounds having

$CH_3-\overset{O}{\overset{||}{C}}-$ or $CH_3-\overset{OH}{\overset{|}{CH}}-$ group.

This test is known as iodoform test.

$$2KI + NaOCl + H_2O \longrightarrow NaCl + I_2 + 2KOH$$

Hence, compound is (I) 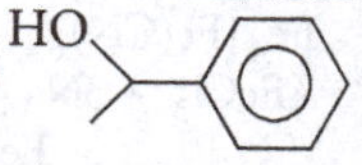


Hence, the correct match is A-III, B-IV, C-II, D-I.

**27.** *(b)* Phenol is acidic in nature as after removing $H^+$ ion, phenoxide ion is stablised by resonance.

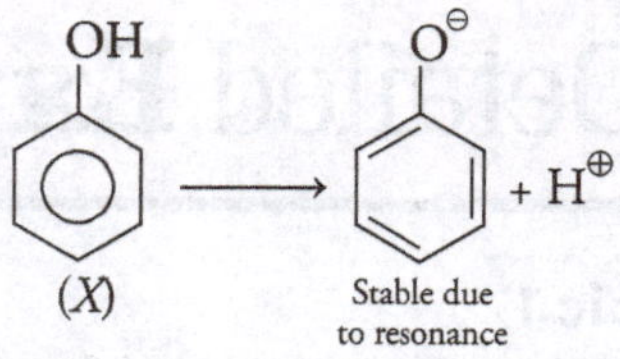


Phenol is although acidic but it is not soluble in $NaHCO_3$ solution. However, it gives violet colour with neutral $FeCl_3$ solution.

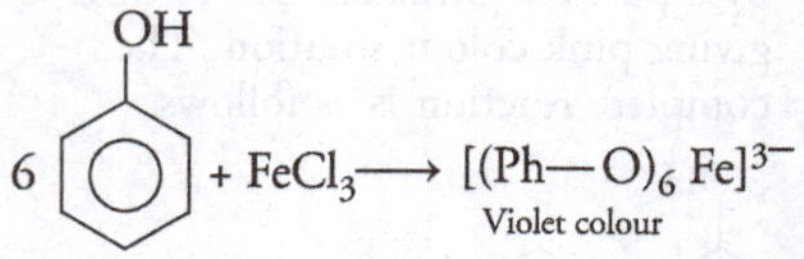


Hence, compound '$X$' is phenol.

**28.** *(a)* Both Statements I and II are correct.

**29.** *(c)* The correct match is I-(D), II-(C), III-(E), IV-(A)

**30.** *(c)* Phenol is a weak acid, it is insoluble in dilute HCl, soluble in NaOH and it decolourises the $Br_2$ water.

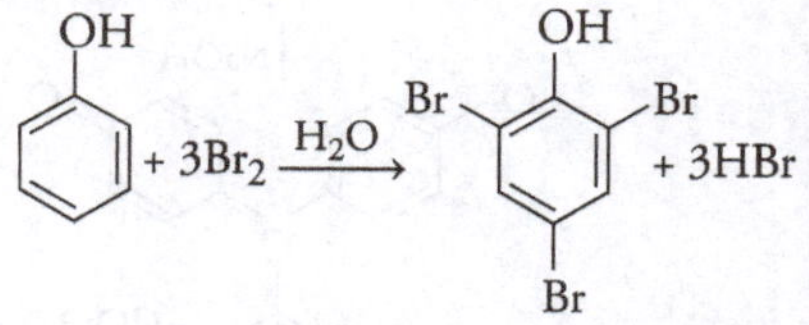


**31.** *(c)* *m*-cresol is the organic compound that shows the following solubility profile.

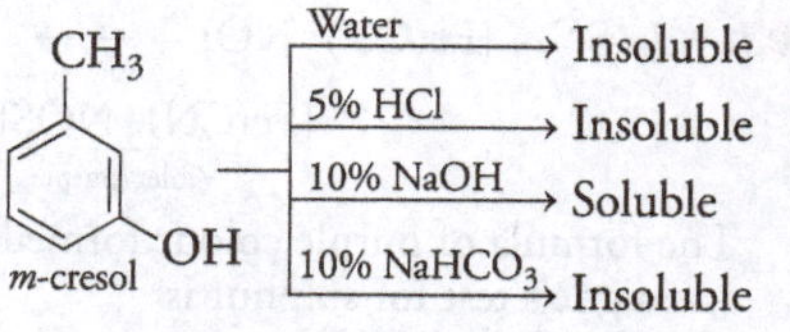


**32.** *(b)* The correct match is (A)-(iii), (B)-(ii), (C)-(i)

**33.** *(2)* The isomers of $C_9H_{12}$ which give negative Baeyer's test are

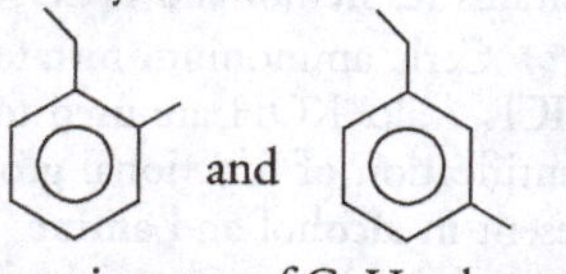


Above two isomers of $C_9H_{12}$ have four different sites for aromatic electrophilic substitution reaction.

## Topic 2

**01.** *(d)* Iron ion ($Fe^{2+}$) undergoes hydrolysis, therefore while preparing aqueous solution of ferrous sulphate and ammonium sulphate in water dilute sulphuric acid is added to prevent hydrolysis of ferrous sulphate. Hence, option (d) is correct.

**02.** *(b)* Compound '*X*' should be (b). Due to presence of —$COCH_3$ group, it gives iodoform test. Due to presence of 3° amine group, it does not undergoes in azo-dye test. On the other hand, presence of $PhCOCH_3$ group, it readily undergoes 2,4-DNP test.

## Topic 3

**01.** *(a)* Both A and R are true and R is the correct explanation of A.

**02.** *(a)* In titration, at the end point methyl orange is present as quinonoid form which is red in colour.

**03.** *(a)* Oxalic acid is used as primary standard for standardising NaOH solution.

**04.** *(d)* When 10 ml, 2M NaOH solution is added to 20 mL of 1M HCl solution following reaction takes place.

$$NaOH + HCl \longrightarrow NaCl + H_2O$$

Initial : $MV = 2 \times 0.1$ $\quad MV = 1 \times 0.2$

$= 0.2$ mole $\quad = 0.2$ mole

Final $\quad 0 \quad 0$

$\therefore$ Resulting solution becomes neutral. Now, when 10 mL of this mixture is poured into volumetric flask of 100 mL containing 2 moles of HCl, then Molarity of HCl will be

$$\frac{2}{100} \times 1000 = 20 \text{ mol/L}$$

**05.** *(a)* Compound '*X*' is carbolic acid and compound '*Y*' is phenolphthalein. When carbolic acid (phenol) reacts with phthalic anhydride in presence of conc. $H_2SO_4$ then forms phenolphthalein which is used as an acid base indicator.

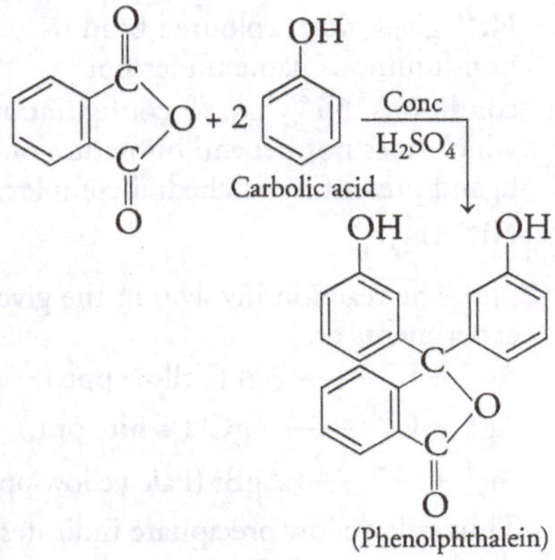

(Phenolphthalein)

**06.** *(d)* Statement I is correct but Statement II is incorrect. Methyl orange is a weak acid by nature. It is yellow in basic medium and acquire more intense red colour in acidic medium due to the formation of quinonoid form.

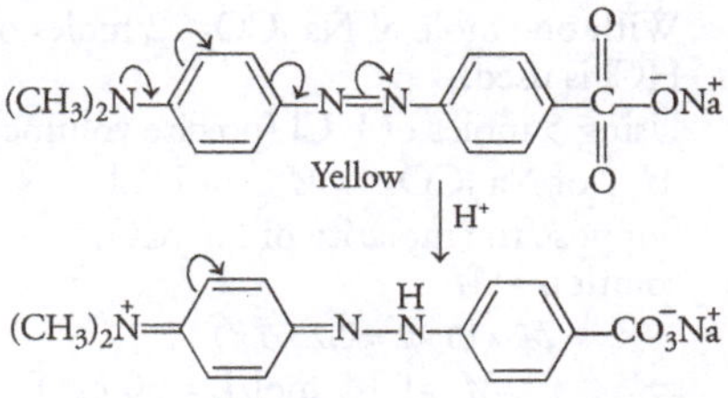


**07.** *(c)* It dissociates in basic medium and forms conjugate base which is pink in colour.

$$HIn + OH^- \rightleftharpoons In^- + H_2O$$

Colourless $\quad$ Pink

Thus, A is true but R is false.

**08.** *(d)* The strength of NaOH measured by titrating against oxalic acid by taking NaOH in the burette and oxalic acid in the conical flask.

**09.** *(b)* The graph that shows the correct change of pH of the titration mixture in the experiment is

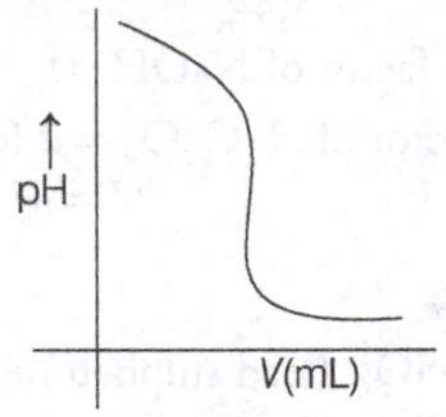


**• Concept Enhancer**

The difference in the volume of NaOH solution between the end point and the equivalence point is not significant for most of the commonly used indicators as there is a large change in the pH value around the equivalence point. Most of them change their colour across this pH change.

**10.** *(c)* According to data given in graph, 2mL NaOH solution is used for neutralisation of HCl and 3 mL NaOH solution is used for neutralisation of $CH_3COOH$.

$\therefore$ Mole of HCl = Mole of NaOH used

$$M \times 40 = 0.1 \times 2$$

$$\Rightarrow \quad M = 0.005$$

$\therefore$ Mole of $CH_3COOH$ = mole of NaOH used

$$M \times 40 = 0.1 \times 3 \Rightarrow M = 0.0075$$

Hence, HCl is strong acid will be neutralised first.

**11.** *(d)* The primary standard is a highly pure stable compound with a known exact composition that can be accurately weighed and dissolved to create solution of known concentration NaOH is hygroscopic and can't be used $FeSO_4 \cdot 6H_2O$ is unstable and can be easily oxidised.
$Na_2Cr_2O_7$ is hygroscopic and can't be used.

**12.** *(d)* Both Statement I and Statement II are true.

$$2MnO_4^- + 5(COO)_2^{2-} + 16H^+ \longrightarrow 10CO_2 + 2Mn^{2+} + 8H_2O$$

This reaction is slow at room temperature but become fast at 60°C. Manganese (II) ions catalyse the reaction, thus the reaction is autocatalytic reaction.

**13.** *(d)* Statements given in C and E are incorrect. While other statements are correct. The correct form of C and E are

- It should not be hygroscopic.
- Both $KMnO_4$ and NaOH cannot be used as a primary standard.

**14.** *(c)* Phenolphthalein is a weak acid that gives pink colour in basic medium due to formation of ions.

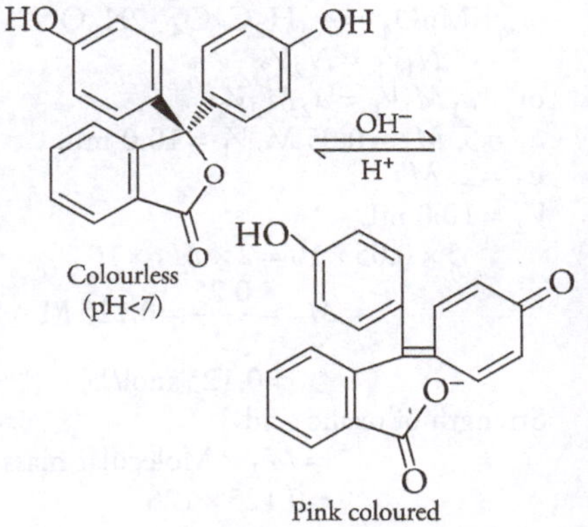

Colourless (pH<7)

Pink coloured (pH > 7)

**15.** *(8)* Equivalent of $KMnO_4$ = Equivalent of $H_2C_2O_4$

So, $M \times 2 \times 5 = 2 \times 20 \times 2$

$\Rightarrow M = 8$ M

**16.** *(1)* When $Ca(OH)_2$ is titrated against standard solution of 0.5 M HCl, 20 mL volume

$$\therefore \quad n_{factor} \times M_1V_1 = n_{factor} \times M_2V_2$$

$2 \times M_1 \times 10 \text{ mL} = 1 \times 0.5 \times 20$
(Since from 25.5 mL to 35.5 mL, only 10 mL of calcium hydroxide is used)

$\therefore \quad [M_1 = 0.5 \text{ M}]$

Now $Ca(OH)_2$ is titrated with 10 mL of $H_2SO_4$

$\therefore \quad 2 \times 0.5 \times 20 = 2 \times M_2 \times 10$

$\therefore \quad [M_2 = 1 \text{ M}]$
20 mL

**17.** *(68)* $2KMnO_4 + 10FeSO_4$
$(NH_4)_2SO_4 \cdot 6H_2O + 8H_2SO_4 \longrightarrow$
$K_2SO_4 + 2MnSO_4 + 5Fe_2(SO_4)_3 + 8H_2O + 10(NH_4)_2SO_4$

With 2 molecule of $KMnO_4$, 8 molecule of $H_2O$ is released.
Also with 10 molecule of $FeSO_4$, 60 molecules of $H_2O$ will also be formed.
Hence total (68) $H_2O$ will be produced.

**18.** *(316)* Let molarity of $KMnO_4 = x$

$KMnO_4 + FeSO_4 \longrightarrow Fe(SO_4)_3 + Mn^{2+}$
$n = 5 \qquad n = 1$

(Equivalent of $KMnO_4$ reacted) = (Equivalent of $FeSO_4$ reacted)

$\Rightarrow \quad 5x \times 10 = 1 \times 0.1 \times 10$
$x = 0.02 \text{ M}$

Molar mass of $KMnO_4 = 158 \text{ g / mol}$
Strength $= x \times 158$
$= 0.02 \times 158$
$= 3.16 \text{ g / L}$
$= 316 \times 10^{-2} \text{ g/L}$

$\therefore \quad x \approx 316.$

**19.** *(1575)* For titration using equation,
$n_{eq} KMnO_4 = n_{eq} H_2C_2O_4 \cdot 2H_2O$
$N_1V_1 = N_2V_2$
or $a_1M_1V_1 = a_2M_2V_2$
$a_1 = 5, M_1 = 0.05 \text{ M}, V_1 = 10.0 \text{ mL},$
$a_2 = 2, M_2 = ?$
$V_2 = 10.0 \text{ mL}$
$5 \times 0.05 \times 10 = 2 \times M_2 \times 10$
$M_2 = \frac{0.25}{2} = 0.125 \text{ M}$
$= 0.125 \text{ mol/L}$

Strength of oxalic acid
$= M_2 \times$ Molecular mass
$= 0.125 \times 126$
$= 15.75 \text{ g/L}$
or $1575 \times 10^{-2} \text{ g/L}$

**20.** *(18)* Normality = Molarity × *n*-factor
$N_{K_2Cr_2O_7} = 0.02 \times 6 = 0.12$
$V_{K_2Cr_2O_7}$ used = 15 mL

Volume of aqueous $Fe^{2+}$ solution used = 10 mL

then, we have $N_1V_1 = N_2V_2$
$0.12 \times 15 = N_2 \times 10$

$N_2 = \frac{1.8}{10} = 0.18 = 18 \times 10^{-2} \text{ N}$
$N_2 = M_2 \times n\text{-factor}$
$M_2 = N_2$ as *n*-factor is 1.

**21.** *(50)* $Na_2CO_3 + 2HCl \longrightarrow 2NaCl + H_2CO_3$

With one mole of $Na_2CO_3$, 2 moles of HCl is used.
Using 5 moles of HCl (precise volume)
$M_{eq}$ of $Na_2CO_3 = M_{eq}$ of HCl
Suppose the molarity of $Na_2CO_3$ solution = $M$

$M \times 10 \times 2 = 0.2 \times 5 \times 1$
$\Rightarrow \quad M = 0.05 \text{ mol/L} = 50 \text{ mM}$

**22.** *(6)* Average burette reading = Volume of NaOH solution $(V_1)$
$= \frac{4.5 + 4.5 + 4.4 + 4.4 + 4.4}{5}$
$= 4.44 \text{ mL}$

Strength of NaOH solution
$= S_1(\text{M}) \text{ (say)} = S_1(\text{N})$

Volume of oxalic acid solution $(V_2)$
$= 10 \text{ mL}$

Strength of oxalic acid solution
$(S_2) = 1.25 \times 2 \text{ N}$

$\Rightarrow S_1 = \frac{V_2S_2}{V_1}$ ($\because$ Law of equivalence)
$= \frac{10 \times (1.25 \times 2)}{4.44} = 5.63 \text{ N} \simeq 6 \text{ M}$

Note *n*-factor of NaOH = 1
*n*-factor of $H_2C_2O_4 = 2$ (oxalic acid)
$N = M \times n$

## Topic 4

**01.** *(a)* $PbSO_4$ (lead sulphate) is white in colour. $PbI_2$ (lead iodide) is a yellow ppt. $(NH_4)_2S$ (ammonium sulphide) is soluble. $(NH_4)_3AsO_4 \cdot 12MoO_3$ (ammonium arsinomolybdate) is yellow.

**02.** *(b)* Dimethyl glyoxime when reacts with nickel ions in basic medium gives bright red precipitate. The complete reaction is as follows

$Ni^{2+} + 2DMG \xrightarrow{\text{Base}} [Ni(DMG)_2]$
Bright red precipitate

**03.** *(d)* In group V, barium ions are precipitated as their carbonates by adding ammonium carbonate to their solution.

**04.** *(b)* $Pb[NO_3]_2$ lead nitrate is a soluble colourless compound. Hence, formation of lead nitrate complex is not used as a confirmatory test. While other complexes given in options gives different precipitates i.e.

$PbCrO_4$ — yellow ppt
$Pb(NO_3)_2$ — soluble
$PbSO_4$ — white ppt
$PbI_2$ — yellow ppt

**05.** *(c)* Among the given options, $Fe^{3+}$ cation does not belongs to group IV in qualitative inorganic analysis. Instead it belongs to group III.

**06.** *(d)* $4Fe^{3+} + 3[Fe(CN)_6]^{4-} \longrightarrow Fe_4[Fe(CN)_6]_3$
(Prussian blue in colour)

**07.** *(b)* Ring is formed due to $Fe[H_2O)_5(NO)]SO_4$ i.e., complex nitrosoferrous sulphate.

**08.** *(c)* AgCl and $Cu_2Cl_2$ are not dissolved in $H_2O$. $ZnCl_2$ and $CuCl_2$ are dissolved in water but $ZnCl_2$ is colourless and $CuCl_2$ is coloured.

**09.** *(a)* $2Cu^{2+} + 4KI \longrightarrow Cu_2I_2 + I_2 + 4K^+$
Copper iodide

**10.** *(d)* The completed reaction is as follows,

$NH_4^+ + NaOH \longrightarrow H_2O + NH_3\uparrow$
$(X)$

$2[HgI_4]^{2-} + 3OH^- + NH_3 \longrightarrow$
$(Y) \qquad (X)$

$HgO.Hg\langle^{NH_2}_{I} + 7I^- + 2H_2O$
Brown ppt.

Thus, compound $X$ and $Y$ are $NH_3$ and $K_2HgI_4 + KOH$ respectively.

**11.** *(d)* $(NH_4)_2S$, $As_2S_3$ and $As_2S_5$ are yellow colour sulphides, while PbS and CuS are black in colour.

**12.** *(b)* Among the given ionic species. $Ni^{2+}$ gives violet coloured bead in non-luminous flame under hot conditions. $Ni^{2+}$ has $d^8$ configuration which does not depend on nature of ligand present in octahedral complex.
$Ni^{2+} : t_{2g}^6 e_g^2$

**13.** *(d)* The reaction involves in the given experiments are

$Ag^+ + I^- \longrightarrow AgI$ (Yellow ppt.)
$Ag^+ + Cl^- \longrightarrow AgCl$ (White ppt.)
$Ag^+ + Br^- \longrightarrow AgBr$ (Pale yellow ppt.)

Thus pale yellow precipitate indicates the presence of $Br^-$.

**14.** (a) Statement I is true but Statement II is false.

The correct form of Statement II is :
The colour of paper turns black because of formation of lead sulphide.
The reaction involved is as follows

$$Na_2S + H_2SO_4 \longrightarrow Na_2SO_4 + H_2S$$

$$\underset{\text{Lead acetate}}{(CH_3COO)_2Pb} + H_2S \longrightarrow \underset{\substack{\text{(Black)}\\ \text{Lead sulphide}}}{PbS} + 2CH_3COOH$$

**15.** (b)

$$BaCl_2 \xrightarrow{SO_3^{2-}} \underset{\text{White ppt.}}{BaSO_3 \downarrow} \xrightarrow{HCl} \underset{\substack{\text{Characteristic}\\ \text{small}}}{SO_2 \uparrow}$$

$\therefore$ $SO_3^{2-}$ ion is present in the inorganic salt.

**16.** (a) $Cu^{2+}$ gives green flame with blue centre.
$Sr^{2+}$ gives crimson red, $Ba^{2+}$ gives apple green and $Ca^{2+}$ gives orange red.

**17.** (c) The addition of NaOH to $Cr^{3+}$ solutions does not form hydroxide but the hydrated oxide is precipitated instead.

$$Cr^{3+} + 3OH^- \longrightarrow Cr(OH)_3 \longrightarrow Cr_2O_3 \cdot (H_2O)_n$$

**18.** (a) Only $Cu^{2+}$ get precipitated when HCl followed by $H_2S$ was added.
$Al^{3+}$ and $Fe^{3+}$ sulphides hydrolyse in water. $Ni^{2+}$ and $Zn^{2+}$ require basic medium with $H_2S$ to form ppt. $Ca^{2+}$ and $Ba^{2+}$ sulphides are soluble.
Hence, we will recieve ppt. of only CuS.

**19.** (b) 1-naphthyl amine and sulphanilic acid is used for the detection of $NO_2^-$ with the help of acetic acid.

**20.** (b) The correct statements are A, B and C

The reaction involved are

$$\text{A. } 2CuSO_4 + K_4[Fe(CN)_6] \xrightarrow{CH_3COOH} \underset{\text{Chocolate brown ppt.}}{Cu_2[Fe(CN)_6]} + 2K_2SO_4$$

$$4FeCl_3 + 3K_4[Fe(CN)_6] \longrightarrow \underset{\text{Prussian blue ppt.}}{Fe_4[Fe(CN)_6]_3} + 12KCl$$

$$3ZnCl_2 + 2K_4[Fe(CN)_6] \xrightarrow{NH_4OH} \underset{\text{White/Bluish ppt.}}{Zn_3[Fe(CN)_6]_2} + 6KCl$$

**21.** (b) The complete reaction occurs as follows.

$$\underset{(A)}{PbS} \xrightarrow{HNO_3} Pb(NO_3)_2 \xrightarrow{H_2SO_4} \underset{(B)}{PbSO_4} \xrightarrow[\text{(ii) Acetic acid, (iii) } K_2CrO_4]{\text{(i) Ammonium acetate}} \underset{\substack{(C)\\ \text{Yellow ppt.}}}{PbCrO_4}$$

Thus, *A*, *B*, *C* respectively are PbS, $PbSO_4$, $PbCrO_4$.

**22.** (b) The correct match is
A-III, B-IV, C-I, D-II

**23.** (a) The correct match is
A-III, B-IV, C-I, D-II

**24.** (a) According to question,

$$CH_3COO^- + 3FeCl_3 + H_2O \longrightarrow \underset{\text{Blood red colour}}{[Fe_3(OH)_2(CH_3COO)_6]^+} \xrightarrow{\Delta} \underset{\text{Red-brown ppt}}{Fe(OH)_2(CH_3COO)\downarrow}$$

$$2NO_3^- + 4H_2SO_4 + 6Fe^{2+} \longrightarrow 6Fe^{3+} + 2NO\uparrow + 4SO_4^{2-} + 4H_2O$$

$$[Fe(H_2O)_6]^{2+} + NO \longrightarrow \underset{\text{Brown}}{[Fe(H_2O)_5(NO)]^{2+}} + H_2O$$

Thus, the mixture contains $CH_3COO^-$ and $NO_3^-$.

**25.** (a) The correct match is A-II, B-I, C-IV, D-III.

**26.** (b) The correct match is A-I, B-III, C-IV, D-II.

**27.** (a) The reagent dimethylglyoxime (*X*) is added to the alkaline solution of nickel salt and a bright red precipitate is formed.

$$Ni^{2+} + \underset{(X)}{DMG^-} \longrightarrow \underset{\text{Bright red precipitate}}{[Ni(DMG)_2]\downarrow}$$

So, here the cation ($Y^{2+}$) is $Ni^{2+}$ and *X* is dimethylglyoxime.

**28.** (a) The correct match is A-(iii), B-(iv), C-(ii), D-(i)

**29.** (c) According to question,

$$\underset{\substack{(X)\\ \text{(Anion)}}}{NO_3^-} + \underset{\text{(conc.)}}{H_2SO_4} + \underset{\text{(Brown fumes)}}{NO_2\uparrow} + H_2O$$

$$FeSO_4 + H_2SO_4 + NO_3^- \longrightarrow \underset{\text{(Dark brown ring)}}{[Fe(H_2O)_5(NO)]SO_4}$$

$$Cu^{2+} + (\text{dil. HCl} + H_2S) \longrightarrow \underset{\substack{\text{Black ppt.}\\ (Y)}}{CuS}$$

$$CuS \xrightarrow{\text{Conc. } HNO_3} Cu(NO_3)_2 + NO_2 + S + H_2O \xrightarrow{\text{Excess } NH_4OH} \underset{\text{Deep blue}}{[Cu(NH_3)_4]^{2+}}$$

Thus, *X* is $Cu(NO_3)_2$.

**30.** (4) Only $Cu^{2+}$, $Fe^{3+}$, $Ca^{2+}$ and $Zn^{2+}$ form precipitate with $K_4[Fe(CN)_6]$

**31.** (5) Among the given salts, the below mention salts are white in colour.

$$\left.\begin{array}{c} SrSO_4 \\ Mg(NH_4)PO_4 \\ Mn(OH)_2 \\ PbSO_4 \\ CaC_2O_4 \end{array}\right\} \text{White colour}$$

Hence, total 5 white coloured salts.

- **Concept Enhancer**

The colour of salt is attributed to the ions present in them, generally salts of cations like sodium, calcium, magnesium, aluminium, zinc, ammonium and anions like chlorides, nitrate, carbonate ion are white in solid state and colourless in their aqueous solution. Most generalised reason for such nature is absence of unpaired electrons in these ions.

**32.** (4) A flame test is a qualitative analysis used to identify the metal and metalloid ion in the sample.
The following metal ions will respond to flame test.
$Sr^{2+}$, $Ba^{2+}$, $Ca^{2+}$, $Cu^{2+}$
Hence, 4 metal ions will respond to flame test.